Expand your learn

Written with your interests and needs in mind, this textbook gives you a solid foundation for success in Precalculus and prepares you for AP® Calculus.

At **Key Math** you'll find engaging activities and resources to support your success in this course.

- Access your **online textbook.** (You'll need to get a ClassPass from your teacher.)
- Investigate **Dynamic Precalculus Explorations.**
- Download **calculator programs** for use with TI-Nspire and TI-83/84 Plus graphing calculators.
- Download **data sets** for use with TI-Nspire, TI-83/84 Plus graphing calculators, Excel, and *Fathom*®.
- Get extra practice with **Supplementary Problems**.
- Expand your understanding of vocabulary with **Bilingual Dictionary Templates.**

Learn more at www.keymath.com/precalc.

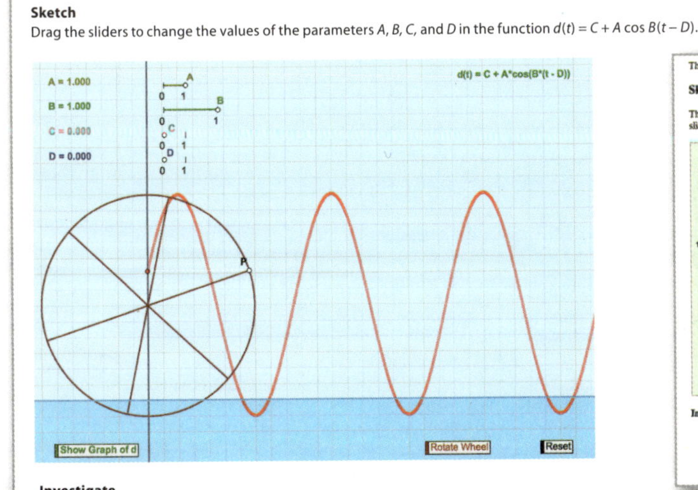

Sketch

Drag the sliders to change the values of the parameters A, B, C, and D in the function $d(t) = C + A\cos B(t - D)$.

Investigate

1. The waterwheel rotates at 6 revolutions per minute (rpm). Two seconds after you start a stopwatch, point P on the rim of the wheel is at its greatest height, $d = 13$ ft, above the surface of the water. The center of the waterwheel is 6 ft above the surface. Based on this information, write an equation expressing the height of point P as a function of time.
2. Drag the sliders for A, B, C, and D so that the function d matches the one you wrote in question 1. Press *Show Graph of d*, and then press *Rotate Wheel* to see if your solution matches the motion of the waterwheel. If not, readjust the sliders to obtain a function that matches.

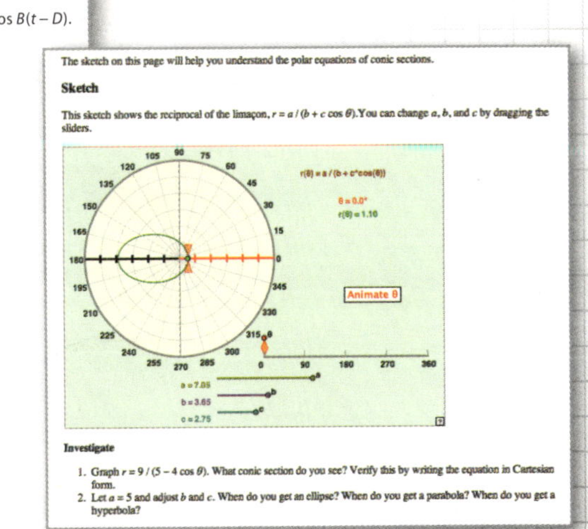

The sketch on this page will help you understand the polar equations of conic sections.

Sketch

This sketch shows the reciprocal of the limaçon, $r = a/(b + c\cos\theta)$. You can change a, b, and c by dragging the sliders.

Investigate

1. Graph $r = 9/(5 - 4\cos\theta)$. What conic section do you see? Verify this by writing the equation in Cartesian form.
2. Let $a = 5$ and adjust b and c. When do you get an ellipse? When do you get a parabola? When do you get a hyperbola?

Dynamic Precalculus Explorations help you investigate and understand a variety of concepts.

PAUL A. FOERSTER

Precalculus
with Trigonometry

CONCEPTS AND APPLICATIONS

Third Edition

Editor
Jocelyn Van Vliet

Project Administrator
Tamar Chestnut

Consulting Editors
Elizabeth DeCarli, Josephine Noah

Accuracy Checker
Cavan Fang

Production Editor
Angela Chen

Copyeditor
Mary Roybal

Editorial Production Supervisor
Kristin Ferraioli

Production Director
Christine Osborne

Text Designer
Marc Publisher Services

Cover Designer
Diana Ghermann

Cover Photo Credits
Getty Images, moodboard

Art Editor and Photo Researcher
Maya Melenchuk

Illustrators
Jon Cannell, Michael Meister, Bill Pasini, Matthew Perry, Charlene Potts, Rose Zgodzinski

Technical Art, Composition, and Prepress
Saferock USA, LLC

Printer
Webcrafters, Inc.

Textbook Product Manager
Elizabeth DeCarli

Executive Editor
Josephine Noah

Publisher
Steven Rasmussen

Key Curriculum Press
1150 65th Street
Emeryville, CA 94608
editorial@keypress.com
www.keypress.com

Printed in the United States of America
10 9 8 7 6 5 4 3 2 1 15 14 13 12 11
ISBN 978-1-60440-044-1

Reviewers & Field Testers

Reviewers, Second Edition

Tim Boykin, Williamsburg, VA
Deborah Davies, The University School of Nashville, Nashville, TN
Daniel Dudley, Verde Valley School, Sedona, AZ
Arnie Hoffman, Acalanes High School, Lafayette, CA
John Quintrell, Montgomery High School, Santa Rosa, CA
Joy Sama, Lake High School, Uniontown, OH
Barbara Uzzell, O'Conner High School, San Antonio, TX

Advisors, First Edition

Bill Medigovich, San Francisco, CA
Mary Anne Molnar, Northern Valley Regional High School, Demarest, NJ
Swapna Mukhopadhyay, San Diego State University, San Diego, CA
Wayne Nirode, Troy High School, Troy, OH
Calvin Rouse, Laney College, Oakland, CA

Field Testers, First Edition

Gary Anderson, Seattle Academy of Arts and Sciences, Seattle, WA
David Badal, Bentley Upper School, Lafayette, CA
Brenda Batten, Thomas Heyward Academy, Ridgeland, SC
Jim Bigelow, Redlands East Valley High School, Redlands, CA
Amarjit Chadda, Los Altos High School, Los Altos, CA
Michelle Clawson, Hamlin High School, Hamlin, TX
Melissa Cleavelin, St. Mary's Academy, Denver, CO
John Crotty, Career High School, New Haven, CT
Deborah Davies, The University School of Nashville, Nashville, TN
Carol DeCuzzi, Audubon Junior/Senior High School, Audubon, NY
Louise Denney, Troup High School, LaGrange, GA
Autie Doerr, Anderson High School, Austin, TX
Paul A. Foerster, Alamo Heights High School, San Antonio, TX
Kevin Fox, Millburn High School, Millburn, NJ
Bill Goldberg, Viewmont High School, Bountiful, UT
Donna Gunter, Anderson High School, Austin, TX
Teresa Ann Guthie-Sarver, Melbourne High School, Melbourne, FL
Karen Hall, Skyview High School, Vancouver, WA
Rick Kempski, Nashoba Regional High School, Bolton, MA
Ken Koppelman, Montgomery High School, Santa Rosa, CA
Stella Leland, Caney Creek High School, Conroe, TX
Dan Lotesto, Riverside University High School, Milwaukee, WI
Bill Miron, Millburn High School, Millburn, NJ
Deb Prantl, Georgetown Day School, Washington, DC
John Quintrell, Montgomery High School, Santa Rosa, CA
Louise Schumann, Redlands High School, Redlands, CA
Luis Shein, Bentley Upper School, Lafayette, CA
Gregory J. Skufca, Melbourne High School, Melbourne, FL
Chris Sollars, Alamo Heights High School, San Antonio, TX
Judy Stringham, Martin High School, Arlington, TX
Nancy Suazo, Española Valley High School, Española, NM
Rickard Swinth, Montgomery High School, Santa Rosa, CA
Susan Thomas, Alamo Heights High School, San Antonio, TX
Anne Thompson, Hawken School, Gates Mills, OH
Julie Van Horn, Manitou Springs High School, Manitou Springs, CO
Barbara Whalen, St. John's Jesuit High School, Toledo, OH
Margaret B. Wirth, J. H. Rose High School, Greenville, NC
Isabel Zsohar, Alamo Heights High School, San Antonio, TX
Richard J. Zylstra, Timothy Christian School, Elmhurst, IL

Author's Acknowledgments

This is dedicated to my students, for whom the text was written, and to my wife, Peggy, who gives me love and encouragement.

Special thanks go to Debbie Davies and her students at The University School in Nashville, Tenn., who gave consistent comments and encouragement by field testing the text in its formative stage, and who drafted and revised much content of the *Teacher's Edition*. Thanks to Debbie Preston of the Keystone School in San Antonio, Texas, for her work on the supplementary materials and for help with in-service for teachers. And thanks to Chris Harrow of The Westminster School in Atlanta, Georgia, for supplying computer-assisted algebra (CAS) materials to supplement the *Teacher's Edition*.

Thanks to other teachers and reviewers whose input helped make earlier editions of the text more useful for the education of our students. In particular, thanks to Gary Anderson, Brenda Batten, Tim Boykin, Daniel Dudley, Bill Goldberg, Karen Hall, Arnie Hoffman, Ken Koppelman, Deb Prantl, John Quintrell, Joy Sama, Louise Schumann, Judy Stringham, Rick Swinth, Anne Thompson, Barbara Uzzell, Barbara Whalen, and Richard Zylstra.

The readability of the text stems from the fact that much of the wording came directly from students as they mastered the ideas. Thanks go to my own students Nancy Carnes, Susan Curry, Brad Foster, Stacey Lawrence, Kelly Sellers, Ashley Travis, and Debbie Wisneski, whose class notes formed the basis for many sections.

Above all, thanks to my colleagues Susan Thomas, Isabel Zsohar, Mercille Wisakowsly, Mickey Bailey, and Chris Sollars, and their students at Alamo Heights High School, who supplied daily input so that the text could be complete, correct, and teachable.

Finally, thanks to the late Richard V. Andree and his wife Josephine for allowing their children, Phoebe Small and Calvin Butterball, to appear in my texts.

Paul A. Foerster

About the Author

Paul A. Foerster has enjoyed teaching mathematics at Alamo Heights High School in San Antonio, Texas, since 1961. He holds a bachelor's degree in chemical engineering, a master's degree in mathematics, and an honorary doctorate in humane letters. Between the first two degrees, he served four years in the U.S. Navy. He has authored five published mathematics textbooks based on problems he wrote for his own and other students, centered on mathematical applications in the real world. In 1983 he received the Presidential Award for Excellence in Mathematics Teaching, the first year of the award. Most recently he was one of six members on the College Board's AP Calculus Review and Alignment Commission. He raised three children with the late Jo Ann Foerster, has two stepchildren (all now grown) through his wife Peggy Foerster, and three grandchildren.

A Note to Students from the Author

Precalculus with Trigonometry is written to give you the thorough understanding of functions that you will need for future courses, including calculus. The text takes advantage of my experience teaching students like you over the past 50 years. This third edition of the text continues to use wording that my students thought up as they wrestled with concepts that are sometimes subtle or elusive. The presentation is organized to take full advantage of technology, particularly graphing calculators, while still giving you experience with pencil-and-paper and mental computations.

Drawing on your experiences in previous courses, you will extend your study of various types of functions that relate one variable quantity to another. You will encounter periodic functions that allow you to predict such things as the time of sunrise on a specific day of the year, logistic functions that model restricted population growth, and parametric and vector equations in which two or three variables all depend on another variable. Also, you will develop methods to help you decide which type of function best fits a set of scattered data. All of these experiences center around the concept that variables really vary, and do not just stand for unknown constants. In the last chapter you will be introduced to the procedure for determining the rate at which a variable varies, which is the basis of the field of calculus.

As a student, you will have the opportunity to learn mathematics four ways—algebraically, numerically, graphically, and verbally. Thus, no matter where your talents lie, you will have an opportunity to excel. For example, if you are a verbal person, you can profit by reading the text, explaining clearly the methods you use, and writing in the journal you will be asked to keep. If your talents are visual, you will have ample opportunity to learn from the shapes of graphs you will plot on the graphing calculator. The calculator will also allow you to do numerical computations, such as regression analysis, that would be too time consuming to do with just pencil and paper.

One thing to keep in mind is that mathematics is not a spectator sport. You must learn by doing, not just by watching others. You will have a chance to participate in cooperative groups, learning from your classmates as you work on explorations that introduce you to new concepts and techniques. The beginning of each problem set includes a Reading Analysis question to help you reflect upon what you read, and ten Quick Review problems to help you quickly recall topics that may need refreshing. Some problems, marked with special icons, ask you to go online to explore concepts using Dynamic Precalculus Explorations. Other problems introduce you to concepts or techniques in the following section. In addition to the sample chapter test at the end of each chapter, there are review problems that correspond to each section of the chapter. You may rehearse for a test on just those topics that have been presented, and check your answers in the back of the book. The concept problems give you a chance to apply your knowledge to new and challenging situations.

Many problems are based on real-world situations, which may serve to motivate you to learn about that particular topic. At other times you may see no immediate use for a topic. Learn the topic anyway and learn it well. Mathematics has a structure that you must discover, and the big picture may become clear only after you have unveiled its various parts. The more you understand mathematics, the more deeply you will be able to understand other subjects, such as history, law, or theology, even though it may seem like there is no obvious connection. Remember, what you know, you may never use. But what you don't know, you'll definitely never use!

In conclusion, let me wish you the best as you embark on this course. Keep a positive attitude and you will find that mastering mathematical concepts and techniques can give you a sense of accomplishment that will make the course seem worthwhile, and even fun.

Paul A. Foerster

Alamo Heights High School

San Antonio, Texas

Contents

Unit 1 Algebraic, Exponential, and Logarithmic Functions

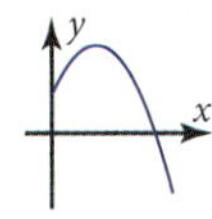

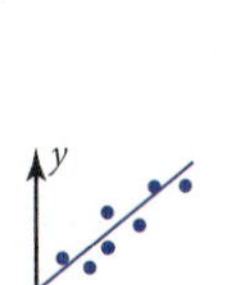

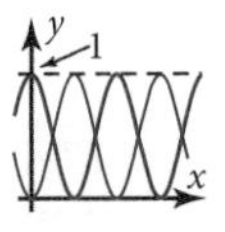

Chapter 7 Trigonometric Function Properties and Identities, and Parametric Functions **343**

Chapter 8 Properties of Combined Sinusoids **385**

Chapter 9 Triangle Trigonometry **441**

Unit 3 Analytic Geometry

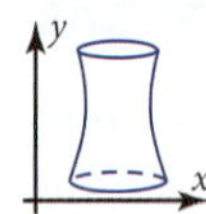

Chapter 10 Conic Sections and Quadric Surfaces 493

Chapter 11 Polar Coordinates, Complex Numbers, and Moving Objects 559

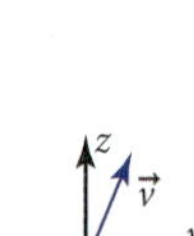

Chapter 12 Three-Dimensional Vectors 601

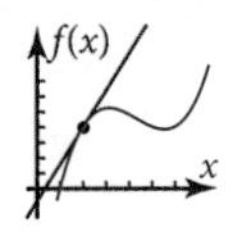
f(x)
x

Chapter 1

Functions and Mathematical Models

If you shoot an arrow into the air, its height above the ground depends on the number of seconds since you released it. In this chapter you will learn ways to express quantitatively the relationship between two variables such as height and time. You will deepen what you have learned in previous courses about functions and the particular relationships that they describe—for example, how height depends on time.

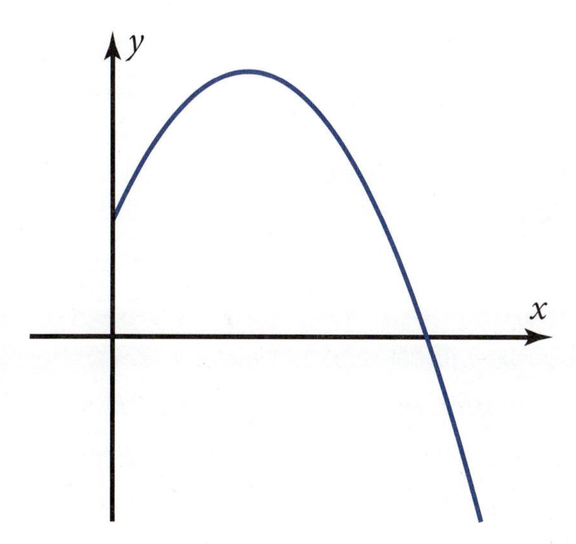

Mathematical Overview

In previous courses you have studied linear functions, quadratic functions, exponential functions, power functions, and others. In precalculus mathematics you will learn general properties that apply to all types of functions. In particular you will learn how to transform a function so that its graph fits real-world data. You will gain this knowledge in four ways.

GRAPHICALLY

The graph at right is the graph of a quadratic function. The y-variable could represent the height of an arrow at various times, x, after its release into the air. For larger time values, the quadratic function shows that y is negative. These values may or may not be reasonable in the real world.

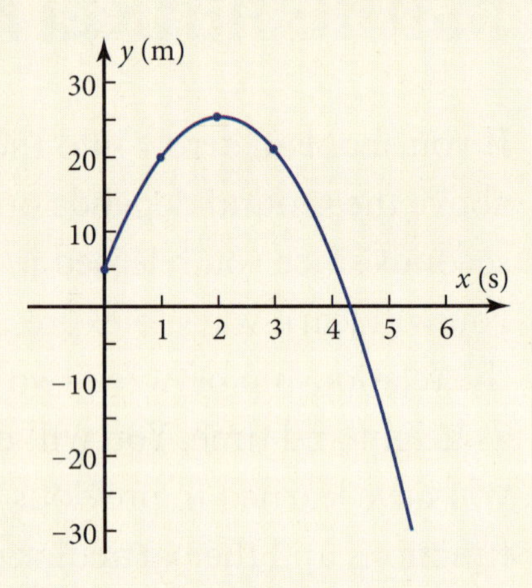

ALGEBRAICALLY

The equation of the function is

$$y = -4.9x^2 + 20x + 5$$

NUMERICALLY

This table shows corresponding x- and y-values that satisfy the equation of the function.

x (s)	y (m)
0	5.0
1	20.1
2	25.4
3	20.9

VERBALLY

When the variables in a function stand for things in the real world, the function is being used as a mathematical model. The coefficients in the equation of the function $y = -4.9x^2 + 20x + 5$ *have a real-world meaning. For example, the coefficient* -4.9 *is a constant that is a result of the gravitational acceleration, 20 is the initial velocity, and 5 reflects the initial height of the arrow.*

1-1 Functions: Graphically, Algebraically, Numerically, and Verbally

If you stack paper cups, the height increases as the number of cups increases. There is *one and only one* height for any given number of cups, so height is called a **function** of the number of cups. In this course you'll refresh your memory about some kinds of functions you have studied in previous courses. You'll also learn some new kinds of functions, and you'll learn properties of functions so that you will be comfortable with them in later calculus courses. In this section you'll see that you can study functions in four ways.

Objective

Work with functions that are defined graphically, algebraically, numerically, or verbally.

In this exploration, you'll find an equation for calculating the height of a stack of paper cups.

EXPLORATION 1-1: Paper Cup Analysis

1. Obtain several paper cups of the same kind. Measure the height of stacks containing 5, 4, 3, 2, and just 1 cup. Record the heights to the nearest 0.1 cm in a copy of this table. State what kind of cup you used.

Number	Height (cm)
1	
2	
3	
4	
5	

2. Plot the points in the table on graph paper. Show the scale you are using on the vertical axis.
3. On average, by how much did the stack height increase for each cup you added? Show how you got your answer.
4. How tall would you expect a 10-cup stack to be? Show how you get your answer. Would this be twice as tall as a 5-cup stack?
5. Let x be the number of cups in a stack, and let y be the height of the stack, measured in centimeters. Write an equation for y as a function of x.
6. What is the name of the kind of function whose equation you wrote in Problem 5?
7. Show that your equation in Problem 5 gives a height close to the measured height for a stack of 3 cups.
8. Use your equation to predict the height of a stack of 35 cups. Round the answer to 1 decimal place.
9. What are the names of the processes of calculating a value *within* the range of the data, as in Problem 7, and *outside* the range of data, as in Problem 8?
10. A cup manufacturer wants to package this kind of cup in boxes that are 45 cm long and hold one stack of cups. What is the maximum number of cups the box could hold? Show how you got your answer.
11. What did you learn as a result of doing this exploration that you did not know before?

If you pour a cup of coffee, it cools more rapidly at first, then less rapidly, finally approaching room temperature. You can show the relationship between coffee temperature and time *graphically.* Figure 1-1a shows the temperature, y, as a function of time, x. At $x = 0$, the coffee has just been poured. The graph shows that as time goes on, the temperature levels off, until it is so close to room temperature, 20°C, that you can't tell the difference.

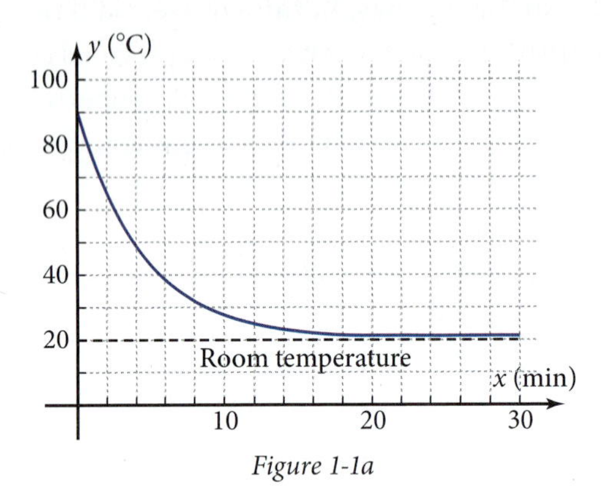

Figure 1-1a

This graph might have come from numerical data, found by experiment. It actually came from an **algebraic** equation, $y = 20 + 70(0.8)^x$.

x (min)	y (°C)
0	90.0
5	42.9
10	27.5
15	22.5
20	20.8

From the equation, you can find **numerical** information. If you enter the equation into your grapher and then use the table feature, you can find these temperatures, rounded to 0.1°C.

Functions that are used to make predictions and interpretations about something in the real world are called **mathematical models.** Temperature is the **dependent variable** because the temperature of the coffee depends on the time it has been cooling. Time is the **independent variable.** You cannot change time simply by changing coffee temperature! Always plot the independent variable on the horizontal axis and the dependent variable on the vertical axis.

The set of values the independent variable of a function can have is called the **domain.** In the coffee cup example, the domain is the set of nonnegative numbers, or $x \geq 0$. The set of values of the dependent variable corresponding to the domain is called the **range** of the function. If you don't drink the coffee (which would end the domain), the range is the set of temperatures between 20°C and 90°C, including 90°C but not 20°C, or $20 < y \leq 90$. The horizontal line at 20°C is called an **asymptote.** The word comes from the Greek *asymptotos*, meaning "not due to coincide." The graph gets arbitrarily close to the asymptote but never touches it. Figure 1-1b shows the domain, range, and asymptote.

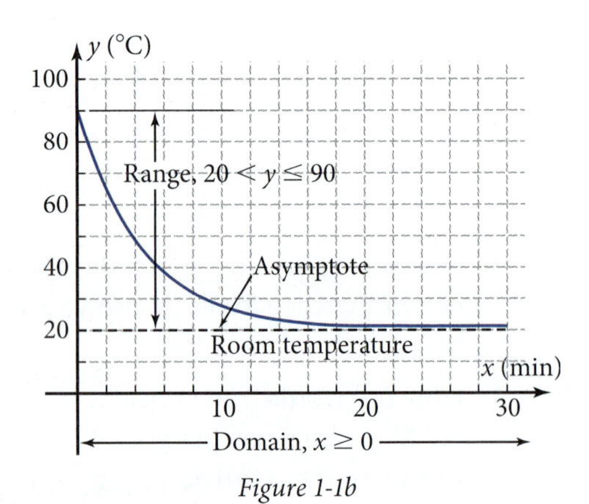

Figure 1-1b

Example 1 shows you how to describe a function *verbally.*

EXAMPLE 1 ➤ The time it takes you to get home from a football game is related to how fast you drive. Sketch a reasonable graph showing how this time and speed are related. Give the domain and range of the function.

SOLUTION It seems reasonable to assume that the time it takes *depends on* the speed you drive. So you must plot time on the *vertical* axis and speed on the *horizontal* axis.

To see what the graph should look like, consider what happens to the time as the speed varies. Pick a speed and plot a point for the corresponding time (Figure 1-1c). Then pick a faster speed. Because the time will be shorter, plot a point closer to the horizontal axis (Figure 1-1d).

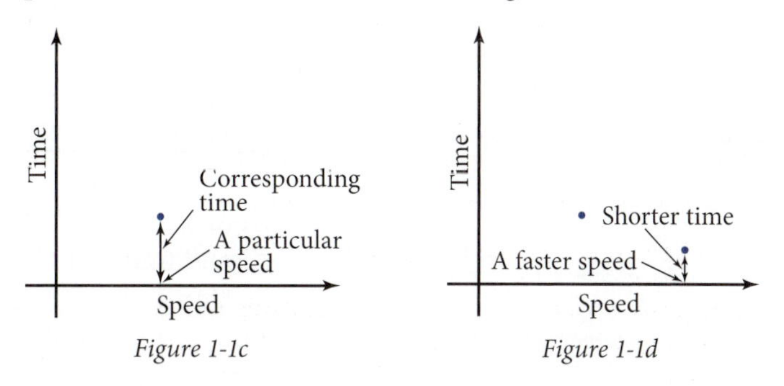

Figure 1-1c *Figure 1-1d*

For a slower speed, the time will be longer. Plot a point farther from the horizontal axis (Figure 1-1e). Finally, connect the points with a smooth curve, because it is possible to drive at any speed within the speed limit. The graph never touches either axis, as Figure 1-1f shows. If the speed were zero, you would never get home. The length of time would be infinite. Also, no matter how fast you drive, it will always take you some time to get home. You cannot arrive home instantaneously.

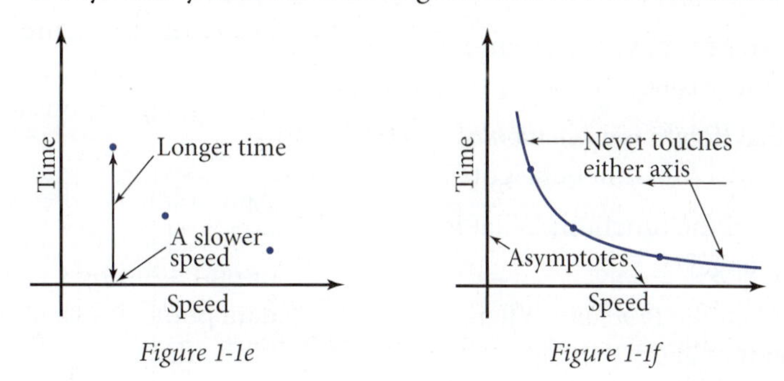

Figure 1-1e *Figure 1-1f*

Domain: $0 \leq$ speed $\leq$ speed limit

Range: time $\geq$ minimum time at speed limit ◄

The problem set will help you see the relationship between variables in the real world and functions in the mathematical world.

Exploratory Problem Set 1-1

1. *Archery Problem:* An archer climbs a tree near the edge of a cliff, then shoots an arrow high into the air. The arrow goes up, then comes back down, going over the cliff and landing in the valley, 30 m below the top of the cliff. The arrow's height, y, in meters above the top of the cliff depends on the time, x, in seconds since the archer released it. Figure 1-1g shows the height as a function of time.

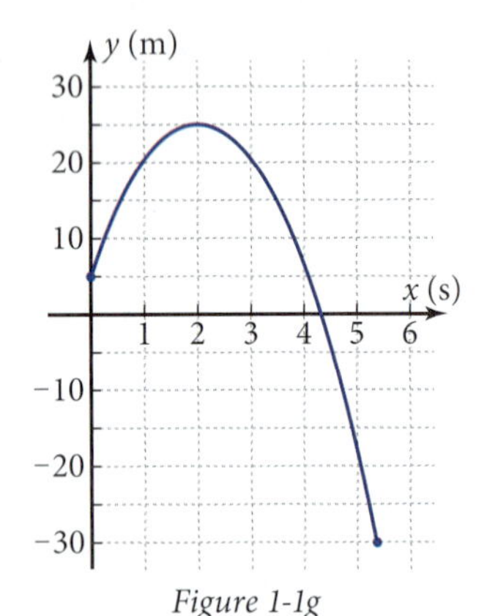

Figure 1-1g

   a. What was the approximate height of the arrow at 1 s? At 5 s? How do you explain the fact that the height is negative at 5 s?

   b. At what *two* times was the arrow 10 m above the ground? At what time does the arrow land in the valley below the cliff?

   c. How high was the archer above the ground when she released the arrow?

   d. Why can you say that height is a *function* of time? Why is time *not* a function of height?

   e. What is the domain of the function? What is the corresponding range?

2. *Gas Temperature and Volume Problem:* When you heat a fixed amount of gas, it expands, increasing its volume. In the late 1700s, French chemist Jacques Charles used numerical measurements of the temperature and volume of a gas to find a quantitative relationship between these two variables. Suppose that these temperatures and volumes had been recorded for a fixed amount of oxygen.

Jacques Charles invented the hydrogen balloon. He participated in the first manned balloon flight in 1783.

   a. On graph paper, plot V as a function of T. Choose scales that go at least from $T = -300$ to $T = 400$, and from $V = 0$ to $V = 35$. You should find that the points lie almost in a straight line. With a ruler, construct the best-fitting line you can for these points. Extend the line to the left until it crosses the T-axis and to the right to $T = 400$.

T (°C)	V (L)
0	9.5
50	11.2
100	12.9
150	14.7
200	16.4
250	18.1
300	19.9

   b. From your graph, read the approximate volumes at $T = 400$ and $T = 30$. Read the approximate temperature at which $V = 0$. How does this temperature compare with *absolute zero,* the temperature at which molecular motion stops?

   c. Finding a value of a variable *beyond* all given data points is called **extrapolation.** *Extra-* means "beyond," and *pol-* comes from "pole," or end. Finding a value *between* two given data points is called **interpolation.** Which of the three values in part b did you find by extrapolation and which by interpolation?

   d. Why can you say that volume is a *function* of temperature? Is temperature also a function of volume? Explain.

   e. Considering volume to be a function of temperature, write the domain and the range of this function.

3. *Mortgage Payment Problem:* People who buy houses usually get a loan to pay for most of the house and make payments on the resulting *mortgage* each month. Suppose you get a \$150,000 loan and pay it back at \$1,074.64 per month with an interest rate of 6% per year (0.5% per month). Your balance, B, in dollars, after n monthly payments is given by the algebraic equation

$$B = 150{,}000(1.005^n) + \frac{1074.64}{0.005}(1 - 1.005^n)$$

   a. Make a table of your balances at the end of each 12 months for the first 10 years of the mortgage. To save time, use the table feature of your grapher to do this.

   b. How many months will it take you to pay off the entire mortgage? Show how you get your answer.

   c. Plot on your grapher the graph of B as a function of n from $n = 0$ until the mortgage is paid off. Sketch the graph on your paper.

   d. True or false: "After half the payments have been made, half the original balance remains to be paid." Show that your conclusion agrees with your graph from part c.

   e. Give the domain and range of this function. Explain why the domain contains only *integers*.

4. WEB *Stopping Distance Problem:* The distance your car takes to stop depends on how fast you are going when you apply the brakes. You may recall from driver's education that it takes more than twice the distance to stop your car if you double your speed.

   a. Sketch a reasonable graph showing your stopping distance as a function of speed.

   b. What is a reasonable domain for this function?

   c. Consult a driver's manual, the Internet, or another reference source to see what the stopping distance is for the maximum speed you stated for the domain in part b.

   d. When police investigate an automobile accident, they estimate the speed the car was going by measuring the length of the skid marks. Which are they considering to be the independent variable, the speed or the length of the skid marks? Indicate how this would be done by drawing arrows on your graph from part a.

5. *Stove Heating Element Problem:* When you turn on the heating element of an electric stove, the temperature increases rapidly at first, then levels off. Sketch a reasonable graph showing temperature as a function of time. Show the horizontal asymptote. Indicate on the graph the domain and range.

6. In mathematics you learn things in four ways—algebraically, graphically, numerically, and verbally.

   a. In which of Problems 1–5 was the function given algebraically? Graphically? Numerically? Verbally?

   b. In which of Problems 1–5 did you go from verbal to graphical? From algebraic to numerical? From numerical to graphical? From graphical to algebraic? From graphical to numerical? From algebraic to graphical?

1-2 Types of Functions

In the previous section you learned that you can describe functions algebraically, numerically, graphically, or verbally. A function defined by an algebraic equation often has a descriptive name. For instance, the function $y = -x^2 + 5x + 3$ is called *quadratic*, from the latin word *quadratum,* meaning square, because the function is a polynomial whose highest power of x is x squared and *quadrangle* is one term for a square. In this section you will refresh your memory about verbal names for algebraically defined functions and see what their graphs look like.

Objective Make connections among the algebraic equation for a function, its name, and its graph.

Definition of Function

If you plot the function $y = -x^2 + 5x + 3$, you get a graph that rises and then falls, as shown in Figure 1-2a. For any x-value you pick, there is only *one* y-value. This is not the case for all graphs. For example, in Figure 1-2b there are places where the graph has more than one y-value for the same x-value. Although the two variables are related, the relation is not a function.

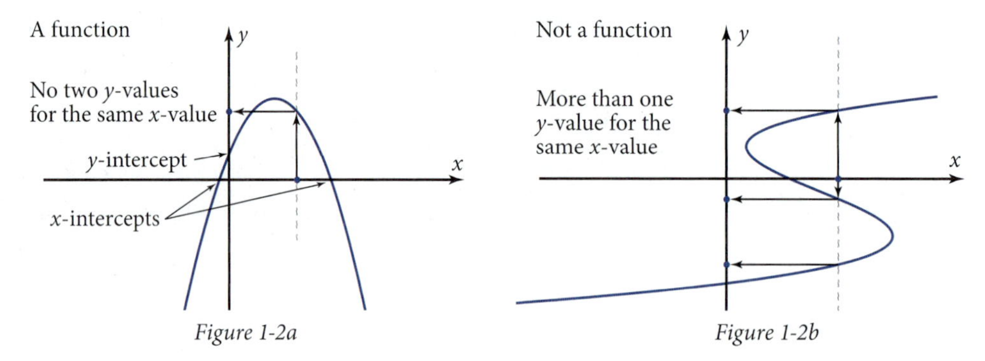

Figure 1-2a *Figure 1-2b*

Each point on a graph corresponds to an *ordered pair* of numbers, (x, y). A **relation** is any set of ordered pairs. A **function** is a set of ordered pairs for which each value of the independent variable (often x) in the domain has only *one* corresponding value of the dependent variable (often y) in the range. So Figures 1-2a and 1-2b are both graphs of relations, but only Figure 1-2a is the graph of a function.

The ***y*-intercept** of a function is the value of y when $x = 0$. It gives the place where the graph crosses the y-axis (Figure 1-2a). An ***x*-intercept** is a value of x for which $y = 0$. Functions can have more than one x-intercept.

f(*x*) Terminology

You should recall $f(x)$ notation from previous courses. It is used for y, the dependent variable of a function. With it, you show what value you substitute for x, the independent variable. For instance, to substitute 4 for x in the quadratic function $f(x) = -x^2 + 5x + 3$, you would write

$$f(4) = -4^2 + 5(4) + 3 = 7$$

The symbol $f(4)$ is pronounced "f of 4" or sometimes "f at 4." You must recognize that the parentheses mean substitution and not multiplication.

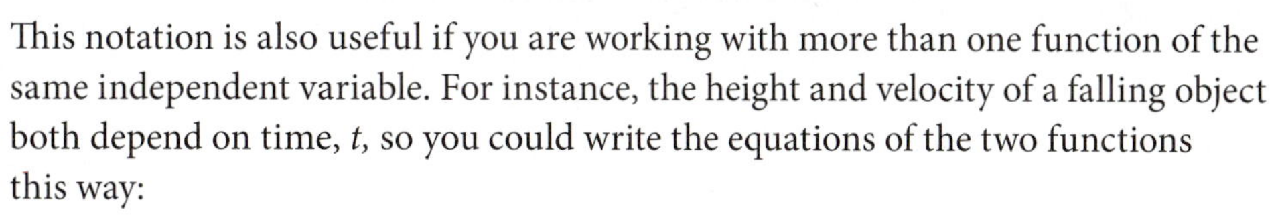

This notation is also useful if you are working with more than one function of the same independent variable. For instance, the height and velocity of a falling object both depend on time, t, so you could write the equations of the two functions this way:

$$h(t) = -4.9t^2 + 10t + 70 \quad \text{(for the height)}$$

$$v(t) = -9.8t + 10 \quad \text{(for the velocity)}$$

In $f(x)$, the variable x or any value substituted for x is called the **argument** of the function. It is important to distinguish between f and $f(x)$. The symbol f is the *name* of the function. The symbol $f(x)$ is the y-value of the function. For instance, if f is the square root function, then $f(x) = \sqrt{x}$ and $f(9) = \sqrt{9} = 3$. Note that the reflexive axiom, $x = x$, requires that you substitute the same number for x everywhere it appears in an expression or equation. It would be improper format to write $f(x) = \sqrt{9}$ if you have substituted 9 for x.

Names of Functions

Functions are named for the operation performed on the independent variable. Here are some types of functions you may recall from previous courses, along with their typical graphs. In these examples, the letters a, b, c, m, and n stand for **constants.** The symbols x and $f(x)$ stand for **variables,** x for the independent variable and $f(x)$ for the dependent variable.

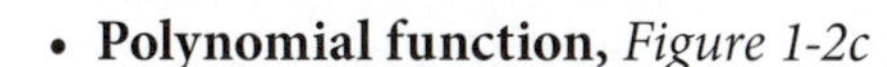

- **Polynomial function,** *Figure 1-2c*

  General equation: $f(x) = a_n x^n + a_{n-1}x^{n-1} + \ldots + a_1 x + a_0$, where n is a nonnegative integer

  Verbally: $f(x)$ is a polynomial function of x. (If $n = 3$, f is a cubic function. If $n = 4$, f is a quartic function.)

  Features: The graph crosses the x-axis up to n times and has up to $n - 1$ vertices (points where the function changes direction). The domain is all real numbers.

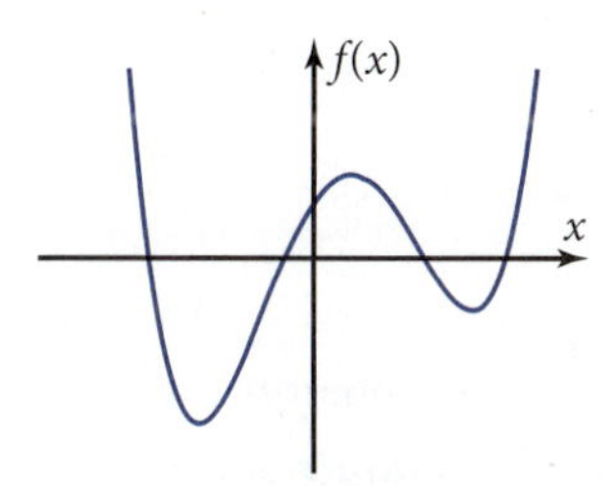

Figure 1-2c

- **Quadratic function,** *Figure 1-2d* (a special case of a polynomial function)

  General equation: $f(x) = ax^2 + bx + c,\ a \neq 0$

  Verbally: $f(x)$ varies quadratically with x, or $f(x)$ is a quadratic function of x.

  Features: The graph changes direction at its one vertex. The domain is all real numbers.

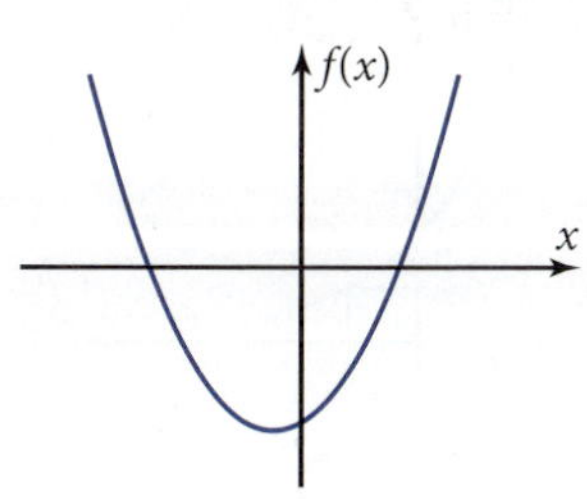

Figure 1-2d

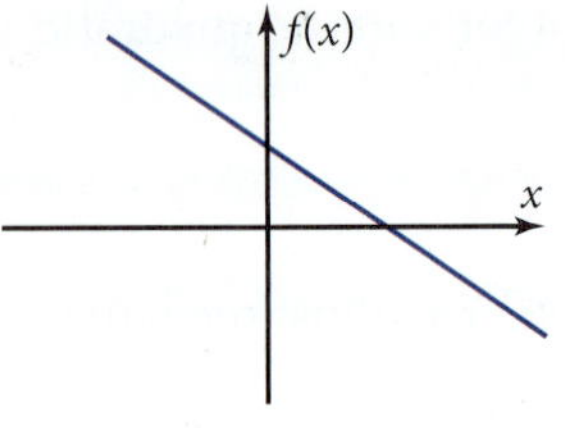

Figure 1-2e

- **Linear function,** *Figure 1-2e* (another special case of a polynomial function)

  General equation: $f(x) = ax + b$ (or $f(x) = mx + b$)

  Verbally: $f(x)$ varies linearly with x, or $f(x)$ is a linear function of x.

  Features: The straight-line graph, $f(x)$, changes at a constant rate as x changes. The domain is all real numbers.

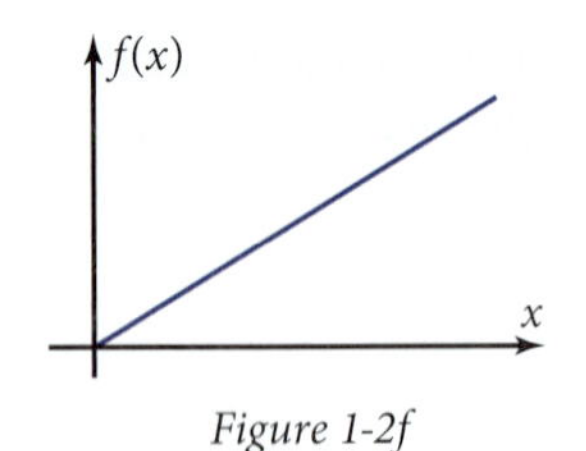

Figure 1-2f

- **Direct variation function,** *Figure 1-2f* (a special case of a linear, power, or polynomial function)

  General equation: $f(x) = ax$ (or $f(x) = mx + 0$, or $f(x) = ax^1$)

  Verbally: $f(x)$ varies directly with x, or $f(x)$ is directly proportional to x.

  Features: The straight-line graph goes through the origin. The domain is all real numbers. However, for most real-world applications, you will use the domain $x \geq 0$ (as shown).

- **Power function,** *Figure 1-2g* (a polynomial function if b is a nonnegative integer)

  General equation: $f(x) = ax^b$ (a *variable* with a *constant* exponent), $a \neq 0$, $b \neq 0$

  Verbally: $f(x)$ varies directly with the bth power of x, or $f(x)$ is directly proportional to the bth power of x.

  Features: The domain depends on the value of b. For positive integer values of b, the domain is all real numbers; for negative integer values of b, the domain is $x \neq 0$. In most real-world applications, the domain is $x \geq 0$ if $b > 0$ and $x > 0$ if $b < 0$.

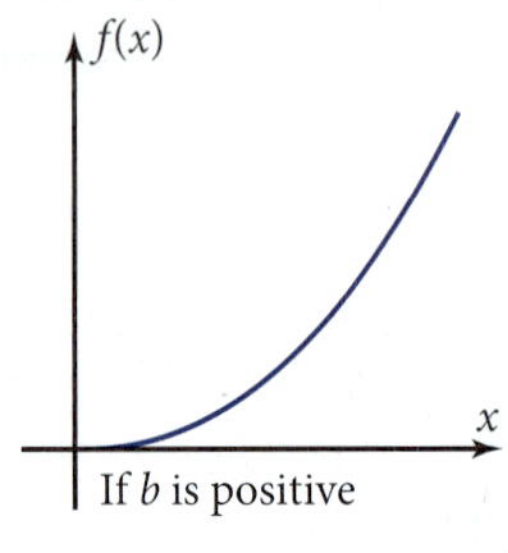

f(x)
x
If b is negative

Figure 1-2g

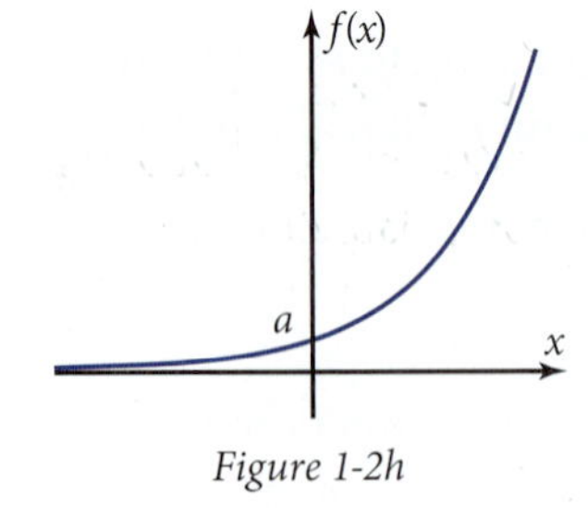

Figure 1-2h

- **Exponential function,** *Figure 1-2h*

  General equation: $f(x) = a \cdot b^x$ (a *constant* with a *variable* exponent), $a \neq 0$, $b > 0$, $b \neq 1$

  Verbally: $f(x)$ varies exponentially with x, or $f(x)$ is an exponential function of x.

  Features: The graph crosses the y-axis at $f(0) = a$ and has the x-axis as an asymptote.

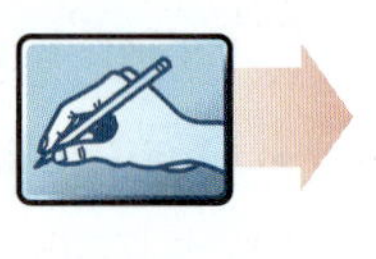

How about if $0 < b < 1$? For example, $y = \left(\frac{1}{2}\right)^x$

y
1
0.5
1
x

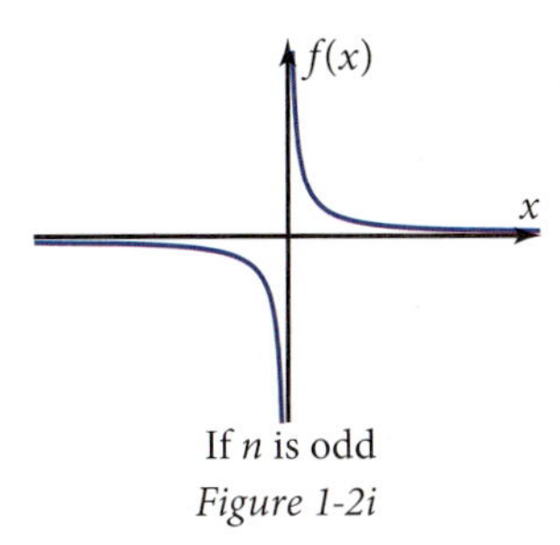

Figure 1-2i

- **Inverse variation function,** *Figure 1-2i*
  (a special case of a power function)

  General equation: $f(x) = \frac{a}{x}$ (or $f(x) = ax^{-1}$)

  $$\text{or } f(x) = \frac{a}{x^n} \text{ (or } f(x) = ax^{-n}\text{)}, a \neq 0, n > 0$$

  Verbally: $f(x)$ varies inversely with x (or with the nth power of x). Alternatively, $f(x)$ is inversely proportional to x (or to the nth power of x).

  Features: Both of the axes are asymptotes. The domain depends on the value of n. For positive integer values of n, the domain is $x \neq 0$. For most real-world applications, the domain is $x > 0$.

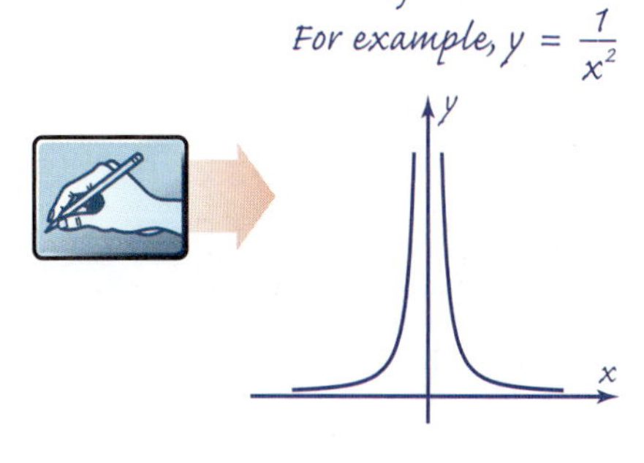

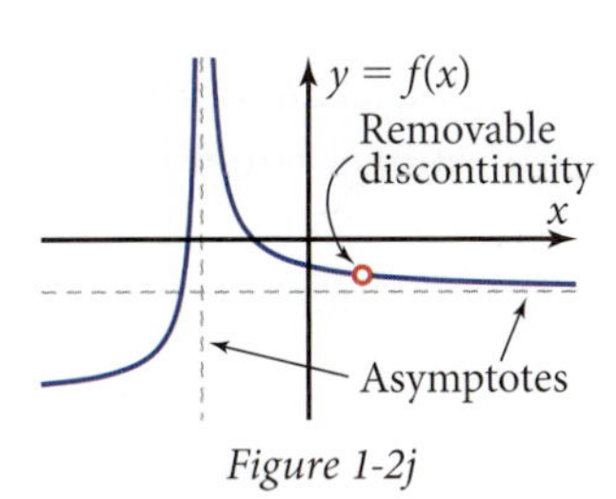

Figure 1-2j

- **Rational algebraic function,** *Figure 1-2j*

  General equation: $f(x) = \frac{n(x)}{d(x)}$, where n and d are polynomial functions

  Verbally: $f(x)$ is a rational function of x.

  Features: A rational function has a discontinuity (asymptote or missing point) where the denominator is zero; it may have horizontal or other asymptotes.

Restricted Domains and Boolean Variables

Suppose that you want to plot a graph using only part of your grapher's window. For instance, let the height of a growing child between ages 3 and 10 be given by $y = 3x + 26$, where x is age in years and y is height in inches. The domain here is $3 \leq x \leq 10$.

Some graphers allow you to enter a restricted domain directly. Other graphers require you to use **Boolean variables.** A Boolean variable, named for George Boole, an Irish logician and mathematician (1815–1864), equals 1 if a given condition is true and 0 if that condition is false. For instance, the compound statement

$$(x \geq 3 \text{ and } x \leq 10)$$

equals 1 if $x = 7$ (which *is* between 3 and 10) and equals 0 if $x = 2$ or $x = 15$ (neither of which is between 3 and 10). To plot a graph in a restricted domain using Boolean variables, divide any term of the equation by the appropriate Boolean variable. For the equation above, enter

$$f_1(x) = 3x + 26 \,/\, (x \geq 3 \text{ and } x \leq 10)$$

If x is between 3 and 10, inclusive, the 26 in $3x + 26$ is divided by 1, which leaves it unchanged. If x is not between 3 and 10, inclusive, the 26 in $3x + 26$ is divided by 0 and the grapher plots nothing.

EXAMPLE 1 ➤ Plot the graph of $f(x) = -x^2 + 5x + 3$ in the domain $0 \le x \le 4$. What kind of function is this? Give the range. Find a pair of real-world variables that could have a relationship described by a graph of this shape.

SOLUTION Enter the equation with restricted domain into your grapher directly. Or, to use Boolean variables, enter

$$f_1(x) = -x^2 + 5x + 3\ /\ (x \ge 0 \text{ and } x \le 4)$$

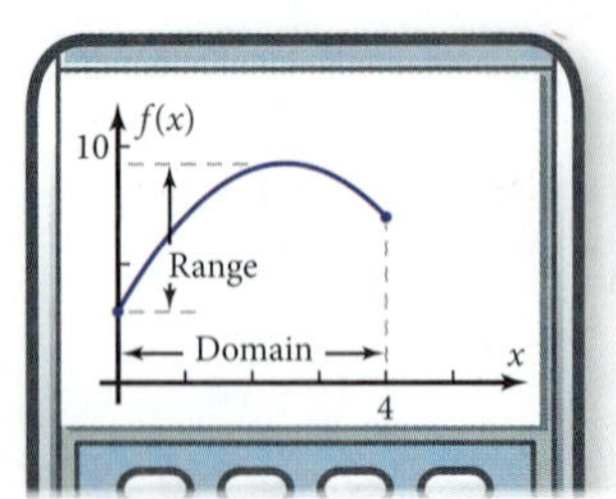

Figure 1-2k

The graph in Figure 1-2k shows the restricted domain.

The function is quadratic because $f(x)$ equals a second-degree polynomial in x.

The range is $3 \le f(x) \le 9.25$. You can find this interval by tracing to the left endpoint of the graph where $f(0) = 3$ and to the high point where $f(2.5) = 9.25$. (At the right endpoint, $f(4) = 7$, which is between 3 and 9.25.)

The function could represent the relationship between something that rises for a while and then falls, such as a punted football's height as a function of time or (if $f(x)$ is multiplied by 10) the grade you could get on a test as a function of the number of hours you study for it. (The grade could be lower for longer times if you stay up too late and thus are sleepy during the test.) ◄

DEFINITION: Boolean Variables

A **Boolean variable** is a variable that has a given condition attached to it. If the condition is true, the variable equals 1. If the condition is false, the variable equals 0.

EXAMPLE 2 ➤ As children grow older, their height and weight are related. Sketch a reasonable graph to show this relation and then describe it. Identify what kind of function has a graph like the one you drew.

SOLUTION Weight depends on height, so weight is on the vertical axis, as shown on the graph in Figure 1-2l. The graph curves upward because doubling the height more than doubles the weight. Extending the graph sends it through the origin, but the domain starts beyond the origin at a value greater than zero, because a person never has zero height or weight. The graph stops at the person's adult height and weight. A power function has a graph like this. ◄

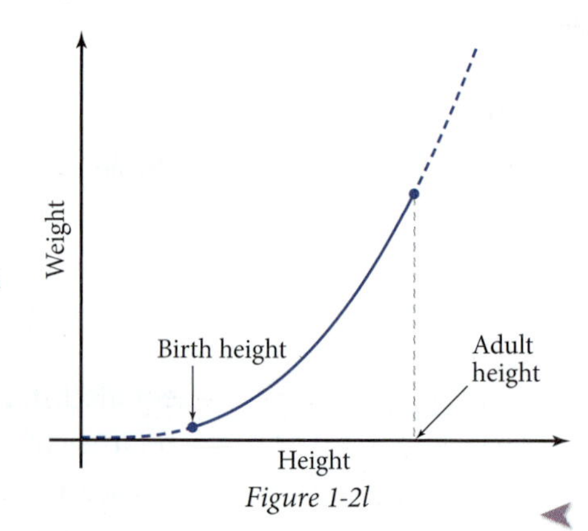

Figure 1-2l

Problem Set 1-2

For Problems 1–4,

a. Plot the graph on your grapher using the given domain. Sketch the result on your paper.

b. Give the range of the function.

c. Name the kind of function.

d. Describe a pair of real-world variables that could be related by a graph of this shape.

1. $f(x) = 2x + 3$ domain: $0 \le x \le 10$

2. $f(x) = 0.2x^3$ domain: $0 \le x \le 4$

3. $g(x) = \frac{12}{x}$ domain: $0 < x \le 10$

4. $h(x) = 5 \cdot 0.6^x$ domain: $-5 \le x \le 5$

For Problems 5–18,

a. Plot the graph using a window set to show the entire graph, when possible. Sketch the result.

b. Give the y-intercept and any x-intercepts and the locations of any asymptotes.

c. Give the range.

5. Quadratic (polynomial) function $f(x) = -x^2 + 4x + 12$ with the domain $0 \le x \le 5$

6. Quadratic (polynomial) function $f(x) = x^2 - 6x + 40$ with the domain $0 \le x \le 8$

7. Cubic (polynomial) function $f(x) = x^3 - 7x^2 + 4x + 12$ with the domain $-1 \le x \le 7$

8. Quartic (polynomial) function $f(x) = x^4 + 3x^3 - 8x^2 - 12x + 16$ with the domain $-3 \le x \le 3$

9. Power function $f(x) = 3x^{2/3}$ with the domain $0 \le x \le 8$

10. Power function $f(x) = 0.3x^{1.5}$ with the domain $0 \le x \le 9$

11. Linear function $f(x) = -0.7x + 4$ with the domain $-3 \le x \le 10$

12. Linear function $f(x) = 3x + 6$ with the domain $-5 \le x \le 5$

13. Exponential function $f(x) = 3 \cdot 1.3^x$ with the domain $-5 \le x \le 5$

14. Exponential function $f(x) = 20 \cdot 0.7^x$ with the domain $-5 \le x \le 5$

15. Inverse-square variation function $f(x) = \frac{25}{x^2}$ with the domain $x > 0$

16. Direct variation function $f(x) = 5x$ with the domain $x \ge 0$

17. Rational function $y = \frac{x - 2}{(x - 4)(x + 1)}$ with the domain $-3 \le x \le 6$, $x \ne 4$, $x \ne -1$

18. Rational function $y = \frac{x^2 - 2x - 2}{x - 3}$ with the domain $-2 \le x \le 6$, $x \ne 3$

For Problems 19–28, name the type of function that has the graph shown.

19.

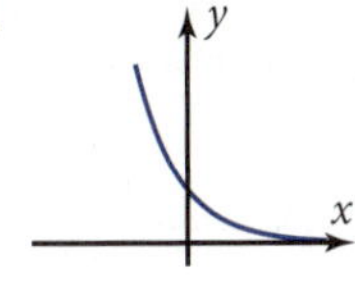

20.

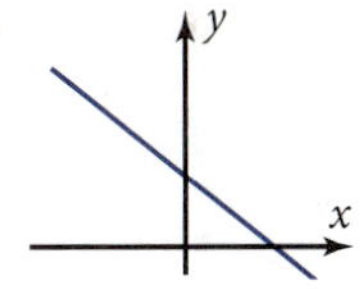

21.

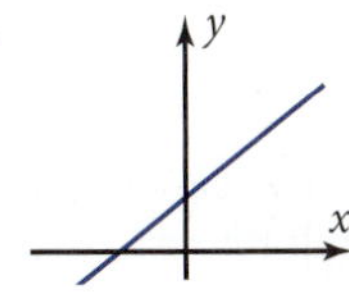

22.

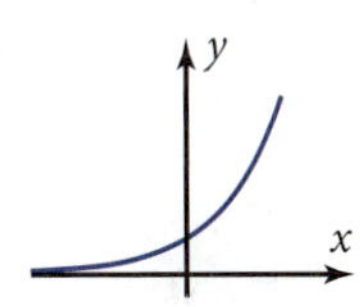

23.

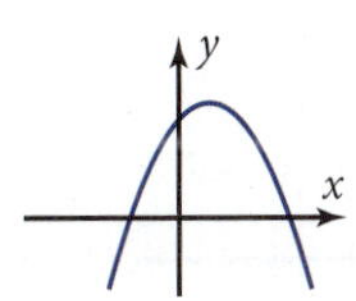

24.

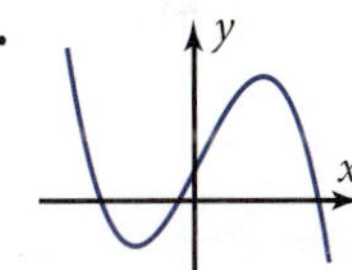

25.

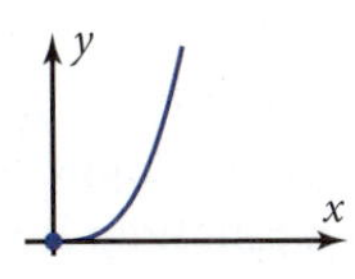

26.

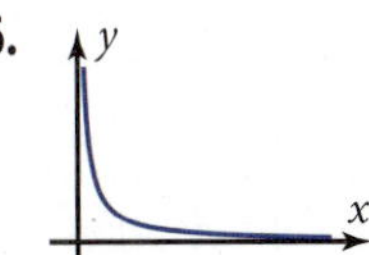

27.

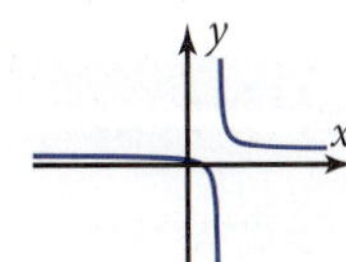

28.

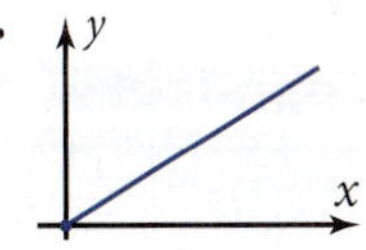

For Problems 29–32,

a. Sketch a reasonable graph showing how the variables are related.

b. Identify the type of function it could be (quadratic, power, exponential, and so on).

29. The weight and length of a dog.

30. The temperature of a cup of coffee and the time since the coffee was poured.

31. The purchase price of a house in a particular neighborhood as a function of the number of square feet of floor space in the house, including a fixed amount for the lot on which the house was built.

32. The height of a punted football as a function of the number of seconds since it was kicked.

For Problems 33–38, tell whether the relation graphed is a function. Explain how you made your decision.

33.

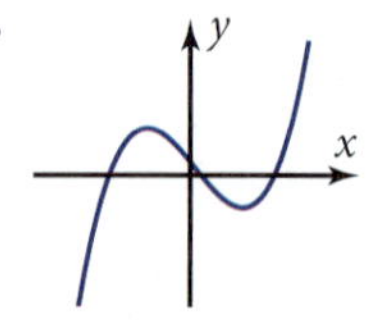

34.

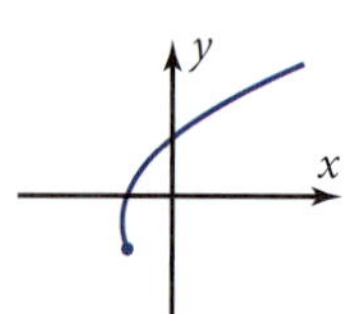

35.

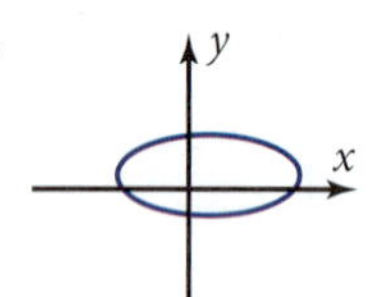

36.

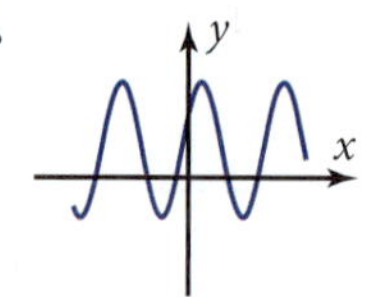

37.

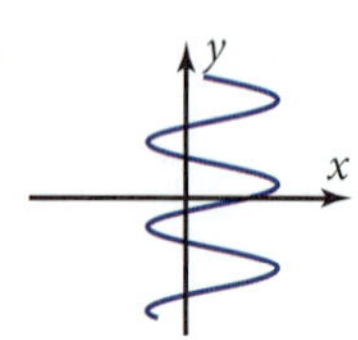

38.

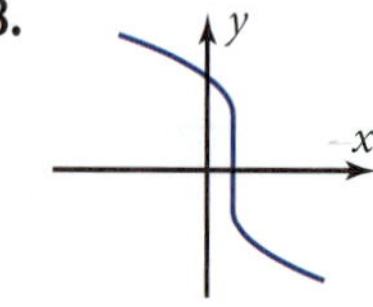

39. *Vertical Line Test Problem:* There is a graphical way to tell whether a relation is a function. It is called the **vertical line test.**

> **PROPERTY: The Vertical Line Test**
>
> If any vertical line cuts the graph of a relation in more than one place, then the relation is not a function.

Figure 1-2m illustrates the test.

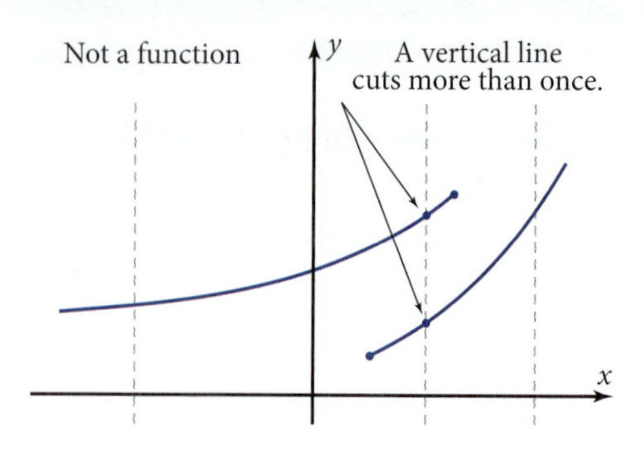

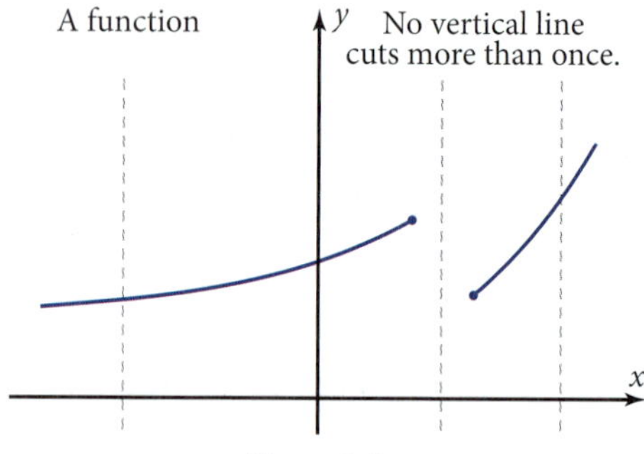

Figure 1-2m

a. Based on the definition of function, explain how the vertical line test distinguishes between relations that are functions and relations that are not functions.

b. Sketch the graphs in Problems 33 and 35. On your sketch, show how the vertical line test tells you that the relation in Problem 33 is a function but the relation in Problem 35 is not.

40. Explain why a function can have more than one x-intercept but only one y-intercept.

41. What is the argument of the function $y = f(x - 2)$?

42. WEB *Research Problem:* Look up George Boole on the Internet or in another reference source. Describe several of Boole's accomplishments that you discover. Include your source.

1-3 Dilation and Translation of Function Graphs

Each of these two graphs shows the unit semicircle and a **transformation** of it. The left graph shows the semicircle *dilated* (magnified) by a factor of 5 in the x-direction and by a factor of 3 in the y-direction. The right graph shows the unit semicircle *translated* by 4 units in the x-direction and by 2 units in the y-direction.

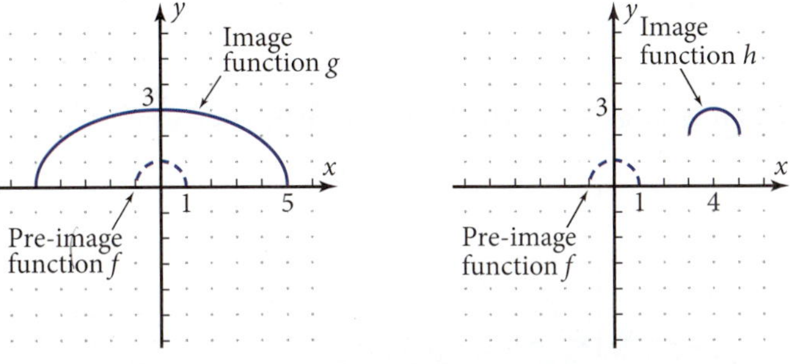

Figure 1-3a

The transformed functions, g and h, in Figure 1-3a are called *images* of the function f. The original function, f, is called the *pre-image*. In this section you will learn how to transform the equation of a function so that its graph will be dilated and translated by given amounts in the x- and y-directions.

Objective

Transform a given pre-image function so that the result is a graph of the image function that has been dilated by given factors and translated by given amounts.

Dilations

To get the vertical dilation in the left graph of Figure 1-3a, multiply each y-coordinate by 3. Figure 1-3b shows the image, $y = 3f(x)$.

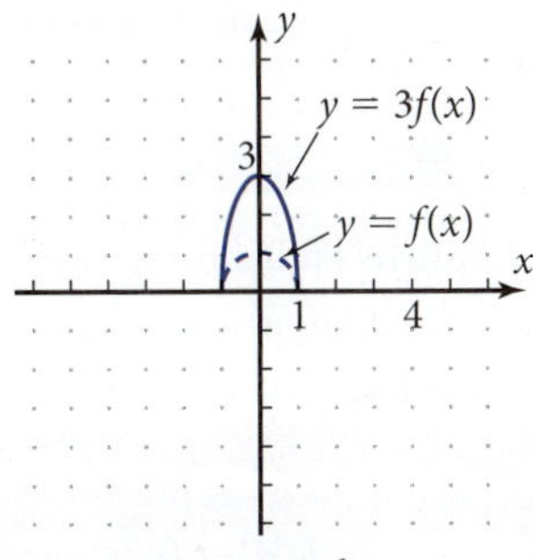

Figure 1-3b

The horizontal dilation is trickier. Each value of the argument must be 5 times what it was in the pre-image to generate the same y-values. Substitute v for the argument of f.

$y = f(v)$ Let v represent the original x-values.

$x = 5v$ The x-values of the dilated image must be 5 times the x-values of the pre-image.

$\frac{1}{5}x = v$ Solve for v.

$y = f\left(\frac{1}{5}x\right)$ Replace v with $\frac{1}{5}x$ for the argument to obtain the equation of the dilated image.

Figure 1-3c shows the graph of the image, $y = f\left(\frac{1}{5}x\right)$

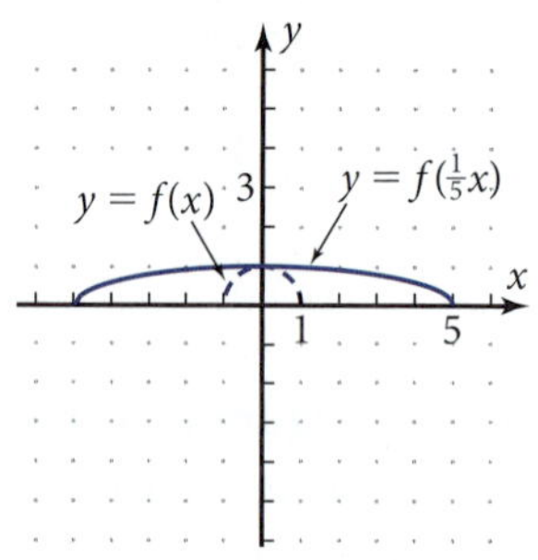

Figure 1-3c

Putting the two transformations together gives the equation for $g(x)$ shown in Figure 1-3a.

$$g(x) = 3f\left(\frac{1}{5}x\right)$$

EXAMPLE 1 ➤ The equation of the pre-image function in Figure 1-3a is $f(x) = \sqrt{1 - x^2}$. Confirm on your grapher that $g(x) = 3f\left(\frac{1}{5}x\right)$ is the transformed image function

a. By direct substitution into the equation

b. By using the grapher's built-in variables feature

SOLUTION

a. $g(x) = 3\sqrt{1 - (x/5)^2}$ Substitute $x/5$ as the argument of f, and multiply the entire expression by 3.

Enter: $f_1(x) = \sqrt{1 - x^2}$

$f_2(x) = 3\sqrt{1 - (x/5)^2}$

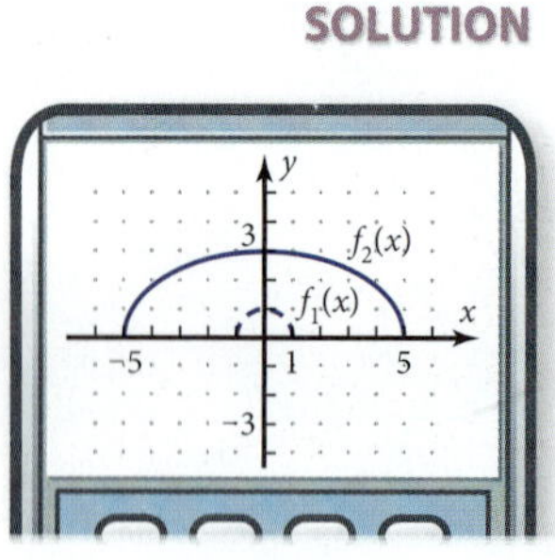

Figure 1-3d

The graph in Figure 1-3d shows a dilation by 5 in the x-direction and by 3 in the y-direction. Use the grid-on feature to make the grid points appear. Use equal scales on the two axes so the graphs have the correct proportions.

b. 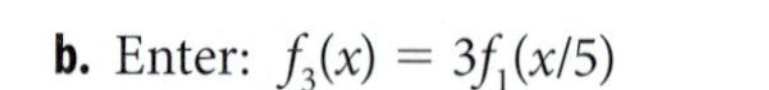
Enter: $f_3(x) = 3f_1(x/5)$ f_1 is the function *name* in this format, not the function value.

This graph is the same as the graph of $f_2(x)$ in Figure 1-3d. ◄

Note that the transformation in $f\left(\frac{1}{5}x\right)$ is applied to the *argument* of function f, *inside* the parentheses. The transformation in $3f(x)$ is applied *outside* the parentheses, to the *value* of the function. For this reason the transformations are given the names **inside transformation** and **outside transformation,** respectively. An inside transformation affects the graph in the horizontal direction, and an outside transformation affects the graph in the vertical direction.

You may ask, "Why do you *multiply* by the y-dilation and *divide* by the x-dilation?" You can see the reason by substituting y for $g(x)$ and dividing both sides of the equation by 3:

$$y = 3f\left(\frac{1}{5}x\right)$$

$$\frac{1}{3}y = f\left(\frac{1}{5}x\right) \qquad \text{Divide both sides by 3 } \left(\text{or multiply by } \frac{1}{3}\right).$$

You actually divide by *both* dilation factors, y by the y-dilation and x by the x-dilation.

Translations

The translations in Figure 1-3a that transform $f(x)$ to $h(x)$ are shown again in Figure 1-3e. To figure out what translation has been done, ask yourself, "To where did the point at the origin move?" As you can see, the center of the semicircle, initially at the origin, has moved to the point (4, 2). So there is a horizontal translation of 4 units and a vertical translation of 2 units.

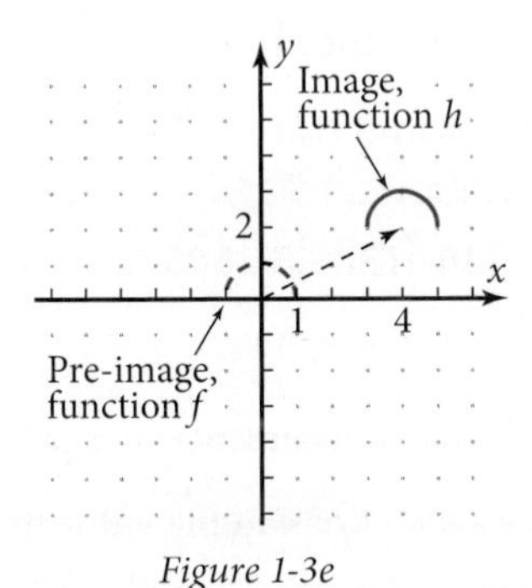

Figure 1-3e

To get a vertical translation of 2 units, add 2 to each y-value:

$$y = 2 + f(x)$$

To get a horizontal translation of 4 units, note that what was happening at $x = 0$ in function f has to be happening at $x = 4$ in function h. Again, substituting v for the argument of f gives

$$h(x) = 2 + f(v)$$

$$x = v + 4$$

$$x - 4 = v$$

$$h(x) = 2 + f(x - 4) \qquad \text{Substitute } x - 4 \text{ as the argument of } f.$$

EXAMPLE 2 ➤ The equation of the pre-image function in Figure 1-3e is $f(x) = \sqrt{1 - x^2}$. Confirm on your grapher that $g(x) = 2 + f(x - 4)$ is the transformed image function by

a. Direct substitution into the equation

b. Using the grapher's built-in variables feature

SOLUTION

a. $g(x) = 2 + \sqrt{1 - (x - 4)^2}$ Substitute $x - 4$ for the argument. Add 2 to the expression.

Enter: $f_1(x) = \sqrt{1 - x^2}$

$f_2(x) = 2 + \sqrt{1 - (x - 4)^2}$

The graph in Figure 1-3f shows an x-translation of 4 units and a y-translation of 2 units.

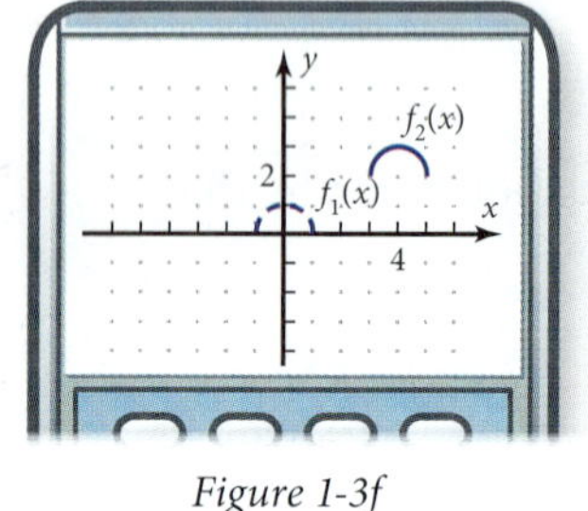

Figure 1-3f

b. Enter: $f_3(x) = 2 + f_1(x - 4)$

The graph is the same as that for $f_2(x)$ in Figure 1-3f. ◄

Again you may ask, "Why do you subtract an x-translation and add a y-translation?" The answer again lies in associating the y-translation with the y-variable. You actually subtract *both* translations:

$y = 2 + f(x - 4)$

$y - 2 = f(x - 4)$ Subtract 2 from both sides.

The reason for writing the transformed equation with y by itself is to make it easier to calculate the dependent variable, either by pencil and paper or on your grapher.

This box summarizes the dilations and translations of a function and its graph.

PROPERTY: Dilations and Translations

The function g given by

$$\frac{1}{a} \cdot g(x) = f\left(\frac{1}{b}x\right) \quad \text{or, equivalently,} \quad g(x) = a \cdot f\left(\frac{1}{b}x\right)$$

represents a **dilation** by a factor of a in the y-direction and by a factor of b in the x-direction.

The function h given by

$$h(x) - c = f(x - d) \quad \text{or, equivalently,} \quad h(x) = c + f(x - d)$$

represents a **translation** by c units in the y-direction and by d units in the x-direction.

Note: If the function is only dilated, the x-dilation is the number you can substitute for x to make the argument equal 1. If the function is only translated, the x-translation is the number to substitute for x to make the argument equal zero.

EXAMPLE 3 ➤ The three graphs in Figure 1-3g show three different transformations of the pre-image graph to image graphs $y = g(x)$. Explain verbally what transformations were done. Write an equation for $g(x)$ in terms of the function f.

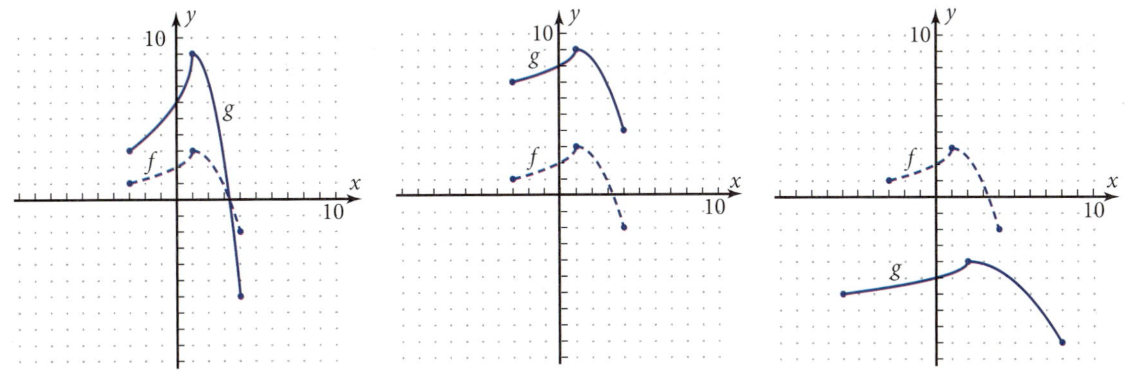

Figure 1-3g

SOLUTION Left graph: Vertical dilation by a factor of 3

Equation: $g(x) = 3f(x)$

Note: Each point on the graph of g is 3 times as far from the x-axis as the corresponding point on the graph of f. Note that the vertical dilation moved points above the x-axis farther up and moved points below the x-axis farther down.

Middle graph: Vertical translation by 6 units

Equation: $g(x) = 6 + f(x)$

Note: The vertical dilation moved all points on the graph of f up by the same amount, 6 units. Also note that the fact that $g(1)$ is three times $f(1)$ is purely coincidental and is not true at other values of x.

Right graph: Horizontal dilation by a factor of 2 and vertical translation by -7 units

Equation: $g(x) = -7 + f\left(\frac{1}{2}x\right)$

Note: Each point on the graph of g is twice as far from the y-axis as the corresponding point on the graph of f. The horizontal dilation moved points to the right of the y-axis farther to the right and moved points to the left of the y-axis farther to the left. ◄

In this exploration, given a pre-image and an image graph, you'll identify the transformation.

EXPLORATION 1-3: Transformations from Graphs

For Problems 1–6, identify the transformation of f (dotted) to get g (solid). Describe the transformation verbally, and give an equation for $g(x)$.

1.

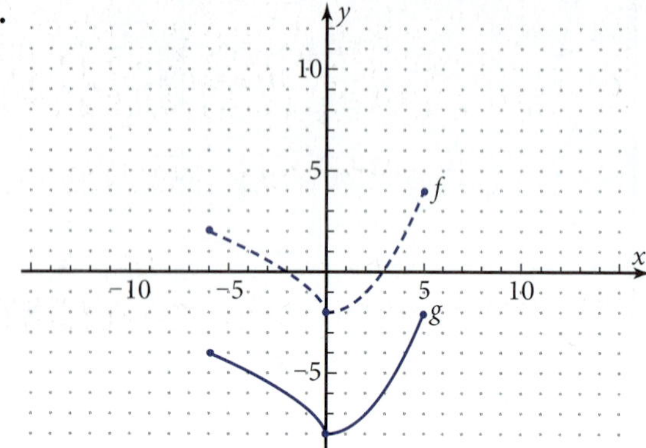

2.

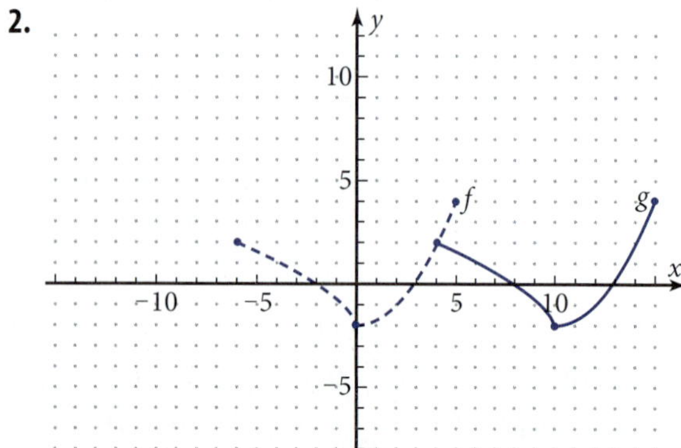

3.

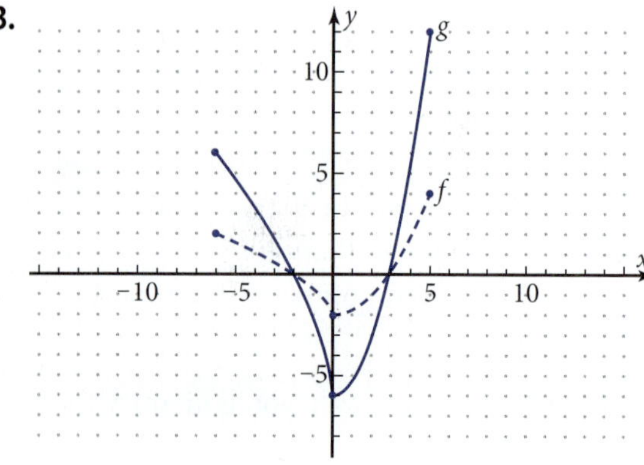

4.

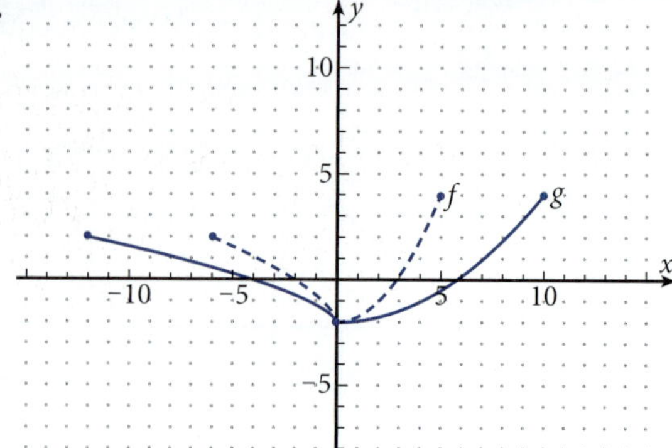

5.

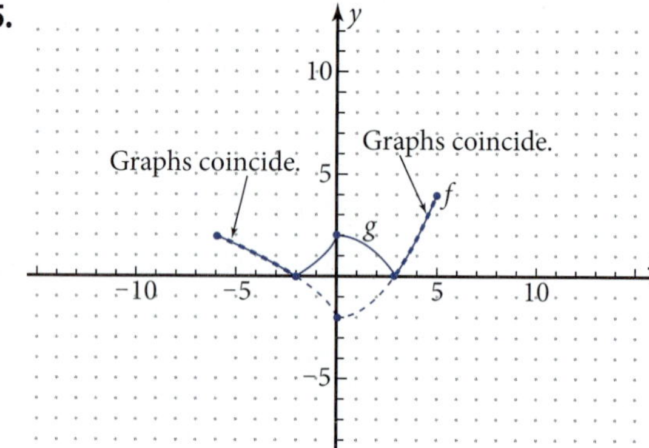

6.

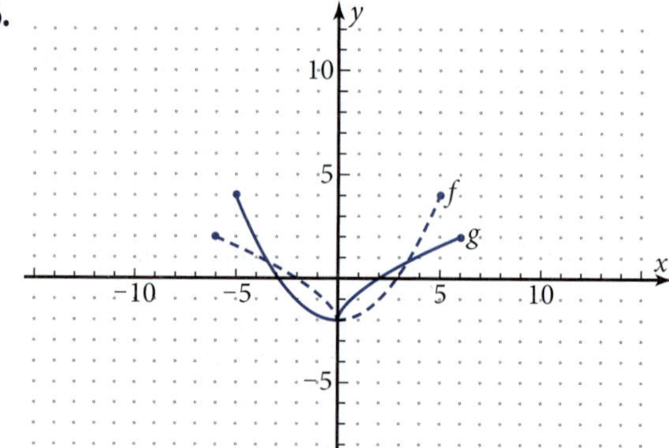

7. What did you learn as a result of doing this exploration that you did not know before?

Problem Set 1-3

Reading Analysis

From now on most problem sets will begin with an assignment that requires you to spend ten minutes or so reading the section and answering some questions to see how well you understand what you have read. This will help develop your ability to read a textbook, a very important skill to have in college. It should also make working the problems in the problem set easier.

From what you have read in this section, what do you consider to be the main idea? What is the major difference on the image graph between a translation and a dilation, and what operation causes each transformation? How can you tell whether a translation or a dilation will be in the x-direction or the y-direction?

Quick Review

From now on there will be ten short problems at the beginning of most problem sets. Some of the problems are intended for review of skills from previous sections or chapters. Others are intended to test your general knowledge. Speed is the key here, not detailed work. Try to do all ten problems in less than five minutes.

Q1. $y = 3x^2 + 5x - 7$ is a particular example of a ___?___ function.

Q2. Write the general equation of a power function.

Q3. Write the general equation of an exponential function.

Q4. Calculate the product: $(x - 7)(x + 8)$

Q5. Expand: $(3x - 5)^2$

Q6. Sketch the graph of a relation that is not a function.

Q7. Sketch the graph of $y = \frac{2}{3}x + 4$.

Q8. Sketch an isosceles triangle.

Q9. Find 30% of 3000.

Q10. Which one of these is not the equation of a function?

A. $y = 3x + 5$ **B.** $f(x) = 3 - x^2$

C. $g(x) = |x|$ **D.** $y = \pm\sqrt{x}$

E. $y = 5x^{2/3}$

For Problems 1–6, let $f(x) = \sqrt{9 - x^2}$.

a. Write the equation for $g(x)$ in terms of x.

b. Plot the graphs of f and g on the same screen. Use a window with integers from about -10 to 10 as grid points. Use the same scale on both axes. Sketch the result.

c. Describe how $f(x)$ was transformed to get $g(x)$, including whether the transformation was an inside or an outside transformation.

1. $g(x) = 2f(x)$ **2.** $g(x) = -3 + f(x)$

3. $g(x) = f(x - 4)$ **4.** $g(x) = f\left(\frac{1}{3}x\right)$

5. $g(x) = 1 + f\left(\frac{1}{2}x\right)$ **6.** $g(x) = \frac{1}{2}f(x + 3)$

For Problems 7–12,

a. Describe how the pre-image function f (dashed) was transformed to get the graph of the image function g (solid).

b. Write an equation for $g(x)$ in terms of function f.

7.

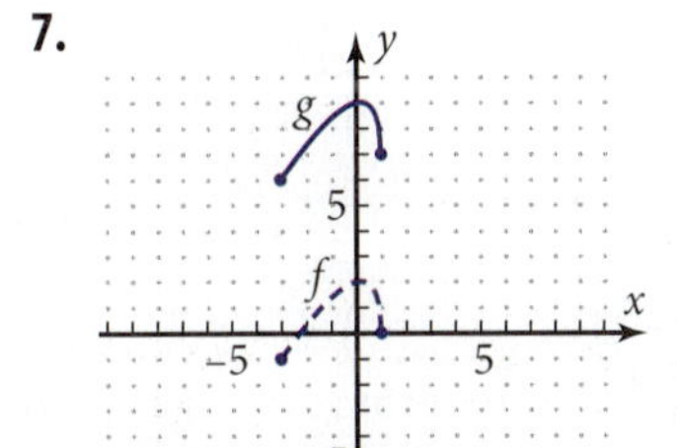

8.

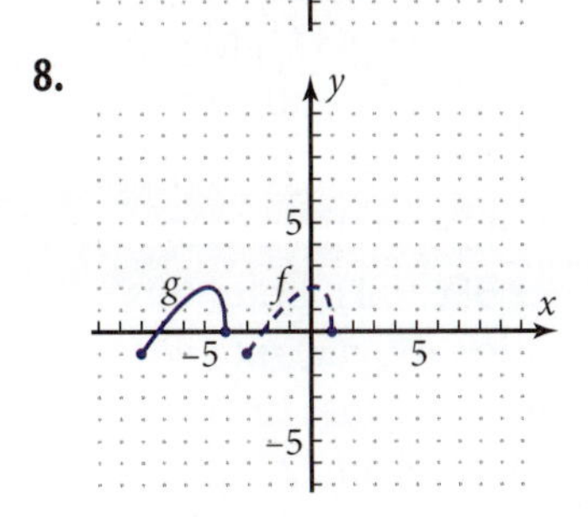

9.

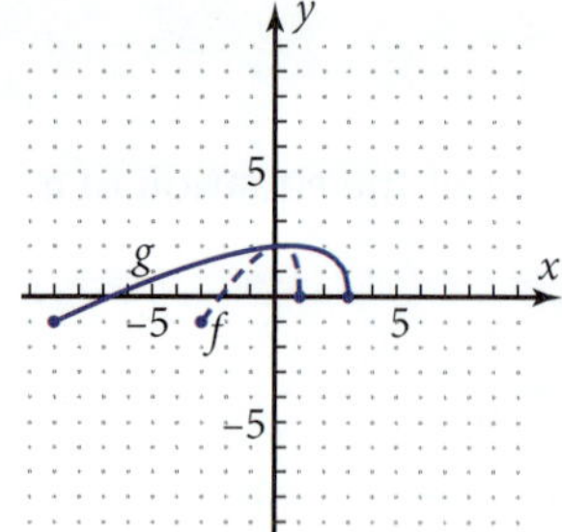

10.

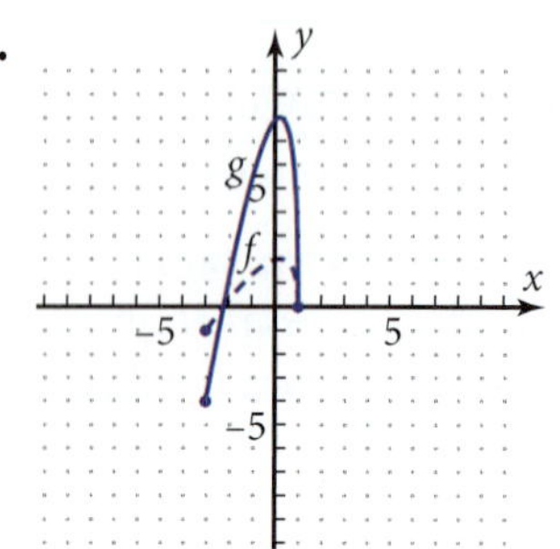

11.

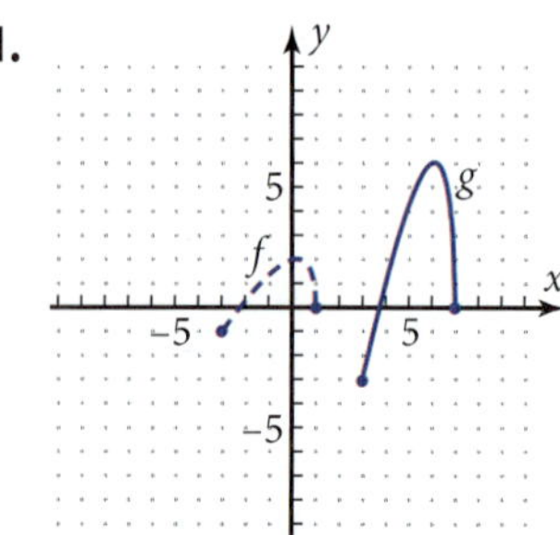

12.

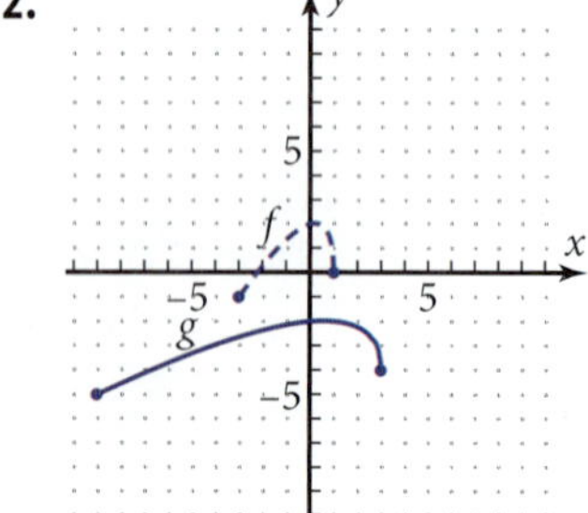

13. The equation of f in Problem 11 is $f(x) = 4.5\sqrt{1 - x} + 2.5(x - 1)$. Enter this equation and the equation for $g(x)$ into your grapher and plot the graphs. Does the result agree with the figure in Problem 11?

14. The equation of f in Problem 12 is $f(x) = 4.5\sqrt{1 - x} + 2.5(x - 1)$. Enter this equation and the equation for $g(x)$ into your grapher and plot the graphs. Does the result agree with the figure in Problem 12?

Figure 1-3h shows the graph of the pre-image function f. For Problems 15–20,

a. Sketch the graph of the image function g on a copy of Figure 1-3h.

b. Identify the transformation(s) that are done.

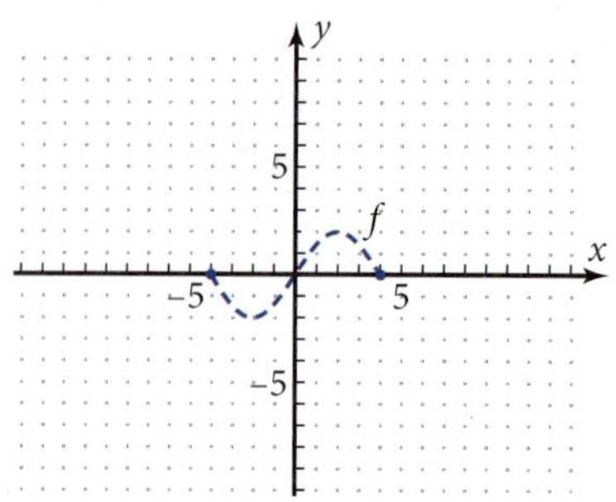

Figure 1-3h

15. $g(x) = f(x + 6)$

16. $g(x) = f\left(\frac{1}{2}x\right)$

17. $g(x) = 5f(x)$

18. $g(x) = 4 + f(x)$

19. $g(x) = 5f(x + 6)$

20. $g(x) = 4 + f\left(\frac{1}{2}x\right)$

21. *Dynamic Transformations Problem:* Go to **www.keymath.com/precalc** and find the Dynamic Precalculus Explorations for Chapter 1. Complete the *Translation* exploration and the *Dilation* exploration, and explain in writing what you learned.

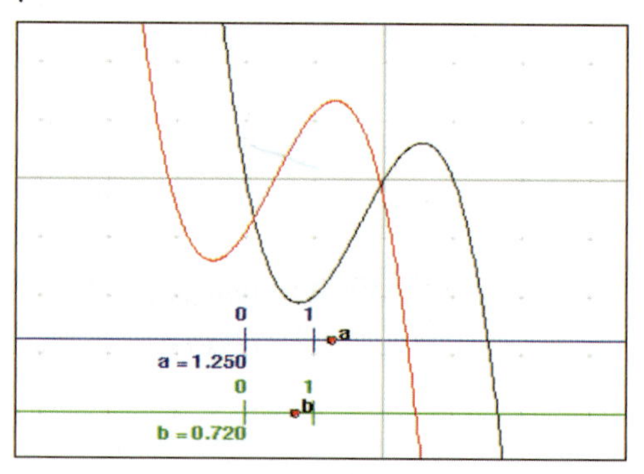

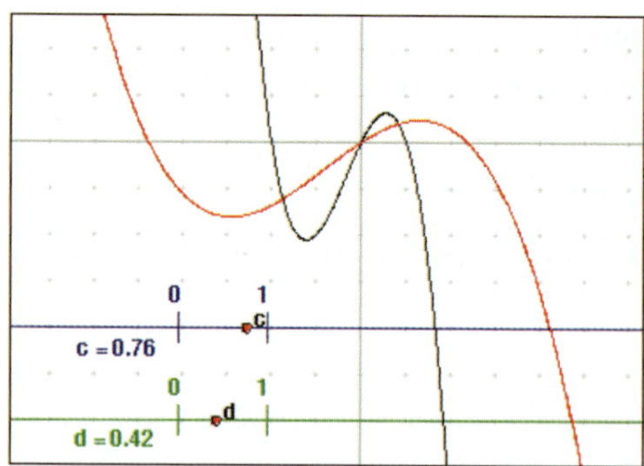

1-4 Composition of Functions

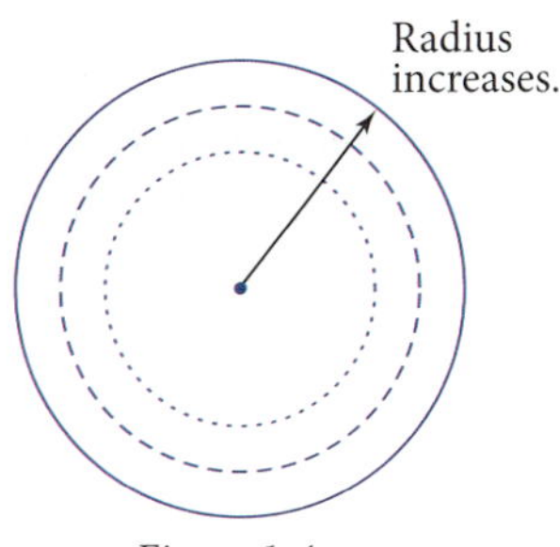

Figure 1-4a

If you drop a pebble into a pond, a circular ripple extends out from the drop point (Figure 1-4a). The radius of the circle is a function of time. The area enclosed by the circular ripple is a function of the radius. Thus area is a function of time through this chain of functions:

- Area depends on radius.
- Radius depends on time.

In this case the area is a **composite function** of time. In this section you will learn some of the mathematics of composite functions.

Objective Given two functions, graph and evaluate the composition of one function with the other.

In this exploration, you'll find the composition of one function with another.

EXPLORATION 1-4: Composition of Functions

1. The figure shows two linear functions, f and g. Write the domain and range of each function.

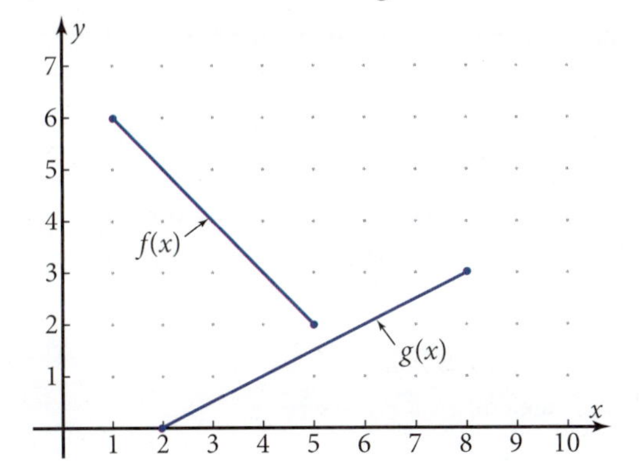

2. Read values of $g(x)$ from the graph and write them in a copy of this table. If the value of x is out of the domain, write "none."

x	g(x)
0	
1	
2	
3	
4	
5	
6	
7	
8	
9	

continued

3. The symbol $f(g(x))$ is read "f of g of x." It means find the value of $g(x)$ first, and then find f of the result. For instance, $g(5) = 1.5$. So $f(g(5)) = f(1.5) = 5.5$. Put another column into the table for values of $f(g(x))$. Write "none" where appropriate.
4. Show in the table an instance where $g(x)$ is defined but $f(g(x))$ is not defined.
5. Plot the values of $f(g(x))$ on a copy of the figure in Problem 1. If the points do not lie in a straight line, go back and check your work.
6. The function in Problem 5 is called the *composition* of f with g, which can be written $f \circ g$. What are the domain and range of $f \circ g$?
7. Write equations for functions f and g.
8. Enter into your grapher the equations of f and g as $f_1(x)$ and $f_2(x)$, respectively. Use Boolean variables or enter the domain directly, depending on your grapher, to make the functions have the proper domains. Then plot the graphs. Does the result agree with the given figure?
9. Enter $f \circ g$ into $f_3(x)$ by entering $f_1(f_2(x))$. Plot this graph. Does it agree with the graph you drew in Problem 5?
10. By suitable algebraic operations on the equations in Problem 7, find an equation for $f(g(x))$. Simplify the equation as much as possible.
11. What did you learn as a result of doing this exploration that you did not know before?

Symbols for Composite Functions

Suppose that the radius of the ripple is increasing at the constant rate of 8 in./s. Then

$$r = 8t$$

where r is the radius in inches and t is the number of seconds. If $t = 5$, then

$$r = 8 \cdot 5 = 40 \text{ in.}$$

The area of the circular region is given by

$$a = \pi r^2$$

where a is the area in square inches and r is the radius in inches. At time $t = 5$, when the radius is 40, the area is given by

$$a = \pi \cdot 40^2 = 1600\pi = 5026.5482... \approx 5027 \text{ in}^2 \text{ or about } 35 \text{ ft}^2.$$

The 5 s is the *input* for the radius function, and the 40 in. is the *output*. Figure 1-4b shows that the output of the radius function becomes the input for the area function. The output of the area function is 5026.5....

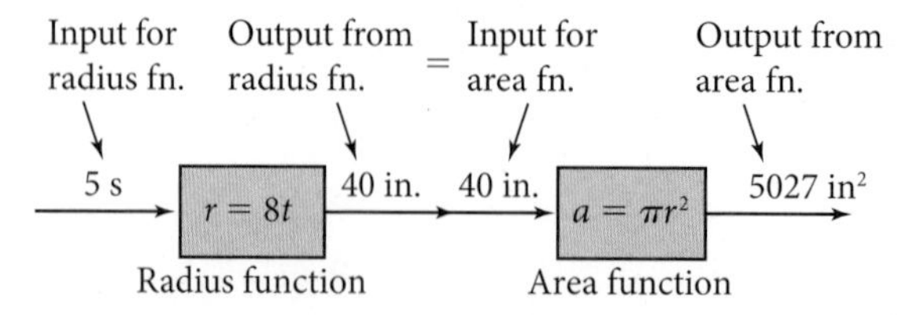

Figure 1-4b

Mathematicians often use $f(x)$ terminology for composite functions. For these radius and area functions, you can write

$r(x) = 8x$ $\quad$ x is the input for function r.

$a(x) = \pi x^2$ $\quad$ x is the input for function a.

The r and a become the *names* of the functions, and the $r(x)$ and $a(x)$ are the *values*, or *outputs*, of the functions. The x simply stands for the *input* of the function. You must keep in mind that the input for function r is the time and the input for function a is the radius.

Combining the symbols leads to this way of writing a composite function:

$$\text{area} = a(r(x))$$

The x is the input for the radius function, and the $r(x)$ is the input for the area function. The notation $a(r(x))$ is pronounced "a of r of x." Function r is called the *inside function* because it appears inside a pair of parentheses. Function a is called the *outside function*. Figure 1-4c shows this symbol and its meanings. The names parallel the terms *inside transformation* and *outside transformation* that you learned in the previous section.

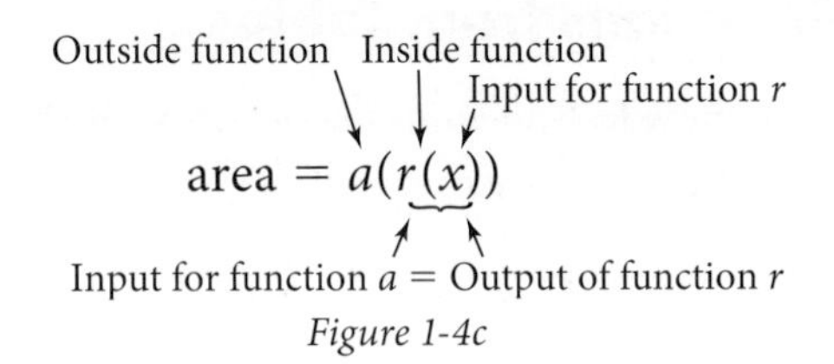

Figure 1-4c

The two function names are sometimes combined this way:

$a \circ r(x)$ $\quad$ or $\quad$ $(a \circ r)(x)$

The symbol $a \circ r$ is pronounced "a composition r." The parentheses in the expression $(a \circ r)$ indicate that $a \circ r$ is the name of the function.

Composite Functions from Graphs

Example 1 shows you how to find a value of the composite function $f(g(x))$ from graphs of the two functions f and g.

EXAMPLE 1 ➤ Functions f and g are graphed in Figure 1-4d. Find $f(g(30))$, showing on copies of the graphs how you found this value.

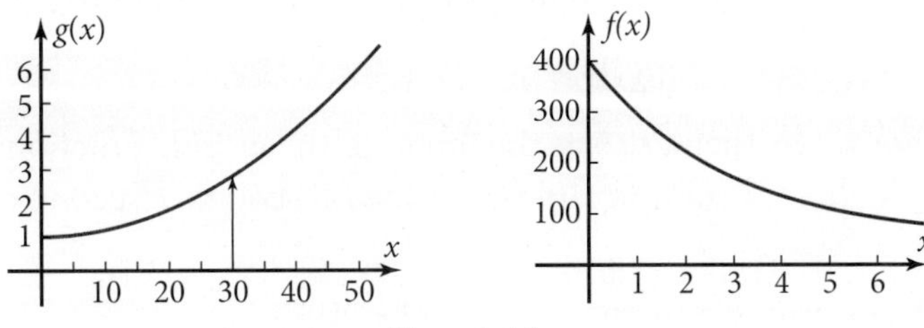

Figure 1-4d

SOLUTION First find the value of the inside function, $g(30)$. As shown on the left in Figure 1-4e,

$$g(30) \approx 2.8$$

Use this output of function g as the input for function f, as shown on the right in Figure 1-4e. Note that the x in $f(x)$ is simply the input for function f and is not the same number as the x in $g(x)$.

$$f(2.8) \approx 180$$

$$\therefore f(g(30)) \approx 180$$

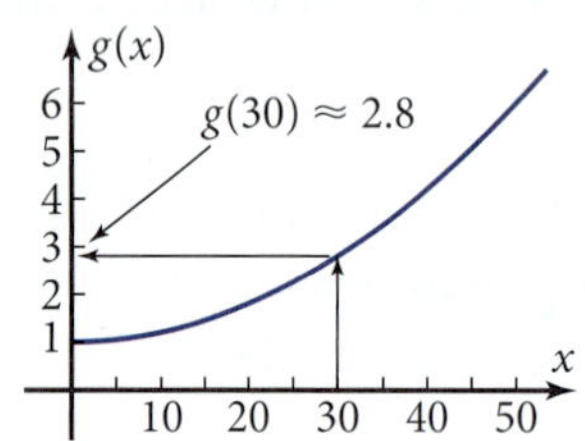

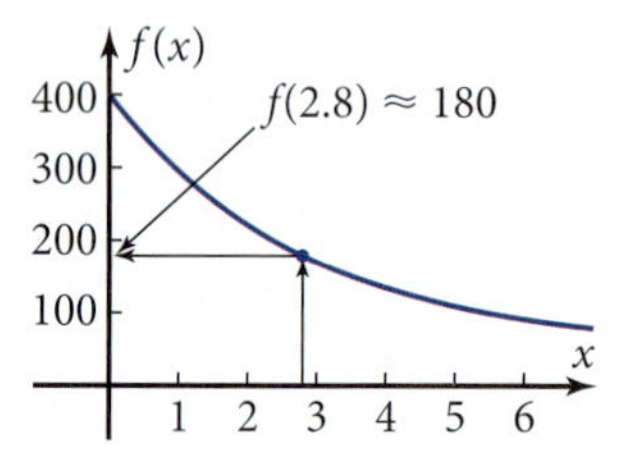

Figure 1-4e

Composite Functions from Tables

Example 2 shows you how to find values of a composite function when the two functions are defined numerically.

EXAMPLE 2 Functions f and g are defined only for the integer values of x in the table.

x	f(x)	g(x)
1	3	5
2	4	3
3	6	2
4	2	1
5	0	7
6	1	4

a. Find $f(g(x))$ for the six values of x in the table.

b. Find $g(f(2))$, and show that it does *not* equal $f(g(2))$.

SOLUTION **a.** To find $f(g(1))$, first find the value of the *inside* function, $g(1)$, by finding 1 in the x-column and $g(1)$ in the $g(x)$ column (third column).

$$g(1) = 5$$

Then use 5 as the input for the *outside* function f by finding 5 in the x-column and $f(5)$ in the $f(x)$ column (second column).

$$f(5) = 0$$

$$\therefore f(g(1)) = 0$$

Find the other values the same way. Here is a compact way to arrange your work.

$$f(g(1)) = f(5) = 0$$
$$f(g2)) = f(3) = 6$$
$$f(g(3)) = f(2) = 4$$
$$f(g(4)) = f(1) = 3$$
$$f(g(5)) = f(7), \text{ which does not exist}$$
$$f(g(6)) = f(4) = 2$$

b. $g(f(2)) = g(4) = 1$, which is not the same as $f(g(2)) = 6$ ◄

Note that in order to find a value of a composite function such as $f(g(x))$, the value of $g(x)$ must be in the domain of the outside function, f. Because $g(5) = 7$ in Example 2 and there is no value for $f(7)$, the value of $f(g(5))$ is undefined.

Composite Functions from Equations

Example 3 shows you how to find values of a composite function if you know the equations of the two functions.

EXAMPLE 3 ➤ Let f be the linear function $f(x) = 3x + 5$, and let g be the exponential function $g(x) = 2^x$.

a. Find $f(g(4))$, $f(g(0))$, and $f(g(-1))$.

b. Find $g(f(-1))$ and show that it is not the same as $f(g(-1))$.

c. Find an equation for $h(x) = f(g(x))$ explicitly in terms of x. Show that $h(4)$ agrees with the value you found for $f(g(4))$.

SOLUTION

a. $g(4) = 2^4 = 16$, and $f(16) = 3 \cdot 16 + 5 = 53$, so $f(g(4)) = 53$

Writing the same steps more compactly for the other two values of x gives

$$f(g(0)) = f(2^0) = f(1) = 3 \cdot 1 + 5 = 8$$
$$f(g(-1)) = f(2^{-1}) = f(0.5) = 3(0.5) + 5 = 6.5$$

b. $g(f(-1)) = g(3 \cdot -1 + 5) = g(2) = 2^2 = 4$, which does not equal 6.5 from part a

c. $h(x) = f(g(x)) = f(2^x) = 3 \cdot 2^x + 5$

The equation is $h(x) = 3 \cdot 2^x + 5$.

So $h(4) = 3 \cdot 2^4 + 5 = 3 \cdot 16 + 5 = 53$, which agrees with part a. ◄

Example 4 shows you that you can compose a function with itself or compose more than two functions.

EXAMPLE 4 ➤ Let f be the linear function $f(x) = 3x + 5$, and let g be the exponential function $g(x) = 2^x$, as in Example 3. Find these values.

a. $f(f(2))$

b. $f(g(f(-3)))$

SOLUTION

a. $f(f(2)) = f(3 \cdot 2 + 5) = f(11) = 3 \cdot 11 + 5 = 38$

b. $f(g(f(-3))) = f(g(3 \cdot -3 + 5)) = f(g(-4)) = f(2^{-4}) = f(0.0625) = 3 \cdot 0.0625 + 5 = 5.1875$

Domain and Range of a Composite Function

In Example 2, you saw that the value of the inside function sometimes is not in the domain of the outside function. Example 5 shows you how to find the domain of a composite function and the corresponding range under this condition.

EXAMPLE 5 ➤ The left graph in Figure 1-4f shows function g with domain $2 \le x \le 7$, and the right graph shows function f with domain $1 \le x \le 5$.

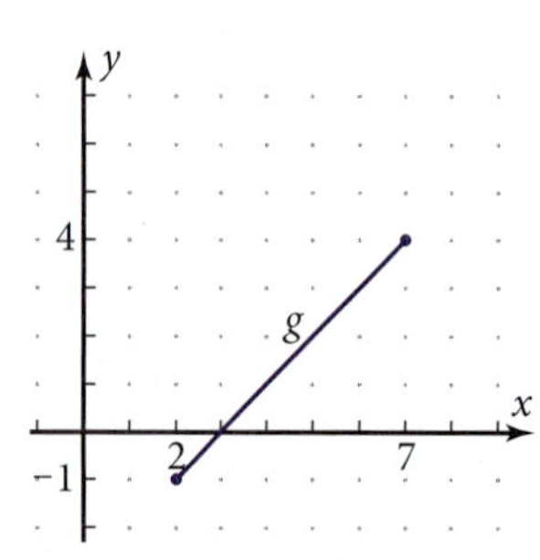

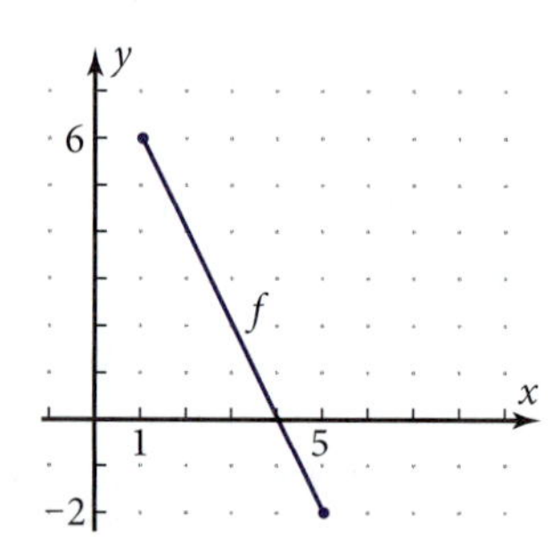

Figure 1-4f

a. Show on copies of these graphs what happens when you try to find $f(g(6))$, $f(g(8))$, and $f(g(2))$.

b. Make a table of values of $g(x)$ and $f(g(x))$ for integer values of x from 1 through 8. If there is no value, write "none." From the table, what does the domain of function $f \circ g$ seem to be?

c. The equations of functions g and f are

$$g(x) = x - 3, \text{ for } 2 \le x \le 7$$

$$f(x) = -2x + 8, \text{ for } 1 \le x \le 5$$

Plot $f(x)$, $g(x)$, and $f(g(x))$ on your grapher, with the grapher's grid showing. Does the domain of $f \circ g$ confirm what you found numerically in part b? What is the range of $f \circ g$?

d. Find the domain of $f \circ g$ algebraically and show that it agrees with part c.

SOLUTION

a. The left graph in Figure 1-4g shows that $g(2) = -1$ and $g(6) = 3$ but $g(8)$ does not exist, because 8 is outside the domain of function g. The right graph shows the two output values of function g, -1 and 3, used as inputs for function f. From the graph, $f(3) = 2$ and $f(-1)$ does not exist, because -1 is outside the domain of function f. Summarize the results:

$f(g(6)) = f(3) = 2$

$f(g(8))$ does not exist because 8 is outside the domain of g.

$f(g(2)) = f(-1)$, which does not exist because -1 is outside the domain of f.

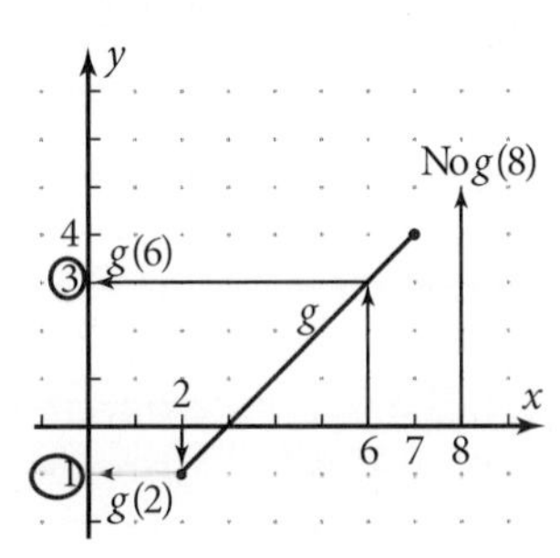

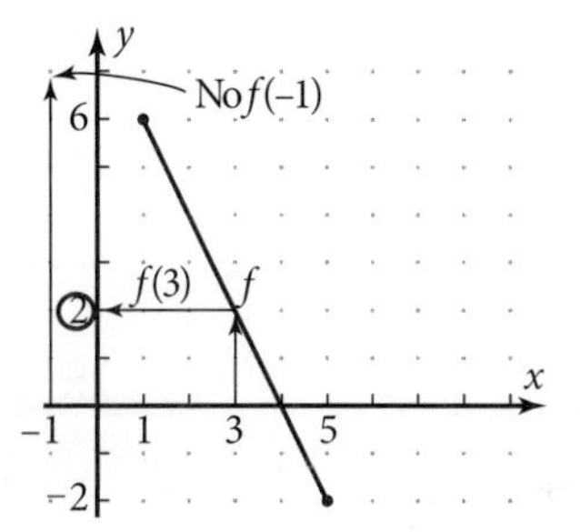

Figure 1-4g

b.

x	**g(x)**	**f(g(x))**
1	none	none
2	−1	none
3	0	none
4	1	6
5	2	4
6	3	2
7	4	0
8	none	none

The domain of $f \circ g$ seems to be $4 \le x \le 7$.

c. Enter: $f_1(x) = x - 3/(x \ge 2 \text{ and } x \le 7)$ for $g(x)$

Use Boolean variables or enter the domain directly, depending on your grapher, to restrict the domain.

Enter: $f_2(x) = -2x + 8/(x \ge 1 \text{ and } x \le 5)$ for $f(x)$

Enter: $f_3(x) = f_2(f_1(x))$ for $f(g(x))$

f_1 and f_2 become function names in this format.

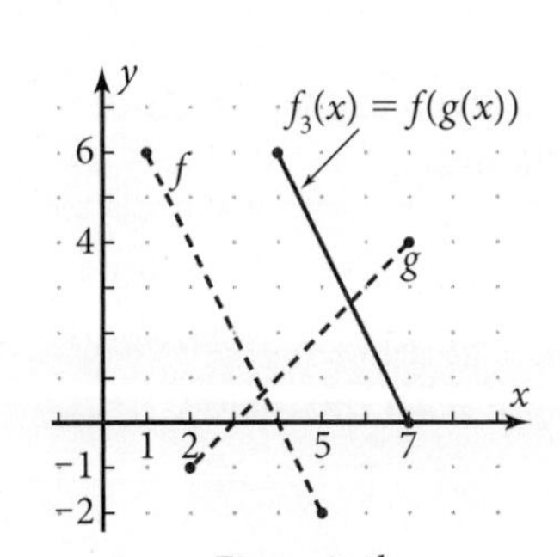

Figure 1-4h

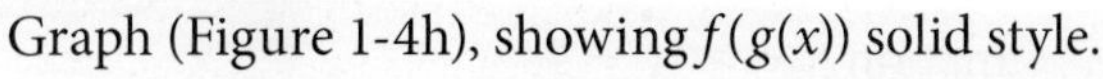

Graph (Figure 1-4h), showing $f(g(x))$ solid style.
The domain of $f \circ g$ is $4 \le x \le 7$, in agreement with part b.
From the graph, the range of $f \circ g$ is $0 \le y \le 6$.

d. To calculate the domain algebraically, first observe that $g(x)$ must be within the domain of f.

$1 \le g(x) \le 5$	Write $g(x)$ in the domain of f.
$1 \le x - 3 \le 5$	Substitute $x - 3$ for $g(x)$.
$4 \le x \le 8$	Add 3 to all three members of the inequality.

Next observe that x must also be in the domain of g, specifically, $2 \le x \le 7$. The domain of $f \circ g$ is the *intersection* of these two intervals. Number-line graphs (Figure 1-4i) will help you visualize the intersection.

$\therefore$ the domain of $f \circ g$ is $4 \le x \le 7$.

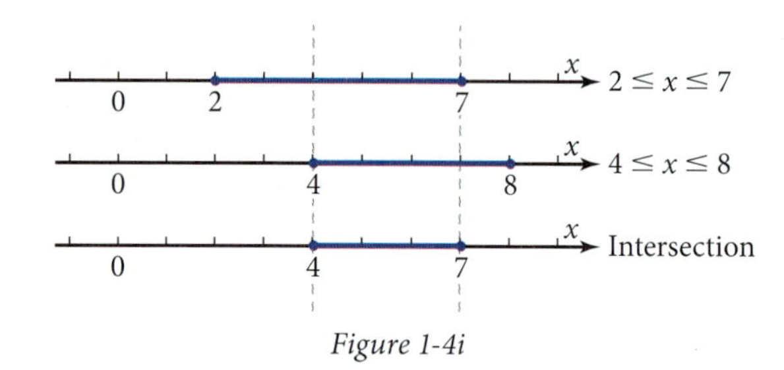

Figure 1-4i

DEFINITION AND PROPERTIES: Composite Function

The composite function $f \circ g$ (pronounced "f composition g" or "f of g") is the function

$$(f \circ g)(x) = f(g(x))$$

Function g, the inside function, is evaluated first, using x as its input. Function f, the outside function, is evaluated next, using $g(x)$ as its input (the output of function g).

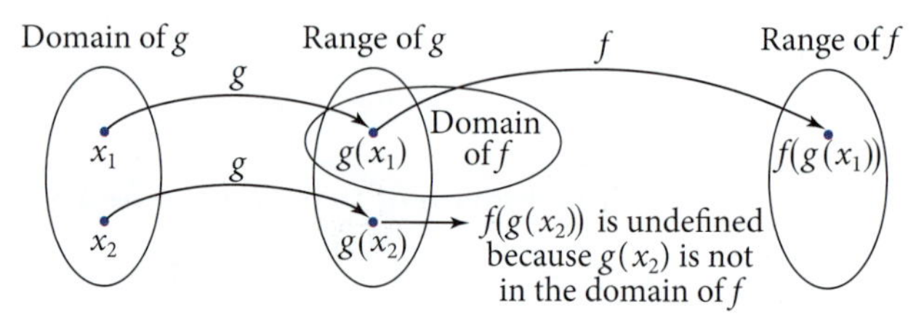

The domain of $f \circ g$ is the set of all values of x in the domain of g for which $g(x)$ is in the domain of f. The figure shows this relationship.

Note: Horizontal dilations and translations are examples of composite functions because they are *inside transformations* applied to x. For instance, the horizontal translation $g(x) = 3(x - 2)$ is actually a composite function with the inside function $f(x) = x - 2$.

Problem Set 1-4

Reading Analysis

From what you have read in this section, what do you consider to be the main idea? Think of a real-world example, other than the one in the text, in which the value of one variable depends on the value of a second variable and the value of the second variable depends on the value of a third variable. In your example, which is the inside function and which is the outside function? For there to be a value of the composite function, what must be true of the value of the inside function?

Quick Review

Q1. What transformation of f is represented by $g(x) = 3f(x)$?

Q2. What transformation of f is represented by $h(x) = 5 + f(x)$?

Q3. If g is a horizontal translation of f by -4 units, then $g(x) =$ __?__.

Q4. If p is a horizontal dilation of f by a factor of 0.2, then $p(x) =$ __?__.

Q5. Why is $y = 3x^5$ not an exponential function, even though it has an exponent?

Q6. Write the general equation of a quadratic function.

Q7. For what value of x will the graph of $y = \frac{x-3}{x-5}$ have a discontinuity?

Q8. Sketch the graph of $f(x) = |x|$.

Q9. Find 40% of 300.

Q10. Which of these is a horizontal dilation by a factor of 2?

A. $g(x) = 2f(x)$ **B.** $g(x) = 0.5f(x)$

C. $g(x) = f(0.5x)$ **D.** $g(x) = f(2x)$

---

1. *Flashlight Problem:* You shine a flashlight, making a circular spot of light on the wall with radius 5 cm. As you back away from the wall, the radius increases at a rate of 7 cm/s.

a. Find the radius at times 4 s and 7 s after you start backing away.

b. Use the radius at times 4 s and 7 s to find the area of the spot of light at these times.

c. Why can it be said that the area is a *composite function* of time?

d. Let t be the number of seconds since you started backing away. Let $r(t)$ be the radius of the spot of light, in centimeters. Let $a(r(t))$ be the area of the spot, in square centimeters. Write an equation for $r(t)$ as a function of t. Write another equation for $a(r(t))$ as a function of $r(t)$. Write a third equation for $a(r(t))$ explicitly in terms of t. Show that the last equation gives the correct area for times $t = 4$ s and $t = 7$ s.

2. *Bacteria Culture Problem*: When you grow a culture of bacteria in a petri dish, the area of the culture is a measure of the number of bacteria present. Suppose that the area of the culture, $A(t)$, in square millimeters, is given by this function of time t, in hours:

$$A(t) = 9(1.1^t)$$

a. Find $A(0)$, $A(5)$, and $A(10)$, the area at times $t = 0$ h, 5 h, and 10 h, respectively.

b. Assume that the bacteria culture is circular. Find the radius of the culture at the three times in part a.

c. Why can it be said that the radius is a *composite function* of time?

d. Let R be the radius function, with input $A(t)$, the area of the culture. Write an equation for $R(A(t))$, the radius as a function of area. Then write an equation for $R(A(t))$ explicitly in terms of t. Show that this equation gives the correct answer for the radius at time $t = 5$ h.

3. *Shoe Size Problem*: The size shoe a person wears, $S(x)$, is a function of the length of the person's foot. The length of the foot, $L(x)$, is a function of the person's age.

a. Sketch reasonable graphs of functions S and L. Label the axes of each graph with the name of the variable represented.

b. What does x represent in function S? What does x represent in function L? What does the composite function $S(L(x))$ represent? What, if any, real-world meaning does $L(S(x))$ have?

c. Let $f(x) = (S \circ L)(x)$. Sketch a reasonable graph of function f. Label the axes of the graph with the name of the variable represented.

4. *Traffic Problem:* The length of time, $T(x)$, it takes you to travel a mile on the freeway depends on the speed at which you travel. The speed, $S(x)$, depends on the number of other cars on that mile of freeway.

   a. Sketch reasonable graphs of functions T and S. Label the axes of each graph with the name of the variable represented.

   b. What does x represent in function T? What does x represent in function S? What does the composite function $T(S(x))$ represent? What, if any, real-world meaning does $S(T(x))$ have?

   c. Let $g(x) = (T \circ S)(x)$. Sketch a reasonable graph of function g. Label the axes of the graph with the name of the variable represented.

5. *Composite Function Graphically, Problem 1:* Functions h and p are defined by the graphs in Figure 1-4j, in the domains shown.

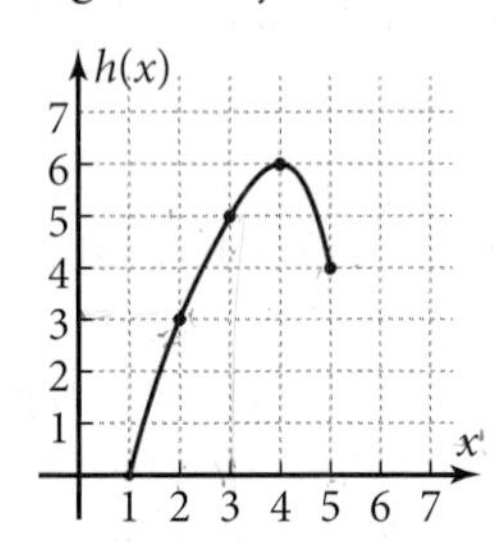

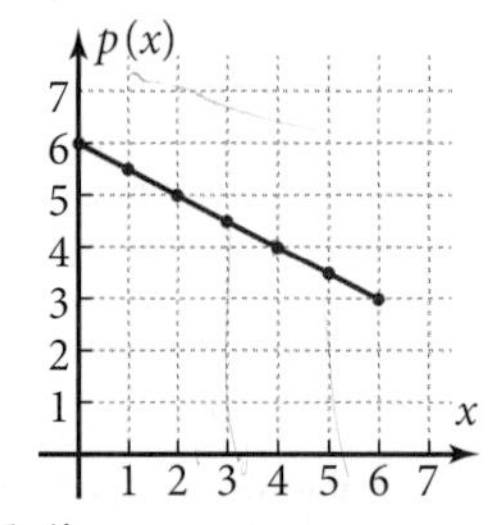

Figure 1-4j

   a. Find $h(3)$. On a copy of the graphs, draw arrows to show how you found this value.

   b. Use the output of $h(3)$ to find $p(h(3))$. Draw arrows to show how you found this value.

   c. Find $p(h(2))$ and $p(h(5))$ by first finding $h(2)$ and $h(5)$ and then using these values as inputs for function p.

   d. Find $h(p(2))$ by first finding $p(2)$ and then using the result as the input for function h. Draw arrows to show how you found this value. Show that $h(p(2)) \neq p(h(2))$.

   e. Explain why there is no value of $h(p(0))$, even though there is a value of $p(0)$.

6. *Composite Function Graphically, Problem 2:* Functions f and g are defined by the graphs in Figure 1-4k, in the domains shown.

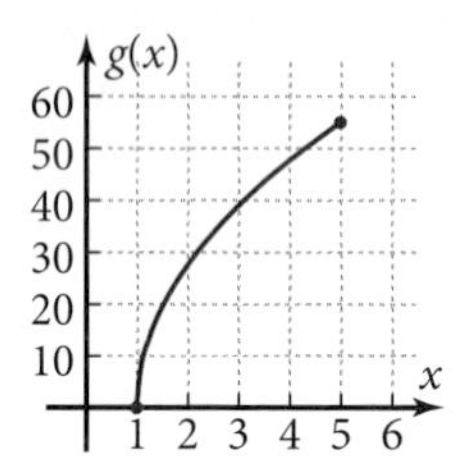

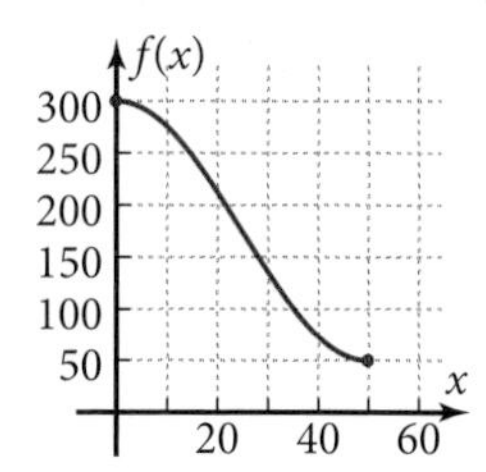

Figure 1-4k

   a. Find the approximate value of $g(4)$. On a copy of the graphs, show how you found this value.

   b. Use the output of $g(4)$ to find the approximate value of $f(g(4))$. Draw arrows to show how you found this value.

   c. Find approximate values of $f(g(3))$ and $f(g(2))$ by first finding $g(3)$ and $g(2)$ and then using these values as inputs for function f.

   d. Explain why there is no value of $f(g(6))$.

   e. Try to find $f(g(5))$ by first finding $g(5)$ and then using the result as the input for function f. Draw arrows to illustrate why there is *no* value of $f(g(5))$.

7. *Composite Function Numerically, Problem 1:* Functions f and g consist of the discrete points in the table, and only these points. Find the values of the composite functions, or explain why no such value exists.

x	f(x)	g(x)
1	3	2
2	5	3
3	4	7
4	2	5
5	1	4

a. Find $g(1)$ and $f(g(1))$.

b. Find $g(2)$ and $f(g(2))$.

c. Find $g(3)$ and $f(g(3))$.

d. Find $f(4)$ and $g(f(4))$.

e. Find $g(f(3))$. **f.** Find $f(f(5))$.

g. Find $g(g(3))$. **h.** Find $f(f(f(1)))$.

8. *Composite Function Numerically, Problem 2:* Functions u and v consist of the discrete points in the table, and only these points. Find the values of the composite functions, or explain why no such value exists.

a. Find $v(2)$ and $u(v(2))$.

b. Find $v(6)$ and $u(v(6))$.

c. Find $v(4)$ and $u(v(4))$.

d. Find $u(4)$ and $v(u(4))$.

e. Find $v(u(10))$.

f. Find $v(v(10))$

g. Find $u(u(6))$.

h. Find $v(v(v(8)))$

x	u(x)	v(x)
2	3	6
4	8	5
6	2	4
8	10	2
10	6	8

9. *Composite Function Algebraically, Problem 1:* Let g and f be defined by

$$f(x) = 9 - x \qquad 4 \le x \le 8$$

$$g(x) = x + 2 \qquad 1 \le x \le 5$$

a. Make a table showing values of $g(x)$ for each integer value of x in the domain of g. In another column, show the corresponding values of $f(g(x))$. If there is no such value, write "none."

b. From your table in part a, what does the domain of the composite function $f \circ g$ seem to be? Confirm (or refute) your conclusion by finding the domain algebraically.

c. Explain why $f(g(6))$ is undefined. Explain why $f(g(1))$ is undefined, even though $g(1)$ is defined.

d. Repeat parts a and b for the composite function $g \circ f$.

e. Figure 1-4l shows functions f and g. Enter the two functions as $f_1(x)$ and $f_2(x)$. Then enter $f(g(x))$ as $f_3(x) = f_1(f_2(x))$, and $g(f(x))$ as $f_2(f_1(x))$. Plot the graphs using the window shown, with the grapher's grid showing and thick style for the two composite function graphs. Sketch the result. Do the domains of the composite functions from the graph agree with your results in parts b and d?

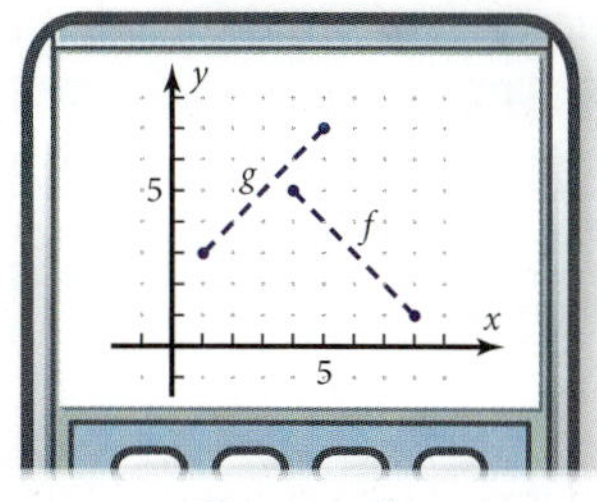

Figure 1-4l

f. Find $f(f(5))$. Explain why $g(g(5))$ is undefined.

10. *Composite Function Algebraically, Problem 2:* Let f and g be defined by

$$f(x) - -x^2 + 8x - 4 \qquad 1 \le x \le 6$$

$$g(x) = 5 - x \qquad 0 \le x \le 7$$

a. Make a table showing values of $g(x)$ for each integer value of x in the domain of g. In another column, show the corresponding values of $f(g(x))$. If there is no such value, write "none."

b. From your table in part a, what does the domain of the composite function $f \circ g$ seem to be? Confirm (or refute) your conclusion by finding the domain algebraically.

c. Show why $f(g(3))$ is defined but $g(f(3))$ is undefined.

d. Figure 1-4m shows the graphs of f and g. Enter these equations as $f_1(x)$ and $f_2(x)$. Then enter $f(g(x))$ as $f_3(x) = f_1(x)(f_2(x))$. Plot the three graphs with the grapher's grid showing. Sketch the result. Does the domain of the composite function agree with your calculation in part b?

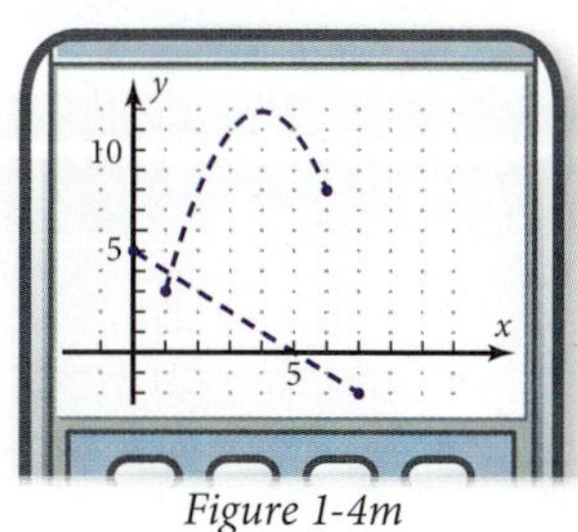

Figure 1-4m

e. Find an equation for $f(g(x))$ explicitly in terms of x. Enter this equation as $f_4(x)$ and plot it on the same screen as the other three functions. What similarities and what differences do you see for $f_4(x)$ and $f_3(x)$?

11. *Square and Square Root Functions:* Let f and g be defined by

$f(x) = x^2$, where x is any real number

$g(x) = \sqrt{x}$, where the values of x make $g(x)$ a real number

a. Find $f(g(3))$, $f(g(7))$, $g(f(5))$, and $g(f(8))$. What do you notice in each case? Make a conjecture: "For all values of x, $f(g(x)) =$ __?__ and $g(f(x)) =$ __?__."

b. Test your conjecture by finding $f(g(-9))$ and $g(f(-9))$. Does your conjecture hold for negative values of x?

c. Plot $f(x)$, $g(x)$, and $f(g(x))$ on the same screen. Use approximately equal scales on both axes, as in Figure 1-4n. Explain why $f(g(x)) = x$, but only for nonnegative values of x.

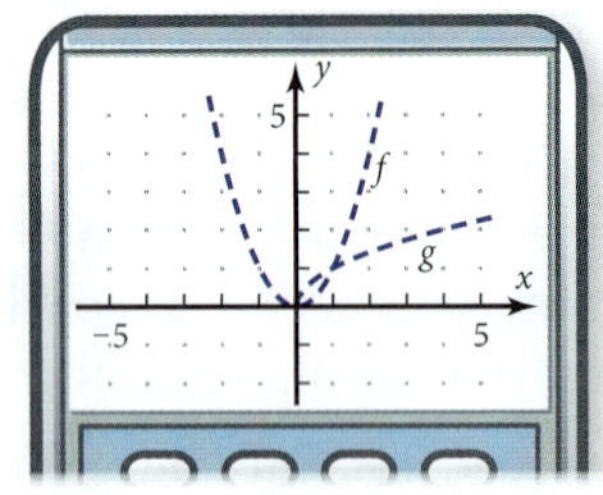

Figure 1-4n

d. Deactivate $f(g(x))$, and plot $f(x)$, $g(x)$, and $g(f(x))$ on the same screen. Sketch the result.

e. Explain why $g(f(x)) = x$ for nonnegative values of x, but $g(f(x)) = -x$ (the opposite of x) for negative values of x. What other familiar function has this property?

12. *Horizontal Translation and Dilation Problem:* Let f, g, and h be defined by

$f(x) = x^2$	$-2 \le x \le 2$
$g(x) = x - 3$	for all real values of x
$h(x) = \frac{1}{2}x$	for all real values of x

a. $f(g(x)) = f(x - 3)$. What transformation is applied to function f by composing it with g?

b. $f(h(x)) = f\left(\frac{1}{2}x\right)$. What transformation is applied to function f by composing it with h?

c. Plot the graphs of f, $f \circ g$, and $f \circ h$. Sketch the results. Do the graphs confirm your conclusions in parts a and b?

For Problems 13 and 14, find what transformation will turn the dashed graph (f) into the solid graph (g).

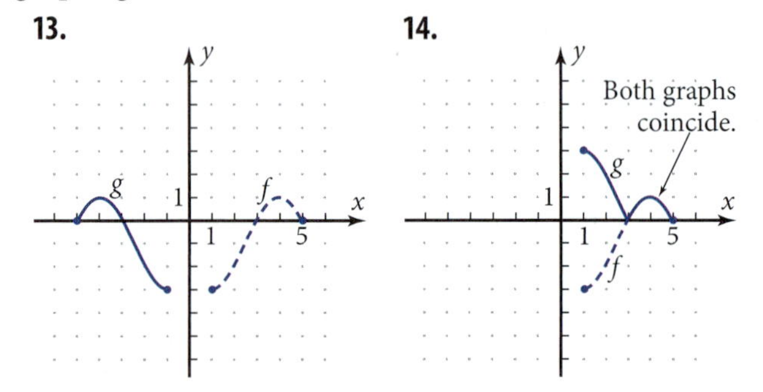

15. *Linear Function and Its Inverse Problem:* Let f and g be defined by

$f(x) = \frac{2}{3}x - 2$ $\qquad$ $g(x) = 1.5x + 3$

a. Find $f(g(6))$, $f(g(-15))$, $g(f(10))$, and $g(f(-8))$. What do you notice in each case?

b. Plot the graphs of f, g, $f \circ g$, and $g \circ f$ on the same screen. How are the graphs of $f \circ g$ and $g \circ f$ related? How are the graphs of $f \circ g$ and $g \circ f$ related to their "parent" graphs, f and g?

c. Show that $f(g(x))$ and $g(f(x))$ both equal x.

d. Functions f and g in this problem are said to be *inverses* of each other. Whatever f does to x, g undoes. Let $h(x) = 5x - 7$. Find an equation for the inverse function of h.

1-5 Inverse Functions and Parametric Equations

The photograph shows a highway crew painting a center stripe. From records of previous work the crew has done, it is possible to predict how much of the stripe the crew will have painted at any time during a normal eight-hour shift. It may also be possible to tell how long the crew has been working by how much stripe has been painted. The input for the distance function is time, and the input for the time function is distance.

If a new relation is formed by interchanging the input and output variables in a given relation, the two relations are called *inverses* of each other. If both relations turn out to be functions, they are called **inverse functions.** If not, the relation and its inverse can still be plotted easily using **parametric equations** in which both x and y are functions of some third variable, such as time.

Objective

Given a function, find its inverse relation, and tell whether the inverse relation is a function. Graph parametric equations both by hand and on a grapher, and use parametric equations to graph the inverse of a function.

Inverse of a Function Numerically

t (h)	*d* (mi)
1	0.2
2	0.6
3	1.0
4	1.4
5	1.8
6	2.2
7	2.6
8	3.0

Suppose that the distance d, in miles, a particular highway crew paints in an eight-hour shift is given numerically by this function of t, in hours, it has been on the job.

Let $d = f(t)$. You can see that $f(1) = 0.2, f(2) = 0.6, \ldots, f(8) = 3.0$. The input for function f is the number of hours, and the output is the number of miles.

As long as the crew does not stop painting during the eight-hour shift, the number of hours it has been painting is a function of the distance. Let $t = g(d)$. You can see that $g(0.2) = 1, g(0.6) = 2, \ldots, g(3) = 8$. The input for function g is the number of miles, and the output is the number of hours. The input and output for functions f and g have been *interchanged,* and thus the two functions are *inverses* of each other.

Symbols for the Inverse of a Function

If function g is the inverse of function f, the symbol f^{-1} is often used for the name of function g. In the highway stripe example, you can write $f^{-1}(0.2) = 1$, $f^{-1}(0.6) = 2, \ldots, f^{-1}(3) = 8$. Note that $f^{-1}(3)$ does not mean the *reciprocal* of $f(3)$.

$$f^{-1}(3) = 8 \qquad \text{and} \qquad \frac{1}{f(3)} = \frac{1}{1} = 1, \text{ not } 8$$

The -1 used with the *name* of a function means the *function* inverse, whereas the -1 used with a *number,* as in 5^{-1}, means the *multiplicative* inverse of that number.

Inverse of a Function Graphically

Figure 1-5a shows a graph of the data for the highway stripe example. Note that the points seem to lie in a straight line. Connecting the points is reasonable if you assume that the crew paints continuously. The line meets the t-axis at about $t = 0.5$ h, indicating that it takes the crew about half an hour at the beginning of the shift to redirect traffic and set up the equipment before it can start painting.

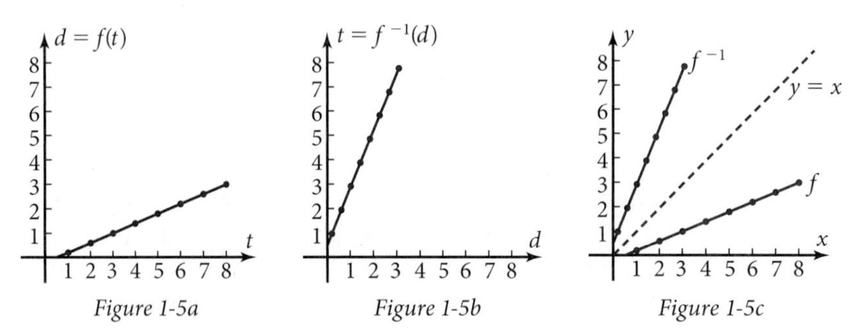

Figure 1-5a *Figure 1-5b* *Figure 1-5c*

Figure 1-5b shows the inverse function, $t = f^{-1}(d)$. Note that every vertical feature on the graph of f is a horizontal feature on the graph of f^{-1}, and vice versa. For instance, the graph of f^{-1} meets the *vertical* axis at 0.5.

Figure 1-5c shows both graphs on the same set of axes. In this figure, x is used for the input variable and y for the output variable. Keep in mind that x for function f represents hours and x for function f^{-1} represents miles. The graphs are **reflections** of each other across the line whose equation is $y = x$.

Inverse of a Function Algebraically

In the highway stripe example, the linear function that fits the graph of function f in Figure 1-5c is

$$y = 0.4(x - 0.5) \qquad \text{or, equivalently,} \qquad y = 0.4x - 0.2$$

slope = 0.4,
x-intercept = 0.5

The linear function that fits the graph of f^{-1} is

$y = 2.5x + 0.5$ slope = 2.5, y-intercept = 0.5

If you know the equation of a function, you can transform it algebraically to find the equation of the inverse relation by first interchanging the variables.

Function: $y = 0.4x - 0.2$

Inverse: $x = 0.4y - 0.2$

The equation of the inverse relation can be solved for y in terms of x.

$x = 0.4y - 0.2$

$y = 2.5x + 0.5$ Solve for y in terms of x.

To distinguish between the function and its inverse, you can write

$f(x) = 0.4x - 0.2$ and $f^{-1}(x) = 2.5x + 0.5$

Bear in mind that the x used as the input for function f is not the same as the x used as the input for function f^{-1}. One is time, and the other is distance.

An interesting thing happens if you take the *composition* of a function and its inverse. In the highway stripe example,

$f(4) = 1.4$ and $f^{-1}(1.4) = 4$

$\therefore f^{-1}(f(4)) = 4$

You get the original input, 4, back again. This result should not be surprising to you. The composite function $f^{-1}(f(4))$ means "How many hours does it take the crew to paint the distance it can paint in four hours?" There is a similar meaning for $f(f^{-1}(x))$. For instance,

$f(f^{-1}(1.4)) = f(4) = 1.4$

In this case 1.4 is the original input of the *inside* function.

Invertibility and the Domain of an Inverse Relation

In the highway stripe example, the length of stripe painted during the first half hour was zero because it took some time at the beginning of the shift for the crew to divert traffic and prepare the equipment. The graph of function f in Figure 1-5d includes times at the beginning of the shift, along with its inverse relation.

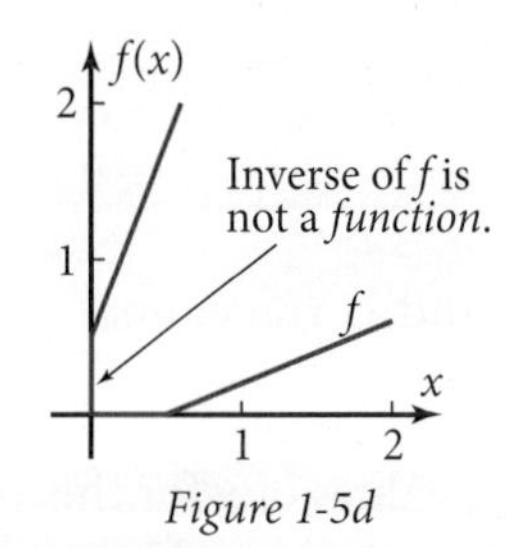

Figure 1-5d

Note that the inverse relation has multiple values of y when x equals zero. Thus the inverse relation is not a *function.* You cannot answer the question "How long has the crew been working when the distance painted is zero?"

If the domain of function f is restricted to times no less than a half hour, the inverse relation *is* a function. In this case, function f is said to be **invertible.** If f is invertible, you are allowed to use the symbol f^{-1} for the inverse function. If the domain of f is $0.5 \leq x \leq 8$, then there is exactly *one* distance for each time and *one* time for each distance. Function f is said to be a **one-to-one function.** Any one-to-one function is invertible. A function that is strictly **increasing,** such as the highway stripe function f, or strictly **decreasing** is a one-to-one function and thus is invertible.

The highway stripe problem gives examples of operations with functions from the real world. Example 1 shows you how to operate with a function and its inverse in a strictly mathematical context.

EXAMPLE 1 ➤ Given $f(x) = 0.5x^2 + 2$

a. Make a table of values for $f(-2), f(-1), f(0), f(1)$, and $f(2)$. From the numbers in the table, explain why you cannot find a unique value of x if $f(x) = 2.5$. How does this result tell you that function f is *not* invertible?

b. Plot the five points in part a on graph paper. Connect the points with a smooth curve. On the same axes, plot the five points for the inverse relation and connect them with another smooth curve. How does the graph of the inverse relation confirm that function f is not invertible?

c. Find an equation for the inverse relation. Plot function f and its inverse on the same screen on your grapher. Show that the two graphs are reflections of each other across the line $y = x$.

SOLUTION

a.

x	$f(x)$
-2	4
-1	2.5
0	2
1	2.5
2	4

If $f(x) = 2.5$, there are two different values of x, -1 and 1. You cannot uniquely determine *the* value of x.

Function f is not invertible because there will be two values of y for the same value of x if the variables are interchanged.

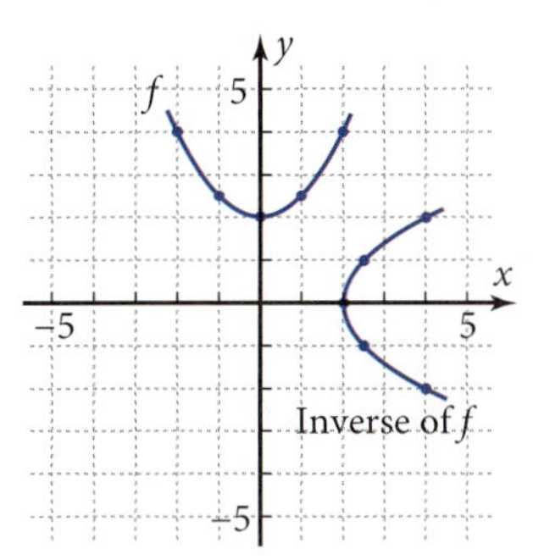

Figure 1-5e

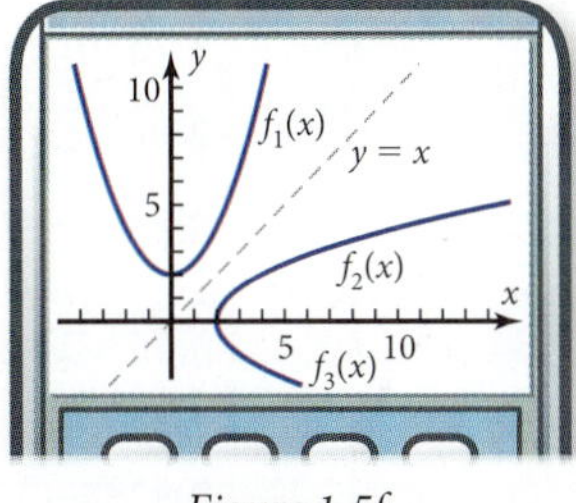

Figure 1-5f

b. Figure 1-5e shows the graphs of function f and its inverse relation. The inverse relation is not a function because there are two values of y for each value of $x > 2$. The inverse relation fails the vertical line test.

c. Function: $y = 0.5x^2 + 2$ — Use y for $f(x)$.

Inverse: $x = 0.5y^2 + 2$ — Interchange x and y.

$y^2 = 2x - 4$

$y = \pm\sqrt{2x - 4}$ — Take the square root of both sides.

$f_1(x) = 0.5x^2 + 2$ — Enter $f(x)$ as $f_1(x)$.

$f_2(x) = \sqrt{(2x - 4)}$

$f_3(x) = -\sqrt{(2x - 4)}$ — Enter the two branches of the inverse relation as $f_2(x)$ and $f_3(x)$.

$f_4(x) = x$ — Enter $y = x$ as $f_4(x)$.

Figure 1-5f shows the graphs of function f and its inverse relation. The graphs are reflections of each other across the line $y = x$.

Parametric Equations

There is a simple way to plot the graph of the inverse of a function with the help of **parametric equations.** Here, x and y are both expressed in terms of some third variable, usually t (because time is often the independent variable in real-world applications).

In this exploration, you will see how to graph a relation specified by parametric equations, both by hand and on your grapher.

EXPLORATION 1-5: Parametric Equations Graph

Let x and y be functions of a third variable, t, as specified by these equations:

$$x = t^2 + 1$$
$$y = t + 2$$

1. Make a table of values of t, x, and y for each integer value of t from -3 to 3.
2. On graph paper, plot the points you found in Problem 1. Connect the points with a smooth curve in the order of increasing values of t.
3. For the relation you plotted in Problem 2, is y a function of x? Explain.
4. Set your grapher to parametric mode. Enter the two equations. Use a window with $-3 \le t \le 3$ and a t-step of 0.1. Set $-10 \le x \le 10$ and $-6 \le y \le 6$. Then plot the graph. Does your grapher's graph agree with your pencil-and-paper graph? If not, make changes until the two graphs agree.
5. What did you learn as a result of doing this exploration that you did not know before?

Example 2 shows you in general what parametric equations are.

EXAMPLE 2 ➤ On graph paper, plot the graph of these parametric equations by first calculating values of x and y for integer values of t from -3 through 7.

$$x_1(t) = |t - 2|$$

$$y_1(t) = \sqrt{t + 2}$$

SOLUTION The table shows values of t, x, and y.

t	x	y
–3	5	None
–2	4	0
–1	3	1
0	2	1.4142...
1	1	1.7320...
2	0	2
3	1	2.2360...
4	2	2.4494...
5	3	2.6457...
6	4	2.8284...
7	5	3

Figure 1-5g shows the graph of the parametric equations. Note that y is not a function of x because there are two values of y for some values of x.

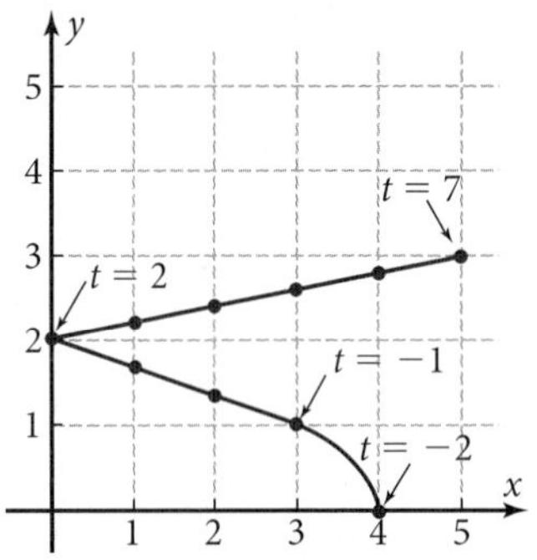

Figure 1-5g

The independent variable t in parametric equations is called the **parameter.** The word comes from the Greek *para-* meaning "alongside," as in "parallel," and *meter,* meaning "measure." The values of t do not show up on the graph in Figure 1-5g unless you write them in.

Your grapher is programmed to plot parametric equations. For the equations in Example 2, use parametric mode and enter

$$x_1(t) = \text{abs}(t - 2)$$

$$y_1(t) = \sqrt{(t + 2)}$$

Use a window with $-3 \le t \le 7$, $0 \le x \le 5$, and $0 \le y \le 5$ and a convenient t-step such as 0.1. The graph will be similar to the graph in Figure 1-5g. ◄

Example 3 shows you how to use parametric equations to plot inverse relations on your grapher.

EXAMPLE 3 ➤ Plot the graph of $y = 0.5x^2 + 2$ for x in the domain $-2 \le x \le 4$ and its inverse using parametric equations. What do you observe about the domain and range of the function and its inverse?

SOLUTION Put your grapher in parametric mode. Then enter

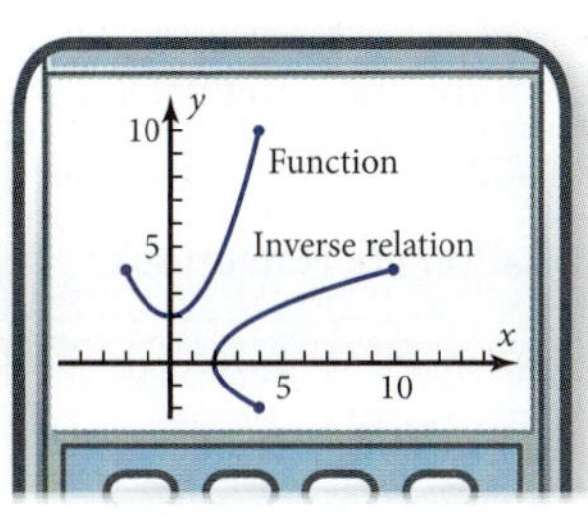

Figure 1-5h

$x_1(t) = t$

$y_1(t) = 0.5t^2 + 2$ Because $x = t$, this is equivalent to $y = 0.5x^2 + 2$.

$x_2(t) = 0.5t^2 + 2$

$y_2(t) = t$ For the inverse, interchange the equations for x and y.

Use a window with $-2 \leq t \leq 4$. Use a convenient t-step, such as 0.1. The result is shown in Figure 1-5h.

The range of the inverse relation is the same as the domain of the function, and vice versa. The range and the domain are interchanged.

Example 4 shows you how to demonstrate algebraically that $f^{-1}(f(x)) = x$ and $f\big(f^{-1}(x)\big) = x$.

EXAMPLE 4 Let $f(x) = 3x + 12$.

a. Find an equation for the inverse of f, and explain how that equation confirms that f is an invertible function.

b. Demonstrate that $f^{-1}(f(x)) = x$ and $f\big(f^{-1}(x)\big) = x$.

SOLUTION

a. Function: $y = 3x + 12$

Inverse: $x = 3y + 12 \Rightarrow y = \frac{1}{3}x - 4$

Because the equation for the inverse relation has the form $y = mx + b$, the inverse is a linear function. Because the inverse relation is a function, f is invertible, so the equation can be written

$$f^{-1}(x) = \frac{1}{3}x - 4$$

b. $f^{-1}(f(x)) = f^{-1}(3x + 12)$ Substitute $3x + 12$ for $f(x)$.

$= \frac{1}{3}(3x + 12) - 4$ Substitute $3x + 12$ as the input for function f^{-1}.

$= x + 4 - 4 = x$ Show that $f^{-1}(f(x))$ equals x.

Also, $f\big(f^{-1}(x)\big) = f\left(\frac{1}{3}x - 4\right)$ Show that $f\big(f^{-1}(x)\big)$ equals x.

$= 3\left(\frac{1}{3}x - 4\right) + 12$

$= x - 12 + 12 = x$

$\therefore f^{-1}(f(x)) = x$ and $f\big(f^{-1}(x)\big) = x$, Q.E.D.

Note: The three-dot mark ∴ stands for "therefore." The letters Q.E.D. stand for the Latin words *quod erat demonstrandum*, meaning "which was to be demonstrated."

The box on the next page summarizes the information of this section regarding inverses of functions.

DEFINITIONS AND PROPERTIES: Function Inverses

- The **inverse** of a relation in two variables is formed by interchanging the two variables.
- The graph of a relation and the graph of its inverse relation are **reflections** of each other across the line $y = x$.
- If the inverse of function f is also a function, then f is **invertible.**
- If f is invertible and $y = f(x)$, then you can write the inverse of f as $y = f^{-1}(x)$.
- To plot the graph of the inverse of a function, either

  Interchange the variables, solve for y, and plot the resulting equation(s), or

  use parametric mode, as in Example 3.
- If f is invertible, then the compositions of f and f^{-1} are

  $f^{-1}(f(x)) = x$, provided x is in the domain of f and $f(x)$ is in the domain of f^{-1}

  $f(f^{-1}(x)) = x$, provided x is in the domain of f^{-1} and $f^{-1}(x)$ is in the domain of f
- A one-to-one function is invertible. Strictly increasing or strictly decreasing functions are one-to-one functions.

Problem Set 1-5

Reading Analysis

From what you have read in this section, what do you consider to be the main idea? Why is it possible to find the inverse of a function even if the function is not invertible? Under what conditions are you permitted to use the symbol f^{-1} for the inverse of function f? How does the meaning of -1 in the function name f^{-1} differ from the meaning of -1 in a numerical expression such as 7^{-1}?

Quick Review

Q1. In the composite function $m(d(x))$, function d is called the __?__ function.

Q2. In the composite function $m(d(x))$, function m is called the __?__ function.

Q3. Give another symbol for $m(d(x))$.

Q4. If $f(x) = 2x$ and $g(x) = x + 3$, find $f(g(1))$.

Q5. Find $g(f(1))$ for the functions in Problem Q4.

Q6. Find $f(f(1))$ for the functions in Problem Q4.

Q7. $|3 - 5| =$ __?__

Q8. Identify the function graphed.

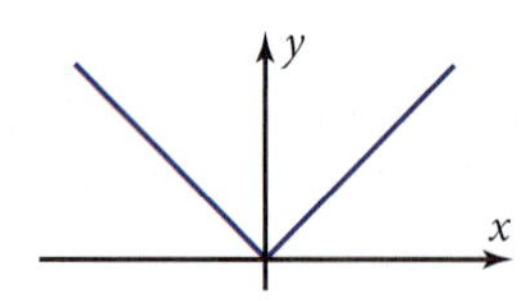

Q9. If $f(x) = 2x$, find $f(0)$.

Q10. If $f(x) = 2x$, find an equation for $g(x)$, a horizontal translation of $f(x)$ by -3 units.

---

1. *Punctured Tire Problem:* Suppose that your car runs over a nail. The table shows the pressure y, in pounds per square inch (psi), of the air inside the tire as a function of x, the number of minutes that have elapsed since the nail punctured the tire.

x (min)	y (psi)
0	36
5	24
10	16
15	10.7
20	7.1
25	4.7
30	0
35	0

a. Let $y = f(x)$. Find $f(5)$, $f(10)$, and $f(15)$.

b. Why is it reasonable to assume that f is an invertible function if x is in the domain $0 \le x \le 25$? Find $f^{-1}(24)$ and $f^{-1}(16)$, and give their real-world meanings.

c. Why is function f not invertible on the whole interval $0 \le x \le 35$? What do you suppose happens between 25 min and 30 min that causes f to not be invertible?

d. Plot the eight given points for function f and the eight corresponding points for the inverse relation. Connect each set of points with a smooth curve. Draw $y = x$ and explain how the two graphs are related to this line.

e. Suppose f is restricted to the domain $0 \le x \le 25$. What is the difference in the meaning of x as an input for function f and x as an input for function f^{-1}?

2. *Cricket Chirping Problem:* The rate at which crickets chirp is a function of the temperature of the air around them. Suppose that the following data have been measured for chirps, *c*, per minute, *y*, at temperatures in degrees Fahrenheit, *x*.

x (°F)	y (c/min)
20	0
30	0
40	5
50	30
60	55
70	80
80	105

a. Let $y = c(x)$. Find $c(40)$, $c(50)$, and $c(60)$.

b. For temperatures of 40°F and above, the chirping rate seems to be a one-to-one function of time. How does this fact imply that function c is invertible for $x \ge 40$? Find the values of $c^{-1}(30)$ and $c^{-1}(80)$. How do these values differ in meaning from $c(30)$ and $c(80)$?

c. Why is function c not invertible for x in the interval $20 \le x \le 80$? What is true in this real-world situation that makes c not invertible?

d. On graph paper, plot the seven given points for function c and the corresponding points for the inverse relation. Connect each set of points with a line or a smooth curve. Draw the line $y = x$ and explain how the two graphs are related to this line.

e. Suppose c is restricted to the domain $40 \le x \le 80$. What is the difference in the meaning of x as an input for function c and x as an input for function c^{-1}?

3. *Punted Football Problem:* Figure 1-5i shows the height of a punted football *y*, in meters, as a function of time *x*, in tenths of a second since it was punted. On a copy of the figure, sketch the graph of the inverse relation and show that the two graphs are reflections across the line $y = x$. How does the graph of the inverse relation reveal that the height function is not invertible?

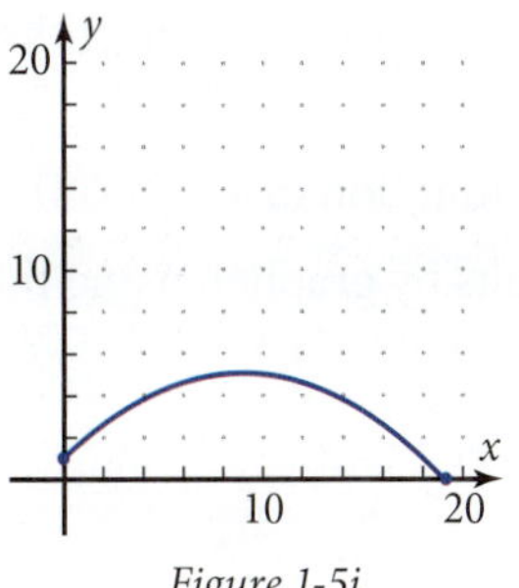

Figure 1-5i

4. *Discrete Function Problem:* Figure 1-5j shows a function that consists of a **discrete** set of points. Show that the function is one-to-one and thus is invertible, even though the function is increasing in some parts of the domain and decreasing in other parts.

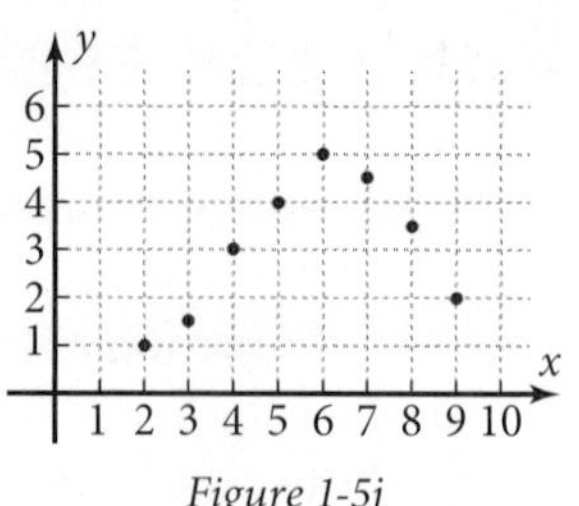

Figure 1-5j

For Problems 5–8, sketch the line $y = x$ and the inverse relation on a copy of the given figure. Be sure that the inverse relation is a reflection of the function graph across the line $y = x$. Tell whether the inverse relation is a function.

5.

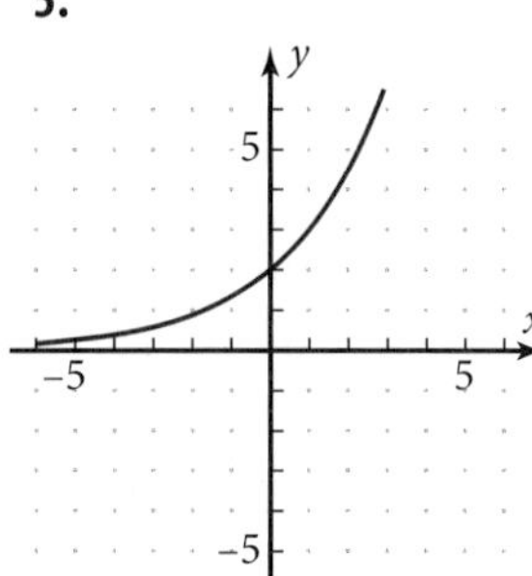

6.

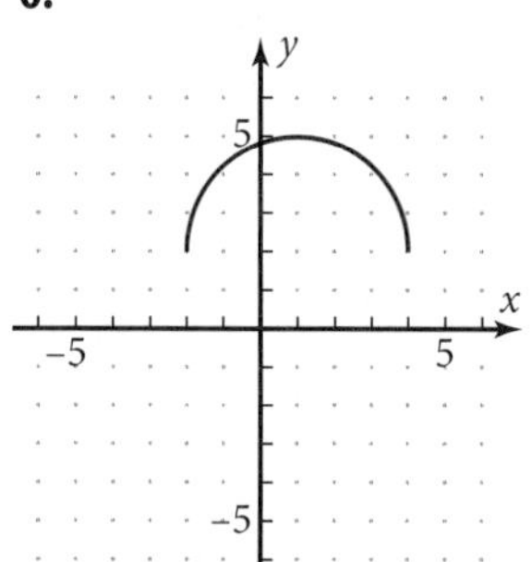

7.

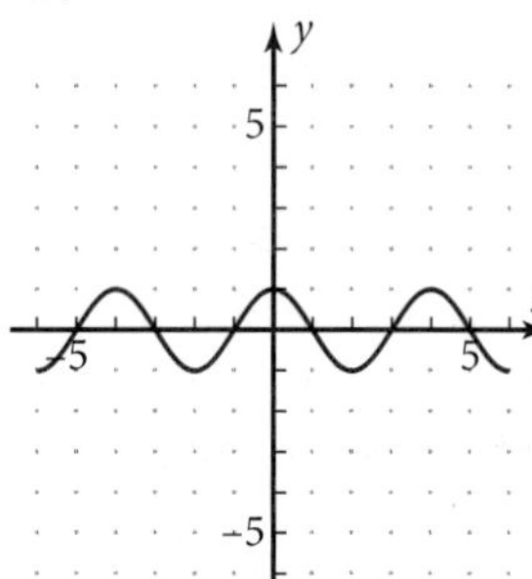

8.

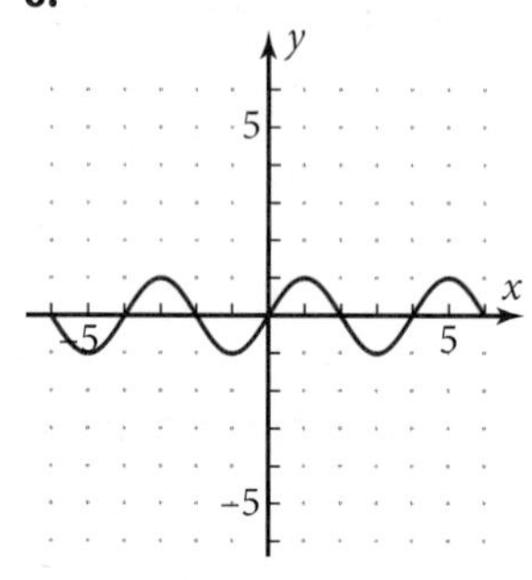

For Problems 9–14,

a. Plot the parametric equations on graph paper using the given domain for t. Connect the points with lines or smooth curves.

b. Tell whether y is a function of x.

c. Confirm your results by grapher, using the given domain for t.

9. $x = |t| - 3 \qquad -5 \leq t \leq 5$
   $y = 2 - t$

10. $x = 5 - |t| \qquad -7 \leq t \leq 4$
    $y = t + 1$

11. $x = 7 - t^2 \qquad -3 \leq t \leq 4$
    $y = t + 2$

12. $x = t - 3 \qquad -1 \leq t \leq 4$
    $y = (t - 2)^2$

13. $x = t + 3 \qquad -1 \leq t \leq 3$
    $y = t^3 - 2t^2$

14. $x = t^2 - 2t + 2 \qquad -1 \leq t \leq 2$
    $y = t^3 - t^2 + t + 1$

15. *Two Paths Problem:* Two particles (small objects) move along the paths shown in Figure 1-5k. The paths are given by these parametric equations, where x and y are distance in meters and t is time in seconds.

    Particle 1: $x = t + 1$, $y = 7 - t^2$ Particle 2: $x = 1.5t + 2$, $y = 1.5t + 6$

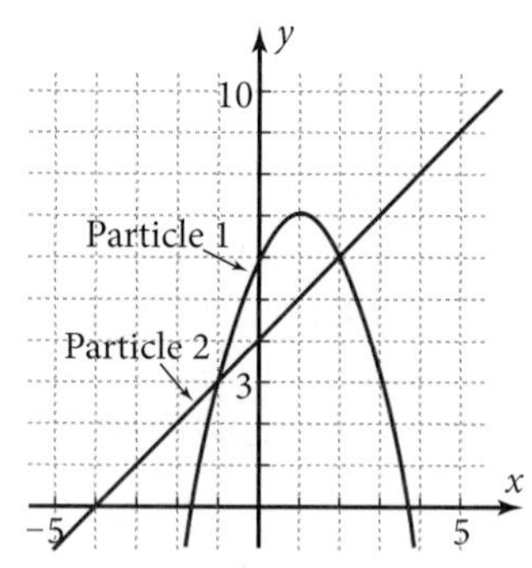

Figure 1-5k

a. The paths intersect at two points. For each point, determine whether the particles reach that point at the same time or at different times. Give numbers to support your conclusion.

b. Confirm your answer to part a graphically by plotting the two sets of parametric equations dynamically on the same screen, setting your grapher to simultaneous mode. Write a sentence or two explaining how your graph confirms your answer to part a.

16. *Two Ships Problem:* At time $t = 0$ h, a freighter is at the point (90, 10) to the east-northeast of a lighthouse located at the origin of a Cartesian coordinate system, where x and y are distance in miles. At time $t = 2$ h, a Coast Guard cutter starts from the lighthouse to intercept the freighter. Figure 1-5l shows the graph of these parametric equations representing the ships' paths:

Freighter: $x = 90 - 10t$, $y = 10 + 5t$ Cutter: $x = 8(t - 2)$, $y = 10(t - 2)$

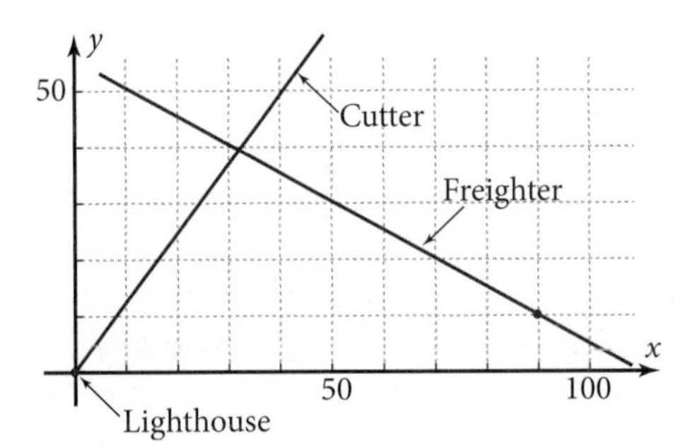

Figure 1-5l

a. Find the value of t at which the y-values of the two paths are equal. At this value of t, are the two x-values equal?

b. Do the two ships arrive at the intersection point at the same time? If so, how can you tell? If not, which ship arrives at the intersection point first?

For Problems 17–28, plot the function in the given domain using parametric mode. On the same screen, plot the inverse relation. Tell whether the inverse relation is a function. Sketch the graphs.

17. $f(x) = 2x - 6 \quad -1 \le x \le 5$
18. $f(x) = -0.4x + 4 \quad -7 \le x \le 10$
19. $f(x) = -x^2 + 4x + 1 \quad 0 \le x \le 5$
20. $f(x) = x^2 - 2x - 4 \quad -2 \le x \le 4$
21. $f(x) = 2^x$ x is any real number.
22. $f(x) = 0.5^x$ x is any real number.
23. $f(x) = -\sqrt{3 - x} \quad -6 \le x \le 3$
24. $f(x) = \sqrt[3]{x} \quad -1 \le x \le 8$
25. $f(x) = \dfrac{1}{x - 3} \quad -2 \le x \le 8$
26. $f(x) = \dfrac{x}{x + 1} \quad -6 \le x \le 4$
27. $f(x) = x^3 \quad -2 \le x \le 1$
28. $f(x) = 0.016x^4 \quad -4 \le x \le 5$

For Problems 29–32, write an equation for the inverse relation by interchanging the variables and solving for y in terms of x. Then plot the function and its inverse on the same screen, using function mode. Sketch the result, showing that the function and its inverse are reflections across the line $y = x$. Tell whether the inverse relation is a function.

29. $y = 2x - 6$
30. $y = -0.4x + 4$
31. $y = -0.5x^2 - 2$
32. $y = 0.4x^2 + 3$
33. Show that $f(x) = \frac{1}{x}$ is its own inverse function.
34. Show that $f(x) = -x$ is its own inverse function.
35. *Cost of Owning a Car Problem:* Suppose that you have fixed costs (car payments, insurance, and so on) of \$500 per month and operating costs of \$0.40 per mile you drive. The monthly cost of owning the car is given by the linear function

$$c(x) = 0.40x + 500$$

where x is the number of miles you drive the car in a given month and $c(x)$ is the number of dollars per month you spend.

a. Find $c(1000)$. Explain the real-world meaning of the answer.

b. Find an equation for $c^{-1}(x)$, where x now stands for the number of dollars you spend instead of the number of miles you drive. Explain why you can use the symbol c^{-1} for the inverse relation. Use the equation of $c^{-1}(x)$ to find $c^{-1}(758)$, and explain its real-world meaning.

c. Plot $f_1(x) = c(x)$ and $f_2(x) = c^{-1}(x)$ on the same screen, using function mode. Use a window with $0 \le x \le 1000$ and use equal scales on the two axes. Sketch the two graphs, showing how they are related to the line $y = x$.

36. *Deer Problem:* The surface area of a deer's body is approximately proportional to the $\frac{2}{3}$ power of the deer's weight. (This is true because the area is proportional to the square of the length and the weight is proportional to the cube of the length.) Suppose that the particular equation for area as a function of weight is given by the power function

$$A(x) = 0.4x^{2/3}$$

where x is the weight in pounds and $A(x)$ is the surface area measured in square feet.

a. Find $A(50)$, $A(100)$, and $A(150)$. Explain the real-world meaning of the answers.

b. True or false: "A deer twice the weight of another deer has a surface area twice that of the other deer." Give numerical evidence to support your answer.

c. Find an equation for $A^{-1}(x)$, where x now stands for area instead of weight.

d. Plot A and A^{-1} on the same screen using function mode. Use a window with $0 \leq x \leq 250$. How are the two graphs related to the line $y = x$?

37. *Braking Distance Problem:* The length of skid marks, $d(x)$ feet, left by a car braking to a stop is a direct square power function of x, the speed in miles per hour when the brakes were applied. Based on information in the *Texas Drivers Handbook* (2002), $d(x)$ is given approximately by

$$d(x) = 0.057x^2 \qquad \text{for } x \geq 0$$

The graph of this function is shown in Figure 1-5m.

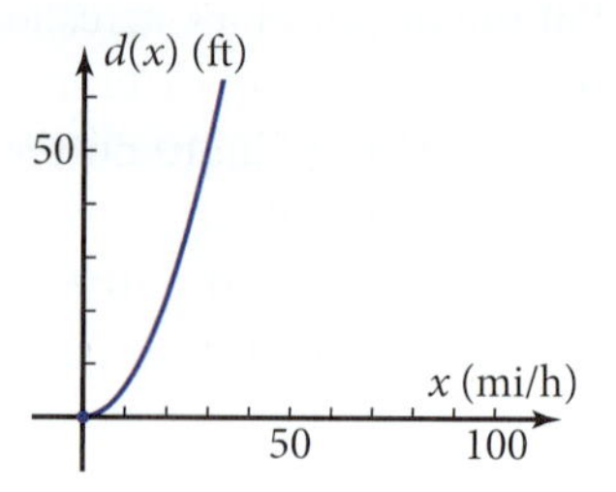

Figure 1-5m

a. When police officers investigate automobile accidents, they use the length of the skid marks to calculate the speed of the car at the time it started to brake. Write an equation for the inverse function, $d^{-1}(x)$, where x is now the length of the skid marks. Explain why you need to take only the positive square root.

b. Find $d^{-1}(200)$. What does this number represent in the context of this problem?

c. Suppose that the domain of function d started at -20 instead of zero. With your grapher in parametric mode, plot the graphs of function d and its inverse relation. Use the window shown in Figure 1-5m with $-20 \leq t \leq 70$. Sketch the result.

d. Explain why the inverse of function d in part c is not a function. What relationship do you notice between the domain and range of d and its inverse?

38. *Horizontal Line Test Problem:* The vertical line test of Section 1-2, Problem 39, helps you see graphically that a relation is a function if no vertical line crosses the graph more than once. The *horizontal line test* allows you to tell whether a function is invertible. Sketch two graphs, one for an invertible function and one for a non-invertible function, that illustrate this test.

PROPERTY: The Horizontal Line Test

If a horizontal line cuts the graph of a function in more than one place, then the function is not invertible because it is not one-to-one.

1-6 Reflections, Absolute Values, and Other Transformations

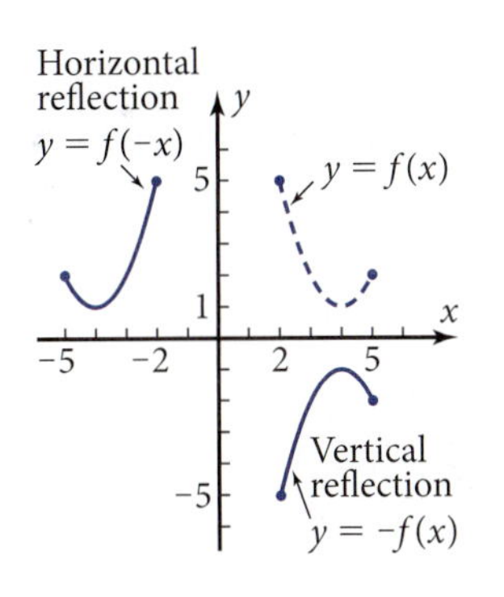

Figure 1-6a

In Section 1-3, you learned that if $y = f(x)$, then multiplying x by a nonzero constant causes a horizontal dilation. Suppose that the constant is -1. Each x-value will be $1/(-1)$ or -1 times what it was in the pre-image. Figure 1-6a shows that the resulting image is a horizontal reflection of the graph across the y-axis. The new graph is the same size and shape, simply a mirror image of the original. Similarly, a vertical dilation by a factor of -1 reflects the graph vertically across the x-axis.

In this section you will learn special transformations of functions that reflect their graphs in various ways. You will also learn what happens when you take the absolute value of a function or of the independent variable x. Finally, you will learn about odd and even functions.

Objective

Given a function, transform it by reflection and by applying *absolute value* to the function or its argument.

Reflections Across the *x*-axis and *y*-axis

Example 1 shows you how to plot the graphs in Figure 1-6a.

EXAMPLE 1 ➤ The pre-image function $y = f(x)$ in Figure 1-6a is $f(x) = x^2 - 8x + 17$, where $2 \le x \le 5$.

a. Write an equation for the reflection of this function across the y-axis.

b. Write an equation for the reflection of this function across the x-axis.

c. Plot the pre-image and the two reflections on the same screen.

SOLUTION

a. A reflection across the y-axis is a horizontal dilation by a factor of -1. So

$y = f(-x) = (-x)^2 - 8(-x) + 17$ — Substitute $-x$ for x.

$y = x^2 + 8x + 17$

Domain: $2 \le -x \le 5$

$-2 \ge x \ge -5$ or $-5 \le x \le -2$ — Multiply all three sides of the inequality by -1. The inequalities reverse.

b. $y = -f(x)$ — For a reflection across the x-axis, find the opposite of $f(x)$.

$y = -x^2 + 8x - 17$ — The domain remains $2 \le x \le 5$.

c. $f_1(x) = x^2 - 8x + 17 \,/\, (x \geq 2 \text{ and } x \leq 5)$ — Divide by a Boolean variable or enter the domain directly, depending on your grapher, to restrict the domain.

$f_2(x) = x^2 + 8x + 17 \,/\, (x \geq -5 \text{ and } x \leq -2)$

$f_3(x) = -x^2 + 8x - 17 \,/\, (x \geq 2 \text{ and } x \leq 5)$

The graphs are shown in Figure 1-6a.

You can check the algebraic solutions by plotting $f_4(x) = f_1(-x)$ and $f_5(x) = -f_1(x)$ using thick style. The graphs should overlay $f_2(x)$ and $f_3(x)$.

PROPERTY: Reflections Across the Coordinate Axes

$g(x) = -f(x)$ is a vertical reflection of function f across the x-axis.

$g(x) = f(-x)$ is a horizontal reflection of function f across the y-axis.

Absolute Value Transformations

Suppose you shoot a basketball. While in the air, it is above the basket level sometimes and below it at other times. Figure 1-6b shows $y = f(x)$, the **displacement** from the level of the basket as a function of time. If the ball is above the basket, its displacement is positive; if the ball is below the basket, its displacement is negative.

Figure 1-6b

Distance, however, is the magnitude (or size) of the displacement, which is never negative.

Distance equals the *absolute value* of the displacement. The solid graph in Figure 1-6c is the graph of $y = g(x) = |f(x)|$. Taking the absolute value of $f(x)$ retains the non-negative values of y and reflects the negative values vertically across the x-axis.

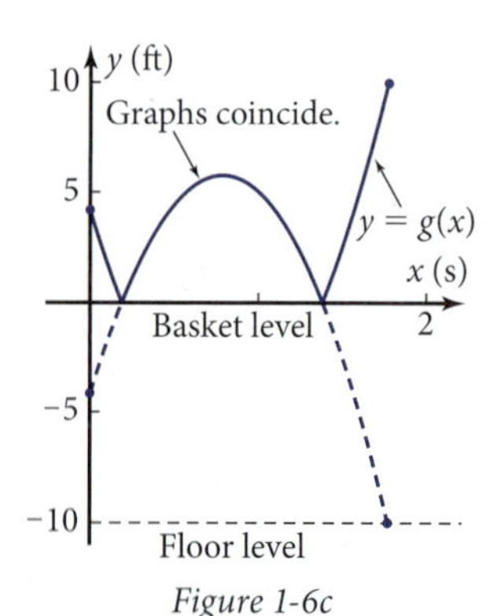

Figure 1-6c

Figure 1-6d shows what happens for $g(x) = f(|x|)$, for which you take the absolute value of the argument (this is a different function f than in the last example). For positive values of x, $|x| = x$, so $g(x) = f(x)$ and the graphs coincide. For negative values of x, $|x| = -x$, so $g(x) = f(-x)$, across the y-axis of the part of function f where $x > 0$. Notice that the graph of f for the negative values of x is not a part of the graph of $f(|x|)$.

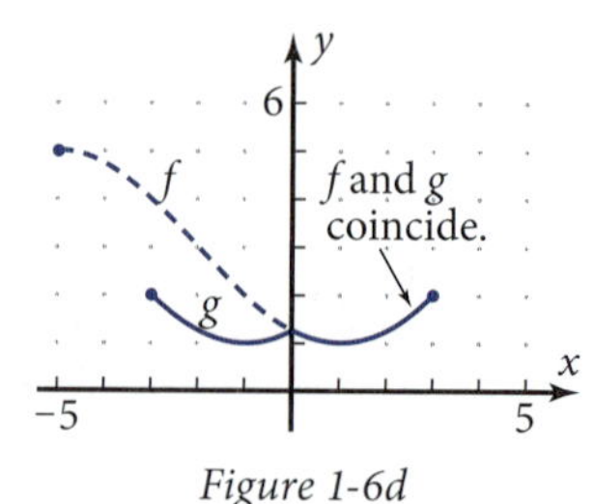

Figure 1-6d

The equation for $g(x)$ can be written this way:

$$g(x) = \begin{cases} f(x) & \text{if } x \geq 0 \\ f(-x) & \text{if } x < 0 \end{cases}$$

Because there are two different rules for $g(x)$ in different "pieces" of the domain, g is called a **piecewise function** of x.

PROPERTY: Absolute Value Transformations

The transformation $g(x) = |f(x)|$
- Reflects f across the x-axis if $f(x)$ is negative
- Leaves f unchanged if $f(x)$ is nonnegative

The transformation $g(x) = f(|x|)$
- Leaves f unchanged for nonnegative values of x
- Reflects the part of the graph for positive values of x to the corresponding negative values of x
- Eliminates the part of f for negative values of x

Even Functions and Odd Functions

Figure 1-6e shows the graph of $f(x) = -x^4 + 5x^2 - 1$, a polynomial function with only even exponents. (The number 1 equals $1x^0$, which has an even exponent.) Figure 1-6f shows the graph of $f(x) = -x^3 + 6x$, a polynomial function with only odd exponents. What symmetries do you observe?

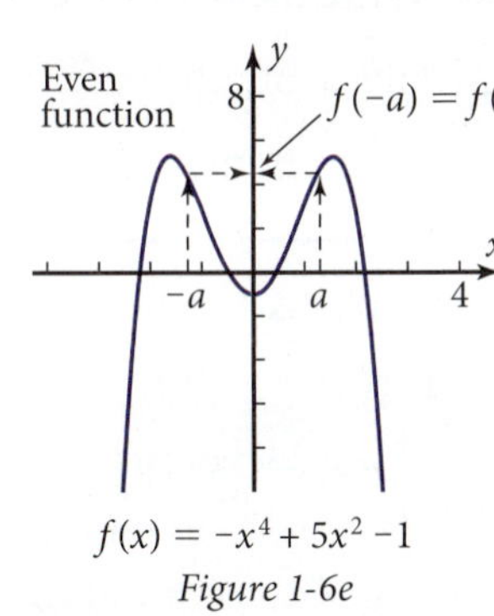

$f(x) = -x^4 + 5x^2 - 1$

Figure 1-6e

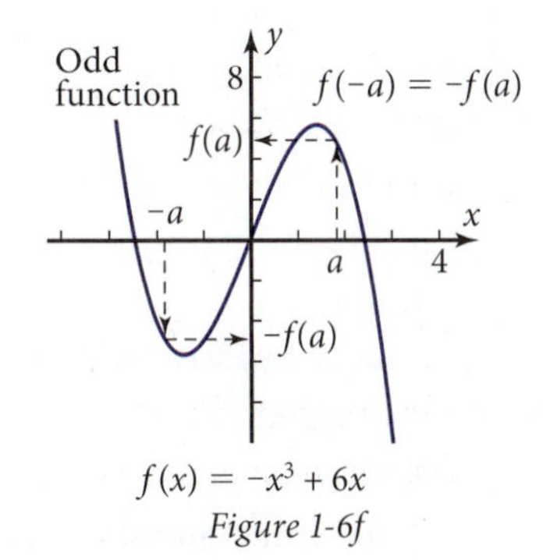

$f(x) = -x^3 + 6x$

Figure 1-6f

Reflecting the graph of the even function $f(x) = -x^4 + 5x^2 - 1$ horizontally across the y-axis leaves the graph unchanged. You can see this algebraically given the property of powers with even exponents.

$$f(-x) = -(-x)^4 + 5(-x)^2 - 1 \quad \text{Substitute } -x \text{ for } x.$$

$$f(-x) = -x^4 + 5x^2 - 1 \quad \text{Negative number raised to an even power.}$$

$$f(-x) = f(x)$$

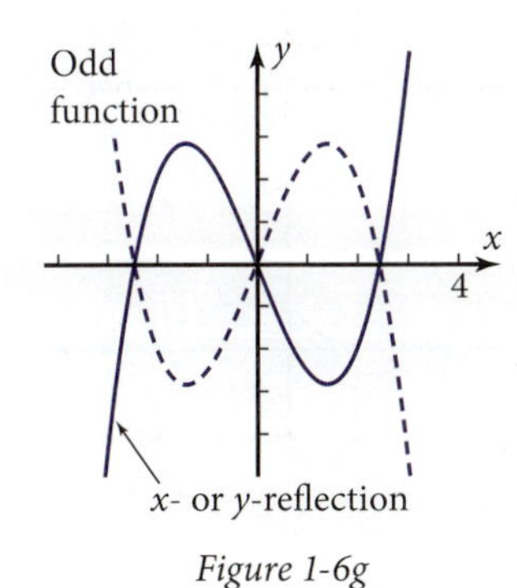

Figure 1-6g

Figure 1-6g shows that reflecting the graph of the odd function $f(x) = -x^3 + 6x$ horizontally across the y-axis has the same effect as reflecting it vertically across the x-axis. Algebraically,

$$f(-x) = -(-x)^3 + 6(-x) \quad \text{Substitute } -x \text{ for } x.$$

$$f(-x) = x^3 - 6x \quad \text{Negative number raised to an odd power.}$$

$$f(-x) = -f(x)$$

Any function having the property $f(-x) = f(x)$ is called an **even function.** Any function having the property $f(-x) = -f(x)$ is called an **odd function.** These names apply even if the equation for the function does not have exponents.

> **DEFINITION: Even Function and Odd Function**
>
> The function f is an **even function** if and only if $f(-x) = f(x)$ for all x in the domain.
>
> The function f is an **odd function** if and only if $f(-x) = -f(x)$ for all x in the domain.

Note: For odd functions, reflection across the y-axis gives the same image as reflection across the x-axis. For even functions, reflection across the y-axis is the same as the pre-image. So odd functions are symmetric about the origin, and even functions are symmetric across the y-axis. Most functions do not possess the property of oddness or evenness.

Problem Set 1-6

Reading Analysis

From what you have read in this section, what do you consider to be the main idea? Reread the paragraph on page 48 that discusses Figure 1-6d. Use numerical values from the graph to guide yourself through this paragraph. Explain in your own words what the sentence about negative values of x means. Why is part of the graph of f "lost" in the graph of $f(|x|)$? Write down specific questions about what you may not understand, and find someone who can answer them.

Quick Review

Q1. If $f(x) = 2x$, then $f^{-1}(x) =$ __?__.

Q2. If $f(x) = x - 3$, then $f^{-1}(x) =$ __?__.

Q3. If $f(x) = 2x - 3$, then $f^{-1}(x) =$ __?__.

Q4. If $f(x) = x^2$, write the equation for the inverse relation.

Q5. Explain why the inverse relation in Problem Q4 is not a function.

Q6. If $f(x) = 2^x$, then $f^{-1}(8) =$ __?__.

Q7. If the inverse relation for function f is also a function, then f is called __?__.

Q8. Write the definition of a one-to-one function.

Q9. Give a number x for which $|x| = x$.

Q10. Give a number x for which $|x| = -x$.

For Problems 1–4, sketch the graphs of

a. $g(x) = -f(x)$ **b.** $h(x) = f(-x)$

c. $a(x) = |f(x)|$ **d.** $v(x) = f(|x|)$

1.

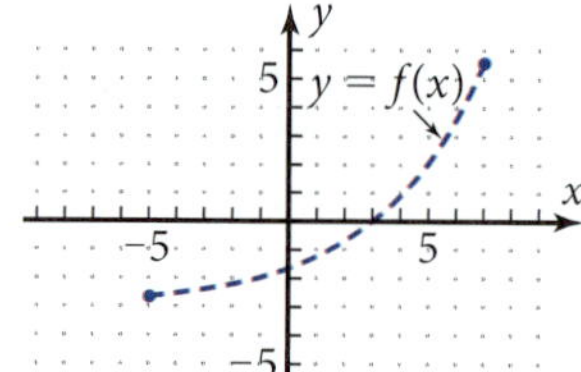

2.

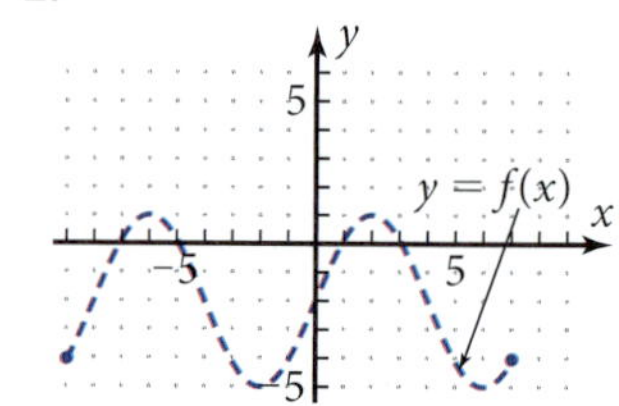

3.

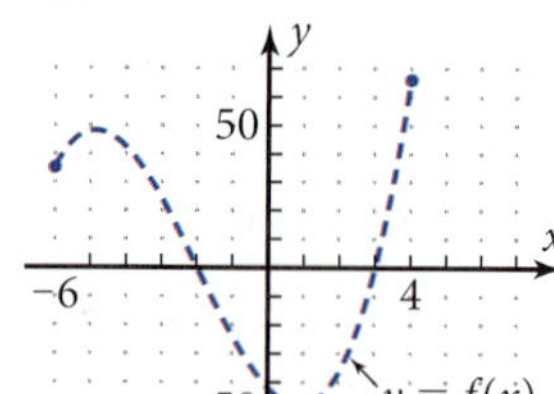

4.

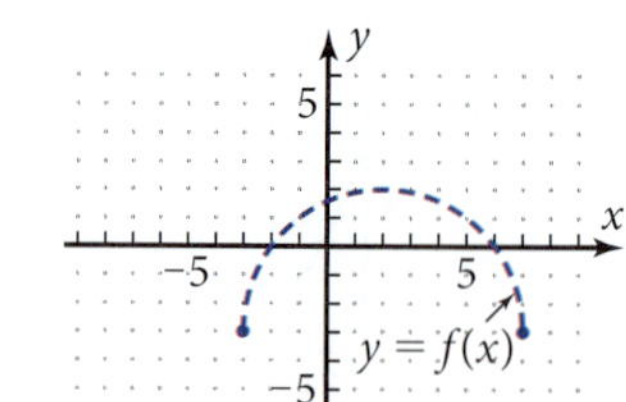

5. The equation for the function in Problem 3 is $f(x) = x^3 + 6x^2 - 13x - 42$ for $-6 \le x \le 4$. Plot the function as $f_1(x)$ on your grapher. Plot $f_2(x) = f_1(|x|)$ using thick style. Does the result confirm your answer to Problem 3, part d?

6. The equation for the function in Problem 4 is $f(x) = -3 + \sqrt{25 - (x - 2)^2}$ for $-3 \le x \le 7$. Plot the function as $f_1(x)$ on your grapher. Plot $f_2(x) = f_1(|x|)$ using thick style. Does the result confirm your answer to Problem 4, part d?

7. *Absolute Value Transformations Problem:* Figure 1-6h shows the graph of $f(x) = 0.5(x - 2)^2 - 4.5$ in the domain $-2 \le x \le 6$.

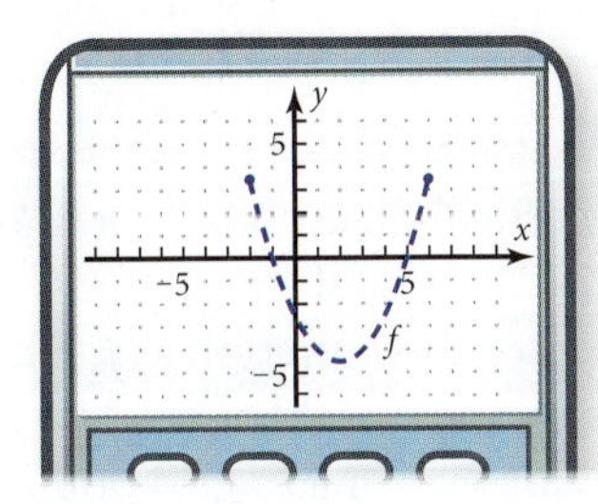

Figure 1-6h

a. Plot the graph of $f_1(x) = f(x)$. On the same screen, plot $f_2(x) = |f(x)|$ using thick style. Sketch the result and describe how this transformation changes the graph of f.

b. Deactivate $f_2(x)$. On the same screen as $f_1(x)$, plot the graph of $f_3(x) = f(|x|)$ using thick style. Sketch the result and describe how this transformation changes the graph of f.

c. Use the equation for function f to find the value of $|f(3)|$ and the value of $f(|-3|)$. Show that both results agree with your graphs in parts a and b. Explain why -3 is in the domain of $f(|x|)$ even though it is not in the domain of f itself.

d. Figure 1-6i shows the graph of a function g, but you don't know the equation for the function. On a copy of this figure, sketch the graph of $y = |g(x)|$, using the conclusion you reached in part a. On another copy of this figure, sketch the graph of $y = g(|x|)$, using the conclusion you reached in part b.

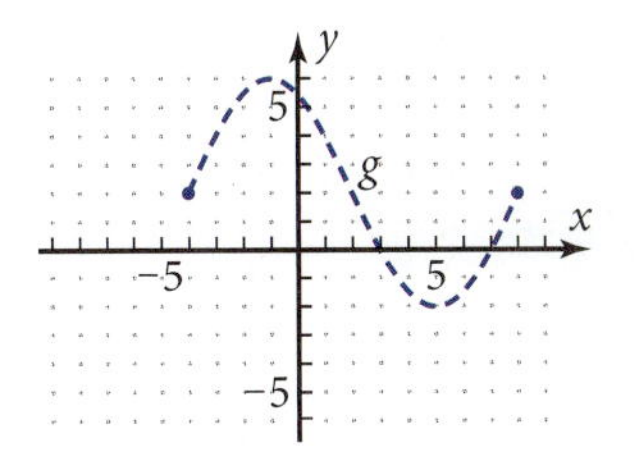

Figure 1-6i

8. *Displacement vs. Distance Absolute Value Problem:* Calvin's car runs out of gas as he is going uphill. He continues to coast uphill for a while, stops, then starts rolling backward without applying the brakes. His displacement, y, in meters, from a gas station on the hill as a function of time, x, in seconds, is given by

$$y = -0.1x^2 + 12x - 250$$

a. Plot the graph of this function. Sketch the result.

b. Find Calvin's displacement at 10 s and at 40 s. What is the real-world meaning of his negative displacement at 10 s?

c. What is Calvin's distance from the gas station at times $x = 10$ s and $x = 40$ s? Explain why both values are positive.

d. Define Calvin's distance from the gas station. Sketch the graph of distance versus time.

e. If Calvin keeps moving as indicated in this problem, when will he pass the gas station as he rolls back down the hill?

9. *Even Function and Odd Function Problem:* Figure 1-6j shows the graph of the even function $f(x) = x^4 - 3x^2 - 4$. Figure 1-6k on the next page shows the graph of the odd function $g(x) = x^5 - 6x^3 + 6x$.

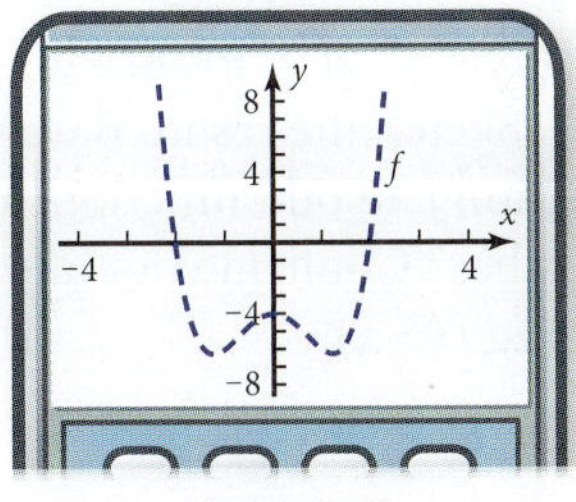

Figure 1-6j

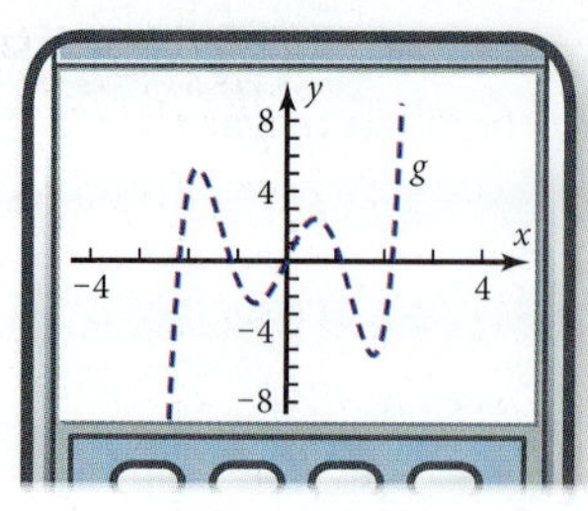

Figure 1-6k

a. On the same screen, plot $f_1(x) = f(x)$ and $f_2(x) = f(-x)$. Use thick style for $f_2(x)$. Based on the properties of negative numbers raised to even powers, explain why the two graphs are identical.

b. Deactivate $f_1(x)$ and $f_2(x)$. On the same screen, plot $f_3(x) = g(x)$, $f_4(x) = g(-x)$, and $f_5(x) = -g(x)$. Use thick style for $f_5(x)$. Based on the properties of negative numbers raised to odd powers, explain why the graphs of $f_4(x)$ and $f_5(x)$ are identical.

c. Even functions have the property $f(-x) = f(x)$. Odd functions have the property $f(-x) = -f(x)$. Figure 1-6l shows two functions, h and j, but you don't know the equation of either function. Tell which function is an even function and which is an odd function.

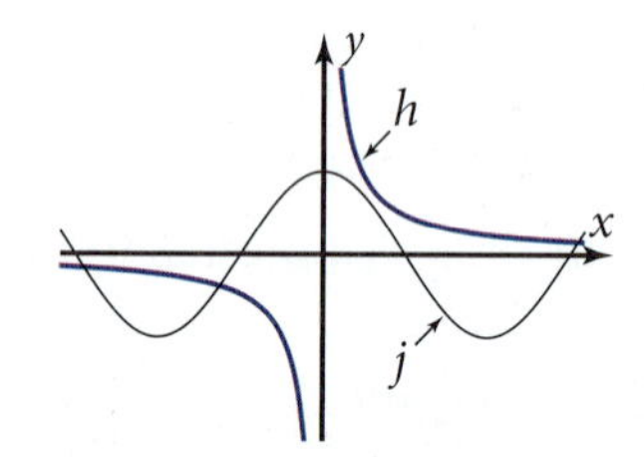

Figure 1-6l

d. Let $e(x) = 2^x$. Sketch the graph. Based on the graph, is function e an odd function, an even function, or neither? Confirm your answer algebraically by finding $e(-x)$.

10. *Absolute Value Function—Odd or Even?* Plot the graph of $f(x) = |x|$. Sketch the result. Based on the graph, is function f an odd function, an even function, or neither? Confirm your answer algebraically by finding $f(-x)$.

11. *Step Discontinuity Problem:* Figure 1-6m shows the graph of

$$f(x) = \frac{|x|}{x}$$

The graph has a *step discontinuity* at $x = 0$, where $f(x)$ jumps instantaneously from -1 to 1.

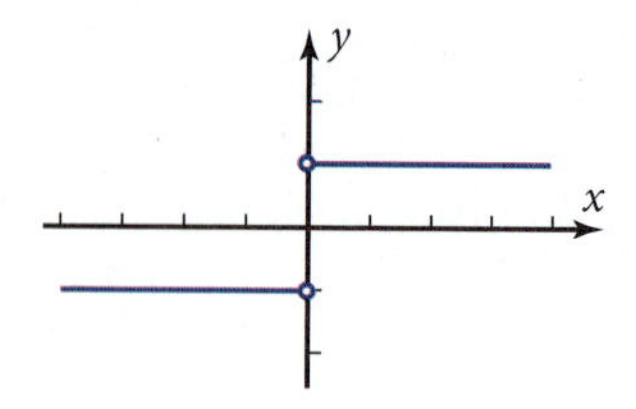

Figure 1-6m

a. Plot the graph of $f_1(x) = f(x)$. Use a window that includes $x = 0$ as a grid point. Does your graph agree with the figure?

b. Figure 1-6n is a vertical dilation of function f with vertical and horizontal translations. Enter an equation for this function as $f_2(x)$, using operations on the variable $f_1(x)$. Use a window that includes $x = 4$ as a grid point. When you have duplicated the graph in Figure 1-6n, write an equation for the transformed function in terms of function f.

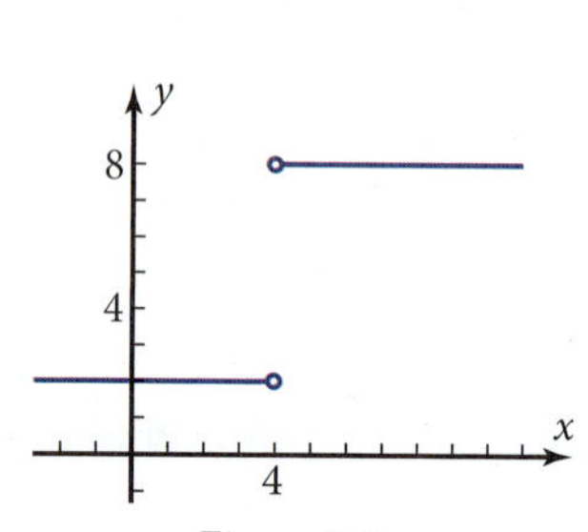

Figure 1-6n

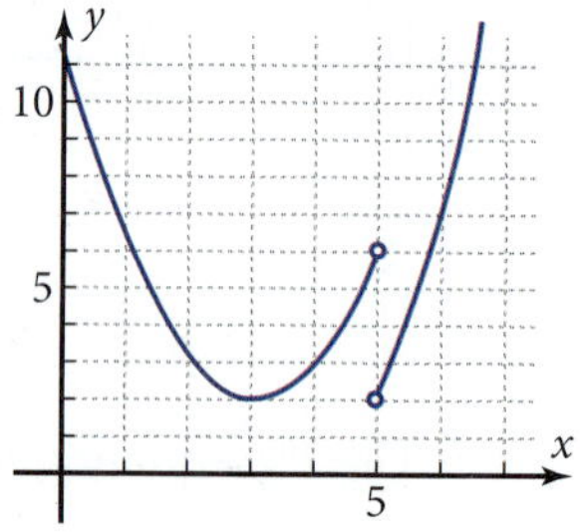

Figure 1-6o

c. Figure 1-6o shows the graph of the quadratic function $y = (x - 3)^2$ to which something has been added or subtracted to give it a step discontinuity of 4 units at $x = 5$. Find an equation of the function. Verify that your equation is correct by plotting it on your grapher.

12. *Step Functions—The Postage Stamp Problem:* Figure 1-6p shows the graph of the **greatest integer function,** $f(x) = \lfloor x \rfloor$. In this function, $\lfloor x \rfloor$ is the greatest integer less than or equal to x. For instance, $\lfloor 3.9 \rfloor = 3$, $\lfloor 5 \rfloor = 5$, and $\lfloor -2.1 \rfloor = -3$.

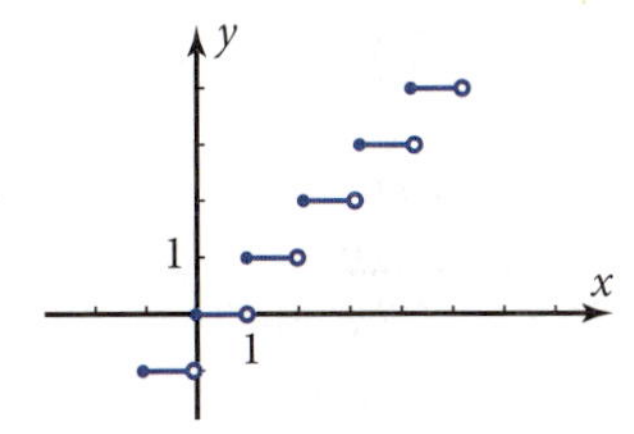

Figure 1-6p

a. Plot the greatest integer function using dot style so that points will not be connected. Most graphers use the symbol int(x) for $\lfloor x \rfloor$. Trace to $x = 2.9$, $x = 3$, and $x = 3.1$. What do you find for the three y-values?

b. In the year 2005, the postage for a first-class letter was 37 cents for weights up to 1 oz and 23 cents more for each additional ounce or fraction of an ounce. Sketch the graph of this function.

c. Using a transformation of the greatest integer function, write an equation for the 2005 postage as a function of the weight. Plot it on your grapher. Does the graph agree with the one you sketched in part b?

d. In 2005, first-class postage rates applied only until the letter reached the weight at which the postage would exceed \$3.13. What is the domain of the function in part c?

e. Check the Internet or another source to find the first-class postage rates *this* year. What differences do you find from the 2005 rates? Cite the source you used.

13. *Piecewise Functions—Weight Above and Below Earth's Surface Problem:* When you are above the surface of Earth, your weight is inversely proportional to the square of your distance from the center of Earth. This is because the farther away you are, the weaker the gravitational force between Earth and you. When you are below the surface of Earth, your weight is directly proportional to your distance from the center. At the center you would be "weightless" because Earth's gravity would pull you equally in all directions.

Figure 1-6q shows the graph of the weight function for a 150-lb person. The radius of Earth is about 4000 mi. The weight is called a *piecewise function* of the distance because it is given by different equations in different "pieces" of the domain. Each piece is called a branch of the function. The equation of the function can be written

$$y = \begin{cases} ax & \text{if } 0 \le x \le 4000 \\ \dfrac{b}{x^2} & \text{if } x \ge 4000 \end{cases}$$

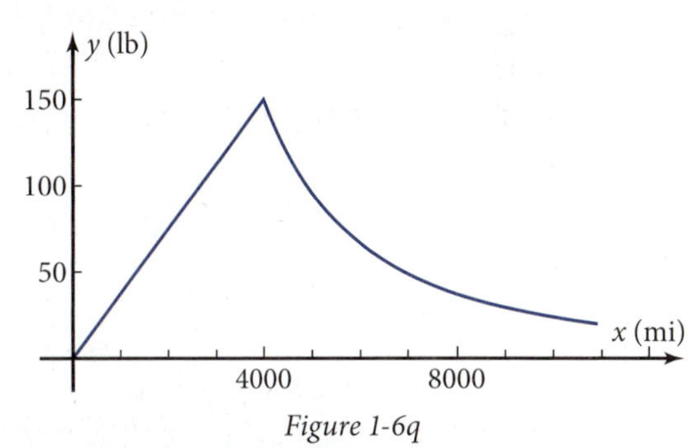

Figure 1-6q

a. Find the values of a and b that make $y = 150$ when $x = 4000$ for each branch.

b. Plot the graph of f. Use piecewise functions or Boolean variables to restrict the domain of the graph.

c. Find y if $x = 3000$ and if $x = 5000$.

d. Find the two distances from the center at which the weight would be 50 lb.

14. *Dynamic Reflection Problem:* Go to **www.keymath.com/precalc** and open the *Dilation* exploration. Set slider c equal to 1 and slider d equal to -1 and describe what you observe. Then set slider c equal to -1 and slider d equal to 1 and describe what you observe. Finally, set both sliders equal to -1 and describe what you observe. Explain how reflections are related to dilations.

1-7 Precalculus Journal

In this chapter you have been learning mathematics graphically, numerically, and algebraically. An important ability to develop for any subject you study is to *verbalize* what you have learned, both orally and in writing. To gain verbal practice, you should start a journal. In it you will record topics you have studied and topics about which you are still unsure. The word *journal* comes from the same word as the French *jour*, meaning "day." *Journey* has the same root and means "a day's travel." Your journal will give you a written record of your travel through mathematics.

Objective

Start writing a journal in which you can record things you have learned about precalculus mathematics and questions you have concerning concepts about which you are not quite clear.

You should use a bound notebook or a spiral notebook with large index cards for pages so that your journal will hold up well under daily use. Researchers use such notebooks to record their findings in the laboratory or in the field. Each entry should start with the date and a title for the topic. A typical entry might look like this sample.

Topic: Inverse of a Function *9/15*

I've learned that you invert a function by interchanging the variables. Sometimes an inverse is a relation that is not a function. If it is a function, the inverse of $y = f(x)$ *is* $y = f^{-1}(x)$. *At first, I thought this meant* $\frac{1}{f(x)}$ *but after losing 5 points on a quiz, I realized that wasn't correct. The graphs of* f *and* f^{-1} *are reflections of each other across the line* $y = x$, *like this:*

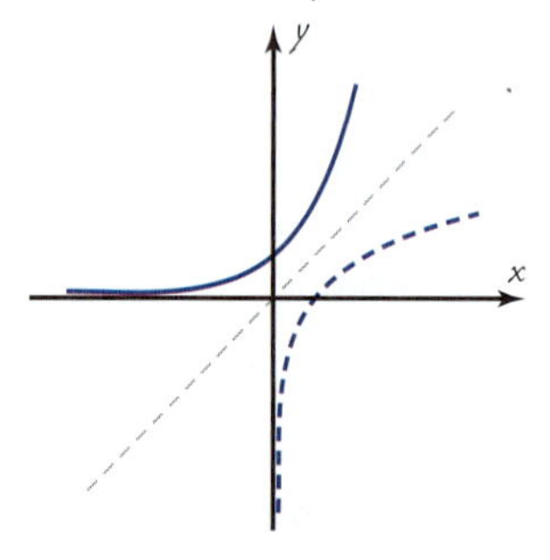

Problem Set 1-7

1. Start a journal for recording your thoughts about precalculus mathematics. The first entry should include things such as these:
   - Sketches of graphs from real-world information
   - Familiar kinds of functions from previous courses
   - How to dilate, translate, reflect, compose, and invert the graph of a function
   - How you feel about what you are learning, such as its potential usefulness to you
   - Any difficulties or misconceptions you had but overcame
   - Any topics about which you are still unsure

1-8 Chapter Review and Test

In this chapter you saw how you can use functions as mathematical models algebraically, graphically, numerically, and verbally. Functions describe a relationship between two variable quantities, such as distance and time for a moving object. Functions defined algebraically are named according to the way the independent variable appears in the equation. If x is an exponent, the function is an exponential function, and so forth. You can transform the graphs of functions by dilating and translating them in the x- and y-directions. Some of these transformations reflect the graph across the x- or y-axis or the line $y = x$. A good understanding of functions will prepare you for later courses in calculus, in which you will learn how to find the rate of change of y as x varies.

You may continue your study of precalculus mathematics either with periodic functions in Chapters 5 through 9, which will probably be quite new to you, or with the fitting of other functions to real-world data in Chapters 2 through 4, which may be more familiar to you from previous courses.

The Review Problems are numbered according to the sections of this chapter. Answers are provided at the back of the book. The Concept Problems allow you to apply your knowledge to new situations. Answers are not provided, and, in some chapters, you may be required to do research to find answers to open-ended problems. The Chapter Test is more like a typical classroom test your instructor might give you. It has a calculator part and a noncalculator part, and the answers are not provided.

Review Problems

R1. *Punctured Tire Problem:* For parts a–d, suppose that your car runs over a nail. The tire's air pressure, y, in pounds per square inch (psi), decreases with time, x, in minutes, as the air leaks out. A graph of pressure versus time is shown in Figure 1-8a.

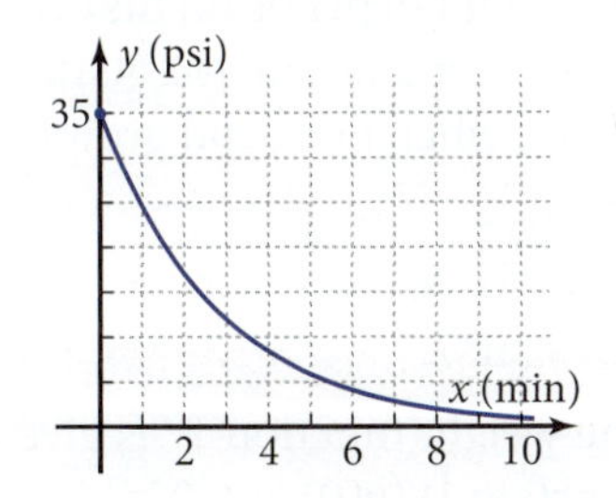

Figure 1-8a

a. Find graphically the pressure after 2 min. Approximately how many minutes can you drive before the pressure reaches 5 psi?

b. The algebraic equation for the function in Figure 1-8a is

$y = 35 \cdot 0.7^x$

Make a table of numerical values of pressure for times of 0, 1, 2, 3, and 4 min.

c. Suppose the equation in part b gives reasonable answers until the pressure drops to 5 psi. At that pressure, the tire comes loose from the rim and the pressure drops to zero. What is the domain of the function described by this equation? What is the corresponding range?

d. The graph in Figure 1-8a gets closer and closer to the x-axis but never quite touches it. What special name is given to the x-axis in this case?

e. *Earthquake Problem:* Earthquakes happen when rock plates slide past each other. The stress between plates that builds up over a number of years is relieved by the quake in a few seconds. Then the stress starts building up again. Sketch a reasonable graph showing stress as a function of time.

In 1989, a magnitude 7.1 earthquake struck Northern California, destroying houses in San Francisco's Marina district.

R2. For parts a–e, name the kind of function for each equation given.

a. $f(x) = 3x + 7$

b. $f(x) = x^3 + 7x^2 - 12x + 5$

c. $f(x) = 1.3^x$

d. $f(x) = x^{1.3}$

e. $f(x) = \dfrac{x - 5}{x^2 - 2x + 3}$

f. Name a pair of real-world variables that could be related by the function in part a.

g. If the domain of the function in part a is $2 \le x \le 10$, what is the range?

h. In a flu epidemic, the number of people currently infected depends on time. Sketch a reasonable graph of the number of people infected as a function of time. What kind of function has a graph that most closely resembles the one you drew?

i. For Figures 1-8b through 1-8d, what kind of function has the graph shown?

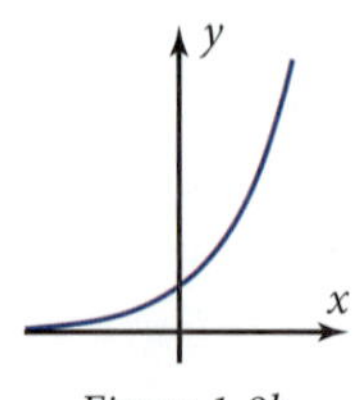

Figure 1-8b

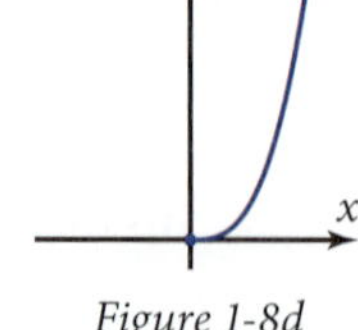

Figure 1-8c

Figure 1-8d

j. Explain how you know that the relation graphed in Figure 1-8e is a function but the relation graphed in Figure 1-8f is not a function.

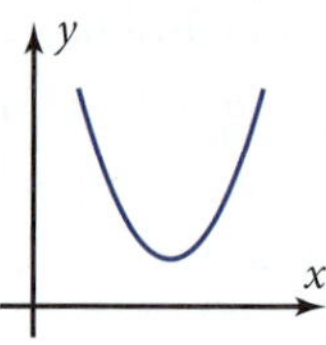

Figure 1-8e

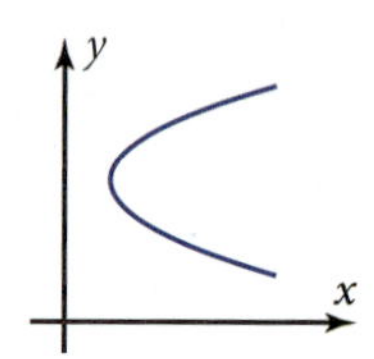

Figure 1-8f

R3. **a.** For functions f and g in Figure 1-8g, identify how the pre-image function f (dashed) was transformed to get the image function g (solid). Write an equation for $g(x)$ in terms of x given that the equation of f is

$f(x) = \sqrt{4 - x^2}$

Confirm the result by plotting the image and the pre-image on the same screen on your grapher.

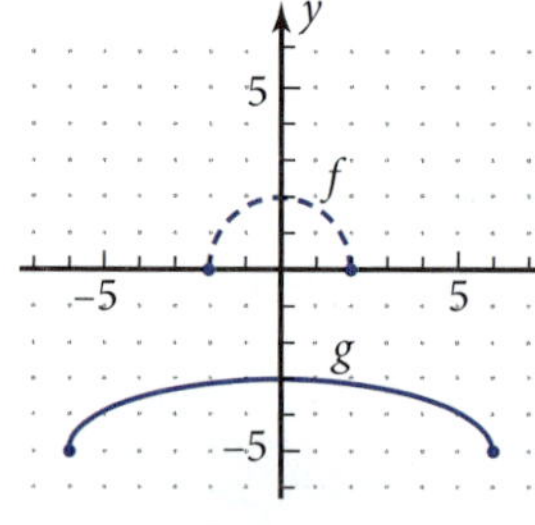

Figure 1-8g

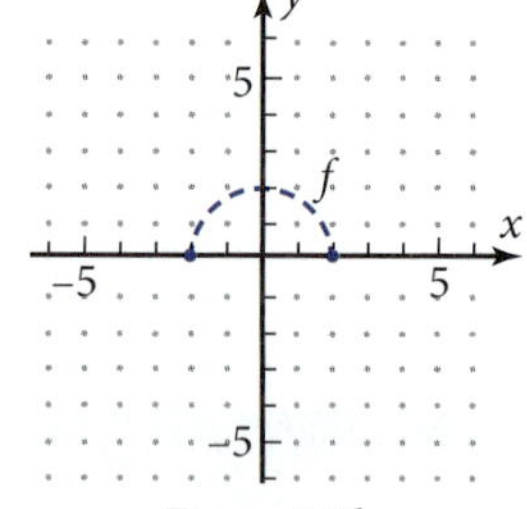

Figure 1-8h

b. If $g(x) = 3f(x - 4)$, explain how function f was transformed to get function g. Using the pre-image in Figure 1-8h, sketch the graph of g on a copy of this figure.

R4. *Height and Weight Problem:* For parts a–e, the weight of a growing child depends on his or her height, and the height depends on age. Assume that the child is 20 in. when born and grows 3 in. per year.

a. Write an equation for $h(t)$ (in inches) as a function of t (in years).

b. Assume that the weight function W is given by the power function $W(h(t)) = 0.004h(t)^{2.5}$. Find $h(5)$, and use the result to calculate the predicted weight of the child at age 5.

c. Plot the graph of $y = W(h(t))$. Sketch the result.

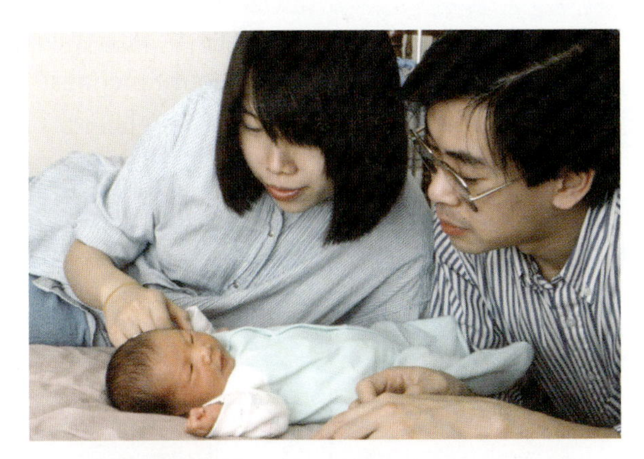

d. Assuming that the height increases at the constant rate of 3 in. per year, does the weight also increase at a constant rate? Explain how you arrived at your answer.

e. What is a reasonable domain for t for the composite function $W \circ h$?

f. *Composite Functions Numerically Problem:* Functions f and g are defined only for the values of x in the table.

x	f(x)	g(x)
3	2	4
4	6	5
5	3	8
6	5	3

Find these values, or explain why they are undefined: $f(g(3))$, $f(g(4))$, $f(g(5))$, $f(g(6))$, $g(f(6))$, $f(f(3))$, and $g(g(3))$.

Two Linear Functions Problem: For parts g–j, let functions f and g be defined by

$$f(x) = x - 2 \qquad 4 \le x \le 8$$

$$g(x) = 2x - 3 \qquad 2 \le x \le 6$$

g. Plot the graphs of f, g, and $f(g(x))$ on the same screen. Sketch the results.

h. Find $f(g(4))$.

i. Show that $f(g(3))$ is undefined, even though $g(3)$ is defined.

j. Calculate the domain of the composite function $f \circ g$ and show that it agrees with the graph you plotted in part g.

R5. Figure 1-8i shows the graph of $f(x) = x^2 + 1$ in the domain $-1 \le x \le 2$.

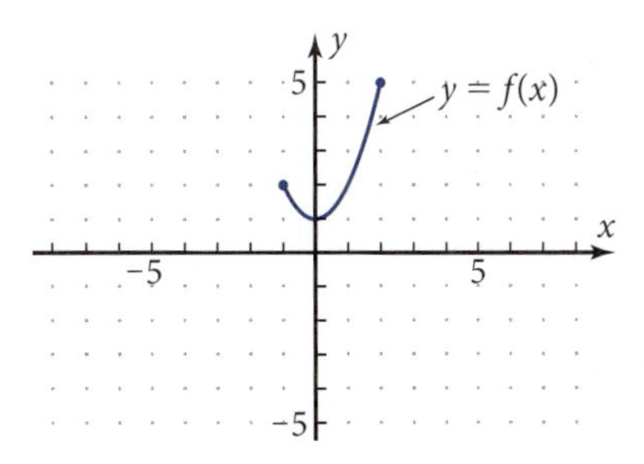

Figure 1-8i

a. On a copy of the figure, sketch the graph of the inverse relation. Explain why the inverse is not a function.

b. Plot the graphs of f and its inverse relation on the same screen using parametric equations. Also plot the line $y = x$. How are the graphs of f and its inverse relation related to the line $y = x$? How are the domain and range of the inverse relation related to the domain and range of function f?

c. Write an equation for the inverse of the function $y = x^2 + 1$ by interchanging the variables. Solve the new equation for y in terms of x. How does this solution reveal that there are two different y-values for some x-values?

d. On a copy of Figure 1-8j, sketch the graph of the inverse relation. What property does the function graph have that allows you to conclude that the function is invertible? What are the vertical lines at $x = -3$ and at $x = 3$ called?

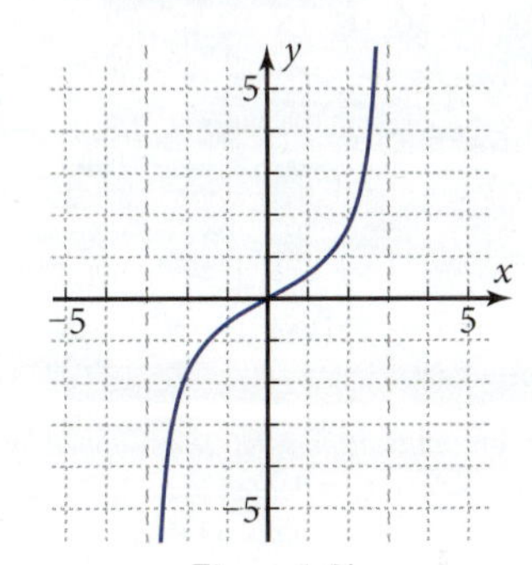

Figure 1-8j

e. Plot these parametric equations on graph paper, using each integer value of t from -3 to 3. Confirm the results by plotting on your grapher. Is y a function of x? Explain.

$$x = 3 + |4 - t^2|$$
$$y = t + 4$$

f. *Spherical Balloon Problem:* The table shows the volume of a spherical balloon, $v(x)$, in cubic meters, as a function of its radius, x, in meters.

x (m)	**v(x) (m³)**
0.2	0.3
0.4	0.27
0.6	0.90
0.8	2.14
1.0	4.19

Plot function v on graph paper by plotting $y = v(x)$ for these points and connecting the points with a smooth curve. What evidence do you have that function v is invertible?

Plot the graph of $y = v^{-1}(x)$ on the same axes. What is the difference in the meaning of x as the input for function v and x as the input for function v^{-1}? Explain why $v^{-1}(v(x))$ equals x.

Echo 1, the first communication satellite developed by NASA, was a giant metal balloon that floated in orbit. It was used to bounce sound signals from one place on Earth to another.

g. Sketch the graph of a one-to-one function. Explain why it is invertible.

R6. a. On four copies of $y = f(x)$ in Figure 1-8k, sketch the graphs of these four functions: $y = -f(x)$, $y = f(-x)$, $y = |f(x)|$, and $y = f(|x|)$.

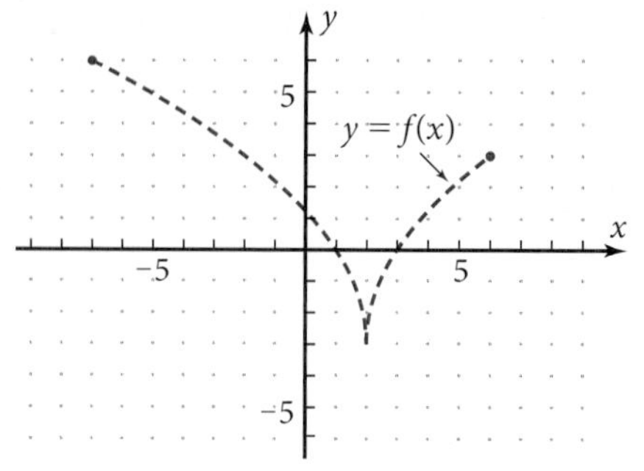

Figure 1-8k

b. Function f in part a is defined piecewise by

$$f(x) = \begin{cases} 3\sqrt{x-2} - 3 & 2 \le x \le 6 \\ 3\sqrt{2-x} - 3 & -7 \le x \le 2 \end{cases}$$

Plot the two branches of this function as $f_1(x)$ and $f_2(x)$ on your grapher. Does the graph agree with Figure 1-8k? Plot $y = f(|x|)$ by plotting $f_3(x) = f_1(x)$ and $f_4(x) = f_2(x)$. Does the graph agree with your result in the corresponding portion of part a?

c. Explain why functions with the property $f(-x) = -f(x)$ are called *odd* functions and functions with the property $f(-x) = f(x)$ are called *even* functions.

d. Plot the graph of

$$f(x) = 0.2x^2 - \frac{|x-3|}{x-3}$$

Use a window that includes $x = 3$ as a grid point. Sketch the result. Name the feature that appears at $x = 3$.

R7. In Section 1-7 you started a precalculus journal. In what ways do you think keeping this journal will help you? How could you use the completed journal at the end of the course? What is your responsibility throughout the year to ensure that writing the journal has been a worthwhile project?

Concept Problems

C1. *Four Transformations Problem:* Figure 1-8l shows a pre-image function f (dashed) and a transformed image function g (solid). Dilations and translations were performed in both directions to get the graph for g. Figure out what the transformations were. Write an equation for $g(x)$ in terms of f. Let $f(x) = x^2$ with domain $-2 \le x \le 2$. Plot the graph of g on your grapher. Does your grapher agree with the figure?

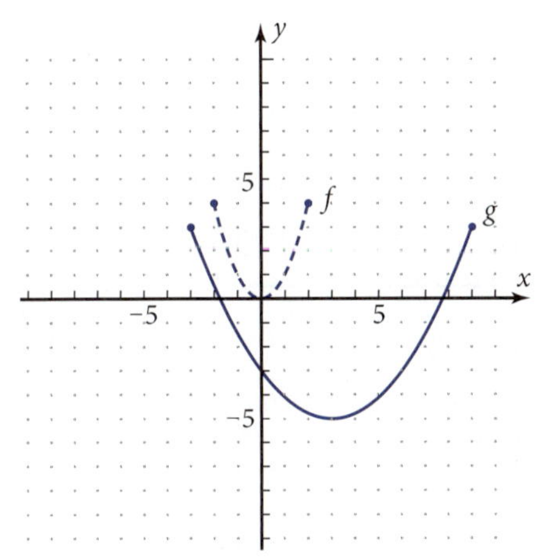

Figure 1-8l

C2. *Sine Function Problem:* If you enter $f_1(x) = \sin(x)$ into your grapher and plot the graph, the result resembles Figure 1-8m. (Your grapher should be in radian mode.) The function is called the **sine function** (pronounced "sign"), which you will study starting in Chapter 5.

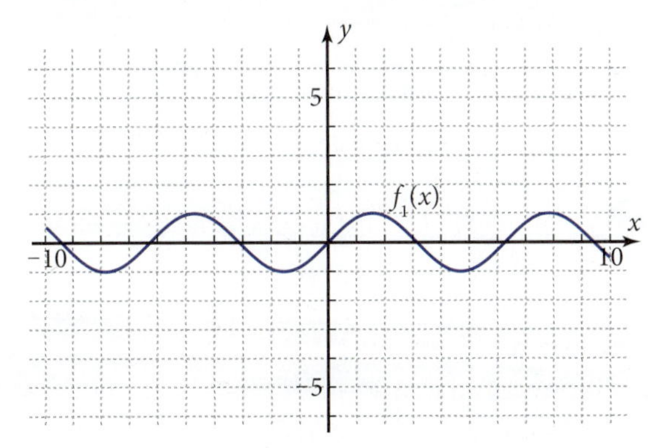

Figure 1-8m

a. The sine function is an example of a **periodic function.** Why do you think this name is given to the sine function?

b. The **period** of a periodic function is the difference in x-values from a point on the graph to the point where the graph first starts repeating itself. Approximately what does the period of the sine function seem to be?

c. Is the sine function an odd function, an even function, or neither? How can you tell?

d. On a copy of Figure 1-8m, sketch a vertical dilation of the sine function graph by a factor of 5. What is the equation of this transformed function? Check your answer by plotting the sine graph and the transformed image graph on the same screen.

e. Figure 1-8n shows a two-step transformation of the sine graph in Figure 1-8m. Name the two transformations. Write an equation for the transformed function, and check your answer by plotting both functions on your grapher.

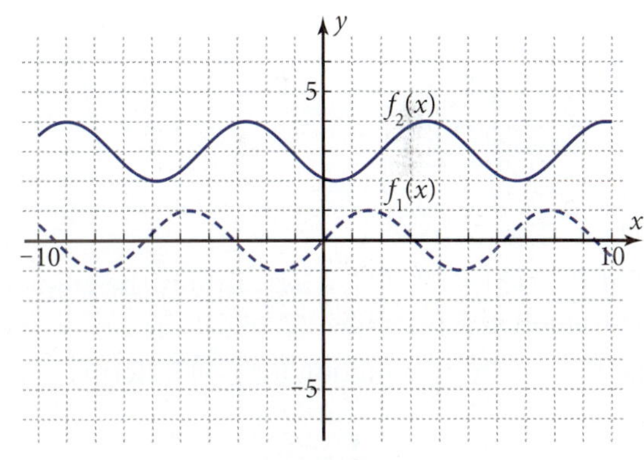

Figure 1-8n

f. Let $f(x) = \sin x$. What transformation would $g(x) = \sin\left(\frac{1}{2}x\right)$ be? Check your answer by plotting both functions on your grapher.

Chapter Test

Part 1: No calculators allowed (T1–T11)

For Problems T1–T4, name the type of function that each graph shows.

T1.

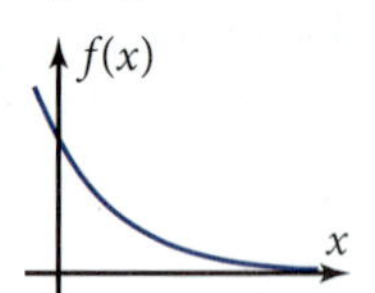

T2.

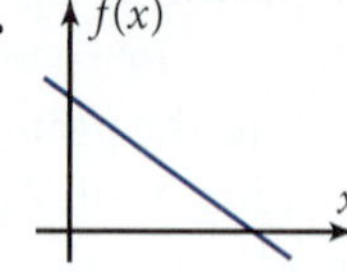

T3.

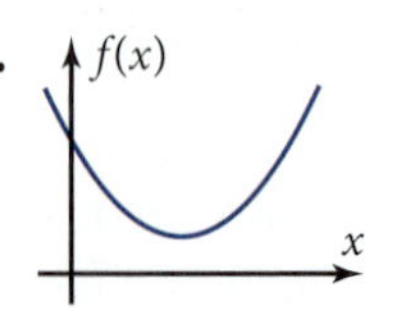

T4.

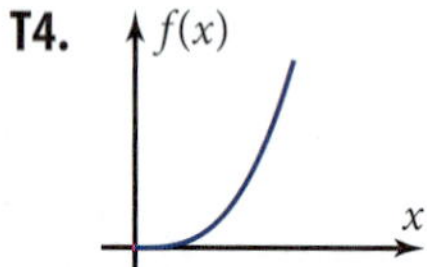

T5. Which of the functions in Problems T1–T4 are one-to-one functions? What conclusion can you make about a function that is not one-to-one?

T6. When you turn on the hot water faucet, the time the water has been running and the temperature of the water are related. Sketch a reasonable graph of this function.

For Problems T7 and T8, tell whether the function is odd, even, or neither.

T7.

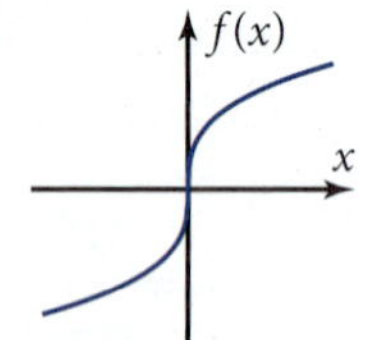

T8.

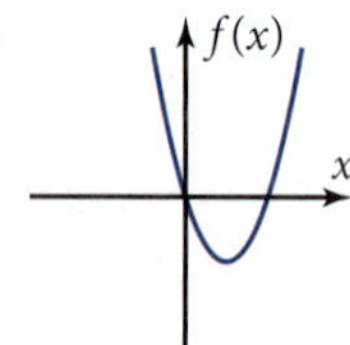

For Problems T9–T11, describe how the graph of f (dashed) was transformed to get the graph of g (solid). Write an equation for $g(x)$ in terms of f.

T9.

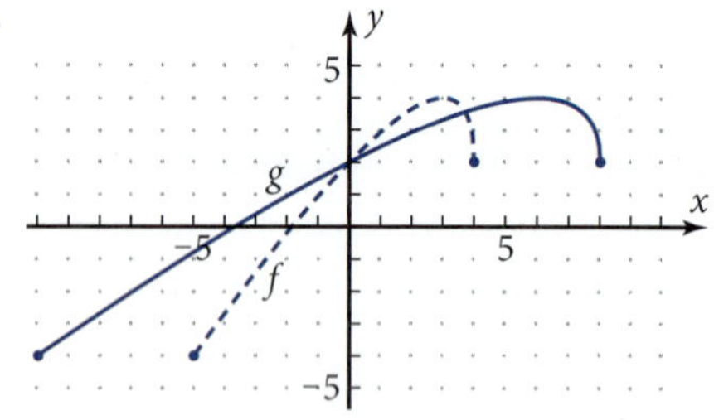

T10.

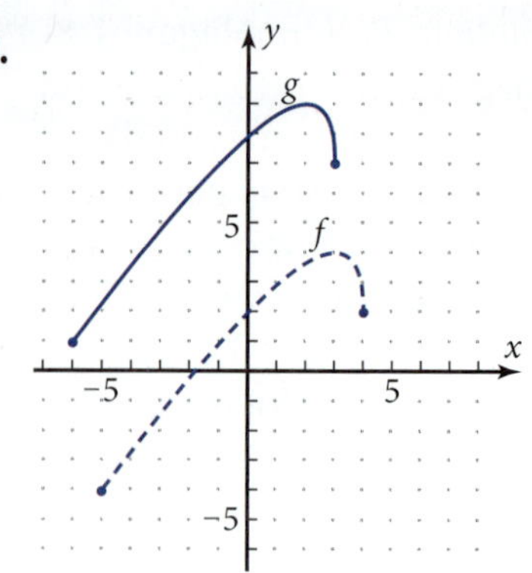

T11.

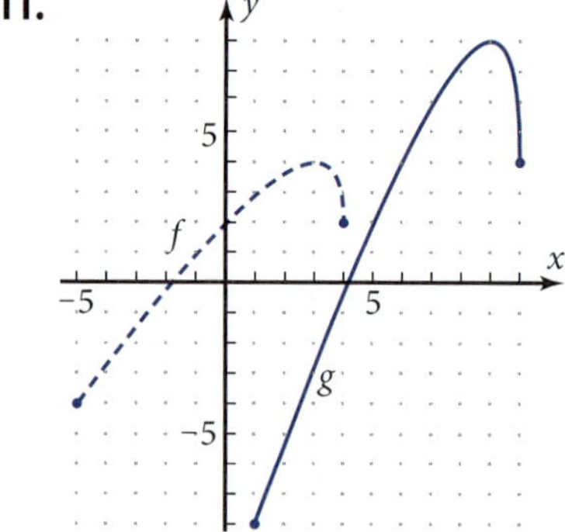

Part 2: Graphing calculators allowed (T12–T29)

T12. Figure 1-8o shows the graph of a function, $y = f(x)$. Give the domain and the range of f.

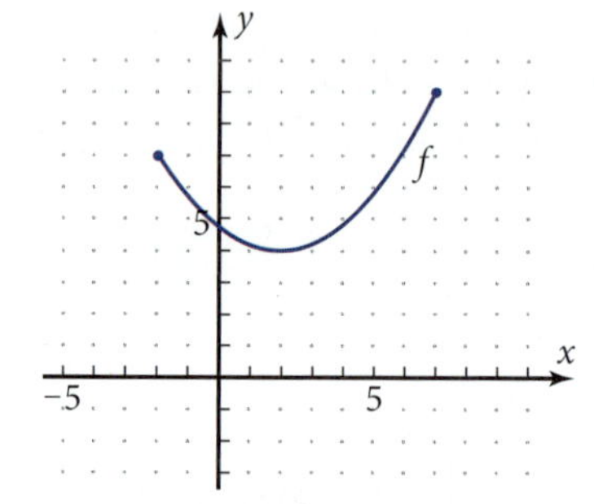

Figure 1-8o

For Problems T13–T16, sketch the indicated transformations on copies of Figure 1-8o. Describe the transformations.

T13. $y = \frac{1}{2}f(x)$

T14. $y = f\left(\frac{2}{3}x\right)$

T15. $y = f(x + 3) - 4$

T16. The inverse relation of $f(x)$

T17. Explain why the inverse relation in Problem T16 is not a function.

T18. Let $f(x) = \sqrt{x}$. Let $g(x) = x^2 - 4$. Find $f(g(3))$. Find $g(f(3))$. Explain why $f(g(1))$ is not a real number, even though $g(1)$ is a real number.

T19. Use the absolute value function to write a single equation for the discontinuous function graphed in Figure 1-8p. Check your answer by plotting it on your grapher.

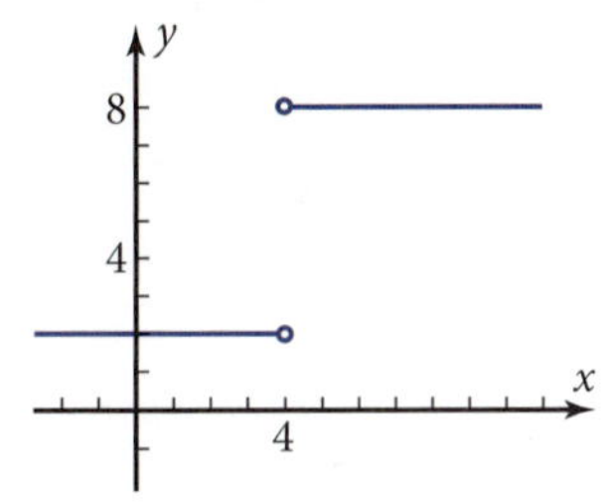

Figure 1-8p

T20. Plot these parametric equations on graph paper, using each integer value of t from -3 to 3. Confirm the results by plotting them on your grapher. Is y a function of x? Explain.

$$x = t + 4$$

$$y = 1 + |4 - t^2|$$

Wild Oats Problem: Problems T21–T28 refer to the competition of wild oats, a kind of weed, with the wheat crop. Based on data in A. C. Madgett's book *Applications of Mathematics: A Nationwide Survey,* the percent loss in wheat crop, $L(x)$, is approximately

$$L(x) = 3.2x^{0.52}$$

where x is the number of wild oat plants per square meter of land.

T21. Describe how $L(x)$ varies with x. What kind of function is L?

T22. Find $L(150)$. Explain verbally what this number means.

T23. Suppose the wheat crop is reduced to 60% of what it would be without the wild oats. How many wild oats per square meter are there?

T24. Let $y = L(x)$. Find an equation for $y = L^{-1}(x)$. For what kind of calculations would the equation $y = L^{-1}(x)$ be more useful than $y = L(x)$?

T25. Find $L^{-1}(100)$. Explain its real-world meaning.

T26. Based on your answer to Problem T25, what would be a reasonable domain and range for L?

T27. Plot $f_1(x) = L(x)$ and $f_2(x) = L^{-1}(x)$ on the same screen. Use equal scales for the two axes. Use the domain and range from Problem T25. Sketch the results along with the line $y = x$.

T28. How can you tell that the inverse relation is a function?

T29. What did you learn as a result of taking this test that you didn't know before?

Chapter 2

Properties of Elementary Functions

If a mother mouse is twice as long as her offspring, then the mother's weight is about eight times the baby's weight. But the mother mouse's skin area is only about four times the baby's skin area. So the baby mouse must eat more than the mother mouse in proportion to its body weight to make up for the heat loss through its skin. In this chapter you'll learn how to use functions to model and explain situations like this.

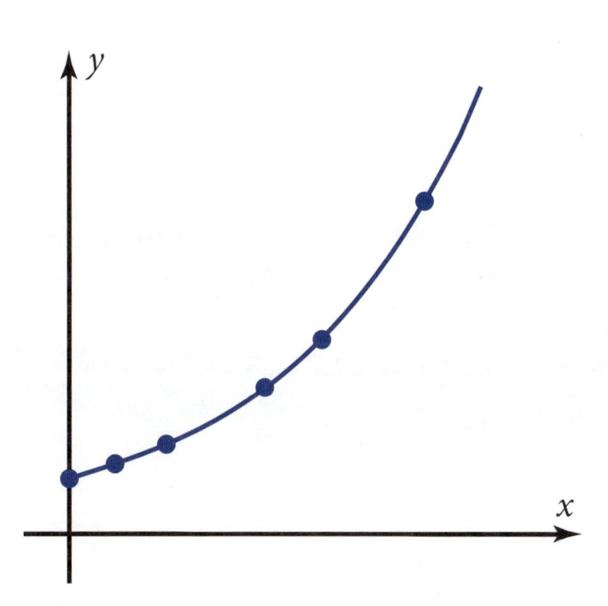

Mathematical Overview

In this chapter you will extend what you have already learned about some of the more familiar functions in algebra, as well as some you may not yet have encountered. These functions are

- Linear
- Quadratic
- Power
- Exponential
- Logarithmic
- Logistic

You will study these functions in four ways.

ALGEBRAICALLY

You can define each of these functions algebraically; for example, the logarithmic function is defined

$$y = \log_b x \qquad \text{if and only if} \qquad b^y = x$$

NUMERICALLY

You can find interesting numerical relationships between the values of variables x and y. Exponential functions exhibit the add–multiply property: As a result of adding a constant to x, the corresponding y-value is multiplied by a constant.

GRAPHICALLY

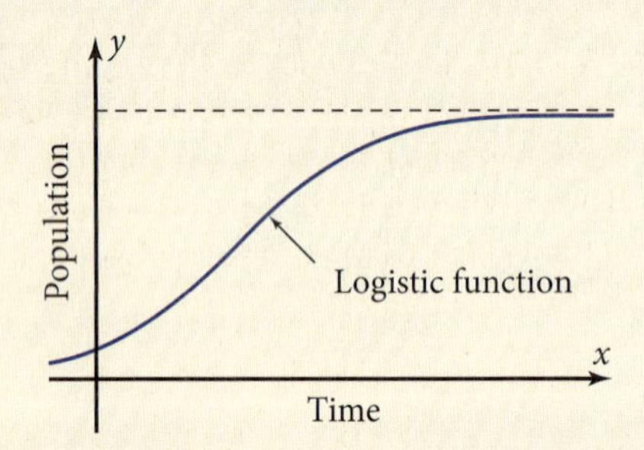

VERBALLY

Exponential functions can describe unrestrained population growth, such as that of rabbits if they have no natural enemies. Logistic functions start out like exponential functions but then level off. Logistic functions can model restrained population growth where there is a maximum sustainable population in a certain region.

2-1 Shapes of Function Graphs

In this chapter you'll learn ways to find a function to fit a real-world situation when the type of function has not been given. You will start by refreshing your memory about graphs of functions you studied in Chapter 1.

Objective Discover patterns in the graphs of linear, quadratic, power, and exponential functions.

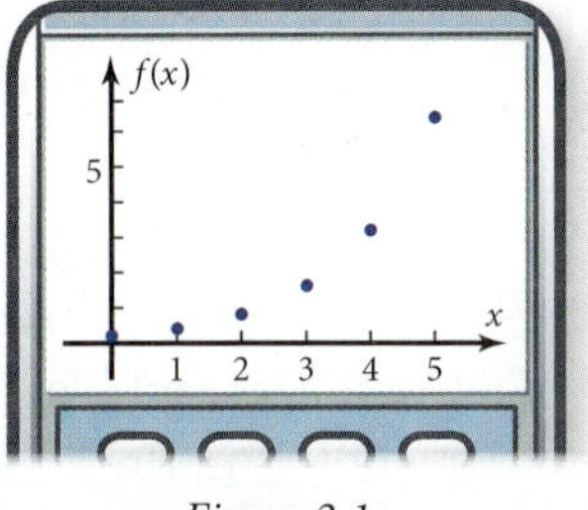

Figure 2-1a

Figure 2-1a shows the plot of points that are values of the exponential function $f(x) = 0.2 \cdot 2^x$. You can make such a plot by storing the x-values in one list and the $f(x)$-values in another and then using the statistics plot feature on your grapher. Figure 2-1b shows that the graph of f contains all the points in the plot. The **concave** side of the graph is up.

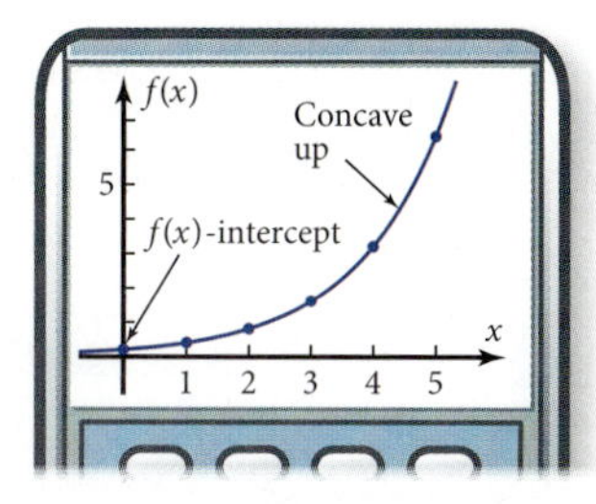

Figure 2-1b

Exploratory Problem Set 2-1

1. *Exponential Function Problem:* In the exponential function $f(x) = 0.2 \cdot 2^x$, $f(x)$ could be the number of thousands of bacteria in a culture as a function of time, x, in hours. Find $f(x)$ for each hour from 0 through 5. Plot the points, and graph the function as in Figure 2-1b. The number of bacteria is increasing as time goes on. How does the concavity of the graph tell you that the *rate* of growth is also increasing?

2. *Power Function Problem:* In the power function $g(x) = 0.1x^3$, $g(x)$ could be the weight in pounds of a snake that is x feet long. Plot the points for each foot from 0 through 6, and graph function g. Because the graphs of f in Problem 1 and g in Problem 2 are both increasing and concave up, what graphical evidence could you use to distinguish between the two types of functions? Is the following statement true or false? "The snake's weight increases by the same amount for each foot it increases in length." Give evidence to support your answer.

3. *Quadratic Function Problem:* In the quadratic function $q(x) = -0.3x^2 + 8x + 7$, $q(x)$ could measure the approximate sales of a new product in the xth week since the product was introduced. Plot the points for every 5 weeks from 0 through 30, and graph function q. Which way is the concave side of the graph, up or down? What feature does the quadratic function graph have that neither the exponential function graph in Problem 1 nor the power function graph in Problem 2 has?

4. *Linear Function Problem:* In the linear function $h(x) = 5x + 27$, $h(x)$ could equal the number of cents you pay for a telephone call of length x, in minutes. Plot the points for every 3 minutes from 0 through 18, and graph function h. What does the fact that the graph is neither concave up nor concave down tell you about the cents per minute you pay for the call?

2-2 Identifying Functions from Graphical Patterns

One way to tell what type of function fits a set of points is by recognizing the properties of the graph of the function.

Objective Given the graph of a function, know whether the function is exponential, power, quadratic, or linear and find the particular equation algebraically.

Here is a brief review of the basic functions used in modeling. Some of these appeared in Chapter 1.

Linear and Constant Functions

General equation: $y = ax + b$ (often written $y = mx + b$), where a (or m) and b stand for constants and the domain is all real numbers. This equation is in the **slope-intercept form** because a (or m) gives the **slope** and b gives the y-intercept. If $a = 0$, then $y = b$; this is a **constant function.**

Parent function: $y = x$

Transformed function: $y = y_1 + a(x - x_1)$, called the **point-slope form** because the graph contains the point (x_1, y_1) and has slope a. The slope, a, is the vertical dilation; y_1 is the vertical translation; and x_1 is the horizontal translation. Note that point-slope form can also be written $y - y_1 = a(x - x_1)$, where the coordinates of the fixed point (x_1, y_1) both appear with a $-$ sign. The form $y = y_1 + a(x - x_1)$ expresses y explicitly in terms of x and thus is easier to enter into your grapher.

Graphical properties: The graph is a straight line. The parent function is shown in the left graph of Figure 2-2a, the slope-intercept form is shown in the middle graph, and the point-slope form is shown in the right graph.

Verbally: For slope-intercept form: "Start at b on the y-axis, run x, and rise ax." For point-slope form: "Start at (x_1, y_1), run $(x - x_1)$, and rise $a(x - x_1)$."

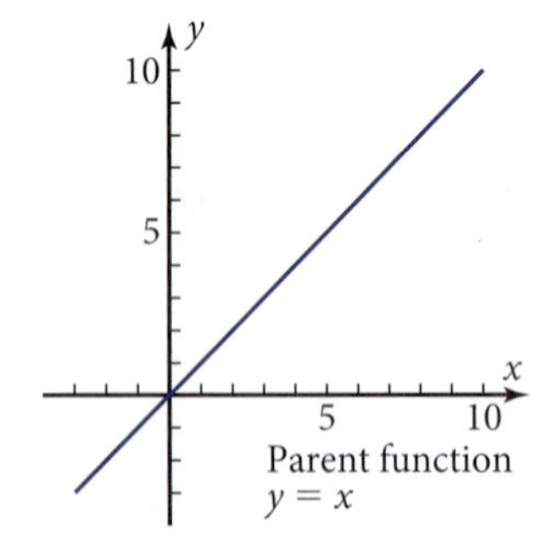

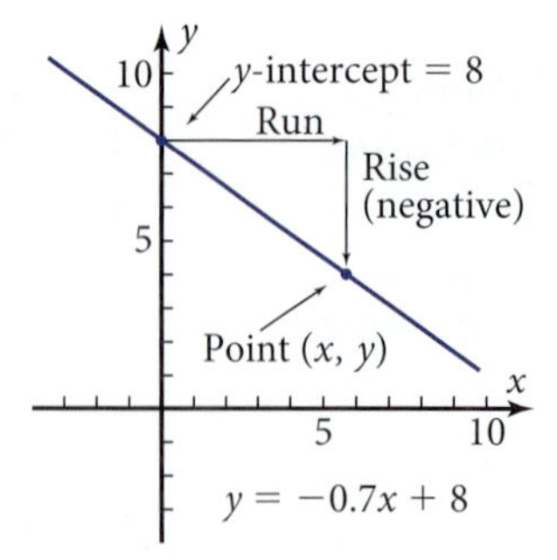

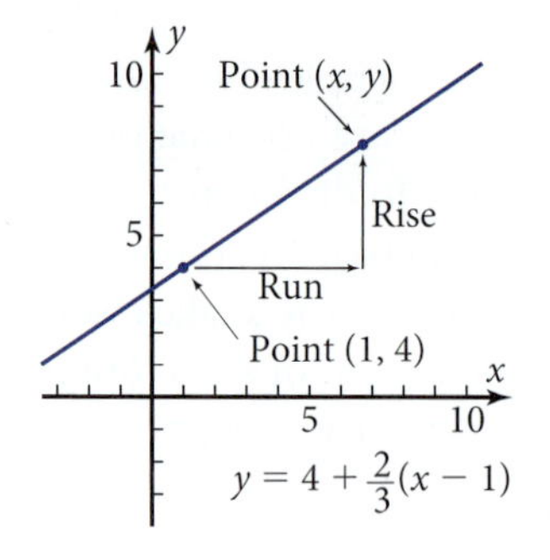

Figure 2-2a: Linear functions

The eruption of Arenal, an active volcano in Costa Rica. The lava particles follow a parabolic path due to gravitational force.

Quadratic Functions

General equation: $y = ax^2 + bx + c$, where $a \neq 0$; a, b, and c stand for c
and the domain is all real numbers

Parent function: $y = x^2$, where the **vertex** is at the origin

Transformed function: $y = k + a(x - h)^2$, called the **vertex form,** with vertex at (h, k). The value k is the vertical translation, h is the horizontal translation, and a is the vertical dilation. Vertex form can also be written $y - k = a(x - h)^2$, but expressing y explicitly in terms of x makes the equation easier to enter into your grapher.

Graphical properties: The graph is a **parabola** (Greek for "along the path of a ball"), as shown in Figure 2-2b. The graph is concave up if $a > 0$ and concave down if $a < 0$.

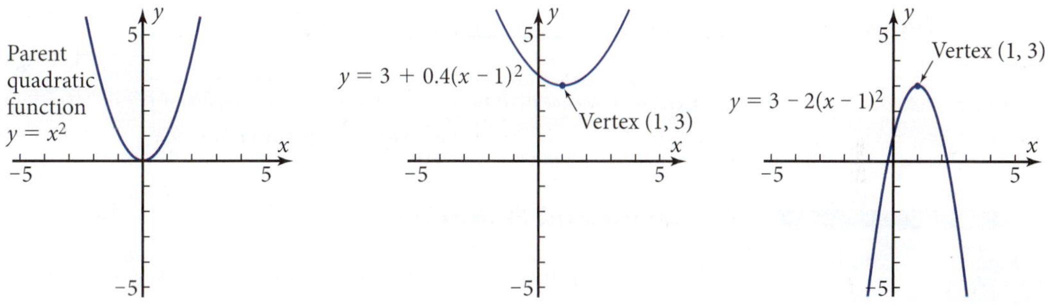

Figure 2-2b: Quadratic functions

Power Functions

General equation: $y = ax^b$, where a and b stand for nonzero constants. If $b > 0$, then the domain can be all real numbers. If $b < 0$, then the domain excludes $x = 0$ to avoid division by zero. If b is not an integer, then the domain usually excludes negative numbers to avoid roots of negative numbers. The domain is also restricted to nonnegative numbers in most applications.

Parent function: $y = x^b$

Verbally: For $y = ax^b$, "if $b > 0$, then y varies directly with the bth power of x, or y is directly proportional to the bth power of x; if $b < 0$, then y varies inversely with the bth power of x, or y is inversely proportional to the bth power of x." The dilation factor a is the **proportionality constant.**

Translated function: $y = d + a(x - c)^b$, where c and d are the horizontal and vertical translation, respectively. Compare the translated form with

$y = y_1 + a(x - x_1)$ for linear functions

$y = k + a(x - h)^2$ for quadratic functions

Unless otherwise stated, "power function" will imply the *untranslated* form, $y = ax^b$.

Graphical properties: Figure 2-2c shows power function graphs for different values of b. In all three cases, $a > 0$. The shape and concavity of the graph depend on the value of b. The graph contains the origin if $b > 0$; it has the axes as asymptotes if $b < 0$. The function is increasing if $b > 0$; it is decreasing if $b < 0$. The graph is concave up if $b > 1$ or if $b < 0$ and concave down if $0 < b < 1$. The concavity of the graph describes the *rate* at which y increases. For $b > 0$, concave up means y is increasing at an *increasing rate,* and concave down means y is increasing at a *decreasing rate.*

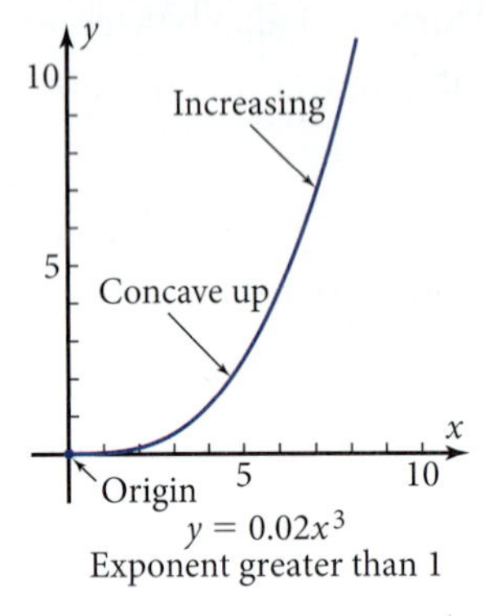

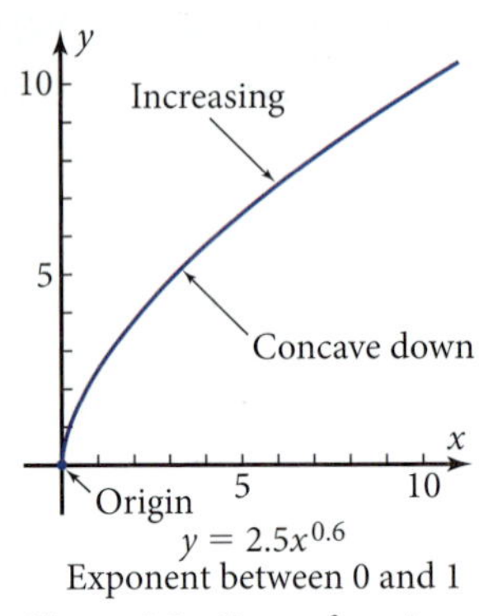

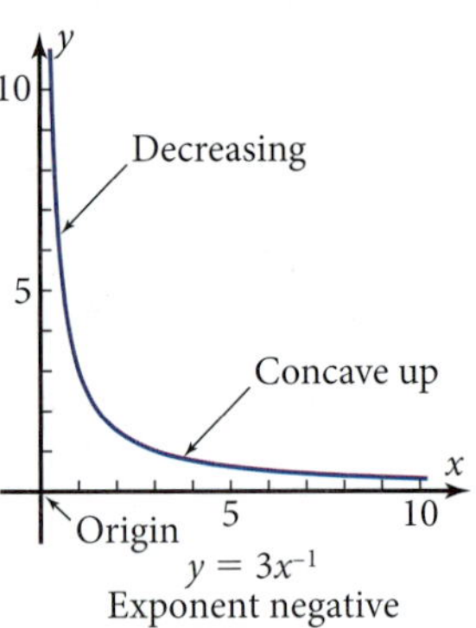

Figure 2-2c: Power functions

Exponential Functions

Marie Curie was awarded the Nobel Prize in chemistry for the discovery of radioactive elements (polonium and radium) in 1911. The breakdown of radioactive elements follows an exponential function.

General equation: $y = ab^x$, where a and b are constants, $a \neq 0$, $b > 0$, $b \neq 1$, and the domain is all real numbers

Parent function: $y = b^x$, where the asymptote is the x-axis

Verbally: In the equation $y = ab^x$, "y varies exponentially with x."

Translated function: $y = ab^x + c$, where the asymptote is the line $y = c$. Unless otherwise stated, "exponential function" will imply the untranslated form, $y = ab^x$.

Graphical properties: Figure 2-2d shows exponential functions for different values of a and b. The constant a is the y-intercept. The function is increasing if $b > 1$ and decreasing if $0 < b < 1$ (provided $a > 0$). If $a < 0$, the opposite is true. The graph is concave up if $a > 0$ and concave down if $a < 0$.

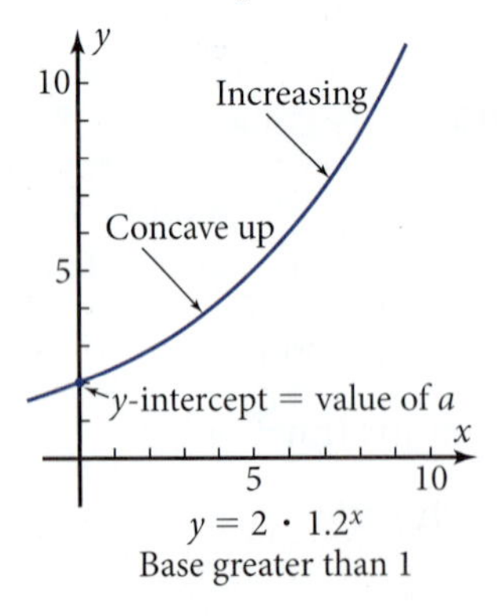

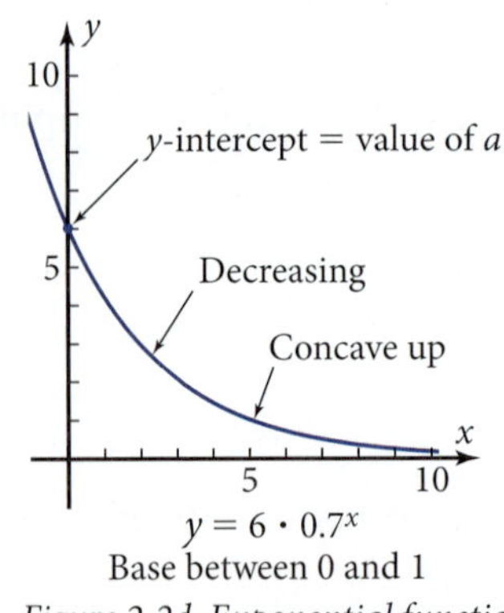

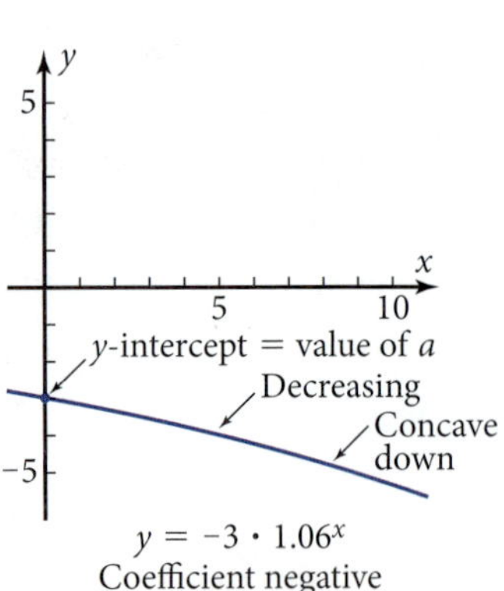

Figure 2-2d: Exponential functions

Mathematicians usually use one of two particular constants for the base of an exponential function: either 10, which is the base of the decimal system, or the naturally occurring number e, which equals 2.71828.... To make the equation more general, multiply the variable in the exponent by a constant. The (untranslated) general equations are given in the box.

DEFINITION: Special Exponential Functions

$y = a \cdot 10^{bx}$ base-10 exponential function

$y = a \cdot e^{bx}$ **natural (base-e) exponential function**

where a and b are constants and the domain is all real numbers.

Note: The equations of these two functions can be generalized by incorporating translations in the x- and y-directions. You'll get $y = a \cdot 10^{b(x-c)} + d$ and $y = a \cdot e^{b(x-c)} + d$.

Base-e exponential functions have an advantage when you study calculus because the rate of change of e^x is equal to e^x.

In this exploration, you'll find the particular equation of a linear, quadratic, power, or exponential function from a given graph.

EXPLORATION 2-2: Graphical Patterns in Functions

1. Identify what kind of function is graphed, and find its particular equation.

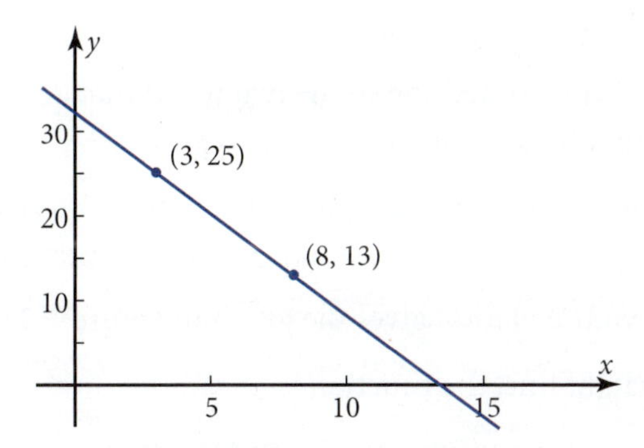

2. Check your answer to Problem 1 graphically. Does your graph agree with the given one?
3. Is the graph in Problem 1 concave up, concave down, or neither?
4. What graphical evidence do you have that the function graphed is an exponential function, not a power function? Find its particular equation.

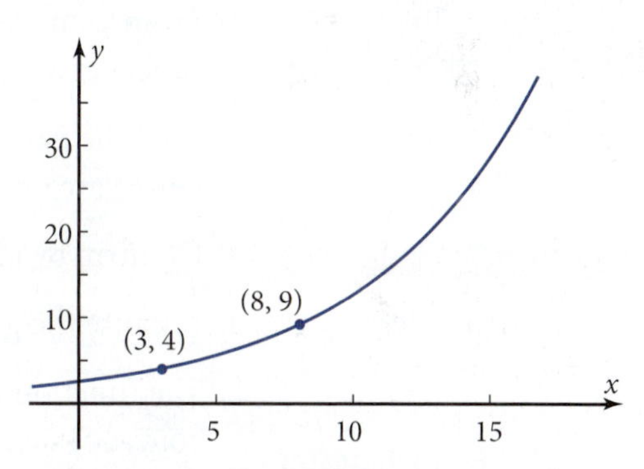

5. Check your answer to Problem 4 graphically. Does your graph agree with the given one?
6. Is the graph in Problem 4 concave up, concave down, or neither?

continued

EXPLORATION, continued

7. What graphical evidence do you have that this function is a power function, not an exponential function? Find its particular equation.

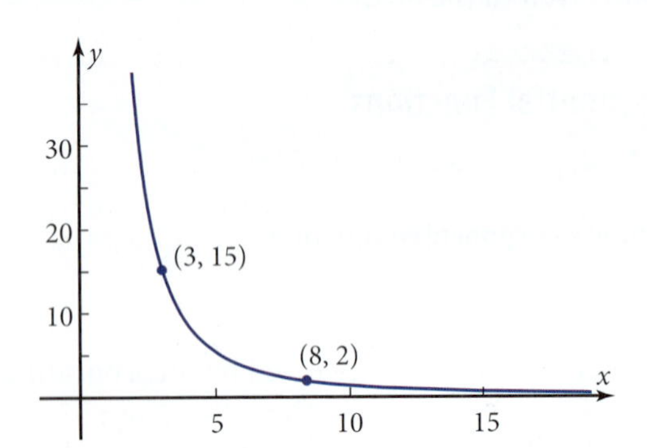

8. Check your answer to Problem 7 graphically. Does your graph agree with the given one?

9. Is the graph in Problem 7 concave up, concave down, or neither?

10. Identify what kind of function is graphed, and find its particular equation.

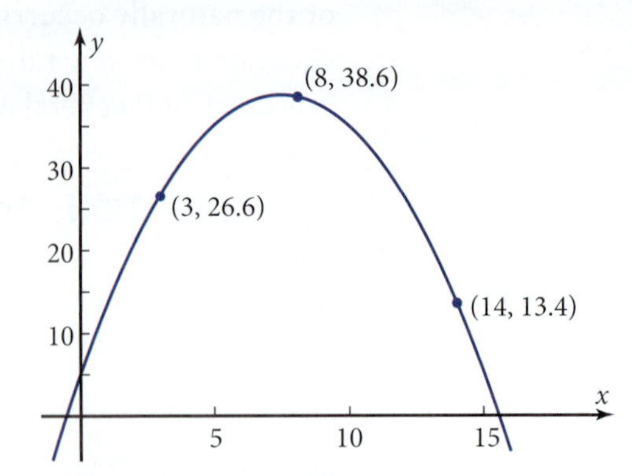

11. Check your answer to Problem 10 graphically. Does your graph agree with the given one?

12. Is the graph in Problem 10 concave up, concave down, or neither?

13. What did you learn as a result of doing this exploration that you did not know before?

EXAMPLE 1 ➤ For the function graphed in Figure 2-2e,

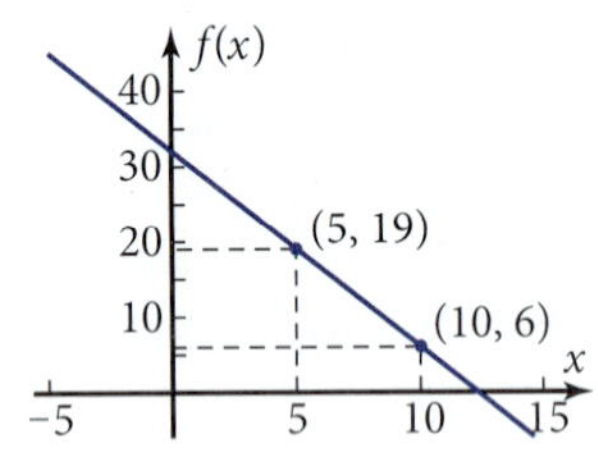

Figure 2-2e

a. Identify the kind of function it is.

b. On what interval or intervals is the function increasing or decreasing? Which way is the graph concave, up or down?

c. From your experience, describe something in the real world that a function with this shape graph could model.

d. Find the particular equation of the function, given that points (5, 19) and (10, 6) are on the graph.

e. Confirm by plotting that your equation gives the graph in Figure 2-2e.

SOLUTION

a. Because the graph is a straight line, the function is linear.

b. The function is decreasing over its entire domain, and the graph is not concave in *either* direction.

c. The function could model anything that decreases at a constant rate. The Quadrant I part of the function could model the number of pages of history text you have left to read as a function of the number of minutes you have been reading.

d. $f(x) = ax + b$ — Write the general equation. Use $f(x)$ as shown on the graph, and use a for the slope.

$$\begin{cases} 19 = 5a + b \\ 6 = 10a + b \end{cases}$$ Substitute the given values of x and y into the equation of f.

$-13 = 5a \Rightarrow a = -2.6$ Subtract the first equation from the second to eliminate b.

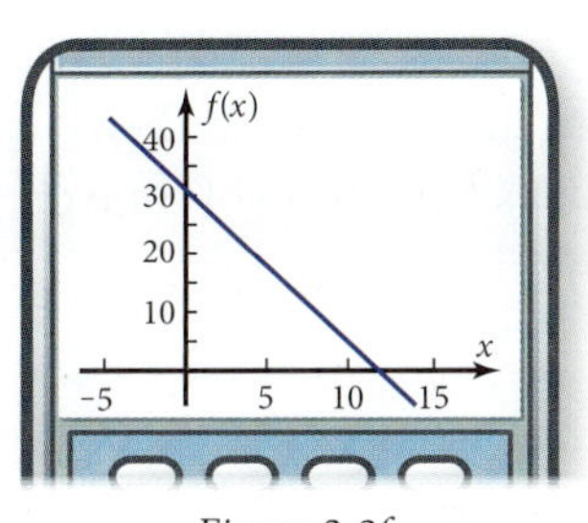

Figure 2-2f

$6 = 10(-2.6) + b \Rightarrow b = 32$ Substitute -2.6 for a in one of the equations.

$\therefore f(x) = -2.6x + 32$ Write the particular equation.

e. Figure 2-2*f* shows the graph of *f*, which agrees with the given graph. Note that the calculated slope, -2.6, is *negative*, which corresponds to the fact that *f(x) decreases as x increases.*

Note that you could have solved the system of equations in Example 1 using matrices.

$$\begin{cases} 5a + b = 19 \\ 10a + b = 6 \end{cases}$$ The given system.

$$\begin{bmatrix} 5 & 1 \\ 10 & 1 \end{bmatrix} \begin{bmatrix} a \\ b \end{bmatrix} = \begin{bmatrix} 19 \\ 6 \end{bmatrix}$$ Write the system in matrix form.

$$\begin{bmatrix} a \\ b \end{bmatrix} = \begin{bmatrix} 5 & 1 \\ 10 & 1 \end{bmatrix}^{-1} \begin{bmatrix} 19 \\ 6 \end{bmatrix}$$ Multiply both sides by the inverse matrix.

$$= \begin{bmatrix} -2.6 \\ 32 \end{bmatrix}$$ Complete the matrix multiplication.

$a = -2.6$ and $b = 32$

You'll study the matrix solution of linear systems more fully in Section 13-2.

EXAMPLE 2 ➤ For the function graphed in Figure 2-2g,

a. Identify the kind of function it could be.

b. On what interval or intervals is the function increasing or decreasing? Which way is the graph concave, up or down?

c. Describe something in the real world that a function with this shape graph could model.

d. Find the particular equation of the function, given that points (1, 76), (2, 89), and (3, 94) are on the graph.

e. Confirm by plotting that your equation gives the graph in Figure 2-2g.

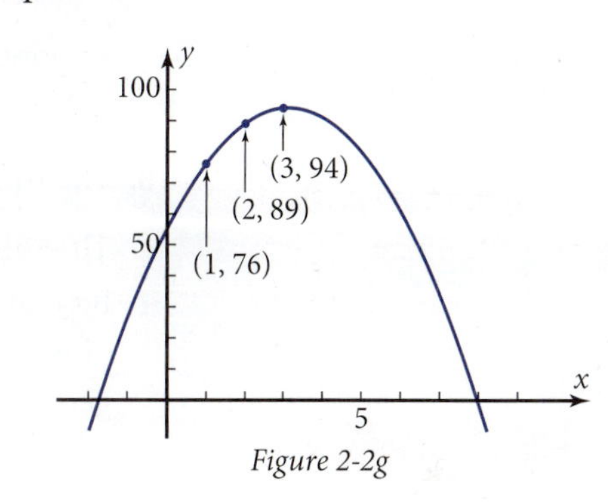

Figure 2-2g

SOLUTION

a. The function could be quadratic because it has a vertex.

b. The function is increasing for $x < 3$ and decreasing for $x > 3$, and it is concave down.

c. The function could model anything that rises to a maximum and then falls back down again, such as the height of a ball as a function of time or the grade you could earn on a final exam as a function of how long you study for it. (Cramming too long might lower your score because of your being sleepy from staying up late!)

d. $y = ax^2 + bx + c$ — Write the general equation.

$$\begin{cases} 76 = a + b + c \\ 89 = 4a + 2b + c \\ 94 = 9a + 3b + c \end{cases}$$ Substitute the given x- and y-values.

$$\begin{bmatrix} 1 & 1 & 1 \\ 4 & 2 & 1 \\ 9 & 3 & 1 \end{bmatrix}^{-1} \begin{bmatrix} 76 \\ 89 \\ 94 \end{bmatrix} = \begin{bmatrix} -4 \\ 25 \\ 55 \end{bmatrix}$$ Solve by matrices.

$\therefore y = -4x^2 + 25x + 55$ — Write the equation.

e. Plotting the graph confirms that the equation is correct. Note that the value of a is negative, which corresponds to the fact that the graph is concave down.

EXAMPLE 3 For the function graphed in Figure 2-2h,

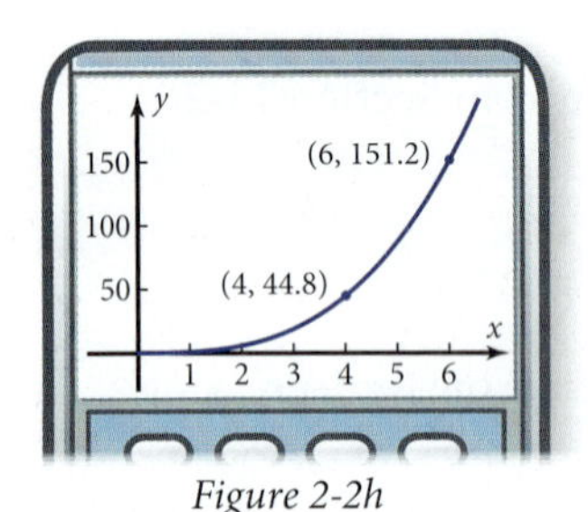

Figure 2-2h

a. Identify the kind of function it could be.

b. On what interval or intervals is the function increasing or decreasing? Which way is the graph concave, up or down?

c. Describe something in the real world that a function with this shape graph could model.

d. Find the particular equation of the function you identified in part a, given that points (4, 44.8) and (6, 151.2) are on the graph.

e. Confirm by plotting that your equation gives the graph in Figure 2-2h.

SOLUTION

a. The function could be a power function or an exponential function, but a power function is chosen because the graph appears to contain the origin. Exponential functions don't contain the origin unless they are translated in the y-direction.

b. The function is increasing and concave up over the entire domain shown.

c. The function could model anything that starts at zero and increases at an increasing rate, such as the power generated by a windmill as a function of wind speed, when the driver applies the brakes, or the volumes of geometrically similar objects as a function of their lengths.

d. $y = ax^b$ — Write the untranslated general equation.

$$\begin{cases} 44.8 = a \cdot 4^b \\ 151.2 = a \cdot 6^b \end{cases}$$
Substitute the given x- and y-values into the equation.

$$\frac{151.2}{44.8} = \frac{a \cdot 6^b}{a \cdot 4^b}$$
Divide the second equation by the first to eliminate a.

$$3.375 = 1.5^b$$
The a's cancel, and $\frac{6^b}{4^b} = \left(\frac{6}{4}\right)^b = 1.5^b$.

$$\log 3.375 = \log 1.5^b$$
Take the logarithm of both sides to get b out of the exponent. See Section 2-4 for a review of logarithms.

$$\log 3.375 = b \log 1.5$$

$$b = \frac{\log 3.375}{\log 1.5} = 3$$

$$44.8 = a \cdot 4^3$$
Substitute 3 for b in one of the equations.

$$a = \frac{44.8}{4^3} = 0.7$$

$$\therefore y = 0.7x^3$$
Write the particular equation.

e. Plotting the graph confirms that the equation is correct. Note that the value of b is between 0 and 1, which corresponds to the fact that the graph is concave up.

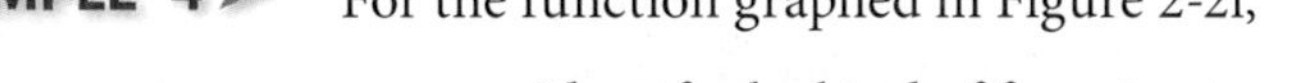

EXAMPLE 4 For the function graphed in Figure 2-2i,

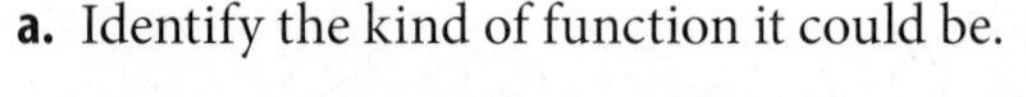

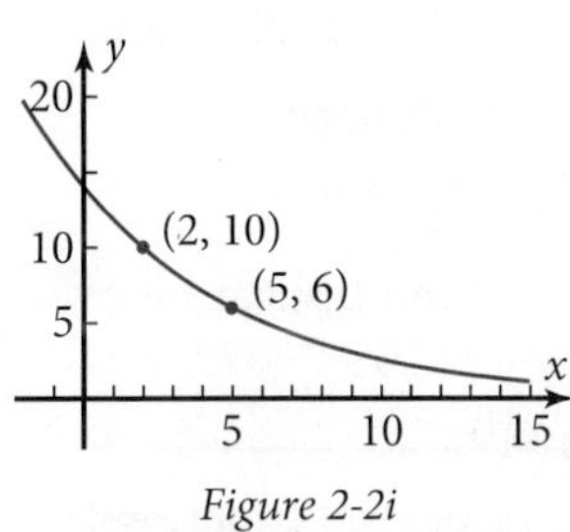

Figure 2-2i

a. Identify the kind of function it could be.

b. On what interval or intervals is the function increasing or decreasing? Which way is the graph concave, up or down?

c. Describe something in the real world that a function with this shape graph could model.

d. Find the particular equation of the function, given that the points (2, 10) and (5, 6) are on the graph.

e. Confirm by plotting that your equation gives the graph in Figure 2-2i.

SOLUTION

a. The function could be exponential or quadratic, but exponential is chosen because the graph appears to approach the x-axis asymptotically.

b. The function is decreasing and concave up over its entire domain.

c. The function could model any situation in which a variable quantity starts at some nonzero value and decreases, gradually approaching zero, such as the number of degrees a cup of coffee is above room temperature as a function of time since it started cooling.

d. $y = ab^x$ — Write the untranslated general equation.

$$\begin{cases} 10 = ab^2 \\ 6 = ab^5 \end{cases}$$ Substitute the given x- and y-values.

$$\frac{6}{10} = \frac{ab^5}{ab^2}$$ Divide the second equation by the first to eliminate a.

$$0.6 = b^3$$

$$0.6^{1/3} = b$$ Raise both sides to the $\frac{1}{3}$ power to eliminate the exponent of b.

$$b = 0.8434\ldots$$ Store without rounding.

$$10 = a(0.8434\ldots)^2$$ Substitute 0.8434... for b in one of the equations.

$$a = 14.0572\ldots$$ Store without rounding.

$$\therefore y = 14.0572\ldots(0.8434\ldots)^x$$ Write the particular equation.

e. Plotting the graph confirms that the equation is correct. Note that the value of b is between 0 and 1, which corresponds to the fact that the function is decreasing.

Problem Set 2-2

Reading Analysis

From what you have read in this section, what do you consider to be the main idea? What is the difference between the parent quadratic function and any other quadratic function? How does the y-intercept of an exponential function differ from the y-intercept of a power function? Sketch the graph of a function that is increasing but concave down.

Quick Review

Q1. If $f(x) = x^2$, find $f(3)$.

Q2. If $f(x) = x^2$, find $f(0)$.

Q3. If $f(x) = x^2$, find $f(-3)$.

Q4. If $g(x) = 2^x$, find $g(3)$.

Q5. If $g(x) = 2^x$, find $g(0)$.

Q6. If $g(x) = 2^x$, find $g(-3)$.

Q7. If $h(x) = x^{1/2}$, find $h(25)$.

Q8. If $h(x) = x^{1/2}$, find $h(0)$.

Q9. If $h(x) = x^{1/2}$, find $h(-9)$.

Q10. What property of real numbers is illustrated by $3(x + 5) = 3(5 + x)$?

A. Associative property of multiplication

B. Commutative property of multiplication

C. Associative property of addition

D. Commutative property of addition

E. Distributive property of multiplication over addition

---

1. Power functions and exponential functions both have exponents. What major algebraic difference distinguishes these two types of functions?

2. What graphical feature do quadratic functions have that linear, exponential, and power functions do not have?

3. Write a sentence or two giving the origin of the word concave and explaining how the word applies to graphs of functions.

4. Explain why direct variation power functions contain the origin but inverse variation power functions do not.

5. Explain why the **reciprocal function** $f(x) = \frac{1}{x}$ is also a power function.

6. In the definition of quadratic function, what is the reason for the restriction $a \neq 0$?

7. The definition of exponential function, $y = ab^x$, includes the restriction $b > 0$. Suppose that $y = (-64)^x$. What would y equal if x were $\frac{1}{2}$? If x were $\frac{1}{3}$? Why do you think there is the restriction $b > 0$ for exponential functions?

8. The vertex form of the quadratic-function equation can be written as

$$y = k + a(x - h)^2 \quad \text{or} \quad y - k = a(x - h)^2$$

Explain why the first form is more useful if you are plotting graphs on your grapher and why the second form is more useful for understanding the translations involved.

9. *Reading Problem:* Clara has been reading her history assignment for 20 min and is now starting page 56 in the text. She reads at a (relatively) constant rate of 0.6 page per minute.

   a. Find the particular equation expressing the page number she is on as a function of minutes, using the point-slope form. Transform your answer to the slope-intercept form.

   b. Which page was Clara on when she started reading the assignment?

   c. The assignment ends at the top of page 63. When would you expect Clara to finish?

10. *Baseball Problem:* Ruth hits a high fly ball to right field. The ball is 4 ft above the ground when she hits it. Three seconds later it reaches its maximum height, 148 ft.

   a. Write an equation in vertex form of the quadratic function expressing the height of the ball explicitly as a function of time.

   b. How high is the ball 5 s after it was hit?

   c. If nobody catches the ball, how many seconds after it was hit will it reach the ground?

For Problems 11–20, the Quadrant I part of a function graph is shown.

   a. Identify the type of function it could represent.

   b. On what interval or intervals is the function increasing or decreasing and which way is the graph concave?

   c. From your experience, what relationship in the real world could be modeled by a function with this shape graph?

   d. Find the particular equation of the function if the given points are on the graph.

   e. Confirm by plotting that your equation gives the graph shown.

11.

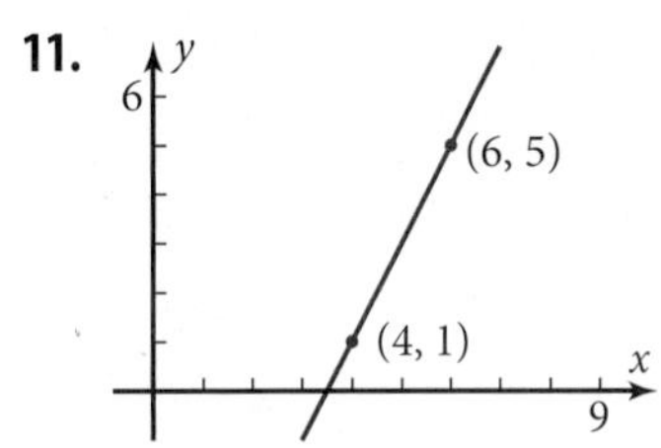

12.

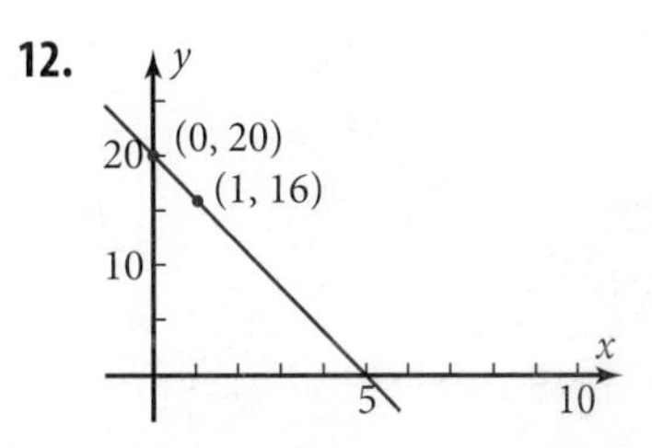

13.

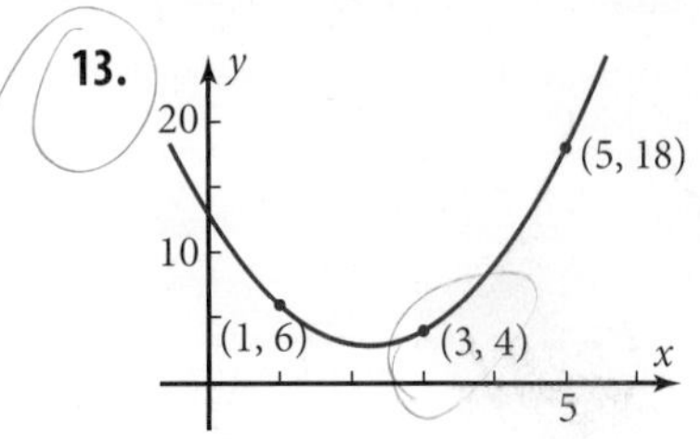

14.

15.

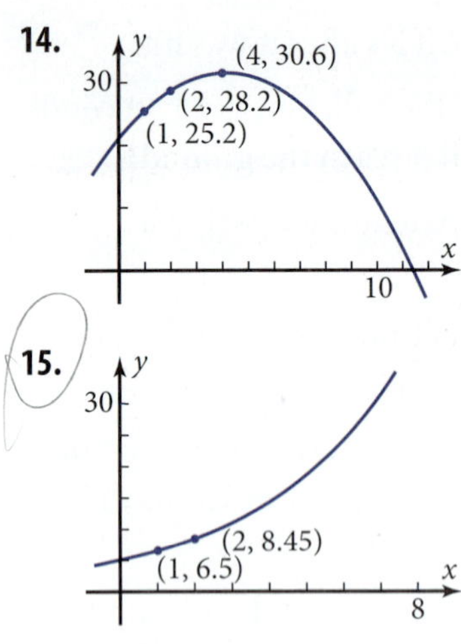

16.

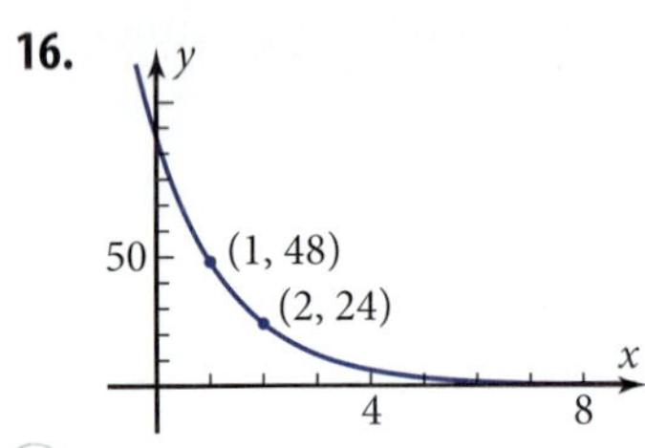

17.

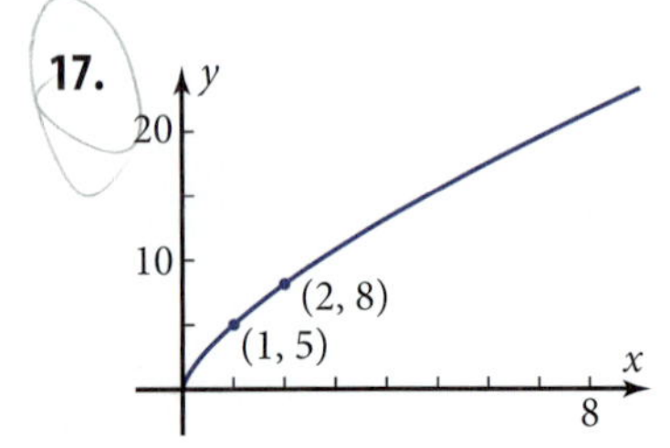

18.

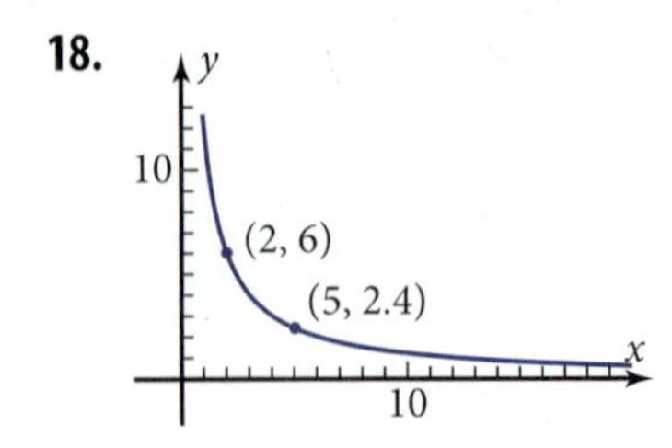

19.

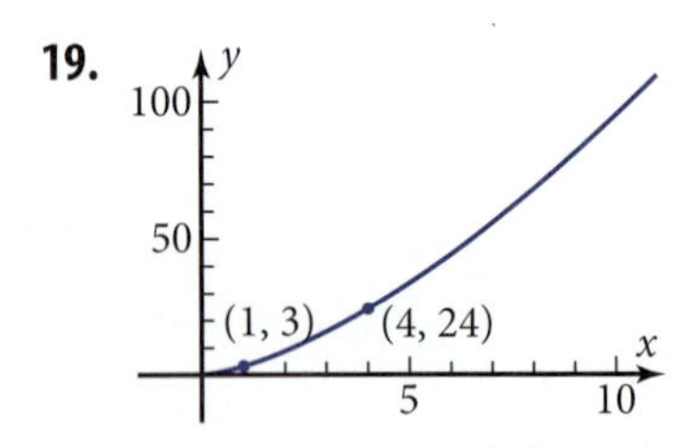

20.

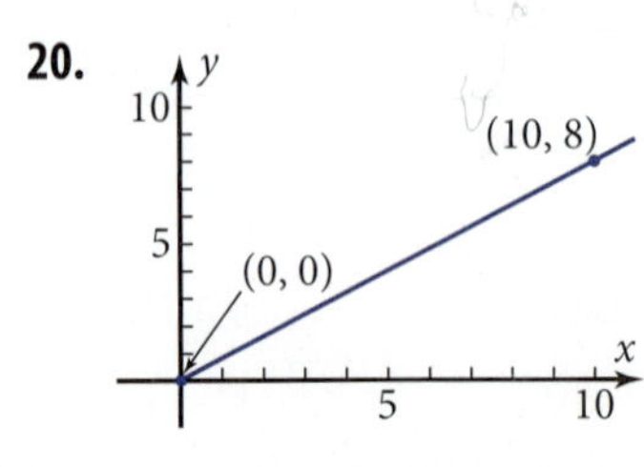

21. Suppose that y increases exponentially with x and that z is directly proportional to the square of x. Sketch the graph of each type of function. In what ways are the two graphs similar to each other? What major graphical difference would allow you to tell which graph is which if they were not labeled?

22. Suppose that y decreases exponentially with x and that z varies inversely with x. Sketch the graph of each type of function. Give at least three ways in which the two graphs are similar to each other. What major graphical difference would allow you to tell which graph is which if they were not labeled?

23. Suppose that y varies directly with x and that z increases linearly with x. Explain why any direct variation function is a linear function but a linear function is not necessarily a direct variation function.

24. Suppose that y varies directly with the square of x and that z is a quadratic function of x. Explain why the direct-square variation function is a quadratic function but the quadratic function is not necessarily a direct-square variation function.

25. *Natural Exponential Function Problem:* Figure 2-2j shows the graph of the natural exponential function $f(x) = 3e^{0.8x}$. Let $g(x) = 3b^x$. Find the value of b for which $g(x) = f(x)$. Show graphically that the two functions are equivalent.

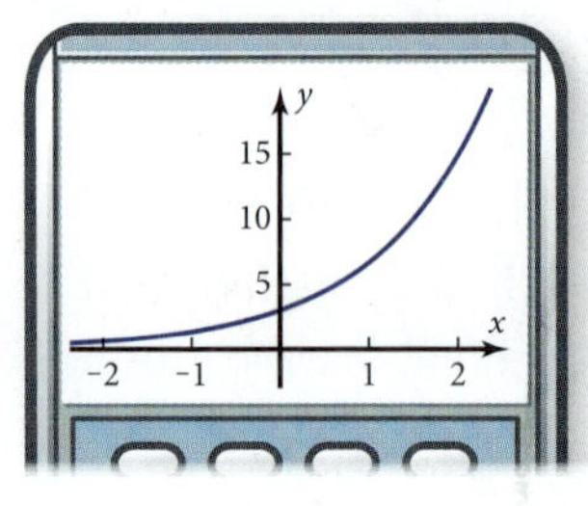

Figure 2-2j

2-3 Identifying Functions from Numerical Patterns

A 16-in. pizza has four times as much area as an 8-in. pizza. A grapefruit whose diameter is 10 cm has eight times the volume of a grapefruit with diameter 5 cm. In general, when you double the linear dimensions of a three-dimensional object, you multiply the surface area by 4 and the volume by 8. This is an example of the multiply–multiply property of power functions. It is similar to the add–add property of linear functions. Every time you add 1000 mi to the distance you have driven your car, you add a constant amount—say, \$300—to the cost of operating that car.

In this section you will use such patterns to identify the type of function that fits a given set of function values. Then you will find more function values, either by following the pattern or by finding the equation of the function.

Objective

- Given a set of regularly spaced *x*-values and the corresponding *y*-values, identify which type of function they fit (linear, quadratic, power, or exponential).
- Find other function values without necessarily finding the particular equation.

The Add–Add Pattern of Linear Functions

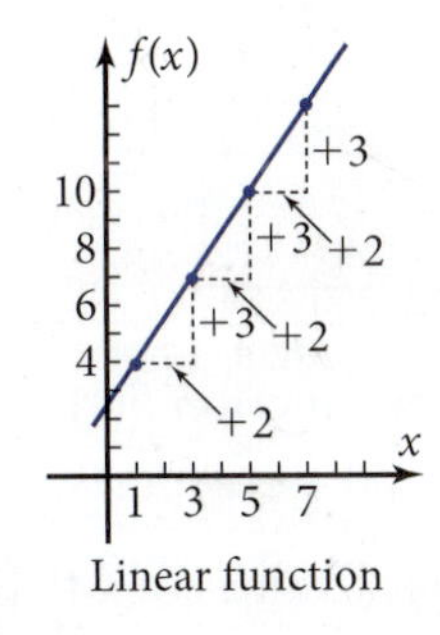

Linear function

x	$f(x)$
1	4
3	7
5	10
7	13
9	16

(Each +2 in x corresponds to +3 in $f(x)$.)

The add-add property

Figure 2-3a

Figure 2-3a shows the graph of the linear function $f(x) = 1.5x + 2.5$. As you can see from the graph and the adjacent table, each time you add 2 to x, y increases by 3. This pattern emerges because a linear function has constant slope. Verbally, you can express this property by saying that every time you add a constant to x, you add a constant (not necessarily the same as the constant added to x) to y. This property is called the **add–add property** of linear functions.

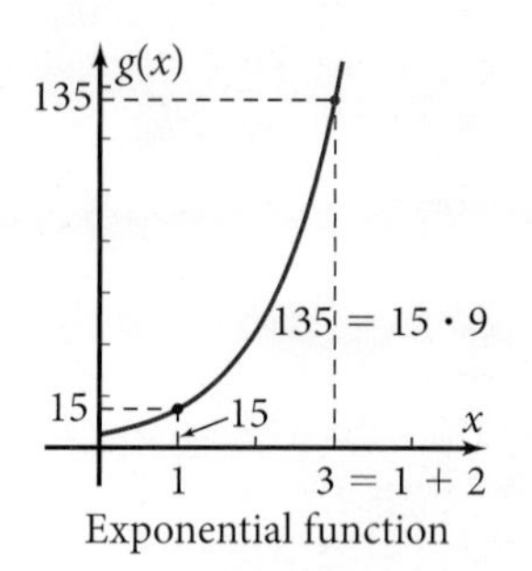

Exponential function

	x	$g(x)$	
	1	15	
+2	3	135	×9
+2	5	1215	×9
+2	7	10935	×9

The add-multiply property

Figure 2-3b

The Add–Multiply Pattern of Exponential Functions

Figure 2-3b shows the graph of the exponential function $g(x) = 5 \cdot 3^x$. This time, adding 2 to x results in the corresponding $g(x)$-values being *multiplied* by the constant 9. This is not coincidental. Here's why the pattern holds.

$$g(1) = 5 \cdot 3^1 = 15$$

$$g(3) = 5 \cdot 3^3 = 135 \qquad \text{(which equals 9 times 15)}$$

You can see algebraically why this is true.

$$g(3) = 5 \cdot 3^3$$

$$= 5 \cdot 3^{1+2} \qquad \text{Write the exponent as 1 increased by 2.}$$

$$= 5 \cdot 3^1 \cdot 3^2 \qquad \text{Product of powers with equal bases property.}$$

$$= (5 \cdot 3^1) \cdot 3^2 \qquad \text{Associate 5 and } 3^1 \text{ to get } g(1) \text{ in the expression.}$$

$$= g(1) \cdot 9$$

The conclusion is that if you add a constant to x, the corresponding y-value is multiplied by the base raised to that constant. This is called the **add–multiply property** of exponential functions.

The Multiply–Multiply Pattern of Power Functions

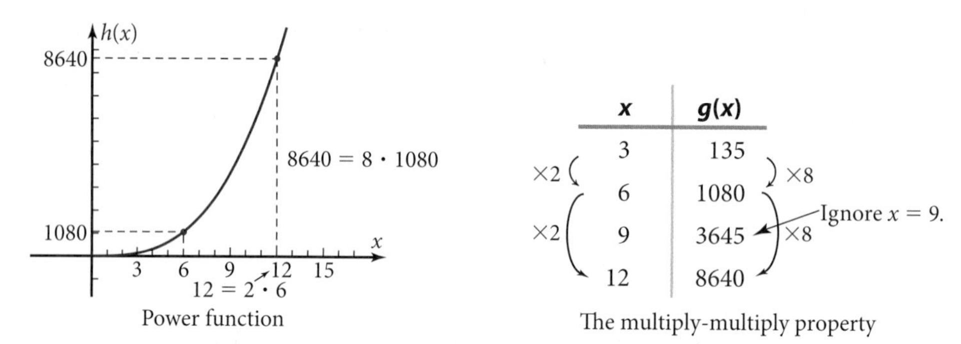

Power function

	x	$g(x)$	
	3	135	
×2	6	1080	×8
	9	3645	
×2	12	8640	×8

The multiply-multiply property

Figure 2-3c

Figure 2-3c shows the graph of the power function $h(x) = 5x^3$. As shown in the table, adding a constant, 3, to x does *not* create a corresponding pattern.

However, a pattern *does* emerge if you pick values of x that change by being *multiplied* by a constant.

$$h(3) = 5 \cdot 3^3 = 135$$

$$h(6) = 5 \cdot 6^3 = 1080 \qquad \text{(which equals 8 times 135)}$$

$$h(12) = 5 \cdot 12^3 = 8640 \qquad \text{(which equals 8 times 1080)}$$

If you double the x-value from 3 to 6 or from 6 to 12, the corresponding y-values are multiplied by 8, or 2^3. You can see algebraically why this is true.

$h(6) = 5 \cdot 6^3$

$= 5(3 \cdot 2)^3$ — Write the x-value 6 as twice 3.

$= (5 \cdot 3^3) \cdot 2^3$ — Distribute the exponent over multiplication and then associate.

$= h(3) \cdot 8$

In conclusion, if you multiply the x-values by 2, the corresponding y-values are multiplied by 8. This is called the **multiply–multiply property** of power functions. Note that extra points may appear in the table, such as (9, 3645) in Figure 2-3c. They do belong to the function, but the x-values do not fit the "multiply" pattern.

In this exploration, you'll find patterns for the y-values in quadratic functions similar to the add–add property of linear functions.

EXPLORATION 2-3: Patterns for Quadratic Functions

1. Show by making a table on your grapher that the points in the table fit the quadratic function

   $q(x) = 0.2x^2 - 1.3x + 14$

x	q(x)
2	12.2
4	12.0
6	13.4
8	16.4
10	21.0

2. Find the differences between consecutive y-values. Then find the *second differences,* that is, the differences between the consecutive differences. What do you notice?
3. Recall that the general equation of a quadratic function is $y = ax^2 + bx + c$, where a, b, and c stand for constants. Substitute the first three ordered pairs from the table in Problem 1 to get three linear equations involving a, b, and c. Solve this system of equations using matrices. Write the particular equation. Does it agree with the equation in Problem 1?
4. On the same screen, plot the graph of $q(x)$ and the five data points. You may use the stat feature on your grapher. Sketch the result.
5. Trace the graph of $q(x)$ to each value of x in the table. Do the five points lie on the graph?
6. Show that a quadratic function fits the data in this table by finding second differences. Find the particular equation, and show that these values satisfy the equation.

x	f(x)
1	12.3
4	34.8
7	44.7
10	42.0
13	26.7

7. What did you learn as a result of doing this exploration that you did not know before?

The Constant-Second-Differences Pattern of Quadratic Functions

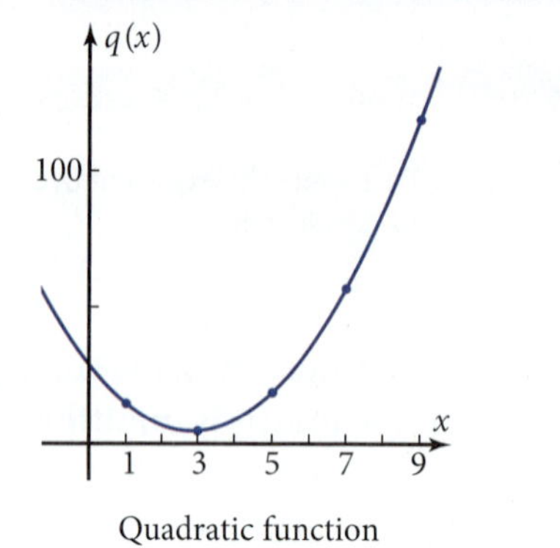

Quadratic function

x	f(x)
1	15
3	5
5	19
7	57
9	119

+2, +2, +2, +2

−10, +14, +38, +62

+24, +24, +24

Constant second differences

First differences

Constant second differences

Figure 2-3d

Figure 2-3d shows the graph of the quadratic function $q(x) = 3x^2 - 17x + 29$. An extension of the add–add property for linear functions applies to quadratics. For equally spaced x-values, the *differences* between the corresponding y-values are equally spaced. Thus the differences between these differences (the *second differences*) are constant. This constant is equal to $2ad^2$, twice the coefficient of the quadratic term times the square of the difference between the x-values.

These four properties are summarized in the box.

PROPERTIES: Patterns for Function Values

Add–Add Property of Linear Functions

If f is a linear function, adding a constant to x results in adding a constant to the corresponding $f(x)$-value. That is,

$$\text{if } f(x) = ax + b \text{ and } x_2 = c + x_1, \text{ then } f(x_2) = ac + f(x_1)$$

Add–Multiply Property of Exponential Functions

If f is an exponential function, adding a constant to x results in multiplying the corresponding $f(x)$-value by a constant. That is,

$$\text{if } f(x) = ab^x \text{ and } x_2 = c + x_1, \text{ then } f(x_2) = b^c \cdot f(x_1)$$

Multiply–Multiply Property of Power Functions

If f is a power function, multiplying x by a constant results in multiplying the corresponding $f(x)$-value by a constant. That is,

$$\text{if } f(x) = ax^b \text{ and } x_2 = cx_1, \text{ then } f(x_2) = c^b \cdot f(x_1)$$

Constant-Second-Differences Property of Quadratic Functions

If f is a quadratic function, $f(x) = ax^2 + bx + c$, and the x-values are spaced d units apart, then the second differences between the $f(x)$-values are constant and equal to $2ad^2$.

EXAMPLE 1 ➤ Identify the pattern in these function values and the kind of function that has this pattern.

x	$f(x)$
4	5
5	7
6	11
7	17
8	25

SOLUTION The values have neither the add–add, add–multiply, nor multiply–multiply property. They do exhibit the constant-second-differences property, as shown in the table. Therefore, a quadratic function fits the data.

	x	$f(x)$		
	4	5		
+1	5	7	+2	
+1	6	11	+4	+2
+1	7	17	+6	+2
+1	8	25	+8	+2

EXAMPLE 2 ➤ If function f has values $f(5) = 12$ and $f(10) = 18$, find $f(20)$ if f is

a. A linear function

b. A power function

c. An exponential function

SOLUTION

a. Linear functions have the add–add property. Notice that you add 5 to the first x-value to get the second one and that you add 6 to the first $f(x)$-value to get the second one. Make a table of values ending at $x = 20$. The answer is $f(20) = 30$.

	x	$f(x)$	
	5	12	
+5	10	18	+6
+5	15	24	+6
+5	20	30	+6

b. Power functions have the multiply–multiply property. When going from the first to the second x- and $f(x)$-values, notice that you multiply 5 by 2 to get 10 and that you multiply 12 by 1.5 to get 18. Make a table of values ending at $x = 20$. The answer is $f(20) = 27$.

	x	$f(x)$	
	5	12	
×2	10	18	×1.5
×2	20	27	×1.5

c. Exponential functions have the add–multiply property. Notice that adding 5 to x results in multiplying the corresponding $f(x)$-value by 1.5. Make a table of values ending at $x = 20$. The answer is $f(20) = 40.5$.

	x	$f(x)$	
	5	12	
+5	10	18	×1.5
+5	15	27	×1.5
+5	20	40.5	×1.5

EXAMPLE 3 ➤ Describe the effect on y of doubling x if

a. y varies directly with x.

b. y varies inversely with the square of x.

c. y varies directly with the cube of x.

SOLUTION a. y is doubled (that is, multiplied by 2^1).

b. y is multiplied by $\frac{1}{4}$ (that is, multiplied by 2^{-2}).

c. y is multiplied by 8 (that is, multiplied by 2^3).

EXAMPLE 4 ➤ Suppose that f is a direct-square power function and that $f(5) = 1000$. Find $f(20)$.

SOLUTION Because f is a power function, it has the multiply–multiply property. Express $x = 20$ as $x = 4 \cdot 5$. Multiplying x by 4 will multiply the corresponding $f(x)$-value by 4^2 because f is a direct square, so

$$f(20) = f(4 \cdot 5) = 4^2 \cdot f(5) = 16 \cdot 1000 = 16{,}000$$

EXAMPLE 5 ➤ *Radioactive Tracer Problem*: The compound 18-fluorodeoxyglucose (18-FDG) is composed of radioactive fluorine (18-F) and a sugar (deoxyglucose). It is used to trace glucose metabolism in the heart. 18-F has a half-life of about 2 h, which means that at the end of each 2-hour time period, only half of the 18-F that was there at the beginning of the time period remains. Suppose a dose of 18-FDG was injected into a patient. Let $f(x)$ be the number of *microcuries* (μCi) of 18-FDG that remains over time x, in hours, as shown in the table.

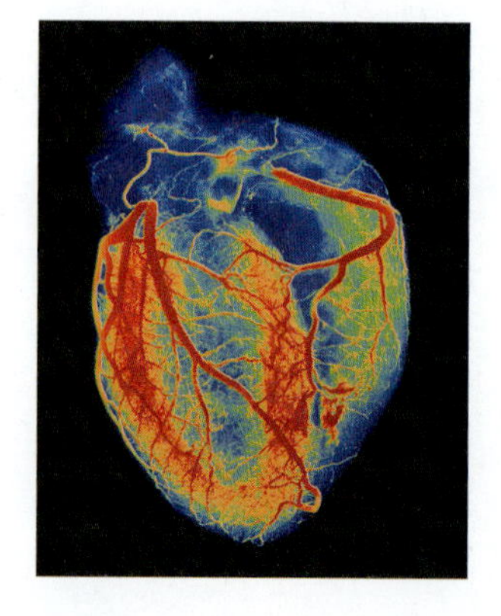

x (h)	$f(x)$ (μCi)
2	5
4	2.5
6	1.25
8	0.652

a. Find the number of microcuries that remains after 12 h.

b. Identify the pattern these data points follow. What type of function shows this pattern?

c. Why can't you use the pattern to find $f(25)$?

d. Find a particular equation of $f(x)$. Show by plotting that all the $f(x)$-values in the table satisfy the equation.

e. Use the equation to calculate $f(25)$. Interpret the solution.

SOLUTION

a. Follow the add pattern in the x-values until you reach 12, and follow the multiply pattern in the corresponding $f(x)$-values.

x (h)	**f (x) (μCi)**
10	0.3125
12	0.15625

$f(12) = 0.15625\ \mu\text{Ci}$

b. The data points have the add–multiply property of exponential functions.

c. 22, 24, 26 — Extend the add pattern in the x-values.

The x-values skip over 25, so $f(25)$ cannot be found using the pattern.

d. $f(x) = ab^x$ — General equation of an exponential function.

$\begin{cases} 5 = ab^2 \\ 2.5 = ab^4 \end{cases}$ — Substitute any two of the ordered pairs.

$\dfrac{2.5}{5} = \dfrac{ab^4}{ab^2}$ — Divide the equations. Have the larger exponent in the numerator.

$0.5 = b^2$ — Simplify.

$0.5^{1/2} = b$ — Raise both sides to the $\frac{1}{2}$ power.

$b = 0.7071\ldots$ — Store without rounding.

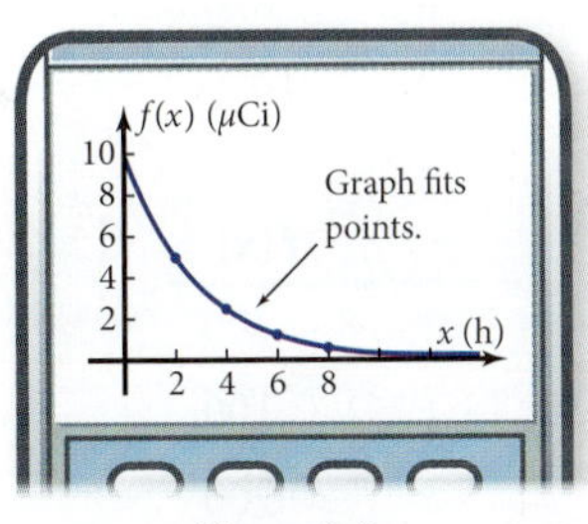

Figure 2-3e

$5 = a(0.7071\ldots)^2$ — Substitute the value for b into one of the equations.

$a = \dfrac{5}{0.7071\ldots^2} = 10$ — Solve for a.

$\therefore f(x) = 10(0.7071\ldots)^x$ — Write the particular equation.

Figure 2-3e shows the graph of f passing through all four given points.

e. $f(25) = 10(0.7071\ldots)^{25} = 0.0017\ldots$

This means that there was about 0.0017 μCi of 18-FDG after 25 h.

Note that part d in Example 5 calls for "a" function that fits the points. It is possible for other functions to fit this set of points, such as the function

$$g(x) = 10(0.7071\ldots)^x + \sin \frac{\pi}{2}x$$

See Chapter 5 for the meaning of the sine function.

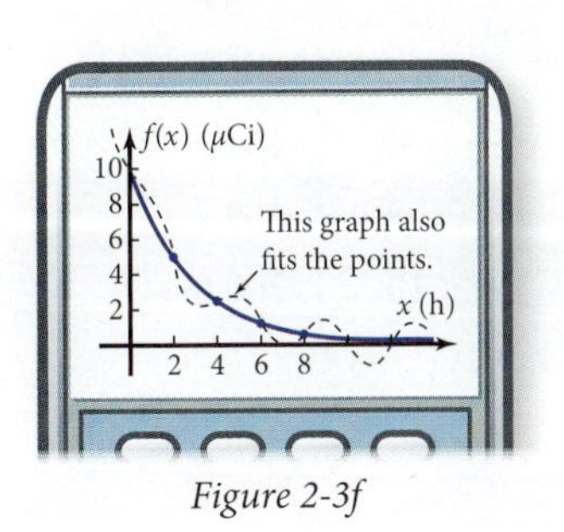

Figure 2-3f

which also fits the given points, as shown in Figure 2-3f. Deciding which function fits better will depend on the situation you are modeling. Also, you can test further to see whether your model is supported by data. For example, to test the second model, you could collect measurements over shorter time intervals and see if the data have a wavy pattern.

Problem Set 2-3

Reading Analysis

From what you have read in this section, what do you consider to be the main idea? How is the add–add property of linear functions consistent with the concept of slope? How do the properties of exponential and power functions differ? If you triple the diameter of a circle, what effect does this have on the circle's area? What type of function has this property? What numerical pattern do quadratic functions have?

Quick Review

Q1. Write the general equation of a linear function.

Q2. Write the general equation of a power function.

Q3. Write the general equation of an exponential function.

Q4. Write the general equation of a quadratic function.

Q5. $f(x) = 3 \cdot x^5$ is the equation of a particular __?__ function.

Q6. $f(x) = 3 \cdot 5^x$ is the equation of a particular __?__ function.

Q7. Name the transformation of $f(x)$ that gives $g(x) = 4 \cdot f(x)$.

Q8. The function $g(x) = 3 + 4 \cdot f(5(x - 6))$ is a vertical dilation of function f by a factor of

A. 3 **B.** 4 **C.** 5 **D.** 6 **E.** -6

Q9. Sketch the graph of a linear function with negative slope and positive y-intercept.

Q10. Sketch the graph of an exponential function with base greater than 1.

For Problems 1–12, determine whether the data have the add–add, add–multiply, multiply–multiply, or constant-second-differences pattern. Identify the type of function that has the pattern.

1.

x	$f(x)$
2	2700
4	2400
6	2100
8	1800
10	1500

2.

x	$f(x)$
2	1500
4	750
6	500
8	375
10	300

3.

x	$f(x)$
2	12
4	48
6	108
8	192
10	300

4.

x	$f(x)$
2	12
4	48
6	192
8	768
10	3072

5.

x	$f(x)$
2	26
4	52
6	78
8	104
10	130

6.

x	$f(x)$
2	4.6
4	6.0
6	7.4
8	8.8
10	10.2

7.

x	$f(x)$
2	1800
4	450
6	200
8	112.5
10	72

8.

x	$f(x)$
2	400
4	100
6	-200
8	-500
10	-800

9.

x	$f(x)$
2	900
4	100
6	111.111...
8	1.2345...
10	0.1371...

10.

x	$f(x)$
2	5.6
4	44.8
6	151.2
8	358.4
10	700.0

11.

x	$f(x)$
1	352
3	136
5	64
7	136
9	352

12.

x	$f(x)$
1	25
5	85
9	113
13	109
17	73

For Problems 13–16, find the indicated function value if f is

a. A linear function

b. A power function

c. An exponential function

13. Given $f(2) = 5$ and $f(6) = 20$, find $f(18)$.

14. Given $f(3) = 80$ and $f(6) = 120$, find $f(24)$.

15. Given $f(10) = 100$ and $f(20) = 90$, find $f(40)$.

16. Given $f(1) = 1000$ and $f(3) = 100$, find $f(9)$.

For Problems 17–20, use the given values to calculate the other values specified.

17. Given f is a linear function with $f(2) = 1$ and $f(5) = 7$, find $f(8)$, $f(11)$, and $f(14)$.

18. Given f is a direct-cube power function with $f(3) = 0.7$, find $f(6)$ and $f(12)$.

19. Given that $f(x)$ varies inversely with the square of x and that $f(5) = 1296$, find $f(10)$ and $f(20)$.

20. Given that $f(x)$ varies exponentially with x and that $f(1) = 100$ and $f(4) = 90$, find $f(7)$, $f(10)$, and $f(16)$.

For Problems 21–24, describe the effect on $f(x)$ if you double the value of x.

21. Direct-square power function

22. Direct fourth-power function

23. Inverse variation power function

24. Inverse-square variation power function

25. *Volume Problem:* The volumes of similarly shaped objects are directly proportional to the cube of a linear dimension.

Baseball

Volleyball

Figure 2-3g

a. Recall from geometry that the volume, V, of a sphere equals $\frac{4}{3}\pi r^3$, where r is the radius. Explain how the formula $V = \frac{4}{3}\pi r^3$ shows that the volume of a sphere varies directly with the cube of the radius. If a baseball has volume 200 cm^3, what is the volume of a volleyball that has three times the radius of the baseball (Figure 2-3g)?

b. King Kong is depicted as having the same proportions as a normal gorilla but as being 10 times as tall. How would his volume (and thus his weight) compare to that of a normal gorilla? If a normal gorilla weighs 400 lb, what would you expect King Kong to weigh? Is this surprising?

c. A great white shark 20 ft long weighs about 4000 lb. Fossilized sharks' teeth from millions of years ago suggest that there were once great whites 100 ft long. How much would you expect such a shark to weigh?

d. Gulliver traveled to Lilliput, where people were $\frac{1}{10}$ as tall as normal people. If Gulliver weighed 200 lb, how much would you expect a Lilliputian to weigh?

Iris Weddell White's illustration The Emperor Visits Gulliver *in Jonathan Swift's* Gulliver's Travels. *(The Granger Collection, New York)*

26. *Area Problem:* The areas of similarly shaped objects are directly proportional to the square of a linear dimension.

a. Give the formula for the area of a circle. Explain why the area varies directly with the square of the radius.

b. If a grapefruit has twice the diameter of an orange, how do the areas of their rinds compare?

c. When Gutzon Borglum designed the reliefs he carved into Mount Rushmore in South Dakota, he started with models $\frac{1}{12}$ the lengths of the actual reliefs. How does the area of each model compare to the area of each of the final reliefs? Explain why a relatively small decrease in the linear dimension results in a relatively large decrease in the surface area to be carved.

d. Gulliver traveled to Brobdingnag, where people were 10 times as tall as normal people. If Gulliver had 2 m^2 of skin, how much skin surface would you expect a Brobdingnagian to have?

27. *Airplane Weight and Area Problem:* In 1896, Samuel Langley successfully flew a model of an airplane he was designing. In 1903, he tried unsuccessfully to fly the full-size airplane. Assume that the full-size plane was 4 times the length of the model (Figure 2-3h).

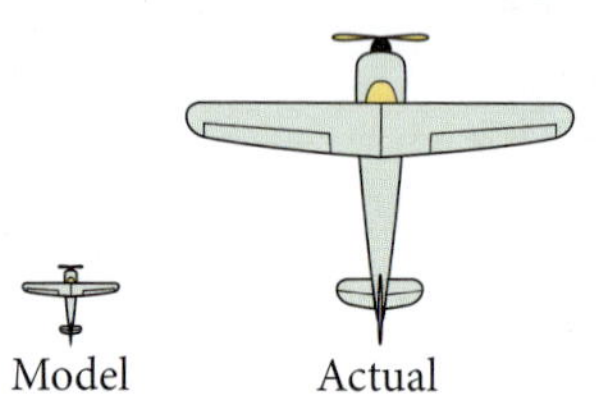

Figure 2-3h

a. The wing area, and thus the lift, of similarly shaped airplanes is directly proportional to the square of the length of each plane. How many times more wing area did the full-size plane have than the model?

b. The volume, and thus the weight, of similarly shaped airplanes is directly proportional to the cube of the length. How many times heavier was the full-size plane than the model?

c. Why do you think the model was able to fly but the full-size plane was not?

28. *Compound Interest Problem:* Money left in a savings account grows exponentially with time. Suppose that you invest \$1000 and find that a year later you have \$1100 in your account.

a. How much will you have after 2 yr? 3 yr? 4 yr?

b. In how many years will your investment double?

29. *Archery Problem:* Ann Archer shoots an arrow into the air. The table lists its height at various times after she shoots it.

Time (s)	Height (ft)
1	79
2	121
3	131
4	109
5	55

a. Show that the second differences between consecutive height values in the table are constant.

b. Use the first three ordered pairs to find the particular equation of the quadratic function that fits these points. Show that the function contains all of the points.

c. Based on the graph you fit to the points, how high was the arrow at 2.3 s? Was it going up or going down? How do you tell?

d. At what two times was the arrow 100 ft high? How do you explain the fact that there were two times?

e. When was the arrow at its highest? How high was that?

f. At what time did the arrow hit the ground?

30. *The Other Function Fit Problem:* It is possible for different functions to fit the same set of discrete data points. Suppose that the data in the table have been given.

x	f(x)
2	12
4	48
6	108
8	192
10	300

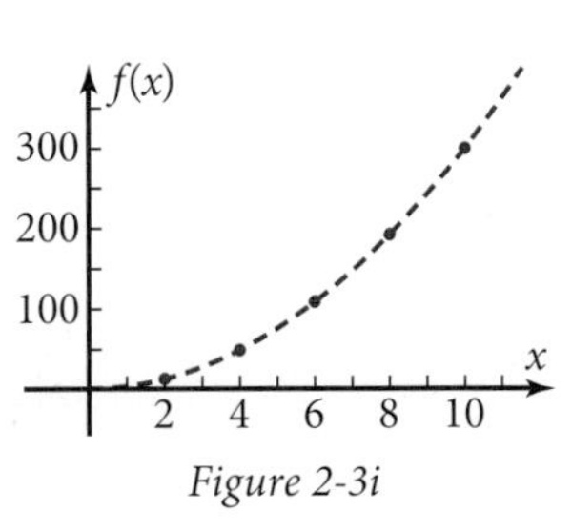

Figure 2-3i

a. Show that the function $f(x) = 3x^2$ fits the data, as shown in Figure 2-3i.

b. Select radian mode, and then plot $f_1(x) = 3x^2$ and $f_2(x) = 3x^2 + 100\sin\left(\frac{\pi}{2}(x)\right)$, where "sin" is the sine function (see Chapter 5). Sketch the result. Does the equation of $f_2(x)$ also fit the given data?

c. Deactivate $f_2(x)$ from part b and plot $f_3(x) = 3x^2\cos(\pi x)$, where "cos" is the cosine function (see Chapter 5). Sketch the result. What do the results tell you about fitting functions to discrete data points?

31. *Incorrect Point Problem:* By considering second differences, show that a quadratic function does *not* fit the values in this table.

x	y
4	5
5	7
6	11
7	17
8	27

What would the last y-value have to be for a quadratic function to fit the values exactly?

32. *Cubic Function Problem:* Figure 2-3j shows the graph of the cubic function

$$f(x) = x^3 - 6x^2 + 5x + 20$$

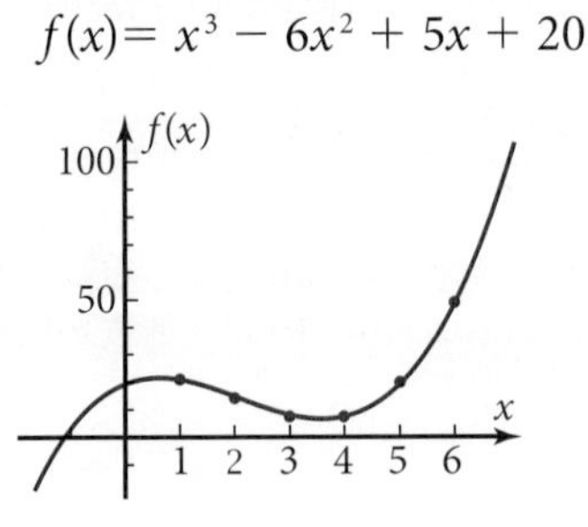

Figure 2-3j

a. Make a table of values of $f(x)$ for each integer value of x from 1 to 6.

b. Show that the *third differences* between the values of $f(x)$ are constant. You can calculate the third differences in a time-efficient way using the list and delta list features of your grapher. If you do it by pencil and paper, be sure to subtract (value $-$ previous value) in each case.

c. Make a conjecture about how you could determine whether a quartic function (fourth degree) fits a set of points.

33. *The Add–Add Property Proof Problem:* Prove that for a linear function, adding a constant to x adds a constant to the corresponding value of $f(x)$. Do this by showing that if $x_2 = x_1 + c$, then $f(x_2)$ equals a constant plus $f(x_1)$. Start by writing the equations of $f(x_1)$ and $f(x_2)$, and then make the appropriate substitutions and algebraic manipulations.

34. *The Multiply–Multiply Property Proof Problem:* Prove that for a power function, multiplying x by a constant multiplies the corresponding value of $f(x)$ by a constant as well. Do this by showing that if $x_2 = cx_1$, then $f(x_2)$ equals a constant times $f(x_1)$. Start by writing the equations of $f(x_1)$ and $f(x_2)$, and then make the appropriate substitutions and algebraic manipulations.

35. *The Add–Multiply Property Proof Problem:* Prove that for an exponential function, adding a constant to x multiplies the corresponding value of $f(x)$ by a constant. Do this by showing that if $x_2 = c + x_1$, then $f(x_2)$ equals a constant times $f(x_1)$. Start by writing the equations of $f(x_1)$ and $f(x_2)$, and then make the appropriate substitutions and algebraic manipulations.

36. *The Constant-Second-Differences Property Proof Problem:* Let $f(x) = ax^2 + bx + c$. Let d be the constant difference between successive x-values. Find $f(x + d)$, $f(x + 2d)$, and $f(x + 3d)$. Simplify. By subtracting consecutive $f(x)$-values, find the three first differences. By subtracting consecutive first differences, show that the two second differences equal the constant $2ad^2$.

2-4 Properties of Logarithms

Any positive number can be written as a power of 10. For instance,

$$3 = 10^{0.4771...}$$

$$5 = 10^{0.6989...}$$

$$15 = 10^{1.1760...}$$

The exponents 0.4771…, 0.6989…, and 1.1760… are called the base-10 **logarithms** of 3, 5, and 15, respectively.

$$\log 3 = 0.4771...$$

$$\log 5 = 0.6989...$$

$$\log 15 = 1.1760...$$

Objective

Learn the properties of base-10 logarithms.

In this exploration you will learn about properties of logarithms. Logarithms give you an algebraic way to solve exponential equations such as $1.06^x = 2$.

EXPLORATION 2-4: Introduction to Logarithms

1. The LOG key on your calculator finds the logarithm of a given number. Evaluate these logarithms:

   $\log 10 =$?

   $\log 100 =$?

   $\log 1000 =$?

   $\log 10000 =$?

   $\log 10^5 =$?

2. From your answers in Problem 1, figure out what $\log x$ means. Write what you discover.
3. Based on your answer to Problem 2, what should $\log 10^{1.8}$ equal?
4. Test your conjecture in Problem 3 by finding the value of $10^{1.8}$ on your calculator and then finding the logarithm of the (unrounded) answer.
5. Test your conjecture again by finding log 347 and then raising 10 to the power of the answer you get.
6. Find log 2 and log 32. Show numerically that log 32 is *five times* log 2.
7. Note that $32 = 2^5$. Thus, $\log 2^5 = 5 \log 2$. Show numerically that $\log 17^{2.34} = 2.34 \log 17$.
8. Complete the property of the *log of a power:* $\log b^x =$?
9. What did you learn as a result of doing this exploration that you did not know before?

A decibel, which measures the relative intensity of sounds, has a logarithmic scale. Prolonged exposure to noise intensity exceeding 85 decibles can lead to hearing loss.

Definition and Properties of Base-10 Logarithms

To gain confidence in the meaning of logarithm, press LOG 3 on your calculator. You will get

$$\log 3 = 0.477121254\ldots$$

Then, without rounding, raise 10 to this power. You will get

$$10^{0.477121254\ldots} = 3$$

The powers of 10 have the normal properties of exponentiation. For instance,

$$15 = (3)(5) = (10^{0.4771\ldots})(10^{0.6989\ldots})$$

$$= 10^{0.4771\ldots+0.6989\ldots} \quad \text{Add the exponents; keep the same base.}$$

$$= 10^{1.1760\ldots}$$

You can check by calculator that $10^{1.1760\ldots}$ really does equal 15.

From this example you can infer that logarithms have the same properties as exponents. This is not surprising, because logarithms *are* exponents. For instance,

$$\log(3 \cdot 5) = \log 3 + \log 5 \quad \text{The log of a product equals the sum of the logs of the factors.}$$

From the values given earlier, you can also show that

$$\log \frac{15}{3} = \log 15 - \log 3 \quad \text{The log of a quotient.}$$

This property is reasonable because you divide powers of equal bases by subtracting the exponents.

$$\frac{15}{3} = \frac{10^{1.1760\ldots}}{10^{0.4771\ldots}} = 10^{1.1760\ldots-0.4771} = 10^{0.6989\ldots} = 5$$

Because a power can be written as a product, you can find the logarithm of a power like this:

$$\log 34 = \log(3 \cdot 3 \cdot 3 \cdot 3) = \log 3 + \log 3 + \log 3 + \log 3$$

$$= 4 \log 3 \quad \text{Combine like terms.}$$

The logarithm of a power equals the exponent of that power times the logarithm of the base. To verify this result, observe that $3^4 = 81$. Press 4 LOG 3 on your calculator, and get 1.9084.... Then press LOG 81. You get the same answer, 1.9084....

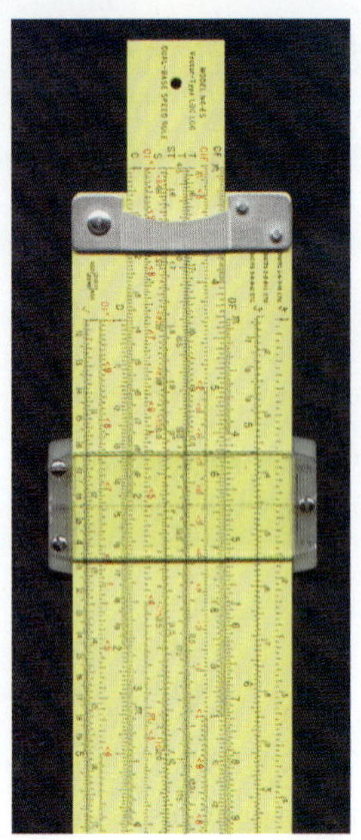

Slide rules, used by engineers in the 19th and early 20th centuries, employ the principle of logarithms for performing complicate calculations.

The definition and three properties of logarithms are summarized in this box.

DEFINITION AND PROPERTIES: Base-10 Logarithms

Definition

$\log x = y$ if and only if $10^y = x$

Verbally: $\log x$ is the exponent in the power of 10 that gives x

Properties

- Log of a Product:

  $\log xy = \log x + \log y$

Verbally: The log of a product equals the sum of the logs of the factors.

- Log of a Quotient:

  $\log \frac{x}{y} = \log x - \log y$

Verbally: The log of a quotient equals the log of the numerator minus the log of the denominator.

- Log of a Power:

  $\log x^y = y \log x$

Verbally: The log of a power equals the exponent times the log of the base.

The reason for the name **logarithm** is historical. Before there were calculators, base-10 logarithms, calculated approximately using infinite series, were recorded in tables. Products with many factors, such as

$$(357)(4.367)(22.4)(3.142)$$

could then be calculated by adding their logarithms (exponents) column-wise in one step rather than by tediously multiplying several pairs of numbers. Englishman Henry Briggs (1561–1630) and Scotsman John Napier (1550–1616) are credited with inventing this "**log**ical way to do **arithm**etic" that you will explore in Problem 47. The word *logarithm* actually comes from the Greek words *logos,* which here means "ratio," and *arithmos*, which means "number."

The most important thing to remember about logarithms is this:

A logarithm is an exponent.

Examples 1 and 2 show you how to verify that a logarithm is an exponent.

EXAMPLE 1 ➤ Find x if $\log 10^{3.721} = x$. Verify your solution numerically.

SOLUTION By definition, the logarithm is the exponent of 10. So $x = 3.721$.

CHECK

$10^{3.721} = 5260.1726\ldots$ By calculator. Do not round.

$\log 5260.1726\ldots = 3.721$

which checks.

EXAMPLE 2 ➤ Find x if $0.258 = 10^x$. Verify your solution numerically.

SOLUTION By definition, x, the exponent of 10, is the logarithm of 0.258.

$x = \log 0.258 = -0.5883...$ By calculator. Do not round.

CHECK $10^{-0.5883...} = 0.258$

which checks. ◄

Examples 3, 4, and 5 show you how to verify numerically the three properties of logarithms.

EXAMPLE 3 ➤ Show numerically that $\log(7 \cdot 9) = \log 7 + \log 9$. Explain how this property agrees with the definition of logarithm.

SOLUTION

$\log(7 \cdot 9) = \log 63 = 1.7993...$

$\log 7 + \log 9 = 0.8451... + 0.9542... = 1.7993...$ Calculate without rounding.

$\therefore \log(7 \cdot 9) = \log 7 + \log 9$

This equality agrees with the definition because

$(7 \cdot 9) = (10^{0.8451}...)(10^{0.9542})$

$= 10^{0.8451...+0.9542}$ Add the exponents. Keep the same base.

$= 10^{1.7993}$

$\therefore \log(7 \cdot 9) = 1.7993...$ The logarithm is the exponent of 10. ◄

EXAMPLE 4 ➤ Show numerically that $\log\frac{51}{17} = \log 51 - \log 17$. Explain how this property agrees with the definition of logarithm.

SOLUTION

$\log\frac{51}{17} = \log 3 = 0.4771...$

$\log 51 - \log 17 = 1.7075... - 1.2304... = 0.4771...$ Calculate without rounding.

$\therefore \log\frac{51}{17} = \log 51 - \log 17$

This equality agrees with the definition because

$\frac{51}{17} = \frac{10^{1.7075}}{10^{1.2304}} = 1^{1.7075...-1.2304}$ Subtract the exponents. Keep the same base.

$= 10^{0.4771...}$

$\therefore \log\frac{51}{17} = 0.4771...$ The logarithm is the exponent of 10. ◄

EXAMPLE 5 ➤ Show numerically that $\log 5^3 = 3 \log 5$. Explain how this property can be derived from the log of a product property.

SOLUTION $\log 5^3 = \log 125 = 2.0969\ldots$

$3 \log 5 = 3 \cdot 0.6989\ldots = 2.0969\ldots$ Calculate without rounding.

$\therefore \log 5^3 = 3 \log 5$ Combine like terms.

This equality derives from the product of a log property because

$\log 5^3 = \log(5 \cdot 5 \cdot 5)$

$= \log 5 + \log 5 + \log 5$ The log of a product equals the sum of the logs of the factors.

$= 3 \log 5$ Combine like terms. ◀

Example 6 shows you how to prove algebraically that the logarithm of the product of two numbers equals the sum of the logarithms of the factors.

EXAMPLE 6 ➤ Prove algebraically that $\log xy = \log x + \log y$.

SOLUTION *Proof*

Let $c = \log x$ and let $d = \log y$.

Then $10^c = x$ and $10^d = y$. By the definition of logarithm.

$xy = 10^c = 10^d$ Multiply x times y.

$xy = 10^{c+d}$ Add the exponents. Keep the same base.

$\log xy = c + d$ The logarithm is the exponent of 10.

$\therefore \log xy = \log x + \log y$, Q.E.D Substitute for c and d. ◀

In Problems 45 and 46, you will prove the other two properties of base-10 logarithms algebraically.

Example 7 shows you how to use the properties of logarithms to simplify expressions that contain logarithms.

EXAMPLE 7 ➤ Use the properties of logarithms to find the number that goes in the blank: $\log 3 + \log 7 - \log 5 = \log$ __?__. Check your answer numerically.

SOLUTION $\log 3 + \log 7 - \log 5 = \log \frac{3 \cdot 7}{5} = \log 4.2$

$\therefore$ 4.2 goes in the blank.

CHECK $\log 3 + \log 7 - \log 5 = 0.6232\ldots$ By calculator.

$\log 4.2 = 0.6232\ldots$

$\therefore \log 3 + \log 7 - \log 5 = \log 4.2$ ◀

Problem Set 2-4

Reading Analysis

From what you have read in this section, what do you consider to be the main idea? Explain the statement "A logarithm is an exponent" and support it with examples.

Quick Review

Q1. In the expression 7^5, the number 7 is called the ___?___.

Q2. In the expression 7^5, the number 5 is called the ___?___.

Q3. The entire expression 7^5 is called a(n) ___?___.

Q4. Write $x^5 \cdot x^7$ as a single exponential expression.

Q5. Write $\frac{x^5}{x^7}$ as a single exponential expression.

Q6. Write the expression $(x^5)^7$ without parentheses.

Q7. For the expression $(xy)^7$, you ___?___ the exponent 7 to get x^7y^7.

Q8. Explain the meaning of 5^{-2}.

Q9. Explain the meaning of $9^{1/2}$.

Q10. The function $y = 5^x$ is called a(n)

- **A.** Power function
- **B.** Exponential function
- **C.** Polynomial function
- **D.** Linear function
- **E.** Inverse of a power function

---

1. What does it mean to say that $-0.1549\ldots$ equals log 0.7?

2. What does it mean to say that $0.9030\ldots$ equals log 8?

3. By the definition of logarithm, if $a = \log b$, then $10^{\underline{?}} = \underline{\ ?\ }$.

4. By the definition of logarithm, if $10^a = b$, then ___?___ = log ___?___.

For Problems 5–8, use the definition of logarithm to write the value of x. Then confirm that your solution is correct by raising 10 to the given power, taking the logarithm of the result, and showing that the final result agrees with your answer.

5. $\log 10^{1.574} = x$ **6.** $\log 10^{2.803} = x$

7. $\log 10^{-0.981} = x$ **8.** $\log 10^{23.58} = x$

For Problems 9–12, use the definition of logarithm to write x as a logarithm. Then evaluate the logarithm by calculator and show that raising 10 to that power gives a result that agrees with the given equality.

9. $57 = 10^x$ **10.** $359 = 10^x$

11. $0.85 = 10^x$ **12.** $0.0321 = 10^x$

For Problems 13–16, find the logarithm by calculator. Then show numerically that raising 10 to that power, gives the argument of the logarithm.

13. log 1066 **14.** log 2001

15. log 0.0596 **16.** log 0.314

For Problems 17–20, evaluate the power of 10. Then show that the logarithm of the answer is equal to the original exponent of 10.

17. $10^{-2.7}$ **18.** $10^{3.5}$

19. $10^{15.2}$ **20.** 10^{-4}

21. Find log 5, log 4, and log 20. Show that $\log 20 = \log 5 + \log 4$. What property of logarithms does this equality illustrate? What property of exponentiation does this property come from?

22. Find log 30, log 4, and log 120. Show that $\log 120 = \log 30 + \log 4$. What property of logarithms does this equality illustrate? What property of exponentiation does this property come from?

23. Find log 35, log 7, and log 5. Show that $\log 5 = \log 35 - \log 7$. What property of logarithms does this equality illustrate? What property of exponentiation does this property come from?

24. Find log 96, log 6, and log 16. Show that log 16 = log 96 − log 6. What property of logarithms does this equality illustrate? What property of exponentiation does this property come from?

25. Find log 32 and log 2. Show that log 32 = 5 log 2. What property of logarithms does this equality illustrate? What property of exponentiation does this property come from?

26. Find log 64 and log 4. Show that log 64 = 3 log 4. What property of logarithms does this equality illustrate? What property of exponentiation does this property come from?

For Problems 27–34, demonstrate numerically the properties of logarithms. Then explain how each result agrees with the definition of logarithm.

27. $\log(0.3 \cdot 0.7) = \log 0.3 + \log 0.7$

28. $\log(7 \cdot 8) = \log 7 + \log 8$

29. $\log(30 \div 5) = \log 30 - \log 5$

30. $\log \frac{2}{8} = \log 2 - \log 8$

31. $\log 2^5 = 5 \log 2$

32. $\log 5^3 = 3 \log 5$

33. $\log \frac{1}{7} = -\log 7$

34. $\log \frac{1}{1000} = -\log 1000$

For Problems 35–44, find the missing number.

35. $\log 7 + \log 3 = \log$ __?__

36. $\log 5 + \log 8 = \log$ __?__

37. $\log 48 - \log 12 = \log$ __?__

38. $\log 4 - \log 20 = \log$ __?__

39. $\log 8 - \log 5 + \log 35 = \log$ __?__

40. $\log 2000 - \log 40 - \log 2 = \log$ __?__

41. $7 \log 2 = \log$ __?__

42. $5 \log 3 = \log$ __?__

43. $\log 125 =$ __?__ $\log 5$

44. $\log 64 =$ __?__ $\log 2$

45. *Logarithm of a Power Property Proof Problem:* Prove algebraically that $\log x^n = n \log x$.

46. *Logarithm of a Quotient Property Proof Problem:* Prove algebraically that $\log \frac{x}{y} = \log x - \log y$.

47. *The Name "Logarithm" Problem:*

   a. Before there were calculators, if you had to multiply $27 \cdot 356 \cdot 43 \cdot 592$, you would have to use long multiplication three times to multiply 27 by 356, that answer by 43, and then that answer by 592. Simulate this process on your calculator by multiplying $27 \cdot 356$ and writing the result, then multiplying that answer by 43 and writing the result, and then multiplying that answer by 592 and writing the final result.

   b. Before there were calculators, if you had to add 27 + 356 + 43 + 592, you could write the numbers in a column and add *without* writing down any intermediate results. Do this addition column-wise, without using a calculator:

$$\begin{array}{r} 27 \\ 356 \\ 43 \\ +\ 592 \\ \hline \end{array}$$

   c. Base-10 logarithms were invented so that strings of numbers could be multiplied by adding their logarithms column-wise. You would look up the logarithms in tables, add these column-wise, and then use the tables backward to find the answer. The computation would look something like this:

   $\log 27 \approx 1.43144$

   $\log 356 \approx 2.5514$

   $\log 43 \approx 1.6335$

   $\log 592 \approx 2.7723$

   Add these logarithms column-wise, without using a calculator. Simulate finding the product by raising 10 to the exponent you found from adding the logarithms and rounding to four significant digits. Does the result agree with your result in part a?

2-5 Logarithms: Equations and Other Bases

In the previous section you learned that a logarithm is an exponent of 10. In this section you will learn that it is possible to find logarithms using other positive numbers as the base. Then you will learn how to use the properties of logarithms to solve an equation for an unknown exponent or to solve an equation involving logarithms.

Objective

Use logarithms with base 10 or other bases to solve exponential or logarithmic equations.

Logarithms with Any Base: The Change-of-Base Property

If $x = 10^y$, then y is the base-10 logarithm of x. Similarly, if $x = 2^y$, then y is the base-2 logarithm of x. The only difference is the number that is the base. To distinguish among logarithms with different bases, the base is written as a subscript after the abbreviation *log.* For instance,

$$3 = \log_2 8 \Leftrightarrow 2^3 = 8$$

$$4 = \log_3 81 \Leftrightarrow 3^4 = 81$$

$$2 = \log_{10} 100 \Leftrightarrow 10^2 = 100$$

The symbol $\log_2 8$ is pronounced "log to the base 2 of 8." The symbol $\log_{10} 100$ is, of course, equivalent to log 100, as defined in the previous section. Note that in all cases *a logarithm is an exponent.*

DEFINITION: Logarithm with Any Base

Algebraically:

$\log_b x = y$ if and only if $b^y = x$, where $b > 0$, $b \neq 1$, and $x > 0$

Verbally:

$\log_b x = y$ means that y is the exponent of b that gives x as the answer.

The way you pronounce the symbol for logarithm gives you a way to remember the definition. Examples 1 and 2 show you how to do this.

EXAMPLE 1 ➤ Write $\log_5 c = a$ in exponential form.

SOLUTION Think this:

- "$\log_5$..." is read "log *base* 5...," so 5 is the base.
- A logarithm is an *exponent*. Because the log equals a, a must be the exponent.
- The "answer" I get for 5^a is the argument of the logarithm, c.

Write only this:

$$5^a = c$$

EXAMPLE 2 ➤ Write $z^4 = m$ in logarithmic form.

SOLUTION $\log_z m = 4$

Two bases of logarithms are used frequently enough to have their own key on most calculators. One is base-10 logarithms, or **common logarithms,** as you saw in the previous section. The other is base-e logarithms, called **natural logarithms,** where $e = 2.71828...$, a naturally occurring number (like π) that you will find advantageous later in your mathematical studies. The symbol $\ln x$ (pronounced "el en of x") is used for the natural logarithms: $\ln x = \log_e x$.

Nautilus shells have a logarithmic spiral pattern.

DEFINITION: Common Logarithm and Natural Logarithm

Common: The symbol $\log x$ means $\log_{10} x$.

Natural: The symbol $\ln x$ means $\log_e x$, where e is a constant equal to 2.71828182845...

To find the value of a base-e logarithm, just press the key on your grapher. For instance,

$$\ln 30 = 3.4011...$$

To show what this answer means, raise e to the 3.4011... power.

$$e^{3.4011...} = 30$$ Use the e^x key. Do not round the 3.4011....

Example 3 shows you how to find a logarithm with a base that is not built into your calculator.

EXAMPLE 3 ➤ Find $\log_5 17$. Check your answer by an appropriate numerical method.

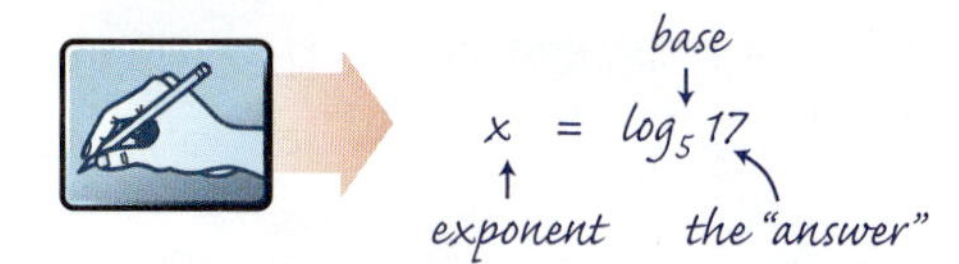

SOLUTION Let $x = \log_5 17$.

$5^x = 17$ Use the definition of logarithm.

$\log_{10} 5^x = \log_{10} 17$ Take $\log_{10}$ of both sides.

$x \log_{10} 5 = \log_{10} 17$ Use the log of a power property to "peel off" the exponent.

$x = \dfrac{\log_{10} 17}{\log_{10} 5} = 1.7603\ldots$ Divide both sides by the coefficient of x.

$\therefore \log_5 17 = 1.7603\ldots$ Substitute for x.

CHECK $5^{1.7603\ldots} = 17$ Do not round the 1.7603.... ◄

In Example 3, note that the base-5 logarithm of a number is directly proportional to the base-10 logarithm of that number. The conclusion of the example can be written this way:

$$\log_5 17 = \frac{1}{\log_{10} 5} \cdot \log_{10} 17 = 1.4306\ldots \log_{10} 17$$

To find the base-5 logarithm of any number, simply multiply its base-10 logarithm by 1.4306... (that is, divide by $\log_{10} 5$).

This proportional relationship is called the *change-of-base property.* From the results of Example 3, you can write

$$\log_5 17 = \frac{\log_{10} 17}{\log_{10} 5}$$

Notice that the logarithm with the desired base is by itself on the left side of the equation and that the two logarithms on the right side have the same base, presumably one available on your calculator. The box shows this property for bases a and b and argument x.

PROPERTY: The Change-of-Base Property of Logarithms

$$\log_a x = \frac{\log_b x}{\log_b a} \quad \text{or} \quad \log_a x = \frac{1}{\log_b a}(\log_b x)$$

EXAMPLE 4 ➤ Find ln 29 using the change-of-base property with base-10 logarithms. Check your answer directly by pressing ln 29 on your calculator.

SOLUTION $\ln 29 = \dfrac{\log 29}{\log e} = \dfrac{1.4623\ldots}{0.4342\ldots} = 3.3672\ldots$ Press log(29)/log(e).

Directly: $\ln 29 = 3.3672\ldots$

which agrees with the answer you got using the change-of-base property. ◄

The spiral arms of galaxies follow a logarithmic pattern.

The properties of base-10 logarithms presented in the previous section are generalized here for any base.

> **Properties of Logarithms**
>
> **The Logarithm of a Power:**
>
> $\log_b x^y = y \log_b x$
>
> *Verbally:* The logarithm of a power equals the product of the exponent and the logarithm of the base
>
> **The Logarithm of a Product:**
>
> $\log_b(xy) = \log_b x + \log_b y$
>
> *Verbally:* The logarithm of a product equals the sum of the logarithms of the factors.
>
> **The Logarithm of a Quotient:**
>
> $\log_b \frac{x}{y} = \log_b x - \log_b y$
>
> *Verbally:* The logarithm of a quotient equals the logarithm of the numerator minus the logarithm of the denominator.

Solving Exponential and Logarithmic Equations

Logarithms provide a way to solve an equation with a variable in the exponent or to solve an equation that already contains logarithms. Examples 5 through 8 show how you can do this.

EXAMPLE 5 ➤ Solve the exponential equation $7^{3x} = 983$ algebraically, using logarithms.

SOLUTION

$7^{3x} = 983$

$\log 7^{3x} = \log 983$ — Take the base-10 logarithm of both sides.

$3x \log 7 = \log 983$ — Use the logarithm of a power property.

$x = \dfrac{\log 983}{3 \log 7}$ — Divide both sides by the coefficient of x.

$x = 1.1803\ldots$ ◀

EXAMPLE 6 ➤ Solve the equation

$$\log_2(x - 1) + \log_2(x - 3) = 3$$

SOLUTION

$\log_2(x - 1) + \log_2(x - 3) = 3$

$\log_2[(x - 1)(x - 3)] = 3$ — Apply the logarithm of a product property.

$2^3 = (x - 1)(x - 3)$ — Use the definition of logarithm.

$8 = x^2 - 4x + 3$ Expand the product.

$x^2 - 4x - 5 = 0$ Reduce one side to zero. Use the symmetric property of equality.

$(x - 5)(x + 1) = 0$ Solve by factoring.

$x = 5 \quad \text{or} \quad x = -1$ Solutions of the quadratic equation.

You need to be cautious here because the solutions in the final step are the solutions of the quadratic equation, and you must make sure they are also solutions of the original logarithmic equation. Check by substituting your solutions into the original equation.

If $x = 5$, then

$$\log_2(5 - 1) + \log_2(5 - 3)$$
$$= \log_2 4 + \log_2 2$$
$$= 2 + 1 = 3$$

If $x = -1$, then

$$\log_2(-1 - 1) + \log_2(-1 - 3)$$
$$= \log_2(-2) + \log_2(-4)$$

which is undefined.

$\therefore x = 5$ is a solution, but $x = -1$ is not.

EXAMPLE 7 Solve the equation and check your solutions.

$$e^{2x} - 3e^x + 2 = 0$$

SOLUTION

$$e^{2x} - 3e^x + 2 = 0$$

$(e^x)^2 - 3e^x + 2 = 0$ Apply the properties of exponents.

You can realize that this is a quadratic equation in the variable e^x. Using the quadratic formula, you get

$$e^x = \frac{+3 \pm \sqrt{9 - 4(2)}}{2} = \frac{3 \pm 1}{2}$$

$$e^x = 2 \quad \text{or} \quad e^x = 1$$

You now have to solve these two equations.

$$e^x = 2 \qquad\qquad e^x = 1$$

$$x = \ln 2 = 0.6931\ldots \qquad\qquad x = 0$$

CHECK

$$e^{2\ln 2} - 3e^{\ln 2} + 2 \qquad\qquad (e^0)^2 - 3e^0 + 2$$

$$= (e^{\ln 2})^2 - 3e^{\ln 2} + 2 \qquad\qquad = 1^2 - 3(1) + 2 = 0$$

$$= 2^2 - 3(2) + 2 = 0$$

Both solutions are correct.

EXAMPLE 8 Solve the logarithmic equation $\ln(x + 3) + \ln(x + 5) = 0$ and check your solution(s).

SOLUTION

$$\ln(x + 3) + \ln(x + 5) = 0$$

$\ln[(x + 3)(x + 5)] = 0$ Use the logarithm of a product property.

$(x + 3)(x + 5) = e^0 = 1$ — Definition of (natural) logarithm.

$x^2 + 8x + 15 = 1$

$x^2 + 8x + 14 = 0$

$x = -2.5857\ldots$ or $x = -5.4142\ldots$ — By the quadratic formula.

CHECK $x = -2.5857\ldots$:

$\ln(-2.5857\ldots + 3) + \ln(-2.5857\ldots + 5)$ — Substitute without rounding.

$= \ln(0.4142\ldots) + \ln(2.4142\ldots)$

$= -0.8813\ldots + 0.8813\ldots = 0$

which checks.

$x = -5.4142\ldots$:

$\ln(-5.4142\ldots + 3) + \ln(-5.4142\ldots + 5)$

$= \ln(-2.4142\ldots) + \ln(-0.4142\ldots)$ — No logarithms of negative numbers.

which is undefined.

The only valid solution is $x = -2.5857\ldots$.

Problem Set 2-5

Reading Analysis

From what you have read in this section, what do you consider to be the main idea? What two bases of logarithms are found on most calculators? How do they differ? How do you find a logarithm with a base other than these? How do you use the logarithm of a power property to solve an equation that has an unknown exponent?

Quick Review

Q1. $\log 5 + \log 8 = \log$ __?__

Q2. $\log 36 - \log 4 = \log$ __?__

Q3. $\log 5^2 = 2 \log$ __?__

Q4. $\log 7^3 =$ __?__ $\log 7$

Q5. $1.5 \log 13 = \log$ __?__

Q6. The name "logarithm" comes from the words __?__ and __?__.

Q7. A logarithm is a(n) __?__.

Q8. $\log 48 = \log 6 + \log$ __?__

Q9. $\log 0.5 = \log 20 - \log$ __?__

Q10. $\log(-2)$ is

A. $\log(1/2)$ **B.** $\dfrac{1}{\log 2}$

C. $-\log 2$ **D.** $-\log(1/2)$

E. Undefined

1. Write the definition of base-b logarithms.
2. State the change-of-base property.
3. Write in exponential form: $\log_7 p = c$
4. Write in exponential form: $\log_v 6 = x$
5. Write in logarithmic form: $k^5 = 9$
6. Write in logarithmic form: $m^d = 13$

For Problems 7–14, find the indicated logarithm. Check your answer by raising the appropriate number to the appropriate power.

7. $\log_7 29$ **8.** $\log_8 352$

9. $\log_3 729$ **10.** $\log_{32} 2$

11. $\log_2 32$ **12.** $\log_5 125$

13. $\log_6 0.3$ **14.** $\log_{15} 0.777$

For Problems 15–34, find the missing values.

15. $\ln 8 + \ln 7 = \ln \underline{\ ?\ }$

16. $\ln 10 + \ln 20 = \ln \underline{\ ?\ }$

17. $\ln 3^5 = \underline{\ ?\ } \ln 3$

18. $2 \ln \underline{\ ?\ } = \ln 81$

19. $\ln 36 - \ln \underline{\ ?\ } = \ln 9$

20. $\ln \underline{\ ?\ } - \ln 7 = \ln 2$

21. $\ln\sqrt{x} = \underline{\ ?\ } \ln x$ **22.** $\ln\sqrt[5]{x} = \underline{\ ?\ } \ln x$

23. $\ln 1 = \underline{\ ?\ }$ **24.** $\ln e = \underline{\ ?\ }$

25. $\log 10 = \underline{\ ?\ }$ **26.** $\log 1 = \underline{\ ?\ }$

27. $\log_7 33 = \dfrac{\log_{10} 33}{?}$ **28.** $\log_{0.07} 53 = \dfrac{\ln 53}{?}$

29. $\dfrac{\log_{0.6} x}{\log_{0.6} 3} = \log_{?} x$ **30.** $\dfrac{\log_{13} n}{\log_{13} 0.5} = \log_{?} n$

31. $\dfrac{\ln x}{\ln 10} = \underline{\ ?\ }\ x$ **32.** $\dfrac{\log x}{\log e} = \underline{\ ?\ }\ x$

33. $\log_k k^3 = \underline{\ ?\ }$

34. If $x = \log_k 2$ and $y = \log_k 5$, then $\log_k 0.4 = \underline{\ ?\ }$.

For Problems 35–48, solve the equation algebraically and check your solution.

35. $\log(3x + 7) = 0$ **36.** $2 \log(x - 3) + 1 = 5$

37. $\log_2(x + 3) + \log_2(x - 4) = 3$

38. $\log_2(2x - 1) - \log_2(x + 2) = -1$

39. $\ln(x - 9)^4 = 8$

40. $\ln(x + 2) + \ln(x - 2) = 0$

41. $5^{3x} = 786$ **42.** $8^{0.2x} = 98.6$

43. $0.8^{0.4x} = 2001$ **44.** $6^{-5x} = 0.007$

45. $3e^{x-4} + 5 = 10$ **46.** $4 - e^{2x-3} = 7$

47. $2e^{2x} + 5e^x - 3 = 0$ **48.** $5 \cdot 2^{2x} - 3 \cdot 2^x - 2 = 0$

49. *Compound Interest Problem:* If you invest \$10,000 in a savings account that pays interest at the rate of 7% APR (annual percentage rate), then the amount M in the account after x years is given by the exponential function

$$M = 10{,}000 \times 1.07^x$$

a. Make a table of values of M for each year from 0 to 6.

b. How can you conclude that the values in the table have the add–multiply property?

c. Suppose that you want to cash in the savings account when the amount M reaches \$27,000. Set M equal to 27,000 and solve the resulting exponential equation algebraically using logarithms. Convert the solution to months, and round appropriately to find how many whole months must elapse before M first exceeds \$27,000.

50. *Population of the United States Problem:* Based on the 1990 and 2000 U.S. censuses, the population increased by an average of 1.24% per year over that time period. That is, the population at the end of any one year was 1.0124 times the population at the beginning of that year.

a. How do you tell that the population function has the add–multiply property?

b. The population in 1990 was about 248.7 million. Write a particular equation expressing population, P, as a function of the n years that have elapsed since 1990.

c. Assume that the population continues to grow at the rate of 1.24% per year. Find algebraically the year in which the population first reached 300 million. In finding the real-world answer, use the fact that the 1990 census was taken as of April 1. How does your prediction compare with that of the U.S. Census Bureau, which placed the date as October 2006?

2-6 Logarithmic Functions

You have already learned about identifying properties of several types of functions.

- Add–add: linear functions
- Add–multiply: exponential functions
- Multiply–multiply: power functions

In this section you'll learn that **logarithmic functions** have the **multiply–add property.**

Objective Show that logarithmic functions have the multiply-add property, and find particular equations algebraically.

Logarithmic Functions

Figures 2-6a and 2-6b show the natural logarithmic function $y = \ln x$ and the common logarithmic function $y = \log x$ (solid graphs). These functions are inverses of the corresponding exponential functions (dashed graphs), as shown by the fact that the graphs are reflections of the graphs of $y = e^x$ and $y = 10^x$ across the line $y = x$. Both logarithmic graphs are concave down. Notice also that the y-values are increasing at a decreasing rate as x increases. In both cases the y-axis is a vertical asymptote for the logarithmic graph. In addition, you can tell that the domain of these basic logarithmic functions is the set of positive real numbers

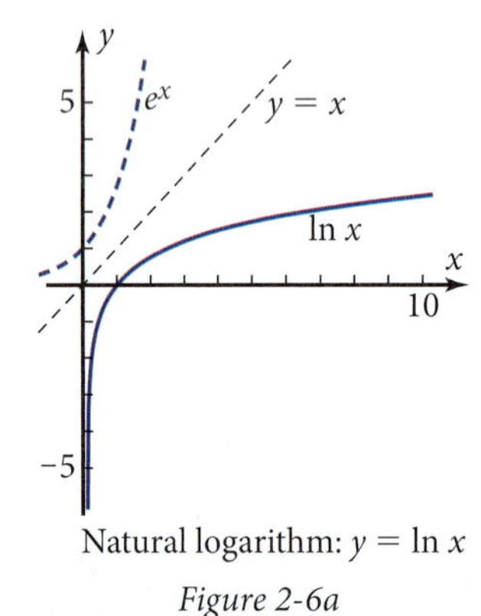

Natural logarithm: $y = \ln x$

Figure 2-6a

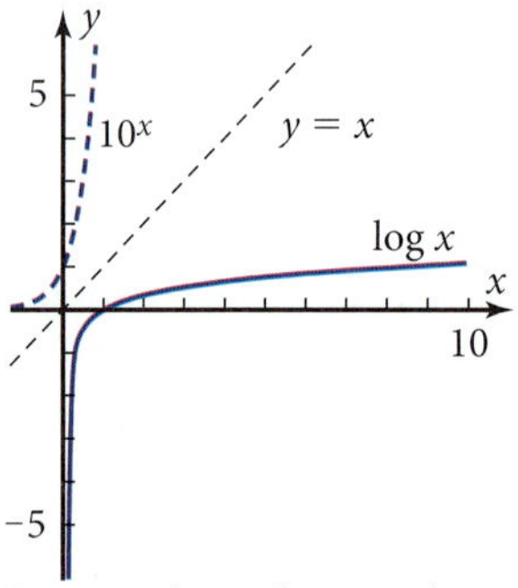

Common logarithm: $y = \log x$

Figure 2-6b

The general equation of a logarithmic function on most graphers has constants to allow for vertical translation and dilation.

DEFINITION: Logarithmic Functions

General equation: $y = a + b \log_c x$ — Base-c logarithmic function

where a, b, and c are constants, with $b \neq 0$, $c > 0$, and $c \neq 1$. The domain is all positive real numbers.

Transformed function: $y = a + b \log_c(x - d)$

where a is the vertical translation, b is the vertical dilation, and d is the horizontal translation.

Note: Remember that log stands for the base-10 logarithm and ln stands for the base-e logarithm.

Multiply–Add Property of Logarithmic Functions

These x- and y-values have the **multiply–add property.** Multiplying x by 3 results in adding 1 to the corresponding y-value.

	x	y	
	6	1	
×3			+1
	18	2	
×3			+1
	54	3	
×3			+1
	162	4	

By interchanging the variables, you can notice that x is an exponential function of y. You can find its particular equation by algebraic calculations.

$$x = 2 \cdot 3^y$$

This equation can be solved for y as a function of x with the help of logarithms.

$$\ln x = \ln(2 \cdot 3^y)$$ Take the natural logarithm (ln) of both sides.

$$\ln x = \ln 2 + y \ln 3$$ Use the product and power properties of logarithms.

$$y \ln 3 = \ln x - \ln 2$$ Isolate the term with y.

$$y = \frac{1}{\ln 3} \ln x - \frac{\ln 2}{\ln 3}$$ Solve for y.

$$y = 0.9102\ldots \ln x - 0.6309\ldots$$ Calculate constants by calculator.

This equation is a logarithmic function with $a = -0.6309\ldots$ and $b = 0.9102\ldots$. So, if a set of points has the multiply–add property, the points represent values of a logarithmic function. Reversing the steps lets you conclude that logarithmic functions have this property in general.

PROPERTY: Multiply–Add Property of Logarithmic Functions

If f is a logarithmic function, then multiplying x by a constant results in adding a constant to the value of $f(x)$. That is,

$$\text{for } f(x) = a + b \log_c x, \text{ if } x_2 = k \cdot x_1, \text{ then } f(x_2) = b \log_c k + f(x_1)$$

Particular Equations of Logarithmic Functions

You can find the particular equation of a logarithmic function algebraically by substituting two points that are on the graph of the function and evaluating the two unknown constants. Example 1 shows you how to do this.

EXAMPLE 1 ➤ Suppose that f is a logarithmic function with values $f(3) = 7$ and $f(6) = 10$.

a. Without finding the particular equation, find $f(12)$ and $f(24)$.

b. Find the particular equation algebraically using natural logarithms.

c. Confirm that your equation gives the value of $f(24)$ found in part a.

SOLUTION

a. Make a table of values using the multiply–add property.

	x	*y*	
	3	7	
×2	6	10	+3
×2	12	13	+3
×2	24	16	+3

$f(12) = 13$ and $f(24) = 16$

b. $f(x) = a + b \ln x$ — Write the general equation.

$\begin{cases} 7 = a + b \ln 3 \\ 10 = a + b \ln 6 \end{cases}$ — Substitute the given points.

$3 = b \ln 6 - b \ln 3$ — Subtract the equations to eliminate a.

$b = \dfrac{3}{\ln 6 - \ln 3} = 4.3280\ldots$ — Factor out b and then divide by $\ln 6 - \ln 3$

$7 = a + 4.3280\ldots \ln 3$ — Substitute $4.3280\ldots$ for b.

$a = 7 - 4.3280\ldots \ln 3 = 2.2451\ldots$ — Store a and b without rounding.

$f(x) = 2.2451\ldots + 4.3280\ldots \ln x$ — Write the particular equation. Graph it on your grapher.

c. By calculator, $f(24) = 16$, which checks. ◄

EXAMPLE 2 Plot the graphs of these functions and identify their domains.

a. $f(x) = 3\log(x - 1)$

b. $f(x) = -\ln(x + 3)$

c. $f(x) = \log_2(x^2 - 1)$

SOLUTION

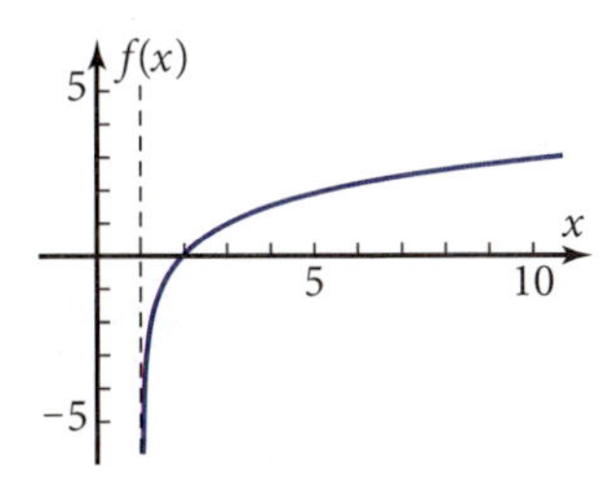

Figure 2-6c

a. You can get the graph of the function $f(x) = 3\log(x - 1)$ through transformations of the parent logarithmic function: a horizontal translation by 1 unit and a vertical dilation by a factor of 3. Figure 2-6c shows the resulting graph.

You know that the domain of a logarithmic function is positive real numbers, so the argument of a logarithmic function has to be positive.

$$x - 1 > 0$$

The domain of the function is $x > 1$. Add 1 to both sides of the inequality.

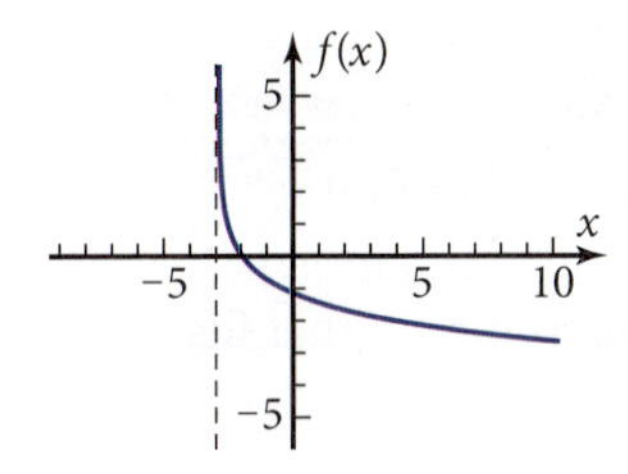

Figure 2-6d

b. Figure 2-6d shows the graph of the function $f(x) = -\ln(x + 3)$. You can get this graph by reflecting the graph of the function $y = \ln x$ across the x-axis and translating it horizontally by -3 units.

Domain:

$$x + 3 > 0$$ Argument of a logarithmic function is positive.

The domain of the function is $x > -3$.

c. In order to graph this function on your grapher, use the change-of-base property.

$$f(x) = \log_2(x^2 - 1) = \frac{\log(x^2 - 1)}{\log 2}$$

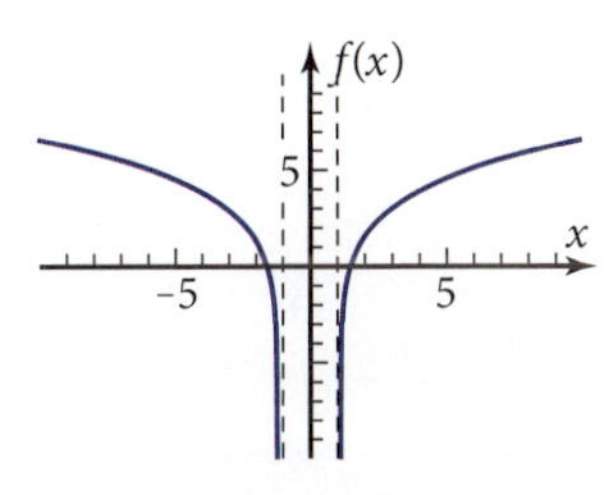

Figure 2-6e

Figure 2-6e shows the resulting graph.

Domain:

$$x^2 - 1 > 0$$ Argument of a logarithmic function is positive.

You can solve this inequality graphically. Graph the quadratic function and look for those values of x for which the function value is greater than zero or the graph is above the x-axis (see Figure 2-6f).

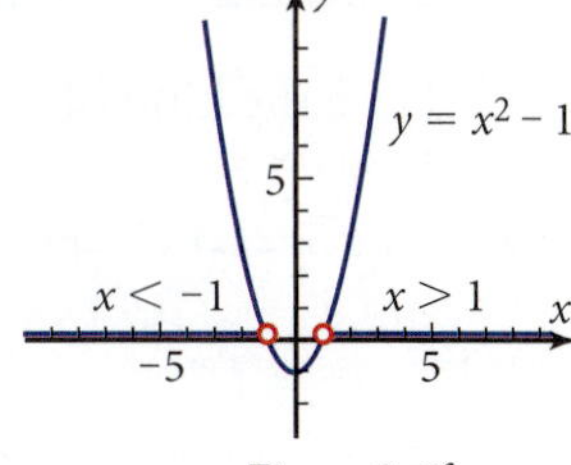

Figure 2-6f

The domain is $x < -1$ or $x > 1$.

Problem Set 2-6

Reading Analysis

From what you have read in this section, what do you consider to be the main idea? What is the general equation of a logarithmic function, and how is a logarithmic function related to an exponential function? What numerical pattern do regularly spaced values of a logarithmic function follow?

Quick Review

Q1. Name the kind of function graphed in Figure 2-6g.

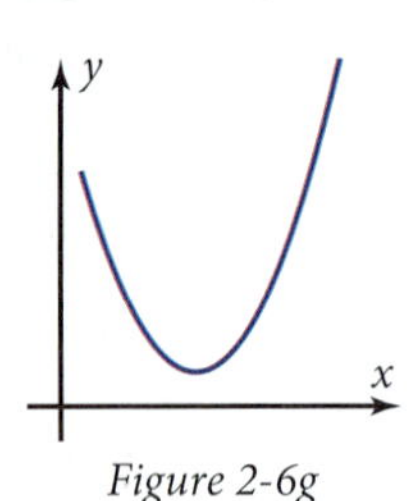

Figure 2-6g

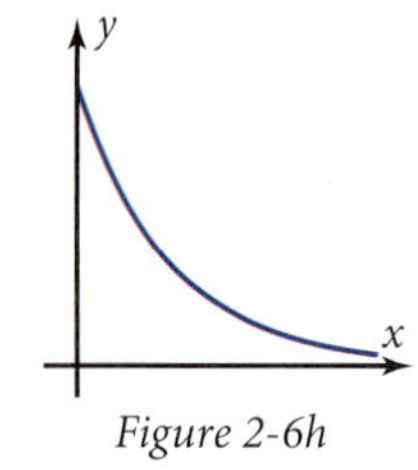

Figure 2-6h

Q2. Name the kind of function graphed in Figure 2-6h.

Q3. Name the kind of function graphed in Figure 2-6i.

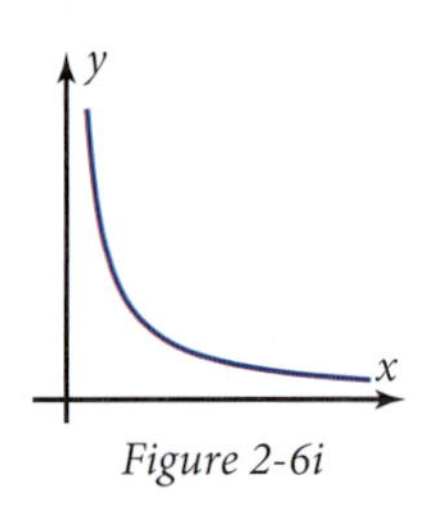

Figure 2-6i

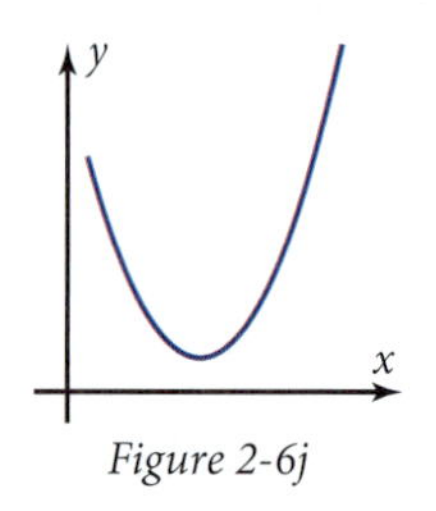

Figure 2-6j

Q4. Name the kind of function graphed in Figure 2-6j.

Q5. Name a real-world situation that could be modeled by the function in Figure 2-6k.

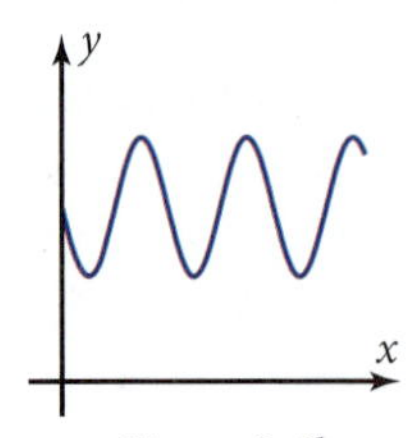

Figure 2-6k

Q6. Sketch a reasonable graph: The population of a city depends on time.

Q7. The graph of a quadratic function is called a __?__.

Q8. Expand the square: $(3x - 7)^2$

Q9. Write the next three terms in this sequence: 3, 6, 12, 24, . . .

Q10. The add–multiply property is a characteristic of __?__ functions.

For Problems 1 and 2,

a. Show that the values in the table have the multiply–add property.

b. Use the first and last points to find algebraically the particular equation of the natural logarithmic function that fits the points.

c. Verify that the equation in part b gives the other points in the table.

1.

x	*y*
3.6	1
14.4	2
57.6	3
230.4	4
921.6	5

2.

x	*y*
1	2
10	3
100	4
1000	5

3. Carbon-14 Dating Problem: The ages of things, such as wood, bone, and cloth, that are made from materials that were once living can be determined by measuring the percentage of original radioactive carbon-14 that remains in them. This table contains data on age as a function of the remaining percentage of carbon-14.

Percentage Remaining	Age (yr)
100	0
90	871
80	1845
70	2949
60	4223
50	5730

The skull of the saber-toothed cat, which lived in the Pleistocene more than 11,000 years ago.

a. Based on theoretical considerations, it is known that the percentage of carbon-14 remaining is an exponential function of the age. How does this fact indicate that the age should be a logarithmic function of the percentage?

b. Using the first and last data points, find the particular equation of the logarithmic function that goes through the points. Show that the equation gives values of other points close to those in the table.

c. You can use your mathematical model to interpolate between the given data points to find fairly precise ages. Suppose that a piece of human bone were found to have carbon-14 content 73.9%. What would you predict its age to be?

d. How old would you predict a piece of wood to be if its carbon-14 content were only 20%?

e. Search on the Internet or in some other resource to find out about early hominid fossil remains and carbon-14 dating. Give the source of your information.

4. *Earthquake Problem:* You can gauge the amount of energy released by an earthquake by its *Richter magnitude,* a scale devised by seismologist Charles F. Richter in 1935. The Richter magnitude is a base-10 logarithmic function of the energy released by the earthquake. These data show the Richter magnitude m for earthquakes that release energy equivalent to the explosion of x tons of TNT (*tri-nitro-toluene*).

x (tons)	m (Richter magnitude)
1,000	4.0
1,000,000	6.0

The director of the National Earthquake Service in Golden, Colorado, studies the seismograph display of a magnitude 7.5 earthquake.

a. Find the particular equation of the common logarithmic function $m = a + b \log x$ that fits the two data points.

1000 = a+b(log4.0)
1000 000 = a+b(log6)

b. Use the equation to predict the Richter magnitude for

- The 1964 Alaska earthquake, one of the strongest on record, which released the energy of 5 billion tons of TNT
- An earthquake that would release the amount of solar energy Earth receives every day, 160 trillion tons of TNT
- Blasting done at a construction site that releases the energy of about 30 lb of TNT

c. The earthquake that caused a tsunami in the Indian Ocean in 2004 had Richter magnitude about 9.0. How many tons of TNT would it take to produce a shock of this magnitude?

d. True or false? "Doubling the energy released by an earthquake doubles the Richter magnitude." Give evidence to support your answer.

e. Look up Richter magnitude on the Internet or in some other resource. Name one thing you learned that is not mentioned in this problem. Give the source of your information.

5. *Logarithmic Function Vertical Dilation and Translation Problem:*

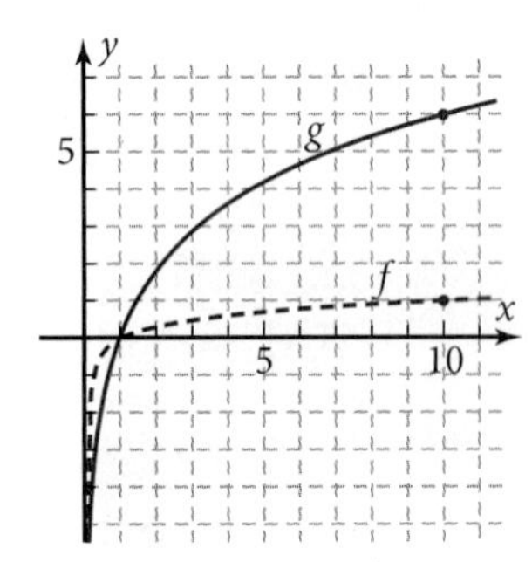

Figure 2-6l

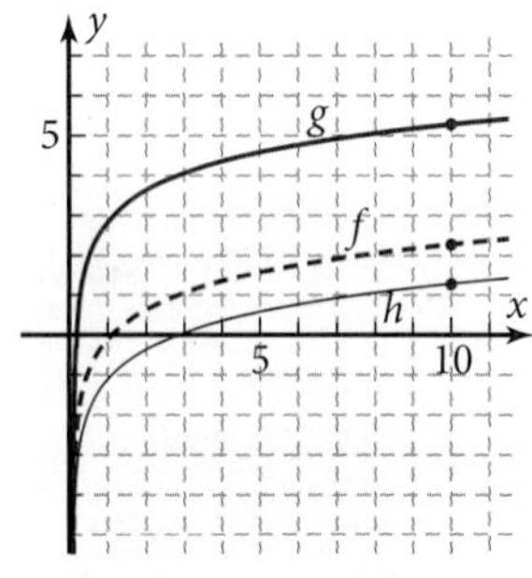

Figure 2-6m

a. Figure 2-6l shows the graph of the common logarithmic function $f(x) = \log_{10} x$ (dashed) and a vertical dilation of this graph by a factor of 6, $y = g(x)$ (solid). Write an equation of $g(x)$, considering it as a vertical dilation. Write another equation of $g(x)$ in terms of $\log_b x$, where b is a number other than 10. Identify the base.

b. Figure 2-6m shows the graph of $f(x) = \ln x$ (dashed). Two vertical translations are also shown, $g(x) = 3 + \ln x$ and $h(x) = -1 + \ln x$. Find algebraically or numerically the x-intercepts of the graphs of g and h.

6. *Logarithmic and Exponential Function Graphs Problem:* Figure 2-6n shows the graph of an exponential function $y = f(x)$ and its inverse function $y = g(x)$.

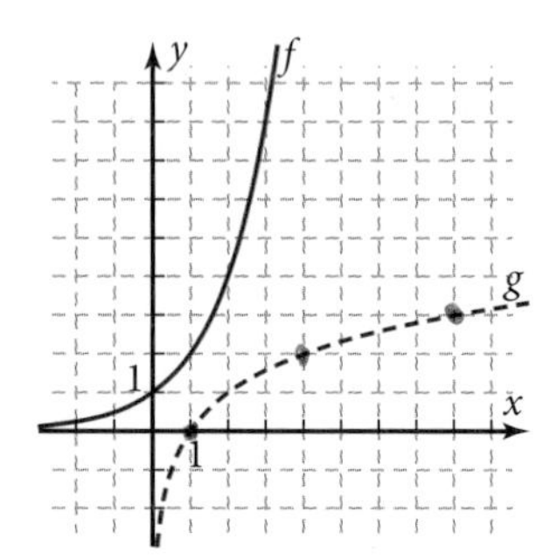

Figure 2-6n

a. The base of the exponential function is an integer. Which integer?

b. Write the particular equation of the inverse function $y = g(x) = f^{-1}(x)$.

c. Confirm that your answers to parts a and b are correct by plotting on your grapher.

d. With your grapher in parametric mode, plot these parametric equations:

$x_1(t) = f(t)$

$y_1(t) = t$

What do you notice about the resulting graph?

e. From your answer to part d, explain how you could plot on your grapher the inverse of any given function. Show that your method works by plotting the inverse of the function $y = x^3 - 9x^2 + 23x - 15$.

For Problems 7–12, graph the functions and identify their domains.

7. $f(x) = -2\log(x + 3)$ **8.** $f(x) = \log(3 - 2x)$

9. $f(x) = \log_3 x^2$ **10.** $f(x) = \ln(x^2 - 4)$

11. $f(x) = \ln 3x$ **12.** $f(x) = 4\log_2(3x + 5)$

This problem prepares you for the next section.

13. *The Definition of e Problem:* Figure 2-6o shows the graph of $y = (1 + 2x)^{1/x}$. If $x = 0$, then y is undefined because of division by zero. If x is close to zero, then $\frac{1}{x}$ is very large. For instance,

$$(1 + 0.0001)^{1/0.0001} = 1.0001^{10000}$$
$$= 2.71814592\ldots$$

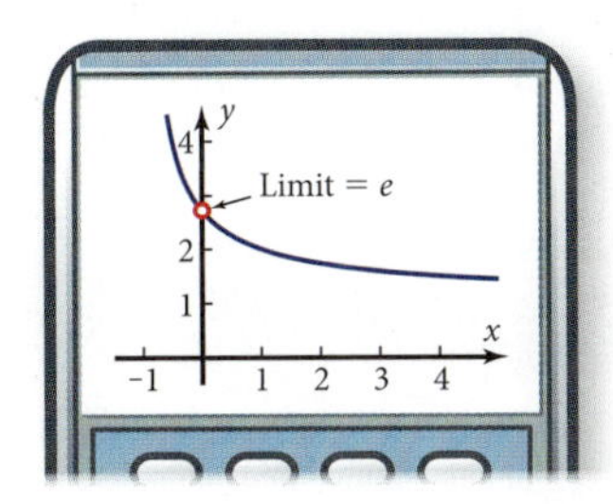

Figure 2-6o

a. Reproduce the graph in Figure 2-6o on your grapher. Use a window that has a grid point at $x = 0$. Trace to values close to zero, and record the corresponding values of y.

b. Two competing properties influence the expression $(1 + x)^{1/x}$ as x approaches zero. A number greater than 1 raised to a large power is very large, but 1 raised to any power is still 1. Which of these competing properties "wins"? Or is there a "compromise" at some number larger than 1?

c. Call up the number e on your grapher. If it does not have an e key, calculate e^1. What do you notice about the answer to part b and the number e?

14. *Research Project*: On the Internet or in some other reference source, find out about Henry Briggs and John Napier and their contributions to the mathematics of logarithms. See if you can find out why natural logarithms are sometimes called *Napierian logarithms.* Give the source of your information.

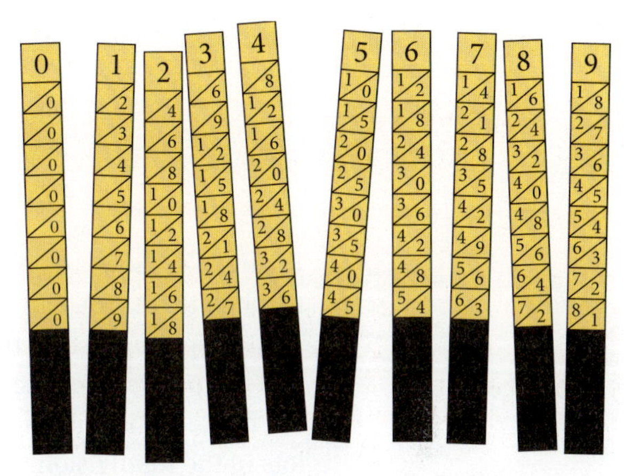

These rods are called "Napier's bones." Invented in the early 1600s, they made multiplication, division, and the extraction of square roots easier.

2-7 Logistic Functions for Restrained Growth

Suppose that the population of a new subdivision is growing rapidly. This table shows monthly population figures.

x (mo)	y (houses)
2	103
4	117
6	132
8	148
10	167

Figure 2-7a shows the plot of points and a smooth (dashed) curve that goes through them. You can tell that it is increasing, is concave up, and has a positive y-intercept, suggesting that an exponential function fits the points. Using the first and last points gives the function $y = 91.2782\ldots(1.0622\ldots)^x$, the curve shown in the figure, which fits the points almost exactly. Suppose that there are only 1000 lots in the subdivision. The actual number of houses will level off, approaching 1000 gradually, as shown in Figure 2-7b.

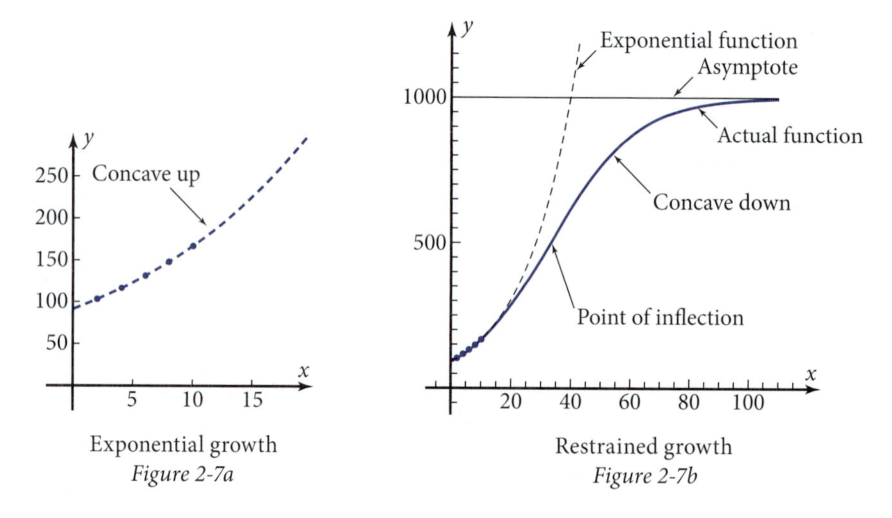

Exponential growth
Figure 2-7a

Restrained growth
Figure 2-7b

In this section you will learn about **logistic functions** that are useful as mathematical models of restrained growth.

Objective

Fit a logistic function to data for restrained growth.

In this exploration, you'll fit the logistic function to restricted population growth.

EXPLORATION 2-7: The Logistic Function for Population Growth

Suppose that this table lists the population of a small community, in thousands of people. The figure shows a scatter plot of the data.

x (years)	y (thousands of people)
1	2
2	3
3	5
4	9
5	13
6	19
7	27
8	32
9	36
10	39

1. Plot the data on your grapher.

2. At first the population seems to be increasing exponentially with time. On a copy of the given graph, sketch the graph of an exponential function that would fit the first six data points reasonably well.

3. Toward the end of the 10-yr period, the function seems to be leveling off. A function that models such population growth is the **logistic function.** Its general equation is

$$y = \frac{c}{1 + ae^{-bx}}$$

where x and y are the variables, e is the base of the natural logarithm, and a, b, and c stand for constants. The community has room for about 43,000 people, meaning $c = 43$. Calculate a and b using the first and the tenth points. Write the particular equation, and plot it on the same screen as the data. Sketch the result.

4. What does the logistic function indicate the population was at time $x = 0$ yr?

5. What graphical evidence do you have that the maximum population in the community is 43,000?

6. Look up the word *logistic* in a dictionary, and find the origin of the word.

7. What did you learn as a result of doing this exploration that you did not know before?

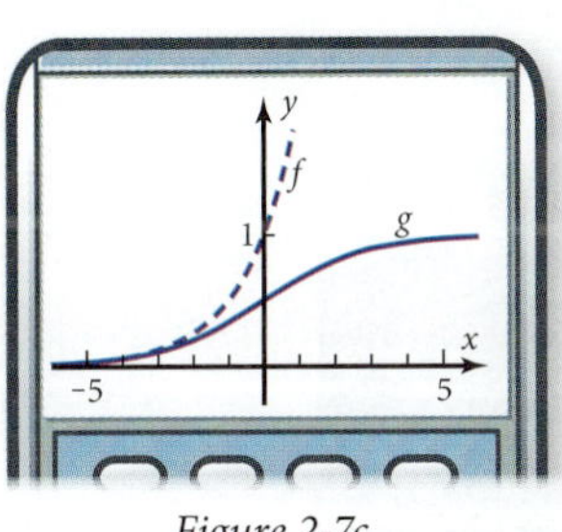

Figure 2-7c

Figure 2-7c shows the graphs of

$$f(x) = 2^x \qquad \text{and} \qquad g(x) = \frac{2^x}{2^x + 1}$$

Function f is an exponential function, and function g is a logistic function. For large positive values of x, the graph of g levels off to $y = 1$. This is because, for large values of x, 2^x is large compared to the 1 in the denominator. So the denominator is not much different from the 2^x in the numerator, and the fraction representing $g(x)$ approaches 1.

$$g(x) \rightarrow \frac{2^x}{2^x} = 1$$

For large negative values of x, the 2^x in the denominator is close to zero. So the denominator is close to 1. Thus the fraction representing $g(x)$ approaches 2^x.

$$g(x) \rightarrow \frac{2^x}{0 + 1} = 2^x$$

As you can see in Figure 2-7c, the logistic function is almost indistinguishable from the exponential function for large negative values of x. But for large positive values, the logistic function levels off, as did the number of occupied houses represented by Figure 2-7b. You can fit logistic functions to data sets by using the same dilations and translations you have used for other types of functions. You'll see how in Example 1.

General Logistic Function

You can transform the equation of function g in Figure 2-7c so that only one exponential term appears.

$$g(x) = \frac{2^x}{2^x + 1}$$

$$= \frac{2^x}{2^x + 1} \cdot \frac{2^{-x}}{2^{-x}} \qquad \text{Multiply by a clever form of 1.}$$

$$= \frac{1}{1 + 2^{-x}}$$

To get a general function of this form, replace the 1 in the numerator with a constant, c, to give the function a vertical dilation by a factor of c. Replace the exponential term 2^{-x} with ab^{-x} or with ae^{-bx} if you want to use the natural exponential function. The result is shown in the box.

DEFINITION: Logistic Function General Equation

$$f(x) = \frac{c}{1 + ae^{-bx}} \qquad \text{or} \qquad f(x) = \frac{c}{1 + ab^{-x}}$$

where a, b, and c are constants and the domain is all real numbers

EXAMPLE 1 ➤ Use the information on the occupied houses from the beginning of the section.

x (months)	y (houses)
2	103
4	117
6	132
8	148
10	167

a. Given that there are 1000 lots in the subdivision, use the points for 2 mo and 10 mo to find the particular equation of the logistic function that satisfies these constraints.

b. Plot the graph of the logistic function from 0 through 100 months. Sketch the result.

c. Make a table showing that the logistic function fits all the points closely.

d. Use the logistic function to predict the number of houses that will be occupied at the value of x corresponding to 2 yr. Which process do you use, extrapolation or interpolation?

e. Find the value of x at the point of inflection. What is the real-world meaning of the fact that the graph is concave up for times before the point of inflection and concave down thereafter?

SOLUTION Start with the second form of the logistic function.

a. $y = \dfrac{1000}{1 + ab^{-x}}$ The vertical dilation is 1000.

$\begin{cases} 103 = \dfrac{1000}{1 + ab^{-2}} \\ 167 = \dfrac{1000}{1 + ab^{-10}} \end{cases}$ Substitititute points (2, 103) and (10, 167)

$\begin{cases} 103 + 103ab^{-2} = 1000 \\ 167 + 167ab^{-10} = 1000 \end{cases}$ Eliminate the fractions.

$\begin{cases} 103ab^{-2} = 897 \\ 167ab^{-10} = 833 \end{cases}$

$\dfrac{167ab^{-10}}{103ab^{-2}} = \dfrac{833}{897}$ Divide to eliminate a and simplify.

$b^{-8} = \dfrac{833}{897} \cdot \dfrac{103}{167}$

$b = \left(\dfrac{833}{897} \cdot \dfrac{103}{167}\right)^{-1/8} = 1.0721\ldots$ Store without rounding.

$103a\,(1.0721\ldots)^{-2} = 897$ Substitute for b.

$a = 10.0106\ldots$ Start without rounding

$y = \dfrac{1000}{1 + 10.0106\ldots(1.0721\ldots)^{-x}}$ Write the particular equation.

b. Figure 2-7b on page 110 shows the graph.

c.

x (months)	y (houses)	Logistic Function	
2	103	103	(exact)
4	117	116.60...	(close)
6	132	131.73...	(close)
8	148	148.50...	(close)
10	167	167	(exact)

d. Trace the function to $x = 24$ for 2 yr.

$y = 347.1053\ldots \approx 347$ houses

The process is extrapolation because $x = 24$ is beyond the range of the given points.

e. The point of inflection is halfway between the x-axis and the asymptote at $y = 1000$. Trace the function to a value that is close to $y = 500$. Use the intersect feature to find $x = 33.0694\ldots$. So the point of inflection occurs at about 33 mo. Before 33 mo, the number of houses is increasing at an increasing rate. After 33 mo, the number is still increasing but at a decreasing rate.

Note that if a, b, and c are all positive, the logistic function will have two horizontal asymptotes, one at the x-axis and one at the line $y = c$. The point of inflection occurs halfway between these two asymptotes.

Properties of the graphs of logistic functions are shown in the box.

PROPERTIES: Logistic Functions

The logistic function is $y = \frac{c}{1 + ae^{-bx}}$ where a, b, and c are constants such that $a > 0$, $b \neq 0$, $c > 0$. The domain is all real numbers. The logistic function has

- Two horizontal asymptotes, one at $y = 0$ and another at $y = c$
- A point of inflection at $y = \frac{c}{2}$

If $b > 0$

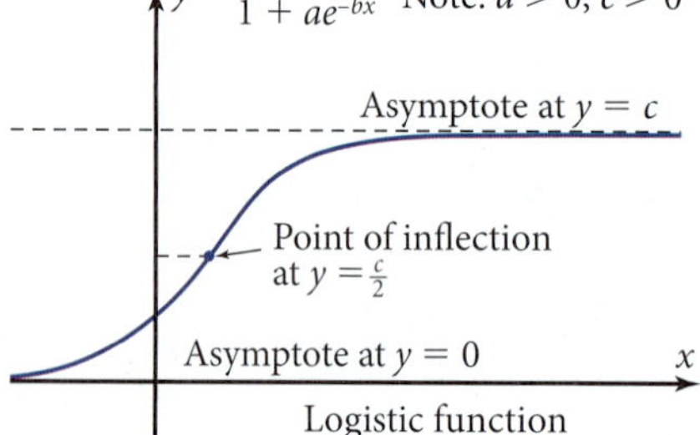

Logistic function

If $b < 0$

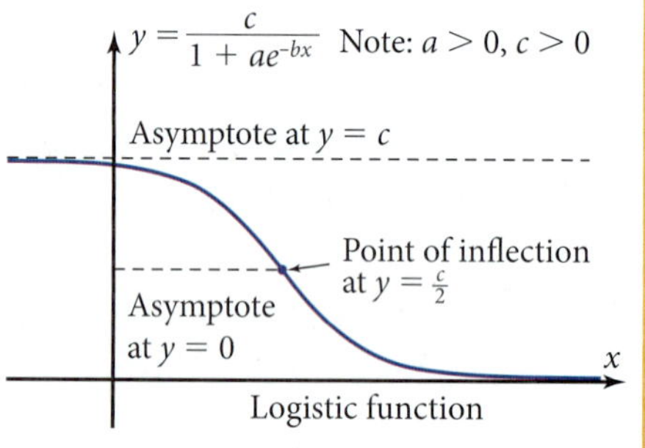

Logistic function

Problem Set 2-7

Reading Analysis

From what you have read in this section, what do you consider to be the main idea? What is the main difference between the graph of a logistic function and the graph of an exponential function? For what kind of real-world situations are logistic functions reasonable mathematical models?

Quick Review

Q1. An exponential function has the ___?___ - ___?___ property.

Q2. A power function has the ___?___ - ___?___ property.

Q3. The equation $y = 3 + 5 \ln x$ defines a ___?___ function.

Q4. The function in Problem Q3 has the __?__ - __?__ property.

Q5. The expression $\ln x$ is a logarithm with the number __?__ as its base.

Q6. Write in exponential form: $h = \log_p m$

Q7. Write in logarithmic form: $c = 5^j$

Q8. If an object rotates at 100 revolutions per minute, how many degrees per second is this?

Q9. Write the general equation of a quadratic function.

Q10. The function $g(x) = 3 + 4 \cdot f(5(x - 6))$ is a horizontal translation of function f by

A. 3 **B.** 4 **C.** 5 **D.** 6 **E.** -6

---

1. Given the exponential function $f(x) = 2.2^x$ and the logistic function $g(x) = \frac{2.2^x}{2.2^x + 1}$,
   - **a.** Plot both graphs on the same screen. Use as a domain $-10 \leq x \leq 10$. Sketch the result.
   - **b.** How do the two graphs compare for large positive values of x? How do they compare for large negative values of x?
   - **c.** Find graphically the approximate x-value of the point of inflection for function g. For what values of x is the graph of function g concave up? Concave down?
   - **d.** Explain algebraically why the logistic function has a horizontal asymptote at $y = 1$.
   - **e.** Transform the equation of the logistic function so that an exponential term appears only *once*. Show numerically that the resulting equation is equivalent to $g(x)$ as given.
2. Figure 2-7d shows the graph of the logistic function

$$f(x) = \frac{3e^{0.2x}}{e^{0.2x} + 4}$$

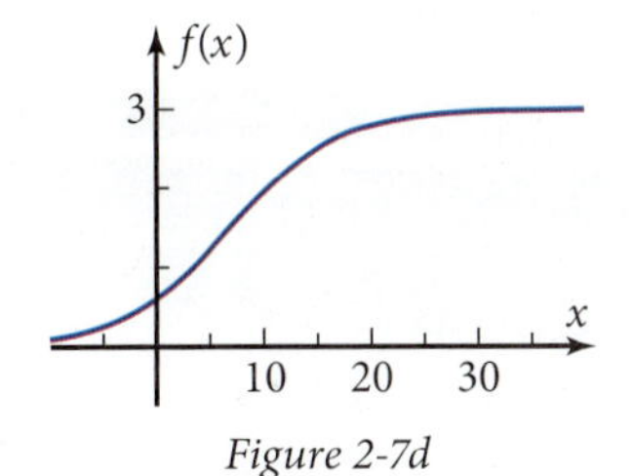

Figure 2-7d

   - **a.** Explain algebraically why the graph has a horizontal asymptote at $y = 3$.
   - **b.** Read the point of inflection from the graph. Find the x-coordinate algebraically.
   - **c.** For what values of x is the graph concave up? Concave down?
   - **d.** Transform the equation so that there is only one exponential term. Confirm by graphing that the resulting equation is equivalent to $f(x)$ as given.
3. *Spreading the News Problem:* You arrive at school and meet your mathematics teacher, who tells you today's test has been canceled! You and your friend spread the good news. The table shows the number of students, y, who have heard the news after time x, in minutes, has passed since you and your friend heard the news.

x (min)	y (students)
0	2
10	5
20	13
30	35
40	90

   - **a.** Plot the points. Imagine a function fit to the points. Is the graph of this function concave up, concave down, or both?
   - **b.** There are 1220 students in the school. Use the numbers of students at time 0 min and at time 40 min to find the equation of the logistic function that meets these constraints.
   - **c.** Plot the graph of the logistic function for the first 3 hours.
   - **d.** Based on the logistic model, how many students have heard the news at 9:00 a.m. if you heard it at 8:00 a.m.? How long will it be until all but ten students have heard the news?

4. *Spreading the News Simulation Experiment:* In this experiment you will simulate the spread of the news in Problem 3. Number each student in your class, starting at 1. Person 1 stands up and then selects two people at random to "tell" the news to. Do this by selecting two random integers between 1 and the number of students in your class, inclusive. (It is not actually necessary to tell any news!) The random number generator on one student's calculator will help make the random selection. The two people with the chosen numbers stand. Thus, after the first iteration, there will probably be three students standing (unless a duplicate random number came up). Each of these (three) people selects two more people to "tell" the news to by selecting a total of six (or four) more random integers. Do this for a total of ten iterations or until the entire class is standing. At each iteration, record the number of iterations and the total number of people who have heard the news. Describe the results of the experiment. Include things such as
   - The plot of the data points.
   - A function that fits the data, and a graph of this function on the plot. Explain why you chose the function you did.
   - A statement of how well the logistic model fits the data.
   - The iteration number at which the good news was spreading most rapidly.

5. WEB *Ebola Outbreak Epidemic Problem:* In the fall of 2000, an epidemic of the Ebola virus broke out in the Gulu district of Uganda. The table shows the total number of people infected from the day the cases were diagnosed as Ebola virus infections. The final number of people who were infected during this epidemic was 396. (Ebola is a virus that causes internal bleeding and is fatal in most cases.)

x (days)	*y* (total infections)
1	71
10	182
15	239
21	281
30	321
50	370
74	394

A Red Cross medical officer instructs villagers about the Ebola virus in Kabede Opong, Uganda.

   a. Plot the data points. Imagine a function that fits the data. Is the graph of this function concave up or concave down?

   b. Use the second and last data points to find the particular equation of a logistic function that fits the data.

   c. Plot on the same screen as the plot in part a the logistic function from part b. Sketch the results.

   d. Where does the point of inflection occur in the logistic model? What is the real-world meaning of this point?

   e. Based on the logistic model, how many people were infected after 40 days?

   f. Find data about other epidemics. Give your source. Try to model the spread of the epidemic for which you found data.

6. *Rabbit Overpopulation Problem:* Figure 2-7e shows two logistic functions represented by the equation

$$y = \frac{1000}{1 + ae^{-x}}$$

Both functions represent the population of rabbits in a particular woods as a function of time x, in years. The value of the constant a is to be determined under two different initial conditions.

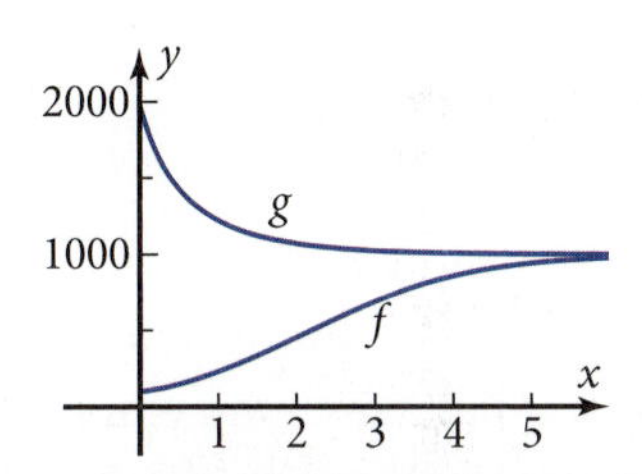

Figure 2-7e

a. For $y = f(x)$ in Figure 2-7e, 100 rabbits were introduced into the woods at time $x = 0$. Find the value of the constant a under this condition. Show that your answer is correct by plotting the graph of f on your grapher.

b. How do you interpret this mathematical model with regard to what happens to the rabbit population under the condition in part a?

c. For $y = g(x)$ in Figure 2-7e, 2000 rabbits were introduced into the woods at time $x = 0$. Find the value of a under this condition. Show that the graph agrees with Figure 2-7e. How does the sign of a represent a generalization of the definition of logistic function?

d. How do you interpret the mathematical model under the condition of part c? What seems to be the implication of trying to stock a region with a greater number of a particular species than the region can support?

7. Given the logistic function

$$f(x) = \frac{c}{1 + ae^{-0.4x}}$$

a. Let $a = 2$. Plot on the same screen the graphs of f for $c = 1$, 2, and 3. Use as a domain $-10 \le x \le 10$. Sketch the results. True or false? "The constant c is a vertical dilation factor."

b. Figure 2-7*f* shows the graph of f with $c = 2$ and with $a = 0.2$, 1, and 5. Which graph is which? How does the value of a transform the graph?

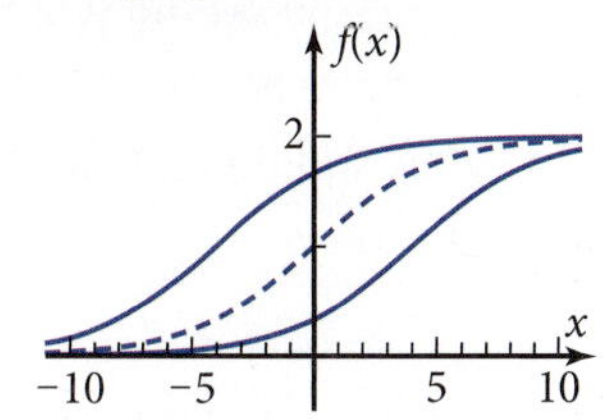

Figure 2-7f

c. Let $g(x) = \dfrac{c}{1 + ae^{-0.4(x-3)}}$.

What transformation applied to f does this represent? Confirm that your answer is correct by plotting f and g on the same screen using $c = 2$ and $a = 1$.

d. What value of a in the equation of $f(x)$ would produce the same transformation as in part c?

2-8 Chapter Review and Test

In this chapter you have learned graphical and numerical patterns for various types of functions:

- Linear
- Quadratic
- Power
- Exponential
- Logarithmic (the inverse of exponential)
- Logistic (for restrained growth)

These patterns allow you to tell which type of function might fit a given real-world situation. Once you have selected a function that has appropriate concavity, increasing–decreasing behavior, and numerical behavior, you can find the particular equation by calculating values of the constants. You can check your work by seeing whether the function fits other given points. Once you have the correct equation, you can use it to interpolate between given values or extrapolate beyond given values to calculate y when you know x, or to calculate x when you know y.

Review Problems

R0. Update your journal with what you have learned in this chapter. Include things such as the definitions, properties, and graphs of the functions listed. Show typical graphs of the various functions, give their domains, and make connections between, for example, the add–multiply property of the exponential functions and the multiply–add property of the logarithmic functions. Show how you can use logarithms and their properties to solve for unknowns in exponential or logarithmic equations, and explain how these equations arise in finding the constants in the particular equation of certain functions. Tell what you have learned about the constant e and where it is used.

R1. This problem concerns these five function values:

x	$f(x)$
2	1.2
4	4.8
6	10.8
8	19.2
10	30.0

a. On the same screen, plot the data points and the graph of $f(x) = 0.3x^2$.

b. Is the function increasing or decreasing? Is the graph concave up or concave down?

c. Name the function in part a. Give an example in the real world that this function might model. Is the y-intercept of f reasonable for this real-world example?

R2. **a.** Find the particular equation of a linear function containing the points (7, 9) and (10, 11). Give an example in the real world that this function could model.

b. Sketch two graphs showing a decreasing exponential function and an inverse variation power function. Give two ways in which the graphs are alike. Give one way in which they are different.

c. How do you tell that the function graphed in Figure 2-8a is an exponential function, not a power function? Find the particular equation of the exponential function. Give an example in the real world that this exponential function could model.

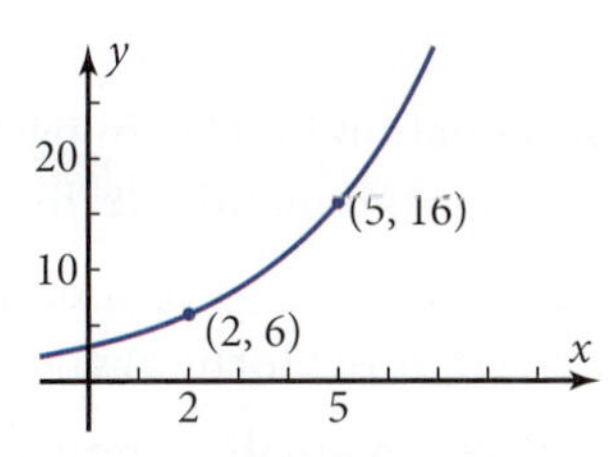

Figure 2-8a

d. Find the particular equation of the quadratic function graphed in Figure 2-8b. How does the equation you find show that the graph is concave down? Give an example in the real world that this function could model.

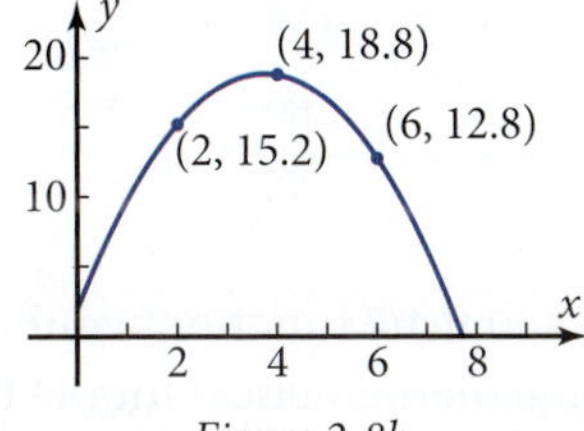

Figure 2-8b

e. A quadratic function has the equation $y - 3 = 2(x - 5)^2$. Where is the vertex of the graph? What is the y-intercept?

R3. For each table of values, tell from the pattern whether the function that fits the points is linear, quadratic, exponential, or power.

a.

x	$f(x)$
3	24
6	12
9	6
12	3

b.

x	$g(x)$
3	24
6	12
9	8
12	6

c.

x	$h(x)$
3	24
6	30
9	36
12	42

d.

x	$q(x)$
3	24
6	12
9	18
12	42

e. Suppose that $f(3) = 90$ and $f(6) = 120$. Find $f(12)$ if the function is

i. An exponential function

ii. A power function

iii. A linear function

f. Demonstrate that the add–multiply property of exponential functions is true for the function $f(x) = 53 \cdot 1.3^x$ by showing algebraically that adding the constant c to x multiplies the corresponding $f(x)$-value by a constant.

R4. **a.** The most important thing to remember about logarithms is that a logarithm is __?__.

b. Write in logarithmic form: $z = 10^p$

c. What does it mean to say that $\log 30 = 1.4771\ldots$?

d. Give numerical examples to illustrate these logarithmic properties:

i. $\log(xy) = \log x + \log y$

ii. $\log \frac{x}{y} = \log x - \log y$

iii. $\log x^y = y \log x$

e. $\log 48 - \log 4 + \log 5 = \log$ __?__

R5. **a.** Write in exponential form: $p = \log_c m$

b. Find $\log_7 30$.

c. $\ln 7 + 2 \ln 3 = \ln$ ___?___

d. Solve the equation:
$\log(x + 1) + \log(x - 2) = 1$

e. Solve the equation: $3^{2x-1} = 7^x$

R6. **a.** On the same screen, plot the graphs of $f_1(x) = \ln x$ and $f_2(x) = e^x$. Use the same scale on both axes. Sketch the results. How are the two graphs related to each other and to the line $y = x$?

b. For the natural exponential function $f(x) = 5e^{-0.4x}$, write the equation in the form $f(x) = ab^x$. For the exponential function $g(x) = 4.3 \cdot 7.4^x$, write the equation as a natural exponential function.

Sunlight Under the Water Problem (*R6c–R6e*): The intensity of sunlight underwater decreases with depth. The table shows the depth, *y*, in feet, below the surface of the ocean you must go to reduce the intensity of light to the given percentage, *x*, of what it is at the surface.

x (%)	**Depth y (ft)**
100	0
50	13
25	26
12.5	39

c. What numerical pattern tells you that a logarithmic function fits the data? Find the particular equation of the function.

d. On the same screen, plot the data and the logarithmic function. Sketch the result.

e. Based on this mathematical model, how deep do you have to go for the light to be reduced to 1% of its intensity at the surface? Do you find this by interpolation or by extrapolation?

R7. **a.** Plot the graphs of these functions on the same screen and sketch the results.

Logistic function: $f(x) = \dfrac{10 \cdot 2^x}{2^x + 10}$

Exponential function: $g(x) = 2^x$

b. Explain why $f(x)$ is very close to $g(x)$ when x is a large negative number. Explain why, when x is a large positive number, $f(x)$ is close to 10 and $g(x)$ is very large.

c. Transform the equation of $f(x)$ in part a so that it has only one exponential term.

d. Transform the equation of $g(x)$ in part a so that it is expressed in the form $g(x) = e^{kx}$.

e. *Population Problem:* A small community is built on an island in the Gulf of Mexico. The population grows steadily, as shown in the table.

x (months)	**y (people)**
6	75
12	153
18	260
24	355

Explain why a logistic function would be a reasonable mathematical model for population as a function of time. If the community has room for 460 residents, find the particular equation of the logistic function that contains the points for 6 mo and for 24 mo. Show that the equation gives approximately the correct solutions for 12 mo and 18 mo. Plot the graph, and sketch the result. When is the population predicted to reach 95% of the capacity?

Concept Problems

C1. *Rise and Run Property of Quadratic Functions Problem:* The sum of consecutive odd counting numbers is always a perfect square. For instance,

$$1 = 1^2$$
$$1 + 3 = 4 = 2^2$$
$$1 + 3 + 5 = 9 = 3^2$$
$$1 + 3 + 5 + 7 = 16 = 4^2$$

This fact can be used to sketch the graph of a quadratic function by a "rise-run" technique similar to that used for linear functions. Figure 2-8c shows that for $y = x^2$, you can start at the vertex and use the pattern "over 1, up 1; over 1, up 3; over 1, up 5; . . . ."

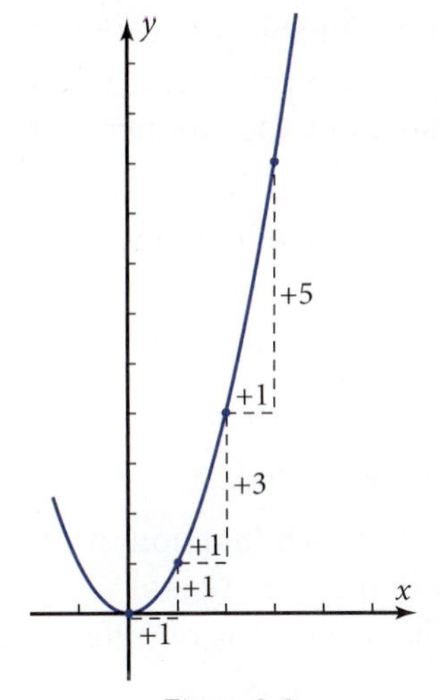

Figure 2-8c

a. On graph paper, plot the graph of $y = x^2$ by using this rise-run technique. Use integer values of x from 0 to 4. Then repeat the pattern for values of x from 0 to -4.

b. The graph of $y = -5 + (x - 2)^2$ is a translation of the graph of $y = x^2$. Locate the vertex, and then plot the graph on graph paper using the rise-run pattern.

c. The graph of $y = -5 + 0.3(x - 2)^2$ is a vertical dilation of the graph in part b. Use the rise-run technique for this function, and then plot its graph on the same axes as in part b.

C2. *Log-log and Semilog Graph Paper Problem:* Let $f(x) = 10000 \cdot 0.65^x$ be the number of bacteria remaining in a culture over time x, in hours. Let $g(x) = 0.09x^2$ be the area of skin, in square centimeters, on a snake of length x, in centimeters. Figure 2-8d shows the graph of the exponential function f plotted on *semilog* graph paper. Figure 2-8e on the next page shows on the graph of the power function g plotted on *log-log* graph paper. On these graphs, one or both axes have scales proportional to the logarithm of the variable's value. Thus the scales are compressed so that a wide range of values can fit on the same sheet of graph paper. For these two functions, the graphs are straight lines.

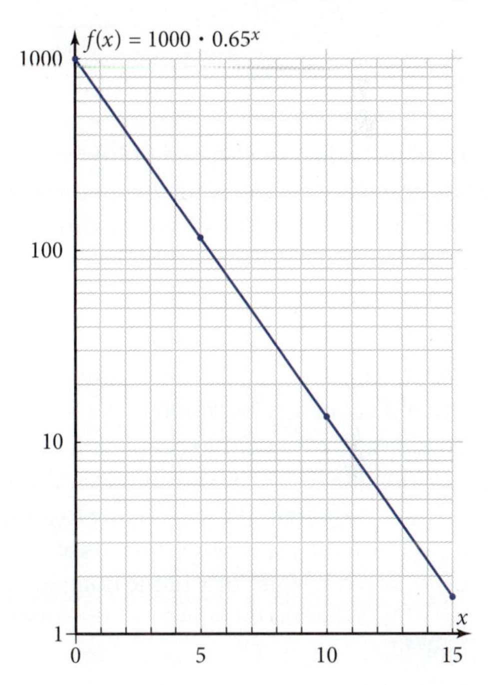

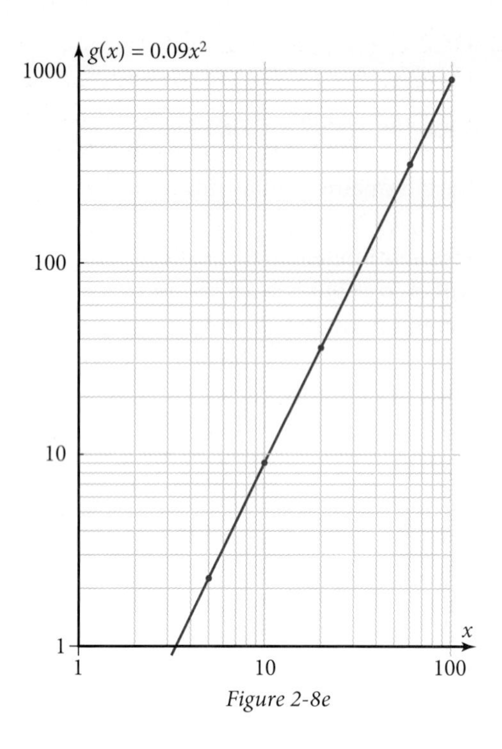

Figure 2-8e

a. Read the values of $f(9)$ and $g(60)$ from the graphs. Then calculate these numbers algebraically using the given equations. If your graphical answers are different from your calculated answers, explain what mistakes you made in reading the graphs.

b. You'll need a sheet of semilog graph paper and a sheet of log-log graph paper for graphing. On the semilog paper, plot the function $h(x) = 2 \cdot 1.5^x$ using several values of x in the domain $[0, 15]$. On the log-log paper, plot the function $p(x) = 700x^{-1.3}$ using several values of x in the domain $[1, 100]$. What do the graphs of the functions look like?

c. Take the logarithm of both sides of the equation $f(x) = 10000 \cdot 0.65^x$. Use the properties of logarithms to show that $\log f(x)$ is a linear function of x. Explain how this fact is connected to the shape of the graph.

d. Take the logarithm of both sides of the equation $g(x) = 0.09x^2$. Use the properties of logarithms to show that $\log g(x)$ is a linear function of $\log x$. How does this fact relate to the graph in Figure 2-8e?

C3. *Slope Field Logistic Function Problem:* The logistic functions you have studied in this chapter model populations that start at a relatively low value and then rise asymptotically to a *maximum sustainable population.* There may also be a *minimum sustainable population.* Suppose that a new variety of tree is planted on a relatively small island. Research indicates that the minimum sustainable population is 300 trees and that the maximum sustainable population is 1000 trees. A logistic function modeling this situation is

$$y = \frac{300C + 1000e^{0.7x}}{C + e^{0.7x}}$$

where y is the number of trees alive at time x, in decades after the trees were planted. The coefficients 300 and 1000 are the minimum and maximum sustainable populations, respectively, and C is a constant determined by the *initial condition,* the number of trees planted at time $x = 0$.

a. Determine the value of C and write the particular equation if, at time $x = 0$,

i. 400 trees are planted.

ii. 1300 trees are planted.

iii. 299 trees are planted.

b. Plot the graph of each function in part a. Use a window with $0 \le x \le 10$ and suitable y-values. What are the major differences among the three graphs?

c. Figure 2-8f shows a *slope field* representing functions with the given equation. The line segment through each grid point indicates the slope the graph would have if it passed through that point. On a copy of Figure 2-8f, plot the three equations from part a. How are the graphs related to the line segments on the slope field?

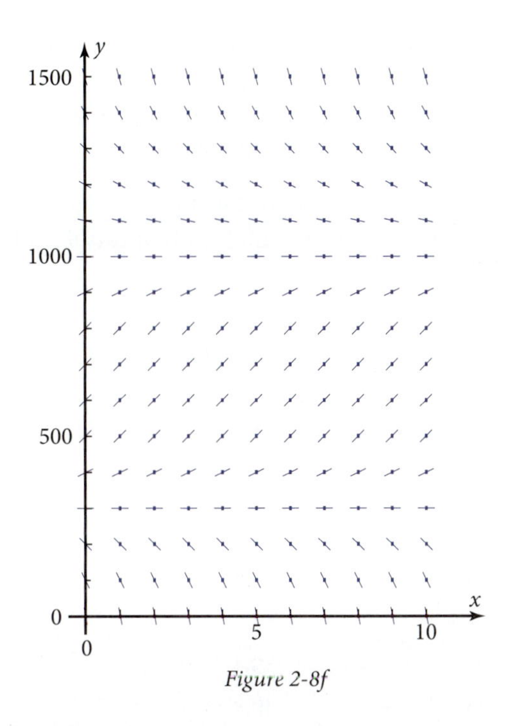

Figure 2-8f

d. Describe the behavior of the tree population for each of the three initial conditions in part a. In particular, explain what happens if too few trees are planted and also what happens if too many trees are planted.

e. Without doing any more computations, sketch on the slope field the graph of the tree population if, at time $x = 0$,

i. 500 trees had been planted.

ii. 1500 trees had been planted.

iii. 200 trees had been planted.

f. How does the slope field allow you to analyze graphically the behavior of many related logistic functions without doing any computations?

Chapter Test

Part 1: No calculators allowed (T1–T9)

T1. Write the general equation of

a. A linear function

b. A quadratic function

c. A power function

d. An exponential function

e. A logarithmic function

f. A logistic function

T2. What type of function could have the graph shown?

a.

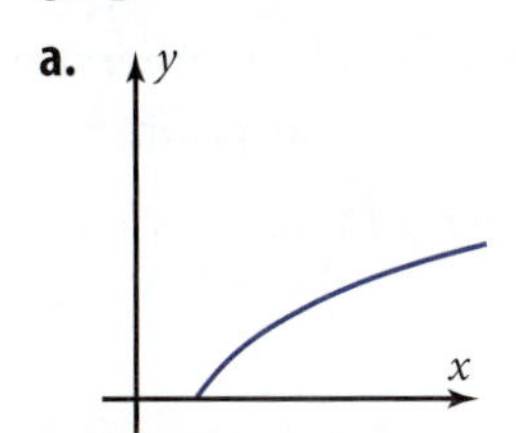

b.

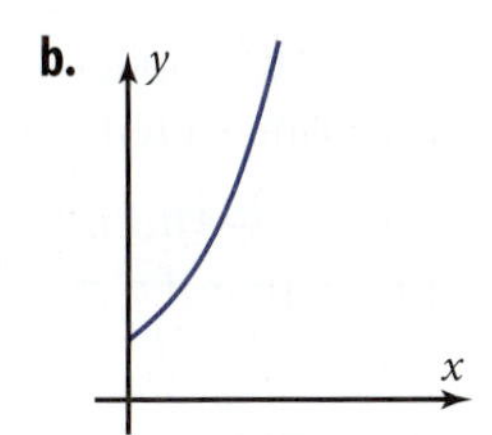

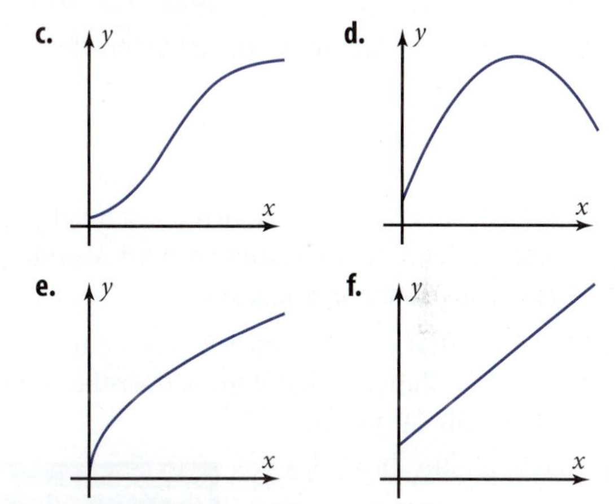

T3. What numerical pattern do regularly spaced data have for

a. A linear function

b. A quadratic function

c. An untranslated power function

d. An untranslated exponential function

e. An untranslated logarithmic function

T4. Write the equation $\log_a b = c$ in exponential form.

T5. Show how to use the logarithm of a power property to simplify $\log 5^x$.

T6. $\ln 80 + \ln 2 - \ln 20 = \ln$ __?__

T7. $\log 5 + 2 \log 3 = \log$ __?__

T8. Solve the equation: $4^x - 3 \cdot 2^x - 4 = 0$

T9. Solve the equation: $\log_2(x - 4) - \log_2(x + 3) = 8$

Part 2: Graphing calculators allowed (T10–T28)

Shark Problem: Suppose that from great white sharks caught in the past, fishermen find these weights and lengths. Use this set of data for Problems T10–T14.

x (ft)	f (x) (lb)
5	75
10	600
15	2025
20	4800

T10. Show that the data set in the table has the multiply–multiply property of power functions

T11. Write the general equation of a power function. Then use the points (5, 75) and (10, 600) to calculate algebraically the two constants in the equation. Store these values without rounding. Write the particular equation.

T12. Confirm that your equation in Problem T11 is correct by showing that it gives the other two data points in the table.

T13. From fossilized shark teeth, naturalists think there were once great white sharks 100 ft long. Based on your mathematical model, how heavy would such a shark be? Is this surprising?

T14. A newspaper report describes a great white shark that weighed 3000 lb. Based on your mathematical model, about how long was the shark? Show the method you use.

Coffee Cup Problem: You pour a cup of coffee. Three minutes after you pour it, you find that it is 94.8°F above room temperature. You record its temperature every 2 minutes thereafter, creating this table of data. Use the data for Problems T15–T18.

x (ft)	g (x) (°F above room temperature)
3	94.8
5	76.8
7	62.2
9	50.4
11	40.8

T15. Plot the information. From the plot, tell whether the graph of the function you can fit to the points is concave up or concave down. Explain why an exponential function would be reasonable for this function but a linear or a power function would not.

T16. Find the particular equation of the exponential function that fits the points at $x = 3$ and $x = 11$. Show that the equation gives approximately the correct values for the other three times.

T17. Extrapolate the exponential function backward to estimate the temperature of the coffee when it was poured.

T18. Use your equation to predict the temperature of the coffee a half-hour after it was poured.

T19. *The Add–Multiply Property Proof Problem:* Prove that if $y = 7(13^x)$, then $\log y$ is a linear function of x.

Model Rocket Problem: A precalculus class launches a model rocket out on the football field. The rocket fires for 2 s. Each second thereafter the class measures the rocket's height, finding the values in this table. Use these data for Problems T20–T22.

t (s)	*h* (ft)
2	166
3	216
4	234
5	220
6	174

T20. Plot the data points. Imagine fitting a function to the data. Is the graph of this function concave up or concave down? What kind of function would be a reasonable mathematical model for this function?

T21. Show numerically that a quadratic function would fit the data by showing that the second differences in the height data are constant.

T22. Use any three of the data points to find the particular equation of the quadratic function that fits the points. Show that the equation gives the correct values for the other two points.

T23. *Logarithmic Function Problem:* A logarithmic function f has $f(2) = 4.1$ and $f(6) = 4.8$. Use the multiply–add property of logarithmic functions to find two more values of $f(x)$. Use the given points to find the particular equation in the form $f(x) = a + b \ln x$.

Population Problem: Problems T24–T27 concern a new subdivision that opens in a small town. The population of the subdivision increases as new families move in. The table lists the population of the subdivision various numbers of months after its opening.

Months	People
2	363
5	481
7	579
11	830

T24. Find the particular equation of the (untranslated) exponential function f that fits the first and last data points. Show that the values of $f(5)$ and $f(7)$ are fairly close to those in the table.

T25. Show that the logistic function g gives values for the population that are also fairly close to the values in the table.

$$g(x) = \frac{3500}{1 + 10.8e^{-0.11x}}$$

T26. On the same screen, plot the four given points, the graph of f, and the graph of g. Use a window with $0 \leq x \leq 70$ and $0 \leq y \leq 5000$. Sketch the result.

T27. Explain why the logistic function g gives more reasonable values for the population than the exponential function f when you extrapolate to large numbers of months.

T28. What did you learn from taking this test that you did not know before?

Chapter 3

Fitting Functions to Data

As a child grows, his or her height is a function of age. But the growth rate is not uniform. There are times when the growth spurts and times when it slows down. However, there might be some mathematical pattern that is true for all children's growth. For example, spurts or slowdowns might occur generally at the same age for boys and for girls. If such a pattern exists, the actual heights will be scattered around some mathematical function that fits the data approximately. The regression techniques you will learn in this chapter can be used to find various types of functions to fit such data and to analyze how good the fit is.

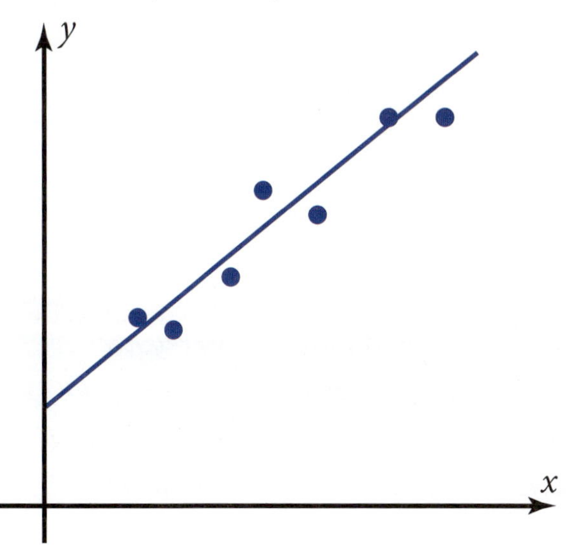

Mathematical Overview

In this chapter you'll learn how to use your grapher to find the best-fitting linear function for a given set of data and learn how to adapt this technique to fit other types of functions. You'll see how you can calculate the correlation coefficient, and you'll learn ways to find the type of function that is most appropriate for a given set of data. You'll apply the techniques you learn to problems such as predicting the increase of carbon dioxide in the atmosphere, a phenomenon that leads to global warming. You'll gain this knowledge in four ways.

GRAPHICALLY

This graph shows a set of data points and the linear function that best fits the data.

ALGEBRAICALLY

The equation of the best-fitting linear function is

$$\hat{y} = 2.1x + 3.4$$

where $\hat{y}$ is the predicted value of y.

NUMERICALLY

In the last column, 17.60 is the sum of the squares of the residuals. For the best-fitting linear function, this number is a minimum.

	x	**y**	$\hat{y}$	$y - \hat{y}$	$(y - \hat{y})^2$
	2	8	7.6	0.4	0.16
	4	10	11.8	−1.8	3.24
	6	19	16.0	3.0	9.00
	8	18	20.2	−2.2	4.84
	10	25	24.4	0.6	0.36
Sums:	30	80		0.0	17.60

VERBALLY

Minimizing the sum of the squares of the residuals determines the regression equation and the correlation coefficient. Even if the correlation coefficient is close to 1, endpoint behavior and residual plots help decide which function fits the data best. It is risky to extrapolate the function too far beyond the range of the given data.

3-1 Introduction to Regression for Linear Data

Moe is recovering from surgery. The table and Figure 3-1a show the number of sit-ups he has been able to do on various days after the surgery. Figure 3-1b shows the best-fitting linear function

$$\hat{y} = 2.1x + 3.4$$

where $\hat{y}$ (pronounced "y hat") is used to distinguish points on this line from the actual data points. The difference $y - \hat{y}$ is called the residual deviation or the **residual.** In this section you'll learn how to use your grapher to find this equation.

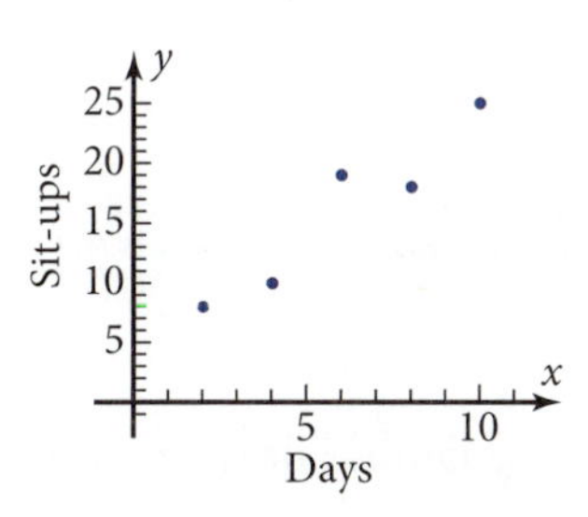

Figure 3-1a

x (days)	y (sit-ups)
2	8
4	10
6	19
8	18
10	25

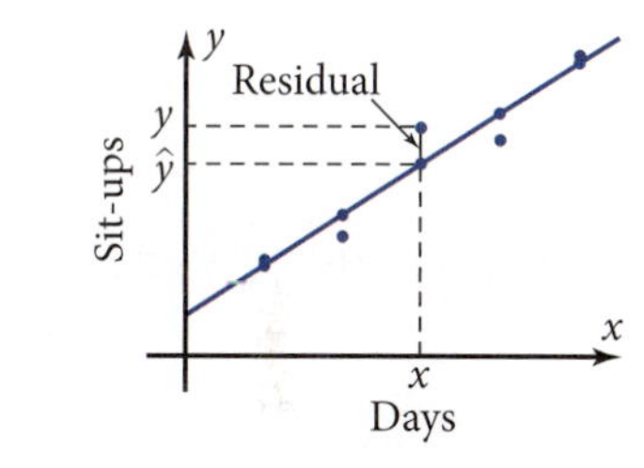

Figure 3-1b

Objective Find the equation of the best-fitting linear function for a set of points by running a linear regression on your grapher, and calculate the sum of the squares of the residuals.

Exploratory Problem Set 3-1

1. Enter the x- and y-values into two lists on your grapher. Then from the statistics menu run a **linear regression** in the form $\hat{y} = ax + b$. Did you find that the equation is $\hat{y} = 2.1x + 3.4$?
2. Plot the given points and the linear equation from Problem 1 on the same screen. Does the result agree with Figure 3-1b?
3. Show how to use the linear function to predict the number of sit-ups Moe could do two weeks after surgery. What real-world reasons could explain why the actual number of sit-ups might be different from the predicted number?
4. Copy the table. Add three new columns:
   - The values of $\hat{y}$ calculated by the equation $\hat{y} = 2.1x + 3.4$
   - The residuals, calculated by the subtraction $y - \hat{y}$
   - The squares of the residuals, $(y - \hat{y})^2$
5. Find the *sum of the squares of the residuals.* This number is abbreviated SS_{res}.
6. The **regression line** is the line that makes SS_{res} *a minimum.* Because SS_{res} is a minimum, the regression line is the best-fitting linear function. Show that SS_{res} would be greater for the function $f_2(x) = 2.1x + 3.5$, which has y-intercept 3.5 instead of 3.4, and also would be greater for the function $f_3(x) = 2.2x + 3.4$, which has slope 2.2 instead of 2.1.

3-2 Deviations, Residuals, and the Correlation Coefficient

In Chapter 2, you found equations of functions that fit given sets of points. If the points represent data measured from the real world, they are usually scattered. For instance, Figure 3-2a shows a scatter plot of the weights and lengths of 43 fish. The scatter of this cloud of points occurs because fish of the same length can have different weights.

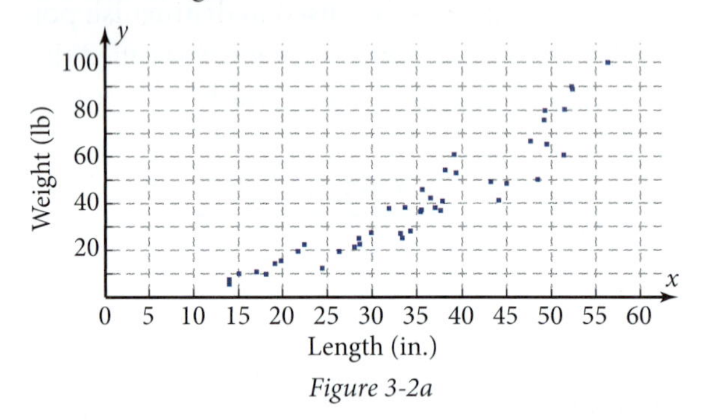

Figure 3-2a

In this section you will learn the reasoning behind doing regression as in Section 3-1 for points that follow a linear pattern. In the next section you will see how to extend regression to fit data that follow a curved pattern, as in Figure 3-2a.

Objective Calculate SS_{res}, the sum of the squares of the residuals, and find out how to determine the equation of the linear function that minimizes SS_{res}.

In Section 3-1, you analyzed data for the number of sit-ups, y, that Moe could do x days after surgery, as listed in this table. Figure 3-2b shows a scatter plot of the data.

	x (days)	**y (sit-ups)**
	2	8
	4	10
	6	19
	8	18
	10	25
Sums:	30	80

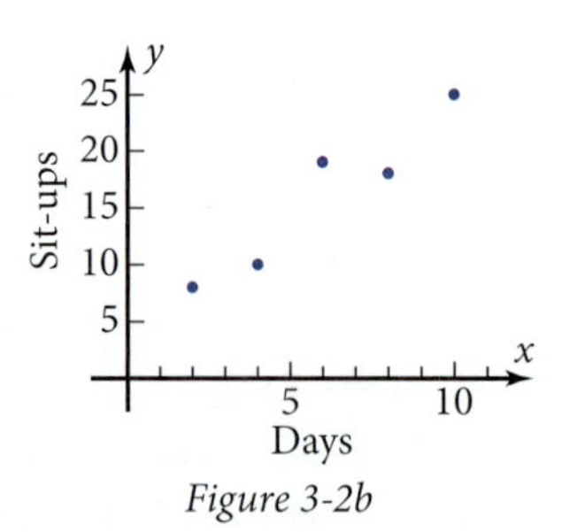

Figure 3-2b

Suppose that you are asked to estimate the number of sit-ups, y, Moe could do, but you are not told the number of days. Your best estimate would be $\bar{y}$ (pronounced "y bar"), the average, or *mean*, of the y-values:

$$\bar{y} = \frac{80}{5} = 16 \text{ sit-ups}$$

Figure 3-2c shows the graph of the constant function $y = \bar{y}$ and the **deviation,** $y - \bar{y}$, of each point from this line.

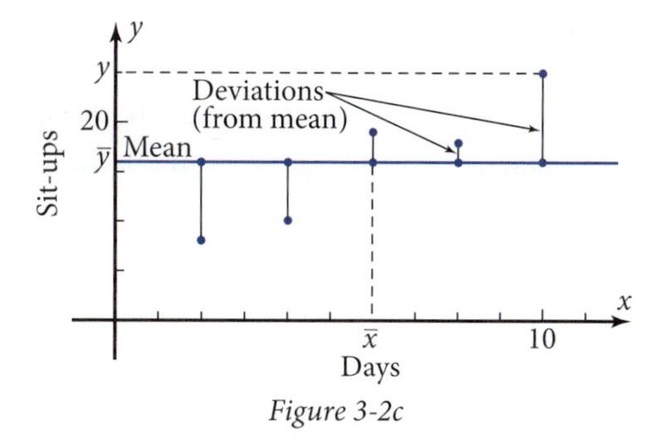

Figure 3-2c

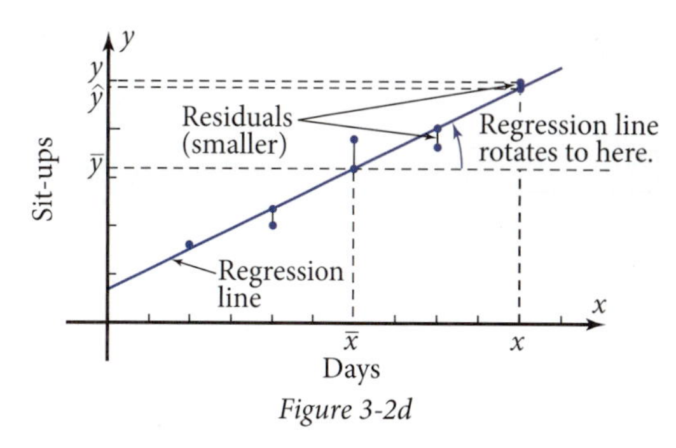

Figure 3-2d

However, if you *do* know for which day to find an estimate, you can find a better estimate of the number of sit-ups for that particular day by assuming that y is a linear function of x. Figure 3-2d shows that making the line slanted instead of horizontal removes much of each deviation. The part of each deviation that remains is called the residual deviation, or residual. If $\hat{y}$ is the value of y for a point on the line, then the residual equals $y - \hat{y}$.

A measure of how well the line fits the point is obtained by squaring each residual, thereby making each value positive or zero, and then summing the results. This sum, SS_{res}, is called the **sum of the squares of the residuals.** Doing the same thing for the deviations from the mean gives a quantity called SS_{dev}, the **sum of the squares of the deviations.** Figures 3-2e and 3-2f show that the squares of the residuals are much smaller than the squares of the deviations from the mean. For a dynamic view of this process, see the *Deviations, Residuals, and the Correlation Coefficient* exploration at **www.keymath.com/precalc.**

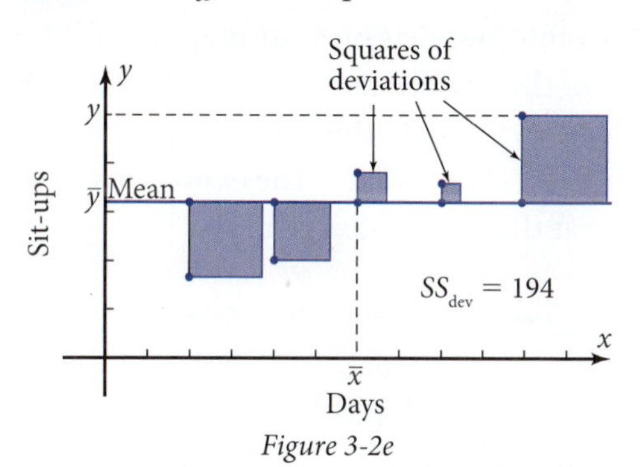

Figure 3-2e

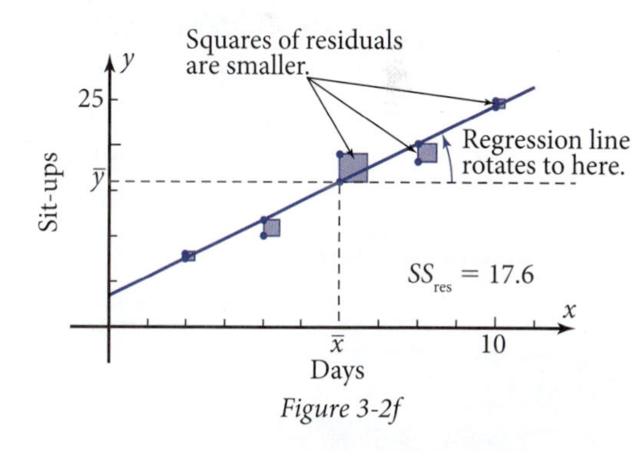

Figure 3-2f

The *regression equation* $\hat{y} = 2.1x + 3.4$ that you found on your grapher in Section 3-1 is the equation of the linear function for which SS_{res} is the minimum of all possible values. The fraction of the original SS_{dev} that has been removed by using the slanted linear function instead of the constant function is

$$r^2 = \frac{SS_{dev} - SS_{res}}{SS_{dev}} = \frac{194 - 17.6}{194} = \frac{176.4}{194} = 0.9092...$$

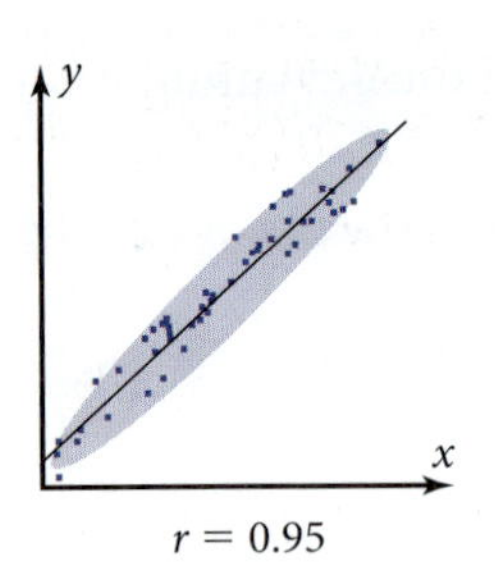

$r = 0.95$

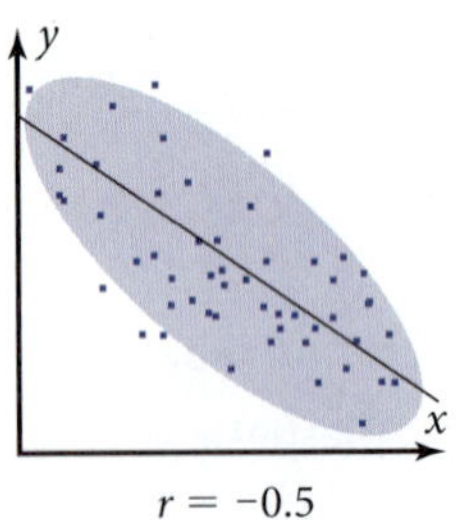

$r = -0.5$

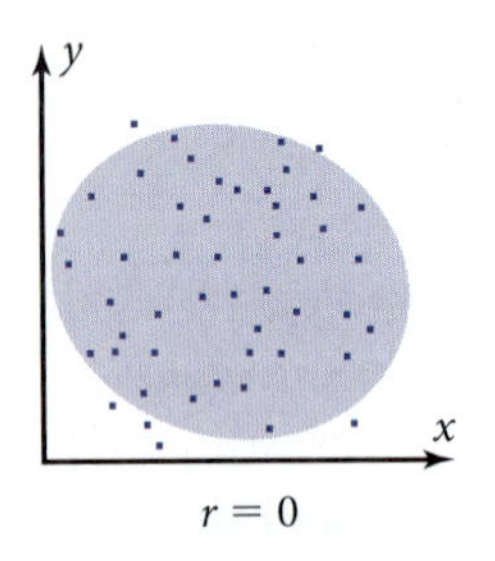

$r = 0$

The quantity r^2 is called the **coefficient of determination.** The symbol r^2 is used because its units are the squares of the y-units. The number r is called the **correlation coefficient.** The correlation coefficient indicates how *well* the best-fitting linear function fits the values. The sign of r shows whether the two variables, x and y, are positively or negatively *associated* with each other. If the y-variable increases as x increases, then the correlation coefficient is positive; if the y-variable decreases as x increases, then the correlation coefficient is negative.

The correlation coefficient also measures the strength of the association. If $r = 1$ or $r = -1$, the function fits the values exactly. This could be the case for a theoretical law in physics or chemistry, where the function that identifies the relationship between the variables follows an exact pattern. Of course, when you collect data, even if there is an underlying law the data will not fit perfectly, due to measurement errors (from equipment or human inexactness).

The closer the correlation coefficient is to 1 or -1, the closer the points cluster around the line and the stronger the association is between the variables. If $r = 0$, there is no relationship between x and y. It is helpful to sketch an ellipse around the cloud of points in a scatter plot. If the ellipse is narrow, the correlation is *strong;* if the ellipse is wide, the correlation is *weak.* The scatter plots at left can help you visualize the strength of the correlation and connect it to the values of the correlation coefficient.

Example 1 shows you how to find SS_{dev} and SS_{res} using lists on your grapher.

EXAMPLE 1 ➤ Find SS_{dev} and SS_{res} for the sit-ups data in this section. Use the results to calculate the coefficient of determination and the correlation coefficient. How do you interpret the value of the coefficient of determination?

SOLUTION Enter the values of x and y into two lists on your grapher, say, L_1 and L_2. In a third list, compute the squares of the deviations, $(y - 16)^2$. In a fourth list, compute the squares of the residuals, $(y - \hat{y})^2$. To do this on your grapher, enter $\hat{y} = 2.1x + 3.4$ as $f_1(x)$, and then enter $\left(L_2 - f_1\left(L_1\right)\right)^2$ into L_4. The expression $f_1\left(L_1\right)$ means the values of $\hat{y}$ (the function in $f_1(x)$) at the values of x in L_1.

x	y	$(y - 16)^2$	$(y - \hat{y})^2$
2	8	64	0.16
4	10	36	3.24
6	19	9	9.00
8	18	4	4.84
10	25	81	0.36

$[8 - (2.1(2) + 3.4)]^2 = [8 - 7.6]^2 = 0.4^2 = 0.16$

$SS_{dev} = 194$ $\quad 17.60 = SS_{res}$ Use the sum command on your grapher.

$$r^2 = \frac{194 - 17.60}{194} = 0.909278...$$

$$r = \sqrt{0.909278...} = 0.9535...$$

Positive square root because the function is increasing.

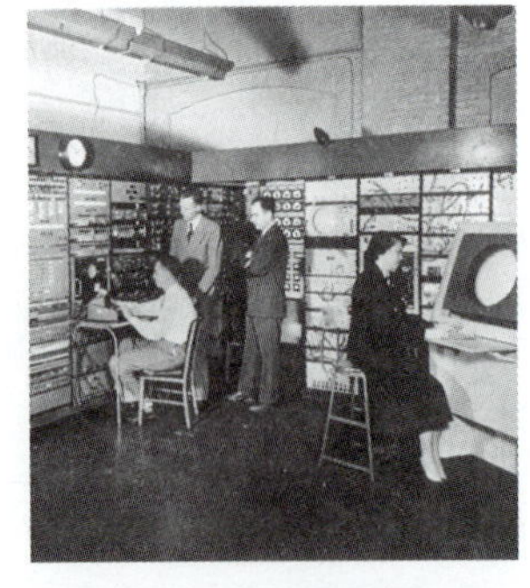

Computer from 1956. Modern computers made statistical data analysis much easier and stimulated the development of various statistical techniques.

The coefficient of determination, $r^2 = 0.9092\ldots$, indicates that 90.92...% of the original deviation has been accounted for by the linear relationship between x and y, and the remaining 9.07...% is due to other fluctuations in the data. The correlation between the variables is fairly strong.

Notes:

- The values of r^2 and r are found when your grapher calculates linear regression, and you can set your grapher to display these values.
- Using any other linear equation besides $\hat{y} = 2.1x + 3.4$ gives a value of SS_{res} larger than 17.60. For instance, $y = 2.2x + 3.4$ gives $SS_{res} = 19.80$.

Here is a summary of the quantities associated with linear regression.

DEFINITIONS: Deviations, Residuals, the Regression Line, and Correlation

The **deviation** of a data point (x, y) is $y - \bar{y}$, the directed distance of its y-value from $\bar{y}$, where $\bar{y}$ is the mean of the y-values.

The **residual** (or residual deviation) of a data point from the line $\hat{y} = mx + b$ is $y - \hat{y}$, the vertical directed distance of its y-value from the line (Figure 3-2g).

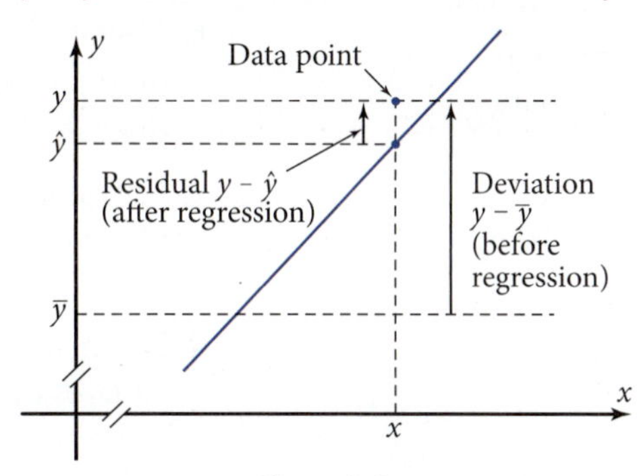

Figure 3-2g

The **sum of the squares of the deviations** is $SS_{dev} = \Sigma(y - \bar{y})^2$, where Σ, the capital Greek letter sigma, means "the sum of the values following the Σ sign."

The **sum of the squares of the residuals** is $SS_{res} = \Sigma(y - \hat{y})^2$.

The **regression line** for a set of data is the line for which SS_{res} is a minimum. The *linear regression equation* is the equation of this line, $\hat{y} = mx + b$.

The **coefficient of determination** is $r^2 = \dfrac{SS_{dev} - SS_{res}}{SS_{dev}}$. It is the fraction of SS_{dev} that has been removed by the linear regression.

The **correlation coefficient,** r, is the positive or negative square root of the coefficient of determination. Use the positive square root if the slope of the line is positive, and use the negative square root if the slope is negative.

Problem Set 3-2

Reading Analysis

From what you have read in this section, what do you consider to be the main idea? What is the difference between the *deviation* of a data point and the *residual* (residual deviation) of that data point? What is true about the residuals of a set of data for the linear function that best fits the data? Why do the correlation coefficients 1 and -1 both indicate a perfect fit of a linear function to a set of data?

Quick Review

Q1. If $\hat{y} = 3x + 5$ and $x = 4$, by how much does $y = 19$ deviate from $\hat{y}$ on the regression line?

Q2. If $\hat{y} = 3x + 5$ and $x = 6$, by how much does $y = 20$ deviate from $\hat{y}$ on the regression line?

Q3. Find the sum of the squares of the residuals if the residuals are 3, -1, -4, and 2.

Q4. How can you tell from the correlation coefficient that a function fits the data perfectly?

Q5. What type of function has the multiply–multiply property?

Q6. For what type of function could $f(x + 3)$ equal $8f(x)$?

Q7. $\log 5 + \log 7 = \log$ __?__

Q8. What is the least common denominator of $\frac{2}{7} + \frac{3}{8}$?

Q9. In a right triangle with hypotenuse m and legs p and r,

A. $m^2 + p^2 = r^2$ **B.** $p^2 + r^2 = m^2$

C. $p^2 - r^2 = m^2$ **D.** $p^2 = m^2 + r^2$

E. $p^2 - m^2 = r^2$

Q10. Expand: $(mx + b)^2$

---

1. *Residuals Problem:* Suppose that these data have been measured for two related variables x and y.

x	y	x	y
5	11	20	29
8	16	23	33
11	19	26	42
14	27	29	44
17	25	32	51

a. Enter the data into two lists on your grapher. (See the data sets at **www.keymath.com/precalc.**) Show by linear regression that the best-fitting linear function is $\hat{y} = 1.4x + 3.8$. Record the correlation coefficient.

b. Make a scatter plot of the data on your grapher. On the same screen, plot $\hat{y}$. How well does the linear function fit the data?

c. Calculate $\bar{x}$ and $\bar{y}$, the means of x and y. Show algebraically that the mean-mean point $(\bar{x}, \bar{y})$ is on the regression line.

d. Define new lists to help you calculate the squares of the deviations, $(y - \bar{y})^2$, and the squares of the residuals, $(y - \hat{y})^2$. By summing these lists, calculate SS_{dev} and SS_{res}. Use the results to calculate the coefficient of determination and the correlation coefficient. Does the correlation coefficient agree with the one you recorded in part a?

e. The line given by the equation $f_2(x) = 1.5x + 1.95$ also contains the mean-mean point $(\bar{x}, \bar{y})$, but it has slope 1.5 instead of 1.4. Plot the line on the same screen as in part b. Can you tell from the graphs which line fits the data better? Explain. Show that SS_{res} for this line is greater than SS_{res} for the regression line.

2. *New Subdivision Problem:* These data represent actual prices of various homes in a new subdivision. The data have been rounded to the nearest 100 square feet and to the nearest 1000 dollars.

Area (ft^2)	Price ($)
1900	155,000
2100	168,000
2400	190,000
2500	189,000
2500	207,000
2600	195,000
2600	199,000
2600	199,000
2700	210,000
2800	220,000

a. Run a linear regression on the data. (See the data sets at **www.keymath.com/precalc.**) Record the correlation coefficient. Plot the regression equation and the data on the same screen. Use a window with $0 \le x \le 3000$. How can you tell that a linear function fits the data reasonably well?

b. Based on the linear model, how much would you expect to pay for a 5000-ft^2 house in this subdivision? How big a house could you buy for a million dollars? What do you call the process of calculating an x- or y-value outside the range of the given data? What do you call the process of estimating an x- or y-value within the range of the given data?

c. What real-world meaning can you give to the y-intercept and the slope?

d. Find $\bar{x}$ and $\bar{y}$. Show that the mean-mean point $(\bar{x}, \bar{y})$ is on the regression line.

e. Find SS_{dev} and SS_{res}. Use the results to calculate the coefficient of determination and the correlation coefficient. Do your answers agree with the results in part a?

f. Why is it reasonable for there to be more than one data point with the same x-value?

3. *Gas Tank Problem:* Lisa Carr fills up her car's gas tank and drives off. The table shows the number of gallons of gas left in the tank at various numbers of miles driven.

Distance Driven (mi)	Gas Left (gal)
6	16.7
22	15.9
44	14.8
50	14.5
60	14.0

a. Run a linear regression on the data. Write down the linear regression equation, along with r^2 and r. How do these numbers tell you that the regression line fits the data perfectly? Why is r negative?

b. Find the coefficient of determination and the correlation coefficient again by calculating SS_{dev} and SS_{res} directly from their definitions. Do the answers agree with the results in part a?

c. Plot the data and the regression equation on the same screen. How can you tell graphically that the regression line fits the data perfectly?

d. According to your mathematical model, how much gas does the gas tank hold? How many miles per gallon does Lisa's car get?

e. Show that your mathematical model predicts that the tank is empty after 340 mi. Because this number is found by extrapolation, how confident are you that the car will actually run out of gas after 340 mi?

4. *Standardized Test Scores Problem:* Figure 3-2h shows a scatter plot of the scores of 1000 12th-graders on the mathematics part of the SAT (Scholastic Aptitude Test) and their high school grade point averages. By regression, the best-fitting linear function is $\hat{y} = 59.0x + 355$, but the coefficient of determination, r^2, is only about 0.14.

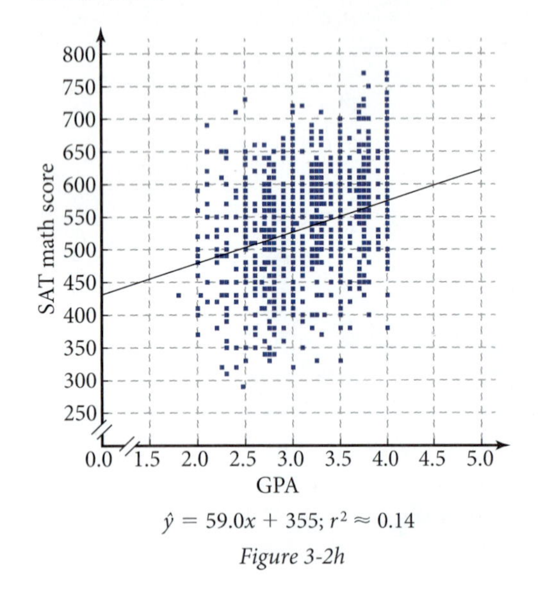

$\hat{y} = 59.0x + 355$; $r^2 \approx 0.14$

Figure 3-2h

a. Suppose that you have a 4.0 grade point average. According to the regression equation, what math score would you be predicted to get on the SAT? How reliable do you think this prediction is?

b. Use the given information to find the correlation coefficient. Explain why you would use the positive square root rather than the negative square root.

5. *Data Cloud Problem:* Figure 3-2i shows an elliptical region in which the "cloud" of data points is expected to lie if the correlation coefficient, r, is -0.95.

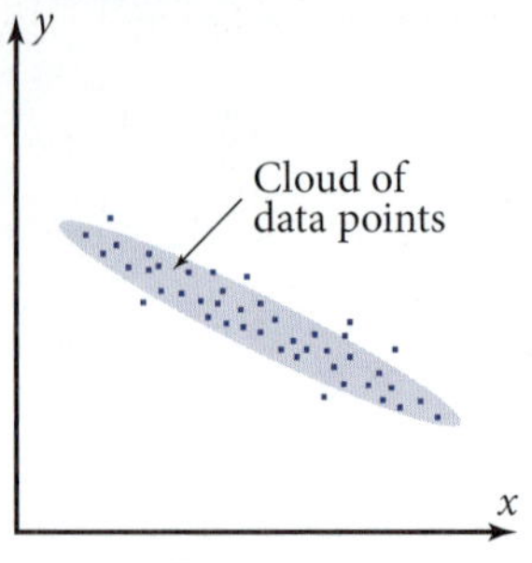

Figure 3-2i

For each given value of r, sketch the cloud you would expect for data with that correlation coefficient.

a. $r = 0.95$ **b.** $r = 0.8$

c. $r = -0.7$ **d.** $r = 0$

6. *Sit-Ups Problem:* Enter the sit-ups data from Example 1 into a data analysis program such as Fathom Dynamic Data™, and make a scatter plot of the data by using the appropriate menu.

a. Add a movable line, and then show squares. The result should resemble Figure 3-2e or 3-2f. Rotate and translate the movable line. Describe what happens to the squares as you move the line. Describe what happens to the quantity *sum of squares.*

b. Without deleting the movable line, add a least-squares line. How do the squares of the residuals for this regression line compare to the squares on the movable line you drew in part a? How does the quantity *sum of squares* for the regression line compare with the values you got as you moved the movable line?

c. What do you understand better about the regression line as a result of working this problem?

3-3 Regression for Nonlinear Data

In Section 3-2, you learned the basis of linear regression. In this section you will learn how to use regression on your grapher to fit other types of functions to data whose scatter plot follows a nonlinear pattern.

Objective

Given a set of data, make a scatter plot, identify the type of function that could model the relationship between the variables, and use regression to find the particular equation that best fits the data.

Figure 3-3a shows the scatter plot of fish weight in pounds, y, versus length in inches, x, that you saw in Section 3-2. The table lists the data for these 43 fish. (See the data sets at **www.keymath.com/precalc.**)

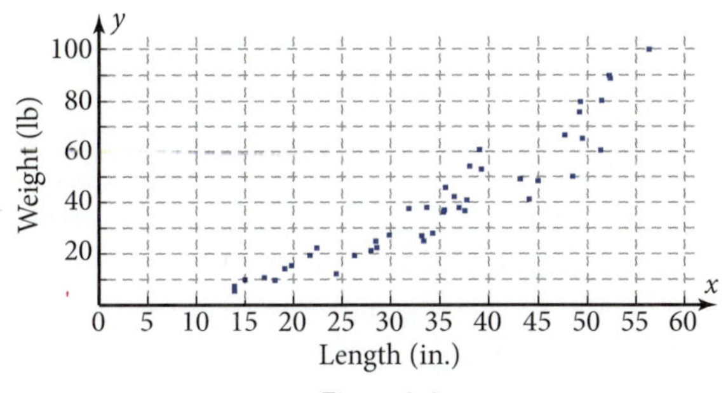

Figure 3-3a

x	y	x	y	x	y	x	y	x	y
14.0	5.5	24.5	12.2	33.8	38.0	38.2	54.1	49.3	75.4
14.0	7.4	26.4	19.3	34.4	28.0	39.2	60.7	49.4	79.4
15.1	9.9	28.1	21.2	35.5	36.2	39.4	52.9	49.6	65.1
17.1	10.7	28.6	24.9	35.6	37.0	43.3	49.1	51.5	60.5
18.2	9.7	28.7	22.4	35.7	45.8	44.2	41.2	51.6	79.9
19.2	14.2	30.0	27.3	36.6	42.2	45.1	48.4	52.4	89.6
19.9	15.4	32.0	37.6	37.1	37.9	47.8	66.4	52.5	88.5
21.8	19.4	33.3	27.0	37.7	36.7	48.6	50.1	56.5	99.8
22.5	22.3	33.5	25.0	37.9	40.9				

Because the points follow a path that is increasing and concave up, a power function or an exponential function might be a reasonable mathematical model for weight of fish as a function of their length. To decide which of the two is more reasonable, consider the **end behavior.** At the lower end of the domain, the graph would contain the origin because a fish of zero length would have zero weight. Because an untranslated power function contains the origin, it would be more reasonable than an untranslated exponential function. Run a **power regression** on your grapher to get

$$\hat{y} = 0.0606\ldots x^{1.7990\ldots}$$

Figure 3-3b shows that the graph of this function fits the points reasonably well.

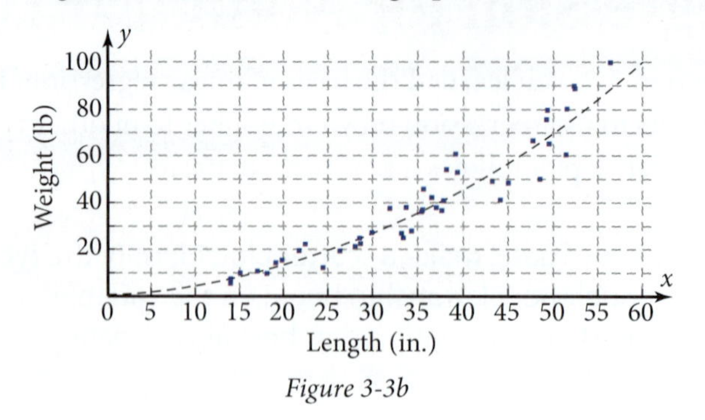

Figure 3-3b

Figure 3-3c shows the result of running **exponential regression** (curved graph) and **linear regression** (straight graph). Neither function has the correct endpoint behavior at $x = 0$.

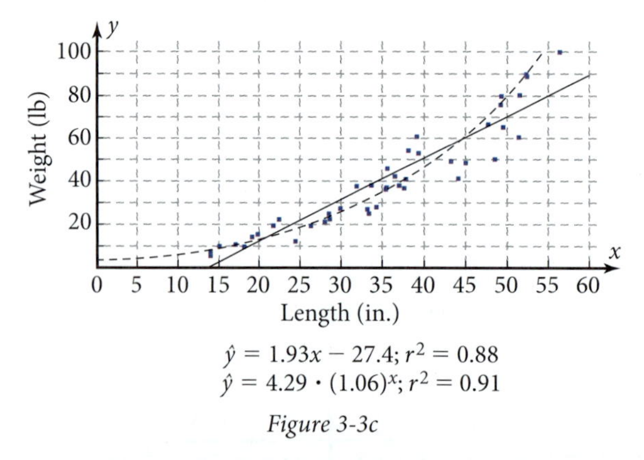

$\hat{y} = 1.93x - 27.4;\ r^2 = 0.88$

$\hat{y} = 4.29 \cdot (1.06)^x;\ r^2 = 0.91$

Figure 3-3c

Your grapher will calculate the correlation coefficient $r = 0.9669\ldots$ for power regression, $r = 0.9557\ldots$ for exponential regression, and $r = 0.9354\ldots$ for linear regression. A correlation coefficient closer to 1 or -1 indicates a better fit. So, in addition to having the wrong endpoint behavior, the exponential and linear functions have a weaker correlation to the data.

Problem Set 3-3

Reading Analysis

From what you have read in this section, what do you consider to be the main idea? What is the primary way you decide what type of function to fit to scattered data that do not follow a linear pattern? If two different functions have graphs with a pattern similar to the scatter plot for a data set, how can you decide which is more reasonable?

Quick Review

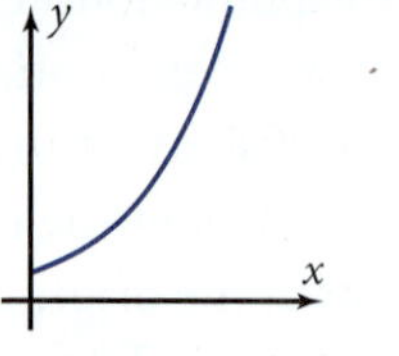

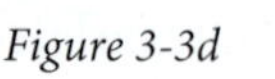

Figure 3-3d

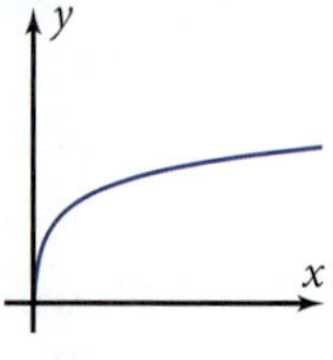

Figure 3-3e

Q1. What type of function is represented by the graph in Figure 3-3d?

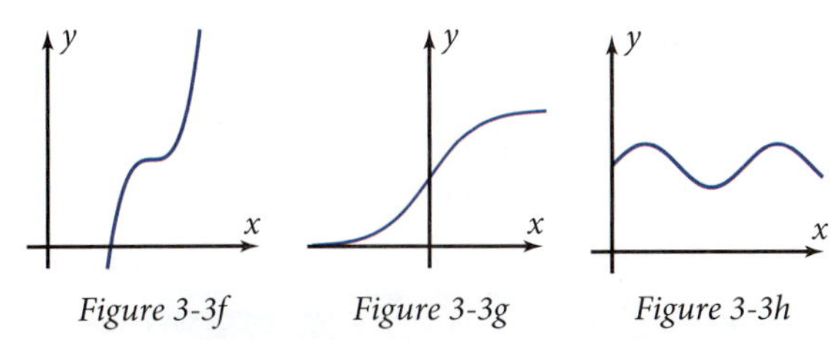

Figure 3-3f *Figure 3-3g* *Figure 3-3h*

Q2. What type of function is represented by the graph in Figure 3-3e?

Q3. What type of function is represented by the graph in Figure 3-3f?

Q4. What type of function is represented by the graph in Figure 3-3g?

Q5. The function in Figure 3-3h is *periodic*. What do you suppose "periodic" means?

Q6. The sales of a particular product depend on the amount spent on advertising. Sketch a reasonable graph.

Q7. The parent quadratic function has equation __?__.

Q8. Expand: $(9x - 4)^2$

Q9. Write the next three terms of the sequence: 192, 96, 48, 24, . . .

Q10. The multiply–multiply property is a characteristic of __?__ functions.

---

1. *Bacteria Problem:* Thirty-six college biology students start bacterial cultures by taking equal-volume samples from a flask in the laboratory. Each student comes back x hours later and measures the number of bacteria, y, in his or her culture. The results are shown in the table and in the scatter plot in Figure 3-3i.

a. Explain why either an untranslated power function or an untranslated exponential function could be a reasonable mathematical model for the number of bacteria as a function of time. Explain why an exponential function would have a more reasonable left endpoint behavior than would a power function.

b. Run an exponential regression on the data. (See the data sets at **www.keymath.com/precalc.**) Write down the equation and the correlation coefficient. Plot the function and the scatter plot on the same screen. Sketch the result on a copy of Figure 3-3i.

c. According to the exponential model, how many bacteria were in each equal sample when the students took them? What do you predict the number of bacteria will be 24 h after the cultures were started?

d. After how many hours do you expect the number of bacteria to reach 100,000?

x (h)	y	x (h)	y
0.6	450	4.6	1963
0.8	446	4.7	1774
1.3	588	4.7	2611
1.5	645	4.9	2853
1.6	718	5.0	1848
1.9	729	5.0	3266
2.0	855	5.1	2229
2.3	1008	5.2	2827
3.1	962	5.3	3776
3.5	1570	5.5	2611
3.7	1620	5.7	3126
3.8	1378	5.7	2928
3.8	1561	5.8	3776
3.8	1580	6.3	4067
3.9	1919	6.3	5393
4.2	2210	6.4	5288
4.5	2212	7.3	7542
4.6	2125	8.6	11042

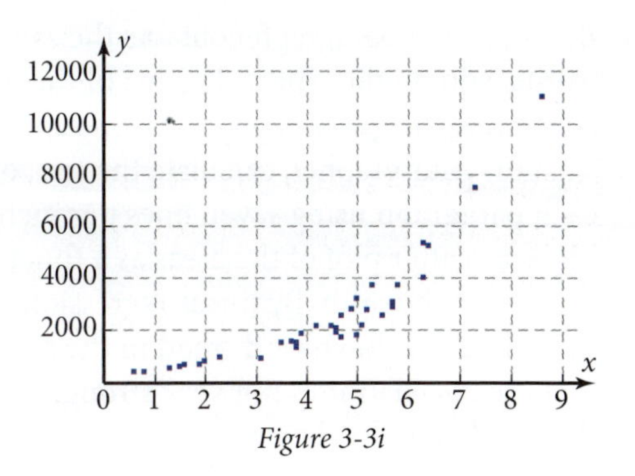

Figure 3-3i

2. *Printed Paragraph Problem:* Ann A. Student types a paragraph on her word processor. Then she adjusts the width of the paragraph by changing the margins. Figure 3-3j and the table show the widths and numbers of lines the paragraph has.

x (in. wide)	y (lines long)
6.5	5
5.5	6
5.0	7
3.75	9
3.0	11
1.5	24
1.0	38

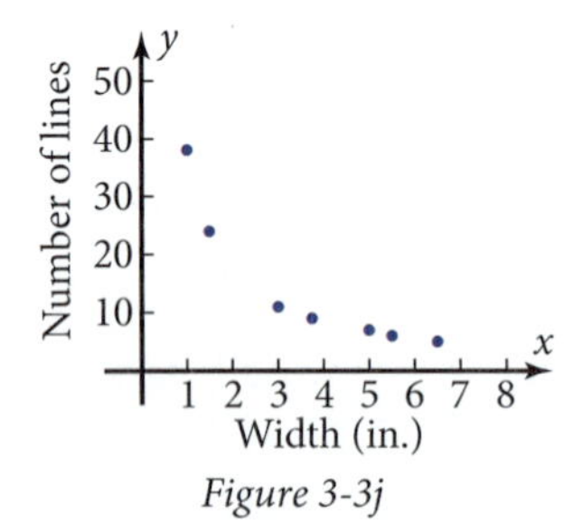

Figure 3-3j

a. Would it be possible to make the width equal zero? Why, then, would a decreasing power function be more reasonable than an exponential function for number of lines as a function of paragraph width? Confirm that a power function fits better by running both power and exponential regressions and comparing the correlation coefficients.

b. Plot the power and exponential functions from part a on the same screen as a scatter plot of the data. How does the result confirm that the power function fits the data better?

c. Because each paragraph contains the same words, you might expect the area of the page covered by the paragraph to be constant. In a list on your grapher, calculate the area of each paragraph using seven lines per inch. Make a scatter plot of the areas as a function of paragraph width. By linear regression, show that there is a downward trend in the areas but that the correlation is not very strong. Sketch the graph and the data points.

3. *Bank Interest Problem:* At the bank, you notice the ad shown that lists the number of years it will take for the balance in your account to reach a certain level if you invest $1000 in one of the bank's accounts.

Dollars	Years to reach
1000.00	0.00
1100.00	1.91
1500.00	8.11
2000.00	13.86
3000.00	21.97

a. Make a scatter plot of the data. Which way is the graph concave? How can you tell from the shape of the graph that a logarithmic function would fit the data?

b. By **logarithmic regression,** find the particular equation of the best-fitting logarithmic function. How does the correlation coefficient confirm that a logarithmic function fits well?

c. Plot the equation from part b on the scatter plot from part a. Sketch the result.

d. Interpolate using your mathematical model to find out how long it takes for $2500 (the mean of $2000 and $3000) to be in the account. Is this number of years equal to the mean of 13.86 and 21.97?

e. If you wanted to leave your money in the account until it reached $5000, how many years would you have to leave it there? Which method do you use, interpolation or extrapolation? Explain.

4. WEB *Planetary Period Problem:* The table shows the period of each planet in years, the mean distance from the Sun in millions of kilometers, and the planets' masses in relation to Earth's mass, as provided by *The New York Times Almanac 2009.*

Name	Period (yr)	Orbit Radius (million km)	Relative Mass
Mercury	0.24	57.9	0.06
Venus	0.62	108.2	0.82
Earth	1	149.6	1
Mars	1.88	227.9	0.11
Jupiter	11.86	778.6	317.83
Saturn	29.42	1433.5	95.16
Uranus	83.75	2872.5	14.54
Neptune	163.72	4495.1	17.15

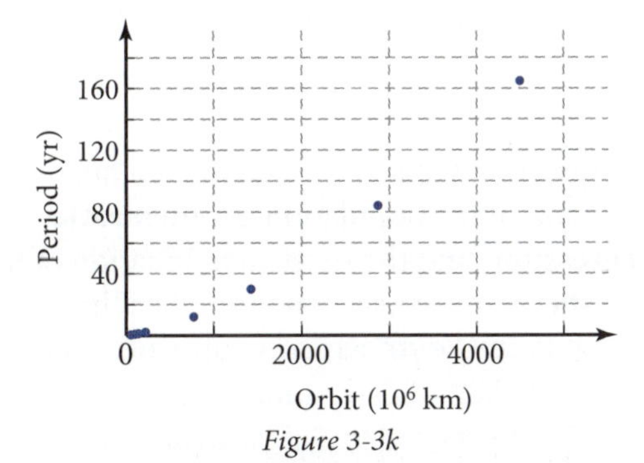

Figure 3-3k

a. If you assume that a power function fits the points in Figure 3-3k, what endpoint behavior are you assuming the period approaches as the distance from the Sun approaches zero? Run both power and exponential regressions for period in years as a function of distance in millions of kilometers. (See the data sets at **www.keymath.com/precalc.**) Give numerical evidence that the untranslated power function fits better.

b. Plot the power function from part a and the scatter plot on the same screen. Does the power function seem to fit the data well?

c. Make a scatter plot of the period versus the relative mass of the planet. Show that there is little or no correlation between these two variables.

d. Most asteroids are located in the "asteroid belt," about 430 million km from the Sun. Some scientists believe that they originated from the breakup of a planet or from material that never coalesced to form a planet. If these asteroids were originally one planet, what would have been the period of that planet?

This view of the asteroid Ida is a composite of five images taken by the Galileo *spacecraft in 1993.*

e. Kepler derived his three laws of planetary motion from careful analysis of data. Look up Kepler's third law on the Internet or in a physics text. How well does the result of your power regression agree with that law?

5. *Roadrunners Problem, Part 1:* Naturalists placed 30 roadrunners in a preserve that formerly had none. The population grows as shown in the table and in Figure 3-3l.

x (yr)	*y* (roadrunners)
0	30
1	44
2	58
3	81
4	110
5	138
6	175
7	203
8	234
9	260
10	276
11	293

Figure 3-3l

a. Give a physical reason and a graphical reason why a logistic function would be a reasonable mathematical model for the roadrunner population as a function of time. By **logistic regression,** find the particular equation of the best-fitting logistic function. (See the data sets at **www.keymath.com/precalc.**) Plot the graph and the data on the same screen. Use a window with $0 \le x \le 15$. Sketch the result.

b. What does your mathematical model predict for the roadrunner population at $x = 20$ yr? What does the model predict for the maximum sustainable population? At approximately what value of x is the point of inflection, the point where the rate of population growth is a maximum?

c. Find $\bar{y}$, the mean of the y-values. Use $\bar{y}$ to calculate SS_{dev}, the sum of the squares of the deviations. Using the regression function from part a, calculate SS_{res}, the sum of the squares of the residuals. Then find the coefficient of determination. How does this coefficient show that the logistic function fits the points quite well?

6. *Roadrunners Problem, Part 2:* The table and the scatter plot in Figure 3-3m show the change in roadrunner population for each year versus the population at the beginning of that year. For instance, the population was 81 at the beginning of the third year and increased by 29, to 110, by the beginning of the fourth year. Thus the rate of increase in year 3 was 29 roadrunners per year.

Year	x (roadrunners)	y (roadrunners/yr)
0	30	14
1	44	14
2	58	23
3	81	29
4	110	28
5	138	37
6	175	28
7	203	31
8	234	26
9	260	16
10	276	17

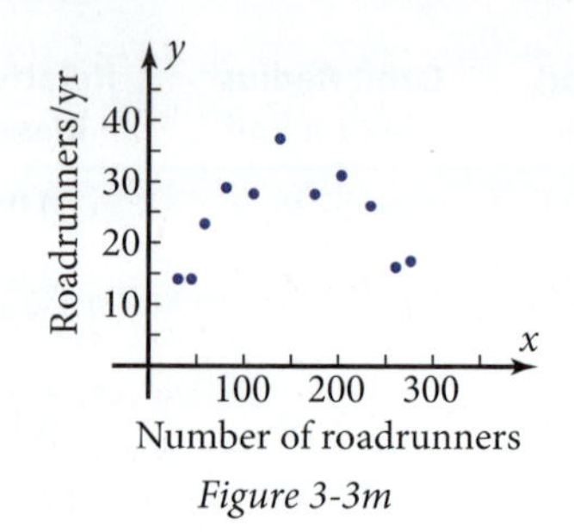

Figure 3-3m

a. The rate of increase of roadrunners is expected to be higher when the population is higher because there are more parents having roadrunner chicks. How, then, can you explain the fact that the population increases at a slower rate when the population is above 150?

b. Assume that the rate of increase is a quadratic function of the population. By **quadratic regression,** find the particular equation of the best-fitting quadratic function. (See the data sets at **www.keymath.com/precalc.**) Record the coefficient of determination, R^2. The lowercase r is used for linear data or data that can be linearized (exponential, power, and logarithmic). The uppercase R is used for data that cannot be linearized (quadratic).

c. Suppose that the naturalists brought in enough roadrunners to increase the population to 400. What does your mathematical model predict for the growth rate? What real-world reason can you think of to explain why this rate is negative?

d. Calculate $\bar{y}$, the mean number of roadrunners per year. Enter lists into your grapher for the squares of the deviations, $(y - \bar{y})^2$ and for the squares of the residuals, $(y - \hat{y})^2$. By summing these lists, calculate SS_{dev} and SS_{res}. Use the results to calculate the coefficient of determination, R^2. Show that the answer is equal to the value you found by regression in part b.

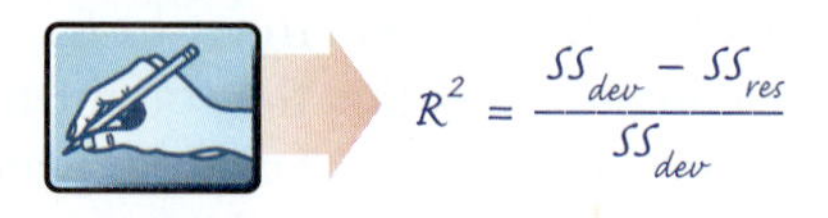

3-4 Linearizing Data and Logarithmic Graph Paper

In the previous section you used exponential, power, or logarithmic regression to find functions that fit nonlinear sets of data. In this section you will see how the calculator does these kinds of regression by using logarithms to transform the data so that a linear function fits the transformed data. You will also learn how to use graph paper on which the scale corresponds to logarithms of the variables.

Objective

Given a set of data that does not follow a linear pattern,

- Transform the data using logarithms, and use linear regression on the transformed data.
- Plot the data on log-log graph paper or semilog graph paper.

In this exploration you will plot graphs of exponential and power functions on graph paper that has one or both axes marked off with logarithmic scales.

EXPLORATION 3-4: Log-Log and Semilog Graph Paper

For Problems 1–3, use these data.

x	$f(x)$
3	4.1
7	24
8	37
11	136
12	212

1. Plot the data on *semilog graph paper*, like that shown here. Note that the distances on the vertical axis correspond to the logarithms of the $f(x)$-values. Do the points follow a linear pattern? If not, go back and correct your work.

2. By regression, find the particular equation for the best-fitting exponential function. Use the equation to calculate predicted values of $f(0)$ and $f(15)$. Plot these points on the graph paper from Problem 1. Do these points fit the linear pattern in Problem 1?

3. Complete the sentence: "An exponential function has a straight-line graph on ___?___ graph paper."

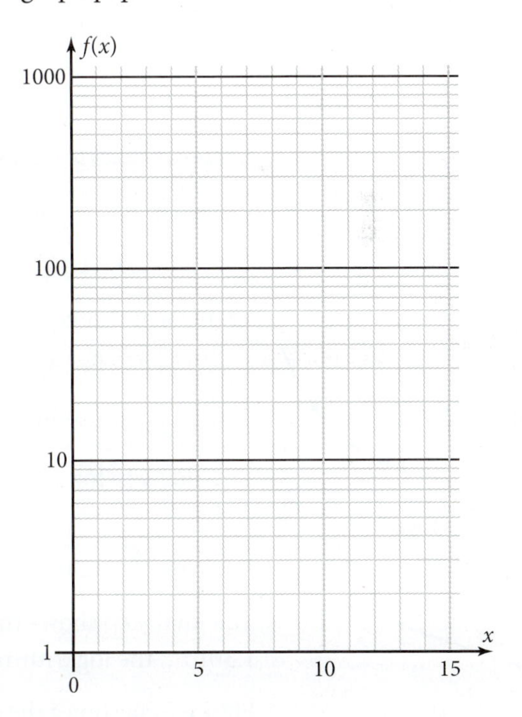

continued

For Problems 4–6, use these data.

x	$g(x)$
2	261
5	87
9	43
20	16.5
70	3.7
100	2.4

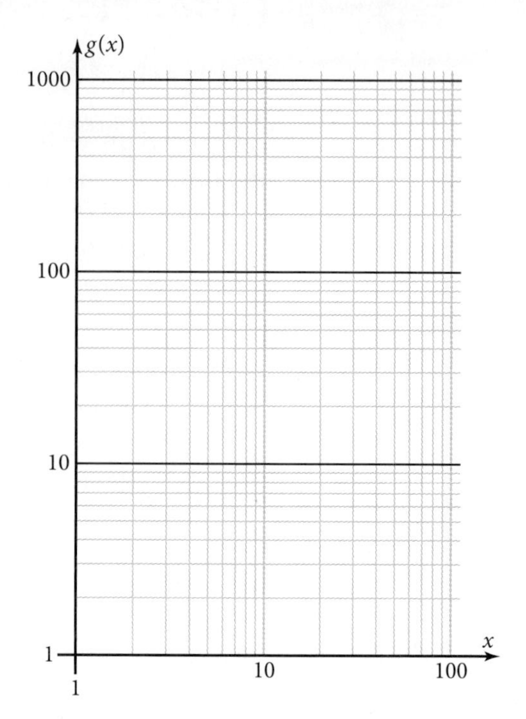

4. Plot the data on a copy of this *log-log graph paper.* Note that both axes have logarithmic scales. Do the points follow a linear pattern? If not, go back and correct your work.

5. By regression, find the particular equation for the best-fitting power function. Use the equation to calculate predicted values of $g(1)$ and $g(50)$. Plot these points on the graph paper from Problem 4. Do these points fit the linear pattern in Problem 4?

6. Complete the sentence: "A power function has a __?__ graph on log-log graph paper."

7. What did you learn as a result of doing this exploration that you did not know before?

Linearizing Exponential Function Data

Consider an exponential function such as

$$y = 20 \cdot 1.45^x$$

Transform this equation by taking the logarithm of both sides.

$\log y = \log(20 \cdot 1.45^x)$	Write "log" in front of both sides of the equation.
$\log y = \log 20 + \log 1.45^x$	The log of a product equals the sum of the logs.
$\log y = \log 20 + x \cdot \log 1.45$	The log of a power equals the exponent times the log of the base.
$\log y = 1.3010... + 0.1613...x$	Evaluate the logs by calculator.

The final step shows that $\log y$ is a linear function of x. The $\log y$-intercept is 1.3010..., the logarithm of 20, and the slope is 0.1613..., the logarithm of 1.45.

Figure 3-4a shows the graph of the original function and the graph of the transformed function. The points on the transformed graph lie on a straight line. The function has been **linearized.**

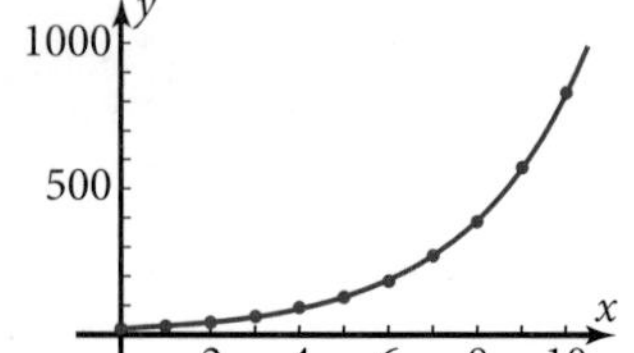

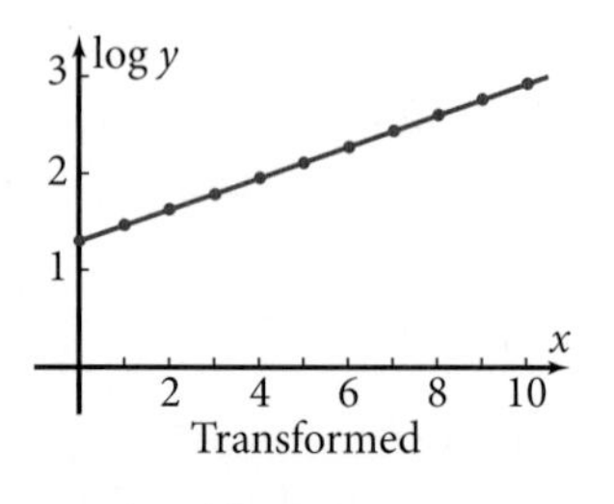

Figure 3-4a

Linearizing Power Function Data

Consider a power function such as

$$y = 800x^{-1.3}$$

Transform this equation by taking the logarithm of both sides.

$$\log y = \log(800x^{-1.3})$$

$$\log y = \log 800 + \log x^{-1.3}$$ The log of a product equals the sum of the logs.

$$\log y = \log 800 - 1.3 \log x$$ The log of a power equals the exponent times the log of the base.

$$\log y = 2.9030... - 1.3 \log x$$ Evaluate the logs by calculator.

The final step shows that log y is a linear function of log x. The log y–intercept is 2.9030..., the logarithm of 800, and the slope is −1.3, the exponent in the power function. Figure 3-4b shows the graph of the original function and the graph of the transformed function. The points on the original graph cover such a wide range that it is almost impossible to tell anything about the function. However, the points on the transformed graph lie on a straight line.

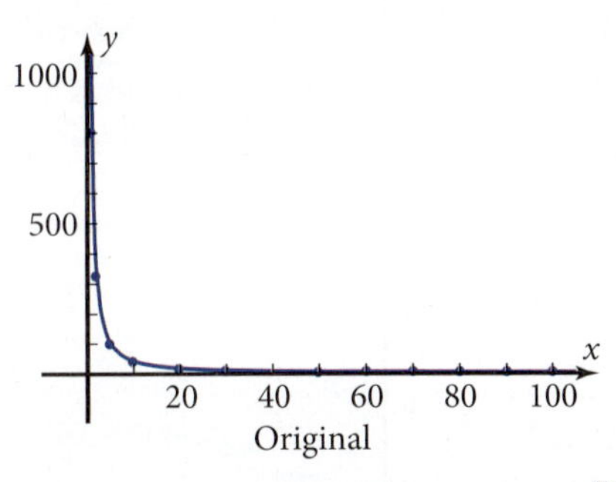

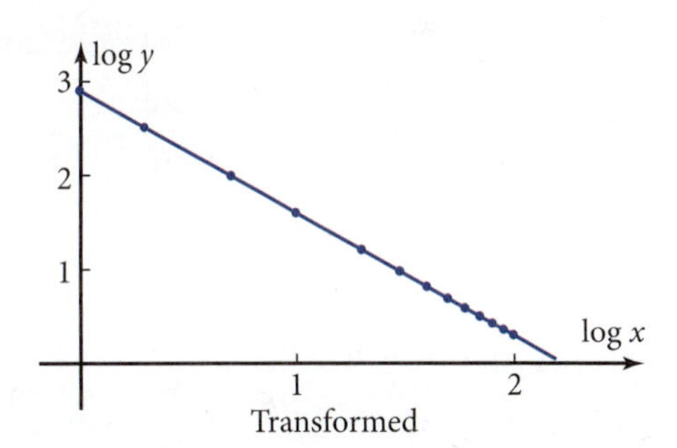

Figure 3-4b

PROPERTY: Linearizing Exponential and Power Functions

Exponential Functions

Algebraically: If $y = ab^x$, then $\log y = \log a + x \log b$.

Verbally: If y is an exponential function of x, then log y is a linear function of x.

Power Functions

Algebraically: If $y = ax^b$, then $\log y = \log a + b \log x$.

Verbally: If y is a power function of x, then log y is a linear function of log x.

Log-Log and Semilog Graph Paper

In many applications of mathematics, real-world data span a large range of values. A graph plotted with ordinary ("arithmetic") scales such as that on the left in Figure 3-4b may be hard to read. In such cases you often see data plotted on graph paper on which one or both axes have scales equal to the *logarithm* of the variable. Figure 3-4c illustrates some features for the function $y = 800x^{-1.3}$, from Figure 3-4b, plotted on *log-log graph paper:* There is more space between 1 and 2 than there is between 2 and 3, there is the same space between 1 and 10 as there is between 10 and 100, and it is possible to read values of x and y directly from the graph. The graph turns out to be a straight line because log y is a linear function of log x for a power function.

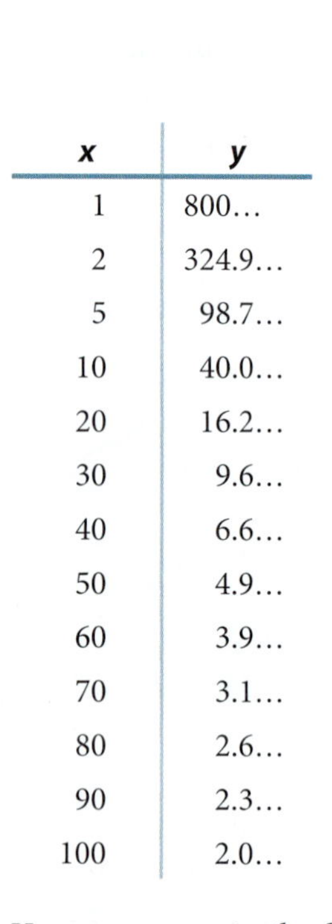

x	y
1	800...
2	324.9...
5	98.7...
10	40.0...
20	16.2...
30	9.6...
40	6.6...
50	4.9...
60	3.9...
70	3.1...
80	2.6...
90	2.3...
100	2.0...

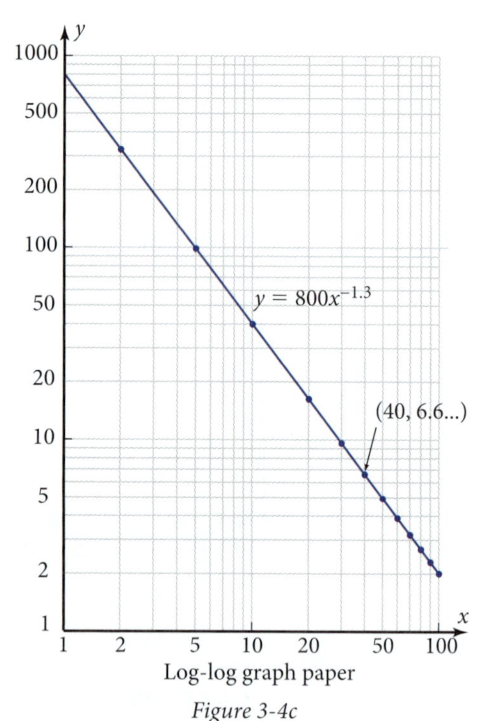

Figure 3-4c

You can use a similar kind of graph paper to plot graphs that have a large range for one variable but a relatively small range for the other variable. Figure 3-4d shows graphs plotted on two kinds of *semilog graph paper.* For add–multiply semilog paper, on the left, the graph of $y = 20 \cdot 1.45^x$ from Figure 3-4a is a straight line, because for exponential functions log y is a linear function of x. For multiply–add semilog paper, on the right, the graph of a logarithmic function such as $y = 14 - 4 \log x$ is a straight line because y is already a linear function of log x.

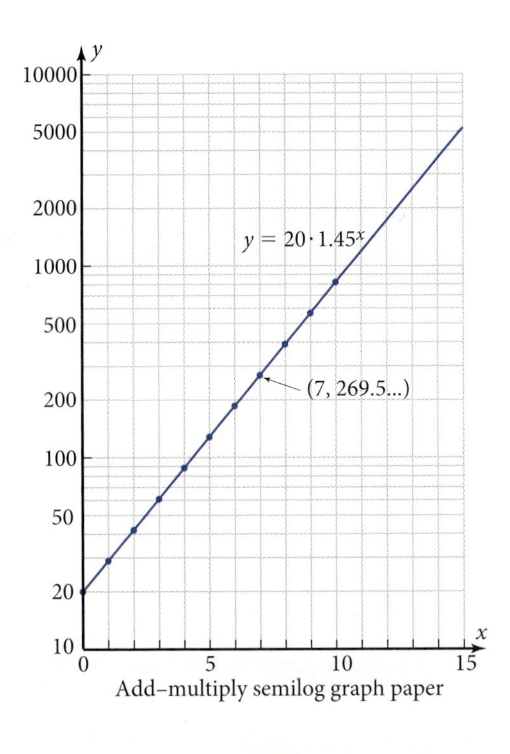

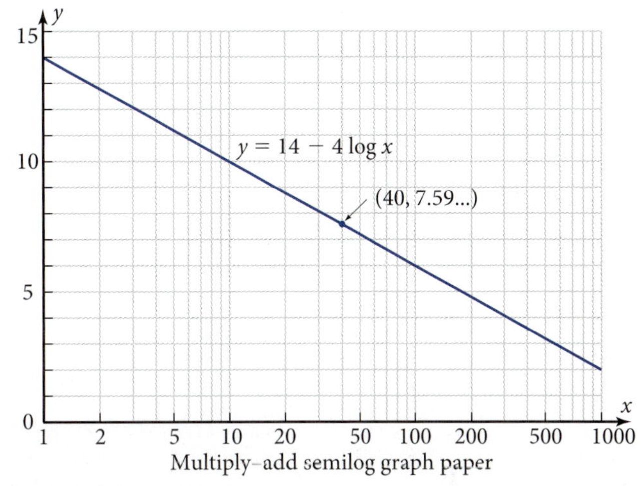

Figure 3-4d

DEFINITIONS: Graph Paper Types

Arithmetic graph paper ("ordinary" graph paper) has linear scales on both axes.

Log-log graph paper has logarithmic scales on both axes.

Add–multiply semilog graph paper has a logarithmic scale on the vertical axis only.

Multiply–add semilog graph paper has a logarithmic scale on the horizontal axis only.

Regression for Nonlinear Data

The table shows the pressure, y, in pounds per square inch (psi) of air remaining in a tire at different times, x, in seconds after the tire was punctured. (See the data sets at **www.keymath.com/precalc.**) Figure 3-4e shows that the data follow a path that is decreasing, and concave up with a positive y-intercept, similar to the graph of an exponential function.

x (s)	y (psi)	x (s)	y (psi)
5	27	30	7
10	21	35	6
15	16	40	4
20	13	45	3
25	9	50	3

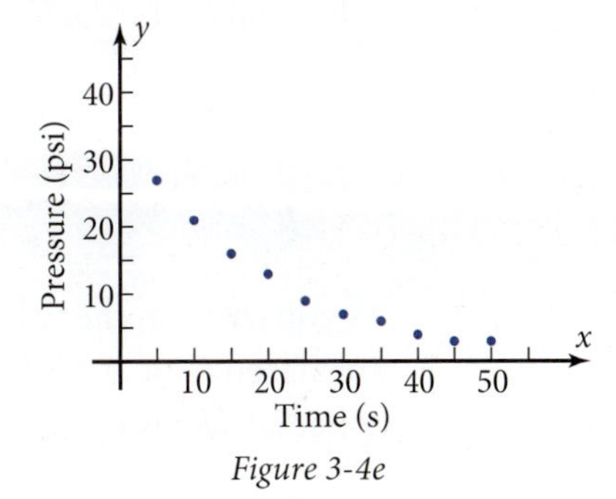

Figure 3-4e

Figure 3-4f and the table show the values of log y for these values of x. The points lie close to a straight line.

x	log y	x	log y
5	1.4313...	30	0.8450...
10	1.3222...	35	0.7781...
15	1.2041...	40	0.6020...
20	1.1139...	45	0.4771...
25	0.9542...	50	0.4711...

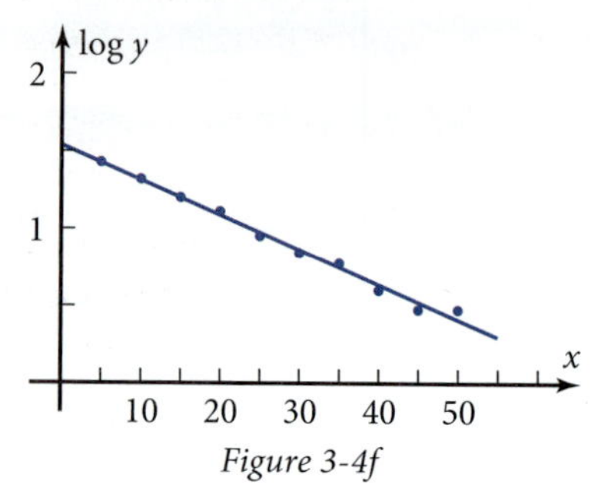

Figure 3-4f

Running a linear regression on log x as a function of x gives

$$\log \hat{y} = -0.0225\ldots x + 1.5415\ldots, \text{ with } r^2 = 0.9905\ldots$$

Transforming this equation back to exponential form gives

$y = 10^{(-0.0225\ldots x + 1.5415\ldots)}$	Definition of logarithm.
$y = (10^{-0.0225\ldots x})(10^{1.5415\ldots})$	Product of powers with equal bases.
$y = (10^{-0.0225\ldots})^x(10^{1.5415\ldots})$	Power of a power property of exponents.
$y = (0.9493\ldots^x)(34.7990\ldots)$	Evaluate the powers of 10.
$y = (34.7990\ldots)(0.9493\ldots^x)$	Commute the two factors.

A revealing fact emerges if you run an exponential regression on the original data. The result is the same as the result of linearizing the data.

$$\hat{y} = (34.7990\ldots)(0.9493\ldots^x), \text{ with } r^2 = 0.9905\ldots$$

The calculator has

- Transformed the original data by taking log y
- Run a linear regression on the transformed data
- Transformed the equation back to exponential form

The coefficient of determination, r^2, is the coefficient for the transformed data, not the original data. The actual coefficient of determination, R^2, for the original data could be found by calculating $\bar{y}$, using the result to find SS_{dev}, then using the regression equation and the original data to find SS_{res}, and finally applying the definition of the coefficient of determination. The results are

$$\bar{y} = \frac{109}{10} = 10.9, SS_{dev} = 606.9, SS_{res} = 1.4971\ldots$$

$$R^2 = \frac{SS_{dev} - SS_{res}}{SS_{dev}} = \frac{606.9 - 1.4971\ldots}{606.9} = 0.9975\ldots$$

As you can see, the value of R^2, 0.9975..., is not equal to the value of r^2 for the linearized data, 0.9905.... The lowercase r^2 is used only for linear data. The uppercase R^2 is used for *multiple regression,* where y is a more complicated function of x. You may have observed that your grapher reports R^2 for quadratic regression. This is because y is a function of both x and x^2. Quadratic functions cannot be linearized by logarithms except in the special case $y = ax^2$, a power function. For functions that can be linearized, r^2 is more meaningful than R^2.

EXAMPLE 1 ➤ Snakes shed their skins periodically. The area of the skin depends on the length of the snake. Suppose that the areas in the table at the left have been measured.

Length (cm)	Area (cm^2)
5	2
10	8
20	40
60	250
100	900

a. Make a scatter plot on your grapher, and use the result to explain why a power function would be a more reasonable mathematical model than either a linear function or an exponential function. Then run a power regression on the data and plot the equation on the same screen as the scatter plot. Record r^2, the coefficient of determination. Sketch the graph.

b. Plot the data points on log-log graph paper. From the regression equation, calculate $\hat{y}$ for $x = 5$ and $x = 100$. Plot these two points on the graph paper and connect them with a straight-line graph. What do you notice about the data points and this straight line, and what does this confirm about power functions?

c. Enter lists into your grapher for log x and log y. Run a linear regression on log y as a function of log x. Write the equation and the value of r^2. What do you notice about the constants in the equation and the value of r^2?

d. Plot the linear equation and the points (log x, log y) on the same screen. Sketch the result.

SOLUTION

a. The scatter plot of the data is shown on the left in Figure 3-4g. The points follow a pattern that is concave up, so a linear function is not reasonable. The x-intercept should be zero because a snake with length zero would have skin area zero, meaning that a power function is more reasonable than an exponential function. Power regression gives

$$\hat{y} = 0.0828\ldots x^{2.0014\ldots}, \text{ with } r^2 = 0.9969\ldots$$

The graph of this equation is shown on the right in Figure 3-4g.

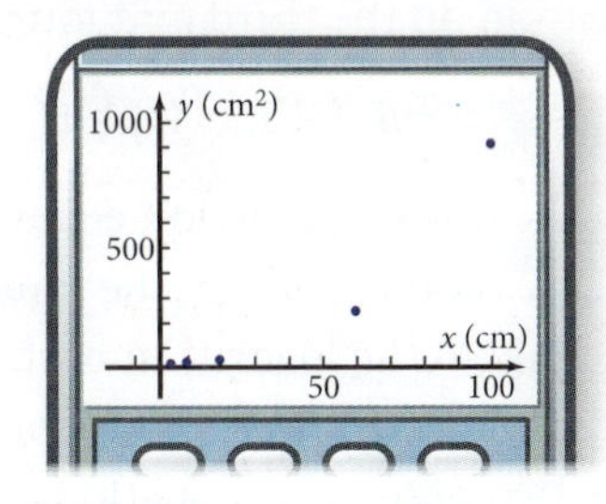

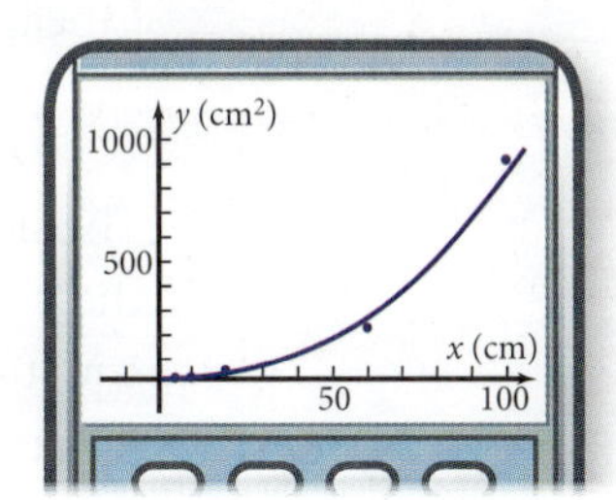

Figure 3-4g

b. $\hat{y}(5) = 2.0767\ldots \approx 2.1$

$\hat{y}(100) = 834.3795\ldots \approx 830$

Figure 3-4h shows that the data points lie close to the regression line, confirming that, for a power function, log y is a linear function of log x.

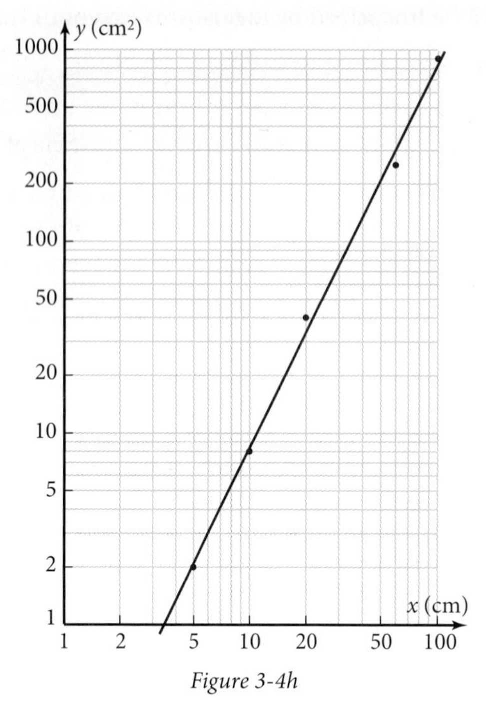

Figure 3-4h

c. On your grapher, calculate the logarithms in two new lists so that they will be available without rounding.

Length, x (cm)	Area, y (cm^2)	log x	log y
5	2	0.6989...	0.3010...
10	8	1	0.9030...
20	40	1.3010...	1.6020...
60	250	1.7781...	2.3979...
100	900	2	2.9542...

Linear regression on the third and fourth columns gives

$$\hat{y} = 2.0014\ldots x - 1.0815\ldots, \text{ with } r^2 = 0.9969\ldots$$

The slope in the linear regression equation is equal to the exponent in the power regression equation. The y-intercept in the linear regression equation is equal to the logarithm of the proportionality constant (dilation factor) in the power regression equation. The values of r^2 are equal for the two regression equations.

d. Figure 3-4i shows that the linear function fits the logarithms of the data, as in Figure 3-4h.

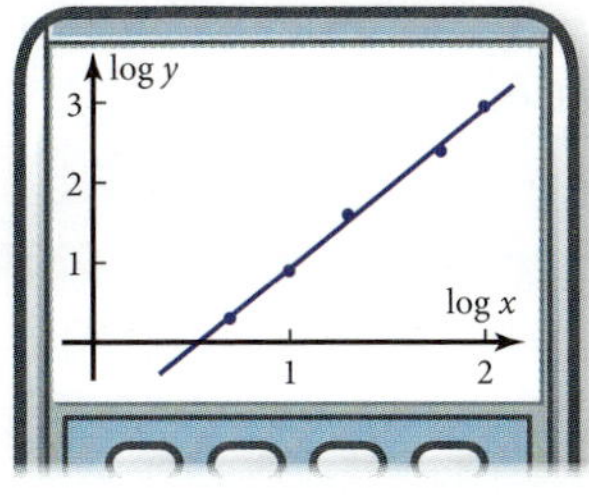

Figure 3-4i

Note: You can plot the graph in Figure 3-4i without actually finding the linear equation. Set your grapher to parametric mode. Then enter log t for x and the log of the power regression equation for y, with x replaced by t.

$x_1(t) = \log(t)$

$y_1(t) = \log(0.08287t\text{^}2.00147\ldots)$

The next example shows you how to graph a logarithmic function on multiply–add semilog graph paper.

EXAMPLE 2 ➤

a. Plot these data on multiply–add semilog graph paper. What evidence do you have that a logarithmic function would fit the data reasonably well?

x	y
2	2.4
10	5.0
50	7.6
300	10.4
900	12.2

b. By logarithmic regression, find the particular equation of the best-fitting logarithmic function. Record the coefficient of determination. Use the equation to calculate $\hat{y}$ for $x = 1$ and $x = 1000$. Plot these two points on the graph in part a and connect them with a straight-line graph. What graphical and numerical evidence do you have that a logarithmic function fits the data reasonably well?

c. Calculate $\hat{y}$ if x is 20. Show that this point lies on the straight-line graph in part b.

SOLUTION Figure 3-4j shows the data and the regression line plotted on multiply–add semilog graph paper.

a. The points lie close to a straight line, indicating that a logarithmic function is reasonable.

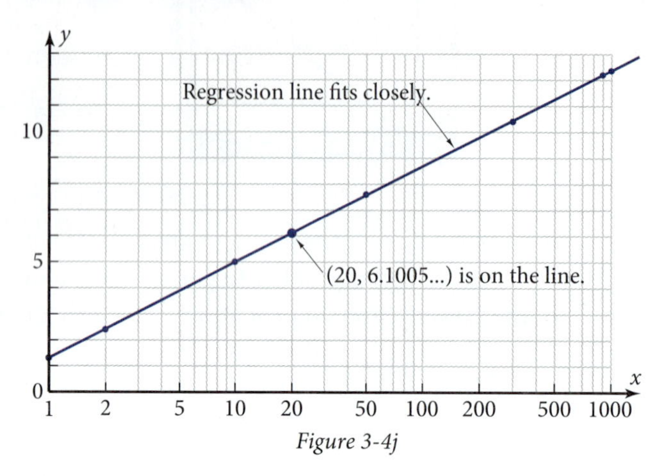

Figure 3-4j

b. The equation is $\hat{y} = 1.3067.... + 1.6001.... \ln x$, with $r^2 = 0.999957....$

$$\hat{y}(1) = 1.3067....$$

$$\hat{y}(1000) = 12.3605....$$

Figure 3-4j shows graphically that the regression line fits the data. Also, $r^2 = 0.999957...$, which is close to 1, confirming numerically that the regression line fits the data.

c. $\hat{y}(20) = 16.1005....$

Figure 3-4j shows that the point is on the line.

Note that most graphers display the results of logarithmic regression in terms of natural logarithms. The values of $\hat{y}$ you get from these regression equations are the same ones you get if you change to base-10 logarithms.

Problem Set 3-4

Reading Analysis

From what you have read in this section, what do you consider to be the main idea? What does your grapher do internally when you run an exponential regression? On what kind of graph paper will a power function graph plot as a straight line, and why is it straight? What is the significance of the slope of this line?

Quick Review

Q1. Sketch the graph of a decreasing exponential function.

Q2. Sketch the graph of a power function with exponent between 0 and 1.

Q3. Sketch the graph of a logarithmic function.

Q4. Sketch the graph of a logistic function.

Q5. Sketch the graph of a quadratic function with a positive y-intercept and a negative coefficient of x^2.

Q6. $\log 5 + \log 7 = \log$ _?_

Q7. $\log 18 - \log 3 = \log$ _?_

Q8. $2 \log 7 = \log$ _?_

Q9. Exponential functions have the _?_ pattern.

Q10. Logarithmic functions have the _?_ pattern.

For the exponential functions in Problems 1 and 2,

a. Calculate the y-values for the given values of x.

b. Plot the points on add–multiply semilog graph paper.

c. Show that the points lie on a straight line.

1. $y = 1.5 \cdot 2^x$; $x = 1, 3, 5, 7, 9$

2. $y = 800 \cdot 0.6^x$; $x = 2, 4, 6, 8, 10$

For the power functions in Problems 3 and 4,

a. Calculate the y-values for the given values of x.

b. Plot the points on log-log graph paper.

c. Show that the points lie on a straight line.

d. Measure the slope with a ruler and show that it equals the exponent in the equation

3. $y = 700x^{-1.3}$; $x = 1, 5, 10, 30, 100$

4. $y = 5x^{0.8}$; $x = 1, 6, 10, 40, 100$

For the logarithmic functions in Problems 5 and 6,

a. Calculate the y-values for the given values of x.

b. Plot the points on multiply–add semilog graph paper.

c. Show that the points lie on a straight line.

5. $y = 2 + 3 \ln x$; $x = 1, 4, 10, 200, 1000$

6. $y = -1 + 2 \log x$; $x = 3, 8, 20, 100, 500$

7. Show that an exponential function graph is *not* a straight line on log-log graph paper by plotting the data from Problem 1 on log-log graph paper.

8. Show that a power function graph is *not* a straight line on add–multiply semilog graph paper by plotting the data from Problem 4 on add–multiply semilog graph paper.

For Problems 9–14,

a. Read the coordinates of the points.

b. Find an equation of the function by running the appropriate regression.

c. Use your regression equation to find the value of $\hat{y}$ for the given value of x, and confirm that the point lies on the graph.

9. $x = 2$

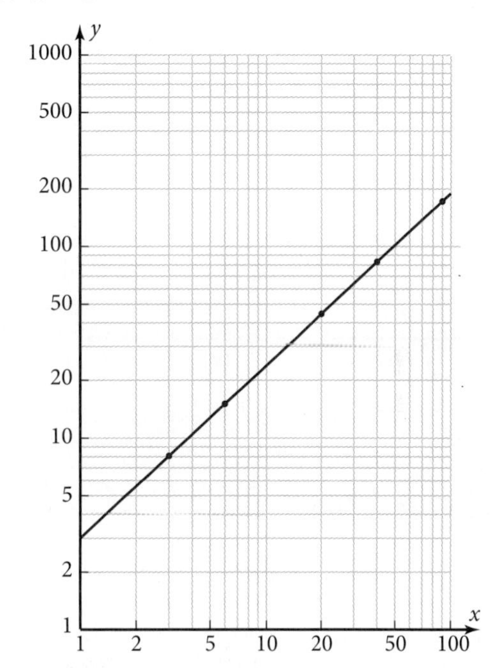

10. $x = 60$

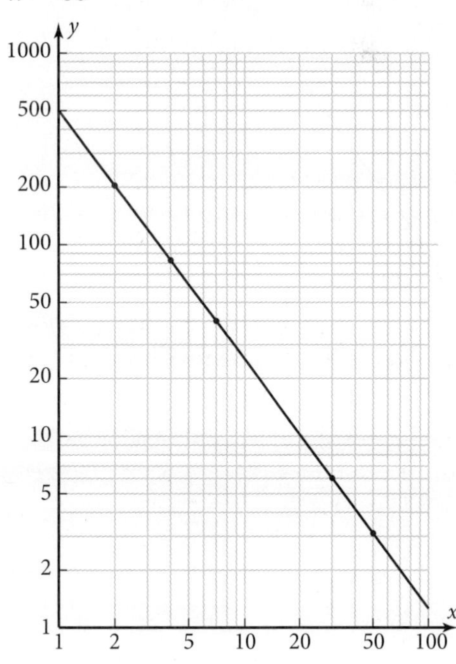

11. $x = 9$

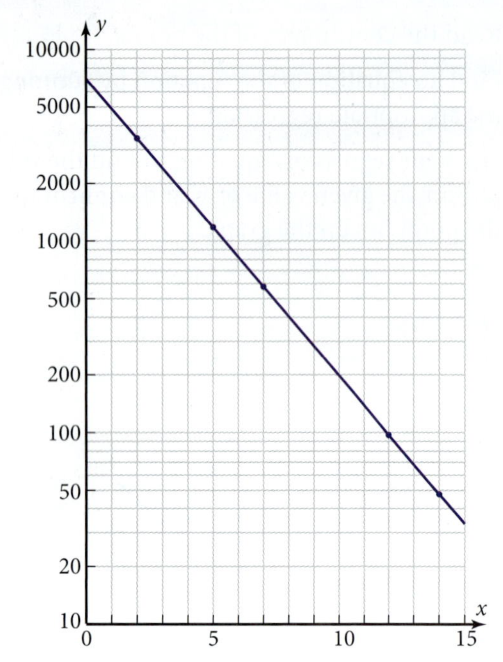

12. $x = 8$

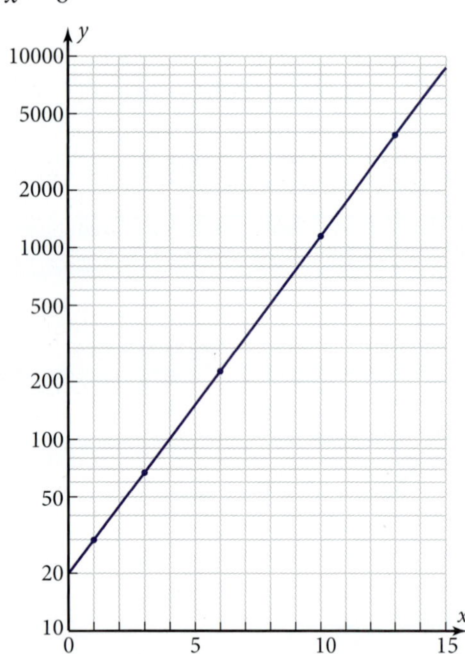

13. $x = 90$

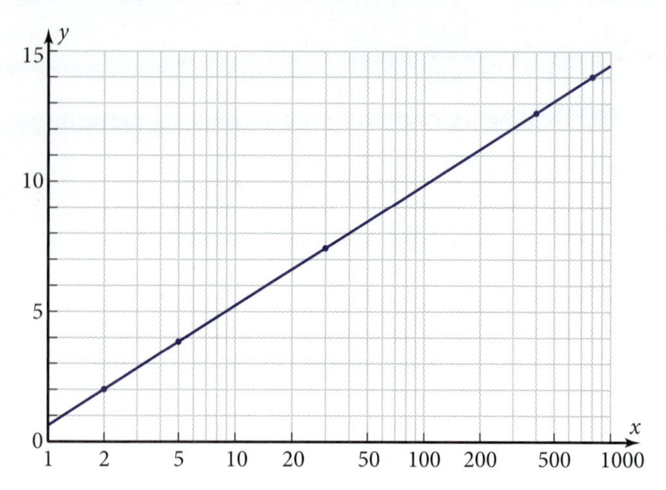

14. $x = 60$

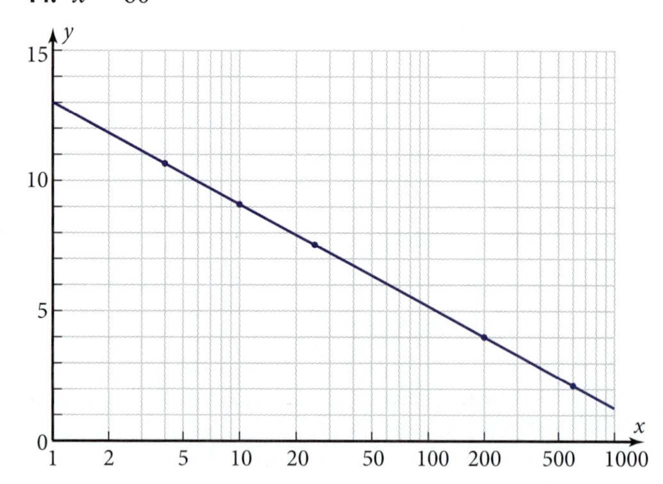

For the exponential functions in Problems 15 and 16, on your grapher plot log y as a function of x. Sketch the results.

15. $y = 5 \cdot 3^x$ **16.** $y = 20 \cdot 0.8^x$

For the power functions in Problems 17 and 18, on your grapher plot log y as a function of log x. You can use parametric mode, as explained in the note following Example 1. Sketch the results.

17. $y = 90x^{-2}$ **18.** $y = 2x^3$

19. *River Basin Problem:* After land masses were formed, rivers formed to drain various regions. The length of the river draining a particular region may be assumed to depend on the area of the region (river basin) drained. Figure 3-4k shows a scatter plot of the data.

River	Drainage Area (10^3 km^2)	Length (km)
Amazon	6915	6296
Yenisey	2580	4438
Congo	3680	4371
Mackenzie	1790	4241
Mississppi	2980	3778
Volga	1380	3687
St. Lawrence	1030	3058
Rio Grande	570	3034
Essequibo	155	970
Mamberamo	79.44	650
Nashua	1.954	90

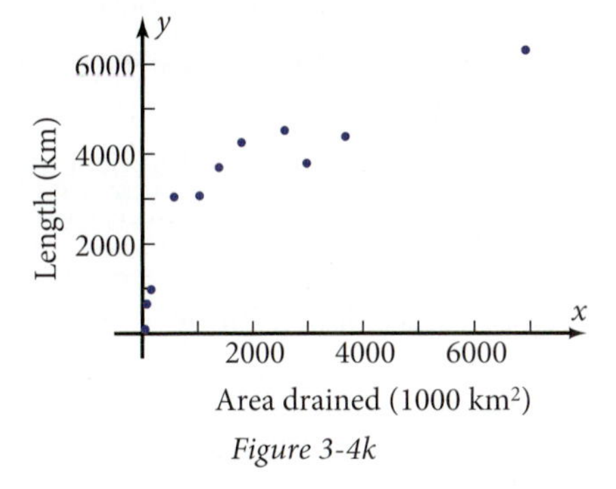

Figure 3-4k

a. The data follow a path that is increasing and concave down, as is typical of increasing logarithmic functions and of power functions with a base between 0 and 1. Explain why a power function would be a more reasonable mathematical model in this case.

b. Enter the data into two lists on your grapher. (See the data sets at **www.keymath.com/precalc.**) By power regression, find the equation of $\hat{y}$, the best-fitting function. Record r^2, the coefficient of determination. Plot the power function and the scatter plot on the same screen, and sketch the result on a copy of Figure 3-4k.

c. Make two more lists containing the logarithms of the area and length. Using parametric mode, plot $\log \hat{y}$ versus $\log x$ on the same screen as a scatter plot of the transformed data. Sketch the result. How does the fit of the linearized function to the transformed data compare to the fit of the power function to the original data?

d. Run a linear regression on $\log y$ as a function of $\log x$. Show that the value you get for r^2 is equal to the value of r^2 you recorded in part b. What does this fact suggest about the way your grapher does power regression?

20. *Water Use Problem:* The booklet *Environmental Mathematics in the Classroom* (MAA, 2003) gives the values in the table for estimated daily water usage (billions of gallons) in the United States for various years starting in 1900. Figure 3-4l shows a scatter plot of these data.

Year	Water Usage (billion gal/day)
1900	40.2
1910	66.4
1920	91.5
1930	110.5
1940	136.4
1950	202.7
1960	322.9
1965	369.6
1970	327.3
1975	420.0
1980	450.0
1985	400.0

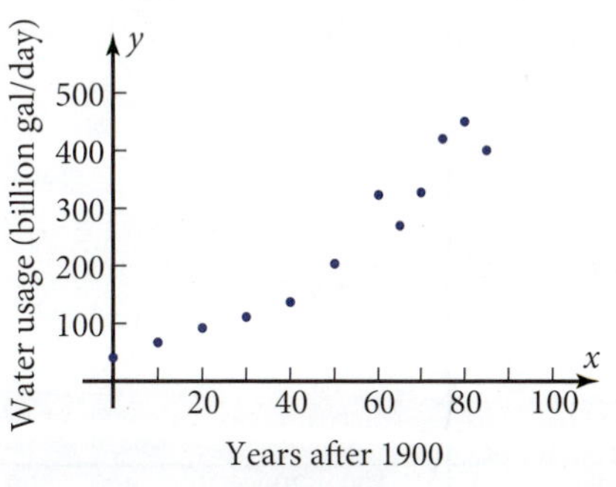

Figure 3-4l

a. The scatter plot shows a pattern that is increasing and concave up, suggesting that either a power function or an exponential function might be a reasonable model. Explain why an exponential function would be more appropriate in this case.

b. By exponential regression, find the equation of $\hat{y}$, the best-fitting function. (See the data sets at **www.keymath.com/precalc.**) Record the value of r^2, the coefficient of determination. Plot the exponential function and the scatter plot on the same screen, and sketch the result on a copy of Figure 3-4l.

c. Plot the data on add–multiply semilog graph paper. Plot the graph of $\hat{y}$ by calculating the values at $x = 0$ and $x = 100$, plotting the points, and connecting them with a straight line. Does the line fit the data reasonably well?

d. Use your regression equation to predict the daily water usage for this year. Did you use extrapolation or interpolation? Based on any patterns you see in the data, how confident are you that your prediction is realistic?

e. Enter another list into your grapher containing the logarithm of the water usage. Run a linear regression on log y as a function of x. How does the value of r^2 compare to the value you recorded for the exponential regression in part b? What does this fact suggest about the way your grapher does exponential regression?

21. *Temperature-Depth Problem:* Suppose that the temperatures listed in the table have been measured at various depths below Earth's surface.

x (°C)	y (km)
30	1.0
60	2.4
100	4.0
200	5.9
500	8.6

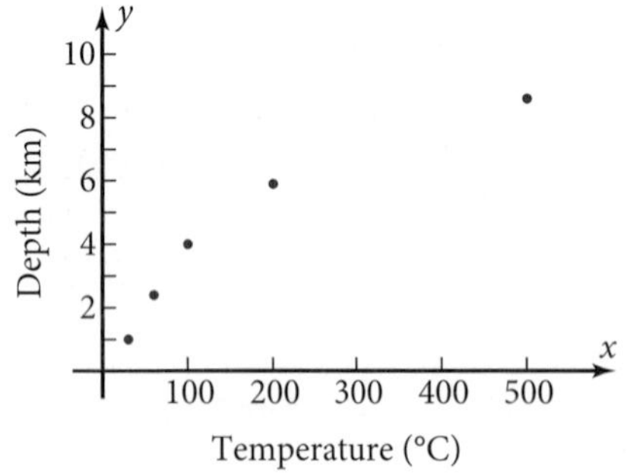

Figure 3-4m

a. The scatter plot in Figure 3-4m shows that the data follow a path that is increasing and concave down, indicating that a logarithmic function or a power function with base between 0 and 1 might fit the data. Explain why a logarithmic function would be more appropriate.

b. By logarithmic regression, find the equation for $\hat{y}$, the best-fitting function. Record r^2, the coefficient of determination. Plot $\hat{y}$ and the data on your grapher, and sketch the result on a copy of Figure 3-4m. Does the graph fit the data reasonably well?

c. Plot the data on multiply–add semilog graph paper. Plot a straight-line graph through the data points and use it to estimate the depth at which the temperature would be 300°C. How close is this graphical value to the value of $\hat{y}$ given by the equation you found in part b?

d. Explain why $\hat{y}$ can be considered to be a linear function of $\ln x$ in the equation you found in part b.

e. Enter a list of values of $\ln x$ into your grapher. Run a linear regression on y as a function of $\ln x$. How does the value of r^2 compare to the value you found by logarithmic regression in part b? What does this fact suggest about the way your grapher does logarithmic regression?

22. *Population Growth Rate Problem:* In the past few censuses, the town of Scorpion Gulch recorded the populations and growth rates listed in the table. Figure 3-4n shows a scatter plot of the data.

Year	Population	Growth Rate (people/yr)
1940	300	19
1950	540	30
1960	950	45
1970	1400	57
1980	2040	53
1990	2400	40
2000	2700	28

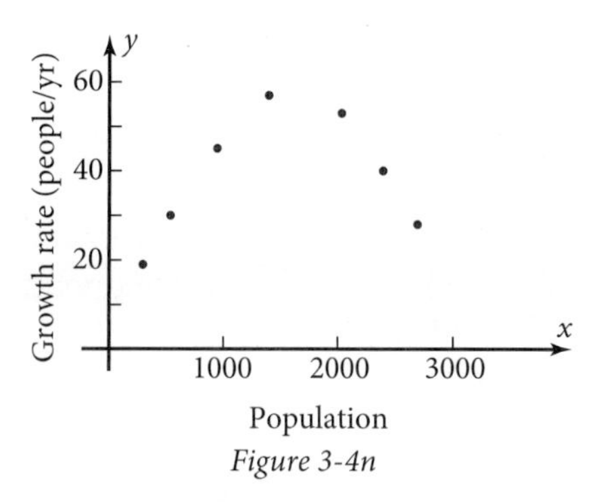

Figure 3-4n

a. The population growth rate follows a path that is increasing, then decreasing, and concave down as the population grows. Why might the rate of population growth slow as the population of the town grows?

b. Run a quadratic regression on the data to find the equation of $\hat{y}$, the best-fitting function. Record the value of R^2. Plot $\hat{y}$ and the data on the same screen, and sketch the result on a copy of Figure 3-4n. Explain what it means that the $\hat{y}$-intercept is so close to zero.

c. Calculate the mean of the growth rate and store it as M, without rounding. Enter into two new lists the squares of the deviations of y from M and the squares of the residual deviations of y from $\hat{y}$. By summing these new lists, find SS_{dev} and SS_{res}. Then use these sums to calculate the coefficient of determination, R^2. Does your result agree with the value of R^2 in part b? Explain why uppercase R is used in this case rather than lowercase r.

23. *Proof Problems:*

a. Prove that for the exponential function $y = 5 \cdot 3^x$, $\log y$ is a linear function of x.

b. Prove in general that for the exponential function $y = ab^x$, $\log y$ is a linear function of x.

c. Prove that for the power function $y = 2x^3$, $\log y$ is a linear function of $\log x$.

d. Prove in general that for the power function $y = ax^b$, $\log y$ is a linear function of $\log x$.

e. Why can you conclude that for a logarithmic function, y is a linear function of $\log x$?

24. *Height-Weight Historical Problem:* Before calculators were available to do regression analysis efficiently, log-log and semilog graph paper were used to help determine what type of function fits a given set of data. The type of paper that gave a straight-line graph indicated which function to use. Suppose that the average weights in the table have been recorded for humans of various heights.

Height (in.)	Weight (lb)
20	8.5
30	21.5
40	41.6
50	69.5
60	105.7

a. Plot the data on add–multiply semilog graph paper. Plot it again on log-log graph paper. Which graph seems to be more nearly a straight line?

b. Based on your answer to part a, which type of function fits the data more closely, exponential or power?

c. Find algebraically the particular equation of the function in part b. Use the first and last data points to find the constants in the equation. (This was the way particular equations were found in the days before calculators had regression analysis built in.)

d. Run an exponential regression and a power regression on the given data. Does the regression analysis confirm your conclusion and equation? How do you tell?

e. Predict the weight of a 90-in.-tall giant using your equation in part c, and again using the regression equation in part d. How closely do the two predictions agree?

3-5 Residual Plots and Mathematical Models

In Section 3-3, you learned how to do regression analysis to fit functions to data by choosing the function that had the shape of the graph and the correct endpoint behavior. In this section you will obtain further evidence of whether a function fits well by making sure that a plot of the residuals follows no particular pattern.

Objective Find graphical evidence of how well a given function fits a set of data by plotting and analyzing the residuals.

In this exploration you will use a residual plot to make conclusions about a cooling cup of coffee.

EXPLORATION 3-5: Coffee Data Residual Plot

You pour a cup of coffee. Rather than drinking it right away, you let it cool, measuring its temperature at 1-minute intervals. This table and graph show y, the number of degrees above room temperature, as a function of x minutes since you poured the coffee.

x (min)	y (°C)	x (min)	y (°C)
2	51.5	7	27.4
3	45.5	8	23.9
4	40.2	9	20.6
5	35.5	10	17.5
6	31.3	11	14.5

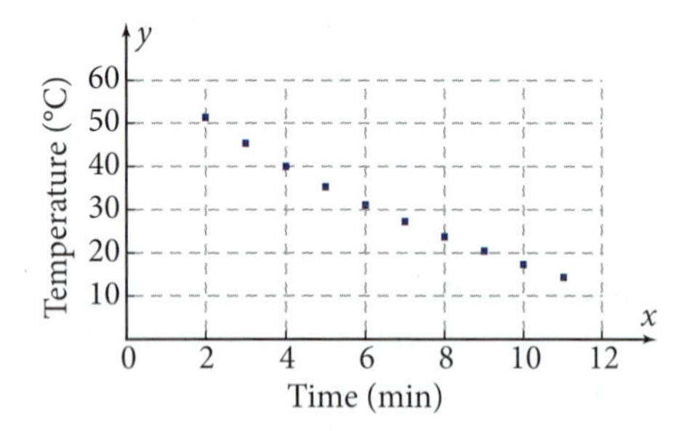

1. Based on the shape and concavity of the graph, explain why both an exponential function and a power function would be reasonable mathematical models for temperature as a function of time.
2. Explain why both exponential and power functions have reasonable endpoint behavior for large values of x but only the exponential function behaves correctly at the left end of the domain.
3. Confirm by exponential and power regression that an exponential function fits the data better than a power function. How do you decide?
4. Plot the exponential function you found in Problem 3 and the data on the same screen. Sketch the result on a copy of the given figure.
5. Calculate the residuals for each point using the exponential function you found in Problem 3. Make a residual plot using a window with $0 \leq x \leq 12$, and a suitable range for the residuals. Sketch the result.
6. What does the fact that the residual plot shows a pattern indicate about how well the exponential function fits the data? What could you conclude if a residual plot showed no discernible pattern?
7. What did you learn as a result of doing this exploration that you did not know before?

Age (wk)	Height (in.)
3	5
4	4
5	7
6	6
7	11
8	11
9	15
10	20
11	21
12	24

A biology class plants one bean each week for 10 weeks. Three weeks after the last bean is planted, the plants have the heights shown in the table and in Figure 3-5a. As you can tell, there is a definite upward trend, but it is not absolutely clear whether the best-fitting function is curved or straight.

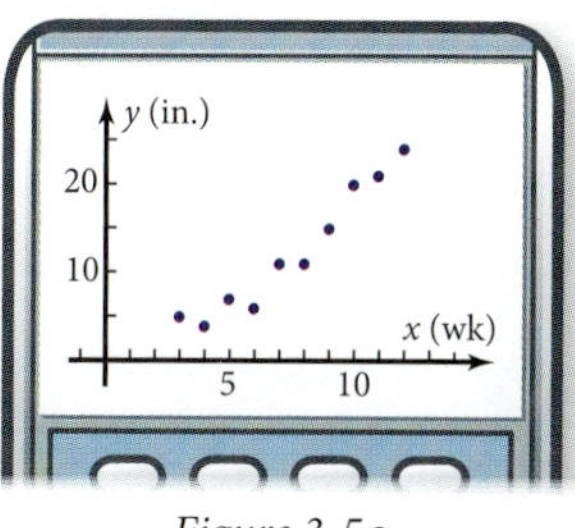

Figure 3-5a

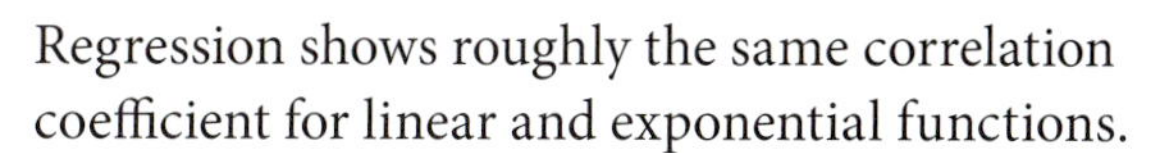

Regression shows roughly the same correlation coefficient for linear and exponential functions.

Linear:

$\hat{y} = 2.3151\ldots x - 4.9636\ldots$

$r = 0.9675\ldots$

Exponential:

$\hat{y} = 2.2539\ldots(1.2267\ldots)^x$

$r = 0.9690\ldots$

Both functions fit reasonably well, as shown in Figures 3-5b and 3-5c.

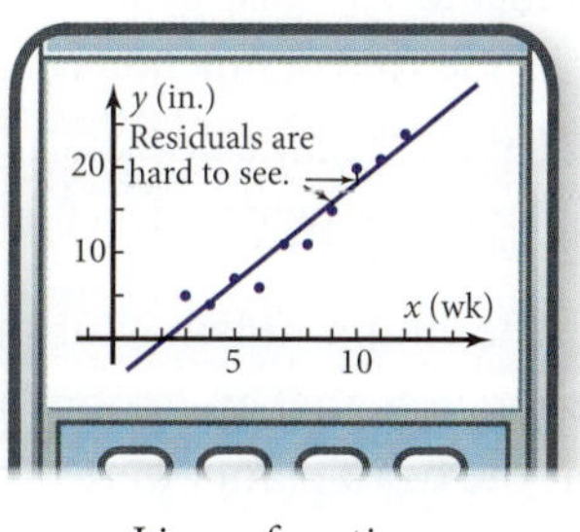

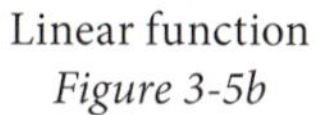
Linear function
Figure 3-5b

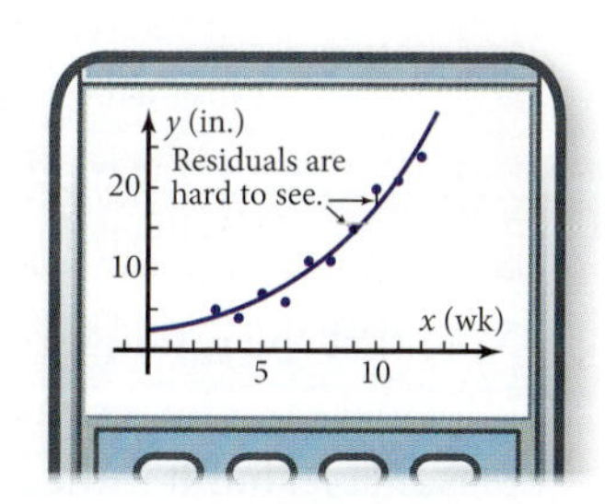

Exponential function
Figure 3-5c

The residuals, $y - \hat{y}$, in these figures are hard to see because they are relatively small. You can see the residuals more easily by making a **residual plot,** a scatter plot of the residuals. Figures 3-5d and 3-5e show residual plots for the linear function and the exponential function, respectively. To make these plots, enter the regression equation for $\hat{y}$ into your grapher as $f_1(x)$. Assuming that the data are in lists L_1 and L_2, you can calculate the residuals like this:

$L_2 - f_1(L_1)$ The expression $f_1(L_1)$ means the values of the function in $f_1(x)$ at the values of x in L_1.

Most graphers have a zoom feature that will set the window automatically to fit the statistical data.

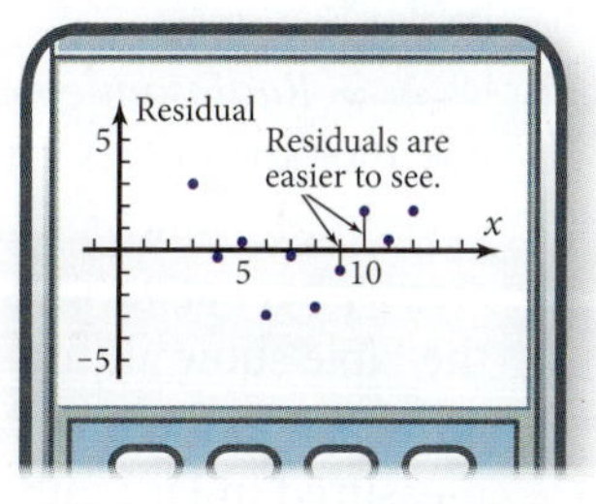

Linear function residuals
Figure 3-5d

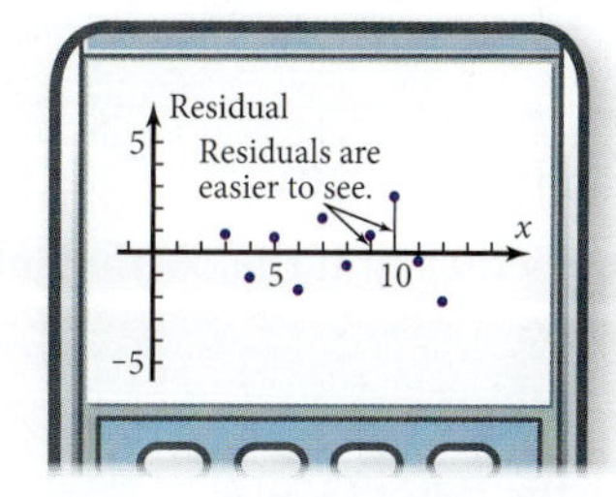

Exponential function residuals
Figure 3-5e

The residuals for the linear function seem to form a pattern: high at both ends and low in the middle. A pattern in the residuals suggests that the data may be nonlinear. The residuals for the exponential function are more random, following no discernible pattern. So the residuals for the exponential function are more likely caused by random variations in the data, such as different growth rates for different bean plants.

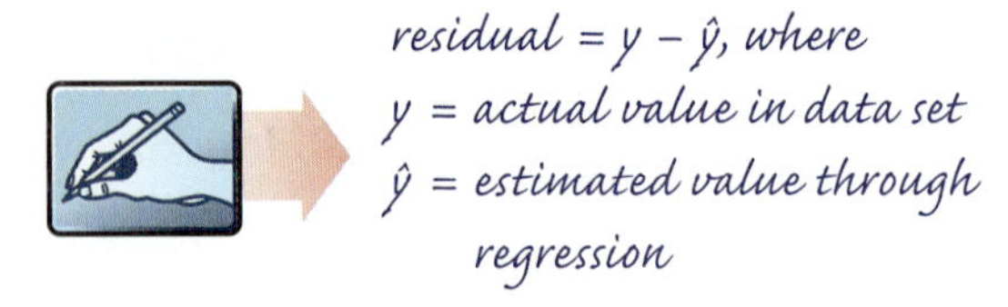

CONCLUSION: Residual Plot Interpretation

If the residual plot follows a regular pattern, then there is a behavior that is not accounted for by the kind of function chosen.

If the residual plot has no identifiable pattern, then the regression equation is likely to account for all but random fluctuations in the data.

Note that the residual plot and the endpoint behavior might give you conflicting information. The residual plot suggests that an exponential function is more reasonable, but it indicates that the bean plants were already sprouted at time $x = 0$, when they were planted.

Problem Set 3-5

Reading Analysis

From what you have read in this section, what do you consider to be the main idea? Why is it that *no* pattern in a residual plot is "good news" and a *definite* pattern is "bad news"? How can a residual plot give you information about how well a function fits a set of data that you cannot get by simply considering the correlation coefficient?

Quick Review

Q1. Find r^2 if $SS_{\text{res}} = 2$ and $SS_{\text{dev}} = 10$.

Q2. Find the correlation coefficient if the coefficient of determination is 0.9.

Q3. Find $\hat{y}$ for the point (3, 9) if $\hat{y} = 2x + 7$.

Q4. Find the residual for the point (3, 9) if $\hat{y} = 2x + 7$.

Q5. Find the deviation for the point (3, 9) if $\bar{x}$ is 5 and $\bar{y}$ is 6.

Q6. Find the $\hat{y}$-intercept if $\hat{y} = 2x + 7$.

Q7. Find the x-coordinate of the vertex of this parabola: $y = 3x^2 + 24x + 71$.

Q8. $2 \log 6 = \log$ __?__

Q9. What transformation of the parent function $y = f(x)$ is the function $y = f(2(x - 3))$?

Q10. Without a calculator, find log 1000.

---

1. *Radiosonde Air Pressure Problem:* Each day meteorologists release weather balloons called radiosondes to measure data about the atmosphere at various altitudes. Figure 3-5f and the table show altitude in meters, pressure in millibars, and temperature in degrees Celsius measured in December 1991 by a radiosonde in Coffeeville, Kansas.

Altitude (m)	Pressure (mbar)	Temperature (°C)
400	965	11.4
620	940	12
800	920	11.9
1220	875	11.7
1611	835	11
2018	795	8.6
2500	750	9
3009	705	6.8
4051	620	0.9
5000	550	−6.8
6048	480	−13
7052	420	−19.8
8075	365	−28.5
9005	320	−35.2
10042	275	−43
11086	235	−48.6
12151	200	−46.8
13024	175	−52.6
14004	150	−58.3
15134	125	−63.1
16197	105	−66.5
17120	90	−71.1
18195	75	−70.7

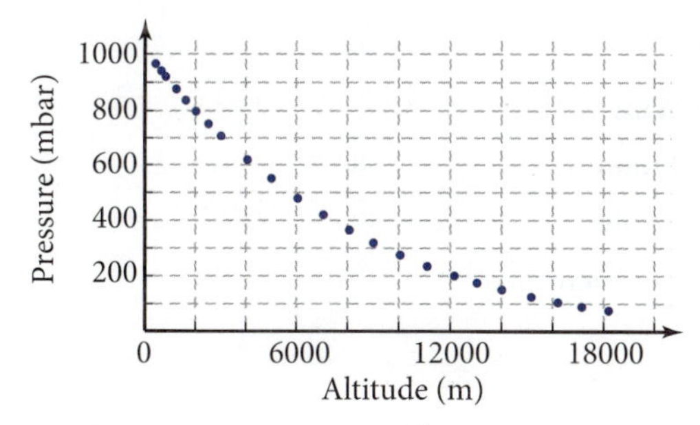

Figure 3-5f

a. From a scatter plot of pressure versus altitude, explain why an exponential function would be expected to fit the data well. Find the best-fitting exponential function. (See the data sets at **www.keymath.com/precalc.**) Record the correlation coefficient and enter the equation into your grapher. Plot the equation on the same screen as the scatter plot. Does the graph fit the points well?

b. Enter another list into your grapher for the residual of each point. Then make a residual plot and sketch the result. Do the residuals follow a definite pattern, or are they randomly scattered? What real-world phenomenon could account for the deviations?

c. Based on your residual plot in part b, could the exponential function be used to make predictions of pressure correct to the nearest millibar, if this accuracy were necessary?

2. *Hot Water Problem:* Tim put some water in a saucepan and then turned the heat on high. Figure 3-5g and the table on page 162 show the temperature of the water in degrees Celsius at various times, in seconds, since he turned on the heat.

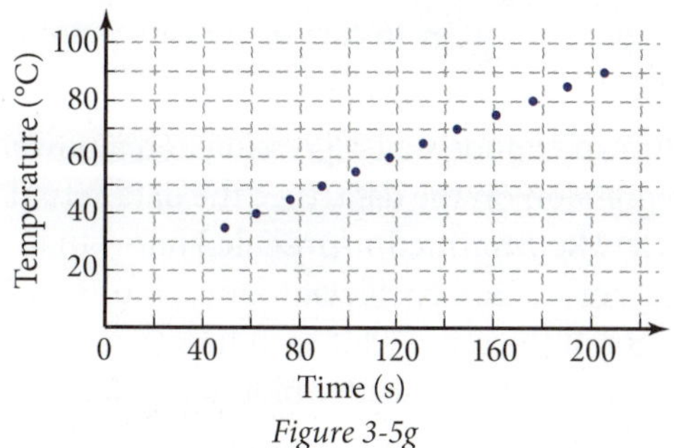

Figure 3-5g

Times (s)	Temperature (°C)
49	35
62	40
76	45
89	50
103	55
117	60
131	65
145	70
161	75
176	80
190	85
205	90

a. Run a linear regression on temperature as a function of time. (See the data sets at **www.keymath.com/precalc.**) Record the correlation coefficient. Plot the regression equation and a scatter plot of the data on the same screen. Does the fit of the line to the data confirm the correlation coefficient is extremely close to 1?

b. Make a residual plot. Sketch the result. How does the residual plot tell you that there is something in the heating of the water that the linear function does not take into account? What real-world reason do you suppose causes this slight nonlinearity?

c. Based on the linear model, at what time would you expect the water to reach 100°C and boil? Based on your observations in part b, would you expect the water to boil sooner than this or later than this? Explain.

3. *Gas Mileage Problem:* Figure 3-5h and the table show the gasoline consumption rate (in miles per gallon) of 16 cars of various weights (in pounds).

a. Run an exponential regression and a power regression on the data. (See the data sets at **www.keymath.com/precalc.**) Show that the correlation coefficient for each function is about the same. Plot each function on the same screen as a scatter plot of the data. Do both functions seem to fit the data well?

b. Make two residual plots, one for the exponential function and one for the power function. Does either residual plot seem to follow a pattern? How do you interpret this result in terms of which function fits the data more closely?

c. Extrapolate using both the exponential function and the power function to predict the gas mileage for a super-compact car weighing only 500 lb. Based on your results, which model—exponential regression or power regression—has a more reasonable endpoint behavior for light cars? Is there a significant difference in endpoint behavior if you extrapolate for very heavy cars?

Model	Weight (lb)	Gasoline Consumption (mi/gal)
Ford Aspire	2140	43
Honda Civic del Sol	2410	36
Honda Civic	2540	34
Ford Escort	2565	34
Honda Prelude	2865	30
Ford Probe	2900	28
Honda Accord	3050	31
BMW 3-series	3250	28
Ford Taurus	3345	25
Ford Mustang	3450	22
Ford Taurus SHO	3545	24
BMW 5-series	3675	23
Lincoln Mark VIII	3810	22
Cadillac Eldorado	3840	19
Cadillac Seville	3935	20
Ford Crown Victoria	4010	22

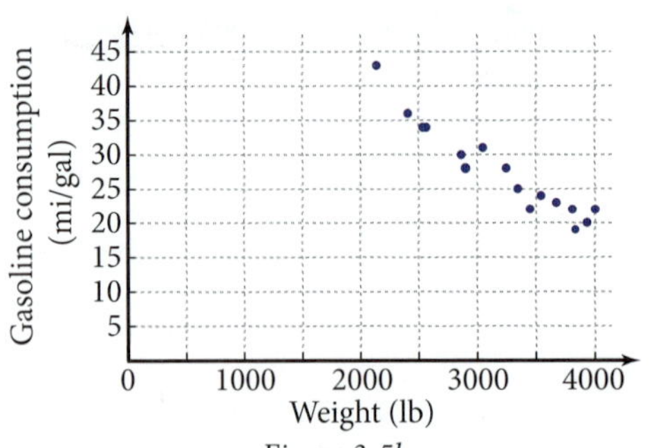

Figure 3-5h

4. *Weed Competition Problem:* In the report *Applications of Mathematics: A Nationwide Survey,* A. C. Madgett reports on the competition of wild oats (a weed) with various crops. The more wild oats that are growing with a crop, the greater the loss in yield of that crop. Figure 3-5i and the table show data from that report.

Wild Oat Plants/m^2	Percent Loss of Wheat Crop
1	3
5	8
10	10
20	17
50	25
100	34
150	40
200	48

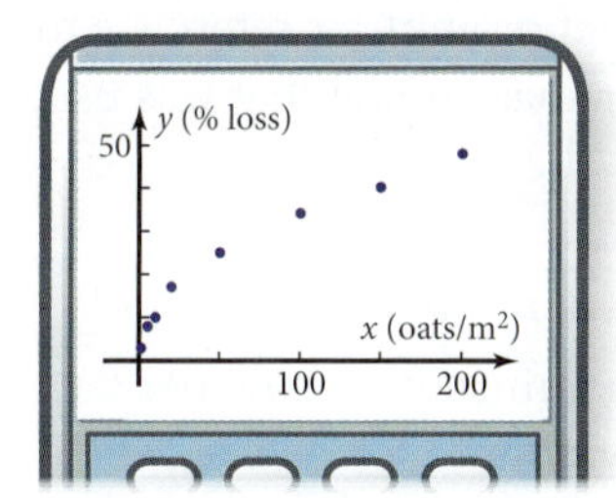

Figure 3-5i

a. From the scatter plot, identify what type of function is most likely to fit the data. By regression, find the particular equation of this kind of function. Plot it and the scatter plot on the same screen. Sketch the result.

b. Make a residual plot. Sketch the result. What information do you get about the way your selected function fits the data from the residual plot? From the correlation coefficient?

c. Based on your mathematical model, what percent crop loss would be expected if there were 500 wild oat plants per square meter? Do you find this number by extrapolation or by interpolation?

d. How many wild oat plants per square meter do you predict it would take to choke out the wheat crop completely? Do you find this number by extrapolation or by interpolation?

5. WEB *Population Problem:* Figure 3-5j and the table show the population of the United States for various years as reported by the U.S. Census Bureau. The variable x is the number of years elapsed since 1930.

Year	x (yr)	Population (million)
1940	10	132.1
1950	20	151.3
1960	30	179.3
1970	40	203.3
1980	50	226.5
1990	60	248.7
2000	70	281.7

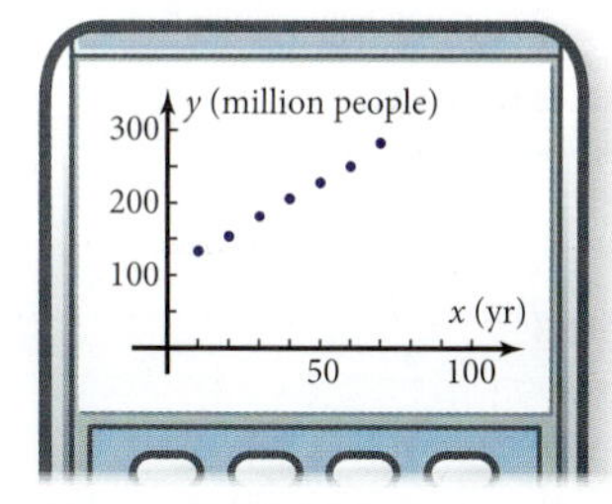

Figure 3-5j

a. Explain why a linear function and an untranslated power function would not have the correct endpoint behavior for dates long before 1930. Explain why an exponential function of the form $y = ab^x$ would not have reasonable endpoint behavior for dates long after the present.

b. Find the particular equation of the best-fitting logistic function. Plot it and the data on the same screen. Use a window that includes dates back to 1830 and forward to 2130. Sketch the result.

c. What does the logistic model predict for the outcome of the 2010 census? How does this prediction compare with the actual 2010 census found on the Internet or from some other source? What does the logistic model predict for the maximum sustainable population of the United States?

d. Suppose that the result of the 2000 census had been 282.4 million, or 1 million higher. Run another logistic regression using the modified data. What difference does a change of 1 million in one data point have on the maximum sustainable population?

6. WEB *Wind Chill Problem:* When the wind is blowing on a cold day, the temperature seems to be colder than it really is. For any given actual temperature with no wind, the equivalent temperature due to wind chill is a function of wind speed. Figure 3-5k and the table show data published by the National Oceanographic and Atmospheric Administration in 2000.

Wind Speed (mi/h)	Equivalent Temperature (°F)
0	35
5	32
10	22
15	16
20	12
25	8
30	6
35	4
40	3

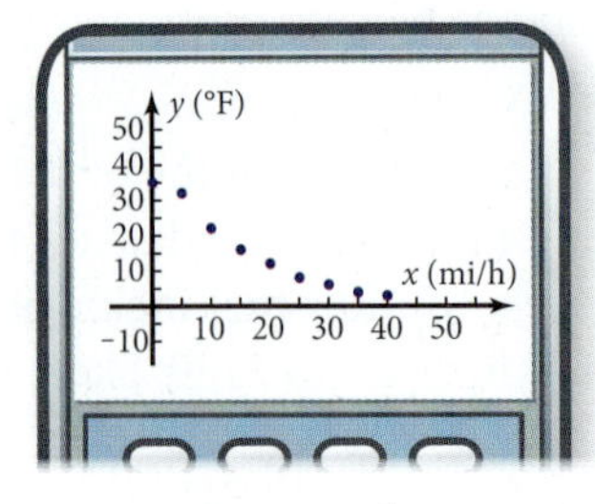

Figure 3-5k

a. Ignoring the data point at 0 mi/h, run a logarithmic regression to find the best-fitting logarithmic function. Plot it and the data on the same screen. Sketch the result.

b. At what wind speed does the logarithmic function predict that the equivalent temperature would be 0°F? Do you find this number by extrapolation or by interpolation? Explain why an untranslated exponential function would predict that the equivalent temperature would never be 0°F. Explain why the logarithmic function does not have the correct endpoint behavior at the lower end of the domain.

c. Make a residual plot for the logarithmic function. Sketch the result. Do the residuals follow a pattern? What does this fact tell you about how well the logarithmic function fits the data?

d. Check the Internet or another reference to find the mathematical model that was used in 2000 and why the model has changed since then.

7. *Calorie Consumption Problem:* The number of calories an animal must consume per day increases with its body mass. However, the number of calories per kilogram of body mass decreases with mass because larger animals have a lower surface-to-volume ratio. In *Studies in Mathematics, Volume XX*, Max Bell reports the data in the table for various mammals.

Animal	Mass (kg)	Consumption-to-Mass Ratio (cal/kg)
Guinea pig	0.7	223
Rabbit	2	58
Human	70	33
Horse	600	22
Elephant	4000	13

a. Because the mass values cover such a wide range, a scatter plot of the given data would offer little useful information. Transform the data by inserting two new columns for log of mass and log of consumption-to-mass ratio, as you did in Section 3-4. Then run a linear regression on the transformed data. Record the value of r^2. Plot the linear function and the transformed data on the same screen, and sketch the result. Do the transformed data seem to follow a linear pattern?

b. By transforming the equation in part a, show that calorie consumption is a power function of mass. Run a power regression on the original data to show that you get the same power function and the same value of r^2.

c. The smallest mammal is the shrew. Predict the calorie consumption per kilogram for a 2-g shrew. (The need to eat so much compared to body mass probably explains why shrews are so mean!)

d. In the referenced article, Bell reports that a 150,000-kg whale consumes about 1.7 cal/kg. What does the power function predict for the number of calories per kilogram if you extrapolate it to the mass of a whale? By what percentage does the answer differ from the given 1.7 cal/kg? Can you think of a reason why the predicted value is so far from the reported value?

8. *Mile Run Record Times:* The table shows that the world record time for the mile run has been decreasing from 1913 through 1993.

Year	Time (s)	Year	Time (s)
1913	254.4	1958	234.5
1915	252.6	1962	234.4
1923	250.4	1964	234.1
1931	249.2	1965	233.6
1933	247.6	1966	231.3
1934	246.8	1967	231.1
1937	246.4	1975	231.0
1942	246.2	1975	229.4
1942	244.6	1979	228.95
1943	242.6	1980	228.80
1944	241.6	1981	228.53
1945	241.4	1981	228.40
1954	239.4	1981	227.33
1954	238.0	1985	226.32
1957	237.2	1993	224.39

a. Make a scatter plot of the data. (See the data sets at **www.keymath.com/precalc.**) Use the last two digits of the year for x and the number of seconds for y. A window with $220 \leq y \leq 260$ will allow you to view the entire data set on the screen. Plot the best-fitting linear function on the same screen. Sketch the result.

b. Find the best-fitting exponential function for the data. Plot it on the same screen as in part a. Can you see any difference between the exponential graph and the linear graph?

c. In 1999, Hicham El Guerrouj of Morocco set a world record of 223.13 s for the mile run. Which function comes closer to predicting this result, the linear or the exponential?

d. Make a residual plot for the linear function. Describe any patterns you see.

e. In 1954, Roger Bannister of Great Britain "broke" the 4-min mile. Until that time, it had been thought that 4 min (240 s) was the quickest a human being could run a mile. What do the data and residual plot suggest happened in the years just before and just after Bannister's feat?

9. *Meatball Problem:* In *Applications of Mathematics: A Nationwide Survey,* A. C. Madgett reports the moisture content of deep-fried meatballs as a function of how long they have been cooked. The data in the table were gathered to determine the effectiveness of adding whey to hamburger meat to make it retain more moisture during cooking, thereby improving its quality.

Cooking Time (min)	Percent Moisture Content	
	No whey	Whey
6	46.4	48.8
8	45.3	—
10	40.6	44.2
12	36.9	41.8
14	33.5	39.2

a. On the same screen, make a scatter plot of both sets of data. Use different symbols for the different data sets. Find the best-fitting linear function for each set of data, and plot the two functions on the same screen as the scatter plots. Sketch the results.

b. Use the linear model to predict the moisture content for the missing data point in the whey data. Which process do you use, extrapolation or interpolation? Explain.

c. Use the linear model to predict how long the meatballs could be cooked without dropping below 30% moisture content, first if they have whey and then if they have no whey. Which process did you use, extrapolation or interpolation? Explain.

d. According to your mathematical models, do the two kinds of meatballs have the same moisture content before they are cooked? Give numbers to support your answer.

10. *Television Set Problem:* The table gives prices of a popular brand of television set for various sizes of screen.

Diagonal (in.)	Price ($)
5	220
12	190
17	230
27	350
36	500

a. Make a scatter plot of the data. Based on the graph, explain why a quadratic function is a more reasonable mathematical model than a linear, a logarithmic, an untranslated power, or an exponential function. Confirm your graphical analysis numerically by finding R^2 or r^2, the coefficient of determination, for each of the five types of functions.

b. Plot the best-fitting quadratic function and the data on the same screen. Sketch the result.

c. If the manufacturer made a 21-in. model and a 50-in. model, how much would you expect to pay for each? For which prediction did you use extrapolation, and for which did you use interpolation?

d. Make a residual plot for the quadratic function. Sketch the result. On the residual plot, indicate which sizes of television sets are slightly overpriced and which are underpriced.

e. Why do prices go *up* for very small television sets?

11. *Journal Problem:* Make an entry in your journal explaining how two different types of regression functions can have the same correlation coefficient although one function is preferred over the other. Show how a residual plot and endpoint behavior can sometimes give conflicting information about which function to use.

3-6 Chapter Review and Test

In this chapter you've learned how to fit various types of functions to data. In choosing the type of function, you considered

- The shape of the scatter plot
- The behavior of the chosen type of function at the endpoints
- Whether the data form a linear pattern on semilog or log-log graph paper
- Whether the residual plot follows a pattern
- The correlation coefficient

You learned how to calculate SS_{dev}, the sum of the squares of the deviations of the y-values from the horizontal line $y = \bar{y}$, the mean of the y-values. Rotating the line to fit the pattern followed by the data reduces the deviations. The linear regression line reduces the deviations enough so that SS_{res}, the sum of the squares of the residuals (residual deviations), is a minimum. The coefficient of determination and the correlation coefficient are calculated from SS_{dev} and SS_{res}. You can use other kinds of regression to fit different types of functions to data, sometimes using logarithms to linearize the data.

Review Problems

R0. Update your journal with what you have learned in this chapter. Include how your knowledge of the shapes of various function graphs guides you in selecting the type of function you'll choose as a model and how regression analysis lets you find the particular equation of the selected type of function. Mention the ways you have of deciding how well the selected function fits, both within the data and possibly beyond the data. Also, explain how the correlation coefficient and the coefficient of determination are calculated from SS_{dev} and SS_{res}, and how taking logarithms of data can linearize the data.

R1. Figure 3-6a shows this set of points and the graph of $\hat{y} = 1.6x + 0.9$.

a. By linear regression, confirm that $\hat{y} = 1.6x + 0.9$ is the correct regression equation.

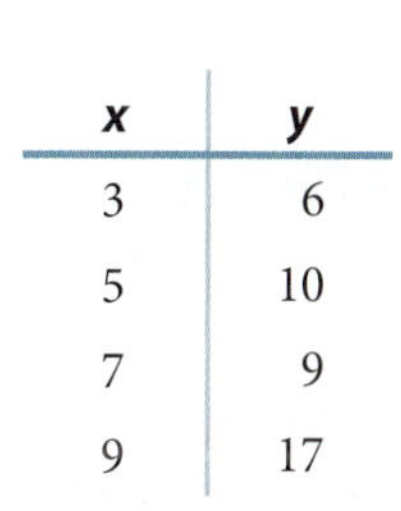

x	y
3	6
5	10
7	9
9	17

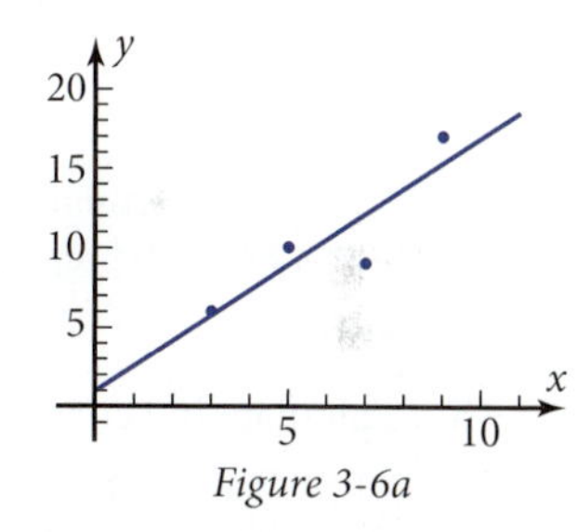

Figure 3-6a

b. Copy the table and add three new columns, one for the values of $\hat{y}$; one for the values of the residuals, $(y - \hat{y})$; and one for the squares of the residuals, $(y - \hat{y})^2$.

c. By calculation, show that the sum of the squares of the residuals, SS_{res}, is 13.8.

d. Show that a slight change in the slope (1.6) or the y-intercept (0.9) leads to a higher value of SS_{res}. For instance, you might try $\hat{y} = 1.5x + 1.0$.

R2. a. Figure 3-6b shows the points from Figure 3-6a with a dashed line at $y = \bar{y}$, the mean of the y-values. Calculate $\bar{y}$ to show that the figure is correct.

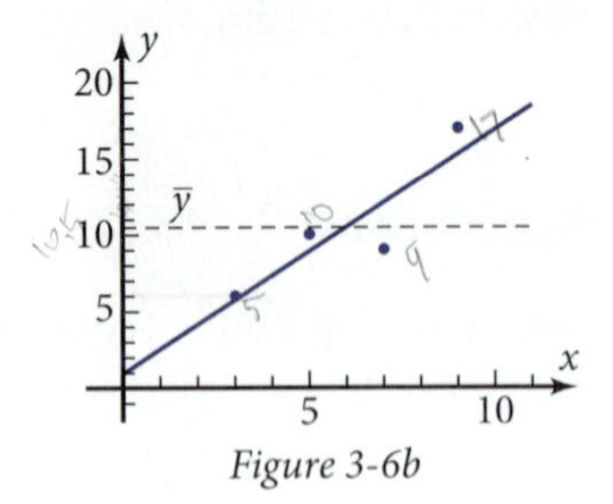

Figure 3-6b

b. Calculate $\bar{x}$. Show algebraically that the point $(\bar{x}, \bar{y})$ is on the regression line $\hat{y} = 1.6x + 0.9$ from Problem R1.

c. On a copy of Figure 3-6b, sketch both the deviation from the mean for the point where x equals 9 and the residual (that is, the residual deviation) for this point.

d. Calculate SS_{dev}, the sum of the squares of the deviations. Why do you suppose that SS_{dev} is so much larger than SS_{res}?

e. Calculate the coefficient of determination,

$$\frac{SS_{dev} - SS_{res}}{SS_{dev}}$$

Then run a linear regression on the points in Problem R1, showing that the fraction equals r^2.

f. Calculate the correlation coefficient from r^2. Show that it agrees with the value from the regression run on your grapher. Why must you choose the *positive* square root?

g. Figures 3-6c through 3-6f show the ellipses in which the clouds of data points lie. For each graph, state whether the correlation coefficient is positive or negative and whether it is closer to 0 or closer to 1 or -1.

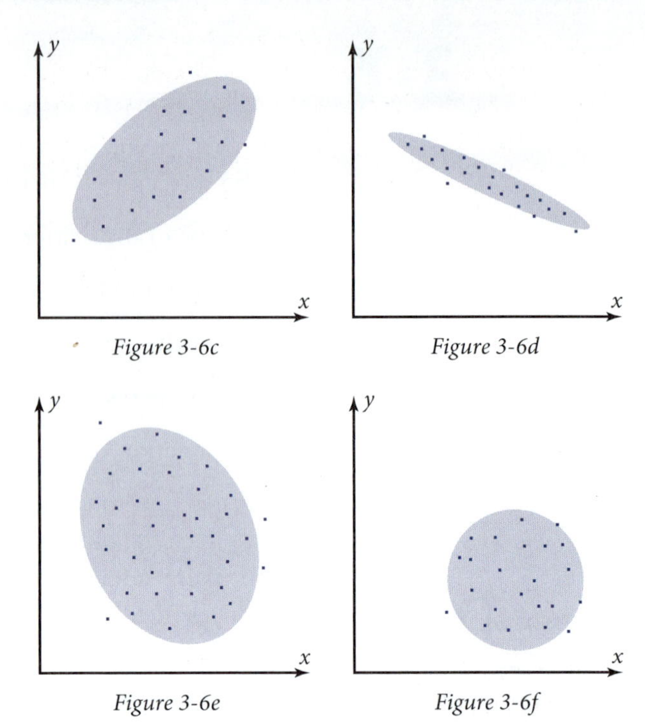

Figure 3-6c *Figure 3-6d*

Figure 3-6e *Figure 3-6f*

R3. *Learning Curve Problem:* When an article is first manufactured, the cost of making each item is relatively high. As the manufacturer gains experience, the cost per item decreases. The function describing how the cost per item varies with the number of items produced is called the learning curve. Suppose that a shoe manufacturer finds the cost per pair of shoes shown in the table and in Figure 3-6g.

Pairs Produced	Cost ($/pair)
100	60
200	46
500	31
700	26
1000	21

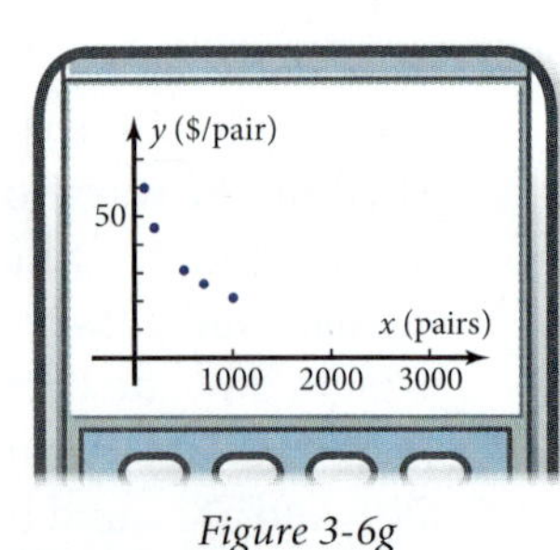

Figure 3-6g

a. By regression analysis, show that a logarithmic function and a power function each fit the data about equally well.

b. It is desired to predict the cost per pair of shoes beyond the upper end of the data. Explain why the power function has a more reasonable endpoint behavior than the logarithmic function for large numbers of pairs produced.

c. Use the power function to predict how many pairs of shoes the shoe manufacturer must produce before the cost per pair drops below \$10.00. Do you find this number by extrapolation or by interpolation? How do you know?

d. According to the power function model, what was the cost of manufacturing the first pair of shoes of this style?

e. The learning curve is sometimes described by saying, "Doubling the number manufactured reduces the cost by __?__ percent." What is the percentage for this kind of shoe? What property of power functions does this fact illustrate?

R4. a. For the power function $y = 7x^5$, show that $\ln y$ is a linear function of $\ln x$.

b. *Hose Problem:* The table and the graph in Figure 3-6h show the rate at which water flows from a garden hose as a function of pressure. The graph has a pattern that is increasing and concave down, suggesting that a power function with base between 0 and 1 would be an appropriate mathematical model. By regression, find the equation of the best-fitting power function. Then on another screen plot both $\log \hat{y}$ versus $\log x$ and the points $(\log x, \log y)$. Sketch the result.

x (psi)	y (gal/min)
1	0.9
5	2.0
10	2.8
15	3.5
25	4.5
40	5.7
50	6.4
70	7.5

Water flow (gal/min)

Pressure (psi)

Figure 3-6h

c. Figure 3-6i shows the best-fitt exponential function for the d table. By exponential regressi equation of this function. Rec of r^2, the coefficient of determination. Then run a linear regression for $\log y$ as a function of x. What do you notice about the coefficient of determination? Transform the exponential equation by taking the log of both sides, thus showing that you get the same equation as the linear equation for $\log y$ as a function of x.

x	y
2	3
15	10
17	16
19	17

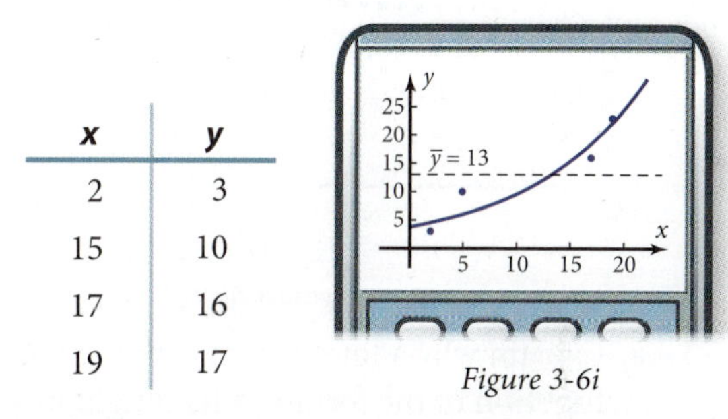

Figure 3-6i

d. Figure 3-6j shows the graph of a function on multiply–add semilog graph paper. Read approximate values of y for $x = 2, 7, 30,$ and 400. By logarithmic regression, find the equation of the best-fitting function for these points. Then use your equation to calculate $\hat{y}$ for $x = 1, 10, 100,$ and 1000. Do the values agree with the given graph?

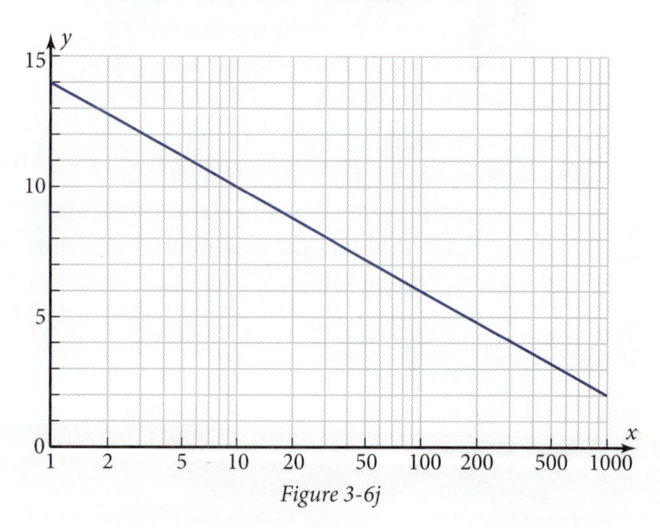

Figure 3-6j

R5. *Carbon Dioxide Problem:* Global warming is caused in part by the increase in the concentration of carbon dioxide in the atmosphere. Suppose that the concentrations in the table were measured monthly over a period of 2 yr. The concentration of carbon dioxide, y, is measured in parts per million (ppm), and x is the number of months.

CO_2 concentration (ppm)

Time (mo)

Figure 3-6k

a. Explain why a linear or an exponential function of the form $y = ab^x$ might fit the data graphed in Figure 3-6k reasonably well, but why an untranslated power function or a logarithmic function would not.

b. Run linear and exponential regressions on the data. (See the data sets at **www.keymath.com/precalc.**) Plot both functions on the same screen as a scatter plot of the data. Show that both functions give close to the actual concentration for $x = 13$ mo but yield a significantly different concentration if you extrapolate to 20 yr.

c. Make a residual plot using the exponential function. Sketch the result. How do you interpret the residual plot in terms of trends in the real world that the exponential function does not account for?

	x (mo)	y (ppm)
January	1	323.8
February	2	324.8
March	3	325.9
April	4	327.2
May	5	328.1
June	6	329.6
July	7	331.5
August	8	333.5
September	9	335.5
October	10	338.0
November	11	340.2
December	12	342.1
January	13	343.5
February	14	344.9
March	15	346.1
April	16	347.4
May	17	348.4
June	18	349.9
July	19	352.0
August	20	354.5
September	21	356.5
October	22	358.9
November	23	360.8
December	24	362.6

Car exhaust contains carbon dioxide, which contributes to global warming.

Concept Problem

C1. *Sinusoidal Regression Problem:* Some graphers are programmed to calculate **sinusoidal regression.** (Sinusoidal functions are covered in Chapter 5.) Suppose the data in the table have been measured.

x	y
2	7.7
4	5.4
6	2.7
8	2.1
10	4.2
12	7.0
14	8.0
16	6.2
18	3.4
20	2.0

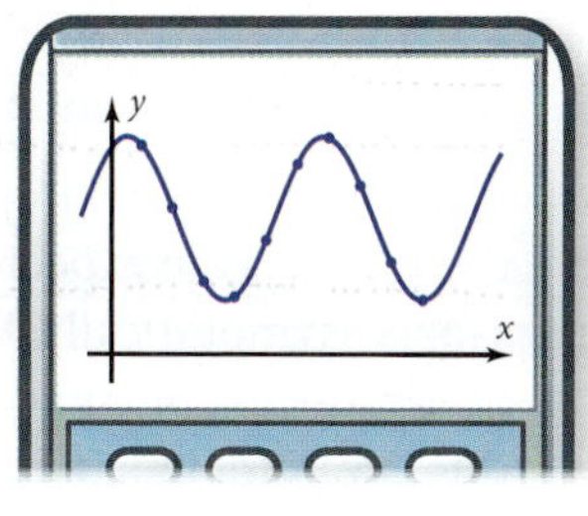

Figure 3-6l

a. Figure 3-6l shows a plot of the data and the best-fitting sinusoidal function. Run a sinusoidal regression on the data. Find the particular equation.

b. Set your grapher to radian mode. Confirm that the graph of your equation goes through the data points, as in Figure 3-6l.

c. Find the approximate values of the period, the amplitude, the phase displacement, and the location of the sinusoidal axis for the sinusoidal function.

Chapter Test

Part 1: No calculators allowed (T1–T10)

For Problems T1–T7, assume that linear regression on a set of data has given the equation $\hat{y} = -2x + 31$.

T1. The mean-mean point, $(\bar{x}, \bar{y})$, is (5, 21). Find the deviation of the data point (7, 15).

T2. Find the residual of the data point (7, 15).

T3. Linear regression gives correlation coefficient -0.95, and exponential regression gives correlation coefficient -0.94. Based on this information, which of these two types of functions fit the data better?

T4. Interpret what is meant by the fact that the correlation coefficient is negative in Problem T3.

T5. Residual plots show the patterns for linear regression (Figure 3-6m) and exponential regression (Figure 3-6n). Based on these plots, which type of function fits the data better? Explain how you decided.

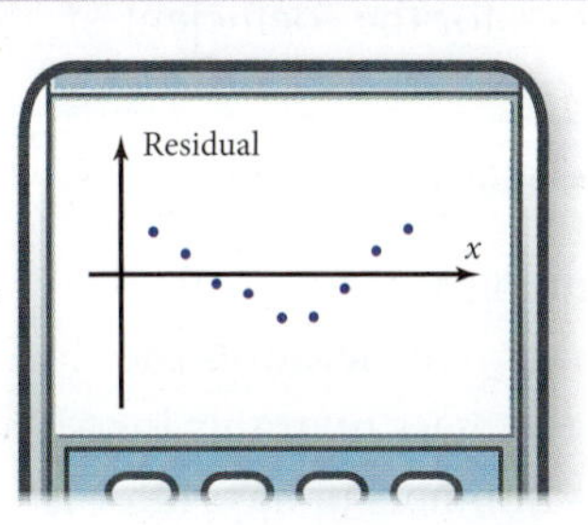

Linear function
Figure 3-6m

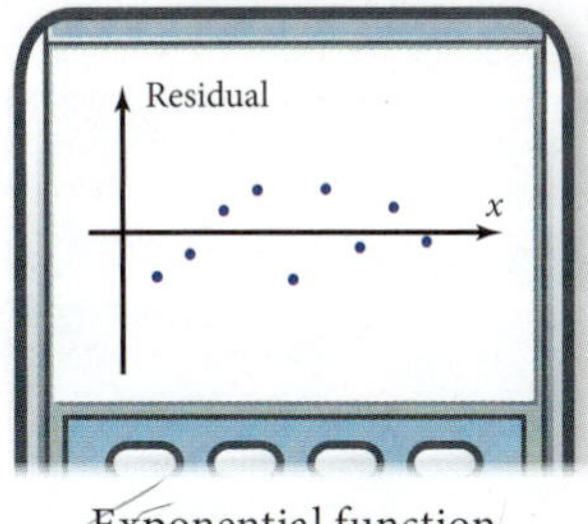

Exponential function
Figure 3-6n

T6. Suppose that in a real-world context you expect to have a positive y-intercept. Which function—the linear, the exponential, both, or neither—has this endpoint behavior?

T7. If the function is extrapolated to large x-values, the y-values are expected to approach the x-axis asymptotically. Which function—the linear, the exponential, both, or neither—has this endpoint behavior?

T8. What type of function has a straight-line graph on log-log graph paper? On add–multiply semilog graph paper? On multiply–add semilog graph paper?

T9. Given SS_{res} and SS_{dev}, how do you calculate the coefficient of determination? How do you calculate the correlation coefficient?

T10. Why does the grapher give the same value of r^2, the coefficient of determination, for power regression and for linear regression with the logarithms of the x-data and y-data? Why does the grapher use R^2 for quadratic regression, instead of r^2, for the coefficient of determination?

Part 2: Graphing calculators allowed (T11–T20)

The Snake Problem: Herpetologist Herbie Tol raises sidewinder rattlesnakes. He measures the length in centimeters and the mass in grams for a baby sidewinder at various numbers of days after it is born. For Problems T11–T19, use these data.

x (days)	y (cm)	m (g)
2	5	1
4	9	5
6	10	9
8	14	20

He runs a linear regression on length as a function of days on his grapher and gets

$$\hat{y} = 1.4x + 2.5$$

He plots a scatter plot and then plots the regression line, as shown in Figure 3-6o.

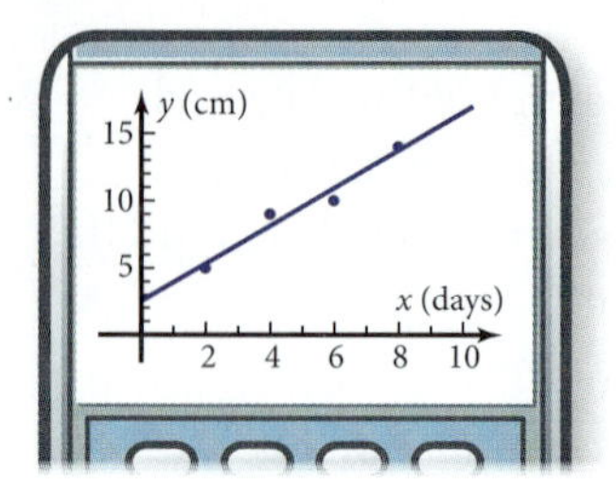

Figure 3-6o

T11. Sketch Figure 3-6o. Demonstrate that you know the meaning of the word *residual* by showing on the sketch the residual for the point where x equals 6.

T12. Add three lists to the data table, one for the value of $\hat{y}$ using the regression equation, one for the residual deviations, and one for the squares of the residuals.

T13. Calculate SS_{res}, the sum of the squares of the residuals.

T14. Calculate the mean-mean point, $(\bar{x}, \bar{y})$. Show algebraically that this point satisfies the regression equation.

T15. Herbie's partner, Peter Doubt, notices that the regression line just misses the points (2, 5) and (8, 14). He calculates the equation containing these points and gets

$$\hat{y}_2 = 1.5x + 2$$

Show that Peter's equation contains the points (2, 5) and (8, 14).

T16. Calculate SS_{res} for Peter's equation. Explain how the answer shows that Peter's equation does not fit the data as well as Herbie's regression equation.

T17. Herbie wants to predict the length of his sidewinder 3 months (91 days) after it is born. Use the linear regression equation $\hat{y} = 1.4x + 2.5$ to make this prediction. Does this prediction involve extrapolation, or does it involve interpolation?

T18. How much different would the predicted length after 3 mo be if Herbie used Peter's equation, $\hat{y}_2 = 1.5x + 2$?

T19. Volumes, and thus masses, of similarly shaped objects are expected to vary directly with the cube of the lengths. By power regression, find the equation of $\hat{m}$, the best-fitting function for mass as a function of length. Calculate $\hat{m}$ for $y = 5$ cm and $y = 50$ cm, and use these values to plot the graph of $\hat{m}$ on log-log graph paper. Plot the four data points $(y, \hat{m})$ on this graph paper. How well does the graph of $\hat{m}$ fit the data points?

T20. What did you learn as a result of taking this test that you did not know before?

Chapter 4

Polynomial and Rational Functions

If you ride the Giant Dipper roller coaster, you'll climb to high points and descend to low points along a continuous track. If the track does not loop or overlap, then its shape can be modeled by a function. In this chapter you will study polynomial functions, and learn how the degree and the coefficients of polynomial functions determine the shape of their graphs. The graphs of polynomial functions often have hills and valleys, and sometimes resemble the track of a roller coaster.

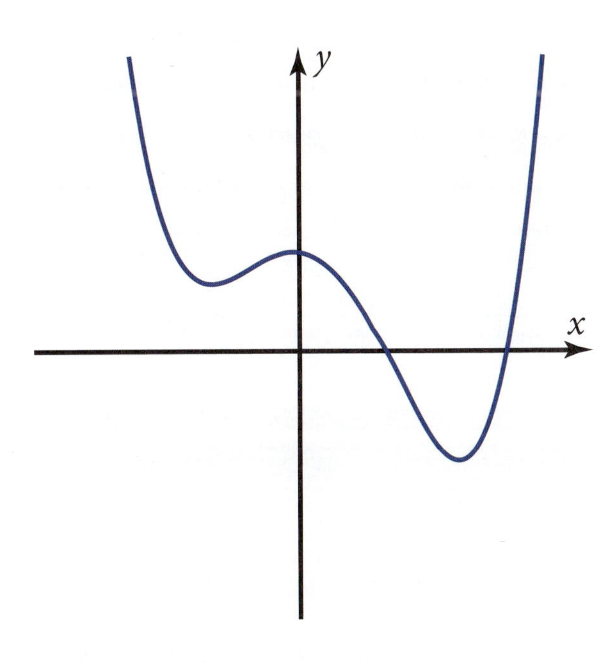

Mathematical Overview

In this chapter you'll learn about polynomial functions. You'll find the zeros of these functions, a generalization of the concept of x-intercept. The techniques you learn will allow you to analyze rational functions, in which $f(x)$ is a ratio of two polynomials. These ratios can represent average rates of change. You'll investigate these concepts in four ways.

GRAPHICALLY

The graph of the polynomial function $f(x) = x^4 - 8x^2 - 8x + 15$ has two x-intercepts.

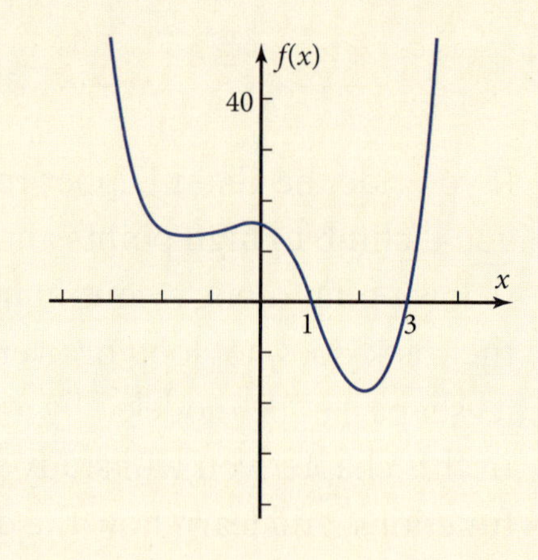

ALGEBRAICALLY

$$f(x) = x^4 + 0x^3 - 8x^2 - 8x + 15$$
$$= (x - 1)(x - 3)(x + 2 + i)(x + 2 - i)$$

NUMERICALLY

$f(1) = 0, f(3) = 0, f(-2 - i) = 0$, and $f(-2 + i) = 0$

Zeros of the function are $x = 1$, $x = 3$, $x = -2 - i$, and $x = -2 + i$.

VERBALLY

A fourth-degree function has exactly four zeros, counting multiple zeros, if the domain of the function includes complex numbers. Only the real-number zeros appear as x-intercepts on the graph, but zeros can also include complex numbers.

4-1 Introduction to Polynomial and Rational Functions

Recall from Section 1-2 that a polynomial function has an equation such as $f(x) = x^2 - 2x - 8$ where no operations other than addition, subtraction, or multiplication are performed on the independent variable x. (Squaring is equivalent to multiplying.) Function f is a quadratic function because the highest power of x is the second power. A **rational function** has an equation such as $g(x) = \frac{x^3 - 5x^2 - 2x + 25}{x - 3}$, where the y-value is a ratio of two polynomials—in this case a cubic polynomial divided by a linear polynomial. Figures 4-1a and 4-1b show the graphs of functions f and g.

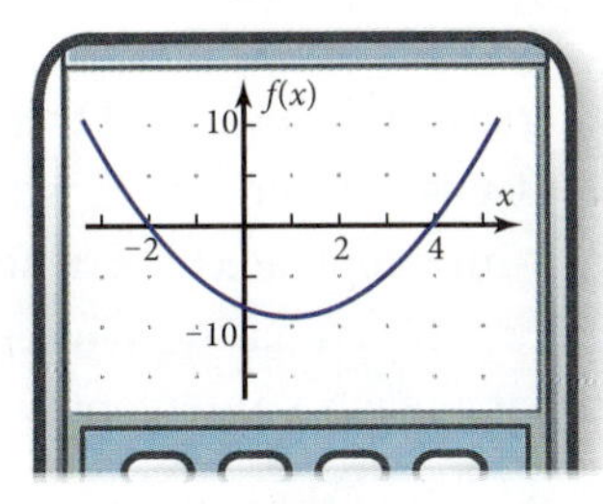

Figure 4-1a

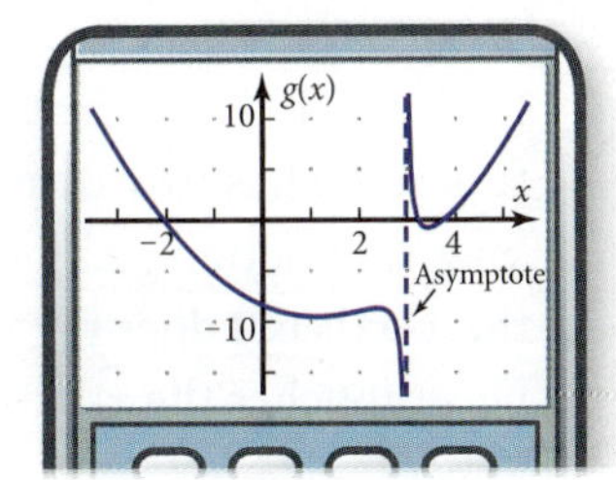

Figure 4-1b

Objective Discover some properties of polynomial and rational functions and their graphs.

Exploratory Problem Set 4–1

1. Function f appears to have x-intercepts at -2 and 4. Confirm by direct substitution that $f(-2)$ and $f(4)$ both equal zero. Then explain why -2 and 4 are also called **zeros** of function f.

2. The x-value -2 appears to be a zero of function g. Confirm or refute this observation numerically.

3. Function g has a vertical asymptote at $x = 3$. Try to find $g(3)$. Then explain why $g(3)$ is undefined. Explain the behavior of $g(x)$ when x is close to 3 by finding $g(3.001)$ and $g(2.999)$.

4. Plot the graphs of functions f and g on the same screen. What do you notice about the two graphs when x is far from 3? When x is close to 3? How could you describe the relationship of the graph of function f to the graph of function g?

5. Let function h be the vertical translation of function f by 10 units. That is, $h(x) = f(x) + 10$. Show by graphing that no zeros of function h are real numbers.

6. Explain why a quadratic function can have no more than two real numbers as zeros.

7. As a challenge, show that the **complex number** $1 + i$ (where $i = \sqrt{-1}$) is a zero of $h(x)$. Explain why a function can have a zero that is not an x-intercept.

8. What did you learn as a result of doing this problem set?

4-2 Quadratic Functions, Factoring, and Complex Numbers

Figure 4-2a shows the graphs of three quadratic functions, f, g, and h. These graphs are parabolas.

$$f(x) = x^2 - 4x - 5$$

$$g(x) = x^2 - 4x + 4$$

$$h(x) = x^2 - 4x + 13$$

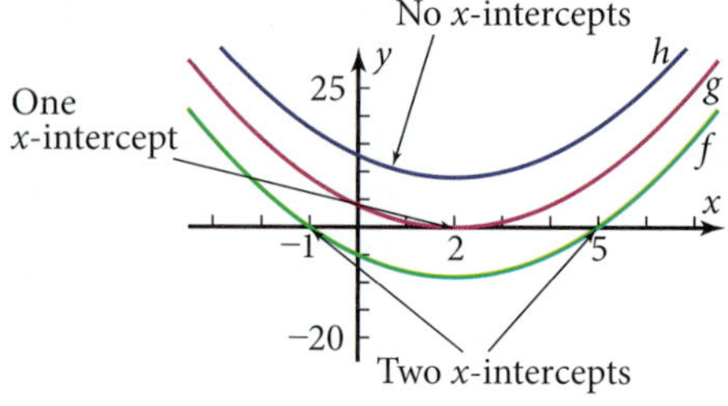

Figure 4-2a

Each graph differs from the others by only a vertical translation. The graph of function f has two x-intercepts, at $x = -1$ and $x = 5$, where the graph crosses the x-axis. Function g has only one x-intercept, at $x = 2$, where the graph touches the x-axis but does not cross it. Function h has no x-intercepts because the graph never touches the x-axis. Note that at each x-intercept, $y = 0$.

In this section you will learn that each of these functions has *exactly two* values of x that make y equal zero. For function f, these values are $x = -1$ and $x = 5$. For function g, they are $x = 2$ and $x = 2$, which are equal to each other. For function h, they are the complex numbers $x = 2 + 3i$ and $x = 2 - 3i$, where i is the unit **imaginary number,** $\sqrt{-1}$. The term **zero of a function** is used to include nonreal values of x that make y equal zero. So every quadratic function has exactly two zeros but may have two, one, or no x-intercepts.

In order to understand the algebra of higher-degree polynomial functions, it will be helpful to review various properties and techniques for quadratic functions.

Objective

Review the properties, graphs, and techniques associated with quadratic functions.

Algebraic Operations with Polynomials

The expression $(3x + 5)(2x - 11)$ is said to be in *factored form* because it is written as a product of two (binomial) factors. To transform the expression to *polynomial form* you use the distributive property to multiply each term in the first binomial by each term in the second.

EXAMPLE 1 ➤ Multiply the binomials and simplify: $(3x + 5)(2x - 11)$

SOLUTION

$(3x + 5)(2x - 11)$	Write the given expression.
$= 6x^2 - 33x + 10x - 55$	Multiply each term in $(3x + 5)$ by each term in $(2x - 11)$.
$= 6x^2 - 23x - 55$	Combine like terms.

The answer in Example 1 is called a quadratic trinomial, *quadratic* meaning "second degree" and *trinomial* meaning "three terms." With practice you can combine the two middle terms mentally and write just the given expression and the answer, like this:

$$(3x + 5)(2x - 11) = 6x^2 - 23x - 55$$

Example 2 shows how to multiply three binomials.

EXAMPLE 2 ➤ Multiply the binomials and simplify: $(x - 5)(x - 2)(x - 1)$

SOLUTION

$(x - 5)(x - 2)(x - 1)$ — Write the given expression.

$= (x - 5)\left[(x - 2)(x - 1)\right]$ — Associate two of the factors.

$= (x - 5)(x^2 - 3x + 2)$ — Multiply the associated factors.

$= x^3 - 3x^2 + 2x - 5x^2 + 15x - 10$ — Multiply each term in $(x - 5)$ by each term in $(x^2 - 3x + 2)$.

$= x^3 - 8x^2 + 17x - 10$ — Combine like terms. ◄

The process of transforming an expression in polynomial form to factored form is called factoring the expression. Example 3 reviews how to factor a quadratic trinomial.

EXAMPLE 3 ➤ Factor: $x^2 - 6x - 40$

SOLUTION

$x^2 - 6x - 40 = (x \qquad)(x \qquad)$ — Write two pairs of parentheses. The first term in each factor will be x.

$= (x - 10)(x + 4)$ — Find two factors of the constant term, -40, that sum to -6. Those factors must be -10 and $+4$ in order to get $x^2 - 6x - 40$ when you multiply the binomials. ◄

You can check the answer to Example 3 by multiplying $(x - 10)(x + 4)$ and showing that the result is the given expression, $x^2 - 6x - 40$.

Solutions of Quadratic Equations

A quadratic equation such as $x^2 - 6x - 23 = 17$ might arise from finding the values of x for the function $y = x^2 - 6x - 23$ that give $y = 17$ (Figure 4-2b).

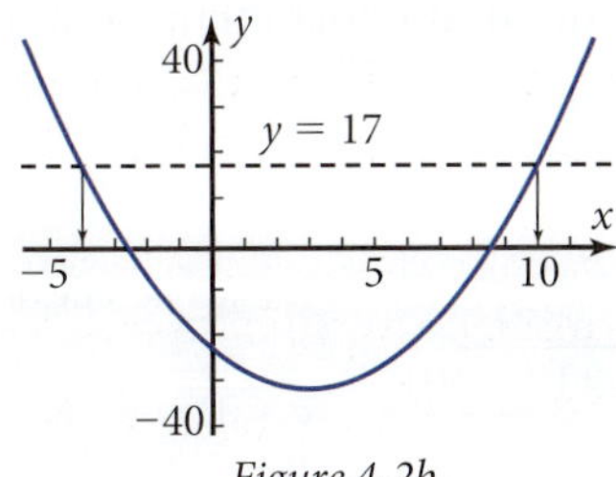

Figure 4-2b

Such an equation can be solved with the help of factoring. The idea is to get a product equal to zero and then set each factor equal to zero.

EXAMPLE 4 ➤ Solve by factoring: $x^2 - 6x - 23 = 17$

SOLUTION

$x^2 - 6x - 23 = 17$

$x^2 - 6x - 40 = 0$ — Make one side of the given equation equal zero.

$(x - 10)(x + 4) = 0$ — Factor the polynomial side.

$x - 10 = 0$ or $x + 4 = 0$ — Set each factor equal to zero.

$x = 10$ or $x = -4$ — Solve the two simpler equations.

Note that the two solutions agree with the graph in Figure 4-2b. ◄

The basis for solving an equation by factoring is the *multiplication property of zero* and its *converse*.

> **PROPERTY: Multiplication Property of Zero and Its Converse**
>
> $a \cdot b = 0$ if and only if $a = 0$ or $b = 0$

It is for this reason that you must first make one side of the equation in Example 4 equal zero before you factor the expression on the other side.

A quadratic equation can also be solved using the quadratic formula. You may recall this formula from earlier courses. The proof is presented in Problems 65 and 66 of this section.

> **PROPERTY: The Quadratic Formula**
>
> *Algebraically:* If $ax^2 + bx + c = 0$ and $a \neq 0$, then $x = \dfrac{-b \pm \sqrt{b^2 - 4ac}}{2a}$.
>
> *Verbally:* If $ax^2 + bx + c = 0$ and $a \neq 0$, then x equals the opposite of b, plus or minus the square root of the quantity $b^2 - 4ac$, all divided by $2a$.

The quantity $b^2 - 4ac$ under the radical sign is called the **discriminant** because it "discriminates" between quadratic equations that have real-number solutions and quadratic equations that don't.

Example 5 shows how to solve the quadratic equation in Example 4 using this formula.

EXAMPLE 5 ➤ Solve using the quadratic formula: $x^2 - 6x - 23 = 17$

SOLUTION To satisfy the "if" part of the quadratic formula, you must make one side of the equation equal zero.

$x^2 - 6x - 23 = 17$

$x^2 - 6x - 40 = 0$ — Make one side of the given equation equal zero.

$x = \dfrac{6 \pm \sqrt{(-6)^2 - 4(1)(-40)}}{2(1)}$ — $a = 1, b = -6, c = -40$

$x = \dfrac{6 \pm \sqrt{196}}{2}$ — Evaluate the discriminant first.

$$x = \frac{6 + 14}{2} \quad \text{or} \quad x = \frac{6 - 14}{2}$$

$$x = 10 \quad \text{or} \quad x = -4$$

This solution agrees with Example 4.

Although using the quadratic formula may take longer than factoring, it works even when the quadratic trinomial will not factor and even when the discriminant, $b^2 - 4ac$, is a negative number.

Imaginary and Complex Numbers

Recall from Figure 4-2a that the function $h(x) = x^2 - 4x + 13$ has no x-intercepts. If you set $h(x)$ equal to zero and apply the quadratic formula to the resulting quadratic equation, $x^2 - 4x + 13 = 0$, you get

$$x = \frac{4 \pm \sqrt{16 - 52}}{2} = \frac{4 \pm \sqrt{-36}}{2}$$

The square root of -36 is not a real number because squares of real numbers are never negative. In order for such a quadratic equation to have solutions, mathematicians define square roots of negative numbers to be *another* kind of number: $\sqrt{-1}$ is defined to be the unit imaginary number, designated by the letter i.

$$\begin{aligned} \sqrt{-36} &= \sqrt{(36) \cdot (-1)} \\ &= \sqrt{36} \cdot \sqrt{-1} \\ &= 6i \end{aligned}$$

Similarly,

$$\sqrt{-9} = \sqrt{(9) \cdot (-1)} = \sqrt{9} \cdot \sqrt{-1} = 3i$$

$$\sqrt{-53} = 7.2801...i$$

$$-\sqrt{-100} = -10i$$

The solutions of the quadratic equation $x^2 - 4x + 13 = 0$ above become

$$x = \frac{4 \pm 6i}{2} = 2 \pm 3i = 2 + 3i \quad \text{or} \quad 2 - 3i$$

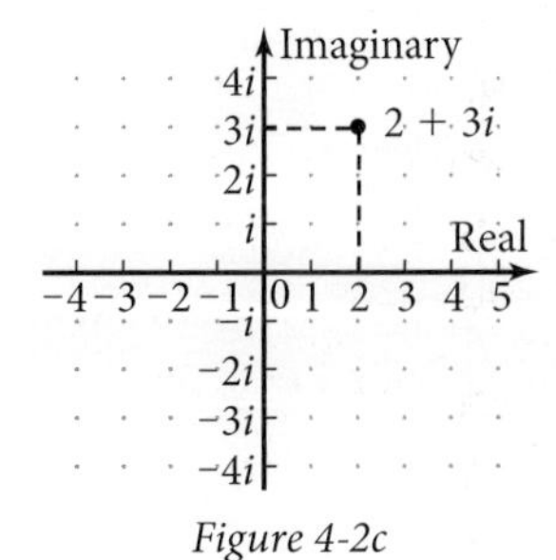

Figure 4-2c

The complex numbers $2 + 3i$ and $2 - 3i$ are the zeros of the function h defined at the beginning of this section. These numbers are not x-intercepts because the graph of function h does not reach the x-axis. The real number 2 is the *real part* of $2 + 3i$, and the real number 3 is the *imaginary part* of $2 + 3i$.

Complex numbers can be plotted on the complex-number plane, sometimes called simply the **complex plane.** The real part is plotted horizontally, and the imaginary part is plotted vertically, as shown in Figure 4-2c.

If the coefficient of i is zero, then only the real part is left. So real numbers are a subset of the complex numbers. You can see this by covering up the regions of the complex plane that are above and below the real-number line.

The definitions and properties of complex numbers appear in the box.

DEFINITIONS: Imaginary and Complex Numbers

$$i = \sqrt{-1} \text{ so that } i^2 = -1$$

If x is a nonnegative real number, then

$$\sqrt{-x} = i\sqrt{x}$$

A *complex number* is a number of the form $a + bi$, where

- The real number a is called the *real part* of $a + bi$.
- The real number b is called the *imaginary part* of $a + bi$.

The complex numbers $a + bi$ and $a - bi$ are called **complex conjugates** of each other.

Complex numbers can be added, subtracted, and multiplied the same way as binomials.

EXAMPLE 6 Multiply and simplify: $(7 + 4i)(9 + 10i)$

SOLUTION

$(7 + 4i)(9 + 10i) = 63 + 70i + 36i + 40i^2$ — Distribute each term in one number to each term of the other.

$= 63 + 70i + 36i - 40$ — Because $i^2 = -1$.

$= 23 + 106i$ — Combine like terms.

The main thing to remember in operating with complex numbers is that i^2 is equal to -1. All the properties and techniques for real numbers apply to complex numbers, including squaring a binomial.

EXAMPLE 7 Check that $x = 2 - 3i$ is a solution to the equation $x^2 - 4x + 13 = 0$ by substituting it into the left side of the equation and showing that the result equals zero.

SOLUTION

$(2 - 3i)^2 - 4(2 - 3i) + 13 \stackrel{?}{=} 0$ — Substitute $2 - 3i$ for x.

$4 - 12i + 9i^2 - 8 + 12i + 13 \stackrel{?}{=} 0$ — Do the squaring and distributing.

$4 - 9 - 8 + 13 \stackrel{?}{=} 0$ — Combine like terms; $9i^2 = -9$.

$0 = 0$ ✓ — The result does equal zero.

Note that complex solutions of a quadratic equation with real coefficients will always be complex conjugates of each other, such as $2 + 3i$ and $2 - 3i$. You can see why this is true by writing the quadratic formula this way:

$$x = \frac{-b}{2a} \pm \frac{\sqrt{b^2 - 4ac}}{2a}$$

The expression $\frac{-b}{2a}$ is a real number. The radical expression will be an imaginary number if the discriminant, $b^2 - 4ac$, is negative.

Vertex Form and Completing the Square

The equation

$$y = 3(x - 5)^2 - 43$$

is said to be in vertex form. As you can see graphically in Figure 4-2d, the vertex is $(5, -43)$. Algebraically, substituting 5 for x makes the expression $3(x - 5)^2$ equal zero, giving -43 for the value of y.

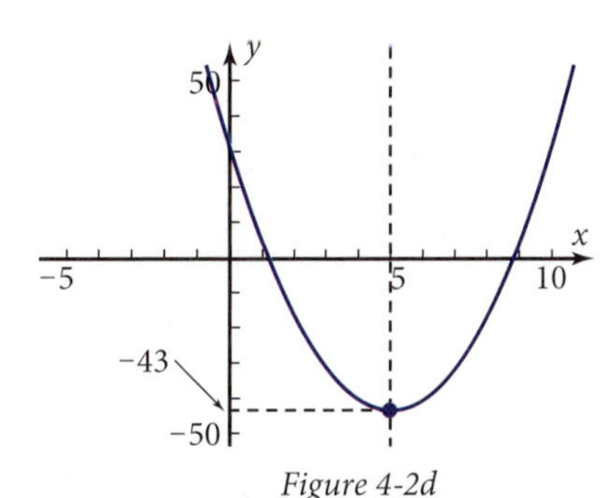

Figure 4-2d

Example 8 shows how to transform an equation in vertex form to polynomial form ($y = ax^2 + bx + c$) by multiplying and combining like terms.

EXAMPLE 8 ➤ Transform the equation $y = 3(x - 5)^2 - 43$ to polynomial form.

SOLUTION

$y = 3(x - 5)^2 - 43$

$y = 3(x^2 - 10x + 25) - 43$ Multiply $(x - 5)(x - 5)$.

$y = 3x^2 - 30x + 75 - 43$ Distribute the 3.

$y = 3x^2 - 30x + 32$ Combine like terms. ◄

If the equation is in polynomial form, like the answer in Example 8, it is easy to find the y-intercept. You can substitute zero for x and get $y = 32$ (in this case), or simply look at the equation and write the constant term, 32. The result agrees with Figure 4-2d. The box on the next page summarizes polynomial form and vertex form.

DEFINITION: Quadratic Function

A quadratic function is a function with the general equation

$y = ax^2 + bx + c$ where a, b, and c stand for constants and $a \neq 0$.

This is known as the *polynomial form* of the equation. (If $a = 0$, the function is linear, not quadratic.)

Vertex form: If the equation of a quadratic function is in the form

$y = a(x - h)^2 + k$ or $y - k = a(x - h)^2$

where a, h, and k stand for constants and $a \neq 0$, then the vertex is at the point (h, k).

Memory aid: h stands for the *horizontal* coordinate of the vertex.
k comes after h in the alphabet, just as y comes after x.

In both forms, the constant a is the vertical dilation of the parent function $y = x^2$. The x-coordinate, h, of the vertex is the average of the two x-values that result from the quadratic formula, specifically, $h = \frac{-b}{2a}$.

A quadratic equation in polynomial form can be transformed to vertex form by **completing the square,** as shown in Example 9.

EXAMPLE 9 ➤ Write the equation $y = 5x^2 + 35x + 17$ in vertex form, and then sketch its graph.

SOLUTION

$y = 5x^2 + 35x + 17$ — Write the given equation.

$y = 5(x^2 + 7x \quad) + 17$ — Factor 5 out of the first two terms to make the x^2-coefficient equal 1.

$y = 5(x^2 + 7x + 3.5^2) + 17 - 5(3.5^2)$ — Complete the square by taking half the x-coefficient, squaring it, adding the resulting 3.5^2 inside the parentheses, and then subtracting $5(3.5^2)$ outside the parentheses.

$y = 5(x + 3.5)^2 - 44.25$ — Write $x^2 + 7x + 3.5^2$ as $(x + 3.5)^2$, and combine the constants.

Vertex: $(-3.5, -44.25)$ — -3.5 is the value of x that makes $(x + 3.5)^2$ equal zero. -44.25 is the value of y when $x = -3.5$.

y-intercept: 17 — Set $x = 0$ in the original polynomial form.

Sketch the graph (Figure 4-2e), showing the vertex, the axis of symmetry, and the y-intercept. (It is not necessary to use the same scales on the two axes.)

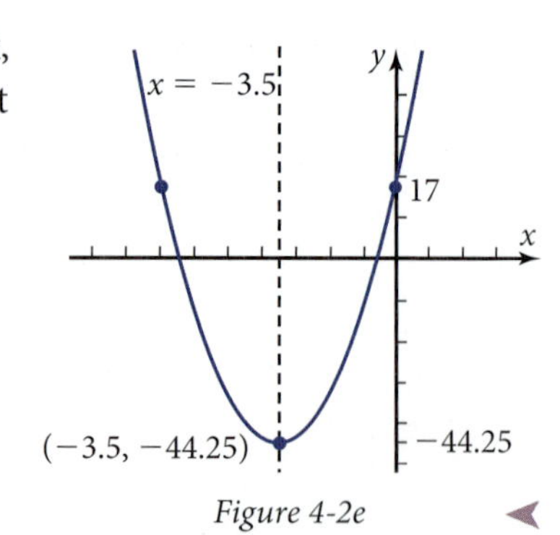

Figure 4-2e

PROCEDURE: Completing the Square

To write a quadratic trinomial $x^2 + bx + c$ as the square of a linear binomial:

- Associate the x^2-term and the x-term.
- Find half the coefficient of x and square the answer.
- Add the result inside the parentheses, and subtract it outside the parentheses.
- Write the trinomial as the square of a binomial plus a constant.

Note: For this procedure to work, the coefficient of x^2 must be 1.

Problem Set 4-2

Reading Analysis

From what you have read in this section, what do you consider to be the main idea? What is meant by a *zero* of a function, and how is it related to an x-intercept? What is the "If . . ." part of the quadratic formula? What are the differences among real numbers, imaginary numbers, and complex numbers? What is meant by the *discriminant* of a quadratic equation? Why does i^2 equal -1? What is meant by the *vertex form* of a quadratic function equation, and what procedure is used to transform a quadratic equation in polynomial form into vertex form?

Quick Review

For Problems Q1–Q4, let $g(x) = 2 + 3f(4x + 5)$. Function g is

Q1. A horizontal translation of function f by __?__ units

Q2. A horizontal dilation of function f by a factor of __?__

Q3. A vertical translation of function f by __?__ units

Q4. A vertical dilation of function f by a factor of __?__

Q5. $f(x) = 3 \cdot 2^x$ is an example of a(n) __?__ function.

Q6. $f(x) = 4 \cdot x^3$ is an example of a(n) __?__ power function.

Q7. $f(x) = 5 \cdot x^{-2}$ is an example of a(n) __?__ power function.

Q8. $f(x) = 6 + 2x$ is an example of a(n) __?__ function.

Q9. A function found by regression fits the data well if the residual plot has a __?__ pattern.

Q10. $(3x - 5)^2 =$

A. $9x^2 - 25$ **B.** $9x^2 + 25$

C. $9x^2 + 30x - 25$ **D.** $9x^2 - 30x + 25$

E. $9x^2 - 30x - 25$

For Problems 1–10, multiply the binomials and simplify.

1. $(4x + 5)(2x - 3)$
2. $(5x - 2)(6x + 3)$
3. $(x - 8)(2x - 9)$
4. $(4x - 1)(x - 10)$
5. $(3x + 7)^2$
6. $(6x - 8)^2$
7. $(5x - 7)(5x + 7)$
8. $(3x + 4)(3x - 4)$
9. $(x + 2)(x - 5)(x - 6)$
10. $(x - 4)(x + 3)(x + 5)$

For Problems 11–20, factor the polynomial.

11. $x^2 + 10x + 21$

12. $x^2 - 5x - 36$

13. $x^2 + 6x - 40$

14. $x^2 - 17x + 60$

15. $x^2 - 36$

16. $x^2 - 121$

17. $x^2 - 20x + 100$

18. $x^2 + 18x + 81$

19. $x^2 + 7x$

20. $x^2 - 9x$

For Problems 21–30, solve the equation by factoring.

21. $x^2 + 10x + 21 = 0$

22. $x^2 - 5x - 36 = 0$

23. $x^2 + 6x - 55 = -15$

24. $x^2 - 17x + 70 = 10$

25. $x^2 - 36 = 0$

26. $x^2 - 121 = 0$

27. $x^2 - 20x + 100 = 0$

28. $x^2 + 18x + 81 = 0$

29. $x^2 + 7x = 0$

30. $x^2 - 9x = 0$

For Problems 31–36, solve the equation by using the quadratic formula.

31. $x^2 + 10x + 21 = 0$

32. $x^2 - 5x - 36 = 0$

33. $3x^2 - 11x + 5 = 0$

34. $5x^2 - 3x - 7 = 0$

35. $6x^2 - 5x = 8$

36. $10x^2 + 1 = -8x$

For Problems 37–42, multiply the complex numbers and simplify.

37. $(3 - 2i)(4 - 5i)$

38. $(5 + 8i)(3 + i)$

39. $(7 + 3i)(7 - 3i)$

40. $(4 - 6i)(4 + 6i)$

41. $(6 + 7i)^2$

42. $(9 - 7i)^2$

For Problems 43–46, solve for complex values of x. Check one of the solutions by substituting it into the given equation.

43. $x^2 - 6x + 34 = 0$

44. $x^2 + 14x + 58 = 0$

45. $9x^2 + 30x + 26 = 0$

46. $25x^2 - 10x + 101 = 0$

For Problems 47–50, one complex solution of the equation is given. Write the other complex solution, and check this second solution by substituting it into the given equation.

47. $x^2 + 22x + 150 = 20; x = -11 - 3i$

48. $x^2 - 16x + 100 = 11; x = 8 - 5i$

49. $25x^2 + 2 = 10x; x = 0.2 + 0.2i$

50. $0.04x^2 + 1.2x = -45; x = -15 + 30i$

For Problems 51–54, find the zeros of the given function. Show that your answer is reasonable by plotting the graph and sketching the result.

51. $f(x) = x^2 - 2x - 15$

52. $f(x) = x^2 + 7x + 6$

53. $f(x) = -x^2 + 6x - 10$

54. $f(x) = -x^2 + 10x - 29$

55. Write the definition of "zero of a function."

56. What is the difference between a zero of a function and an x-intercept of a function?

For Problems 57–60,

a. Transform the equation to polynomial form.

b. Write the coordinates of the vertex.

c. Use the vertex, the y-intercept, and the reflection of the y-intercept across the axis of symmetry to sketch the graph.

57. $f(x) = -0.2(x - 3)^2 + 5$

58. $f(x) = 3(x - 2)^2 - 8$

59. $f(x) = 5(x + 4)^2 + 7$

60. $f(x) = -0.5(x + 1)^2 - 3$

For Problems 61–64,

a. Transform the equation in polynomial form to vertex form by completing the square.

b. Write the coordinates of the vertex.

c. Use the vertex, the y-intercept, and the reflection of the y-intercept across the axis of symmetry to sketch the graph.

61. $f(x) = x^2 + 6x + 11$

62. $f(x) = x^2 - 8x + 21$

63. $f(x) = 3x^2 - 24x + 17$

64. $f(x) = 8x^2 + 40x + 37$

65. *Solving Quadratics by Completing the Square Problem:* If you did not know the quadratic formula, you could solve a quadratic equation by completing the square. These steps show how it can be done.

$5x^2 + 30x + 7 = 0$
Write the given equation.

$x^2 + 6x + 1.4 = 0$
Divide by 5 to make the x2-coefficient equal 1.

$x^2 + 6x \qquad = -1.4$
Move the constant term to the right side, and leave a space to complete the square.

$x^2 + 6x + 9 = 9 - 1.4$
Add 9 to both sides to complete the square on the left.

$(x + 3)^2 = 7.6$
Factor on the left, and simplify on the right.

$x + 3 = \pm\sqrt{7.6}$
Take the square root of both sides.

$x = -3 \pm\sqrt{7.6}$
$= -0.2431...$ or $-5.7568...$
Simplify.

Solve these equations by completing the square.

a. $x^2 + 6x + 4 = 0$

b. $x^2 - 10x + 21 = 0$

c. $x^2 - 7x - 4 = 0$

d. $7x^2 + 14x + 3 = 0$

e. $2x^2 - 10x + 20 = 9$

66. *Derivation of the Quadratic Formula Problem:* In this problem you will see how the quadratic formula is derived by completing the square to solve the general quadratic equation as in Problem 65. Write each step, parts a–g, on your paper, and supply the reason for each step.

a. $ax^2 + bx + c = 0$ ___?___

b. $x^2 + \frac{b}{a}x + \frac{c}{a} = 0$ ___?___

c. $x^2 + \frac{b}{a}x \qquad = -\frac{c}{a}$ ___?___

d. $x^2 + \frac{b}{a}x + \frac{b^2}{(2a)^2} = \frac{b^2}{(2a)^2} - \frac{c}{a}$ ___?___

e. $\left(x + \frac{b}{2a}\right)^2 = \frac{b^2 - 4ac}{(2a)^2}$ ___?___

f. $x + \frac{b}{2a} = \frac{\pm\sqrt{b^2 - 4ac}}{2a}$ ___?___

g. $x = \frac{-b \pm\sqrt{b^2 - 4ac}}{2a}$ ___?___

4-3 Graphs and Zeros of Higher-Degree Functions

In Section 4-2, you reviewed quadratic functions—polynomial functions of degree 2. In this section you will learn to identify cubic, quartic, quintic, and higher-degree functions from their equations or graphs, to find their real and complex zeros, and to find their x-intercepts.

Objective Given a polynomial function,

- Determine from the graph what degree it might be, and vice versa.
- Find the zeros from the equation or the graph.

In this exploration you will explore the zeros of cubic functions and the behavior of their graphs.

EXPLORATION 4-3: Graphs and Zeros of Cubic Functions

1. The figure shows the graphs of $f(x) = x^3$ (dashed)—the parent cubic function—and $g(x) = x^3 - x$, a vertical translation of function f by the *variable* amount $-x$. Set $g(x) = 0$ and show that function g has the three distinct real zeros shown in the graph by factoring the polynomial on the right side of the resulting equation.

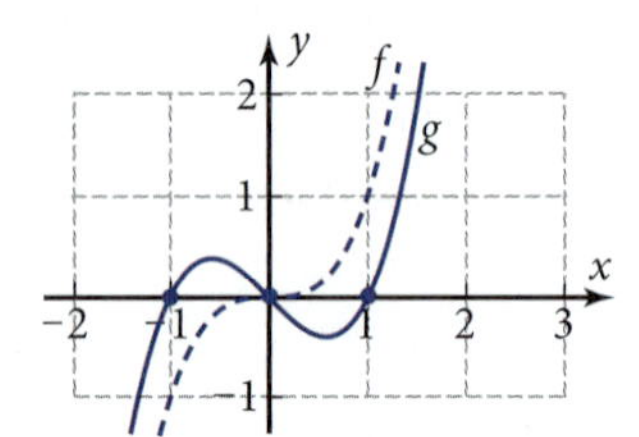

2. The parent function $f(x) = x^3$ can be factored as $f(x) = x \cdot x \cdot x$. Use this factorization to explain why function f also has three real zeros but not three *distinct* real zeros.

3. The figure shows the graphs of $f(x) = x^3$ and function v, a vertical translation of function f by 1 unit.

$v(x) = x^3 + 1$ (polynomial form)

$v(x) = (x + 1)(x^2 - x + 1)$ (factored form)

Multiply the factors to show that the factored form is equivalent to the polynomial form.

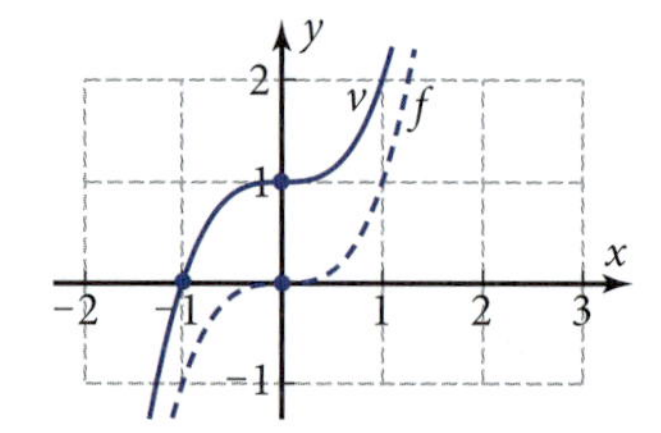

4. Use the factored form of function v from Problem 3 to show that function v also has three zeros, but two of the zeros are nonreal complex numbers.

5. The figure shows the graphs of $f(x) = x^3$ and $h(x) = x^3 - x^2$, a vertical translation of function f by the variable amount $-x^2$. By first factoring $h(x)$, show that function h has three real zeros, not all of which are distinct.

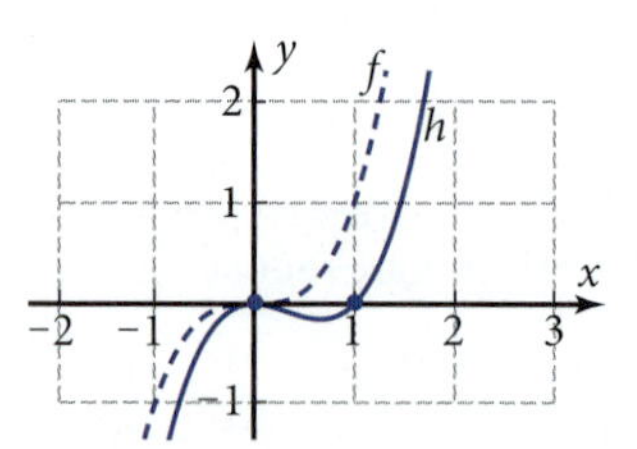

6. From the graphs of functions f and h above, describe the behavior of the graph at a *single zero* such as function h has at $x = 1$, a *double zero* such as function h has at $x = 0$, and a *triple zero*, such as function f has at $x = 0$.

continued

7. The graphs of $f(x) = x^3$ and $p(x) = (x - 2)^3 + 1$ are shown in the figure. Name the two transformations of function f that were done to get function p. Show that $x = 1$ is a zero of function p.

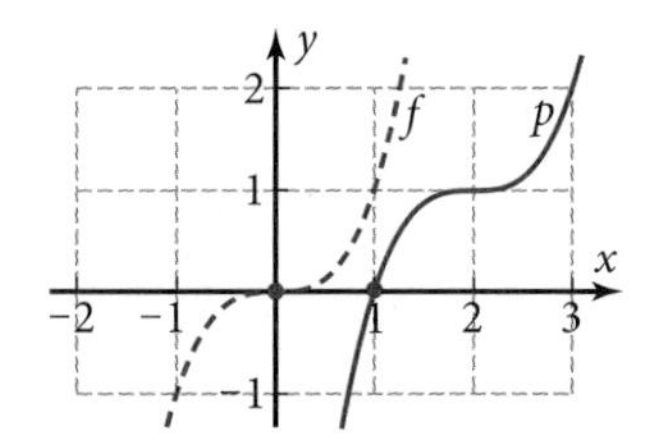

One factor of $p(x)$ is $(x - 1)$. For a challenge, see if you can find the other factor, and use the result to find the two nonreal complex zeros of $p(x)$.

8. From what you have seen in this exploration, how many zeros does a cubic function have? Give two reasons why a cubic function might have fewer than this number of x-intercepts.

9. What did you learn as a result of doing this exploration that you did not know before?

Graphs of Polynomial Functions

Recall the general equation of a polynomial function from Chapter 1. Here are some examples of polynomial functions.

$f(x) = 4x^2 - 7x + 3$	Quadratic function, 2nd degree
$f(x) = 2x^3 - 5x^2 + 4x + 7$	Cubic function, 3rd degree
$f(x) = x^4 + 6x^3 - 3x^2 + 5x - 8$	Quartic function, 4th degree
$f(x) = -6x^5 - x^3 + 2x$	Quintic function, 5th degree
$f(x) = 5x^9 + 4x^8 - 11x^3 + 63$	9th-degree function (no special name)

In each case, $f(x)$ is a **polynomial** expression. The only operations performed on the variable in a polynomial are the **polynomial operations:** addition, subtraction, and multiplication. Nonnegative integer powers are allowed in a polynomial because they are defined in terms of multiplication. Polynomials are **continuous** for all real-number values of x; that is, there are no discontinuities caused by roots of negative numbers or division by zero. The graphs are also *smooth,* which means that they have no abrupt changes or sharp corners such as those caused by absolute values.

The **degree** of a polynomial is the greatest number of variables multiplied together in any one term. For a one-variable polynomial, the degree is the same as the highest exponent on the variable. The coefficient of the term with the highest degree is called the **leading coefficient.** Figure 4-3a on page 190 shows examples of a quadratic, a cubic, and a quartic function.

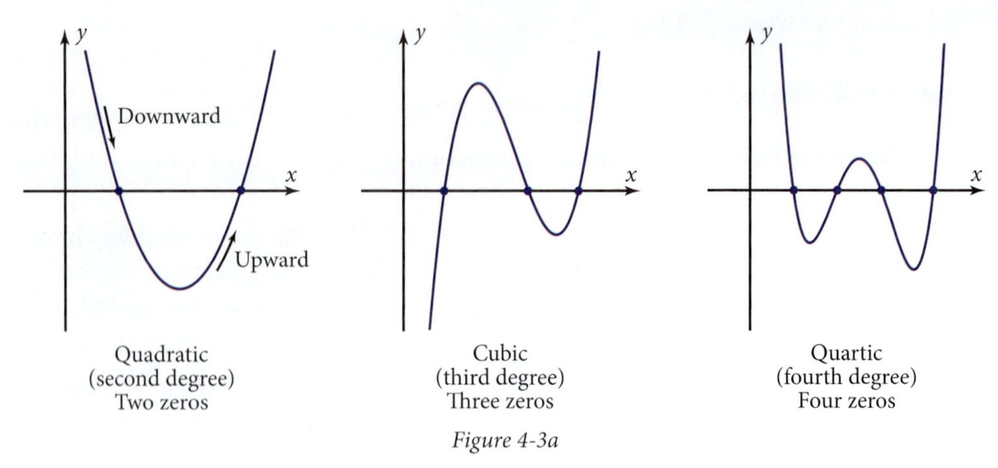

Figure 4-3a

The **end behavior** of a function f refers to what happens to $f(x)$ as x gets large in either the positive or the negative direction. As shown in Figure 4-3a, all three example functions become large and positive as x gets large in the positive direction. But the odd-degree function (cubic) becomes large and negative as x gets large in the negative direction. This happens because both positive and negative numbers raised to even powers are positive, whereas negative numbers raised to odd powers are negative. If the leading coefficient is negative, as in $y = -5x^3$, then this end behavior is reflected across the x-axis.

As x gets large, the highest-degree term eventually dominates the other terms, and the graph looks like that of the highest-degree term. Figure 4-3b shows the graph of the cubic function $f(x) = x^3 - 7x^2 + 10x + 2$ and the graph of the cubic term, $g(x) = x^3$. In the narrow viewing window on the left, the two graphs are clearly different. But if you zoom out, as in the graph on the right, the graphs look quite similar.

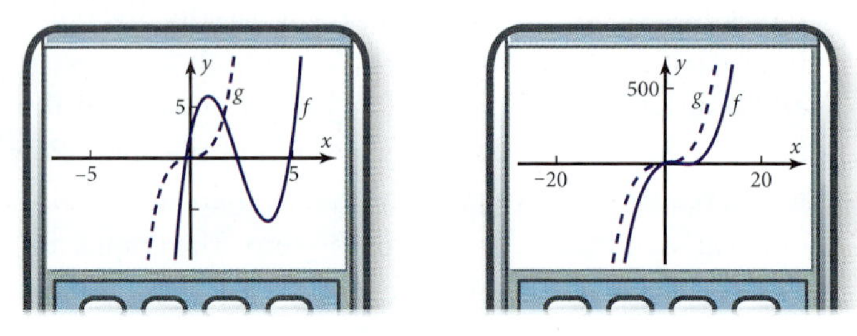

Figure 4-3b

Refer back to Figure 4-3a. A quadratic function graph has two branches, one downward and one upward. As a result there is one **extreme point** (or vertex, or **critical point**) and as many as two x-intercepts. Similarly, a cubic function graph can have three branches, a quartic function graph can have four branches, and a quintic function graph can have five branches. The number of x-intercepts can be as high as the degree of the function, and the number of extreme points can be up to one less than the degree. As you saw in the exploration at the beginning of this section, the number of zeros of a polynomial function is equal to the degree of the function if you count multiple zeros and complex zeros.

> **DEFINITIONS: Zeros and x-Intercepts of a Function**
>
> A zero of a function f is a number c, real or complex, for which $f(c) = 0$.
>
> An x-intercept of a function f is a real zero of that function.

As you saw in Exploration 4-3, it is possible for two or more zeros to occur at the same value of x. The function

$$f(x) = (x - 2)(x - 3)^2(x - 1)^3 = (x - 2)[(x - 3)(x - 3)][(x - 1)(x - 1)(x - 1)]$$

in Figure 4-3c has a single zero at $x = 2$, where the graph crosses the x-axis at an angle. There is a double zero at $x = 3$, where the graph only touches the x-axis but does not cross it. And there is a triple zero at $x = 1$, where the graph levels off and crosses the x-axis tangent to it. Algebraically, you can see from the factored form of $f(x)$ on the far right that the *multiplicity* of a zero is the number of identical linear factors that are zero for that particular value of x.

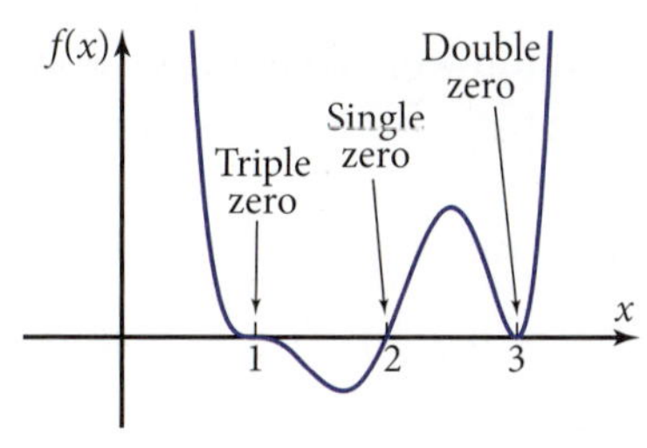

Figure 4-3c

Finding Zeros of a Polynomial Function: Synthetic Substitution

Synthetic substitution is a quick pencil-and-paper method for evaluating a polynomial function. Suppose that $f(x) = x^3 - 9x^2 - x + 105$ and that you want to find $f(6)$ by synthetic substitution.

- Write the x-value, 6, and the coefficients of $f(x)$ like this:

$$\begin{array}{r|rrrr} 6 & 1 & -9 & -1 & 105 \\ & & & & \\ \hline & & & & \end{array}$$

Leave space here.

Leave space here.

- Bring down the leading coefficient, 1, below the line and multiply it by the 6, writing the answer in the next column, under the -9.

$$\begin{array}{r|rrrr} 6 & 1 & -9 & -1 & 105 \\ & & 6 & & \\ \hline & 1 & & & \end{array}$$

- Add the −9 and the 6, and write the answer, −3, below the line. Multiply −3 by 6, and write the answer, −18, above the line. Repeat the steps, adding and multiplying. The final result is

$$\begin{array}{r|rrrr} 6 & 1 & -9 & -1 & 105 \\ & & 6 & -18 & -114 \\ \hline & 1 & -3 & -19 & \underline{-9} \end{array}$$

$\therefore f(6) = -9$ The value of the function $f(6)$ is the last number below the line.

Check:

$$f(6) = 6^3 - 9(6^2) - 1(6) + 105 = -9$$

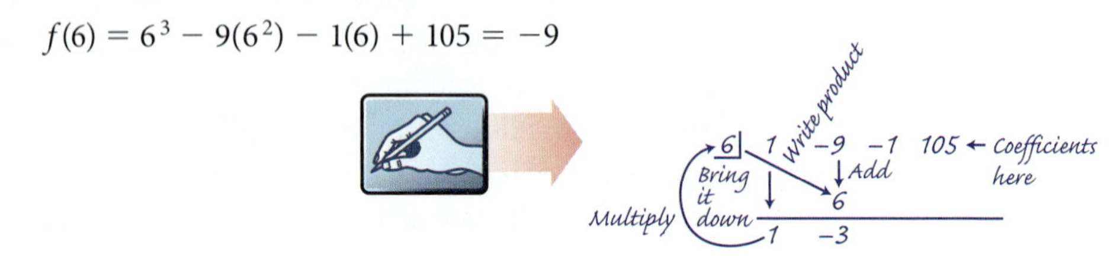

To see why synthetic substitution works, factor the polynomial into nested form:

$$f(x) = x^3 - 9x^2 - x + 105$$
$$= (1x - 9)x^2 - x + 105$$
$$= [(1x - 9)x - 1]\,x + 105 \qquad \text{Nested form.}$$

In this form you can evaluate the polynomial by repeating these steps:

Multiply by x.

Add the next coefficient.

Synthetic substitution is closely related to long division of polynomials. To see why, divide $f(x)$ by $(x - 6)$, a linear binomial that equals zero when x is 6. First, divide the x from the term $(x - 6)$ into the x^3 from the polynomial. Write the answer, x^2, above the x^2-term of the polynomial.

$$\begin{array}{rll} & x^2 - 3x - 19 & \text{Quotient} \\ x - 6 \big) & x^3 - 9x^2 - x + 105 & \\ & \underline{x^3 - 6x^2} & \text{Multiply } x^2 \text{ and } (x-6). \\ & -3x^2 - x & \text{Subtract and bring } -x \text{ down.} \\ & \underline{-3x^2 + 18x} & \text{Multiply } -3x \text{ and } (x-6). \\ & -19x + 105 & \text{Subtract and bring the next term down.} \\ & \underline{-19x + 114} & \text{Multiply } -19 \text{ and } (x-6). \\ & -9 & \text{Remainder.} \end{array}$$

$$\therefore \frac{x^3 - 9x^2 - x + 105}{x - 6} = x^2 - 3x - 19 + \frac{-9}{x - 6} \qquad \text{“Mixed-number” form.}$$

Note: The term *“mixed-number” form* is used here because of the similarity of this form to the result of whole-number division when there is a remainder. For example, when you divide 13 by 4, the quotient is 3 and the remainder is 1, so you can write $\frac{13}{4}$ as $3\frac{1}{4}$.

Notice that the coefficients of the quotient, 1, −3, and −19, are the values below the line in the synthetic substitution process. Thus synthetic substitution gives you a way to do long division of a polynomial by a linear binomial expression. Just substitute the value of x for which the linear binomial equals zero.

The fact that the *remainder* after division by $(x - 6)$ equals the *value* of $f(6)$ is an example of the *remainder theorem.*

> **PROPERTY: The Remainder Theorem**
>
> If $p(x)$ is a polynomial, then $p(c)$ equals the remainder when $p(x)$ is divided by the quantity $(x - c)$.

> **COROLLARY: The Factor Theorem**
>
> The quantity $(x - c)$ is a factor of the polynomial $p(x)$ if and only if $p(c) = 0$.

The corollary is true because if the remainder equals zero, then $(x - c)$ divides $p(x)$ evenly, which means $(x - c)$ is a *factor* of $p(x)$.

EXAMPLE 1 ➤ Let $f(x) = x^3 - 4x^2 - 3x + 2$.

Let $g(x) = x^3 - 4x^2 - 3x + 18$.

Let $h(x) = x^3 - 4x^2 - 3x + 54$.

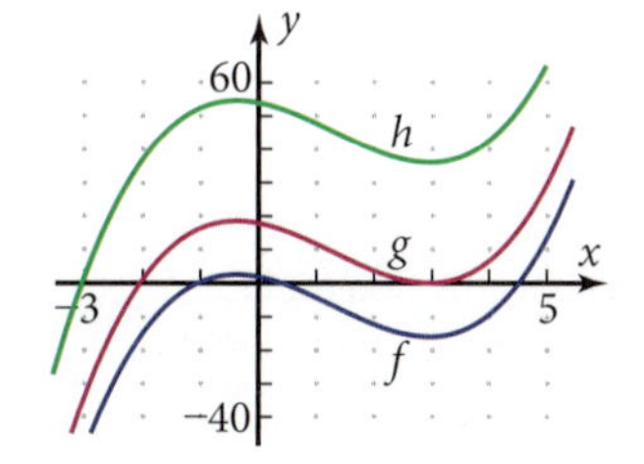

Figure 4-3d

The graphs are shown in Figure 4-3d.

a. Show that −1 is a zero of $f(x)$. Find the other two zeros, and check by graphing.

b. Show that −2 is a zero of $g(x)$. Find the other two zeros, and check by graphing.

c. Show that −3 is a zero of $h(x)$. Find the other two zeros, and check by graphing.

SOLUTION For each part of the problem, use synthetic substitution to show that the remainder is zero.

a.

$$\begin{array}{r|rrrr} -1 & 1 & -4 & -3 & 2 \\ & & -1 & 5 & -2 \\ \hline & 1 & -5 & 2 & \underline{0} \end{array}$$

Synthetically substitute −1 for x.

The remainder is zero.

Therefore, −1 is a zero of $f(x)$. $f(-1) = 0$ because $f(-1)$ equals the remainder.

$f(x) = (x + 1)(x^2 - 5x + 2)$ — $(x + 1)$ is a factor of $f(x)$ because $f(x)$ equals zero when $x = -1$; coefficients of the other factor appear in the bottom row of the synthetic substitution.

$x^2 - 5x + 2 = 0$ — Set the other factor equal to zero.

$x = 4.5615...$ and $x = 0.4384...$ are also zeros of $f(x)$. — Solve using the quadratic formula.

The graph of f in Figure 4-3d crosses the x-axis at about -1, 0.4, and 4.6, values that agree with the algebraic solutions.

b.

-2	1	-4	-3	18	Synthetically substitute -2 for x.
		-2	12	-18	
	1	-6	9	0	The remainder is zero.

Therefore, -2 is a zero of $g(x)$. — $g(-2) = 0$ because it equals the remainder.

$g(x) = (x + 2)(x^2 - 6x + 9)$ — $(x + 2)$ is a factor of $g(x)$ because $g(x)$ equals zero when $x = -2$.

$g(x) = (x + 2)(x - 3)(x - 3)$ — The second factor can itself be factored.

-2, 3, and 3 are zeros of $g(x)$. — 3 is called a *double zero* of $g(x)$.

The graph of g in Figure 4-3d crosses the x-axis at -2 and touches the x-axis at 3, in agreement with the algebraic solutions.

c.

-3	1	-4	-3	54	Synthetically substitute -3 for x.
		-3	21	-54	
	1	-7	18	0	The remainder is zero.

Therefore, -3 is a zero of $h(x)$. — $h(-3) = 0$ because it equals the remainder.

$h(x) = (x + 3)(x^2 - 7x + 18)$ — $(x + 3)$ is a factor of $h(x)$ because $h(x)$ equals zero when $x = -3$.

$x^2 - 7x + 18 = 0$ — Set the other factor equal to zero.

$$x = \frac{7 \pm \sqrt{7^2 - 4(1)(18)}}{2(1)}$$ Use the quadratic formula.

$$= 3.5 \pm 0.5\sqrt{-23}$$

$$= 3.5 + 2.3979...i \quad \text{or} \quad 3.5 - 2.3979...i$$

Complex solutions are a conjugate pair.

The graph of h in Figure 4-3d crosses the x-axis only at -3, which agrees with the algebraic solutions.

Notes:

- In part b of Example 1, $x = 3$ is called a double zero of $f(x)$. It appears twice, once for each factor $(x - 3)$. As you can see from Figure 4-3d, the graph of g just touches the x-axis and does not cross it at $x = 3$. So there are three zeros, -2, 3, and 3, although there are only two distinct x-intercepts.

- In part c of Example 1, there are three zeros of $h(x)$ but only one x-intercept, as you can see in Figure 4-3d. The other two zeros are nonreal complex numbers. As you saw in Section 4-2, the two complex zeros are *complex conjugates* of each other.

The results of Example 1 illustrate the *fundamental theorem of algebra* and its corollaries.

PROPERTY: The Fundamental Theorem of Algebra and Its Corollaries

A polynomial function has at least one zero in the set of complex numbers.

Corollary
An nth-degree polynomial function has exactly n zeros in the set of complex numbers, counting multiple zeros.

Corollary
If a polynomial has only real coefficients, then any nonreal complex zeros appear in conjugate pairs.

Be sure you understand what the theorem says. As you saw in Figure 4-2c, the real numbers are a subset of the complex numbers, so the zeros of the function could be nonreal or real complex numbers. The x-intercepts correspond to the real-number zeros of the function. From now on, these real-number zeros will be referred to as *real zeros* and nonreal complex zeros wil be referred to as *complex zeros.*

EXAMPLE 2 ➤ Identify the degree and the number of real and nonreal complex zeros that the polynomial function in Figure 4-3e could have. State whether the leading coefficient is positive or negative.

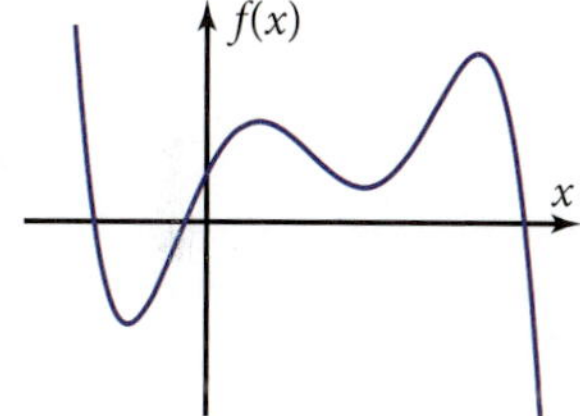

Figure 4-3e

SOLUTION The function could be 5th or higher degree because it has five branches—down, up, down, up, down—resulting in four extreme points in the domain shown. The degree of the polynomial must be odd, because for large negative values of x, $f(x)$ becomes larger in the positive direction, whereas for large positive values of x, $f(x)$ becomes larger in the negative direction.

There are three real zeros because there are three x-intercepts, none of which is a repeated zero. If the degree is 5, then there are two complex zeros because the total number of zeros must be five.

The leading coefficient is negative because $f(x)$ becomes very large in the negative direction as x becomes larger in the positive direction. ◄

Sums and Products of Zeros

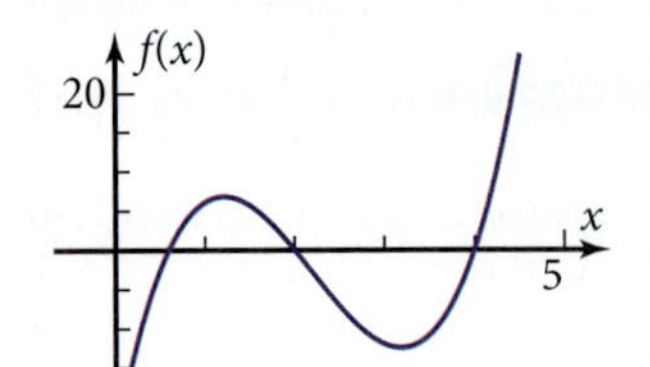

Figure 4-3f

Figure 4-3f shows the graph of the cubic function

$$f(x) = 5x^3 - 33x^2 + 58x - 24$$

The three zeros are $x = z_1 = 0.6$, $x = z_2 = 2$, and $x = z_3 = 4$. If you factor out the leading coefficient, 5, the sum and product of these zeros appear in the equation.

$$f(x) = 5\left(x^3 - \frac{33}{5}x^2 + \frac{58}{5}x - \frac{24}{5}\right) = 5(x^3 - 6.6x^2 + 11.6x - 4.8)$$

$$z_1 + z_2 + z_3 = 0.6 + 2 + 4 = 6.6$$ The opposite of the quadratic coefficient.

$$z_1z_2z_3 = (0.6)(2)(4) = 4.8$$ The opposite of the constant term.

The sum of the products of the zeros taken two at a time equals the linear coefficient.

$$z_1z_2 + z_1z_3 + z_2z_3 = (0.6)(2) + (0.6)(4) + (2)(4) = 11.6$$

Equal to the linear coefficient.

To see why these properties hold, start with $f(x)$ in factored form, expand it, and combine like terms *without* completing the calculations.

$$f(x) = 5(x - 0.6)(x - 2)(x - 4)$$

$$= 5[x^3 - 0.6x^2 - 2x^2 - 4x^2 + (0.6)(2)x + (0.6)(4)x + (2)(4)x - (0.6)(2)(4)]$$

$$= 5[x^3 \underbrace{- (0.6 + 2 + 4)x^2}_{\text{Opposite of sum}} + \underbrace{[(0.6)(2) + (0.6)(4) + (2)(4)]x}_{\text{Sum of pairwise products}} \underbrace{- (0.6)(2)(4)}_{\text{Opposite of product}}]$$

This property is true, in general, for any cubic function. In Problems 38 and 39, you will extend this property to quadratic functions and to higher-degree functions.

PROPERTY: Sums and Products of the Zeros of a Cubic Function

If the cubic function $p(x) = ax^3 + bx^2 + cx + d$ has zeros z_1, z_2, and z_3, then

$$z_1 + z_2 + z_3 = -\frac{b}{a}$$ Sum of the zeros.

$$z_1z_2 + z_1z_3 + z_2z_3 = \frac{c}{a}$$ Sum of the pairwise products of the zeros.

$$z_1z_2z_3 = -\frac{d}{a}$$ Product of the zeros.

This property allows you to determine something about the zeros of the function without actually calculating them. It also gives you a way to find the particular equation of a cubic function when you know its zeros.

EXAMPLE 3 ➤ Find the zeros of the function $f(x) = x^3 - 13x^2 + 59x - 87$.

Then show that the sum of the zeros, the sum of their pairwise products, and the product of all three zeros correspond to the coefficients of the equation that defines function f.

SOLUTION Plot the graph to help you find that $x = 3$ is a zero (Figure 4-3g).

Find the other factors by synthetic substitution.

$$\begin{array}{r|rrrr} 3 & 1 & -13 & 59 & -87 \\ & & 3 & -30 & 87 \\ \hline & 1 & -10 & 29 & \underline{0} \end{array}$$

Figure 4-3g

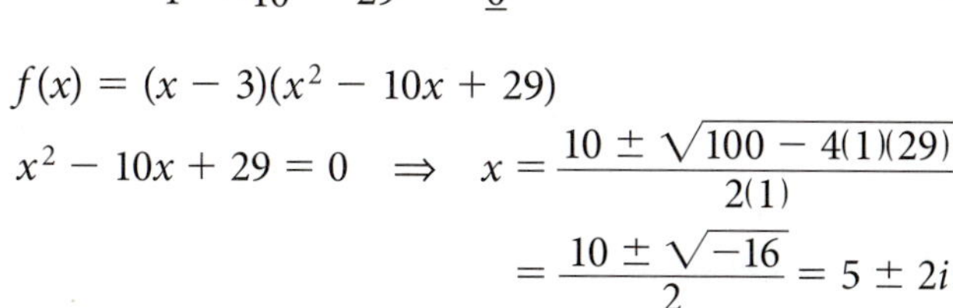

$$f(x) = (x - 3)(x^2 - 10x + 29)$$

$$x^2 - 10x + 29 = 0 \quad \Rightarrow \quad x = \frac{10 \pm \sqrt{100 - 4(1)(29)}}{2(1)}$$

$$= \frac{10 \pm \sqrt{-16}}{2} = 5 \pm 2i$$

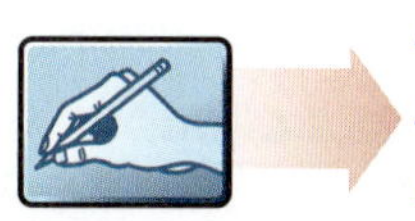

Sum $= -\frac{b}{a}$

Sum of pairwise products $= \frac{c}{a}$

Product $= -\frac{d}{a}$

Sum: $3 + (5 + 2i) + (5 - 2i) = 13$ — The opposite of the quadratic coefficient.

Pairwise-product sum:

$3(5 + 2i) + 3(5 - 2i) + (5 + 2i)(5 - 2i)$ — Recall that $i^2 = -1$.

$= 15 + 6i + 15 - 6i + 25 + 4$

$= 59$ — The linear coefficient.

Product: $3(5 + 2i)(5 - 2i) = 3(25 + 4) = 87$

The opposite of the constant term.

EXAMPLE 4 Find the particular equation for a cubic function with integer coefficients if the function's zeros have the given sum, product, and sum of pairwise products. Confirm these properties after finding the zeros of the function.

Sum: $-\frac{5}{3}$ Sum of pairwise products: $-\frac{58}{3}$ Product: $\frac{40}{3}$

SOLUTION The particular equation for one possible function is

$$y = x^3 + \frac{5}{3}x^2 - \frac{58}{3}x - \frac{40}{3}$$

The simplest particular equation for a function with integer coefficients is

$$f(x) = 3x^3 + 5x^2 - 58x - 40$$

By the graph (Figure 4-3h), the zeros are $x = -5$, $x = -\frac{2}{3}$, and $x = 4$.

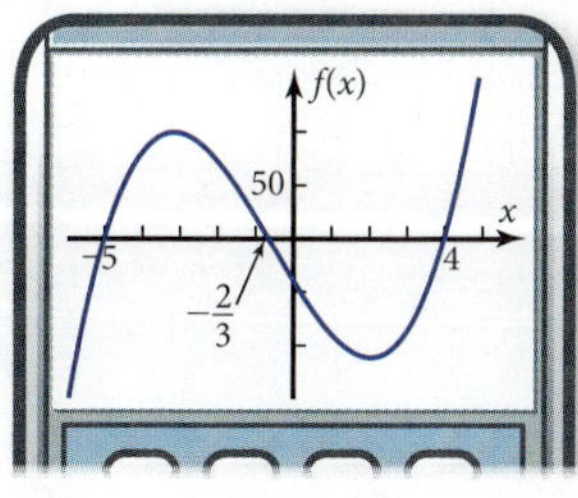

Figure 4-3h

Check:

Sum: $(-5) + \left(-\frac{2}{3}\right) + 4 = -\frac{5}{3}$ (Correct)

Sum of the pairwise products:

$(-5)\left(-\frac{2}{3}\right) + (-5)(4) + \left(-\frac{2}{3}\right)(4) = -\frac{58}{3}$ (Correct)

Product: $(-5)\left(-\frac{2}{3}\right)(4) = \frac{40}{3}$ (Correct)

Problem Set 4-3

Reading Analysis

From what you have read in this section, what do you consider to be the main idea? How can you determine the degree of a polynomial function by looking at its graph? What are the similarities and differences between zeros of a function and x-intercepts? How is the factor theorem related to the remainder theorem? How can you find the sum and the product of the zeros of a cubic function simply by looking at the equation?

Quick Review

Q1. Sketch the graph of $y = x^2$.

Q2. Sketch the graph of $y = x^3$.

Q3. Sketch the graph of $y = x^4$.

Q4. Sketch the graph of $y = 2^x$.

Q5. Multiply the complex numbers and simplify: $(3 + 2i)(5 + 6i)$

Q6. Solve the equation using the quadratic formula: $x^2 - 14x + 54 = 0$

Q7. Solve the equation using the quadratic formula: $x^2 - 14x + 58 = 0$

Q8. Multiply and simplify: $(4 - 7i)^2$

Q9. What type of function has the add–multiply property?

Q10. The vertex of the graph of the quadratic function $f(x) = 3(x - 4)^2 - 5$ is at

A. $(3, 4)$

B. $(4, 5)$

C. $(4, -5)$

D. $(-4, 5)$

E. $(-4, -5)$

---

1. Given the function $p(x) = x^3 - 5x^2 + 2x + 8$,

a. Plot the graph using an appropriate domain. How many branches (increasing or decreasing) does the graph have? How is this number related to the degree of $p(x)$?

b. Find graphically the three real zeros of $p(x)$.

c. Show by synthetic substitution that $(x + 1)$ is a factor of $p(x)$. Use the result to write the other factor, and factor it further if you can.

d. Explain the relationship between the zeros of $p(x)$ and the factors of $p(x)$.

2. Given the function $p(x) = x^3 - 3x^2 + 9x + 13$,

a. Plot the graph using an appropriate domain. From the graph, find the real zero of $p(x)$.

b. Show by synthetic substitution that $(x + 1)$ is a factor of $p(x)$. Write the other factor.

c. Find the other two zeros. Substitute one of the two complex zeros into the equation for $p(x)$ to show that it really is a zero.

d. Explain the relationship between the zeros of $p(x)$ and the graph of $p(x)$.

For the polynomial functions in Problems 3–8, give the minimum possible degree, the number of real zeros the function could have (counting multiple zeros), the number of complex zeros the function could have, and the sign of the leading coefficient.

3.

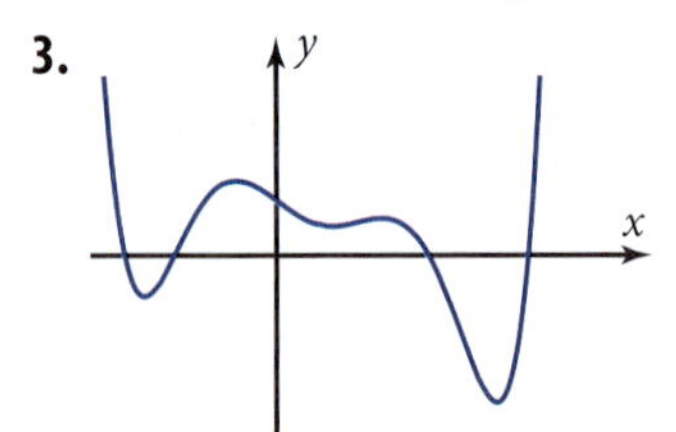

4.

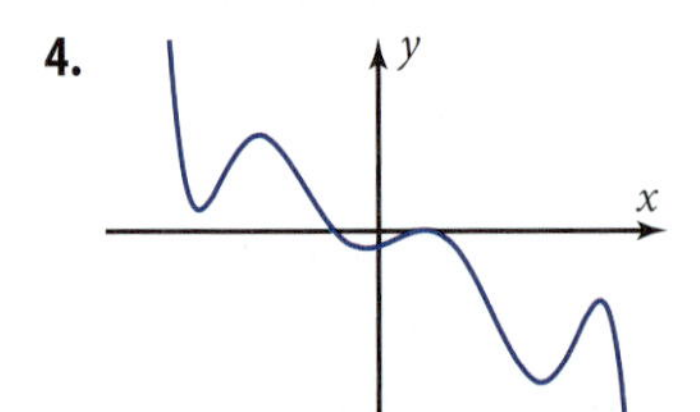

5.

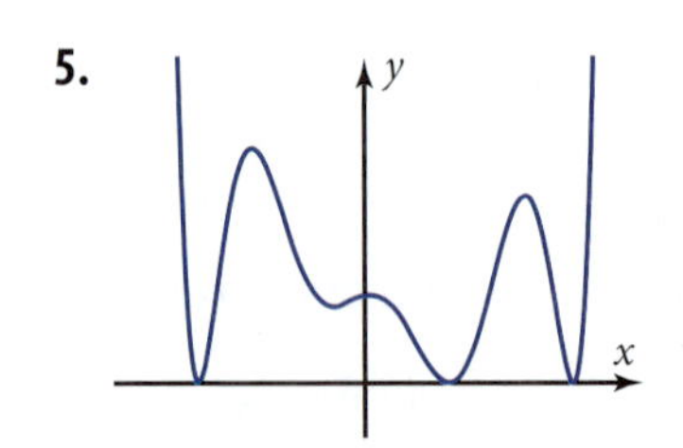

6.

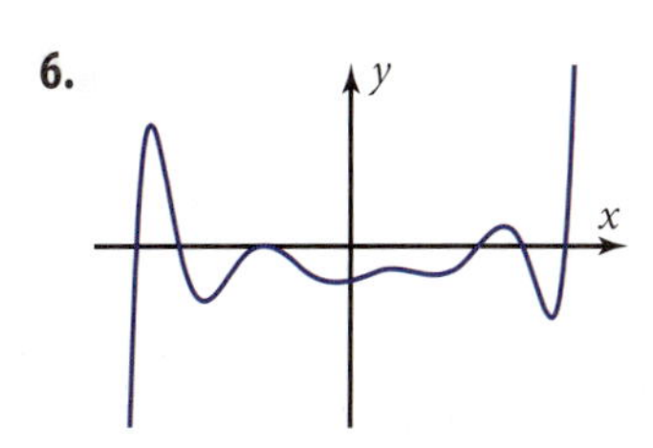

7.

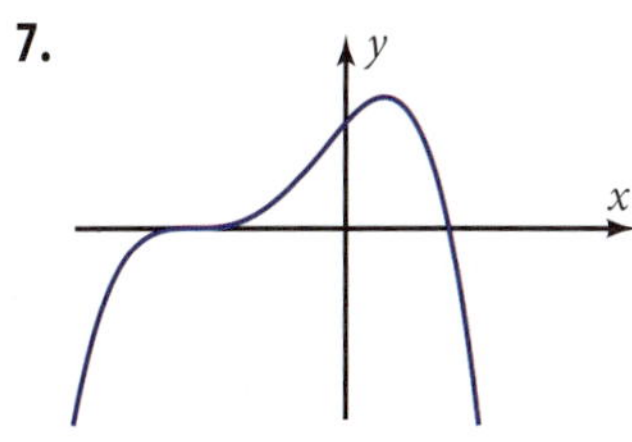

8.

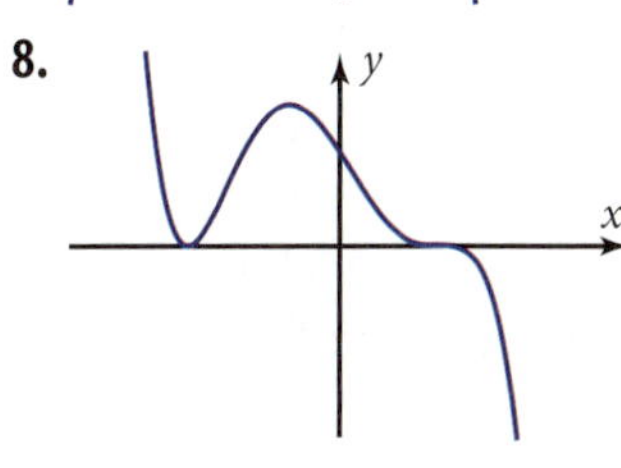

For Problems 9–20, sketch the graph of the polynomial function described, or explain why no such function can exist.

9. Cubic function with two distinct negative zeros, one positive zero, and a positive y-intercept

10. Cubic function with a negative double zero, a positive zero, and a negative leading coefficient

11. Cubic function with one real zero, two complex zeros, and a positive leading coefficient

12. Cubic function with no real zeros

13. Cubic function with no extreme points

14. Quartic function with no extreme points

15. Quartic function with no real zeros

16. Quartic function with two distinct positive zeros, two distinct negative zeros, and a negative y-intercept

17. Quartic function with two double zeros

18. Quartic function with two distinct real zeros and two complex zeros

19. Quartic function with five distinct real zeros

20. Quintic function with five distinct real zeros

For Problems 21–24, find quickly the sum, the product, and the sum of the pairwise products of the zeros from the coefficients, using the properties in this section. Then find the zeros and confirm that your answers satisfy the properties.

21. $f(x) = x^3 - x^2 - 22x + 40$

22. $f(x) = x^3 + x^2 - 7x - 15$

23. $f(x) = -5x^3 - 18x^2 + 7x + 156$

24. $f(x) = 2x^3 - 9x^2 - 8x + 15$

For Problems 25–28, find a particular equation for the cubic function, with zeros as described and leading coefficient equal to 1. Then find the zeros and confirm that your answers satisfy the given properties.

25. Sum: 4; sum of the pairwise products: -11; product: -30

26. Sum: 9; sum of the pairwise products: 26; product: 24

27. Sum: 8; sum of the pairwise products: 29; product: 52

28. Sum: -5; sum of the pairwise products: 4; product: 10

For Problems 29 and 30,

a. By synthetic substitution, find $p(c)$.

b. Write $\frac{p(x)}{x-c}$ in mixed-number form.

29. $p(x) = x^3 - 7x^2 + 5x + 4$, $c = 2$ and $c = -3$

30. $p(x) = x^3 - 9x^2 + 2x - 5$, $c = 3$ and $c = -2$

31. State the remainder theorem.

32. State the factor theorem.

33. State the fundamental theorem of algebra.

34. State the two corollaries of the fundamental theorem of algebra.

35. *Spaceship Problem:* Ella Vader (Darth's other daughter) is approaching Alderaan in her spaceship. Her distance from the surface, $d(x)$, in kilometers, at time x, in minutes, after she starts a maneuver, is given by the function

$$d(x) = x^4 - 22x^3 + 158x^2 - 414x + 405$$

a. State the degree of the function, and give the number of terms.

b. Plot the graph of function d and sketch the result. Use a window with $0 \le x \le 12$ and an appropriate range for y.

c. How far was Ella from Alderaan when she started the maneuver? What part of the function tells you this?

d. Function d has an extreme point at a time soon after Ella started the maneuver. Find Ella's distance from Alderaan at this point and the time at which she reached this point. Describe the method you use.

e. Function d has a double zero. Find this zero. Describe what happens to Ella's spaceship at this zero.

f. Function d has two complex zeros. One of these zeros is $x = 2 + i$. What is the other complex zero? Show that $2 + i$ really is a zero of function d.

g. Find the sum of the zeros of $d(x)$, counting the double zero twice. Where does this number appear in the equation of the function?

36. *Bent Board Problem:* A group of students conducts a project in which they construct a bridge out of a board 10 ft long. They nail down the board at distances $x = 0$ ft, $x = 4$ ft, and $x = 10$ ft from the left end of the board. Then one of the group members stands on the board at distance $x = 3$ ft. As a result, the board bends into the shape shown in Figure 4-3i. The vertical scale has been expanded for clarity.

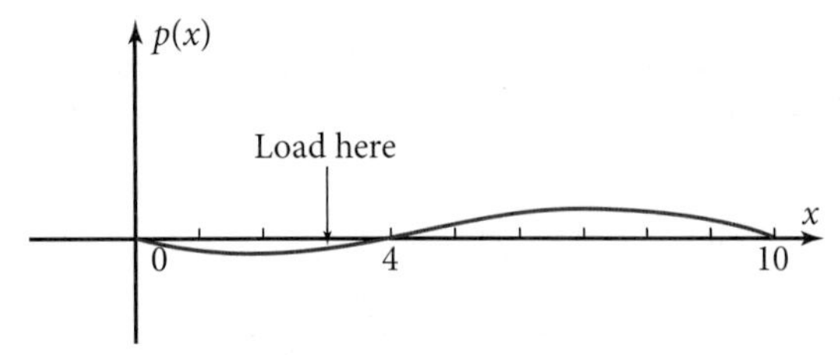

Figure 4-3i

a. Assume that the board bends into the shape of the graph of a cubic function with zeros at the three support points. Find the equation for $p(x)$, a cubic polynomial function with these three zeros and a leading coefficient equal to 1.

b. The actual deflection, $d(x)$, in inches, of the board from its rest position is a vertical dilation of function p by a constant factor of a. That is, $d(x) = a \cdot p(x)$. If $d(3) = -0.7$, find the value of a.

c. Explain why the equation for $d(x)$ does not give meaningful answers for values of x less than 0 or greater than 10, and thus why the domain of x is $0 \le x \le 10$.

d. At what value of x within the domain does the board reach its maximum positive deflection? How far above the board's rest position (the x-axis) is the board at this value of x?

37. *Bungee Jumping Problem:* Lucy Lastic bungee jumps from a high tower. On the way down she passes an elevated walkway on which her friends are standing. Her distance, $f(x)$, in feet above the walkway, is given by the cubic function $f(x) = -x^3 + 13x^2 - 50x + 56$ where x is the number of seconds that have elapsed since she jumped. Part of the graph of function f is shown in Figure 4-3j.

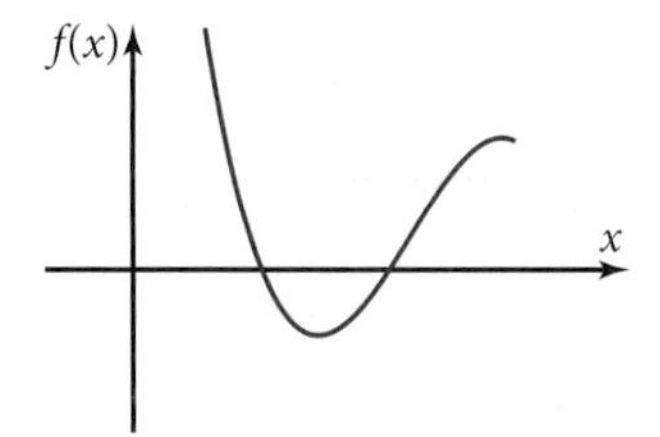

Figure 4-3j

a. How high above the platform was Lucy when she jumped? What part of the mathematical model tells you this?

b. By synthetic substitution, show that 2 is a zero of $f(x)$ and thus that Lucy was passing the walkway at time $x = 2$ s. Was she going down or coming up at this time?

c. Find the other zeros of $f(x)$. Show that the sum, the product, and the sum of the pairwise products of these zeros agree with the properties you have learned in this section.

d. Explain why $f(x)$ has the wrong end behavior for large positive values of x.

38. *Quadratic Function Sum and Product of Zeros Problem:* By the quadratic formula, the zeros of the general quadratic function $f(x) = ax^2 + bx + c$ are

$$z_1 = \frac{-b + \sqrt{b^2 - 4ac}}{2a} \quad \text{and}$$

$$z_2 = \frac{-b - \sqrt{b^2 - 4ac}}{2a}$$

By finding $z_1 + z_2$ and $z_1 z_2$, show that there is a property of the sum and product of zeros of a quadratic function that is similar to the corresponding property for cubic functions.

39. *Quartic Function Sum and Product of Zeros Problem:* By repeated synthetic substitution or using your grapher, find the zeros of the function

$$f(x) = 2x^4 + 3x^3 - 14x^2 - 9x + 18$$

Then find these quantities:

- Sum of the zeros ("products" of the zeros taken one at a time)
- Sum of all possible products of zeros taken two at a time
- Sum of all possible products of zeros taken three at a time
- Product of the zeros ("sum" of the products taken all four at a time)

From the results of your calculations, make a conjecture about how the property of the sums and products of the zeros can be extended to functions of degree higher than 3.

40. *Reciprocals of the Zeros Problem:* Prove that if the function

$$p(x) = ax^3 + bx^2 + cx + d$$

has zeros z_1, z_2, and z_3, then the function

$$q(x) = dx^3 + cx^2 + bx + a$$

has zeros $\frac{1}{z_1}$, $\frac{1}{z_2}$, and $\frac{1}{z_3}$.

41. *Horizontal Translation and Zeros Problem:* Let $f(x) = x^3 - 5x^2 + 7x - 12$. Let $g(x)$ be a horizontal translation of $f(x)$ by 1 unit in the positive direction. Find the equation for $g(x)$. Show algebraically that each zero of $g(x)$ is 1 unit larger than the corresponding zero of $f(x)$.

4-4 Fitting Polynomial Functions to Data

U.S. speed skater Shani Davis competed in the men's 1500 m at the 2006 Winter Olympic Games in Torino, Italy, winning the silver medal.

Figure 4-4a shows four cubic functions that could model the position of a moving object as a function of time. From left to right in the figure, the graphs show an object that

- Passes a reference point ($y = 0$) while slowing down (decreasing slope) and then speeds up again (increasing slope)
- Stops momentarily (zero slope) and then continues forward
- Reverses direction (negative slope) and then continues forward
- Approaches the reference point ($y = 0$) going in the negative direction

In all four cases, there is only one x-intercept, indicating that the other two zeros are nonreal complex numbers.

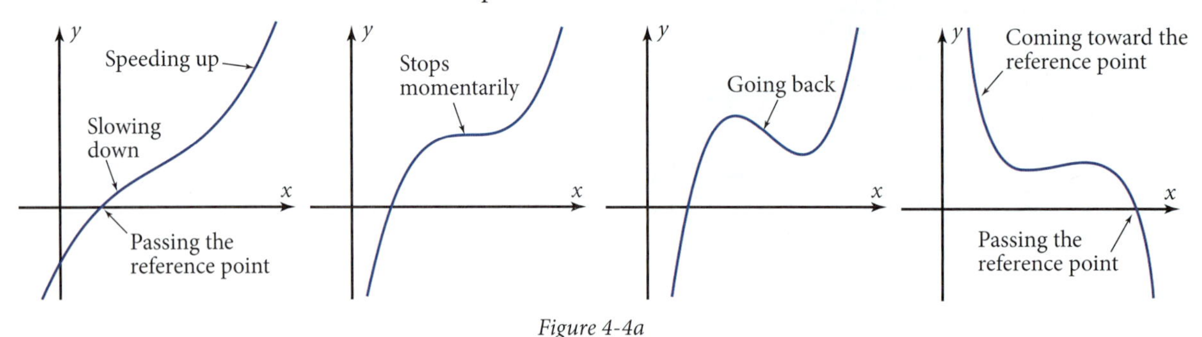

Figure 4-4a

In this section you will apply curve-fitting techniques to polynomial functions.

Objective Given a set of points, find the equation for the polynomial function that fits the data exactly or fits best for a given degree.

Concavity and Points of Inflection

A smoothly curved graph can have a concave (hollowed out) side and a convex (bulging) side. For reasons you will learn when you study calculus, mathematicians usually refer to the concave side. Figure 4-4b shows regions where the graph is concave up or concave down. The point at which the direction of concavity changes is called a **point of inflection.** The word *inflection* originates from the British spelling, *inflexion,* meaning "not flexed."

Half-coconut
Convex side
Concave side

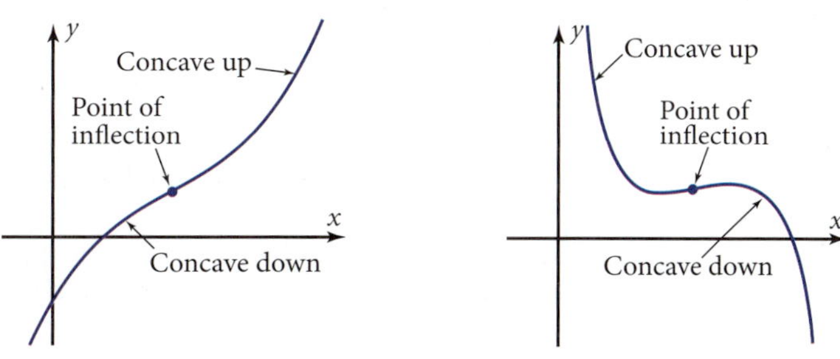

Figure 4-4b

Comparing the graphs on the left in Figure 4-4b and Figure 4-4a, notice that if a function is increasing and concave up, then y is increasing at an increasing rate. If the function is increasing but concave down, then y is still increasing, but at a decreasing rate.

A cubic function has exactly one point of inflection. Its x-coordinate is given by the formula in the box, a property you will prove in subsequent calculus courses.

> **PROPERTY: Horizontal Coordinate of a Cubic Function's Inflection Point**
>
> For the cubic function $f(x) = ax^3 + bx^2 + cx + d$, the horizontal coordinate of the point of inflection is
>
> $$x = \frac{-b}{3a}$$

Note: The vertex of the quadratic function $f(x) = ax^2 + bx + c$ is at $x = -\frac{b}{2a}$.

EXAMPLE 1 ➤ A cubic function f contains the points (6, 38), (5, 74), (2, 50), and (−1, 80).

a. Find the equation for f algebraically. Check by cubic regression.

b. Verify the answer in part a by plotting and tracing on the graph.

c. Find the x-coordinate of the point of inflection and show that it agrees with the graph.

SOLUTION

a. $f(x) = ax^3 + bx^2 + cx + d$ — Write the general equation.

$$\begin{cases} 216a + 36b + 6c + d = 38 \\ 125a + 25b + 5c + d = 74 \\ 8a + 4b + 2c + d = 50 \\ -a + b - c + d = 80 \end{cases}$$

Substitute 6 for x and 38 for $f(x)$, and so forth, to get a system of equations.

$$\begin{bmatrix} 216 & 36 & 6 & 1 \\ 125 & 25 & 5 & 1 \\ 8 & 4 & 2 & 1 \\ -1 & 1 & -1 & 1 \end{bmatrix}^{-1} \begin{bmatrix} 38 \\ 74 \\ 50 \\ 80 \end{bmatrix} = \begin{bmatrix} -2 \\ 15 \\ -19 \\ 44 \end{bmatrix}$$

Solve the system using matrices.

$\therefore f(x) = -2x^3 + 15x^2 - 19x + 44$ — Cubic regression gives the same equation, with $R^2 = 1$.

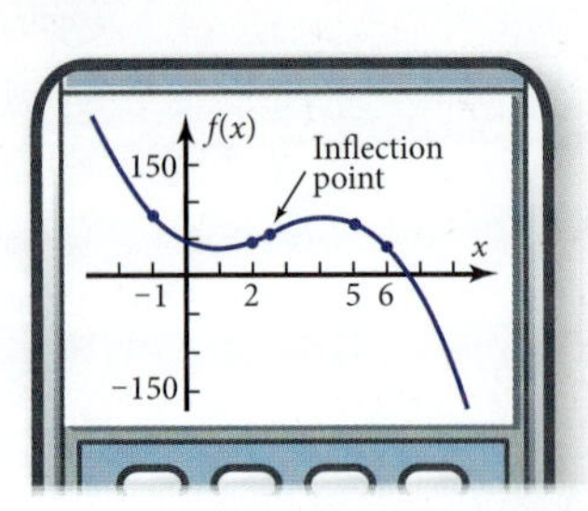

Figure 4-4c

b. The graph is shown in Figure 4-4c. Tracing to $x = 6$, $x = 5$, $x = 2$, and $x = -1$ confirms that the given points are on the graph. Note that for large positive values of x, the values of $f(x)$ get larger in the negative direction, consistent with the fact that the leading coefficient is negative.

c. $x = \dfrac{-b}{3a} = \dfrac{-15}{3(-2)} = 2.5$, which agrees with the graph in Figure 4-4c

Note: If you do not recall the matrix solution of linear systems from previous courses, see Section 13-2 for a refresher.

Recall the constant-second-differences property of quadratic functions. For a cubic function, the *third* differences between the y-values are constant. The table shows the result for the function $f(x) = -2x^3 + 15x^2 - 19x + 44$ from Example 1.

x	f(x)			
−1	80			
		−36		
0	44		30	
		−6		−12
1	38		18	
		12		−12
2	50		6	
		18		−12
3	68		−6	
		12		−12
4	80		−18	
		−6		−12
5	74		−30	
		−36		
6	38			

PROPERTY: Constant-*n*th-Differences Property

For an nth-degree polynomial function, if the x-values are equally spaced, then the $f(x)$-values have constant nth differences.

You will prove this property in Problem 14.

EXAMPLE 2 ➤ An object moving in a straight line passes a reference point at time $x = 2$ s. It slows down, stops, reverses direction, and passes the reference point going backward. Then it stops and reverses direction again, passing the reference point a third time. The table shows its displacements, $f(x)$, in meters, at various times.

x (s)	f(x) (m)
2	0
3	27
4	24
5	13
6	−4
7	−11
8	6
9	32

a. Make a scatter plot of the data. Explain why a cubic function would be a reasonable mathematical model for displacement as a function of time.

b. Write the equation for the best-fitting cubic function. Plot the graph of the function on the scatter plot in part a.

c. Use the equation in part b to calculate the approximate time the object passes the reference point going backward.

d. Show that a quartic function gives a coefficient of determination closer to 1 but has the wrong behavior for the given information.

SOLUTION

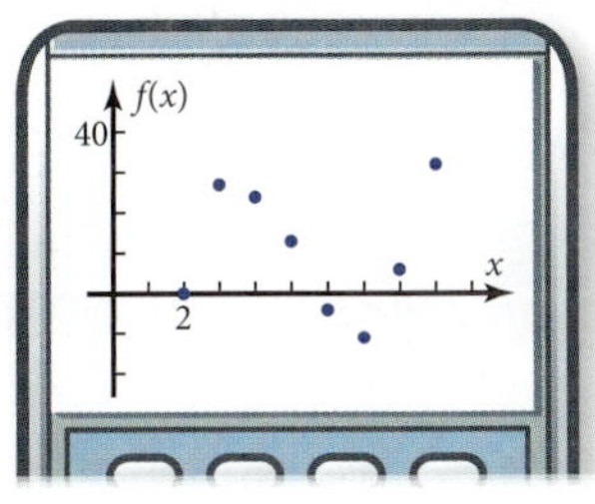

Figure 4-4d

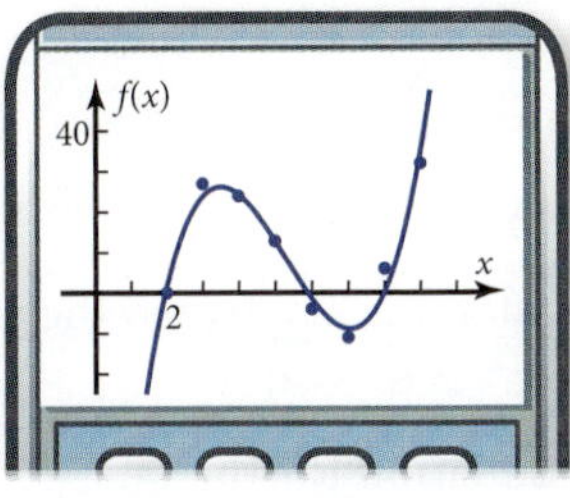

Figure 4-4e

a. Figure 4-4d shows the scatter plot. A cubic function is a reasonable mathematical model because it can reverse direction twice (it has two extreme points), as shown by the scatter plot.

b. By cubic regression, the equation is

$$f(x) = 1.5782...x^3 - 25.0119...x^2 + 117.2669...x - 145.4761...$$

$R^2 = 0.9611...$, indicating a reasonably good fit because it is close to 1. The graph of f in Figure 4-4e shows that it is a reasonably good fit.

c. From the graph or the table, the x-value when the object passes the reference point going backward is close to 6. Use the zeros, intersect, or solver feature on your grapher.

$$x = 5.8908... \approx 5.9 \text{ s}$$

d. Quartic regression gives a coefficient of determination $R^2 = 0.9861...$, which is closer to 1 than the $R^2 = 0.961...$ from the cubic regression. However, the quartic function would have a third extreme point and would cross the x-axis a fourth time, which wasn't mentioned in the statement of the problem.

Problem Set 4-4

Reading Analysis

From what you have read in this section, what do you consider to be the main idea? How is the constant-second-differences property of quadratic functions extended to higher-degree functions? How would you go about finding the equation for the quartic function that best fits a given set of data points algebraically? Numerically?

Quick Review

Q1. Multiply: $(x - 3)(x - 5)$

Q2. Multiply and simplify: $(3 + 2i)(5 + 4i)$

Q3. Expand: $(x - 7)^2$

Q4. Expand and simplify: $(5 + 3i)^2$

Q5. What is the maximum number of extreme points a quintic function graph can have?

Q6. One zero of a particular cubic function with real-number coefficients is $-7 + 4i$. What is another zero?

Q7. Sketch the graph of a cubic function with a positive double zero and x^3-coefficient -2.

Q8. Find the sum of the zeros of the function

$$f(x) = 2x^3 + 7x^2 - 5x + 13$$

Q9. If polynomial $p(x)$ has remainder 7 when divided by $(x - 5)$, find $p(5)$.

Q10. If polynomial $p(x)$ has $p\left(\frac{-3}{2}\right) = 0$, then a factor of $p(x)$ is

A. $x - 3$ **B.** $x + 3$ **C.** $x - 2$
D. $x + 2$ **E.** $2x + 3$

---

1. Given $p(x) = x^3 - 5x^2 + 2x + 10$,

a. Plot the graph using an appropriate domain. Sketch the result.

b. How many zeros does the function have? How many extreme points does the graph have? How are these numbers related to the degree of $p(x)$?

c. Make a table of values of $p(x)$ for each integer value of x from 4 to 9. Show that the third differences between the $p(x)$-values are constant.

d. Find the x-coordinate of the point of inflection, and mark this point on the graph.

2. Given $p(x) = -x^4 + 6x^3 + 6x^2 - 12x + 11$,

   a. Plot the graph using an appropriate domain. Sketch the result.

   b. How many zeros does the function have? How many extreme points does the graph have? How are these numbers related to the degree of $p(x)$?

   c. Make a table of values of $p(x)$ for each integer value of x from -2 to 4. Show that the fourth differences between the $p(x)$-values are constant.

   d. Darken the parts of the graph that are concave down.

3. The values in the table give the coordinates of points that are on the graph of function f:

x	$f(x)$
2	25.4
3	13.1
4	−3.8
5	−23.5
6	−44.2
7	−64.1

   a. Make a scatter plot of the points.

   b. Show that the third differences between the $f(x)$-values are constant.

   c. Find algebraically the equation for the cubic function that fits the first four points. Show that running a cubic regression on all six points gives the same equation. Plot the equation for the function on the same screen as the scatter plot from part a.

   d. Find the x-coordinate of the point of inflection, and mark this point on the graph.

4. The values in the table give the coordinates of points that are on the graph of function g:

x	$g(x)$
2	−3
3	−25
4	−31
5	27
6	221
7	647
8	1425

   a. Make a scatter plot of the points.

   b. Show that the fourth differences between the $g(x)$-values are constant.

   c. Find algebraically the equation for the quartic function that fits the first five points. Show that running a quartic regression on all seven points gives the same equation. Plot the equation for the function on the same screen as the scatter plot from part a.

   d. Darken the parts of the graph that are concave up.

5. *Diving Board Problem:* Figure 4-4f shows a diving board deflected by a person standing on it. Theoretical results on strength of materials indicate that the deflection of such a *cantilever beam* below its horizontal rest position at any point x from the built-in end of the beam is a cubic function of x. Suppose that these deflections, $f(x)$, are measured in thousandths of an inch, where x is measured in feet.

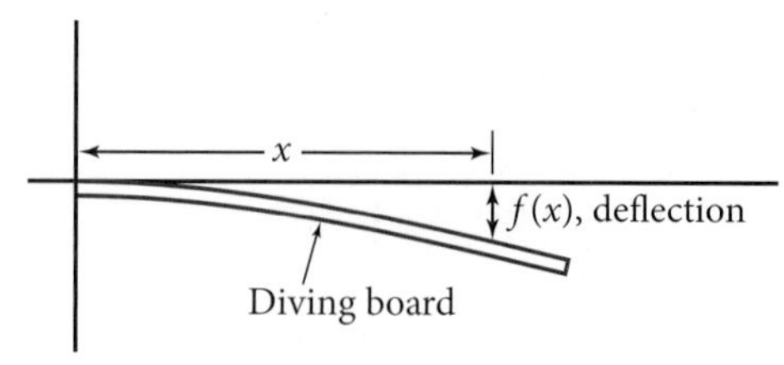

Figure 4-4f

x (ft)	f(x) (0.001 in.)
0	0
1	116
2	448
3	972

a. Write the equation for the cubic function that fits the data in the table. Show that the linear and constant coefficients are zero.

b. How far does the diving board deflect when the person is standing at its end, $x = 10$ ft?

c. Show that the function has another zero but that it is out of the domain of the function.

d. Sketch the graph of the function, showing both vertices. Darken the part of the graph that is in the domain determined by the 10-ft-long diving board.

6. *Two-Stage Rocket Problem:* A two-stage rocket is fired straight up. After the first stage finishes firing, the rocket slows down until the second stage starts firing. Its altitudes, $h(x)$, in feet above the ground, at each 10 s after firing, are given in the table.

x (s)	h(x) (ft)
10	1750
20	3060
30	3510
40	3700
50	4230
60	5700

a. Show that the $h(x)$-values have constant third differences.

b. What type of function will fit the data exactly? Write its equation.

c. Plot the graph of h. Based on the graph, does the rocket start coming back down before the second stage fires? How can you tell?

d. The first stage of the rocket was fired at time $x = 0$. How do you explain the fact that the function in part b has a zero at time $x = 3$ s?

7. *Cylinder in a Paraboloid Problem:* Figure 4-4g shows the paraboloid formed by rotating about the y-axis the parabola portion of the graph $y = 16 - x^2$ that is above the x-axis. A cylinder is inscribed in the paraboloid with its upper rim touching the parabola at the sample point (x, y) in the first quadrant. Both x and y are measured in centimeters.

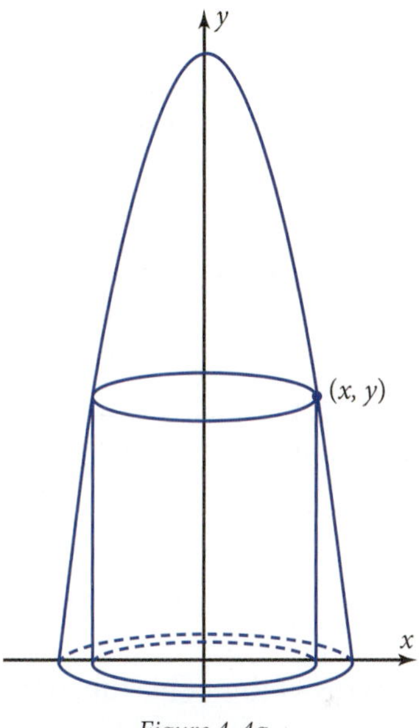

Figure 4-4g

a. Find the volume of the cylinder as a function of x and y. (Recall that the volume of a cylinder equals the area of the base multiplied by the altitude.) Then substitute $16 - x^2$ for y to get an equation for $v(x)$, the volume of the cylinder as a function of x alone. Explain why function v is a *quartic* function of x.

b. Show that function v has four real zeros, two of which are a double zero. Explain why one of the real zeros is outside the real-world domain of this function. A graph might help.

c. Find the value of x that gives the cylinder its maximum volume. What is the maximum volume?

8. *Christmas Tree Problem:* A charity organization sells Christmas trees of varying heights. Some prices they charge are shown in the table.

x (ft)	$p(x)$ (dollars)
4	58
5	78
6	100
7	130
8	174

a. Use the first four data points to find algebraically the equation for the cubic function that fits these data. Then show that the price of an 8-ft tree fits the mathematical model.

b. Plot the graph of function p. Use a window with $0 \leq x \leq 10$ and $-50 \leq y \leq 300$. Sketch the result.

c. How much would you expect to pay for a 20-ft tree? Is this surprising?

d. According to your cubic model, trees below a certain height are worthless. At what height does this happen? What part of the mathematical model tells you this?

e. Find the three zeros of this cubic function. Use the result to explain how you know there are no x-intercepts other than the one you found in part d.

f. The price of relatively short trees increases at a decreasing rate. The price of relatively tall trees increases at an increasing rate. At what height does this behavior change? Name the feature the graph has at this point.

9. *Television Set Pricing:* A retail store has various sizes of television sets made by the same manufacturer. The price of a television set is a function of the screen size, measured in inches along the diagonal. The prices are listed in the table.

x (in.)	$p(x)$ (dollars)
2	160
5	100
7	120
12	250
17	220
21	200
27	340
32	680
35	1100

a. Make a scatter plot of the data. Based on the scatter plot, explain why a quartic function is a more appropriate mathematical model than a cubic function.

b. Write the equation for the best-fitting quartic function, $p(x)$. Plot function p on the same screen as the scatter plot.

c. Based on the quartic model, which size television set is most overpriced?

d. What real-world reason can you think of to explain why the 17-in. and 21-in. sets are less expensive than the smaller, 12-in. set?

10. *Pilgrim's Bean Crop Problem:* When the Pilgrims arrived in America, they brought along seeds from which to grow crops. If they had planted beans, the number of bean plants would have increased rapidly, leveled off, and then decreased with the approach of winter. The next spring, a new crop would have come up. Suppose that the number of bean plants, $B(x)$, as a function of x, in weeks since the planting, is given by this table:

x	$B(x)$
3	59
4	113
5	160
6	203
7	240
8	272

Squanto Teaching Pilgrims by Charles Jefferys (The Granger Collection, New York)

a. Write the equation for the best-fitting cubic function. Plot the equation and the data on the same screen. Sketch the result.

b. According to the cubic model, what is the maximum number of bean plants the Pilgrims had in the first year? After how many weeks did the number of bean plants reach this maximum? Did any plants survive through the next winter? If so, what was the smallest number of plants, and when did the number reach this minimum? If not, when did the last plant die, and when did the first plant emerge the next spring?

c. Show that if $B(8)$ had been 273 instead of 272, the conclusions of part b would be much different. (This phenomenon is called *sensitive dependence on initial conditions.*)

d. In what year did the first Pilgrims arrive in America? What did they name the place where they landed?

11. *River Bend Problem:* A river meanders back and forth across Route 66. Three crossings are 1.7 mi, 3.8 mi, and 5.5 mi east of the intersection of Route 66 and Farm Road 13, or FM 13 (Figure 4-4h).

a. Use the sum and product of zeros properties to find quickly an equation of the cubic function $y = x^3 + bx^2 + cx + d$ that has 1.7, 3.8, and 5.5 as zeros. What is the y-intercept?

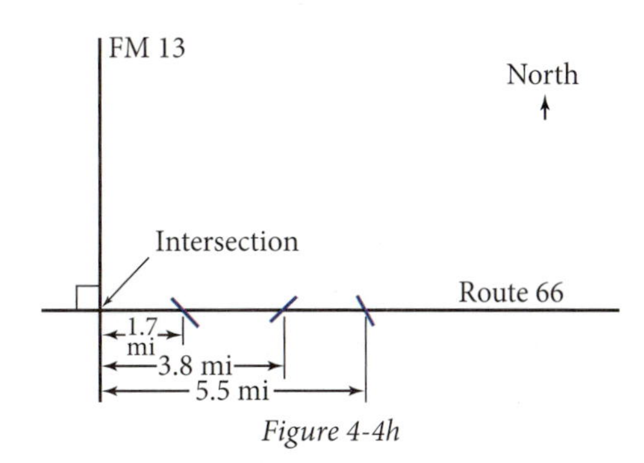

Figure 4-4h

b. Let $f(x)$ be the number of miles north of Route 66 for a point on the river that is distance x, in miles, east of FM 13. Suppose that the river crosses FM 13 at a point 4.1 mi north of the intersection. What *negative* vertical dilation of the equation in part a would give a cubic function with the correct $f(x)$-intercept as well as the correct x-intercepts? Write the particular equation for $f(x)$.

c. Plot function f. Sketch the result. What is the graphical significance of the fact that the leading coefficient is negative?

d. Based on the cubic model, what is the farthest south of Route 66 that the river goes between the 1.7-mi crossing and the 3.8-mi crossing? What is the farthest north of Route 66 that the river goes between the 3.8-mi crossing and the 5.5-mi crossing? How far east of the 5.5-mi crossing would you have to go for the river to be 10 mi south of Route 66?

12. *Airplane Payload Problem:* The number of kilograms of "payload" an airplane can carry equals the number of kilograms the wings can lift minus the mass of the airplane, minus the mass of the crew and their equipment. Use these facts to write an equation for the payload as a function of the airplane's length:

- The plane's mass is directly proportional to the cube of the plane's length.
- The plane's lift is directly proportional to the square of the plane's length.

a. Assume that a plane of a particular design and length $L = 20$ m can lift 2000 kg and has mass 800 kg. Write an equation for the lift and an equation for the mass as functions of L.

b. Assume that the crew and their equipment have mass 400 kg. Write the equation for $P(L)$, the payload the plane can carry in kilograms.

c. Make a table of values of $P(L)$ for each 10 m from 0 m to 50 m.

d. Function P is cubic and thus has three zeros. Find these three zeros, and explain what each represents in the real world.

13. *Behavior of Polynomial Functions for Large Values of x:* Figure 4-4i shows

$$f(x) = x^3 - 7x^2 + 10x + 2 \text{ (solid) and}$$
$$g(x) = x^3 \text{ (dashed)}$$

The graph on the right is zoomed out by a factor of 4 in the x-direction and by a factor of 64 (equal to 4^3) in the y-direction.

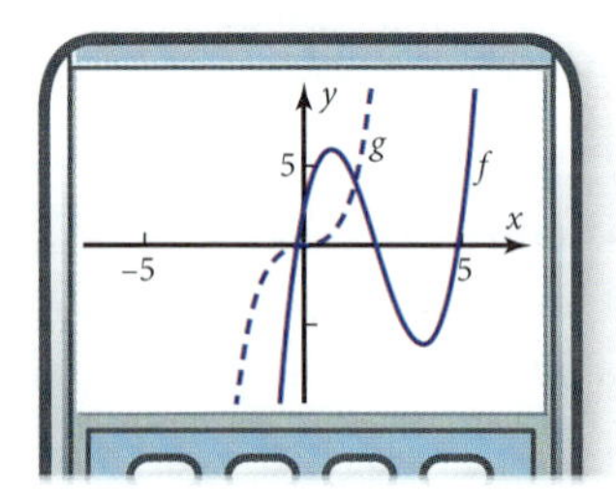

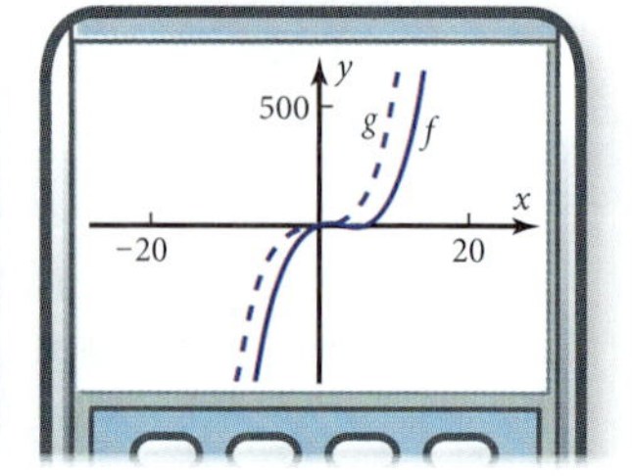

Figure 4-4i

a. Plot the two graphs on your grapher with a window as shown in the graph on the left. Then change the window to widen the domain by a factor of 4 and the range by a factor of 64. Does the result resemble the graph on the right in Figure 4-4i?

b. Zoom out again by the same factors. Sketch the resulting graphs.

c. What do you notice about the shapes of the two graphs as you zoom out farther and farther? Can you still see the intercepts and vertices of the graph of f? What do you think is the reason for saying that the highest-degree term *dominates* the function for large positive and negative values of x?

14. *Constant-nth-Differences Proof Project:* Let $f(x) = ax^3 + bx^2 + cx + d$.

a. Find algebraically four consecutive values of $f(x)$ for which the x-values are k units apart. That is, find $f(x), f(x + k), f(x + 2k)$, and $f(x + 3k)$. Expand the powers.

b. Show that the third differences between the values in part a are independent of x and are equal to $6ak^3$.

c. Let $g(x) = 5x^3 - 11x^2 + 13x - 19$. Find $g(3)$, $g(10)$, $g(17)$, $g(24)$, and $g(31)$. By finding the third differences between consecutive values, show numerically that the conclusion in part b is correct.

15. *Coefficient of Determination Review Problem:*

a. Enter the data from Example 2 into your grapher and run the cubic regression. Confirm that the coefficient of determination is 0.9611..., as shown in the example.

b. Using the appropriate list features on your grapher, find SS_{res}, the sum of the squares of the residual deviations of each data point from the regression curve.

c. Find the mean of the given $f(x)$-values. Find SS_{dev}, the sum of the squares of the deviations of each data point from this mean value.

d. Recall that the coefficient of determination is defined to be the fraction of SS_{dev} that is removed by the regression. That is,

$$R^2 = \frac{SS_{dev} - SS_{res}}{SS_{dev}}$$

Confirm that this formula gives 0.9611..., the value found by regression.

16. *Journal Problem:* Enter into your journal what you have learned so far about higher-degree polynomial functions. Include things such as the shapes of the graphs and constant differences and their relationship to the degree of the polynomial.

4-5 Rational Functions: Asymptotes and Discontinuities

In Section 4–1, you were introduced to rational algebraic functions such as $f(x) = \frac{x^3 + 5x^2 + 8x - 6}{x - 3}$. A rational algebraic function is defined to be a function that is a ratio of two polynomials. In this section you will investigate the behavior of such functions at values of x that make the denominator equal zero, such as at $x = 3$ for function f above.

Objective

Discover some properties of polynomial and rational functions and their graphs.

In this exploration you will analyze transformations of the parent reciprocal function.

EXPLORATION 4-5: Transformations of the Parent Reciprocal Function

1. The figure shows the graph of the parent reciprocal function, $f(x) = \frac{1}{x}$. Find $f(1)$, $f(2)$, $f(3)$, $f\left(\frac{1}{2}\right)$, and $f(-2)$. Are these values consistent with the graph? Find $f(1000)$ and $f(0.001)$. Based on the answers, explain why the y-axis is a vertical asymptote and the x-axis is a horizontal asymptote.

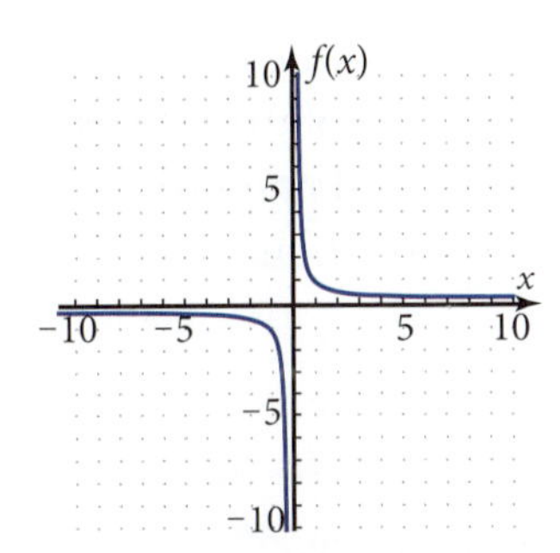

2. The figure shows the graph of function g, a translation of the parent reciprocal function in both the x- and y-directions. Where are the vertical and horizontal asymptotes for function g? What is the magnitude of each translation? Write the particular equation for $g(x)$. Show that the value of $g(4)$ given by the equation agrees with the graph.

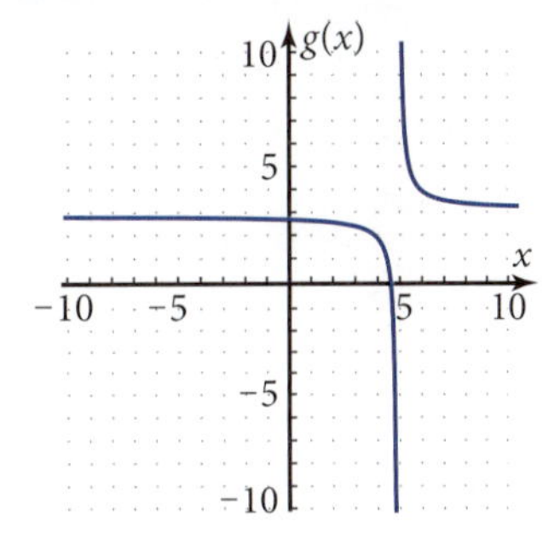

3. The figure shows the graph of function h, a dilation of the parent reciprocal function. Considering the dilation to be vertical, write the equation for $h(x)$. Considering the dilation to be horizontal, write another equation for $h(x)$. Show that the value of $h(-6)$ given by each equation agrees with the graph.

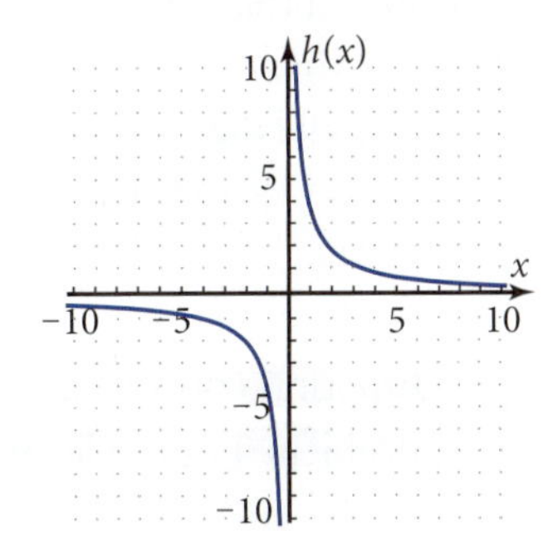

continued

4. The figure shows the graph of function c, which is a vertical dilation and a vertical and horizontal translation of the parent reciprocal function. Use the asymptotes to find the magnitude of each translation. Find the vertical dilation factor, and explain how you calculated it. Write the equation for $c(x)$, and show that the value of $c(-1)$ agrees with the graph.

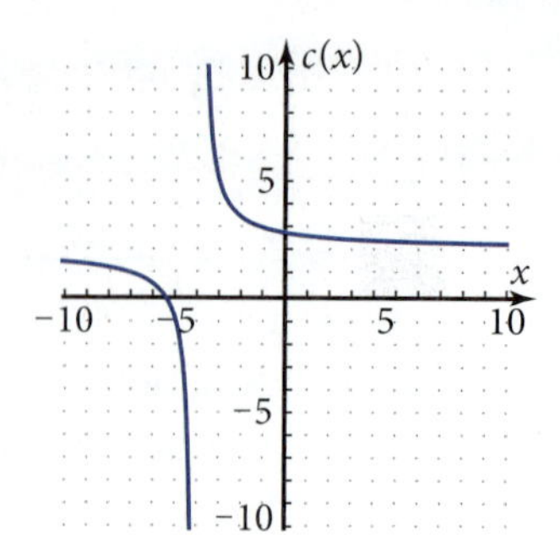

5. What did you learn as a result of doing this exploration that you did not know before?

DEFINITION: Rational Algebraic Function

A rational algebraic function (or, more simply, a "rational function") is a function for which $f(x)$ can be expressed as a rational algebraic fraction, that is, as the ratio of two polynomials. The general equation is

$$f(x) = \frac{n(x)}{d(x)}\text{t}$$

where the numerator, $n(x)$, and the denominator, $d(x)$, are both polynomial expressions.

The definition of a rational algebraic fraction is similar to the definition of a rational number: "a number that can be expressed as a ratio of two integers." Just as not all ratios of two numbers are rational numbers, not all ratios of expressions are rational algebraic fractions. For example, $\frac{\sqrt{7}}{\log 13}$ is not a rational number, and $\frac{|x|}{x^{1.5}}$ is not a rational algebraic fraction.

Proper Rational Algebraic Fractions

Recall that a fraction like $\frac{3}{7}$ is called a proper fraction because the numerator is less than the denominator and therefore its value is less than 1. Similarly, if a rational algebraic fraction has a numerator of degree less than the degree of the denominator, then the expression is called a *proper rational algebraic fraction.* For example, $\frac{3x+9}{x^2+7x+12}$ is a proper rational algebraic fraction. If the equation for a rational function f contains an expression of the form $\frac{\text{constant expression}}{\text{linear expression}}$, then function f may be a transformation of the parent reciprocal function, as you saw in Problem 4 of Exploration 4-5.

Example 1 shows two interesting things that can happen to a function whose equation contains a proper rational algebraic fraction.

EXAMPLE 1 ➤ For the function $f(x) = \dfrac{3x + 9}{x^2 + 7x + 12} + 2$,

a. Simplify the fraction. Describe what happens at values of x that make the denominator equal zero.

b. Use the result from part a to sketch the graph.

SOLUTION **a.** First, factor the numerator and the denominator:

$$f(x) = \frac{3(x + 3)}{(x + 3)(x + 4)} + 2$$

If $x = -3$ or $x = -4$, the denominator is zero, so there is no value for $f(-3)$ or $f(-4)$. Function f is said to have a *discontinuity* at these values of x.

If you cancel the $(x + 3)$ factors, you get

$$f(x) = \frac{3}{(x + 4)} + 2, \text{ provided } x \neq -3 \qquad x \neq -3 \text{ because } f(-3) \text{ is undefined.}$$

b. The graph in Figure 4-5a shows a horizontal asymptote at $y = 2$, a vertical asymptote at $x = -4$, and a gap (circled) in the graph at $x = -3$.

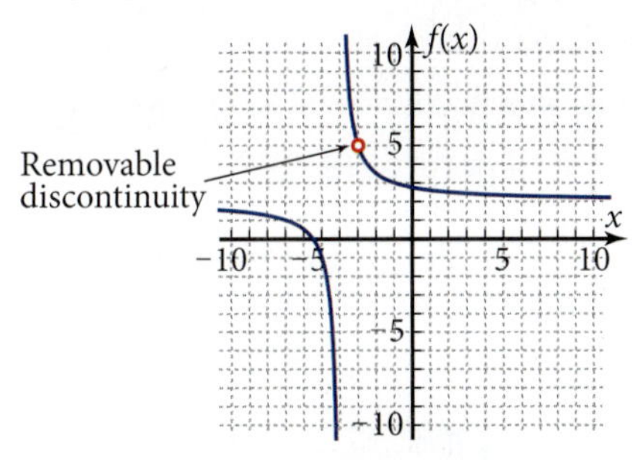

Figure 4-5a

Example 1 shows that two kinds of discontinuity can occur in a function graph at a value of x that makes the denominator zero. The box gives the definition of these kinds of discontinuity.

> **DEFINITION: Discontinuities**
>
> If the denominator of the equation for a function f contains the factor $(x - c)$, then $f(c)$ is undefined because of division by zero, and the graph has a discontinuity at $x = c$.
>
> If the factor $(x - c)$ is canceled out in simplifying the equation, then the discontinuity is a *removable discontinuity,* and the graph has a gap at that one point.
>
> If the factor $(x - c)$ is still present in the denominator after all possible simplifying has been done, then $f(x)$ becomes infinitely large (in the positive or negative direction, or both) as x gets close to c, and the graph has a vertical asymptote at $x = c$.

Some proper rational algebraic fractions have denominators that never equal zero. In this case there are no discontinuities in the corresponding function graph, as shown in Example 2.

EXAMPLE 2 ➤ For the rational function $f(x) = \frac{5}{(x-3)^2+1} - 2$,

a. Explain why there are no discontinuities in the graph.

b. Explain why the graph has a horizontal asymptote at $y = -2$.

c. What feature will the graph of function f have at $x = 3$?

d. Sketch the graph of function f. Verify that your sketch is reasonable by plotting the graph of function f on your grapher. Is the graph concave up or concave down at $x = 3$?

SOLUTION

a. Function f has no discontinuities because the denominator $(x-3)^2 + 1$ is never less than 1, so it never equals zero.

b. The graph of function f has a horizontal asymptote at $y = -2$ because the denominator gets very large as x gets large in either the positive or the negative direction. The algebraic fraction gets very small, causing values of $f(x)$ to get closer and closer to -2.

c. The denominator has its minimum value, 1, at $x = 3$. At this value of x, $f(x) = 3$. All other values of x give a value of $f(x)$ less than 3, so the graph has a high point at (3, 3).

d. Sketch the graph as in Figure 4-5b. Plotting the graph on your grapher confirms that your sketch is reasonable. The graph is concave down at $x = 3$.

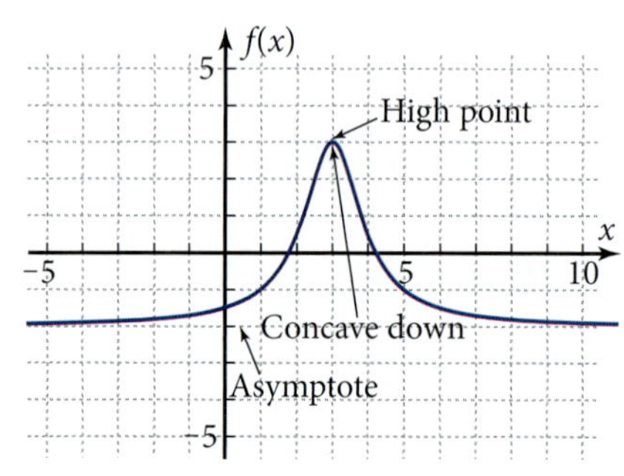

Figure 4-5b

Improper Rational Algebraic Fractions

Functions f and g are examples of *improper rational algebraic fractions:*

$$f(x) = \frac{x^3 - 5x^2 + 8x - 6}{x - 3} \qquad g(x) = \frac{x^3 - 5x^2 + 8x - 5}{x - 3}$$

In an improper rational algebraic fraction, the numerator is of equal or higher degree than the denominator. The equations of $f(x)$ and $g(x)$ are the same except that the numerators differ by 1. However, as shown in Figure 4-5c, the graphs of the two functions are quite different. Both functions have a discontinuity at $x = 3$, but the discontinuity for function f is a removable discontinuity while the discontinuity for function g is a vertical asymptote.

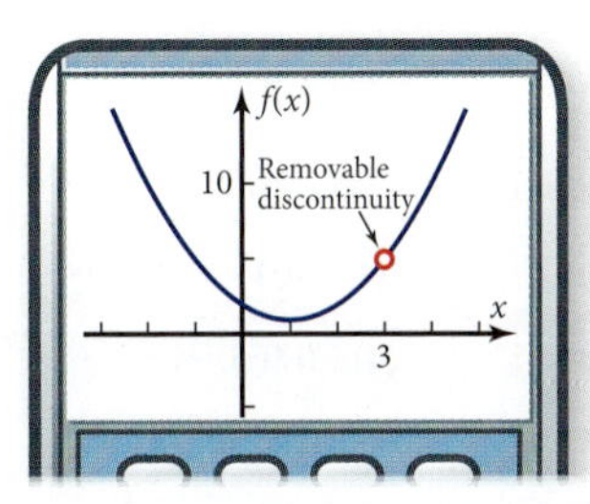

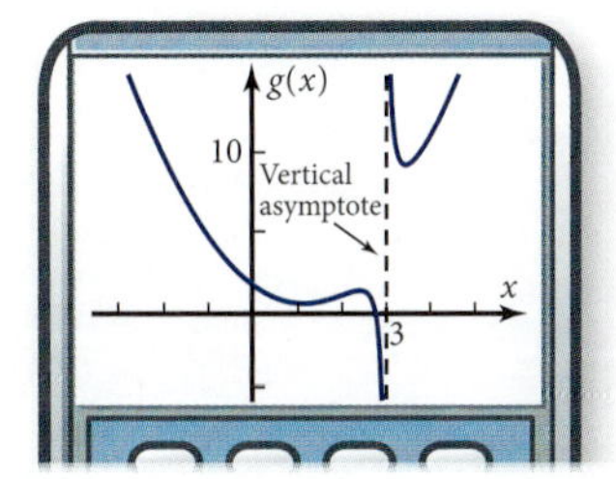

Figure 4-5c

Example 3 shows you how to analyze functions f and g algebraically to see why their graphs have such different behavior at the discontinuity.

EXAMPLE 3 ➤ Refer to functions f and g above.

a. Transform functions f and g to mixed-number form using synthetic substitution.

b. Use the results from part a to explain why function f has a removable discontinuity at $x = 3$ but function g has a vertical asymptote at $x = 3$.

c. Make a table showing the values of $f(x)$ and $g(x)$ for $x = 2, 2.9, 2.99, 3, 3.01, 3.1$, and 4. From the results, describe the difference in behavior of the two functions for values of x close to 3.

d. Find the exact coordinates of the removable discontinuity in function f.

SOLUTION

a. $f(x) = \dfrac{x^3 - 5x^2 + 8x - 6}{x - 3}$ $\qquad$ $g(x) = \dfrac{x^3 - 5x^2 + 8x - 5}{x - 3}$

$$\begin{array}{c|cccc} 3 & 1 & -5 & 8 & -6 \\ & & 3 & -6 & 6 \\ \hline & 1 & -2 & 2 & 0 \end{array} \qquad \begin{array}{c|cccc} 3 & 1 & -5 & 8 & -5 \\ & & 3 & -6 & 6 \\ \hline & 1 & -2 & 2 & 1 \end{array}$$

$f(x) = x^2 - 2x + 2$, provided $x \neq 3$ $\qquad$ $g(x) = x^2 - 2x + 2 + \dfrac{1}{x - 3}$

b. Function f has a removable discontinuity at $x = 3$ because $(x - 3)$ is a factor of both the numerator and the denominator. (Because there is still no value for $f(3)$, you must state $x \neq 3$.)

Function g has a vertical asymptote at $x = 3$ because $(x - 3)$ is still in the denominator after the fraction has been simplified.

c. The table shows that $f(x)$ stays close to 5 when x is close to 3 but that $g(x)$ becomes very large in the negative direction when x is close to 3 on the negative side and very large in the positive direction when x is close to 3 on the positive side.

x	f(x)	g(x)
2	2	1
2.9	4.61	−5.39
2.99	4.9601	−95.0399
3	(none)	(none)
3.01	5.0401	105.0401
3.1	5.41	15.41
4	10	11

d. Substitute 3 for x in the simplified form of $f(x)$:

$$y = (3)^2 - 2(3) + 2 = 5$$

You are not allowed to say $f(3) = 5$ because function f is undefined at $x = 3$.

The removable discontinuity is at (3, 5). The table in part c shows that values of $f(x)$ are close to 5 if x is close to 3.

Note that an interesting feature of function g shows up if you plot the graph of g and the graph of $y = x^2 - 2x + 2$, the polynomial part of the mixed-number form, on the same screen. As shown in Figure 4-5d, the graph of $y = x^2 - 2x + 2$ is a curved asymptote for the graph of g.

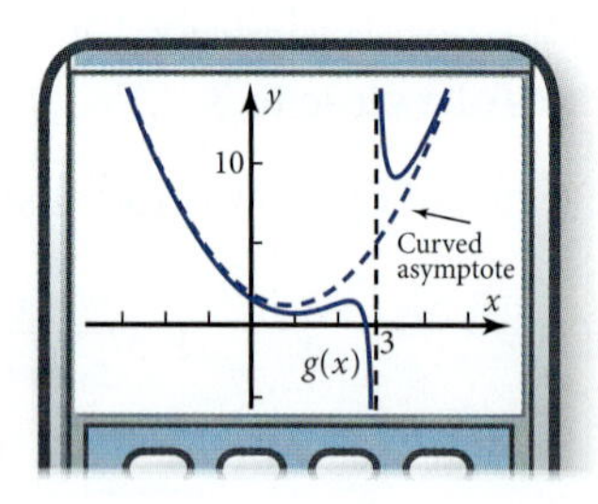

Figure 4-5d

Example 4 shows the behavior of a rational function in which the numerator and the denominator are polynomials of the same degree.

EXAMPLE 4 ➤ Plot the graph of the rational function $f(x)$, showing any discontinuities and stating which kind they are.

$$f(x) = \frac{3x^2 - 25x - 50}{x^2 - 16x + 60}$$

Describe the end behavior (the behavior when x is far away from zero) and explain how this behavior is related to the coefficients in the rational expression.

SOLUTION

$$f(x) = \frac{3x^2 - 25x - 50}{x^2 - 16x + 60}$$

$$f(x) = \frac{(3x + 5)(x - 10)}{(x - 10)(x - 6)}$$ Factor the numerator and denominator.

$$f(x) = \frac{3x + 5}{x - 6}, \text{ provided } x \neq 10$$ $x \neq 10$ because $f(10)$ is undefined.

Figure 4-5e shows the graph, with a vertical asymptote at $x = 6$ and a removable discontinuity at $x = 10$.

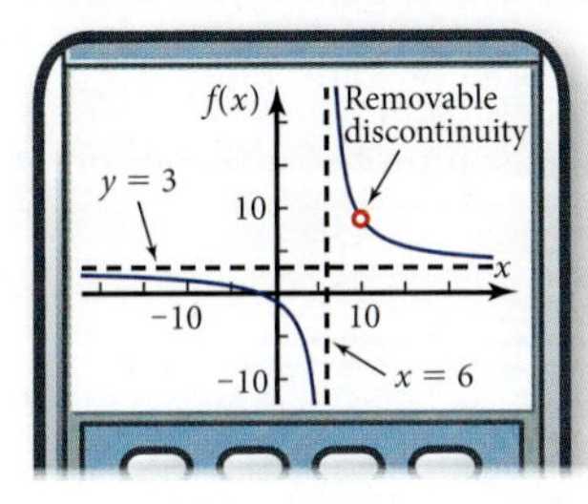

Figure 4-5e

The end behavior shows the graph approaching the horizontal asymptote $y = 3$ as x gets large in both the positive and the negative directions. The 3 is equal to the ratio of the leading coefficients in the numerator and denominator, in both the simplified and the unsimplified fractions. ◄

The **end behavior** of a function refers to the behavior of a function near the positive and negative ends of its domain. For the rational functions in this section, the domain extends to infinity in both directions. The box on the next page summarizes the end behavior of rational functions.

PROPERTY: End Behavior and Nonvertical Asymptotes

A rational function with an *improper* rational algebraic expression $f(x) = \frac{n(x)}{d(x)}$, where the degree of $n(x)$ is greater than or equal to the degree of $d(x)$, can be transformed to the form $f(x) = p(x) + \frac{r(x)}{d(x)}$, where $p(x)$, remainder $r(x)$, and denominator $d(x)$ are polynomials and $\frac{r(x)}{d(x)}$ is a *proper* rational algebraic fraction (that is, $r(x)$ is of lower degree than $d(x)$). The remainder fraction, $\frac{r(x)}{d(x)}$, approaches zero as x gets large in the positive or negative direction.

The graph of $y = p(x)$ forms a *non-vertical asymptote* for the graph of function f. If $y = p(x)$ is a nonconstant linear equation, it is called a slant asymptote. The graph of function f approaches the graph of this asymptote as x gets large in the positive or negative direction.

If the degrees of $n(x)$ and $d(x)$ are the same, then $f(x)$ can be transformed to $f(x) = k + \frac{r(x)}{d(x)}$, where k is a constant. Then k is a vertical translation and the line $y = k$ is a horizontal asymptote for the graph of function f.

Problem Set 4-5

Reading Analysis

From what you have read in this section, what do you consider to be the main idea? What causes a function to have a discontinuity at a particular value of x? What makes a discontinuity removable, and what makes a discontinuity a vertical asymptote? What operation can you perform on the equation of a rational function to reveal a nonvertical asymptote?

Quick Review

Q1. Sketch the graph of a quadratic function.

Q2. Sketch the graph of a cubic function with a double zero at $x = 3$.

Q3. Sketch the graph of a cubic function with two nonreal zeros.

Q4. Sketch the graph of a quartic function with four distinct real zeros and a negative leading coefficient.

Q5. How many extreme points can the graph of a 7th-degree function have?

Q6. Write the zeros of the function $f(x) = (3x - 5)(2x + 7)(x - 4)$.

Q7. How many times does the graph of $g(x) = (3x - 5)(x^2 + 5x + 2)$ cross the x-axis?

Q8. What transformation of function f gives function g if $g(x) = f\left(\frac{1}{3}x\right)$?

Q9. Multiply and simplify: $(3 + 5i)^2$

Q10. Which of these is not a rational number?

A. 7 **B.** $\frac{3}{4}$ **C.** $\sqrt{25}$

D. 2^5 **E.** All are rational numbers.

---

1. Explain the similarities and differences between a rational number and a rational algebraic expression.

2. Explain the similarities and differences between a numerical improper fraction and an improper algebraic expression.

3. What causes a rational algebraic function to have a discontinuity at a particular value of x? What two kinds of discontinuities have you encountered so far?

4. Write the equation for $g(x)$, a transformation of the parent reciprocal function that is a vertical dilation by a factor of 5, a vertical translation by -4 units, and a horizontal translation by 3 units.

For Problems 5–8, function f is a transformation of the parent reciprocal function. Write the vertical dilation and translation and the horizontal translation. Calculate the values of $f(x)$ 1 unit to the left of the vertical asymptote and 1 unit to the right. Sketch the graph of function f, showing these features.

5. $f(x) = 3 - \dfrac{2}{x - 4}$

6. $f(x) = -2 + \dfrac{3}{x + 4}$

7. $f(x) = \dfrac{4}{x + 1} - 3$

8. $f(x) = \dfrac{-1}{x - 2} + 3$

For Problems 9–22, $f(x)$ equals an improper rational algebraic expression with numerator and denominator of the same degree.

a. Write $f(x)$ in the form $p(x) + \frac{r(x)}{d(x)}$.

b. Find the x- and y-coordinates of any removable discontinuities and the location of any vertical asymptotes.

c. Show that the horizontal asymptote is located at the value of y equal to the ratio of the leading coefficients of the numerator and the denominator.

d. Find the x- and y-intercepts.

e. Sketch the graph.

9. $f(x) = \dfrac{x - 3}{x + 2}$

10. $f(x) = \dfrac{x + 1}{x - 3}$

11. $f(x) = \dfrac{2 - x}{x - 4}$

12. $f(x) = \dfrac{3 - x}{x + 5}$

13. $f(x) = \dfrac{3x + 2}{x - 1}$

14. $f(x) = \dfrac{1 - 2x}{x + 3}$

15. $f(x) = \dfrac{(x - 2)(x + 7)}{(x + 7)(x + 5)}$

16. $f(x) = \dfrac{(x + 2)(x - 3)}{(x - 3)(x - 4)}$

17. $f(x) = \dfrac{x^2 + 3x - 4}{x^2 + 2x - 3}$

18. $f(x) = \dfrac{x^2 - 6x + 8}{x^2 - 9x + 20}$

19. $f(x) = \dfrac{3x^2 + 5x - 28}{x^2 - x - 20}$

20. $f(x) = \dfrac{-2x^2 + 3x + 20}{x^2 - 16}$

21. $f(x) = \dfrac{x^2 - 4}{x^2 + 4}$

22. $f(x) = \dfrac{x^2 - 9}{x^2 - 4x + 10}$

For Problems 23–34, function f is an improper rational algebraic expression with a numerator of higher degree than the denominator.

a. Write $f(x)$ in mixed-number form or polynomial form.

b. Find the x- and y-coordinates of any removable discontinuities and the location of any vertical asymptotes.

c. Write the equation of any nonvertical asymptotes.

d. Find the x- and y-intercepts.

e. Sketch the graph.

23. $f(x) = \dfrac{x^2 + 2x - 8}{x + 1}$

24. $f(x) = \dfrac{x^2 - 6x + 8}{x - 1}$

25. $f(x) = \dfrac{x^2 - 4}{x - 3}$

26. $f(x) = \dfrac{x^2 - 2x - 3}{x - 2}$

27. $f(x) = \dfrac{2x^2 - 11x + 12}{x - 4}$

28. $f(x) = \dfrac{8 + 2x - x^2}{x + 2}$

29. $f(x) = \dfrac{x^3 - x^2 - 7x + 13}{x - 2}$

30. $f(x) = \dfrac{x^3 - 5x^2 + 2x + 10}{x - 3}$

31. $f(x) = \dfrac{x^3 + 5x^2 + 4x - 16}{x + 3}$

32. $f(x) = \dfrac{x^3 - x^2 - 13x - 18}{x + 2}$

33. $f(x) = \dfrac{x^3 - 10x^2 + 17x + 28}{x - 4}$

34. $f(x) = \dfrac{x^3 - 4x^2 - 16x - 11}{x + 1}$

35. Functions f, g, and h have the same numerator but different denominators. Use appropriate algebra to describe the features of each graph in Figures 4-5f to 4-5h, including the location and type of each discontinuity, the x- and y-intercepts, and the location of any slant asymptotes.

a. $f(x) = \dfrac{(x-4)(x-1)(x+3)}{(x-4)(x+3)}$

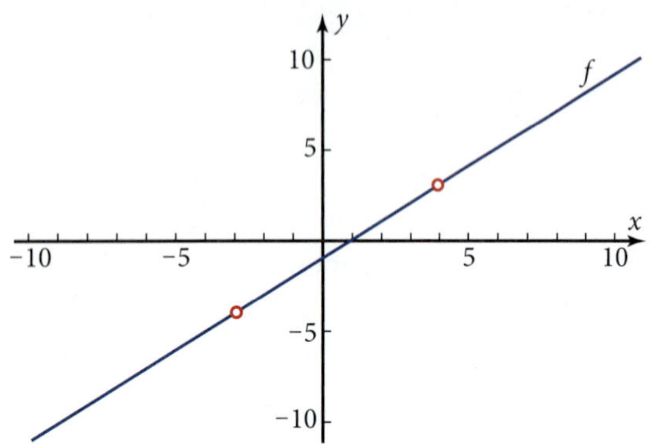

Figure 4-5f

b. $g(x) = \dfrac{(x-4)(x-1)(x+3)}{(x-3)(x+2)}$

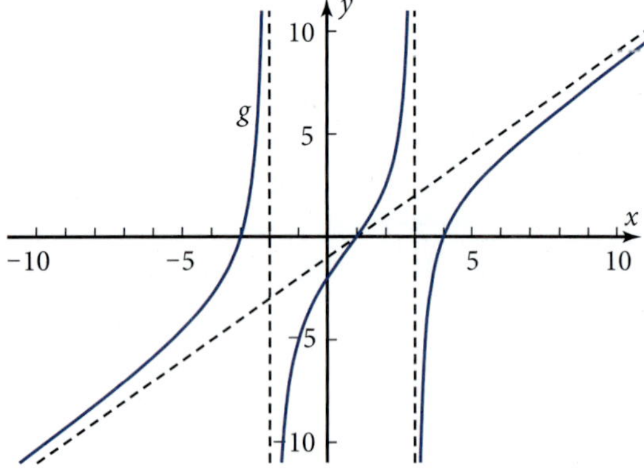

Figure 4-5g

c. $h(x) = \dfrac{(x-4)(x-1)(x+3)}{(x-4)(x+2)}$

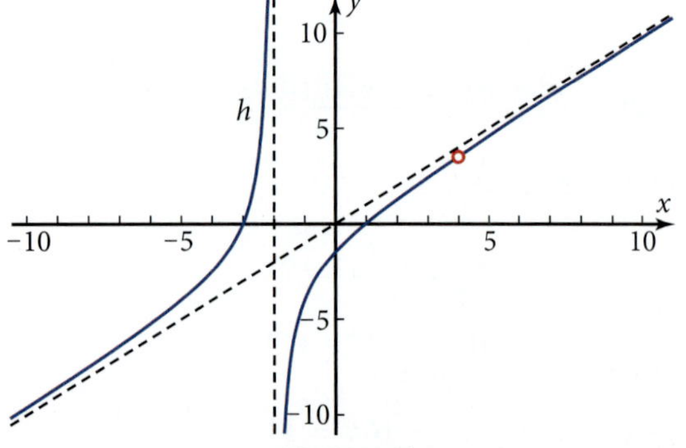

Figure 4-5h

36. Functions f, g, and h have the same numerator but different denominators. Use appropriate algebra to describe the features of each graph in Figures 4-5i to 4-5k, including the location and type of each discontinuity, the x- and y-intercepts, and the location of any slant asymptotes.

a. $f(x) = \dfrac{x^3 + 3x^2 - 13x - 15}{x^2 + 6x + 5}$

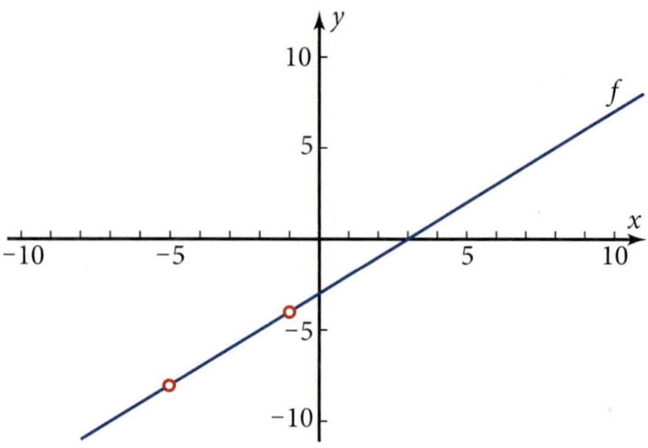

Figure 4-5i

b. $g(x) = \dfrac{x^3 + 3x^2 - 13x - 15}{x^2 - 16}$

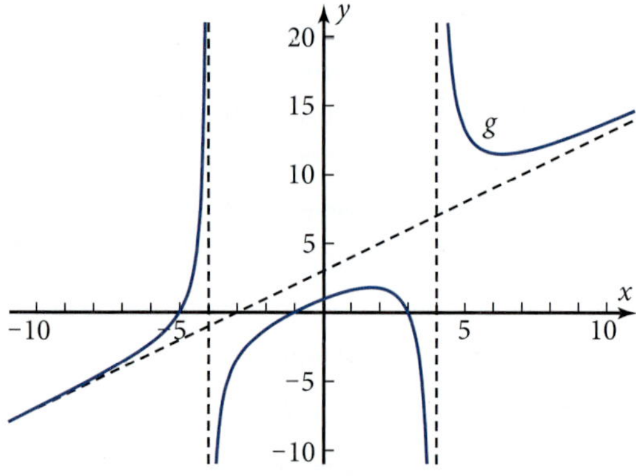

Figure 4-5j

c. $h(x) = \dfrac{x^3 + 3x^2 - 13x - 15}{x^2 + 16}$

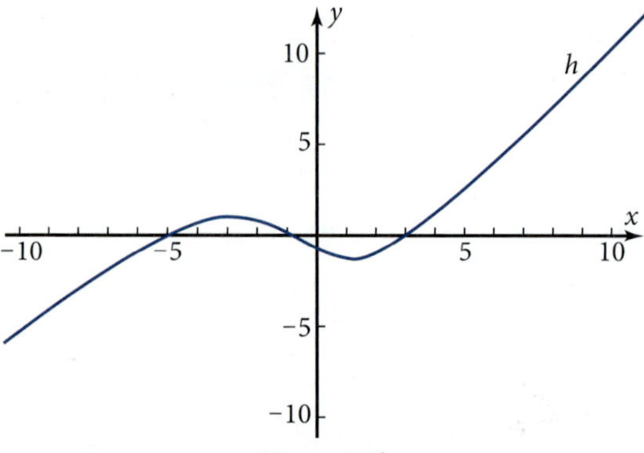

Figure 4-5k

4-6 Partial Fractions and Operations with Rational Expressions

From previous courses, recall how to multiply, divide, add, and subtract fractions:

Multiply: $\frac{2}{3} \cdot \frac{5}{7} = \frac{2 \cdot 5}{3 \cdot 7} = \frac{10}{21}$ Top times top, over bottom times bottom.

Divide: $\frac{2}{3} \div \frac{5}{7} = \frac{2}{3} \cdot \frac{7}{5} = \frac{14}{15}$ Multiply by the reciprocal of the second fraction.

Add: $\frac{2}{3} + \frac{5}{7} = \frac{2}{3} \cdot \frac{7}{7} + \frac{5}{7} \cdot \frac{3}{3} = \frac{14}{21} + \frac{15}{21} = \frac{29}{21}$ Get a common denominator.

Subtract: $\frac{2}{3} - \frac{5}{7} = \frac{14}{21} - \frac{15}{21} = -\frac{1}{21}$ Get a common denominator.

In this section you will perform these operations on *algebraic* fractions. You will also learn about partial fractions, in which you start with the answer to an addition problem and find the fractions that were added. You will use these techniques in the next section to solve fractional equations.

Objective

Multiply, divide, add, and subtract rational expressions, and resolve proper algebraic fractions into the sum of two or more partial fractions.

Multiplication and Division

Two properties of real numbers and a definition allow you to multiply and divide rational expressions and simplify the result. These properties, stated in the box, can be proved from the field axioms in Appendix A.

PROPERTIES: Products of Fractions

Reciprocal of a Product Property

Algebraically: If $x \neq 0$ and $y \neq 0$, then $\frac{1}{xy} = \frac{1}{x} \cdot \frac{1}{y}$.

Verbally: The reciprocal of a product of two nonzero numbers equals the product of the two reciprocals.

Multiplication Property of Fractions

Algebraically: If $x \neq 0$ and $y \neq 0$, then $\frac{ab}{xy} = \frac{a}{x} \cdot \frac{b}{y}$.

Verbally: A quotient of two products can be split into a product of two quotients. Or, the other way around: To multiply two fractions, multiply their numerators and multiply their denominators.

DEFINITION: Division of Fractions

Algebraically: If $y \neq 0$, then $\frac{x}{y} = x \cdot \frac{1}{y}$.

Verbally: Dividing a number by y means multiplying that number by the reciprocal of y.

Applied to two fractions, this definition becomes

$$\frac{a}{x} \div \frac{b}{y} = \frac{a}{x} \cdot \frac{y}{b}, \text{ provided } x \neq 0, y \neq 0, \text{ and } b \neq 0$$

Example 1 shows how to apply these properties to a product of two rational expressions and simplify the result.

EXAMPLE 1 ➤ Multiply and simplify: $\frac{x^2 + 5x + 6}{x^2 - x - 20} \cdot \frac{x^2 + 3x - 4}{x^2 + x - 2}$

SOLUTION

$$\frac{x^2 + 5x + 6}{x^2 - x - 20} \cdot \frac{x^2 + 3x - 4}{x^2 + x - 2}$$

$$= \frac{(x^2 + 5x + 6) \cdot (x^2 + 3x - 4)}{(x^2 - x - 20) \cdot (x^2 + x - 2)}$$ Numerator times numerator over denominator times denominator.

$$= \frac{(x + 2)(x + 3)(x - 1)(x + 4)}{(x + 4)(x - 5)(x - 1)(x + 2)}$$ Factor the numerator and the denominator.

$$= \frac{(x + 3)(x - 1)(x + 2)(x + 4)}{(x - 5)(x - 1)(x + 2)(x + 4)}$$ Commute factors appropriately.

$$= \frac{x + 3}{x - 5} \cdot \frac{(x - 1)(x + 2)(x + 4)}{(x - 1)(x + 2)(x + 4)}$$ Use the multiplication property of fractions.

$$= \frac{x + 3}{x - 5}, x \neq 1, x \neq -2, x \neq -4$$ Because the second fraction equals 1. ◀

Note that the restrictions on x must be stated for the factors that cancel, as they were for removable discontinuities. Of course, x cannot equal 5 either, but this is clear from the fact that the term $(x - 5)$ still appears in the denominator.

The process can be simplified if the numerator and denominator involve only multiplication of factors. If you realize this, you can cancel by striking out the common factors in the numerator and denominator without doing the commuting and breaking the rational expression into two fractions.

$$\frac{\cancel{(x + 2)}(x + 3)\cancel{(x - 1)}\cancel{(x + 4)}}{\cancel{(x + 4)}(x - 5)\cancel{(x - 1)}\cancel{(x + 2)}} = \frac{x + 3}{x - 5}, x \neq 1, x \neq -2, x \neq -4$$

For this purpose it is helpful to define canceling in a fraction.

DEFINITION: Canceling in a Fraction

Canceling in a fraction means dividing the numerator and denominator by the same common factor.

This definition will help you avoid mistakes such as canceling the x's in a fraction like $\frac{x + 3}{x - 5}$. Because canceling means division, which distributes over addition and subtraction, if you cancel the x's the result is $\frac{1 + \frac{3}{x}}{1 - \frac{5}{x}}$, which is by no means simpler!

Addition and Subtraction

To add or subtract two or more rational expressions, such as $\frac{3}{x+4} + \frac{5}{x-7}$, you must first find a common denominator. Then add the numerators. Example 2 shows how to do this.

EXAMPLE 2 Add by first finding a common denominator: $\frac{3}{x+4} + \frac{5}{x-7}$

SOLUTION

$$\frac{3}{x+4} + \frac{5}{x-7}$$

$$= \frac{(x-7)}{(x-7)} \cdot \frac{3}{x+4} + \frac{5}{x-7} \cdot \frac{(x+4)}{(x+4)}$$ Multiply each fraction by a "clever" form of 1.

$$= \frac{(x-7)(3)}{(x-7)(x+4)} + \frac{(5)(x+4)}{(x-7)(x+4)}$$ Multiply the numerators. Multiply the denominators.

$$= \frac{(x-7)(3) + (5)(x+4)}{(x-7)(x+4)}$$ Add the numerators; keep the common denominator.

$$= \frac{3x - 21 + 5x + 20}{(x-7)(x+4)}$$ Multiply in the numerator.

$$= \frac{8x-1}{x^2 - 3x - 28}$$ Combine like terms. Multiply in the denominator.

Partial Fractions

The process of adding or subtracting two rational expressions can be reversed to find the two simpler proper fractions that were added or subtracted. Finding these two fractions from the answer is called *resolving into partial fractions.*

To resolve $\frac{8x-1}{x^2-3x-28}$ into partial fractions, first factor the denominator:

$$\frac{8x-1}{x^2-3x-28} = \frac{8x-1}{(x+4)(x-7)}$$

So the two target fractions will have denominators $(x + 4)$ and $(x - 7)$. Observe that the first fraction is a proper rational algebraic expression—the numerator is of lower degree than the denominator. So the two target fractions will also be proper fractions, meaning that their numerators will be of degree zero—that is, constant. Picking letters such as A and B to stand for these constant numerators, write:

$$\frac{8x-1}{(x+4)(x-7)} = \frac{A}{x+4} + \frac{B}{x-7}$$

Next follow these steps:

$$\frac{8x-1}{(x-7)} = A + \left(\frac{B}{x-7}\right)(x+4)$$ Multiply both sides of the equation by $(x + 4)$ to remove the discontinuity on the left and leave the constant A by itself on the right, with no denominator.

$$\frac{8(-4)-1}{-4-7} = A$$ Substitute -4 for x to make the second term on the right equal zero.

$$A = \frac{-33}{-11} = 3$$ Simplify.

Following similar steps but multiplying both sides by $(x - 7)$ instead of $(x + 4)$ allows you to conclude that $B = 5$. So

$$\frac{8x - 1}{(x + 4)(x - 7)} = \frac{3}{x + 4} + \frac{5}{x - 7}$$

This process is known as *Heaviside's method* after Oliver Heaviside (1850–1925). A verbal description of a shortcut to this method is "To find the numerator of the fraction containing $(x + 4)$, cover up the $(x + 4)$ in the original fraction with your finger and substitute -4 for x (the value of x that makes $(x + 4)$ equal zero) in what remains." The result is the numerator of the partial fraction with denominator $(x + 4)$. Example 3 shows how to apply this shortcut.

EXAMPLE 3 ➤ Resolve into partial fractions: $\frac{8x - 1}{(x + 4)(x - 7)}$

SOLUTION Write $\frac{8x - 1}{(x + 4)(x - 7)} = \frac{\quad}{x + 4} + \frac{\quad}{x - 7}$.

Cover the $(x + 4)$ with your finger and substitute -4 for x in what is left.

$$\frac{8x - 1}{(\quad)(x - 7)} = \frac{3}{x + 4} + \frac{\quad}{x - 7} \qquad [8(-4) - 1]/(-4 - 7) = 3$$

Cover the $(x - 7)$ with your finger and substitute 7 for x in what is left.

$$\frac{8x - 1}{(x + 4)(\quad)} = \frac{3}{x + 4} + \frac{5}{x - 7} \qquad [8(7) - 1]/(7 + 4) = 5$$

$$\frac{8x - 1}{(x + 4)(x - 7)} = \frac{3}{x + 4} + \frac{5}{x - 7}$$

Problem Set 4-6

Reading Analysis

From what you have read in this section, what do you consider to be the main idea? What property of fractions allows you to multiply two fractions by multiplying the numerators and multiplying the denominators? Why can't you add two fractions by adding the numerators and adding the denominators? What process allows you to take the answer to the sum of two algebraic fractions and break it into the original fractions that were added?

Quick Review

Q1. Multiply: $(x - 3)(x + 7)$

Q2. Factor: $x^2 - 7x - 8$

For Problems Q3–Q8, $f(x) = \frac{(x - 1)(x + 2)}{(x - 1)(x - 2)(x + 3)}$.

Q3. Find the zeros of the denominator.

Q4. Find the zeros of the numerator.

Q5. Find the x-coordinate(s) of any removable discontinuities.

Q6. Find the x-coordinate(s) of any vertical asymptotes.

Q7. Find any x-intercepts.

Q8. Find the y-intercept.

Q9. Where does the graph of $f(x) = \frac{5x^2 + 3x - 7}{2x^2 + 11x - 1}$ have a horizontal asymptote?

Q10. The x-coordinate of the vertex of $y = 5x^2 - 30x + 17$ is

A. 6 **B.** 5 **C.** 3

D. -3 **E.** -5

For Problems 1–12, multiply or divide and simplify.

1. $\frac{2}{x-2} \cdot \frac{x^2-4}{4}$
2. $\frac{x^2+7x+12}{12} \cdot \frac{4}{x+4}$
3. $\frac{x^2+4x+3}{5x} \div \frac{x+1}{x+5}$
4. $\frac{x^2-64}{x^2-16} \div \frac{x+8}{x+4}$
5. $\frac{x^2+6x}{6} \cdot \frac{x^2+6}{x^3+6x^2}$
6. $\frac{x^2-4}{2x-4} \cdot \frac{2}{x+2}$
7. $\frac{x^2+x-2}{x^2-4x-12} \cdot \frac{x^2-5x-6}{x^2-2x+1}$
8. $\frac{x^2+3x-10}{x^2-7x+6} \cdot \frac{x^2+2x-3}{x^2+x-6}$
9. $\frac{x^2-7x+12}{x^2-x-6} \div \frac{x^2-16}{x^2+x-2}$
10. $\frac{x^2-6x+8}{x^2-5x+6} \div \frac{x^2-7x+12}{x^2-4x+4}$
11. $\frac{x+3}{x-7} \div \frac{x-7}{x+5} \cdot \frac{x+5}{x+3}$
12. $\frac{x+9}{x-10} \cdot \frac{x+8}{x+9} \div \frac{x-10}{x+8}$

Problems 13–16 involve complex fractions, which are fractions with one or more fractions in the numerator, the denominator, or both. Multiply the fraction by a clever form of 1 that eliminates the "minor" denominators. Then do any other possible simplifying.

13. $\frac{x+3+\frac{2}{x}}{1-\frac{4}{x^2}}$ (Multiply by $\frac{x^2}{x^2}$.)
14. $\frac{x-5+\frac{6}{x}}{1-\frac{9}{x^2}}$
15. $\frac{x-3+\frac{12}{x+5}}{x-8+\frac{42}{x+5}}$ (Multiply by $\frac{x+5}{x+5}$.)
16. $\frac{x+3+\frac{5}{x-3}}{x+2+\frac{4}{x-3}}$

For Problems 17–28, add or subtract and simplify.

17. $\frac{1}{x+1} + \frac{1}{x-1}$
18. $\frac{3}{x-1} + \frac{1}{1-x}$
19. $\frac{2x-1}{x+1} - \frac{2x-1}{x-1}$
20. $\frac{x+3}{x-3} - \frac{x-3}{x+3}$
21. $x+2-\frac{x^2+x-6}{x-3}$
22. $2x+5-\frac{x^2+2x-15}{x-3}$
23. $\frac{1}{x+3} - \frac{1}{x-3} + \frac{2x}{x^2-9}$
24. $\frac{3}{x+6} - \frac{4x}{x^2-36} + \frac{2}{x-6}$
25. $\frac{3x+13}{x^2-3x-10} - \frac{16}{x^2-6x+5}$
26. $\frac{6}{x^2-7x+12} + \frac{5x+9}{x^2-2x-3}$
27. $\frac{x-2}{x^2-x-2} + \frac{x-4}{x^2-5x+4}$
28. $\frac{x+4}{x^2-3x-28} - \frac{x-5}{x^2+2x-35}$

For Problems 29–38, resolve the algebraic fraction into partial fractions.

29. $\frac{11x-15}{x^2-3x+2}$
30. $\frac{7x+25}{x^2-7x-8}$
31. $\frac{5x-11}{x^2-2x-8}$
32. $\frac{3x-12}{x^2-5x-50}$
33. $\frac{21}{x^2+7x+10}$
34. $\frac{10x}{x^2-9x-36}$
35. $\frac{9x^2-25x-50}{(x+1)(x-7)(x+2)}$
36. $\frac{7x^2+22x-54}{(x-2)(x+4)(x-1)}$
37. $\frac{4x^2+15x-1}{x^3+2x^2-5x-6}$
38. $\frac{-3x^2+22x-31}{x^3-8x^2+19x-12}$

Unfactorable Quadratics

Heaviside's method does not work for an algebraic fraction such as

$$\frac{7x^2 - 4x}{(x^2 + 1)(x - 2)}$$

which has an unfactorable quadratic factor, $(x^2 + 1)$, in the denominator (unless you are willing to use imaginary numbers). Because the numerators of the corresponding partial fractions need be only one degree lower than the degree of their denominators, the quadratic term can have a linear numerator. Thus the partial fractions would be

$$\frac{7x^2 - 4x}{(x^2 + 1)(x - 2)} = \frac{Ax + B}{x^2 + 1} + \frac{C}{x - 2}$$

where A, B, and C stand for constants. Adding the fractions on the right, then simplifying, gives

$$\frac{7x^2 - 4x}{(x^2 + 1)(x - 2)} = \frac{Ax + B}{x^2 + 1} + \frac{C}{x - 2}$$

$$= \frac{(Ax + B)(x - 2) + C(x^2 + 1)}{(x^2 + 1)(x - 2)}$$

$$= \frac{Ax^2 - 2Ax + Bx - 2B + Cx^2 + C}{(x^2 + 1)(x - 2)}$$

So $Ax^2 + Cx^2 = 7x^2$, $-2Ax + Bx = -4x$, and $-2B + C = 0$. Solving the system

$$\begin{cases} 7 = A + C \\ -4 = -2A + B \\ 0 = -2B + C \end{cases}$$

gives $A = 3$, $B = 2$, and $C = 4$. Therefore,

$$\frac{7x^2 - 4x}{(x^2 + 1)(x - 2)} = \frac{3x + 2}{x^2 + 1} + \frac{4}{x - 2}$$

For Problems 39 and 40, resolve the rational expression into partial fractions.

39. $\dfrac{4x^2 + 6x + 11}{(x^2 + 1)(x + 4)}$

40. $\dfrac{4x^2 - 15x - 1}{x^3 - 5x^2 + 3x + 1}$

Repeated Linear Factors

If a power of a linear factor appears in the denominator, the fraction could have come from adding partial fractions with that power or any lower power. For example, the algebraic fraction

$$\frac{x^2 - 4x + 18}{(x + 4)(x - 1)^2}$$

has a repeated linear factor, $(x - 1)$, in the denominator. In this case the partial fractions will include one fraction with denominator $(x - 1)^2$. But they could also include a fraction with denominator $(x - 1)$. So the fractions would be

$$\frac{A}{x + 4} + \frac{B}{x - 1} + \frac{C + Dx}{(x - 1)^2}$$

Because the numerator has only three coefficients, one of the four constants, A, B, C, or D, is arbitrary. To get the simplest partial fractions, it is helpful to let D be the arbitrary constant, and set D equal to zero. The partial fractions thus have the form

$$\frac{x^2 - 4x + 18}{(x + 4)(x - 1)^2} = \frac{A}{x + 4} + \frac{B}{x - 1} + \frac{C}{(x - 1)^2}$$

Adding the fractions on the right, then simplifying, gives

$$= \frac{A(x - 1)^2 + B(x + 4)(x - 1) + C(x + 4)}{(x + 4)(x - 1)^2}$$

$$= \frac{Ax^2 - 2Ax + A + Bx^2 + 3Bx - 4B + Cx + 4C}{(x + 4)(x - 1)^2}$$

So $Ax^2 + Bx^2 = x^2$, $-2Ax + 3Bx + Cx = -4x$, and $A - 4B + 4C = 18$. Solving the system

$$\begin{cases} 1 = A + B \\ -4 = -2A + 3B + C \\ 18 = A - 4B + 4C \end{cases}$$

gives $A = 2$, $B = -1$, and $C = 3$. Therefore,

$$\frac{x^2 - 4x + 18}{(x + 4)(x - 1)^2} = \frac{2}{x + 4} + \frac{-1}{x - 1} + \frac{3}{(x - 1)^2}$$

For Problems 41 and 42, resolve the rational expression into partial fractions.

41. $\dfrac{4x^2 + 18x + 6}{(x + 5)(x + 1)^2}$

42. $\dfrac{3x^2 - 53x + 245}{x^3 - 14x^2 + 49x}$

4-7 Fractional Equations and Extraneous Solutions

For each kind of function you have studied so far, you have evaluated $f(x)$ for various given values of x and solved for x if a value of $f(x)$ is given. That is,

- Given x, find y.
- Given y, find x.

In this section you will solve for x if function f is a rational algebraic function.

Objective Given a rational algebraic function f, find x for a given value of $f(x)$.

Suppose you are given the rational algebraic function

$$f(x) = \frac{x^2 - 5x + 10}{x - 3}$$

and you are to find x if $f(x) = 6$. Figure 4-7a shows the graph of f and a line plotted at $y = 6$.

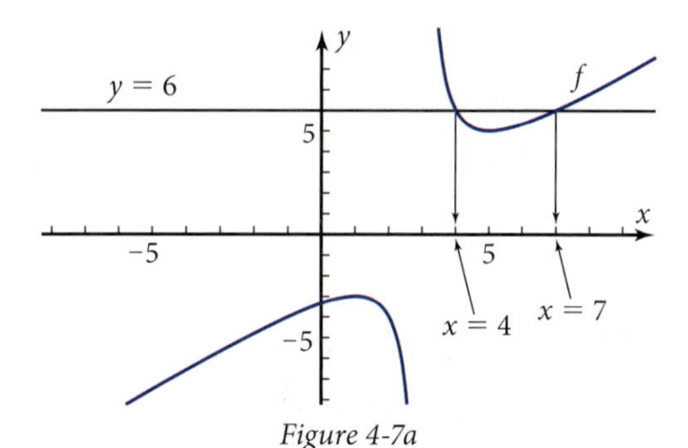

Figure 4-7a

It is relatively easy to find that $x = 4$ or $x = 7$ *numerically*, by making a table, or *graphically*, using the intersect or solver feature of your grapher. Example 1 shows how to find these values *algebraically*.

EXAMPLE 1 ➤ For $f(x) = \dfrac{x^2 - 5x + 10}{x - 3}$, find x algebraically if $f(x) = 6$.

SOLUTION

$x \neq 3$	Write any domain restrictions.
$6 = \dfrac{x^2 - 5x + 10}{x - 3}$	Substitute 6 for $f(x)$.
$6(x - 3) = x^2 - 5x + 10$	Multiply both sides by $(x - 3)$ to eliminate the fraction.
$6x - 18 = x^2 - 5x + 10$	
$0 = x^2 - 11x + 28$	Make one side equal zero.
$0 = (x - 4)(x - 7)$	Factor the quadratic trinomial (or solve using the quadratic formula).
$x = 4$ or $x = 7$	These values of x make the right side equal zero.

Neither value of x is excluded from the domain, so both 4 and 7 are solutions. ◄

The equation $6 = \frac{x^2 - 5x + 10}{x - 3}$ in the solution to Example 1 is called a *fractional equation* because there is a variable in a denominator. An equation such as $\frac{1}{2}x + 8 = 13$ is *not* called a fractional equation, even though it has a fraction in it, because there are no variables in a denominator.

Suppose you are asked to find x for function f in Example 1 if $f(x) = 4$. Figure 4-7b shows that a horizontal line drawn at $y = 4$ does not intersect the graph of function f. Example 2 shows how to verify this algebraically.

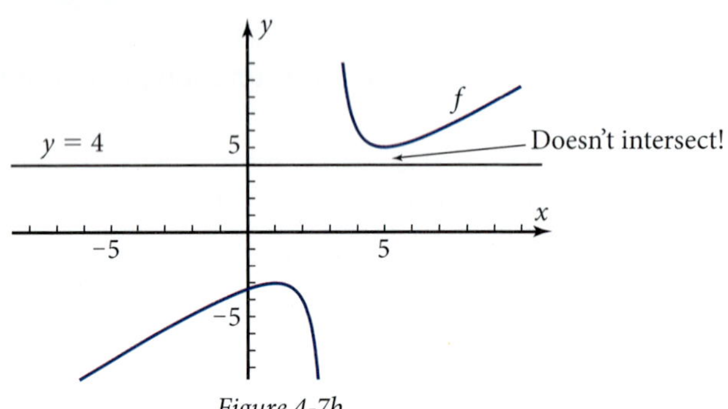

Figure 4-7b

EXAMPLE 2 ➤ For $f(x) = \frac{x^2 - 5x + 10}{x - 3}$, find x algebraically if $f(x) = 4$.

SOLUTION

$x \neq 3$ — Write any domain restrictions.

$$4 = \frac{x^2 - 5x + 10}{x - 3}$$

$$4(x - 3) = x^2 - 5x + 10$$

$$4x - 12 = x^2 - 5x + 10$$

$$0 = x^2 - 9x + 22$$

$b^2 - 4ac = (-9)^2 - 4(1)(22) = -7$ — Find the discriminant.

No real solutions. — The discriminant is negative. ◄

Interesting things sometimes happen if you try to find the points at which the graphs of two rational algebraic functions intersect. Suppose

$$f(x) = \frac{x}{x - 2} + \frac{2}{x + 3} \quad \text{and} \quad g(x) = \frac{10}{x^2 + x - 6}$$

Figure 4-7c shows the graphs of $f(x)$ and $g(x)$ plotted on the same axes.

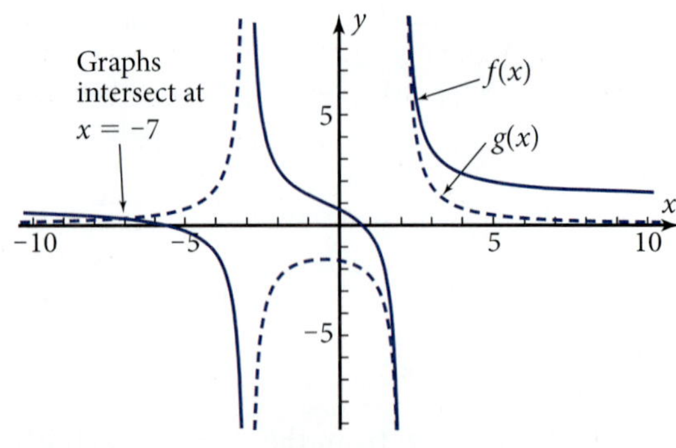

Figure 4-7c

Both graphs have asymptotes at $x = -3$ and $x = 2$. Although it is a bit difficult to see, the graphs intersect at $x = -7$. It is not clear whether the graphs intersect where they come together near the asymptote at $x = 2$. The algebraic solution allows you to resolve this ambiguity.

EXAMPLE 3 ➤ Find the intersections of the graphs of $f(x) = \frac{x}{x-2} + \frac{2}{x+3}$ and $g(x) = \frac{10}{x^2 + x - 6}$ algebraically.

SOLUTION

$x \neq 2, x \neq -3$ — Write any domain restrictions.

$$\frac{x}{x-2} + \frac{2}{x+3} = \frac{10}{x^2 + x - 6}$$ Set $f(x)$ equal to $g(x)$.

$$\frac{x}{x-2} + \frac{2}{x+3} = \frac{10}{(x-2)(x+3)}$$ Factor the denominator on the right side.

To get rid of the fractions, you can multiply both sides of the equation by $(x - 2)(x + 3)$, which is the least common multiple of the three denominators.

$$(x-2)(x+3)\left(\frac{x}{x-2} + \frac{2}{x+3}\right) = \left(\frac{10}{(x-2)(x+3)}\right)(x-2)(x+3)$$

On the right side, the denominator cancels immediately, leaving 10. On the left side, the $[(x - 2)(x + 3)]$ must be distributed to *both* fractions before canceling.

$$(x-2)(x+3)\left(\frac{x}{x-2}\right) + (x-2)(x+3)\left(\frac{2}{x+3}\right) = 10$$

$$(x+3)(x) + (x-2)(2) = 10$$ Cancel.

$$x^2 + 5x - 14 = 0$$ Distribute, and combine like terms.

$$(x-2)(x+7) = 0$$ Factor.

$$x = 2 \quad \text{or} \quad x = -7$$

The value $x = 2$ is one of the values excluded from the domain. A solution of the transformed equation $x^2 + 5x - 14 = 0$ that does not work in the original equation is called an *extraneous solution*. Strike through $x = 2$ and write the word "extraneous."

extraneous
~~$x = 2$~~ or $x = -7$

$x = -7$ is the only solution. — Write the solution. ◄

Note: The extraneous solution in Example 3 occurs where both graphs have a common asymptote. It is possible for the graphs of two rational functions to cross each other near a common asymptote. The most reliable way to be sure of the behavior of the two functions is to find an algebraic solution.

Problem Set 4-7

Reading Analysis

From what you have read in this section, what do you consider to be the main idea? How does this main idea correspond to operations you have performed on other kinds of functions? Why is not every equation with fractions in it called a fractional equation? What surprising type of solution might you find when you solve a fractional equation?

Quick Review

Q1. Find x if $5x - 7 = 13$.

Q2. Find x if $(2x - 5)(x + 3) = 0$.

Q3. Find x if $x^2 - 5x - 14 = 0$.

Q4. Find x if $|x - 8| = 10$.

Q5. Find x if $\frac{2}{3}x = 24$.

Q6. Find x if $x^{1/3} = 8$.

Q7. Find x if $3^x = 200$.

Q8. Find x if $\frac{3}{x} = 12$.

Q9. Multiply: $(3x - 7)^2$

Q10. "$(x - a)$ is a factor of $P(x)$ if and only if $P(a) = 0$" is a statement of

- **A.** The multiplication property of zero
- **B.** The factor theorem
- **C.** The multiplicative identity axiom
- **D.** The synthetic substitution property
- **E.** The converse of the multiplication property of zero

---

1. *Extraneous Solutions Problem:* Start with the equation $x = 3$.

- **a.** Multiply each side by $(x - 4)$. Explain why the transformed equation is true when $x = 4$ but the original equation is not.
- **b.** What name is given to the solution $x = 4$ for the transformed equation?
- **c.** Multiplying both sides of an equation by an expression such as $(x - 4)$ that can equal zero is called an *irreversible step.* Why do you suppose this name is used?

2. *Depressed Equation Problem:* Start with the equation $x^2 = 3x$.

- **a.** Explain why $x = 0$ and $x = 3$ are both solutions of the equation.
- **b.** Divide each side of the equation by x. The transformed equation is called a "depressed equation" because its degree is lower than the degree of the original equation. Write all solutions of the depressed equation.
- **c.** Why is it more dangerous to divide both members by an expression that can equal zero than it is to multiply by such an expression?

3. Given $f(x) = \frac{-x^2 + 6x - 11}{x - 2}$,

- **a.** Find algebraically the values of x for which $f(x) = -2$.
- **b.** Show algebraically that $f(x)$ never equals 4.
- **c.** Does $f(x)$ ever equal 7? Justify your answer.
- **d.** Confirm the results of parts a, b, and c by plotting the graph of function f on your grapher and sketching the result.

4. Given $f(x) = \frac{x^2 + 10x + 32}{x + 5}$,

- **a.** Find algebraically the values of x for which $f(x) = 8$.
- **b.** Show algebraically that $f(x)$ never equals 5.
- **c.** Does $f(x)$ ever equal -5? Justify your answer.
- **d.** Confirm the results of parts a, b, and c by plotting the graph of function f on your grapher and sketching the result.

5. Given $f(x) = \frac{x^2 - 4x - 1}{x - 3}$,

- **a.** Find algebraically the values of x for which $f(x) = 5$.
- **b.** Find graphically the values of x for which $f(x) = -6$.
- **c.** Sketch the graph, showing the answers to parts a and b.
- **d.** Is the range of function f the set of all real numbers? Explain.

6. Given $f(x) = \frac{-x^2 - 6x - 3}{x + 4}$,

   a. Find algebraically the values of x for which $f(x) = 6$.

   b. Find graphically the values of x for which $f(x) = -3$.

   c. Sketch the graph, showing the answers to parts a and b.

   d. Is the range of function f the set of all real numbers? Explain.

7. Given $f(x) = x + \frac{x}{x - 2}$ and $g(x) = \frac{2}{x - 2}$,

   a. Find algebraically all values of x for which $f(x) = g(x)$. Discard any extraneous solutions.

   b. Plot the graphs of functions f and g on your grapher. Sketch the result. Show that the graphs confirm your answer to part a.

8. Given $f(x) = x + \frac{2x}{x - 1}$ and $g(x) = \frac{3 - x}{x - 1}$,

   a. Find algebraically all values of x for which $f(x) = g(x)$. Discard any extraneous solutions.

   b. Plot the graphs of functions f and g on your grapher. Sketch the result. Show that the graphs confirm your answer to part a.

For Problems 9–28, state any domain restrictions and solve the equation algebraically. Show the step where you discard any extraneous solutions.

9. $\frac{x}{x - 3} - \frac{7}{x + 5} = \frac{24}{x^2 + 2x - 15}$

10. $\frac{x}{x + 2} + \frac{7}{x - 5} = \frac{14}{x^2 - 3x - 10}$

11. $\frac{3x}{x + 4} + \frac{4x}{x - 3} = \frac{84}{x^2 + x - 12}$

12. $\frac{4x}{x - 1} - \frac{5x}{x - 2} = \frac{2}{x^2 - 3x + 2}$

13. $\frac{3}{x - 3} + \frac{4}{x - 4} = \frac{25}{x^2 - 7x + 12}$

14. $\frac{11x}{x + 20} + \frac{24}{x} = 11 + \frac{88}{x(x + 20)}$

15. $\frac{x + 2}{x - 3} + \frac{x - 2}{x - 6} = 2$

16. $\frac{3x + 2}{x - 1} + \frac{2x - 4}{x + 2} = 5$

17. $\frac{2}{x + 2} + \frac{x}{x - 2} = \frac{x^2 + 4}{x^2 - 4}$

18. $\frac{x}{x + 4} + \frac{4}{x - 4} = \frac{x^2 + 16}{x^2 - 16}$

19. $\frac{1}{1 - x} = 1 - \frac{x}{x - 1}$

20. $\frac{x}{x - 1} + \frac{2}{x^2 - 1} = \frac{8}{x + 1}$

21. $\frac{x + 3}{2x - 3} = \frac{18x}{4x^2 - 9}$

22. $3 - \frac{22}{x + 5} = \frac{6x - 1}{2x + 7}$

23. $\frac{4x}{x^2 - 9} - \frac{x - 1}{x^2 - 6x + 9} = \frac{2}{x + 3}$

24. $\frac{x}{x^2 - 2x + 1} = \frac{2}{x + 1} + \frac{4}{x^2 - 1}$

25. $\frac{3x}{x - 2} + \frac{2x}{x + 3} = \frac{30}{x^2 + x - 6}$

26. $\frac{5}{x - 6} - \frac{4}{x + 3} = \frac{x + 39}{x^2 - 3x - 18}$

27. *River Barge Problem:* A tugboat pushes a string of barges 50 mi up the Mississippi River against a current of 4 mi/h, stops for an hour to pick up a string of empty barges, and then returns to the starting point.

   a. If x is the number of miles per hour the tugboat travels through the water, explain why the total time, $f(x)$, for the round trip is given by the function $f(x) = \frac{50}{x - 4} + 1 + \frac{50}{x + 4}$.

   b. How long will the round trip take if the tugboat goes 20 mi/h? 10 mi/h? 5 mi/h? Does the round trip take 4 times as long at 5 mi/h than it does at 20 mi/h, more than 4 times as long, or less than 4 times as long? Do you find this surprising?

   c. The tugboat owner wants to complete the round trip in 8 h so that she will not have to pay the crew for overtime. Find, algebraically, the minimum speed at which the tugboat can travel through the water to accomplish this.

4-8 Chapter Review and Test

In this chapter you extended your knowledge of quadratic functions to include higher-degree polynomial functions. A quadratic function may have two, one, or no x-intercepts and has exactly two *zeros*. Zeros that have real-number values are x-intercepts, while *complex* zeros are not x-intercepts. By the *fundamental theorem of algebra* and its corollaries, an nth-degree polynomial function has exactly n zeros, counting complex values and repeated zeros. You learned that while polynomial functions are defined for all real values of x, *rational algebraic functions* are undefined at values of x that make a denominator zero. The graph of a rational function may have a *vertical asymptote* or a *removable discontinuity* at any such x-value and may have horizontal, slant, or curved asymptotes if the numerator is of equal or higher degree than the denominator. You have also reviewed operations with algebraic fractions and solutions to fractional equations.

Review Problems

R0. Update your journal with what you have learned in this chapter. Include things such as

- The most important thing you have learned as a result of studying this chapter
- The new terms you have learned and what they mean
- Significant features of polynomial and rational function graphs
- The types of real-world situations polynomial and rational functions can model
- The similarities and differences between multiplying and adding rational expressions, and between adding rational expressions and resolving a rational expression into partial fractions
- How to solve a fractional equation and why a fractional equation may have extraneous solutions

R1. Figure 4-8a shows the graphs of

$f(x) = x^2 - x - 6$ (solid) and

$g(x) = \dfrac{x^3 - 5x^2 - 2x + 29}{x - 4}$ (dashed).

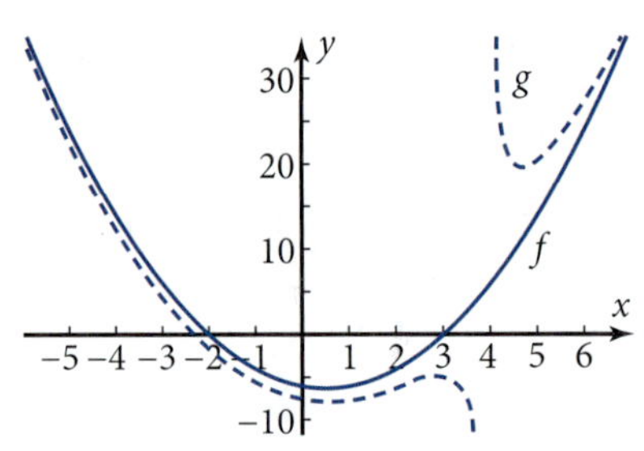

Figure 4-8a

a. Explain why there is a value of $f(4)$ but no value of $g(4)$.

b. What feature does the graph of function g have at $x = 4$?

c. Show that $x = -2$ and $x = 3$ are zeros of function f.

d. Show numerically that $g(6)$ is close to $f(6)$ but $g(4.01)$ is not close to $f(4.01)$.

e. Based on your answer to part d, explain why the graph of function f is a curved asymptote for function g.

R2. **a.** Multiply and simplify: $(4x + 7)(2x - 5)$

b. Expand and simplify: $(3x - 10)^2$

c. Factor the polynomial: $x^2 - 9x - 10$

d. Factor the polynomial: $x^2 - 121$

e. Solve the equation by factoring: $x^2 - 7x = -12$

f. Solve the equation using the quadratic formula: $5x^2 - 7x - 13 = 0$

g. Solve the equation using the quadratic formula: $x^2 + 10x + 34 = 0$

h. Show by direct substitution that $x = 2 + 3i$ is a solution to the equation $x^2 - 4x = -13$.

i. Quick! Write the other solution to the equation in part h.

j. Write the coordinates of the vertex: $y = 3(x - 5)^2 - 7$

k. Transform into polynomial form: $y = 3(x - 5)^2 - 7$

l. Complete the square to transform to vertex form: $y = 5x^2 + 80x + 57$

R3. **a.** Figure 4-8b shows the graph of a polynomial function f.

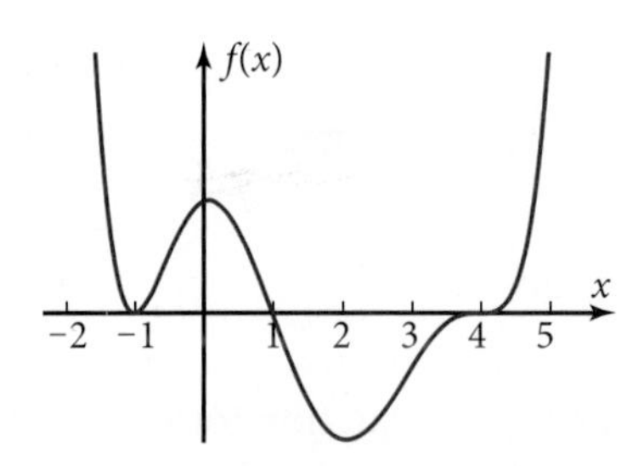

Figure 4-8b

i. Write the single zero.

ii. Write the double zero.

iii. Write the triple zero.

iv. Write a possible equation for $f(x)$ in factored form.

v. Write the smallest possible degree of function f.

vi. Give the sign of the leading coefficient of $f(x)$.

b. Given the polynomial function $f(x) = 2x^3 - 43x^2 + 271x - 440$,

i. Find the sum of the zeros (without actually calculating the zeros).

ii. Find the sum of the pairwise products of the zeros.

iii. Find the product of the zeros.

iv. Confirm your answers in parts i–iii by finding the zeros using your grapher.

c. Sketch the graph of a quintic function that has three distinct real zeros and two complex zeros.

d. **i.** State the remainder theorem.

ii. Use the remainder theorem to find the remainder if $x^{10} - 723$ is divided by $x - 2$.

iii. Explain why the factor theorem is a special case of the remainder theorem.

R4. **a.** Figure 4-8c shows the graph of the cubic function $f(x) = -x^3 + 8x^2 - 29x + 52$.

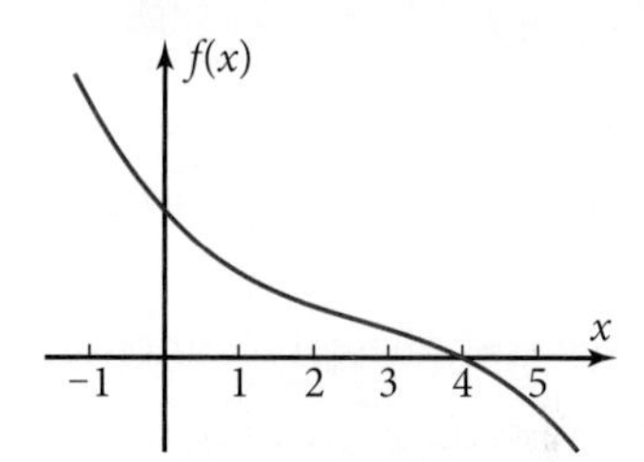

Figure 4-8c

i. From the graph, at approximately what value of x is the point of inflection?

ii. Find algebraically the exact x-coordinate of the point of inflection.

iii. For what values of x is the graph concave up?

iv. Find the three zeros of function f.

b. *Stock Market Problem:* The stock for a new biotechnology company goes on the market. It is an immediate success, and the price of the stock rises dramatically. The price per share of the stock is listed in the table.

Weeks	Dollars per Share
1	14.00
2	26.00
3	60.00
4	110.00
5	170.00
6	234.00

i. Find algebraically the equation for the cubic function that fits the first four data points. Show the steps leading to your answer.

ii. Show that the fifth and sixth data points agree with the cubic function in part i.

iii. Run a cubic regression on the data. Explain how the result confirms that the cubic function in part i fits all six data points.

iv. Plot the graph of the cubic function in part i. Sketch the result.

v. If you use this mathematical model to extrapolate a few months into the future, what does it predict will eventually happen to the price of the stock? Think of a real-world reason why this end behavior might occur.

R5. a. Write the definition of a rational algebraic function.

b. What is the difference between a *proper* algebraic fraction and an *improper* algebraic fraction?

c. For the rational algebraic function $f(x) = \frac{3x - 2}{x - 2}$,

i. Write the equation in the form $f(x) = p(x) + \frac{r(x)}{d(x)}$.

ii. What are the three transformations of the parent reciprocal function, $y = \frac{1}{x}$, that appear in function f?

iii. Sketch the graph. Show the horizontal and vertical asymptotes.

iv. Explain how to find the horizontal asymptote without transforming the equation to the form $f(x) = p(x) + \frac{r(x)}{d(x)}$.

d. Given the rational functions

$$f(x) = \frac{x^3 - 13x^2 + 57x - 81}{x - 3} \quad \text{and}$$

$$g(x) = \frac{x^3 - 13x^2 + 57x - 80}{x - 3}$$

i. Plot both graphs on your grapher. Sketch the result, identifying which function is which.

ii. Function f has a removable discontinuity at $x = 3$. Remove the discontinuity by canceling. Find the y-coordinate of the discontinuity.

iii. Function g has a vertical asymptote at $x = 3$. Write an equation for $g(x)$ in the form $p(x) + \frac{r(x)}{d(x)}$, and explain why the discontinuity at $x = 3$ cannot be removed.

iv. Function g also has a curved asymptote. Write the equation for this asymptote and describe how the asymptote relates to the graphs of $f(x)$ and $g(x)$.

e. Figure 4-8d shows the graph of $h(x) = \frac{10x - 30}{x^3 - 3x^2 + 5x - 15}$.

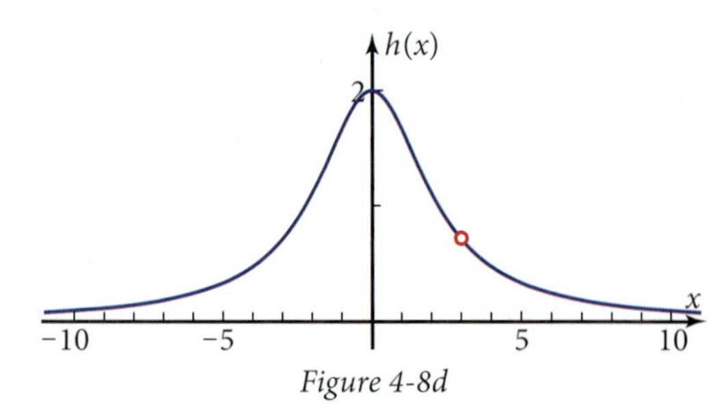

Figure 4-8d

i. Simplify the fraction by factoring and canceling.

ii. Explain algebraically why there is a removable discontinuity at $x = 3$.

iii. Explain algebraically why the x-axis is a horizontal asymptote but there are no vertical asymptotes.

R6. **a.** State the multiplication property of fractions algebraically and verbally.

b. State the definition of division algebraically and verbally.

c. Divide and simplify. State any domain restrictions.

$$\frac{x^2 - 5x + 6}{x^2 - 4x - 5} \div \frac{x^2 - 2x - 3}{x^2 - 9x + 20}$$

d. Add and simplify. State any domain restrictions.

$$\frac{16}{x^2 - 2x - 3} + \frac{3x - 13}{x^2 - 5x + 6}$$

e. Resolve into partial fractions:

$$\frac{6x^2 - x - 31}{(x - 2)(x + 1)(x - 3)}$$

R7. **a.** Starting with the equation $x = 4$, multiply both sides by $(x - 3)$. What value of x is a solution of the transformed equation that was not a solution of the original equation? What special name is given to this new solution? What name is given to the step where you multiplied both sides of the equation by $(x - 3)$, an expression that can equal zero?

b. Solve by factoring: $x^2 - 5x - 14 = 0$

c. Given $f(x) = \frac{x^2 - 2x - 15}{x - 5}$,

i. Substitute 2 for $f(x)$, and solve for x by first multiplying both sides by $(x - 5)$. Show that one solution is valid and one solution is extraneous.

ii. Substitute 8 for $f(x)$, and solve for x as in part i.

iii. Sketch the graph of function f. Show the results of parts i and ii.

d. Each of these equations illustrates a different situation that could occur when you solve a fractional equation. Solve each equation by multiplying both sides by the least common multiple of the denominators. Describe what happens.

i. $\frac{x}{x - 2} - \frac{1}{x + 4} = \frac{12}{x^2 + 2x - 8}$

ii. $\frac{5}{x - 2} - \frac{5}{x + 4} = \frac{30}{x^2 + 2x - 8}$

iii. $\frac{x}{x - 2} - \frac{2}{x + 4} = \frac{12}{x^2 + 2x - 8}$

iv. $\frac{x}{x - 2} - \frac{2}{x + 4} = \frac{19}{x^2 + 2x - 8}$

e. *Airplane Flight Time Problem:* Suppose you are a route planner for an airline company that is planning a daily shuttle service between Chicago and San Francisco. You want to know how fast an airplane would have to travel in order to complete a round trip in less than 24 hours, so that only one plane is needed to provide the shuttle service. Let x be the plane's airspeed in miles per hour, and let $f(x)$ be the number of hours it takes to travel round trip.

i. You find that the two airports are about 2000 mi apart. The prevailing wind is from west to east at about 50 mi/h, so the plane's ground speed from Chicago to San Francisco is $(x - 50)$ mi/h, and its ground speed for the return trip is $(x + 50)$ mi/h. Recalling that $time = \frac{distance}{rate}$, write an equation for $f(x)$, including 2 h "turn-around" time at each airport.

ii. Assuming that no intermediate stops are needed for refueling or maintenance, how long would it take a Boeing 737 flying at 500 mi/h to travel round trip? A blimp flying at 60 mi/h? Is this surprising?

iii. Let $f(x) = 24$ h, and solve the resulting equation to find the minimum speed a plane would have to fly in order to travel round trip in no more than 24 h.

Concept Problems

C1. *Graphs of Complex Zeros Problem:* Figure 4-8e shows the graph of the cubic function $f(x) = x^3 - 18x^2 + 105x - 146$ and the line containing the points (2, 0) and $(8, f(8))$. You will learn a way to use this line to find the complex zeros of a cubic function.

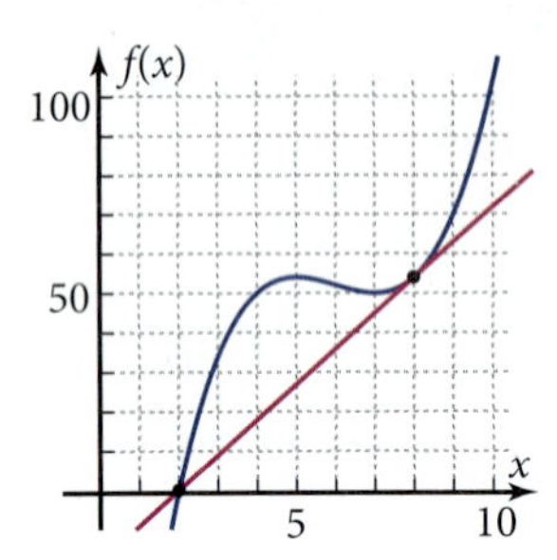

Figure 4-8e

a. Find the two nonreal zeros of $f(x)$.

b. Show algebraically that the line is tangent to the graph of function f at the point $(8, f(8))$ by finding the equation for the line and setting it equal to $f(x)$. Show that $x = 8$ is a double zero of the resulting equation and thus the graph touches the line but does not cross it at $x = 8$.

c. Show that the real part, a, of the nonreal zeros equals the x-coordinate of the point of tangency. Show that the imaginary part, b, is equal to the square root of the absolute value of the slope of the tangent line.

d. Figure 4-8f shows the graph of another cubic function, g, but the equation for $g(x)$ is not given. On a copy of the figure, construct a tangent line to the graph of function g at the appropriate point $(x, g(x))$ that also contains the x-intercept. Use the technique of part c to find graphically the nonreal complex zeros.

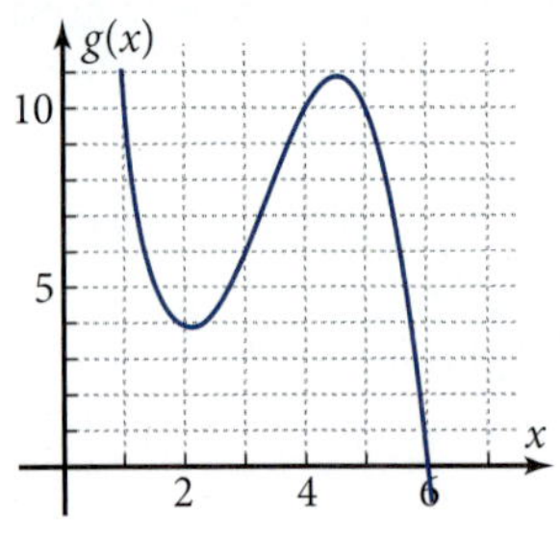

Figure 4-8f

e. For integer values of x, the values of $g(x)$ in Figure 4-8f are also integers. Use four of these integer points to find an equation for $g(x)$. Then calculate the three zeros and confirm that your graphical solution is correct.

f. Find Paul J. Nahin's book *An Imaginary Tale: The Story of* $\sqrt{-1}$, published by Princeton University Press in 1998. Consult pages 27–30 for an in-depth analysis of the property you have learned in this problem.

C2. *The Rational Root Theorem:*

a. By synthetic substitution, show that -1 is a zero of the function $f(x) = 6x^3 + 17x^2 - 24x - 35$. Use the results to show that $f(x)$ factors into three linear factors:

$$f(x) = (x + 1)(2x + 7)(3x - 5)$$

What are the other two zeros of $f(x)$?

b. The box contains a statement of the *rational root theorem* from algebra. Show that the zeros of $f(x)$ in part a agree with the conclusion of this theorem.

> **PROPERTY: Rational Root Theorem**
>
> If the rational number $\frac{n}{d}$ is a zero of the polynomial function p with integer coefficients (or a root of the polynomial equation $p(x) = 0$), then n is a factor of the constant term of the polynomial and d is a factor of the leading coefficient.

c. Show that these two polynomial functions agree with the rational root theorem:

$$g(x) = 3x^3 - 19x^2 + 13x + 35$$

$$h(x) = 6x^3 - 35x^2 - 31x + 280$$

C3. *Rational Function Surprise Problem:* Figure 4-8g shows the graph of the rational function

$$f(x) = \frac{x^3 + 6x^2 - 9x - 46}{x^2 + 2x - 15}$$

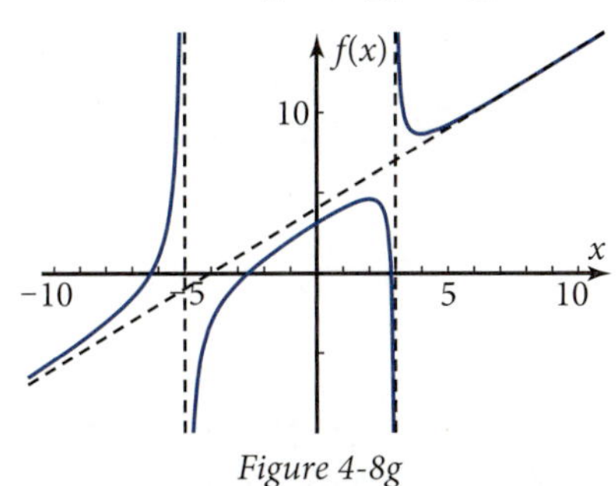

Figure 4-8g

a. Transform the equation for $f(x)$ to the form $p(x) + \frac{r(x)}{d(x)}$, and use the result to write equations for the three asymptotes. Explain why there are no removable discontinuities.

b. The diagonal asymptote appears to intersect the graph of function f somewhere between $x = -10$ and $x = 10$. Using appropriate algebra, either find the point(s) where the asymptote intersects the graph or show that the asymptote does not intersect the graph.

Chapter Test

Part 1: No calculators allowed (T1–T11)

T1. Multiply and simplify: $(x - 5)(3x - 7)$

T2. Factor: $x^2 - 7x - 60$

T3. Solve by factoring: $x^2 - 7x - 60 = 0$

T4. Expand and simplify: $(5 - 3i)^2$

T5. Write the zeros of the function $f(x) = (x - 5)(x - 2)(x + 1)$.

T6. Write the x-intercepts of function f in Problem T5.

T7. Figure 4-8h shows the graph of the function $g(x) = x^3 + x^2 - 7x - 15$.

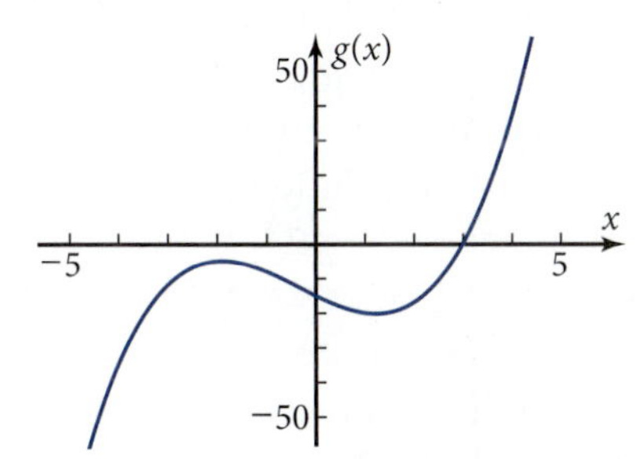

Figure 4-8h

a. Explain graphically why function g has two nonreal zeros.

b. Use synthetic substitution to show that $x = 3$ is really a zero of function g.

c. Find the nonreal zeros of function g.

d. Show that the sum of the pairwise products of the zeros equals the coefficient of the x-term in the given equation.

T8. For the polynomial function f in Figure 4-8i, which has no nonreal zeros,

a. Write the zeros.

b. Write the degree of the function.

c. If the leading coefficient equals -1, write the equation for $f(x)$ in factored form.

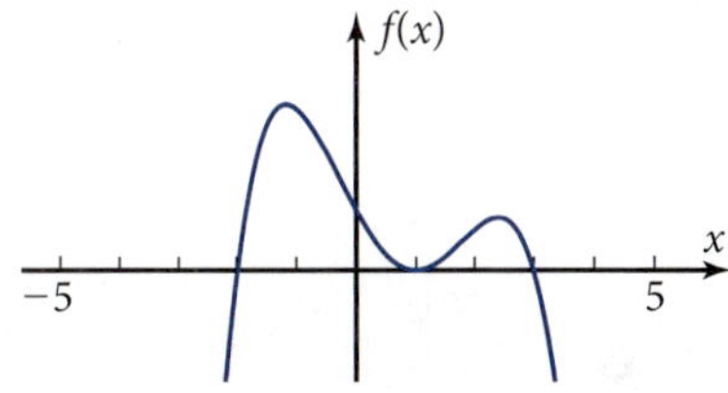

Figure 4-8i

T9. Function h in Figure 4-8j is a transformation of the parent reciprocal function, $y = \frac{1}{x}$.

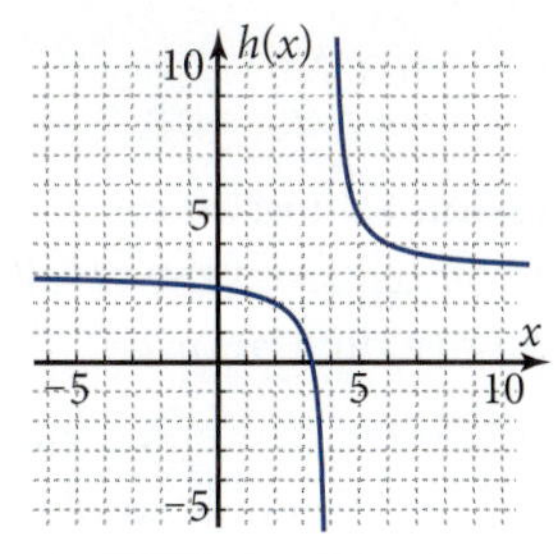

Figure 4-8j

a. Name the three transformations.

b. Write an equation for $h(x)$.

T10. Resolve $f(x) = \frac{4}{(x - 1)(x - 3)}$ into partial fractions.

T11. The equation $x - 3 = 5$ has $x = 8$ as its only solution. If you multiply both sides by $(x - 2)$, the resulting equation has another solution. What is this other solution? What name is given to this new solution?

Part 2: Graphing calculators allowed (T12–T19)

T12. Factor $x^2 - 22x - 720$ by first finding the zeros either graphically or numerically.

T13. Find the complex-number solutions of $x^2 + 10x + 44 = 10$ using the quadratic formula. Substitute one of the solutions into the given equation and show that the quadratic expression on the left really equals 10 for this value of x.

T14. Transform $y = 3x^2 - 24x + 37$ into vertex form by completing the square; then state the coordinates of the vertex. Show that the value of h, the horizontal coordinate of the vertex, agrees with the formula $h = \frac{-b}{(2a)}$.

T15. *Driving Problem:* Hezzy Tate slows down as her car approaches an intersection, then speeds up again as she passes through it. Figure 4-8k and the table show her distance, $d(t)$ in feet, from the beginning of the intersection at time t, in seconds.

t	$d(t)$
3	13
4	20
5	27
6	40
7	65

Figure 4-8k

a. Run a cubic regression to find an equation for $d(t)$. How can you tell from the result that the equation fits all five data points?

b. At what time, t, did Hezzy first enter the intersection? Show that the cubic model has no other real zeros by finding the other zeros of function d.

c. The slope of the graph indicates Hezzy's speed. At what time, t, was she going the slowest? What special name is given to this point on the graph of function d?

d. At time $t = 3$ s, is the graph concave down or concave up?

T16. Let $f(x) = \frac{x - 2}{x - 3}$ and $g(x) = \frac{x - 3}{x + 2}$.

a. Find $f(x) \cdot g(x)$. **b.** Find $f(x) \div g(x)$.

c. Find $f(x) - g(x)$.

T17. Figure 4-8l shows the graph of the rational function $f(x) = \frac{x^3 - 3x^2 - 25x + 75}{x^2 + 3x - 10}$. Simplify the algebraic fraction, and then transform it to the form $p(x) + \frac{r(x)}{d(x)}$. Explain how the simplified fraction allows you to find the two asymptotes, the removable discontinuity, and the x-intercepts.

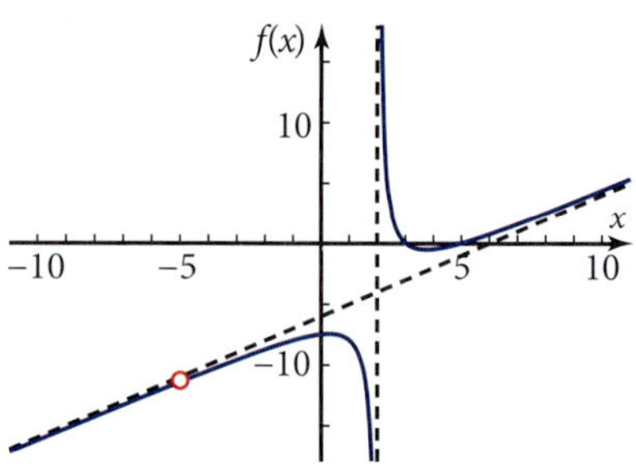

Figure 4-8l

T18. Figure 4-8m shows $f(x) =$ *left side of fractional equation* (dashed) and $g(x) =$ *right side* (solid). Solve this equation by multiplying both sides of the equation by the least common multiple of the denominators to eliminate the fractions and solving the resulting equation. How does the valid solution agree with the graphs of functions f and g? What feature do the graphs have at the extraneous solution of x?

$$\frac{x}{x - 3} - \frac{7}{x + 5} = \frac{24}{x^2 + 2x - 15}$$

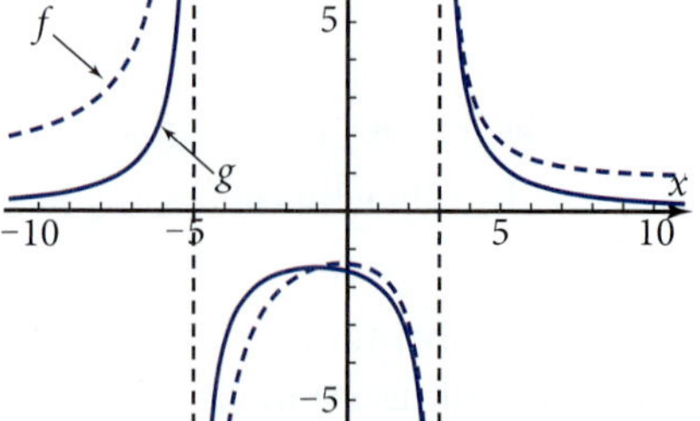

Figure 4-8m

T19. What did you learn as a result of taking this test that you did not know before?

4-9 Cumulative Review, Chapters 1–4

These problems constitute a 2- to 3-hour "rehearsal" for your examination on Chapters 1–4.

Review Problems

Part 1: No calculators allowed (1–29)

In Chapter 1, you studied functions and transformations that can be performed on functions.

1. What is the difference between a relation that is a function and a relation that is not a function?
2. Name the transformations applied to function f to get function g, if
   a. $g(x) = 5f(3x)$
   b. $g(x) = 4 + f(x + 7)$
3. In Figure 4-9a, name the transformations applied to function f to get function h. Write an equation for $h(x)$ in terms of function f.

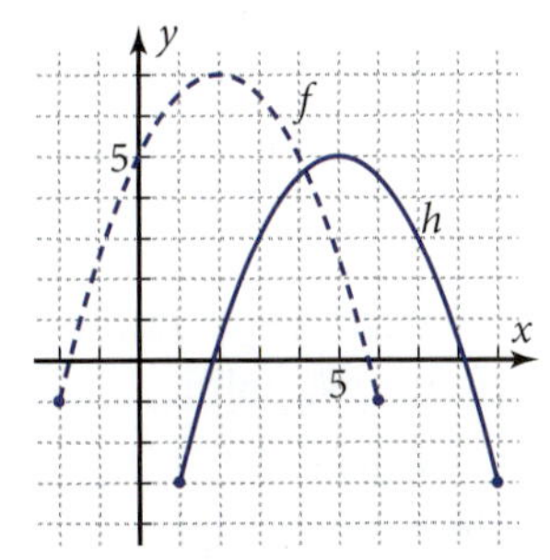

Figure 4-9a

4. On a copy of Figure 4-9b, sketch the graph of $y = f^{-1}(x)$ and the line $y = x$. Explain how the graphs of functions f and f^{-1} are related to the line.

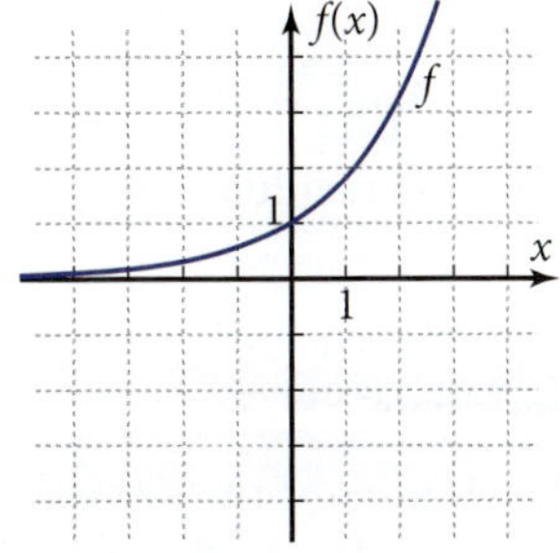

Figure 4-9b

5. If $f(-x) = -f(x)$ for all values of x in the domain, then function f is a(n) __?__ function.
6. Write an equation for $g(x)$ in terms of $f(x)$ that has *all* these features:
   - A horizontal dilation by a factor of 2
   - A vertical dilation by a factor of 3
   - A horizontal translation by 4 units
   - A vertical translation by 5 units
7. Name the kind of function specified by the equations

$$x = t^3 - 7t$$
$$y = 9 - 2t^2$$

8. Figure 4-9c shows the graph of a function f. On a copy of the figure, draw the graph of
   a. $g(x) = |f(x)|$
   b. $h(x) = f(|x|)$

   Indicate which graph is which.

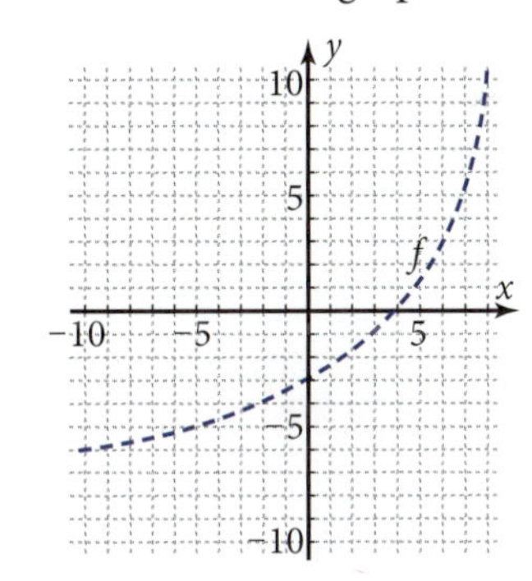

Figure 4-9c

In Chapter 2, you studied properties of elementary functions.

9. Write the general equation for a(n)
   a. Linear function
   b. Quadratic function
   c. Logarithmic function
   d. Exponential function
   e. Power function

10. Sketch graphs of these functions.

a. $f(x) = -2(x - 3)^2 + 5$

b. $g(x) = e^{x+2}$

c. $h(x) = \log_{10} x - 1$

11. Name the pattern followed by the y-values of functions with regularly spaced x-values for

a. Logarithmic functions

b. Power functions

c. Quadratic functions

12. Name the type of function shown by each graph.

a.

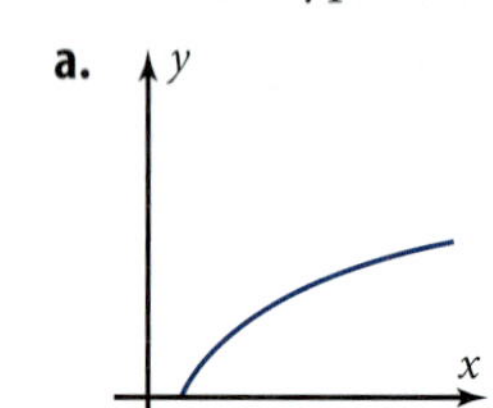

b.

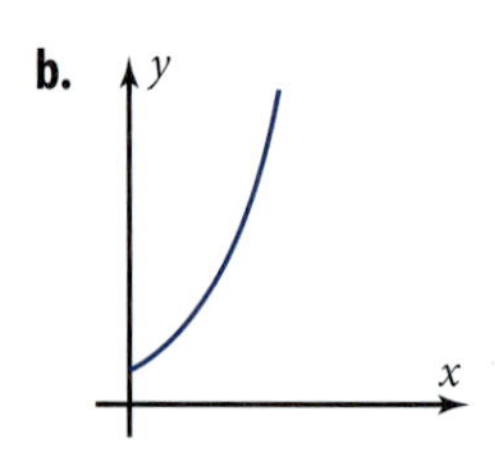

c.

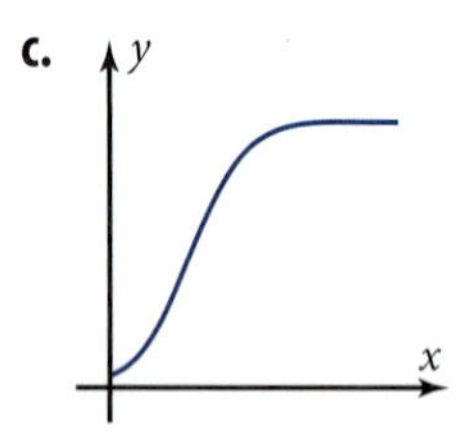

d.

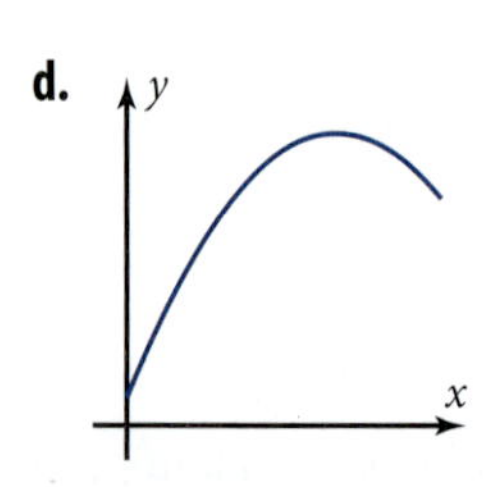

e.

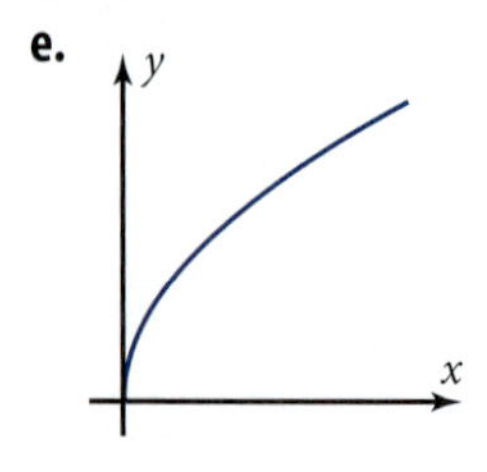

f.

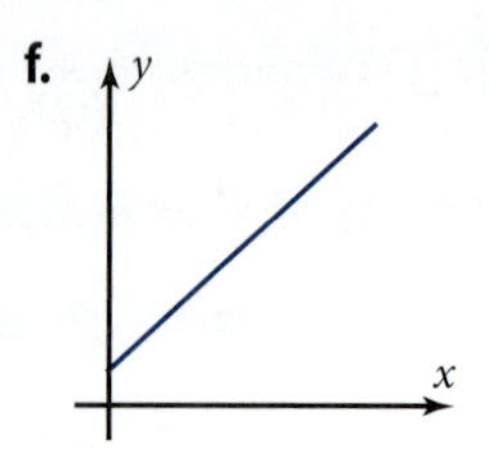

13. Use the appropriate property or definition to complete the statement.

a. $p = \log_c m$ if and only if ___?___

b. $\log 7 + \log 8 = \log$ ___?___

c. $\log 32 =$ ___?___ $\log 2$

In Chapter 3, you fit equations of functions to scattered data.

For Problems 14–16, Figure 4-9d shows a set of points with a dashed horizontal line drawn at $y = \bar{y}$, where $\bar{y}$ is the mean of the y-values.

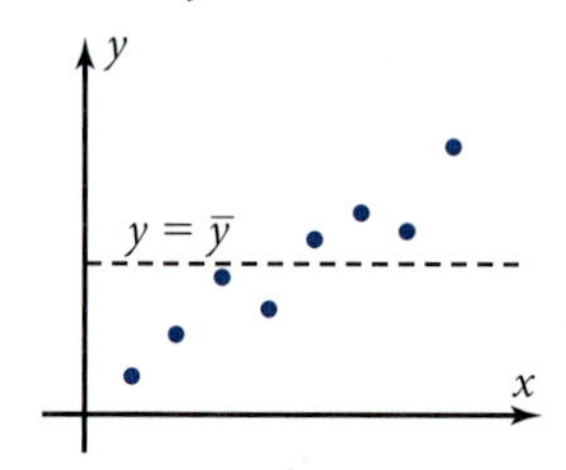

Figure 4-9d

14. On a copy of the figure, sketch the deviation from the mean for the rightmost point.

15. Sketch what you think is the best-fitting linear function. Show the residual for the rightmost point. How does the size of the residual compare with the size of the deviation?

16. Based on what you know about the mean-mean point, $(\bar{x}, \bar{y})$, show where $\bar{x}$ would be on the graph in Problem 14.

17. Suppose the regression equation for a set of data is $\hat{y} = 3x + 5$. What does the residual equal for the data point (4, 15)?

18. For a regression line, what is true about the sum of the squares of the residuals, SS_{res}?

19. Suppose the sum of the squares of the deviations is $SS_{dev} = 100$ and $SS_{res} = 36$. What does the coefficient of determination, r^2, equal?

20. Suppose you run a power regression on a set of data. The correlation coefficient, r, is -0.99, and the residual plot shows a definite pattern. What do these two facts tell you about how well the power function fits the data?

For Problems 21–23, Figure 4-9e shows the graph of $y = 5000(0.72^x)$ on a special kind of graph paper.

21. What kind of graph paper is this? Why is this kind of graph paper useful if there is a wide range of y-values for the function being graphed?

22. Find the approximate value of y for $x = 8$. Find the approximate value of x for $y = 80$.

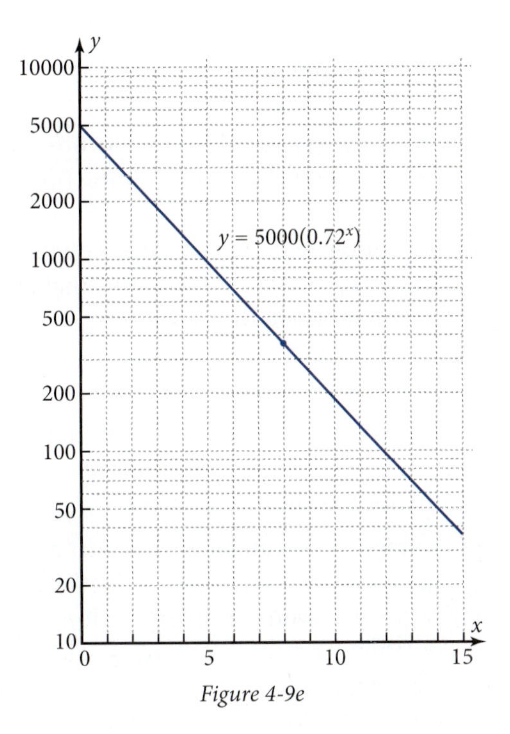

Figure 4-9e

23. Take the logarithm of both sides of the given equation. Use the result to explain why the graph of this function is a straight line on this kind of graph paper.

In Chapter 4, you studied polynomial and rational functions and how imaginary numbers are useful in analyzing the behavior of these functions.

24. Multiply: $(3x - 2)(5x + 7)$

25. Factor: $x^2 - 11x - 12$

26. Expand and simplify: $(7 + 3i)^2$

27. For the function

$$f(x) = (x - 2)(x + 6)(x + 6),$$

a. Write the degree of $f(x)$.

b. Write the x-intercepts of function f.

c. Write the zeros of function f.

d. Find quickly the coefficient of the x^2-term if $f(x)$ is expanded into polynomial form.

28. For the rational function

$$f(x) = \frac{(x - 3)(x + 2)}{(x + 2)(x + 5)(x - 7)},$$

a. Write the x-coordinate(s) of any removable discontinuities.

b. Write the equation(s) of any vertical asymptotes.

c. Write any x-intercept(s).

29. Resolve $\frac{18}{(x - 4)(x + 2)}$ into partial fractions.

Part 2: Graphing calculators allowed (30–37)

30. Figure 4-9f shows the graph of the function $f(x) = x^2 - 7$.

a. With your grapher in parametric mode, plot the graph of function f and its inverse relation. Sketch the graph of the inverse relation on a copy of the given figure.

b. Find an equation for the inverse of function f. Explain why you are not permitted to use the symbol $f^{-1}(x)$ for this inverse relation.

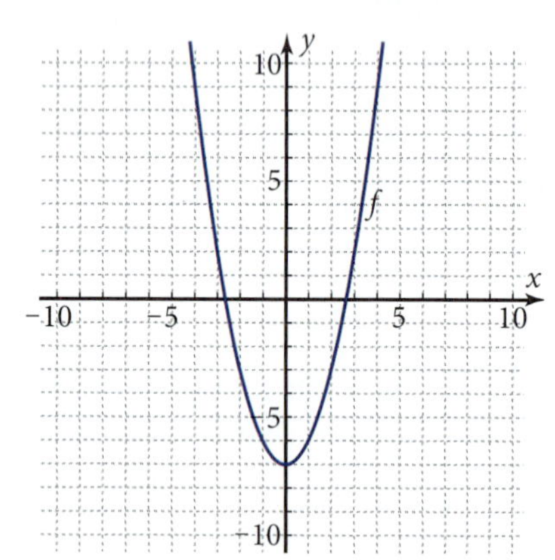

Figure 4-9f

31. Figure 4-9g shows the graph of the parametric function

$$x = t^3 - 7t$$
$$y = 9 - 2t^2$$

Find the coordinates of the points (x, y) for $t = -1$, $t = 2$, and $t = 3$. Do these points lie on the given graph?

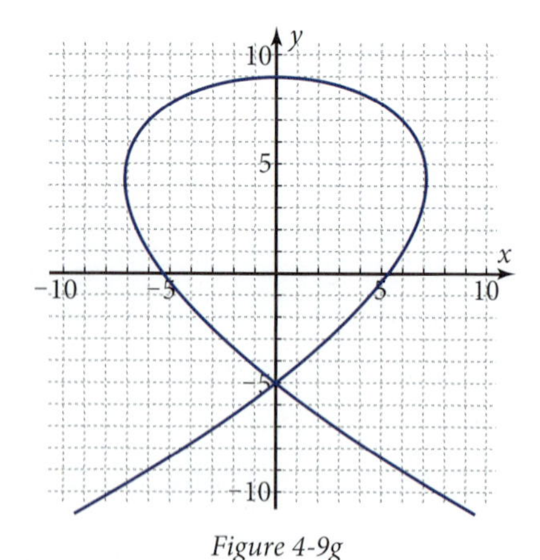

Figure 4-9g

32. *Light Intensity Problem:* The table shows the intensity, y, of light beneath the water surface as a function of distance, x, in meters, below the surface.

x	y
0	100
3	50
6	25
8	16
11	8
17	2

a. What pattern do the first three data points follow? What type of function has this pattern?

b. Find the particular equation for the function in part a algebraically by substituting the second and third points into the general equation. Show that the particular equation gives values for the last three points that are close to the values in the table.

c. Run the appropriate kind of regression to find the function of the type in part a that best fits all six data points. Write the correlation coefficient, and explain how it indicates that the function fits the data quite well.

d. Use the regression equation from part c to predict the light intensity at a depth of 14 m. Which do you use, interpolation or extrapolation, to find this intensity? How do you decide?

33. *Spindletop Problem:* On January 10, 1901, the first oil-well gusher in Texas happened at Spindletop near Beaumont. In the following months, many more wells were drilled. Assume that the data in the table are number of wells, y, as a function of number of months, x, after January 10, 1901.

x (months)	y (wells)
0	1
1	3
2	10
3	27
4	75
5	150

a. Run a regression to find the particular equation for the best-fitting logistic function for these data.

b. Plot the logistic function on your grapher. Sketch the result on a copy of Figure 4-9h. Show the point of inflection.

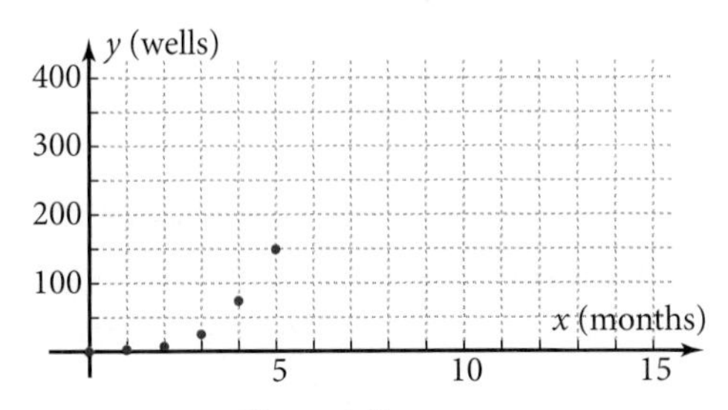

Figure 4-9h

c. How many wells does your model predict were ultimately drilled at Spindletop?

d. Why is a logistic model more reasonable than an exponential model for this problem?

34. *Cricket Problem:* The frequency at which crickets chirp increases as the temperature increases. Let x be the temperature, in degrees Fahrenheit, and let y be the number of chirps per second. Suppose the data in the table have been measured. The data are graphed in Figure 4-9i.

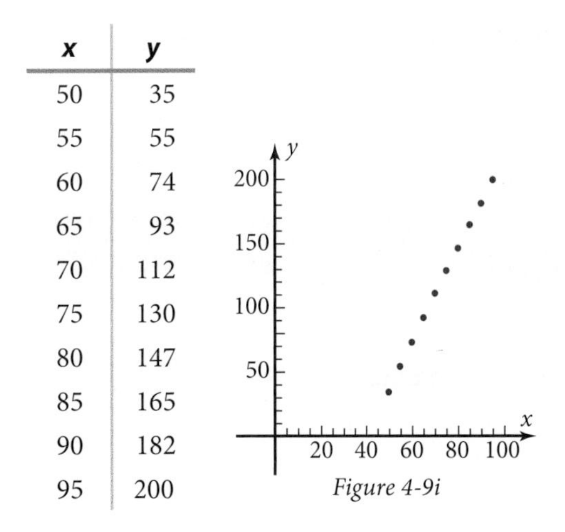

x	y
50	35
55	55
60	74
65	93
70	112
75	130
80	147
85	165
90	182
95	200

Figure 4-9i

a. A linear function appears to fit the data. Write the linear regression equation, and give numerical evidence from the regression result that a linear function fits very well.

b. If you extrapolate to temperatures below 50°F, what does the linear function indicate will eventually happen to the crickets? What, then, would be a reasonable lower bound for the domain of the linear function?

c. If you extrapolate to temperatures above 95°F, what does the linear function indicate will eventually happen to the crickets? Is this a reasonable endpoint behavior?

d. Enter a list into your grapher to calculate the residuals, $y - \hat{y}$. Use the results to make a residual plot. Sketch the residual plot. What information do you get from the residual plot concerning how well the linear function fits the data?

35. *Logarithmic Function Problem:*

a. Use the definition of logarithm to evaluate $\log_9 53$.

b. Use the log-of-a-power property to solve the exponential equation $3^{4x} = 93$.

c. Use the change-of-base property to evaluate $\log_5 47$ using natural logarithms.

d. Plot the graph of $f(x) = \ln x$. Sketch the result. Explain why $f(1) = 0$. Give numerical evidence to show how $f(5)$ and $f(7)$ are related to $f(35)$.

36. *Tree Problem:* Ann R. Burr has a tree nursery in which she stocks various sizes of live oak trees. Figure 4-9j shows the price, $f(x)$, in dollars, she charges for a tree of height x, in feet, planted on a customer's property. For short trees, the price per foot decreases as the height increases. For taller trees, the price per foot increases because the trees are harder to move and plant.

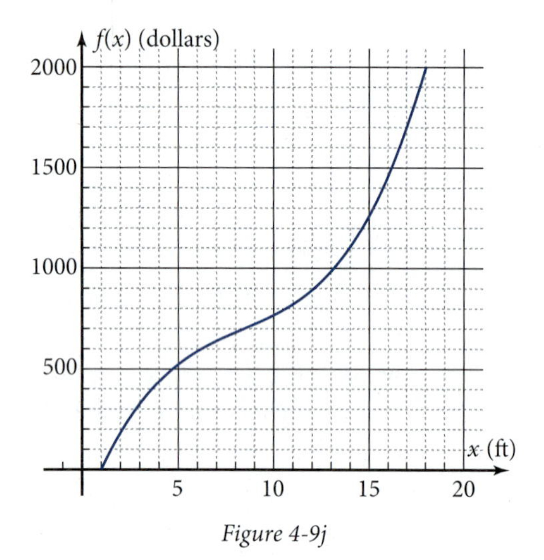

Figure 4-9j

a. The particular equation for this function is the cubic function

$$f(x) = x^3 - 25x^2 + 249x - 225$$

Show by synthetic substitution that $x = 1$ is a zero of this function. What is the real-world meaning of this fact?

b. Find the other two zeros of function f. How does the fact that these are nonreal zeros agree with the given graph?

c. Add the three zeros of function f. Where does this sum appear in the equation for $f(x)$?

d. The graph goes off the scale. If the vertical scale were extended far enough, how much would a 20-ft tree cost?

e. The slope of the graph changes from decreasing to increasing at the point of inflection. Find the x-coordinate of this point. Does this value agree with the graph?

f. The **average rate of change** in price from 10 ft to x ft is given by the rational function $r(x) = \frac{f(x) - f(10)}{x - 10}$. Find an equation for $r(x)$ explicitly in terms of x. Using appropriate algebra, show that the graph of function r has a removable discontinuity at $x = 10$. Find the y-value of this discontinuity.

g. The y-value you found in part f is the **instantaneous rate of change** in price at $x = 10$ ft. On a copy of the given graph, plot a line through the point $(10, f(10))$ with slope equal to this instantaneous rate. Describe the relationship of this line to the given graph.

37. What is the most important thing you have learned as a result of completing this Cumulative Review?

Chapter 5

Periodic Functions and Right Triangle Problems

Ice-skater Michelle Kwan rotates through many degrees during a spin. Her extended hands come back to the same position at the end of each rotation. Thus the position of her hands is a periodic function of the angle through which she rotates. (A periodic function is a function that repeats at regular intervals.) In this chapter you will learn about some special periodic functions that can be used to model situations like this.

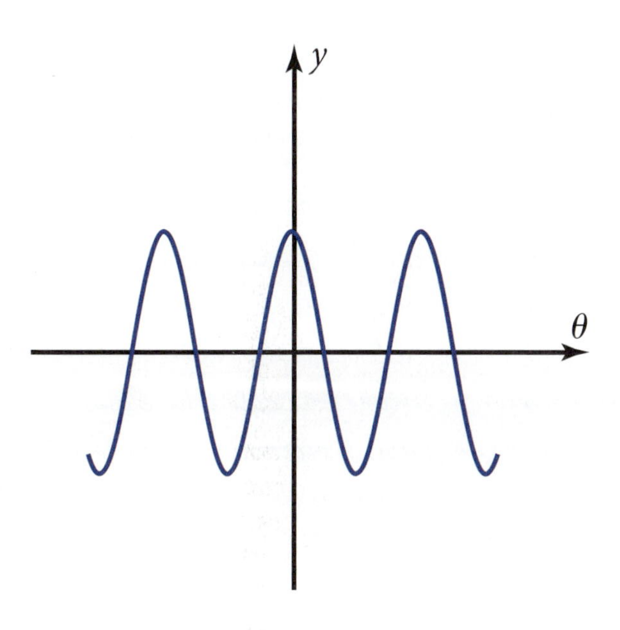

Mathematical Overview

In Chapters 1–4, you studied various types of functions and how these functions can be mathematical models of the real world. In this chapter you will study functions for which the y-values repeat at regular intervals. You will study these periodic functions in four ways.

ALGEBRAICALLY

$$\cos\theta = \frac{u}{r} = \frac{\text{displacement of adjacent leg}}{\text{length of hypotenuse}}$$

(θ is the Greek letter theta.)

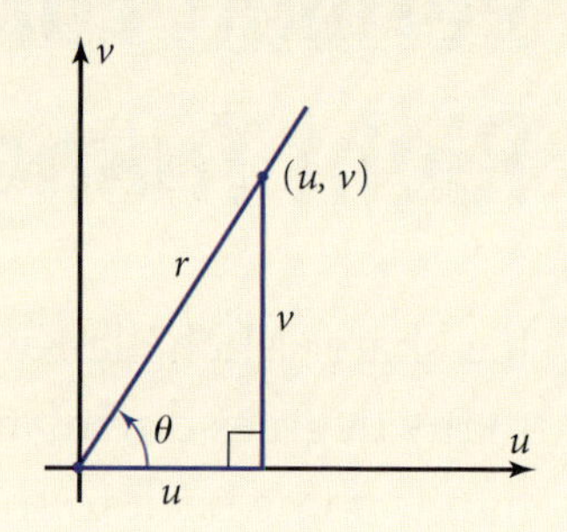

NUMERICALLY

θ	$y = \cos\theta$
0°	1
30°	0.8660...
60°	0.5
90°	0

GRAPHICALLY

This is the graph of a cosine function. Here y depends on the angle, θ, which can take on negative values and values greater than 180°.

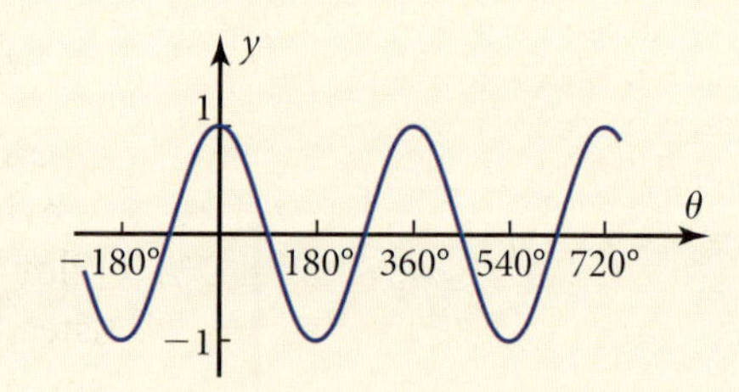

VERBALLY

The trigonometric functions cosine, sine, tangent, cotangent, secant, and cosecant are initially defined as ratios of sides of a right triangle. The definitions are extended to positive and negative angles measuring rotation by forming a reference right triangle whose legs are positive or negative displacements and whose hypotenuse is the radius of the circle formed as the angle increases. The resulting functions are periodic as the angle increases beyond 360°.

5-1 Introduction to Periodic Functions

As you ride a Ferris wheel, your distance from the ground depends on the number of degrees the wheel has rotated (Figure 5-1a). Suppose you start measuring the number of degrees when the seat is on a horizontal line through the axle of the wheel. The Greek letter θ (theta) often stands for the measure of an angle through which an object rotates. A wheel rotates through 360° each revolution, so θ is not restricted. If you plot θ, in degrees, on the horizontal axis and the height above the ground, y, in meters, on the vertical axis, the graph looks like Figure 5-1b. Notice that the graph has repeating y-values corresponding to each revolution of the Ferris wheel.

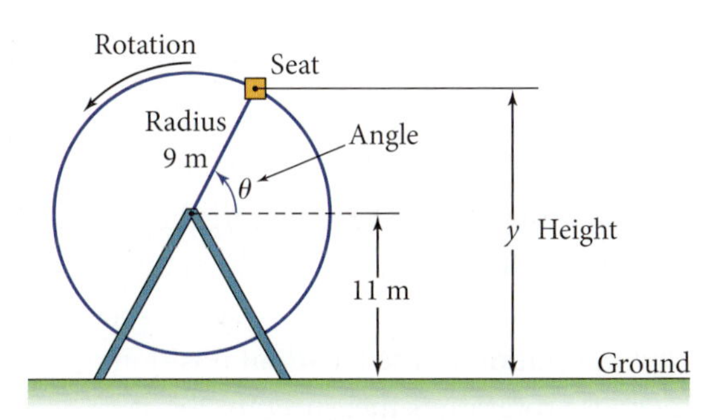

Figure 5-1a

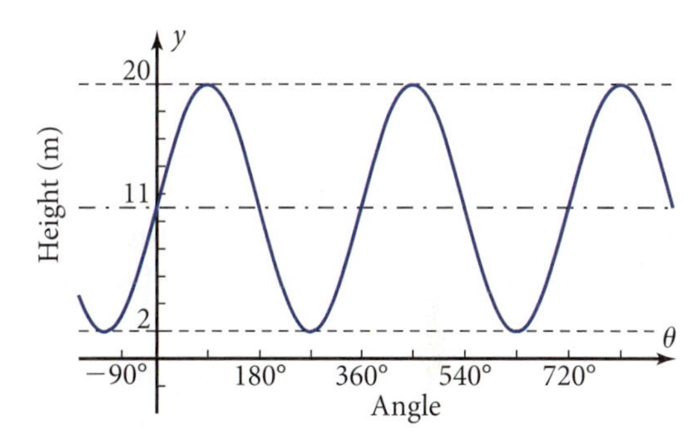

Figure 5-1b

Objective Find the function that corresponds to the graph of a sinusoid and graph it on your grapher.

Exploratory Problem Set 5-1

1. The graph in Figure 5-1c is the sine function (pronounced "sign"). Its abbreviation is sin, and it is written $\sin(\theta)$ or $\sin\theta$. Plot $f_1(x) = \sin(x)$ on your grapher (using x instead of θ). Use the window shown, and make sure your grapher is in degree mode. Does your graph agree with the figure?

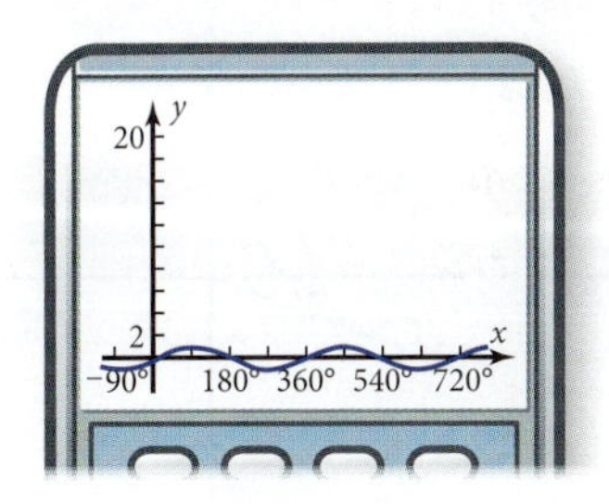

Figure 5-1c

2. The graphs in Figures 5-1b and 5-1c are called **sinusoids** (pronounced like "sinus," a skull cavity). What two transformations must you perform on the parent sine function in Figure 5-1c to get the sinusoid in Figure 5-1b?

3. Enter into your grapher an appropriate equation for the sinusoid in Figure 5-1b as $f_2(x)$. Verify that your equation gives the correct graph.

4. Explain how an angle can have a measure greater than 180°. Explain the real-world significance of the negative values of θ and x in Figures 5-1b and 5-1c.

5-2 Measurement of Rotation

In the Ferris wheel problem of Section 5-1, you saw that you can use an angle to measure an amount of rotation. In this section you will extend the concept of an angle to angles whose measures are greater than 180° and to angles whose measures are negative. You will learn why functions such as your height above the ground are periodic functions of the angle through which the Ferris wheel turns.

Objective Given an angle of any measure, draw a picture of that angle.

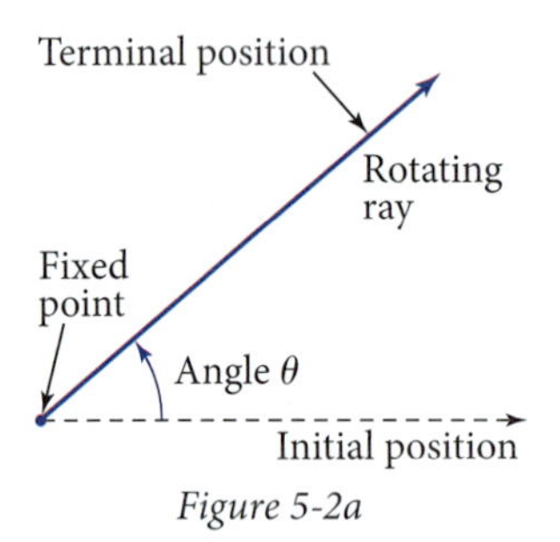

Figure 5-2a

An angle as a measure of rotation can be as large as you like. For instance, a figure skater might spin through an angle of thousands of degrees. To put this idea into mathematical terms, consider a ray with a fixed starting point. Let the ray rotate through a certain number of degrees, θ, and come to rest in a terminal (or final) position, as in Figure 5-2a.

So that the terminal position is uniquely determined by the angle measure, a **standard position** is defined. The initial position of the rotating ray is along the positive horizontal axis in a coordinate system, with its starting point at the origin. Counterclockwise rotation to the terminal position is measured in positive degrees, and clockwise rotation is measured in negative degrees.

DEFINITION: Standard Position of an Angle

An angle is in **standard position** in a Cartesian coordinate system if

- Its vertex is at the origin.
- Its initial side is along the positive horizontal axis.
- It is measured *counterclockwise* from the horizontal axis if the angle measure is positive and *clockwise* from the horizontal axis if the angle measure is negative.

Figure 5-2b shows a rotating ray in several positions in a uv-coordinate system (v for vertical) with a point (u, v) on the ray at a fixed distance from the origin. The angle θ in standard position measures the location of the ray. (The customary variables x and y will be used later for other purposes.)

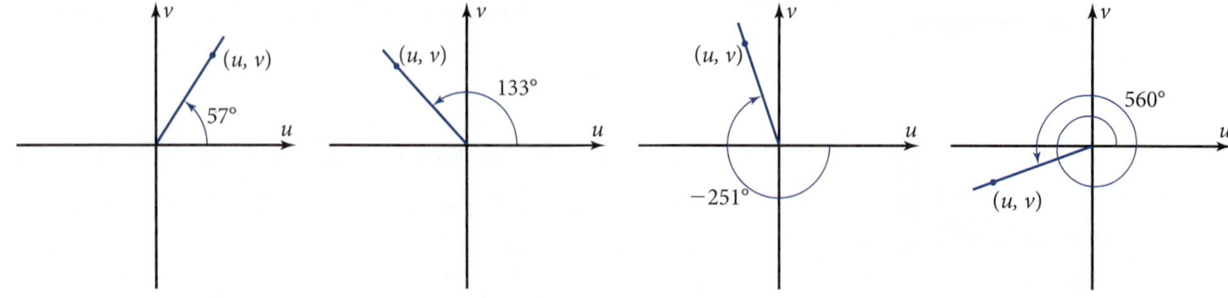

Figure 5-2b

The same position can have several corresponding angle measures. For instance, the 493° angle terminates in the same position as the 133° angle after one full revolution (360°) more. The −227° angle terminates there as well, by rotating clockwise instead of counterclockwise. Figure 5-2c shows these three **coterminal angles.**

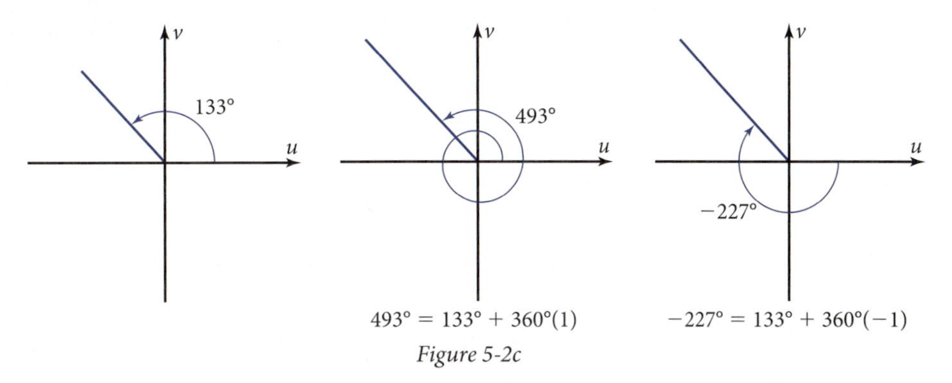

$493° = 133° + 360°(1)$ $\qquad$ $-227° = 133° + 360°(-1)$

Figure 5-2c

Letters such as θ may be used for the measure of an angle or for the angle itself. Other Greek letters are often used as well: α (alpha), β (beta), γ (gamma), ϕ (phi) (pronounced "fye" or "fee"), and ω (omega).

You might recognize some of the Greek letters on this subway sign in Athens, Greece.

DEFINITION: Coterminal Angles

Two angles in standard position are **coterminal** if and only if their degree measures differ by a multiple of 360°. That is, ϕ and θ are coterminal if and only if

$$\phi = \theta + 360°n$$

where n stands for an integer.

Note: Coterminal angles have *terminal* sides that *coincide,* hence the name.

To draw an angle in standard position, you can find the measure of the positive acute angle between the horizontal axis and the terminal side. This angle is called the **reference angle.**

> **DEFINITION: Reference Angle**
>
> The **reference angle** of an angle in standard position is the *positive acute angle* between the horizontal axis and the terminal side.

Note: Reference angles are always measured *counterclockwise.* Angles whose terminal sides fall on one of the axes do not have reference angles.

In this exploration you will apply this definition to find the measures of several reference angles.

EXPLORATION 5-2: Reference Angles

1. The figure shows an angle, $\theta = 152°$, in standard position. The reference angle, θ_{ref}, is measured *counterclockwise* between the terminal side of θ and the nearest side of the horizontal axis. Show that you know what *reference angle* means by drawing θ_{ref} and calculating its measure.

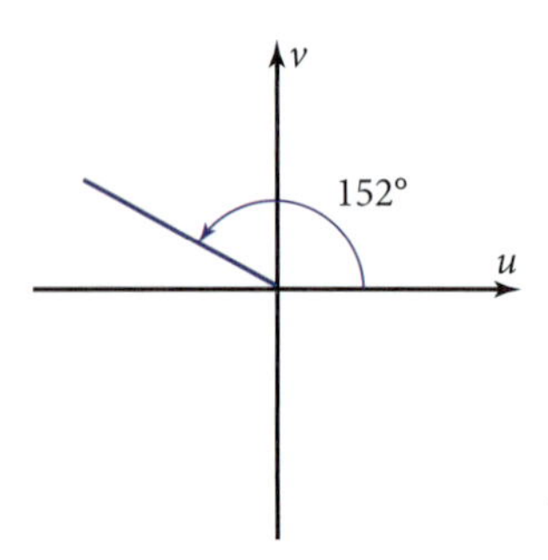

2. The figure shows $\theta = 250°$. Sketch the reference angle and calculate its measure.

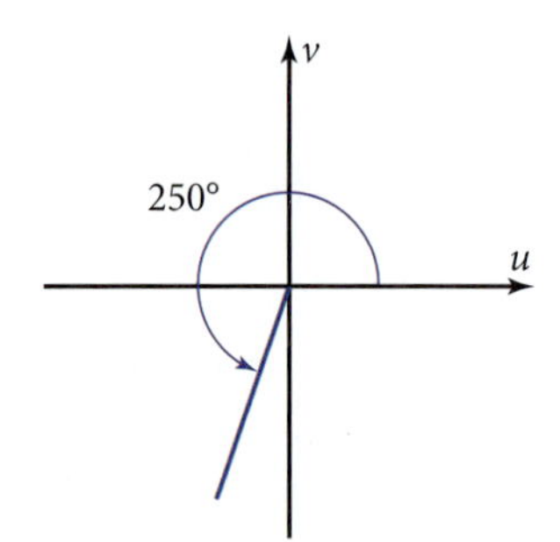

3. You should have drawn arrowheads on the arcs for the reference angles in Problems 1 and 2. If you haven't, draw them now. Explain why the arc for 152° goes from the terminal side to the u-axis but the arc for 250° goes from the u-axis to the terminal side.

4. Amos Take thinks the reference angle for 250° should go to the v-axis because the terminal side is closer to it than the u-axis. Tell Amos why his conclusion does not agree with the definition of *reference angle* in Problem 1.

5. Sketch an angle of 310° in standard position. Sketch its reference angle and find the measure of the reference angle.

6. Sketch an angle whose measure is between 0° and 90°. What is the reference angle of this angle?

7. The figure shows an angle of −150°. Sketch the reference angle and find its measure.

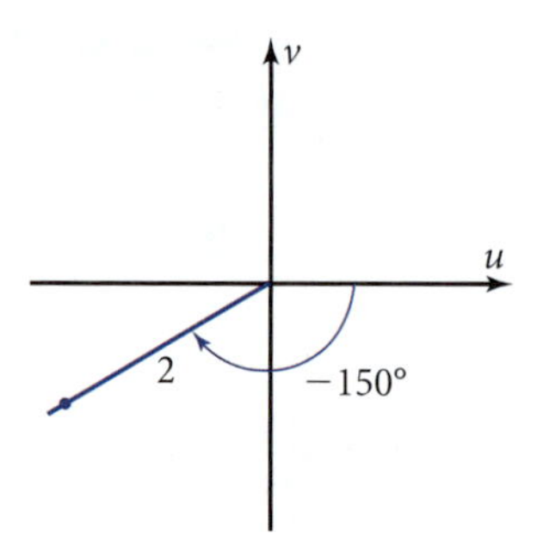

8. The figure in Problem 7 shows a point 2 units from the origin and on the terminal side of the angle. Draw a segment from this point perpendicular to the u-axis, thus forming a right triangle whose hypotenuse is 2 units long. Use what you recall from geometry to find the lengths of the two legs of the triangle.

9. What did you learn as a result of doing this exploration that you did not know before?

Example 1 shows how to find reference angles for angles terminating in each of the four quadrants.

EXAMPLE 1 ➤ Sketch angles of 71°, 133°, 254°, and 317° in standard position and calculate the measure of each reference angle.

SOLUTION To calculate the measure of the reference angle, sketch an angle in the appropriate quadrant; then look at the geometry to figure out what to do.

Figure 5-2d shows the four angles along with their reference angles. For an angle between 0° and 90° (in Quadrant I), the angle and the reference angle are the same. For angles in other quadrants, you have to calculate the positive acute angle between the u-axis and the terminal side of the angle.

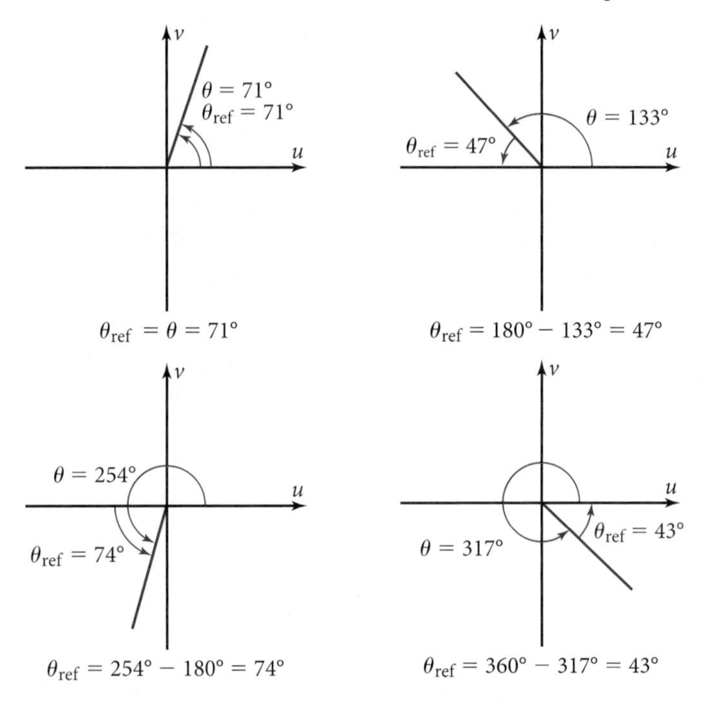

Figure 5-2d

Note that if the angle is not between 0° and 360°, you can first find a coterminal angle that is between these values. It then becomes an "old" problem like Example 1.

EXAMPLE 2 ➤ Sketch an angle of 4897° in standard position and calculate the measure of the reference angle.

SOLUTION $\frac{4897}{360} = 13.6027...$ Divide 4897 by 360 to find the number of whole revolutions.

This number tells you that the terminal side makes 13 whole revolutions plus another 0.6027... revolution. To find out which quadrant the angle falls in, multiply the decimal part of the number of revolutions by 360 to find the number of degrees. The answer is θ_c, a coterminal angle to θ between 0° and 360°.

$\theta_c = (0.6027...)(360) = 217°$ Compute without rounding.

Sketch the 217° angle in Quadrant III, as in Figure 5-2e.

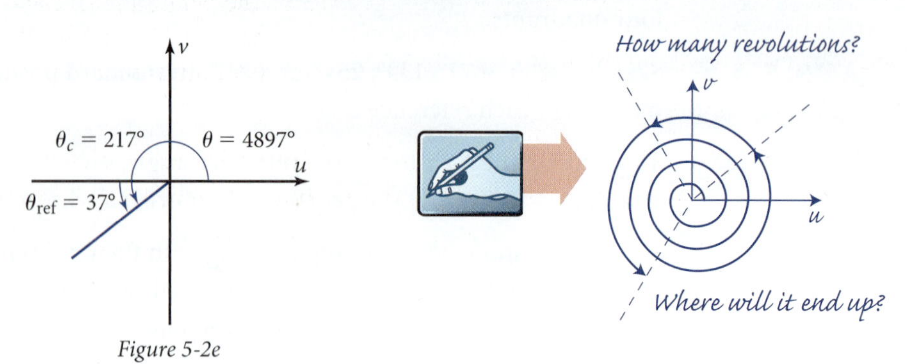

Figure 5-2e

From the figure, you should be able to see that

$$\theta_{ref} = 217° - 180° = 37°$$

As you draw the reference angle, remember that it is always between the terminal side and the *horizontal* axis (never the vertical axis). The reference angle sometimes goes from the axis to the terminal side and sometimes from the terminal side to the axis. To figure out which way it goes, recall that the reference angle is *positive*. Thus it always goes in the *counterclockwise* direction.

Problem Set 5-2

Reading Analysis

From what you have read in this section, what do you consider to be the main idea? How can an angle have a measure greater than 180° or a negative measure? If the terminal side of an angle is drawn in standard position in a *uv*-coordinate system, why can there be more than one value for the measure of the angle but only one value for the measure of the reference angle?

Quick Review

Q1. A function that repeats its values at regular intervals is called a __?__ function.

In Problems Q2–Q5, describe the transformation.

Q2. $g(x) = 5f(x)$

Q3. $g(x) = f(3x)$

Q4. $g(x) = 4 + f(x)$

Q5. $g(x) = f(x - 2)$

Q6. If $f(x) = 2x + 6$, then $f^{-1}(x) =$ __?__.

Q7. How many degrees are there in two revolutions?

Q8. Sketch the graph of $y = 2^x$.

Q9. 40 is 20% of what number?

Q10. $\dfrac{x^{20}}{x^5} =$

A. x^{15} **B.** x^4

C. x^{25} **D.** x^{100}

E. None of these

For Problems 1–20, sketch the angle in standard position, mark the reference angle, and find its measure.

1. 130° **2.** 198°

3. 259° **4.** 147°

5. 342° **6.** 21°

7. 54° **8.** 283°

9. −160° **10.** −220°

11. −295° **12.** −86°

13. 98.6°

14. 57.3°

15. −154.1°

16. −273.2°

17. 5481°

18. 7321°

19. −2746°

20. −3614°

For Problems 21–26, the angles are measured in degrees, minutes, and seconds. There are 60 minutes (60′) in a degree and 60 seconds (60″) in a minute. To find 180° − 137°24′, you calculate 179°60′ − 137°24′. Sketch each angle in standard position, mark the reference angle, and find its measure.

21. 145°37′

22. 268°29′

23. 213°16′

24. 121°43′

25. 308°14′51″

26. 352°16′44″

For Problems 27 and 28, sketch a reasonable graph of the function, showing how the dependent variable is related to the independent variable.

27. A student jumps up and down on a trampoline. Her distance from the ground depends on time.

28. The pendulum in a grandfather clock swings back and forth. The distance from the end of the pendulum to the left side of the clock depends on time.

For Problems 29 and 30, write an equation for the image function, g (solid), in terms of the pre-image function, f (dashed).

29.

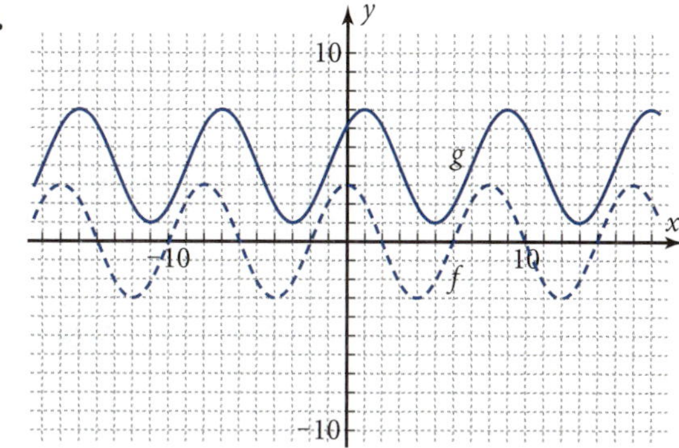

30.

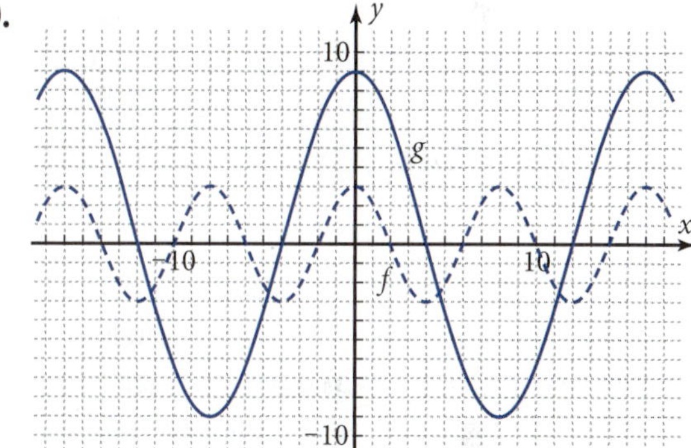

5-3 Sine and Cosine Functions

From previous mathematics courses, you may recall working with sine, cosine, and tangent for angles in a right triangle. Your grapher has these functions in it. If you plot the graphs of $y = \sin\theta$ and $y = \cos\theta$, you get the periodic functions shown in Figure 5-3a. Each graph is called a sinusoid, as you learned in Section 5-1. To get these graphs, you may enter the equations in the form $y = \sin x$ and $y = \cos x$ and use degree mode.

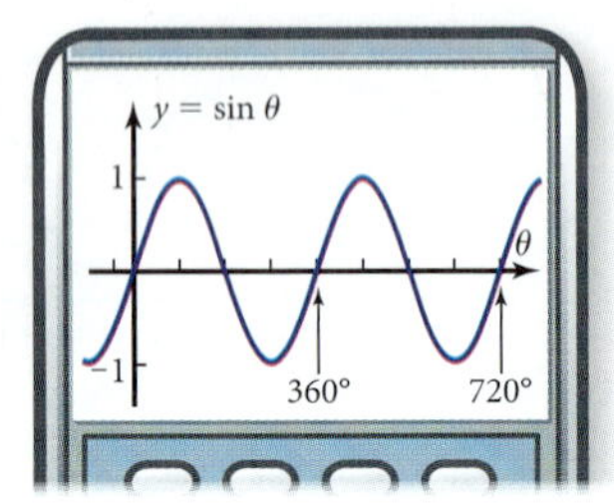

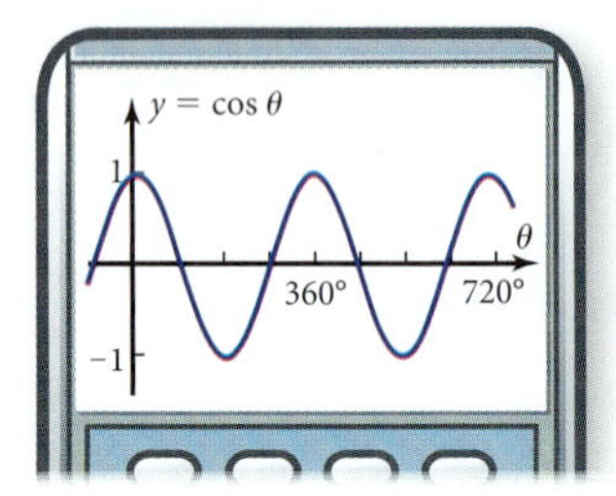

Figure 5-3a

In this section you will see how the reference angles of Section 5-2 let you extend the right triangle definitions of sine and cosine to include angles of any measure. You will also see how these definitions lead to sinusoids.

Objective Extend the definitions of sine and cosine to any angle.

A **periodic function** is a function whose values repeat at regular intervals. The graphs in Figure 5-3a are examples. The part of the graph from any point to the point where the graph starts repeating itself is called a **cycle.** For a periodic function, the **period** is the difference between the horizontal coordinates corresponding to one cycle, as shown in Figure 5-3b. The sine and cosine functions complete a cycle every 360°, as you can see in Figure 5-3a. So the period of these functions is 360°. A horizontal translation of one period makes the pre-image and image graphs identical.

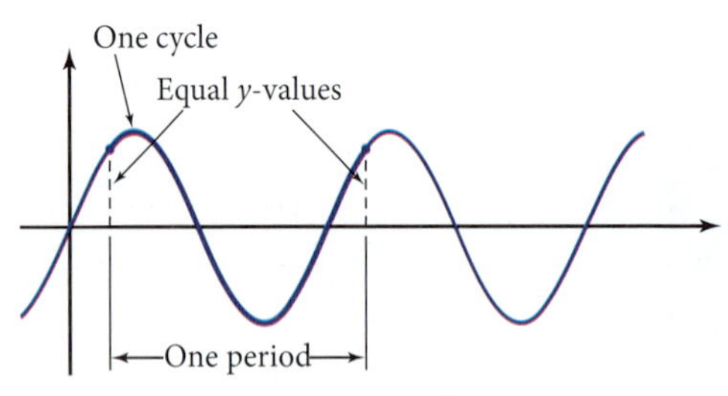

Figure 5-3b

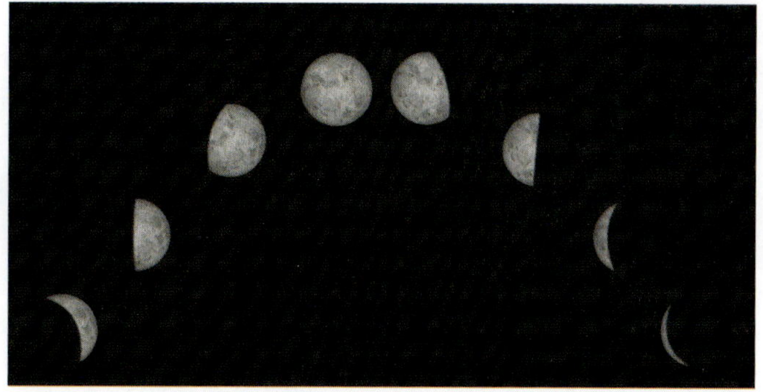

Periodicity is common. The phases of the moon are one example of a periodic phenomenon.

DEFINITION: Periodic Function

The function f is a **periodic function** of x if and only if there is a number $p > 0$ for which $f(x - p) = f(x)$ for all values of x in the domain.

If p is the smallest such number, then p is called the **period** of the function.

Definition of Sine and Cosine for Any Angle

To understand why sine and cosine are periodic functions, consider a ray rotating about the origin in a uv-coordinate system, forming a variable angle θ in standard position. The left side of Figure 5-3c shows the ray terminating in Quadrant I. A point on the ray r units from the origin traces a circle of radius r as the ray rotates. The coordinates (u, v) of the point vary, but r stays constant.

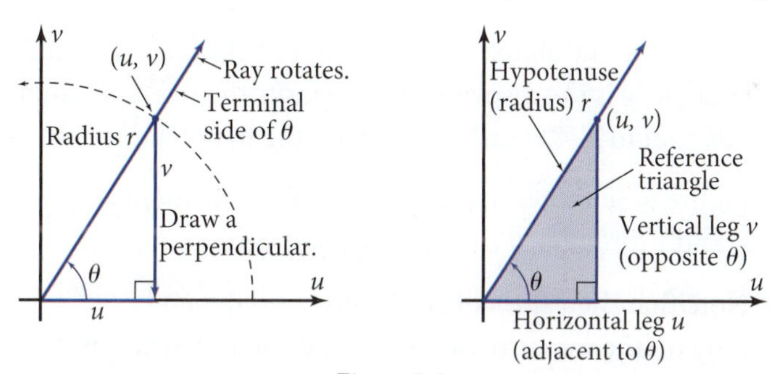

Figure 5-3c

Drawing a perpendicular from point (u, v) to the u-axis forms a right triangle with θ as one of the acute angles. This triangle is called the **reference triangle.** As shown on the right in Figure 5-3c, the coordinate u is the length of the leg adjacent to angle θ, and the coordinate v is the length of the leg opposite angle θ. The radius, r, is the length of the hypotenuse.

The right triangle definitions of the sine and cosine functions are

$$\sin\theta = \frac{\text{opposite leg}}{\text{hypotenuse}} \qquad \cos\theta = \frac{\text{adjacent leg}}{\text{hypotenuse}}$$

These functions are called **trigonometric functions,** from the Greek roots for "triangle" (*trigon-*) and "measurement" (*-metry*).

As θ increases beyond 90°, the values of u, v, or both become negative. As shown in Figure 5-3d, the reference triangle appears in different quadrants, with the reference angle at the origin. Because u and v can be negative, consider them to be **displacements** of the point (u, v) from the respective axes. The value of r stays positive because it is the radius of the circle. The definitions of sine and cosine for any angle use these displacements, u and v.

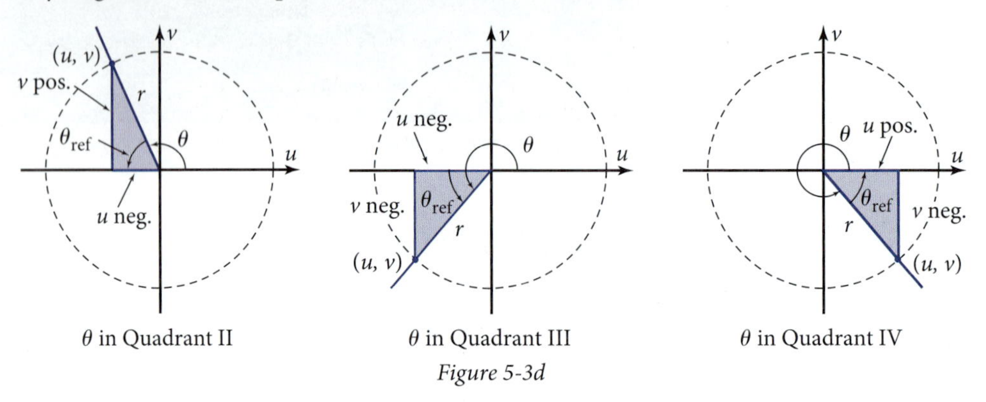

Figure 5-3d

DEFINITION: Sine and Cosine Functions for Any Angle

Let (u, v) be a point r units from the origin on a rotating ray. Let θ be the angle to the ray, in standard position. Then

$$\sin\theta = \frac{v}{r} = \frac{\text{vertical displacement}}{\text{radius}} \qquad \cos\theta = \frac{u}{r} = \frac{\text{horizontal displacement}}{\text{radius}}$$

You can remember these definitions by thinking "v as in *vertical*" and "u comes before v in the alphabet, like x comes before y." With respect to the reference triangle, v is the displacement *opposite* the reference angle, and u is the displacement *adjacent* to it. The radius is always the hypotenuse of the reference triangle.

Figure 5-3e shows the signs of the displacements u and v for angle θ in the four quadrants.

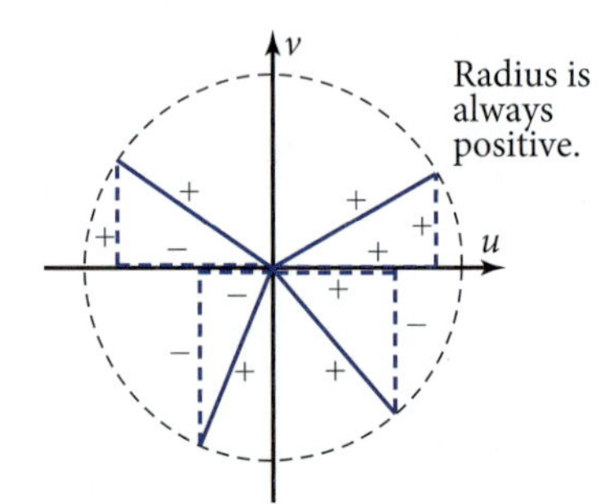

Figure 5-3e

Note that the values of $\sin\theta$ and $\cos\theta$ depend only on the measure of angle θ, not on the location of the point on the terminal side.

As shown in Figure 5-3f, reference triangles for the same angle are similar. Thus

$$\sin\theta = \frac{v_1}{r_1} = \frac{v_2}{r_2} \qquad \text{and} \qquad \cos\theta = \frac{u_1}{r_1} = \frac{u_2}{r_2}$$

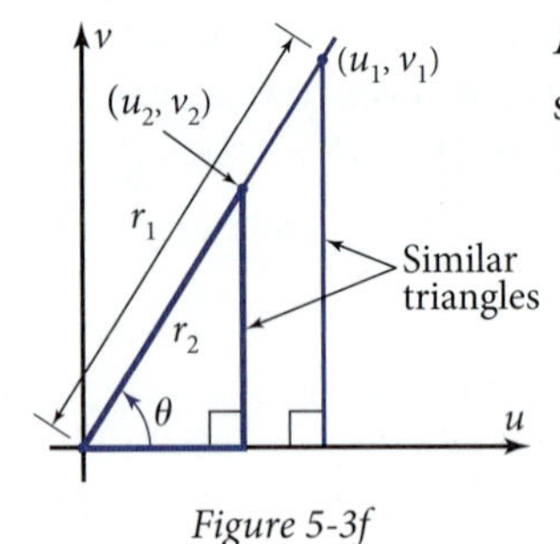

Figure 5-3f

Periodicity of Sine and Cosine

Figure 5-3g shows why the sine function is periodic. On the left, θ is shown as an angle in standard position in a uv-coordinate system. As θ increases from 0°, v is positive and increasing until it equals r, when θ reaches 90°. Then v decreases, becoming negative as θ passes 180° until it equals $-r$, when θ reaches 270°. From 270° to 360°, v is negative but increasing until it equals 0. Beyond 360°, the pattern repeats. Recall that $\sin \theta = \frac{v}{r}$. Thus $\sin \theta$ starts at 0, increases to 1, decreases to -1, and then increases to 0 as θ increases from 0° through 360°. The right side of Figure 5-3g shows θ as the horizontal coordinate in a θy-coordinate system, with $y = \sin \theta$. This is the first sinusoid of Figure 5-3a.

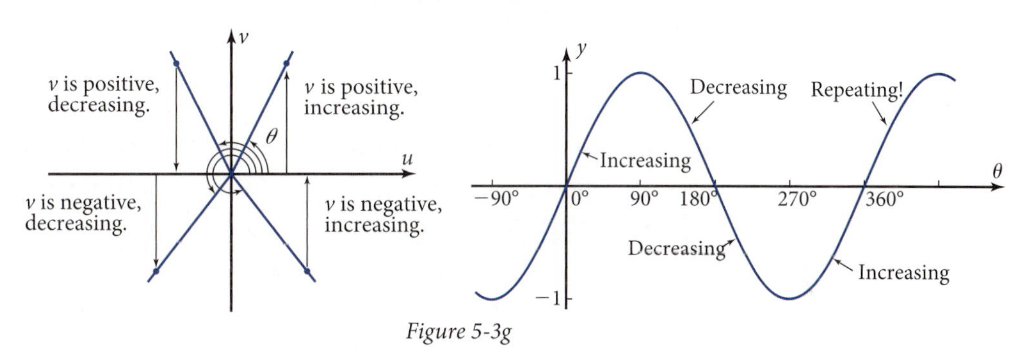

Figure 5-3g

You can see how the rotating ray in a uv-coordinate system generates a sinusoid by viewing the *Sine Wave Tracer* exploration at **www.keymath.com/precalc.** You can also create your own animation using geometry software such as The Geometer's Sketchpad®.

EXAMPLE 1 ➤ Draw angle θ equal to 147° in standard position in a uv-coordinate system. Draw the reference triangle and show the measure of θ_{ref}, the reference angle. Find $\cos 147°$ and $\cos \theta_{\text{ref}}$, and explain the relationship between the two cosine values.

SOLUTION Draw the 147° angle and its reference angle, as in Figure 5-3h. Pick a point on the terminal side of the angle and draw a perpendicular to the horizontal axis, forming the reference triangle.

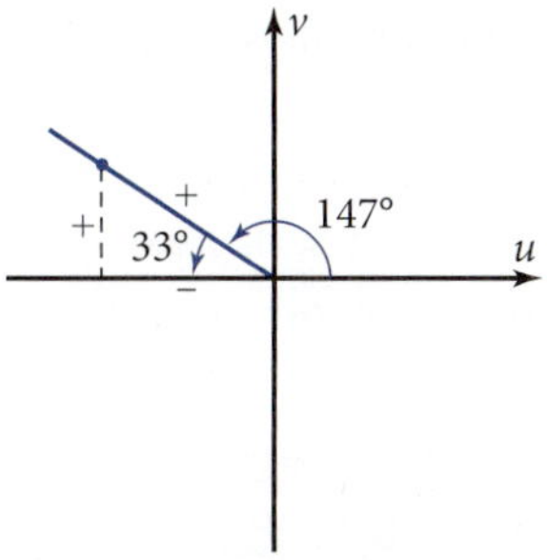

$\theta_{\text{ref}} = 180° - 147° = 33°$ — Because θ_{ref} and 147° must sum to 180°.

$\cos 147° = -0.8386...$

$\cos 33° = 0.8386...$ — By calculator.

Figure 5-3h

Both cosine values have the same magnitude. Cos 147° is negative because the horizontal coordinate of a point in Quadrant II is negative. The radius, r, is always considered to be positive because it is the radius of a circle traced as the ray rotates. ◄

Note: When you write the cosine of an angle in degrees, such as cos 147°, you must write the degree symbol. Writing cos 147 without the degree symbol has a different meaning, as you will see when you learn about angles in radians in the next chapter.

EXAMPLE 2 ➤ The terminal side of angle θ contains the point $(u, v) = (8, -5)$. Sketch the angle in standard position. Use the definitions of sine and cosine to find $\sin \theta$ and $\cos \theta$.

SOLUTION As shown in Figure 5-3i, draw the point $(8, -5)$ and draw angle θ with its terminal side passing through the point. Draw a perpendicular from the point $(8, -5)$ to the horizontal axis, forming the reference triangle. Label the displacements $u = 8$ and $v = -5$. Calculate the radius, r, using the Pythagorean theorem, and show it on your sketch.

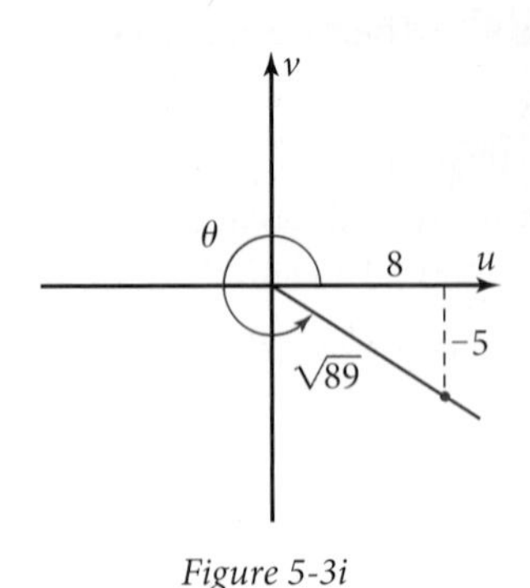

Figure 5-3i

$r = \sqrt{8^2 + (-5)^2} = \sqrt{89}$ Show $\sqrt{89}$ on the figure.

$\sin \theta = \dfrac{-5}{\sqrt{89}} = 0.5299...$ Sine is $\dfrac{\text{opposite displacement}}{\text{hypotenuse}}$.

$\cos \theta = \dfrac{8}{\sqrt{89}} = 0.8479...$ Cosine is $\dfrac{\text{adjacent displacement}}{\text{hypotenuse}}$.

Figure 5-3j shows the parent sine function, $y = \sin \theta$. You can plot sinusoids with other proportions and locations by transforming this parent graph. Example 3 shows you how to do this.

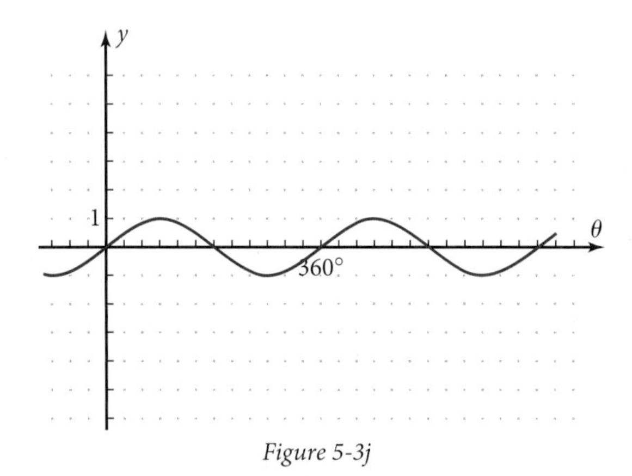

Figure 5-3j

EXAMPLE 3 ➤ Let $y = 4 \sin \theta$. What transformation of the parent sine function is this? On a copy of Figure 5-3j, sketch the graph of this image sinusoid. Check your sketch by plotting the parent sinusoid and the transformed sinusoid on the same screen.

SOLUTION The transformation is a vertical dilation by a factor of 4.

Find places where the pre-image function has high points, low points, or zeros. Multiply the y-values by 4 and plot the resulting points (Figure 5-3k on the next page, left side). Sketch a smooth curve through the critical points (Figure 5-3k, right side). Your grapher will confirm that your sketch is correct.

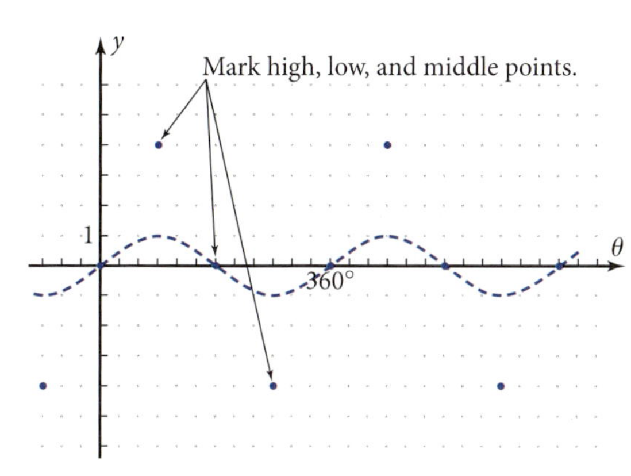

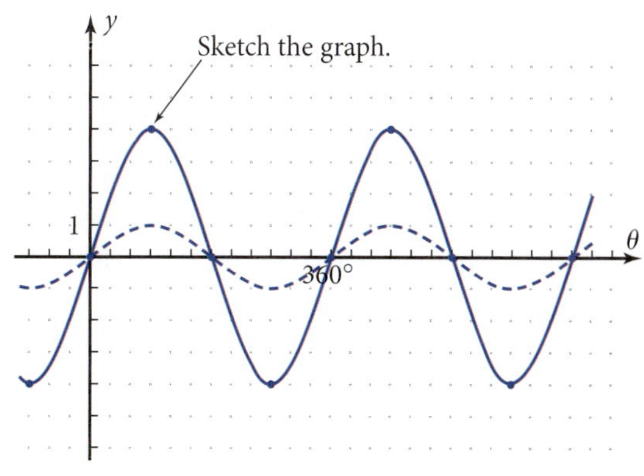

Figure 5-3k

Problem Set 5-3

Reading Analysis

From what you have read in this section, what do you consider to be the main idea? Give a real-world example of what it means for a function to be *periodic*. How are the definitions of sine and cosine extended from right triangles, with angle measures limited to between 0° and 90°, to angles of any measure, positive or negative? What is the difference between the way θ appears in a uv-coordinate system and the way it appears in a θy-coordinate system?

Quick Review

Q1. Write the general equation for an exponential function.

Q2. The equation $y = 3x^{1.2}$ represents a particular __?__ function.

Q3. Find the reference angle for a 241° angle.

Q4. Name these Greek letters: α, β, γ, ϕ.

Q5. What transformation of the pre-image function $y = x^5$ is the image function $y = (x - 3)^5$?

Q6. Find x if $5 \log 2 = \log x$.

Q7. Sketch a reasonable graph showing the height of your foot above the pavement as a function of the distance your bicycle has traveled.

Q8. $3.7^0 =$ __?__ (3.7 with a zero exponent, not 3.7 degrees)

Q9. What is the value of 5! (five factorial)?

Q10. What percent of 300 is 60?

For Problems 1–6, sketch the angle in standard position in a uv-coordinate system. Draw the reference triangle, showing the measure of the reference angle. Find the sine or cosine of the angle and its reference angle. Explain the relationship between them.

1. sin 250°

2. sin 320°

3. cos 140°

4. cos 200°

5. cos 300°

6. sin 120°

For Problems 7–14, use the definitions of sine and cosine to write $\sin\theta$ and $\cos\theta$ for angles whose terminal side contains the given point.

7. $(7, 11)$ **8.** $(4, 1)$

9. $(-2, 5)$ **10.** $(-6, 9)$

11. $(4, -8)$ **12.** $(8, -3)$

13. $(-24, -7)$ (What do you notice about r?)

14. $(-3, -4)$ (What do you notice about r?)

Figure 5-3l shows the parent function graphs $y = \sin\theta$ and $y = \cos\theta$. For Problems 15–20, give the transformation of the parent function represented by the equation. Sketch the transformed graph on a copy of Figure 5-3l. Confirm your sketch by plotting both graphs on the same screen on your grapher.

15. $y = \sin(\theta - 60°)$ **16.** $y = 4 + \sin\theta$

17. $y = 3\cos\theta$ **18.** $y = \cos\frac{1}{2}\theta$

19. $y = 3 + \cos 2\theta$ **20.** $y = 4\cos(\theta + 60°)$

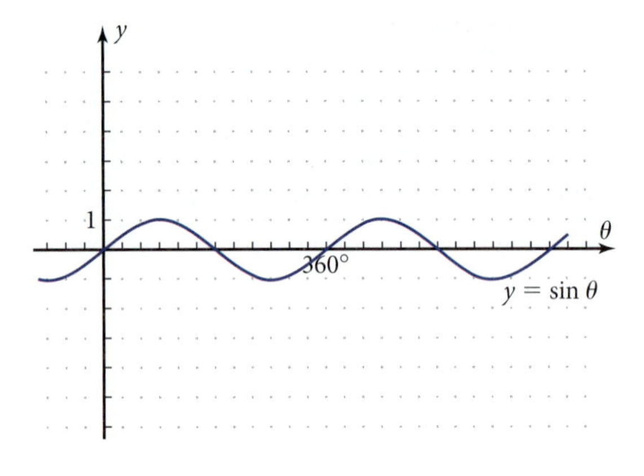

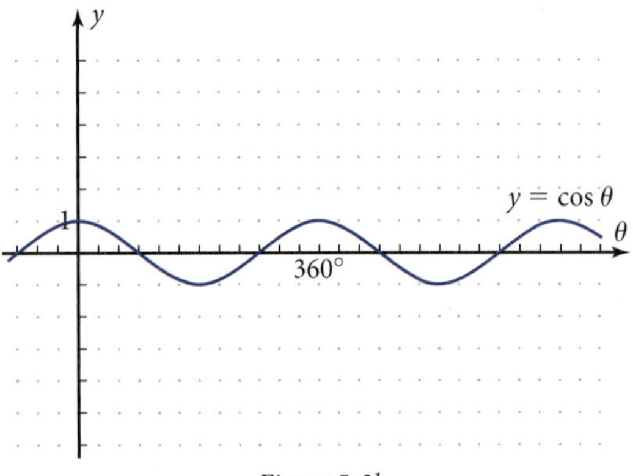

Figure 5-3l

21. Draw the uv-coordinate system. In each quadrant, put a + sign or a − sign to show whether $\cos\theta$ is positive or negative when angle θ terminates in that quadrant.

22. Draw the uv-coordinate system. In each quadrant, put a + sign or a − sign to show whether $\sin\theta$ is positive or negative when angle θ terminates in that quadrant.

23. *Functions of Reference Angles Problem:* This property relates the sine and cosine of an angle to the sine and cosine of the reference angle. Give numerical examples to show that the property is true for both sine and cosine.

> **PROPERTY: Sine and Cosine of a Reference Angle**
>
> $\sin\theta_{\text{ref}} = |\sin\theta|$ and $\cos\theta_{\text{ref}} = |\cos\theta|$

24. *Construction Problem:* For this problem use pencil and paper or a computer graphing program such as The Geometer's Sketchpad. Construct a right triangle with one horizontal leg with length 8 cm and an acute angle of measure 35° with its vertex at one end of the 8-cm leg. Measure the hypotenuse and the other leg. Use these measurements to calculate the values of sin 35° and cos 35° from the definitions of sine and cosine. How well do the answers agree with the values you get directly by calculator? While keeping the angle measure equal to 35°, increase the lengths of the sides of the right triangle. Calculate the values of sin 35° and cos 35° in the new triangle. What do you find?

5-4 Values of the Six Trigonometric Functions

In Section 5-3, you recalled the definitions of the sine and cosine of an angle and saw how to extend these definitions to include angles beyond the range of 0° to 90°. With the extended definitions, $y = \sin\theta$ and $y = \cos\theta$ are periodic functions whose graphs are called sinusoids. In this section you will define four other trigonometric functions. You will learn how to evaluate all six trigonometric functions approximately, by calculator, and exactly in special cases, using the definitions. In the next section you will use what you have learned to calculate unknown side lengths and angle measures in right triangles.

Objective Be able to find values of the six trigonometric functions approximately, by calculator, for any angle and exactly for certain special angles.

Sine and cosine have been defined for any angle as ratios of the coordinates (u, v) of a point on the terminal side of the angle and, equivalently, as ratios of the displacements in the reference triangle.

$$\sin\theta = \frac{v}{r} = \frac{\text{vertical displacement}}{\text{radius}} = \frac{\text{opposite}}{\text{hypotenuse}}$$

$$\cos\theta = \frac{u}{r} = \frac{\text{horizontal displacement}}{\text{radius}} = \frac{\text{adjacent}}{\text{hypotenuse}}$$

In this exploration you will explore the values of sine and cosine for various angles.

EXPLORATION 5-4: Values of the Sine and Cosine Functions

1. This figure shows an angle θ in standard position in a uv-coordinate system. Write the sine and cosine of θ in terms of the coordinates (u, v) and the distance r from the origin to the point.

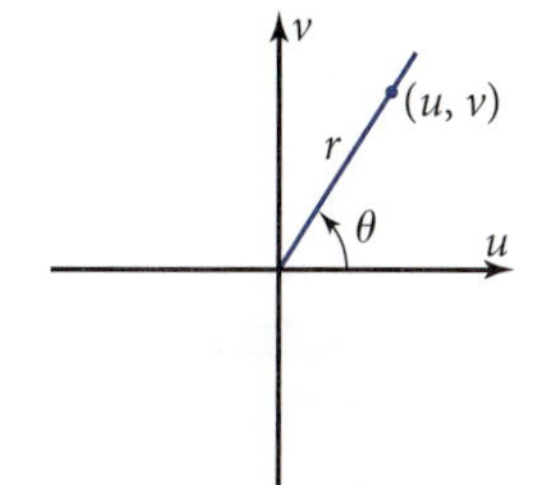

2. This figure shows an angle of 123° in standard position in a uv-coordinate system. By calculator, find sin 123° and cos 123°. Write the answers in **ellipsis format.** Explain why sin 123° is positive but cos 123° is negative.

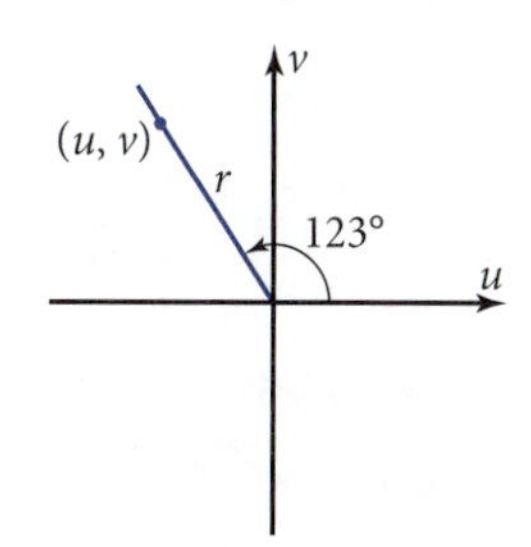

continued

EXPLORATION, continued

3. This figure shows an angle θ in standard position. The terminal side contains the point $(-3, -7)$. Write $\sin \theta$ and $\cos \theta$ exactly, using fractions and radicals.

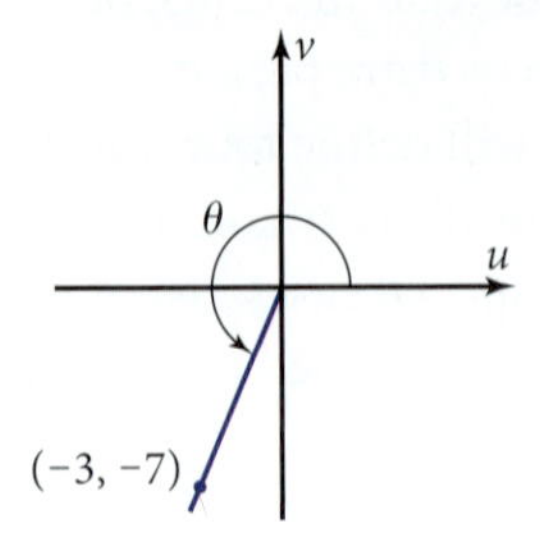

4. This figure shows an angle of 300° in standard position. Choose a convenient point on the terminal side. Use properties of special triangles from geometry to find u, v, and r for the point you picked. Then write sin 300° and cos 300° exactly, using radicals if necessary. Explain why sin 300° is negative but cos 300° is positive.

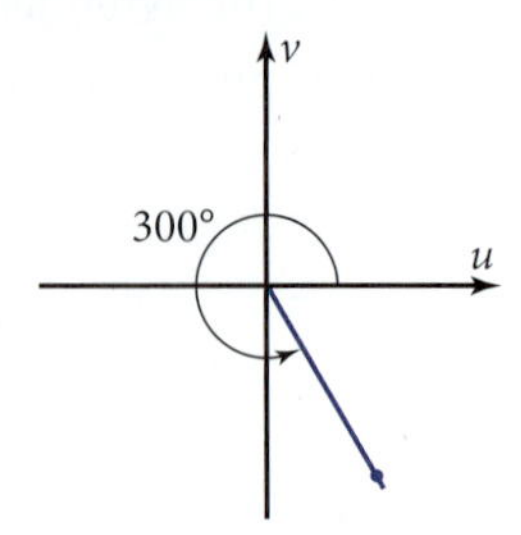

5. What did you learn as a result of doing this exploration that you did not know before?

Four other ratios can be made using u, v, and r. Their names are **tangent, cotangent, secant,** and **cosecant,** and their definitions are given in the box on the next page.

The right triangle definition of tangent for an acute angle is extended to the ratio of v to u for a point on the terminal side of any angle.

$$\tan \theta = \frac{v}{u} = \frac{\text{vertical displacement}}{\text{horizontal displacement}} = \frac{\text{opposite}}{\text{adjacent}}$$

The cotangent, secant, and cosecant functions are reciprocals of the tangent, cosine, and sine functions, respectively. The relationship between each pair of functions, such as cotangent and tangent, is called the **reciprocal property** of trigonometric functions, which you will explore fully in Chapter 7.

When you write the functions in a column in the order $\sin \theta$, $\cos \theta$, $\tan \theta$, $\cot \theta$, $\sec \theta$, and $\csc \theta$, the functions and their reciprocals have this pattern:

DEFINITIONS: The Six Trigonometric Functions

Let (u, v) be a point r units from the origin on the terminal side of a rotating ray. If θ is the angle to the ray, in standard position, then the following definitions hold.

Right Triangle Form

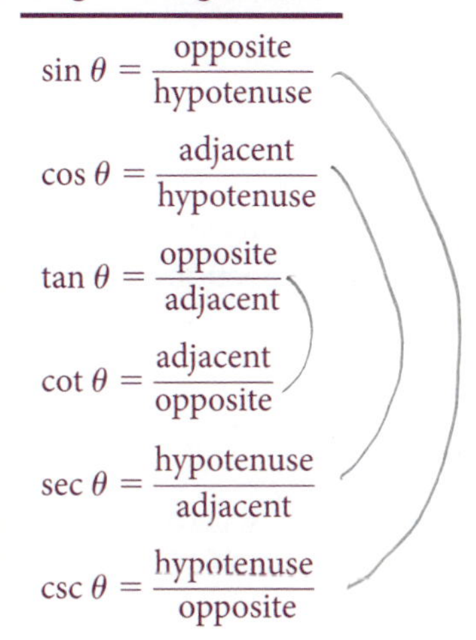

$$\sin\theta = \frac{\text{opposite}}{\text{hypotenuse}}$$

$$\cos\theta = \frac{\text{adjacent}}{\text{hypotenuse}}$$

$$\tan\theta = \frac{\text{opposite}}{\text{adjacent}}$$

$$\cot\theta = \frac{\text{adjacent}}{\text{opposite}}$$

$$\sec\theta = \frac{\text{hypotenuse}}{\text{adjacent}}$$

$$\csc\theta = \frac{\text{hypotenuse}}{\text{opposite}}$$

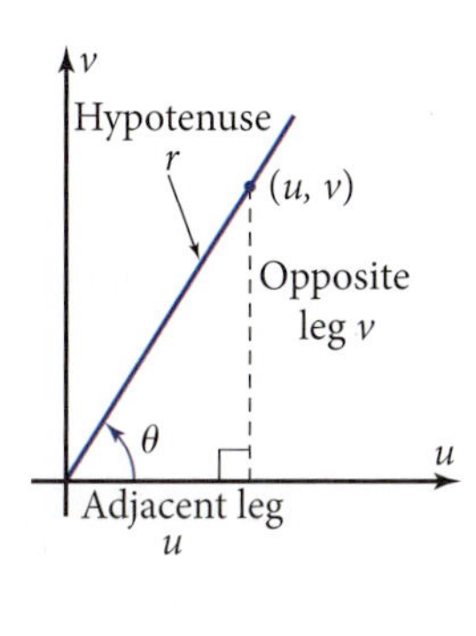

Coordinate Form

$$\sin\theta = \frac{\text{vertical coordinate}}{\text{radius}} = \frac{v}{r}$$

$$\cos\theta = \frac{\text{horizontal coordinate}}{\text{radius}} = \frac{u}{r}$$

$$\tan\theta = \frac{\text{vertical coordinate}}{\text{horizontal coordinate}} = \frac{v}{u}$$

$$\cot\theta = \frac{\text{horizontal coordinate}}{\text{vertical coordinate}} = \frac{u}{v}$$

$$\sec\theta = \frac{\text{radius}}{\text{horizontal coordinate}} = \frac{r}{u}$$

$$\csc\theta = \frac{\text{radius}}{\text{vertical coordinate}} = \frac{r}{v}$$

Note: The coordinates u and v are also the horizontal and vertical displacements of the point (u, v) in the reference triangle.

The Names Tangent and Secant

To see why the names tangent and secant are used, look at Figure 5-4a.

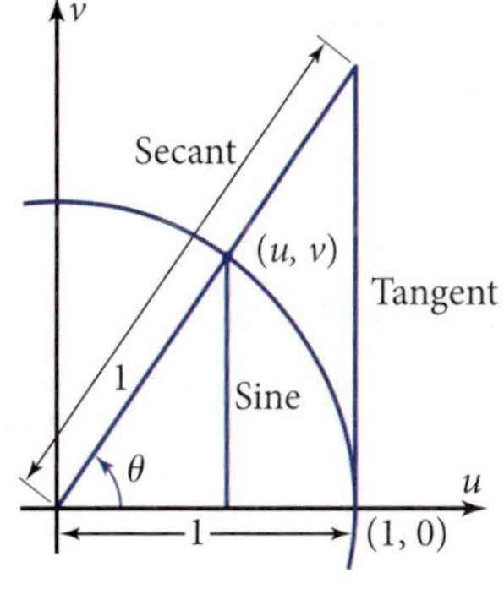

Figure 5-4a

The point (u, v) has been chosen on the terminal side of angle θ where $r = 1$ unit. The circle traced by (u, v) is called the **unit circle** (a circle with radius 1 unit). The value of the sine is given by

$$\sin\theta = \frac{v}{r} = \frac{v}{1} = v$$

Thus the sine of an angle is equal to the vertical coordinate of a point on the unit circle. Similarly, $\cos\theta = \frac{u}{1} = u$, the horizontal coordinate of a point on the unit circle.

If you draw a vertical line tangent to the circle at the point $(1, 0)$, another reference triangle is formed, with the adjacent displacement equal to 1 unit. In this larger triangle,

$$\tan\theta = \frac{\text{opposite}}{\text{adjacent}} = \frac{\text{length of tangent segment}}{1} = \text{length of tangent segment}$$

Hence the name tangent is used.

The hypotenuse of this larger reference triangle is part of a secant line, a line that cuts through the circle. In the larger triangle,

$$\sec \theta = \frac{\text{hypotenuse}}{\text{adjacent}} = \frac{\text{length of secant segment}}{1} = \text{length of secant segment}$$

Hence the name secant is used. By the properties of similar triangles, for any point (u, v) on the terminal side

$$\tan \theta = \frac{v}{u} \quad \text{and} \quad \sec \theta = \frac{r}{u}$$

You have seen why the names tangent and secant are used for these ratios. You will explore the reason for the prefix *co-* in the names cosine, cotangent, and cosecant when you do Problem 43 in the problem set for this section. Section 8-3 explains in detail how the reciprocal functions are related to *complementary angles*.

Approximate Values by Calculator

Example 1 shows how you can find approximate values of all six trigonometric functions by calculator.

EXAMPLE 1 ➤ Evaluate by calculator the six trigonometric functions of 58.6°. Round to four decimal places.

SOLUTION You can find sine, cosine, and tangent directly by calculator.

$\sin 58.6° = 0.8535507... \approx 0.8536$ Note preferred usage of the ellipsis and the $\approx$ sign.

$\cos 58.6° = 0.5210096... \approx 0.5210$

$\tan 58.6° = 1.6382629... \approx 1.6383$

The other three functions are the reciprocals of the sine, cosine, and tangent functions. Notice that the reciprocals follow the pattern described earlier.

$$\cot 58.6° = \frac{1}{\tan 58.6°} = 0.6104026... \approx 0.6104$$

$$\sec 58.6° = \frac{1}{\cos 58.6°} = 1.9193503... \approx 1.9194$$

$$\csc 58.6° = \frac{1}{\sin 58.6°} = 1.1715764... \approx 1.1716$$

◄

Exact Values by Geometry

If you know a point on the terminal side of an angle, you can calculate the values of the trigonometric functions exactly. Example 2 shows you the steps.

EXAMPLE 2 ➤ The terminal side of angle θ contains the point $(-5, 2)$. Find *exact* values of the six trigonometric functions of θ. Use radicals if necessary, but no decimals.

SOLUTION

- Sketch the angle in standard position (Figure 5-4b).
- Pick a point on the terminal side, $(-5, 2)$ in this instance, and draw a perpendicular to the horizontal axis.

- Use the Pythagorean theorem to mark the displacements on the reference triangle.

$$r = \sqrt{(-5)^2 + 2^2} = \sqrt{29}$$

$$\sin\theta = \frac{\text{vertical}}{\text{radius}} = \frac{2}{\sqrt{29}}$$

$$\cos\theta = \frac{\text{horizontal}}{\text{radius}} = \frac{-5}{\sqrt{29}} = -\frac{5}{\sqrt{29}}$$

$$\tan\theta = \frac{\text{vertical}}{\text{horizontal}} = \frac{2}{-5} = -\frac{2}{5}$$

$$\cot\theta = \frac{1}{\tan\theta} = -\frac{5}{2}$$

$$\sec\theta = \frac{1}{\cos\theta} = -\frac{\sqrt{29}}{5}$$

$$\csc\theta = \frac{1}{\sin\theta} = \frac{\sqrt{29}}{2}$$

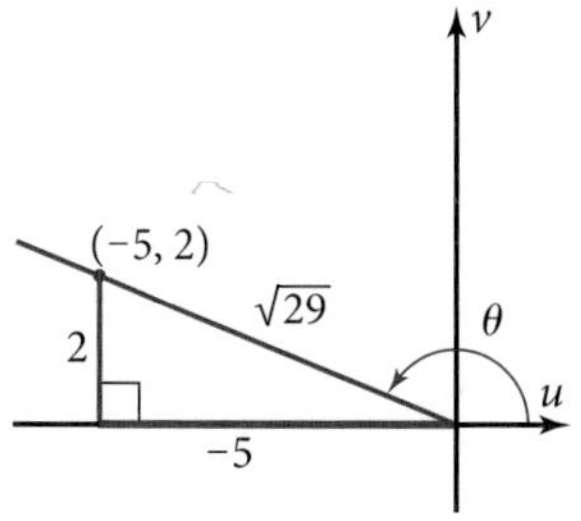

Figure 5-4b

Note: You can use the proportions of the side lengths in the 30°–60°–90° triangle and the 45°–45°–90° triangle to find exact function values for angles whose reference angle is a multiple of 30° or 45°. Figure 5-4c shows these proportions.

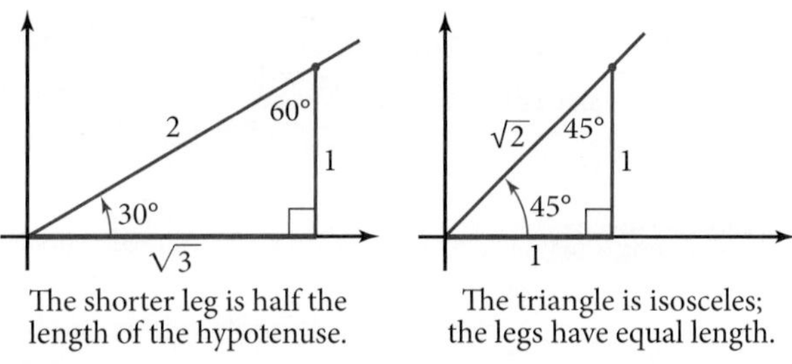

Figure 5-4c

EXAMPLE 3 ➤ Find *exact* values (no decimals) of the six trigonometric functions of 300°.

SOLUTION Sketch an angle terminating in Quadrant IV and a reference triangle with $u = 1$ (Figure 5-4d).

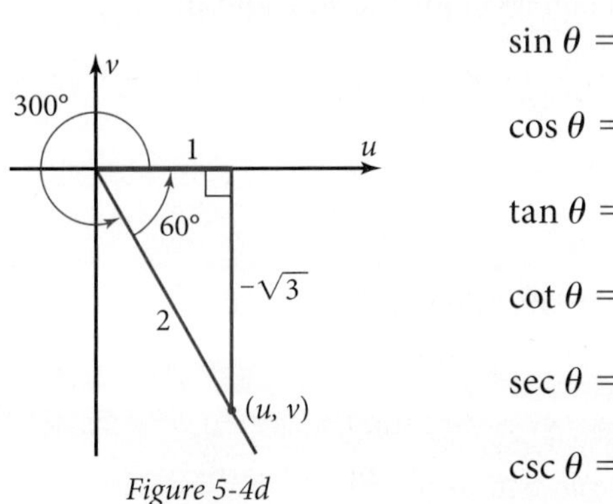

Figure 5-4d

$$\sin\theta = \frac{-\sqrt{3}}{2} = -\frac{\sqrt{3}}{2}$$ Use the *negative* square root because v is negative.

$$\cos\theta = \frac{1}{2}$$

$$\tan\theta = -\frac{\sqrt{3}}{1} = -\sqrt{3}$$ Simplify.

$$\cot\theta = \frac{1}{\tan\theta} = -\frac{1}{\sqrt{3}}$$ Use the reciprocal relationship.

$$\sec\theta = \frac{1}{\cos\theta} = \frac{2}{1} = 2$$

$$\csc\theta = \frac{1}{\sin\theta} = -\frac{2}{\sqrt{3}}$$

To avoid errors in placing the 1, 2, and, $\sqrt{3}$ on the reference triangle, remember that the hypotenuse is the longest side of a right triangle and that 2 is greater than $\sqrt{3}$.

Example 4 shows how to find the function values for an angle that terminates on a quadrant boundary.

EXAMPLE 4 ➤ Without using a calculator, evaluate the six trigonometric functions for an angle of 180°.

SOLUTION Figure 5-4e shows an angle of 180° in standard position. The terminal side falls on the negative side of the horizontal axis. Pick any point on the terminal side, such as (−3, 0). Note that although the *u*-coordinate of the point is negative, the distance *r* from the origin to the point is positive because it is the radius of a circle. The vertical coordinate, *v*, is 0.

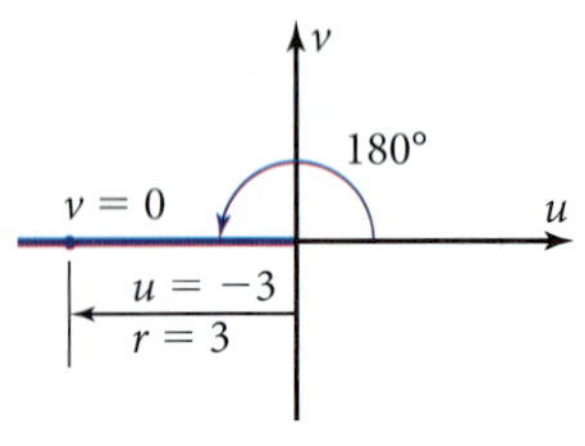

Figure 5-4e

$\sin 180° = \dfrac{\text{vertical}}{\text{radius}} = \dfrac{0}{3} = 0$ Use the "vertical, radius" rather than "opposite, hypotenuse."

$\cos 180° = \dfrac{\text{horizontal}}{\text{radius}} = \dfrac{-3}{3} = -1$

$\tan 180° = \dfrac{\text{vertical}}{\text{horizontal}} = \dfrac{0}{-3} = 0$

$\cot 180° = \dfrac{1}{\tan 180°} = \dfrac{1}{0}$ No value. Undefined because of division by zero.

$\sec 180° = \dfrac{1}{\cos 180°} = \dfrac{1}{-1} = -1$

$\csc 180° = \dfrac{1}{\sin 180°} = \dfrac{1}{0}$ No value. ◄

Problem Set 5-4

Reading Analysis

From what you have read in this section, what do you consider to be the main idea? In what way does the reference triangle help you explain why certain trigonometric functions have negative values in certain quadrants of a *uv*-coordinate system? How do geometric properties of right triangles allow you to find the exact values of trigonometric functions of certain angles?

Quick Review

Problems Q1–Q5 concern the right triangle in Figure 5-4f.

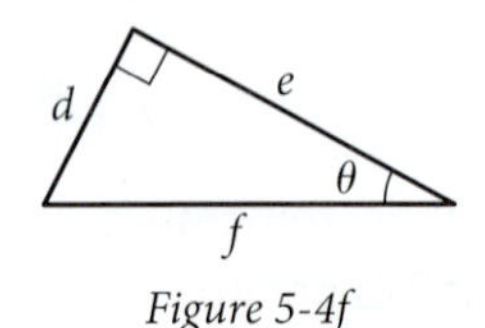

Figure 5-4f

Q1. Which side is the leg opposite angle θ?

Q2. Which side is the leg adjacent to angle θ?

Q3. Which side is the hypotenuse?

Q4. $\cos\theta =$ ___?___

Q5. $\sin\theta =$ ___?___

Q6. Write an equation for the sinusoid in Figure 5-4g.

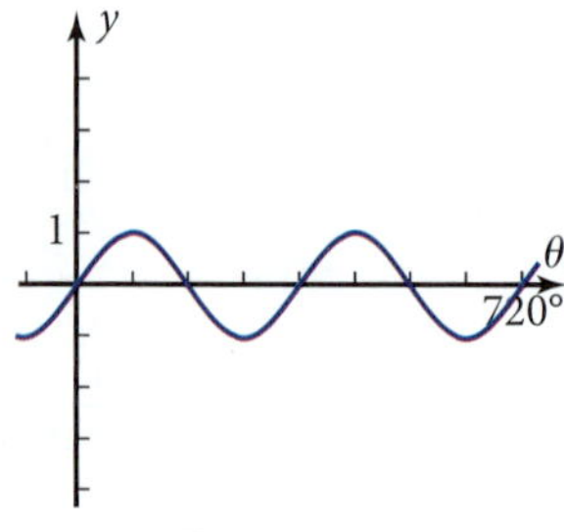

Figure 5-4g

Q7. Write an equation for the sinusoid in Figure 5-4h.

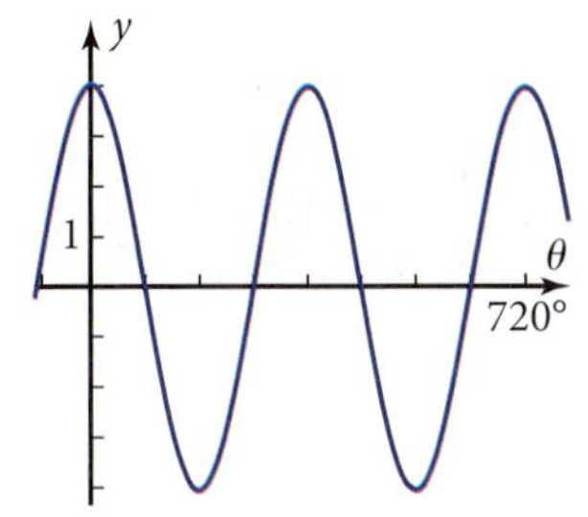

Figure 5-4h

Q8. How was the parent cosine function transformed to get the sinusoid in Figure 5-4h?

Q9. Sketch the graph of $y = -x^2$.

Q10. A one-to-one function is

A. Always increasing

B. Always decreasing

C. Always positive

D. Always negative

E. Always invertible

For Problems 1–6, find a decimal approximation of the given function value. Round the answer to four decimal places.

1. cot 38° | **2.** cot 140°

3. sec 238° | **4.** sec(−53°)

5. csc(−179°) | **6.** csc 180° (Surprising?)

For Problems 7–10, find the exact values (no decimals) of the six trigonometric functions of an angle θ in standard position whose terminal side contains the given point.

7. $(4, -3)$ | **8.** $(-12, 5)$

9. $(-5, -7)$ | **10.** $(2, 3)$

For Problems 11–14, if angle θ terminates in the given quadrant and has the given function value, find the exact values (no decimals) of the six trigonometric functions of θ.

11. Quadrant II, $\sin\theta = \frac{4}{5}$

12. Quadrant III, $\cos\theta = -\frac{1}{3}$

13. Quadrant IV, $\sec\theta = 4$

14. Quadrant I, $\csc\theta = \frac{13}{12}$

For Problems 15–20, find the exact values of the six trigonometric functions of the given angle.

15. 60° | **16.** 135°

17. −315° | **18.** 330°

19. 180° | **20.** −270°

For Problems 21–32, find the exact value (no decimals) of the given function. Try to do this quickly, from memory or by visualizing the figure in your head.

21. sin 180° | **22.** sin 225°

23. cos 240° | **24.** cos 120°

25. tan 315° | **26.** tan 270°

27. cot 0° | **28.** cot 300°

29. sec 150° | **30.** sec 0°

31. csc 45° | **32.** csc 330°

33. Find all values of θ from 0° through 360° for which

a. $\sin\theta = 0$ | **b.** $\cos\theta = 0$

c. $\tan\theta = 0$ | **d.** $\cot\theta = 0$

e. $\sec\theta = 0$ | **f.** $\csc\theta = 0$

34. Find all values of θ from 0° through 360° for which

a. $\sin\theta = 1$ | **b.** $\cos\theta = 1$

c. $\tan\theta = 1$ | **d.** $\cot\theta = 1$

e. $\sec\theta = 1$ | **f.** $\csc\theta = 1$

For Problems 35–42, find the exact value (no decimals) of the given expression. Note that the expression $\sin^2 \theta$ means $(\sin \theta)^2$. You may check your answers by calculator.

35. $\sin 30° + \cos 60°$

36. $\tan 120° + \cot(-30°)$

37. $\sec^2 45°$

38. $\cot^2 30°$

39. $\sin 240° \csc 240°$

40. $\cos 120° \sec 120°$

41. $\tan^2 60° - \sec^2 60°$

42. $\cos^2 210° + \sin^2 210°$

43. Recall that complementary angles sum to 90°.

a. If $\theta = 23°$, what is the complement of θ?

b. Find cos 23° and find sin(complement of 23°). What relationship do you notice?

c. Based on what you've discovered, what do you think the prefix *co-* stands for in the names cosine, cotangent, and cosecant?

44. *Pattern in Sine Values Problem:* Find the exact values of sin 0°, sin 30°, sin 45°, sin 60°, and sin 90°. Make all the denominators equal to 2 and all the numerators radicals. Describe the pattern you see.

45. *Sine Wave Tracer Project:* Figure 5-4i shows the unit circle in a *uv*-coordinate system and a sinusoid whose *y*-values equal the *v*-values of the point *P* where the rotating ray cuts the unit circle. Create this sketch using dynamic geometry software such as The Geometer's Sketchpad so that when you move point *P* around the circle, point *Q* traces the sinusoid. You can also find the *Sine Wave Tracer* exploration at **www.keymath.com/precalc** that provides this sketch.

Explore this sketch. Write a paragraph telling what you learned. In particular, explain the relationship between θ plotted as an angle in a *uv*-coordinate system and θ plotted horizontally in a θy-coordinate system.

46. *Journal Problem:* Update your journal, writing about concepts you have learned since the last entry and concepts about which you are still unsure.

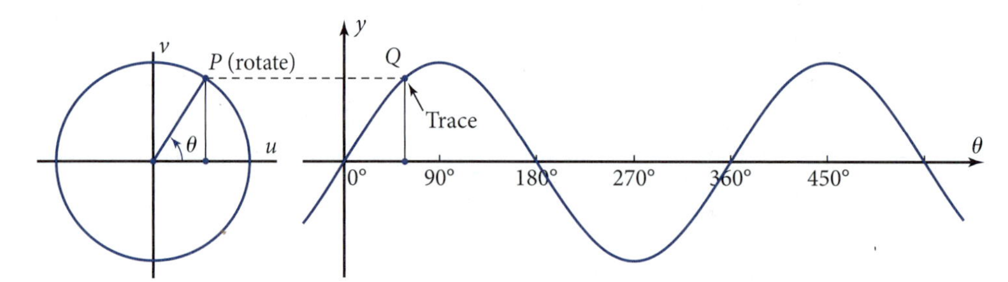

Figure 5-4i

5-5 Inverse Trigonometric Functions and Triangle Problems

You have learned how to evaluate trigonometric functions for specific angle measures. Next you'll learn to use function values to find the angle measures. You'll also learn how to find unknown side and angle measures in a right triangle.

Objective Given two sides of a right triangle or a side and an acute angle, find measures of the other sides and angles.

Inverses of Trigonometric Functions

In order to find the measure of an angle when its function value is given, you could press the appropriate inverse function keys on your calculator.

The symbol $\cos^{-1}$ is the familiar inverse function terminology of Chapter 1, read as "inverse cosine." Note that it does *not* mean the -1 power of cos, which is the reciprocal of cosine:

$$(\cos x)^{-1} = \frac{1}{\cos x}$$

Trigonometric functions are periodic, so they are not one-to-one functions. There are many angles whose cosine is 0.8 (Figure 5-5a). However, for each trigonometric function there is a **principal branch** of the function that is one-to-one and includes angles between 0° and 90°. The calculator is programmed to give the one angle on the principal branch. The symbol $\cos^{-1} 0.8$ means the one angle on the principal branch whose cosine is 0.8. The inverse of the cosine function on the principal branch is a function denoted $\cos^{-1} x$.

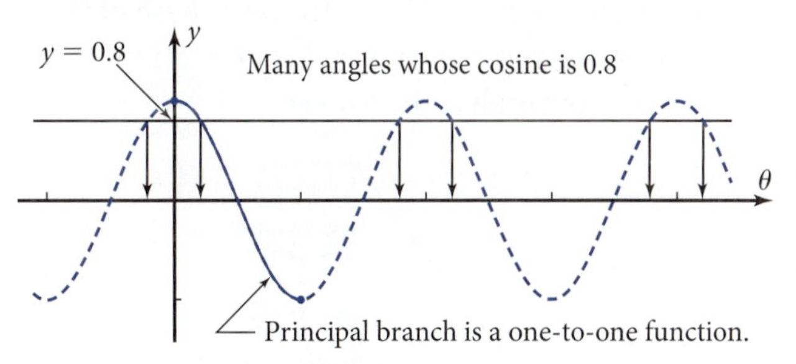

Figure 5-5a

In Section 7-6, you will learn more about the principal branches of all six trigonometric functions. For the triangle problems of this section, all the angles will be acute, so the value the calculator gives you is the value of the angle you want.

The definitions of the inverse trigonometric functions are given in the box.

DEFINITIONS: Inverse Trigonometric Functions

If x is the value of a trigonometric function at angle θ, then the **inverse trigonometric functions** can be defined on limited domains.

$\theta = \sin^{-1} x$ means $\sin \theta = x$ and $-90° \leq \theta \leq 90°$.

$\theta = \cos^{-1} x$ means $\cos \theta = x$ and $0° \leq \theta \leq 180°$.

$\theta = \tan^{-1} x$ means $\tan \theta = x$ and $-90° < \theta < 90°$.

Notes:

- Words: "Angle θ is the angle on the principal branch whose sine (and so on) is x."
- Pronunciation: "Inverse sine of x" and so on, never "sine to the negative 1."
- The symbols $\sin^{-1} x$, $\cos^{-1} x$, and $\tan^{-1} x$ are used only for the value the calculator gives you, not for other angles that have the same function value. The symbols $\cot^{-1} x$, $\sec^{-1} x$, and $\csc^{-1} x$ are similarly defined.
- The symbol $\sin^{-1} x$ does *not* mean the reciprocal of $\sin x$.

Right Triangle Problems

Trigonometric functions and inverse trigonometric functions often come up in applications, such as right triangle problems

EXAMPLE 1 ➤ Suppose you have the job of measuring the height of the local water tower. Climbing makes you dizzy, so you decide to do the whole job at ground level. You find that from a point 47.3 m from the base of the tower you must aim a laser pointer at an angle of 53° (angle of elevation) to hit the top of the tower (Figure 5-5b). How high is the tower?

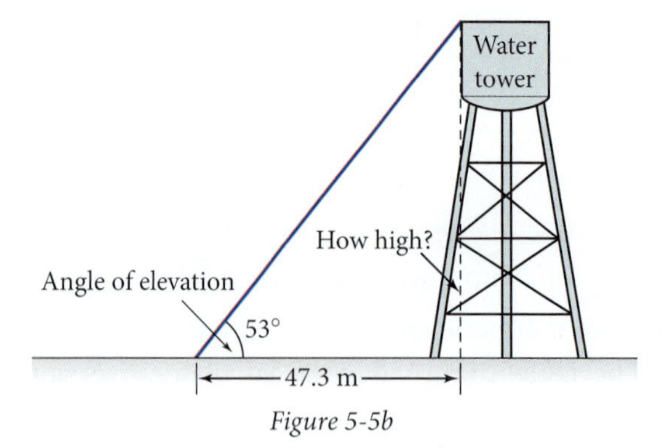

Figure 5-5b

SOLUTION Sketch an appropriate right triangle, as in Figure 5-5c. Label the known and unknown information.

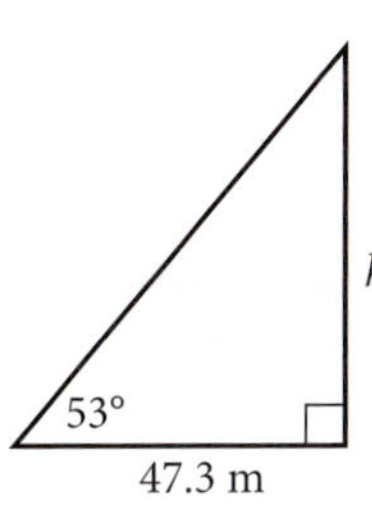

Figure 5-5c

$$\frac{h}{47.3} = \tan 53°$$ Write a ratio for tangent.

$$h = 47.3 \tan 53° = 62.7692...$$ Solve for h.

The tower is about 62.8 m high. Write the real-world answer.

Note that the angle must always have the degree sign, even during the computations. The symbol tan 53 has a different meaning, as you will learn when you study radians in the next chapter.

EXAMPLE 2 A ship is passing through the Strait of Gibraltar along a straight path. At its closest point of approach, Gibraltar radar determines that the ship is 2400 m away. Later, the radar determines that the ship is 2650 m away (Figure 5-5d).

a. By what angle θ did the ship's bearing from Gibraltar change?

b. How far did the ship travel between the two observations?

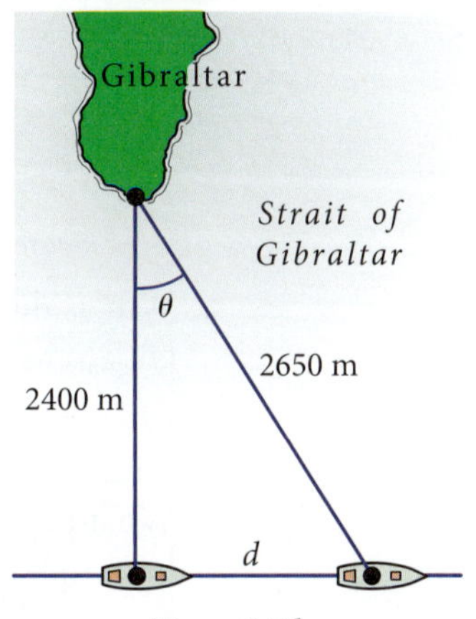

Figure 5-5d

SOLUTION

a. Draw the right triangle and label the unknown angle θ. By the definition of cosine,

$$\cos \theta = \frac{\text{adjacent}}{\text{hypotenuse}} = \frac{2400}{2650}$$

$$\theta = \cos^{-1} \frac{2400}{2650} = 25.0876...°$$ Take the inverse cosine to find θ.

The angle measure is about 25.09°.

b. Label the unknown side d, for distance. By the definition of sine,

$$\frac{d}{2650} = \sin 25.0876...°$$ Use the unrounded angle measure that is in your calculator.

$$d = 2650 \sin 25.0876...° = 1123.6102...$$

The ship traveled about 1124 m.

Problem Set 5-5

Reading Analysis

From what you have read in this section, what do you consider to be the main idea? In what way does the inverse of a trigonometric function compare with the inverse of a function that you studied in Section 1-5? How can you use the inverse of a trigonometric function to find an unknown angle measure in a right triangle?

Quick Review

Problems Q1–Q6 refer to the right triangle in Figure 5-5e.

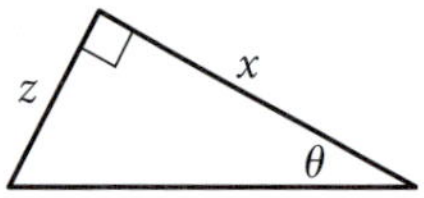

Q1. $\sin\theta =$ __?__

Q2. $\cos\theta =$ __?__

Q3. $\tan\theta =$ __?__

Q4. $\cot\theta =$ __?__

Q5. $\sec\theta =$ __?__

Q6. $\csc\theta =$ __?__

z x θ y

Figure 5-5e

Q7. $\sin 60° =$ __?__ (No decimals!)

Q8. $\cos 135° =$ __?__ (No decimals!)

Q9. $\tan 90° =$ __?__ (No decimals!)

Q10. The graph of the periodic function $y = \cos\theta$ is called a __?__

For Problems 1–6, evaluate the inverse trigonometric function for the given value. (Note the domain restrictions for inverse trigonometric functions, given on page 270.)

1. Find $\sin^{-1} 0.3$. Explain what the answer means.
2. Find $\cos^{-1} 0.2$. Explain what the answer means.
3. Find $\tan^{-1} 7$. Explain what the answer means.
4. Explain why $\sin^{-1} 2$ is undefined.
5. Find $\cos(\sin^{-1} 0.8)$. Explain, based on the Pythagorean theorem, why the answer is a rational number.
6. Find $\sin(\cos^{-1} 0.28)$. Explain, based on the Pythagorean theorem, why the answer is a rational number.

7. *Principal Branches of Sine and Cosine Problem:* Figure 5-5f shows the principal branch of the function $y = \sin\theta$ as a solid line on the sine graph. Figure 5-5g shows the principal branch of the function $y = \cos\theta$ as a solid line on the cosine graph.

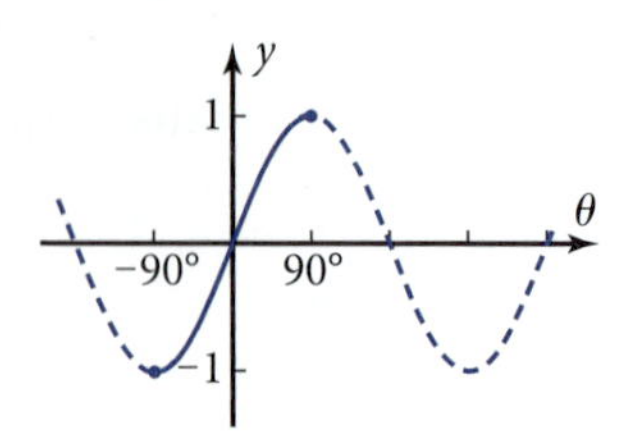

Figure 5-5f

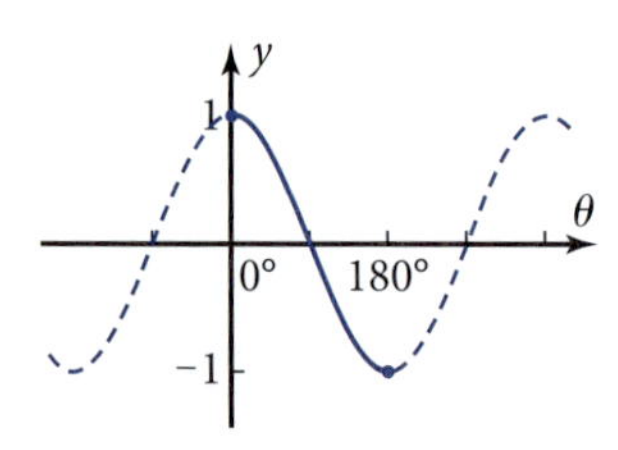

Figure 5-5g

   a. Why is neither the entire sine function nor the entire cosine function invertible?

   b. What are the domains of the principal branches of the cosine and sine functions? What property do these principal branches have that makes them invertible functions?

   c. Find $\theta = \sin^{-1}(-0.9)$. Explain why the answer is a negative number.

8. *Construction Problem:* Draw a right triangle to scale, with one leg 8 cm long and the adjacent acute angle 34°. Draw on paper with ruler and protractor or on the computer with a program such as The Geometer's Sketchpad.

   a. Measure the lengths of the opposite leg and the hypotenuse to the nearest 0.1 cm.

   b. Calculate the lengths of the opposite leg and the hypotenuse using appropriate trigonometric functions. Show that your measured values and the calculated values agree within 0.1 cm.

The Berlin Wall, long a symbol of the Cold War, was demolished in 1989. This man is chiseling away a souvenir piece of the wall.

9. *Ladder Problem:* Suppose you have a ladder 2.7 m long like the one in the picture above.

a. If the ladder makes an angle of 63° with the level ground when you lean it against a vertical wall, how high up the wall is the top of the ladder?

b. Your cat is trapped on a tree branch 2.6 m above the ground. If you lean the top of the ladder against the branch, what angle does the bottom of the ladder make with the level ground?

10. *Flagpole Problem:* You must order a new rope for the flagpole. To find out what length of rope is needed, you observe that the pole casts a shadow 11.6 m long on the ground. The angle of elevation of the Sun is 36° at this time of day (Figure 5-5h). How tall is the flagpole?

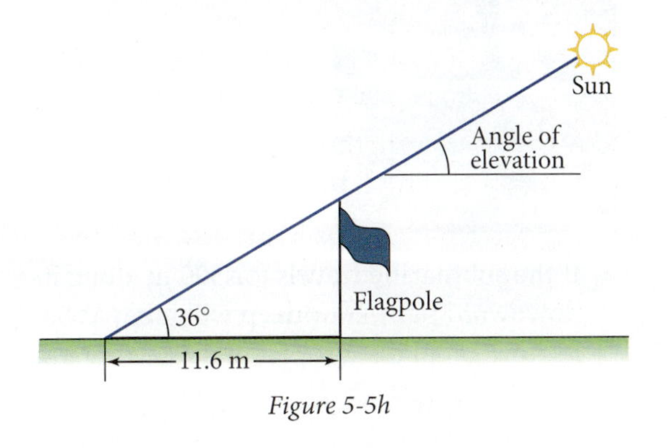

Figure 5-5h

11. *The Grapevine Problem:* Interstate 5 in California enters the San Joaquin Valley through a mountain pass called the Grapevine. The road descends from an elevation of 3500 ft above sea level to 1500 ft above sea level in a slant distance of 6.5 mi.

a. Approximately what angle does the roadway make with the horizontal?

b. What assumption must you make about how the road slopes?

12. *Grand Canyon Problem:* From a point on the North Rim of the Grand Canyon, a surveyor measures an angle of depression of 1.3° to a point on the South Rim (Figure 5-5i). From an aerial photograph she determines that the horizontal distance between the two points is 10 mi. How many feet is the South Rim below the North Rim?

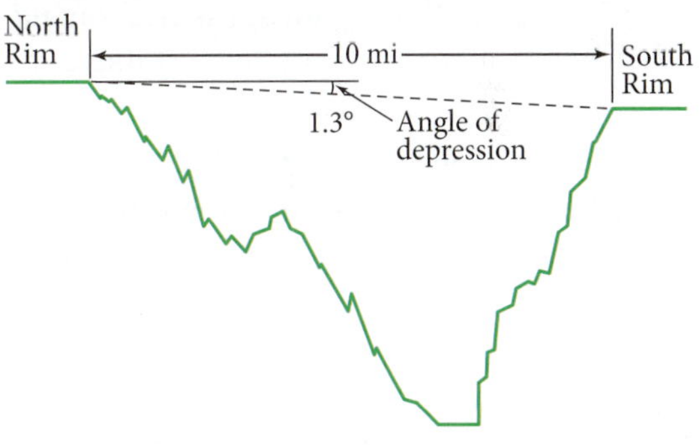

Figure 5-5i

13. *Tallest Skyscraper Problem:* In 2005, Taipei 101 in Taiwan was the world's tallest skyscraper, reaching 509 m above the ground. Suppose that at a particular time the building casts a shadow on the ground 1070 m long. What is the Sun's angle of elevation at this time?

14. *Airplane Landing Problem:* Commercial airliners fly at an altitude of about 10 km. Pilots start descending toward the airport when they are far away so that the airplane will not have to dive at a steep angle.

 a. If the pilot wants the plane's path to make an angle of 3° with the ground, at what horizontal distance from the airport must she start descending?

 b. If she starts descending when the plane is at a horizontal distance of 300 km from the airport, what angle will the plane's path make with the horizontal?

 c. Sketch the actual path of the plane just before and just after it touches the ground.

15. *Radiotherapy Problem:* A doctor plans to use a beam of gamma rays to treat a tumor that is 5.7 cm beneath the patient's skin. To avoid damaging an organ, the radiologist moves the source over 8.3 cm (Figure 5-5j).

 a. At what angle to the patient's skin must the radiologist aim the source to hit the tumor?

 b. How far will the beam travel through the patient's body before reaching the tumor?

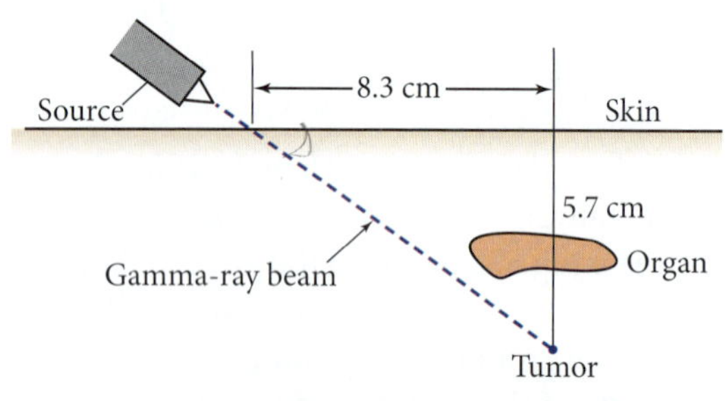

Figure 5-5j

16. *Triangular Block Problem:* A block bordering Market Street is a right triangle (Figure 5-5k). You take 125 paces on Market Street and 102 paces on Pine Street as you walk around the block.

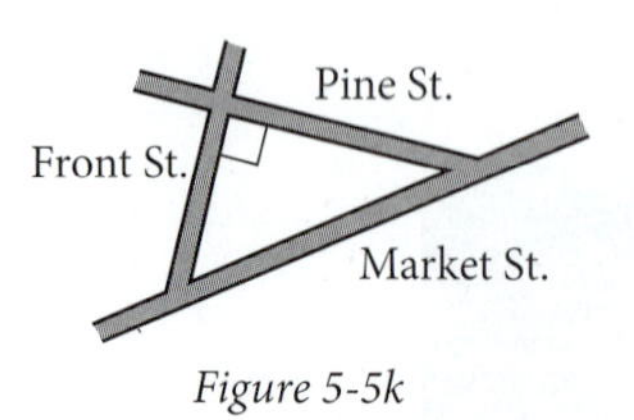

Figure 5-5k

 a. At what angle do Pine and Market Streets intersect?

 b. How many paces must you take on Front Street to complete the trip?

17. *Surveying Problem:* When surveyors measure land that slopes significantly, the slant distance they measure is longer than the horizontal distance they must draw on the map. Suppose that the distance from the top edge of the Cibolo Creek bed to the edge of the water is 37.8 m (Figure 5-5l). The land slopes downward at an angle of 27.6° to the horizontal.

 a. What is the horizontal distance from the top of the creek bed to the edge of the creek?

 b. How far below the level of the surrounding land is the surface of the water in the creek?

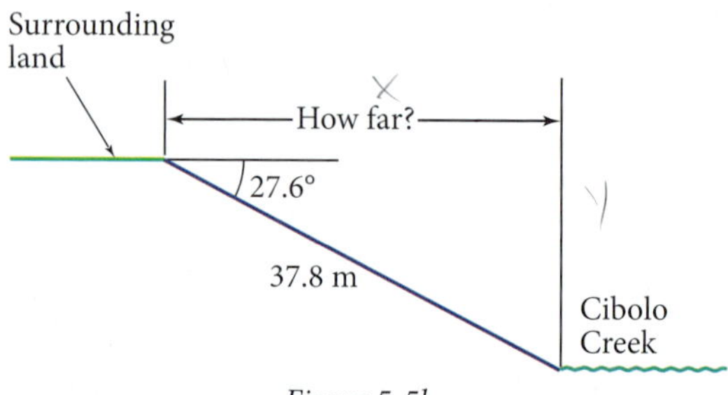

Figure 5-5l

18. *Submarine Problem:* As a submarine at the surface of the ocean makes a dive, its path makes an angle of 21° with the surface.

 a. If the submarine travels for 300 m along its downward path, how deep will it be? What is its horizontal distance from its starting point?

 b. How many meters must it go along its downward path to reach a depth of 1000 m?

19. *Highland Drive Problem:* One of the steeper streets in the United States is the 500 block of Highland Drive on Queen Anne Hill in Seattle. To measure the slope of the street, Tyline held a builder's level so that one end touched the pavement. The pavement was 14.4 cm below the level at the other end. The level itself was 71 cm long (Figure 5-5m).

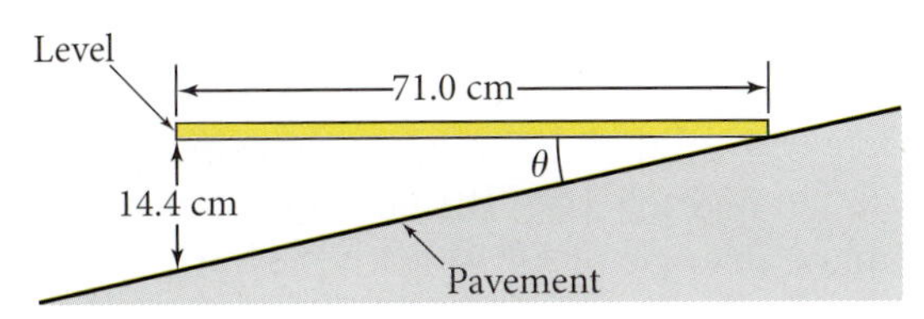

Figure 5-5m

a. What angle does the pavement make with the level?

b. A map of Seattle shows that the horizontal length of this block of Highland Drive is 365 ft. How much longer than 365 ft is the slant distance up this hill?

c. How high does the street rise in this block?

20. WEB *Planet Diameter Problem:* You can find the approximate diameter of a planet by measuring the angle between the lines of sight to the two sides of the planet (Figure 5-5n).

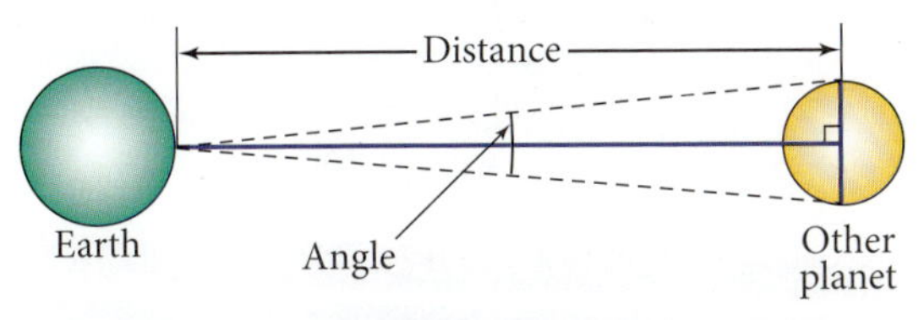

Figure 5-5n

a. When Venus is closest to Earth (25,000,000 mi), the angle is 0°1′2.5″ (0 degrees, 1 minute, 2.5 seconds). Find the approximate diameter of Venus.

b. When Jupiter is closest to Earth (390,000,000 mi), the angle is 0°0′46.9″. Find the approximate diameter of Jupiter.

c. Check an encyclopedia, an almanac, or the Internet to see how close your answers are to the accepted diameters.

21. *Window Problem:* Suppose you want the windows of your house built so that the eaves completely shade them from the sunlight in the summer and the sunlight completely fills the windows in the winter. The eaves have an overhang of 3 ft (Figure 5-5o).

a. How far below the eaves should the top of a window be placed for the window to receive full sunlight in midwinter, when the Sun's noontime angle of elevation is 25°?

b. How far below the eaves should the bottom of a window be placed for the window to receive no sunlight in midsummer, when the Sun's angle of elevation is 70°?

c. How tall will the windows be if they meet both requirements?

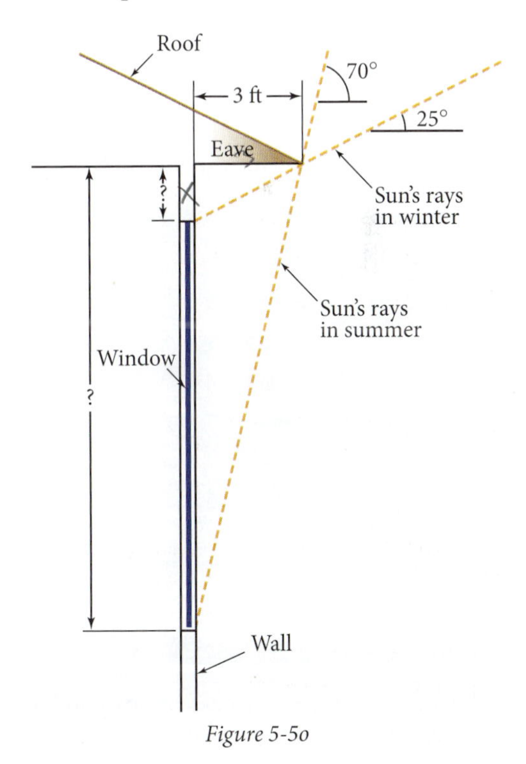

Figure 5-5o

22. *Grand Piano Problem:* A 28-in. prop holds open the lid on a grand piano. The base of the prop is 55 in. from the lid's hinge.

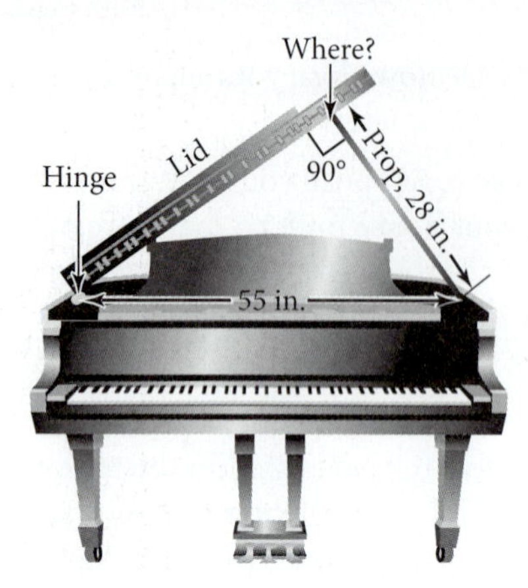

a. What angle does the lid make with the piano top when the prop is placed perpendicular to the lid?

b. Where on the lid should the prop be placed to make the right angle in part a?

c. The piano also has a shorter (13-in.) prop. Where on the lid should this prop be placed to make a right angle with the lid?

23. WEB *Handicap Ramp Project:* In this project you will measure the angle a typical handicap access ramp makes with the horizontal.

a. Based on handicap access ramps you have seen, make a conjecture about the angle these ramps make with the horizontal. Each member of your group should make his or her own conjecture.

b. Find a convenient handicap access ramp. With a level and ruler, measure the run and the rise of the ramp. Use these numbers to calculate the angle the ramp makes with the horizontal.

c. Check on the Internet or use another reference source to find the maximum angle a handicap access ramp is allowed to make with the horizontal.

24. WEB *Pyramid Problem:* The Great Pyramid of Cheops in Egypt has a square base measuring 230 m on each side. The faces of the pyramid make an angle of 51°50′ with the horizontal (Figure 5-5p).

a. How tall is the pyramid?

b. What is the shortest distance you would have to climb to get to the top?

c. Suppose you decide to make a model of the pyramid by cutting four isosceles triangles out of cardboard and gluing them together. How large should you make the base angles of these isosceles triangles?

d. Show that the ratio of the distance you calculated in part b to one-half the length of the base of the pyramid is very close to the golden ratio, $\frac{\sqrt{5}+1}{2}$.

e. Check on the Internet or use another reference source to find other startling relationships among the dimensions of this pyramid. Also visit Public Broadcasting System's website and look for Nova: Secrets of Lost Empires for photographs and information about the pyramids.

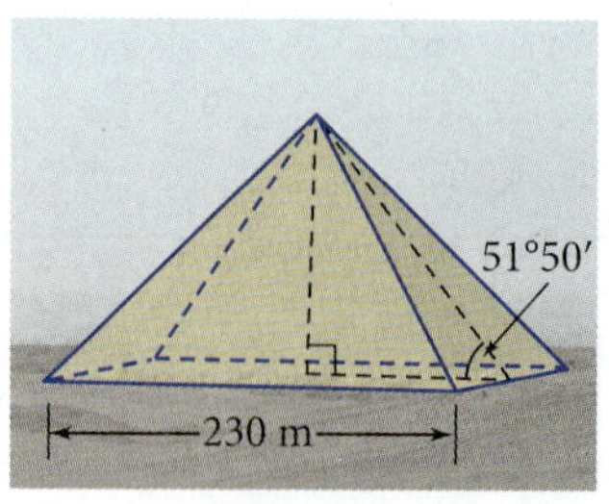

Figure 5-5p

5-6 Chapter Review and Test

This chapter introduced you to periodic functions. You saw how these functions occur in the real world. For instance, your distance from the ground as you ride a Ferris wheel changes periodically. The sine and cosine functions are periodic. By dilating and translating these functions, you can get sinusoids of different proportions, with their critical points located at different places. You also learned about the other four trigonometric functions and their relationship to the sine and cosine functions. Finally, you learned how to use these six functions and their inverses to find unknown side lengths and angle measures in right triangles.

Review Problems

R0. Update your journal with what you have learned since the last entry. Include things such as

- How angles can have measures that are negative or greater than 180°, and what reference angles are
- The definitions of sine, cosine, tangent, cotangent, secant, and cosecant
- Why sine and cosine graphs are periodic
- Inverse trigonometric functions used to find angle measures
- Applications to right triangle problems

R1. *Hose Reel Problem:* You unwind a hose by turning the crank on a hose reel (Figure 5-6a). As you crank, the distance your hand is above the ground is a periodic function of the angle through which the reel has rotated (Figure 5-6b, solid graph). The distance, y, is measured in feet, and the angle, θ, is measured in degrees.

a. The dashed graph in Figure 5-6b is the pre-image function $y = \sin\theta$. Plot this parent sine function graph on your grapher. Does the result agree with Figure 5-6b?

b. The solid graph in Figure 5-6b is a dilation and translation of the pre-image function $y = \sin\theta$. Write the full names of these transformations, and write an equation for the function. When you plot the transformed graph on your grapher, does the result agree with Figure 5-6b?

c. What is the name for the periodic graphs in Figure 5-6b?

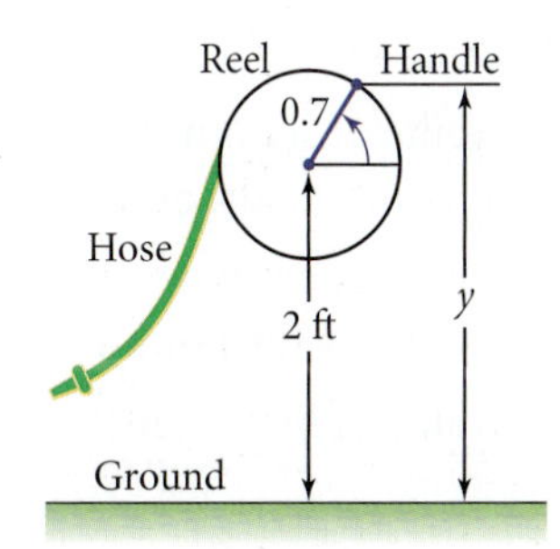

Figure 5-6a

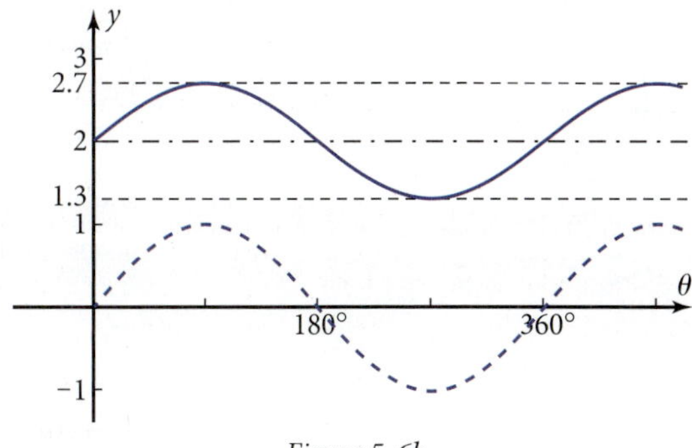

Figure 5-6b

R2. For each angle measure, sketch an angle in standard position. Mark the reference angle and find its measure.

a. $110°$

b. $-79°$

c. $2776°$

R3. a. Find $\sin\theta$ and $\cos\theta$ given that the terminal side of angle θ contains the point $(-5, 7)$.

b. Find decimal approximations of $\sin 160°$ and $\cos 160°$. Draw an angle of $160°$ in standard position and mark the reference angle. Based on displacements in the reference triangle, explain why $\sin 160°$ is positive but $\cos 160°$ is negative.

c. Sketch the graphs of the parent sinusoids $y = \cos\theta$ and $y = \sin\theta$.

d. In which two quadrants on a *uv*-coordinate system is $\sin\theta$ negative?

e. For the function $y = 4 + \cos 2\theta$, what are the transformations of the parent function graph $y = \cos\theta$? Sketch the graph of the transformed function.

R4. a. Find a decimal approximation of $\csc 256°$.

b. Find exact values (no decimals) of the six trigonometric functions of $150°$.

c. Find the exact value of $\sec\theta$ if $\theta_{\text{ref}} = 45°$ and angle θ terminates in Quadrant III.

d. Find the exact value of $\cos\theta$ if the terminal side of angle θ contains the point $(-3, 5)$.

e. Find the exact value of $\sec(-120°)$.

f. Find the exact value of $\tan^2 30° - \csc^2 30°$.

g. Explain why $\tan 90°$ is undefined.

R5. a. Find a decimal approximation of $\theta = \cos^{-1} 0.6$. What does the answer mean?

b. *Sunken Ship Problem:* Imagine that you are on a salvage ship in the Gulf of Mexico. Your sonar system has located a sunken ship at a slant distance of 683 m from your ship, with an angle of depression of 28°.

i. How deep is the water at the location of the sunken ship?

ii. How far must your ship go to be directly above the sunken ship?

iii. Your ship moves horizontally toward the sunken ship. After 520 m, what is the angle of depression?

iv. How could the crew of a fishing vessel use the techniques you used to solve this problem while searching for schools of fish?

Concept Problems

C1. *Tide Problem:* The average depth of the water at the beach varies with time due to the motion of the tides. Figure 5-6c shows the graph of depth, in feet, versus time, in hours, for a particular beach. Find the four transformations of the parent cosine graph that would give the sinusoid shown. Write an equation for this particular sinusoid, assuming 1 degree represents 1 h. Confirm your answer on your grapher.

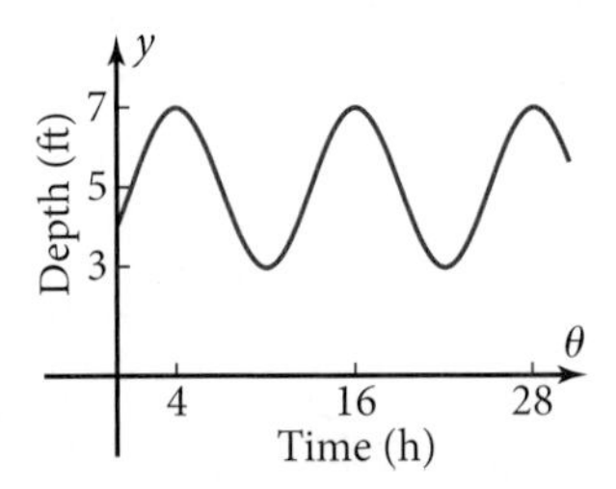

Figure 5-6c

C2. Figure 5-6d shows three cycles of the sinusoid function $y = 10 \sin \theta$. The horizontal line $y = 3$ cuts each cycle at two points.

a. Estimate graphically the six values of θ where the line intersects the sinusoid.

b. Calculate the six points in part a numerically, using the intersect feature of your grapher.

c. Calculate the six points in part a algebraically, using the inverse sine function.

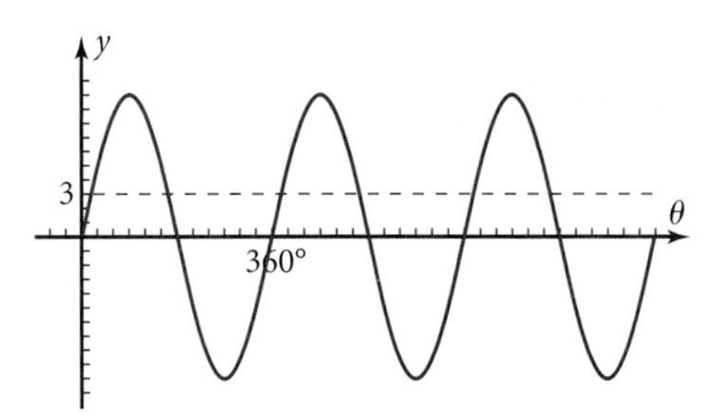

Figure 5-6d

C3. On your grapher, make a table of values of $\cos^2 \theta + \sin^2 \theta$ for each 10 degrees, starting at $\theta = 0°$. What is the pattern in the answers? Explain why this pattern applies to any value of θ.

C4. A ray from the origin of a uv-coordinate system starts along the positive u-axis and rotates around and around the origin. The slope of the ray depends on the angle through which the ray has rotated.

a. Sketch a reasonable graph of the slope as a function of the angle of rotation.

b. What function on your grapher is the same as the one you sketched in part a?

Chapter Test

Part 1: No calculators allowed (T1–T8)

T1. Sketch an angle θ in standard position whose terminal side contains the point $(3, -4)$. Sketch the reference triangle, showing the reference angle and the displacements u, v, and r. Find the exact values of the six trigonometric functions of θ.

T2. Sketch a 120° angle in standard position. Sketch the reference triangle. Show the reference angle and its measure and the displacements u, v, and r. Find the exact values of the six trigonometric functions of 120°.

T3. Sketch a 225° angle in standard position. Sketch the reference triangle. Show the reference angle and its measure and the displacements u, v, and r. Find the exact values of the six trigonometric functions of 225°.

T4. Sketch a 180° angle in standard position. Pick a point on the terminal side, and write the horizontal coordinate, vertical coordinate, and radius. Use these numbers to find the exact values of the six trigonometric functions of 180°.

T5. The number of hairs on a person's head and his or her age are related. Sketch a reasonable graph.

T6. The distance between the tip of the "second" hand on a clock and the floor depends on time. Sketch a reasonable graph.

T7. Only one of the functions in Problems T5 and T6 is periodic. Which one is that?

T8. Figure 5-6e shows the graph of $y = \sin\theta$ (dashed) and its principal branch (solid). Explain why the function $y = \sin\theta$ is not invertible but the function defined by its principal branch is invertible.

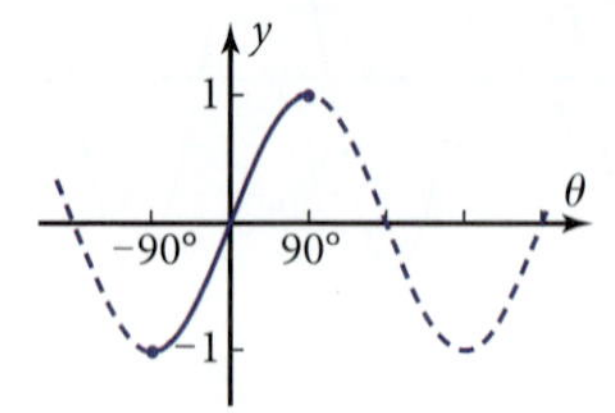

Figure 5-6e

Part 2: Graphing calculators allowed (T9–T22)

For Problems T9–T12, use your calculator to find each value.

T9. sec 39° **T10.** cot 173° **T11.** csc 191°

T12. $\tan^{-1} 0.9$. Explain the meaning of the answer.

T13. Calculate the length of side *x*.

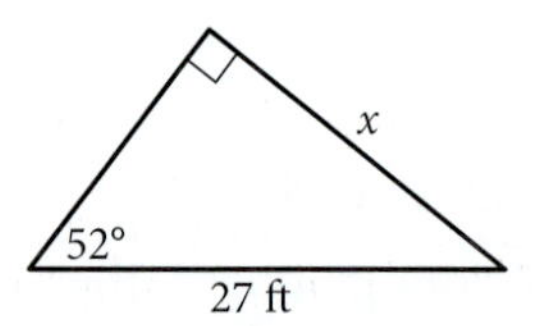

T14. Calculate the length of side *y*.

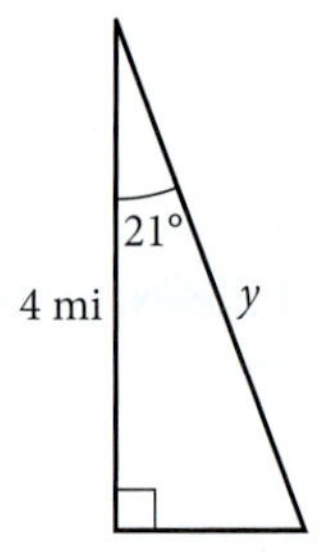

T15. Calculate the measure of angle *B*.

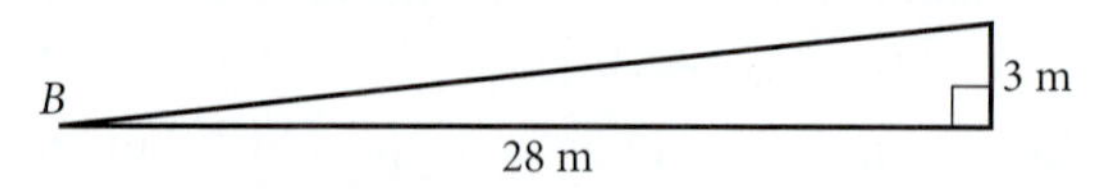

T16. Calculate the length of side *z*.

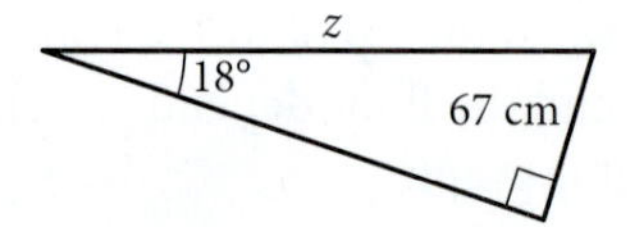

T17. Calculate the measure of angle *A*.

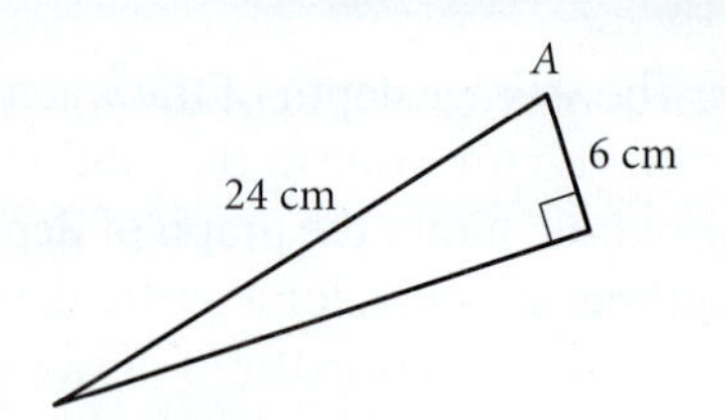

Buried Treasure Problem: For Problems T18–T20, use Figure 5-6f, the diagram of a buried treasure. From the point on the ground at the left of the figure, sonar detects the treasure at a slant distance of 19.3 m, at an angle of 33° with the horizontal.

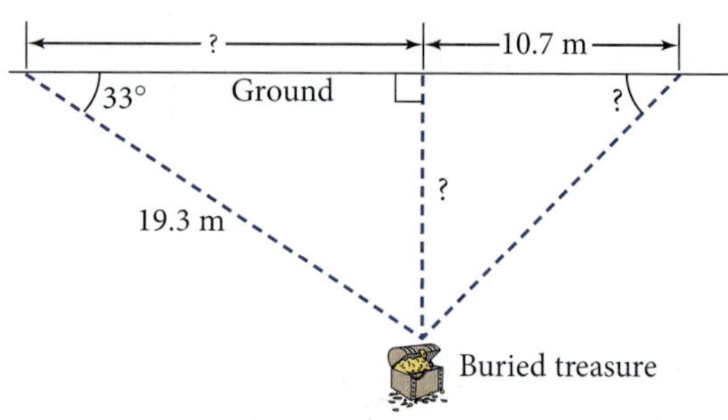

Figure 5-6f

T18. How far must you go from the point on the left to be directly over the treasure?

T19. How deep below the ground is the treasure?

T20. If you keep going to the right 10.7 m from the point directly above the treasure, at what angle would you have to dig to reach the buried treasure?

T21. In Figure 5-6g, the solid graph shows the result of *three* transformations applied to the parent function $y = \cos\theta$ (dashed). Write the equation for the transformed function. Check your results on your grapher.

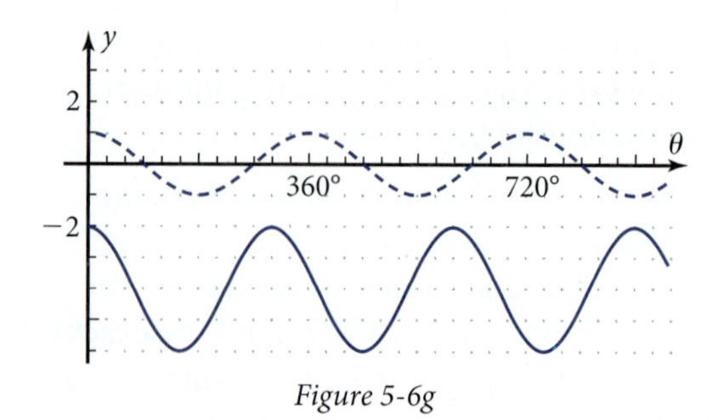

Figure 5-6g

T22. What did you learn as a result of taking this test that you didn't know before?

Chapter 6

Applications of Trigonometric and Circular Functions

Stresses in the earth compress rock formations and cause them to buckle into sinusoidal shapes. It is important for geologists to be able to predict the depth of a rock formation at a given point. Such information can be very useful for structural engineers as well. In this chapter you'll learn about the circular functions, which are closely related to the trigonometric functions. Geologists and engineers use these functions as mathematical models to perform calculations for such wavy rock formations.

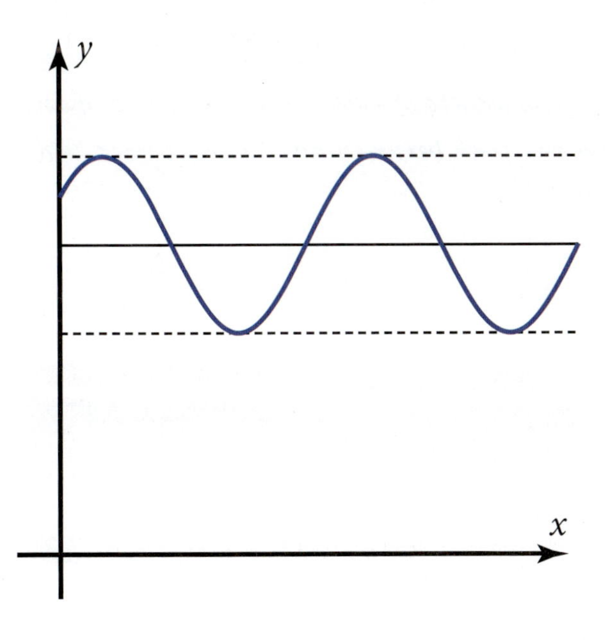

Mathematical Overview

You have learned that the graphs of $y = \cos\theta$ and $y = \sin\theta$ are sinusoids. You have also learned about horizontal and vertical dilations, and horizontal and vertical translations of functions. Now it is time to apply these four transformations to sinusoids so that you can fit them to many real-world situations where the y-values vary periodically.

GRAPHICALLY

The graph is a sinusoid that is a cosine function transformed through vertical and horizontal translations and dilations. The independent variable here is x rather than θ so that you can fit sinusoids to situations that do not involve angles.

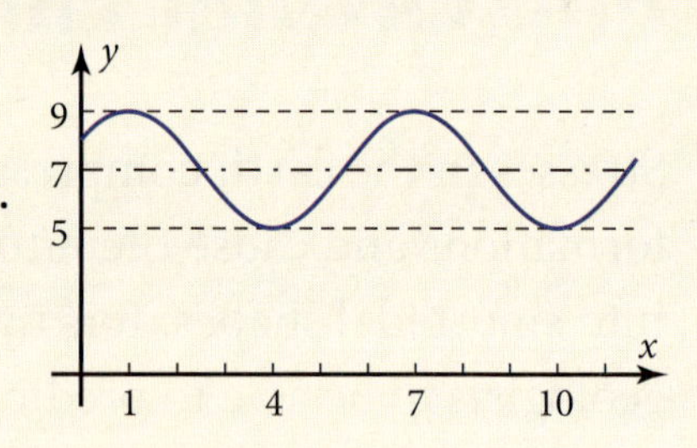

ALGEBRAICALLY

Particular equation: $y = 7 + 2\cos\frac{\pi}{3}(x - 1)$

NUMERICALLY

x	y
1	9
2	8
3	6
4	5

VERBALLY

The circular functions are just like the trigonometric functions except that the independent variable is an arc of a unit circle instead of an angle. Angles in radians form the link between angles in degrees and numbers of units of arc length.

6-1 Sinusoids: Amplitude, Period, and Cycles

Figure 6-1a shows a dilated and translated sinusoid and some of its graphical features. In this section you will learn how these features relate to transformations you've already learned.

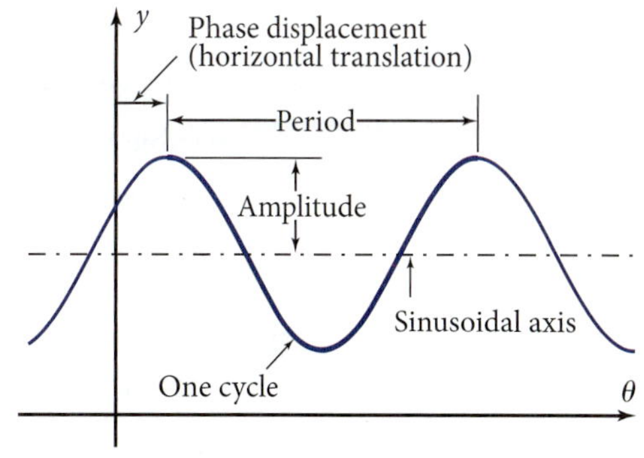

Figure 6-1a

Objective Learn the meanings of *amplitude, period, phase displacement,* and *cycle* of a sinusoidal graph.

Exploratory Problem Set 6-1

1. Sketch one **cycle** of the graph of the parent sinusoid $y = \cos\theta$, starting at $\theta = 0°$. What is the **amplitude** of this graph?
2. Plot the graph of the transformed cosine function $y = 5\cos\theta$. What is the amplitude of this graph? What is the relationship between the amplitude and the vertical dilation of a sinusoid?

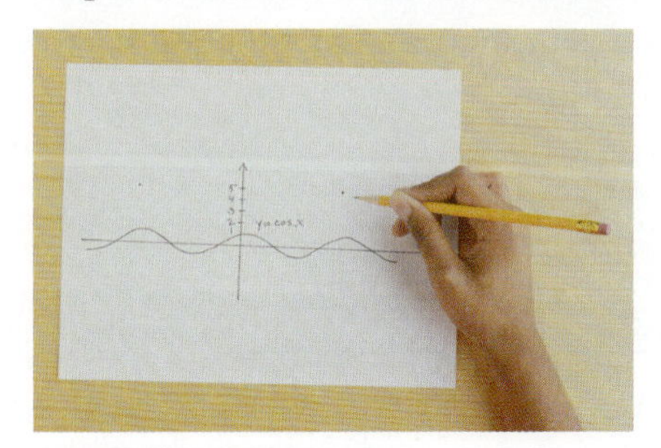

3. What is the period of the transformed function in Problem 2? What is the period of the parent function $y = \cos\theta$?
4. Plot the graph of $y = \cos 3\theta$. What is the period of this transformed function graph? How is the 3 related to the transformation? How could you calculate the period using the 3?
5. Plot the graph of $y = \cos(\theta - 60°)$. What transformation is caused by the 60°?
6. The $(\theta - 60°)$ in Problem 5 is called the *argument* of the cosine. The **phase displacement** is the value of θ that makes the argument equal zero. What is the phase displacement for this function? How is the phase displacement related to the horizontal translation?
7. Plot the graph of $y = 6 + \cos\theta$. What transformation is caused by the 6?
8. The **sinusoidal axis** runs along the middle of the graph of a sinusoid. It is the dashed centerline in Figure 6-1a. What transformation of the function $y = \cos x$ does the location of the sinusoidal axis indicate?
9. What are the amplitude, period, phase displacement, and sinusoidal axis location of the graph of $y = 6 + 5\cos 3(\theta - 60°)$? Check by plotting on your grapher.

6-2 General Sinusoidal Graphs

In Section 6-1, you encountered the terms *period, amplitude, cycle, phase displacement,* and *sinusoidal axis.* They are often used to describe horizontal and vertical translation and dilation of sinusoids. In this section you'll make the connection between the new terms and these transformations so that you will be able to fit an equation to any given sinusoid. This in turn will help you use sinusoidal functions as mathematical models for real-world applications such as the variation of average daily temperature with the time of year.

Objective

Given any one of these sets of information about a sinusoid, find the other two:

- The equation
- The graph
- The amplitude, period or frequency, phase displacement, and sinusoidal axis

In this exploration you will apply a transformed sinusoid to model the average daily high temperature at a particular location as a function of time.

EXPLORATION 6-2: Periodic Daily Temperatures

This graph shows the average daily high temperatures in San Antonio, by month, for 24 months. The dots are connected by a smooth curve resembling a sinusoid. Month 1 is January, month 2 is February, and so forth.

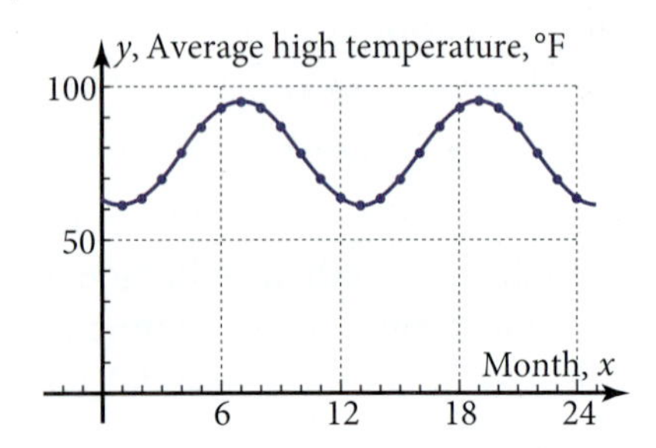

1. The graph of $y = \cos\theta$ completes a cycle each 360° (angle, not temperature!). Shown here is a transformed sinusoid that completes a cycle each 12°—a horizontal dilation by a factor of $\frac{1}{30}$. Write the equation of this sinusoid and plot it on your grapher. Does the result agree with this figure?

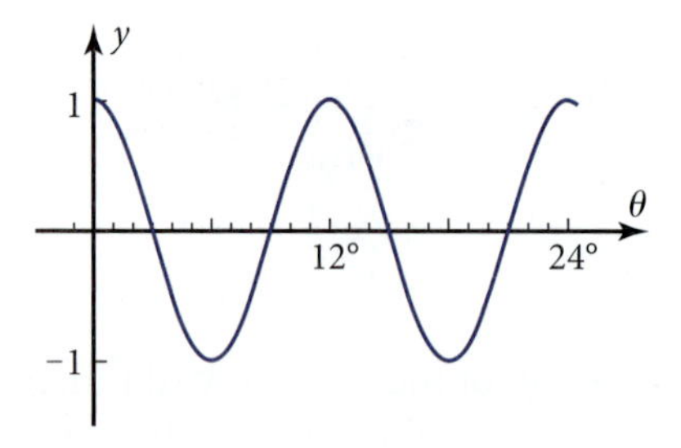

continued

EXPLORATION, continued

2. The greatest average high temperature occurs in July, month 7. Write an equation for a horizontal translation by 7 for the function in Problem 1 and plot it on your grapher. Does the result agree with the graph shown here?

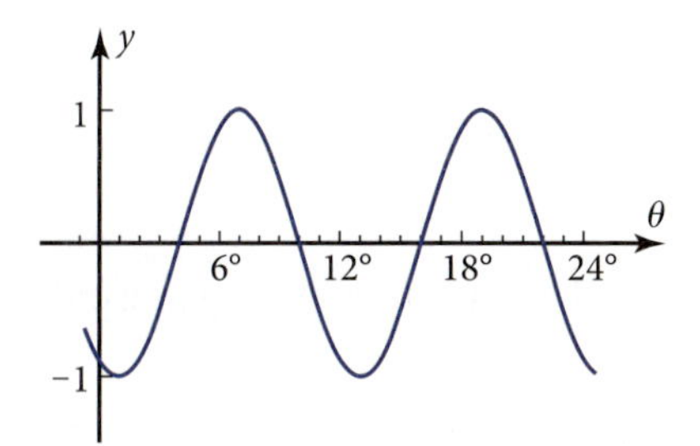

3. The July average high temperature is 95°F. The lowest average high temperature is 61°F in January. So the vertical dilation of the sinusoid is $\frac{(95-61)}{2}$, or 17. Transform the equation in Problem 2 so that it has a vertical dilation by a factor of 17 and plot it on your grapher. Does the result agree with this graph?

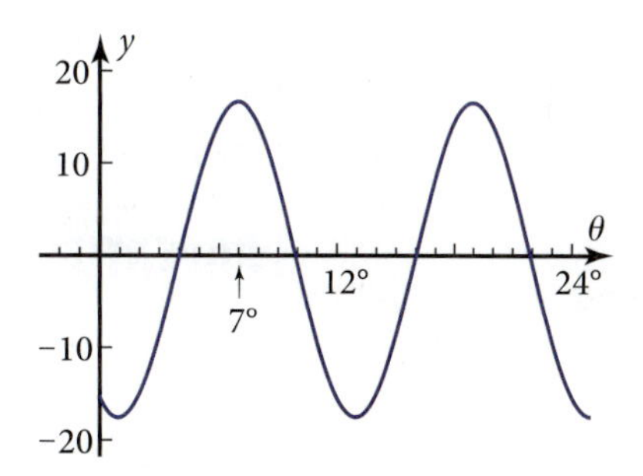

4. The graph before Problem 1 is a vertical translation of the graph in Problem 3 by 95 − 17 (or 61 + 17), which equals 78. Write an equation for this translation of the function in Problem 3 and plot it on your grapher. Does the result agree with the graph before Problem 1?
5. Predict the average high temperature in San Antonio for March, month 3. Do you get the same answer for March, month 15?
6. What did you learn as a result of doing this exploration that you did not know before?

Recall from Chapter 5 that the period of a sinusoid is the number of degrees per cycle. The reciprocal of the period, or the number of cycles per degree, is called the **frequency.** It is convenient to use the frequency when the period is very short. For instance, the alternating electrical current in the United States has a frequency of 60 cycles per second, meaning that the period is 1/60 second per cycle.

You can see how the general sinusoidal equations allow for all four transformations.

DEFINITION: General Sinusoidal Equation

$y = C + A\cos B(\theta - D)$ or $y = C + A\sin B(\theta - D)$, where

- $|A|$ is the amplitude (A is the vertical dilation, which can be positive or negative).
- B is the reciprocal of the horizontal dilation.
- C is the location of the sinusoidal axis (vertical translation).
- D is the phase displacement (horizontal translation).

The period can be calculated from the value of B. Because $\frac{1}{B}$ is the horizontal dilation and because the parent cosine and sine functions have period 360°, the period of a sinusoid equals $\frac{1}{|B|}(360°)$. Dilations can be positive or negative, so you must use the absolute value symbol.

PROPERTIES: Period and Frequency of a Sinusoid

For general equations $y = C + A \cos B(\theta - D)$ or $y = C + A \sin B(\theta - D)$

$$\text{period} = \frac{1}{|B|}(360°) \qquad \text{and} \qquad \text{frequency} = \frac{1}{\text{period}} = \frac{|B|}{360°}$$

Next you'll use these properties and the general equation to graph sinusoids and find their equations.

Concavity, Points of Inflection, and Bounds of Sinusoids

In Section 4-4 you learned about concavity and points of inflection. As you can see from Figure 6-2a, sinusoids have regions that are concave upward and other regions that are concave downward. The concavity changes at points of inflection.

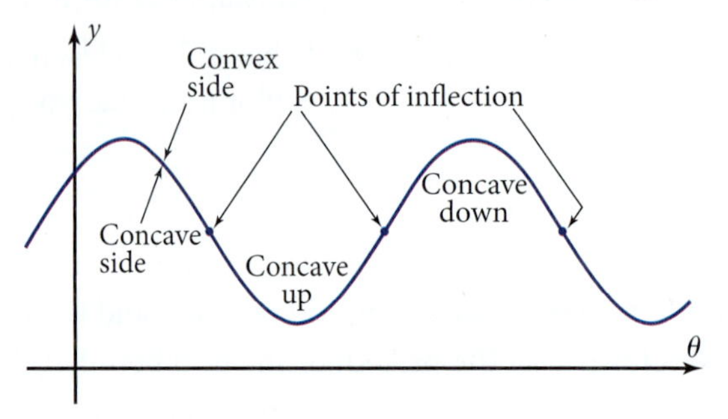

Figure 6-2a

As you can see from Figure 6-2b, the sinusoidal axis goes through the points of inflection. The lines through the high points and the low points are called the *upper bound* and the *lower bound*, respectively. The high points and low points are called *critical points* because they have a "critical" influence on the size and location of the sinusoid. Note that it is a quarter-cycle between a critical point and the next point of inflection.

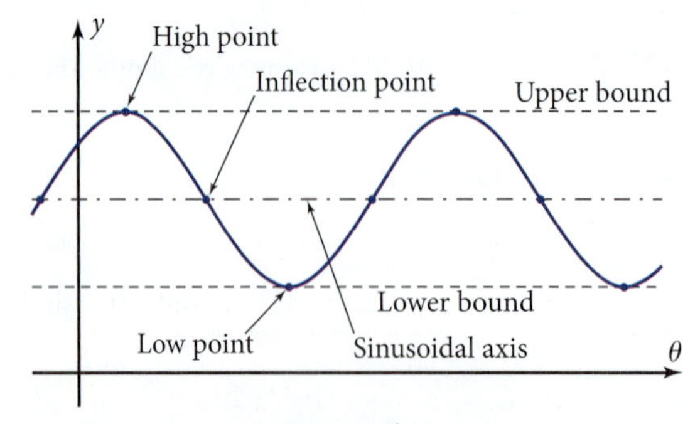

Figure 6-2b

EXAMPLE 1 ➤ Suppose that a sinusoid has period 12° per cycle, amplitude 7 units, phase displacement $-4°$ with respect to the parent cosine function, and a sinusoidal axis 5 units below the θ-axis. Without using your grapher, sketch this sinusoid and then find an equation for it. Verify with your grapher that your equation and the sinusoid you sketched agree with each other.

SOLUTION First draw the sinusoidal axis at $y = -5$, as in Figure 6-2c. (The long-and-short dashed line is used by draftspersons for centerlines.) Use the amplitude, 7, to draw the upper and lower bounds 7 units above and 7 units below the sinusoidal axis.

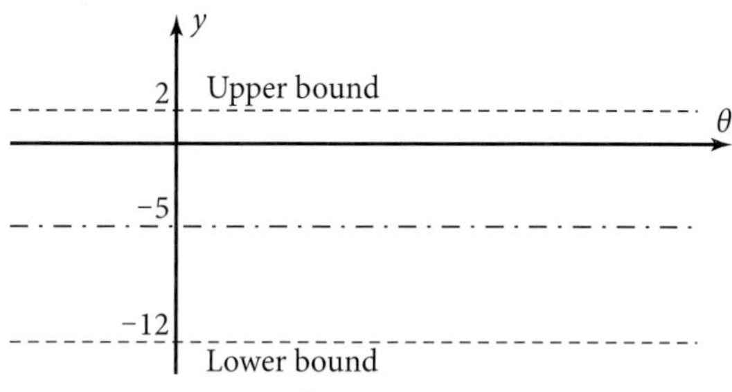

Figure 6-2c

Next find some critical points on the graph (Figure 6-2d). Start at $\theta = -4°$, because that is the phase displacement, and mark a high point on the upper bound. (The cosine function starts a cycle at a high point because $\cos 0° = 1$.) Then use the period, 12°, to plot the ends of the next two cycles.

$$-4° + 12° = 8° \qquad\qquad -4° + 2(12°) = 20°$$

Mark some low critical points halfway between consecutive high points.

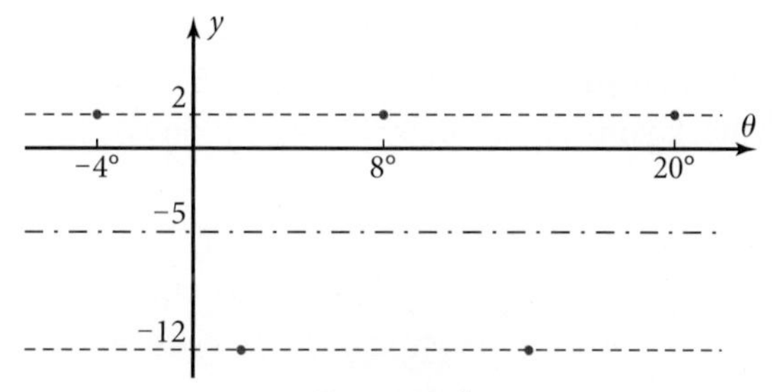

Figure 6-2d

Now mark the points of inflection (Figure 6-2e). They lie on the sinusoidal axis, halfway between consecutive high and low points.

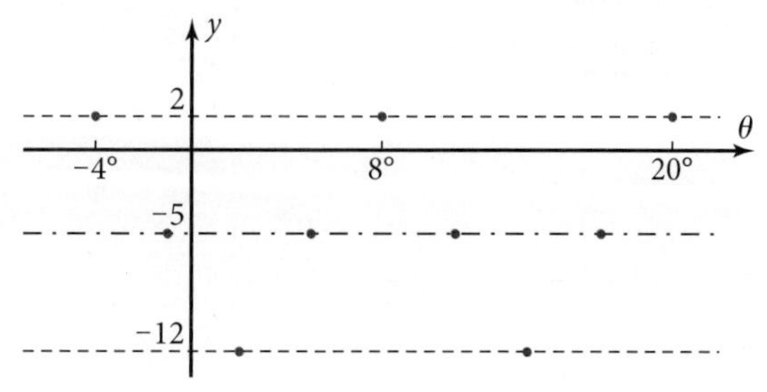

Figure 6-2e

Finally, sketch the graph in Figure 6-2f by connecting the critical points and points of inflection with a smooth curve. Be sure that the graph is rounded at the critical points and that it changes concavity at the points of inflection.

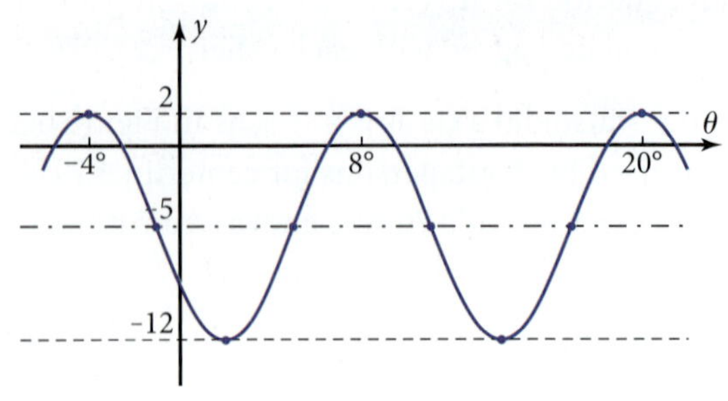

Figure 6-2f

Because the period of this sinusoid is 12° and the period of the parent cosine function is 360°, the horizontal dilation is

$$\text{dilation} = \frac{12°}{360°} = \frac{1}{30}$$

The coefficient B in the sinusoidal equation is the reciprocal of $\frac{1}{30}$, namely, 30. The horizontal translation is $-4°$. Thus a particular equation is

$$y = -5 + 7\cos 30(\theta + 4°)$$

Plotting the graph on your grapher confirms that this equation produces the correct graph (Figure 6-2g).

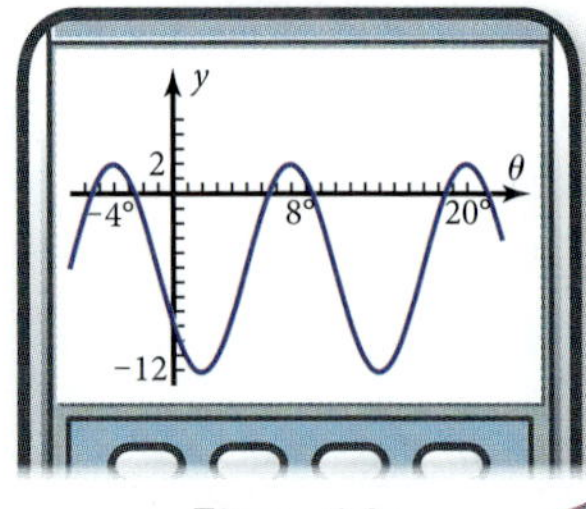

Figure 6-2g

EXAMPLE 2 ➤ For the sinusoid in Figure 6-2h, give the period, frequency, amplitude, phase displacement, and sinusoidal axis location. Write a particular equation for the sinusoid. Check your equation by plotting it on your grapher.

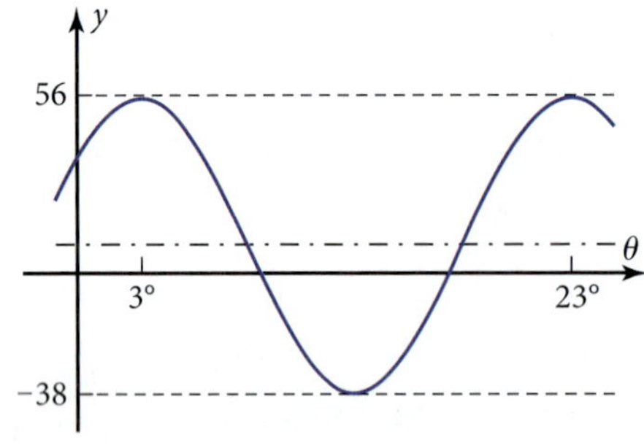

Figure 6-2h

SOLUTION As you will see later, you can use either the sine or the cosine as the pre-image function. Here, use the cosine function, because its "first" cycle starts at a high point and two high points are known.

- To find the period, look at the cycle shown in Figure 6-2h. It starts at 3° and ends at 23°, so the period is $23° - 3°$, or 20°.
- The frequency is the reciprocal of the period, or $\frac{1}{20}$ cycle per degree.
- The sinusoidal axis is halfway between the upper and lower bounds, so $y = \frac{1}{2}(-38 + 56)$, or 9.
- The amplitude is the distance between the upper or lower bound and the sinusoidal axis.

$$A = 56 - 9 = 47$$

- Using the cosine function as the parent function, the phase displacement is 3°. (You could also use 23° or −17°.)
- The horizontal dilation is $\frac{20°}{360°}$, so $B = \frac{360°}{20°}$, or 18 (the reciprocal of the horizontal dilation). So a particular equation is

$$y = 9 + 47 \cos 18(\theta - 3°)$$

Plotting the corresponding graph on your grapher confirms that the equation is correct.

You can find an equation of a sinusoid when only part of a cycle is given. The next example shows you how to do this.

EXAMPLE 3 ➤ Figure 6-2i shows a quarter-cycle of a sinusoid. Write a particular equation and check it by plotting it on your grapher.

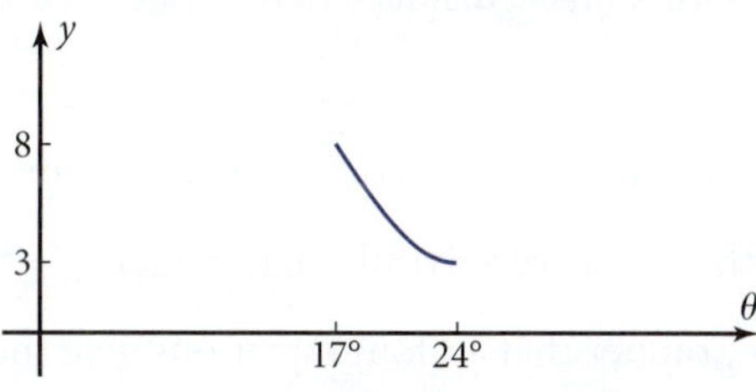

Figure 6-2i

SOLUTION Imagine the entire cycle from the part of the graph that is shown. You can tell that a low point is at $\theta = 24°$ because the graph appears to level out there. So the lower bound is at $y = 3$. The point at $\theta = 17°$ must be an inflection point on the sinusoidal axis at $y = 8$ because the graph is a quarter-cycle. So the amplitude is $8 - 3$, or 5. Sketch the lower bound, the sinusoidal axis, and the upper bound. Next locate a high point.

Each quarter-cycle covers (24° − 17°), or 7°, so the critical points and points of inflection are spaced 7° apart. Thus a high point is at $\theta = 17° - 7°$, or 10°. Sketch at least one complete cycle of the graph (Figure 6-2j).

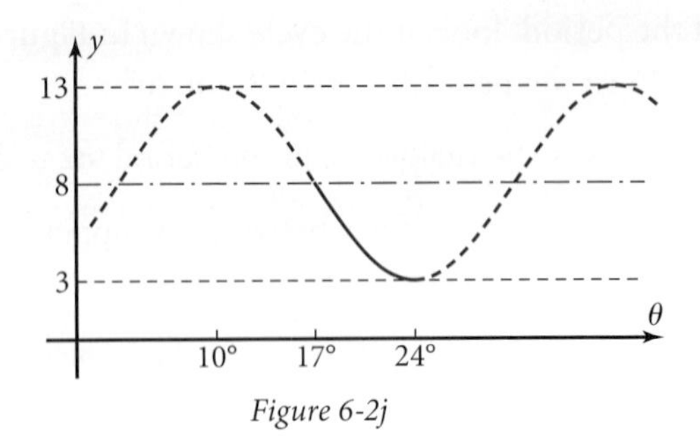

Figure 6-2j

The period is 4(7°), or 28°, because a quarter of the period is 7°. The horizontal dilation is $\frac{28°}{360°}$, or $\frac{7}{90}$.

The coefficient B in the sinusoidal equation is the reciprocal of this horizontal dilation. If you use the techniques of Example 2, a particular equation is

$$y = 8 + 5\cos\frac{90}{7}(\theta - 10°)$$

Plotting the graph on your grapher shows that the equation is correct.

Note that in all the examples so far *a* particular equation is used, not *the*. There are many equivalent forms of the equation, depending on which cycle you pick for the "first" cycle and whether you use the parent sine or parent cosine function. The next example shows some possibilities.

EXAMPLE 4 ➤ For the sinusoid in Figure 6-2k, write a particular equation using

a. Cosine, with a phase displacement other than 10°

b. Sine

c. Cosine, with a negative vertical dilation factor

d. Sine, with a negative vertical dilation factor

Confirm on your grapher that all four equations give the same graph.

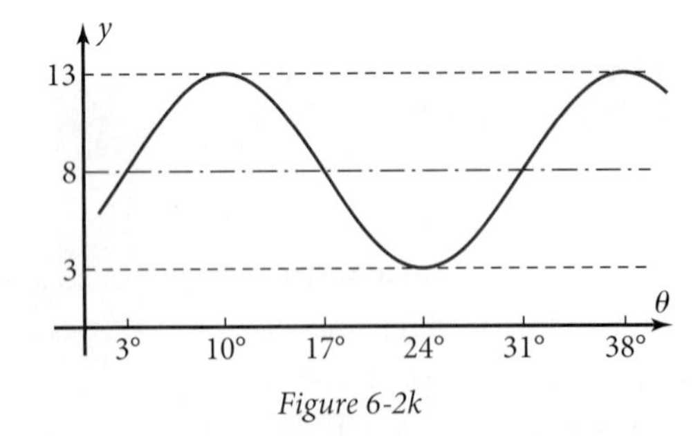

Figure 6-2k

SOLUTION

a. Notice that the sinusoid is the same one as in Example 3. To find a different phase displacement, look for another high point. A convenient one is at $\theta = 38°$. All the other constants remain the same. So another particular equation is

$$y = 8 + 5\cos\frac{90}{7}(\theta - 38°)$$

b. The graph of the parent sine function starts at a point of inflection on the sinusoidal axis while *going up.* Two possible starting points appear in Figure 6-2k, one at $\theta = 3°$ and another at $\theta = 31°$.

$$y = 8 + 5\sin\frac{90}{7}(\theta - 3°) \quad \text{or} \quad y = 8 + 5\sin\frac{90}{7}(\theta - 31°)$$

c. Changing the vertical dilation factor from 5 to -5 causes the sinusoid to be reflected across the sinusoidal axis. If you use -5, the "first" cycle starts as a *low* point instead of a high point. The most convenient low point in this case is at $\theta = 24°$.

$$y = 8 - 5\cos\frac{90}{7}(\theta - 24°)$$

d. With a negative dilation factor, the sine function starts a cycle at a point of inflection while going *down.* One such point is shown in Figure 6-2k at $\theta = 17°$.

$$y = 8 - 5\sin\frac{90}{7}(\theta - 17°)$$

Plotting these four equations on your grapher reveals only one image. The graphs are superimposed on one another.

Problem Set 6-2

Reading Analysis

From what you have read in this section, what do you consider to be the main idea? How are the *period, frequency,* and *cycle* related to one another in connection with sinusoids? What is the difference between the way θ appears on the graph of a sinusoid and the way it appears in a *uv*-coordinate system, as in Chapter 5? How can there be more than one particular equation for a given sinusoid?

Quick Review

Problems Q1–Q5 refer to Figure 6-2l.

Q1. How many cycles are there between $\theta = 20°$ and $\theta = 80°$?

Q2. What is the amplitude?

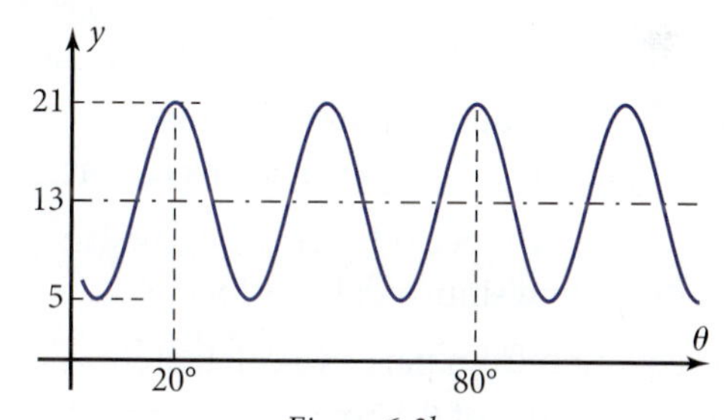

Figure 6-2l

Q3. What is the period?

Q4. What is the vertical translation?

Q5. What is the horizontal translation (for cosine)?

Q6. Find the exact value (no decimals) of sin 60°.

Q7. Find the approximate value of sec 71°.

Q8. Find the approximate value of $\cot^{-1} 4.3$.

Q9. Find the measure of the larger acute angle of a right triangle with legs of lengths 11 ft and 9 ft.

Q10. Expand: $(3x - 5)^2$

For Problems 1–4, find the amplitude, period, phase displacement, and sinusoidal axis location. Without using your grapher, sketch the graph by locating critical points. Then check your graph using your grapher.

1. $y = 7 + 4\cos 3(\theta + 10°)$
2. $y = 3 + 5\cos \frac{1}{4}(\theta - 240°)$
3. $y = -10 + 20\sin \frac{1}{2}(\theta - 120°)$
4. $y = -8 + 10\sin 5(\theta + 6°)$

For Problems 5–8,

a. Find a particular equation for the sinusoid using cosine or sine, whichever seems easier.

b. Give the amplitude, period, frequency, phase displacement, and sinusoidal axis location.

c. Use the equation from part a to calculate y for the given values of θ. Show that the result agrees with the given graph for the first value.

5. $\theta = 60°$ and $\theta = 1234°$

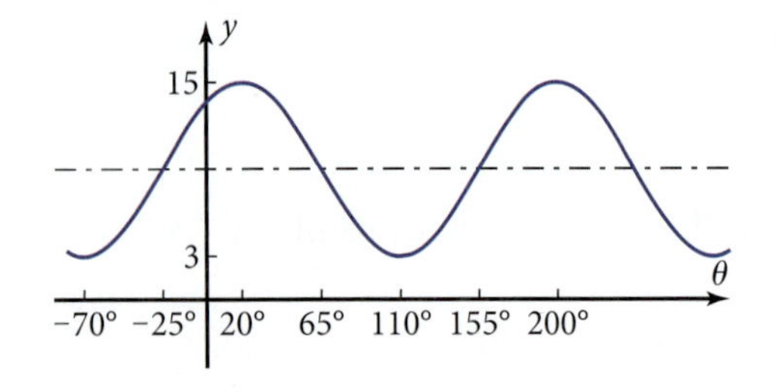

6. $\theta = 10°$ and $\theta = 453°$

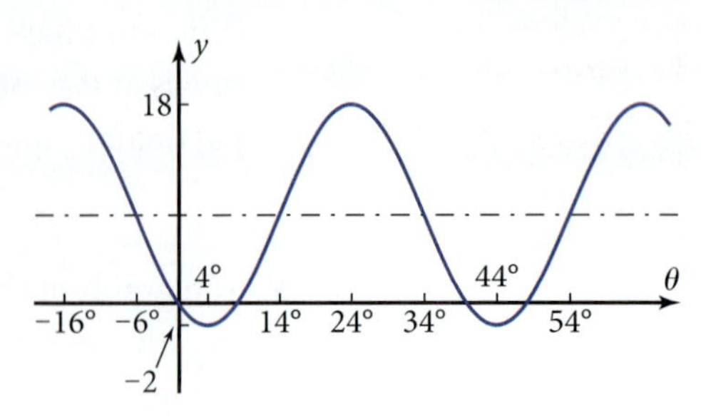

7. $\theta = 70°$ and $\theta = 491°$

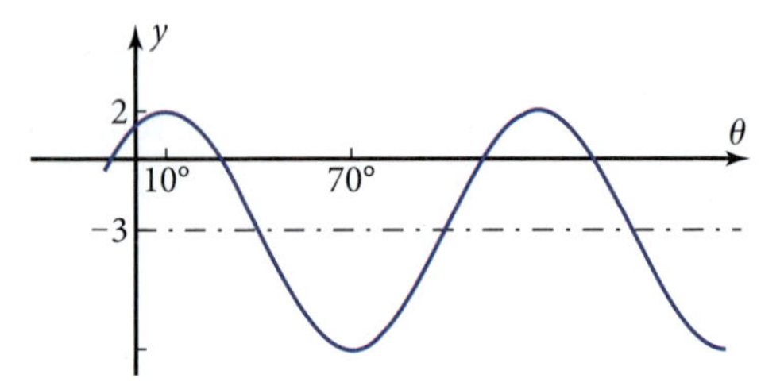

8. $\theta = 8°$ and $\theta = 1776°$

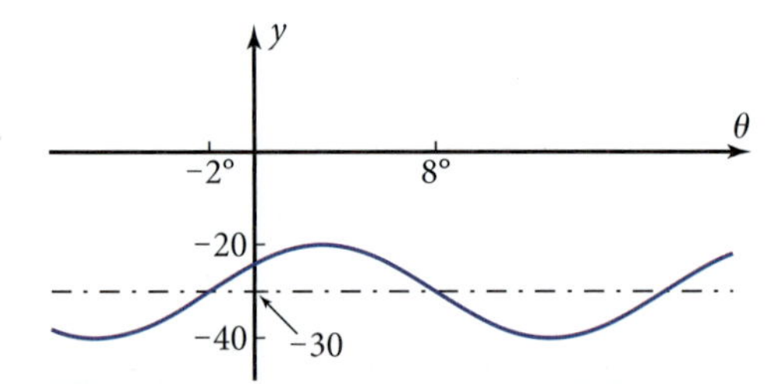

For Problems 9–14, find a particular equation for the sinusoid that is graphed.

9.

10.

11.

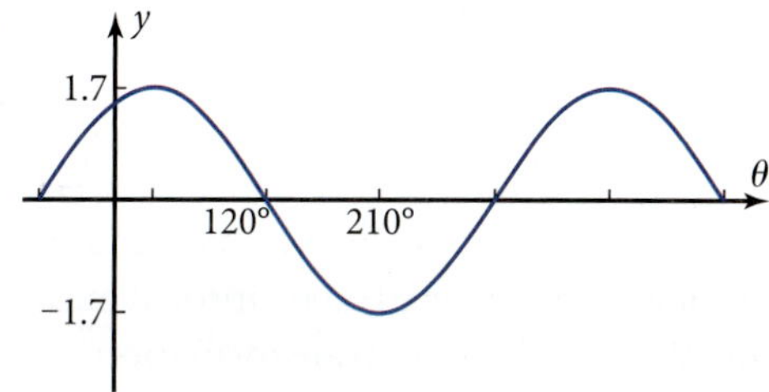

12.

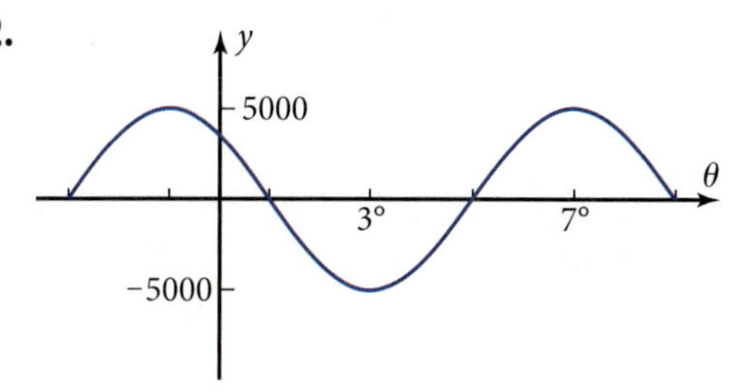

13.

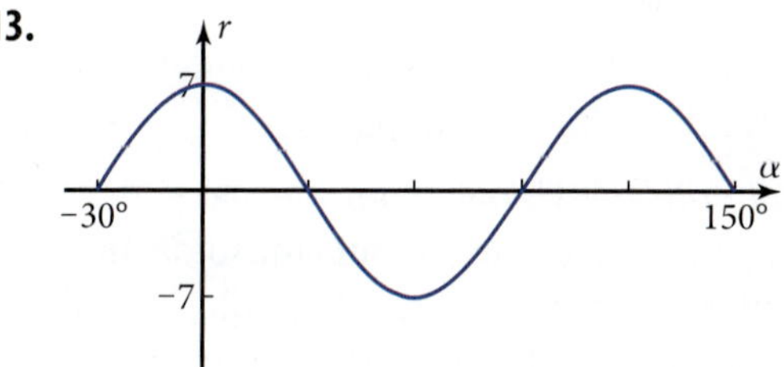

14.

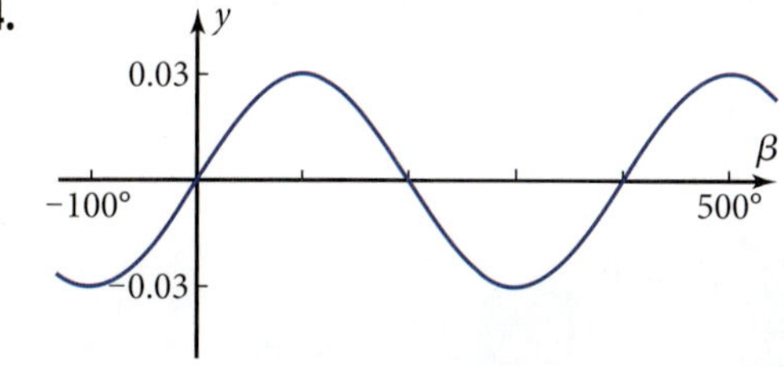

In Problems 15 and 16, a half-cycle of a sinusoid is shown. Find a particular equation for the sinusoid.

15.

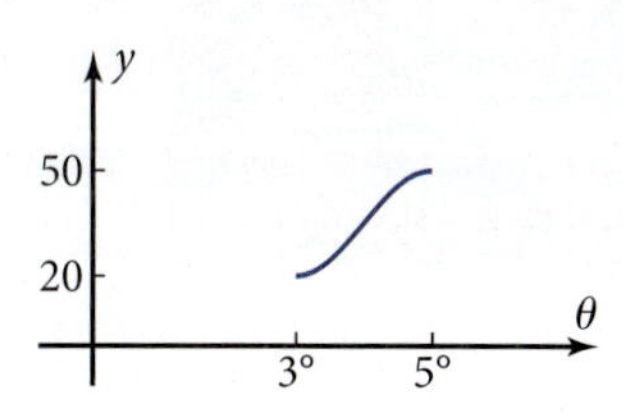

16.

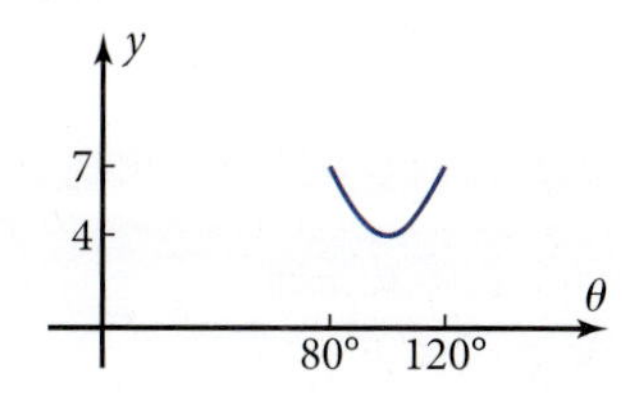

In Problems 17 and 18, a quarter-cycle of a sinusoid is shown. Find a particular equation for the sinusoid.

17.

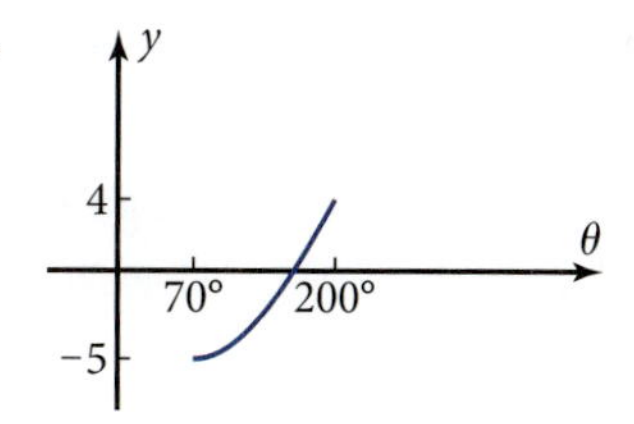

18.

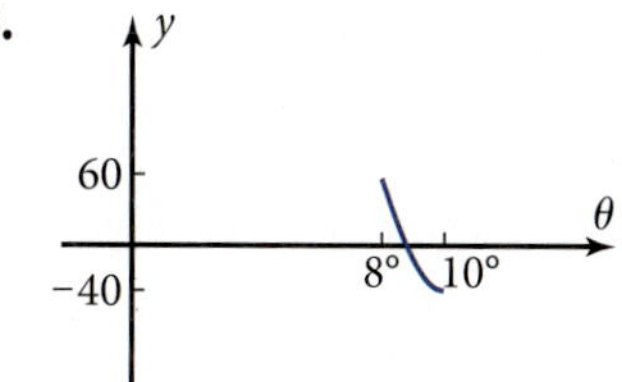

19. If the sinusoid in Problem 17 is extended to $\theta = 300°$, what is the value of y? If the sinusoid is extended to $\theta = 5678°$, is the point on the graph above or below the sinusoidal axis? How far?

20. If the sinusoid in Problem 18 is extended to the left to $\theta = 2.5°$, what is the value of y? If the sinusoid is extended to $\theta = 328°$, is the point on the graph above or below the sinusoidal axis? How far?

For Problems 21 and 22, sketch the sinusoid described and write a particular equation for it. Check the equation on your grapher to make sure it produces the graph you sketched.

21. The period equals 72°, the amplitude is 3 units, the phase displacement (for $y = \cos\theta$) equals 6°, and the sinusoidal axis is at $y = 4$ units.

22. The frequency is $\frac{1}{10}$ cycle per degree, the amplitude equals 2 units, the phase displacement (for $y = \cos\theta$) equals $-3°$, and the sinusoidal axis is at $y = -5$ units.

For Problems 23 and 24, write four different particular equations for the given sinusoid, using

a. Cosine as the parent function with positive vertical dilation

b. Cosine as the parent function with negative vertical dilation

c. Sine as the parent function with positive vertical dilation

d. Sine as the parent function with negative vertical dilation

Plot all four equations on the same screen on your grapher to confirm that the graphs are the same.

23.

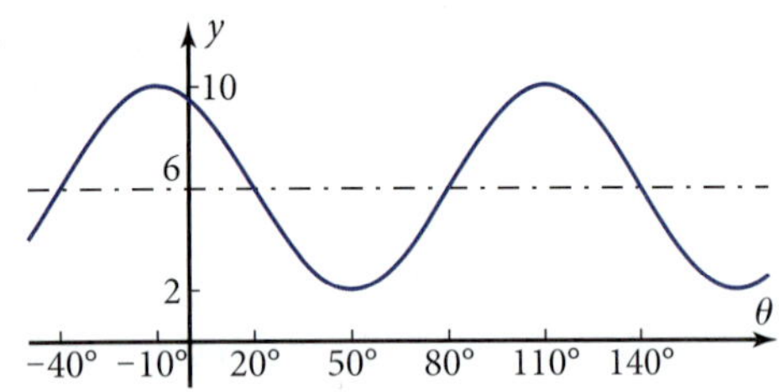

24.

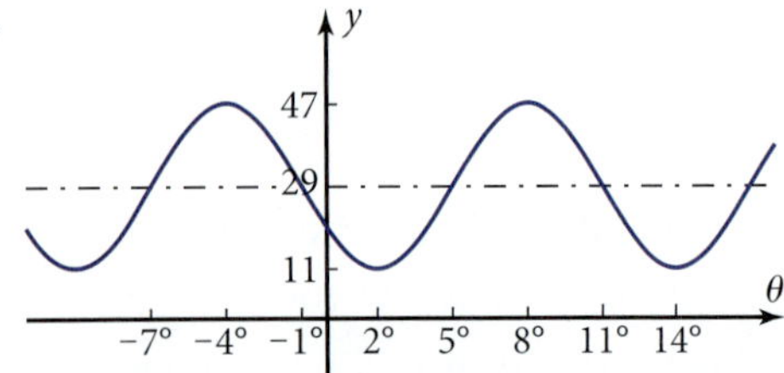

25. *Frequency Problem:* The unit for the period of a sinusoid is degrees per cycle. The unit for the frequency is cycles per degree.

a. Suppose a sinusoid has period $\frac{1}{60}$ degree/cycle. What would the frequency be? Why might people prefer to speak of the frequency of such a sinusoid rather than the period?

b. For $y = \cos 300\theta$, what is the period? What is the frequency? How can you calculate the frequency quickly, using the 300?

26. *Inflection Point Problem:* Sketch the graph of a function that has high and low critical points. On the sketch, show

a. A point of inflection

b. A region where the graph is concave up

c. A region where the graph is concave down

27. *Horizontal vs. Vertical Transformations Problem:* In the function

$$y = 3 + 4\cos 2(\theta - 5^\circ)$$

the 3 and the 4 are the vertical transformations, but the 2 and the -5° are the reciprocal and opposite of the horizontal transformations.

a. Show that you can transform the given equation to

$$\frac{y-3}{4} = \cos\left(\frac{\theta - 5^\circ}{1/2}\right)$$

b. Examine the equation in part a for the transformations that are applied to the x- and y-variables. What is the form of these transformations?

c. Why is the original form of the equation more useful than the form in part a?

28. *Journal Problem:* Update your journal with things you have learned about sinusoids. In particular, explain how the amplitude, period, phase displacement, frequency, and sinusoidal axis location are related to the four constants in the general sinusoidal equation. What is meant by *critical points, concavity,* and *points of inflection*?

6-3 Graphs of Tangent, Cotangent, Secant, and Cosecant Functions

If you enter tan 90° into your calculator, you will get an error message because tangent is defined as a quotient. On the unit circle, a point on the terminal side of a 90° angle has horizontal coordinate zero and vertical coordinate 1. Division of any number by zero is undefined, which you'll see leads to **vertical asymptotes** at angle measures for which division by zero would occur. In this section you'll also see that the graphs of the tangent, cotangent, secant, and cosecant functions are **discontinuous** where the function value would involve division by zero.

Objective Plot the graphs of the tangent, cotangent, secant, and cosecant functions, showing their behavior when the function value is undefined.

You can plot cotangent, secant, and cosecant by using the fact that they are reciprocals of tangent, cosine, and sine, respectively.

$$\cot\theta = \frac{1}{\tan\theta} \qquad \sec\theta = \frac{1}{\cos\theta} \qquad \csc\theta = \frac{1}{\sin\theta}$$

Figure 6-3a shows the graphs of $y = \tan\theta$ and $y = \cot\theta$, and Figure 6-3b shows the graphs of $y = \sec\theta$ and $y = \csc\theta$, all as they might appear on your grapher. If you use a window that includes multiples of 90° as grid points, you'll see that the graphs are discontinuous. Notice that the graphs go off to infinity (positive or negative) at odd or even multiples of 90°, exactly those places where the functions are undefined.a

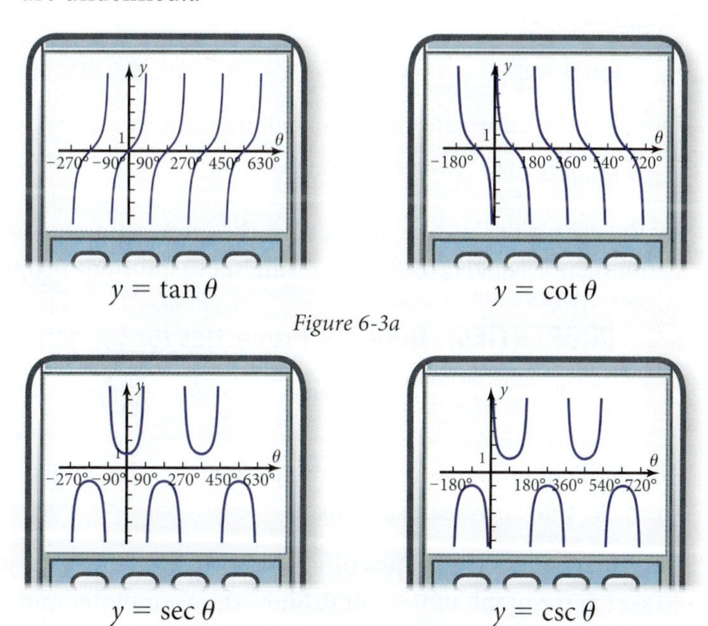

$y = \tan\theta$ $y = \cot\theta$

Figure 6-3a

$y = \sec\theta$ $y = \csc\theta$

Figure 6-3b

To see why the graphs have these shapes, it is helpful to look at transformations performed on the parent cosine and sine graphs.

EXAMPLE 1 ➤ Sketch the graph of the parent sine function, $y = \sin \theta$. Use the fact that $\csc \theta = \frac{1}{\sin \theta}$ to sketch the graph of the cosecant function. Show how the asymptotes of the cosecant function are related to the graph of the sine function.

SOLUTION Sketch the sine graph as in Figure 6-3c. Where the value of the sine function is zero, the cosecant function will be undefined because of division by zero. Draw vertical asymptotes at these values of θ.

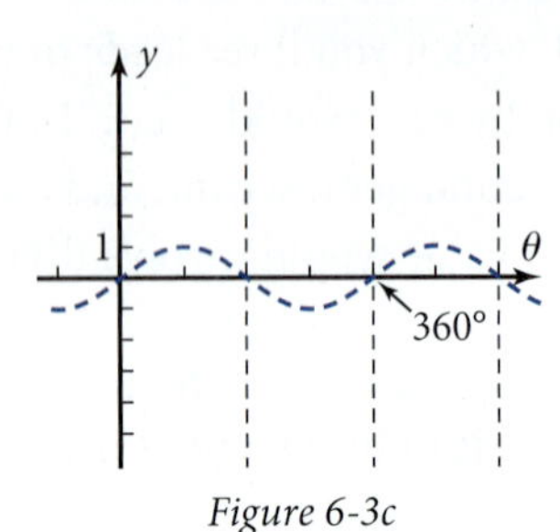

Figure 6-3c

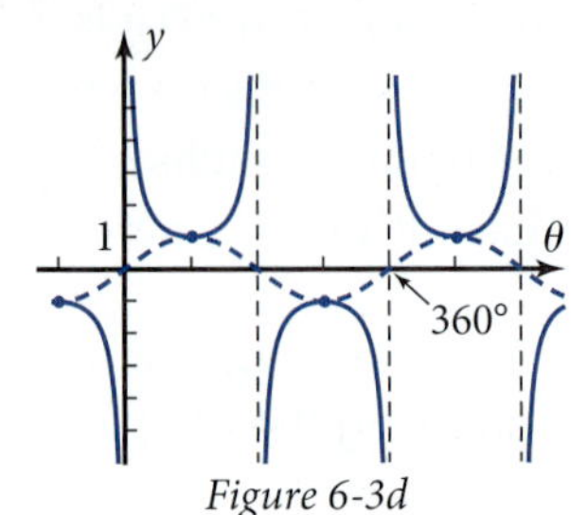

Figure 6-3d

Where the sine function equals 1 or -1, so does the cosecant function, because the reciprocal of 1 is 1 and the reciprocal of -1 is -1. Mark these points as in Figure 6-3d. As the sine gets smaller, the cosecant gets larger, and vice versa. For instance, the reciprocal of 0.2 is 5, and the reciprocal of 0.1 is 10. Sketch the graph consistent with these facts, as in Figure 6-3d. ◄

To understand why the graphs of the tangent and cotangent functions have the shapes in Figure 6-3a, it is helpful to examine how these functions are related to the sine and cosine functions. If θ is the standard position angle of a ray from the origin to point (u, v) in a uv-coordinate system, and r is the distance from the origin to (u, v), then by definition,

$$\tan \theta = \frac{v}{u}$$

Dividing the numerator and the denominator by r gives

$$\tan \theta = \frac{v/r}{u/r}$$

By the definitions of sine and cosine, the numerator equals $\sin \theta$ and the denominator equals $\cos \theta$. As a result, these **quotient properties** are true.

PROPERTIES: Quotient Properties for Tangent and Cotangent

$$\tan \theta = \frac{\sin \theta}{\cos \theta} \quad \text{and} \quad \cot \theta = \frac{\cos \theta}{\sin \theta}$$

The quotient properties allow you to construct the tangent and cotangent graphs from the sine and cosine graphs.

EXAMPLE 2 ➤ On paper, sketch the graphs of $y = \sin \theta$ and $y = \cos \theta$. Use the quotient property to sketch the graph of $y = \cot \theta$. Show the asymptotes and the points where the graph crosses the θ-axis.

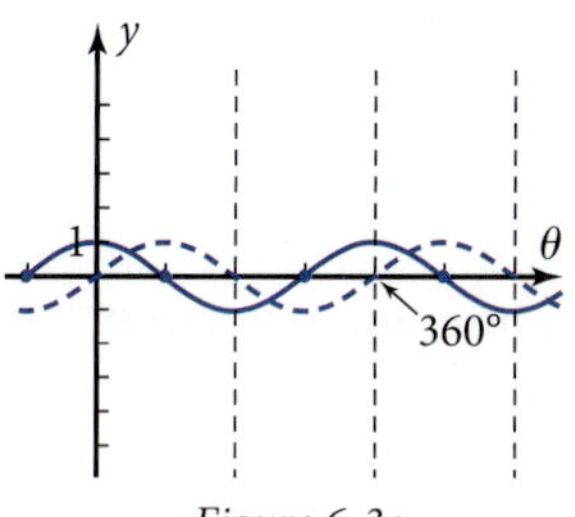

Figure 6-3e

SOLUTION Draw the graphs of the sine and cosine functions (dashed and solid, respectively) as in Figure 6-3e. Because $\cot\theta = \frac{\cos\theta}{\sin\theta}$, show the asymptotes where $\sin\theta = 0$, and show the θ-intercepts where $\cos\theta = 0$.

At $\theta = 45°$, and wherever else the graphs of the sine and the cosine functions intersect each other, $\frac{\cos\theta}{\sin\theta}$ will equal 1. Wherever sine and cosine are opposites of each other, $\frac{\cos\theta}{\sin\theta}$ will equal -1. Mark these points as in Figure 6-3f. Then sketch the cotangent graph through the marked points, consistent with the asymptotes. The final graph is shown in Figure 6-3g.

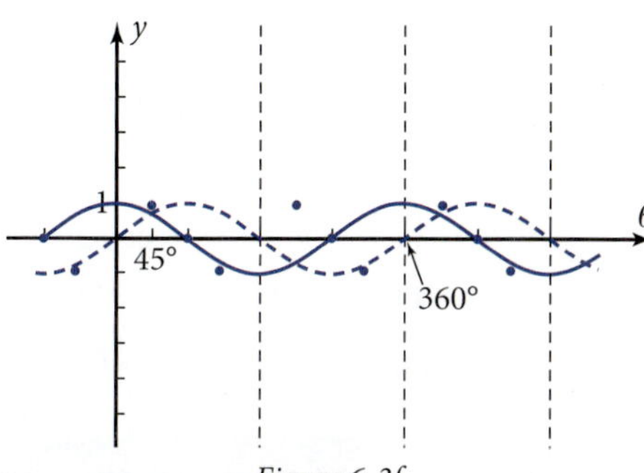

Figure 6-3f

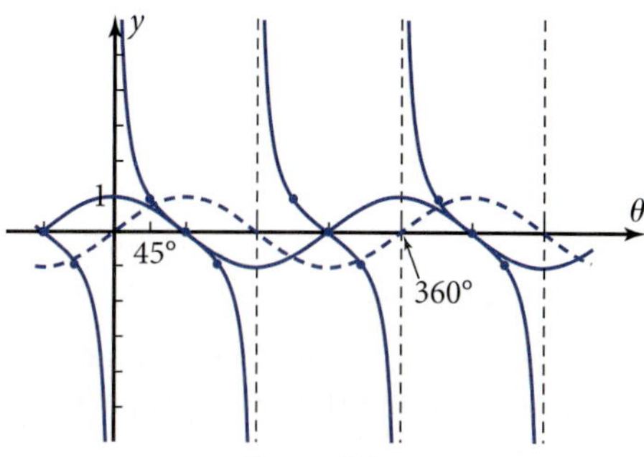

Figure 6-3g

Problem Set 6-3

Reading Analysis

From what you have read in this section, what do you consider to be the main idea? What feature do the graphs of the tangent, cotangent, secant, and cosecant functions have that sinusoids do not have, and why do they have this feature? What algebraic properties allow you to sketch the graph of the tangent or cotangent function from two sinusoids?

Quick Review

Problems Q1–Q7 refer to the equation $y = 3 + 4\cos 5(\theta - 6°)$.

Q1. The graph of the equation is called a __?__.

Q2. The amplitude is __?__.

Q3. The period is __?__.

Q4. The phase displacement with respect to $y = \cos\theta$ is __?__.

Q5. The frequency is __?__.

Q6. The sinusoidal axis is at $y =$ __?__.

Q7. The lower bound is at $y =$ __?__.

Q8. What kind of function is $y = x^{5.2}$?

Q9. What kind of function is $y = 5^x$?

Q10. The "If..." part of the statement of a theorem is called the

A. Conclusion **B.** Hypothesis
C. Converse **D.** Inverse
E. Contrapositive

---

1. *Secant Function Problem*

a. Sketch two cycles of the parent cosine function, $y = \cos\theta$. Use the fact that $\sec\theta = \frac{1}{\cos\theta}$ to sketch the graph of $y = \sec\theta$.

b. How can you locate the asymptotes in the secant graph by looking at the cosine graph? How does your graph compare with the secant graph in Figure 6-3b?

c. Does the secant function have critical points? If so, find some of them. If not, explain why not.

d. Does the secant function have points of inflection? If so, find some of them. If not, explain why not.

2. *Tangent Function Problem*

   a. Sketch two cycles of the parent function $y = \cos\theta$ and two cycles of the parent function $y = \sin\theta$ on the same axes.

   b. Explain how you can use the graphs in part a to locate the θ-intercepts and the vertical asymptotes of the graph of $y = \tan\theta$.

   c. Mark the asymptotes, intercepts, and other significant points on your sketch in part a. Then sketch the graph of $y = \tan\theta$. How does the result compare with the tangent graph in Figure 6-3a?

   d. Does the tangent function have critical points? If so, find some of them. If not, explain why not.

   e. Does the tangent function have points of inflection? If so, find some of them. If not, explain why not.

3. *Quotient Property for Tangent Problem:* Plot these three graphs on the same screen on your grapher. Explain how the result confirms the quotient property for tangent.

   $f_1(x) = \sin\theta$

   $f_2(x) = \cos\theta$

   $f_3(x) = f_1(x)/f_2(x)$

4. *Quotient Property for Cotangent Problem:* On the same screen on your grapher, plot these three graphs. Explain how the result confirms the quotient property for cotangent.

   $f_1(x) = \sin\theta$

   $f_2(x) = \cos\theta$

   $f_3(x) = f_2(x)/f_1(x)$

5. Without referring to Figure 6-3a, quickly sketch the graphs of $y = \tan\theta$ and $y = \cot\theta$.

6. Without referring to Figure 6-3b, quickly sketch the graphs of $y = \sec\theta$ and $y = \csc\theta$.

7. Explain why the period of the functions $y = \tan\theta$ and $y = \cot\theta$ is only 180°, instead of 360° like the periods of the other four trigonometric functions.

8. Explain why it is meaningless to talk about the amplitude of the tangent, cotangent, secant, and cosecant functions.

9. What is the domain of the function $y = \sec\theta$? What is its range?

10. What is the domain of the function $y = \tan\theta$? What is its range?

For Problems 11–14, what are the dilation and translation caused by the constants in the equation? Plot the graph on your grapher and show that these transformations are correct.

11. $y = 2 + 5\tan 3(\theta - 5°)$

12. $y = -1 + 3\cot 2(\theta - 30°)$

13. $y = 4 + 6\sec\frac{1}{2}(\theta + 50°)$

14. $y = 3 + 2\csc 4(\theta + 10°)$

15. *Rotating Lighthouse Beacon Problem:* Figure 6-3h shows a lighthouse located 500 m from the shore.

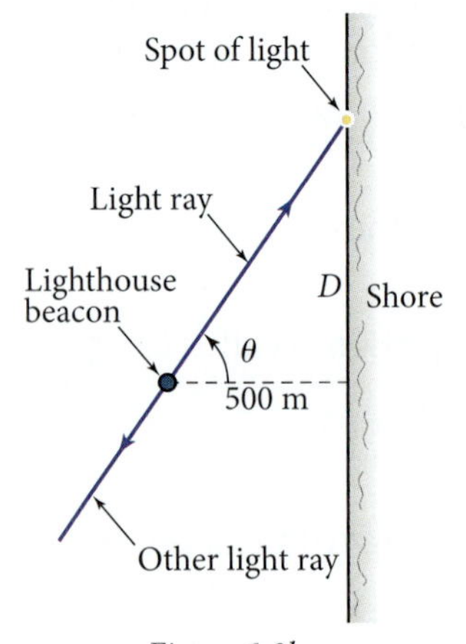

Figure 6-3h

A rotating light on top of the lighthouse sends out rays of light in opposite directions. As the beacon rotates, the ray at angle θ makes a spot of light that moves along the shore. As θ increases beyond 90°, the other ray makes the spot of light. Let D be the displacement of the spot of light from the point on the shore closest to the beacon, with the displacement positive to the right and negative to the left as you face the beacon from the shore.

a. Plot the graph of D as a function of θ. Use a window with $0° \le \theta \le 360°$ and $-2000 \le D \le 2000$. Sketch the result.

b. Where does the spot of light hit the shore when $\theta = 55°$? When $\theta = 91°$?

c. What is the first positive value of θ for which D equals 2000? For which D equals -1000?

d. Explain the physical significance of the asymptote at $\theta = 90°$.

16. *Variation of Tangent and Secant Problem:* Figure 6-3i shows the unit circle in a uv-coordinate system and a ray from the origin, O, at an angle, θ, in standard position. The ray intersects the circle at point P. A line is drawn tangent to the circle at point P, intersecting the u-axis at point A and the v-axis at point B. A vertical segment from point P intersects the u-axis at point C, and a horizontal segment from point P intersects the v-axis at point D.

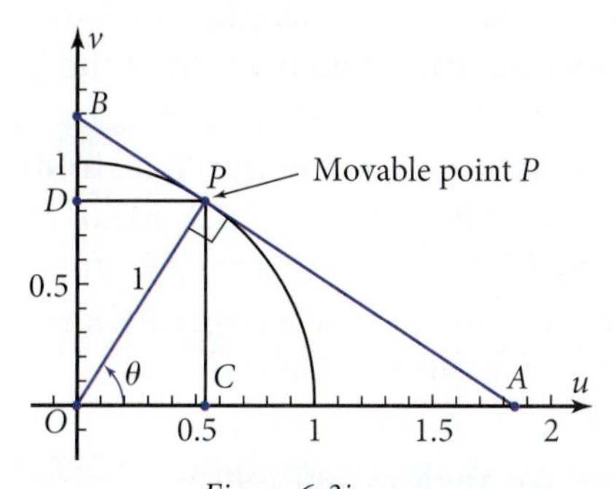

Figure 6-3i

a. Use the properties of similar triangles to explain why these segment lengths are equal to the six corresponding function values:

$PA = \tan\theta$

$PB = \cot\theta$

$PC = \sin\theta$

$PD = \cos\theta$

$OA = \sec\theta$

$OB = \csc\theta$

b. The angle between the ray and the v-axis is the *complement* of angle θ; that is, its measure is $90° - \theta$. Show that in each case the *co*function of θ is equal to the function of the *complement* of θ.

c. Construct Figure 6-3i using dynamic geometry software such as The Geometer's Sketchpad, or use the *Variation of Tangent and Secant* exploration at **www.keymath.com/precalc.** Observe what happens to the six function values as θ changes. Describe how the sine and cosine vary as θ is made larger or smaller. Based on the figure, explain why the tangent and secant become infinite as θ approaches 90° and why the cotangent and cosecant become infinite as θ approaches 0°.

6-4 Radian Measure of Angles

With your calculator in degree mode, press sin 60°. You get

$$\sin 60° = 0.866025403...$$

Now change to **radian** mode and press $\sin\left(\frac{\pi}{3}\right)$. You get the same answer!

$$\sin\left(\frac{\pi}{3}\right) = 0.866025403...$$

In this section you will learn what radians are and how to convert angle measures between radians and degrees. The radian measure of angles allows you to expand on the concept of trigonometric functions, as you'll see in the next section. Through this expansion of trigonometric functions, you can model real-world phenomena in which independent variables represent distance, time, or any other quantity, not just an angle measure in degrees.

Objective Given an angle measure in degrees, convert it to radians, and vice versa. Given an angle measure in radians, find trigonometric function values.

In this exploration you will explore the radian as a unit of angular measure, and you'll see how this unit arises naturally from the radius of a circle.

EXPLORATION 6-4: Introduction to Radians

This graph shows a circle of radius r units in a uv-coordinate system. The radius has been marked off in units of tenths of a radius. An arc of the circle with a curved length exactly one radius unit long has been marked off starting at the positive u-axis. A central angle has been drawn subtending (cutting off) this arc.

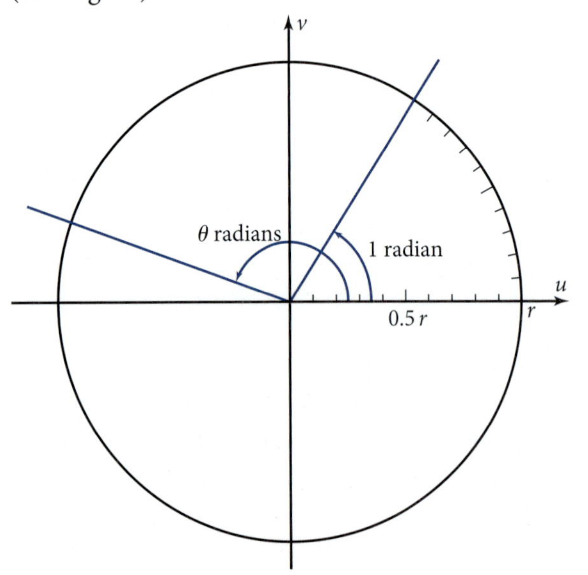

1. Mark a flexible ruler (an index card will do) with the scale shown on the u-axis. Then measure the arc on the circle opposite the angle marked 1 radian. (See the illustration in Problem 2 of the following Problem Set for ideas on how to do this.) Does the arc have length r? How does this fact tell you that the angle measures 1 radian?
2. Mark off two more radius-lengths on your flexible ruler starting from one end of the scale you have already marked. Then measure around the circumference of the circle from the positive u-axis to the terminal side of the angle marked θ. Based on this measurement and the definition of radians (page 302), what is the radian measure of angle θ?
3. What is the radian measure of one full revolution? An angle of 180°? 90°?
4. What is the degree measure of an angle of 1 radian? Show how you found your answer.
5. What did you learn as a result of doing this exploration that you did not know before?

Excerpt from an old Babylonian cuneiform text

The degree as a unit of angular measure came from ancient mathematicians, probably Babylonians. It is assumed that they divided a revolution into 360 parts we call degrees because there were approximately 360 days in a year and they used the base-60 (sexagesimal) number system. There is another way to measure angles, called radian measure. This mathematically more natural unit of angular measure is derived by wrapping a number line around the unit circle (a circle of radius 1 unit) in a *uv*-coordinate system, as in Figure 6-4a. Each point on the number line corresponds to a point on the perimeter of the circle.

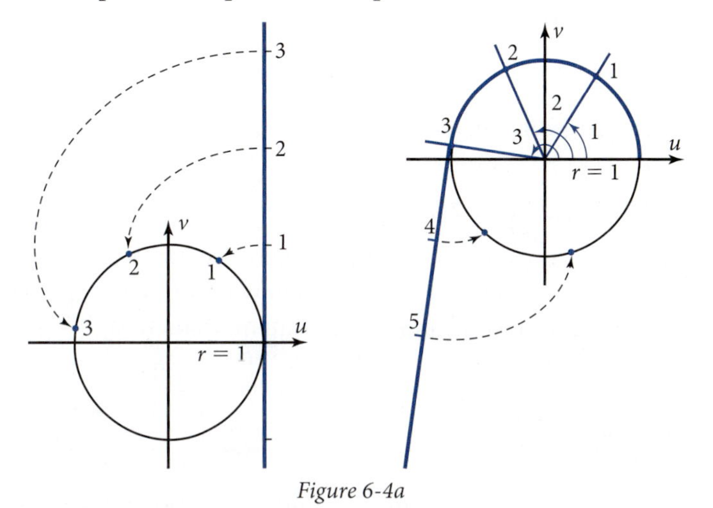

Figure 6-4a

If you draw rays from the origin to the points 1, 2, and 3 on the circle (right side of Figure 6-4a), the corresponding central angles have radian measures 1, 2, and 3, respectively.

But, you may ask, what happens if the same angle is in a larger circle? Would the same radian measure correspond to it? How would you calculate the radian measure in this case? Figures 6-4b and 6-4c answer these questions. Figure 6-4b shows an angle of measure 1, in radians, and the arcs it *subtends* (cuts off) on circles of radius 1 unit and *x* units. The arc subtended on the unit circle has length 1 unit. By the properties of similar geometric figures, the arc subtended on the circle of radius *x* has length *x* units. So 1 radian subtends an arc of length equal to the radius of the circle.

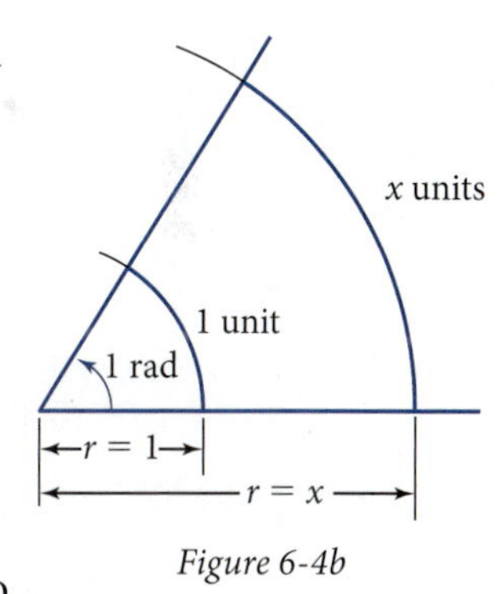

Figure 6-4b

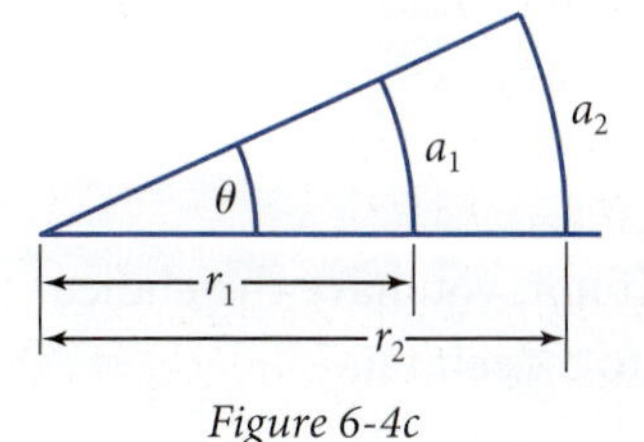

Figure 6-4c

For any angle measure, the arc length and the radius are proportional $\left(\frac{a_1}{r_1} = \frac{a_2}{r_2}\right.$, as shown in Figure 6-4c$\left.\right)$, and their quotient is a unitless number that uniquely corresponds to and describes the angle. So, in general, the **radian measure** of an angle equals the length of the subtended arc divided by the radius.

DEFINITION: Radian Measure of an Angle

$$\text{radian measure} = \frac{\text{arc length}}{\text{radius}}$$

For the work that follows, it is important to distinguish between the name of the angle and the measure of that angle. Measures of angle θ will be written this way:

θ is the *name* of the angle.

$m^{\circ}(\theta)$ is the degree measure of angle θ.

$m^{R}(\theta)$ is the radian measure of angle θ.

Because the circumference of a circle is $2\pi r$ and because r for the unit circle is 1, the wrapped number line in Figure 6-4a divides the circle into 2π units (a little more than six parts). So there are 2π radians in a complete revolution. There are also 360° in a complete revolution. You can convert degrees to radians, or the other way around, by setting up these proportions:

$$\frac{m^{R}(\theta)}{m^{\circ}(\theta)} = \frac{2\pi}{360^{\circ}} = \frac{\pi}{180^{\circ}} \qquad \text{or} \qquad \frac{m^{\circ}(\theta)}{m^{R}(\theta)} = \frac{360^{\circ}}{2\pi} = \frac{180^{\circ}}{\pi}$$

Solving for $m^{R}(\theta)$ and $m^{\circ}(\theta)$, respectively, gives

$$m^{R}(\theta) = \frac{\pi}{180^{\circ}}\, m^{\circ}(\theta) \qquad \text{and} \qquad m^{\circ}(\theta) = \frac{180^{\circ}}{\pi}\, m^{R}(\theta)$$

These equations lead to a procedure for accomplishing the objective of this section.

PROCEDURE: Radian–Degree Conversion

To find the radian measure of θ, multiply the degree measure by $\frac{\pi}{180^{\circ}}$.

To find the degree measure of θ, multiply the radian measure by $\frac{180^{\circ}}{\pi}$.

EXAMPLE 1 ➤ Convert 135° to radians.

SOLUTION In order to keep the units straight, write each quantity as a fraction with the proper units. If you have done the work correctly, certain units will cancel, leaving the proper units for the answer.

$$\left.\begin{matrix} 135^{\circ} & x \\ 180^{\circ} & \pi \end{matrix}\right\} \Rightarrow \frac{135}{180} = \frac{x}{\pi}$$

$$m^{R}(\theta) = \frac{135\ \cancel{\text{degrees}}}{1} \cdot \frac{\pi \text{ radians}}{180\ \cancel{\text{degrees}}} = \frac{3}{4}\pi = 2.3561\ldots \text{ radians}$$

Notes:

- If the *exact* value is called for, leave the answer as $\frac{3}{4}\pi$. If not, you have the choice of writing the answer as a multiple of π or converting to a decimal.
- The procedure for canceling units used in Example 1 is called *dimensional analysis.* You will use this procedure throughout your study of mathematics.

EXAMPLE 2 ➤ Convert 5.73 radians to degrees.

SOLUTION

$$\frac{5.73 \cancel{\text{radians}}}{1} \cdot \frac{180 \text{ degrees}}{\pi \cancel{\text{radians}}} = 328.3048...^\circ$$

EXAMPLE 3 ➤ Find tan 3.7.

SOLUTION Unless the argument of a trigonometric function has the degree symbol, it is assumed to be a measure in radians. (That is why it has been important for you to include the degree symbol until now.) Set your calculator to radian mode and enter tan 3.7.

$$\tan 3.7 = 0.6247...$$

EXAMPLE 4 ➤ Find the radian measure and the degree measure of an angle whose sine is 0.3.

SOLUTION

$\sin^{-1} 0.3 = 0.3046...$ radian — Set your calculator to radian mode.

$\sin^{-1} 0.3 = 17.4576...^\circ$ — Set your calculator to degree mode.

To check whether these answers are in fact equivalent, you could convert one to the other.

$$0.3046...\text{radian} \cdot \frac{180 \text{ degrees}}{\pi \cancel{\text{radians}}} = 17.4576...^\circ$$

Use the 0.3046... already in your calculator, without rounding.

Radian Measures of Some Special Angles

It will help you later in calculus to be able to recall quickly the radian measures of certain special angles, such as those whose degree measures are multiples of 30° and 45°.

By the technique of Example 1,

$$30^\circ \rightarrow \frac{\pi}{6} \text{ radian, or } \frac{1}{12} \text{ revolution}$$

$$45^\circ \rightarrow \frac{\pi}{4} \text{ radian, or } \frac{1}{8} \text{ revolution}$$

If you remember these two, you can find others quickly by multiplication. For instance,

$$60^\circ \rightarrow 2(\pi/6) = \frac{\pi}{3} \text{ radians, or } \frac{1}{6} \text{ revolution}$$

$$90^\circ \rightarrow 3(\pi/6) \text{ or } 2(\pi/4) = \frac{\pi}{2} \text{ radians, or } \frac{1}{4} \text{ revolution}$$

$$180^\circ \rightarrow 6(\pi/6) \text{ or } 4(\pi/4) = \pi \text{ radians, or } \frac{1}{2} \text{ revolution}$$

For 180°, you can simply remember that a full revolution is 2π radians, so half a revolution is π radians.

Figure 6-4d shows the radian measures of some special first-quadrant angles. Figure 6-4e shows radian measures of larger angles that are $\frac{1}{4}$, $\frac{1}{2}$, $\frac{3}{4}$, and 1 revolution. The box summarizes this information.

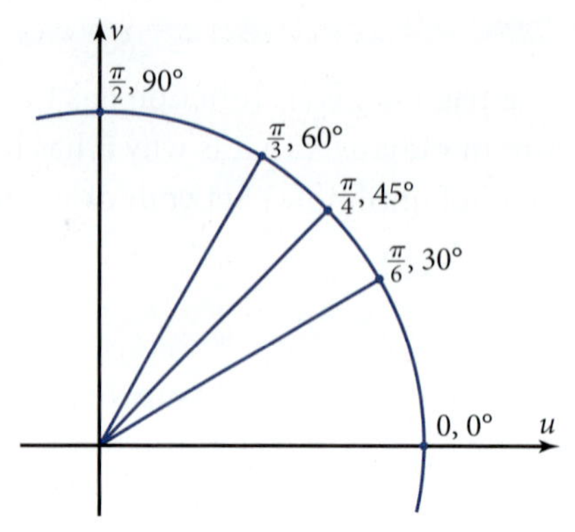

Figure 6-4d

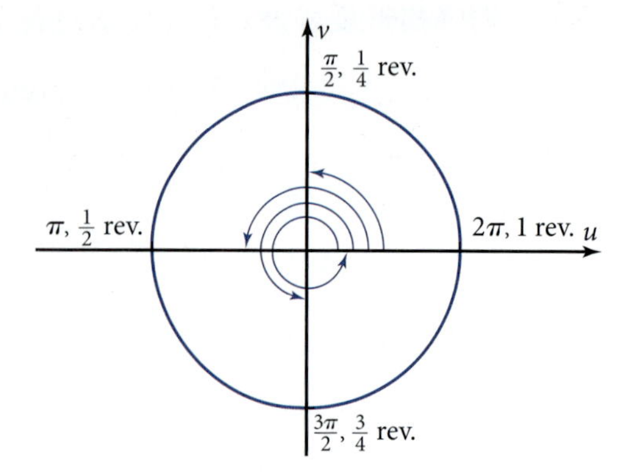

Figure 6-4e

PROPERTY: Radian Measures of Some Special Angles

Degrees	30°	45°	60°	90°	180°	360°
Radians	$\pi/6$	$\pi/4$	$\pi/3$	$\pi/2$	π	2π
Revolutions	$\frac{1}{12}$	$\frac{1}{8}$	$\frac{1}{6}$	$\frac{1}{4}$	$\frac{1}{2}$	1

EXAMPLE 5 ➤ Find the exact value of $\sec \frac{\pi}{6}$.

SOLUTION $\sec \frac{\pi}{6} = \sec 30° = \frac{1}{\cos 30°} = \frac{1}{\sqrt{3}/2} = \frac{2}{\sqrt{3}}$

Recall how to use the reference triangle to find the exact value of cos 30°.

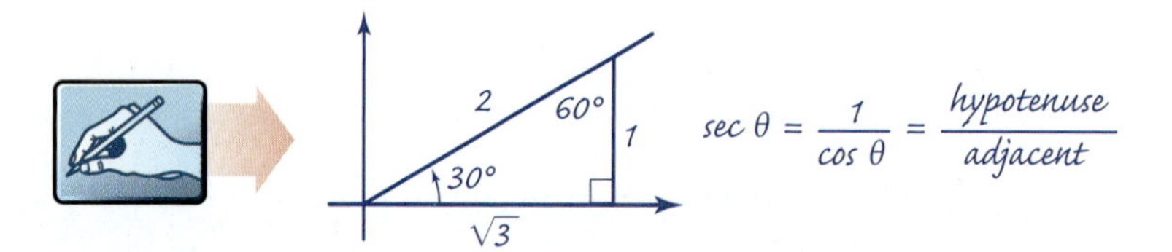

Problem Set 6-4

Reading Analysis

From what you have read in this section, what do you consider to be the main idea? Is a radian large or small compared to a degree? How do you find the radian measure of an angle if you know its degree measure? How can you remember that there are 2π radians in a full revolution?

Quick Review

Q1. Sketch the graph of $y = \tan \theta$.

Q2. Sketch the graph of $y = \sec \theta$.

Q3. What is the first positive value of θ at which the graph of $y = \cot \theta$ has a vertical asymptote?

Q4. What is the first positive value of θ for which the graph of $\csc \theta = 0$?

Q5. What is the exact value of tan 60°?

Q6. What transformation of function f is represented by $g(x) = 3f(x)$?

Q7. What transformation of function f is represented by $h(x) = f(10x)$?

Q8. Write the general equation for a quadratic function.

Q9. $3^{2015} \div 3^{2011} = \underline{\ \ ?\ \ }$

Q10. The "then" part of the statement for a theorem is called the

A. Converse **B.** Inverse

C. Contrapositive **D.** Conclusion

E. Hypothesis

---

1. *Wrapping Function Problem:* Figure 6-4f shows the unit circle in a uv-coordinate system. Suppose you want to use the angle measure in radians as the independent variable. Imagine the x-axis from an xy-coordinate system placed tangent to the circle. Its origin, $x = 0$, is at the point $(u, v) = (1, 0)$. Then the x-axis is wrapped around the circle.

a. Show where the points $x = 1$, 2, and 3 on the number line map onto the circle.

b. On your sketch from part a, show angles of 1, 2, and 3 radians in standard position.

c. Explain how the length of the arc of the unit circle subtended by a central angle of the circle is related to the radian measure of that angle.

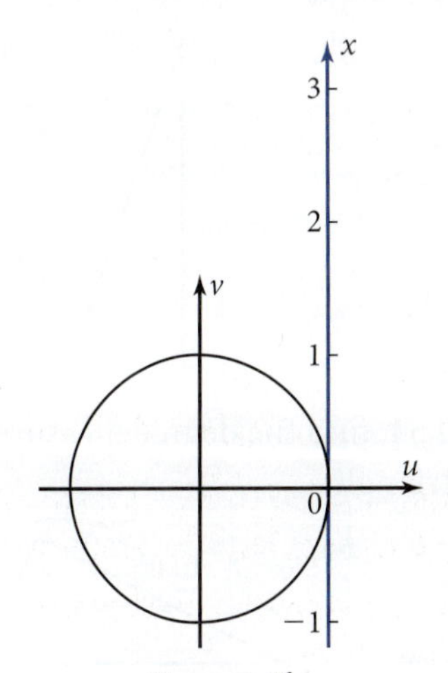

Figure 6-4f

2. *Arc Length and Angle Problem:* As a result of the definition of radian, you can calculate the arc length as the product of the angle in radians and the radius of the circle. Figure 6-4g shows arcs of three circles subtended by a central angle of 1.3 radians. The radii of the circles have lengths 1, 2, and 3 cm.

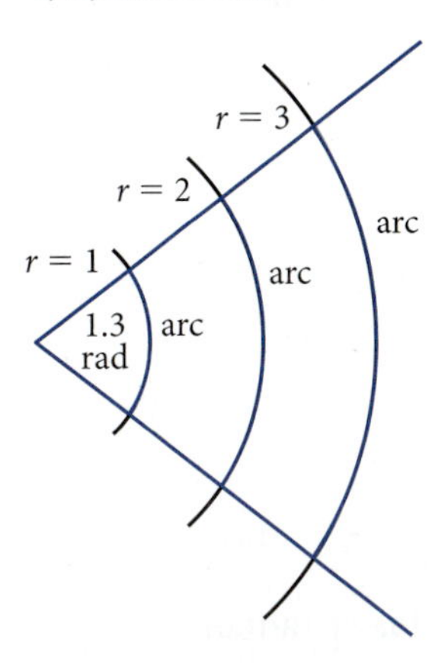

Figure 6-4g

a. How long would the arc of the 1-cm circle be if you measured it with a flexible ruler?

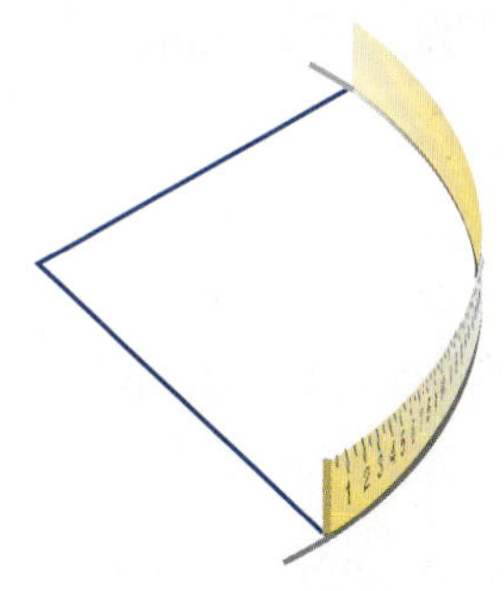

b. Find the lengths of the arcs on the 2-cm circle and the 3-cm circle using the properties of similar geometric figures.

c. On a circle of radius r meters, what is the length of an arc that is subtended by an angle of 1.3 radians?

d. How could you *quickly* find the length a of an arc of a circle of radius r meters that is subtended by a central angle of θ radians? Write a formula representing the arc length.

For Problems 3–10, find the *exact* radian measure of the angle (no decimals).

3. 60° **4.** 45°
5. 30° **6.** 180°
7. 120° **8.** 450°
9. −225° **10.** 1080°

For Problems 11–14, find the radian measure of the angle in decimal form.

11. 37° **12.** 58°
13. 123° **14.** 258°

For Problems 15–24, find the exact degree measure of the angle given in radians (no decimals). Use the most time-efficient method.

15. $\frac{\pi}{10}$ radian **16.** $\frac{\pi}{2}$ radians
17. $\frac{\pi}{6}$ radian **18.** $\frac{\pi}{4}$ radian
19. $\frac{\pi}{12}$ radian **20.** $\frac{2\pi}{3}$ radians
21. $\frac{3\pi}{4}$ radians **22.** π radians
23. $\frac{3\pi}{2}$ radians **24.** $\frac{5\pi}{6}$ radians

For Problems 25–30, find the degree measure in decimal form of the angle given in radians.

25. 0.34 radian **26.** 0.62 radian
27. 1.26 radians **28.** 1.57 radians
29. 1 radian **30.** 3 radians

For Problems 31–34, find the function value (in decimal form) for the angle in radians.

31. sin 5 **32.** cos 2
33. tan(−2.3) **34.** sin 1066

For Problems 35–38, find the radian measure (in decimal form) of the angle.

$\tan^{-1} 5 = x$
⇓
$\tan x = 5$

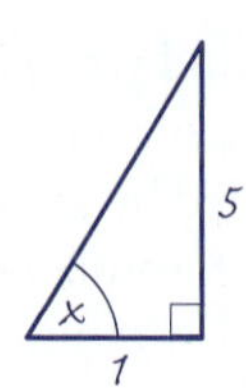

35. $\sin^{-1} 0.3$ **36.** $\tan^{-1} 5$
37. $\cot^{-1} 3$ **38.** $\csc^{-1} 1.001$

For Problems 39–44, find the *exact* value of the indicated function (no decimals). Note that because the degree sign is not used, the angle is assumed to be in radians.

39. $\sin \frac{\pi}{3}$ **40.** $\cos \pi$
41. $\tan \frac{\pi}{6}$ **42.** $\cot \frac{\pi}{2}$
43. $\sec 2\pi$ **44.** $\csc \frac{\pi}{4}$

For Problems 45–48, find the *exact* value of the expression (no decimals).

45. $\sin \frac{\pi}{2} + 6 \cos \frac{\pi}{3}$ **46.** $\csc \frac{\pi}{6} \sin \frac{\pi}{6}$
47. $\cos^2 \pi + \sin^2 \pi$ **48.** $\tan^2 \frac{\pi}{3} - \sec^2 \frac{\pi}{3}$

For Problems 49 and 50, write a particular equation for the sinusoid graphed.

49.

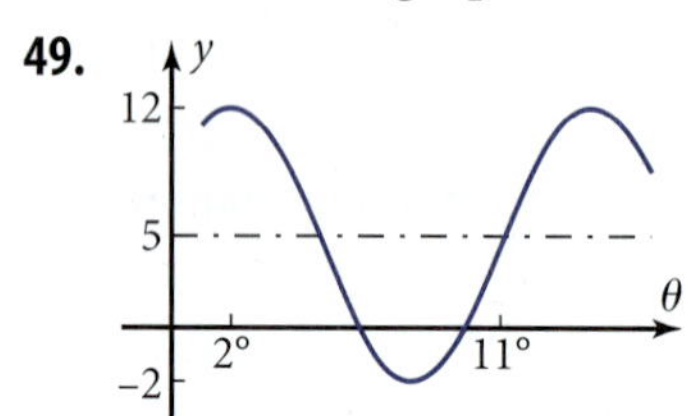

50.

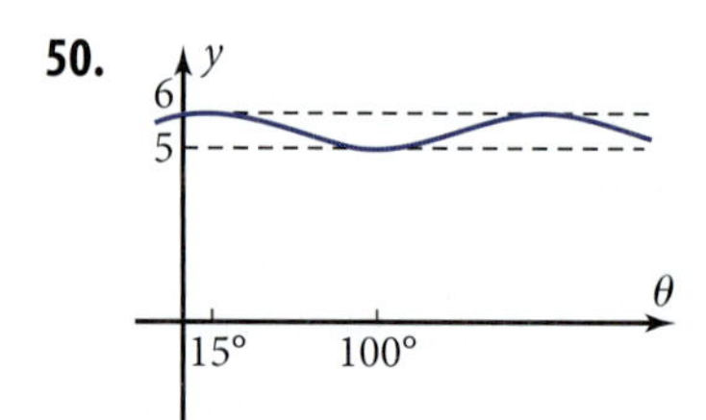

For Problems 51 and 52, find the length of the side labeled x in the right triangle.

51.

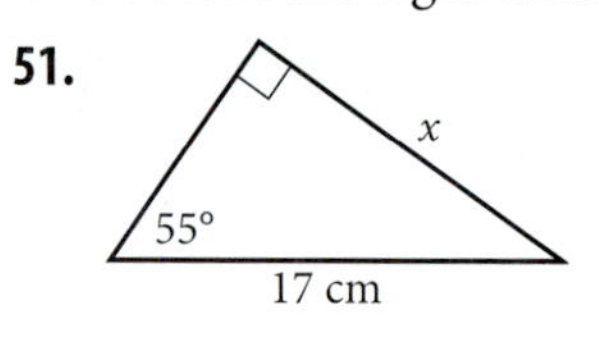

52.

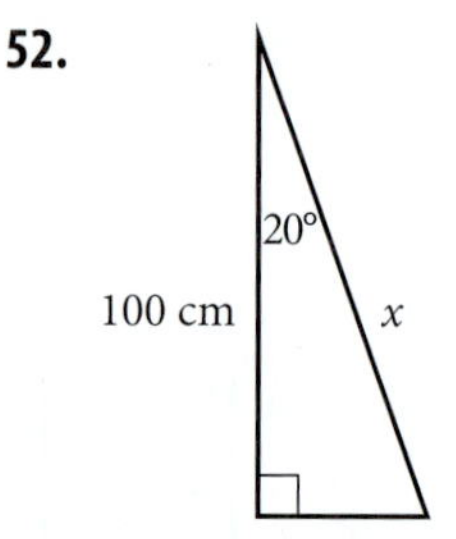

For Problems 53 and 54, find the degree measure of angle θ in the right triangle.

53.

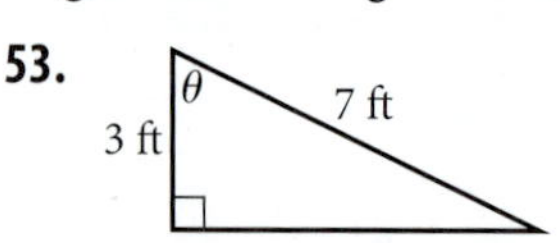

54.

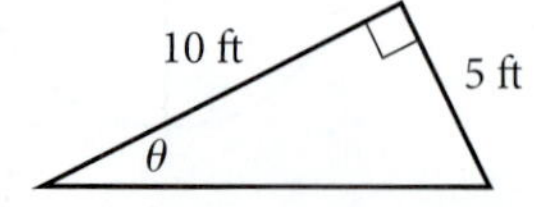

6-5 Circular Functions

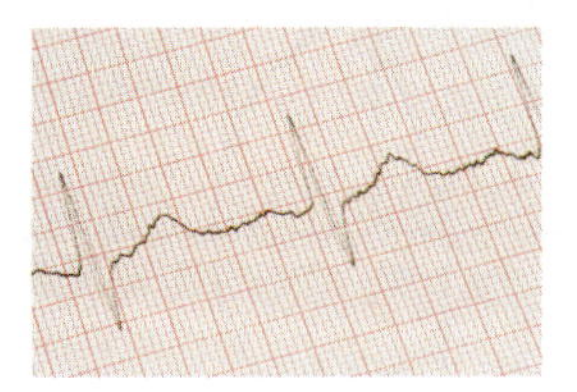

The normal human EKG (electrocadiogram) is periodic.

In many real-world situations, the independent variable of a periodic function is time or distance, with no angle evident. For instance, the normal daily high temperature varies periodically with the day of the year. In this section you will learn about **circular functions,** periodic functions whose independent variable is a real number without any units. These functions, as you will see, are identical to trigonometric functions in every way except for their argument. Circular functions are more appropriate for real-world applications. They also have some advantages in later courses in calculus, for which this course is preparing you.

Objective Learn about the circular functions and their relationship to trigonometric functions.

Two cycles of the graph of the parent cosine function are completed in 720° (Figure 6-5a, left) or in 4π units (Figure 6-5a, right), because 4π radians correspond to two revolutions.

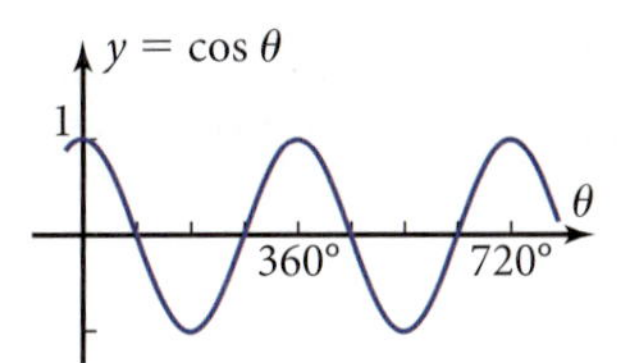

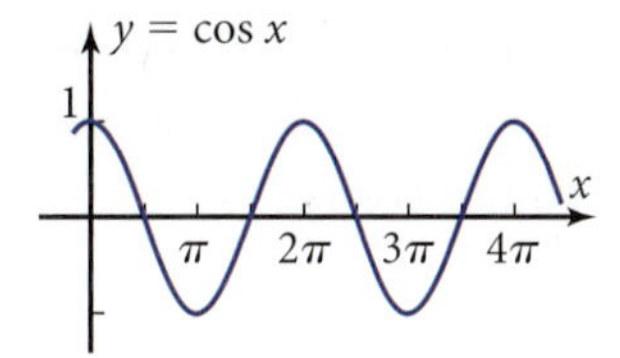

Figure 6-5a

To see how the independent variable can represent a real number, imagine the x-axis from an xy-coordinate system lifted out and placed vertically tangent to the unit circle in a uv-coordinate system with its origin at the point $(u, v) = (1, 0)$, as on the left side in Figure 6-5b. Then wrap the x-axis around the unit circle. As shown on the right side in Figure 6-5b, $x = 1$ maps onto an angle of 1 radian, $x = 2$ maps onto 2 radians, $x = \pi$ maps onto π radians, and so on.

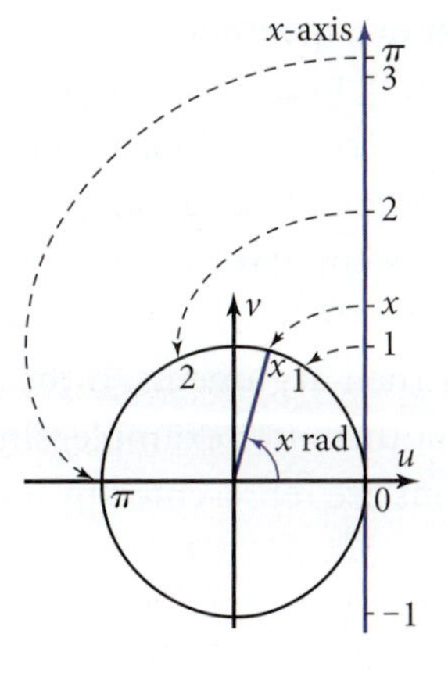

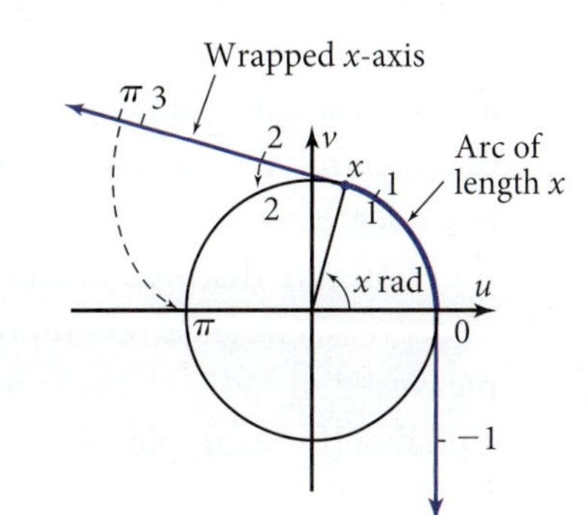

Figure 6-5b

The distance x on the x-axis is equal to the arc length on the unit circle. This arc length is equal to the radian measure for the corresponding angle. Thus the functions $\sin x$ and $\cos x$ for a number x on the x-axis are the same as the sine and cosine of an angle of x radians.

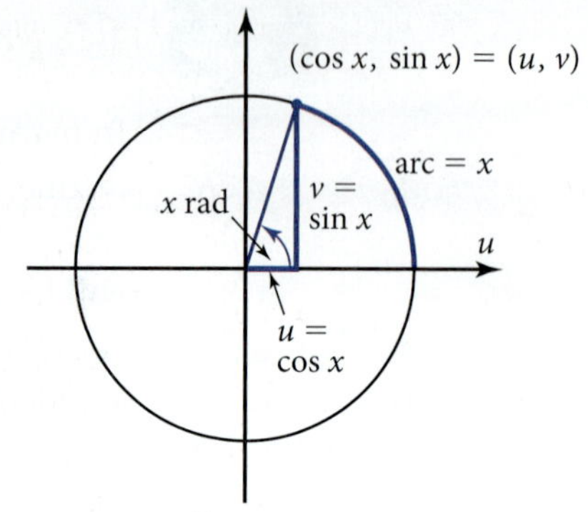

Figure 6-5c

Figure 6-5c shows an arc of length x on the unit circle, with the corresponding angle. The arc is in **standard position** on the unit circle, with its initial point at (1, 0) and its terminal point at (u, v). The sine and cosine of x are defined in the same way as for the trigonometric functions:

$$\cos x = \frac{\text{horizontal coordinate}}{\text{radius}} = \frac{u}{1} = u$$

$$\sin x = \frac{\text{vertical coordinate}}{\text{radius}} = \frac{v}{1} = v$$

The name *circular function* comes from the fact that x equals the length of an arc on the unit circle. The other four circular functions are defined as ratios of sine and cosine.

DEFINITION: Circular Functions

If (u, v) is the terminal point of an arc of length x in standard position on the unit circle, then the **circular functions** of x are defined as

$\sin x = v$ $\qquad$ $\cos x = u$

$\tan x = \dfrac{\sin x}{\cos x} = \dfrac{v}{u}$ $\qquad$ $\cot x = \dfrac{\cos x}{\sin x} = \dfrac{u}{v}$

$\sec x = \dfrac{1}{\cos x} = \dfrac{1}{u}$ $\qquad$ $\csc x = \dfrac{1}{\sin x} = \dfrac{1}{v}$

Circular functions are equivalent to trigonometric functions in radians. This equivalency provides an opportunity to expand the concept of trigonometric functions. You have seen trigonometric functions first defined using the angles of a right triangle and later expanded to include all angles. From now on, the concept of trigonometric functions includes circular functions, and the functions can have both degrees and radians as arguments. The way the two kinds of trigonometric functions are distinguished is by their arguments. If the argument is measured in degrees, Greek letters represent them (for example, $\sin \theta$). If the argument is measured in radians, the functions are represented by letters from the Roman alphabet (for example, $\sin x$).

EXAMPLE 1 ➤ Plot the graph of $y = 4 \cos 5x$ on your grapher, in radian mode. Find the period graphically and algebraically. Compare your results.

SOLUTION Figure 6-5d shows the graph.

Tracing the graph, you find that the first high point beyond $x = 0$ is between $x = 1.25$ and $x = 1.3$. So graphically the period is between 1.25 and 1.3.

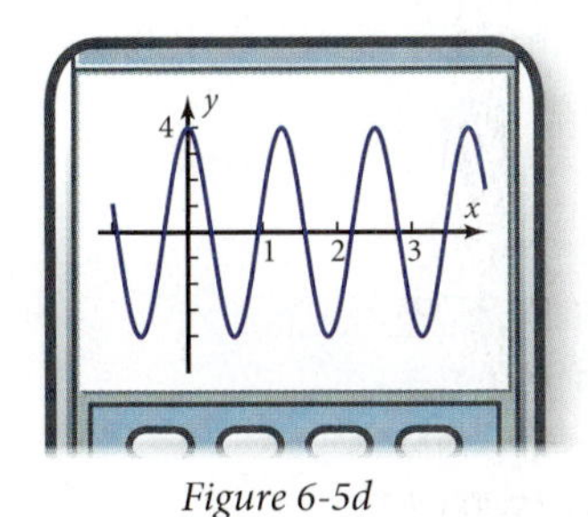

Figure 6-5d

To find the period algebraically, recall that the 5 in the argument of the cosine function is the reciprocal of the horizontal dilation. The period of the parent cosine function is 2π, because there are 2π radians in a complete revolution. Thus the period of the given function is

$$\frac{1}{5}(2\pi) = 0.4\pi = 1.2566...$$

The answer found graphically is close to this exact answer. ◄

Note: Using the MAXIMUM feature of your grapher confirms that the high point is at $x = 1.2566....$

EXAMPLE 2 ➤ Find a particular equation for the sinusoid function graphed in Figure 6-5e. Notice that the horizontal axis is labeled x, not θ, indicating that the angle is measured in radians. Confirm your answer by plotting the equation on your grapher.

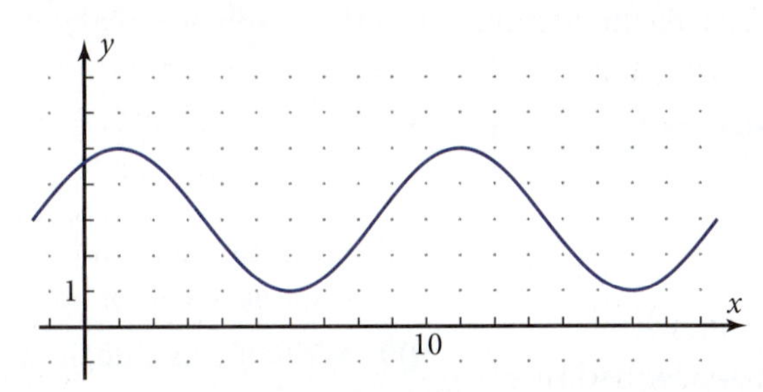

Figure 6-5e

SOLUTION

$y = C + A \cos B(x - D)$ — Write the general sinsoidal equation, using x instead of θ.

- Sinusoidal axis is at $y = 3$, so $C = 3$. — Find A, B, C, and D using information from the graph.
- Amplitude is 2, and the cycle starts at a high point, so $A = 2$.
- Period is 10. — From one high point to the next is $11 - 1$.
- Dilation is $\frac{10}{2\pi}$ or $\frac{5}{\pi}$, so $B = \frac{\pi}{5}$. — B is the reciprocal of the horizontal dilation.
- Phase displacement is 1 (for the parent function $\cos x$), so $D = 1$. — Cosine starts a cycle at a high point.

$y = 3 + 2 \cos \frac{\pi}{5}(x - 1)$ — Write the particular equation.

Plotting this equation in radian mode confirms that it is correct. ◄

EXAMPLE 3 ➤ Sketch the graph of $y = \tan \frac{\pi}{6}x$.

SOLUTION In order to graph the function, you need to identify its period, the locations of its inflection points, and its asymptotes.

$$\text{Period} = \frac{6}{\pi} \cdot \pi = 6$$ Horizontal dilation is the reciprocal of $\frac{\pi}{6}$; the period of the parent tangent function is π.

For this function, the points of inflection are also the x-intercepts, or the points where the value of the function equals zero. So

$$\frac{\pi}{6}x = 0, \pm\pi, \pm 2\pi, \ldots$$

$$x = 0, \pm 6, \pm 12, \ldots$$

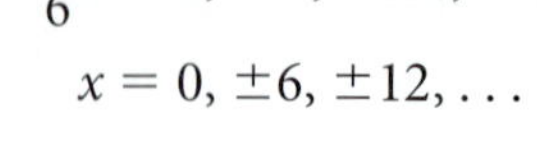

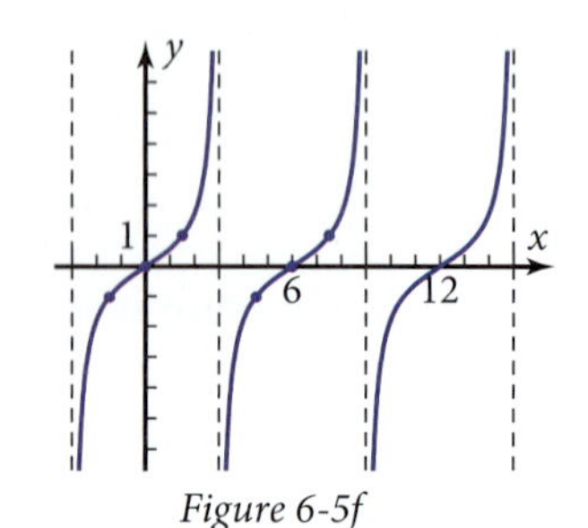

Figure 6-5f

Asymptotes are at values where the function is undefined. So

$$\frac{\pi}{6}x = -\frac{\pi}{2}, \frac{\pi}{2}, \frac{3\pi}{2}, \frac{5\pi}{2}, \ldots$$

$$x = -3, 3, 9, 15, \ldots$$

Recall that halfway between a point of inflection and an asymptote the tangent equals 1 or −1. The graph in Figure 6-5f illustrates these features. ◄

Note that in the graphs of circular functions the number π appears either in the equation as a coefficient of x or in the graph as a scale mark on the x-axis.

Problem Set 6-5

Reading Analysis

From what you have read in this section, what do you consider to be the main idea? As defined in this text, what are the differences and the similarities between a *circular* function and a *trigonometric* function? How do angle measures in radians link the circular functions to the trigonometric functions?

Quick Review

Q1. How many radians are in 180°?

Q2. How many degrees are in 2π radians?

Q3. How many degrees are in 1 radian?

Q4. How many radians are in 34°?

Q5. Find sin 47°.

Q6. Find sin 47.

Q7. Find the period of $y = 3 + 4\cos 5(\theta - 6°)$.

Q8. Find the upper bound for y for the sinusoid in Problem Q7.

Q9. How long does it take you to go 300 mi at an average speed of 60 mi/h?

Q10. Write 5% as a decimal.

For Problems 1–4, find the exact arc length on the unit circle subtended by the given angle (no decimals).

1. 30°

2. 60°

3. 90°

4. 45°

For Problems 5–8, find the exact degree measure of the angle that subtends the given arc length on the unit circle.

5. $\frac{\pi}{3}$ units

6. $\frac{\pi}{6}$ unit

7. $\frac{\pi}{4}$ unit

8. $\frac{\pi}{2}$ units

For Problems 9–12, find the exact arc length on the unit circle subtended by the given angle in radians.

9. $\frac{\pi}{2}$ radians

10. π radians

11. 2 radians

12. 1.467 radians

For Problems 13–16, evaluate the circular function in decimal form.

13. $\tan 1$ 　　**14.** $\sin 2$

15. $\sec 3$ 　　**16.** $\cot 4$

For Problems 17–20, find the inverse circular function in decimal form.

17. $\cos^{-1} 0.3$ 　　**18.** $\tan^{-1} 1.4$

19. $\csc^{-1} 5$ 　　**20.** $\sec^{-1} 9$

For Problems 21–24, find the exact value of the circular function (no decimals).

21. $\sin \frac{\pi}{3}$ 　　**22.** $\cos \frac{\pi}{4}$

23. $\tan \frac{\pi}{6}$ 　　**24.** $\csc \pi$

For Problems 25–28, find the period, amplitude, phase displacement, and sinusoidal axis location. Use these features to sketch the graph. Confirm your graph by plotting the sinusoids on your grapher.

25. $y = 3 + 2 \cos \frac{\pi}{5}(x - 4)$

26. $y = -4 + 5 \sin \frac{2\pi}{3}(x + 1)$

27. $y = 2 + 6 \sin \frac{\pi}{4}(x + 1)$

28. $y = 5 + 4 \cos \frac{\pi}{3}(x - 2)$

For Problems 29–32, find the period, asymptotes, and critical points or points of inflection, and then sketch the graph.

29. $y = \cot \frac{\pi}{4}x$ 　　**30.** $y = \tan 2\pi x$

31. $y = 2 + \sec x$ 　　**32.** $y = 3 \csc x$

For Problems 33–42, find a particular equation for the circular function graphed.

33.

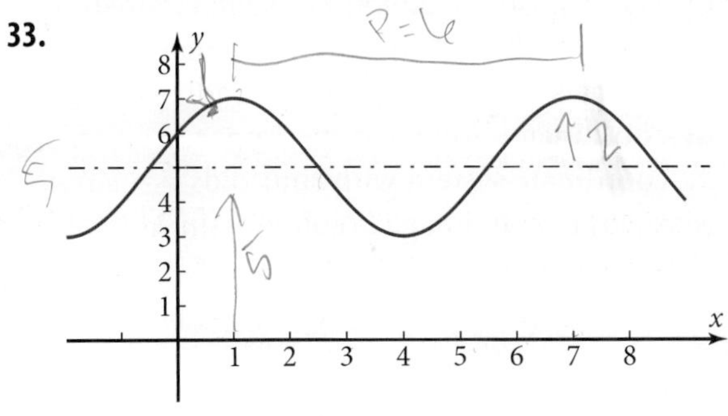

34.

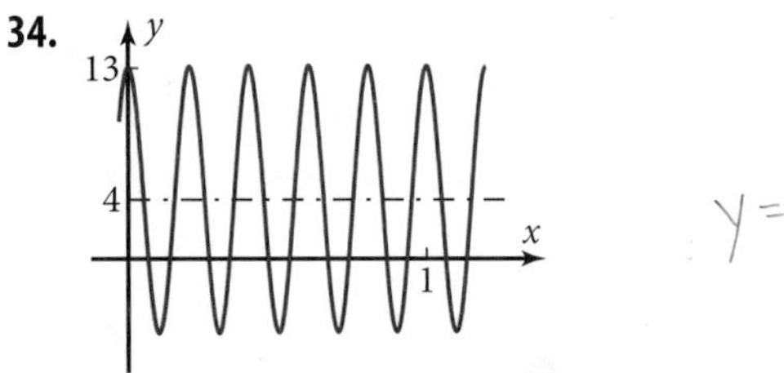

35.

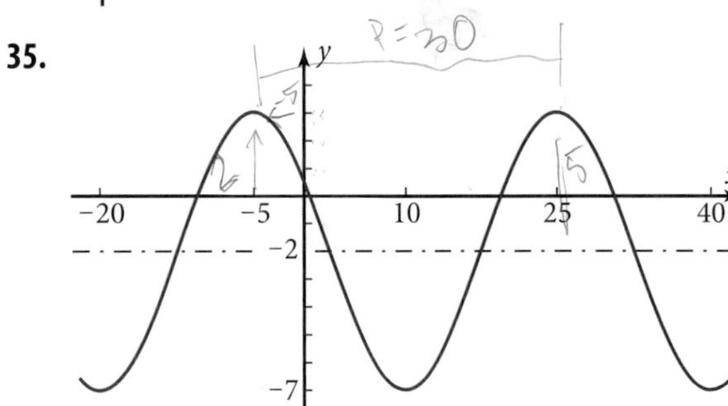

36.

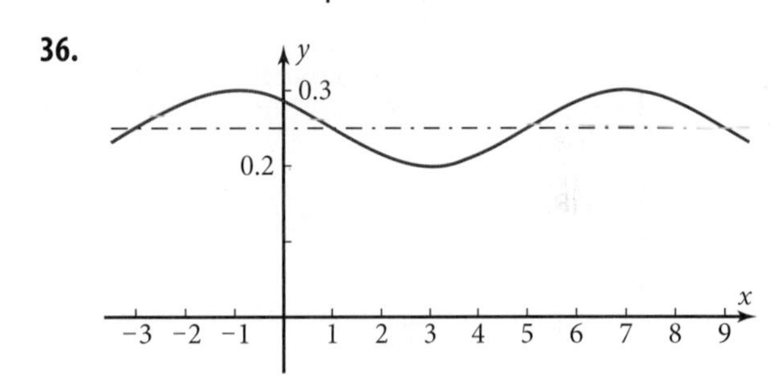

37.

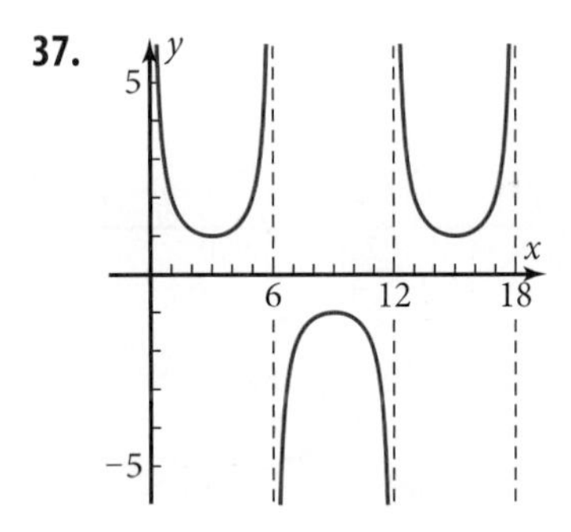

38.

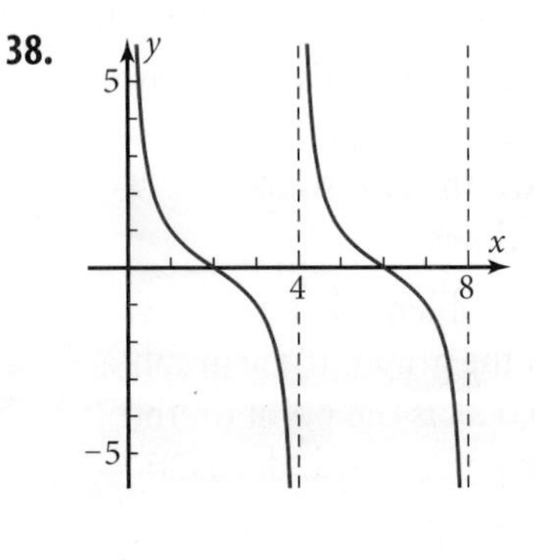

39.

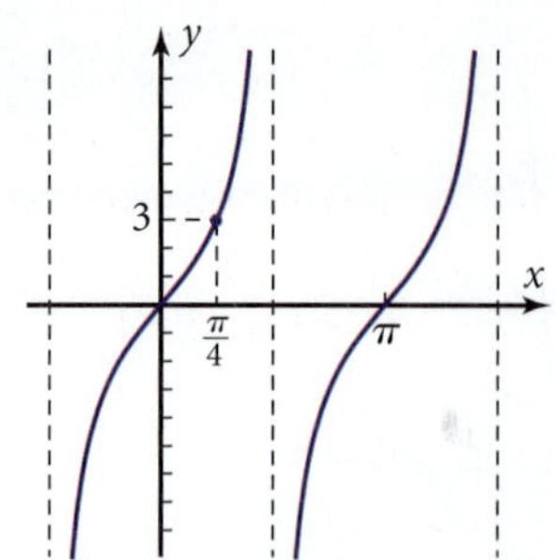

40.

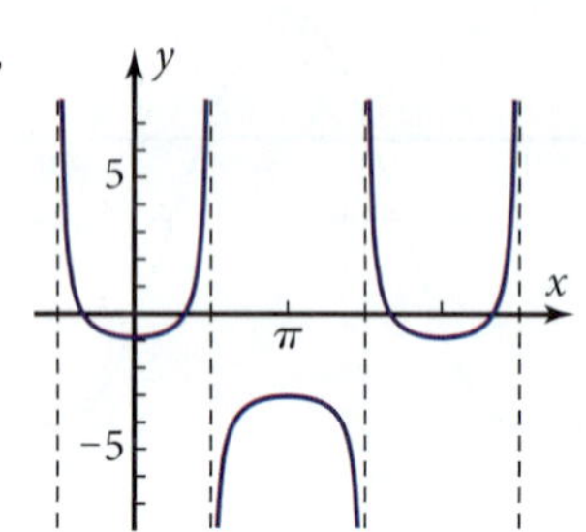

41.

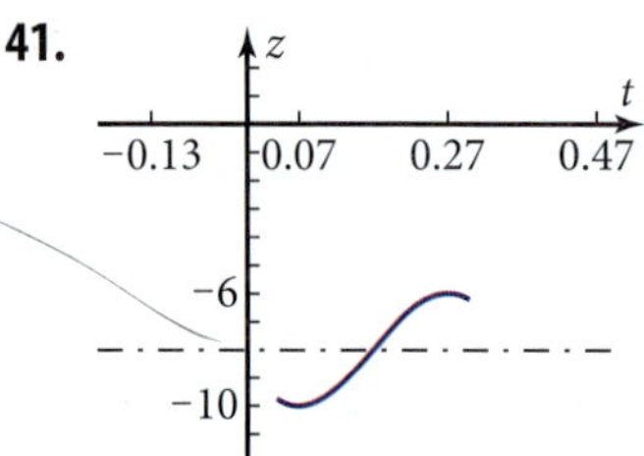

42.

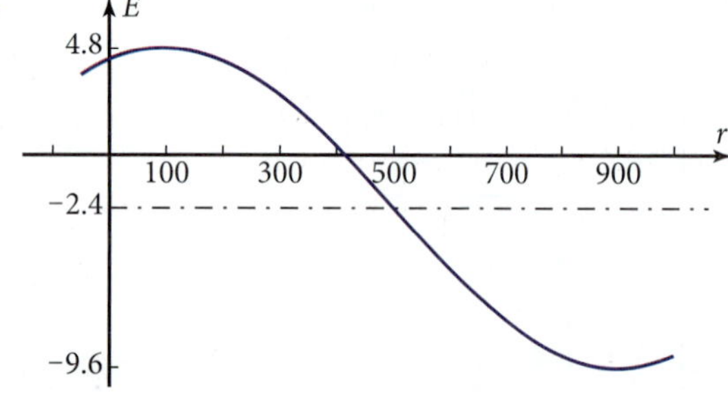

43. For the sinusoid in Problem 41, find the value of z at $t = 0.4$ on the graph. If the graph is extended to $t = 50$, is the point on the graph above or below the sinusoidal axis? How far above or below?

44. For the sinusoid in Problem 42, find the value of E at $r = 1234$ on the graph. If the graph is extended to $r = 10{,}000$, is the point on the graph above or below the sinusoidal axis? How far above or below?

45. *Sinusoid Translation Problem:* Figure 6-5g shows the graphs of $y = \cos x$ (dashed) and $y = \sin x$ (solid). Note that the graphs are congruent to each other (if superimposed, they coincide), differing only in horizontal translation.

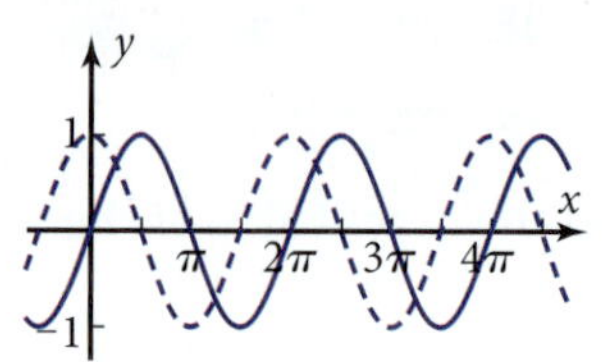

Figure 6-5g

a. What translation would make the cosine graph coincide with the sine graph? Complete the equation: $\sin x = \cos(\underline{\quad ? \quad})$.

b. Let $y = \cos(x - 2\pi)$. What effect does this translation have on the cosine graph?

c. Name a positive and a negative translation that would make the sine graph coincide with itself.

d. Explain why $\sin(x - 2\pi n) = \sin x$ for any integer n. How is the 2π related to the sine function?

e. Using dynamic geometry software such as The Geometer's Sketchpad, plot two sinusoids with different colors illustrating the concept of this problem, or use the *Sinusoid Translation* exploration at **www.keymath.com/precalc.** One sinusoid should be $y = \cos x$ and the other $y = \cos(x - k)$, where k is a slider or parameter with values between -2π and 2π. Describe what happens to the transformed graph as k varies.

46. *Sinusoid Dilation Problem*: Figure 6-5h shows the unit circle in a uv-coordinate system with angles of measure x and $2x$ radians. The uv-coordinate system is superimposed on an xy-coordinate system with sinusoids $y = \sin x$ (dashed) and an image graph $y = \sin 2x$ (solid).

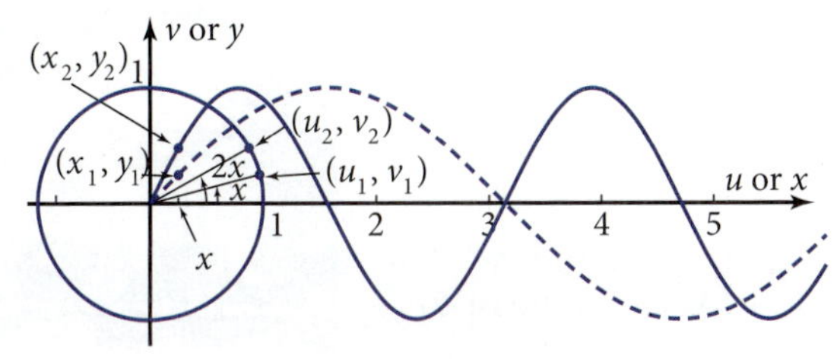

Figure 6-5h

a. Explain why the value of v for each angle is equal to the value of y for the corresponding sinusoid.

b. Create Figure 6-5h with dynamic geometry software such as Sketchpad, or go to **www.keymath.com/precalc** and use the *Sinusoid Dilation* exploration. Show the whole unit circle, and extend the x-axis to $x = 7$. Use a slider or parameter to vary the value of x. Is the second angle measure double the first one as x varies? Do the moving points on the two sinusoids have the same value of x?

c. Replace the 2 in sin $2x$ with a variable factor, k. Use a slider or parameter to vary k. What happens to the period of the (solid) image graph as k increases? As k decreases?

47. *Circular Function Comprehension Problem:* For circular functions such as cos x, the independent variable, x, represents the length of an arc of the unit circle. For other functions you have studied, such as the quadratic function $y = ax^2 + bx + c$, the independent variable, x, stands for a distance along a horizontal number line, the x-axis.

a. Explain how the concept of wrapping the x-axis around the unit circle links the two kinds of functions.

b. Explain how angle measures in radians link the circular functions to the trigonometric functions.

48. *The Inequality sin x < x < tan x Problem:* In this problem you will examine the inequality $\sin x < x < \tan x$ for $0 < x < \frac{\pi}{2}$. Figure 6-5i shows angle AOB in standard position, with subtended arc AB of length x on the unit circle.

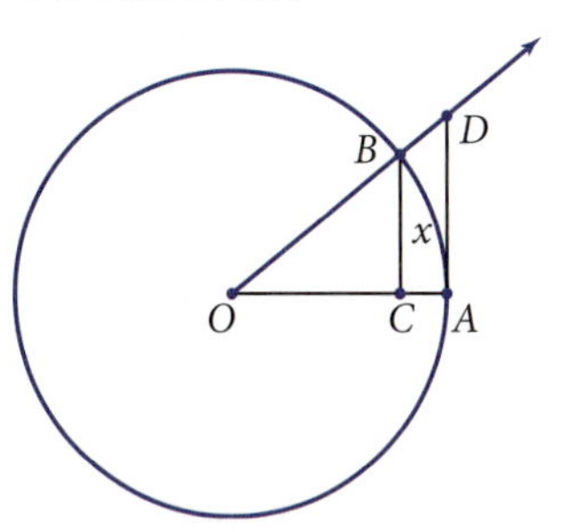

Figure 6-5i

a. Based on the definition of radians, explain why x is also the radian measure of angle AOB.

b. Based on the definitions of sine and tangent, explain why BC and AD equal sin x and tan x, respectively.

c. From Figure 6-5i it appears that $\sin x < x < \tan x$. Make a table of values to show numerically that this inequality is true even for values of x very close to zero.

d. Construct Figure 6-5i with dynamic geometry software such as Sketchpad, or go to **www.keymath.com/precalc** and use the *Inequality sin x < x < tan x* exploration. On your sketch, display the values of x and the ratios $(\sin x)/x$ and $(\tan x)/x$. What do you notice about the relative sizes of these values when angle AOB is in the first quadrant? What value do the two ratios seem to approach as angle AOB gets close to zero?

49. *Journal Problem:* Update your journal with things you have learned about the relationship between trigonometric functions and circular functions.

6-6 Inverse Circular Relations: Given y, Find x

Radar speed guns use inverse relations to calculate the speed of a car from time measurements.

A major reason for finding the particular equation for a sinusoid is to use it to evaluate y for a given x-value or to calculate x when you are given y. Functions are used this way to make predictions in the real world. For instance, you can express the time of sunrise as a function of the day of the year. With this equation, you can predict the time of sunrise on a given day by simply evaluating the expression. Predicting the day(s) on which the Sun rises at a given time is more complicated. In this section you will learn graphical, numerical, and algebraic ways to find x for a given value of y.

Objective Given the equation of a circular or trigonometric function and a particular value of y, find specified values of x or θ graphically, numerically, and algebraically.

The Inverse Cosine Relation

The symbol $\cos^{-1} 0.3$ means the inverse cosine function evaluated at 0.3, a particular arc or angle whose cosine is 0.3. By calculator, in radian mode,

$$\cos^{-1} 0.3 = 1.2661...$$

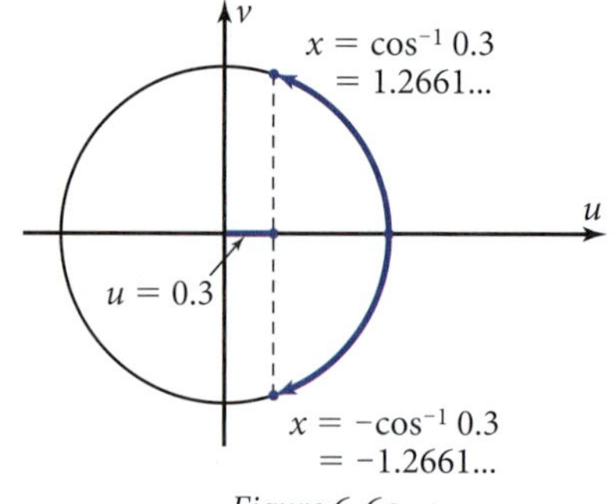

Figure 6-6a

The inverse cosine relation includes all arcs or angles whose cosine is a given number. The term that you'll use in this text is *arccosine,* abbreviated arccos. So arccos 0.3 means any arc or angle whose cosine is 0.3, not just the function value. Figure 6-6a shows that both 1.2661... and −1.2661... have cosines equal to 0.3. So −1.2661... is also a value of arccos 0.3.

The general solution for the arccosine of a number is written this way:

$$\arccos 0.3 = \pm\cos^{-1} 0.3 + 2\pi n \qquad \text{General solution for arccos 0.3.}$$

where n stands for an integer. The $\pm$ sign tells you that both the value from the calculator and its opposite are values of arccos 0.3. The $2\pi n$ tells you that any arc that is an integer number of revolutions added to these values is also a value of arccos 0.3. If n is a negative integer, a number of revolutions is being subtracted from these values. Note that there are infinitely many such values.

The arcsine and arctangent relations will be defined in Section 7-4 in connection with solving more general equations.

DEFINITION: Arccosine, the Inverse Cosine Relation

$$\arccos x = \pm\cos^{-1} x + 2\pi n \quad \text{or} \quad \arccos x = \pm\cos^{-1} x + 360°n$$

where n is an integer

Verbally: Inverse cosines come in opposite pairs with all their coterminal angles.

Note: The function value $\cos^{-1} x$ is called the **principal value** of the inverse cosine relation. This is the value the calculator is programmed to give. In Section 7-6, you will learn why certain quadrants are picked for these inverse function values.

EXAMPLE 1 ➤ Find the first five positive values of arccos(−0.3).

SOLUTION Assume that the inverse *circular* function is being asked for.

$$\begin{aligned} \arccos(-0.3) &= \pm\cos^{-1}(-0.3) + 2\pi n \\ &= \pm 1.8754... + 2\pi n && \text{By calculator.} \\ &= 1.8754..., 1.8754... + 2\pi, 1.8754... + 4\pi && \text{Use } \cos^{-1}(-0.3). \\ \text{or} \quad & -1.8754... + 2\pi, \ -1.8754... + 4\pi && \text{Use } -\cos^{-1}(-0.3). \\ &= 1.8754..., 8.1586..., 14.4418... \\ & \quad \text{or } 4.4076..., 10.6908... \\ &= 1.8754..., 4.4076..., 8.1586..., \\ & \quad 10.6908..., 14.4418... && \text{Arrange in ascending order.} \end{aligned}$$

Note: Do not round the value of $\cos^{-1}(-0.3)$ before adding the multiples of 2π.

Finding *x* When You Know *y*

Figure 6-6b shows a sinusoid with a horizontal line drawn at $y = 5$. The horizontal line cuts the part of the sinusoid shown at six different points. Each point corresponds to a value of x for which $y = 5$. The next examples show how to find the values of x by three methods.

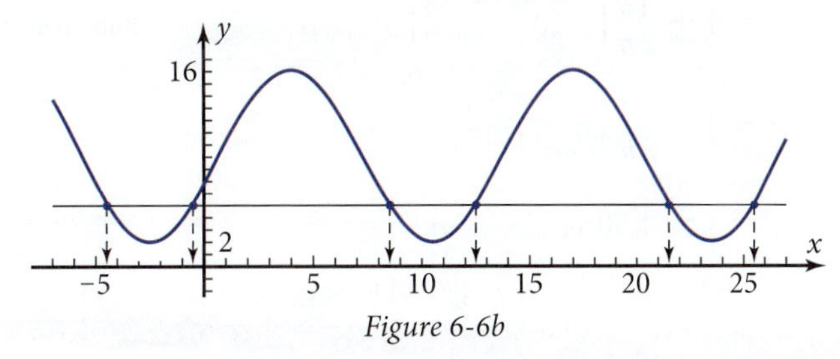

Figure 6-6b

EXAMPLE 2 ➤ Find *graphically* the six values of x for which $y = 5$ for the sinusoid in Figure 6-6b.

SOLUTION On the graph, draw vertical lines from the intersection points down to the x-axis (Figure 6-6b). The values are

$$x \approx -4.5, -0.5, 8.5, 12.5, 21.5, 25.5$$

EXAMPLE 3 ➤ Find *numerically* the six values of x in Example 2. Show that the answers agree with those found graphically in Example 2.

SOLUTION

$f_1(x) = 9 + 7\cos\frac{2\pi}{13}(x - 4)$ — Write the particular equation using the techniques of Section 6-5.

$f_2(x) = 5$ — Plot a horizontal line at $y = 5$.

$x \approx 8.5084...$ and $x \approx 12.4915...$ — Use the intersect or solver feature on your grapher to find two adjacent x-values.

$x \approx 8.5084... + 13(-1) \approx -4.4915...$ — Add multiples of the period to find other x-values.

$x \approx 12.4915... + 13(-1) \approx -0.5085...$

$x \approx 8.5084... + 13(1) \approx 21.5084...$

$x \approx 12.4915... + 13(1) \approx 25.4915...$

These answers agree with the answers found graphically in Example 2. ◄

Note that the $\approx$ sign is used for answers found numerically because the solver or intersect feature on most calculators gives only approximate answers.

EXAMPLE 4 ➤ Find *algebraically* (by calculation) the six values of x in Example 2. Show that the answers agree with those in Examples 2 and 3.

SOLUTION

$9 + 7\cos\frac{2\pi}{13}(x - 4) = 5$ — Set the two functions equal to each other.

$\cos\frac{2\pi}{13}(x - 4) = -\frac{4}{7}$ — Simplify the equation by isolating the cosine expression (start "peeling" constants away from x).

$\frac{2\pi}{13}(x - 4) = \arccos\left(-\frac{4}{7}\right)$ — Take the arccosine of both sides.

$x = 4 + \frac{13}{2\pi}\arccos\left(-\frac{4}{7}\right)$ — Rearrange the equation to isolate x (finish "peeling" constants away from x).

$x = 4 + \frac{13}{2\pi}\left(\pm\cos^{-1}\left(-\frac{4}{7}\right) + 2\pi n\right)$ — Substitute for arccosine.

$x = 4 \pm \frac{13}{2\pi}\cos^{-1}\left(-\frac{4}{7}\right) + 13n$ — Distribute the $\frac{13}{2\pi}$ over both terms.

$x = 4 \pm 4.5084... + 13n$

$x = 8.5084... + 13n$ or $-0.5084... + 13n$

$x = -4.4915..., -0.5084..., 8.5084...,$
$12.4915..., 21.5084..., 25.4915...$ — Let n be $0, \pm 1, \pm 2$.

These answers agree with the graphical and numerical solutions in Examples 2 and 3. ◄

Notes:

- In the term $13n$, the 13 is the period. The $13n$ in the general solution for x means that you need to add multiples of the period to the values of x you find for the inverse function.
- You can enter $8.5084... + 13n$ and $-0.5084... + 13n$ into your grapher and make a table of values. For most graphers you will have to use x in place of n.
- The algebraic solution gets all the values at once rather than one at a time numerically.

Problem Set 6-6

Reading Analysis

From what you have read in this section, what do you consider to be the main idea? Why does the arccosine of a number have more than one value while $\cos^{-1}$ of that number has only one value? What do you have to do to the inverse cosine value you get on your calculator in order to find other values of arccosine? Explain the sentence "Inverse cosines come in opposite pairs with all their coterminal angles" that appears in the definition box for arccosine.

Quick Review

Q1. What is the period of the circular function $y = \cos 4x$?

Q2. What is the period of the trigonometric function $y = \cos 4\theta$?

Q3. How many degrees are in $\frac{\pi}{6}$ radian?

Q4. How many radians are in 45°?

Q5. Sketch the graph of $y = \sin \theta$.

Q6. Sketch the graph of $y = \csc \theta$.

Q7. Find the smaller acute angle in a right triangle with legs of length 3 mi and 7 mi.

Q8. $x^2 + y^2 = 9$ is the equation of a(n) __?__.

Q9. What is the general equation for an exponential function?

Q10. Functions that repeat themselves at regular intervals are called __?__ functions.

For Problems 1–4, find the first five positive values of the inverse circular relation.

1. arccos 0.9 **2.** arccos 0.4

3. arccos(−0.2) **4.** arccos(−0.5)

For the circular sinusoids graphed in Problems 5–10,

a. Estimate graphically the x-values shown for the indicated y-value.

b. Find a particular equation for the sinusoid.

c. Find the x-values in part a numerically, using the equation from part b.

d. Find the x-values in part a algebraically.

e. Find the first value of x greater than 100 for which the sinusoid has the given y-value.

5. $y = 6$

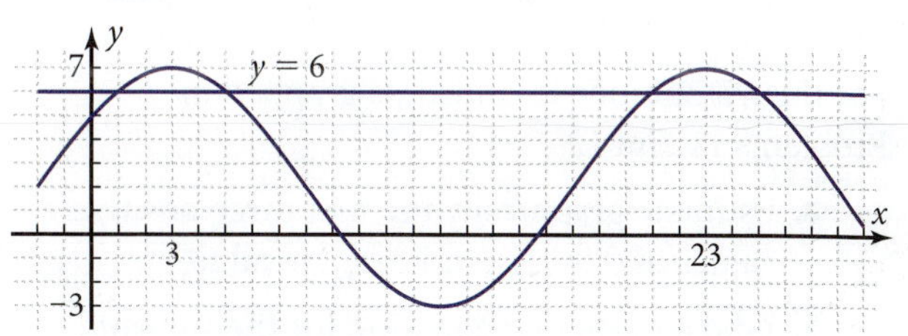

6. $y = 5$

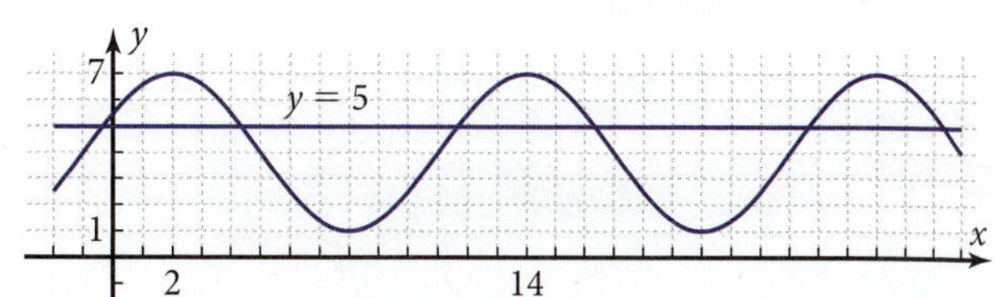

7. $y = -1$

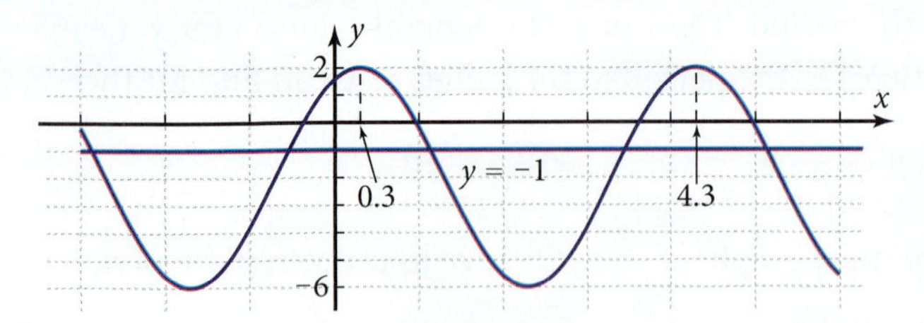

8. $y = -2$

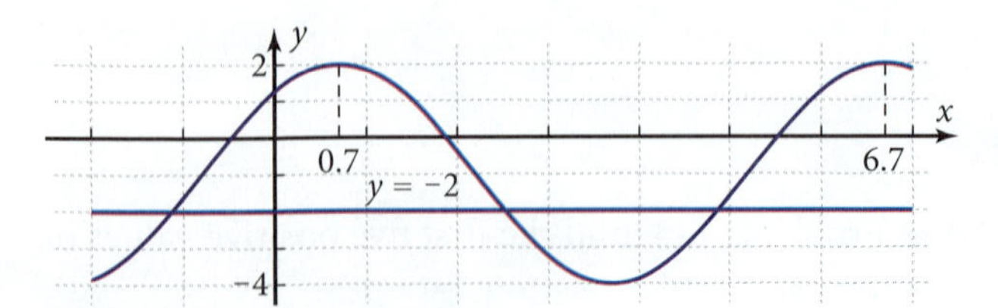

9. $y = 1.5$

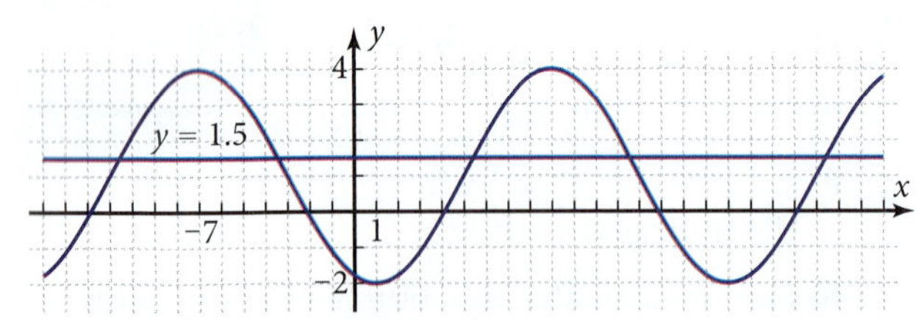

10. $y = -4$

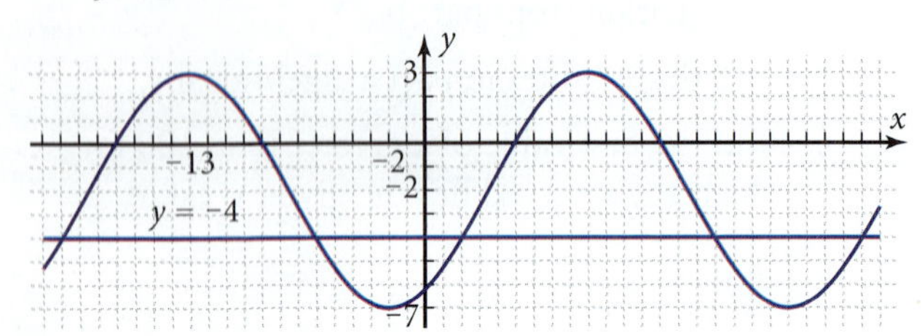

For the trigonometric sinusoids graphed in Problems 11 and 12,

a. Estimate graphically the first three positive values of θ for the indicated y-value.

b. Find a particular equation for the sinusoid.

c. Find the θ-values in part a numerically, using the equation from part b.

d. Find the θ-values in part a algebraically.

11. $y = 3$

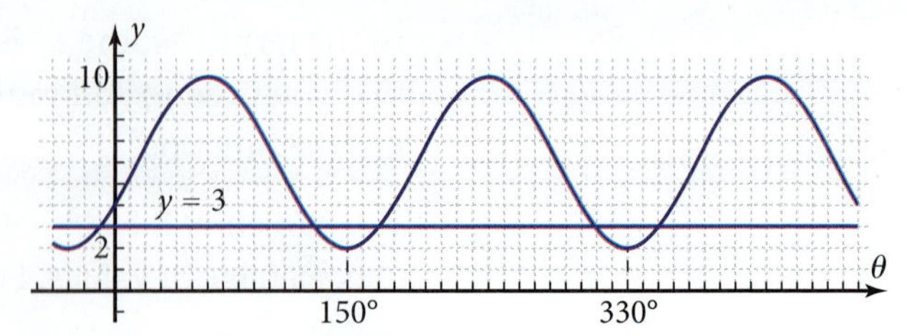

12. $y = 5$

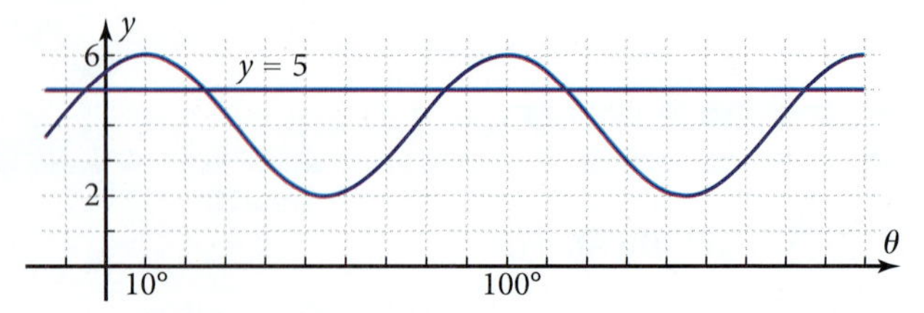

13. Figure 6-6c shows the graph of the parent cosine function, $y = \cos x$.

a. Find algebraically the six values of x shown on the graph for which $\cos x = -0.9$.

b. Find algebraically the first value of x greater than 200 for which $\cos x = -0.9$.

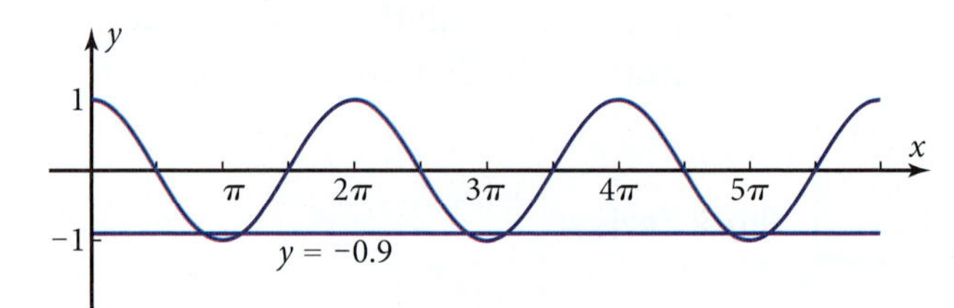

Figure 6-6c

6-7 Sinusoidal Functions as Mathematical Models

A chemotherapy treatment destroys red blood cells along with cancer cells. The red cell count goes down for a while and then comes back up again. If a treatment is taken every three weeks, then the red cell count resembles a periodic function of time (Figure 6-7a). If such a function is regular enough, you can use a sinusoidal function as a mathematical model.

In this section you'll start with a verbal description of a periodic phenomenon, interpret it graphically, find an algebraic equation from the graph, and use the equation to calculate numerical answers.

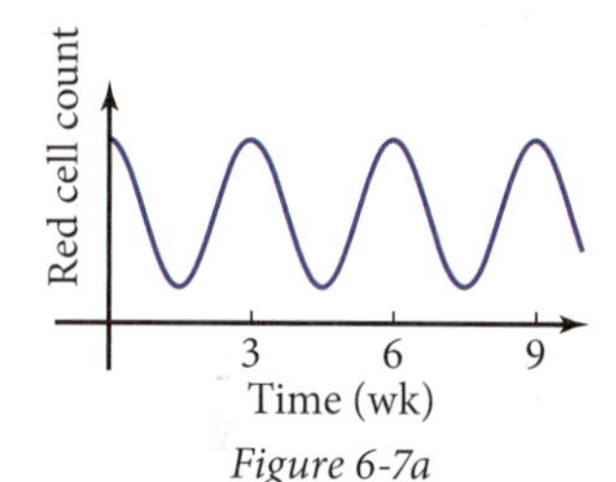

Figure 6-7a

Objective

Given a verbal description of a periodic phenomenon, write an equation using the sine or cosine function and use the equation as a mathematical model to make predictions and interpretations about the real world.

In this exploration you will use sinusoids to predict events in the real world.

EXPLORATION 6-7: Chemotherapy Problem

1. Figure 6-7a shows the red blood cell count for a patient taking chemotherapy treatments each 3 weeks. Suppose that a treatment is given at time $t = 0$ weeks, when the count is $c = 800$ units. The count drops to a low of 200, and then rises back to 800 when $t = 3$, at which time the next treatment is given. Write the equation for this sinusoidal function.
2. The patient feels "good" if the count is above 700. Use the equation in Problem 1 to predict the count 17 days after the treatment, and thus conclude whether she has started feeling good again at this time.
3. For what interval of times around 9 weeks will the patient feel good? For how many days does the good feeling last? Show how you got your answers.
4. What did you learn as a result of doing this exploration that you did not know before?

EXAMPLE 1 ➤ *Waterwheel Problem:* Suppose that the waterwheel in Figure 6-7b rotates at 6 revolutions per minute (rev/min). Two seconds after you start a stopwatch, point P on the rim of the wheel is at its greatest height, $d = 13$ ft, above the surface of the water. The center of the waterwheel is 6 ft above the surface.

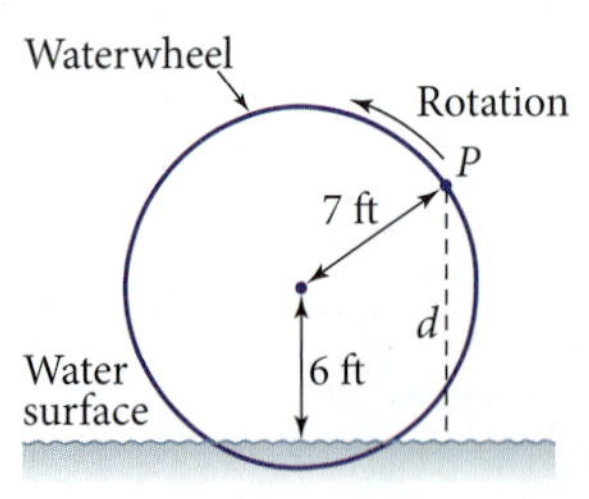

Figure 6-7b

a. Sketch the graph of d as a function of time t, in seconds, since you started the stopwatch.

b. Assuming that d is a sinusoidal function of t, write a particular equation. Confirm by grapher that your equation gives the graph in part a.

c. How high above or below the water's surface will point P be at time $t = 17.5$ s? At that time, will P be going up or down?

d. At what positive time t does point P first emerge from the water?

SOLUTION

a. From what's given, you can tell the location of the sinusoidal axis, the "high" and "low" points, and the period.

Sketch the sinusoidal axis at $d = 6$ as shown in Figure 6-7c.

Sketch the upper bound at $d = 6 + 7$, or 13, and the lower bound at $d = 6 - 7$, or -1.

Sketch a high point at $t = 2$. Because the waterwheel rotates at 6 rev/min, the period is $\frac{60}{6}$, or 10 s. Mark the next high point at $t = 2 + 10$, or 12.

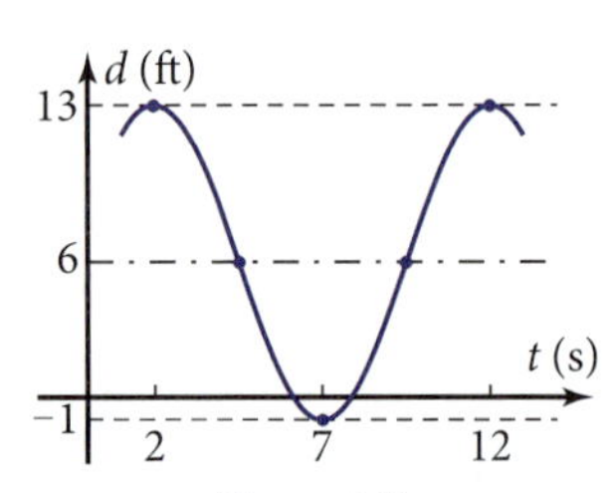

Figure 6-7c

Mark a low point halfway between the two high points, and mark the points of inflection on the sinusoidal axis halfway between each consecutive high and low.

Sketch the graph through the critical points and the points of inflection. Figure 6-7c shows the finished sketch.

b. $d = C + A\cos B(t - D)$ — Write the general equation. Use d and t for the variables.

From the graph, $C = 6$ and $A = 7$.

$D = 2$ — Cosine starts a cycle at a high point.

Horizontal dilation: $\frac{10}{2\pi} = \frac{5}{\pi}$ — The period of this sinusoid is 10; the period of the cosine function is 2π.

$B = \frac{\pi}{5}$ — B is the reciprocal of the horizontal dilation.

$\therefore d = 6 + 7\cos\frac{\pi}{5}(t - 2)$ — Write the particular equation.

Plotting on your grapher confirms that the equation is correct (Figure 6-7d).

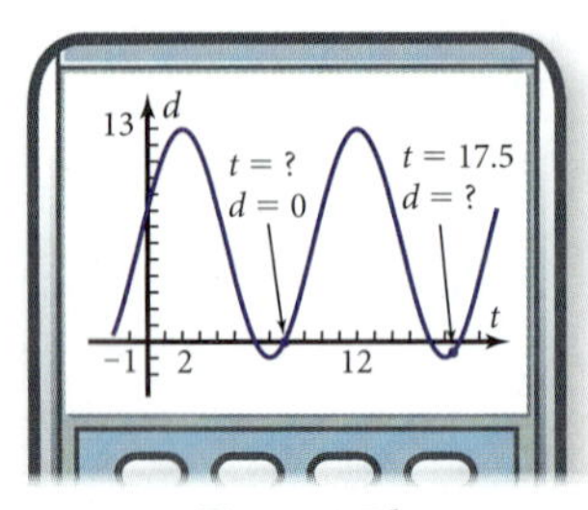

Figure 6-7d

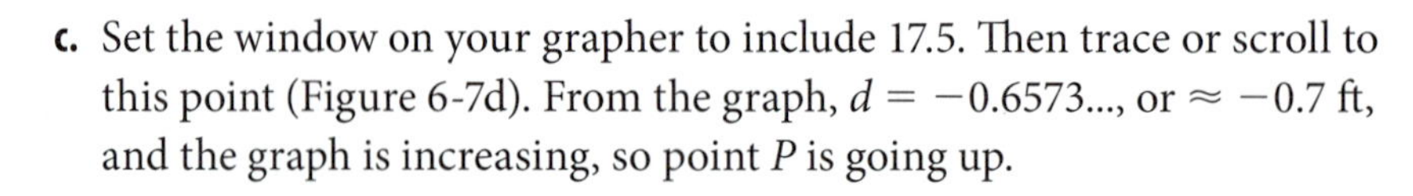

c. Set the window on your grapher to include 17.5. Then trace or scroll to this point (Figure 6-7d). From the graph, $d = -0.6573...$, or ≈ -0.7 ft, and the graph is increasing, so point P is going up.

d. Point P is either submerging into or emerging from the water when $d = 0$. At the first zero for positive t-values, shown in Figure 6-7d, the point is going into the water. At the next zero, the point is emerging. Using the intersect, zeros, or solver feature of your grapher, you'll find that the point is at

$$t = 7.8611... \approx 7.9 \text{ s}$$

If you go to **www.keymath.com/precalc,** you can view the *Waterwheel* exploration for a dynamic view of the waterwheel and the graph of d as a function of t.

Note that it is usually easier to use the cosine function for these problems, because its graph starts a cycle at a high point.

Problem Set 6-7

Reading Analysis

From what you have read in this section, what do you consider to be the main idea? What is the first step in solving a sinusoidal model problem that takes it out of the real world and puts it into the mathematical world? After you have taken this step, how does your work in this chapter allow you to answer questions about the real-world situation?

Quick Review

Problems Q1–Q8 concern the circular function $y = 4 + 5\cos\frac{\pi}{6}(x - 7)$.

Q1. The amplitude is __?__.

Q2. The period is __?__.

Q3. The frequency is __?__.

Q4. The sinusoidal axis is at $y =$ __?__.

Q5. The phase displacement with respect to the parent cosine function is __?__.

Q6. The upper bound is at $y =$ __?__.

Q7. If $x = 9$, then $y =$ __?__.

Q8. The first three positive x-values at which low points occur are __?__, __?__, and __?__.

Q9. Two values of $x = \arccos 0.5$ are __?__ and __?__.

Q10. If $y = 5 \cdot 3^x$, adding 2 to the value of x multiplies the value of y by __?__.

1. *Steamboat Problem:* Mark Twain sat on the deck of a river steamboat. As the paddle wheel turned, a point on the paddle blade moved so that its distance, d, in feet, from the water's surface was a sinusoidal function of time t, in seconds. When Twain's stopwatch read 4 s, the point was at its highest, 16 ft above the water's surface. The wheel's diameter was 18 ft, and it completed a revolution every 10 s.

a. Sketch the graph of the sinusoid.

b. What is the lowest the point goes? Why is it reasonable for this value to be negative?

c. Find a particular equation for distance as a function of time.

d. How far above the surface was the point when Mark's stopwatch read 17 s?

e. What is the first positive value of t at which the point was at the water's surface? At that time, was the point going into or coming out of the water? How can you tell?

f. "Mark Twain" is a pen name used by Samuel Clemens. What is the origin of that pen name? Give the source of your information.

2. *Fox Population Problem:* Naturalists find that populations of some kinds of predatory animals vary periodically with time. Assume that the population of foxes in a certain forest varies sinusoidally with time. Records started being kept at time $t = 0$ yr. A minimum number of 200 foxes appeared at $t = 2.9$ yr. The next maximum, 800 foxes, occurred at $t = 5.1$ yr.

   a. Sketch the graph of this sinusoid.

   b. Find a particular equation expressing the number of foxes as a function of time.

   c. Predict the fox population when $t = 7, 8, 9,$ and 10 yr.

   d. Suppose foxes are declared a vulnerable species when their population drops below 300. Between what two nonnegative values of t did the foxes first become vulnerable?

   e. Show on your graph in part a that your answer to part d is correct.

3. *Bouncing Spring Problem:* A weight attached to the end of a long spring is bouncing up and down (Figure 6-7e). As it bounces, its distance from the floor varies sinusoidally with time. You start a stopwatch. When the stopwatch reads 0.3 s, the weight first reaches a high point 60 cm above the floor. The next low point, 40 cm above the floor, occurs at 1.8 s.

   a. Sketch the graph of this sinusoidal function.

   b. Find a particular equation for distance from the floor as a function of time.

   c. What is the distance from the floor when the stopwatch reads 17.2 s?

   d. What was the distance from the floor when you started the stopwatch?

   e. What is the first positive value of time when the weight is 59 cm above the floor?

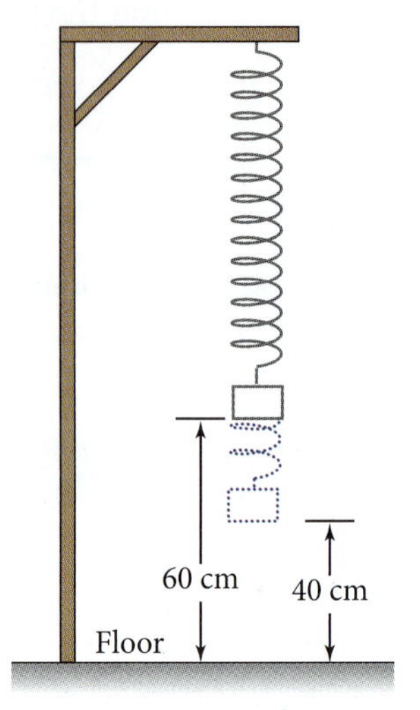

Figure 6-7e

4. *Rope Swing Problem:* Zoey is at summer camp. One day she is swinging on a rope tied to a tree branch, going back and forth alternately over land and water. Nathan starts a stopwatch. At time $x = 2$ s, Zoey is at one end of her swing, at a distance $y = -23$ ft from the riverbank (see Figure 6-7f). At time $x = 5$ s, she is at the other end of her swing, at a distance $y = 17$ ft from the riverbank. Assume that while she is swinging, y varies sinusoidally with x.

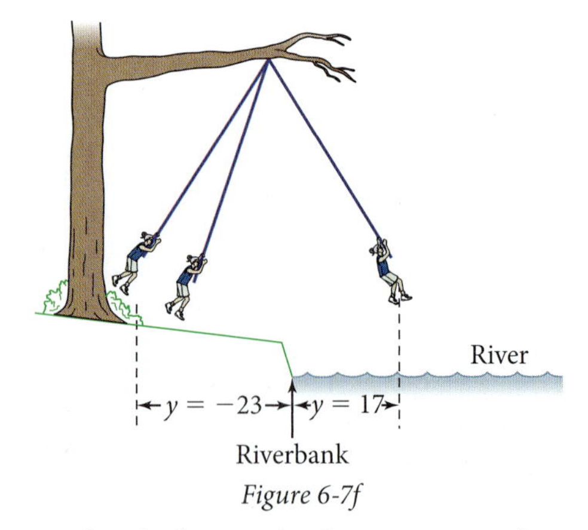

Figure 6-7f

   a. Sketch the graph of y versus x and write a particular equation.

b. Find y when $x = 13.2$ s. Was Zoey over land or over water at this time?

c. Find the first positive time when Zoey was directly over the riverbank ($y = 0$).

d. Zoey lets go of the rope and splashes into the water. What is the value of y for the end of the rope when it comes to rest? What part of the mathematical model tells you this?

5. *Roller Coaster Problem:* A theme park is building a portion of a roller coaster track in the shape of a sinusoid (Figure 6-7g). You have been hired to calculate the lengths of the horizontal and vertical support beams.

a. The high and low points of the track are separated by 50 m horizontally and 30 m vertically. The low point is 3 m below the ground. Let y be the distance, in meters, a point on the track is above the ground. Let x be the horizontal distance, in meters, a point on the track is from the high point. Find a particular equation for y as a function of x.

b. The vertical support beams are spaced 2 m apart, starting at the high point and ending just before the track goes below the ground. Make a table of values of the lengths of the beams.

c. The horizontal beams are spaced 2 m apart, starting at ground level and ending just below the high point. Make a table of values of horizontal beam lengths.

d. The builder must know how much support beam material to order. In the most time-efficient way, find the total length of the vertical beams and the total length of the horizontal beams.

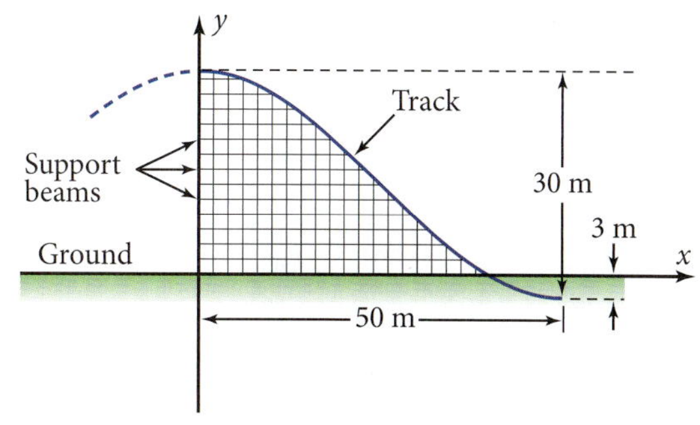

Figure 6-7g

6. *Buried Treasure Problem:* Suppose you seek a treasure that is buried in the side of a mountain. The mountain range has a sinusoidal vertical cross section (Figure 6-7h). The valley to the left is filled with water to a depth of 50 m, and the top of the mountain is 150 m above the water level. You set up an x-axis at water level and a y-axis 200 m to the right of the deepest part of the water. The top of the mountain is at distance $x = 400$ m.

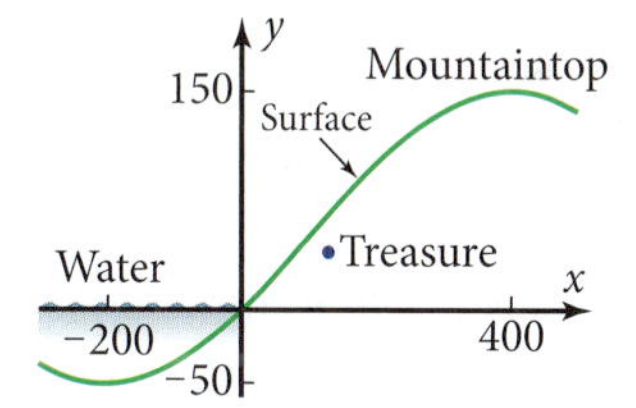

Figure 6-7h

a. Find a particular equation expressing y for points on the *surface* of the mountain as a function of x.

b. Show algebraically that the sinusoid in part a contains the origin, (0, 0).

c. The treasure is located beneath the surface at the point (130, 40), as shown in Figure 6-7h. Which would be the shorter way to dig to the treasure, a horizontal tunnel or a vertical tunnel? Show your work.

7. WEB *Sunspot Problem:* For several hundred years, astronomers have kept track of the number of sunspots that occur on the surface of the Sun. The number of sunspots in a given year varies periodically, from a minimum of about 10 per year to a maximum of about 110 per year. Between 1750 and 1948, there were exactly 18 complete cycles.

a. What is the period of a sunspot cycle?

b. Assume that the number of sunspots per year is a sinusoidal function of time and that a maximum occurred in 1948. Find a particular equation expressing the number of sunspots per year as a function of the year.

c. How many sunspots will there be in the year 2020? This year?

d. What is the first year after 2020 in which there will be about 35 sunspots? What is the first year after 2020 in which there will be a maximum number of sunspots?

e. Find out how closely the sunspot cycle resembles a sinusoid by looking on the Internet or in another reference.

8. *Tide Problem:* Suppose you are on the beach at Port Aransas, Texas, on August 2. At 2:00 p.m., at high tide, you find that the depth of the water at the end of a pier is 1.5 m. At 7:30 p.m., at low tide, the depth of the water is 1.1 m. Assume that the depth varies sinusoidally with time.

a. Find a particular equation expressing depth as a function of the time that has elapsed since 12:00 a.m. August 2.

b. Use your mathematical model to predict the depth of the water at 5:00 p.m. on August 3.

c. At what time does the first low tide occur on August 3?

d. What is the earliest time on August 3 that the water depth will be 1.27 m?

e. A high tide occurs because the Moon is pulling the water away from Earth slightly, making the water a bit deeper at a given point. How do you explain the fact that there are *two* high tides each day at most places on Earth? Provide the source of your information.

9. *Shock Felt Round the World Problem:* Suppose that one day all 300+ million people in the United States climb up on tables. At time $t = 0$, they all jump off. The resulting shock wave starts Earth vibrating at its fundamental period, 54 min. The surface first moves down from its normal position and then moves up an equal distance above its normal position (Figure 6-7i). Assume that the amplitude is 50 m.

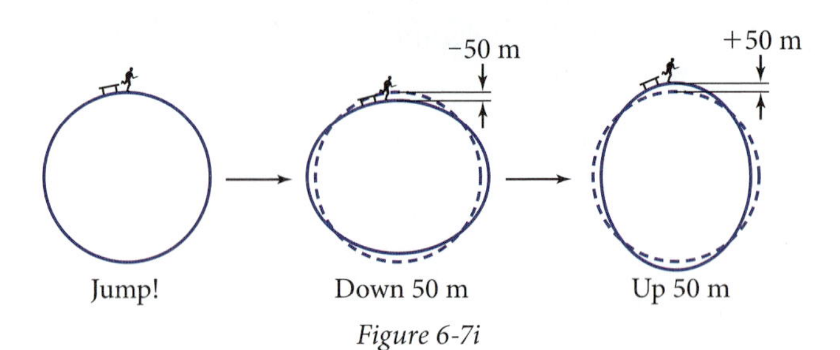

Figure 6-7i

a. Sketch the graph of the displacement of the surface from its normal position as a function of time elapsed since the people jumped.

b. At what time will the surface be farthest above its normal position?

c. Write a particular equation expressing displacement above normal position as a function of time elapsed since the jump.

d. What is the displacement at time $t = 21$ min?

e. What are the first three positive times at which the displacement is -37 m?

10. *Island Problem:* Ona Nyland owns an island several hundred feet from the shore of a lake. Figure 6-7j shows a vertical cross section through the shore, lake, and island. The island was formed millions of years ago by stresses that caused the earth's surface to warp into the sinusoidal pattern shown. The highest point on the shore is at distance $x = -150$ ft. From measurements on and near the shore (solid part for the graph), topographers find that an equation for the sinusoid is

$$y = -70 + 100 \cos \frac{\pi}{600}(x + 150)$$

where x and y are distance in feet. Ona consults you to make predictions about the rest of the graph (dashed).

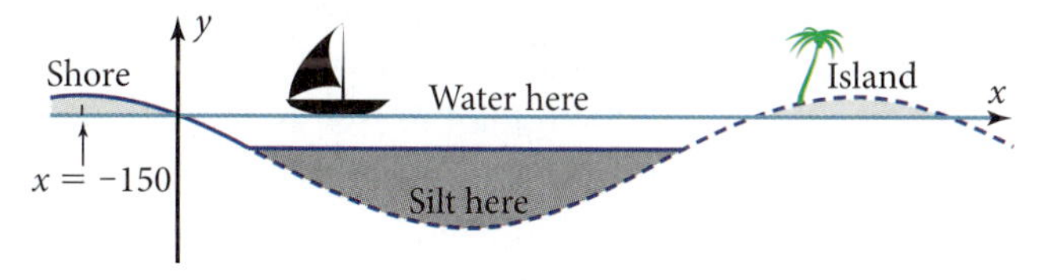

Figure 6-7j

a. What is the highest the island rises above the water level in the lake? How far from the y-axis is this high point? Show how you got your answers.

b. What is the deepest the sinusoid goes below the water level in the lake? How far from the y-axis is this low point? Show how you got your answers.

c. Over the centuries silt has filled the bottom of the lake so that the water is only 40 ft deep. That is, the silt line is at $y = -40$ ft. Plot the graph using a suitable domain and range. Then find graphically the interval of x-values between which Ona would expect to find silt if she goes scuba diving in the lake.

d. If Ona drills an offshore well at distance $x = 700$ ft, through how much silt would she drill before she reaches the sinusoid? Show how you got your answer.

e. The sinusoid appears to go through the origin. Does it actually do so, or does it just miss? Justify your answer.

f. Find algebraically the interval of x-values between which the island is at or above the water level. How wide is the island, from the water on one side to the water on the other?

11. *Pebble in the Tire Problem:* As you stop your car at a traffic light, a pebble becomes wedged between the tire treads. When you start moving again, the distance between the pebble and the pavement varies sinusoidally with the distance you have gone. The period is the circumference of the tire. Assume that the diameter of the tire is 24 in.

a. Sketch the graph of this sinusoidal function.

b. Find a particular equation for the function. (It is possible to get an equation with zero phase displacement.)

c. What is the pebble's distance from the pavement when you have gone 15 in.?

d. What are the first two distances you have gone when the pebble is 11 in. from the pavement?

12. *Oil Well Problem:* Figure 6-7k shows a vertical cross section through a piece of land. The y-axis is drawn coming out of the ground at the fence bordering land owned by your boss, Earl Wells. Earl owns the land to the left of the fence and is interested in acquiring land on the other side to drill a new oil well. Geologists have found an oil-bearing formation below Earl's land that they believe to be sinusoidal in shape. At distance $x = -100$ ft, the top surface of the formation is at its deepest, $y = -2500$ ft. A quarter-cycle closer to the fence, at distance $x = -65$ ft, the top surface is only 2000 ft deep. The first 700 ft of land beyond the fence is inaccessible. Earl wants to drill at the first convenient site beyond $x = 700$ ft.

a. Find a particular equation expressing y as a function of x.

b. Plot the graph on your grapher. Use a window with $-100 \leq x \leq 900$. Describe how the graph confirms that your equation is correct.

c. Find graphically the first interval of x-values in the available land for which the top surface of the formation is no more than 1600 ft deep.

d. Find algebraically the values of x at the ends of the interval in part c. Show your work.

e. Suppose the original measurements were slightly inaccurate and the value of y shown at -65 ft is actually at $x = -64$ ft. Would this fact make much difference in the answer to part c? Use the most time-efficient method to arrive at your answer. Explain what you did.

13. *Sound Wave Problem:* The hum you hear on some radios when they are not tuned to a station is a sound wave of 60 cycles per second.

Bats navigate and communicate using ultrasonic sounds with frequencies of 20–100 kilohertz (kHz), which are undetectable by the human ear. A kilohertz is 1000 cycles per second.

a. Is 60 cycles per second the period, or is it the frequency? If it is the period, find the frequency. If it is the frequency, find the period.

b. The *wavelength* of a sound wave is defined as the distance the wave travels in a time interval equal to one period. If sound travels at 1100 ft/s, find the wavelength of the 60-cycle-per-second hum.

c. The lowest musical note the human ear can hear is about 16 cycles per second. In order to play such a note, a pipe on an organ must be exactly half as long as the wavelength. What length organ pipe would be needed to generate a 16-cycle-per-second note?

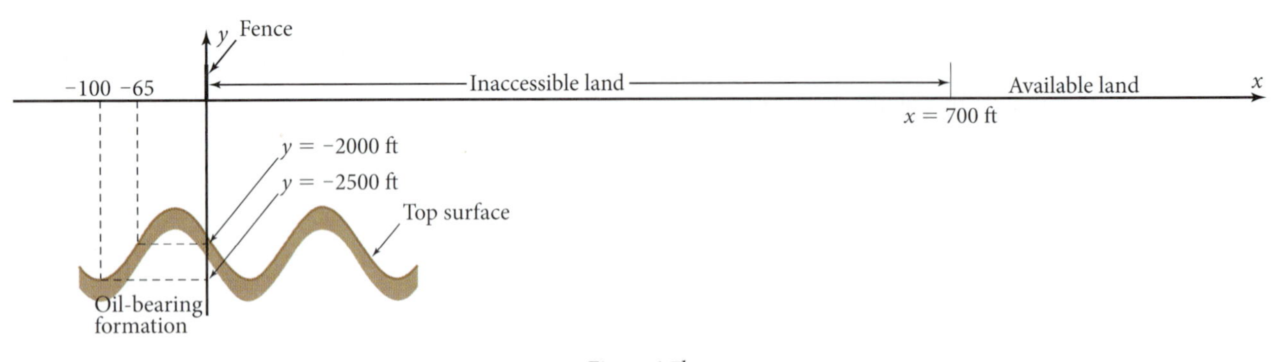

Figure 6-7k

14. WEB *Sunrise Project:* Assume that the time of sunrise varies sinusoidally with the day of the year. Let t be the time of sunrise. Let d be the day of the year, starting with $d = 1$ on January 1.

a. On the Internet or from an almanac, find for your location the time of sunrise on the longest day of the year, June 21, and on the shortest day of the year, December 21. If you choose, you can use the data for San Antonio, 5:34 a.m. and 7:24 a.m., CST, respectively. The phase displacement for cosine will be the value of d at which the Sun rises the latest. Use the information to find a particular equation expressing time of sunrise as a function of the day number.

b. Calculate the time of sunrise today at the location you used in part a. Compare the answer to your data source.

c. What is the time of sunrise on your birthday, taking daylight saving time into account?

d. What is the first day of the year on which the Sun rises at 6:07 a.m. in the location in part a?

e. In the northern hemisphere, Earth moves faster in wintertime, when it is closer to the Sun, and slower in summertime, when it is farther from the Sun. As a result, the actual high point of the sinusoid occurs later than predicted, and the actual low point occurs earlier than predicted (Figure 6-7l). A representation of the actual graph can be plotted by putting in a phase displacement that *varies*. See if you can duplicate the graph in Figure 6-7l on your grapher. Is the modified graph a better fit for the actual sunrise data for the location in part a?

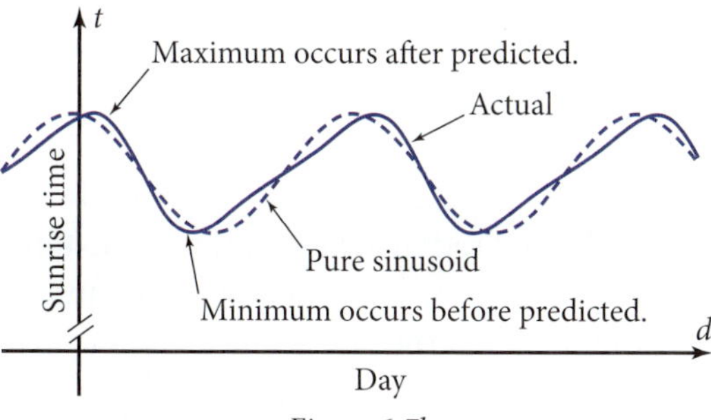

Figure 6-7l

15. *Variable Amplitude Pendulum Project:* If there were no friction, the displacement of a pendulum from its rest position would be a sinusoidal function of time,

$$y = A \cos Bt$$

To account for friction, assume that the amplitude, A, decreases exponentially with time:

$$A = a \cdot b^t$$

Make a pendulum by tying a weight to a string hung from the ceiling or some other convenient place (see Figure 6-7m).

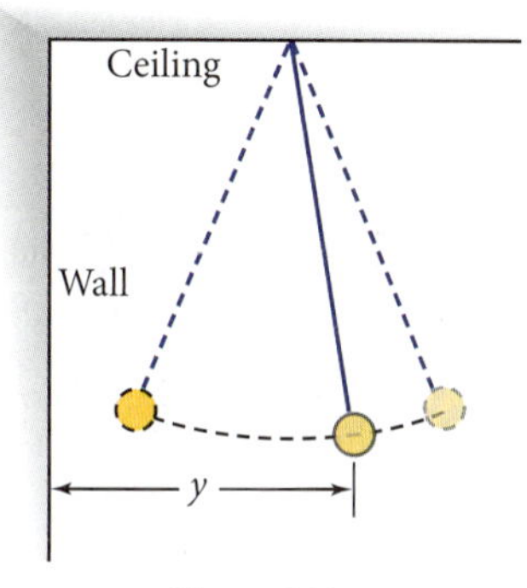

Figure 6-7m

Find its period by measuring the time for 10 swings and dividing by 10. Record the amplitude when you first start the pendulum, and measure it again after 30 s. From these measurements, find the constants a, b, and B and write a particular equation expressing the position of the pendulum as a function of time. Test your equation by using it to predict the displacement of the pendulum at time $t = 10$ s and seeing if the pendulum really is where you predicted it to be at that time. Write an entry in your journal describing this experiment and your results.

6-8 Rotary Motion

When you ride a merry-go-round, you go faster when you sit nearer the outside. As the merry-go-round rotates through a certain angle, you travel farther in the same amount of time when you sit closer to the outside (Figure 6-8a).

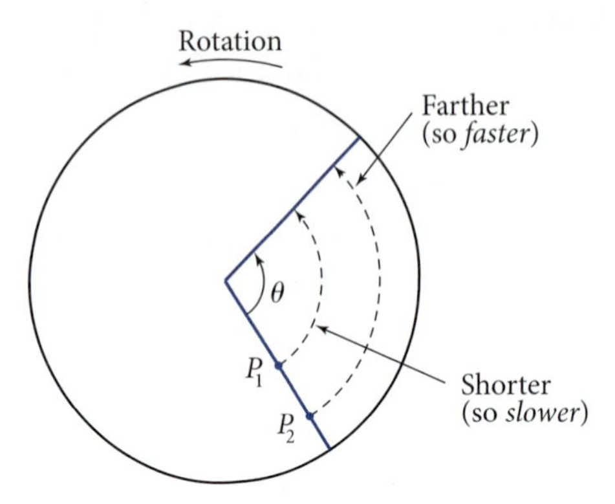

Figure 6-8a

However, all points on the merry-go-round turn through the *same* number of degrees per unit of time. So there are two different kinds of speed, or velocity, associated with a point on a rotating object. The **angular velocity** is the number of degrees or radians per unit of time, and the **linear velocity** is the distance per unit of time.

Objective Given information about a rotating object or connected rotating objects, find linear and angular velocities of points on the objects.

In this exploration you will practice computing and interpreting linear and angular velocities.

EXPLORATION 6-8: Angular and Linear Velocity

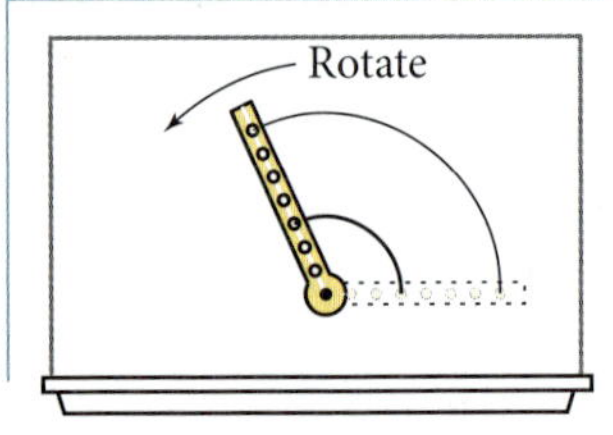

The figure shows a rotating ruler attached by suction cup to a dry-erase board. Marking pens are put into holes in the ruler, and the ruler and pens are rotated slowly from the initial position (dotted), reaching the final position (solid) after 5 s. Each pen leaves an arc on the board.

1. With a flexible ruler you find that the curved lengths of the arcs are 60 cm (outer) and 24 cm (inner). At what average speed (cm/s) did each pen travel in the 5 s? Which pen moved faster?
2. Suppose the angle from the initial position to the final position is 115°. At what number of degrees per second did each pen move? Did one pen move more degrees per second than the other?
3. Suppose the inner pen is 12 cm from the center of rotation and the outer pen is 30 cm from the center of rotation. An angle of 115° is about 2 radians. Show that the distance each pen moved can be found by multiplying the radius of the arc by the angle measure in radians.

continued

4. The numbers of cm/s in Problem 1 are *linear velocities.* The number of deg/s in Problem 2 is an *angular velocity.* What is the angular velocity in radians per second?
5. Show that you can find each linear velocity in Problem 1 by multiplying the radius of the arc by the angular velocity in radians per second.
6. Write the definitions of linear velocity and angular velocity. Then write some conclusions about angular velocity and linear velocity of points on the same rotating object, and how the two kinds of velocity are related.
7. What did you learn as a result of doing this exploration that you did not know before?

To reduce rotary motion to familiar algebraic terms, certain symbols are usually used for radius, arc length, angle measure, linear velocity, angular velocity, and time (Figure 6-8b). They are

r Radius from the center of rotation to the point in question

a Number of units of arc length through which the point moves

θ Angle through which the point rotates (usually in radians, but not always)

v Linear velocity, in distance per unit of time

ω Angular velocity (often in radians per unit of time; Greek "omega")

t Length of time to rotate through a particular angle θ

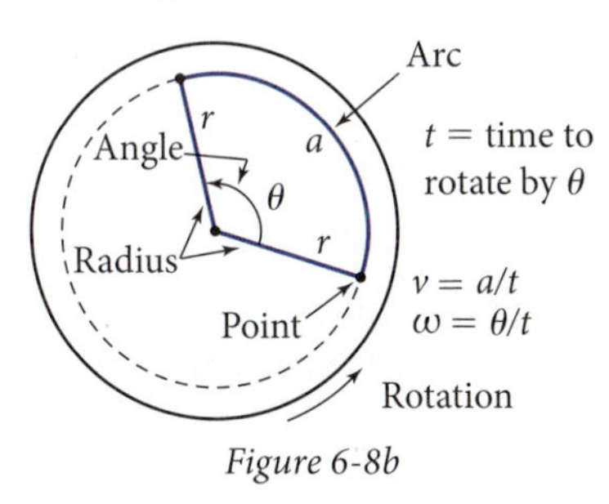

Figure 6-8b

These definitions relate the variables.

DEFINITIONS: Angular Velocity and Linear Velocity

The **angular velocity,** ω, of a point on a rotating object is the number of degrees (radians, revolutions, and so on) through which the point turns per unit of time.

The **linear velocity,** v, of a point on a rotating object is the distance the point travels along its circular path per unit of time.

Algebraically: $\omega = \frac{\theta}{t}$ and $v = \frac{a}{t}$

Properties of linear and angular velocity help you accomplish this section's objective. First, by the definition of radians, the length of an arc of a circle is equal to the radius multiplied by the radian measure of the central angle.

In physics, θ is used for angles, even if the angle is measured in radians. Because you might study rotary motion elsewhere, you'll see the same notation here.

$$a = r\theta \qquad \theta \text{ must be in radians.}$$

$$\frac{a}{t} = \frac{r\theta}{t} = r \cdot \frac{\theta}{t} \qquad \text{Divide both sides of the equation by time.}$$

By definition, the left side equals the linear velocity, v, and the right side is r multiplied by the angular velocity, ω. So you can write the equation

$$v = r\omega \qquad \omega \text{ must be in radians per unit of time.}$$

PROPERTIES: Linear Velocity and Angular Velocity

If θ is in radians and ω is in radians per unit of time, then

$$a = r\theta$$

$$v = r\omega$$

Analysis of a Single Rotating Object

EXAMPLE 1 ➤ An old LP ("long play") record, as in Figure 6-8c, rotates at $33\frac{1}{3}$ rev/min.

a. Find the angular velocity in radians per second.

b. Find the angular and linear velocities of the record (per second) at the point at which the needle is located when it is just starting to play, 14.5 cm from the center.

Figure 6-8c

c. Find the angular and linear velocities (per second) at the center of the turntable.

SOLUTION

a. The $33\frac{1}{3}$ rev/min is already an angular velocity because it is a number of revolutions (angle) per unit of time. All you need to do is change to the desired units. For this purpose, it is helpful to use dimensional analysis. There are 2π radians in one revolution and 60 seconds in 1 minute. Write the conversion factors this way:

$$\omega = \frac{33\frac{1}{3}\text{ rev}}{\text{min}} \cdot \frac{2\pi \text{ rad}}{\text{rev}} \cdot \frac{1 \text{ min}}{60 \text{ s}} = 1\frac{1}{9}\pi = 3.4906... \approx 3.49 \text{ rad/s}$$

Notice that the revolutions and minutes cancel, leaving radians in the numerator and seconds in the denominator.

b. All points on the same rotating object have the same angular velocity. So the point 14.5 cm from the center is also rotating at $\omega = 1\frac{1}{9}\pi$ radians per second. The computation of linear velocity is

$$v = r\omega = \frac{14.5 \text{ cm}}{\text{rad}} \cdot \frac{1\frac{1}{9}\pi \text{ rad}}{\text{s}} = 50.6145... \approx 50.6 \text{ cm/s}$$

Note that for the purpose of dimensional analysis, the radius has the units cm/rad. A point 14.5 cm from the center moves 14.5 cm along the arc for each radian the record rotates.

c. The turntable and record rotate as a single object. So all points on the turntable have the same angular velocity as the record, even the point that is the center of the turntable. The radius to the center is, of course, zero. So

$$\omega = 1\frac{1}{9}\pi \approx 3.49 \text{ rad/s}$$
$$v = r\omega = (0)(1\frac{1}{9}\pi) = 0 \text{ cm/s}$$

Interestingly, the center of a rotating object has zero linear velocity, but it still rotates with the same angular velocity as all other points on the object.

Analysis of Connected Rotating Objects

Figure 6-8d shows the back wheel of a bicycle. A small sprocket is connected to the axle of the wheel. This sprocket is connected by a chain to the large sprocket to which the pedals are attached. So there are several rotating objects whose motions are related to one another. Example 2 shows you how to analyze the motion.

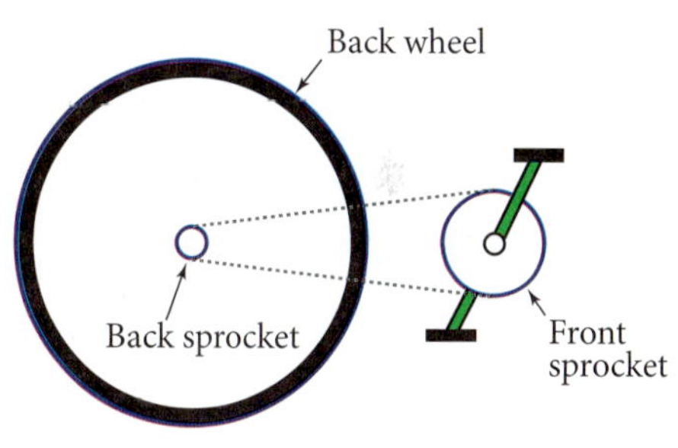

Figure 6-8d

EXAMPLE 2 ➤ A cyclist turns the pedals of her bicycle (Figure 6-8d) at 8 rad/s. The front sprocket has diameter 20 cm and is connected by the chain to the back sprocket, which has diameter 6 cm. The rear wheel has radius 35 cm and is connected to the back sprocket.

a. What is the angular velocity of the front sprocket?

b. What is the linear velocity of points on the chain?

c. What is the linear velocity of points on the rim of the back sprocket?

d. What is the angular velocity of the center of the back sprocket?

e. How fast is the bicycle going in kilometers per hour?

SOLUTION

a. $\omega = 8$ rad/s

Because the pedals and the front sprocket are connected at their axles, they rotate as one object. All points on the same rotating object have the same angular velocity.

b. $v = r\omega = \dfrac{10 \text{ cm}}{\text{rad}} \cdot \dfrac{8 \text{ rad}}{\text{s}} = 80 \text{ cm/s}$

The linear velocity of points on the chain is the same as the linear velocity of points on the rim of the front sprocket. The radius of the front sprocket is $\frac{20}{2}$, or 10 cm.

c. $v = 80$ cm/s

The back sprocket's rim has the same linear velocity as the chain and the front sprocket's rim.

d. $v = r\omega \Rightarrow \omega = \dfrac{v}{r} = \dfrac{80 \text{ cm}}{\text{s}} \cdot \dfrac{\text{rad}}{3 \text{ cm}} = 26\frac{2}{3}$ rad/s

The angular velocity is the same at every point on the same rotating object, even at the center. So the angular velocity at the center of the back sprocket is the same as at the rim. You can calculate this angular velocity using the equation $v = r\omega$. The radius is 3 cm, half the diameter.

e. $v = r\omega = \dfrac{35 \text{ cm}}{\text{rad}} \cdot \dfrac{26\frac{2}{3} \text{ rad}}{\text{s}} \cdot \dfrac{3{,}600 \text{ s}}{\text{h}} \cdot \dfrac{1 \text{ km}}{100{,}000 \text{ cm}} = 33.6$ km/h

The wheel is connected by an axle to the back sprocket, so it rotates with the same angular velocity as the sprocket. Unless the wheel is skidding, the speed the bicycle goes is the same as the linear velocity of points on the rim of the wheel. You can calculate this linear velocity using the equation $v = r\omega$.

From Example 2, you can draw some general conclusions about rotating objects connected either at their rims or by an axle.

CONCLUSIONS: Connected Rotating Objects

Two rotating objects connected by an axle have the same angular velocity.

Two rotating objects connected at their rims have the same linear velocity at their rims.

Problem Set 6-8

Reading Analysis

From what you have read in this section, what do you consider to be the main idea? Give a real-world example involving rotary motion. What is the difference between linear velocity and angular velocity? Explain why it is possible for one type of velocity to equal zero when the other does not equal zero.

Quick Review

Q1. A runner goes 1000 m in 200 s. What is her average speed?

Q2. A skater rotates 3000° in 4 s. How fast is he rotating?

Q3. If one value of $\theta = \arccos x$ is 37°, then another value of θ for $0° \le \theta \le 360°$ is __?__.

Q4. If one value of $y = \arccos x$ is 1.2 radians, then the first negative value of y is __?__.

Q5. What is the period of the function $y = 7 + 4 \cos 2(x - 5)$?

Q6. What transformation of function f is $g(x) = f(0.2x)$?

Q7. Sketch a right triangle with hypotenuse 8 cm and one leg 4 cm. How long is the other leg?

Q8. What are the measures of the angles of the triangle in Problem Q7?

Q9. Factor: $x^2 - 11x + 10$

Q10. Find the next term in the geometric sequence 3, 6, ...

1. *Hammer Throw Problem:* An athlete spins around in the hammer throw event to propel the hammer. In order for the hammer to land where he wants, it must leave his hand at a speed of 60 ft/s. Assume that the hammer is 4 ft from his center of rotation.

   a. How many radians per second must he rotate to achieve his objective?

   b. How many revolutions per minute must he rotate?

2. *Ship's Propeller Problem:* The propeller on a freighter has a radius of about 4 ft (Figure 6-8e). At full speed, the propeller turns at 150 rev/min.

Figure 6-8e

   a. What is the angular velocity of the propeller in radians per second at the tip of the blades? At the center of the propeller?

   b. What is the linear velocity in feet per second at the tip of the blades? At the center of the propeller?

3. *Lawn Mower Blade Problem:* The blade on a rotary lawn mower is 19 in. long. The cutting edges begin 6 in. from the center of the blade (Figure 6-8f). In order for a lawn mower blade to cut grass, it must strike the grass at a speed of at least 900 in./s.

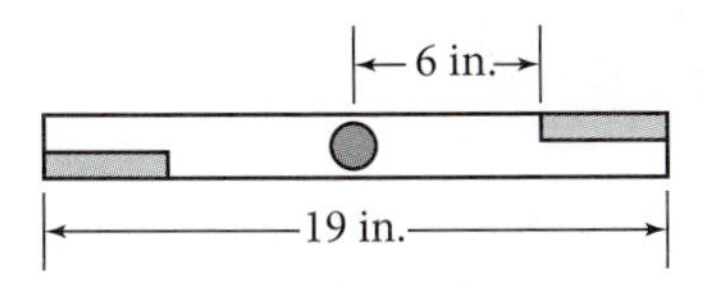

Figure 6-8f

   a. If you want the innermost part of the cutting edge to cut grass, how many radians per second must the blade turn? How many revolutions per minute is this?

   b. What is the linear velocity of the outermost tip of the blade while it is turning as in part a?

   c. If the outermost tip of the blade strikes a stone while it is turning as in part a, how fast could the stone be propelled from the mower? How many miles per hour is this?

4. *Bicycle Problem:* Rhoda rides a racing bike at a speed of 50.4 km/h. The wheels have diameter 70 cm.

   a. What is the linear velocity of the points farthest out on the wheels?

   b. Find the angular velocity of the wheels in radians per second.

   c. Find the angular velocity of the wheels in revolutions per minute.

5. *Dust Problem:* A speck of dust is sitting 4 cm from the center of a turntable. Phoebe spins the turntable through an angle of 120°.

   a. Through how many radians does the speck of dust turn?

   b. What distance does it travel?

   c. If Phoebe rotates the turntable 120° in 0.5 s, what is the dust speck's angular velocity? What is its linear velocity?

6. *Seesaw Problem:* Stan and his older brother Ben play on a seesaw. Stan sits at a point 8 ft from the pivot. On the other side of the seesaw, Ben, who is heavier, sits just 5 ft from the pivot. As Ben goes up and Stan goes down, the seesaw rotates through an angle of 37° in 0.7 s.

   a. What are Ben's angular velocity in radians per second and linear velocity in feet per second?

   b. What are Stan's angular and linear velocities?

7. *Figure Skating Problem:* Ima N. Aspin goes figure skating. She goes into a spin with her arms outstretched, making four complete revolutions in 6 s.

   a. How fast is she rotating in revolutions per second?

   b. Find Ima's angular velocity in radians per second.

   c. Ima's outstretched fingertips are 70 cm from the central axis of her body (around which she rotates). What is the linear velocity of her fingertips?

d. As Ima spins there are points on her body that have *zero* linear velocity. Where are these points? What is her angular velocity at these points?

e. Ima pulls her arms in close to her body, just 15 cm from her axis of rotation. As a result, her angular velocity increases to 10 rad/s. Are her fingertips going faster or slower than they were in part c? Justify your answer.

8. *Paper Towel Problem:* In 0.4 s, Dwayne pulls from the roll three paper towels with total length 45 cm.

a. How fast is he pulling the paper towels?

b. The roll of towels has diameter 14 cm. What is the linear velocity of a point on the outside of the roll?

c. What is the angular velocity of a point on the outside of the roll?

d. At how many revolutions per minute is the roll of towels spinning?

e. The next day Dwayne pulls the last few towels off the roll with the same linear velocity as before. This time the roll's diameter is only 4 cm. What is the angular velocity now?

9. *Pulley Problem:* Two pulleys are connected by a pulley belt (Figure 6-8g).

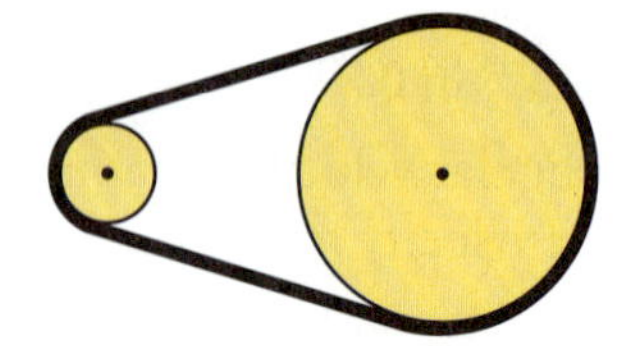

Figure 6-8g

a. The small pulley has diameter 10 cm and rotates at 100 rev/min. Find its angular velocity in radians per second.

b. Find the linear velocity of a point on the rim of the 10-cm pulley.

c. Find the linear velocity of a point on the belt connecting the two pulleys.

d. Find the linear velocity of a point on the rim of the large pulley, which has diameter 30 cm.

e. Find the angular velocity of a point on the rim of the 30-cm pulley.

f. Find the angular and linear velocities of a point at the center of the 30-cm pulley.

10. *Gear Problem:* A gear with diameter 30 cm is revolving at 45 rev/min. It drives a smaller gear that has diameter 8 cm (similar to Figure 6-8h).

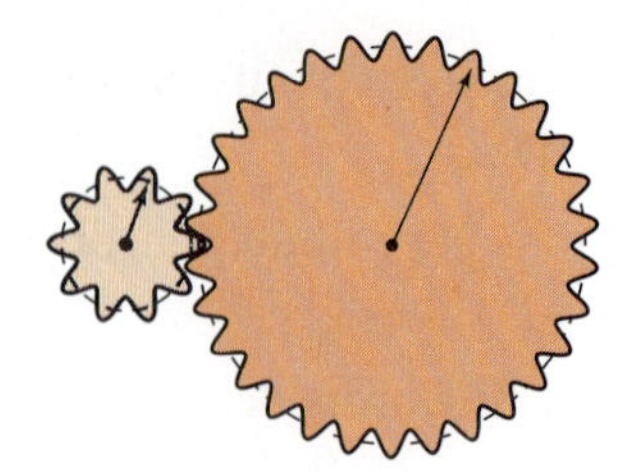

Figure 6-8h

a. How fast is the large gear turning in radians per minute?

b. What is the linear velocity of the teeth on the large gear?

c. What is the linear velocity of the teeth on the small gear?

d. How fast is the small gear turning in radians per minute?

e. How fast is the small gear turning in revolutions per minute?

f. If you double an angular velocity by using gears, what is the ratio of the diameters of the gears? Which gear does the driving, the large gear or the small gear?

11. *Tractor Problem:* The rear wheels of a tractor (Figure 6-8i) are 4 ft in diameter and are turning at 20 rev/min.

Figure 6-8i

a. How fast is the tractor going in feet per second? How fast is this in miles per hour?

b. The front wheels have a diameter of only 1.8 ft. How fast are the tread points moving in feet per second around the wheel? Is this an angular velocity or a linear velocity?

c. How fast in revolutions per minute are the front wheels turning? Is this an angular velocity or a linear velocity?

12. *Wheel and Grindstone Problem:* A waterwheel with diameter 12 ft turns at 0.3 rad/s (Figure 6-8j).

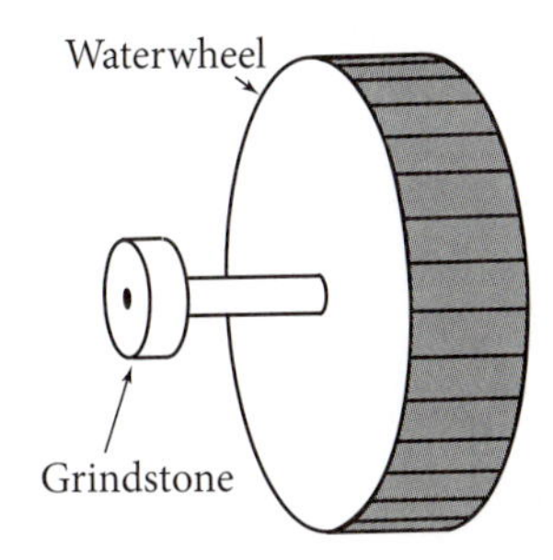

Figure 6-8j

a. What is the linear velocity of points on the rim of the waterwheel?

b. The waterwheel is connected by an axle to a grindstone with diameter 3 ft. What is the angular velocity of points on the rim of the grindstone?

c. What is the fastest velocity of any point on the grindstone? Where are these points?

13. *Three Gear Problem:* Three gears are connected as depicted schematically (without showing their teeth) in Figure 6-8k.

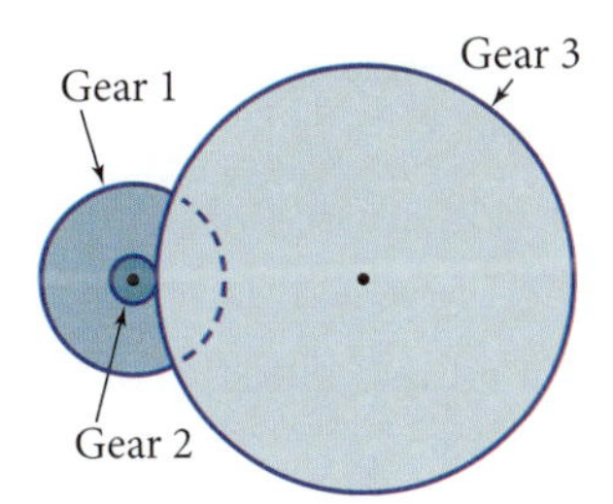

Figure 6-8k

a. Gear 1 rotates at 300 rev/min. Its radius is 8 in. What is its angular velocity in radians per second?

b. Gear 2 is attached to the same axle as Gear 1 but has radius 2 in. What is its angular velocity?

c. What is the linear velocity at a point on the teeth of Gear 2?

d. Gear 3 is driven by Gear 2. What is the linear velocity of the teeth on Gear 3?

e. Gear 3 has radius 18 in. What is the angular velocity of its teeth?

f. What are the linear and angular velocities at the center of Gear 3?

14. *Truck Problem:* In the 1930s, some trucks used a chain to transmit power from the engine to the wheels (Figure 6-8l). Suppose the drive sprocket had diameter 6 in., the wheel sprocket had diameter 20 in., and the drive sprocket rotated at 300 rev/min.

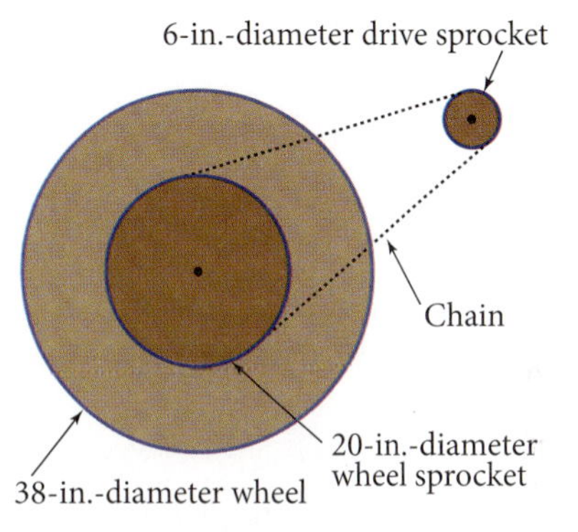

Figure 6-8l

a. Find the angular velocity of the drive sprocket in radians per second.

b. Find the linear velocity of the wheel sprocket in inches per minute.

c. Find the angular velocity of the wheel in radians per minute.

d. If the wheel has diameter 38 in., find the speed the truck is going, to the nearest *mile per hour.*

15. *Marching Band Formation Problem:* Suppose a marching band executes a formation in which some members march in a circle 50 ft in diameter and others in a circle 20 ft in diameter. The band members in the small circle march in such a way that they mesh with the members in the big circle without bumping into each other. Figure 6-8m on the next page shows the formation. The members in the big circle march at a normal pace of 5 ft/s.

a. What is the angular velocity of the big circle in radians per second?

b. What is the angular velocity of the big circle in revolutions per minute?

c. Which is the same about the two circles, their linear or their angular velocities at the rims?

d. What is the angular velocity of the small circle?

e. How many times faster does the small circle revolve? How can you find this factor using only the two diameters?

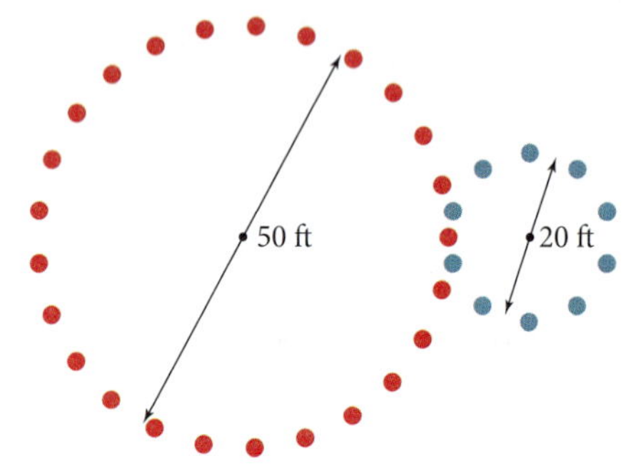

Figure 6-8m

16. *Four Pulley Problem:* Four pulleys are connected to one another as shown in Figure 6-8n. Pulley 1 is driven by a motor at an angular velocity of 120 rev/min. It is connected by a belt to Pulley 2. Pulley 3 is on the same axle as Pulley 2. It is connected by another belt to Pulley 4. The dimensions of the pulleys are

Pulley 1: radius = 10 cm

Pulley 2: radius = 2 cm

Pulley 3: radius = 12 cm

Pulley 4: radius = 3 cm

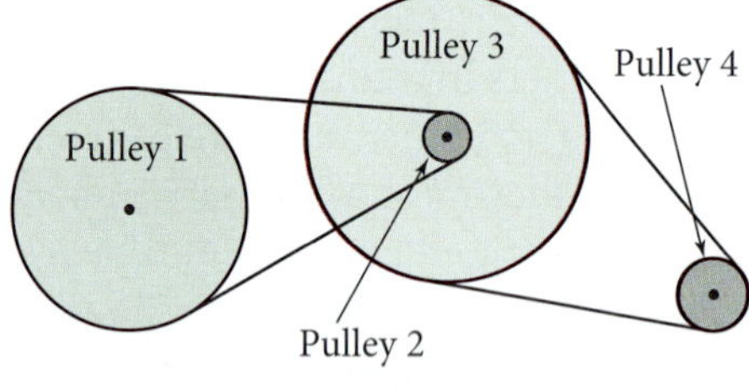

Figure 6-8n

a. What is the angular velocity of Pulley 1 in radians per minute?

b. What is the linear velocity of the rim of Pulley 1?

c. Find the linear and angular velocities of the rims of Pulley 2, Pulley 3, and Pulley 4.

d. Find the linear and angular velocities of the center of Pulley 4.

e. Find the angular velocity of Pulley 4 in revolutions per minute.

f. How many times faster than Pulley 1 is Pulley 4 rotating? How can you find this factor simply from the radii of the four pulleys?

17. *Gear Train Problem:* When something that rotates fast, like a car's engine, drives something that rotates slower, like the car's wheels, a gear train is used. In Figure 6-8o, Gear 1 is rotating at 2700 rev/min. The teeth on Gear 1 drive Gear 2, which is connected by an axle to Gear 3. The teeth on Gear 3 drive Gear 4. The sizes of the gears are

Gear 1: radius = 2 cm

Gear 2: radius = 15 cm

Gear 3: radius = 3 cm

Gear 4: radius = 18 cm

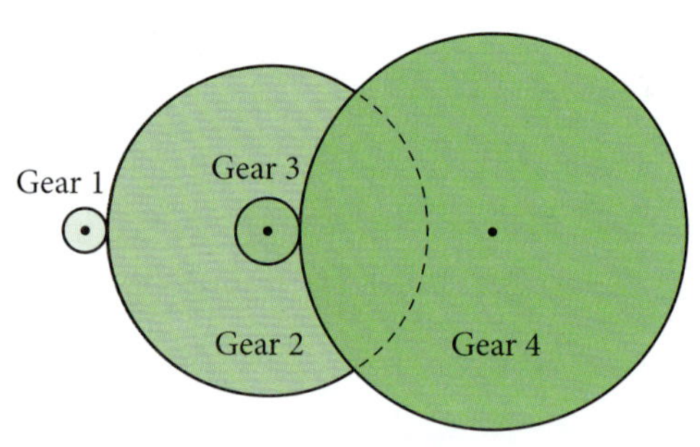

Figure 6-8o

a. What is the angular velocity of Gear 1 in radians per second?

b. Find the linear and angular velocities of the teeth on the rims of Gear 2, Gear 3, and Gear 4.

c. Find the linear and angular velocities at the center of Gear 4.

d. Find the angular velocity of Gear 4 in revolutions per minute.

e. The *reduction ratio* is the ratio of the angular velocity of the fastest gear to the angular velocity of the slowest gear. What is the reduction ratio for the gear train in Figure 6-8o? Calculate this reduction ratio without working parts a–d of this problem.

6-9 Chapter Review and Test

In this chapter you learned how to graph trigonometric functions. The sine and cosine functions are continuous sinusoids, while other trigonometric functions are discontinuous, having vertical asymptotes at regular intervals. You also learned about circular functions, which you can use to model real-world phenomena mathematically, and you learned how radians provide a link between these circular functions and the trigonometric functions. Radians also provide a way to calculate linear and angular velocity in rotary motion problems.

Review Problems

R0. Update your journal with what you have learned since the last entry. Include things such as

- The one most important thing you have learned as a result of studying this chapter
- The graphs of the six trigonometric functions
- How the transformations of sinusoidal graphs relate to the function transformations in Chapter 1
- How the circular and trigonometric functions are related
- Why circular functions usually are more appropriate as mathematical models than are trigonometric functions

R1. a. Sketch the graph of a sinusoid. On the graph, show the difference in meaning between a cycle and a period. Show the amplitude, the phase displacement, and the sinusoidal axis.

b. In the equation $y = 3 + 4\cos 5(\theta - 10°)$, what name is given to the quantity $5(\theta - 10°)$?

R2. a. Without using your grapher, show that you understand the effects of the constants in a sinusoidal equation by sketching the graph of $y = 3 + 4\cos 5(\theta - 10°)$. Give the amplitude, period, sinusoidal axis location, and phase displacement.

b. Using the cosine function, find a particular equation for the sinusoid in Figure 6-9a. Find another particular equation using the sine function. Show that the equations are equivalent to each other by plotting them on the same screen. What do you observe about the two graphs?

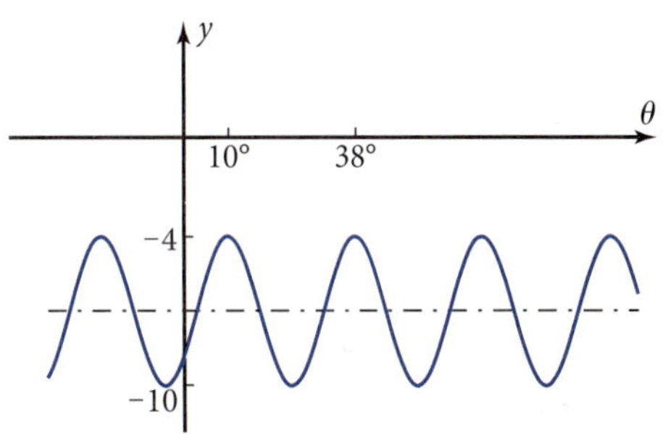

Figure 6-9a

c. A quarter-cycle of a sinusoid is shown in Figure 6-9b. Find a particular equation of the sinusoid.

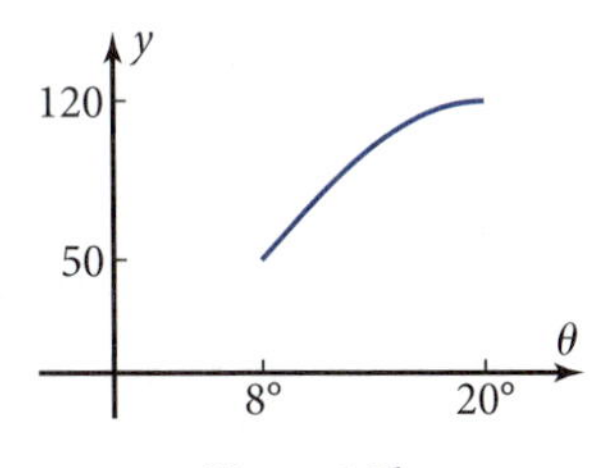

Figure 6-9b

d. At what value of θ shown in Figure 6-9b does the graph have a point of inflection? At what point does the graph have a critical point?

e. Find the frequency of the sinusoid in Figure 6-9b.

R3. **a.** Sketch the graph of $y = \tan \theta$.

b. Explain why the period of the tangent function is 180° rather than 360° like the sine and cosine functions.

c. Plot the graph of $y = \sec \theta$ on your grapher. Explain how you did this.

d. Use the relationship between sine and cosecant to explain why the cosecant function has vertical asymptotes at $\theta = 0°$, 180°, 360°,....

e. Explain why the graph of the cosecant function has high and low points but no points of inflection. Explain why the graph of the cotangent function has points of inflection but no high or low points.

f. For the function $y = 2 + 0.4 \cot \frac{1}{3}(\theta - 40°)$, give the vertical and horizontal dilations and the vertical and horizontal translations of the parent cotangent function. Then plot the graph to confirm that your answers are correct. What is the period of this function? Why is it not meaningful to talk about its amplitude?

R4. **a.** How many radians are in 30°? In 45°? In 60°? Give the answers exactly, in terms of π.

b. How many degrees are in an angle of 2 radians? Write the answer as a decimal.

c. Find cos 3 and cos 3°.

d. Find the radian measure of $\cos^{-1} 0.8$ and $\csc^{-1} 2$.

e. How long is the arc of a circle subtended by a central angle of 1 radian if the radius of the circle is 17 units?

R5. **a.** Draw the unit circle in a uv-coordinate system. In this coordinate system, draw an x-axis vertically with its origin at the point $(u, v) = (1, 0)$. Show where the points $x = 1$ unit, 2 units, and 3 units map onto the unit circle as the x-axis is wrapped around it.

b. How long is the arc of the unit circle subtended by a central angle of 60°? Of 2.3 radians?

c. Find sin 2° and sin 2.

d. Find the value of the inverse trigonometric function $\cos^{-1} 0.6$.

e. Find the exact values (no decimals) of the circular functions $\cos \frac{\pi}{6}$, $\sec \frac{\pi}{4}$, and $\tan \frac{\pi}{2}$.

f. Sketch the graphs of the parent circular functions $y = \cos x$ and $y = \sin x$.

g. Explain how to find the period of the circular function $y = 3 + 4 \sin \frac{\pi}{10}(x - 2)$ from the constants in the equation. Sketch the graph. Confirm by plotting on your grapher that your sketch is correct.

h. Find a particular equation for the circular function sinusoid for which a half-cycle is shown in Figure 6-9c.

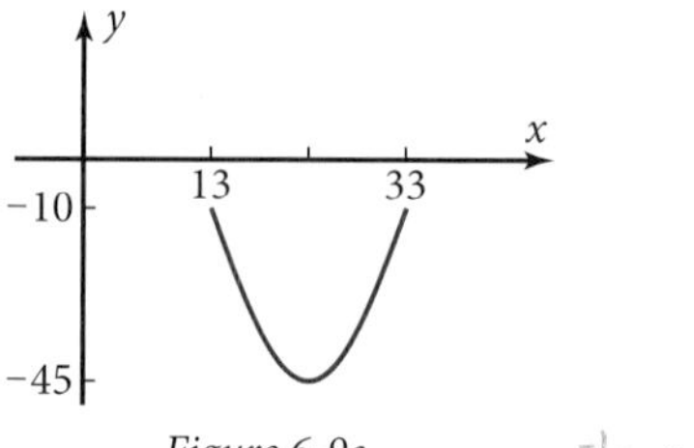

Figure 6-9c

R6. **a.** Find the general solution of the inverse circular relation arccos 0.8.

b. Find the first three positive values of the inverse circular relation arccos 0.8.

c. Find the least value of arccos 0.1 that is greater than 100.

d. For the sinusoid in Figure 6-9d, find the four values of x shown for which $y = 2$

- Graphically, to one decimal place
- Numerically, by finding a particular equation and plotting the graph
- Algebraically, using the particular equation

e. What is the next positive value of x for which $y = 2$, beyond the last positive value shown in Figure 6-9d?

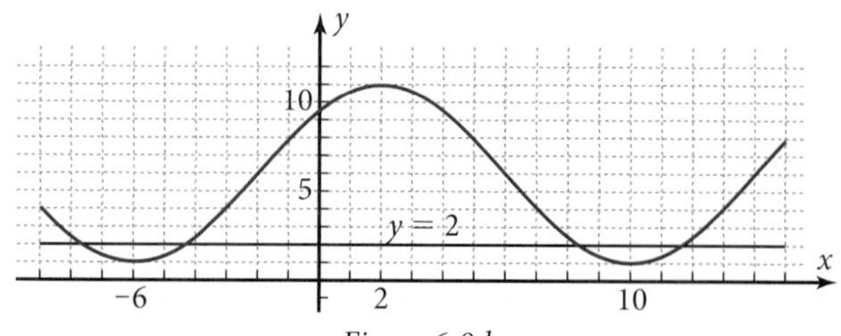

Figure 6-9d

R7. *Porpoising Problem:* Assume that you are aboard a research submarine doing submerged training exercises in the Pacific Ocean. At time $t = 0$, you start porpoising (going alternately deeper and shallower). At time $t = 4$ min, you are at your deepest, $y = -1000$ m. At time $t = 9$ min, you next reach your shallowest, $y = -200$ m. Assume that y varies sinusoidally with time.

- **a.** Sketch the graph of y versus t.
- **b.** Write an equation expressing y as a function of t.
- **c.** Your submarine can't communicate with ships on the surface when it is deeper than $y = -300$ m. At time $t = 0$, could your submarine communicate? How did you arrive at your answer?
- **d.** Between what two nonnegative times is your submarine first unable to communicate?

R8. *Clock Problem:* The "second" hand on a clock rotates through an angle of 120° in 20 s.

- **a.** What is its angular velocity in degrees per second?
- **b.** What is its angular velocity in radians per second?
- **c.** How far does a point on the tip of the hand, 11 cm from the axle, move in 20 s? What is the linear velocity of the tip of the hand? How can you calculate this linear velocity quickly from the radius and the angular velocity?

Three Wheel Problem: Figure 6-9e shows Wheel 1 with radius 15 cm, turning with an angular velocity of 50 rad/s. It is connected by a belt to Wheel 2, with radius 3 cm. Wheel 3, with radius 25 cm, is connected to the same axle as Wheel 2.

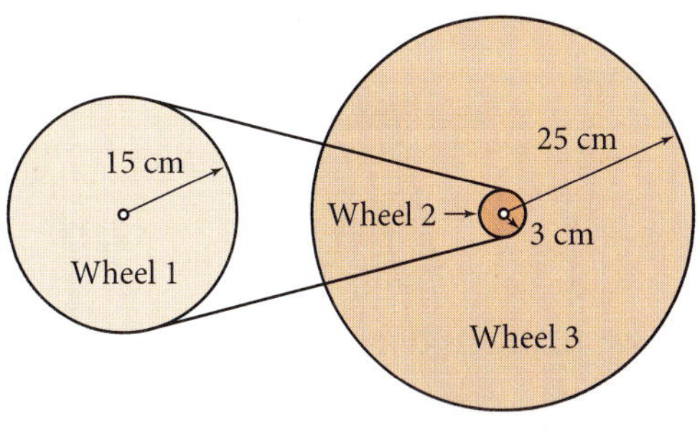

Figure 6-9e

- **d.** Find the linear velocity of points on the belt connecting Wheel 1 to Wheel 2.
- **e.** Find the linear velocity of points on the rim of Wheel 2.
- **f.** Find the linear velocity of a point at the center of Wheel 2.
- **g.** Find the angular velocity of Wheel 2.
- **h.** Find the angular velocity of Wheel 3.
- **i.** Find the linear velocity of points on the rim of Wheel 3.
- **j.** If Wheel 3 is touching the ground, how fast (in kilometers per hour) would the vehicle connected to the wheel be moving?

Concept Problems

C1. *Pump Jack Problem:* An oil well pump jack is shown in Figure 6-9f on the next page. As the motor turns, the walking beam rocks up and down, pulling the rod out of the well and letting it go back into the well. The connection between the rod and the walking beam is a steel cable that wraps around the cathead. The distance d from the ground to point P, where the cable connects to the rod, varies periodically with time.

- **a.** As the walking beam rocks, the angle θ it makes with the ground varies sinusoidally with time. The angle goes from a minimum of -0.2 radian to a maximum of 0.2 radian. How many degrees correspond to this range of angle θ?
- **b.** The radius of the circular arc on the cathead is 8 ft. What arc length on the cathead corresponds to the range of angles in part a?

c. The distance, d, between the cable-to-rod connector and the ground varies sinusoidally with time. What is the amplitude of the sinusoid?

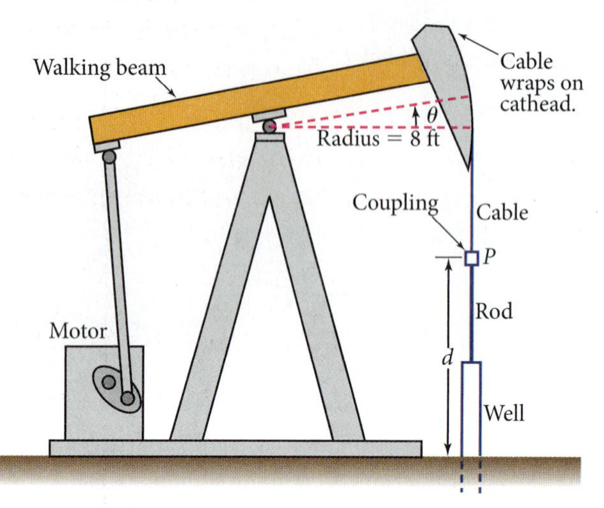

Figure 6-9f

d. Suppose the pump is started at time $t = 0$. One second later, P is at its highest point above the ground. P is at its next low point 2.5 s after that. When the walking beam is horizontal, point P is 7 ft above the ground. Sketch the graph of this sinusoid.

e. Find a particular equation expressing d as a function of t.

f. How far above the ground is point P at time $t = 9$ s?

g. How long does point P stay more than 7.5 ft above the ground on each cycle?

h. True or false? "The angle is always the independent variable in a periodic function."

C2. *Inverse Circular Relation Graphs:* In this problem you'll investigate the graphs of the inverse sine and inverse cosine functions and the general inverse sine and inverse cosine relations from which they come.

a. On your grapher, plot the inverse circular function $y = \sin^{-1} x$. Use a window with $-10 \leq x \leq 10$ that includes $x = 1$ and $x = -1$ as grid points. Use the same scales on both the x- and y-axes. Sketch the result.

b. The graph in part a is only for the inverse sine *function*. You can plot the entire inverse sine *relation*, $y = \arcsin x$, by putting your grapher in parametric mode. In this mode, both x and y are functions of a third variable, usually t. Enter the parametric equations this way:

$x = \sin t$

$y = t$

Plot the graph, using a window with the t-values the same as the x-values in part a. Sketch the graph.

c. Describe how the graphs in part a and part b are related to each other.

d. Explain algebraically how the parametric functions in part b and the function $y = \sin^{-1} x$ are related.

e. Find a way to plot the ordinary sine function, $y = \sin x$, on the same screen as in part b. Use a different style for this graph so that you can distinguish it from the other one. The result should look like the graphs in Figure 6-9g.

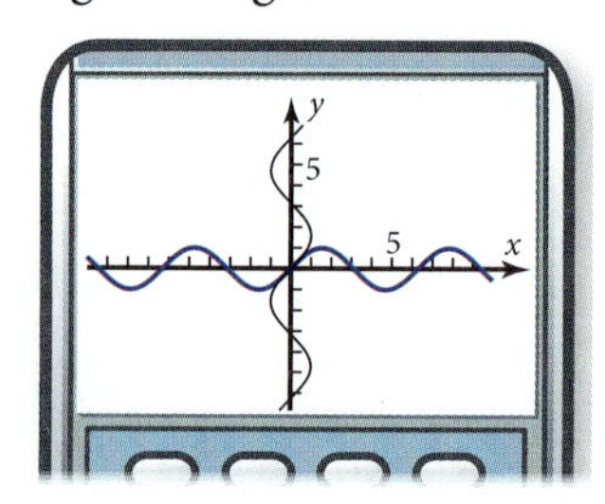

Figure 6-9g

f. How are the two graphs in Figure 6-9g related to each other? Find a geometric transformation of the sine graph that gives the arcsine graph.

g. Explain why the arcsine graph in Figure 6-9g is *not* a function graph but the graph you plotted in part a of the principal values of the inverse sine is a function graph.

h. Using the same scales as in part b, plot the graphs of the cosine function, $y = \cos x$, and the inverse cosine relation. Sketch the result. Do the two graphs have the same relationship as those in Figure 6-9g?

i. Repeat part h for the inverse tangent function.

j. Write an entry in your journal about what you have learned from this problem.

C3. *Merry-Go-Round Problem:* A merry-go-round rotates at a constant angular velocity while rings of seats rotate at a different (but constant) angular velocity (Figure 6-9h). Suppose the seats rotate at 30 rev/min counterclockwise with respect to the ground while the merry-go-round is rotating at 12 rev/min counterclockwise.

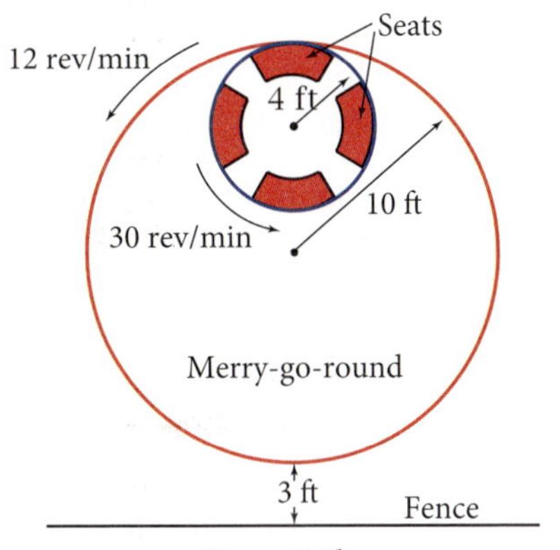

Figure 6-9h

a. Find your linear velocity, in feet per second, due to the *combined* rotations of the seats and the merry-go-round when your seat is

- Farthest from the center of the merry-go-round.
- Closest to the center of the merry-go-round.

b. In what direction are you actually moving when your seat is closest to the center of the merry-go-round?

c. As your seat turns, your distance from the fence varies sinusoidally with time. As the merry-go-round turns, the axis of this sinusoid also varies sinusoidally with time, but with a different period and amplitude. Suppose that at time $t = 0$ your seat is at its farthest distance from the fence, 23 ft. Write an equation expressing your distance from the fence as a function of time, t, in seconds.

d. Plot the graph of the function in part c. Sketch the result.

e. Use the answers in parts a–d to explain why many people don't feel well after riding on this type of ride.

Chapter Test

Part 1: No calculators allowed (T1–T9)

T1. Figure 6-9i shows an x-axis drawn tangent to the unit circle in a uv-coordinate system. On a copy of this figure, show approximately where the point $x = 2.3$ maps onto the unit circle when the x-axis is wrapped around the circle.

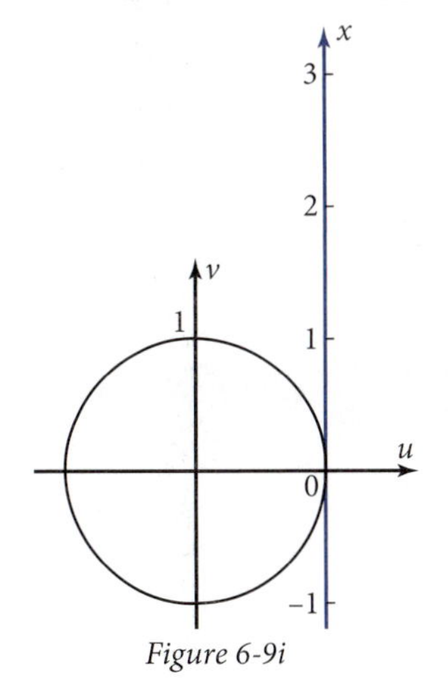

Figure 6-9i

T2. Sketch an angle of 2.3 radians on the copy of Figure 6-9i.

T3. What are the steps needed to find a decimal approximation of the degree measure of an angle of 2.3 radians? In what quadrant would this angle terminate?

T4. Give the exact number of radians in 120° (no decimals).

T5. Give the exact number of degrees in $\frac{\pi}{5}$ radian (no decimals).

T6. Give the period, amplitude, sinusoidal axis, and phase displacement of the circular function

$$f(x) = 3 + 4\cos\frac{\pi}{5}(x - 1)$$

T7. Sketch at least two cycles of the sinusoid in Problem T6.

T8. An object rotates with angular velocity $\omega = 3$ rad/s. What is the linear velocity of a point 20 cm from the axis of rotation?

T9. A gear with radius 5 in. rotates so that its teeth have angular velocity 40 in./s. Its teeth mesh with a larger gear with radius 10 in. What is the linear velocity of the teeth on the larger gear?

Part 2: Graphing calculators allowed (T10–T24)

T10. A long pendulum hangs from the ceiling. As it swings back and forth, its distance from the wall varies sinusoidally with time. At time $x = 1$ s it is at its closest point, $y = 50$ cm. Three seconds later it is at its farthest point, $y = 160$ cm. Sketch the graph.

T11. Figure 6-9j shows a half-cycle of a circular function sinusoid. Find a particular equation for this sinusoid.

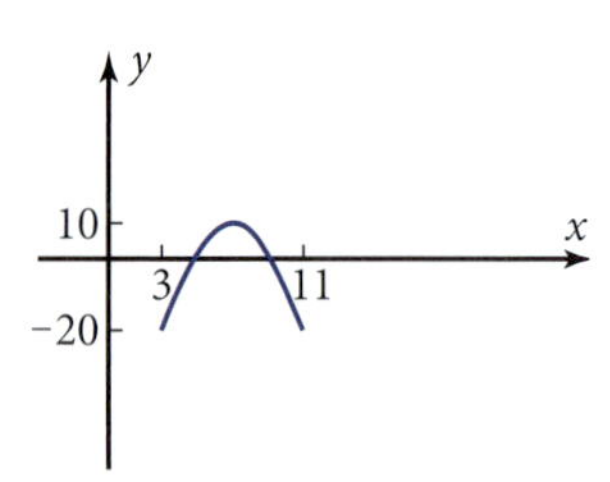

Figure 6-9j

For Problems T12–T18, Figure 6-9k shows the depth of the water at a point near the shore as it varies due to the tides. A particular equation relating d, in feet, to t, in hours after midnight on a given day, is

$$d = 3 + 2\cos\frac{\pi}{5.6}(t - 4)$$

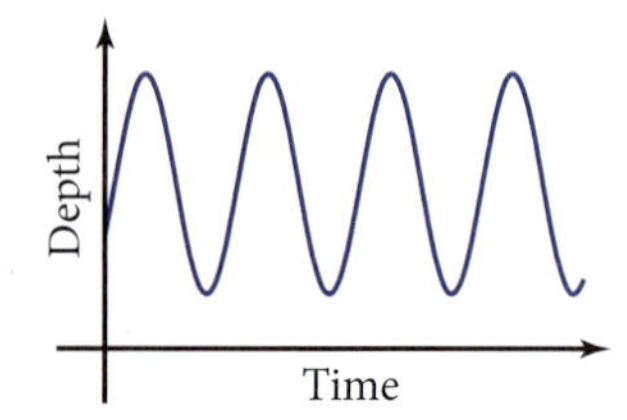

Figure 6-9k

T12. Find a time at which the water is deepest. How deep is it at that time?

T13. After the time you found in Problem T12, when is the water next at its shallowest? How deep is it at that time?

T14. What does t equal at 3:00 p.m.? How deep is the water at that time?

T15. Plot the graph of the sinusoid in Figure 6-9k on your grapher. Use a window with x (actually t) about $0 \le x \le 50$ and an appropriate window for y (actually d).

T16. By tracing your graph in Problem T15, find, approximately, the first interval of nonnegative times for which the water is less than 4.5 ft deep.

T17. Set your grapher's table mode to begin at the later time from Problem T16, and set the table increment at 0.01. Find to the nearest 0.01 h the latest time at which the water is still less than 4.5 ft deep.

T18. Solve algebraically for the first positive time at which the water is exactly 4.5 ft deep.

Bicycle Problem: For Problems T19–T23, Anna Racer is riding her bike. She turns the pedals at 120 rev/min. The dimensions of the bicycle are shown in Figure 6-9l.

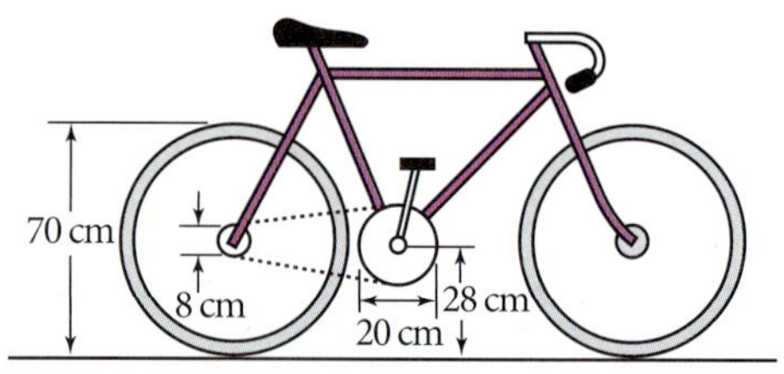

Figure 6-9l

T19. What is the angular velocity of the pedals in radians per second?

T20. What is the linear velocity of the chain in centimeters per second?

T21. What is the angular velocity of the back wheel?

T22. How fast is Anna's bike going, in kilometers per hour?

T23. The pedals are 24 cm from the axis of the large sprocket. Sketch a graph showing the distance of Anna's right foot from the pavement as a function of the number of seconds since her foot was at a high point. Show the upper and lower bounds, the sinusoidal axis, and the location of the next three high points.

T24. What did you learn as a result of taking this test that you did not know before?

Chapter 7

Trigonometric Function Properties and Identities, and Parametric Functions

The Foucault pendulum shown in the photograph provided physical proof in the middle of the 19th century that Earth is rotating about its axis. Pendulums can have many different paths in different planes. A pendulum's path is a two-dimensional curve that you can describe by x- and y-displacements from its rest position as functions of time. You can predict a pendulum's position at any given time using parametric equations. Pythagorean properties of trigonometric functions can be used to model periodic relationships and allow you to conclude whether the path of a pendulum is an ellipse or a circle.

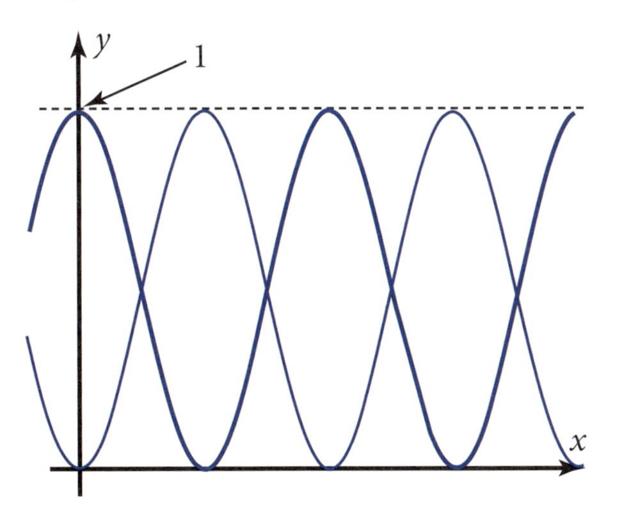

Mathematical Overview

Three kinds of algebraic properties relate different trigonometric functions that have the same argument. For example, the graphs of $y = \cos^2 x$ and $y = \sin^2 x$ are sinusoids. If you add these two functions, the result is always 1. In this chapter you'll learn properties you can apply to proving identities and solving trigonometric equations. You'll gain this knowledge in four ways.

ALGEBRAICALLY

Pythagorean property: $\cos^2 x + \sin^2 x = 1$

GRAPHICALLY

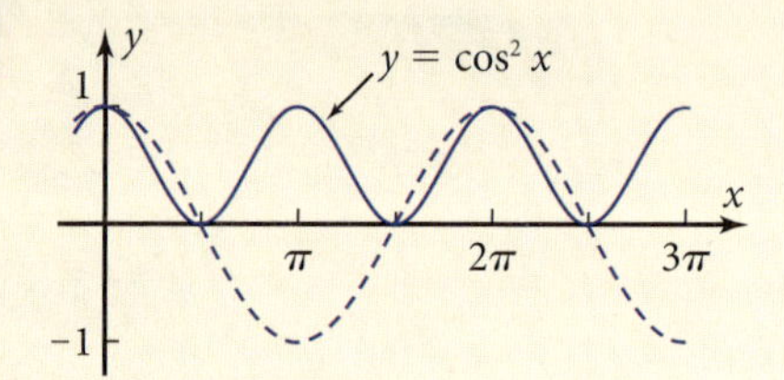

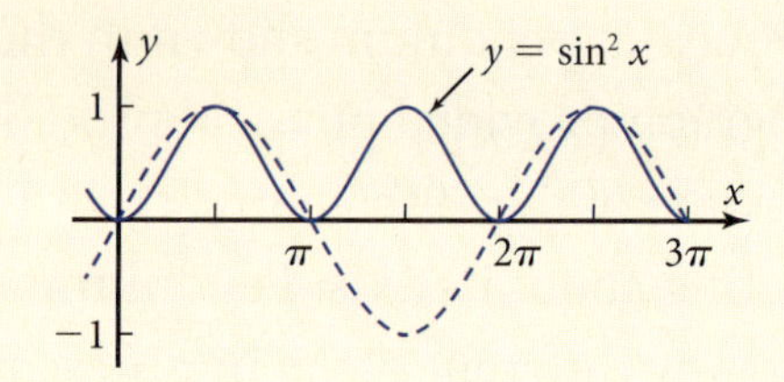

NUMERICALLY

$\cos^2 \frac{\pi}{5} + \sin^2 \frac{\pi}{5} = 0.6545... + 0.3454... = 1$

VERBALLY

The Pythagorean, reciprocal, and quotient properties transform one trigonometric expression into another form. Results can be checked graphically by plotting the original expression and the transformed one or numerically by making a table of values for both expressions.

7-1 Introduction to the Pythagorean Property

Figure 7-1a shows the graphs of $y = \cos^2 x$ (on the left) and $y = \sin^2 x$ (on the right). Both graphs are sinusoids, as you will see in the next chapter. In this section you'll learn that the sum of the two functions always equals 1.

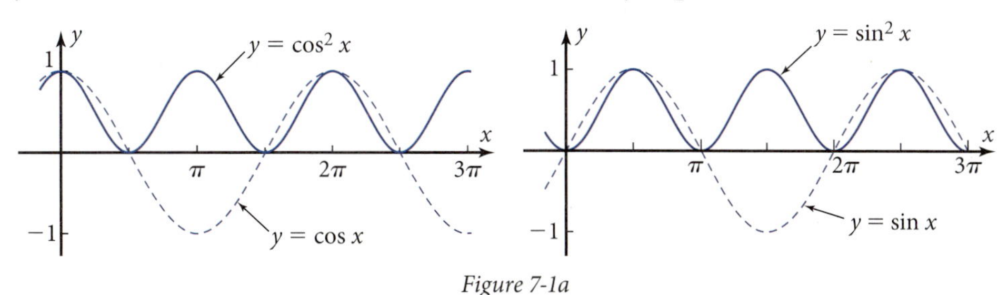

Figure 7-1a

Objective

Investigate the sum of the squares of the cosine and sine of the same argument.

Exploratory Problem Set 7-1

1. If you enter $\cos^2 0.7$ and $\sin^2 0.7$ into your calculator, you get these numbers:

   $\cos^2 0.7 \approx 0.5849835715$

   $\sin^2 0.7 \approx 0.4150164285$

   Without using your calculator, add the numbers. What do you notice?

2. Enter $f_1(x) = (\cos(x))^2$ and $f_2(x) = (\sin(x))^2$ into your grapher. (This is how your grapher recognizes $\cos^2 x$ and $\sin^2 x$.) Enter $f_3(x) = f_1(x) + f_2(x)$ and then make a table of values of the three functions for each 0.1 radian, starting at 0. What do you notice about $f_3(x)$?

3. Plot the three functions on the same screen. Do the graphs of $f_1(x)$ and $f_2(x)$ agree with those in Figure 7-1a? How does the relationship between $f_1(x)$ and $f_2(x)$ give you graphical evidence that $\cos^2 x + \sin^2 x$ is equal to 1, no matter what x is?

4. Remake the table of Problem 2 with your grapher in degree mode. Does your conclusion in Problem 3 apply to trigonometric functions independent of whether x is measured in degrees or radians?

5. Figure 7-1b shows the unit circle in a uv-coordinate system and an angle of 50° in standard position. Use the definitions of cosine and sine to explain why $\cos 50° = u$ and $\sin 50° = v$.

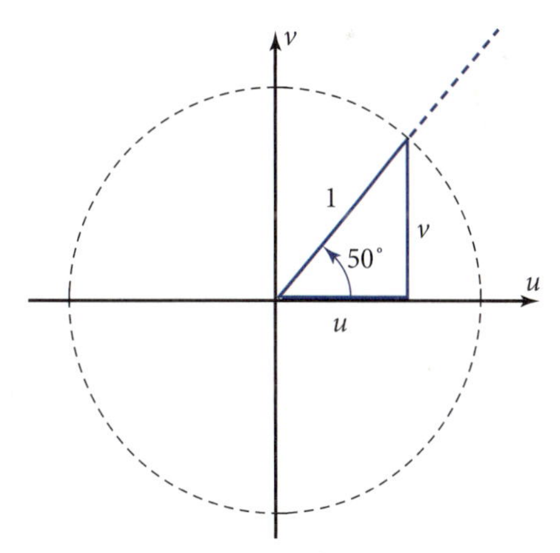

Figure 7-1b

6. Show that $\cos^2 50° + \sin^2 50° = 1$ numerically. Explain graphically why this **Pythagorean property** is true.

7-2 Pythagorean, Reciprocal, and Quotient Properties

In Section 7-1, you discovered the Pythagorean property

$$\cos^2 x + \sin^2 x = 1$$

for all values of x. You also know that secant, cosecant, and cotangent are reciprocals of cosine, sine, and tangent, respectively. In this section you will prove these properties algebraically, along with the quotient properties, such as

$$\tan x = \frac{\sin x}{\cos x}$$

Objective Derive algebraically three kinds of properties expressing relationships among trigonometric functions.

In this exploration you will use the definitions of the six trigonometric functions to find relationships among them.

EXPLORATION 7-2: Properties of Trigonometric Functions

This figure shows an angle of x radians in standard position and the reference triangle with legs u and v and hypotenuse r.

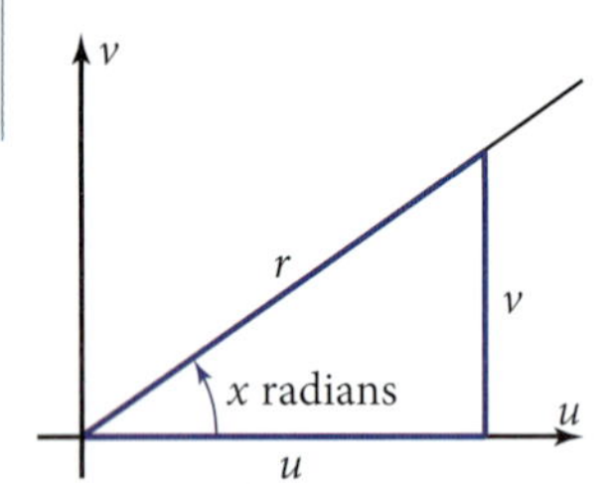

1. Write the definitions of $\sin x$, $\cos x$, $\tan x$, $\cot x$, $\sec x$, and $\csc x$ as ratios of the three sides of the reference triangle, u, v, and r.
2. Use the definitions in Problem 1 to write *reciprocal properties* expressing $\sec x$, $\csc x$, and $\cot x$ as reciprocals of other trigonometric functions.
3. Use the definitions in Problem 1 to write $\frac{\sin x}{\cos x}$ in terms of u, v, and r. Simplify. What other trigonometric function does this quotient equal? Write an equation for the resulting *quotient property* expressing this function as the quotient $\sin x$ divided by $\cos x$.
4. Write another quotient property, expressing $\cot x$ in terms of $\sin x$ and $\cos x$.
5. Use the reciprocal and quotient properties to derive two quotient properties, one for $\tan x$ and one for $\cot x$, in terms of $\sec x$ and $\csc x$.
6. The property $\cos^2 x + \sin^2 x = 1$ at the beginning of this section is called a *Pythagorean property.* Divide both sides of this equation by $\cos^2 x$. Then use the reciprocal and quotient properties to write a Pythagorean property involving the squares of tangent and secant.
7. Derive a Pythagorean property relating $\csc x$ and $\cot x$.
8. Transform the expression $\csc x \tan x$ into $\sec x$. Write the given expression. Then substitute using appropriate properties and simplify.
9. The expression $\csc x \tan x$ in Problem 8 involves two functions. The result, $\sec x$, involves only one function. What could be your thought process in deciding how to start this problem?
10. Transform $\csc x \tan x \cos x$ into 1.
11. What did you learn as a result of doing this exploration that you did not know before?

Because the properties you'll learn in this section apply to all trigonometric functions, the argument x will be used both for degrees and for radians.

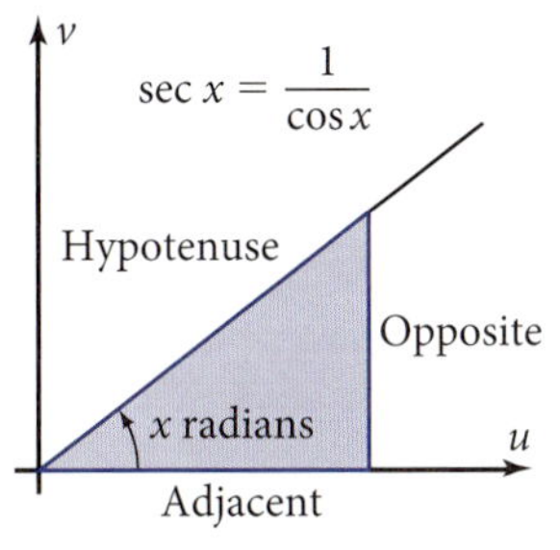

Figure 7-2a

Reciprocal Properties

In Chapter 5, in order to find values of the secant, cosecant, and tangent functions, you took advantage of the fact that each is the reciprocal of one of the functions on your grapher. For instance,

$$\sec x = \frac{1}{\cos x}$$

because, in the reference triangle (Figure 7-2a),

$$\sec x = \frac{\text{hypotenuse}}{\text{adjacent leg}} \quad \text{and} \quad \cos x = \frac{\text{adjacent leg}}{\text{hypotenuse}}$$

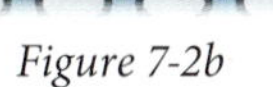

Figure 7-2b

This relationship between secant and cosine is called a **reciprocal property**. As you can see from the graphs in Figure 7-2b, each y-value for the secant graph is the reciprocal of the corresponding y-value for the cosine graph. For instance, because $\cos\left(\frac{\pi}{3}\right) = \frac{1}{2}$, it follows that $\sec\left(\frac{\pi}{3}\right) = 2$. As you saw in Section 6-3, the asymptotes for the graph of the secant function occur at

$$x = \frac{\pi}{2}, \frac{3\pi}{2}, \frac{5\pi}{2}, \ldots$$

where the value of the cosine function is zero.

If $\cos x = 0$, $\sec x = \frac{1}{\cos x} = \frac{1}{0}$, which is undefined!

This box summarizes the three reciprocal properties.

PROPERTIES: The Reciprocal Properties

$$\sec x = \frac{1}{\cos x} \qquad \csc x = \frac{1}{\sin x} \qquad \cot x = \frac{1}{\tan x}$$

The domain excludes those values of x that produce a denominator equal to zero.

Quotient Properties

If you divide $\sin x$ by $\cos x$, you get an interesting result.

$$\frac{\sin x}{\cos x} = \frac{\dfrac{\text{opposite leg}}{\text{hypotenuse}}}{\dfrac{\text{adjacent leg}}{\text{hypotenuse}}}$$ Definition of sine and cosine.

$$= \frac{\text{opposite leg}}{\text{hypotenuse}} \cdot \frac{\text{hypotenuse}}{\text{adjacent leg}}$$ Multiply the numerator by the reciprocal of the denominator.

$$= \frac{\text{opposite leg}}{\text{adjacent leg}}$$ Simplify.

$$= \tan x$$ Definition of tangent.

$$\therefore \tan x = \frac{\sin x}{\cos x}$$ Transitivity and symmetry.

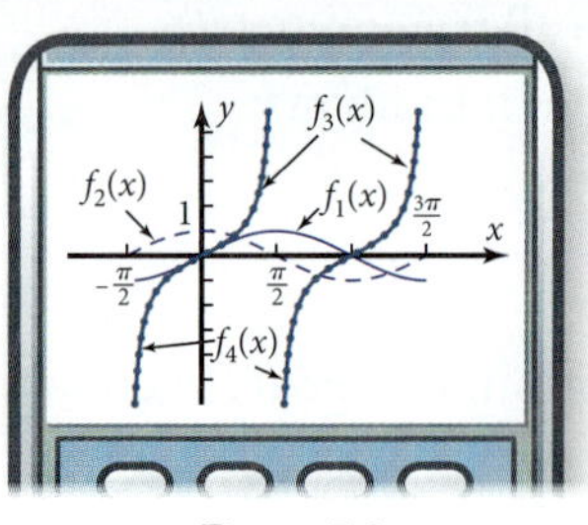

Figure 7-2c

This relationship is called a **quotient property.** If you plot

$$f_1(x) = \sin x$$

$$f_2(x) = \cos x$$

$$f_3(x) = f_1(x)/f_2(x)$$

$$f_4(x) = \tan x$$

the graphs of $f_3(x)$ and $f_4(x)$ will coincide (Figure 7-2c).

Because cotangent is the reciprocal of tangent, another quotient property is

$$\cot x = \frac{\cos x}{\sin x}$$

Each of these quotient properties can be expressed in terms of secant and cosecant. For instance,

$$\tan x = \frac{\sin x}{\cos x}$$

$$= \frac{\frac{1}{\csc x}}{\frac{1}{\sec x}} \qquad \text{Use the reciprocal properties for sine and cosine.}$$

$$= \frac{\sec x}{\csc x} \qquad \text{Simplify.}$$

$$\therefore \tan x = \frac{\sec x}{\csc x}$$

Using the reciprocal property for cotangent gives

$$\cot x = \frac{\csc x}{\sec x}$$

This box records the two quotient properties in both of their forms. The properties apply unless a denominator equals zero.

PROPERTIES: The Quotient Properties

$$\tan x = \frac{\sin x}{\cos x} = \frac{\sec x}{\csc x} \qquad \text{Domain: } x \neq \frac{\pi}{2} + \pi n\text{, where } n \text{ is an integer.}$$

$$\cot x = \frac{\cos x}{\sin x} = \frac{\csc x}{\sec x} \qquad \text{Domain: } x \neq \pi n\text{, where } n \text{ is an integer.}$$

Pythagorean Properties

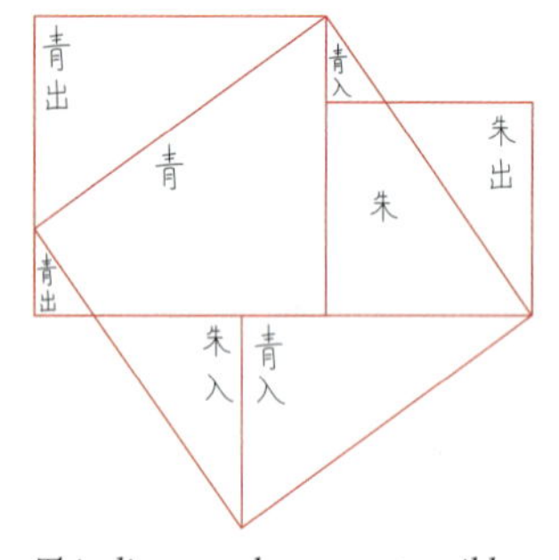

This diagram shows one possible reconstruction of the geometric figure in Liu Hui's 3rd-century-A.D. proof of the Pythagorean theorem.

Figure 7-2d on the next page shows an arc of length x in standard position on the unit circle in a uv-coordinate system. By the Pythagorean theorem, point (u, v) at the endpoint of arc x has the property

$$u^2 + v^2 = 1$$

This property is true even if x terminates in a quadrant where u or v is negative, because squares of negative numbers are the same as the squares of their absolute values.

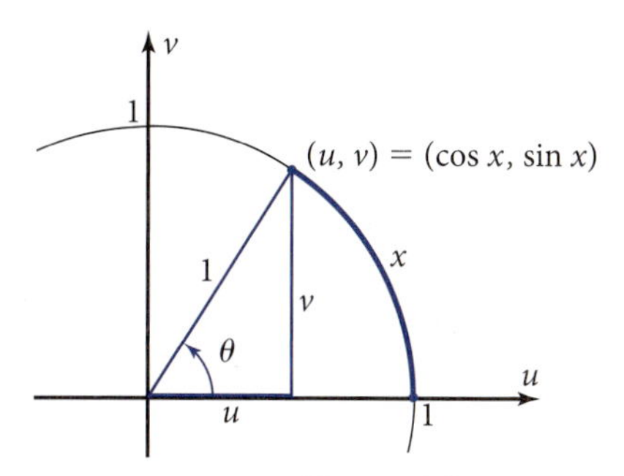

Figure 7-2d

By the definitions of cosine and sine,

$$\cos x = \frac{u}{1} = u \qquad \text{and} \qquad \sin x = \frac{v}{1} = v$$

Substitution into the equation $u^2 + v^2 = 1$ gives the Pythagorean property for sine and cosine.

$$\cos^2 x + \sin^2 x = 1$$

Two other Pythagorean properties can be derived from this one.

$$\cos^2 x + \sin^2 x = 1$$ Start with the Pythagorean property for cosine and sine.

$$\frac{\cos^2 x}{\cos^2 x} + \frac{\sin^2 x}{\cos^2 x} = \frac{1}{\cos^2 x}$$ Divide both sides of the equation by $\cos^2 x$.

$$1 + \tan^2 x = \sec^2 x$$ $\frac{\sin^2 x}{\cos^2 x} = \left(\frac{\sin x}{\cos x}\right)^2 = \tan^2 x$ and $\frac{1}{\cos^2 x} = \sec^2 x$

Dividing by $\sin^2 x$ instead of by $\cos^2 x$ results in the property

$$\cot^2 x + 1 = \csc^2 x$$

This box records the three Pythagorean properties.

PROPERTIES: The Three Pythagorean Properties

$\cos^2 x + \sin^2 x = 1$ Domain: All real values of x.

$1 + \tan^2 x = \sec^2 x$ Domain: $x \neq \frac{\pi}{2} + \pi n$, where n is an integer.

$\cot^2 x + 1 = \csc^2 x$ Domain: $x \neq \pi n$, where n is an integer.

Problem Set 7-2

Reading Analysis

From what you have read in this section, what do you consider to be the main idea? Name the three kinds of properties, and give an example of each kind. If (u, v) is a point at which the terminal side of an angle θ in standard position intersects the unit circle centered at the origin, explain why $\sin\theta = v$ and $\cos\theta = u$.

Quick Review

Q1. What is the exact value of cos 30°?

Q2. What is the exact value of $\sin\left(\frac{\pi}{4}\right)$?

Q3. What is the exact value of tan 60°?

Q4. What is the exact value of $\cot\left(\frac{\pi}{2}\right)$?

Q5. Write cos 57° in decimal form.

Q6. Write $\sin 33°$ in decimal form.

Q7. Write $\sec 81°$ in decimal form.

Q8. Write $\csc 9°$ in decimal form.

Q9. From the answers to Problems Q5–Q8, what relationship exists between the values of sine and cosine? What about between the values of secant and cosecant?

Q10. The period of the circular function $y = 3 + 4\cos 5(x - 6)$ is

A. 3 **B.** 4 **C.** 5

D. 6 **E.** None of these

---

1. What is the reciprocal property for $\sec x$?
2. Explain why $\cot x \tan x = 1$.
3. Write $\tan x$ in terms of $\sin x$ and $\cos x$.
4. Show how you can transform the reciprocal property $\cot x = \frac{\cos x}{\sin x}$ algebraically to express $\cot x$ in terms of $\sec x$ and $\csc x$.
5. Explain geometrically why the property $\cos^2 x + \sin^2 x = 1$ is called a Pythagorean property.
6. By appropriate operations on the Pythagorean property $\cos^2 x + \sin^2 x = 1$, derive the Pythagorean property $\cot^2 x + 1 = \csc^2 x$.
7. Sketch the graph of the trigonometric function $y = \sin\theta$. On the same axes, sketch the graph of the function $y = \csc\theta$ using the fact that the y-value for $\csc\theta$ is the reciprocal of the corresponding y-value for $\sin\theta$. Where do the asymptotes occur in the graph of the cosecant function?
8. On your grapher, make a table with columns showing the values of the trigonometric expressions $\tan^2\theta$ and $\sec^2\theta$ for values of θ of $0°, 15°, 30°, \ldots$. What relationship do you notice between the two columns? How do you explain this relationship? How do you explain what happens at $\theta = 90°$?
9. Show algebraically that $\sin^2 x = 1 - \cos^2 x$.
10. Show algebraically that $\cot^2 x = \csc^2 x - 1$.
11. Many trigonometric properties involve the number 1. Use these properties to write six trigonometric expressions that equal 1.
12. Use the Pythagorean properties to write expressions equivalent to each of these expressions.

    a. $\sin^2 x$ **b.** $\cos^2 x$ **c.** $\tan^2 x$

    d. $\cot^2 x$ **e.** $\sec^2 x$ **f.** $\csc^2 x$

13. *Duality Property of Trigonometric Functions:* The cosine of an angle is the sine of the *complement* of that angle. Two functions that satisfy this property are called **cofunctions** of each other. For instance, $\cos 70° = \sin 20°$, as you can check by calculator. Each property of this section has a *dual,* a property in which each function in the original property has been replaced by its cofunction. For example,

    $$\tan x = \frac{\sin x}{\cos x} \rightarrow \cot x = \frac{\cos x}{\sin x}$$

    Show that each property in this section has a dual that is also a valid property. Explain how this duality property can help you memorize the properties.

14. *Other Quadrants Problem:* The text at the bottom of page 348 states that $u^2 + v^2 = 1$ is true even if angle θ terminates in another quadrant. Sketch a copy of Figure 7-2d with the angle terminating in Quadrant II, where the displacement u is negative. Would $\cos\theta$ equal u, or would it equal $-u$ in this case? Why would it still be true that $u^2 + v^2 = 1$?

15. *Dynamic Unit Circle Properties Problem:* Open the *Unit Circle Properties* exploration at **www.keymath.com/precalc.** Explore the relationships you notice as you move point P around the circle. Use the action buttons that illustrate the Pythagorean properties. Explain in writing what you learned from doing this dynamic exploration.

7-3 Identities and Algebraic Transformation of Expressions

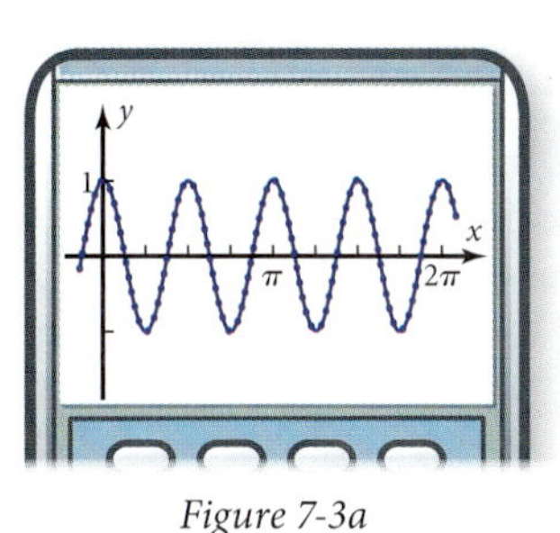

Figure 7-3a

Figure 7-3a shows the graphs of these functions:

$$f_1(x) = \cos^2 x - \sin^2 x$$

$$f_2(x) = 1 - 2\sin^2 x$$

The first graph is a thin, solid line, and the second is a thick, dotted line. The two graphs are identical. The equation

$$\cos^2 x - \sin^2 x = 1 - 2\sin^2 x$$

is called an **identity** because the two sides of the equation represent identical numbers for all values of x for which the expressions are defined. In this section you will gain fluency with the properties from the previous section by using them to transform one trigonometric expression to another one, such as the left side of the identity to the right side.

Objective

Given a trigonometric expression, transform it into an equivalent expression whose form is perhaps simpler or more useful.

Transformations

Here are examples of transforming one expression into another.

EXAMPLE 1 ➤ Transform $\sin x \cot x$ into $\cos x$.

SOLUTION Your thought process should be: "The product $\sin x \cot x$ has two factors, and the result has only one factor. Can I convert one of the factors into a fraction and cancel?"

$\sin x \cot x$	Start by writing the given expression.
$= \sin x \cdot \dfrac{\cos x}{\sin x}$	Substitute using the quotient properties to get $\cos x$ into the expression.
$= \cos x$	Simplify.
$\therefore \sin x \cot x = \cos x$, Q.E.D.	Use the transitive property for completeness.

◄

EXAMPLE 2 ➤ Transform $\cos^2 x - \sin^2 x$ into $1 - 2\sin^2 x$.

SOLUTION Your thought process should be:

- The result has only sine in it, so I need to get rid of cosine.
- The expressions involve *squares* of functions, so I'll think of the *Pythagorean* properties.

- I can write the Pythagorean property $\cos^2 x + \sin^2 x = 1$ as $\cos^2 x = 1 - \sin^2 x$.

$\cos^2 x - \sin^2 x$	Start by writing the given expression.
$= (1 - \sin^2 x) - \sin^2 x$	Substitute $1 - \sin^2 x$ for $\cos^2 x$ using the Pythagorean property.
$= 1 - 2\sin^2 x$	Combine like terms.
$\therefore \cos^2 x - \sin^2 x = 1 - 2\sin^2 x$, Q.E.D.	Use the transitive property.

Identities

To prove that a given trigonometric equation is an identity, start with the expression on one side of the equation and transform it into the other. You can pick either side of the equation to work on.

EXAMPLE 3 Prove algebraically that $(1 + \cos x)(1 - \cos x) = \sin^2 x$ is an identity.

SOLUTION Proof:

$(1 + \cos x)(1 - \cos x)$	Start with one member of the equation, usually the more complicated one.
$= 1 - \cos^2 x$	Complete the multiplication.
$= \sin^2 x$	Look for familiar expressions. Because the functions are squared, think Pythagorean!
$\therefore (1 + \cos x)(1 - \cos x) = \sin^2 x$, Q.E.D.	Use the transitive property.

Notes:

- Start by writing "Proof." This word tells the reader of your work that you have stopped *stating* the problem and started *solving* it. Writing "Proof" also gets your pencil moving! Sometimes you don't see how to prove something until you actually start doing it.

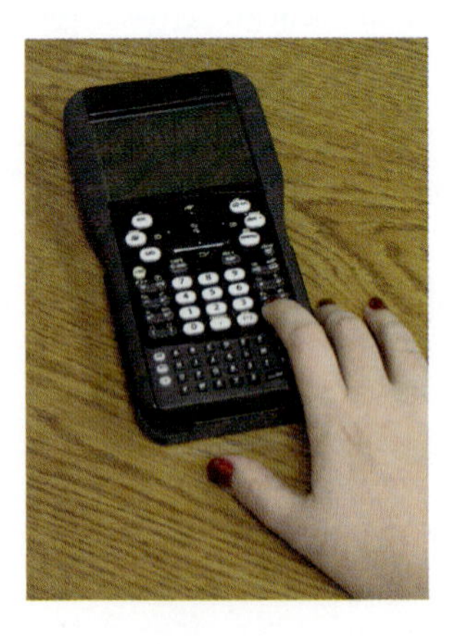

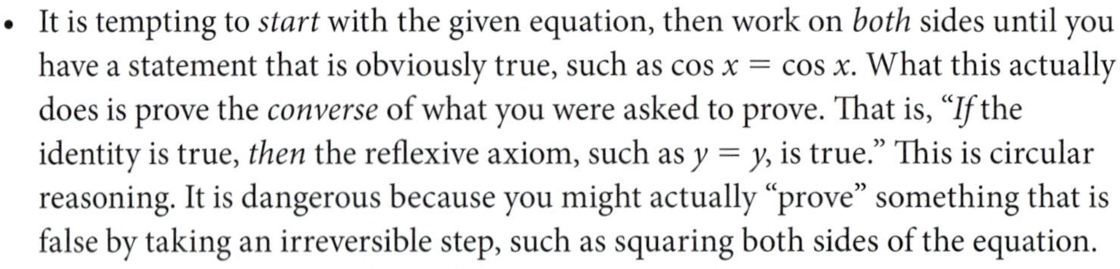

- It is tempting to *start* with the given equation, then work on *both* sides until you have a statement that is obviously true, such as $\cos x = \cos x$. What this actually does is prove the *converse* of what you were asked to prove. That is, "*If* the identity is true, *then* the reflexive axiom, such as $y = y$, is true." This is circular reasoning. It is dangerous because you might actually "prove" something that is false by taking an irreversible step, such as squaring both sides of the equation.
- You can never prove graphically or numerically that an identity is true for all values of x. However, you can confirm the validity of an identity for a set of values graphically by plotting both sides and showing that the graphs coincide, or numerically by generating a table of values. Figure 7-3a at the beginning of this section is an example of a graphical verification.

EXAMPLE 4 ➤ Prove algebraically that $\cot A + \tan A = \csc A \sec A$ is an identity.

SOLUTION Proof:

$\cot A + \tan A$	Pick one member of the equation to start with.
$= \dfrac{\cos A}{\sin A} + \dfrac{\sin A}{\cos A}$	The result has only one term. Try writing fractions to add together.
$= \dfrac{\cos^2 A + \sin^2 A}{\sin A \cos A}$	Find a common denominator and add the fractions.
$= \dfrac{1}{\sin A \cos A}$	Recognize the Pythagorean property, and use it to simplify the numerator.
$= \dfrac{1}{\sin A} \cdot \dfrac{1}{\cos A}$	The result has *two* factors, so *make* two factors.
$= \csc A \sec A$	Use the reciprocal properties to get the csc *A* and sec *A* that appear in the result.
$\therefore \cot A + \tan A = \csc A \sec A$, Q.E.D.	Use the transitive property.

Note: Avoid the temptation to use a shortcut by writing only cos or sec. These are the *names* of the functions, not the values of the functions. Equality applies to numbers, not to names.

EXAMPLE 5 ➤ Prove algebraically that $\frac{1 - \cos B}{\sin B} = \frac{\sin B}{1 + \cos B}$ is an identity. Confirm it graphically for a reasonable interval.

SOLUTION Proof:

$\dfrac{\sin B}{1 + \cos B}$	Start with the more complicated side of the equation (binomial denominator).
$= \dfrac{\sin B}{1 + \cos B} \cdot \dfrac{1 - \cos B}{1 - \cos B}$	Multiply by a clever form of 1 (see the note following this example).
$= \dfrac{\sin B(1 - \cos B)}{1 - \cos^2 B}$	Distribute in the denominator but not in the numerator. You want $(1 - \cos B)$ in your result.
$= \dfrac{\sin B(1 - \cos B)}{\sin^2 B}$	Recognize the Pythagorean property, and use it to get a denominator with one term.
$= \dfrac{1 - \cos B}{\sin B}$	Cancel the sin *B* in the numerator with one sin *B* in the denominator.
$\therefore \dfrac{1 - \cos B}{\sin B} = \dfrac{\sin B}{1 + \cos B}$, Q.E.D.	Use the transitive property.

Enter $f_1(x) = (1 - \cos(x))/\sin(x)$ and $f_2(x) = \sin(x)/(1 + \cos(x))$ into your grapher. Plot the graphs using different styles, such as dashed for one and solid or path style for the other. Figure 7-3b shows the result.

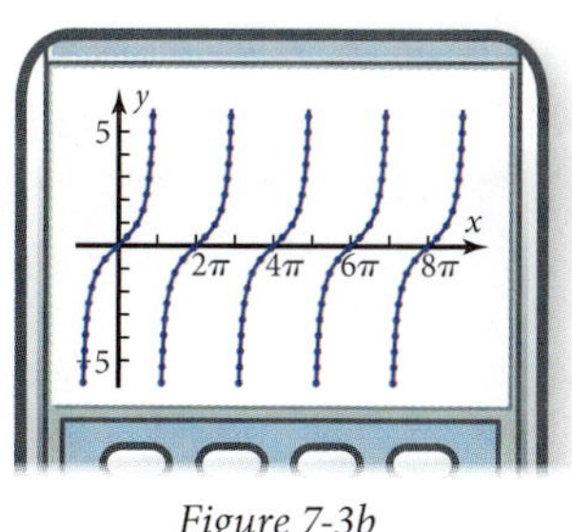

Figure 7-3b

Note: There are two ways to think up the form of 1 to multiply by in the second line of Example 5. First, the expressions $(1 + \cos B)$ and $(1 - \cos B)$ are **conjugate binomials,** or conjugates. When you multiply conjugates, you get a difference of two squares (no middle term). This allows you to use the Pythagorean property in the next step. Second, you want the quantity $(1 - \cos B)$ in the result, so you *put* it there by multiplying by a form of 1 that contains it.

The box summarizes useful techniques from the examples in this section.

PROCEDURE: Transforming Trigonometric Expressions and Proving Identities

1. Start by writing the given expression or, for an identity, by picking the side of the equation you wish to start with and writing it down. Usually it is easier to start with the more complicated side.
2. Look for *algebraic* things to do.
   - **a.** If there are two terms and you want only one term, then
     - **i.** Add fractions, or
     - **ii.** Factor something out.
   - **b.** Multiply by a clever form of 1 in order to
     - **i.** Multiply a numerator or denominator by its conjugate binomial, or
     - **ii.** Get a desired expression into the numerator or denominator.
   - **c.** Perform any obvious calculations (distribute, square, multiply polynomials, and so on).
   - **d.** Factor out an expression you want to appear in the result.
3. Look for *trigonometric* things to do.
   - **a.** Look for familiar trigonometric expressions you can transform.
   - **b.** If there are *squares* of functions, think of Pythagorean properties.
   - **c.** Reduce the number of different functions, transforming them into ones you want in the result.
   - **d.** Leave unchanged any expressions you want in the result.
4. Keep looking at the result and thinking of ways you can get closer to it.

Problem Set 7-3

Reading Analysis

From what you have read in this section, what do you consider to be the main idea? What are the similarities and differences between transforming one expression into another expression and proving that an equation is an identity? Why can't you start proving that an equation is an identity by writing down the given equation?

Quick Review

Q1. Write the Pythagorean property for cosine and sine.

Q2. Write the quotient property for tangent in terms of sine and cosine.

Q3. Write the quotient property for tangent in terms of secant and cosecant.

Q4. Write the reciprocal property for secant.

Q5. Why does $\tan x \cot x$ equal 1?

Q6. Sketch the graph of the parent cosine function $y = \cos x$.

Q7. Sketch the graph of the parent sine function $y = \sin \theta$.

Q8. What is the vertical dilation for $y = 2 + 3\cos 4(x - 5)$?

Q9. The reference angle for 260° is __?__.

Q10. $y = 3(1.06^x)$ is an example of a(n) __?__ function.

For Problems 1–26, show the steps in transforming the expression on the left into the one on the right.

1. $\cos x \tan x$ to $\sin x$
2. $\csc x \tan x$ to $\sec x$
3. $\sec A \cot A \sin A$ to 1
4. $\csc B \tan B \cos B$ to 1
5. $\sin^2 \theta \sec \theta \csc \theta$ to $\tan \theta$
6. $\cos^2 \alpha \csc \alpha \sec \alpha$ to $\cot \alpha$
7. $\cot R + \tan R$ to $\csc R \sec R$
8. $\cot D \cos D + \sin D$ to $\csc D$
9. $\csc x - \sin x$ to $\cot x \cos x$
10. $\sec x - \cos x$ to $\sin x \tan x$
11. $(\tan x)(\cot x \cos x + \sin x)$ to $\sec x$
12. $(\cos x)(\sec x + \cos x \csc^2 x)$ to $\csc^2 x$
13. $(1 + \sin B)(1 - \sin B)$ to $\cos^2 B$
14. $(\sec E - 1)(\sec E + 1)$ to $\tan^2 E$
15. $(\cos \phi - \sin \phi)^2$ to $1 - 2\cos \phi \sin \phi$
16. $(1 - \tan \phi)^2$ to $\sec^2 \phi - 2\tan \phi$
17. $(\tan n + \cot n)^2$ to $\sec^2 n + \csc^2 n$
18. $(\cos k - \sec k)^2$ to $\tan^2 k - \sin^2 k$
19. $\dfrac{\csc^2 x - 1}{\cos x}$ to $\cot x \csc x$
20. $\dfrac{1 - \cos^2 x}{\tan x}$ to $\sin x \cos x$
21. $\dfrac{\sec^2 \theta - 1}{\sin \theta}$ to $\tan \theta \sec \theta$
22. $\dfrac{1 + \cot^2 \theta}{\sec^2 \theta}$ to $\cot^2 \theta$

23. $\dfrac{\sec A}{\sin A} - \dfrac{\sin A}{\cos A}$ to $\cot A$
24. $\dfrac{\csc B}{\cos B} - \dfrac{\cos B}{\sin B}$ to $\tan B$
25. $\dfrac{1}{1 - \cos x} + \dfrac{1}{1 + \cos x}$ to $2\csc^2 x$
26. $\dfrac{1}{\sec D - \tan D} + \dfrac{1}{\sec D + \tan D}$ to $2\sec D$

For Problems 27–36, prove algebraically that the given equation is an identity.

27. $(\sec x)(\sec x - \cos x) = \tan^2 x$

28. $(\tan x)(\cot x + \tan x) = \sec^2 x$

29. $(\sin x)(\csc x - \sin x) = \cos^2 x$

30. $(\cos x)(\sec x - \cos x) = \sin^2 x$

31. $\csc^2 \theta - \cos^2 \theta \csc^2 \theta = 1$

32. $\cos^2 \theta + \tan^2 \theta \cos^2 \theta = 1$

33. $(\sec \theta + 1)(\sec \theta - 1) = \tan^2 \theta$

34. $(1 + \sin \theta)(1 - \sin \theta) = \cos^2 \theta$

35. $(2 \cos x + 3 \sin x)^2 + (3 \cos x - 2 \sin x)^2 = 13$

36. $(5 \cos x - 4 \sin x)^2 + (4 \cos x + 5 \sin x)^2 = 41$

37. Confirm that the equation in Problem 33 is an identity by plotting the two graphs on the same screen.

38. Confirm that the equation in Problem 34 is an identity by plotting the two graphs on the same screen.

39. Confirm that the equation in Problem 35 is an identity by making a table of values.

40. Confirm that the equation in Problem 36 is an identity by making a table of values.

41. Prove that the equation $\cos x = 1 - \sin x$ is *not* an identity.

42. Prove that the equation $\tan^2 x - \sec^2 x = 1$ is *not* an identity.

Problems 43–54 involve more complicated algebraic techniques. Prove that each equation is an identity.

43. $\sec^2 A + \tan^2 A \sec^2 A = \sec^4 A$

44. $\cos^4 t - \sin^4 t = 1 - 2 \sin^2 t$

45. $\dfrac{1}{\sin x \cos x} - \dfrac{\cos x}{\sin x} = \tan x$

46. $\dfrac{\sin x}{\csc x} + \dfrac{\cos x}{\sec x} = 1$

47. $\dfrac{1}{1 + \cos p} = \csc^2 p - \csc p \cot p$

48. $\dfrac{\cos x}{\sec x - 1} - \dfrac{\cos x}{\tan^2 x} = \cot^2 x$

49. $\dfrac{1 + \sin x}{1 - \sin x} = 2 \sec^2 x + 2 \sec x \tan x - 1$

50. $\sin^3 z \cos^2 z = \sin^3 z - \sin^5 z$

51. $\sec^2 \theta + \csc^2 \theta = \sec^2 \theta \csc^2 \theta$

52. $\sec \theta + \tan \theta = \dfrac{1}{\sec \theta - \tan \theta}$

53. $\dfrac{1 - 3 \cos x - 4 \cos^2 x}{\sin^2 x} = \dfrac{1 - 4 \cos x}{1 - \cos x}$

54. $\dfrac{\sec^2 x - 6 \tan x + 7}{\sec^2 x - 5} = \dfrac{\tan x - 4}{\tan x + 2}$

55. *Journal Problem:* Update your journal with what you have learned recently about transforming trigonometric expressions algebraically. In particular, show how you can use the three kinds of properties from Section 7-2 to transform an expression into a different form.

7-4 Arcsine, Arctangent, Arccosine, and Trigonometric Equations

In Section 6-6, you learned how to solve equations that reduce to the form

$$\cos x = a, \qquad \text{where } a \text{ is a constant}$$

You learned that the general solution is

$$x = \arccos a = \pm\cos^{-1} a + 2\pi n$$

where n is an integer representing a number of cycles or revolutions. In this section you will solve trigonometric equations involving sine or tangent rather than just cosine, where the argument may be in degrees or in radians.

Objective Find algebraically or numerically the solutions to equations involving circular or trigonometric sines, cosines, and tangents of one argument.

Arcsine, Arctangent, and Arccosine

You recall from Section 6-6 that arccos x means any of the angles whose cosine is x. Arcsin x and arctan x have the analogous meaning for sine and tangent. Within any one revolution there are two values of the inverse trigonometric relation for any given argument. Figure 7-4a shows how to find the values of arcsin $\frac{3}{5}$, arccos $\frac{3}{5}$, and arctan $\frac{3}{5}$. Sketch a reference triangle with appropriate side lengths 3 and 5, then look for a reference triangle in another quadrant for which the sides have displacements in the ratio $\frac{3}{5}$. You find the general solution by adding integer numbers of revolutions, $360°n$ or $2\pi n$ radians.

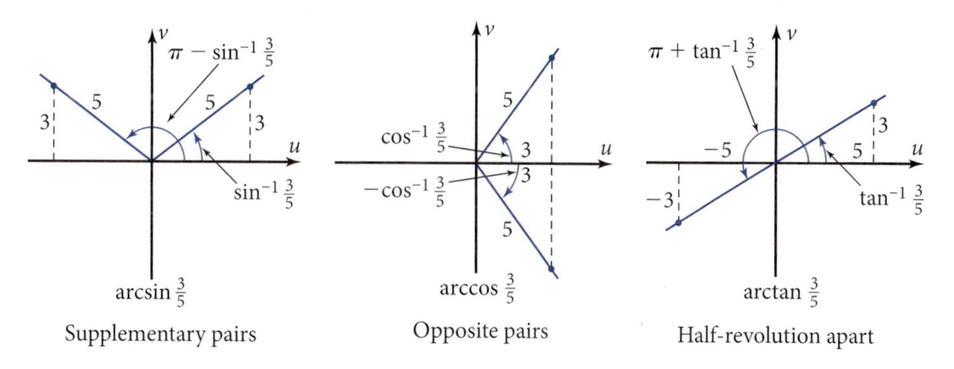

Figure 7-4a

The second angle involves a reflection of the reference triangle across an axis. To remember *which* axis, recall the definitions of the trigonometric functions.

arcsine:	$\sin x = \dfrac{\text{vertical coordinate}}{\text{radius}} = \dfrac{v}{r}$	Reflect across the vertical axis (Figure 7-4a, left).
arccosine:	$\cos x = \dfrac{\text{horizontal coordinate}}{\text{radius}} = \dfrac{u}{r}$	Reflect across the horizontal axis (Figure 7-4a, middle).
arctangent:	$\tan x = \dfrac{\text{vertical coordinate}}{\text{horizontal coordinate}} = \dfrac{v}{u}$	Reflect across *both* axes (Figure 7-4a, right).

EXAMPLE 1 ➤ Solve the equation $10\sin(x - 0.2) = -3$ algebraically for x in the domain $0 \le x \le 4\pi$. Verify the solutions graphically.

SOLUTION

$10\sin(x - 0.2) = -3$ — Write the given equation.

$\sin(x - 0.2) = -0.3$ — Reduce to the form $f(\text{argument}) = \text{constant}$.

$x - 0.2 = \arcsin(-0.3)$ — Take the arcsine of both sides.

$x = 0.2 + \arcsin(-0.3)$ — Isolate x.

$x = 0.2 + \sin^{-1}(-0.3) + 2\pi n$ — Substitute supplementary pairs for arcsine.

$\quad \text{or } 0.2 + [\pi - \sin^{-1}(-0.3)] + 2\pi n$

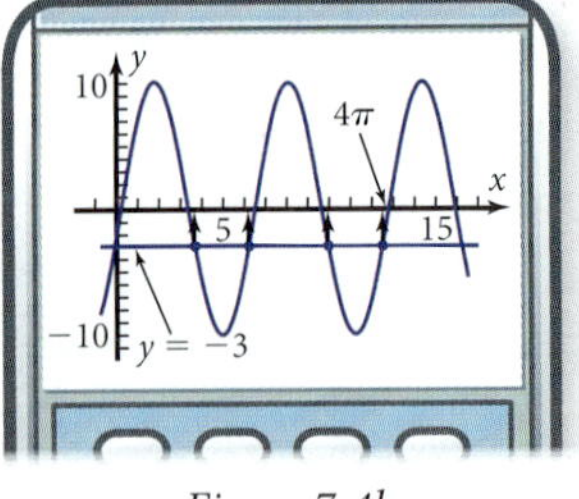

Figure 7-4b

$x = -0.1046... + 2\pi n \quad \text{or} \quad x = 3.6462... + 2\pi n$

$x = 3.6462..., 6.1784..., 9.9294..., 12.4616...$ — Choose the values of n that give solutions in the given domain.

The graph in Figure 7-4b shows $f_1(x) = 10\sin(x - 0.2)$ with the line $f_2(x) = -3$. Use the intersect feature of your grapher to show that the lines do intersect at the points in the solution. (Some intersections are out of the domain.) ◄

Notes:

- You can enter $f_1(x) = 0.2 + \sin^{-1}(-0.3) + 2\pi n$ and $f_2(x) = 0.2 + [\pi - \sin^{-1}(-0.3)] + 2\pi n$ into your grapher, using x in place of n, and use the table feature to find the particular values.
- The function value $\sin^{-1}(-0.3) = -0.3046\ldots$ terminates in Quadrant IV. The other value is the supplement of this number. Figure 7-4c shows that subtracting $-0.3046...$ from π gives an angle in Quadrant III, where the other value must be if its sine is negative.

$$\pi - (-0.3046...) = \pi + 0.3046... = 3.4462...$$

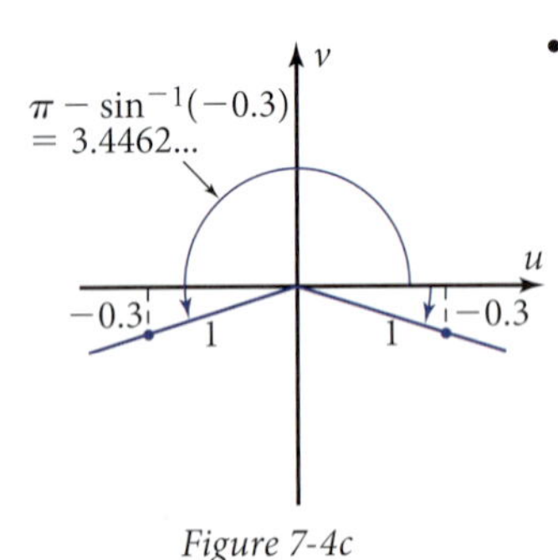

Figure 7-4c

Interval Notation

A compact way to write a domain such as $0 \le x \le 4\pi$ is $[0, 4\pi]$. This set of values of x is called the **closed interval** from $x = 0$ to $x = 4\pi$. The **open interval** from $x = 0$ to $x = 4\pi$ is written $(0, 4\pi)$ and means $0 < x < 4\pi$. The symbol $\in$ from set terminology is used to show that x is an "element of" or is in a given interval. So you can write the domain for the closed interval

$$x \in [0, 4\pi]$$

which is pronounced "x is an element of the closed interval from 0 to 4π." The various interval notations are summarized in this box.

DEFINITIONS: Interval Notation

Written	Meaning	Name	Visually
$x \in [0, 4\pi]$	$0 \le x \le 4\pi$	Closed interval	●—● 0 to 4π
$x \in (0, 4\pi)$	$0 < x < 4\pi$	Open interval	○—○ 0 to 4π
$x \in [0, 4\pi)$	$0 \le x < 4\pi$	Half-open interval	●—○ 0 to 4π
$x \in (0, 4\pi]$	$0 < x \le 4\pi$	Half-open interval	○—● 0 to 4π

EXAMPLE 2 Solve the equation $4 \tan 2\theta = -5$ algebraically for the first three positive values of θ. Verify the solutions graphically.

SOLUTION

$4 \tan 2\theta = -5$

$\tan 2\theta = -1.25$

$2\theta = \arctan(-1.25) = \tan^{-1}(-1.25) + 180°n$ — Write the general solution for 2θ, angles a half-revolution apart (Figure 7-4d).

$\theta = \frac{1}{2}\tan^{-1}(-1.25) + 90°n$ — Solve for θ; divide *both* terms on the right side of the equation by 2.

$\theta = -25.6700...° + 90°n$

$\theta = 64.3299...°, 154.3299...°, 244.3299...°$ — Choose the values of n that give the first three positive answers.

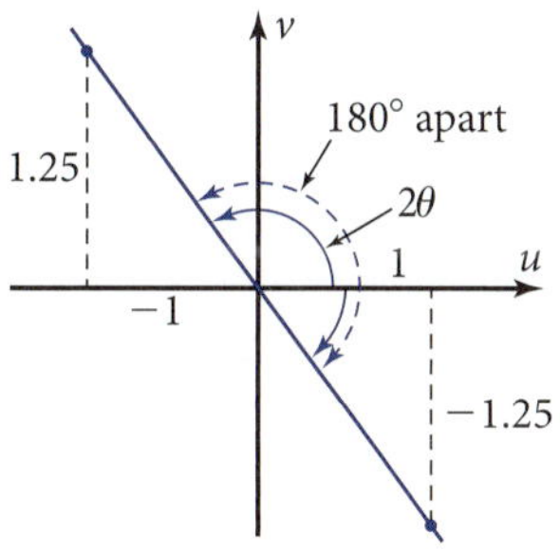

Figure 7-4d

The graph in Figure 7-4e shows $y = 4 \tan 2\theta$ and $y = -5$, with intersections at the three positive values that are in the solution.

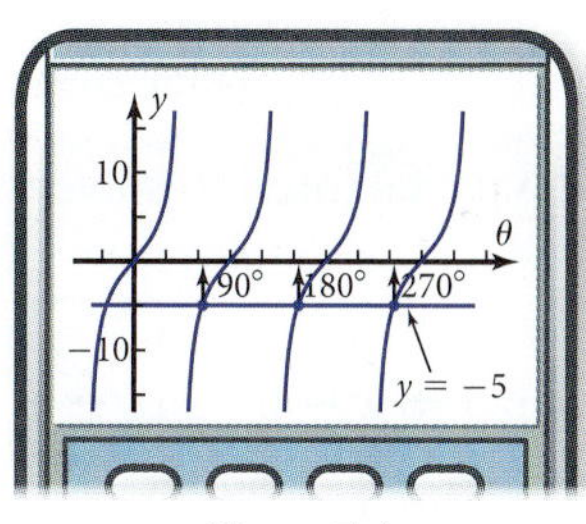

Figure 7-4e

PROPERTIES: General Solutions for Arcsine, Arccosine, and Arctangent

Let A stand for the argument of the inverse sine, cosine, or tangent, and let n represent an integer.

$$\theta = \arcsin A = \sin^{-1} A + 360°n \quad \text{or} \quad (180° - \sin^{-1} A) + 360°n$$

$$x = \arcsin A = \sin^{-1} A + 2\pi n \quad \text{or} \quad (\pi - \sin^{-1} A) + 2\pi n$$

Verbally: Inverse sines come in supplementary pairs (plus coterminals).
Graphically: Reflect the reference triangle across the *vertical* axis.

$$\theta = \arccos A = \pm\cos^{-1} A + 360°n$$

$$x = \arccos A = \pm\cos^{-1} A + 2\pi n$$

Verbally: Inverse cosines come in opposite pairs (plus coterminals).
Graphically: Reflect the reference triangle across *horizontal* axis.

$$\theta = \arctan A = \tan^{-1} A + 180°n$$

$$x = \arctan A = \tan^{-1} A + \pi n$$

Verbally: Inverse tangents come in pairs a half-revolution apart.
Graphically: Reflect the reference triangle across *both* axes.

Quadratic Forms

You may need to use the quadratic formula or factoring to solve algebraically an equation that has squares of trigonometric functions.

EXAMPLE 3 ➤ Solve algebraically $\cos^2\theta + \sin\theta + 1 = 0$, $\theta \in [-90°, 270°)$.

SOLUTION

$\cos^2\theta + \sin\theta + 1 = 0$

$(1 - \sin^2\theta) + \sin\theta + 1 = 0$ — Use the Pythagorean property to change all terms to *one* trigonometric function.

$\sin^2\theta - \sin\theta - 2 = 0$ — Write the equation in $ax^2 + bx + c = 0$ form and multiply by -1.

No solution: ~~$\sin\theta = 2$~~ or $\sin\theta = -1$ — Use the quadratic formula. Discard impossible solutions.

$\theta = \arcsin(-1) = -90° + 360°n$ — $180° - (-90°) = 270°$ is coterminal with $-90°$.

$\theta = -90°$ — The angle measure $270°$ when $n = 1$ is out of the domain (half-*open* interval). ◄

Note that in this case you could have factored to solve the quadratic equation.

$(\sin\theta - 2)(\sin\theta + 1) = 0$

$\sin\theta - 2 = 0 \quad \text{or} \quad \sin\theta + 1 = 0$

$\sin\theta = 2 \quad \text{or} \quad \sin\theta = -1$

If a product is zero, then one of its factors has to be zero!

Numerical Solutions

Some trigonometric equations cannot be solved algebraically. This is often true if the variable appears both transcendentally (in the argument of a non-algebraic function such as $y = \sin x$) and in an algebraic expression. For example,

$$0.2x + \sin x = 2$$

There is no algebraic solution because you cannot transform the equation to the form $f(\text{argument}) = \text{constant}$. In other cases, the algebraic solution may be difficult to find. In such cases, a numerical solution with the help of graphs is appropriate.

EXAMPLE 4 Solve $0.2x + \sin x = 2$ for all real values of x.

SOLUTION The graph in Figure 7-4f shows $f_1(x) = 0.2x + \sin x$ and $f_2(x) = 2$ intersecting at the three points $x \approx 7.0$, 9.3, and 12.1.

$x = 6.9414...,\ 9.2803...,\ 12.1269...$ Use the intersect or solver feature on your grapher.

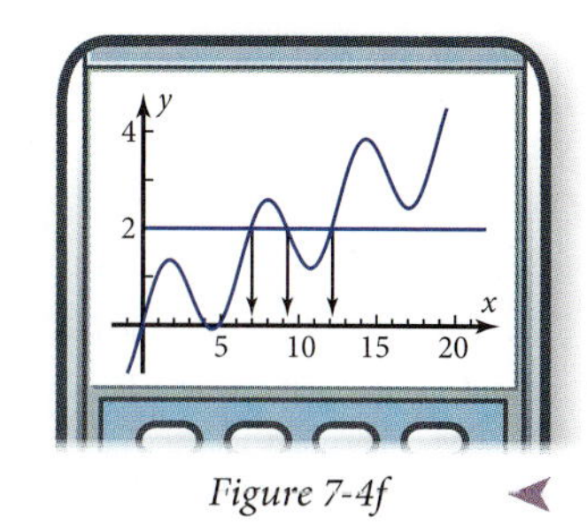

Figure 7-4f

Problem Set 7-4

Reading Analysis

From what you have read in this section, what do you consider to be the main idea? Why can you write $\arccos A = \pm\cos^{-1} A$ but not $\arctan A = \pm\tan^{-1} A$? What is the difference between a closed interval and an open interval? Give an example of a trigonometric equation that cannot be solved algebraically.

Quick Review

Q1. Write the particular equation for the sinusoid with amplitude 2, period 120°, sinusoidal axis at $y = 5$, and phase displacement 17° with respect to $y = \cos x$.

Q2. Sketch a reasonable graph depicting the time of sunset as a function of the day of the year.

Q3. Sketch the graph of $y = \sec x$.

Q4. What is the exact value (no decimals) of $\cos 30°$?

Q5. What is the exact value (no decimals) of $\sin \frac{\pi}{4}$?

Q6. Sketch the reference angle for 260°.

Q7. Right triangle XYZ has right angle Y. Side x is opposite angle X, and so on. Find $\csc X$.

Q8. Find the degree measure of the acute angle $\cot^{-1} 3$.

Q9. What is the value of n if $\log 32 = n \log 2$?

Q10. The graph of $y = 3x^2 + 2x - 7$ is called a(n) __?__.

For Problems 1–10,

a. Find the general solution for θ or x.

b. Find the particular solutions that are in the given interval.

1. $\theta = \arcsin 0.7$ $\quad \theta \in [0°, 720°]$

2. $\theta = \arcsin(-0.6)$ $\quad \theta \in [0°, 720°]$

3. $x = \arcsin(-0.2)$ $\quad x \in [0, 4\pi]$

4. $x = \arcsin 0.9$ $\quad x \in [0, 4\pi]$

5. $\theta = \arctan(-4)$ $\quad \theta \in [0°, 720°]$

6. $\theta = \arctan 0.5$ $\quad \theta \in [0°, 720°]$

7. $x = \arctan 10$ $\quad x \in [0, 4\pi]$

8. $x = \arctan(-0.9)$ $\quad x \in [0, 4\pi]$

9. $\theta = \arccos 0.2 \qquad \theta \in [0°, 720°]$

10. $x = \arccos(-0.8) \qquad x \in [0, 4\pi]$

11. Confirm graphically that the solutions to Problem 1b are correct.

12. Confirm graphically that the solutions to Problem 8b are correct.

13. Explain why there are *no* solutions to $x = \arccos 2$ but there *are* solutions to $x = \arctan 2$.

14. Explain why there are *no* solutions to $\theta = \arcsin 3$ but there *are* solutions to $\theta = \arctan 3$.

15. If one value of arctan A is 37°, find another value of arctan A in the interval [0°, 360°).

16. If one value of arctan A is $\frac{\pi}{3}$, find another value of arctan A in the interval $[0, 2\pi)$.

17. If one value of arcsin A is $\frac{5\pi}{6}$, find another value of arcsin A in the interval $[0, 2\pi)$.

18. If one value of arcsin A is 143°, find another value of arcsin A in the interval [0°, 360°).

19. If one value of arccos A is 2, find another value of arccos A in the interval $[0, 2\pi)$.

20. If one value of arccos A is −50°, find a value of arccos A in the interval [0°, 360°).

For Problems 21–26, solve the equation in the given domain.

21. $\tan\theta + \sqrt{3} = 0 \qquad \theta \in [0°, 720°]$

22. $2\cos\theta + \sqrt{3} = 0 \qquad \theta \in [0°, 720°]$

23. $2\sin(\theta + 47°) = 1 \qquad \theta \in [-360°, 360°]$

24. $\tan(\theta - 81°) = 1 \qquad \theta \in [-180°, 540°]$

25. $3\cos\pi x = 1 \qquad x \in [0, 6]$

26. $5\sin\pi x = 2 \qquad x \in [-2, 4]$

27. Confirm graphically that the solutions to Problem 23 are correct.

28. Confirm graphically that the solutions to Problem 24 are correct.

29. Figure 7-4g shows the graph of $y = 2\cos^2\theta - \cos\theta - 1$. Calculate algebraically the θ-intercepts in the domain $0° \le \theta \le 720°$, and show that they agree with the graph.

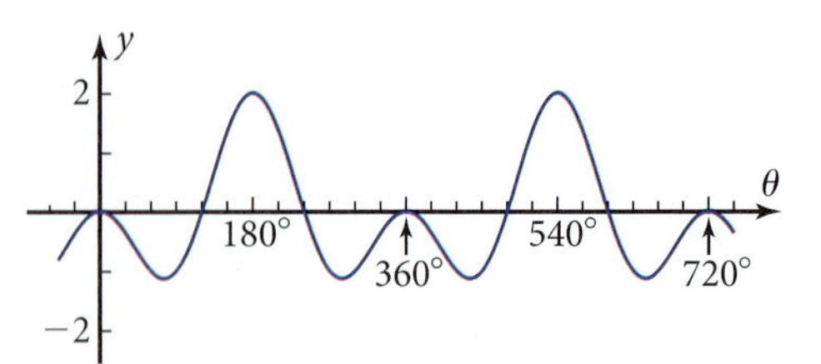

Figure 7-4g

30. Figure 7-4h shows the graph of $y = 2\sin^2\theta - 3\sin\theta + 1$. Calculate algebraically the θ-intercepts in the domain $0° \le \theta \le 720°$, and show that they agree with the graph.

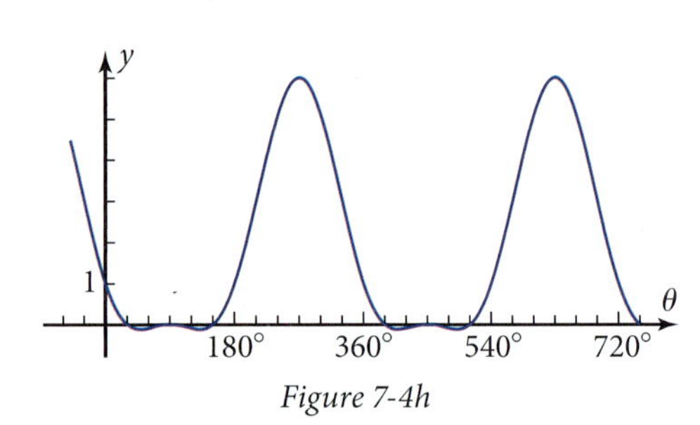

Figure 7-4h

31. Figure 7-4i shows the graph of $y = 2\sin^2\theta - 3\sin\theta - 2$. Calculate algebraically the θ-intercepts in the domain $0° \le \theta \le 720°$, and show that they agree with the graph.

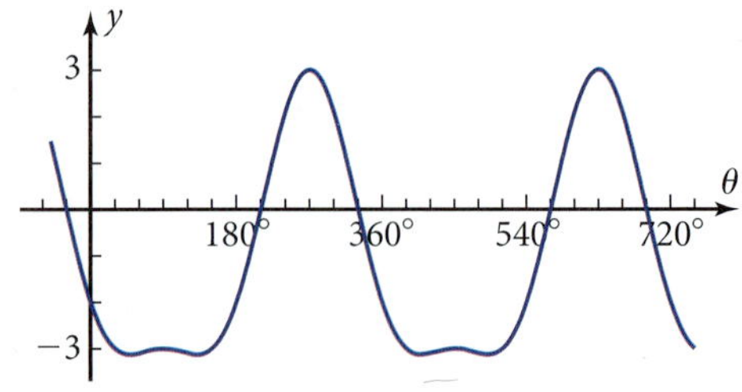

Figure 7-4i

32. Figure 7-4j shows the graph of $y = \cos^2\theta + 5\cos\theta + 6$. Calculate the θ-intercepts algebraically. Tell why the results you got agree with the graph.

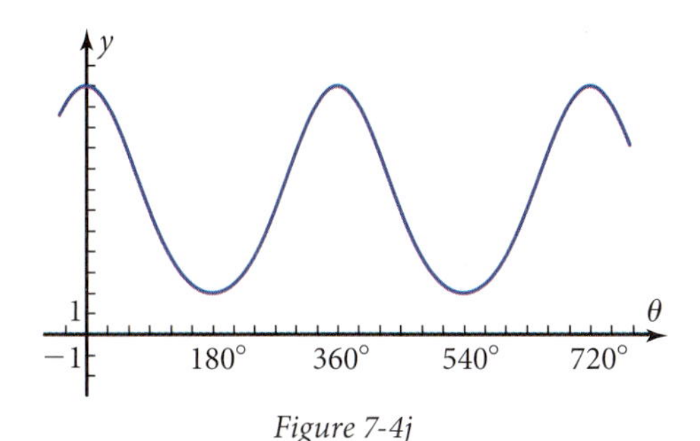

Figure 7-4j

For Problems 33–36,

a. Solve the equation graphically using the intersect feature of your grapher.

b. Solve the equation algebraically, confirming the graphical solution.

33. $3\cos^2\theta = 2\cos\theta$ $\quad\theta \in [0°, 360°)$

34. $\tan^2\theta = 2\tan\theta$ $\quad\theta \in [0°, 360°)$

35. $4\cos^2 x + 2\sin x = 3$ $\quad x \in [0, 2\pi)$

36. $5\sin^2 x - 3\cos x = 4$ $\quad x \in [0, 2\pi)$

37. *Rotating Beacon Problem:* Figure 7-4k shows a rotating beacon on a lighthouse 500 yd offshore. The beam of light shines out of both sides of the beacon, making a spot of light that moves along the beach with a displacement y, measured in yards, from the point on the beach that is closest to the lighthouse.

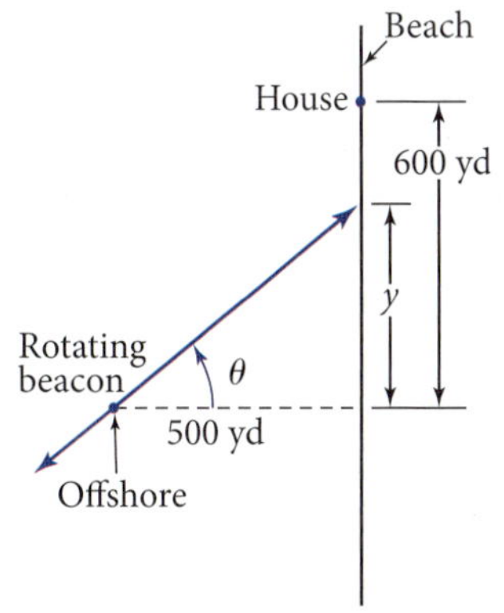

Figure 7-4k

a. Write an equation for y in terms of θ.

b. The beacon rotates with an angular velocity of 5 deg/s. Let t be the time, in seconds, since the beam was perpendicular to the beach (that is, $y = 0$). By appropriate substitution, write an equation for y as a function of t.

c. A house on the beach is at a displacement $y = 600$ yd. Find the first four positive values of t when the spot of light illuminates the house.

38. *Numerical Solution of Equation Problem 1:* Figure 7-4l shows the graphs of $y = x$ and $y = \cos x$ as they might appear on your grapher.

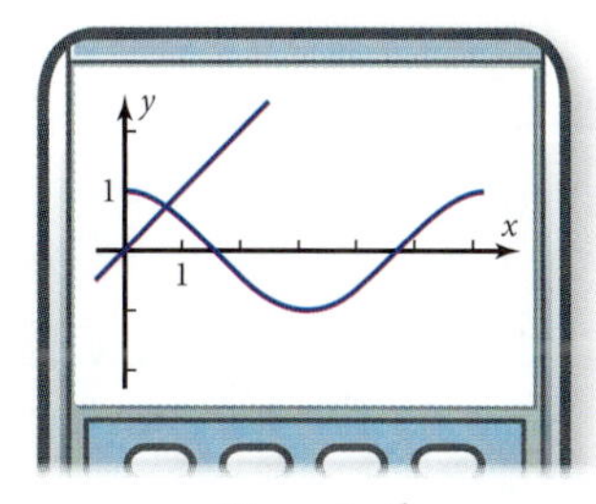

Figure 7-4l

a. Read from the graph a value of x for which $\cos x \approx x$.

b. Solve numerically to find a more precise value of x in part a.

c. Are there other values of x for which $\cos x \approx x$? How did you reach your conclusion?

d. Explain why the equation $\cos x = x$ cannot be solved algebraically.

39. *Numerical Solution of Equation Problem 2:* Figure 7-4m shows the graphs of $y = x$ and $y = \tan \pi x$.

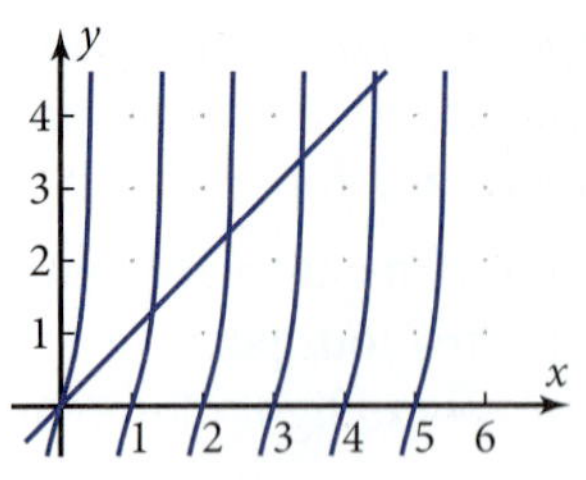

Figure 7-4m

a. Read from the graph the first three values of x for which $\tan \pi x = x$.

b. Solve numerically to find the three precise values in part a.

c. Explain why the equation $\tan \pi x = x$ cannot be solved algebraically.

40. *Numerical Solution of Equation Problem 3:* Figure 7-4n shows the graphs of $y = x$ and $y = 5 \sin \frac{\pi}{2}x$.

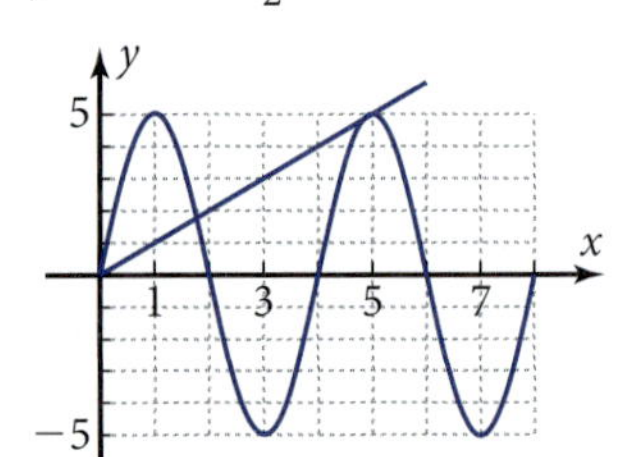

Figure 7-4n

a. Find numerically the greatest value of x for which $5 \sin \frac{\pi}{2}x = x$.

b. Find numerically the next-to-greatest value of x for which $5 \sin \frac{\pi}{2}x = x$. The zoom feature on your grapher may help.

c. Explain why the equation $5 \sin \frac{\pi}{2}x = x$ cannot be solved algebraically.

41. *Trigonometric Inequality Problem 1:* Figure 7-4o shows the region of points that satisfy the trigonometric inequality

$$y \leq 2 + 3 \cos \frac{\pi}{6}x$$

a. Duplicate the figure on your grapher. Use the appropriate style to shade the region below the boundary curve.

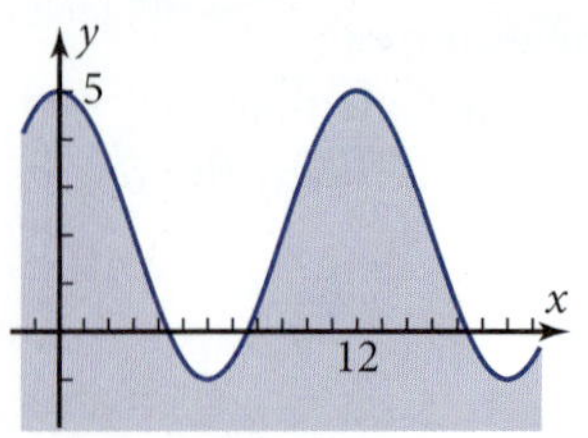

Figure 7-4o

b. On the same screen, plot the graph of $y \geq 0$. Use the appropriate style so that the grapher will shade the region. Sketch the intersection of the two regions.

c. Find the interval of x-values centered at $x = 12$ in which both inequalities are satisfied.

42. *Trigonometric Inequality Problem 2:* Figure 7-4p shows the region of points that satisfy the trigonometric inequality

$$y \leq 5 \sin \frac{\pi}{4}x$$

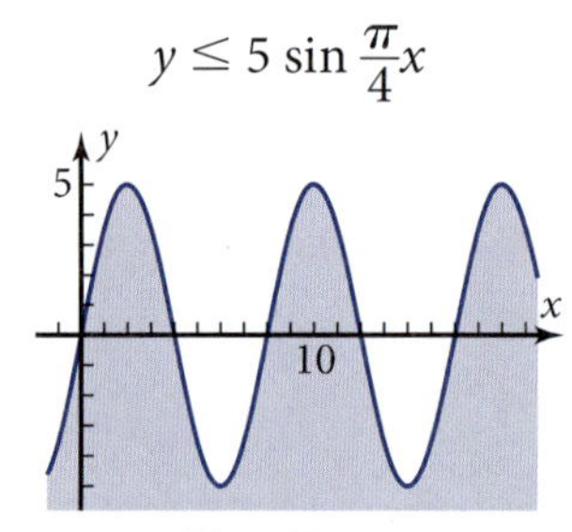

Figure 7-4p

a. Duplicate the figure on your grapher. Use the appropriate style to shade the region.

b. On the same screen, plot the region of points that satisfy the inequality $y \geq 0.3x$. Sketch the intersection of the two regions for $x \geq 0$.

c. Find all intervals of x-values for $x \geq 0$ for which both inequalities are satisfied.

43. *Surprise Problem:* Try solving this equation algebraically. Show how to interpret the results graphically. In particular, what do the graphs of the two sides of the equation look like?

$$\frac{1 - \sin x}{\cos x} = \frac{\cos x}{1 + \sin x}$$

7-5 Parametric Functions

In Section 1-5, you used parametric equations in connection with plotting the graph of the inverse of a function. In Section 7-6, you will use parametric equations to graph inverses of the six trigonometric functions. In this section you will use parametric equations to represent functions for which both x and y depend on a third variable, such as time. The pendulum in Figure 7-5a represents such a function.

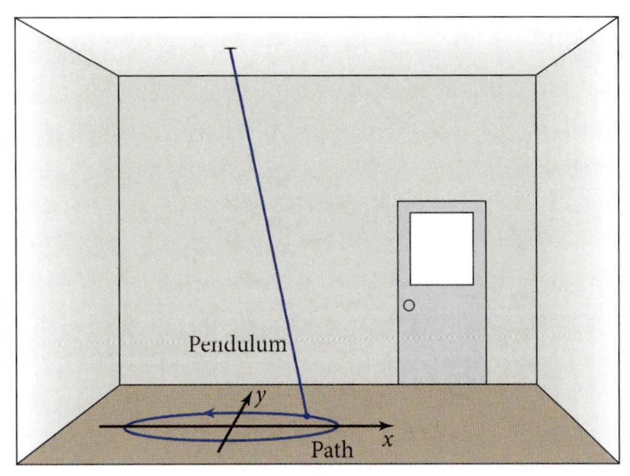

Figure 7-5a

Objective Given equations for a parametric function, plot the graph and make conclusions about the geometric figure that results.

In this exploration you will predict the position, (x, y), of a pendulum moving in both the x- and y-directions at time t.

EXPLORATION 7-5: Parametric Function Pendulum Problem

A pendulum hangs from the ceiling, as shown in Figure 7-5a. When the pendulum bob is at rest, it hangs above the origin, $(0, 0)$, of an xy-plane on the horizontal floor.

1. The pendulum is pulled to a displacement, x, of 30 cm and released at time $t = 0$ s, causing it to swing back and forth between $x = 30$ cm and $x = -30$ cm. By stopwatch you find that the period for one complete back-and-forth swing is 3.8 s. Assuming that displacement x varies sinusoidally with time, sketch the graph and write an equation for x as a function of t.

2. The pendulum is restarted by giving it a push from the origin in the y-direction at time $t = 0$ s. Again, it swings back and forth, this time swinging first to $y = 20$ cm, then back to $y = -20$ cm, with each complete back-and-forth swing again taking 3.8 s. Assuming that displacement y varies sinusoidally with time, sketch the graph and write a simple equation for y as a function of t.

continued

3. The pendulum is started a third time, this time by pulling it out to $x = 30$ cm, then giving it a push in the y-direction just hard enough to make it trace an elliptical path, as shown in Figure 7-5a and plotted in the figure below. The equations you found in Problems 1 and 2 express the position of the pendulum bob as a *parametric function* of time t. With your grapher in parametric mode, plot three cycles of the graph of this function. Use path style so that you can see when the graph is being retraced. Does your graph agree with the figure?

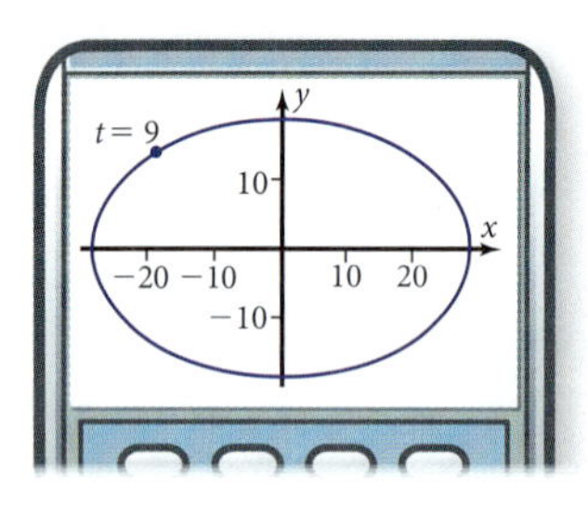

4. Trace to $t = 9$ s and write the coordinates of the point on the graph. Keep at least three decimal places. Does your answer agree with the point shown in the figure?
5. What special name is given to a variable, such as t in this problem, upon which two or more other variables depend?
6. What special name is given to functions in which two or more variables depend on the same independent variable?
7. Just for fun, see if you can transform the two parametric equations by eliminating the *parameter*, t, to get one equation involving only x and y. How does this equation confirm that the path really is an ellipse?
8. What did you learn as a result of doing this exploration that you did not know before?

In Exploration 7-5, you saw an example in which two related variables, x and y, depend on an independent third variable, t. The set of points (x, y) can be thought of as a dependent "variable" in a **parametric function.** The independent variable t is called the *parameter*. (The prefix *para-* is a Greek word meaning "beside" or "near," as in *parallel*, and the suffix *-meter* means "measure.") The letter t is usually used for the parameter because variables x and y often depend on time. Example 1 shows you how to plot a parametric function using degree mode.

EXAMPLE 1 ➤ Plot the graph of this parametric function in degree mode.

$$x = 5\cos t$$
$$y = 7\sin t$$

SOLUTION Set your grapher to parametric mode and enter the two equations. Choose a window that uses equal scales on both axes. Because the amplitudes of x and y are 5 and 7, respectively, the window will have to be at least $-5 \le x \le 5$ and $-7 \le y \le 7$. Use an interval of at least $0° \le t \le 360°$ and a t-step of 5°. Figure 7-5b shows the graph. ◄

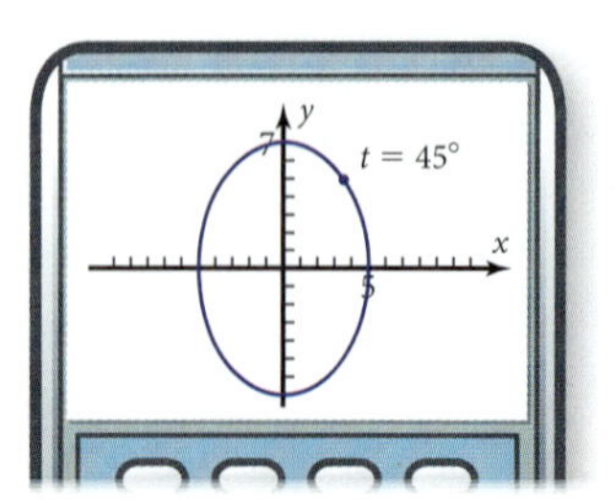

Figure 7-5b

Note that the parameter t is *not* an angle in standard position. Figure 7-5b shows that when $t = 45°$, the angle measure is considerably larger than 45°. In Chapter 11, you'll learn graphical properties of parametric functions that reveal how the angle is related to points on the path.

Pythagorean Properties to Eliminate the Parameter

You can sometimes discover properties of a graph by eliminating the parameter, thereby reducing the function to a single Cartesian equation with only two variables, x and y. The next example shows you how.

EXAMPLE 2 For the parametric function $x = 5 \cos t$, $y = 7 \sin t$ in Example 1, eliminate the parameter to get a Cartesian equation relating x and y. Describe the graph.

SOLUTION Because $\cos^2 t + \sin^2 t = 1$, you can eliminate the parameter by solving the given equations for $\cos t$ and $\sin t$, squaring both sides of each equation, and then adding.

$$x = 5 \cos t \Rightarrow \frac{x}{5} = \cos t \Rightarrow \left(\frac{x}{5}\right)^2 = \cos^2 t$$

$$y = 7 \sin t \Rightarrow \frac{y}{7} = \sin t \Rightarrow \left(\frac{y}{7}\right)^2 = \sin^2 t$$

$$\left(\frac{x}{5}\right)^2 + \left(\frac{y}{7}\right)^2 = \cos^2 t + \sin^2 t$$

Add the two equations, left side to left side, right side to right side.

$$\left(\frac{x}{5}\right)^2 + \left(\frac{y}{7}\right)^2 = 1$$

Use the Pythagorean property for cosine and sine.

The graph is an ellipse.

The unit circle $x^2 + y^2 = 1$ is dilated horizontally by 5 and vertically by 7.

EXAMPLE 3 Figure 7-5c shows the graph of the parametric equations

$$x = 6 + 5 \cos t$$
$$y = -3 + 7 \sin t$$

Describe the effect of the constants 6 and -3 on the graph.

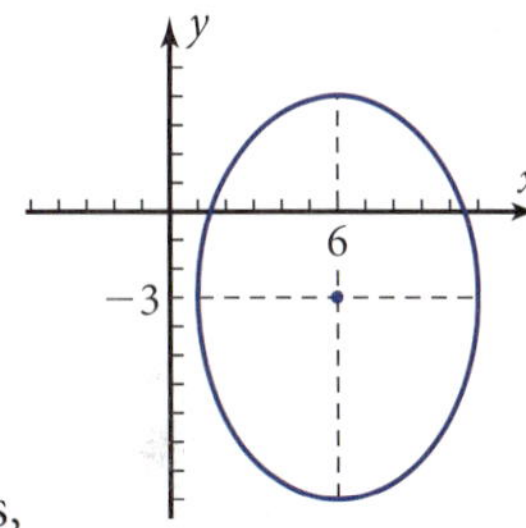

Figure 7-5c

SOLUTION Comparing Figure 7-5c to Figure 7-5b, you can see that the constants 6 and -3 are horizontal and vertical translations, respectively, and give the coordinates of the center of the ellipse.

Parametric Equations from Graphs

From the previous examples you can generalize the parametric equations for an ellipse with axes parallel to the coordinate axes.

PROPERTY: Parametric Equations for an Ellipse

The general parametric equations for an ellipse with axes parallel to the coordinate axes are

$$x = h + a \cos t$$
$$y = k + b \sin t$$

where $|a|$ and $|b|$ are called the x- and y-radii, respectively, and h and k are the coordinates of the center. If $|a| = |b|$, the figure is a circle.

Note: The coefficients a and b are the horizontal and vertical dilations of the **unit circle**

$$x = \cos t$$
$$y = \sin t$$

Also, the constants h and k are the horizontal and vertical translations, respectively, of the center of the unit circle.

If you know this property, you can use parametric functions to plot solid three-dimensional figures such as cones and cylinders on your grapher. Example 4 shows you how to do this.

EXAMPLE 4 ➤ Figure 7-5d shows the outlines of a cylinder. Duplicate this figure on your grapher by finding parametric equations for ellipses to represent the bases and then drawing lines to represent the walls.

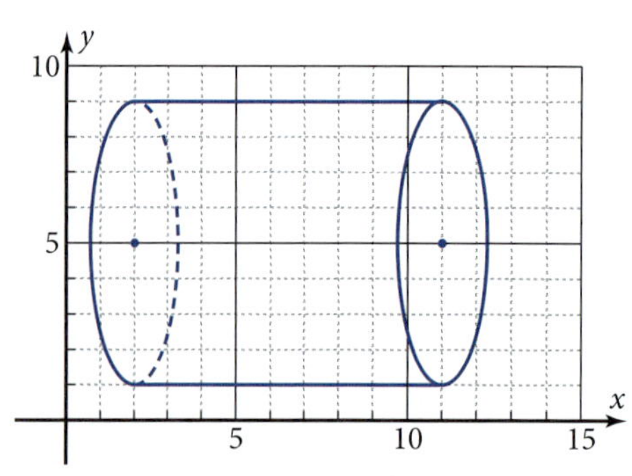

Figure 7-5d

SOLUTION The bases are ellipses because the cylinder is shown in perspective. The right ellipse is centered at the point (11, 5), with x-radius about 1.3 and y-radius 4. The left ellipse is congruent with the right one, with the center at the point (2, 5). Half of the left ellipse is hidden by the cylinder.

$$x_1(t) = 11 + 1.3 \cos t$$
$$y_1(t) = 5 + 4 \sin t$$

Parametric equations representing the right ellipse.

$$x_2(t) = 2 + 1.3 \cos t \,/\, (90 \leq t \text{ and } t \leq 270)$$
$$y_2(t) = 5 + 4 \sin t$$

Solid half of the left ellipse.

$$x_3(t) = 2 + 1.3 \cos t \,/\, (t \leq 90 \text{ or } t \geq 270)$$
$$y_3(t) = 5 + 4 \sin t$$

Dashed half of the left ellipse. Use dot style.

Plot these functions on your grapher in parametric mode, using degrees, with equal scales on the two axes. Use an interval of 0° to 360° for t to get a complete revolution for each ellipse. Draw lines from the point (2, 9) to the point (11, 9) and from the point (2, 1) to the point (11, 1), representing the walls of the cylinder. ◄

Problem Set 7-5

Reading Analysis

From what you have read in this section, what do you consider to be the main idea? The last statement in the solution to Example 2 defines an ellipse in terms of transformations on the unit circle. Look back at the discussion of dilations in Section 1-3, and then explain in your own words why the Cartesian equation at the end of Example 2 defines an ellipse.

Quick Review

Q1. What is the Pythagorean property for cosine and sine?

Q2. What is the Pythagorean property for secant and tangent?

Q3. If $\cos^{-1} x = 1.2$, what is the general solution for $\arccos x$?

Q4. If $\sin^{-1} x = 56°$, what is the general solution for $\arcsin x$?

Q5. For right triangle ABC, if angle B is the right angle, then $\sin A =$ __?__.

Q6. For right triangle ABC in Q5, side $a^2 =$ __?__ in terms of sides b and c.

Q7. If $y = \cos B\theta$ has period 180°, what does B equal?

Q8. What is the period of the parent sine function, $y = \sin x$?

Q9. If an angle has measure $\frac{\pi}{6}$ radian, what is its degree measure?

Q10. The exact value of $\cos \frac{\pi}{4}$ is

A. 0 **B.** $\frac{1}{\sqrt{2}}$ **C.** $\frac{1}{2}$

D. $\frac{\sqrt{3}}{2}$ **E.** 1

Problems 1 and 2 show the relationship among x, y, and t in parametric functions. For each problem,

a. Make a table of x- and y-values for a range of t-values. Include negative values of t.

b. Plot the points (x, y) on graph paper and connect them with a line or a smooth curve.

c. Confirm that your graph is correct by plotting it on your grapher using parametric mode.

1. $x = 3t + 1$
$y = 2t - 1$

2. $x = 1 + t^2$
$y = t + 2$

For Problems 3–6,

a. Plot the graph on your grapher. Sketch the result.

b. Use the Pythagorean property for cosine and sine to eliminate the parameter t.

c. Explain how you know that the graph is an ellipse or a circle.

3. $x = 3 \cos t$
$y = 5 \sin t$

4. $x = 6 \cos t$
$y = 6 \sin t$

5. $x = 5 + 7 \cos t$
$y = 2 + 3 \sin t$

6. $x = 4 + 3 \cos t$
$y = -1 + 6 \sin t$

The truncated cylindrical tower of the Museum of Modern Art in San Francisco, California, has an elliptical face.

Problems 7–14 show solid three-dimensional figures. The ellipses represent circular bases of the solids. The dashed lines represent hidden edges.

a. Write parametric equations for the ellipses.

b. Plot the figure on your grapher by using parametric equations and drawing lines.

7. Cone

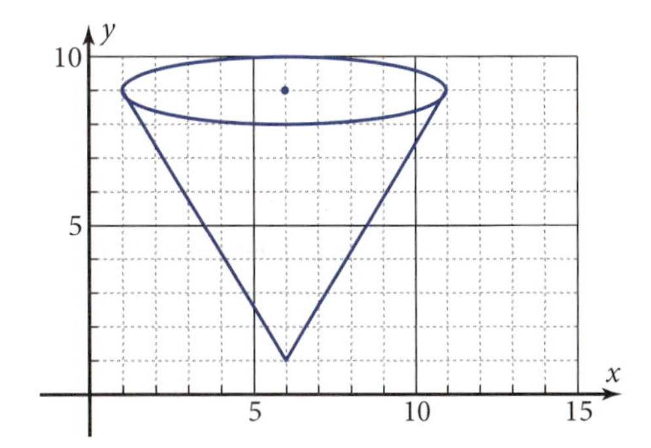

8. Cone

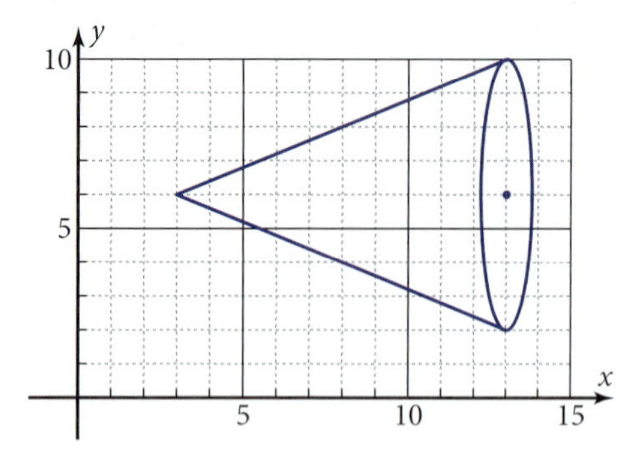

9. Cylinder

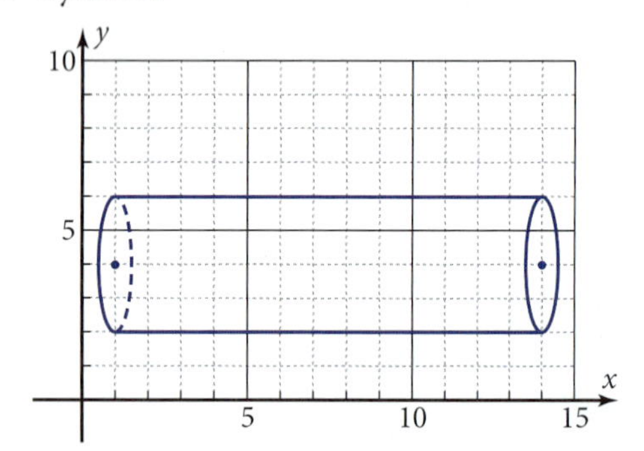

10. Cylinder

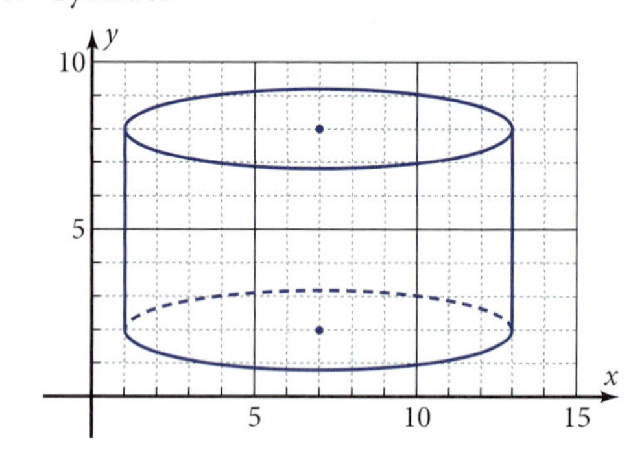

11. Frustum of a cone

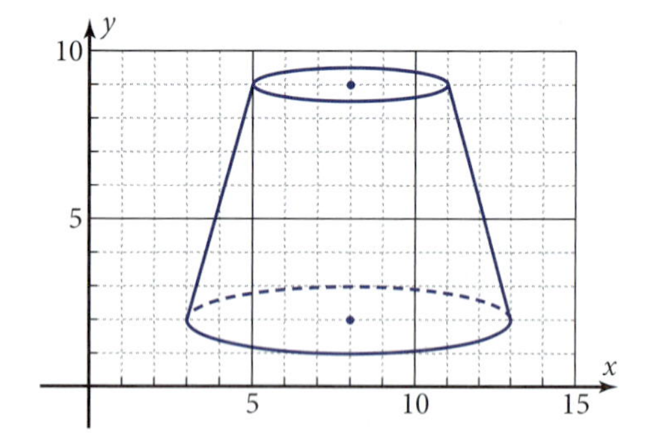

12. Two-napped cone

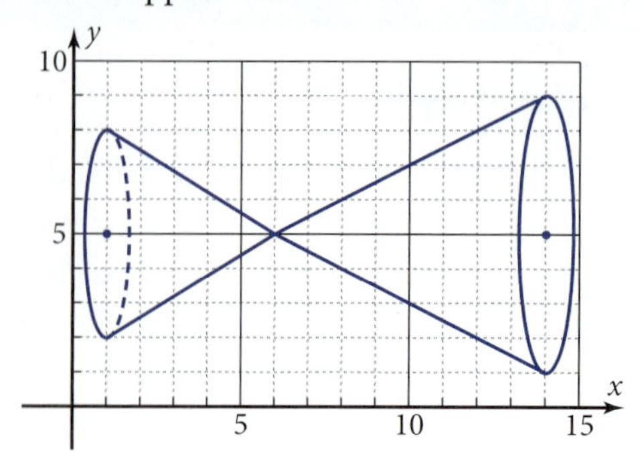

13. Hemisphere
Include the equation for the semicircle.

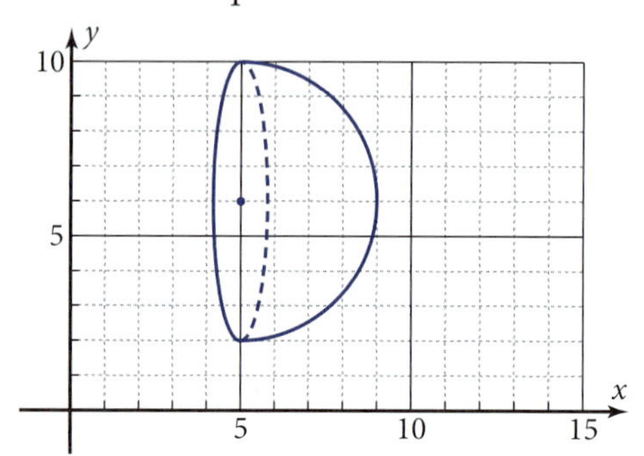

14. Hemisphere
Include the equation for the semicircle.

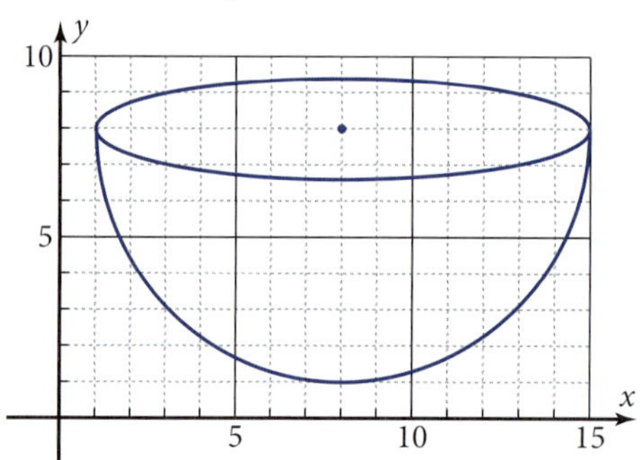

15. *Projectile Problem:* If a ball is thrown through the air, its motion in the horizontal and vertical directions is modeled by two different physical laws. Horizontally, the ball moves at a constant rate if you ignore air resistance. Vertically, the ball accelerates downward due to gravity. Let x be the ball's horizontal displacement, in meters, from its starting point, and let y be the vertical displacement, in meters, above its starting point. Suppose a ball is thrown with a horizontal velocity of 20 m/s and an initial upward velocity of 40 m/s. The parametric equations for its position (x, y) at time t, in seconds, are

$$x = 20t$$
$$y = 40t - 4.9t^2$$

The graph of x and y as functions of t is shown in Figure 7-5e.

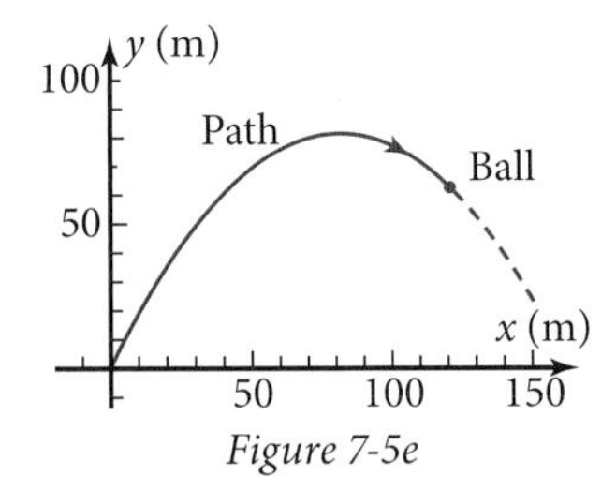

Figure 7-5e

a. What is the position of the ball at time $t = 3$ s?

b. When is the ball at a horizontal displacement, x, of 100 m? How high is it at that time?

c. At what *two* times is the ball 30 m above its starting height? Find x at these times.

d. A fence 2 m high is at $x = 160$ m. According to this parametric function, will the ball go over the fence, hit the fence, or hit the ground before reaching the fence? How can you tell?

e. Eliminate the parameter t, showing that y is a quadratic function of x.

16. *Parametric Function Domain Problem:* Sometimes when you eliminate the parameter, the Cartesian function has a domain different from that of the parametric function. In this problem you will investigate the parametric function

$$x_1(t) = 3\cos^2 t$$
$$y_1(t) = 2\sin^2 t$$

a. Set your grapher to radian mode. Plot the graph using $[-2\pi, 2\pi]$ for t and a window that includes positive and negative values of x and y. Sketch the result.

b. Based on your graph, make a conjecture about what geometric figure the graph is.

c. Eliminate the parameter t with the help of the Pythagorean properties. Solve the equation for y in terms of x; that is, find $y = f(x)$. Does the Cartesian equation confirm or refute your conjecture in part b?

d. You can plot the Cartesian equation in part c in parametric mode this way:

$$x_2(t) = t$$
$$y_2(t) = f(t)$$

The equation $y = f(t)$ is the Cartesian equation you found in part c, with t in place of x. Plot the graph. Compare the x-domain and y-range of the Cartesian equation to those of the parametric equations. Describe your observations.

Problems 17–20 involve parametric functions that have interesting graphs. Plot the graphs on your grapher and sketch the results. Use radian mode.

17. *Involute of a Circle Problem:*

$$x = \cos t + t\sin t$$
$$y = \sin t - t\cos t$$

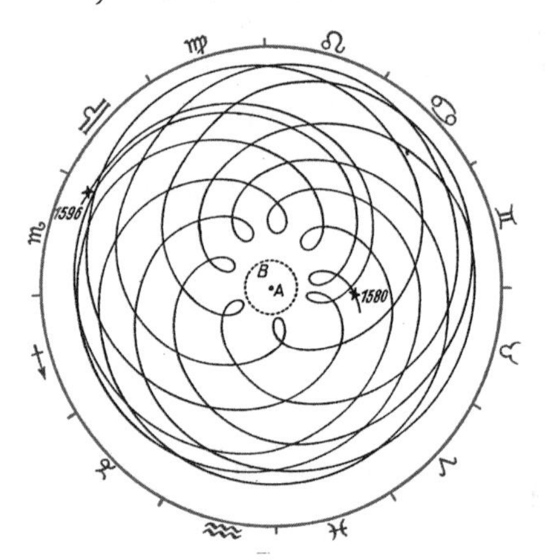

The motion of Mars. Adapted from Johannes Kepler's Astronomia Nova *(1609)*

18. *Asteroid Problem:* $x = 8\cos^3 t$
$x = 8\sin^3 t$

This curve is also called a hypocycloid of four cusps.

19. *Cycloid Problem:* $x = t + \sin t$
$y = 1 - \cos t$

20. *Conchoid of Nicomedes Problem:*

$$x = \tan t + 5\sin t$$
$$y = 1 + 5\cos t$$

21. *Sine Curve Tracer Problem:* Figure 7-5f shows two parametric functions plotted in radian mode,

$x_1(t) = \cos t \qquad x_2(t) = t$
$y_1(t) = \sin t \qquad y_2(t) = \sin t$

The first is a unit circle centered at the origin, and the second is the parent sine function.

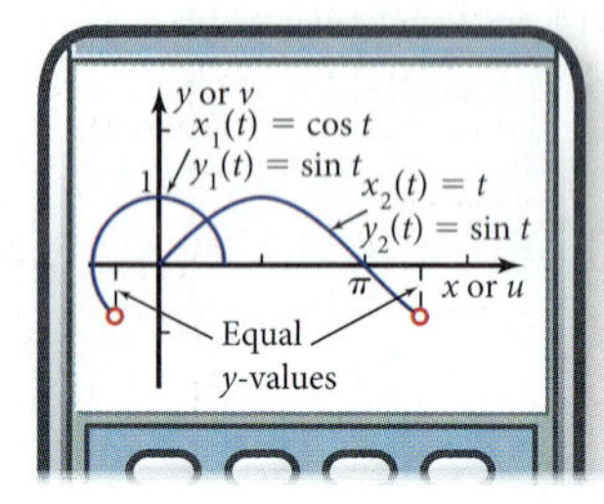

Figure 7-5f

a. Enter these functions into your grapher. Use simultaneous mode so that both graphs will be plotted at the same time. Use a window with $[-1, 2\pi]$ for x and equal scales on the x- and y-axes. Use $[0, 2\pi]$ for t and a small t-step, such as 0.02, so that the graphs are plotted slowly and you can observe the relationship between the graphs as they develop. Does your graph resemble Figure 7-5f at some time as it is being plotted? If not, go back and check your work.

b. In what way does observing the graph being plotted clarify in your mind the relationship between $\sin x$ for x as an angle in standard position in a uv-coordinate system and $\sin x$ plotted in an xy-coordinate system?

c. What changes would you need to make to the parametric equations in order to trace the cosine curve?

Problem 22 prepares you for the next section.

22. *Graphs of Inverse Trigonometric Relations by Parametrics:* Figure 7-5g shows the graph of the relation $y = \arcsin x$. Note that this is *not* a function, because there is more than one value of y corresponding to the same value of x. In this problem you will learn how to duplicate the graph on your grapher.

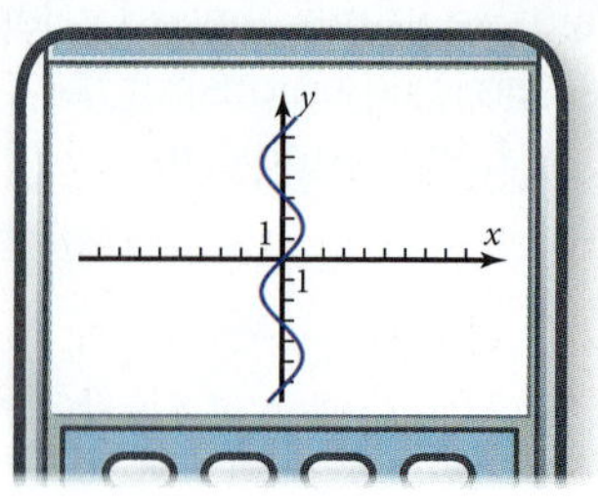

Figure 7-5g

a. With your grapher in function mode, plot $y = \sin^{-1} x$ (radian mode). Use a window with at least $-7 \leq y \leq 7$ and approximately equal scales on both axes. Why does the grapher show only part of the graph in Figure 7-5g?

b. Set your grapher to parametric mode. Enter the parametric equations

$x_1(t) = \sin t$
$y_1(t) = t$

Use a t-interval as large as the y-window. Describe the results. Based on the definition of arcsine, explain why these parametric equations generate the graph of the entire inverse sine relation.

c. With the first parametric equations still active, enter the equations

$x_2(t) = t$
$y_2(t) = \sin t$

Use a t-interval as large as the x-window. Sketch the resulting graphs. How are the two graphs related to each other?

d. Repeat parts b and c for $y = \arccos x$ and $y = \cos x$.

e. Repeat parts b and c for $y = \arctan x$ and $y = \tan x$.

7-6 Inverse Trigonometric Relation Graphs

You have learned that an inverse trigonometric relation, such as arcsin 0.4, has many values; but when you enter $\sin^{-1} 0.4$ into your calculator, it gives you only one of those values. In this section you'll learn which value the calculator has been programmed to give. You'll also learn how to calculate exact values of inverse trigonometric functions.

Objective

- Plot graphs of inverse trigonometric functions and relations.
- Find exact values of inverse trigonometric functions.

In this exploration you will investigate the graph of the inverse sine and inverse cosine relations.

EXPLORATION 7-6: Graphs of Inverse Trigonometric Relations

The figure shows the graph of $y = \arcsin x$, the *inverse relation* for $y = \sin x$. In this exploration you will plot this graph on your grapher along with $y = \sin x$ and make some conclusions about how to restrict the range of the relation in order to get the *inverse function* for $y = \sin x$, $y = \sin^{-1} x$.

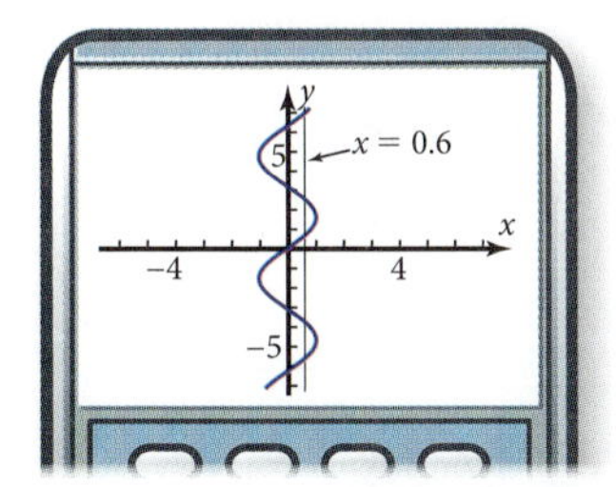

1. The figure shows the vertical line $x = 0.6$. From the graph, read to one decimal place the five values of y for $x = 0.6$. How does this indicate that $y = \arcsin x$ is not a function?

2. Plot $y = \arcsin x$ on your grapher using parametric mode. Enter

   $x_1(t) = \sin t$
   $y_1(t) = t$

   Use a window with $[-7, 7]$ for y, equal scales on the two axes, and a t-interval of $-7 \le t \le 7$. Does your graph agree with the figure?

3. The inverse sine *function*, $y = \sin^{-1} x$, is defined by limiting the range of $y = \arcsin x$ using these criteria:

   - The graph must be a *function*.
   - The graph must use the entire domain, $[-1, 1]$, for x.
   - The graph should be one *continuous* piece.
   - The graph should be centrally located, near the origin.

   What interval of y-values meets all these criteria?

4. On the same screen as in Problem 2, plot the graph of $y = \sin^{-1} x$, using thick style to distinguish the two graphs. Enter

   $x_2(t) = t$
   $y_2(t) = \sin^{-1} t$

   Does the range you chose in Problem 3 agree with the range of $y = \sin^{-1} x$ on your grapher?

5. Deactivate $y = \sin^{-1} x$ from Problem 4, and then plot $y = \sin x$ on the same screen as $y = \arcsin x$ in Problem 2. Use thick style to distinguish the two graphs. How are the two graphs related to the diagonal line $y = x$? Sketch the result on a copy of the figure.

continued

6. Plot the graph of $y = \arccos x$ and $y = \cos x$ on your grapher. Use the same window as in the previous problems. Sketch the result and the line $y = x$ on graph paper.
7. What range of y-values would meet all the criteria in Problem 3 to define the inverse cosine function, $y = \cos^{-1} x$? Confirm or refute your answer by plotting the graph of $y = \cos^{-1} x$ on your grapher.
8. What did you learn as a result of doing this exploration that you did not know before?

Graphs and Principal Branches

Figure 7-6a shows the graph of $y = \tan^{-1} x$, the inverse tangent *function.* It is a reflection of one branch of the graph of $y = \tan x$ across the line $y = x$ (Figure 7-6b).

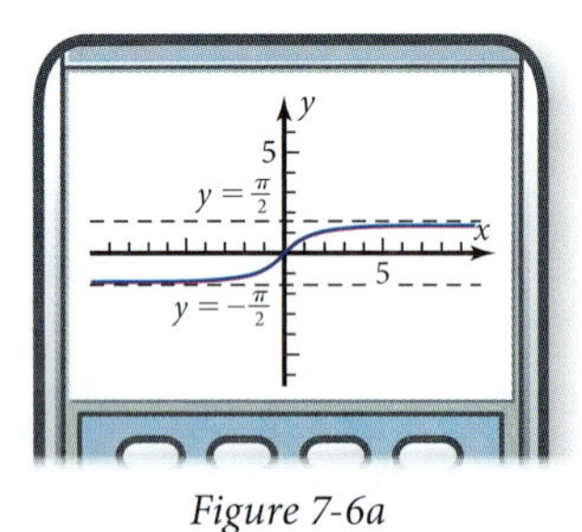

Figure 7-6a

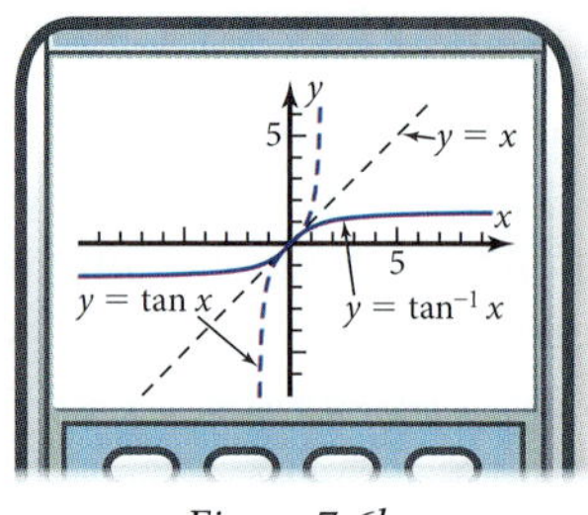

Figure 7-6b

The inverse tangent function, $y = \tan^{-1}x$, is called the **principal branch** of the inverse tangent relation, $y = \arctan x$. Using parametric mode, you can plot the graph of $y = \arctan x$. Enter

$$x = \tan t$$
$$y = t$$

The definition of arctangent tells you that $y = \arctan x$ if and only if $x = \tan y$. The graph will have all branches of $y = \arctan x$ that are in the window you have chosen (Figure 7-6c).

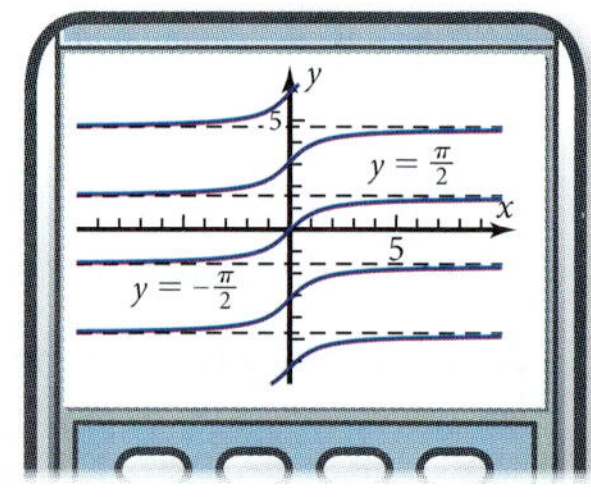

Figure 7-6c

The function $y = \tan^{-1} x$ is defined by designating one branch of arctangent to be the principal branch. Do this by restricting the range of the arctangent to meet the criteria in the box on the next page.

Criteria for Selecting Principal Branches of Inverse Trigonometric Functions

1. The principal branch must be a *function.*
2. It must use the *entire domain* of the inverse trigonometric relation.
3. It should be one *continuous* graph, if possible.
4. It should be *centrally located*, near the orgin.
5. If there is a choice between two possible branches, use the *postive* one.

Figure 7-6d shows the results of applying these criteria to all six inverse trigonometric relations. The highlighted portion of each graph shows the inverse trigonometric function. The rest of each graph shows more of the inverse trigonometric relation. Notice the ranges of y that give the principal branches.

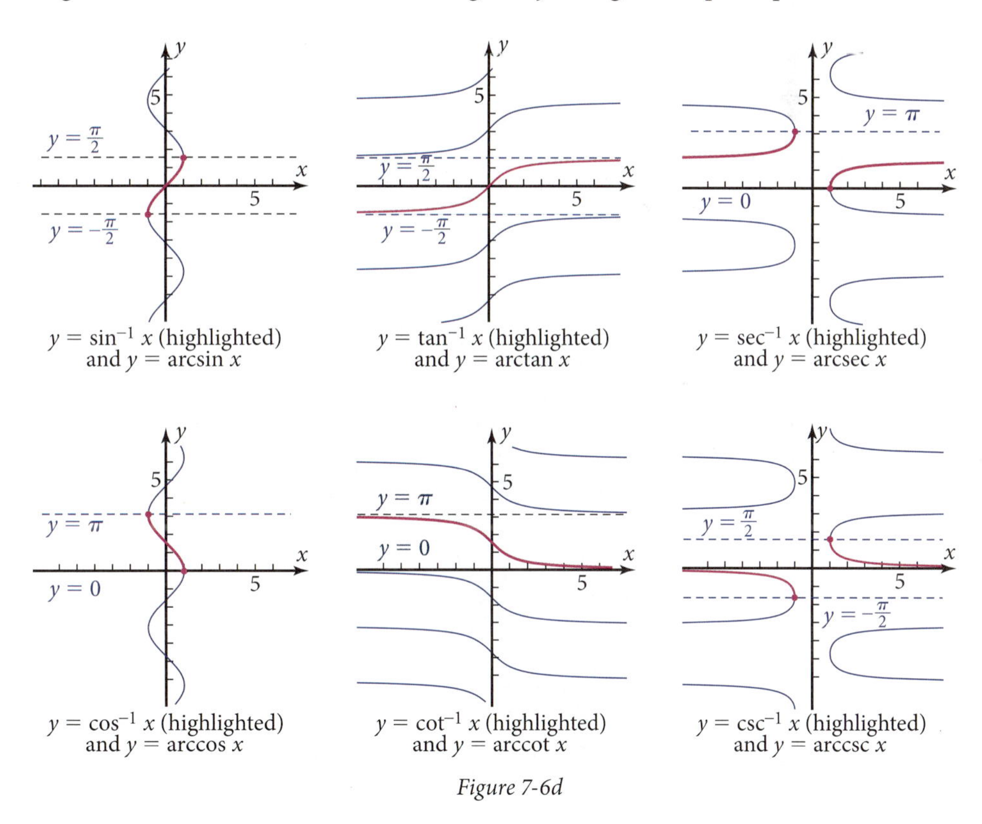

Figure 7-6d

DEFINITIONS: Ranges and Domains of Inverse Trigonometric Functions

Function	Range (Numerically)	Range (Graphically)	Domain
$y = \sin^{-1} x$	$y \in \left[-\frac{\pi}{2}, \frac{\pi}{2}\right]$	Quadrants I and IV	$x \in [-1, 1]$
$y = \cos^{-1} x$	$y \in [0, \pi]$	Quadrants I and II	$x \in [-1, 1]$
$y = \tan^{-1} x$	$y \in \left(-\frac{\pi}{2}, \frac{\pi}{2}\right)$	Quadrants I and IV	$x \in (-\infty, \infty)$
$y = \cot^{-1} x$	$y \in (0, \pi)$	Quadrants I and II	$x \in (-\infty, \infty)$
$y = \sec^{-1} x$	$y \in [0, \pi]$ and $y \neq \frac{\pi}{2}$	Quadrants I and II	$\lvert x\rvert \geq 1$
$y = \csc^{-1} x$	$y \in \left[-\frac{\pi}{2}, \frac{\pi}{2}\right]$ and $y \neq 0$	Quadrants I and IV	$\lvert x\rvert \geq 1$

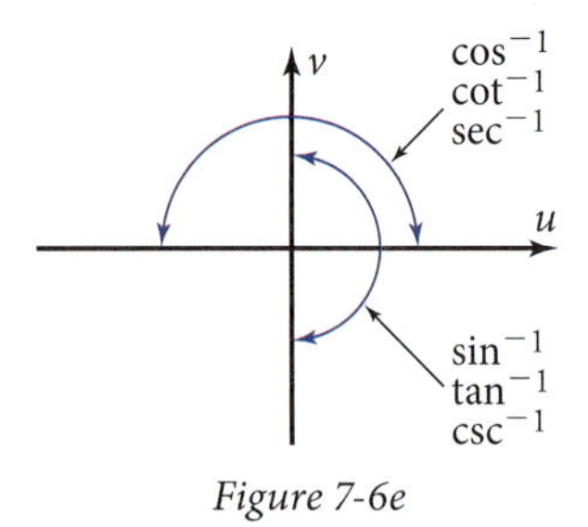

Figure 7-6e

To remember these ranges, it might help you to think of y as an angle in radians in a uv-coordinate system (Figure 7-6e). If the argument is positive, the arc or angle terminates in Quadrant I. If the argument is negative, the arc or angle terminates in Quadrant II or Quadrant IV, depending on which inverse function it is. None of the inverse functions terminates in Quadrant III.

If you could perpendicularly project an image of the railing in this spiral staircase onto the wall behind it, you would get a curve that resembles the graph of arccosine. (Compare to lower-left graph in Figure 7-6d.)

Exact Values of Inverse Circular Functions

Recall that it is possible to find exact trigonometric and circular function values of certain special angles or arcs. For instance, $\cos\left(\frac{\pi}{6}\right) = \frac{\sqrt{3}}{2}$. It is also possible to find exact values of expressions involving inverse trigonometric functions.

EXAMPLE 1 ➤ Evaluate $\tan\left(\sin^{-1}\left(-\frac{2}{3}\right)\right)$ geometrically to find the exact value. Check your answer numerically.

SOLUTION Draw an angle in standard position whose sine is $-\frac{2}{3}$. The angle terminates in Quadrant IV because the range of the inverse sine function is Quadrants I and IV. Draw the reference triangle, as shown in Figure 7-6f, and find the length of the third side. Then use the definition of tangent.

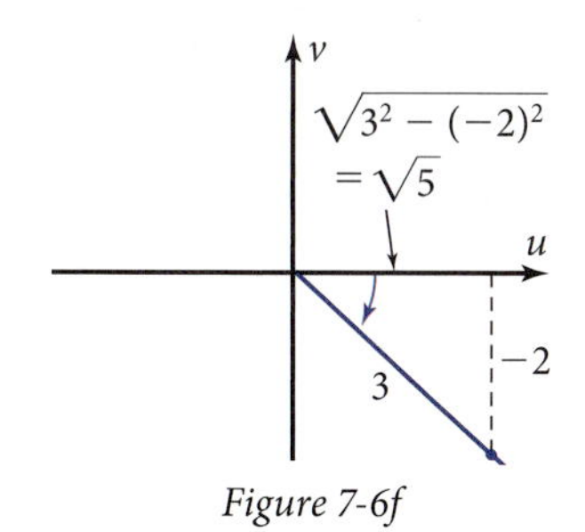

Figure 7-6f

$$\tan\left(\sin^{-1}\left(-\frac{2}{3}\right)\right) = \frac{-2}{\sqrt{5}}$$

Check: When you evaluate $\frac{-2}{\sqrt{5}}$, you get $-0.8944\ldots$,

which agrees with $\tan\left(\sin^{-1}\left(-\frac{2}{3}\right)\right) = \tan(-0.7297\ldots) = -0.8944\ldots$.

EXAMPLE 2 ➤ Evaluate $y = \sin(\cos^{-1} x)$ graphically to find the value in radical form. Set this value equal to y, and plot it together with the original equation on the same screen to confirm that your answer is correct.

SOLUTION Figure 7-6g shows the two possible quadrants for $\cos^{-1} x$.

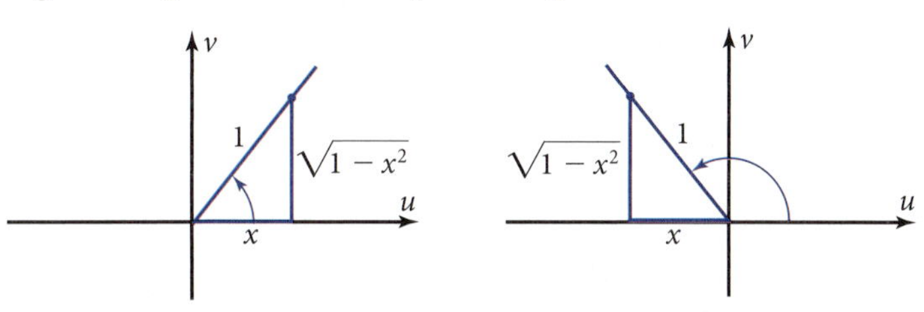

Figure 7-6g

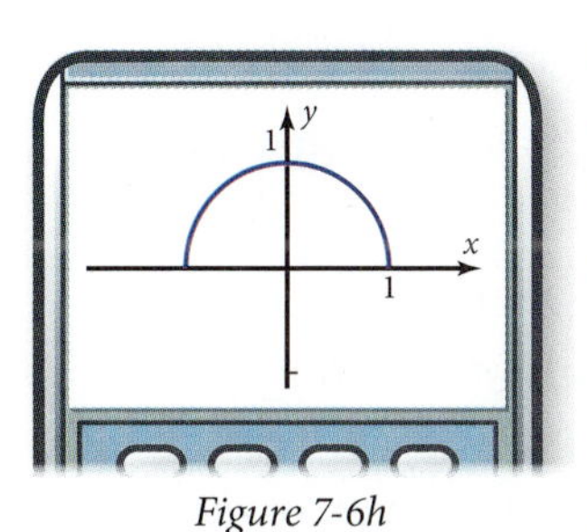

Figure 7-6h

By the definition of cosine, you can label the horizontal leg of the reference triangle x and the radius 1. The third side is given by the Pythagorean theorem. Note that, in both Quadrants I and II, the third side is *positive*. So you use the positive square root in both cases. By the definition of sine,

$$y = \sin(\cos^{-1} x) = \frac{\sqrt{1 - x^2}}{1} = \sqrt{1 - x^2}$$

Figure 7-6h shows that the graph is a semicircle with radius 1.

The Composite of a Function and Its Inverse Function

If you apply the techniques of Examples 1 and 2 to a function and its inverse function, an interesting property reveals itself. Example 3 shows you how to do this.

EXAMPLE 3 ➤ Evaluate $\cos(\cos^{-1} x)$. Explain why the answer is reasonable. Set your answer equal to y, and plot it together with $y = \cos(\cos^{-1} x)$ on the same screen to verify that the answer is correct.

SOLUTION $\cos^{-1} x$ means "the angle whose cosine is x." So by definition,

$$y = \cos(\cos^{-1} x) = x$$

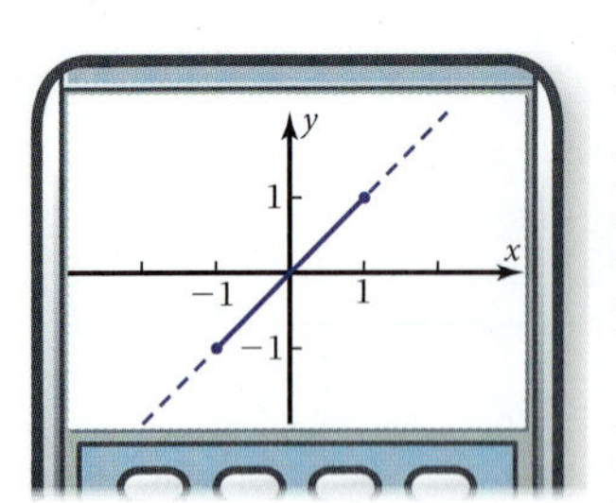

Figure 7-6i

Draw a *uv*-diagram as in the earlier examples if you need further convincing.

Figure 7-6i shows both the graph of $y = x$ and the graph of $y = \cos(\cos^{-1} x)$. Note that the graph of $\cos(\cos^{-1} x)$ has domain $[-1, 1]$ because the inverse cosine function is defined only for those values of x.

Example 3 illustrates a general property relating a function and its inverse function, which you may recall from Section 1-5.

PROPERTY: The Composite of a Function and Its Inverse Function

$$f(f^{-1}(x)) = x \quad \text{and} \quad f^{-1}(f(x)) = x$$

provided x is in the *range* of the outside function and in the *domain* of the inside function.

In Problem 23, you will prove this property. To illustrate the restrictions in the box,

$$\cos^{-1}(\cos 10) = \cos^{-1}(-0.8390...)$$
$$= 2.5663..., \text{ not } 10 \qquad \text{10 is not in the range of } \cos^{-1}.$$

$$\cos(\cos^{-1} 3) \text{ is undefined, not 3.} \qquad \text{3 is not in the domain of } \cos^{-1}.$$

In the first case, 10 is not in the range of the inverse cosine function (principal branch). In the second case, 3 is not in the domain of the inverse cosine function.

Problem Set 7-6

Reading Analysis

From what you have read in this section, what do you consider to be the main idea? Why does the graph of $y = \arctan x$ have more than one branch, and how do you decide which of these branches is the principal branch? How is the principal branch related to the value of $\tan^{-1} x$ your calculator gives you?

Quick Review

Q1. The function $y = 5 + 6\cos 7(x - 8)$ is a horizontal translation of $y = \cos x$ by __?__.

Q2. The sinusoid in Q1 is a vertical translation of $y = \cos x$ by __?__ units.

Q3. The sinusoid in Q1 is a vertical dilation of $y = \cos x$ by a factor of __?__.

Q4. The sinusoid in Q1 is a horizontal dilation of $y = \cos x$ by a factor of __?__.

Q5. The period of the sinusoid in Q1 is __?__.

Q6. If $f(x) = x^3$, then the inverse function $f^{-1}(x) =$ __?__.

Q7. What geometric figure is the graph of the parametric function defined by $x = 3\cos t$ and $y = 5\sin t$?

Q8. Write the Pythagorean property that involves tangent.

Q9. Without your grapher, evaluate $\cos \pi$.

Q10. Given $A = \arcsin x$, write the general solution for A in terms of $\sin^{-1} x$.

1. With your grapher in function mode, plot the graphs of $y = \sin^{-1} x$, $y = \cos^{-1} x$, and $y = \tan^{-1} x$. How do the graphs compare with the graphs in Figure 7-6d? Specifically, does each graph have the same y-range as shown for the principal branch?

2. With your grapher in parametric mode, plot the graphs of $y = \arcsin x$, $y = \arccos x$, and $y = \arctan x$. Use equal scales on both axes. To make the graph fill the screen, the t-interval should be the same as the y-range. How do the graphs compare with the graphs in Figure 7-6d?

For Problems 3 and 4, with your grapher in parametric mode, plot the two graphs on the same screen. Use a window with at least $[-7, 7]$ for y and a window for x that makes the scales on the two axes the same. Use different styles for the two graphs. Describe what you can do to show that these two graphs are reflections across the line $y = x$.

3. $y = \arcsin x$ and $y = \sin x$

4. $y = \arctan x$ and $y = \tan x$

For Problems 5–14, calculate the exact value of the function geometrically. Assume the principal branch in all cases. Check your answers by direct calculation.

5. $\tan\left(\cos^{-1}\frac{4}{5}\right)$

6. $\cos\left(\tan^{-1}\frac{4}{3}\right)$

7. $\sin\left(\tan^{-1}\frac{5}{12}\right)$

8. $\sec\left(\sin^{-1}\frac{15}{17}\right)$

9. $\cos\left(\sin^{-1}\left(-\frac{8}{17}\right)\right)$

10. $\cot\left(\csc^{-1}\left(-\frac{13}{12}\right)\right)$

11. $\sec\left(\cos^{-1}\frac{2}{3}\right)$ (Surprised?)

12. $\tan(\cot^{-1} 4)$ (Surprised?)

13. $\cos(\cos^{-1} 3)$

14. $\sec(\sec^{-1} 0)$

15. Explain why $\cos(\cos^{-1} 3)$ in Problem 13 does *not* equal 3.

16. Explain why $\sec(\sec^{-1} 0)$ in Problem 14 does *not* equal 0.

For Problems 17–22, evaluate the function geometrically to find the answer in radical form. Set your answer equal to y, and plot this equation together with the original equation on the same screen to show that your answer is correct.

17. $y = \cos(\sin^{-1} x)$

18. $y = \tan(\sin^{-1} x)$

19. $y = \sin(\tan^{-1} x)$

20. $y = \cos(\tan^{-1} x)$

21. $y = \sin(\sin^{-1} x)$

22. $y = \tan(\tan^{-1} x)$

23. *Composite of a Function and Its Inverse Problem:* In Problems 21 and 22, you found that $\sin(\sin^{-1} x) = x$ and that $\tan(\tan^{-1} x) = x$. These are examples of a general property of functions and their inverse functions, to which you were introduced in Chapter 1. In this problem you will prove the property. Let $f(x)$ be an invertible function.

    a. Prove that $f^{-1}(f(x)) = x$ by letting $y = f(x)$, applying the definition of f^{-1}, and using a clever substitution.

    b. Prove that $f(f^{-1}(x)) = x$ by letting $y = f^{-1}(x)$, applying the definition of f^{-1}, and using a clever substitution.

24. *Interpretation Problem—Composite of a Function and Its Inverse:* In Problem 23, you proved that the composite function of an invertible function and its inverse function is equal to x. In this problem you will see some surprises!

    a. Explain why the graph of $y = \tan(\tan^{-1} x)$ in Problem 22 is equivalent to the graph of $y = x$ for all values of x but the graph of $y = \sin(\sin^{-1} x)$ in Problem 21 is equivalent to the graph of $y = x$ for only certain values of x.

    b. Figure 7-6j shows the result of plotting on your grapher

    $$y = \sin^{-1}(\sin x)$$

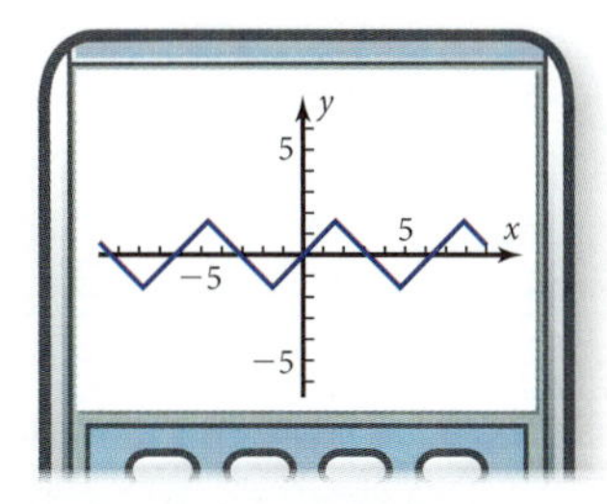

Figure 7-6j

Why is the graph "saw-toothed" instead of linear?

c. Plot $y = \cos^{-1}(\cos x)$. Sketch the result. Explain why the graph is *not* equivalent to the graph of $y = x$.

d. Plot $y = \tan^{-1}(\tan x)$. Use dot style rather than solid style. Sketch the result. Explain why the graph is *not* equivalent to the graph of $y = x$.

25. *Tunnel Problem:* Scorpion Gulch and Western Railway are preparing to build a new line through Rolling Mountains. They have hired you to do some calculations for tunnels and bridges needed on the line.

You set up a Cartesian coordinate system with its origin at the entrance to a tunnel through Bald Mountain. Your surveying crew finds that the mountain rises 250 m above the level of the track and that the next valley descends 50 m below the level of the track. The cross section of the mountain and valley is roughly sinusoidal, with a horizontal distance of 700 m from the top of the mountain to the bottom of the valley (Figure 7-6k).

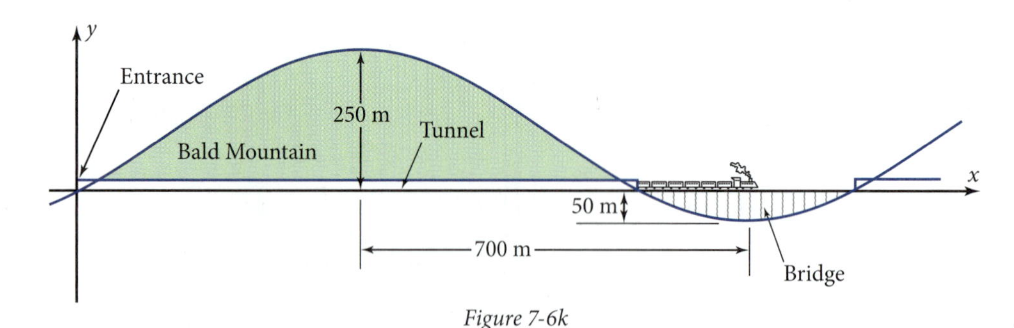

Figure 7-6k

a. Write the particular equation expressing the vertical displacement y, in meters, from the track to the surface of the mountain or valley as a function of distance x, in meters, from the tunnel entrance. You can find the constants A, B, and C from the given information. Finding the phase displacement D requires that you substitute the other three constants and the coordinates $(0, 0)$ for (x, y), then solve for D.

b. How long will the tunnel be? How long will the bridge be?

c. The railway company thinks it might be cheaper to build the line if it is raised by 20 m. The tunnel will be shorter, and the bridge will be longer. Find the new values of x at the beginning and end of the tunnel and at the beginning and end of the bridge. How long will each section be under these conditions?

26. *Journal Problem:* Update your journal with things you have learned since the last entry. Include things such as

- How to plot graphs of *inverse trigonometric relations*
- How the ranges of the *inverse trigonometric functions* are chosen
- How to calculate values of inverse trigonometric functions geometrically

7-7 Chapter Review and Test

In this chapter you've learned how to transform trigonometric expressions and solve equations using the Pythagorean, quotient, and reciprocal properties. The Pythagorean properties help show that certain parametric function graphs are circles or ellipses. The parametric functions let you plot graphs of inverse trigonometric relations. Analyzing these graphs and identifying the principal branches gives more meaning to the values the calculator gives for the inverse trigonometric functions.

Review Problems

R0. Update your journal with what you have learned in this chapter. Include topics such as

- Statements of the three kinds of properties
- How to prove that a trigonometric equation is an identity
- How to solve conditional trigonometric equations algebraically, numerically, and graphically
- What a parametric function is, how to graph it, and how to eliminate the parameter to get a Cartesian equation
- How to graph inverse trigonometric relations and find ranges for inverse trigonometric functions

R1. Figure 7-7a shows the unit circle and an angle θ in standard position.

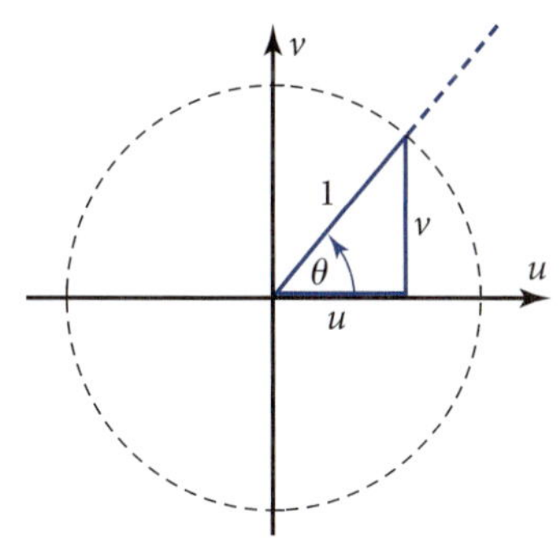

Figure 7-7a

a. Explain why $u^2 + v^2 = 1$.

b. Explain why $u = \cos\theta$ and $v = \sin\theta$.

c. Explain why $\cos^2\theta + \sin^2\theta = 1$.

d. Give a numerical example that confirms the property in part c.

e. Plot on the same screen $f_1(\theta) = \cos^2\theta$ and $f_2(\theta) = \sin^2\theta$. Sketch the graphs. How do the graphs support the Pythagorean property $\cos^2\theta + \sin^2\theta = 1$?

R2. a. Write equations expressing $\tan x$ and $\cot x$ in terms of $\sin x$ and $\cos x$.

b. Write equations expressing $\tan x$ and $\cot x$ in terms of $\sec x$ and $\csc x$.

c. Write three equations in which the product of two trigonometric functions equals 1.

d. Make a table of values showing numerically that $\cos^2 x + \sin^2 x = 1$.

e. Write equations expressing

i. $\sin^2 x$ in terms of $\cos x$

ii. $\tan^2 x$ in terms of $\sec x$

iii. $\csc^2 x$ in terms of $\cot x$

f. Sketch the graph of the parent function $y = \cos x$. On the same set of axes, sketch the graph of $y = \sec x$ using the fact that secant is the reciprocal of cosine.

R3. a. Transform $\tan A \sin A + \cos A$ into $\sec A$. What values of A are excluded from the domain?

b. Transform $(\cos B + \sin B)^2$ into $1 + 2\cos B \sin B$. What values of B are excluded from the domain?

c. Tranform $\frac{1}{1 + \sin C} + \frac{1}{1 - \sin C}$ into $2\sec^2 C$. What values of C are excluded from the domain?

d. Prove that this equation is an identity: $\csc D(\csc D - \sin D) = \cot^2 D$. What values of D are excluded from the domain?

e. Prove that this equation is an identity: $(3 \cos E + 5 \sin E)^2 + (5 \cos E - 3 \sin E)^2 = 34$.

f. Show that the two expressions in part b are equivalent by plotting each on your grapher.

g. Make a table of values to show that the equation in part e is an identity.

R4. a. Find the general solution for $\theta = \arcsin 0.3$.

b. Solve $1 + \tan 2\pi(x + 0.6) = 0$ algebraically for the first four positive values of x. Confirm graphically that your solutions are correct.

c. Solve $(2 \cos \theta - 1)(2 \sin \theta + \sqrt{3}) = 0$ algebraically in the domain $\theta \in [0°, 540°]$. Confirm graphically that your solutions are correct.

R5. a. Plot the graph of this parametric function on your grapher. Sketch the result.

$$x = -2 + 5 \cos t$$
$$y = 1 + 3 \sin t$$

b. Use the Pythagorean property for cosine and sine to eliminate the parameter in part a.

c. How can you conclude from the answer to part b that the graph is an ellipse? Where is the center of the ellipse? What are the x- and y-radii?

d. Figure 7-7b shows a solid cone in perspective. Write parametric equations for the ellipse that represents the circular base of the cone. Draw the cone on your grapher.

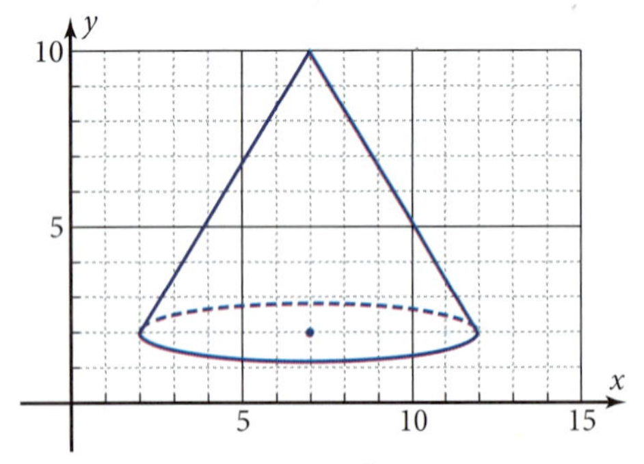

Figure 7-7b

R6. a. Using parametric mode on your grapher, duplicate the graph of the circular relation $y = \arccos x$ shown in Figure 7-7c.

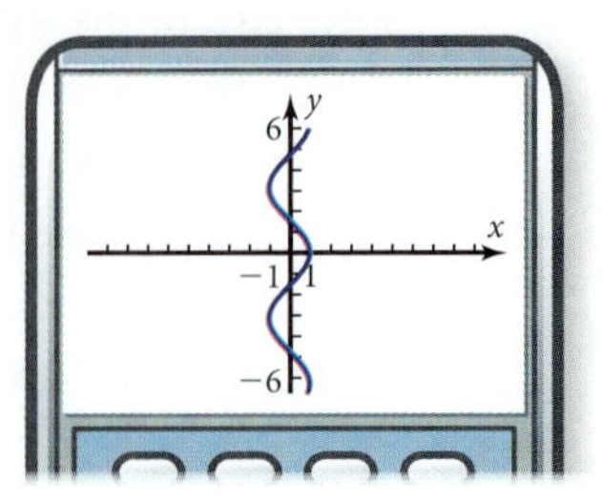

Figure 7-7c

b. Sketch the graph of the function $y = \cos^{-1} x$, the principal branch of the relation $y = \arccos x$. Explain the specifications used in selecting this principal branch. What is the range of this inverse cosine function?

c. How is the graph of $y = \arccos x$ related to the graph of $y = \cos x$?

d. Find geometrically the exact value (no decimals) of $\sin(\tan^{-1} 2)$. Check the answer by direct calculation.

e. Write an equation for $y = \tan(\cos^{-1} x)$ that does not involve trigonometric or inverse trigonometric functions. Confirm your answer by plotting this equation together with the given function on the same screen. Sketch the result.

f. Prove that $\cos(\cos^{-1} x) = x$.

g. Show on a uv-diagram the range of values of the functions $\sin^{-1}$ and $\cos^{-1}$.

h. Explain why the prefix *arc-* is appropriate in the names *arccos*, *arcsin*, and so on.

Concept Problems

C1. *Pendulum Problem:* Figure 7-7d shows a pendulum hanging from the ceiling. The pendulum bob traces out a counterclockwise circular path with radius 20 cm (which appears elliptical because it is drawn in perspective). At any time t, in seconds, since the pendulum was started in motion, it is over the point (x, y) on the floor, where x and y are in centimeters. The pendulum makes a complete cycle in 3 s.

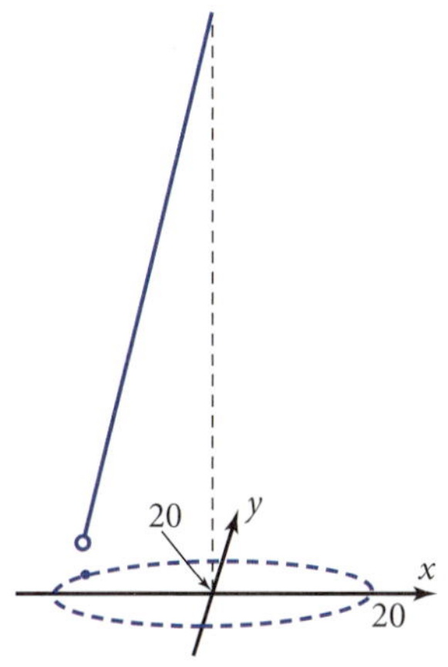

Figure 7-7d

a. Assuming that the pendulum bob was at the point $(20, 0)$ at time $t = 0$ s, write parametric equations for the circular path it traces.

Léon Foucault demonstrates his pendulum and the rotation of Earth at the Pantheon in Paris (1851).

b. Where is the pendulum at time $t = 5$ s?

c. Find the first three times when the pendulum bob has y-coordinate 10 cm. What are the x-coordinates at each of these times?

d. Explain how this problem ties together all the topics in this chapter.

C2. Prove that each equation is an identity.

a. $\dfrac{1 + \sin x + \cos x}{1 + \sin x - \cos x} = \dfrac{1 + \cos x}{\sin x}$

b. $\dfrac{1 + \sin x + \cos x}{1 - \sin x + \cos x} = \dfrac{1 + \sin x}{\cos x}$

C3. *Square of a Sinusoid Problem:* Figure 7-7e shows the graphs of $f_1(x) = \cos x$ (dashed) and $f_2(x) = \cos^2 x$ (solid). The squared graph seems to be sinusoidal.

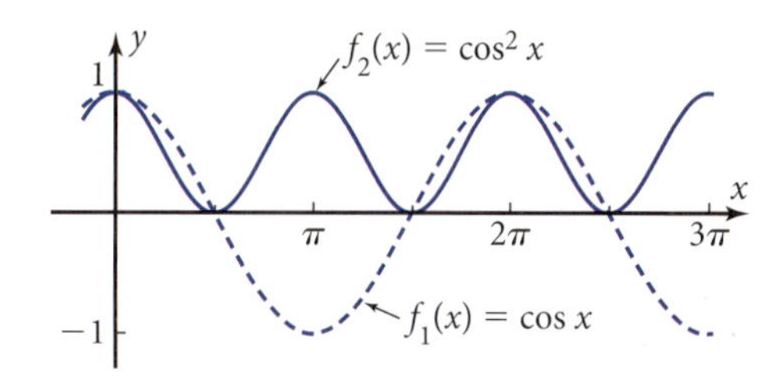

Figure 7-7e

a. Assuming that the graph of $f_2(x) = \cos^2 x$ is a sinusoid, find its period, amplitude, sinusoidal axis location, and phase displacement from the graph of the function $y = \cos x$.

b. Write the particular equation for the sinusoid you described in part a.

c. Give numerical and graphical evidence that the sinusoid in part b is identical to $f_2(x) = \cos^2 x$.

Chapter Test

Part 1: No calculators allowed (T1–T8)

T1. Write the Pythagorean property for cosine and sine.

T2. Write a quotient property involving cosine and sine.

T3. Write the reciprocal property for cotangent.

T4. Write the reciprocal property for secant.

T5. The value of $\sin^{-1} 0.5$ is 30°. Write the general solution for $\theta = \arcsin 0.5$.

T6. The value of $\tan^{-1}\sqrt{3}$ is $\frac{\pi}{3}$. Write the general solution for $x = \arctan\sqrt{3}$.

T7. Explain why the range of $y = \cos^{-1} x$ is $[0, \pi]$ but the range of $y = \sin^{-1} x$ is $\left[-\frac{\pi}{2}, \frac{\pi}{2}\right]$.

T8. Find geometrically the exact value of $\cos(\tan^{-1} 2)$.

Part 2: Graphing calculators allowed (T9–T20)

T9. Transform $(1 + \sin A)(1 - \sin A)$ into $\cos^2 A$. What values of A are excluded from the domain?

T10. Prove that $\tan B + \cot B = \csc B \sec B$ is an identity. What values of B must be excluded from the domain?

T11. Multiply the numerator and denominator of

$$\frac{\sin C}{1 + \cos C}$$

by the conjugate of the denominator. Show that the result is equivalent to

$$\frac{1 - \cos C}{\sin C}$$

T12. Plot the graphs of both expressions in Problem T11 to confirm that the two expressions are equivalent. Sketch the graphs. What values of C are excluded from the domain?

T13. With your calculator in degree mode, find the value of $\cos^{-1} 0.6$. Show the angle in a uv-coordinate system.

T14. Find another angle between 0° and 360° whose cosine is 0.6. Show it on the uv-coordinate system in Problem T13.

T15. Write the general solution of the inverse trigonometric relation $\theta = \arccos 0.6$. Show how you can use the $\pm$ sign to simplify writing this solution.

T16. Find the fifth positive value of θ for which $\cos\theta = 0.6$. How many revolutions, n, are needed to get to that value of θ in the uv-coordinate system?

T17. Find algebraically the general solution of

$$4\tan(\theta - 25°) = 7$$

T18. Write parametric equations for the ellipse in Figure 7-7f.

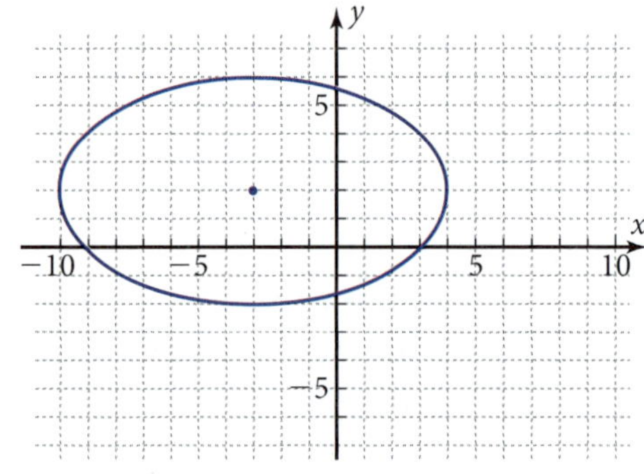

Figure 7-7f

T19. Write the parametric equations you use to plot $y = \arctan x$ (Figure 7-7g).

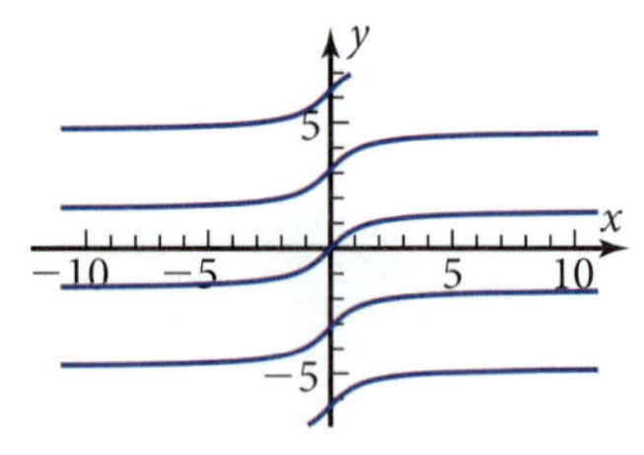

Figure 7-7g

T20. What did you learn as a result of this test that you did not know before?

Chapter 8

Properties of Combined Sinusoids

When two vehicles are going nearly the same speed on the highway, the combined sound of their engines sometimes seems to pulsate. The same thing happens when two airplane engines are going at slightly different speeds. The phenomenon is called *beats*. Using the concept of beats, a vibrato sound can be generated on a piano by tuning two strings for the same note at slightly different frequencies. In this chapter you'll learn about combinations of sinusoids so that you can analyze these harmonic phenomena.

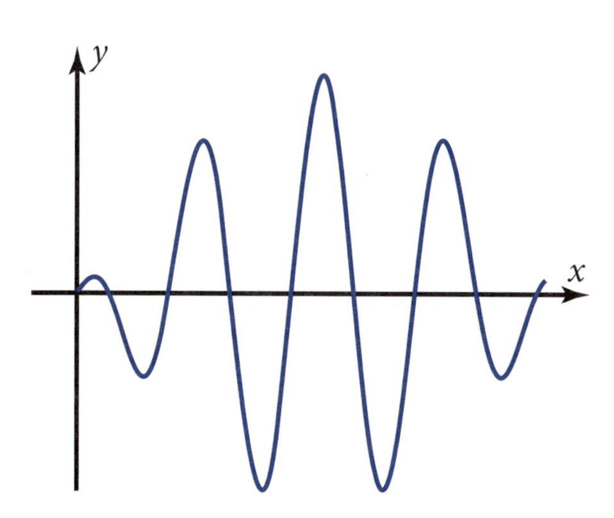

Mathematical Overview

In Chapter 7 you learned the Pythagorean, quotient, and reciprocal properties of the trigonometric functions. Each of these properties involves functions of *one* argument. In this chapter you'll learn properties of functions in which *different* arguments appear. These properties allow you to analyze more complicated periodic functions that are sums or products of sinusoids. You'll learn this in four ways.

GRAPHICALLY

A variable-amplitude periodic function

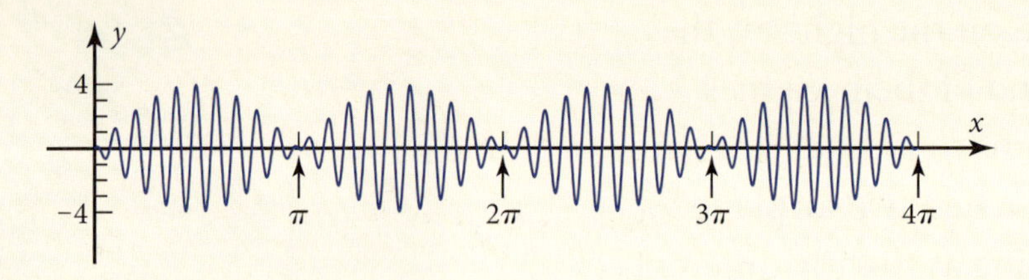

ALGEBRAICALLY

You can represent this graph as either a product of two sinusoids or a sum (or difference) of two sinusoids.

$$y = 4 \sin x \cdot \cos 20x \qquad \text{or} \qquad y = 2 \sin 21x - 2 \sin 19x$$

NUMERICALLY

If $x = 1$, the y-value of either function equals 1.3735....

VERBALLY

If the amplitude varies, the combined graph is a product of two sinusoids or a sum of two sinusoids. For the product, the sinusoids have much different periods. For the sum, they have nearly equal periods. The sum and product properties can transform one expression to the other. This property explains how AM radio and FM radio work.

8-1 Introduction to Combinations of Sinusoids

Music, like any other sound, is transmitted by waves. A "pure" musical note can be represented by a sine or cosine graph. The frequency of the note is represented by the period of the graph, and the loudness of the note is represented by the amplitude of the graph. Figure 8-1a shows two musical notes of the same frequency that are played at the same time, represented by $f_1(\theta) = 3 \cos \theta$ (green) and $f_2(\theta) = 4 \sin \theta$ (magenta). The blue graph represents the sound wave formed by adding these two sounds, $f_3(\theta) = 3 \cos \theta + 4 \sin \theta$. In this section you will explore this combined wave graph and show that it, too, is a sinusoid.

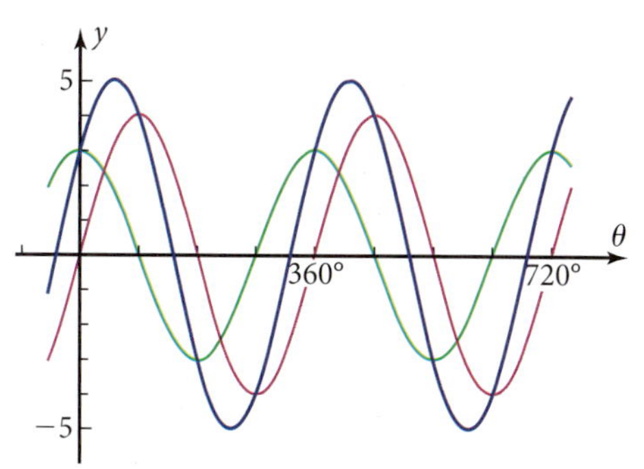

Figure 8-1a

Objective Investigate graphs formed by sums of sines and cosines.

Exploratory Problem Set 8-1

1. Plot the graphs of $f_1(\theta) = 3 \cos \theta$, $f_2(\theta) = 4 \sin \theta$, and $f_3(\theta) = f_1(\theta) + f_2(\theta)$ on the same screen. Do your graphs agree with the ones in Figure 8-1a?
2. The graph of $f_3(\theta)$ in Problem 1 seems to be a sinusoid. Estimate graphically its period, amplitude, and phase displacement with respect to $y = \cos x$.
3. Find numerically the amplitude and phase displacement of $f_3(\theta)$ by using the maximum feature of your grapher to find the first high point. Do the results confirm your estimates in Problem 2?
4. Plot the sinusoid $f_4(\theta) = A \cos(\theta - D)$, where A and D are the amplitude and phase displacement you found in Problem 3. Use a different style for this graph so that you can distinguish it from the graph of $f_3(\theta)$. Does the graph of $f_3(\theta)$ really seem to be a sinusoid?
5. Make a table of values of $f_3(\theta)$ and $f_4(\theta)$ for various values of θ. Do the values confirm or refute the conjecture that $f_3(\theta)$ is a sinusoid?
6. See if you can find the constants A and D in Problem 4 algebraically, using the factors 3 and 4 from the equations for $f_1(\theta)$ and $f_2(\theta)$.
7. Substitute two different angle measures for θ and D and show that $\cos(\theta - D)$ does not equal $(\cos \theta - \cos D)$.
8. Based on your observation in Problem 7, what property of multiplication and subtraction does *not* apply to the operations cosine and subtraction?

8-2 Composite Argument and Linear Combination Properties

If you add two sinusoids such as $f_1(\theta) = 3\sin\theta$ and $f_2(\theta) = 4\sin\theta$, you get another sinusoid with amplitude equal to the sum of the two amplitudes (Figure 8-2a, left side). These sinusoids are said to be **in phase** because their high and low points occur at the same values of θ.

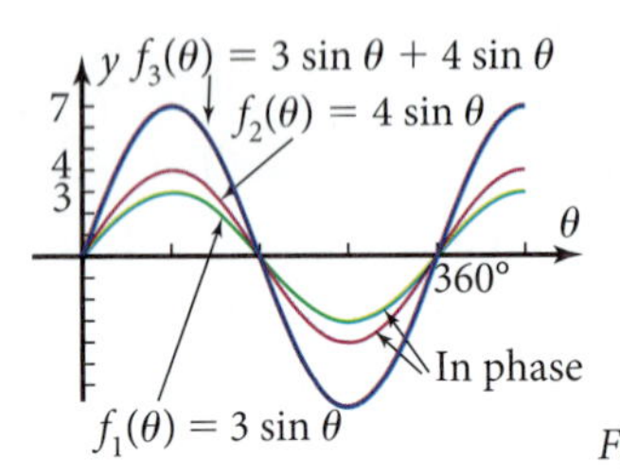

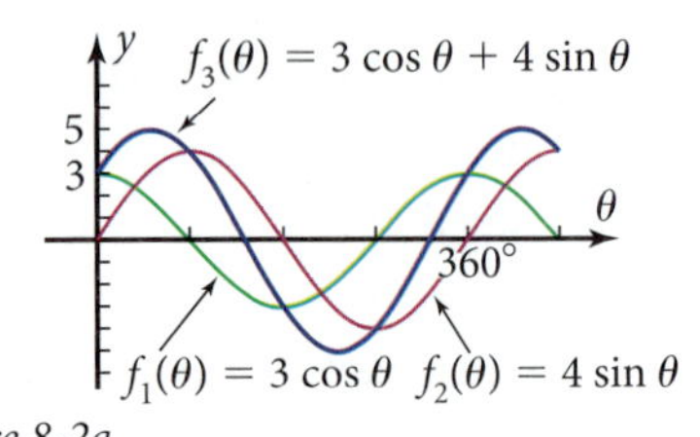

Figure 8-2a

If you add two sinusoids with the same period that are *not* in phase (Figure 8-2a, right side), the result is also a sinusoid, with amplitude less than the sum, as you found in Section 8-1. In this section you will learn algebraic ways to find the amplitude and phase displacement of a **linear combination** of cosine and sine having the same argument, that is, an equation in the form $y = a\cos\theta + b\sin\theta$.

With the help of the **composite argument property,** by which you can express $\cos(A - B)$ in terms of cosines and sines of A and B, you will prove that any linear combination of cosine and sine with equal periods is another sinusoid with the same period. This principle governs the sound produced by musical instruments. Two instruments playing middle C produce the sound of middle C whether or not the sound waves are in phase.

Objective

Derive a composite argument property expressing $\cos(A - B)$ in terms of cosines and sines of A and B, and use it to express a linear combination of cosine and sine as a single cosine with a phase displacement.

In this exploration you will write $\cos(x - y)$ in terms of $\cos x$, $\cos y$, $\sin x$, and $\sin y$.

EXPLORATION 8-2: Cosine of a Difference Discovery

1. Prove by counterexample that the operation cosine does *not* distribute over subtraction by showing numerically that

   $\cos(58° - 20°) \neq \cos 58° - \cos 20°$

2. It is possible to write the exact value of $\cos(58° - 20°)$ using just cosines and sines of 58° and 20°.

   Let $A = \cos 58°$.
   Let $B = \cos 20°$.
   Let $C = \sin 58°$.
   Let $D = \sin 20°$.

   Store these values in your calculator without rounding. By experimenting, find a combination of these four values that gives 0.7880…, the value of $\cos(58° - 20°)$.

continued

3. Make a conjecture: "If x and y are angle measures, then $\cos(x - y) = \underline{\ \ ?\ \ }$."
4. Select several pairs of values for x and y and test your conjecture. Does the conjecture work for all the values? If not, go back and modify your conjecture, and then try again.
5. When you feel reasonably sure that your conjecture is correct, write a sentence explaining how you can write the cosine of the difference of two angles in terms of the cosines and sines of the angles. Start by writing "cosine (*first angle* − *second angle*) = ..."
6. Apply the composite argument property in Problem 5 to $f_1(\theta) = 6\cos(\theta - 70°)$. The result should be a *linear combination* of $\cos\theta$ and $\sin\theta$. That is,

   $f_2(\theta) = b\cos\theta + c\sin\theta$, where b and c stand for constants
7. Plot the graphs of $f_1(\theta)$ and $f_2(\theta)$ from Problem 6 on the same screen. Use a window with $[-90°, 450°]$ for x. Is $f_2(\theta)$ really a sinusoid with amplitude 6 and phase displacement 70° with respect to the parent cosine graph? How can you tell?
8. What did you learn as a result of doing this exploration that you did not know before?

Linear Combination Property

As you will prove in Problem 32, the graph of $y = 3\cos\theta + 4\sin\theta$ is a sinusoid that you can express in the form

$$y = A\cos(\theta - D)$$

where A is the amplitude and D is the phase displacement with respect to $y = \cos x$. If you plot the graph and then use the maximum feature on your grapher, you will find that

When you "mix" sound, the principles of linear combination of sound waves apply.

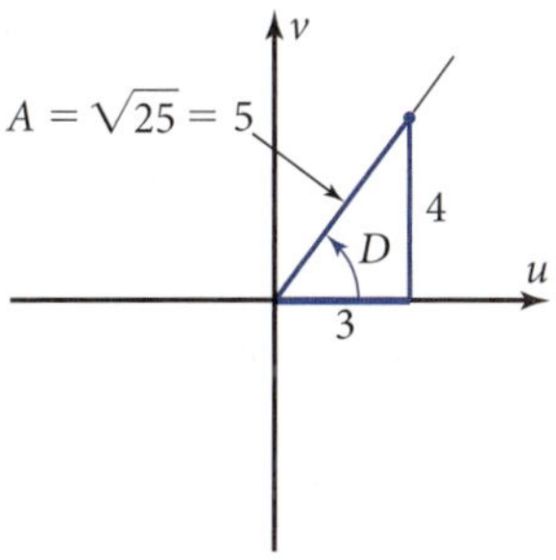

$$A = 5 \qquad \text{and} \qquad D = 53.1301\ldots°$$

Figure 8-2b

Actually, D is the angle in standard position with $u = 3$ (the coefficient of cosine in $y = 3\cos\theta + 4\sin\theta$) and $v = 4$ (the coefficient of sine), as shown in Figure 8-2b. You can see this relationship in the *Linear Combination* exploration at **www.keymath.com/precalc.** Once you derive the composite argument property mentioned in the objective, you can prove that the amplitude A is the length of the hypotenuse of the reference triangle for angle D. You can find the length of A by using the Pythagorean theorem and the measure of angle D by finding the arctangent.

$$A = \sqrt{3^2 - 4^2} = 5$$

$$D = \arctan\frac{4}{3} = 53.1301\ldots° + 180°n = 53.1301\ldots°$$

Choose $n = 0$ so that D terminates in Quadrant I.

So $y = 5\cos(\theta - 53.1301\ldots°)$ is equivalent to $y = 3\cos\theta + 4\sin\theta$. The graphical solution confirms this.

EXAMPLE 1 ➤ Express $y = -8\cos\theta + 3\sin\theta$ as a single cosine with a phase displacement.

SOLUTION Sketch angle D in standard position with $u = -8$ and $v = 3$ (the coefficients of cosine and sine, respectively), as shown in Figure 8-2c.

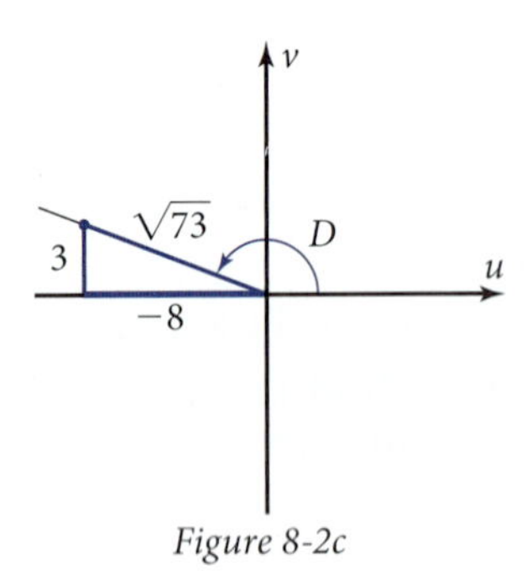

Figure 8-2c

$A = \sqrt{(-8)^2 + 3^2} = \sqrt{73}$ Find A by the Pythagorean theorem.

$D = \arctan\dfrac{3}{-8} = -20.5560...° + 180°n$ Find D using the definition of arctangent.

$= 159.4439...°$ Choose $n = 1$ to place D in the correct quadrant.

$\therefore y = \sqrt{73}\cos(\theta - 159.4439...°)$ ◀

> **PROPERTY: Linear Combination of Cosine and Sine with Equal Periods**
>
> $$b\cos x + c\sin x = A\cos(x - D)$$
>
> where
>
> $$A = \sqrt{b^2 + c^2} \qquad \text{and} \qquad D = \arctan\frac{c}{b}$$
>
> The quadrant for $D = \arctan\frac{c}{b}$ depends on the signs of b and c and may be determined by sketching angle D in standard position. The length of the hypotenuse of the reference triangle is A.

Composite Argument Property for Cosine ($A - B$)

In Exploration 8-2, you found by counterexample that the cosine function does *not* distribute over addition or subtraction:

$$\cos(58° - 20°) = \cos 38° = 0.7880\ldots$$

$$\cos 58° - \cos 20° = 0.5299\ldots - 0.9396\ldots = -0.4097\ldots$$

$$\therefore \cos(58° - 20°) \neq \cos 58° - \cos 20°$$

However, you *can* express $\cos(58° - 20°)$ exactly in terms of sines and cosines of 58° and 20°. The result is

$$\cos(58° - 20°) = \cos 58° \cos 20° + \sin 58° \sin 20°$$

Both sides equal 0.7880....

Next you'll see how to generalize the results for any angles A and B. The left side of Figure 8-2d shows angles A and B in standard position and shows their difference, angle $(A - B)$. The coordinates of the points at which the sides of angle $(A - B)$ cut the unit circle are

$$(\cos A, \sin A) \qquad \text{and} \qquad (\cos B, \sin B)$$

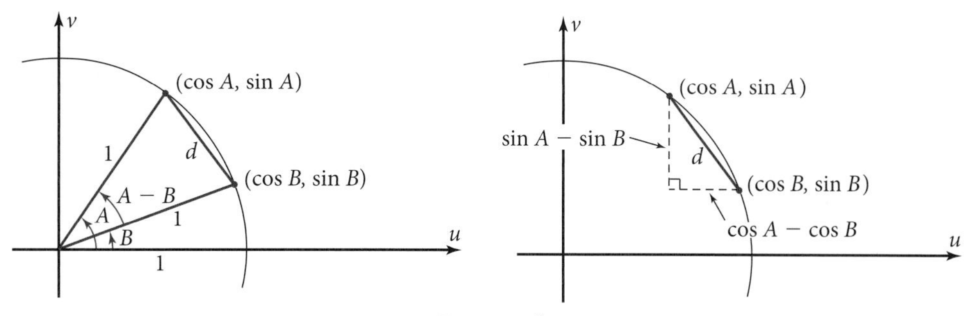

Figure 8-2d

The chord between the initial and terminal points has length d, which can be calculated by the Pythagorean theorem. As shown on the right side of Figure 8-2d, d is the hypotenuse of a right triangle with side lengths $(\cos A - \cos B)$ and $(\sin A - \sin B)$. You may recall this as the **distance formula** from earlier courses you have taken.

$$d^2 = (\cos A - \cos B)^2 + (\sin A - \sin B)^2$$ By the Pythagorean theorem.

$$= \cos^2 A - 2\cos A\cos B + \cos^2 B + \sin^2 A - 2\sin A\sin B + \sin^2 B$$ Expand the squares.

$$= (\cos^2 A + \sin^2 A) + (\cos^2 B + \sin^2 B) - 2\cos A\cos B - 2\sin A\sin B$$ Commute and associate the squared terms.

$$= 1 + 1 - 2\cos A\cos B - 2\sin A\sin B$$ Use the Pythagorean property for cosine and sine.

$$\therefore d^2 = 2 - 2\cos A\cos B - 2\sin A\sin B$$

Now consider Figure 8-2e, which shows angle $(A - B)$ rotated into standard position. The coordinates of the terminal and initial points of the angle in this position are

$$(\cos(A - B), \sin(A - B)) \quad \text{and} \quad (1, 0)$$

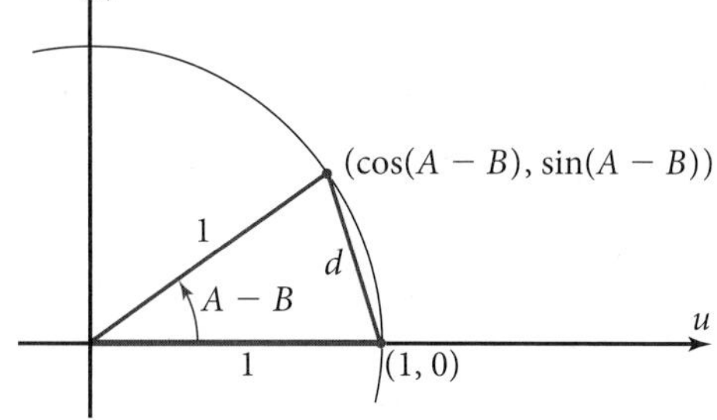

Figure 8-2e

The chord still has length d. By the distance formula and subsequent algebra,

$$d^2 = [\cos(A - B) - 1]^2 + [\sin(A - B) - 0]^2$$

$$= \cos^2(A - B) - 2\cos(A - B) + 1 + \sin^2(A - B)$$

Expand the squares.

$$= [\cos^2(A - B) + \sin^2(A - B)] + 1 - 2\cos(A - B)$$

Associate the square terms.

$$\therefore d^2 = 2 - 2\cos(A - B)$$

Use the Pythagorean property.

Equate the two expressions for d^2 to get

$$2 - 2\cos(A - B) = 2 - 2\cos A \cos B - 2\sin A \sin B$$

$$-2\cos(A - B) = -2\cos A \cos B - 2\sin A \sin B$$

$$\cos(A - B) = \cos A \cos B + \sin A \sin B$$

This is the property that was illustrated by numerical example with $A = 58°$ and $B = 20°$.

PROPERTY: Composite Argument Property for cos($A - B$)

$$\cos(A - B) = \cos A \cos B + \sin A \sin B$$

Verbally: Cosine of the difference of two angles is equal to cosine of first angle times cosine of second angle, plus sine of first angle times sine of second angle.

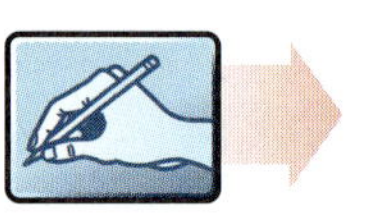

Cosine of first minus second is cosine first times cosine second plus sine first times sine second.

EXAMPLE 2 ➤ Express $7\cos(\theta - 23°)$ as a linear combination of $\cos\theta$ and $\sin\theta$.

SOLUTION

$$7\cos(\theta - 23°) = 7(\cos\theta \cos 23° + \sin\theta \sin 23°)$$

Apply the composite argument property.

$$= 7\cos\theta \cos 23° + 7\sin\theta \sin 23°$$

$$= (7\cos 23°)\cos\theta + (7\sin 23°)\sin\theta$$

Associate the constant factors.

$$\therefore 7\cos(\theta - 23°) = 6.4435...\cos\theta + 2.7351...\sin\theta$$

A linear combination of $\cos\theta$ and $\sin\theta$. ◄

Algebraic Solution of Equations

You can use the linear combination property to solve certain trigonometric equations algebraically.

EXAMPLE 3 ➤ Solve $-2\cos x + 3\sin x = 2$ for $x \in [-2\pi, 2\pi]$. Verify the solution graphically.

SOLUTION

$-2\cos x + 3\sin x = 2$ Write the given equation.

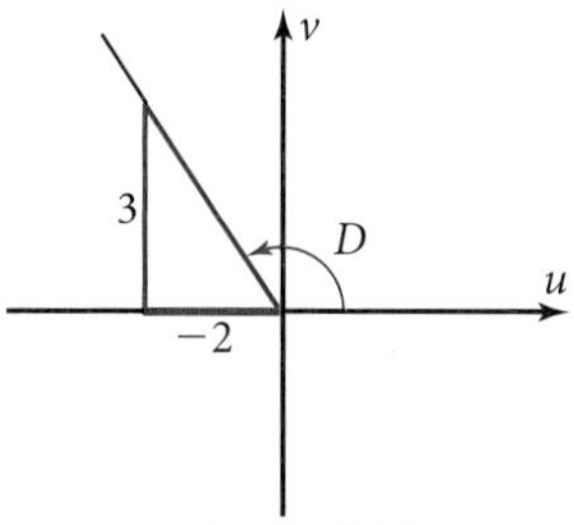

Transform the left side, $-2\cos x + 3\sin x$, into the form $A\cos(x - D)$. Start by drawing angle D in standard position (Figure 8-2f).

Figure 8-2f

$A = \sqrt{(-2)^2 + (3)^2} = \sqrt{13}$ Use the Pythagorean theorem to calculate A.

$D = \arctan\dfrac{3}{-2} = -0.9827... + \pi n = 2.1587...$

Use $n = 1$ for the proper arctangent value.

$\therefore \sqrt{13}\cos(x - 2.1587...) = 2$ Rewrite the equation using $A\cos(x - D)$.

$$\cos(x - 2.1587...) = \frac{2}{\sqrt{13}}$$

$$x - 2.1587... = \arccos\left(\frac{2}{\sqrt{13}}\right)$$

$$x = 2.1587... \pm 0.9827... + 2\pi n$$

Rewrite the equation and evaluate the arccosine.

$$x = 3.1415... + 2\pi n \quad \text{or} \quad 1.1760... + 2\pi n$$

Evaluate 2.1587... + 0.9827... and 2.1587... − 0.9827.

$$x = -5.1071..., -3.1415..., 1.1760..., 3.1415...$$

Pick values of n to get x in the domain.

Figure 8-2g shows the graph of $y = -2\cos x + 3\sin x$ and the line $y = 2$. Note that the graph is a sinusoid, as you discovered algebraically while solving the equation. By using the intersect feature, you can see that the four solutions are correct and that they are the only solutions in the domain $[-2\pi, 2\pi]$. The graph also shows the phase displacement 2.1587... and the amplitude $\sqrt{13}$, approximately 3.6.

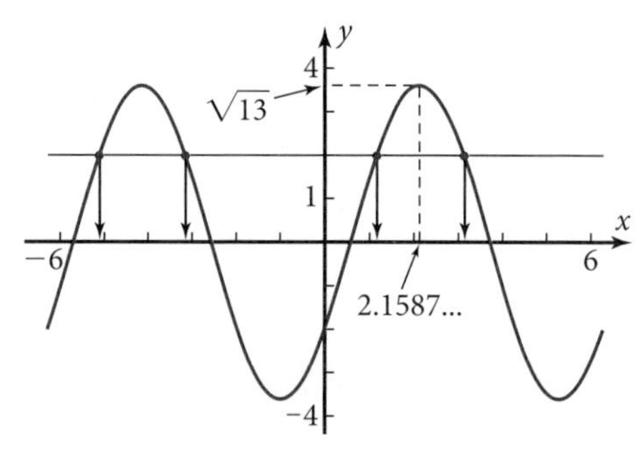

Figure 8-2g

Problem Set 8-2

Reading Analysis

From what you have read in this section, what do you consider to be the main idea? Explain how you know that the cosine function does not distribute over subtraction, and tell what the cosine of the difference of two angles *does* equal in terms of functions of the two angles. What does this property allow you to prove about a linear combination of two sinusoids with equal periods?

Quick Review

Q1. State the Pythagorean property for secant and tangent.

Q2. State the reciprocal property for cosecant.

Q3. State the quotient property for cotangent in terms of sine and cosine.

Q4. Is $\cos^2 x = 1 - \sin^2 x$ an identity?

Q5. Is $\cot x \tan x = 1$ an identity?

Q6. Is $\cos x \sin x = 1$ an identity?

Q7. Find the exact value (no decimals) of $\cos \frac{\pi}{4}$.

Q8. Find the exact value (no decimals) of tan 30°.

Q9. Find the first three positive angles for $\theta = \arccos 0.5$.

Q10. Factor: $x^2 - 5x - 6$

For Problems 1–12, write the linear combination of cosine and sine as a single cosine with a phase displacement.

1. $y = 12 \cos \theta + 5 \sin \theta$

2. $y = 4 \cos \theta + 3 \sin \theta$

3. $y = -7 \cos \theta + 24 \sin \theta$

4. $y = -15 \cos \theta + 8 \sin \theta$

5. $y = -8 \cos \theta - 11 \sin \theta$

6. $y = -7 \cos \theta - 10 \sin \theta$

7. $y = 6 \cos \theta - 6 \sin \theta$

8. $y = \cos \theta - \sin \theta$

9. $y = \sqrt{3} \cos \theta + \sin \theta$

10. $y = (\sqrt{6} + \sqrt{2}) \cos \theta + (\sqrt{6} - \sqrt{2}) \sin \theta$
Is the result surprising?

11. $y = -3 \cos x + 4 \sin x$ (radian mode)

12. $y = -5 \cos x - 12 \sin x$ (radian mode)

13. Confirm by graphing that your answer to Problem 1 is correct.

14. Confirm by graphing that your answer to Problem 2 is correct.

15. Express the circular function $y = \cos 3x + \sin 3x$ as a single cosine with a phase displacement. What effect does the 3 have on your work?

M. C. Escher's Rippled Surface *shows the reflection of branches in the water waves, which have a sinusoidal pattern. (M. C. Escher's* Rippled Surface *© 2006 Cordon Art B.V.-Baarn-Holland. All rights reserved.)*

16. Figure 8-2h shows a cosine graph and a sine graph. Find equations for these two sinusoids. Then find an equation for the sum of the two sinusoids as a single cosine with a phase displacement. Verify your answers by plotting the equations on your grapher.

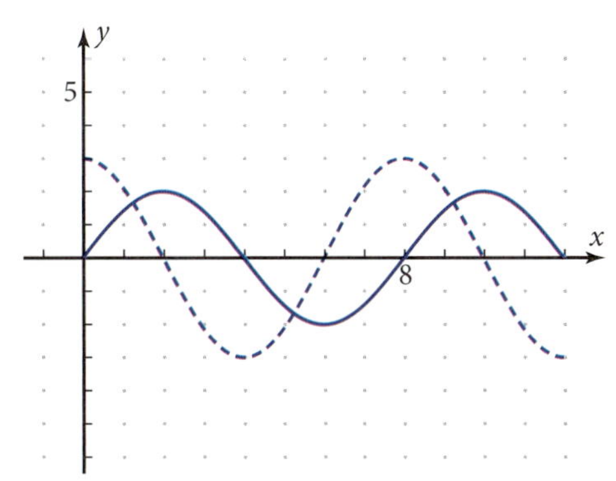

Figure 8-2h

17. Prove by counterexample that cosine does not distribute over subtraction. That is, give a numerical example to show that $\cos(A - B) \neq \cos A - \cos B$.

18. Make a table of values to show numerically that $\cos(A - B) = \cos A \cos B + \sin A \sin B$.

For Problems 19–22, express each equation as a linear combination of cosine and sine.

19. $y = 10 \cos(\theta - 30°)$. Confirm graphically that your answer is correct.

20. $y = 20 \cos(\theta - 60°)$. Confirm graphically that your answer is correct.

21. $y = 5 \cos(3\theta - 150°)$

22. $y = 8 \cos(2\theta - 120°)$

For Problems 23–26, solve the equation algebraically. Use the domain $x \in [0, 2\pi]$ or $\theta \in [0°, 360°]$.

23. $5 \cos \theta + 7 \sin \theta = 3$

24. $2 \cos x + 5 \sin x = 4$

25. $-8 \cos x - 3 \sin x = 5$

26. $7 \cos \theta - 4 \sin \theta = 6$

27. Use the composite argument property to show that this equation is an identity:

$$\cos 2\theta = \cos 5\theta \cos 3\theta + \sin 5\theta \sin 3\theta$$

Use the result to solve this equation for $\theta \in [0°, 360°]$:

$$\cos 5\theta \cos 3\theta + \sin 5\theta \sin 3\theta = 0.3$$

28. *Musical Note Problem:* The Nett sisters, Cora and Clara, are in a band. Each is playing the note A. Their friend Tom is standing at a place where the notes arrive exactly a quarter cycle out of phase. If x is time, in seconds, the function equations for Cora's and Clara's notes are

Cora: $y = 100 \cos 440\pi x$

Clara: $y = 150 \sin 440\pi x$

a. The sound Tom hears is the sum of Cora's and Clara's sound waves. Write an equation for this sound as a single cosine with a phase displacement.

b. The amplitudes 100 and 150 measure the loudness of the two notes Cora and Clara are playing. Is this statement true or false? "Tom hears a note 250 units loud, the sum of 100 and 150." Explain how you arrived at your answer.

c. The frequency of the note A being played by Cora and Clara is 220 cycles per second. Explain how you can figure this out from the two equations. Is this statement true or false? "The note Tom hears also has a frequency of 220 cycles per second."

Problems 29 and 30 prepare you for the next section.

29. *Cofunction Property for Cosines and Sines Problem:*

a. Show that $\cos 70° = \sin 20°$.

b. Use the composite argument property and the definition of complementary angles to show in general that $\cos(90° - \theta) = \sin \theta$.

c. What does the prefix *co-* mean in the name *cosine*?

30. *Even Property of Cosine Problem:*

a. Show that $\cos(-54°) = \cos 54°$.

b. You can write $\cos(-54°)$ as $\cos(0° - 54°)$. Use the composite argument property to show algebraically that $\cos(-\theta) = \cos \theta$.

c. Recall that functions with the property $f(-x) = f(x)$ are called *even* functions. Show why this name is picked by letting $f(x) = x^6$ and showing that $f(-x) = f(x)$.

31. *Composite Argument Property Derivation Problem:* Derive the property

$$\cos(A - B) = \cos A \cos B + \sin A \sin B$$

Try to do this on your own, looking at the text only long enough to get yourself started again if you get stuck.

32. *Linear Combination of Cosine and Sine Derivation Problem:* In this problem you'll see how to prove the linear combination property.

a. Use the composite argument property to show that

$$A \cos(\theta - D)$$
$$= (A \cos D) \cos \theta + (A \sin D) \sin \theta$$

b. Let $A \cos D = b$, and let $A \sin D = c$. Square both sides of each equation to get

$$A^2 \cos^2 D = b^2$$

$$A^2 \sin^2 D = c^2$$

Explain why $A^2 = b^2 + c^2$.

c. Explain why $D = \arccos \frac{b}{A}$ and $D = \arcsin \frac{c}{A}$, and thus why $D = \arctan \frac{c}{b}$.

33. *Journal Problem:* Update your journal with what you have learned since the last entry. In particular, explain what the composite argument property is and how you can use it to prove that a sum of cosine and sine with equal periods is a single cosine with the same period and a phase displacement.

8-3 Other Composite Argument Properties

In Section 8-2, you used the composite argument property for cosine to show that a linear combination of cosine and sine with equal periods is a single cosine with the same period but with a different amplitude and a phase displacement. In this section you'll learn composite argument properties for sine and tangent involving $(A + B)$ as well as $(A - B)$. You'll learn the cofunction properties and odd–even function properties that allow you to derive these new composite argument properties quickly from $\cos(A - B)$.

Objective

For trigonometric functions f, derive and learn properties for

- $f(-x)$ in terms of $f(x)$
- $f(90° - \theta)$ in terms of functions of θ, or $f\left(\frac{\pi}{2} - x\right)$ in terms of functions of x
- $f(A + B)$ and $f(A - B)$ in terms of functions of A and functions of B

The Odd–Even Properties

If you take the functions of opposite angles or arcs, interesting patterns emerge.

$\sin(-20°) = -0.3420\ldots$ and $\sin 20° = 0.3420\ldots$

$\cos(-20°) = 0.9396\ldots$ and $\cos 20° = 0.9396\ldots$

$\tan(-20°) = -0.3639\ldots$ and $\tan 20° = 0.3639\ldots$

These numerical examples illustrate the fact that sine and tangent are odd functions and cosine is an even function. Figure 8-3a shows graphically why these properties apply for any value of θ.

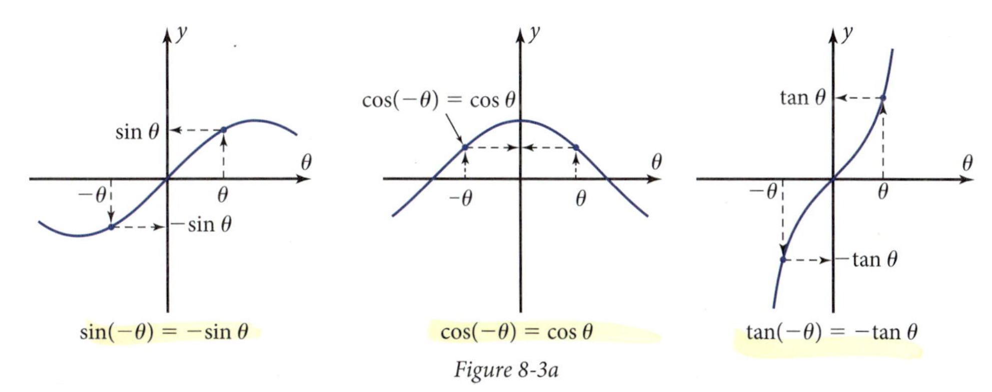

Figure 8-3a

The reciprocals of the functions have the same *parity* (oddness or evenness) as the original functions.

PROPERTIES: Odd and Even Functions

Cosine and its reciprocal are *even* functions. That is,

$$\cos(-x) = \cos x \qquad \text{and} \qquad \sec(-x) = \sec x$$

Sine and tangent, and their reciprocals, are *odd* functions. That is,

$$\sin(-x) = -\sin x \qquad \text{and} \qquad \csc(-x) = -\csc x$$

$$\tan(-x) = -\tan x \qquad \text{and} \qquad \cot(-x) = -\cot x$$

The Cofunction Properties: Functions of $(90° - \theta)$ or $\left(\frac{\pi}{2} - x\right)$

The angles 20° and 70° are *complementary angles* because they sum to 90°. (The word comes from "complete," because the two angles *complete* a right angle.) The angle 20° is the complement of 70°, and the angle 70° is the complement of 20°. An interesting pattern shows up if you take the function and the cofunction of complementary angles.

$$\cos 70° = 0.3420\ldots \qquad \text{and} \qquad \sin 20° = 0.3420\ldots$$

$$\cot 70° = 0.3639\ldots \qquad \text{and} \qquad \tan 20° = 0.3639\ldots$$

$$\csc 70° = 1.0641\ldots \qquad \text{and} \qquad \sec 20° = 1.0641\ldots$$

You can verify these patterns by using the right triangle definitions of the trigonometric functions. Figure 8-3b shows a right triangle with acute angles that measure 70° and 20°. The opposite leg for 70° is the adjacent leg for 20°.

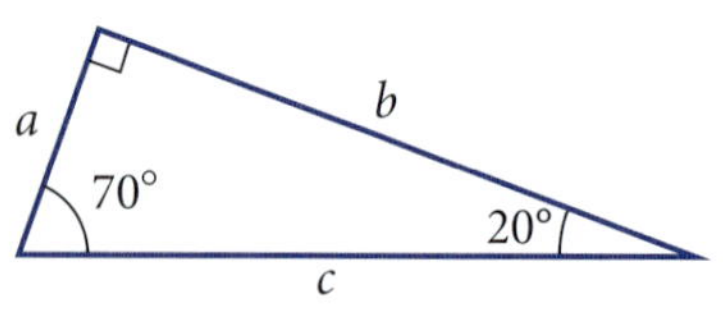

Figure 8-3b

$$\cos 70° = \frac{\text{adjacent leg}}{\text{hypotenuse}} = \frac{a}{c} \qquad \text{and} \qquad \sin 20° = \frac{\text{opposite leg}}{\text{hypotenuse}} = \frac{a}{c}$$

$$\therefore \cos 70° = \sin 20°$$

The prefix *co-* in the names cosine, cotangent, and cosecant comes from the word *complement.* In general, the cosine of an angle is the *sine* of the *complement* of that angle. The same property is true for cotangent and cosecant, as you can verify with the help of Figure 8-3b.

The cofunction properties are true regardless of the measure of the angle or arc. For instance, if θ is 234°, then the complement of θ is $90° - 234°$, or $-144°$.

$$\cos 234° = -0.5877... \qquad \text{and}$$

$$\sin(90° - 234°) = \sin(-144°) = -0.5877...$$

$$\therefore \cos 234° = \sin(90° - 234°)$$

Note that it doesn't matter which of the two angles you consider to be "the angle" and which you consider to be "the complement." It is just as true, for example, that

$$\sin 20° = \cos(90° - 20°)$$

The cofunction properties of the trigonometric functions are summarized verbally as

The cosine of an angle equals the *sine* of the *complement* of that angle.

The cotangent of an angle equals the *tangent* of the *complement* of that angle.

The cosecant of an angle equals the *secant* of the *complement* of that angle.

PROPERTIES: Cofunction Properties of Trigonometric Functions

When working with degrees:

$$\cos\theta = \sin(90° - \theta) \quad \text{and} \quad \sin\theta = \cos(90° - \theta)$$

$$\cot\theta = \tan(90° - \theta) \quad \text{and} \quad \tan\theta = \cot(90° - \theta)$$

$$\csc\theta = \sec(90° - \theta) \quad \text{and} \quad \sec\theta = \csc(90° - \theta)$$

When working with radians:

$$\cos x = \sin\left(\frac{\pi}{2} - x\right) \quad \text{and} \quad \sin x = \cos\left(\frac{\pi}{2} - x\right)$$

$$\cot x = \tan\left(\frac{\pi}{2} - x\right) \quad \text{and} \quad \tan x = \cot\left(\frac{\pi}{2} - x\right)$$

$$\csc x = \sec\left(\frac{\pi}{2} - x\right) \quad \text{and} \quad \sec x = \csc\left(\frac{\pi}{2} - x\right)$$

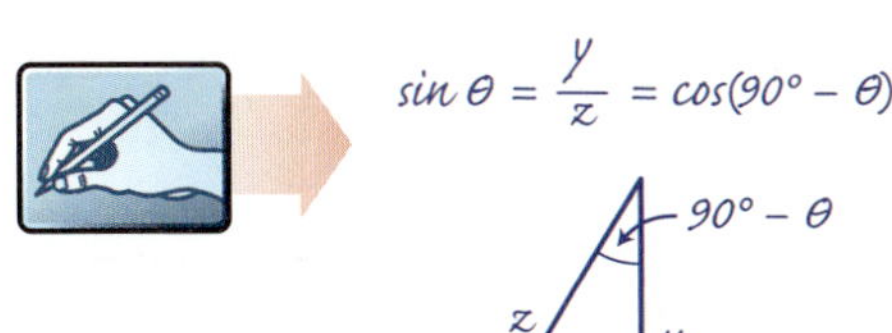

The Composite Argument Property for cos(*A* + *B*)

You can write the cosine of a *sum* of two angles in terms of functions of those two angles. You can transform the cosine of a sum to a cosine of a difference with some insightful algebra and the odd–even properties.

$$\cos(A + B)$$

$$= \cos[A - (-B)]$$ Change the sum to a difference.

$$= \cos A \cos(-B) + \sin A \sin(-B)$$ Use the composite argument property for cos(*first* − *second*).

$$= \cos A \cos B + \sin A(-\sin B)$$ Cosine is an even function. Sine is an odd function.

$$= \cos A \cos B - \sin A \sin B$$

$$\therefore \cos(A + B) = \cos A \cos B - \sin A \sin B$$

The only difference between this property and the one for $\cos(A - B)$ is the sign between the terms on the right side of the equation.

The Composite Argument Properties for sin(*A* − *B*) and sin(*A* + *B*)

You can derive composite argument properties for $\sin(A - B)$ with the help of the cofunction property.

$$\sin(A - B) = \cos[90° - (A - B)]$$ Transform into a cosine using the cofunction property.

$$= \cos[(90° - A) + B]$$ Distribute the minus sign, then associate $(90° - A)$.

$$= \cos(90° - A) \cos B - \sin(90° - A) \sin B$$ Use the composite argument property for cos(*first* + *second*).

$$= \sin A \cos B - \cos A \sin B$$ Use the cofunction property the other way around.

$$\therefore \sin(A - B) = \sin A \cos B - \cos A \sin B$$

$\cos[\underbrace{(90° - A)}_{\text{first}} + \underbrace{B}_{\text{second}}] = \cos(\text{first}) \cdot \cos(\text{second}) - \sin(\text{first}) \cdot \sin(\text{second})$

The composite argument property for $\sin(A + B)$ is

$$\sin(A + B) = \sin A \cos B + \cos A \sin B$$

You can derive it by writing $\sin(A + B)$ as $\sin[A - (-B)]$ and using the same reasoning as for $\cos(A + B)$.

The Composite Argument Properties for tan(*A* − *B*) and tan(*A* + *B*)

You can write the tangent of a composite argument in terms of tangents of the two angles. This requires factoring out a "common" factor that isn't actually there!

$$\tan(A - B) = \frac{\sin(A - B)}{\cos(A - B)}$$

Use the quotient property for tangent to "bring in" sines and cosines.

$$= \frac{\sin A \cos B - \cos A \sin B}{\cos A \cos B + \sin A \sin B}$$

Use the composite argument properties for $\sin(A - B)$ and $\cos(A - B)$.

$$= \frac{\cos A \cos B\left(\frac{\sin A \cos B}{\cos A \cos B} - \frac{\cos A \sin B}{\cos A \cos B}\right)}{\cos A \cos B\left(\frac{\cos A \cos B}{\cos A \cos B} + \frac{\sin A \sin B}{\cos A \cos B}\right)}$$

Factor out $(\cos A \cos B)$ in the numerator and denominator to put cosines in the minor denominators.

$$= \frac{\frac{\sin A}{\cos A} - \frac{\sin B}{\cos B}}{1 + \frac{\sin A \sin B}{\cos A \cos B}}$$

Cancel all common factors.

$$= \frac{\tan A - \tan B}{1 + \tan A \tan B}$$

Use the quotient property to get only tangents.

$$\therefore \tan(A - B) = \frac{\tan A - \tan B}{1 + \tan A \tan B}$$

$\sin A \cos B = \frac{\cos A \cos B}{\cos A \cos B} \cdot \sin A \cos B = \cos A \cos B \cdot \frac{\sin A \cos B}{\cos A \cos B}$

Multiply by 1

Associate the lower $\cos A \cos B$ with $\sin A \cos B$

You can derive the composite argument property for $\tan(A + B)$ by writing $\tan(A + B)$ as $\tan[A - (-B)]$ and then using the fact that tangent is an odd function. The result is

$$\tan(A + B) = \frac{\tan A + \tan B}{1 - \tan A \tan B}$$

This box summarizes the composite argument properties for cosine, sine, and tangent. As with the composite argument properties for $\cos(A - B)$ and $\cos(A + B)$, notice that the signs between the terms change when you compare $\sin(A - B)$ and $\sin(A + B)$ or $\tan(A - B)$ and $\tan(A + B)$.

PROPERTIES: Composite Argument Properties for Cosine, Sine, and Tangent

$$\cos(A - B) = \cos A \cos B + \sin A \sin B$$

$$\cos(A + B) = \cos A \cos B - \sin A \sin B$$

$$\sin(A - B) = \sin A \cos B - \cos A \sin B$$

$$\sin(A + B) = \sin A \cos B + \cos A \sin B$$

$$\tan(A - B) = \frac{\tan A - \tan B}{1 + \tan A \tan B}$$

$$\tan(A + B) = \frac{\tan A + \tan B}{1 - \tan A \tan B}$$

Algebraic Solution of Equations

You can use the composite argument properties to solve certain trigonometric equations algebraically.

EXAMPLE 1 ➤ Solve the equation for $x \in [0, 2\pi]$. Verify the solutions graphically.

$$\sin 5x \cos 3x - \cos 5x \sin 3x = \frac{1}{2}$$

SOLUTION

$\sin 5x \cos 3x - \cos 5x \sin 3x = \frac{1}{2}$ — Write the given equation.

$\sin(5x - 3x) = \frac{1}{2}$ — Use the composite argument property for $\sin(A - B)$.

$\sin 2x = \frac{1}{2}$

$2x = \arcsin \frac{1}{2}$

$= 0.5235... + 2\pi n$ or $(\pi - 0.5235...) + 2\pi n$ — Use the definition of arcsine to write the general solution.

$x = 0.2617... + \pi n$ or $1.3089... + \pi n$ — Divide by 2.

$x = 0.2617..., 1.3089..., 3.4033..., 4.4505...$ — Use $n = 0$ and $n = 1$ to get the solutions in the domain.

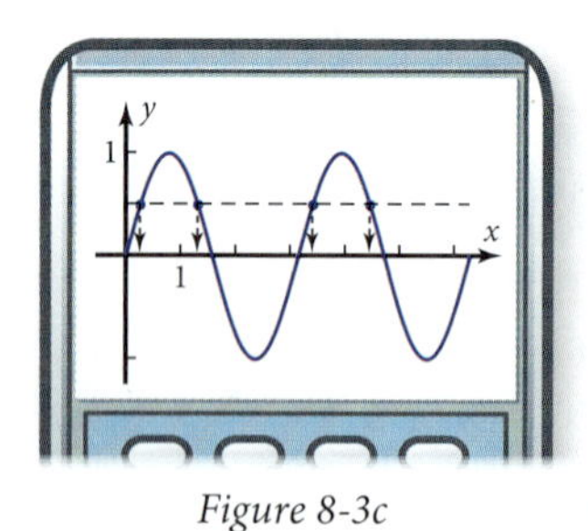

Figure 8-3c

In this case, the answers turn out to be simple multiples of π. See if you can figure out why $x = \frac{\pi}{12}, \frac{5\pi}{12}, \frac{13\pi}{12}$, and $\frac{17\pi}{12}$.

Figure 8-3c shows the graph of $y = \sin 5x \cos 3x - \cos 5x \sin 3x$ and the line $y = 0.5$. Note that the graph of $y = \sin 5x \cos 3x - \cos 5x \sin 3x$ is equivalent to the sinusoid $y = \sin 2x$. By using the intersect feature, you can see that the four solutions are correct and that they are the only solutions in the domain $x \in [0, 2\pi]$.

Problem Set 8-3

Reading Analysis

From what you have read in this section, what do you consider to be the main idea? What is the difference between an *odd* function and an *even* function? How can the composite argument property for cos(*first* − *second*) be used to derive a similar property for cos(*first* + *second*)? What is the meaning of the prefix *co-* in the name *cosine*?

Quick Review

Q1. If one value of arcsin x is 30°, find another positive value of arcsin x less than 360°.

Q2. The value 30 is what percentage of 1000?

Q3. $2^{10} =$ ___?___

Q4. $\cos 7 \cos 3 + \sin 7 \sin 3 = \cos$ ___?___

Q5. What is the amplitude of the sinusoid $y = 8 \cos \theta + 15 \sin \theta$?

Q6. If $A \cos(\theta - D) = 8 \cos \theta + 15 \sin \theta$, then D could equal ___?___.

Q7. $\tan^2 47° - \sec^2 47° =$ ___?___

Q8. $\log 3 + \log 4 = \log$ ___?___

Q9. $\tan x = \frac{\sec x}{\csc x}$ is called a ___?___ property.

Q10. Sketch the graph of an exponential function with base between 0 and 1.

---

1. Prove by counterexample that $\sin(A + B) \neq \sin A + \sin B$.
2. Prove by counterexample that tangent does not distribute over addition.
3. Show by numerical example that $\tan(A - B) = \frac{\tan A - \tan B}{1 + \tan A \tan B}$.

4. Show by numerical example that $\sin(A - B) = \sin A \cos B - \cos A \sin B$.

5. Make a table of values to confirm that $\cos(-x) = \cos x$.

6. Make a table of values to confirm that $\tan(-x) = -\tan x$.

7. Confirm graphically that $\cot \theta = \tan(90° - \theta)$.

8. Confirm graphically that $\cos \theta = \sin(90° - \theta)$.

9. *Odd–Even Property Geometrical Proof Problem:* Figure 8-3d shows angles θ and $-\theta$ in standard position in a uv-coordinate system. The u-coordinates of the points where the angles cut the unit circle are the same. The v-coordinates of these points are opposites of each other.

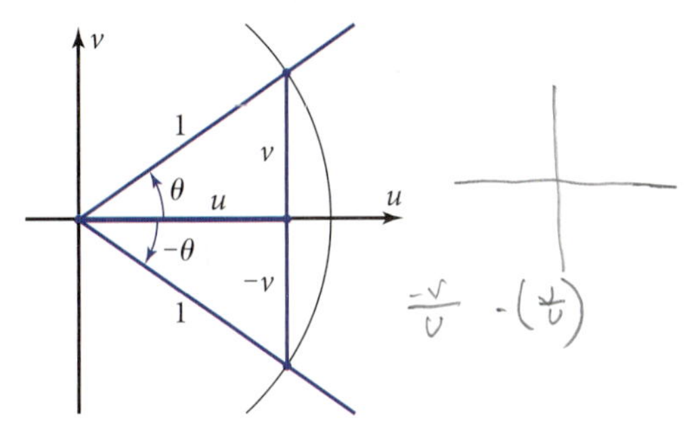

Figure 8-3d

a. Based on the definition of sine, explain why $\sin(-\theta) = -\sin \theta$.

b. Based on the definition of cosine, explain why $\cos(-\theta) = \cos \theta$.

c. Based on the definition of tangent, explain why $\tan(-\theta) = -\tan \theta$.

d. Based on the reciprocal properties, explain why secant, cosecant, and cotangent have the same odd–even properties as cosine, sine, and tangent, respectively.

10. *Odd–Even Property Proof:* Recall that the graph of $y = -f(x)$ is a vertical reflection of $y = f(x)$ across the x-axis and that the graph of $y = f(-x)$ is a horizontal reflection across the y-axis.

a. Figure 8-3e shows the graph of $y = \sin x$. Sketch the graph resulting from a reflection of $y = \sin x$ across the y-axis, and sketch another graph resulting from a reflection of $y = \sin x$ across the x-axis. Based on the results, explain why sine is an *odd* function.

b. Figure 8-3f shows the graph of $y = \cos x$. Sketch the graph resulting from a reflection of $y = \cos x$ across the y-axis. From the result, explain why cosine is an *even* function.

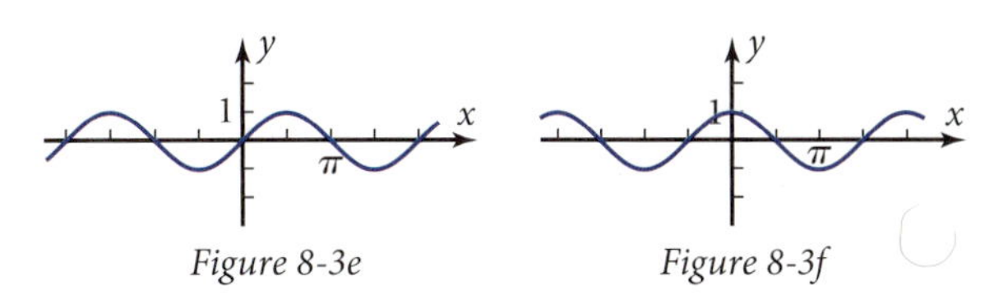

Figure 8-3e *Figure 8-3f*

For Problems 11–20, use the composite argument properties to show that the given equation is an identity.

11. $\cos(\theta - 90°) = \sin \theta$

12. $\cos\left(x - \frac{\pi}{2}\right) = \sin x$

13. $\sin\left(x - \frac{\pi}{2}\right) = -\cos x$

14. $\sec(\theta - 90°) = \csc \theta$ (Be clever!)

15. $\sin(\theta + 60°) - \cos(\theta + 30°) = \sin \theta$

16. $\sin(\theta + 30°) + \cos(\theta + 60°) = \cos \theta$

17. $\sqrt{2} \cos\left(x - \frac{\pi}{4}\right) = \cos x + \sin x$

18. $(\cos A \cos B - \sin A \sin B)^2 + (\sin A \cos B + \cos A \sin B)^2 = 1$

19. $\sin 3x \cos 4x + \cos 3x \sin 4x = \sin 7x$

20. $\cos 10x \cos 6x + \sin 10x \sin 6x = \cos 4x$

For Problems 21–26, use the composite argument properties to transform the left side of the equation to a *single* function of a composite argument. Then solve the equation algebraically to get

a. The general solution for x or θ

b. The particular solutions for x in the domain $x \in [0, 2\pi)$ or for θ in the domain $\theta \in [0°, 360°)$

21. $\cos x \cos 0.6 - \sin x \sin 0.6 = 0.9$

22. $\sin \theta \cos 35° + \cos \theta \sin 35° = 0.5$

23. $\sin 3\theta \cos \theta - \cos 3\theta \sin \theta = 0.5\sqrt{2}$

24. $\cos 3x \cos x + \sin 3x \sin x = -1$

25. $\dfrac{\tan 2x - \tan x}{1 + \tan 2x \tan x} = \sqrt{3}$

26. $\dfrac{\tan \theta + \tan 27°}{1 - \tan \theta \tan 27°} = 1$

Exact Function Value Problems: Figure 8-3g shows angles A and B in standard position in a uv-coordinate system. For Problems 27–32, use the information in the figure to find *exact* values (no decimals!). Check your answers by calculating A and B, adding or subtracting them, and finding the function values directly.

27. $\cos(A - B)$ **28.** $\cos(A + B)$

29. $\sin(A - B)$ **30.** $\sin(A + B)$

31. $\tan(A - B)$ **32.** $\tan(A + B)$

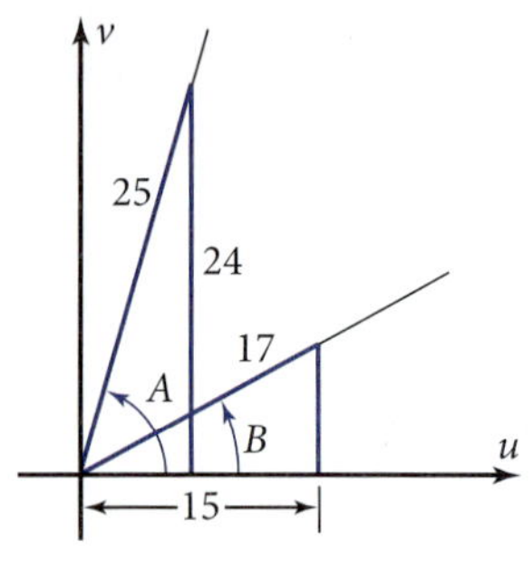

Figure 8-3g

For Problems 33 and 34, use the composite argument properties with exact values of functions of special angles (such as 30°, 45°, 60°) to show that these numerical expressions are exact values of sin 15° and cos 15°. Confirm numerically that the values are correct.

33. $\sin 15^\circ = \dfrac{\sqrt{6} - \sqrt{2}}{4}$

34. $\cos 15^\circ = \dfrac{\sqrt{6} + \sqrt{2}}{4}$

For Problems 35 and 36, use the exact values of sin 15° and cos 15° from Problems 33 and 34 and the cofunction properties to find exact values (no decimals) of the expressions.

35. $\sin 75^\circ$ **36.** $\cos 75^\circ$

For Problems 37 and 38, use the values of sin 15° and cos 15° from Problems 33 and 34, with appropriate simplification, to show that the numerical expressions are exact values of tan 15° and cot 15°.

37. $\tan 15^\circ = 2 - \sqrt{3}$ **38.** $\cot 15^\circ = 2 + \sqrt{3}$

39. *Cofunction Property for the Inverse Sine Function Problem*: In this problem you will prove that $\cos^{-1} x$ is the complement of $\sin^{-1} x$.

a. Let $\theta = 90^\circ - \sin^{-1} x$. Use the composite argument property to prove that $\cos\theta = x$.

b. From part a, it follows that $\theta = \arccos x$, the inverse trigonometric *relation*. Use the fact that $-90^\circ \le \sin^{-1} x \le 90^\circ$ to show that θ is in the interval $[0^\circ, 180^\circ]$.

c. How does part b allow you to conclude that θ is $\cos^{-1} x$, the inverse trigonometric *function*?

40. *Cofunction Properties for the Inverse Circular Functions Problem:* Use the cofunction properties for the inverse circular functions to calculate these values. Show that each answer is in the range of the inverse cofunction.

a. $\cos^{-1}(-0.4)$

b. $\cot^{-1}(-1.5)$

c. $\csc^{-1}(-2)$

PROPERTIES: Cofunction Properties for the Inverse Circular Functions

$$\cos^{-1} x = \frac{\pi}{2} - \sin^{-1} x$$

$$\cot^{-1} x = \frac{\pi}{2} - \tan^{-1} x$$

$$\csc^{-1} x = \frac{\pi}{2} - \sec^{-1} x$$

Triple Argument Properties Problems: The composite argument properties have sums of *two* angle or arc measures. It is possible to derive *triple argument properties* for three angle or arc measures. For Problems 41 and 42, derive properties expressing the given function in terms of $\sin A$, $\sin B$, $\sin C$, $\cos A$, $\cos B$, and $\cos C$ (start by associating two of the three angles).

41. $\cos(A + B + C)$ **42.** $\sin(A + B + C)$

8-4 Composition of Ordinates and Harmonic Analysis

In Sections 8-1 and 8-2, you learned that a sum of two sinusoids with equal periods is another sinusoid with the same period, as shown in Figure 8-4a. A product of sinusoids with equal periods is also a sinusoid, as shown in Figure 8-4b.

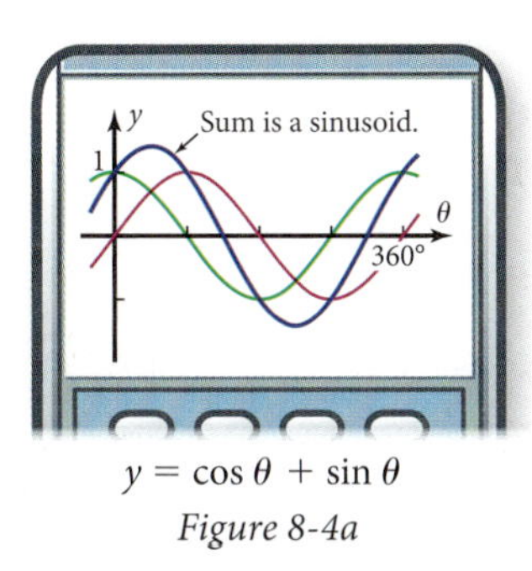

$y = \cos\theta + \sin\theta$

Figure 8-4a

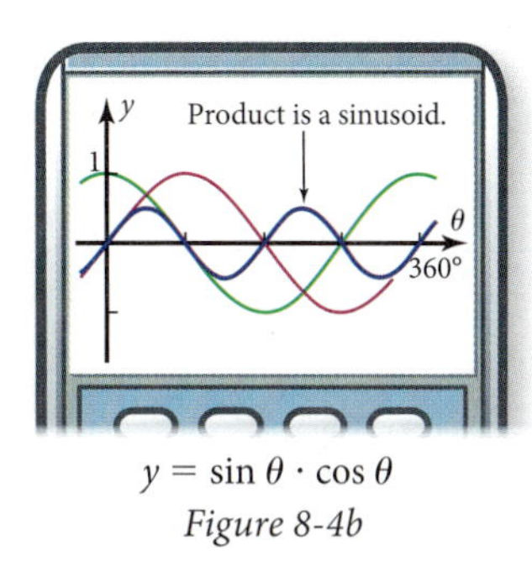

$y = \sin\theta \cdot \cos\theta$

Figure 8-4b

Figure 8-4c shows the result of adding two sinusoids with unequal periods, which might happen, for example, if two musical notes of different frequencies are played at the same time.

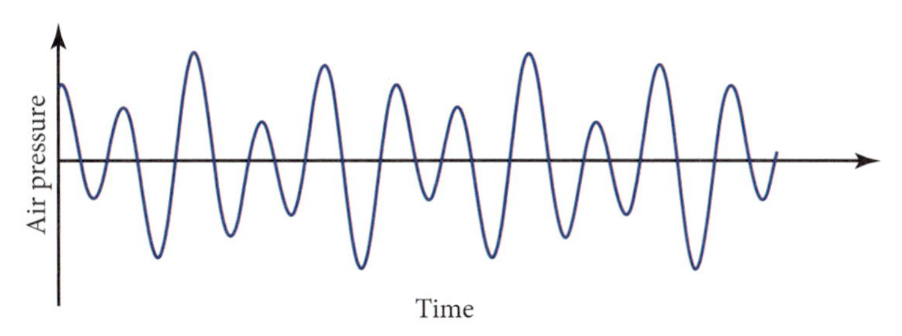

Figure 8-4c

In this section you will learn about **composition of ordinates,** by which sinusoids are added or multiplied, and **harmonic analysis,** by which you reverse the process to find the parent sinusoids.

Objective

- Given two sinusoids, form a new graph by adding or multiplying ordinates (y-coordinates).
- Given a graph formed by adding or multiplying two sinusoids, find the equations for the two sinusoids.

In this exploration you will draw conclusions about graphs composed of a product or a sum of two sinusoids with unequal periods.

EXPLORATION 8-4: Sum or Product of Sinusoids with Unequal Periods

In this exploration you will investigate the product and the sum of these two functions:

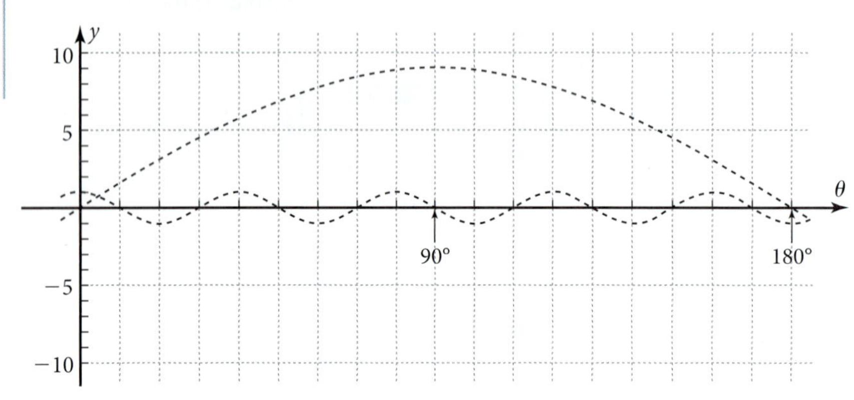

1. The figure shows the graphs of

$f_1(\theta) = 9 \sin \theta$ and $f_2(\theta) = \cos 9\theta$

Indicate which graph is which.

2. Without first plotting on your grapher, on a copy of the figure draw by composition of ordinates the graph of

$y = 9 \sin \theta \cdot \cos 9\theta$

(To compose ordinates, find the y-coordinates of $f_1(\theta)$ and $f_2(\theta)$ and perform the indicated operation.)

3. Plot $f_1(\theta)$ and $f_2(\theta)$ on your grapher. On the same screen, plot $f_3(\theta) = 9 \sin \theta \cdot \cos 9\theta$. Is your composed graph in Problem 2 close to this actual graph?

4. On the same screen, plot the graph of $f_4(\theta) = -f_1(\theta)$. Explain why the graphs of $f_1(\theta)$ and $f_4(\theta)$ can be said to form an *envelope* for the composed graph.

5. Recall that if $y = A \cos 9\theta$, then A is the vertical dilation (the amplitude) of the graph. From your work in Problems 1–3, write a few sentences describing the effect on the graph of making the dilation variable, $A = 9 \sin \theta$, instead of constant.

6. By composition of ordinates, draw the graph of $y = 9 \sin \theta + \cos 9\theta$ on a copy of the figure.

7. Plot $y = 9 \sin \theta + \cos 9\theta$ on your grapher. How closely does your graph in Problem 6 resemble the actual graph?

8. Recall that if $y = C + \cos 9\theta$, then C is a (constant) vertical translation of the sinusoid. Describe the effect of making C a variable, $9 \sin \theta$, instead of simply a constant.

9. Write conclusions:

a. *Adding* sinusoids with very different periods produces . . .

b. *Multiplying* sinusoids with very different periods produces . . .

10. What did you learn as a result of doing this exploration that you did not know before?

Sum of Two Sinusoids with Unequal Periods

In Exploration 8-4, you found that if you add two sinusoids with different periods, the result is a periodic function with a *variable sinusoidal axis*. Example 1 shows you how to sketch such a graph on paper if the two *auxiliary graphs* are given.

EXAMPLE 1 ➤ Figure 8-4d shows the graphs of $f_1(\theta) = 3\cos\theta$ and $f_2(\theta) = \sin 4\theta$.

On a copy of this figure, sketch the graph of

$$y = 3\cos\theta + \sin 4\theta$$

Then plot the function on your grapher. How well does your sketch match the actual graph?

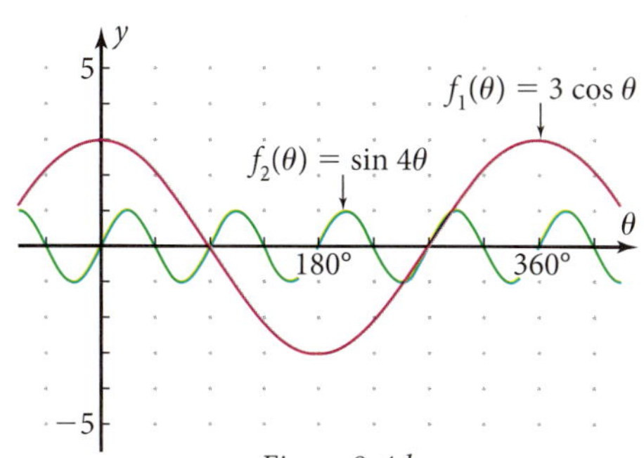

Figure 8-4d

SOLUTION Look for high points, low points, and zeros on the two graphs. At each θ-value you chose, estimate the ordinate of each graph and add them. Put a dot on the graph paper at that value of θ with the appropriate ordinate. Figure 8-4e shows how you might estimate the ordinate at one particular point.

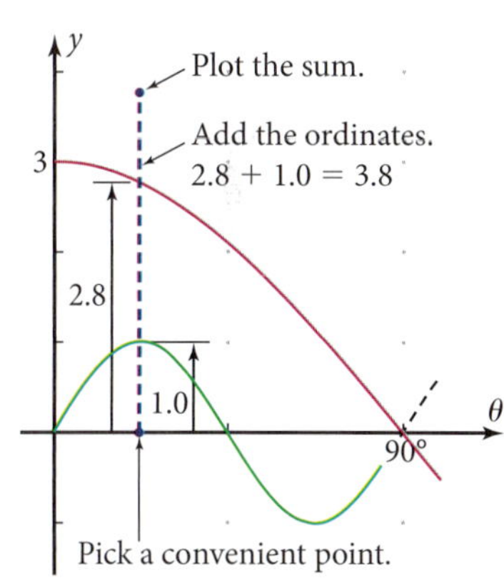

Figure 8-4e

Figure 8-4f shows the dots plotted at places where the auxiliary graphs have critical points or zeros. Once you see the pattern, you can connect the dots with a smooth curve, as in Figure 8-4g.

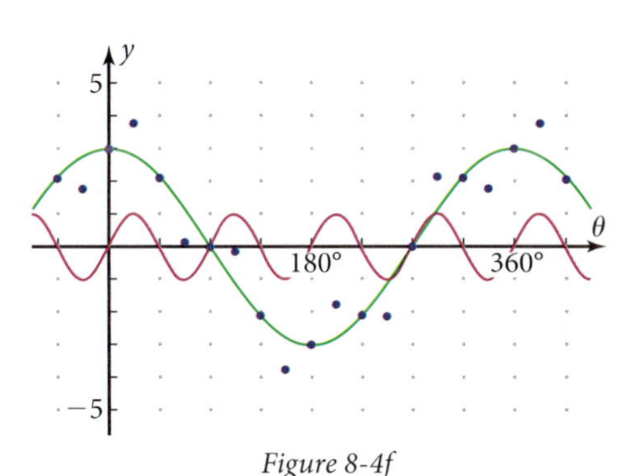

Figure 8-4f

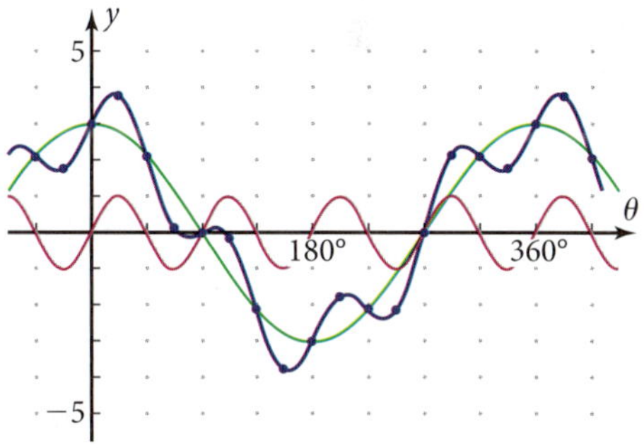

Figure 8-4g

Plotting the equation on your grapher confirms that your sketch is correct. ◄

Product of Two Sinusoids with Unequal Periods

Example 2 shows you how to combine the two sinusoids of Example 1 by multiplying instead of adding.

EXAMPLE 2 ➤ Figure 8-4d shows the graphs of $f_1(\theta) = 3 \cos \theta$ and $f_2(\theta) = \sin 4\theta$.

On a copy of Figure 8-4d, sketch the graph of

$$y = 3 \cos \theta \cdot \sin 4\theta$$

Then plot the function on your grapher. How well does your sketch match the actual graph?

SOLUTION The thought process is the same as for Example 1, but this time you multiply the ordinates instead of adding them. Figure 8-4h shows the results.

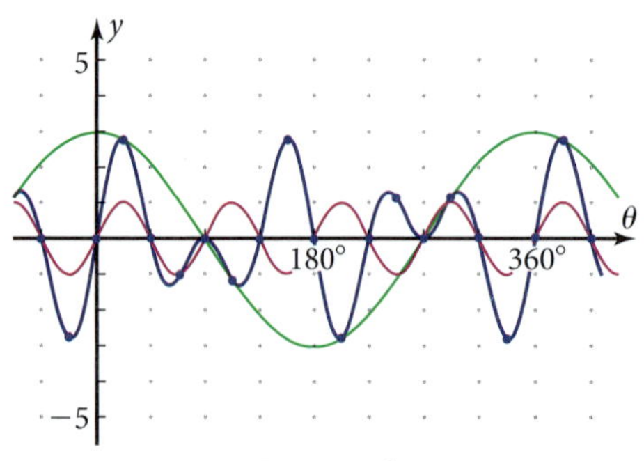

Figure 8-4h

Plotting the equation on your grapher confirms that your sketch is correct. ◄

In Example 2, multiplying $\sin 4\theta$ by $3 \cos \theta$ gives $\sin 4\theta$ a different dilation at different points. The function $y = 3 \cos \theta \sin 4\theta$ seems to have a *variable amplitude*. Because $y = \sin 4\theta$ has amplitude 1, the composed graph touches the graph of $y = 3 \cos \theta$ wherever the graph of $y = \sin 4\theta$ has a high point. The graphs of $y = 3 \cos \theta$ and its opposite, $y = -3 \cos \theta$, form an *envelope* for the composed graph. Note that at each place where either graph crosses the θ-axis, the composed graph also crosses the axis.

PROPERTIES: Sums and Products of Sinusoids with Unequal Periods

If two sinusoids have greatly different periods, then

- Adding the two sinusoids produces a function with a variable sinusoidal axis.
- Multiplying the two sinusoids produces a function with a variable amplitude.

Harmonic Analysis: The Reverse of Composition of Ordinates

You can use the properties of sums and products of sinusoids with different periods to help you "decompose" a complicated graph into the two sinusoids that formed it. The procedure is used, for example, by technicians when their ship's sonar detects a complicated wave pattern and they want to find out if the sounds are being generated by another ship or by a whale.

EXAMPLE 3 ➤ The function in Figure 8-4i is a sum or a product of two sinusoids. Find the particular equation, and confirm your answer by plotting the equation on your grapher.

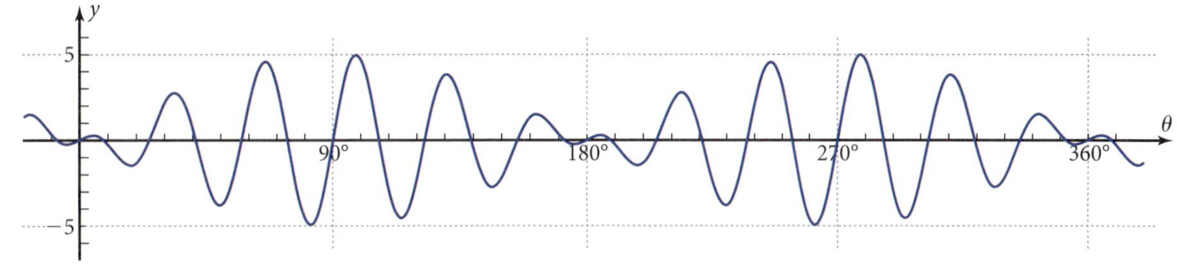

Figure 8-4i

SOLUTION Your thought process should be

1. The *amplitude* varies, so it is a *product* of sinusoids.
2. Sketch the larger-period sinusoid, forming an *envelope curve* that is tangent to the graph near the high points, as shown in Figure 8-4j. These points indicate the ends of cycles of the shorter-period sinusoid. Where the envelope curve is negative, it is tangent to the given curve near the *low* points. If the envelope curve crosses the given curve abruptly, such as at 180° in Figure 8-4j, it is *not* tangent and thus does *not* indicate the end of a smaller-sinusoid cycle.

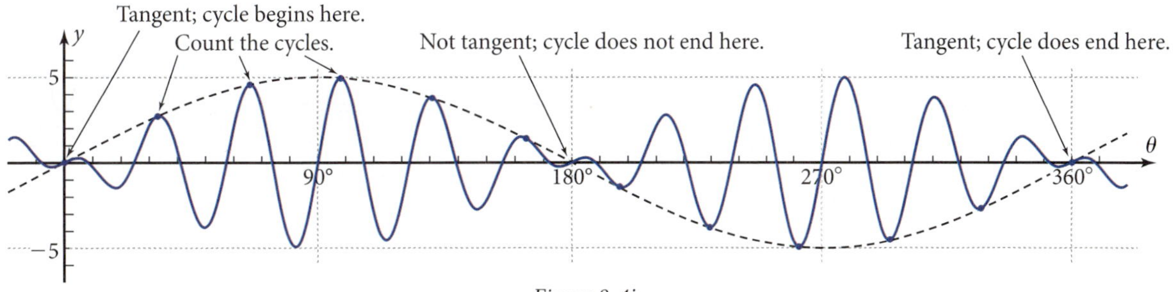

Figure 8-4j

3. For convenience, assume the larger-period sinusoid has the equation $y = 5 \sin \theta$.
4. Let the amplitude of the smaller-period sinusoid be 1. Count the number of cycles the given graph makes in one cycle of the longer-period envelope, 11 in this case. To help you count cycles, look for points where the given curve is tangent to the envelope curve.
5. The smaller-period sinusoid appears to be a cosine. For very small values of θ, the composed graph is very close to the larger-period graph, as happens when you multiply by a number close to 1 ($\cos 0°$) rather than a number close to 0 ($\sin 0°$). So its equation is $y = \cos 11\theta$.
6. Write the equation for the composed function: $y = 5 \sin \theta \cdot \cos 11\theta$.
7. Plot the graph (Figure 8-4k) to check your answer. Use a window, such as $[-20, 200]$ for x, that is small enough to separate the cycles and includes the ends where characteristic behavior occurs. If the plotted graph doesn't agree with the given graph, go back and check your work.

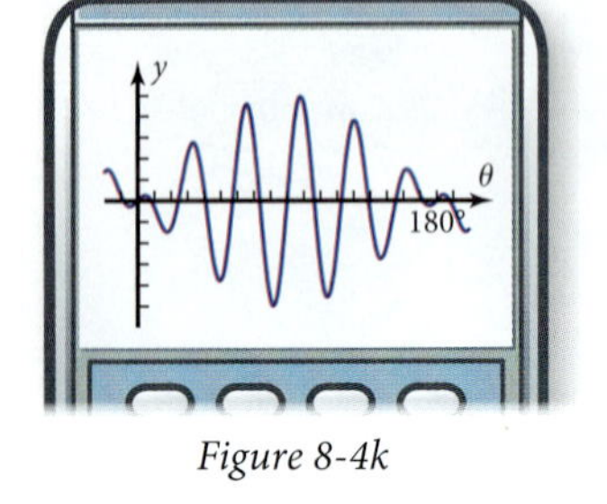

Figure 8-4k

The next example shows how to perform the harmonic analysis if the argument is in radians.

EXAMPLE 4 The function in Figure 8-4l is a sum or a product of two sinusoids. Find the particular equation, and confirm your answer by plotting the equation on your grapher.

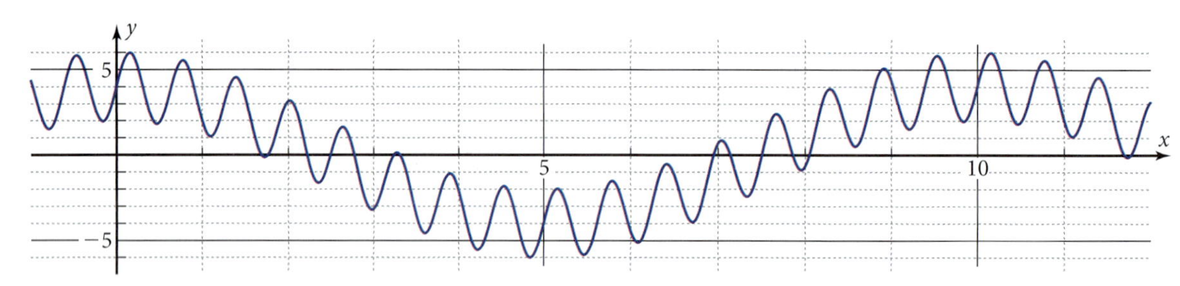

Figure 8-4l

SOLUTION Your thought process should be

1. The *sinusoidal axis* varies, so it is a *sum* of sinusoids.
2. Sketch the larger-period sinusoid as a sinusoidal axis through the points of inflection halfway between high points and low points (Figure 8-4m).

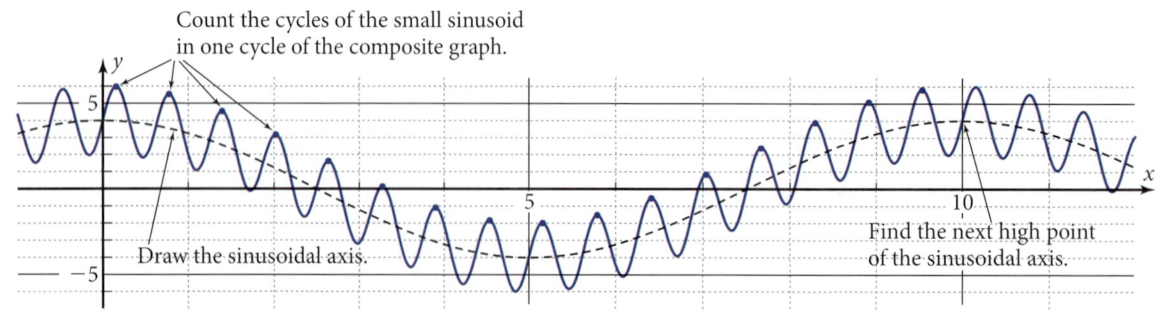

Figure 8-4m

3. The larger-period sinusoid has amplitude 4 and period 10 and starts at a high point, so its equation is

$$y = 4\cos\frac{\pi}{5}x \qquad \text{The coefficient of } x \text{ is } 2\pi \text{ divided by the period.}$$

4. The smaller-period sinusoid has amplitude 2. Because it is on the varying sinusoidal axis at $x = 0$, its function is sine. It makes 16 cycles in one period of the larger-period sinusoid, $0 \le x \le 10$. So its equation is

$$y = 2\sin\frac{16\pi}{5}x \qquad \text{The period is } 10 \cdot \frac{1}{16}.$$

5. The equation for the composed function is $y = 4\cos\frac{\pi}{5}x + 2\sin\frac{16\pi}{5}x$.
6. Plotting the equation on your grapher in radian mode confirms that the equation is correct.

Problem Set 8-4

Reading Analysis

From what you have read in this section, what do you consider to be the main idea? How do you decide whether a combination of two sinusoids with much different periods is a *sum* or a *product*? What is the difference in meaning between composition of ordinates and harmonic analysis?

Quick Review

Q1. If $\cos A = 0.6$, $\sin A = 0.8$, $\cos B = \frac{1}{\sqrt{2}}$, and $\sin B = \frac{-1}{\sqrt{2}}$, then $\cos(A - B) =$ __?__.

Q2. In general, $\cos(x + y) =$ __?__ in terms of cosines and sines of x and y.

Q3. $\sin 5 \cos 3 + \cos 5 \sin 3 = \sin($__?__$)$

Q4. $\sin(90° - \theta) = \cos($__?__$)$

Q5. How do you tell that sine is an *odd* function?

Q6. $\cos(x + x) =$ __?__ in terms of $\cos x$ and $\sin x$.

Q7. If the two legs of a right triangle are 57 and 65, find the tangent of the smallest angle.

Q8. The period of the sinusoid $y = 5 + 7\cos\frac{\pi}{4}(x - 6)$ is __?__.

Q9. $x = 3 + 2\cos t$ and $y = 5 + 4\sin t$ are parametric equations for a(n) __?__.

Q10. If $g(x) = f(x - 7)$, what transformation is applied to function f to get function g?

Problems 1 and 2 refer to Figure 8-4n, which shows the graphs of

$$f_1(\theta) = 6\sin\theta \qquad \text{and} \qquad f_2(\theta) = \cos 6\theta$$

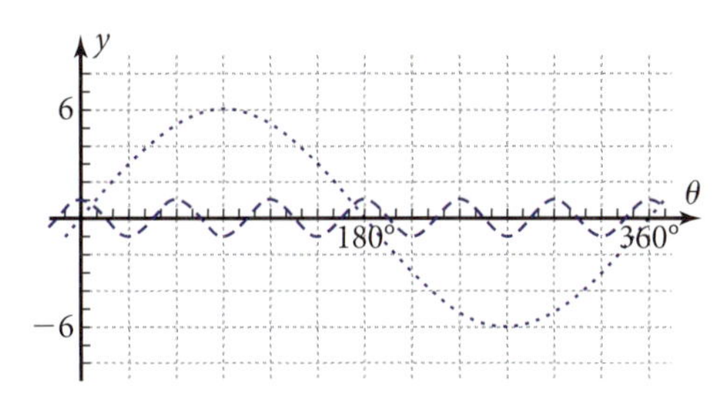

Figure 8-4n

1. *Product of Two Sinusoids Problem:*
   a. Which graph corresponds to each function?
   b. Without using your grapher, sketch on a copy of Figure 8-4n the graph of

   $$y = 6\sin\theta \cdot \cos 6\theta$$

   c. On your grapher, plot the graph of $f_3(\theta) = f_1(\theta) \cdot f_2(\theta)$. Use a window with at least $[0°, 360°]$ for θ and an appropriate y-range. How closely does your composed graph in part b resemble the actual graph on your grapher?
   d. The graph of $y = A\cos 6\theta$ is a sinusoid with a fixed amplitude, A. How would you describe the graph of $y = 6\sin\theta \cdot \cos 6\theta$?

2. *Sum of Two Sinusoids Problem:*

   a. Without using your grapher, sketch on a copy of Figure 8-4n the graph of

   $$y = 6\sin\theta + \cos 6\theta$$

   b. On your grapher, plot the graph of $f_3(\theta) = f_1(\theta) + f_2(\theta)$. Use a window with at least $[0°, 360°]$ for θ and an appropriate y-range. How closely does your composed graph in part a resemble the actual graph on your grapher?

   c. The graph of $y = C + \cos 6\theta$ is a sinusoid with a fixed vertical translation, C. How would you describe the graph of $y = 6\sin\theta + \cos 6\theta$?

The Tacoma Narrows Bridge in Tacoma, Washington, collapsed on November 7, 1940, due to wind that caused vibrations of increasing amplitude.

Harmonic Analysis Problems: For Problems 3–12, find the particular equation for the graph shown. Each is the sum or the product of sinusoids with very different periods. Make note of whether the argument is in degrees or radians.

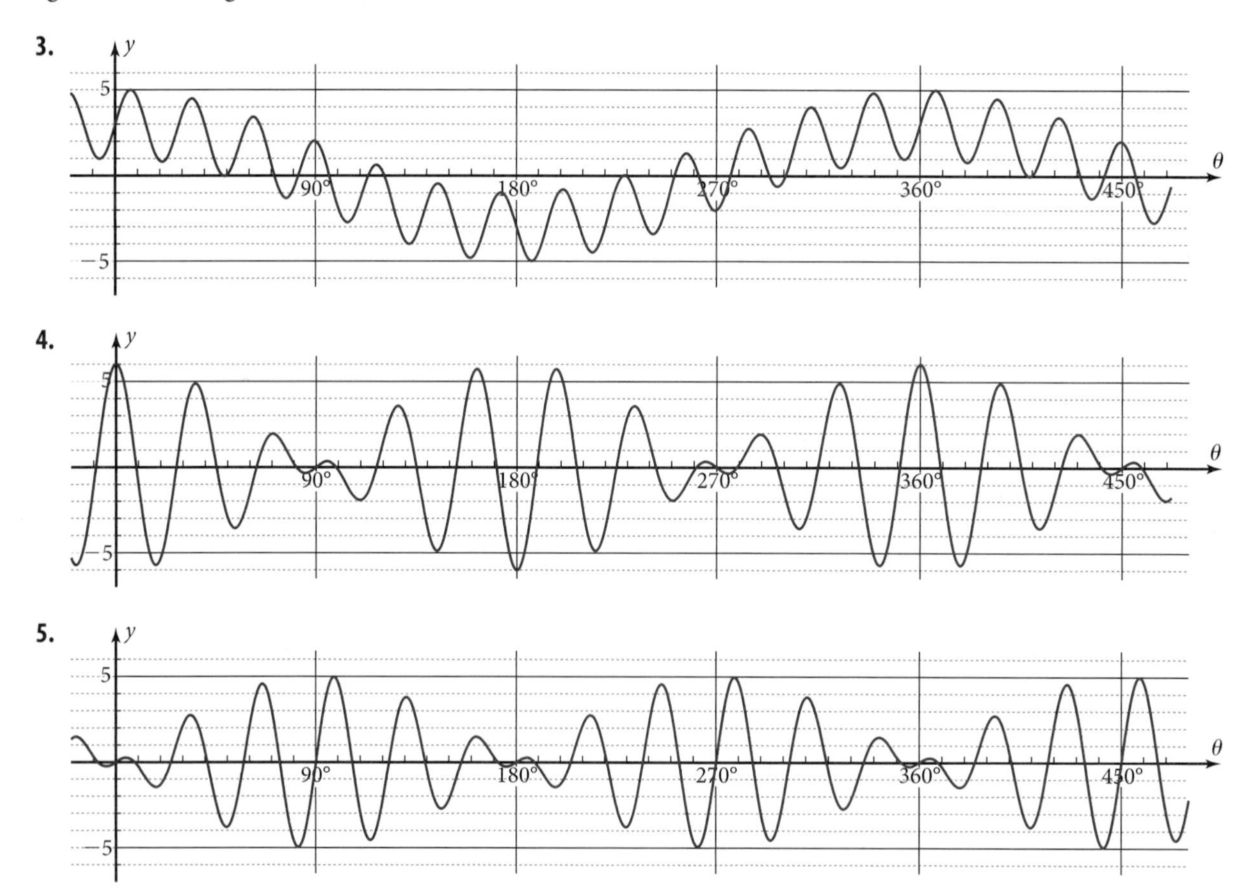

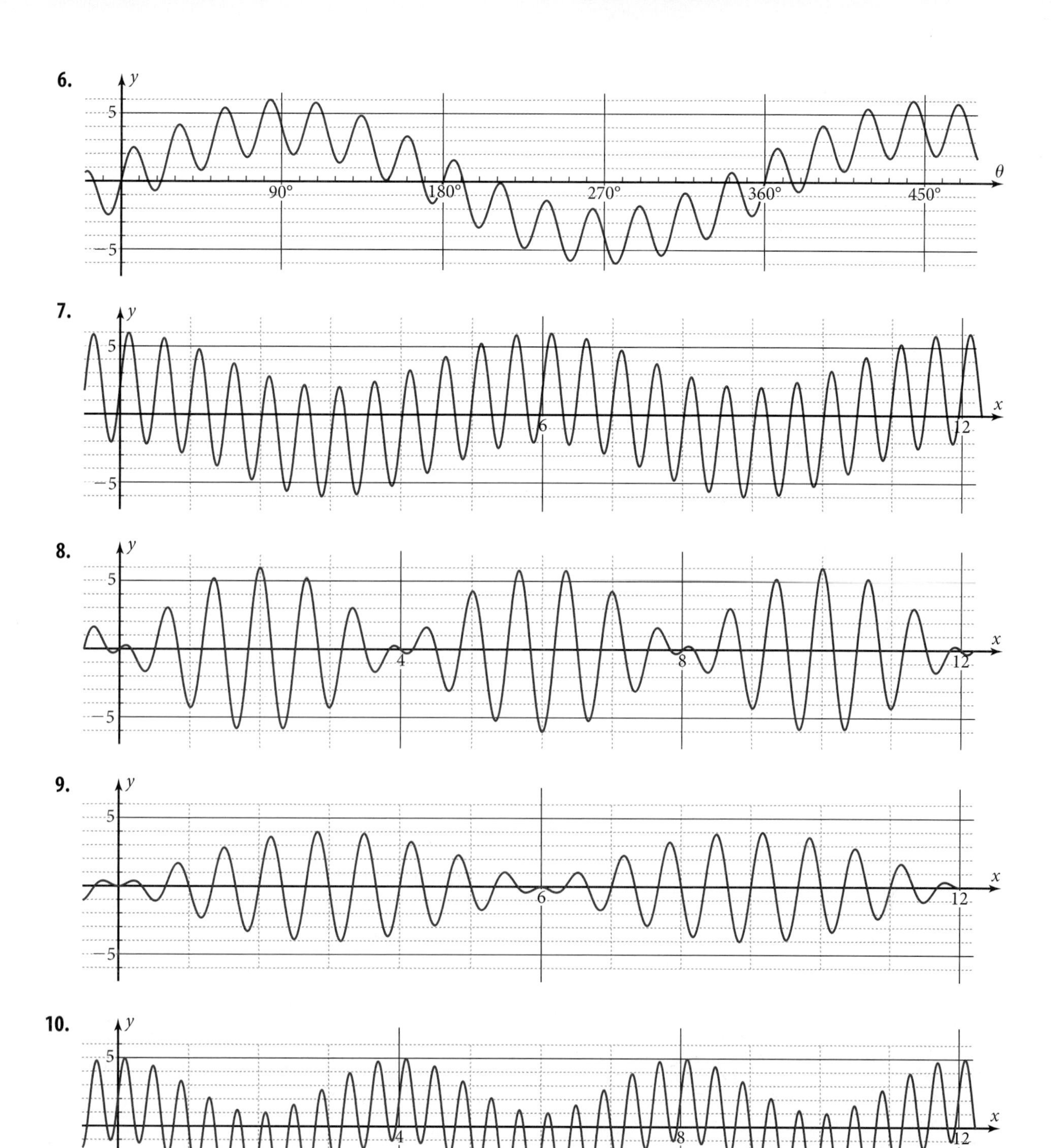

6.
y
5
−5
90°
180°
270°
360°
450°
θ
7.
y
5
−5
6
12
x
8.
y
5
−5
4
8
12
x
9.
y
5
−5
6
12
x
10.
y
5
−5
4
8
12
x

11.

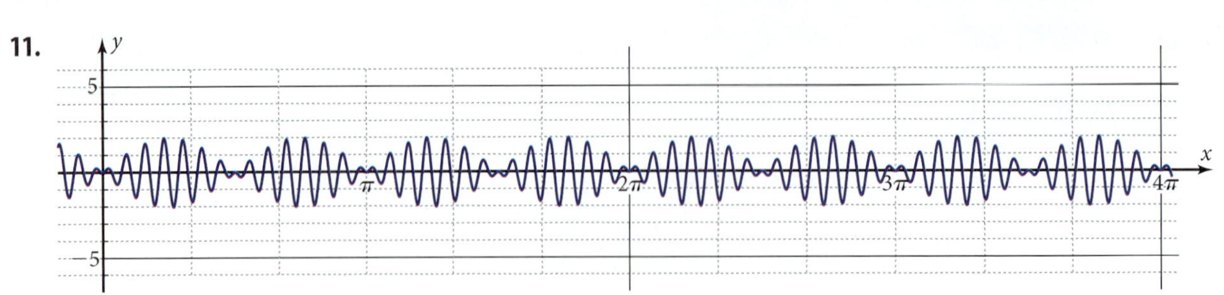

12. A combination of *three* sinusoids!

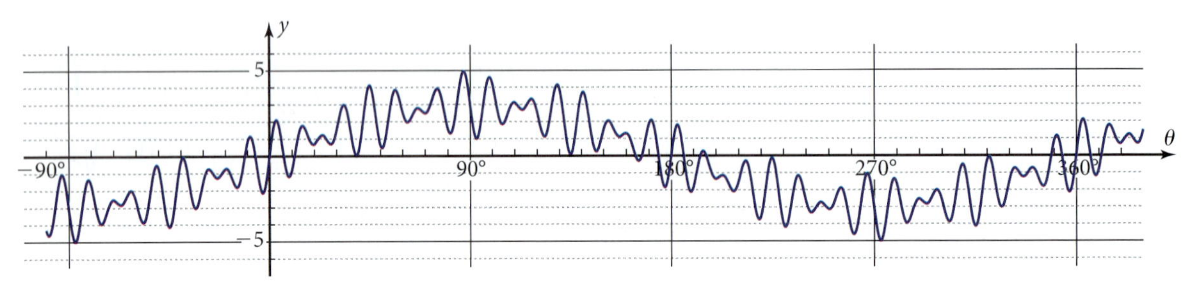

13. *Submarine Sonar Problem:* The sonar on a surface ship picks up sound being generated by equipment on a submarine. By performing harmonic analysis on the sound wave, it is possible to identify the national origin of the submarine.

a. Figure 8-4o shows the pattern of a sound wave, where x is time, in seconds. Find an equation for this graph. Observe that the longer-wave sinusoidal axis completes *three* cycles before the wave pattern starts repeating itself.

b. What is the period of the longer sinusoid? What is the period of the shorter sinusoid?

c. Recall that the frequency of a sinusoid is the reciprocal of the period. What are the frequencies of the two sinusoids in part b?

d. United States submarines have electrical generators that rotate at 60 cycles per second and other electrical generators that rotate at 400 cycles per second. Based on your results in part c, could the sound have been coming from the generators on a U.S. submarine?

14. *Sunrise Project:* The angle of elevation of the Sun at any time of day on any day of the year is a sum of two sinusoids. The first sinusoid is caused by the rotation of Earth about its own axis. In San Antonio this sinusoid has period 1 day and amplitude 61° (the complement of the 29° latitude on which San Antonio lies). The sinusoid reaches a minimum at midnight on any day. The other sinusoid is caused by the rotation of Earth around the Sun. That sinusoid has period 365.25 days and amplitude 23.5° (equal to the tilt of Earth's axis with respect to the ecliptic plane). This second sinusoid reaches a minimum at day -10 (December 21). Part of the graph of this composed function is shown in Figure 8-4p. On this graph, y is the angle of elevation on day x. Note that $x = 1$ at the *end* of day 1, $x = 2$ at the *end* of day 2, and so on.

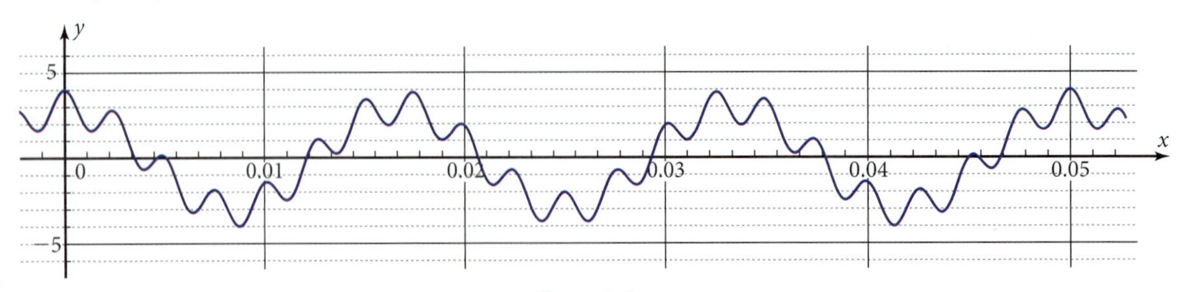

Figure 8-4o

a. Write the particular equation for this composed graph. Use radians as the argument of the function. Use the equation to show that the maximum angle of elevation on day 1 ($x = 0.5$) is lower than the maximum angle of elevation on day 30. Confirm that this is true by direct measurement on Figure 8-4p.

b. Find the maximum angle of elevation on the longest day of the year, June 21.

c. Sunrise occurs at the time when the angle of elevation equals zero, as it is increasing. Find the time of sunrise on day 1 by solving numerically for the value of x close to 0.25 when $y = 0$. Then convert your answer to hours and minutes. Show that the Sun rises earlier on day 30 than it does on day 1. How much earlier?

d. How do you interpret the parts of the graph that are *below* the x-axis?

This is a mariners' astrolabe from around 1585. An astrolabe measured the elevation of the Sun or another star to predict the time of sunrise and sunset.

15. *Journal Problem:* Update your journal with things you have learned since the last entry. In particular, mention how you decide whether a composed graph is a sum or a product of sinusoids. You might also mention how comfortable you are becoming with the amplitudes and periods of sinusoidal graphs and whether and how your confidence in working with these concepts has grown since you first encountered them in Chapter 5.

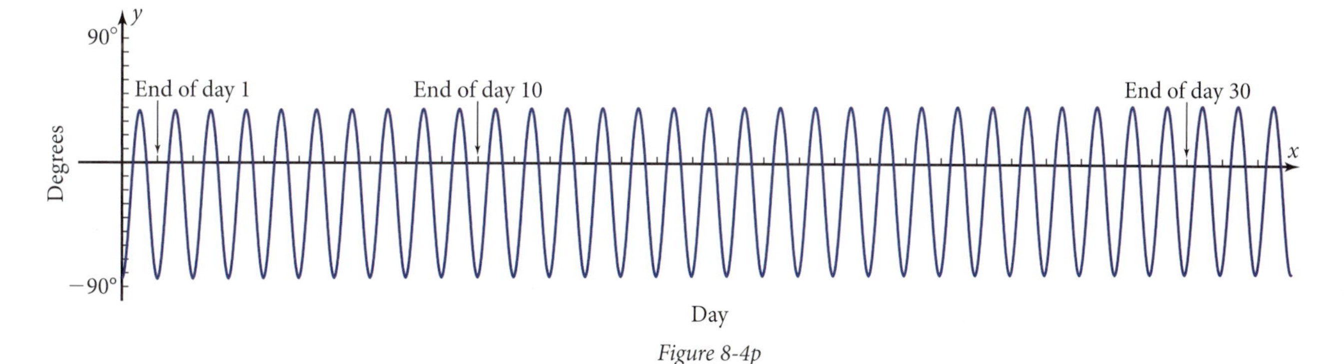

Figure 8-4p

8-5 The Sum and Product Properties

Figure 8-5a shows the graphs of two sinusoids with very different periods. Figure 8-5b shows two sinusoids with nearly equal periods. The product of the two sinusoids in Figure 8-5a is the variable-amplitude function shown in Figure 8-5c. Surprisingly, the *sum* of the two nearly equal sinusoids in Figure 8-5b also produces the wave pattern in the third figure! The y-values of the sinusoids in Figure 8-5b add up where the waves are in phase (i.e., two high points coincide) and cancel out where the waves are out of phase. This phenomenon is heard as what are called *beats*. For example, two piano strings for the same note can be tuned at slightly different frequencies to produce a vibrato effect. This is what piano tuners listen for.

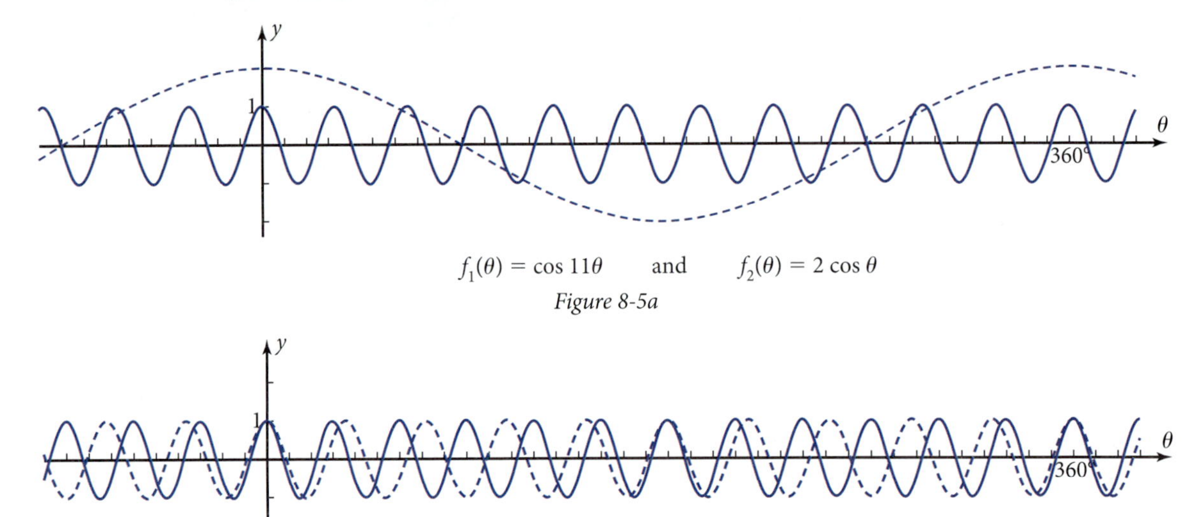

$f_1(\theta) = \cos 11\theta$ and $f_2(\theta) = 2\cos\theta$

Figure 8-5a

$f_1(\theta) = \cos 12\theta$ and $f_2(\theta) = \cos 10\theta$

Figure 8-5b

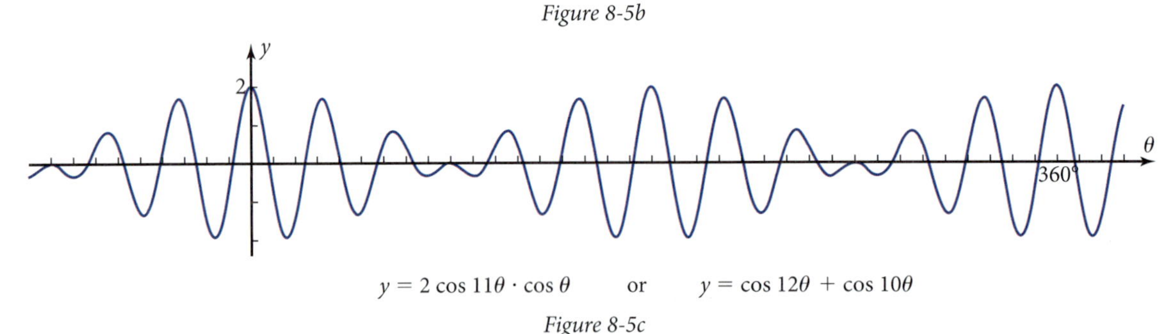

$y = 2\cos 11\theta \cdot \cos\theta$ or $y = \cos 12\theta + \cos 10\theta$

Figure 8-5c

In this section you will learn algebraic properties that allow you to prove that the sum graph and the product graph are equivalent.

Objective Transform a sum of two sinusoids to a product of two sinusoids, and vice versa.

Product of Sum

From Figure 8-5c, it appears that

$$\cos 12\theta + \cos 10\theta = 2 \cos 11\theta \cdot \cos \theta$$

To see why, write 12θ as $(11\theta + \theta)$ and 10θ as $(11\theta - \theta)$ and use composite argument properties.

$$\cos 12\theta = \cos(11\theta + \theta) = \cos 11\theta \cos \theta - \sin 11\theta \sin \theta$$

$$\cos 10\theta = \cos(11\theta - \theta) = \cos 11\theta \cos \theta + \sin 11\theta \sin \theta$$

Add the two equations to get

$$\cos 12\theta + \cos 10\theta = 2 \cos 11\theta \cos \theta$$

If you subtract the equations, you get

$$\cos 12\theta - \cos 10\theta = -2 \sin 11\theta \sin \theta$$

Using similar steps, you can derive these properties in general. If you use A and B in place of 11θ and θ, you'll get **sum and product properties** expressing the sum or difference of two cosines as a product of two cosines or two sines.

$$\cos(A + B) + \cos(A - B) = 2 \cos A \cos B$$

$$\cos(A + B) - \cos(A - B) = -2 \sin A \sin B$$

Two other sum and product properties come from adding or subtracting the composite argument properties for sine.

$$\sin(A + B) = \sin A \cos B + \cos A \sin B$$

$$\sin(A - B) = \sin A \cos B - \cos A \sin B$$

$$\sin(A + B) + \sin(A - B) = 2 \sin A \cos B$$ By adding the two equations.

$$\sin(A + B) - \sin(A - B) = 2 \cos A \sin B$$ By subtracting the two equations.

These properties can also be expressed by reversing the two sides of the equation. This box summarizes the four properties.

PROPERTIES: Sum and Product Properties—Product to Sum

$$2 \cos A \cos B = \cos(A + B) + \cos(A - B)$$

$$2 \sin A \sin B = -\cos(A + B) + \cos(A - B)$$

$$2 \sin A \cos B = \sin(A + B) + \sin(A - B)$$

$$2 \cos A \sin B = \sin(A + B) - \sin(A - B)$$

It is probably easier to derive these properties as you need them than it is to memorize them.

EXAMPLE 1 ➤ Transform $2 \sin 13° \cos 48°$ into a sum (or difference) of functions with *positive* arguments. Demonstrate numerically that the result is correct.

SOLUTION Sine multiplied by cosine appears in the *sine* composite argument properties. So the answer will be the sum (or difference) of two sines. If you have not memorized the sum and product properties, you would write

$$\sin(13° + 48°) = \sin 13° \cos 48° + \cos 13° \sin 48°$$

$$\underline{\sin(13° - 48°) = \sin 13° \cos 48° - \cos 13° \sin 48°}$$

$$\sin 61° + \sin(-35°) = 2 \sin 13° \cos 48°$$ *Add* the equations.

$$2 \sin 13° \cos 48° = \sin 61° - \sin 35°$$ Use the symmetric property of equality. Also, sine is an *odd* function. ◄

Check: 0.3010... = 0.3010...

Sum to Product

Example 2 shows you how to reverse the process and transform a sum of two sinusoids into a product.

EXAMPLE 2 ➤ Transform $\cos 7\theta - \cos 3\theta$ into a product of functions with positive arguments.

SOLUTION First think of writing $\cos 7\theta$ and $\cos 3\theta$ as cosines with composite arguments. Then use appropriate calculations to find out what those two arguments are.

Let $\cos 7\theta = \cos(A + B)$ and let $\cos 3\theta = \cos(A - B)$.

$$A + B = 7\theta$$

$$\underline{A - B = 3\theta}$$

$$2A = 10\theta$$

$$A = 5\theta$$

Substituting 5θ for A in either equation, you get $B = 2\theta$. Now, substitute these values for A and B in the composite argument properties for cosine.

$$\cos 7\theta = \cos(5\theta + 2\theta) = \cos 5\theta \cos 2\theta - \sin 5\theta \sin 2\theta$$

$$\underline{\cos 3\theta = \cos(5\theta - 2\theta) = \cos 5\theta \cos 2\theta + \sin 5\theta \sin 2\theta}$$

$$\cos 7\theta - \cos 3\theta = -2 \sin 5\theta \sin 2\theta$$

The arguments have no negative signs, so you need no further transformations. ◄

From the algebraic steps in Example 2, you can see that A equals half the sum of the arguments. You can also tell that B equals half the difference of the arguments. So a general property expressing a difference of two cosines as a product is

$$\cos x - \cos y = -2 \sin \tfrac{1}{2}(x + y) \sin \tfrac{1}{2}(x - y)$$

You can also write the other three sum and product properties in this form. The results are in the box. Again, do not try to memorize the properties. Instead, derive them from the composite argument properties, as in Example 2, or look them up when you need to use them.

PROPERTIES: Sum and Product Properties—Sum to Product

$$\sin x + \sin y = 2 \sin \tfrac{1}{2}(x + y) \cos \tfrac{1}{2}(x - y)$$
$$\sin x - \sin y = 2 \cos \tfrac{1}{2}(x + y) \sin \tfrac{1}{2}(x - y)$$
$$\cos x + \cos y = 2 \cos \tfrac{1}{2}(x + y) \cos \tfrac{1}{2}(x - y)$$
$$\cos x - \cos y = -2 \sin \tfrac{1}{2}(x + y) \sin \tfrac{1}{2}(x - y)$$

Note: Both functions on the "sum" side are always the same function.

EXAMPLE 3 ➤ Figure 8-5d shows the graph of a periodic trigonometric function with a variable sinusoidal axis. Using harmonic analysis, find the particular equation for this function as a sum of two sinusoids. Then transform the sum into a product. Confirm graphically that both equations produce the function graphed in Figure 8-5d.

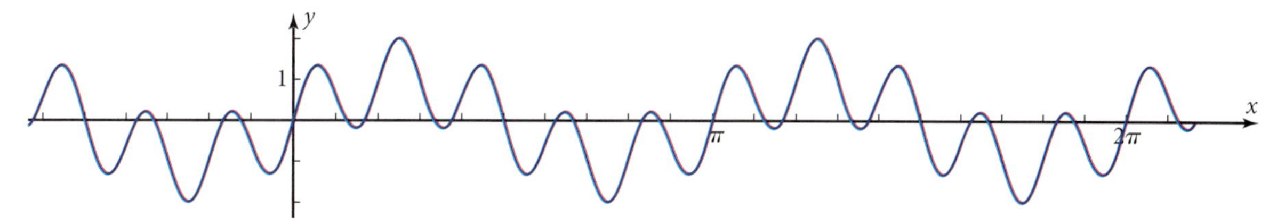

Figure 8-5d

SOLUTION On a copy of Figure 8-5d, sketch the sinusoidal axis. Figure 8-5e shows the result.

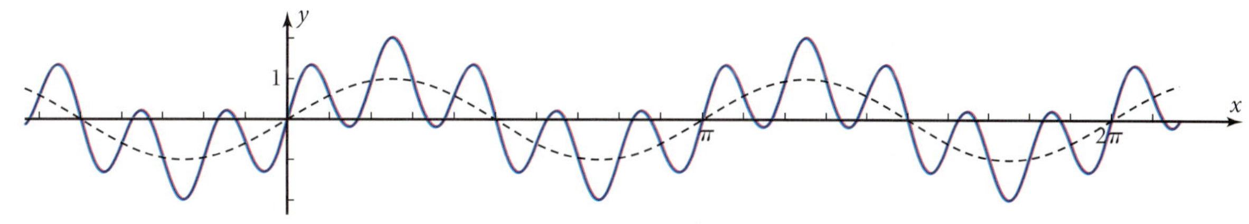

Figure 8-5e

By the techniques of Section 8-4, $y = \sin 2x + \sin 10x$

To transform this equation to a product, write

$$y = 2 \sin \tfrac{1}{2}(2x + 10x) \cos \tfrac{1}{2}(2x - 10x) \quad \text{Use the property from the box, or derive it.}$$
$$y = 2 \sin 6x \cos(-4x)$$
$$y = 2 \sin 6x \cos 4x \quad \text{Cosine is an } \textit{even} \text{ function.}$$

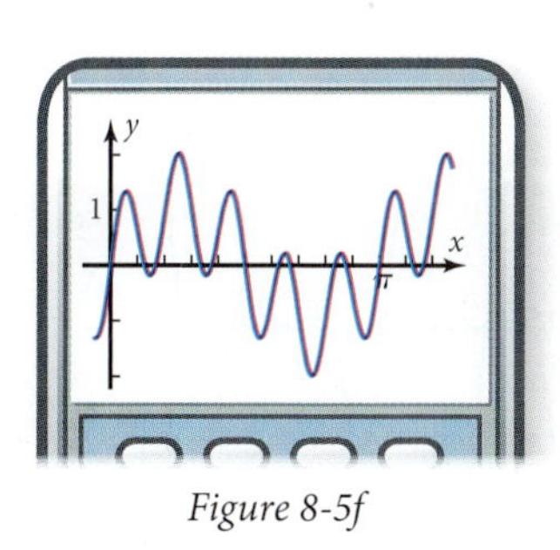

Figure 8-5f

Both equations give the graph shown in Figure 8-5f, which agrees with Figure 8-5d. ◄

Problem Set 8-5

Reading Analysis

From what you have read in this section, what do you consider to be the main idea? How is it possible for a sum of two sinusoids with nearly equal periods to produce the same wave pattern as a product of two sinusoids with very different periods? What does the text say about the advisability of memorizing the sum and product properties?

Quick Review

Q1. Find values of A and B if $A + B = 20$ and $A - B = 12$.

Q2. Find values of A and B if $A + B = x$ and $A - B = y$.

Q3. Sketch the graph of the parent trigonometric sinusoid $y = \cos\theta$.

Q4. Sketch the graph of the parent circular sinusoid $y = \sin x$.

Q5. How many degrees are in $\frac{\pi}{4}$ radian?

Q6. How many radians are in 180°?

Q7. What is the exact value (no decimals) of $\cos\left(\frac{\pi}{6}\right)$?

Q8. What is the exact value (no decimals) of tan 30°?

Q9. $\cos(3x + 5x) =$ ___?___ in terms of functions of $3x$ and $5x$.

Q10. Find the value of θ in decimal degrees if $\theta = \cot^{-1}\left(\frac{3}{7}\right)$.

Transformation Problems: For Problems 1–8, transform the product into a sum or difference of sines or cosines with *positive* arguments.

1. 2 sin 41° cos 24°

2. 2 cos 73° sin 62°

3. 2 cos 53° cos 49°

4. 2 sin 29° sin 16°

5. 2 cos 3.8 sin 4.1

6. 2 cos 2 cos 3

7. $2 \sin 3x \sin 7.2$

8. $2 \sin 8x \cos 2x$

For Problems 9–16, transform the sum or difference to a product of sines and/or cosines with *positive* arguments.

9. cos 46° + cos 12°

10. cos 56° − cos 24°

11. sin 2 + sin 6

12. sin 3 − sin 8

13. cos 2.4 − cos 4.4

14. sin 1.8 + sin 6.4

15. $\sin 3x - \sin 8x$

16. $\cos 9x + \cos 11x$

Graphing Problems: For Problems 17–20, use harmonic analysis to find an equation of the given graph as a product or sum of sinusoids. Then transform the product into a sum or the sum into a product. Confirm graphically that both of your equations produce the given graph.

17.

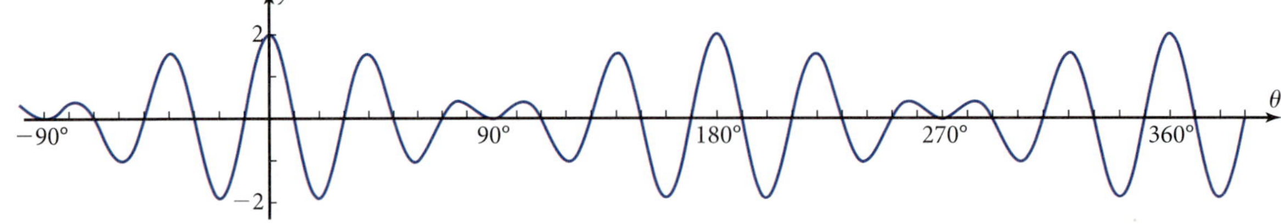

18.

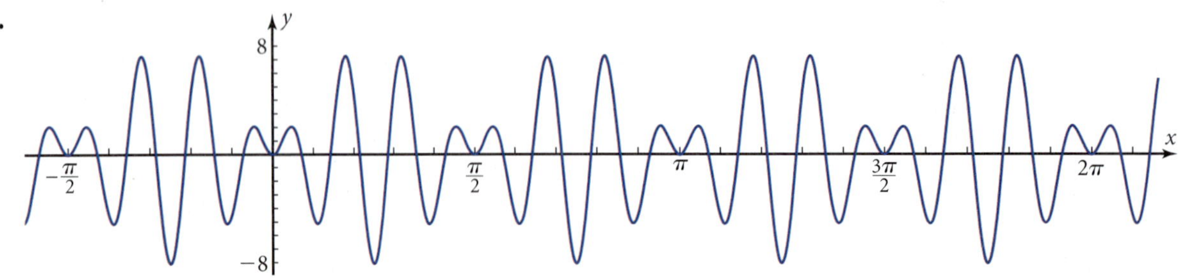

19.

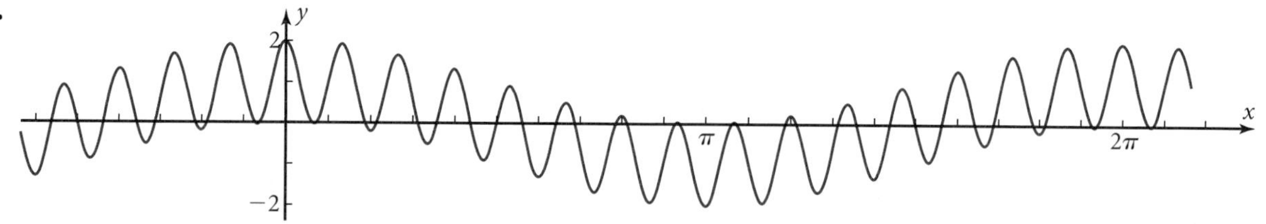

20.

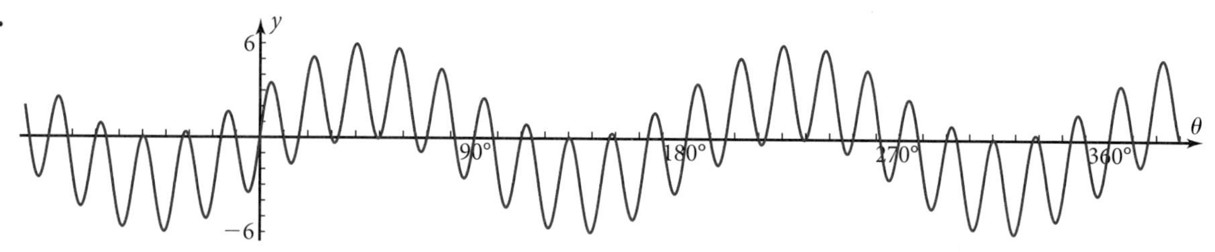

Algebraic Solution of Equations Problems: You can use the sum and product properties to find algebraically the exact solutions of certain equations. For Problems 21–24, solve the equation by first transforming it into a product equal to zero and then setting each factor equal to zero. Use the domain $\theta \in [0°, 360°]$ or $x \in [0, 2\pi]$.

21. $\sin 3x - \sin x = 0$ **22.** $\sin 3\theta + \sin \theta = 0$

23. $\cos 5\theta + \cos 3\theta = 0$ **24.** $\cos 5x - \cos x = 0$

Identities Problems: For Problems 25–30, use the sum and product properties to prove that the given equation is an identity.

25. $\cos x - \cos 5x = 4 \sin 3x \sin x \cos x$

26. $\dfrac{\sin 5x + \sin 7x}{\cos 5x + \cos 7x} = \tan 6x$

27. $\cos x + \cos 2x + \cos 3x = (\cos 2x)(1 + 2 \cos x)$

28. $\sin(x + y) \sin(x - y) = \sin^2 x - \sin^2 y$

29. $\cos(x + y) \cos(x - y) = \cos^2 x - \sin^2 y$

30. $\sin(x + y) \cos(x - y) = \frac{1}{2}\sin 2x + \frac{1}{2}\sin 2y$

31. *Piano Tuning Problem:* Note A on the piano has frequency 220 cycles/s. Inside the piano there are three strings for this note. When the note is played, the hammer attached to the A key hits all three strings and starts them vibrating. Suppose that two of the A strings are tuned to 221 cycles/s and 219 cycles/s, respectively.

a. The combined sound of these two notes is the sum of the two sound waves. Write an equation for the combined sound wave, where the independent variable is time t, in seconds. Use the (undisplaced) cosine function for each sound wave, and assume that each has amplitude 1.

b. Transform the sum in part a into a product of two sinusoids.

c. Explain why the combined sound is equivalent to a sound with frequency 220 cycles/s and an amplitude that varies. What is the frequency of the variable amplitude? Describe how the note sounds.

32. *Car and Truck Problem:* Suppose that you are driving an 18-wheeler tractor-trailer truck along the highway. A car pulls up alongside your truck and then moves ahead very slowly. As it passes, the combined sound of the car and truck engines pulsates louder and softer. The combined sound produced when the truck has a higher frequency than the car can be either a sum of the two sinusoids or a product of them.

a. The tachometer says that your truck's engine is turning at 3000 revolutions per minute (rev/min), which is equivalent to 50 revolutions per second (rev/s). The pulsations come and go once a second, which means that the amplitude sinusoid has period *two* seconds. Write an equation for the sound intensity as a product of two sinusoids. Use cosine for each, with independent variable t, in seconds. Use 2 for the amplitude of the larger sinusoid and 1 for that of the smaller.

b. Transform the product in part a into a sum of two sinusoids.

c. At what rate is the car's engine rotating?

d. Explain why the period of the amplitude sinusoid is *two* seconds, not one second.

33. 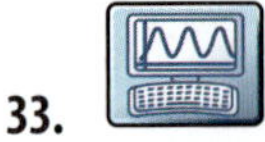  *Dynamic Sinusoidal Sums and Products Project:* Using dynamic geometry software such as The Geometer's Sketchpad, create a sketch with a slider k to plot the function

$$f(x) = \cos 6x \sin kx$$

Go to **www.keymath.com/precalc** if you need information on sliders or to use the *Sinusoidal Sums and Products* exploration. Use the sketch to answer these questions.

a. Set k equal to 0.5 so that the periods of the two sinusoids are very different. What kind of wave pattern do you notice? Then slide the value of k up to 5. What differences do you notice taking place in the wave pattern as the periods of the two sinusoids get closer to each other? What happens to the wave pattern when k equals 6?

b. Change the equation to

$$f(x) = \cos 6x + \sin kx$$

Again set k equal to 0.5 and describe the wave pattern. What happens this time as you slide k up closer to 6? Equal to 6?

c. Describe how this problem illustrates the sum and product properties and what these properties tell you about wave patterns for sums of sinusoids and for products of sinusoids.

34. *AM Radio Project:* AM ("amplitude modulation") radio works by having a sound wave of a relatively low frequency (long period) cause variations in the amplitude of a "carrier wave" that has a very high frequency (VHF). So the sound wave is multiplied by the carrier wave. An example of the resulting wave pattern is shown in Figure 8-5g. In this project you will find equations of the two waves that were multiplied to form this graph. Then you will see how you can form the same wave pattern by *adding* two waves of nearly equal frequency.

a. By harmonic analysis, find the equations of the two circular function sinusoids that were multiplied to form the graph in Figure 8-5g. Which equation is the sound wave, and which is the carrier wave?

b. You can add two sinusoids with nearly equal periods to form the same wave pattern as in Figure 8-5g. Use the sum and product properties to find the equations of these two sinusoids.

c. Confirm by graphing that the product of the sinusoids in part a and the sum of the sinusoids in part b give the same wave pattern as in Figure 8-5g.

d. The scale on the t-axis in Figure 8-5g is in milliseconds. Find the frequency of the carrier wave in kilocycles per second. (This number, divided by 10, is what appears on an AM radio dial.)

e. Research to find the name used in radio waves for "cycles per second."

Problem 35 prepares you for the next section.

35. *Power-Reducing Identities Problem:* Derive the formulas shown by applying the product to sum properties, using the same angle for A and B. Why do you think these are called power-reducing identities?

$$\sin^2 x = \frac{1}{2}(1 - \cos 2x) \qquad \cos^2 x = \frac{1}{2}(1 + \cos 2x)$$

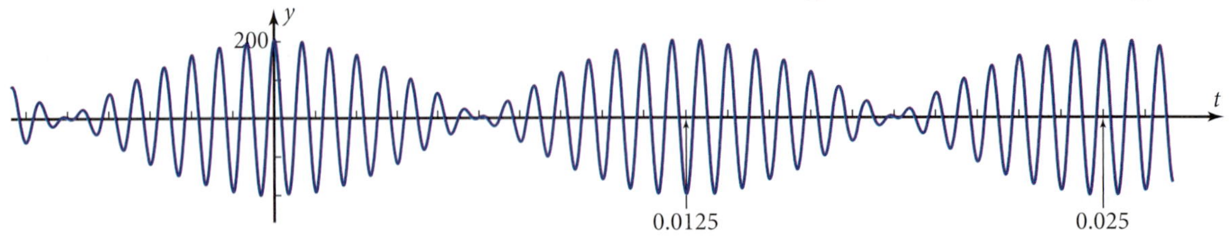

Figure 8-5g

8-6 Double and Half Argument Properties

If you write $\cos 2x$ as $\cos(x + x)$ and use the composite argument property, you can get a **double argument property** expressing $\cos 2x$ in terms of sines and cosines of x. By performing algebraic operations on the double argument properties, you can derive similar **half argument properties.** In this section you'll learn these properties and show that the product of two sinusoids with equal periods is a sinusoid with half the period and half the product of the amplitudes of the two orginal sinusoids.

Objective

- Prove that a product of sinusoids with equal periods is also a sinusoid.
- Derive formulas for $\cos 2A$, $\sin 2A$, and $\tan 2A$ in terms of functions of A.
- Derive formulas for $\cos \frac{1}{2}A$, $\sin \frac{1}{2}A$, and $\tan \frac{1}{2}A$ in terms of functions of A.

Product of Sinusoids with Equal Periods

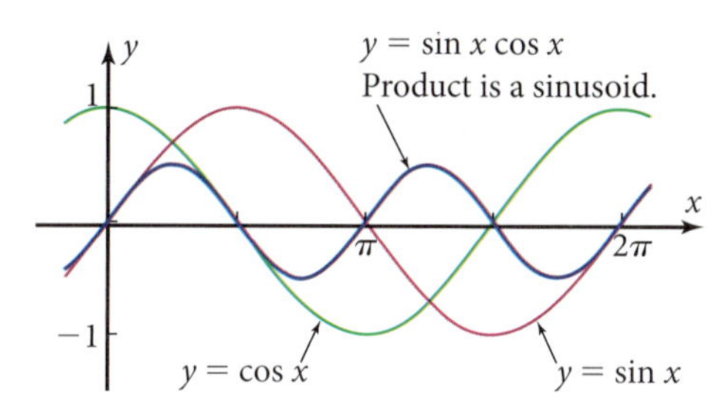

Figure 8-6a

Figure 8-6a shows that the graph of $y = \sin x \cos x$ appears to be a sinusoid with half the period and half the product of amplitudes of the parent sine and cosine functions. Its equation would be $y = \frac{1}{2}\sin 2x$. This calculation shows that this is true.

$$\frac{1}{2}\sin 2x = \frac{1}{2}\sin(x + x)$$ Write $2x$ as $x + x$.

$$= \frac{1}{2}(\sin x \cos x + \cos x \sin x)$$ Use the composite argument property.

$$= \frac{1}{2}(2 \sin x \cos x)$$

$$= \sin x \cos x$$

$$\therefore \sin x \cos x = \frac{1}{2}\sin 2x$$ The product of sine and cosine with equal arguments is a sinusoid.

The square of cosine and the square of sine have a similar property. Figure 8-6b gives graphical evidence.

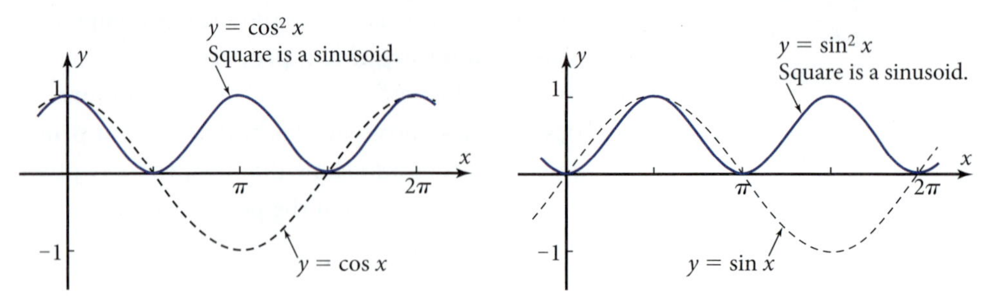

Figure 8-6b

Both $y = \cos^2 x$ and $y = \sin^2 x$ are sinusoids with amplitude $\frac{1}{2}$, period π, and sinusoidal axis $y = \frac{1}{2}$. Notice that the two graphs are a half-cycle out of phase. (This is consistent with the fact that $\cos^2 x + \sin^2 x = 1$.) This box summarizes the conclusions, fulfilling the first of this section's objectives. The second and third properties in the box are also referred to as *power-reducing identities.*

PROPERTIES: Products and Squares of Cosine and Sine

Product of Sine and Cosine Property

$$\sin x \cos x = \frac{1}{2} \sin 2x$$

Square of Cosine Property

$$\cos^2 x = \frac{1}{2} + \frac{1}{2} \cos 2x$$

Square of Sine Property

$$\sin^2 x = \frac{1}{2} - \frac{1}{2} \cos 2x$$

Note: The product of two sinusoids with equal periods, equal amplitudes, and zero vertical translation is a sinusoid with half the period and half the product of the amplitudes.

Double Argument Properties

Multiplying both sides of the equation $\sin x \cos x = \frac{1}{2} \sin 2x$ by 2 gives an equation expressing $\sin 2x$ in terms of the sine and cosine of x.

$$\sin 2x = 2 \sin x \cos x \qquad \text{Double argument property for sine.}$$

Here's how the double argument property for cosine can be derived.

$$\cos 2x = \cos(x + x)$$

$$= \cos x \cos x - \sin x \sin x$$

$$= \cos^2 x - \sin^2 x$$

This box summarizes the double argument properties. You will prove the double argument property for tangent in Problem 31.

PROPERTIES: Double Argument Properties

Double Argument Property for Sine

$\sin 2A = 2 \sin A \cos A$

Double Argument Properties for Cosine

$\cos 2A = \cos^2 A - \sin^2 A$

$\cos 2A = 2 \cos^2 A - 1$ (using $\sin^2 A = 1 - \cos^2 A$)

$\cos 2A = 1 - 2 \sin^2 A$ (using $\cos^2 A = 1 - \sin^2 A$)

Double Argument Property for Tangent

$\tan 2A = \dfrac{2 \tan A}{1 - \tan^2 A}$

EXAMPLE 1 ➤ If $\cos x = 0.3$, find the exact value of $\cos 2x$. Check your answer numerically by finding the value of x, doubling it, and finding the cosine of the resulting argument.

SOLUTION

$\cos 2x = 2 \cos^2 x - 1$ — Double argument property for cosine in terms of cosine alone.

$= 2(0.3)^2 - 1$

$= -0.82$

Check: $x = \cos^{-1} 0.3 = 1.2661...$

$\cos(2 \cdot 1.2661...) = \cos 2.5322... = -0.82$, which checks ◄

Note that for the properties to apply, all that matters is that one argument is twice the other argument. Example 2 shows this.

EXAMPLE 2 ➤ Write an equation expressing $\cos 10x$ in terms of $\sin 5x$.

SOLUTION

$\cos 10x = \cos(2 \cdot 5x)$ — Transform to double argument.

$= 1 - 2 \sin^2 5x$ — Use the double argument property for cosine involving only sine.

$\therefore \cos 10x = 1 - 2 \sin^2 5x$ ◄

EXAMPLE 3 ➤ Write the equation in Example 2 directly from the composite argument property for cosine.

SOLUTION

$\cos 10x = \cos(5x + 5x)$ — Transform into composite argument.

$= \cos 5x \cos 5x - \sin 5x \sin 5x$ — Use the composite argument property for cosine.

$= \cos^2 5x - \sin^2 5x$

$= (1 - \sin^2 5x) - \sin^2 5x$ — Use the Pythagorean property.

$= 1 - 2 \sin^2 5x$ ◄

Half Argument Properties

The square of cosine property and the square of sine property state that

$$\cos^2 x = \frac{1}{2} + \frac{1}{2}\cos 2x \qquad \text{and} \qquad \sin^2 x = \frac{1}{2} - \frac{1}{2}\cos 2x$$

The argument x on the left in the equations is half the argument $2x$ on the right. Let $A = 2x$. Then

$$\cos^2 \tfrac{1}{2}A = \tfrac{1}{2}(1 + \cos A) \qquad \text{and} \qquad \sin^2 \tfrac{1}{2}A = \tfrac{1}{2}(1 - \cos A)$$

Taking the square roots gives half argument properties for $\cos \frac{1}{2}A$ and $\sin \frac{1}{2}A$.

$$\cos \tfrac{1}{2}A = \pm\sqrt{\tfrac{1}{2}(1 + \cos A)} \qquad \text{and} \qquad \sin \tfrac{1}{2}A = \pm\sqrt{\tfrac{1}{2}(1 - \cos A)}$$

The way you can determine whether to choose the positive or negative sign of the ambiguous $\pm$ sign is by looking at the quadrant in which $\frac{1}{2}A$ terminates (*not* the quadrant in which A terminates). For instance, if $A = 120°$, then $\frac{1}{2}A$ is 60°, which terminates in the first quadrant, as shown in the left graph of Figure 8-6c. In this case, both $\sin \frac{1}{2}A$ and $\cos \frac{1}{2}A$ are positive. If $A = 480°$ (which is coterminal with 120°), then $\frac{1}{2}A$ is 240°, which terminates in Quadrant III, as shown in the right graph of Figure 8-6c. In this case, you'll choose the negative sign for $\cos \frac{1}{2}A$ and $\sin \frac{1}{2}A$. The signs of sine and cosine don't have to be the same for the same argument. If $\frac{1}{2}A$ falls in Quadrant II or Quadrant IV, the signs of $\cos \frac{1}{2}A$ and $\sin \frac{1}{2}A$ are opposites.

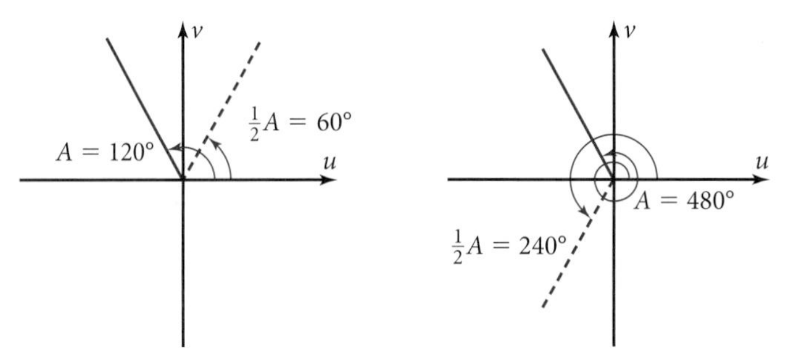

Figure 8-6c

You can derive a half argument property for $\tan \frac{1}{2}A$ by dividing the respective properties for sine and cosine. In Problem 32, you will see how this is done. The result is remarkable because you can drop the radical sign and the ambiguous $\pm$ sign.

The half argument properties are summarized in this box.

PROPERTIES: Half Argument Properties

Half Argument Property for Sine

$$\sin \tfrac{1}{2}A = \pm\sqrt{\tfrac{1}{2}(1 - \cos A)}$$

Half Argument Property for Cosine

$$\cos \tfrac{1}{2}A = \pm\sqrt{\tfrac{1}{2}(1 + \cos A)}$$

Half Argument Properties for Tangent

$$\tan \tfrac{1}{2}A = \pm\sqrt{\frac{1 - \cos A}{1 + \cos A}} = \frac{1 - \cos A}{\sin A} = \frac{\sin A}{1 + \cos A}$$

Note: The ambiguous $\pm$ sign is determined by the quadrant in which $\frac{1}{2}A$ terminates.

Example 4 shows you how to calculate the functions of half an angle and twice an angle using the properties and how to verify the result by direct computation.

EXAMPLE 4 ➤ If $\cos A = \frac{15}{17}$ and A is in the open interval $(270°, 360°)$,

a. Find the exact value of $\cos \frac{1}{2}A$.

b. Find the exact value of $\cos 2A$.

c. Verify your answers numerically by calculating the values of $2A$ and $\frac{1}{2}A$ and finding the cosines.

SOLUTION Sketch angle A in standard position, as shown in Figure 8-6d. Pick a point on the terminal side with horizontal coordinate 15 and radius 17. Draw the reference triangle. By the Pythagorean theorem and by noting that angle A terminates in Quadrant IV, you can determine that the vertical displacement of the point is -8.

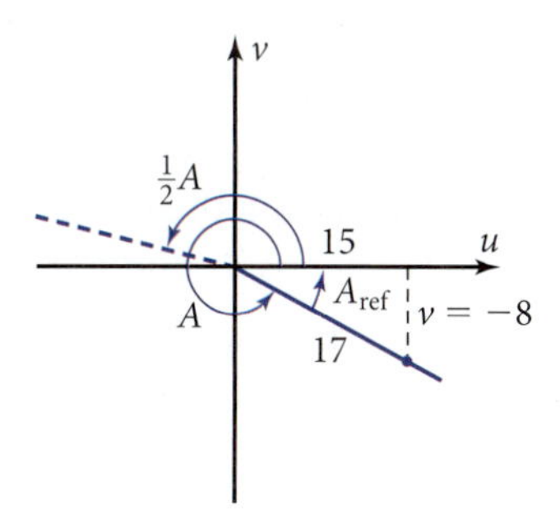

Figure 8-6d

a. $270° < A < 360°$ — Write the given interval as an inequality.

$135° < \frac{1}{2}A < 180°$ — Divide by 2 to find the quadrant in which $\frac{1}{2}A$ terminates.

Therefore, $\frac{1}{2}A$ terminates in Quadrant II, where cosine is negative.

$\cos \frac{1}{2}A = -\sqrt{\frac{1}{2}\left(1 + \frac{15}{17}\right)} = -\frac{4}{\sqrt{17}}$ Use the half argument property with the minus sign.

b. $\cos 2A = 2\cos^2 A - 1 = \frac{450}{289} - 1 = \frac{161}{289}$ Use the form of cos 2*A* involving the given function, cos *A*.

c. $A = \arccos \frac{15}{17} = \pm 28.0724...° + 360°n$

Use the definition of arccosine to write the general solution for *A*.

$A = -28.0724...° + 360° = 331.9275...°$

Write the particular solution. See Figure 8-6d.

$\frac{1}{2}A = 165.9637...°$ and $2A = 663.8550...°$

$\cos 165.9637...° = -0.9701... = -\frac{4}{\sqrt{17}}$

$\cos 663.8550...° = 0.5570... = \frac{161}{289}$

The answers are correct.

Problem Set 8-6

Reading Analysis

From what you have read in this section, what do you consider to be the main idea? Write the double argument property for sine and show how it is derived from the composite argument property. How does this property allow you to conclude that the product of a sine and a cosine with equal periods is another sinusoid? What does the text say about how you determine whether to use the + sign or the − sign in the half argument properties for cosine and sine?

Quick Review

Q1. By the composite argument properties, $\cos(x - y) =$ __?__.

Q2. By the composite argument properties, $\sin x \cos y - \cos x \sin y =$ __?__.

Q3. True or false: $\tan(x + y) = \tan x + \tan y$

Q4. $\log x + \log y = \log$ (__?__)

Q5. The equation $3(x + y) = 3x + 3y$ is an example of the __?__ property of multiplication over addition.

Q6. Find the amplitude of the sinusoid $y = 2\cos\theta + 5\sin\theta$.

Q7. Find the phase displacement with respect to $y = \cos\theta$ of $y = 2\cos\theta + 5\sin\theta$.

Q8. Find the measure of the smaller acute angle of a right triangle with legs 13 cm and 28 cm.

Q9. What is the period of the circular function $y = \sin 5x$?

Q10. The graph of the parametric function $x = 5\cos t$ and $y = 4\sin t$ is a(n) __?__.

---

1. Explain why $\cos 2x$ does *not* equal $2\cos x$ by considering the differences in the graphs of $f_1(x) = \cos 2x$ and $f_2(x) = 2\cos x$.

2. Prove by a numerical counterexample that $\tan \frac{1}{2}x$ does *not* equal $\frac{1}{2}\tan x$.

For Problems 3–6, illustrate by numerical example that the double argument property is true by making a table of values.

3. $\sin 2x = 2\sin x \cos x$

4. $\cos 2x = \cos^2 x - \sin^2 x$

5. $\cos 2x = 2\cos^2 x - 1$

6. $\tan 2x = \frac{2\tan x}{1 - \tan^2 x}$

For Problems 7–10, illustrate by numerical example that the half argument property is true by making a table of values.

7. $\sin \frac{1}{2}A$ for $A \in [0, 180°]$

8. $\cos \frac{1}{2}A$ for $A \in [0°, 180°]$

9. $\cos \frac{1}{2}A$ for $A \in [360°, 540°]$

10. $\sin \frac{1}{2}A$ for $A \in [360°, 540°]$

Sinusoid Problems: For Problems 11–16, the graph of each function is a sinusoid.

- **a.** Plot the graph of the given function.
- **b.** From the graph, find the equation for the sinusoid.
- **c.** Verify algebraically that the equation is sinusoidal.

11. $y = 6 \sin x \cos x$

12. $y = 8 \cos^2 x$

13. $y = 10 \sin^2 x$

14. $y = \cos x + \sin x$

15. $y = \cos^2 3x$

16. $y = 12 \cos 5x \sin 5x$

17. *Sinusoid Conjecture Problem 1:* In this section you proved that certain graphs that look like sinusoids really are sinusoids. Figure 8-6e shows the graphs of

$$f_1(x) = 2 + \cos x$$
$$f_2(x) = 4 + \sin x$$
$$f_3(x) = (2 + \cos x)(4 + \sin x)$$

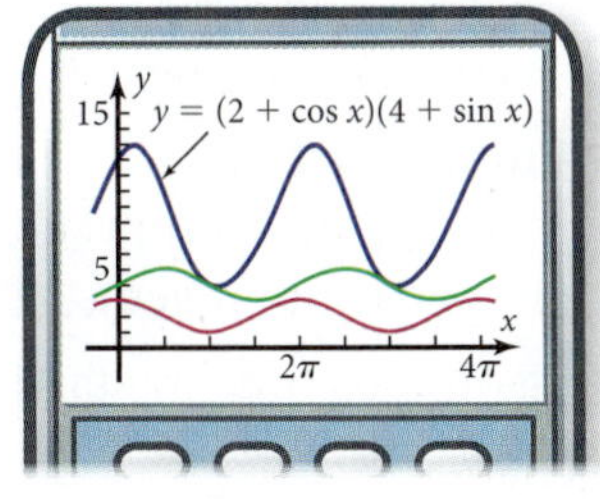

Figure 8-6e

The graphs of $f_1(x)$ and $f_2(x)$ are sinusoids with vertical displacements. Is the graph of $f_3(x)$ a sinusoid? If so, find the particular equation for the graph. If not, explain why not.

18. *Sinusoid Conjecture Problem 2:* Figure 8-6f shows the graphs of

$$f_1(\theta) = \cos \theta$$
$$f_2(\theta) = \sin(\theta - 30°)$$
$$f_3(\theta) = (\cos \theta)[\sin (\theta - 30°)]$$

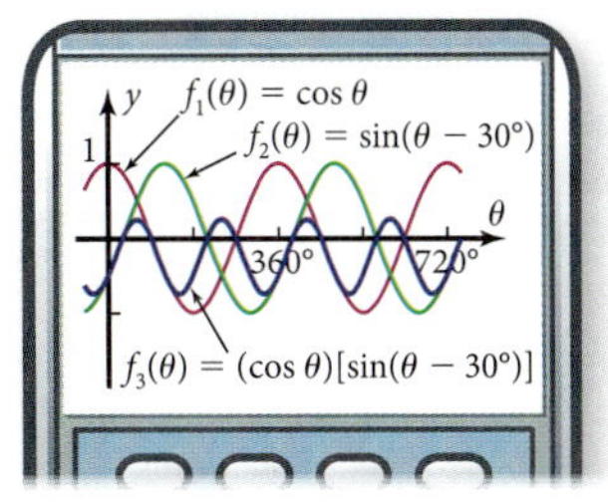

Figure 8-6f

The graph of $f_2(\theta)$ is a sinusoid with a horizontal displacement. Is the graph of $f_3(\theta)$ a sinusoid? If so, find the particular equation for the graph. If not, explain why not.

19. *Half Argument Interpretation Problem:* Figure 8-6g shows the graphs of

$$f_1(\theta) = \cos \frac{1}{2}\theta$$
$$f_2(\theta) = \sqrt{\frac{1}{2}(1 + \cos \theta)}$$

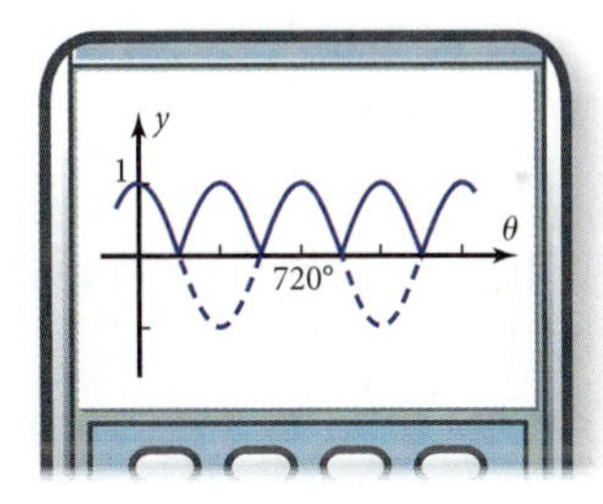

Figure 8-6g

- **a.** Which graph is $f_1(\theta)$ and which is $f_2(\theta)$?
- **b.** The half argument property for cosine contains the ambiguous $\pm$ sign. For what intervals of θ should you use the positive sign? The negative sign?
- **c.** What one transformation can you apply to the $f_1(\theta)$ equation to make its graph identical to the graph of $f_2(\theta)$?
- **d.** Explain why $\sqrt{n^2}$ equals $|n|$, not just n. Use the result, and the way you derived the half argument property for cosine, to explain the origin of the $\pm$ sign in the half argument property for cosine.

20. *Terminal Position of $\frac{1}{2}A$ Problem:* You have seen that the half argument properties can involve a different sign for different coterminal angles. Is the same thing true for the double argument properties? For instance, 30° and 390° are coterminal, as shown in Figure 8-6h, but $\frac{1}{2}(30°)$ and $\frac{1}{2}(390°)$ are not coterminal. Show that the same thing does *not* happen for 2(30°) and 2(390°). Show that if any two angles A and B are coterminal, then $2A$ and $2B$ are also coterminal.

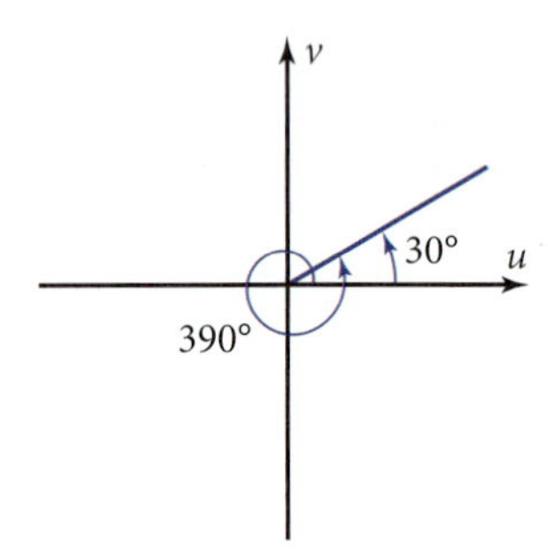

Figure 8-6h

Exact Values Problems: For Problems 21–26,

a. Use the double and half argument properties to find the *exact* values of the functions, using radicals and fractions if necessary.

b. Show that your answers are correct by finding the measure of angle A and then evaluating the functions directly.

21. If $\cos A = \frac{3}{5}$ and $A \in (0°, 90°)$, find $\sin 2A$ and $\cos \frac{1}{2}A$.

22. If $\cos A = \frac{3}{5}$ and $A \in (270°, 360°)$, find $\cos 2A$ and $\sin \frac{1}{2}A$.

23. If $\cos A = -\frac{3}{5}$ and $A \in (180°, 270°)$, find $\sin 2A$ and $\cos \frac{1}{2}A$.

24. If $\cos A = -\frac{3}{5}$ and $A \in (90°, 180°)$, find $\cos 2A$ and $\sin \frac{1}{2}A$.

25. If $\cos A = \frac{3}{5}$ and $A \in (630°, 720°)$, find $\sin 2A$ and $\cos \frac{1}{2}A$.

26. If $\cos A = -\frac{3}{5}$ and $A \in (450°, 540°)$, find $\cos 2A$ and $\sin \frac{1}{2}A$.

27. *Sine Double Argument Property Derivation Problem:* Starting with $\sin 2x = \sin(x + x)$, derive the property $\sin 2x = 2 \sin x \cos x$.

28. *Cosine Double Argument Properties Derivation Problem:*

a. Starting with $\cos 2x = \cos(x + x)$, derive the property $\cos 2x = \cos^2 x - \sin^2 x$.

b. Using the Pythagorean properties, prove that $\cos 2x = 2 \cos^2 x - 1$.

c. Using the Pythagorean properties, prove that $\cos 2x = 1 - 2 \sin^2 x$.

29. *Sine Times Cosine Is a Sinusoid Problem:* Using the double argument properties, prove algebraically that the graph of $y = \sin x \cos x$ is a sinusoid.

30. *Squares of Cosine and Sine Are Sinusoids Problem:* Using the double argument properties, prove algebraically that the graphs of $y = \cos^2 x$ and $y = \sin^2 x$ are sinusoids.

31. *Double Argument Property for Tangent:* The table of properties in this section lists this property expressing $\tan 2A$ in terms of $\tan A$:

$$\tan 2A = \frac{2 \tan A}{1 - \tan^2 A}$$

a. Show graphically that the property is true on an interval of your choice. Write a few sentences explaining what you did and your results. List any domain restrictions on the argument A.

b. Derive this double argument property algebraically by starting with the appropriate composite argument property for tangent.

c. Derive this property again, this time by starting with the quotient property for $\tan 2A$ and substituting the double argument properties for sine and cosine. You will need to use some insightful algebraic operations to transform the resulting sines and cosines back into tangents.

32. *Half Argument Property for Tangent:*

a. Based on the quotient properties, tell why

$$\tan \frac{1}{2}A = \pm\sqrt{\frac{1 - \cos A}{1 + \cos A}}$$

b. Starting with the property in part a, derive another form of the half argument property for tangent given in the box on page 427,

$$\tan \frac{1}{2}A = \frac{\sin A}{1 + \cos A}$$

First, multiply under the radical sign by 1 in the form

$$\frac{1 + \cos A}{1 + \cos A}$$

Explain what happens to the $\pm$ sign.

c. Confirm graphically that the result in part b is true. Write a few sentences explaining what you did and your results. Based on the graphs, explain why only the positive sign applies and never the negative sign.

Algebraic Solution of Equations Problems: For Problems 33–38, solve the equation algebraically, using the double argument or half argument properties appropriately to transform the equation into a suitable form.

33. $4 \sin x \cos x = \sqrt{3}, x \in [0, 2\pi]$

34. $\cos^2 \theta - \sin^2 \theta = -1, \theta \in [0°, 360°]$

35. $\cos^2 \theta = 0.5, \theta \in [0°, 360°]$

36. $\dfrac{2 \tan x}{1 - \tan^2 x} = \sqrt{3}, x \in [0, 2\pi]$

37. $\sqrt{\frac{1}{2}(1 + \cos x)} = \frac{1}{2}\sqrt{3}, x \in [0, 4\pi]$

38. $\sqrt{\frac{1}{2}(1 - \cos \theta)} = 1, \theta \in [0°, 720°]$

Identity Problems: For Problems 39–44, prove that the given equation is an identity.

39. $\sin 2x = \dfrac{2 \tan x}{1 + \tan^2 x}$

40. $\cos 2y = \dfrac{1 - \tan^2 y}{1 + \tan^2 y}$

41. $\sin 2\phi = 2 \cot \phi \sin^2 \phi$

42. $\tan \beta = \dfrac{1 - \cos 2\beta}{\sin 2\beta}$

43. $\sin^2 5\theta = \frac{1}{2}(1 - \cos 10\theta)$

44. $\cos^2 3x = \frac{1}{2}(1 + \cos 6x)$

8-7 Chapter Review and Test

In this chapter you have extended the study of trigonometric function properties that you started in Chapter 7. Specifically, you have learned properties such as $\cos(x - y) = \cos x \cos y + \sin x \sin y$ that apply to functions of more than one argument. You saw that these properties allow you to analyze graphs that are composed of sums and products of sinusoids. Sometimes the composed graph was another sinusoid, and sometimes it was a periodic function with a varying amplitude or a varying sinusoidal axis. Finally, you applied these properties to derive double and half argument properties that express, for instance, $\sin 2A$ and $\sin \frac{1}{2}A$ in terms of functions of A.

For your future reference, the box on page 439 lists the ten kinds of properties you have learned in Chapters 7 and 8.

Review Problems

R0. Update your journal with what you have learned in this chapter. For example, include

- Products and sums of sinusoids with equal periods
- Products and sums of sinusoids with much different periods
- Harmonic analysis of graphs composed of two sinusoids
- Transformations between sums and products of sinusoids
- Double argument and half argument properties to prove that certain products are sinusoids

R1. Figure 8-7a shows the graph of $y = 5 \cos \theta + 12 \sin \theta$.

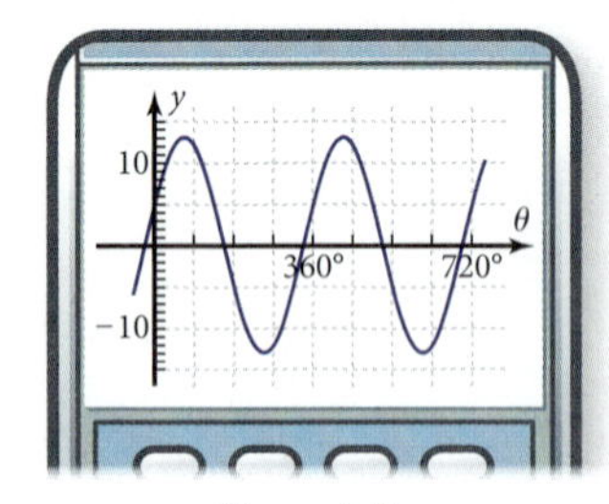

Figure 8-7a

a. The graph in Figure 8-7a is a sinusoid. Calculate its amplitude and its phase displacement with respect to the parent cosine curve, and then write an equation for the displaced sinusoid.

b. Plot your equation in part a and the equation $y = 5 \cos \theta + 12 \sin \theta$ on the same screen. Do the two graphs agree with each other and with Figure 8-7a?

R2. Figure 8-7b shows the graphs of $f_1(\theta) = \cos(\theta - 60°)$ and $f_2(\theta) = \cos \theta - \cos 60°$ as they might appear on your grapher.

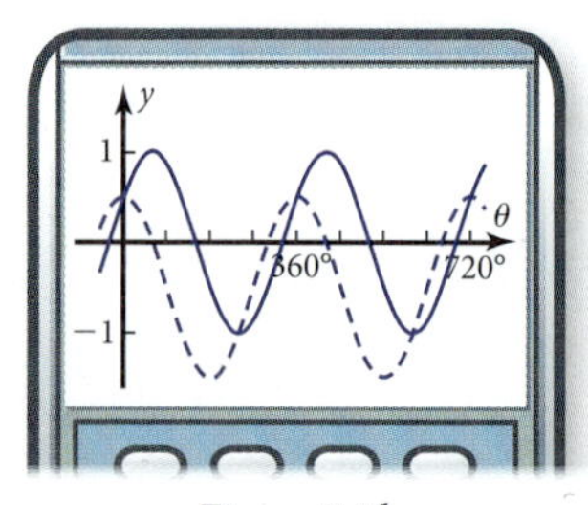

Figure 8-7b

a. Without actually plotting the graphs, identify which is $f_1(\theta)$ and which is $f_2(\theta)$. Explain how you chose them. From the graphs, how can you conclude that the cosine function does *not* distribute over subtraction?

b. Write $\cos(\theta - 60°)$ as a linear combination of sine and cosine. Verify graphically that your answer is correct.

c. Write the trigonometric expression $8 \cos \theta + 15 \sin \theta$ as a single sinusoid with a phase displacement with respect to $y = \cos \theta$. Plot the graphs of the given expression and your answer. Explain how the two graphs confirm that your answer is correct.

d. Write the circular function expression $-9 \cos x + 7 \sin x$ as a single sinusoid with a phase displacement with respect to $y = \cos x$. Make a table of values for the given expression and your answer. Explain how the numbers in the table indicate that your answer is correct.

e. Figure 8-7c shows the graphs of $f_1(x) = 4 \cos x + 3 \sin x$ and $f_2(x) = 2$ as they might appear on your grapher. Solve the equation $4 \cos x + 3 \sin x = 2$ algebraically for $x \in [0, 2\pi]$. Show that the solutions agree with the graph.

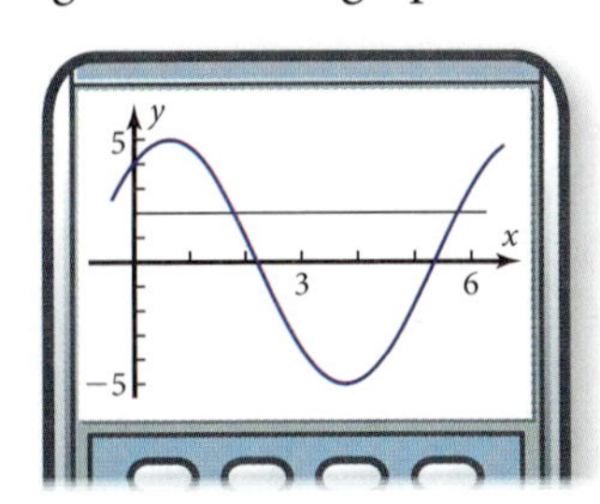

Figure 8-7c

R3. a. Express $\sin(-x)$, $\cos(-x)$, and $\tan(-x)$ in terms of the same function of (positive) x.

b. Express $\sin(x + y)$ in terms of sines and cosines of x and y.

c. Express $\cos(x + y)$ in terms of sines and cosines of x and y.

d. Express $\tan(x - y)$ in terms of $\tan x$ and $\tan y$.

e. Express $\cos(90° - \theta)$ in terms of $\sin \theta$.

f. Express $\cot\left(\frac{\pi}{2} - x\right)$ in terms of $\tan x$. What restrictions are there on the domain?

g. Express $\csc\left(\frac{\pi}{2} - x\right)$ in terms of $\sec x$. What restrictions are there on the domain?

h. Sketch the graph of $y = \tan x$, and thus show graphically that $\tan(-x) = -\tan x$. What restrictions are there on the domain?

i. Figure 8-7d shows the graph of $y = \cos 3x \cos x + \sin 3x \sin x$ and the line $y = 0.4$. Solve the equation $\cos 3x \cos x + \sin 3x \sin x = 0.4$ algebraically for $x \in [0, 2\pi]$. Explain how the graph confirms your solution.

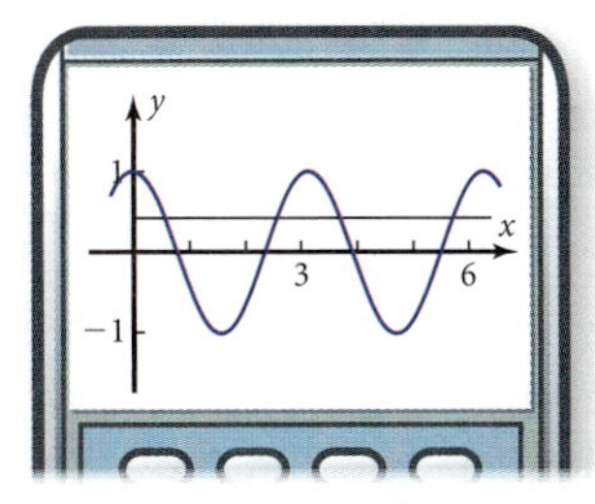

Figure 8-7d

R4. Figure 8-7e shows a sinusoid and a linear function.

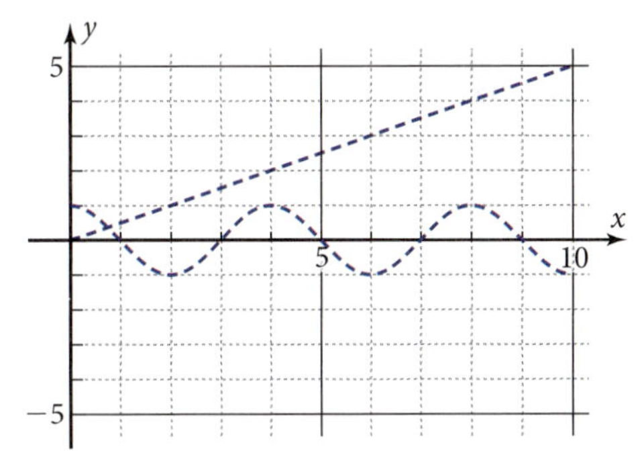

Figure 8-7e

a. On a copy of the figure, sketch the graph of the function composed by *adding* the ordinates.

b. On another copy of Figure 8-7e, sketch the graph composed by *multiplying* the ordinates.

c. Confirm your answers to parts a and b by finding equations of the two parent graphs, plotting them and the composed graphs on your grapher, and comparing them with your sketches.

d. Describe verbally the two composed graphs in part c.

e. By harmonic analysis, find the particular equation for the composed function graph in Figure 8-7f. Check your answer by plotting on your grapher.

f. By harmonic analysis, find the particular equation for the composed function graph in Figure 8-7g. Check your answer by plotting on your grapher.

R5. a. Transform $\cos 13° \cos 28°$ into a sum (or difference) of sines or cosines with positive arguments.

b. Transform $\sin 5 - \sin 8$ into a product of sines and cosines with positive arguments.

c. Figure 8-7h shows the graph of $y = 4 \sin x \sin 11x$. Transform the expression on the right side of this equation into a sum (or difference) of sinusoids whose arguments have positive coefficients. Check graphically that your answer and the given equation both agree with Figure 8-7h.

d. Solve $2 \sin 3\theta + 2 \sin \theta = 0$ algebraically for $\theta \in [0°, 360°]$.

e. Use the sum and product properties to prove that $\cos(x + \frac{\pi}{3}) \cos(x - \frac{\pi}{3}) = \cos^2 x - \frac{3}{4}$ is an identity.

R6. Figure 8-7i shows the graph of $y = \sin^2 3x$.

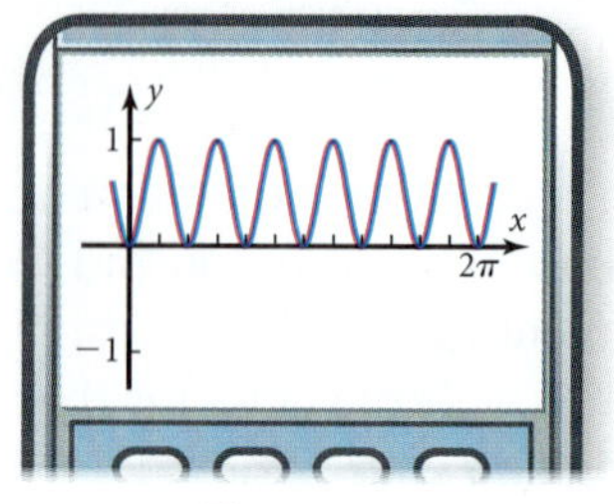

Figure 8-7i

a. The function is a sinusoid. Find its equation from the graph. Confirm algebraically that your equation is correct by using the double argument properties.

b. Give a numerical counterexample to prove that $\cos 2x \neq 2 \cos x$.

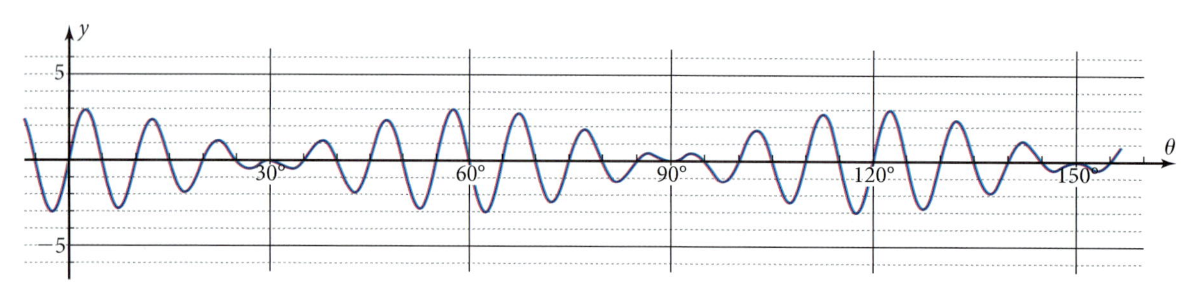

Figure 8-7f

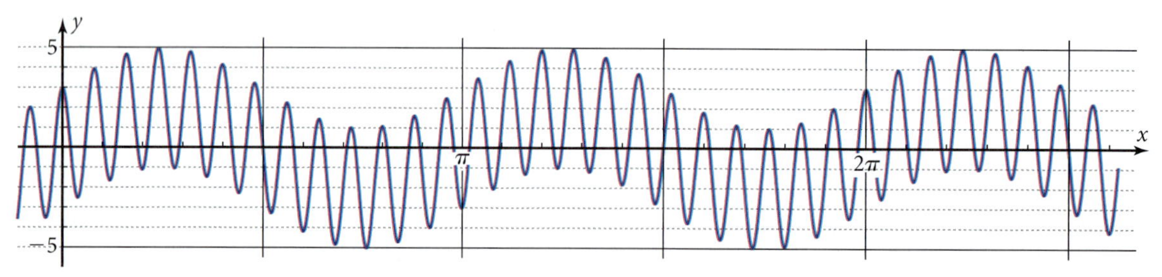

Figure 8-7g

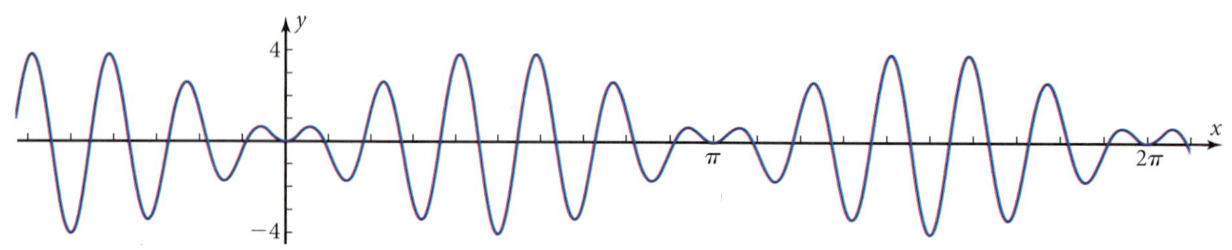

Figure 8-7h

c. Write an equation expressing $\cos 2x$ in terms of $\cos x$ alone. Make a table of values to demonstrate that the equation is an identity.

d. Write an equation expressing $\tan 2x$ in terms of $\tan x$ alone. What restrictions are there on the domain of x?

e. Suppose that $A \in (360°, 450°)$ and $\sin A = \frac{24}{25}$. Find the exact values (no decimals) of $\cos 2A$ and $\cos \frac{1}{2}A$. Then calculate the measure of angle A (no round-off), and find decimal values of $\cos 2A$ and $\cos \frac{1}{2}A$ directly. How do these values compare with the exact values?

f. Prove that $\sin 2A = 2 \tan A \cos^2 A$ is an identity for all the values where both sides are defined. What is the domain of this identity?

g. Find an algebraic solution of the equation

$$\sqrt{\frac{1}{2}(1 - \cos \theta)} = 0.5 \qquad \theta \in [0°, 720°]$$

by applying the half argument properties. Then find a graphical solution by finding where the left and right members of the equation intersect. Do the solutions agree? Explain.

Concept Problems

C1. *Exact Value of sin 18° Project:* You have learned how to find exact values of functions of multiples of 30° and 45°. By the composite argument property, you found an exact value of sin 15° by writing it as sin(45° − 30°). In this problem you will combine trigonometric properties with algebraic techniques and some ingenuity to find an exact value of sin 18°.

a. Use the double argument property for sine to write an equation expressing sin 72° in terms of sin 36° and cos 36°.

b. Transform the equation in part a so that sin 72° is expressed in terms of sin 18° and cos 18°. You should find that the *sine* form of the double argument property for cos 36° works best.

c. Recall by the cofunction property that sin 72° = cos 18°. Replace sin 72° in your equation from part b with cos 18°. If you have done everything correctly, cos 18° should disappear from the equation, leaving a *cubic* (third-degree) equation in sin 18°.

d. Solve the equation in part c for sin 18°. It may help to let $x = \sin 18°$ and solve for x. If you rearrange the equation so that the right side is 0, you should find that $(2x - 1)$ is a factor of the left side. You can find the other factor by long division or synthetic substitution. To find the exact solutions, recall the multiplication property of zero and the quadratic formula.

e. You should have *three* solutions for the equation in part d. Only one of these solutions is possible. *Which* solution?

f. A pattern shows up for some exact values of $\sin \theta$:

$$\sin 15° = \frac{\sqrt{6} - \sqrt{2}}{4}$$

$$\sin 18° = \frac{\sqrt{5} - 1}{4} = \frac{\sqrt{5} - \sqrt{1}}{4}$$

$$\sin 30° = \frac{1}{2} = \frac{2}{4} = \frac{\sqrt{4} - \sqrt{0}}{4}$$

See if you can extend this pattern to sines of other angles.

C2. *A Square Wave Function and Fourier Series Project:* Figure 8-7j shows the graph of

$$y = \cos x - \frac{1}{3}\cos 3x$$

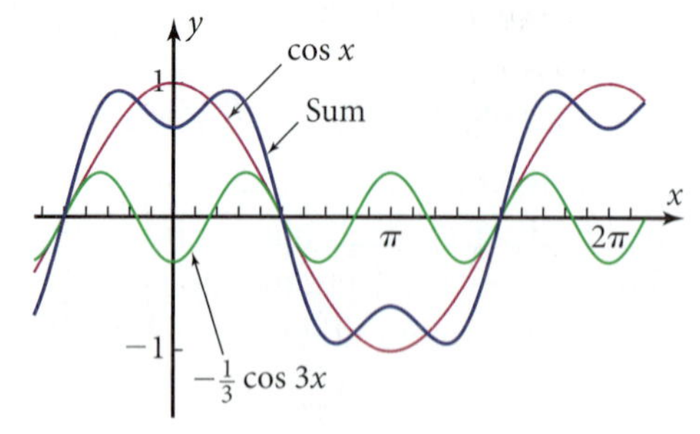

Figure 8-7j

The smaller sinusoid pulls the larger one closer to the x-axis at some places and pushes it farther away at others. The terms in the given equation form this **partial sum** of a *Fourier series:*

$$y = \cos x - \frac{1}{3}\cos 3x + \frac{1}{5}\cos 5x - \frac{1}{7}\cos 7x + \frac{1}{9}\cos 9x - \frac{1}{11}\cos 11x + \ldots$$

The more terms that are added to the partial sum, the more the result looks like a *square wave.* For instance, the graph of the partial sum with 11 terms is shown in Figure 8-7k.

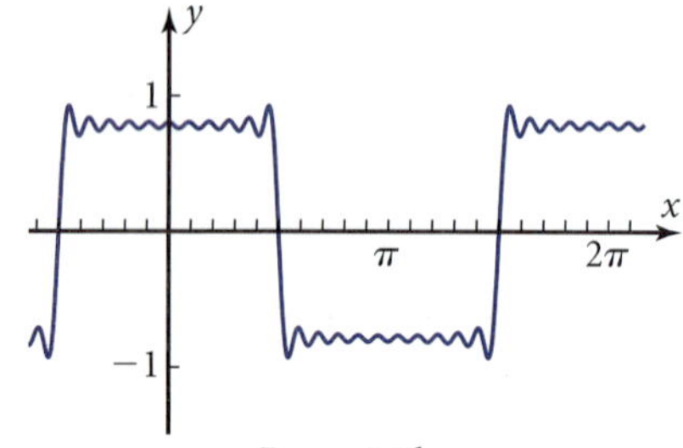

Figure 8-7k

a. Plot the graph of the partial sum with 11 terms and show that it is similar to the graph in Figure 8-7k. Find the sum and sequence commands that will allow you to write the equation without entering all the terms.

b. If you plot the partial sum with 12 terms, some of the high points and low points in Figure 8-7k are reversed. So if you average the 11-term and the 12-term sums, you get a better square-wave pattern. Do this on your grapher. (If you perform some clever calculations first, you can find a relatively easy way to add one term to the 11-term sum that gives the average without having to compute the 12-term sum.)

Electrical generators can produce alternating currents with the square-wave pattern shown in Figure 8-7k.

c. Plot the average of the 50-term sum and the 51-term sum. How close does the graph come to a "square wave"?

d. When the square wave is on the high portion of its graph, there is an "axis" about which it oscillates. Find the y-value of this axis. Explain how you found your answer. How is the answer related to π?

e. Plot the 10th partial sum of the Fourier sine series

$$y = \sin x + \frac{1}{2}\sin 2x + \frac{1}{3}\sin 3x + \ldots + \frac{1}{10}\sin 10x$$

From your graph, figure out why the result is called a *sawtooth wave pattern.*

Chapter Test

Part 1: No calculators (T1–T9)

T1. What are the amplitude and period of the sinusoid $y = 6 \cos x + 7 \sin x$?

T2. The graph of $y = \cos(x - 2)$ is a sinusoid with a phase displacement of 2. Given that $\cos 2 \approx -0.42$ and $\sin 2 \approx 0.91$, write an equation for y as a linear combination of $\cos x$ and $\sin x$.

T3. What are the amplitude and period of the sinusoid $y = 2 \sin \theta \cos \theta$?

T4. The graph of $y = 5 \cos x + \sin 8x$ is periodic but not a sinusoid. Based on the form of the equation, describe in words what the graph would look like.

T5. The process of finding the two sinusoids that have been added or multiplied to form a combined graph is called __?__.

T6. Is tangent an odd or an even function? Write the odd–even property in algebraic form.

T7. According to the cofunction property, $\cos 13° =$ __?__.

T8. The expression $\cos 9 + \cos 5$ can be transformed to $2 \cos x \cos y$. What do x and y equal?

T9. You can write the double argument property for cosine in the form

$$\cos x = 1 - 2 \sin^2 \frac{1}{2}x$$

Show algebraically how you can transform this equation into the half argument property for sine.

Part 2: Graphing calculators allowed (T10–T19)

T10. Figure 8-7l shows the graph of the linear combination of two sinusoids with equal periods

$$y = -4 \cos \theta + 3 \sin \theta$$

Write the equation as a single cosine with a phase displacement. Show that the calculated phase displacement agrees with the graph in the figure.

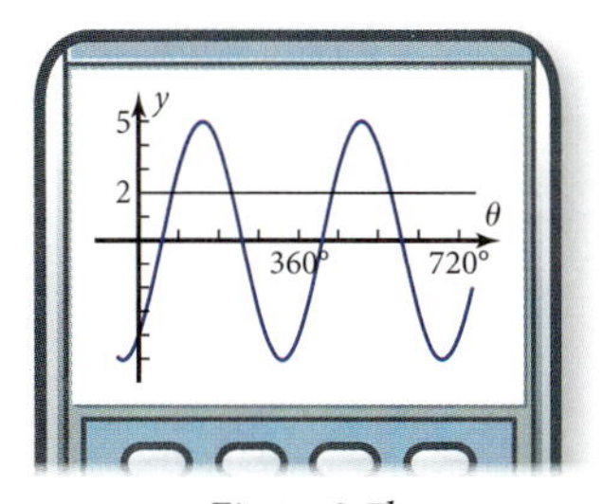

Figure 8-7l

T11. Solve the equation $-4 \cos \theta + 3 \sin \theta = 2$ algebraically for $\theta \in [0°, 720°]$. Show that the solutions agree with the four shown in Figure 8-7l.

T12. In Problem T2 you transformed

$$y = \cos(x - 2) \qquad \text{into}$$

$$y = \cos x \cos 2 + \sin x \sin 2$$

Plot both graphs on the same screen. Explain in writing how your graphs confirm that the two equations are equivalent.

T13. Figure 8-7m shows the graph of $y = \cos^2 \theta$ as it might appear on your grapher. From the graph, find the equation for this sinusoid.

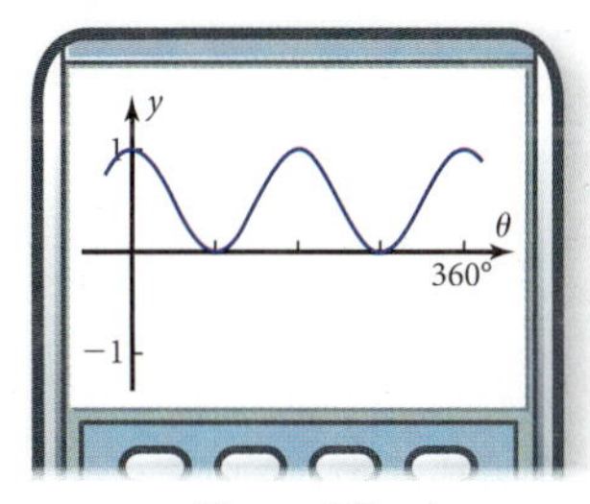

Figure 8-7m

Use the double argument properties to prove algebraically that your equation is correct.

T14. Figure 8-7n shows the graphs of two sinusoids with different periods. On a copy of the figure, sketch the graph of the *sum* of these two sinusoids. Then write equations for each sinusoid and plot the result on your grapher. How well does the sketch agree with the plot?

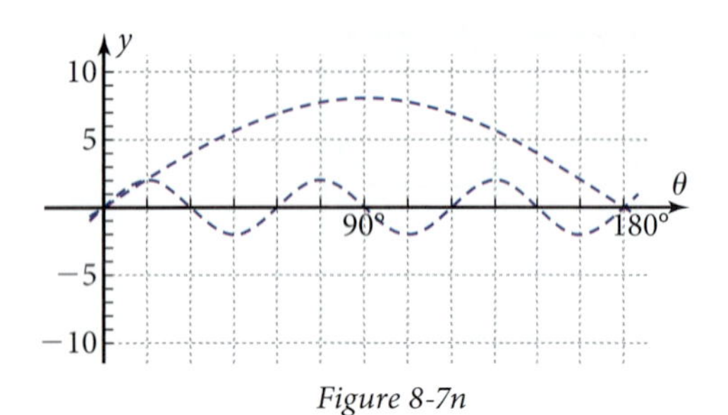

Figure 8-7n

T15. Figure 8-7o shows a periodic function whose graph is a product of two sinusoids with unequal periods. Write the particular equation for the function. Confirm your answer by plotting it on your grapher.

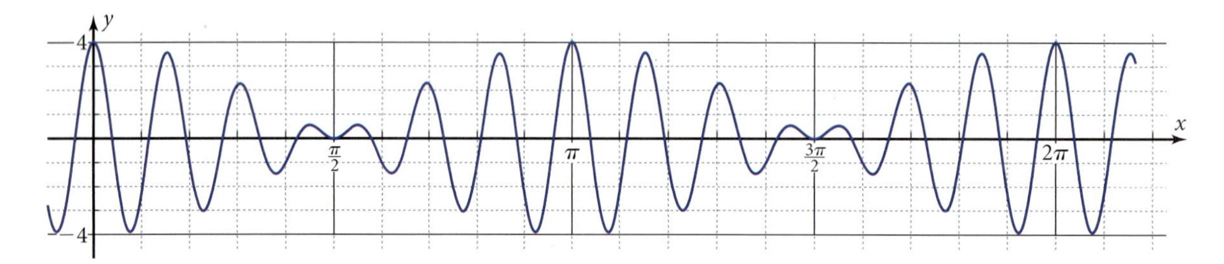

Figure 8-7o

T16. A periodic function has the equation $y = 4 \cos x \cos 11x$. Transform the equation so that the expression on the right side is a sum of two sinusoids. Confirm numerically that your equation is correct by making a table of values.

T17. Suppose that A is an angle between 0° and 90° and that $\cos A = \frac{15}{17}$. What does $\sin(90° - A)$ equal? What property can you use to find this answer quickly?

T18. For angle A in Problem T17, find the exact value (no decimals) of $\sin 2A$ and $\cos \frac{1}{2}A$. Then find the measure of angle A, and calculate $\sin 2A$ and $\cos \frac{1}{2}A$ directly. How do the answers compare?

T19. What did you learn as a result of taking this test that you did not know before?

PROPERTIES: Summary of Trigonometric Function Properties

Reciprocal Properties

$\cot x = \dfrac{1}{\tan x}$ or $\tan x \cot x = 1$

$\sec x = \dfrac{1}{\cos x}$ or $\cos x \sec x = 1$

$\csc x = \dfrac{1}{\sin x}$ or $\sin x \csc x = 1$

Quotient Properties

$\tan x = \dfrac{\sin x}{\cos x} = \dfrac{\sec x}{\csc x}$

$\cot x = \dfrac{\cos x}{\sin x} = \dfrac{\csc x}{\sec x}$

Pythagorean Properties

$\cos^2 x + \sin^2 x = 1$

$1 + \tan^2 x = \sec^2 x$

$\cot^2 x + 1 = \csc^2 x$

Odd–Even Function Properties

$\sin(-x) = -\sin x$ (odd function)

$\cos(-x) = \cos x$ (even function)

$\tan(-x) = -\tan x$ (odd function)

$\cot(-x) = -\cot x$ (odd function)

$\sec(-x) = \sec x$ (even function)

$\csc(-x) = -\csc x$ (odd function)

Cofunction Properties

$\cos(90° - \theta) = \sin \theta$, $\cos\left(\frac{\pi}{2} - x\right) = \sin x$

$\cot(90° - \theta) = \tan \theta$, $\cot\left(\frac{\pi}{2} - x\right) = \tan x$

$\csc(90° - \theta) = \sec \theta$, $\csc\left(\frac{\pi}{2} - x\right) = \sec x$

Linear Combination of Cosine and Sine

$b \cos x + c \sin x = A \cos(x - D)$, where

$A = \sqrt{b^2 + c^2}$ and $D = \arctan \frac{c}{b}$

Composite Argument Properties

$\cos(A - B) = \cos A \cos B + \sin A \sin B$

$\cos(A + B) = \cos A \cos B - \sin A \sin B$

$\sin(A - B) = \sin A \cos B - \cos A \sin B$

$\sin(A + B) = \sin A \cos B + \cos A \sin B$

$\tan(A - B) = \dfrac{\tan A - \tan B}{1 + \tan A \tan B}$

$\tan(A + B) = \dfrac{\tan A + \tan B}{1 - \tan A \tan B}$

Sum and Product Properties

$2 \cos A \cos B = \cos(A + B) + \cos(A - B)$

$2 \sin A \sin B = -\cos(A + B) + \cos(A - B)$

$2 \sin A \cos B = \sin(A + B) + \sin(A - B)$

$2 \cos A \sin B = \sin(A + B) - \sin(A - B)$

$\cos x + \cos y = 2 \cos \frac{1}{2}(x + y) \cos \frac{1}{2}(x - y)$

$\cos x - \cos y = -2 \sin \frac{1}{2}(x + y) \sin \frac{1}{2}(x - y)$

$\sin x + \sin y = 2 \sin \frac{1}{2}(x + y) \cos \frac{1}{2}(x - y)$

$\sin x - \sin y = 2 \cos \frac{1}{2}(x + y) \sin \frac{1}{2}(x - y)$

Double Argument Properties

$\sin 2x = 2 \sin x \cos x$

$\cos 2x = \cos^2 x - \sin^2 x = 1 - 2 \sin^2 x$

$= 2 \cos^2 x - 1$

$\tan 2x = \dfrac{2 \tan x}{1 - \tan^2 x}$

$\sin^2 x = \frac{1}{2}(1 - \cos 2x)$

$\cos^2 x = \frac{1}{2}(1 + \cos 2x)$

Half Argument Properties

$\sin \frac{1}{2}x = \pm\sqrt{\frac{1}{2}(1 - \cos x)}$

$\cos \frac{1}{2}x = \pm\sqrt{\frac{1}{2}(1 + \cos x)}$

$\tan \frac{1}{2}x = \pm\sqrt{\dfrac{1 - \cos x}{1 + \cos x}}$

$= \dfrac{\sin x}{1 + \cos x} = \dfrac{1 - \cos x}{\sin x}$

Chapter 9

Triangle Trigonometry

Tracts of land are often made up of irregular shapes created by geographical features such as lakes and rivers. Surveyors often measure these irregularly shaped tracts of land by dividing them into triangles. To calculate the area, side lengths, and angle measures of each triangle, they must extend their knowledge of right triangle trigonometry to include triangles that have no right angle. In this chapter you'll learn how to do these calculations.

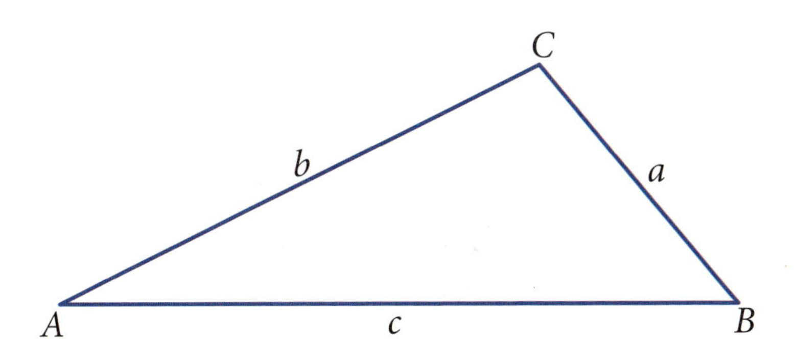

Mathematical Overview

The Pythagorean theorem describes how to find the length of the hypotenuse of a right triangle if you know the lengths of the two legs. If the angle formed by the two given sides is not a right angle, you can use the law of cosines (an extension of the Pythagorean theorem) to find side lengths or angle measures. If two angles and a side opposite one of the angles or two sides and an angle opposite one of the sides are given, you can use the law of sines. The area can also be calculated from side and angle measures. These techniques give you a way to analyze vectors, which are quantities (such as velocity) that have both direction and magnitude. You will learn about these techniques in four ways.

GRAPHICALLY

Make a scale drawing of the triangle using the given information, and measure the length of the third side.

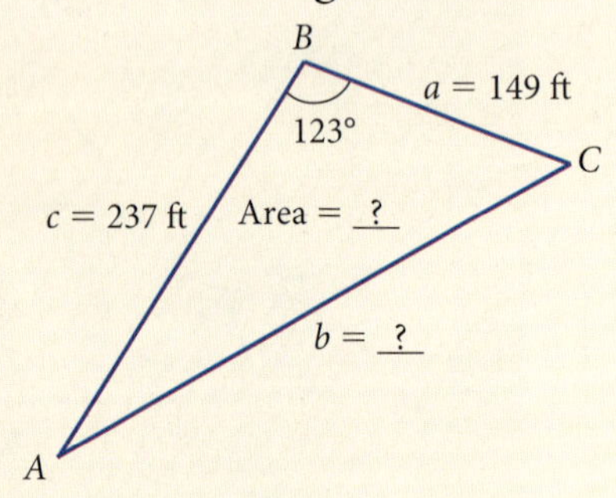

ALGEBRAICALLY

Law of cosines: $b^2 = a^2 + c^2 - 2ac \cos B$

Area of a triangle: $A = \frac{1}{2}ac \sin B$

NUMERICALLY

The length of side b: $b = \sqrt{149^2 + 237^2 - 2(149)(237) \cos 123°}$

$= 341.8123\ldots$ ft

The area of the triangle: $A = \frac{1}{2}(237)(149) \sin 123° = 14{,}807.9868\ldots$

$\approx 14{,}808 \text{ ft}^2$

VERBALLY

Given two sides of a triangle and the included angle, the law of cosines can be used to find the length of the third side and the sine can be used to find the area. If all three sides are given, the law of cosines can be used in reverse to find any angle measure.

9-1 Introduction to Oblique Triangles

You already know how to find unknown side lengths and angle measures in right triangles by using trigonometric functions. In this section you'll be introduced to a way of calculating the same kind of information if none of the angles of the triangle is a right angle. Such triangles are called **oblique triangles.**

Objective Given two sides and the included angle of a triangle, find by direct measurement the length of the third side of the triangle.

Exploratory Problem Set 9-1

1. Figure 9-1a shows five triangles. Each has sides of length 3 cm and 4 cm. They differ in the measure of the angle included between the two sides. Measure the sides and angles. Do you agree with the given measurements in each case?
2. Measure *a,* the third side of each triangle. Find the length of the third side if *A* were 180° and if *A* were 0°. Record your results in table form.
3. Store the data from Problem 2 in lists on your grapher. Make a connected plot of the data on your grapher.
4. The plot looks like a half-cycle of a sinusoid. Find the equation of the sinusoid that has the same low and high points and plot it on the same screen. Do the data really seem to follow a sinusoidal pattern?
5. By the Pythagorean theorem, $a^2 = 3^2 + 4^2$ if A is 90°. If A is *less* than 90°, side a is less than 5, so it seems you must *subtract* something from $3^2 + 4^2$ to get the value of a^2. See if you can find *what* is subtracted!
6. What did you learn from doing this problem set that you did not know before?

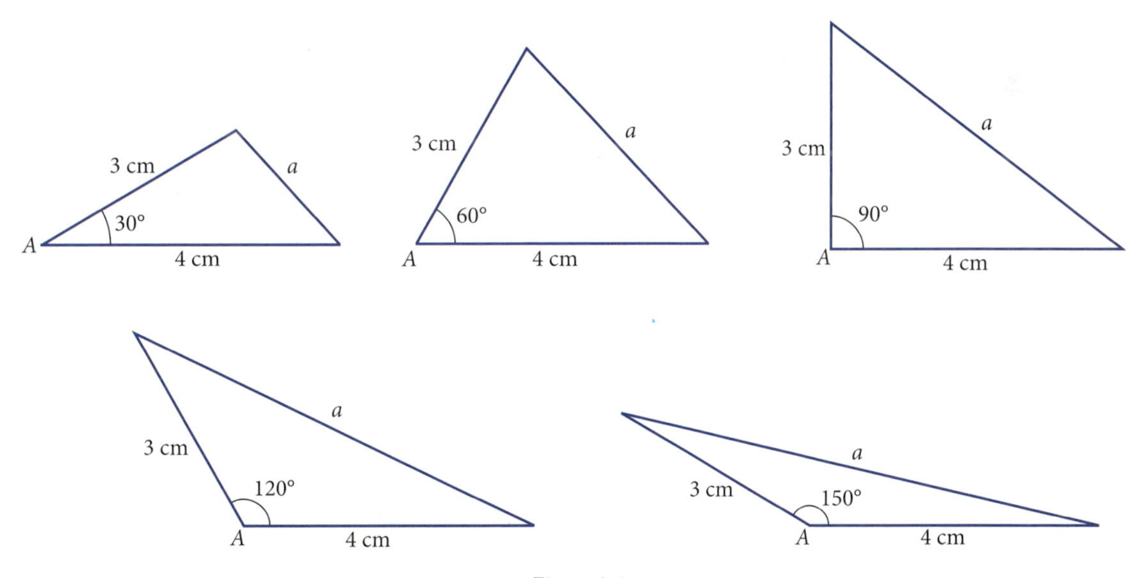

Figure 9-1a

9-2 Oblique Triangles: The Law of Cosines

In Section 9-1, you measured the third side of triangles for which two sides and the included angle were known. Think of the three triangles with included angles 60°, 90°, and 120°.

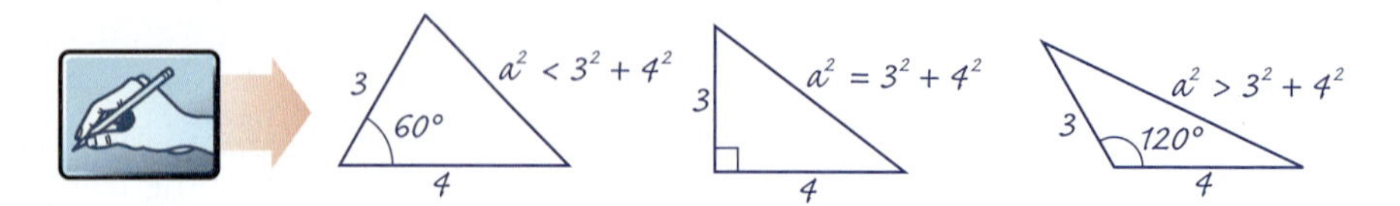

For the right triangle in the middle, you can find the third side, a, using the Pythagorean theorem.

$$a^2 = 3^2 + 4^2$$

For the 60° triangle on the left, the value of a^2 is less than $b^2 + c^2$. For the 120° triangle on the right, a^2 is greater than $b^2 + c^2$.

The equation you'll use to find the exact length of the third side from the measures of two sides and the included angle is called the **law of cosines** (because it involves the cosine of the angle). In this section you'll see why the law of cosines is true and how to use it.

Objective

- Given two sides and the included angle of a triangle, derive and use the law of cosines to find the length of the third side.
- Given three sides of a triangle, find an angle measure.

In this exploration you will demonstrate by measurement that the law of cosines gives the correct value for the third side of a triangle if two sides and the included angle are given.

EXPLORATION 9-2: Derivation of the Law of Cosines

The figure shows $\triangle ABC$. Angle A has been placed in standard position in a uv-coordinate system.

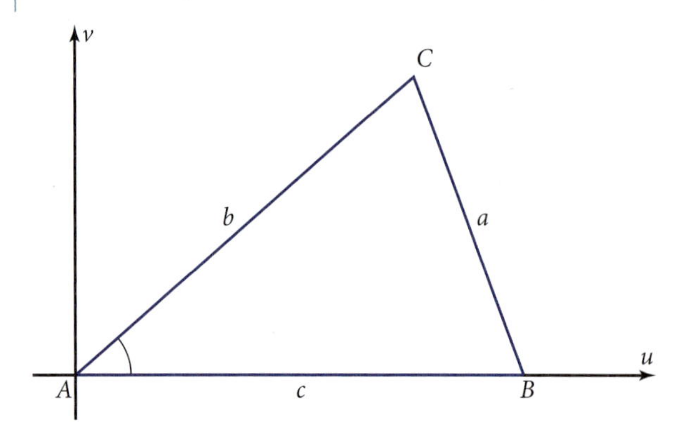

1. The sides that include angle A have lengths b and c. Write the coordinates of points B and C using b, c, and functions of angle A.

   B: $(u, v) = (\underline{\ ?\ }, \underline{\ ?\ })$

   C: $(u, v) = (\underline{\ ?\ }, \underline{\ ?\ })$

2. Use the distance formula to write the square of the length of the third side, a^2, in terms of b, c, and functions of angle A.
3. Simplify the equation in Problem 2 by expanding the square. Use the Pythagorean property for cosine and sine to simplify the terms containing $\cos^2 A$ and $\sin^2 A$.

continued

4. The equation in Problem 3 is called the law of cosines. Show that you understand what the law of cosines says by using it to calculate the length of the third side of this triangle.

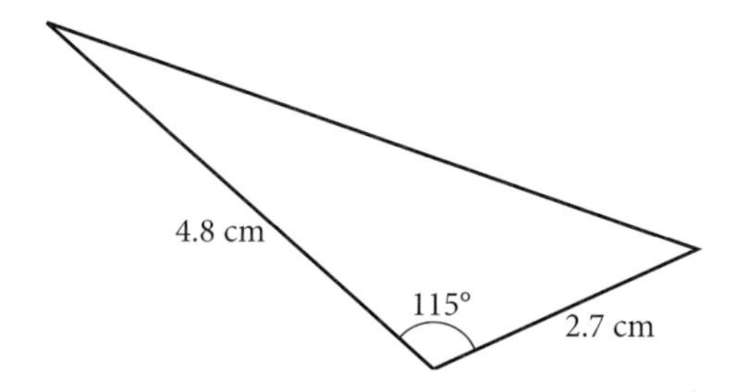

5. Measure the given sides and angle of the triangle in Problem 4. Do you agree with the given measurements? Measure the third side. Does it agree with your calculated value?
6. Describe how the unknown side in the law of cosines is related to the given sides and their included angle. Start by writing, "Given two sides and the included angle . . .".
7. What have you learned as a result of doing this exploration that you did not know before?

Derivation of the Law of Cosines

In Exploration 9-2, you demonstrated that the square of side a of a triangle can be found by subtracting a quantity from the Pythagorean expression $b^2 + c^2$. Here is why this property is true. Suppose that the lengths of two sides, b and c, of $\triangle ABC$ are known, as is the measure of the included angle, A (Figure 9-2a, left).

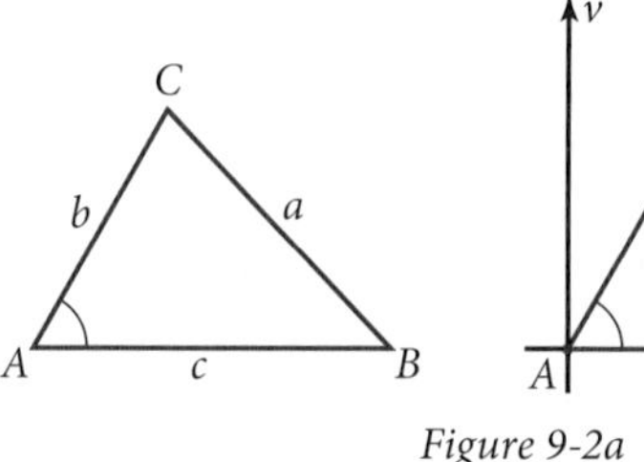

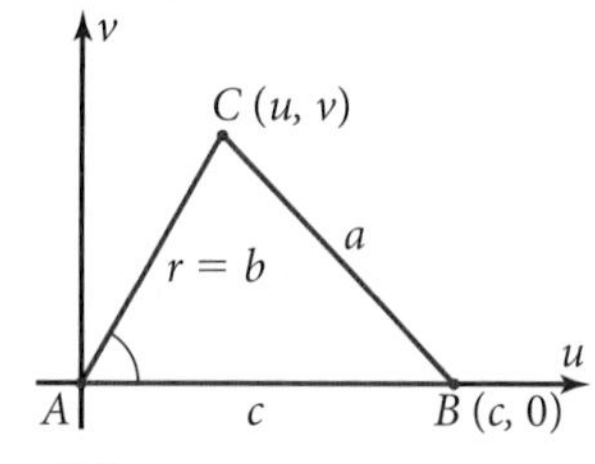

Figure 9-2a

If you construct a uv-coordinate system with angle A in standard position, as on the right in Figure 9-2a, then vertices B and C have coordinates $B(c, 0)$ and $C(u, v)$. By the distance formula,

$$a^2 = (u - c)^2 + (v - 0)^2$$

By the definitions of cosine and sine,

$$\frac{u}{b} = \cos A \Rightarrow u = b\cos A$$

$$\frac{v}{b} = \sin A \Rightarrow v = b\sin A$$

Substituting these values for u and v and completing the appropriate algebraic operations gives

$$a^2 = (u - c)^2 + (v - 0)^2$$

$$a^2 = (b \cos A - c)^2 + (b \sin A - 0)^2$$

$$a^2 = b^2 \cos^2 A - 2bc \cos A + c^2 + b^2 \sin^2 A$$ Calculate the squares.

$$a^2 = b^2(\cos^2 A + \sin^2 A) - 2bc \cos A + c^2$$ Factor b^2 from the first and last terms.

$$a^2 = b^2 + c^2 - 2bc \cos A$$ Use the Pythagorean property.

PROPERTY: The Law of Cosines

In triangle ABC with sides a, b, and c,

$$a^2 = b^2 + c^2 - 2bc \cos A$$

side² + side² − 2(side)(side)(cosine of included angle) = (third side)²

Notes:

- If the angle measures 90°, the law of cosines reduces to the Pythagorean theorem, because cos 90° is zero.
- If angle A is obtuse, $\cos A$ is negative. So you are subtracting a negative number from $b^2 + c^2$, giving the larger value for a^2, as you found in Section 9-1.
- You should not jump to the conclusion that the law of cosines gives an easy way to *prove* the Pythagorean theorem. Doing so would involve circular reasoning, because the Pythagorean theorem (in the form of the distance formula) was used to *derive* the law of cosines.
- A capital letter is used for the vertex, the angle at that vertex, or the measure of that angle, whichever is appropriate. If confusion results, you can use the symbols from geometry, such as $m\angle A$ for the measure of angle A.

Applications of the Law of Cosines

You can use the law of cosines to calculate the measure of either a side or an angle. In each case, different parts of a triangle are given. Watch for what these "givens" are.

EXAMPLE 1 In $\triangle PMF$, $M = 127°$, $p = 15.78$ ft, and $f = 8.54$ ft. Find the measure of the third side, m.

SOLUTION First, sketch the triangle and label the sides and angles, as shown in Figure 9-2b. (It does not need to be accurate, but it must have the right relationship among sides and angles.)

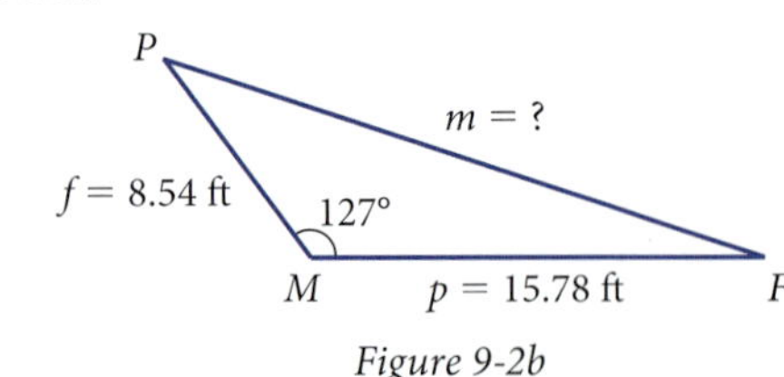

Figure 9-2b

$$m^2 = 8.54^2 + 15.78^2 - 2(8.54)(15.78)\cos 127°$$

Use the law of cosines for side m.

$$m^2 = 484.1426...$$

$$m = 22.0032... \approx 22.0 \text{ ft}$$

EXAMPLE 2 In $\triangle XYZ$, $x = 3$ m, $y = 7$ m, and $z = 9$ m. Find the measure of the largest angle.

SOLUTION Make a sketch of the triangle and label the sides, as shown in Figure 9-2c.

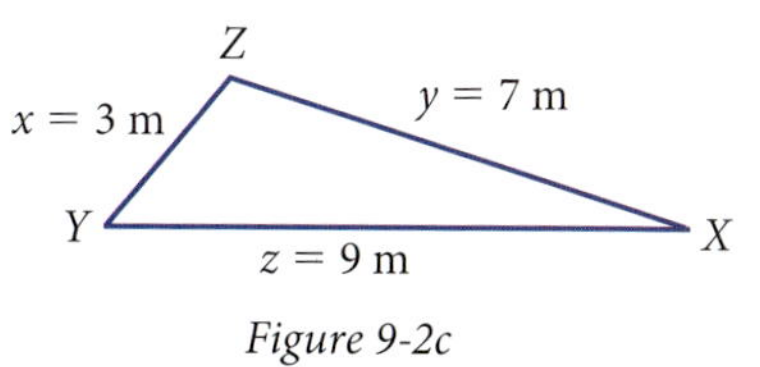

Figure 9-2c

Recall from geometry that the largest side is opposite the largest angle, in this case, Z. Use the law of cosines with this angle and the two sides that include it.

$$9^2 = 7^2 + 3^2 - 2 \cdot 7 \cdot 3 \cos Z$$

$$81 = 49 + 9 - 42 \cos Z$$

$$\frac{81 - 49 - 9}{-42} = \cos Z$$

$$-0.5476... = \cos Z$$

$$Z = \arccos(-0.5476...) = \cos^{-1}(-0.5476...) = 123.2038... \approx 123.2°$$

Note that $\arccos(-0.5476...) = \cos^{-1}(-0.5476...)$ in Example 2 because there is only one value of an arccosine between 0° and 180°, the range of angles possible in a triangle.

EXAMPLE 3 Suppose that the lengths of the sides in Example 2 had been $x = 3$ m, $y = 7$ m, and $z = 11$ m. What would the measure of angle Z be in this case?

SOLUTION Write the law of cosines for side z, the side that is opposite angle Z.

$$11^2 = 7^2 + 3^2 - 2 \cdot 7 \cdot 3 \cos Z$$

$$121 = 49 + 9 - 42 \cos Z$$

$$\frac{121 - 49 - 9}{-42} = \cos Z$$

$$-1.5 = \cos Z$$

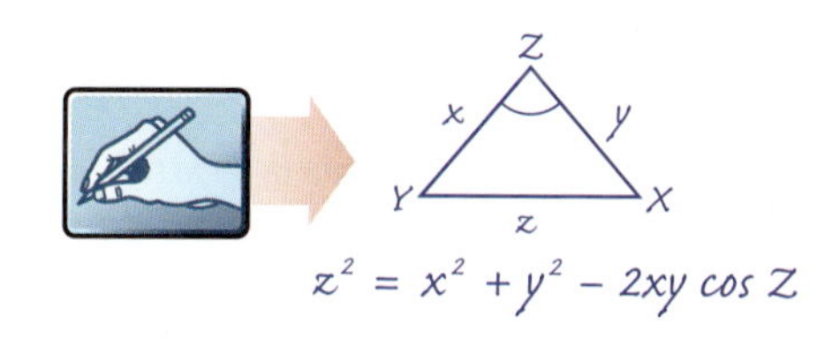

There is no such triangle.

$\cos Z$ must be in the range $[-1, 1]$.

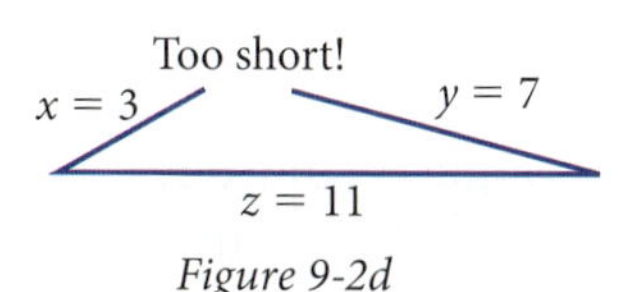

Figure 9-2d

The geometric reason why there is no solution in Example 3 is that no two sides of a triangle can sum to less than the third side. Figure 9-2d illustrates this fact. The law of cosines signals this inconsistency algebraically by giving a cosine value outside the interval $[-1, 1]$.

Problem Set 9-2

Reading Analysis

From what you have read in this section, what do you consider to be the main idea? How is the law of cosines related to the Pythagorean theorem? What three parts of a triangle should you know in order to use the law of cosines in its "frontward" form, and what part can you calculate using the law? How can you use the law of cosines to calculate the measure of an angle?

Quick Review

Problems Q1–Q6 refer to right triangle *QUI* (Figure 9-2e).

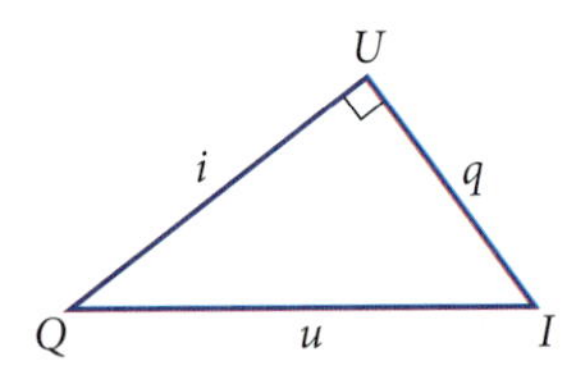

Figure 9-2e

Q1. $\cos Q = $ _?_

Q2. $\tan I = $ _?_

Q3. $\sin U = $ _?_

Q4. In terms of side u and angle I, what does i equal?

Q5. In terms of sides u and q, what does i equal?

Q6. In terms of the inverse tangent function, $Q = $ _?_.

Q7. The graph of $y = 5 \cos \theta + \sin 12\theta$ is periodic with a varying _?_.

Q8. In terms of cosines and sines of 53° and 42°, $\cos(53° - 42°) = $ _?_.

Q9. What transformation of $y = \cos x$ is expressed by $y = \cos 5x$?

Q10. Express $\sin 2x$ in terms of $\sin x$ and $\cos x$.

For Problems 1–4, find the length of the specified side.

1. Side r in $\triangle RPM$, if $p = 4$ cm, $m = 5$ cm, and $R = 51°$
2. Side d in $\triangle CDE$, if $c = 7$ in., $e = 9$ in., and $D = 34°$
3. Side r in $\triangle PQR$, if $p = 3$ ft, $q = 2$ ft, and $R = 138°$
4. Side k in $\triangle HJK$, if $h = 8$ m, $j = 6$ m, and $K = 172°$

For Problems 5–12, find the measure of the specified angle.

5. Angle U in $\triangle UMP$, if $u = 2$ in., $m = 3$ in., and $p = 4$ in.
6. Angle G in $\triangle MEG$, if $m = 5$ cm, $e = 6$ cm, and $g = 8$ cm
7. Angle T in $\triangle BAT$, if $b = 6$ km, $a = 7$ km, and $t = 12$ km
8. Angle E in $\triangle PEG$, if $p = 12$ ft, $e = 22$ ft, and $g = 16$ ft
9. Angle Y in $\triangle GYP$, if $g = 7$ yd, $y = 5$ yd, and $p = 13$ yd
10. Angle N in $\triangle GON$, if $g = 6$ mm, $o = 3$ mm, and $n = 12$ mm
11. Angle O in $\triangle NOD$, if $n = 1475$ yd, $o = 2053$ yd, and $d = 1428$ yd
12. Angle Q in $\triangle SQR$, if $s = 1504$ cm, $q = 2465$ cm, and $r = 1953$ cm
13.  *Accurate Drawing Project:*
    a. Using computer software such as The Geometer's Sketchpad, or using a ruler and protractor, construct $\triangle RPM$ from Problem 1. Then measure side r. Does the measured value agree with the calculated value in Problem 1 within ±0.1 cm?
    b. Using Sketchpad or a ruler, compass, and protractor, construct $\triangle MEG$ from Problem 6. Construct the longest side, 8 cm, first. Then draw an arc or circle of radius 5 cm from one endpoint and an arc of radius 6 cm from the other endpoint. The third vertex is the point where the arcs intersect. Measure angle G. Does the measured value agree with the calculated value in Problem 6 within ±1°?

14. *Fence Problem:* Mattie works for a fence company. She has the job of pricing a fence to go across a triangular lot at the corner of Alamo and Heights Streets, as shown in Figure 9-2f. The streets intersect at a 65° angle. The lot extends 200 ft from the intersection along Alamo and 150 ft from the intersection along Heights.

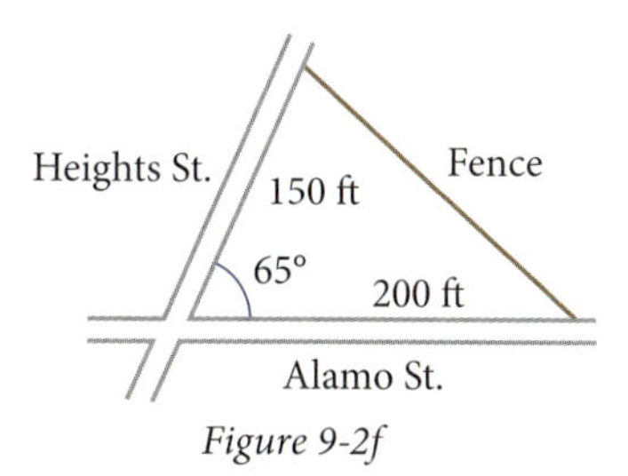

Figure 9-2f

a. How long will the fence be?

b. How much will it cost her company to build the fence if fencing costs \$3.75 per foot?

c. What price should she quote to the customer if the company is to make a 35% profit?

15. *Flight Path Problem:* Sam flies a helicopter to drop supplies to stranded flood victims. He will fly from the supply depot, *S*, to the drop point, *P*. Then he will return to the helicopter's base at *B*, as shown in Figure 9-2g. The drop point is 15 mi from the supply depot. The base is 21 mi from the drop point. It is 33 mi between the supply depot and the base. Because the return flight to the base will be made after dark, Sam wants to know in what direction to fly. What is the angle between the two paths at the drop point?

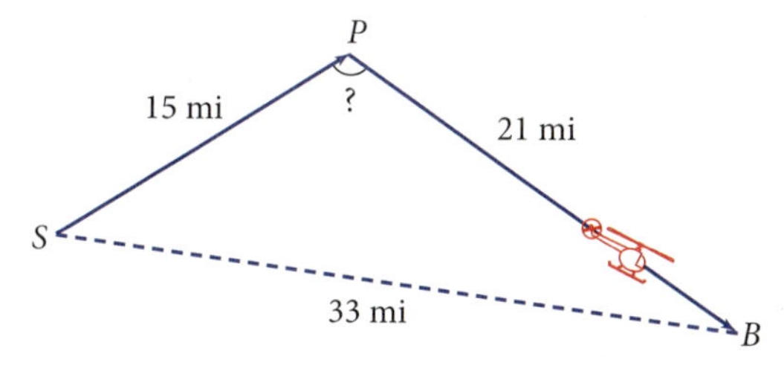

Figure 9-2g

16. *Geometrical Derivation of the Law of Cosines Problem:* Open the *Law of Cosines* exploration at **www.keymath.com/precalc.** Explain in writing how this sketch provides a visual verification of the law of cosines.

17. *Derivation of the Law of Cosines Problem:* Figure 9-2h shows $\triangle XYZ$ with angle Z in standard position. The sides that include angle Z are 4 units and 5 units long, as shown. Find the coordinates of points X and Y in terms of 4, 5, and angle Z. Then use the distance formula, appropriate algebra, and trigonometry to show that

$$z^2 = 5^2 + 4^2 - 2 \cdot 5 \cdot 4 \cdot \cos Z$$

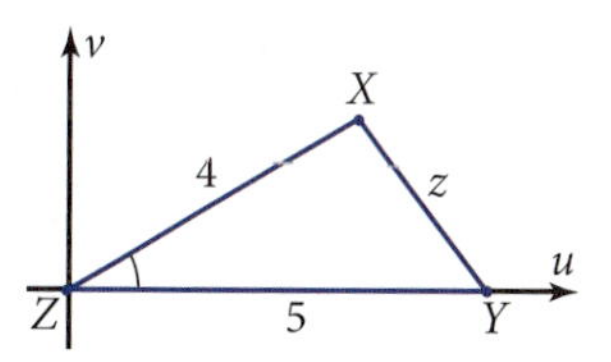

Figure 9-2h

18. *Acute, Right, or Obtuse Problem:* The law of cosines states that in $\triangle XYZ$

$$x^2 = y^2 + z^2 - 2yz \cos X$$

a. Explain how the law of cosines allows you to make a quick test to see whether angle X is acute, right, or obtuse, as shown in this box:

> **PROPERTY: Test for the Size of an Angle in a Triangle**
>
> In $\triangle XYZ$:
>
> If $x^2 < y^2 + z^2$, then angle X is an acute angle.
>
> If $x^2 = y^2 + z^2$, then angle X is a right angle.
>
> If $x^2 > y^2 + z^2$, then angle X is an obtuse angle.

b. Without using your calculator, find whether angle X is acute, right, or obtuse if $x = 7$ cm, $y = 5$ cm, and $z = 4$ cm.

9-3 Area of a Triangle

Recall from earlier math classes that the area of a triangle equals half the product of the base and the altitude. In this section you'll learn how to find this area from two side lengths and the included angle measure. This is the same information you use in the law of cosines to calculate the length of the third side.

Objective

Given the measures of two sides and the included angle, or the measures of all three sides, find the area of the triangle.

In this exploration you will discover a quick method for calculating the area of a triangle from the measures of two sides and the included angle.

EXPLORATION 9-3: Area of a Triangle and Hero's Formula

For Problems 1–3, $\triangle XYZ$ has sides $y = 8$ cm and $z = 7$ cm and included angle X with measure 38°.

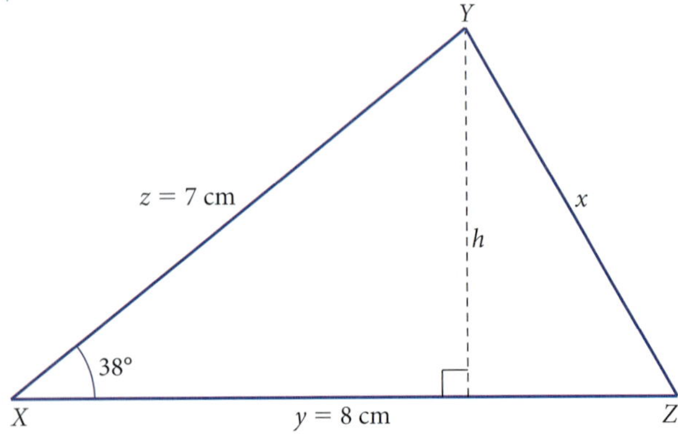

1. Do you agree with the given measurement for y? for z? for $\angle X$?
2. Use the given measurements to calculate altitude h. Measure h. Does it agree with the calculation?
3. Recall from geometry that the area of a triangle is $\frac{1}{2}$(base)(altitude). Find the area of $\triangle XYZ$.
4. By substituting $z \sin X$ for the altitude in Problem 3 you get

$$\text{Area} = \frac{1}{2}yz \sin X$$

or, in general,

$$\text{Area} = \frac{1}{2}\text{(side)(side)(sine of included angle)}$$

Sketch a triangle with sides 43 m and 51 m and included angle 143°. Use this area formula to find the area of this triangle.

For Problems 5–8, $\triangle ABC$ has sides $a = 8$, $b = 7$, and $c = 11$.

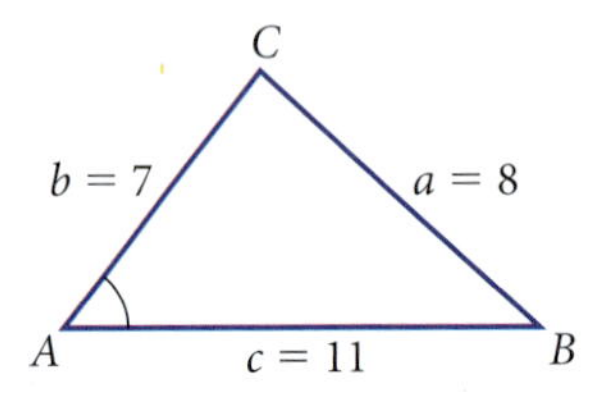

5. Find the measure of angle A using the law of cosines. Store the answer without rounding.
6. Use the unrounded value of A and the area formula of Problem 4 to find the area of $\triangle ABC$.
7. Calculate the *semiperimeter* (half the perimeter) of the triangle, $s = \frac{1}{2}(a + b + c)$.
8. Evaluate the quantity $\sqrt{s(s-a)(s-b)(s-c)}$. What do you notice about the answer?
9. Use *Hero's formula*, namely,

$$\text{Area} = \sqrt{s(s-a)(s-b)(s-c)}$$

to find the area of this triangle.

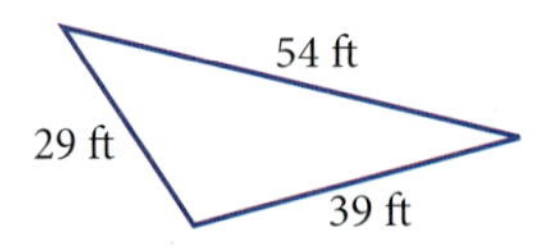

10. What did you learn as a result of doing this exploration that you did not know before?

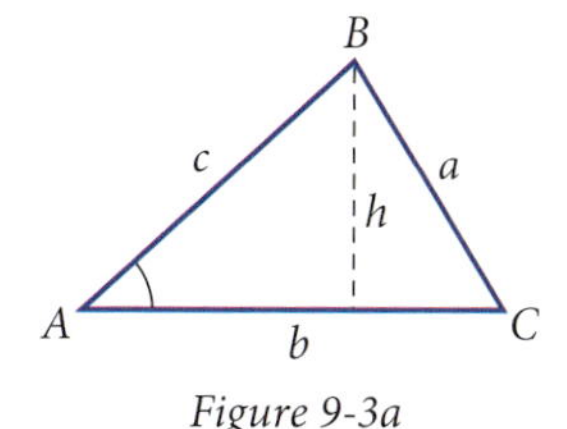

Figure 9-3a

The following is a derivation of the area formula you discovered in Exploration 9-3. Figure 9-3a shows $\triangle ABC$ with base b and altitude h.

$\text{Area} = \frac{1}{2}bh$ — From geometry, area equals half base times altitude.

$\text{Area} = \frac{1}{2}b(c \sin A)$ — Because $\sin A = \frac{h}{c}$.

$\text{Area} = \frac{1}{2}bc \sin A$

PROPERTY: Area of a Triangle

In $\triangle ABC$,

$$\text{Area} = \frac{1}{2}bc \sin A$$

Verbally: The area of a triangle equals half the product of two of its sides and the sine of the included angle.

EXAMPLE 1 ➤ In $\triangle ABC$, $a = 13$ in., $b = 15$ in., and $C = 71°$. Find the area of the triangle.

SOLUTION Sketch the triangle to be sure you're given two sides and the *included* angle (Figure 9-3b).

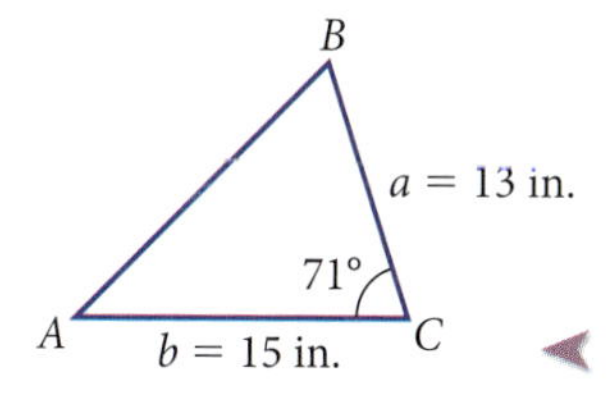

Figure 9-3b

$$\text{Area} = \frac{1}{2}(13)(15)\sin 71°$$
$$= 92.1880... \approx 92.19 \text{ in.}^2$$ ◄

EXAMPLE 2 ➤ Find the area of $\triangle JDH$ if $j = 5$ cm, $d = 7$ cm, and $h = 11$ cm.

SOLUTION Sketch the triangle to give yourself a picture of what has to be done (Figure 9-3c).

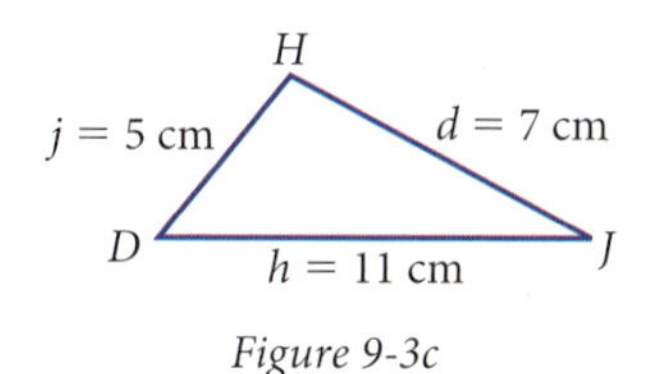

Figure 9-3c

$h^2 = j^2 + d^2 - 2jd \cos H$ — Use the law of cosines to calculate an angle measure.

$\cos H = \dfrac{j^2 + d^2 - h^2}{2jd} = \dfrac{5^2 + 7^2 - 11^2}{2(5)(7)} = -0.6714...$ — Solve for $\cos H$.

$H = \arccos(-0.6714...) = \cos^{-1}(-0.6714...) = 132.1774...°$ — Store without rounding.

$\text{Area} = \frac{1}{2}(5)(7)\sin 132.1774...° = 12.9687... \approx 12.97 \text{ cm}^2$ ◄

Hero's Formula

Hero of Alexandria

It is possible to find the area of a triangle directly from the lengths of three sides without going through the angle calculations of Example 2. The method uses **Hero's formula,** named after Hero of Alexandria, who lived around 100 B.C.E.

PROPERTY: Hero's Formula

In $\triangle ABC$, the area is given by

$$\text{Area} = \sqrt{s(s-a)(s-b)(s-c)}$$

where s is the semiperimeter (half the perimeter), $\frac{1}{2}(a+b+c)$.

EXAMPLE 3 ➤ Find the area of $\triangle JDH$ in Example 2 using Hero's formula. Confirm that you get the same answer as in Example 2.

SOLUTION

$$s = \tfrac{1}{2}(5+7+11) = 11.5$$

$$\text{Area} = \sqrt{11.5(11.5-5)(11.5-7)(11.5-11)} = \sqrt{168.1875}$$

$$= 12.9687... \approx 12.97 \text{ cm}^2$$ Agrees with Example 2. ◄

Problem Set 9-3

Reading Analysis

From what you have read in this section, what do you consider to be the main idea? In what way is the area formula Area $= \frac{1}{2}bc \sin A$ related to the formula Area $= \frac{1}{2}(\textit{base})(\textit{height})$? What formula allows you to calculate the area of a triangle *directly,* from three given side lengths?

Quick Review

Problems Q1–Q5 refer to Figure 9-3d.

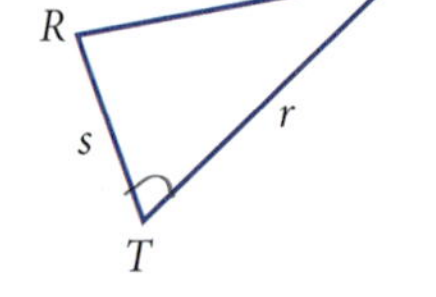

Figure 9-3d

Q1. State the law of cosines using angle R.

Q2. State the law of cosines using angle S.

Q3. State the law of cosines using angle T.

Q4. Express $\cos T$ in terms of sides r, s, and t.

Q5. Why do you need only the function $\cos^{-1}$, not the relation arccos, when using the law of cosines to find an angle?

Q6. When you multiply two sinusoids with very different periods, you get a function with a varying __?__.

Q7. What is the first step in proving that a trigonometric equation is an identity?

Q8. Which trigonometric functions are *even* functions?

Q9. If angle θ is in standard position, then $\frac{\text{horizontal coordinate}}{\text{radius}}$ is the definition of __?__.

Q10. In the composite argument properties, $\cos(x+y) =$ __?__.

For Problems 1–4, find the area of the indicated triangle.

1. $\triangle ABC$, if $a = 5$ ft, $b = 9$ ft, and $C = 14°$
2. $\triangle ABC$, if $b = 8$ m, $c = 4$ m, and $A = 67°$
3. $\triangle RST$, if $r = 4.8$ cm, $t = 3.7$ cm, and $S = 43°$
4. $\triangle XYZ$, if $x = 34.19$ yd, $z = 28.65$ yd, and $Y = 138°$

For Problems 5–7, use Hero's formula to calculate the area of the triangle.

5. $\triangle ABC$, if $a = 6$ cm, $b = 9$ cm, and $c = 11$ cm
6. $\triangle XYZ$, if $x = 50$ yd, $y = 90$ yd, and $z = 100$ yd
7. $\triangle DEF$, if $d = 3.7$ in., $e = 2.4$ in., and $f = 4.1$ in.
8. *Comparison of Methods Problem:* Reconsider Problems 1 and 7.
   a. For $\triangle ABC$ in Problem 1, calculate the length of the third side using the law of cosines. Store the answer without rounding. Then find the area using Hero's formula. Do you get the same answer as in Problem 1?
   b. For $\triangle DEF$ in Problem 7, calculate the measure of angle D using the law of cosines. Store the answer without rounding. Then find the area using the area formula as in Example 2. Do you get the same answer as in Problem 7?

9. *Hero's Formula and Impossible Triangles Problem:* Suppose someone tells you that $\triangle ABC$ has sides $a = 5$ cm, $b = 6$ cm, and $c = 13$ cm.

   a. Explain why there is no such triangle.

   b. Apply Hero's formula to the given information. How does Hero's formula allow you to detect that there is no such triangle?

10. *Lot Area Problem:* Sean works for a real estate company. The company has a contract to sell the triangular lot at the corner of Alamo and Heights Streets (Figure 9-3e). The streets intersect at a 65° angle. The lot extends 200 ft from the intersection along Alamo and 150 ft from the intersection along Heights.

   a. Find the area of the lot.

   b. Land in this neighborhood is valued at \$35,000 per acre. An acre is 43,560 ft^2. How much is the lot worth?

   c. The real estate company will earn a commission of 6% of the sales price. If the lot sells for what it is worth, how much will the commission be?

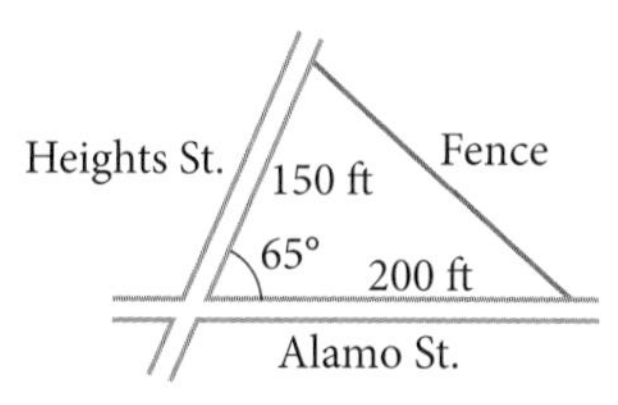

Figure 9-3e

11. *Variable Triangle Problem:* Figure 9-3f shows angle θ in standard position in a uv-coordinate system. The fixed side of the angle is 3 units long, and the rotating side is 4 units long. As θ increases, the area of the triangle shown in the figure is a function of θ.

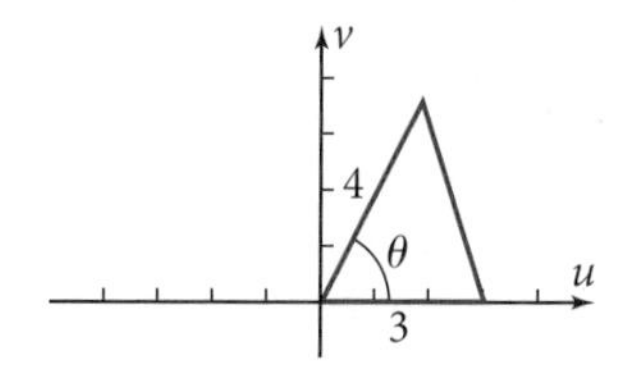

Figure 9-3f

   a. Write the area as a function of θ.

   b. Make a table of values of area for each 15° from 0° through 180°.

   c. Is this statement true or false? "The area is an increasing function of θ for all angles from 0° through 180°." Give evidence to support your answer.

   d. Find the domain of θ for which this statement is true: "The area is a sinusoidal function of θ." Explain why the statement is false outside this domain.

12. *Unknown Angle Problem:* Suppose you need to construct a triangle with one side 14 cm, another side 11 cm, and a given area.

   a. What *two* possible values of the included angle will produce an area 50 cm^2?

   b. Show that there is only *one* possible value of the included angle if the area is 77 cm^2.

   c. Show algebraically that there would be *no* possible value of the angle if the area were 100 cm^2.

13. *Comparing Formulas Problem:* Demonstrate numerically that the area formula and Hero's formula both yield the same results for a 30°-60°-90° triangle with hypotenuse 2 cm. Can you show it without using decimals?

14. *Derivation of the Area Formula Problem:* Figure 9-3g shows $\triangle XYZ$ with angle Z in standard position. The two sides that include angle Z are 4 and 5 units long. Find the altitude h in terms of length 4 and angle Z. Then show that the area of the triangle is given by

   Area $= \frac{1}{2}(5)(4)\sin Z$

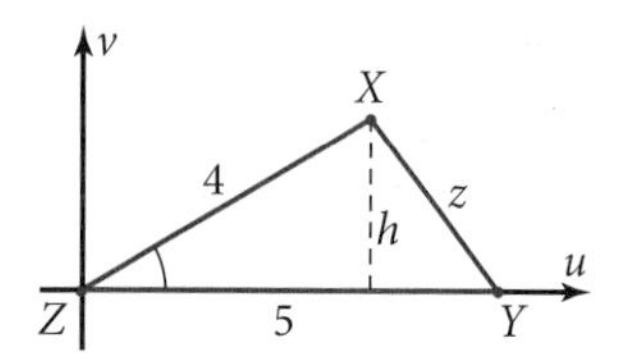

Figure 9-3g

9-4 Oblique Triangles: The Law of Sines

Because the law of cosines involves all three sides of a triangle, you must know at least two of the sides to use it. In this section you'll learn the law of sines, which lets you calculate a side length of a triangle if only one side and two angles are given.

Objective Given the measure of an angle, the length of the side opposite this angle, and one other piece of information about a triangle, find the other side lengths and angle measures.

In this exploration you will use the ratio of a side length to the sine of the opposite angle to find the measures of other parts of a triangle.

EXPLORATION 9-4: The Law of Sines

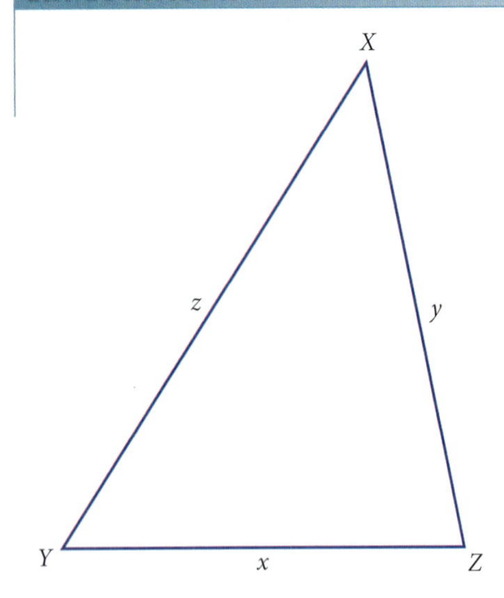

1. In $\triangle XYZ$, are the following measurements correct?

   $y = 6.0$ cm $\quad z = 7.0$ cm

   $Y = 57°$ $\quad Z = 78°$

2. Assuming that the measurements in Problem 1 are correct, calculate these ratios:

   $$\frac{y}{\sin Y} \qquad \frac{z}{\sin Z}$$

3. The law of sines states that within a triangle, the ratio of the length of a side to the sine of the opposite angle is constant. Do the calculations in Problem 2 seem to confirm this property?

4. Measure angle X.

5. Assuming that the law of sines is correct,

   $$\frac{x}{\sin X} = \frac{y}{\sin Y}$$

   Use this information and the measured value of X to calculate length x.

6. Measure side x. Does your measurement agree with the calculated value in Problem 5?

The law of sines can be derived algebraically.

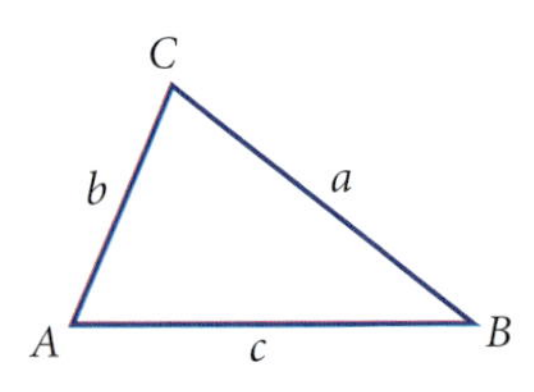

7. For $\triangle ABC$, use the area formula to write the area *three* ways:
   a. Involving angle A
   b. Involving angle B
   c. Involving angle C

8. The area of a triangle is independent of the way you calculate it, so all three expressions in Problem 7 are equal to each other. Write a three-part equation expressing this fact.

9. Divide all three "sides" of the equation in Problem 8 by whatever is necessary to leave only the sines of the angles in the numerators. Simplify.

continued

EXPLORATION, continued

10. The equation you should have gotten in Problem 9 is the law of sines. Explain why it is equivalent to the law of sines as written in Problem 5.

11. What did you learn as a result of doing this exploration that you did not know before?

In Exploration 9-4, you demonstrated that the law of sines is correct and used it to find an unknown side length of a triangle from information about other sides and angles. The law of sines can be proved with the help of the area formula from Section 9-3.

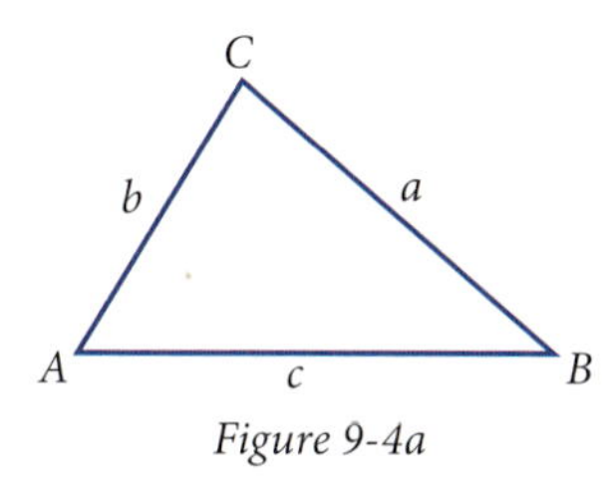

Figure 9-4a

Figure 9-4a shows $\triangle ABC$. In the previous section you found that the area is equal to $\frac{1}{2}bc \sin A$. The area is constant no matter which pair of sides and included angle you use.

$$\tfrac{1}{2}bc \sin A = \tfrac{1}{2}ac \sin B = \tfrac{1}{2}ab \sin C \qquad \text{Set the areas equal.}$$

$$bc \sin A = ac \sin B = ab \sin C \qquad \text{Multiply by 2.}$$

$$\frac{bc \sin A}{abc} = \frac{ac \sin B}{abc} = \frac{ab \sin C}{abc} \qquad \text{Divide by } abc.$$

$$\frac{\sin A}{a} = \frac{\sin B}{b} = \frac{\sin C}{c}$$

This final relationship is called the **law of sines.** If three nonzero numbers are equal, then their reciprocals are equal. So you can write the law of sines in another algebraic form:

$$\frac{a}{\sin A} = \frac{b}{\sin B} = \frac{c}{\sin C}$$

PROPERTY: The Law of Sines

In $\triangle ABC$,

$$\frac{\sin A}{a} = \frac{\sin B}{b} = \frac{\sin C}{c} \quad \text{and} \quad \frac{a}{\sin A} = \frac{b}{\sin B} = \frac{c}{\sin C}$$

Verbally: Within any given triangle, the ratio of the sine of an angle to the length of the side opposite that angle is constant.

Because of the different combinations of sides and angles for any given triangle, it is convenient to revive some terminology from geometry. The initials SAS stand for "side, angle, side." This means that as you go around the perimeter of the triangle, you are given the length of a side, the measure of an angle, and the length of a side, in that order. SAS is equivalent to knowing two sides and the included angle, the same information used in the law of cosines and in the area formula. Similar meanings are attached to ASA, AAS, SSA, and SSS.

Given AAS, Find the Other Sides

Example 1 shows you how to calculate two side lengths given the third side and two angles.

EXAMPLE 1 ➤ In $\triangle ABC$, $B = 64°$, $C = 38°$, and $b = 9$ ft. Find the lengths of sides a and c.

SOLUTION First, draw a picture, as in Figure 9-4b.

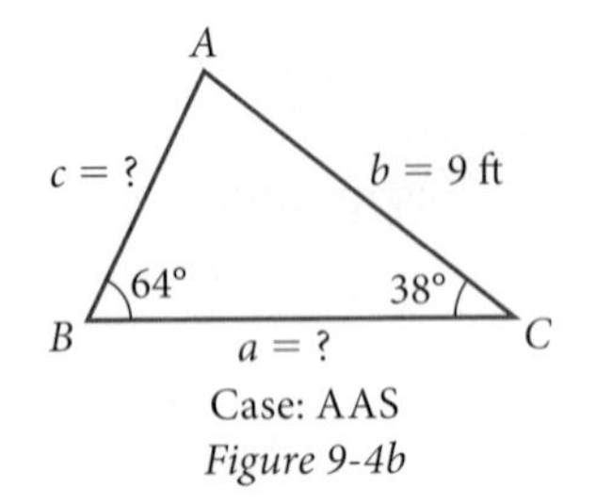

Case: AAS

Figure 9-4b

Because you know the angle opposite side c but not the angle opposite side a, it's easier to start with finding the length of side c.

$$\frac{c}{\sin 38°} = \frac{9}{\sin 64°}$$ Use the law of sines. Put the unknown in the numerator on the left side.

$$c = \frac{9 \sin 38°}{\sin 64°} = 6.1648\ldots \text{ ft}$$ Multiply both sides by sin 38° to isolate c on the left.

To find a by the law of sines, you need the measure of A, the opposite angle.

$$A = 180° - (38° + 64°) = 78°$$ The sum of the interior angles in a triangle is 180°.

$$\frac{a}{\sin 78°} = \frac{9}{\sin 64°}$$ Use the appropriate parts of the law of sines with a in the numerator.

$$a = \frac{9 \sin 78°}{\sin 64°} = 9.7946\ldots \text{ ft}$$

$$\therefore a \approx 9.79 \text{ ft} \quad \text{and} \quad c \approx 6.16 \text{ ft}$$

Given ASA, Find the Other Sides

Example 2 shows you how to calculate side lengths if the given side is included between the two given angles.

EXAMPLE 2 ➤ In $\triangle ABC$, $a = 8$ m, $B = 64°$, and $C = 38°$. Find the lengths of sides b and c.

SOLUTION First, draw a picture (Figure 9-4c). The picture reveals that in this case you do not know the angle opposite the given side. So you calculate this angle measure first. From there on, it is a familiar problem, similar to Example 1.

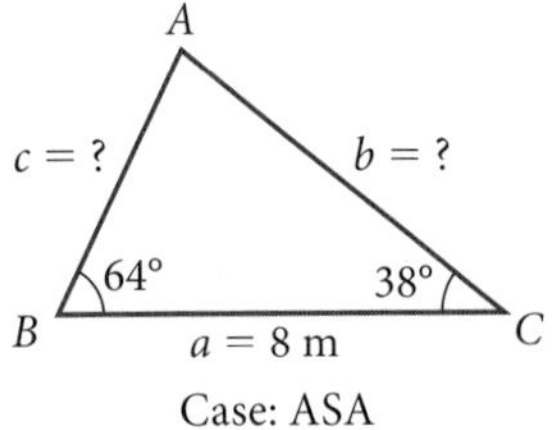

Case: ASA

Figure 9-4c

$$A = 180° - (38° + 64°) = 78°$$

$$\frac{b}{\sin 64°} = \frac{8}{\sin 78°}$$ Use the appropriate parts of the law of sines.

$$b = \frac{8 \sin 64°}{\sin 78°} = 7.3509\ldots \text{ m}$$

$$\frac{c}{\sin 38°} = \frac{8}{\sin 78°}$$ Use the appropriate parts of the law of sines.

$$c = \frac{8 \sin 38°}{\sin 78°} = 5.0353\ldots \text{ m}$$

$$\therefore b \approx 7.35 \text{ m} \quad \text{and} \quad c \approx 5.04 \text{ m}$$

The Law of Sines for Angles

You can use the law of sines to find an unknown angle of a triangle. However, you must be careful because there are *two* values of the inverse sine relation between 0° and 180°, either of which could be the answer. For instance, arcsin 0.8 = 53.1301...° or 126.8698...°; both could be angles of a triangle. Problem 11 shows you what to do in this situation.

Problem Set 9-4

Reading Analysis

From what you have read in this section, what do you consider to be the main idea? Based on the verbal statement of the law of sines, why is it necessary to know at least one angle in the triangle to use the law? In the solution to Example 1, why is it advisable to put the unknown side length in the numerator on the left side of the equation? Why can you be led to an incorrect answer if you try to use the law of sines to find an angle measure?

Quick Review

Q1. State the law of cosines for $\triangle PAF$ involving angle P.

Q2. State the formula for the area of $\triangle PAF$ involving angle P.

Q3. Write two values of $\theta = \arcsin 0.5$ that lie between 0° and 180°.

Q4. If $\sin \theta = 0.3726\ldots$, then $\sin(-\theta) =$ __?__.

Q5. $\cos \frac{\pi}{6} =$

A. $\frac{1}{\sqrt{3}}$ **B.** $\frac{1}{2}$ **C.** $\frac{2}{\sqrt{3}}$

D. $\frac{\sqrt{3}}{2}$ **E.** $\sqrt{3}$

Q6. A(n) __?__ triangle has no equal sides and no equal angles.

Q7. A(n) __?__ triangle has no right angle.

Q8. State the Pythagorean property for cosine and sine.

Q9. $\cos 2x = \cos(x + x) =$ __?__ in terms of cosines and sines of x.

Q10. The amplitude of the sinusoid $y = 3 + 4 \cos 5(x - 6)$ is __?__.

---

1. In $\triangle ABC$, $A = 52°$, $B = 31°$, and $a = 8$ cm. Find the lengths of side b and side c.

2. In $\triangle PQR$, $P = 13°$, $Q = 133°$, and $q = 9$ in. Find the lengths of side p and side r.

3. In $\triangle AHS$, $A = 27°$, $H = 109°$, and $a = 120$ yd. Find the lengths of side h and side s.

4. In $\triangle BIG$, $B = 2°$, $I = 79°$, and $b = 20$ km. Find the lengths of side i and side g.

5. In $\triangle PAF$, $P = 28°$, $f = 6$ m, and $A = 117°$. Find the lengths of side a and side p.

6. In $\triangle JAW$, $J = 48°$, $a = 5$ ft, and $W = 73°$. Find the lengths of side j and side w.

7. In $\triangle ALP$, $A = 85°$, $p = 30$ ft, and $L = 87°$. Find the lengths of side a and side l.

8. In $\triangle LOW$, $L = 2°$, $o = 500$ m, and $W = 3°$. Find the lengths of side l and side w.

9. *Island Bridge Problem:* Suppose that you work for a construction company that is planning to build a bridge from the land to a point on an island in a lake (Figure 9-4d). The only two places on the land to start the bridge are point X and point Y, 1000 m apart. Point X has better access to the lake but is farther from the island than point Y. To help decide between X and Y, you need the precise lengths of the two possible bridges. From point X you measure a 42° angle to the point on the island, and from point Y you measure a 58° angle.

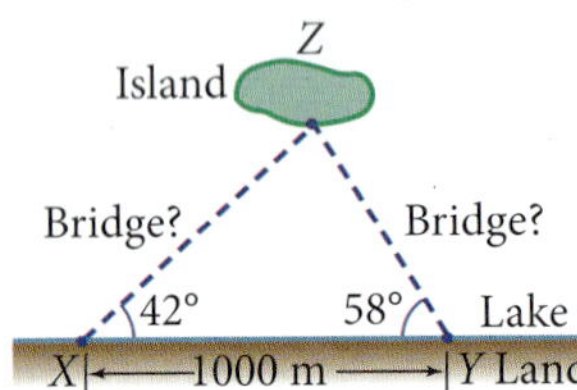

Figure 9-4d

a. How long would each bridge be?

b. If constructing the bridge costs \$370 per meter, how much could be saved by constructing the shorter bridge?

c. How much could be saved by constructing the shortest possible bridge (if that were okay)?

10. *Walking Problem:* Amos walks 800 ft along the sidewalk next to a field. Then he turns at an angle of 43° to the sidewalk and heads across the field (Figure 9-4e). When he stops, he looks back at the starting point, finding a 29° angle between his path across the field and the direct route back to the starting point.

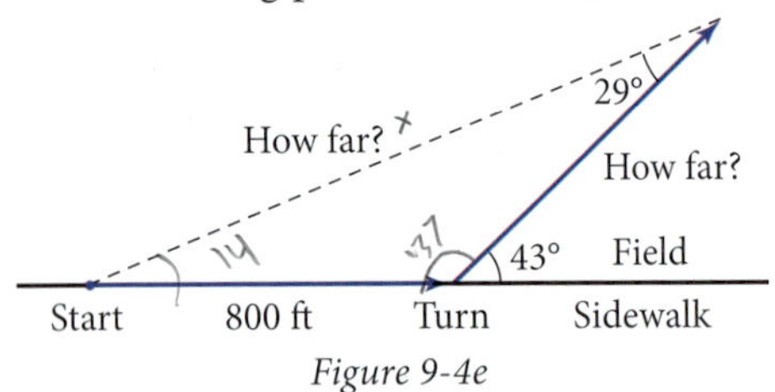

Figure 9-4e

a. How far across the field did Amos walk?

b. How far does he have to walk to go directly back to the starting point?

c. Amos walks 5 ft/s on the sidewalk but only 3 ft/s across the field. Which way is quicker for him to return to the starting point—by going directly across the field or by retracing the original route?

11. *Law of Sines for Angles Problem:* You can use the law of sines to find an unknown angle measure, but the technique is risky. Suppose that $\triangle ABC$ has sides 4 cm, 7 cm, and 10 cm, as shown in Figure 9-4f.

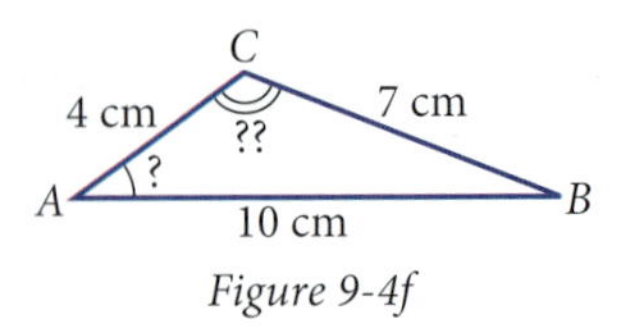

Figure 9-4f

a. Use the law of cosines to find the measure of angle A.

b. Use the answer to part a (don't round off) and the law of *sines* to find the measure of angle C.

c. Find the measure of angle C again, using the law of *cosines* and the given side lengths.

d. Your answers to parts b and c probably do not agree. Show that you can get the correct answer from your work with the law of sines in part b by considering the *general* solution for arcsine.

e. Why is it dangerous to use the law of sines to find an angle measure but not dangerous to use the law of cosines?

12. *Accurate Drawing Problem:* Using computer software such as The Geometer's Sketchpad, or using a ruler and protractor with pencil and paper, construct a triangle with base 10.0 cm and base angles 40° and 30°. Measure the length of the side opposite the 30° angle. Then calculate its length using the law of sines. Your measured value should be within ±0.1 cm, of the calculated value.

13. *Geometric Derivation of the Law of Sines Problem:* Open the *Law of Sines* exploration at **www.keymath.com/precalc.** Explain in writing how this sketch provides a visual verification of the law of sines.

14. *Algebraic Derivation of the Law of Sines Problem:* Derive the law of sines algebraically. If you cannot do it from memory, consult the text long enough to get started. Then try finishing on your own.

9-5 The Ambiguous Case

From one end of a long segment, you draw an 80-cm segment at a 26° angle. From the other end of the 80-cm segment, you draw a 50-cm segment, completing a triangle. Figure 9-5a shows the two possible triangles you might create.

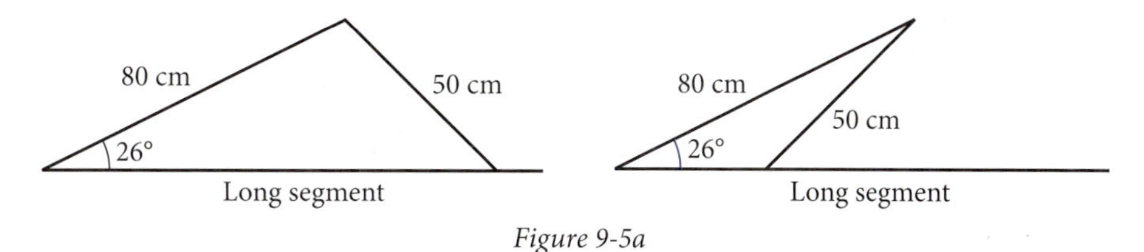

Figure 9-5a

Figure 9-5b shows why there are two possible triangles. A 50-cm arc drawn from the upper vertex cuts the long segment in two places. Each point could be the third vertex of the triangle.

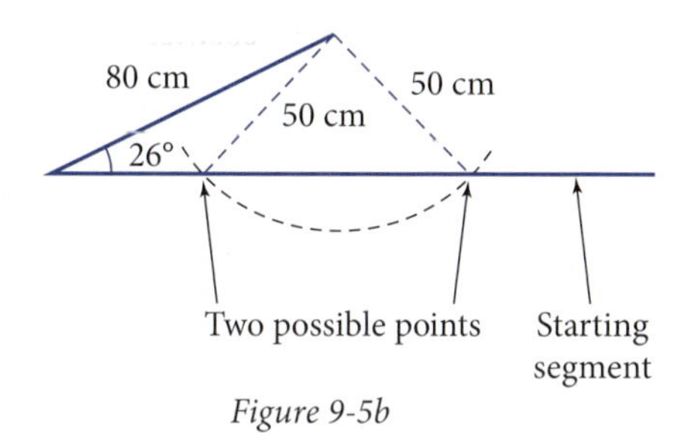

Figure 9-5b

As you go around the perimeter of the triangle in Figure 9-5b, the given information is a side, another side, and an angle (SSA). Because there are two possible triangles that have these specifications, SSA is called the *ambiguous case.*

Objective Given two sides and a non-included angle, calculate the possible lengths of the third side.

EXAMPLE 1 ➤ In $\triangle XYZ$, $x = 50$ cm, $z = 80$ cm, and $X = 26°$, as in Figure 9-5a. Find the possible lengths of side y.

SOLUTION Sketch a triangle and label the given sides and angle (Figure 9-5c).

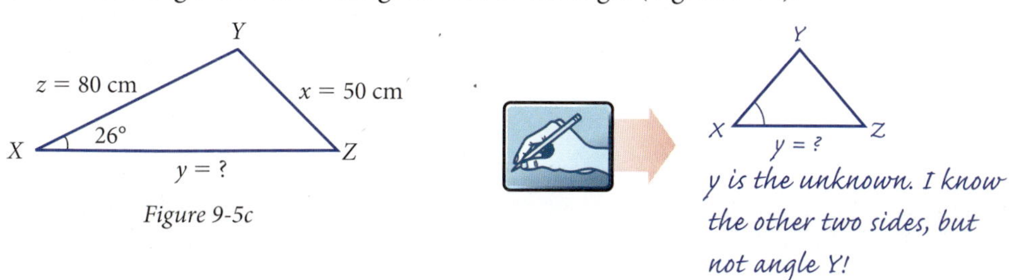

Figure 9-5c

Using the law of sines to find y would require several steps. Here is a shorter method, using the law of cosines.

$$50^2 = y^2 + 80^2 - 2 \cdot y \cdot 80 \cdot \cos 26°$$

Write the law of cosines for the known angle, $X = 26°$.

This is a quadratic equation in the variable y. You can solve it using the quadratic formula.

$$y^2 - (160 \cos 26°)y + 6400 - 2500 = 0$$

Make one side equal zero.

$$y^2 + (-160 \cos 26°)y + 3900 = 0$$

Get the form $ay^2 + by + c = 0$.

$$y = \frac{160 \cos 26° \pm \sqrt{(-160 \cos 26°)^2 - 4 \cdot 1 \cdot 3900}}{2 \cdot 1}$$

Use the quadratic formula: $y = \frac{-b \pm \sqrt{b^2 - 4ac}}{2a}$.

$$y = 107.5422... \text{ or } 36.2648...$$

$$y \approx 107.5 \text{ cm or } 36.3 \text{ cm}$$

You may be surprised if you use different lengths for side x in Example 1. Figures 9-5d and 9-5e show this side as 90 cm and 30 cm, respectively, instead of 50 cm. In the first case, there is only *one* possible triangle. In the second case, there is none.

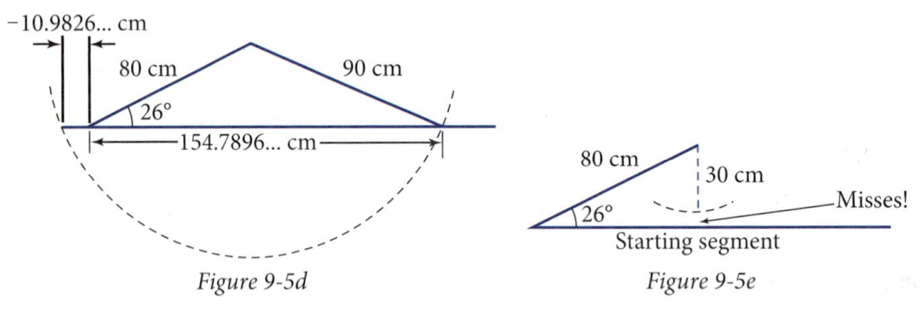

Figure 9-5d

Figure 9-5e

The quadratic formula technique of Example 1 detects both of these results. For 30 cm, the discriminant, $b^2 - 4ac$, equals $-1319.5331...$, meaning there are no real solutions to the equation and thus no triangle. For 90 cm,

$$y = 154.7896... \text{ or } -10.9826...$$

Although $-10.9826...$ cannot be a side measure of a triangle, it does equal the *displacement* (the *directed* distance) to the point where the arc would cut the starting segment if this segment were extended in the other direction.

Problem Set 9-5

Reading Analysis

From what you have read in this section, what do you consider to be the main idea? Sketch a triangle with two given sides and a given non-included angle that illustrates that there can be two different triangles with the same given information. How can the law of cosines be applied in the ambiguous case to find both possible lengths of the third side with the same computation?

Quick Review

Problems Q1–Q6 refer to the triangle in Figure 9-5f.

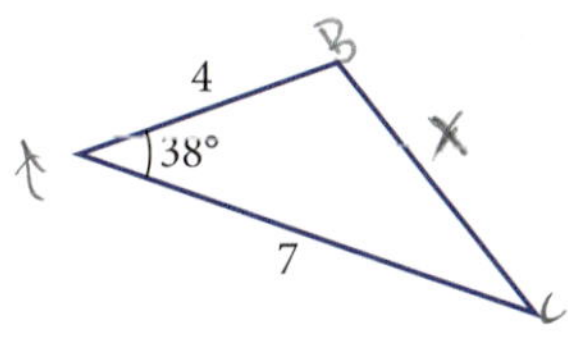

Figure 9-5f

Q1. The initials SAS stand for __?__.

Q2. Find the length of the third side.

Q3. What method did you use in Problem Q2?

Q4. Find the area of this triangle.

Q5. The largest angle in this triangle is opposite the __?__ side.

Q6. The sum of the angle measures in this triangle is __?__.

Q7. Find the amplitude of the sinusoid $y = 4 \cos x + 3 \sin x$.

Q8. The period of the circular function $y = 3 + 7 \cos \frac{\pi}{8}(x - 1)$ is

A. 16 **B.** 8 **C.** $\frac{\pi}{8}$ **D.** 7 **E.** 3

Q9. The value of the inverse circular function $x = \sin^{-1} 0.5$ is __?__.

Q10. A value of the inverse circular relation $x = \arcsin 0.5$ between $\frac{\pi}{2}$ and 2π is __?__.

For Problems 1–8, find the possible lengths of the indicated side.

1. In $\triangle ABC$, $B = 34°$, $a = 4$ cm, and $b = 3$ cm. Find c.
2. In $\triangle XYZ$, $X = 13°$, $x = 12$ ft, and $y = 5$ ft. Find z.
3. In $\triangle ABC$, $B = 34°$, $a = 4$ cm, and $b = 5$ cm. Find c.
4. In $\triangle XYZ$, $X = 13°$, $x = 12$ ft, and $y = 15$ ft. Find z.
5. In $\triangle ABC$, $B = 34°$, $a = 4$ cm, and $b = 2$ cm. Find c.
6. In $\triangle XYZ$, $X = 13°$, $x = 12$ ft, and $y = 60$ ft. Find z.
7. In $\triangle RST$, $R = 130°$, $r = 20$ in., and $t = 16$ in. Find s.
8. In $\triangle OBT$, $O = 170°$, $o = 19$ m, and $t = 11$ m. Find b.
9. *Radio Station Problem:* Radio station KROK plans to broadcast rock music to people on the beach near Ocean City (O.C. in Figure 9-5g). Measurements show that Ocean City is 20 mi from KROK, at an angle 50° north of west. KROK's broadcast range is 30 mi.

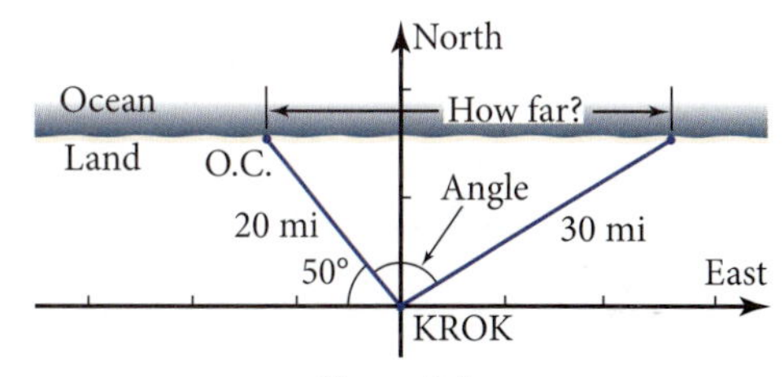

Figure 9-5g

a. Use the law of cosines to calculate how far along the beach to the east of Ocean City people can hear KROK.

b. There are two answers to part a. Show that *both* answers have meaning in the real world.

c. KROK plans to broadcast only in an angle between a line from the station through Ocean City and a line from the station through the point on the beach farthest to the east of Ocean City that people can hear the station. What is the measure of this angle?

For Problems 10–13, use the law of sines to find the indicated angle measure. Determine beforehand whether there are two possible angles or just one.

10. In $\triangle ABC$, $A = 19°$, $a = 25$ mi, and $c = 30$ mi. Find C.

11. In $\triangle HSC$, $H = 28°$, $h = 50$ mm, and $c = 20$ mm. Find S.

12. In $\triangle XYZ$, $X = 58°$, $x = 9.3$ cm, and $z = 7.5$ cm. Find Z.

13. In $\triangle BIG$, $B = 110°$, $b = 1000$ yd, and $g = 900$ yd. Find G.

14. *Six SSA Possibilities Problem:* Parts a through f show six possibilities of $\triangle XYZ$ if angle X and sides x and y are given. For each case, explain the relationship among x, y, and the quantity $y \sin X$

a.

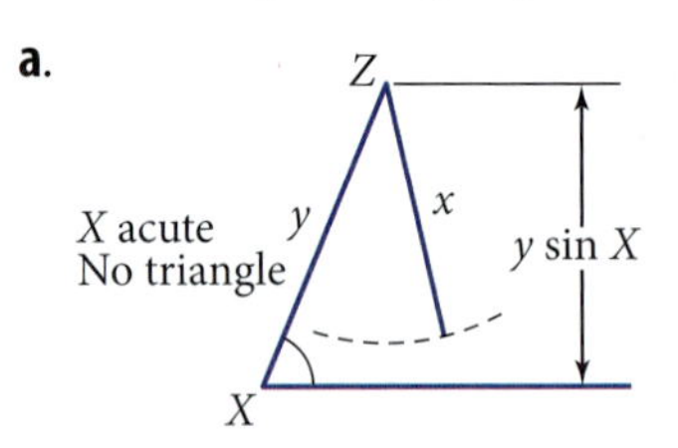

b.

Z
X acute
One triangle
y
x
X z Y

c.

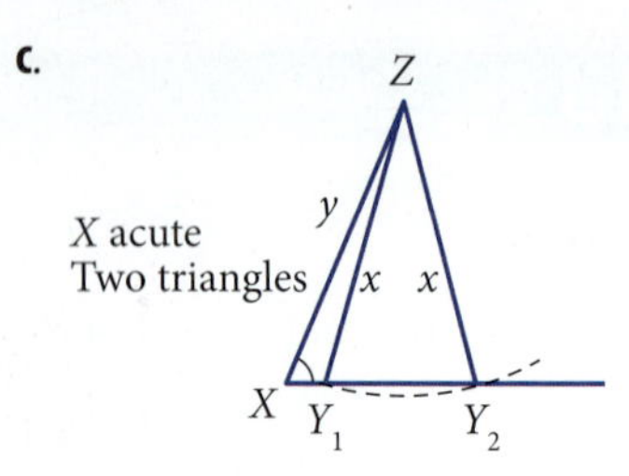

d

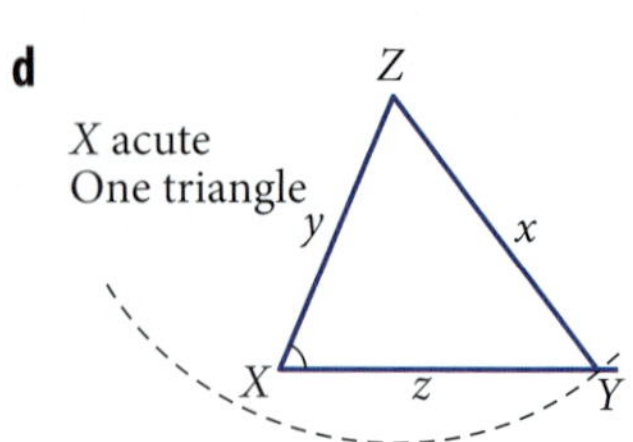

e.

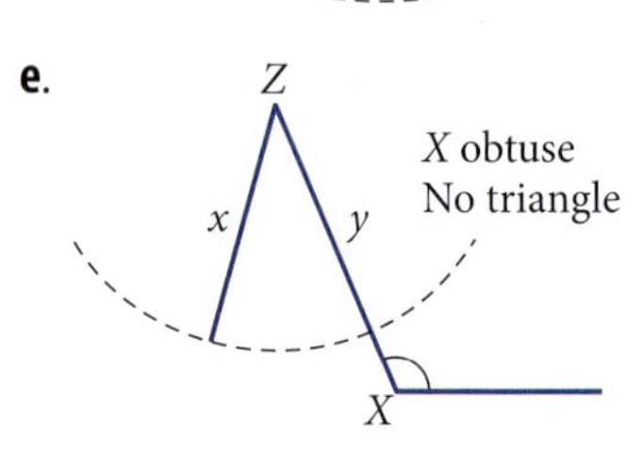

f.

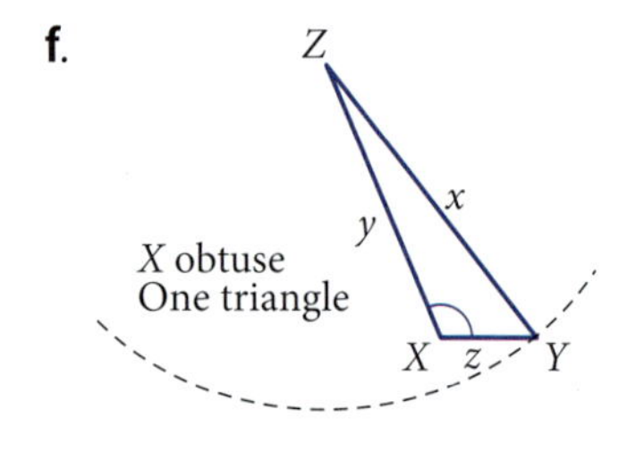

9-6 Vector Addition

Suppose you start at the corner of a room and walk 10 ft at an angle of 70° to one of the walls (Figure 9-6a). Then you turn 80° clockwise and walk another 7 ft. If you had walked straight from the corner to your stopping point, how far and in what direction would you have walked?

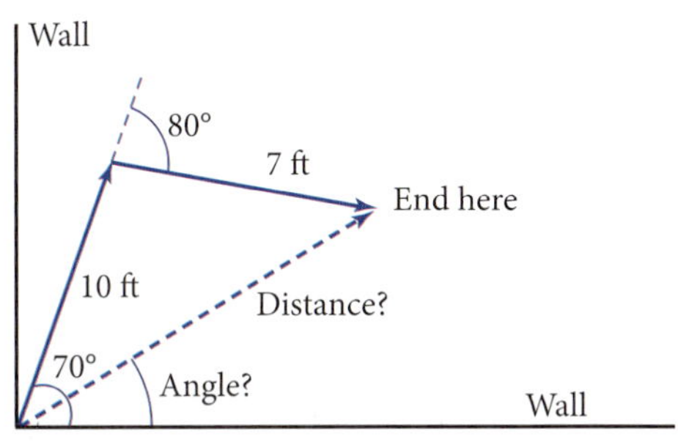

Figure 9-6a

The two motions described are called **displacements.** They are **vector quantities** that have both *magnitude* (size) and *direction* (angle). Vector quantities are represented by directed line segments called **vectors.** A quantity such as distance, time, or volume that has no direction is called a **scalar** quantity.

Objective Given two vectors, add them to find the resultant vector.

In this exploration you will use the properties of triangles to add vectors.

EXPLORATION 9-6: Sum of Two Displacement Vectors

1. The figure shows two vectors starting from the origin. One ends at the point (4, 7), and the other ends at the point (5, 3). Copy the figure on graph paper and translate one of the two vectors so that the beginning of the translated vector is at the end of the other vector. Then draw the resultant vector—the sum of the two vectors.

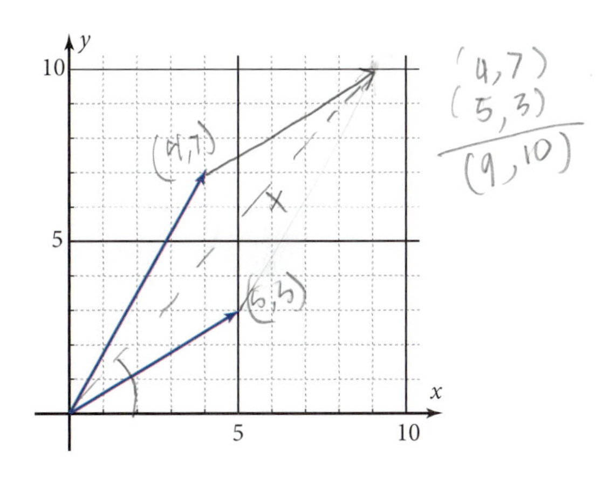

2. Calculate the length of the resultant vector in Problem 1 and the angle it makes with the x-axis.
3. The two given vectors and the resultant vector form a triangle. Calculate the measure of the largest angle in this triangle.
4. Calculate the measure of the angle between the two vectors when they are placed tail-to-tail, as they were given in Problem 1.
5. In Problem 1, you translated one of the vectors. Show on your copy of the figure that you would have gotten the *same* resultant vector if you had translated the *other* vector. Use a different color than you used in Problem 1.

continued

6. The vectors in Problem 1 have *components* in the *x*-direction and in the *y*-direction. These components are a horizontal vector and a vertical vector that can be added together to equal the given vector. On your graph from Problem 1, show how the components of the longer vector can be added to give that vector.

7. Give an *easy* way to get the components of the resultant vector of the two given vectors in Problem 1.

8. What did you learn as a result of doing this exploration that you did not know before?

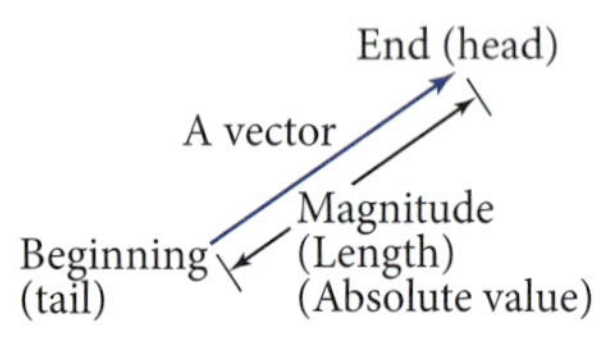

Figure 9-6b

The *length* of a directed line segment represents the *magnitude* of the vector quantity, and the *direction* of the segment represents the vector's direction. An arrowhead on a vector distinguishes the end (its *head*) from the beginning (its *tail*), as shown in Figure 9-6b.

A typhoon's wind speed can reach up to 150 mi/h.

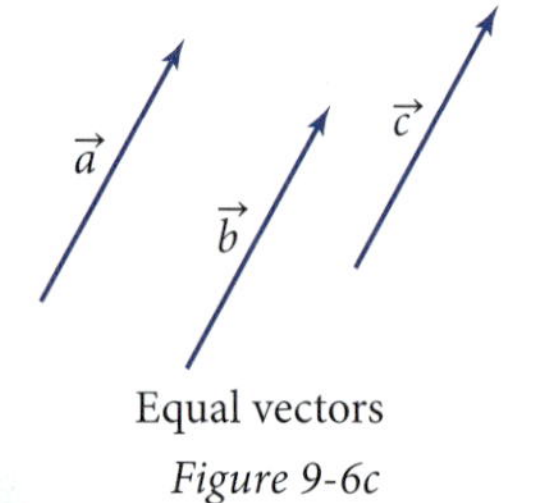

Equal vectors

Figure 9-6c

A variable used for a vector has a small arrow over the top of it, like this, $\vec{x}$, to distinguish it from a scalar. The magnitude of the vector is also called its *absolute value* and is written $|\vec{x}|$. Vectors are *equal* if they have the same magnitude and the same direction. Vectors $\vec{a}, \vec{b}, \vec{c}$ in Figure 9-6c are equal vectors, even though they start and end at different places. So you can translate a vector without changing its value.

DEFINITION: Vector

A **vector,** $\vec{v}$, is a directed line segment.

The absolute value, or magnitude, of a vector, $|\vec{v}|$, is a scalar quantity equal to its length.

Two vectors are equal if and only if they have the same magnitude and the same direction.

EXAMPLE 1 ➤ You start at the corner of a room and walk as shown in Figure 9-6a. Find the displacement that results from the two motions.

SOLUTION Draw a diagram showing the two given vectors and the displacement that results, $\vec{x}$ (Figure 9-6d). They form a triangle with sides 10 ft and 7 ft and included angle 100° (180° − 80°).

$$|\vec{x}|^2 = 10^2 + 7^2 - 2(10)(7)\cos 100° = 173.3107... \quad \text{Use the law of cosines.}$$

$$|\vec{x}| = 13.1647...\text{ ft} \quad \text{Store without rounding for use later.}$$

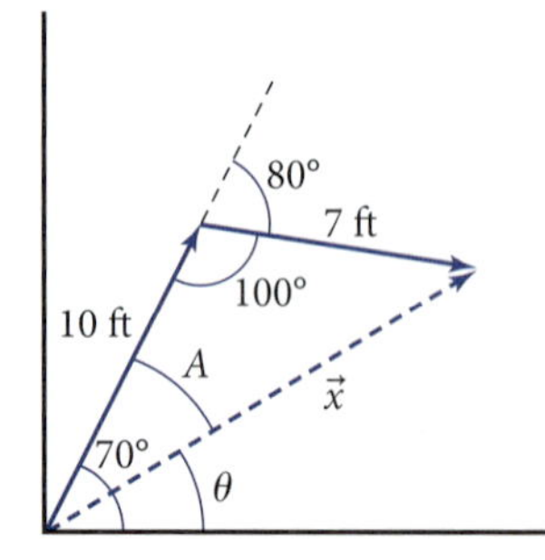

Figure 9-6d

$$7^2 = 10^2 + 13.1647...^2 - 2(10)(13.1647...)\cos A$$

Use the law of cosines to find A.

$$\cos A = \frac{10^2 + 13.1647...^2 - 7^2}{2(10)(13.1647...)} = 0.8519...$$

$$A = 31.5770...°$$

$$\theta = 70° - 31.5770...° = 38.4229...°$$

The vector representing the resultant displacement is approximately 13.2 ft at an angle of 38.4° to the wall.

This example leads to the graphical definition of vector addition. If the tail of one vector is placed at the head of another vector, the *sum* of two vectors goes from the beginning of the first vector to the end of the second, representing the resultant displacement. Because of this, the sum of two vectors is also called the **resultant vector.**

DEFINITION: Vector Addition

The *sum* $\vec{a} + \vec{b}$ is the vector from the beginning of $\vec{a}$ to the end of $\vec{b}$ if the tail of $\vec{b}$ is placed at the head of $\vec{a}$.

Example 2 shows how to add two vectors that are not yet head-to-tail, using *velocity* vectors, for which the magnitude is the scalar *speed.*

EXAMPLE 2 ➤ A ship near the coast is going 9 knots at an angle of 130° to a current of 4 knots. (A knot, kt, is a nautical mile per hour, slightly faster than a regular mile per hour.) What is the ship's resultant velocity with respect to the shore?

SOLUTION Draw a diagram showing two vectors 9 and 4 units long, tail-to-tail, making an angle of 130° with each other, as shown in Figure 9-6e. Translate one of the vectors so that the two vectors are head-to-tail. Draw the resultant vector, $\vec{v}$, from the beginning (tail) of the first to the end (head) of the second.

Translate head-to-tail.

4 kt

50°

Resultant, $\vec{v}$

9 kt

A

θ

130°

4 kt

Figure 9-6e

In its new position, the 4-kt vector is parallel to its original position. The 9-kt vector is a transversal cutting two parallel lines. So the angle between the vectors forming the triangle shown in Figure 9-6e is the supplement of the given 130° angle, namely, 50°. From here on the problem is like Example 1.

$$|\vec{v}| = 7.1217... \text{ kt}$$

$$A = 25.4838...°$$

$$\theta = 130° - 25.4838...° = 104.5161...°$$

$$\therefore \vec{v} \approx 7.1 \text{ kt at about } 104.5° \text{ to the current}$$

Vector Addition by Components

Suppose that an airplane is climbing with a horizontal velocity of 300 mi/h and a vertical velocity of 170 mi/h (Figure 9-6f). Let $\vec{i}$ and $\vec{j}$ be **unit vectors** in the horizontal and vertical directions, respectively. This means that each vector has magnitude 1 mi/h.

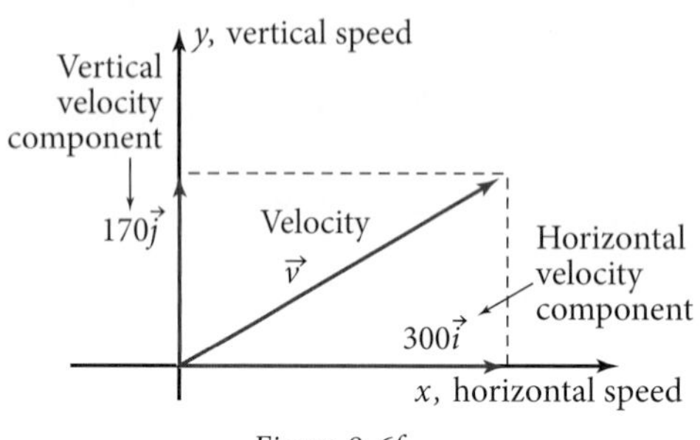

Figure 9-6f

You can write the resultant velocity vector, $\vec{v}$, as the sum

$$\vec{v} = 300\vec{i} + 170\vec{j}$$

The $300\vec{i}$ and $170\vec{j}$ are called the horizontal and vertical **components** of $\vec{v}$. The product of a scalar, 300, and the unit vector $\vec{i}$ is a vector in the same direction as the unit vector but 300 times as long.

EXAMPLE 3 ➤ Vector $\vec{a}$ has magnitude 3 and direction 143° from the horizontal (Figure 9-6g). Resolve $\vec{a}$ into horizontal and vertical components.

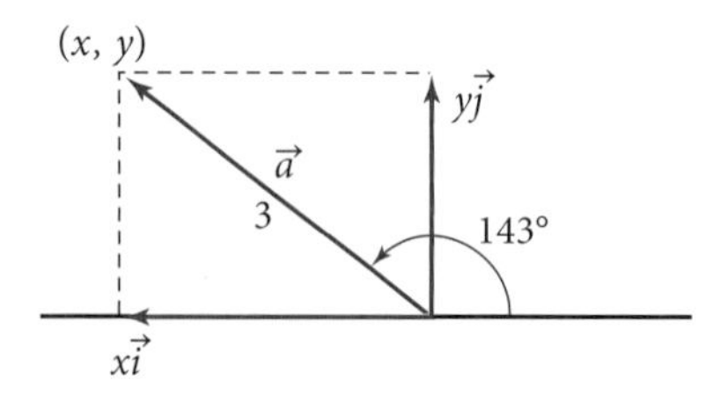

Figure 9-6g

SOLUTION Let (x, y) be the point at the head of $\vec{a}$. Using a reference triangle for 143°,

$$\frac{x}{3} = \cos 143° \qquad \text{and} \qquad \frac{y}{3} = \sin 143°$$

$$\therefore x = 3\cos 143° = -2.3959... \qquad \text{and} \qquad y = 3\sin 143° = 1.8054...$$

$$\therefore \vec{a} \approx -2.396\vec{i} + 1.805\vec{j}$$

Note that multiplying a vector by a *negative* number, such as $-2.396\vec{i}$ in Example 3, gives a vector that points in the *opposite* direction.

Example 3 demonstrates the following property.

> **PROPERTY: Components of a Vector**
>
> If $\vec{v}$ is a vector in the direction θ in standard position, then
>
> $$\vec{v} = x\vec{i} + y\vec{j}$$
>
> where $x = |\vec{v}|\cos\theta$ and $y = |\vec{v}|\sin\theta$.

Components make it easy to add two vectors. As shown in Figure 9-6h, if $\vec{r}$ is the resultant vector of $\vec{a}$ and $\vec{b}$, then the components of $\vec{r}$ are the sums of the components of $\vec{a}$ and $\vec{b}$. Because the two horizontal components have the same direction, you can add them simply by adding their coefficients. The same is true for the vertical components.

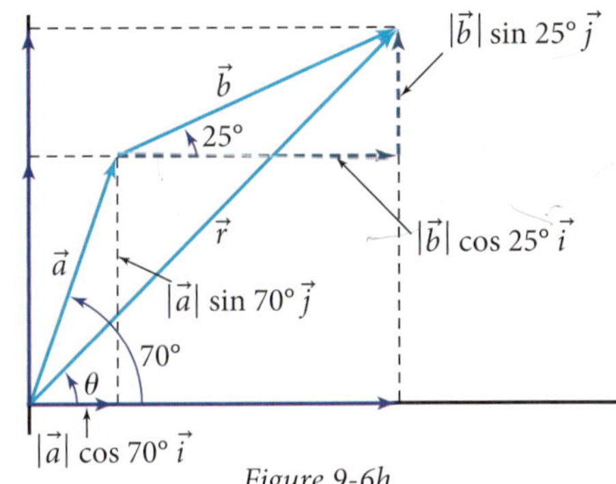

Figure 9-6h

EXAMPLE 4 ➤ Vector $\vec{a}$ has magnitude 5 at 70°, and $\vec{b}$ has magnitude 6 at 25° (Figure 9-6h). Find the resultant vector, $\vec{r}$, as

a. The sum of two components

b. A magnitude and a direction angle

SOLUTION

a. $\vec{r} = \vec{a} + \vec{b}$

$= (5\cos 70°)\vec{i} + (5\sin 70°)\vec{j} + (6\cos 25°)\vec{i} + (6\sin 25°)\vec{j}$ — Write the components.

$= (5\cos 70° + 6\cos 25°)\vec{i} + (5\sin 70° + 6\sin 25°)\vec{j}$ — Combine like terms.

$= 7.1479...\vec{i} + 7.2341...\vec{j}$

$\approx 7.15\vec{i} + 7.23\vec{j}$ — Round the *final* answer.

b. $|\vec{r}| = \sqrt{(7.1479...)^2 + (7.2341...)^2} = 10.1698...$ — By the Pythagorean theorem.

$\theta = \arctan\dfrac{7.2341...}{7.1479...} = 45.3435...° + 180°n = 45.3435...°$ — Pick $n = 0$.

$\therefore \vec{r} \approx 10.17$ at 45.34° — Round the *final* answer. ◄

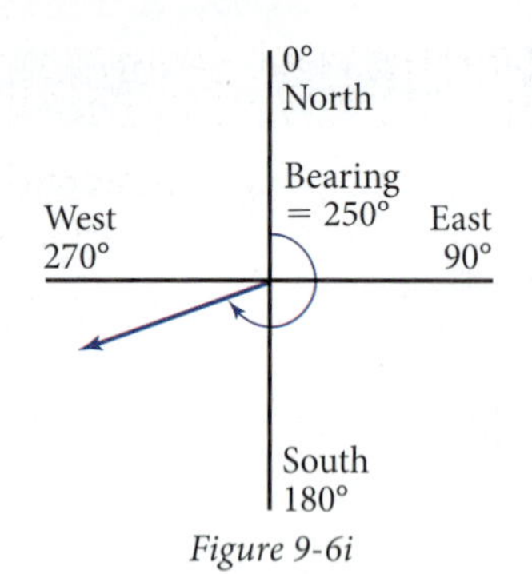

Figure 9-6i

Navigation Problems

A **bearing,** an angle measured clockwise from north, is used universally by navigators for a velocity or a displacement vector. Figure 9-6i shows a bearing of 250°.

These sailors continue the tradition of one of the first seafaring people. Pacific Islanders read the waves and clouds to determine currents and predict weather.

EXAMPLE 5 ➤ Victoria walks 90 m due south (bearing 180°), then turns and walks 40 m more along a bearing of 250° (Figure 9-6j).

a. Find her resultant displacement vector from the starting point.

b. What is the starting point's bearing from the place where Victoria stops?

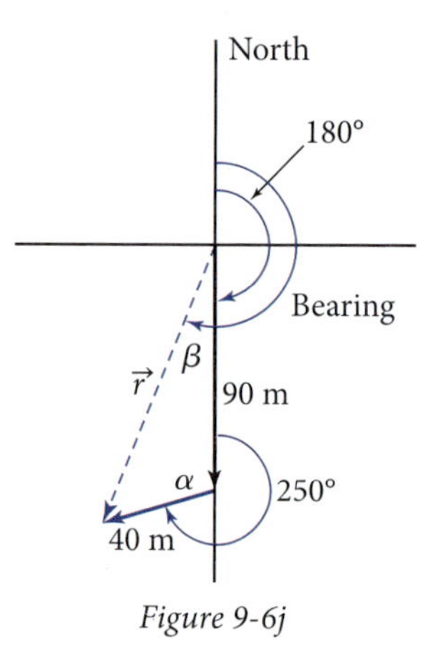

Figure 9-6j

SOLUTION

a. The resultant vector, $\vec{r}$, goes from the beginning of the first vector to the end of the second. Angle α is an angle in the resulting triangle.

$\alpha = 360° - 250° = 110°$

$|\vec{r}|^2 = 90^2 + 40^2 - 2(90)(40)\cos 110° = 12162.5450...$ Use the law of cosines.

$|\vec{r}| = 110.2839...\text{ m}$ Store without rounding.

To find the bearing, first calculate the measure of angle β in the resulting triangle.

$$\cos\beta = \frac{90^2 + (110.2839...)^2 - 40^2}{2(90)(110.2839...)} = 0.9401...$$ Use the law of cosines.

$\beta = 19.9272...°$

Bearing $= 180° + 19.9272...° = 199.9272...°$ See Figure 9-6j.

$\therefore \vec{r} \approx 110.3\text{ m at a bearing of }199.9°$

b. The bearing from the ending point to the starting point is the opposite of the bearing from the starting point to the ending point. To find the opposite, add 180° to the original bearing.

$$\text{Bearing} = 199.9272...° + 180° = 379.9272...°$$

Because this bearing is greater than 360°, find a coterminal angle by subtracting 360°.

$$\text{Bearing} = 379.9272...° \approx 19.9°$$

Problem Set 9-6

Reading Analysis

From what you have read in this section, what do you consider to be the main idea? What is the difference between a vector quantity and a scalar quantity? What is the difference between a vector and a vector quantity? What is meant by the unit vectors $\vec{i}$ and $\vec{j}$, and how can they be used to write the components of a vector in an xy-coordinate system?

Quick Review

Q1. $\cos 90° =$

A. 1 **B.** 0 **C.** -1 **D.** $\frac{1}{2}$ **E.** $\frac{\sqrt{3}}{2}$

Q2. $\tan \frac{\pi}{4} =$

A. 1 **B.** 0 **C.** -1 **D.** $\frac{1}{2}$ **E.** $\frac{\sqrt{3}}{2}$

Q3. In $\triangle FED$, the law of cosines states that $f^2 =$ __?__.

Q4. A triangle has sides 5 ft and 8 ft and included angle 30°. What is the area of the triangle?

Q5. For $\triangle MNO$, $\sin M = 0.12$, $\sin N = 0.3$, and side $m = 24$ cm. How long is side n?

Q6. Finding equations of two sinusoids that are combined to form a graph is called __?__.

Q7. If $\sin \theta = \frac{5}{13}$ and angle θ is in Quadrant II, what is $\cos \theta$?

Q8. If $\theta = \csc^{-1}\left(\frac{11}{7}\right)$, then $\theta = \sin^{-1}$ (__?__).

Q9. The equation $y = 3 \cdot 5^x$ represents a particular __?__ function.

Q10. What transformation is applied to $f(x)$ to get $g(x) = f(3x)$?

For Problems 1–4, translate one vector so that the two vectors are head-to-tail, and then use appropriate triangle trigonometry to find $|\vec{a} + \vec{b}|$ and the angle the resultant vector makes with $\vec{a}$ (Figure 9-6k).

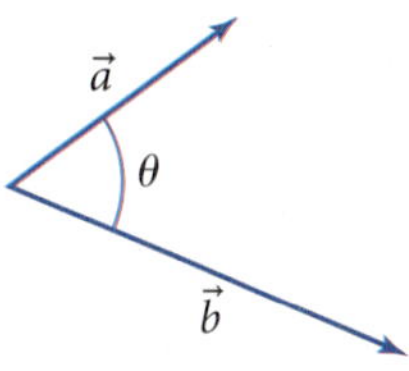

Figure 9-6k

1. $|\vec{a}| = 7$ cm, $|\vec{b}| = 11$ cm, and $\theta = 73°$
2. $|\vec{a}| = 8$ ft, $|\vec{b}| = 2$ ft, and $\theta = 41°$
3. $|\vec{a}| = 9$ in., $|\vec{b}| = 20$ in., and $\theta = 163°$
4. $|\vec{a}| = 10$ mi, $|\vec{b}| = 30$ mi, and $\theta = 122°$
5. *Displacement Vector Problem:* Lucy walks on a bearing of 90° (due east) for 100 m and then on a bearing of 180° (due south) for 180 m.
   a. What is her bearing from the starting point?
   b. What is the starting point's bearing from where she stops?
   c. How far along the bearing in part b must Lucy walk in order to go directly back to the starting point?

6. *Velocity Vector Problem:* A plane flying with an air velocity of 400 mi/h crosses the jet stream, which is blowing at 150 mi/h. The angle between the two velocity vectors is 42° (Figure 9-6l). The plane's actual velocity with respect to the ground is the vector sum of these two velocities.

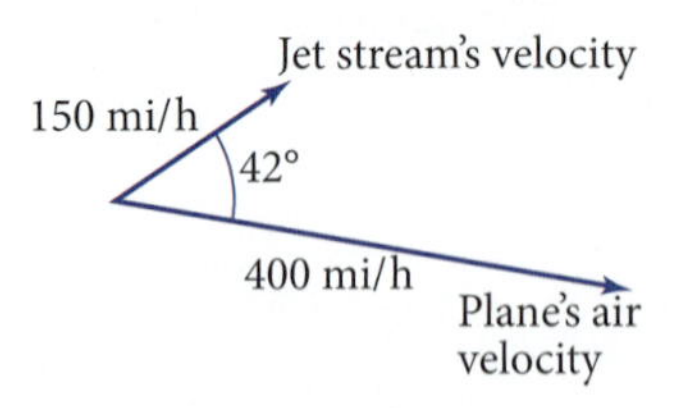

Figure 9-6l

   a. What is the plane's actual velocity with respect to the ground? Why is it *less* than 400 mi/h + 150 mi/h?

   b. What angle does the plane's ground velocity vector make with its 400-mi/h air velocity vector?

7. *Force Vector Problem:* Abe and Bill cooperate to pull a tree stump out of the ground. They think it will take a force of 350 lb to do the job. They tie ropes around the stump. Abe pulls his rope with a force of 200 lb, and Bill pulls his rope with a force of 150 lb. The force vectors make an angle of 40°, as shown in Figure 9-6m.

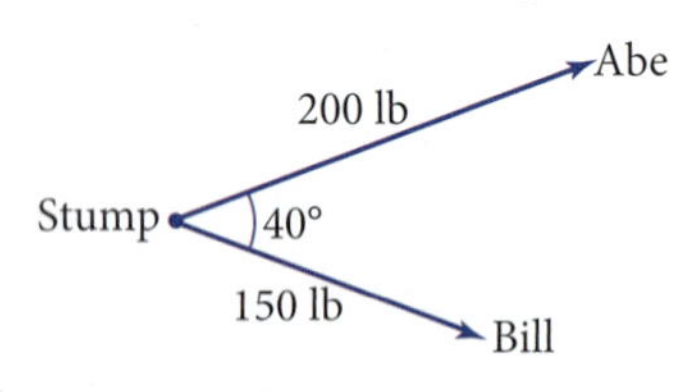

Figure 9-6m

   a. Find the magnitude of the resultant force vector and the angle the resultant vector makes with Abe's vector.

   b. What false assumption about vectors did Abe and Bill make?

8. *Swimming Problem:* Suppose that you swim across a stream that has a 5-km/h current.

   a. Find your actual velocity vector if you swim perpendicular to the current at 3 km/h.

   b. Find your speed through the water if you swim perpendicular to the current but your resultant velocity makes an angle of 34° with the direction you are heading.

   c. If you swim at 3 km/h, can you make it straight across the stream? Explain.

For Problems 9–12, resolve the vector into horizontal and vertical components.

9.

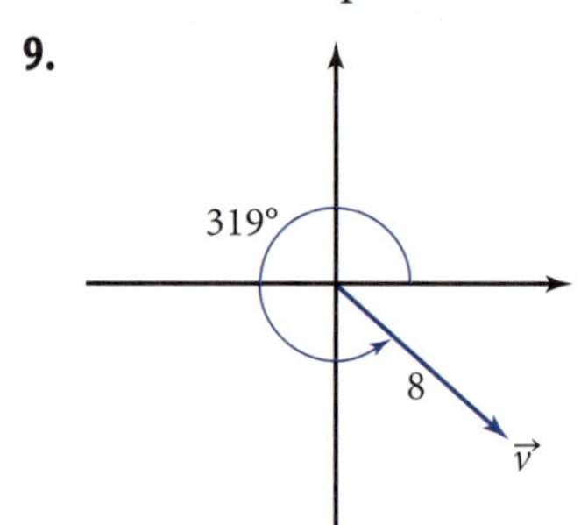

10.

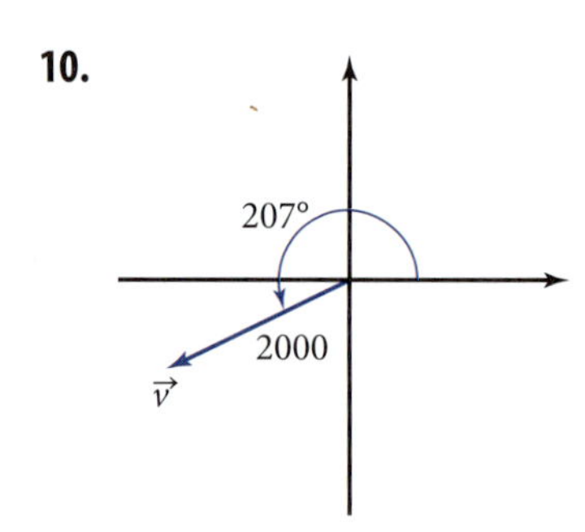

11.

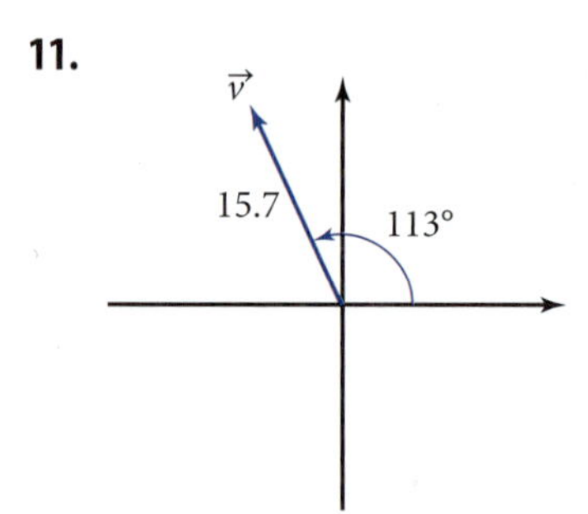

12.

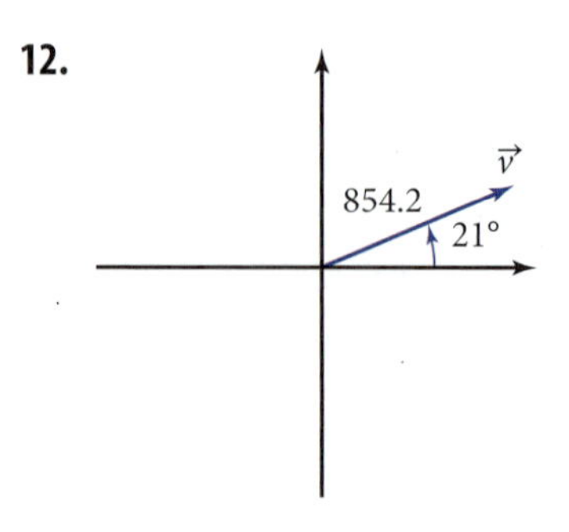

13. *Airplane Vector Components Problem:* A jet plane flying with a velocity of 500 mi/h through the air is climbing at an angle of 35° to the horizontal (Figure 9-6n).

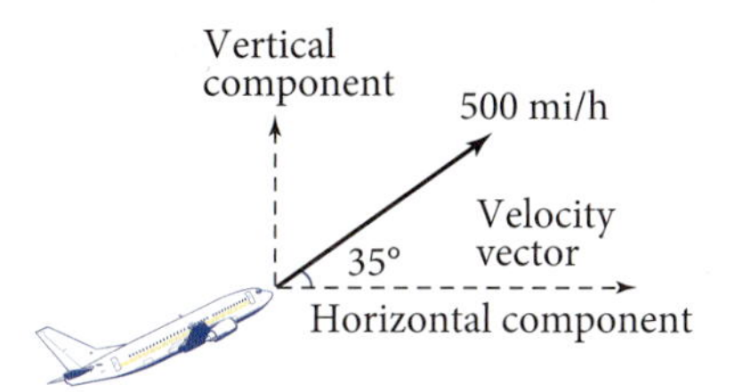

Figure 9-6n

a. The magnitude of the horizontal component of the velocity vector represents the plane's ground speed. Find this ground speed.

b. The magnitude of the vertical component of the velocity vector represents the plane's climb rate. How many feet per second is the plane climbing? (Recall that a mile is 5280 ft.)

14. *Baseball Vector Components Problem:* At time $t = 0$ s, a baseball is hit with a velocity of 150 ft/s at an angle of 25° to the horizontal (Figure 9–7o). At time $t = 3$ s, the ball has slowed to 100 ft/s and is going downward at an angle of 12° to the horizontal.

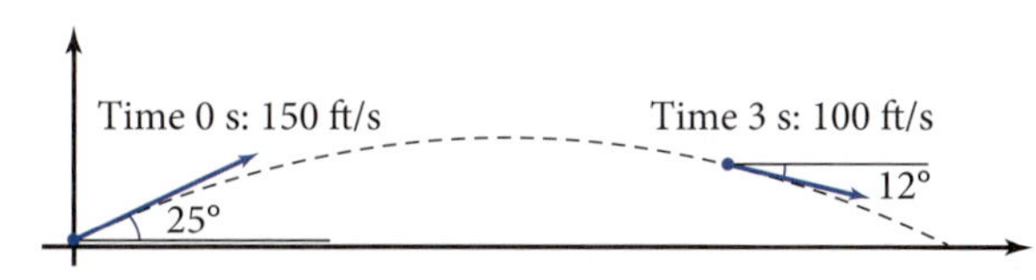

Figure 9-6o

a. Find the magnitudes of the horizontal and vertical components of the velocity vector at time $t = 0$ s. What information do these components give you about the motion of the baseball?

b. How fast is the baseball dropping at time $t = 3$ s? What mathematical quantity reveals this information?

15. If $\vec{r} = 21$ units at 70° and $\vec{s} = 40$ units at 120°, find $\vec{r} + \vec{s}$

a. As a sum of two components

b. As a magnitude and direction

16. If $\vec{u} = 12$ units at 60° and $\vec{v} = 8$ units at 310°, find $\vec{u} + \vec{v}$

a. As a sum of two components

b. As a magnitude and direction

17. A ship sails 50 mi on a bearing of 20° and then turns and sails 30 mi on a bearing of 80°. Find the resultant displacement vector as a distance and a bearing.

18. A plane flies 30 mi on a bearing of 200° and then turns and flies 40 mi on a bearing of 10°. Find the resultant displacement vector as a distance and a bearing.

19. A plane flies 200 mi/h on a bearing of 320°. The air is moving with a wind speed of 60 mi/h on a bearing of 190°. Find the plane's resultant velocity vector (speed and bearing) by adding these two velocity vectors.

20. A scuba diver swims 100 ft/min on a bearing of 170°. The water is moving with a current of 30 ft/min on a bearing of 115°. Find the diver's resultant velocity (speed and bearing) by adding these two velocity vectors.

21. *Spaceship Problem:* A spaceship is moving in the plane of the Sun, the Moon, and Earth. It is being acted upon by three forces (Figure 9-6p). The Sun pulls with a force of 90 newtons at 40°. The Moon pulls with a force of 50 newtons at 110°. Earth pulls with a force of 70 newtons at 230°. What is the resultant force as a sum of two components? What is the magnitude of this force? In what direction will the spaceship move as a result of these forces?

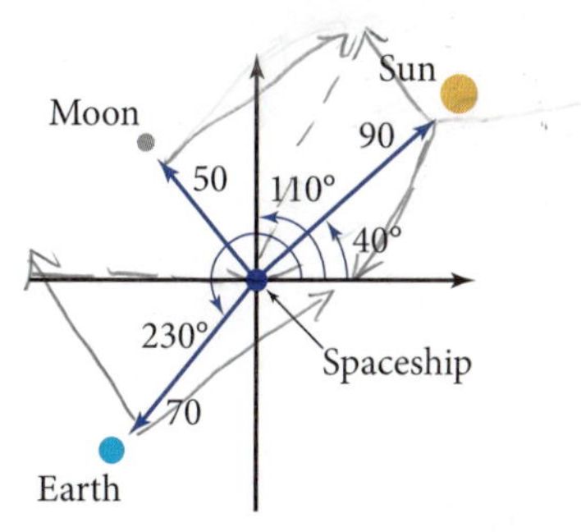

Figure 9-6p

Problems 22–26 refer to vectors $\vec{a}$, $\vec{b}$, and $\vec{c}$ in Figure 9-6q.

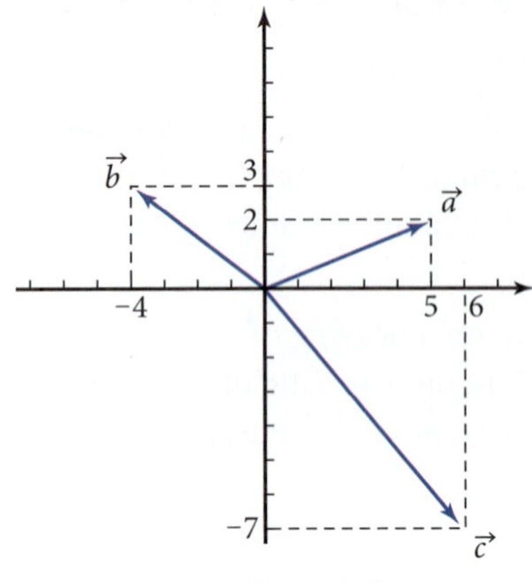

Figure 9-6q

22. *Commutativity Problem:*

a. On graph paper, plot $\vec{a} + \vec{b}$ by translating $\vec{b}$ so that its tail is at the head of $\vec{a}$.

b. On the same axes, plot $\vec{b} + \vec{a}$ by translating $\vec{a}$ so that its tail is at the head of $\vec{b}$.

c. How does your figure show that vector addition is *commutative*?

23. *Associativity Problem:* Show that vector addition is *associative* by plotting on graph paper $(\vec{a} + \vec{b}) + \vec{c}$ and $\vec{a} + (\vec{b} + \vec{c})$.

24. *Zero Vector Problem:* Plot on graph paper the sum $\vec{a} + (-\vec{a})$. What is the magnitude of the resultant vector? Can you assign a direction to the resultant vector? Why is the resultant called the *zero vector*?

25. *Closure Under Addition Problem:* How can you conclude that the set of vectors is *closed* under addition? Why is the existence of the zero vector necessary to ensure closure?

26. *Closure Under Multiplication by a Scalar Problem:* How can you conclude that the set of vectors is closed under multiplication by a scalar? Is the existence of the zero vector necessary to ensure closure in this case?

27. Look up the origin of the word *scalar.* Give the source of your information.

9-7 Real-World Triangle Problems

Previously in this chapter you encountered some real-world triangle problems in connection with learning the law of cosines, the law of sines, the area formula, and Hero's formula. You were able to tell which technique to use by the section of the chapter in which the problem appeared. In this section you will encounter such problems without having those external clues.

Objective

Given a real-world problem, identify a triangle and use the appropriate technique to calculate unknown side lengths and angle measures.

To accomplish this objective, it helps to formulate some conclusions about which method is appropriate for a given set of information. Some of these conclusions are contained in this box.

Surveying instrument

PROCEDURES: Triangle Techniques

Law of Cosines

- Usually you use it to find the length of the third side from two sides and the included angle (SAS).
- You can also use it in reverse to find an angle measure if you know three sides (SSS).
- You can use it to find *both* lengths of the third side in the ambiguous SSA case.
- You *can't* use it if you know only *one* side because it involves all three sides.

Law of Sines

- Usually you use it to find a side length when you know an angle, the opposite side, and another angle (ASA or AAS).
- You can also use it to find an angle measure, but there are *two* values of arcsine between 0° and 180° that could be the answer.
- You *can't* use it for the SSS case because you must know at least one angle.
- You *can't* use it for the SAS case because the side *opposite* the angle is unknown.

Area Formula

- You can use it to find the area from two sides and the included angle (SAS).

Hero's Formula

- You can use it to find the area from three sides (SSS).

Problem Set 9-7

Reading Analysis

From what you have read in this section, what do you consider to be the main idea? Under what condition could you *not* use the law of cosines for a triangle problem? Under what conditions could you *not* use the law of sines for a triangle problem? In each case tell why you couldn't.

Quick Review

Q1. For $\triangle ABC$, write the law of cosines involving angle B.

Q2. For $\triangle ABC$, write the law of sines involving angles A and C.

Q3. For $\triangle ABC$, write the area formula involving angle A.

Q4. Sketch $\triangle XYZ$ given x, y, and angle X, showing how you can draw *two* possible triangles.

Q5. Draw a sketch showing a vector sum.

Q6. Draw a sketch showing the components of $\vec{v}$.

Q7. Write $\vec{a} + \vec{b}$ if $\vec{a} = 4\vec{i} + 7\vec{j}$ and $\vec{b} = -6\vec{i} + 8\vec{j}$.

Q8. $\cos \pi =$

A. 1 **B.** 0 **C.** -1 **D.** $\frac{1}{2}$ **E.** $\frac{\sqrt{3}}{2}$

Q9. By the composite argument properties, $\sin(A - B) =$ __?__.

Q10. What is the phase displacement of $y = 7 + 6\cos 5(\theta + 37°)$ with respect to the parent cosine function?

---

1. *Mountain Height Problem:* A surveying crew has the job of measuring the height of a mountain (Figure 9-7a). From a point on level ground they measure an angle of elevation of 21.6° to the top of the mountain. They move 507 m closer horizontally and find that the angle of elevation is now 35.8°. How high is the mountain? (You might have to calculate some other information along the way!)

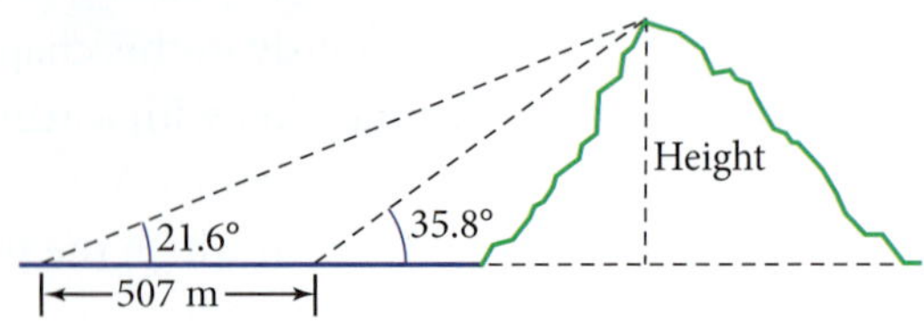

Figure 9-7a

2. *Studio Problem:* A contractor plans to build an artist's studio with a roof that slopes differently on the two sides (Figure 9-7b). On one side, the roof makes an angle of 33° with the horizontal. On the other side, which has a window, the roof makes an angle of 65° with the horizontal. The walls of the studio are planned to be 22 ft apart.

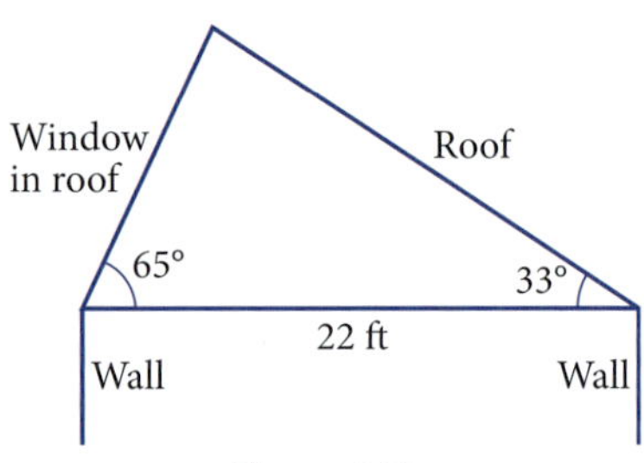

Figure 9-7b

   a. Calculate the lengths of the two parts of the roof.

   b. How many square feet will need to be painted for each triangular end of the roof?

3. *Detour Problem:* Suppose that you are the pilot of an airliner. You find it necessary to detour around a group of thundershowers, as shown in Figure 9-7c. You turn your plane at an angle of 21° to your original path, fly for a while, turn, and then rejoin your original path at an angle of 35°, 70 km from where you left it.

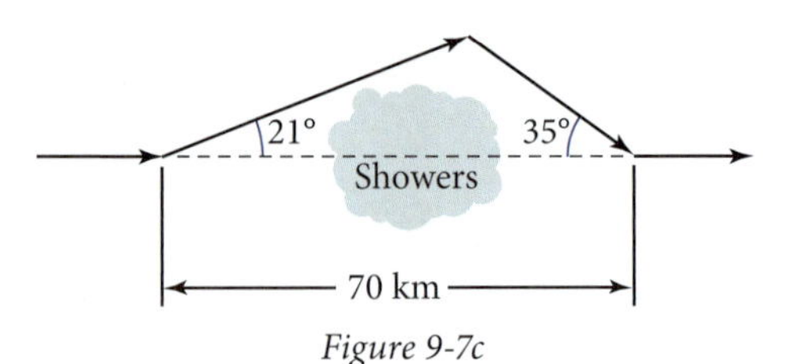

Figure 9-7c

   a. How much farther did you have to fly because of the detour?

b. What is the area of the region enclosed by the triangle?

4. *Pumpkin Sale Problem:* Scorpion Gulch Shelter is having a pumpkin sale for Halloween. The pumpkins will be displayed on a triangular region in the parking lot, with sides 40 ft, 70 ft, and 100 ft. Each pumpkin takes about 3 ft^2 of space.

   a. About how many pumpkins can the shelter display?

   b. Find the measure of the middle-size angle.

5. *Underwater Research Lab Problem:* A ship is sailing on a path that will take it directly over an occupied research lab on the ocean floor. Initially, the lab is 1000 yd from the ship on a line that makes an angle of 6° with the surface (Figure 9-7d). When the ship's slant distance has decreased to 400 yd, the ship can contact people in the lab by underwater telephone. Find the *two* distances from the starting point at which the ship is at a slant distance of 400 yd from the lab.

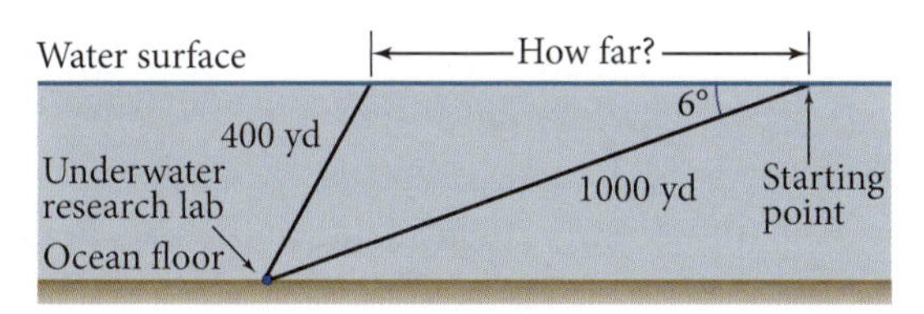

Figure 9-7d

6. *Truss Problem:* A builder has specifications for a triangular truss to hold up a roof. The horizontal side of the triangle will be 30 ft long. An angle at one end of this side will be 50°. The side to be constructed at the other end will be 20 ft long. Use the law of sines to find the angle measure opposite the 30-ft side. Interpret the results.

7. *Rocket Problem:* An observer 2 km from the launchpad observes a rocket ascending vertically. At one instant, the angle of elevation is 21°. Five seconds later, the angle has increased to 35°.

Space shuttle on launchpad at Cape Canaveral, Florida

   a. How far did the rocket travel during the 5-s interval?

   b. Find its average speed during this interval.

   c. If the rocket keeps going vertically at the same average speed, what will be the angle of elevation 15 s after the *first* sighting?

8. *Grand Piano Problem:* The lid on a grand piano is held open by a 28-in. prop. The base of the prop is 55 in. from the lid's hinges, as shown in Figure 9-7e. At what possible distances along the lid could you place the end of the prop so that the lid makes a 26° angle with the piano?

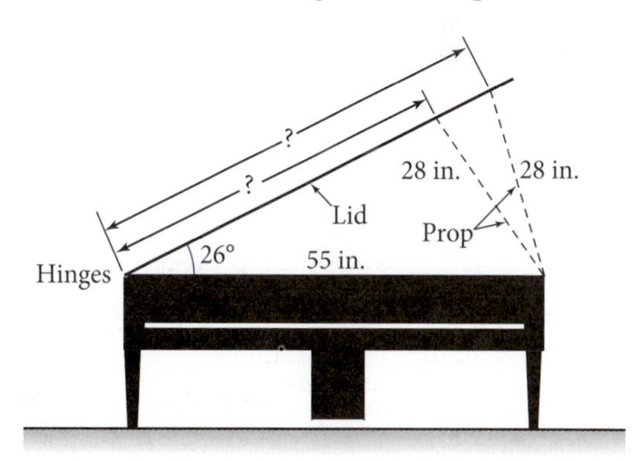

Figure 9-7e

9. *Airplane Velocity Problem:* A plane is flying through the air at a speed of 500 km/h. At the same time, the air is moving at 40 km/h with respect to the ground at an angle of 23° with the plane's path. The plane's ground speed is the magnitude of the vector sum of the plane's air velocity and the wind velocity. Find the plane's ground speed if it is flying

a. Against the wind

b. With the wind

10. *Airplane Lift Problem:* When an airplane is in flight, the air pressure creates a force vector, called the *lift,* that is perpendicular to the wings. When the plane banks for a turn, this lift vector may be resolved into horizontal and vertical components. The vertical component has magnitude equal to the plane's weight (this is what holds the plane up). The horizontal component is a *centripetal* force that makes the plane go on its curved path. Suppose that a jet plane weighing 500,000 lb banks at an angle θ (Figure 9-7f).

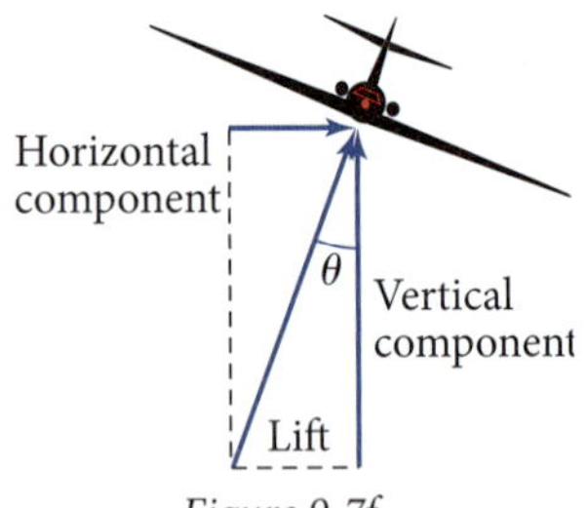

Figure 9-7f

a. Make a table of magnitudes of lift and horizontal component for each 5° from 0° through 30°.

b. Based on your table in part a, why can a plane turn in a *smaller* circle when it banks at a *greater* angle?

c. Why does a plane fly *straight* when it is *not* banking?

d. If the maximum lift the wings can sustain is 600,000 lb, what is the maximum angle at which the plane can bank?

e. What might happen if the plane tried to bank at an angle greater than in part d?

11. *Canal Barge Problem:* In the past, it was common to pull a barge with tow ropes on opposite sides of a canal (Figure 9-7g). Assume that one person exerts a force of 50 lb at an angle of 20° with the direction of the canal. The other person pulls at an angle of 15° with respect to the canal with just enough force so that the resultant vector is directly along the canal. Find the force, in pounds, with which the second person must pull and the magnitude of the resultant force vector.

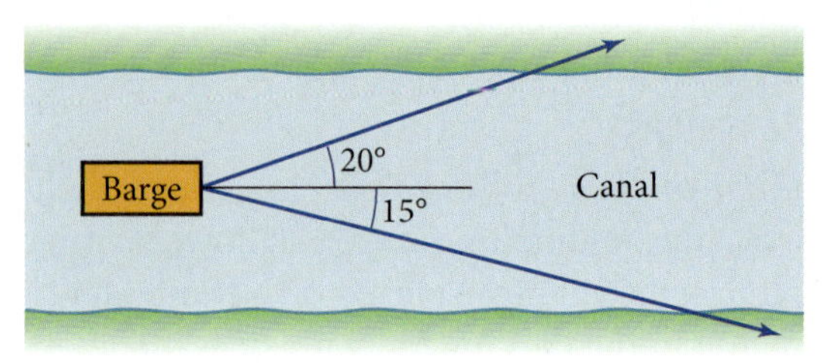

Figure 9-7g

12. WEB *Sailboat Force Vector Problem:* Figure 9-7h represents a sailboat with one sail, set at a 30° angle with the axis of the boat. The wind exerts a force vector of 300 lb that acts on the mast in a direction perpendicular to the sail.

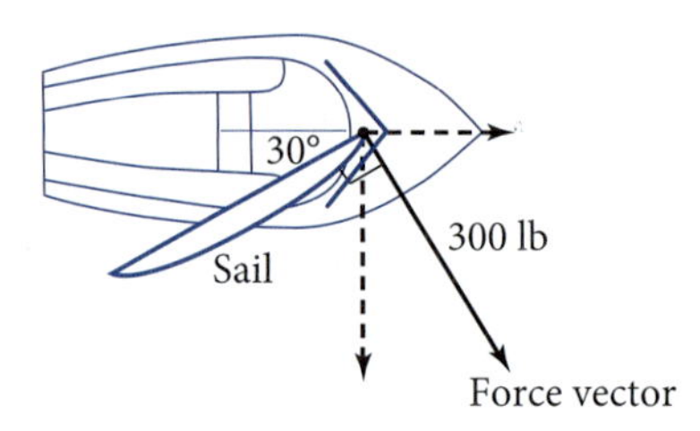

Figure 9-7h

a. Find the absolute value of the component of the force vector along the axis of the boat. (This force makes the boat move forward.)

b. How hard is the wind pushing the boat in the direction perpendicular to the axis of the boat? (The keel minimizes the effect of this force in pushing the boat sideways.)

c. On the Internet or in some other reference source, look up the physics of sailboats to find out why two sails more than double the forward force produced by one sail. Give the source of your information.

Figure 9-7i

13. *Truck on a Hill Problem:* One of the steepest streets in the United States is Marin Street, in Berkeley, California. In some blocks the street makes a 13° angle with the horizontal. Suppose that a truck is parked on such a street (Figure 9-7i). The 40,000-lb weight vector of the truck can be resolved into components perpendicular to the street surface (the *normal component*) and parallel to the street surface.

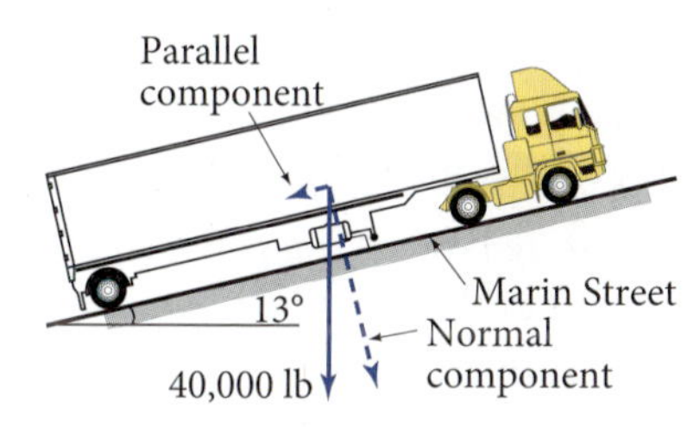

Figure 9-7i

a. Find the magnitude of the normal component. Show that it is not much less than the 40,000-lb weight of the truck.

b. Find the magnitude of the parallel component. Is this surprising? This is the force the brakes must exert to keep the truck from rolling down the hill.

14. WEB *Sliding Friction Force Problem:* Figure 9-7j, left, shows a 100-lb box being pulled across a level floor. Figure 9-7j, right, shows the same box being pulled up a ramp (an *inclined plane,* as physicists call it) that makes a 27° angle with the horizontal. In this problem you will learn how to calculate the force needed to pull the box up the ramp.

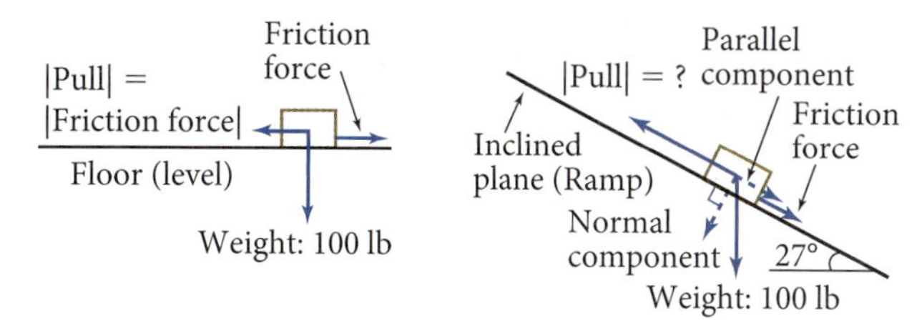

Figure 9-7j

a. To pull the box along the level floor, all you need to do is overcome the force of friction. The magnitude of this friction force is directly proportional to the magnitude of the force acting perpendicular to the floor (the *normal force*), in this case the weight of the box. Suppose that it takes a 60-lb force to pull a 100-lb box across the floor. Let x be the magnitude of the friction force, and let y be the magnitude of the normal force. Write the particular equation expressing y as a function of x. (The proportionality constant in this equation is called the *coefficient of friction.*)

b. When the box is being pulled up the ramp, the normal force is the component of the weight in the direction perpendicular to the ramp (Figure 9-7j, right). Assuming that the coefficient of friction for the ramp is the same as for the floor, use your equation from part a to calculate the magnitude, y, of the force needed to overcome friction for the 100-lb box.

c. The total force needed to move the box up the ramp is the sum of the force needed to overcome friction (part b) and the component of the weight parallel to the ramp. How hard must you pull on the box to move it up the ramp?

d. On the Internet or in another reference source, find the difference between *static* coefficient of friction and *dynamic* coefficient of friction. Give the source of your information.

15. *Hanging Weight Problem 1:* Figure 9-7k shows a 10-lb weight hanging on a string 20 in. long. You pull the weight sideways with a force of magnitude x, in pounds, making the string form an angle θ with the vertical. In this problem you will find the measure of angle θ as a function of how hard you pull and the resulting tension force in the string.

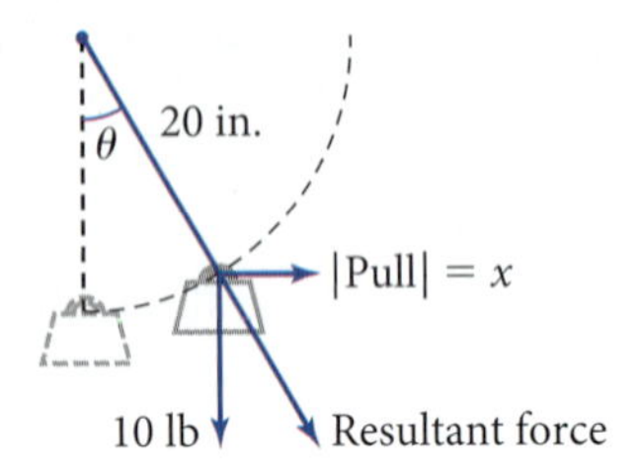

Figure 9-7k

a. The resultant force exerted on the string by the block is the vector sum of the 10-lb weight of the block and the x-lb force, and it acts in the direction of the string. With what force must you pull to make $\theta = 30°$? What will be the tension in the string (the magnitude of the resultant vector)?

b. Write an equation expressing θ as a function of x. Sketch the graph of this function. What happens to the angle measure as x becomes very large?

c. Write another equation expressing the tension in the string as a function of x. Sketch the graph of this function. What happens to this tension as x becomes very large?

16. *Hanging Weight Problem 2:* Figure 9-7l shows an object weighing 50 lb supported by two cables connected to walls 65 ft apart on opposite sides of an alley. Tension vectors $\vec{t}_1$ and $\vec{t}_2$ in the cables make angles of 20° and 40°, respectively, with the horizontal. The resultant vector of these tension vectors is the 50-lb vector pointed straight up, in a direction opposite to the weight vector. In this problem you will calculate the magnitudes of the two tension vectors.

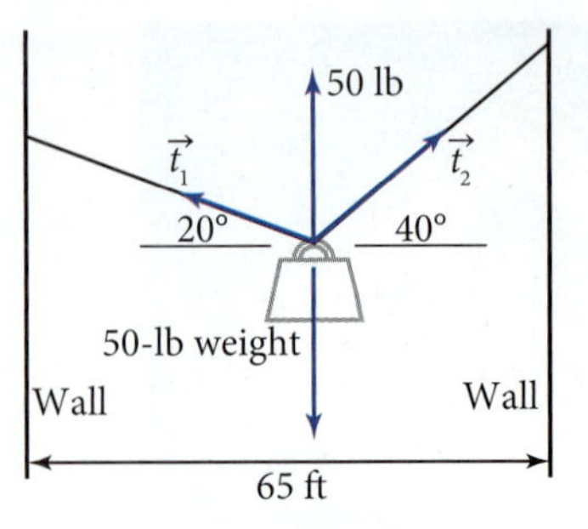

Figure 9-7l

a. The horizontal components of vectors $\vec{t}_1$ and $\vec{t}_2$ have opposite directions but equal magnitudes. (Otherwise the object would move sideways!) Write an equation involving these magnitudes that expresses this fact.

b. The vertical components of $\vec{t}_1$ and $\vec{t}_2$ sum to the upward-pointing 50-lb vector. Write another equation involving the magnitudes of these tension vectors that expresses this fact.

c. Solve the system of equations in parts a and b to find the magnitudes of $\vec{t}_1$ and $\vec{t}_2$. Store the results without rounding.

d. Demonstrate numerically that the magnitudes of the horizontal components of $\vec{t}_1$ and $\vec{t}_2$ are equal and that the magnitudes of the vertical components sum to 50 lb.

e. Which tension vector bears more of the 50-lb weight, the one with the larger angle to the horizontal or the one with the smaller angle?

17. *Hanging Weight by Law of Sines Problem:* Figure 9-7m shows the two tension vectors $\vec{t}_1$ and $\vec{t}_2$ from Figure 9-7l drawn head-to-tail, with the 50-lb sum vector starting at the tail of $\vec{t}_1$ and ending at the head of $\vec{t}_2$. Use the law of sines to find the magnitudes of $\vec{t}_1$ and $\vec{t}_2$.

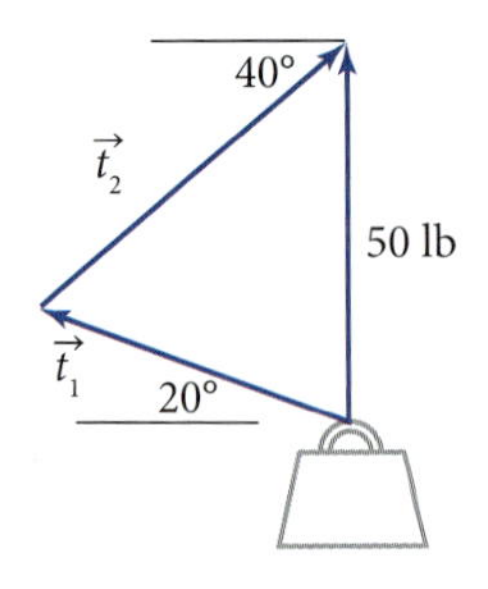

Figure 9-7m

18. *Ship's Velocity Problem:* A ship is sailing through the water in the English Channel with velocity 22 knots on a bearing of 157°, as shown in Figure 9-7n. The current has velocity 5 knots on a bearing of 213°. The actual velocity of the ship is the vector sum of the ship's velocity and the current's velocity. Find the ship's actual velocity.

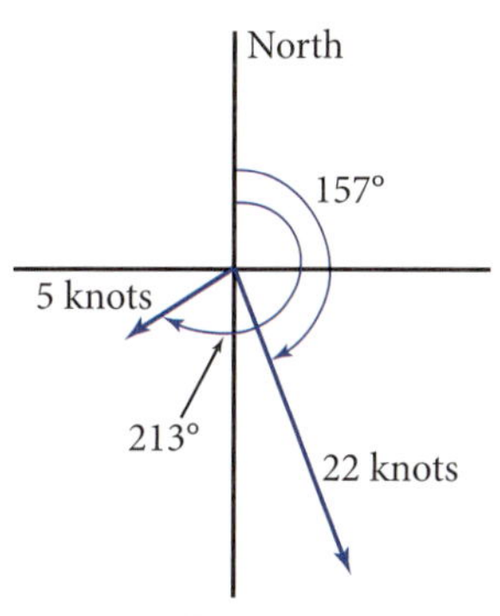

Figure 9-7n

19. *Wind Velocity Problem:* A navigator on an airplane knows that the plane's velocity through the air is 250 km/h on a bearing of 237°. By observing the motion of the plane's shadow across the ground, she finds to her surprise that the plane's ground speed is only 52 km/h and that its direction is along a bearing of 15°. She realizes that the ground velocity is the vector sum of the plane's velocity and the wind velocity. What wind velocity would account for the observed ground velocity?

20. *Space Station Problem:* Ivan is in a space station orbiting Earth. He has the job of observing the motion of two communications satellites.

a. As Ivan approaches the two satellites, he finds that one of them is 8 km away, the other is 11 km away, and the angle between the two (with Ivan at the vertex) is 120°. How far apart are the satellites?

b. A few minutes later, Satellite 1 is 5 km from Ivan and Satellite 2 is 7 km from him. At this time, the two satellites are 10 km apart. At which of the three space vehicles does the largest angle of the resulting triangle occur? What is the measure of this angle? What is the area of the triangle?

c. Several orbits later, only Satellite 1 is visible, while Satellite 2 is near the opposite side of Earth (Figure 9-7o). Ivan determines that the measure of angle A is 37.7°, the measure of angle B is 113°, and the distance between him and Satellite 1 is 4362 km. To the nearest kilometer, how far apart are Ivan and Satellite 2?

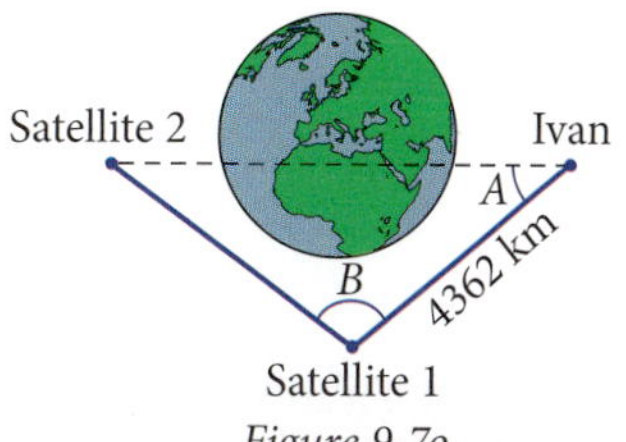

Figure 9-7o

The International Space Station is a joint project of the United States, the Russian Federation, Japan, the European Union, Canada, and Brazil. Construction began in 1998 and continues today through the efforts of astronauts who live aboard the station for many months at a time.

21. *Visibility Problem:* Suppose that you are aboard a plane destined for Hawaii. The pilot announces that your altitude is 10 km. You decide to calculate how far away the horizon is. You draw a sketch as in Figure 9-7p and realize that you must calculate an *arc length.* You recall from geography that the radius of Earth is about 6400 km. How far away is the horizon along Earth's curved surface? Is this surprising?

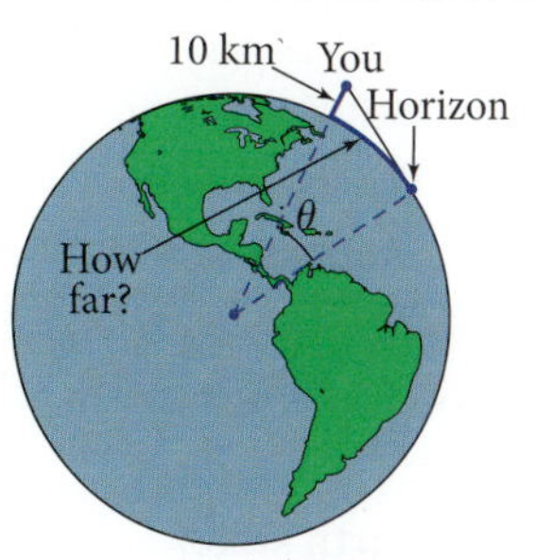

Figure 9-7p

22. *Hinged Rulers Problem:* Figure 9-7q shows a meterstick (100-cm ruler) with a 60-cm ruler attached to one end by a hinge. The other ends of both rulers rest on a horizontal surface. The hinge is pulled upward so that the meterstick makes an angle θ with the surface.

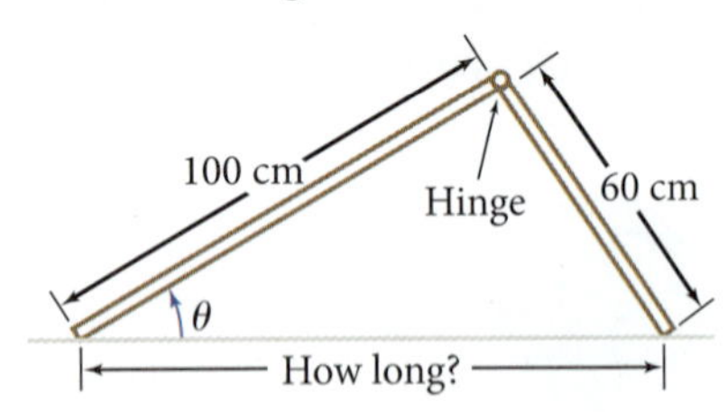

Figure 9-7q

a. Find the two possible distances between the ruler ends if $\theta = 20°$.

b. Show that there is *no* possible triangle if $\theta = 50°$.

c. Find the value of θ that gives just *one* possible distance between the ends.

23. *Surveying Problem 1:* A surveyor measures the three sides of a triangular field and gets lengths 114 m, 165 m, and 257 m.

a. What is the measure of the largest angle of the triangle?

b. What is the area of the field?

24. *Surveying Problem 2:* A field has the shape of a quadrilateral that is *not* a rectangle. Three sides measure 50 m, 60 m, and 70 m, and two angles measure 127° and 132° (Figure 9-7r).

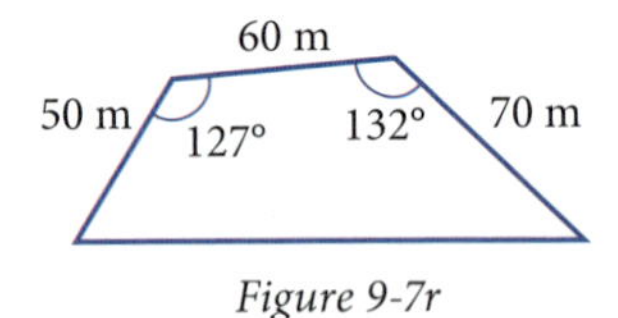

Figure 9-7r

a. By dividing the quadrilateral into two triangles, find its area.

b. Find the length of the fourth side.

c. Find the measures of the other two angles.

25. *Surveying Problem 3:* Surveyors find the area of an irregularly shaped tract of land by taking "field notes." These notes consist of the length of each side and information for finding each angle measure. For this problem, starting at one vertex, the tract is divided into triangles. For the first triangle, two sides and the included angle are known (Figure 9-7s), so you can calculate its area. To calculate the area of the next triangle, you must recognize that one of its sides is also the *third* side of the *first* triangle and that one of its angles is an angle of the polygon (147° in Figure 9-7s) *minus* an angle of the first triangle. By calculating the measures of this side and angle and using the next side of the polygon (15 m in Figure 9-7s), you can calculate the area of the second triangle. The areas of the remaining triangles are calculated in the same manner. The area of the tract is the sum of the areas of the triangles.

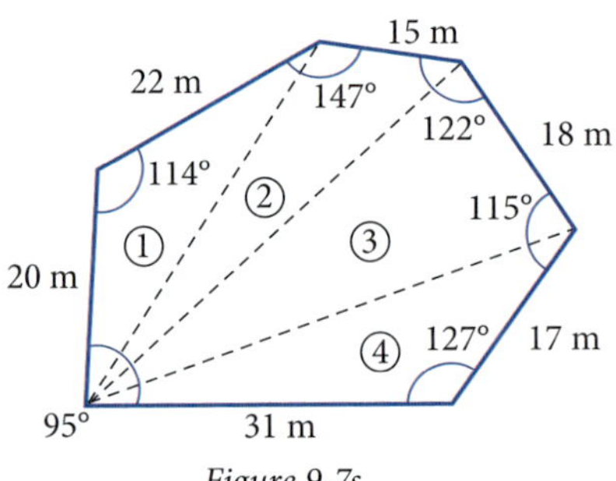

Figure 9-7s

a. Write a program for calculating the area of a tract using the technique described. The input should be the measures of the sides and angles of the polygon. The output should be the area of the tract.

b. Use your program to calculate the area of the tract in Figure 9-7s. If you get approximately 1029.69 m^2, you can assume that your program is working correctly.

c. Show that the last side of the polygon has length 30.6817... m, which is close to the measured value, 31 m.

d. The polygon in Figure 9-7s is a *convex polygon* because none of the angles measure more than 180°. Explain why your program might give wrong answers if the polygon were *not* convex.

9-8 Chapter Review and Test

In this chapter you returned to the analysis of triangles started in Chapter 5. You expanded your knowledge of trigonometry to include oblique triangles as well as right triangles. You learned techniques to find side lengths and angle measures for various sets of given information. These techniques are useful for real-world problems, including analyzing vectors.

Review Problems

R0. Update your journal with things you learned in this chapter. Include topics such as the laws of cosines and sines, the area formulas, how these are derived, and when it is appropriate to use them. Also include how triangle trigonometry is applied to vectors.

R1. Figure 9-8a shows triangles with sides 4 cm and 5 cm, with a varying included angle θ. The length of the third side (dashed) is a function of θ. The five values of θ shown are 30°, 60°, 90°, 120°, and 150°.

a. Measure the length of the third side (dashed) for each triangle.

b. How long would the third side be if the angle were 180°? If it were 0°?

c. If $\theta = 90°$, you can calculate the length of the dashed line by means of the Pythagorean theorem. Does your measured length in part a agree with this calculated length?

d. If y is the length of the dashed line, the law of cosines states that

$$y = \sqrt{5^2 + 4^2 - 2 \cdot 5 \cdot 4 \cdot \cos\theta}$$

Plot the data from parts a and b and this equation for y on the same screen. Do the data seem to fit the law of cosines? Does the graph seem to be part of a sinusoid? Explain.

R2. **a.** Make a sketch of a triangle with sides 50 ft and 30 ft and included angle 153°. Find the length of the third side.

b. Make a sketch of a triangle with sides 8 m, 5 m, and 11 m. Calculate the measure of the largest angle.

c. Suppose you want to construct a triangle with sides 3 cm, 5 cm, and 10 cm. Explain why this is geometrically impossible. Show how computation of an angle using the law of cosines leads to the same conclusion.

d. Sketch $\triangle DEF$ with angle D in standard position in a uv-coordinate system. Find the coordinates of points E and F in terms of sides e and f and angle D. Use the distance formula to prove that you can calculate d using

$$d^2 = e^2 + f^2 - 2ef\cos D$$

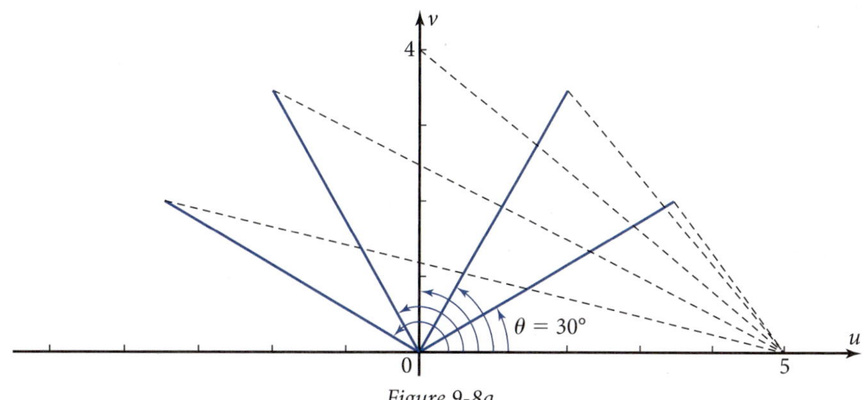

Figure 9-8a

R3. a. Make a sketch of a triangle with sides 50 ft and 30 ft and included angle 153°. Find the area of the triangle.

b. Make a sketch of a triangle with sides 8 mi, 11 mi, and 15 mi. Find the measure of one angle and use it to find the area of the triangle. Calculate the area again using Hero's formula. Show that the results are the same.

c. Suppose that two sides of a triangle have lengths 10 yd and 12 yd and that the area is 40 yd^2. Find the two possible measures of the included angle between these two sides.

d. Sketch $\triangle DEF$ with side d horizontal. Draw the altitude from vertex D to side d. What does this altitude equal in terms of side e and angle F? By appropriate geometry, show that the area of the triangle is

$$\text{Area} = \frac{1}{2}de \sin F$$

R4. a. Make a sketch of a triangle with one side 6 in., the angle opposite that side 39°, and another angle, 48°. Calculate the length of the side opposite the 48° angle.

b. Make a sketch of a triangle with one side 5 m and its two adjacent angles measuring 112° and 38°. Find the length of the longest side of the triangle.

c. Make a sketch of a triangle with one side 7 cm, a second side 5 cm, and the angle opposite the 5-cm side 31°. Find the *two* possible measures of the angle opposite the 7-cm side.

d. Sketch $\triangle DEF$ and show sides d, e, and f. Write the area three ways: in terms of angle D, in terms of angle E, and in terms of angle F. Equate the areas and then perform calculations to derive the three-part equation expressing the law of sines.

R5. Figure 9-8b shows a triangle with sides 5 cm and 8 cm and angles θ and ϕ, not included by these sides.

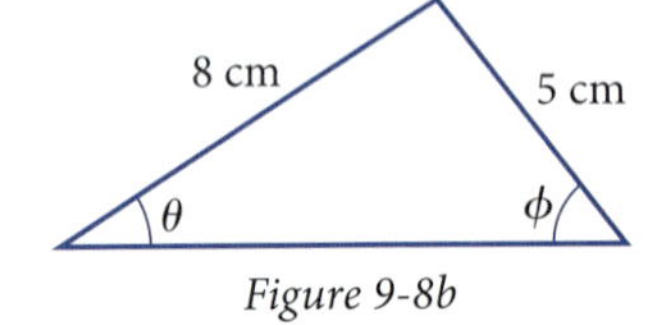

Figure 9-8b

a. If $\theta = 22°$, calculate the *two* possible values of the length of the third side.

b. If $\theta = 85°$, show algebraically that there is *no* possible triangle.

c. Calculate the value of θ for which there is exactly *one* possible triangle.

d. If $\phi = 47°$, calculate the *one* possible length of the third side of the triangle.

R6. a. Vectors $\vec{a}$ and $\vec{b}$ make a 174° angle when placed tail-to-tail (Figure 9-8c). The magnitudes of the vectors are $|\vec{a}| = 6$ and $|\vec{b}| = 10$. Find the magnitude of the resultant vector $\vec{a} + \vec{b}$ and the angle this resultant vector makes with $\vec{a}$ when they are placed tail-to-tail.

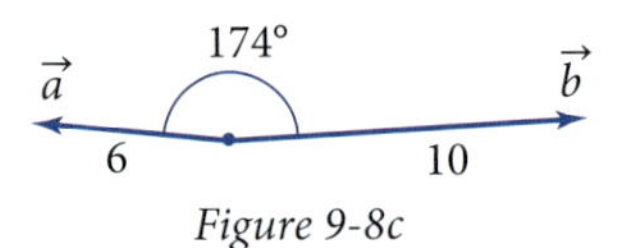

Figure 9-8c

b. Suppose that $\vec{a} = 5\vec{i} + 3\vec{j}$ and $\vec{b} = 7\vec{i} - 6\vec{j}$. Find the resultant vector $\vec{a} + \vec{b}$ as sums of components. Then find the vector again as a magnitude and an angle in standard position.

c. A ship moves west (bearing of 270°) for 120 mi and then turns and moves on a bearing of 130° for another 200 mi. How far is the ship from its starting point? What is the ship's bearing relative to its starting point?

d. A plane flies through the air at 300 km/h on a bearing of 220°. Meanwhile, the air is moving at 60 km/h on a bearing of 115°. Find the plane's resultant ground velocity as a sum of two components, where unit vector $\vec{i}$ points north and $\vec{j}$ points east. Then find the plane's resultant ground speed and the bearing on which it is actually moving.

e. *Calvin's Roof Vector Problem:* Calvin does roof repairs. Figure 9-8d shows him sitting on a roof that makes an angle θ with the horizontal. The parallel component of his 160-lb weight vector acts to pull him down the roof. The frictional force vector counteracts the parallel component with magnitude μ (Greek letter *mu*) times the magnitude of the normal component of the force vector. Here μ is the coefficient of friction, a nonnegative constant which is usually less than or equal to 1. If $\mu = 0.9$ and $\theta = 40°$, will Calvin be able to sit on the roof without sliding? What is the steepest roof Calvin can sit on without sliding? Why could Calvin never be held by friction alone on a roof with $\theta > 45°$?

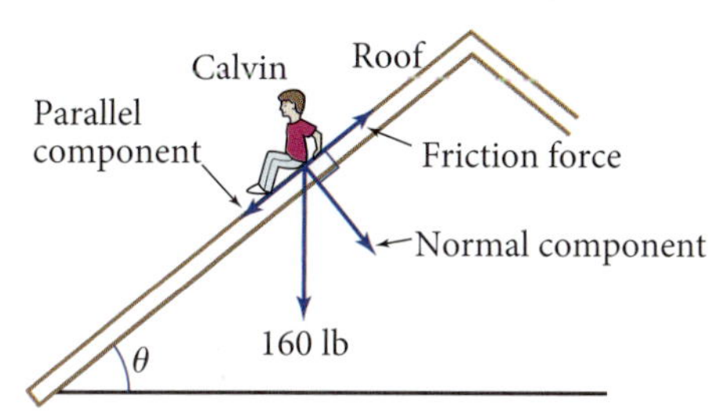

Figure 9-8d

R7. *Airport Problem (parts a–f):* Figure 9-8e shows Nagoya Airport and Tokyo Airport 260 km apart. The ground controllers at Tokyo Airport monitor planes within a 100-km radius of the airport.

a. Plane 1 is 220 km from Nagoya Airport at an angle of 32° to the straight line between the airports. How far is Plane 1 from Tokyo Airport? Is it really out of range of Tokyo Ground Control, as suggested by Figure 9-8e?

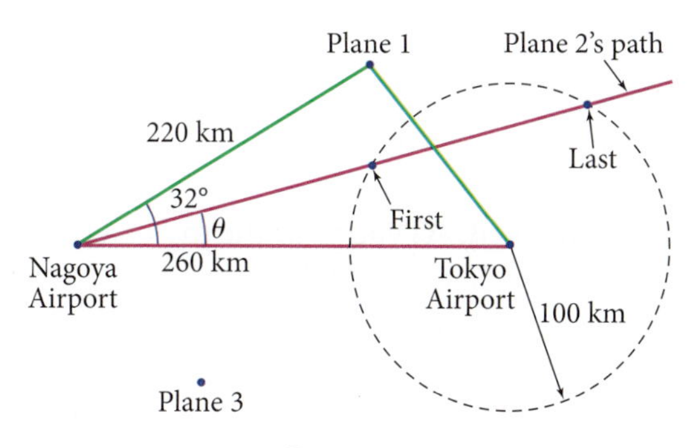

Figure 9-8e

b. Plane 2 is going to take off from Nagoya Airport and fly past Tokyo Airport. Its path will make an angle θ with the line between the airports. If $\theta = 15°$, how far will Plane 2 be from Nagoya Airport when it first comes within range of Tokyo Ground Control? How far from Nagoya Airport is it when it is last within range? Store both of these distances in your calculator, without rounding.

c. Show that if $\theta = 40°$, Plane 2 is never within range of Tokyo Ground Control.

d. Calculate the value of θ for which Plane 2 is within range of Tokyo Ground Control at just *one* point. How far from Nagoya Airport is this point? Store the distance in your calculator, without rounding.

e. Show numerically that the square of the distance in part d is exactly equal to the product of the two distances in part b. What theorem from geometry expresses this result?

f. Plane 3 (Figure 9-8e) reports that it is being forced to land on an island at sea! Nagoya Airport and Tokyo Airport report that the angle measures between Plane 3's position and the line between the airports are 35° and 27°, respectively. Which airport is Plane 3 closer to? How much closer?

Helicopter Problem (parts g–i): The rotor on a helicopter creates an upward force vector (Figure 9-8f). The vertical component of this force (the lift) balances the weight of the helicopter and keeps it in the air. The horizontal component (the thrust) makes the helicopter move forward. Suppose that the helicopter weighs 3000 lb.

g. At what angle will the helicopter have to tilt forward to create a thrust of 400 lb?

h. What will be the magnitude of the total force vector?

i. Explain why the helicopter can hover over the same spot by judicious choice of the tilt angle.

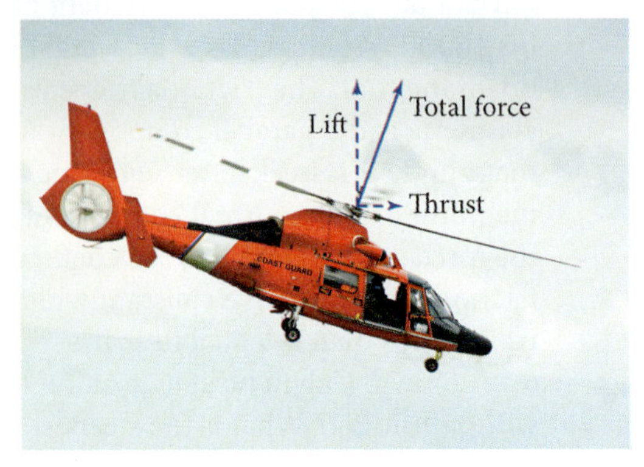

Figure 9-8f

Concept Problems

C1. WEB *Essay Project:* Research the contributions of different cultures to trigonometry. Use these resources or others you might find on the Web or in your local library: Eli Maor, *Trigonometric Delights* (Princeton: Princeton University Press, 1998); David Blatner, *The Joy of* π (New York: Walker Publishing Co., 1997). Write an essay about what you have learned.

C2. *Reflex Angle Problem:* Figure 9-8g shows quadrilateral *ABCD,* in which angle *A* is a *reflex angle* measuring 250°. The resulting figure is called a *nonconvex polygon.* Note that the diagonal from vertex *B* to *D* lies outside the figure.

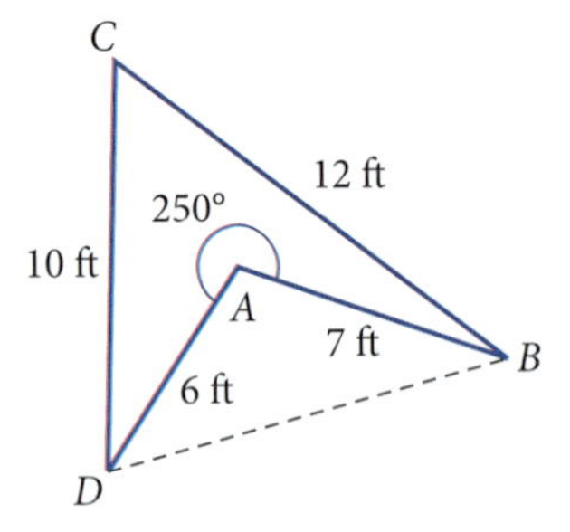

Figure 9-8g

a. Find the measure of angle *A* in $\triangle ABD$. Next, calculate the length *DB* using the side lengths 6 ft and 7 ft shown in Figure 9-8g. Then calculate *DB directly,* using the 250° measure of angle *A*. Do you get the same answer? Explain why or why not.

b. Calculate the area of $\triangle ABD$ using the nonreflex angle you calculated in part a. Then calculate the area of this triangle directly using the 250° measure of angle *A*. Do you get the same answer for the area? Explain why or why not.

c. Use the results in part a to find the area of $\triangle BCD$. Then find the area of quadrilateral *ABCD.* Explain how you can find this area *directly* using the 250° measure of angle *A*.

C3. *Angle of Elevation Experiment:* Construct an *inclinometer* that you can use to measure angles of elevation. One way to do this is to hang a piece of wire, such as a straightened paper clip, from the hole in a protractor, as shown in Figure 9-8h. Then tape a straw to the protractor so that you can sight a distant object more accurately. As you view the top of a building or tree along the straight edge of the protractor, gravity holds the paper clip vertical, allowing you to determine the angle of elevation.

Use your apparatus to measure the height of a tree or building using the techniques of this chapter.

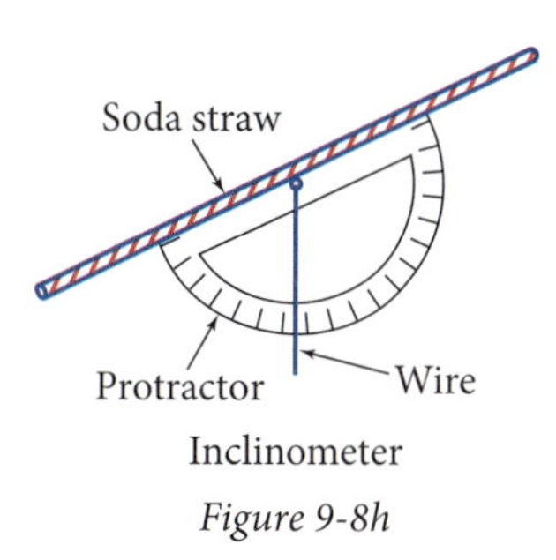

Inclinometer

Figure 9-8h

C4. *Euclid's Problem:* This problem comes from Euclid's *Elements.* Figure 9-8i shows a circle with a secant line and a tangent line.

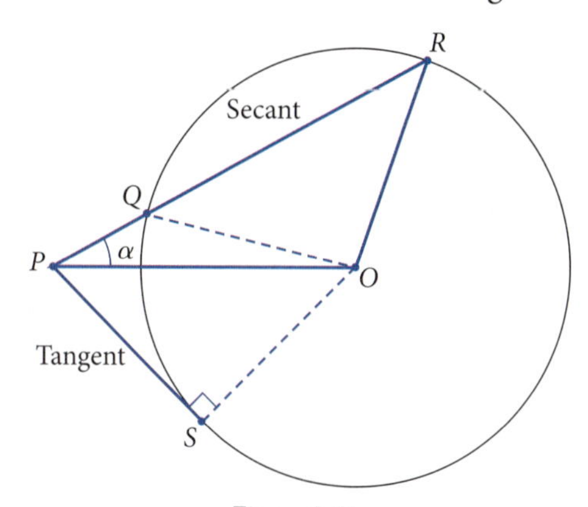

Figure 9-8i

a. Sketch a similar figure using a dynamic geometry program, such as The Geometer's Sketchpad, and measure the lengths of the secant segments, $\overline{PQ}$ and $\overline{PR}$, and the tangent segment $\overline{PS}$. By varying the radius of the circle and the angle QPO, see if it is true that

$$PS^2 = PQ \cdot PR$$

b. Using the trigonometric laws and identities you've learned, prove that the equation in part a is a true statement.

Euclid of Alexandria

C5. *Dot (Scalar) Product of Two Vectors Problem:* Figure 9-8j shows two vectors in standard position:

$$\vec{a} = 3\vec{i} + 4\vec{j}$$
$$\vec{b} = 7\vec{i} + 2\vec{j}$$

The **dot product,** written $\vec{a} \cdot \vec{b}$, is defined to be

$$\vec{a} \cdot \vec{b} = |\vec{a}|\,|\vec{b}| \cos \theta$$

where θ is the angle between the two vectors when they are placed tail-to-tail. Find the measure of the angle between $\vec{a}$ and $\vec{b}$, and store it without rounding. Use the result and the exact lengths of $\vec{a}$ and $\vec{b}$ to calculate $\vec{a} \cdot \vec{b}$. You should find that the answer is an integer! Figure out a way to calculate $\vec{a} \cdot \vec{b}$ using only the coefficients of the unit vectors: 3, 4, 7, and 2. Why do you suppose the dot product is also called the **scalar product** of the two vectors?

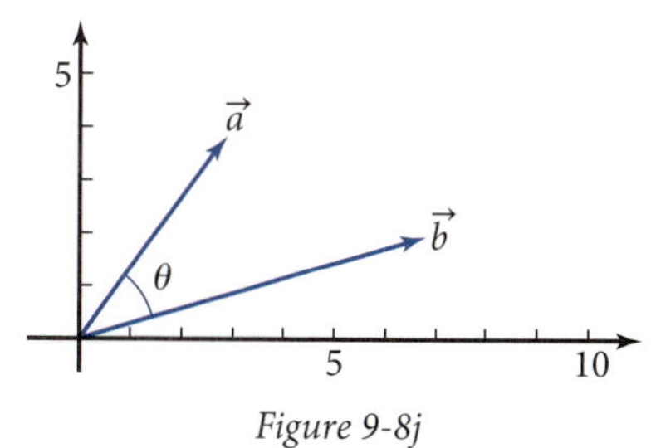

Figure 9-8j

Chapter Test

Part 1: No calculators (T1–T9)

To answer Problems T1–T3, refer to Figure 9-8k.

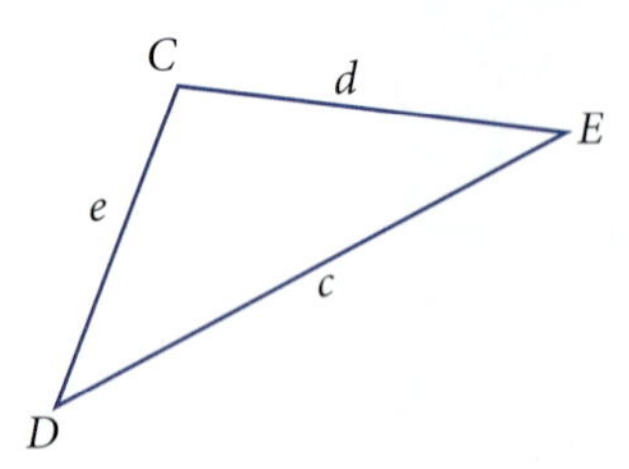

Figure 9-8k

T1. Write the law of cosines involving angle D.

T2. Write the law of sines (either form).

T3. Write the area formula involving sides d and e.

T4. Explain why you cannot use the law of cosines for the triangle in Figure 9-8l.

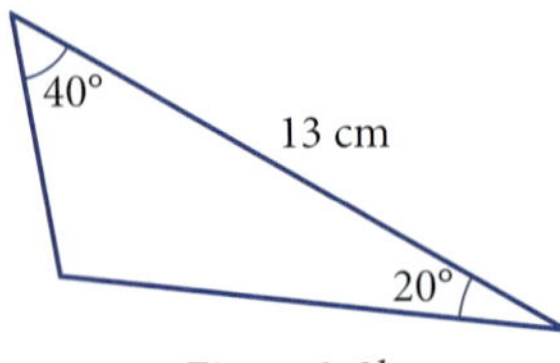

Figure 9-8l

T5. Explain why you cannot use the law of sines for the triangle in Figure 9-8m.

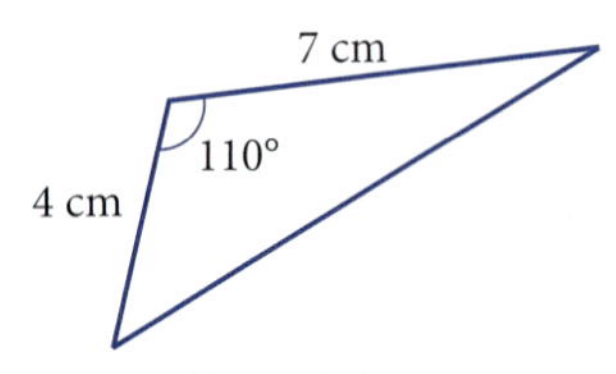

Figure 9-8m

T6. Explain why there is *no* triangle with the side lengths given in Figure 9-8n.

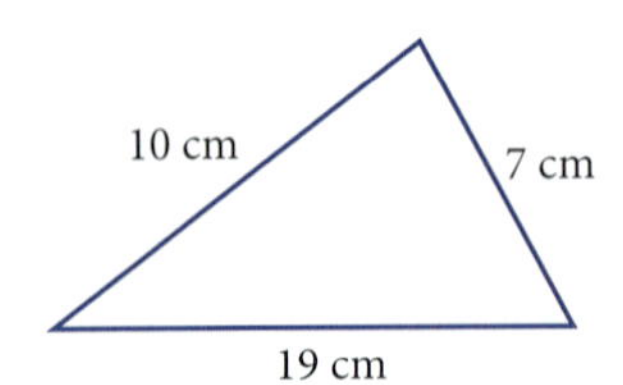

Figure 9-8n

T7. Explain why you can use the inverse cosine *function*, $\cos^{-1}$, when you are finding an angle of a triangle by the law of cosines but must use the inverse sine *relation*, arcsin, when you are finding an angle of a triangle by the law of sines.

T8. Sketch the vector sum $\vec{a} + \vec{b}$ (Figure 9-8o).

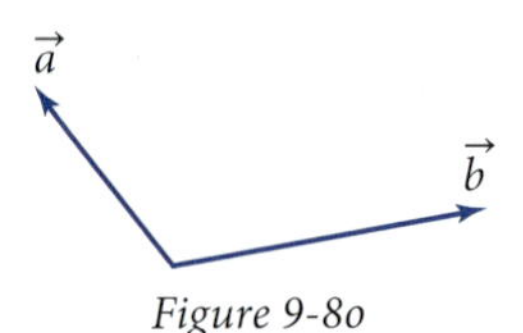

Figure 9-8o

T9. Sketch vector $\vec{v} = 3\vec{i} - 5\vec{j}$ and its components in the x- and y-directions.

Part 2: Graphing calculators are allowed (T10–T22)

T10. Construct a triangle with sides 7 cm and 5 cm and an included angle 24°. Measure the third side.

T11. Calculate the length of the third side in Problem T10. Does the measurement in Problem T10 agree with this calculated value?

T12. Make a sketch of a triangle with base 50 ft and base angles 38° and 47°. Calculate the measure of the third angle.

T13. Calculate the length of the shortest side of the triangle in Problem T12.

T14. Sketch a triangle. Make up lengths for the three sides that give a *possible* triangle. Calculate the measure of the largest angle. Store the answer without rounding.

T15. Find the area of the triangle in Problem T14. Use the angle measure you calculated in Problem T14. Store the answer without rounding.

T16. Use Hero's formula to calculate the area of your triangle in Problem T14. Does it agree with your answer to Problem T15?

T17. Figure 9-8p shows a circle of radius 3 cm. Point P is 5 cm from the center. From point P, a secant line is drawn at an angle of 26° to the line connecting the center to P. Use the law of cosines to calculate the two unknown lengths labeled a and b in the figure.

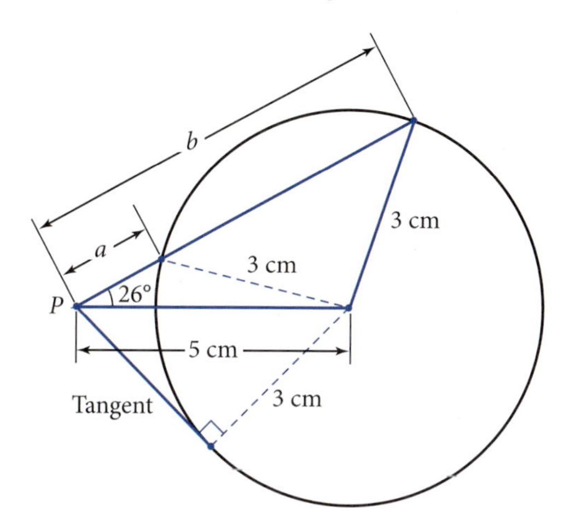

Figure 9-8p

T18. Recall that the radius of a circle drawn to the point of tangency is perpendicular to the tangent. Use this fact to calculate the length of the tangent segment from point P in Figure 9-8p.

T19. Show numerically that the product of the two lengths you found in Problem T17 equals the square of the tangent length you found in Problem T18. This geometrical property appears in Euclid's *Elements.*

T20. For $\vec{v} = 3\vec{i} - 5\vec{j}$, calculate the magnitude. Calculate the direction as an angle in standard position.

T21. *Vector Difference Problem:* Figure 9-8q shows position vectors

$$\vec{a} = 3\vec{i} + 4\vec{j}$$
$$\vec{b} = 7\vec{i} + 2\vec{j}$$

By subtracting components, find the difference vector, $\vec{d} = \vec{a} - \vec{b}$. On a copy of Figure 9-8q, show that $\vec{d}$ is equal to the **displacement vector** from the head of $\vec{b}$ to the head of $\vec{a}$. Explain how this interpretation of a vector difference is analogous to the way you determine how far your car has gone by subtracting the beginning odometer reading from the ending odometer reading.

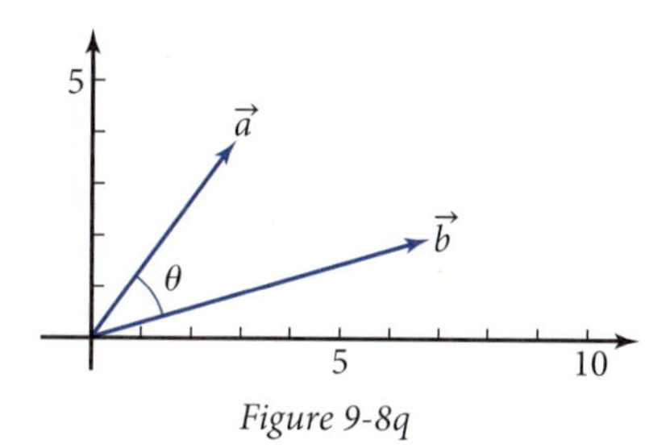

Figure 9-8q

T22. What did you learn as a result of taking this test that you did not know before?

9-9 Cumulative Review, Chapters 5–9

These problems constitute a 2- to 3-hour "rehearsal" for your examination on the trigonometric functions unit, Chapters 5–9. You began by studying **periodic functions.**

Review Problems

1. *Satellite Problem 1:* A satellite is in a circular orbit around Earth. From where you are on Earth's surface, the straight-line distance to the satellite (*through* Earth, at times) is a periodic function of time. Sketch a reasonable graph.

To write equations for periodic functions such as the one in Problem 1, you generalized the **trigonometric functions** from geometry by allowing angles to be negative or greater than 180°.

2. Sketch a $-213°$ angle in standard position. Draw the reference triangle and find the measure of the reference angle.
3. The terminal side of angle θ contains the point $(12, -5)$ in the uv-coordinate system. Write the *exact* values (no decimals) of the six trigonometric functions of θ.
4. Write the exact value (no decimals) of $\sin 240°$.
5. Draw 180° in standard position. Explain why $\cos 180° = -1$.

If θ is allowed to take on any real number of degrees, the trigonometric functions become periodic functions of θ.

6. Sketch the graph of the parent sine function, $y = \sin \theta$.
7. What special name is given to the kind of periodic function you graphed in Problem 6?

Periodic functions such as the one in Problem 1 have independent variables that can be time or distance, not an angle measure. So you learned about **circular functions** whose independent variable is x, not θ. The **radian** is the link between trigonometric functions and circular functions.

8. How many radians are in 360°? 180°? 90°? 45°?
9. How many degrees are in 2 radians?
10. Sketch a graph showing the unit circle centered at the origin of a uv-coordinate system. Sketch an x-axis tangent to the circle, going vertically through the point $(u, v) = (1, 0)$. If the x-axis is wrapped around the unit circle, show that the point $(2, 0)$ on the x-axis corresponds to angle measure 2 radians.
11. Sketch the graph of the parent circular sinusoidal function $y = \cos x$.

Translation and dilation transformations also apply to circular function sinusoids.

12. For $y = 3 + 4\cos 5(x + 6)$, find
    a. The horizontal dilation
    b. The vertical dilation
    c. The horizontal translation
    d. The vertical translation
13. For sinusoids, list the special names given to
    a. The horizontal dilation
    b. The vertical dilation
    c. The horizontal translation
    d. The vertical translation

To use sinusoids as **mathematical models,** you learned to write a particular equation from the graph.

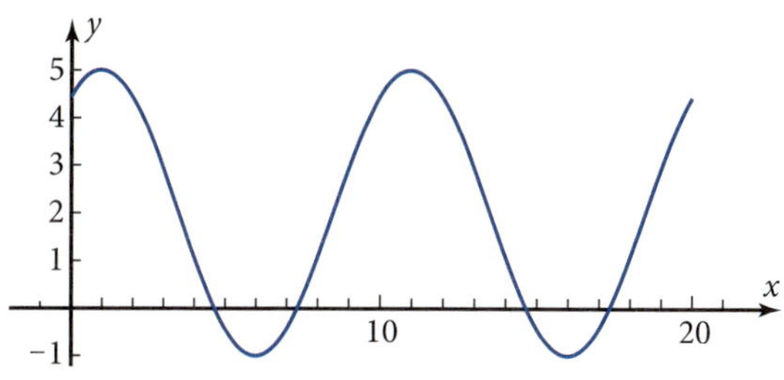

Figure 9-9a

14. Write the particular equation for the sinusoid in Figure 9-9a.

15. If the graph in Problem 14 were plotted on a wide-enough domain, predict y for $x = 342.7$.

16. For the sinusoid in Problem 14, find algebraically the first three positive values of x if $y = 4$.

17. Show graphically that the three values you found in Problem 16 are correct.

18. *Satellite Problem 2:* Assume that in Problem 1, the satellite's distance varies sinusoidally with time. Suppose that the satellite is closest, 1000 mi from you, at time $t = 0$ min. Half a period later, at $t = 50$ min, it is at its maximum distance from you, 9000 mi. Write a particular equation for distance, in thousands of miles, as a function of time.

Radians gave you a convenient way to analyze the motion of two or more rotating objects.

19. Figure 9-9b shows a 5-cm-radius gear on a machine tool driving a 12-cm-radius gear. The design engineers want the smaller gear's teeth to have linear velocity 120 cm/s.

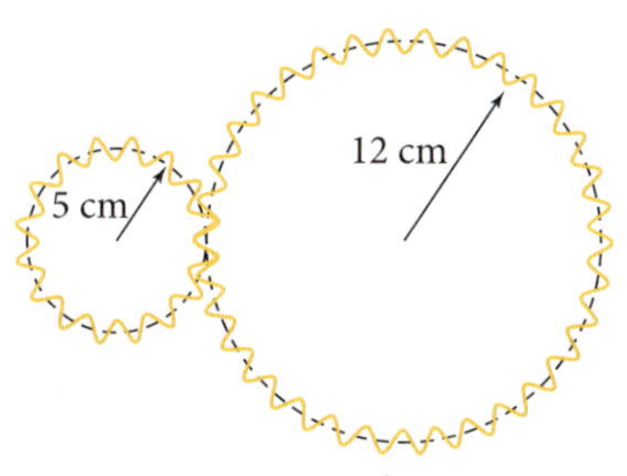

Figure 9-9b

a. What will be the angular velocity of the smaller gear in radians per second? In revolutions per minute?

b. What will be the linear velocity of the larger gear's teeth?

c. At how many revolutions per minute will the larger gear rotate?

20. *Satellite Problem 3:* Figure 9-9c shows the satellite of Problems 1 and 18 in an orbit with radius 5000 mi around Earth. Earth is assumed to have radius 4000 mi. As in Problem 18, assume that it takes 100 min for the satellite to make one complete orbit around Earth.

a. What is the satellite's angular velocity in radians per minute?

b. How fast is it going in miles per hour?

c. What interesting connection do you notice between the angular velocity in part a and the sinusoidal equation in Problem 18?

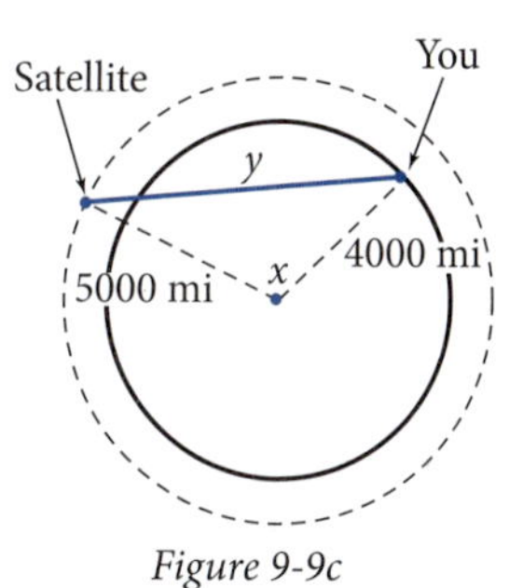

Figure 9-9c

Next you learned some properties of trigonometric and circular functions.

21. There are three kinds of properties that involve just *one* argument. Write the name of each kind of property, and give an example of each.

22. Use the properties in Problem 21 to prove that this equation is an identity. What restrictions are there on the domain of x?

$$\sec^2 x \sin^2 x + \tan^4 x = \frac{\sin^2 x}{\cos^4 x}$$

23. Other properties involve functions of a composite argument. Write the composite argument property for $\cos(x - y)$. Then express this property verbally.

24. Show numerically that $\cos 34° = \sin 56°$.

25. Use the property in Problem 23 to prove that the equation $\cos(90° - \theta) = \sin \theta$ is an identity. How does this explain the result in Problem 24?

The properties can be used to explain why certain combinations of graphs come out the way they do.

26. Show that the function

$$y = 3 \cos \theta + 4 \sin \theta$$

is a sinusoid by finding algebraically the amplitude and phase displacement with respect to $y = \cos \theta$ and writing y as a single sinusoid.

27. The function

$$y = 12 \sin \theta \cos \theta$$

is equivalent to the sinusoid $y = 6 \sin 2\theta$. Prove algebraically that this is true by applying the composite argument property to $\sin 2\theta$.

28. Write the double argument property expressing $\cos 2x$ in terms of $\sin x$ alone. Use this property to show algebraically that the graph of $y = \sin^2 x$ is a sinusoid.

Sums and products of sinusoids with different periods have interesting wave patterns. By using **harmonic analysis,** you can write equations of the two sinusoids that were added or multiplied.

29. Find the particular equation for the function in Figure 9-9d.

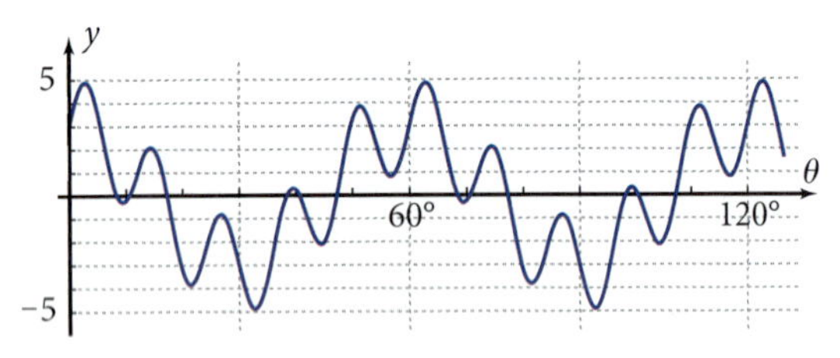

Figure 9-9d

30. Find the particular equation for the function in Figure 9-9e.

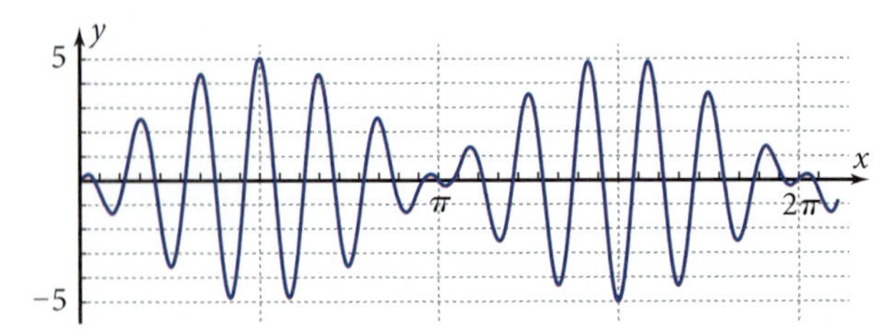

Figure 9-9e

A product of sinusoids with very different periods can be transformed to a sum of sinusoids with nearly equal periods.

31. Transform the function

$$y = 2 \cos 20\theta \cos \theta$$

into a sum of two cosine functions.

32. Find the periods of the two sinusoids in the equation given in Problem 31 and the periods of the two sinusoids in the answer. What can you tell about relative sizes of the periods of the two sinusoids in the given equation and about relative sizes of the periods of the sinusoids in the answer?

Trigonometric and circular functions are periodic, so there are many values of θ or x that give the same value of y. Thus, the inverses of these functions are not functions.

33. Find the (one) value of the inverse trigonometric function $\theta = \tan^{-1} 5$.

34. Find the general solution of the inverse trigonometric relation $x = \arcsin 0.4$.

Parametric functions make it possible to plot the graphs of inverse circular relations.

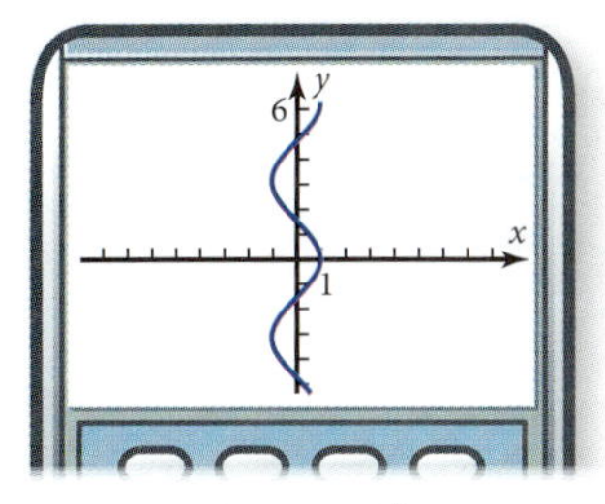

Figure 9-9f

35. Use parametric functions to create the graph of $y = \arccos x$, as shown in Figure 9-9f.

36. The inverse trigonometric function $y = \cos^{-1} x$ is the principal branch of $y = \arccos x$. Define the domain and range of $y = \cos^{-1} x$.

37. Find the first four positive values of θ if $\theta = \arctan 2$.

Last, you studied triangle and vector problems.

38. State the law of cosines.

39. State the law of sines.

40. State the area formula for a triangle given two sides and the included angle.

41. If a triangle has sides 6 ft, 7 ft, and 12 ft, find the measure of the largest angle.

42. Find the area of the triangle in Problem 41 using Hero's formula.

43. Given $\vec{a} = -3\vec{i} + 4\vec{j}$ and $\vec{b} = 5\vec{i} + 12\vec{j}$,

a. Find the resultant vector, $\vec{a} + \vec{b}$, in terms of its components.

b. Find the magnitude and angle in standard position of the resultant vector.

c. Sketch a figure to show $\vec{a} + \vec{b}$ added geometrically, head-to-tail.

d. Is this true or false?

$$|\vec{a} + \vec{b}| = |\vec{a}| + |\vec{b}|$$

Explain why your answer is reasonable.

The triangle properties can be used to show that periodic functions that *look* like sinusoids may not actually *be* sinusoids.

44. *Satellite Problem 4:* In Problem 18, you assumed that the distance between you and the satellite was a sinusoidal function of time. In this problem you will get a more accurate mathematical model.

a. Use the law of cosines and the distances in Figure 9-9g to find y as a function of angle x, in radians.

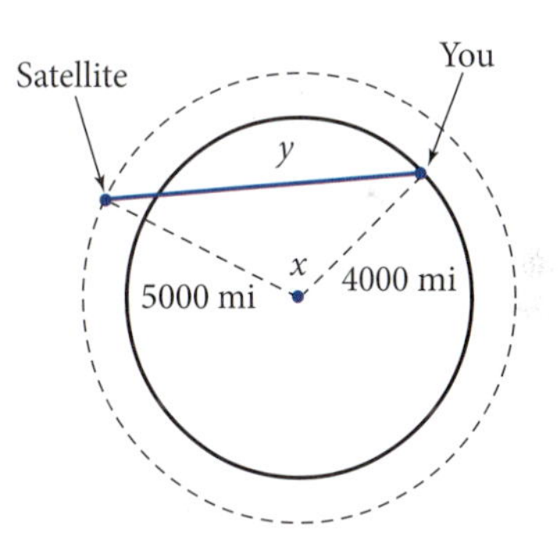

Figure 9-9g

b. Use the fact that it takes 100 min for the satellite to make one orbit to write the equation for y as a function of time t. Assume that $x = 0$ at time $t = 0$ min.

c. Plot the equation from part b and the equation from Problem 18 on the same screen, thus showing that the functions have the same high points, low points, and period but that the equation from part b is *not* a sinusoid.

45. *Three Force Vectors Problem:* Figure 9-9h shows three force vectors, $\vec{a}$, $\vec{b}$, and $\vec{c}$, acting on a point at the origin.

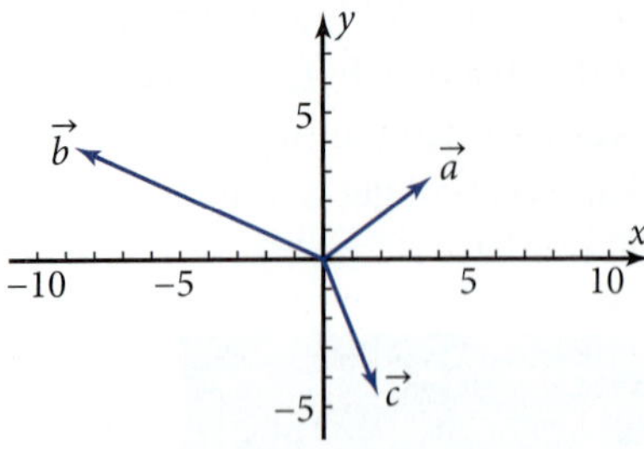

Figure 9-9h

a. Write each of the three vectors in terms of the unit vectors $\vec{i}$ and $\vec{j}$.

b. Find the resultant vector, $\vec{d} = \vec{a} + \vec{b} + \vec{c}$, in terms of the unit vectors $\vec{i}$ and $\vec{j}$.

c. If the forces are measured in newtons, write the resultant force vector as a magnitude and a direction angle.

d. On a copy of Figure 9-9h, draw the three vectors head-to-tail in the order $(\vec{a} + \vec{b}) + \vec{c}$. Show that the resultant vector agrees with your answer to part b. Measure the magnitude and angle of the resultant vector with a ruler and protractor. Show that the results agree with part c.

It is important for you to be able to state *verbally* the things you have learned.

46. What do you consider to be the *one* most important thing you have learned so far as a result of studying precalculus?

Chapter 10

Conic Sections and Quadric Surfaces

Three-dimensional figures that model objects such as the Kobe Port Tower in Japan, a parabolic antenna, and an egg are generated by rotating conic sections—hyperbolas, parabolas, or ellipses—about an axis of symmetry. These figures have rich geometric and algebraic properties. The general name *conic sections* comes from the fact that each figure can be generated geometrically by a plane slicing through (sectioning) a cone. The type of conic section formed is determined by the angle of the plane.

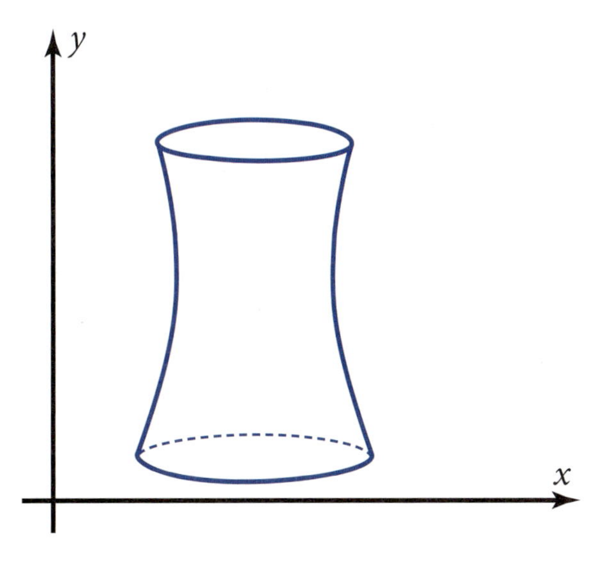

Mathematical Overview

In this chapter you will learn how to apply algebraic techniques to analyze ellipses, hyperbolas, parabolas, and circles, all of which are formed by a plane sectioning a cone. Fixed focal points and directrix lines help define these *conic sections*. The shapes of spaceship and comet paths are conic sections, and reflective surfaces are formed by rotating the conic section shapes about their axes. You will study conic sections in four ways.

GRAPHICALLY

Ellipse

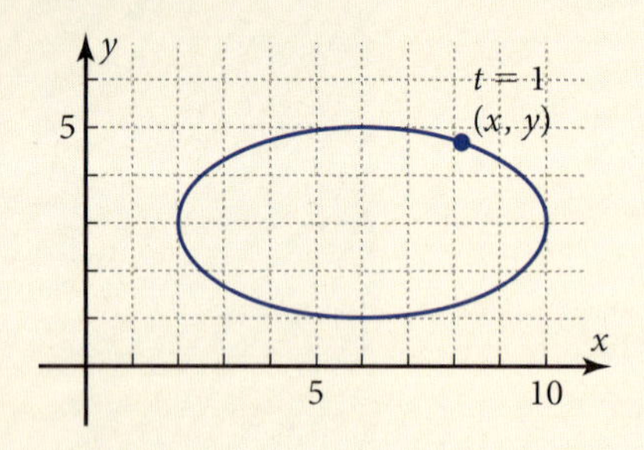

ALGEBRAICALLY

Cartesian equation of an ellipse:

$$x^2 + 4y^2 - 12x - 24y + 56 = 0$$

Parametric equations of an ellipse:

$$x = 6 + 4\cos t$$

$$y = 3 + 2\sin t$$

NUMERICALLY

$t = 1$: $x = 6 + 4\cos 1 = 8.1612...$

$y = 3 + 2\sin 1 = 4.6829...$

VERBALLY

To plot an ellipse on a grapher, either use parametric equations or solve the Cartesian equation $x^2 + 4y^2 - 12x - 24y + 56 = 0$ for y. Both methods generate the same graph. Algebraic techniques can transform the parametric equations to Cartesian form, and vice versa.

10-1 Quadratic Relations and Conic Sections

Figures 10-1a to 10-1d show the graphs of four *quadratic relations*. The equations for these relations can have y^2- and xy-terms as well as x^2-terms. These are not functions because there are x-values that have more than one y-value. These graphs are called **conic sections**—in short, conics—because they are formed by a plane cutting, or sectioning, a cone.

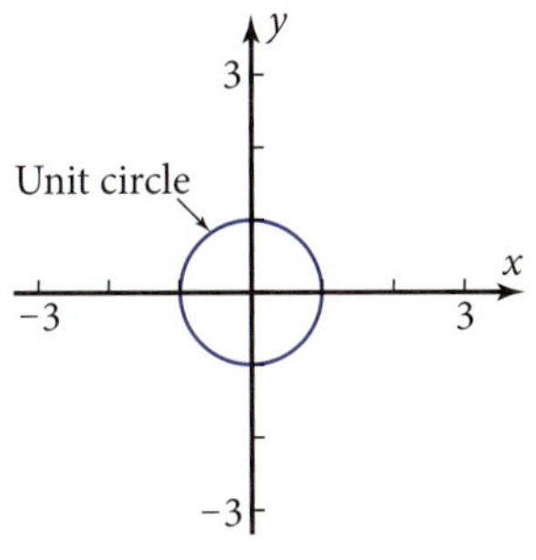

Figure 10-1a

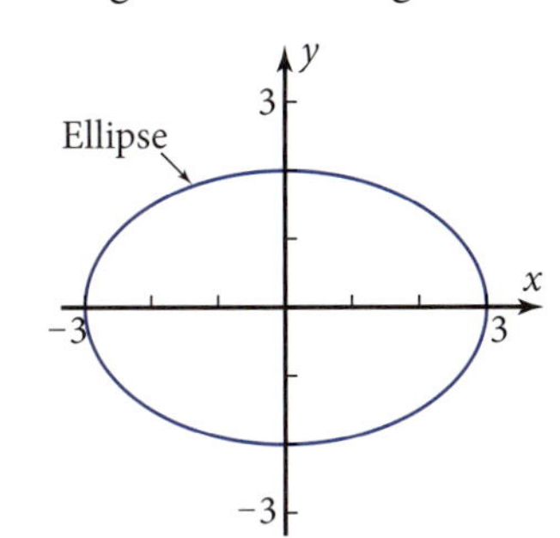

Figure 10-1b

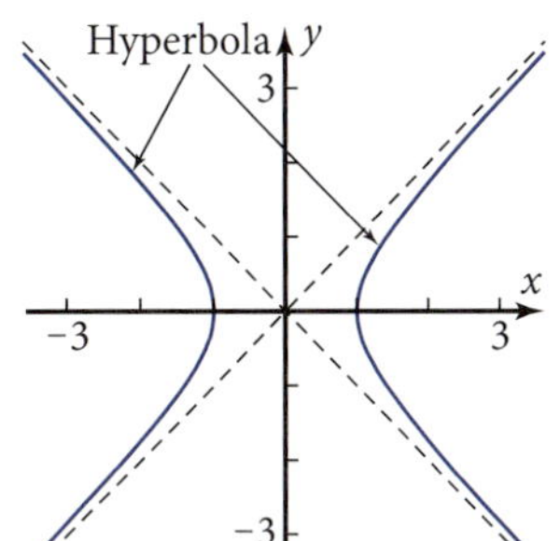

Figure 10-1c

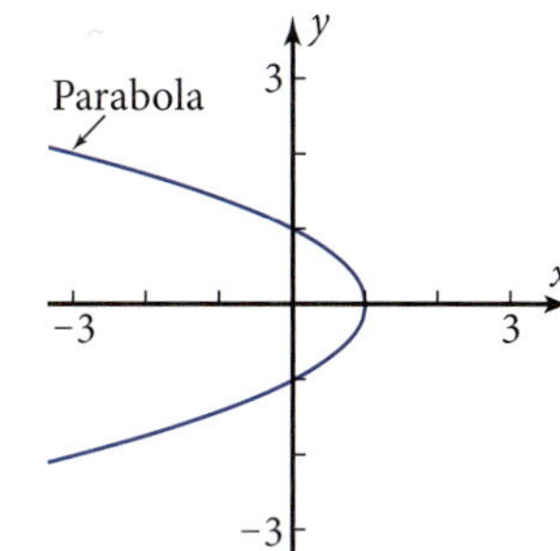

Figure 10-1d

Objective Given a quadratic equation with two variables, plot its graph and formulate conclusions.

Exploratory Problem Set 10-1

1. Plot $x^2 + y^2 = 1$ by solving for y in terms of x. Enter the two solutions as $f_1(x)$ and $f_2(x)$. (One is a positive square root, and the other is a negative square root.) Use a window that has equal scales on the two axes. Based on the Pythagorean theorem, explain why the graph is the unit circle in Figure 10-1a.

2. Plot $4x^2 + 9y^2 = 36$ by first solving for y in terms of x. Show that the result is the ellipse in Figure 10-1b.

3. The ellipse in Problem 2 is a dilation of the unit circle by a factor of 3 units in the x-direction and by a factor of 2 units in the y-direction. By making the right side equal 1, transform the given equation to this equivalent form:

$$\left(\frac{x}{3}\right)^2 + \left(\frac{y}{2}\right)^2 = 1$$

This form is sometimes called the **center-radius form** of the equation of the ellipse. Where do the two dilation factors show up in the transformed equation?

4. Plot $x^2 - y^2 = 1$ by solving for y in terms of x. Show that the result is the **hyperbola** in Figure 10-1c. Plot the two lines $y = x$ and $y = -x$. How are these lines related to the graph?

5. Plot the hyperbola $4x^2 - 9y^2 = 36$. Use a window with $[-10, 10]$ for x and use equal scales on the two axes. Show that the asymptotes now have slopes $\pm\frac{2}{3}$ instead of ± 1.

6. Transform the equation in Problem 5 to make the right side equal 1, as in Problem 3. Show that the hyperbola in Problem 5 is a dilation of the hyperbola in Figure 10-1c with x-dilation 3 units and y-dilation 2 units. Where do these dilation factors appear in the transformed equation?

7. The equation $x + y^2 = 1$ has only one squared term. Solve the equation for y in terms of x and plot the two solutions as $f_1(x)$ and $f_2(x)$. Show that the graph is the parabola in Figure 10-1d.

8. How could you tell from an equation before it is transformed whether its graph will be a circle, an ellipse, a hyperbola, or a parabola?

10-2 Cartesian Equations for Conic Sections

The paths of satellites or comets traveling in space under the action of gravity are circles, ellipses, parabolas, or hyperbolas. Figure 10-2a shows how these conic sections are formed by a plane sectioning one or both *nappes* of a cone at various angles. The graphs of quadratic equations in two variables are conic sections, as you saw in Section 10-1.

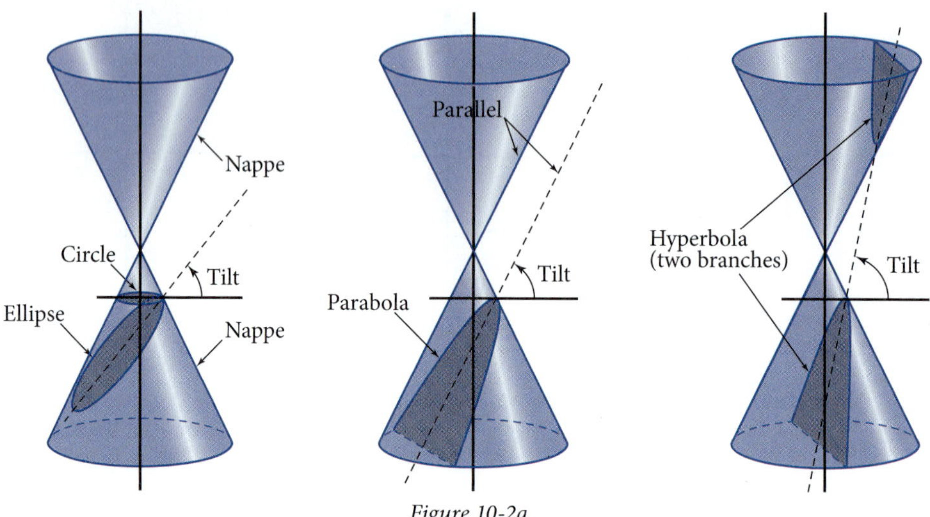

Figure 10-2a

You may have seen shadows cast by a cone of light from a floor lamp that is cut by the planes of walls and ceilings, as in Figure 10-2b. The wall cuts both nappes of the cone, forming a hyperbola. The horizontal ceiling cuts the cone at right angles to its axis, forming a circle. The sloped ceiling can form an ellipse, a parabola, or a hyperbola, depending on the slope.

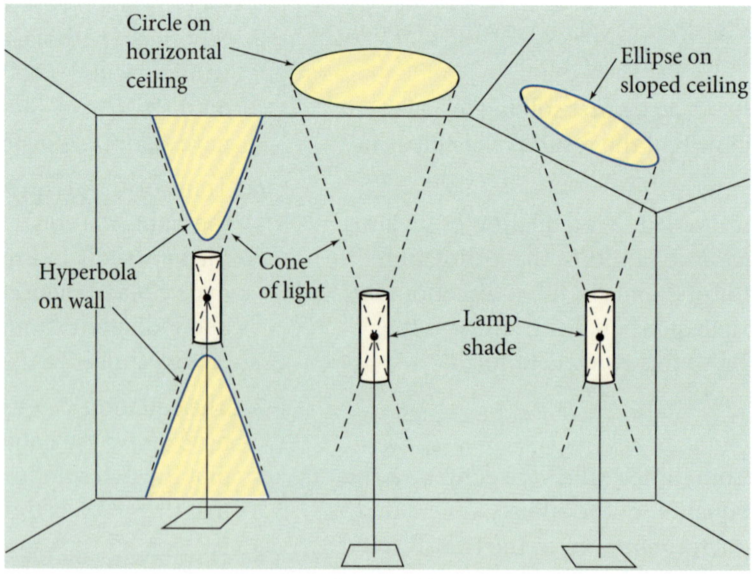

Figure 10-2b

Objective Given a Cartesian equation for a conic section, transform it to other equivalent forms and sketch or plot the graph; given the graph of a conic section, write a Cartesian equation.

Quadratic Relation Equations for Conic-Section Graphs

The general equation for a quadratic relation has terms of second, first, and zero degree:

$$Ax^2 + Bxy + Cy^2 + Dx + Ey + F = 0$$

where A, B, C, D, E, and F are constants. In Section 10-1, you found that by selecting various values of A, B, and C, the graphs of these relations could be circles, ellipses, hyperbolas, or parabolas. The graphs and equations of the parent relations are shown in Figure 10-2c. Note that a hyperbola has two diagonal asymptotes, separating two disconnected branches.

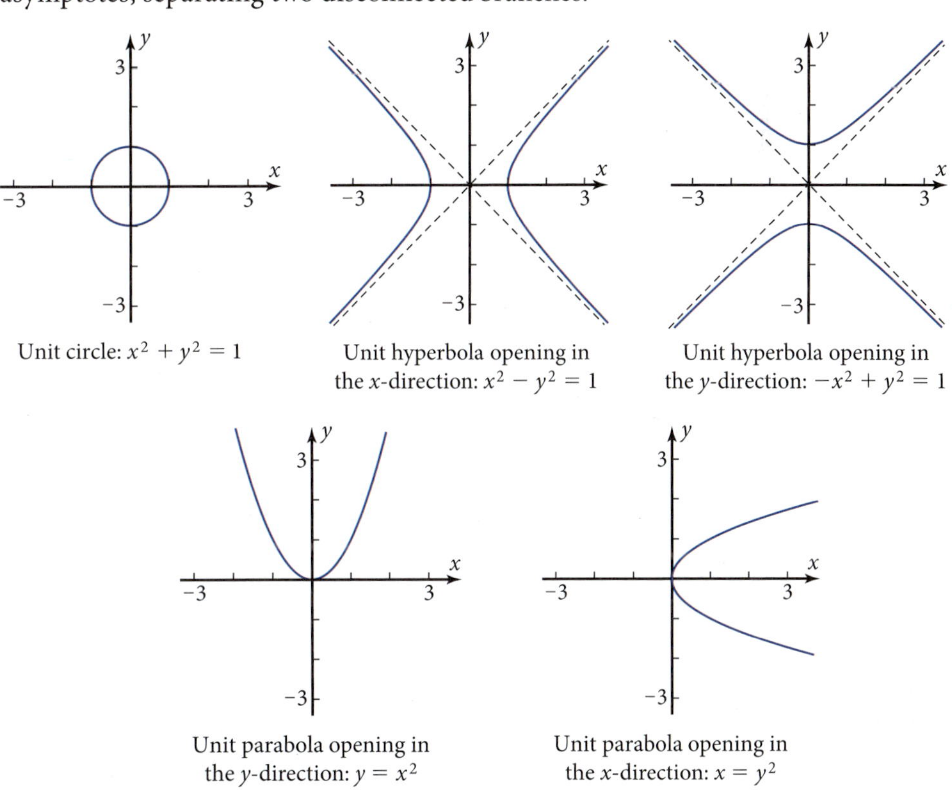

Figure 10-2c

If there is no xy-term ($B = 0$ in the general equation), you can identify the graph from the coefficients of the x^2- and y^2-terms, A and C, respectively, in the general equation. The box on the next page summarizes the properties you discovered in Section 10-1.

PROPERTIES: Recognition of Conic Sections from Equations

If a quadratic equation in two variables has no xy-term, then the graph is

- A circle if x^2 and y^2 have equal nonzero coefficients
- An ellipse if x^2 and y^2 have unequal coefficients but the same sign
- A hyperbola if x^2 and y^2 have opposite signs
- A parabola if only one of the two variables is squared

Transformed Conic Sections

You can form any conic section with its axis of symmetry parallel to the x- or y-axis by dilating and translating its parent graph. (In Section 10-6, you will learn that the xy-term produces conic sections that are *rotated* about the origin.) For instance, the ellipse in Figure 10-2d is formed by dilating the unit circle, $x^2 + y^2 = 1$, by a factor of 2 in the x-direction and by a factor of 3 in the y-direction, giving the ellipse x-radius 2 and y-radius 3.

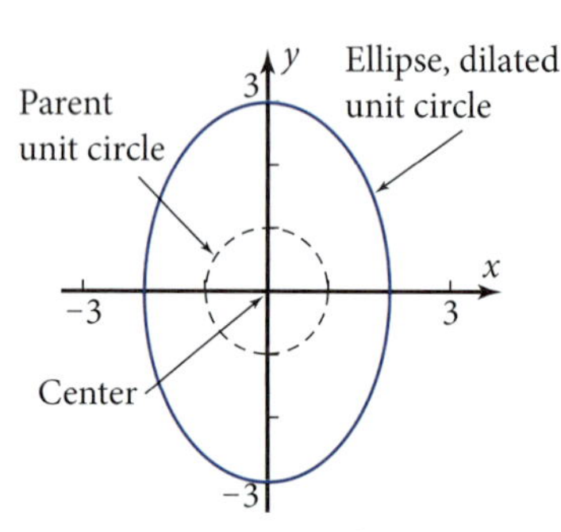

Figure 10-2d

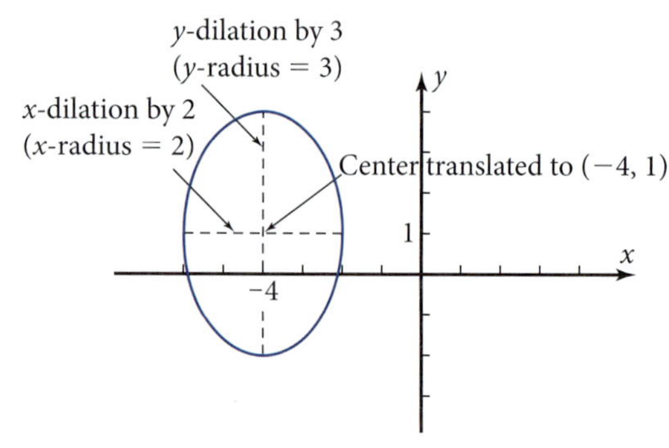

Figure 10-2e

Recall from previous work that you dilate a graph in a particular direction by *dividing* the respective variable by the dilation factor. So the equation of the ellipse in Figure 10-2d is

$$\left(\frac{x}{2}\right)^2 + \left(\frac{y}{3}\right)^2 = 1$$

You translate a graph in either direction by *subtracting* a constant from the respective variable. Figure 10-2e shows the ellipse of Figure 10-2d translated by -4 units in the x-direction and by 1 unit in the y-direction, thus moving the center to the point (–4, 1). The particular equation for the ellipse is

$$\left(\frac{x+4}{2}\right)^2 + \left(\frac{y-1}{3}\right)^2 = 1$$

As before, the translations are the values of x and y that make the quantities inside the parentheses equal zero. The box presents the general equations of parent, dilated, and translated circles, ellipses, hyperbolas, and parabolas. When $B = 0$, these equations are given the name **center-radius form** or **vertex form.** An equation such as $3x^2 + 5xy - 4y^2 + 11x + 8y - 97 = 0$ is said to be in *polynomial form.*

PROPERTIES: Equations of Parent and Transformed Conic Sections

General quadratic relation equation in polynomial form:

$Ax^2 + Bxy + Cy^2 + Dx + Ey + F = 0$ where A, B, C, D, E, and F are constants and at least one of A, B, and C is not zero.

Circle

$x^2 + y^2 = 1$ — Parent equation, unit circle

$(x - h)^2 + (y - k)^2 = r^2$ or $\left(\frac{x - h}{r}\right)^2 + \left(\frac{y - k}{r}\right)^2 = 1$ — Center (h, k), radius r

Ellipse

$x^2 + y^2 = 1$ — Parent equation, unit ellipse, same as unit circle

$\left(\frac{x - h}{d_x}\right)^2 + \left(\frac{y - k}{d_y}\right)^2 = 1$ — Center (h, k), x- and y-dilations d_x and d_y

Hyperbola

Opening in x-direction, x^2-coefficient is positive:

$x^2 - y^2 = 1$ — Parent equation, unit hyperbola. Asymptote slopes are ± 1

$\left(\frac{x - h}{d_x}\right)^2 - \left(\frac{y - k}{d_y}\right)^2 = 1$ — Center (h, k), x - and y-dilations d_x and d_y. Asymptote slopes are $\pm\frac{y\text{-dilation}}{x\text{-dilation}}$.

Opening in y-direction, y^2-coefficient is positive:

$-x^2 + y^2 = 1$ — Parent equation, unit hyperbola. Asymptote slopes are ± 1

$-\left(\frac{x - h}{d_x}\right)^2 + \left(\frac{y - k}{d_y}\right)^2 = 1$ — Center (h, k), x- and y-dilations d_x and d_y. Asymptote slopes are $\pm\frac{y\text{-dilation}}{x\text{-dilation}}$.

Parabola

Opening in y-direction, equation has "$y =$":

$y = x^2$ — Parent equation, unit parabola

$y = k + a(x - h)^2$ — Vertex (h, k), dilation factor a

Opening in x-direction, equation has "$x =$":

$x = y^2$ — Parent equation, unit parabola

$x = h + a(y - k)^2$ — Vertex (h, k), dilation factor a

Pencil-and-Paper Sketches

Examples 1, 2, and 3 show you how to sketch the graph of a conic section if the equation is given in center-radius form.

EXAMPLE 1 ➤ Sketch the graph of $\left(\frac{x-2}{5}\right)^2 + \left(\frac{y-4}{3}\right)^2 = 1.$

SOLUTION Analysis:

- The graph will be an ellipse. — x^2 and y^2 have unequal coefficients with the same sign.
- The center is at the point (2, 4). — The x- and y-translations are 2 and 4 units, respectively.
- The x-dilation is 5 units and the y-dilation is 3 units. — x-radius $= 5$, y-radius $= 3$

First plot the center point, (2, 4). Then plot points ± 5 units from the center in the x-direction and ± 3 units from the center in the y-direction. These points are ends of the **major axis** (the longer axis) and the **minor axis** (Figure 10-2f, left). The endpoints of the major axis are called **vertices** (plural of *vertex*). Sketch the ellipse by connecting the four points you have plotted with a smooth curve, as shown in Figure 10-2f, right.

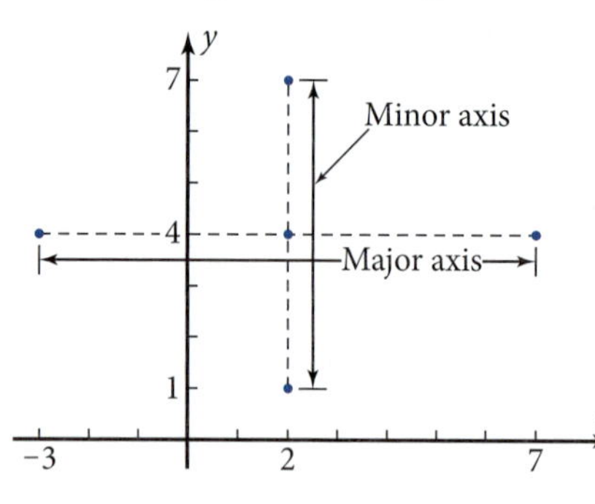

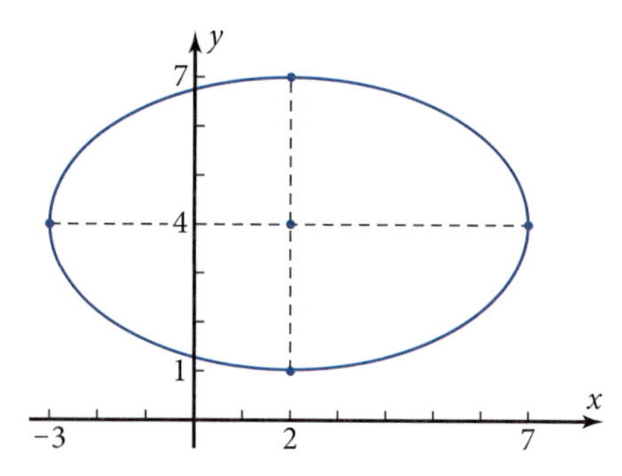

Figure 10-2f

EXAMPLE 2 ➤ Sketch the graph of $-\left(\frac{x+5}{2}\right)^2 + \left(\frac{y-4}{3}\right)^2 = 1.$

Translations give the center, (−5, 4).

$$-\left(\frac{x-(-5)}{2}\right)^2 + \left(\frac{y-4}{3}\right)^2 = 1$$

Opposite signs mean hyperbola.

SOLUTION Analysis:

- The graph is a hyperbola. — The squared terms have opposite signs.
- It opens in the y-direction. — The y-containing term is positive.
- The center is at the point $(-5, 4)$. — These values make the x- and y-terms zero, respectively.
- The asymptotes have slopes $\pm\frac{3}{2}$. — Slope is $\pm\frac{y\text{-dilation}}{x\text{-dilation}}$.

Sketch the center, vertices, and asymptotes (Figure 10-2g, left). A quick way to draw the asymptotes is by plotting the center, then sketching a box that extends to the left and right by the x-dilation, and up and down by the y-dilation. Draw the asymptotes through the corners of this box. The asymptotes cross at the center and have slopes given by

$$\text{Slope} = \pm\frac{y\text{-dilation}}{x\text{-dilation}}$$

Then sketch the graph (Figure 10-2g, right). Be sure the branches of the graph get closer and closer to the asymptotes and do not curve away.

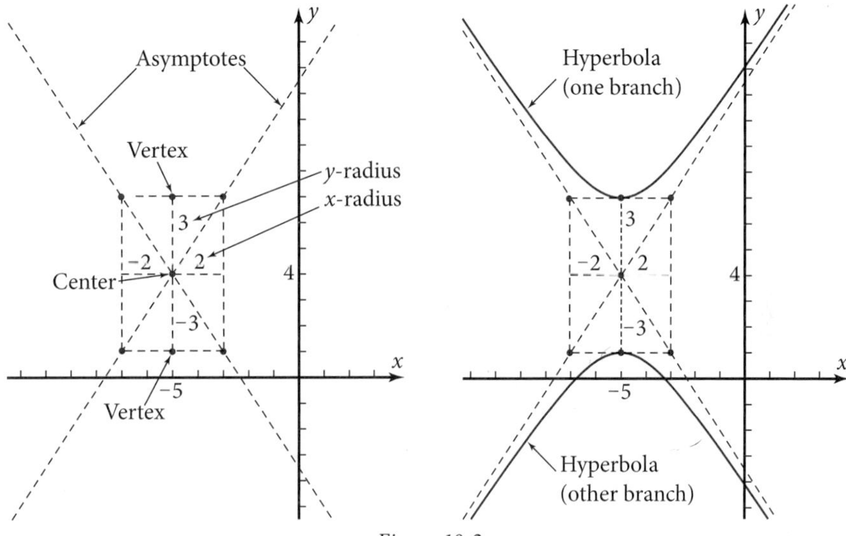

Figure 10-2g

EXAMPLE 3 ➤ Sketch the graph of $x = 6 - \frac{1}{2}(y - 2)^2$.

SOLUTION Analysis:

- The graph is a parabola. — The equation has only one squared term.
- The axis of symmetry is horizontal. — The equation is $x = \ldots$.
- It opens in the negative x-direction. — The dilation factor is negative.
- The vertex is at the point (6, 2). — Substituting 2 for y makes the squared term equal zero. When the squared term is zero, x is 6.

First plot the vertex (Figure 10-2h, left). Then through the vertex draw the horizontal **axis of symmetry,** $y = 2$.

Find another point, say, the x-intercept. Set $y = 0$ and evaluate x.

$$x = 6 - \frac{1}{2}(0 - 2)^2 = 6 - 2 = 4$$

Then draw the x-intercept and its reflection across the axis of symmetry.

Finally, sketch the graph containing the three points (Figure 10-2h, right).

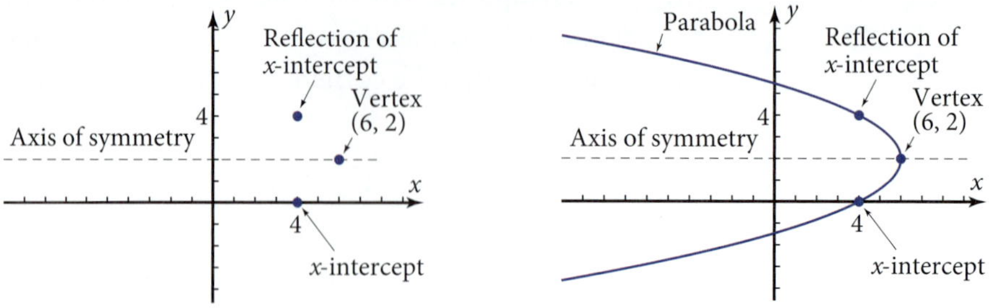

Figure 10-2h

Note: It is helpful to draw the axis of symmetry through the vertex *before* plotting the x-intercept so that you will be less likely to mistake the x-intercept for the vertex.

Examples 4 and 5 show you how to reverse the process and find an equation for a conic section from a given graph.

EXAMPLE 4 Find the particular equation in center-radius form for the hyperbola in Figure 10-2i (left).

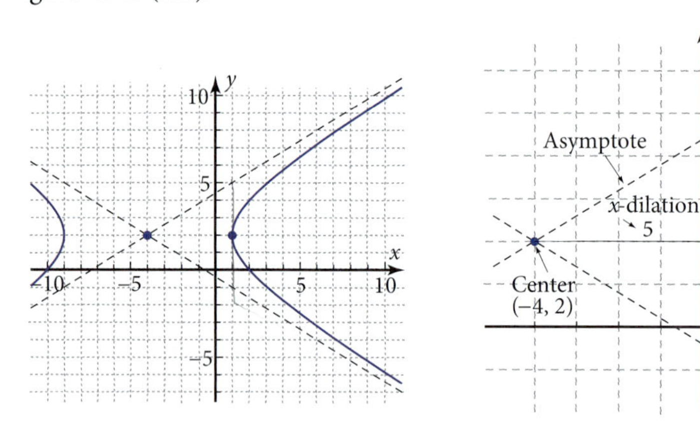

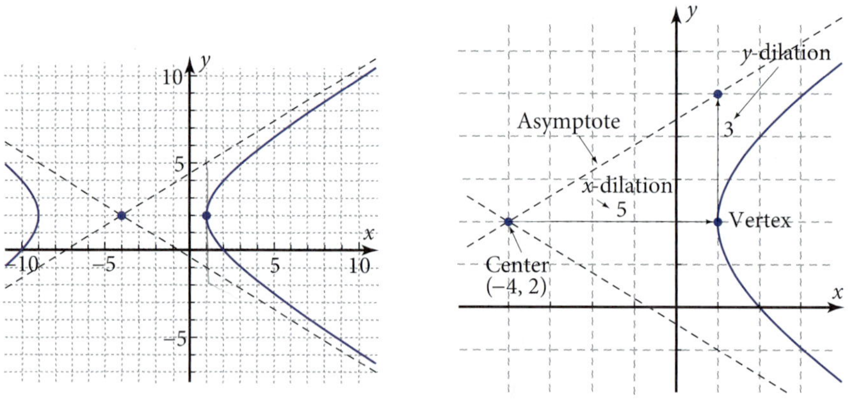

Figure 10-2i

SOLUTION From the graph, the center is at the point $(-4, 2)$. So the x-translation is -4 units and the y-translation is 2 units.

Because it is 5 units from the center to a vertex, the x-dilation is 5, as shown on the right in Figure 10-2i. Because it is 3 units from the vertex up to the asymptote, the y-dilation is 3.

Because the hyperbola opens in the x-direction, the term containing x has a $+$ sign in front of it, and the term containing y has a $-$ sign in front of it. (Memory aid: "$-$" in front implies "Doesn't open in this direction.")

The equation is $\left(\frac{x+4}{5}\right)^2 - \left(\frac{y-2}{3}\right)^2 = 1.$

EXAMPLE 5 ➤ Find the particular equation in vertex form for the parabola in Figure 10-2j.

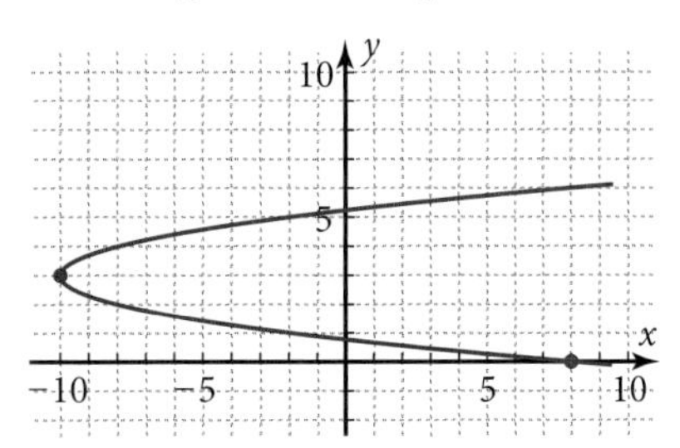

Figure 10-2j

SOLUTION The vertex is at the point $(-10, 3)$ and the parabola opens in the x-direction, so the equation has the form

$$x = -10 + a(y - 3)^2$$

To find the dilation factor, a, substitute the point $(8, 0)$ that is marked on the graph.

$$8 = -10 + a(0 - 3)^2$$

$$18 = 9a$$

$$2 = a$$

The equation is $x = -10 + 2(y - 3)^2$. ◄

Transforming an Equation to or from Polynomial Form

Example 6 shows you how to transform an equation from center-radius form to polynomial form. Example 7 shows you how to reverse the process and transform an equation from polynomial form to center-radius form.

EXAMPLE 6 ➤ Transform the equation in Example 1 into the form

$$Ax^2 + Bxy + Cy^2 + Dx + Ey + F = 0$$

where A, B, C, D, E, and F are constants.

SOLUTION

$$\left(\frac{x-2}{5}\right)^2 + \left(\frac{y-4}{3}\right)^2 = 1$$

$$\frac{(x-2)^2}{25} + \frac{(y-4)^2}{9} = 1$$

$$9(x-2)^2 + 25(y-4)^2 = 225$$ Multiply both sides by (25)(9) to eliminate the fractions.

$$9(x^2 - 4x + 4) + 25(y^2 - 8y + 16) = 225$$

$$9x^2 - 36x + 36 + 25y^2 - 200y + 400 = 225$$

$$9x^2 + 25y^2 - 36x - 200y + 211 = 0$$ Commute and associate the terms and make the right side equal zero. ◄

Note that the x^2- and y^2-terms in Example 6 have the same sign but unequal coefficients, indicating that the graph is an ellipse. Note also that there is no xy-term. In Section 10-6, you will learn that an xy-term rotates the graph.

EXAMPLE 7 ➤ Transform the equation $9x^2 - 4y^2 + 90x - 32y + 197 = 0$ to center-radius form by completing the square.

SOLUTION

$$9x^2 - 4y^2 + 90x - 32y + 197 = 0$$

$$9(x^2 + 10x + \quad) - 4(y^2 + 8y + \quad) = -197$$

Subtract 197 from both sides. Factor 9 from the x-terms and -4 from the y-terms. Leave space to complete the squares.

$$9(x^2 + 10x + 25) - 4(y^2 + 8y + 16) = -197 + 9(25) - 4(16)$$

To complete the square, take half the linear coefficient, square it, and add the result inside the parentheses. Add 9(25) and $-4(16)$ to the right side to balance the equation.

$$9(x + 5)^2 - 4(y + 4)^2 = -36$$

Factor on the left. Combine on the right.

$$\frac{9(x + 5)^2}{-36} - \frac{4(y + 4)^2}{-36} = \frac{-36}{-36}$$

Divide by -36 to get 1 on the right.

$$-\frac{(x + 5)^2}{4} + \frac{(y + 4)^2}{9} = 1$$

Reduce the fractions on the left. Cancel on the right.

$$-\left(\frac{x + 5}{2}\right)^2 + \left(\frac{y + 4}{3}\right)^2 = 1$$

Write as squares of fractions on the left. ◀

Note: The reason for getting 1 on the right in the next-to-last step in Example 7 is that center-radius form involves transformations of the *unit* hyperbola.

Conic Sections by Grapher from the Cartesian Equation

The general Cartesian equation of a conic section is

$$Ax^2 + Bxy + Cy^2 + Dx + Ey + F = 0$$

where A, B, C, D, E, and F are constants. You can plot the graph by first writing the equation as a quadratic function of y and then using the quadratic formula:

$$Cy^2 + (Bx + E)y + (Ax^2 + Dx + F) = 0$$

$$y = \frac{-(Bx + E) \pm \sqrt{(Bx + E)^2 - 4(C)(Ax^2 + Dx + F)}}{2C}$$

By the quadratic formula.

You can write or download a program from **www.keymath.com/precalc** to use this result. The input would be the six coefficients, A, B, C, D, E, and F. The program should generate and plot two functions, one with the + sign and one with the − sign.

EXAMPLE 8 ➤ Use a grapher program to plot the graph of $9x^2 + 25y^2 - 36x - 200y + 211 = 0$. Sketch the result.

SOLUTION Use a window with at least $[-3, 7]$ for x that has the integers as grid points and that has equal scales for both axes.

Run the program. Input 9 for A, 0 for B, 25 for C, -36 for D, -200 for E, and 211 for F.

The result is shown in Figure 10-2k. Use the zoom square feature to make equal scales on both axes. Note that if you use low-resolution graphics, the two branches of the ellipse may appear not to connect, as in Figure 10-2l. If this is the case, your sketch should show them connected, as in Figure 10-2k.

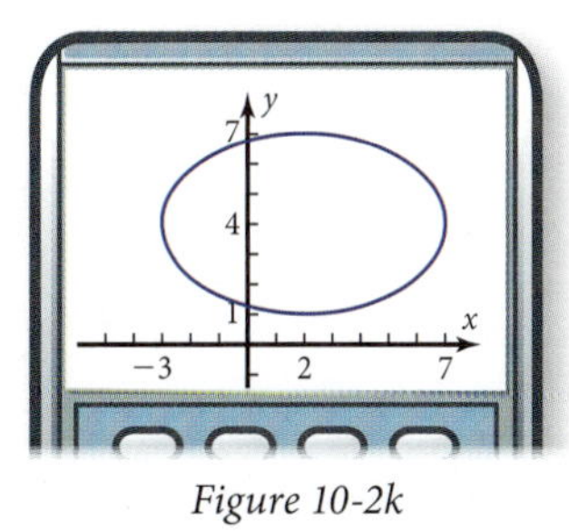

Figure 10-2k

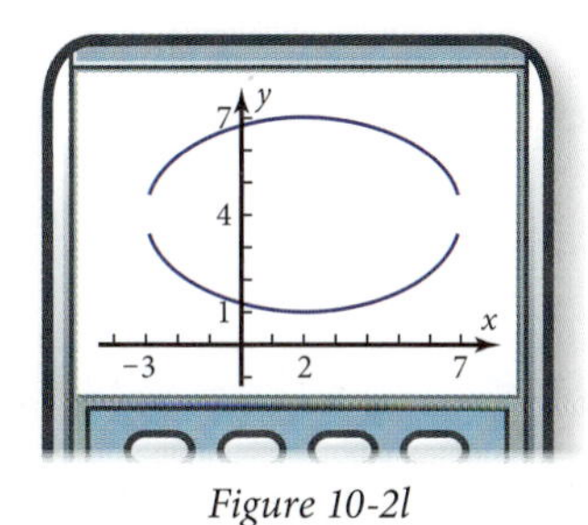

Figure 10-2l

Problem Set 10-2

Reading Analysis

From what you have read in this section, what do you consider to be the main idea? Why are parabolas, ellipses, circles, and hyperbolas called *conic sections*? Why does a hyperbola have two branches, while a parabola has only one? How can you form an ellipse by dilating and translating the unit circle? What is the major difference between an equation in *polynomial* form and an equation for the same conic section in *center-radius* form or *vertex* form?

Quick Review

For Problems Q1–Q4, identify the transformations applied to the parent cosine function to get the circular function $y = 3 + 4\cos 5(x - 6)$.

Q1. What is the horizontal translation?

Q2. What is the vertical translation?

Q3. What is the horizontal dilation?

Q4. What is the vertical dilation?

Q5. What is the difference between a *vector* quantity and a *scalar* quantity?

Q6. If $\vec{a} = 3\vec{i} + 5\vec{j}$ and $\vec{b} = 4\vec{i} - 6\vec{j}$, find $\vec{a} + \vec{b}$.

Q7. How do you translate two vectors so that they are in position to be added?

Q8. When the vectors are placed as in Problem Q7, where does the resultant vector go?

Q9. Write the equation for the parent quadratic function.

Q10. Write the Pythagorean property involving cosine and sine.

For Problems 1–4, sketch the graph.

1. $x^2 + y^2 = 1$

2. $x^2 - y^2 = 1$

3. $-x^2 + y^2 = 1$

4. $x = y^2$

For Problems 5–12,

a. Name the conic section simply by looking at the Cartesian equation.

b. Sketch the graph.

c. Transform the given equation into polynomial form, $Ax^2 + Bxy + Cy^2 + Dx + Ey + F = 0$.

d. Plot your answer to part c using the program you used in Example 8. Does the plotted graph agree with your sketch in part b?

5. $\left(\frac{x-3}{2}\right)^2 + \left(\frac{y-1}{4}\right)^2 = 1$

6. $\left(\frac{x+2}{7}\right)^2 + \left(\frac{y-4}{3}\right)^2 = 1$

7. $-\left(\frac{x-2}{5}\right)^2 + \left(\frac{y+1}{3}\right)^2 = 1$

8. $\left(\frac{x+3}{4}\right)^2 - \left(\frac{y+3}{2}\right)^2 = 1$

9. $\left(\frac{x+1}{6}\right)^2 + \left(\frac{y-2}{6}\right)^2 = 1$

10. $\left(\frac{x-4}{10}\right)^2 + \left(\frac{y-2}{10}\right)^2 = 1$

11. $y = 6 - 0.2(x-1)^2$

12. $x = -6 + 1.5(y-3)^2$

For Problems 13–18, name the conic section and write the particular equation in center-radius form or vertex form.

13.

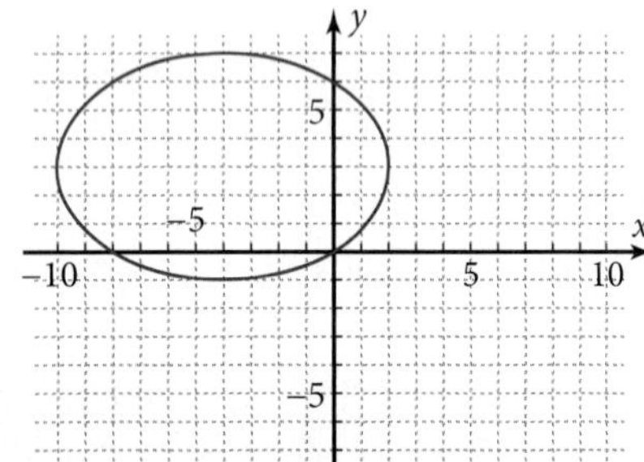

14.

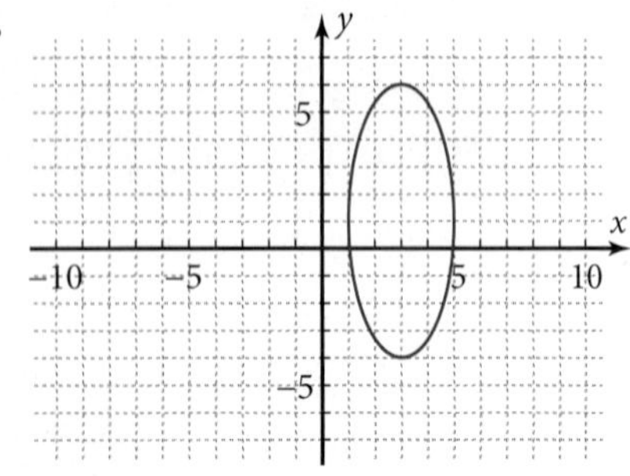

15.

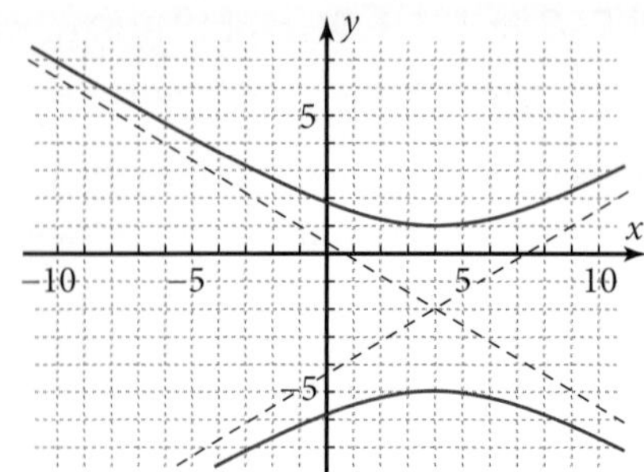

16.

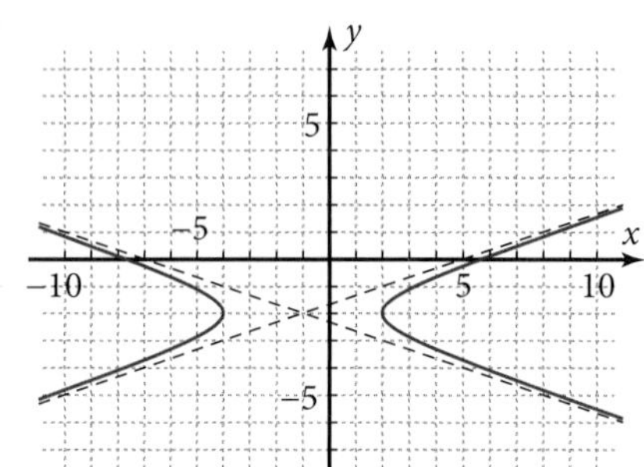

17.

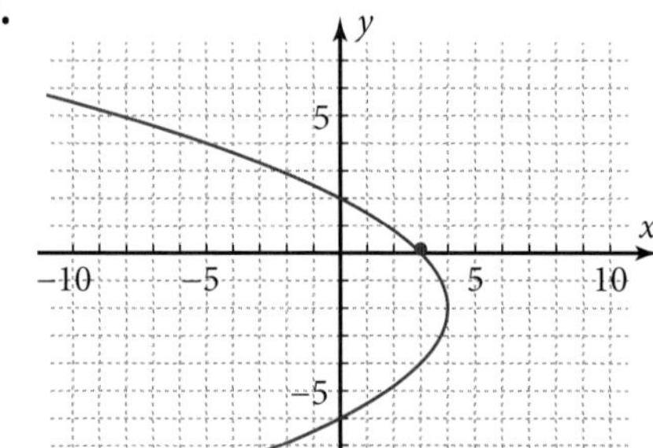

18.

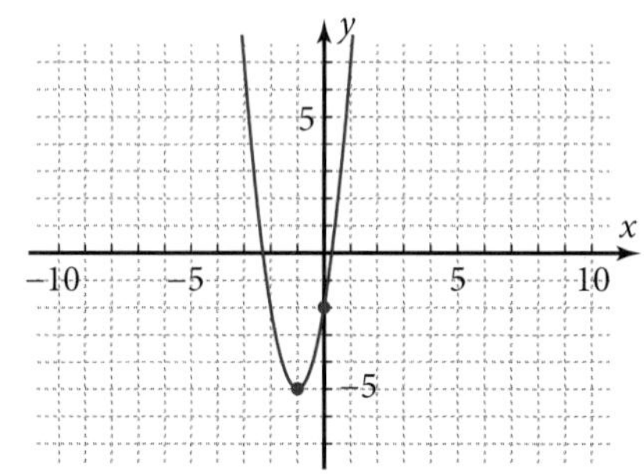

For Problems 19–34,

a. Identify the conic section by inspecting the squared-term coefficients.

b. Transform the equation to center-radius form or vertex form by completing the square, and describe the characteristics of the graph.

c. Plot your answer to part b using the program you used in Example 8. Does it agree with your conclusions in parts a and b?

19. $x^2 + y^2 - 10x + 8y + 5 = 0$

20. $x^2 + y^2 + 12x - 2y + 21 = 0$

21. $4x^2 + 9y^2 - 16x + 90y + 205 = 0$

22. $25x^2 + 4y^2 - 150x + 32y + 189 = 0$

23. $25x^2 + 9y^2 + 50x - 36y - 164 = 0$

24. $x^2 + 9y^2 + 12x + 54y + 81 = 0$

25. $16x^2 + 25y^2 - 300y + 500 = 0$

26. $25x^2 + 9y^2 = 900$

27. $25x^2 - 16y^2 - 100x - 96y - 444 = 0$

28. $4x^2 - 9y^2 + 16x + 108y - 344 = 0$

29. $25x^2 - 9y^2 + 300x - 126y + 684 = 0$

30. $4x^2 - 36y^2 - 40x + 216y - 80 = 0$

31. $x = -3y^2 - 12y - 5$

32. $x = -2y^2 + 12y - 10$

33. $x = 0.5y^2 + 3y + 4$

34. $y = 0.2x^2 + 2x - 2.2$ (Be observant!)

35. *Spaceship Problem*: A spaceship orbits in an elliptical path close to the Sun, which is at the origin. Let x and y be distance in millions of miles. The equation for the orbit is

$$\left(\frac{x-12}{13}\right)^2 + \left(\frac{y}{5}\right)^2 = 1$$

a. Sketch the elliptical orbit.

b. What is the closest the spaceship is to the Sun? The farthest?

c. If $x = 20$, what are the two possible values of y?

d. At the points in part c, how far is the spaceship from the Sun?

36. *Comet Problem:* As a comet approaches Earth, gravity pulls it out of its straight-line path into a hyperbolic path. After it passes Earth (provided it misses!) it approaches another straight-line path as the effect of Earth's gravity becomes weaker. The two straight lines are the asymptotes of the hyperbola. Suppose tracking instruments find that the path of the particular comet shown in Figure 10-2m is given by this equation, where x and y are distance in thousands of miles:

$$\frac{(x+20)^2}{225} - \frac{y^2}{175} = 1$$

a. Use the information in the equation to find the vertex of the hyperbola. Earth's radius is 4000 mi. How close will the comet be to Earth's surface when it is at the vertex?

b. Transform the equation to polynomial form. Plot the hyperbola and the two asymptotes on your grapher using an appropriate window. Does the graph agree with Figure 10-2m?

c. Use the equation to find the value of y when the comet is at the point shown in Figure 10-2m, where $x = 20$. How far from the center of Earth is the comet at this point? From the surface of Earth?

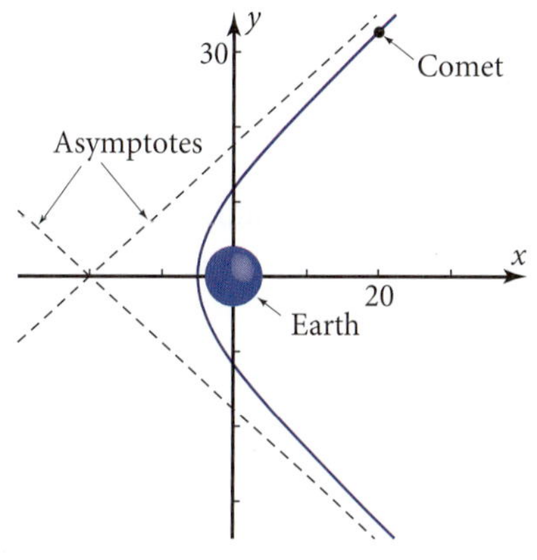

Figure 10-2m

37. *xy-Term Problem:* Figure 10-2n shows the graphs of

$$9x^2 + 25y^2 = 225 \quad \text{and}$$
$$9x^2 - 20xy + 25y^2 = 225$$

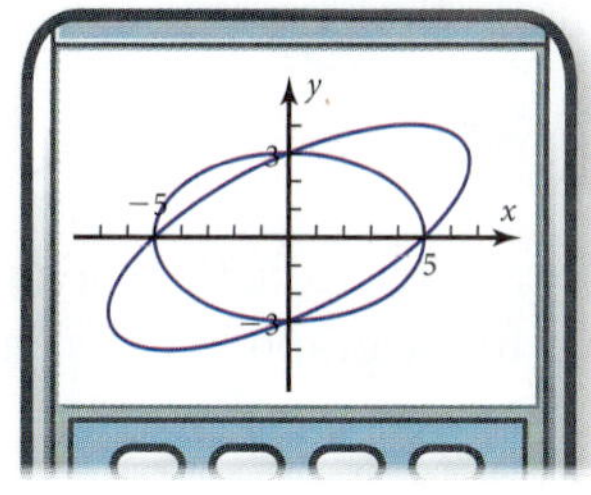

Figure 10-2n

a. Plot these graphs using the program of Example 8. (With low-resolution graphics, the ends of the second graph may not close.)

b. What changes and what does not change when the xy-term, $-20xy$, is added to the equation?

10-3 Parametric Equations for Conic Sections

In Section 7-5, you learned that parametric equations can be used to graph circles and ellipses. In this section you will learn that hyperbolas and "sideways" parabolas can also be plotted parametrically.

Objective

Given parametric equations of a conic section, identify which conic section the graph will be, plot the graph, and confirm your identification by transforming the parametric equation into a Cartesian equation.

In this exploration you will show graphically and algebraically that some parametric function graphs are conic sections.

EXPLORATION 10-3: Introduction to Parametric Equations for Conic Sections

1. With your grapher in parametric mode and radian mode, use $[0, 2\pi]$ for t, a window with $[-10, 10]$ for x, and equal scales on the two axes. Plot these parametric equations and sketch the result.

   $x = \cos t$
   $y = \sin t$

2. Square both sides of both equations in Problem 1. Then add the two equations, left side to left side and right side to right side. Use the Pythagorean property for cosine and sine to show that the result is equivalent to the *unit circle* with equation $x^2 + y^2 = 1$.

3. Plot these parametric equations. Describe verbally how the resulting *ellipse* is related to the unit circle in Problem 2.

   $x = 5 \cos t$
   $y = 3 \sin t$

4. Plot these parametric equations. How is the graph related to the graph in Problem 3?

   $x = 2 + 5 \cos t$
   $y = -1 + 3 \sin t$

5. Plot the *unit hyperbola* with these parametric equations, and sketch the result.

   $x = \sec t$
   $y = \tan t$

6. Square both sides of both equations in Problem 5. Then combine the squared equations to show that $x^2 - y^2 = 1$.

7. Plot the unit hyperbola with these parametric equations and explain how it is related to the unit hyperbola in Problem 5.

   $x = \tan t$
   $y = \sec t$

8. Plot the hyperbola that has these parametric equations. Sketch the result on graph paper.

   $x = -4 + 3 \tan t$
   $y = 1 + 2 \sec t$

9. What did you learn as a result of doing this exploration that you did not know before?

In Exploration 10-3, you recalled the parametric equations for a circle and an ellipse and encountered parametric equations for hyperbolas. The parametric equations for a parabola come from setting one of the variables x or y equal to t, and setting the other variable equal to a quadratic function of t. The box summarizes these equations.

PROPERTIES: Parametric Equations of Conic Sections

Circle	$x = \cos t$ $y = \sin t$	Parent equation, unit circle
	$x = h + r\cos t$ $y = k + r\sin t$	Center (h, k), radius r
Ellipse	$x = \cos t$ $y = \sin t$	Parent equation, unit ellipse, same as unit circle
	$x = h + d_x\cos t$ $y = k + d_y\sin t$	Center (h, k), x- and y-dilations d_x and d_y
Hyperbola	Opening in x-direction:	
	$x = \sec t$ $y = \tan t$	Parent equation, unit hyperbola Asymptote slopes are ± 1
	$x = h + d_x\sec t$ $y = k + d_y\tan t$	Center (h, k), x- and y-dilations d_x and d_y Asymptote slopes are $\pm\frac{y\text{-dilation}}{x\text{-dilation}}$.
	Opening in y-direction:	
	$x = \tan t$ $y = \sec t$	Parent equation, unit hyperbola Asymptote slopes are ± 1
	$x = h + d_x\tan t$ $y = k + d_y\sec t$	Center (h, k), x- and y-dilations d_x and d_y Asymptote slopes are $\pm\frac{y\text{-dilation}}{x\text{-dilation}}$.
Parabola	Opening in x-direction:	
	$x = t^2$ $y = t$	Parent equation, unit parabola
	$x = h + a(t - k)^2$ $y = t$	Vertex (h, k), dilation factor a
	Opening in y-direction:	
	$x = t$ $y = t^2$	Parent equation, unit parabola
	$x = t$ $y = k + a(t - h)^2$	Vertex (h, k), dilation factor a

Examples 1 and 2 on the next page show you how to write parametric equations for a given conic-section graph.

EXAMPLE 1 ➤ Write parametric equations for the parabola in Figure 10-3a. Confirm your answer by plotting the parametric equations on your grapher.

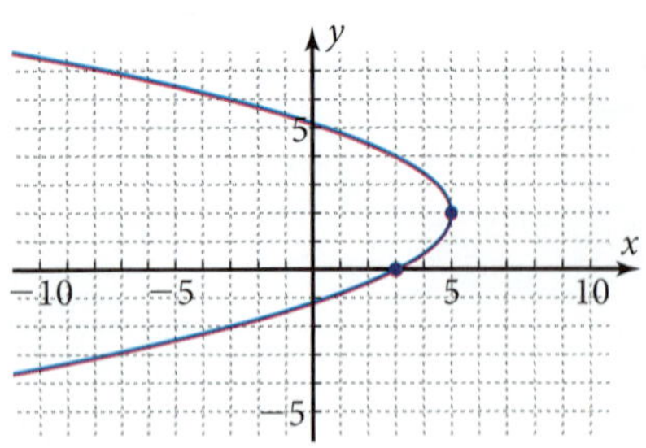

Figure 10-3a

SOLUTION Analysis:

- Opens in the x-direction, so $x = h + a(t - k)^2$ and $y = t$.
- Vertex is at the point (5, 2), so $x = 5 + a(t - 2)^2$.
- Point $(x, y) = (3, 0)$ is given, so substitute this point to calculate a.

$$3 = 5 + a(0 - 2)^2$$ $t = y$, so substitute 0 for t in the other parametric equation.

$$-2 = 4a$$

$$-0.5 = a$$

The parametric equations are

$$x = 5 - 0.5(t - 2)^2$$
$$y = t$$

The plotted graph agrees with the given figure. ◄

EXAMPLE 2 ➤ Write parametric equations for the hyperbola in Figure 10-3b. Confirm your answer by plotting the parametric equations on your grapher.

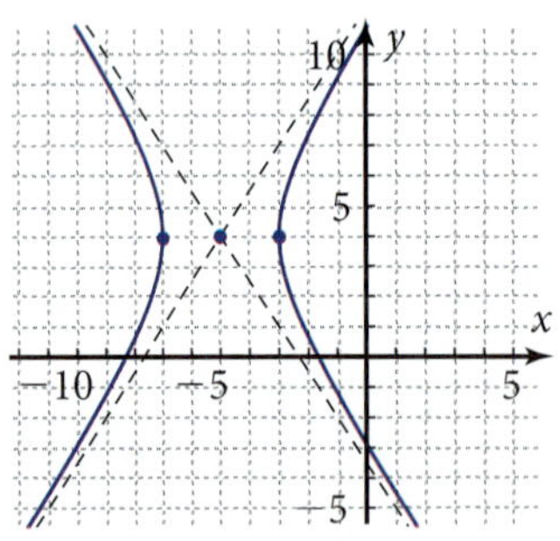

Figure 10-3b

SOLUTION Analysis:

- The graph opens in the x-direction, so the secant goes with x.
- x-translation by -5 units, y-translation by 4 units (coordinates of the center)
- x-dilation by a factor of 2 (center to vertex)
- y-dilation by a factor of 3 (vertex to asymptote)

The parametric equations are

$$x = -5 + 2 \sec t$$
$$y = 4 + 3 \tan t$$

The plotted graph agrees with the given figure. You might need to enter $x = -5 + 2/\cos t$.

Example 3 shows you how to transform the parametric equations in Example 2 to Cartesian form.

EXAMPLE 3 ➤ Transform these parametric equations to an equivalent Cartesian equation by eliminating the parameter t, thus confirming that the graph is the hyperbola in Figure 10-3b.

$$x = -5 + 2 \sec t$$
$$y = 4 + 3 \tan t$$

SOLUTION

$$x = -5 + 2 \sec t$$
$$y = 4 + 3 \tan t$$

Write the given equations.

$$\frac{x+5}{2} = \sec t$$
$$\frac{y-4}{3} = \tan t$$

Isolate $\sec t$ and $\tan t$.

$$\left(\frac{x+5}{2}\right)^2 = \sec^2 t$$
$$\left(\frac{y-4}{3}\right)^2 = \tan^2 t$$

Square both sides of each equation for the eventual use of Pythagorean properties.

$$\left(\frac{x+5}{2}\right)^2 - \left(\frac{y-4}{3}\right)^2 = \sec^2 t - \tan^2 t$$

Subtract the bottom equation from the top equation.

$$\left(\frac{x+5}{2}\right)^2 - \left(\frac{y-4}{3}\right)^2 = 1$$

Pythagorean property for secant and tangent.

The equation is in center-radius form, confirming agreement with Figure 10-3b:

- a hyperbola opening in the x-directions because the $-$ sign is in front of the y-term
- x- and y-translations of -5 units and 4 units, because these values make the numerators equal zero
- x-dilation by a factor of 2 and y-dilation by a factor of 3, the denominators of the terms

Problem Set 10-3

Reading Analysis

From what you have read in this section, what do you consider to be the main idea? How do you determine from the parametric equations whether the graph will be an ellipse or a hyperbola without plotting the graph? How are the Pythagorean properties of the trigonometric functions useful in eliminating the parameter to find a Cartesian equation from given parametric equations?

Quick Review

Problems Q1–Q5 refer to the equation

$$\left(\frac{x-7}{9}\right)^2 + \left(\frac{y+5}{16}\right)^2 = 1$$

Q1. Which conic section will the graph be?

Q2. Which way is the major axis oriented, in the x-direction or the y-direction?

Q3. What is the x-dilation factor?

Q4. What is the y-translation?

Q5. What is the significance of the 1 on the right side of the equation?

Q6. Write the Pythagorean property for cosine and sine.

Q7. Write the Pythagorean property for secant and tangent.

Q8. $\frac{\pi}{3}$ radians is __?__ degrees.

Q9. Write arccos x in terms of $\cos^{-1} x$.

Q10. $-\frac{\sqrt{3}}{2} =$

A. $\cos\frac{\pi}{6}$ **B.** $\cos\frac{\pi}{3}$ **C.** $\cos\frac{\pi}{2}$

D. $\cos\frac{2\pi}{3}$ **E.** $\cos\frac{5\pi}{6}$

For Problems 1–8,

a. Name the conic section simply by looking at the parametric equations.

b. Sketch the graph.

c. Plot the parametric equations on your grapher. Does the plotted graph agree with your sketch in part b?

1. $x = \cos t$, $y = \sin t$

2. $x = \sec t$, $y = \tan t$

3. $x = 3 \tan t$, $y = 2 \sec t$

4. $x = 3 \cos t$, $y = 2 \sin t$

5. $x = 4 + 5 \sec t$, $y = 3 + 2 \tan t$

6. $x = -2 + 3 \tan t$, $y = 1 + 4 \sec t$

7. $x = -6 + 5 \cos t$, $y = -2 + 5 \sin t$

8. $x = 3 + 4 \cos t$, $y = -2 + 4 \sin t$

9. Transform the parametric equations in Problem 1 to an equivalent Cartesian equation. How does the Cartesian equation support your answer to Problem 1, part a?

10. Transform the parametric equations in Problem 2 to an equivalent Cartesian equation. How does the Cartesian equation support your answer to Problem 2, part a?

11. Transform the parametric equations in Problem 5 to an equivalent Cartesian equation. How does the Cartesian equation support your answer to Problem 5, part a?

12. Transform the parametric equations in Problem 8 to an equivalent Cartesian equation. How does the Cartesian equation support your answer to Problem 8, part a?

For Problems 13–18, write parametric equations for the given conic section.

13.

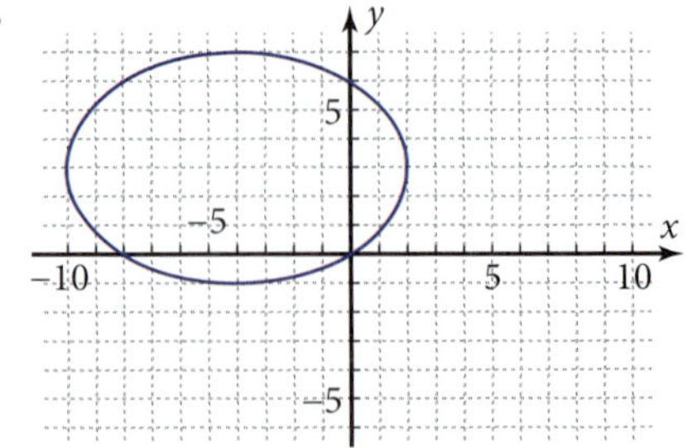

14.

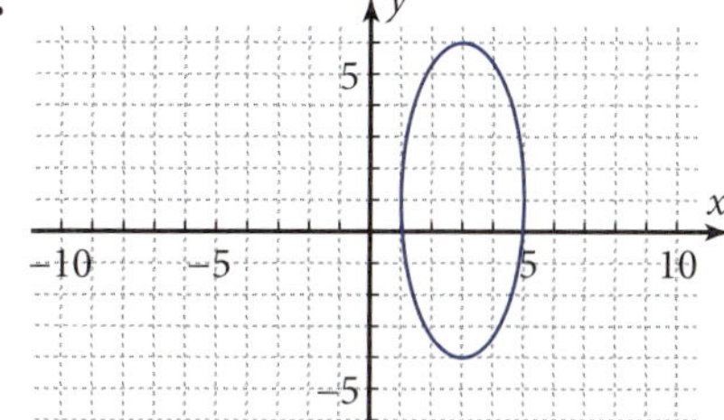

15.

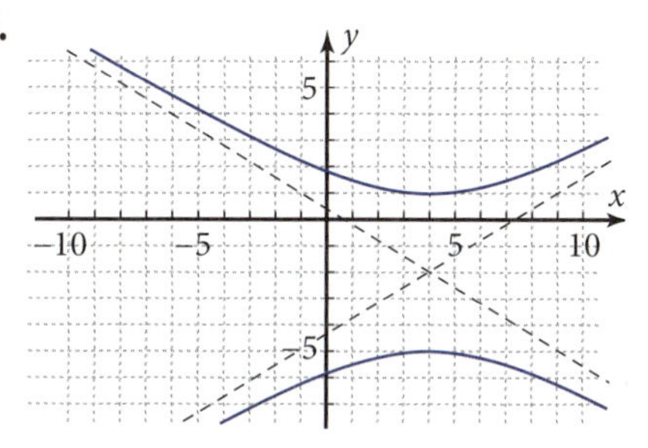

16.

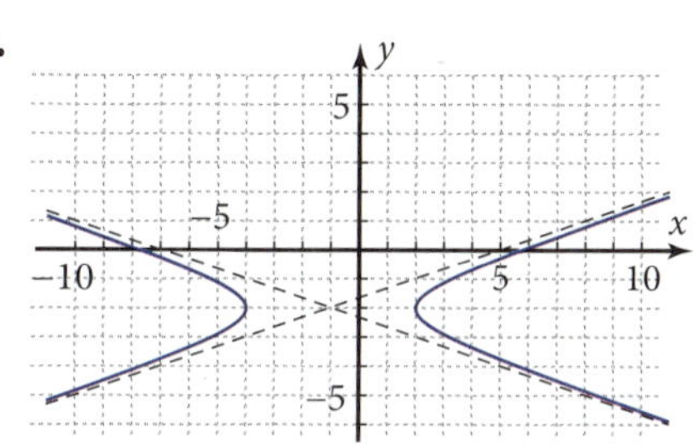

17.

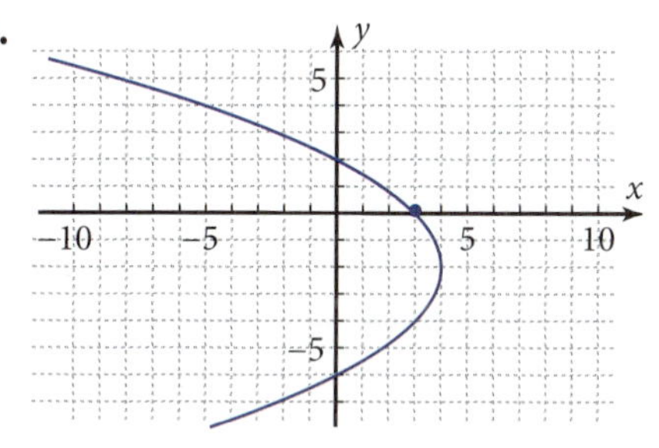

18.

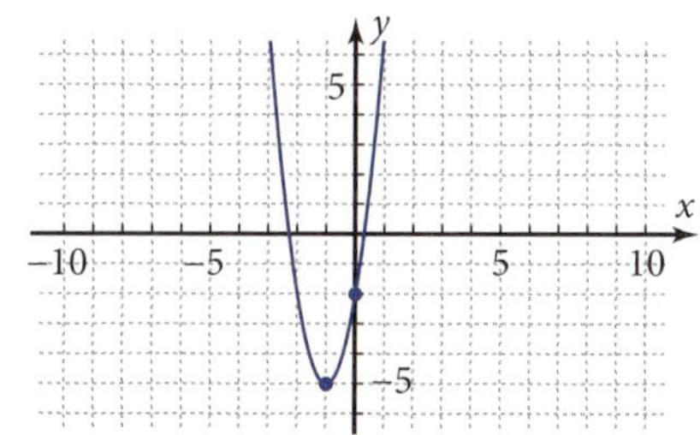

19. *Meteor Problem:* Astronomers detect a meteor approaching Earth. They determine that its path is the branch of the hyperbola given by

$$x = -50 + 40 \sec t$$
$$y = 30 \tan t$$

with $[-0.5\pi, 0.5\pi]$ for t. The center of Earth is at the origin, and x and y are distances in thousands of miles.

a. Plot the branch of the hyperbola. Sketch the result.

b. When $t = -1$ radian, what are the x- and y-coordinates of the meteor? How far is it from the center of Earth at this time?

c. At what value of t is the meteor closest to Earth? At that time, how far is the meteor from the surface of Earth? (Earth's diameter is about 7920 mi.)

d. Before Earth's gravity deflected the meteor into its curved path, it was traveling straight along one asymptote of the hyperbola. What is the Cartesian equation for this asymptote?

e. What do you suppose is the physical significance of the other asymptote?

The Willamette Meteorite in the American Museum of Natural History in New York. For the Clackamas tribe, this meteorite represented a union of sky, earth, and water.

10-4 Quadric Surfaces and Inscribed Figures

A paraboloid surface reflects the parallel electromagnetic rays of television broadcasts into its focal point.

Figure 10-4a shows a parabola in the xy-plane in a three-dimensional coordinate system. The axis of symmetry is along the y-axis. Figure 10-4b shows the three-dimensional surface generated by rotating this parabola about its axis of symmetry. The surface is called a **paraboloid.** The suffix *-oid* means "like," so a paraboloid is "parabola-like."

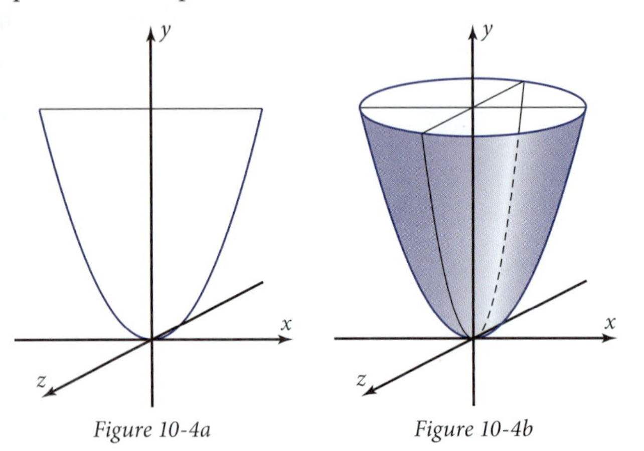

Figure 10-4a *Figure 10-4b*

Real-world objects such as reflectors in flashlights and microphones that pick up quarterbacks' voices at football games have this shape. In general, the three-dimensional analog of a conic section is a **quadric surface.** Quadric surfaces in general are defined by quadratic equations in three variables. In this section you'll encounter some special quadric surfaces that are generated by rotating conic sections about their axes, such as a paraboloid. You'll also study plane and solid figures that can be inscribed inside these surfaces.

Objective

Given the equation for a conic section, sketch the surface generated by rotating it about one of its axes, and find the area or volume of a figure inscribed either in the plane region bounded by the graph or in the solid region bounded by the surface.

The parabola in Figure 10-4a was rotated about its axis of symmetry. The axes of symmetry of ellipses and hyperbolas are given different names to distinguish between them. Figure 10-4c shows that for an ellipse the names are *major axis* and *minor axis*. The names refer to the relative sizes of the two axes, not to the directions in which they point. For a hyperbola, the names are **transverse axis** (from vertex to vertex) and **conjugate axis** (perpendicular to the transverse axis). The latter name comes from the conjugate hyperbola, which has the same asymptotes and dilations but opens in the other direction.

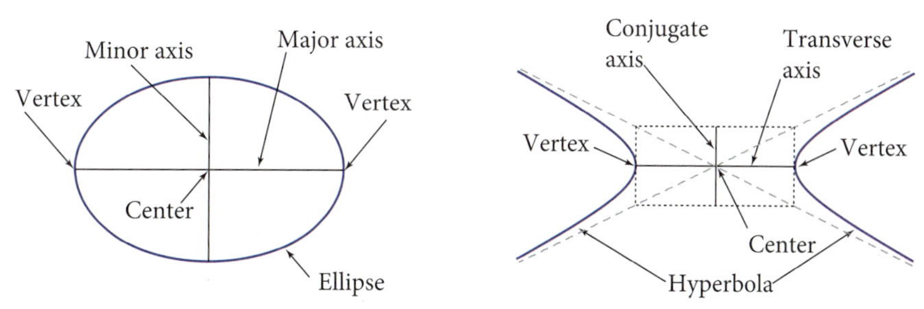

Figure 10-4c

The somewhat flattened ball is akin to an oblate spheroid.

Rotating an ellipse about one of its axes generates an **ellipsoid.** If the rotation is about the major axis, the ellipsoid is called a *prolate spheroid,* reminiscent of a football or an egg. If the rotation is about the minor axis, the ellipsoid is called an *oblate spheroid,* similar to a round pillow. Figure 10-4d shows these shapes.

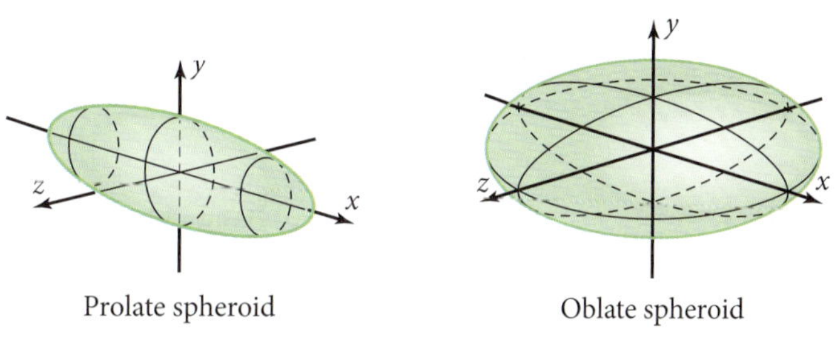

Figure 10-4d

Rotating a hyperbola about one of its axes generates a **hyperboloid.** If the rotation is about the transverse axis, the two branches of the hyperbola form two disconnected surfaces, called a *hyperboloid of two sheets.* If the rotation is about the conjugate axis, the surface is connected and is called a *hyperboloid of one sheet.* Figure 10-4e shows these shapes.

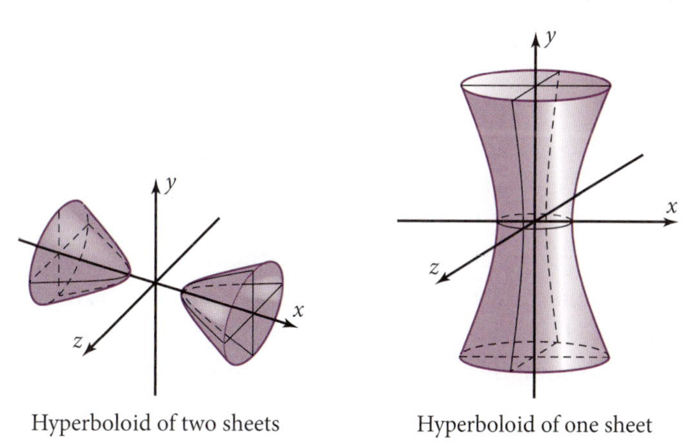

Figure 10-4e

Hyperboloids of one sheet have the remarkable property that they can be generated by rotating a line about an axis that is skew to the line. The decorative table in Figure 10-4f shows this property. The power plant cooling towers in the figure take advantage of the fact that they can be built with straight reinforcing materials, without any internal support structure.

Figure 10-4f

EXAMPLE 1 ➤ Sketch the hyperboloid formed by rotating about the y-axis the hyperbola $-9x^2 + y^2 = 9$ from $x = 0$ to $x = 2$.

SOLUTION

$$-9x^2 + y^2 = 9$$

$$-\left(\frac{x}{1}\right)^2 + \left(\frac{y}{3}\right)^2 = 1$$

Figure 10-4g shows the hyperbola centered at the origin, opening in the y-direction, with x-dilation 1 and y-dilation 3, and asymptotes having slopes ± 3. Show the circular cross sections in perspective, using dashed lines where the cross section is hidden.

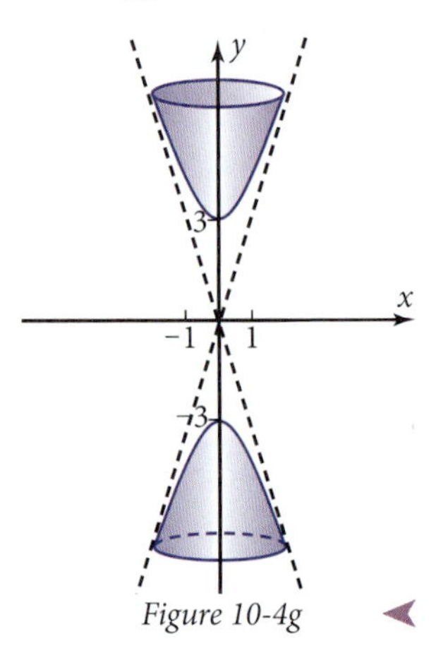

Figure 10-4g

EXAMPLE 2 ➤ Rectangles of various proportions are inscribed in the region under the half-ellipse $64x^2 + 25y^2 = 1600$, $y \geq 0$, as shown in Figure 10-4h. A vertex of the rectangle is the sample point (x, y) that lies on the ellipse. The rectangle can be tall and skinny, short and wide, or somewhere in between, depending on where the sample point is placed on the ellipse. The area of the rectangle depends on the location of the sample point. Place the sample point in Quadrant I.

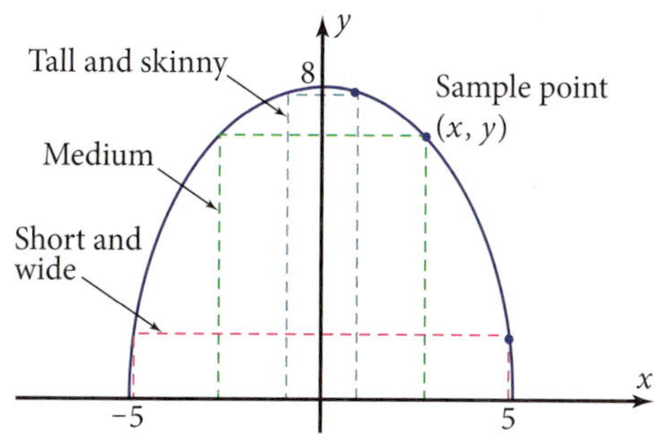

Figure 10-4h

a. Write the area of a representative rectangle in terms of the values of x and y at the sample point.

b. Transform the equation in part a so that the area is a function of x alone.

c. Make a table of values of area as a function of x for each 1 unit from $x = 0$ to $x = 5$. Based on your table, approximately what value of x in this interval seems to give the maximum area?

d. Plot the graph of area as a function of x. Use the maximum feature on your grapher to find the value of x that produced the maximum area and to find that maximum.

SOLUTION Let A represent the area.

a. $A = 2xy$ — The width of the rectangle is $2x$, not x.

b. $64x^2 + 25y^2 = 1600, \qquad y \geq 0$ — Transform the ellipse equation to get y in terms of x.

$$25y^2 = 1600 - 64x^2$$

$$y^2 = \frac{64}{25}(25 - x^2)$$

$$y = \frac{8}{5}\sqrt{25 - x^2}$$

Use only the *positive* square root, because $y \geq 0$.

$$\therefore A = 3.2x\sqrt{25 - x^2}$$

$A = 2xy$

c.

x	A
0	0
1	15.6767...
2	29.3284...
3	38.4
4	38.4
5	0

The maximum area seems to occur at a value of x between 3 and 4.

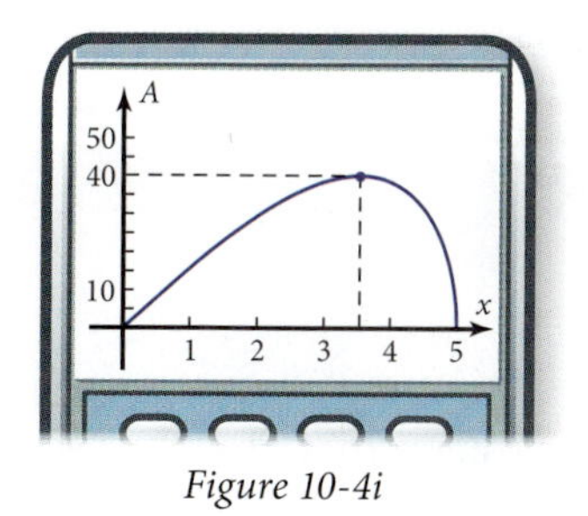

Figure 10-4i

d. Figure 10-4i shows the graph of area as a function of x. The maximum, 40, occurs at $x = 3.5355....$

EXAMPLE 3 ➤ The part of the parabola $y = 4 - x^2$ in the first quadrant is rotated about the y-axis to form a paraboloid. A cylinder is inscribed in the region under this surface with the center of its lower base at the origin and the points on the circumference of its upper base on the paraboloid.

a. Sketch the paraboloid and the cylinder.

b. Find the volume of the cylinder as a function of its radius.

c. Find the maximum volume the cylinder can have, and then find the radius of this maximal cylinder.

SOLUTION First sketch the parabola as in Figure 10-4j.

a. Pick a sample point (x, y) on the parabola in Quadrant I. Then draw the paraboloid by considering what happens as the parabola rotates about its axis of symmetry. The upper base of the cylinder will be traced by the sample point. The two circular bases will appear as ellipses in the figure because they are being viewed in perspective.

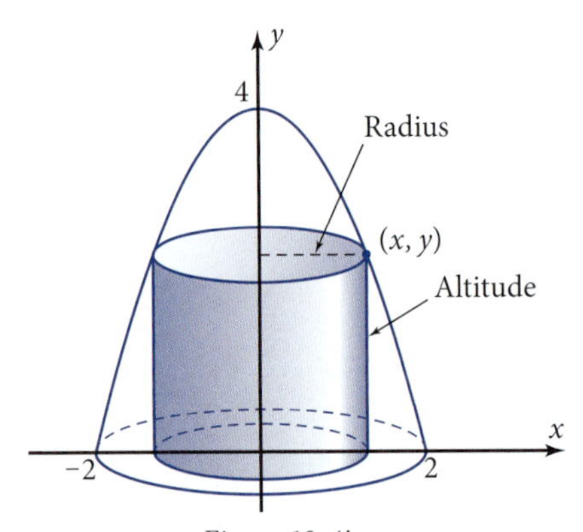

Figure 10-4j

b. Let V represent the volume of the cylinder.

$V = \pi x^2 y$ Volume = π(radius^2(altitude)

$V = \pi x^2(4 - x^2)$ Substitute for y.

c. Plot the volume function as shown in Figure 10-4k.

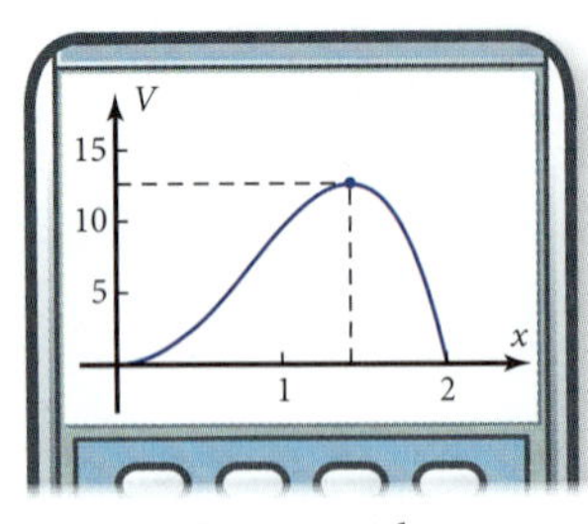

Figure 10-4k

The maximum feature of your grapher gives the maximum volume, 12.5663... cubic units, at $x = 1.4142...$ (which is the square root of 2). ◀

Recall that the *lateral area* of a solid is its surface area, excluding the area of the bases. These formulas from geometry will help you with the problems in Problem Set 10-4.

PROPERTIES: Geometry Formulas

Figure	Volume	Surface Area	
Cylinder	$V = \pi r^2 h$	$L = 2\pi rh$	Lateral area
		$S = 2\pi rh + 2\pi r^2$	Total area
Cone	$V = \frac{1}{3}\pi r^2 h$	$L = \pi rl$	Lateral area (l = slant height)
		$S = \pi rl + \pi r^2$	Total area
Sphere	$V = \frac{4}{3}\pi r^3$	$S = 4\pi r^2$	

Problem Set 10-4

Reading Analysis

From what you have read in this section, what do you consider to be the main idea? What is the difference between a prolate spheroid and an oblate spheroid? Why do hyperboloids have two sheets when the rotation is about the transverse axis of the hyperbola, but only one sheet when the rotation is about the conjugate axis?

Quick Review

For the equations in Problems Q1–Q7, name the conic section.

Q1. $x^2 + 4y^2 + 5x + 6y = 100$

Q2. $x^2 - 4y^2 + 5x + 6y = 100$

Q3. $-x^2 + 4y^2 + 5x + 6y = 100$

Q4. $4x^2 + 4y^2 + 5x + 6y = 100$

Q5. $4x^2 + 5x + 6y = 100$

Q6. $4y^2 + 5x + 6y = 100$

Q7. $x = 3 + 5\cos t$
$y = 4 + 2\sin t$

Q8. Complete the square: $5x^2 + 30x + 58$

Q9. Complete the square: $y^2 + 10y + 10$

Q10. Which operation causes dilation of a figure, multiplication or addition?

For Problems 1–10, sketch the quadric surface.

1. Paraboloid formed by rotating the part of the graph of $y = x^2$ from $x = 0$ to $x = 3$ about the y-axis

2. Paraboloid formed by rotating the part of the graph of $y = 9 - x^2$ that lies in the first quadrant about the y-axis

3. Ellipsoid formed by rotating the graph of $4x^2 + y^2 = 16$ about the x-axis

4. Ellipsoid formed by rotating the graph of $4x^2 + y^2 = 16$ about the y-axis

5. Hyperboloid formed by rotating the part of the graph of $4x^2 - y^2 = 4$ from $x = 1$ to $x = 2$ about the y-axis

6. Hyperboloid formed by rotating the part of the graph of $-x^2 + y^2 = 9$ from $y = 3$ to $y = 6$ about the x-axis

7. Hyperboloid formed by rotating the part of the graph of $x^2 - 4y^2 = 4$ from $x = -5$ to $x = 5$ about the x-axis

8. Hyperboloid formed by rotating the part of the graph of $-x^2 + y^2 = 9$ from $x = -6$ to $x = 6$ about the y-axis

9. Cone formed by rotating the part of the line $y = 3x$ from $x = -2$ to $x = 2$ about the y-axis

10. Cone formed by rotating the part of the line $y = 0.5x$ from $x = -6$ to $x = 6$ about the x-axis

11. *Ellipsoids in Sketchpad Problem:* Go to **www.keymath.com/precalc** and open the sketch Surfaces.gsp. (Note that this requires the use of The Geometer's Sketchpad software.) By double-clicking the parameters A and B, change their values to graph the ellipse $4x^2 + y^2 = 16$ from Problems 3 and 4. Click the `Paint x Surface` button and observe the result. Then click the `Clear Traces` button, and then the `Paint y Surface` button. Write a paragraph telling what you observed and describing how the Sketchpad surfaces compare with your sketches in Problems 3 and 4. For instance, what do you notice about the orientations of the three axes? How can you make Sketchpad's surfaces look the same as your sketches?

12. *Hyperboloids in Sketchpad Problem:* Following the directions given in Problem 11, plot the hyperboloid from Problem 5 formed by rotating the graph of $4x^2 - y^2 = 4$ about the y-axis. Then clear the trace and rotate this hyperbola about the x-axis. Note the differences between the two hyperboloids. Finally, plot the hyperboloid from Problem 7 formed by rotating the graph of $x^2 - 4y^2 = 4$ about the x-axis. Write a paragraph telling what you observed and describing how the computer graphs compare with the graphs you sketched in Problems 5 and 7. What do the results of this problem help you understand about quadric surfaces?

Frei Otto used hyperbolic-paraboloid surfaces in the architectural design of the tensile structure in Olympia Park for the 1972 Summer Olympics in Munich, Germany.

13. *Triangle in Parabola Problem:* A triangle is inscribed in the region bounded by the graph of the parabola $y = 9 - x^2$ and the x-axis (Figure 10-4l). A vertex of the triangle is at the origin, and the opposite side is parallel to the x-axis. Another vertex touches the parabola at the sample point (x, y) in the first quadrant. Plot the area of the triangle as a function of x, and sketch the result. Find the value of x that maximizes the area of the triangle, and find the area of this maximal triangle.

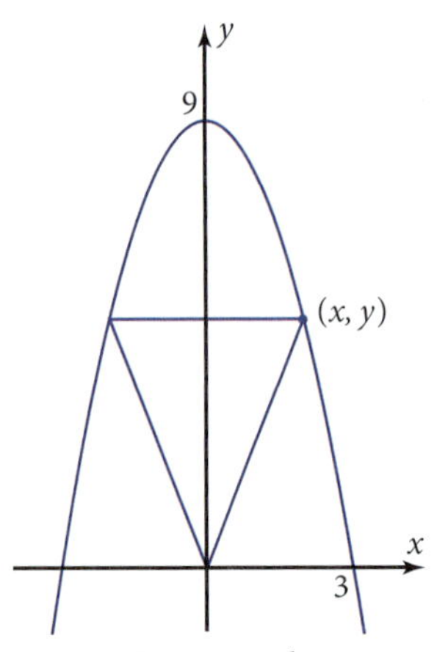

Figure 10-4l

14. *Rectangle in Ellipse Problem:* A rectangle is inscribed in the ellipse $9x^2 + 25y^2 = 225$. The sides of the rectangle are parallel to the coordinate axes. Sketch the ellipse and the rectangle. Then find the area of the rectangle in terms of a sample point at which a vertex of the rectangle touches the ellipse in the first quadrant. Plot the area of the rectangle as a function of x, and sketch the result. Find the value of x that maximizes the area of the rectangle, and find this maximum area.

15. *Cylinder in Sphere Volume Problem:* The circle $x^2 + y^2 = 25$ is rotated about the y-axis to form a sphere (Figure 10-4m). A cylinder is inscribed in the sphere, with its axis along the y-axis. Write an equation expressing the volume of the cylinder in terms of a sample point (x, y) in the first quadrant where the upper base of the cylinder touches the circle. Plot the volume of the cylinder as a function of x, and sketch the result. Find the value of x that gives the maximum volume, and find this maximum volume.

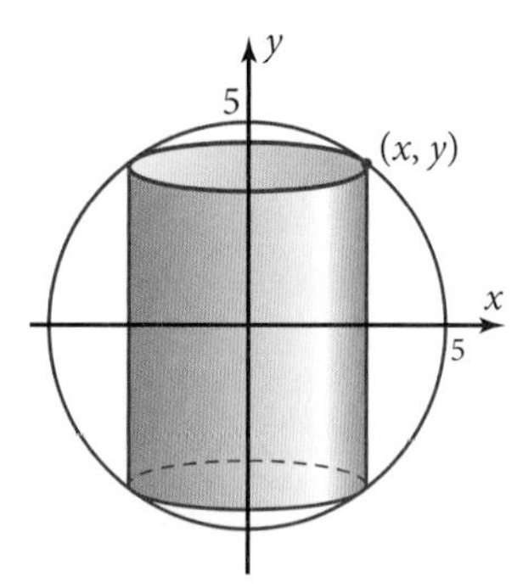

Figure 10-4m

16. *Cylinder in Ellipsoid Problem:* The ellipse $x^2 + 4y^2 = 4$ is rotated about the x-axis to form an ellipsoid. A cylinder is inscribed in the ellipsoid, with its axis along the x-axis. The right base of the cylinder touches the ellipse at the sample point (x, y) in the first quadrant. Sketch the ellipsoid and the cylinder. Then plot the volume of the cylinder as a function of x. Sketch the graph. Find the value of x that maximizes the volume of the cylinder, and find this maximum volume.

17. *Cylinder in Sphere Area Problem:* A cylinder is inscribed in a sphere with radius 5 units, as in Figure 10-4m.

a. Find the radius and altitude of the cylinder with maximum lateral area.

b. Find the radius and altitude of the cylinder with maximum total area.

c. Does the cylinder with maximum lateral area also have the maximum total area?

d. Does the cylinder with maximum total area also have maximum volume, as found in Problem 15?

18. *Cylinder in Ellipsoid Area Problem:* A cylinder is inscribed in the ellipsoid in Problem 16.

a. Find the radius and altitude of the cylinder with maximum lateral area.

b. Find the radius and altitude of the cylinder with maximum total area.

c. Does the cylinder with maximum lateral area also have the maximum total area?

d. Does the cylinder with maximum total area also have maximum volume, as found in Problem 16?

19. *Submarine Problem:* The bow of a submarine has the shape of the half-ellipsoid formed by rotating about the x-axis the right half of the ellipse

$$225x^2 + 900y^2 = 202{,}500$$

where x and y are length in feet. The ellipsoid (a doubly curved surface) is to be shaped from thin metal that is relatively easy to mold. The pressure hull, made of thick metal, is in the form of a cylinder (a singly curved surface) inscribed in the ellipsoid (Figure 10-4n). How should the cylinder be constructed to give it the maximum volume? How much of the heavy steel plate will be needed to form the curved walls of the cylinder?

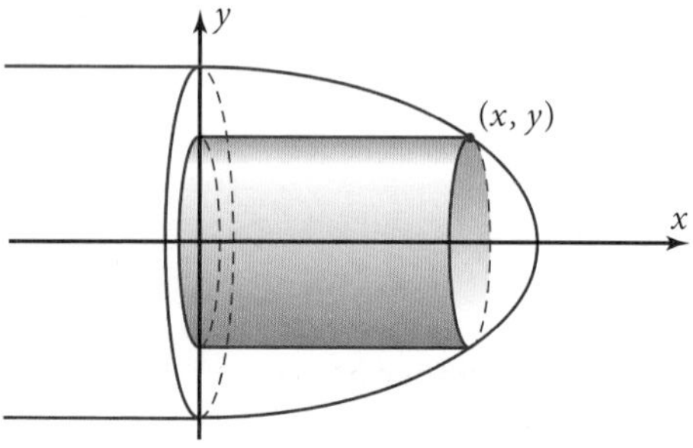

Figure 10-4n

10-5 Analytic Geometry of the Conic Sections

So far in this chapter you have studied algebraic properties of conic sections. Now you will analyze some geometric properties of conic sections, a subject that carries the name **analytic geometry.**

Objective Given the equation for a conic section, find the foci, the directrix, and the eccentricity, and vice versa.

In this exploration you will learn about the *focus, directrix,* and *eccentricity* of an ellipse.

EXPLORATION 10-5: Focus, Directrix, and Eccentricity of an Ellipse

The figure shows an ellipse centered at the point (0, 0) with these features:

- One *vertex* is at the point (6, 0).
- One *directrix* is the vertical line $x = 10$.
- One *focus* is at the point (3.6, 0); the other focus is at the point (−3.6, 0).
- The *eccentricity* is $e = 0.6$.

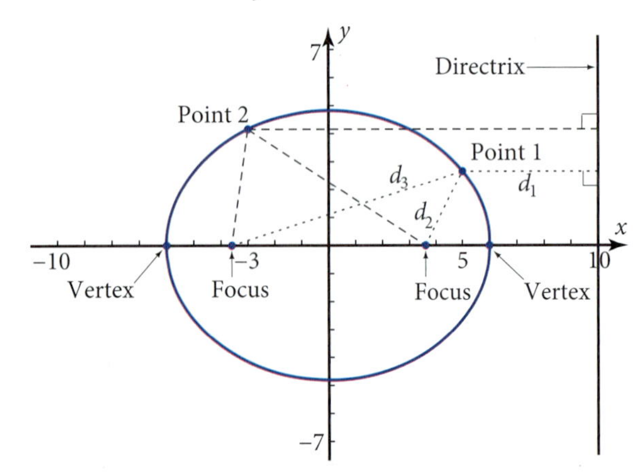

1. Point 1 on the ellipse is at $x = 5$. Its distance from the directrix is d_1, its distance from the right focus is d_2, and its distance from the left focus is d_3, as shown by the short-dashed line segments. Measure these distances using the scales on the given axes. You can draw marks on a piece of paper to measure the slant distances.
2. Show that $d_2 \approx 0.6d_1$ (where 0.6 is the eccentricity).
3. Point 2 on the ellipse is at $x = -3$. Its distances d_1, d_2, and d_3 (not labeled) are indicated by the long-dashed line segments. Measure these three distances for Point 2. Show that, again, $d_2 \approx 0.6d_1$.
4. Show that for both Point 1 and Point 2, $d_2 + d_3 \approx 12$, where 12 is the length of the *major axis* (the line segment connecting the two vertices).
5. The equation for the ellipse is

$$\left(\frac{x}{6}\right)^2 + \left(\frac{y}{4.8}\right)^2 = 1$$

Substitute 5 for x and calculate y for Point 1. Use this y-value and the Pythagorean theorem to calculate lengths d_2 and d_3. Use the results to show that $d_2 + d_3$ is *exactly* equal to 12. Show that d_2 is *exactly* equal to $0.6d_1$.

6. Read ahead to find the *focus-directrix property* of an ellipse. Write a statement of this property.
7. Read ahead to find the *two-foci property* of an ellipse. Write a statement of this property.
8. What did you learn as a result of doing this exploration that you did not know before?

The Old Senate Chamber in the U.S. Capitol in Washington, D.C., is a famous whispering chamber.

For each vertex of a conic section, there is a special fixed point on the concave side called the **focus,** as shown in Figure 10-5a. For ellipses, light or sound rays starting at one focus are reflected toward the other focus, making the ellipse a useful shape for auditoriums and "whispering chambers." For parabolas, rays starting at the focus are reflected parallel to the axis of symmetry, making the parabola a useful shape for headlight reflectors and dish TV antennas. For hyperbolas, the reflected rays diverge, but when extended backward they pass through the other focus.

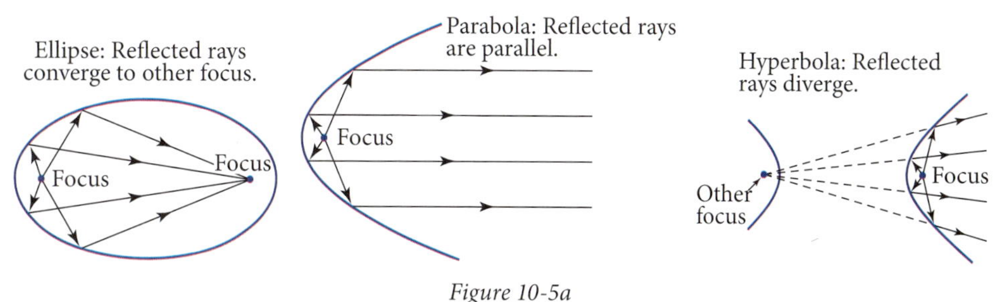

Figure 10-5a

For each focus a of parabola, ellipse, or hyperbola, there is a fixed line on the convex side, called the **directrix** (plural = directrices) perpendicular to the axis of symmetry. For each point on the graph, its distance from the focus is directly proportional to its distance from the corresponding directrix. The proportionality constant is called the **eccentricity,** written e. As shown in Figure 10-5b, if d_1 is the distance from a point on the graph to the directrix and d_2 is the distance from that point to the focus, then

$$d_2 = ed_1$$

This is the *focus-directrix property.*

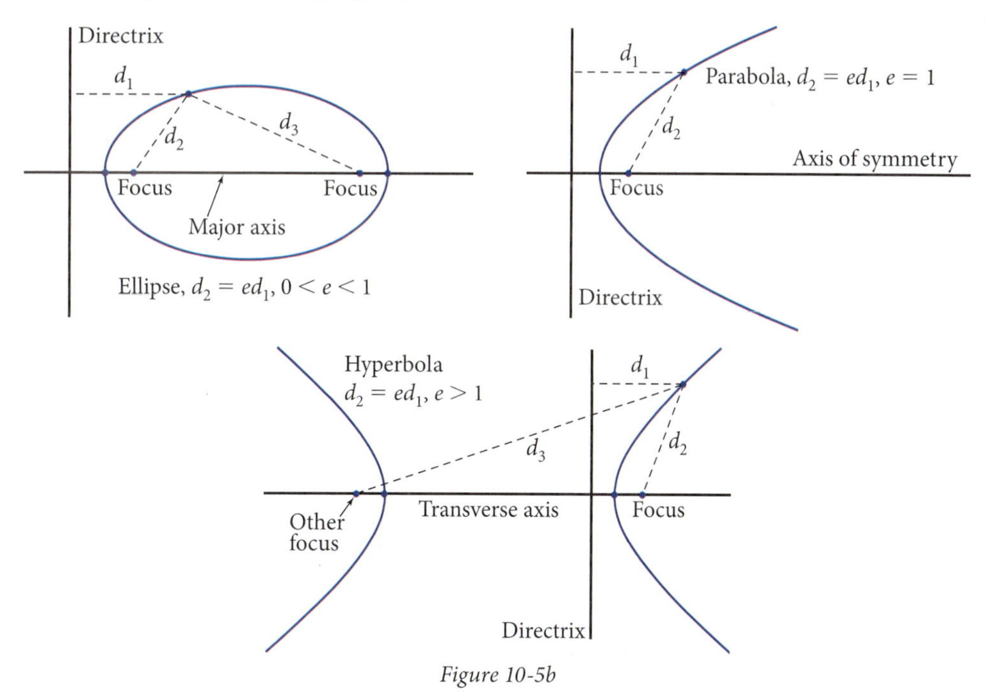

Figure 10-5b

The closer e is to 1, the longer and more "eccentric" the ellipse is. If the eccentricity is 1, the graph is a parabola. If the eccentricity is greater than 1, the graph is a hyperbola. You can see dynamically how varying the eccentricity produces different conic sections by viewing the *Unified Conics* exploration at **www.keymath.com/precalc.**

Some properties of conic sections are summarized in this box.

PROPERTIES: Focus, Directrix, and Eccentricity of a Conic Section

If d_1 is the distance from the point (x, y) on a conic section to one of its directrices and d_2 is the distance from (x, y) to the corresponding focus, then

$$d_2 = ed_1 \qquad \text{or, equivalently} \qquad e = \frac{d_2}{d_1}$$

Verbally: The distance to the focus is e times the distance to the directrix.

The eccentricity is the ratio: $\dfrac{\text{distance from point to focus}}{\text{distance from point to directrix}}$.

$e > 1 \Rightarrow$ Hyperbola

$e = 1 \Rightarrow$ Parabola

$0 < e < 1 \Rightarrow$ Ellipse

$e = 0 \Rightarrow$ Circle The "directrix" is infinitely far away.

Radii of an Ellipse and of a Hyperbola

An ellipse has four constant "radii," identified with letters in Figure 10-5c.

- a, **major radius,** from the center to a vertex, along the major axis
- b, **minor radius,** from the center to one end of the minor axis
- c, **focal radius,** from the center to a focus, along the major axis
- d, **directrix radius,** from the center to a directrix, along the major axis

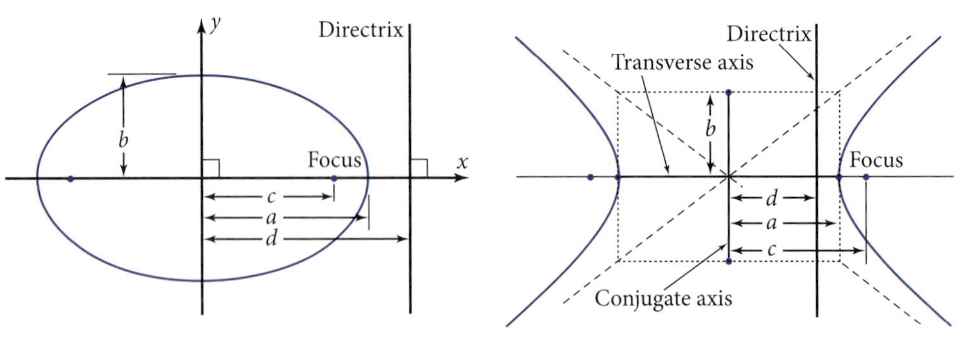

Figure 10-5c *Figure 10-5d*

The radii for a hyperbola are labeled with the same letters in Figure 10-5d. Each focus is on the concave side of the corresponding vertex, and the directrix is on the convex side. Recall that the transverse axis goes from vertex to vertex and that the conjugate axis goes through the center, perpendicular to the transverse axis.

- a, **transverse radius**, from the center to a vertex, along the transverse axis
- b, **conjugate radius,** from the center to one endpoint of the conjugate axis
- c, **focal radius,** from the center to a focus, along the transverse axis
- d, **directrix radius,** from the center to a directrix, along the transverse axis

You'll learn how the transverse, conjugate, and focal radii are related later in this section.

Focal Radii for an Ellipse or a Hyperbola

Figure 10-5b shows distances d_2 and d_3 from a point on an ellipse or a hyperbola to the two foci. An ellipse has the property that the sum of these distances is constant. For a hyperbola, the difference between these distances is constant. These properties allow you to define the ellipse and the hyperbola geometrically.

DEFINITIONS: Ellipse and Hyperbola

An **ellipse** is the set of points P in a plane for which the sum of the distances from point P to two fixed points (the foci) is constant.

A **hyperbola** is the set of all points P in a plane for which the absolute value of the difference of the distances from point P to two fixed points (the foci) is constant.

Figure 10-5e shows how you can demonstrate this property for ellipses. Tie two pins to a string so that, when stretched out, the pins are 10 cm apart. Then stick the pins into a piece of cardboard, foam board, or wood so that the pins are 8 cm apart. Set up a coordinate system with the pins at the fixed points $(-4, 0)$ and $(4, 0)$, as shown in the figure. Place a pencil as shown and draw a curve, keeping the string taut. For any point (x, y) on the resulting ellipse, the sum of its distances from the two foci equals the constant length of the string, 10 cm in this case. This distance is the length of the major axis, $2a$, twice the major radius.

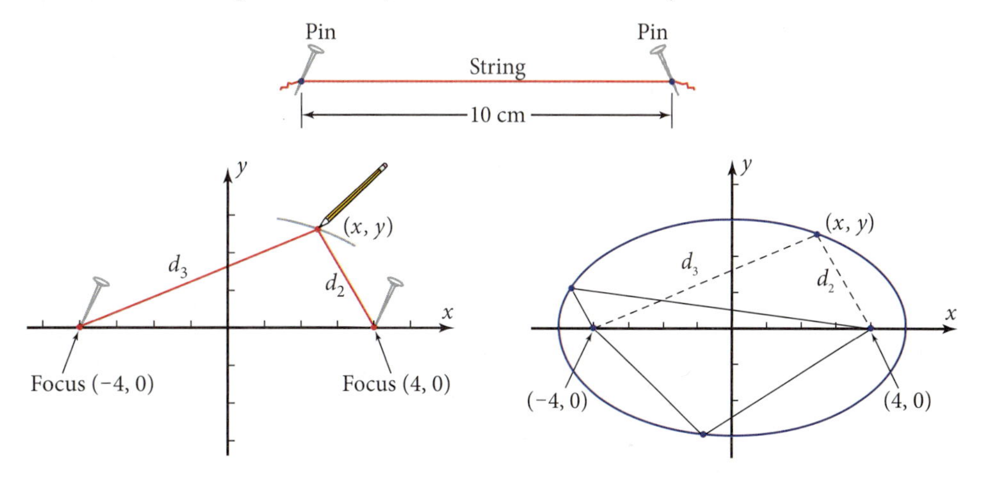

Figure 10-5e

PROPERTIES: Two-Foci Properties for Ellipses and Hyperbolas

If d_2 and d_3 are the distances from point (x, y) on an ellipse or a hyperbola to the two foci and a is the major radius or transverse radius, then

$d_2 + d_3 = 2a$ for ellipses

$|d_2 - d_3| = 2a$ for hyperbolas

Pythagorean Property of an Ellipse or a Hyperbola

The major, minor, and focal radii of an ellipse are related by a Pythagorean property. Placing the pencil of Figure 10-5e on page 525 at the end of the minor axis (Figure 10-5f) shows that the major radius, a (half the length of the string), equals the hypotenuse of a right triangle whose legs are the minor radius, b, and the focal radius, c. By the Pythagorean theorem, $a^2 = b^2 + c^2$.

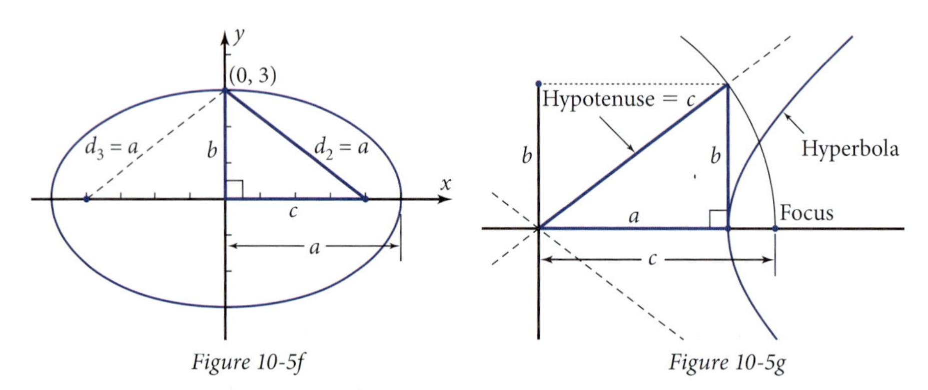

Figure 10-5f *Figure 10-5g*

Figure 10-5g illustrates that for hyperbolas the focal radius equals the hypotenuse of a right triangle whose legs are the transverse radius, a, and the conjugate radius, b. Thus, $c^2 = a^2 + b^2$.

PROPERTIES: Pythagorean Properties of Ellipses and Hyperbolas

For an ellipse, if a is the major radius, b is the minor radius, and c is the focal radius, then

$a^2 = b^2 + c^2$

For a hyperbola, if a is the transverse radius, b is the conjugate radius, and c is the focal radius, then

$c^2 = a^2 + b^2$

Radii and Eccentricity Properties of an Ellipse or a Hyperbola

The major (or transverse) radius, focal radius, and directrix radius of an ellipse or a hyperbola are related by the eccentricity, e. Figure 10-5h shows a point, P, on the ellipse at the end of the minor axis. Here, the distance d_1 to the directrix equals d, the directrix radius, and the distance d_2 to the focus equals a, the major radius. Substituting a for d_2 and d for d_1 in the equation $d_2 = ed_1$, you get

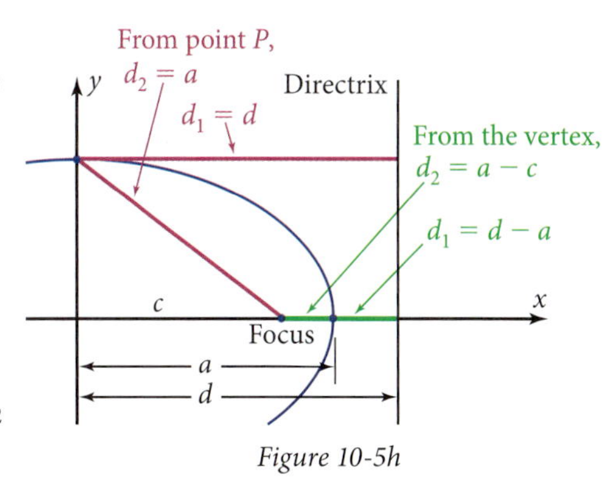

Figure 10-5h

$$a = ed$$ Major (transverse) radius equals e times the directrix radius.

If the point is at the vertex of the ellipse, then $d_1 = d - a$ and $d_2 = a - c$. Substituting $a - c$ for d_2 and $d - a$ for d_1 in the equation $d_2 = ed_1$, you get

$$a - c = e(d - a)$$

Knowing that $a = ed$, you can transform this equation to

$$c = ea$$ Focal radius equals e times the major (transverse) radius.

Figure 10-5i shows a similar relationship for a hyperbola. The equations $a = ed$ and $c = ea$ are also true for hyperbolas.

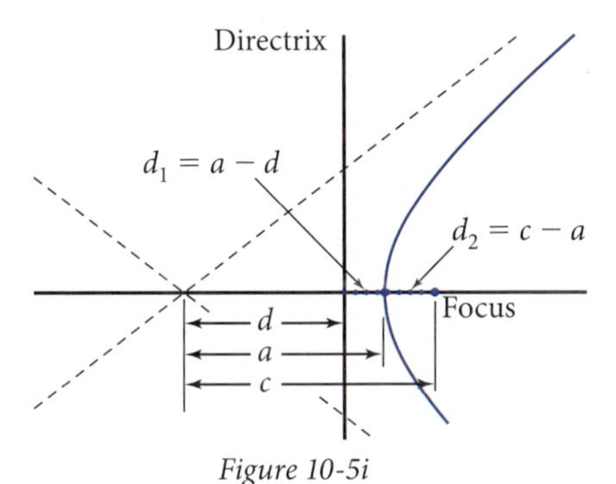

Figure 10-5i

> **PROPERTIES: Radii and Eccentricity of an Ellipse or a Hyperbola**
>
> If e is the eccentricity of an ellipse or a hyperbola, then
>
> $c = ea$ — Focal radius equals the eccentricity times the major (transverse) radius.
>
> $a = ed$ — Major (transverse) radius equals the eccentricity times the directrix radius.
>
> $e = \frac{c}{a}$ and $e = \frac{a}{d}$ — Eccentricity $= \frac{\text{focal radius}}{\text{major radius}} = \frac{\text{major radius}}{\text{directrix radius}}$

The geometric properties of ellipses and hyperbolas are summarized in this box.

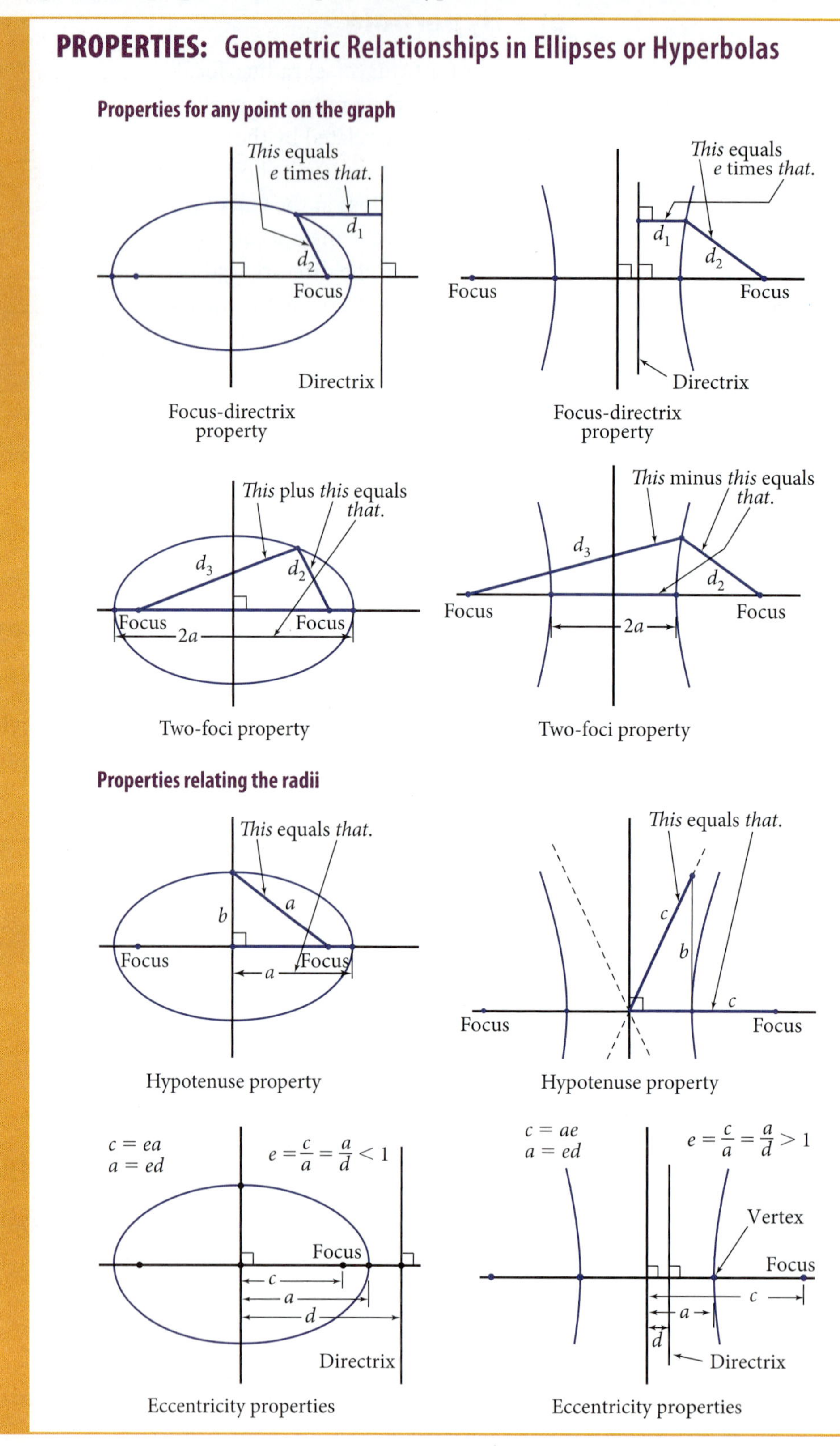

EXAMPLE 1 ➤ Consider the ellipse $49x^2 + 16y^2 = 784$.

a. Sketch the graph. Show the two foci and the two directrices.

b. Find the major, minor, and focal radii, the eccentricity, and the directrix radius.

c. Calculate y for $x = 3$. Show that the distance from the point (x, y) to one focus is e times its distance to the corresponding directrix.

SOLUTION First transform the equation to find the two dilation factors.

a. $49x^2 + 16y^2 = 784$ — Write the given equation.

$\left(\frac{x}{4}\right)^2 + \left(\frac{y}{7}\right)^2 = 1$ — Make the right side equal 1 to find the two dilation factors.

Sketch the graph (Figure 10-5j, left) with x-radius 4 and y-radius 7. Show the foci on the concave side of the corresponding vertices and the directrices on the convex side.

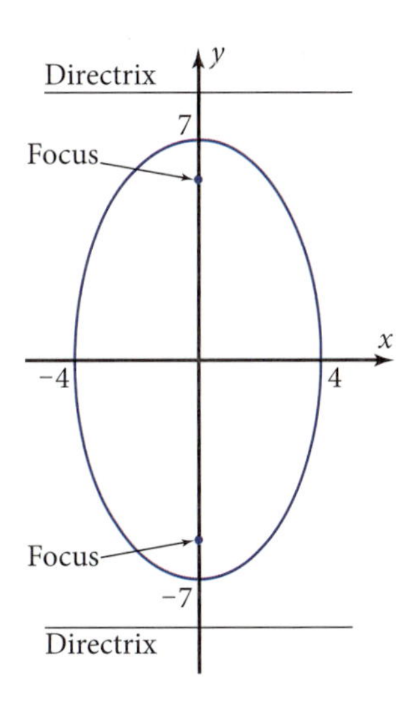

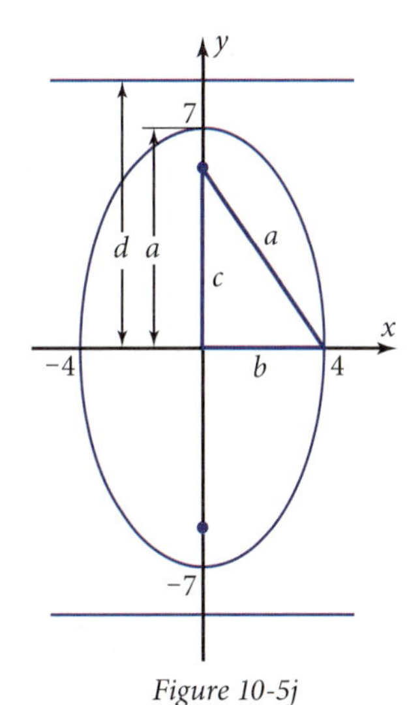

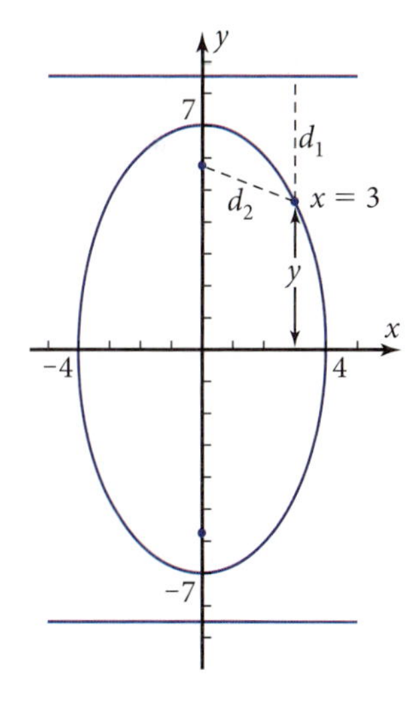

Figure 10-5j

b. Major radius: $a = 7$ — The major radius is the longer radius (Figure 10-5j, middle).

Minor radius: $b = 4$

Focal radius: $c^2 = a^2 - b^2 = 7^2 - 4^2 = 33$ — Use the Pythagorean property: Hypotenuse = major radius.

$$c = \sqrt{33} = 5.7445...$$

Store as c in your grapher.

Eccentricity: $e = \frac{c}{a} = \frac{\sqrt{33}}{7} = 0.8206...$ — e is between 0 and 1, as is true for ellipses. Store as e.

Directrix radius:

$$e = \frac{a}{d} \Rightarrow d = \frac{a}{e} = \frac{7}{\frac{\sqrt{33}}{7}} = \frac{49}{\sqrt{33}} = 8.5298...$$

d is the distance from the center to the directrix (the directrix radius).

Note that $d = 8.5298...$ is greater than 7, which means that the directrix is on the convex side of the vertex, and that $c = 5.7445...$ is less than 7, which means that the focus is on the concave side.

c. $49(3^2) + 16y^2 = 784$ — Substitute 3 for x in the given equation.

$$y = \pm\sqrt{\frac{343}{16}} = \pm 4.6300...$$

The right ellipse in Figure 10-5j shows the point (3, 4.6300...) and distances d_1 and d_2.

$$d_1 = d - y = 8.5298... - 4.6300... = 3.8997...$$

$$d_2 = \sqrt{3^2 + (c - y)^2} = \sqrt{3^2 + 1.1144...^2} = 3.2003...$$

Use the Pythagorean theorem.

$$ed_1 = (0.8206...)(3.8997...) = 3.2003..., \text{ which equals } d_2$$

EXAMPLE 2 ➤ Consider a conic section with eccentricity 1.25 and foci (6, 2) and (−4, 2).

a. Identify the conic and find its equation.

b. Sketch the graph, showing the foci and directrices. You may first plot the conic parametrically or using a program like the one in Section 10-2.

SOLUTION Start by making a sketch of the foci (Figure 10-5k).

a. The conic is a hyperbola because e is greater than 1. The hyperbola opens in the x-direction because the transverse axis (through the foci) is horizontal.

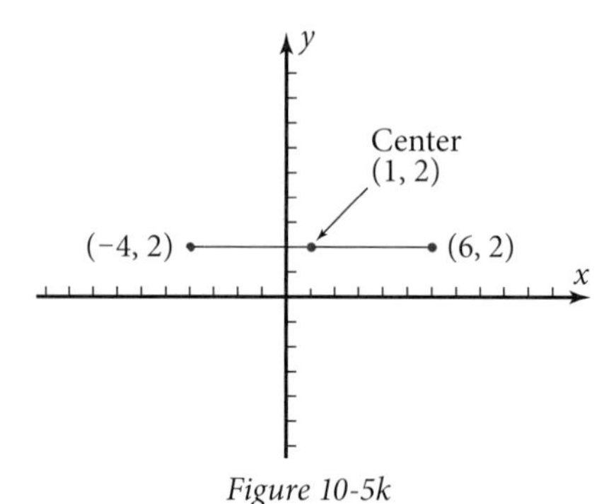

Figure 10-5k

The center is at the point (1, 2). — Average −4 and 6 to find the x-coordinate.

$c = 6 - 1 = 5$ — Focal radius goes from the center to the focus.

$e = \frac{c}{a} \Rightarrow a = \frac{c}{e} = \frac{5}{1.25} = 4$ — Use the eccentricity to find the transverse radius.

$b^2 = c^2 - a^2 = 25 - 16 = 9 \Rightarrow b = 3$ — Use the Pythagorean property to find the conjugate radius.

The equation is $\left(\frac{x-1}{4}\right)^2 - \left(\frac{y-2}{3}\right)^2 = 1.$ — a goes under the positive term; b goes under the negative term.

b. Use the eccentricity to find the directrix radius.

$$e = \frac{a}{d} \Rightarrow d = \frac{a}{e} = \frac{4}{1.25} = 3.2$$

See the graph in Figure 10-5l. Each focus is on the concave side of the vertex, and each directrix is on the convex side.

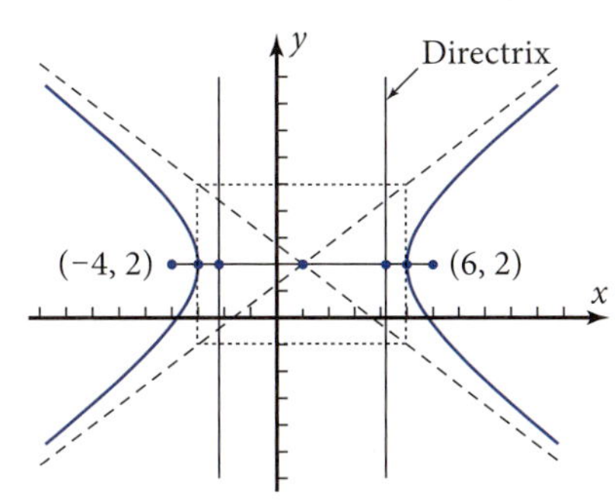

Figure 10-5l

Parametrically,

$x = 1 + 4 \sec t$ Secant goes with the direction in which the

$y = 2 + 3 \tan t$ hyperbola opens.

To use the program in Section 10-2, first transform the Cartesian equation to polynomial form:

$$9x^2 - 16y^2 - 18x + 64y - 199 = 0$$

EXAMPLE 3 ➤ Find the Cartesian equation for the conic with focus $(-1, 2)$, directrix $x = 3$, and eccentricity $e = \frac{3}{4}$. How is the result consistent with the eccentricity property of conics? Plot the graph on your grapher using the program in Section 10-2, Example 8. Sketch the result.

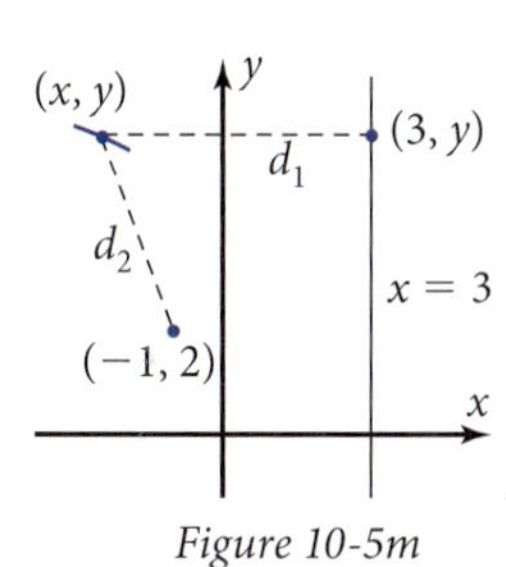

Figure 10-5m

SOLUTION Draw a sketch showing the given directrix and focus and a point (x, y) representing a point on the graph (Figure 10-5m). Show d_1 from point (x, y) to the directrix and d_2 from point (x, y) to the focus.

$d_2 = ed_1$ Focus-directrix property of conics.

$\sqrt{(x + 1)^2 + (y - 2)^2} = \frac{3}{4} \cdot |x - 3|$ By the distance formula.

$16(x^2 + 2x + 1 + y^2 - 4y + 4) = 9(x^2 - 6x + 9)$

Square both sides and simplify.

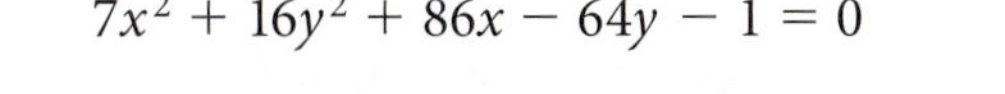

$7x^2 + 16y^2 + 86x - 64y - 1 = 0$ Make the right side equal zero.

The squared terms have the same sign but unequal coefficients, indicating an ellipse. This is consistent with the fact that e is between 0 and 1.

Figure 10-5n

Figure 10-5n shows the ellipse with the given focus and directrix.

Focal Distance of a Parabola

The radius properties given previously do not apply to parabolas because a parabola has only one vertex and no center. For a parabola, $d_2 = d_1$ because the eccentricity equals 1. This property lets you define parabolas geometrically.

> **DEFINITION: Parabola**
>
> A **parabola** is the set of all points P in a plane for which point P's distance to a fixed point (the focus) is equal to its distance to a fixed line that does not pass through the focus (the directrix).

Using this property you can find the equation of a parabola with vertex at the origin in terms of the distance p from the vertex to the focus or directrix (Figure 10-5o). This distance, p, is called the **focal distance** of a parabola.

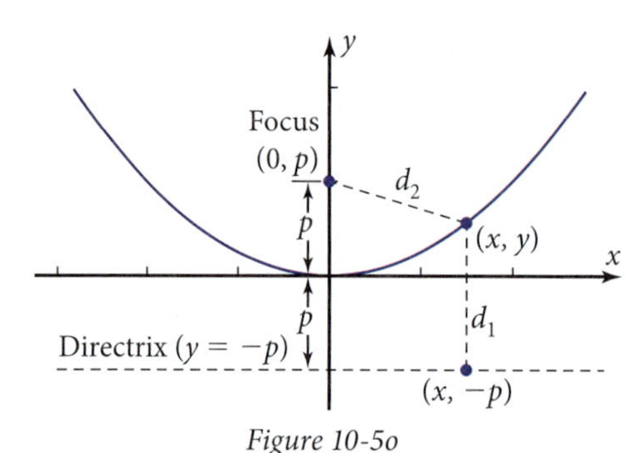

Figure 10-5o

Pick a point (x, y) on the parabola (Figure 10-5o).

$$d_2 = d_1 \qquad \text{The eccentricity of a parabola is 1.}$$

$$\sqrt{x^2 + (y - p)^2} = |y + p| \qquad \text{By the distance formula.}$$

$$x^2 + y^2 - 2py + p^2 = y^2 + 2py + p^2 \qquad \text{Square both sides and expand.}$$

$$x^2 = 4py$$

$$y = \frac{1}{4p}x^2$$

> **PROPERTY: Focus-Directrix Equation of Parabolas**
>
> The equation of a parabola with vertex at the origin and axis of symmetry along a coordinate axis is
>
> $y = \frac{1}{4p}x^2$ or $x = \frac{1}{4p}y^2$ — Opening in the positive direction
>
> $y = -\frac{1}{4p}x^2$ or $x = \frac{1}{4p}y^2$ — Opening in the negative direction
>
> where p is the distance from the vertex to the focus or to the directrix.

EXAMPLE 4 ➤ Find the particular equation for the parabola with focus at the point (2, 3) and directrix $y = 5$. Plot the equation on your grapher and sketch the result. Calculate the two x-intercepts, and show that the graph agrees with your calculations.

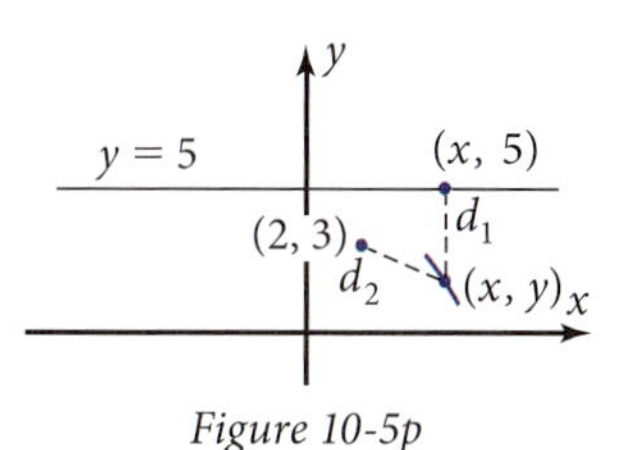

Figure 10-5p

SOLUTION Sketch the focus and directrix as shown in Figure 10-5p.

Pick (x, y) representing a point on the parabola, and draw on the sketch the distances d_1 from the point (x, y) to the directrix and d_2 from the point (x, y) to the focus. Because the graph is a parabola, the eccentricity equals 1. Thus, $d_1 = d_2$.

$$d_2 = d_1$$ Write an equation.

$$\sqrt{(x-2)^2 + (y-3)^2} = |y-5|$$ Substitute for the distances in terms of x and y.

$$x^2 - 4x + 4 + y^2 - 6y + 9 = y^2 - 10y + 25$$ Square both sides, and expand.

$$4y = -x^2 + 4x + 12$$ Solve for y as a function of x.

$$y = -0.25x^2 + x + 3$$

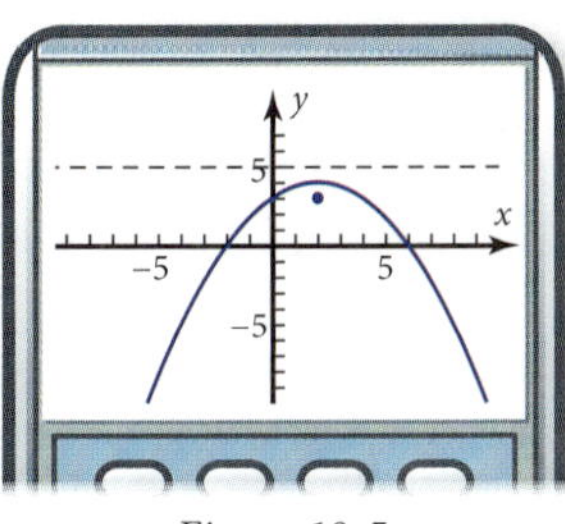

Figure 10-5q

The graph in Figure 10-5q shows the vertex of the parabola halfway between the focus and the directrix.

$$0 = -0.25x^2 + x + 3$$ Set $y = 0$ to find the x-intercepts.

$$x = 6 \quad \text{or} \quad x = -2$$ By the quadratic formula.

The graph agrees with these values. ◄

EXAMPLE 5 ➤ Find the focal distance for the parabola in Example 4.

SOLUTION The focal distance, p, is the distance from the vertex to the focus. The focus is the point (2, 3). The equation of the parabola is $y = -0.25x^2 + x + 3$.

The x-value of the vertex is $x = 2$. The x-value of the vertex is the same as the x-value of the focus.

$$y = -0.25(2)^2 + (2) + 3 = (-1) + 2 + 3 = 4$$ Find the value of y when $x = 2$.

The vertex is the point (2, 4).

$$p = 4 - 3 = 1$$ Since the x-coordinates are the same, the distance between the vertex and the focus is the difference in their y-coordinates.

The focal distance for the parabola is 1 unit. ◄

Note: The dilation factor, a, of the parabola in Examples 4 and 5 is -0.25. The focus-directrix equation of a parabola establishes that $a = \pm\frac{1}{4p}$. So,

$$-0.25 = \pm\frac{1}{4p} \Rightarrow \pm 4(-0.25) = p \Rightarrow \pm 1 = p$$

Since p is defined as a distance, it will always be positive, so $p = 1$ and the equation can be written as $|a| = \frac{1}{4p}$.

Problem Set 10-5

Reading Analysis

From what you have read in this section, what do you consider to be the main idea? How does the eccentricity of a conic section act as a proportionality constant relating the distances from a point on the curve to two fixed objects? Explain why the major radius of an ellipse can also be depicted as the hypotenuse of a right triangle within the ellipse. Why does a parabola have only one focus and one vertex, while both ellipses and hyperbolas have two?

Quick Review

For Problems Q1–Q6, name the quadric surface you get by rotating as stated.

Q1. An ellipse about its major axis

Q2. An ellipse about its minor axis

Q3. A hyperbola about its transverse axis

Q4. A hyperbola about its conjugate axis

Q5. A circle about its diameter

Q6. A parabola about its axis of symmetry

Q7. How do you recognize a hyperbola from its Cartesian equation?

Q8. How do you distinguish between a circle and an ellipse from the Cartesian equation?

Q9. How do you tell which way a parabola opens from its Cartesian equation?

Q10. What is the origin of the word *ellipse*?

---

1. *Hyperbola Problem:* Figure 10-5r shows the hyperbola $16x^2 - 9y^2 = 144$. Its foci are at the points $(-5, 0)$ and $(5, 0)$, and a directrix is the line $x = 1.8$. Its eccentricity, e, is $\frac{5}{3}$.

a. A point on the hyperbola in Quadrant I has x-coordinate 7. Calculate y for this point. Does the value agree with the graph? Store the answer as y in your grapher.

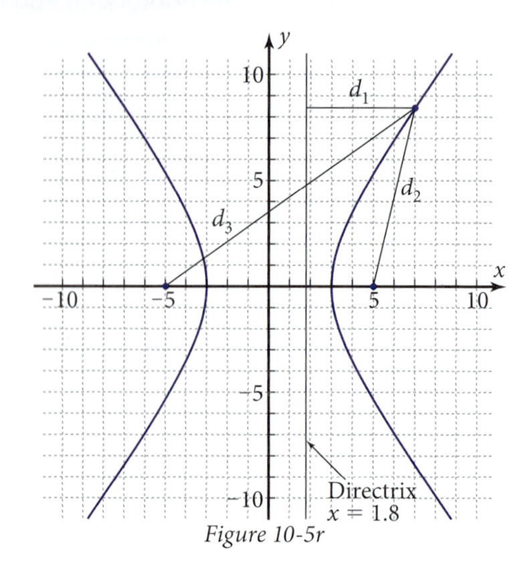

Figure 10-5r

b. Use the Pythagorean theorem and the result of part a to calculate these distances:

- d_1 from the point $(7, y)$ to the directrix
- d_2 from the point $(7, y)$ to the focus $(5, 0)$
- d_3 from the point $(7, y)$ to the focus $(-5, 0)$

c. Show that $d_2 = ed_1$.

d. Show that $|d_2 - d_3| = 6$, the length of the transverse axis (between the vertices).

e. Find the x- and y-dilations. Which of these is the transverse radius, a, and which is the conjugate radius, b?

f. As shown in Figure 10-5r, the focal radius is $c = 5$. Show that $c^2 = a^2 + b^2$, the Pythagorean property of hyperbolas.

g. Show that the directrix radius, $d = 1.8$, satisfies the equation $a = ed$ and that the focal radius, $c = 5$, satisfies the equation $c = ea$.

2. *Ellipse Problem:* Figure 10-5s shows the ellipse $25x^2 + 9y^2 = 225$. Its foci are at the points $(0, -4)$ and $(0, 4)$, and a directrix is the line $y = 6.25$. Its eccentricity, e, is 0.8.

a. A point on the ellipse in Quadrant I has y-coordinate 3. Calculate x for this point. Does your answer agree with the graph? Store the answer as x in your grapher.

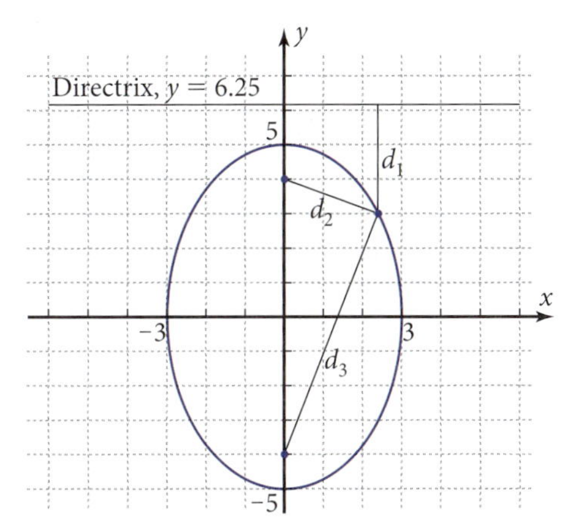

Figure 10-5s

b. Use the Pythagorean theorem and the result of part a to calculate these distances:

- d_1 from the point $(x, 3)$ to the directrix
- d_2 from the point $(x, 3)$ to the focus $(0, 4)$
- d_3 from the point $(x, 3)$ to the focus $(0, -4)$

c. Show that $d_2 = ed_1$.

d. Show that $d_2 + d_3 = 10$, the length of the major axis.

e. Find the x- and y-dilations. Which of these is the major radius, a, and which is the minor radius, b?

f. As shown in Figure 10-5s, the focal radius is $c = 4$. Show that $a^2 = b^2 + c^2$, the Pythagorean property of ellipses.

g. Show that the directrix radius, $d = 6.25$, satisfies the equation $a = ed$ and that the focal radius, $c = 4$, satisfies the equation $c = ea$.

The Coliseum in Rome, Italy, is an elliptical amphitheater that once seated 50,000 people.

3. *Parabola Problem:* Figure 10-5t shows the parabola $y = \frac{1}{8}x^2$. Its focus is at the point $(0, 2)$, and its directrix is the line $y = -2$.

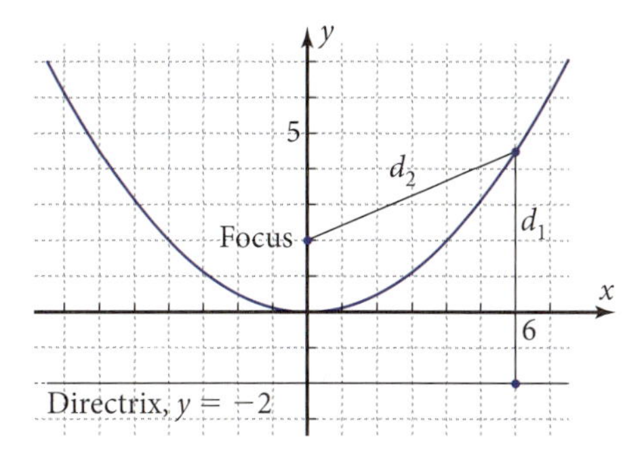

Figure 10-5t

a. The vertex is at the origin, $(0, 0)$. Explain how this fact confirms that the parabola's eccentricity is 1.

b. The point shown in the first quadrant has x-coordinate 6. Calculate y for this point. Does the value agree with the graph?

c. Calculate these distances:

- d_1 from the point $(6, y)$ to the directrix
- d_2 from the point $(6, y)$ to the focus

d. Show that $d_2 = d_1$ and that this fact is consistent with eccentricity $= 1$ for a parabola.

The paraboloid mirror behind the lightbulb in a headlight's focus reflects its light in parallel rays.

4. *Circle Problem:* The circle $x^2 + y^2 = 25$ can be considered an ellipse with major and minor radii equal to each other.

   a. Find the major and minor radii.

   b. Based on the Pythagorean property of ellipses, explain why the focal radius of a circle is zero. Where, then, are the foci of a circle?

   c. The eccentricity of an ellipse, e, is $\frac{c}{a}$, where c and a are the focal radius and major radius, respectively. Explain why the eccentricity of a circle is zero. Why is the name *eccentricity* appropriate in this case?

   d. The eccentricity of an ellipse is also equal to $\frac{a}{d}$, where d is the directrix radius. Based on the answer to part c, explain why the directrix of a circle is infinitely far from its center.

5. *Conic Construction Problem 1:* Plot on graph paper the conic with focus (0, 0), directrix $x = -6$, and eccentricity $e = 2$. Put the x-axis near the middle of the graph paper and the y-axis just far enough from the left side to fit the directrix on the paper. Plot the points for which the distance d_1 from the directrix equals 2, 4, 6, 8, and 10 units. Connect the points with a smooth curve. Which conic section have you graphed?

6. *Conic Construction Problem 2:* Plot on graph paper the conic with focus (0, 0), directrix $x = -6$, and eccentricity $e = 1$. Plot points for which the distance d_1 from the directrix equals 3, 6, 10, and 20 units. Connect the points with a smooth curve. Which conic section have you graphed?

7. *Computer Construction Problem:* Use geometry software such as The Geometer's Sketchpad to plot the conics in Problems 5 and 6. Sketch the results. How do the graphs confirm your conclusion about the kind of conic section plotted?

8. *Dynamic Conics Problem:* Figure 10-5u shows a fixed directrix and a fixed focus, with conics of varying eccentricity. Open the *Unified Conics* exploration at **www.keymath.com/precalc** and drag point E to view each conic shown in Figure 10-5u. Then return to the ellipse.

   a. Click the Distances button. Explain why the eccentricity is equal to the distance from the vertex, point E, to the focus, point F, divided by the distance from point E to point A on the directrix.

   b. Drag the point P on the ellipse. What happens to the distances between point P and the focus and between point P and the directrix? What happens to the ratio of these distances? Why is it correct to say that the eccentricity of a given ellipse is constant?

   c. Click the Hyperbola button. Describe the changes in the figure and in the eccentricity. Drag point P on the hyperbola. Does the eccentricity remain constant as the distances between point P and the focus and between point P and the directrix change?

   d. Click the Parabola button. What happens to the second branch of the hyperbola? What happens to the eccentricity? Drag point P on the parabola and describe what you notice.

   e. Click the Animate E button. Describe the changes that occur in the graph and in the eccentricity as point E moves closer to the focus and as it moves closer to the directrix.

   f. What do you understand better about conic sections as a result of working this problem?

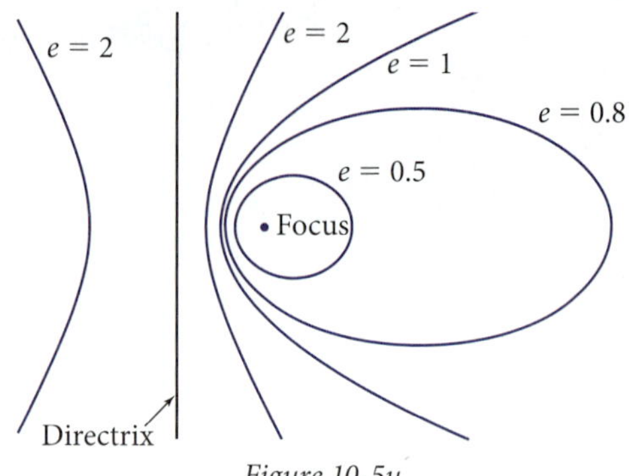

Figure 10-5u

9. *Mars Orbit Problem*: Mars is in an elliptical orbit around the Sun, with the Sun at one focus. The *aphelion* (the point farthest from the Sun) and the *perihelion* (the point closest to the Sun) are 155 million miles and 128 million miles from the Sun, respectively, as shown in Figure 10-5v.

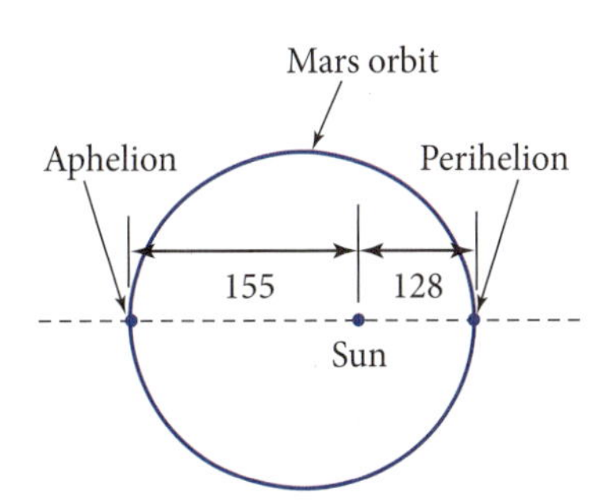

Figure 10-5v

a. How long is the major axis of the ellipse? What is the major radius?

b. Find the focal radius and the minor radius of the ellipse.

c. Write a Cartesian equation for the ellipse, with the center at the origin and the major axis along the x-axis.

d. At the two equinoxes (times of equal day and night), the angle at the Sun between the major axis and Mars is 90°. At these times, what is the value of x? How far is Mars from the Sun?

e. Find the eccentricity of the ellipse.

f. How far from the Sun is the closer directrix of the ellipse?

g. Write parametric equations for the ellipse. Plot the graph using parametric mode. Zoom appropriately to make equal scales on the two axes.

h. The ellipse you plotted in part g looks almost circular. How do the major and minor radii confirm this? How does the eccentricity confirm this?

10. *Comet Path Problem:* Figure 10-5w shows the path of a comet approaching Earth. The path is a conic section with eccentricity $e = 1.1$ and directrix radius $d = 100{,}000$ mi.

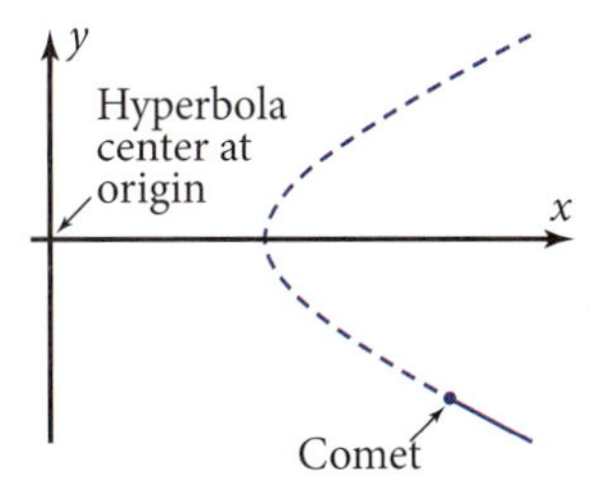

Figure 10-5w

a. How can you tell from the given information that the path is a hyperbola?

b. The center of the hyperbola is at the origin in Figure 10-5w, and the transverse axis is along the x-axis. On a sketch of the figure, show the focus and the directrix.

c. The center of Earth is at one focus of the hyperbola. Find the coordinates of the focus.

d. The comet is closest to Earth when it is at one vertex. How close does it come to the center of Earth? How close does it come to the surface of Earth, 4000 mi from the center?

e. Write parametric equations for the hyperbola. What range of t-values will generate the branch shown in Figure 10-5w?

f. When the comet is at the point shown, x equals 200,000 mi. At this time, what does the parameter t equal? What does y equal? How far is the comet from the center of Earth?

For Problems 11–20,

a. Identify the conic section.

b. Calculate the four radii (ellipses and hyperbolas) or the focal distance (parabolas) and the eccentricity.

c. Plot the graph. Sketch the result.

11. $\frac{x^2}{9} + \frac{y^2}{25} = 1$

12. $\frac{x^2}{289} + \frac{y^2}{64} = 1$

13. $-\frac{x^2}{36} + \frac{y^2}{9} = 1$

14. $\frac{x^2}{9} - \frac{y^2}{16} = 1$

15. $\left(\frac{x-1}{4}\right)^2 + \left(\frac{y+2}{3}\right)^2 = 1$

16. $-\left(\frac{x+1}{3}\right)^2 + \left(\frac{y-2}{16}\right)^2 = 1$

17. $5x^2 - 3y^2 = -30$

18. $16x^2 + 25y^2 = 1600$

19. $x = -\frac{1}{4}y^2 + 3$

20. $x = \frac{1}{8}y^2 + 1$

For Problems 21–32,

a. Sketch the conic section showing the given information.

b. Find the particular equation (Cartesian or parametric).

c. Plot the graph on your grapher. Does your sketch in part a agree with the graph?

21. Focus $(0, 0)$, directrix $y = 3$, eccentricity $e = 2$. Identify the conic section.

22. Focus $(0, 0)$, directrix $x = 5$, eccentricity $e = \frac{3}{4}$. Identify the conic section.

23. Focus $(0, 0)$, directrix $y = -4$, eccentricity $e = 1$. Identify the conic section.

24. Focus $(0, 0)$, directrix $x = \frac{1}{2}$, eccentricity $e = 1$. Identify the conic section.

25. Focus $(2, -3)$, directrix $y = 0$, eccentricity $e = \frac{1}{2}$. Identify the conic section.

26. Focus $(3, 1)$, directrix $x = 2$, eccentricity $e = 4$. Identify the conic section.

27. Ellipse with foci $(12, 0)$ and $(-12, 0)$ and constant sum of distances equal to 26

28. Hyperbola with foci $(0, 5)$ and $(0, -5)$ and constant difference of distances equal to 8

29. Hyperbola with vertices $(-1, 3)$ and $(5, 3)$ and slope of asymptotes $\pm\frac{2}{3}$

30. Ellipse with vertices $(4, -2)$ and $(4, 8)$ and minor radius 3

31. Parabola with focus $(2, 3)$ and directrix $y = 5$

32. Parabola with focus $(4, 5)$ and vertex $(4, 2)$

33. *Latus Rectum Problem:* The *latus rectum* of a conic section is the chord through a focus parallel to the directrix (Figure 10-5x). By appropriate substitution into the equations, find the length of the latus rectum for the conics in Problems 9, 11, and 17.

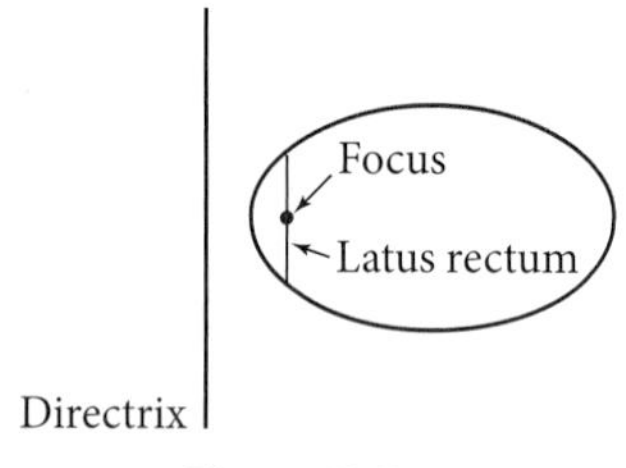

Figure 10-5x

10-6 Parametric and Cartesian Equations for Rotated Conics

Previously in this chapter you learned how to write parametric equations for circles, ellipses, hyperbolas, and parabolas. In this section you will learn how you can transform the parametric equations to rotate the conic through a given angle. The Cartesian equations for these rotated conics turn out to contain an xy-term as well as x^2- or y^2-terms.

Objective

- Plot a conic section rotated by a specified angle to the coordinate axes.
- Identify a rotated conic from its Cartesian equation.
- Plot a rotated conic using its Cartesian equation or parametric equations.

In this exploration you will learn how to rotate a figure by multiplying its parametric equations by a rotation matrix.

EXPLORATION 10-6: Rotation of a Figure by Matrix Multiplication

1. Let $[A]$ be the 2×2 matrix $[A] = \begin{bmatrix} 2 & 5 \\ 6 & 7 \end{bmatrix}$.

Let $[B]$ be the 2×1 matrix $[B] = \begin{bmatrix} 4 \\ 3 \end{bmatrix}$.

The *product matrix* $[A] \cdot [B]$ is found by multiplying the elements in a *row* of the first matrix by the corresponding elements in a *column* of the second matrix. The first element in the product matrix is calculated this way:

$$[A] \cdot [B] = \begin{bmatrix} 2 & 5 \\ 6 & 7 \end{bmatrix}\begin{bmatrix} 4 \\ 3 \end{bmatrix} = \begin{bmatrix} 23 \\ \ \end{bmatrix}$$

$(2)(4) + (5)(3) = 23$

Follow this pattern to multiply the second row of the first matrix by the column of the second matrix. Write the resulting 2×1 product matrix. Then check your answer by entering $[A]$ and $[B]$ into your grapher and multiplying.

2. Matrix $[B]$ in Problem 1 can represent the point (4, 3) shown in the figure. Matrix $[A]$ causes a transformation of that point to a new location. Replace $[A]$ in your grapher with the following matrix. Then multiply $[A]$ times $[B]$ and write the product matrix.

$$[A] = \begin{bmatrix} \cos 25° & \cos(25° + 90°) \\ \sin 25° & \sin(25° + 90°) \end{bmatrix}$$

A 25° counterclockwise rotation matrix.

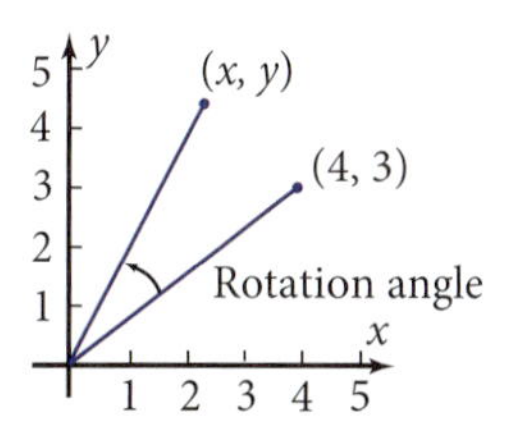

3. The product matrix in Problem 2 gives the coordinates of the point (x, y) in the figure. Does your calculation agree with the point plotted in the figure? If not plotted, check your work.

4. In Problem 2, the 25° angle in $[A]$ is the counterclockwise angle through which the line segment from the origin to the point (4, 3) is rotated. Measure this rotation angle with a protractor. Is it really 25°?

5. Show that the rotation in Problem 2 has not changed the length of the line segment.

6. Write a rotation matrix for

a. A counterclockwise rotation by 100°

b. A clockwise rotation by 70°

7. What did you learn as a result of doing this exploration that you did not know before?

Parametric Equations for Rotated Conics

In Exploration 10-6, you learned that by multiplying a certain 2×2 *rotation matrix* by a 2×1 matrix representing a point (x, y), you get a 2×1 product matrix representing the *image* of the point (x, y) rotated about the origin. In Section 13-3 you will learn the basis for this kind of rotation. For now, the box shows you how to form the rotation matrix.

PROPERTY: Rotation Matrix

If $[A] = \begin{bmatrix} \cos\alpha & \cos(\alpha + 90°) \\ \sin\alpha & \sin(\alpha + 90°) \end{bmatrix}$ and $[B] = \begin{bmatrix} x \\ y \end{bmatrix}$, then the matrix product

$$[A][B] = \begin{bmatrix} \cos\alpha & \cos(\alpha + 90°) \\ \sin\alpha & \sin(\alpha + 90°) \end{bmatrix} \begin{bmatrix} x \\ y \end{bmatrix}$$

represents a rotation of the point (x, y) through an angle of α degrees counterclockwise about the origin.

If you substitute the parametric equations $x(t)$ and $y(t)$ of a conic section into the 2×1 matrix, you get parametric equations for the conic section rotated through an angle of α degrees. Example 1 shows you how to do this.

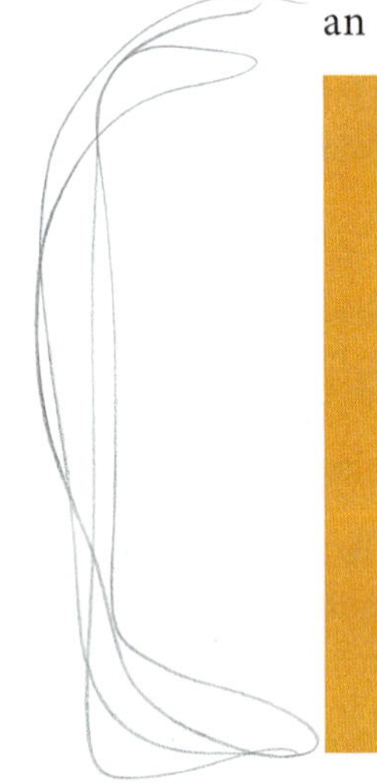

PARAMETRIC EQUATIONS OF CONIC SECTIONS: Center at Origin

Where a is the major radius, the transverse radius, or, in a parabola, the squared-term coefficient, and b is the minor radius or conjugate radius

	***x*-axis**	***y*-axis**	
Circle or ellipse:	$x = a\cos t$ $y = b\sin t$	$x = b\cos t$ $y = a\sin t$	$a \geq b$
Hyperbola:	$x = a\sec t$ $y = b\tan t$	$x = b\tan t$ $y = a\sec t$	
Parabola:	$x = at^2$ $y = t$	$x = t$ $y = at^2$	Vertex at origin.

EXAMPLE 1 ➤ Write parametric equations for the hyperbola centered at the point (1, 3) and with transverse radius 5 and conjugate radius 2 if the transverse radius makes an angle of 25° (counterclockwise) with respect to the x-axis. Plot the graph on your grapher and sketch the result.

SOLUTION First write parametric equations for the unrotated hyperbola centered at the origin, shown in the graph on the left in Figure 10-6a. The x-dilation equals the transverse radius, and the y-dilation equals the conjugate radius.

$x = 5\sec t$
$y = 2\tan t$

The hyperbola opens in the x-direction, so secant goes with x.

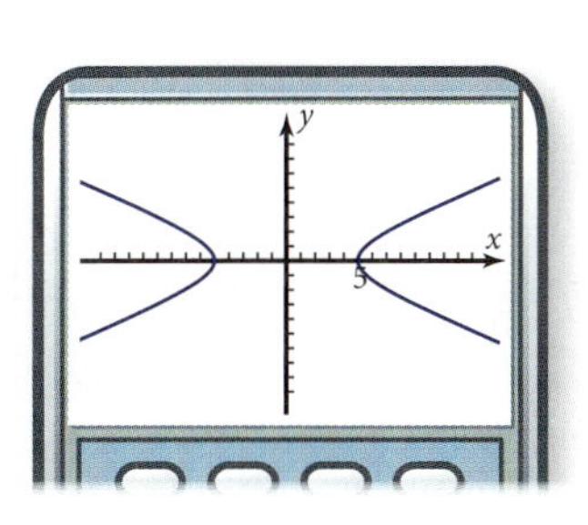

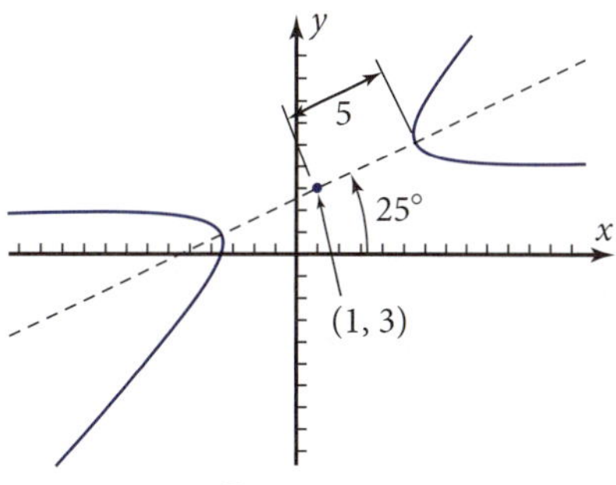

Figure 10-6a

Next, rotate the hyperbola 25° about the origin.

$$\begin{bmatrix} \cos 25° & \cos 115° \\ \sin 25° & \sin 115° \end{bmatrix} \begin{bmatrix} 5 \sec t \\ 2 \tan t \end{bmatrix} \qquad 115° = 90° + 25°$$

$$= \begin{bmatrix} 5 \cos 25° \sec t + 2 \cos 115° \tan t \\ 5 \sin 25° \sec t + 2 \sin 115° \tan t \end{bmatrix}$$

$$= \begin{bmatrix} 4.5315... \sec t - 0.8452... \tan t \\ 2.1130... \sec t + 1.8126... \tan t \end{bmatrix}$$

Finally, write the parametric equations for x and y with a horizontal translation by 1 unit and a vertical translation by 3 units to put the center at (1, 3).

$$x = 1 + 4.5315... \sec t - 0.8452... \tan t$$
$$y = 3 + 2.1130... \sec t + 1.8126... \tan t$$

With your grapher in parametric mode and degree mode, plot these parametric equations. Use [0°, 360°] for t. The graph should look like the graph on the right in Figure 10-6a. Be sure to use equal scales on the two axes.

This box summarizes the three-step process in Example 1.

TECHNIQUE: Parametric Equations for a Rotated Conic Section

- Write parametric equations for the desired conic with its center or vertex at the origin.
- Rotate that conic with respect to the origin.
- Translate the rotated conic by the desired amount.

Cartesian Equation with *xy*-Term

Beginning in Section 10-2, you have been using a program to plot conics in function mode if the equation has the form

$$Ax^2 + Bxy + Cy^2 + Dx + Ey + F = 0$$

You can use this program to explore the graphs of conics for which the xy-term is not zero.

EXAMPLE 2 ➤ Plot the graph of the equation $9x^2 - 40xy + 25y^2 - 8x + 2y - 28 = 0$. Which conic section does the graph appear to be? Does this conclusion agree with what you have learned about the coefficients of the x^2- and y^2-terms?

SOLUTION Graph the conic using the program in Section 10-2, Example 8. The graph is shown in Figure 10-6b. The result appears to be a hyperbola whose axes are tilted at an angle to the coordinate axes. Based on the coefficients of x^2 and y^2, the graph would be expected to be an ellipse, not a hyperbola.

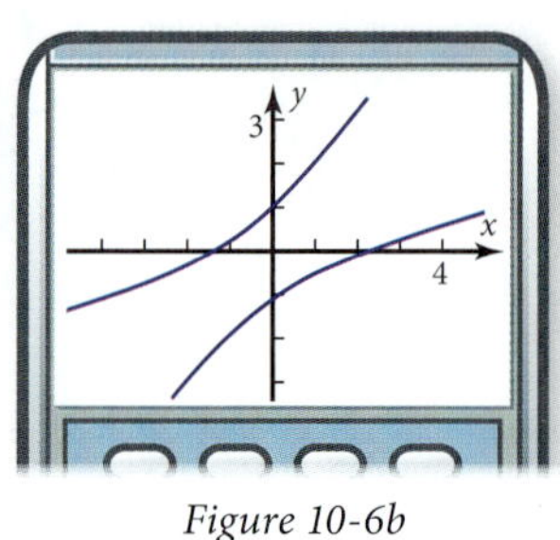

Figure 10-6b

EXAMPLE 3 ➤ Plot the graph of the equation $9x^2 - 20xy + 25y^2 - 8x + 2y - 28 = 0$ on the same screen as the hyperbola in Figure 10-6b. Describe the similarities and differences.

SOLUTION The graph is shown in green in Figure 10-6c, superimposed on the hyperbola in Example 2. With $-20xy$ instead of $-40xy$, the graph is now a rotated ellipse. The ellipse and the hyperbola have the same x- and y-intercepts because the xy-term equals zero if either x or y is zero.

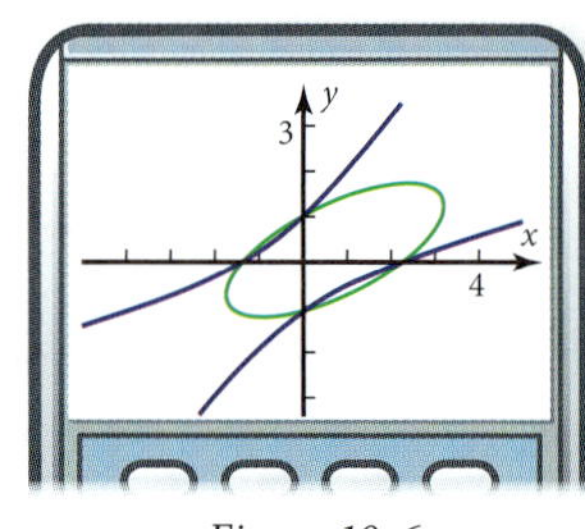

Figure 10-6c

EXAMPLE 4 ➤ Plot the graph of the equation $9x^2 - 30xy + 25y^2 - 8x + 2y - 28 = 0$ on the same screen as the hyperbola and ellipse in Figure 10-6c. Describe the similarities and differences.

SOLUTION The new graph is a parabola, shown in magenta in Figure 10-6d. An xy-term with the right coefficient makes the rotated conic a parabola. The x- and y-intercepts of the graphs are the same as before.

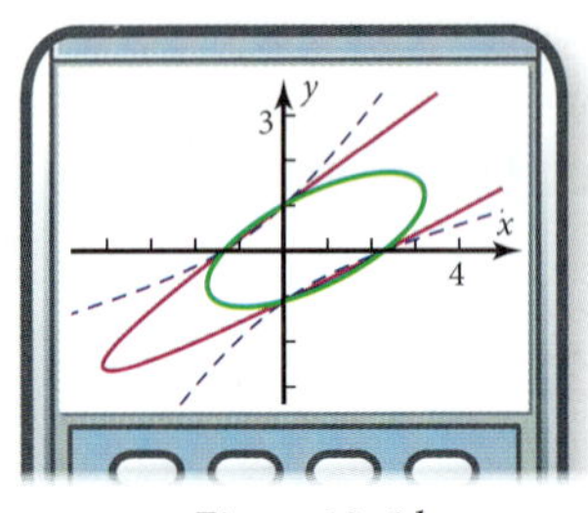

Figure 10-6d

Identifying a Conic by the Discriminant

In Examples 2, 3, and 4, the xy-coefficient takes on different values. As the absolute value of the coefficient increases, the graph changes from an ellipse to a parabola to a hyperbola. It is possible to tell which figure the graph will be from the coefficients of the three quadratic terms, x^2, xy, and y^2. To see what that relationship is, start with the equation of the hyperbola in Example 2, and solve it for y in terms of x using the quadratic formula.

$$9x^2 - 40xy + 25y^2 - 8x + 2y - 28 = 0$$ Write the equation from Example 2.

$$25y^2 + (-40x + 2)y + (9x^2 - 8x - 28) = 0$$ Write the equation as a quadratic in y.

$$y = \frac{-(-40x + 2) \pm \sqrt{(-40x + 2)^2 - 4(25)(9x^2 - 8x - 28)}}{2(25)}$$

$$y = \frac{20x - 1 \pm \sqrt{175x^2 + 160x + 701}}{25}$$ Use the quadratic formula and simplify.

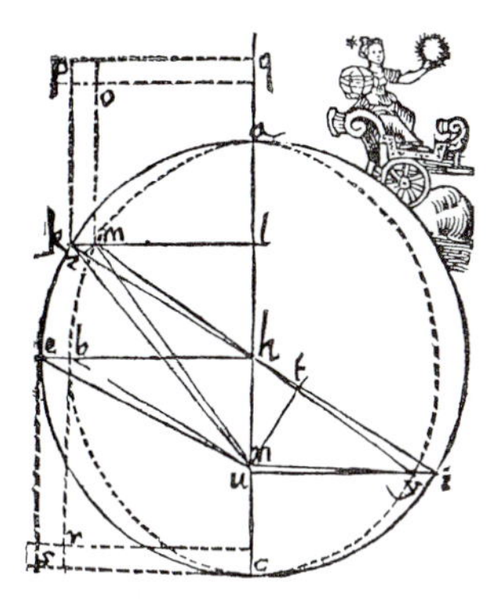

Johannes Kepler's diagram of the elliptical path of planet m around the Sun at point n—from his book Astronomia Nova *(1609).*

This transformation is the basis for the program you have been using since Section 10-2. The grapher plots two equations for y, one with the $+$ sign and one with the $-$ sign. Repeating the calculation for the equations in Examples 3 and 4 gives these results:

Ellipse, $-20xy$: $\quad y = \dfrac{10x - 1 \pm \sqrt{-125x^2 + 180x + 701}}{25}$

Parabola, $-30xy$: $\quad y = \dfrac{15x - 1 \pm \sqrt{0x^2 + 170x + 701}}{25}$

Hyperbola, $-40xy$: $\quad y = \dfrac{20x - 1 \pm \sqrt{175x^2 + 160x + 701}}{25}$

The graph will be an ellipse if the coefficient of x^2 under the radical sign is negative, a parabola if the x^2-coefficient is zero, and a hyperbola if the x^2-coefficient is positive. To find out what this coefficient is in general, repeat the steps given for the hyperbola in Example 2 using A, B, C, D, E, and F for the six coefficients.

$$Ax^2 + Bxy + Cy^2 + Dx + Ey + F = 0$$ Write the general equation.

$$Cy^2 + (Bx + E)y + (Ax^2 + Dx + F) = 0$$ Write the equation as a quadratic in y.

$$y = \frac{-(Bx + E) \pm \sqrt{(Bx + E)^2 - 4(C)(Ax^2 + Dx + F)}}{2C}$$ Use the quadratic formula.

Expand the expression under the radical sign and regroup.

$$B^2x^2 + 2BEx + E^2 - 4ACx^2 - 4CDx - 4CF$$

$$= (B^2 - 4AC)x^2 + (2BE - 4CD)x + (E^2 - 4CF)$$

The x^2-coefficient, $B^2 - 4AC$, is called the **discriminant,** like $b^2 - 4ac$ in the familiar quadratic formula. This box summarizes the results so far.

PROPERTY: Discriminant of a Conic Section

For the general equation $Ax^2 + Bxy + Cy^2 + Dx + Ey + F = 0$, the **discriminant** is $B^2 - 4AC$.

$B^2 - 4AC > 0 \quad \Rightarrow \quad$ Hyperbola

$B^2 - 4AC = 0 \quad \Rightarrow \quad$ Parabola

$B^2 - 4AC < 0 \quad \Rightarrow \quad$ Circle or ellipse
(The graph is a circle only if $B = 0$ and $A = C \neq 0$.)

EXAMPLE 5 ➤ For the equation $2x^2 - 5xy + 8y^2 + 5x - 56y + 120 = 0$, use the discriminant to identify which conic the equation represents. Confirm by plotting on your grapher.

SOLUTION $B^2 - 4AC = 25 - 4(2)(8) = -39 < 0 \quad \Rightarrow \quad$ Graph will be an ellipse.

The graph in Figure 10-6e confirms that the conic is an ellipse.

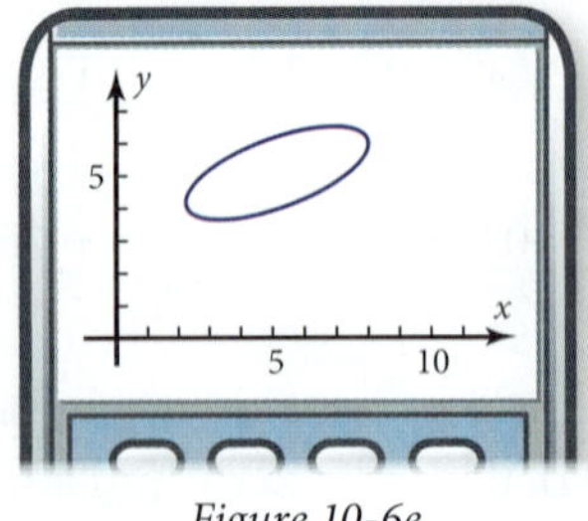

Figure 10-6e

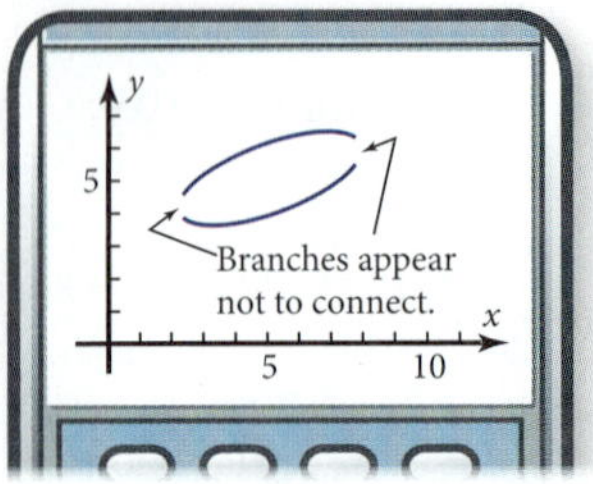

Branches appear not to connect.

Figure 10-6f

Note that if you are using a low-resolution grapher, such as a handheld graphing calculator, the two branches of the ellipse might not meet. Figure 10-6f shows you what the graph in Example 5 might look like. When you sketch the graph, you should show it closing, as in Figure 10-6e.

Problem Set 10-6

Reading Analysis

From what you have read in this section, what do you consider to be the main idea? What two major changes are caused by introduction of an xy-term into the Cartesian equation of a conic section? What feature does *not* change when an xy-term is introduced? How can you use matrices to rotate a conic section by a given number of degrees without changing its proportions?

Quick Review

Q1. Find the major radius of this ellipse:

$$\left(\frac{x}{8}\right)^2 + \left(\frac{y}{17}\right)^2 = 1$$

Q2. Find the focal radius of the ellipse in Problem Q1.

Q3. Find the eccentricity of the ellipse in Problem Q1.

Q4. Find the directrix radius of the ellipse in Problem Q1.

Q5. Find the conjugate radius of this hyperbola:

$$\left(\frac{x}{3}\right)^2 - \left(\frac{y}{4}\right)^2 = 1$$

Q6. Find the focal radius of the hyperbola in Problem Q5.

Q7. Find the eccentricity of the hyperbola in Problem Q5.

Q8. Find the directrix radius of the hyperbola in Problem Q5.

Q9. How far apart are the vertices of the hyperbola in Problem Q5?

Q10. What is the eccentricity of the parabola $y = 0.1x^2$?

1. *Rotation Problem 1:* Figure 10-6g shows the point (6, 8) connected to the origin, and the image of this point and segment rotated counterclockwise by 50°.

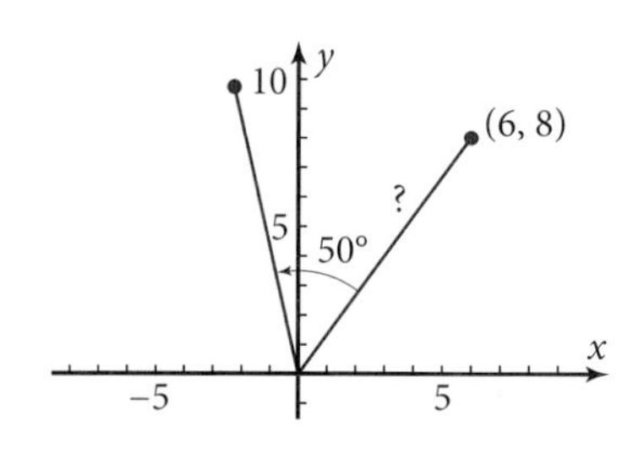

Figure 10-6g

   a. Write a matrix to represent this 50° rotation and multiply it by the matrix representing the point (6, 8). Write the resulting product matrix and the ordered pair for the image point. Does this ordered pair agree with Figure 10-6g? Measure the angle with a protractor and show that it really is 50°.

   b. Show that the rotation in part a did not change the length of the line segment connecting the point to the origin.

   c. Write a matrix to represent a clockwise rotation by 90° and simplify. Multiply this rotation matrix by the matrix representing the point (6, 8). Give evidence that the line segment from the origin to the resulting image point really is perpendicular to the segment ending at point (6, 8).

2. *Rotation Problem 2:* Write a rotation matrix for

   a. A counterclockwise rotation by 123°

   b. A clockwise rotation by 39°

   c. A counterclockwise rotation by 90° (Simplify)

For Problems 3–8, find the parametric equations for the conic section shown, and confirm that your equations are correct by plotting the graph on your grapher.

3. Ellipse

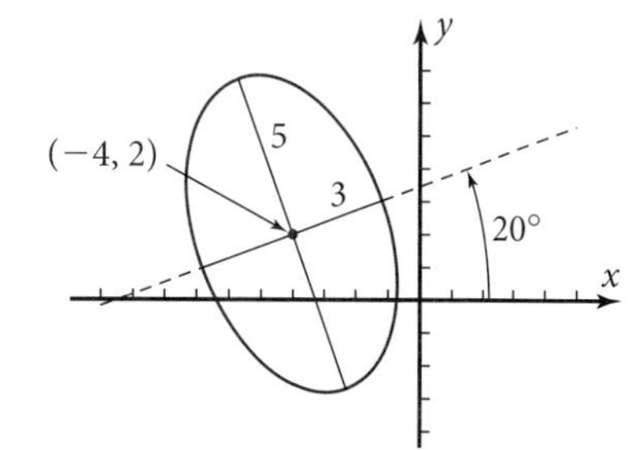

4. Ellipse

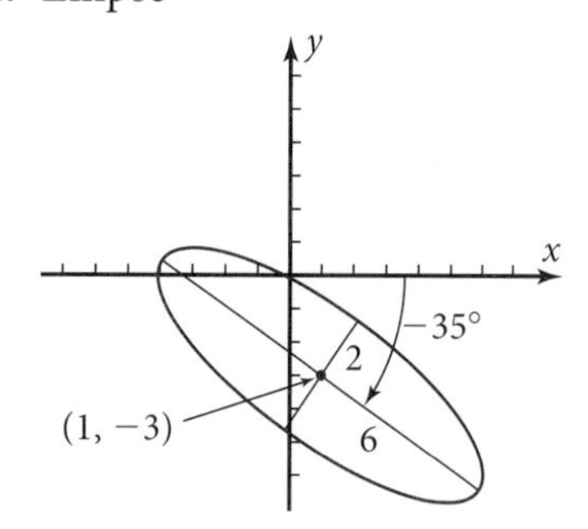

5. Hyperbola

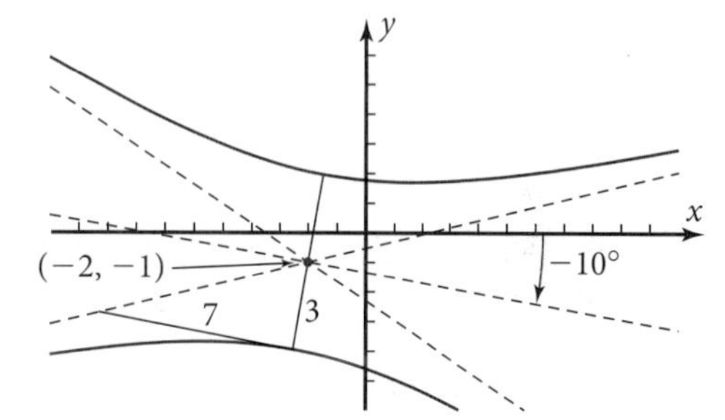

6. Hyperbola

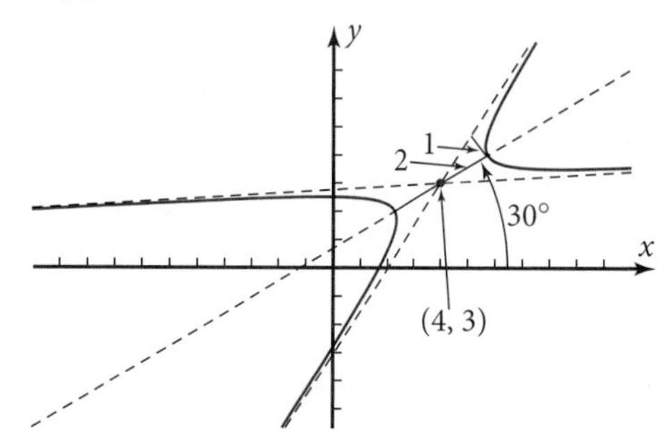

7. Parabola (Use $[-10, 10]$ for t.)

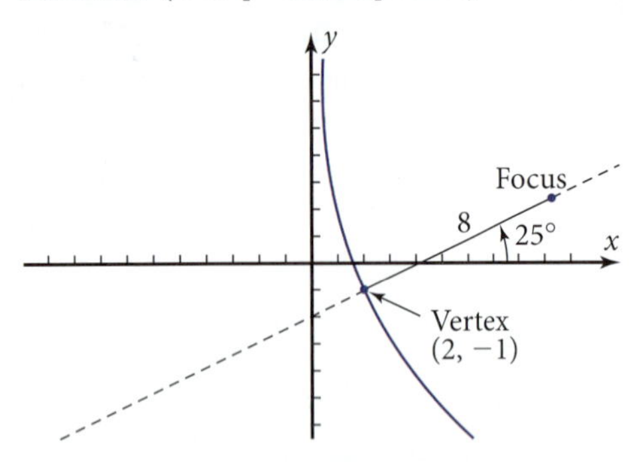

8. Parabola (Use $[-10, 10]$ for t.)

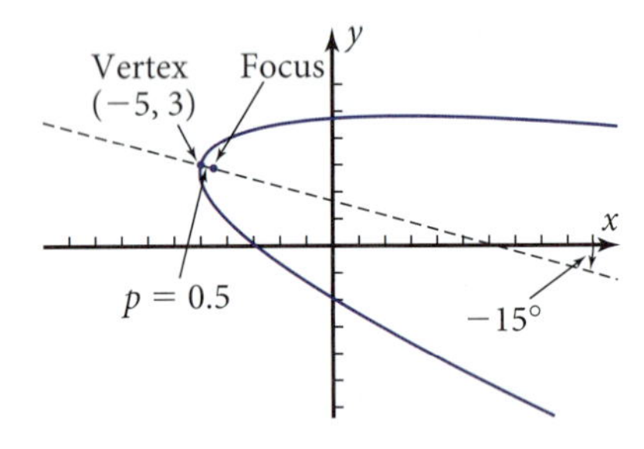

For Problems 9–14, find the parametric equations for the conic section described. Plot the graph on your grapher and sketch the result.

9. Ellipse with center (8, 5), eccentricity 0.96, and major radius 25 at an angle of $-30°$ to the x-axis. Use a window with $[-30, 30]$ for x and equal scales on the two axes.

10. Ellipse with center $(6, -2)$, eccentricity 0.8, and major radius 5 at an angle of 70° to the x-axis. Use a window with $[-10, 10]$ for x and equal scales on the two axes.

11. Hyperbola with center (0, 0), eccentricity 1.25, and transverse radius 4 at an angle of 15° to the x-axis. Use a window with $[-10, 10]$ for x and equal scales on the two axes.

12. Hyperbola with center $(-5, 10)$, eccentricity $\frac{25}{24}$, and transverse radius 24 at an angle of $-20°$ to the x-axis. Use a window with $[-50, 50]$ for x and equal scales on the two axes.

13. Parabola with vertex at the point $(-8, 5)$, focus $\frac{1}{4}$ unit from the vertex, and axis of symmetry at an angle of $-30°$ to the x-axis, opening to the lower right. Use $[-5, 5]$ for t, a window with $[-10, 10]$ for x, and equal scales on the two axes.

14. Parabola with vertex at the origin, focus in the first quadrant 10 units from the vertex, and axis of symmetry at an angle of 45° to the x-axis. Use $[-10, 10]$ for t, a window with $[-10, 10]$ for x, and equal scales on the two axes.

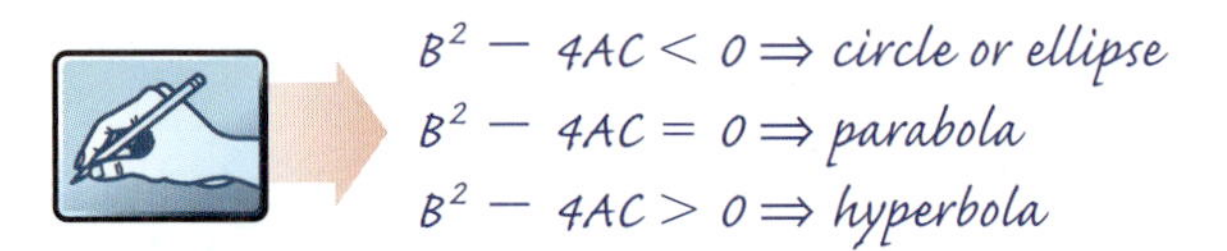

$B^2 - 4AC < 0 \Rightarrow$ *circle or ellipse*
$B^2 - 4AC = 0 \Rightarrow$ *parabola*
$B^2 - 4AC > 0 \Rightarrow$ *hyperbola*

For Problems 15–20, use the discriminant to determine which conic section the graph will be. Confirm your conclusion by plotting the graph. Sketch the result.

15. $3x^2 - 5xy + 9y^2 - 21x + 35y - 50 = 0$

16. $10x^2 - 4xy + 2y^2 + 87x - 13y + 100 = 0$

17. $3x^2 - 10xy + 6y^2 - 12x + 4y + 10 = 0$

18. $8x^2 + 40xy + 2y^2 - 20x - 10y - 31 = 0$

19. $x^2 + 6xy + 9y^2 - 3x - 4y - 10 = 0$

20. $16x^2 - 40xy + 25y^2 + 80x - 150y - 200 = 0$

21. *Discriminant of Unrotated Conics Problem:* If there is no xy-term, you can identify a conic section from the signs of the squared terms. For instance, the conic is an ellipse if x^2 and y^2 have the same sign. Show that the property you have learned about identifying a conic from the sign of the discriminant is consistent with the properties you have learned for identifying an unrotated conic from the signs of the squared terms.

22. *Inverse Variation Function Problem:* An inverse variation power function has the particular equation $y = \frac{12}{x}$. Prove that the graph is a hyperbola. What angle does its transverse axis make with the x-axis?

23. *Rotation and Dilation from Parametric Equations Problem:* Here are the general parametric equations for an ellipse centered at the origin with x- and y-radii a and b, respectively, that has been rotated counterclockwise through an angle v. Assume $[0, 2\pi]$ for t.

$$x = (a \cos v) \cos t + \left[b \cos\left(v + \frac{\pi}{2}\right)\right] \sin t$$

$$y = (a \sin v) \cos t + \left[b \sin\left(v + \frac{\pi}{2}\right)\right] \sin t$$

a. Use a rotation matrix to show how these equations follow from the unrotated ellipse equations

$$x = a \cos t$$
$$y = b \sin t$$

b. A particular ellipse has parametric equations

$$x = 3 \cos t - 2 \sin t$$
$$y = 5 \cos t + 1.2 \sin t$$

Calculate algebraically the major and minor radii and the angle v that one of the axes makes with the x-axis. Show by graphing that your answers are correct.

24. *Parabolic Lamp Reflector Project:* Figure 10-6h, left, shows in perspective a paraboloid that forms the reflector for a table lamp. The reflector is 12 in. long in the x-direction and has radius 5 in. The circular lip with 5-in. radius is shown in perspective as an ellipse with major radius 5 in. and minor radius 2 in.

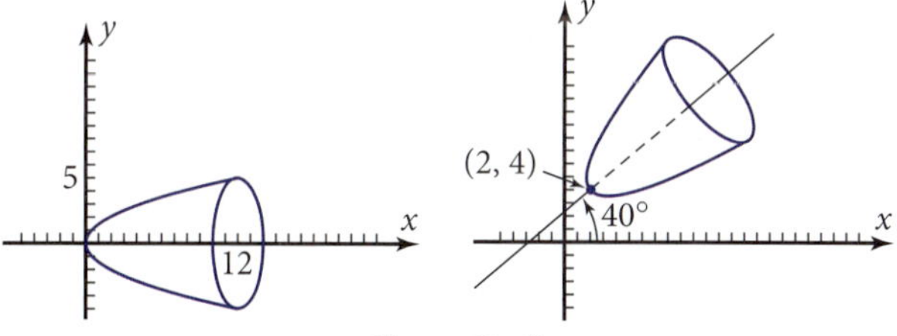

Figure 10-6h

a. Write parametric equations for the parabola shown. Pick a suitable t-interval and plot the graph. Does the parabola start and end at the points shown in the figure?

b. Write parametric equations for the ellipse. Plot the ellipse on the same screen as the parabola. If your grapher does not allow you to pick different t-intervals for the two curves, you will have to be clever about setting the period for the cosine and sine in the parametric equations so that the entire ellipse will be plotted.

c. Figure 10-6h, right, shows the same lamp reflector rotated and translated so that its axis makes an angle of 40° with the x-axis. Write parametric equations for the rotated parabola and rotated ellipse. Plot the two graphs on the same screen. If the result does not look like the given figure, keep working on your equations until your graph is correct.

25. *Variable xy-Term Problem:* Go to **www.keymath.com/precalc** and open the *General Conics* exploration. Plot the equation

$$x^2 + y^2 - 10x - 8y + 16 = 0$$

by changing the sliders for A, B, C, D, E, and F to the appropriate values.

a. Describe the graph that appears. How is the result consistent with what you have learned about the relationships between the graph of a conic section and its equation?

b. Move the slider to vary B from -4 to 4. Describe what happens to the graph as B is varied. What stays the same about the graph, and what changes?

c. Move the slider to $B - 2$. Give numerical and graphical evidence to support the fact that the graph is a parabola. Is it still a parabola if $B = -2$?

d. How has working this problem clarified your understanding of the properties of conic sections?

10-7 Applications of Conic Sections

You have learned how to find equations for ellipses, parabolas, hyperbolas, and circles from their analytic properties. These figures appear in the real world as paths of thrown objects, orbits of planets and comets, and shapes of bridges. They even appear in solutions of business problems. In this section you will consolidate your knowledge of conic sections by applying it to some real-world problems.

This giant parabolic mirror is a solar furnace on the side of a building harvesting solar power in Odeillo, France.

Objective Given a situation from the real world in which conic sections appear, create a mathematical model and use it to make predictions and interpretations.

Problem Set 10-7

Quick Review

Q1. Why are parabolas, ellipses, hyperbolas, and circles called *conic sections*?

Q2. Why does a hyperbola have two branches, while a parabola has only one branch?

Q3. Write a Cartesian equation for a unit hyperbola opening in the y-direction.

Q4. Write parametric equations for a unit hyperbola opening in the y-direction.

Q5. Complete the square: $x^2 - 10x + \underline{\ ?\ }$

Q6. Sketch a paraboloid.

Q7. Write the definition of *eccentricity* in terms of the distances d_1 from the directrix and d_2 from the focus to a point on the graph.

Q8. Sketch an ellipse, showing the approximate location of the two foci.

Q9. What is special about the distances from the two foci to a point on an ellipse?

Q10. If an ellipse has major radius 7 and minor radius 3, what is the focal radius?

1. *Coffee Table Problem:* A furniture manufacturer wishes to make elliptical tops for coffee tables 52 in. long and 26 in. wide, as shown in Figure 10-7a. A pattern is to be cut from plywood so that the outline of the tabletops can easily be marked on the tabletop's surface. Give detailed instructions for a rapid way to mark the ellipse on the plywood. What will be the eccentricity of the tabletops?

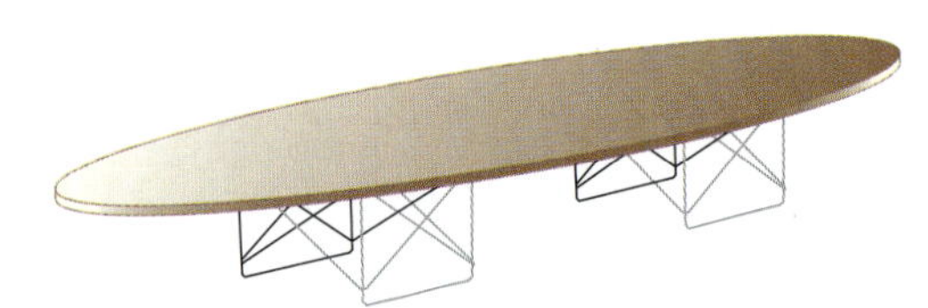

Figure 10-7a

2. *Stadium Problem:* The plan for a new football stadium calls for the stands to be in a region defined by two concentric ellipses (Figure 10-7b). The outer ellipse is to be 240 yd long and 200 yd wide. The inner ellipse is to be 200 yd long and 100 yd wide. A football field of standard dimensions, 120 yd by 160 ft, will be laid out in the center of the inner ellipse.

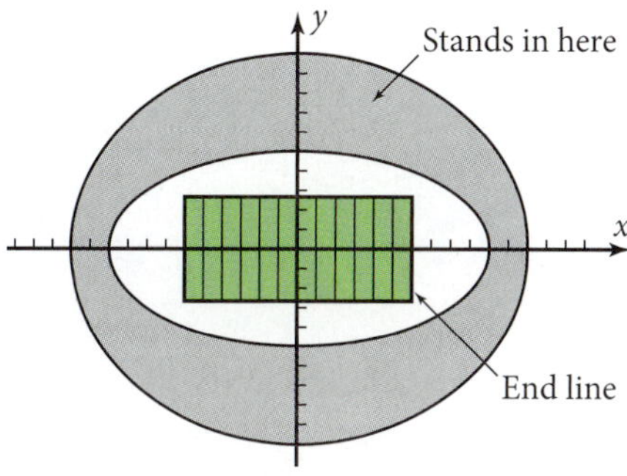

Figure 10-7b

a. Find particular equations for the two ellipses.

b. Find the eccentricities of the two ellipses.

c. How much clearance will there be between the corner of the field and the inner ellipse in the direction of the end line?

d. The formula for the area of an ellipse is πab, where a and b are the major and minor radii, respectively. To the nearest square yard, what is the area of the stands? If each seat takes about 0.8 yd^2, what will be the approximate seating capacity of the stadium?

e. Show that the familiar formula for the area of a circle is a special case of the formula for the area of an ellipse.

3. *Bridge Problem:* Figure 10-7c is a photograph of Bixby Bridge in Big Sur, California. A similar bridge under construction is shown in Figure 10-7d. The span of the bridge is to be 1000 ft, and it is to rise 250 ft at the vertex of the parabola. The roadway is horizontal and will pass 20 ft above the vertex. Vertical columns extend between the parabola and the roadway, spaced every 50 ft horizontally.

Figure 10-7c

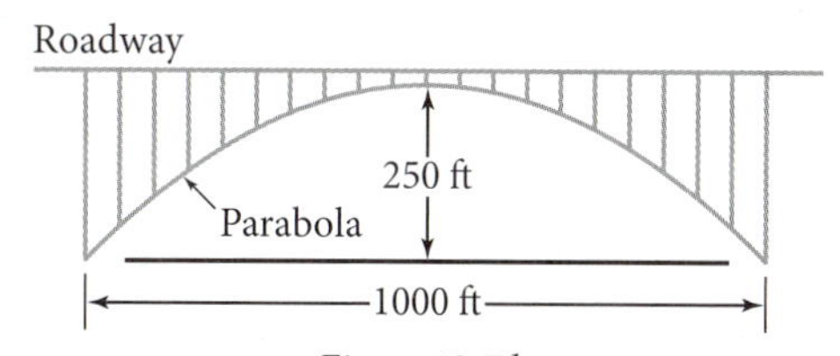

Figure 10-7d

a. Using convenient axes, find the particular equation for the parabola.

b. The construction company that builds the bridge must know how long to make each vertical column. Make a table of values showing these lengths.

c. To order enough concrete and steel to make the vertical columns, the construction company must know the total length of the columns. By appropriate operations on the values in the table in part b, calculate this total length. Observe that there is a row of columns on both sides of the bridge.

4. *Halley's Comet Problem:* Halley's comet moves in an elliptical orbit around the Sun, with the Sun at one focus. It passes within about 50 million miles of the Sun once every 76 years. The other end of its orbit is about 5000 million miles from the Sun, beyond the orbit of Uranus.

a. Find the eccentricity of the comet's orbit.

b. Find the particular equation for the orbit. Put the origin at the Sun and the x-axis along the major axis.

c. The length of the orbit's major axis is 5050 million miles. What is its the length of its minor axis?

d. How far is the comet from the Sun when the line from its position to the Sun is perpendicular to the major axis?

e. What is the distance between the Sun and the directrix closest to the Sun?

5. *Meteor Tracking Problem 1:* Suppose that you have been hired by Palomar Observatory near San Diego. Your mission is to track incoming meteors to predict whether they will strike Earth. Because Earth has a circular cross section, you set up a coordinate system with its origin at Earth's center (Figure 10-7e). The equation for Earth's surface is

$$x^2 + y^2 = 40$$

where x and y are distances in thousands of kilometers.

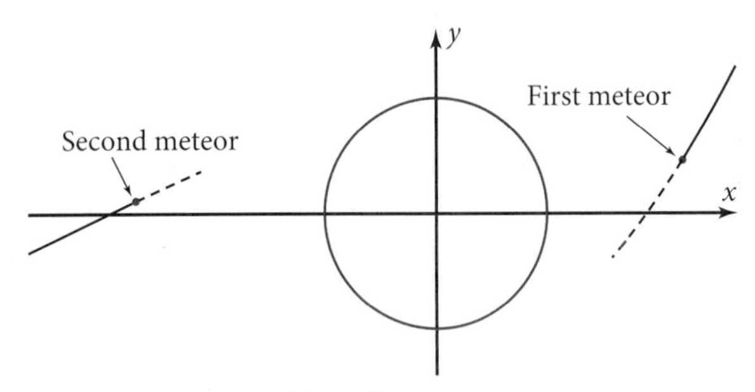

Figure 10-7e

a. The first meteor you observe is moving along a path whose equation is

$$x^2 - 18y = 144$$

What geometric figure is the path? Determine graphically whether the meteor's path intersects Earth's surface.

b. Confirm your conclusion in part a algebraically by solving the system of equations

$$x^2 + y^2 = 40$$
$$x^2 - 18y = 144$$

To do this, try eliminating x algebraically and solving the resulting equation for y.

c. The second meteor you observe is moving along a path whose equation is

$$x^2 - 4y^2 + 80y = 340$$

What geometric figure is the path? Confirm graphically that the path *does* intersect Earth's surface. Find numerically the point at which the meteor will strike Earth.

d. Find algebraically the point at which the meteor in part c will strike Earth.

The Palomar Observatory near San Diego, California houses five telescopes used nightly for a variety of astronomical research programs.

6. *Meteor Tracking Problem 2:* A meteor originally moving along a straight path will be deflected by Earth's gravity into a path that is a conic section. If the meteor is moving fast enough, the path will be a hyperbola with the original straight-line path as an asymptote. Assume that a meteor is approaching the vicinity of Earth along a hyperbola with general equations

$$x = a \sec t$$
$$y = b \tan t$$

a. Suppose that at $t = -1.4$, the position of the meteor, (x, y), is $(29.418, -11.596)$, where x and y are displacements in millions of miles. Find the particular values of a and b.

b. Plot the branch of the hyperbola in the t-interval $\left[-\frac{\pi}{2}, \frac{\pi}{2}\right]$. Sketch the result.

c. Earth is at the focus of the hyperbola that is closest to the vertex in part b. What is the focal radius? How far is the meteor from Earth when it is at the vertex of the hyperbola?

d. Assuming that the meteor does not hit Earth, its path as it leaves Earth's vicinity will approach the other asymptote of the hyperbola. What is the equation of this other asymptote? Show both asymptotes on your sketch in part b.

7. *Overpass Problem:* An overpass is to be built for train tracks to go across a highway. For aesthetic reasons, the designers want the underneath to have the shape of a semi-ellipse, as shown in Figure 10-7f. The ellipse will have vertices (x, y) at the points $(100, 0)$ and $(300, 0)$, where x and y are distances in feet. The roadway

underneath the overpass is 140 ft wide, extending from the point (130, 0) to the point (270, 0). So that large trucks can pass under, the clearance must be at least 20 ft at the points where the underneath of the overpass is closest to the roadway.

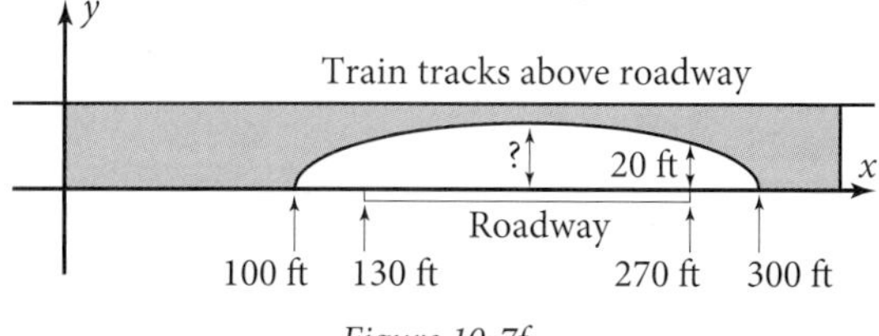

Figure 10-7f

a. Write the particular equation for the ellipse.

b. How much clearance will there be at the center of the overpass, where the ellipse is farthest from the roadway?

c. If a load on a truck is 25 ft high, between what two values of x must the truck stay in order for it to not scrape the overpass?

8. *I-Beam Problem:* Two steel I-beams being used in the construction of a chemical manufacturing plant are to cross at an angle of 70°, as shown in Figure 10-7g. For added strength at the junction of the I-beams, a gusset is to be welded between them. The gusset is made from a flat plate of steel that fits into the angle between the I-beams. For the proper distribution of stress, the gusset is to be cut in the shape of a hyperbola with its asymptotes along the I-beams. The vertex of the hyperbola is 10 in. from the intersection of the I-beams, and the hyperbola ends 25 in. from the intersection of the I-beams.

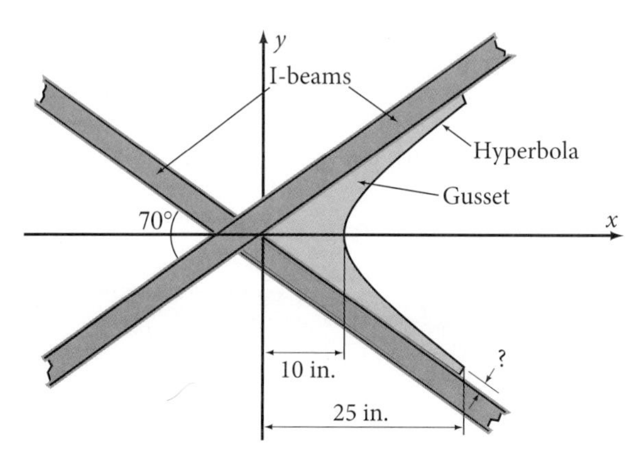

Figure 10-7g

a. Write the particular equation for the hyperbola.

b. In order to make the gusset, the metal workers must know the y-values of the hyperbola at various values of x. Make a table showing these values for each 2 inches from 10 in. to 22 in.

c. The gusset will be its narrowest at the end where $x = 25$ (see Figure 10-7g). How narrow is it there?

d. Why is an I-beam called an "I-beam"?

9. *Marketing Problem 1:* A customer located at point (x, y) can purchase goods from Supplier 1 located at the point (0, 0) or from Supplier 2 located at the point (6, 0), where x and y are distance in miles (Figure 10-7h). The charge for delivery of the goods is \$10.00 per mile from Supplier 1 and \$20.00 per mile from Supplier 2.

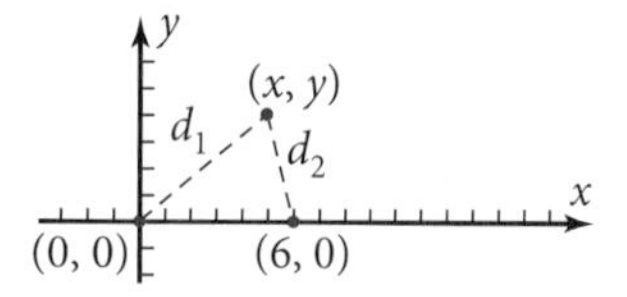

Figure 10-7h

a. To decide which supplier to use for a given point (x, y), it would help the customer to know all points for which the total cost of shipping is the same for either supplier. These points will satisfy the equation

$$10d_1 = 20d_2$$

where d_1 and d_2 are the distances from Supplier 1 and Supplier 2, respectively, to point (x, y). Find the particular Cartesian equation for this set of points. Show that the graph is a circle.

b. If point (x, y) is close enough to Supplier 2, it is cheaper to pay the higher cost per mile. On a copy of the figure, shade the region in which it is cheaper to receive goods from Supplier 2.

c. If a customer is located at the point (15, 0), which supplier is closer? Which supplier's shipping charges would be less? Is this surprising?

10. *Marketing Problem 2:* Suppose that the shipping charge from both suppliers in Problem 9 is \$10.00 but that the purchase price of the goods is \$980 from Supplier 1 and \$1000 from Supplier 2.

a. Find the particular equation for the set of points (x, y) for which the total cost (shipping plus goods) is the same for both suppliers.

b. Explain why the graph of the set of points in part a is a hyperbola.

c. Explain why one branch of the hyperbola in part a is not relevant to this application. Sketch the other branch.

d. From which supplier should you purchase the goods if you are located at the point $(7, 20)$? Justify your answer.

11. *Hyperboloid Project:* A hyperboloid of one sheet is graphed in Figure 10-7i. The top and bottom ellipses in the figure represent the circular bases of the hyperboloid in perspective. Find parametric equations for the ellipses and hyperbola, and use the results to plot the hyperboloid. If your grapher does not allow you to set different t-intervals for different graphs, find a clever way to graph the figure correctly.

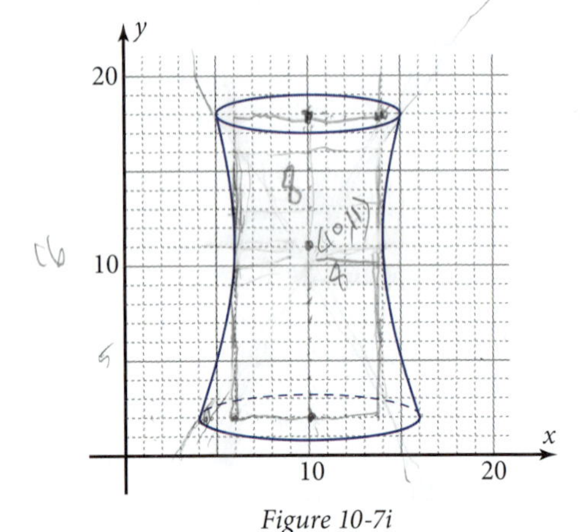

Figure 10-7i

12. *Elliptical Pendulum Project:* Make a pendulum by tying a small weight to the end of a string, as shown in Figure 10-7j. Tie the other end of the string to the ceiling or an overhead doorway so that it clears the floor by 1 to 2 cm. Place two metersticks perpendicular to each other to represent the x- and y-axes. The origin should be under the rest position of the pendulum. Measure the period of the pendulum by pulling it out to $x = 30$ cm on the x-axis and counting the time for ten swings. Confirm that the period is the same if you start the pendulum from $y = 20$ cm on the y-axis. Then set the pendulum in motion by holding it at $x = 30$ on the x-axis and pushing it in the y-direction just hard enough so that it crosses the y-axis at $y = 20$. Write parametric equations for the elliptical path that the pendulum traces in terms of the parameter t, the time, in seconds, since the pendulum was started. Using the equations, calculate the expected position of the pendulum at time $t = 7$ s. Then set the pendulum in motion again. Start timing at $t = 0$ when the pendulum crosses the positive x-axis the second time. Does the pendulum pass through the point you calculated for $t = 7$ s?

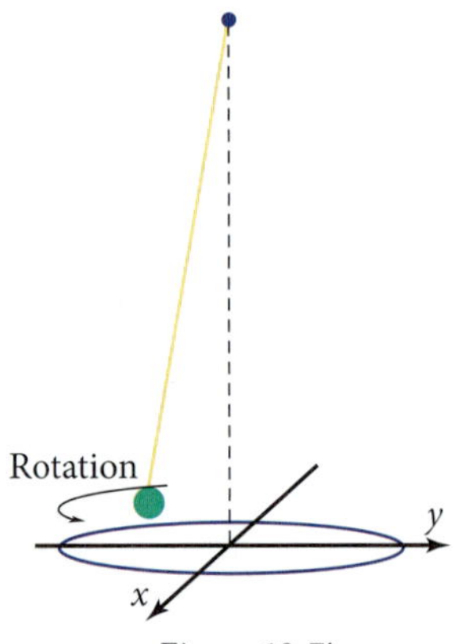

Figure 10-7j

10-8 Chapter Review and Test

In this chapter you have studied ellipses, circles, parabolas, and hyperbolas. These conic sections get their names from plane slices of cones. Each one has geometric or analytic properties, such as foci, directrices, eccentricity, and axes of symmetry. You have analyzed the graphs algebraically, both in Cartesian form and in parametric form. Rotation of conic section graphs about their axes of symmetry produces three-dimensional quadric surfaces. You have also seen how conic sections appear in the real world, such as in the shapes of bridges and the paths of celestial objects.

Review Problems

R0. Update your journal with what you have learned in this chapter. Include things such as

- The way conic sections are formed by slicing a cone with a plane
- The origins of the names *ellipse, parabola,* and *hyperbola*
- Cartesian equations for the conic sections
- Parametric equations for the conic sections
- Dilations, translations, and rotations of the conic sections
- Quadric surfaces formed by rotating conic sections about axes of symmetry
- Focus, directrix, and eccentricity properties of conic sections
- Some real-world situations where conic sections appear

R1. Without plotting, tell whether the graph will be a circle, an ellipse, a parabola, or a hyperbola.

a. $x^2 + y^2 = 36$

b. $x^2 + 9y^2 = 36$

c. $4x^2 - y^2 = 36$

d. $4x^2 - 4y^2 = 36$

e. $4x^2 + y = 36$

R2. a. Write the Cartesian equation for the specified parent relation.

i. The unit circle

ii. The unit hyperbola opening in the y-direction

iii. The unit parabola opening in the x-direction

b. Consider the ellipse with equation

$$\left(\frac{x-2}{7}\right)^2 + \left(\frac{y+3}{4}\right)^2 = 1$$

i. Sketch the graph.

ii. Transform the equation to polynomial form, $Ax^2 + Bxy + Cy^2 + Dx + Ey + F = 0$.

iii. Plot the graph using the program in Section 10-2, Example 8. Does the plotted graph agree with your sketch in part i?

c. For the equation $16x^2 - 25y^2 - 96x - 100y + 444 = 0$ in polynomial form,

i. How can you tell without plotting that the graph will be a hyperbola?

ii. Complete the squares to transform the equation to center-radius form.

iii. What is the significance of making the right side of the equation equal 1?

iv. Sketch the graph, showing the asymptotes and vertices.

d. For the parabola in Figure 10-8a,

i. Write a Cartesian equation.

ii. Plot the graph using the program in Section 10-2, Example 8. Does the plotted graph agree with Figure 10-8a?

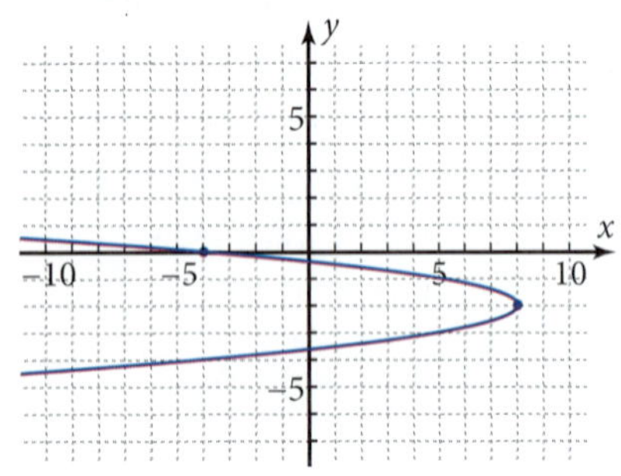

Figure 10-8a

R3. a. Write parametric equations for the specified parent relation.

i. The unit circle

ii. The unit hyperbola opening in the x-direction

iii. The unit parabola opening in the y-direction

b. For the ellipse $\left(\frac{x-2}{7}\right)^2 + \left(\frac{y+3}{4}\right)^2 = 1$ in Problem R2, part b,

i. Write equivalent parametric equations for x and y.

ii. Plot the graph on your grapher. Sketch the result.

c. For the hyperbola in Figure 10-8b,

i. Write parametric equations.

ii. Plot the parametric equations in part i on your grapher. Does the plotted graph agree with Figure 10-8b?

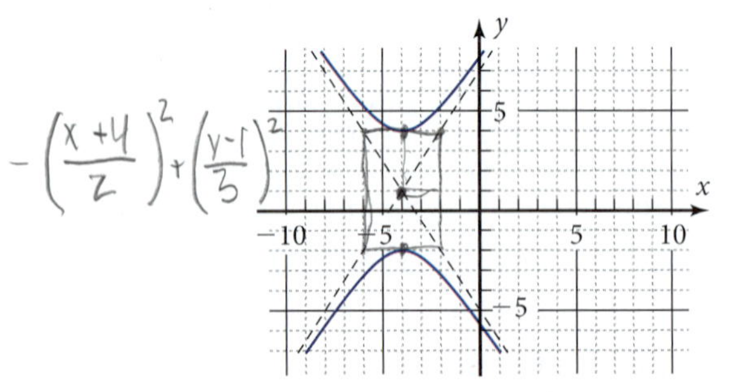

Figure 10-8b

d. For the parabola in Figure 10-8a in Problem R2, part d,

i. Write parametric equations.

ii. Plot the parametric equations on your grapher. Does the plotted graph agree with Figure 10-8a?

e. Transform these parametric equations to an equivalent Cartesian equation by eliminating the parameter t.

$$x = 3 \sec t$$
$$y = 5 \tan t$$

R4. a. Sketch the quadric surface.

i. Ellipsoid formed by rotating the graph of $4x^2 + y^2 = 16$ about the y-axis

ii. Hyperboloid formed by rotating the graph of $x^2 - 9y^2 = 9$ about the x-axis, using $[-6, 6]$ for x

iii. Hyperboloid formed by rotating the graph of $4x^2 - y^2 = 4$ about the y-axis, using $[1, 2]$ for x

b. A paraboloid is formed by rotating about the x-axis the part of the parabola $x = 4 - y^2$ that lies in the first quadrant. A cylinder is inscribed in the paraboloid, with its axis along the x-axis, its left base containing the origin, and its right base touching the paraboloid (Figure 10-8c). Find the radius and altitude of the cylinder with maximum volume. Find this maximum volume.

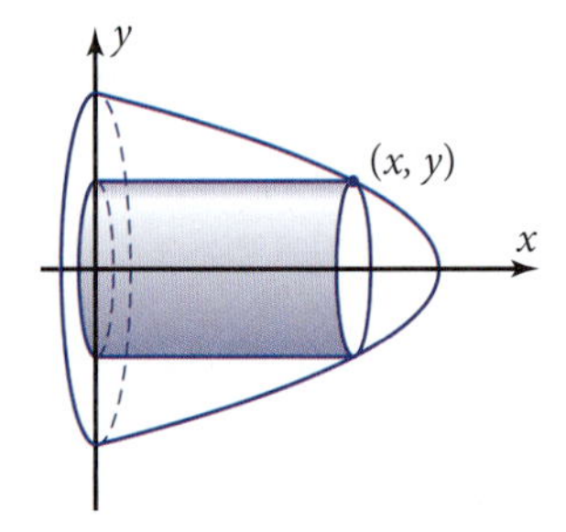

Figure 10-8c

R5. For parts a and b, Figure 10-8d shows an ellipse with eccentricity 0.8, foci at the points $(-6.4, 0)$ and $(6.4, 0)$, and one directrix $x = 10$. The point shown on the ellipse has x-coordinate 3.

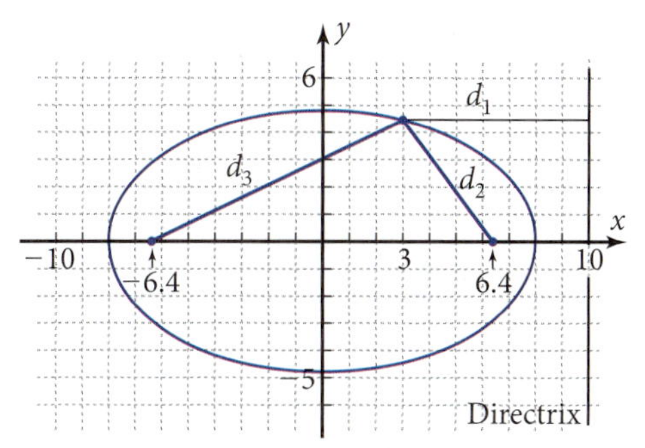

Figure 10-8d

a. Measure the three distances d_1, d_2, and d_3 using the scales shown. Confirm that $d_2 = ed_1$. Confirm that $d_2 + d_3 = 16$, the major diameter.

b. Write a Cartesian equation for the ellipse. Use the equation to calculate y for the point with x-coordinate 3. Use the value of y and the Pythagorean theorem to calculate d_2 and d_3 exactly. Do your measurements in part a agree with the calculated values? Does $d_2 + d_3$ equal 16 exactly?

c. *Billiards Table Problem:* Figure 10-8e shows that an elliptical billiards table is to be built with eccentricity 0.9 and foci at the points $(-81, 0)$ and $(81, 0)$, where x and y are in centimeters. Find the major and minor radii. Write parametric equations for the ellipse and check them by plotting the equations on your grapher and comparing the plotted graph to Figure 10-8e. Measure the angles formed between the tangent line to the ellipse at the point where $t = 1$ and the lines connecting that point to each focus. What does the result tell you about where the billiard ball will go if it is spotted at one focus and shot straight to the point where $t = 1$?

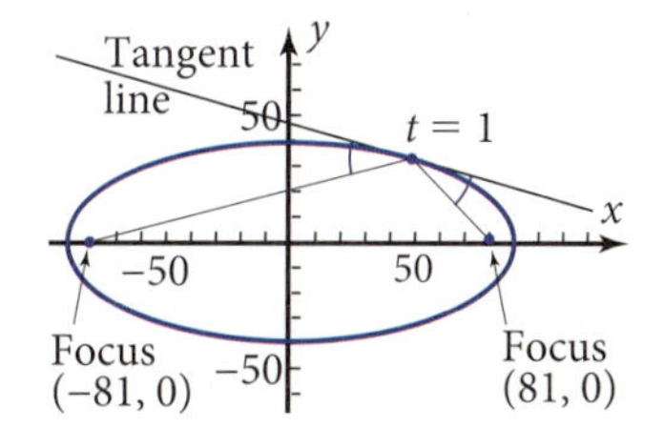

Figure 10-8e

d. Identify the conic section whose equation is given here. Sketch the graph. Find the four radii and the eccentricity.

$$\left(\frac{x}{4}\right)^2 - \left(\frac{y}{3}\right)^2 = 1$$

e. A parabola has directrix $x = 4$ and focus at the point $(0, 0)$. Sketch this information. Write the particular Cartesian equation. Plot the graph and sketch the result.

f. Write parametric equations for the hyperbola with eccentricity $\frac{5}{3}$ and foci at the points $(2, 1)$ and $(2, 7)$. Plot the graph and sketch the result.

R6. a. Write parametric equations for the ellipse in Figure 10-8f. Confirm that your equations are correct by plotting them on your grapher.

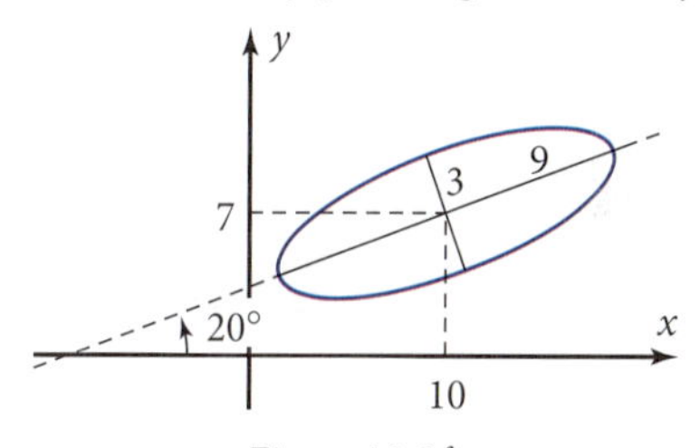

Figure 10-8f

b. Write parametric equations for the hyperbola with eccentricity 2 centered at the point $(3, -4)$ if its transverse radius is 6 and its transverse axis makes an angle of 35° with the x-axis. Confirm that your equations are correct by plotting them on your grapher.

c. Write parametric equations for the parabola with vertex at the point $(1, 2)$ and focus at the point $(4, 5)$. Confirm that your equations are correct by plotting them on your grapher.

d. Use the discriminant to identify the conic. Confirm by plotting the graph. Sketch the result.

i. $4x^2 + 2xy + 9y^2 + 15x - 13y - 19 = 0$

ii. $4x^2 + 12xy + 9y^2 + 15x - 13y - 19 = 0$

iii. $4x^2 + 22xy + 9y^2 + 15x - 13y - 19 = 0$

e. What graphical feature do all three conics in part d have in common?

R7. *Parabolic Antenna Problem:* A satellite dish antenna is to be constructed in the shape of a paraboloid (Figure 10-8g). The paraboloid is formed by rotating about the x-axis the parabola with focus at the point (25, 0) and directrix $x = -25$, where x and y are in inches. The diameter of the antenna is to be 80 in.

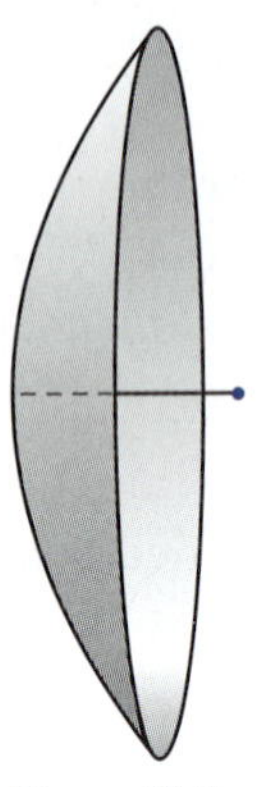

Figure 10-8g

a. Find the equation of the parabola and the domain of x.

b. Sketch the graph of the parabola, showing the focus and the directrix.

c. A receiver is to be placed at the focus. Figure 10-8g suggests that the receiver would touch the ground if the antenna were placed "facedown." Determine algebraically whether this observation is correct.

Concept Problems

C1. *Reflecting Property of a Parabola Problem:* Figure 10-8h shows the parabolic cross section of the satellite dish antenna shown in Figure 10-8g. The equation for the parabola is $x = 0.01y^2$.

a. Confirm that the focus of the parabola is at the point (25, 0) as shown in Figure 10-8h.

b. A television signal ray comes in parallel to the x axis at $y = 20$. Find the coordinates of the point at which the ray strikes the parabola.

c. Calculate the measure of angle A between the incoming ray and a line to the focus from the point where the incoming ray strikes the parabola.

d. An incoming ray and its reflected ray make angles of equal measure with a line tangent to the curved surface. Use your answer to part c to calculate the measures of angles B and C if these angles have equal measure. With a protractor, measure A, B, and C in Figure 10-8h to confirm your calculations.

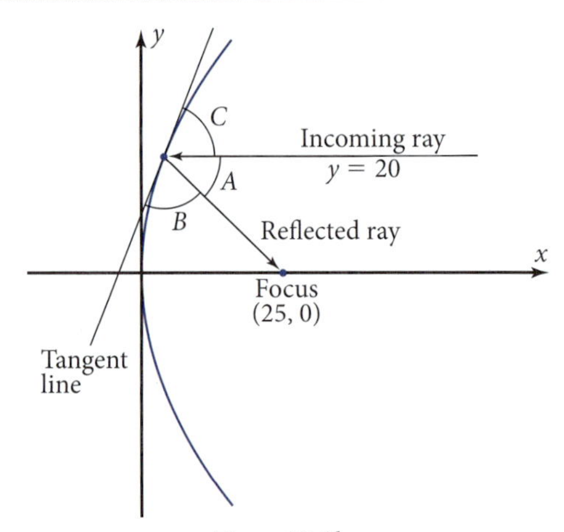

Figure 10-8h

e. Find the particular equation for the tangent line. Use angle C to find its slope. Plot the parabola and the tangent line on your grapher. Zoom in on the point of tangency. What do you notice about the tangent line and the graph as you zoom in?

f. Pick another incoming ray. Calculate the measure of angle A. Let C be half the supplement of angle A, as in part d. Is the line at angle C with the incoming ray tangent to the parabola at the point where the incoming ray strikes the graph? How do you know?

g. Write a conjecture about the direction the reflected ray takes whenever the incoming ray is parallel to the axis of the parabola. Your conjecture should give you insight into why the name *focus* is used and why television satellite antennas and other listening devices are made in the shape of a paraboloid.

C2. *Systems of Quadratic Equations Problem:* Figure 10-8i shows an ellipse, a hyperbola, and a line. The equations for each pair of these graphs form a system of equations. In this problem you will solve the systems by finding the points at which these graphs intersect each other.

a. Solve the ellipse and hyperbola system graphically, to one decimal place.

b. Solve the ellipse and line system graphically, to one decimal place.

c. Solve the hyperbola and line system graphically, to one decimal place.

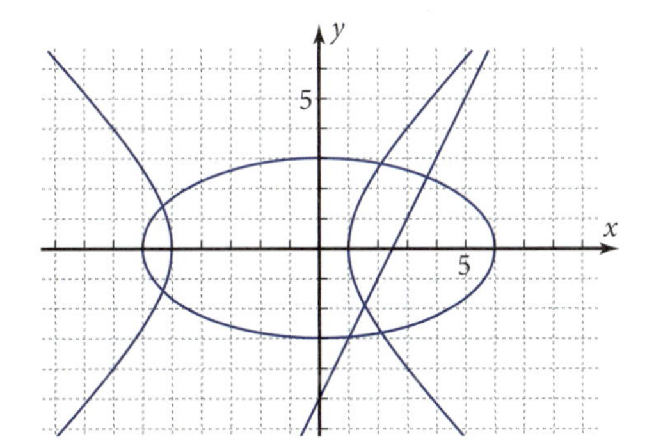

Figure 10-8i

d. The equations for the ellipse and hyperbola in Figure 10-8i are

$$x^2 - y^2 + 4x - 5 = 0$$

$$x^2 + 4y^2 - 36 = 0$$

Quickly state which equation is which. How do you know?

e. Solve the ellipse and hyperbola system algebraically. To do this, first eliminate y by adding a multiple of the first equation to the second. Solve the resulting quadratic equation for x. Finally, substitute the two resulting x-values into one of the original equations and calculate y. What do you notice about two of the four solutions? How does this observation agree with the graphs?

f. Solve the ellipse and hyperbola system numerically. Do this by plotting the two graphs on your grapher and then using the intersect feature. Does the result agree with your solutions in parts a and e?

g. The equation for the line in Figure 10-8i is

$$2x - y = 5$$

Solve the ellipse and line system algebraically. To do this, first solve the linear equation for y in terms of x and then substitute the result for y in the ellipse equation. After you solve the resulting quadratic equation, substitute the two values of x into the *linear* equation to find y. Does the result agree with your solution in part b?

h. Solve the hyperbola and line system algebraically. Which solution does *not* appear in the graphical solution of part c?

Chapter Test

Part 1: No calculators allowed (T1–T10)

For Problems T1–T6, identify the conic section.

T1. $4x^2 + 9y^2 + 24x + 36y - 72 = 0$ ellipse

T2. $9x^2 - 25y^2 + 36x + 200y - 589 = 0$ hyperbola

T3. $x^2 - 14x - 36y + 13 = 0$

T4. $x^2 + 3xy + 4y^2 - 400 = 0$

T5. $x^2 + 4xy + 4y^2 + 5x - 400 = 0$

T6. $x^2 + 5xy + 4y^2 + 5x - 400 = 0$

T7. Sketch a hyperboloid of one sheet.

T8. Give another name for an ellipsoid formed by rotating an ellipse about its major axis. Name something in the real world that has (approximately) this shape.

T9. Sketch an ellipse. Show the approximate locations of the foci and the directrices. Pick a point on the ellipse and draw its distances to a directrix and to the corresponding focus. How are the two distances related to each other?

T10. Draw a sketch showing an ellipse as a section of a cone.

Part 2: Graphing calculators allowed (T11–T20)

T11. Sketch the graph of

$$-\left(\frac{x-3}{6}\right)^2 + \left(\frac{y-1}{2}\right)^2 = 1$$

T12. Write parametric equations for the graph in Problem T11. Plot the graph. Does your sketch in Problem T11 agree with the plotted graph?

T13. Transform the equation in Problem T11 to the form $Ax^2 + Cy^2 + Dx + Ey + F = 0$.

T14. How can you tell by looking at the equation that the graph of the conic section

$$4x^2 + 9y^2 - 40x + 36y - 8 = 0$$

is an ellipse? Transform the equation to center-radius form by completing the squares. Write the x-dilation factor and the y-translation.

T15. How can you tell by looking at the parametric equations that the graph of

$$x = 5 + \cos t$$
$$y = -3 + \sin t$$

is a translated unit circle? Write the coordinates of the center. Transform these equations to an equivalent Cartesian equation by eliminating the parameter t.

T16. An ellipsoid is formed by rotating the ellipse $x^2 + 4y^2 = 16$ about the y-axis. A cylinder is inscribed in the ellipsoid with its axis along the y-axis and its two bases touching the ellipsoid (Figure 10-8j). Plot the graph of the volume of the cylinder as a function of half its height, y. Find numerically the radius and altitude of the cylinder with maximum volume and the approximate value of this maximum volume.

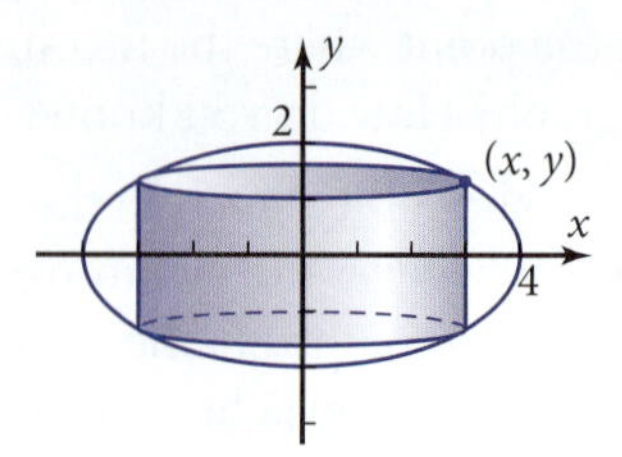

Figure 10-8j

T17. *Satellite Problem:* A satellite is in elliptical orbit around Earth, as shown in Figure 10-8k. The major radius of the ellipse is 51,000 mi, and the focal radius is 45,000 mi. The center of Earth is at one focus of the ellipse.

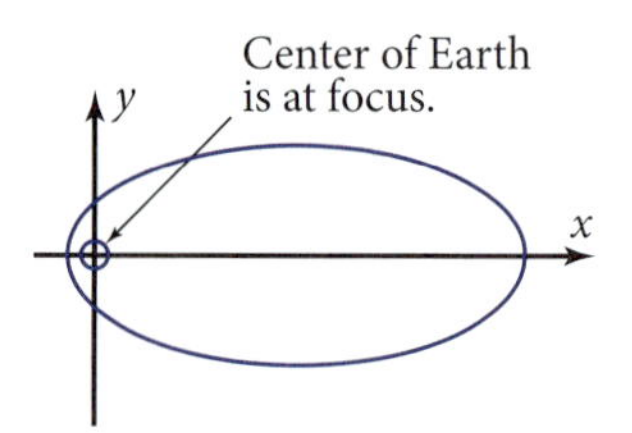

Figure 10-8k

a. Where is the center of the ellipse? What is the minor radius? What is the eccentricity?

b. Write the Cartesian equation for the ellipse. Use the equation to find the y-intercepts of the ellipse.

c. The satellite is closest to Earth when it is at a vertex of the ellipse. The radius of Earth is about 4000 mi. What is the closest the satellite comes to the surface of Earth?

T18. Find the particular equation for the parabola with focus at the point $(-4, -5)$ and directrix $y = 2$.

T19. Figure 10-8l shows an ellipse centered at the point $(1, -2)$, with major radius 7 units making an angle of $-25°$ to the x-axis and minor radius 3 units. Find parametric equations for the ellipse. Confirm your equations by plotting the graph.

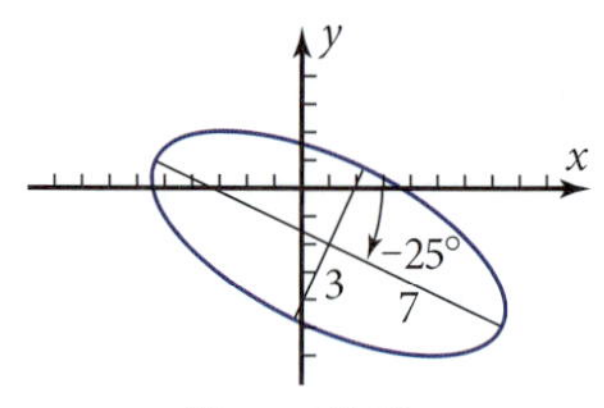

Figure 10-8l

T20. What did you learn as a result of taking this test that you did not know before?

Chapter 11

Polar Coordinates, Complex Numbers, and Moving Objects

Paths traced by rotating objects can be modeled by coordinates in which the independent variable is an angle measure and the dependent variable is a directed distance from the origin. These *polar coordinates,* along with the parametric functions of Chapter 7, give a way to find equations for things such as the *involute of a circle,* which determines the shape of the surfaces of the gear teeth seen in the photograph. This shape allows one gear to transmit its motion to another in a smooth manner.

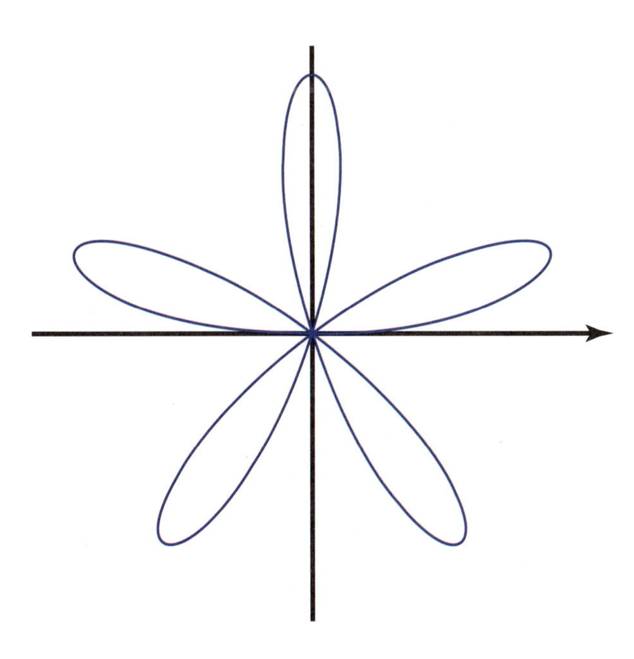

Mathematical Overview

In this chapter you will learn about polar coordinates, in which r and θ are the variables instead of x and y. Polar coordinates allow you to plot complicated graphs such as the five-leaved rose in the graph below. These coordinates have surprising connections to complex numbers and to the parametric functions you studied in earlier chapters. You will study polar coordinates in four ways.

GRAPHICALLY

Five-leaved rose

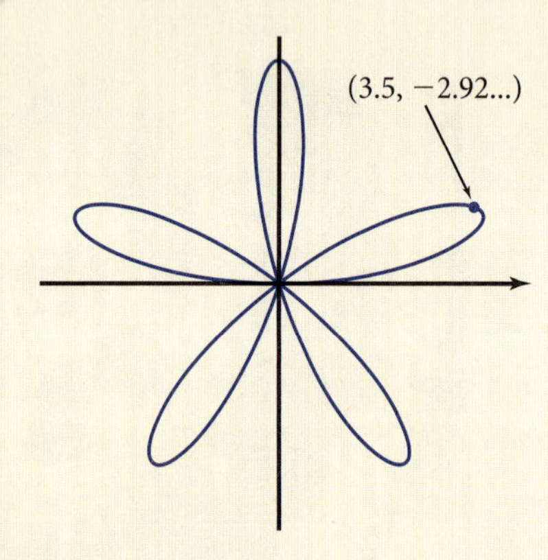

ALGEBRAICALLY

$r = 3 \sin 5\theta$ Polar equation of a rose.

NUMERICALLY

$\theta = 3.5$ radians: $r = -2.9268\ldots$ units The r-value can be negative!

VERBALLY

To plot a point (r, θ) in polar coordinates, you first rotate to the angle θ. If r is positive, go that many units in the same direction as angle θ. If r is negative, go in the direction opposite angle θ. Thus, an angle in the third quadrant corresponds to a point in the first quadrant if the radius is negative.

11-1 Introduction to Polar Coordinates

The graphs of the trigonometric functions you have plotted so far have been in the familiar Cartesian coordinate system, in which points are located by x- and y-coordinates. A more natural way to plot some graphs is to locate points by an angle θ in standard position and a displacement r from the origin. Such graphs are said to be plotted in **polar coordinates.**

Objective Given a point with coordinates (r, θ), plot it on polar coordinate graph paper.

Exploratory Problem Set 11-1

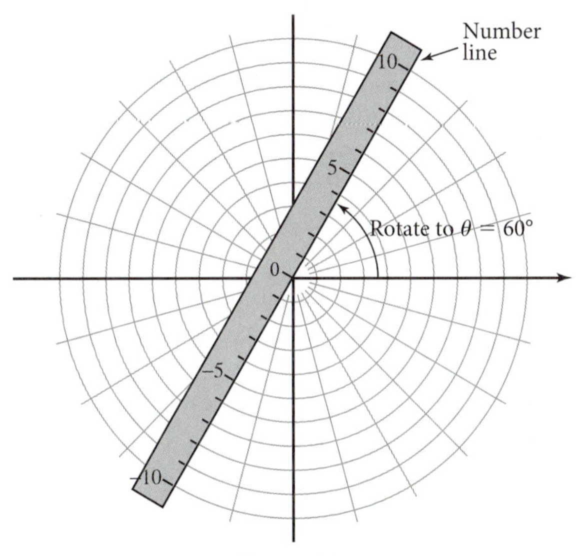

Figure 11-1a

1. Obtain a piece of polar coordinate graph paper. As shown in Figure 11-1a, make a "number-line" ruler out of paper or an index card and mark it off with the same scale as the polar coordinate paper. Put the origin (zero) of the number line at the **pole** (the origin on the polar coordinate paper). Then rotate the number line counterclockwise to $\theta = 60°$ in standard position. On the polar coordinate paper, draw a ray at 60° and mark the angle. On this ray, mark the point $(r, \theta) = (7, 60°)$ by going 7 units in the positive direction along the number line. Write $(7, 60°)$ at this point.
2. With the number-line ruler at the same place, go to -8 units. Extend the ray in Problem 1 backward using a dashed line, and mark the point $(r, \theta) = (-8, 60°)$.
3. Rotate the number-line ruler to $\theta = 240°$. Plot the points $(r, \theta) = (8, 240°)$ and $(-7, 240°)$. What do you notice about these two points?
4. Rotate the number-line ruler to $\theta = 130°$. Mark the points $(r, \theta) = (5, 130°)$ and $(r, \theta) = (-6, 130°)$ and their coordinates on the polar coordinate paper. Draw a solid ray representing the terminal side of the 130° angle and a dashed ray extending in the opposite direction for the negative value of r.
5. Rotate the number-line ruler clockwise far enough for its negative side to pass through the point $(r, \theta) = (5, 130°)$ from Problem 4. Then write another ordered pair for this point using the appropriate negative angle.
6. Write a third ordered pair (r, θ) for the point in Problem 5 using a negative angle and a positive value of r.
7. Based on the way you plotted the points in this exploration, which is the independent variable, r or θ? Does this agree or disagree with the custom of putting the independent variable first in an ordered pair?
8. What did you learn as a result of doing this problem set that you did not know before?

11-2 Polar Equations for Conics and Other Curves

Figure 11-2a shows the graph of the **polar equation** $r = 2 + 8\cos\theta$. The figure is called a **limaçon of Pascal.** (The *ç* in *limaçon* is pronounced like an *s*.) *Limaçon* is a French word for "snail." Graphs of polar functions may also be conic sections, such as the ellipse $r = \frac{9}{5 - 4\cos\theta}$ graphed in Figure 11-2b. In this section you will use algebra to see why the graphs of some polar functions are conic sections.

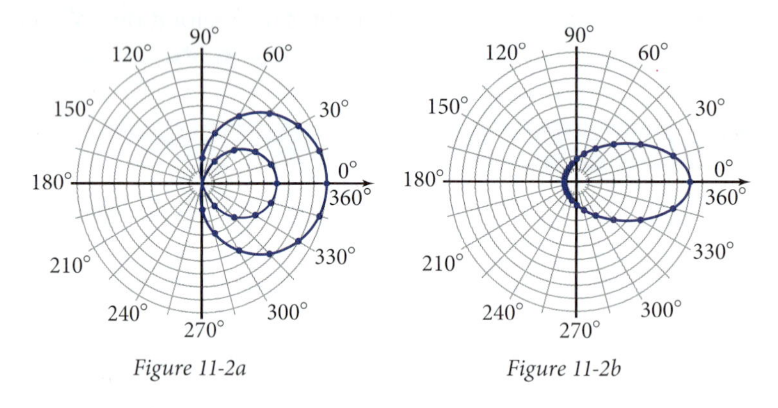

Figure 11-2a *Figure 11-2b*

Objective

- Given a polar equation, plot the graph.
- Given the polar equation for a conic section, transform it into Cartesian coordinates.

In this exploration you will plot polar curves on paper and on your grapher.

EXPLORATION 11-2: Limaçon in Polar Coordinates

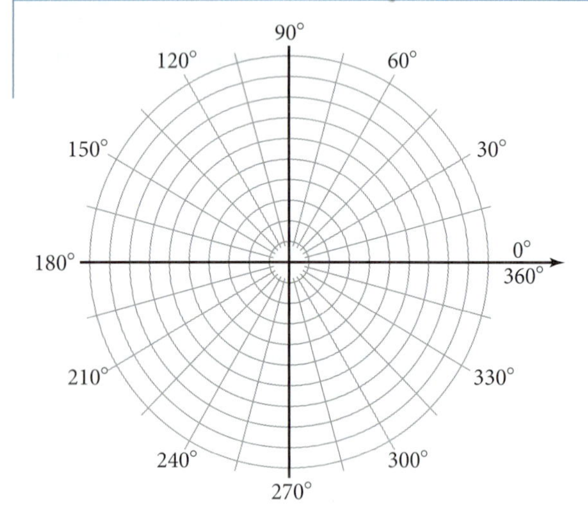

1. Here is a table of values of r and θ for a curve in polar coordinates. On polar coordinate paper, plot the points and connect them with a smooth curve.

θ	r	θ	r	θ	r
0°	3.0	135°	7.9	255°	−3.8
15°	4.8	150°	6.5	270°	−4.0
30°	6.5	165°	4.8	285°	−3.8
45°	7.9	180°	3.0	300°	−3.1
60°	9.1	195°	1.2	315°	−1.9
75°	9.8	210°	−0.5	330°	−0.5
90°	10.0	225°	−1.9	345°	1.2
105°	9.8	240°	−3.1	360°	3.0
120°	9.1				

continued

EXPLORATION, continued

2. Explain how you plot points for which r is negative.
3. The equation for the curve in Problem 1 is

   $$r = 3 + 7\sin\theta$$

   Do you agree that r-values from this equation, rounded to one decimal place, are the same as the values in the table in Problem 1?
4. Plot the graph in Problem 2 on your grapher. Use [0°, 360°] for θ and a θ-step of 5°. Use equal scales on the two axes. Does the graph on your grapher confirm the graph you plotted in Problem 1?
5. What is the name of the geometric figure in Problems 1 and 4?
6. Set r equal to 0 in the equation from Problem 3 and solve the resulting equation for θ. Write the general solution.
7. Give the *two* values of θ in [0°, 360°] at which the graph goes through the pole.
8. Draw a ray on your graph in Problem 1 at each of the two values of θ in Problem 7. How are the rays related to the graph?
9. What did you learn as a result of doing this exploration that you did not know before?

Background on Polar Coordinates

In Figure 11-2c (top graph), a point is labeled with its Cartesian coordinates (x, y). In the middle graph, the same point is labeled with its polar coordinates (r, θ). In the polar coordinate system, the origin is called the pole and the positive horizontal axis is called the **polar axis.** The graph on the bottom shows the relationships among x, y, r, and θ.

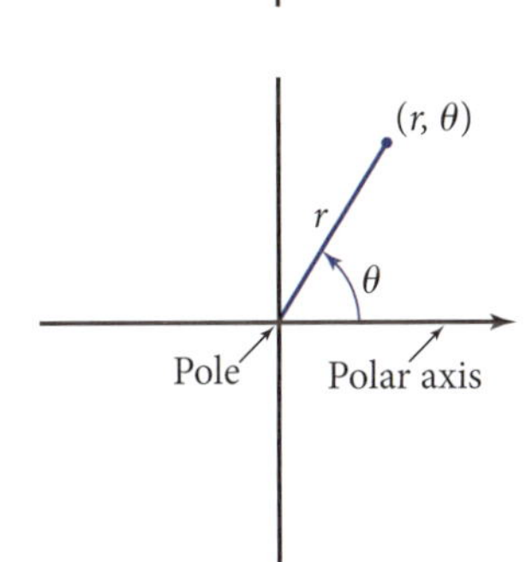

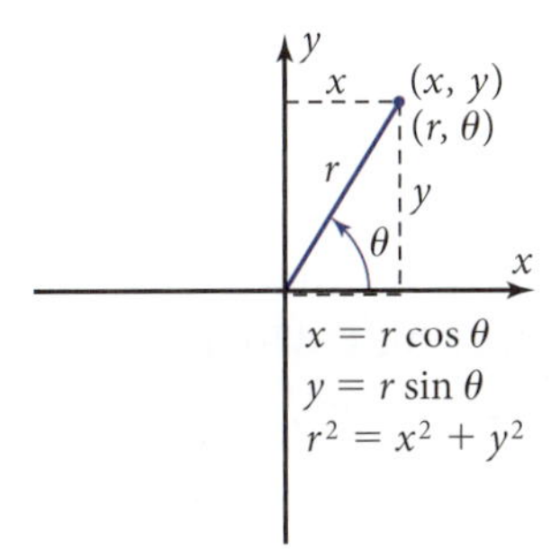

Figure 11-2c

Airplanes use a form of polar coordinates for navigation where north is the polar axis. The direction of an object, called the azimuth, is the angle measured clockwise around the horizon from north.

DEFINITIONS: Polar Coordinates

The pole in the polar coordinate system is the same as the origin in the Cartesian coordinate system. The polar axis is shown in the same position as the positive x-axis in the Cartesian coordinate system.

A point in polar coordinates is written as an ordered pair (r, θ), where

- θ is the measure of an angle in standard position whose terminal side contains the point.
- r (for "radius") is the displacement (directed distance) from the pole to the point in the direction of the terminal ray of θ.

Note: Although the angle θ is the independent variable, it is customary to write ordered pairs as (r, θ) instead of (θ, r). Also, it is customary to use θ for the angle whether it is in degrees or radians.

PROPERTIES: Polar and Cartesian Coordinates

If a point with Cartesian coordinates (x, y) has polar coordinates (r, θ) with $r \neq 0$, then

$r^2 = x^2 + y^2$ A Pythagorean property.

$\frac{x}{r} = \cos\theta$ or equivalently $x = r\cos\theta$

$\frac{y}{r} = \sin\theta$ or equivalently $y = r\sin\theta$

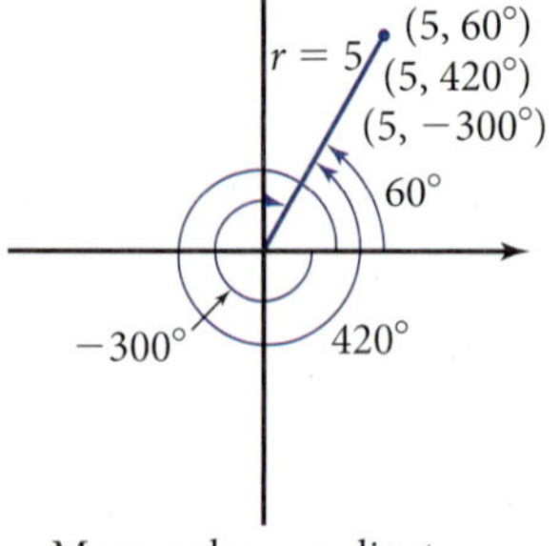

Many polar coordinates for the same point
Figure 11-2d

Because the terminal rays of an infinite number of coterminal angles pass through any given point, a point in polar coordinates may be represented by an infinite number of ordered pairs. Figure 11-2d shows three ordered pairs representing the same point.

Points in polar coordinates can have negative values of r. To plot the point $(-7, 130°)$, imagine a number line rotating about the pole to an angle of $130°$ (Figure 11-2e). Because $r = -7$, go 7 units in the negative direction along the rotated number line.

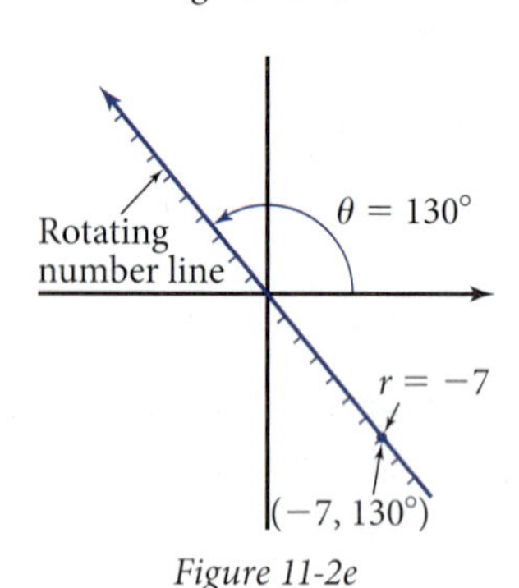

Figure 11-2e

Graphs of Polar Equations

To plot the graph of a polar equation such as

$$r = 1 + 2\cos\theta$$

set your grapher in polar mode. Use a window with equal scales for the x- and y-axes. You must also select a domain for the independent variable θ. Usually $[0°, 360°]$ with a θ-step of $5°$, or $[0, 2\pi]$ with a θ-step of 0.1 radian (or $\frac{\pi}{48}$ radian) will give a fairly smooth graph in a reasonable length of time.

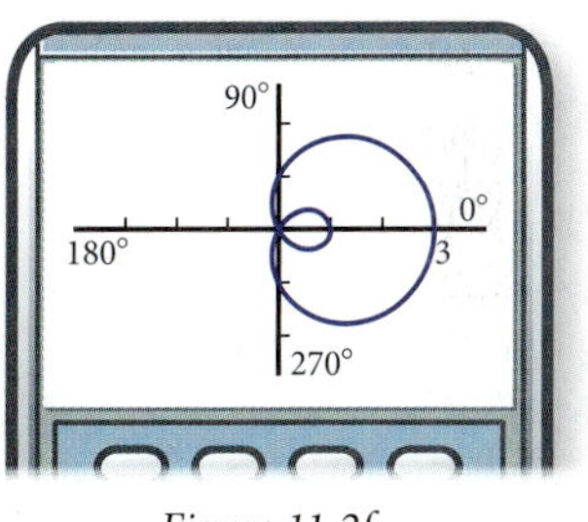

Figure 11-2f

Like the graph you plotted in Exploration 11-2, the graph of $r = 1 + 2\cos\theta$ in Figure 11-2f is a limaçon of Pascal. If you trace the graph, you'll find that the inner loop corresponds to θ-values between 120° and 240°. To see why, it helps to change mode and plot the **auxiliary Cartesian graph** (Figure 11-2g) of the equation,

$$y = 1 + 2\cos\theta$$

The Cartesian graph reveals that when θ is between 120° and 240°, r is negative. So the points are plotted in the opposite direction (Figure 11-2h) for θ in this range. Note also that where the graph goes through the pole, it is tangent to the lines $\theta = 120°$ and $\theta = 240°$.

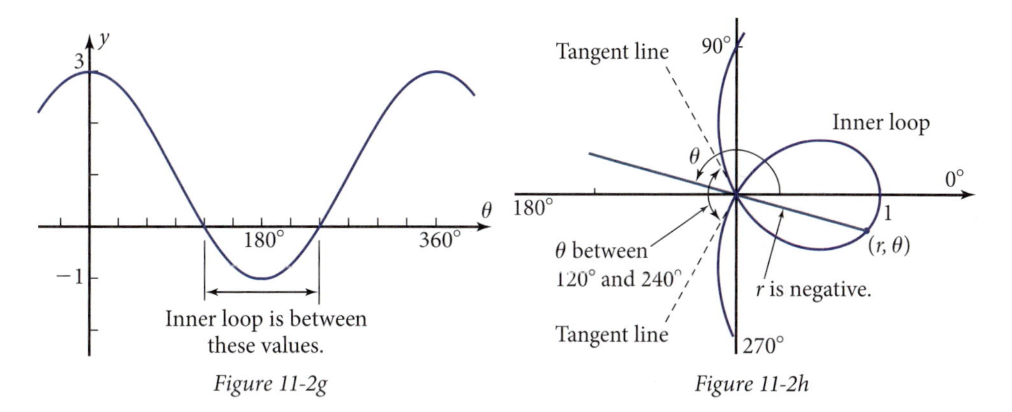

Figure 11-2g

Figure 11-2h

You can see dynamically how polar graphs and Cartesian graphs are related by viewing the *Polar Graphs* exploration at **www.keymath.com/precalc.** Use the sliders to enter the equation for the limaçon discussed. Then drag the slider that changes the angle θ or click the ANIMATE button. Observe how the location of the moving point on the auxiliary Cartesian graph is related to its location on the polar graph, and notice how the point is plotted in the direction opposite that of angle θ when r is negative, forming the inner loop.

DEFINITION: Limaçon in Polar Coordinates

A **limaçon** is a figure with polar equation

$$r = a + b\cos\theta \quad \text{or} \quad r = a + b\sin\theta, \quad a \neq 0, b \neq 0$$

If $|a| < |b|$, then the limaçon has an inner loop.

If $|a| > |b|$, then the limaçon has no inner loop.

If $|a| = |b|$, then the limaçon has a **cusp** at the pole and is called a **cardioid** (which means "heartlike").

A **cusp** is a point at which the graph of a relation changes direction abruptly.

EXAMPLE 1 ➤ Plot the graph of the five-leaved rose $r = 6\sin 5\theta$. Find numerically the first interval of positive θ-values for which r is negative. Confirm your answer by plotting this part of the graph.

SOLUTION The graph is shown in Figure 11-2i. Notice the five "leaves."

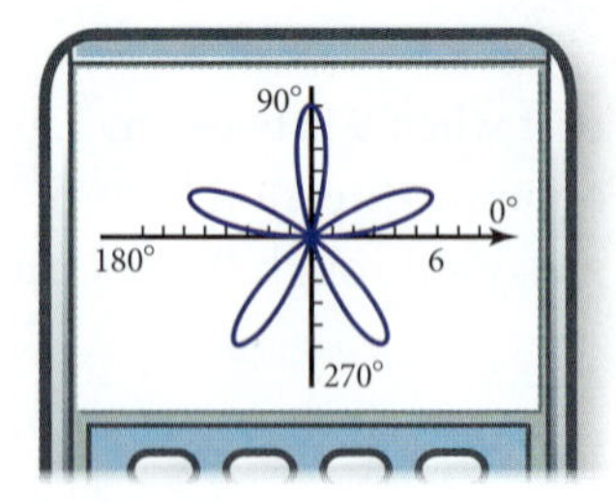

Figure 11-2i

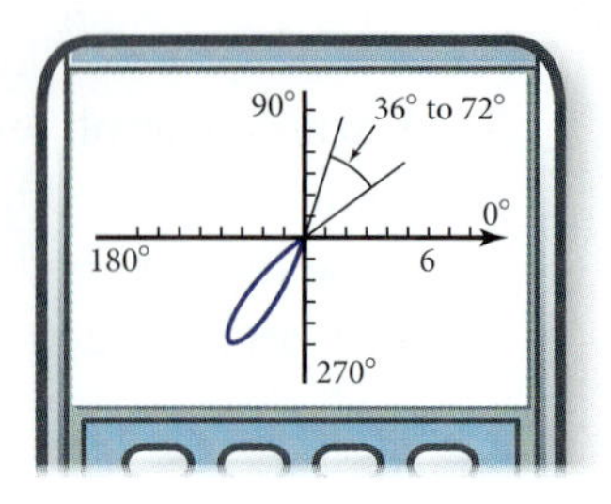

Figure 11-2j

θ	r
0°	0
15°	−5.7955...
30°	3
45°	−4.2426...
60°	−5.1961...
75°	1.5529...

Make a table of values for θ and r starting at 0° and stepping by 15°. Negative values first occur at 45° and 60°. By narrowing your search, you will find that r changes from positive to negative at 36° and from negative back to positive at 72°. So $36° < \theta < 72°$ is the first interval of positive θ-values for which r is negative.

You can confirm the answer graphically (Figure 11-2j) or algebraically by solving the equation $6\sin 5\theta = 0$ to find $\theta = 0°, 36°, 72°, 108°, \ldots$. ◄

Conics in Polar Coordinates

If you graph the reciprocal of the polar equation for a limaçon, the result is, surprisingly, a conic section! Figure 11-2k shows two examples.

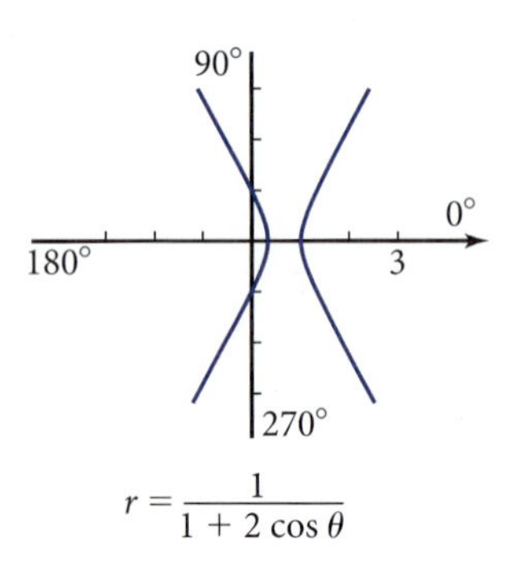

$r = \dfrac{1}{1 + 2\cos\theta}$

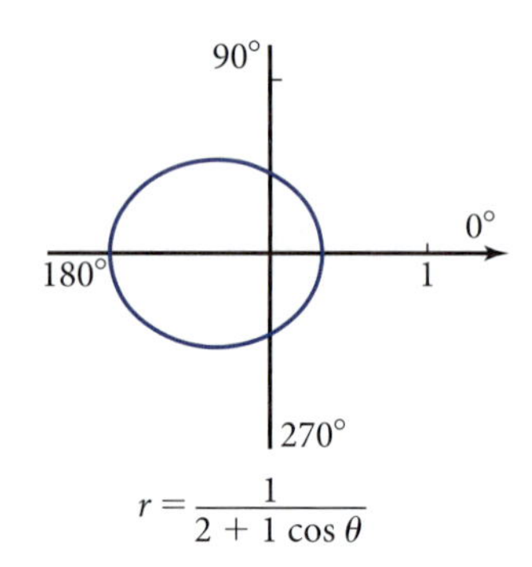

$r = \dfrac{1}{2 + 1\cos\theta}$

Figure 11-2k

EXAMPLE 2 ➤ Plot the graph of the equation $r = \frac{9}{5 - 4\cos\theta}$. Show algebraically that the graph is an ellipse.

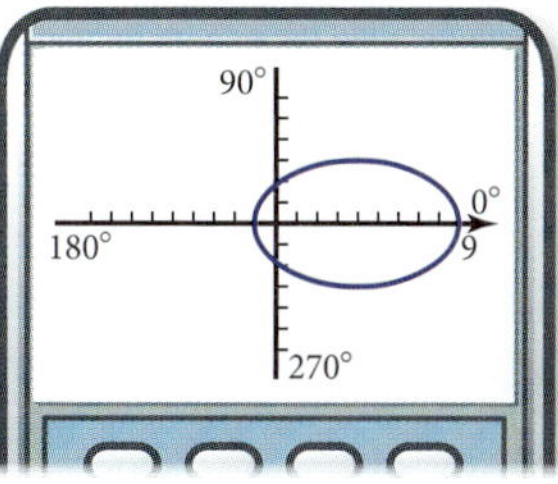

Figure 11-2l

SOLUTION The graph (Figure 11-2l) has an elliptical shape. For a dynamic, high-resolution graph, use the *Polar Conics* exploration at **www.keymath.com/precalc.** To verify that the graph is an ellipse, transform the equation into Cartesian form.

$5r - 4r\cos\theta = 9$ — Eliminate the fraction by multiplying both sides by $5 - 4\cos\theta$.

$5\sqrt{x^2 + y^2} - 4x = 9$ — Substitute $\sqrt{x^2 + y^2}$ for r, and x for $r\cos\theta$.

$5\sqrt{x^2 + y^2} = 4x + 9$ — Isolate the radical term on one side of the equation.

$25x^2 + 25y^2 = 16x^2 + 72x + 81$ — Square both sides to eliminate the radical term.

$9x^2 + 25y^2 - 72x - 81 = 0$

Therefore, the graph is an ellipse. — x^2 and y^2 have the same sign but unequal coefficients. ◄

One focus of the ellipse in Example 2 is at the pole. The eccentricity is $\left|\frac{-4}{5}\right|$, or 0.8, the absolute value of the ratio of the coefficients in the denominator of the polar equation. The general properties of conics with one focus at the pole are summarized in the box.

PROPERTIES: Conic Sections in Polar Coordinates

The general polar equation for a conic section with one focus at the pole is

$$r = \frac{k}{a + b\cos\theta} \quad \text{or} \quad r = \frac{k}{a + b\sin\theta}$$

The eccentricity of the conic is $e = \left|\frac{b}{a}\right|$.

$|k| = |aep|$, where p is the distance between the focus and the directrix.

- If $|b| < |a|$, then the graph is an ellipse ($e < 1$).
- If $|b| > |a|$, then the graph is a hyperbola ($e > 1$).
- If $|b| = |a|$, then the graph is a parabola ($e = 1$).

This box summarizes the relationship between conic sections and limaçons.

PROPERTY: Relationship Between Conics and Limaçons

The r-values for a conic section are reciprocals of the r-values for a limaçon.

- If the limaçon has a loop, then the conic is a hyperbola.
- If the limaçon has no loop and no cusp, then the conic is an ellipse.
- If the limaçon is a cardioid, then the conic is a parabola.

Polar Equations for Special Circles and Lines

Figure 11-2m shows a circle with diameter 9 units centered on the polar axis and passing through the pole. Point (r, θ) is on the circle. The triangle shown is a right triangle because the angle at the point (r, θ) is inscribed in the semicircle. By the definition of cosine, $\frac{r}{9} = \cos\theta$. So a polar equation for the circle is $r = 9\cos\theta$.

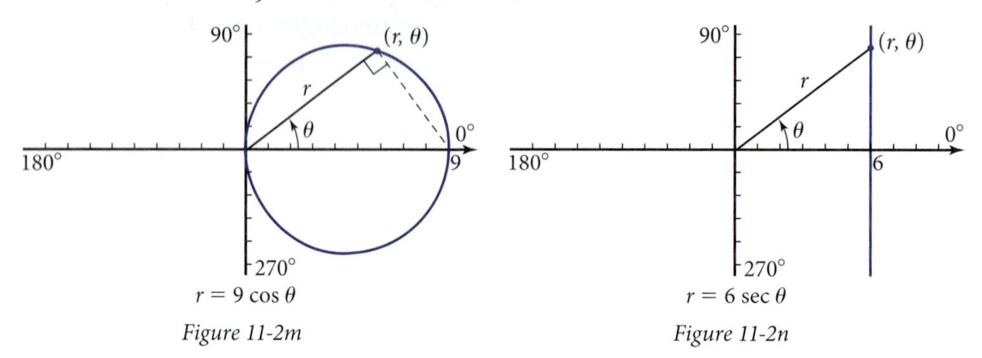

Figure 11-2m *Figure 11-2n*

Similarly, the graph of $r = 9\sin\theta$ is a circle passing through the pole and centered on the line $\theta = 90°$.

Figure 11-2n shows a point (r, θ) on the line perpendicular to the polar axis and 6 units from the pole. In the right triangle formed by r, the line, and the polar axis, $\frac{r}{6} = \sec\theta$. So a polar equation for the line is $r = 6\sec\theta$.

The graph of $r = 6\csc\theta$ is a line parallel to the polar axis and 6 units from the pole. Note that $r = a \sec\theta$ is equivalent to $r\cos\theta = a$, and the relationship between polar and Cartesian coordinates establishes that $x = r\cos\theta$. So, $x = a$. A similar relationship emerges for $r = a\csc\theta$. For a dynamic, high-resolution graph, use the *Polar Lines* exploration at **www.keymath.com/precalc.** The box contains a summary of the equations of these special circles and lines.

PROPERTIES: Polar Equations of Special Circles and Lines

The equations apply to the circles and lines shown.

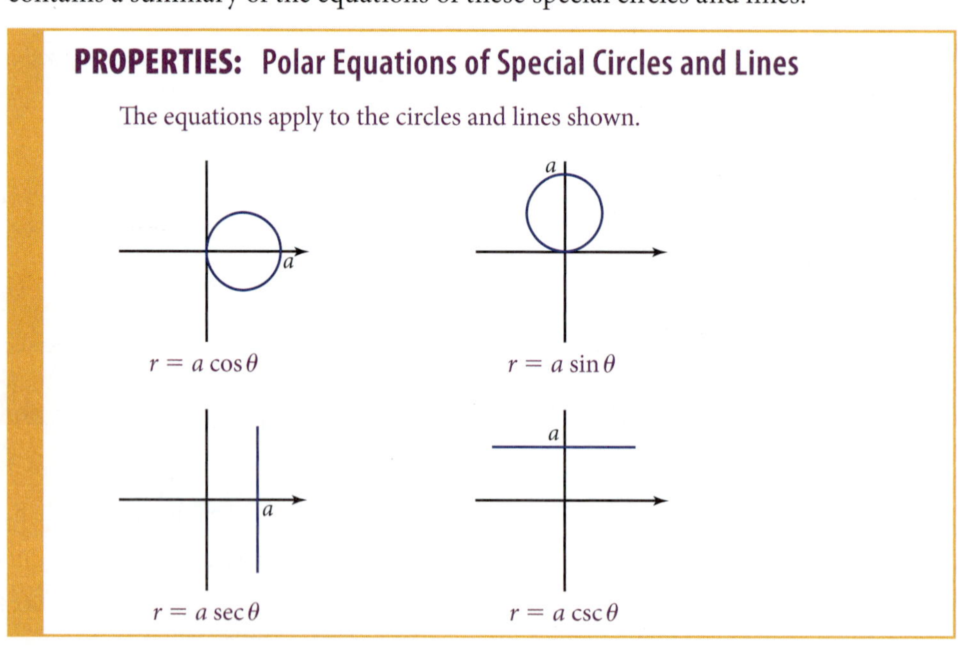

Problem Set 11-2

Reading Analysis

From what you have read in this section, what do you consider to be the main idea? Can you think of a real-world situation in which it might be more advantageous to use polar coordinates than Cartesian coordinates? Why are there many different ordered pairs for the same point in polar coordinates? How do you interpret a negative r-value? How is the polar equation for a limaçon related to the polar equation for a conic section?

Quick Review

For Q1–Q7, refer to Figure 11-2o.

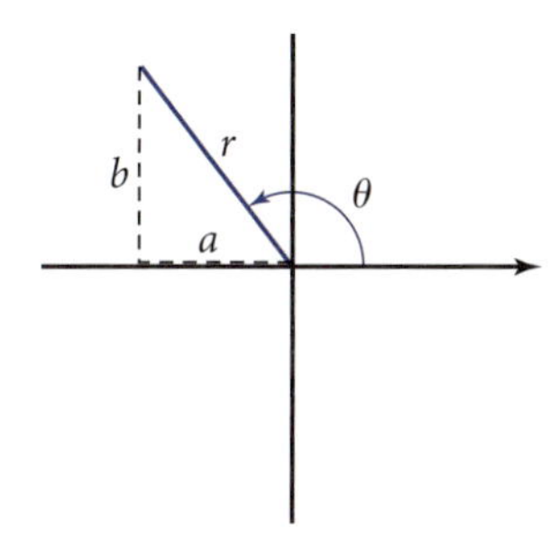

Figure 11-2o

Q1. What does $\sin\theta$ equal?

Q2. What does $\cos\theta$ equal?

Q3. What does $\tan\theta$ equal?

Q4. What does $\cot\theta$ equal?

Q5. What does $\sec\theta$ equal?

Q6. What does $\csc\theta$ equal?

Q7. Which of the six trigonometric functions of angle θ will be negative?

Q8. What is the exact value of $\cos 150°$?

Q9. What is the exact value of $\sin\frac{\pi}{4}$?

Q10. If one value of $\arcsin x$ is 15°, then the other value between 0° and 360° is

A. 75° **B.** 105° **C.** 165° **D.** 195°

E. None of these

For Problems 1 and 2, plot the points on polar coordinate paper and connect them in order with a smooth curve.

1.

θ	r	θ	r
0°	0	105°	−4.8
15°	1.3	120°	−7.5
30°	4.3	135°	−7.1
45°	7.1	150°	−4.3
60°	7.5	165°	−1.3
75°	4.8	180°	0
90°	0		

2.

θ	r	θ	r
0°	0	105°	−7.2
15°	0.1	120°	−3.0
30°	0.6	135°	−1.4
45°	1.4	150°	−0.6
60°	3.0	165°	−0.1
75°	7.2	180°	0
90°	(infinite)		

3. Plotting the points in Problem 1 gives the *bifolium* $r = 20\cos\theta\sin^2\theta$. Plot the graph in the domain [0°, 360°], or view the *Bifolium* exploration at **www.keymath.com/precalc.** Trace to $\theta = 225°$ and sketch the result. What is happening to the graph as θ goes from 180° to 360°? Look up *bifolium* on the Internet or in some other source and tell what you find.

4. Plotting the points in Problem 2 gives the *cissoid of Diocles* $r = 2\sin\theta\tan\theta$. Plot the graph in the domain [0°, 360°], or view the *Cissoid* exploration at **www.keymath.com/precalc.** Trace to $\theta = 240°$ and sketch the result. What is the root of the word *cissoid*? Look up *Diocles* on the Internet or in some other source and tell what you find.

For most of the remaining problems in this section, you can create dynamic, high-resolution graphs by using the Dynamic Precalculus Explorations for Chapter 11 at **www.keymath.com/precalc.** Because the equations in these explorations are expressed in terms of cosine, you need to change the value of d to 90° for problems expressed in terms of sine.

5. Plot the *three-leaved rose* $r = 10\sin 3\theta$. Find the first interval of positive θ-values for which r is negative. Confirm your answer by plotting this part of the graph.

6. Plot the *four-leaved rose* $r = 5\cos 2\theta$. Find the first interval of positive θ-values for which r is negative. Confirm your answer by plotting this part of the graph.

7. Plot the circle $r = 6\sin\theta$. Use a domain of $[0°, 360°]$. Trace to $\theta = 300°$. Draw a sketch to help you explain why the point is in the second quadrant even though $\theta = 300°$ is a fourth-quadrant angle.

8. Plot the line $r = 2\csc\theta$. Use a domain of $[0°, 360°]$. Trace to $\theta = 240°$. Draw a sketch to help you explain why the point is in the first quadrant even though $\theta = 240°$ is a third-quadrant angle.

9. Show algebraically that the graph of $r = 6\sin\theta$ in Problem 7 is a circle through the pole.

10. Show algebraically that the graph of $r = 2\csc\theta$ in Problem 8 is the line $y = 2$.

11. Figure 11-2p shows the graph of $r = 9\cos\frac{\theta}{2}$. Plot the graph using two revolutions. Find the first interval of positive θ-values for which r is negative. Sketch this part of the graph.

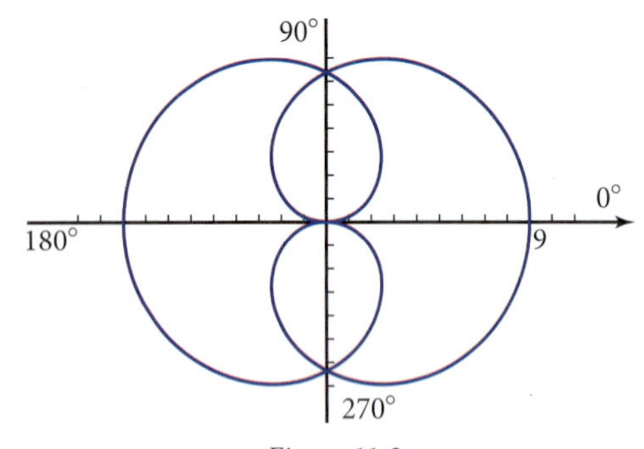

Figure 11-2p

12. Figure 11-2q shows a *lemniscate of Bernoulli*, $r = \sqrt{9\cos 2\theta}$. Plot the graph on your grapher. Describe the behavior of the graph as it goes to the pole. Explain why there are values of θ for which there is no graph. Explain why there are no negative values of r. Look up the name *Bernoulli* on the Internet or in some other source. See how many different members of the Bernoulli family you can find. Tell approximately when each person lived and what they studied.

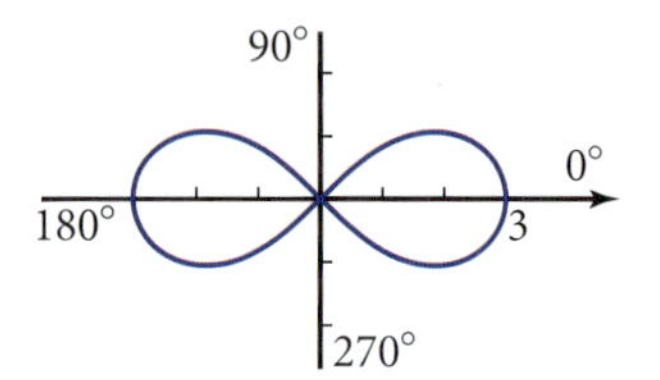

Figure 11-2q

13. Figure 11-2r shows the first three revolutions of the *spiral* $r = \frac{2\theta}{\pi}$, where angle θ is measured in radians. Plot the graph on your grapher in the domain $[0, 6\pi]$. At what three positive values of r does the graph cross the polar axis? Extend the domain to $[-6\pi, 6\pi]$. Describe what you see on your grapher.

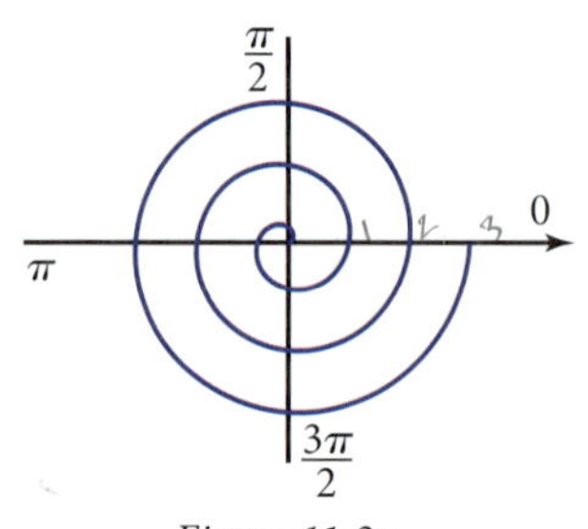

Figure 11-2r

14. WEB Figure 11-2s shows a *conchoid of Nicomedes*, $r = 8 + 3\csc\theta$. Plot the graph on your grapher. Use $[0°, 360°]$ for θ with a θ-step of 5°. What happens to the graph as θ approaches 180° and 360°? At the point P shown in the figure, is r positive or negative? What interval of θ-values generates the loop below the horizontal axis? Look up the name *Nicomedes* on the Internet or in some other source. What did you learn?

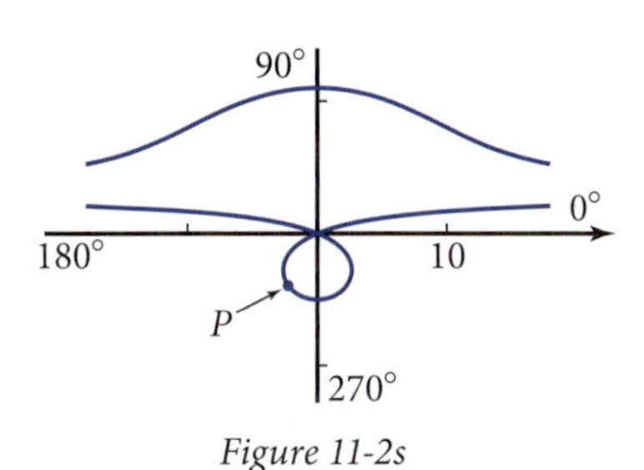

Figure 11-2s

15. *Circles Problem:* Each of these graphs is a circle of diameter 1 unit passing through the pole. Write the polar equation for each circle.

a.

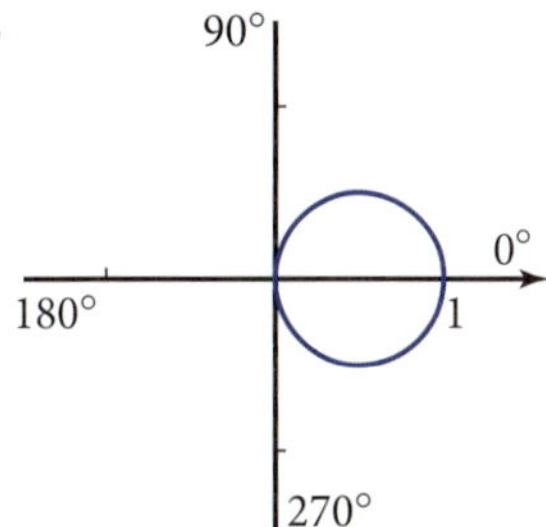

b.

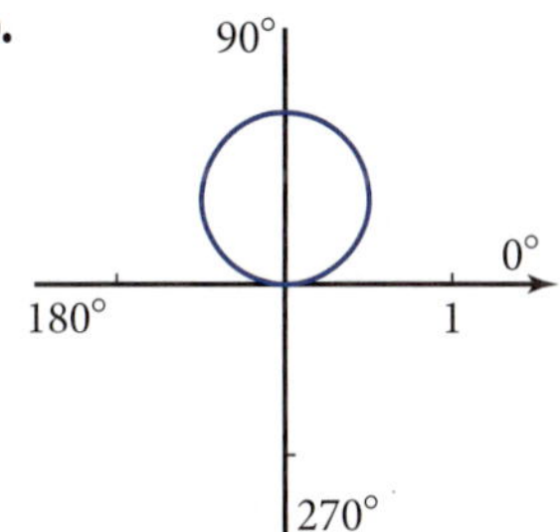

c.

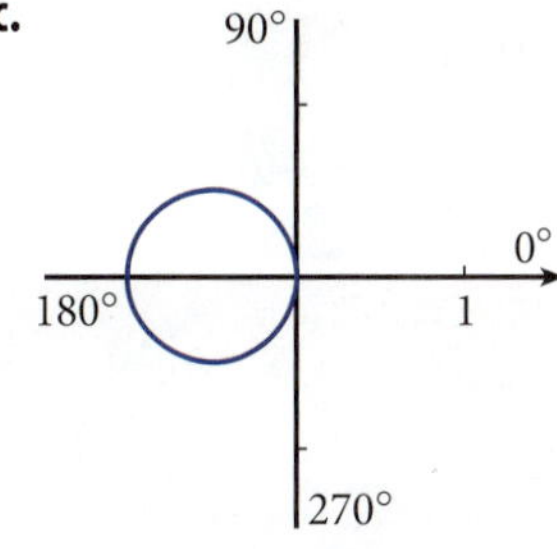

d.

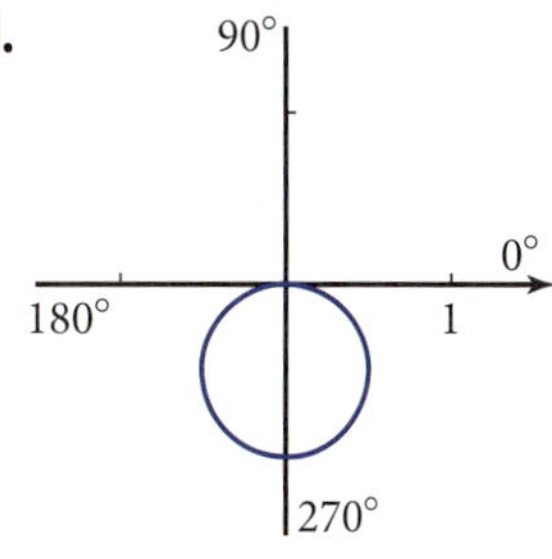

16. *Lines Problem:* Each of these graphs is a line 1 unit from the pole. Write the polar equation for each line.

a.

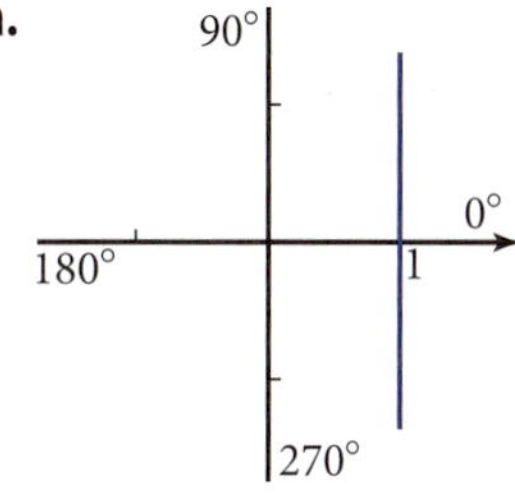

b.

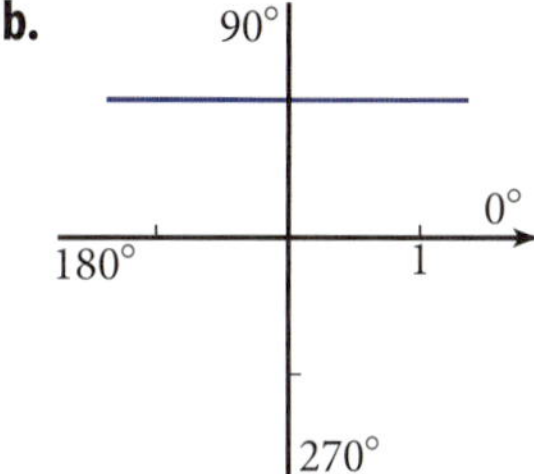

c.

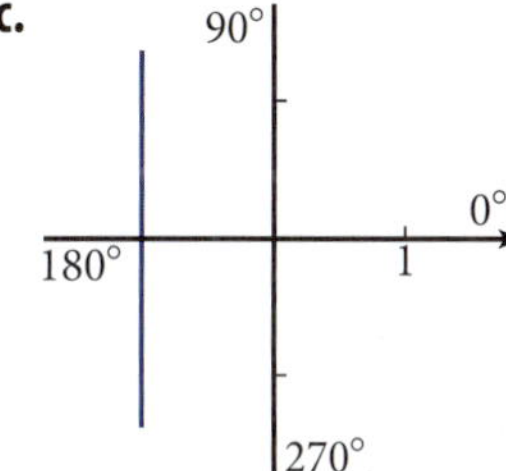

d.

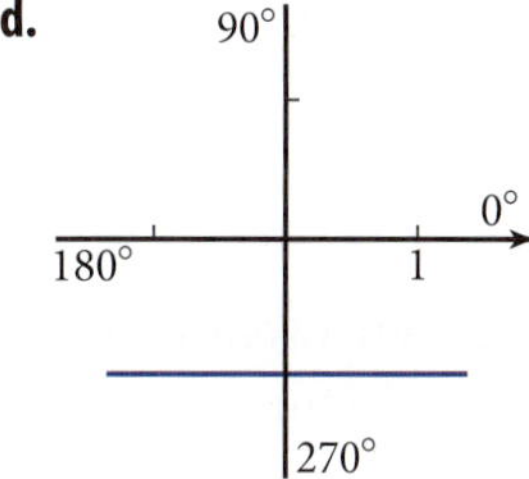

17. *Hyperbola Problem:* Consider the polar equation $r = \frac{8}{3 + 5\cos\theta}$.

a. Plot the graph. Sketch the result.

b. Show algebraically that the graph is a hyperbola by transforming the equation into Cartesian form.

c. Where is one focus of the hyperbola? What is the eccentricity?

18. *Ellipse Problem:* Consider the polar equation $r = \frac{10}{3 + 2\sin\theta}$.

a. Plot the graph. Sketch the result.

b. Show algebraically that the graph is an ellipse by transforming the equation into Cartesian form.

c. Where is one focus of the ellipse? What is the eccentricity?

19. *Parabola Problem:* Consider the polar equation $r = \frac{6}{1 + \cos\theta}$.

a. Plot the graph. Sketch the result.

b. Show algebraically that the graph is a parabola by transforming the equation into Cartesian form.

c. Where is the focus? How can you tell from the equation that the eccentricity equals 1?

20. *Rotated Polar Graphs Problem:* Figure 11-2t shows two ellipses,

$$r = \frac{19}{10 - 9\cos\theta} \quad \text{and}$$

$$r = \frac{19}{10 - 9\cos(\theta - 30°)}$$

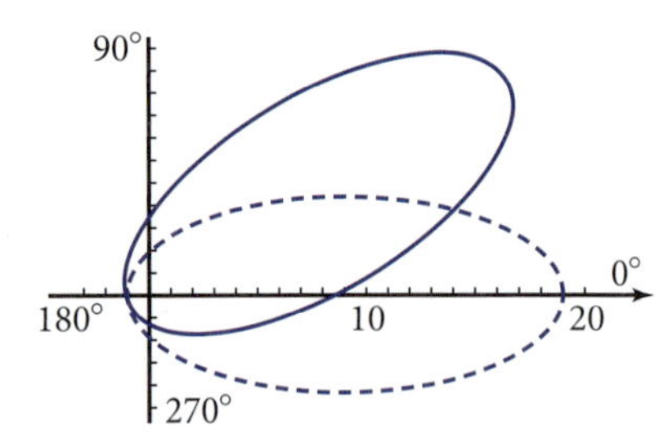

Figure 11-2t

a. Which graph is which? What is the effect of subtracting 30° from θ in the second equation?

b. Write an equation that would rotate the dashed graph 90° counterclockwise. Confirm that your equation is correct by plotting the graph.

c. Explain the difference in the effect of subtracting a constant from θ in polar coordinates and subtracting a constant from θ in Cartesian coordinates.

d. The graph of $r = 3\sec\theta$ is a line perpendicular to the polar axis 3 units from the pole. Write an equation that would rotate the line 60° counterclockwise. Plot the graph. Where does the line cross the polar axis? Is the line still a perpendicular distance of 3 units from the pole? How can you tell?

21. *Roller Skating Problem:* Figure 11-2u shows a roller skating loop as it appears in a manual of the Roller Skating Rink Operations of America. The figure is composed of arcs of circles that are easy to mark on the rink floor. The finished figure resembles a limaçon with a loop. Write a polar equation for the limaçon. Confirm your answer by plotting the limaçon on your grapher.

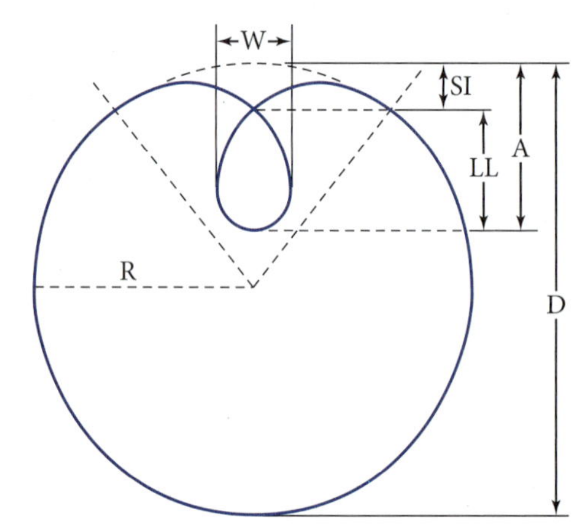

Figure 11-2u

Loop Dimensions

D	240 cm
SI $= \frac{1}{8}$D	30 cm
LL $= \frac{1}{4}$D	60 cm
A	90 cm
W $= \frac{1}{6}$D	40 cm

22. *Rose Problem:* The general equation for a rose in polar coordinates is

$$r = k\cos n\theta$$

where n is an integer.

a. Plot the four-leaved rose $r = 9\cos 2\theta$.

b. Plot the three-leaved rose $r = 9\cos 3\theta$.

c. Find the relationship between n and the number of leaves in the rose by plotting other roses with both even and odd coefficients of θ.

d. Write the equation for the five-leaved rose that appears on page 559. Use a distance from the pole to the tip of a leaf of 7 units. Plot the graph on your grapher. Does your graph agree with the figure?

23. *Comparing Graphs Problem:* Figure 11-2v shows the graph of $y = 2 + 4\sin x$ in the domain $[-360°, 390°]$. Figure 11-2w shows the graph of $r = 2 + 4\sin\theta$.

a. Find the rectangular coordinates of the five marked points in Figure 11-2v.

b. Find the polar coordinates of the five marked points in Figure 11-2w.

c. Explain how the marked points on the rectangular coordinate graph correspond to the points marked on the polar coordinate graph.

d. Which parts of the graph in rectangular coordinates corresponds to the inner loop on the limaçon?

e. Explain why $r = 2 + 4\sin x$ is a function even though its polar graph fails the vertical line test.

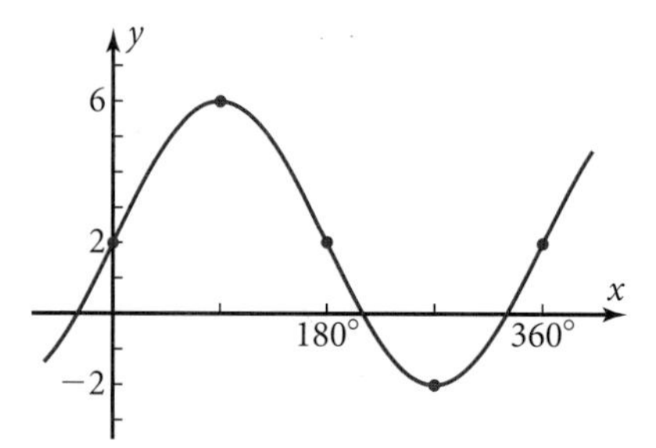

Figure 11-2v

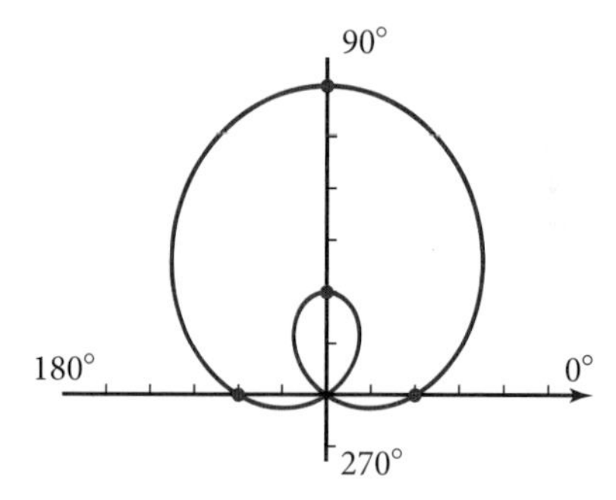

Figure 11-2w

11-3 Intersections of Polar Curves

The limaçon $r_1(\theta) = 3 + 2\cos\theta$ and the four-leaved rose $r_2(\theta) = 5\sin 2\theta$ (Figure 11-3a) appear to intersect at eight points, for instance, P_1 and P_2. Normally, a point of intersection represents a solution to a system of equations, but in this section you will discover that P_1 is a solution to this system of equations but P_2 is not. In fact, only four of the eight points are solutions. The other four intersections are points where the graphs cross, but for different values of θ.

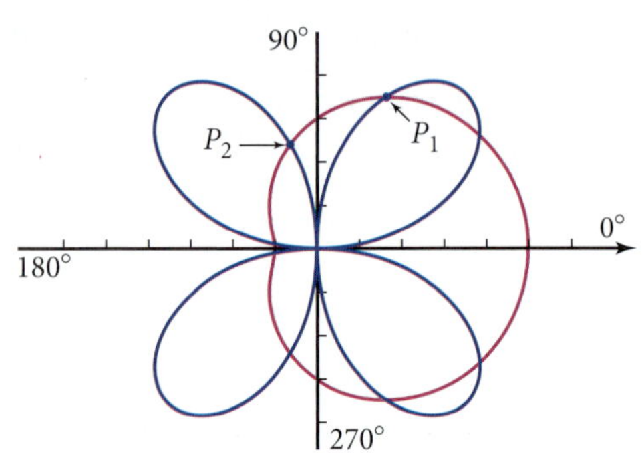

Figure 11-3a

Objective Given two polar equations, find the solutions to the system of equations and relate them to the intersections of the polar curves.

In this exploration you will plot two polar curves on your grapher and determine which of the intersections represent solutions to the system of equations.

EXPLORATION 11-3: Intersections of Polar Curves

The figure shows the polar curves

$r_1(\theta) = 5 - 2\cos\theta$ and
$r_2(\theta) = -4\tan\theta$

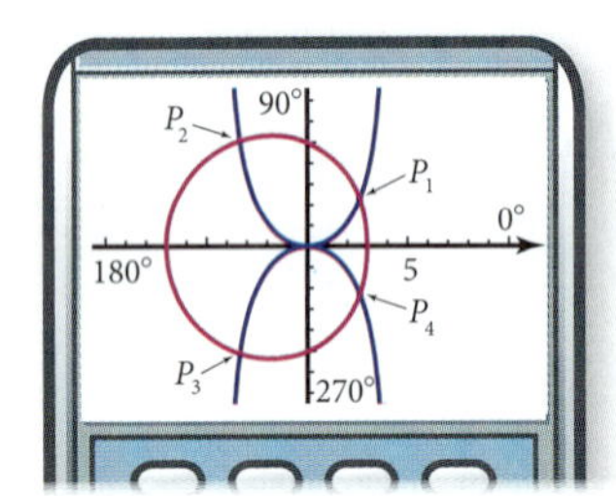

1. Plot the two graphs on your grapher. Use simultaneous mode, path style, and a fairly small θ-step so that the graphs plot relatively slowly. Pause the plotting when the graph of $r_1(\theta)$ reaches point P_1 on the figure. What do you notice about the graph of $r_2(\theta)$ for this value of θ?
2. Resume the plotting, pausing again where the graph of $r_1(\theta)$ is at point P_2 on the figure. Do both graphs pass through point P_2 for the same value of θ?
3. If two polar curves pass through the same point for the same value of θ, then the point represents a solution to the system of equations. Otherwise the intersection is not a solution. Do points P_3 and P_4 on the figure represent solutions? How do you know?
4. Curve $r_1(\theta)$ passes through point P_1 at $\theta = 41.1344...°$. Trace $r_1(\theta)$ to this value of θ. Does the result coincide with point P_1? Write the value of $r_1(\theta)$ for this value of θ, using ellipsis format.
5. Trace $r_2(\theta)$ to $\theta = 41.1344...°$. What interesting thing do you notice about the value of $r_2(\theta)$ for this value of θ? Sketch $r_1(\theta)$ and $r_2(\theta)$ for this value of θ on a copy of the figure.
6. With your grapher in function mode, plot the auxiliary Cartesian graphs

   $f_1(x) = 5 - 2\cos x$ and
   $f_2(x) = -4\tan x$

   Sketch the result on graph paper.

continued

7. On the same screen as your graph from Problem 6, use the intersect feature to find the coordinates of the two solutions to the system in the domain [0°, 360°].

8. What did you learn as a result of doing this exploration that you did not know before?

If you plot the equations $r_1(\theta) = 3 + 2\cos\theta$ and $r_2(\theta) = 5\sin 2\theta$ with your grapher in simultaneous mode, the two graphs will be drawn at the same time.

Figure 11-3b shows the effect of pausing at about 65°. Both graphs are at point P_1 for a value of θ close to 65°. But if you continue plotting to about 105° (Figure 11-3c), the limaçon is at point P_2 and the rose is at point P_3. Because the rose has a negative r-value and the limaçon has a positive r-value at this angle θ, point P_2 is not a solution to this system of equations.

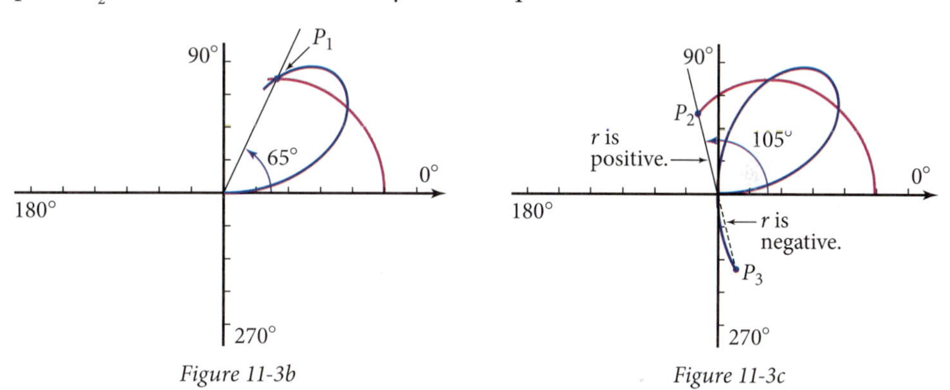

Figure 11-3b *Figure 11-3c*

EXAMPLE 1 ➤ Find the solution to the system consisting of the limaçon $r_1(\theta) = 3 + 2\cos\theta$ and the four-leaved rose $r_2(\theta) = 5\sin 2\theta$ in the domain [0°, 360°].

SOLUTION To see which points are solutions, it helps to plot auxiliary Cartesian graphs. Figure 11-3d shows the graphs of $f_1(\theta) = 3 + 2\cos\theta$ and $f_2(\theta) = 5\sin 2\theta$. The solutions occur where the auxiliary graphs intersect. You can find the precise values using the intersect feature on your grapher.

(4.6529..., 34.2630...°), (3.8511..., 64.8126...°), (1.0103..., 185.8289...°), (2.4855..., 255.0953...°)

These solution points are circled in Figure 11-3e.

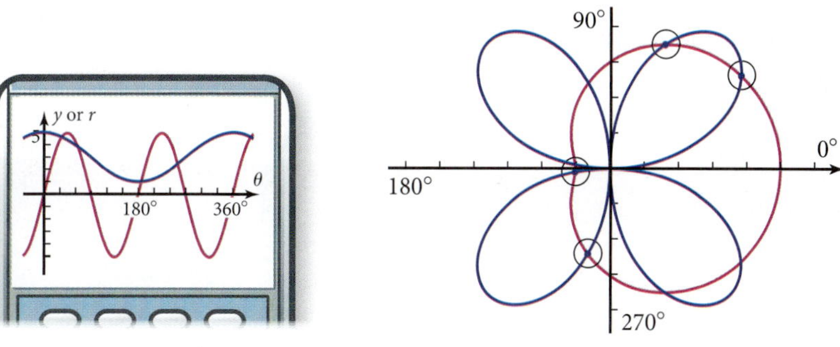

Figure 11-3d *Figure 11-3e*

Problem Set 11-3

Reading Analysis

From what you have read in this section, what do you consider to be the main idea? How is it possible for an intersection of two polar curves to not be a solution to the system of equations? How can you use auxiliary Cartesian graphs to find the solution points?

Quick Review

Q1. Find the Cartesian coordinates of the polar point $(r, \theta) = (6, 30°)$.

Q2. Find polar coordinates of the Cartesian point $(x, y) = (3, 7)$.

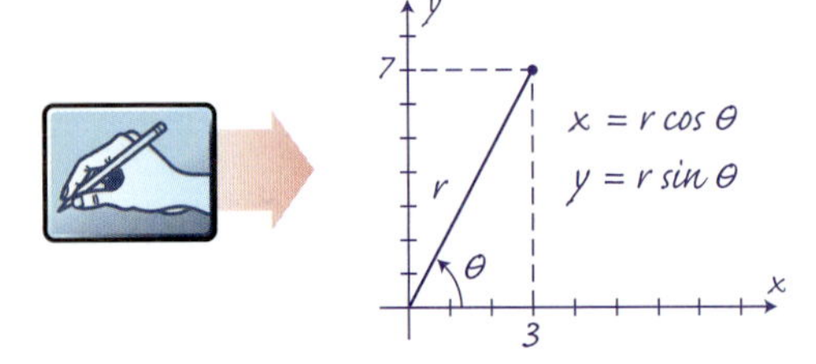

For Problems Q3–Q6, write another ordered pair for the polar point (6, 30°) with the given specifications.

Q3. r and θ are both positive.

Q4. r is positive and θ is negative.

Q5. r is negative and θ is positive.

Q6. r and θ are both negative.

Q7. What geometric figure has the polar equation $r = 5(3 + 4\cos\theta)$?

Q8. What geometric figure has the polar equation $r = \frac{5}{3 + 4\cos\theta}$?

Q9. Write a polar equation for the circle in Figure 11-3f.

Q10. Write parametric equations for the circle in Figure 11-3f.

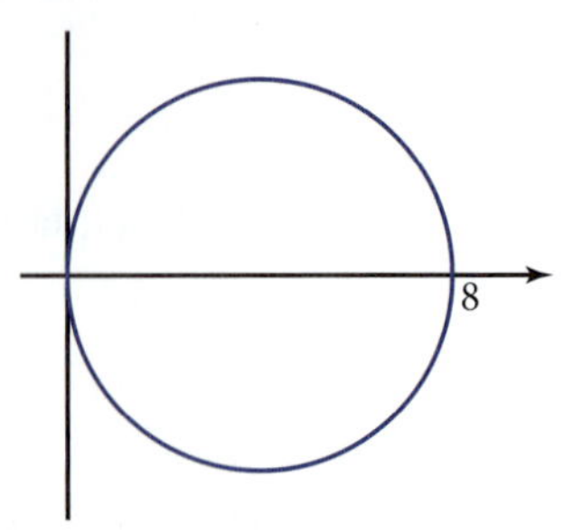

Figure 11-3f

For Problems 1–8, find the solutions to the system of equations and mark the solution points on a sketch of the graphs.

1. Ellipse $r_1(\theta) = \dfrac{5}{3 - 2\cos\theta}$ and hyperbola $r_2(\theta) = \dfrac{5}{2 + 3\cos\theta}$ (Figure 11-3g)

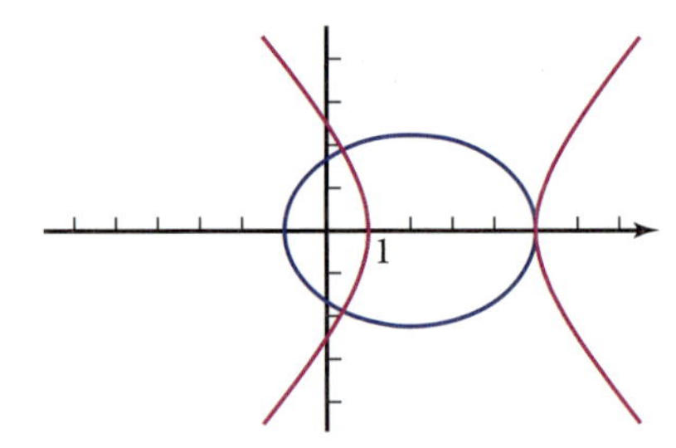

Figure 11-3g

2. Limaçon $r_1(\theta) = 2 + 3\cos\theta$ and ellipse $r_2(\theta) = \dfrac{5}{3 + 2\cos\theta}$ (Figure 11-3h)

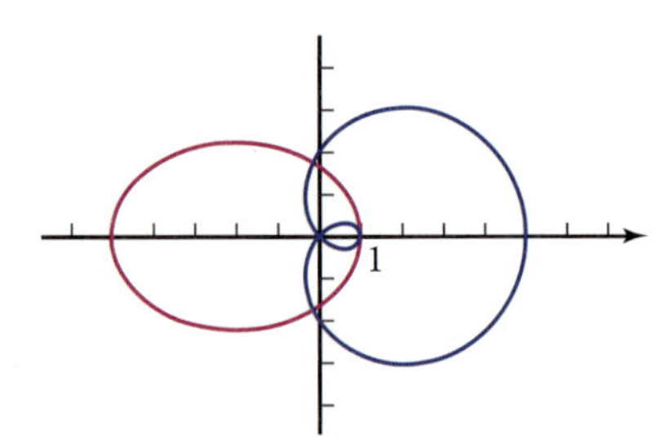

Figure 11-3h

3. Limaçon $r_1(\theta) = -1 - 5\cos(\theta - 180°)$ and limaçon $r_2(\theta) = 3 + 2\cos\theta$ (Figure 11-3i)

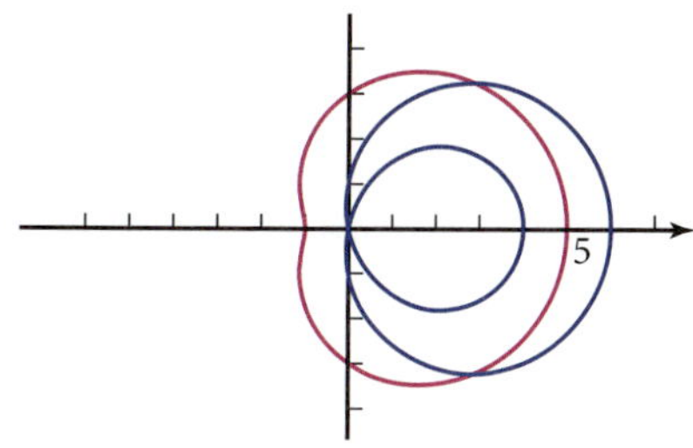

Figure 11-3i

4. Limaçon $r_1(\theta) = 1 + 5\cos\theta$ and line $r_2(\theta) = 3\sec(\theta - 30°)$ (Figure 11-3j)

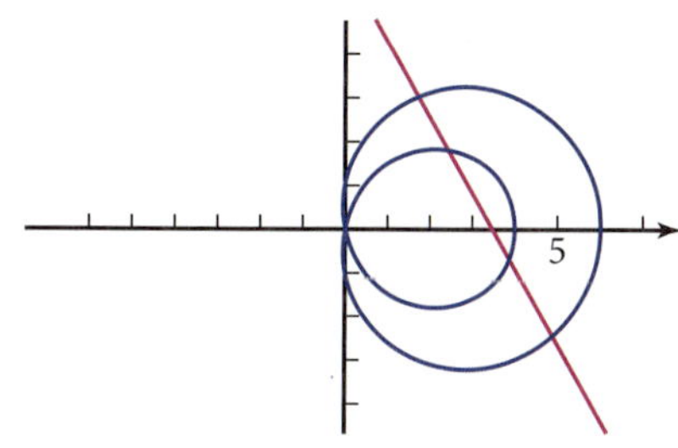

Figure 11-3j

5. Conchoid $r_1(\theta) = 3 + \csc\theta$ and ellipse $r_2(\theta) = \dfrac{5}{3 - 2\sin\theta}$ (Figure 11-3k)

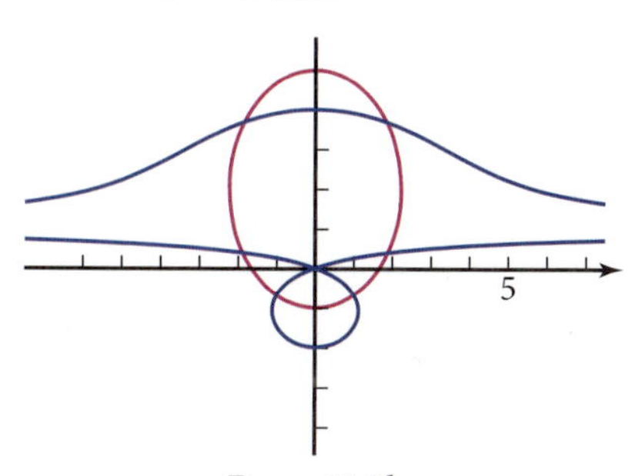

Figure 11-3k

6. Limaçon $r_1(\theta) = 3 - 2\sin\theta$ and rose $r_2(\theta) = 5\cos 2\theta$ (Figure 11-3l)

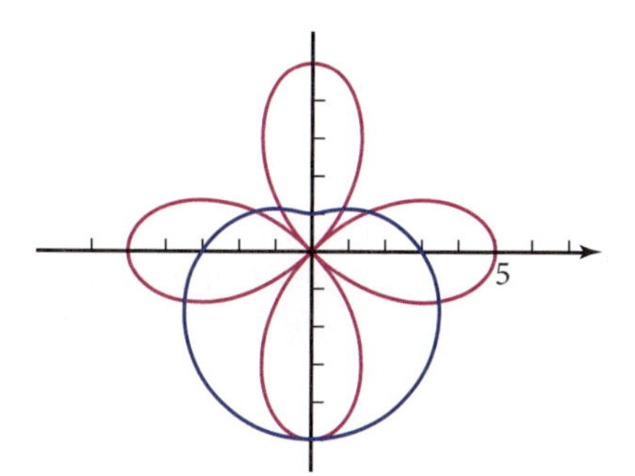

Figure 11-3l

7. Spiral $r_1(\theta) = 0.5\theta$ and rose $r_2(\theta) = 5\cos 2\theta$ (Figure 11-3m). Use radian mode and positive values of θ.

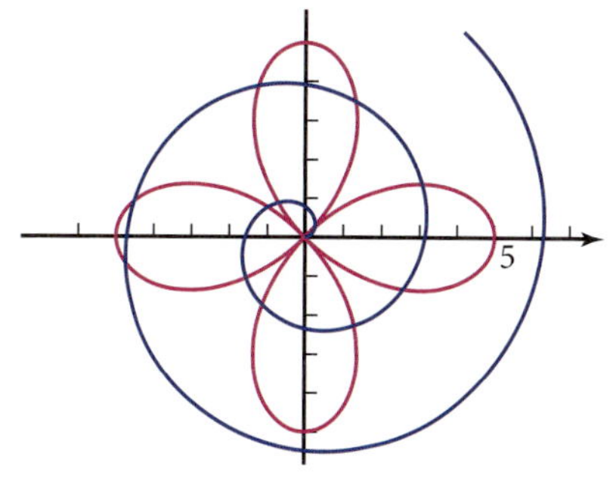

Figure 11-3m

8. Spiral $r_1(\theta) = 0.5\theta$ and circle $r_2(\theta) = 5\cos\theta$ (Figure 11-3n). Use radian mode and positive values of θ.

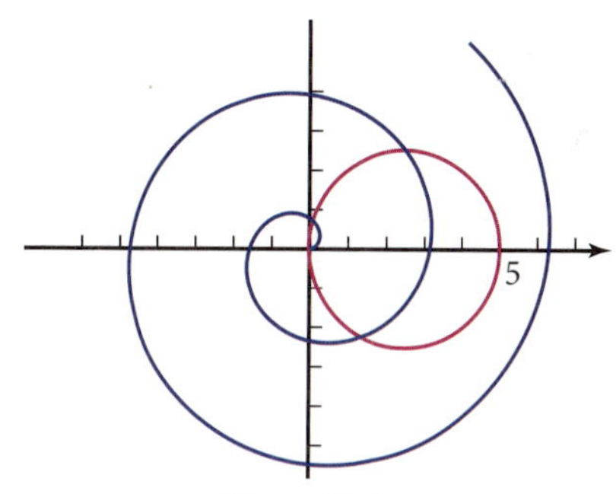

Figure 11-3n

9. *Finding Other Intersection Points Problem:* Plot on the same screen the limaçons

$r_1(\theta) = 3 + 2\sin\theta$ and

$r_2(\theta) = -3 - 2\sin(\theta - 180°)$

Use simultaneous mode and path style. Describe what you observe about the points at which each graph is being plotted and the final graphs. Tracing to a particular value of θ on the final graphs might help. Then plot the circle $r_3(\theta) = 4$. Show that the intersections of $r_3(\theta)$ with $r_2(\theta)$ do not represent solutions to the system of those two equations, but do represent the solutions to the system of $r_3(\theta)$ and $r_1(\theta)$. From what you have observed, write instructions that could be used to find the intersections of two polar curves that are not solutions to the system of equations.

11-4 Complex Numbers in Polar Form

An **imaginary number** is a square root of a negative number. Using i for the *unit imaginary number* $\sqrt{-1}$, you can write

$$\sqrt{-9} = i\sqrt{9} = 3i$$

A **complex number** is the sum of a real number and an imaginary number. For instance, the complex number $z = 4 + 3i$ is the sum of the real number 4 and the imaginary number $3i$.

You can represent complex numbers by points in a Cartesian coordinate system called the **complex plane,** as shown in Figure 11-4a. The horizontal coordinate of a point is the real part of the number, and the vertical coordinate is the imaginary part.

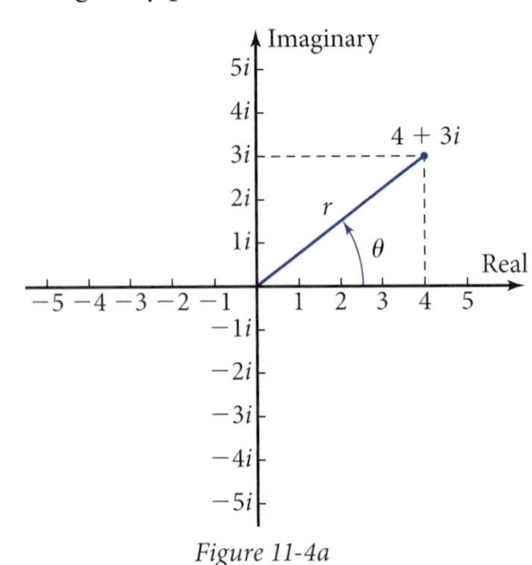

Figure 11-4a

You can write a complex number in *polar form* by writing the real and imaginary parts in terms of the polar coordinates of the point. In this section you will make remarkable discoveries about products and roots of complex numbers by writing the numbers in polar form.

Objective Operate with complex numbers in polar form.

In this exploration you will review products of complex numbers and draw some conclusions.

EXPLORATION 11-4: Review of Complex Numbers

1. The imaginary number i is defined to be $\sqrt{-1}$, the square root of -1. Explain why $i^2 = -1$.
2. The product of two complex numbers such as $(4 + 3i)(5 + 12i)$ is found the same way as multiplying any two binomials. Multiply each term in one binomial by each term in the other, and then combine like terms. Multiply these numbers and simplify the answer as much as possible, recalling from Problem 1 that $i^2 = -1$.

continued

EXPLORATION, continued

3. The *modulus* or *absolute value* of a complex number equals the distance from the origin of the complex plane to the point representing the number. For example, if $z_1 = 4 + 3i$ as in Figure 11-4a, then by the Pythagorean theorem

   $$|z_1| = \sqrt{4^2 + 3^2} = 5$$

   Let $z_2 = 5 + 12i$. Find $|z_2|$.

4. From your answer to Problem 2, find the absolute value of the product. That is, find $|z_1z_2|$. What relationship do you notice between this number and $|z_1|$ and $|z_2|$?

5. Figure 11-4a shows the angle θ in standard position for a complex number. Find the measure of angle θ_1 for $z_1 = 4 + 3i$ and the measure of angle θ_2 for $z_2 = 5 + 12i$. Then find the measure of angle θ_3 for the product z_1z_2 you found in Problem 2. What relationship do you notice between the angle measure for the product and the angle measures for the two factors in the product of these complex numbers?

6. What did you learn as a result of doing this exploration that you did not know before?

These definitions apply to complex numbers in Cartesian form.

DEFINITIONS: Imaginary and Complex Numbers

$i = \sqrt{-1}$	The unit imaginary number.
$i^2 = -1$	Squaring a square root removes the radical sign.
$z = a + bi$	General form of a complex number in Cartesian form. The real number a is called the *real part* of z. The real number b is called the *imaginary part* of z.
$a - bi$	The **complex conjugate** of $a + bi$.
$\lvert z\rvert = \lvert a + bi\rvert = \sqrt{a^2 + b^2}$	The **modulus** or magnitude of a complex number.

You operate with complex numbers in the same way you operate with other binomials. For example, to add or subtract complex numbers, combine like terms:

$$(4 + 3i) - (5 - 2i) = 4 + 3i - 5 + 2i$$
$$= -1 + 5i$$

To multiply complex numbers, expand and combine like terms:

$$(4 + 3i)(5 - 2i) = 20 - 8i + 15i - 6i^2$$
$$= 20 + 7i - 6(-1)$$
$$= 26 + 7i$$

Figure 11-4b shows the number $z = a + bi$ in the complex plane. The polar coordinates of z are (r, θ).

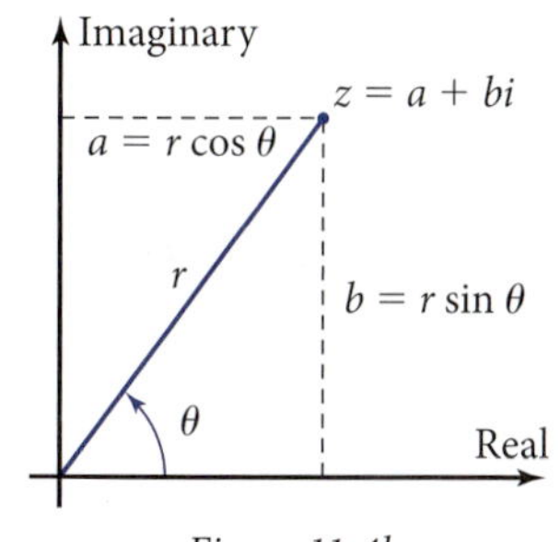

Figure 11-4b

By the definitions of cosine and sine,

$$a = r\cos\theta \qquad \text{and} \qquad b = r\sin\theta$$

Therefore, z can be written

$$z = (r\cos\theta) + i(r\sin\theta) \qquad \text{or} \qquad z = r(\cos\theta + i\sin\theta)$$ Factor out r.

The expression $(\cos\theta + i\sin\theta)$ is written "cis θ," pronounced "sis" as in "sister." The *c* comes from cosine, the *i* from the unit imaginary number, and the *s* from *sine*. Thus, any complex number can be written

$$z = r\operatorname{cis}\theta$$

DEFINITION: Polar Form of a Complex Number

$$z = r\operatorname{cis}\theta = r(\cos\theta + i\sin\theta)$$

where r is called the *modulus* (or magnitude) of z and θ is called the *argument* of z (either degrees or radians).

The *absolute value* of a complex number is the modulus of that number:

$$|z| = r$$

These relationships let you convert between the polar and Cartesian forms of a complex number:

PROPERTY: Relationships Between Polar and Cartesian Forms

If $z = a + bi = r\operatorname{cis}\theta$, then

$$a = r\cos\theta$$

$$b = r\sin\theta$$

$$r = \sqrt{a^2 + b^2}$$

EXAMPLE 1 ➤ Transform the complex number $z = -5 + 7i$ into polar form.

SOLUTION

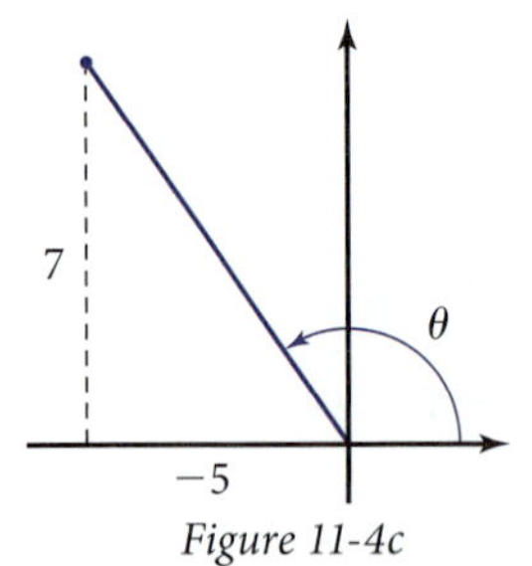

Figure 11-4c

Sketch the number in the complex plane (Figure 11-4c).

$$r = \sqrt{(-5)^2 + 7^2} = \sqrt{74}$$ Use only the coefficient 7, not $7i$.

$$\theta = \arctan\frac{7}{-5} = -54.4623...^\circ + 180^\circ n = 125.5376...^\circ$$

θ is in Quadrant II.

$$\therefore z = \sqrt{74}\operatorname{cis}(125.5376...^\circ)$$

EXAMPLE 2 ➤ Transform $z = 5 \operatorname{cis} 144°$ into Cartesian form.

SOLUTION

$$z = 5(\cos 144° + i \sin 144°) \quad \text{Definition of cis } \theta.$$
$$= -4.0450... + (2.9389...)i$$
$$\approx -4.05 + 2.94i$$

EXAMPLE 3 ➤ If $z_1 = 3 \operatorname{cis} 83°$ and $z_2 = 2 \operatorname{cis} 41°$, find the product z_1z_2.

SOLUTION **Long Way**

$$z_1z_2 = 3(\cos 83° + i \sin 83°) \cdot 2(\cos 41° + i \sin 41°) \quad \text{Definition of cis } \theta.$$
$$= (3)(2)[\cos 83° \cos 41° + i(\sin 83° \cos 41°) + i(\cos 83° \sin 41°) + i^2(\sin 83° \sin 41°)]$$
$$= 6[(\cos 83° \cos 41° - \sin 83° \sin 41°) + i(\sin 83° \cos 41° + \cos 83° \sin 41°)]$$
$$= 6[\cos(83° + 41°) + i \sin(83° + 41°)] \quad \text{Composite argument properties from Chapter 8.}$$
$$= 6 \operatorname{cis} 124°$$

Short Way

$$z_1z_2 = (3 \operatorname{cis} 83°)(2 \operatorname{cis} 41°) = 6 \operatorname{cis} 124°$$

Multiply the moduli (plural of modulus); add the arguments.

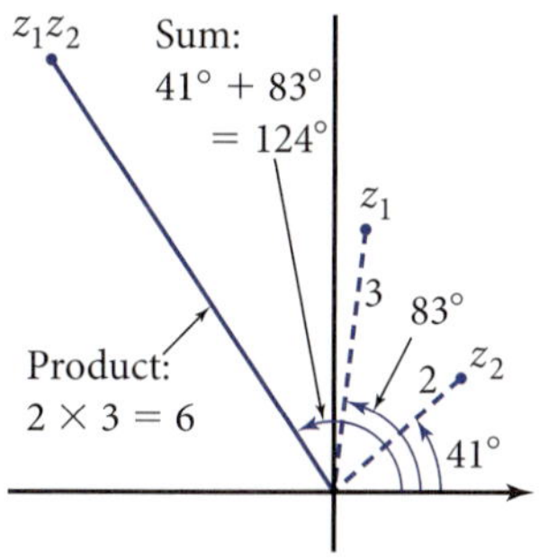

Figure 11-4d

> **PROPERTY: Product of Two Complex Numbers in Polar Form**
>
> If $z_1 = r_1 \operatorname{cis} \theta_1$ and $z_2 = r_2 \operatorname{cis} \theta_2$, then
>
> $$z_1z_2 = r_1r_2 \operatorname{cis}(\theta_1 + \theta_2)$$
>
> *Verbally:* Multiply the moduli; add the arguments.

Figure 11-4d illustrates this property using the complex numbers from Example 3.

EXAMPLE 4 ➤ Find the reciprocal of $z = 2 \operatorname{cis} 29°$.

SOLUTION **Long Way**

$$\frac{1}{z} = \frac{1}{2(\cos 29° + i \sin 29°)} \quad \text{Definition of cis } \theta.$$
$$= \frac{1}{2} \cdot \frac{1}{\cos 29° + i \sin 29°} \cdot \frac{\cos 29° - i \sin 29°}{\cos 29° - i \sin 29°} \quad \text{Multiply by a clever form of 1.}$$
$$= \frac{1}{2} \cdot \frac{\cos 29° - i \sin 29°}{\cos^2 29° - i^2 \sin^2 29°} \quad \text{Product of conjugate binomials in the denominator.}$$
$$= \frac{1}{2} \cdot \frac{\cos 29° - i \sin 29°}{\cos^2 29° + \sin^2 29°} \quad i^2 = -1$$
$$= \frac{1}{2}(\cos 29° - i \sin 29°) \quad \text{Pythagorean property of cosine and sine.}$$
$$= \frac{1}{2}[\cos(-29°) + i \sin(-29°)] \quad \text{Odd–even properties of cosine and sine.}$$
$$= \frac{1}{2} \operatorname{cis}(-29°) \quad \text{Definition of cis } \theta.$$

Short Way

$$\frac{1}{z} = \frac{1}{2\operatorname{cis} 29^\circ} = \frac{1}{2}\operatorname{cis}(-29^\circ)$$

Take the reciprocal of the modulus and the opposite of the argument.

PROPERTY: Reciprocal of a Complex Number in Polar Form

If $z = r \operatorname{cis} \theta$ and $z \neq 0$, then

$$\frac{1}{z} = \frac{1}{r}\operatorname{cis}(-\theta)$$

Verbally: Take the reciprocal of the modulus and the opposite of the argument.

EXAMPLE 5 If $z_1 = 5\operatorname{cis} 71^\circ$ and $z_2 = 2\operatorname{cis} 29^\circ$, find $\frac{z_1}{z_2}$.

SOLUTION **Long Way**

$$\frac{z_1}{z_2} = \frac{5\operatorname{cis} 71^\circ}{2\operatorname{cis} 29^\circ}$$

$$= (5\operatorname{cis} 71^\circ)\left[\frac{1}{2}\operatorname{cis}(-29^\circ)\right]$$ Apply the reciprocal property.

$$= \frac{5}{2}\operatorname{cis} 42^\circ$$ Apply the multiplication property.

Short Way

$$\frac{z_1}{z_2} = \frac{5\operatorname{cis} 71^\circ}{2\operatorname{cis} 29^\circ} = \frac{5}{2}\operatorname{cis} 42^\circ$$

Divide the moduli; subtract the arguments.

PROPERTY: Quotient of Two Complex Numbers in Polar Form

If $z_1 = r_1 \operatorname{cis} \theta_1$ and $z_2 = r_2 \operatorname{cis} \theta_2$, then

$$\frac{z_1}{z_2} = \frac{r_1}{r_2}\operatorname{cis}(\theta_1 - \theta_2)$$

Verbally: Divide the moduli; subtract the arguments.

EXAMPLE 6 If $z = 2\operatorname{cis} 29^\circ$, find z^5.

SOLUTION **Long Way**

$$z^5 = (2\operatorname{cis} 29^\circ)^5$$

$$= (2\operatorname{cis} 29^\circ)(2\operatorname{cis} 29^\circ)(2\operatorname{cis} 29^\circ)(2\operatorname{cis} 29^\circ)(2\operatorname{cis} 29^\circ)$$

$$= 2^5 \operatorname{cis}(5 \cdot 29^\circ)$$ Apply the product property repeatedly.

$$= 32\operatorname{cis} 145^\circ$$

$(2\ cis\ 29^\circ)(2\ cis\ 29^\circ)(2\ cis\ 29^\circ)\ldots$

$4\ cis\ 58^\circ$

$8\ cis\ 87^\circ$

Short Way

$$z^5 = 2^5 \operatorname{cis}(5 \cdot 29^\circ) = 32\operatorname{cis} 145^\circ$$

Raise the modulus to the power; multiply the argument by the exponent.

The "short way" to raise a complex number to a power is known as **De Moivre's theorem.**

> **PROPERTY: De Moivre's Theorem**
>
> If $z = r \operatorname{cis} \theta$, then
>
> $$z^n = r^n \operatorname{cis} n\theta$$
>
> *Verbally:* Raise the modulus to the power, and multiply the argument by the exponent.

Because De Moivre's theorem is true for fractional exponents, you can use it to find *roots* of a complex number in polar form. There is a surprise, as you will see in Example 7.

EXAMPLE 7 ➤ If $z = 8 \operatorname{cis} 60°$, find the cube roots of z, $\sqrt[3]{z}$.

SOLUTION There are multiple polar coordinates for a given point. In general, $z = 8 \operatorname{cis}(60° + 360° k)$, where k stands for an integer. So there are multiple cube roots of a complex number:

$$\sqrt[3]{z} = z^{1/3}$$

$$= (8 \operatorname{cis} 60° + 360° k)^{1/3}$$

$$= 8^{1/3} \operatorname{cis} \frac{1}{3}(60° + 360° k) \quad \text{By De Moivre's theorem.}$$

$$= 2 \operatorname{cis}(20° + 120° k)$$

$$= 2 \operatorname{cis} 20°, 2 \operatorname{cis} 140°, 2 \operatorname{cis} 260°, 2 \operatorname{cis} 380°, \ldots$$

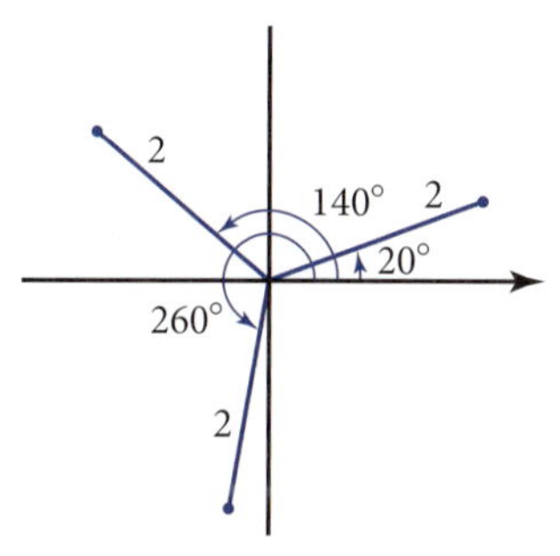

Figure 11-4e

But $2 \operatorname{cis} 380°$ is coterminal with $2 \operatorname{cis} 20°$, so only the first three results are *distinct* cube roots.

$$\sqrt[3]{z} = 2 \operatorname{cis} 20°, 2 \operatorname{cis} 140°, \text{ or } 2 \operatorname{cis} 260°$$ ◄

In general, a complex number has exactly n distinct nth roots. Figure 11-4e shows that the three cube roots of $z = 8 \operatorname{cis} 60°$ are equally spaced around the pole in the complex plane.

Problem Set 11-4

Reading Analysis

From what you have read in this section, what do you consider to be the main idea? What is the definition of i, and why does i^2 equal -1? What does $\operatorname{cis} \theta$ mean, and why is this form of a complex number useful for multiplication? What is the graphical relationship among the three cube roots of a complex number?

Quick Review

Q1. Write Cartesian coordinates for the polar point (3, 90°).

Q2. Write polar coordinates for the Cartesian point (−3, 3).

Q3. What geometric figure is the graph of $r = 10 \cos \theta$?

Q4. If the limaçon $r = a + b\cos\theta$ has a loop, how are a and b related?

Q5. If the limaçon $r = a + b\cos\theta$ is a cardioid, how are a and b related?

Q6. How do you decide whether a point where two polar curves cross is a true intersection?

Q7. How are the graphs of $r_1(\theta) = 2\cos\theta$ and $r_2(\theta) = 10\cos\theta$ related?

Q8. How are the graphs of $r_1(\theta) = 2\cos\theta$ and $r_3(\theta) = 2\cos(\theta - 50°)$ related?

Q9. Find the discriminant: $3x^2 + 5x + 11 = 0$

Q10. The polar graph of $r = \dfrac{10}{2 + 3\cos\theta}$ is a(n)

A. Hyperbola **B.** Ellipse **C.** Parabola
D. Circle **E.** None of these

For Problems 1–12, write the complex number in polar form, $r\operatorname{cis}\theta$.

1. $-1 + i$
2. $1 - i$
3. $\sqrt{3} - i$
4. $1 + i\sqrt{3}$
5. $-4 - 3i$
6. $-3 + 4i$
7. $5 + 7i$
8. $-11 - 2i$
9. 1
10. i
11. $-i$
12. -8

For Problems 13–22, write the complex number in Cartesian form, $a + bi$.

13. $8\operatorname{cis}34°$
14. $11\operatorname{cis}247°$
15. $6\operatorname{cis}120°$
16. $8\operatorname{cis}150°$
17. $\sqrt{2}\operatorname{cis}225°$
18. $3\sqrt{2}\operatorname{cis}45°$
19. $5\operatorname{cis}180°$
20. $9\operatorname{cis}90°$
21. $3\operatorname{cis}270°$
22. $2\operatorname{cis}0°$

For Problems 23–26,

a. Find z_1z_2.
b. Find $\dfrac{z_1}{z_2}$.
c. Find $z_1^{\,2}$.
d. Find $z_2^{\,3}$.

23. $z_1 = 3\operatorname{cis}47°$, $z_2 = 5\operatorname{cis}36°$

24. $z_1 = 2\operatorname{cis}154°$, $z_2 = 3\operatorname{cis}27°$

25. $z_1 = 4\operatorname{cis}238°$, $z_2 = 2\operatorname{cis}51°$

26. $z_1 = 6\operatorname{cis}19°$, $z_2 = 4\operatorname{cis}96°$

For Problems 27–36, find the indicated roots and sketch the answers on the complex plane.

27. Cube roots of $27\operatorname{cis}120°$
28. Cube roots of $8\operatorname{cis}15°$
29. Fourth roots of $16\operatorname{cis}80°$
30. Fourth roots of $81\operatorname{cis}64°$
31. Square roots of i
32. Square roots of $-i$
33. Cube roots of 8
34. Cube roots of -27
35. Sixth roots of -1
36. Tenth roots of 1

Complex numbers are used to analyze the flow of alternating current in electrical circuits.

37. *Triple Argument Properties Problem:* By De Moivre's theorem,

$$(\cos\theta + i\sin\theta)^3 = \cos 3\theta + i\sin 3\theta$$

Expand the expression on the left. By equating the real parts and the imaginary parts on the left and right sides of the resulting equation, derive *triple argument properties* expressing $\cos 3\theta$ and $\sin 3\theta$ in terms of sines and cosines of θ.

Abraham De Moivre (1667–1754) introduced complex numbers in trigonometry. (The Granger Collection, New York)

38. WEB *Research Project:* From the Internet or some other source, find out about De Moivre. For instance, learn about his major mathematical contributions, the books he wrote, and his life.

39. *Journal Problem:* Update your journal. Include things such as the use of polar coordinates to represent complex numbers and how polar coordinates make it relatively easy to find products, quotients, roots, and powers.

11-5 Parametric Equations for Moving Objects

So far in this chapter you have seen how you can plot ellipses, limaçons, and other graphs relatively easily using polar coordinates, and how polar coordinates lead to a way of analyzing complex numbers. In this section you will return to the polar graphs you studied at the beginning of the chapter and analyze them with the help of parametric functions and vectors. Many of the graphs you will encounter in this section come from moving objects, such as a wheel rolling along a line or around a circle, a string unwinding from a circle, or a projectile moving through the air.

Objective Given a geometric description of the path followed by a moving object, write parametric equations to describe the path and plot it on your grapher.

EXAMPLE 1 ➤ A ship moves with an eastward velocity of 21 km/h and a northward velocity of 13 km/h. At time $t = 0$ h the ship is at the point $P_0(-43, 19)$, where the distances are in kilometers from a lighthouse (Figure 11-5a).

a. Write parametric equations for the ship's path, using t, in hours, as the parameter.

b. Confirm that your equations are correct by plotting them on your grapher.

c. Predict the time when the ship will be 60 km north of the lighthouse. How far east or west of the lighthouse will it be at this time?

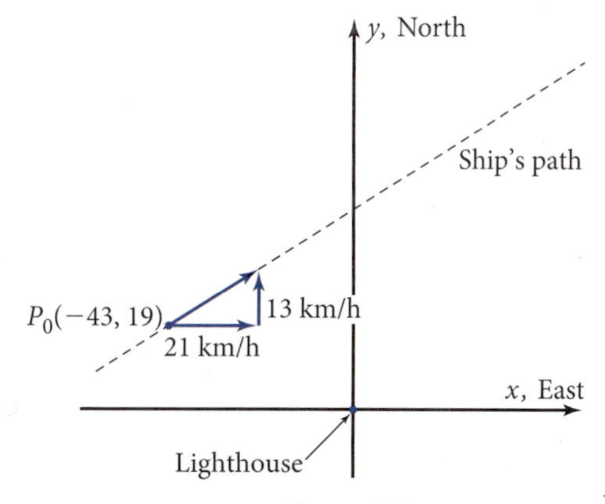

Figure 11-5a

SOLUTION **a.** $x = -43 + 21t$
$y = 19 + 13t$ Distance = (rate)(time)

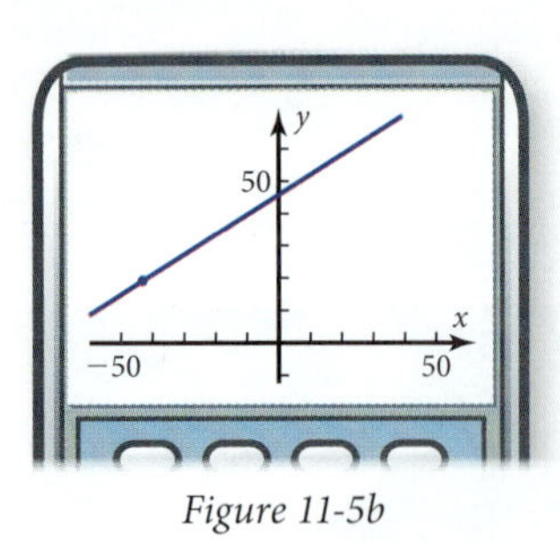

Figure 11-5b

b. The graph (Figure 11-5b) confirms that the equations represent the given path. Use equal scales on the two axes.

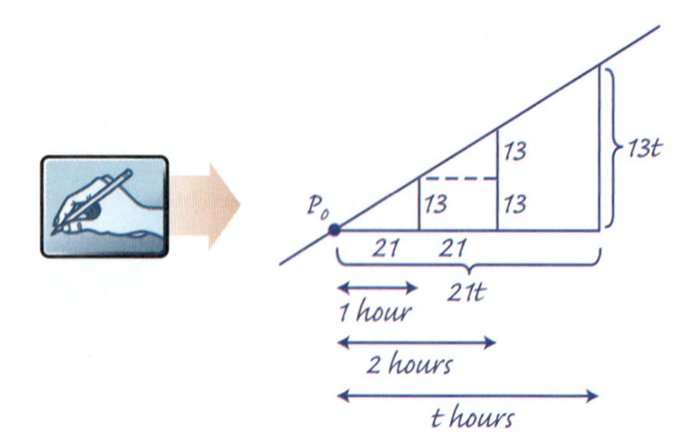

c. $60 = 19 + 13t \Rightarrow t = 3.1538... \approx 3.15 \text{ h} \approx 3 \text{ h } 9 \text{ min}$

Set y equal to 60 and solve for t.

$x = -43 + 21(3.1538...) = 23.2307...$ Substitute the solution into the x-equation.

The ship will be approximately 23.23 km east of the lighthouse. ◄

EXAMPLE 2 ➤ As a wheel rolls along a straight-line path, a fixed point on the rim of the wheel traces a curve called a *cycloid*. The wheel in Figure 11-5c has radius 6 cm and rolls in a positive direction along the x-axis. Let the parameter t represent the number of radians the wheel has rolled since a point $P(x, y)$ on its rim was at the origin. Write parametric equations for the cycloid traced by point P. Check your equation by graphing.

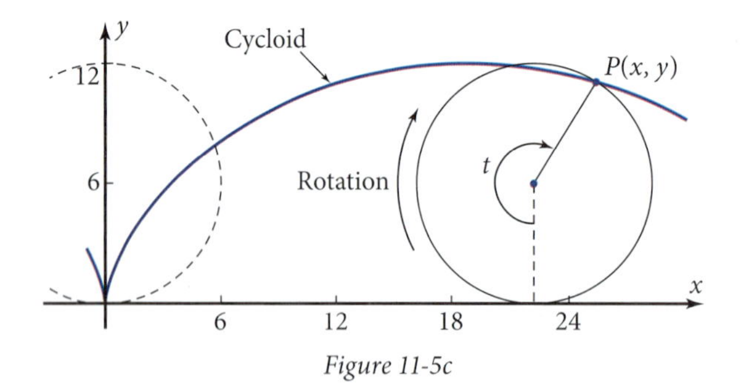

Figure 11-5c

SOLUTION Let $\vec{r}$ be the vector from the origin to the point $P(x, y)$ on the circle after it has rolled t radians.

Figure 11-5d shows that you can write $\vec{r}$ as the sum of three other vectors:

$$\vec{r} = \vec{v}_1 + \vec{v}_2 + \vec{v}_3$$

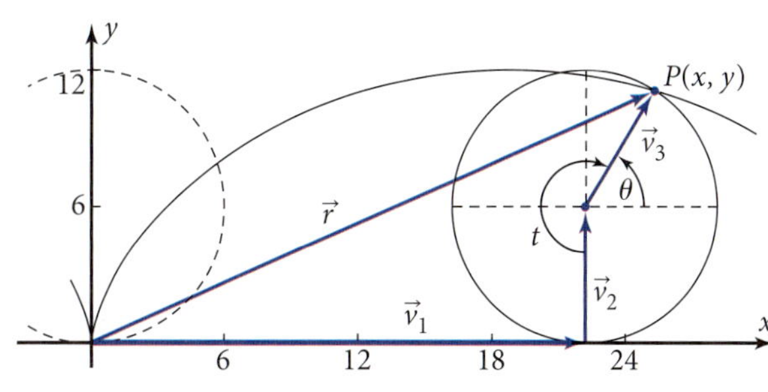

Figure 11-5d

Vector $\vec{v}_1$ is horizontal, and its magnitude is the distance the wheel has rolled.

$$|\vec{v}_1| = 6t \Rightarrow \vec{v}_1 = (6t)\vec{i}$$

Arc length = (radius)(central angle in radians) (Note that it's important to use radians rather than degrees because arc length, $6t$, is a distance, and radians are defined as a distance, whereas degrees are not.)

Vector $\vec{v}_2$ is vertical, with constant magnitude 6 units.

$$\vec{v}_2 = 6\vec{j}$$

Vector $\vec{v}_3$ extends from the center of the wheel to the point $P(x, y)$. It has constant magnitude 6 units, but it rotates clockwise. Let θ be the angle in standard position for $\vec{v}_3$. Then

$$\vec{v}_3 = (6\cos\theta)\vec{i} + (6\sin\theta)\vec{j}$$

Definitions of cosine and sine.

To get θ in terms of t, observe that θ starts at 1.5π radians (270°) when $t = 0$, and decreases as t increases. Therefore,

$$\theta = 1.5\pi - t$$

$$\vec{v}_3 = [6\cos(1.5\pi - t)]\vec{i} + [6\sin(1.5\pi - t)]\vec{j}$$

$$\therefore \vec{r} = (6t)\vec{i} + 6\vec{j} + [6\cos(1.5\pi - t)]\vec{i} + [6\sin(1.5\pi - t)]\vec{j}$$

Add the three vectors.

$$\vec{r} = [6t + 6\cos(1.5\pi - t)]\vec{i} + [6 + 6\sin(1.5\pi - t)]\vec{j}$$

Combine like terms.

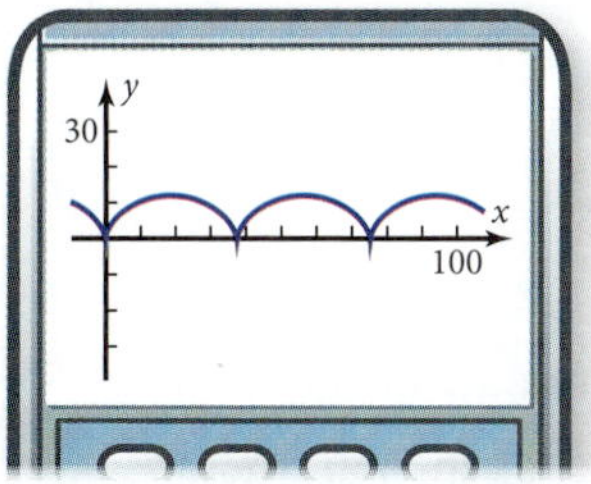

Figure 11-5e

This is called a *vector equation* of the cycloid. The parametric equations are the x- and y-components of the vector equation.

$$x = 6t + 6\cos(1.5\pi - t)$$
$$y = 6 + 6\sin(1.5\pi - t)$$

The graph is shown in Figure 11-5e. Use radian mode and equal scales on both axes. For a dynamic demonstration, view the *Cycloid* exploration at

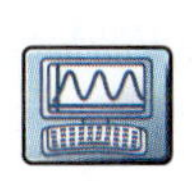

www.keymath.com/precalc. Note that although y is a periodic function of x, it is not a sinusoid. The low points are cusps rather than rounded curves.

You can use the composite argument properties from Chapter 8 to simplify the cosine and sine terms in Example 2. Because $\cos 1.5\pi$ equals 0 and $\sin 1.5\pi$ equals -1,

$$\cos(1.5\pi - t) = \cos 1.5\pi \cos t + \sin 1.5\pi \sin t = -\sin t$$
$$\sin(1.5\pi - t) = \sin 1.5\pi \cos t - \cos 1.5\pi \sin t = -\cos t$$

Thus the parametric equations are

$$x = 6t - 6\sin t$$
$$y = 6 - 6\cos t$$

The general parametric equations of a cycloid are listed in the box.

PROPERTY: Parametric Equations of a Cycloid

As a wheel of radius a rolls along the x-axis, a point on the rim of the wheel that starts at the origin traces the path of a cycloid whose parametric equations are

$$x = a(t - \sin t)$$
$$y = a(1 - \cos t)$$

where t is the number of radians the wheel has rolled since the point was at the origin.

Problem Set 11-5

Reading Analysis

From what you have read in this section, what do you consider to be the main idea? What is the relationship between the vector equation for the path of a moving object and the parametric equations for the path? Give two examples of paths of moving objects that can be modeled by parametric equations or by a vector.

Quick Review

Q1. Add: $(3 - 7i) + (5 + 6i)$

Q2. Add: $(3\vec{i} - 7\vec{j}) + (5\vec{i} + 6\vec{j})$

Q3. Add: $20 \operatorname{cis} 80° + 10 \operatorname{cis} 50°$

Q4. Multiply: $(20 \operatorname{cis} 80°)(10 \operatorname{cis} 50°)$

Q5. Divide: $(20 \operatorname{cis} 80°) \div (10 \operatorname{cis} 50°)$

Q6. Expand: $(2 \operatorname{cis} 20°)^3$

Q7. Write $20 \operatorname{cis} 80°$ in the form $a + bi$.

Q8. Write $3 - 7i$ in polar form.

Q9. Write the composite argument property for $\cos(A - B)$.

Q10. Which two of the six trigonometric functions are even functions?

---

1. *Airplane's Path Problem:* An airplane is flying at a velocity of 300 km/h west and 100 km/h north. At time $t = 0$ h the plane is at the point (473, 155), where the distances are in kilometers from a Federal Aviation Agency station located at the origin (Figure 11-5f).

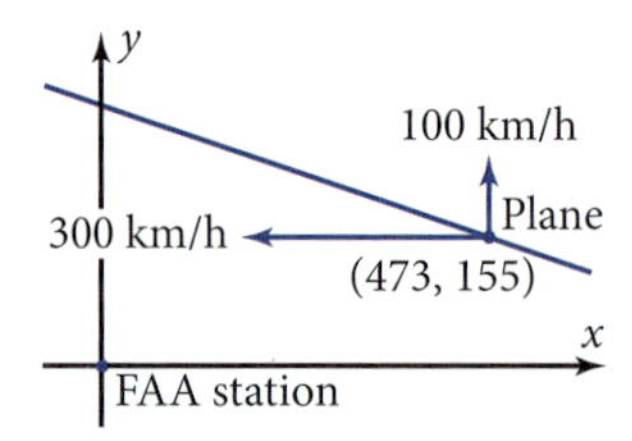

Figure 11-5f

a. Write parametric equations for the plane's path, using t, in hours, as the parameter.

b. If the plane continues on this path, when will it be due north of the FAA station? How far north of the station will it be?

c. What is the plane's actual speed?

2. *Walking Problem:* Calvin is walking at a speed of 6 ft/s along a path that makes an angle of 55° with the x-axis. At time $t = 0$ s he is at the point (263, 107), where the distances are in feet from a particular traffic light (Figure 11-5g).

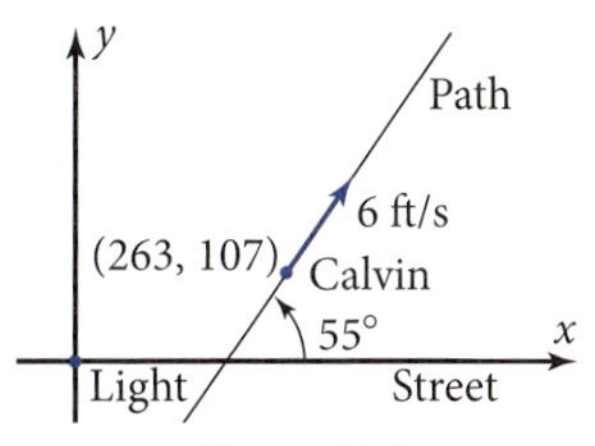

Figure 11-5g

a. What are Calvin's speeds in the x- and y-directions?

b. Write parametric equations for his position as a function of the parameter t, in seconds.

c. A street goes along the x-axis. Assuming Calvin was walking at his 6 ft/s pace before time $t = 0$ s, at what time t did he cross the street?

d. How far from the light does the path cross the street?

3. WEB *Projectile Motion Problem:* Sir Francis Drake's ship fires a cannonball at an enemy galleon. At time $t = 0$ s the cannonball has an initial velocity of 200 ft/s and a 20° angle of elevation (Figure 11-5h). The cannonball's position is represented by $P(x, y)$. Assume the point (0, 10) represents the point from which the cannonball was fired.

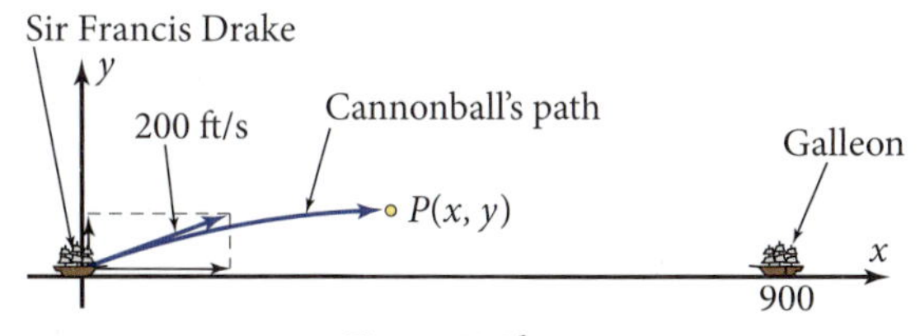

Figure 11-5h

a. Write parametric equations for the cannonball's path. To do this, assume there is no air friction, so the horizontal velocity remains what it was at time $t = 0$ s. The vertical position equals the height the cannonball is above the water plus the distance the cannonball would travel at the initial upward velocity at time $t = 0$ s, minus $16t^2$ to account for the effects of gravity.

b. Plot the parametric equations. Use a window with an interval for x large enough to show the point where the cannonball hits the water. Sketch the result.

c. The galleon is at distance $x = 900$ ft from Sir Francis Drake's ship. The tops of the sails are 40 ft above the surface of the water. Will the cannonball fall short of the galleon, pass over the galleon, or hit it somewhere between the waterline and the tops of the sails? Show how you get your answer.

d. To be most effective, the cannonball should hit the galleon right at the waterline ($y = 0$). At what angle of elevation should the cannonball be fired to accomplish this objective?

e. Using the Internet or some other reference source, find out who Sir Francis Drake was, when he lived, and to which country the enemy galleon might have belonged.

4. WEB *Ship Collision Project:* Two ships are steaming through the fog. At time $t = 0$ min, their positions and velocities are

Ship A: Point $(x, y) = (2000, 600)$,
velocity 500 m/min at an angle of 140°

Ship B: Point $(x, y) = (200, 300)$,
velocity 400 m/min at an angle of 80°

a. Write parametric equations representing the position (x, y) of each ship, with x and y in meters and the parameter t in minutes.

b. Plot the parametric equations for both ships on the same screen. Use simultaneous mode so that you can see where each ship is with respect to the other as the graphs are being plotted. Based on what you observe, do the ships collide, almost collide, or miss each other by a significant distance?

c. Use the distance formula to help you write a Cartesian equation for the distance between the ships as a function of time. Plot this function on another screen and sketch the result.

d. By appropriate operations on the function in part c, find numerically the time the ships are closest to each other and how close they get. Based on your answer, should the ships have changed their courses to avoid a collision, or will they miss each other by a safe distance?

e. In July 1956, there was a serious collision between the ships *Andrea Doria* and *Stockholm*. Using the Internet or some other reference source, find out what happened.

5. *Ellipse from Geometric Properties Problem:* Figure 11-5i shows concentric circles of radii 3 units and 5 units centered at the origin of a Cartesian coordinate system. A ray from the center at an angle of t radians to the x-axis cuts the two circles at points A and B, respectively. From point A a horizontal line is drawn, and from point B a vertical line is drawn. These two lines intersect at point P on the graph of a curve.

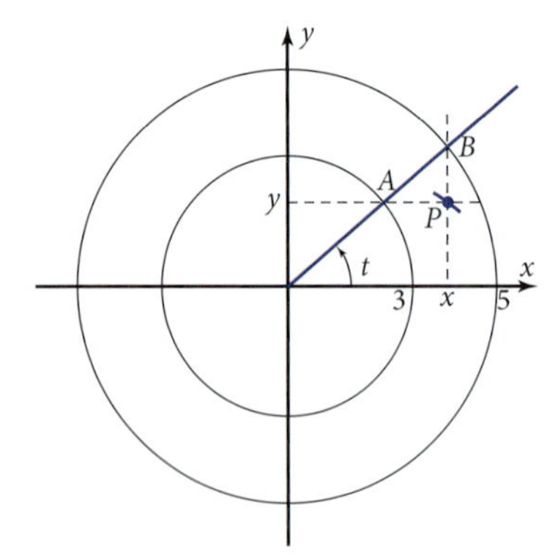

Figure 11-5i

a. On a copy of Figure 11-5i, pick other values of the angle t and plot more points using the given specifications. Connect the resulting points with a smooth curve. Does the graph seem to be an ellipse?

b. Write parametric equations for the point $P(x, y)$ in terms of the parameter t using the given geometric description. Is the result the same as the parametric equations for an ellipse from Section 7-5?

c. Plot the parametric equations on your grapher. Use equal scales on the two axes. Also plot the two circles. Do the circles have the same relationship to the curve as in your sketch?

d. The parameter t can be eliminated from the two parametric equations to get a single equation involving only the variables x and y. Clever use of the Pythagorean properties will allow you to do this. Write the Cartesian equation of this curve. How can you tell whether the equation represents an ellipse?

6. *Serpentine Curve Problem:* Figure 11-5j shows the *serpentine curve*, so called for its snakelike shape. A fixed circle of radius 5 units has its center on the x-axis and passes through the origin. A variable line from the origin makes an angle of t radians with the x-axis. It intersects the circle at point A, and it intersects the fixed line $y = 5$ at point B. A horizontal line from point A and a vertical line from point B intersect at the point $P(x, y)$ on the serpentine curve.

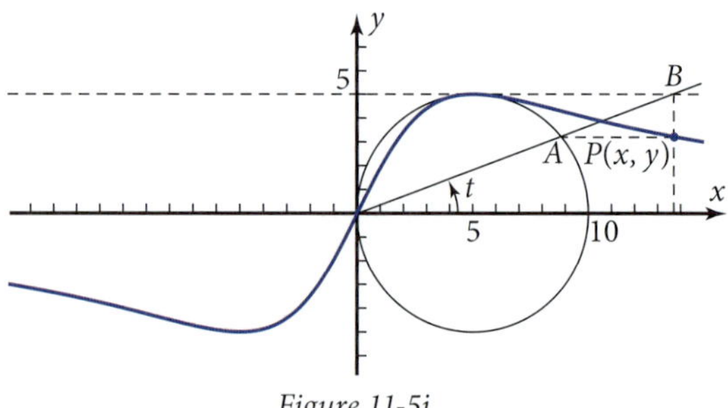

Figure 11-5j

a. On a copy of Figure 11-5j, pick a different value of t between 0 and $\frac{\pi}{2}$ and plot the corresponding point P as described. Plot another point for a t-value between $\frac{\pi}{2}$ and π. Show that the resulting points really are on the serpentine curve.

b. Write parametric equations expressing x and y in terms of the parameter t. (To find y, first find the distance from the origin to point A. You can do this by recalling the polar equation for a circle or by drawing a right triangle inscribed in the semicircle with right angle at point A and hypotenuse 10 units.)

c. Confirm that your parametric equations in part b are correct by plotting them on your grapher. Use a window with an interval for x at least as large as the one shown, and use equal scales on both axes.

d. The point P in Figure 11-5j corresponds to $t = 0.35$ radian. Confirm that this is correct by showing that the values of x and y you find from the equation agree with the values in the figure.

7. *Flanged Wheel Prolate Cycloid Problem:* Train wheels have flanges that project beyond the rims to keep the wheels from slipping off the track. A point P on the flange traces a *prolate cycloid* as the wheel turns. Figure 11-5k shows an example. The radius of the flange has been exaggerated so that you can see more clearly what a prolate cycloid looks like. Assume that the wheel has radius 50 cm, the flange has radius 70 cm, and the wheel has rotated t radians since the point $P(x, y)$ was farthest below the track.

- Let $\vec{r}$ (not shown) be the position vector to the point $P(x, y)$ on the flange.
- Let $\vec{v}_1$ be the vector from the origin to the point where the wheel touches the track.
- Let $\vec{v}_2$ be the vector from that point to the center of the wheel.
- Let $\vec{v}_3$ be the vector from the center of the wheel to the point on the flange.

a. Explain why $\vec{r} = \vec{v}_1 + \vec{v}_2 + \vec{v}_3$.

b. The length of $\vec{v}_1$ equals the distance the wheel has rolled. Vector $\vec{v}_2$ is a constant vector in the vertical direction. Write $\vec{v}_1$ and $\vec{v}_2$ in terms of their components.

c. Vector $\vec{v}_3$ goes from the center of the wheel to the point $P(x, y)$. Write $\vec{v}_3$ as a function of t. Use the result to write $\vec{r}$ as a vector function of t.

d. Plot the graph of $\vec{r}$ using parametric mode. Does your graph look like Figure 11-5k?

e. How far does point P move in the x-direction between $t = 0$ radians and $t = 0.1$ radian? How do you explain the fact that the displacement is *negative,* even though the wheel is going in the positive x-direction?

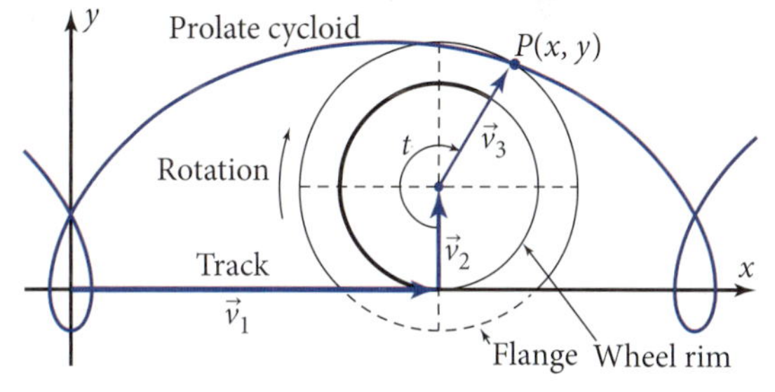

Figure 11-5k

8. *Epicycloid Problem:* Figure 11-5l shows the *epicycloid* traced by a point on the rim of a wheel of radius 2 cm as it rotates, without slipping, around the outside of a circle of radius 6 cm. The wheel starts with point $P(x, y) = (6, 0)$. The parameter t is the number of radians from the positive x-axis to a line from the origin through the center of the wheel.

- Let $\vec{r}$ (not shown) be the position vector to the point $P(x, y)$.
- Let $\vec{v}_1$ be the vector from the origin to the center of the wheel.
- Let $\vec{v}_2$ be the vector from the center of the wheel to the point $P(x, y)$.

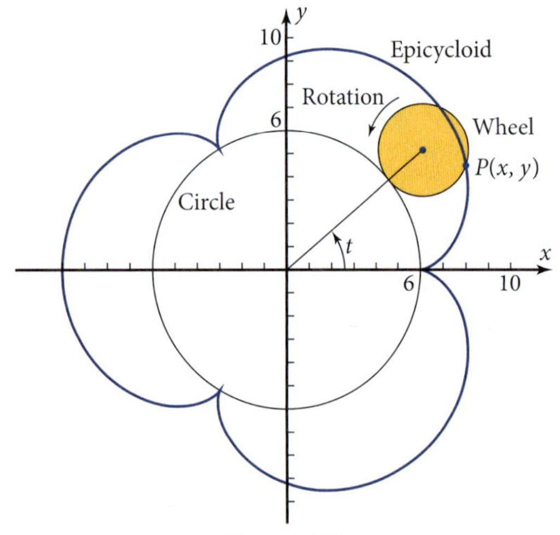

Figure 11-5l

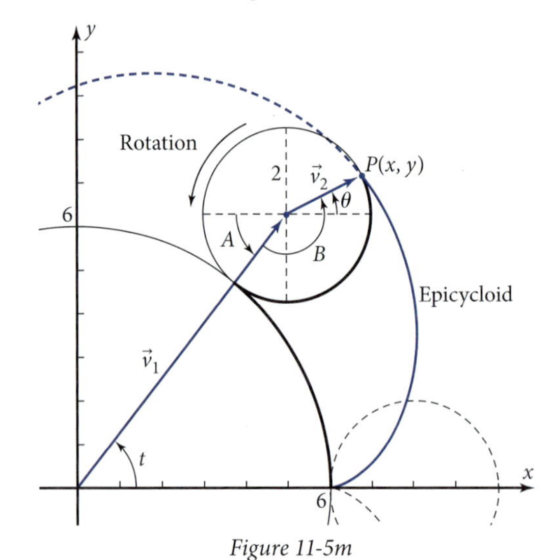

Figure 11-5m

a. Find $\vec{v}_1$ in terms of t and the unit vectors $\vec{i}$ and $\vec{j}$. Find $\vec{v}_2$ in terms of angle θ in Figure 11-5m and the unit vectors $\vec{i}$ and $\vec{j}$.

b. Write θ in terms of t by observing that θ starts at $-\pi$ radians when $t = 0$. Thus, θ is given by the sum $-\pi + A + B$, where angle A and angle B are as shown in Figure 11-5m. Express angles A and B in terms of t. Note that the arc of the wheel subtended by angle B equals the arc of the circle subtended by angle t, because the wheel rotates without slipping. Note also that the length of an arc of a circle equals the central angle in radians times the radius.

c. Write a vector equation for $\vec{r}$ as a function of t by observing that $\vec{r} = \vec{v}_1 + \vec{v}_2$. Use the result to plot the epicycloid using parametric mode. Use a t-domain large enough to get one complete cycle, and use equal scales on both axes. Does the graph agree with Figure 11-5m? If not, go back and check your work.

d. Plot the 6-cm circle on the same screen as in part c. Does the result agree with the figure?

9. *Involute of a Circle Problem:* Figure 11-5n shows an *involute of a circle.* A string is wrapped around a circle of radius 5 units. A pen is tied to the string at the point (5, 0). Then the string is unwound in the counterclockwise direction. The involute is the spiral path followed by the pen as the string unwinds. The curve is interesting because gear teeth made with their surfaces in this shape transmit the rotation smoothly from one gear to the next.

a. The parameter t is the radian measure of the angle between the positive x-axis and the line from the origin to the point of tangency of the string. Vector $\vec{v}_1$ goes from the center of the circle to the point of tangency. Vector $\vec{v}_2$ goes along the string from the point of tangency to the point $P(x, y)$ on the involute. Write a vector equation for the position vector $\vec{r}$ (not shown) to point P in terms of the parameter t. Note that $\vec{v}_1$ and $\vec{v}_2$ are perpendicular, because $\vec{v}_1$ is a radius to the point of tangency. Confirm your equation by plotting it on your grapher. Use three revolutions for the t-domain.

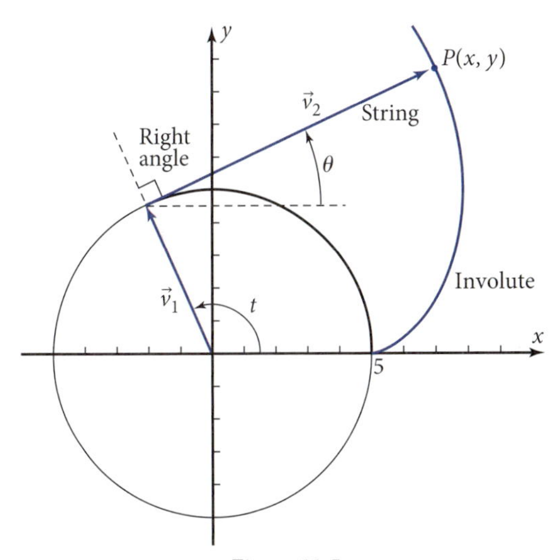

Figure 11-5n

b. Construct an involute by wrapping a string around a roll of tape, tying a pencil or pen to the end of the string, and tracing the path as you unwind the string. Does the involute you plotted in part a agree with this actual involute?

10. *Roller Coaster Problem:* Figure 11-5o shows part of a roller coaster track whose shape is a prolate cycloid (see Problem 7). This shape can be generated through this method: Imagine a circle of radius 5 ft rolling along the underside of the line $y = 20$ (Figure 11-5p). The prolate cycloid results from tracing point $P(x, y)$, located at 12 ft from the center of the rolling circle as it rotates through an angle of t radians. Point $P(x, y)$ is at its top position, directly above the origin, when angle $t = 0$ radians. The position vector $\vec{r}$ (not shown) to the point $P(x, y)$ is the sum of four other vectors starting at the origin, $\vec{v}_1 + \vec{v}_2 + \vec{v}_3 + \vec{v}_4$. Write a vector equation for the track in terms of t. By plotting on your grapher, show that your equation is correct.

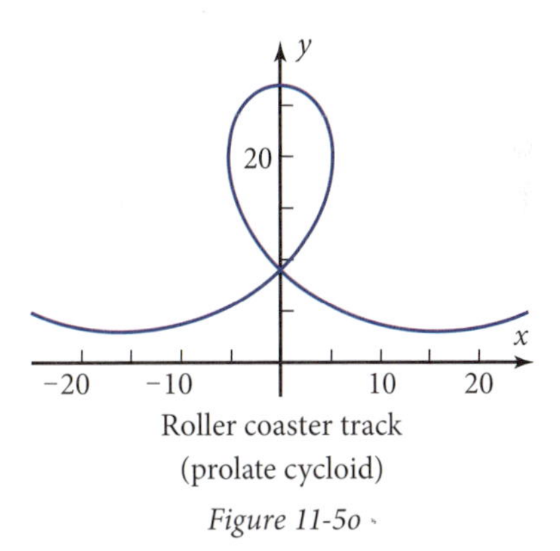

Roller coaster track (prolate cycloid)

Figure 11-5o

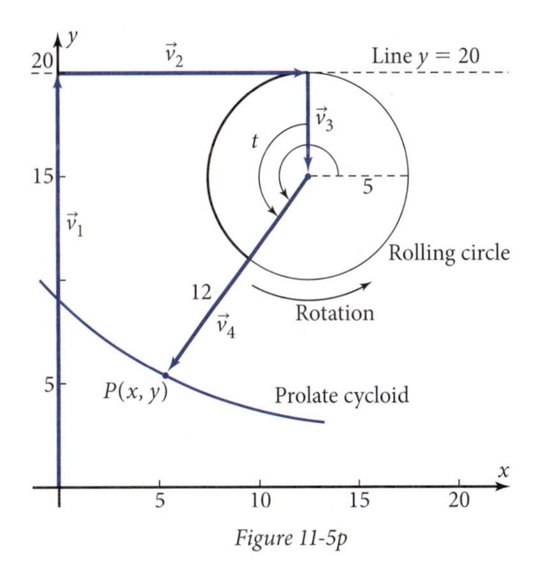

Figure 11-5p

11. *Parametric Equations of Polar Curves Problem:* Figure 11-5q shows the ellipse with this polar equation, superimposed on a rectangular coordinate grid:

$$r = \frac{9}{5 - 4\cos\theta}$$

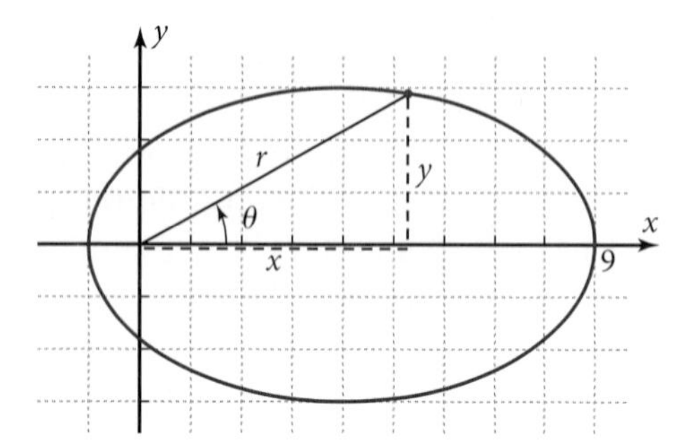

Figure 11-5q

a. Confirm on your grapher that the equation gives the graph in Figure 11-5q. Note that the value of θ shown in the figure is 0.5 radian.

b. Use the definitions of cosine and sine and the value of r from the polar equation to write the parametric equations for the ellipse using t in place of θ. Confirm that your parametric equations give the same ellipse and that tracing to $t = 0.5$ radian gives the point shown.

c. The ellipse in Figure 11-5q has center at the point (4, 0), x-radius 5, and y-radius 3. Use this information to write another set of parametric equations for the ellipse in terms of the parameter t.

d. With your grapher in simultaneous mode, plot the parametric equations of part c on the same screen as the parametric equations of part b. Use path style so that you can see where each function is being plotted as t increases from 0 to 2π. Do the two finished graphs coincide? Does the parameter t in part c represent the same angle as t in part b?

11-6 Chapter Review and Test

In this chapter you started by plotting graphs in polar coordinates. Polar coordinates allowed you to write complex numbers in polar form. You then saw that the product of two complex numbers has a magnitude equal to the product of the two magnitudes and an angle measure equal to the sum of the two angle measures. By De Moivre's theorem you were able to find powers and roots of complex numbers. Vectors and parametric functions gave you a way to describe the path followed by a point on a moving object.

Review Problems

R0. Update your journal with what you have learned in this chapter. Include things such as

- Multiple ways to write polar coordinates for the same point
- Graphs such as limaçons and conic sections in polar coordinates
- The fact that r can be negative in polar coordinates
- False and true intersections of polar curves
- How a complex number can be written in polar form
- De Moivre's theorem and products, quotients, and roots of complex numbers
- Parametric equations for cycloids and other paths of moving objects

R1. Plot these points on polar coordinate paper. Show especially what happens when r is negative. Connect the points with a smooth curve.

θ	r
45°	7.1
60°	5.0
75°	2.6
90°	0
105°	−2.6
120°	−5.0
135°	−7.1

R2. **a.** Plot the three-leaved rose $r = 10\cos 3\theta$. Sketch the graph.

b. Plot the limaçon $r = 3 - 5\cos\theta$. Sketch the graph.

c. Plot the hyperbola $r = \frac{8}{3 - 5\cos\theta}$. Sketch the graph.

d. Plot the circle $r = 2\cos(\theta - 60°)$. Sketch the graph.

e. Plot the line $r = 2\sec(\theta - 60°)$. Sketch the graph.

f. Prove algebraically that the graph in part c is a hyperbola by transforming the equation into Cartesian form. Where is a focus of the hyperbola? What is its eccentricity?

R3. Find the solutions to the system $r_1(\theta) = 4 + 6\cos\theta$ and $r_2(\theta) = 5 - 3\cos\theta$, shown in Figure 11-6a. At what other point(s) do the graphs cross?

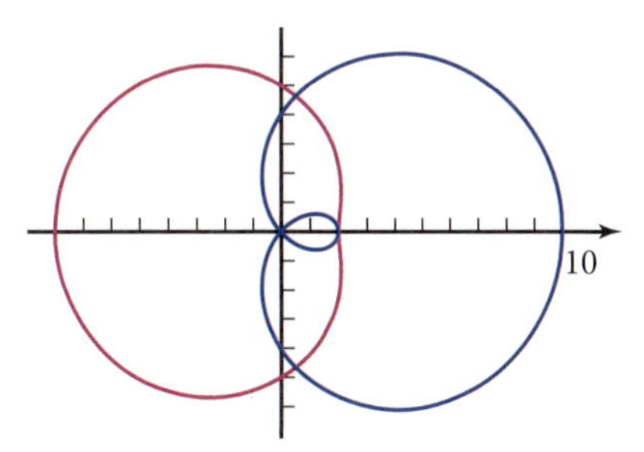

Figure 11-6a

R4. **a.** Write in polar form: $-5 + 12i$

b. Write in Cartesian form: $7 \text{ cis } 234°$

c. Find the product: $(2 \text{ cis } 52°)(5 \text{ cis } 38°)$

d. Find the quotient: $(51 \text{ cis } 198°) \div (17 \text{ cis } 228°)$

e. Raise to the power: $(2 \text{ cis } 27°)^5$

f. Raise to the power: $(8 \text{ cis } 120°)^{1/3}$

g. Sketch the three answers to part f on the complex plane.

h. Express this sum in Cartesian form:

$$10 \text{ cis } 43° + 7 \text{ cis } 130° - 5 \text{ cis } 215°$$

i. Write the answer to part h as a complex number in polar form.

R5. **a.** *Parametric Line Problem:* As you travel eastward on Highway I-90 from Cleveland to Erie, the highway makes an angle of 24° north of east. Suppose you start at Cleveland at time $t = 0$ h and drive at 60 mi/h. Write parametric equations expressing your position $P(x, y)$ as a function of t, in hours, where x and y are the number of miles east and north of Cleveland, respectively. The highway passes through Erie, which is 50 mi east of Cleveland. How far north of Cleveland is Erie?

b. *Quarter and Dime Epicycloid Problem:* A quarter is placed with its center at the origin of a Cartesian coordinate system. A dime is placed with its center on the x-axis, as shown in Figure 11-6b. The quarter is held fixed while the dime rotates counterclockwise around it without slipping. Figure 11-6c shows the dime rotated to the place where a line from the origin through its center makes an angle of t radians with the x-axis. Point P on the dime traces an epicycloid. The quarter has radius 12 mm, and the dime has radius 9 mm.

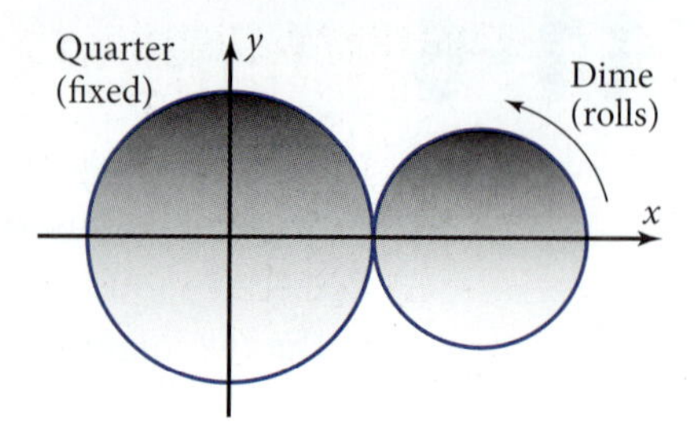

Figure 11-6b

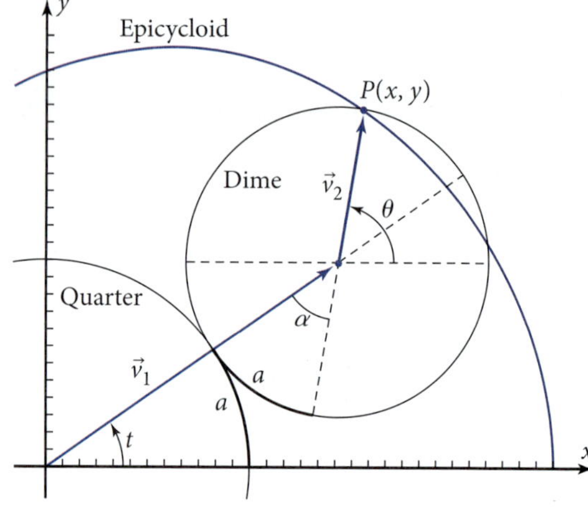

Figure 11-6c

i. Do you agree that the circles in Figure 11-6b really are the size of a quarter and a dime? Do you agree that the radii are 12 mm and 9 mm, respectively?

ii. Derive a vector equation for the point $P(x, y)$ on the edge of the rolling dime in terms of t. To help you do this, notice that the position vector $\vec{r}$ to point P (not shown in Figure 11-6c) is equal to $\vec{v}_1 + \vec{v}_2$, where $\vec{v}_1$ extends from the origin to the center of the dime and $\vec{v}_2$ extends from the center of the dime to point P. Once you find $\vec{v}_2$ in terms of θ in Figure 11-6c, you can find $\theta = \frac{7}{3}t$ by realizing that arc a on the quarter equals arc a on the dime, thus letting you write α in terms of t.

iii. Plot one complete cycle of the epicycloid with your grapher in parametric mode. Note that it takes more than one revolution for the graph to close.

Concept Problems

C1. *Planetary Motion Science Fiction Problem:* Figure 11-6d shows a small planet with a 3-mi radius orbiting a black hole. The orbit is circular with radius 10 mi. As it orbits the black hole, the planet rotates counterclockwise. Vector $\vec{v}_1$ goes from the center of the black hole to the center of the planet. Vector $\vec{v}_2$ goes from the center of the planet to the point $P(x, y)$ on the surface of the planet. Angles A and B are in standard position at the centers of the black hole and the planet, respectively. Vector $\vec{r}$ (not shown) is the position vector to the point $P(x, y)$.

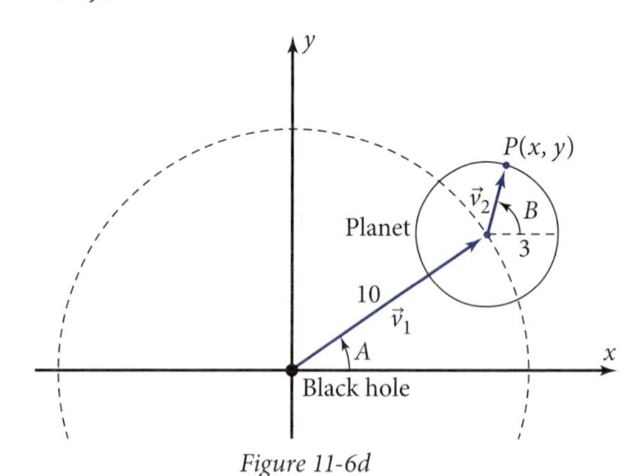

Figure 11-6d

a. Write $\vec{r}$ in terms of the unit vectors $\vec{i}$ and $\vec{j}$ and angles A and B.

b. At time $t = 0$ h both angle A and angle B measure 0 radians. The planet rotates counterclockwise at 12 radians per hour, and it orbits the black hole counterclockwise at 2 radians per hour. Write equations expressing A and B in terms of t. Use the results to write $\vec{r}$ as a function of t.

c. Plot the path of point P on your grapher. Use radian mode and equal scales on both axes. Sketch the result.

d. Explain why there are only *five* loops in the graph, in spite of the fact that the angular velocity of the planet is *six* times the angular velocity of the orbit.

e. Plot the graph of the path of point P under these conditions.

i. The planet slows from 12 radians per hour to 8 radians per hour.

ii. The planet rotates clockwise at 12 radians per hour instead of counterclockwise.

iii. The planet rotates at exactly the right angular velocity to make cusps instead of loops in the path.

C2. *Gear Tooth Problem:* The surfaces of gear teeth are made to model the shape of an involute of a circle (Figure 11-6e). This form is used because it allows the motion of one gear to be transmitted uniformly to the motion of another. An involute is the path traced by the end of a string as it is unwound from around a circle. In this problem you will see how the polar coordinates of a point on an involute are related to the Cartesian coordinates that can be found from parametric equations. You will do this by using vectors in the form of complex numbers.

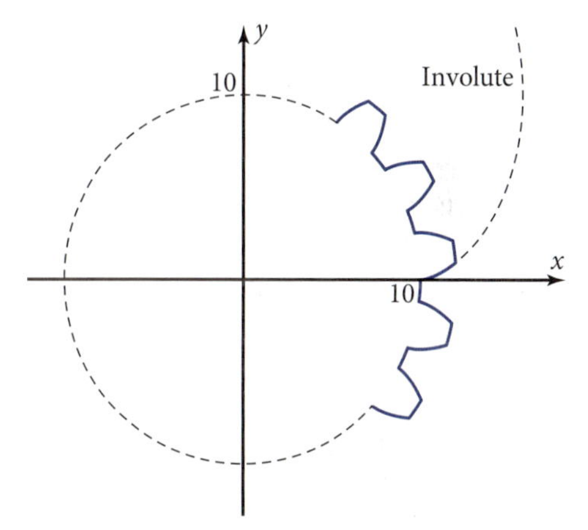

Figure 11-6e

a. Suppose a gear is to have radius 10 cm (to the inside of the teeth). The curved part of the gear tooth surface has the shape of the involute formed by unwrapping a string from this circle. Vector $\vec{v}_1$ (Figure 11-6f on the next page) goes from the center of the circle to the point of tangency of the string. Write $\vec{v}_1$ as a complex number in polar form, in terms of the angle t, in radians.

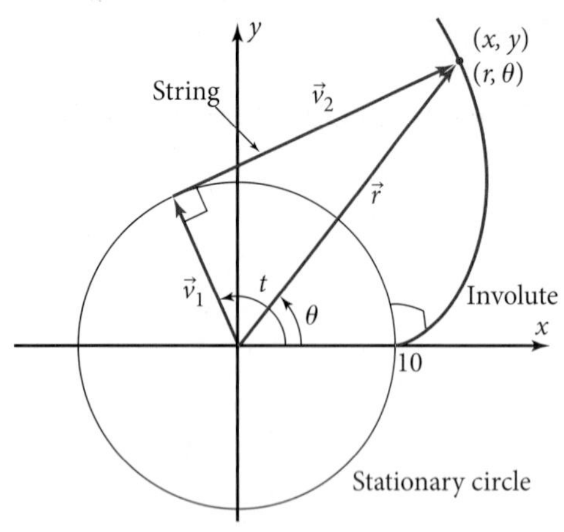

Figure 11-6f

b. Vector $\vec{v}_2$ goes along the string from the point of tangency to the point on the involute. Explain why its length is the same as that of the arc of the circle subtended by angle t. Write $\vec{v}_2$ as a complex number in polar form. Observe that, with respect to the horizontal axis, the angle for $\vec{v}_2$ is $\frac{\pi}{2}$ radians less than t because $\vec{v}_2$ is perpendicular to $\vec{v}_1$. Use the appropriate properties to write $\vec{v}_2$ in terms of functions of t.

c. Vector $\vec{r}$ is the position vector to the point $P(x, y)$ on the involute. Show that

$$\vec{r} = 10(\cos t + t\sin t) + 10i(\sin t - t\cos t)$$

d. Show that r (the length of $\vec{r}$) is given by the equation $r = 10\sqrt{1 + t^2}$.

e. The gear teeth are to be 2 cm deep, which means that the outer radius of the gear will be 12 cm. If the inside of the tooth shown in Figure 11-6f is at $t = 0$, what will be the value of t at the outside of the tooth? What will be the value of θ for this value of t?

f. Machinists who make the gear need to know the degree measure of angle θ in part e. Find this measure in degrees and minutes, to the nearest minute.

C3. *General Polar Equation of a Circle Problem:* The general polar equation of a circle that does not pass through the pole can be found with the help of the law of cosines. Suppose a circle of radius a is centered at the point with polar coordinates (k, α), as in

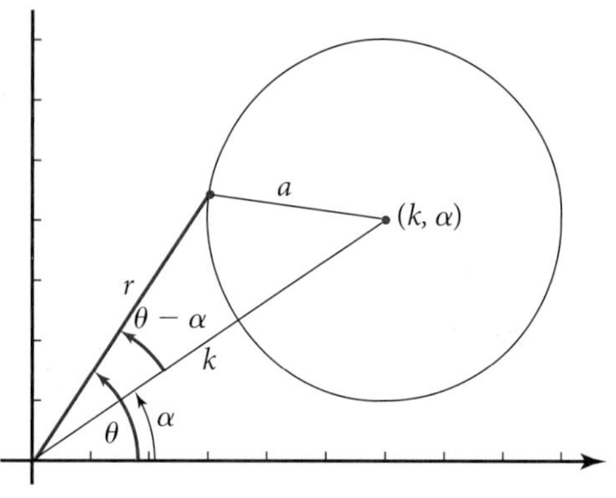

Figure 11-6g

a. Use the law of cosines to write an equation relating r, a, k, and the angle $(\theta - \alpha)$. Solve the equation for r with the help of the quadratic formula.

b. The quadratic formula has an ambiguous $\pm$ sign in it. Use the form of the solution with the $+$ sign to plot the circle of radius 3 centered at the point $(7, 40°)$. Use $[0°, 360°]$ for θ. Does the grapher plot the graph as one continuous circle?

c. Use the form of the solution with the $-$ sign. How does the graph relate to the graph in part b?

d. Find the two values of r if θ equals $50°$.

e. Show algebraically that there are no real values of r if θ equals $90°$.

Chapter Test

Part 1: No calculators allowed (T1–T9)

T1. Plot these points on polar coordinate paper. Connect the points with a smooth curve.

θ	r
150°	−8.4
165°	−5.6
180°	−5.0
195°	−5.6
210°	−8.4

T2. Write the polar equation for the circle in Figure 11-6h.

T3. Write the polar equation for the line in Figure 11-6.

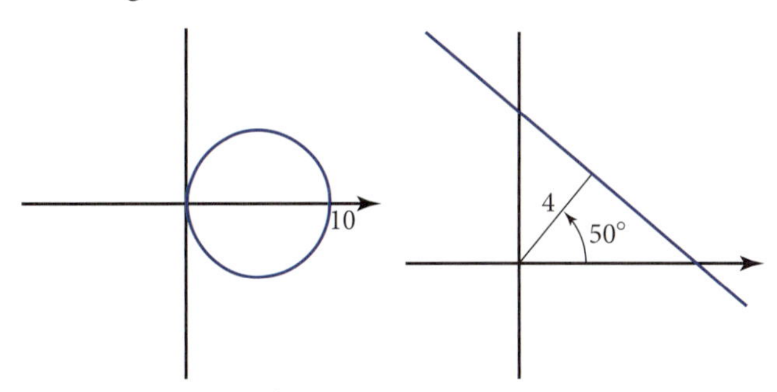

Figure 11-6h *Figure 11-6i*

T4. Write the definition of $r\operatorname{cis}\theta$.

T5. Find the product: $(5\operatorname{cis}37°)(3\operatorname{cis}54°)$

T6. Find the quotient: $(3\operatorname{cis}100°) \div (12\operatorname{cis}20°)$

T7. Raise to the power: $(4\operatorname{cis}50°)^3$

T8. Write i as a complex number in polar form. Use the result and De Moivre's theorem to find the two square roots of i. Plot the answers on a sketch of the complex plane.

T9. Write two other polar ordered pairs for the point (7, 30°), one with a positive r-value and a positive value of θ and the other with a negative r-value and a positive value of θ.

Part 2: Graphing calculators allowed (T10–T17)

T10. The polar equation of the graph in Problem T1 is

$$r = \frac{5}{2 + 3\cos\theta}$$

Eliminate the fraction by multiplying both sides by the denominator of the right side. Then derive a Cartesian equation, thus showing that the graph really is a hyperbola.

T11. The graphs of $r = 3\sin 4\theta$ and $r = 3\sin 5\theta$ are both roses. Plot each equation on your grapher. How many "leaves" are on each rose? How can you tell from the coefficient of θ how many leaves will be in a rose graph?

T12. Write $24 - 7i$ as a complex number in polar form.

T13. Write $6\operatorname{cis}300°$ as a complex number in rectangular form.

T14. Figure 11-6j shows the two polar curves

$$r_1(\theta) = 5 + 4\cos\theta \quad \text{and} \quad r_2(\theta) = 1 + 6\sin\theta$$

a. Which graph is which? What special name is given to each graph?

b. The graphs cross when $x = 0$ and $y = 5$. Is this a true intersection? Explain.

c. The graphs cross at a point in the first quadrant. Show that this point is a true intersection, and find its polar coordinates.

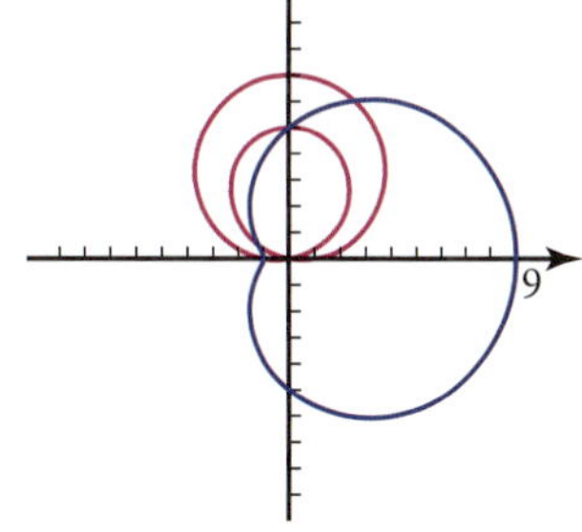

Figure 11-6j

T15. *Airplane Looping Problem:* A stunt pilot is doing a loop with her plane. As shown in Figure 11-6k, three forces are acting on the plane:

Wing lift: 2500 lb at 127°

Propeller thrust: 700 lb at 37° (perpendicular to the lift)

Gravitational force: 2000 lb straight down

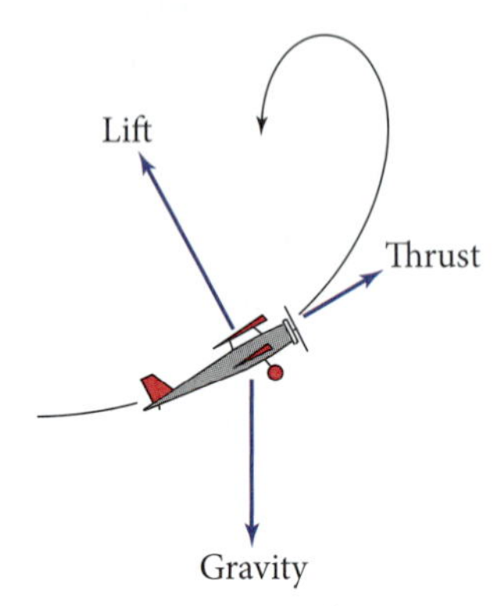

Figure 11-6k

Write the sum of these three force vectors as a complex number in polar form.

T16. *Car Wheel Curtate Cycloid Problem:* Figure 11-6l shows the *curtate cycloid* path traced by the valve stem (where you put in the air) on a car tire as the car moves. (*Curtate* comes from the Latin word *curtus,* which means "shortened." The words *curt* and *curtail* have the same origin.) The wheel rotates an angle *t,* in radians, from a point where the valve stem was at its lowest.

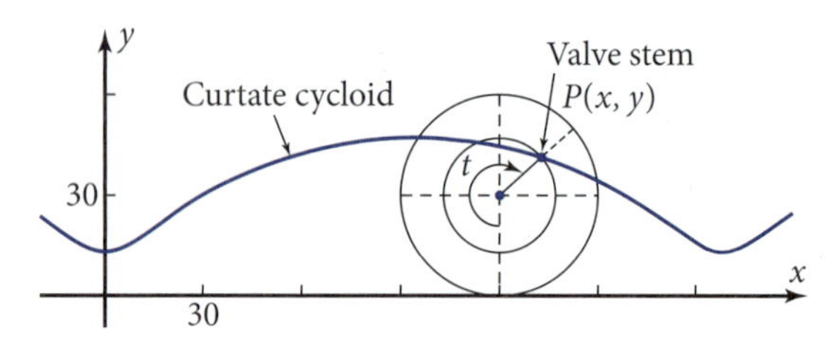

Figure 11-6l

a. The wheel has radius 30 cm. The valve stem is 17 cm from the center of the wheel. Write the position vector to point *P* on the curtate cycloid as a sum of three other vectors: $\vec{v}_1$ from the origin to the point where the tire touches the road, $\vec{v}_2$ from the head of $\vec{v}_1$ to the center of the wheel, and $\vec{v}_3$ from the center of the wheel to the valve stem. Use the result to write parametric equations for *x* and *y* in terms of the parameter *t.*

b. If you have not already done so, simplify your equations in part a using the composite argument properties to get equations that have only *t* as the argument.

c. Confirm that your equations are correct by plotting three cycles of the curtate cycloid. Use equal scales on the two axes.

d. On the same screen, plot a sinusoid with the same period, amplitude, and high and low points as the curtate cycloid. Sketch the results. Tell how you can distinguish between the two graphs.

T17. What did you learn as a result of taking this test that you did not know before?

Chapter 12

Three-Dimensional Vectors

When a building is constructed, it is important for the builders to know relationships between various planes, such as walls, ceilings, and roofs. These relationships contribute to building something that is both structurally sound and aesthetically pleasing to the eye. In this chapter you will learn to write vector equations that represent three-dimensional vectors, which can be used to calculate distances, angle measures, and intersections of lines and planes in space—quantities that would be hard to find graphically.

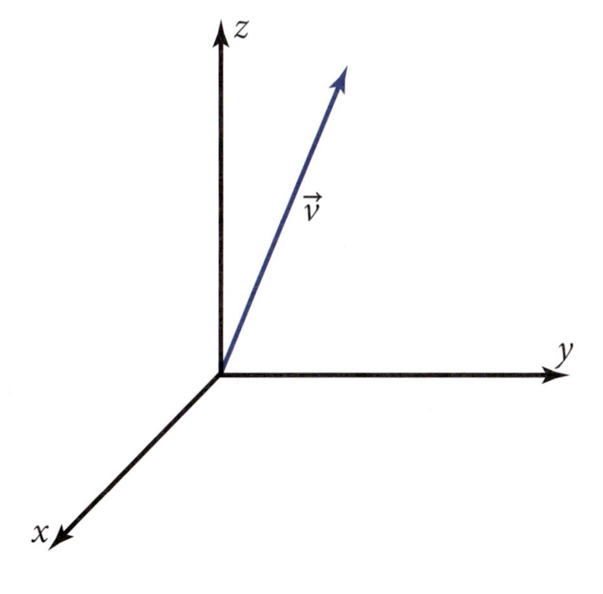

Mathematical Overview

In this chapter you will extend what you learned about two-dimensional vectors to vectors in space. You can perform some operations, such as vector addition and subtraction, by simply adding a third component to the representation of a vector. Other operations require new techniques—for example, there are two different kinds of multiplication for vectors. The payoff is the ability to calculate things such as the place where a line in space intersects a plane in space so that parts of three-dimensional objects will fit together properly when they are constructed. You will gain this knowledge in four ways.

NUMERICALLY

Dot product:

$$\left(4\vec{i} + 3\vec{j} + 5\vec{k}\right) \cdot \left(2\vec{i} + 6\vec{j} + 8\vec{k}\right) = (4)(2) + (3)(6) + (5)(8) = 66$$

Cross product:

$$\left(4\vec{i} + 3\vec{j} + 5\vec{k}\right) \times \left(2\vec{i} + 6\vec{j} + 8\vec{k}\right) = \begin{vmatrix} i & j & k \\ 4 & 3 & 5 \\ 2 & 6 & 8 \end{vmatrix}$$

$$= -6\vec{i} - 22\vec{j} + 18\vec{k}$$

GRAPHICALLY

A line intersecting a plane

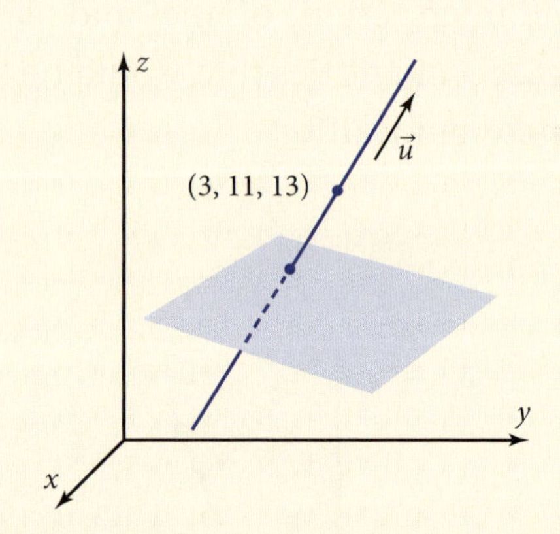

ALGEBRAICALLY

Equation for a line in three-dimensional space:

$$\vec{r} = \left(5 - \tfrac{2}{3}d\right)\vec{i} + \left(7 + \tfrac{1}{3}d\right)\vec{j} + \left(-1 + \tfrac{2}{3}d\right)\vec{k}$$

VERBALLY

A dot product results in a number, called a scalar, which is why it is often referred to as a "scalar product." A cross product results in a vector, which is why it is often referred to as a "vector product."

12-1 Review of Two-Dimensional Vectors

In Chapter 9, you learned that vectors are directed line segments. You used them as mathematical models of vector quantities, such as displacement, velocity, and force, that have direction as well as magnitude. In this chapter you will extend your knowledge to vectors in space. First, you will refresh your memory about two-dimensional vectors.

Objective Given two vectors, find the resultant vector by adding or subtracting them.

Exploratory Problem Set 12-1

1. Figure 12-1a shows two vectors, $\vec{a}$ and $\vec{b}$. They have magnitude and direction but no fixed location. You find the vector sum $\vec{a} + \vec{b}$ by translating $\vec{b}$ so that its beginning (or tail) is at the end (or head) of $\vec{a}$. The resultant vector goes from the beginning of $\vec{a}$ to the end of the translated $\vec{b}$. Show that you understand how to add two vectors by drawing a sketch showing $\vec{a} + \vec{b}$.

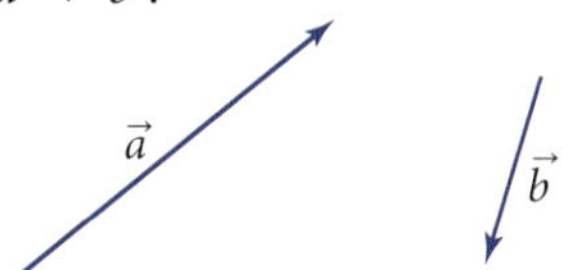

Figure 12-1a

2. You find the vector difference $\vec{a} - \vec{b}$ by adding $-\vec{b}$, the opposite of $\vec{b}$, to $\vec{a}$. The opposite of $\vec{b}$ is a vector of the same length pointing in the opposite direction. Show that you understand vector subtraction by sketching $\vec{a} - \vec{b}$.
3. The **position vector** for a point (x, y) in a coordinate system has a fixed location. It goes from the origin to the point. Figure 12-1b shows the position vector $\vec{p}$ for the point (4, 3). You can think of this vector as the sum of a vector in the x-direction and a vector in the y-direction. On a copy of Figure 12-1b, sketch these two components of $\vec{p}$.

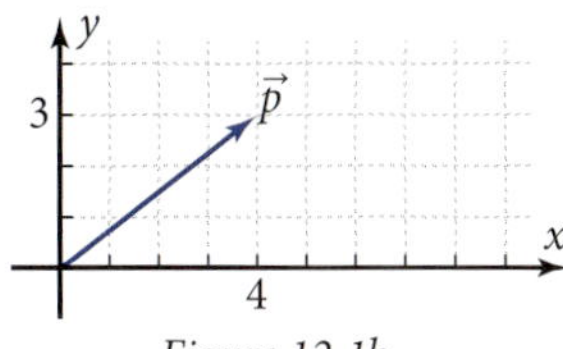

Figure 12-1b

4. In Problem 3, the vector in the x-direction is 4 units long. Vectors $\vec{i}$ and $\vec{j}$ are **unit vectors** in the x- and y-directions, respectively. A unit vector is a vector that is 1 unit long. So the x-component of $\vec{p}$ is the vector $4\vec{i}$, a product of a **scalar** and a vector. A scalar is a quantity that has only magnitude, not direction. What is the y-component of $\vec{p}$ in Figure 12-1b?
5. Suppose that $\vec{v} = 6\vec{i} + 8\vec{j}$. Sketch $\vec{v}$ as a position vector. Find the length of $\vec{v}$ (the *magnitude*) by means of the Pythagorean theorem. Find the measure of the angle $\vec{v}$ makes with the x-axis.
6. Suppose that $\vec{a} = 6\vec{i} + 2\vec{j}$ and $\vec{b} = 3\vec{i} + 5\vec{j}$. Draw $\vec{a}$ on graph paper as a position vector. Draw $\vec{b}$ with its tail at the head of $\vec{a}$. Draw the resultant vector, $\vec{a} + \vec{b}$. Write this sum vector as a position vector. What simple method can you think of for adding two vectors if you know their components? What is the length of $\vec{a} + \vec{b}$?
7. Suppose $\vec{c} = -9\vec{i} + 4\vec{j}$. Sketch $\vec{c}$ as a position vector. The measure of angle θ in standard position is a value of $\arctan\left(-\frac{4}{9}\right)$. Is θ equal to the value of $\tan^{-1}\left(-\frac{4}{9}\right)$ given by your calculator?
8. Make a list of all the important words in this section. Put a check mark by the ones you understand and a question mark by the ones you don't quite understand.

12-2 Two-Dimensional Vector Practice

In this section you will consolidate your knowledge of two-dimensional vectors from Chapter 9 and from Section 12-1. In the next section you will extend these concepts to vectors in three-dimensional space.

Objective

- Given the components of a two-dimensional position vector, find its length, a unit vector in its direction, a scalar multiple of it, and its direction angle.
- Given two two-dimensional position vectors, find their sum and their difference.

The box on the next page summarizes the definitions and properties of vectors. Figure 12-2a illustrates some of these definitions. The top figure shows a vector $\vec{v}$. The bottom left figure shows two equal vectors, $\vec{v}$ and $\vec{a}$, that are translations of each other. The bottom center figure shows a vector $\vec{v}$ and its opposite, $\vec{a} = -\vec{v}$. The bottom right figure shows a vector $\vec{v}$ that is 5 units long and a unit vector $\vec{u}$ in the same direction.

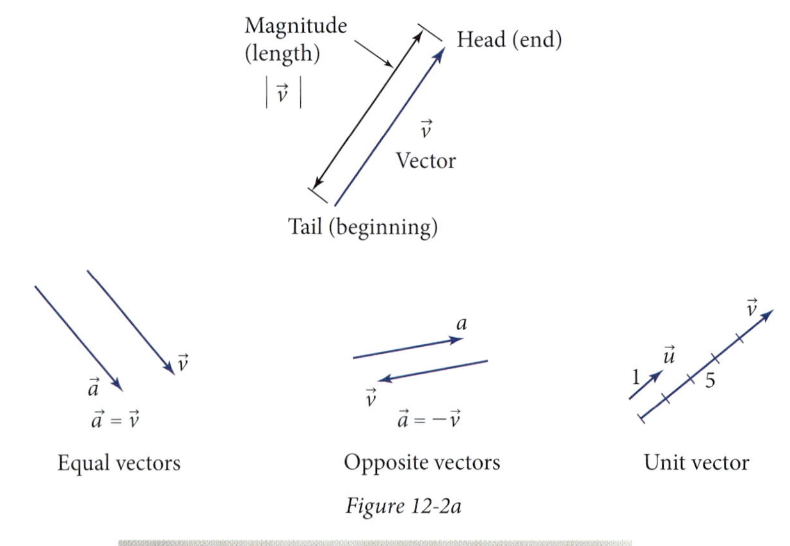

Figure 12-2a

Vectors play an important role in aerodynamics. Here airplane co-inventor Wilbur Wright glides down the north slope of Big Kill Devil Hill near Kitty Hawk in a double-ruddered glider.

DEFINITIONS AND PROPERTIES RELATING TO VECTORS

- A **vector quantity** is a quantity, such as force, velocity, or displacement, that has both magnitude (size) and direction.
- A **scalar** is a quantity, such as time, speed, or volume, that has only magnitude, no direction.
- A **vector** is a directed line segment that represents a vector quantity. Symbol: $\vec{v}$
- The *tail* of a vector is the point where it begins. The *head* of a vector is the point where it ends. An arrowhead is drawn at the head of a vector.
- The *magnitude,* or *absolute value,* of a vector is its length. Symbol: $|\vec{v}|$
- Two vectors are *equal* if they have the same magnitude and the same direction. So you can translate a vector without changing it, but you can't rotate or dilate it.
- The *opposite* of a vector is a vector of the same length in the opposite direction. Symbol: $-\vec{v}$
- A **unit vector** is a vector that is 1 unit long. Vectors $\vec{i}$ and $\vec{j}$ are unit vectors in the x- and y-directions, respectively. A unit vector $\vec{u}$ in the direction of a given vector $\vec{v}$ is found by dividing the vector by its length:

  $$\vec{u} = \frac{\vec{v}}{|\vec{v}|}$$

- A **position vector,** $\vec{v} = x\vec{i} + y\vec{j}$, starts at the origin and ends at the point (x, y).
- If $\vec{v} = x\vec{i} + y\vec{j}$, then its magnitude is $|\vec{v}| = \sqrt{x^2 + y^2}$ by the Pythagorean theorem.
- A **displacement vector** is the difference between an object's initial and final positions.

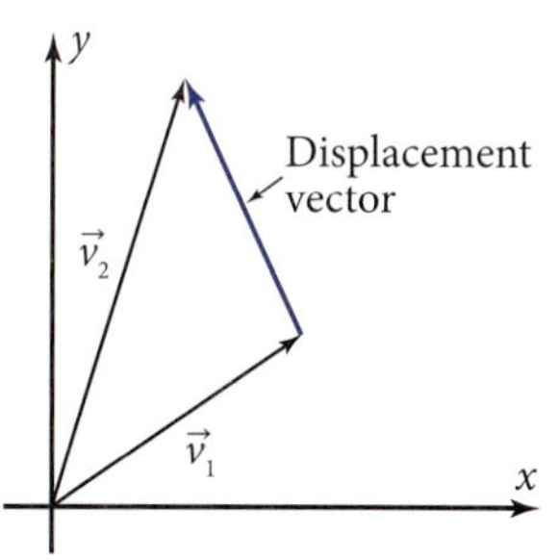

Operations on Vectors

In Section 9-6, you learned that to add two vectors $\vec{a} + \vec{b}$ you translate the beginning of the second vector to the end of the first vector. The sum, called the **resultant vector,** is the vector that goes from the beginning of the first vector to the end of the second, as in Figure 12-2b.

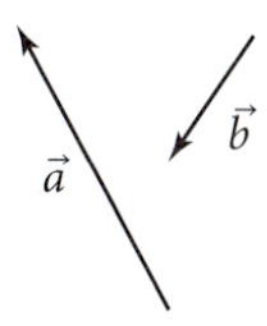

Two vectors

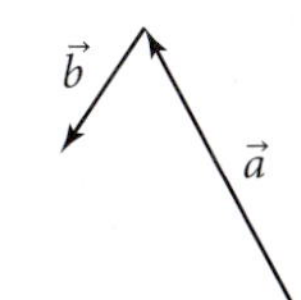

Translate head to tail.

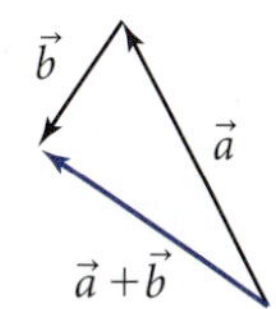

Sum goes from beginning of first vector to end of last.

Vector addition
Figure 12-2b

Vector subtraction is defined the same way as subtraction of real numbers—adding the opposite. Just as $5 - 3 = 5 + (-3)$, $\vec{a} - \vec{b} = \vec{a} + \left(-\vec{b}\right)$. Figure 12-2c illustrates this definition.

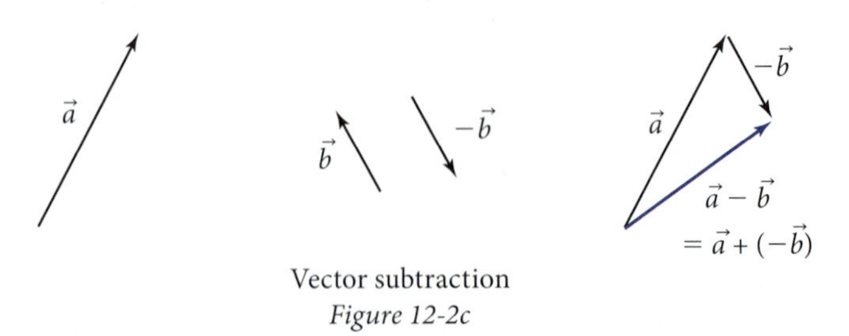

Vector subtraction
Figure 12-2c

A geometric property of the difference of two vectors is useful in many applications. If $\vec{a}$ and $\vec{b}$ in Figure 12-2c are placed *tail-to-tail,* the difference vector, $\vec{a} - \vec{b}$, is the *displacement vector* that goes from the head of $\vec{b}$ to the head of $\vec{a}$, as in Figure 12-2d. You can remember this property using an analogy to a car's odometer reading. To find out how far you have traveled on a trip, take "where you end minus where you began."

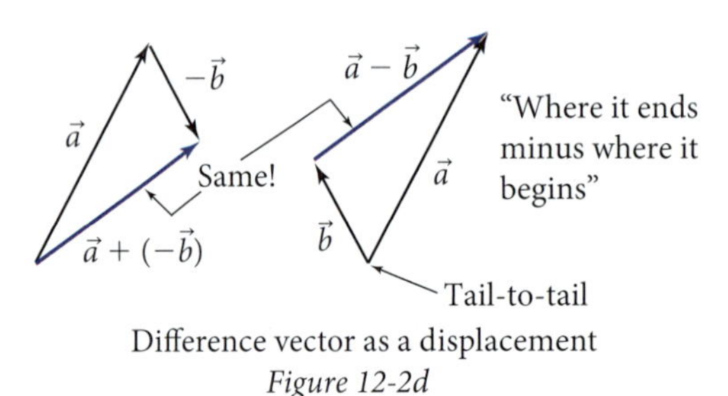

Difference vector as a displacement
Figure 12-2d

You can add or subtract two vectors easily if you write them in terms of their components in the x- and y-directions. As shown in Figure 12-2e, $\vec{v}$ is the position vector to the point $(x, y) = (-5, 3)$. Vectors $\vec{i}$ and $\vec{j}$ are unit vectors in the x- and y-directions, respectively. So $\vec{v}$ can be written

$$\vec{v} = -5\vec{i} + 3\vec{j}$$

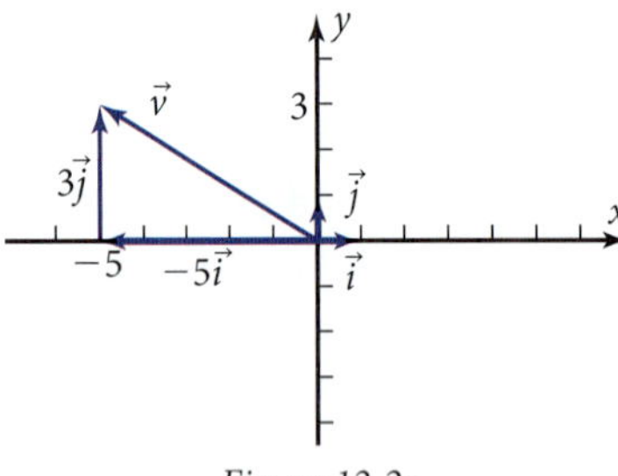

Figure 12-2e

These examples show how to operate on vectors that are written in terms of their components.

EXAMPLE 1 ➤ Figure 12-2f shows two vectors, $\vec{v}$ and $\vec{w}$.

a. Write these vectors in terms of their components.

b. On a copy of the figure, translate $\vec{w}$ so that its tail is at the head of $\vec{v}$. Then draw the resultant vector, $\vec{r} = \vec{v} + \vec{w}$. Find $\vec{r}$ numerically by adding the components of $\vec{v}$ and $\vec{w}$, and show that the answer agrees with your drawing.

c. Explain how you would find $\vec{w} + \vec{v}$. Why is the answer equivalent to $\vec{v} + \vec{w}$?

d. Find $|\vec{v}|$, $|\vec{w}|$, and $|\vec{v} + \vec{w}|$. Based on the graph, explain why $|\vec{v} + \vec{w}| < |\vec{v}| + |\vec{w}|$.

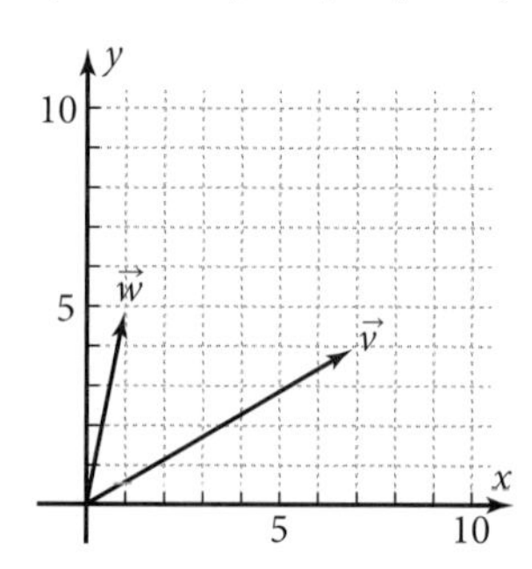

Figure 12-2f

SOLUTION

a. $\vec{v} = 7\vec{i} + 4\vec{j}$, $\vec{w} = 1\vec{i} + 5\vec{j}$

b. Figure 12-2g shows $\vec{w}$ translated so that $\vec{v}$ and $\vec{w}$ are head-to-tail, along with the resultant vector $\vec{r} = \vec{v} + \vec{w}$ going from the beginning of $\vec{v}$ to the end of $\vec{w}$.

$$\vec{r} = \left(7\vec{i} + 4\vec{j}\right) + \left(1\vec{i} + 5\vec{j}\right) = 8\vec{i} + 9\vec{j}$$

This answer agrees with Figure 12-2g, which shows $\vec{r}$ ending at the point (8, 9).

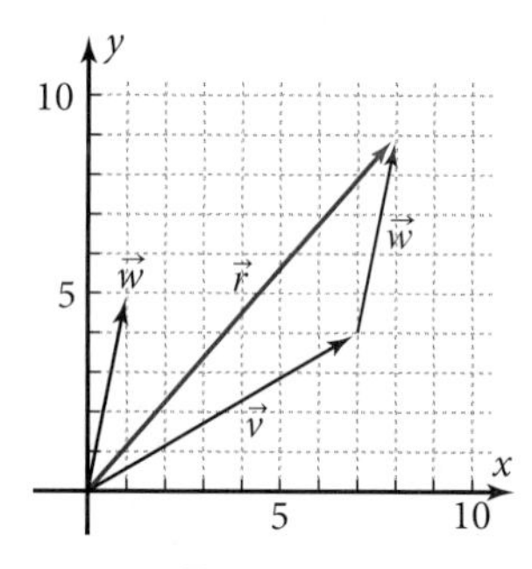

Figure 12-2g

c. To find $\vec{w} + \vec{v}$, translate $\vec{v}$ so that its tail is at the head of $\vec{w}$. The resultant vector goes from the beginning of $\vec{w}$ to the end of $\vec{v}$, which is also at the point (8, 9) shown in Figure 12-2g. Numerically,

$$\vec{w} + \vec{v} = \left(1\vec{i} + 5\vec{j}\right) + \left(7\vec{i} + 4\vec{j}\right) = 8\vec{i} + 9\vec{j}$$

which equals $\vec{v} + \vec{w}$, showing that vector addition is a *commutative* operation.

d. $\left|\vec{v}\right| = \sqrt{7^2 + 4^2} = \sqrt{65} = 8.0622...$

$\left|\vec{w}\right| = \sqrt{1^2 + 5^2} = \sqrt{26} = 5.0990...$

$\left|\vec{v} + \vec{w}\right| = \sqrt{8^2 + 9^2} = \sqrt{145} = 12.0415...$

$\left|\vec{v}\right| + \left|\vec{w}\right| = 8.0622... + 5.0990... = 13.1612...$, so

$\left|\vec{v} + \vec{w}\right| < \left|\vec{v}\right| + \left|\vec{w}\right|$.

This is reasonable because the third side of the triangle formed by the three vectors in Figure 12-2g is shorter than the sum of the lengths of the other two sides.

EXAMPLE 2 Consider vectors $\vec{v}$ and $\vec{w}$ in Example 1.

a. Draw $-\vec{w}$, the opposite of $\vec{w}$, as a position vector (starting at the origin). Then translate $-\vec{w}$ so that its tail is at the head of $\vec{v}$. Using the definition of vector addition, draw $\vec{v} + (-\vec{w})$. Explain why $\vec{v} - \vec{w}$ is equivalent to $\vec{v} + (-\vec{w})$.

b. Draw a displacement vector from the head of $\vec{w}$ to the head of $\vec{v}$. Explain why this vector is equivalent to $\vec{v} - \vec{w}$ from part a.

c. Find $\vec{v} - \vec{w}$ numerically from the coordinates of $\vec{v}$ and $\vec{w}$, and show that the answer agrees with your drawings in parts a and b.

SOLUTION

a. Figure 12-2h shows $-\vec{w}$ as a position vector and $-\vec{w}$ translated so that $\vec{v}$ and $-\vec{w}$ are head-to-tail. The sum $\vec{v} + (-\vec{w})$ goes from the beginning of $\vec{v}$ to the end of $-\vec{w}$. By the definition of vector subtraction, $\vec{v} - \vec{w} = \vec{v} + (-\vec{w})$.

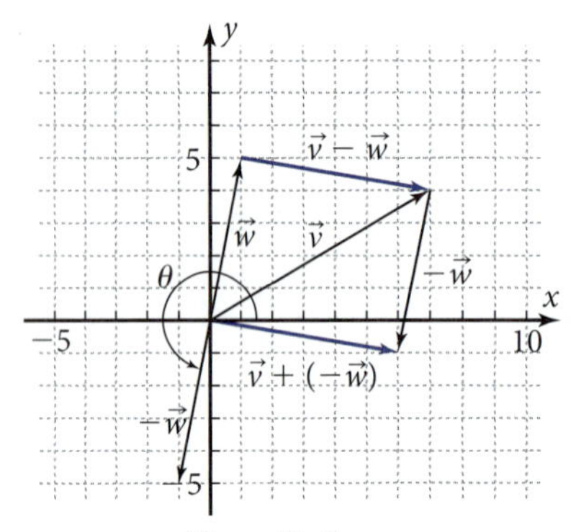

Figure 12-2h

b. Figure 12-2h also shows that the displacement vector from the head of $\vec{w}$ to the head of $\vec{v}$ has the same length and direction as $\vec{v} - \vec{w}$.

Displacement vector $\vec{v} - \vec{w}$ is "where it ends minus where it begins."

c. $\vec{v} - \vec{w} = \left(7\vec{i} + 4\vec{j}\right) - \left(1\vec{i} + 5\vec{j}\right) = 6\vec{i} - 1\vec{j}$, which agrees with both $\vec{v} - \vec{w}$ and $\vec{v} + (-\vec{w})$ in Figure 12-2h.

EXAMPLE 3 ➤ In Figure 12-2h, $\vec{v} = 7\vec{i} + 4\vec{j}$ and $\vec{w} = 1\vec{i} + 5\vec{j}$.

a. Find the linear combination $-3\vec{v} + 6\vec{w}$.

b. Find a unit vector, $\vec{u}$, in the direction of $\vec{v}$. Use the result to find a vector 20 units long in the direction of $\vec{v}$.

c. Find the measure of θ, the angle from the positive x-axis to $-\vec{w}$.

SOLUTION **a.**

$$\begin{aligned} & -3\vec{v} + 6\vec{w} \\ &= -3\left(7\vec{i} + 4\vec{j}\right) + 6\left(1\vec{i} + 5\vec{j}\right) \\ &= -21\vec{i} - 12\vec{j} + 6\vec{i} + 30\vec{j} \\ &= -15\vec{i} + 18\vec{j} \end{aligned}$$

b. $|\vec{v}| = \sqrt{7^2 + 4^2} = \sqrt{65} = 8.0622...$

$$\vec{u} = \frac{\vec{v}}{|\vec{v}|} = \frac{7}{\sqrt{65}}\vec{i} + \frac{4}{\sqrt{65}}\vec{j} = 0.8682...\vec{i} + 0.4961...\vec{j}$$

Divide the vector by its length.

$$20\vec{u} = 20\left(0.8682...\vec{i} + 0.4961...\vec{j}\right) = 17.3648...\vec{i} + 9.9227...\vec{j}$$

c. $\theta = \arctan\left(\frac{-5}{-1}\right) = 78.6900...° + 180°n \approx 258.7°$, using $n = 1$ ◀

EXAMPLE 4 ➤ Given points $C(8, 25)$ and $D(17, 3)$,

a. Find $\overrightarrow{CD}$, the vector pointing from C to D.

b. Find the position vector of the point $\frac{3}{4}$ of the way from C to D.

SOLUTION **a.** Sketch points C and D and position vectors $\vec{c}$ and $\vec{d}$ to these points, as in Figure 12-2i.

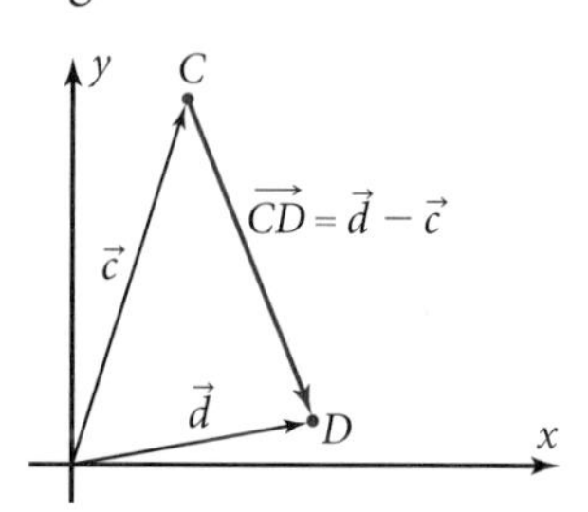

Figure 12-2i

Vector $\overrightarrow{CD}$ starts at C and ends at D. Therefore,

$$\overrightarrow{CD} = \vec{d} - \vec{c}$$

Where it ends minus where it begins.

$$= \left(17\vec{i} + 3\vec{j}\right) - \left(8\vec{i} + 25\vec{j}\right)$$

Write the position vectors to the two points.

$$= 9\vec{i} - 22\vec{j}$$

b. Sketch a vector starting at C and going $\frac{3}{4}$ of the way to D, as in Figure 12-2j. This vector will be $\frac{3}{4}\overrightarrow{CD}$. Because this vector and $\vec{c}$ are head-to-tail, the position vector $\vec{p}$ will be the vector sum.

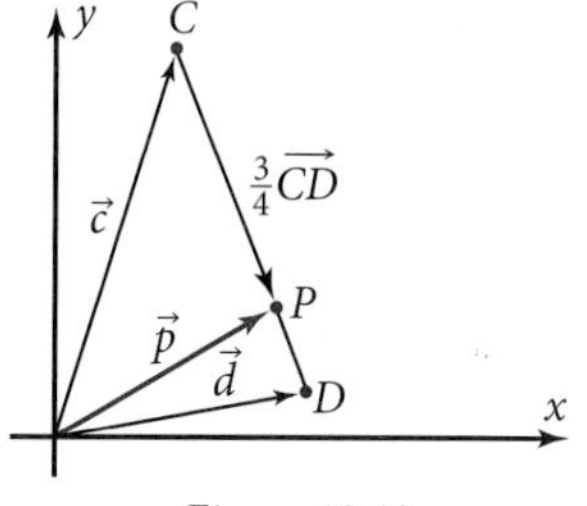

Figure 12-2j

$$\begin{aligned} \vec{p} &= \vec{c} + \tfrac{3}{4}\overrightarrow{CD} \\ &= \left(8\vec{i} + 25\vec{j}\right) + \tfrac{3}{4}\left(9\vec{i} - 22\vec{j}\right) \\ &= 14.75\vec{i} + 8.5\vec{j} \end{aligned}$$ ◀

Problem Set 12-2

Reading Analysis

From what you have read in this section, what do you consider to be the main idea? How do you find the resultant of two vectors geometrically? Numerically from the vectors' components? In what way is subtraction of two vectors similar to subtraction of two real numbers? What is the difference between a position vector and a displacement vector?

Quick Review

For Problems Q1–Q6, express the given values for right triangle ABC in Figure 12-2k.

Q1. $\sin A$

Q2. $\cos C$

Q3. $\tan C$

Q4. $a^2 + c^2$

Q5. The area

Q6. Angle A as an inverse tangent

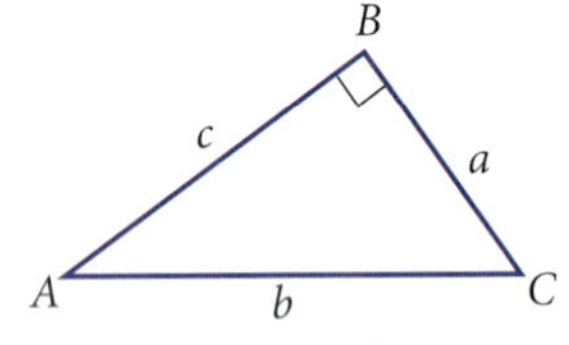

Figure 12-2k

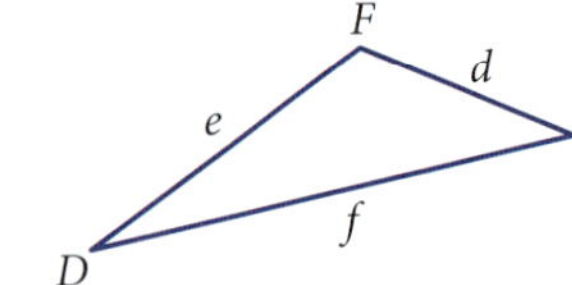

Figure 12-2l

Problems Q7–Q9 refer to oblique triangle DEF in Figure 12-2l.

Q7. Find f^2 by the law of cosines.

Q8. $\dfrac{d}{\sin D} = \underline{\quad ? \quad}$ by the law of sines

Q9. Find the area in terms of two sides and angle D.

Q10. $\cos 180° =$

A. 1 **B.** $\frac{1}{2}$ **C.** 0 **D.** $-\frac{1}{2}$ **E.** -1

1. Figure 12-2m shows two vectors, $\vec{a}$ and $\vec{b}$.

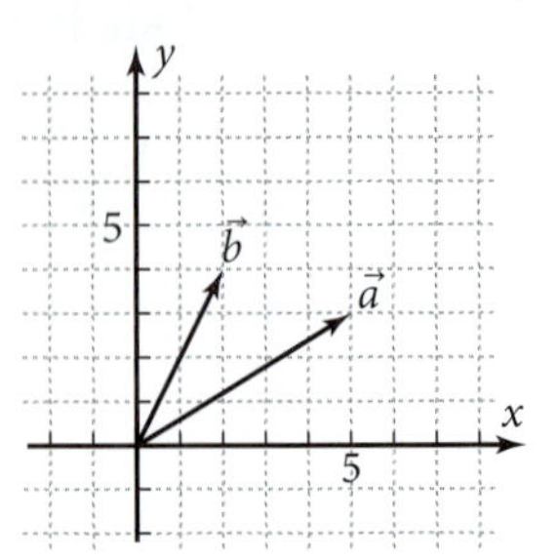

Figure 12-2m

a. Write $\vec{a}$ and $\vec{b}$ in terms of their components.

b. On a copy of Figure 12-2m, translate $\vec{b}$ so that its tail is at the head of $\vec{a}$. Then draw the resultant vector $\vec{r} = \vec{a} + \vec{b}$. Find $\vec{r}$ numerically by adding the components of $\vec{a}$ and $\vec{b}$, and show that the answer agrees with your drawing.

c. Explain how you would find $\vec{b} + \vec{a}$. Why is the answer equivalent to $\vec{a} + \vec{b}$?

d. Find $|\vec{a}|$, $|\vec{b}|$, and $|\vec{a} + \vec{b}|$. Based on the graph, explain why $|\vec{a} + \vec{b}| < |\vec{a}| + |\vec{b}|$.

2. Figure 12-2n shows two vectors, $\vec{c}$ and $\vec{d}$.

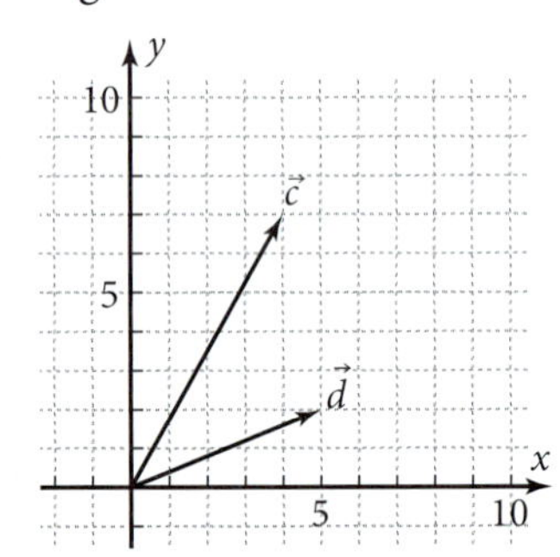

Figure 12-2n

a. Write $\vec{c}$ and $\vec{d}$ in terms of their components.

b. On a copy of Figure 12-2n, translate $\vec{d}$ so that its tail is at the head of $\vec{c}$. Then draw the resultant vector $\vec{r} = \vec{c} + \vec{d}$. Find $\vec{r}$ numerically by adding the components of $\vec{c}$ and $\vec{d}$, and show that the answer agrees with your drawing.

c. Explain how you would find $\vec{d} + \vec{c}$. Why is the answer equivalent to $\vec{c} + \vec{d}$?

d. Find $|\vec{c}|$, $|\vec{d}|$, and $|\vec{c} + \vec{d}|$. Based on the graph, explain why $|\vec{c} + \vec{d}| < |\vec{c}| + |\vec{d}|$.

3. Refer to $\vec{a}$ and $\vec{b}$ in Figure 12-2m.

a. On another copy of the figure, draw $-\vec{b}$, the opposite of $\vec{b}$, as a position vector (starting at the origin). Then translate $-\vec{b}$ so that its tail is at the head of $\vec{a}$. Using the definition of vector addition, draw $\vec{a} + (-\vec{b})$. Use components to demonstrate that $\vec{a} - \vec{b}$ is equivalent to $\vec{a} + (-\vec{b})$.

b. Draw a displacement vector from the head of $\vec{b}$ to the head of $\vec{a}$. Explain why this vector is equivalent to $\vec{a} - \vec{b}$ from part a. Explain verbally how you can tell whether $\vec{a} - \vec{b}$ goes from $\vec{a}$ to $\vec{b}$ or from $\vec{b}$ to $\vec{a}$.

c. Find $\vec{a} - \vec{b}$ numerically from the coordinates of $\vec{a}$ and $\vec{b}$, and show that the answer agrees with your drawings in parts a and b.

4. Refer to $\vec{c}$ and $\vec{d}$ in Figure 12-2n.

a. On another copy of the figure, draw $-\vec{d}$, the opposite of $\vec{d}$, as a position vector (starting at the origin). Then translate $-\vec{d}$ so that its tail is at the head of $\vec{c}$. Using the definition of vector addition, draw $\vec{c} + (-\vec{d})$. Use components to demonstrate that $\vec{c} - \vec{d}$ is equivalent to $\vec{c} + (-\vec{d})$.

b. Draw a displacement vector from the head of $\vec{d}$ to the head of $\vec{c}$. Explain why this vector is equivalent to $\vec{c} - \vec{d}$ from part a. Explain verbally how you can tell whether $\vec{c} - \vec{d}$ goes from $\vec{c}$ to $\vec{d}$ or from $\vec{d}$ to $\vec{c}$.

c. Find $\vec{c} - \vec{d}$ numerically from the coordinates of $\vec{c}$ and $\vec{d}$, and show that the answer agrees with your drawings in parts a and b.

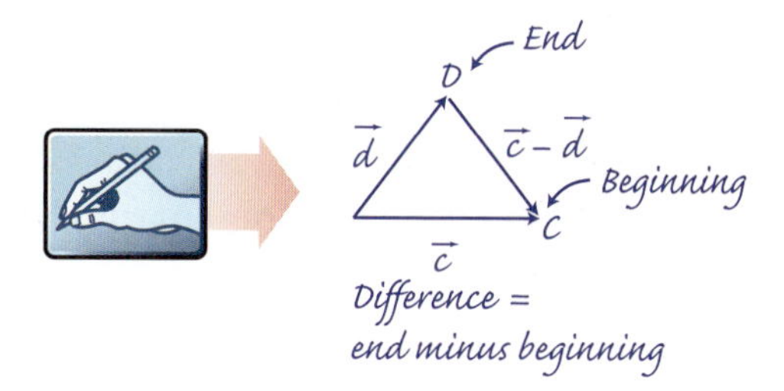

5. In Figure 12-2m, $\vec{a} = 5\vec{i} + 3\vec{j}$ and $\vec{b} = 2\vec{i} + 4\vec{j}$.

a. Find the linear combination $7\vec{a} + 6\vec{b}$.

b. Find a unit vector, $\vec{u}$, in the direction of $\vec{a}$. Use the result to find a vector 10 units long in the direction of $\vec{a}$.

c. Find the measure of the angle from the positive x-axis to $-\vec{b}$.

6. In Figure 12-2n, $\vec{c} = 4\vec{i} + 7\vec{j}$ and $\vec{d} = 5\vec{i} + 2\vec{j}$.

a. Find the linear combination $3\vec{c} - 8\vec{d}$.

b. Find a unit vector, $\vec{u}$, in the direction of $\vec{c}$. Use the result to find a vector 7 units long in the direction of $\vec{c}$.

c. Find the measure of the angle from the positive x-axis to $\vec{d}$.

For Problems 7–10, find the vector.

7. $\overrightarrow{AB}$ for points $A(3, 4)$ and $B(2, 7)$

8. $\overrightarrow{CD}$ for points $C(4, 1)$ and $D(3, 5)$

9. $\overrightarrow{BA}$ for points $A(7, 3)$ and $B(5, -1)$

10. $\overrightarrow{DC}$ for points $C(-2, 3)$ and $D(4, -3)$

11. *Highway Rest Stop Problem:* A highway rest stop will be built 40% of the way from the town of Artesia, A, to the town of Brooks, B, in Figure 12-2o. The two towns are located at points $A(20, 73)$ and $B(45, 10)$. Big City is located at the origin. The distances are in kilometers.

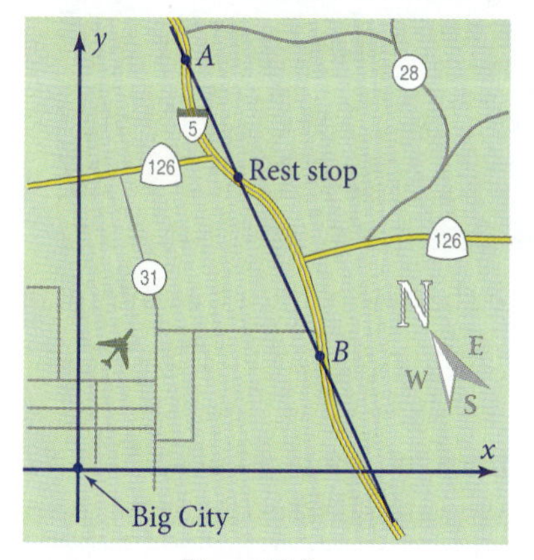

Figure 12-2o

a. Write the position vectors for Artesia and Brooks.

b. Write the displacement vector from Artesia to Brooks.

c. Write the displacement vector from Artesia to the rest stop.

d. Write the position vector for the rest stop.

e. It has been proposed to supply electricity to the rest stop directly from Big City. How long would the electric lines have to be? At what angle to the x-axis would the lines have to run?

f. How long would the electric lines have to be if they came from the closer of the two towns, Artesia or Brooks?

12. *Archaeology Problem:* Archaeologists often cut a trial trench through an archaeological site to reveal the different layers under the topsoil. This stratigraphy helps with dating the artifacts they unearth and with identifying any geological movements that might have disturbed the original position of objects. They usually lay a grid on the site, much like a coordinate system. Assume that they dig the trial trench from point $C(200, -300)$ to point $D(400, 500)$, where the distances are in yards.

a. Make a sketch showing the given information.

b. Write the position vectors for C and D and the vector from C to D.

c. The crew finds the remnants of a wall 65% of the way from C to D. Write the vector from C to this point.

d. How long is the trench from C to the wall?

e. How far is the wall from the origin?

For Problems 13–16, use the techniques of Problems 11 and 12 to find the vector or point.

13. Find the position vector for the point $\frac{1}{3}$ of the way from point $A(2, 7)$ to point $B(14, 5)$.

14. Find the position vector for the point $\frac{2}{3}$ of the way from point $C(11, 5)$ to point $D(2, 17)$.

15. Find the midpoint of the segment connecting points $E(6, 2)$ and $F(10, -4)$. From the result, give a quick way to find the midpoint of the segment connecting two given points.

16. Find the midpoint of the segment connecting points $G(5, 7)$ and $H(-3, 13)$. From the result, give a quick way to find the midpoint of the segment connecting two given points.

17. *Vector Properties Problem:* Prove these properties of vectors. A sketch may help. Express the vectors as the sum of their components, and prove the properties algebraically.

a. Vector addition is commutative.

b. Vector addition is associative.

c. Vector subtraction is *not* commutative.

d. Multiplication by a scalar distributes over vector addition.

e. The set of vectors is closed under addition. (Why is it necessary for there to be a zero vector in order for this closure property to be true?)

18. *Triangle Inequality Problem:*

a. Sketch two nonparallel vectors $\vec{a}$ and $\vec{b}$ head-to-tail. Then draw $\vec{a} + \vec{b}$. What can you say about $|\vec{a} + \vec{b}|$ compared to $|\vec{a}| + |\vec{b}|$?

b. Sketch two parallel vectors $\vec{a}$ and $\vec{b}$ head-to-tail, pointing in the same direction. Then draw $\vec{a} + \vec{b}$. What can you say about $|\vec{a} + \vec{b}|$ compared to $|\vec{a}| + |\vec{b}|$?

c. Use the appropriate theorems and postulates from geometry to prove that this *triangle inequality* is true.

PROPERTY: Triangle Inequality for Vectors

$$|\vec{a} + \vec{b}| \leq |\vec{a}| + |\vec{b}|$$

12-3 Vectors in Space

Suppose that a helicopter rises 700 ft and then moves to a point 300 ft east and 400 ft north of its original ground position. As shown in Figure 12-3a, the position vector has three components rather than just two. The techniques you learned for analyzing two-dimensional vectors in Section 12-2 carry over to three-dimensional vectors. You simply add a third component perpendicular to the other two.

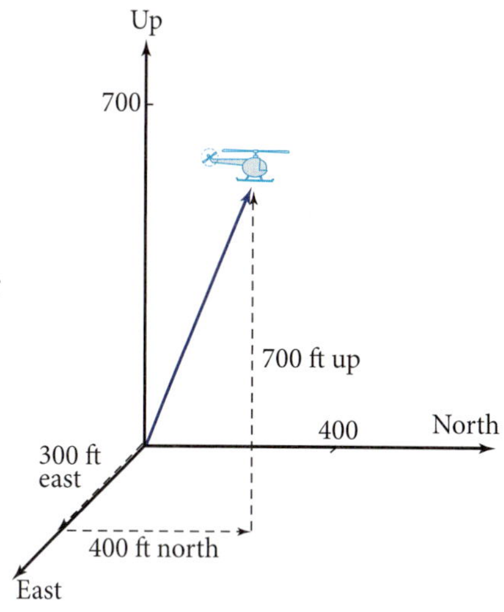

Figure 12-3a

Objective

- Given two three-dimensional vectors, find their lengths, add them, subtract them, and use the results to analyze real-world problems.
- If a position vector terminates in the first octant, sketch it on graph paper.

In this exploration you will plot and perform operations on three-dimensional vectors by using their components.

EXPLORATION 12-3: Introduction to Three-Dimensional Vectors

1. On a sheet of graph paper, mark off positive parts of x-, y-, and z-axes as shown. Write the scale numbers 5 and 10 on each axis, noting that the x-axis is shortened so that 10 occurs near the arrowhead labeled x. Then draw

$$\vec{v} = 5\vec{i} + 7\vec{j} + 10\vec{k}$$

as a position vector by starting at the origin and counting 5 units in the x-direction, then 7 units in the y-direction, and finally 10 units in the z-direction. Draw an arrow from the origin to the head of $\vec{v}$. Show a "box" surrounding $\vec{v}$ that makes the vector look three-dimensional.

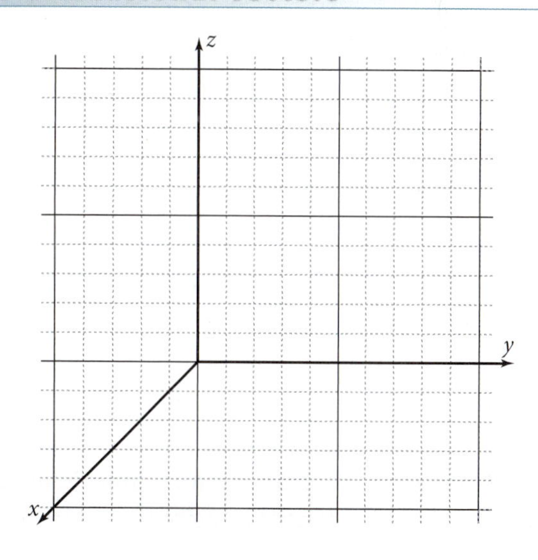

2. Show that you know what the three unit vectors $\vec{i}$, $\vec{j}$, and $\vec{k}$ mean by sketching them starting from the origin on your graph in Problem 1.

continued

3. Find the length of $\vec{v}$ in Problem 1 using the three-dimensional Pythagorean theorem $d = \sqrt{x^2 + y^2 + z^2}$.
4. Write $3\vec{v}$ as a sum of components.
5. Show by means of the three-dimensional Pythagorean theorem that $3\vec{v}$ really is three times as long as $\vec{v}$.
6. Write as a sum of components a *unit* vector in the direction of $\vec{v}$.
7. Let $\vec{a} = 4\vec{i} - 3\vec{j} + 8\vec{k}$. Quickly find $\vec{v} + \vec{a}$, where $\vec{v} = 5\vec{i} + 7\vec{j} + 10\vec{k}$, as in Problem 1.
8. Find the displacement vector $\vec{b}$ from the point (1, 5, 2) to the point (7, 3, 11).
9. Find the position vector of the point 0.3 of the way from the point (1, 5, 2) to the point (7, 3, 11).
10. What did you learn as a result of doing this exploration that you did not know before?

Figure 12-3b shows a two-dimensional representation of a three-dimensional coordinate system. In 3-D, the z-axis points upward, so you draw it going up on your paper. The x-axis is drawn obliquely down to the left, and the y-axis is drawn horizontally to the right. Imagine looking down from somewhere above the first quadrant of the xy-plane. Notice that the x- and y-axes have the same orientation with respect to each other as they do in two dimensions. To assist you visually, the tick marks on the x-axis are drawn horizontally, and unit intervals are drawn shorter than on the other two axes.

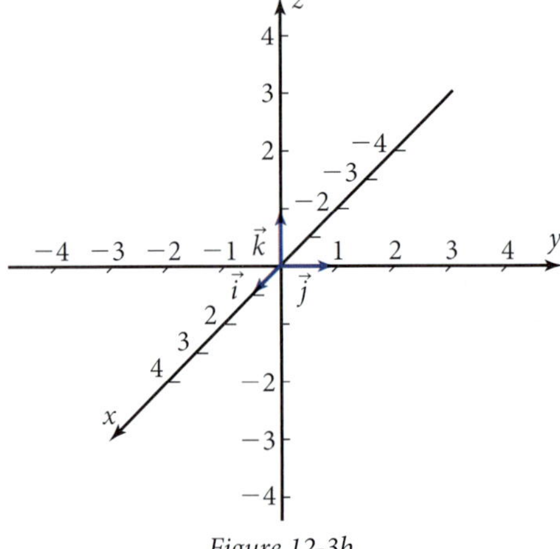

Figure 12-3b

The xy-plane, yz-plane, and xz-plane divide space into eight regions called *octants*. The region in which all three variables are positive is called the *first octant*. The other octants are not usually named.

The symbols $\vec{i}$, $\vec{j}$, and $\vec{k}$ are used for **unit vectors** in the x-, y-, and z-directions, respectively.

EXAMPLE 1 ➤ On graph paper, draw a sketch of the position vector $\vec{p} = 3\vec{i} + 5\vec{j} + 7\vec{k}$. Write the coordinates of the point P at the end (the head) of $\vec{p}$. Find the length of $\vec{p}$.

SOLUTION Draw the three axes, with the x-axis along the diagonal as in Figure 12-3c. To get the desired perspective on the x-axis, mark two units on the diagonal for each grid line. Because $\vec{p}$ is a position vector, it starts at the origin. Starting there, draw a vector 3 units long in the x-direction. From its head, draw a second vector 5 units long in the y-direction by counting spaces. From the head of the second vector, draw a third vector 7 units long in the z-direction. Because the vectors are head-to-tail, the sum goes from the beginning of the first vector to the end of the last vector.

If you draw in the dashed lines as shown, they help make the drawing look three-dimensional. They form a box for which $\vec{p}$ is the main diagonal.

Because $\vec{p}$ is the position vector of a point (x, y, z), the point has coordinates (3, 5, 7).

You can find the length of the vector using the three-dimensional Pythagorean theorem. It is an extension of the two-dimensional formula, with the square of the third coordinate appearing under the radical sign as well.

$$|\vec{p}| = \sqrt{3^2 + 5^2 + 7^2} = \sqrt{83} = 9.1104...$$

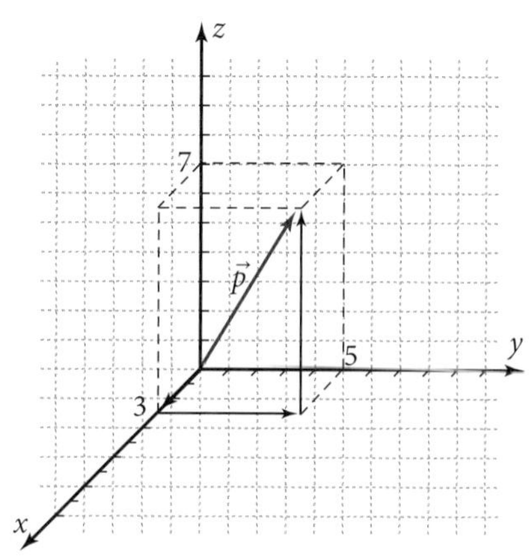

Figure 12-3c

EXAMPLE 2 ➤ Find the displacement vector from point $A(8, 2, 13)$ to point $B(3, 10, 4)$. Use the result to find the distance between the two points.

SOLUTION Sketch two points in three-dimensional coordinates as in Figure 12-3d. It is not necessary to draw them to scale. Draw position vectors $\vec{a}$ and $\vec{b}$ to the two points. The vector from A to B is equal to $\vec{b} - \vec{a}$.

$$\overrightarrow{AB} = \vec{b} - \vec{a} = (3\vec{i} + 10\vec{j} + 4\vec{k}) - (8\vec{i} + 2\vec{j} + 13\vec{k}) = (-5\vec{i} + 8\vec{j} - 9\vec{k})$$

Displacement vector from point A to point B.

The distance between A and B is the length of $\overrightarrow{AB}$.

$$|\overrightarrow{AB}| = \sqrt{(-5)^2 + 8^2 + (-9)^2} = \sqrt{170} = 13.0384... \text{ units}$$

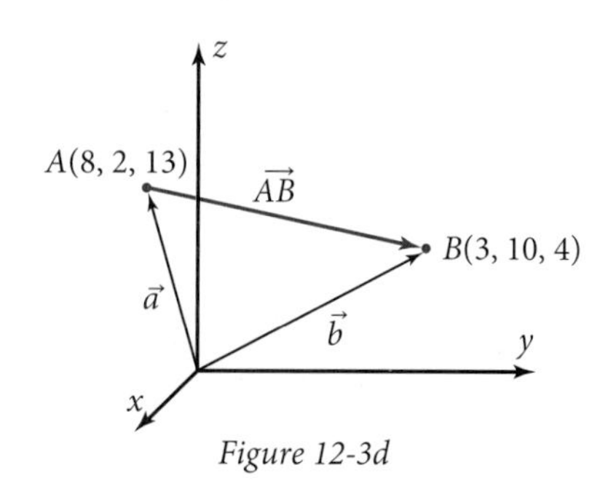

Figure 12-3d

EXAMPLE 3 ➤ Find the position vector to the point 70% of the way from point $A(8, 2, 13)$ to point $B(3, 10, 4)$ in Example 2. Write the coordinates of the point.

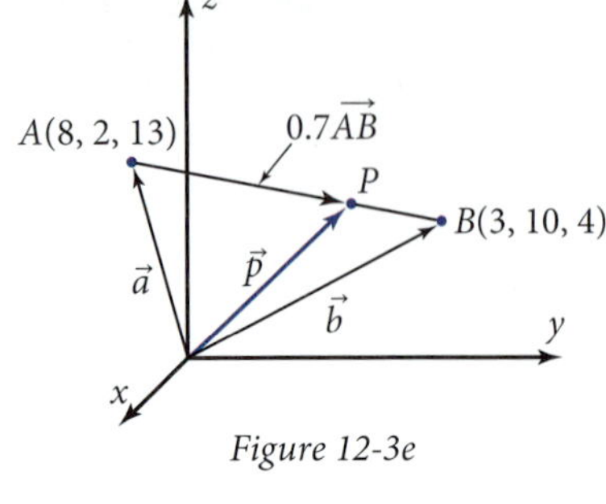

Figure 12-3e

SOLUTION On the diagram you drew for Example 2, sketch a vector starting at A and ending at a point 70% of the way from A to B, as in Figure 12-3e.

This vector, $0.7\overrightarrow{AB}$, added to $\vec{a}$ is the position vector of point P, as you can see from the fact that the two vectors are in position for adding (that is, head-to-tail). Using $\vec{p}$ for the position vector,

$$\begin{aligned}\vec{p} &= \vec{a} + 0.7\overrightarrow{AB} = 8\vec{i} + 2\vec{j} + 13\vec{k} + 0.7\left(-5\vec{i} + 8\vec{j} - 9\vec{k}\right)\\ &= 8\vec{i} + 2\vec{j} + 13\vec{k} + \left(-3.5\vec{i} + 5.6\vec{j} - 6.3\vec{k}\right)\\ &= 4.5\vec{i} + 7.6\vec{j} + 6.7\vec{k} \qquad \text{Position vector to point } P.\end{aligned}$$

The coordinates of point P are (4.5, 7.6, 6.7).

Problem Set 12-3

Reading Analysis

From what you have read in this section, what do you consider to be the main idea? When you sketch a vector in xyz-coordinates, why are the tick marks closer together on the x-axis than they are on the other two axes? How do you find the length of a three-dimensional vector? How do you find the displacement vector from one point to another point?

Quick Review

Q1. Where a vector starts is called its __?__, and where it ends is called its __?__.

Q2. A vector may be translated to another position without changing its __?__ or __?__.

Q3. How do you translate two vectors to add them geometrically?

Q4. When you have translated two vectors as in Problem Q3, the sum goes from __?__ to __?__.

Q5. How do you translate two vectors to subtract them geometrically?

Q6. When you have translated two vectors as in Problem Q5, the difference goes from __?__ to __?__.

Q7. Where does a position vector always start?

Q8. The absolute value of a vector, $|\vec{v}|$, is the same as its __?__.

Q9. True or false? The difference of two vectors is less than their sum.

Q10. What makes a vector a *unit* vector?

For Problems 1–4, draw the position vector on graph paper. Show the circumscribed "box" that makes the vector look three-dimensional. Write the coordinates of the point at the head of the vector.

1. $\vec{p} = 5\vec{i} + 9\vec{j} + 6\vec{k}$
2. $\vec{p} = 8\vec{i} + 2\vec{j} + 7\vec{k}$
3. $\vec{p} = 3\vec{i} + 8\vec{j} + 4\vec{k}$
4. $\vec{p} = 10\vec{i} + 7\vec{j} + 3\vec{k}$
5. Let $\vec{a} = 4\vec{i} + 2\vec{j} - 3\vec{k}$ and $\vec{b} = 7\vec{i} - 5\vec{j} + \vec{k}$.
   a. Find $\vec{a} + \vec{b}$, $\vec{a} - \vec{b}$, and $\vec{b} - \vec{a}$.
   b. Find $3\vec{a}$ and $6\vec{a} - 5\vec{b}$.
   c. Find $|\vec{a} + \vec{b}|$ and $|\vec{a}| + |\vec{b}|$. Does $|\vec{a} + \vec{b}|$ equal $|\vec{a}| + |\vec{b}|$?
   d. Find a unit vector in the direction of $\vec{b}$. Find a vector 20 units long in the direction of $\vec{b}$.

6. Let $\vec{c} = -4\vec{i} + 6\vec{j} + 3\vec{k}$ and $\vec{d} = 9\vec{i} + 8\vec{j} - 2\vec{k}$.

   a. Find $\vec{c} + \vec{d}$, $\vec{c} - \vec{d}$, and $\vec{d} - \vec{c}$.

   b. Find $-(\vec{c} + \vec{d})$ and $3\vec{c} - 4\vec{d}$.

   c. Find $|\vec{c}| + |\vec{d}|$ and $|\vec{c} + \vec{d}|$. Does $|\vec{c} + \vec{d}|$ equal $|\vec{c}| + |\vec{d}|$?

   d. Find a unit vector in the direction of $\vec{d}$. Find a vector 10 units long in the direction of the opposite of $\vec{d}$.

For Problems 7–10, find the indicated displacement vector. Use the answer to find the distance between the two points.

7. $\overrightarrow{RS}$ for points $R(5, 6, 12)$ and $S(8, 13, 6)$

8. $\overrightarrow{PQ}$ for points $P(6, 8, 14)$ and $Q(10, 16, 9)$

9. $\overrightarrow{BA}$ for points $A(9, 13, -4)$ and $B(3, 6, -10)$

10. $\overrightarrow{DC}$ for points $C(2, 9, 0)$ and $D(1, 4, 8)$

11. *Tree House Problem:* Elmer builds a tree house in his level backyard for the children to play in. He uses one corner of the yard as the origin of a three-dimensional coordinate system. The x- and y-axes run along the ground, and the z-axis is vertical. The tree house is at the point (30, 55, 17), where the dimensions are in feet.

   a. Sketch the coordinate axes and the point (30, 55, 17).

   b. Write the position vector $\vec{h}$ to the tree house. How high is the tree house above the ground? How far is the tree house from the origin?

   c. A wire is to be stretched from the tree house to the point (10, 0, 8) at the top of the back door of Elmer's house so that the children can slide messages down it. Write a vector representing the displacement from the tree house to the back door.

   d. How long will the wire in part c need to be?

   e. The children slide a message down the wire. It gets stuck when it is only 30% of the way from the tree house to the back door. Find a vector representing the displacement from the tree house to the stuck message. How far along the wire did the message go before it got stuck?

   f. Find the position vector of the stuck message. How high above the ground is it?

12. *Space Station Problem:* Two communications satellites are in geosynchronous orbit around Earth. (Geosynchronous satellites orbit Earth with a period of 1 day, so they don't appear to move with respect to an observer on the ground.) From a point on the ground, the position vectors to the two satellites are

   Satellite 1: $\vec{p}_1 = 18\vec{i} + 5\vec{j} + 12\vec{k}$

   Satellite 2: $\vec{p}_2 = 15\vec{i} + 9\vec{j} + 14\vec{k}$

   The distances are in thousands of miles. These vectors are shown schematically (not to scale) in Figure 12-3f. A space station is to be located at point P on the line between the two satellites, 40% of the way from Satellite 1 to Satellite 2.

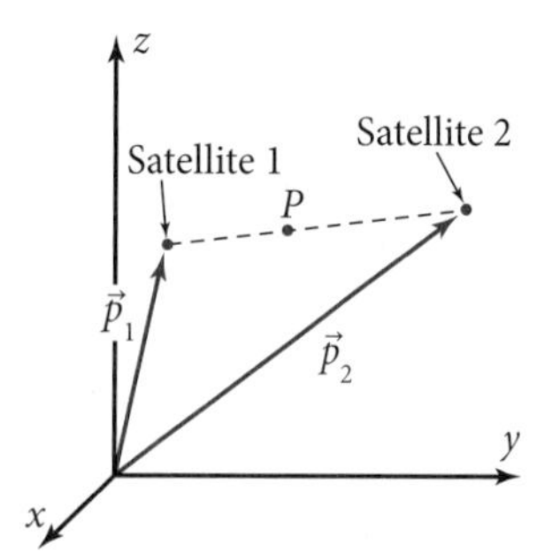

Figure 12-3f

   a. Find the displacement vector from Satellite 1 to Satellite 2.

   b. Find the position vector of point P, where the space station will be located.

   c. How far will the space station be from Satellite 1? How far will the space station be from the point on the ground?

For Problems 13–16, find the position vector of the indicated point.

13. $\frac{2}{3}$ of the way from (7, 8, 11) to (34, 32, 14)

14. $\frac{1}{3}$ of the way from (5, 1, 23) to (26, 13, 14)

15. 130% of the way from (2, 9, 7) to (4, −3, 1)

16. 270% of the way from (3, 8, 5) to (7, 1, −10)

17. *Perspective Problem:* Prove that if the x-axis is drawn obliquely on a piece of graph paper, as in Figure 12-3b, and two units are marked off for each grid line crossed, then the distances along the x-axis are about 70% of the distances along the y- and z-axes.

18. *Three-Dimensional Distances Problem:* Figure 12-3g shows point P with coordinates (x, y, z) at distance d from the origin. Segments x and y are the legs of a right triangle with hypotenuse a, and segments a and z are the legs of a right triangle with hypotenuse d. By applying the two-dimensional Pythagorean theorem to these two triangles, prove the *three-dimensional Pythagorean theorem:*

$$d = \sqrt{x^2 + y^2 + z^2}$$

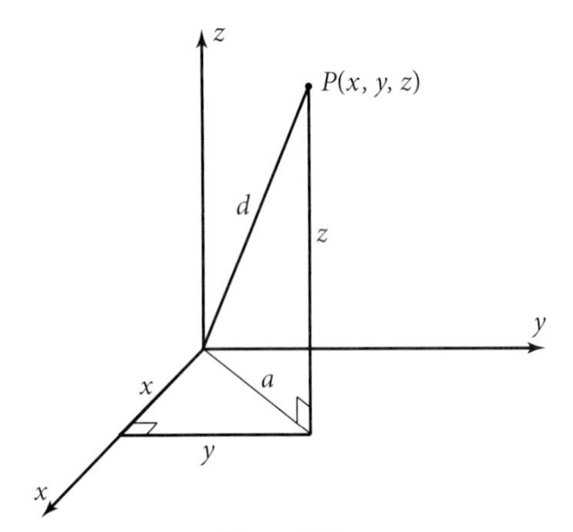

Figure 12-3g

19. *Four-Dimensional Vector Problem:* In Einstein's theory of time and space, time is a fourth dimension. Although it is impossible to draw a vector with more than three dimensions, the techniques you have learned make it possible to analyze them algebraically. The definitions and techniques for adding, subtracting, and finding lengths can be extended to higher-dimensional vectors. It is convenient to drop $\vec{i}$, $\vec{j}$, and $\vec{k}$ and use ordered quadruples, ordered quintuples, and so on to represent the vectors. Let

$$\vec{a} = (3, 5, 2, 7) \text{ and}$$
$$\vec{b} = (5, 11, 7, 1)$$

a. Find $|\vec{a}|$ and $|\vec{b}|$.

b. Find $\vec{a} + \vec{b}$.

c. Find $\vec{a} - \vec{b}$.

d. If $\vec{a}$ and $\vec{b}$ are considered to be position vectors, write the displacement vector from the head of $\vec{a}$ to the head of $\vec{b}$.

e. Find the position vector of the point 40% of the way from the head of $\vec{a}$ to the head of $\vec{b}$ in part d.

Albert Einstein (1879–1955) published the theory of special relativity in 1905 and the theory of general relativity in 1916. He received the Nobel Prize in physics in 1921. Hermann Minkowski (1864–1909) laid the mathematical foundations for Einstein's theory of relativity.

12-4 Scalar Products and Projections of Vectors

You have learned how to add and subtract two vectors and how to multiply a vector by a scalar. In this section you will learn about the **dot product** of two vectors, one way of multiplying vectors, which is useful for finding the measure of an angle between two vectors numerically and for determining the effect a vector such as a force vector has on the direction of a different vector. The dot product is also called the **scalar product** or **inner product,** for reasons you will see in this section.

Objective

Given two vectors, find their dot product. Use the result to find the measure of an angle between the vectors and the projection of one vector on the other.

Dot Products (Scalar or Inner Products)

The symbol for the dot product of $\vec{a}$ and $\vec{b}$ is $\vec{a} \cdot \vec{b}$. It is read "vector *a* dot vector *b*." If you translate the two vectors so that they are tail-to-tail, as in Figure 12-4a, you find the dot product by multiplying the magnitudes of the vectors and the cosine of the angle between them.

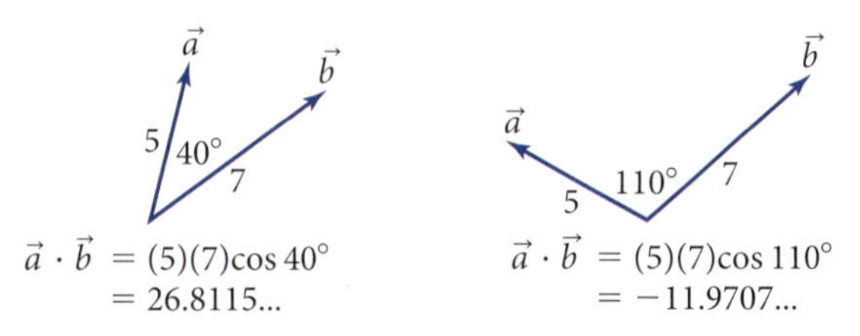

Figure 12-4a

Note that the answer is not a vector. It is a scalar. This is why the dot product is sometimes called the scalar product. Note also that if the angle between the vectors is acute, the dot product is positive. If the angle is obtuse, the dot product is negative. Three special cases are shown in Figure 12-4b. If the vectors point in the same direction, the angle between them has measure 0°. Because cos 0° equals 1, the dot product is the product of the two magnitudes. Similarly, if the vectors point in opposite directions, the angle has measure 180°. The dot product is the opposite of the product of the magnitudes because cos 180° equals −1. Finally, if the vectors are perpendicular, the dot product is zero because cos 90° equals 0.

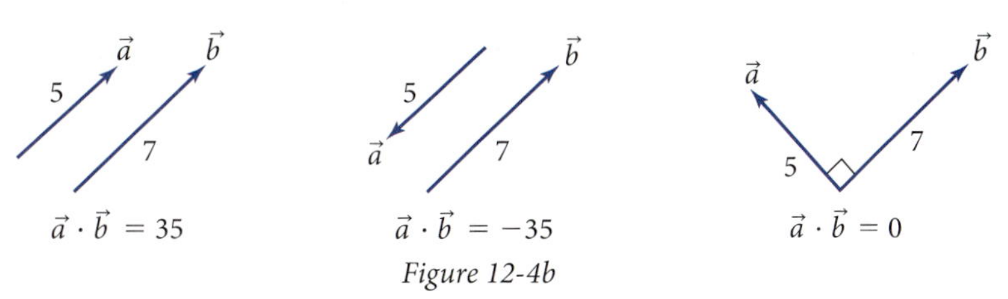

Figure 12-4b

Here is the formal definition of dot product.

DEFINITION: Dot Product (Scalar Product, Inner Product)

$$\vec{a} \cdot \vec{b} = |\vec{a}|\,|\vec{b}|\cos\theta$$

where θ is the angle between the two vectors when they are translated tail-to-tail (Figure 12-4c)

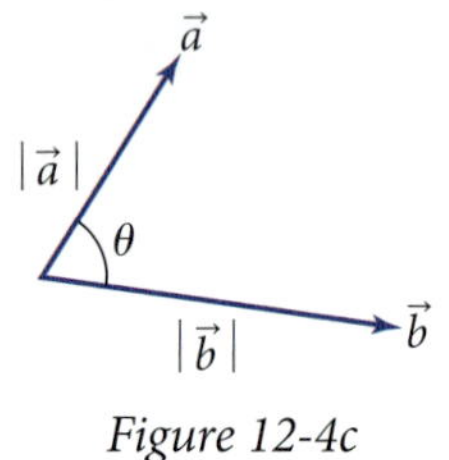

Figure 12-4c

Unfortunately, the definition is not very useful for finding dot products of three-dimensional vectors given by components because you don't know the angle between the vectors. So you must seek another way to do the calculation.

Let $\vec{a} = 2\vec{i} + 5\vec{j} + 7\vec{k}$.

Let $\vec{b} = 9\vec{i} + 3\vec{j} + 4\vec{k}$.

$\vec{a} \cdot \vec{b} = \left(2\vec{i} + 5\vec{j} + 7\vec{k}\right) \cdot \left(9\vec{i} + 3\vec{j} + 4\vec{k}\right)$ Substitute for the two vectors.

$$\begin{aligned} &= 18\vec{i} \cdot \vec{i} + 6\vec{i} \cdot \vec{j} + 8\vec{i} \cdot \vec{k} \\ &+ 45\vec{j} \cdot \vec{i} + 15\vec{j} \cdot \vec{j} + 20\vec{j} \cdot \vec{k} \\ &+ 63\vec{k} \cdot \vec{i} + 21\vec{k} \cdot \vec{j} + 28\vec{k} \cdot \vec{k} \end{aligned}$$

The dot product distributes over addition. Distribute each term in the first vector to each term in the second.

Note that $\vec{i}$ is a unit vector, so $\vec{i} \cdot \vec{i} = (1)(1)\cos 0° = 1$. The same is true for $\vec{j} \cdot \vec{j}$ and $\vec{k} \cdot \vec{k}$. However, $\vec{i} \cdot \vec{j} = (1)(1)\cos 90° = 0$. Each of the preceding dot products with perpendicular unit vectors is equal to zero. Therefore,

$$\begin{aligned} \vec{a} \cdot \vec{b} &= 18 + 0 + 0 \\ &+ 0 + 15 + 0 \\ &+ 0 + 0 + 28 \\ &= 61 \end{aligned}$$

The answer is 61, a scalar.

This calculation reveals a reason for calling the dot product the inner product. The numbers that contribute to the dot product are "inside" in the array shown above.

The calculation also reveals a quick way to find a dot product from its components.

$$\vec{a} \cdot \vec{b} = (2)(9) + (5)(3) + (7)(4) = 61$$

You multiply the x-coefficients, the y-coefficients, and the z-coefficients and then sum the products.

TECHNIQUE: Computation of Dot Product

If $\vec{a} = x_1\vec{i} + y_1\vec{j} + z_1\vec{k}$ and $\vec{b} = x_2\vec{i} + y_2\vec{j} + z_2\vec{k}$ then

$$\vec{a} \cdot \vec{b} = x_1x_2 + y_1y_2 + z_1z_2$$

Verbally: The dot product of two three-dimensional vectors equals the sum of the respective products of the coefficients of the $\vec{i}$, $\vec{j}$, and $\vec{k}$ unit vectors.

EXAMPLE 1 ➤ Find the dot product $\vec{c} \cdot \vec{d}$ if

$$\vec{c} = 4\vec{i} - 6\vec{j} + 9\vec{k}$$

$$\vec{d} = 2\vec{i} + 5\vec{j} - 3\vec{k}$$

SOLUTION

$$\vec{c} \cdot \vec{d} = (4)(2) + (-6)(5) + (9)(-3) = -49$$

To calculate the dot product, multiply x_1 by x_2, y_1 by y_2, z_1 by z_2, then add.

EXAMPLE 2 ➤ Use the dot product in Example 1 to find the measure of angle θ between $\vec{c}$ and $\vec{d}$.

SOLUTION

$\vec{c} \cdot \vec{d} = -49$ — From Example 1.

$|\vec{c}|\,|\vec{d}| \cos\theta = -49$ — Use the definition of dot product.

$|\vec{c}| = \sqrt{4^2 + (-6)^2 + 9^2} = \sqrt{133}$ — Find the lengths of $\vec{c}$ and $\vec{d}$.

$|\vec{d}| = \sqrt{2^2 + 5^2 + (-3)^2} = \sqrt{38}$

$\sqrt{133}\sqrt{38}\cos\theta = -49$ — Substitute for the magnitudes of the vectors.

$$\cos\theta = \frac{-49}{\sqrt{133}\sqrt{38}} = -0.6892...$$

$$\theta = 133.5709...^\circ$$

Projections of Vectors

Suppose you drag a box across the floor by pulling it with a force, $\vec{F}$, as shown in the photo. The component of $\vec{F}$ in the direction of the displacement influences things such as the change in speed of the box and the amount of work done by the force in moving the box. This component is called the **vector projection** of the force vector on the displacement vector.

Force vector, $\vec{F}$

Displacement vector

Force component (vector projection)

Figure 12-4d shows $\vec{p}$, the vector projection of $\vec{a}$ on $\vec{b}$, and the reason for the name "projection." Light rays shining perpendicular to $\vec{b}$ would "project" a shadow of $\vec{a}$ onto $\vec{b}$, corresponding to $\vec{p}$. To calculate the magnitude and direction of $\vec{p}$, note that its magnitude equals the magnitude of $\vec{a}$ times the cosine of θ, the angle between $\vec{a}$ and $\vec{b}$ when they are placed tail-to-tail. So $\vec{p}$ equals this magnitude times a unit vector in the direction of $\vec{b}$.

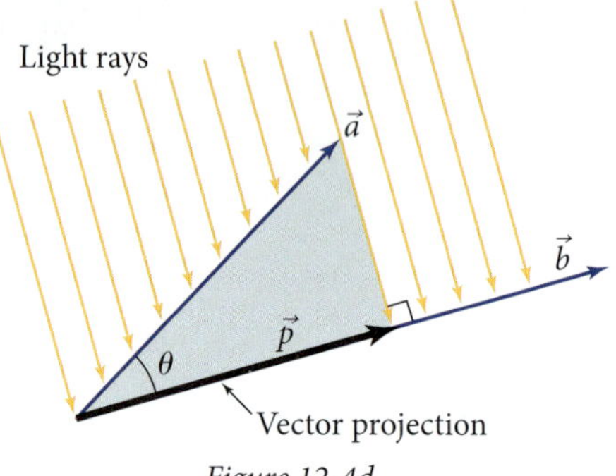

Figure 12-4d

$$\vec{p} = (|\vec{a}|\cos\theta)\frac{\vec{b}}{|\vec{b}|}$$

Example 3 shows you how to use this fact to find a vector projection.

EXAMPLE 3 ➤ If $\vec{b} = 8\vec{i} - 5\vec{j} + 3\vec{k}$ and $\vec{a}$ is 10 units long at an angle $\theta = 70°$ to $\vec{b}$, find $\vec{p}$, the vector projection of $\vec{a}$ on $\vec{b}$.

SOLUTION

$|\vec{b}| = \sqrt{8^2 + (-5)^2 + 3^2} = \sqrt{98}$ — Find the length of $\vec{b}$.

Unit vector $\vec{u} = \dfrac{\vec{b}}{|\vec{b}|} = \dfrac{8}{\sqrt{98}}\vec{i} - \dfrac{5}{\sqrt{98}}\vec{j} + \dfrac{3}{\sqrt{98}}\vec{k}$ — Find a unit vector in the direction of $\vec{b}$.

Length $= |\vec{a}|\cos\theta = 10\cos 70° = 3.4202...$ — Find the length of the vector projection of $\vec{a}$ on $\vec{b}$.

$\vec{p} = 3.4202...\left(\dfrac{8}{\sqrt{98}}\vec{i} - \dfrac{5}{\sqrt{98}}\vec{j} + \dfrac{3}{\sqrt{98}}\vec{k}\right)$ — Multiply the unit vector by the length.

$\vec{p} = 2.7639...\vec{i} - 1.7274...\vec{j} + 1.0364...\vec{k}$ ◀

The quantity $|\vec{a}|\cos\theta$ in Example 3 is called the **scalar projection** of $\vec{a}$ on $\vec{b}$. The letter p (without the vector symbol) will be used for the scalar projection.

$p = |\vec{a}|\cos\theta$ — Scalar projection of $\vec{a}$ on $\vec{b}$.

If angle θ is acute, the scalar projection equals the magnitude of $\vec{p}$. If angle θ is obtuse, the scalar projection is negative and thus is the opposite of the magnitude of $\vec{p}$. Figure 12-4e shows the two cases. If angle θ is obtuse, the vector projection points in the direction opposite $\vec{b}$.

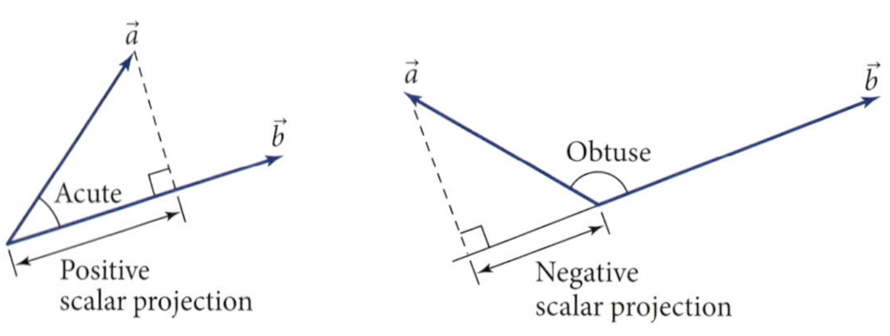

Figure 12-4e

DEFINITIONS: Projections of Vectors

If θ is the angle between $\vec{a}$ and $\vec{b}$ when they are placed tail-to-tail, then the **scalar projection** of $\vec{a}$ on $\vec{b}$ is

$$p = |\vec{a}| \cos\theta$$

If $\vec{u}$ is a unit vector in the direction of $\vec{b}$, then the **vector projection** of $\vec{a}$ on $\vec{b}$ is

$$\vec{p} = p\vec{u}$$

EXAMPLE 4 ➤ Consider $\vec{a}$ and $\vec{b}$, where $\vec{a} = -4\vec{i} + 5\vec{j} + 9\vec{k}$ and $\vec{b} = 6\vec{i} - 8\vec{j} + \vec{k}$.

a. Find the scalar projection of $\vec{a}$ on $\vec{b}$.

b. Find the vector projection of $\vec{a}$ on $\vec{b}$.

SOLUTION

a. $\vec{a} \cdot \vec{b} = (-4)(6) + (5)(-8) + (9)(1) = -55$

To find θ, first find the dot product and the two lengths.

$|\vec{a}| = \sqrt{(-4)^2 + 5^2 + 9^2} = \sqrt{122}$

$|\vec{b}| = \sqrt{6^2 + (-8)^2 + 1^2} - \sqrt{101}$

$\sqrt{122}\sqrt{101} \cos\theta = -55$

Use the definition of dot product.

$$\cos\theta = \frac{-55}{\sqrt{122}\sqrt{101}} = -0.4954...$$

$$\theta = 119.7011...°$$

$\therefore p = \sqrt{122} \cos 119.7011...° = -5.4727...$

Use the definition of scalar projection.

b. $\vec{u} = \dfrac{\vec{b}}{|\vec{b}|} = \dfrac{6\vec{i} - 8\vec{j} + \vec{k}}{\sqrt{101}}$

Find a unit vector in the direction of $\vec{b}$.

$\therefore \vec{p} = p\vec{u} = -5.4727...\vec{u} = -3.2673...\vec{i} + 4.3564...\vec{j} - 0.5445...\vec{k}$ ◄

Problem Set 12-4

Reading Analysis

From what you have read in this section, what do you consider to be the main idea? What is the definition of dot product, and why is it also called a scalar product? In what relatively simple way can you compute a dot product without having to find the measure of the angle between the two vectors? What is meant by the projection of one vector on another?

Quick Review

Q1. Give two names for the symbol $|\vec{a}|$.

Q2. What is the y-component of $\vec{v} = 3\vec{i} - 5\vec{j} + 2\vec{k}$?

Q3. Find the length of $\vec{v}$ in Problem Q2.

Q4. True or false? $1\vec{i} + 1\vec{j} + 1\vec{k}$ is a unit vector.

Q5. Find the position vector to the point (5, 8, 6).

Q6. Find the displacement vector from the point (5, 8, 6) to the point (11, 3, 7).

Q7. Find the coefficient of determination if $SS_{res} = 5$ and $SS_{dev} = 100$.

Q8. For the polar curve $r = 3 \cos \theta$, find r if $\theta = 2$ radians.

Q9. Which functions have constant second differences in y-values for equally spaced x-values?

Q10. Without using your calculator, find $\cos \pi$.

For Problems 1–6, use the definition of dot product to find $\vec{a} \cdot \vec{b}$, where θ is the measure of the angle between $\vec{a}$ and $\vec{b}$ when they are placed tail-to-tail.

1. $|\vec{a}| = 30$, $|\vec{b}| = 25$, and $\theta = 37°$

2. $|\vec{a}| = 17$, $|\vec{b}| = 8$, and $\theta = 23°$

3. $|\vec{a}| = 29$, $|\vec{b}| = 50$, and $\theta = 127°$

4. $|\vec{a}| = 40$, $|\vec{b}| = 53$, and $\theta = 126°$

5. $|\vec{a}| = 51$, $|\vec{b}| = 27$, and $\theta = 90°$

6. $|\vec{a}| = 43$, $|\vec{b}| = 29$, and $\theta = 180°$

For Problems 7–12, use the definition of dot product to find the measure of the angle between $\vec{a}$ and $\vec{b}$ if the two vectors are placed tail-to-tail.

7. $|\vec{a}| = 20$, $|\vec{b}| = 30$, and $\vec{a} \cdot \vec{b} = 100$

8. $|\vec{a}| = 8$, $|\vec{b}| = 9$, and $\vec{a} \cdot \vec{b} = 24$

9. $|\vec{a}| = 11$, $|\vec{b}| = 17$, and $\vec{a} \cdot \vec{b} = -123$

10. $|\vec{a}| = 300$, $|\vec{b}| = 500$, and $\vec{a} \cdot \vec{b} = -100{,}000$

11. $|\vec{a}| = 60$, $|\vec{b}| = 80$, and $\vec{a} \cdot \vec{b} = 4800$

12. $|\vec{a}| = 29$, $|\vec{b}| = 31$, and $\vec{a} \cdot \vec{b} = 0$

For Problems 13–18, find $\vec{a} \cdot \vec{b}$ and the measure of the angle between $\vec{a}$ and $\vec{b}$ when they are tail-to-tail.

13. $\vec{a} = 2\vec{i} + 5\vec{j} + 3\vec{k}$
$\vec{b} = 7\vec{i} - \vec{j} + 4\vec{k}$

14. $\vec{a} = 3\vec{i} + 2\vec{j} - 4\vec{k}$
$\vec{b} = 8\vec{i} + 5\vec{j} - 2\vec{k}$

15. $\vec{a} = -3\vec{i} + 5\vec{j} + 2\vec{k}$
$\vec{b} = 6\vec{i} - 3\vec{j} + \vec{k}$

16. $\vec{a} = 4\vec{i} - 3\vec{j} - 7\vec{k}$
$\vec{b} = \vec{i} + 5\vec{j} + 3\vec{k}$

17. $\vec{a} = 8\vec{i} + 9\vec{j} - 2\vec{k}$
$\vec{b} = 3\vec{i} - 4\vec{j} - 6\vec{k}$

18. $\vec{a} = \vec{i} + 3\vec{j} - 5\vec{k}$
$\vec{b} = -7\vec{i} + 4\vec{j} + \vec{k}$

19. *Sailboat Force Problem:* Two ropes from the sail of a sailboat are both attached to the same cleat on the deck. The force vectors created by the ropes are

$$\vec{F}_1 = 15\vec{i} + 70\vec{j} + 10\vec{k}$$
$$\vec{F}_2 = 30\vec{i} + 50\vec{j} + 5\vec{k}$$

where the forces are in pounds. The vectors are shown (not to scale) in Figure 12-4f.

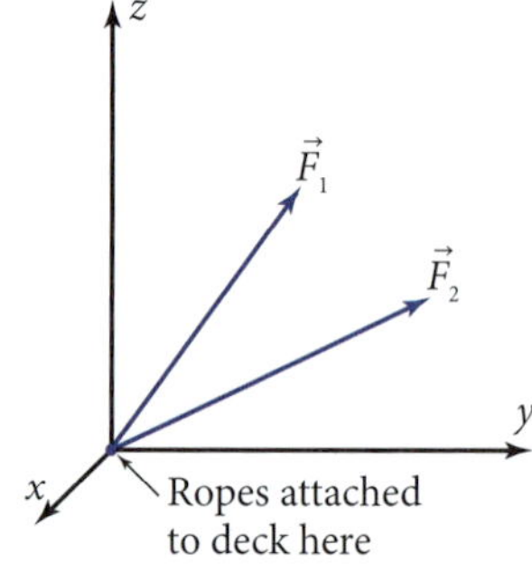

Figure 12-4f

a. Find the resultant force vector.

b. The y-axis runs along the length of the sailboat. The force in the y-direction is what makes the sailboat move forward. How many pounds does the resultant force exert in the y-direction?

c. The x-axis runs across the sailboat. The force in the x-direction makes the ship heel over. How many pounds does the resultant force exert in the x-direction?

d. The z-axis goes perpendicular to the deck. The force in the z-direction tends to pull the cleat out of the deck! How many pounds does the resultant force exert in the z-direction?

e. How many pounds, total, does the resultant force exert on the cleat?

f. What is the magnitude of each of the two forces? Do these magnitudes sum to the magnitude of the resultant force in part e?

g. Find the dot product of $\vec{F}_1$ and $\vec{F}_2$. Use the answer to find the measure of the angle the two forces make with each other.

20. *Hip Roof Problem:* A house is to be built with a hip roof. The triangular end of the roof is shown in Figure 12-4g. An xyz-coordinate system is set up with its origin at a bottom corner of the roof at the back of the house. The position vector $\vec{h}$ to the front bottom corner where the angle is marked and the position vector $\vec{v}$ to the peak of the roof are

$$\vec{h} = 20\vec{i} + 45\vec{j} + 0\vec{k}$$
$$\vec{v} = 10\vec{i} + 35\vec{j} + 8\vec{k}$$

The dimensions are in feet.

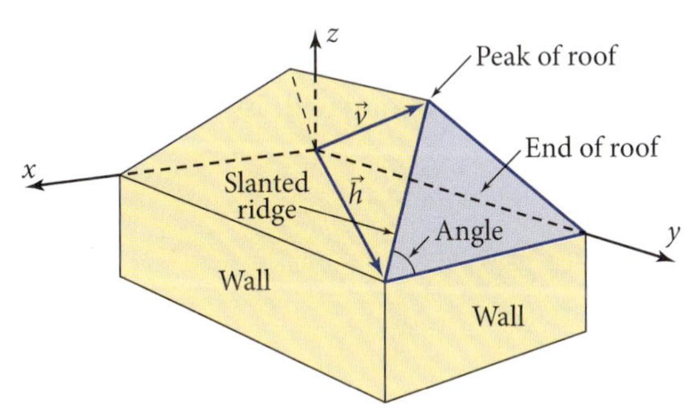

Figure 12-4g

a. The floor of the attic is along the xy-plane. How far will the peak of the roof be above this floor? How can you tell?

b. How long and how wide will the house be? How can you tell?

c. Builders need to know how long to make the rafter that goes from the corner with the marked angle to the peak of the roof along the slanted ridge. How long will the rafter be?

d. The roof tiles that come next to the slanted ridge will have to be cut at an angle, as shown in the figure. What measure will this angle have?

e. Is the triangular end of the roof an isosceles triangle? How can you tell?

21. *Dynamic Vector Projection Problem:* Go to **www.keymath.com/precalc** and complete the *Vector Projection* exploration in which you can drag the endpoints of $\vec{a}$ and $\vec{b}$. Use the buttons to view the vector projection of $\vec{a}$ on $\vec{b}$ and vice versa, as well as the dot products $\vec{a} \cdot \vec{b}$ and $\vec{b} \cdot \vec{a}$. Observe the effects on the vector projections and dot products of dragging the two vectors, and answer the questions asked. Then explain in writing what you learned from doing this activity.

22. In Figure 12-4h, $\vec{v}$ is 10 units long and makes a 28° angle with $\vec{a} = 7\vec{i} + 3\vec{j} + 4\vec{k}$. Find the vector projection of $\vec{v}$ on $\vec{a}$.

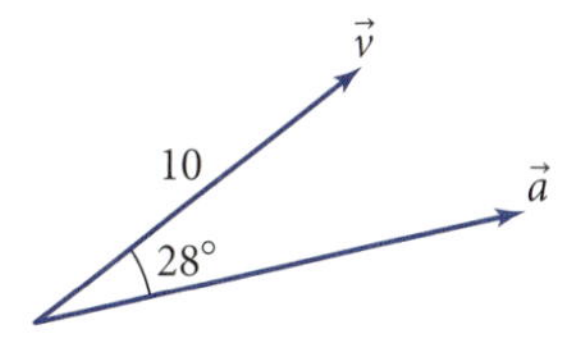

Figure 12-4h

23. In Figure 12-4i, $\vec{v}$ is 100 units long and makes a 145° angle with $\vec{b} = 50\vec{i} - 60\vec{j} + 40\vec{k}$. Find the vector projection of $\vec{v}$ on $\vec{b}$.

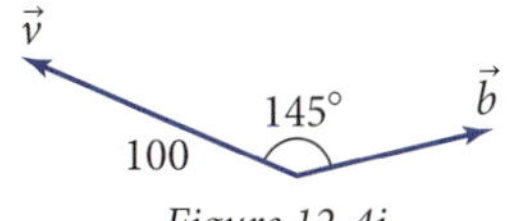

Figure 12-4i

24. *Shortcuts for Projections Problem:* Show that these formulas give the scalar and vector projections of $\vec{a}$ on $\vec{b}$.

> **TECHNIQUES: Formulas for Scalar and Vector Projections**
>
> The scalar projection of $\vec{a}$ on $\vec{b}$ is given by
>
> $$p = \frac{\vec{a} \cdot \vec{b}}{|\vec{b}|}$$
>
> The vector projection of $\vec{a}$ on $\vec{b}$ is given by
>
> $$\vec{p} = \frac{\vec{a} \cdot \vec{b}}{|\vec{b}|^2}\vec{b}$$

For Problems 25–28,

a. Find the scalar projection of $\vec{r}$ on $\vec{s}$.

b. Find the vector projection of $\vec{r}$ on $\vec{s}$.

25. $\vec{r} = \vec{i} + 4\vec{j} - 7\vec{k}$
$\vec{s} = 5\vec{i} - 2\vec{j} - 3\vec{k}$

26. $\vec{r} = 3\vec{i} + 2\vec{j} + 5\vec{k}$
$\vec{s} = 7\vec{i} - \vec{j} - 3\vec{k}$

27. $\vec{r} = 6\vec{i} - \vec{j} - 7\vec{k}$
$\vec{s} = \vec{i} - 5\vec{j} + 3\vec{k}$

28. $\vec{r} = 4\vec{i} - 3\vec{j} + 3\vec{k}$
$\vec{s} = -2\vec{i} + 5\vec{j} + \vec{k}$

29. *Vocabulary Problem:* Give three names commonly used for $\vec{a} \cdot \vec{b}$.

30. *Cube Problem:* Figure 12-4j shows a cube with one corner at the origin of a three-dimensional coordinate system.

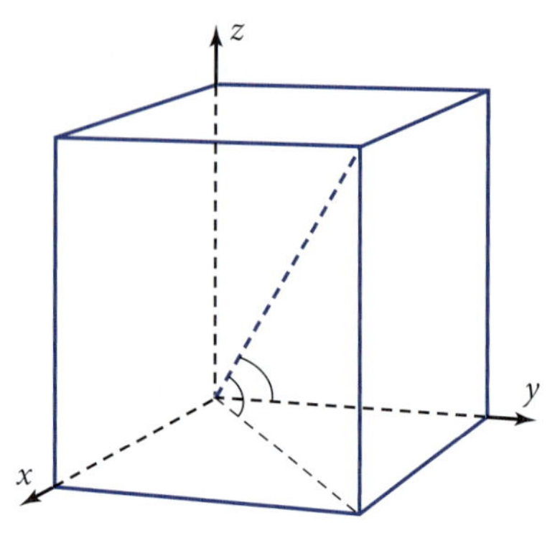

Figure 12-4j

a. Find the measure of the angle the main diagonal of the cube makes with one of the edges of the cube.

b. Find the measure of the angle the main diagonal makes with a diagonal in a face of the cube.

31. *Journal Problem:* Update your journal with what you have learned since the last entry. Include the definition, computational technique, and uses of the dot product, along with its two other names.

12-5 Planes in Space

Figure 12-5a shows the plane surface of a tilted underground rock formation. By finding an equation relating the x-, y-, and z-coordinates of points on this plane, you can calculate how far you would have to drill down to reach the plane. From algebra you may remember that the general equation for a plane in space has the form

$$5x + 7y - 4z = 19 \qquad \text{or, in general,} \qquad Ax + By + Cz = D$$

where x, y, and z are coordinates of a point on the plane and A, B, C, and D stand for constants. In this section you will see how to derive an equation for a plane from a vector **normal** (perpendicular) to the plane.

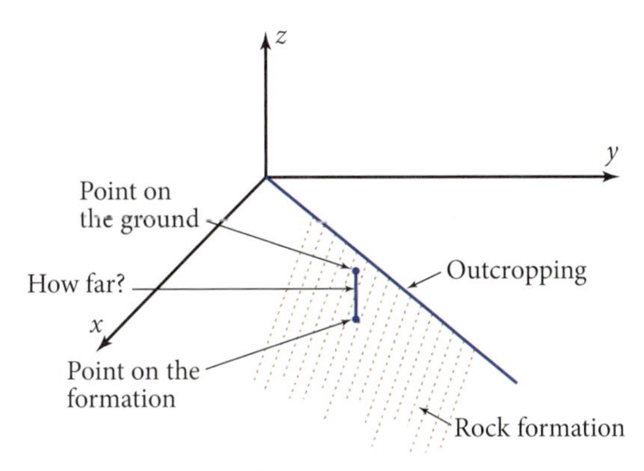

Figure 12-5a

Objective

Given a point in a plane and a vector perpendicular to the plane, find a particular equation for the plane and use it to find other points in the plane.

Equation for a Plane

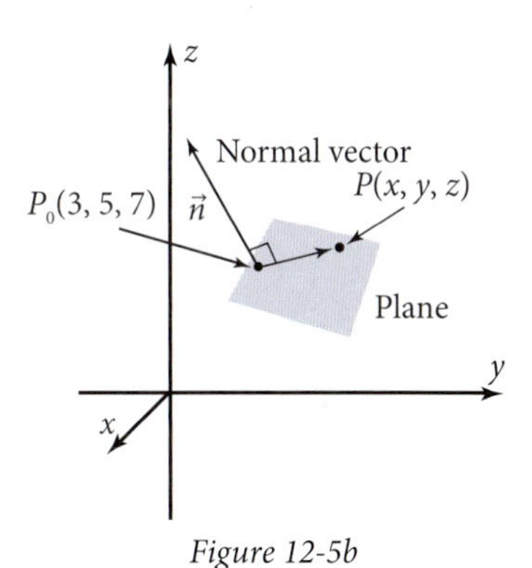

Figure 12-5b

Figure 12-5b shows a plane in space with a vector $\vec{n}$ normal to it. Suppose that

$$\vec{n} = 11\vec{i} + 2\vec{j} + 13\vec{k}$$

and that point P_0 in the plane has coordinates (3, 5, 7). Point $P(x, y, z)$ is a variable point in the plane. You need to find an equation for the plane relating x, y, and z.

The displacement vector $\overrightarrow{P_0P}$ goes from P_0 to P. Subtracting coordinates, this vector is

$$\overrightarrow{P_0P} = (x - 3)\vec{i} + (y - 5)\vec{j} + (z - 7)\vec{k}$$

Because $\overrightarrow{P_0P}$ is in the plane and $\vec{n}$ is perpendicular to the plane, the dot product of these two vectors equals zero. Some calculations lead to an equation of the plane.

$$\vec{n} \cdot \overrightarrow{P_0P} = 0$$

$$\left(11\vec{i} + 2\vec{j} + 13\vec{k}\right) \cdot \left[(x-3)\vec{i} + (y-5)\vec{j} + (z-7)\vec{k}\right] = 0$$

Substitute for the two vectors.

$$11(x-3) + 2(y-5) + 13(z-7) = 0$$

Evaluate the dot product.

$$11x + 2y + 13z - 134 = 0$$

$$11x + 2y + 13z = 134$$

This is an equation of the plane in space.

By comparing the answer with the given information, you see that the coefficients in the particular equation for the plane are the same as the coefficients of the given normal vector. This is true in general, and you can use this property to find an equation quickly, as shown in Example 1. (In Problems 15 and 16, you will prove the property.)

Notice that if the plane contains the origin, (0, 0, 0), then the equation in the third step from the end reduces to

$$11x + 2y + 13z = 0$$

The equation $11(x-3) + 2(y-5) + 13(z-7) = 0$ indicates a translation of this plane by 3 units in the x-direction, 5 units in the y-direction, and 7 units in the z-direction.

EXAMPLE 1 ➤ Find an equation for the plane containing the point (3, 5, 7) with normal vector $\vec{n} = 11\vec{i} + 2\vec{j} + 13\vec{k}$.

SOLUTION

$$11x + 2y + 13z = D$$

Substitute the coefficients of the components of $\vec{n}$ into the equation $Ax + By + Cz = D$.

$$11(3) + 2(5) + 13(7) = D$$

Substitute the given point for (x, y, z).

$$134 = D$$

$\therefore$ an equation for the plane is $11x + 2y + 13z = 134$. ◀

EXAMPLE 2 ➤ Find a vector $\vec{n}$ normal to the plane $7x - 3y + 8z = -51$.

SOLUTION

$$\vec{n} = 7\vec{i} - 3\vec{j} + 8\vec{k}$$ ◀

Note that any nonzero multiple of $\vec{n}$ is also a solution to Example 2. In particular, the opposite of $\vec{n}$, $-7\vec{i} + 3\vec{j} - 8\vec{k}$, is also a normal vector to the plane $7x - 3y + 8z = -51$.

EXAMPLE 3 ➤ Find an equation for the plane perpendicular to the segment connecting points $P_1(3, 8, -2)$ and $P_2(7, -1, 6)$ and passing through the point 30% of the way from point P_1 to point P_2. Figure 12-5c illustrates the problem in general.

SOLUTION The displacement vector $\overrightarrow{P_1P_2}$ is normal to the plane, so you can write

$$\begin{aligned}\vec{n} &= (7-3)\vec{i} + (-1-8)\vec{j} + [6-(-2)]\vec{k} \\ &= 4\vec{i} - 9\vec{j} + 8\vec{k}\end{aligned}$$

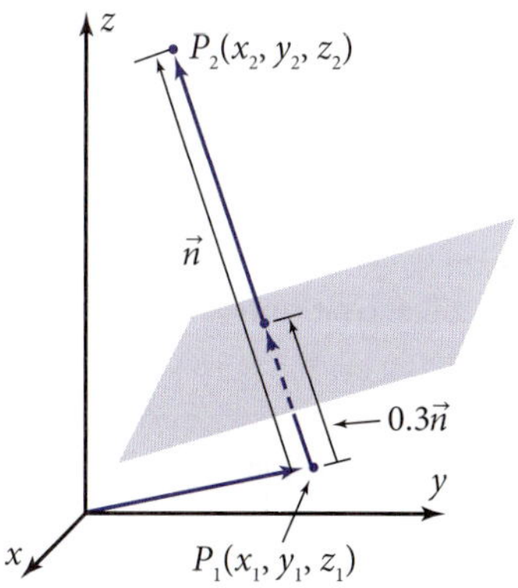

The position vector $\vec{p}$ to the point on the plane equals the position vector to point P_1 plus 0.3 times the normal vector $\vec{n}$.

$$\begin{aligned}\vec{p} &= \left(3\vec{i} + 8\vec{j} - 2\vec{k}\right) + 0.3\left(4\vec{i} - 9\vec{j} + 8\vec{k}\right) \\ &= 4.2\vec{i} + 5.3\vec{j} + 0.4\vec{k}\end{aligned}$$

Figure 12-5c

Thus, a point in the plane is (4.2, 5.3, 0.4).

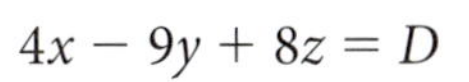

$$4x - 9y + 8z = D$$

The coefficients in the plane's equation are the coefficients of the components of the normal vector.

$$4(4.2) - 9(5.3) + 8(0.4) = D$$

$$-27.7 = D$$

$\therefore$ an equation for the plane is $4x - 9y + 8z = -27.7$. ◄

This box summarizes the technique for finding an equation of a plane in space.

TECHNIQUE: Equation for a Plane in Space

1. Use the given information to find a normal vector and a point in the plane.
2. Substitute the coefficients of the components of the normal vector for A, B, and C in the general equation.

   $Ax + By + Cz = D$
3. Substitute the coordinates of the given point for (x, y, z) to calculate the value of D.
4. Write the equation.

EXAMPLE 4 ➤ If an equation for a plane is $-7x + 8y + 4z = 200$, find the z-coordinate of point $P(3, 5, z)$ in the plane.

SOLUTION

$$-7(3) + 8(5) + 4z = 200$$

Substitute 3 for x and 5 for y in the equation.

$$4z = 181$$

$$z = 45.25$$ ◄

Problem Set 12-5

Reading Analysis

From what you have read in this section, what do you consider to be the main idea? What is the general equation for a plane in space, and how are the coefficients in this equation related to a vector normal to the plane? Why does the dot product of a vector in a plane and a vector normal to that plane always equal zero?

Quick Review

Q1. Give a name for $\vec{a} \cdot \vec{b}$.

Q2. Give a second name for $\vec{a} \cdot \vec{b}$.

Q3. Give the third name for $\vec{a} \cdot \vec{b}$.

Q4. How can you find the scalar projection of $\vec{a}$ on $\vec{b}$?

Q5. If p is the scalar projection of $\vec{a}$ on $\vec{b}$ and $\vec{u}$ is a unit vector in the direction of $\vec{b}$, what does the vector projection of $\vec{a}$ on $\vec{b}$ equal?

Q6. How do you tell from the dot product whether two vectors are perpendicular?

Q7. If $\vec{a} = 3\vec{i} + 2\vec{j} + 1\vec{k}$ and $\vec{b} = 4\vec{i} - 3\vec{j} - 5\vec{k}$, find $\vec{a} \cdot \vec{b}$.

Q8. Find the supplementary angle of 84°.

Q9. Find the supplementary angle of an angle with measure 1 radian.

Q10. What does $\cos^2 A$ equal in terms of $\sin A$?

For Problems 1 and 2, find two normal vectors to the plane, pointing in opposite directions.

1. $3x + 5y - 7z = -13$ **2.** $4x - 7y + 2z = 9$

For Problems 3–8, find a particular equation for the plane described.

3. Perpendicular to $\vec{n} = 3\vec{i} - 5\vec{j} + 4\vec{k}$, containing the point $(6, -7, -2)$

4. Perpendicular to $\vec{n} = -\vec{i} + 3\vec{j} - 2\vec{k}$, containing the point $(4, 7, 5)$

5. Perpendicular to the line segment connecting the points $(3, 8, 5)$ and $(11, 2, -3)$ and passing through the midpoint of the segment

6. Parallel to the plane $3x - 7y + 2z = 11$ and containing the point $(8, 11, -3)$

7. Parallel to the plane $5x - 3y - z = -4$ and containing the point $(4, -6, 1)$

8. Perpendicular to $\vec{n} = 4\vec{i} + 3\vec{j} - 2\vec{k}$ and having x-intercept 5 (The *x-intercept of a plane* is the value of x when the other two variables equal zero.)

9. A plane has the equation $3x - 7y + 5z = 54$. Points $P_1(6, 2, z_1)$ and $P_2(4, -3, z_2)$ are on the plane. Find the z-coordinates of the two points. How far apart are the points? What is the y-intercept of the plane (the value of y when the other two variables equal zero)?

10. A plane has the equation $4x + 2y - 10z = 300$. Points $P_1(x_1, 4, 5)$ and $P_2(7, y_2, 8)$ are on the plane. Find x_1 and y_2. Calculate the distance between the two points. What is the z-intercept of the plane (the value of z when the other two variables equal zero)?

11. *Geology Problem:* Figure 12-5d shows an underground rock formation that slants up and outcrops at ground level along a line in a field. The x- and y-axes run along perpendicular fence lines. A point on the outcropping has coordinates (200, 300, 0), where distances are in meters. A vector normal to the plane of the underground formation is

$$\vec{n} = 30\vec{i} - 17\vec{j} + 11\vec{k}$$

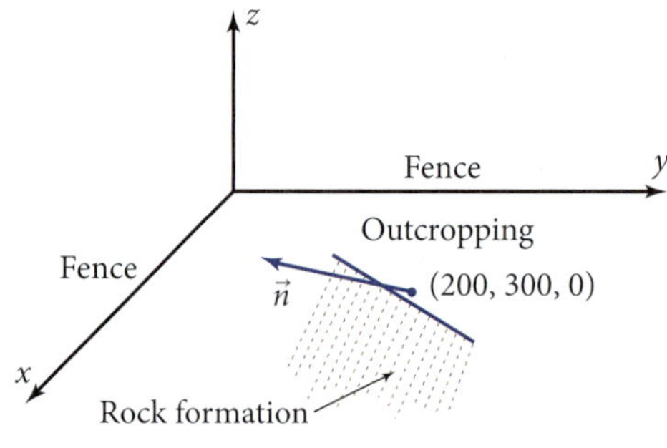

Figure 12-5d

a. Find an equation for the plane surface of the underground rock formation.

b. If you follow the outcropping line to the fences, which axis will it cross first, the x-axis or the y-axis? At what point will it meet the fence?

c. If a well is drilled vertically starting at the point (70, 50, 0), how deep will it be when it first encounters the rock formation?

d. The angle between the plane of the rock formation and the plane of the ground is called a *dihedral angle*. Geologists call this angle the *dip* of the formation. It is equal to the angle between the normal vectors to the two planes or to the supplement of this angle. Find the acute dip angle.

12. *Roof Valley Problem:* Figure 12-5e shows an L-shaped house that is to be built. Roof 1 and Roof 2 will have normal vectors

$$\vec{n}_1 = 0\vec{i} + 6\vec{j} + 12\vec{k}$$
$$\vec{n}_2 = 6\vec{i} + 0\vec{j} + 12\vec{k}$$

The two roofs will meet at a "valley." Point (30, 30, 10) is at the lower end of the valley. The dimensions are in feet.

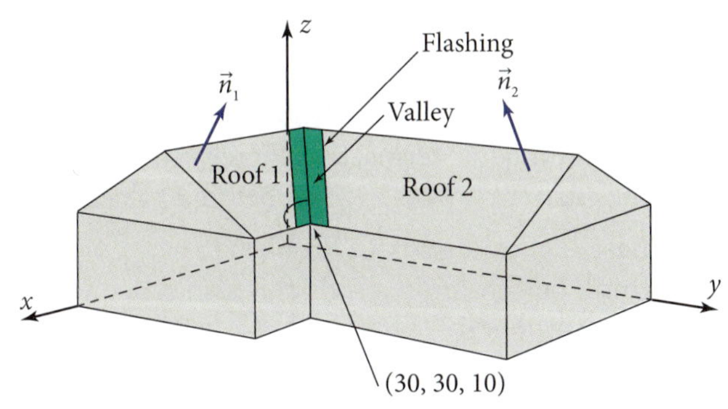

Figure 12-5e

a. Find particular equations for the two roof planes.

b. The top end of the valley is at the point $(15, 15, z)$. Use the equation for Roof 1 to calculate the value of z. Show that the point satisfies the equation for Roof 2, and give the real-world meaning of this fact.

c. How high will the ridge of the roof rise above the top of the walls?

d. Write the displacement vector from the bottom of the valley to the top.

e. Find the measure of the obtuse angle the valley makes with the bottom edge of Roof 1.

f. A piece of sheet metal flashing is to be fitted into the valley to go underneath the shingles. How long will the valley be?

g. The two roof sections form a dihedral angle equal to the angle between the two normal vectors or to the supplement of this angle. The flashing must be bent to fit this angle. Calculate the obtuse dihedral angle between the two roof sections.

h. Why do you think builders put flashing in roof valleys?

13. Prove that these two planes are perpendicular.

$$2x - 5y + 3z = 10$$
$$7x + 4y + 2z = 17$$

14. Find the value of A that makes these two planes perpendicular.

$$Ax + 3y - 2z = -8$$
$$4x - 5y + z = 7$$

15. *Plane's Equation Proof Problem:* Prove that if $\vec{n} = A\vec{i} + B\vec{j} + C\vec{k}$ is a normal vector to a plane, then a particular equation for the plane is $Ax + By + Cz = D$, where D stands for a constant.

16. *Normal Vector Proof Problem:* Prove the converse of the property in Problem 15. Specifically, prove that if $Ax + By + Cz = D$, where D stands for a constant, then a normal vector to the plane is $\vec{n} = A\vec{i} + B\vec{j} + C\vec{k}$.

12-6 Vector Product of Two Vectors

The dot product of two vectors is a scalar. In this section you will learn about the **cross product** of $\vec{a}$ and $\vec{b}$, written $\vec{a} \times \vec{b}$ and read "vector *a* cross vector *b*." In this course you will learn some geometric uses of the cross product. In later courses cross products are used in fields such as alternating electric current theory and accelerated rotary motion.

Objective

Be able to calculate cross products of two vectors and use cross products for geometric computations.

In this exploration you will discover the meaning of and a way of computing the cross product of two vectors.

EXPLORATION 12-6: Introduction to the Cross Product

Let $\vec{a} = 2\vec{i} + 3\vec{j} + 6\vec{k}$.
Let $\vec{b} = 4\vec{i} + 5\vec{j} + 20\vec{k}$.

The vector $\vec{c} = 30\vec{i} - 16\vec{j} - 2\vec{k}$ is the cross product of $\vec{a}$ and $\vec{b}$, written $\vec{a} \times \vec{b}$.

1. Find $|\vec{a}|$, $|\vec{b}|$, and $|\vec{a} \times \vec{b}|$. Does the length of the cross product vector equal the product of the lengths of the two factors?
2. Find the measure of the angle θ between $\vec{a}$ and $\vec{b}$ when they are placed tail-to-tail. Show that

$$|\vec{a} \times \vec{b}| = |\vec{a}|\,|\vec{b}|\sin\theta$$

3. Find $(\vec{a} \times \vec{b}) \cdot \vec{a}$ and $(\vec{a} \times \vec{b}) \cdot \vec{b}$. From the answers, what do you conclude about the direction of the cross product with respect to the direction of the two vectors being cross multiplied?
4. Look up the right-hand rule later in this section. When you understand it, demonstrate it to a classmate.
5. Read the formal definition of *cross product* on the next page and confirm that it is consistent with your answers to Problems 1–3.
6. Explain why $\vec{i} \times \vec{i}$, $\vec{j} \times \vec{j}$, and $\vec{k} \times \vec{k}$ all equal the zero vector.
7. Explain why $\vec{i} \times \vec{j} = \vec{k}$, $\vec{j} \times \vec{k} = \vec{i}$, and $\vec{k} \times \vec{i} = \vec{j}$.
8. Explain why $\vec{j} \times \vec{i}$ is the *opposite* of $\vec{i} \times \vec{j}$.
9. Calculate $\vec{a} \times \vec{b}$ by distributing each term in the first factor to each term in the second factor. Write the nine terms in the answer on three lines, as you did the first time you calculated a dot product. Simplify the result by taking advantage of the results of Problems 6, 7, and 8, but leave the nine terms in the same three lines.
10. Simplify the answer to Problem 9 by combining like terms, and thereby show that the answer really is $\vec{c}$ given before Problem 1.
11. Based on your work in Problem 9, why do you think that the cross product is also called the outer product?
12. Why is the cross product also called the vector product?
13. The coefficients of $\vec{a}$ and $\vec{b}$ can be written as a determinant with the three unit vectors as the top row, like this:

$$\begin{vmatrix} \vec{i} & \vec{j} & \vec{k} \\ 2 & 3 & 6 \\ 4 & 5 & 20 \end{vmatrix}$$

Evaluate this determinant, thereby showing that the answer is $\vec{a} \times \vec{b}$. See page 635 if you need help expanding the determinant.

continued

14. Find $\left(4\vec{i} - 5\vec{j} + 8\vec{k}\right) \times \left(3\vec{i} + 6\vec{j} - 7\vec{k}\right)$ quickly.

15. What did you learn as a result of doing this exploration that you did not know before?

Here is the formal definition of cross product.

DEFINITION: Cross Product (Vector Product, Outer Product)

The cross product of two vectors, $\vec{a} \times \vec{b}$, is a vector with these properties:

1. $\vec{a} \times \vec{b}$ is perpendicular to the plane containing $\vec{a}$ and $\vec{b}$.
2. The magnitude of $\vec{a} \times \vec{b}$ is

$$\left|\vec{a} \times \vec{b}\right| = \left|\vec{a}\right| \left|\vec{b}\right| \sin\theta$$

   where θ is the angle between the two vectors when they are placed tail-to-tail.
3. The direction of $\vec{a} \times \vec{b}$ is determined by the "right-hand rule." Put the fingers of your right hand so that they curl in the shortest direction *from* the *first* vector *to* the *second* vector. The cross product is in the same direction your thumb points (Figure 12-6a).

Note that the right-hand rule leads you to conclude that $\vec{b} \times \vec{a}$ is the *opposite* of $\vec{a} \times \vec{b}$. As shown in Figure 12-6b, curling your fingers from $\vec{b}$ to $\vec{a}$ makes your thumb point in the opposite direction. So cross multiplication of vectors isn't commutative. The cross product $\vec{b} \times \vec{a}$ equals $-\left(\vec{a} \times \vec{b}\right)$, not $\vec{a} \times \vec{b}$.

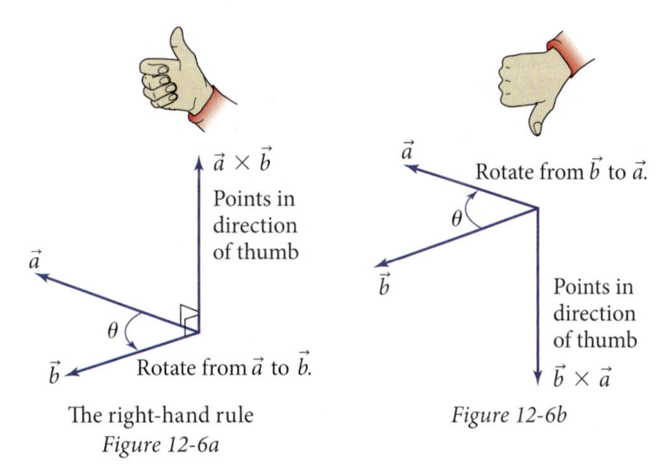

The right-hand rule
Figure 12-6a

Figure 12-6b

Computation of Cross Products from the Definition

The unit vectors $\vec{i}$, $\vec{j}$, and $\vec{k}$ have special properties when they are cross multiplied. Because the angle between a vector and itself is 0° and because $\sin 0° = 0$, the cross product of a vector by itself is the zero vector. For instance, $\vec{i} \times \vec{i} = \vec{0}$. Also, $\vec{j} \times \vec{j} = \vec{0}$ and $\vec{k} \times \vec{k} = \vec{0}$.

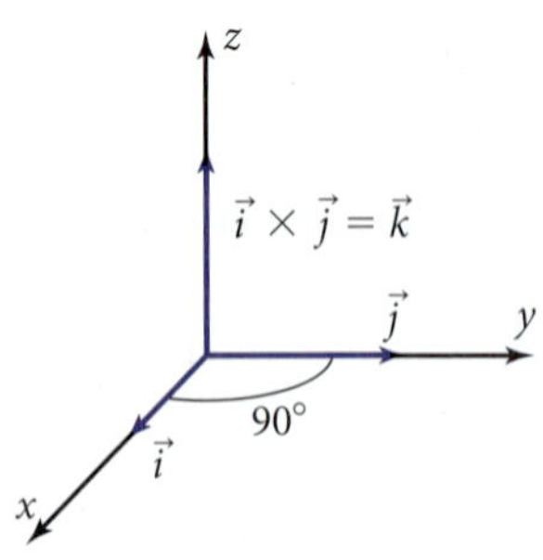

Figure 12-6c

Figure 12-6c shows that $\vec{i} \times \vec{j} = \vec{k}$. Because $\vec{i}$ and $\vec{j}$ are perpendicular,

$$|\vec{i} \times \vec{j}| = |\vec{i}||\vec{j}|\sin 90° = (1)(1)(1) = 1$$

By the right-hand rule, as $\vec{i}$ rotates toward $\vec{j}$, your thumb points in the z-direction. So $\vec{i} \times \vec{j} = \vec{k}$, a unit vector in the z-direction. Similarly, $\vec{j} \times \vec{k} = \vec{i}$ and $\vec{k} \times \vec{i} = \vec{j}$. These and other special cases are summarized in the box.

PROPERTIES: Cross Products of the Unit Coordinate Vectors

$\vec{i} \times \vec{i} = \vec{0}$ $\quad$ $\vec{i} \times \vec{j} = \vec{k}$ $\quad$ $\vec{j} \times \vec{i} = -\vec{k}$

$\vec{j} \times \vec{j} = \vec{0}$ $\quad$ $\vec{j} \times \vec{k} = \vec{i}$ $\quad$ $\vec{k} \times \vec{j} = -\vec{i}$

$\vec{k} \times \vec{k} = \vec{0}$ $\quad$ $\vec{k} \times \vec{i} = \vec{j}$ $\quad$ $\vec{i} \times \vec{k} = -\vec{j}$

Note that you can remember the cross products in the middle column because they involve $\vec{i}$ to $\vec{j}$ to $\vec{k}$ and $\vec{k}$ back to $\vec{i}$, in alphabetical order. This fortunate memory aid occurs because of the *right-handed coordinate system* that is being used. It is for this reason that the x-axis is shown going to the left and the y-axis is shown going to the right, rather than the more intuitive but less useful way of drawing the x-axis to the right. Note also that if you reverse the order, as in the farthest right column in the box, the cross products are the opposites of the unit vectors.

With these special cases in mind, you can use algebra to compute a cross product.

EXAMPLE 1 ➤ Find $\vec{a} \times \vec{b}$ if

$$\vec{a} = 3\vec{i} + 5\vec{j} + 7\vec{k} \quad \text{and} \quad \vec{b} = 11\vec{i} + 2\vec{j} + 13\vec{k}$$

SOLUTION

$\vec{a} \times \vec{b} = (3\vec{i} + 5\vec{j} + 7\vec{k}) \times (11\vec{i} + 2\vec{j} + 13\vec{k})$ — Substitute for $\vec{a}$ and $\vec{b}$.

$= 33\vec{i} \times \vec{i} + 6\vec{i} \times \vec{j} + 39\vec{i} \times \vec{k} + 55\vec{j} \times \vec{i} + 10\vec{j} \times \vec{j} + 65\vec{j} \times \vec{k} + 77\vec{k} \times \vec{i} + 14\vec{k} \times \vec{j} + 91\vec{k} \times \vec{k}$ — The cross product distributes over addition. Cross each term in the first vector with each term in the second vector.

$= \vec{0} + 6\vec{k} - 39\vec{j} - 55\vec{k} + \vec{0} + 65\vec{i} + 77\vec{j} - 14\vec{i} + \vec{0}$ — Use the special cross products.

$= 51\vec{i} + 38\vec{j} - 49\vec{k}$ — Combine like terms. ◄

Notice that there are zeros down the main diagonal in the next-to-last step of Example 1. The only terms that contribute to the cross product are the "outer" terms in the array, leading to the name *outer product* for cross product. The name *vector product* is also used because the result is a vector.

Computation of Cross Products by Means of Determinants

The method for computing cross products in Example 1 can seem tedious because you must make sure you have the correct unit vector when you calculate the cross product of unit vectors. Fortunately, there is a more easily remembered technique.

This square array of numbers, formed using the vectors from Example 1, is called a third-order **determinant.** You form the determinant by writing the three unit vectors along the top row, the coefficients of the first vector in the middle row, and the coefficients of the second vector in the bottom row.

$$\begin{vmatrix} \vec{i} & \vec{j} & \vec{k} \\ 3 & 5 & 7 \\ 11 & 2 & 13 \end{vmatrix}$$

You may have encountered determinants in algebra in conjunction with inverting a matrix or solving a system of linear equations.

To expand this determinant along the top row, you find the first term by writing the top-left element, $\vec{i}$, mentally crossing out its row and its column, and multiplying by the second-order determinant that remains. You do the same for $\vec{j}$ and $\vec{k}$. The signs of the expanded determinant alternate, so the second term has a $-$ sign and the third term has a $+$ sign.

$$\vec{i}\begin{vmatrix} 5 & 7 \\ 2 & 13 \end{vmatrix} - \vec{j}\begin{vmatrix} 3 & 7 \\ 11 & 13 \end{vmatrix} + \vec{k}\begin{vmatrix} 3 & 5 \\ 11 & 2 \end{vmatrix}$$

Remember the $-$ sign for the middle term!

To expand the second-order determinants, multiply the top-left and bottom-right numbers, and then subtract the product of the top-right and bottom-left numbers.

$$\vec{i}\,[(5)(13) - (7)(2)] - \vec{j}\,[(3)(13) - (7)(11)] + \vec{k}\,[(3)(2) - (5)(11)]$$
$$= 51\vec{i} + 38\vec{j} - 49\vec{k}$$

This is the same as the cross product found in Example 1.

Geometric Applications of Cross Products

You can use the cross product to find a vector normal to two other vectors. You can use the result to find the equation of the plane containing the two vectors, as you did in Section 12-5. Example 2 on the next page shows you how.

EXAMPLE 2 ➤ Find a particular equation for the plane containing the points $P_1(-5, 5, 5)$, $P_2(-3, 2, 7)$, and $P_3(1, 12, 6)$.

SOLUTION Sketch a plane and the three points, as in Figure 12-6d. It helps to show the coordinates of the points on your sketch.

Figure 12-6d

Write the displacement vectors from one point to the other two.

$$\overrightarrow{P_1P_2} = (-3+5)\vec{i} + (2-5)\vec{j} + (7-5)\vec{k} = 2\vec{i} - 3\vec{j} + 2\vec{k}$$

$$\overrightarrow{P_1P_3} = (1+5)\vec{i} + (12-5)\vec{j} + (6-5)\vec{k} = 6\vec{i} + 7\vec{j} + \vec{k}$$

Find the cross product to find a normal vector, $\vec{n}$.

$$\vec{n} = \overrightarrow{P_1P_2} \times \overrightarrow{P_1P_3} = \begin{vmatrix} \vec{i} & \vec{j} & \vec{k} \\ 2 & -3 & 2 \\ 6 & 7 & 1 \end{vmatrix}$$

$$= \vec{i}\begin{vmatrix} -3 & 2 \\ 7 & 1 \end{vmatrix} - \vec{j}\begin{vmatrix} 2 & 2 \\ 6 & 1 \end{vmatrix} + \vec{k}\begin{vmatrix} 2 & -3 \\ 6 & 7 \end{vmatrix}$$

$$= \vec{i}(-3-14) - \vec{j}(2-12) + \vec{k}(14+18)$$ Watch for double negatives!

$$\vec{n} = -17\vec{i} + 10\vec{j} + 32\vec{k}$$

$$\therefore -17x + 10y + 32z = D$$ The coefficients in the equation equal the coefficients of the components of the normal vector.

$$-17(-5) + 10(5) + 32(5) = D$$ Substitute one of the given points.

$$295 = D$$

The equation is $-17x + 10y + 32z = 295$. ◄

Note that Example 2 is the same kind of problem as in Section 12-5. The only difference is the way you calculate the normal vector.

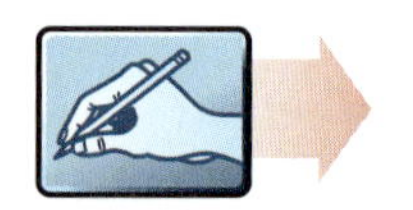

The cross product of two vectors is perpendicular to the original vectors.

Geometric Meaning of $|\vec{a} \times \vec{b}|$

The magnitude of the cross product of two vectors has a geometric meaning. Figure 12-6e shows a parallelogram with $\vec{a}$ and $\vec{b}$ as two adjacent sides.

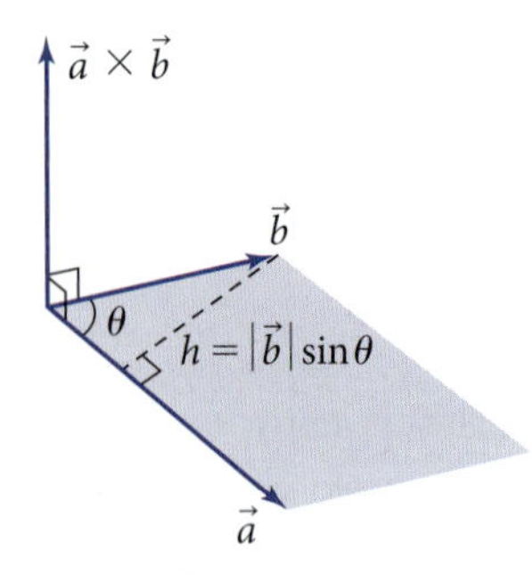

Figure 12-6e

The altitude of the parallelogram is

$$h = |\vec{b}|\sin\theta$$

The base of the parallelogram is $|\vec{a}|$. The area formula for a parallelogram is base times altitude, so

$$\text{Area} = |\vec{a}||\vec{b}|\sin\theta$$

You should recognize that $|\vec{a}||\vec{b}|\sin\theta$ is also defined to be the magnitude of the cross product.

You can find the area of a triangle with two vectors as sides by the same technique because its area is half the area of the corresponding parallelogram. Example 3 shows you how to do this.

EXAMPLE 3 ➤ Find the area of the triangle with vertices at the points $P_1(-5, 5, 5)$, $P_2(-3, 2, 7)$, and $P_3(1, 12, 6)$.

SOLUTION Figure 12-6f shows the three points (the same as in Example 2), along with the triangle having $\overrightarrow{P_1P_2}$ and $\overrightarrow{P_1P_3}$ as adjacent sides.

$\vec{n} = \overrightarrow{P_1P_2} \times \overrightarrow{P_1P_3}$

$P_2(-3, 2, 7)$ $P_1(-5, 5, 5)$ $P_3(1, 12, 6)$

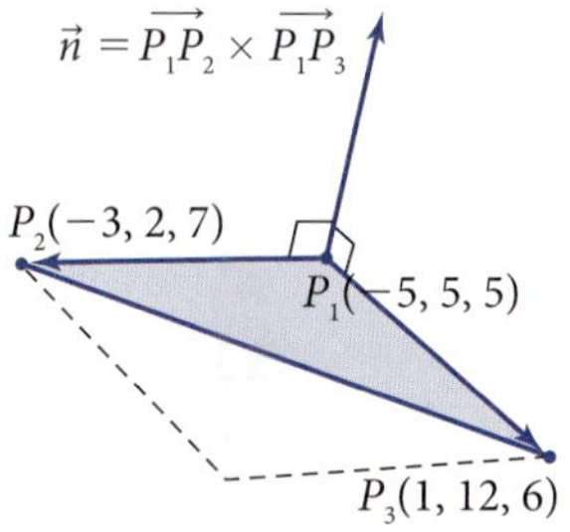

Figure 12-6f

From Example 2, $\overrightarrow{P_1P_2} \times \overrightarrow{P_1P_3} = -17\vec{i} + 10\vec{j} + 32\vec{k}$.

The area of the triangle is half the area of the parallelogram having $\overrightarrow{P_1P_2}$ and $\overrightarrow{P_1P_3}$ as adjacent sides. The area of this parallelogram is equal to the magnitude of $\overrightarrow{P_1P_2} \times \overrightarrow{P_1P_3}$, which you can calculate using the three-dimensional Pythagorean theorem.

$$\left|\overrightarrow{P_1P_2} \times \overrightarrow{P_1P_3}\right| = \left|-17\vec{i} + 10\vec{j} + 32\vec{k}\right| = \sqrt{(-17)^2 + 10^2 + 32^2}$$

$$= \sqrt{1413} = 37.5898...$$

$$\therefore \text{area of triangle} = \frac{1}{2}\sqrt{1413} = 18.7949...$$

The box summarizes these geometric properties.

PROPERTY: Area of a Parallelogram or Triangle from the Cross Product

The area of the parallelogram having $\vec{a}$ and $\vec{b}$ as adjacent sides is

$$\text{Area of parallelogram} = \left|\vec{a} \times \vec{b}\right|$$

The area of the triangle having $\vec{a}$ and $\vec{b}$ as adjacent sides is

$$\text{Area of triangle} = \frac{1}{2}\left|\vec{a} \times \vec{b}\right|$$

Problem Set 12-6

Reading Analysis

From what you have read in this section, what do you consider to be the main idea? What is the main difference between a cross product and a dot product of two vectors? How is the formula for the magnitude of a cross-product vector related to the formula for the dot product?

Quick Review

Q1. If $\vec{a}$ is 7 units long and $\vec{b}$ is 3 units long, what can you say about $\left|\vec{a} + \vec{b}\right|$?

Q2. If $\vec{a} \cdot \vec{b} = 25$, $\left|\vec{a}\right| = 5$, and $\left|\vec{b}\right| = 10$, find the measure of the angle between the vectors.

Q3. Find the magnitude of the vector $\vec{i} + \vec{j} + \vec{k}$.

Q4. Find the dot product $\left(3\vec{i} + 2\vec{j} - \vec{k}\right) \cdot \left(\vec{i} + 2\vec{j} + 10\vec{k}\right)$.

Q5. Is the angle between the vectors in Problem Q4 obtuse or acute?

Q6. Sketch two nonzero vectors whose sum is zero.

Q7. Sketch two nonzero vectors whose dot product is zero.

Q8. In $\triangle ABC$, if $\sin A = 0.6$, $\sin B = 0.2$, and $a = 60$, how long is side b?

Q9. Sketch the graph of a logistic function.

Q10. Find the amplitude of the sinusoid $y = 3 + 4\cos 5\pi(x - 6)$.

For Problems 1–4, find the cross product using determinants.

1. $(3\vec{i} + 4\vec{j} + 2\vec{k}) \times (5\vec{i} + 6\vec{j} + \vec{k})$

2. $(7\vec{i} + 2\vec{j} + 3\vec{k}) \times (6\vec{i} + \vec{j} + 5\vec{k})$

3. $(4\vec{i} - 3\vec{j} - \vec{k}) \times (2\vec{i} - \vec{j} + \vec{k})$

4. $(-3\vec{i} + 8\vec{j} + 2\vec{k}) \times (\vec{i} + 7\vec{j} + 6\vec{k})$

For Problems 5 and 6, find the dot product.

5. $(2\vec{i} + 7\vec{j} - 5\vec{k}) \cdot (9\vec{i} + 3\vec{j} + \vec{k})$

6. $(8\vec{i} - 4\vec{j} - 2\vec{k}) \cdot (5\vec{i} + 6\vec{j} - 7\vec{k})$

7. *Program for Cross Products Problem:* Write a program to calculate the cross product of two vectors. The program should prompt you to enter the three coefficients of each vector. Then it should calculate and display the coefficients of the cross product. Test your program by using it on the vectors in Problem 1. The correct answer is $-8\vec{i} + 7\vec{j} - 2\vec{k}$.

8. *Multipliers of Zero Problem:* The multiplication property of zero states that, for real numbers x and y, "if $x = 0$ or $y = 0$, then $xy = 0$." Its converse is "if $xy = 0$, then $x = 0$ or $y = 0$."

a. Show that there is a multiplication property of zero for cross products.

b. Show by counterexample that the converse of the property in part a is *false*. That is, find two nonzero vectors whose cross product equals zero.

For Problems 9–11, find a particular equation for the plane containing the given points.

9. (3, 5, 8), (−2, 4, 1), and (−4, 7, 3)

10. (5, 7, 3), (4, −2, 6), and (2, −6, 1)

11. (0, 3, −7), (5, 0, −1), and (4, 3, 9)

12. The cross product of the normal vectors to two planes is a vector that points in the direction of the line of intersection of the planes. Find a particular equation for the plane containing the point (−3, 6, 5) and normal to the line of intersection of the planes $3x + 5y + 4z = -13$ and $6x - 2y + 7z = 8$.

For Problems 13–16, find the area of the given figure.

13. The parallelogram determined by $\vec{a} = 2\vec{i} + 3\vec{j} + 6\vec{k}$ and $\vec{b} = 3\vec{i} - 4\vec{j} + 12\vec{k}$

14. The parallelogram determined by $\vec{c} = 4\vec{i} + 4\vec{j} - 7\vec{k}$ and $\vec{d} = -2\vec{i} + 5\vec{j} - 14\vec{k}$

15. The triangle with vertices (3, 7, 5), (2, −1, 7), and (−4, 6, 10)

16. The triangle with vertices (7, 8, 11), (−4, 2, 1), and (3, 8, 2)

17. *Awning Problem:* Figure 12-6g shows a triangular awning for the corner of a building. The vertices of the awning are to be at the points (10, 0, 8), (0, 15, 8), and (0, 0, 13), where the dimensions are in feet.

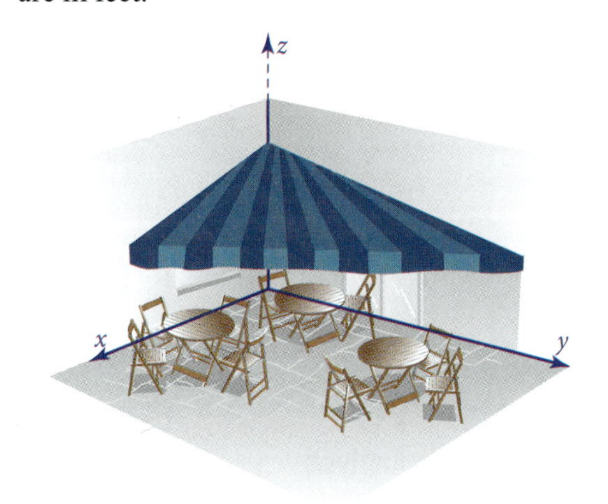

Figure 12-6g

a. Find the displacement vectors from the vertex on the z-axis to the other two vertices.

b. Find a vector normal to the plane.

c. Find the area of the awning when it is completed.

d. Find the lengths of the three sides of the awning and the measures of the three vertex angles so that the people who make the awning will know how to cut the canvas.

e. Find an equation for the plane. Use the equation to find out whether the point (5, 6, 9) is above the awning or below it.

18. *Torque Problem:* Figure 12-6h shows a wrench on a nut and bolt. As you tighten the nut, you exert a force $\vec{F} = 5\vec{i} + 2\vec{j} + 0\vec{k}$, where the magnitude of the force vector is in pounds. The displacement vector from the center of the bolt to where your hand applies the force is $\vec{d} = 7\vec{i} + 10\vec{j} + 0\vec{k}$, where the magnitude of the displacement is in inches. The force exerts a **torque** on the nut that twists it tight. The torque is defined to be the cross product of the force vector and the displacement vector. Find the torque vector. In which direction does the torque vector act?

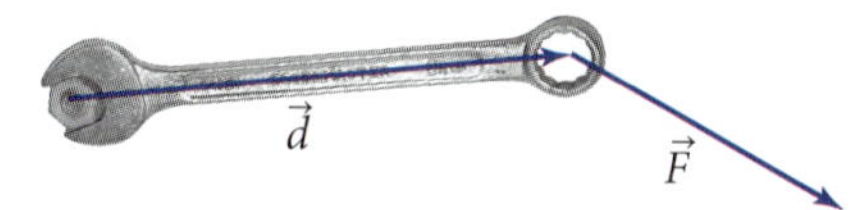

Figure 12-6h

19. Consider $\vec{a}$ and $\vec{b}$, where $\vec{a} = 5\vec{i} - 2\vec{j} + 3\vec{k}$ and $\vec{b} = 4\vec{i} + 7\vec{j} - 6\vec{k}$.

a. Find $\vec{a} \times \vec{b}$.

b. Use dot products to show that $\vec{a} \times \vec{b}$ really is normal to both $\vec{a}$ and $\vec{b}$.

c. Calculate $\vec{a} \cdot \vec{b}$, and use it to find the measure of the angle θ between $\vec{a}$ and $\vec{b}$.

d. Use θ from part c to show that $|\vec{a} \times \vec{b}|$ equals $|\vec{a}||\vec{b}|\sin\theta$.

20. Consider $\vec{e}$ and $\vec{f}$, where $\vec{e} = 2\vec{i} + 5\vec{j} - 3\vec{k}$ and $\vec{f} = 7\vec{i} - 4\vec{j} - 2\vec{k}$.

a. Prove that $\vec{e}$ and $\vec{f}$ are perpendicular.

b. Show that $|\vec{e} \times \vec{f}|$ equals $|\vec{e}||\vec{f}|$.

c. Explain why the result of part b is consistent with the definition of cross product.

21. Consider $\vec{g}$ and $\vec{h}$, where $\vec{g} = -3\vec{i} + 6\vec{j} - 12\vec{k}$ and $\vec{h} = 5\vec{i} - 10\vec{j} + 20\vec{k}$.

a. Prove that $\vec{g}$ and $\vec{h}$ are parallel.

b. Show that $|\vec{g} \times \vec{h}|$ equals 0.

c. Explain why the result of part b is consistent with the definition of cross product.

22. A plane is determined by the points (2, 1, 7), (3, 4, 9), and (6, −4, 5). Find its x-, y-, and z-intercepts.

23. If $\vec{u} = 3\vec{i} + 4\vec{j} + 6\vec{k}$ and $\vec{v} = x\vec{i} + \vec{j} + z\vec{k}$, find the values of x and z that make the cross product $\vec{u} \times \vec{v}$ equal $2\vec{i} + 24\vec{j} - 17\vec{k}$.

24. *Pythagorean Quadruples Problem:* You may have observed that the length of a vector sometimes turns out to be an integer. For example,

$$|8\vec{i} + 9\vec{j} + 12\vec{k}| = \sqrt{8^2 + 9^2 + 12^2} = \sqrt{289} = 17$$

Four positive integers a, b, c, and d for which

$$a^2 + b^2 + c^2 = d^2$$

form a *Pythagorean quadruple.* Write a program for your grapher or other computer to find all Pythagorean quadruples for values of a, b, and c up to 20. See if you can get the grapher to give only the *primitive* Pythagorean quadruples by eliminating those that are multiples of others, such as 2, 4, 4, 6, which is twice 1, 2, 2, 3.

Planes in space are visible in this geometric art by Jason Luz titled Elevator No. 14.

12-7 Direction Angles and Direction Cosines

A beam for a building under construction is to be held in place by three triangular gussets, as shown in Figure 12-7a. So that the beam will point in the correct direction, the gussets must make the correct angles with the three coordinate axes. In this section you will learn how to calculate these **direction angles.** The cosines of these angles are called the **direction cosines.**

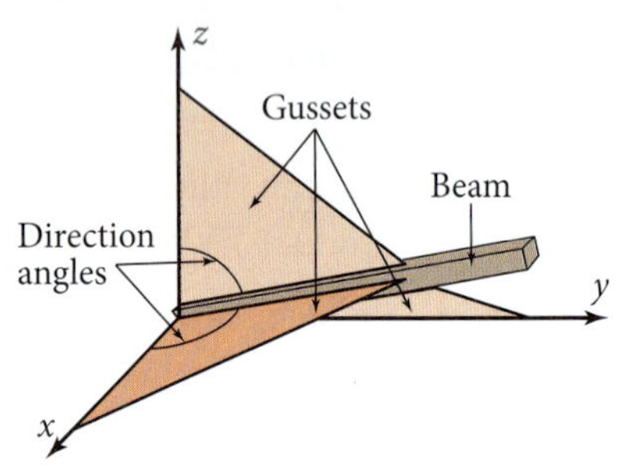

Figure 12-7a

Objective Given a vector, find its direction angles and direction cosines, and vice versa.

Figure 12-7b shows a vector $\vec{v}$ and its direction angles. The first three letters of the Greek alphabet, α (alpha), β (beta), and γ (gamma), are commonly used for the three direction angles. The letters are used in alphabetical order, corresponding to the x-, y-, and z-axes.

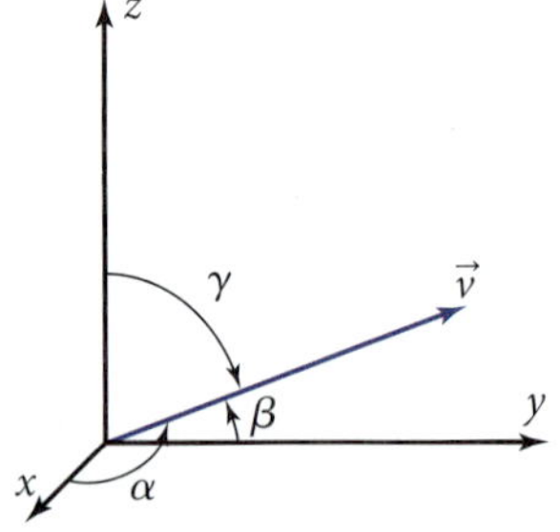

Figure 12-7b

DEFINITIONS: Direction Angles and Direction Cosines

The direction angles of a position vector are

α, from the x-axis to the vector

β, from the y-axis to the vector

γ, from the z-axis to the vector

The direction cosines of a position vector are the cosines of the direction angles.

$c_1 = \cos \alpha$

$c_2 = \cos \beta$

$c_3 = \cos \gamma$

EXAMPLE 1 ➤ Find the direction cosines and the direction angles of $\vec{v} = 3\vec{i} + 7\vec{j} + 5\vec{k}$.

SOLUTION Find the dot products $\vec{v} \cdot \vec{i}$, $\vec{v} \cdot \vec{j}$, and $\vec{v} \cdot \vec{k}$. Then use these to find the angle measures.

$$\vec{v} \cdot \vec{i} = \left(3\vec{i} + 7\vec{j} + 5\vec{k}\right) \cdot \left(1\vec{i} + 0\vec{j} + 0\vec{k}\right) = 3 \quad \text{Equal to the coefficient of } \vec{i} \text{ in } \vec{v}.$$

$$\vec{v} \cdot \vec{j} = \left(3\vec{i} + 7\vec{j} + 5\vec{k}\right) \cdot \left(0\vec{i} + 1\vec{j} + 0\vec{k}\right) = 7 \quad \text{Equal to the coefficient of } \vec{j} \text{ in } \vec{v}.$$

$$\vec{v} \cdot \vec{k} = \left(3\vec{i} + 7\vec{j} + 5\vec{k}\right) \cdot \left(0\vec{i} + 0\vec{j} + 1\vec{k}\right) = 5 \quad \text{Equal to the coefficient of } \vec{k} \text{ in } \vec{v}.$$

$$|\vec{v}| = \sqrt{3^2 + 7^2 + 5^2} = \sqrt{83}$$

$$|\vec{i}| = |\vec{j}| = |\vec{k}| = 1 \quad \text{They are } \textit{unit} \text{ vectors.}$$

$$\sqrt{83}(1)\cos\alpha = 3 \Rightarrow \cos\alpha = \frac{3}{\sqrt{83}} \Rightarrow \alpha = 70.7741\ldots^\circ \quad \text{Use the definition of dot product.}$$

$$\sqrt{83}(1)\cos\beta = 7 \Rightarrow \cos\beta = \frac{7}{\sqrt{83}} \Rightarrow \beta = 39.7940\ldots^\circ$$

$$\sqrt{83}(1)\cos\gamma = 5 \Rightarrow \cos\gamma = \frac{5}{\sqrt{83}} \Rightarrow \gamma = 56.7138\ldots^\circ$$

There is nothing special about the sum of α, β, and γ. In Example 1, for instance, $\alpha + \beta + \gamma = 167.2820\ldots^\circ$. However, there is a remarkable property related to the sum of the squares of the direction cosines.

$$\cos^2\alpha + \cos^2\beta + \cos^2\gamma = \frac{9}{83} + \frac{49}{83} + \frac{25}{83} = \frac{83}{83} = 1$$

The sum of the squares of the direction cosines of a position vector is always 1. This Pythagorean property holds because the three dot products are equal to the three coefficients of the components of $\vec{v}$. When you square the numerators in Example 1 and sum them, you get 83, the same as the radicand in the magnitude of $\vec{v}$. The ratio will always be 1.

The Pythagorean property explains another property related to direction cosines. Find a unit vector in the direction of $\vec{v}$ in Example 1.

$$\vec{u} = \frac{\vec{v}}{|\vec{v}|}$$

$$= \frac{3\vec{i} + 7\vec{j} + 5\vec{k}}{\sqrt{83}}$$

$$= \frac{3}{\sqrt{83}}\vec{i} + \frac{7}{\sqrt{83}}\vec{j} + \frac{5}{\sqrt{83}}\vec{k}$$

Notice that the coefficients of the components of the unit vector are the direction cosines. These two properties are summarized in the box on the next page.

PROPERTIES: Direction Cosines

Pythagorean Property of Direction Cosines

If α, β, and γ are the direction angles of a position vector and $c_1 = \cos\alpha$, $c_2 = \cos\beta$, and $c_3 = \cos\gamma$ are its direction cosines, then

$$\cos^2\alpha + \cos^2\beta + \cos^2\gamma = 1 \qquad \text{or} \qquad c_1^{\,2} + c_2^{\,2} + c_3^{\,2} = 1$$

Unit Vector Property of Direction Cosines

$\vec{u} = c_1\vec{i} + c_2\vec{j} + c_3\vec{k}$ is a unit vector in the direction of the given vector.

EXAMPLE 2 ➤ Find the direction angles and the direction cosines of $\vec{v} = 13\vec{i} - 6\vec{j} + 18\vec{k}$ quickly. Use the result to write a unit vector in the direction of $\vec{v}$.

SOLUTION

$$|\vec{v}| = \sqrt{13^2 + 6^2 + 18^2} = \sqrt{529} = 23$$

$$\cos\alpha = \frac{13}{23} \Rightarrow \alpha = 55.5826...^\circ$$

$$\cos\beta = \frac{-6}{23} \Rightarrow \beta = 105.1216...^\circ$$

$$\cos\gamma = \frac{18}{23} \Rightarrow \gamma = 38.4999...^\circ$$

The unit vector is $\vec{u} = \frac{13}{23}\vec{i} - \frac{6}{23}\vec{j} + \frac{18}{23}\vec{k}$. The direction cosines are the coefficients of $\vec{i}$, $\vec{j}$, and $\vec{k}$. ◄

EXAMPLE 3 ➤ If $\alpha = 152^\circ$ and $\beta = 73^\circ$, find γ.

SOLUTION

$\cos^2 152^\circ + \cos^2 73^\circ + c_3^{\,2} = 1$ Pythagorean property of direction cosines.

$c_3^{\,2} = 1 - \cos^2 152^\circ - \cos^2 73^\circ = 0.1349...$

$c_3 = \pm 0.3673...$ Don't forget the $\pm$ sign!

$\gamma = 68.4497...^\circ$ or $111.5502...^\circ$ There are two possible values of γ, one acute and the other obtuse. ◄

Example 3 shows algebraically that there are two possibilities for the third direction angle if the other two angles are given. Figure 12-7c illustrates this fact geometrically. One angle is acute. The other is its supplement, which is obtuse.

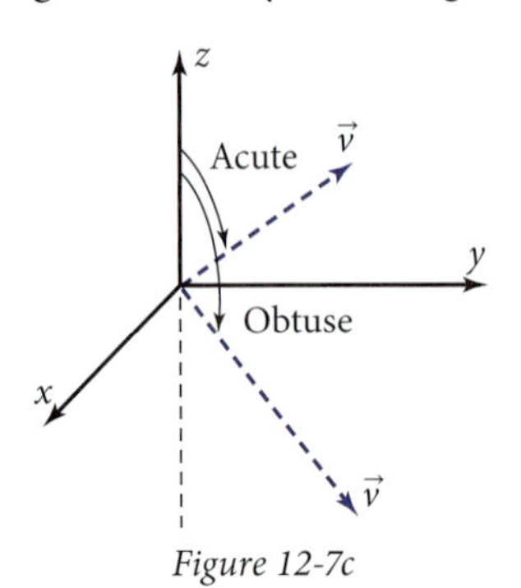

Figure 12-7c

Problem Set 12-7

Reading Analysis

From what you have read in this section, what do you consider to be the main idea? Given $\vec{v} = 2\vec{i} + 3\vec{j} + 5\vec{k}$, find $\vec{v} \cdot \vec{i}$ and tell how you can use the answer to find $\cos \alpha$. If you know two direction angles of a vector, explain why the third direction angle may have *two* possible values.

Quick Review

Q1. Write the definition of $\vec{a} \cdot \vec{b}$.

Q2. In the definition of cross product, what does $|\vec{a} \times \vec{b}|$ equal?

Q3. How are the directions of $\vec{a}$ and $\vec{b}$ related to the direction of $\vec{a} \times \vec{b}$?

Q4. $\vec{a} \cdot (\vec{a} \times \vec{b}) =$ ___?___

Q5. How is $\vec{b} \times \vec{a}$ related to $\vec{a} \times \vec{b}$?

Q6. If $|\vec{a} \times \vec{b}| = 50$, find the area of the triangle determined by $\vec{a}$ and $\vec{b}$.

Q7. Find a normal vector for the plane $3x + 4y - 5z = 37$.

Q8. Find another normal vector for the plane in Problem Q7.

Q9. The equation for the linear function that best fits a set of data is called the ___?___ equation.

Q10. The equation $a^2 = b^2 + c^2 - 2bc \cos A$ is a statement of the ___?___.

For Problems 1 and 2, sketch the vector and show its direction angles.

1. $\vec{v} = 4\vec{i} + 10\vec{j} + 3\vec{k}$

2. $\vec{v} = 5\vec{i} + 4\vec{j} + 9\vec{k}$

For Problems 3–6, find the direction cosines and direction angles of the position vector to the given point.

3. $(2, -5, 3)$

4. $(5, 7, -1)$

5. $(-4, 8, 19)$

6. $(10, -15, 6)$

For Problems 7–10, find the direction cosines of the vector from the first point to the second.

7. $(-3, 7, 1)$ to $(4, 8, -2)$

8. $(6, 9, 4)$ to $(-2, 10, 1)$

9. $(2, 9, 4)$ to $(11, 1, 16)$

10. $(4, 2, -9)$ to $(-7, 10, 7)$

For Problems 11 and 12, find a unit vector in the direction from the first point to the second point, and write its direction cosines.

11. $(3, 7, -2)$ to $(11, 23, -9)$

12. $(-5, 3, 2)$ to $(3, 2, 6)$

For Problems 13 and 14, prove that the vector is a unit vector, and find its direction cosines and direction angles.

13. $\frac{7}{9}\vec{i} + \frac{4}{9}\vec{j} - \frac{4}{9}\vec{k}$

14. $\frac{1}{9}\vec{i} - \frac{4}{9}\vec{j} + \frac{8}{9}\vec{k}$

For Problems 15 and 16, find the third direction cosine and the two possible values of the third direction angle.

15. $c_1 = \frac{18}{23}, c_2 = -\frac{13}{23}$

16. $c_1 = -\frac{12}{17}, c_3 = \frac{8}{17}$

For Problems 17–20, α, β, and γ are direction angles of different position vectors. Find the possible values of γ.

17. $\alpha = 120°, \beta = 60°$

18. $\alpha = 110°, \beta = 70°$

19. $\alpha = 17°, \beta = 12°$

20. $\alpha = 173°, \beta = 168°$

21. *Circus Cannon Problem:* In the circus, a dummy clown is shot from a cannon. Its velocity vector as it leaves the cannon is

$$\vec{v} = 5\vec{i} + 11\vec{j} + 7\vec{k}$$

where 5, 11, and 7 are speeds in the x-, y-, and z-directions, respectively, measured in feet per second. Figure 12-7d shows the *azimuth angle* and the angle of elevation of the cannon.

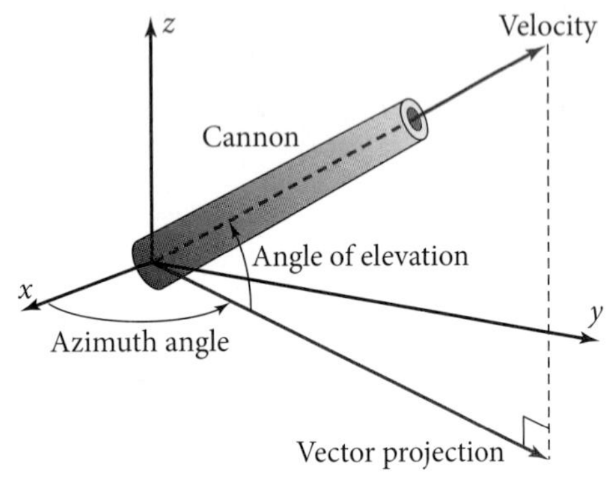

Figure 12-7d

a. At what speed will the dummy leave the cannon?

b. Find the three direction angles of $\vec{v}$.

c. Use the direction angles to find quickly the angle of elevation.

d. Find $\vec{p}$, the vector projection of $\vec{v}$ on the (horizontal) xy-plane.

e. The azimuth angle is the angle in the xy-plane from the x-axis to the vector projection $\vec{p}$. Find the measure of this angle.

f. If the angle of elevation were increased by 5°, would this change affect the other two direction angles? Would this change affect the azimuth angle?

22. *Shoe Box Construction Project:* As shown in Figure 12-7e, run a stiff wire (such as coat hanger wire) or a thin stick on the main diagonal of a shoe box. (The front face of the shoe box is not shown so that you can see into the interior.) Let $\vec{v}$ be the vector running along the wire from the bottom corner to the diagonally opposite top corner. Measure the three dimensions of the shoe box, and use the results to write $\vec{v}$ in terms of its coordinates. Then calculate the three direction angles. Cut out three triangular pieces of cardboard, each with one angle equal to a direction angle. Do the three triangles fit the direction angles? If so, tape the cardboard pieces in place and check your project with your instructor. If not, repeat your computations and measurements until the pieces of cardboard do fit. Write up this project in your journal, describing what you have learned as a result of the calculations and the construction.

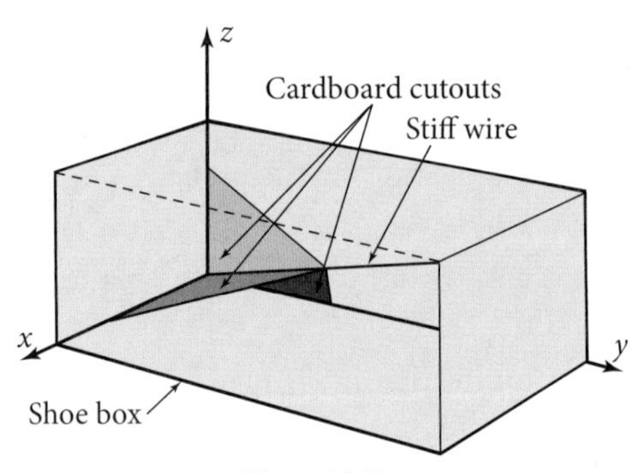

Figure 12-7e

23. *Proof of the Pythagorean Property of Direction Cosines:* Prove that if $\vec{v} = A\vec{i} + B\vec{j} + C\vec{k}$ and c_1, c_2, and c_3 are the direction cosines of $\vec{v}$, then $c_1^2 + c_2^2 + c_3^2 = 1$.

24. *Journal Problem:* Update your journal with what you have learned about vectors since your last entry. In particular, explain the difference between the definitions of cross product and dot product, and describe how you calculate these products. Then give some uses of dot product and cross product.

12-8 Vector Equations for Lines in Space

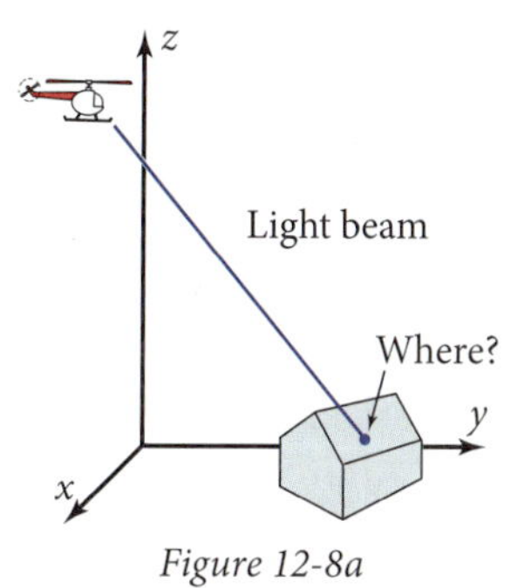

Figure 12-8a

A rescue helicopter tries to locate people stranded on the roof of a house. It shines a light beam on the roof, as shown in Figure 12-8a. By finding an equation relating the x-, y-, and z-coordinates of points on the beam, you can predict where the spot of light will appear on the roof and other information to help with the task. Unfortunately, you can not define a line in space using just one linear Cartesian equation as you can for a plane. In this section you will derive a *vector* equation that gives position vectors for points on a line.

Objective Given information about a line in space, find a vector equation for the line and use it to calculate coordinates of points on the line.

Figure 12-8b shows a line in space. Vector $\vec{r}$ is the position vector to a variable point $P(x, y, z)$ on the line. Point $P_0(5, 11, 13)$ is a fixed point on the line. The unit vector $\vec{u}$ points along the line. Let d be the directed distance from P_0 to P. Thus, the displacement vector from P_0 to P is

$$\overrightarrow{P_0P} = d\vec{u}$$

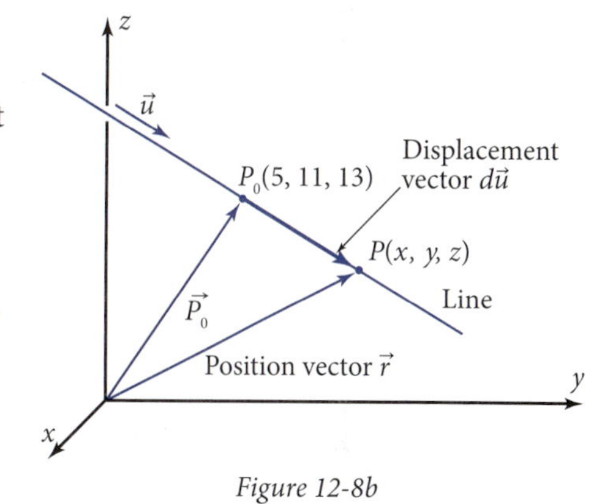

Figure 12-8b

Note that the position vector $\vec{r}$ to the variable point is the sum of $d\vec{u}$ and the position vector $\vec{P_0}$ to the fixed point.

$$\vec{r} = \vec{P_0} + d\vec{u}$$ General vector equation of a line in space.

To calculate points (x, y, z) on the line, you must find a particular equation by substituting for the fixed point and for the unit vector.

EXAMPLE 1 ➤ Find a particular equation for the line that contains the fixed point $P_0(5, 11, 13)$ and is parallel to the unit vector

$$\vec{u} = \frac{3}{7}\vec{i} + \frac{6}{7}\vec{j} + \frac{2}{7}\vec{k}$$

SOLUTION Make a general sketch of the line, as shown in Figure 12-8c.

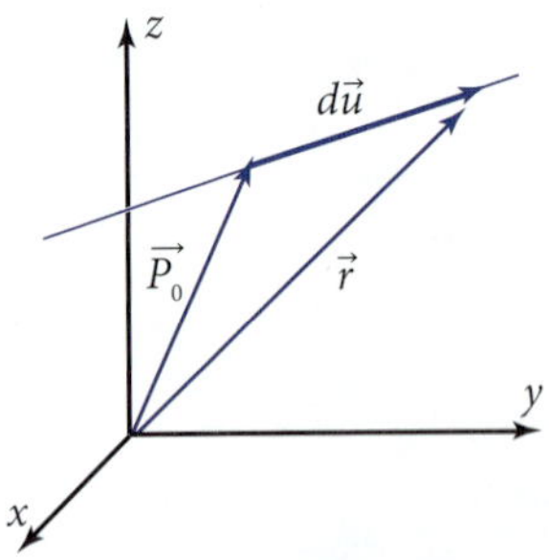

The general equation is $\vec{r} = \vec{P_0} + d\vec{u}$. $\quad$ $\vec{r}$ is the resultant of $\vec{P_0}$ and $d\vec{u}$.

$\vec{P_0} = 5\vec{i} + 11\vec{j} + 13\vec{k}$ $\quad$ $\vec{P_0}$ is the position vector to point P_0.

Therefore, the particular equation is

$$\vec{r} = \left(5\vec{i} + 11\vec{j} + 13\vec{k}\right) + d\left(\tfrac{3}{7}\vec{i} + \tfrac{6}{7}\vec{j} + \tfrac{2}{7}\vec{k}\right) \quad \text{Substitute for } \vec{P_0} \text{ and } \vec{u}.$$

$$\vec{r} = \left(5 + \tfrac{3}{7}d\right)\vec{i} + \left(11 + \tfrac{6}{7}d\right)\vec{j} + \left(13 + \tfrac{2}{7}d\right)\vec{k} \quad \text{Combine like terms.}$$

Figure 12-8c

The variable d that represents the directed distance from the fixed point to the variable point is a parameter similar to those you have encountered previously.

PROPERTY: General Vector Equation of a Line in Space

$$\vec{r} = \vec{P_0} + d\vec{u}$$

where $\vec{r}$ is the position vector to variable point $P(x, y, z)$ on the line,

$\vec{P_0}$ is the position vector to a fixed point $P_0(x_0, y_0, z_0)$ on the line,

d is a parameter equal to the directed distance from fixed point P_0 to P, and

$\vec{u}$ is a unit vector in the direction of the line.

Given unit vector $\vec{u} = c_1\vec{i} + c_2\vec{j} + c_3\vec{k}$, the equation is

$$\vec{r} = \left(x_0 + c_1 d\right)\vec{i} + \left(y_0 + c_2 d\right)\vec{j} + \left(z_0 + c_3 d\right)\vec{k}$$

Once you have the particular equation, you can use it in various ways to find points on the line. For instance, if you want to find the coordinates of a point at a particular distance from point P_0, substitute the appropriate value of d and complete the calculations.

EXAMPLE 2 Find the point on the line in Example 1 that is at a directed distance of −21 units from P_0.

SOLUTION

$$\vec{r} = \left[5 + \tfrac{3}{7}(-21)\right]\vec{i} + \left[11 + \tfrac{6}{7}(-21)\right]\vec{j} + \left[13 + \tfrac{2}{7}(-21)\right]\vec{k} \quad \text{Substitute } -21 \text{ for } d.$$

$$= (5 - 9)\vec{i} + (11 - 18)\vec{j} + (13 - 6)\vec{k}$$

$$= -4\vec{i} - 7\vec{j} + 7\vec{k}$$

The point is $(-4, -7, 7)$.

To find a point that has a particular value of x, y, or z, you must first calculate the value of d, then proceed as in Example 2. Example 3 shows how to do this.

EXAMPLE 3 Find the point where the line in Example 1 intersects the xy-plane.

SOLUTION For any point in the xy-plane, $z = 0$. So you set the z-coordinate of the line equal to zero.

$$13 + \frac{2}{7}d = 0 \Rightarrow d = -\frac{91}{2}$$ Calculate the value of d.

$$\therefore \vec{r} = \left[5 + \frac{3}{7}\left(-\frac{91}{2}\right)\right]\vec{i} + \left[11 + \frac{6}{7}\left(-\frac{91}{2}\right)\right]\vec{j} + \left[13 + \frac{2}{7}\left(-\frac{91}{2}\right)\right]\vec{k}$$

Substitute $-\frac{91}{2}$ for d.

$$= (5 - 19.5)\vec{i} + (11 - 39)\vec{j} + (13 - 13)\vec{k}$$

$$= -14.5\vec{i} - 28\vec{j} + 0\vec{k}$$

The point is $(-14.5, -28, 0)$. As a check, the z-coordinate really does equal 0.

By extending the technique of Example 3, you can find the point where a given line intersects a given plane.

EXAMPLE 4 A line containing the point $(5, 3, -1)$ has direction cosines

$$c_1 = \frac{6}{11}, \quad c_2 = -\frac{2}{11}, \quad \text{and} \quad c_3 = \frac{9}{11}$$

a. Write a particular equation for the line.

b. Find the point where the line intersects the plane $7x + 4y - 2z = 39$.

SOLUTION

a. Make a general sketch for the line and plane, as shown in Figure 12-8d.

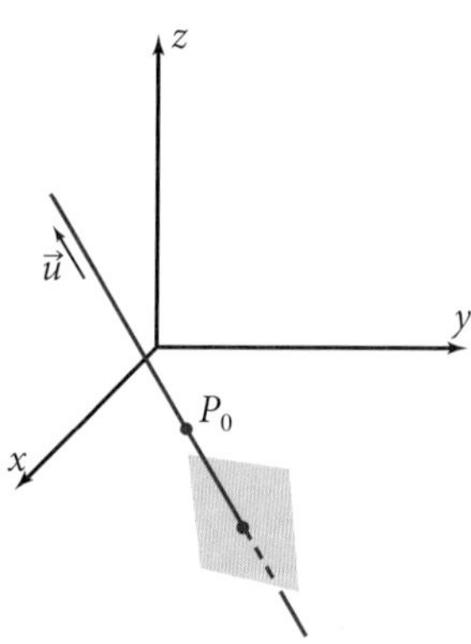

Figure 12-8d

The general equation for the line is $\vec{r} = \vec{P_0} + d\vec{u}$.

$$\vec{P_0} = 5\vec{i} + 3\vec{j} - \vec{k}$$ The coefficients of the position vector are the coordinates of the fixed point.

$$\vec{u} = \frac{6}{11}\vec{i} - \frac{2}{11}\vec{j} + \frac{9}{11}\vec{k}$$ The coefficients of the components of the unit vector are the direction cosines.

$$\vec{r} = \left(5\vec{i} + 3\vec{j} - \vec{k}\right) + d\left(\frac{6}{11}\vec{i} - \frac{2}{11}\vec{j} + \frac{9}{11}\vec{k}\right)$$ Substitute into the general equation.

$$\therefore \vec{r} = \left(5 + \frac{6}{11}d\right)\vec{i} + \left(3 - \frac{2}{11}d\right)\vec{j} + \left(-1 + \frac{9}{11}d\right)\vec{k}$$ The particular equation of the line.

b. Assume that the line intersects the plane at the point (x, y, z), so this point is both on the line and on the plane. This means that the coordinates x, y, and z will satisfy both the equation for the line and the equation for the plane. So you can substitute x, y, and z from the equation for the line in part a into the given equation for the plane and use the result to calculate d.

$$7\left(5 + \frac{6}{11}d\right) + 4\left(3 - \frac{2}{11}d\right) - 2\left(-1 + \frac{9}{11}d\right) = 39$$

$$35 + \frac{42}{11}d + 12 - \frac{8}{11}d + 2 - \frac{18}{11}d = 39$$

$$\frac{16}{11}d = -10$$

$$d = -\frac{55}{8}$$

$$\therefore \vec{r} = \left[5 + \frac{6}{11}\left(-\frac{55}{8}\right)\right]\vec{i} + \left[3 - \frac{2}{11}\left(-\frac{55}{8}\right)\right]\vec{j} + \left[-1 + \frac{9}{11}\left(-\frac{55}{8}\right)\right]\vec{k}$$

Substitute $-\frac{55}{8}$ for d.

$$= 1.25\vec{i} + 4.25\vec{j} - 6.625\vec{k}$$

The point is $(1.25, 4.25, -6.625)$.

Problem Set 12-8

Reading Analysis

From what you have read in this section, what do you consider to be the main idea? What is the major difference between the equation for a plane in space and the equation for a line in space? Where do direction cosines appear in the equation for a line in space?

Quick Review

Q1. If $\cos \alpha = \frac{1}{2}$, find α.

Q2. For direction angles α, β, and γ, if $\cos^2 \alpha = 0.3$ and $\cos^2 \beta = 0.2$, find $\cos^2 \gamma$.

Q3. Find the magnitude of $\vec{v} = 1\vec{i} - 2\vec{j} + 2\vec{k}$.

Q4. Find the direction cosine c_2 for $\vec{v}$ in Problem Q3.

Q5. Give one major difference between a dot product and a cross product of two vectors.

Q6. If p is the scalar projection of $\vec{a}$ on $\vec{b}$, then the vector projection is $\vec{p} =$ __?__.

Q7. Find $\vec{i} \cdot \vec{j}$.

Q8. Find $\vec{i} \times \vec{j}$.

Q9. If $|\vec{a} \times \vec{b}| = 70$, find the area of the parallelogram with adjacent sides $\vec{a}$ and $\vec{b}$.

Q10. What transformation on $f(x)$ is represented by $y = f(3x)$?

For Problems 1 and 2, given the equation of a line,

a. Identify the coordinates of the fixed point on the given line.

b. Identify a unit vector, $\vec{u}$, in the direction of the line.

c. Confirm that $\vec{u}$ really is a *unit* vector.

1. $\vec{r} = \left(5 + \frac{9}{17}d\right)\vec{i} + \left(-3 + \frac{12}{17}d\right)\vec{j} + \left(4 + \frac{8}{17}d\right)\vec{k}$

2. $\vec{r} = \left(6 - \frac{1}{9}d\right)\vec{i} + \left(7 + \frac{8}{9}d\right)\vec{j} + \left(-5 + \frac{4}{9}d\right)\vec{k}$

3. Find the point on the line in Problem 1 for which $d = 34$.

4. Find the point on the line in Problem 2 for which $d = 27$.

5. Find the point where the line in Problem 1 intersects the xy-plane.

6. Find the point where the line in Problem 2 intersects the yz-plane.

For Problems 7 and 8, show that $\vec{u}$ is a unit vector. Then write a vector equation for the line parallel to $\vec{u}$ containing the given point.

7. $\vec{u} = \frac{2}{7}\vec{i} + \frac{6}{7}\vec{j} - \frac{3}{7}\vec{k}$, $P_0 = (5, -1, 4)$

8. $\vec{u} = \frac{11}{15}\vec{i} - \frac{2}{15}\vec{j} + \frac{2}{3}\vec{k}$, $P_0 = (-3, 4, 7)$

For Problems 9 and 10, find the direction cosines of $\vec{v}$. Then write a vector equation for the line parallel to $\vec{v}$ containing the given point.

9. $\vec{v} = 2\vec{i} - 3\vec{j} + 4\vec{k}$, $P_0 = (1, -8, -5)$

10. $\vec{v} = \vec{i} + 2\vec{j} - 5\vec{k}$, $P_0 = (-6, 3, -4)$

For Problems 11 and 12, find a vector equation for the line from the first point to the second.

11. $(5, 1, -4)$ to $(14, 21, 8)$

12. $(6, -2, 7)$ to $(10, 6, 26)$

For Problems 13 and 14, find the point where the given line intersects the given plane.

13. Line: $\vec{r} = \left(3 + \frac{2}{3}d\right)\vec{i} + \left(4 + \frac{2}{3}d\right)\vec{j} + \left(3 - \frac{1}{3}d\right)\vec{k}$
Plane: $7x - 3y + 5z = -20$

14. Line: $\vec{r} = \left(4 + \frac{1}{3}d\right)\vec{i} + \left(1 + \frac{2}{3}d\right)\vec{j} + \left(7 + \frac{2}{3}d\right)\vec{k}$
Plane: $x + 4y - 3z = 35$

15. *Forensic Bullet Path Problem:* A bullet has pierced the wall and ceiling of a small house (Figure 12-8e), and it may have lodged in the roof. You have studied vectors, so the investigators call on you to calculate the point in the roof where the bullet is expected to be found. The dimensions of the house are in feet.

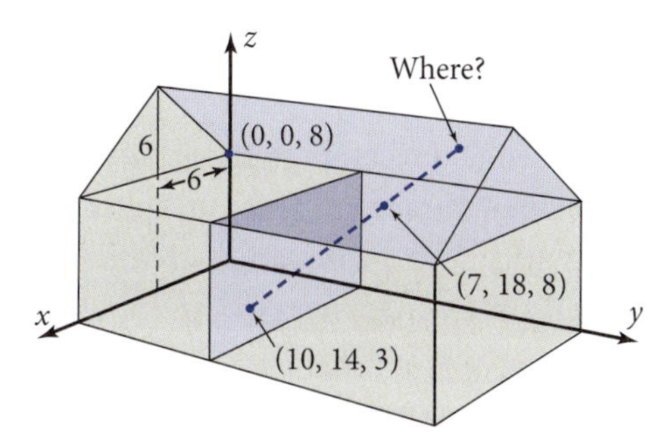

Figure 12-8e

a. You set up a three-dimensional coordinate system as shown in Figure 12-8e, with the origin at the floor in the back corner of the house. The bullet pierced the wall at the point (10, 14, 3) and then pierced the ceiling at the point (7, 18, 8). Find a unit vector in the direction of the bullet's path.

b. Using the point (10, 14, 3) as the fixed point, write a vector equation for the line followed by the bullet.

c. How tall is the interior of the house, from floor to ceiling? How can you tell?

d. Figure 12-8e shows that the point (0, 0, 8) is at the back corner of the slanted roof. If you run horizontally in the x-direction 6 ft from this point and then rise vertically in the z-direction 6 ft, you reach the crest of the roof. Explain why $\vec{n} = -6\vec{i} + 0\vec{j} + 6\vec{k}$ is a vector normal to the plane of the roof section that is shaded in Figure 12-8e.

e. Find a Cartesian equation for the plane of the roof in part d.

f. Find the point in the roof at which investigators may expect to find the bullet.

g. What is the meaning of the word *forensic,* and why is the word appropriate in the title of this problem?

16. *Flood Control Tunnel Problem:* Suppose that you work for a construction company that has been hired to dig a drainage tunnel under a city. The tunnel will carry excess water to the other side of the city during heavy rains, thus preventing flooding (Figure 12-8f). The tunnel is to start at ground level and then slant downward until it reaches a point 100 ft below the surface. Then it will become horizontal, extending far enough to reach the other side of the city (not shown). Your job is to analyze the slanted part of the tunnel.

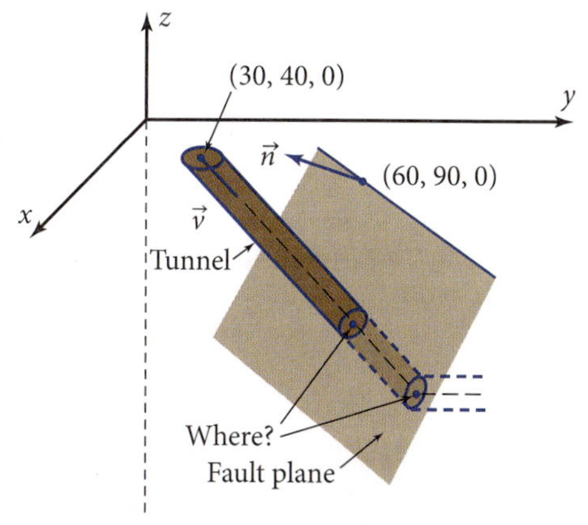

Figure 12-8f

a. The Engineering Department has determined that the tunnel will slant downward in the direction of the vector $\vec{v} = 9\vec{i} + 12\vec{j} - 20\vec{k}$. The centerline of the tunnel starts at the point (30, 40, 0) on the surface. The measurements are in feet. Find a particular vector equation for the centerline.

b. How far along the centerline must the construction crews dig to reach the end of the slanted part of the tunnel, 100 ft below the ground? What are the coordinates of this endpoint?

c. Construction crews must be careful when they reach a fault plane that is in the path of the slanted part of the tunnel. The Geology Department has determined that the point (60, 90, 0) is on the fault plane where it outcrops at ground level and that the vector $\vec{n} = 2\vec{i} - 4\vec{j} + \vec{k}$ is normal to the plane (Figure 12-8f). Find a particular equation for the plane.

d. How far along the centerline of the tunnel must the construction crews dig in order to reach the fault plane? What are the coordinates of the point at which the centerline intersects the plane? How far beneath ground level is this point?

12-9 Chapter Review and Test

In this chapter you have extended your knowledge of vectors to three-dimensional space. You extended the operations of addition, subtraction, and multiplication by a scalar to three dimensions by simply giving a third component to the vectors. Two new operations, dot product and cross product, allow you to "multiply" two vectors. Dot multiplication gives a scalar for the answer. Cross multiplication gives a vector perpendicular to the two factor vectors. These techniques have allowed you to find the measure of the angle between two vectors, to project one vector onto another vector, to write equations of planes and lines in space, and to find areas of triangles and parallelograms in space.

Review Problems

R0. Update your journal with what you have learned in Chapter 12. Include things such as

- The difference between a position vector and a displacement vector
- The difference between a dot product and a cross product
- The difference between a scalar projection and a vector projection
- The difference between how you *calculate* a dot product or cross product and what dot and cross products *mean*
- How dot products are used to find equations for planes in space
- How cross products are used to find areas of triangles and parallelograms in space

R1. Figure 12-9a shows two-dimensional vectors $\vec{a} = 3\vec{i} + 4\vec{j}$ and $\vec{b} = 7\vec{i} + 2\vec{j}$.

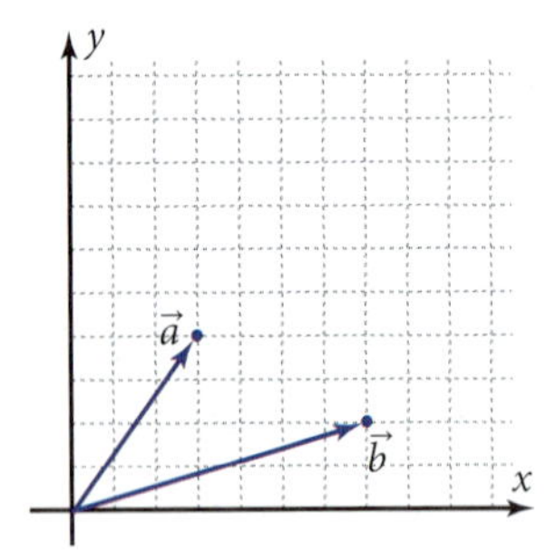

Figure 12-9a

a. Explain why these vectors, as shown, are position vectors.

b. On a copy of Figure 12-9a, sketch $\vec{r} = \vec{a} + \vec{b}$, $\vec{d} = \vec{a} - \vec{b}$, and $\vec{v} = 2\vec{a}$.

c. Write the displacement vector $\vec{d} = \vec{a} - \vec{b}$ in terms of its components.

d. Find the length of the resultant vector $\vec{r} = \vec{a} + \vec{b}$.

e. Find the measure of the angle $\vec{a}$ makes with the x-axis.

R2. a. For position vectors $\vec{a}$ and $\vec{b}$ in Problem R1, find the displacement vector $\vec{d}$ from the head of $\vec{a}$ to the head of $\vec{b}$.

b. Find the displacement vector from the head of $\vec{a}$ to a point 40% of the way from the head of $\vec{a}$ to the head of $\vec{b}$.

c. Find the position vector for the point 40% of the way from the head of $\vec{a}$ to the head of $\vec{b}$.

d. Write the coordinates of the point 40% of the way from the head of $\vec{a}$ to the head of $\vec{b}$.

R3. a. Draw $\vec{v} = 5\vec{i} + 9\vec{j} + 4\vec{k}$ as a position vector. Show the "box" that makes it look three-dimensional. Indicate the unit vectors $\vec{i}$, $\vec{j}$, and $\vec{k}$ on the drawing.

b. If $\vec{a} = 6\vec{i} - 5\vec{j} + 2\vec{k}$ and $\vec{b} = 3\vec{i} + 4\vec{j} - 7\vec{k}$, find $3\vec{a} - 2\vec{b}$.

c. Find $|\vec{a}|$ for $\vec{a}$ in part b.

d. Find a unit vector in the direction of $\vec{a}$ in part b.

e. If $\vec{a}$ and $\vec{b}$ in part b are position vectors, find the displacement vector from the head of $\vec{a}$ to the head of $\vec{b}$.

f. For $\vec{a}$ and $\vec{b}$ in part b, find the position vector to the point 70% of the way from the head of $\vec{a}$ to the head of $\vec{b}$.

R4. a. Write the definition of dot product. Give two other names for dot product.

b. If $|\vec{a}| = 7$, $|\vec{b}| = 8$, and $\theta = 155°$, find $\vec{a} \cdot \vec{b}$.

c. If $|\vec{a}| = 10$, $|\vec{b}| = 20$, and $\vec{a} \cdot \vec{b} = -35$, find θ.

For parts d–h, let $\vec{a} = 6\vec{i} - 5\vec{j} + 2\vec{k}$ and $\vec{b} = 3\vec{i} + 4\vec{j} - 7\vec{k}$.

d. Find $|\vec{a}|$, $|\vec{b}|$, and $\vec{a} \cdot \vec{b}$.

e. Find the measure of the angle between $\vec{a}$ and $\vec{b}$ when they are placed tail-to-tail.

f. Find a unit vector in the direction of $\vec{b}$.

g. Find the scalar projection of $\vec{a}$ on $\vec{b}$.

h. Find the vector projection of $\vec{a}$ on $\vec{b}$.

R5. a. Use a dot product to prove that if the vector $\vec{n} = A\vec{i} + B\vec{j} + C\vec{k}$ is normal to a plane, then the equation for the plane is $Ax + By + Cz = D$, where D stands for a constant.

b. Find two normal vectors for the plane $3x - 7y + z = 5$, pointing in opposite directions.

c. Find the equation for the plane containing the point $(6, 2, -1)$, with normal vector $\vec{n} = 2\vec{i} - 7\vec{j} - 3\vec{k}$. Use the equation to find z for the point $P(10, 20, z)$ on the plane.

d. Find the equation for the plane perpendicular to the segment with endpoints $(5, 7, 2)$ and $(8, 13, 11)$ if the x-intercept of the plane is $x = 15$.

R6. a. Write the definition of $\vec{a} \times \vec{b}$. Give three names for $\vec{a} \times \vec{b}$.

b. If $|\vec{a}| = 7$, $|\vec{b}| = 8$, and $\theta = 155°$, find $|\vec{a} \times \vec{b}|$.

For parts c–e, let $\vec{a} = 3\vec{i} + 2\vec{j} - \vec{k}$ and $\vec{b} = -4\vec{i} + 3\vec{j} + 5\vec{k}$.

c. Find $\vec{a} \times \vec{b}$ and $\vec{b} \times \vec{a}$.

d. Find $\vec{a} \cdot \vec{b}$ and $\vec{b} \cdot \vec{a}$.

e. Find the area of the triangle determined by $\vec{a}$ and $\vec{b}$.

f. Find the equation for the plane containing the points $(2, 5, 8)$, $(3, 7, 4)$, and $(-1, 9, 6)$.

R7. a. Sketch a position vector and show its three direction angles.

b. Find the direction cosines and the direction angles of $\vec{v} = 6\vec{i} - 8\vec{j} + 5\vec{k}$.

c. Vector $\vec{a}$ has direction cosines $c_1 = 0.2$ and $c_2 = -0.3$. Find the two possible values of c_3 and the two possible values of the third direction angle, γ.

d. Show algebraically, using the Pythagorean property, that there is no vector with direction angles $\alpha = 30°$ and $\beta = 40°$. Explain geometrically why such a vector cannot exist.

R8. For parts a–d, the position vector $\vec{r}$ to a point on a line is given by the vector equation

$$\vec{r} = \left(6 + \tfrac{7}{9}d\right)\vec{i} + \left(3 + \tfrac{4}{9}d\right)\vec{j} + \left(2 - \tfrac{4}{9}d\right)\vec{k}$$

a. Write the fixed point and the fixed vector that appear in the equation. Show that the vector is a unit vector.

b. Show that you understand the meaning of the independent variable d in the equation by finding the coordinates of the point on the line that is a directed distance -18 units from the fixed point. Explain the significance of the fact that d is negative in this case.

c. Find the coordinates of the point where the line intersects the xy-plane.

d. Find the coordinates of the point where the line intersects the plane $3x - 7y + z = 5$.

e. Find a vector equation for the line containing the points $(2, 8, 4)$ and $(11, 13, 7)$.

Concept Problems

C1. *Distance Between a Point and a Line Problem:* Figure 12-9b shows a line and a point P_1 not on the line. Vector $\vec{v}$ is parallel to the line, and d is the perpendicular distance between P_1 and the line.

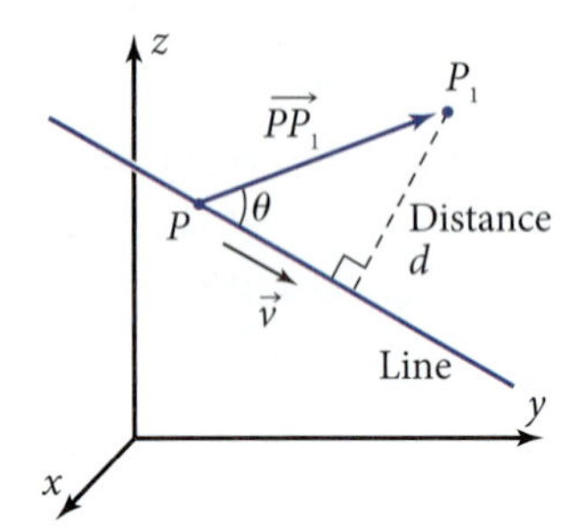

Figure 12-9b

Now suppose that the equation for the line is

$$\vec{r} = \left(5 + \frac{6}{11}t\right)\vec{i} + \left(3 - \frac{2}{11}t\right)\vec{j} + \left(-1 + \frac{9}{11}t\right)\vec{k}$$

and the point off the line is $P_1(4, 7, 6)$.

a. By trigonometry, $d = |\overrightarrow{PP_1}| \cdot \sin\theta$. Multiply the right side of this equation by 1 in the form $\frac{|\vec{v}|}{|\vec{v}|}$. Take advantage of the fact that the numerator on the right now equals $|\overrightarrow{PP_1} \times \vec{v}|$ to find d without first finding θ.

b. *Ladder Problem:* Figure 12-9c shows a 25-ft ladder leaning against a wall to reach a high window. To miss the flower bed, the ladder is moved over so that its left foot is at the point (7, 9, 0). The top of the ladder is 24 ft up the wall. Find a vector equation for the line along the left side of the ladder. Given that the rungs on the ladder are 1 ft apart, use the equation to find the rung that is closest to the upper-left corner of the window, at the point (0, 5, 18). Find the *perpendicular* distance from the left side of the ladder to the point (0, 5, 18), taking advantage of the results of part a,

$$d = \frac{|\overrightarrow{PP_1} \times \vec{v}|}{|\vec{v}|}$$

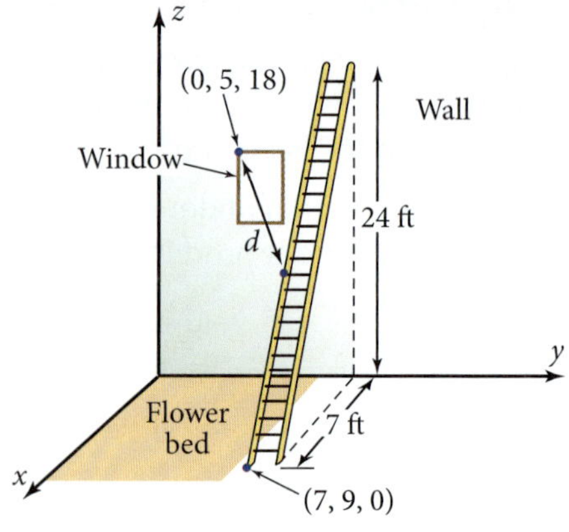

Figure 12-9c

C2. a. *Distance Between Skew Lines Problem:* Figure 12-9d shows Line 1 containing the point $P_1(3, 8, 5)$ and parallel to $\vec{v}_1 = 6\vec{i} + 3\vec{j} + 5\vec{k}$, and Line 2 containing the point $P_2(5, 2, 7)$ and parallel to $\vec{v}_2 = 9\vec{i} + 7\vec{j} + 1\vec{k}$. Lines 1 and 2 are *skew* lines because they are not parallel yet do not intersect. The cross product $\vec{v}_1 \times \vec{v}_2$ is perpendicular to both lines. The perpendicular distance d between the two lines is the absolute value of the scalar projection of $\overrightarrow{P_1P_2}$ on this cross product. Find this distance.

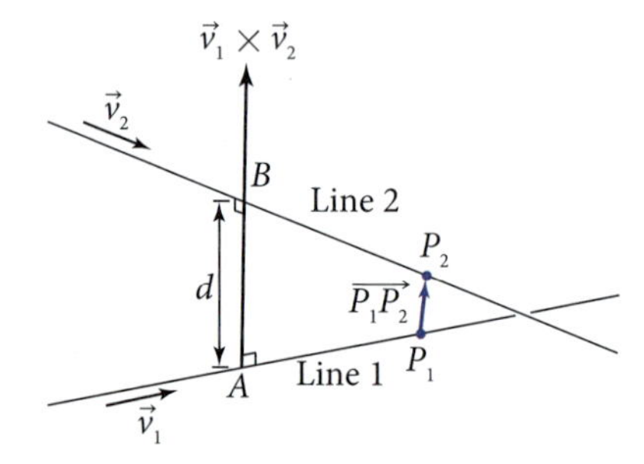

Figure 12-9d

b. *Airplane Near-Miss Velocity Vector Problem:* Flight 007 took off from the point $P_1(3000, 2000, 0)$ on one runway at an airport, where distances are in feet. At the same instant, Flight 1776 was at the point $P_2(1000, 500, 300)$ preparing to land on another runway that crosses the first one, as shown in Figure 12-9e. Computers in the control tower found that the velocity vectors for the two flights were

Flight 007: $\vec{v}_1 = -100\vec{i} + 50\vec{j} + 20\vec{k}$

Flight 1776: $\vec{v}_2 = 40\vec{i} + 200\vec{j} - 15\vec{k}$

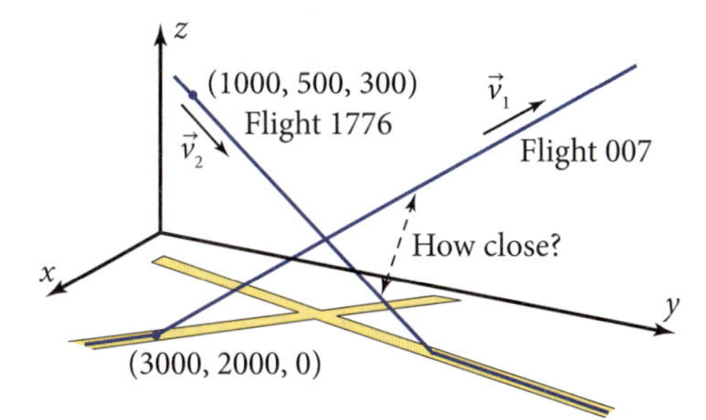

Figure 12-9e

Find the closest the two planes' paths came to each other using the results of part a, namely,

$$d = \frac{\left|(\vec{v}_1 \times \vec{v}_2) \cdot \overrightarrow{P_1P_2}\right|}{\left|\vec{v}_1 \times \vec{v}_2\right|}$$

c. The speeds in part b are in feet per second. Thus, the position vectors of the planes are

Flight 007:

$\vec{r}_1 = (3000 - 100t)\vec{i} + (2000 + 50t)\vec{j} + (0 + 20t)\vec{k}$

Flight 1776:

$\vec{r}_2 = (1000 + 40t)\vec{i} + (500 + 200t)\vec{j} + (300 - 15t)\vec{k}$

where the parameter t is time in seconds. Write the displacement vector from Flight 1776 to Flight 007 as a function of time. Using appropriate algebraic or numerical techniques, find the time t at which the flights were closest to each other by finding the value of t at which the length of the displacement vector was a minimum. Explain why the closest the flights came to each other is *not* the same as the closest the two paths came to each other. Is there cause for concern about how close the flights came to each other?

Chapter Test

Part 1: No calculators allowed (T1–T9)

T1. On a copy of Figure 12-9f,

a. Show the direction angles α, β, and γ for $\vec{v}$.

b. Mark the three unit vectors $\vec{i}$, $\vec{j}$, and $\vec{k}$.

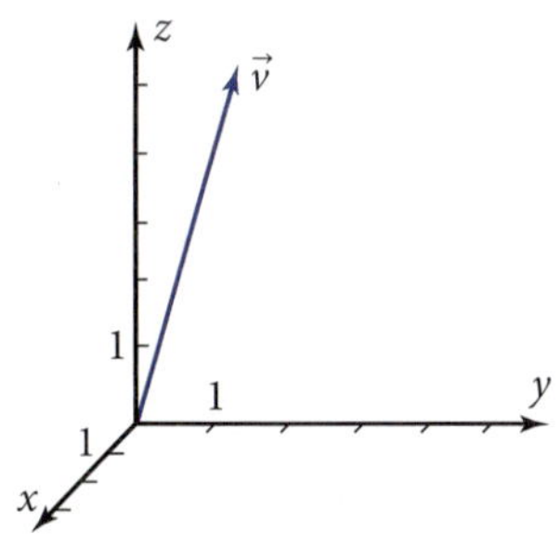

Figure 12-9f

T2. On a copy of Figure 12-9g,

a. Sketch the cross product vector $\vec{a} \times \vec{b}$.

b. Sketch $\vec{p}$, the vector projection of $\vec{a}$ on $\vec{b}$.

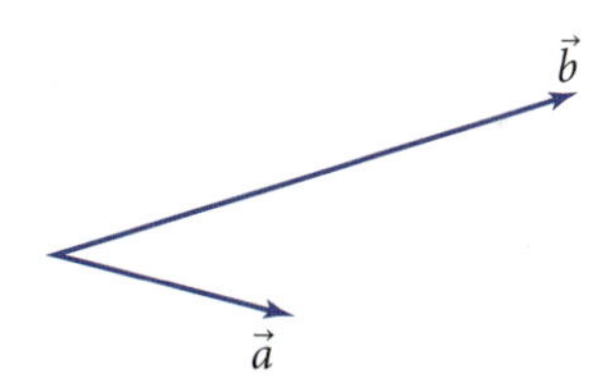

Figure 12-9g

T3. What is the major difference in meaning between $\vec{a} \cdot \vec{b}$ and $\vec{a} \times \vec{b}$?

T4. Write the definition of $\vec{a} \cdot \vec{b}$.

T5. Given: $\vec{a} = 3\vec{i} + 2\vec{j} + 4\vec{k}$
$\vec{b} = 1\vec{i} + 5\vec{j} + 2\vec{k}$

a. Find $\vec{a} \cdot \vec{b}$.

b. Find $\vec{a} \times \vec{b}$.

T6. What does it mean to say that one vector is normal to another vector?

T7. How can you tell quickly whether one three-dimensional vector is perpendicular to another three-dimensional vector?

T8. Find a vector normal to the plane $-13x + 10y - 5z = 22$.

T9. What makes a vector a unit vector?

Part 2: Graphing calculators allowed (T10–T27)

Problems T10–T17 refer to the vectors

$$\vec{a} = 5\vec{i} + 2\vec{j} + 9\vec{k}$$
$$\vec{b} = 3\vec{i} + 8\vec{j} + 4\vec{k}$$

T10. Find the resultant of $\vec{a}$ and $\vec{b}$.

T11. If $\vec{a}$ and $\vec{b}$ are placed tail-to-tail, find the displacement vector from the head of $\vec{b}$ to the head of $\vec{a}$.

T12. Find $|\vec{a}|$ and $|\vec{b}|$.

T13. Find a unit vector in the direction of $\vec{b}$.

T14. Find the measure of the angle between $\vec{a}$ and $\vec{b}$ if they are placed tail-to-tail.

T15. Find a vector perpendicular to both $\vec{a}$ and $\vec{b}$.

T16. Find the area of the triangle with $\vec{a}$ and $\vec{b}$ as two of its sides.

T17. Find the scalar projection of $\vec{a}$ on $\vec{b}$.

Problems T18–T22 refer to the line with vector equation

$$\vec{r} = \left(3 + \tfrac{8}{9}d\right)\vec{i} + \left(5 + \tfrac{1}{9}d\right)\vec{j} + \left(8 + \tfrac{4}{9}d\right)\vec{k}$$

T18. Find the coordinates of the fixed point that appears in the equation.

T19. Prove that the vector $\vec{u} = \frac{8}{9}\vec{i} + \frac{1}{9}\vec{j} + \frac{4}{9}\vec{k}$ that appears in the equation is a *unit* vector.

T20. Find the point on the line that is at a directed distance 27 units from the fixed point.

T21. Find the directed distance from the fixed point on the line to the point where the line intersects the xy-plane.

T22. Find γ, the direction angle the line makes with the z-axis.

Awning Problem: For Problems T23–T26, an awning is to be built in the corner of a building, as shown in Figure 12-9h. A vertical column on the left of the awning starts on the x-axis and ends at the point (10, 0, 7) on the awning. The dimensions are in feet. A normal vector to the plane of the awning is

$$\vec{n} = 7\vec{i} + 5\vec{j} + 10\vec{k}$$

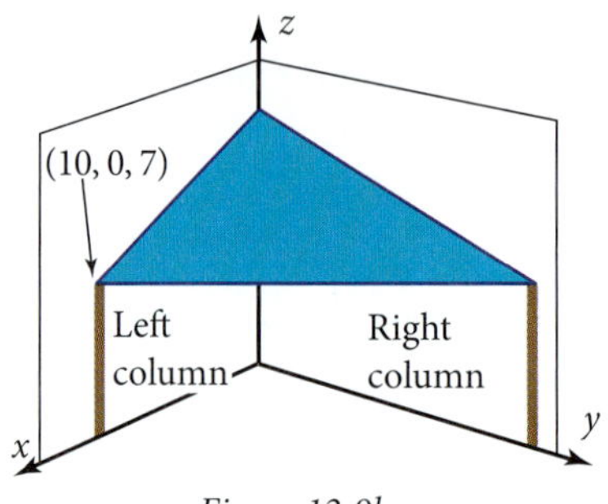

Figure 12-9h

T23. Write a particular equation for the plane.

T24. How long will the column on the right be, where $x = 0$ ft and $y = 12$ ft?

T25. How high will the awning be at the back corner, where the walls meet?

T26. A light fixture is to be located at the point (4, 6, 9). Find the vertical distance between the light and the awning. Is the light above the awning or below it?

T27. What did you learn as a result of taking this test that you did not know before?

Chapter 13

Matrix Transformations and Fractal Figures

Each part of the chambered nautilus shell is a transformation—a duplicate of the original shape, with a different size and location. This type of pattern is also found in pinecones and sunflowers. Matrix operations can be used to apply such transformations. When several simple operations are performed iteratively (over and over, starting with the previous result), they can produce an amazingly complicated object with self-similar parts.

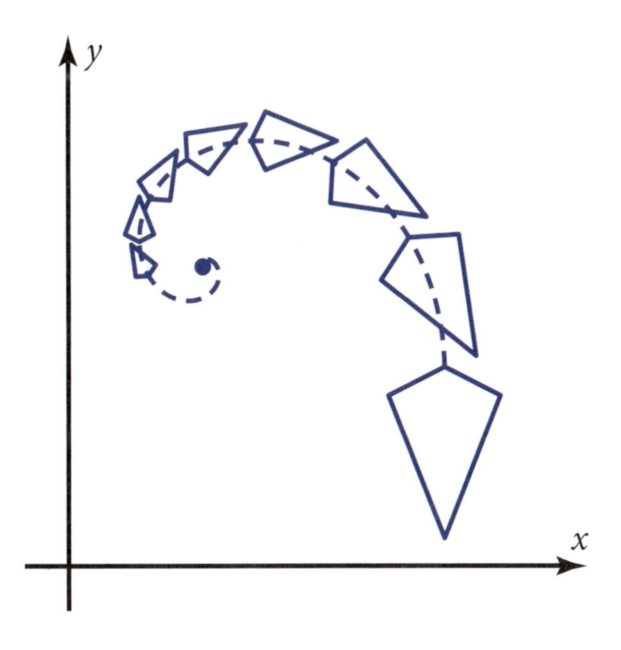

Mathematical Overview

In this chapter you will learn how to use matrices (the plural of *matrix*) to transform two-dimensional figures into complex images. A single transformation, iterated many times, transforms a simple trapezoid into the snail-like figure below. Iterating several matrix transformations can produce figures so complex that they have fractional dimensions. You will study such *fractal* figures in four ways.

NUMERICALLY

$$\begin{bmatrix} 0.9\cos 30^\circ & 0.9\cos 120^\circ & 6 \\ 0.9\sin 30^\circ & 0.9\sin 120^\circ & 2 \\ 0 & 0 & 1 \end{bmatrix}\begin{bmatrix} 2 & -2 & -5 & 5 \\ 5 & 5 & -5 & -5 \\ 1 & 1 & 1 & 1 \end{bmatrix}$$

$$= \begin{bmatrix} 5.3088... & 2.1911... & 4.3528... & 12.1471... \\ 6.7971... & 4.9971... & -4.1471... & 0.3528... \\ 1 & 1 & 1 & 1 \end{bmatrix}$$

GRAPHICALLY

Spiraling trapezoids

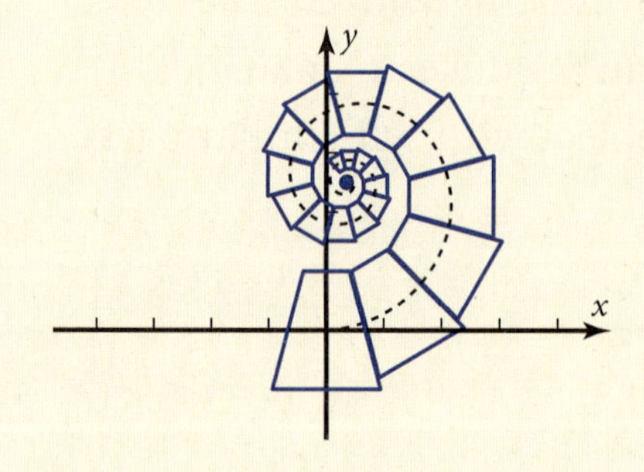

ALGEBRAICALLY

50th iteration: [Image] = $[\text{transformation}]^{50}$[pre-image]

VERBALLY

Surprisingly, images can have fractional dimensions. Matrices can be used recursively to generate such images, called "fractals," and the fractional dimensions can be calculated by analyzing the result of repeating this process infinitely many times.

13-1 Introduction to Iterated Transformations

In this section you will explore what happens to a two-dimensional figure when you perform the same transformation over and over, each time applying the transformation to the result of the previous transformation. This process is called **iteration.** The result of each transformation is also called an iteration.

Objective See what happens to the perimeter and area of a square when you perform the same set of transformations repeatedly (iteratively).

Exploratory Problem Set 13-1

The left diagram in Figure 13-1a shows a 10-cm by 10-cm square. To create the middle diagram, the original, or *pre-image,* square was transformed into four similar squares, each with sides whose length is 40% of the original side length. These image squares were then translated so that each has a corner at one of the corners of the pre-image. The right diagram shows the result of applying the same transformation to each of the four squares from the first iteration. In this problem set you will explore the perimeter and area of various iterations.

1. Find the perimeter and area of the pre-image square. Find the *total* perimeter and area of the four squares in the first iteration. Find the *total* perimeter and area of the 16 squares in the second iteration. Display the answers in a table with these column headings: Iteration number, Side length, Total perimeter, and Total area.
2. What pattern do you notice that relates the total perimeter to the iteration number? What pattern relates the total area to the iteration number? What pattern relates the total area to the total perimeter?
3. Using the patterns you observed in Problem 2, find the total perimeter and the total area of the third and fourth iterations.
4. Calculate the total perimeter and the total area of the 20th iteration.
5. If the iterations could be performed infinitely many times, the images would approach a figure sometimes called *Sierpiński's carpet*. What would be the total perimeter of this figure? What would be the total area? Do the answers surprise you? (In this chapter you will encounter other surprises, such as the fact that this figure is less than two-dimensional but more than one-dimensional!)

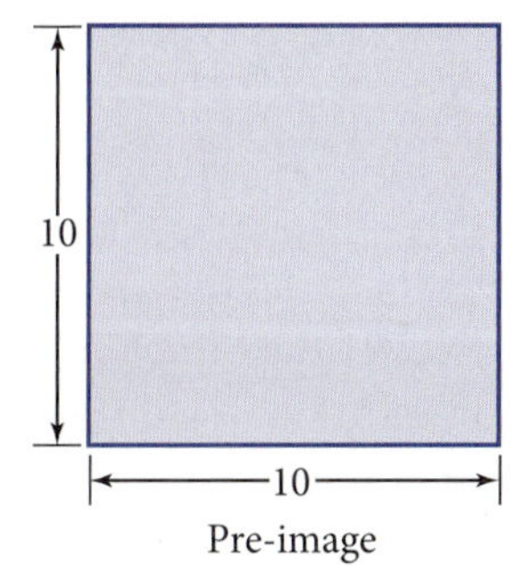

Pre-image

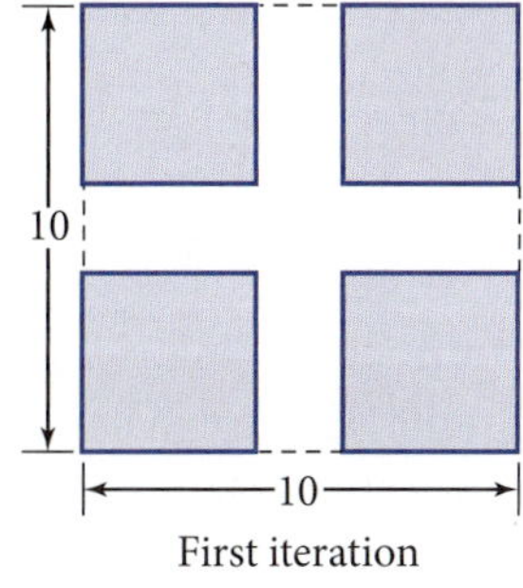

First iteration

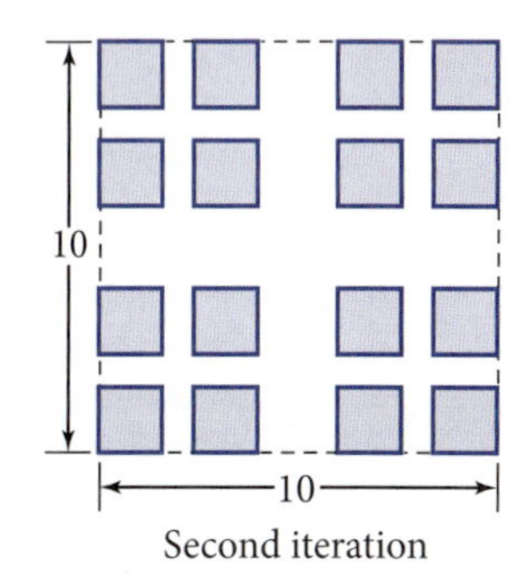

Second iteration

Figure 13-1a

13-2 Matrix Operations and Solutions of Linear Systems

In Section 13-1, a pre-image square was duplicated, dilated, and translated to create four new squares. Iterating these transformations many times creates images that approach a figure with an infinite number of pieces, zero area, and infinite perimeter. In this section you will refresh your memory about matrices, which you may have studied in previous courses. In the rest of the chapter you will see how you can use matrices to perform these geometric transformations algebraically.

Objective

- Given two matrices, find their sum and product.
- Given a square matrix, find its multiplicative inverse.
- Use matrices to solve a system of linear equations.

A **matrix** is a rectangular array of numbers.

$$\begin{bmatrix} 2 & 5 & 3 \\ -1 & 4 & -2 \end{bmatrix} \quad \begin{bmatrix} 5 & 1 \\ 7 & 3 \\ 2 & -4 \end{bmatrix} \quad \begin{bmatrix} 9 \\ 7 \\ 1 \\ 3 \end{bmatrix} \quad \begin{bmatrix} 2 \\ -5 \end{bmatrix} \quad \begin{bmatrix} 2 & 3 & -4 \\ -1 & 5 & 7 \\ 9 & -8 & -6 \end{bmatrix}$$

2×3 matrix 3×2 matrix 4×1 matrix 2×1 matrix 3×3 (square) matrix

The numbers in a matrix are called **elements.** When the dimensions of a matrix are stated, the number of rows is always given first. So the first matrix above is called a 2×3 (read "two by three") matrix because it has 2 *rows* and 3 *columns.* A *square matrix* has the same number of rows and columns. In this section you will see how to perform operations on matrices.

Addition and Subtraction

You add or subtract two matrices with the same dimensions by adding or subtracting their corresponding elements. For instance,

$$\begin{bmatrix} 5 & 2 & 7 \\ 1 & 3 & 9 \end{bmatrix} + \begin{bmatrix} 4 & 6 & 8 \\ 5 & 1 & 3 \end{bmatrix} = \begin{bmatrix} 9 & 8 & 15 \\ 6 & 4 & 12 \end{bmatrix}$$

Likewise,

$$\begin{bmatrix} 6 & 5 \\ 2 & 3 \end{bmatrix} - \begin{bmatrix} 4 & 7 \\ 5 & 2 \end{bmatrix} = \begin{bmatrix} 2 & -2 \\ -3 & 1 \end{bmatrix}$$

You can add or subtract two matrices only if they have exactly the same dimensions. Such matrices are called *commensurate* foar addition or subtraction. If you try adding or subtracting incommensurate matrices on your calculator, you will get an error message.

Multiplication by a Scalar

To multiply a matrix by a scalar (a number), multiply each element of the matrix by that scalar. For instance,

$$5\begin{bmatrix} 2 & 3 & 4 \\ 6 & 1 & -7 \end{bmatrix} = \begin{bmatrix} 10 & 15 & 20 \\ 30 & 5 & -35 \end{bmatrix}$$

Multiplication by a scalar is equivalent to repeated addition. For example, multiplying a matrix by 5 is equivalent to adding five of the same matrices.

Multiplication of Two Matrices

Traffic engineers often use matrices to analyze traffic flow.

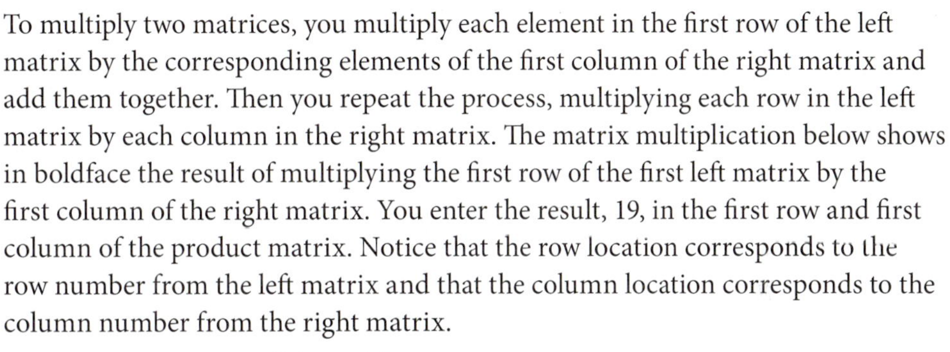

To multiply two matrices, you multiply each element in the first row of the left matrix by the corresponding elements of the first column of the right matrix and add them together. Then you repeat the process, multiplying each row in the left matrix by each column in the right matrix. The matrix multiplication below shows in boldface the result of multiplying the first row of the first left matrix by the first column of the right matrix. You enter the result, 19, in the first row and first column of the product matrix. Notice that the row location corresponds to the row number from the left matrix and that the column location corresponds to the column number from the right matrix.

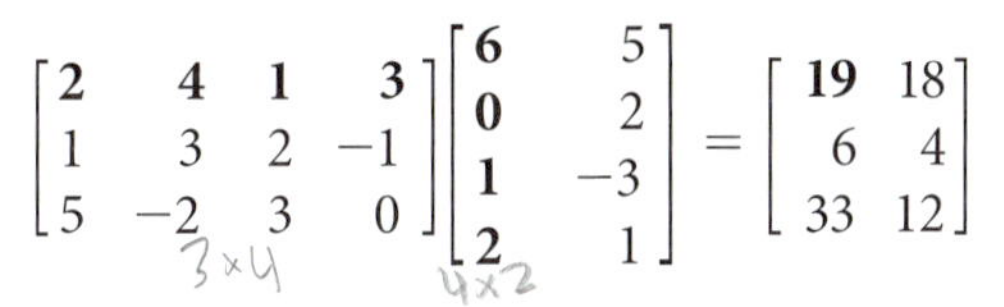

$$\begin{bmatrix} \mathbf{2} & \mathbf{4} & \mathbf{1} & \mathbf{3} \\ 1 & 3 & 2 & -1 \\ 5 & -2 & 3 & 0 \end{bmatrix}\begin{bmatrix} \mathbf{6} & 5 \\ \mathbf{0} & 2 \\ \mathbf{1} & -3 \\ \mathbf{2} & 1 \end{bmatrix} = \begin{bmatrix} \mathbf{19} & 18 \\ 6 & 4 \\ 33 & 12 \end{bmatrix}$$

$$(\text{row } 1) \cdot (\text{column } 1) = (2)(6) + (4)(0) + (1)(1) + (3)(2) = 19$$

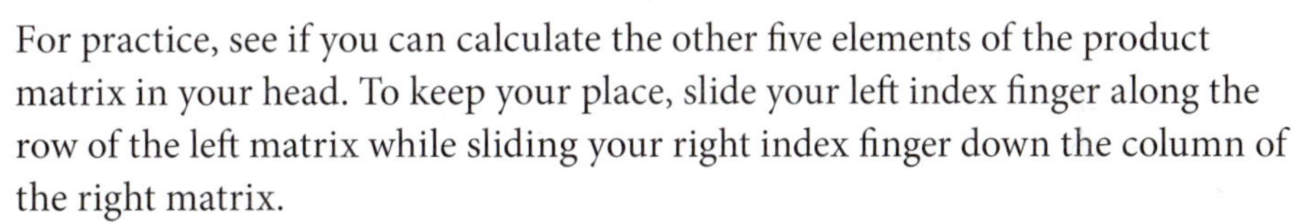

For practice, see if you can calculate the other five elements of the product matrix in your head. To keep your place, slide your left index finger along the row of the left matrix while sliding your right index finger down the column of the right matrix.

Make note of these important statements about matrix multiplication.

- To be commensurate for multiplication, each *row* of the left matrix must have the same number of elements as each *column* of the right matrix.

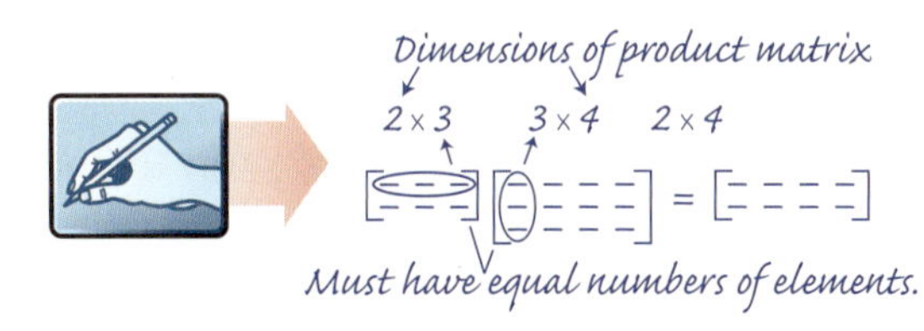

- The product matrix has the same number of *rows* as the left matrix and the same number of *columns* as the right matrix.
- Commuting two matrices $[A]$ and $[B]$ might make them *incommensurate* for multiplication. Even when the commuted matrices are commensurate, the product $[B][A]$ can be different from the product $[A][B]$. Thus matrix multiplication is *not commutative.*

Identities and Inverses for Multiplication of Square Matrices

The 3×3 **identity matrix** is the square matrix

$$[I] = \begin{bmatrix} 1 & 0 & 0 \\ 0 & 1 & 0 \\ 0 & 0 & 1 \end{bmatrix}$$

An identity matrix is a square matrix with 1s along the *main diagonal* (the diagonal from the upper-left corner to the lower-right corner) and 0s everywhere else. Multiplying a square matrix $[A]$ by the identity matrix in either order leaves $[A]$ unchanged, as you can check either manually or on your grapher.

$$\begin{bmatrix} 1 & 0 & 0 \\ 0 & 1 & 0 \\ 0 & 0 & 1 \end{bmatrix}\begin{bmatrix} 4 & -7 & 5 \\ 3 & 6 & 9 \\ 2 & 8 & -1 \end{bmatrix} = \begin{bmatrix} 4 & -7 & 5 \\ 3 & 6 & 9 \\ 2 & 8 & -1 \end{bmatrix} \quad \text{and}$$

$$\begin{bmatrix} 4 & -7 & 5 \\ 3 & 6 & 9 \\ 2 & 8 & -1 \end{bmatrix}\begin{bmatrix} 1 & 0 & 0 \\ 0 & 1 & 0 \\ 0 & 0 & 1 \end{bmatrix} = \begin{bmatrix} 4 & -7 & 5 \\ 3 & 6 & 9 \\ 2 & 8 & -1 \end{bmatrix}$$

If the product of two square matrices is the identity matrix, then the two matrices are **inverses** of each other. The inverse of a matrix $[M]$ is denoted $[M]^{-1}$. For instance,

$$\begin{bmatrix} 3 & 2 \\ 8 & 7 \end{bmatrix}\begin{bmatrix} \frac{7}{5} & -\frac{2}{5} \\ -\frac{8}{5} & \frac{3}{5} \end{bmatrix} = \begin{bmatrix} 1 & 0 \\ 0 & 1 \end{bmatrix} \quad \text{so} \quad \begin{bmatrix} 3 & 2 \\ 8 & 7 \end{bmatrix}^{-1} = \begin{bmatrix} \frac{7}{5} & -\frac{2}{5} \\ -\frac{8}{5} & \frac{3}{5} \end{bmatrix}$$

Note that the numerators in the inverse matrix are the elements of the original matrix but rearranged and, for two values, changed in sign. The denominators are all 5, which is the **determinant** of the first matrix, written det $[M]$ or $|M|$. You can find the determinant of a matrix by using the built-in features on your grapher. For a 2×2 matrix, you can calculate the determinant as

det $[M]$ = (upper left times lower right) − (upper right times lower left)

$$\det\begin{bmatrix} 3 & 2 \\ 8 & 7 \end{bmatrix} = (3)(7) - (2)(8) = 5$$

An easy way to find the inverse of a 2×2 matrix is to interchange the top-left and bottom-right elements, change the signs of the other two elements, and multiply by the reciprocal of the determinant of the matrix. That is,

$$\text{If } [M] = \begin{bmatrix} a & b \\ c & d \end{bmatrix} \quad \text{then} \quad [M]^{-1} = \frac{1}{\det [M]}\begin{bmatrix} d & -b \\ -c & a \end{bmatrix}$$

The matrix $\begin{bmatrix} d & -b \\ -c & a \end{bmatrix}$ is called the *adjugate* or *adjoint* of $[M]$, abbreviated adj $[M]$. You can express the inverse of any square matrix as

$$[M]^{-1} = \frac{1}{\det [M]} \cdot \text{adj } [M]$$

In Exploration 13-2 on the next page, you will learn how to compute the inverse of a 3×3 matrix with pencil and paper. However, the examples and problems in this chapter assume that you will use your grapher to find the inverse of a matrix.

EXAMPLE 1 ➤ Consider the matrix $[C] = \begin{bmatrix} 2 & 3 & 4 \\ 5 & 1 & 2 \\ 6 & 8 & 7 \end{bmatrix}$.

a. Find $[C]^{-1}$, and show that $[C]^{-1}[C]$ equals the identity matrix.

b. Find det $[C]$ and adj $[C]$.

SOLUTION First, enter $[C]$ into your grapher using the matrix menu.

a. To find the inverse, enter $[C]^{-1}$.

$$[C]^{-1} = \begin{bmatrix} -0.1836... & 0.2244... & 0.0408... \\ -0.4693... & -0.2040... & 0.3265... \\ 0.6938... & 0.0408... & -0.2653... \end{bmatrix}$$

To multiply the result by $[C]$, enter ans*$[C]$ or $[C]^{-1}[C]$.

$$[C]^{-1}[C] = \begin{bmatrix} 1 & 0 & 0 \\ 0 & 1 & 0 \\ 0 & 0 & 1 \end{bmatrix}$$

This is the 3×3 identity matrix.

b. det $[C] = 49$

Enter det $[C]$ on your grapher.

Most graphers do not calculate the adjoint of a matrix directly. However,

$$[C]^{-1} = \frac{1}{\det [C]} \cdot \text{adj } [C]$$

so you can find adj $[C]$ this way:

$$\text{adj } [C] = 49\,[C]^{-1} = \begin{bmatrix} -9 & 11 & 2 \\ -23 & -10 & 16 \\ 34 & 2 & -13 \end{bmatrix}$$

Note that if the elements of $[C]$ are integers, then the elements of adj $[C]$ are also integers.

James Clerk Maxwell (1831–1879), a Scottish physicist, developed equations that quantify the relationship between electricity, magnetism, and electromagnetic waves. These equations can be expressed in a set of matrix equations.

Matrix Solution of a Linear System

You can use the inverse of a matrix to solve a system of linear equations, such as

$$\begin{aligned} 2x + 3y + 4z &= 14 \\ 5x + y + 2z &= 24 \\ 6x + 8y + 7z &= 29 \end{aligned}$$

You can write the left sides of these three equations as the product of two matrices:

$$\begin{bmatrix} 2 & 3 & 4 \\ 5 & 1 & 2 \\ 6 & 8 & 7 \end{bmatrix} \begin{bmatrix} x \\ y \\ z \end{bmatrix} = \begin{bmatrix} 2x + 3y + 4z \\ 5x + y + 2z \\ 6x + 8y + 7z \end{bmatrix}$$

So you can write the system as

$$\begin{bmatrix} 2 & 3 & 4 \\ 5 & 1 & 2 \\ 6 & 8 & 7 \end{bmatrix} \begin{bmatrix} x \\ y \\ z \end{bmatrix} = \begin{bmatrix} 14 \\ 24 \\ 29 \end{bmatrix}$$

Let $[C]$ stand for the coefficient matrix on the far left, $[V]$ for the 3×1 matrix containing the variables, and $[A]$ for the 3×1 matrix on the right containing the "answers." You can write this system (or any other system of n linear equations in n variables) in matrix form:

$$[C][V] = [A]$$

To find the values of the variables in $[V]$, you can eliminate $[C]$ by left-multiplying both sides of the equation by $[C]^{-1}$:

$$[C]^{-1}([C][V]) = [C]^{-1}[A]$$

Associating $[C]^{-1}$ and $[C]$ gives the identity matrix, $[I]$.

$$([C]^{-1}[C])[V] = [C]^{-1}[A]$$

$$[I][V] = [C]^{-1}[A]$$

$$[V] = [C]^{-1}[A] \qquad \text{or} \qquad [C]^{-1}[A] = [V]$$

For these equations, the calculation is

$$\begin{bmatrix} x \\ y \\ z \end{bmatrix} = \begin{bmatrix} 2 & 3 & 4 \\ 5 & 1 & 2 \\ 6 & 8 & 7 \end{bmatrix}^{-1} \begin{bmatrix} 14 \\ 24 \\ 29 \end{bmatrix} = \begin{bmatrix} 4 \\ -2 \\ 3 \end{bmatrix}$$

The solution is $x = 4$, $y = -2$, and $z = 3$.

In this exploration you will learn how to invert a 3×3 matrix "by hand" and confirm your answer by grapher.

EXPLORATION 13-2: Inverse of a 3 × 3 Matrix

You learned that the inverse of a matrix $[M]$ is given by

$$[M]^{-1} = \frac{1}{\det [M]} \cdot \text{adj } [M]$$

where $\det [M]$ is the determinant of $[M]$ and adj $[M]$ is the adjugate or adjoint of $[M]$. In this exploration, you will use this fact to calculate the inverse of a 3×3 matrix.

1. Let $[C] = \begin{bmatrix} 2 & 3 & 4 \\ 5 & 1 & 2 \\ 6 & 8 & 7 \end{bmatrix}$

The *transpose* of $[C]$ is $[C]^T = \begin{bmatrix} 2 & 5 & 6 \\ 3 & 1 & 8 \\ 4 & 2 & 7 \end{bmatrix}$

Explain what you must do to $[C]$ to get $[C]^T$.

2. Each element of $[C]^T$ has a *minor determinant* that is found by covering up the row and column of that element. For instance, the minor of 2 in the upper-left corner of $[C]^T$ is

$$\begin{vmatrix} 1 & 8 \\ 2 & 7 \end{vmatrix} = -9$$

Show that the minor of 5 in the first row, second column of $[C]^T$ is equal to -11.

continued

3. Each element of $[C]^T$ also has a *cofactor,* equal to the minor determinant with this alternating sign pattern applied:

$$\begin{matrix} + & - & + \\ - & + & - \\ + & - & + \end{matrix}$$

Use the minors in Problem 2 to write the cofactors of 2 and 5 in $[C]^T$.

4. The adjugate or adjoint of $[C]$, written adj $[C]$, is the matrix of the cofactors of $[C]^T$. By evaluating the other seven cofactors, show that

$$\text{adj }[C] = \begin{bmatrix} -9 & 11 & 2 \\ -23 & -10 & 16 \\ 34 & 2 & -13 \end{bmatrix}$$

5. Show that the determinant of $[C]$ is det $[C] = 49$.

6. The inverse of $[C]$ is

$$[C]^{-1} = \frac{1}{49}\begin{bmatrix} -9 & 11 & 2 \\ -23 & -10 & 16 \\ 34 & 2 & -13 \end{bmatrix}$$

$$= \begin{bmatrix} -\frac{9}{49} & \frac{11}{49} & \frac{2}{49} \\ -\frac{23}{49} & -\frac{10}{49} & \frac{16}{49} \\ \frac{34}{49} & \frac{2}{49} & -\frac{13}{49} \end{bmatrix}$$

Check this by finding $[C]^{-1}$ on your grapher and then multiplying the answer by 49. Does the result equal adj $[C]$?

7. What did you learn as a result of doing this exploration that you did not know before?

Problem Set 13-2

Reading Analysis

From what you have read in this section, what do you consider to be the main idea? The word order "rows, then columns" appears frequently in connection with matrices. Explain how this word order appears in naming the dimensions of a matrix, in determining that two matrices are commensurate for multiplication, and in writing the product of two matrices. How can you tell that two matrices are inverses of each other? How can you solve a system of linear equations using matrices?

Quick Review

Q1. A square dilated to 40% of its original length has __?__% of the original area.

Q2. After two iterations, Sierpiński's square has a total area that is __?__% of the pre-image area.

Q3. After two iterations, Sierpiński's square has a total perimeter that is __?__% of the pre-image perimeter.

Q4. What type of function has the add–multiply property for regularly spaced x-values?

Q5. Write the general equation for a power function.

Q6. Find the slope of the line perpendicular to the graph of the equation $3x + 7y = 41$.

Q7. How many degrees are there in an angle with measure $\frac{\pi}{3}$ radians?

Q8. Expand the square: $(3x - 5)^2$

Q9. Find 2% of 3000.

Q10. Find $(3\vec{i} - 2\vec{j} - 7\vec{k}) \cdot (8\vec{i} + 6\vec{j} - \vec{k})$.

For Problems 1–10, perform the given operation by hand. Use your grapher to confirm that your answers are correct.

1. $\begin{bmatrix} 3 & 5 \\ -2 & 4 \\ 7 & 1 \end{bmatrix} + \begin{bmatrix} -5 & 8 \\ 2 & 6 \\ -7 & 10 \end{bmatrix}$

2. $\begin{bmatrix} 5 & 7 & -4 \\ 10 & 0 & -2 \\ 11 & -3 & 12 \end{bmatrix} - \begin{bmatrix} 4 & 5 & -7 \\ 6 & -5 & -8 \\ 4 & -11 & 3 \end{bmatrix}$

3. $4[-8\ \ 5\ \ 3] - 2[-5\ \ -1\ \ 7]$

4. $7\begin{bmatrix} 2 & 8 \\ -4 & 1 \end{bmatrix} + 3\begin{bmatrix} -5 & 1 \\ 2 & -6 \end{bmatrix}$

5. $[-2\ \ 3\ \ 5]\begin{bmatrix} 1 & 4 \\ 7 & -3 \\ -1 & -5 \end{bmatrix}$

6. $\begin{bmatrix} 2 & 4 & -3 \\ 5 & 1 & 2 \\ -1 & 3 & 4 \end{bmatrix}\begin{bmatrix} -1 & 3 & 1 \\ 2 & 4 & 3 \\ 1 & 0 & 2 \end{bmatrix}$

7. $\begin{bmatrix} 4 & 7 & 5 \\ 3 & 2 & -1 \end{bmatrix}\begin{bmatrix} 6 & 8 \\ 3 & -6 \end{bmatrix}$

8. $\begin{bmatrix} 1 & 0 & 0 \\ 0 & 1 & 0 \\ 0 & 0 & 1 \end{bmatrix}\begin{bmatrix} 1 & 4 & 7 \\ 2 & 5 & 8 \\ 3 & 6 & 9 \end{bmatrix}$

9. $\begin{bmatrix} 4 & 4 \\ 5 & 3 \end{bmatrix}\begin{bmatrix} 1 & 0 \\ 0 & 1 \end{bmatrix}$

10. $\begin{bmatrix} 2 & 3 & 4 \\ 5 & 1 & 2 \\ 6 & 5 & 7 \end{bmatrix}\begin{bmatrix} -3 & -1 & 2 \\ -23 & -10 & 16 \\ 19 & 8 & -13 \end{bmatrix}$

11. *Investment Income Problem:* A brokerage company has investments in four states: California, Arkansas, Texas, and South Dakota. The investments are bonds, mortgages, and other loans. $[M]$ shows the number of millions of dollars in each investment in each state.

$$[M] = \begin{array}{c} \text{CA} \quad \text{AR} \quad \text{TX} \quad \text{SD} \\ \begin{bmatrix} 32 & 8 & 15 & 2 \\ 15 & 20 & 17 & 9 \\ 14 & 22 & 23 & 7 \end{bmatrix} \end{array} \begin{array}{l} \text{Bonds} \\ \text{Mortgages} \\ \text{Loans} \end{array}$$

The percentages of annual income that the investments yield are bonds, 6%; mortgages, 9%; loans, 11%. These numbers are represented by a yield matrix, $[Y]$.

$$[Y] = [0.06\ \ \ 0.09\ \ \ 0.11]$$

a. Find the product $[Y][M]$. Use the product matrix to find the annual income the company earns from investments in Texas. How much of this comes from mortgages?

b. Explain why you cannot find the real-world product $[M][Y]$.

c. Explain why it is impossible in the mathematical world to find the product $[M][Y]$.

12. *Virus Problem:* A virus sweeps through a high school, infecting 30% of the 11th graders and 20% of the 12th graders, as represented by $[P]$.

$$[P] = \begin{array}{c} \text{11th} \quad \text{12th} \\ \begin{bmatrix} 0.3 & 0.2 \\ 0.7 & 0.8 \end{bmatrix} \end{array} \begin{array}{l} \text{Ill} \\ \text{Well} \end{array}$$

There are 100 11th-grade boys, 110 11th-grade girls, 120 12th-grade boys, and 130 12th-grade girls, as represented by $[S]$.

$$[S] = \begin{array}{c} \text{Boys} \quad \text{Girls} \\ \begin{bmatrix} 100 & 110 \\ 120 & 130 \end{bmatrix} \end{array} \begin{array}{l} \text{11th grade} \\ \text{12th grade} \end{array}$$

a. Show that the product $[P][S]$ does not equal $[S][P]$.

b. Identify the real-world quantities that the elements of the product $[P][S]$ represent.

c. Identify the real-world quantities that the elements of the product $[S][P]$ represent.

For Problems 13 and 14,

a. Find $[M]^{-1}$. Show that $[M]^{-1}[M] = [I]$ and that $[M][M]^{-1} = [I]$.

b. Find det $[M]$. Find adj $[M]$ and show that all the elements of adj $[M]$ are integers.

13. $[M] = \begin{bmatrix} 3 & 5 & 2 \\ 4 & 7 & 7 \\ 5 & 8 & 9 \end{bmatrix}$

14. $[M] = \begin{bmatrix} 3 & 7 & 1 & -2 \\ 4 & 5 & -1 & 6 \\ 2 & 3 & 8 & 1 \\ 5 & 4 & 9 & 7 \end{bmatrix}$

For Problems 15 and 16, find det $[M]$. Explain why your grapher gives you an error message when you try to find $[M]^{-1}$. Then state what you think a determinant "determines."

15. $[M] = \begin{bmatrix} 6 & 3 \\ 8 & 4 \end{bmatrix}$

16. $[M] = \begin{bmatrix} 1 & 2 & 3 \\ 4 & 5 & 6 \\ 7 & 8 & 9 \end{bmatrix}$

For Problems 17 and 18, solve the system of equations using the inverse of a matrix.

17.
$$\begin{aligned} 5x + 3y - 7z &= 3 \\ 10x - 4y + 6z &= 5 \\ 15x + y - 8z &= -2 \end{aligned}$$

18.
$$\begin{aligned} w - 5x + 2y - z &= -18 \\ 3w + x - 3y + 2z &= 17 \\ 4w - 2x + y - z &= -1 \\ -2w + 3x - y + 4z &= 11 \end{aligned}$$

19. *Quadratic Function Problem:* Recall that a quadratic function has the general equation $y = ax^2 + bx + c$. To find the equation for the particular function that contains the points (4, 13), (6, 29), and (8, 49), substitute each pair of x- and y-values into the general equation to get three linear equations with the three unknown constants a, b, and c. Solve the equations as a system to find a particular quadratic function that contains these three points. Use the equation to predict the value of y when x equals 20.

20. *Quartic Function Problem:* The general equation for a quartic (fourth-degree) function is $y = ax^4 + bx^3 + cx^2 + dx + e$, where a, b, c, d, and e stand for constants. Find a particular equation for the quartic function that contains the points (1, 15), (2, 19), (3, 75), (4, 273), and (5, 751). Use the equation to predict the value of y when x equals -3.

21. Show that matrix multiplication is not commutative by showing that

$$\begin{bmatrix} 2 & 3 \\ 4 & 5 \end{bmatrix}\begin{bmatrix} 6 & 7 \\ 8 & 9 \end{bmatrix} \neq \begin{bmatrix} 6 & 7 \\ 8 & 9 \end{bmatrix}\begin{bmatrix} 2 & 3 \\ 4 & 5 \end{bmatrix}$$

22. *Multipliers of Zero Problem:* For real numbers, the zero-product property states that if the product of two factors is 0, then at least one of the factors is 0. Show that this property is *false* for matrix multiplication by finding two 2×2 matrices whose product is the zero matrix (the matrix in which each element is 0) but for which no element of either matrix is 0. (Find a matrix whose determinant is 0, and multiply it by its adjoint matrix.) The two matrices you find are called *multipliers of zero.*

13-3 Rotation and Dilation Matrices

Matrices have many uses both in the real world and in the mathematical world. One use is transforming geometric figures. In this section you will see how to write a matrix that will rotate or dilate a figure.

Objective Write a matrix that will apply a given dilation and rotation when it is multiplied by a matrix representing a geometric figure.

Dilations

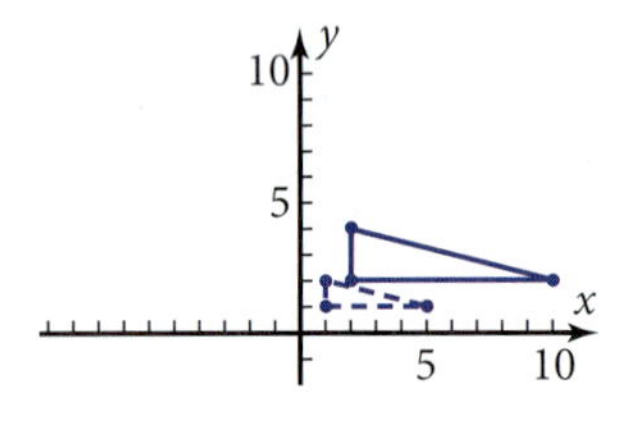

Figure 13-3a

You can represent a figure in the plane by a matrix with two rows. For instance, you can represent the smaller (dashed) triangle in Figure 13-3a by this matrix:

$$[M] = \begin{bmatrix} 1 & 5 & 1 \\ 1 & 1 & 2 \end{bmatrix}$$

Each column represents an ordered pair corresponding to one of the vertices. The top element is the x-coordinate, and the bottom element is the y-coordinate.

To dilate this triangle, you can multiply $[M]$ by a scalar. Multiplying $[M]$ by 2 multiplies each element of the matrix by 2, dilating the triangle by a factor of two.

When applying more than one transformation to a figure (such as a dilation and a rotation), it is helpful to use a matrix to dilate a figure. You know that multiplying a matrix by the identity matrix multiplies each element in that matrix by one, resulting in the original matrix.

If you multiply the 2×2 identity matrix by 2, you get

$$[T] = 2[I] = 2\begin{bmatrix} 1 & 0 \\ 0 & 1 \end{bmatrix} = \begin{bmatrix} 2 & 0 \\ 0 & 2 \end{bmatrix}$$

which, if multiplied by $[M]$, will multiply each element of $[M]$ by a factor of 2. You can use $[T]$ as a **transformation matrix.** A transformation matrix is always the matrix on the left in a matrix multiplication. The product $[T][M]$ is the *image matrix.*

$$[T][M] = \begin{bmatrix} 2 & 10 & 2 \\ 2 & 2 & 4 \end{bmatrix}$$

If you plot the ordered pairs represented by this matrix, you get the vertices of the larger (solid) triangle in Figure 13-3a. The sides of this **image** triangle are twice as long as those of the original triangle (the **pre-image**). Each point on the image is twice as far from the origin as the corresponding point on the pre-image. The transformation represented by $[T]$ dilates any point in the Cartesian plane by a factor of 2, doubling the side lengths of any figure in the plane.

In general, you dilate by a factor of k by multiplying by the general dilation matrix.

PROPERTY: General Dilation Matrix

Matrix $[T]$ dilates a figure by a factor of k with respect to the origin:

$$[T] = \begin{bmatrix} k & 0 \\ 0 & k \end{bmatrix}$$

Rotations

You can write the identity matrix this way:

$$\begin{bmatrix} 1 & 0 \\ 0 & 1 \end{bmatrix} = \begin{bmatrix} \cos 0° & \cos 90° \\ \sin 0° & \sin 90° \end{bmatrix}$$

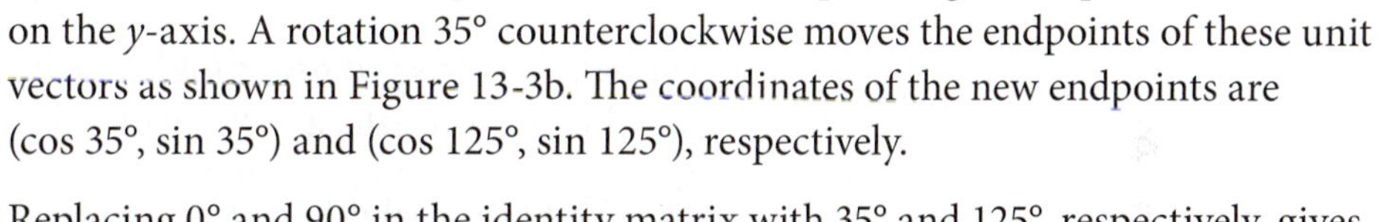

Figure 13-3b

The 1 and the 0 in the first column are the coordinates of the endpoint of a unit vector along the positive x-axis. Similarly, the 0 and the 1 in the second column are the coordinates of the endpoint of a unit vector pointing in the positive direction on the y-axis. A rotation 35° counterclockwise moves the endpoints of these unit vectors as shown in Figure 13-3b. The coordinates of the new endpoints are (cos 35°, sin 35°) and (cos 125°, sin 125°), respectively.

Replacing 0° and 90° in the identity matrix with 35° and 125°, respectively, gives a matrix $[T]$ that rotates a figure 35° counterclockwise with respect to the origin.

$$[T] = \begin{bmatrix} \cos 35° & \cos 125° \\ \sin 35° & \sin 125° \end{bmatrix}$$

The angle measure in the first column is the number of degrees by which the figure is rotated. The angle measure in the second column is 90° more than the rotation angle. The general rotation matrix for an angle θ is shown in the box.

PROPERTY: General Rotation Matrix

Matrix $[T]$ rotates a figure in the plane counterclockwise with respect to the origin by an angle θ.

$$[T] = \begin{bmatrix} \cos\theta & \cos(\theta + 90°) \\ \sin\theta & \sin(\theta + 90°) \end{bmatrix}$$

EXAMPLE 1 ➤ Write a transformation matrix that will rotate a figure clockwise 70° and dilate it by a factor of 1.6. Use it to transform the small pre-image triangle in Figure 13-3a. Plot the pre-image and the image on graph paper.

SOLUTION The desired transformation matrix is the product of the dilation matrix and the rotation matrix. Note that the rotation is clockwise, so the rotation angle is −70°.

$$[T] = \begin{bmatrix} 1.6 & 0 \\ 0 & 1.6 \end{bmatrix} \begin{bmatrix} \cos(-70°) & \cos 20° \\ \sin(-70°) & \sin 20° \end{bmatrix}$$

20° is 90° + (−70°).

$$= \begin{bmatrix} 1.6\cos(-70°) & 1.6\cos 20° \\ 1.6\sin(-70°) & 1.6\sin 20° \end{bmatrix}$$

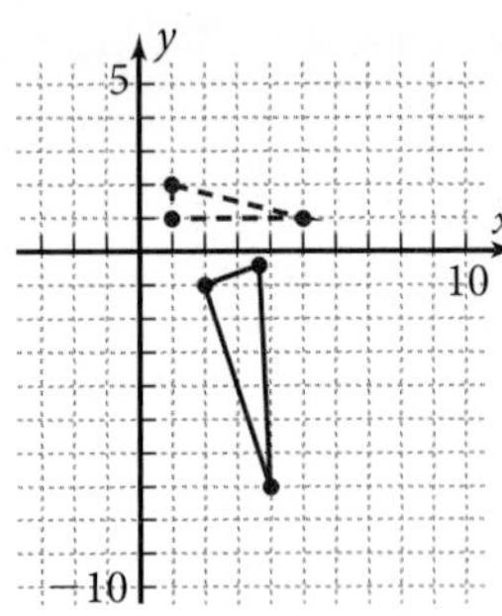

Figure 13-3c

To find the image matrix, multiply $[T]$ by the earlier pre-image matrix, $[M]$:

$$[T][M] = \begin{bmatrix} 1.6\cos(-70°) & 1.6\cos 20° \\ 1.6\sin(-70°) & 1.6\sin 20° \end{bmatrix}\begin{bmatrix} 1 & 5 & 1 \\ 1 & 1 & 2 \end{bmatrix}$$

$$= \begin{bmatrix} 2.0507... & 4.2396... & 3.5542... \\ -0.9562... & -6.9703... & -0.4090... \end{bmatrix}$$

Figure 13-3c shows the pre-image (dashed) and the image (solid) plotted on graph paper. You can confirm with a protractor that the image has been rotated clockwise 70°. You can confirm with a ruler that each point on the image is 1.6 times as far from the origin as the corresponding point on the pre-image; thus, the lengths of the sides in the image have been multiplied by 1.6.

Iterated Transformations

Iterating a transformation means performing the transformation over and over, each time operating on the image that resulted from the previous transformation.

EXAMPLE 2 Write a transformation matrix $[A]$ that will rotate a figure counterclockwise 60° and dilate it by a factor of 0.9. Starting with the pre-image triangle $[M]$ from Example 1, apply $[A]$ iteratively for four iterations. Plot the pre-image and each of the four images on graph paper. Describe the pattern formed by the triangles.

SOLUTION

$$[A] = \begin{bmatrix} 0.9\cos 60° & 0.9\cos 150° \\ 0.9\sin 60° & 0.9\sin 150° \end{bmatrix}$$

$$[A][M] \approx \begin{bmatrix} -0.3 & 1.5 & 1.1 \\ 1.2 & 4.3 & -1.7 \end{bmatrix}$$ Display only one decimal place to make plotting easier.

$$[A]\text{ Ans} \approx \begin{bmatrix} -1.1 & -2.7 & -1.8 \\ 0.3 & 3.1 & -0.1 \end{bmatrix}$$ Multiply $[A]$ by the unrounded result of the previous iteration; this is the second iteration.

$$[A]\text{ Ans} \approx \begin{bmatrix} -0.7 & -3.6 & -0.7 \\ -0.7 & -0.7 & -1.5 \end{bmatrix}$$ Third iteration.

$$[A]\text{ Ans} \approx \begin{bmatrix} 0.2 & -1.1 & 0.8 \\ -0.9 & -3.2 & -1.2 \end{bmatrix}$$ Fourth iteration.

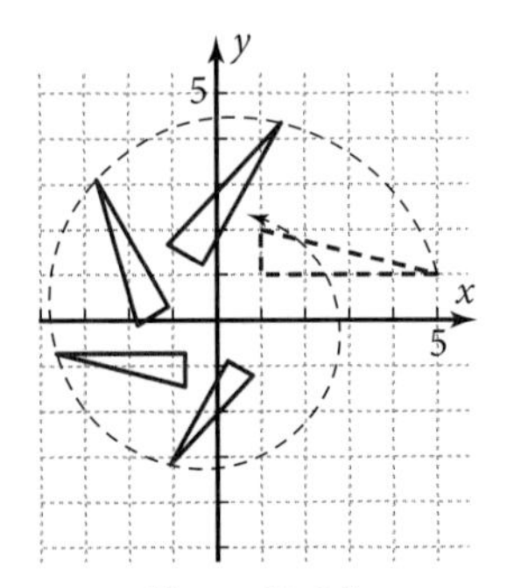

Figure 13-3d

Plot the images on graph paper. As shown in Figure 13-3d, the images get smaller, and they spiral toward the origin as if they were "attracted" to it.

In Problem 19, you will write a program to calculate and plot image matrices iteratively on your grapher.

Problem Set 13-3

Reading Analysis

From what you have read in this section, what do you consider to be the main idea? Write the 2×2 identity matrix and show how it can be written using cosines and sines. How would this form of the identity matrix be changed so that it rotates an image counterclockwise by 27°? How would this rotation matrix be changed so that it also dilates the image by a factor of 0.7?

Quick Review

Q1. Find $\begin{bmatrix} 2 & 5 \\ 3 & 1 \end{bmatrix} + \begin{bmatrix} 4 & 7 \\ 1 & 6 \end{bmatrix}$.

Q2. Find $\begin{bmatrix} 4 & 7 \\ 1 & 6 \end{bmatrix} + \begin{bmatrix} 2 & 5 \\ 3 & 1 \end{bmatrix}$.

Q3. Does $[A] + [B]$ always equal $[B] + [A]$?

Q4. Based on your answer to Problem Q3, matrix addition is a(n) __?__ operation.

Q5. Find $\begin{bmatrix} 2 & 5 \\ 3 & 1 \end{bmatrix}\begin{bmatrix} 4 & 7 \\ 1 & 6 \end{bmatrix}$.

Q6. Find $\begin{bmatrix} 4 & 7 \\ 1 & 6 \end{bmatrix}\begin{bmatrix} 2 & 5 \\ 3 & 1 \end{bmatrix}$.

Q7. Does $[A][B]$ always equal $[B][A]$?

Q8. Find det $\begin{bmatrix} 4 & 7 \\ 1 & 6 \end{bmatrix}$.

Q9. Find $\begin{bmatrix} 4 & 7 \\ 1 & 6 \end{bmatrix}^{-1}$.

Q10. The position vector for the point (3, −7, 5) is __?__.

For Problems 1–6, draw the pre-image represented by the matrix on the right. Assume that the points are connected in the order they appear to form a closed figure. Then multiply and plot the image. Describe the transformation.

1. $\begin{bmatrix} 2 & 0 \\ 0 & 2 \end{bmatrix}\begin{bmatrix} 2 & 3 & 1 \\ 1 & 2 & 3 \end{bmatrix}$

2. $\begin{bmatrix} 3 & 0 \\ 0 & 3 \end{bmatrix}\begin{bmatrix} 1 & 3 & 2 \\ 1 & 1 & 4 \end{bmatrix}$

3. $\begin{bmatrix} \frac{1}{2} & 0 \\ 0 & \frac{1}{2} \end{bmatrix}\begin{bmatrix} 2 & 2 & -4 & -4 \\ 6 & -3 & -3 & 6 \end{bmatrix}$

During World War II, Navajo Code Talkers created a code based on the Navajo language—a code that the Germans and Japanese could not break. Matrices are often used to create and break secret codes.

4. $\begin{bmatrix} \frac{1}{3} & 0 \\ 0 & \frac{1}{3} \end{bmatrix}\begin{bmatrix} 9 & -3 & -3 & 9 \\ 0 & 0 & 12 & 12 \end{bmatrix}$

5. $\begin{bmatrix} 0.8 & -0.6 \\ 0.6 & 0.8 \end{bmatrix}\begin{bmatrix} 3 & 3 & 6 & 6 \\ 1 & 2 & 2 & 1 \end{bmatrix}$

6. $\begin{bmatrix} 0.8 & 0.6 \\ -0.6 & 0.8 \end{bmatrix}\begin{bmatrix} 1 & 1 & 2 \\ 2 & 5 & 5 \end{bmatrix}$

For Problems 7–12, write a transformation matrix, and then use it to transform the given figure. Plot the pre-image and the image, confirming that your transformation matrix is correct.

7. Dilate this triangle by a factor of 3.

$$\begin{bmatrix} 1 & 3 & 4 \\ 1 & 1 & 5 \end{bmatrix}$$

8. Dilate this dart by a factor of 2.

$$\begin{bmatrix} 1 & 3 & 5 & 3 \\ 1 & 2 & 1 & 6 \end{bmatrix}$$

9. Rotate the pre-image triangle in Problem 7 clockwise 50°.

10. Rotate the pre-image dart in Problem 8 counterclockwise 70°.

11. Dilate the pre-image triangle in Problem 7 by a factor of 3 and rotate it clockwise 50°.

12. Dilate the pre-image dart in Problem 8 by a factor of 2 and rotate it counterclockwise 70°.

For Problems 13 and 14, write a matrix for the given pre-image, describe the effect the transformation matrix $[A]$ will have, and then iterate the transformation four times. Plot each image on a copy of the figure.

13. $[A] = \begin{bmatrix} 0.8\cos 20^\circ & 0.8\cos 110^\circ \\ 0.8\sin 20^\circ & 0.8\sin 110^\circ \end{bmatrix}$
applied to Figure 13-3e

14. $[A] = \begin{bmatrix} 0.7\cos(-40^\circ) & 0.7\cos(50^\circ) \\ 0.7\sin(-40^\circ) & 0.7\sin(50^\circ) \end{bmatrix}$
applied to Figure 13-3f

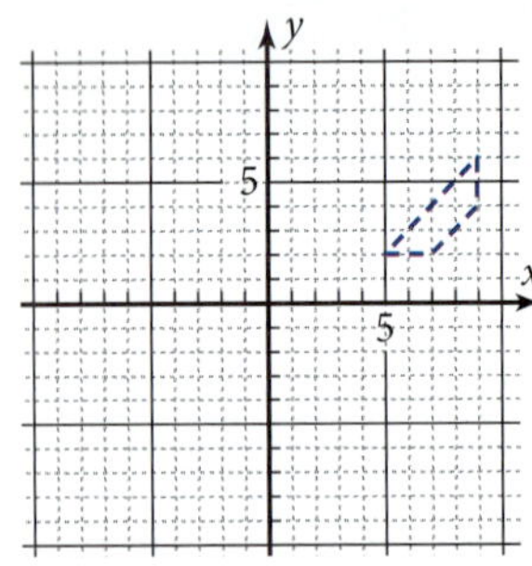

Figure 13-3e

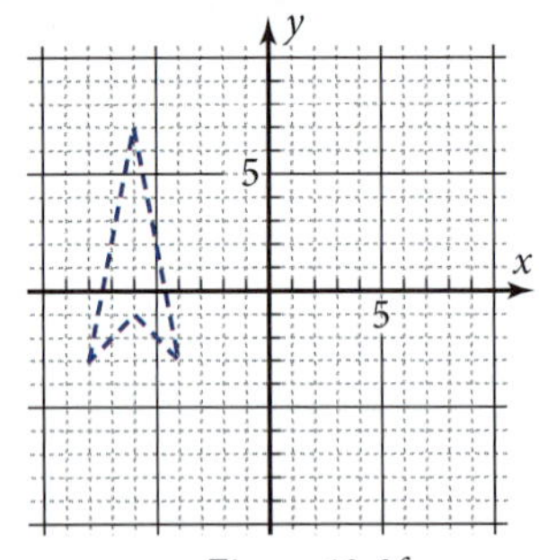

Figure 13-3f

For Problems 15–18, write a transformation matrix $[A]$ for the transformation described.

15. Rotation counterclockwise 90°

16. Rotation 180°

17. Dilation by a factor of 5 with respect to the origin

18. Dilation by a factor of 0.9 with respect to the origin

19. *Grapher Program for Iterative Transformations:* Write or download a program from **www.keymath.com/precalc** to perform iterative transformations. The program should allow you to store a transformation matrix as $[A]$ and a pre-image matrix as $[D]$. When you run the program, the grapher should first store a copy of $[D]$ as $[E]$ and plot the pre-image on the screen. When you press ENTER, the grapher should multiply $[A]$ by $[E]$, store the result back in $[E]$, plot the image, and then pause until the ENTER key is pressed again. Check your program using the transformation and pre-image in Example 2.

20. *Grapher Program Test:* Run your program from Problem 19 using the transformation and pre-image in Problem 14. Sketch the path followed by the uppermost point in the pre-image. To what fixed point do the images seem to be attracted?

21. *Dynamic Matrix Transformations Problem:* Go to **www.keymath.com/precalc** and complete the *Matrix Transformations* exploration to view the effects of multiplying the pre-image by the dilation matrix and by the rotation matrix. Answer the questions asked, and explain in writing what you learned from this problem.

22. *Journal Problem:* Write in your journal the most important thing you have learned as a result of studying matrix transformations.

13-4 Translation with Rotation and Dilation Matrices

In the previous section you saw how to use a transformation matrix to rotate and dilate a figure in the plane. In this section you will learn how to use a matrix to translate a figure to a new position without changing its size or orientation. Then you will explore the combined effects of rotating, dilating, and translating a figure iteratively. Surprisingly, the iterated images are attracted to a fixed point if the dilation reduces the images in size.

Objective Given a desired dilation, rotation, and translation, write a matrix that will apply the transformation when it is multiplied by a pre-image matrix, and find the fixed point to which the images are attracted.

In this exploration you will perform matrix operations that translate a given figure as well as rotate and dilate it.

EXPLORATION 13-4: Combined Translation, Rotation, and Dilation

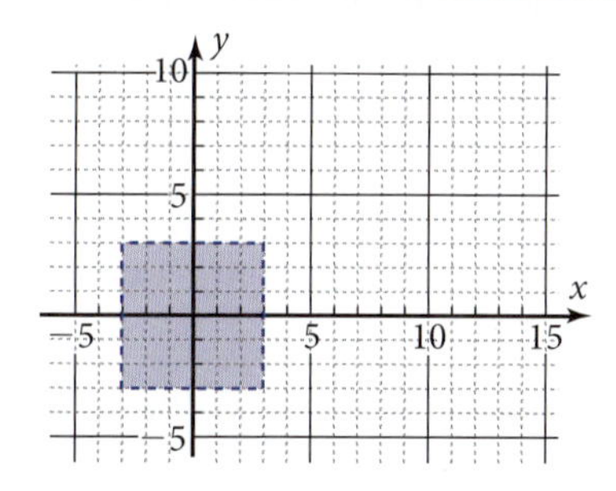

1. The figure shows the square specified by the matrix

$$[D] = \begin{bmatrix} 3 & 3 & -3 & -3 \\ 3 & -3 & -3 & 3 \end{bmatrix}$$

On graph paper, draw the image formed by translating this square by 5 units in the x-direction and 4 units in the y-direction. Write the matrix for the image.

2. Insert a row of 1s in $[D]$. Multiply it by the transformation matrix $[A]$, as shown.

$$[A][D] = \begin{bmatrix} 1 & 0 & 5 \\ 0 & 1 & 4 \\ 0 & 0 & 1 \end{bmatrix}\begin{bmatrix} 3 & 3 & -3 & -3 \\ 3 & -3 & -3 & 3 \\ 1 & 1 & 1 & 1 \end{bmatrix}$$

3. What do you notice about the first two rows of the image matrix in Problem 2?

4. Explain how the 5 and the 4 in the third column of $[A]$, along with the 1s in the bottom row of $[D]$, accomplish the translations. (Try multiplying by hand to see!)

5. Describe what the row 0, 0, 1 at the bottom of $[A]$ does.

6. What special 2 × 2 matrix is in the upper-left corner of $[A]$ in Problem 2? How would you modify $[A]$ so that it accomplishes a 70% reduction and a 30° counterclockwise rotation as well as the translations?

7. Run your iterative transformation program (from Problem 19 on page 670) with $[A]$ as modified in Problem 6 and $[D]$ as in Problem 2. Sketch the first image and the path of the centers of the images on your graph.

8. If your work in Problem 7 is correct, you should have found that the squares are being attracted to a fixed point. Find *graphically* the approximate coordinates of this "attractor."

continued

EXPLORATION, continued

9. Find the fixed point *numerically* by finding $[A]^{50}[D]$.

10. Change the pre-image matrix to

$$[D] = \begin{bmatrix} 16 & 16 & 10 & 10 \\ 2 & 8 & 8 & 2 \\ 1 & 1 & 1 & 1 \end{bmatrix}$$

Run the program again. Iterate at least 20 times. Sketch the first image and the path of the centers of the images on your graph from Problem 1. Do the images converge to the same fixed point as in Problem 8, to a different fixed point, or to no point?

11. What did you learn as a result of doing this exploration that you did not know before?

Translations

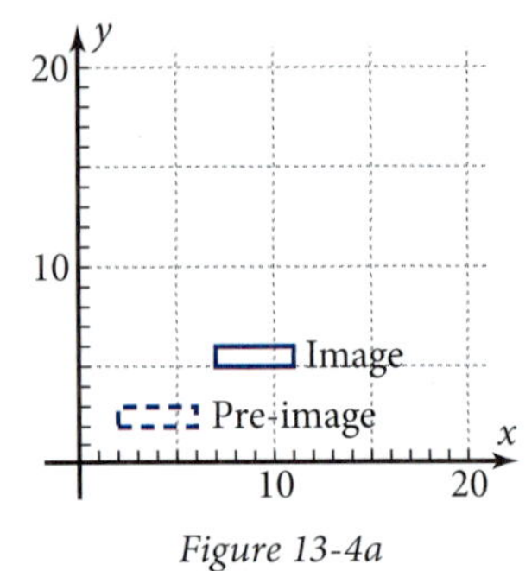

Figure 13-4a

Figure 13-4a shows the (dashed) rectangle represented by this pre-image matrix:

$$[M] = \begin{bmatrix} 2 & 6 & 6 & 2 \\ 2 & 2 & 3 & 3 \end{bmatrix}$$

The (solid) image shows the rectangle **translated** by 5 units in the x-direction and 3 units in the y-direction. You accomplish this translation algebraically by adding 5 to each x-coordinate in the top row of $[M]$ and adding 3 to each y-coordinate in the bottom row. You could perform this translation using matrix addition:

$$\begin{bmatrix} 5 & 5 & 5 & 5 \\ 3 & 3 & 3 & 3 \end{bmatrix} + \begin{bmatrix} 2 & 6 & 6 & 2 \\ 2 & 2 & 3 & 3 \end{bmatrix} = \begin{bmatrix} 7 & 11 & 11 & 7 \\ 5 & 5 & 6 & 6 \end{bmatrix}$$

However, there is a way to accomplish the translation by multiplying by a transformation matrix rather than by adding.

First, insert a third row into $[M]$ containing all 1s. Then write a 3×3 transformation matrix $[T]$ containing the identity matrix in the upper-left corner, the translations 5 and 3 in the third column, and the row 0, 0, 1 across the bottom.

$$[T][M] = \begin{bmatrix} 1 & 0 & 5 \\ 0 & 1 & 3 \\ 0 & 0 & 1 \end{bmatrix} \begin{bmatrix} 2 & 6 & 6 & 2 \\ 2 & 2 & 3 & 3 \\ 1 & 1 & 1 & 1 \end{bmatrix}$$

If you multiply these matrices, the image matrix is

$$[T][M] = \begin{bmatrix} 7 & 11 & 11 & 7 \\ 5 & 5 & 6 & 6 \\ 1 & 1 & 1 & 1 \end{bmatrix}$$

The translated figure's coordinates appear along the top two rows, and the bottom row of the image matrix contains all 1s. If you multiply the matrices by hand, you will see what has happened. The 1s in the bottom row of the pre-image matrix, together with the third column of the transformation matrix, cause the translations by 5 units and 3 units. The row 0, 0, 1 in the transformation matrix causes 1, 1, 1, 1 to appear in the bottom row of the image matrix.

Combined Translations, Dilations, and Rotations

The preceding transformation matrix $[T]$, as mentioned, has the identity matrix in its upper-left corner:

$$[T] = \begin{bmatrix} 1 & 0 & 5 \\ 0 & 1 & 3 \\ 0 & 0 & 1 \end{bmatrix}$$

If you replace this embedded identity matrix with a rotation and dilation matrix, multiplying a pre-image matrix by $[T]$ applies all the transformations.

EXAMPLE 1 ➤ Write a transformation matrix $[T]$ to rotate a figure counterclockwise 30°, dilate it by a factor of 0.8, and translate it by 5 units in the positive x-direction and 3 units in the positive y-direction. Apply the transformation to the kite specified by matrix $[M]$. Plot the pre-image and the image.

$$[M] = \begin{bmatrix} 8 & 10 & 12 & 10 \\ 7 & 2 & 7 & 8 \end{bmatrix}$$

SOLUTION First add a row of 1s to the bottom of $[M]$.

$$[M] = \begin{bmatrix} 8 & 10 & 12 & 10 \\ 7 & 2 & 7 & 8 \\ 1 & 1 & 1 & 1 \end{bmatrix}$$

The transformation matrix has the rotation and dilation matrix in the upper-left corner. The translations appear in the third column.

$$[T] = \begin{bmatrix} 0.8 \cos 30^\circ & 0.8 \cos 120^\circ & 5 \\ 0.8 \sin 30^\circ & 0.8 \sin 120^\circ & 3 \\ 0 & 0 & 1 \end{bmatrix}$$

The image is

$$[T][M] = \begin{bmatrix} 7.7425... & 11.1282... & 10.5138... & 8.7282... \\ 11.0497... & 8.3856... & 12.6497... & 12.5425... \\ 1 & 1 & 1 & 1 \end{bmatrix}$$

$$\approx \begin{bmatrix} 7.7 & 11.1 & 10.5 & 8.7 \\ 11.0 & 8.4 & 12.6 & 12.5 \\ 1 & 1 & 1 & 1 \end{bmatrix}$$

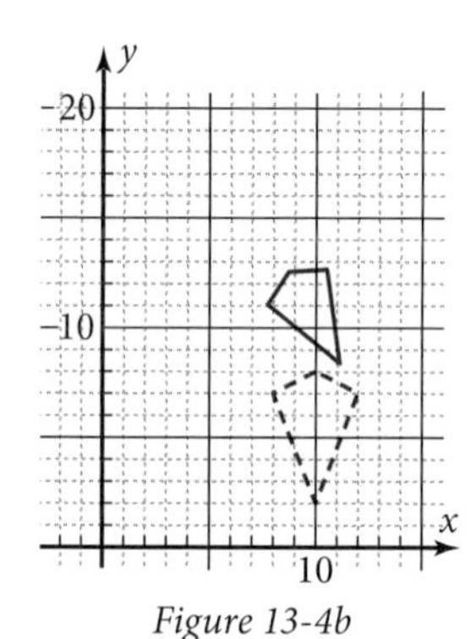

Figure 13-4b

The pre-image and image are shown in Figure 13-4b. If you are plotting on paper, you will find it easier if you set your grapher to round to one decimal place. ◄

If the transformation of Example 1 is applied iteratively, the images spiral around and are attracted to a fixed point. Unlike the pure dilation and rotation transformations of the previous section, however, the attractor is not the origin. You can use the program from Problem Set 13-3, Problem 19, to plot the iterated images and then find the approximate coordinates of this fixed point.

EXAMPLE 2 ➤ Apply the transformation of Example 1 iteratively. Sketch the resulting images. Estimate the coordinates of the point to which the images are attracted.

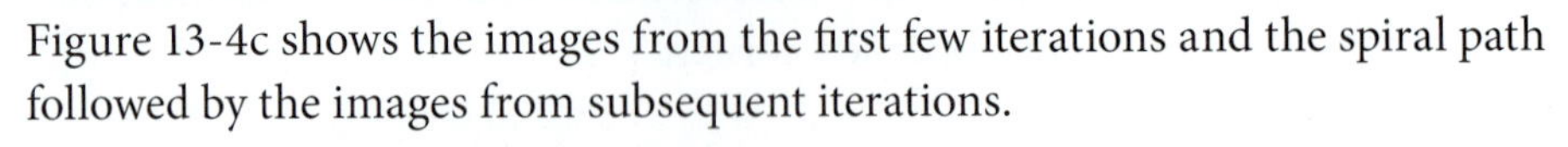

SOLUTION Figure 13-4c shows the images from the first few iterations and the spiral path followed by the images from subsequent iterations.

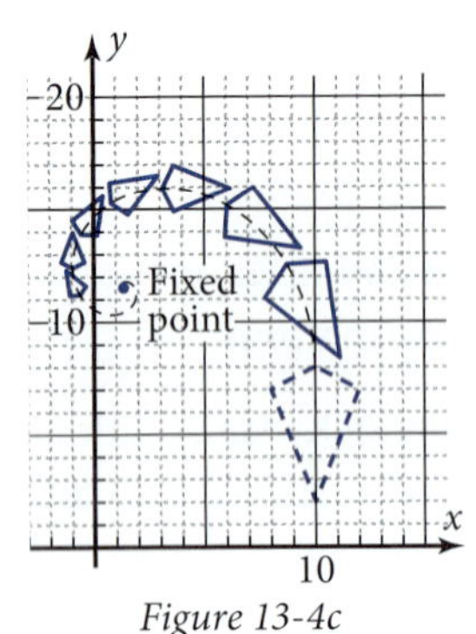

Figure 13-4c

Graphically, the images seem to be attracted to the point at about (1.3, 11.5). To estimate the coordinates of the point numerically, you can display the matrix where the program stores the images (matrix $[E]$). After 30 iterations, the image matrix is

$$\begin{bmatrix} 1.3122... & 1.3098... & 1.3073... & 1.3098... \\ 11.4914... & 11.4976... & 11.4914... & 11.4901... \\ 1 & 1 & 1 & 1 \end{bmatrix}$$

As you can see, each point in the image has been attracted to a point close to the point (1.31, 11.49). This point is called a **fixed point attractor** or simply a **fixed point.** If you apply the transformation to this point, it does not move—that is, it remains fixed. ◄

Fixed Point Attractors and Limits

The fixed point attractor is the **limit** of the image points as the number of iterations approaches infinity. The fixed point depends only on the transformation matrix, not on the pre-image. Using (x_0, y_0) for a pre-image point, (x_1, y_1) for the image of the point after the first iteration, (x_2, y_2) for the image after the second iteration, and so forth, and using (X, Y) for the fixed point, you can write the limit as described in the box.

DEFINITION: Fixed Point Attractor

If the images approach a fixed point (X, Y) when a transformation $[T]$ is applied iteratively, then

$$(X, Y) = \lim_{n\to\infty}(x_n, y_n)$$

where (x_n, y_n) is the image of a point (x_0, y_0) after n iterations.

Verbally: (X, Y) is the limit of (x_n, y_n) as n approaches infinity.

It is possible to calculate a fixed point algebraically and numerically.

EXAMPLE 3 ➤ Calculate algebraically the fixed point in Example 2.

SOLUTION If the fixed point is (X, Y), then applying the transformation $[T]$ to (X, Y) will give (X, Y) as the image. Write the fixed point, (X, Y), as a 3×1 matrix and multiply it by the transformation matrix.

$$\begin{bmatrix} 0.8\cos 30° & 0.8\cos 120° & 5 \\ 0.8\sin 30° & 0.8\sin 120° & 3 \\ 0 & 0 & 1 \end{bmatrix} \begin{bmatrix} X \\ Y \\ 1 \end{bmatrix} = \begin{bmatrix} X \\ Y \\ 1 \end{bmatrix}$$

The image is the same as the pre-image.

$$\begin{bmatrix} (0.8\cos 30°)X + (0.8\cos 120°)Y + 5 \\ (0.8\sin 30°)X + (0.8\sin 120°)Y + 3 \\ 1 \end{bmatrix} = \begin{bmatrix} X \\ Y \\ 1 \end{bmatrix}$$

Multiply the matrices on the left.

$$(0.8\cos 30°)X + (0.8\cos 120°)Y + 5 = X$$
$$(0.8\sin 30°)X + (0.8\sin 120°)Y + 3 = Y$$

Equate the top two rows of the matrices.

$$(0.8\cos 30° - 1)X + (0.8\cos 120°)Y = -5$$
$$(0.8\sin 30°)X + (0.8\sin 120° - 1)Y = -3$$

Get X and Y on the left and the constants on the right.

$$\begin{bmatrix} 0.8\cos 30° - 1 & 0.8\cos 120° \\ 0.8\sin 30° & 0.8\sin 120° \end{bmatrix}^{-1} \begin{bmatrix} X \\ Y \end{bmatrix} = \begin{bmatrix} -5 \\ -3 \end{bmatrix}$$

Write the system in matrix form.

$$\begin{bmatrix} X \\ Y \end{bmatrix} = \begin{bmatrix} 0.8\cos 30° - 1 & 0.8\cos 120° \\ 0.8\sin 30° & 0.8\sin 120° \end{bmatrix}^{-1} \begin{bmatrix} -5 \\ -3 \end{bmatrix} = \begin{bmatrix} 1.3205... \\ 11.4858... \end{bmatrix}$$

Solve.

The fixed point is (1.3205..., 11.4858...), which confirms the approximate values found graphically.

If you apply transformation $[T]$ iteratively to matrix $[M]$, the images are

Iteration 1: $[T][M]$

Iteration 2: $[T]([T][M]) = ([T][T])[M] = [T]^2[M]$

Iteration 3: $[T]([T]([T][M])) = ([T][T][T])[M] = [T]^3[M]$

These equations are true because matrix multiplication is associative. To find the transformation matrix for the 30th iteration, you would calculate

$$[T]^{30} = \begin{bmatrix} -0.0012... & 0.0000... & 1.3222... \\ 0.0000... & -0.0012... & 11.5000... \\ 0 & 0 & 1 \end{bmatrix}$$

The four elements in the rotation and dilation part of the matrix are close to zero because the dilation, 0.8, is less than 1 and is being raised to a high power. The two elements in the translation part of the matrix are close to the coordinates of the fixed point. If the rotation and dilation elements were equal to zero, the resulting matrix would translate any point (a, b) to the fixed point.

$$\begin{bmatrix} 0 & 0 & X \\ 0 & 0 & Y \\ 0 & 0 & 1 \end{bmatrix} \begin{bmatrix} a \\ b \\ 1 \end{bmatrix} = \begin{bmatrix} X \\ Y \\ 1 \end{bmatrix}$$

Example 4 shows you how to take advantage of this fact to find the fixed point numerically.

EXAMPLE 4 ➤ Calculate the fixed point in Example 2 numerically in a time-efficient way.

SOLUTION

$$[T]^{100} = \begin{bmatrix} 0.0000... & 0.0000... & 1.3205... \\ 0.0000... & 0.0000... & 11.4858... \\ 0 & 0 & 0 \end{bmatrix}$$

Raise $[T]$ to a high power. Be sure the rotation-dilation is close to zero.

The fixed point is $(X, Y) \approx (1.3205..., 11.4858...)$.

Fixed point is in the translation part of the matrix.

The information in Examples 1 through 4 is summarized in the box.

PROPERTY: General Rotation, Dilation, and Translation Matrix

When applied to a matrix of the form

$$[M] = \begin{bmatrix} x_1 & x_2 & \dots \\ y_1 & y_2 & \dots \\ 1 & 1 & \dots \end{bmatrix}$$

the transformation matrix

$$[T] = \begin{bmatrix} d\cos A & d\cos(A + 90°) & h \\ d\sin A & d\sin(A + 90°) & k \\ 0 & 0 & 1 \end{bmatrix}$$

- Dilates by a factor of d
- Rotates counterclockwise $A°$
- Translates by h units in the x-direction and k units in the y-direction.

The techniques for finding the fixed point are summarized in the box below.

PROCEDURE: Fixed Point of a Linear Transformation

A linear transformation has a fixed point attractor (X, Y) if and only if

$$(X, Y) = \lim_{n\to\infty}(x_n, y_n)$$

where (x_n, y_n) is the image of a point (x_0, y_0) after n iterations of transformation matrix $[T]$.

To find the fixed point (X, Y) algebraically:

1. Write the equation $[T]\begin{bmatrix} X \\ Y \\ 1 \end{bmatrix} = \begin{bmatrix} X \\ Y \\ 1 \end{bmatrix}$.
2. Multiply the matrices on the left side of the equation.
3. Equate the elements in the first and second rows of the resulting matrix on the left to X and Y in the matrix on the right.
4. Solve the resulting system of equations for X and Y.

To find the fixed point (X, Y) numerically:

1. Raise $[T]$ to a high power.
2. Make sure the rotation and dilation elements are close to zero.
3. Write the fixed point from the translation part of the resulting matrix.

Problem Set 13-4

Reading Analysis

From what you have read in this section, what do you consider to be the main idea? Translating an image is a process of *addition*. Explain how a translation can be accomplished by matrix *multiplication* by a clever choice of the transformation matrix. What happens to the images under an iterative application of a transformation matrix that reduces the size of the image, translates, and rotates? Why is the word *limit* used to describe what happens?

Quick Review

Q1. What is the dilation factor for an 80% reduction?

Q2. Find the image of the point (1, 0) under a rotation counterclockwise 30°.

Q3. Find the dimensions of $[A][B]$ if $[A]$ is 3×5 and $[B]$ is 5×2.

Q4. If $[M] = \begin{bmatrix} 7 & 2 \\ 8 & 4 \end{bmatrix}$, find det $[M]$.

Q5. Find $[M]^{-1}$ for $[M]$ in Problem Q4.

Q6. Find $[M]^{-1}[M]$ for $[M]$ in Problem Q4.

Q7. Find $[M]^2$ for $[M]$ in Problem Q4.

Q8. Explain why $[D] = \begin{bmatrix} 3 & 5 \\ 3 & 5 \end{bmatrix}$ has no multiplicative inverse.

Q9. Find the dot product $(3\vec{i} + 7\vec{j} - 2\vec{k}) \cdot (4\vec{i} - \vec{j} + 5\vec{k})$.

Q10. Is the angle between the two vectors in Problem Q9 acute or obtuse?

---

1. Consider the rectangle with vertices (3, 2), (7, 2), (7, 4), and (3, 4).

a. Plot the rectangle on graph paper. Write a transformation matrix $[A]$ to rotate this rectangle counterclockwise 20°, dilate it by a factor of 0.9, and translate it by 6 units in the x-direction and -1 unit in the y-direction. Write a matrix $[M]$ for the pre-image rectangle, apply the transformation, and plot the image on graph paper.

b. Enter the matrices into your grapher and apply the transformation in part a iteratively, plotting the images using the program from Section 13-3, Problem 19. Sketch the path followed by the images. To what fixed point do the images seem to be attracted?

c. Find the approximate location of the fixed point numerically by finding $[A]^{100}[M]$. Does it agree with the answer you found graphically in part b?

d. Find the location of the fixed point algebraically. Show that your answer agrees with the answers you found graphically and numerically in parts b and c.

2. Consider the dart with vertices (7, 1), (9, 2), (11, 1), and (9, 5).

a. Plot the dart on graph paper. Write a transformation matrix $[A]$ to rotate the dart counterclockwise 40°, dilate it by a factor of 0.8, and translate it by -3 units in the x-direction and 4 units in the y-direction. Write a matrix $[M]$ for the pre-image figure, apply the transformation, and plot the image on the graph.

b. Enter the matrices into your grapher and apply the transformation in part a iteratively, plotting the images using the program from Section 13-3, Problem 19. Sketch the path followed by the images. To what fixed point do the images seem to be attracted?

c. Find the approximate location of the fixed point numerically by finding $[A]^{100}[M]$. Does it agree with the answer you found graphically in part b?

d. Find the location of the fixed point algebraically. Show that your answer agrees with the answers you found graphically and numerically in parts b and c.

Three-dimensional video images are created from iterated transformations of a pre-image.

3. *Fixed Point Problem:* Figure 13-4d shows a rectangle that is to be the pre-image for a set of linear transformations. In this problem you will find out which matrix determines the fixed point: the transformation matrix or the pre-image matrix.

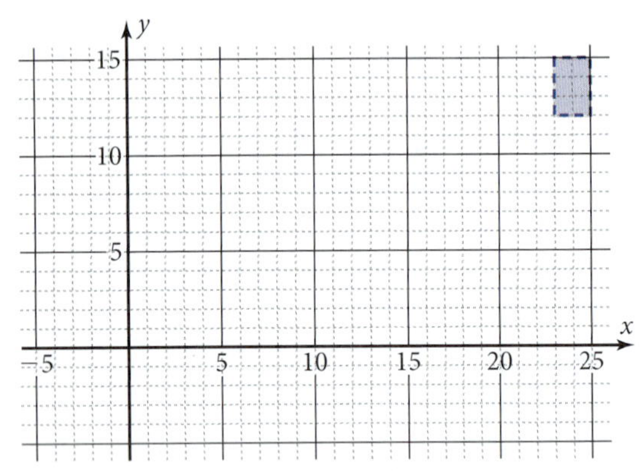

Figure 13-4d

a. Describe the transformations applied by $[T_1]$.

$$[T_1] = \begin{bmatrix} 0.8\cos(-20°) & 0.8\cos(70°) & 2 \\ 0.8\sin(-20°) & 0.8\sin(70°) & 6 \\ 0 & 0 & 1 \end{bmatrix}$$

b. Write a matrix $[M_1]$ for the rectangle in Figure 13-4d. Apply $[T_1]$ iteratively to $[M_1]$ using your grapher program. To what fixed point do the images converge? Show this fixed point on a copy of Figure 13-4d, along with the path the images follow to reach this point.

c. Apply $[T_1]$ iteratively to another rectangle, $[M_2]$ below. Are the images attracted to the same fixed point as the images of $[M_1]$? Sketch the pre-image and the path the images follow.

$$[M_2] = \begin{bmatrix} 2 & -2 & -2 & 2 \\ 10 & 10 & 0 & 0 \\ 1 & 1 & 1 & 1 \end{bmatrix}$$

d. Write a transformation matrix $[T_2]$ that performs this set of transformations:

- A 30% reduction; that is, a dilation by a factor of 0.7
- A rotation counterclockwise 35°
- A translation by 7 units in the x-direction and -3 units in the y-direction

Apply $[T_2]$ iteratively to the rectangle represented by $[M_1]$. Are the images attracted to the same fixed point as the images generated by $[T_1]$? On a copy of Figure 13-4d, sketch the path the images follow to reach the fixed point.

e. Based on your results in parts b–d, which determines the location of the fixed point attractor, the transformation matrix or the pre-image matrix? Does applying $[T_2]$ to $[M_2]$ support your conclusion?

4. *Third Row Problem*: Multiply the given matrices "by hand." From the results, explain the effect of the elements 1, 1 in the third row of the pre-image matrix $[M]$. Explain the effect of the elements 0, 0, 1 in the third row of the transformation matrix $[T]$ and why it is important for these elements to have this effect.

$$[T][M] = \begin{bmatrix} 2 & 0 & 7 \\ 0 & 2 & 3 \\ 0 & 0 & 1 \end{bmatrix} \begin{bmatrix} 8 & 5 \\ 4 & 9 \\ 1 & 1 \end{bmatrix}$$

13-5 Strange Attractors for Several Iterated Transformations

In the previous section you saw how an iterated transformation can cause images to be attracted to a fixed point. In this section you will see what happens when several different transformations are applied iteratively. Instead of being attracted to a single point, the images are attracted to a figure of remarkable complexity. Sometimes these **strange attractors** have shapes that look like trees, ferns, snowflakes, or islands.

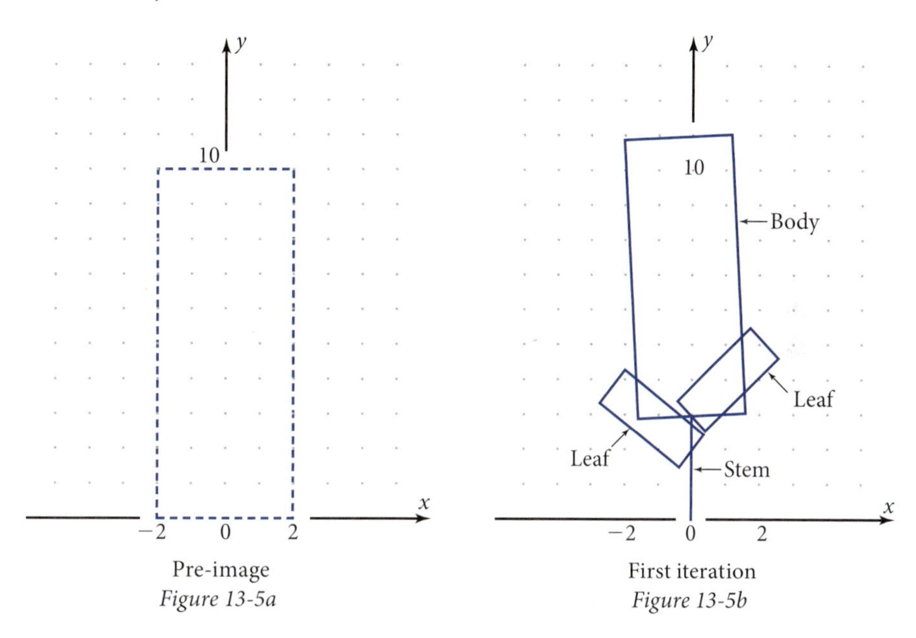

Pre-image
Figure 13-5a

First iteration
Figure 13-5b

Objective Given several different transformations, apply them iteratively, starting with a pre-image, and plot the resulting images.

Strange Attractors Graphically

Figure 13-5a shows a rectangular pre-image whose matrix is

$$[M] = \begin{bmatrix} 2 & 2 & -2 & -2 \\ 0 & 10 & 10 & 0 \\ 1 & 1 & 1 & 1 \end{bmatrix}$$

Figure 13-5b shows the four images that result from applying four transformations—$[A]$, $[B]$, $[C]$, and $[D]$—to the pre-image in Figure 13-5a.

The image matrices representing these transformations are

$$[A][M] = \begin{bmatrix} 0.8\cos 3^\circ & 0.8\cos 93^\circ & 0 \\ 0.8\sin 3^\circ & 0.8\sin 93^\circ & 3 \\ 0 & 0 & 1 \end{bmatrix}[M] \approx \begin{bmatrix} 1.6 & 1.2 & -2.0 & -1.6 \\ 3.1 & 11.1 & 10.9 & 2.9 \\ 1 & 1 & 1 & 1 \end{bmatrix}$$

$$[B][M] = \begin{bmatrix} 0.3\cos 52^\circ & 0.3\cos 142^\circ & 0 \\ 0.3\sin 52^\circ & 0.3\sin 142^\circ & 2 \\ 0 & 0 & 1 \end{bmatrix}[M] \approx \begin{bmatrix} 0.4 & -2.0 & -2.7 & -0.4 \\ 2.5 & 4.3 & 3.4 & 1.5 \\ 1 & 1 & 1 & 1 \end{bmatrix}$$

$$[C][M] = \begin{bmatrix} 0.3\cos(-46^\circ) & 0.3\cos 44^\circ & 0 \\ 0.3\sin(-46^\circ) & 0.3\sin 44^\circ & 3 \\ 0 & 0 & 1 \end{bmatrix}[M] \approx \begin{bmatrix} 0.4 & 2.6 & 1.7 & -0.4 \\ 2.6 & 4.7 & 5.5 & 3.4 \\ 1 & 1 & 1 & 1 \end{bmatrix}$$

$$[D][M] = \begin{bmatrix} 0 & 0 & 0 \\ 0 & 0.3 & 0 \\ 0 & 0 & 1 \end{bmatrix}[M] \approx \begin{bmatrix} 0 & 0 & 0 & 0 \\ 0 & 3 & 3 & 0 \\ 1 & 1 & 1 & 1 \end{bmatrix}$$

The first three transformations are dilations, rotations, and translations. $[D]$ is a dilation by 0.3 unit in the y-direction and by 0 unit in the x-direction, shrinking the rectangle to a line segment along the y-axis.

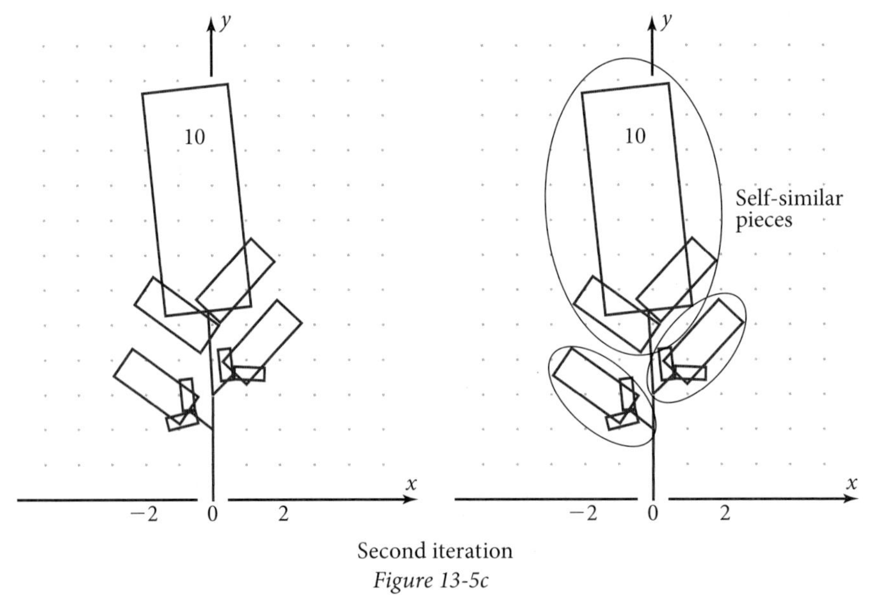

Second iteration

Figure 13-5c

Figure 13-5c shows the 16 images that result from applying each of the four transformations to each of the four images from the first iteration. Note that the figure created by this second iteration has three pieces that are similar to each other and to the image from the first iteration. Each piece has a linear "stem," a large rectangle coming out of the stem (body), and two small rectangles at the base of the large rectangle (leaves). A fourth piece is a long stem extending vertically from the origin—it is actually composed of four overlapping line segments formed by the four transformations $[D][A][M]$, $[D][B][M]$, $[D][C][M]$, and $[D][D][M]$. The longest of these segments extends up to $y = 3.3218....$

With a third iteration, each of the four transformations is applied to the 16 images from the second iteration, giving 64 new images. The 20th iteration has 4^{20}, or 1,099,511,627,776, images and resembles the fern leaf in Figure 13-5d below.

Strange Attractors Numerically—Barnsley's Method

As you have seen, plotting images created by applying several transformations iteratively can be tedious. In 1988, Michael Barnsley published a method for plotting such images more efficiently. Instead of starting with a pre-image figure, you start with *one point*. You then select *one* of the four transformations at random, apply it to that point, and plot the image point. Then you again select one of the four transformations at random, apply it to the first image, and plot the new point. As more and more points are plotted, there are regions to which the points are attracted and regions that they avoid. The resulting image is an approximation of the image you would get by carrying out the iterations in Figure 13-5c an infinite number of times. The more points you plot, the better the approximation. Figure 13-5d shows this strange attractor plotted with 200 points, 1000 points, and 5000 points.

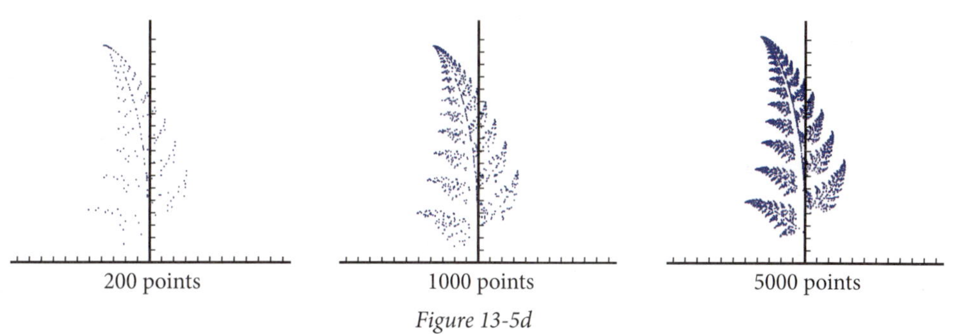

Figure 13-5d

If you were to magnify any one of the branches of the fern leaf by the proper amount, it would look exactly like the whole leaf. Figures with this quality are said to be **self-similar.** In Section 13-6, you will learn about such images, called **fractals.** They are so "fractured" that their dimensions turn out to be fractions, as you'll learn in the next section.

The technique of plotting a strange attractor pointwise is called **Barnsley's method.** To give the images reasonable point densities, you must select a probability for each transformation. For Figure 13-5d there was an 80% probability of picking transformation $[A]$, which produces the main form of the leaf; a 9% probability for each of $[B]$ and $[C]$, which produce the two side branches; and a 2% probability for $[D]$, which draws the stem of the fern leaf.

Self-similarity is readily apparent in cauliflower.

PROCEDURE: Barnsley's Method

To plot a strange attractor pointwise:

1. Write a transformation matrix for each transformation. (The dilation factors must be less than 1.).
2. Select a probability for each matrix.
3. Select any point to be the pre-image.
4. Pick a transformation at random using the appropriate probability, apply the transformation to the point, and plot the image.
5. Repeat step 4 on the image from the previous iteration. Do this for as many iterations as you choose.

Problem Set 13-5

Reading Analysis

From what you have read in this section, what do you consider to be the main idea? If a strange attractor is constructed by applying three different transformations iteratively starting with one pre-image, how do you find the number of images in the second, third, and fourth iterations? How does Barnsley's method of constructing a strange attractor differ from constructing all images for each iteration?

Quick Review

Q1. Write a 2×2 matrix to dilate a figure by a factor of 2.

Q2. Write a 2×2 matrix to rotate a figure clockwise 12°.

Q3. Write a 3×3 matrix for a 40% reduction, a rotation counterclockwise 23°, an x-translation by 4 units, and a y-translation by -3 units.

Q4. After a 40% reduction, a segment 10 cm long is __?__ cm long.

Q5. A 40% reduction transforms a 100-cm^2 rectangle to a rectangle with area __?__

Q6. If a rotation takes the x-axis to $\theta = 40°$, then it takes the y-axis to $\theta =$ __?__.

Q7. Write cos 30° exactly, in radical form.

Q8. By the cofunction property, what does $\sin\left(\frac{\pi}{2} - x\right)$ equal?

Q9. How many degrees are there in $\frac{\pi}{3}$ radians?

Q10. $\vec{a}$ is 7 units long, $\vec{b}$ is 8 units long, and the angle between them when they are placed tail-to-tail is 38°. How long is $\vec{a} \times \vec{b}$?

---

1. *Sierpiński's Triangle Problem:* Figure 13-5e shows a triangular pre-image whose matrix is

$$[M] = \begin{bmatrix} 15 & -0 & -15 \\ -10 & 20 & -10 \\ 1 & -1 & 1 \end{bmatrix}$$

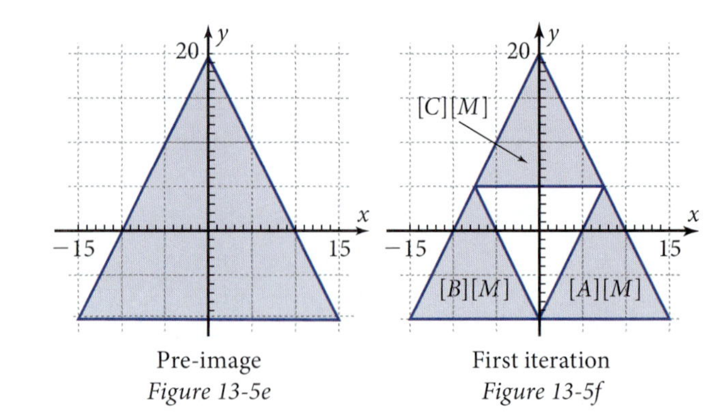

Pre-image
Figure 13-5e

First iteration
Figure 13-5f

Figure 13-5f shows the first iteration of three transformations. Each transformation reduces the triangle by 50%. Transformation $[A]$ translates the dilated image so that its lower-right vertex coincides with the lower-right vertex of the pre-image. Transformation $[B]$ translates the dilated image so that its lower-left vertex coincides with the lower-left vertex of the pre-image. Transformation $[C]$ translates the dilated image so that its upper vertex coincides with the upper vertex of the pre-image.

a. Write the transformation matrices $[A]$, $[B]$, and $[C]$.

b. Apply the nine transformations required for the second iteration. That is, calculate the images

$[A][A][M]$ $\quad$ $[A][B][M]$
$[A][C][M]$ $\quad$ $[B][A][M]$
$[B][B][M]$ $\quad$ $[B][C][M]$
$[C][A][M]$ $\quad$ $[C][B][M]$
$[C][C][M]$

c. On graph paper, plot the nine images from the second iteration in part b.

d. How many images will be in the third iteration? The 20th iteration?

e. Find the area of the pre-image triangle. Find the total area of the three triangles in the first iteration. Use what you observe in these calculations to find a formula for the total area of the nth image.

f. If the iterations are performed infinitely, the figure is called *Sierpiński's triangle* or Sierpiński's gasket. What is the area of Sierpiński's triangle? Surprising?

2. *Sierpiński's Carpet Problem:* Figure 13-5g shows the square pre-image whose matrix is

$$[M] = \begin{bmatrix} 20 & 20 & 0 & 0 \\ 20 & 0 & 0 & 20 \\ 1 & 1 & 1 & 1 \end{bmatrix}$$

Figure 13-5h shows the first iteration of four transformations. Each transformation reduces the square by 40%. The four transformations then translate the reduced images to the four corners of the original pre-image.

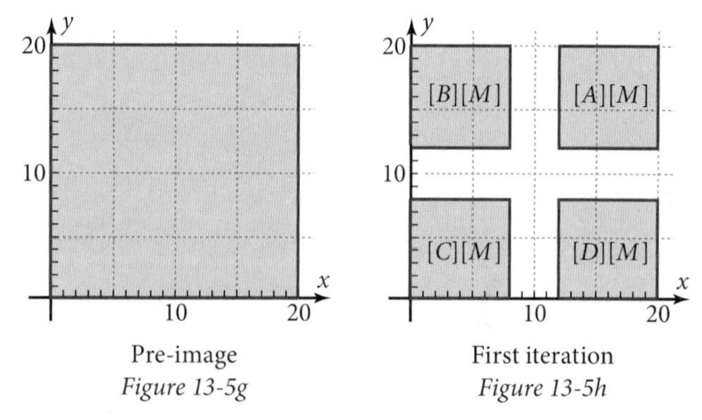

Pre-image
Figure 13-5g

First iteration
Figure 13-5h

a. Write a transformation matrix for each of the four transformations.

b. Apply the 16 transformations required for the second iteration. Plot the 16 images on graph paper.

c. How many images will there be in the third iteration? The 20th iteration?

d. Find the perimeter of the pre-image. Find the total perimeter of the squares in the first iteration. Find the total perimeter of the squares in the third iteration. By extending the pattern you observe in your answers, find the total perimeter of the squares in the 20th iteration. What happens to the total perimeter as the number of iterations becomes very large?

e. As a research project, find out about Waclaw Sierpiński, the man for whom Sierpiński's triangle and carpet are named.

3.  *Barnsley's Method Program:* Write or download a program that will plot a strange attractor using Barnsley's method. You can find both a program for your grapher and a sketch for use with The Geometer's Sketchpad at **www.keymath.com/precalc.** Both are named *Barnsley.* Before running the program, store up to four transformation matrices and the probability associated with each matrix. The program should allow you to input a starting pre-image point and the number of points to plot. Then the program should iteratively select a transformation at random, apply it to the preceding image, and plot the new image.

4. *Barnsley's Method Program Debugging:* Test your program for Barnsley's method by plotting the fern-shaped strange attractor in Figure 13-5d. Use the transformation matrices $[A]$, $[B]$, $[C]$, and $[D]$ described on pages 679–680, with probabilities 0.8, 0.09, 0.09, and 0.02, respectively. Use (1, 1) as the initial pre-image point, and plot a sufficient number of points to get a reasonably good image. When your program is working, run it again using a different pre-image point. Does the pre-image you select seem to change the final image? Run the program again using five times as many points. Describe the similarities and differences in the final image created by using more points.

5. Use your Barnsley's method program to plot Sierpiński's triangle from Problem 1. Use probability $\frac{1}{3}$ for each of the three transformations. Only three transformations are involved, so use probability 0 for the fourth transformation. You should get an image similar to that in Figure 13-5i.

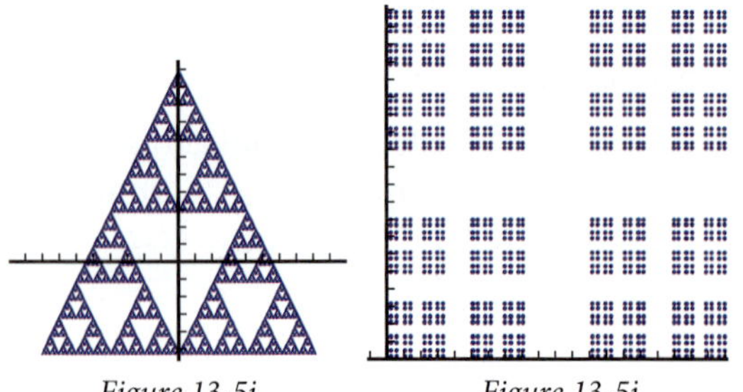

Figure 13-5i *Figure 13-5j*

6. Use your Barnsley's method program to plot Sierpiński's carpet from Problem 2. Use probability $\frac{1}{4}$ for each of the four transformations. You should get an image similar to that in Figure 13-5j.

7. Change the matrices for Sierpiński's carpet in Problem 6 so that the dilation factor is 0.5 instead of 0.4. You must also change the translations so that the upper-right square's upper-right corner still goes to the point (20, 20), and so forth. Explain why the pattern of points in Figure 13-5j disappears when the dilation factor is changed to 0.5.

8. Change the matrices for Sierpiński's carpet in Problem 6 so that the dilation factor is 0.6 instead of 0.4. You must also change the translations so that the upper-right square's upper-right corner still goes to the point (20, 20), and so forth. Does any pattern seem to appear in the points?

9. *Foerster's Tree Problem:* Figure 13-5k shows a vertical segment 10 units high, starting at the origin. This pre-image is to be transformed into a "tree" with three pieces, each 6 units long, as shown in Figure 13-5l. The three pieces satisfy these conditions:

   - The left branch is rotated $+30°$ from the trunk and starts at $y = 5$ on the y-axis.
   - The trunk starts at the origin.
   - The right branch is rotated $-30°$ from the trunk and starts at $y = 4$ on the y-axis.

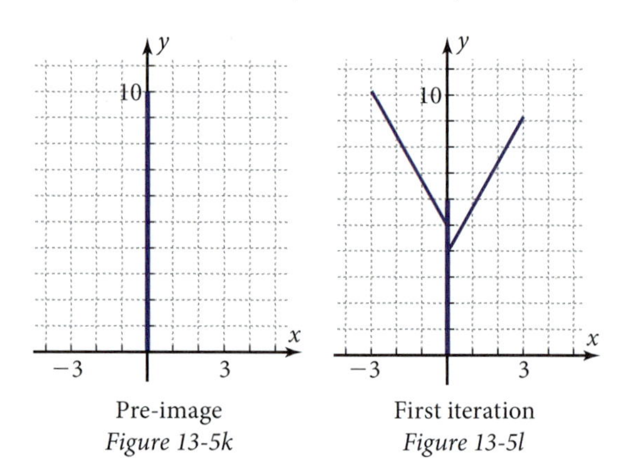

Pre-image
Figure 13-5k

First iteration
Figure 13-5l

   a. Write the pre-image on the left as a 3×2 matrix, $[D]$, with two 1s in the bottom row.

b. Write three 3 × 3 transformation matrices to perform these tasks:

$[A]$ should transform the pre-image to the left branch.

$[B]$ should transform the pre-image to the trunk.

$[C]$ should transform the pre-image to the right branch.

c. Figure 13-5l shows the three images in the first iteration. To get the nine images in the second iteration, you multiply each image by $[A]$, $[B]$, and $[C]$. Calculate the nine images $[A][A][D]$, $[A][B][D]$, $[A][C][D]$, $[B][A][D]$, $[B][B][D]$, $[B][C][D]$, $[C][A][D]$, $[C][B][D]$, and $[C][C][D]$. Round the entries of the image matrices to one decimal place. Plot the nine images on graph paper.

d. Use your Barnsley's method program to see what the tree would look like if the iterations were performed infinitely many times. Use probability $\frac{1}{3}$ for each transformation.

e. The tree in this problem "attracts" the points. What special name is given to such an attractor?

f. Calculate the sum of the lengths of the images in the first, second, third, and 100th iterations of this tree. If the iterations were performed infinitely many times, what would the sum of the lengths of the images approach? Does the answer surprise you?

g. Calculate the fixed points for each of the three transformations in this problem. How do these points relate to points on the graph in part d?

10. *Koch's Snowflake Problem:* Figure 13-5m shows the line segment represented by the pre-image matrix

$$[M] = \begin{bmatrix} 12 & 12 \\ 6 & -6 \\ 1 & 1 \end{bmatrix}$$

Figure 13-5n shows the images of these four transformations applied to the pre-image matrix:

$$[A] = \begin{bmatrix} \frac{1}{3} & 0 & 8 \\ 0 & \frac{1}{3} & 4 \\ 0 & 0 & 1 \end{bmatrix}$$

$$[B] = \begin{bmatrix} \frac{1}{3}\cos 60° & \frac{1}{3}\cos 150° & 11.7320... \\ \frac{1}{3}\sin 60° & \frac{1}{3}\sin 150° & -2.4641... \\ 0 & 0 & 1 \end{bmatrix}$$

$$[C] = \begin{bmatrix} \frac{1}{3}\cos(-60°) & \frac{1}{3}\cos 30° & 11.7320... \\ \frac{1}{3}\sin(-60°) & \frac{1}{3}\sin 30° & 2.4621... \\ 0 & 0 & 1 \end{bmatrix}$$

$$[D] = \begin{bmatrix} \frac{1}{3} & 0 & 8 \\ 0 & \frac{1}{3} & -4 \\ 0 & 0 & 1 \end{bmatrix}$$

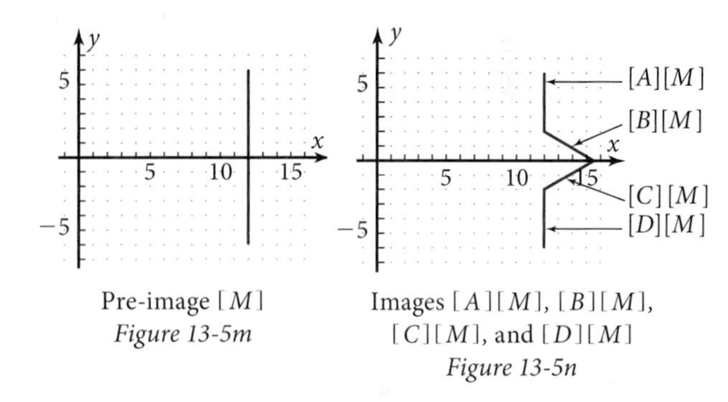

Pre-image $[M]$
Figure 13-5m

Images $[A][M]$, $[B][M]$, $[C][M]$, and $[D][M]$
Figure 13-5n

a. Draw a sketch showing how $[A]$ dilates $[M]$ by a factor of $\frac{1}{3}$ and then translates the dilated image so that its top point is at the top of the pre-image.

b. Show algebraically that the rotation and dilation part of $[B]$ moves the point (12, 6) at the top of the pre-image to the point $(4\cos 60° + 2\cos 150°, 4\sin 60° + 2\sin 150°)$. Then show how the translation part of $[B]$ moves this point to the point (12, 2) at the bottom of image $[A][M]$.

c. Based on your answers to parts a and b, describe the effects transformations $[C]$ and $[D]$ have on the pre-image segment.

d. In the second iteration, the four transformations are applied to each of the four images. Write matrices for the 16 images in the second iteration, with elements rounded to one decimal place. Plot these images on dot paper or graph paper.

e. If the transformations are applied infinitely many times, the result is part of Helge von Koch's *snowflake curve* shown in Figure 13-5q.

The first and second iterations of that curve are shown, respectively, in Figure 13-5o and Figure 13-5p. The result of many iterations is shown in Figure 13-5q. On a copy of Figure 13-5q, circle two parts of the snowflake curve of different sizes that show that the snowflake curve is self-similar.

f. The length of the pre-image in Figure 13-5m is 12 units. The first iteration in Figure 13-5n has total length 16 units because it has four segments that are each 4 units long. What is the total length of the second iteration? The third iteration? The fourth iteration? Find the length of the 100th iteration. What would be the total length of the final snowflake curve? Is this surprising?

g. Use Barnsley's method to show that you get the same strange attractor when you start with *one* point as a pre-image and apply the four transformations iteratively, at random, on the resulting images. Use probability $\frac{1}{4}$ for each transformation.

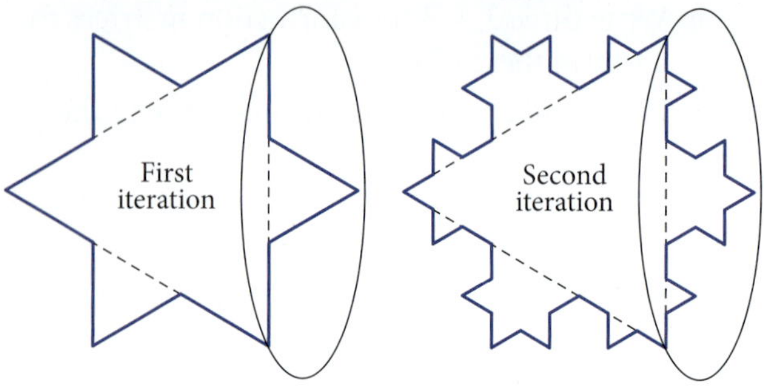

Figure 13-5o *Figure 13-5p*

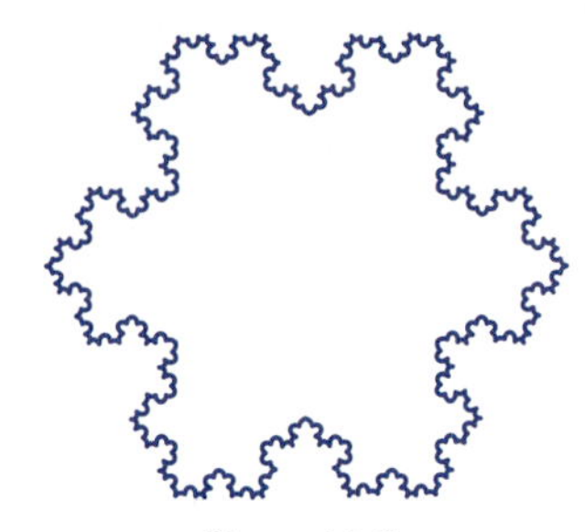

Figure 13-5q

11. *Fixed Points in a Strange Attractor:* Transformation $[A]$ for the fern image in Figure 13-5d is

$$[A] = \begin{bmatrix} 0.8\cos 3^\circ & 0.8\cos 93^\circ & 0 \\ 0.8\sin 3^\circ & 0.8\sin 93^\circ & 3 \\ 0 & 0 & 1 \end{bmatrix}$$

a. Find the fixed point for this transformation. To what part of the fern image does this fixed point correspond?

b. Make a conjecture about the approximate locations of the fixed points for the transformations $[B]$, $[C]$, and $[D]$ applied to the fern image in Figure 13-5d.

c. Compute the fixed points in part b numerically by raising the transformation matrices to a high power. Does this confirm or refute your conjecture?

12. *Sketchpad Fractal Project:* Download the sketch Barnsley.gsp at **www.keymath.com/precalc** and explore the fractal images presented by changing the parameters. To investigate how this sketch was made, download Fractal Fern.pdf and Fractal Fern.gsp. Describe what you learned from doing this project.

13-6 Fractal Dimensions

In Section 13-5, you had the chance to explore Koch's snowflake curve and Sierpiński's triangle. These are shown in Figure 13-6a. You may have found in your explorations that the lengths of successive iterations of the snowflake curve follow an increasing geometric sequence: 12, 16, 21.3333..., 28.4444..., . . . , and the areas of successive iterations of Sierpiński's triangle form a decreasing geometric sequence: 450, 337.5, 253.125, 189.84375, ... (See Chapter 15 for more on geometric sequences.) So the length of the snowflake curve approaches infinity and the area of Sierpiński's triangle approaches zero.

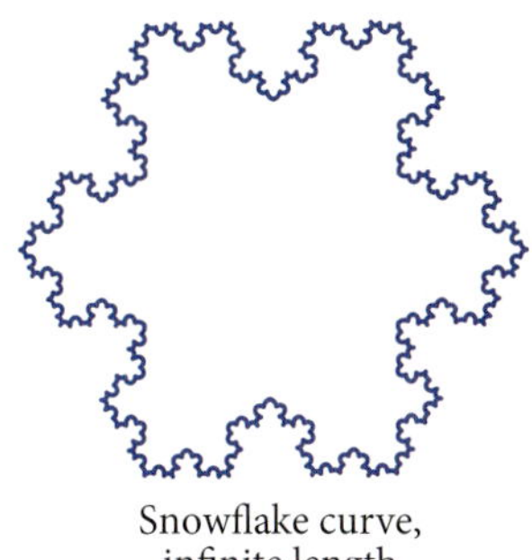

Snowflake curve, infinite length

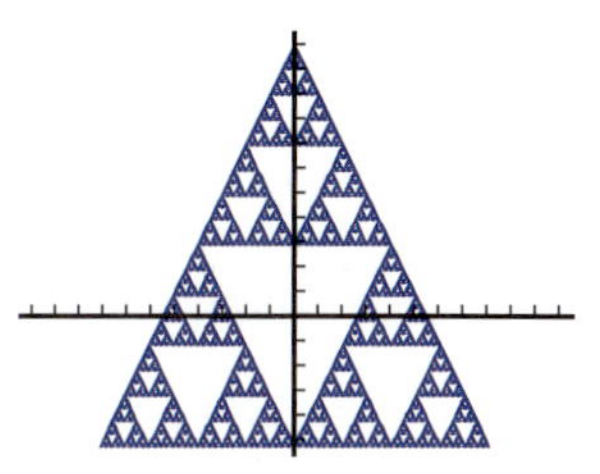

Sierpiński's triangle, zero area

Figure 13-6a

In this section you will make sense out of these seeming contradictions by learning a precise definition of the dimension of a figure. Both figures in Figure 13-6a are called fractals because their dimensions are fractions. As you'll see, the dimensions of both these fractals are between 1 and 2.

Objective Given a figure formed by iteration of several transformation matrices, determine its fractal dimension.

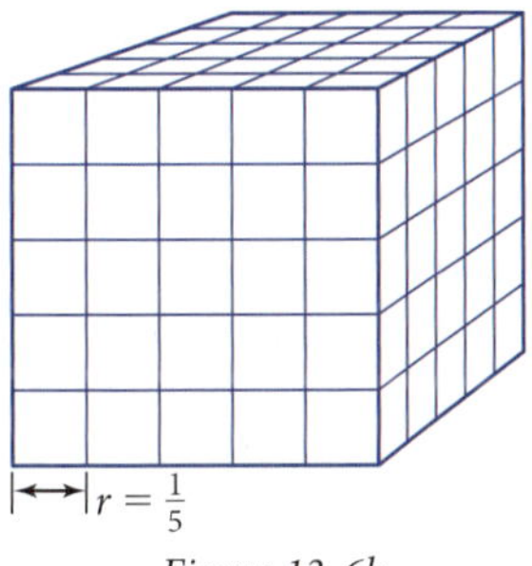

Figure 13-6b

A solid (three-dimensional) cube is a *self-similar* object because it can be broken into smaller cubes that are similar to one another and to the original cube. Figure 13-6b shows a cube with edge length 1 unit divided into smaller cubes, each with edge length $\frac{1}{5}$ unit. Note that each of the smaller cubes is also self-similar.

There are $N = 5^3$, or 125, small cubes. The exponent 3 in this equation is the dimension of the cube. With the help of logarithms, you can isolate the exponent 3.

$$\log 5^3 = \log N$$

$$3 \log 5 = \log N$$

$$3 = \frac{\log N}{\log 5}$$

The 5 in the equation is equal to $\frac{1}{r}$, where r, in this case $\frac{1}{5}$, is the ratio of the edge length of one small cube to the length of the original, pre-image cube. Substituting this information into the last equation gives

$$3 = \frac{\log N}{\log \frac{1}{r}}$$

This equation is the basis for the definition of dimension credited to Felix Hausdorff, a German mathematician who lived from 1868 to 1942.

DEFINITION: Hausdorff Dimension

If an object is subdivided into N self-similar pieces, the ratio of the length of each piece to the length of the original object is r, and the subdivisions can be done infinitely many times, then the dimension D of the object is

$$D = \frac{\log N}{\log \frac{1}{r}}$$

This is called the **Hausdorff dimension.**

To see how this definition applies to the snowflake curve, consider the pre-image and first iteration of any one segment in the curve. As Figure 13-6c shows, the segment is transformed into four self-similar segments, each of which is one-third as long as the original segment, so $N = 4$ and $r = \frac{1}{3}$.

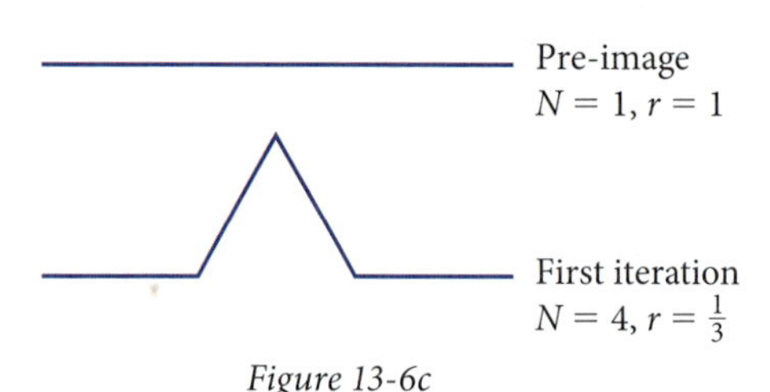

Figure 13-6c

The iterations are performed infinitely many times in the same pattern, so Hausdorff's definition of dimension applies. Thus the dimension of the snowflake curve is

$$D = \frac{\log 4}{\log \frac{1}{\frac{1}{3}}} = \frac{\log 4}{\log 3} = 1.2618...$$

So the snowflake curve is 1.2618...-dimensional! The dimension of a fractal measures the fractal's "space-filling ability." The curve's infinite length helps explain why it has a fractional dimension. If it had finite length, it could be "straightened out" into line segments and would therefore be one-dimensional. However, because the snowflake curve has infinite length, it can never be straightened out all the way; there will always be "spikes" that extend into the second dimension.

EXAMPLE 1 ➤ At each iteration in the generation of the snowflake curve, any one segment in the preceding iteration is divided into four self-similar segments, each of which is one-third as long as the previous segment.

a. For iterations 0 through 4, make a table of values showing the iteration number (n), the number of segments (N), the ratio of the length of each segment to the length of the pre-image (r), and $\frac{1}{r}$.

b. Calculate the dimension D of the snowflake using N and r from iteration 1. Show that you get the *same* value of D using N and r from iteration 4. How does the name used for this sort of figure reflect the fact that the Hausdorff dimension is not an integer?

c. Perform a linear regression on $\log N$ as a function of $\log\left(\frac{1}{r}\right)$. Plot the points and the equation on the same screen. Show numerically that the slope of the line equals the dimension of the snowflake.

d. If the pre-image has length 12 units, calculate the total length of the images at each iteration, 1 through 4. Use the pattern you observe to calculate the total length of the images at the 50th iteration. Explain why the total length of the snowflake approaches infinity as the iterations continue.

a. The total-length values are computed in part d.

Iteration, n	N	r	$\frac{1}{r}$	Total Length, L
0	1	1	1	$12(1)(1) = 12$
1	4	$\frac{1}{3}$	3	$12(4)\left(\frac{1}{3}\right) = 16$
2	16	$\frac{1}{9}$	9	$12(16)\left(\frac{1}{9}\right) = 21.3333\ldots$
3	64	$\frac{1}{27}$	27	$12(64)\left(\frac{1}{27}\right) = 28.4444\ldots$
4	256	$\frac{1}{81}$	81	$12(256)\left(\frac{1}{81}\right) = 37.9259\ldots$

The coastline of Cape Cod, Massachusetts. The total length of a coastline increases as you use measure with greater precision, giving the length the characteristics of a fractal.

b.

$$D = \frac{\log 4}{\log \frac{1}{\frac{1}{3}}} = \frac{\log 4}{\log 3} = 1.2618...$$

$$D = \frac{\log 256}{\log \frac{1}{\frac{1}{81}}} = \frac{\log 256}{\log 81} = 1.2618..., \text{ which is the same value}$$

It's called a *fractal*, indicating that its Hausdorff dimension is a fraction.

c.

$$\log N = (1.2618...) \log \frac{1}{r} + 0$$

Fit is exact because the correlation coefficient r equals 1.

The slope 1.2618... equals the dimension from part b, as shown in Figure 13-6d.

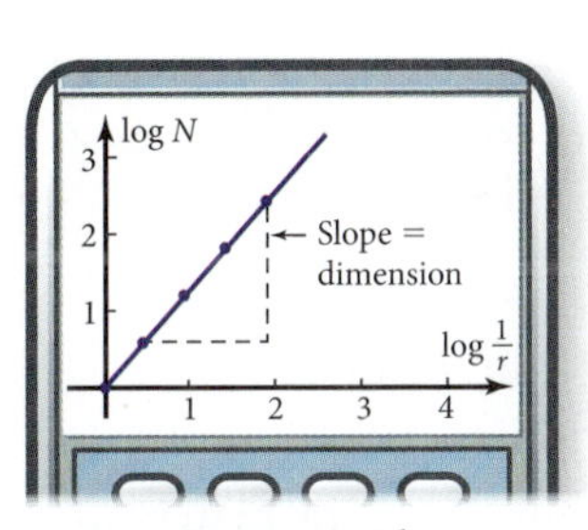

Figure 13-6d

d. The total lengths are given in the table in part a. The total length at each iteration is the original length, 12, multiplied by the number of segments N times the ratio of the length of each segment to the original length. From this pattern, the equation for the length L as a function of the iteration number n is

$$L = 12(4)^n\left(\frac{1}{3}\right)^n = 12\left(\frac{4}{3}\right)^n$$

Substituting 50 for n in this equation gives

$$L = 12\left(\frac{4}{3}\right)^{50} = 21{,}189{,}371.5590...$$

The values of L form a geometric sequence with common ratio $\frac{4}{3}$. Thus the values of L are unbounded as n increases. The number of iterations performed to generate the snowflake curve is infinite, so the length is infinite.

Note that the data for N in Example 1, part a, lie on a straight line when plotted on log-log graph paper, as shown in Figure 13-6e. This result confirms that N is a power function of $\frac{1}{r}$.

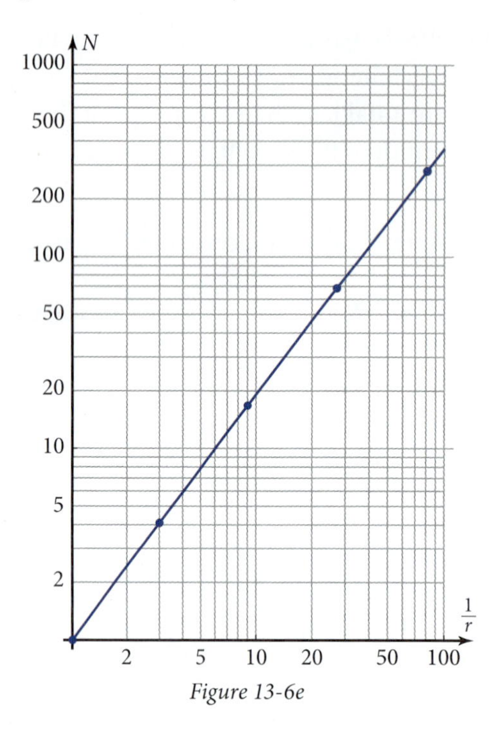

Figure 13-6e

Problem Set 13-6

Reading Analysis

From what you have read in this section, what do you consider to be the main idea? Write the formula for the Hausdorff dimension, and explain the meaning of each of the three variables that appear in the formula.

Quick Review

Q1. Find the common ratio of this geometric sequence: 100, 90, 81, 72.9, . . .

Q2. If you apply one linear transformation iteratively, the images can be attracted to a __?__.

Q3. If you apply several linear transformations iteratively, the images can be attracted to a __?__.

Q4. What rotation results from the application of this matrix?

$$\begin{bmatrix} 0.7\cos(-35°) & 0.7\cos 55° & -4 \\ 0.7\sin(-35°) & 0.7\sin 55° & 2 \\ 0 & 0 & 1 \end{bmatrix}$$

Q5. What dilation results from the matrix in Problem Q4?

Q6. What x-translation results from the matrix in Problem Q4?

Q7. What y-translation results from the matrix in Problem Q4?

Q8. Explain the purpose of the elements 0, 0, 1 in the bottom row of the matrix in Problem Q4.

Q9. What type of function has the "multiply–multiply" property?

Q10. If $g(x) = f\left(\frac{1}{3}x\right)$, what transformation is applied to $f(x)$ to get $g(x)$?

1. *Dimension Definition Applied to a Square Problem:* Figure 13-6f shows a square region divided into 25 self-similar squares, each with side length one-fifth the side length of the original square.

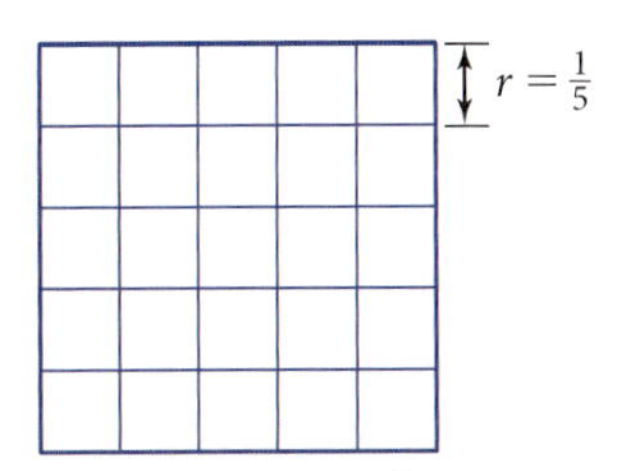

Figure 13-6f

   a. Show that the Hausdorff dimension leads you to conclude that a square is two-dimensional.

   b. If the square were cut into smaller squares with sides r equal to 0.01 times the length of the original sides, show that the Hausdorff dimension would still lead to the conclusion that a square is two-dimensional.

   c. What allows you to conclude that the Hausdorff dimension really does apply to the square when it is cut into smaller self-similar squares?

2. *Dimension of Sierpiński's Triangle Problem:* Figure 13-6g shows the pre-image and the first iteration of Sierpiński's triangle. The original pre-image triangle is transformed into three self-similar triangles, each of which has side lengths one-half the length of the sides in the pre-image.

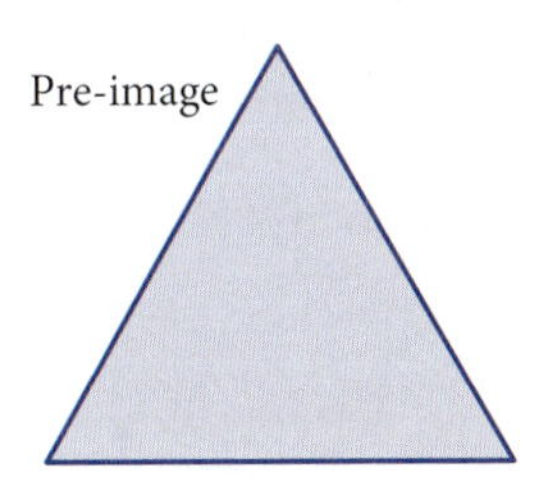

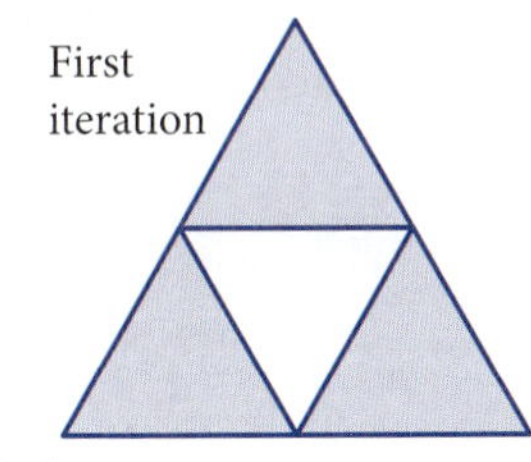

Figure 13-6g

   a. For iterations 0 through 4, make a table of values showing the iteration number (n), the number of triangles (N), the ratio of the side length of each triangle to the side length of the pre-image (r), and $\frac{1}{r}$.

   b. On log-log graph paper, plot N as a function of $\frac{1}{r}$. What does the fact that the points lie in a straight line tell you about the type of function that relates the points? Measure the slope of the line with a ruler, and record the result.

   c. Perform a power regression on N as a function of $\frac{1}{r}$. Write the regression equation. How does the exponent in the equation compare with the slope you measured in part b?

   d. Calculate the dimension D of Sierpiński's triangle using iterations 1 and 4. How does the result compare with the slope you measured in part b and with the exponent of the power function in part c?

   e. If the pre-image is an equilateral triangle with sides 16 cm long, calculate the total perimeter of the images in each iteration, 1 through 4. Use the pattern you observe to calculate the total perimeter of the images in the 50th iteration. Explain why the total perimeter of Sierpiński's triangle approaches infinity as the iterations continue.

3. *Dimension of Sierpiński's Carpet Problem:* Figure 13-6h shows the pre-image and the first iteration of Sierpiński's carpet. The original pre-image is divided into four self-similar squares, each of which has side lengths that are 40% of the side lengths in the pre-image.

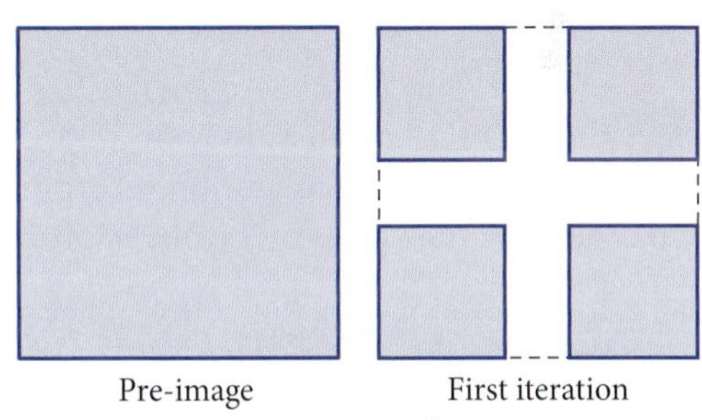

Figure 13-6h

   a. The complete Sierpiński's carpet is formed by iterating infinitely many times. What is the dimension of the final Sierpiński's carpet?

   b. Calculate the total area of the final Sierpiński's carpet. How does the result correspond to the dimension of the carpet?

   c. Calculate the total perimeter of the final Sierpiński's carpet. How does the result correspond to the dimension of the square?

d. Suppose that the self-similar squares at each iteration had side lengths that were 50% of the side lengths of the preceding image. How would this change the dimension of the square? Why is it *not* correct to call Sierpiński's carpet a fractal in this case?

e. Suppose that the self-similar squares at each iteration had side lengths that were 60% of the side lengths of the preceding image. How would this change the dimension of the square? How would it affect the total area of the square?

4. *Coastline Length Problem:* Lengths of the west coast of Great Britain as measured on a map by rulers of varying length are listed in the table. Shorter rulers can measure around smaller features so the total length appears to be greater (Figure 13-6i). The numbers are based on data collected by Lewis F. Richardson and reported in Benoit Mandelbrot's book *Fractals: Form, Chance, and Dimension.*

Ruler (mi)	Coastline (mi)
10.2	3020
30	2090
100	1580
500	1000
950	950

a. Enter the data into two lists in your grapher. Use your grapher to calculate and record in two other lists the values of $\frac{1}{r}$ (the reciprocal of the ratio of ruler length to 950 mi, the longest ruler) and the number N of ruler lengths in the coastline. Plot N as a function of $\frac{1}{r}$ on log-log graph paper, and draw a line that you think best fits the data. Measure the slope of the line and record the result.

b. Assuming that the coastline follows the same pattern for shorter and shorter rulers, calculate the Hausdorff dimension of the west coast of Great Britain. Use the values of $\frac{1}{r}$ and N for the shortest ruler. How does the result compare with the slope you measured in part a?

c. On your grapher, calculate and record in two other lists the values of $\log \frac{1}{r}$ and $\log N$. Perform a linear regression on N as a function of $\log \frac{1}{r}$. How does the slope of this linear function compare with the slope you measured in part a and with the dimension you calculated in part b?

d. Use power regression to find an equation for N as a function of $\frac{1}{r}$. What do you notice about the exponent in this equation? Use the equation to predict what the total length of the coastline would measure if a 1-in. ruler were used. Does the answer surprise you?

5. *Conclusions Problem:* At the beginning of this section you read that the snowflake curve has infinite length and that Sierpiński's triangle has zero area. How is the fractal dimension of a figure related to its total length and area?

6. *Journal Problem:* In your journal describe the most significant things you have learned about iterated transformations and fractals.

Figure 13-6i

13-7 Chapter Review and Test

In this chapter you have seen how you can use the concept of *matrix* to solve systems of linear equations and to apply linear transformations to geometric figures. Applied iteratively, a linear transformation can cause the images to be attracted to a fixed point. If several such transformations are applied iteratively, the images can be attracted to a figure of great complexity, sometimes resembling an object in nature, such as a fern or a tree. These strange attractors are called *fractals* because they have fractional Hausdorff dimensions rather than integer dimensions.

Applying linear transformations to a simple starting shape can produce images of startling realism—evident in this computer-generated landscape.

Review Problems

R0. Update your journal with what you have learned in this chapter. Include things such as the most important thing you have learned as a result of studying this chapter, the new terms you have learned and what they mean, the ways matrix transformations can change an image, how Barnsley's method for generating fractal images differs from repeating transformations on all images in the previous iteration, and Hausdorff's definition of dimension.

R1. Figure 13-7a shows a line segment 1 unit long. In the first iteration, an image is formed by removing the middle third of the segment. In the second iteration, an image is formed by removing the middle third of each segment in the first iteration. If the iterations are performed infinitely many times, the image is called the *Cantor set,* after Georg Cantor, a German mathematician who lived from 1845 to 1918. How many segments are there in the 10th iteration? What is the total length of the segments in the 10th iteration? What does the total length approach as the number of iterations increases without bound?

Pre-image, 1 unit long
First iteration
Second iteration

Figure 13-7a

R2. **a.** Evaluate: $9\begin{bmatrix} 5 & 2 \\ 7 & -1 \end{bmatrix} - 6\begin{bmatrix} 3 & 8 \\ 5 & 4 \end{bmatrix}$

b. Evaluate: $\begin{bmatrix} 3 & -5 & 2 \\ -1 & 4 & 3 \end{bmatrix}\begin{bmatrix} 6 \\ 2 \\ -3 \end{bmatrix}$

c. Evaluate: $\det\begin{bmatrix} 3 & 8 \\ 5 & 4 \end{bmatrix}$

d. Solve this system using matrices.

$$3x - 5y + 2z = -7$$
$$4x + y - 6z = 33$$
$$9x - 8y - 7z = 38$$

R3. **a.** Describe the transformations produced by $[T]$.

$$[T] = \begin{bmatrix} 0.6\cos 30° & 0.6\cos 120° \\ 0.6\sin 30° & 0.6\sin 120° \end{bmatrix}$$

b. Plot the pre-image triangle specified by $[M]$.

$$[M] = \begin{bmatrix} 5 & 8 & 7 \\ 1 & 2 & 3 \end{bmatrix}$$

c. Plot the image that results from the third iteration, $[T][T][T][M]$, on the same grid as the pre-image triangle.

d. By how many degrees has the image in part c been rotated from the original pre-image?

e. Show that the image vertex closest to the origin in part c is exactly 0.6^3 times as far from the origin as the corresponding vertex in the pre-image.

R4. **a.** Describe the transformations produced by $[A]$.

$$[A] = \begin{bmatrix} 0.6\cos 30° & 0.6\cos 120° & 5 \\ 0.6\sin 30° & 0.6\sin 120° & 2 \\ 0 & 0 & 1 \end{bmatrix}$$

b. Plot the pre-image rectangle specified by $[M]$.

$$[M] = \begin{bmatrix} 0 & 10 & 10 & 0 \\ 0 & 0 & 3 & 3 \\ 1 & 1 & 1 & 1 \end{bmatrix}$$

c. Plot the image at the third iteration, $[A][A][A][M]$.

d. To what fixed point are the images attracted when transformation $[A]$ is applied iteratively many times? How do you calculate this point numerically in a time-efficient way?

e. Calculate the fixed point in part d algebraically, and show that it agrees with the result you found numerically in part d.

f. Explain the purpose of the row 1, 1, 1, 1 in $[M]$ and of the row 0, 0, 1 in $[A]$.

R5. Figure 13-7b shows the rectangular pre-image

$$[M] = \begin{bmatrix} -3 & 3 & 3 & -3 \\ -1 & -1 & 1 & 1 \\ 1 & 1 & 1 & 1 \end{bmatrix}$$

It also shows the images $[A][M]$ and $[B][M]$, where $[A]$ is the transformation in Problem R4, part a, and $[B]$ rotates and dilates the same as $[A]$ but translates in the opposite direction.

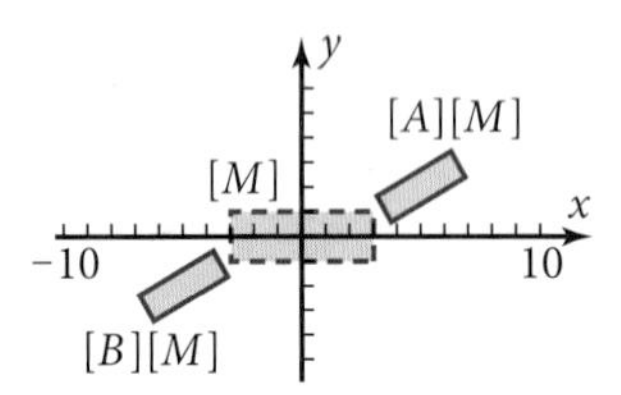

Figure 13-7b

a. Write a matrix for transformation $[B]$.

b. The two image rectangles in Figure 13-7b form the first iteration for a more complicated geometric figure. In the second iteration, transformations $[A]$ and $[B]$ are applied to both images in the first iteration. Apply these transformations and plot the resulting four images on graph paper.

c. The figure that results from applying $[A]$ and $[B]$ iteratively infinitely many times can be plotted approximately using Barnsley's method. Use probability 50% for each transformation. The result is shown in Figure 13-7c. Confirm on your grapher that the attractor in Figure 13-7c is correct. Write a paragraph describing the procedure used in Barnsley's method.

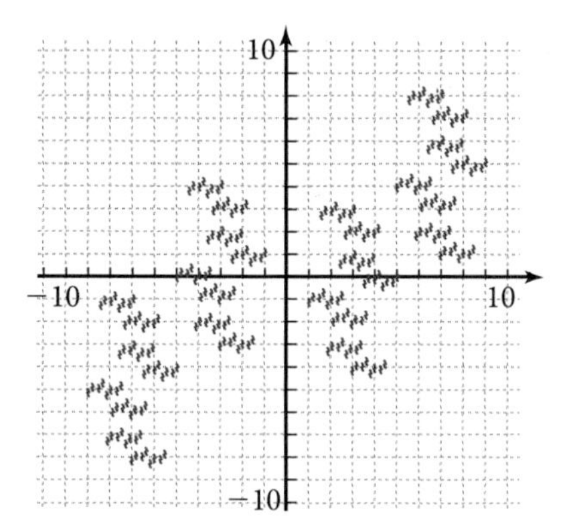

Figure 13-7c

d. On a copy of Figure 13-7c, plot the fixed point you found for transformation $[A]$ in Problem R4, part d. How does this fixed point seem to correspond to the fractal image?

e. For the pre-image and the first three iterations, make a table showing the iteration number, the number of rectangles, the perimeter of each rectangle, and the total perimeter of the figure. From the pattern in the table, find the total perimeter in the 50th iteration. If the iterations were performed infinitely many times, the result would be the strange attractor in Figure 13-7c. What is the total perimeter of the strange attractor?

R6. **a.** State the definition of the Hausdorff dimension.

b. Show that the fractal in Figure 13-7c is more than one-dimensional but less than two-dimensional.

c. How does the result in part b agree with the total perimeter of the fractal in Figure 13-7c?

d. Find the area of the pre-image rectangle and the total areas of the rectangles in the first, second, and third iterations in Problem R5. What number does the area of the fractal in Figure 13-7c approach as the number of iterations increases without bound? Does this number agree with the dimension you calculated in part b?

e. Suppose that the dilation factors for matrices $[A]$ and $[B]$ from Problems R4 and R5 were changed from 0.6 to 0.5. Use Barnsley's method to plot the resulting figure. Use probability 50% for each matrix.

f. Calculate the dimension of the figure in part e. Calculate the limit of the total perimeter of the rectangles as the number of iterations approaches infinity. How do the results of these calculations explain the change in the figure caused by reducing the dilation factor to 0.5?

g. If the dilation factor in transformations $[A]$ and $[B]$ were reduced to 0.4, what would be the dimension of the resulting figure? What number would the total perimeter of the figure approach? How does this number correspond to the dimension of the figure?

Concept Problems

C1. WEB *Matrix Inverse Research Problem:* Let $[A]$ be the 4×4 matrix

$$[A] = \begin{bmatrix} 1 & 2 & 3 & 5 \\ 2 & 3 & 5 & 6 \\ 4 & 6 & 7 & 8 \\ 9 & 8 & 6 & 4 \end{bmatrix}$$

Using the Internet or another resource, look up "adjoint matrix" or "adjugate matrix." Use what you find to answer these questions.

a. Write the source of your information.

b. Evaluate the cofactor of the element 1 in row 1, column 1 of the matrix.

c. Evaluate the cofactor of the element 5 in row 1, column 4. Describe how you determine the signs of the cofactors.

d. Find the adjugate (or adjoint) of $[A]$ using your calculator in an appropriate manner and the fact that

$$[A]^{-1} = \frac{1}{\det [A]} \cdot \text{adj } [A]$$

Do the cofactors you found in parts a and b agree with the entries in adj $[A]$? Explain.

e. Why would it be tedious to find the inverse of a 4×4 matrix completely "by hand," with no assistance from the built-in matrix features of your grapher?

C2. *Why Barnsley's Method Works Problem:* Consider a simple case involving Sierpiński's triangle to see why Barnsley's method works. Figure 13-7d shows the nine images in the second iteration from Problem Set 13-5, Problem 1. The transformation matrices are

$$[A] = \begin{bmatrix} 0.5 & 0 & 7.5 \\ 0 & 0.5 & -5 \\ 0 & 0 & 1 \end{bmatrix}$$

$$[B] = \begin{bmatrix} 0.5 & 0 & -7.5 \\ 0 & 0.5 & -5 \\ 0 & 0 & 1 \end{bmatrix}$$

$$[C] = \begin{bmatrix} 0.5 & 0 & 0 \\ 0 & 0.5 & 10 \\ 0 & 0 & 1 \end{bmatrix}$$

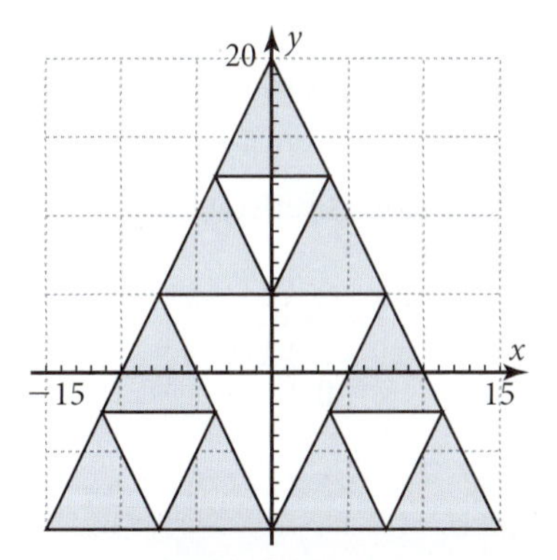

Figure 13-7d

a. Show that the fixed points of the three transformations $[A]$, $[B]$, and $[C]$ are the three vertices of the largest triangle.

b. On a copy of Figure 13-7d, mark the point $(10, -10)$, which is on the boundary of the shaded region. Write this point as a pre-image matrix and then apply transformation $[C]$ to it. Mark the image point on the copy of the figure, showing that it is on the boundary of another shaded region but has skipped over two of the unshaded regions.

c. The transformation in part b took the point $(10, -10)$ halfway from where it was to the fixed point $(0, 20)$ at the top vertex of the largest triangle. Make a conjecture about where the image point in part b will go if you apply transformation $[C]$ to it. Then confirm (or refute) your conjecture by applying the transformation.

d. Apply transformation $[B]$ to the image from part b. Show that the new image is halfway from the image in part b to the fixed point at the lower left vertex of the largest triangle, is on a boundary of a shaded region, and has skipped over an unshaded region.

e. Demonstrate that Barnsley's method really works by showing that it produces a self-similar fractal. Specifically for Sierpiński's triangle, show that if point P is anywhere in or on the largest triangle, the result of any one of the specified transformations applied to point P is a point in or on one of the three large corner triangles (those with base length 15). Why does this demonstration that the largest triangle is reproduced into the three corner triangles ensure that Sierpiński's triangle will have regions in which there are no points?

Chapter Test

Part 1: No calculators allowed (T1–T6)

T1. Multiply: $\begin{bmatrix} 3 & 5 & -2 \\ 4 & 1 & 7 \end{bmatrix} \begin{bmatrix} 6 & 8 \\ -1 & 9 \\ 0 & 3 \end{bmatrix}$

T2. Explain why the matrices in Problem T1 are commensurate for multiplication.

T3. In what way is matrix multiplication similar to the multiplication of two vectors?

T4. Find det $\begin{bmatrix} 3 & 8 \\ 2 & 7 \end{bmatrix}$. Use the result to find $\begin{bmatrix} 3 & 8 \\ 2 & 7 \end{bmatrix}^{-1}$.

Show that the product of the matrix and its inverse is equal to the identity matrix.

T5. What transformation is represented by this matrix?

$$\begin{bmatrix} 0.9 \cos 15° & 0.9 \cos 105° & 3 \\ 0.9 \sin 15° & 0.9 \sin 105° & 2 \\ 0 & 0 & 1 \end{bmatrix}$$

T6. A line segment is transformed by iterative matrix multiplication. The matrices for the pre-image and the images of the first and second iterations are

$$\begin{bmatrix} 5 & 5 \\ 2 & 6 \\ 1 & 1 \end{bmatrix} \quad \begin{bmatrix} 6.7 & 5.1 \\ 6.4 & 9.2 \\ 1 & 1 \end{bmatrix} \quad \begin{bmatrix} 6.1 & 3.8 \\ 10.1 & 11.4 \\ 1 & 1 \end{bmatrix}$$

Pre-image First iteration Second iteration

Plot these three images on graph paper or dot paper. Explain what will happen to the images as more and more iterations are performed.

Part 2: Graphing calculators allowed (T7–T19)

For Problems T7–T18, the graph in Figure 13-7e shows a vertical segment 10 units high, starting at the origin. This pre-image is to be transformed into a "tree" made from three segments, each 5 units long. The three segments satisfy these conditions:

- The left branch starts at the point $(0, 5)$ and is rotated $+20°$ from the trunk.
- The trunk extends from the point $(0, 0)$ to the point $(0, 5)$.
- The right branch starts at the point $(0, 5)$ and is rotated $-30°$ from the trunk.

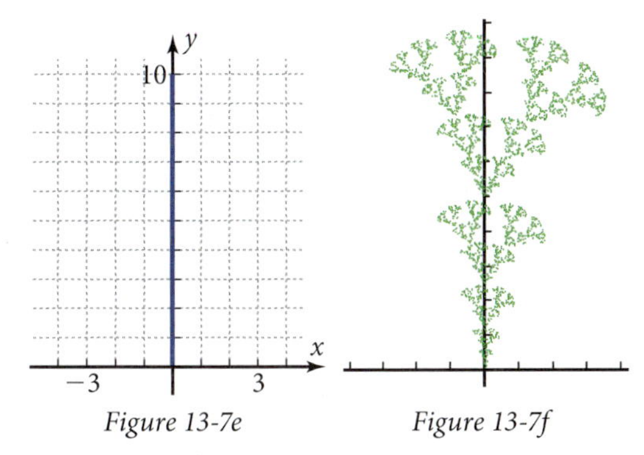

Figure 13-7e *Figure 13-7f*

T7. Write three 3×3 transformation matrices to perform these tasks:

$[A]$ should transform the pre-image to the left branch.

$[B]$ should transform the pre-image to the trunk.

$[C]$ should transform the pre-image to the right branch.

T8. Multiply the pre-image matrix $[D]$ by each of $[A]$, $[B]$, and $[C]$. Write the three image matrices, rounding the entries to one decimal place. Plot the three images on graph paper.

T9. The three images in Problem T8 are the results of the first iteration. If the three transformations are applied to each of these three images, the nine resulting images form the second iteration. Calculate the image $[A][C][D]$ of the second iteration, and plot it on the same axes as the images in Problem T8.

T10. If the iterations are performed infinitely many times, the resulting tree is a fractal. If the transformations are applied at random to a single point, each time using the previous image, the points are attracted to the *same* fractal figure. Use Barnsley's method with 1000 points to plot this strange attractor.

T11. Figure 13-7f shows the strange attractor from Problem T10 plotted with 4000 points. On a copy of Figure 13-7f, illustrate self-similarity by circling two parts of the figure of different size, each of which is similar to the whole tree.

T12. The first iteration has three images, each 5 units long. The second iteration has nine images, each 2.5 units long. Calculate the sum of the lengths of the images at iterations 0, 1, 2, 3, and 100.

T13. If the iterations were performed infinitely many times, what would the sum of the lengths of the images approach?

T14. Each iteration divides each previous segment into three self-similar pieces, each 0.5 times as long as the previous segment. Let N be the number of pieces, and let r be the ratio of the length of each piece to the length of the original pre-image. Complete a table for iterations 0 through 5 listing values of n, r, $\frac{1}{r}$, and N.

T15. Complete the statement: "Each time $\frac{1}{r}$ is multiplied by 2, N is multiplied by __?__."

T16. Write Hausdorff's definition of dimension.

T17. Calculate the dimension of the tree that would result if the iterations were performed infinitely many times.

T18. Strange attractors such as the one in Figure 13-7f result from iterating several different transformations. If just *one* transformation is iterated, the images are attracted to a *single* fixed point. Find the fixed point to which the images are attracted if $[A]$ is applied iteratively to the pre-image $[D]$. On a copy of Figure 13-7f, mark this fixed point. How does it relate to the fractal figure?

T19. What did you learn as a result of taking this test that you did not know before?

13-8 Cumulative Review, Chapters 10–13

This problem set is a final exam on these topics:

- Conic sections and quadric surfaces
- Polar coordinates
- Complex numbers
- Parametric equations for moving objects
- Three-dimensional vectors
- Matrix transformations and fractal figures

If you are thoroughly familiar with these topics, you should be able to finish the problem set in about three hours.

Review Problems

Annie's Conic Section Problems: (Problems 1–8) Annie takes a test on conic sections.

1. How can Annie tell that this equation in center-radius form is the equation for a hyperbola? Sketch the graph opening in the proper direction, showing the asymptotes and the vertices.

$$-\left(\frac{x-4}{3}\right)^2 + \left(\frac{y-1}{5}\right)^2 = 1$$

2. Transform the equation in Problem 1 into polynomial form. How does the result confirm the fact that the graph is a hyperbola?

3. This equation in polynomial form appears to be for a hyperbola because the squared terms have opposite signs.

$$25x^2 - 9y^2 - 200x + 18y + 391 = 0$$

However, the graph is actually a *degenerate hyperbola.* Plot the graph using your conic section plotting program (from Section 10-2) and sketch the result. Then complete the squares to transform the equation into center-radius form. What do you notice about the right side of the transformed equation that would suggest that the hyperbola is degenerate?

4. Figure 13-8a shows the ellipse $9x^2 + 25y^2 = 225$. Find the focal radius, eccentricity, and directrix radius.

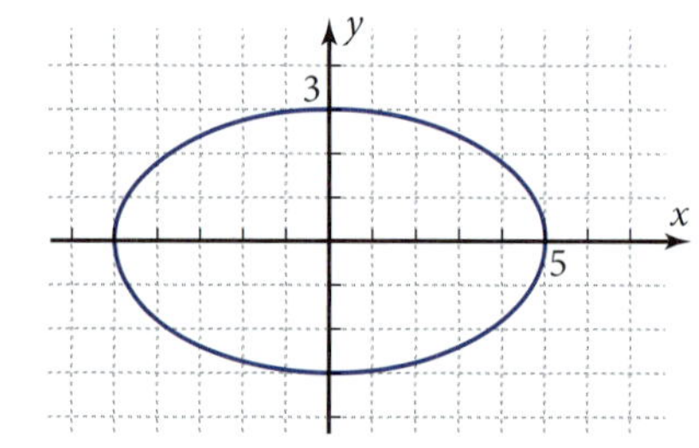

Figure 13-8a

5. On a copy of Figure 13-8a, pick a point (x, y) on the ellipse. Measure the distances from the point (x, y) to the two foci. How do these distances compare with the length of the major axis? Measure the perpendicular distance from the point (x, y) to the directrix on the right side. How is this distance related to the distance from the point (x, y) to the focus on the right side?

6. If the equation in Problem 4 included a nonzero xy-term, as in the equation $9x^2 + 10xy + 25y^2 = 225$, how could you tell from the discriminant that the graph would still be an ellipse? What would be the major change in the graph?

7. Write parametric equations for the ellipse in Figure 13-8a. Then write parametric equations for the ellipse after it is translated 2 units in the positive x-direction and 1 unit in the positive y-direction. Confirm that your equations are correct by plotting them on your grapher.

8. Figure 13-8b shows the ellipsoid formed by rotating the ellipse of Figure 13-8a about the x-axis. A cylinder is inscribed in this ellipsoid, with its axis along the x-axis. Find the volume of the cylinder in terms of the sample point (x, y) where the cylinder touches the ellipse. Use the result to find the radius and altitude (length) of the cylinder with maximum volume.

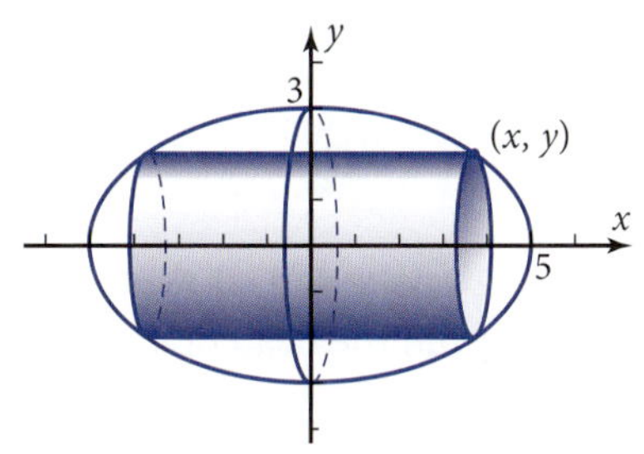

Figure 13-8b

Polar Coordinates Problems: (Problems 9–11)

9. Figure 13-8c shows part of the polar graph of $r = 1 - 7\cos\theta$. Calculate r if $\theta = 30°$. Sketch the graph and show the 30° angle and the point on the graph. Explain why the point is in the *third* quadrant despite the fact that 30° is a *first*-quadrant angle.

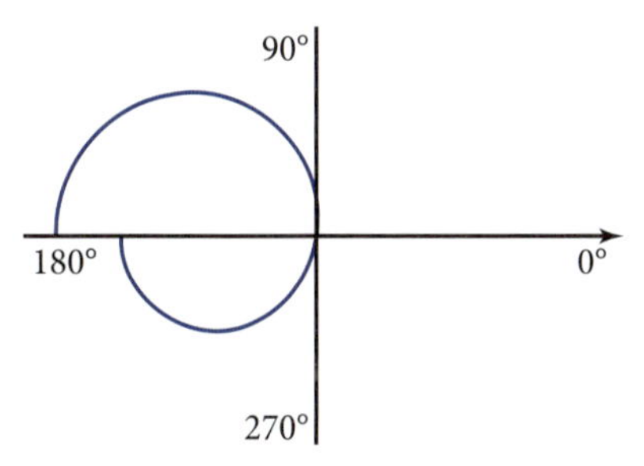

Figure 13-8c

10. Plot the cardioid $r = 24 - 24\cos\theta$. Use a window with $[-50, 50]$ for the horizontal axis and equal scales on both axes. Sketch the result. Why do you think the figure is called a *cardioid*?

11. Figure 13-8d shows the graph of the circle with equation $r = 3960$, in polar coordinates, representing Earth, with radius 3960 mi. It also shows part of the elliptical path of a spaceship given by the polar equation

$$r = \frac{5600}{1 + 0.5\cos\theta}$$

If the spaceship continues on its present path, will it hit Earth's surface? If so, what are the polar coordinates of the point where it will hit? If not, explain how you know it will not.

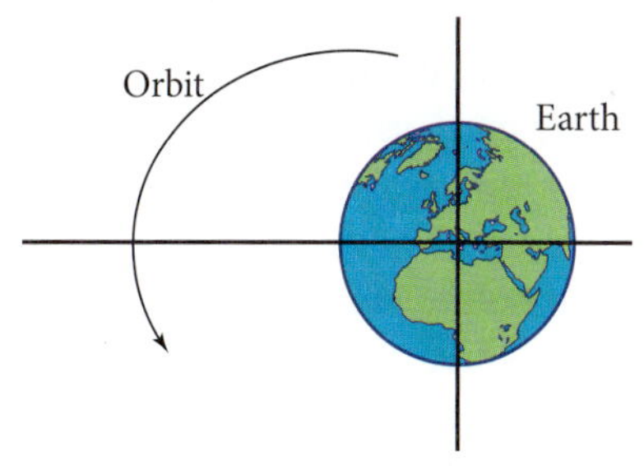

Figure 13-8d

Complex Numbers Problems: (Problems 12–14)

12. Square and simplify: $(3 - 5i)^2$

13. Multiply the complex numbers 5 cis 70° and 8 cis 40°. Write the answer as a complex number in the form $r(\cos\theta + i\sin\theta)$. Transform the answer into $a + bi$ form.

14. Write the three cube roots of 64 as complex numbers in polar form.

Parametric Function Problems: (Problems 15 and 16)

15. Figure 13-8e shows the graph of the parametric function given by the equations

$$x = t^3$$
$$y = t^2$$

Calculate the coordinates of the point (x, y) if $t = 2$, and describe how you can check your answer graphically. Eliminate the parameter t, thus transforming the equations into a single Cartesian equation with y as a function of x. Plot the Cartesian equation on your grapher. Does the graph agree with the given figure?

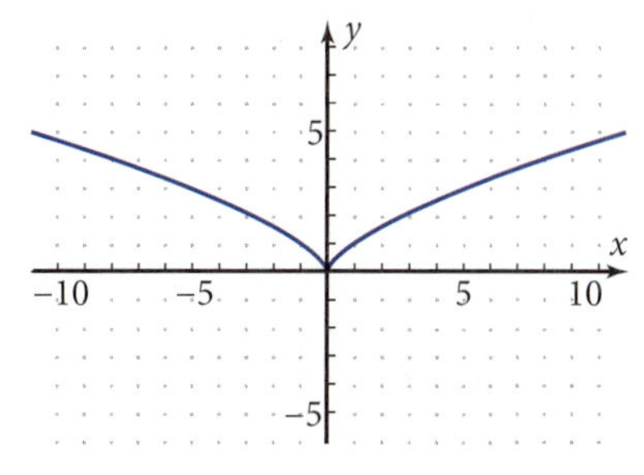

Figure 13-8e

16. Figure 13-8f shows a quarter (radius 12 mm) centered at the origin of a Cartesian coordinate system. Another quarter rolls around it (without slipping). A point on the moving quarter traces an *epicycloid of one cusp.* Write parametric equations for the path. It will help if you sketch the rolling quarter in another position, draw a vector from the origin to the center of the rolling quarter, and then draw another vector from that point to the point on the graph. Confirm that your equations are correct by plotting them on your grapher. How does the graph relate to the cardioid in Problem 10?

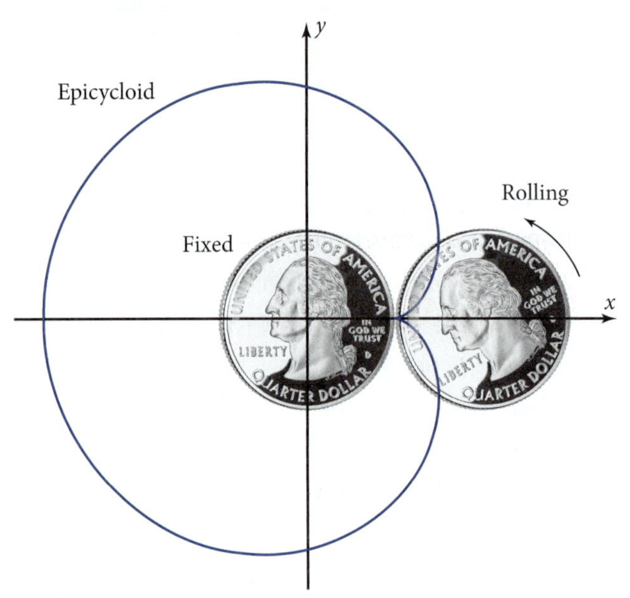

Figure 13-8f

Air Show Problems: (Problems 17–24) A pilot is doing stunts with her plane at an air show.

At times $t = 0$ s and $t = 10$ s, her position vectors from the control tower are

$$\vec{a} = 7\vec{i} + 4\vec{j} + 3\vec{k} \qquad \text{at } t = 0$$
$$\vec{b} = 10\vec{i} + 20\vec{j} + 5\vec{k} \qquad \text{at } t = 10$$

where $\vec{i}$ and $\vec{j}$ are unit vectors in the horizontal xy-plane and $\vec{k}$ is a vertical unit vector in the direction of the z-axis. The magnitude of each unit vector is 100 m.

17. Sketch vectors $\vec{a}$ and $\vec{b}$ tail-to-tail. Show the displacement vector from the head of $\vec{a}$ to the head of $\vec{b}$.

18. Calculate the displacement vector in Problem 17.

19. At $t = 10$ s, how far was the pilot from her position at $t = 0$ s?

20. Was the plane higher up or lower down at $t = 10$ s than it was at $t = 0$ s? By how many meters?

21. What is the measure of the angle between $\vec{a}$ and $\vec{b}$?

22. Find the scalar projection of $\vec{a}$ on $\vec{b}$.

23. The cross product of $\vec{a}$ and $\vec{b}$ is

$$\vec{a} \times \vec{b} = -40\vec{i} - 5\vec{j} + 100\vec{k}$$

Show that $\vec{a} \times \vec{b}$ is perpendicular to $\vec{a}$.

24. Find the area of the triangle formed by $\vec{a}$ and $\vec{b}$.

Matrix Operations Problems: (Problems 25–29)

25. For this matrix product show how 50, the lower-right element of the product matrix, is calculated.

$$\begin{bmatrix} 6 & 8 & 3 \\ 2 & 7 & 4 \end{bmatrix} \begin{bmatrix} 3 & 2 \\ 5 & 6 \\ 4 & 1 \end{bmatrix} = \begin{bmatrix} 70 & 63 \\ 57 & 50 \end{bmatrix}$$

26. Show how the determinant −91 is calculated.

$$\det\begin{bmatrix}70 & 63\\57 & 50\end{bmatrix} = -91$$

27. Show how the determinant in Problem 26 is used to calculate the inverse matrix

$$\begin{bmatrix}70 & 63\\57 & 50\end{bmatrix}^{-1} = \begin{bmatrix}-0.5494... & 0.6923...\\0.6263... & -0.7692...\end{bmatrix}$$

28. Show that the product $\begin{bmatrix}70 & 63\\57 & 50\end{bmatrix}\begin{bmatrix}70 & 63\\57 & 50\end{bmatrix}^{-1}$ equals the identity matrix.

29. How is multiplying two matrices similar to finding the dot product of two vectors?

Doug's Iterative Transformation Problems: (Problems 30–38) Doug is an archaeologist. He unearths a set of paving stones that follow a spiral pattern generated by this matrix:

$$[A] = \begin{bmatrix}0.45 & -0.78 & 20\\0.78 & 0.45 & 10\\0 & 0 & 1\end{bmatrix}$$

30. Given that the dilation factor is 0.9, what is the angle of rotation?

31. The pre-image matrix is

$$[D] = \begin{bmatrix}2 & -2 & -5 & 5\\5 & 5 & -5 & -5\\1 & 1 & 1 & 1\end{bmatrix}$$

Use your iterative program (from Section 13-3, Problem 19) to apply transformation $[A]$ for 40 iterations. Use a window with $[-50, 50]$ for x and $[-20, 50]$ for y. Find numerically to one decimal place the coordinates of the fixed point to which the images are attracted.

32. Show that you understand how matrix multiplication is done by finding this product:

$$\begin{bmatrix}0.45 & -0.78 & 20\\0.78 & 0.45 & 10\\0 & 0 & 1\end{bmatrix}\begin{bmatrix}x\\y\\1\end{bmatrix}$$

33. At the fixed point, the image of the point (x, y) must equal (x, y). Thus

$$\begin{bmatrix}0.45 & -0.78 & 20\\0.78 & 0.45 & 10\\0 & 0 & 1\end{bmatrix}\begin{bmatrix}x\\y\\1\end{bmatrix} = \begin{bmatrix}x\\y\\1\end{bmatrix}$$

Use this information to calculate algebraically the coordinates of the fixed point.

34. Doug finds a pattern on a wall that resembles the fractal image in Figure 13-8g.

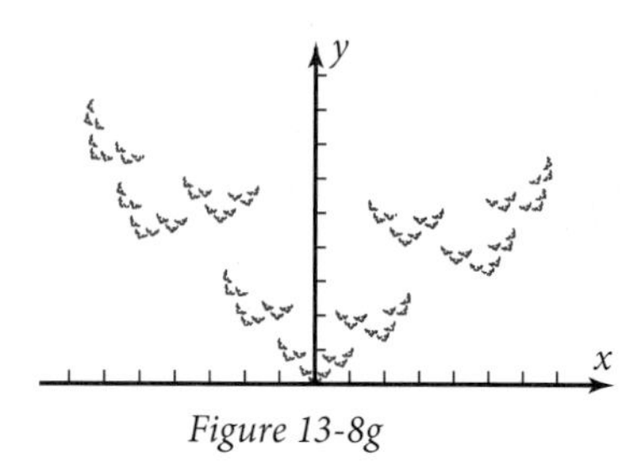

Figure 13-8g

Circle two parts of different size, each of which is similar to the entire figure.

35. One of the three transformations that generated Figure 13-8g is

$$[A] = \begin{bmatrix}0.4\cos 20° & 0.4\cos 110° & 5\\0.4\sin 20° & 0.4\sin 110° & 3\\0 & 0 & 1\end{bmatrix}$$

Find numerically the fixed point for this transformation and explain how it relates to the image in Figure 13-8g.

36. Write another transformation matrix $[B]$ to do all these things:

- Rotate 20° clockwise
- Dilate by a factor of 0.4
- Translate by −5 units in the x-direction
- Translate by 4 units in the y-direction

37. Write a third transformation matrix $[C]$ to dilate by a factor of 0.4 without rotating or translating.

38. Check your answers to Problems 36 and 37 by running your Barnsley program (from Section 13-5. Problem 3) with 1000 points. Use probabilities $\frac{1}{3}, \frac{1}{3}, \frac{1}{3}$, and 0 for the four transformations the program is expecting.

The Great Wave by Katsushika Hokusai. The waves in this painting resemble a fractal pattern.

Fractal Dimension Problems: (Problems 39–43) Figure 13-8h shows a 9×9 square pre-image and the first iteration of four transformations applied to the pre-image. Each transformation dilates the pre-image to $\frac{1}{3}$ its original length but translates by different amounts.

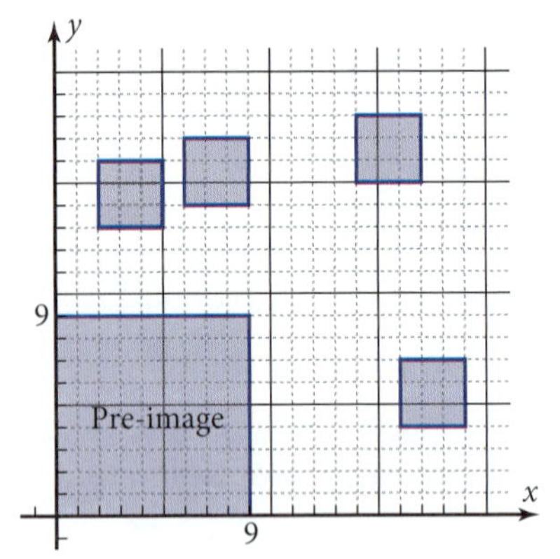

Figure 13-8h

39. How many images, N, will there be in the second iteration and in the third iteration? What is the total perimeter of the images in the first iteration? The second iteration? The third iteration?

Iteration	N	Perimeter of Each	Total Perimeter
0	1	36	36
1	4	12	
2			
3			

40. Following the pattern in the table, what will be the total perimeter of the 50th iteration?

41. What is the total area of the first iteration? The second iteration? The third iteration?

42. If the transformations of this problem are applied using Barnsley's method, what will be the dimension of the resulting fractal image?

43. What limit does the total area approach as the number of iterations approaches infinity? What limit does the total perimeter approach? Explain why these answers are consistent with the dimension you calculated in Problem 42.

44. What is the most important thing you learned as a result of completing this cumulative review?

Chapter

14

Probability, and Functions of a Random Variable

To protect themselves against losing their life savings in a fire, homeowners purchase fire insurance. Each homeowner pays an insurance company a relatively small premium each year. From that money, the insurance company pays to replace the few homes that burn. Actuaries at the insurance company use the probability that a given house will burn to calculate the premiums to charge each year so that the company will have enough money to cover insurance claims and to pay employee salaries and other expenses. In this chapter you will learn some of the mathematics involved in these calculations.

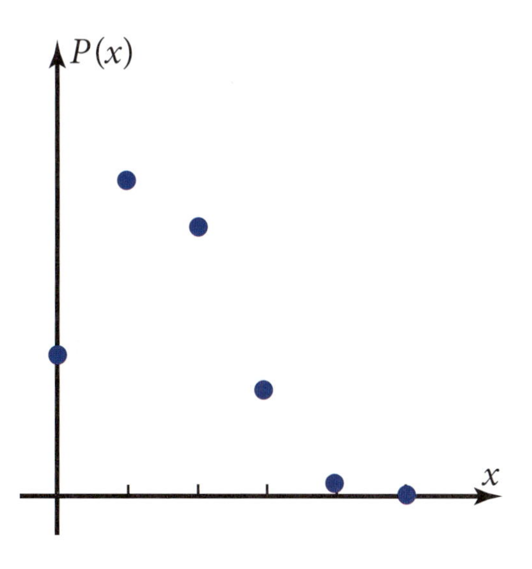

Mathematical Overview

In this chapter you'll learn about random experiments and events. Results of random experiments constitute events. You will also learn the concepts of probability and mathematical expectation. The probabilities of various events are values that give you the likelihood of a particular event among all possible events in an experiment. You will gain this knowledge in four ways.

NUMERICALLY

$$\text{Probability} = \frac{\text{number of favorable outcomes}}{\text{total number of possible outcomes}}$$

ALGEBRAICALLY

If the probability that a thumbtack will land point up on any one flip is 0.3, then $P(x)$, the probability that it will land point up exactly x times in five flips, is this function of a random variable:

$$P(x) = {}_5C_x \cdot 0.6^{5-x} \cdot 0.3^x$$

GRAPHICALLY

This is the graph of $P(x)$. The graph shows the probability that the thumbtack will land point up 0, 1, 2, 3, 4, or 5 times in five flips.

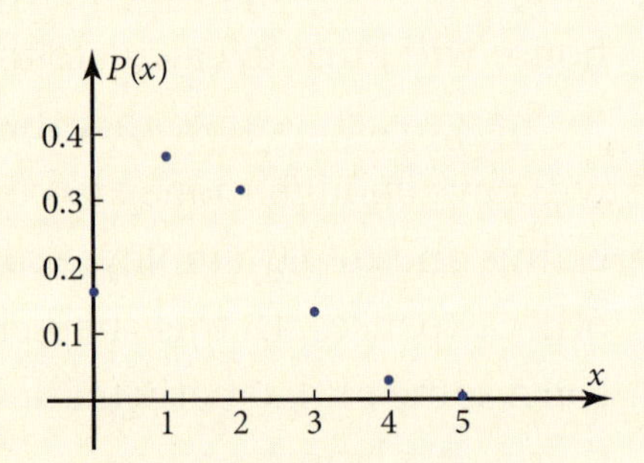

VERBALLY

To find a function of a random variable, find the probability of each event in some random experiment, such as getting "point up" x times when a thumbtack is flipped five times. Then plot the graph of the probability as a function of x. Don't connect the dots, because the number of point-ups has to be a whole number. If the thumbtack experiment is done for money, the average winnings or losses (called the mathematical expectation) are calculated by multiplying the probability of each outcome by the result of that outcome.

14-1 Introduction to Probability

Suppose two dice are rolled, one white and one black. Figure 14-1a shows the 36 possible outcomes.

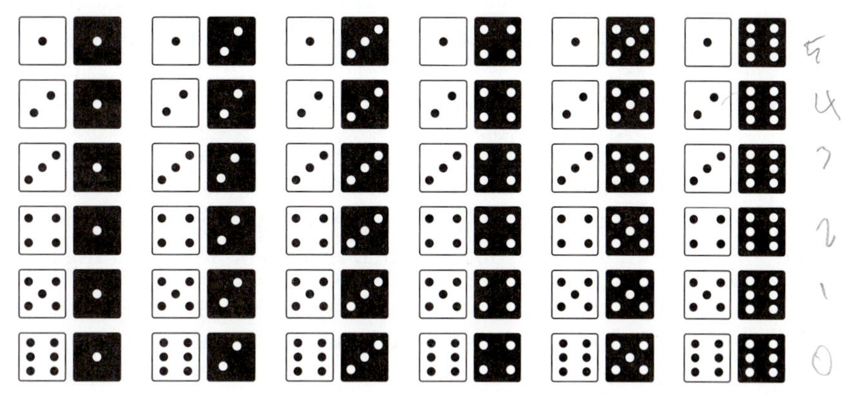

Figure 14-1a

There are five outcomes for which the total on the dice is 6:

Because each outcome is equally likely, you would expect that in many rolls of the dice the total would be 6 roughly $\frac{5}{36}$ of the time. This number, $\frac{5}{36}$, is called the **probability** of rolling a 6. In this section you will find the probabilities of other events in the dice-rolling experiment.

Objective Find the probability of various events in a dice-rolling experiment.

Exploratory Problem Set 14-1

Two dice are rolled, one white and one black. Find the probability of each of these events.

1. The total is 10.
2. The total is at least 10.
3. The total is less than 10.
4. The total is at most 10.
5. The total is 7.
6. The total is 2.
7. The total is between 3 and 7, inclusive.
8. The total is between, but does not include, 3 and 7.
9. The total is between 2 and 12, inclusive.
10. The total is 13.
11. The numbers are 2 and 5.
12. The black die shows 2 and the white die shows 5.
13. The black die shows 2 or the white die shows 5.

14-2 Words Associated with Probability

You have heard statements such as "It will probably rain today." Mathematicians give the word *probably* a precise meaning by attaching *numbers* to it, such as "The probability that it will rain is 30%." To understand and use probability, you need to learn the definitions of a few important terms.

Objective Distinguish among various words used to describe probability.

For the dice-rolling experiment of Section 14-1, the act of rolling the dice is called a **random experiment.** Each time you roll the dice is called a **trial.** The word *random* lets you know that there is no way of telling beforehand how any roll is going to come out.

Each way the dice could come up, such as

is called an **outcome** or a **simple event.** Outcomes are results of a random experiment. In this experiment, the outcomes are *equally likely.* That is, each has the same chance of occurring.

An **event** is a set of outcomes. For example, the event "the total on two dice is 6" is the five-element set

The set of *all* outcomes of an experiment is called the **sample space.** The sample space for the dice-rolling experiment of Section 14-1 is the set of all 36 outcomes in Figure 14-1a.

The **probability** of an event may now be defined numerically.

DEFINITION: Probability

If the outcomes of a random experiment are equally likely, then the **probability** that a particular event will occur is

Verbally: $\text{Probability} = \dfrac{\text{number of outcomes in the event}}{\text{number of outcomes in the sample space}}$

Symbolically: $P(E) = \dfrac{n(E)}{n(S)}$

where $n(E)$ is the number of outcomes in event E and $n(S)$ is the number of outcomes in the sample space, S.

The symbols in the definition of probability are variations of $f(x)$ notation. Using these symbols makes sense because the probability of an event *depends* on the event.

Note that all probabilities are between 0 and 1, inclusive. An event that is *certain* to occur has probability 1 because $n(E) = n(S)$. An event that cannot possibly occur has probability 0 because $n(E) = 0$.

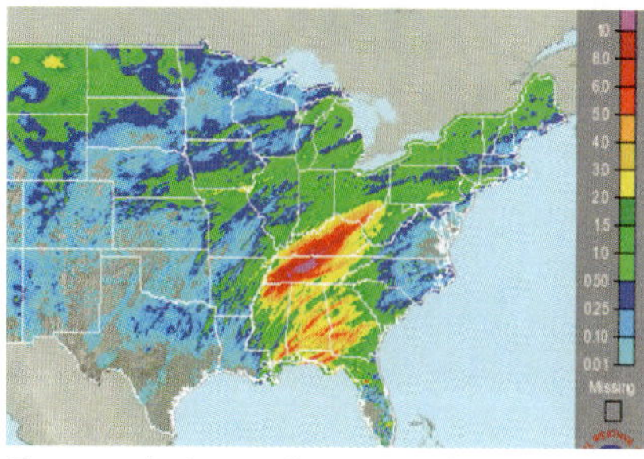

Tomorrow's chance of rain is rarely 0% or 100%.

Problem Set 14-2

Quick Review

Q1. If you flip a coin, what is the probability that the result will be heads?

Q2. If you flip the coin again, what is the probability that the second flip will be heads?

Q3. Does the result of the second flip depend on the result of the first flip?

Q4. What is $\frac{3}{5}$ expressed as a percent?

Q5. If i is the imaginary number $\sqrt{-1}$, then $i^2 = $ __?__.

Q6. $\begin{bmatrix} 3 & 1 & 2 & 9 & 6 \\ 7 & 1 & 4 & 0 & 8 \end{bmatrix}$ is a __?__ × __?__ matrix.

Q7. In the expression ab, the numbers a and b are called __?__.

Q8. In the expression $a + b$, the numbers a and b are called __?__.

Q9. True or false: $(ab)^2 = a^2b^2$

Q10. True or false: $(a + b)^2 = a^2 + b^2$

---

1. A card is drawn at random from a standard 52-card deck. (See Figure 14-3a on page 710 if you are unfamiliar with a deck of cards.)

a. What term is used in probability for the act of drawing the card?

b. How many outcomes are in the sample space?

c. How many outcomes are in the event "the card is a face card"?

d. Calculate P(the card is a face card).

e. Calculate P(the card is black).

f. Calculate P(the card is an ace).

g. Calculate P(the card is between 3 and 7, inclusive).

h. Calculate P(the card is the ace of clubs).

i. Calculate P(the card belongs to the deck).

j. Calculate P(the card is a joker).

2. A penny, a nickel, and a dime are flipped at the same time. Each coin can land either heads up (H) or tails up (T).

a. What term is used in probability for the act of flipping the coins?

b. One possible outcome is THT. List all eight outcomes in the sample space.

c. How many outcomes are in the event "exactly two of the coins show heads"?

d. Calculate P(HHT).

e. Calculate P(exactly two heads).

f. Calculate P(at least two heads).

g. Calculate P(penny and nickel are tails).

h. Calculate P(penny or nickel is tails).

i. Calculate P(none are tails).

j. Calculate P(zero, one, two, or three heads).

k. Calculate P(four heads).

3. WEB *Historical Search Project:* Check the Internet or other sources for information about early contributors to the field of mathematical probability. Describe the dice problem investigated by Blaise Pascal and Pierre de Fermat that led to the foundations of probability theory. Give the source of your information.

14-3 Two Counting Principles

Counting the outcomes in an event or sample space can be difficult. For example, suppose a CD player is programmed to play eight songs in random order and you want to find the probability that your two favorite songs will play in a row. The sample space for this experiment contains over 40,000 outcomes! In this section you will learn ways of computing numbers of outcomes without actually counting them.

Objective Calculate the number of outcomes in an event or sample space.

Independent and Mutually Exclusive Events

Counting outcomes sometimes involves considering two or more events. To find the number of outcomes in a situation involving two events, you need to consider whether *both* events occur or whether *either* one event *or* the other occurs, but not both.

For example, suppose a summer camp offers four outdoor activities and three indoor activities:

Outdoor	Indoor
swimming	pottery
canoeing	computers
volleyball	music
archery	

On Monday, each camper is assigned an outdoor activity in the morning and an indoor activity in the afternoon. In how many ways can the two activities be chosen?

In this situation, the two events are "an outdoor activity is chosen" and "an indoor activity is chosen." You want to count the number of ways *both* events could occur. You could find the answer by making an organized list of all the possible pairs:

swimming–pottery	swimming–computers	swimming–music
canoeing–pottery	canoeing–computers	canoeing–music
volleyball–pottery	volleyball–computers	volleyball–music
archery–pottery	archery–computers	archery–music

You could also reason like this: There are four choices for the outdoor activity. For each of these choices, there are three choices for the indoor activity. So there are $4 \cdot 3$ or 12 ways of choosing both activities. Note that the events in this situation are said to be **independent** because the way one occurs does not affect the ways the other could occur.

On the day of the camp talent show, there is time for only one activity. Each camper is assigned *either* an outdoor activity *or* an indoor activity (not both).

In this situation, the events are **mutually exclusive,** meaning that the occurrence of one of them excludes the possibility that the other will occur. If a camper is assigned an outdoor activity, he or she cannot also be assigned an indoor activity, and vice versa. Because there are four ways of choosing an outdoor activity and three ways of choosing an indoor activity, there are $4 + 3$ or 7 ways of choosing one type of activity or the other.

These examples illustrate two counting principles, summarized in the box.

PROPERTIES: Two Counting Principles

1. Let A and B be two events that occur independently. Then the number of ways *both* events can occur is given by

   $n(A \text{ and } B) = n(A) \cdot n(B)$ *Multiply* the number of ways.

2. Let A and B be mutually exclusive events. Then the number of ways one event *or* the other event can occur is given by

   $n(A \text{ or } B) = n(A) + n(B)$ *Add* the number of ways.

Dependent and Overlapping Events

What if two events are **dependent,** not independent? For instance, suppose that on Tuesday the outdoor activities are canceled due to rain, so each camper is assigned one indoor activity in the morning and a different indoor activity in the afternoon. Event A, "Select a morning activity," can occur three ways. But event B, "Select an afternoon activity," can occur only *two* ways because the activity cannot be the same as the morning one. So there are only $3 \cdot 2$ or 6 ways of choosing A and B. The number of ways B can occur *depends* on whether or not A has already occurred.

What if two events are overlapping events and thus are not mutually exclusive? For instance, suppose that you draw one card from a standard deck of 52 cards and want to know the number of ways the card could be a heart or a face card. As shown in Figure 14-3a, $n(\text{heart}) = 13$ and $n(\text{face}) = 12$. Adding 13 and 12, as in the second counting principle, gives the wrong answer. The three cards in the intersection (those that are both hearts *and* face cards) have been counted twice. An easy way to get the correct answer is to subtract the number of outcomes that are both hearts and face cards from the sum $13 + 12$. That is,

$$n(\text{heart } or \text{ face}) = n(\text{heart}) + n(\text{face}) - n(\text{heart } and \text{ face})$$

This results in the correct numerical solution $13 + 12 - 3$, or 22. You can verify this result by counting.

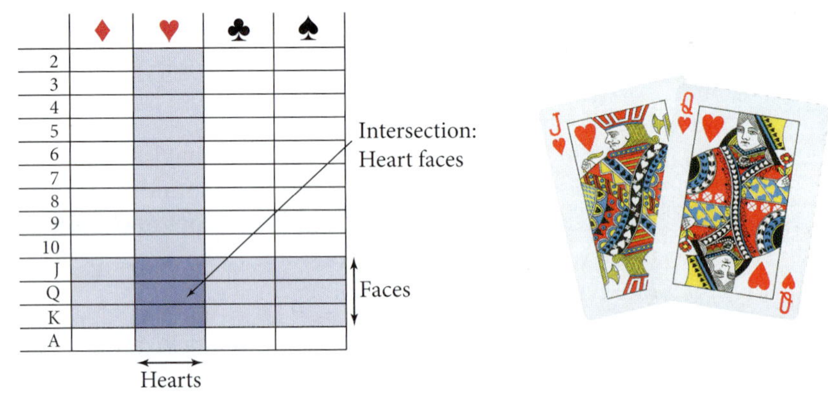

Figure 14-3a

The box shows generalizations of the two counting principles for dependent events and for overlapping events.

PROPERTIES: Two Counting Principles, Generalized

1. Let A and B be two dependent events that can occur in sequence. Then the number of ways *both* events can occur is given by

$n(A \text{ and then } B) = n(A) \cdot n(B \mid A)$ *Multiply* the numbers of ways.

where $n(B \mid A)$ is the number of ways B could occur given A has already occurred.

2. Let A and B be two overlapping events that are not mutually exclusive. Then the number of ways one event *or* the other event (or possibly both) can occur is given by

$n(A \text{ or } B) = n(A) + n(B) - n(A \text{ and } B)$ *Add* the numbers of individual ways, and then *subtract* the number of ways that were counted twice.

Notes:

- If A and B are independent, then $n(B|A) = n(B)$, and the first counting principle reduces to $n(A \text{ and } B) = n(A) \cdot n(B)$, as in the box on page 709.
- If A and B are two mutually exclusive events, then $n(A \text{ and } B) = 0$, and the second counting principle reduces to $n(A \text{ or } B) = n(A) + n(B)$, as in the box on page 709.

Problem Set 14-3

Reading Analysis

From what you have read in this section, what do you consider to be the main idea? Under what condition would you *multiply* two numbers of ways two events could occur, and under what condition would you *add* the numbers? Under what condition would adding the two numbers of ways need to be adjusted by subtraction to give the correct answer?

Quick Review

Q1. Simplify the fraction $\frac{12}{36}$.

Q2. Evaluate $1 \cdot 2 \cdot 3 \cdot 4 \cdot 5$

Q3. If $n(A) = 71$ and there are 300 possible outcomes in the sample space, then $P(A) = \underline{\quad ? \quad}$.

Q4. Multiply: $\left(\frac{2}{7}\right)\left(\frac{3}{4}\right)$

Q5. Add: $\frac{3}{8} + \frac{1}{4}$

Q6. The exact value (no decimals) of $\cos \frac{\pi}{6}$ is $\underline{\quad ? \quad}$.

Q7. How well does a regression equation fit the data if the correlation coefficient is -1?

Q8. $x = 3t^2$ and $y = \cos t$ are equations for a $\underline{\quad ? \quad}$ function.

Q9. Factor: $x^2 - 3x - 4$

Q10. The dot product $\vec{a} \cdot \vec{b} = (3\vec{i} + 4\vec{j} + 7\vec{k}) \cdot (5\vec{i} + 2\vec{j} + 1\vec{k})$ equals

A. $\sqrt{14} \cdot \sqrt{8}$ **B.** $\sqrt{74} \cdot \sqrt{30}$
C. $15\vec{i} + 8\vec{j} + 7\vec{k}$ **D.** 112 **E.** 30

---

1. A salesperson has 7 customers in Denver and 13 customers in Reno. In how many different ways could she telephone

a. A customer in Denver and then a customer in Reno

b. A customer in Denver or a customer in Reno, but not both

2. A pizza establishment offers 12 vegetable toppings and 5 meat toppings. Find the number of different ways you could select

a. A meat topping or a vegetable topping

b. A meat topping and a vegetable topping

3. A reading list consists of 11 novels and 5 biographies. Find the number of different ways a student could select

a. A novel or a biography

b. A novel and then a biography

c. A biography and then another biography

4. A convoy of 20 cargo ships and 5 escort vessels approaches the Suez Canal. In each scenario, in how many different ways could these vessels begin to go through the canal?

a. A cargo ship and then an escort vessel

b. A cargo ship or an escort vessel

c. A cargo ship and then another cargo ship

5. The menu at Paesano's lists 7 salads, 11 entrees, and 9 desserts. How many different salad–entree–dessert meals could you select? (Meals are considered to be different if any one thing is different.)

6. Admiral Motors manufactures cars with 5 different body styles, 11 different exterior colors, and 6 different interior colors. A dealership wants to display one of each possible variety of car in its showroom. Explain to the manager of the dealership why the plan would be impractical.

7. Consider the letters in the word LOGARITHM.
   a. In how many different ways could you select a vowel or a consonant?
   b. In how many different ways could you select a vowel and then a consonant?
   c. How many different three-letter "words" (for example, ORL, HLG, AOI) could you make using each letter no more than once in any one word? (There are three events: "select the first letter," "select the second letter," and "select the third letter." Find the number of ways each event can occur, and then figure out what to do with the three results.)

8. Lee brought two jazz CDs and five rap CDs to play at the class picnic.
   a. In how many different ways could he choose a jazz CD and then a rap CD?
   b. In how many different ways could he choose a jazz CD or a rap CD?
   c. Lee's CD player allows him to load four CDs at once. The CDs will play in the order he loads them. How many different orderings of four CDs are possible? (See Problem 7 for a hint.)

9. There are 20 girls on the basketball team. Of these, 17 are over 16 years old, 12 are taller than 170 cm, and 9 are both older than 16 and taller than 170 cm. How many of the girls are older than 16 or taller than 170 cm?

10. Lyle's DVD collection includes 37 classic films and 29 comedies. Of these, 21 are classic comedies. How many DVDs does Lyle have that are classics or comedies?

11. The library has 463 books dealing with science and 592 books of fiction. Of these, 37 are science fiction books. How many books are either science or fiction?

12. The senior class has 367 girls and 425 students with brown hair. Of the girls, 296 have brown hair. In how many different ways could you select a girl or a brown-haired student from the senior class?

13. *Seating Problem:* There are ten students in a class and ten chairs, numbered 1 through 10.
   a. In how many different ways could a student be selected to occupy chair 1?
   b. After someone is seated in chair 1, how many different ways are there of seating someone in chair 2?
   c. In how many different ways could chairs 1 and 2 be filled?
   d. If two of the students are sitting in chairs 1 and 2, in how many different ways could chair 3 be filled?
   e. In how many different ways could chairs 1, 2, and 3 be filled?
   f. In how many different ways could all ten chairs be filled? Do you find this surprising?

14. *Baseball Team Problem:* Nine people on a baseball team are trying to decide who will play each position.

a. In how many different ways could they select a person to be pitcher?

b. After someone has been selected as pitcher, in how many different ways could they select someone to be catcher?

c. In how many different ways could they select a pitcher and a catcher?

d. After the pitcher and catcher have been selected, in how many different ways could they select a first-base player?

e. In how many different ways could they select a pitcher, catcher, and first-base player?

f. In how many different ways could all nine positions be filled? Do you find this surprising?

15. *License Plate Problem:* Many states use car license plates that have six characters. Some states use two letters followed by a number from 1 to 9999. Others use three letters followed by a number from 1 to 999.

a. Which of these two plans allows more possible license plates? How many more?

b. How many different license plates could there be if the state allowed either two letters and four digits or three letters and three digits?

c. Assuming there are about 250 million motor vehicles in the United States, would it be possible to have a national license plate program using the plan in part b? Explain.

16. *Telephone Number Problem:* When ten-digit telephone numbers were introduced into the United States and Canada in the 1950s, certain restrictions were placed on the groups of numbers:

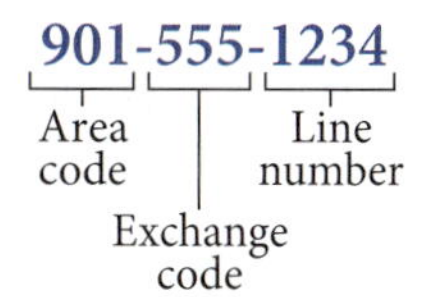

Area code: three digits; the first must not be 0 or 1, and the second must be 0 or 1.

Exchange code: three digits; the first and second must not be 0 or 1.

Line number: four digits; at least one must not be 0.

a. Find the possible numbers of area codes, exchange codes, and line numbers.

b. How many valid numbers could there be under this numbering scheme?

c. How many ten-digit numbers could there be if there were no restrictions on the three groups of numbers?

d. What is the probability that a ten-digit number dialed at random would be a valid number under the original restrictions?

e. The total population of the United States and Canada is currently about 350 million. In view of the fact that there are now area codes and exchange codes that do not conform to the original restrictions, what assumption can you make about the number of telephones per person in the United States and Canada?

17. *Journal Problem:* Update your journal with things you have learned about probability and about counting outcomes.

14-4 Probabilities of Various Permutations

Many counting problems involve finding the number of different ways to *arrange,* or *order,* things. For example, the three letters A, B, and C can be arranged in six different ways:

ABC ACB BAC BCA CAB CBA

But the ten letters A, B, C, D, E, F, G, H, I, and J can be arranged in more than 3 *million* different ways! In this section you will learn a time-efficient way to calculate the number of arrangements, or **permutations,** of a set of objects. As a result you will be able to calculate relatively quickly the probability that a permutation selected at random will have certain characteristics.

Objective Given a description of a permutation, find the probability of getting that permutation if an arrangment is selected at random.

Here is a formal definition of permutation.

> **DEFINITION: Permutation**
>
> A **permutation** of a set of objects is an arrangement in a definite order of some or all of the elements in that set.

EXAMPLE 1 In how many different ways could you arrange three books on a shelf if you have seven books from which to choose?

SOLUTION The process of selecting an arrangement (a permutation) can be divided into three events:

A—Choose a book to go in the first position (Figure 14-4a).

B—Choose a book to go in the second position.

C—Choose a book to go in the third position.

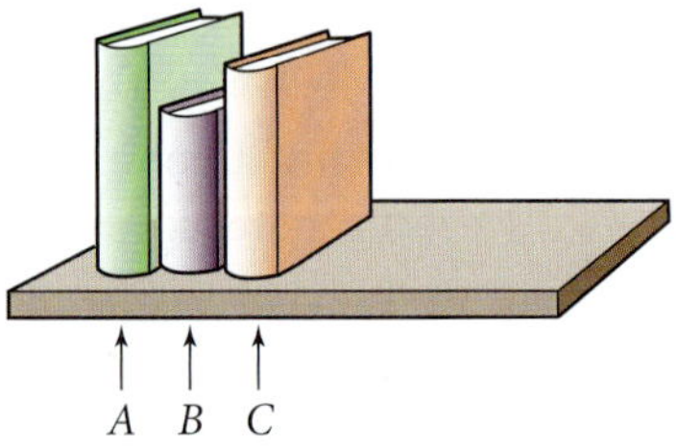

A B C

Figure 14-4a

Let n be the number of permutations.

$n = \_\ \_\ \_$ Mark three spaces for the three events.

$n = \underline{7}\ \underline{6}\ \underline{5}$ Seven ways to select the first book; six ways to select the second; five ways to select the third.

$n = \underline{7} \cdot \underline{6} \cdot \underline{5} = 210$ ways Apply the counting principle for sequential events.

In Example 1, the answer 210 is "the number of permutations of seven elements taken three at a time."

EXAMPLE 2 ➤ A permutation is selected at random from letters in the word SEQUOIA. What is the probability that it has letter Q in the fourth position and ends with a vowel?

SOLUTION Let E be the set of all favorable outcomes. First, find the number of outcomes (permutations) in E.

$n(E) = \_\ \_\ \_\ \_\ \_\ \_\ \_$	Mark seven spaces on which to record the number of ways of selecting each letter.
$n(E) = \_\ \_\ \_\ \underline{1}\ \_\ \_\ \_$	Write 1 in the fourth space, because there is only one Q to go there.
$n(E) = \_\ \_\ \_\ \underline{1}\ \_\ \_\ \underline{5}$	Write 5 in the last space, because there are five ways to select a vowel.
$n(E) = \underline{5}\ \underline{4}\ \underline{3}\ \underline{1}\ \underline{2}\ \underline{1}\ \underline{5}$	There are five letters left, so there are 5, 4, 3, 2, and 1 ways to select the remaining letters.
$n(E) = \underline{5} \cdot \underline{4} \cdot \underline{3} \cdot \underline{1} \cdot \underline{2} \cdot \underline{1} \cdot \underline{5} = 600$	Apply the counting principle for sequential events.

Next, find the number of outcomes in the sample space.

$$n(S) = \underline{7} \cdot \underline{6} \cdot \underline{5} \cdot \underline{4} \cdot \underline{3} \cdot \underline{2} \cdot \underline{1} = 5040$$

Finally, find the probability, using the definition.

$$P(E) = \frac{n(E)}{n(S)} = \frac{600}{5040} = \frac{5}{42} = 0.1190... \approx 12\%$$

Note that in Example 2 the fourth position is a *fixed position*, because there is only one way it can be filled. The last position is a *restricted position*; more than one letter can go there, but the choices are limited to vowels.

Note also that the number of outcomes in the sample space,

$$7 \cdot 6 \cdot 5 \cdot 4 \cdot 3 \cdot 2 \cdot 1$$

is the product of consecutive positive integers ending with 1. This is called a **factorial.** The factorial symbol is the exclamation mark, ! . So

$$7! = 7 \cdot 6 \cdot 5 \cdot 4 \cdot 3 \cdot 2 \cdot 1 = 5040$$ Pronounced "7 factorial."

Most calculators have a built-in factorial function.

DEFINITION: Factorial

For any positive integer n, ***n* factorial** ($n!$) is given by

$$n! = 1 \cdot 2 \cdot 3 \cdot \cdots \cdot n$$

or, equivalently,

$$n! = n \cdot (n-1) \cdot (n-2) \cdot \cdots \cdot 2 \cdot 1$$

0! is defined to be equal to 1.

Problem Set 14-4

Reading Analysis

From what you have read in this section, what do you consider to be the main idea? What is a *permutation* of several objects, and what is the basic method for finding the number of possible permutations without actually listing and counting them?

Quick Review

Q1. If the outcomes of a random experiment are equally likely, then the probability of an event is defined to be __?__.

Q2. For events A and B, $n(A \text{ and then } B) =$ __?__.

Q3. If independent events A and B are mutually exclusive, then $n(A \text{ or } B) =$ __?__.

Q4. If events A and B are *not* mutually exclusive, then $n(A \text{ or } B) =$ __?__.

Q5. The set of all possible outcomes of a random experiment is called the __?__.

Q6. __?__ functions have the multiply–multiply property.

Q7. The slope of the linear function $4x + 5y = 40$ is __?__.

Q8. The "If" part of a theorem is called the __?__.

Q9. An equation that is true for *all* values of the variable is called a(n) __?__.

Q10. 4% of 700 is __?__.

---

1. The Hawaiian alphabet has 12 letters. How many permutations could be made using

a. Two different letters

b. Four different letters

c. Twelve different letters

2. Fran Tick takes a ten-problem precalculus test. The problems may be worked in any order.

a. In how many different orders could she work all ten problems?

b. In how many different orders could she choose the first seven of the ten problems she will work?

3. Triangles are often labeled by placing a different letter at each vertex. In how many different ways could a given triangle be labeled using any of the 26 letters of the alphabet?

4. Alma, Bella, and Cristina each draw two cards from a standard 52-card deck and place them face up in a row. The cards are not replaced. Alma goes first. Find the number of different two-card sets which

a. Alma could draw

b. Bella could draw *after* Alma has already drawn

c. Cristina could draw *after* Alma and Bella have drawn theirs

5. Frost Bank has seven vice presidents, but only three spaces in the parking lot are labeled "Vice President." In how many different ways could these spaces be occupied by the vice presidents' cars?

6. A professor says to her class, "You may work these six problems in any order you choose." There are 100 students in the class. Is it possible for each student to work the problems in a different order? Explain.

7. A six-letter permutation is selected at random from the letters in the word NIMBLE.

a. How many permutations are possible?

b. How many of these permutations begin with the letter M?

c. What is the probability that the permutation begins with the letter M?

d. Express the probability in part c as a percent.

e. What is the probability that the permutation is NIMBLE?

8. A six-letter permutation is selected at random from the letters in the word NIMBLE. Find the probability of each event.

a. The third letter is I and the last letter is B.

b. The second letter is a vowel and the third letter is a consonant.

c. The second and third letters are both vowels.

d. The second letter is a consonant and the last letter is E.

e. The second letter is a consonant and the last letter is L.

9. *Baseball Team Problem 1:* Nine people try out for the nine positions on a baseball team.

   a. In how many different ways could the positions be filled if there are no restrictions on who plays which position?

   b. In how many different ways could the positions be filled if Fred must be the pitcher but the other eight people can take any of the remaining eight positions?

   c. If the positions are selected at random, what is the probability that Fred will be the pitcher?

   d. What is the probability in part c expressed as a percent?

10. *Soccer Team Problem 1:* Eleven girls try out for the 11 positions on a soccer team.

   a. In how many different ways could the 11 positions be filled if there are no restrictions on who plays which position?

   b. In how many different ways could the positions be filled if Mabel must be the goalkeeper?

   c. If the positions are selected at random, what is the probability that Mabel will be the goalkeeper?

   d. What is the probability in part c expressed as a percent?

11. *Baseball Team Problem 2:* Nine people try out for the nine positions on a baseball team. If the players are selected at random for the positions, find the probability of each event.

   a. Fred, Mike, or Jason is the pitcher.

   b. Fred, Mike, or Jason is the pitcher, and Sam or Paul plays first base.

   c. Fred, Mike, or Jason is the pitcher, Sam or Paul plays first base, and Bob is the catcher.

12. *Soccer Team Problem 2:* Eleven girls try out for the 11 positions on the varsity soccer team. If the players are selected at random, find the probability of each event.

   a. Mabel, Keisha, or Diedra is the goalkeeper.

   b. Mabel, Keisha, or Diedra is the goalkeeper, and Alice or Phyllis is the center forward.

   c. Mabel, Keisha, or Diedra is the goalkeeper, Alice or Phyllis is the center forward, and Bea is the left fullback.

13. Eight children line up at random for a fire drill (Figure 14-4b).

Figure 14-4b

   a. How many possible arrangements are there?

   b. In how many of these arrangements are Calvin and Phoebe next to each other? (*Clue:* Arrange *seven* things—the Calvin and Phoebe pair and the other six children. Then arrange Calvin and Phoebe.)

   c. What is the probability that Calvin and Phoebe will be next to each other?

14. The ten digits, 0, 1, 2, 3, . . . , 9, are arranged at random with no repeats. Find the probability that the numeral formed represents

   a. A number greater than 6 billion

   b. An even number greater than 6 billion (There are two cases to consider: "first digit is odd" and "first digit is even.")

Permutations with Repeated Elements: The word CARRIER has seven letters. But there are fewer than 7! permutations, because in any arrangement of these seven letters the three Rs are interchangeable. If these Rs were distinguishable, there would be 3!, or 6, ways of arranging them. This implies that only $\frac{1}{6}$ $\left(\text{that is, } \frac{1}{3!}\right)$ of the 7! permutations are actually different. So the number of permutations is $\frac{7!}{3!} = 840$.

There are four Is, four Ss, and two Ps in the word MISSISSIPPI, so the number of different permutations of its letters is $\frac{11!}{4!\,4!\,2!} = 34{,}650$.

15. Find the number of different permutations of the letters in each word.

a. FREELY **b.** BUBBLES

c. LILLY **d.** MISSISSAUGA

e. HONOLULU **f.** HAWAIIAN

16. Nine pennies are lying on a table. Five show heads and four show tails. In how many different ways, such as HHTHTTHHT, could the coins be lined up if you consider all the heads to be identical and all the tails to be identical?

Circular Permutations: In Figure 14-4c, the letters A, B, C, and D are arranged in a circle. Though these may seem to be different permutations, they are considered the same permutation because the letters have the same position *with respect to one another.* That is, each of the four letters has the same letter to its left and the same letter to its right. An easy way to calculate the number of different *circular permutations* of n elements is to fix the position of one element and then arrange the other $(n - 1)$ elements with respect to it (Figure 14-4d). So, for the letters A, B, C, and D, the number of circular permutations is $n = \underline{1} \cdot \underline{3} \cdot \underline{2} \cdot \underline{1} = \underline{6}$.

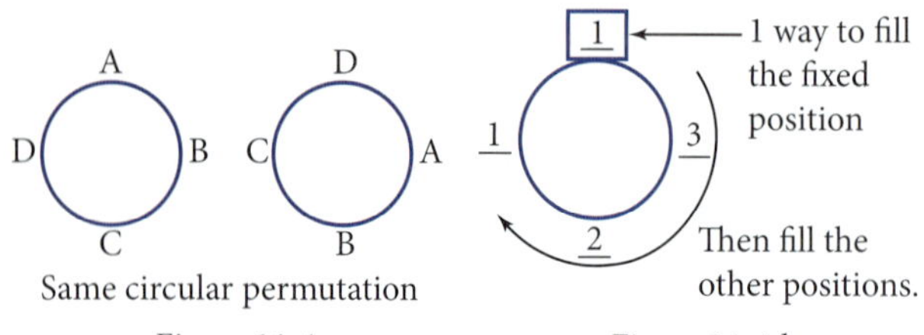

Figure 14-4c

Figure 14-4d

17. How many different circular permutations could be made with these letters?

a. ABCDE **b.** QLMXTN

c. LOGARITHM

18. In how many different ways could King Arthur's 12 knights be seated around the Round Table?

19. Four girls and four boys sit on a merry-go-round.

a. In how many different ways can they be arranged with respect to one another so that boys and girls alternate?

b. If they seat themselves at random, what is the probability that boys and girls will alternate?

20. Suppose that you are concerned only with which elements come *between* other elements in a circular permutation, not with which elements are to the left and to the right. Then there are two circular permutations that would be considered the same, a clockwise one and a counterclockwise one (Figure 14-4e). In this case, there would be only half the number of circular permutations as calculated earlier.

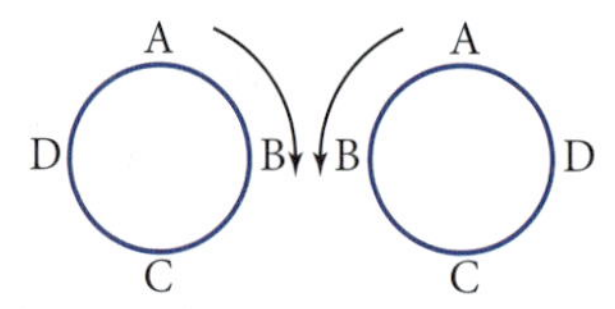

The same-"betweenness" property

Figure 14-4e

Find the number of different ways you could arrange

a. Seven different beads to form a bracelet if you consider only which bead is *between* which other beads

b. Five keys on a key ring if you consider only which key is *between* which other keys

14-5 Probabilities of Various Combinations

In this section you will learn about *combinations* of various numbers of elements in a given set and how combinations differ from permutations. In this exploration you will practice reading mathematics by doing structured reading of some new material and answering questions about what you have read.

EXPLORATION 14-5: Reading about Combinations

1. Skim this section. Summarize what you think is the major point of the section.
2. Read the section again. Explain why AEI and IAE are the same combination of three vowels but different permutations.
3. State the definition of *combination*.
4. There are 60 different three-letter permutations of the five vowels, AEIOU. How many different three-letter combinations can be made?
5. Write one symbol for the number of combinations of five elements taken three at a time.
6. Read Example 2. Use what you learn to write $10 \cdot 9 \cdot 8 \cdot 7$ as a ratio of two factorials.
7. Write ${}_{10}C_4$ as a fraction involving factorials.
8. Evaluate the fraction in Problem 7.
9. Evaluate ${}_{10}C_4$ using the built-in feature of your grapher. Show that the answer is the same as in Problem 8.
10. Expand: $(a + b)^2$
11. Expand $(a + b)^3$ by writing it as $(a + b)(a + b)^2$ and using the result of Problem 10.
12. Expand $(a + b)^4$ by writing it as $(a + b)(a + b)^3$ and using the result of Problem 11.
13. Show that the coefficients 1, 4, 6, 4, and 1 in Problem 12 are equal to ${}_4C_0$, ${}_4C_1$, ${}_4C_2$, ${}_4C_3$, and ${}_4C_4$, respectively.
14. What did you learn as a result of doing this exploration that you did not know before?

There are 24 different three-letter "words" that can be made from the four letters A, B, C, and D. These are

ABC	ACB	BAC	BCA	CAB	CBA	One combination.
ABD	ADB	BAD	BDA	DAB	DBA	A second combination.
ACD	ADC	CAD	CDA	DAC	DCA	A third combination.
BCD	BDC	CBD	CDB	DBC	DCB	A fourth combination.

Because these words are *arrangements* of the letters in a definite order, each is a *permutation* of four elements taken three at a time.

Suppose you are concerned only with *which* letters appear in the word, not with the order in which they appear. For instance, you would consider ADC and DAC to be the same because they have the same three letters. Each different group of three letters is called a **combination** of the letters A, B, C, and D. There are 24 different permutations but only four different combinations of the letters A, B, C, and D taken three at a time. In this section you will learn a time-efficient way to calculate the number of combinations of the elements in a set.

Objective Calculate the number of different combinations containing *r* elements taken from a set containing *n* elements.

Here is the formal definition of combination.

DEFINITION: Combination

A **combination** of elements in a set is a *subset* of those elements, without regard to the order in which the elements are arranged.

In the example following Exploration 14-5, you can see that for every *one* combination, there are *six* possible permutations. So the total number of combinations is equal to the total number of permutations *divided by* 6. That is,

$$\text{Number of combinations} = \frac{24}{6} = 4$$

This idea allows you to calculate a number of combinations by dividing two numbers of permutations.

PROPERTY: Computation of the Number of Combinations

$$\text{Number of combinations} = \frac{\text{total number of permutations}}{\text{number of permutations of } \textit{one} \text{ combination}}$$

Before you proceed with examples, it is helpful to define some symbols.

DEFINITIONS: Symbols for Numbers of Combinations and Permutations

${}_nC_r$ = number of different combinations of *n* elements taken *r* at a time
Pronounced: "*n, C, r*"

${}_nP_r$ = number of different permutations of *n* elements taken *r* at a time
Pronounced: "*n, P, r*"

Example 1 shows you how to calculate a number of combinations with the help of these symbols.

EXAMPLE 1 ➤ Calculate ${}_4C_3$.

SOLUTION

$${}_4C_3 = \frac{{}_4P_3}{{}_3P_3} = \frac{4 \cdot 3 \cdot 2}{3 \cdot 2 \cdot 1} = 4$$

◄

EXAMPLE 2 ➤ Write ${}_9P_4$ as a ratio of factorials. Interpret the answer in terms of the number of elements in the set and the number of elements selected for the permutation.

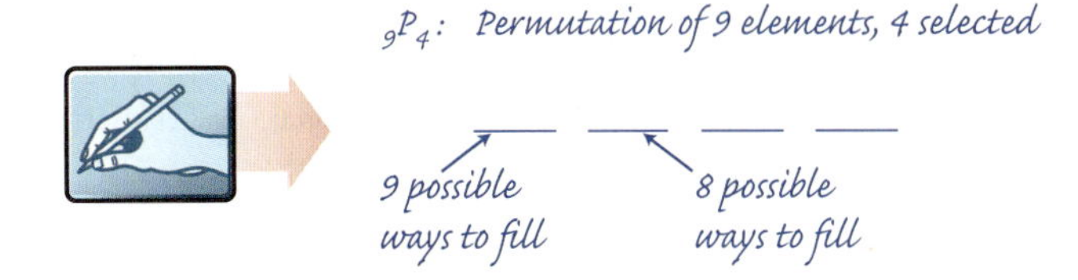

SOLUTION

$${}_9P_4 = \underline{9} \cdot \underline{8} \cdot \underline{7} \cdot \underline{6}$$

$$= 9 \cdot 8 \cdot 7 \cdot 6 \cdot \frac{5 \cdot 4 \cdot 3 \cdot 2 \cdot 1}{5 \cdot 4 \cdot 3 \cdot 2 \cdot 1}$$ Multiply by a clever form of 1.

$$= \frac{9 \cdot 8 \cdot 7 \cdot 6 \cdot 5 \cdot 4 \cdot 3 \cdot 2 \cdot 1}{5 \cdot 4 \cdot 3 \cdot 2 \cdot 1}$$

$$= \frac{9!}{5!}$$

The 9 in the numerator is the total number of elements in the set from which elements are selected for the permutation. The 5 in the denominator is the number of elements *not* selected for the permutation. ◀

EXAMPLE 3 ➤ Write ${}_9C_4$ as a ratio of a factorial to a product of factorials. Interpret the answer in terms of the number of elements in the set and the number of elements selected for the combination.

SOLUTION

$${}_9C_4 = \frac{{}_9P_4}{{}_4P_4}$$ Use the preceding property.

$$= \frac{9!/5!}{4!}$$ From the solution to Example 2.

$$= \frac{9!}{4!\,5!}$$

The 9 in the numerator is the total number of elements in the set from which elements are selected for the combination. The 4 and 5 in the denominator are the number of elements *selected* for the combination and the number of elements *not* selected for the combination, respectively. ◀

From Examples 2 and 3, you can find relatively simple patterns to use for calculating a number of combinations or permutations.

TECHNIQUE: Calculation of the Number of Permutations or Combinations

Example: Permutations

Same

$${}_9P_4 = \frac{9!}{5!}$$

Number of selected elements; Number *not* selected

Example: Combinations

Same

$${}_9C_4 = \frac{9!}{4!\,5!}$$

Same; Sum to numeral in numerator.

With the techniques on page 721 in mind, you are ready to solve problems in which you must find the probability that a specified combination occurs.

EXAMPLE 4 ➤ In how many different ways could you form a committee of three people from a group of seven people? Explain how you know that a number of combinations is being asked for, not a number of permutations.

SOLUTION Committees with the same members are different only if the people on the committees have special roles. If there are no special roles, how the committee members are arranged does not matter. So the answer is a number of combinations, not a number of permutations. Let n(3 people) stand for the number of different three-person committees.

$$n(\text{3 people}) = {}_7C_3 = \frac{7!}{3!\,4!} = 35 \text{ committees}$$ Use the pattern for combinations.

EXAMPLE 5 ➤ If a committee of five is selected at random from a group of six women and three men, find the probability that it will include

a. Eileen and Ben (two of the nine people)

b. Exactly three women and two men

c. At least three women

SOLUTION The sample space for all these probabilities is the set of all possible five-member committees.

$$n(\text{sample space}) = {}_9C_5 = \frac{9!}{5!\,4!} = 126 \text{ committees}$$

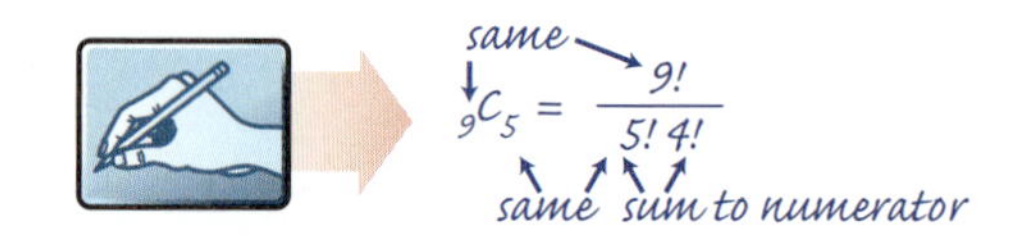

a. $P(\text{Eileen and Ben}) = \dfrac{n(\text{Eileen and Ben})}{n(\text{sample space})}$ Definition of probability.

To find n(Eileen and Ben), first select Eileen and Ben (one way), then select the other three committee members from the seven people who remain. There are ${}_7C_3$ ways to select the three committee members.

$$n(\text{Eileen and Ben}) = 1 \cdot {}_7C_3 = \frac{7!}{3!\,4!} = 35$$ By the counting principle for sequential events.

$$\therefore P(\text{Eileen and Ben}) = \frac{35}{126} = 0.2777... \approx 28\%$$ The definition of probability.

b. To find the number of three-woman, two-man committees, notice that people are being selected from two different groups. You can divide the "hard" problem of selecting the committee into two "easy" problems, selecting the women and selecting the men. So, by the counting principle for sequential events,

$$n(\text{3 women and 2 men}) = n(\text{3 women}) \cdot n(\text{2 men})$$

$$= {}_6C_3 \cdot {}_3C_2$$ "3 women" and "2 men" are *independent* events.

$$= \frac{6!}{3!\,3!} \cdot \frac{3!}{2!\,1!}$$

$$= 20 \cdot 3 = 60$$

$$\therefore P(\text{3 women and 2 men}) = \frac{60}{126} = 0.4761... \approx 48\%$$

c. If the committee has at least three women, it could have three women *or* four women *or* five women. In each case, the remainder of the committee consists of men. Turn this problem into three easier problems.

$$n(3W, 2M) = 60$$ From part b.

$$n(4W, 1M) = {}_6C_4 \cdot {}_3C_1 = \frac{6!}{4!\,2!} \cdot \frac{3!}{1!\,2!} = 15 \cdot 3 = 45$$

$$n(5W, 0M) = {}_6C_5 \cdot {}_3C_0 = \frac{6!}{5!\,1!} \cdot \frac{3!}{0!\,3!} = 6 \cdot 1 = 6$$ Recall that $0! = 1$.

Because these are mutually exclusive events, you can add the numbers of ways.

$$n(\text{at least 3 women}) = 60 + 45 + 6 = 111$$

$$\therefore P(\text{at least 3 women}) = \frac{111}{126} = 0.8809... \approx 88\%$$

Note that most graphers have built-in functions to calculate numbers of permutations and combinations directly. For instance, to calculate ${}_9C_4$, you might enter 9 nCr 4. The answer would be 126, the same as $\frac{9!}{4!\,5!}$.

Problem Set 14-5

Reading Analysis

From what you have read in this section, what do you consider to be the main idea? What question can you ask yourself to decide whether a particular problem involves combinations or permutations? How is the symbol ${}_9C_4$ pronounced, and how can you calculate this quantity using factorials?

Quick Review

Q1. $4! = \underline{\quad ? \quad}$

Q2. $\frac{4!}{4} = \underline{\quad ? \quad}!$

Q3. $\frac{3!}{3} = \underline{\quad ? \quad}!$

Q4. $\frac{2!}{2} = \underline{\quad ? \quad}!$

Q5. $\frac{1!}{1} = \underline{\quad ? \quad}!$

Q6. Why does 0! equal 1, not 0?

Q7. Write ${}_5P_5$ as a factorial.

Q8. Write ${}_nP_n$ as a factorial...

Q9. Express 0.4385... as a percent rounded to the nearest integer.

Q10. The *exact* value (no decimals) of $\tan \frac{\pi}{3}$ is $\underline{\quad ? \quad}$.

For Problems 1–12, evaluate the number of combinations or permutations two ways:

a. Using factorials, as in the examples of this section

b. Directly, using your grapher

1. ${}_5C_3$ **2.** ${}_6C_4$ **3.** ${}_{27}C_{19}$

4. ${}_{44}C_{24}$ **5.** ${}_{10}C_{10}$ **6.** ${}_{100}C_{100}$

7. ${}_{10}C_0$ **8.** ${}_{100}C_0$ **9.** ${}_6P_4$

10. ${}_{11}P_5$ **11.** ${}_{47}P_{30}$ **12.** ${}_{50}P_{20}$

13. Twelve people apply to go on a biology field trip, but there is room in the car for only five of them. In how many different ways could the group of five making the trip be chosen? How can you tell that a number of combinations is being asked for, not a number of permutations?

14. Seven people come to an evening bridge party. Only four people can play bridge at any one time, so they decide to play as many games as it takes to use every possible foursome once. How many games would have to be played? Could all of these games be played in one evening?

15. A donut franchise sells 34 varieties of donuts. Suppose one of the stores decides to make sample boxes with six different donuts in each box. How many different sample boxes could be made? Would it be practical to stock one of each kind of box?

16. Just before each Supreme Court session, each of the nine justices shakes hands once with every other justice. How many handshakes take place?

17. Horace Holmsley bought blueberries, strawberries, a watermelon, grapes, plums, and peaches. Find the number of different fruit salads he could make if he uses

a. Three ingredients

b. Four ingredients

c. Three ingredients or four ingredients

d. All six ingredients

18. A pizzeria offers 11 different toppings. Find the number of different kinds of pizza it could make using

a. Three toppings

b. Five toppings

c. Three toppings or five toppings

d. All 11 toppings

19. A standard deck of playing cards has 52 cards.

a. How many different five-card poker hands could be formed from a standard deck?

b. How many different 13-card bridge hands could be formed?

c. How can you tell that numbers of combinations are being asked for, not numbers of permutations?

20. The diagonals of a convex polygon are made by connecting the vertices two at a time. However, some of the combinations are *sides,* not diagonals (Figure 14-5a). How many diagonals are there in each convex figure?

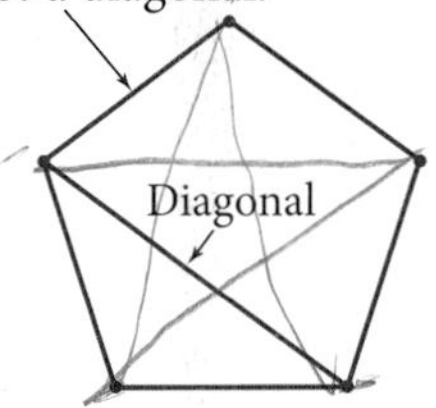

Figure 14-5a

a. Pentagon (five sides)

b. Decagon (ten sides)

c. n-gon (n sides). From the answer, write a simple formula for the number of diagonals.

21. A set has ten distinct elements. Find the number of subsets that contain exactly

a. Two elements

b. Five elements

c. Eight elements. Explain the relationship between this answer and the answer to part a.

22. A set has five elements, {❀, ❁, ✌, ✢, ⚯}.

a. Find the number of different subsets that contain

i. One element

ii. Two elements

iii. Three elements

iv. Four elements

v. All five elements

vi. No elements

b. How many subsets are there altogether? What relationship does this number have to the number of elements in the set?

c. Based on your answer to part b, how many subsets would a ten-element set have? A 100-element set?

23. *Review Problem 1:* You draw a 5-card hand from a standard 52-card deck and then arrange the cards from left to right.

a. After the cards have been selected, in how many different ways could you arrange them?

b. How many different five-card hands could be selected without considering arrangement?

c. How many different five-card arrangements could be formed from the deck?

d. Which part(s) of this problem involve permutations and which involve combinations?

24. *Review Problem 2:* At South High School, 55 students entered an essay contest. From these students, 10 are selected as finalists.

a. After the finalists have been selected, in how many different ways could they be ranked from 1st to 10th?

b. In how many different ways could the ten finalists be selected?

c. How many different ten-student rankings could be made from the 55 entrants?

d. Which part(s) of this problem involve permutations and which involve combinations?

25. Charlie has 13 socks in his drawer, 7 blue and 6 green. He selects 5 socks at random. Find the probability that he selects

a. Two blue socks and three green socks

b. Three blue socks and two green socks

c. Two blue socks and three green socks, or three blue socks and two green socks

d. The one sock that has a hole in it

26. In a group of 15 people, 6 are left-handed and the rest are right-handed. If 7 people are selected at random from this group, find the probability that

a. Three are left-handed and four are right-handed

b. All are right-handed

c. All are left-handed

d. Harry and Peg, two of the left-handers, are selected

27. Three baseball cards are selected at random from a group of seven cards. Two of the cards are rookie cards.

a. What is the probability that exactly one of the three selected cards is a rookie card?

b. What is the probability that at least one of the three cards is a rookie card?

c. What is the probability that none of the three cards is a rookie card?

d. What is the relationship between the answers to parts b and c?

28. Emma, who is three years old, tears the labels off all ten cans of soup on her mother's shelf (Figure 14-5b). Her mother knows that there are two cans of tomato soup and eight cans of vegetable soup. She selects four cans at random.

Figure 14-5b

a. What is the probability that exactly one of the four cans contains tomato soup?

b. What is the probability that at least one of the cans contains tomato soup?

c. What is the probability that none of the four cans contain tomato soup?

d. What is the relationship between the answers to parts b and c?

29. *Light Bulb Problem:* Light bulb manufacturers like to be assured that their bulbs will work. Because testing every bulb is impractical, a random sample of bulbs is tested. Suppose the quality control department at a light bulb factory decides to test a random sample of 5 bulbs for every 100 that are made.

a. In how many different ways could a 5-bulb sample be taken from 100 bulbs?

b. To check the quality control process, the manager of the factory puts 2 defective bulbs in with 98 working bulbs before the sample is taken. In how many different ways could a sample of 5 of these 100 bulbs be selected that contains at least 1 of the defective bulbs?

c. If the 100 bulbs include 2 defective bulbs, what is the probability that the sampling process will reveal at least 1 defective bulb?

d. Based on your answers to parts b and c, do you think that the 5-bulb-in-100 sampling plan is sufficiently effective?

30. A standard 52-card deck of playing cards has four suits, with 13 cards in each suit. In a particular game, each of the four players is dealt 13 cards at random.

a. Find the probability that such a 13-card hand has

i. Exactly five spades

ii. Exactly three clubs

iii. Exactly five spades and three clubs

iv. Exactly five spades, three clubs, and two diamonds

b. Which is more probable, getting four aces or getting 13 cards of the same suit? Give numbers to support your answer.

31. *Journal Problem:* Update your journal with things you have learned since the last entry. In particular, tell how large numbers of outcomes can be calculated, rather than counted, using the concepts of factorials, combinations, and permutations.

14-6 Properties of Probability

In the preceding sections you learned how to calculate the probability of an event using the definition

$$P(E) = \frac{n(E)}{n(S)}$$

That is, you divided the number of outcomes in an event by the number of outcomes in the sample space. In this section you will learn some properties of probability that will allow you to calculate probabilities more quickly.

Objective

Given events A and B, calculate

- $P(A \text{ and } B)$, the probability of the *intersection* of A and B
- $P(A \text{ or } B)$, the probability of the *union* of A and B
- $P(\text{not } A)$ and $P(\text{not } B)$, the probabilities of the *complement* of A and the *complement* of B

Intersection of Events

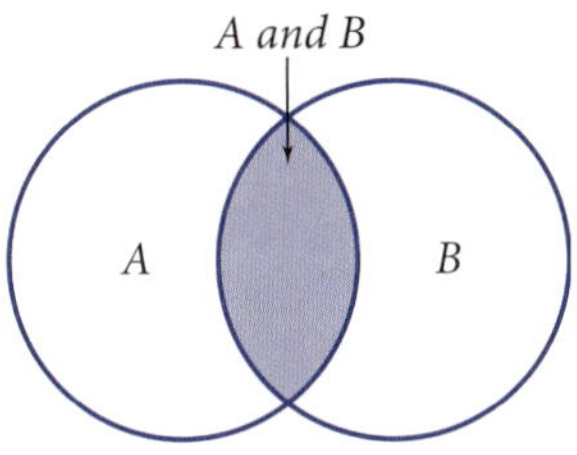

A and B

A B

The shaded region is $A \cap B$.

Figure 14-6a

If A and B are two events, then the **intersection** of A and B, $A \cap B$, is the set of all outcomes in event A *and* event B (Figure 14-6a).

For example, suppose you draw 2 cards from a standard deck of 52 playing cards without replacing the first card before you draw the second. What is the probability that both cards will be black? Here, you are looking for the probability of the intersection of the events "the first card is black" and "the second card is black."

There are 52 ways to choose the first card and 51 ways to choose the second card after the first has already been chosen. So the number of outcomes in the sample space is

$$n(S) = 52 \cdot 51 = 2652$$

There are 26 ways the first card could be black. After the first black card has been drawn, there are only 25 ways the second card could be black. So

$$n(\text{both black}) = 26 \cdot 25 = 650$$

By the counting principle for sequential events.

$$\therefore P(\text{both black}) = \frac{650}{2652} = 0.2450...$$

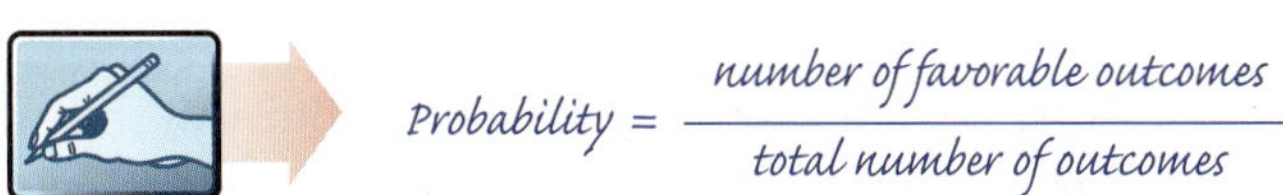

A pattern appears if you do *not* simplify the numbers of outcomes:

$$P(\text{both black}) = \frac{26 \cdot 25}{52 \cdot 51}$$

$$= \frac{26}{52} \cdot \frac{25}{51}$$ Multiplication property of fractions.

$$= P(\text{first card is black}) \cdot P(\text{second card is black after the first card is black})$$

In this example, the events are *not* independent because the result of the first draw affects the choices for the second draw. However, if the first card were *replaced* before the second card was drawn, then the events would be independent. In this case,

$$P(\text{2nd is black after 1st is black}) = P(\text{2nd is black}) = \frac{26}{52}$$

So, when the two cards are drawn with replacement,

$$P(\text{both are black}) = P(\text{1st is black}) \cdot P(\text{2nd is black})$$

$$= \frac{26}{52} \cdot \frac{26}{52} = \frac{676}{2704} = 0.25$$

PROPERTY: Probability of the Intersection of Two Events

If $P(B \mid A)$ is the probability that B occurs given that event A occurs, then

$$P(A \text{ and } B) = P(A \cap B) = P(A) \cdot P(B \mid A)$$

If A and B are independent events, then

$$P(A \text{ and } B) = P(A \cap B) = P(A) \cdot P(B)$$

Note that this property corresponds to the counting principle for sequential events, $n(A \text{ and } B) = n(A) \cdot n(B \mid A)$.

Union of Events

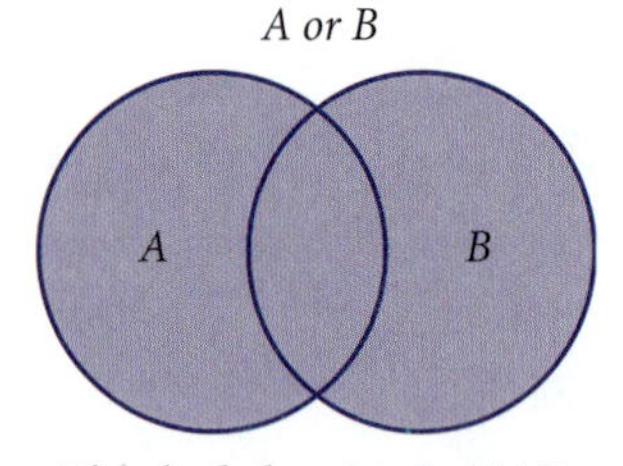

The shaded region is $A \cup B$.

Figure 14-6b

If A and B are two events, then the **union** of A and B, $A \cup B$, is the set of all outcomes in event A *or* event B (Figure 14-6b).

For example, suppose a bag contains 7 chocolate chip cookies, 11 macadamia nut cookies, 12 oatmeal cookies, 4 gingersnaps, and 9 oatmeal-chocolate cookies. If you select 1 cookie at random, what is the probability that it will contain oatmeal or chocolate?

Here, you are looking for the probability of the union of the events "the cookie contains chocolate" and "the cookie contains oatmeal."

The sample space is all the cookies, so

$$n(S) = 7 + 11 + 12 + 4 + 9 = 43$$

The events overlap—that is, there are cookies that contain both oatmeal and chocolate. Use the counting principle for non-mutually exclusive events to count the favorable outcomes:

$$\begin{aligned} n(\text{chocolate or oatmeal}) &= n(\text{chocolate}) + n(\text{oatmeal}) \\ &\quad - n(\text{chocolate} \cap \text{oatmeal}) \\ &= 16 + 21 - 9 \\ &= 28 \end{aligned}$$

$$\therefore P(\text{chocolate or oatmeal}) = \frac{28}{43} = 0.6511...$$

Here again, a pattern appears if you resist the temptation to simplify first:

$$\begin{aligned} P(\text{chocolate or oatmeal}) &= \frac{16 + 21 - 9}{43} \\ &= \frac{16}{43} + \frac{21}{43} - \frac{9}{43} \\ &= P(\text{chocolate}) + P(\text{oatmeal}) \\ &\quad - P(\text{chocolate} \cap P\text{oatmeal}) \end{aligned}$$

If two events are mutually exclusive, then their intersection is empty. In this case, you can find the probability of the union of the events simply by summing the probabilities of the two events. For example, there are no cookies that contain both ginger and macadamia nuts, so

$$\begin{aligned} P(\text{ginger or macadamia}) &= P(\text{ginger}) + P(\text{macadamia}) \\ &= \frac{4}{43} + \frac{11}{43} \\ &= \frac{15}{43} = 0.3488... \end{aligned}$$

PROPERTY: Probability of the Union of Two Events

If events A and B are not mutually exclusive, then

$$P(A \text{ or } B) = P(A \cup B) = P(A) + P(B) - P(A \cap B)$$

If events A and B are mutually exclusive, then

$$P(A \text{ or } B) = P(A \cup B) = P(A) + P(B)$$

Note that the first form of the property reduces to the second form when the events are mutually exclusive.

Complementary Events

You can accomplish the third section objective using the property for the union of mutually exclusive events:

Let $P(A)$ be the probability that event A occurs.

Let $P(\text{not } A)$ be the probability that event A does *not* occur.

The events A and "not A" are mutually exclusive, and one or the other is certain to occur. Thus,

$P(A \text{ or not } A) = P(A) + P(\text{not } A) = 1$ — Probability is 1 if the event is certain to occur.

$\therefore\ P(\text{not } A) = 1 - P(A)$

Together, the events A and "not A" complete all the possibilities. Therefore, they are called **complementary events.**

> **PROPERTY: Complementary Events**
>
> The probability that event A will not occur is
>
> $P(\text{not } A) = 1 - P(A)$

EXAMPLE 1 ➤ Calvin and Phoebe volunteer in the children's ward of a hospital. The probability that Calvin catches mumps as the result of a visit to the ward is 13%, and the probability that Phoebe catches mumps is 7%. Find the probability of each event.

a. Both catch mumps.

b. Calvin does not catch mumps.

c. Phoebe does not catch mumps.

d. Neither Calvin nor Phoebe catches mumps.

e. At least one of them catches mumps.

SOLUTION Using probability notation, $P(\text{C}) = 0.13$ and $P(\text{Ph}) = 0.07$.

a. $P(\text{C and Ph}) = P(\text{C}) \cdot P(\text{Ph})$ — Assuming C and Ph are independent events.

$= 0.13 \cdot 0.07$

$= 0.0091$, or 0.91%

b. $P(\text{not C}) = 1 - P(\text{C})$ — C and "not C" are complementary events.

$= 1 - 0.13$

$= 0.87$, or 87%

c. $P(\text{not Ph}) = 1 - P(\text{Ph}) = 1 - 0.07$

$= 0.93$, or 93%

d. $P(\text{not C and not Ph}) = P(\text{not C}) \cdot P(\text{not Ph})$

$= 0.87 \cdot 0.93$ — From parts b and c.

$= 0.8091$, or 80.91%

e. $P(\text{at least 1}) = 1 - P(\text{not C and not Ph})$ — Complementary events.

$= 1 - 0.8091$ — From part d.

$= 0.1909$, or 19.09%

Alternate solution for part e:

$P(\text{at least 1}) = P(\text{C}) + P(\text{Ph}) - P(\text{C and Ph})$ — C and Ph are not mutually exclusive events.

$= 0.13 + 0.07 - 0.0091$ — From the given probabilities and part a.

$= 0.1909$, or 19.09%

Problem Set 14-6

Reading Analysis

From what you have read in this section, what do you consider to be the main idea? In what way do the properties of probability in this section correspond to the counting principles in Section 14-3? How do you find the probability that a particular event will *not* happen?

Quick Review

Q1. If A and B are mutually exclusive, then $n(A \text{ or } B) =$ __?__.

Q2. If A and B are *not* mutually exclusive, then $n(A \text{ or } B) =$ __?__.

Q3. $n(A \text{ and } B) = n(A) \cdot n(B|A)$, where $n(B|A)$ is the number of ways B can happen __?__.

Q4. The number of combinations of five objects taken three at a time is equal to __?__.

Q5. The number of permutations of five objects taken three at a time is equal to __?__.

Q6. Why is a number of permutations greater than the corresponding number of combinations?

Q7. What is the definition of residual deviation?

Q8. Write the exact value (no decimals) of $\sin^{-1} 0.5$.

Q9. $y = 5 \cdot 3^x$ is a particular equation for a(n) __?__ function.

Q10. The area of a triangle with sides 5 cm and 3 cm and included angle 30° is __?__.

---

1. *Calculator Components Problem:* The "heart" of a calculator is one or more chips on which thousands of components are etched. Chips are mass-produced and have a fairly high probability of being defective. Suppose a particular brand of calculator uses two kinds of chips. Chip A has probability 70% of being defective, and chip B has probability 80% of being defective. If one chip of each kind is randomly selected, find the probability that

a. Both chips are defective

b. Chip A is not defective

c. Chip B is not defective

d. Neither chip is defective

e. At least one chip is defective

2. *Car Breakdown Problem:* Suppose you plan to drive on a long trip. The probability that your car will have a flat tire is 0.1, and the probability that it will have engine trouble is 0.05. If these probabilities are independent, what is the probability of

   a. No flat tire b. No engine trouble

   c. Neither flat tire nor engine trouble

   d. Both a flat tire and engine trouble

   e. At least one, either a flat tire or engine trouble

3. *Traffic Light Problem:* Two traffic lights on Broadway operate independently. Your probability of being stopped at the first light is 40%. Your probability of being stopped at the second is 70%. Find the probability of being stopped at

   a. Both lights b. Neither light

   c. The first light but not the second

   d. The second light but not the first

   e. Exactly one of the lights

4. *Visiting Problem:* Eileen and Ben are away at college. They visit home on random weekends. The probability that Eileen will visit on any given weekend is 20%. The probability that Ben will visit is 25%. On a given weekend, find the probability that

   a. Both of them will visit

   b. Neither will visit

   c. Eileen will visit but Ben will not

   d. Ben will visit but Eileen will not

   e. Exactly one of them will visit

5. *Hide-and-Seek Problem:* The Katz brothers, Bob and Tom, are hiding in the cellar. If either one sneezes, he will reveal their hiding place. Bob's probability of sneezing is 0.6, and Tom's probability is 0.7. What is the probability that at least one brother will sneeze?

6. *Backup System Problem:* Vital systems such as electric power generating systems have "backup" components in case one component fails. Suppose two generators each have a 98% probability of working. The system will continue to operate as long as at least one of the generators is working. What is the probability that the system will continue to operate?

7. *Basketball Problem:* Three basketball teams from Lowe High each play on Friday night. The probabilities that the teams will win are 70% for varsity, 60% for junior varsity, and 80% for freshmen. Find the probability that

   a. All three teams win

   b. All three teams lose

   c. At least one team wins

   d. The varsity team wins and the other two teams lose

8. *Grade Problem:* Terry Tory has these probabilities of passing various courses: Humanities, 90%; Speech, 80%; and Latin, 95%. Find the probability of

   a. Passing all three b. Failing all three

   c. Passing at least one d. Passing exactly one

9. *Spaceship Problem:* Complex systems such as spaceships have many components. Unless the system has backup components, the failure of any one component could cause the entire system to fail. Suppose a spaceship has 1000 such vital components and is designed without backups.

   a. If each component is 99.9% reliable, what is the probability that all 1000 components work and the spaceship does not fail? Surprising?

   b. What is the minimum reliability needed for each component to ensure a 90% probability that all 1000 components will work?

At least one generator must work.

10. *Silversword Problem:* The silversword is a rare relative of the sunflower that grows only on Haleakala volcano on Maui, Hawaii. The seeds have only a small probability of germinating, but if enough are planted there is a fairly good chance of getting a new plant. Suppose the probability that any one seed will germinate is 0.004.

    a. What is the probability that any one seed will *not* germinate?

    b. If 100 seeds are planted, find the probability that

        i. None will germinate

        ii. At least one will germinate

    c. What is the fewest number of seeds that would need to be planted to ensure a 99% probability that at least one will germinate?

11. *Football Plays Problem:* Backbay Polytechnic Institute's quarterback selects passing and running plays at random. By analyzing previous records, an opposing team finds these probabilities:

    - The probability that he will pass on first down is 0.4.
    - If he passes on first down, the probability that he will pass on second down is 0.3.
    - If he selects a running play on first down, the probability that he will pass on second down is 0.8.

    a. Find the probability that he will pass on

        i. First down and second down

        ii. First down but not second down

        iii. Second down but not first down

        iv. Neither first down nor second down

    b. Sum the four probabilities you have calculated. How do you explain the answer?

12. *Measles and Chicken Pox Problem:* Suppose that in any one year a child has probability 0.12 of catching measles and probability 0.2 of catching chicken pox.

    a. If these events are *independent* of each other, what is the probability that a child will get *both* diseases in a given year?

    b. Suppose statistics show that the probability of getting measles and then chicken pox in the same year is 0.006, and the probability of getting chicken pox and then measles in the same year is 0.018.

        i. If you know a child had measles, calculate the probability of that child getting chicken pox after having measles in the same year.

        ii. If you know a child had chicken pox, calculate the probability of that child getting measles after having chicken pox in the same year.

    c. Based on the given probabilities and your answers to part b, what could you conclude about the effects of the two diseases on each other?

13. *Airplane Engine Problem:* One reason airplanes are designed with more than one engine is to increase the planes' reliability. Usually a twin-engine plane can make it to an airport on just one engine should the other engine fail during flight. Suppose that for a twin-engine plane the probability that any one engine will fail during a given flight is 3%.

    a. If the engines operate independently, what is the probability that *both* engines will fail during a flight?

    b. Suppose flight records indicate that the probability that both engines will fail during a given flight is actually 0.6%. What is the probability that the second engine fails after the first has already failed?

    c. Based on your answer to part b, do the engines actually operate independently? Explain.

14-7 Functions of a Random Variable

Suppose you conduct the random experiment of flipping a coin five times. The coin is bent so that the probability of heads on any one flip is only 0.4. What is the probability of the event that exactly two of the outcomes are heads and the other three are tails? In this section you'll learn how to calculate the probabilities of all possible events for a random experiment. In the coin-flipping experiment, each probability depends on the number of heads in the event. Thus, the probability is a *function* of the number of heads. Such a function is called a *function of a random variable.*

Objective

Given a random experiment, find and graph the probabilities of all possible events.

To find $P(3\text{T}, 2\text{H})$ for the random experiment just described, it helps to start by looking at a simpler event, $P(\text{TTTHH})$, the probability of three tails and two heads *in that order:*

$$P(\text{TTTHH}) = P(\text{T}) \cdot P(\text{T}) \cdot P(\text{T}) \cdot P(\text{H}) \cdot P(\text{H})$$

$$= (0.6)(0.6)(0.6)(0.4)(0.4) \qquad P(\text{T}) = 1 - P(\text{H}) = 1 - 0.4 = 0.6$$

$$= 0.6^3 \cdot 0.4^2$$

There are ten possible outcomes that have exactly two heads:

HHTTT	HTHTT	HTTHT	HTTTH	THHTT
THTHT	THTTH	TTHHT	TTHTH	TTTHH

Ten is the number of ways of selecting a *group* of two of the five flips to be heads. But this is also the number of *combinations* of five elements taken two at a time, or ${}_5C_2$. So

$$P(3\text{T}, 2\text{H}) = 10 \cdot 0.6^3 \cdot 0.4^2$$

$$= {}_5C_2 \cdot 0.6^3 \cdot 0.4^2 \qquad \text{By calculator or by computing factorials, } {}_5C_2 = 10.$$

The expression ${}_5C_2 \cdot 0.6^3 \cdot 0.4^2$ is a term in the **binomial series** that comes from expanding

$$(0.6 + 0.4)^5$$

If x stands for the number of times *heads* appears in five flips, then

$$P(x) = {}_5C_x \cdot 0.6^{5-x} \cdot 0.4^x$$

As you can see, the probability $P(x)$ is a *function* of the random variable x. Because this function tells how the 100% probability is *distributed* among the various possible events, it is called a **probability distribution.** In this case, because the probabilities are terms in a binomial series, it is called a **binomial distribution.**

A random experiment, such as the coin-flipping experiment, in which the trials are repeated a number of times and in which there are only two possible outcomes for each trial, is called a **binomial experiment.**

PROPERTY: Binomial Probability Distribution

Suppose a random experiment consists of repetitions of the same action and that the action has only two possible results. Let E be one of the two possible results.

Let b be the probability that event E occurs in any one repetition.
Let a be the probability that event E does *not* occur in any one repetition.
Let x be the number of times event E occurs in n repetitions.

Then

$$P(x) = {}_nC_x \cdot a^{n-x} \cdot b^x$$

That is, $P(x)$ has the value of the term with b^x as a factor in the binomial series $(a + b)^n$.

EXAMPLE 1 ➤ A bent coin is flipped five times (as described previously). The probability of getting heads on any one toss is 40%.

a. Find all terms in the probability distribution.

b. Show that the total of the probabilities equals 1, and explain the significance of this fact.

c. Plot the graph. Sketch the result.

d. Calculate the probability that the coin lands heads up at least two of the five times.

SOLUTION Let $P(x)$ be the probability that there are exactly x heads in five flips.

a. Because the probability of getting heads on any one flip is 40%, or 0.4, the probability of *not* getting heads (that is, of getting tails) is $1 - 0.4$, or 0.6.

$$P(0) = {}_5C_0 \cdot 0.6^5 \cdot 0.4^0 = 1 \cdot 0.6^5 \cdot 0.4^0 = 0.07776$$

$$P(1) = {}_5C_1 \cdot 0.6^4 \cdot 0.4^1 = 5 \cdot 0.6^4 \cdot 0.4^1 = 0.2592$$

$$P(2) = {}_5C_2 \cdot 0.6^3 \cdot 0.4^2 = 10 \cdot 0.6^3 \cdot 0.4^2 = 0.3456$$

$$P(3) = {}_5C_3 \cdot 0.6^2 \cdot 0.4^3 = 10 \cdot 0.6^2 \cdot 0.4^3 = 0.2304$$

$$P(4) = {}_5C_4 \cdot 0.6^1 \cdot 0.4^4 = 5 \cdot 0.6^1 \cdot 0.4^4 = 0.0768$$

$$P(5) = {}_5C_5 \cdot 0.6^0 \cdot 0.4^5 = 1 \cdot 0.6^0 \cdot 0.4^5 = 0.01024$$

A time-efficient way to compute all the probabilities is to put the six possible values of x—the integers 0 through 5—into one list and then put the formula for $P(x)$ into a second list.

b. The sum of the six probabilities in part a is exactly 1. This indicates that one of the six events listed is certain to happen and that there are no other possible events in this random experiment.

c. Make a scatter plot using the values in part a (Figure 14-7a). Note that only integers are in the domain of this function.

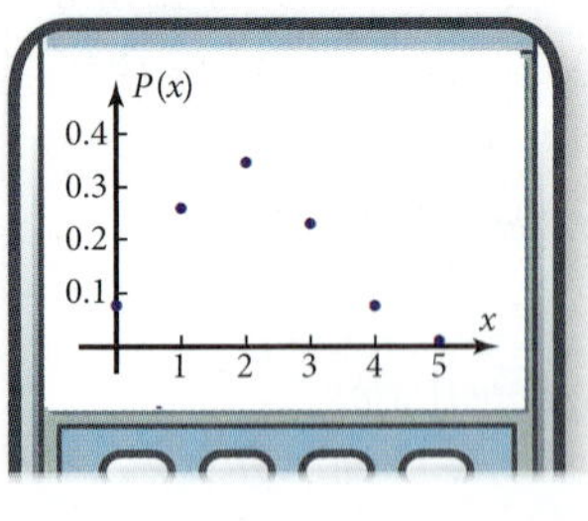

Figure 14-7a

d. $P(x \geq 2) = P(2) + P(3) + P(4) + P(5)$ — The results are mutually exclusive.

$= 0.3456 + 0.2304 + 0.0768 + 0.01024$ — From part a.

$= 0.66304 \approx 66\%$

Alternate solution to part d:

$P(x \geq 2) = 1 - P(0) - P(1)$

$= 1 - 0.07776 - 0.2592$

$= 0.66304 \approx 66\%$

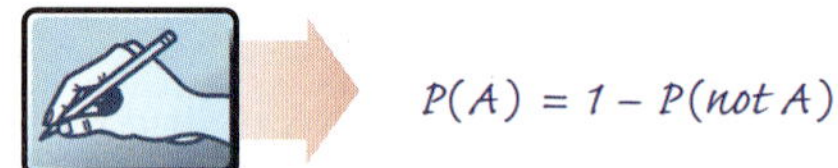

Problem Set 14-7

Reading Analysis

From what you have read in this section, what do you consider to be the main idea? In a function of a random variable, what is the independent variable and what is the dependent variable? If a random experiment is performed repeatedly and there are only two possible outcomes for each repetition, why is the resulting probability distribution called a *binomial* distribution?

Quick Review

Q1. If A and B are mutually exclusive, then $P(A \text{ or } B) =$ __?__.

Q2. If A and B are *not* mutually exclusive, then $P(A \text{ or } B) =$ __?__.

Q3. If A and B happen in that order, then $P(A \text{ and } B) = P(A) \cdot P(B|A)$, where $P(B|A)$ is __?__.

Q4. If a fair coin is flipped twice, $P(\text{TT}) =$ __?__.

Q5. If a fair coin is flipped twice, $P(\text{at least one is heads}) =$ __?__.

Q6. If a fair coin is flipped three times, $P(\text{TTT}) =$ __?__.

Q7. If a fair coin is flipped three times, $P(\text{THT in that order}) =$ __?__. Is this surprising?

Q8. If $g(x) = f(x - 2)$, then g is a __?__ transformation of f.

Q9. If $g(x) = f(2x)$, then g is a __?__ transformation of f.

Q10. Sketch the graph of a logistic function.

---

1. *Heredity Problem:* If a dark-haired mother and father have a particular combination of genes, each of their babies have a $\frac{1}{4}$ probability of having light hair.

a. What is the probability of any one baby having dark hair?

b. If they have three babies, calculate $P(0)$, $P(1)$, $P(2)$, and $P(3)$, the probabilities of having zero, one, two, and three dark-haired babies, respectively.

c. Find the sum of the probabilities in part b. How do you interpret the answer?

d. Plot the graph of this probability distribution.

e. What special name is given to this kind of probability distribution?

2. *Multiple-Choice Test Problem:* A short multiple-choice test has four questions. Each question has five choices, exactly one of which is right. Willie Passitt has not studied for the test, so he guesses answers at random.

a. What is the probability that his answer on a particular question is right? What is the probability that it is wrong?

b. Calculate his probabilities of guessing 0, 1, 2, 3, and 4 answers right.

c. Perform a calculation that shows that your answers in part b are reasonable.

d. Plot the graph of this probability distribution.

e. Willie will pass the test if he answers at least three of the four questions right. What is his probability of passing?

f. This binomial probability distribution is an example of a function of __?__.

3. *Thumbtack Problem:* If you flip a thumbtack, it can land either point up or point down (Figure 14-7b). Suppose the probability that any one flip will land point up is 0.7 and the tack is flipped ten times. Let $P(x)$ be the probability that x of ten flips land point up.

a. Show how $P(3)$ is calculated.

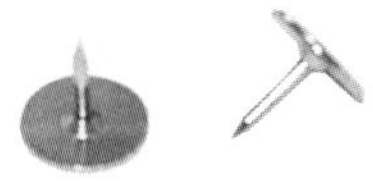

b. Calculate $P(x)$ for each of the 11 possible values of x. Make a scatter plot of the probability distribution, and sketch the result.

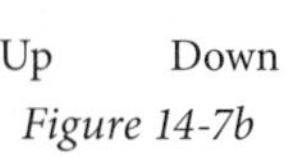

Up Down

Figure 14-7b

c. Which is more probable, that the thumbtack will land point up more than five times or that the thumbtack will land point up at most five times? Show results that support your answer.

4. *Traffic Light Problem:* Three widely separated traffic lights on U.S. Route 1 operate independently of one another. The probability that you will be stopped at any one light is 40%.

a. Show how to calculate the probability of being stopped at exactly two of the three lights.

b. Which is more probable, being stopped at more than one light or being stopped at one or fewer lights? Show results that support your answer.

c. Suppose you make the trip four times, encountering a total of 12 lights. Make a scatter plot of the probability distribution. Sketch the result.

5. *Color Blindness Problem:* Statistics show that about 8% of all males are color-blind. Interestingly, women are less likely to have this condition. Suppose 20 males are selected at random. Let $P(x)$ be the probability that x of the 20 men are color-blind.

a. Compute the probability distribution and plot its graph. Use a window that makes the graph fill most of the screen. Sketch the pattern followed by the points on the graph.

b. From your output in part a, find $P(0)$, $P(1)$, $P(2)$, and $P(3)$.

c. In a time-efficient way, calculate the probability that at least 4 of the 20 males are color-blind. Show the method you used for the computation.

6. *Eighteen-Wheeler Problem:* Large tractor-trailer trucks usually have 18 tires. Suppose the probability that any one tire will blow out on a given cross-country trip is 0.03.

a. What is the probability that any one tire will *not* blow out?

b. Find the probability that

i. None of the 18 tires blows out

ii. Exactly one of the tires blows out

iii. Exactly two of the tires blow out

iv. More than two tires blow out

c. If a trucker wants to have a 95% probability of making the trip without a blowout, what must be the reliability of each tire? That is, what would the probability that any one tire blows out have to be?

7. *Perfect Solo Problem:* Clara Nett plays a musical solo. She guesses that her probability of playing any one note right is 99%. The solo has 60 notes.

   a. Find the probability that

      i. She plays every note right

      ii. She makes exactly one mistake

      iii. She makes exactly two mistakes

      iv. She makes at least two mistakes

      v. She makes more than two mistakes

   b. What must be Clara's probability of playing any one note right if she wants to have a 95% probability of playing all 60 notes right?

8. *Airplane Engine Problem:* One reason commercial airplanes have more than one engine is to reduce the consequences should an engine fail during flight (Figure 14-7c). Under certain circumstances, some counterintuitive things happen when the number of engines is increased. Assume that the probability that any one engine will fail on a given flight is 0.1 (this is high, but assume it anyway).

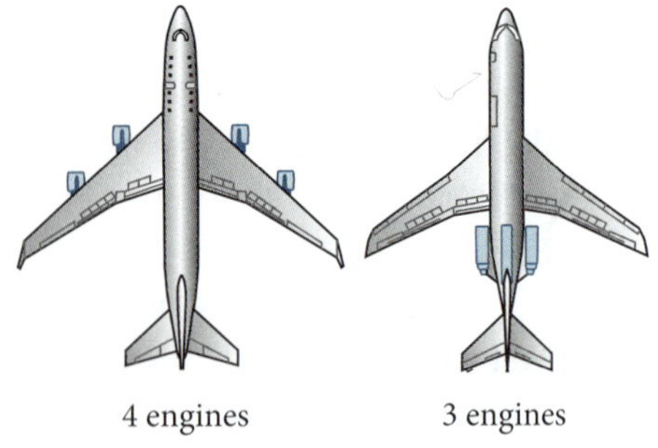

Which is safer?

Figure 14-7c

   a. For a plane that has four engines, calculate the probabilities that zero, one, two, three, and all four engines fail during the given flight. Show that the probabilities sum to 1, and explain the significance of this fact.

   b. If the plane will keep flying as long as no more than one engine fails, what is the probability that the four-engine plane keeps flying?

   c. Suppose a different kind of plane has three engines of the same reliability and it, too, will keep flying if no more than one engine fails. What is the probability that the three-engine plane keeps flying?

   d. Based on your computations in this problem, which is safer, the four-engine plane or the three-engine plane?

9. *World Series Problem:* Suppose the Dodgers and the Yankees are in the World Series. A baseball team must win four games to win the World Series. From their season records, you predict that the Dodgers have probability 0.6 of beating the Yankees in any one game. Assume that this probability is independent of which team has won a preceding game in this World Series.

   a. Find the probability that the Dodgers win the series by winning the first four games.

   b. Find the probability that the Yankees win the series by winning the first four games.

   c. For a team to win the series in exactly five games, they must win exactly three of the first four games, then win the fifth game. Find the probability that the Dodgers win the series in five games.

   d. Find the probability that the Yankees win the series in five games.

   e. Find the probability of each of these events.

      i. The Dodgers win the series in six games.

      ii. The Yankees win the series in six games.

      iii. The Dodgers win the series in seven games.

      iv. The Yankees win the series in seven games.

   f. Find the probability that the Dodgers win the series.

   g. What is the most probable length of the series—four, five, six, or seven games?

Problems 10–13 involve probability distributions other than binomial distributions.

10. *Dice Problem:* A random experiment consists of rolling two dice, one black and one white.

a. Plot the probability distribution for each of these random variables. You can count the outcomes in Figure 14-1a.

i. x is the sum of the numbers on the two dice.

ii. x is the number on the black die minus the number on the white die.

iii. x is the absolute value of the difference between the number on the black die and the number on the white die.

b. For each probability distribution in part a, find the most probable value of x.

11. *Proper Divisors Problem:* An integer from 1 through 10 is selected at random. Let x be the number of proper divisors the integer has. (A **proper divisor** of an integer n is a positive integer less than n that divides n exactly. For example, 12 has five proper divisors: 1, 2, 3, 4, and 6.)

a. List the proper divisors and the number of proper divisors for each integer from 1 through 10.

b. For each possible value of x, identify how many of the integers from 1 through 10 have that number of proper divisors.

c. Let $P(x)$ be the probability that an integer from 1 through 10 has x proper divisors. Calculate $P(x)$ for each value of x in the domain.

d. Plot the graph of the probability distribution in part c on your grapher. Do you see any pattern followed by the points on the graph?

12. *First Girl Problem:* Eva and Paul want to have a baby girl. They know that the probability of having a girl on any single birth is 0.5.

a. Let x be the number of babies they have, and $P(x)$ the probability that the xth baby is the *first* girl. Then $P(1) = 0.5$. $P(2)$ is the probability that the first baby is *not* a girl and that the second baby *is* a girl. Calculate $P(2)$, $P(3)$, and $P(4)$.

b. Plot the graph of $P(x)$. Sketch the graph, showing what happens as x becomes large.

c. Besides being called a probability distribution, what other special kind of function is this?

d. Show that the sum of the values of $P(x)$ approaches 1 as x becomes very large.

13. *Same Birthday Problem:* A group of students compares birthdays.

a. What is the probability that Shawn's birthday is *not* the same as Mark's?

b. If Shawn and Mark have different birthdays, what is the probability that Frieda's birthday is not the same as Shawn's or Mark's?

c. What is the probability that Shawn and Mark have different birthdays *and* that Frieda has a birthday different from both of theirs?

d. Using the pattern you observe in part c, find the probability that a group of ten students will all have different birthdays. Give a decimal approximation of the result.

e. What is the probability that in a group of ten students at least two have the same birthday (that is, *not* all ten have different birthdays)?

f. Write a program to compute a list of probabilities that in a group of x people at least two people have the same birthday. Store the output in lists of x and $P(x)$ for use in subsequent graphing. Use the program to make a list of $P(x)$ for 2 through 60 people.

g. Plot the graph of the probability distribution in part f on your grapher. Use the data in the lists outputted by the program without any further computation. Sketch the result.

h. From the graphical or numerical data, determine how many people must be in a group to have a probability that at least two people will have the same birthday equal to

i. 50%

ii. 99%

14. *Journal Problem:* Update your journal with things you have learned since the last entry. In particular, explain how the properties of probability and the concept of function lead to functions of a random variable.

14-8 Mathematical Expectation

One of the main uses of probability is in calculating an expected value without actually conducting a random experiment. Insurance companies can use an expected value to calculate the insurance costs and expected profit from a particular policy. In this section you will study **mathematical expectation,** a value you can calculate based on the outcomes for each event in a random experiment.

Objective Calculate the mathematical expectation for a given random experiment.

In this exploration you will calculate the payoff expected for a random experiment.

EXPLORATION 14-8: Mathematical Expectation

Grades Problem (Problems 1–3): Ernie DeGrades is good in mathematics and languages but not as good in history and chemistry. He figures that these are his probabilities of earning various grades in the five courses he is taking.

	A	B	C
Mathematics	95%	5%	
English	90%	10%	
Spanish	80%	15%	5%
Chemistry	70%	20%	10%
History	55%	30%	15%

1. Ernie gets 4 grade points for an A in any subject, 3 for a B, and 2 for a C. What is Ernie's mathematically expected number of grade points for the mathematics course?
2. Calculate Ernie's mathematically expected number of grade points in each of the other courses.
3. Ernie's GPA (grade point average) is the total number of grade points earned divided by the total number of courses. What is his mathematically expected GPA?

Bent Coin Problem (Problems 4 and 5): You "pay" ten tokens for a game. You flip a bent coin five times. The probability of heads on any one flip is 40%.

4. Calculate your probabilities of getting heads exactly 0, 1, 2, 3, 4, and 5 times in the five flips.
5. Your payoff for five heads is 100 tokens. Your payoff for four heads is 50 tokens and for three heads is 20 tokens. For any other number of heads, you win nothing. Calculate the mathematical expectation for this random experiment and explain its real-world meaning. Remember to subtract the 10 tokens you paid to play the game.
6. What did you learn as a result of doing this exploration that you did not know before?

At a school carnival, students are awarded points for winning games. At the end of the evening, they can trade in their points for prizes. For a particular game, students start out paying 50 points to roll a single die. The payoffs for the game are

- Roll a 6: Win 100 points (and get your 50 points back).
- Roll a 2 or a 4: Win 10 points (and get your 50 points back).
- Roll an odd number: Win nothing (and lose your 50 points).

Because each outcome is equally likely, you would "expect" to get each number *once* in six rolls of the die. (You probably won't, but that is what you expect to happen on average if you roll the die many times.) If you did roll each number exactly once, your winnings would be

Number	Points Won
1	−50
2	10
3	−50
4	10
5	−50
6	100
Total:	−30

Because you would *lose* 30 points in six rolls, your average winnings would be

$$\text{Average winnings} = \frac{-30}{6} = -5 \text{ points per roll}$$

This average is your mathematical expectation. If you play the game thousands of times, you would expect to lose about 5 points per roll, on average.

A pattern shows up if you do not carry out the addition when calculating the mathematical expectation in the preceding random experiment. Let E stand for the mathematical expectation:

$$E = \frac{-50 + 10 - 50 + 10 - 50 + 100}{6}$$

Sum the six values, one for each outcome, and divide by 6.

$$= \frac{3(-50) + 2(10) + 1(100)}{6}$$

Combine "like terms."

$$= \frac{3(-50)}{6} + \frac{2(10)}{6} + \frac{1(100)}{6}$$

Division distributes over addition.

$$= \frac{3}{6}(-50) + \frac{2}{6}(10) + \frac{1}{6}(100)$$

Properties of fractions.

The fraction $\frac{3}{6}$ is the probability of getting an odd number, and −50 is the value (sometimes called the "payoff") associated with getting an odd number. Similarly, $\frac{2}{6}$ and $\frac{1}{6}$ are the probabilities associated with 10 and 100, respectively, which are the values for the other two events. So you can calculate the mathematical expectation of a random experiment by multiplying the probability and the value for each mutually exclusive event and then summing the results. This fact leads to the algebraic definition of mathematical expectation.

DEFINITION: Mathematical Expectation

Algebraically: The **mathematical expectation,** E, of a random experiment is the sum

$$E = P(A_1)a_1 + P(A_2)a_2 + P(A_3)a_3 + \cdots + P(A_n)a_n \text{ or}$$

$$E = \sum_{k=1}^{n} P(A_k)a_k$$

for the n mutually exclusive events $A_1, A_2, A_3, \ldots, A_n$ in the experiment. The values $a_1, a_2, a_3, \ldots, a_n$ correspond to the outcomes of $A_1, A_2, A_3, \ldots, A_n$.

Verbally: The mathematical expectation is the *weighted average value (payoff)* for a random experiment each time it is run.

EXAMPLE 1 ➤ The basketball toss at an amusement park costs 50¢ to play. To play, you shoot three balls. If you make no baskets, you win nothing (and lose your 50¢). If you make just one basket, you win a key chain worth 5¢. If you make two of the three baskets, you win a stuffed animal worth 60¢. If you make all three baskets, you win a doll worth $2.50. The basketball hoop is small, so your probability of making any one basket is only 30%. What is the mathematical expectation for the game? How do you interpret the answer?

SOLUTION Let $P(x)$ be your probability of making x baskets. Your probability of missing any one basket is 100 − 30, or 70% (0.7, as a decimal). Therefore,

$$P(0) = {}_3C_0 \cdot 0.7^3 \cdot 0.3^0 = 0.343$$

$$P(1) = {}_3C_1 \cdot 0.7^2 \cdot 0.3^1 = 0.441$$

$$P(2) = {}_3C_2 \cdot 0.7^1 \cdot 0.3^2 = 0.189$$

$$P(3) = {}_3C_3 \cdot 0.7^0 \cdot 0.3^3 = \underline{0.027}$$

Check: Total = 1.000

The payoff for each event is found by subtracting the 50¢ "admission fee" from the amount you win:

x	Payoff (in cents)
0	0 − 50 = −50
1	5 − 50 = −45
2	60 − 50 = 10
3	250 − 50 = 200

$$E = (0.343)(-50) + (0.441)(-45) + (0.189)(10) + (0.027)(200)$$

Definition of expectation.

$$= -29.705$$

So on average you would expect to *lose* about 30¢ per game if you played many times. (This is the way amusement parks make money on such games!)

Once you understand how mathematical expectation is calculated, you can use a spreadsheet or list operations on your grapher to compute the values.

x	$P(x)$	Payoff	$P(x) \cdot$ Payoff
0	0.343	−50	−17.15
1	0.441	−45	−19.845
2	0.189	10	1.89
3	0.027	200	5.4
Totals:	1.000		−29.705

$\therefore E = -29.705$, or a loss of about 30¢ each time you play the game

Problem Set 14-8

Reading Analysis

From what you have read in this section, what do you consider to be the main idea? Assume that a random experiment has only two outcomes, A and "not A," where the probability of A is 0.6. If the payoff for A is 20 and the payoff for "not A" is 30, what is the mathematical expectation for the experiment?

Quick Review

Q1. Expand: $(a + b)^2$

Q2. Expand: $(a + b)^3$

Q3. Expand: $(a + b)^4$

Q4. Evaluate ${}_4C_1$ and ${}_4C_3$.

Q5. How are the answers to Problem Q4 related to the answer to Problem Q3?

Q6. If $a = 0.3$ and $b = 0.7$, then ${}_4C_3\, a^1b^3 =$ __?__.

Q7. How is the answer to Problem Q6 related to a binomial probability distribution?

Q8. Which equals $\sec^2 x - \tan^2 x$: $\cos^2 x + \sin^2 x$ or $\cos^2 x - \sin^2 x$?

Q9. What is the mathematical expectation of a spinner that ends on +5 two-thirds of the time and on −16 one-third of the time?

Q10. Whose name is associated with finding the area of a triangle directly from its side lengths?

1. *Uranium Fission Problem:* When a uranium atom splits ("fissions"), it releases 0, 1, 2, 3, or 4 neutrons. Let $P(x)$ be the probability that x neutrons are released. Assume that the probability distribution is

x	$P(x)$
0	0.05
1	0.2
2	0.25
3	0.4
4	0.1

   a. What is the mathematically expected number of neutrons released per fission?

   b. The number of neutrons released in any one fission must be an integer. How do you explain the fact that the mathematically expected number in part a is not an integer?

2. *Archery Problem:* An expert archer has the probabilities listed in the table of hitting various rings on a target (Figure 14-8a).

Figure 14-8a

Color	Probability	Points
Gold	0.20	9
Red	0.36	7
Blue	0.23	5
Black	0.14	3
White	0.07	1

a. What is her mathematically expected number of points on any one shot?

b. If she shoots 48 arrows, what would her expected score be?

3. *Sales Incentive Problem:* Calvin is a salesperson at a car dealership that wants to sell last year's car models to make space for new cars. The manager wants minivans to clear out fastest, station wagons next, and other models last, so she puts together two choices of incentives for the sales staff.

Option A: Receive a \$100 bonus for each vehicle sold of last year's models.

Option B: Receive a \$2000 bonus for selling four minivans, two station wagons, one pickup truck, and one sedan of last year's models.

Calvin estimates the probabilities that he will sell the given number of each model in time to make space for next year's cars:

$P(\text{four minivans}) = 50\%$

$P(\text{two station wagons}) = 70\%$

$P(\text{one pickup truck}) = 80\%$

$P(\text{one sedan}) = 90\%$

a. Suppose Calvin chooses Option A. What is the mathematical expectation for Calvin's bonus if he sells four minivans? Two station wagons? One pickup truck? One sedan? Use the results to find the mathematical expectation for his bonus if he sells these eight cars.

b. Calculate Calvin's probability of being able to sell four minivans, two station wagons, one pickup truck, and one sedan.

c. Calculate the mathematical expectation for Calvin's bonus if he chooses Option B.

d. Based on your answers to parts a–c, which option should he choose?

4. *Seed Germination Problem:* A package of seeds for an exotic tropical plant states that the probability that any one seed will germinate is 80%. Suppose you plant four of the seeds.

a. Find the probabilities that exactly 0, 1, 2, 3, and 4 of the seeds will germinate.

b. Find the mathematically expected number of seeds that will germinate.

5. *Batting Average Problem:* Jackie Robinson's highest major league batting average for one season was .342, which means that his probability of getting a hit at any one official at-bat was 0.342. Suppose Robinson had five official at-bats during a game.

a. Calculate the probabilities that he got 0, 1, 2, 3, 4, and 5 hits.

b. What was Robinson's mathematically expected number of hits for this game?

6. *Expectation for a Binomial Experiment:* Suppose you conduct a random experiment that has a binomial probability distribution. Suppose the probability that outcome C occurs on any one repetition is 0.4. Let $P(x)$ be the probability that outcome C occurs x times in five repetitions.

   a. Calculate $P(x)$ for each value of x in the domain.

   b. Find the mathematically expected value of x. (*Hint:* The value if C occurs x times is x.)

   c. Show that the mathematically expected value of x is equal to 0.4 (the probability that C occurs on *one* repetition) times 5 (the total number of repetitions).

   d. If the probability that C occurs on any one repetition is b and the probability that C does not occur on any one repetition is $a = 1 - b$, prove that in five trials the expected value of x is $5b$.

   e. From what you have observed in this problem, make a conjecture about the mathematically expected value of x in n repetitions, if the probability that C occurs on any one repetition is b.

   f. If you plant 100 seeds, each of which has probability 0.71 of germinating, how many seeds would you expect to germinate?

7. *Multiple-Choice Test Problem:* Suppose you are taking your college entrance exams. You answer all the questions you know and have some time left over. So you decide to guess the answers to the rest of the questions.

   a. Each question is multiple choice with five choices. If you guess at random, what is the probability of getting an answer right? Of getting an answer wrong?

   b. When the testing service grades your paper, it gives you 1 point if the answer is right and subtracts $\frac{1}{4}$ point if the answer is wrong. What is your mathematically expected score on any question for which you guess at random?

   c. Suppose that, on one question, you can eliminate one choice you know is wrong and then randomly guess among the other four. What is your mathematically expected score on this question? Is it surprisingly low?

   d. Calculate your mathematically expected score on a question for which you can eliminate two of the choices and then on a question for which you can eliminate three of the choices.

   e. Based on your answers in this problem, do you think it is worthwhile guessing answers on a multiple-choice test?

8. *Accident/Illness Insurance Problem:* Some of the highest-paid mathematicians are *actuaries,* who figure out what you should pay for various types of insurance. Suppose an insurance company has an accident/illness policy that pays \$500 if you become ill during any one year, \$1000 if you have an accident, and \$6000 if you both become ill and have an accident. The premium, or payment, for this policy is \$100 per year. One of your friends, who has studied actuarial science, tells you that your probability of becoming ill in any one year is 0.05 and that your probability of having an accident is 0.03. Becoming ill and having an accident are independent events.

   a. Find the probability of each event.

      i. Becoming ill and having an accident

      ii. Becoming ill and not having an accident

      iii. Not becoming ill but having an accident

      iv. Not becoming ill and not having an accident

   b. What is the customer's mathematical expectation for this policy?

   c. An insurance policy is *actuarially* sound if the insurance company is expected to make a profit from it. Based on the probabilities assumed, is this policy actuarially sound?

9. *Life Insurance Problem 1:* Functions of random variables are used as mathematical models in the insurance business. The numbers in the table were taken from a *mortality table.* The table shows the probability, $P(x)$, that a person who is alive on his or her xth birthday will die before he or she reaches age $x + 1$.

Age, x	$P(x)$
15	0.00146
16	0.00154
17	0.00162
18	0.00169
19	0.00174
20	0.00179

A group of 10,000 15-year-olds gets together to form its own life insurance company. For a premium of $40 per year, the members agree to pay $20,000 to the family of anyone in the group who dies while he or she is 15 through 20 years old.

a. Calculate $D(15)$, the number out of 10,000 expected to die while they are age 15. Round to an integer.

b. Calculate $A(16)$, the number out of 10,000 expected to be alive on their 16th birthday.

c. Calculate $D(16)$. Round to an integer.

d. Make a table of x, $P(x)$, $A(x)$, and $D(x)$ for each value of x from 15 through 20.

e. Put columns into the table of part d for $I(x)$ and $O(x)$, the income from the $40 premiums and the amount paid out from the $20,000 death benefits. Take into account that a person who dies no longer pays premiums the following years.

f. Calculate $NI(x) = I(x) - O(x)$, the net income of the company each year. Explain why $NI(x)$ *decreases* each year. (There are two reasons!)

g. On average, how much would the company expect to make per year? Would this be enough to pay a full-time employee to operate the company?

10. *Life Insurance Problem 2:* A group of 10,000 people, each now 55 years old, is to be insured as described in Problem 9. Upon the death of the insured person at any age from 55 through 59, his or her survivors receive $20,000. Your job is to calculate the annual premium that should be charged for this policy. Here is the portion of the mortality table that applies to this age group.

Age, x	$P(x)$
55	0.01300
56	0.01421
57	0.01554
58	0.01700
59	0.01859

a. Make a table showing $D(x)$, $A(x)$, and $O(x)$, the number of deaths, the number still alive, and the amount paid out in death benefits, respectively.

b. An administrator is to be paid $30,000 a year to operate the program. Calculate the total paid out by the company over the five-year period, including the administrator's salary and the death benefits.

c. Calculate the total number of premiums received during the five-year period, taking into consideration the fact that the number of premiums received each year decreases after the first year as the insured persons die. From the result, calculate the premium that must be charged per person per year for the company to break even.

d. Why is the premium in part c so much higher than the premium of $40 per year in Problem 9?

14-9 Chapter Review and Test

In this chapter you analyzed functions in which the independent variable takes on random values. The dependent variable is the probability that a particular value of the random variable occurs. You used the definition of probability as a ratio of numbers of outcomes of a random experiment to derive properties that allow you to calculate probabilities algebraically. You learned that the binomial probability distribution has many real-world applications. Such functions are useful in finding mathematical expectation, which is the potential payoff for a random experiment.

Review Problems

R0. Update your journal with what you have learned in this chapter. Include things such as the definitions of random variable, probability, outcome, event, sample space, permutations, combinations, functions of a random variable, binomial probability distributions, and mathematical expectation. Show how what you have learned allows you to compute numbers of outcomes algebraically rather than by actually counting.

R1. *Quarter, Dime, and Nickel Problem:* A quarter, a dime, and a nickel are marked with 1 on the tails side and 2 on the heads side. All three coins are flipped. The eight possible outcomes are shown in Figure 14-9a. Find the probability of each event.

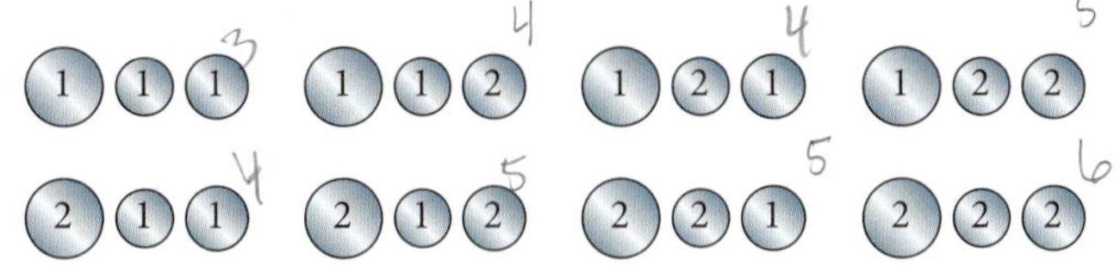

Figure 14-9a

a. The total is 4.

b. The total is 5.

c. The total is 6.

d. The total is 7.

e. The total is odd.

f. The total is between 4 and 6, inclusive.

g. The total is between 3 and 6, inclusive.

h. The quarter shows 2 and the nickel shows 1.

i. The quarter shows 2 or the nickel shows 1.

R2. *Numbered Index Card Problem:* Twenty-five index cards are numbered from 1 through 25. The cards are placed number side down on the table, and one card is drawn at random.

a. How many outcomes are in the sample space?

b. How many outcomes are in the event "the number is odd"?

c. What is the difference between an outcome and an event?

d. Find the probability that

- **i.** The number is odd
- **ii.** The number is divisible by 3
- **iii.** The number has two digits
- **iv.** The number is less than 30
- **v.** The number is at least 30

R3. a. An ice cream shop has 20 flavors of ice cream and 11 flavors of sherbet. Find the number of different ways you could select

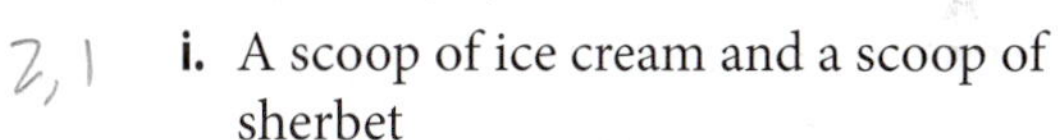

- **i.** A scoop of ice cream and a scoop of sherbet
- **ii.** A scoop of ice cream or a scoop of sherbet

b. Using the letters in EXACTING, find the number of different ways you could select

- **i.** A consonant and then a vowel
- **ii.** A consonant and then a different consonant

R4. **a.** The Russian alphabet has 33 characters. Find the number of different permutations that can be made

i. Using 3 different characters

ii. Using 33 different characters

b. How many different three-letter "words" can be made from the letters in PRECAL?

c. Find the probability that a permutation of all the letters in the word CHAPTER begins with a consonant and ends with a consonant.

d. Find the number of different permutations of all the letters in

HUMUHUMUNUKUNUKU

which are the first 16 letters of the name of Hawaii's state fish.

R5. **a.** Evaluate ${}_7C_3$ using factorials.

b. What is the difference between a permutation and a combination?

c. The 12th-grade class at Scorpion Gulch High School has 100 students: 53 girls and 47 boys. In how many different ways could they select the following?

i. A group of four students to be class officers

ii. A president, a vice president, a secretary, and a treasurer

iii. A seven-member debate team consisting of four boys and three girls

d. If a seven-member debate team is selected at random, what is the probability that it will have four boys and three girls as in part c.iii?

R6. **a.** *Car Trouble Problem:* Mr. Rhee's car has a 70% probability of starting, and Ms. Rhee's car has an 80% probability of starting. Find the probability of each event.

i. Neither car will start.

ii. Both cars will start.

iii. Either both cars will start or neither car will start.

iv. Exactly one of the cars will start.

b. *Basketball Game Problem:* High school basketball teams often play each other twice during the season. Suppose Central High has a 60% probability of winning its first game against Tech. If Central wins the first game, it has an 85% probability of winning the second game. If Central loses the first game, it has a 45% probability of winning the second game. Find the probability of each event.

i. Central wins both games.

ii. Central wins the first game and loses the second game.

iii. Central loses the first game and wins the second game.

iv. Central loses both games.

v. Show by calculation that the answers in parts b.i–iv are reasonable.

R7. *Candle Lighter Problem:* A butane candle lighter does not always light when you pull the trigger. Suppose a lighter has a 60% probability of lighting on any one pull. You pull the trigger six times. Let $P(x)$ be the probability that it lights exactly x times.

a. Show the method used to calculate $P(4)$.

b. In a time-efficient way, calculate $P(x)$ for each value of x in the domain.

c. Plot the graph of $P(x)$ as a scatter plot on your grapher. Sketch the result.

d. Find the probability that the lighter lights at least half the time.

e. Why is this random experiment called a *binomial* experiment?

R8. *Airline Overbooking Problem:* For parts a–d, a small commuter airline charges $100 for tickets on a particular flight. The plane holds 20 people, so the total revenue for a full flight is $2000. The airline expects a higher revenue by booking 21 passengers and taking its chances that one or more passengers will not show up (the tickets are nonrefundable). Company records indicate that, on average, there is a 10% probability that any one passenger will not show up. However, if everyone does show up, one passenger must be "bumped" and given a $300 payment. (The $100 is not refunded because the ticket can be used on a later flight.)

a. What is the probability that all 21 passengers show up and the airline makes only $1800 ($2100 − $300)? What is the probability that 20 or fewer passengers show up and the airline makes the full $2100 on that flight? What is the airline's mathematically expected revenue if it books 21 passengers?

b. If the airline books 22 passengers, the revenue is $1600 for zero no-shows, $1900 for one no-show, and $2200 for two or more no-shows. What is the airline's mathematically expected revenue if it books 22 passengers?

c. Calculate the mathematically expected revenue if the airline books 23 passengers. Is this more or less than the $2000 it would make if it did no overbooking?

d. What things besides a possible loss of money might make the airline limit the amount of overbooking?

Weighted Average Problem: For parts e and f, a college professor gives students a *weighted average.* Test 1 counts as 10% of the grade. Tests 2, 3, and 4 count as 20% each. The final exam counts as 30% of the grade.

e. To receive a grade of B or above, a student must have a weighted average of at least 80. Suppose Nita B. Topaz gets scores of 72, 86, 83, 77, and 81 on the five tests, in that order. Will she receive at least a B? Show numbers to support your answer.

f. Explain why the mathematics involved in finding a weighted average is the same as that used to find the mathematical expectation of a random experiment.

Concept Problem

C1. *Nuclear Reactor Project:* When a uranium atom inside the reactor of a nuclear power plant is hit by a neutron, it splits (fissions), releasing energy and some new neutrons (Figure 14-9b). The mathematically expected number of new neutrons per fission is 2.3.

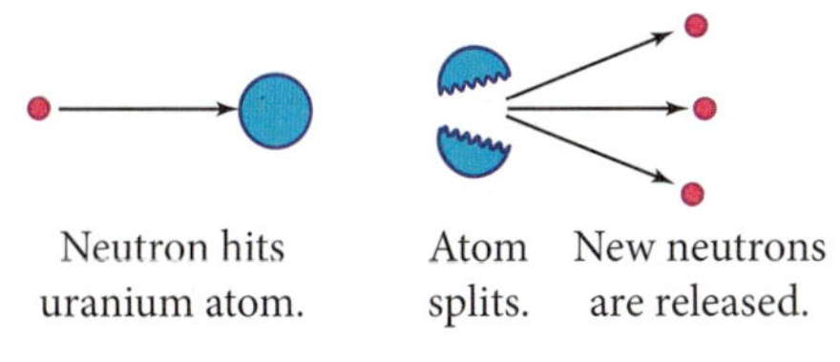

Figure 14-9b

a. Suppose there are 100 neutrons in the reactor initially. If all these neutrons cause fissions, how many neutrons would you expect there to be after this first "generation" of fissions?

b. If all the neutrons from the first generation cause fissions, how many neutrons would you expect after two generations? After three generations? After four generations? What kind of sequence do these numbers form?

c. If each generation takes 0.001 s, how many neutrons would you expect there to be after 1 s? Does this answer surprise you? This is what makes atomic bombs explode!

d. Not all of the neutrons from one generation actually cause fissions in the next generation. Some leak out of the reactor, some are captured by atoms other than uranium, and some that are captured by uranium atoms do not cause fission.

Assume that

$P(\text{leaking}) = 0.36$

$P(\text{capture by other atom}) = 0.2$

$P(\text{nonfission capture}) = 0.15$

Calculate the probability that *none* of these things happens and thus that the neutron *does* cause a fission in the next generation.

e. Use the probability in part d and the fact that there are 2.3 new neutrons per fission to calculate k, the expected number of new neutrons in the second generation caused by one neutron in the first generation.

f. How many neutrons would you expect there to be after 1 s under the conditions in part e if each generation still takes 0.001 s as in part c? Would the reactor explode like a bomb?

g. Why can you say that the number of neutrons is *increasing exponentially* with time?

h. The constant k in part e is called the *multiplication factor.* The *chain reaction* in a nuclear reactor is controlled by moving control rods out or in to absorb fewer or more neutrons. If k is slightly more than 1, the power level increases. If k is less than 1, the power level decreases. If k equals 1, the power level remains constant and the reactor is said to be *critical.* What would $P(\text{capture by other atom})$, mentioned in part d, have to equal to make the reactor critical?

Chapter Test

Part 1: No calculators allowed (T1–T8)

T1. Calculate the number of permutations of seven objects taken three at a time. Show your method.

T2. Calculate the number of combinations of six objects taken four at a time. Show how this number is calculated using factorials.

T3. What is the difference between a permutation and a combination?

T4. If A and B are independent events and if $P(A) = 0.8$ and $P(B) = 0.9$, find $P(A \text{ and } B)$.

T5. If A and B are independent events and if $P(A) = 0.8$ and $P(B) = 0.9$, find $P(A \text{ or } B)$.

T6. Suppose that, in each repetition of a random experiment, the probability that event A occurs is 0.8. Find the probability that A occurs in exactly two out of three repetitions.

T7. Explain why the random experiment in Problem T6 is called a *binomial* experiment.

T8. Suppose C, D, and E are three mutually exclusive events of a random experiment and $P(C)$, $P(D)$, and $P(E)$ are 0.5, 0.3, and 0.2, respectively. If the payoffs are \$10, \$6, and −\$100 for C, D, and E, respectively, find the mathematical expectation for the random experiment.

Part 2: Graphing calculators allowed (T9–T28)

Pick-Three Problem: Problems T9–T13 concern the 11th-grade class, which decides to run a lottery to help finance the prom. A person pays \$1 and picks three different digits. If all three digits match the winning digits, the class pays the person \$100 (but keeps the \$1). If not, the class keeps the \$1.

T9. The sample space for this random experiment contains ${}_{10}C_3$ outcomes. What is the probability that any one pick is the winning combination?

T10. What is the probability that any one pick is *not* the winning combination?

T11. What is the 11th-grade class's payoff if the pick is the winning combination? What is the class's payoff if the pick is not the winning combination?

T12. What is the class's mathematical expectation for any one pick?

T13. How much would the class expect to make from the sale of 1000 picks?

Multiple-Choice Test Problem: Problems T14–T17 refer to a typical statewide exit-level test with four different choices for each question. A test taker receives 1 point for each correct answer, but for each wrong answer $\frac{1}{3}$ point is subtracted.

T14. If you guess at random on a question, what is the probability of guessing the right answer? Of guessing a wrong answer?

T15. What is thc mathematically expected number of points for any question for which you guess the answer?

T16. Suppose you know that one of the four choices is incorrect. What is the probability of guessing the right answer from the remaining three choices? What is the probability of guessing a wrong answer?

T17. What is your mathematically expected number of points for a question for which you can eliminate one of the four choices?

Punctuality Problem: Problems T18–T23 concern Hezzy Tate, who has a 30% probability of being late to class on any one day.

T18. What is his probability of *not* being late on any one day?

T19. Show how to calculate Hezzy's probability of being late on exactly two of the five days in a week.

T20. Make a list on your grapher of his probabilities of being late on zero through five days. Write the result.

T21. Perform a calculation that shows that your answers in Problem T20 are reasonable.

T22. What is the special name of the probability distribution in Problem T20?

T23. Plot a graph of the probability distribution in Problem T20 on your grapher. Sketch the graph.

Cup and Saucer Problem: Problems T24–T27 concern Wanda, who washes dishes at a restaurant. Her probability of breaking a cup on any one shift is 8%, and her probability of breaking a saucer is 6%. Calculate the probabilities of these events.

T24. P(cup and saucer)

T25. P(cup and not saucer)

T26. P(saucer and not cup)

T27. P(not cup and not saucer)

T28. What did you learn as a result of doing this chapter test that you did not know before?

Chapter 15

Sequences and Series

Positioning spaceships requires highly accurate computations because slight errors can make a difference of many miles in the landing point or cause an unacceptable reentry angle. Using a series of powers, you can calculate cosines and sines to as many decimal places as desired, simply by using the operations of addition, subtraction, and multiplication many times. In this chapter you will learn about these power series as well as other series that can be used to calculate things such as compound interest on money in a savings account and the cumulative effects of repeated doses of medication.

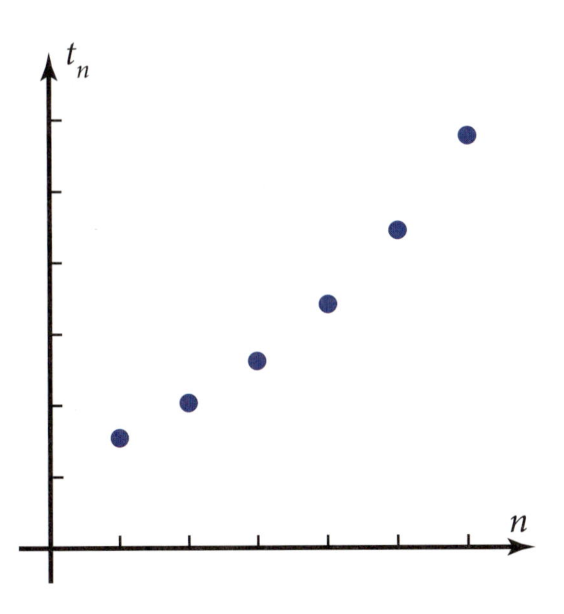

Mathematical Overview

In this chapter you will learn about sequences of numbers and about series, which are sums of the terms of sequences. Geometric and arithmetic series are logical mathematical models for functions such as compound interest, where the amount of money in an account increases by jumps each month rather than rising continuously. You will look at sequences and series in four ways.

NUMERICALLY

Sequence: 3, 6, 12, 24, . . .
Series: $3 + 6 + 12 + 24 + \cdots$
Partial sum: $S_{10} = 3 + 6 + 12 + 24 + \cdots + 1536 = 3069$

ALGEBRAICALLY

$$S_{10} = 3 \cdot \frac{1 - 2^{10}}{1 - 2} = 3069$$

GRAPHICALLY

The figure shows a graph of the terms of the geometric sequence $1.2 \cdot 1.3^n$ as a function of n, the term number. The dotted line indicates the continuous function graph that fits the discrete values of the sequence.

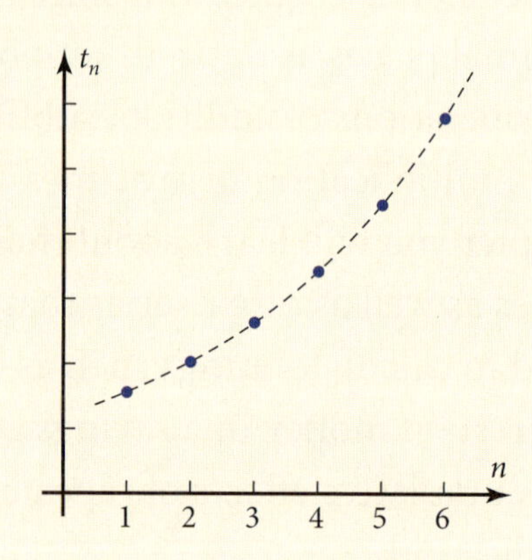

VERBALLY

Arithmetic and geometric sequences are similar. In arithmetic sequences the terms progress by adding a constant, and in geometric sequences the terms progress by multiplying by a constant. They are analogous to linear and exponential functions.

15-1 Introduction to Sequences and Series

Most of the functions you have studied so far have been **continuous**—their graphs have no discontinuities. Many of these graphs have been smooth curves. Where discrete data points have been measured, you looked for the continuous function that best fit these points. In this chapter you will study **sequences** of numbers, such as

$$5, 7, 9, 11, 13, \ldots$$

and **series** formed by summing the terms of a sequence, such as

$$5 + 7 + 9 + 11 + 13 + \cdots$$

Objective

- Given a few terms in a sequence or series of numbers, find more terms.
- Given a series, find the sum of a specified number of terms.

Exploratory Problem Set 15-1

1. The infinite set of numbers 5, 7, 9, 11, . . . is an **arithmetic sequence.** It progresses by adding 2 to one term to get the next term. What does the tenth term equal? How many 2s would you have to add to the first term, 5, to get the tenth term? How could you get the tenth term quickly? Find the 100th term quickly.
2. Enter the first ten terms of the sequence in Problem 1 into a list on your grapher, and enter the term numbers 1, 2, 3, 4, . . . , 10 into another list. Make a point plot of term value as a function of term number. Sketch the plot.
3. What type of continuous function contains all the points in the plot of Problem 2?
4. The infinite sum $5 + 7 + 9 + 11 + \cdots$ is an **arithmetic series.** Because the series has an infinite number of terms, you cannot add them all. But you can add *part* of the series. Find the tenth **partial sum** by adding the first ten terms.
5. Find the average of the first and tenth terms in the series of Problem 4. Multiply this number by 10. What do you notice about the answer? Use the pattern you observe to find the 100th partial sum of the series. Show how you found it.
6. Calculate the first ten partial sums of the series in Problem 4 and enter them into a third data list. Make a point plot of partial sum as a function of term number, using the term numbers in the data list of Problem 2. Sketch the result.
7. Run regressions to find out which type of continuous function exactly fits the partial sums in Problem 6. Write its particular equation. Use the result to find quickly the 100th partial sum of the series.
8. The infinite set of numbers 6, 12, 24, 48, . . . is a **geometric sequence.** How do the terms progress from one to the next? Find the tenth term of the sequence.
9. The infinite sum $6 + 12 + 24 + 48 + \cdots$ is a **geometric series.** Find the tenth partial sum of the series.
10. What did you learn as a result of doing this problem set that you did not know before?

15-2 Arithmetic, Geometric, and Other Sequences

(The Granger Collection, New York)

Suppose you have $40 that you are saving in a piggy bank to spend on a special project. You take on a part-time job that pays $13 per day. Each day you put this cash into the piggy bank. The number of dollars in the bank is a function of the number of days you have worked.

Days (term numbers, n)	1	2	3	4	5	. . .
Dollars (terms, t_n)	53	66	79	92	105	. . .

This is a **discrete function** rather than a **continuous function.** A discrete function is a series of disconnected points, whereas a continuous function has no gaps or discontinuities. After $3\frac{1}{2}$ days you still have the same $79 that you had after 3 days. A function like this, whose domain is a set of consecutive integers, is called a **sequence.** In this section you will look for patterns in sequences that allow you to calculate a term from its term number or to find the term number of a given term.

Objective

- Represent sequences explicitly and recursively.
- Find a term in a sequence given its term number.
- Find the term number of a given term in a sequence.

In this exploration you will use the first few terms of a sequence to discover a pattern and find more terms or term numbers.

EXPLORATION 15-2: Patterns in Sequences

1. An *arithmetic sequence* progresses by adding a constant to the preceding term to get the next term. Could these sequences be arithmetic sequences?

a. 5, 10, 20, . . .

b. 5, 10, 15, . . .

c. 5, 10, 40, . . .

2. The constant added to a term of an arithmetic sequence to get the next term is called the *common difference.* What is the common difference for the arithmetic sequence in Problem 1? Why do you think it is called a "difference"?

3. A *geometric sequence* progresses by multiplying the preceding term by a constant to get the next term. Could these sequences be geometric sequences?

a. 5, 10, 20, . . .

b. 5, 10, 15, . . .

c. 5, 10, 40, . . .

4. The constant that multiplies a term of a geometric sequence to get the next term is called the *common ratio.* What is the common ratio for the geometric sequence in Problem 3? Why do you think it is called a "ratio"?

continued

EXPLORATION, continued

Problems 5–9 pertain to this sequence:

n:	1	2	3	4	5	6	7 . . .
t_n:	4,	10,	18,	28,	40,	54,	70, . . .

5. By discovering a pattern in the sequence, write the next two terms.

6. Figure out what operation(s) can be performed on the term numbers, n, to get the values of t_n. That is, figure out a formula for t_n as a function of n.

7. On graph paper, plot the graph of the sequence. Connect the points with a dashed line to show the pattern.

8. Use the formula from Problem 6 to calculate t_{53}.

9. The number 23,868 is a term in the sequence. Calculate its term number.

10. What did you learn as a result of doing this exploration that you did not know before?

EXAMPLE 1 ➤ For the sequence of dollars 53, 66, 79, 92, 105, . . . at the beginning of this section:

a. Sketch the graph of the first few terms of the sequence.

b. Find t_{100}, the 100th term of the sequence.

c. Write an equation for t_n, the nth term of the sequence, as a function of n.

SOLUTION

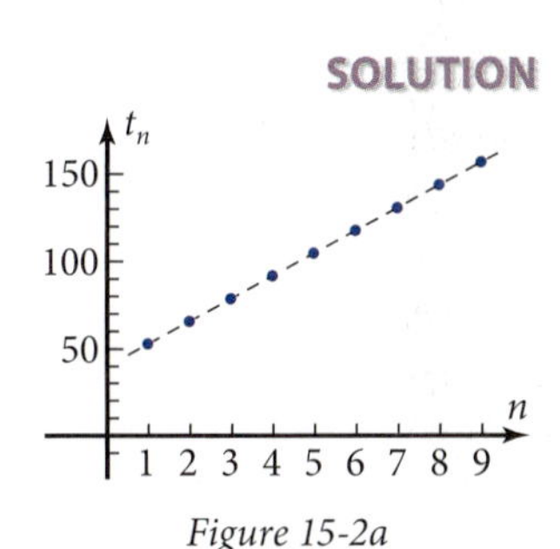

Figure 15-2a

a. The graph in Figure 15-2a shows discrete points. You can connect the points with a dashed line to show the pattern, but don't make it a solid line because sequences are defined only on the set of positive integers. There are no points between consecutive terms in a sequence.

b. To find a pattern, write the term number, n, in one column and the term, t_n, in another column. Then show the 13s being added to the preceding terms to get the next terms.

n	t_n	
1	53	+13
2	66	+13
3	79	+13
4	92	

To get the *fourth* term, you start with 53 and add 13 *three* times. So to get the *100th* term, you start with 53 and add 13 *ninety-nine* times.

$$t_{100} = 53 + 99(13) = 1340$$

c. $t_n = 53 + 13(n - 1)$ or $t_n = 40 + 13n$ ◀

The sequence in Example 1 is called an **arithmetic sequence.** You get each term by adding the same constant to the preceding term. You can also say that the difference of consecutive terms is a constant. This constant is called the **common difference.**

The pattern "add 13 to the previous term to get the next term" in Example 1 is a *recursive* pattern for the sequence—each term is calculated using the previous term. You can write an algebraic **recursion formula:**

$$t_n = t_{n-1} + 13$$

Of course, in this example it is necessary to specify the value of the first term, $t_1 = 53$. The sequence mode on your grapher makes it easy to calculate terms recursively. Here's how you would enter the equation into a typical grapher:

$n\text{Min} = 1$	Enter the beginning value of the term number, n.
$u(n) = u(n-1) + 13$	Enter the recursion formula; $u(n)$ stands for t_n.
$u(n\text{Min}) = \{53\}$	Enter the first term.

The table view gives the values shown at right.

The pattern $t_n = 53 + 13(n-1)$ you saw in Example 1, part c, is called an **explicit formula** for the sequence. It "explains" how to calculate any desired term without finding the terms before it.

n	*u(n)*
1	53
2	66
3	79
.	.
.	.
.	.

EXAMPLE 2 ➤ When you leave money in a savings account, the interest is *compounded*. This means that interest is paid on the previously earned interest as well as on the amount originally deposited. If the interest rate is 6% per year, compounded once a year, the amount at the beginning of any year is 1.06 times the amount at the beginning of the previous year. Suppose parents invest $1000 in an account on their baby's first birthday.

a. Find recursively the first four terms, t_1, t_2, t_3, and t_4, in the sequence of amounts.

b. Make a point plot of t_n as a function of n for the first 18 birthdays.

c. Calculate explicitly the value of t_{18}, the amount on the 18th birthday.

d. Write an explicit formula for the amount, t_n, as a function of the birthday number, n.

e. If the money is left in the account and the interest rate stays the same, when will the amount first exceed $11,000?

SOLUTION

a.

Birthday, *n*	Dollars, t_n	
1	1000.00	
		×1.06
2	1060.00	
		×1.06
3	1123.60	
		×1.06
4	1192.02	

Calculate *without* rounding; round the *answers*.

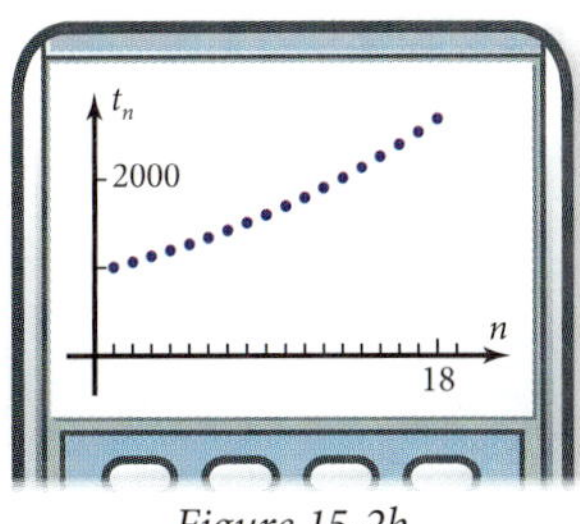

Figure 15-2b

b. With your grapher in sequence mode, enter the recursion formula. Figure 15-2b shows the point plot.

$$n\text{Min} = 1$$
$$u(n) = u(n-1) * 1.06$$
$$u(n\text{Min}) = \{1000\}$$

c. For the *fourth* birthday, you multiply 1000 by *three* factors of 1.06. So for the 18th birthday you multiply 1000 by 17 factors of 1.06:

$$t_{18} = 1000 \cdot 1.06^{17} = 2692.7727... \approx \$2692.77$$

d. $t_n = 1000 \cdot 1.06^{n-1}$ Multiply 1000 by $(n-1)$ factors of 1.06.

e. Algebraic solution:

$$1000 \cdot 1.06^{n-1} > 11{,}000$$
$$1.06^{n-1} > 11$$
$$(n-1)\log 1.06 > \log 11$$
$$n - 1 > \frac{\log 11}{\log 1.06} = 41.1522... \Rightarrow n > 42.1522...$$

The amount would first exceed \$11,000 on the 43rd birthday.

43 because n is greater than 42.1522....

Numerical solution: With your grapher in sequence mode, make a table and scroll down to $n = 43$, the first year in which t_n exceeds 11,000. ◄

The sequence in Example 2 is called a **geometric sequence.** You get each term by multiplying the previous term by the same constant. You can also say that the ratio of consecutive terms is a constant. This constant is called the **common ratio.** Notice that this pattern is the same as the add-multiply property of exponential functions, which you learned about in Chapter 2.

EXAMPLE 3 ➤ Consider the sequence 6, 12, 20, 30, 42, 56, 72, ...

a. Write a recursion formula for t_n as a function of t_{n-1}. Use it to find the next few terms.

b. Write an explicit formula for t_n as a function of n. Use it to find t_{100}.

SOLUTION

a. Make a table showing the term number and the corresponding term value.

The terms progress by adding amounts that increase by 2 each time. For instance, to get term 5, you add 12, which is 2 times the quantity $(n + 1)$. So

$$t_n = t_{n-1} + 2(n+1)$$

n	t_n		
1	6		2×3
		+6	
2	12		3×4
		+8	
3	20		4×5
		+10	
4	30		5×6
		+12	
5	42		6×7
		+14	
6	56		7×8
		+16	
7	72		8×9

Enter this recursion formula into your grapher with nMin = 1 and $t_1 = u(n\text{Min}) = 6$. Scroll down the table to find the next few values of t_n. Note that the value 72 for t_7 confirms that your formula is correct.

n	t_n
7	72
8	90
9	110
10	132
.	.
.	.
.	.

b. The terms are also the products of consecutive integers, term number plus 1 and term number plus 2, as shown in the table in part a. So an explicit formula is

$$t_n = (n + 1)(n + 2)$$

$$t_{100} = (101)(102) = 10{,}302$$

Example 3 illustrates that the recursion formula is useful for finding the next few terms but the explicit formula for t_n in terms of n lets you find terms farther along in the sequence without having to find all the intermediate terms.

The following definitions pertain to sequences.

DEFINITION: Sequences

A **sequence** is a function whose domain is a set of integers. The independent variable is the term number, n, and the dependent variable is the term value, t_n.

A **recursion formula** for a sequence specifies the initial term and defines t_n as a function of one or more preceding terms, such as t_{n-1}.

An **explicit formula** for a sequence specifies t_n as a function of n.

Notes:

- Although a sequence usually has an infinite number of terms, it could have a finite number if the domain is restricted.
- A recursion formula gives an easy way to find the next few terms in a sequence.
- An explicit formula is useful for calculating terms later in the sequence or for calculating the term number of a given term value.

DEFINITIONS: Arithmetic and Geometric Sequences

An **arithmetic sequence** is a sequence in which each term is formed recursively by *adding* a constant to the previous term. The constant added is called the **common difference.**

A **geometric sequence** is a sequence in which each term is formed recursively by *multiplying* the previous term by a constant. The constant multiplier is called the **common ratio.**

Notes:

- An arithmetic sequence is a *linear function* of the term number.
- A geometric sequence is an *exponential function* of the term number.

There are techniques for finding a specified term and for finding the term number of a given term value.

TECHNIQUES: Terms, Term Numbers, and Graphs of Sequences

To find more terms in a sequence, make a table of term numbers and terms and then:

- Find a recursive pattern and follow the pattern to the desired term, or
- Write an explicit formula for t_n in terms of n and substitute a value for n.

To find the value of n for a given term:

- Follow the recursive pattern until you reach the given term, or
- Substitute the given term value into the explicit formula and solve for n.

To plot the graph of a sequence:

- Make a table of n and t_n on your grapher, then plot the points (n, t_n), or
- Set your grapher in sequence mode, enter the formula for t_n, and then graph.

Problem Set 15-2

Reading Analysis

From what you have read in this section, what do you consider to be the main idea? How is a sequence related to other types of functions you have studied in this course? What is the difference between a recursion formula and an explicit formula? An arithmetic sequence has the add–add property, and a geometric sequence has the add–multiply property. What other types of functions that you have studied have these properties?

Quick Review

Q1. What type of function has the add–multiply property?

Q2. What type of function has the multiply–add property?

Q3. What type of function has the add–add property?

Q4. What type of function has the multiply–multiply property?

Q5. If y is a direct-cube power function of x, then what does doubling x do to y?

Q6. What is an integer?

Q7. Solve: $x^2 + 7x + 6 = 0$

Q8. Add the vectors $3\vec{i} + 4\vec{j}$ and $2\vec{i} - 5\vec{j}$.

Q9. Add the complex numbers $3 + 4i$ and $2 - 5i$.

Q10. Write polar coordinates of the point $(-5, 30°)$ using a *positive* value of r.

For Problems 1–12,

a. State whether the sequence is arithmetic, geometric, or neither.

Arithmetic sequences: $\underbrace{\underbrace{t_1 + d}_{t_2} + d}_{t_3} + \cdots$ ...

Geometric sequences: $\underbrace{\underbrace{t_1 \cdot r}_{t_2} \cdot r}_{t_3} \cdot \cdots$ ...

b. Write the next two terms for the first ellipsis.

c. Find t_{100}.

d. Find the term number of the term after the first ellipsis.

1. 27, 36, 48, . . . , 849490.0219... , . . .

2. 27, 31, 35, . . . , 783, . . .

3. 58, 45, 32, . . . , −579, . . .

4. 100, 90, 81, . . . , 3.0903... , . . .

5. 54.8, 137, 342.5, . . . , 3266334.5336... , . . .

6. 67.3, 79, 90.7, . . . , 38490.1, . . .

7. 50, −45, 40.5, . . . , −15.6905... , . . .

8. −1234, −1215.7, −1197.4, . . . , 2426, . . .

9. 0, 3, 8, 15, 24, 35, 48, 63, 80, 99, . . . , 3248, . . .

10. 4, 10, 18, 28, 40, 54, 70, . . . , 178504, . . .

11. $x, 2x - a, 3x - 2a, \ldots, 240x - 239a, \ldots$

12. $5, 5\sqrt{2}, 10, \ldots, 20480, \ldots$

13. *Grains of Rice Problem:* A story is told that the person who invented chess centuries ago was to be rewarded by the king. The inventor gave the king a simple request: "Place one grain of rice on the first square of a chessboard, place two grains on the second, then four, eight, and so forth, till all 64 squares are filled." What type of sequence do the numbers of grains form? On which square would the number of grains first exceed 1000? How many grains would be on the last square? Why do you think the king was upset about having granted the inventor's request?

14. *George Washington's Will Problem:* Suppose you find that when George Washington died in 1799, he left $1000 in his will to your ancestors. The money has been in a savings account ever since, earning interest. The amounts 1 yr, 2 yr, and 3 yr after Washington died were $1050.00, $1102.50, and $1157.63, respectively. Show that these numbers form a geometric sequence (allowing for round-off, if necessary). When will (or did) the total in the account first exceed $1 million? How much would be in the account this year? Why do you think banks have rules limiting the number of years money can be left in a dormant account before they stop paying interest on it?

15. *Depreciation Problem:* The Internal Revenue Service (IRS) assumes that an item that can wear out, such as a house, car, or computer, depreciates by a constant number of dollars per year. (If the item is used in a business, the owner is allowed to subtract the amount of the depreciation from the business's income before figuring taxes.) Suppose that an office building is originally valued at $1,300,000.

a. If the building depreciates by $32,500 per year, write the first few terms of the sequence of values of the building after 1, 2, 3, . . . yr. What type of sequence do these numbers form? How much will the building be worth after 30 yr? How long will it be until the building is *fully depreciated?* Why does the IRS call this *straight-line depreciation?*

b. Suppose the IRS allows the business to take *accelerated depreciation,* each year deducting 10% of the building's value at the beginning of the year. Write the first few terms in the sequence of values in each year of the building's life. How much will the business get to deduct the first, second, and third years of the building's life? How old will the building be when the business can deduct less than $32,500, which is the amount using straight-line depreciation?

16. *Piggy Bank Problem:* Suppose you decide to save money by putting $5 into a piggy bank the first week, $7 the second week, $9 the third week, and so forth.

a. What type of sequence do the deposits form? How much will you deposit at the end of the tenth week? In what week will you deposit $99?

b. Find the total you would have in the bank at the end of the tenth week. Show that you can calculate this total by averaging the first and the tenth deposits and then multiplying this average by the number of weeks.

c. What is the total amount you would have in the bank at the end of a year? (Do the computation in a time-efficient way.)

17. *Laundry Problem:* An item of clothing loses a certain percentage of its color with each washing. Suppose a pair of blue jeans loses 9% of its color with each washing. What percentage remains after the first, second, and third washings? What type of sequence do these numbers form? What percentage of the original color would be left after 20 washings? How many washings would it take until only 10% of the original color remains?

18. *Ancestors Problem:* Your ancestors in the first, second, and third generations back are your biological parents, grandparents, and great-grandparents, respectively.

a. Write the number of ancestors you have (living or dead) in the first, second, and third generations back. What type of sequence do these numbers form? How many ancestors do you have in the 10th generation back? In the 20th generation back?

b. As the number of generations back gets larger, the calculated number of ancestors increases without limit and eventually will exceed the population of the world. What do you conclude must be true to explain this seeming contradiction?

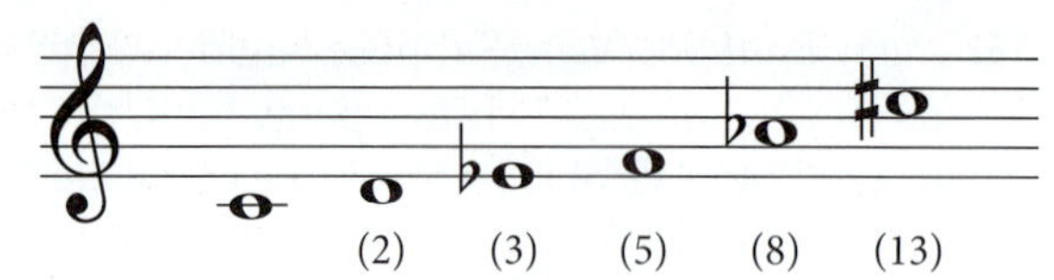

The successive tones in one of Béla Bartók's musical scales increase in a Fibonacci sequence of halftones.

19. WEB *Fibonacci Sequence Problem:* These numbers form the **Fibonacci sequence:**

1, 1, 2, 3, 5, 8, 13, 21, 34, 55, . . .

a. Figure out the recursion pattern followed by these *Fibonacci numbers*. Write the next two terms of the sequence. Enter the recursion formula into your grapher. You will need to enter $u(n\text{Min}) = \{1, 1\}$ to show that the first two terms are given. Make a table of Fibonacci numbers and scroll down to find the 20th term of the sequence.

b. Find the first ten ratios, r_n, of the Fibonacci numbers, where

$$r_n = \frac{t_{n+1}}{t_n}$$

Show that these ratios get closer and closer to the *golden ratio*,

$$r = \frac{\sqrt{5} + 1}{2} = 1.61803398\ldots$$

c. Find a pinecone, a pineapple, or a sunflower, or a picture of one of these. Each has sections formed by intersections of two spirals, one in one direction and another in the opposite direction. Count the number of spirals in each direction. What do you notice about these numbers?

d. Look up Leonardo Fibonacci (also known as Leonardo of Pisa) on the Internet or in another reference source. Find out when and where he lived. See if you can find out how he related the sequence to the growth of a population of rabbits and why, therefore, his name is attached to the sequence.

20. *Factorial Sequence Problem:* These numbers form the sequence of factorials:

1, 2, 6, 24, 120, 720, . . .

a. Figure out a recursive pattern in the sequence, and use it to write the next two factorials.

b. Recall from Chapter 14 that you use the exclamation mark, !, to designate a factorial. For example, 6! = 720. Write an explicit formula and use it to find 10! and 20!. What do you notice about the magnitude of the values? Think of a possible reason why the exclamation mark is used for factorials.

21. *Staircase Problem:* Debbie can take the steps of a staircase one at a time or two at a time. She wants to find out how many different ways she can go up staircases with different numbers of steps. She realizes that there is one way she can go up a staircase of 1 step and two ways she can go up a staircase of 2 steps (one step and one step, or both steps simultaneously).

a. Explain why the numbers of ways she can go up 3- and 4-step staircases are three and five, respectively.

b. If Debbie wants to get to the 14th step of a staircase, she can reach it by taking either one step from the 13th step or two steps from the 12th step. So the number of ways to get to step 14 is the number of ways to get to step 13 plus the number of ways to get to step 12. Let n be the number of steps in the staircase, and let t_n be the number of different ways Debbie can go up that staircase. Write a recursion formula for t_n as a function of t_{n-1} and t_{n-2}. Use the recursion formula to find the number of ways she could go up a 20-step staircase. On your grapher, you must enter $u(n\text{Min}) = \{2, 1\}$ to show that $t_2 = 2$ and $t_1 = 1$.

c. How does the number of ways of climbing stairs relate to the Fibonacci sequence in Problem 19?

d. In how many different ways could Debbie go up the 91 steps to the top of the pyramid in Chichén Itzá, Mexico? Do you find this surprising?

22. *Mortgage Payment Problem:* Suppose a family borrows $150,000 to purchase a house. It agrees to pay back this *mortgage* at $1074.65 per month. But part of that payment goes to pay the interest for the month on the balance remaining. The interest rate is 6% per year, so the family pays 0.5% per month. The balance b_n remaining after month n is given by the recursion formula

$$b_n = b_{n-1} + 0.005b_{n-1} - 1074.65$$

a. Explain the meaning of each of the three terms on the right side of the recursion formula.

b. Find b_{12}, the balance remaining at the end of the first year of the mortgage. How much money did the family pay for the year? How much of this amount went to pay interest, and how much went to reducing the balance of the mortgage?

c. After how many months will the balance drop to zero and the mortgage be paid off?

23. *Credit Card Problem:* A credit card company charges 18% interest per year (1.5% per month) on your unpaid balance. Suppose you have a balance of $3000 at month 1, and the minimum payment specified on the monthly bill is $100. The new balance at month 2 will be $3000 plus 1.5% interest, minus the $100 payment. These balances form the sequence

$$u(1) = 3000, u(2) = u(1) + 0.015u(1) - 100$$

If this pattern of payments continues, how long will it be until the balance drops below the $100 minimum payment? How much, total, will you have paid when the balance is paid off? Do you find the answers surprising?

24. *Arithmetic and Geometric Means Problem:* **Arithmetic means** and **geometric means** between two numbers are terms between the two numbers that form an arithmetic or geometric sequence with the two numbers.

a. Insert three arithmetic means between 47 and 84 so that the sequence 47, __?__, __?__, __?__, 84 is part of an arithmetic sequence.

b. Insert three geometric means between 3 and 48 so that the sequence 3, __?__, __?__, __?__, 48 is part of a geometric sequence.

c. There are two different sets of geometric means in part b, one involving only positive numbers and another involving both positive and negative numbers. Find the set of means you didn't find in part b.

25. *Office Building Problem:* Suppose you are responsible for estimating the cost of constructing a new office building. From previous records you find that the cost of constructing similar buildings was $1200/m^2 for the first floor and $1500/m^2 for the fifth floor.

a. Assuming that the numbers of dollars per square meter form an arithmetic sequence, find the costs per square meter for the second, third, and fourth floors. What mathematical name is given to these numbers?

b. The 48th floor is to have area 1000 m^2. Assuming that the arithmetic sequence continues, what is the total cost of constructing this floor?

26. WEB *Musical Scale Problem:* Figure 15-2c shows the keys on a piano around middle C. The lower A has frequency 220 Hz (Hertz, or cycles per second), and the higher A has a frequency twice as high. The 11 notes (counting black keys) in between have frequencies that form a geometric sequence between 220 and 440.

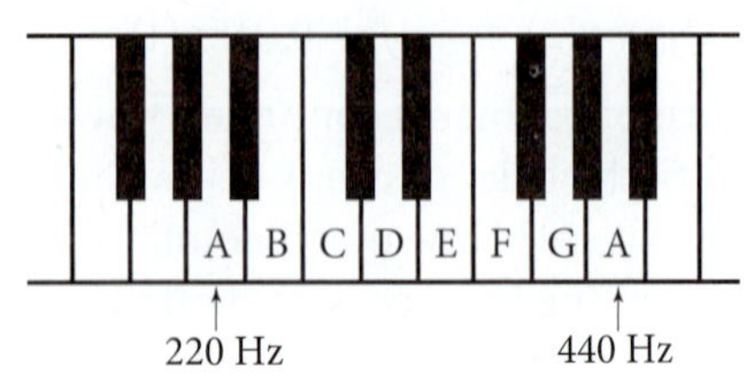

Figure 15-2c

a. Calculate the frequencies of the 11 notes between 220 Hz and 440 Hz. Check "equal temperament scale" on the Internet or in some other source to see if your answers are correct. Cite the source of your information. What mathematical name is given to these 11 numbers?

b. Calculate the frequencies of the highest and lowest notes on the piano keyboard, 51 notes above and 36 notes below the 220 Hz A, respectively.

27. *Logarithm Sequence Problem:* Given the sequence of natural logarithms

$$\ln 3, \ln 6, \ln 12, \ln 24, \ldots$$

a. Write the next three terms.

b. What kind of sequence do the arguments of the logarithms form?

c. Show that the logarithms themselves form an arithmetic sequence. What is the common difference? Use the properties of logarithms to express this difference as a logarithm.

28. *Formulas for Sequences Problem:* Let t_n be the nth term of a given sequence. Based on the patterns followed by arithmetic and geometric sequences, explain how the formulas in the box are derived.

PROPERTIES: Formulas for Arithmetic and Geometric Sequences

For an arithmetic sequence with first term t_1 and common difference d,

$$t_n = t_1 + (n - 1)d$$

For a geometric sequence with first term t_1 and common ratio r,

$$t_n = t_1 \cdot r^{n-1}$$

15-3 Series and Partial Sums

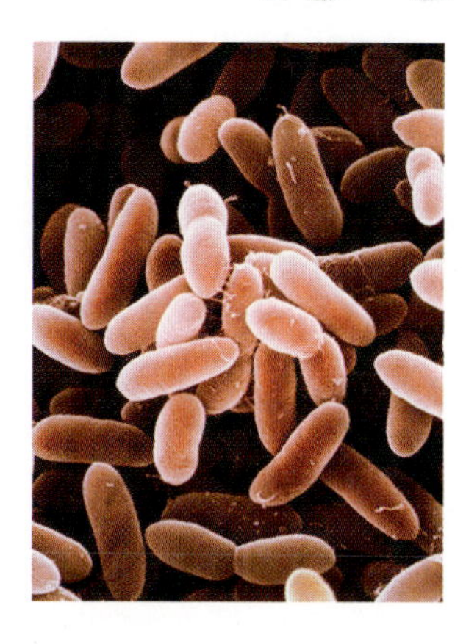

A population of bacteria grows by subdividing. Suppose the number of new bacteria in any one generation is a term in the sequence

$$5, 12, 21, 32, 45, \ldots$$

The total number of bacteria present at any number of generations n is the sum of the terms in the sequence,

$$5 + 12 + 21 + 32 + 45 + \cdots + t_n$$

The indicated sum of the terms of a sequence is called a series, as defined in Section 15-1. The total number of bacteria at the fifth generation, for instance, is the fifth partial sum of the series,

$$5 + 12 + 21 + 32 + 45 = 115$$

At the fifth generation there are 45 new bacteria, for a total of 115 bacteria.

In this section you will learn ways to calculate partial sums of series. You will also revisit **binomial series** that arise from raising a binomial to a power, such as $(a + b)^{10}$.

Objective

- Given a series, find a specified partial sum, or find the number of terms if the partial sum is given.
- Use sigma notation to write partial sums.
- Given a power of a binomial, expand it as a binomial series.

Numerical Computation of Partial Sums of Series: Sigma Notation

This example shows you how to calculate a partial sum of a series directly by summing the terms on your grapher.

EXAMPLE 1 ➤ For the series $5 + 12 + 21 + 32 + 45 + \cdots$, calculate S_{100}, the 100th partial sum.

SOLUTION The terms in the given series are products of integers, as shown in this table:

n	t_n	Pattern
1	5	5×1
2	12	6×2
3	21	7×3
4	32	8×4
⋮	⋮	⋮
n	$(n + 4)(n)$	

Write a program to compute partial sums or download the Series program provided at **www.keymath.com/precalc.** The program should use a sequence formula stored in the grapher to calculate the term values and should allow you to put in the desired number of terms. Then it should enter a loop that calculates the term values one at a time and accumulates them by adding each term to a variable such as S (for sum). At each iteration the program should have your grapher display the current partial sum. The final output should be the last partial sum. Your program should give

$$S_{100} = 358{,}550$$

A partial sum can be written compactly using sigma notation. The symbol Σ, the uppercase Greek letter sigma, is often used to indicate a sum.

$$S_{100} = \sum_{n=1}^{100} (n + 4)(n)$$

The expression on the right side of the equation is read "the sum from n equals 1 to 100 of $(n + 4)(n)$." It means to substitute $n = 1, 2, 3, \ldots, 100$ into the formula, perform the computations, and sum the results. The variable n is called the **term index.** You may recall sigma notation from your work in Chapter 3.

In this exploration you will find several partial sums of a series for which a formula for t_n is known.

EXPLORATION 15-3: Introduction to Series

1. Write out the indicated terms of this series:

$$\sum_{k=1}^{6} 3k + 5$$

2. Evaluate the partial sum in Problem 1 by summing the terms.

3. Use the Series program described in Example 1 to confirm the answer you got in Problem 2.

4. Use your Series program to evaluate S_{100} for the series in Problem 1. That is, find

$$\sum_{k=1}^{100} 3k + 5$$

5. Figure out a formula for t_n for this next series. Then use your Series program to find the fifth partial sum, S_5, for the series. Confirm that the program gives you the correct answer by actually summing the first five terms:

$$2 + 5 + 10 + 17 + 26 + \cdots$$

6. Use your program to calculate the 50th partial sum for the series in Problem 5.

7. Write out the first four terms of the geometric series with first term 1000 and common ratio 1.06. Calculate the fourth partial sum by summing these terms.

8. Write a formula for the nth term of the geometric series in Problem 7. Confirm that your formula is correct by using the Series program to find the fourth partial sum you calculated manually in Problem 7.

9. Use your program to find the 30th partial sum of the geometric series in Problem 7. (This number is the amount of money you would have at the end of 30 years if you invested $1000 a year in a savings account that pays an interest rate of 6% per year, compounded annually.)

10. Calculate the partial sums S_{10}, S_{20}, S_{50}, S_{100}, and S_{200}, of the geometric series with first term 800 and common ratio 0.9.

continued

11. The partial sums in Problem 10 *converge to 8000.* What do you think this means?

12. What did you learn as a result of doing this exploration that you did not know before?

Finding Partial Sums of Arithmetic Series Algebraically

Suppose you put \$7 in a piggy bank the first day, \$10 the second day, \$13 the third day, and so forth. The amounts you put in follow the arithmetic sequence

$$7, 10, 13, 16, 19, \ldots$$

The total in the piggy bank on any one day is a partial sum of the arithmetic series that results from summing the appropriate number of terms in the preceding sequence.

Day 1: 7

Day 2: $7 + 10 = 17$

Day 3: $7 + 10 + 13 = 30$

Day 4: $7 + 10 + 13 + 16 = 46$

$\vdots$

Suppose you want to find S_{10}, the tenth partial sum of the series:

$$S_{10} = 7 + 10 + 13 + 16 + 19 + 22 + 25 + 28 + 31 + 34 = 205$$

A pattern shows up if you add the first and last terms, the second and next-to-last terms, and so forth.

$$\begin{aligned} S_{10} &= (7 + 34) + (10 + 31) + (13 + 28) + (16 + 25) + (19 + 22) \\ &= 41 + 41 + 41 + 41 + 41 \\ &= 5(41) \\ &= 205 \end{aligned}$$

So a time-efficient way to find the partial sum algebraically is to add the first and last terms and then multiply the result by the number of pairs of terms:

$$S_{10} = \frac{10}{2}(7 + 34) = 5(41) = 205$$

By associating the 2 in the denominator with (7 + 34), you can see that

$$S_{10} = 10 \cdot \frac{7 + 34}{2} = 10 \cdot 20.5 = 205$$

So the nth partial sum is the same as the sum of n terms, each of which is equal to the *average* of the first and last terms. This fact allows you to see why the pattern works for an odd number of terms as well as for an even number.

EXAMPLE 2 ➤ Find algebraically the 100th partial sum of the arithmetic series $53 + 60 + 67 + \cdots$. Check the answer by calculating the partial sum numerically, as in Example 1.

SOLUTION

$t_{100} = 53 + 99(7) = 746$ — Add 99 common differences to 53 to get the 100th term.

$S_{100} = \frac{100}{2}(53 + 746) = 39{,}950$ — There are $\frac{100}{2}$ pairs, each equal to the sum of the first term and the last term.

Check: Enter $f_1(x) = 53 + (x - 1)(7)$. Then run your grapher program to get

$$S_{100} = 39{,}950$$

which agrees with the algebraic solution. ◄

Finding Partial Sums of Geometric Series Algebraically

If you make regular deposits (for example, monthly) into a savings account, the total you have in the account at any given time is a partial sum of a geometric series. It is possible to calculate such partial sums algebraically.

Consider this sixth partial sum of a geometric series:

$$S_6 = 7 + 21 + 63 + 189 + 567 + 1701$$

The first term is 7, and the common ratio is 3. If you multiply both sides of the equation by −3 and add the result to the original sum, you get

$$\begin{array}{rl} S_6 &= 7 + 21 + 63 + 189 + 567 + 1701 \\ -3S_6 &= \quad -21 - 63 - 189 - 567 - 1701 - 5103 \\ \hline S_6 - 3S_6 &= 7 + 0 + 0 + 0 + 0 + 0 - 5103 \end{array}$$

Middle terms "telescope."

The middle terms *telescope*, or cancel out, leaving only the first term of the top equation and the last term of the bottom equation. So

$S_6 - 3S_6 = 7 - 7 \cdot 3^6$ — 5103 is $7 \cdot 3^6$.

$S_6(1 - 3) = 7(1 - 3^6)$ — Factor out S_6 on the left and 7 on the right.

$S_6 = 7 \cdot \frac{1 - 3^6}{1 - 3}$ — Divide both sides by $(1 - 3)$.

In this form of the equation, 7 is the first term, t_1. The 3 in the numerator and denominator is the common ratio, r, and the exponent 6 is the number, n, of terms to be added, demonstrating that, in general,

$$S_n = t_1 \cdot \frac{1 - r^n}{1 - r}$$

Verbally, you can remember this result by saying "First term times a fraction. The fraction is 1 minus r^n, divided by 1 minus r."

EXAMPLE 3 ➤ If you deposit $100 a month into an account that pays 6% interest per year, compounded monthly, then each deposit earns 0.5% interest per month. For instance, after four deposits, the first deposit has earned three months' interest, the second has earned two months' interest, the third has earned one month's interest, and the last has earned no interest. Thus, the total is

$$S_4 = 100 + 100(1.005) + 100(1.005)^2 + 100(1.005)^3$$

a. How much will be in the account after ten years (120 deposits)? How much of this is interest?

b. How long will it take until the total in the account first exceeds $50,000?

SOLUTION

a. Algebraically: The series is geometric, with first term 100 and common ratio 1.005.

$$S_{120} = 100 \cdot \frac{1 - 1.005^{120}}{1 - 1.005} = 16{,}387.9346... \approx \$16{,}387.93$$

Numerically: Enter $y = 100(1.005)^{x-1}$ into your grapher and run the Series program.

$$S_{120} = 16{,}387.9346... \approx \$16{,}387.93$$

The amount of interest is $16,387.93 minus $12,000, or $4,387.93.

b. Algebraically:

$$100 \cdot \frac{1 - 1.005^n}{1 - 1.005} > 50{,}000 \quad \text{Make } S_n \text{ greater than 50,000.}$$

$$1 - 1.005^n < -2.5$$

$$1.005^n > 3.5$$

$$n \log 1.005 > \log 3.5$$

$$n > \frac{\log 3.5}{\log 1.005} = 251.1784\ldots$$

It will take 252 months for the amount in the account to exceed $50,000.

Numerically: Store $100(1.005)^{x-1}$ in your grapher and run the Series program until the amount first exceeds 50,000. This will be $50,287.41 at 252 months. ◄

Convergent and Divergent Geometric Series

Assume that a person who is 200 cm from another person is allowed to take steps each of which is half the remaining distance. So the steps will be of lengths

$$100, 50, 25, 12.5, 6.25, \ldots$$

The total distance traveled will be given by the geometric series

$$100 + 50 + 25 + 12.5 + 6.25 + \ldots$$

The person in motion will never go the entire 200 cm, but the partial sums of the series **converge** to 200 as a limit.

$$S_5 = 193.75$$

$$S_{10} = 199.8046...$$

$$S_{20} = 199.999809...$$

$$S_{30} = 199.9999998...$$

The algebraic formula for S_n shows you why this happens:

$$S_n = 100 \cdot \frac{1 - 0.5^n}{1 - 0.5}$$

The value 0.5^n approaches zero as n becomes large. So you can write

$$\lim_{n \to \infty} S_n = 100 \cdot \frac{1 - 0}{1 - 0.5} = 100 \cdot 2 = 200$$

The symbol in front of S_n is read "the limit as n approaches infinity." A geometric series will converge to a limit if and only if the common ratio r satisfies the inequality $|r| < 1$. If $|r| \geq 1$ and $t_1 \neq 0$, then the terms of the series do not go to zero, and the series **diverges.** The partial sums do not approach a limit.

EXAMPLE 4 To what limit does the geometric series $50 + 45 + 40.5 + \ldots$ converge? How many terms must be added in order for the partial sums to be within 1 unit of this limit?

SOLUTION

$$r = \frac{45}{50} = 0.9$$

$$\therefore S_n = 50 \cdot \frac{1 - 0.9^n}{1 - 0.9} = 500(1 - 0.9^n)$$ $50/(1 - 0.9) = 50/0.1 = 500$

$$\lim_{n \to \infty} S_n = 500(1 - 0) = 500$$ 0.9^n approaches zero as n approaches infinity.

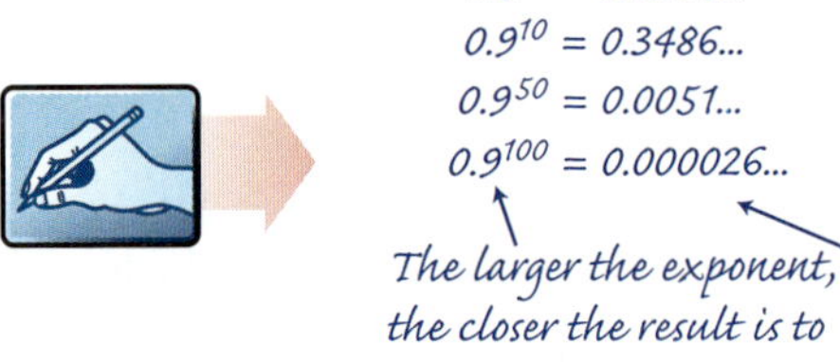

$$500(1 - 0.9^n) > 499$$

Make S_n greater than $(500 - 1)$.

$$1 - 0.9^n > 0.998$$

$$0.9^n < 0.002$$

$$n \log 0.9 < \log 0.002$$

Take the log of both sides.

$$n > \frac{\log 0.002}{\log 0.9} = 58.9842...$$

Solve. Reverse the inequality because $\log 0.9$ is a negative number.

To be within 1 unit of the limit, 59 terms must be added.

Round to the next higher term.

Binomial Series

If you expand a power of a binomial expression, you get a series with a finite number of terms. For instance,

$$(a + b)^5 = a^5 + 5a^4b + 10a^3b^2 + 10a^2b^3 + 5ab^4 + b^5$$

Such a series is called a **binomial series** or **binomial expansion.** You can see several patterns in the binomial series that result from expanding $(a + b)^n$.

- There are $(n + 1)$ terms.
- Each term has degree n because the exponents sum to n.
- The powers of a start at a^n and decrease by 1 with each term. The powers of b start at b^0 and increase by 1 with each term.
- The coefficients are symmetrical with respect to the ends of the series.
- The coefficients form a row of *Pascal's triangle*:

$(a + b)^0$	1
$(a + b)^1$	1 1
$(a + b)^2$	1 2 1
$(a + b)^3$	1 3 3 1
$(a + b)^4$	1 4 6 4 1
$(a + b)^5$	1 5 10 10 5 1
⋮	⋮

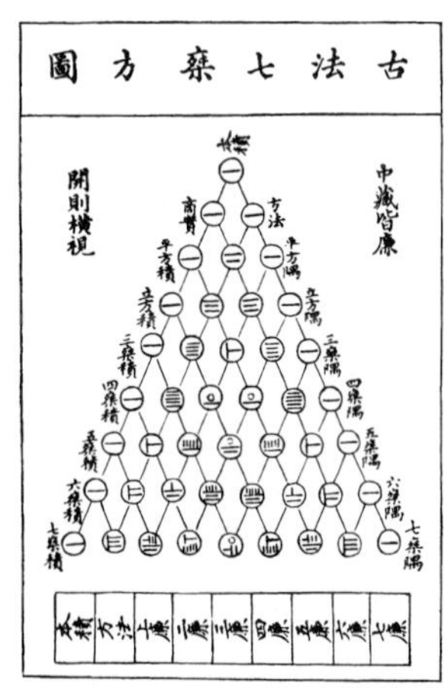

Jia Xian, a Chinese mathematician, discovered the relationship between the numbers in consecutive rows of this triangle approximately 500 years earlier than Pascal did.

Each number in the interior of the triangle is the sum of the numbers to its left and right in the previous row. The first and last numbers in each row are 1.

- The coefficients can be calculated using parts of the preceding term.

$$\frac{\text{(coefficient of term)(exponent of } a\text{)}}{\text{term number}} = \text{coefficient of next term}$$

For example, $5a^4b$ is the *second* term in the binomial series given earlier. To find the coefficient of the third term, write

$$\frac{(5)(4)}{2} = 10$$

- The coefficients can also be calculated *algebraically*. They are equal to numbers of combinations of n objects taken r at a time, where r is the exponent of b. From Section 14-5 you may recall, for example, that

$$_5C_2 = \frac{5!}{3!\,2!} = \frac{5 \cdot 4 \cdot 3 \cdot 2 \cdot 1}{(3 \cdot 2 \cdot 1)(2 \cdot 1)} = 10$$

So the term containing b^2 is

$$\frac{5!}{3!\,2!}a^3b^2$$

The factorials in the denominator are the same as the exponents of a and b. The factorial in the numerator is the sum of these exponents, which is equal to the original exponent of the binomial. Because they appear as coefficients of the terms in a binomial series, expressions such as $\frac{5!}{3!\,2!}$ are sometimes called **binomial coefficients.** In general, the coefficient of the term containing b^r is $\frac{n!}{(n-r)!\,r!}$. Using combination notation, you can write this expression as $_nC_r$. Another common way to write the same expression is $\binom{n}{r}$, which is read "n choose r." Recall that 0! is defined to equal 1. So $\binom{n}{0} = 1$ and $\binom{n}{n} = 1$. Using this notation, you can write the **binomial formula** for finding the terms of a binomial series (or "expanding a binomial").

Binomial Formula (or Binomial Theorem)

For any postive integer n and numbers a and b,

$$(a+b)^n = a^n + \binom{n}{1}a^{n-1}b + \binom{n}{2}a^{n-2}b^2 + \ldots + \binom{n}{n-2}a^2b^{n-2} + \binom{n}{n-1}ab^{n-1} + b^n$$

You can also write the binomial formula with sigma notation:

$$(a+b)^n = \sum_{r=0}^{n}\binom{n}{r}a^{n-r}\,b^r$$

EXAMPLE 5 ➤ Use the binomial formula to expand the binomial $(x - 2y)^4$.

SOLUTION

$$\begin{aligned}(x-2y)^4 &= x^4 + \binom{4}{1}x^3(-2y) + \binom{4}{2}x^2(-2y)^2 + \binom{4}{3}x(-2y)^3 + (-2y)^4\\ &= x^4 + 4x^3(-2y) + 6x^2(-2y)^2 + 4x(-2y)^3 + (-2y)^4\\ &= x^4 - 8x^3y + 24x^2y^2 - 32xy^3 + 16y^4\end{aligned}$$

EXAMPLE 6 ➤ Find the eighth term of the binomial expansion of $(3 - 2x)^{12}$.

SOLUTION For the eighth term, $r = 7$ and $n = 12$. Therefore,

$$\text{8th term} = \frac{12!}{5!\,7!}(3)^5(-2x)^7 = 792(243)(-128x^7) = -24{,}634{,}368x^7$$

This box summarizes the properties of series introduced in this section.

PROPERTIES: Formulas for Arithmetic, Geometric, and Binomial Series

Arithmetic Series

The terms of an arithmetic sequence progress by *adding* a common difference d.

$$t_n = t_1 + (n - 1)d$$ Add $(n - 1)$ common differences to the first term.

The nth partial sum of the arithmetic series is

$$S_n = \frac{n}{2}(t_1 + t_n)$$ Add $\frac{n}{2}$ terms, each equal to the sum of the first and last terms.

Geometric Series

The terms of a geometric sequence progress by *multiplying* by a common ratio r.

$$t_n = t_1 \cdot r^{n-1}$$ Multiply the first term by $(n - 1)$ common ratios.

The nth partial sum of the geometric series is

$$S_n = t_1 \cdot \frac{1 - r^n}{1 - r}$$ Multiply the first term by a fraction.

The sum of an infinite geometric series with $|r| < 1$ is

$$\lim_{n \to \infty} S_n = t_1 \cdot \frac{1}{1 - r}$$

Binomial Series

The terms of a binomial series come from expanding a binomial $(a + b)^n$:

$$\text{Term with } b^r = \frac{n!}{(n - r)!\, r!} a^{n-r} b^r = \binom{n}{r} a^{n-r} b^r$$

The sum of a binomial series, or the binomial formula, is

$$(a + b)^n = \sum_{r=0}^{n} \binom{n}{r} a^{n-r} b^r$$

Problem Set 15-3

Reading Analysis

From what you have read in this section, what do you consider to be the main idea? What is the major difference between a sequence and a series? What is a partial sum of a series? What does it mean to say that a geometric series *converges*? What is meant by a binomial series?

Quick Review

Q1. Write the next two terms of this arithmetic sequence: 10, 20, . . .

Q2. Write the next two terms of this geometric sequence: 10, 20, . . .

Q3. Write the next two terms of this harmonic sequence: $\frac{1}{3}, \frac{1}{4}, \frac{1}{5}, \frac{1}{6}, \ldots$

Q4. Write the next two terms of this factorial sequence: 1, 2, 6, . . .

Q5. Find the 101st term of the arithmetic sequence with first term 20 and common difference 3.

Q6. Find t_{101} for the geometric sequence with $t_1 = 20$ and $r = 1.1$.

Q7. Find the product and simplify: $(2 + 3i)(2 - 3i)$.

Q8. Find the dot product of these vectors: $(2\vec{i} + 3\vec{j}) \cdot (2\vec{i} - 3\vec{j})$

Q9. Which conic section is the graph of the equation $x^2 - y^2 + 3x - 5y = 100$?

Q10. Find the product of these matrices:

$$\begin{bmatrix} 3 & 2 \\ 5 & 1 \end{bmatrix}\begin{bmatrix} 4 & 3 \\ 6 & 2 \end{bmatrix}$$

---

1. *Arithmetic Series Problem:* A series has a partial sum

$$S_{10} = \sum_{n=1}^{10} [3 + (n - 1)(5)]$$

a. Write out the terms of the partial sum. How can you tell that the series is arithmetic?

b. Evaluate S_{10} three ways: numerically, by summing the 10 terms; algebraically, by averaging the first and last terms and multiplying by the number of terms; and numerically, by entering the formula for t_n and using your grapher program. Are the answers the same?

c. Evaluate S_{100} for this series. Which method did you use?

2. *Geometric Series Problem:* A series has a partial sum

$$S_6 = \sum_{n=1}^{6} 5 \cdot 3^{n-1}$$

a. Write out the terms of the partial sum. How can you tell that the series is geometric?

b. Evaluate S_6 three ways: numerically, by summing the six terms; algebraically, by using the pattern (first term) times (fraction involving r); and numerically, by entering the formula for t_n and using your grapher program. Are the answers the same?

c. Evaluate S_{20} for this series. Which method did you use?

3. *Convergent Geometric Series Pile Driver Problem:* A pile driver pounds a piling (a column) into the ground for a new building that is being constructed (Figure 15-3a). Suppose that on the first impact the piling is driven 100 cm into the ground. On the second impact the piling is driven another 80 cm into the ground. Assume that the distances the piling is driven with each impact form a geometric sequence.

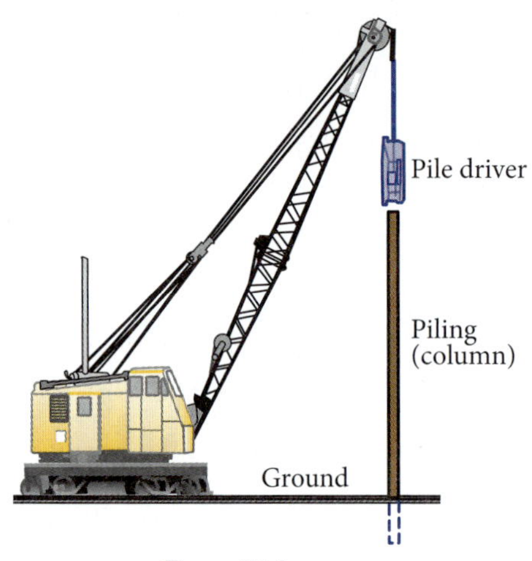

Figure 15-3a

a. How far will the piling be driven on the tenth impact? How deep in the ground will it be after ten impacts?

b. Run the Series program using $n = 100$. What do you notice about the partial sums as the grapher displays each sum? What does it mean to say that the partial sums are "converging to 500"? What is the real-world meaning of this limit to which the series converges?

c. Show how to calculate algebraically that the limit to which the partial sums converge is 500.

4. *Harmonic Series Divergence Problem:* If you stack a deck of cards so that they just barely balance, the top card overhangs by $\frac{1}{2}$ the deck length, the second card overhangs by $\frac{1}{4}$ the deck length, the third card overhangs by $\frac{1}{6}$ the deck length, and so on (Figure 15-3b).

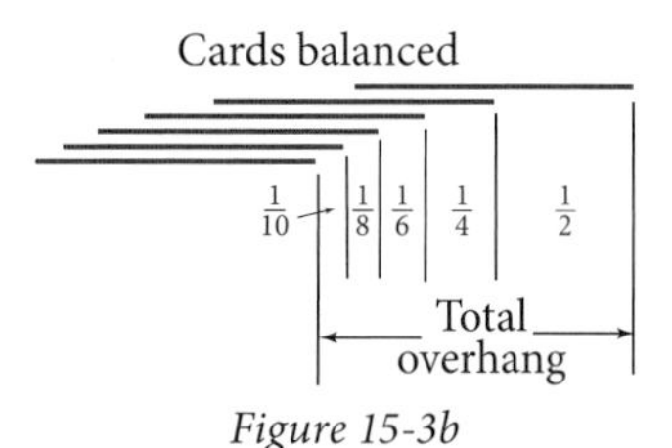

Figure 15-3b

The total overhang for n cards is thus a partial sum of this **harmonic series:**

$$\frac{1}{2}+\frac{1}{4}+\frac{1}{6}+\frac{1}{8}+\frac{1}{10}+\frac{1}{12}+\cdots$$

a. The figure indicates that the total overhang for four cards is greater than the length of the deck. Show numerically that this is true.

b. How many cards would you have to stack in order for the total overhang to exceed two deck lengths?

c. What would the total overhang be for a standard 52-card deck hanging over the edge of a table? Is this surprising?

d. Factor out $\frac{1}{2}$ from the terms, then group the terms this way:

$$\frac{1}{2}\left[1+\frac{1}{2}+\left(\frac{1}{3}+\frac{1}{4}\right)+\left(\frac{1}{5}+\frac{1}{6}+\frac{1}{7}+\frac{1}{8}\right)\right.$$
$$\left.+\text{ (next 8 terms)} + \text{(next 16 terms)} + \cdots\right]$$

Show that each group of terms inside the brackets is greater than or equal to $\frac{1}{2}$. How does this fact allow you to conclude that the partial sums of a harmonic series *diverge* and can get larger than any real number?

5. *Geometric Series for Compound Interest Problem:* Suppose you invest \$100 in an investment account that pays 12% interest per year, compounded monthly. The amount during any one month is 1.01 times the amount the month before. The amounts follow the geometric sequence shown in the table.

Month	Dollars
1	100
2	$100(1.01)$
3	$100(1.01)^2$
4	$100(1.01)^3$
5	$100(1.01)^4$
6	$100(1.01)^5$
...	...

If you make regular \$100 deposits each month, each deposit follows the geometric sequence shown in the table below.

a. Explain why the amount in the account at the fifth month is the fifth *partial sum* of a geometric series. Calculate this amount using a time-efficient method.

b. If you continue the regular \$100 monthly deposits, how much will be in the account at the end of ten years? How much of this will be interest?

c. How many months would it take until the total first exceeds \$100,000?

Month	1	2	3	4	5	6	...
Dollars	100	$100(1.01)$	$100(1.01)^2$	$100(1.01)^3$	$100(1.01)^4$	$100(1.01)^5$	...
		100	$100(1.01)$	$100(1.01)^2$	$100(1.01)^3$	$100(1.01)^4$	...
			100	$100(1.01)$	$100(1.01)^2$	$100(1.01)^3$	...
				100	$100(1.01)$	$100(1.01)^2$	...
					100	$100(1.01)$	...
						100	...

6. *Present Value Compound Interest Problem:* Suppose money is invested in a savings account at 3% annual interest, compounded monthly. Because the interest rate is 0.25% per month, the amounts in the account each month form a geometric sequence with common ratio 1.0025.

   a. Find the amount you would have to invest now to have $10,000 at the end of ten years. This amount is called the *present value* of $10,000.

   b. If you invest x dollars a month in this account, the total at the end of each month is a partial sum of a geometric series with x as the first term and common ratio 1.0025. How much would you have to invest each month in order to have $10,000 at the end of ten years?

7. *Geometric Series Mortgage Problem:* Suppose someone takes out a $200,000 mortgage (loan) to buy a house. The interest rate, I, is 0.5% or 0.005 per month (6% per year), and the payments, P, are $1050.00 per month. Most of the monthly payment goes to pay the interest for that month, with the rest going to pay on the *principal,* thus reducing the *balance,* B, owed on the loan. The table shows payment, interest, principal, and balance for the first few months.

Month, n	Payment, P	Interest	Principal	Balance, B_n
0				200,000.00
1	1050.00	1000.00	50.00	199,950.00
2	1050.00	999.75	50.25	199,899.75
3	1050.00	999.50	50.50	199,849.25

   a. Show that you understand how the table is constructed by calculating the row entries for month 4.

   b. The balance, B_1, after one month is given by the equation

   $$B_1 = B_0 + B_0I - P = B_0(1 + I) - P$$

   Show that the balance after four months can be written

   $$B_4 = B_0(1 + I)^4 - P(1 + I)^3 - P(1 + I)^2 - P(1 + I) - P$$

   Then use the formula for the partial sum of a geometric series to show that

   $$B_4 = B_0(1 + I)^4 - P\frac{1 - (1 + I)^4}{1 - (1 + I)}$$
   $$= B_0(1 + I)^4 + \frac{P}{I}[1 - (1 + I)^4]$$

   c. The formula in part b can be generalized to find B_n by replacing the 4's with n's. Use this information to calculate the number of months it takes to pay off the mortgage; that is, find the value of n for which $B_n = 0$.

   d. Plot the graph of B_n as a function of n from $n = 0$ to the time the mortgage is paid off. Sketch a smooth curve showing the pattern followed by the points. True or false: "Halfway through the duration of the mortgage, half of the mortgage has been paid off." Explain your reasoning.

8. *Geometric Series by Long Division Problem:* The limit, S, of the partial sums of a convergent geometric series is given by

   $$S = t_1 \cdot \frac{1}{1 - r}$$

   where t_1 is the first term of the sequence and r is the common ratio.

   a. Use long division to divide $(1 - r)$ into 1. Show that the result is the geometric series

   $$S = t_1 + t_1r + t_1r^2 + t_1r^3 + t_1r^4 + t_1r^5 + \cdots$$

   b. The result illustrates the way you can do mathematical problems "backward" as well as "forward." Give another instance in which you can use this forward-and-backward feature.

9. *Thumbtack Binomial Series Problem:* If you flip a thumbtack five times, there are six possible numbers of "point-ups" you could get: 0, 1, 2, 3, 4, and 5. If the probability of "point-up" on any one flip is 60% (0.6), then the probabilities of each of the six outcomes are terms in the binomial series you get from expanding

   $$(0.4 + 0.6)^5$$

   The probability of exactly three "point-ups" is the term that contains 0.6^3.

   a. Calculate the six terms of the binomial series.

b. What is the probability of exactly three "point-ups"?

c. Explain why the probability of no more than three "point-ups" is a partial sum of this binomial series. Calculate this probability.

d. Calculate the probability of no more than six "point-ups" in ten flips of the thumbtack. Is the answer the same as the probability of no more than three "point-ups" in five flips?

10. *Snowflake Curve Series Problem:* Figure 15-3c shows Koch's snowflake curve, which you may have encountered in Section 13-5.

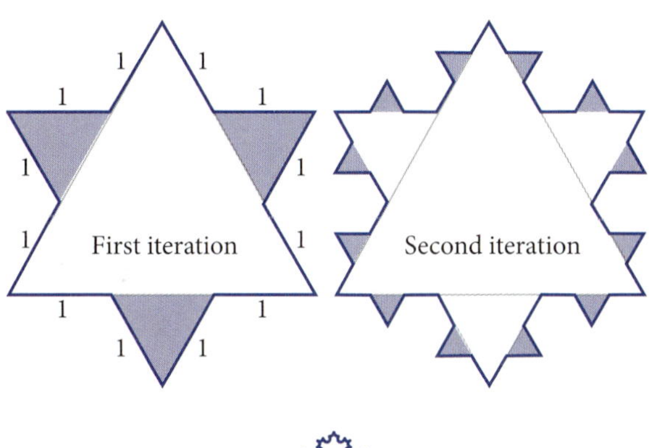

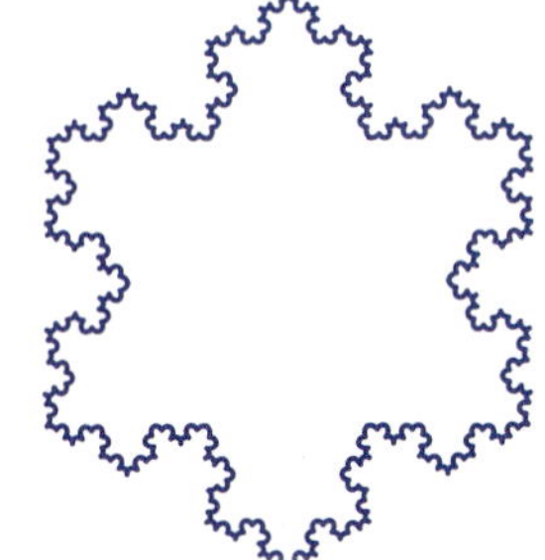

Figure 15-3c

In the first iteration, segments 1 unit long are marked on the sides of an equilateral triangle with sides 3 units, and three equilateral triangles (shaded) are drawn. In the second iteration, equilateral triangles with sides $\frac{1}{3}$ unit are constructed on each side of the figure from the first iteration. The iterations are carried on this way infinitely. The snowflake curve is the boundary of the resulting figure.

a. Find the total perimeter of the first iteration. Find the total perimeter of the second iteration. What type of sequence do the lengths of the iterations form? How do you conclude that the perimeter of the completed snowflake curve is *infinite*?

b. What is the total area of the shaded triangles in the first iteration? What area is added to this by the shaded triangles in the second iteration? What type of series do the areas of the iterations form? Does this series converge? If so, to what number? If not, show why not.

For Problems 11–14, write out the terms of the partial sum and add them.

11. $S_5 = \sum_{n=1}^{5} 2n + 7$ **12.** $S_7 = \sum_{n=1}^{7} n^2$

13. $S_6 = \sum_{n=1}^{6} 3^n$ **14.** $S_6 = \sum_{n=1}^{6} n!$

For Problems 15–22, each series is either geometric or arithmetic. Find the indicated partial sum algebraically.

15. For $2 + 10 + 50 + \cdots$, find S_{11}.

16. For $97 + 131 + 165 + \cdots$, find S_{37}.

17. For $24 + 31.6 + 39.2 + \cdots$, find S_{54}.

18. For $36 + 54 + 81 + \cdots$, find S_{29}.

19. For $1000 + 960 + 920 + \cdots$, find S_{78}.

20. For $1000 + 900 + 810 + \cdots$, find S_{22}.

21. For $50 - 150 + 450 - \cdots$, find S_{10}.

22. For $32.5 - 52 + 83.2 - \cdots$, find S_{41}.

For Problems 23–28, the series is either arithmetic or geometric. Find *n* for the given partial sum.

23. For $32 + 43 + 54 + \cdots$, find n if $S_n = 4407$.

24. For $13 + 26 + 52 + \cdots$, find n if $S_n = 425{,}971$.

25. For $18 + 30 + 50 + \cdots$, find n if $S_n \approx 443{,}061$.

26. For $97 + 101 + 105 + \cdots$, find n if $S_n = 21{,}663$.

27. For $97 + 91 + 85 + \cdots$, find n if $S_n = 217$. (Is this surprising?)

28. For $60 + 54 + 48.6 + \cdots$, find n if $S_n \approx 462.74$.

For Problems 29–36, state whether the geometric series converges. If it does converge, find the limit to which it converges.

29. $100 + 90 + 81 + \cdots$

30. $25 + 20 + 16 + \cdots$

31. $40 + 50 + 62.5 + \cdots$

32. $200 - 140 + 98 - \cdots$

33. $300 + 90 + 27 + \cdots$

34. $20 + 60 + 180 + \cdots$

35. $1000 - 950 + 902.5 - \cdots$

36. $360 + 240 + 160 + \cdots$

For Problems 37–42, expand as a binomial series and simplify.

37. $(x - y)^3$

38. $(4m - 5n)^2$

39. $(2x - 3)^5$

40. $(3a + 2)^4$

41. $(x^2 + y^3)^6$

42. $(a^3 - b^2)^5$

For Problems 43–52, find the indicated term in the binomial series.

43. $(x + y)^8$, y^5-term

44. $(p + j)^{11}$, j^4-term

45. $(p - j)^{15}$, j^{11}-term

46. $(c - d)^{19}$, d^{15}-term

47. $(x^3 - y^2)^{13}$, x^{18}-term

48. $(x^3 - y^2)^{24}$, x^{30}-term

49. $(3x + 2y)^8$, y^5-term

50. $(3x + 2y)^7$, y^4-term

51. $(r - q)^{15}$, 12th term

52. $(a - b)^{17}$, 8th term

53. *Dynamic Geometric Series Problem:* Go to **www.keymath.com/precalc** and open the *Geometric Series* exploration. Explain in writing how this exploration results in the derivation of the formula for the sum of an infinite geometric series. What effect does the value of r have on this result?

54. *Partial Sum Proof Problem:*

a. Prove that for an arithmetic series with first term t_1, nth term t_n, and common difference d, the nth partial sum is given by the formula

$$S_n = \frac{n}{2}(t_1 + t_n)$$

b. Prove that for a geometric series with first term t_1, nth term t_n, and common ratio r, the nth partial sum is given by

$$S_n = t_1 \cdot \frac{1 - r^n}{1 - r}$$

55. *Journal Problem:* Update your journal with things you have learned in this chapter. Include the difference between a sequence and a series and what makes a sequence or series arithmetic or geometric. Explain how partial sums of series can be calculated numerically and how calculations for arithmetic, geometric, and binomial series can be performed algebraically.

15-4 Chapter Review and Test

The length of each cycle in this French braid is a term in a series, and the total length of the braid is a partial sum.

In this chapter you studied sequences of numbers. You can consider these sequences to be functions in which the independent variable is an integer (the term number) and the dependent variable is the term itself. For some sequences you can find the term values either recursively as a function of the preceding term or explicitly as a function of the term number. Then you learned about series, which are sums of the terms of sequences. You can calculate partial sums of series numerically using your grapher. For arithmetic and geometric series, you can also calculate partial sums algebraically.

Review Problems

R0. Update your journal with what you have learned in this chapter. Include things such as

- The difference between a sequence and a series
- The similarities and differences between arithmetic and geometric series
- The meaning and computation of terms in a binomial series
- Numerical ways to compute term values and partial sums of various kinds of series
- Algebraic ways to compute terms and partial sums of arithmetic and geometric series
- An algebraic way to find the sum of an infinite geometric series with $|r| < 1$

R1. **a.** Find the next two terms of the arithmetic sequence 5, 8, 11, 14, ...

b. Find the sixth partial sum of the arithmetic series $5 + 8 + 11 + 14 + \cdots$.

c. Show that the sixth partial sum in part b equals 6 times the average of the first and last terms.

d. Find the next two terms of the geometric sequence 5, 10, 20, 40, ...

e. Find the sixth partial sum of the geometric series $5 + 10 + 20 + 40 + \cdots$.

R2. **a.** Is the sequence 23, 30, 38, . . . arithmetic, geometric, or neither?

b. Find t_{200}, the 200th term of the arithmetic sequence 52, 61, 70, ...

c. 3571 is a term of the sequence in part b. What is its term number?

d. Find t_{100}, the 100th term of the geometric sequence with $t_1 = 200$ and $r = 1.03$.

e. $t_n = 5644.6417...$ is a term of the sequence in part d. What does n equal?

f. Write a recursion formula and the next three terms for the sequence 0, 3, 8, 15, 24, ...

g. Write an explicit formula for the sequence in part f and use it to calculate the 100th term.

h. *Monthly Interest Problem:* Suppose you invest \$3000 in an account that pays 6% interest per year (0.5% per month), compounded monthly. At any time during the first month, you have \$3000 in the account. At any time during the second month, you have \$3000 plus the interest for the first month, and so on.

i. Show that the month number and the amount in the account during that month have the add–multiply property of exponential functions. Explain why a sequence is a more appropriate mathematical model than a continuous exponential function.

ii. In what month will the amount first exceed \$5000?

iii. If you leave the money in the account until you retire 50 years from now, how much will be in the account?

R3. a. Write the terms of the sixth partial sum of this series:

$$\sum_{k=1}^{n} [2 + (k - 1)(3)]$$

b. How do you know that the series in part a is an arithmetic series?

c. Show that the partial sum in part a equals 6 times the average of the first and last terms.

d. 418,440 is a partial sum of the series in part a. Which partial sum is it?

e. A geometric series has $t_1 = 4000$ and $r = 0.95$. Find the 200th partial sum of this series numerically using your grapher.

f. Find the 200th partial sum in part e again, this time algebraically using the formula for S_n.

g. 78,377.8762... is a partial sum of the series in part e. Which partial sum is it?

h. To what limit do the partial sums of the series in part e converge?

i. Write the term containing b^7 in the binomial expansion of $(a - b)^{13}$.

j. *Vincent and Maya's Walking Problem:* Vincent and Maya start walking at the same time from the same point and in the same direction. Maya starts with a 12-in. step and increases her stride by $\frac{1}{2}$ in. each step. She goes 21 steps, then stops. Vincent starts with a 36-in. step, and each subsequent step is 90% as long as the preceding one.

i. What kind of sequence do Maya's steps follow? What kind of sequence do Vincent's steps follow?

ii. How long is Maya's last step? How long is Vincent's 21st step?

iii. After each has taken 21 steps, who is ahead? Show how you reached your conclusion.

iv. If Vincent keeps walking in the same manner, will he ever get to where Maya stopped? Explain.

k. *Vitamin C Dosage Problem:* When you take a dose of medication, the amount of medication in your body jumps up immediately to a higher value and then decreases as the medication is used up and expelled. Suppose you take 500-mg doses of vitamin C each 6 h to fight a cold. Assume that by the time of the second dose only 60% of the first dose remains. How much vitamin C will you have just after you take the second dose? Just after the third dose? Show that the total amount of vitamin C in your system is a partial sum of a geometric series. How much vitamin C will you have in your system at the end of the fourth day, when you have just taken the 16th dose? What limit does the amount of vitamin C approach as the number of doses approaches infinity?

Concept Problems

C1. *Tree Problem:* A treelike figure is drawn as shown in Figure 15-4a. The first year, the tree grows a trunk 2 m long. The next year, two branches, each 1 m long, grow at right angles to each other from the top of the trunk, symmetrical to the line of the trunk. In subsequent years, each branch grows two new branches, each half as long as the preceding branch.

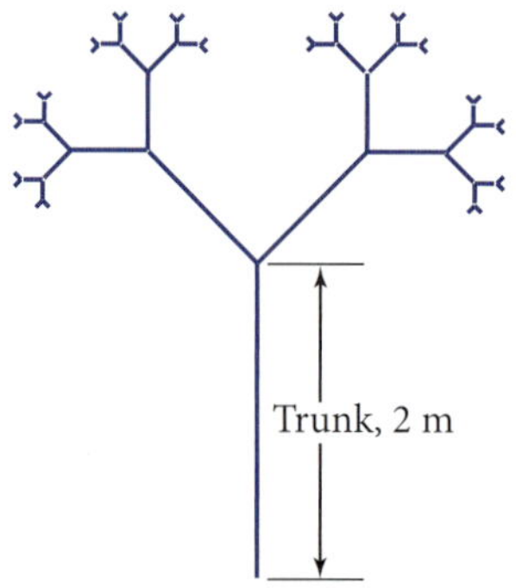

Trunk, 2 m

Figure 15-4a

a. Show that the lengths of the branches form a geometric sequence.

b. Find the height of the tree after 2 yr, 3 yr, and 4 yr.

c. Show that the height of the tree is the partial sum of *two* geometric series. What is the common ratio of each?

d. Suppose the tree keeps growing like this forever.

i. What limit will the height approach?

ii. What limit will the width approach?

iii. What limit will the length of each branch approach?

iv. What limit will the total length of all branches approach?

v. How close to the ground will the lowest branches come?

C2. WEB *Bode's Law Problem:* In 1766, Johann Titius, a German astronomer, discovered that the distances of some planets from the Sun are proportional to the terms of a rather simple sequence:

Term Number	Name	Term Value
1	Mercury	4
2	Venus	7
3	Earth	10
4	Mars	16
5	Ceres	28
6	Jupiter	52
7	Saturn	100

(Mercury, at 4, does not fit the sequence. Ceres is a dwarf planet in the asteroid belt.)

a. Describe the pattern followed by the term values. Use this pattern to find the term value for Uranus, planet 8.

b. Planets 9 and 10, Neptune and Pluto (a dwarf planet), have distances corresponding to the numbers 301 and 395, respectively. Are these numbers terms of the sequence? Justify your answer.

c. Find a formula for t_n, the term value, in terms of n, the planet number. For what values of n is the formula valid?

d. The dwarf planet Ceres was found by looking at a distance from the Sun calculated by Bode's law. If you were to look beyond Pluto for planet 11, how far from the Sun would Bode's law suggest that you look? (Earth is 93 million miles from the Sun.)

e. Consult the Internet or some other reference source to see why Titius's discovery is called Bode's law.

C3. *Binomial Series with Noninteger Exponent Problem:* If you raise a binomial to a noninteger power, such as $(a+b)^{1.8}$, you can still find the coefficient of the next term from the pattern

$$\frac{(\text{coefficient})(\text{exponent of } a)}{(\text{term number})}$$

a. Write the first five terms of the binomial expansion of $(a+b)^{1.8}$.

b. The series in part a has an infinite number of terms. A binomial series for a positive integer exponent may also be considered to have an infinite number of terms. Use the coefficient pattern in this problem to show what happens beyond the b^5-term in the expansion of $(a+b)^5$ and why such series seem to have a finite number of terms.

C4. *Power Series Problem:* In these three **power series,** each term involves a nonnegative integer power of x. (Such series are also called *Maclaurin series* or *Taylor series.*)

$$f(x) = 1 + x + \frac{1}{2!}x^2 + \frac{1}{3!}x^3 + \cdots$$

$$g(x) = x - \frac{1}{3!}x^3 + \frac{1}{5!}x^5 - \frac{1}{7!}x^7 + \cdots$$

$$h(x) = 1 - \frac{1}{2!}x^2 + \frac{1}{4!}x^4 - \frac{1}{6!}x^6 + \cdots$$

a. Evaluate the seventh partial sum of the series for $f(0.6)$ (terms through x^6). Show that the answer is close to the value of $e^{0.6}$, where e is the base of the natural logarithms.

b. Evaluate the fourth partial sum of the series for $g(0.6)$. Show that the answer is close to the value of sin 0.6. Show that the fifth partial sum is even closer to the value of sin 0.6.

c. Evaluate the fourth partial sum of the series for $h(0.6)$. Which function on your grapher does $h(x)$ seem to be close to?

d. Figure 15-4b shows the graphs of $f_1(x)$—a cubic function equal to the second partial sum of $g(x)$—and $f_2(x) = \sin x$. Graph these functions on your grapher. Then add the next three terms of the series for $g(x)$ to the equation of $f_1(x)$. You'll get a 9th-degree function. Plot the two graphs again. Sketch the result, and describe what you observe.

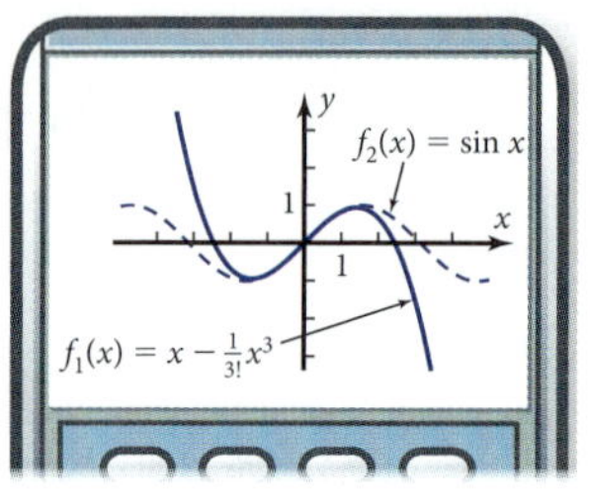

Figure 15-4b

Chapter Test

Part 1: No calculators allowed (T1–T9)

T1. A geometric series has first term $t_1 = 6$ and common ratio $r = 2$. Write the first five terms of the series. Find the fifth partial sum, S_5, numerically by summing the terms of the series.

T2. An arithmetic series has first term $t_1 = 7$ and common difference $d = 3$. Write the first six terms of the series. Find the sixth partial sum, S_6, numerically by summing the terms and algebraically by using the sum of the first and last terms in an appropriate way.

T3. Write an algebraic formula for S_n, the nth partial sum of an arithmetic series, in terms of n, t_1, and t_n.

T4. Is this series arithmetic, geometric, or neither? Give numerical evidence to support your answer.

$$3 + 7 + 12 + 18 + 25 + \cdots$$

T5. Evaluate the partial sum numerically, by writing out and summing the terms.

$$S_4 = \sum_{k=1}^{4} 7 \cdot 3^{k-1}$$

T6. Is the series in Problem T5 arithmetic, geometric, or neither? Give numerical evidence to support your answer.

T7. Write the term containing b^9 in the binomial expansion of $(a - b)^{15}$. Leave the answer in factorial form.

T8. Write a recursive formula for t_n for the arithmetic sequence 17, 21, 25, ...

T9. Write an explicit formula for t_n for the geometric sequence 7, 14, 28, ...

Part 2: Graphing calculators allowed (T10–T24)

T10. The arithmetic series in Problem T2 is $7 + 10 + 13 + \cdots$. Calculate the 200th term, t_{200}. Use the answer to calculate the 200th partial sum, S_{200}, algebraically, with the help of the pattern for partial sums of an arithmetic series. Confirm that your answer is correct by summing the terms numerically using the Series program on your grapher.

Bouncing Ball Problem: Imagine that you're bouncing a ball. Each time the ball bounces, it comes back up to 80% of its previous height. For Problems T11–T13, assume that the starting height of the ball was 5 ft.

T11. What kind of sequence describes the ball's successive maximum heights after each bounce? What is the length of the *total* vertical path the ball covers before the tenth bounce?

T12. Find the formula for the ball's maximum height after its nth bounce. Find the formula for the *total* length of the path of the ball before the nth bounce.

T13. The series in Problem T12 converges to a certain number. Based on your answers, to what number does it converge? Show algebraically that this is correct.

Push-ups Problem: For Problems T14 and T15, Emma starts an exercise program. At the first workout she does five push-ups. The next workout she does eight push-ups. She decides to let the number of push-ups in each workout be a term in an arithmetic sequence.

T14. Find algebraically the number of push-ups she does on the tenth workout and the total number of push-ups she has done after ten workouts.

T15. If Emma were able to keep up the arithmetic sequence of push-ups, which of the terms in the sequence would be 101? Calculate this term number algebraically.

Medication Problem: For Problems T16–T19, Natalie takes 50 mg of allergy medicine each day. By the next day, some of the medicine has decomposed, but the rest is still in her body. Natalie finds that the amount still in her body after n days is given by this partial sum, where k is an integer:

$$S_n = \sum_{k=1}^{n} 50(0.8^{k-1})$$

T16. Demonstrate that you know what sigma notation means by writing out the first three terms of this series. What does S_3 equal?

T17. Run the Series program on your grapher to find S_{40} numerically.

T18. After how many days does the amount in Natalie's body first exceed 200 mg?

T19. Does the amount of active medicine in Natalie's body seem to be converging to a certain number, or does it simply keep getting larger without limit? How can you tell?

Loan Problem: For Problems T20 and T21, Leonardo borrows $200.00 from his parents to buy a new calculator. They require him to pay back 10% of his unpaid balance at the end of each month.

T20. Find Leonardo's unpaid balances at the end of 0, 1, 2, and 3 mo. Is the sequence of unpaid balances arithmetic, geometric, or neither? How do you know?

T21. Leonardo must pay off the rest of the loan when his unpaid balance has dropped below $5.00. After how many months will this have happened? Indicate the method you use. (It is not enough to say "I used my calculator" or "guess and check.")

T22. Find the eighth term of the binomial series $(a - b)^{12}$.

T23. *Genetics Problem:* If a dark-haired mother and father have a particular combination of genes, each of their babies has a 25% probability of having light hair (and a 75% probability of having dark hair). If they have five babies, the probability that exactly x of the babies have light hair is the term with 0.25^x in the binomial expansion of $(0.25 + 0.75)^5$. Find the probability that exactly three of the five babies have light hair.

T24. What did you learn as a result of taking this test that you did not know before?

Chapter 16

Introduction to Limits, Derivatives, and Integrals

A bee flies back and forth past a flower. Its distance from the flower is a function of time. The bee doesn't fly at a constant speed, so you can't find the distance it has traveled simply by multiplying rate by time. In this chapter you'll explore the motion of an object with varying speed. You'll learn to use the concept of limit (which you saw in connection with geometric series), to find such an object's instantaneous rate (its rate at a given moment in time), and to estimate the total distance it travels.

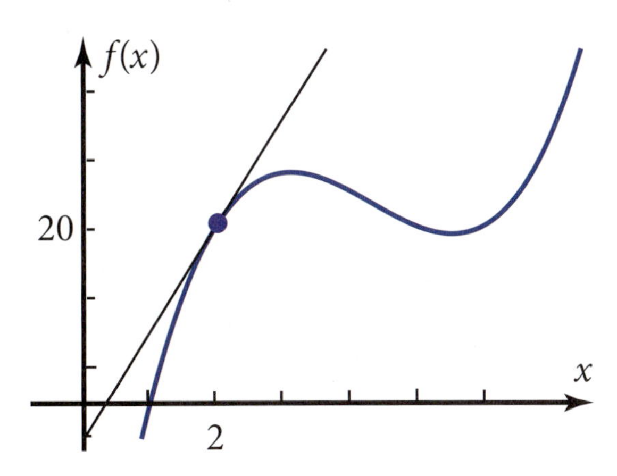

Mathematical Overview

In this chapter you will tie together familiar concepts and techniques to take a first look at calculus. Instantaneous rates of change involve limits of average rates of change, such as displacement divided by time. Accumulated amounts of change, such as rate multiplied by time, where the rate is variable, involve taking the limit of a series of distances or other quantities found for short time intervals. You will investigate these concepts in four ways.

GRAPHICALLY

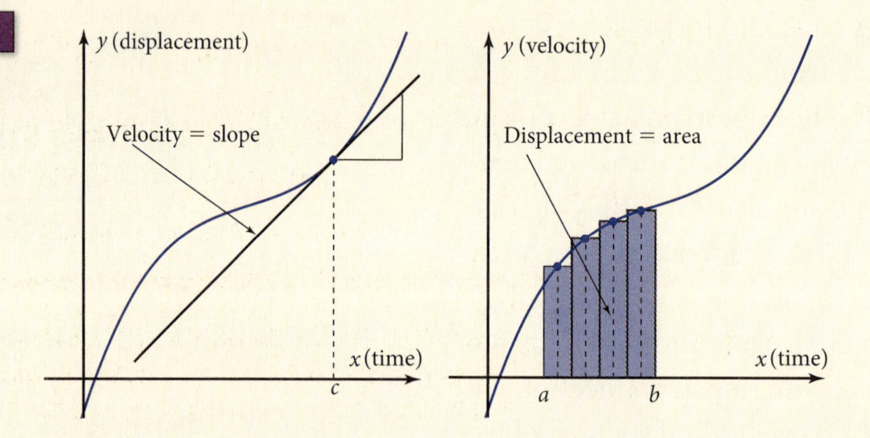

In the graph on the left, velocity equals the slope of the tangent line. In the graph on the right, displacement equals approximately the area of the shaded region.

ALGEBRAICALLY

$$\text{Instantaneous rate of change of } f(x) \text{ at } c = \lim_{x \to c} \frac{f(x) - f(c)}{x - c}$$

NUMERICALLY

If $f(x) = 0.1x^3 - 0.9x^2 + 3x - 0.6$ represents velocity, then the change in displacement from $x = 1$ s to $x = 3$ s is approximately

$$f(1.25)(0.5) + f(1.75)(0.5) + f(2.25)(0.5) + f(2.75)(0.5) = 5.0125$$

VERBALLY

The instantaneous velocity equals the limit of the average velocity.

16-1 Exploring Limits, Derivatives, and Integrals

Figure 16-1a shows the displacement of an object, $f(x)$, in feet, from a particular point as a function of time x, in seconds, since the object started to be timed. Figure 16-1b shows the velocity $g(x)$, in centimeters per second, of a second moving object as a function of time x, in seconds, since the object started to be timed.

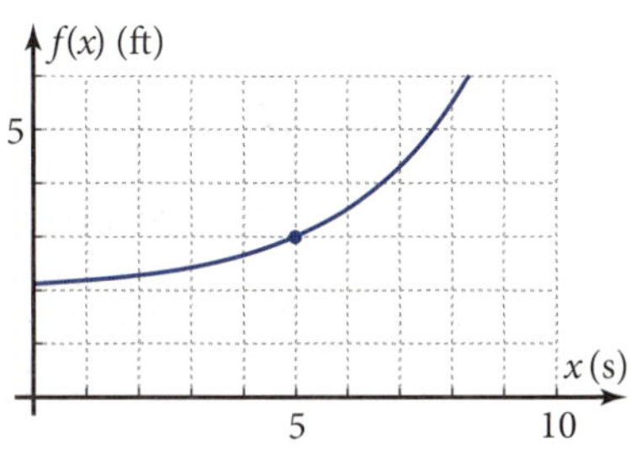

Figure 16-1a

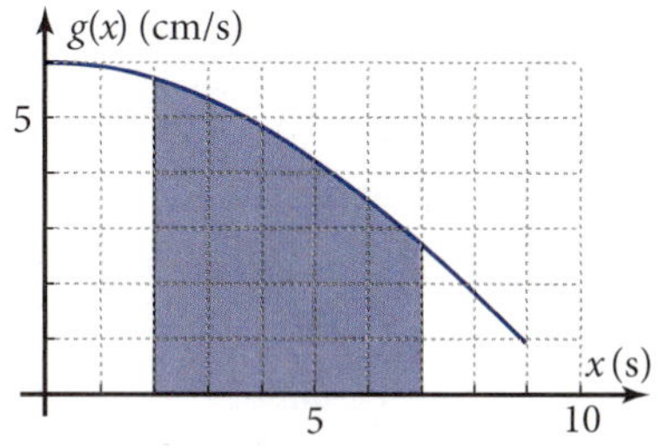

g(x) (cm/s)

5

x (s)

5 10

Figure 16-1b

Objective Given a time–distance graph, approximate the instantaneous rate of change, and given a time–velocity graph, approximate the distance traveled in a given time interval.

Exploratory Problem Set 16-1

1. The equation for function f in Figure 16-1a is $f(x) = 2 + 2^{0.6(x-5)}$ where $f(x)$ is the displacement, in feet, from the origin at time x, in seconds. How far did the object travel between $x = 5$ s and $x = 6$ s? Find the average rate of change of $f(x)$ for this interval of time.
2. Find the average rate of change of $f(x)$ for the time interval from $x = 5$ s to $x = 5.1$ s and from $x = 5$ s to $x = 5.01$ s.
3. The instantaneous rate of change of $f(x)$ at $x = 5$ is the limit of the average rates of change between $x = 5$ and $x = c$ as as c gets closer and closer to 5. Make a conjecture: Instantaneous rate $\approx$ __?__
4. On a copy of Figure 16-1a, draw a line through the point (5, 3) with slope equal to the instantaneous rate of change in Problem 3. How is the line related to the curved graph of function f?
5. Because distance is equal to rate multiplied by time, the area of the shaded region in Figure 16-1b is equal to the distance traveled by the object from time $x = 2$ s to $x = 7$ s. Approximate the area by counting squares.
6. On a copy of Figure 16-1b, draw five rectangles with width 1 unit and heights $g(2.5)$, $g(3.5)$, $g(4.5)$, $g(5.5)$, and $g(6.5)$. Each rectangle should intersect the function graph at the midpoint of its top side. The leftmost rectangle will have vertices (2, 0), (2, $g(2.5)$), (3, $g(2.5)$), and (3, 0). Given that $g(x) = 6 \cos \frac{\pi}{20} x$, where $g(x)$ is velocity in centimeters per second and x is time in seconds, find the sum of the areas of these rectangles. Is the result approximately equal to the area you estimated graphically in Problem 5?
7. How could you use limits to define the exact area of the region in Figure 16-1b?
8. What did you learn as a result of doing this problem set that you did not know before?

16-2 Limits

The rational function in Figure 16-2a,

$$f(x) = \frac{x^3 - 9x^2 + 31x - 34}{x - 2}$$

has a discontinuity at $x = 2$ because $(x - 2)$ in the denominator equals zero when $x = 2$. Functions like this arise in finding instantaneous rates of change, as you will learn in the next section. The concept of **limit** gives you a way to examine such discontinuities.

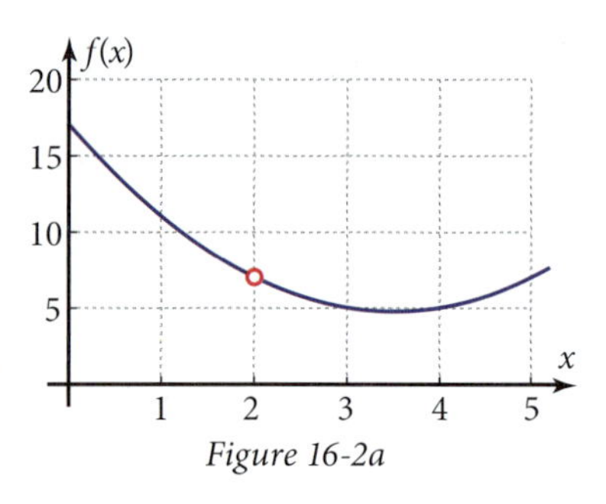

Figure 16-2a

Objective

Given function f with a discontinuity at a particular value of $x = c$, use limits to find the number $f(x)$ stays close to when x is kept close to c but x is not equal to c.

In this exploration you will find the limit of a rational algebraic function as x approaches a value that makes the denominator equal zero.

EXPLORATION 16-2: Limits

Use the rational function f graphed in Figure 16-2a to answer these questions.

1. What is the meaning of the word *rational* in the name "rational algebraic function"?
2. Try to find $f(2)$. What form does $f(x)$ take when 2 is substituted for x? Why is $f(2)$ undefined? What feature does the graph of function f have at $x = 2$?
3. Although there is no value for $f(2)$, there is a number L that $f(x)$ stays close to if x is kept close to 2 but is not equal to 2. From the graph, what does this number appear to be?
4. The number L you wrote in Problem 3 is the *limit* of $f(x)$ as x approaches 2. Show that $f(1.9)$ and $f(2.1)$ are close to this number. Show that $f(1.99)$ and $f(2.01)$ are even closer to this number.
5. $f(x)$ can be made to stay within 0.02 unit of L when x is kept close enough to 2 (but is not equal to 2). Find this corresponding interval of x-values by simplifying the equation for $f(x)$, and then substituting $L - 0.02$ and $L + 0.02$ for $f(x)$.
6. Choose a value of x in the interval you found in Problem 5 and show that f(that value) really is within 0.02 unit of L.
7. Substitute 2 for x in the simplified form of $f(x)$. What is remarkable about the result? What does it mean to say that function f has a "removable" discontinuity at $x = 2$?
8. Why is the letter L appropriate for the number $f(x)$ stays close to when x is kept close enough to 2 (but is not equal to 2)?
9. What did you learn as a result of doing this exploration that you did not know before?

The box on the next page gives a definition of limit. As you work through the examples and problems in this section, the meaning of each word in the definition will become clearer to you. Meanwhile, commit this definition to memory so that you can say and write it correctly without having to look at the text.

DEFINITION: Limit

Verbally: L is the **limit** of $f(x)$ as x approaches c
if and only if
L is the *one* number you can keep $f(x)$ arbitrarily close to
just by keeping x close enough to c on both sides, but not equal to c.

Algebraically: $L = \lim_{x \to c} f(x)$

Figure 16-2b helps you understand the definition of a limit. If you pick an interval of numbers that includes the limit L, then $f(x)$ can be kept in this interval by keeping x in a suitably small interval around $x = c$, but not at c itself. If this is true no matter how arbitrarily small an interval you pick around L, then L is the limit of $f(x)$ as x approaches c.

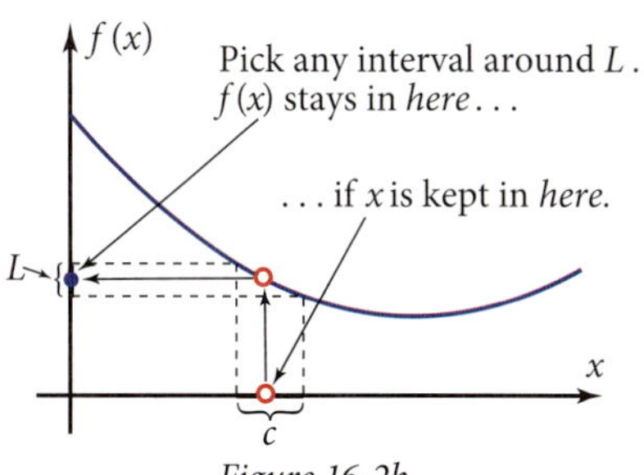

Figure 16-2b

Example 1 furthers your understanding of the definition of limit by analyzing the graph of a function that has distinctive features.

EXAMPLE 1 Figure 16-2c shows the graph of a piecewise function, f.

Find each limit and use the definition of limit to explain your answer.

a. $\lim_{x \to 1} f(x)$

b. $\lim_{x \to 2} f(x)$

c. $\lim_{x \to 3} f(x)$

d. $\lim_{x \to 4} f(x)$

e. $\lim_{x \to 5} f(x)$

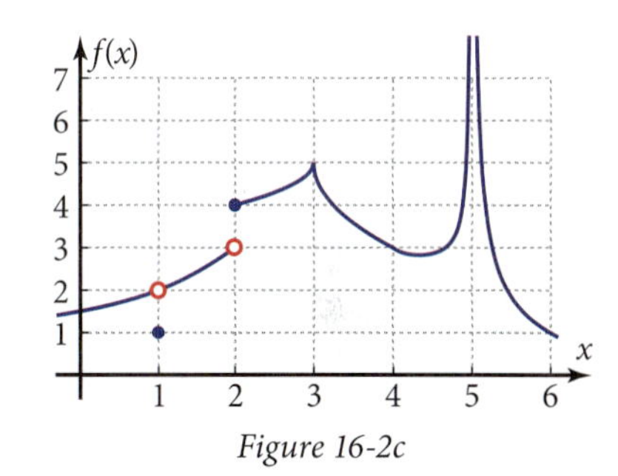

Figure 16-2c

SOLUTION

a. $\lim_{x \to 1} f(x) = 2$

$f(x)$ stays close to 2 if x is kept close to 1 but is not *equal* to 1. The fact that $f(1) = 1$, not 2, is irrelevant because "but x not equal to c" from the definition of limit means you can ignore what happens at $x = 1$.

b. $\lim_{x \to 2} f(x)$ does not exist.

The definition says, "L is the *one* number . . . ." $f(x)$ is close to 3 if x is close to 2 on the negative side, but it is close to 4 if x is close to 2 on the positive side.

c. $\lim_{x \to 3} f(x) = 5$

$f(x)$ stays close to 5 if x is kept close to 3 but is not equal to 3. The fact that $f(3)$ also equals 5 is irrelevant because "but x not equal to c" from the definition of limit means you can ignore what happens at $x = 3$. The fact that there is a *corner* at $x = 3$ is also irrelevant.

d. $\lim_{x \to 4} f(x) = 3$

$f(x)$ stays close to 3 if x is kept close to 4 but is not equal to 4. The fact that the graph is smooth and has no discontinuity at $x = 4$ is irrelevant.

e. $\lim_{x \to 5} f(x)$ does not exist.

Because there is a vertical asymptote at $x = 5$, $f(x)$ becomes infinite there. Thus there is no number $f(x)$ stays close to if x is kept close to 5 (but is not equal to 5).

Example 2 shows you how to calculate algebraically an interval of x-values that will keep $f(x)$ arbitrarily close to the limit.

EXAMPLE 2 Figure 16-2d shows the graph of the rational function

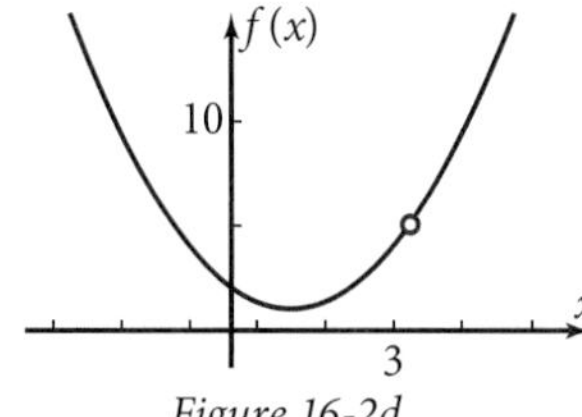

Figure 16-2d

$$f(x) = \frac{x^3 - 5x^2 + 8x - 6}{x - 3}$$

a. Simplify the fraction in the equation for $f(x)$. By substituting 3 into the simplified equation, find $\lim_{x \to 3} f(x)$. Does the result agree with the graph?

b. In what interval must x be kept in order for $f(x)$ to stay within 0.1 unit of the limit?

c. Choose a value of x in the interval in part b and show that $f(x)$ really is within 0.1 unit of the limit for this value of x.

d. Would the procedure in parts b and c work for *any* interval containing L, no matter how small?

SOLUTION

a. Factor the numerator by synthetically substituting 3 for x (the value that makes the denominator equal zero).

3	1	−5	8	−6
		3	−6	6
	1	−2	2	0

$$f(x) = \frac{(x - 3)(x^2 - 2x + 2)}{x - 3} = x^2 - 2x + 2,\ x \neq 3$$

The limit, L, is the vertical coordinate of the discontinuity. Substitute 3 for x in the simplified form to find this value. Note that this is not the value of $f(3)$ because $f(3)$ is undefined.

$$L = \lim_{x \to 3} f(x) = 3^2 - 2(3) + 2 = 5$$

The answer agrees with the graph.

b. Let $f(x) = 5.1$. — 0.1 unit above 5.

$$x^2 - 2x + 2 = 5.1 \Rightarrow x^2 - 2x - 3.1 = 0$$

Substitute 5.1 for $f(x)$. Why can you use the simplified form of $f(x)$?

By the quadratic formula, $x = 3.0248...$. — The other result of the quadratic formula, $-1.0248...$, is not close to 3.

Let $f(x) = 4.9$. — 0.1 unit below 5.

$$x^2 - 2x + 2 = 4.9 \Rightarrow x^2 - 2x - 2.9 = 0$$

By the quadratic formula, $x = 2.9748...$. — $-0.9748...$ is not close to 3.

The interval is $2.9748... < x < 3.0248...$, $x \neq 3$.

c. Choose $x = 3.01$, for example, in the interval in part b.

$$f(3.01) = 3.01^2 - 2(3.01) + 2 = 5.0401$$

which is within 0.1 unit of 5.

d. The procedure in parts b and c would work no matter how small the interval around 5. Thus $f(x)$ stays *arbitrarily* close to 5 when x is close to 3 but is not equal to 3.

Example 3 shows you a function with an equation similar to $f(x)$ in Example 2 but with much different behavior at $x = 3$.

EXAMPLE 3 ➤ Figure 16-2e shows the graph of the rational function

$$g(x) = \frac{x^3 - 5x^2 + 8x - 5}{x - 3}$$

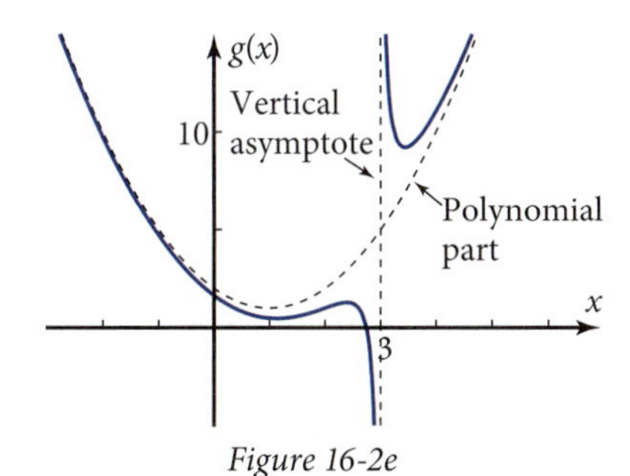

Figure 16-2e

a. Write the fraction in the equation for $g(x)$ in mixed-number form.

b. How does the remainder fraction in the mixed-number form of $g(x)$ reveal that there is a vertical asymptote at $x = 3$ and thus no value of $\lim_{x \to 3} g(x)$?

c. How is the polynomial part of the mixed-number form of $g(x)$ related to the graph of function g?

SOLUTION

a. Synthetically substitute 3 (the value that makes the denominator equal zero) for x.

$$\begin{array}{c|rrrr} 3 & 1 & -5 & 8 & -5 \\ & & 3 & -6 & 6 \\ \hline & 1 & -2 & 2 & 1 \end{array}$$

$$g(x) = x^2 - 2x + 2 + \frac{1}{x - 3}$$

b. The fraction $y = \frac{1}{x - 3}$ becomes infinite as x approaches 3, indicating a vertical asymptote.

c. The polynomial part, $y = x^2 - 2x + 2$, is a curved asymptote for the graph of function g, as shown in Figure 16-2e.

Example 4 shows that the limit of a function might equal the function value if the function value exists (that is, if the function is *continuous*). It also demonstrates how to find the limit of a function that is not a polynomial.

EXAMPLE 4 ➤ Figure 16-2f shows the graph of

$$h(x) = 2^x$$

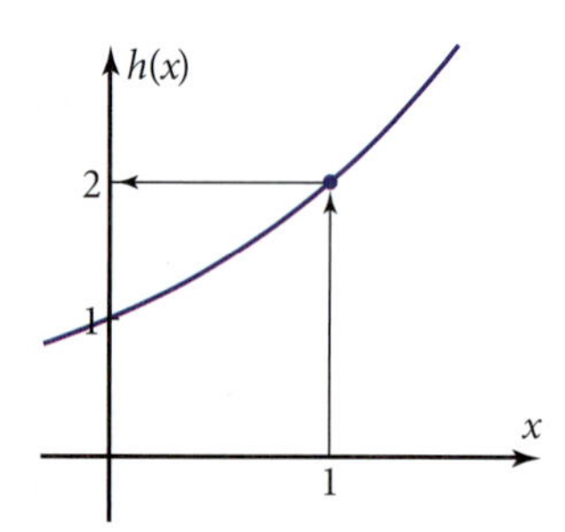

Figure 16-2f

Given that $\lim_{x \to 1} h(x) = h(1)$, find an interval of x-values for which $h(x)$ stays within 0.1 unit of $h(1)$ when x is kept in this interval but x is not equal to 1. Illustrate your answer on a copy of Figure 16-2f. Choose a value of x in this interval and demonstrate numerically that this value of x gives a value of $h(x)$ that really is within 0.1 unit of $h(1)$.

SOLUTION $h(1) = 2^1 = 2$

Let $h(x) = 2.1$. 2.1 is 0.1 unit above $h(1) = 2$.

$2^x = 2.1$

$\log 2.1 = x \log 2$

$$x = \frac{\log 2.1}{\log 2} = 1.0703...$$

Let $h(x) = 1.9$. 1.9 is 0.1 unit below $h(1) = 2$.

$2^x = 1.9$

$$x = \frac{\log 1.9}{\log 2} = 0.9259...$$

Keeping x in the interval $0.9259... < x < 1.0703...$ makes $h(x)$ stay within 0.1 unit of 2, as shown in Figure 16-2g.

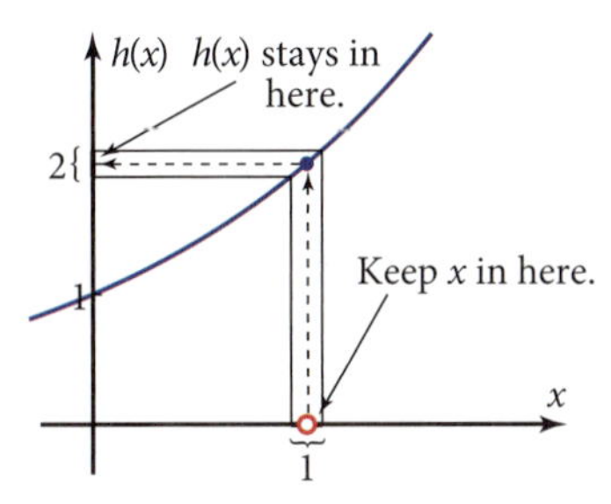

Figure 16-2g

Let $x = 1.04$, which is in the interval.

$$f(1.04) = 2^{1.04} = 2.0562...$$

which is within 0.1 unit of 2.

Problem Set 16-2

Reading Analysis

From what you have read in this section, what do you consider to be the main idea? Is the limit of a function a horizontal coordinate, a vertical coordinate, or an $f(x)$-value? If the limit of $f(x)$ as x approaches 3 is 5, then what happens to $f(x)$ when x is kept close to 3 but does not equal 3? How can $f(x)$ have a limit as x approaches 3 even if there is no value of $f(3)$? Why can a function have a limit $L = 5$ as x approaches 3 even if $f(3) = 6$?

Quick Review

Q1. If $f(x) = x^3 + 5x^2 - 7x + 13$, then function f is a(n) __?__ function.

Q2. If $f(x) = \dfrac{x^2 - 5x + 6}{x - 2}$, then function f is a(n) __?__ function.

Q3. Why is $f(x)$ in Problem Q2 undefined at $x = 2$?

Q4. What result do you get if you substitute 2 for x in Problem Q2?

Q5. What feature will the graph of function f in Problem Q2 have where $x = 2$?

Q6. Is 2, 5, 8, 11, . . . a sequence or a series?

Q7. Is 2, 5, 8, 11, . . . geometric, arithmetic, or neither?

Q8. If A and B are not mutually exclusive events, then $P(A \text{ or } B) =$ ___?___.

Q9. How is $\vec{a} \times \vec{b}$ related to $\vec{a}$ and $\vec{b}$?

Q10. How is arccos x related to $\cos^{-1} x$?

---

1. Write the definition of limit without looking at the definition in the text. Then check to make sure your definition is complete and correct. If it is not, try again until you can write it perfectly.

2. What is the reason for the clause, "but not equal to c" in the definition of limit?

For Problems 3–8, write the limits as x approaches a, b, and c, or give a reason why the limit does not exist.

3.

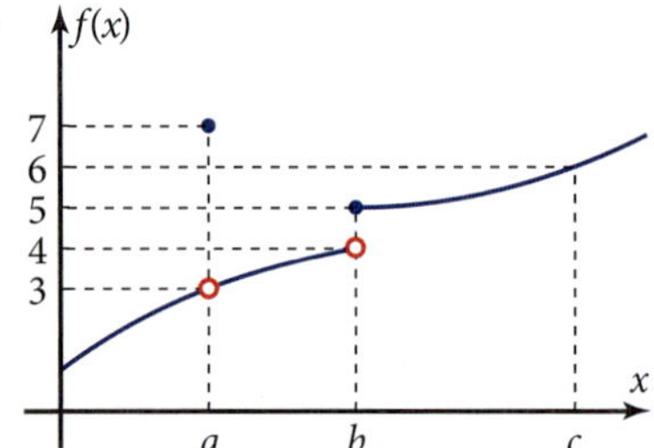

4.

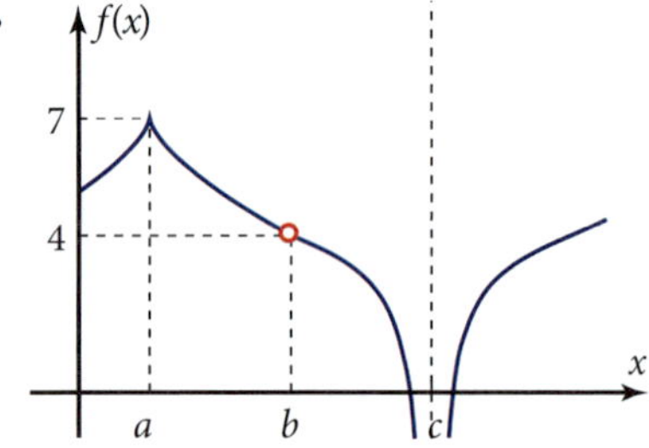

5.

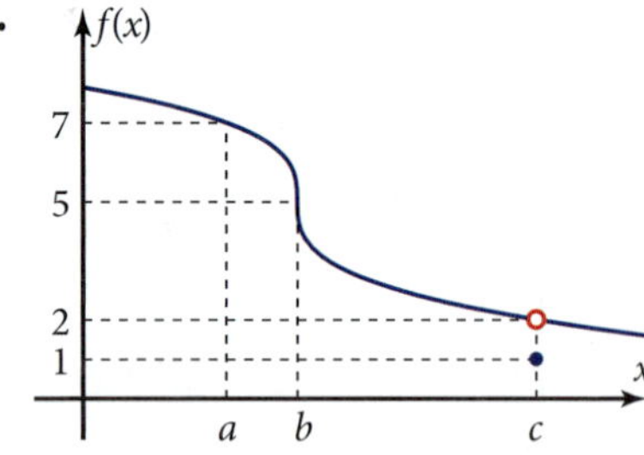

6.

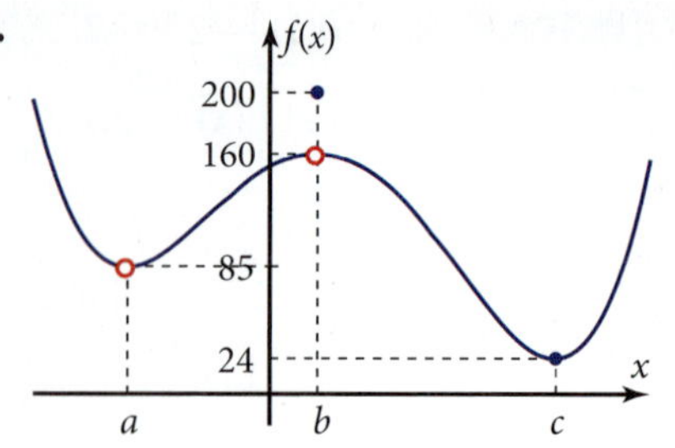

7.

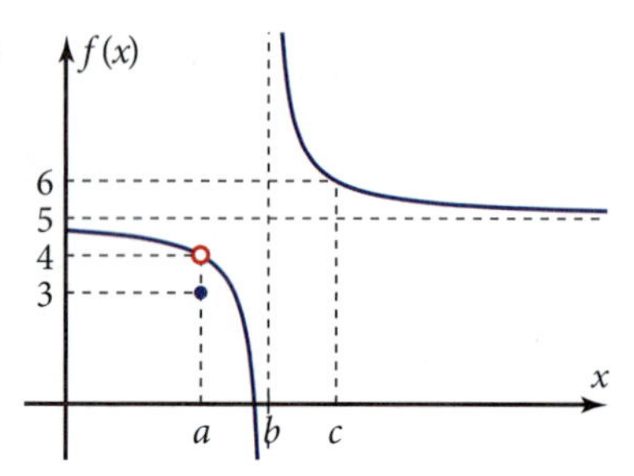

8.

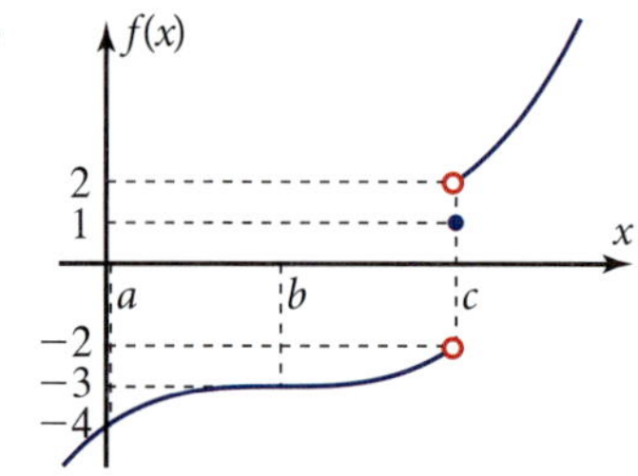

For Problems 9–12, simplify the fraction, then find the limit of $f(x)$ as x approaches the value that makes the denominator equal zero. Sketch the graph, showing the removable discontinuity.

9. $f(x) = \dfrac{-2x^2 + 21x - 52}{x - 4}$

10. $f(x) = \dfrac{x^2 + 2x - 15}{2x - 6}$

11. $f(x) = \dfrac{-x^3 + 11x^2 - 37x + 36}{x - 4}$

12. $f(x) = \dfrac{x^3 - 20x^2 + 152x - 408}{x - 6}$

13. In Problem 9, $f(x) = -2x + 13$ if $x \neq 4$. The graph of function f is shown in Figure 16-2h.

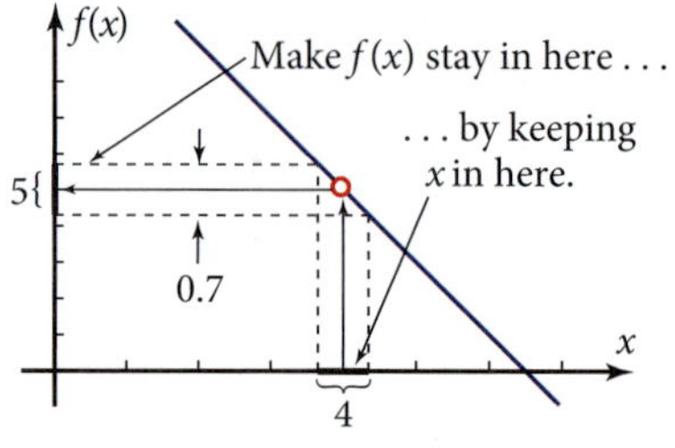

Figure 16-2h

a. Find the interval of x-values close to $x = 4$ that will make $f(x)$ stay within 0.7 unit of 5 when x is kept in this interval.

b. Pick an x-value from the interval in part a and show that $f(x)$ really is within 0.7 unit of 5.

c. How can the reasoning in parts a and b be generalized to show that 5 really is the limit of $f(x)$ as x approaches 4?

14. In Problem 10 you should have found that $f(x) = 0.5x + 2.5$ if $x \neq 3$.

a. Find the interval of x-values close to $x = 3$ that will make $f(x)$ stay within 0.3 unit of 4 when x is kept in this interval.

b. Pick an x-value, c, from the interval in part a and show that $f(c)$ really is within 0.3 unit of 4.

c. How can the reasoning in parts a and b be generalized to show that 4 really is the limit of $f(x)$ as x approaches 3?

15. In Problem 11, $f(x) = -x^2 + 7x - 9$ if $x \neq 4$ (Figure 16-2i). Find an interval about $x = 4$ that makes $f(x)$ stay within 0.2 unit of 3 when x is kept in that interval. How could this reasoning be generalized to show that 3 really is the limit of $f(x)$ as x approaches 4?

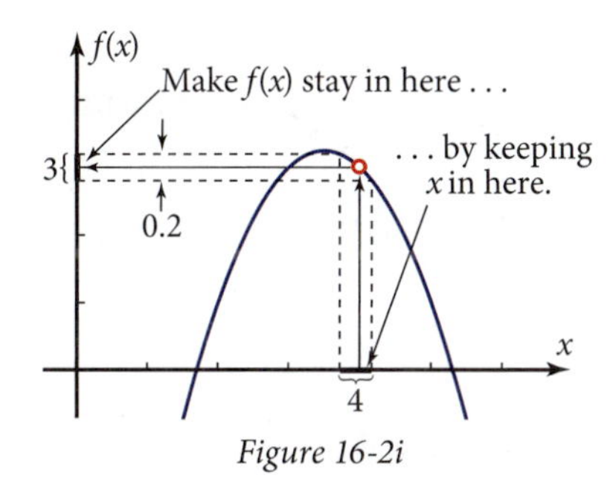

Figure 16-2i

16. In Problem 12 you should have found that $f(x) = x^2 - 14x + 68$ if $x \neq 6$. (Figure 16-2j). Find an interval about $x = 6$ that makes $f(x)$ stay within 0.9 unit of 20 when x is kept in that interval. How can this reasoning be generalized to show that 20 really is the limit of $f(x)$ as x approaches 6?

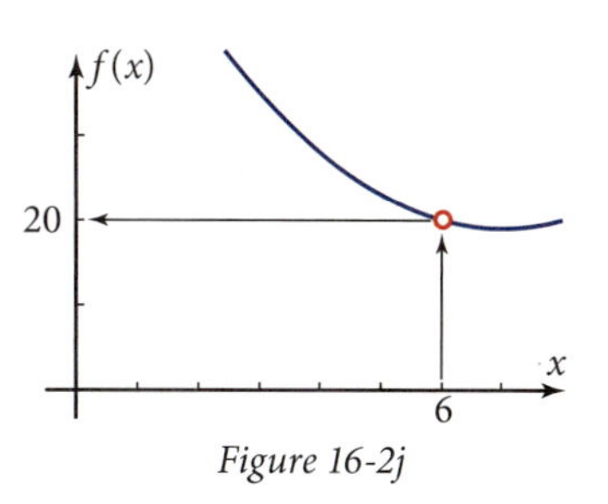

Figure 16-2j

17. Figure 16-2k shows the graph (solid curve) of the rational function

$$f(x) = \frac{-x^3 + 10x^2 - 23x + 15}{x - 2}$$

Write the equation in mixed-number form. Based on the result, explain why there is no limit of $f(x)$ as x approaches 2. How is the graph of the polynomial part of $f(x)$ (dashed curve) related to the graph of function f?

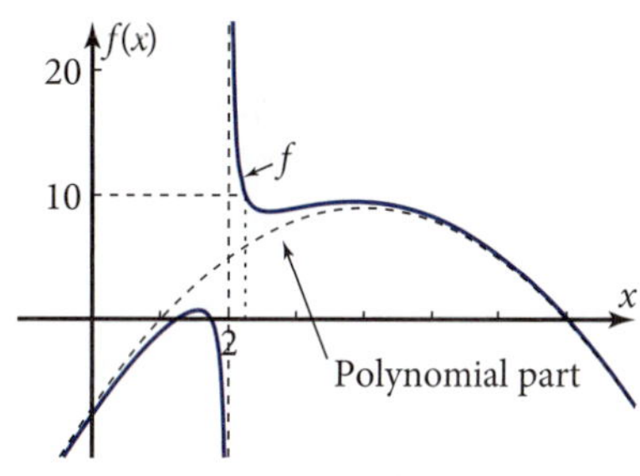

Figure 16-2k

18. Figure 16-2l shows the graph of the trigonometric function

$$f(x) = 3 + 2\cos\frac{\pi}{6}(x - 7)$$

Find an interval about $x = 5$ that makes $f(x)$ stay within 0.1 unit of $f(5)$ if x is kept in that interval. Explain how this calculation could be generalized to show that $\lim_{x \to 5} f(x) = f(5)$.

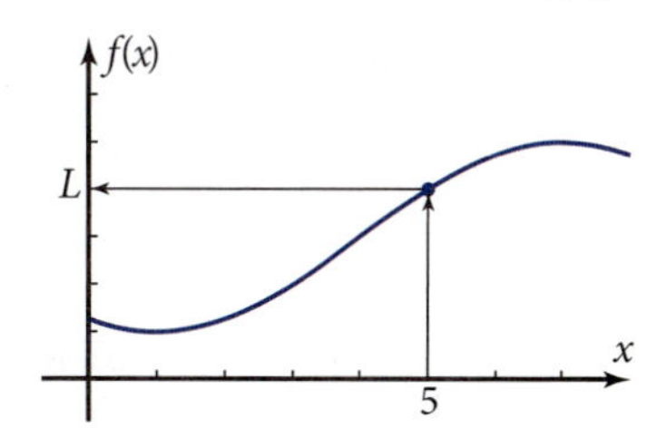

Figure 16-2l

16-3 Rate of Change of a Function: The Derivative

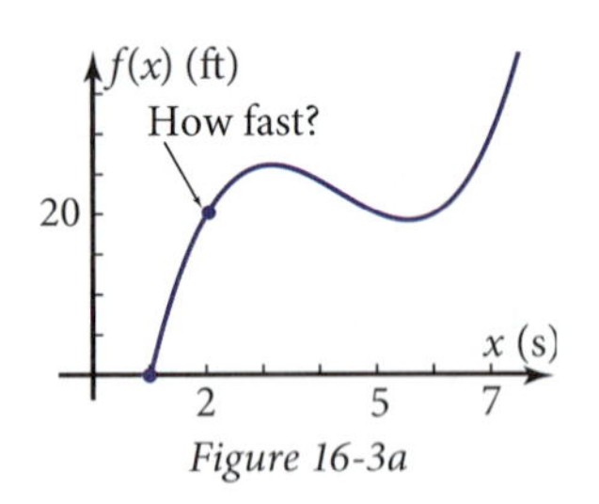

Figure 16-3a

Suppose a bird takes off from ground level at time $x = 1$ s. It climbs for a while, then dives for a while, and then swoops back up again. Figure 16-3a shows what its height, $f(x)$, in feet, might be as a function of time, x, in seconds.

From the graph you can tell that the bird is still climbing at time $x = 2$ s. You might ask, "At what *rate* is the bird climbing at the instant $x = 2$ s?" In this section you will use limits to find such an **instantaneous rate,** and you will get a preview of one of the major concepts in calculus, the **derivative,** which gives the instantaneous rate of change of a function at a particular value of x.

Objective Given an equation for a function, use limits to find the derivative, or instantaneous rate of change, of the function at a given point in the domain.

About 300 years ago, Sir Isaac Newton (1643–1727) in England and Gottfried Wilhelm Leibniz (1646–1716) in Germany were credited with solving the problem of finding an instantaneous rate of change.

The Derivative (Instantaneous Rate) Numerically

Rate equals distance divided by time. An "instant" is 0 s long. In 0 s, the bird mentioned earlier would travel 0 ft. So the instantaneous rate takes the form $\frac{0}{0}$, called an **indeterminate form.**

Suppose the bird's height, depicted in Figure 16-3a, is given by the cubic function

$$f(x) = x^3 - 13x^2 + 52x - 40$$

where x is time in seconds and $f(x)$ is distance in feet. To estimate the instantaneous rate at which the bird is climbing at $x = 2$ s, first find the **average rate** over a small time interval. Using your grapher, you find

$$f(2) = 20 \quad \text{and} \quad f(2.1) = 21.131$$

So the bird climbed $f(2.1) - f(2)$ or 1.131 ft in 0.1 s, for an average rate of

$$\frac{1.131}{0.1} = 11.31 \text{ ft/s}$$

To get a better estimate of the instantaneous rate, use smaller time intervals.

The average rate from $x = 2$ s to $x = 2.01$ s is

$$\frac{f(2.01) - f(2)}{2.01 - 2} = \frac{0.119301}{0.01} = 11.9301 \text{ ft/s} \qquad f(2.01) = 20.119301$$

The average rate from $x = 2$ s to $x = 2.001$ s is

$$\frac{f(2.001) - f(2)}{2.001 - 2} = \frac{0.011993001}{0.001} = 11.993001 \text{ ft/s}$$

$f(2.001) = 20.011993001$

The instantaneous rate of change at $x = 2$ s is the *limit* of the average rates of change as the later time, x, approaches 2 s. The average rates of change seem to be getting closer to 12 as x gets closer to 2. Thus a reasonable conjecture for the derivative of function f at $x = 2$, written $f'(2)$ and pronounced "f prime of 2," would be $f'(2) = 12$ ft/s.

The Derivative (Instantaneous Rate) Algebraically

In the bird flight example, the distance the bird rose between 2 s and x s was found by subtraction,

$$\text{Distance risen} = (\text{later height}) - (\text{height at 2 s}) = f(x) - f(2)$$

The time it took is also found by subtraction:

$$\text{Time taken} = (\text{later time}) - (2 \text{ s}) = x - 2$$

So the average rate of change, $r(x)$ ft/s, is given by the function

$$r(x) = \frac{f(x) - f(2)}{x - 2}$$

$\text{Rate} = \frac{\text{distance}}{\text{time}}$

Substituting for $f(x)$ and using the limit techniques of the previous section, the derivative is

$$f'(2) = \lim_{x \to 2} r(x) = \lim_{x \to 2} \frac{(x^3 - 13x^2 + 52x - 40) - 20}{x - 2}$$

Substitute for $f(x)$ and $f(2)$.

$$= \lim_{x \to 2} \frac{x^3 - 13x^2 + 52x - 60}{x - 2}$$

$$= \lim_{x \to 2} \frac{(x - 2)(x^2 - 11x + 30)}{x - 2}$$

Factor by synthetic substitution.

$$= \lim_{x \to 2} (x^2 - 11x + 30) = 12$$

The derivative at $x = 2$ is exactly 12, which is the number conjectured above.

Figure 16-3b shows that the value of the derivative (the *instantaneous* rate of change) is the vertical coordinate of the removable discontinuity in the average rate function, $r(x)$, at $x = 2$.

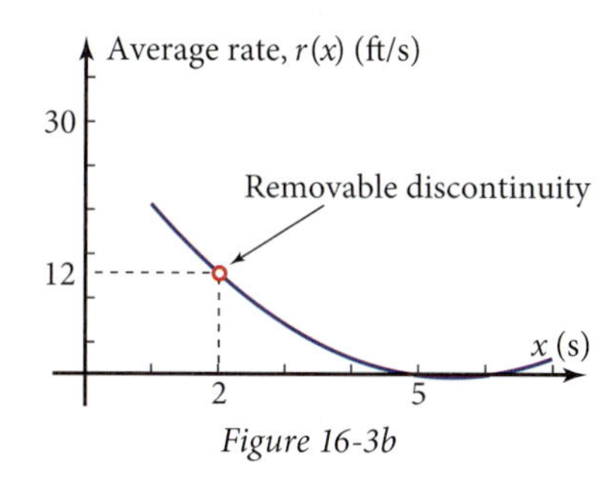

Figure 16-3b

The box shows the formal definition of derivative. (You will refine this definition further when you study calculus.)

> **DEFINITION: Derivative of a Function at a Point**
>
> *Verbally:* The **derivative** of function f at $x = c$ is the limit of the average rates of change of $f(x)$ as x approaches c.
>
> *Algebraically:* $f'(c) = \lim_{x \to c} \dfrac{f(x) - f(c)}{x - c}$

The Derivative (Instantaneous Rate) Graphically

In the bird flight example, the average rate of change of the bird's height from $x = 2$ s to $x = 3$ s is given by

$$\frac{f(x) - f(2)}{x - 2} = \frac{f(3) - f(2)}{3 - 2} = \frac{26 - 20}{1} = 6 \text{ ft/s}$$

From geometry, recall that a secant line to a circle is a line that intersects the circle at two points. Figure 16-3c shows a secant line to the graph of function f. The secant line contains the points (2, 20) and (3, 26) on the graph of function f. The slope of this secant line, 6 ft/s, equals the average rate of change of the bird's height from $x = 2$ s to $x = 3$ s.

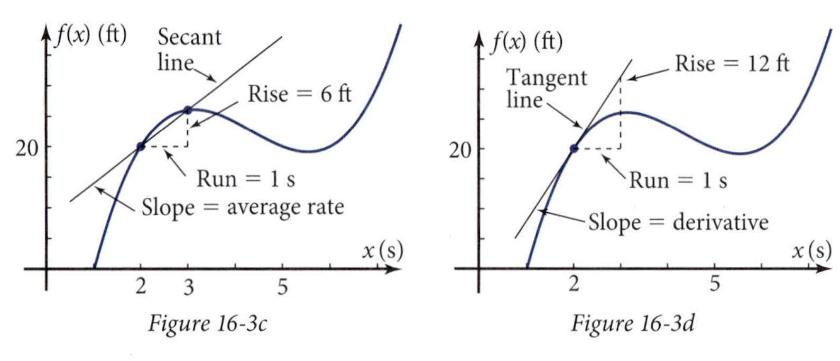

Figure 16-3c

Figure 16-3d

If you take shorter and shorter time intervals from 2 s to x s, the secant lines have slopes that get closer and closer to the instantaneous rate, 12 ft/s. Figure 16-3d shows that 12 is the slope of the **tangent line** to the graph of function f at the point where $x = 2$. Example 1 shows you how to verify graphically that the line with slope equal to the derivative, $f'(2)$, really is tangent to the graph of function f at $x = 2$.

EXAMPLE 1 ➤ Find a particular equation of the line tangent to the graph of $f(x) = x^3 - 13x^2 + 52x - 40$ at the point where $x = 2$. Plot the graph of function f and the tangent line on the same screen. Zoom in on the graphs at the point where $x = 2$ and record what you observe.

SOLUTION

$f(2) = 20$ — Found previously in this section.

$f'(2) = \lim_{x \to 2} \frac{f(x) - f(2)}{x - 2} = 12$ — Found previously in this section.

The equation of the line is

$y = 20 + 12(x - 2)$ — Use point-slope form.

which simplifies to $y = 12x - 4$.

Figure 16-3e shows that the line appears to be tangent to the graph at $x = 2$.

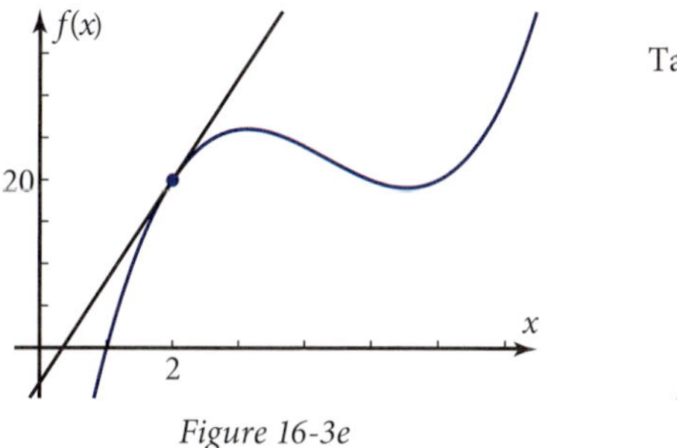

Figure 16-3e

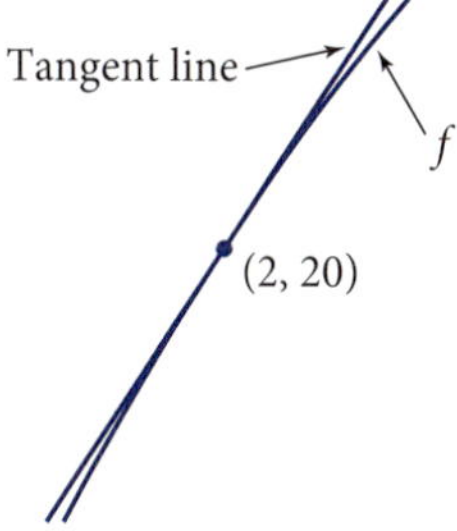

Figure 16-3f

Figure 16-3f zooms in on the point (2, 20) by a factor of 10. The tangent line and the graph of f seem to be very close to each other but still touch only at the point (2, 20). Note that the x- and y-axes in Figure 16-3f lie off the screen. ◀

The box summarizes the concepts of average and instantaneous rates of change.

PROPERTIES: Average and Instantaneous Rates of Change

- The **average rate of change,** $r(x)$, of a function f over an interval starting at $x = c$ is the change in the y-values of the function divided by the corresponding change in the x-values. It is given by the rational function

  $r(x) = \frac{f(x) - f(c)}{x - c}$ — Average rate of change from c to x.

  Graphically: The average rate of change equals the slope of the *secant line* to the graph of function f that contains the points $(c, f(c))$ and $(x, f(x))$.

- The **instantaneous rate of change** of $f(x)$ at $x = c$, called the **derivative** of function f at $x = c$, is the limit of the average rates of change as x approaches c:

  $f'(c) = \lim_{x \to c} \frac{f(x) - f(c)}{x - c}$ — Instantaneous rate of change at $x = c$.

  Graphically: The instantaneous rate equals the slope of the *tangent line* to the graph of function f at the point $(c, f(c))$.

Algebraic Derivative of the Power Function

Suppose you are asked to find the derivative of the power function $f(x) = 7x^4$ at the point where $x = 5$. From the definition of derivative,

$$f'(5) = \lim_{x \to 5} \frac{f(x) - f(5)}{x - 5}$$

The derivative is the limit of the average rates of change.

By substituting an unspecified constant, say c, for 5 you can derive a formula for the instantaneous rate of change that can be used for *any* point c in the domain, without having to go through the entire limit process each time.

$$f'(c) = \lim_{x \to c} \frac{f(x) - f(c)}{x - c} = \lim_{x \to c} \frac{7x^4 - 7c^4}{x - c} = \lim_{x \to c} \frac{7(x^4 - c^4)}{x - c}$$

Factor $(x - c)$ out of the polynomial $x^4 - c^4$ by synthetic substitution.

$$\begin{array}{r|rrrrr} c & 1 & 0 & 0 & 0 & -c^4 \\ & & c & c^2 & c^3 & c^4 \\ \hline & 1 & c & c^2 & c^3 & 0 \end{array}$$

$$f'(c) = \lim_{x \to c} \frac{7(x - c)(x^3 + cx^2 + c^2x + c^3)}{x - c} = \lim_{x \to c} 7(x^3 + cx^2 + c^2x + c^3)$$

$$= 7(c^3 + c^3 + c^3 + c^3) = 7 \cdot 4c^3$$

Substitute c for x to evaluate the limit.

So the formula for $f'(c)$ is $f'(c) = 7 \cdot 4c^3$. To make this formula apply to any value of x, substitute x for c:

If $f(x) = 7x^4$, then $f'(x) = 7 \cdot 4x^3$ or $f'(x) = 28x^3$.

You can see now why the instantaneous rate of change is called the "derivative": The formula for the derivative is *derived* from the equation for $f(x)$. The formula can be used to find an instantaneous rate of change quickly and thus to find the slope of the tangent line at any value of x.

$f'(5) = 28 \cdot 5^3 = 3500$ — The slope of the tangent line at $x = 5$ is 3500.

$f'(1) = 28 \cdot 1^3 = 28$ — The slope of the tangent line at $x = 1$ is 28.

$f'(-2) = 28 \cdot (-2)^3 = -224$ — The slope of the tangent line at $x = -2$ is -224.

Several patterns show up in the equation for $f'(x)$ that allow you to write the derivative of any power function by inspection, without having to go through the limit process at all.

For $f(x) = 7x^4 \Rightarrow f'(x) = 7 \cdot 4x^3$:

- The vertical dilation factor 7 in $f(x)$ appears the same way in $f'(x)$.
- The exponent 4 in $f(x)$ appears as a coefficient in $f'(x)$.
- The exponent 3 in $f'(x)$ is one less than the exponent in $f(x)$.

These patterns occur in the derivative of any power function.

Example 2 shows you how to find the derivative of another power function.

EXAMPLE 2 ➤ Given $f(x) = 0.3x^6$, find an equation for $f'(x)$ and use it to find $f'(2)$. Write an equation for the tangent line at $x = 2$ and plot the tangent line and function f on the same screen, thereby verifying that the line really is tangent to the graph of function f at that point.

SOLUTION

$f(x) = 0.3x^6 \Rightarrow f'(x) = 0.3 \cdot 6x^5$ — Keep the 0.3 dilation factor. Multiply by the old exponent. Decrease the exponent by 1.

$f'(x) = 1.8x^5$ — Simplify.

$f'(2) = 1.8(2^5) = 57.6$ — Find the slope of the tangent line at $x = 2$.

$f(2) = 0.3(2^6) = 19.2$ — Find the y-coordinate at $x = 2$.

The equation of the tangent line is

$y = 19.2 + 57.6(x - 2)$ — Use point-slope form.

Figure 16-3g shows that the line is tangent to the graph at the point where $x = 2$.

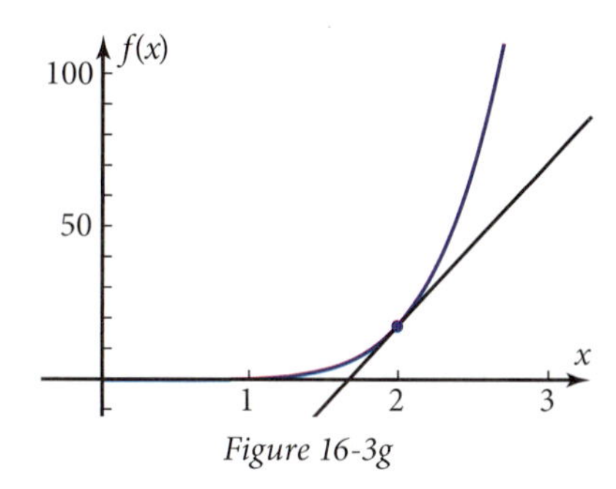

Figure 16-3g

Example 3 shows you how to find the derivative of a polynomial function, the sum of several monomial power functions. The example also shows you how to use the derivative to find a high point or a low point on the graph of a function.

EXAMPLE 3 ➤ Figure 16-3h shows the polynomial function $f(x) = x^3 - 11x^2 + 36x - 26$.

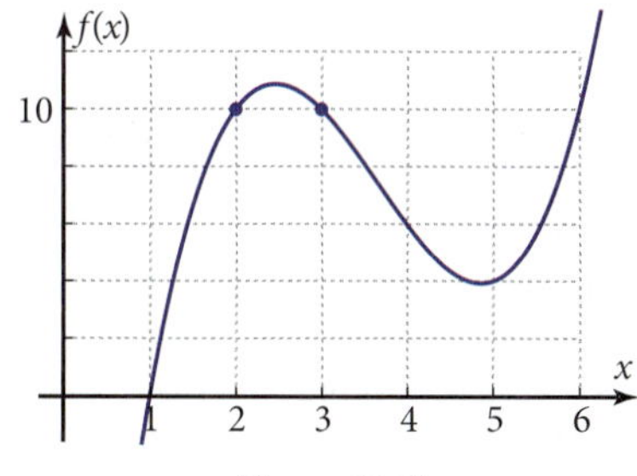

Figure 16-3h

a. Assuming that the derivative of the sum (or difference) of several power functions is the sum (or difference) of the derivatives of the individual terms, find an equation for $f'(x)$.

b. Find $f'(2)$ and $f'(3)$. Show that these values are consistent with the graph.

c. From Figure 16-3h, estimate to one decimal place the x-coordinates of the two extreme points (vertices).

d. At each extreme point in part c, the tangent line will be a horizontal line with slope zero. Thus the derivative will also equal zero. Find the x-coordinates of these two points by setting $f'(x)$ equal to 0 and solving algebraically. Are your estimates in part c close to these exact values?

SOLUTION

a. $f(x) = x^3 - 11x^2 + 36x - 26$ — Write the given function.

$f'(x) = 3x^2 - 22x + 36$ — The derivative of $36x$ is 36. The derivative of -26 is 0. (Why?)

b. $f'(2) = 3(2^2) - 22(2) + 36 = 4$ — Consistent because the graph is sloping up at $x = 2$.

$f'(3) = 3(3^2) - 22(3) + 36 = -3$ — Consistent because the graph is sloping down at $x = 3$.

c. High point at $x \approx 2.5$

Low point at $x \approx 4.9$

d. $3x^2 - 22x + 36 = 0$ — Set the derivative equal to zero.

$x = 2.4648...$ or $x = 4.8685...$ — By the quadratic formula.

The estimates in part c are close to these exact answers.

The property you have been using is called the **power rule for derivatives.** Be careful! The power rule works only for power functions, as you will see in the problem set.

PROPERTIES: The Power and Sum Rules for Derivatives

The Power Rule for Derivatives

If $f(x) = kx^n$, where n stands for a nonnegative integer and k is a constant dilation factor, then $f'(x) = k \cdot nx^{n-1}$.

Verbally: To find the derivative of a power function, multiply by the original exponent and decrease the exponent by 1.

The Sum Rule for Derivatives

If $p(x)$ is a polynomial function $p(x) = f(x) + g(x)$ and if $f(x)$ and $g(x)$ are monomials, then $p'(x) = f'(x) + g'(x)$.

Verbally: To find the derivative of a polynomial function, find the derivative of each term and sum the results.

Problem Set 16-3

Reading Analysis

From what you have read in this section, what do you consider to be the main idea? How are the concepts average rate, instantaneous rate, limit, and derivative related to one another? What kind of line represents the average rate of change of a function? What kind of line represents the instantaneous rate of change of a function? How is the derivative of a function related to a line tangent to the graph of the function? What is the rule for finding the derivative of a power function?

Quick Review

For Problems Q1–Q4, a rational algebraic function has the equation $r(x) = \frac{p(x)}{q(x)}$. What do you know about $p(x)$ and $q(x)$ for each given condition?

Q1. There is a discontinuity at $x = 4$.

Q2. The discontinuity at $x = 4$ is removable.

Q3. The discontinuity at $x = 4$ yields a vertical asymptote.

Q4. $r(5) = 0$

Q5. What is the form $\frac{0}{0}$ called?

Q6. Find: $\lim_{x \to c} \frac{x^2 - 81}{x - 9}$.

Q7. What is the slope of the linear function $y = 3x + 5$?

Q8. What is the slope of the linear function for which $f(2) = 7$ and $f(5) = 19$?

Q9. Write a linear factor of the polynomial function f if $f(3) = 0$.

Q10. If $(3 - 5i)$ is a zero of a polynomial with real-number coefficients, what is another complex zero?

---

1. Consider the function

$$f(x) = -x^3 + 10x^2 - 22x + 23$$

where $f(x)$ is the displacement, in feet, of a moving object from a fixed point at time x, in minutes (Figure 16-3i).

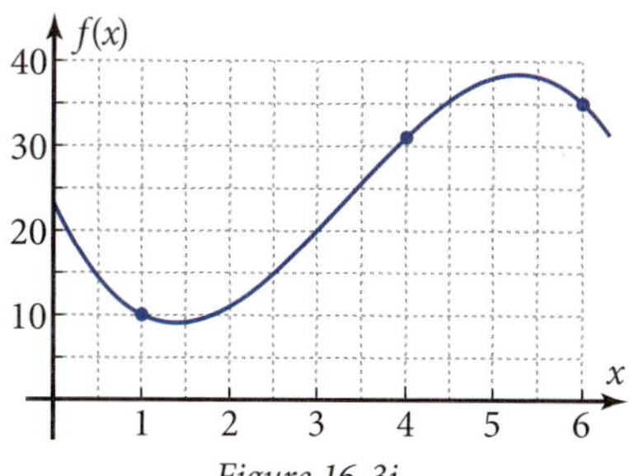

Figure 16-3i

a. Find the average rate of change of $f(x)$ from $x = 4$ to $x = 6$. On a copy of Figure 16-3i, plot a line showing the graphical representation of this rate. How is the line related to the graph of function f?

b. Let $r(x)$ be the average rate of change of $f(x)$ from 4 to x. Write an equation for $r(x)$ and simplify it. Does the simplified equation for $r(x)$ give you the same value of $r(6)$ that you found in part a? If not, correct your work.

c. Find the instantaneous rate of change of $f(x)$ at $x = 4$ by taking the limit of $r(x)$ as x approaches 4. Write an equation of the line containing the point $(4, f(4))$ with slope equal to this instantaneous rate. Plot the graph of this line on your copy of the figure from part a. How is this line related to the graph of function f?

d. Use the sum and power rules for derivatives to find an equation for $f'(x)$. Use this equation to find $f'(4)$. How is your answer related to the instantaneous rate of change in part c?

e. Find $f'(1)$. Write an equation of the line containing $(1, f(1))$ that has slope equal to the derivative at this point. Plot the graph of this line and function f on the same screen. Zoom in on the point $(1, f(1))$ and explain how the result confirms that the line is tangent to the graph of function f at this point.

2. Figure 16-3j shows the position of a moving object as a function of time.

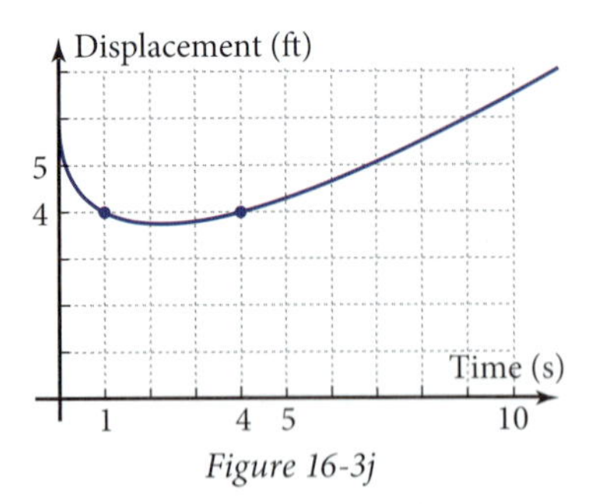

Figure 16-3j

a. On a copy of the figure, draw lines tangent to the graph at times 1 s and 4 s.

b. Estimate the instantaneous rate of change of the displacement at these two times.

c. Is the displacement increasing or decreasing at these two times? At what rate?

3. *Tim and Lum's Board Pricing Problem:* Tim Burr and his brother Lum own a lumber company. Figure 16-3k shows the price, in cents, they charge for boards of varying length, in feet.

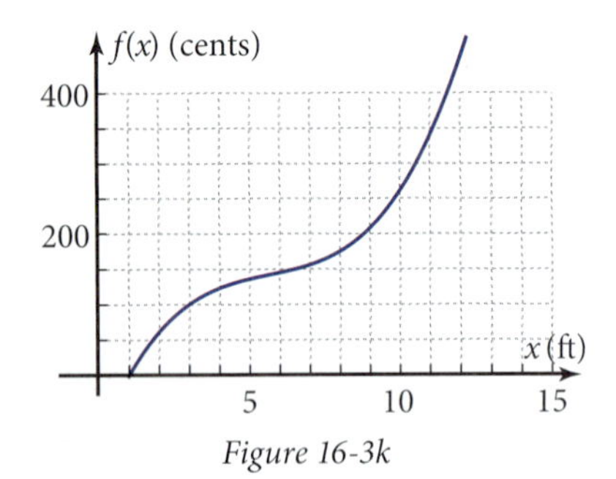

Figure 16-3k

a. A particular equation of this function is the cubic function

$$f(x) = x^3 - 17x^2 + 105x - 89$$

where x is the length of the board in feet, and $f(x)$ is the price of the board in cents. Show by synthetic substitution that $x = 1$ is a zero of this function. What is the real-world meaning of this fact? Find the other two zeros of this function, and interpret their meaning with regard to this mathematical model.

b. What price would you expect to pay for a 20-ft board? How long is a board that costs exactly $10.00?

c. Find the average rate of change of the price for the intervals [10, 10.1], [10, 10.01], and [10, 10.001]. What limit do the average rates seem to be approaching as the width of the intervals gets closer to zero?

d. Find an equation for $f'(x)$. Use the equation to find the instantaneous rate of change of the price at $x = 10$ ft. Use the result to write an equation of the line tangent to the graph at $x = 10$. Use your grapher to confirm that the line really is tangent to the graph by plotting the tangent line and function f on the same screen. Sketch the result.

e. When you buy larger quantities of a product, you expect to pay less per unit for "buying in quantity." Show that this benefit does *not* apply to this mathematical model for cents per foot by showing that the instantaneous rate of change of the price at 10 ft is greater than the instantaneous rate at 5 ft. Give a real-world reason for this behavior.

4. *Door Closer Problem:* Figure 16-3l shows the top view of a door. Suppose the door is equipped with an automatic closer so that when you give the door a push it swings open, slows down, starts swinging back again, and slows down again as it nears its closed position. Assume that while the door is in motion, the number of degrees, $d(t)$, from the closed position as a function of time, t, in seconds, since you pushed it is given by $d(t) = 200t \cdot 2^{-t}$.

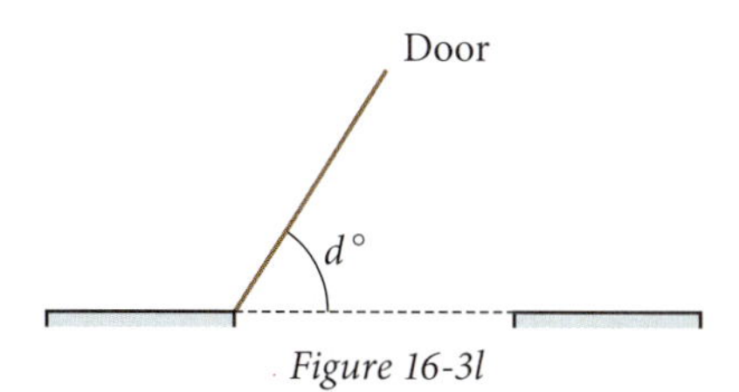

Figure 16-3l

a. Find the average rate of change (in degrees per second) of the door's position from $t = 1$ s to $t = 1.5$ s.

b. Find an estimate of the instantaneous rate of change of the door's position at time $t = 1$ s by finding the average rates of change for these time intervals:

- From 1 s to 1.1 s
- From 1 s to 1.01 s
- From 1 s to 1.001 s

How can you tell from this instantaneous rate of change that the door is still opening at $t = 1$ s?

c. Open the *Instantaneous Rate* exploration at **www.keymath.com/precalc**. Figure 16-3m shows the graph of $d(t)$ as it appears in this Dynamic Precalculus Exploration. Set t_1 equal to 1 s by dragging point t_1 or using the T → 1 button, and then click the SHOW T2 and SHOW RATE buttons. Set Δt equal to 0.5 s by dragging the Δt slider or using the 0.5 button, setting t_2 equal to 1.5. Does the rate $(d_2 - d_1)/(t_2 - t_1)$ equal the value you calculated in part a? What is the graphical interpretation of this average rate of change?

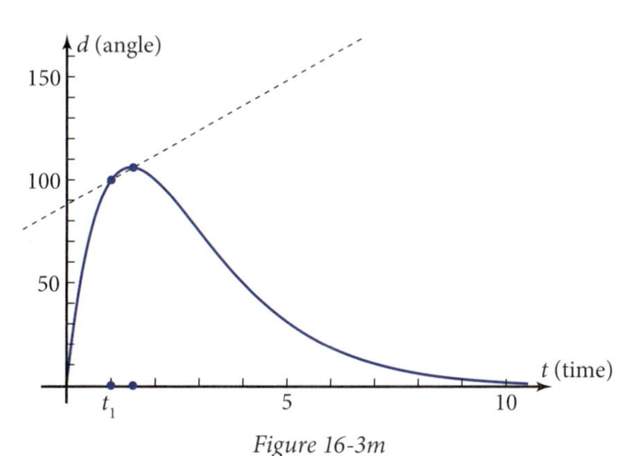

Figure 16-3m

d. Click the buttons that set Δt equal to 0.1 s, 0.01 s, and 0.001 s. How do the average rates of change compare to the values you calculated in part b? Click successive buttons to set Δt to the smallest value shown, 0.00001 s. Describe what happens to the values of the average rate of change and the line through the points t_1 and t_2 as Δt approaches zero.

e. Reset Δt to 0.5 s. Then click the OPEN DOOR button. Describe what happens to the line connecting the two points on the graph. Click the OPEN DOOR button again, and describe what happens numerically to the rate $(d_2 - d_1)/(t_2 - t_1)$. How do you interpret the negative rates for larger values of t?

f. What do you understand better about instantaneous rates of change as a result of working this problem?

For Problems 5–12, find an equation for the derivative, $f'(x)$.

5. $f(x) = x^7$

6. $f(x) = x^9$

7. $f(x) = 8x^6$

8. $f(x) = 12x^{10}$

9. $f(x) = 9x^3 - 5x^2 + 2x - 16$

10. $f(x) = 11x^3 - 3x^2 - 13x + 37$

11. $f(x) = x^6 - 3^6$

12. $f(x) = x^5 + 4^5$

For Problems 13 and 14, find the derivative, $f'(x)$. Use the fact that the derivative is zero at an extreme point (maximum or minimum) of a smooth curve to find the x-coordinates of all extreme points. Confirm your answer graphically.

13. $f(x) = x^3 - 12x^2 + 36x + 17$

14. $f(x) = x^3 - 3x^2 - 9x + 7$

15. *Mistake Problem:* Amos Take has to find the derivative of $f(x) = (2x - 7)^3$. He writes the old exponent, 3, and reduces the exponent by 1, getting $3(2x - 7)^2$. Show Amos that he has made a mistake by first expanding $(2x - 7)^3$, then finding the derivative of each term, and finally comparing this correct answer with Amos's answer. Why can't Amos use the power rule?

16. *Error Problem:* Mae Danerror thinks that because the derivative of a sum of two functions equals the sum of the derivatives, the derivative of a product should equal the product of the derivative. Show Mae that she has made an error by showing her that the derivative of $f(x) = x^3 \cdot x^5$ does not equal $3x^2 \cdot 5x^4$.

17. *Derivative of the Sine Function Problem:* Figure 16-3n shows the graph of $f(x) = \sin x$. In this problem you will make a conjecture about an equation for $f'(x)$.

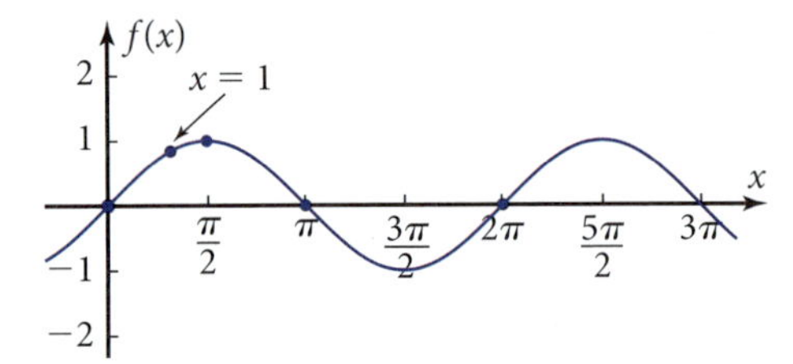

Figure 16-3n

a. Find the average rate of change of $f(x)$ from $x = 0$ to $x = 0.001$. Based on your answer, make a conjecture about the limit of this average rate of change as x approaches zero.

b. Find the average rate of change of $f(x)$ from $x = 1$ to $x = 1.0001$. Show that the answer is close to cos 1.

c. If $f(x) = \sin x$, make a conjecture about an equation for the **derivative function,** $f'(x)$. Show that the conjecture is consistent with the graph of $f(x) = \sin x$ for $x = \frac{\pi}{2}$, $x = \pi$, and $x = 2\pi$.

18. *Derivative of an Exponential Function Problem:* Figure 16-3o shows the graph of the exponential function $f(x) = 2^x$.

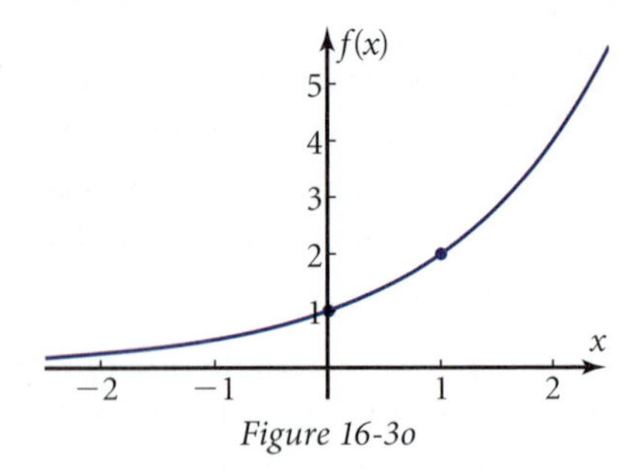

Figure 16-3o

a. Find an estimate of $f'(0)$ by finding the average rate of change from $x = -0.001$ to $x = 0.001$.

b. Mae Danerror thinks the power rule tells us that the derivative of $f(x) = 2^x$ is $f'(x) = x \cdot 2^{x-1}$. Tell Mae what mistake she is making and then show Mae that your answer for part a is inconsistent with the answer Mae got using her incorrect formula.

c. The exact answer in part a is the natural logarithm of a number you may recognize. Find the exact value of $f'(0)$.

d. Find approximations of $f'(1)$ and $f'(3)$ using the method of part a. Each is a multiple of the natural logarithm you found in part b. Use this information to make a conjecture about an equation for the derivative of the exponential function $f(x) = 2^x$.

e. Test your conjecture in part c by using it to find $f'(5)$ for $f(x) = 1.3^x$ and showing that the result is consistent with the average rate of change from $x = 4.999$ to $x = 5.001$.

16-4 Accumulated Rates: The Definite Integral

Recall that distance equals rate multiplied by time. This simple formula assumes that the rate is *constant.* In most real-world situations this formula does not apply because the rate *varies.* In this section you will get a preview of the other major concept of calculus, the *definite integral,* which can be used to find the distance traveled by an object moving at a varying rate.

Objective Given a function that describes the rate of a moving object, find the distance traveled over a specified time period.

Suppose Portia is driving at a rate of 60 ft/s and must slow to 20 ft/s. Figure 16-4a shows a graph of what her velocity, $f(x)$, in feet per second, might look like as a function of time, x, in seconds.

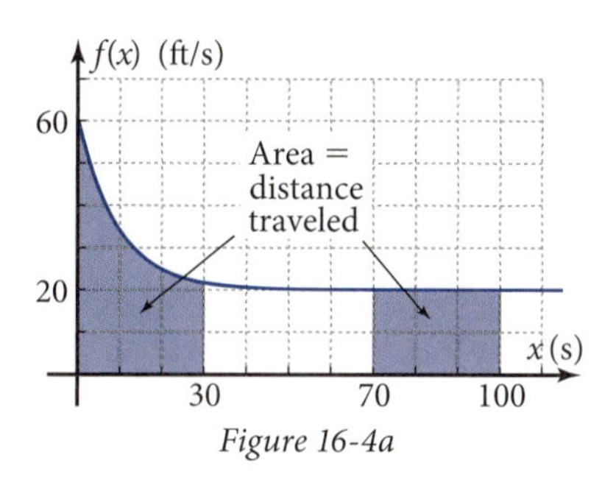

Figure 16-4a

Between 70 s and 100 s, Portia's velocity has leveled off to about 20 ft/s. So during this 30-s time period the distance she has traveled is approximately (20 ft/s)(30 s) or 600 ft. Graphically, this distance equals the area of the rectangular region under the graph of function f between $x = 70$ s and $x = 100$ s in Figure 16-4a.

Between 0 s and 30 s, Portia's velocity is changing. However, the distance is still equal to the area of the region under the graph, in this case from $x = 0$ s to $x = 30$ s. You can estimate that there are about 9.6 grid squares in this region by counting them.

Each grid square represents

$$10\,\frac{\text{ft}}{\text{s}} \cdot 10\text{ s} = 100\text{ ft}$$ Note that the seconds cancel out, leaving ft.

The total distance traveled between 0 s and 30 s is approximately (9.6)(100) or 960 ft. This value is called the **definite integral** of $f(x)$ with respect to x. A definite integral is evaluated by finding the product of two factors, in which one factor may vary.

Counting grid squares is a graphical method of finding the approximate value of a definite integral. The method relies on having an accurate graph of the function and on your skill and patience in counting partial grid squares. Example 1 shows you a numerical method for finding a definite integral.

EXAMPLE 1 ➤ The equation of the function in Figure 16-4a is

$$f(x) = 60 - 40(1 - 0.9^x)$$

Find the approximate value of the definite integral of $f(x)$ with respect to x from $x = 0$ s to $x = 30$ s by approximating the region with rectangles and summing their areas.

SOLUTION Figure 16-4b shows the region under the graph of f divided into six strips of equal width. The strips divide the interval [0, 30] into six 5-s subintervals. Figure 16-4c shows rectangles drawn corresponding to the six subintervals. The base of each rectangle is 5 units, and each height equals the value of $f(x)$ at the midpoint of the subinterval.

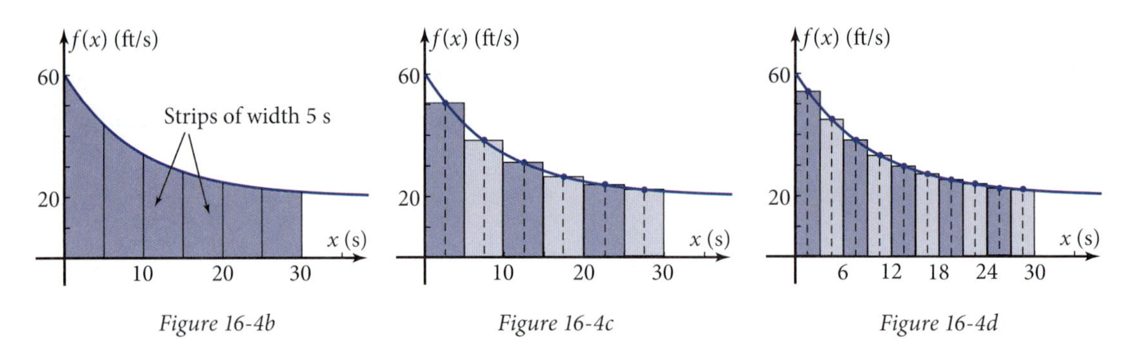

Figure 16-4b *Figure 16-4c* *Figure 16-4d*

The areas of the rectangles in Figure 16-4c are close to the areas of the respective strips in Figure 16-4b. So summing the areas of the rectangles gives an answer close to the area of the region, and this area equals the value of the definite integral—the distance traveled by the car from 0 s to 30 s.

$$\begin{aligned}\text{Integral} &\approx f(2.5)\cdot 5 + f(7.5)\cdot 5 + f(12.5)\cdot 5 + f(17.5)\cdot 5 \\ &\quad + f(22.5)\cdot 5 + f(27.5)\cdot 5 \\ &\approx 5(50.7373... + 38.1500... + 30.7174... + 26.3285... \\ &\quad + 23.7369... + 22.2066...) \\ &\approx 5(191.8769...) \approx 959.3849... \approx 959 \text{ ft}\end{aligned}$$

This is close to the 960 ft found by counting grid squares. ◄

The sum in the solution of Example 1 is called a *midpoint Riemann sum.* The word *midpoint* indicates that the x-values used to find the $f(x)$-values are at the midpoints of the subintervals. The technique is named in honor of German mathematician Bernhard Riemann (1826–1866). An advantage of using a Riemann sum to estimate an integral, as in Example 1, is that it is less subjective than counting grid squares. A more significant advantage is that more accurate approximations of the definite integral can be found simply by dividing the interval, [0, 30], into more subintervals. Figure 16-4d shows a midpoint Riemann sum with 10 subintervals. Note that these rectangles have areas even closer to the areas of the strips in Figure 16-4b.

Example 2 shows you what happens when you increase the number of subintervals numerically.

EXAMPLE 2 ➤ Use a grapher or computer program to find approximations of the definite integral of $f(x) = 60 - 40(1 - 0.9^x)$ from $x = 0$ to $x = 30$ using midpoint Riemann sums with $n = 10$, 100, and 500 subintervals. Use the results to make a conjecture about the exact value of the integral, and hence the exact distance Portia traveled by car as described earlier in this section.

SOLUTION Write or download a program from **www.keymath.com/precalc** for your grapher or computer that will evaluate midpoint Riemann sums. Before running the program, enter the equation for the function into your grapher. When you run the program, it should ask you for the lower and upper bounds of integration (0 and 30 in this instance) and the number of subintervals to be used. Then the program should calculate the width of each subinterval $\left(\frac{(30-0)}{6} = 5 \text{ in Example 1}\right)$ and enter a loop that calculates and accumulates (adds) successive values of $f(x)$ for the stored function. Finally, the program should calculate the Riemann sum by multiplying the accumulated sum by the width of the subinterval (5 in Example 1) and display the answer.

Let R_n stand for the midpoint Riemann sum with n increments. Running the program should give these results:

$R_6 = 959.3849...$

Check that your program gives the correct result for Example 1.

$R_{10} = 962.0460...$

$R_{100} = 963.5399...$

$R_{500} = 963.5545...$

Conjecture:

- Integral ≈ 963.6, because the Riemann sums are getting larger.
- Distance traveled ≈ 963.6 ft, which is close to the previous estimates. ◄

Extending the ideas in Example 2 leads to a way to define a definite integral. Take the limit of the midpoint Riemann sums, R_n, as n becomes infinitely large. If this limit exists, then it is equal to the definite integral. In calculus, you will learn algebraic ways to calculate the limit of a Riemann sum. The exact result for the definite integral in Example 2 is

$$600 + \frac{40(0.9^{30} - 0.9^{0})}{\ln 0.9} = 963.5551...$$

A helpful way to remember the symbol that is used for a definite integral is to take the letter S (for "sum") and stretch it out (as in taking a limit) so that it becomes $\int$. Then write the starting and ending x-values as a subscript and a superscript and apply the symbol to the function.

$$\int_0^{30} 60 - 40(1 - 0.9^x)\,dx$$

The dx is pronounced "dee x," or "with respect to x," and means that x is the independent variable.

DEFINITION: Definite Integral

Verbally: The definite integral of $f(x)$ from a to b with respect to x is equal to the limit of the Riemann sums for the region bounded by the graph of function f and the x-axis between $x = a$ and $x = b$ as the number, n, of subintervals of $[a, b]$ becomes infinite and the width, h (for "horizontal"), of each subinterval approaches zero.

Algebraically: $\int_a^b f(x)\,dx = \lim_{n\to\infty} R_n = \lim_{h\to 0} R_n$

If function f represents a rate of change of another function, such as the velocity of a moving object (the rate of change of displacement), then the definite integral *accumulates* the rates to give the displacement, explaining the reason for the title of this section.

Note: In this section you have learned how to use definite integrals to approximate the distance traveled when an object moves at a varying rate, but always in one direction (so the rate is always positive). In calculus, you will learn to use definite integrals to calculate the distance traveled when both the rate at which an object travels and its direction may vary.

Problem Set 16-4

Reading Analysis

From what you have read in this section, what do you consider to be the main idea? How is a definite integral of a function related to the graph of that function? What is meant by a Riemann sum for a function, and how is a definite integral related to a Riemann sum?

Quick Review

Q1. Write an equation for the derivative of the power function $f(x) = 7x^3$.

Q2. What is the graphical meaning of the derivative of a function at a particular point?

Q3. What is the real-world meaning of the derivative of a function at a particular point?

Q4. If $f'(3) = -7$, what do you know about the behavior of function f at $x = 3$?

Q5. If function f has a vertex at $x = 4$ and there is a value for $f'(4)$, what does $f'(4)$ equal?

Q6. Find $\lim_{n\to\infty} S_n$ for the partial sums of the geometric series $100 + 80 + 64 + \cdots$.

Q7. Multiply: $\begin{bmatrix} 2 & 4 \\ 3 & 7 \end{bmatrix}\begin{bmatrix} 5 & 1 \\ 6 & 9 \end{bmatrix}$

Q8. Find the dot product: $(3\vec{i} + 2\vec{j} + 5\vec{k}) \cdot (4\vec{i} + 6\vec{j} + 1\vec{k})$

Q9. Sketch the point $(r, \theta) = (-3, 40°)$ in polar coordinates.

Q10. In the equations $\begin{cases} x = \cos t \\ y = \tan t \end{cases}$, the variable t is called a(n) __?__.

For Problems 1 and 2, estimate the definite integral graphically.

1. $\int_2^{14} f(x)\,dx$

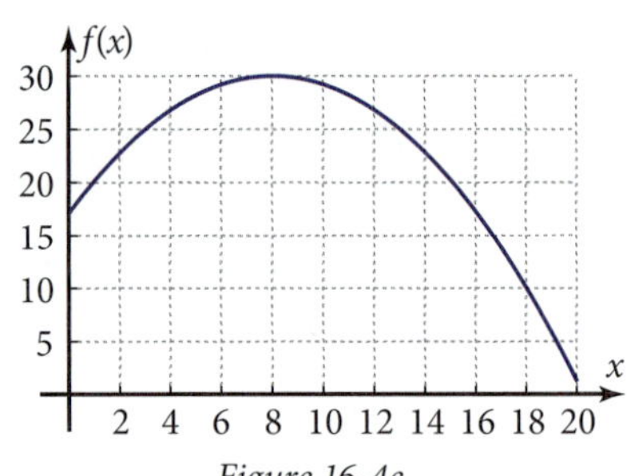

Figure 16-4e

2. $\int_0^4 f(x)\,dx$

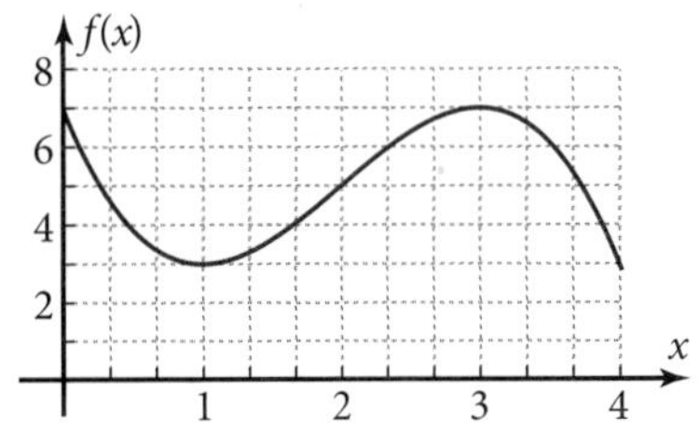

Figure 16-4f

In Problems 3 and 4, use Riemann sums to estimate the definite integral representing the distance traveled at a variable rate.

3. *Distance Problem 1:* Ron S. Miles sprints for 10 s, speeding up at first and then slowing down as he gets tired. His velocity, $f(x)$, in meters per second, is given by $f(x) = 11x \cdot 1.4^{-x}$ where time, x, is in seconds (Figure 16-4g).

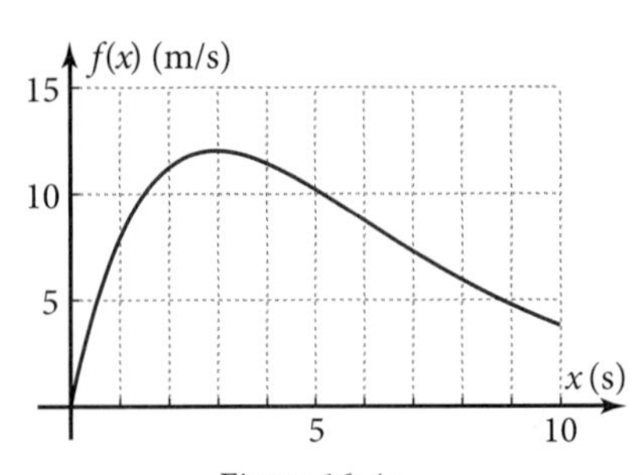

Figure 16-4g

a. Explain why the distance he runs between $x = 0$ s and $x = 10$ s must be found as a definite integral, not just as rate multiplied by time. Write this integral using proper mathematical notation.

b. On a copy of Figure 16-4g, sketch the five rectangles for the midpoint Riemann sum R_5 for the integral in part a. Write the five terms of R_5 and evaluate the sum without using a grapher or computer program.

c. Use your program from Example 2 or download the Riemann program from **www.keymath.com/precalc** to evaluate R_5, and show that it gives the same result as your answer in part b. Then use your program to find R_{10}, R_{20}, and R_{100}.

d. Ron's exact total distance for the 10-s time period is the limit of R_n as n becomes infinitely large. What does this distance seem to be?

4. *Distance Problem 2:* As Iona Carr drives away from a stop sign, her sister Lisa determines that the car's velocity, in feet per second, from time $x = 1$ s to $x = 9$ s is given by $f(x) = 3\sqrt{x}$.

a. Graph $f(x)$ and identify the quantity that represents the total distance they travel in this time interval.

b. Write a definite integral for this total distance and estimate the distance using midpoint Riemann sums with increasing numbers of increments. Give the values of the sums you use to lead you to your conclusion.

Problems 5 and 6 show that definite integrals can represent quantities other than distance.

5. *Bacteria Problem:* Figure 16-4h shows the rate of growth of bacteria in a laboratory culture, in bacteria per day. The rate of growth is given by $r(x) = 200 \cdot 1.2^x$, where x is the number of days since the culture was started. Between day 3 and day 9 the number of bacteria in the culture increased by an amount equal to

$$\int_3^9 r(x)\,dx$$

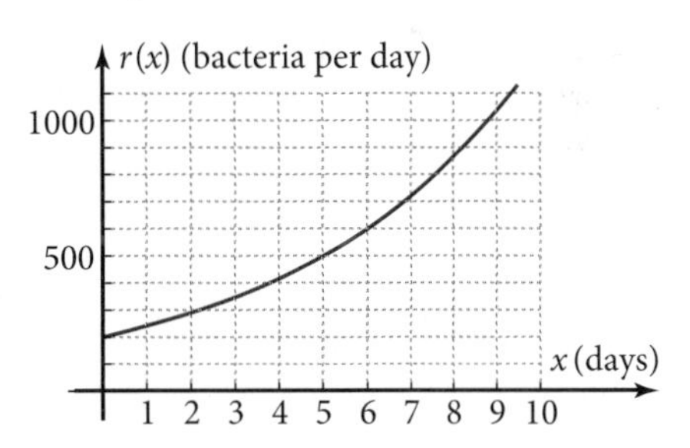

Figure 16-4h

a. On a copy of Figure 16-4h, sketch the three rectangles for the midpoint Riemann sum R_3 for this integral. Evaluate R_3 without using a grapher or computer program. State the units of the answer, and explain why these units are consistent with the real-world meaning of this integral.

b. Use the Riemann program to evaluate R_3, and show that it gives the same result as your answer in part a. Then use the program to find R_{10}, R_{20}, and R_{100}. Make a conjecture about the limit of the Riemann sums as the number of increments becomes infinitely large. What is the real-world meaning of this sum?

6. *Football Volume Problem:* A football is about 12 in. long. Between the two ends, the cross-sectional area (Figure 16-4i) increases, reaches a maximum halfway between the ends, and then decreases. Suppose the cross-sectional area, in square inches, is given by

$$A(x) = 40 \sin \frac{\pi}{12} x$$

where x is the number of inches from one end of the ball to the point where the cross-sectional area is being measured.

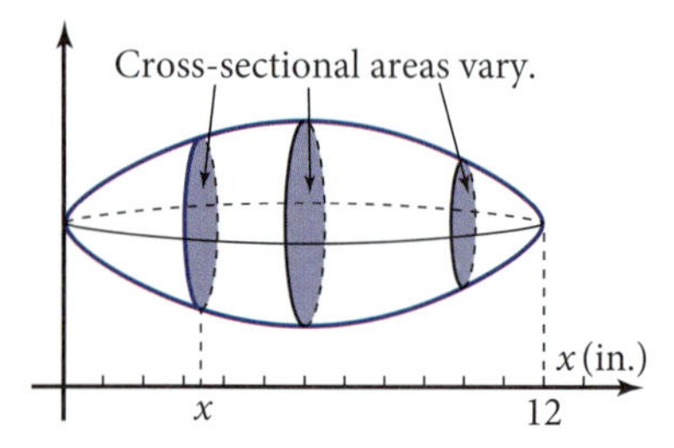

Figure 16-4i

a. Sketch the graph of function A with the appropriate domain.

b. The volume of a solid object equals its cross-sectional area times its length. Explain why the volume of the football must be found by evaluating a definite integral, not by simple multiplication.

c. Write a definite integral to describe the volume of the football. Explain why the integral has units of volume. Estimate the volume by finding midpoint Riemann sums with increasing numbers of increments.

7. WEB *Historical Research Problem:* On the Internet or in another reference source, look up Sir Isaac Newton and Gottfried Wilhelm Leibniz. Write a paragraph or two about each person. Include, if possible, their contributions to the subject of calculus, the branch of mathematics that is concerned with derivatives and integrals of functions.

Gottfried Wilhelm Leibniz (1646–1716), a German mathematician, made significant contributions to the development of calculus, including the power rule for derivatives. (The Granger Collection, New York)

16-5 Chapter Review and Test

In this chapter you have been introduced to the concepts of calculus. The derivative measures the instantaneous rate of change in the y-value of a function with respect to the x-value. The definite integral measures the product of the x-value and the y-value where the y-value is variable. The concept of limit is used in evaluating both derivatives and integrals.

Review Problems

R0. Update your journal with what you have learned in this chapter. Include things such as the most important thing you have learned as a result of studying this chapter and the meaning of limit, derivative, and integral.

R1. Figure 16-5a shows the graph of $f(x) = \sin x$.

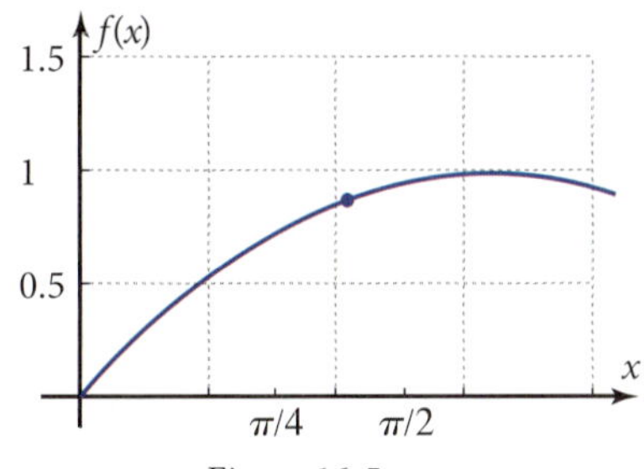

Figure 16-5a

a. The point shown on the graph is at $x = \frac{\pi}{3}$. The instantaneous rate of change of $f(x)$ with respect to x at this point is exactly 0.5. (You will learn how to find this value when you study calculus.) Show that the average rate of change of $f(x)$ from $x = \frac{\pi}{3}$ to $x = \left(\frac{\pi}{3} + 0.01\right)$ is close to 0.5. Show that the average rate of change from $x = \frac{\pi}{3}$ to $x = \left(\frac{\pi}{3} + 0.001\right)$ is even closer to 0.5.

b. On a copy of Figure 16-5a, plot a line containing the point $\left(\frac{\pi}{3}, f\left(\frac{\pi}{3}\right)\right)$ and having slope 0.5 (the instantaneous rate of change). How is this line related to the graph of function f?

c. On your copy of Figure 16-5a, shade the region under the graph of function f from $x = 0$ to $x = 1.5$. By counting squares, estimate the area of this region. Show that your estimate is close to the exact area, which equals $(1 - \cos 1.5)$.

R2. **a.** Write the definition of $L = \lim_{x \to c} f(x)$.

b. For the piecewise function in Figure 16-5b, write the limits of $f(x)$ as x approaches a, b, and c, or state that the limit does not exist. Give a reason for each answer.

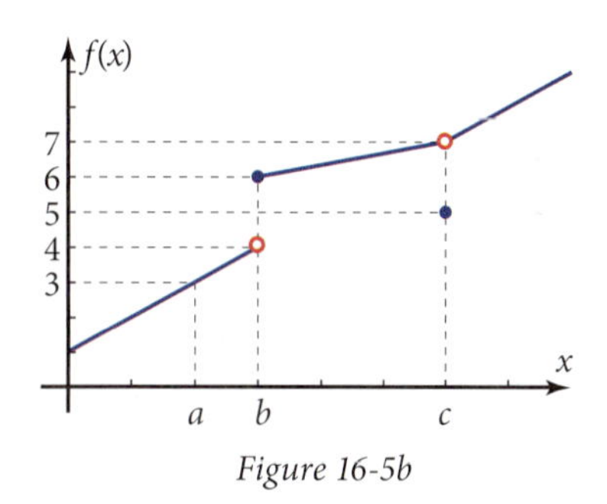

Figure 16-5b

c. Figure 16-5c shows the graph of the rational function

$$g(x) = \frac{x^3 - 13x^2 + 57x - 81}{x - 3}$$

Simplify the fraction in the equation of the function. Use the result to show that $\lim_{x \to 3} g(x) = 6$. Find an interval of x-values close to 3 such that keeping x in this interval, but not equal to 3, makes $g(x)$ stay within 0.7 unit of 6. How could this reasoning be generalized to show that 6 really is the limit of $g(x)$ as x approaches 3?

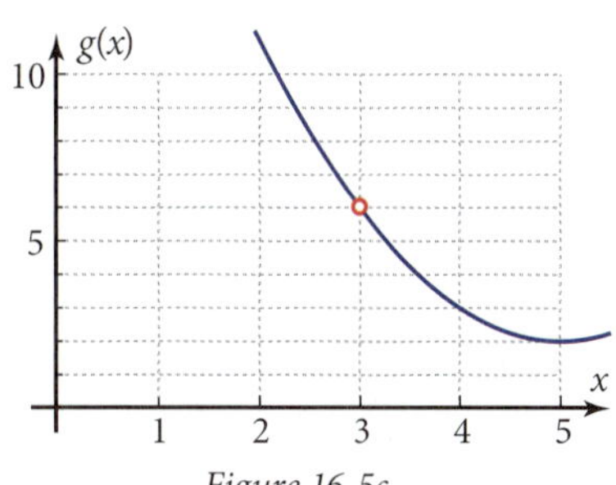

Figure 16-5c

d. Figure 16-5d shows the graph (solid curve) of the rational function

$$h(x) = \frac{x^3 - 5x^2 - 2x + 29}{x - 4}$$

Write the equation for function h in mixed-number form. From the result, how can you tell that there is a vertical asymptote at $x = 4$, not a removable discontinuity? Plot the graph of function h and the polynomial part of function h on the same screen. Does your graph match the dashed parabola in Figure 16-5d? How is the parabola related to the graph of function h?

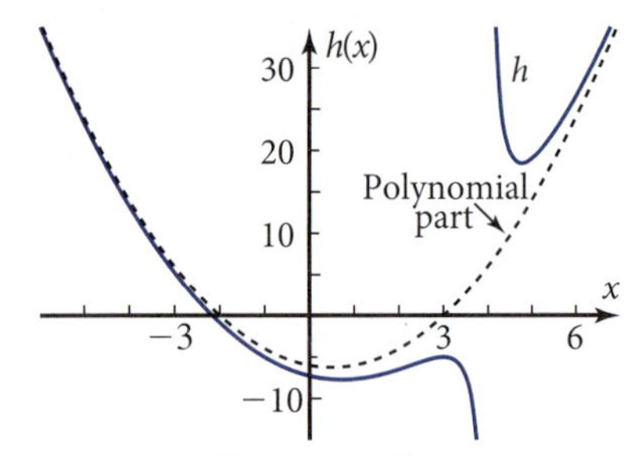

Figure 16-5d

R3. *Train Problem:* A train headed north stops for some switching operations. As it does these, the locomotive goes back and forth across a roadway crossing. The displacement, $d(x)$, in feet, of the front of the locomotive north of the crossing at time x, in minutes, is given by this cubic function (Figure 16-5e):

$$d(x) = 120x^3 - 1200x^2 + 3480x - 2400$$

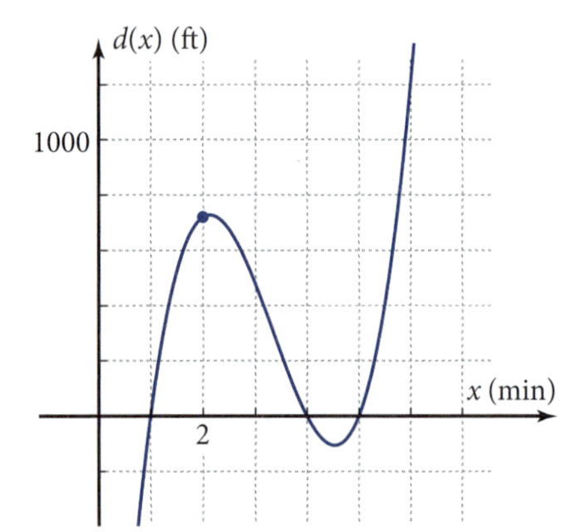

Figure 16-5e

a. The average rate of change (or velocity) of $d(x)$ for the time interval $[2, x]$ is

$$r(x) = \frac{d(x) - d(2)}{x - 2} \qquad \text{Rate} = \frac{\text{displacement}}{\text{time}}$$

Find $r(2.01)$ and $r(2.001)$. From the results, make a conjecture about the instantaneous velocity of the locomotive at time $x = 2$ min.

b. Substitute the expression for $d(x)$ and the value of $d(2)$ into the equation for $r(x)$. Then simplify the rational function by using synthetic substitution. Find the instantaneous velocity by taking the limit of $r(x)$ as x approaches 2. Does your conjecture in part a agree with this instantaneous velocity? Is the train going north or south at time $x = 2$ min? Explain.

c. Use the power rule to find the derivative of $d(x)$. Substitute to find $d'(2)$. Do you get the same value as the instantaneous velocity you found in part b?

d. Plot the graph of the average rate of change, $r(x)$. What is the vertical coordinate of the removable discontinuity at $x = 2$? How is this number related to earlier parts of this problem?

e. On a copy of Figure 16-5e, plot a line containing the point $(2, d(2))$ with slope equal to the derivative, $d'(2)$. How is this line related to the graph of function d?

f. By setting $d'(x)$ equal to zero, find the time between $x = 4$ min and $x = 5$ min when the locomotive has stopped and changes direction.

R4. *Dam Problem:* Water flows out of a lake behind a dam at a rate, $f(x)$, in thousands of cubic feet per hour, that decreases exponentially with time according to the equation $f(x) = 50 \cdot 0.8^x$, where x is the number of hours that have elapsed since the flood gates were opened (Figure 16-5f).

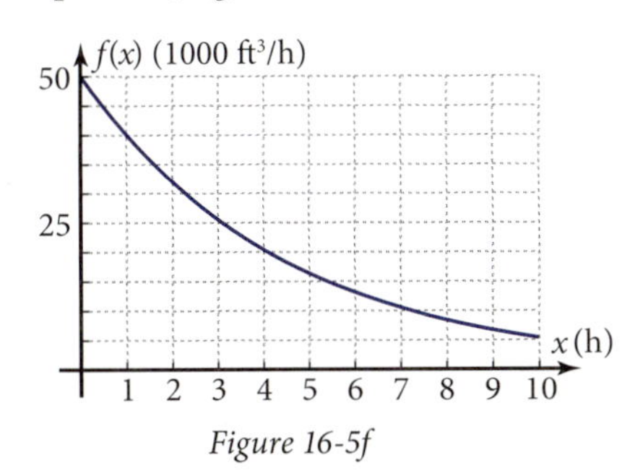

Figure 16-5f

a. Find $\int_1^9 f(x)\,dx$ approximately, by counting squares.

b. The symbol in part a is pronounced __?__. What real-world quantity does the answer in part a represent?

c. On a copy of Figure 16-5f, sketch a midpoint Riemann sum, R_4, with four rectangles of equal width. Evaluate this Riemann sum numerically, showing your work.

d. Use the Riemann program to evaluate R_{20} and R_{100}. What limit do the Riemann sums seem to be approaching as the number of rectangles becomes infinitely large?

Concept Problem

C1. *Exact Value of an Integral Problem:* Figure 16-5g shows the graph of $f(x) = 3x^2$.

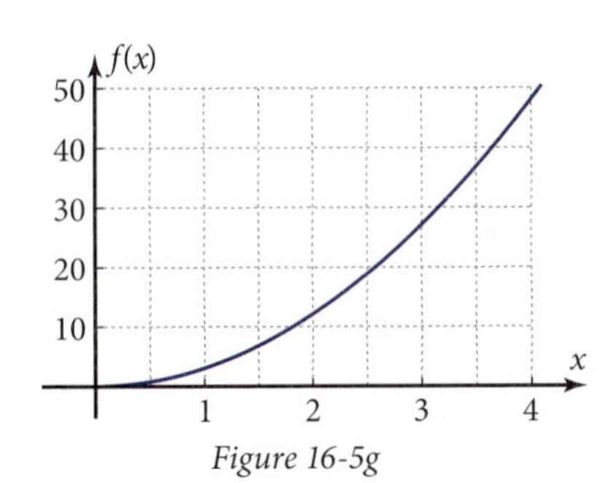

Figure 16-5g

a. The definite integral of $f(x)$ with respect to x from $x = 2$ to $x = 4$ turns out to be an integer. Using your Riemann sum program, find this integer. Show the values of the sums you use.

b. Confirm your answer in part a by counting squares.

c. Let $g(x) = x^3$. Find $g(4) - g(2)$. What is interesting about the answer? How is function f related to function g?

d. In calculus you will learn that the derivative of $g(x) = -\cos x$ is $g'(x) = \sin x$. Use the pattern you observed in part c to find the exact value of the definite integral of $\sin x$ with respect to x from $x = 0$ to $x = \pi$. Show that a midpoint Riemann sum with 100 increments is very close to this exact value.

Chapter Test

Part 1: No calculators allowed (T1–T6)

T1. Write the definition of $L = \lim_{x \to c} f(x)$.

T2. For the piecewise function in Figure 16-5h, find the limit of $f(x)$ as x approaches a, b, and c, or state that the limit does not exist. Give a reason for each answer.

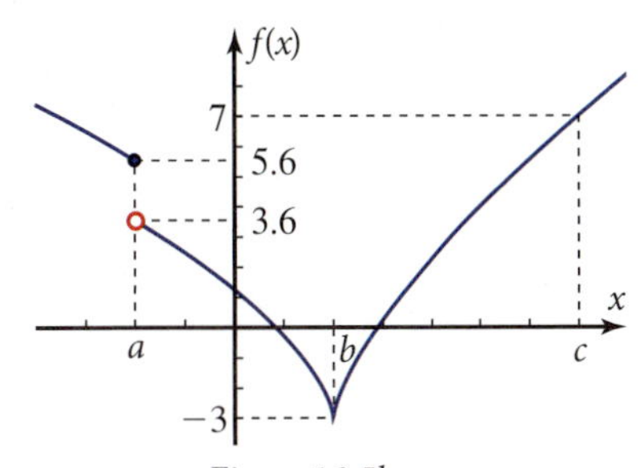

Figure 16-5h

For Problems T3–T6, Figure 16-5i shows the graph of a function f and two lines, one green and one magenta.

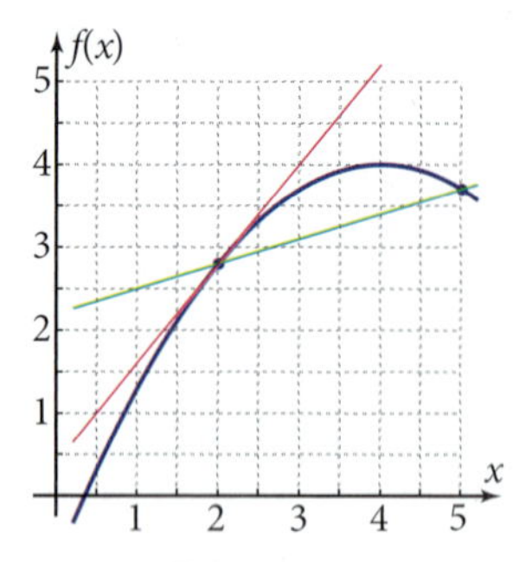

Figure 16-5i

T3. Which line corresponds to the average rate of change of $f(x)$ from $x = 2$ to $x = 5$? This line is called a __?__ line. Find the approximate value of this rate of change.

T4. The line that corresponds to the instantaneous rate of change of $f(x)$ at $x = 2$ is called a __?__ line. Find the approximate value of this rate of change. The instantaneous rate of change of a function is called the __?__ of the function.

T5. Find the approximate value of the definite integral of $f(x)$ with respect to x from $x = 2$ to $x = 5$.

T6. If $f(x)$ is the velocity of a moving object in centimeters per second and x is time in seconds, what quantity does the definite integral represent?

Part 2: Graphing calculators allowed (T7–T17)

For Problems T7–T10, Figure 16-5j shows the graph of the rational function

$$f(x) = \frac{-x^3 + 11x^2 - 37x + 36}{x - 4}$$

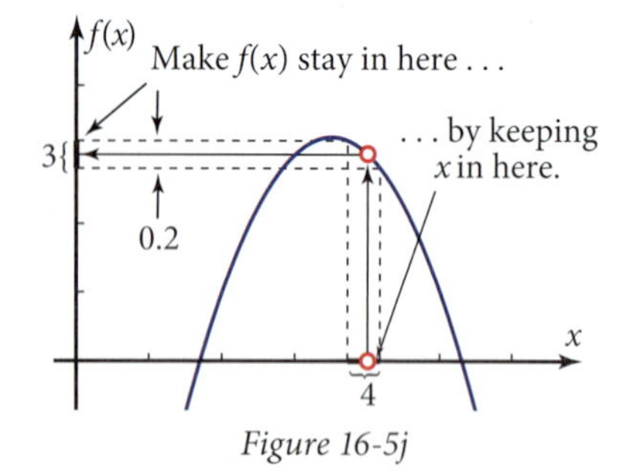

Figure 16-5j

T7. Explain why there is no value of $f(4)$.

T8. Simplify the fraction and use the result to show that $\lim_{x \to 4} f(x) = 3$.

T9. Find the largest interval around $x = 4$ in which you could keep x to make $f(x)$ stay within 0.2 unit of 3. Why must this interval exclude 4 itself?

T10. How could the reasoning of Problem T9 be generalized to show that 3 really is the limit of $f(x)$ as x approaches 4?

For Problems T11–T13, Figure 16-5k shows the graph of the rational function

$$f(x) = \frac{x^2 - 4x - 9}{2x - 10}$$

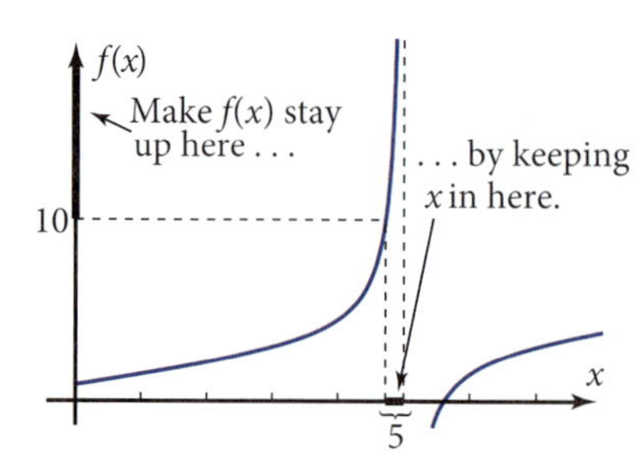

Figure 16-5k

T11. Write the equation for $f(x)$ in mixed-number form. Use the result to show that the discontinuity at $x = 5$ is not removable and thus there is no limit of $f(x)$ as x approaches 5.

T12. Find an interval on the negative side of $x = 5$ that will make $f(x)$ stay greater than 10 when x is kept in this interval. Pick an x-value in this interval and show that $f(x)$ really is greater than 10. How could this reasoning be generalized to show that $f(x)$ can be made greater than any positive number, no matter how large, simply by keeping x close enough to 5 on the negative side?

T13. Plot the graph of function f and the polynomial part of the simplified equation for function f on the same screen. Sketch the result. Describe the relation of the polynomial graph to the graph of function f.

For Problems T14–T16, Figure 16-5l shows the graph of $f(x) = -0.3x^2 + 2.4x - 0.8$.

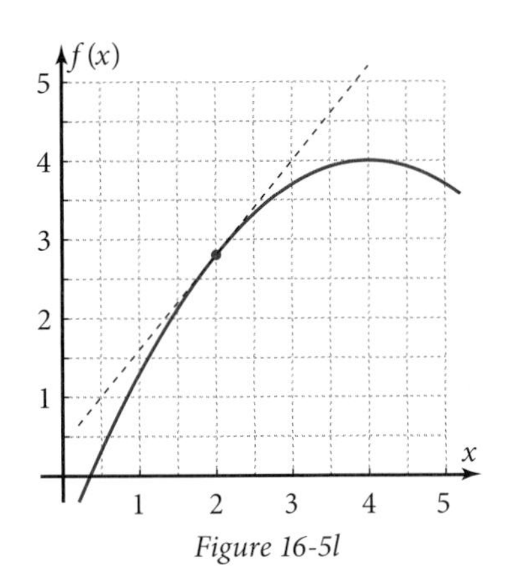

Figure 16-5l

T14. Find an equation for the derivative, $f'(x)$. Use the equation to find the exact value of $f'(2)$. On a copy of Figure 16-5l, draw a rise and a run between appropriate points to show that $f'(2)$ is the slope of the dashed line tangent to the graph at $x = 2$ in the given figure.

T15. Find a midpoint Riemann sum with three rectangles of equal width for $\int_2^5 f(x)\,dx$ without using a grapher or computer program. Show the calculations that led to your answer.

T16. By finding Riemann sums with more subintervals, make a conjecture about the exact value of the definite integral in Problem T15. Show the Riemann sums you calculated to arrive at your conclusion.

T17. What did you learn as a result of taking this test that you did not know before?

16-6 Cumulative Review, Chapters 14–16

These problems constitute a 2- to 3-hour "rehearsal" for your examination on these topics:

- Probability, and functions of a random variable
- Sequences and series
- Limits, derivatives, and integrals

Review Problems

Part 1: No calculators allowed (1–19)

For Problems 1–3, two independent events A and B have probabilities $P(A) = 0.8$ and $P(B) = 0.9$. Find each probability.

1. $P(A \text{ and } B)$
2. $P(A \text{ or } B)$
3. $P(\text{not } A)$
4. Find the number of permutations of five objects taken three at a time.
5. Using factorials, write the number of combinations of ten objects taken six at a time.
6. The probability that a bent coin lands heads up on any one flip is 60%. If the coin lands heads up, you win 10 points. If it lands tails up, you lose 20 points. What is your mathematical expectation for any one flip?

For Problems 7–12, a series starts $100 + 90 + \cdots$.

7. Write the common ratio if the series is geometric.
8. Write the third term if the series is geometric.
9. Find the number to which the series converges if the series is geometric.
10. Write the third term if the series is arithmetic.
11. Find the third partial sum if the series is arithmetic.
12. Find four arithmetic means between 100 and 90.

For Problems 13–19, Figure 16-6a shows the graph of a function f.

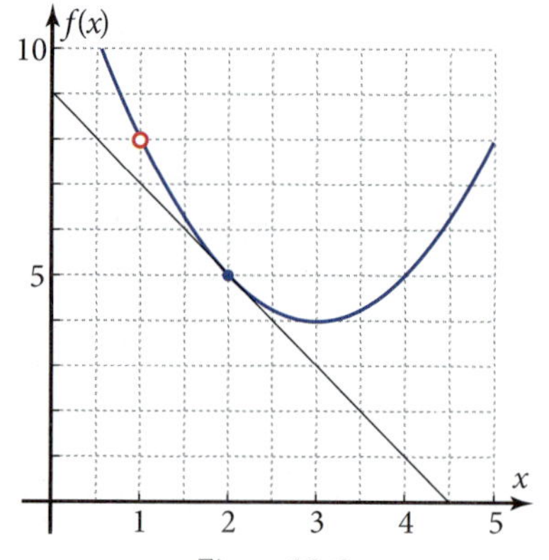

Figure 16-6a

13. What feature does the graph of function f have at $x = 1$?
14. Find $\lim_{x \to 1} f(x)$.
15. Find the average rate of change of $f(x)$ with respect to x from $x = 2$ to $x = 5$.
16. The line is tangent to the graph of function f at the point (2, 5). Find $f'(2)$.
17. What quantity does $f'(2)$ in Problem 16 represent?
18. Find the approximate value of $\int_2^5 f(x)\,dx$.
19. The symbol in Problem 18 is pronounced __?__.

Part 2: Graphing calculators allowed (20–49)

Probability Distribution Problem (Problems 20–26): You flip a thumbtack several times. The probability that it lands "point up" on any one flip is 0.4.

20. What is the probability that the thumbtack lands point down?

21. The probability that flip 2 is the first time the thumbtack lands point up is the probability that it does not land point up on flip 1 and does land point up on flip 2. Let $P(x)$ be the probability that the first time the thumbtack lands point up is on flip x. Find $P(1)$, $P(2)$, $P(3)$, $P(4)$, and $P(5)$. Plot these values on graph paper.

22. Find the probability that the thumbtack first lands point up on flip 1 or flip 2 or flip 3.

23. The probability that the thumbtack first lands point up on one of the first n flips is S_n, the nth partial sum of a geometric series. Use the formula for S_n to find the probability that the first point up is on one of the first ten flips.

24. Find the limit of S_n in Problem 23 as n becomes infinitely large. Interpret the answer in terms of this random experiment.

25. You flip the thumbtack six times. What is the probability that it lands point up exactly two times? What special name is given to the kind of probability distribution that occurs in this random thumbtack experiment?

26. Show how to calculate the number of combinations of six objects taken two at a time. How is this number related to the probability in Problem 25?

Indy 500 Problem (Problems 27–31): Manny Moore has probability 70% of finishing in one of the top ten places in the Indianapolis 500 car race. His wife, Annie, has probability 80% of finishing in the top ten.

27. What is the probability that Manny and Annie both finish in the top ten? What must you assume in order to make this computation, given no additional information about these probabilities?

28. The probability that Manny finishes in the top ten but Annie does not is 0.14. Show how you could calculate this number.

29. What is the probability that Annie finishes in the top ten but Manny does not?

30. What is the probability that neither Manny nor Annie finishes in the top ten?

31. Manny and Annie decide to spend any winnings on various vacations. They will spend these amounts, depending on who wins.

M and A:	Hawaii, $8000
M, not A:	California, $3000
A, not M:	Florida, $4000
No Moore:	Stay home, $0

Calculate the mathematically expected amount they will spend.

Sequences and Series Problems (Problems 32–40)

32. Give the first four terms of the arithmetic *sequence* with first term 7 and common difference 5.

33. Give the fourth partial sum of the geometric *series* with first term 8 and common ratio 3.

34. You put $12 into a piggy bank the first week, $15 the second week, $18 the third week, and so on. How much will you put into the piggy bank the tenth week? What total amount will be in the piggy bank after the tenth deposit?

35. After a long time, you break open the piggy bank in Problem 34 and invest $500 of the money in a bank account that pays an interest rate of 3% per year, compounded monthly. If you make no withdrawals, how much, total, will you have in the account at the end of the 36th month? Why is a sequence a more reasonable mathematical model for this problem than a continuous function?

36. Wildlife conservationists find that the number of catfish in a particular lake at the end of any one year is only 0.9 times what it was at the beginning of that year. Assume that at time $t = 0$ yr there were 100 catfish in the lake. The conservationists decide to add 30 more catfish at the end of each year. Write a recursion formula and use it to predict the number of catfish at $t = 1, 2, 3, 4$, and 5 yr. If this recursion formula is followed for many years, will the catfish population level off asymptotically, or will it increase without limit? How do you know?

37. For the sequence 2, 5, 10, 17, 26, 37, 50, . . . , there is a relatively simple pattern relating the term values to the term numbers. Find this pattern and then write an explicit formula for t_n as a function of n. Use the formula to find t_{100}.

38. One of the terms in the sequence in Problem 37 is $t_n = 5042$. Use the formula you found in Problem 37 to calculate n algebraically.

39. The series $\sum_{n=1}^{\infty} \frac{1}{n^{1.2}}$ is called a p-series because the term index in the denominator is raised to a power. Write the first four terms of the series. Find S_{100}, the 100th partial sum.

40. In the binomial series that results from expanding $(a - b)^{10}$, find the term containing b^7.

Tree Problem (Problems 41–44): Ann R. Burr has a tree nursery in which she stocks various sizes of live oak trees. Figure 16-6b shows the price, $f(x)$, in dollars, she charges for a tree of height x, in feet, planted on the customer's property, given by the cubic function

$$f(x) = x^3 - 25x^2 + 249x - 225$$

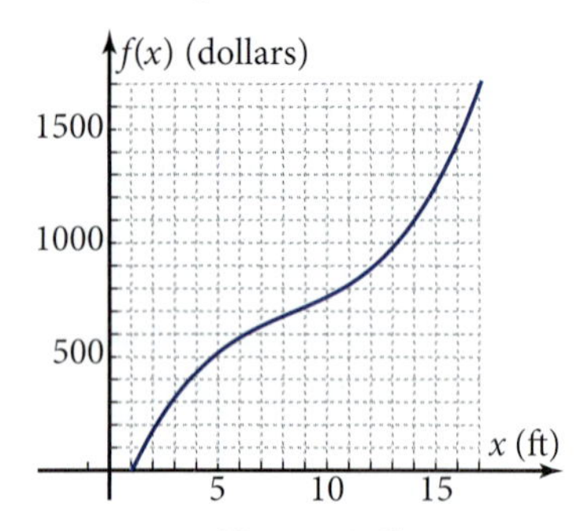

Figure 16-6b

41. Find the average rate of change of price, in dollars per foot, from $x = 10$ ft to $x = 15$ ft. On a copy of Figure 16-6b, plot a secant line whose slope is this average rate.

42. The average rate of change of price from 10 ft to x ft is given by the rational function

$$r(x) = \frac{f(x) - f(10)}{x - 10}$$

Substitute for $f(x)$ and $f(10)$. Simplify the resulting rational expression by using synthetic substitution. Then find the instantaneous rate of change of the price at $x = 10$ ft by taking the limit of $r(x)$ as x approaches 10.

43. Use the power rule for derivatives to find an equation for $f'(x)$. Use the result to show that $f'(10)$ equals the instantaneous rate of change you found in Problem 42. On your copy of Figure 16-6b, plot a line containing the point $(10, f(10))$ with slope equal to $f'(10)$. Describe the relationship of this line to the given graph.

44. Show that the instantaneous rate of change of the price at 16 ft is greater than that at 10 ft. What is a possible real-world reason for the greater rate?

Bathtub Problem (Problems 45–48): The rate, $r(x)$, in liters per minute, at which water flows out of a particular bathtub at time x, in minutes after the drain is opened, is given by the function (Figure 16-6c)

$$r(x) = x^2 - 20x + 84$$

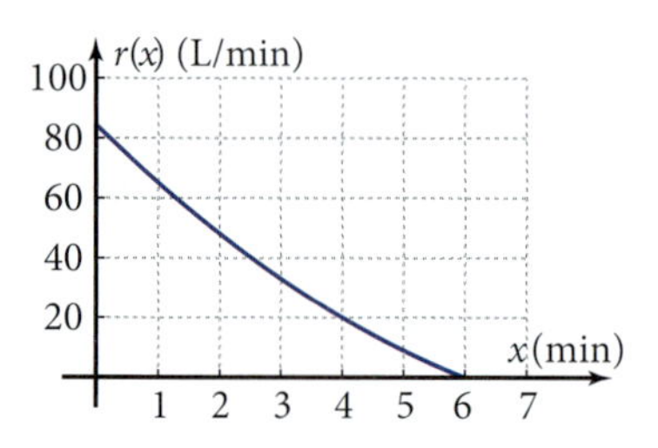

Figure 16-6c

45. By counting squares, estimate the value of the definite integral of $r(x)$ with respect to x from $x = 0$ min, when the drain is opened, to $x = 6$ min, when the tub is empty. Use the appropriate symbol to represent this integral.

46. On a copy of Figure 16-6c, sketch the three rectangles of R_3, the midpoint Riemann sum with three rectangles of equal width, for the integral in Problem 45. Evaluate R_3, showing your work. Is the answer close to your answer in Problem 45?

47. The limit of the Riemann sums as the number of rectangles becomes infinite is the exact value of the definite integral, an integer in this case. Use the Riemann program to find R_{10} and R_{100}. Use the results to make a conjecture about the value of the integral.

48. Find the units of the definite integral. What is the real-world meaning of this integral?

49. Summarize what you have learned as a result of doing this cumulative review.

Types of Numbers, Axioms, and Other Properties

In algebra you learned names for various types of numbers, such as real numbers, imaginary numbers, and rational numbers. You also learned commutative axioms, closure axioms, transitive axioms, and so on, as well as other properties such as the multiplication property of zero that can be proved from these axioms. In this appendix you will refresh your memory about the names of these numbers, axioms, and other properties, and some definitions. You'll also see, in the examples, how to use axioms to prove properties.

Types of Numbers

Complex numbers are numbers of the form $a + bi$, where a and b are **real numbers** and $i = \sqrt{-1}$. The real numbers form a subset of the complex numbers, for which the number b equals 0. You can group real numbers several ways: as positive or negative, rational or irrational, algebraic or transcendental, and so forth. Figure A-1a on the next page shows the set of complex numbers and some of its subsets.

An important thing to realize is that a **rational number** is a number that *can be* expressed as a ratio of two integers. It does not have to be written that way. For instance,

23 is rational because it can be written $\frac{23}{1}$, $\frac{46}{2}$, and so on.
$\sqrt{9}$ is rational because it can be written as 3, which equals $\frac{3}{1}$.
$2\frac{3}{7}$ is rational because it can be written $\frac{17}{7}$.
3.87 is rational because it can be written $\frac{387}{100}$.
5.3333... (repeating) is rational because it can be written $5\frac{1}{3}$, which equals $\frac{16}{3}$.

Irrational numbers, on the other hand, *cannot* be expressed as the ratio of two integers. Irrational numbers include the nth root of any integer that is not an exact nth power. For instance, you can prove by contradiction that $\sqrt{3}$ is irrational. To do this, assume that $\sqrt{3}$ is a rational number, so it could be written as

$$\sqrt{3} = \frac{a}{b}$$

where a and b are relatively prime integers. But then

$$(\sqrt{3})^2 = 3 = \frac{a^2}{b^2}$$

which is impossible because a and b have no common factors to cancel. So you've arrived at a contradiction, which could be caused only by the faulty assumption that $\sqrt{3}$ is a rational number.

An **algebraic number** is a number that is a solution to a polynomial equation with rational-number coefficients. You can think of it as a number that can be expressed using a finite number of algebraic operations, including taking the roots.

Transcendental numbers are numbers that are not algebraic. They "transcend," or go beyond, the algebraic operations. The most notable examples you may know of are π and e. Most trigonometric and circular function values, as well as most logarithm values, are also transcendental.

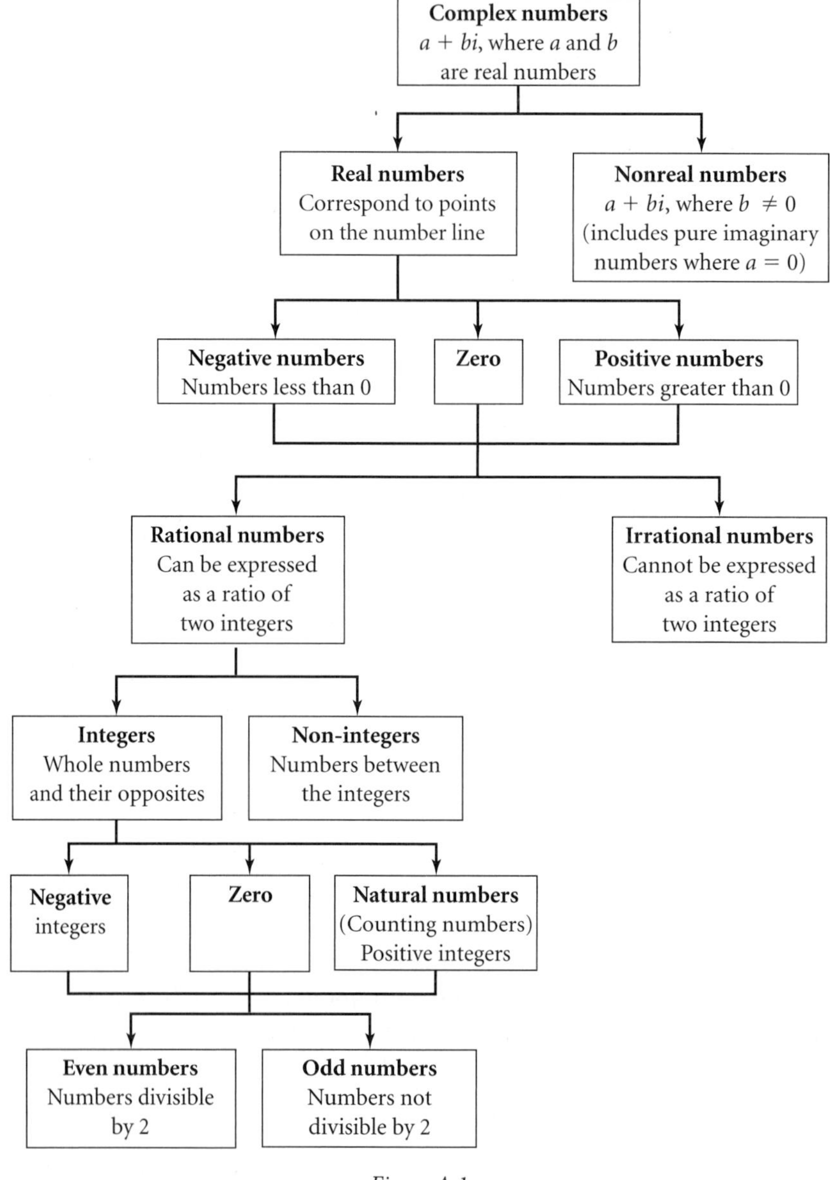

Figure A-1a

Axioms for Addition and Multiplication

There are 11 basic axioms for real numbers that apply to the operations of addition and multiplication. These axioms are called the **field axioms**. An *axiom* is a property that is assumed to be true so that it can be used as the basis for a mathematical system. Generally speaking, mathematicians prefer to have as few axioms as possible and to prove other properties using these axioms. The box lists the field axioms and what they state.

PROPERTIES: The Field Axioms

If x, y, and z are real numbers, then the following statements are true.

1. **Closure Under Addition**

   $x + y$ is a unique real number.

   "You can't get out of the set of real numbers when you add two real numbers."

2. **Closure Under Multiplication**

   xy is a unique real number.

   "You can't get out of the set of real numbers when you multiply two real numbers."

3. **Additive Identity**

   $$0 + x = x + 0 = x$$

   "Adding zero does not change a number."

4. **Multiplicative Identity**

   $$1 \cdot x = x \cdot 1 = x$$

   "Multiplying by 1 does not change a number."

5. **Additive Inverses**

   Every real number x has a unique additive inverse $-x$ such that $x + (-x) = 0$.

   "You can undo the addition of a number by adding its opposite."

6. **Multiplicative Inverses**

   Every real number x (except 0) has a unique multiplicative inverse $\frac{1}{x}$ such that $x \cdot \frac{1}{x} = 1$.

   "You can undo multiplication by a number by multiplying by its reciprocal."

(continued)

Appendix A

Properties: The Field Axioms, continued

7. **Commutative Axiom for Addition**

$$x + y = y + x$$

"You can commute two terms in a sum without changing the answer."

8. **Commutative Axiom for Multiplication**

$$xy = yx$$

"You can commute two factors in a product without changing the answer."

9. **Associative Axiom for Addition**

$$(x + y) + z = x + (y + z)$$

"You can associate terms in a sum differently without changing the answer."

10. **Associative Axiom for Multiplication**

$$(xy)z = x(yz)$$

"You can associate factors in a product differently without changing the answer."

11. **Distributive Axiom for Multiplication Over Addition**

$$x(y + z) = xy + xz$$

"You can distribute multiplication over addition without changing the answer."

Subtraction and division can be defined in terms of addition and multiplication with the aid of the inverse axioms.

DEFINITIONS: Subtraction and Division

$$x - y = x + (-y)$$

"Subtracting a number means adding its opposite."

$$x \div y = x \cdot \frac{1}{y}$$

"Dividing by a number means multiplying by its reciprocal."

Appendix A

Axioms for Equality and Order

There are three axioms for equality and three axioms for order (inequality), all of which state facts related to the = sign, the < sign, and the > sign. The term *order* pertains to the order in which numbers appear on the number line.

PROPERTIES: Axioms for Equality and for Order

If x, y, and z stand for real numbers, then the following statements are true.

Reflexive Axiom for Equality

$x = x$

"A real number is equal to itself, so a variable stands for the same number wherever it appears in an expression."

Symmetric Axiom for Equality

If $x = y$, then $y = x$.

"You can reverse the sides of an equation without affecting the equality."

Transitive Axioms for Equality and Order

If $x = y$ and $y = z$, then $x = z$.

If $x < y$ and $y < z$, then $x < z$.

If $x > y$ and $y > z$, then $x > z$.

"If the first number equals the second number and the second number equals the third number, then the first number equals the third number." (And so on.)

"Equality *goes through* (hence the name 'transit…') from first to last number."

Trichotomy Axiom (or Comparison Axiom)

For any two given numbers x and y, exactly *one* of these is true:

$x < y$

$x = y$

$x > y$

"A number y cuts the number line into three pieces (hence the name *trichotomy*): numbers less than it, numbers equal to it, and numbers greater than it."

Properties That Can Be Proved from the Axioms

The other familiar properties of real numbers can be proved from the axioms. Four examples are shown here. Other provable properties are listed after Example 4.

EXAMPLE 1 ➤ **Substitution into Sums and Products**

$x + y$ and $z + y$ stand for the same number, provided $x = z$.

xy and zy stand for the same number, provided $x = z$.

"You can substitute equal quantities for equal quantities in a sum or a product."

PROOF By the closure axioms, $x + y$ and xy each stand for a *unique* real number. Thus, it does not matter what symbol is used for x if $x = z$; both have the same value. Given that $y = y$, from the reflexive axiom, $x + y = z + y$ and $xy = zy$. Q.E.D. ◄

EXAMPLE 2 ➤ **Addition Property of Equality**

If $x = y$, then $x + z = y + z$.

"You can add the same number to both sides of an equation without affecting the equality."

PROOF

$x + z = x + z$	Reflexive axiom: A number equals itself.
$x = y$	Given.
$\therefore\ x + z = y + z$, Q.E.D.	Substitution into a sum. ◄

The **converse** of a property is the statement you get by interchanging the hypothesis (the "if" part) and the conclusion (the "then" part). The converse of a property may or may not be true. For instance, "If you run marathons, then you are in good shape" is true. But the converse, "If you are in good shape, then you run marathons," is false. So converses must also be proved.

EXAMPLE 3 ➤ **Converse of the Addition Property of Equality**

If $x + z = y + z$, then $x = y$.

"You can cancel equal terms on both sides of the equal sign."

PROOF

$x + z = y + z$	Given.
$(x + z) + (-z) = (y + z) + (-z)$	Addition property of equality: Add the opposite of z to both sides.
$x + [z + (-z)] = y + [z + (-z)]$	Associative axiom for addition.
$x + 0 = y + 0$	Additive inverse axiom.
$\therefore\ x = y$, Q.E.D.	Additive identity axiom. ◄

Note that the proof may seem to involve an excessive number of steps, especially because you are so familiar with "adding the opposite to both sides." However, to constitute a proof, each step must be justified by an axiom or by a previously proved property.

EXAMPLE 4 ➤ **Combining Like Terms**

Prove that $2x + 3x = 5x$.

PROOF

$2x + 3x = (2 + 3)x$	Write one side of the desired equation. Use the distributive axiom (read in reverse).
$= 5x$	Arithmetic.
$\therefore\ 2x + 3x = 5x$, Q.E.D.	Transitive axiom for equality.

The box shows these four properties as well as others. Each one can be proved from the axioms and from those properties that appear before it in the box.

PROPERTIES: Real Number Properties Provable from the Axioms

If x, y, and z are real numbers, then the following statements are true.

1. Substitution into Sums and Products

$x + y$ and $z + y$ stand for the same number, provided $x = z$.

xy and zy stand for the same number, provided $x = z$.

"You can substitute equal quantities for equal quantities in a sum or a product."

2. Addition Property of Equality

If $x = y$, then $x + z = y + z$.

"You can add the same number to both sides of an equation without affecting the equality."

3. Converse of the Addition Property of Equality

If $x + z = y + z$, then $x = y$.

"You can cancel equal terms on both sides of the equal sign."

4. Combining Like Terms

Example: $2x + 3x = 5x$

"You can combine like terms by adding their coefficients."

5. Multiplication Property of Equality

If $x = y$, then $xz = yz$.

"You can multiply both sides of an equation by the same number without affecting the equality."

(continued)

Properties: Real Number Properties Provable from the Axioms, continued

6. Cancellation Property of Equality for Multiplication

If $xz = yz$ and $z \neq 0$, then $x = y$.

"You can divide both sides of an equation by the same nonzero number."

7. Opposite of an Opposite

$-(-x) = x$

"The opposite of the opposite of a number is the original number."

8. Reciprocal of a Reciprocal

If $x \neq 0$, then $\frac{1}{\frac{1}{x}} = x$

"The reciprocal of the reciprocal of a nonzero number is the original number."

9. Reciprocal of a Product

If $x \neq 0$ and $y \neq 0$, then $\frac{1}{xy} = \frac{1}{x} \cdot \frac{1}{y}$.

"The reciprocal of a product can be split into the product of the two reciprocals."

10. Multiplication Property of Fractions

If $x \neq 0$ and $y \neq 0$, then $\frac{ab}{xy} = \frac{a}{x} \cdot \frac{b}{y}$.

"A quotient of two products can be split into a product of two fractions."

11. Multiplication Property of Zero

For any real number x, $x \cdot 0 = 0$.

"Zero times any real number is zero."

12. Converse of the Multiplication Property of Zero

If $xy = 0$, then $x = 0$ or $y = 0$.

"The only way a product can equal zero is for one of the factors to equal zero."

13. Multiplication Property of −1

$-1 \cdot x = -x$

"-1 times a number equals the opposite of that number."

(continued)

Appendix A

Properties: Real Number Properties Provable from the Axioms, continued

14. Product of Two Opposites

For two positive numbers, x and y,

$$(-x)(-y) = xy$$

"Negative times negative is positive."

15. Opposites of Equal Numbers

If $x = y$, then $-x = -y$.

"If two real numbers are equal, then their opposites are equal. Therefore, you can take the opposite of both sides of an equation without affecting the equality."

16. Reciprocals of Equal Numbers

If $x = y \neq 0$, then $\frac{1}{x} = \frac{1}{y}$.

"If two nonzero numbers are equal, then their reciprocals are equal. Therefore, you can take the reciprocal of both sides of an equation, unless it involves taking the reciprocal of zero, without affecting the equality."

17. The Square of a Real Number

$x^2 \geq 0$ for any real number x.

"The square of a real number is never negative."

18. Distributive Property for Subtraction

$$x(y - z) = xy - xz$$

"Multiplication distributes over subtraction."

19. Distributive Property for Division

$$\frac{x + y}{z} = \frac{x}{z} + \frac{y}{z} \qquad z \neq 0$$

"Division distributes over addition."

(Reading this property from right to left explains why you can add fractions that have a common denominator.)

$$\frac{x - y}{z} = \frac{x}{z} - \frac{y}{z} \qquad z \neq 0$$

"Division distributes over subtraction."

(continued)

Appendix A

Properties: Real Number Properties Provable from the Axioms, continued

20. Division of a Number by Itself

$\frac{n}{n} = 1 \qquad n \neq 0$

"A nonzero number divided by itself equals 1."

21. Reciprocal of 1

$\frac{1}{1} = 1$

"1 is its own reciprocal."

22. Dividing by 1

$\frac{n}{1} = 1$

"Any number divided by 1 equals that number."

23. Dividing Numbers with Opposite Signs

$\frac{-x}{y} = -\frac{x}{y}$

"A negative number divided by a positive number is negative."

$\frac{x}{-y} = -\frac{x}{y}$

"A positive number divided by a negative number is negative."

24. Opposite of Sum and Difference

$-(x + y) = -x + (-y)$

"The opposite of a sum equals the sum of the opposites."

$-(x - y) = y - x$

"$x - y$ and $y - x$ are opposites of each other."

Mathematical Induction

You recall that the distributive axiom states that

$$a(x_1 + x_2) = ax_1 + ax_2$$

In words, "Multiplication distributes over a sum of *two* terms." It seems reasonable that multiplication distributes over sums of three terms, four terms, and so forth. In general,

$$a(x_1 + x_2 + x_3 + \cdots + x_n) = ax_1 + ax_2 + ax_3 + \cdots + ax_n$$

In this appendix you will learn about *mathematical induction*, a technique by which you can prove that this *extended distributive property* is true for *any* number of terms, no matter how large.

Unfortunately, the field axioms are not sufficient to prove this extended distributive property. Another axiom, the **well-ordering axiom,** allows the proof to be done.

PROPERTY: The Well-Ordering Axiom

Any nonempty set of positive integers has a *least* element.

The truth of this axiom, like that of most axioms, should be obvious to you. The name comes from the fact that a set is said to be "well-ordered" if its elements can be arranged in order, starting with a least element. The set of positive real numbers and the set of all integers do not have this property. There is a restriction to nonempty sets because the empty set has no elements at all and thus cannot have a least element.

You can prove the extended distributive property by contradiction. You assume that it is false and then show that this assumption leads to a contradiction. Example 1 shows how to do this.

EXAMPLE 1 ➤ **Extended Distributive Property**

Prove that $a(x_1 + x_2 + x_3 + \cdots + x_n) = ax_1 + ax_2 + ax_3 + \cdots + ax_n$ is true for any integer $n \geq 2$.

Appendix B

PROOF (by contradiction)

Assume that the property is false. Then there is a positive integer $n = p$ for which

$$a(x_1 + x_2 + x_3 + \cdots + x_n) \neq ax_1 + ax_2 + ax_3 + \cdots + ax_p$$

By the distributive axiom, the property is true for $n = 2$. That is,

$$a(x_1 + x_2) = ax_1 + ax_2$$

Draw Venn diagrams for two sets of positive integers (Figure B-1a),

T = {positive integers n for which the property is true}

F = {positive integers n for which the property is false}

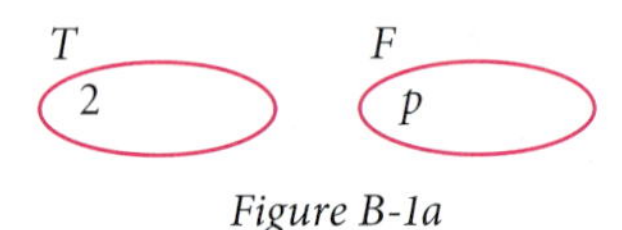

Figure B-1a

The integer 2 is an element of T, and (by assumption) p is an element of F. Write these integers in the Venn diagram of Figure B-1a.

Because F is a nonempty set of positive integers, the well-ordering axiom allows you to conclude that it has a least element, ℓ. Because ℓ is the *least* element of F, $\ell - 1$ is not an element of F. Because 2 is in T, $\ell \geq 3$, and thus both ℓ and $\ell - 1$ are positive integers. Because $\ell - 1$ is a positive integer and is not in F, it must be an element of T. Write $\ell - 1$ and ℓ in the Venn diagram (Figure B-1b).

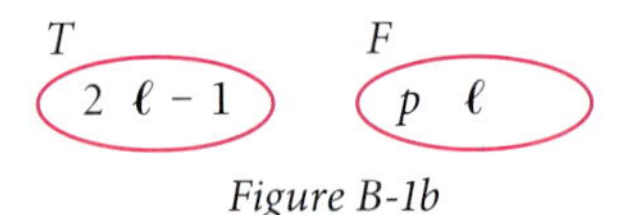

Figure B-1b

Because $\ell - 1$ is in T, the property is true for $n = \ell - 1$. Because ℓ is in F, the property is false for $n = \ell$. Write the (true) statement of the property if $n = \ell - 1$.

$$a(x_1 + x_2 + x_3 + \cdots + x_{\ell-1}) \neq ax_1 + ax_2 + ax_3 + \cdots + ax_{\ell-1}$$

The distributive property if $n = \ell - 1$.

Now, start with $n = \ell$.

$a(x_1 + x_2 + x_3 + \ldots + x_{\ell-1} + x_\ell)$

$= a((x_1 + x_2 + x_3 + \cdots + x_{\ell-1}) + x_\ell)$ — Associate the first $\ell - 1$ terms.

$= a(x_1 + x_2 + x_3 + \cdots + x_{\ell-1}) + ax_\ell$ — Multiplication distributes over a sum of two terms.

$= ax_1 + ax_2 + ax_3 + \cdots + ax_{\ell-1} + ax_\ell$ — Substitute from the statement of the distributive property if $n = \ell - 1$.

Therefore, the property is true for $n = \ell$ terms. But this statement contradicts the statement that the property is false for $n = \ell$ terms. The only place this contradiction could have arisen is the assumption that there is a positive integer $n = p$ for which the property is false. Thus, there is no positive integer n for which the property is false, and it is true for all integers $n \geq 2$. Q.E.D.

Once you understand the process, the proof may be shortened. All you need to do is (1) prove that assuming the property is true for *one* value of n implies that it is true for the *next* value of n, and (2) prove that there is one value of n for which it actually *is* true. These two ideas combine to form the **induction principle.**

PROPERTY: The Induction Principle

If

(1) there is a positive integer n_0 for which a property is true, and

(2) for any integer $k \geq n_0$, assuming the property is true for $n = k$ allows you to conclude it is also true for $n = k + 1$,

then the property is true for *any* integer $n \geq n_0$.

Note that the step where you demonstrate that the property is true for one value of n is called the *anchor.* This step "anchors" the induction. The step in which you assume that the property is true for $n = k$ is called the *induction hypothesis.*

Example 2 shows how the proof of the extended distributive property can be shortened with the help of the induction principle. Proofs done this way are said to be done by *mathematical induction,* which is the topic of this appendix.

EXAMPLE 2 ➤ Extended Distributive Property, Again

Prove that $a(x_1 + x_2 + x_3 + \cdots + x_n) = ax_1 + ax_2 + ax_3 + \cdots + ax_n$ is true for any integer $n \geq 2$.

PROOF (by induction on n)

Anchor: $a(x_1 + x_2) = ax_1 + ax_2$ by the distributive axiom. Therefore, the property is true for $n = 2$.

Induction Hypothesis: Assume the property is true for some positive integer $n = k$, where $k \geq 2$. That is, assume that

$$a(x_1 + x_2 + x_3 + \cdots + x_k) = ax_1 + ax_2 + ax_3 + \cdots + a_k$$

Demonstration for $n = k + 1$:

For $n = k + 1$,

$a(x_1 + x_2 + x_3 + \cdots + x_k + x_{k+1})$

$= a((x_1 + x_2 + x_3 + \cdots + x_k) + x_{k+1})$ — Associate the first k terms.

$= a(x_1 + x_2 + x_3 + \cdots + x_k) + ax_{k+1}$ — By the anchor (distribute over *two* terms).

$= ax_1 + ax_2 + ax_3 + \cdots + ax_k + ax_{k+1}$ — Substitute, using the induction hypothesis.

Conclusion: Because (1) the property is true for one value of n, namely, $n = 2$, and because (2) assuming it is true for $n = k$ implies that it is true for $n = k + 1$, you can conclude that

$$a(x_1 + x_2 + x_3 + \cdots + x_n) = ax_1 + ax_2 + ax_3 + \cdots + ax_n$$

is true for any integer $n \geq 2$. Q.E.D.

You may have detected one weakness in the proof of the extended distributive property. Without ever stating it, you have assumed an extended *associative property* and an extended *transitive property*. The proof of the extended associative property is a bit tricky; it is presented as Example 3.

EXAMPLE 3 ➤ Extended Associative Property for Addition

Prove that

$$x_1 + x_2 + x_3 + \cdots + x_{n-1} + x_n = (x_1 + x_2 + x_3 + \cdots + x_{n-1}) + x_n$$

for any integer $n \geq 3$.

PROOF The associative property, $(x_1 + x_2) + x_3 = x_1 + (x_2 + x_3)$, guarantees a unique sum of three numbers. So you can write $x_1 + x_2 + x_3 = (x_1 + x_2) + x_3$. This alternative form of the associative property is the starting point you'll use.

Anchor: For $n = 3$,

$$x_1 + x_2 + x_3 = (x_1 + x_2) + x_3 \qquad \text{Agreed-upon order of operations.}$$

Induction Hypotheses: Assume that for $n = k$, where $k \geq 3$:

$$x_1 + x_2 + x_3 + \cdots + x_{k-1} + x_k = (x_1 + x_2 + x_3 + \cdots + x_{k-1}) + x_k$$

Demonstration for $n = k + 1$: If there are $k + 1$ terms, then

$$x_1 + x_2 + x_3 + \cdots + x_{k-1} + x_k + x_{k+1}$$

$$= ((\cdots((x_1 + x_2) + x_3) + \cdots + x_{k-1}) + x_k) + x_{k+1} \qquad \text{Order of operations, sum of } k + 1 \text{ terms.}$$

$$= ((x_1 + x_2 + x_3 + \cdots + x_{k-1}) + x_k) + x_{k+1} \qquad \text{Order of operations, sum of } k - 1 \text{ terms.}$$

$$= (x_1 + x_2 + x_3 + \cdots + x_{k-1} + x_k) + x_{k+1} \qquad \text{Induction hypothesis.}$$

Therefore, the property is true for a sum of $k + 1$ terms.

Conclusion:

$$\therefore x_1 + x_2 + x_3 + \cdots + x_{n-1} + x_n = (x_1 + x_2 + x_3 + \cdots + x_{n-1}) + x_n$$

for any integer $n \geq 3$. Q.E.D.

Induction is useful for proving that the formulas for series work for *any* finite number of terms. Example 4 shows you how to prove that the formula for the partial sum S_n of a geometric series is true for any value of n, no matter how large.

EXAMPLE 4 ➤ **Partial Sum of a Geometric Series Property**

Prove that $S_n = a + ar + ar^2 + ar^3 + \cdots + ar^{n-1} = \dfrac{a(1- r^n)}{1 - r}$ for all integers $n \geq 1$.

(Note that a is used for t_1 for simplicity of writing and that it is written in the numerator of the partial sum formula instead of out in front of the fraction.)

PROOF (by induction on n)

Anchor: If $n = 1$, then $S_1 = a$. The formula gives $\dfrac{a(1 - r^1)}{1 - r} = a$, which anchors the induction.

Induction Hypothesis: Assume that the formula works for $n = k$. That is, assume that

$$S_k = a + ar + ar^2 + ar^3 + \cdots + ar^{k-1} = \frac{a(1 - r^k)}{1 - r}$$

Demonstration for $n = k + 1$:

S_{k+1}

$= a + ar + ar^2 + ar^3 + \cdots + ar^{k-1} + ar^k$	Definition of geometric series.
$= (a + ar + ar^2 + ar^3 + \cdots + ar^{k-1}) + ar^k$	Extended associative property.
$= \dfrac{a(1- r^k)}{1 - r} + ar^k$	Induction hypothesis.
$= \dfrac{a(1- r^k) + ar^k(1 - r)}{1 - r}$	Find a common denominator and add fractions.
$= \dfrac{a - ar^k + ar^k - ar^{k+1}}{1 - r}$	Distribute.
$= \dfrac{a(1 - r^{k+1})}{1 - r}$	Combine like terms, then factor out a.

which is the formula with $k + 1$ substituted for n.

Conclusion:

$$\therefore S_n = a + ar + ar^2 + ar^3 + \cdots + ar^{n-1} = \frac{a(1 - r^k)}{1 - r}$$

for all integers $n \geq 1$. Q.E.D. ◄

You should be careful not to read too much into the conclusion of an induction proof. The proof is good only for any *finite* number of terms. If there is an infinite number of terms, the sum may or may not converge to a real number, depending on the value of r.

EXAMPLE 5 ➤ **Partial Sum for the Series of Squares**

Prove that $S_n = 1 + 4 + 9 + 16 + \cdots + n^2 = \dfrac{n(n+1)(2n+1)}{6}$ for all integers $n \geq 1$.

PROOF *Anchor:* If $n = 1$, then $S_1 = 1$. The formula gives $\dfrac{1(2)(3)}{6} = 1$ as well, which anchors the induction.

Induction Hypothesis: Assume that the formula is correct for $n = k$. That is, assume that

$$S_k = 1 + 4 + 9 + 16 + \cdots + k^2 = \frac{k(k+1)(2k+1)}{6}$$

Demonstration for $n = k + 1$:

$$S_{k+1} = 1 + 4 + 9 + \cdots + k^2 + (k+1)^2$$

Definition of the sum of square series.

$$= (1 + 4 + 9 + \cdots + k^2) + (k+1)^2$$

Extended associative property.

$$= \frac{k(k+1)(2k+1)}{6 + (k+1)^2}$$

Induction hypothesis.

$$= \frac{k(k+1)(2k+1) + 6(k+1)^2}{6}$$

Write the quotient as one fraction with common denominator.

$$= \frac{(k+1)(k(2k^2+1) + 6(k+1))}{6}$$

Factor out $(k + 1)$ from the two terms in the numerator.

$$= \frac{(k+1)(2k^2 + k + 6k + 6)}{6}$$

Multiply the factors in the second parentheses.

$$= \frac{(k+1)(2k^2 + 7k + 6)}{6}$$

Combine like terms in the second parentheses.

$$= \frac{(k+1)(k+2)(2k+3)}{6}$$

Factor the second factor in the numerator.

$$= \frac{(k+1)((k+1)+1)(2(k+1)+1)}{6}$$

Set apart the $(k + 1)$ terms inside the parentheses

which is the formula with $(k + 1)$ substituted for n.

Conclusion:

$$S_n = 1 + 4 + 9 + 16 + \cdots + n^2 = \frac{n(n+1)(2n+1)}{6}$$

is true for all integers $n \geq 1$. Q.E.D. ◀

You can use mathematical induction to prove other extended field, equality, and order axioms, properties of exponentiation and logarithms, formulas for sequences and series, and some interesting properties of numbers.

Answers to Selected Problems

Chapter 1

Problem Set 1-1

1a. 20 m; −17.5 m; it is below the top of the cliff.

1b. ≈0.3 s; ≈3.8 s; ≈5.3 s

1c. 5 m

1d. There is only one altitude for any given time; some altitudes correspond to more than one time.

1e. Domain: $0 \le x \le \approx 5.3$; range: $-30 \le y \le 25$.

3a.

n	B
0	150,000
12	145,995
24	141,744
36	137,230
48	132,438
60	127,350
72	121,948
84	116,213
96	110,125
108	103,661
120	96,798

3b. Changing ΔTbl to 1 shows that the balance becomes negative at the end of month 241, so the balance will become 0 during month 241.

3c., 3d. False

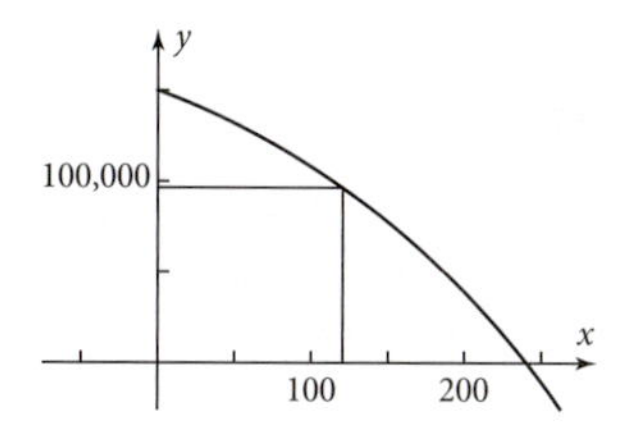

3e. Domain: $0 \le x \le 241$, x is an integer; range: $0 \le y \le 150{,}000$.

5. Answers will vary.

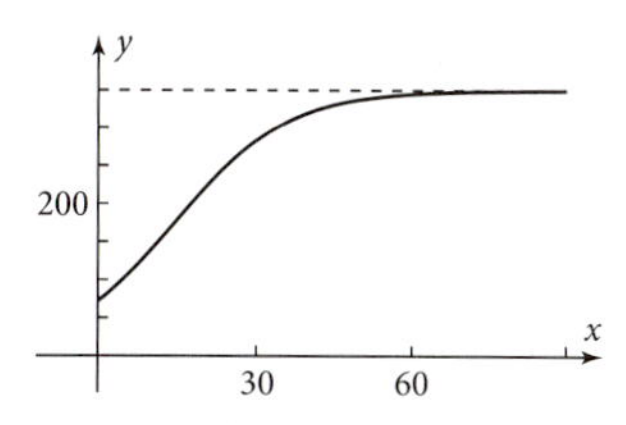

Domain: $x \ge 0$ s; range: $72°\text{F} \le y \le 350°\text{F}$

Problem Set 1-2

1a.

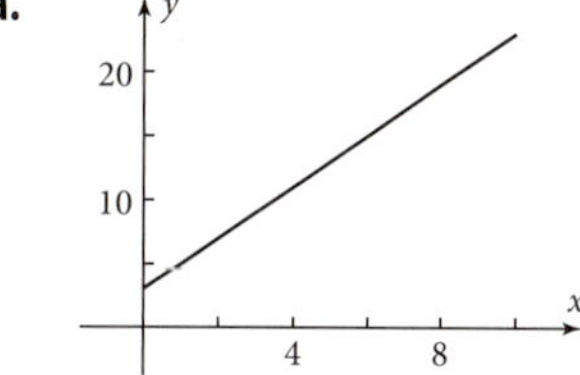

1b. $3 \le f(x) \le 23$

1c. Linear

1d. Answers will vary.

3a.

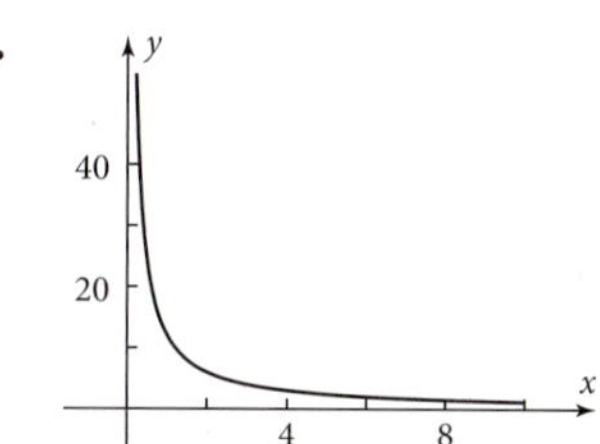

3b. $g(x) \ge 1.2$

3c. Inverse variation

3d. Answers will vary.

5a.

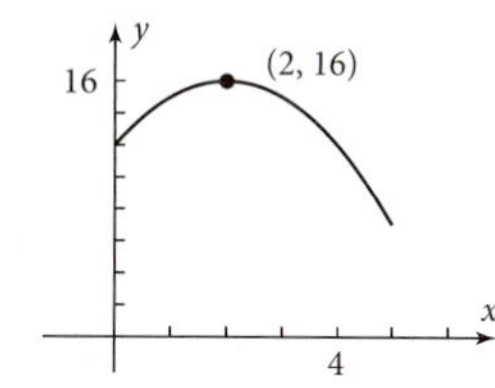

5b. y-intercept at $y = 12$; no x-intercepts; no asymptotes

5c. $7 \le y \le 16$

7a.

7b. y-intercept at $y = 12$; x-intercepts at $x = -1$, $x = 2$ and $x = 6$; no asymptotes

7c. $-20.7453\ldots \le y \le 40$

9a.

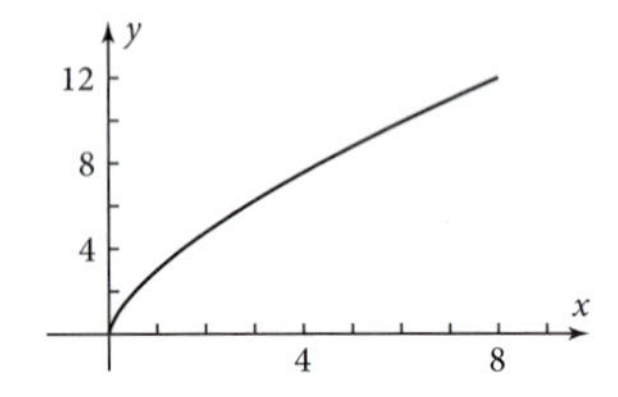

9b. y-intercept at $y = 0$; x-intercept at $x = 0$; no asymptotes

9c. $0 \le y \le 12$

11a.

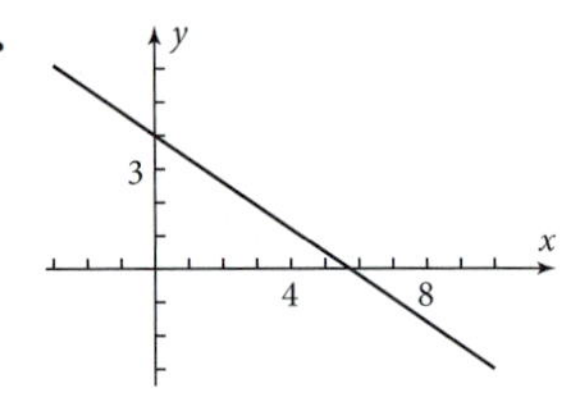

11b. y-intercept at $y = 4$; x-intercept at $x = 5\frac{5}{7}$; no asymptotes

11c. $-3 \le y \le 6.1$

13a.

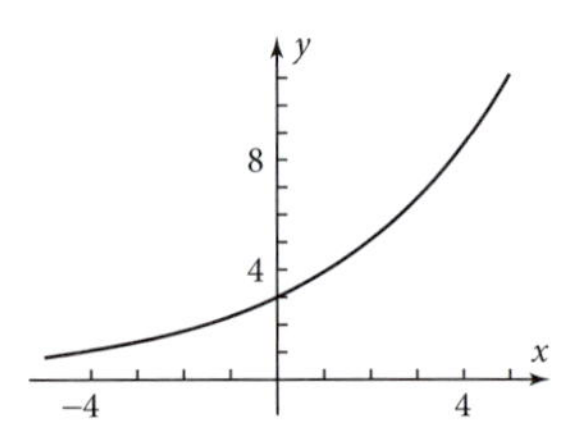

13b. y-intercept at $y = 3$; no x-intercepts; asymptote $y = 0$ (the x-axis)

13c. $0.8079\ldots \le y \le 11.1387\ldots$

15a.

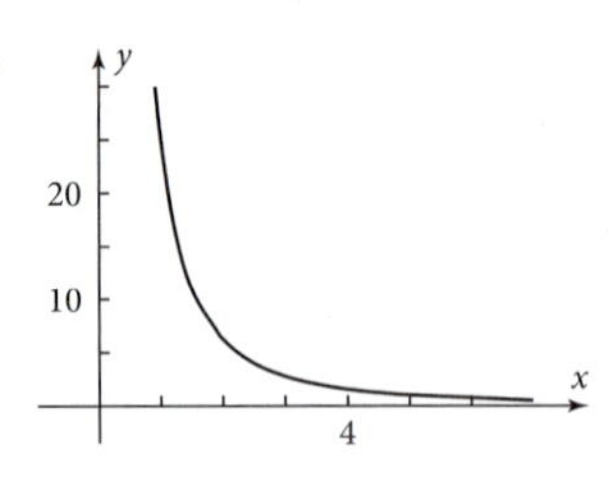

15b. No y-intercept; no x-intercept; asymptotes $x = 0$ (the y-axis) and $y = 0$ (the x-axis)

15c. $y > 0$

17a.

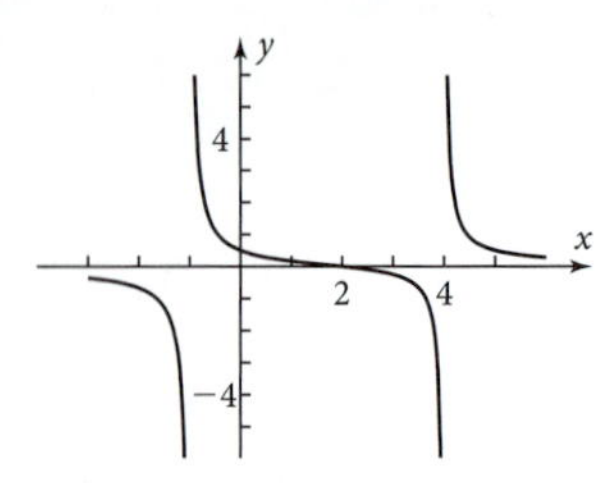

17b. y-intercept at $y = \frac{1}{2}$; x-intercept at $x = 2$; asymptotes $x = -1$, $x = 4$, and $y = 0$ (the x-axis)

17c. Range: all real numbers

19. Exponential

21. Linear

23. Quadratic

25. Power

27. Rational

29a.

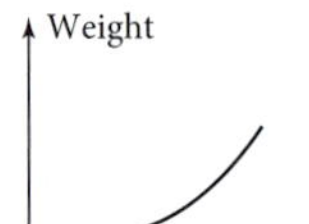

29b. Power (cubic)

31a.

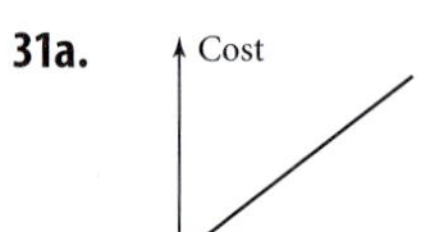

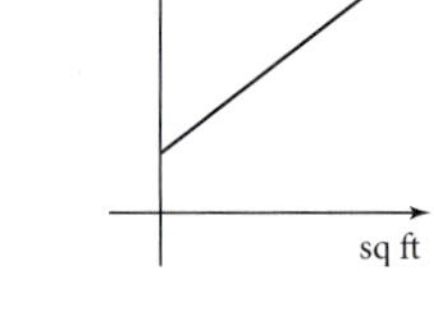

31b. Linear

33. Function; no x-value has more than one corresponding y-value.

35. Not a function; there is at least one x-value with more than one corresponding y-value.

37. Not a function; there is at least one x-value with more than one corresponding y-value.

39a. A vertical line through a given x-value crosses the graph at the y-values that correspond to that x-value. So, if a vertical line crosses the graph more than once, it means that that x-value has more than one y-value.

39b. In Problem 33, any vertical line crosses the graph at most once, but in Problem 35, any vertical line between the two endpoints crosses the graph twice.

41. $x - 2$

Problem Set 1-3

1a. $g(x) = 2\sqrt{9 - x^2}$

1b.

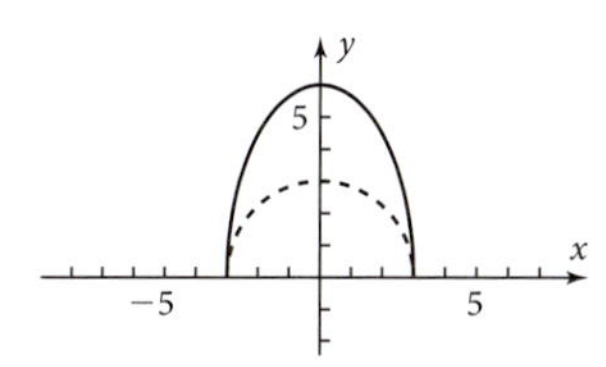

1c. y-dilation by 2 (outside transformation)

3a. $g(x) = \sqrt{9 - (x - 4)^2}$

3b.

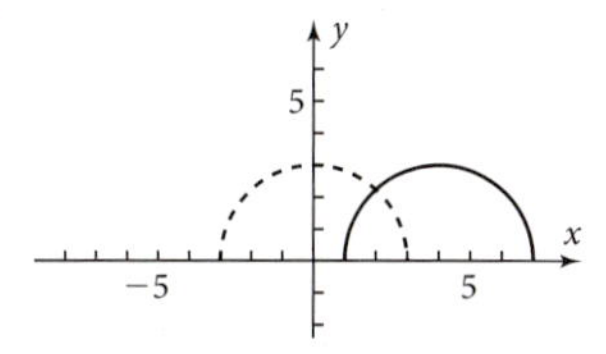

3c. x-translation by 4 (inside transformation)

5a. $g(x) = 1 + \sqrt{9 - \left(\frac{x}{2}\right)^2}$

5b.

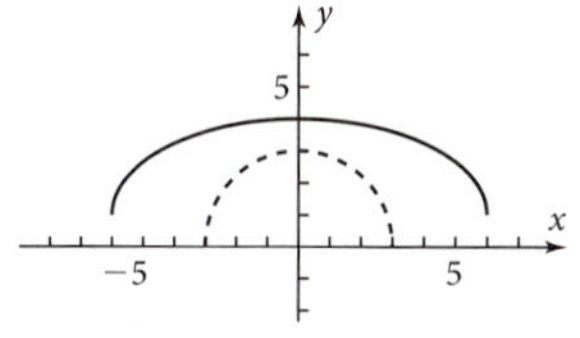

5c. x-dilation by 2 (inside transformation), y-translation by 1 (outside transformation)

7a. y-translation by 7

7b. $g(x) = 7 + f(x)$

9a. x-dilation by 3

9b. $g(x) = f\left(\frac{x}{3}\right)$

11a. x-translation by 6, y-dilation by 3

11b. $g(x) = 3 \cdot f(x - 6)$

13. No. The domain of $f(x)$ is $x \leq 1$, but the domain of the graph is $-3 \leq x \leq 1$.

15a.

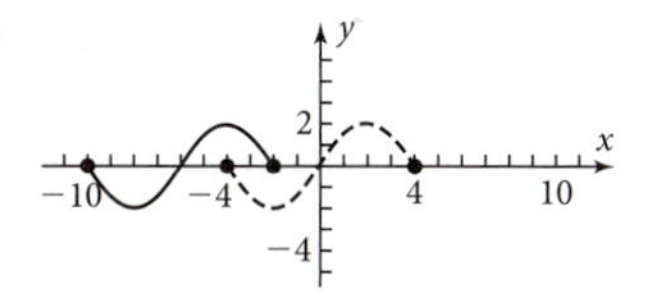

15b. x-translation by -6

17a.

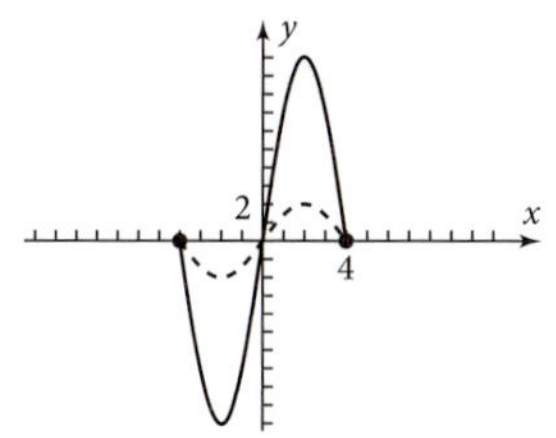

17b. y-dilation by 5

19a.

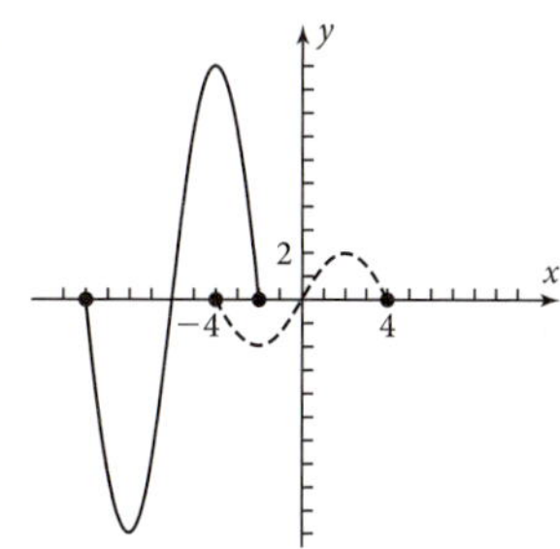

19b. y-dilation by 5, x-translation by -6

21. Answers will vary.

Problem Set 1-4

1a. 33 cm; 54 cm

1b. 3421.1943… cm²; 9160.8841… cm²

1c. The area depends on the radius, which in turn depends on the time.

1d. $r(t) = 5 + 7t$; $a(r(t)) = \pi(r(t))^2$;
$a(r(t)) = \pi(5 + 7t)^2$;
$a(4) = 3421.1943\ldots\ \text{cm}^2$;
$a(7) = 9160.8841\ldots\ \text{cm}^3$

3a. Answers will vary. Note that shoe size is a discrete graph, because shoe sizes come only in half units.

Sample answer:

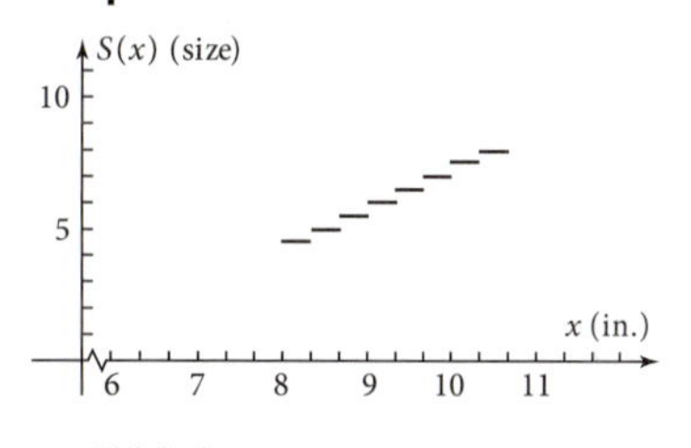

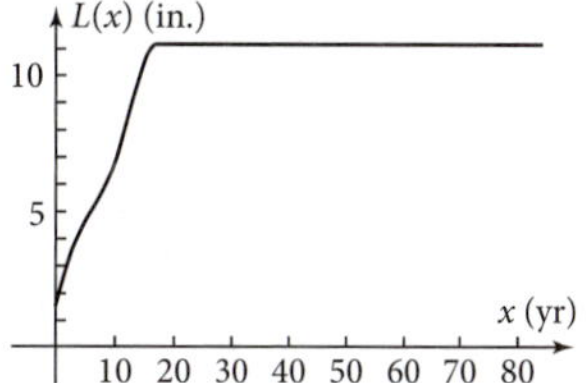

3b. In $S(x)$, x represents foot length (in inches, for the preceding graph). In $L(x)$, x represents age (in years). The composite function $S(L(x))$ gives shoe size as a function of age (x represents age). $L(S(x))$ would be meaningless with the given functions L and S. Because x is substituted into S, x must represent foot length. S then gives shoe size. But this is substituted into L, which expects to have an age, not a shoe size, substituted into it.

3c. Answers will vary. Sample Answer:

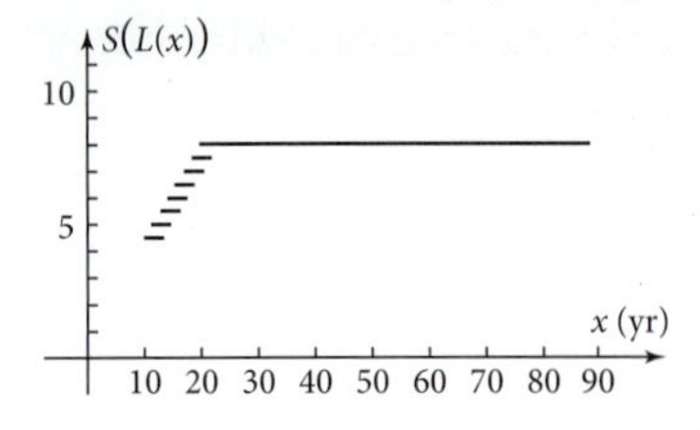

5a. $h(3) = 5$

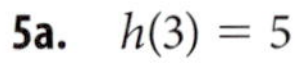

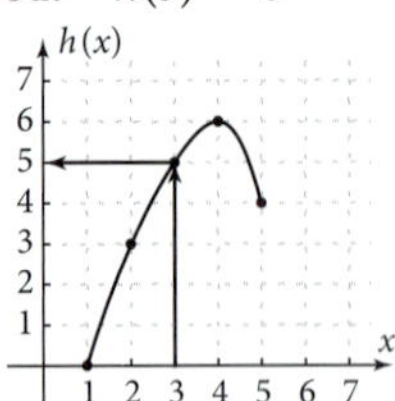

5b. $p(h(3)) = p(5) = 3.5$

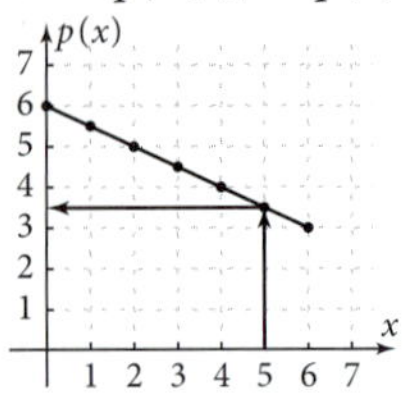

5c. $p(h(2)) = 4.5; p(h(5)) = 4$

5d. $h(p(2)) = 4$, which is different from $p(h(2)) = 4.5$.

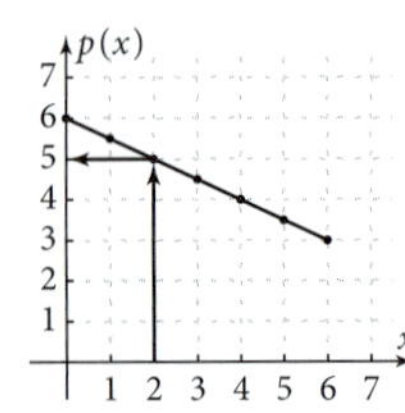

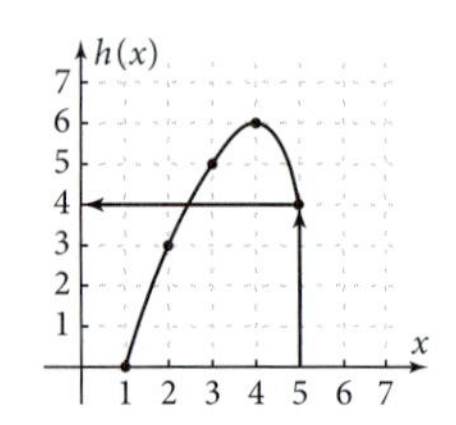

5e. $h(p(0)) = h(6)$, which is undefined because 6 is not in the domain of h.

7a. $g(1) = 2; f(g(1)) = 5$

7b. $g(2) = 3; f(g(2)) = 4$

7c. $g(3) = 7; f(g(3))$ is undefined.

7d. $f(4) = 2; g(f(4)) = 3$

7e. $g(f(3)) = 5$ **7f.** $f(f(5)) = 3$

7g. $g(g(3))$ is undefined.

7h. $f(f(f(1))) = 2$

9a.

x	g(x)	f(g(x))
1	3	none
2	4	5
3	5	4
4	6	3
5	7	2

9b. $2 \le x \le 5$

9c. 6 is not in the domain of g, so $g(x)$ is undefined. $g(1) = 3$, but 3 is not in the domain of f.

9d.

x	f(x)	g(f(x))
4	5	7
5	4	6
6	3	5
7	2	4
8	1	3

Domain: $4 \le x \le 8$

9e.

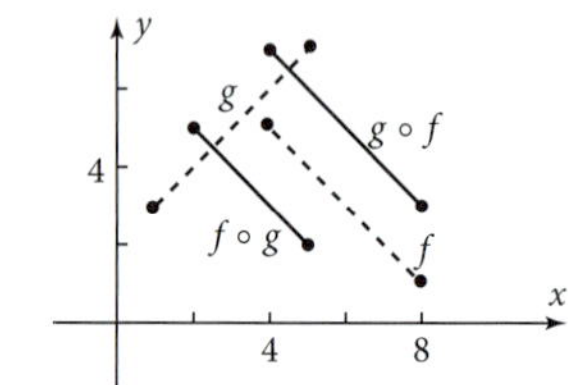

The domains of the composite functions match the calculations in parts b and d.

9f. $f(f(5)) = f(4) = 5; g(5) = 7$, and 7 is not in the domain of g.

11a. $f(g(3)) = 3; f(g(7)) = 7; g(f(5)) = 5; g(f(8)) = 8;$ Conjecture: For all values of x, $f(g(x)) = g(f(x)) = x$.

11b. $f(g(-9))$ is undefined. $g(f(-9)) = 9 \neq -9$. No.

11c.

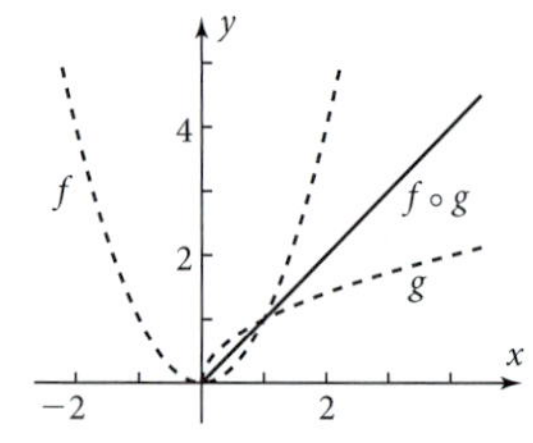

g is defined only for nonnegative x, so $f \circ g$ is defined only for nonnegative x.

11d.

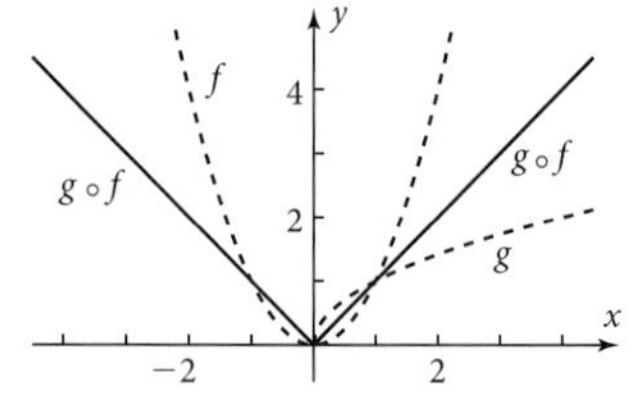

11e. $g(f(x)) = g(x^2) = \sqrt{x^2}$

$$= \begin{cases} x & \text{if } x \ge 0 \\ -x & \text{if } x < 0 \end{cases} = |x|$$

Answers to Selected Problems

13. If the dotted graph is $f(x)$, $1 \leq x \leq 5$, then the solid graph is $g(x) = f(-x)$, $-5 \leq x \leq -1$. In terms of composition of functions, the solid graph is $g(x) = f(h(x))$, where $h(x) = -x$.

15a. $f(g(6)) = 6$; $f(g(-15)) = -15$; $g(f(10)) = 10$; $g(f(-8)) = -8$; $f(g(x)) = g(f(x)) = x$

15b.

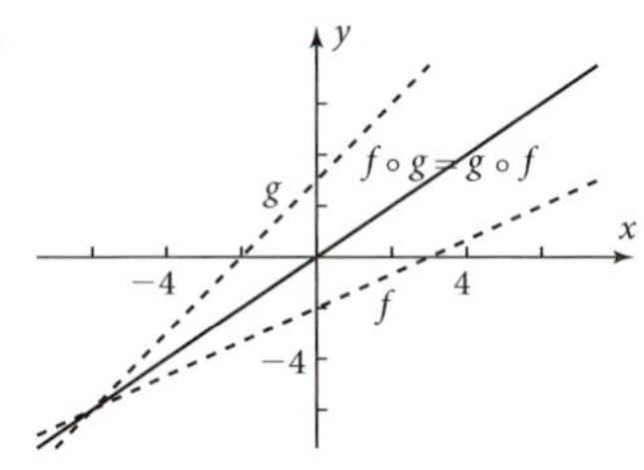

$f(g(x))$ and $g(f(x))$ coincide with each other, and with the line $y = x$. $f(x)$ and $g(x)$ are each other's reflections across that line.

15c. $f(g(x)) = f(1.5x + 3) = \frac{2}{3}(1.5x + 3) - 2$
$= \frac{2}{3} \cdot \frac{3}{2}x + \frac{2}{3} \cdot 3 - 2 = x + 2 - 2 = x;$
$g(f(x)) = g\left(\frac{2}{3}x - 2\right) = 1.5\left(\frac{2}{3}x - 2\right) + 3$
$= \frac{3}{2} \cdot \frac{2}{3}x + \frac{3}{2}(-2) + 3 = x - 3 + 3 = x$

15d. Find $j(x)$ such that $h(j(x)) = x$;
$j(x) = \frac{x + 7}{5} = \frac{1}{5}x + \frac{7}{5}.$

Problem Set 1-5

1a. $f(5) = 24$ psi; $f(10) = 16$ psi; $f(15) = 10.7$ psi

1b. The air leaks out of the tire as time passes, so the pressure is constantly getting lower. Thus, f is a decreasing function and hence is invertible.
$f^{-1}(24) = 5$ min, which answers the question "At what time was the pressure 24 psi?" $f^{-1}(16) = 10$ min, which answers the question "At what time was the pressure 16 psi?"

1c. Somewhere between $x = 25$ and $x = 30$ min, all the air goes out of the tire, and the pressure remains zero. So it is not possible to give a unique time corresponding to a pressure of 0 psi; $f^{-1}(0)$ cannot be defined.

1d. The graph of the inverse relation is dotted. The two graphs are reflections of each other over the line $y = x$. (They coincidentally happen to be very close over most of their length.)

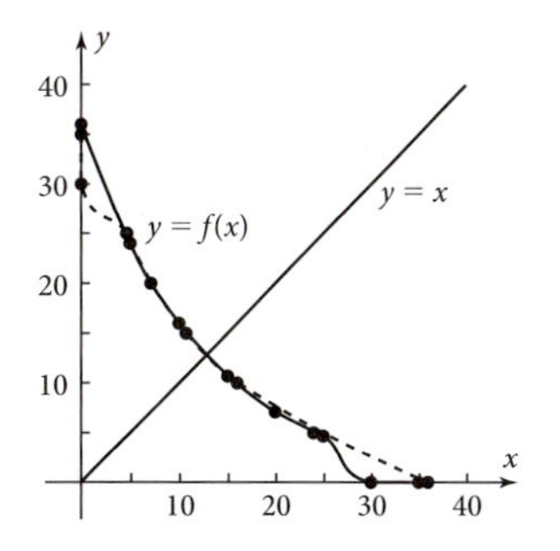

1e. As an input for f, x represents time in minutes. As an input for f^{-1}, it represents pressure in psi.

3.

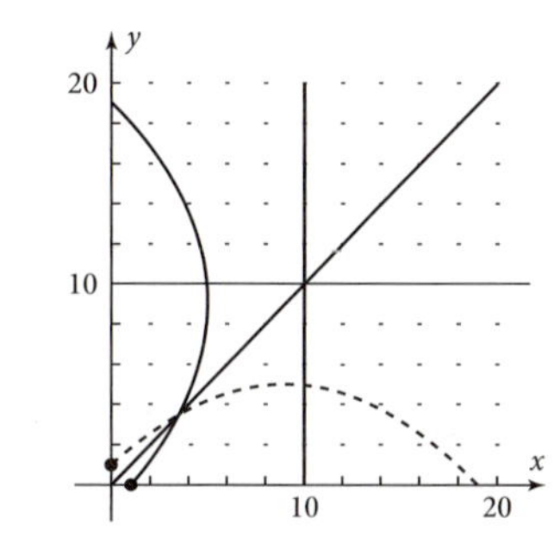

Throughout most of its domain, the inverse relation has two y-values for every x-value.

5. Function

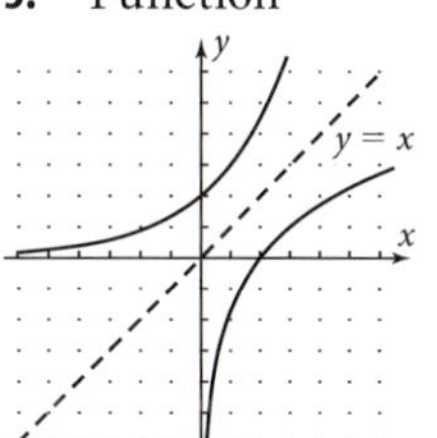

7. Not a function

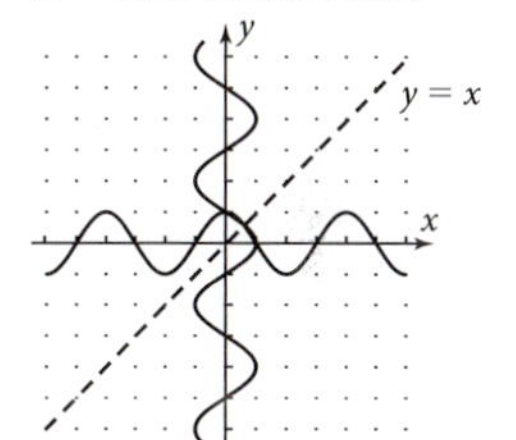

9a.

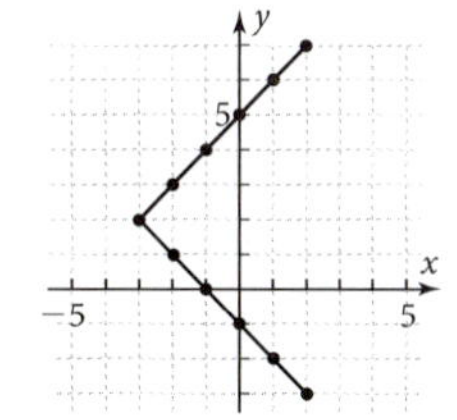

9b. Not a function

9c. Grapher graph agrees with graph on paper.

Answers to Selected Problems

11a.

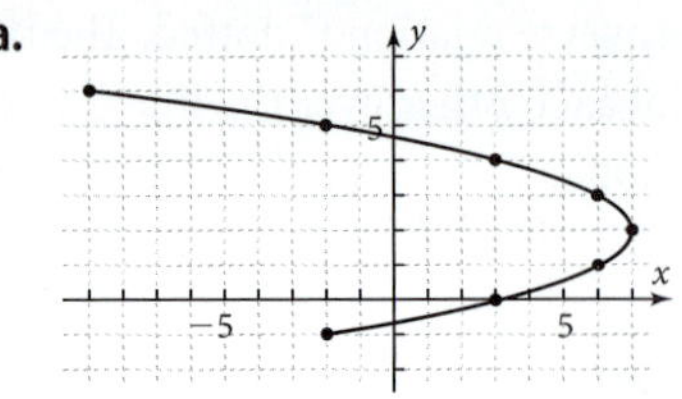

11b. Not a function

11c. Grapher graph agrees with graph on paper.

13a.

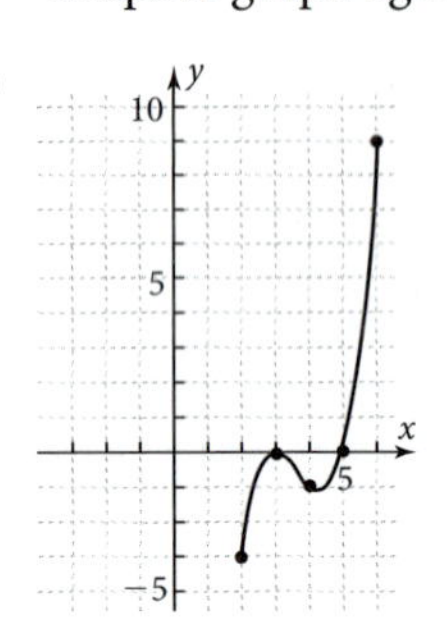

13b. Function

13c. Grapher graph agrees with graph on paper.

15a. Paths intersect simultaneously at point $(-1, 3)$ when $t = -2$ s. Paths intersect at point $(2, 6)$ but not simultaneously.

15b. Grapher graph confirms that the paths intersect simultaneously only at point $(-1, 3)$ when $t = -2$ s.

17. Function

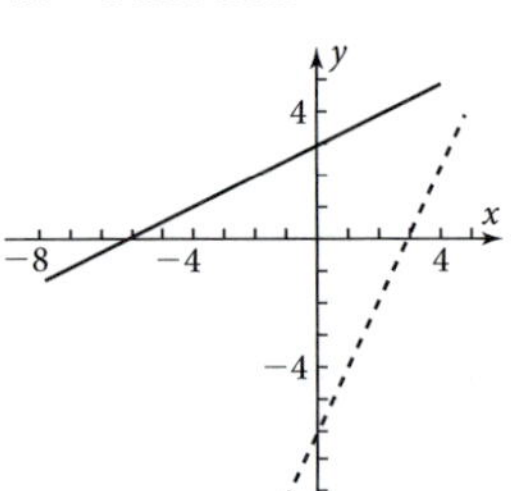

19. Not a function

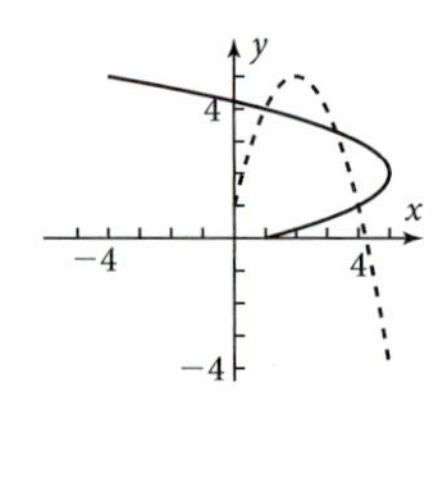

21. Function

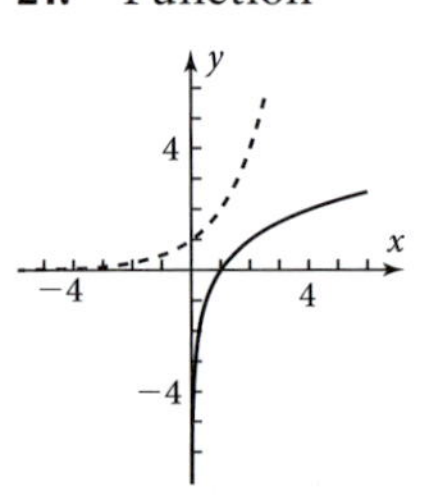

23. Function

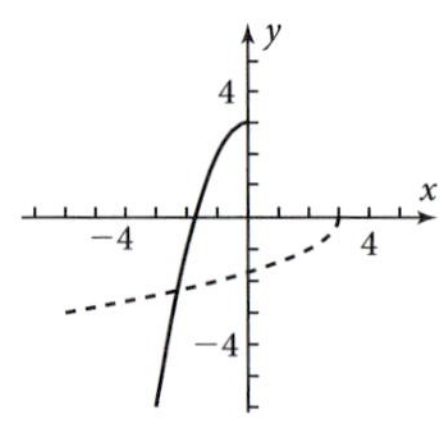

25. Function

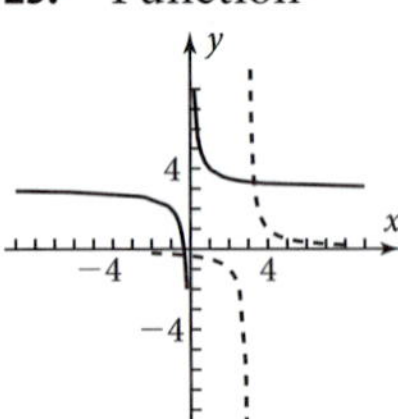

27. Function

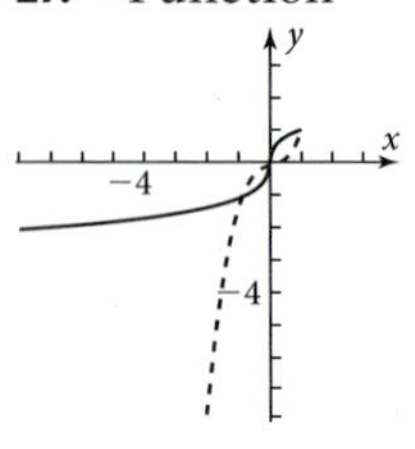

29. $y = f^{-1}(x) = \frac{1}{2}x + 3$

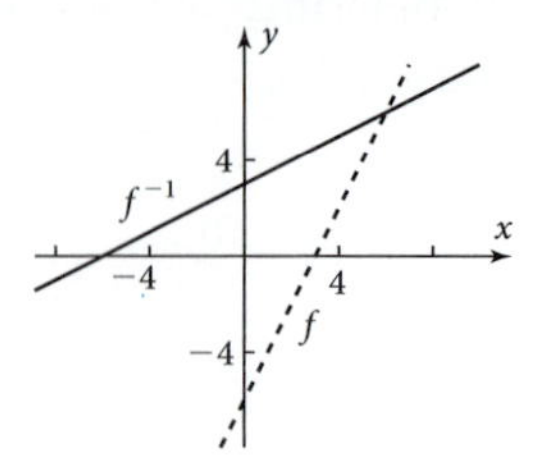

The inverse relation is a function.

31. $y = \pm\sqrt{-2x - 4}$

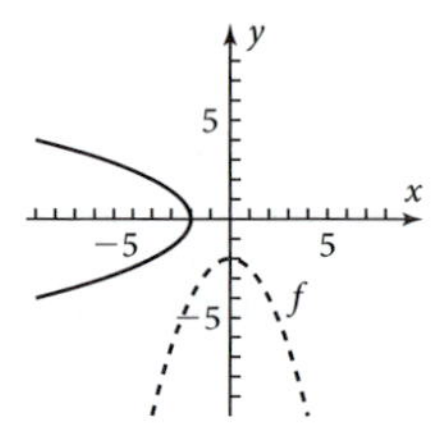

The inverse relation is not a function.

33. $f(f(x)) = \dfrac{1}{f(x)} = \dfrac{1}{(1/x)} = x, x \neq 0$

35a. $c(1000) = 900$. If you drive 1000 mi in a month, your monthly cost is \$900.

35b. $c^{-1}(x) = 2.5x - 1250$. $c^{-1}(x)$ is a function because no input produces more than one output. $c^{-1}(758) = 645$. You would have a monthly cost of \$758 if you drove 645 mi in a month.

35c.

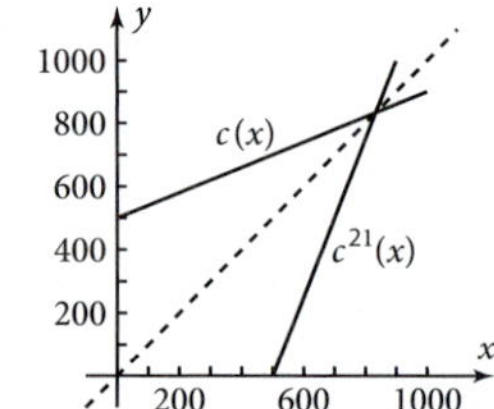

37a. $d^{-1}(x) = \sqrt{\dfrac{x}{0.057}}$. Because the domain of d is $x \geq 0$, the range of d^{-1} is $d^{-1}(x) \geq 0$.

37b. $d^{-1}(200) = 59.234\ldots$ This means that a 200-ft skid mark is caused by a car moving at a speed of about 59 mi/h.

37c.

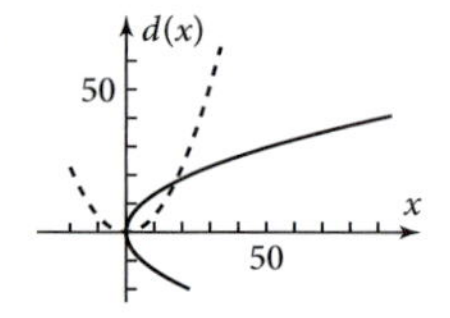

37d. Because the domain of d now contains negative numbers, the range of the inverse relation contains negative numbers. Now, because the range of the inverse relation contains negative numbers, $y = \pm\sqrt{\dfrac{x}{0.057}}$, which is not a function.

Problem Set 1-6

1a.

1b.

1c.

1d.

3a.

3b.

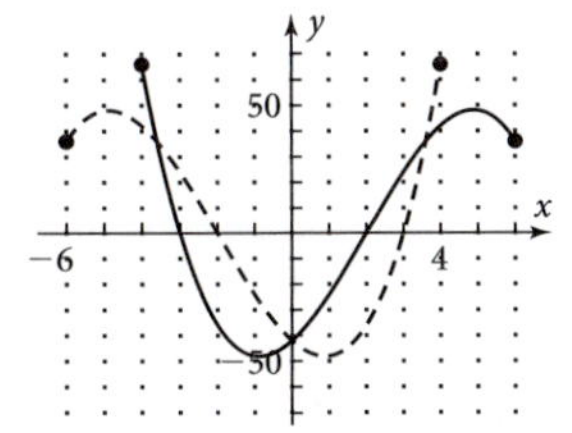

3c.

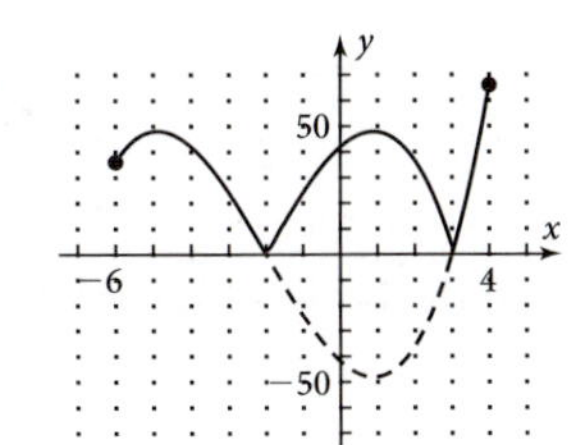

3d.

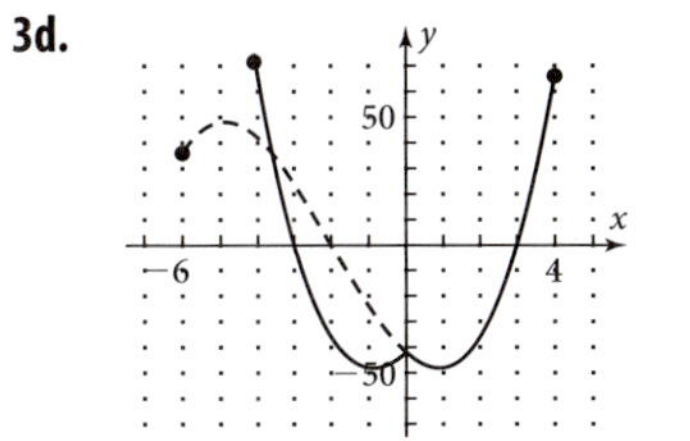

5. The graphs match.

7a.

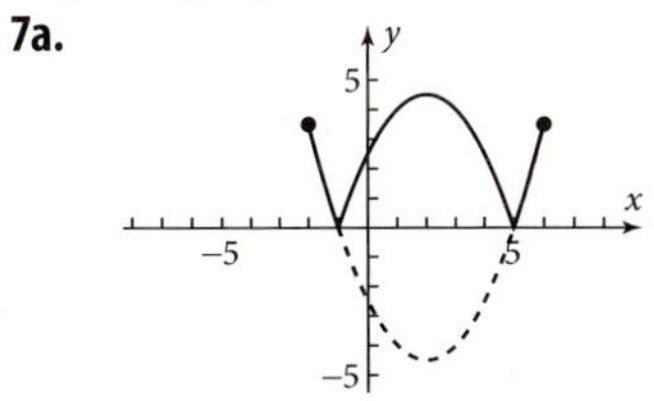

This transformation reflects all the points on the graph below the x-axis across the x-axis.

7b.

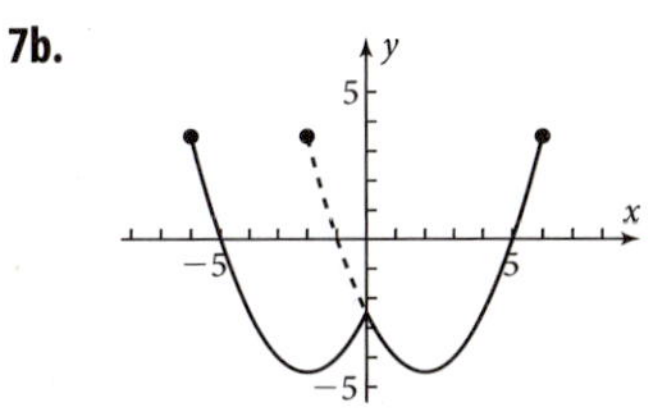

This transformation reflects $f(x)$, for positive values of x, across the y-axis.

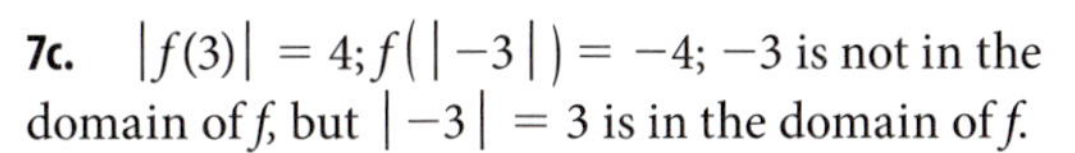

7c. $|f(3)| = 4; f(|-3|) = -4; -3$ is not in the domain of f, but $|-3| = 3$ is in the domain of f.

7d.

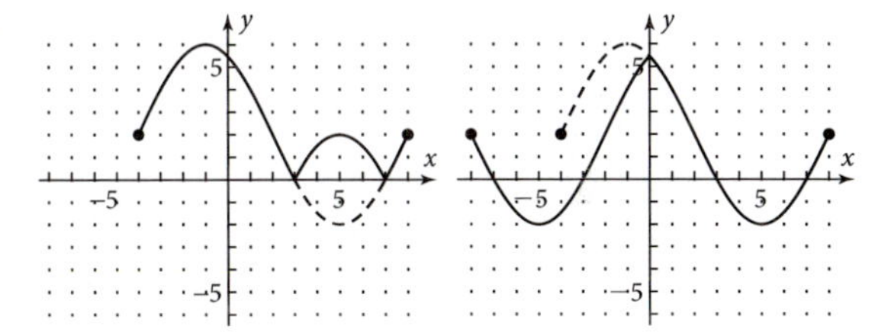

9a.

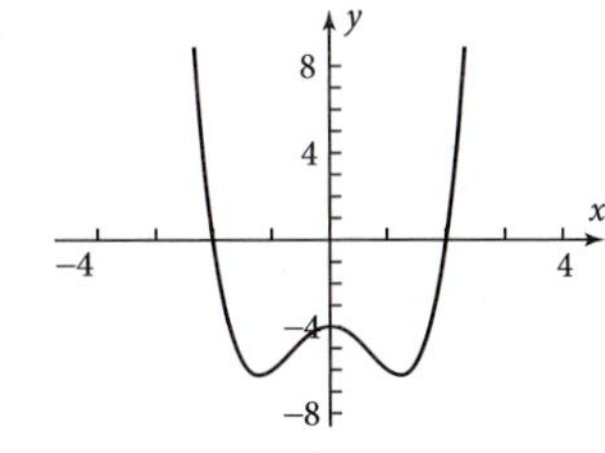

The polynomial function $f(x)$ is the sum of even powers of x. A negative number raised to an even power is equal to the absolute value of that number raised to the same power. So, for $\pm x$, the same corresponding y-value occurs, and therefore $f(x) = f(-x)$.

9b.

A negative number raised to an odd power is equal to the opposite of the absolute value of that number raised to the same power. Because each term in $g(x)$ is a monomial in x raised to an odd power, $g(-x)$ has the same effect on $g(x)$ as $-g(x)$.

9c. Function h is odd; function j is even.

9d.

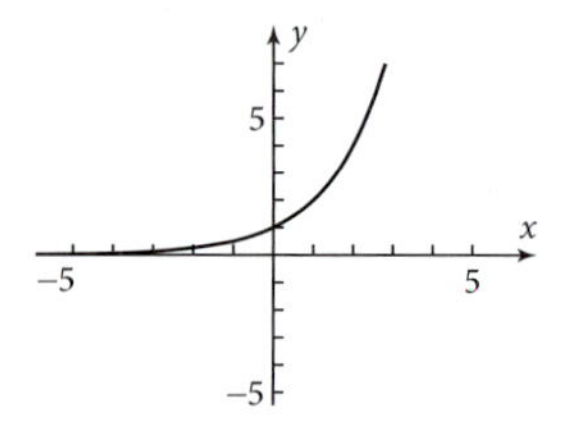

The function $e(x)$ is neither odd nor even. $e(-x) \neq -e(x)$

11a. The graphs match.

11b. $g(x) = 3\dfrac{|x-4|}{x-4} + 5;$
$g(x) = 3f(x-4) + 5$

11c. $f(x) = (x-3)^2 - 2 \cdot \dfrac{|x-5|}{x-5}$
The graphs match.

13a. $a = 0.0375;\ b = 2{,}400{,}000{,}000$

13b.

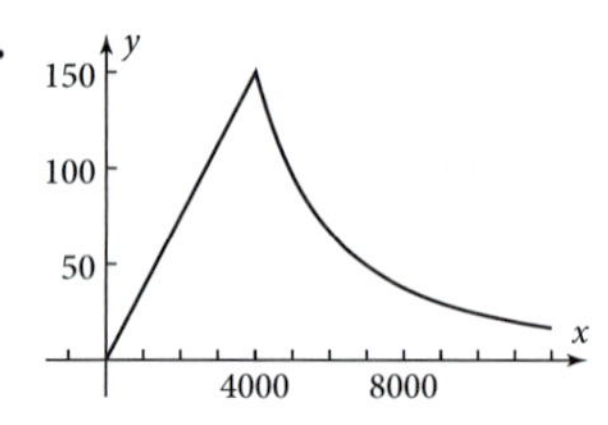

13c. $y(3000) = 112.5$ lb; $y(5000) = 96$ lb

13d. $f_1(x) = 50 \Rightarrow x = 1.333.\overline{3}$ mi;
$f_2(x) = 50 \Rightarrow x = 6928.2032\ldots$ mi

Problem Set 1-8

R1a. 17.15 psi; 5.4 min

R1b.

x	y
0	35
1	24.5
2	17.15
3	12.005
4	8.4035

R1c. Domain: $0 \leq x \leq \approx 5.5$; range $5 \leq y \leq 35$.

R1d. Asymptote

R1e.

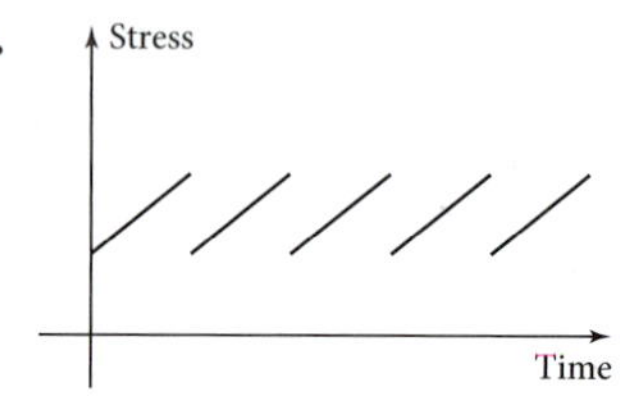

R3a. Horizontal dilation by a factor of 3, vertical translation by -5;
$g(x) = \sqrt{4 - \left(\frac{x}{3}\right)^2} - 5$

R3b. Horizontal translation by $+4$, vertical dilation by a factor of 3

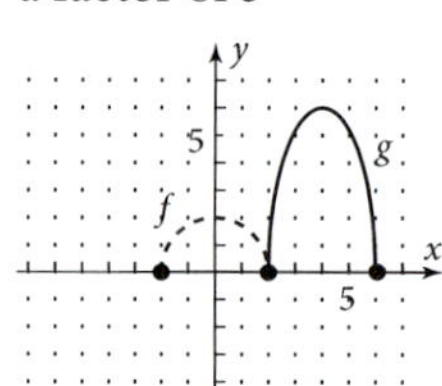

R5a. The inverse does not pass the vertical line test.

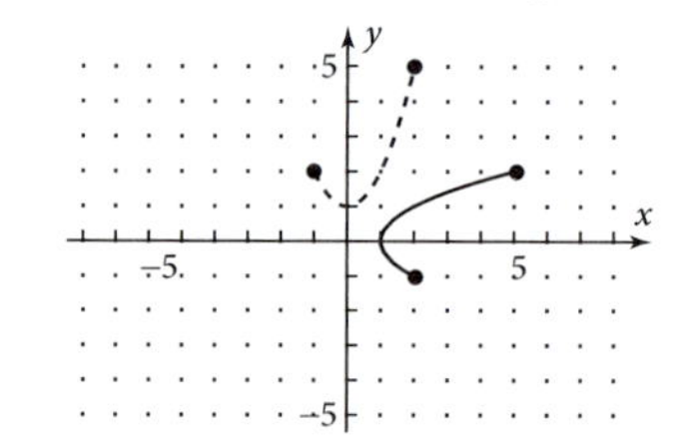

R5b.

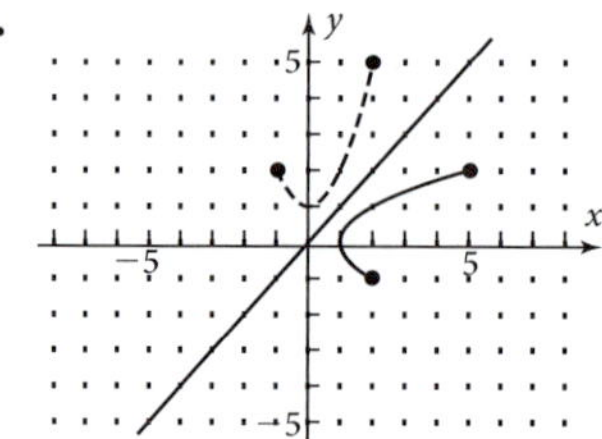

The graphs are each other's reflections across the line. The domain of f corresponds to the range of the inverse relation. The range of f corresponds to the domain of the inverse relation.

R5c. $x = y^2 + 1 \Rightarrow y = \pm\sqrt{x-1}$. The $\pm$ reveals that there are two different y-values for some x-values.

R5d.

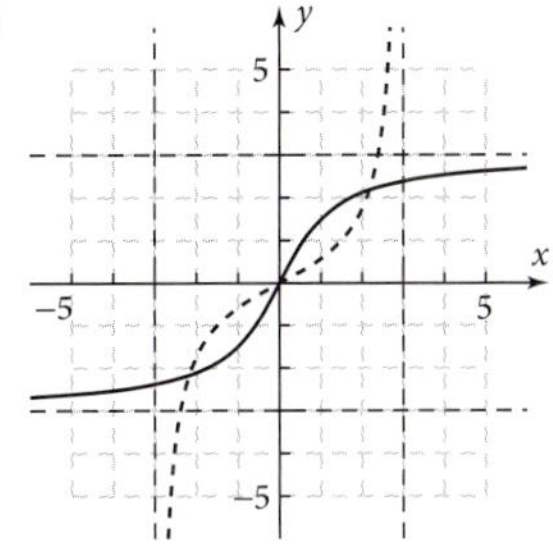

It passes the horizontal line test; asymptotes.

R5e.

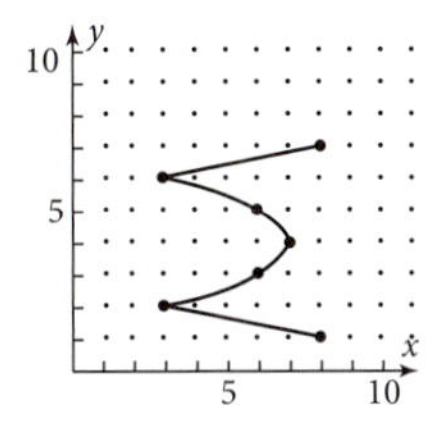

Grapher graph agrees with graph on paper; Not a function because every x in the domain has multiple values of y.

R5f.

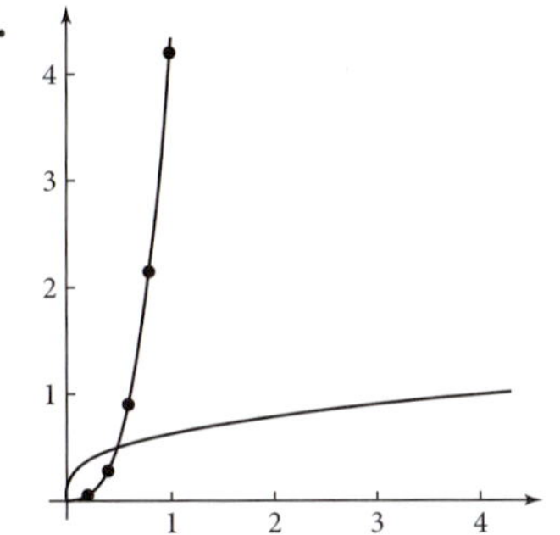

The curve is invertible because it is increasing. As the input to v, x represents radius in meters. As the input to v^{-1}, it represents volume in cubic meters. If x_0 is a particular input to v, then $(x_0, v(x_0))$ is a point on the graph of $v(x)$. Plugging the output, $v(x_0)$, into v^{-1} gives the point $(v(x_0), v^{-1}(v(x_0)))$ on the graph of $v^{-1}(x)$. But the graph of $v^{-1}(x)$ is just the graph of $v(x)$ with all the x- and y-values exchanged, so this point is actually $(v(x_0), x_0)$. Thus, $v^{-1}(v(x_0)) = x_0$.

R5g. Since no y corresponds to more than one x in the original function, no x corresponds to more than one y in the inverse relation, so the inverse relation is a function. Sample graph:

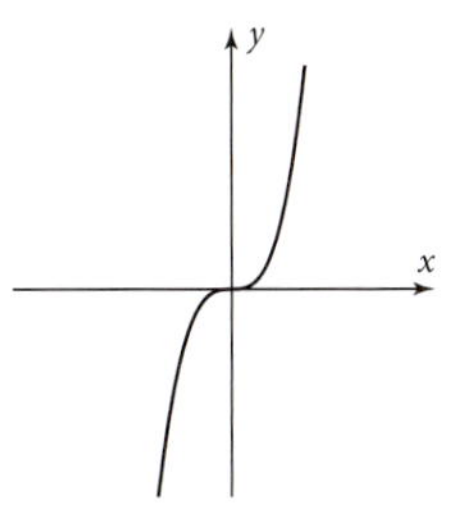

R7. Answers will vary.

C1. Horizontal dilation by 3, vertical dilation by 2, horizontal translation by +3, vertical translation by −5; $g(x) = 2\left(\frac{x-3}{3}\right)^2 - 5$

T1. Exponential

T3. Polynomial (quadratic)

T5. All except T3; Functions that are not one-to-one are not invertible; that is, their inverses are not functions.

T7. Odd

T9. Horizontal dilation by 2; $g(x) = f\left(\frac{x}{2}\right)$

T11. Horizontal translation by +6, vertical dilation by 2; $g(x) = 2 \cdot f(x-6)$

T13. Vertical dilation by $\frac{1}{2}$

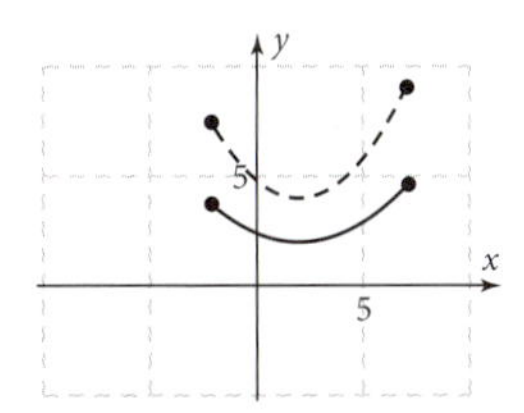

T15. Horizontal translation by −3, vertical translation by −4

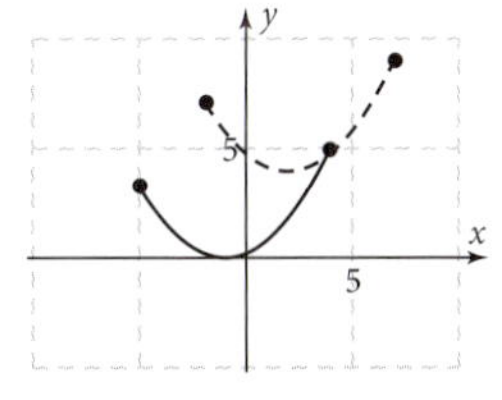

T17. The graph fails the vertical line test. (The pre-image graph fails the horizontal line test—it is not one-to-one.)

T19. Horizontal translation by +4, vertical translation by +5, and vertical dilation by 3 of $\frac{x}{|x|}$;
$y = 3 \cdot \frac{x-4}{|x-4|} + 5$

T21. $L(x)$ varies proportionately to the 0.52 power of x. Power function.

T23. About 129 plants per square meter

T25. $L^{-1}(100) = \left(\dfrac{100}{3.2}\right)^{\frac{1}{0.52}} = 749.3963\ldots$

If the crop loss is 100% (i.e., the total crop is lost), there must have been about 750 wild oat plants (or more) per square meter.

T27.

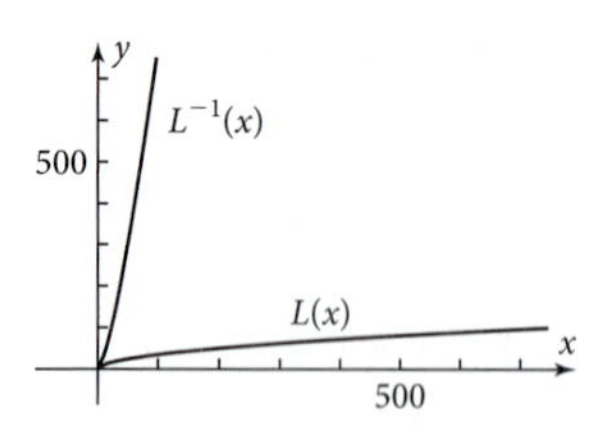

T29. Answers will vary.

Chapter 2

Problem Set 2-1

1.

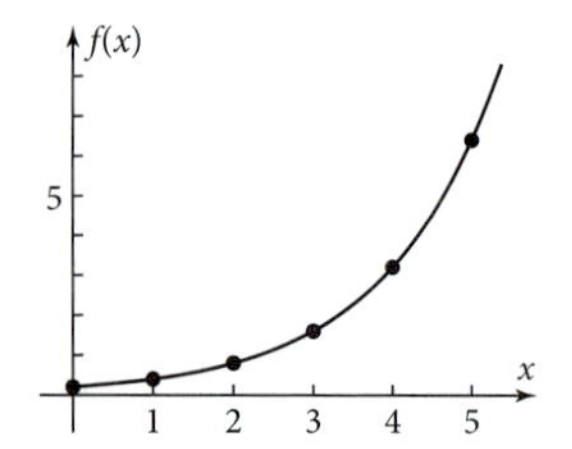

The "hollow" section is upward. The bacteria are growing faster and faster.

3.

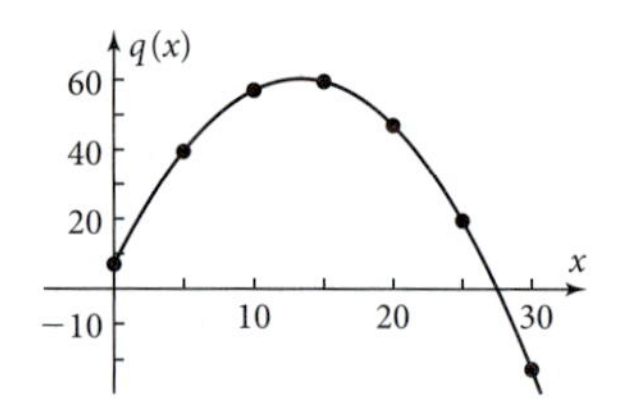

Concave down; a maximum (high point)

Problem Set 2-2

1. In power functions, the exponent is constant and the independent variable is in the base. In exponential functions, the base is constant and the independent variable is in the exponent.

3. Answers will vary.

5. $\frac{1}{x} = x^{-1}$

7. $(-64)^{1/2}$ is undefined but $(-64)^{1/3} = -4$. The restriction allows the function to be defined for all values of x.

9a. $y = 56 + 0.6(x - 20) = 0.6x + 44$

9b. Page 44

9c. $11\frac{2}{3}$ min from now.

11a. Linear

11b. Increasing for all real-number values of x, not concave

11c. Answers will vary.

11d. $y = 2x - 7$

11e. The graphs match.

13a. Quadratic

13b. Decreasing for $x < 2.25$ and increasing for $x > 2.25$, concave up

13c. Answers will vary.

13d. $y = 2x^2 - 9x + 13$

13e. The graphs match.

15a. Exponential

15b. Increasing for all real-number values of x, concave up

15c. Answers will vary.

15d. $y = 5 \cdot (1.3)^x$

15e. The graphs match.

17a. Power

17b. Increasing for $x \geq 0$, concave down

17c. Answers will vary.

17d. $y = 5x^{\log_2 1.6}$

17e. The graphs match.

19a. Power

19b. Increasing for $x \geq 0$, concave up

19c. Answers will vary.

19d. $y = 3x^{3/2}$

19e. The graphs match.

21.

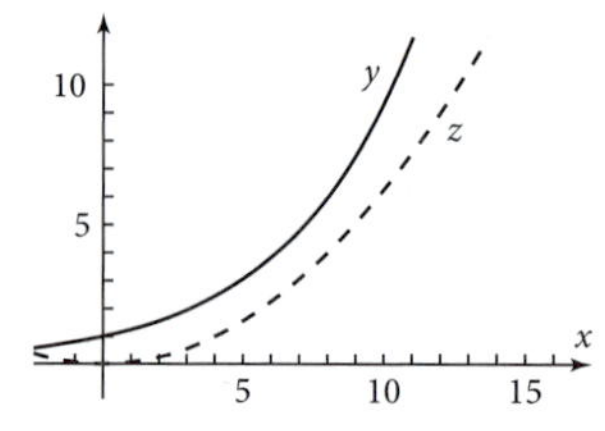

Both graphs are concave up and do not change their concavity, and each becomes infinite on one side of the vertical axis. But the graph proportional to the square of x passes through the origin and becomes infinite on both sides of the vertical axis, whereas the exponential function does not pass through the origin and becomes infinite only on the positive side of the vertical axis.

23. A direct variation function can be written in the linear form $y = ax + b$ with $b = 0$. But you cannot write a linear function $y = ax + b$ with $b \neq 0$ as a direct variation function $y = ax$.

25. $b = e^{0.8} = 2.2255...$. The graphs are equivalent.

Problem Set 2-3

1. Add–add property: linear

3. Multiply–multiply property: power; and constant-second-differences property: quadratic

5. Add–add property: linear; multiply–multiply property: power

7. Multiply–multiply property: power, inverse variation

9. Add–multiply property: exponential

11. Constant-second-differences property: quadratic

13a. 65 **13b.** 80

13c. 1280

15a. 70 **15b.** 81

15c. 72.9

17. $f(8) = 13, f(11) = 19, f(14) = 25$

19. $f(10) = 324, f(20) = 81$

21. Multiply y by 4.

23. Divide y by 2.

25a. $V(r)$ has the form $V = ar^3$ where $a = \frac{4}{3}\pi$; 5400 cm^3.

25b. 400,000 lb = 200 tons

25c. 500,000 lb

25d. 0.2 lb

27a. 16 times more wing area

27b. 64 times heavier

27c. The full-sized plane had four times as much weight per unit of wing area as the model.

29a. $[H(3) - H(2)] - [H(2) - H(1)] = -32$ ft; $[H(4) - H(3)] - [H(3) - H(2)] = -32$ ft; $[H(5) - H(4)] - [H(4) - H(3)] = -32$ ft

29b. $H(t) = -16t^2 + 90t + 5; H(4) = 109; H(5) = 55$

29c. $H(2.3) = 127.36$ ft; going up; the height seems to peak at about $t = 3$ s.

29d. 1.4079... s (going up) or 4.2170... s (coming down); there are two solutions to the equation at $y = 100$.

29e. 2.8125 s; 131.5625 ft

29f. 5.6800... s

31. If $y(8)$ were 25, then a quadratic function would fit.

33. If $f(x) = ax + b$, then $f(x_2) = f(x_1 + c) + b = ax_1 + ac + b = (ax_1 + b) + ac = f(x_1) + ac$.

35. If $f(x) = ax^b$, then $f(x_2) = f(c + x_1) = ab^{c+x_1} = a(b^c \cdot b^{x_1}) = b^c \cdot ab^{x_1} = b^c \cdot f(x_1)$.

Problem Set 2-4

1. $10^{-0.1549...} = 0.7$

3. $10^a = b$

5. $x = 1.574; 10^{1.574} = 37.4973..., \log 37.4973... = 1.574$

7. $x = -0.981; 10^{-0.981} = 0.1044..., \log 0.1044... = -0.981$

9. $x = \log 57 = 1.7558...; 10^{1.7558...} = 57$

11. $x = \log 0.85 = -0.0705...; 10^{-0.0705...} = 0.85$

13. $3.0277...; 10^{3.0277...} = 1066$

15. $-1.2247...; 10^{-1.2247...} = 0.0596$

17. $0.001995...; \log 0.001995... = -2.7$

19. $1.5848... \times 10^{15}; \log(1.5848... \times 10^{15}) = 15.2$

21. $\log(5 \cdot 4) = \log 20 = 1.3010... = 0.6989... + 0.6020... = \log 5 + \log 4$; $\log xy = \log x + \log y$; $b^c \cdot b^d = b^{c+d}$

23. $\log(35 \div 7) = \log 5 = 0.6989... = 1.5440... - 0.8450... = \log 35 - \log 7$; $\log \frac{x}{y} = \log x - \log y$; $\frac{b^c}{b^d} = b^{c-d}$

25. $\log(2^5) = \log 32 = 1.5051... = 5(0.3010...) = 5 \log 2$; $\log b^x$; $x \log b$; $(b^c)^d = b^{cd}$

27. $\log 0.21 = -0.6777... = -0.5228... + (-0.1549...) = \log 0.3 + \log 0.7$; $0.21 = 10^{-0.6777...} = 10^{-0.5228... + (-0.1549...)} = 10^{-0.5228...} \cdot 10^{-0.1549...} = 0.3 \cdot 0.7$

29. $\log 6 = 0.7781... = 1.4771... - 0.6989... = \log 30 - \log 5$; $6 = 10^{0.7781...} = 10^{1.4771... - 0.6989...} = 10^{1.4771...} \div 10^{0.6989...} = 30 \div 5$

31. $\log 32 = 1.5051... = 5(0.3010...) = 5 \log 2$; $32 = 10^{1.5051...} = 10^{5(0.3010...)} = (10^{0.3010...})^5 = 2^5$

33. $\log \frac{1}{7} = -0.8450... = -\log 7;$
$\frac{1}{7} = 10^{-0.8450...} = \frac{1}{10^{0.8450...}} = \frac{1}{7}$

35. 21

37. 4

39. 56

41. 128

43. 3

45. Let $c = \log x$, so $x = 10^c$. Then $x^n = (10^c)^n = 10^{cn}$, so $\log x^n = cn = nc = n \log x$.

47a.

$$\begin{array}{r} 27 \\ \times 356 \\ \hline 9{,}612 \\ \times 43 \\ \hline 415{,}316 \\ \times 592 \\ \hline 244{,}683{,}072 \end{array}$$

47b. 1018

47c. 8.3886; $10^{8.3886} \approx 245{,}700{,}000$, which agrees (to four significant digits) with the answer from part a.

Problem Set 2-5

1. $\log_b x = y$ if and only if $b^y = x$ for $x > 0$, $b > 0$, $b \neq 1$

3. $7^c = p$

5. $\log_k 9 = 5$

7. 1.7304...

9. 6

11. 5

13. $-0.6719...$

15. 56

17. 5

19. 4

21. $\frac{1}{2}$

23. 0

25. 1

27. $\log_{10} 7$

29. 3

31. log (or $\log_{10}$)

33. 3

35. $x = -2$

37. $x_1 = 5, x_2 = -4$
The equation is undefined for $x = -4$.

39. $x - 9 = 16.3890...$

41. $x = 1.3808...$

43. $x = -85.1626...$

45. $x = 4.5108...$

47. $x = -0.6931...$

49a.

x	M
0	10,000
1	10,700
2	11,449
3	12,250
4	13,108
5	14,026
6	15,007

49b. Whenever you add 1 to x, you multiply M by 1.07.

49c. 177 mo, or 14 yr 9 mo

Problem Set 2-6

1a. $\frac{14.4}{3.6} = \frac{57.6}{14.4} = \frac{230.4}{57.6} = \frac{921.6}{230.4} = 4$

1b. $y = 1 - \frac{4 \ln 3.6}{\ln 256} + \frac{4}{\ln 256} \ln x$
$= 0.0760... + 0.7213... \ln x$

1c. The equation fits the data.

3a. The inverse of an exponential function is a logarithmic function.

3b. $y = -38{,}069.2959... - 8{,}266.6425... \ln p$;
$r = -0.999...$, very close to -1

3c. $y(73.9) = 2500.3068... \approx 2500$ years old

3d. $y(20) = 13{,}304.6479... \approx 13{,}300$ years old

3e. Answers will vary.

5a. $g(x) = 6 \log_{10} x$; $\log_{6\sqrt{10}} x = \log_{1.4677...} = x$

5b. The x-intercept of g is e^{-3}, and the x-intercept of h is e.

7. Domain: $x > -3$

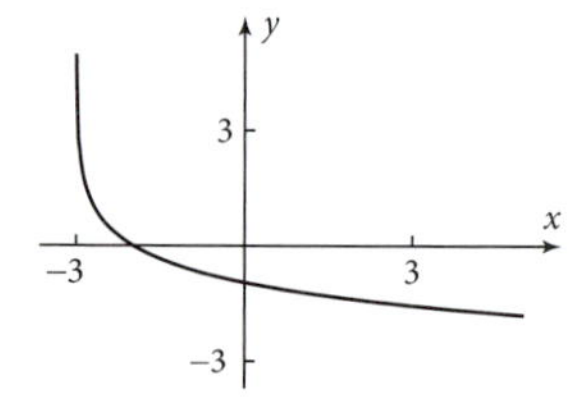

9. Domain: $x \neq 0$

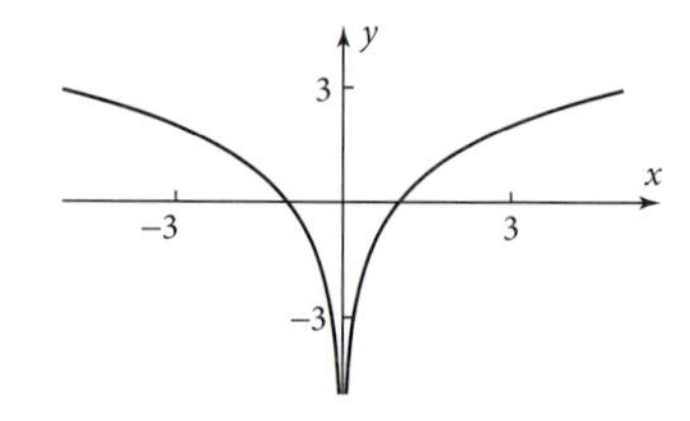

11. Domain: $x > 0$

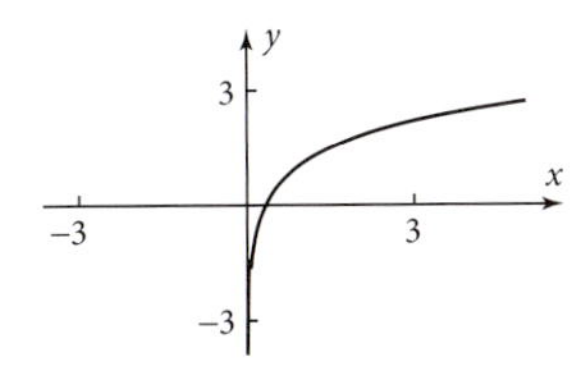

13a.

x (years)	*y* (dollars)
−0.10000	2.8679...
0.10000	2.5937...
−0.01000	2.7319...
0.01000	2.7048...
−0.00100	2.7196...
0.00100	2.7169...
−0.00010	2.7184...
0.00010	2.7181...
−0.00001	2.7182...
0.00001	2.7182...

13b. The two properties balance out, so that as x approaches 0, y approaches 2.7182....

13c. $e = 2.7182\ldots$; they are the same.

Problem Set 2-7

1a.

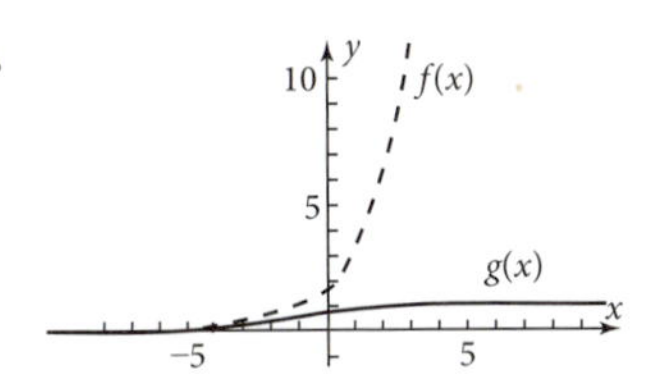

1b. The graphs are almost the same for large negative values of x, but widely different for large positive values of x.

1c. $x = 0$; concave up for $x < 0$ and concave down for $x > 0$

1d. As x grows very large, the 1 in the denominator becomes insignificant in comparison to the 2.2^x, so $g(x) = \dfrac{2.2^x}{2.2^x + 1} \approx \dfrac{2.2^x}{2.2^x} = 1$

1e. $g(x) = \dfrac{1}{1 + 2.2^{-x}}$. A table of values shows that the expressions are equivalent.

3a. Concave up

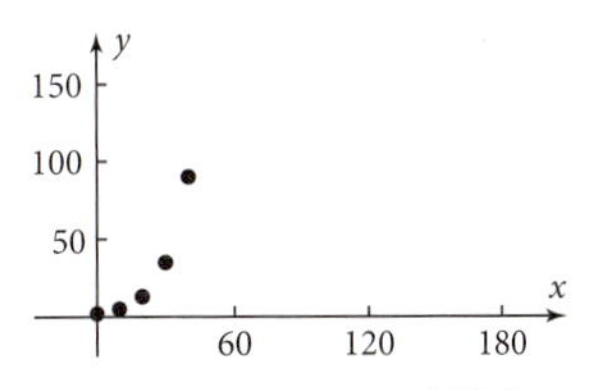

3b. $y = \dfrac{1220}{1 + (609)(1.1019\ldots)^{-x}}$

3c.

3d. ≈435 students; 115.4930... min

5a. Concave down

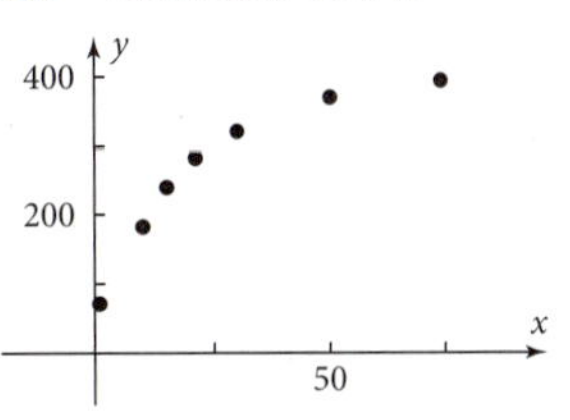

5b. $y = \dfrac{396}{1 + (2.7532\ldots)(1.0888\ldots)^{-x}}$

5c.

5d. The point of inflection occurs at (11.9037..., 198). Before approximately 12 days passed, the rate of new infection was increasing; after that, the rate was decreasing.

5e. Approximately 363 people were infected.

5f. Answers will vary.

7a. True.

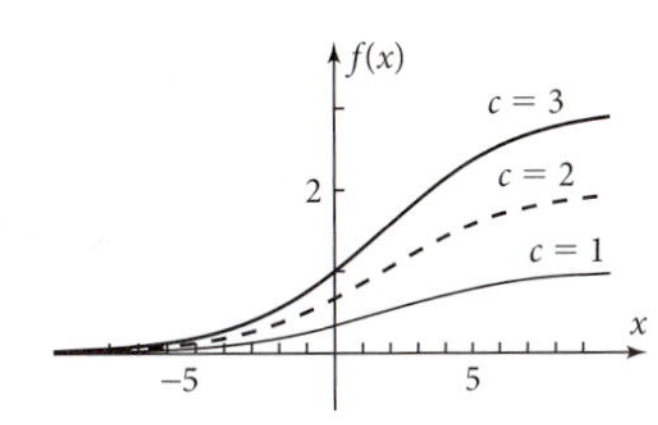

7b. Changing a seems to translate the graph horizontally.

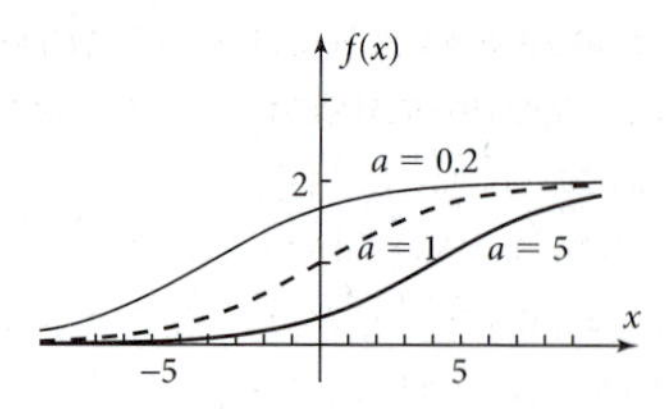

7c. Horizontal translation by 3

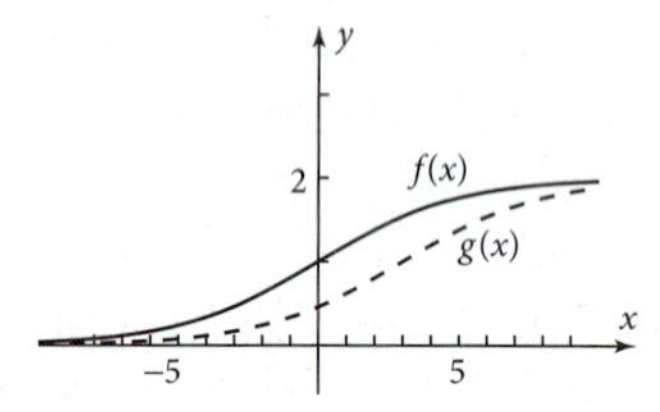

7d. $a = e^{1.2} = 3.3201....$

Problem Set 2-8

R1a.

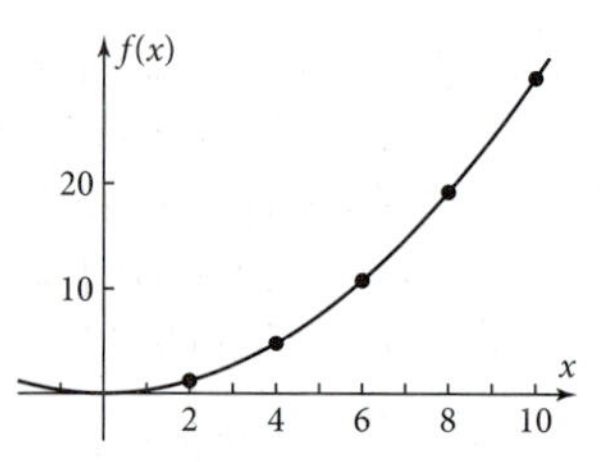

R1b. Increasing for $x > 0$, decreasing for $x < 0$, concave up

R1c. Quadratic power function. Real-world interpretations may vary.

R3a. Exponential

R3b. Power (inverse variation)

R3c. Linear **R3d.** Quadratic

R3e. i. $f(12) = 213\frac{1}{3}$

R3e. ii. $f(12) = 160$

R3e. iii. $f(12) = 180$

R3f. $f(x + c) = 53 \cdot 1.3^{x+c} = 53 \cdot 1.3^x \cdot 1.3^c$
$= 1.3^c \cdot f(x)$

R5a. $c^p = m$

R5b. $\log_7 30 = 1.7478...$

R5c. 63

R5d. $x = 4$, the equation is undefined if $x = -3$.

R5e. $x = \dfrac{1}{2 - \dfrac{\log 7}{\log 3}} = 4.3714...$

R7a.

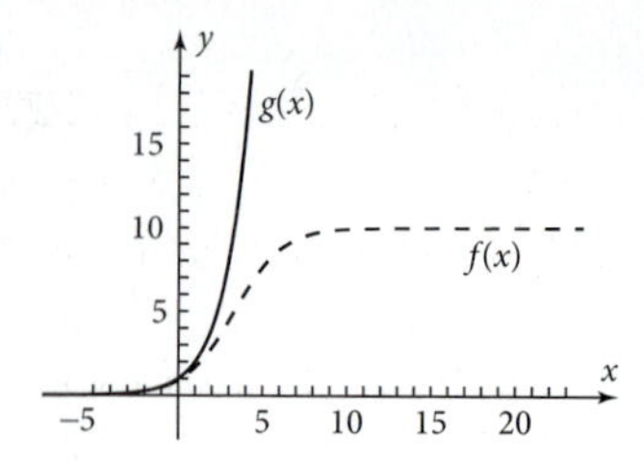

R7b. When x is a large negative number, the denominator of $f(x)$ is essentially equal to 10, so for large negative x, $f(x) = \frac{10 \cdot 2^x}{2^x + 10} \approx \frac{10 \cdot 2^x}{10} = 2^x = g(x)$. But for large positive x, the 10 in the denominator of $f(x)$ is negligible compared with the 2^x; so $f(x) = \frac{10 \cdot 2^x}{2^x + 10} \approx \frac{10 \cdot 2^x}{2^x}$ $= 10$.

R7c. $f(x) = \dfrac{10}{1 + 10 \cdot 2^{-x}}$

R7d. $g(x) = e^{(\ln 2)x}$

R7e. The size of the population would be limited by the capacity of the island.

$f(x) = \dfrac{460}{1 + (13.2906...)(1.1718...)^{-x}}$

$f(12) = 154.2335...$

$f(18) = 260.5072...$

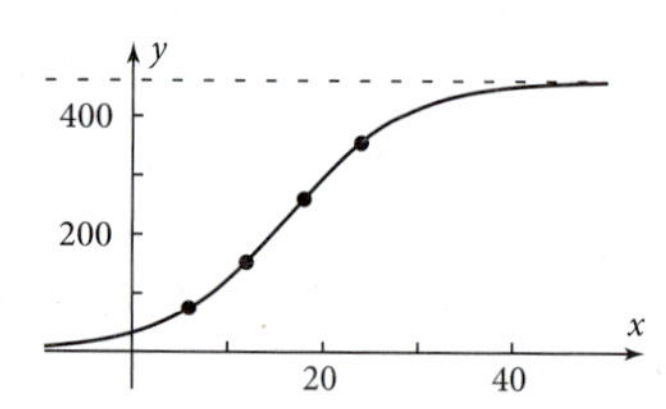

34.8878... months

C1a.

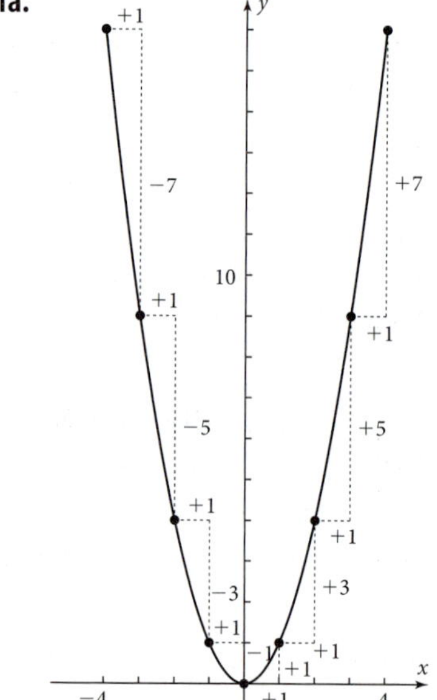

C1b. Vertex at (2, −5)

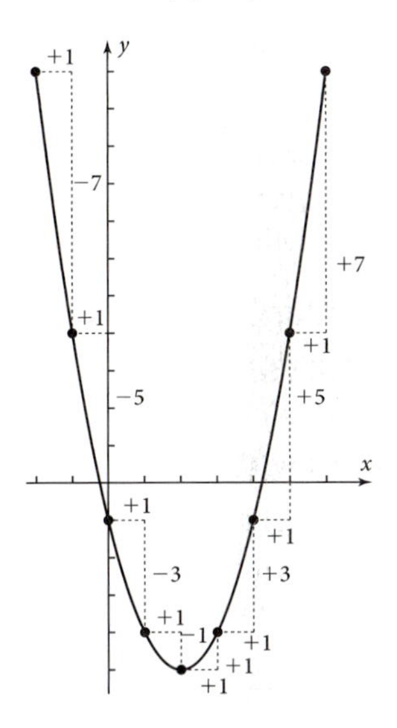

C1c.

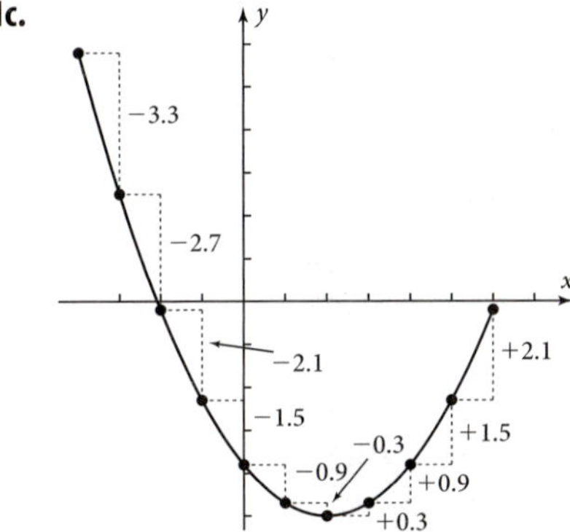

C3a. i. $y = \dfrac{1800 + 1000e^{0.7x}}{6 + e^{0.7x}}$

C3a. ii. $y = \dfrac{-90 + 1000e^{0.7x}}{-0.3 + e^{0.7x}}$

C3a. iii. $y = \dfrac{-210{,}300 + 1000e^{0.7x}}{-701 + e^{0.7x}}$

C3b.

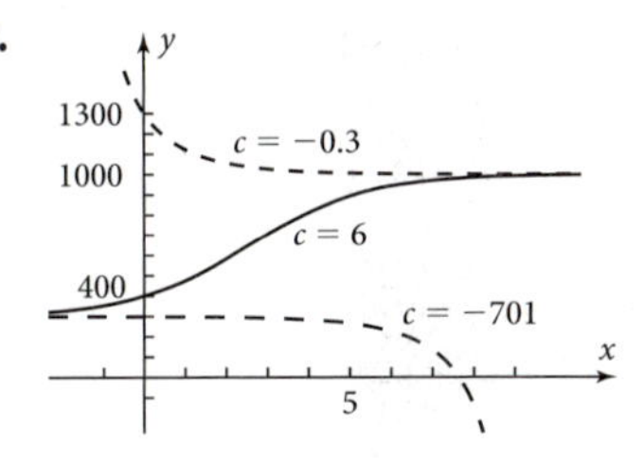

C3c. The graphs follow the direction of the line segments.

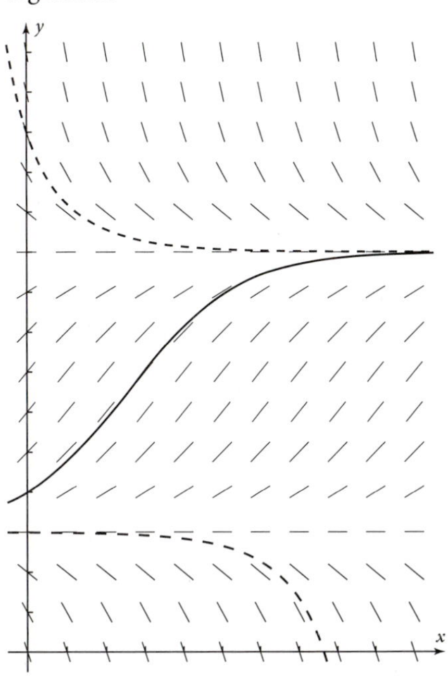

C3d. If 400 trees are planted, the population increases at first and then levels off at 1000. If 1300 (too many) trees are planted, the population decreases to level off at 1000. If 299 (too few) trees are planted, the population dwindles until all trees are dead.

C3e.

C3f. You can draw the graph following the direction of the line segments to get an idea of what happens at different initial conditions.

T1a. $y = ax + b$

T1b. $y = ax^2 + bx + c, a \neq 0$

T1c. $y = ax^b, a \neq 0$

T1d. $y = ae^{bx}$ or $y = ab^x$, $a, b \neq 0, b > 0$ and $b \neq 1$ in the case of $y = ab^x$

T1e. $y = a + b \log_c x, b \neq 0$ and $c > 0, c \neq 1$

T1f. $y = \dfrac{c}{1 + ae^{-bx}}$ or $y = \dfrac{c}{1 + ab^{-x}}$, $a, b, c \neq 0$, $b > 0$ and $b \neq 1$ in the case of $y = \dfrac{c}{1 + ab^{-x}}$

T3a. Add–add

T3b. Constant-second differences

T3c. Multiply–multiply

T3d. Add–multiply

T3e. Multiply–add

T5. $\log 5^x = x \cdot \log 5$

T7. 45

T9. No solutions.

T11. $f(x) = ax^b$; $f(x) = 0.6x^3$

T13. $f(100) = 600{,}000$ lb $= 300$ tons

T15.

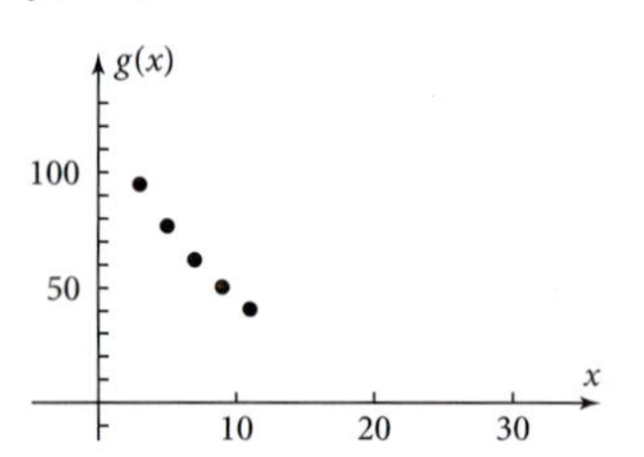

Graph will be concave up. The function appears to start at a positive number, decrease rapidly, and then level off as x grows large. A linear function cannot work, because the graph appears to be concave. An inverse variation power function cannot work, because it appears that the graph will intersect the vertical axis.

T17. 130.0510... °F above room temperature.

T19. $\log y = \log 7 + (\log 13) \cdot x$

T21. The second differences are all -32.

T23. $f(18) = 5.5, f(54) = 6.2$; $y = 3.6583\ldots + 0.6371\ldots \ln x$

T25. $g(2) = 362.0488\ldots; g(5) = 484.0232\ldots$; $g(7) = 583.2807\ldots; g(11) = 829.2796\ldots$

T27. The town can hold only a limited number of people.

Chapter 3

Problem Set 3-1

1. Yes

3. $\hat{y}(14) = 32.8$; ≈ 33 sit-ups. Explanations may vary.

5. $SS_{res} = 17.60$

Problem Set 3-2

1a. A graphing calculator gives $\hat{y} = 1.4x + 3.8$, with $r = 0.9842\ldots$

1b.

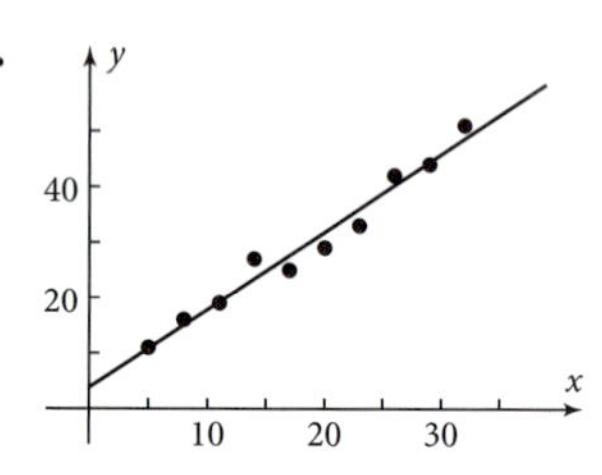

The line fits the data well.

1c. $\bar{x} = 18.5; \bar{y} = 29.7; \hat{y}(18.5) = 29.7$

1d. $SS_{dev} = 1502.10, SS_{res} = 46.80; r^2 = 0.9688\ldots$; $r = 0.9842\ldots$, which agrees with part a.

1e.

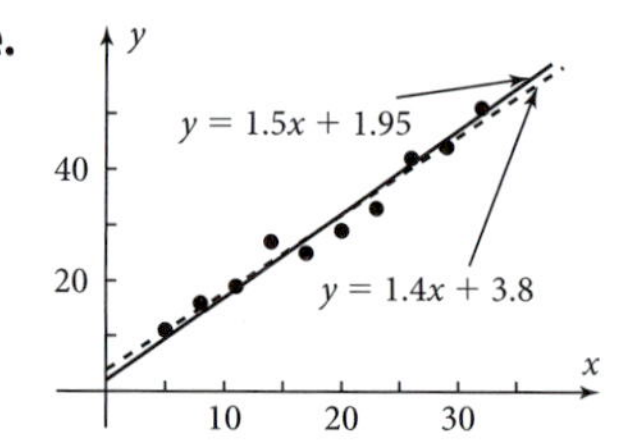

It is hard to tell which line fits better. $SS_{res} = 54.2250$, which is larger than SS_{res} for the regression line.

3a. $\hat{y} = -0.05x + 17, r^2 = 1, r = -1$, which means a perfect fit; r is negative because the remaining gas decreases as the distance driven increases.

3b. $\bar{y} = 15.18$ gal; $SS_{dev} = 4.828, SS_{res} = 0$; $r^2 = 1, r = -1$, which agrees with part a.

3c.

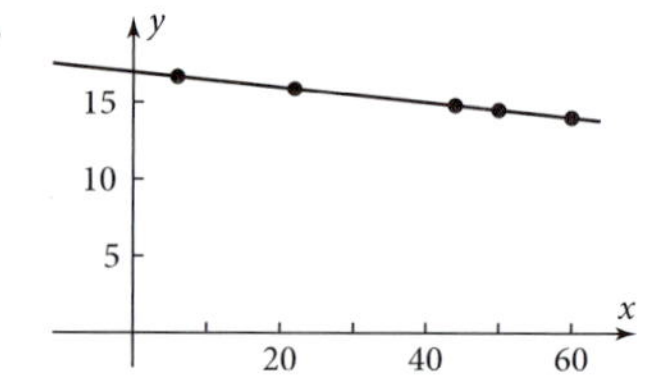

Data points are all on the line.

3d. 17 gal; 20 mi/gal

3e. $\hat{y}(340) = 0$ gal; not very confident, because driving conditions could change.

5a.

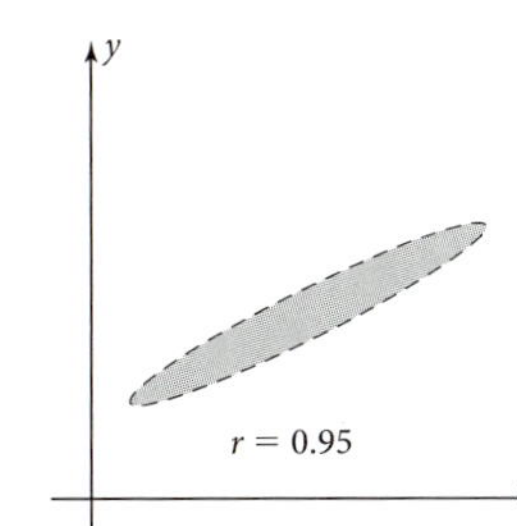

5b.

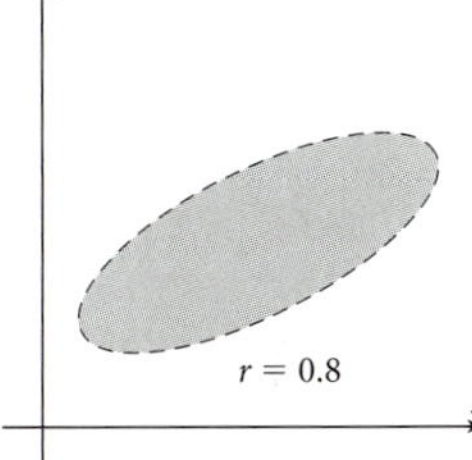

5c.

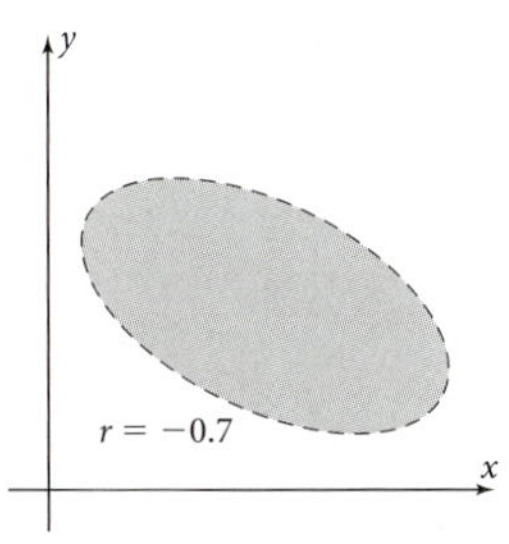

5d.

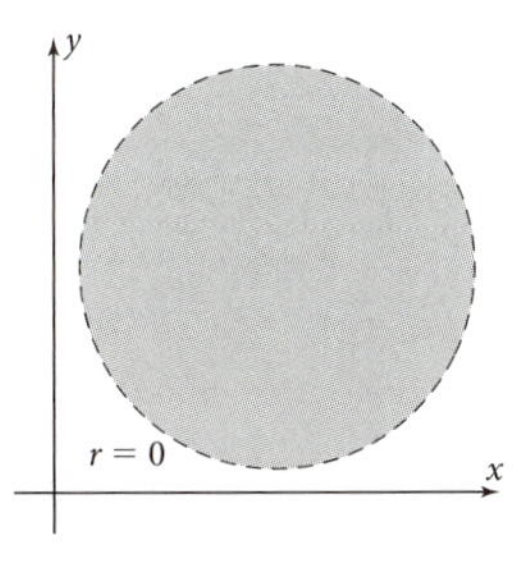

Problem Set 3-3

1a. Both functions have the proper right endpoint behavior: increasing to infinity. Only an exponential function has the correct left endpoint behavior: being nonzero.

1b. $\hat{y} = 346.9291\ldots \cdot 1.4972\ldots^x$, $r = 0.9818$

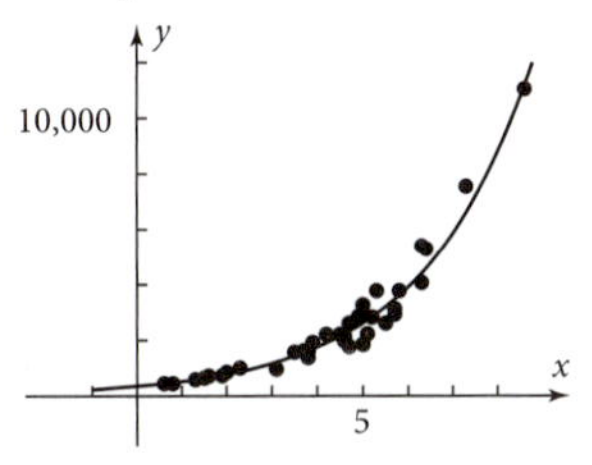

1c. $\hat{y}(0) \approx 347$ bacteria; $\hat{y}(24) \approx 5.6$ million bacteria

1d. ≈ 14.0 h

3a.

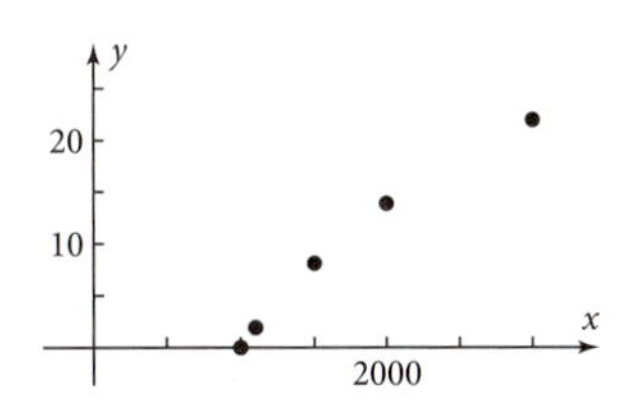

Concave down. The graph decreases more steeply (presumably to $-\infty$) toward $x = 0$, has a positive x-intercept, and increases less steeply to the right.

3b. $\hat{y} = -138.1230\ldots + 19.9956\ldots \ln x$; $r = 0.9999999799\ldots$, which is nearly 1.

3c.

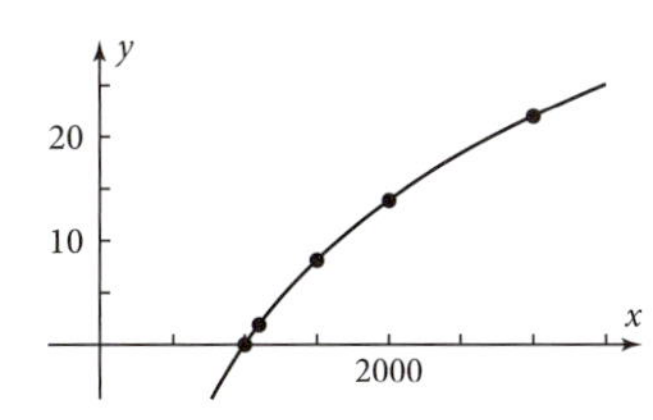

3d. $\hat{y}(2500) \approx 18.32 \text{ yr} \neq \dfrac{13.86 + 21.97}{2} = 17.915 \text{ yr}$

3e. ≈ 32.18 yr

Extrapolation, because $5000 > 3000$.

5a. Growth is basically exponential, but physical limits eventually make the population level off. A logistic function fits data that have asymptotes at both endpoints but are exponential in the middle.

$$\hat{y} = \frac{327.5140\ldots}{1 + 10.0703\ldots e^{-0.4029\ldots x}}$$

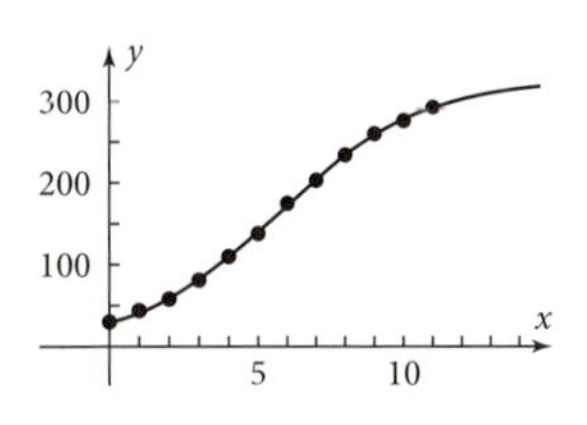

5b. $y \approx 326$ roadrunners; ≈ 328 roadrunners as $x \to \infty$; $x \approx 5.5$ yr.

5c. $\bar{y} = 158.5$ roadrunners

$SS_{dev} = 98{,}653.00$; $SS_{res} = 25.4205\ldots$

$r^2 = 0.9997\ldots$, which is very close to 1.

Problem Set 3-4

1a.

x	y
1	3
3	12
5	48
7	192
9	768

1b., 1c.

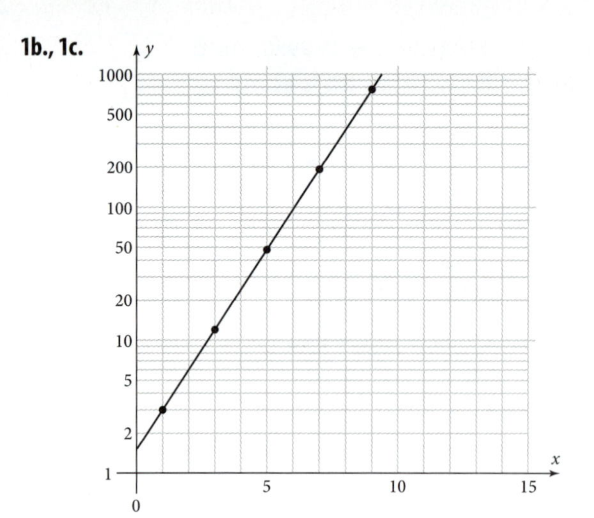

3a.

x	y
1	700
5	86.3847…
10	35.0831…
30	8.4108…
100	1.7583…

3b., 3c.

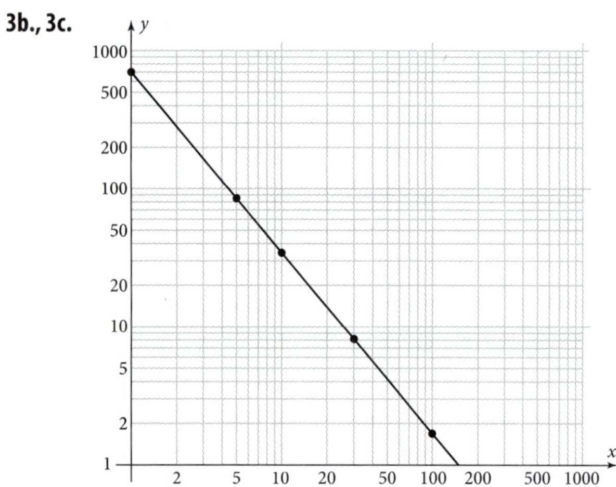

3d. Slope = −1.3

5a.

x	y
1	2
4	6.1588…
10	8.9077…
200	17.8949…
1000	22.7232…

5b., 5c.

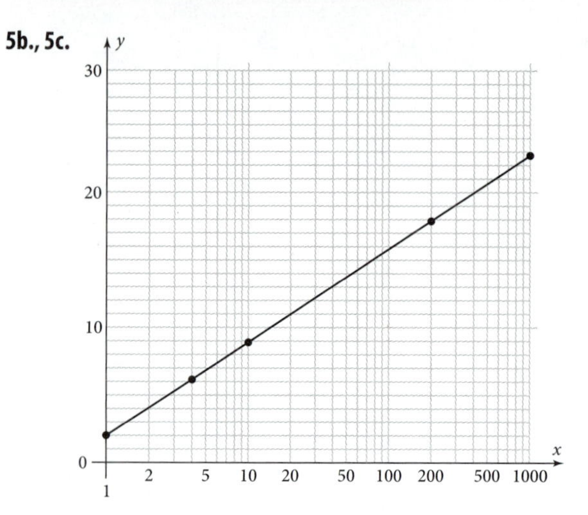

7.

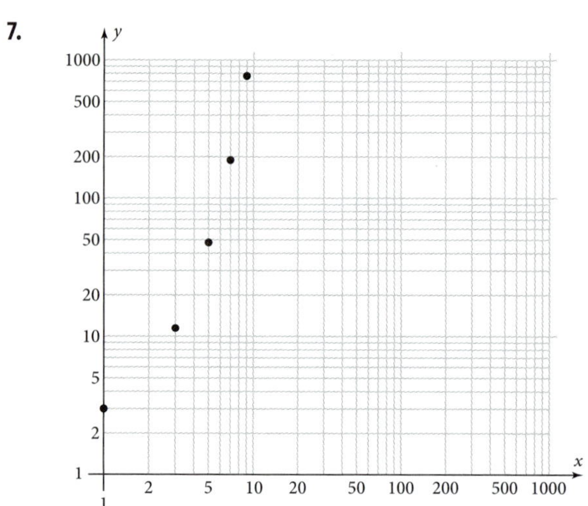

9a. Approximately (3, 8), (6, 15), (20, 44), (40, 82), and (90, 170)

9b. $\hat{y}(x) \approx 3x^{0.9}$

9c. $\hat{y}(2) \approx 5.6$, and (2, 5.6) is on the graph.

11a. Approximately (2, 3300), (5, 1200), (7, 570), (12, 99), and (14, 48)

11b. $\hat{y}(x) \approx 6814(0.7)^x$

11c. $\hat{y}(9) \approx 283.6$, and (9, 284) is on the graph.

13a. Approximately (2, 2), (5, 3.8), (30, 7.4), (400, 12.6), and (800, 14)

13b. $\hat{y}(x) \approx 2 \ln x + 0.6$

13c. $\hat{y}(90) \approx 9.6$, and (90, 9.6) is on the graph.

15.

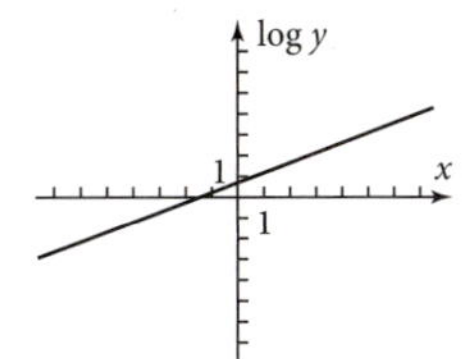

17.

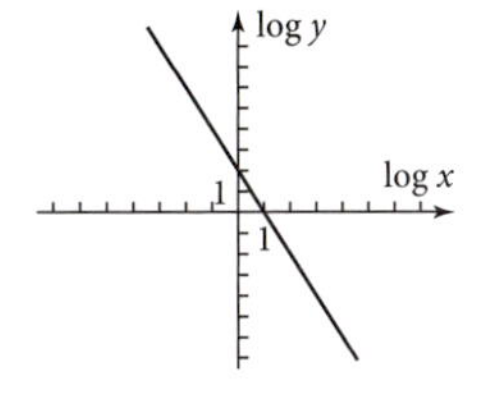

19a. If the drainage area is 0, the length should be 0 also. A power function contains the point (0, 0); a logarithmic function does not.

19b. $\hat{y}(x) = 70.2930\ldots x^{0.5274\ldots}$; $r^2 = 0.9760\ldots$

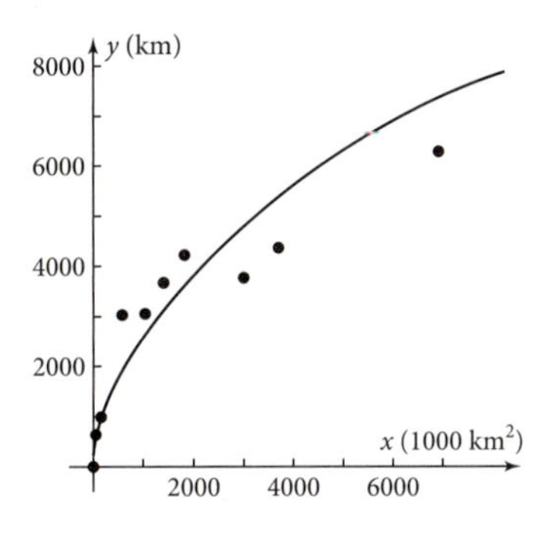

19c.

log (Drainage Area)	log (Length)
3.8397...	3.7990...
3.4116...	3.6471...
3.5658...	3.6405...
3.2528...	3.6274...
3.4742...	3.5772...
3.1398...	3.5666...
3.0128...	3.4854...
2.7558...	3.4820...
2.1903...	2.9867...
1.9000...	2.8129...
0.2909...	1.9542...

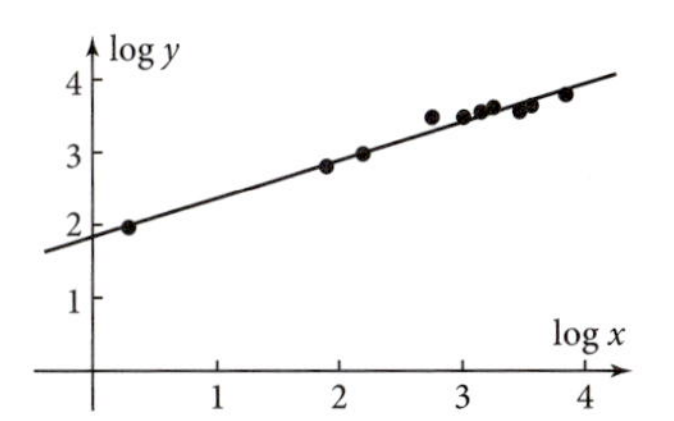

The linearized function appears to fit the transformed data better than the power function fits the original data.

19d. $\log \hat{y}(x) = 1.8469\ldots + 0.5274\ldots \log x$; $r^2 = 0.9760\ldots$; the grapher does power regression by doing linear regression on the linearized data, then changing back into power form.

21a. A power function would have 0°C at 0 km, that is, at Earth's surface. A logarithmic function correctly gives a positive temperature at the surface.

21b. $\hat{y}(x) = -8.5753\ldots + 2.7431\ldots \ln x$; $r^2 = 0.9958\ldots$; yes

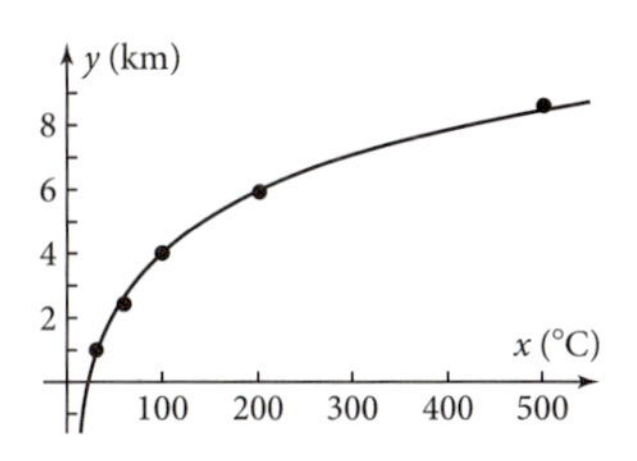

21c.

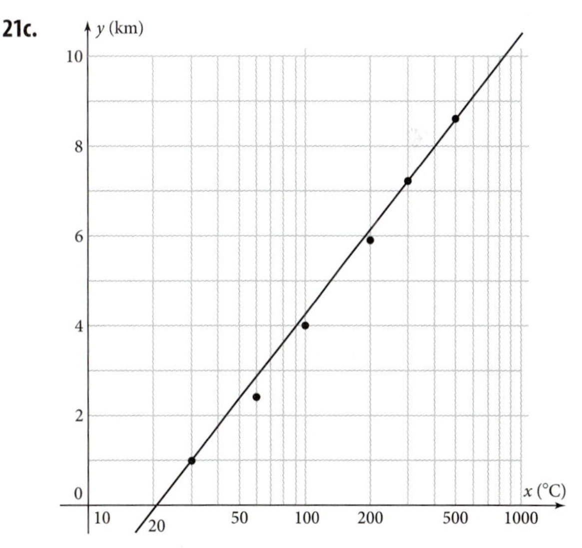

Answers will vary. The line shown here connects (30, 1.0) and (500, 8.6), and predicts that the temperature will be 300°C when the depth is about 7.2 km. The equation from part b predicts $\hat{y}(300) = 7.0712\ldots$ km.

21d. It has the form $\hat{y} = a + b \ln x$, so it is linear in the "variable" $\ln x$.

21e.

$\ln x$	y
3.4011...	1.0
4.0943...	2.4
4.6051...	4.0
5.2983...	5.9
6.2146...	8.6

$\hat{y}(\ln x) = -8.5753... + 2.7431... \ln x$; $r^2 = 0.9958...$; this indicates that the grapher does logarithmic regression by doing linear regression on the linearized data, then transforming back into logarithmic form.

23a. $y = 5 \cdot 3^x \Rightarrow \log y = \log(5 \cdot 3^x) = \log 5 + (\log 3)x$

23b. $y = ab^x \Rightarrow \log y = \log(ab^x) = \log a + (\log b)x$

23c. $y = 2x^3 \Rightarrow \log y = \log(2x^3) = \log 2 + 3 \log x$

23d. $y = ax^b \Rightarrow \log y = \log(ax^b) = \log a + b \log x$

23e. $y = a + c \log_b x = a + c \cdot \dfrac{\log x}{\log b}$
$= a + \dfrac{c}{\log b} \cdot \log x$

Problem Set 3-5

1a. The scatter plot is decreasing, is concave up, has a finite value for $x = 0$, and seems to approach 0 as $x \rightarrow \infty$. $\hat{y} = 107.9102... \cdot 0.998...^x$, $r = -0.9985...$

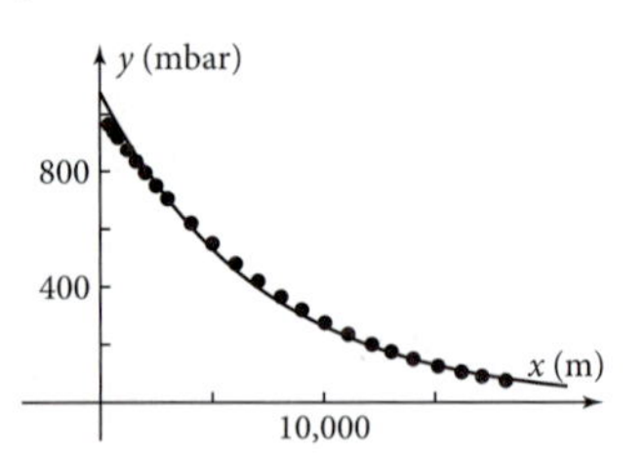

The points seem to lie very near the graph on the right, somewhat less near on the left and in the middle.

1b.

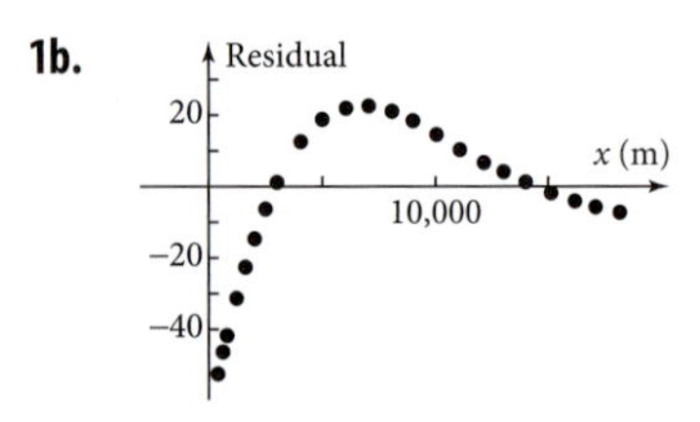

The residuals have a very definite pattern.

1c. No. The residuals are as great as ≈ -52.7 millibars (the mean absolute residual is ≈ 16.9 millibars).

3a. Exponential: $\hat{y} = 91.7362... \cdot 0.9996...^x$, $r = -0.9646...$

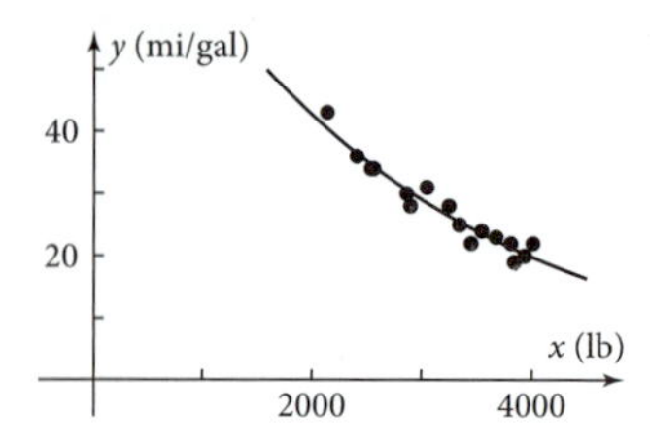

Power: $\hat{y} = 338{,}947.0156...x^{-1.1721...}$, $r = -0.9672...$

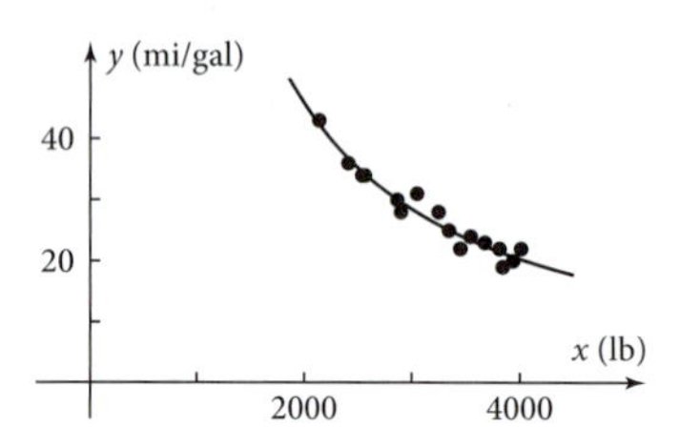

The functions appear to fit equally well, both graphically and by their r-values.

3b. Exponential:

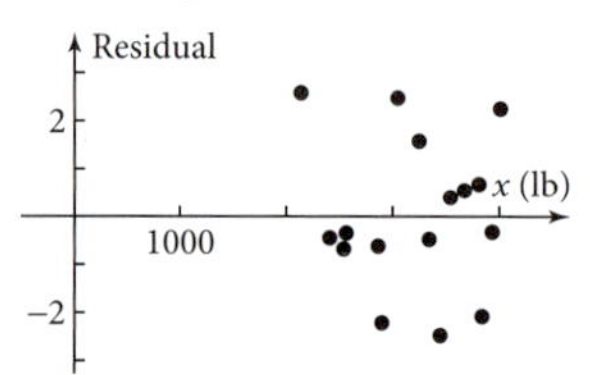

Power:

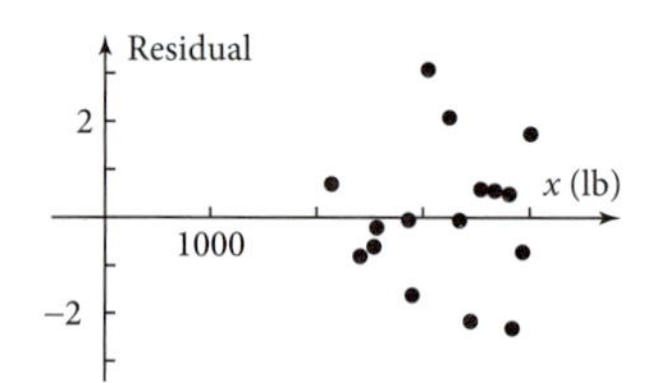

Both residual plots are fairly random. It is unclear which fits better.

3c. Exponential: ≈ 76 mi/gal.
Power: ≈ 233 mi/gal.
The exponential model is much more reasonable. The right endpoint behaviors are not significantly different.

5a. Both a linear and a power function would give a population of 0 at some recent time in the past. The exponential function would have even more rapid growth as years go by.

5b. 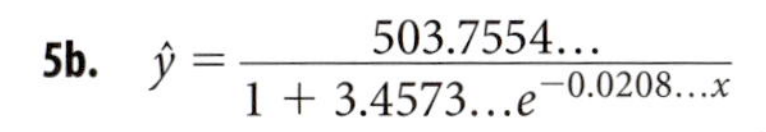
$\hat{y} = \dfrac{503.7554\ldots}{1 + 3.4573\ldots e^{-0.0208\ldots x}}$

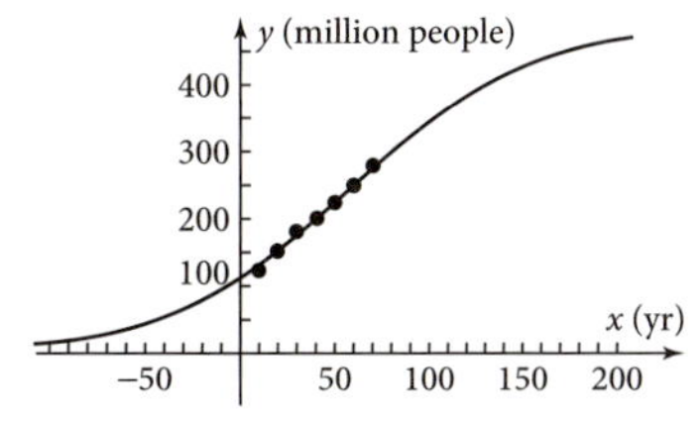

5c. $\hat{y}(80) \approx 304.6$ million; $\hat{y} \to 503.8$ million as $x \to \infty$

5d. $\hat{y} = \dfrac{524.4171\ldots}{1 + 3.6272\ldots e^{-0.0203\ldots x}}$

$\hat{y} \to \approx 524.4$ million as $x \to \infty$

7a.

log (mass)	log (cal/kg)
−0.1549...	2.3483...
0.3010...	1.7634...
1.8450...	1.5185...
2.7781...	1.3424...
3.6020...	1.1139...

$\log \hat{y} = -0.2774\ldots \log x + 2.0817\ldots$, $r^2 = 0.8773\ldots$

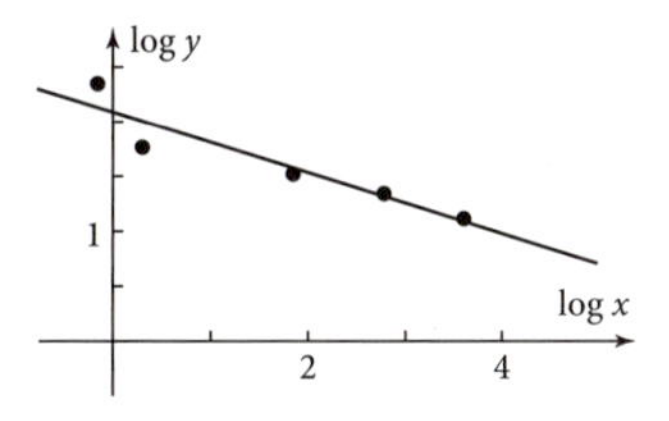

The pattern is roughly linear.

7b. $\log \hat{y} = 120.7251\ldots x^{-0.2774\ldots}$, which is the same equation as found by power regression, with $r^2 = 0.8773\ldots$, as in part a.

7c. $\hat{y}(0.002) = 676.9192\ldots \approx 677$ cal/kg

7d. $\approx$4.4 cal/kg, approximately 160% larger than or 260% of the actual value. You are extrapolating to an x-value quite far out of the data range.

9a. $\hat{y}_{\text{no whey}} = -1.71x + 57.64$
$\hat{y}_{\text{whey}} = -1.1942\ldots x + 56.04$

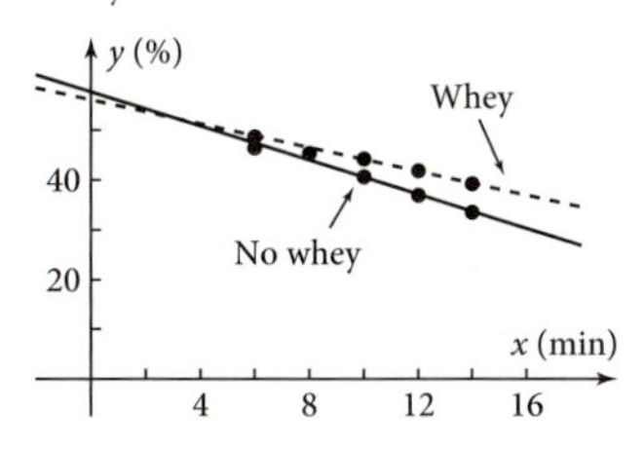

9b. $\hat{y}_{\text{whey}}(8) \approx 46.5\%$; Interpolation

9c. If whey is used: $x \approx 22$ min;
If no whey is used: $x \approx 16$ min; Extrapolation in both cases

9d. $\hat{y}_{\text{no whey}}(0) \approx 57.6\%$,
$\hat{y}_{\text{whey}}(0) \approx 56.0\%$. They are close.

11. Journal entries will vary.

Problem Set 3-6

R1a. A graphing calculator confirms that $\hat{y} = 1.6x + 0.9$.

R1b.

$\hat{y}$	$(y - \hat{y})$	$(y - \hat{y})^2$
5.7	0.3	0.09
8.9	1.1	1.21
12.1	−3.1	9.61
15.3	1.7	2.89

R1c. $\Sigma(y - \hat{y})^2 = 13.80$

R1d. For $\hat{y} = 1.5x + 1.0$: $\Sigma(y - \hat{y})^2 = 15.00$

R3a. Logarithmic: $\hat{y} = 136.6412\ldots - 16.8782\ldots \ln x$, $r = -0.9979\ldots$

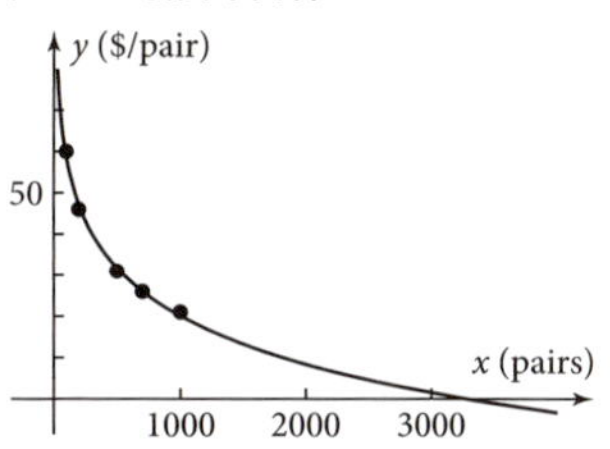

Power: $\hat{y} = 488.0261\ldots x^{-0.4494\ldots}$, $r = -0.9970\ldots$

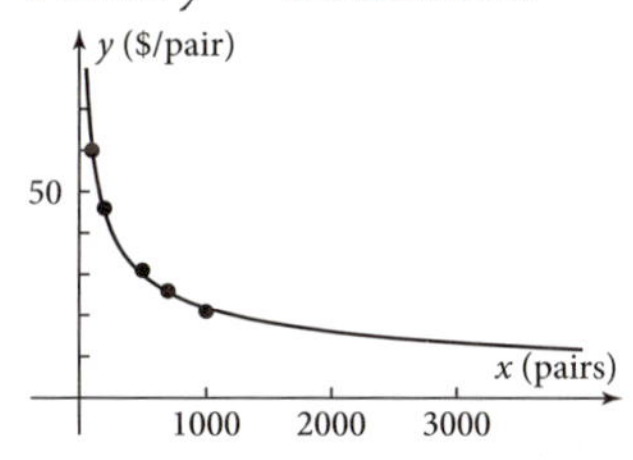

R3b. The logarithmic function predicts that the price of a pair eventually reaches $0.00 (at $x \approx 3280$ pairs), and then even becomes negative. The power function never predicts a price of $0.00 or lower (it approaches $0.00 asymptotically).

R3c. 5715 pairs by extrapolation

R3d. $\approx$$488.03 per pair

R3e. The cost is reduced by $\approx$27%; multiply–multiply

R5a. A power function would have $\hat{y}(0) = 0$ and a logarithmic function would have $\hat{y}(0) \to -\infty$ but both the exponential function and linear function would have $\hat{y}(0) \approx 325$ ppm.

R5b. Linear: $\hat{y} = 1.7115\ldots x + 320.5101\ldots, r = 0.9974\ldots$

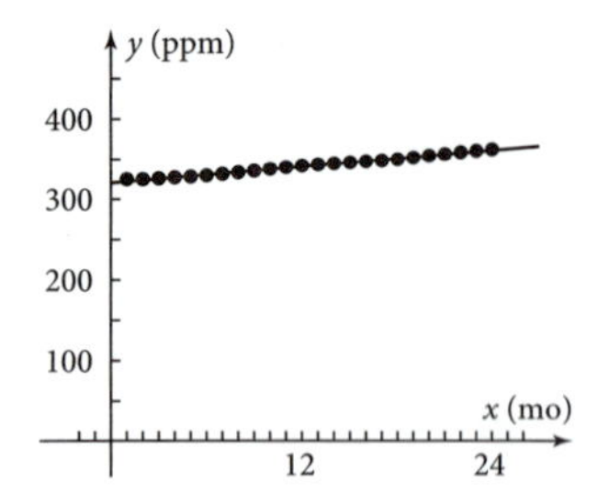

Exponential: $\hat{y} = 320.9749\ldots \cdot 1.0050\ldots^x, r = 0.9977\ldots$

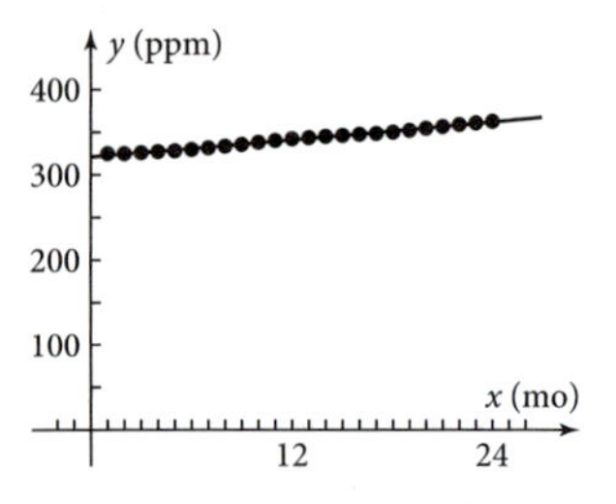

Linear: $\hat{y}(13) \approx 342.8$ ppm
Exponential: $\hat{y}(13) \approx 342.6$ ppm
Actual: 343.5 ppm
Linear: $\hat{y}(20 \cdot 12) \approx 731.3$ ppm
Exponential: $\hat{y}(20 \cdot 12) \approx 1067.0$ ppm

R5c.

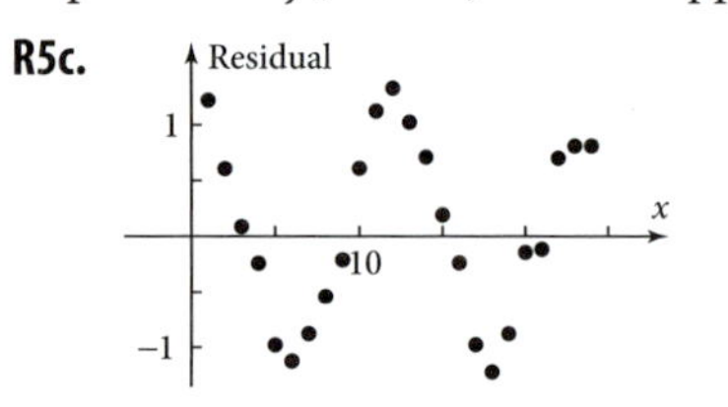

There is a repeating pattern, with a maximum in the winter and a minimum in the summer.

C1a. $\hat{y} = 3.0080...\sin(0.4990\ldots x + 1.0188...) + 5.0046...$

C1b. A table of values of $\hat{y}(x)$ agrees with the actual y-values to within round-off errors. The residuals are very small:

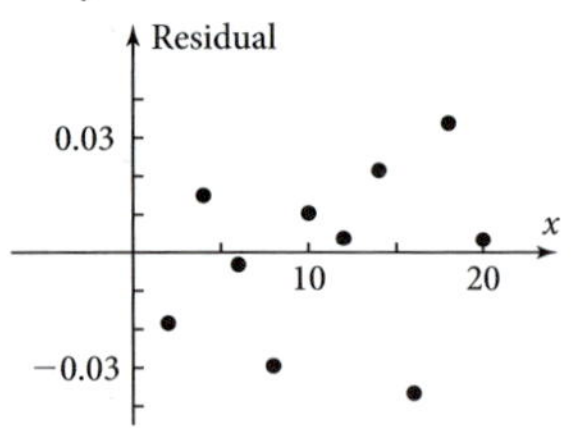

C1c. Period = 12.5915..., Amplitude = 3.0080..., Phase shift = −2.0418..., Sinusoidal axis $y = 5.0046...$

T1. −6

T3. Linear

T5. Exponential, because there is no clear pattern in the residuals for the exponential function

T7. Exponential

T9. Coefficient of determination, $R^2 = \dfrac{SS_{dev} - SS_{res}}{SS_{dev}}$

Correlation coefficient, $R = \pm\sqrt{R^2}$

T11.

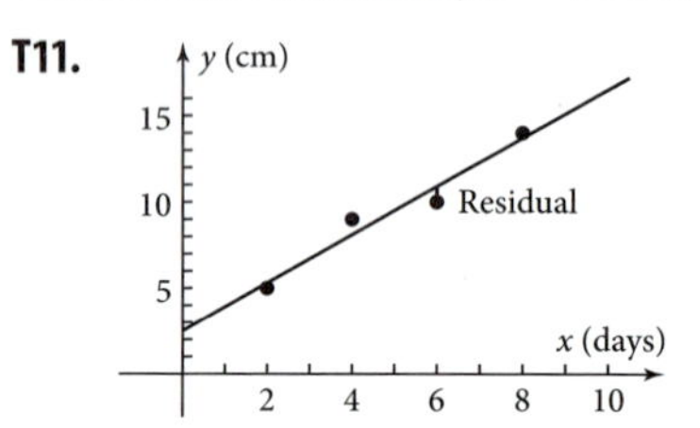

T13. $SS_{res} = 1.8$

T15. $\hat{y}_2(2) = 1.5(2) + 2 = 5$; $\hat{y}_2(8) = 1.5(8) + 2 = 14$

T17. 129.9 cm;
Extrapolation: 91 days > 8 days

T19. $\hat{m}(y) = 0.0087...y^{2.9455...}$; $\hat{m}(5) = 0.9991...$g; $\hat{m}(50) = 881.3357...$g

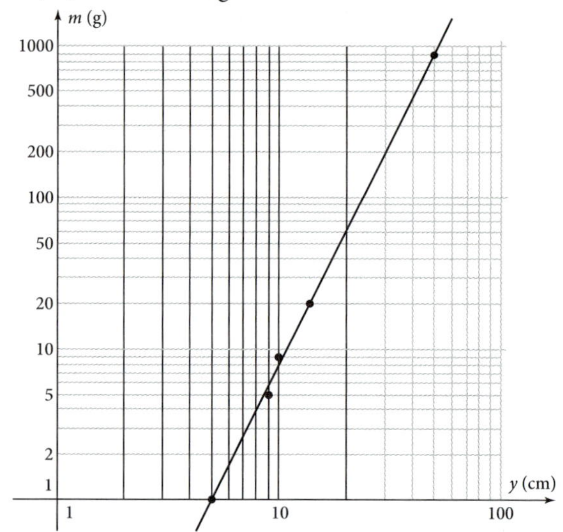

The graph of $\hat{m}$ appears to fit the data fairly well.

Chapter 4

Problem Set 4-1

1. $f(-2) = 0; f(4) = 0$; -2 and 4 are called *zeros of the function* because when substituted for x, they make the function value equal zero.

3. $g(3)$ takes the form $\frac{1}{0}$, which is undefined because of division by zero; $g(3.001) = 995.004001$, $g(2.999) = -1005.003999$; $g(x)$ is very large in the positive direction when x is close to 3 on the positive side, and very large in the negative direction when x is close to 3 on the negative side.

5. The graph of function h does not cross the x-axis, and thus there are no real zeros for function h.

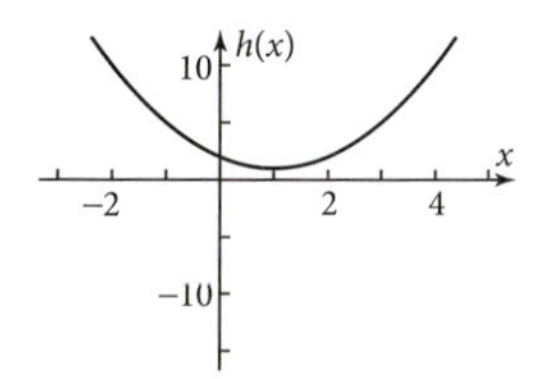

Problem Set 4-2

1. $8x^2 - 2x - 15$

3. $2x^2 - 25x + 72$

5. $9x^2 + 42x + 49$

7. $25x^2 - 49$

9. $x^3 - 9x^2 + 8x + 60$

11. $(x + 3)(x + 7)$

13. $(x - 4)(x + 10)$

15. $(x - 6)(x + 6)$

17. $(x - 10)^2$

19. $x(x + 7)$

21. $x = -3, -7$

23. $x = 4, -10$

25. $x = 6, -6$

27. $x = 10$

29. $x = 0, -7$

31. $x = -3, -7$

33. $x = 3.1350..., 0.5316...$

35. $x = 1.6442..., -0.8109...$

37. $2 - 23i$

39. 58

41. $-13 + 84i$

43. $x = 3 \pm 5i$

45. $x = -\frac{5}{3} \pm \frac{1}{3}i$

47. $x = -11 + 3i$

49. $x = 0.2 - 0.2i$

51. $x = -3, x = 5$

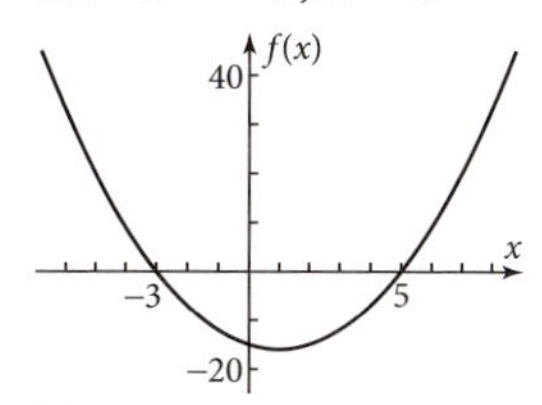

53. $x = 3 + i, x = 3 - i$; the graph confirms that the vertex is at $x = 3$ and there are no real zeros.

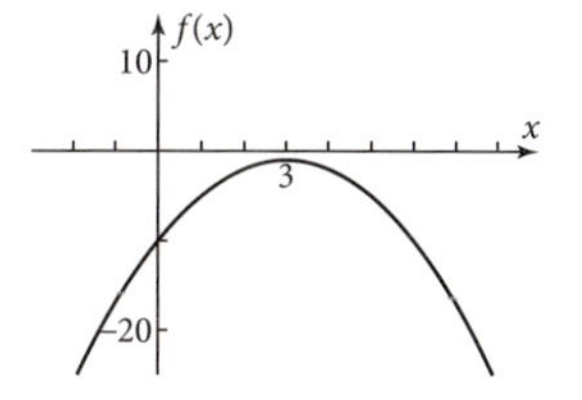

55. A zero of a function f is a value of x (real or complex) that makes $f(x) = 0$.

57a. $f(x) = -0.2x^2 + 1.2x + 3.2$

57b. $(3, 5)$

57c.

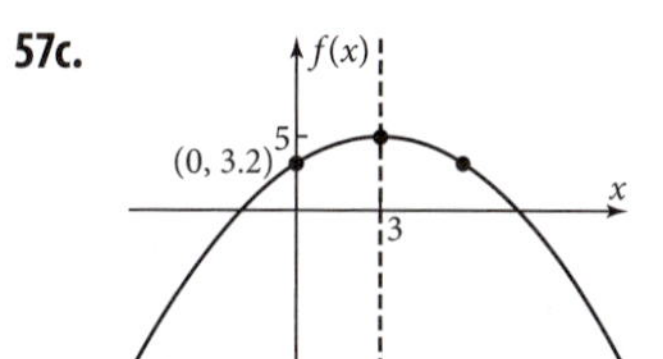

59a. $f(x) = 5x^2 + 40x + 87$

59b. $(-4, 7)$

59c.

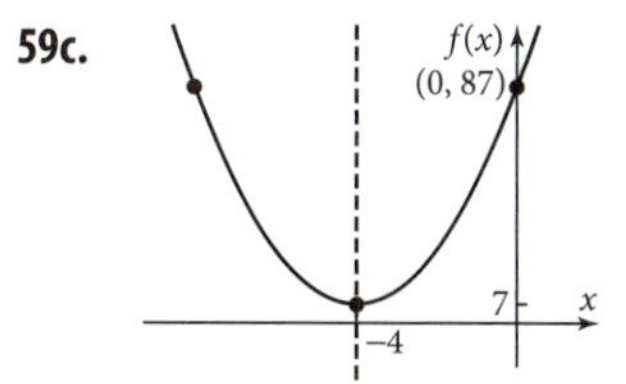

61a. $f(x) = (x + 3)^2 + 2$

61b. $(-3, 2)$

61c.

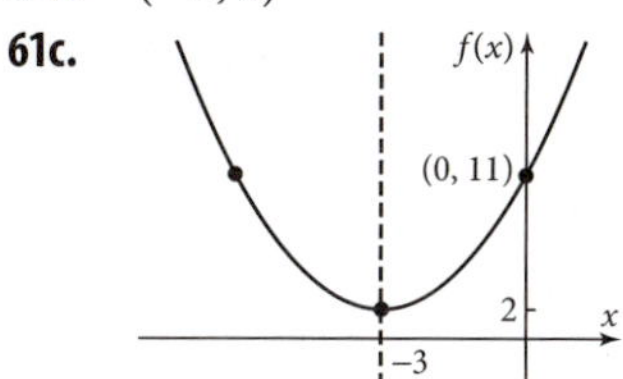

63a. $f(x) = 3(x - 4)^2 - 31$

63b. $(4, -31)$

63c.

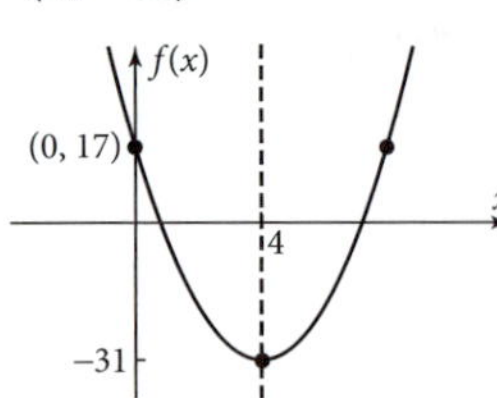

65a. $x = -3 \pm \sqrt{5} = -0.7639..., -5.2360...$

65b. $x = 7, 3$

65c. $x = 3.5 \pm \sqrt{16.25} = 7.5311..., -0.5311...$

65d. $x = -1 \pm \sqrt{\frac{4}{7}} = -0.2440..., -1.7559...$

65e. $x = 2.5 \pm \sqrt{0.75} = 3.3660..., 1.6339...$

Problem Set 4-3

1a.

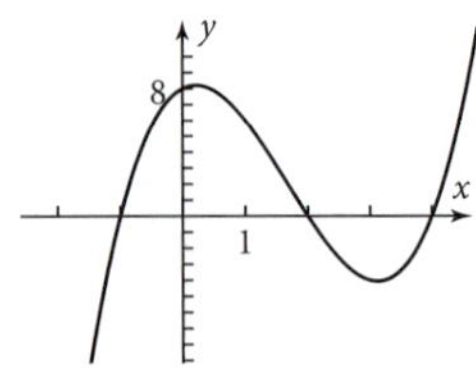

Three (two up and one down), the same as the degree of the polynomial

1b. $x \approx -1, 2, 4$

1c. $p(x) = (x + 1)(x - 2)(x - 4)$

1d. If c is a zero, then $x - c$ is a factor.

3. Sixth-degree; four real distinct zeros; two nonreal complex zeros; positive leading coefficient

5. Eighth-degree; six real zeros (including three double zeros); two nonreal complex zeros; positive leading coefficient

7. Fourth-degree; four real zeros (including a triple zero); no nonreal complex zeros; negative leading coefficient

9. Example: $f(x) = -(x + 3)(x + 1)(x - 1)$

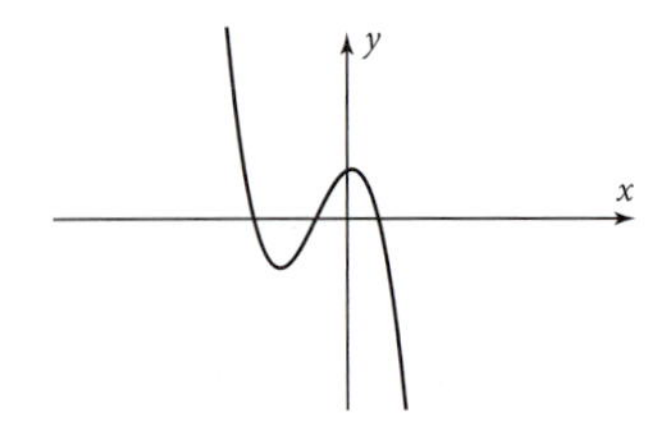

11. Example: $f(x) = (x + 1)(x^2 - 4x + 5)$

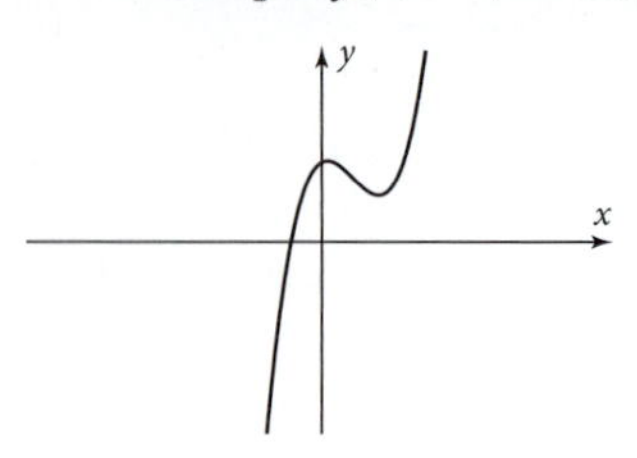

13. Example: $f(x) = x^3 + x$

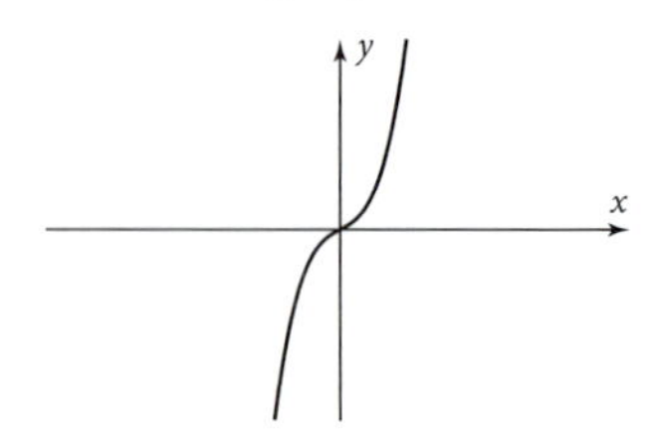

15. Example: $f(x) = x^4 + 1$

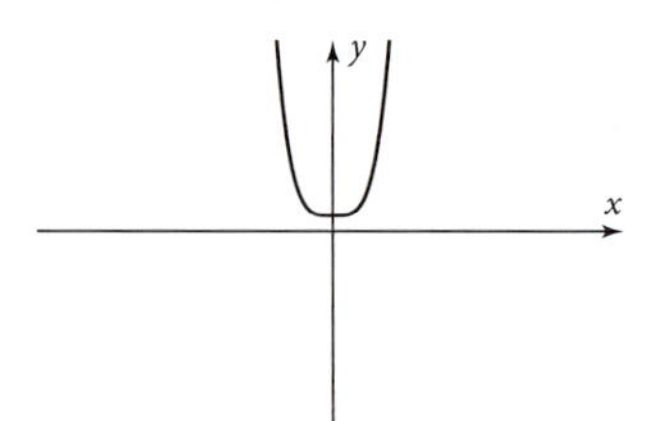

17. Example: $f(x) = (x + 1)^2(x - 2)^2$

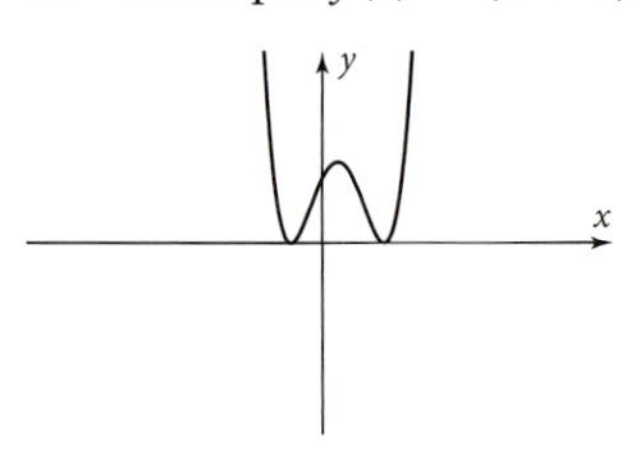

19. No such polynomial exists; a polynomial cannot have more zeros than its degree.

21. Sum = 1, product = −40; sum of pairwise products = −22, $x = -5, 2, 4$

23. Sum = $-\frac{18}{5}$, product = $\frac{156}{5}$; sum of pairwise products = $-\frac{7}{5}$, $x = \frac{12}{5}, -3 \pm 2i$

25. $f(x) = x^3 - 4x^2 - 11x + 30 = (x + 3)(x - 2)(x - 5)$

27. $f(x) = x^3 - 8x^2 + 29x - 52$
$= (x - 4)(x - 2 - 3i)(x - 2 + 3i)$

29a. $p(2) = -6$; $p(-3) = -101$

29b. $\dfrac{p(x)}{x - 2} = x^2 - 5x - 5 - \dfrac{6}{x - 2}$;
$\dfrac{p(x)}{x + 3} = x^2 - 10x + 35 - \dfrac{101}{x + 3}$

31. If $p(x)$ is a polynomial, then $p(c)$ equals the remainder when $p(x)$ is divided by the quantity $(x - c)$.

33. A polynomial function has at least one zero in the set of complex numbers.

35a. Quartic function; five terms

35b.

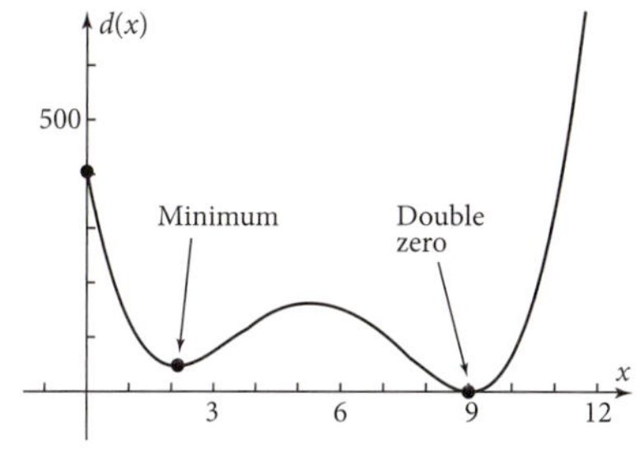

35c. 405 km as shown by the y-intercept.

35d. The extreme point is (2.1492..., 47.9782...). So Ella turned around and started moving away from Alderaan at about 2.15 minutes after she started the maneuver, when she was at about 48.0 km from the surface.

35e. $x = 9$ minutes; at this point, the spaceship just touches the surface of Alderaan, and then starts moving away.

35f. $x = 2 - i$

35g. 22; the opposite of the coefficient of the second term.

37a. 56 ft as indicated by the y-intercept.

37b. Synthetic substitution shows $x = 2$ is a zero of $f(x)$. Lucy was going down at this time.

37c. Zeros are $x = 2$, $x = 4$, and $x = 7$; the sum of the zeros is 13; $-\left(\frac{b}{a}\right) = -\left(\frac{13}{-1}\right) = 13$. The product of the zeros is 56; $-\left(\frac{d}{a}\right) = -\left(\frac{56}{-1}\right) = 56$. The sum of the pairwise products is 50; $\frac{c}{a} = \frac{-50}{-1} = 50$.

37d. For large positive values of x, $f(x)$ keeps going downward as x increases, whereas Lucy actually comes up again and continues bounces of decreasing amplitude as time goes on.

39. $x = -3, -\frac{3}{2}, 1, 2$; (one at a time) $-\frac{3}{2}$; (two at a time) $\frac{-14}{2}$; (three at a time) $-\frac{-9}{2}$; (four at a time) $\frac{18}{2}$. Conjecture: For a degree n polynomial $a_n x^n + a_{n-1}x^{n-1} + \ldots + a_1 x + a_0$, the coefficient a_k of the degree k term is $a_n \cdot (-1)^{n-k} \cdot$ (the sum of the product of roots taken k at a time).

41. $g(x) = x^3 - 8x^2 + 20x - 25$;
zeros of $f(x)$: $x = 4, \frac{1}{2} \pm \frac{1}{2}i\sqrt{11}$;
zeros of $g(x)$: $x = 5, \frac{3}{2} \pm \frac{1}{2}i\sqrt{11}$

Problem Set 4-4

1a., 1d.

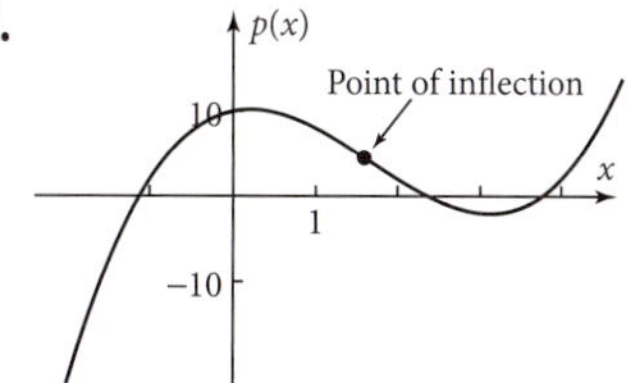

1b. The function has three zeros (equaling the degree of the function) and two extreme points (one less than the degree of the function).

1c. The third differences all equal 6.

1d. $x = 1.6666...$

3a.

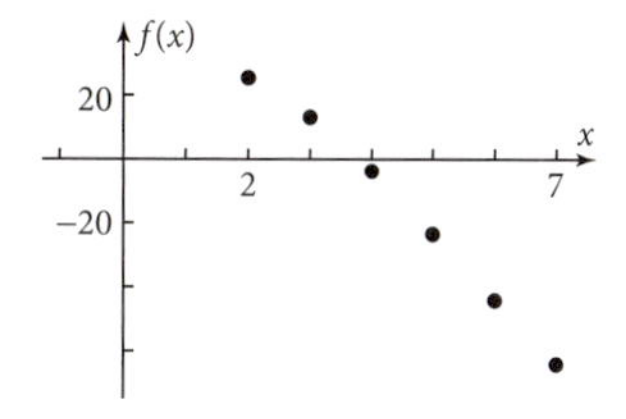

3b. The third differences all equal 1.8.

3c. $f(x) = 0.3x^3 - 5x^2 + 7x + 29$; Cubic regression gives the same result, with $R^2 = 1$.

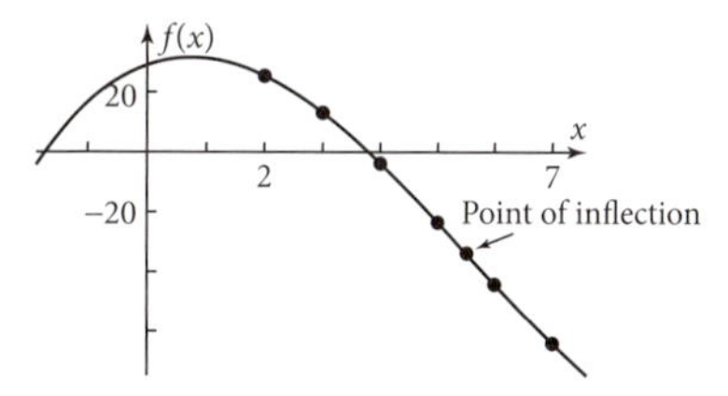

3d. The point of inflection is at $x = 5.5555...$

5a. $f(x) = -4x^3 + 120x^2$; Cubic regression gives the same result, with $R^2 = 1$.

5b. 8 in.

5c. Zeros are $x = 0$ ft or 30 ft

5d.

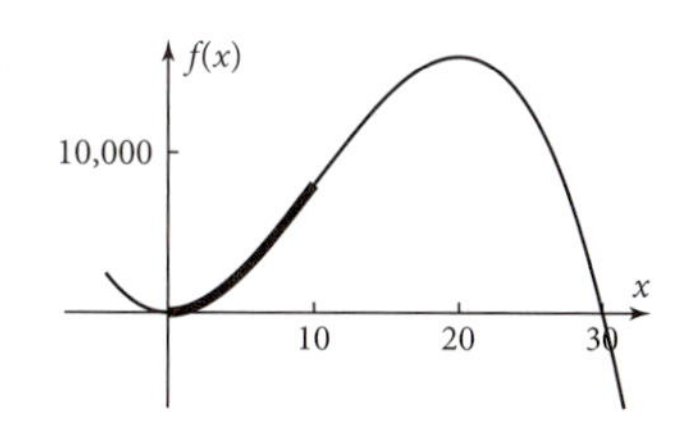

7a. Volume $= \pi x^2 y$, so $v(x) = \pi(16x^2 - x^4)$; The function is quartic because of the x^4.

7b. $v(x) = \pi \cdot x \cdot x \cdot (4 - x)(4 + x)$, so zeros are $x = 0$, 0, 4, or -4. $x = 0$ is a double zero, $x = -4$ is a zero out of the domain.

7c. About 201.06 cm^3 at $x \approx 2.83$ cm.

9a.

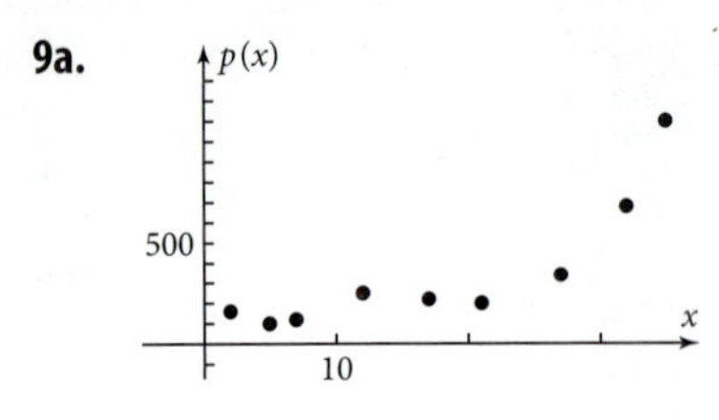

The function appears to have three vertices, so it must have degree at least four.

9b. $p(x) = 0.0058...x^4 - 0.3289...x^3 + 6.1803...x^2 - 36.6237...x + 193.0963...$ with $R^2 = 0.9931...$

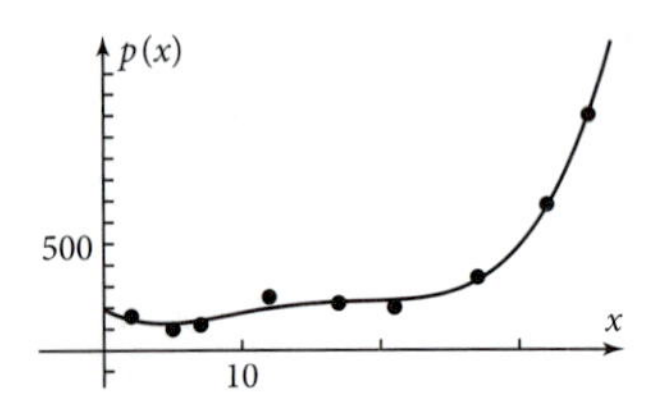

9c. The 12-in.-model is the most overpriced because its residual is greatest.

9d. Answers will vary.

11a. $y = x^3 - 11x^2 + 36.71x - 35.53$; the y-intercept is -35.53 mi.

11b. $-\left(\frac{4.1 \text{ mi}}{35.53 \text{ mi}}\right)$; $y = \frac{4.1}{35.53}x^3 + \frac{45.1}{35.53}x^2 - \frac{150.511}{35.53}x + 4.1$

11c.

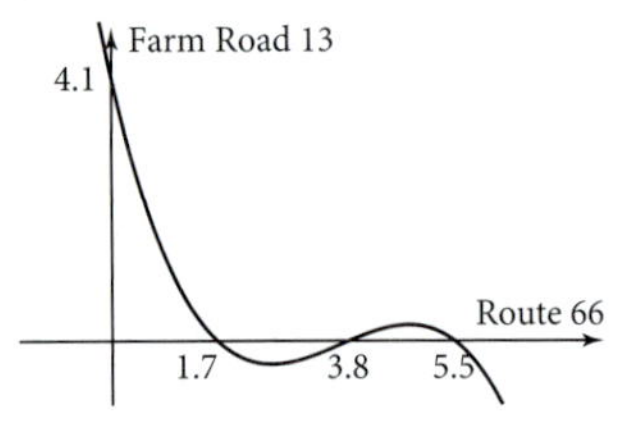

The river goes south before the first crossing and after the last.

11d. Farthest south is 0.3618... mi ($y = -0.3618...$) at 2.5676... mi.
Farthest north is 0.2508... mi at 4.7656... mi.
$-2.8568...$ mi east of the zero at 5.5 mi.

13a. The graphs match.

13b.

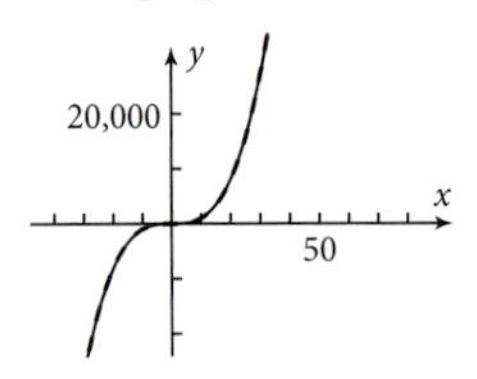

13c. Both graphs look similar to $y = x^3$. The vertices and intercepts of the f graph are hard to see. The terms of lower degree do not significantly affect the graph for large x.

15a. $R^2 = 0.9611...$

15b. $SS_{res} = 67.0324...$

15c. $\bar{y} = 10.875$; $SS_{dev} = 1724.875$

15d. $R^2 = \dfrac{1724.875 - 67.0324...}{1724.875} = 0.9611...$

Problem Set 4-5

1. A rational number is a number that can be written as a ratio of two integers. A rational algebraic expression is an expression that can be written as a ratio of two polynomials.

3. A rational algebraic function has a discontinuity at any value of x that makes the denominator equal zero. It can be a removable discontinuity or a vertical asymptote.

5. Vertical dilation by a factor of -2, vertical translation by 3, horizontal translation by 4; $f(3) = 5$, $f(5) = 1$

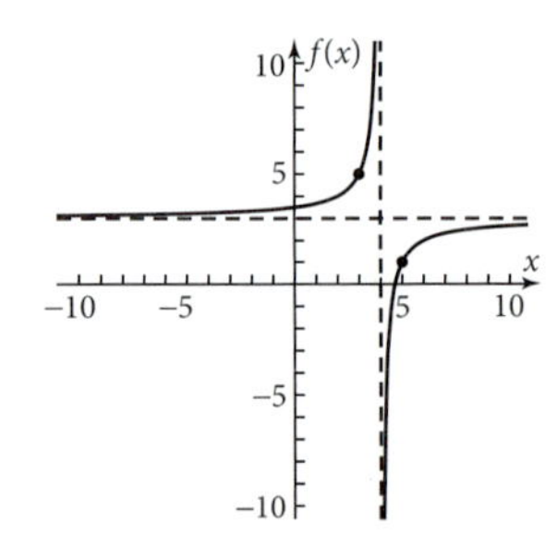

7. Vertical dilation by a factor of 4, vertical translation by -3, horizontal translation by -1; $f(-2) = -7$, $f(0) = 1$

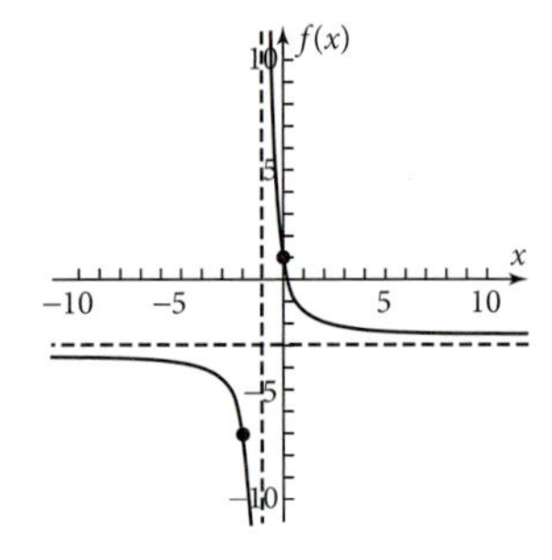

9a. $f(x) = 1 + \dfrac{-5}{x+2}$

9b. No removable discontinuities, vertical asymptote: $x = -2$.

9c. Horizontal asymptote: $y = 1$, which equals the ratio of the leading coefficients

9d. x-intercept: $x = 3$; y-intercept: $f(0) = 1.5$

9e.

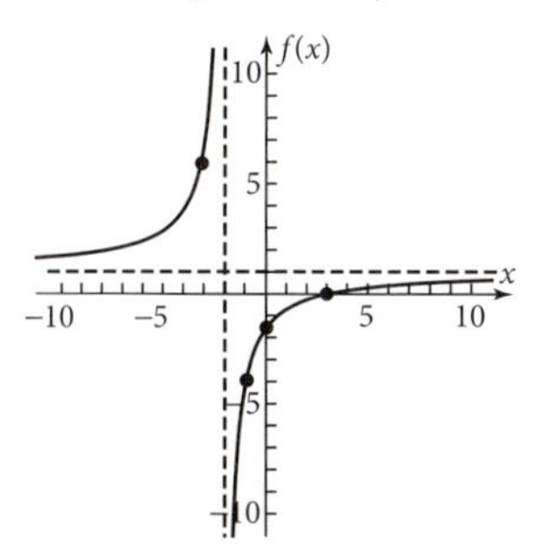

11a. $f(x) = -1 + \dfrac{-2}{x-4}$

11b. No removable discontinuities, vertical asymptote: $x = 4$

11c. Horizontal asymptote: $y = -1$, which equals the ratio of the leading coefficients

11d. x-intercept: $x = 2$; y-intercept: $f(0) = -0.5$

11e.

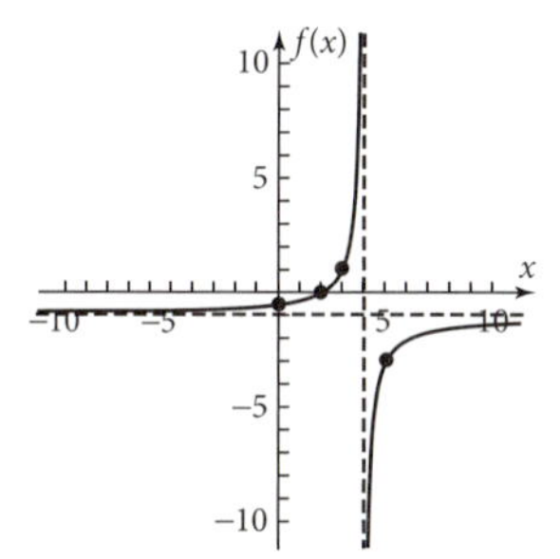

13a. $f(x) = 3 + \dfrac{5}{x-1}$

13b. No removable discontinuities, vertical asymptote: $x = 1$

13c. Horizontal asymptote: $y = 3$, which equals the ratio of the leading coefficients

13d. x-intercept: $x = -\frac{2}{3}$; y-intercept: $f(0) = -2$

13e.

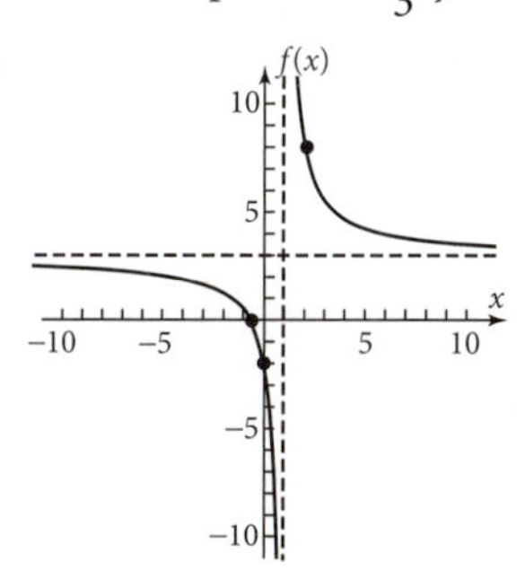

15a. $f(x) = 1 + \dfrac{-7}{x+5}, x \neq -7$

15b. Removable discontinuity at $(-7, 4.5)$; vertical asymptote: $x = -5$

15c. Horizontal asymptote: $y = 1$, which equals the ratio of the leading coefficients

15d. x-intercept: $x = 2$; y-intercept: $f(0) = -0.4$

15e.

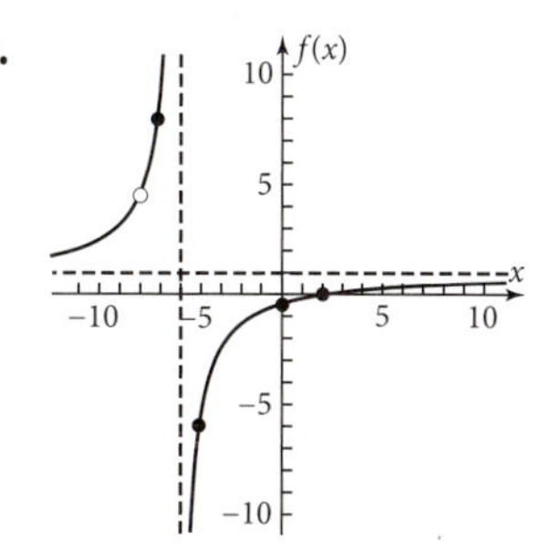

17a. $f(x) = 1 + \dfrac{1}{x+3}, x \neq 1$

17b. Removable discontinuity at $(1, 1.25)$; vertical asymptote: $x = -3$

17c. Horizontal asymptote: $y = 1$, which equals the ratio of the leading coefficients

17d. x-intercept: $x = -4$; y-intercept: $f(0) = \frac{4}{3}$

17e.

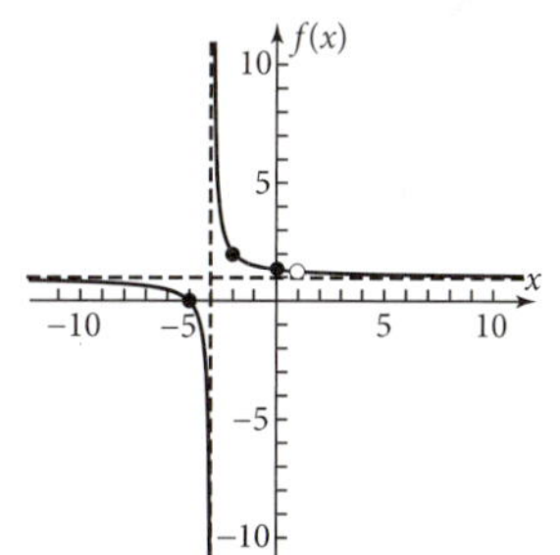

19a. $f(x) = 3 + \dfrac{8}{x-5}, x \neq -4$

19b. Removable discontinuity at $\left(-4, \frac{19}{9}\right)$; vertical asymptote: $x = 5$

19c. Horizontal asymptote: $y = 3$, which equals the ratio of the leading coefficients

19d. x-intercept: $x = \frac{7}{3}$; y-intercept: $f(0) = 1.4$

19e.

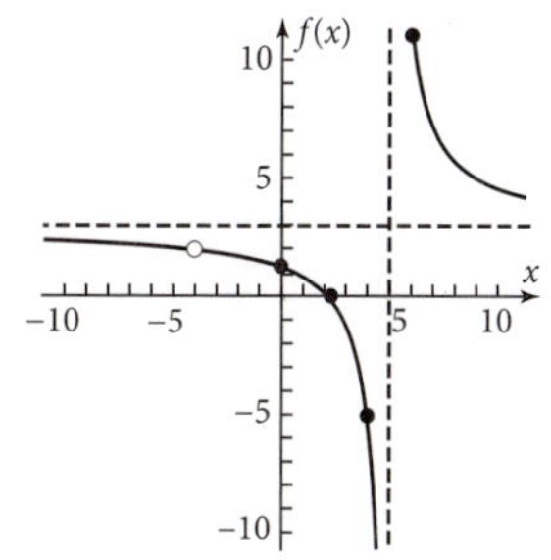

21a. $f(x) = 1 + \dfrac{-8}{x^2 + 4}$

21b. No removable discontinuities; no vertical asymptotes

21c. Horizontal asymptote: $y = 1$, which equals the ratio of the leading coefficients

21d. x-intercepts: $x = \pm 2$; y-intercept: $f(0) = -1$

21e.

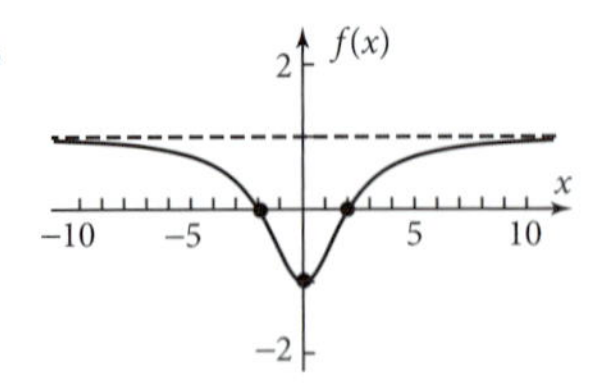

23a. $f(x) = x + 1 + \dfrac{-9}{x + 1}$

23b. No removable discontinuities, vertical asymptote: $x = -1$

23c. Diagonal asymptote: $y = x + 1$

23d. x-intercepts: $x = -4$ or $x = 2$; y-intercept: $f(0) = -8$

23e.

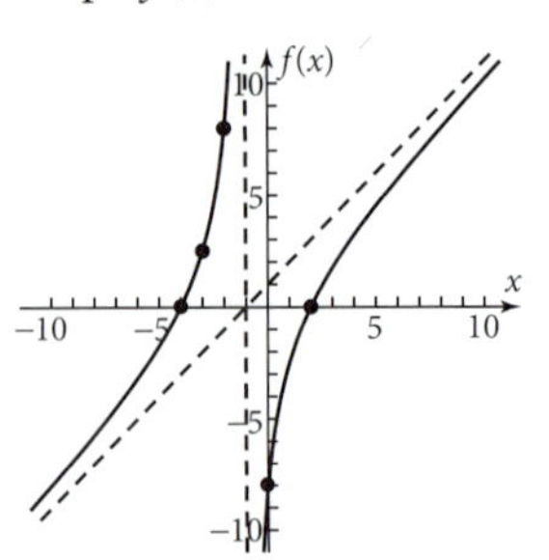

25a. $f(x) = x + 3 + \dfrac{5}{x - 3}$

25b. No removable discontinuities; vertical asymptote: $x = 3$

25c. Diagonal asymptote: $y = x + 3$

25d. x-intercepts: $x = -2$ or $x = 2$; y-intercept: $f(0) = \frac{4}{3}$

25e.

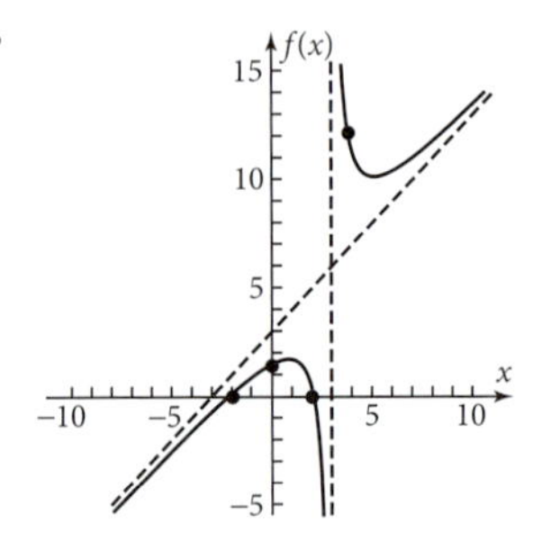

27a. $f(x) = 2x - 3, x \neq 4$

27b. Removable discontinuity at (4, 5); no vertical asymptotes

27c. No non-vertical asymptotes

27d. x-intercepts: $x = 1.5$; y-intercept: $f(0) = -3$

27e.

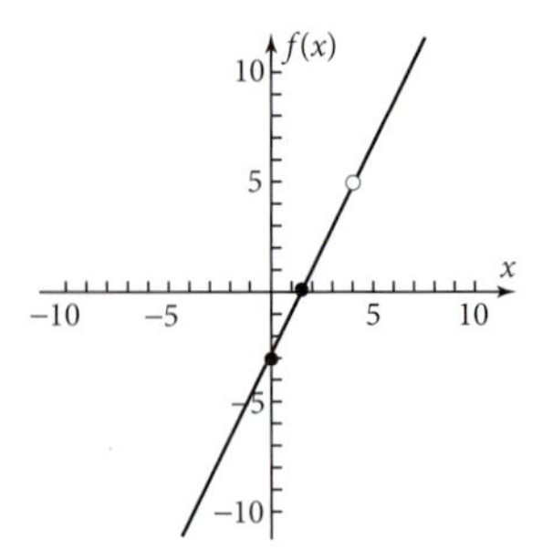

29a. $f(x) = x^2 + x - 5 + \dfrac{3}{x - 2}$

29b. No removable discontinuities; vertical asymptote: $x = 2$

29c. Curved asymptote: $y = x^2 + x - 5$

29d. x-intercepts: $x = -2.9206\ldots$; y-intercept: $f(0) = -6.5$

29e.

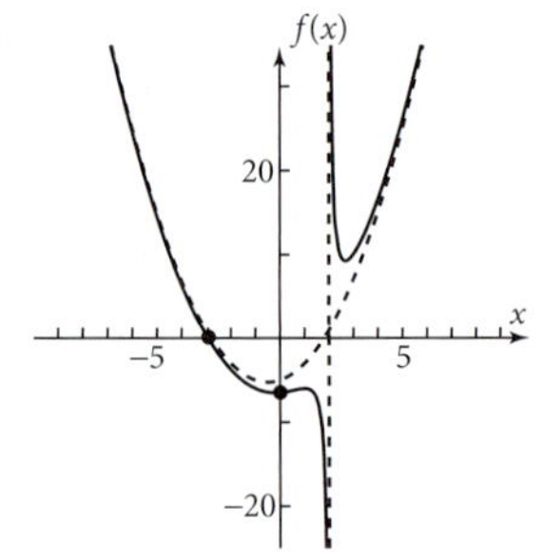

31a. $f(x) = x^2 + 2x - 2 + \dfrac{-10}{x + 3}$

31b. No removable discontinuities; vertical asymptote: $x = -3$

31c. Curved asymptote: $y = x^2 + 2x - 2$

31d. x-intercepts: $x = 1.3069\ldots$; y-intercept: $f(0) = -\frac{16}{3}$

31e.

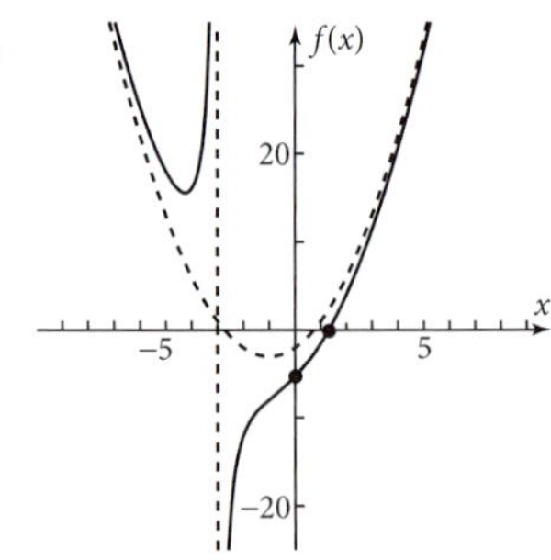

33a. $f(x) = x^2 - 6x - 7, x \neq 4$

33b. Removable discontinuity at $(4, -15)$; no vertical asymptotes

33c. No other asymptotes

33d. x-intercepts: $x = -1, 7$; y-intercept: $f(0) = -7$

33e.

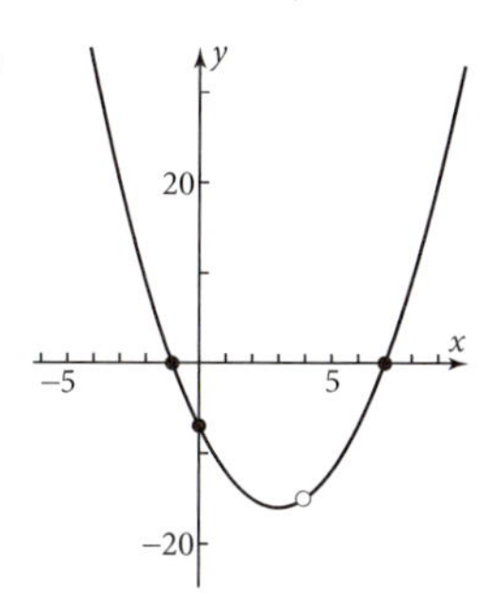

35a. The graph of f is the line $y = x - 1$ with removable discontinuities at $x = 4$ and $x = -3$, x-intercept 1, and y-intercept -1.

35b. There are vertical asymptotes at $x = 3$ and $x = -2$, and x-intercepts 4, 1, and -3. The y-intercept is $g(0) = -2$. $g(x) = x - 1 + \frac{-6x - 6}{x^2 - x - 6}$, so the diagonal asymptote has equation $y = x - 1$.

35c. There is a vertical asymptote at $x = -2$, a removable discontinuity at $x = 4$, and x-intercepts 1 and -3. The y-intercept is $h(0) = -1.5$. $h(x) = x - \frac{3}{x + 2}$, if $x \neq 4$, so the diagonal asymptote has equation $y = x$.

Problem Set 4-6

1. $\frac{x + 2}{2}, x \neq 2$

3. $\frac{(x + 3)(x + 5)}{5x}$, or $\frac{x^2 + 8x + 15}{5x}, x \neq -1, x \neq -5$

5. $\frac{x^2 + 6}{6x}, x \neq -6$

7. $\frac{x + 1}{x - 1}, x \neq 6, x \neq -2$

9. $\frac{x - 1}{x + 4}, x \neq 3, x \neq -2, x \neq 4$

11. $\frac{(x + 5)^2}{(x - 7)^2}$, or $\frac{x^2 + 10x + 25}{x^2 - 14x + 49}, x \neq -5, x \neq -3$

13. $\frac{x^2 + x}{x - 2}, x \neq 0, x \neq -2$

15. $\frac{x + 3}{x - 2}, x \neq -5, x \neq 1$

17. $\frac{2x}{(x - 1)(x + 1)}$

19. $\frac{2 - 4x}{x^2 - 1}$

21. $\frac{-2x}{x - 3}$

23. $\frac{2}{x + 3}, x \neq 3$

25. $\frac{3x + 9}{(x + 2)(x - 1)}, x \neq 5$

27. $\frac{2x}{x^2 - 1}, x \neq 2, x \neq 4$

29. $\frac{4}{x - 1} + \frac{7}{x - 2}$

31. $\frac{3.5}{x + 2} + \frac{1.5}{x - 4}$

33. $\frac{7}{x + 2} + \frac{-7}{x + 5}$

35. $\frac{2}{x + 1} + \frac{3}{x - 7} + \frac{4}{x + 2}$

37. $\frac{3}{x - 2} + \frac{2}{x + 1} + \frac{-1}{x + 3}$

39. $\frac{x + 2}{x^2 + 1} + \frac{3}{x + 4}$

41. $\frac{1}{x + 5} + \frac{3}{x + 1} + \frac{-2}{(x + 1)^2}$

Problem Set 4-7

1a. $x(x - 4) = 3(x - 4)$; The transformed equation is true for $x = 4$ because both sides equal 0 if $x = 4$.

1b. Extraneous solution

1c. Multiplying by $(x - 4)$ is an irreversible step because you cannot divide by $(x - 4)$ without risking division by 0.

3a. $x = 3, 5$

3b. $4 = \frac{-x^2 + 6x - 11}{x - 2}, x \neq 2 \Rightarrow x^2 - 2x + 3 = 0$; $b^2 - 4ac = -8$; The discriminant is negative, so there are no real values of x for which $f(x) = 4$.

3c. $7 = \frac{-x^2 + 6x - 11}{x - 2}, x \neq 2 \Rightarrow x^2 + x - 3 = 0$; $b^2 - 4ac = 13$; The discriminant is positive, so there are real values of x for which $f(x) = 7$.

3d.

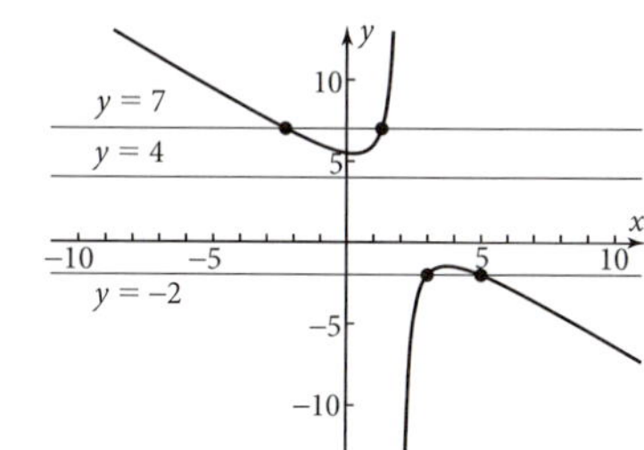

Answers to Selected Problems

5a. $x = 2, 7$

5b., 5c. The graph shows that $f(x) = -6$ if $x = -5.4721\ldots$ or $x = 3.4721\ldots$.

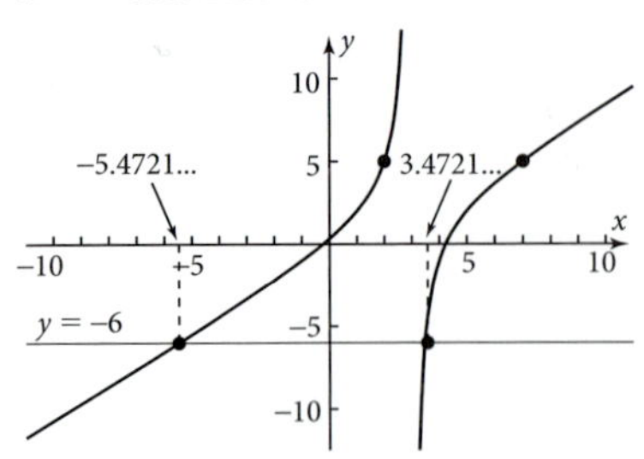

5d. The range is all real numbers; any horizontal line will intersect the graph of f (at two places).

7a. $x = -1$ or $\overset{\text{extraneous}}{\cancel{x = 2}} \Rightarrow x = -1$

7b. The graph shows that $f(x) = g(x)$ only at $x = -1$, and there is an asymptote at $x = 2$, where the extraneous solution is.

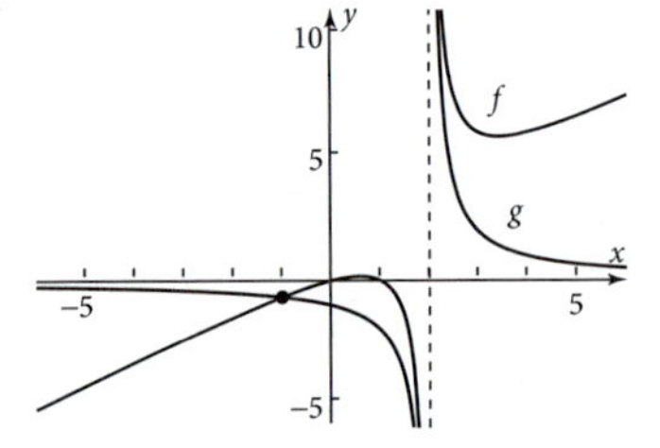

9. $x \neq 3, x \neq -5; x = -1$ or $\cancel{x = 3} \Rightarrow x = -1$

11. $x \neq -4, x \neq 3;$ $\overset{\text{extraneous}}{\cancel{x = -4}}$ or $\overset{\text{extraneous}}{\cancel{x = 3}} \Rightarrow$ No solutions

13. $x \neq 3, x \neq 4; x = 7$

15. $x \neq 3, x \neq 6; x = \frac{14}{3} = 4\frac{2}{3}$.

17. $x \neq -2, x \neq 2;$ $\overset{\text{extraneous}}{\cancel{x = 2}} \Rightarrow$ No solutions

19. $x \neq 1$
All real numbers are solutions except 1.

21. $x \neq \frac{3}{2}, x \neq -\frac{3}{2}; x = 3$ or $\overset{\text{extraneous}}{\cancel{x = \frac{3}{2}}} \Rightarrow x = 3$

23. $x \neq 3, x \neq -3; x = 5$ or $\overset{\text{extraneous}}{\cancel{x = -3}} \Rightarrow x = 5$

25. $x \neq 2, x \neq -3;$ $\overset{\text{extraneous}}{\cancel{x = 2}}$ or $\overset{\text{extraneous}}{\cancel{x = -3}} \Rightarrow$ No solutions

27a. *distance* = *rate* · *time*, so *time* = $\frac{\textit{distance}}{\textit{rate}}$. The rate upstream is $(x - 4)$ mi/h, the rate downstream is $(x + 4)$ mi/h, and the round trip distance is 50 mi. The waiting time is 1 hour.
$\therefore f(x) = \dfrac{50}{x - 4} + 1 + \dfrac{50}{x + 4}$

27b. $f(20) \approx 6.2$ hours; $f(10) \approx 12.9$ hours; $f(5) \approx 56.6$ hours
At 5 mi/h, which is $\frac{1}{4}$ of 20 mi/h, the round trip takes more than 4 times as long; in fact, it takes almost 10 times as long!

27c. The minimum speed is about 15.3 mi/h for a less-than-8-hour round trip.

Problem Set 4-8

R1a. $f(4) = 6$, which is a real number. $g(4)$ would involve division by zero, so there is no value for $g(4)$.

R1b. A vertical asymptote at $x = 4$

R1c. $f(3) = 0$ and $f(-2) = 0$

R1d. $f(6) = 24$ and $g(6) = 26.5$ (Close!); $f(4.01) = 6.0701$ and $g(4.01) = 506.0701$ (Not close!!)

R1e. The graph of function g gets arbitrarily close to the graph of function f as x gets farther away from zero, showing that the graph of f is a curved asymptote for the graph of g.

R3a. i. $x = 1$

R3a. ii. $x = -1$

R3a. iii. $x = 4$

R3a. iv. $f(x) = (x + 1)^2(x - 1)(x - 4)^3$

R3a. v. 6th degree

R3a. vi. Positive

R3b. i. 21.5

R3b. ii. 135.5

R3b. iii. 220

R3b. iv. Zeros are 2.5, 8, and 11; $2.5 + 8 + 11 = 21.5$; $(2.5)(8) + (2.5)(11) + (8)(11) = 135.5$; $(2.5)(8)(11) = 220$

R3c. Graphs will vary. The sample graph is $y = x^5 + 4x^4 - 3x^3 - 12x^2 - 4x - 16$, with zeros -4, -2, i, $-i$, and 2.

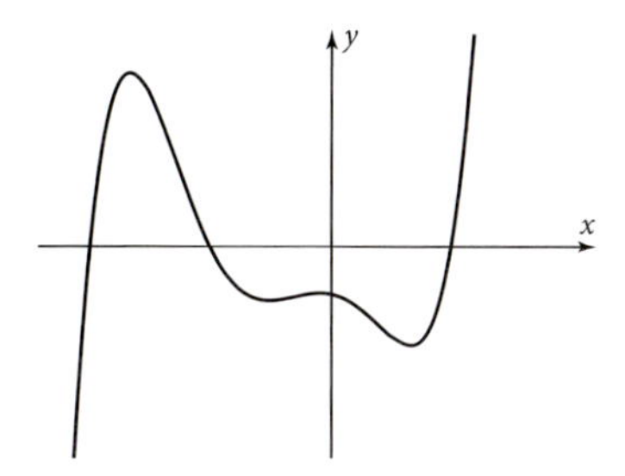

R3d. i. If $p(x)$ is a ploynomial, then $p(c)$ equals the remainder when $p(x)$ is divided by te quantity $(x - c)$

R3d. ii. 301

R3d. iii. The factor theorem is equivalent to the remainder theorem with the remainder equal to 0. So $(x - c)$ is a factor of $p(x)$ if and only if $f(c) = 0$.

R5a. A rational algebraic function is a function with a general equation that can be expressed as $f(x) = \frac{n(x)}{d(x)}$, where $n(x)$ and $d(x)$ are polynomials.

R5b. A proper algebraic fraction has a numerator of lower degree than the denominator. An improper algebraic fraction has a numerator of equal or greater degree than the denominator.

R5c. i. $f(x) = 3 + \dfrac{4}{x - 2}$

R5c. ii. Vertical translation by 3 units, vertical dilation by a factor of 4, and horizontal translation by 2 units.

R5c. iii.

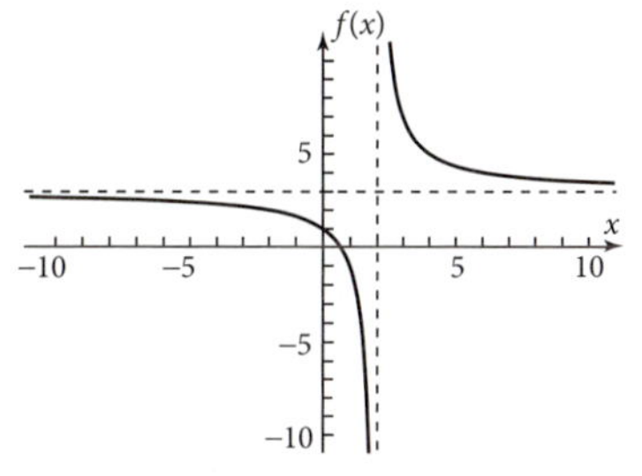

R5c. iv. The ratio of the leading coefficients of the numerator to denominator, 3 to 1, gives the horizontal asymptote: $y = 3$.

R5d. i. The graph of $f(x)$ is dashed; the graph of $g(x)$ is solid.

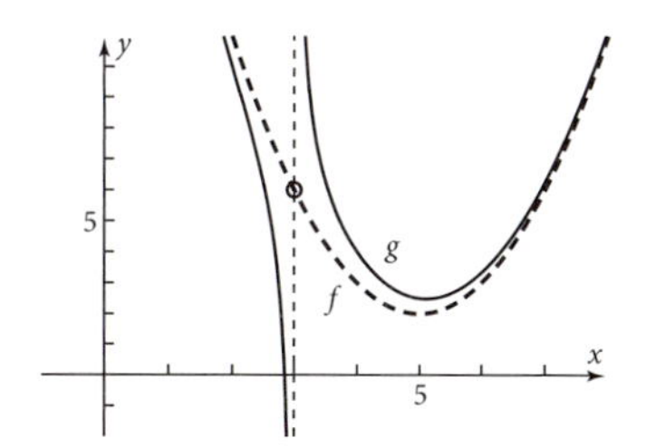

R5d. ii. $f(x) = x^2 - 10x + 27$, $x \neq 3$; y-coordinate 6

R5d. iii. $g(x) = x^2 - 10x + 27 + \frac{1}{x - 3}$; The term $\frac{1}{x - 3}$ is infinite if $x = 3$, so the discontinuity is a vertical asymptote and cannot be removed.

R5d. iv. The curved asymptote is $y = x^2 - 10x + 27$, the polynomial part of the equation for function g, and the same graph as function f.

R5e. i. $h(x) = \dfrac{10}{x^2 + 5}$, $x \neq 3$

R5e. ii. There is a removable discontinuity at $x = 3$ because the factor $(x - 3)$ cancels from the denominator.

R5e. iii. The denominator is of higher degree than the numerator, so the values of $h(x)$ approach zero as x gets far from the origin, making the x-axis a horizontal asymptote. There are no vertical asymptotes because the denominator $(x^2 + 5)$ never equals zero.

R7a. $x(x - 3) = 4(x - 3)$; The value $x = 3$ is now a solution.
The new solution is an *extraneous* solution. The step is an *irreversible* step. (You can't divide by $(x - 3)$ without risking division by zero.)

R7b. $x = 7, -2$

R7c. i. $x = -1$ or ~~$x = 5$~~ (extraneous); The valid solution is $x = -1$.

R7c. ii. ~~$x = 5$~~ (extraneous)
No valid solutions

R7c. iii. The graph shows that $y = 2$ intersects the graph of f at $x = -1$ only, and the graph of $y = 8$ intersects the graph of f at $x = 5$, where there is a removable discontinuity.

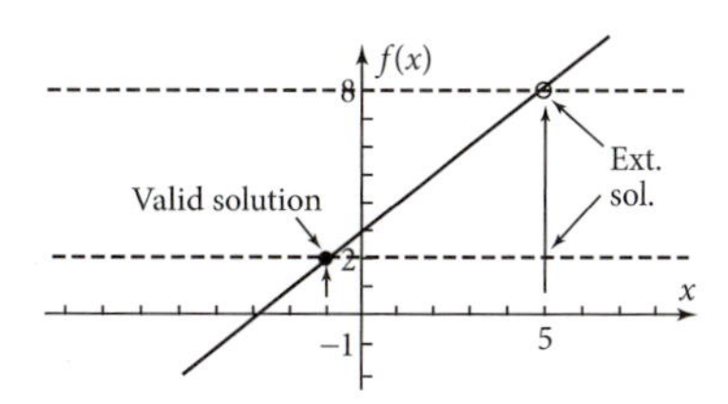

R7d. i. $x = -5$ or ~~$x = 2$~~ (extraneous) $\Rightarrow x = -5$ (One solution is extraneous.)

R7d. ii. All real numbers are solutions except $x = 2$ and $x = -4$.

R7d. iii. ~~$x = 2$~~ (extraneous) or ~~$x = -4$~~ (extraneous) $\Rightarrow$ No valid solutions

R7d. iv. $x = 3$ or $x = -5$ (Both solutions are valid.)

R7e. i. $f(x) = \dfrac{2000}{x - 50} + \dfrac{2000}{x + 50} + 4$

R7e. ii. ≈ 12.1 h for 737; ≈ 222.2 h for the blimp!

R7e. iii. The plane must fly at least about 212 mi/h.

C1a. $x = 8 \pm 3i$

C1b. $f(8) = 54$; The equation of the line is $y = 9x - 18$.
$x^3 - 18x^2 + 105x - 146 = 9x - 18$
$\Rightarrow (x - 2)(x - 8)(x - 8) = 0$
$\Rightarrow x = 2, 8, 8$. So 8 is a double zero, and the line just touches the graph at $x = 8$ without crossing, meaning that the line is tangent to the graph.

C1c. The complex zeros are $a \pm bi = 8 \pm 3i$. $a = 8$ is the x-coordinate of the point of tangency. The slope of the tangent line is 9, so $b = \pm 3$ is the square root of the absolute value of the slope of the tangent line.

C1d. The graph shows the tangent line through (2, 4) and (6, 0), with slope -1. Nonreal complex zeros are $2 \pm i$.

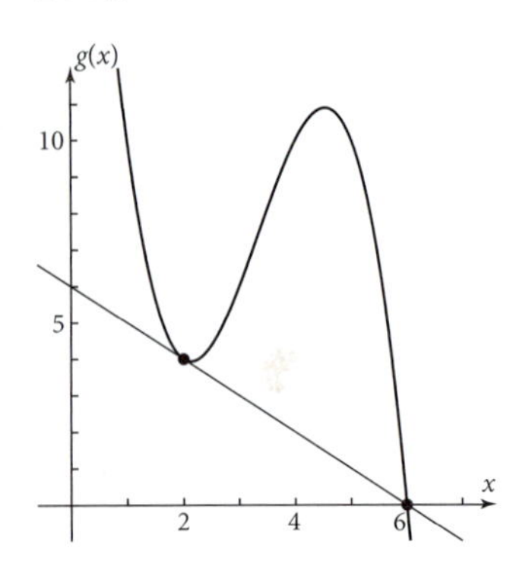

C1e. The integer points on the graph are shown in the table.
$g(x) = -x^3 + 10x^2 - 29x + 30$.
The three zeros are $x = 6, 2 \pm i$, which agrees with the graphical solution.

x	$g(x)$
1	10
2	4
3	6
4	10
5	10
6	0

C1f. Student research problem.

C3a. $f(x) = x + 4 + \dfrac{-2x + 14}{(x + 5)(x - 3)}$. Vertical asymptotes: $x = -5$ and $x = 3$, diagonal asymptote: $y = x + 4$. There are no removable discontinuities because the factors $(x + 5)$ and $(x - 3)$ cannot be factored out.

C3b. The asymptote intersects the graph at the point (7, 11).

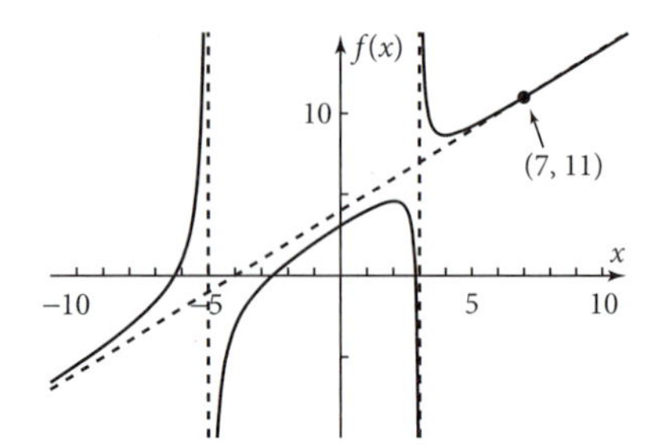

T1. $x^2 - 22x + 35$

T3. $x = 12, -5$

T5. $x = 5, 2$, and -1

T7a. There are two nonreal complex zeros because the vertex on the left approaches but does not cross the x-axis.

T7b.

3	1	1	−7	−15
		3	12	15
	1	4	5	0

Therefore $x = 3$ is a real-number zero of $g(x)$.

T7c. $x = -2 \pm i$

T7d. The coefficient of the x-term is -7.
$3(-2 + i) + 3(-2 - i) + (-2 + i)(-2 - i) = -7$

T9a. Vertical translation by 3 units, vertical dilation by a factor of 2, and horizontal translation by 4 units

T9b. $h(x) = \dfrac{2}{x - 4} + 3$

T11. 2; extraneous solution

T13. $-5 \pm 3i$
Check $x = -5 + 3i$: $(-5 + 3i)^2 + 10(-5 + 3i) + 44$
$= 25 - 30i + 9i^2 - 50 + 30i + 44 = 25 - 9 - 50 + 44$
$= 10$

T15a. $d(t) = t^3 - 12t^2 + 54t - 68$; $R^2 = 1$.

T15b. $t = 2$ s; the other zeros are $t = 5 \pm 3i$, so $t = 2$ is the only real zero

T15c. She was going the slowest at $t = 4$ s; the point of inflection.

T15d. Concave down

T17. $f(x) = \frac{(x-3)(x-5)}{(x-2)}, x \neq -5$

By synthetic substitution, $f(x) = x - 6 + \frac{3}{x-2}$. There is a removable discontinuity at $x = -5$ because the $(x + 5)$ factor in the denominator cancels. There is a vertical asymptote at $x = 2$ because the $(x - 2)$ factor in the denominator does not cancel. The diagonal asymptote is $y = x - 6$, the polynomial part of the mixed-number form of the equation. The graph has x-intercepts $x = 3$ and $x = 5$ because the factors $(x - 3)$ and $(x - 5)$ in the numerator do not cancel.

T19. Answers will vary.

Problem Set 4-9

1. A function is a relation for which there is never a value of x that has more than one value of y.

3. Horizontal translation by 3 units and vertical translation by -2 units; $h(x) = -2 + f(x - 3)$

5. Odd

7. Parametric function

9a. $y = mx + b$

9b. $y = ax^2 + bx + c, a \neq 0$

9c. $y = a + b\ln x$ or $y = a + b\log x, b \neq 0$

9d. $y = ab^x, a \neq 0, b > 0$

9e. $y = ax^b, a \neq 0, b \neq 0$

11a. Multiply–add

11b. Multiply–multiply

11c. Constant-second-differences

13a. $p = \log_c m$ if and only if $c^p = m$

13b. $\log 7 + \log 8 = \log 56$

13c. $\log 32 = 5 \log 2$

15. The residual is smaller than the deviation.

17. -2

19. $r^2 = 0.64$

21. Add–multiply semilog. Taking the logarithm of the y-values compresses them into a narrower range.

23. $\log y = (\log 5000) + (\log 0.72)x$, which shows that $\log y$ is linear in x.

25. $(x - 12)(x + 1)$

27a. Cubic (third degree)

27b. $x = 2, -6$

27c. 2, -6, and -6 (a double zero)

27d. 10

29. $\frac{3}{x-4} + \frac{-3}{x+2}$

31. When $t = -1$, the coordinates are (6, 7).
When $t = 2$, the coordinates are $(-6, 1)$.
When $t = 3$, the coordinates are $(6, -9)$.
The three points lie on the graph.

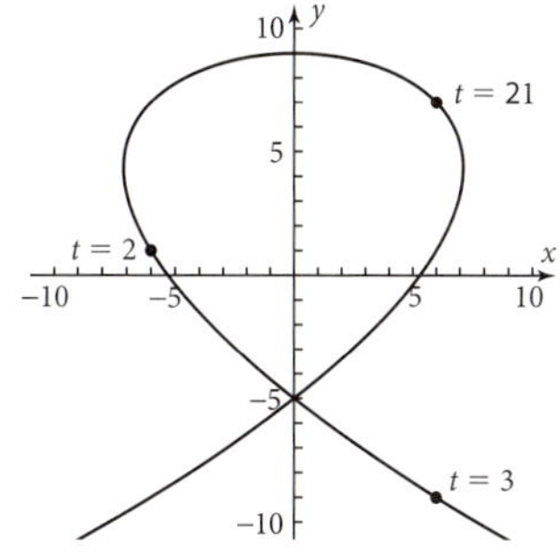

33a. $\hat{y} = \frac{263.8637...}{1 + 314.7597...e^{-1.2056...x}}$

33b. Actual point of inflection is (4.7707..., 131.9318...).

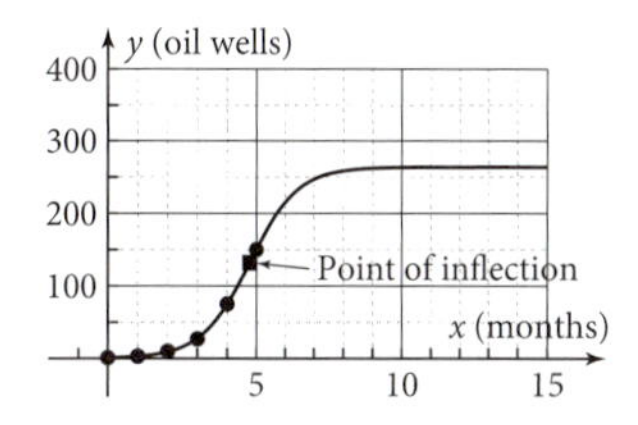

33c. About 264 wells

33d. An exponential model predicts that the number of oil wells would grow without bound. The logistic model shows that the number of wells will level off because of overcrowding.

35a. 1.8069...

35b. $x = 1.0314...$

35c. 2.3922...

35d.

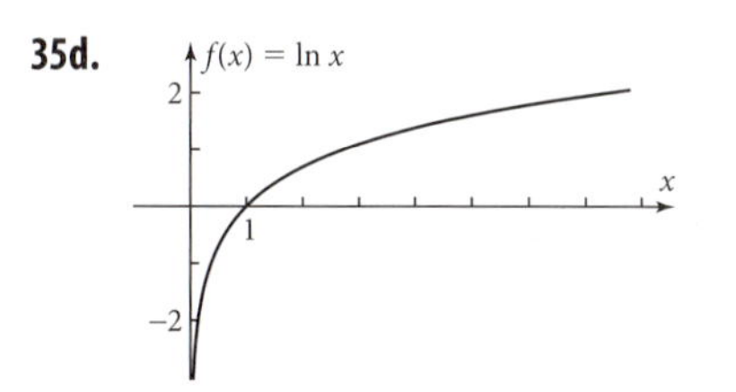

$f(1) = 0$ because $\ln 1 = 0$, because $\ln 1 = \log_e 1$, and $e^0 = 1$.
$f(5) = \ln 5 = 1.6094...$;
$f(7) = \ln 7 = 1.9459...$;
$f(35) = \ln 35 = 3.5553... = 1.6094... + 1.9459...$
$= f(5) + f(7)$

37. Answers will vary.

Chapter 5

Problem Set 5-1

1. The graph should match Figure 5-1c.
3. $f_2(x) = 11 + 9\sin(x)$

Problem Set 5-2

1. $\theta_{ref} = 50°$

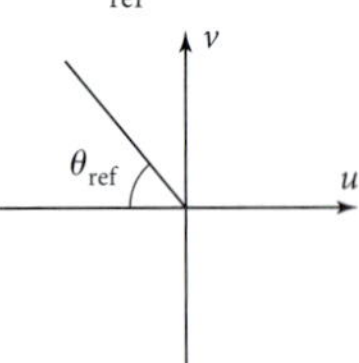

3. $\theta_{ref} = 79°$

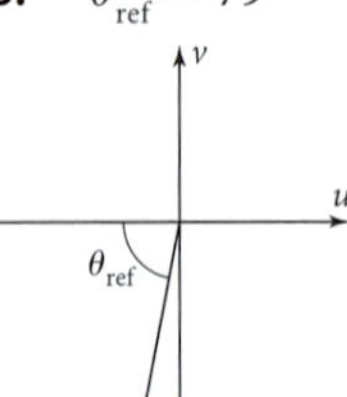

5. $\theta_{ref} = 18°$

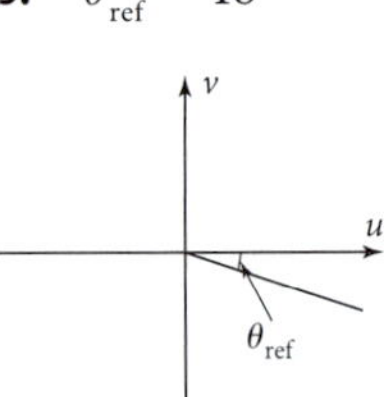

7. $\theta_{ref} = 54°$

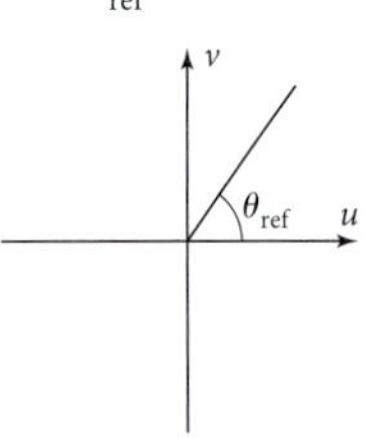

9. $\theta_{ref} = 20°$

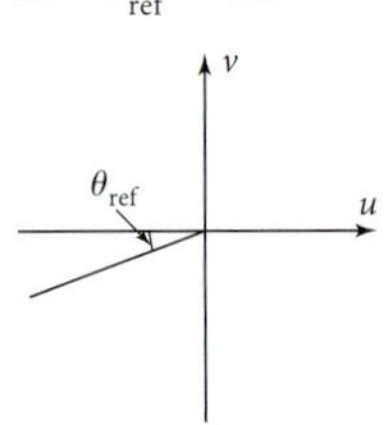

11. $\theta_{ref} = 65°$

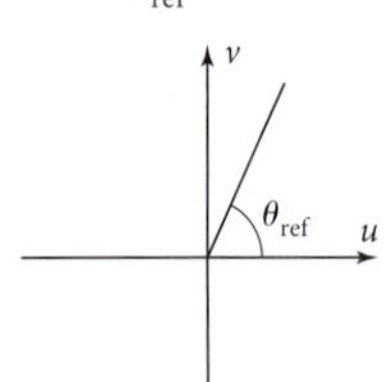

13. $\theta_{ref} = 81.4°$

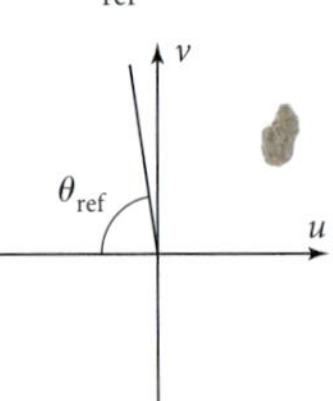

15. $\theta_{ref} = 25.9°$

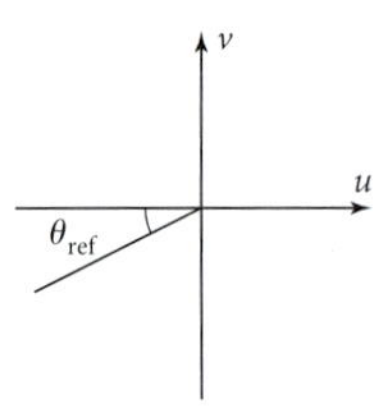

17. $\theta_{ref} = 81°$

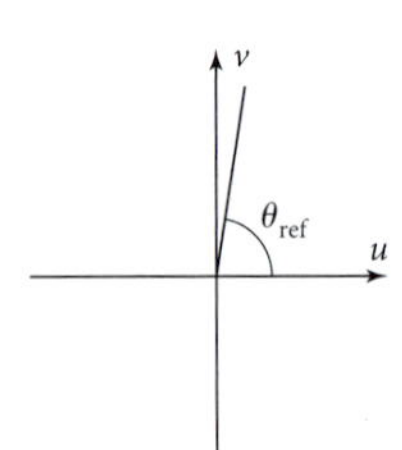

19. $\theta_{ref} = 46°$

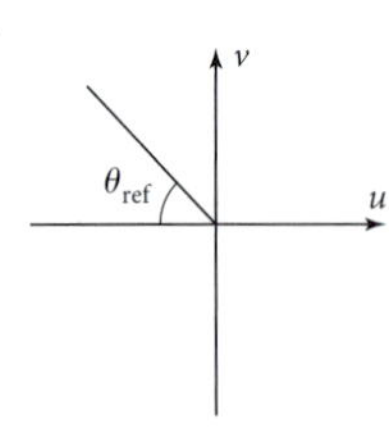

21. $\theta_{ref} = 34°23'$

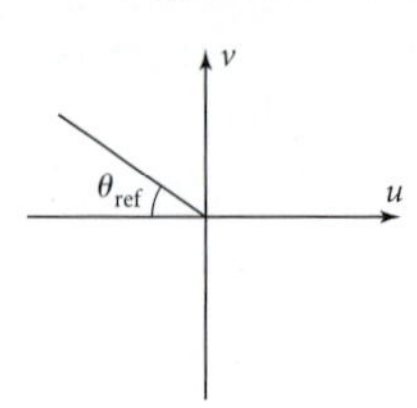

23. $\theta_{ref} = 33°16'$

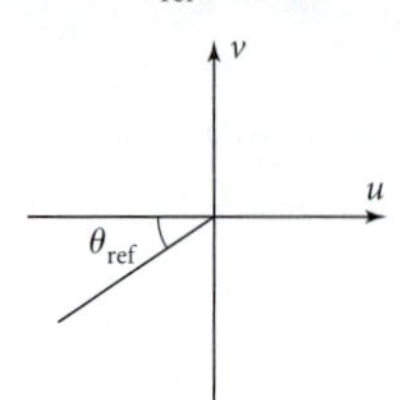

25. $\theta_{ref} = 51°45'9''$

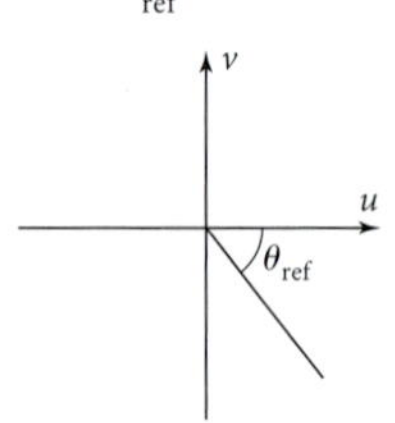

27.

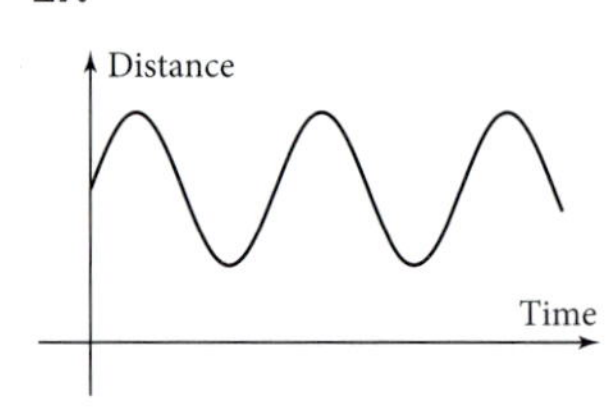

29. $g(x) = 4 + f(x - 1)$

Problem Set 5-3

1. $\theta_{ref} = 70°$

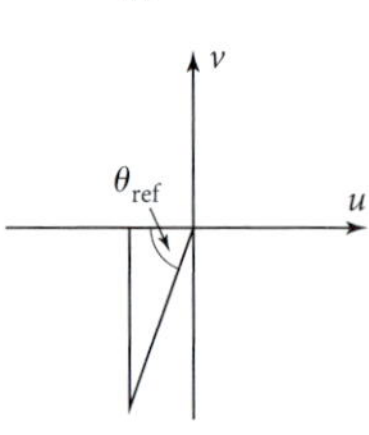

$\sin 250° = -0.9396...$, $\sin 70° = 0.9396...$,
$\sin 250° = -\sin 70°$

3. $\theta_{ref} = 40°$

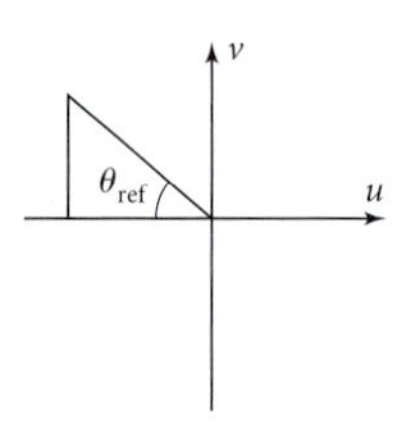

$\cos 140° = -0.7660...$,
$\cos 40° = 0.7660...$, $\cos 140° = -\cos 40°$

5. $\theta_{ref} = 60°$

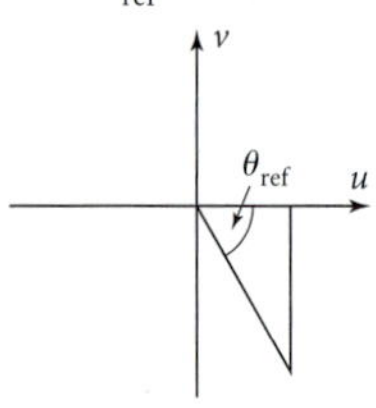

$\cos 300° = 0.5$, $\cos 60° = 0.5$, $\cos 300° = \cos 60°$

7. $\sin\theta = \frac{11}{\sqrt{170}} = 0.8436...;\ \cos\theta = \frac{7}{\sqrt{170}}$ $= 0.5368...$

9. $\sin\theta = \frac{5}{\sqrt{29}} = 0.9284...;\ \cos\theta = \frac{-2}{\sqrt{29}} = -0.3713...$

11. $\sin\theta = \frac{-8}{4\sqrt{5}} = \frac{-2}{\sqrt{5}} = -0.8944...;$
$\cos\theta = \frac{4}{4\sqrt{5}} = \frac{1}{\sqrt{5}} = 0.4472...$

13. $\sin\theta = \frac{-7}{25} = -0.28;\ \cos\theta = \frac{-24}{25} = -0.96;$
r is an integer

15. θ-translation of $y = \sin\theta$ by $+60°$

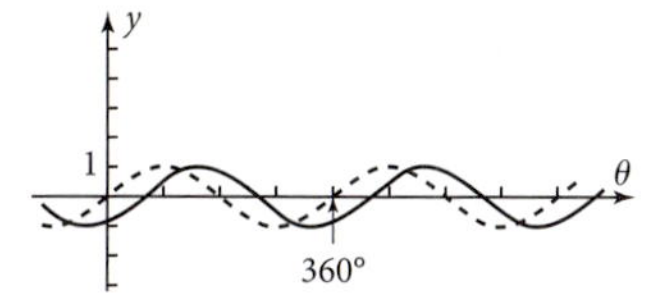

17. y-dilation of $y = \cos\theta$ by 3

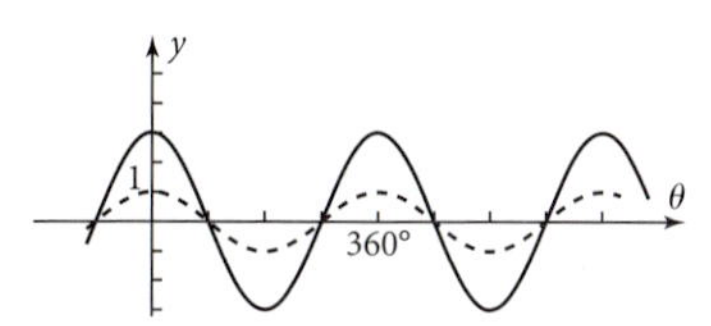

19. θ-dilation of $y = \cos\theta$ by $\frac{1}{2}$, y-translation by $+3$

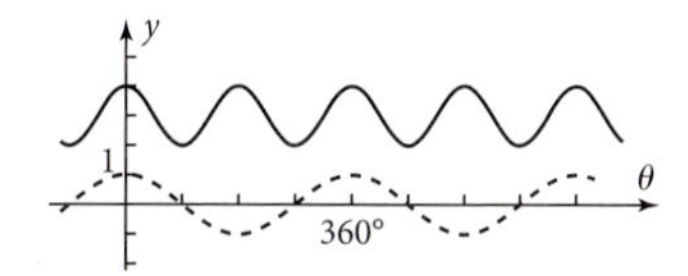

21.

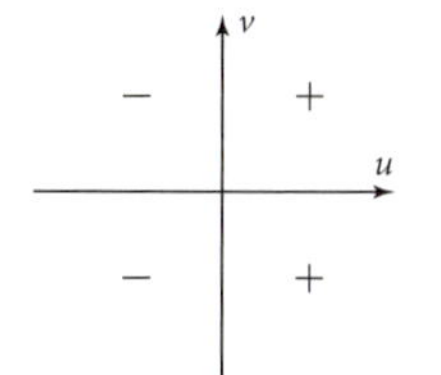

23. Examples will vary.

Problem Set 5-4

1. ≈ 1.2799

3. ≈ -1.8871

5. ≈ -57.2987

7. $\sin\theta = -\frac{3}{5};\ \cos\theta = \frac{4}{5};\ \tan\theta = -\frac{3}{4};$
$\cot\theta = -\frac{4}{3};\ \sec\theta = \frac{5}{4};\ \csc\theta = -\frac{5}{3}$

9. $\sin\theta = -\frac{7}{\sqrt{74}};\ \cos\theta = -\frac{5}{\sqrt{74}};$
$\tan\theta = \frac{7}{5};\ \cot\theta = \frac{5}{7};\ \sec\theta = -\frac{\sqrt{74}}{5};$
$\csc\theta = -\frac{\sqrt{74}}{7}$

11. $\sin\theta = \frac{4}{5};\ \cos\theta = -\frac{3}{5};\ \tan\theta = -\frac{4}{3};$
$\cot\theta = -\frac{3}{4};\ \sec\theta = -\frac{5}{3};\ \csc\theta = \frac{5}{4}$

13. $\sin\theta = -\frac{\sqrt{15}}{4};\ \cos\theta = \frac{1}{4};\ \tan\theta = -\sqrt{15};$
$\cot\theta = -\frac{1}{\sqrt{15}};\ \sec\theta = 4;\ \csc\theta = -\frac{4}{\sqrt{15}}$

15. $\sin 60° = \frac{\sqrt{3}}{2};\ \cos 60° = \frac{1}{2};\ \tan 60° = \sqrt{3};$
$\cot 60° = \frac{1}{\sqrt{3}};\ \sec 60° = 2;\ \csc 60° = \frac{2}{\sqrt{3}}$

17. $\sin(-315°) = \frac{1}{\sqrt{2}};\ \cos(-315°) = \frac{1}{\sqrt{2}};$
$\tan(-315°) = 1;\ \cot(-315°) = 1;\ \sec(-315°) = \sqrt{2};$
$\csc(-315°) = \sqrt{2}$

19. $\sin 180° = 0;\ \cos 180° = -1;\ \tan 180° = 0;$
$\cot 180°$ is undefined; $\sec 180° = -1$; $\csc 180°$ is undefined.

21. 0

23. $-\frac{1}{2}$

25. -1

27. undefined

29. $-\frac{2}{\sqrt{3}}$

31. $\sqrt{2}$

33a. $\theta = 0°, 180°, 360°$

33b. $\theta = 90°, 270°$

33c. $\theta = 0°, 180°, 360°$

33d. $\theta = 90°, 270°$

33e. $\sec\theta \neq 0$ for all θ

33f. $\csc\theta \neq 0$ for all θ

35. 1

37. 2

39. 1

41. -1

43a. 67°

43b. $\cos 23° = 0.93205...$, $\sin 67° = 0.9205...$;
They are equal.

43c. "Complement"

45. In the uv-coordinate plane, θ is the angle from the origin to the point (u, v) in standard position. In the θy-coordinate plane, θ represents the same angle, but is now used as the independent variable.

Problem Set 5-5

1. 17.4576...°, because $\sin 17.4576...° = 0.3$

3. 81.8698...°, because $\tan 81.8698...° = 7$

5. $\cos(\sin^{-1} 0.8) = 0.6$; $\theta = \sin^{-1} 0.8$ represents an angle of a right triangle with sides 3, 4, and 5.

7a. They are not one-to-one.

7b. Sine: $-90° \le \theta \le 90°$. Cosine: $0° \le \theta \le 180°$. They are one-to-one.

7c. $\sin^{-1}(-0.9) = -64.1580...°$. On the principal branch, only negative angles correspond to negative values of the sine.

9a. ≈ 2.4 m

9b. ≈ 74.4J

11a. $\approx 3.6°$

11b. Assume that the slope of the road is constant.

13. $\approx 25.4°$

15a. $\approx 34.5°$

15b. ≈ 10.1 cm

17a. ≈ 33.5 m

17b. ≈ 17.5 m

19a. $\approx 11.5°$

19b. ≈ 7 ft

19c. ≈ 74 ft

21a. ≈ 1 ft 5 in.

21b. ≈ 8 ft 3 in.

21c. ≈ 6 ft 10 in.

23. Answers will vary.

Problem Set 5-6

R0. Journal entries will vary.

R1a. The graphs match.

R1b. y-dilation by 0.7, y-translation by +2; $y = 2 + 0.7 \sin\theta$; The result agrees with the graph.

R1c. Sinusoid

R3a. $\sin\theta = \frac{7}{\sqrt{74}}$; $\cos\theta = -\frac{5}{\sqrt{74}}$

R3b. $\sin 160° = 0.3420...$; $\cos 160° = -0.9396...$; $\theta_{\text{ref}} = 20°$

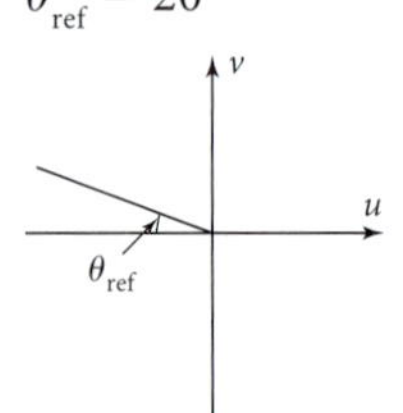

The angle 160° terminates within Quadrant II, above the x-axis (so $\sin 160° > 0$), and to the left of the y-axis (so $\cos 160° < 0$).

R3c.

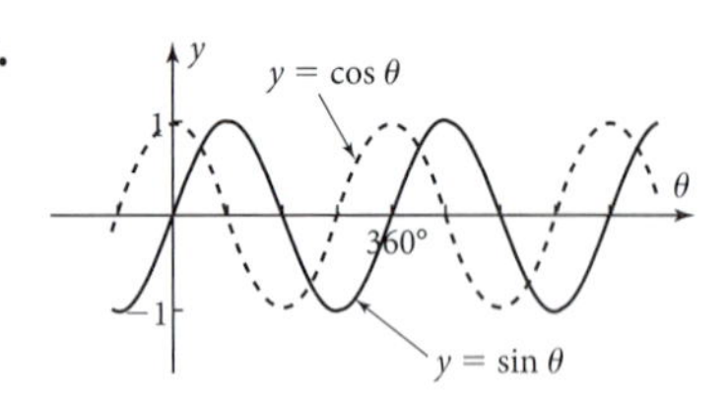

R3d. Quadrants III and IV

R3e. y-translation of +4, θ-dilation of $\frac{1}{2}$

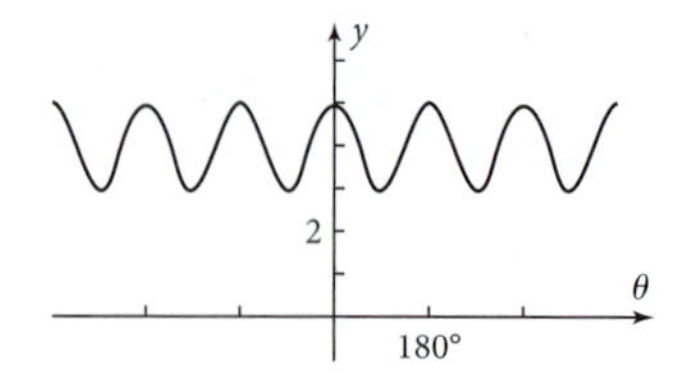

R5a. $53.1301...°$; this means that $\cos 53.1301...° = 0.6$.

R5b. i. ≈ 321 m

R5b. ii. ≈ 603 m

R5b. iii. $\approx 75°$

R5b. iv. Fishing crews could use this technique to find the slant distance and depth of a school of fish.

C1. x-dilation of $\frac{1}{30}$ h/deg; x-translation of +4 h; y-dilation of 2; y-translation of +5 ft; $y = 5 + 2 \cdot \cos 30(x - 4)$

C2a., C2b., C2c. $\theta = 17.4576...°, 162.5423...°, 377.4576...°, 522.5423...°, 737.4576...°, 882.5423...°$

C3. $\cos^2\theta + \sin^2\theta = 1$ for all θ. If $\cos\theta$ represents the u-coordinate of a point on the unit circle, and $\sin\theta$ represents the v-coordinate of this point, then $\cos^2\theta + \sin^2\theta$ is the square of this point to the origin. Because the point is on the unit circle, the distance is 1.

T1.

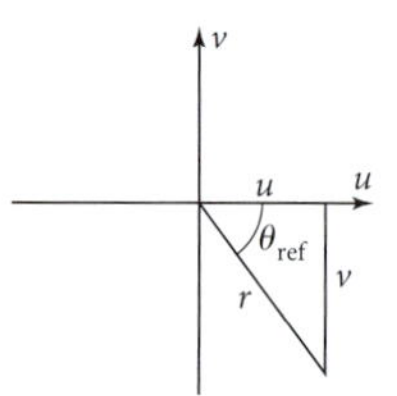

$\sin\theta = -\frac{4}{5}$; $\cos\theta = \frac{3}{5}$; $\tan\theta = -\frac{4}{3}$;
$\cot\theta = -\frac{3}{4}$; $\sec\theta = \frac{5}{3}$; $\csc\theta = -\frac{5}{4}$

T3. $\theta_{\text{ref}} = 45°$

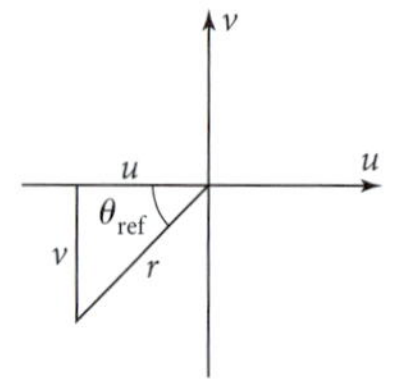

$\sin 225° = -\frac{1}{\sqrt{2}}$; $\cos 225° = -\frac{1}{\sqrt{2}}$; $\tan 225° = 1$;
$\cot 225° = 1$; $\sec 225° = -\sqrt{2}$; $\csc 225° = -\sqrt{2}$

T5.

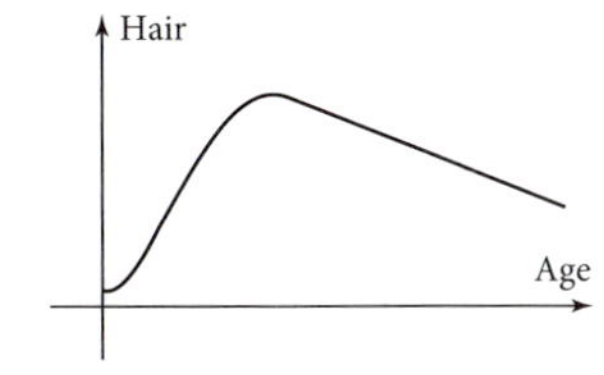

T7. The function in Problem T6 is periodic.

T9. 1.2867...

T11. −5.2408...

T13. 21.2762... ft

T15. 6.1155...°

T17. 75.5224...°

T19. 10.5115... m

T21. $y = -4 + 2\cos\left(\frac{6}{5}\theta\right)$

Chapter 6

Problem Set 6-1

1. Amplitude = 1

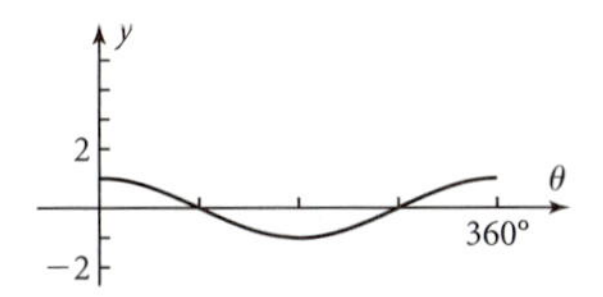

3. 360° for both functions

5.

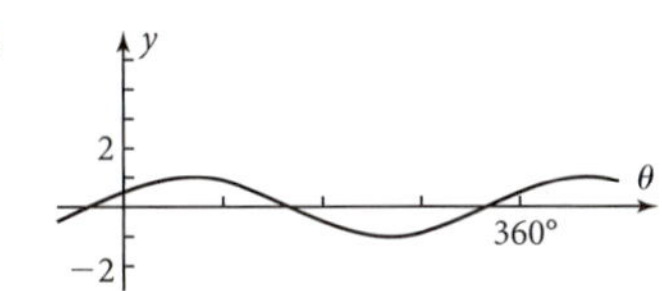

Horizontal translation by +60°

7.

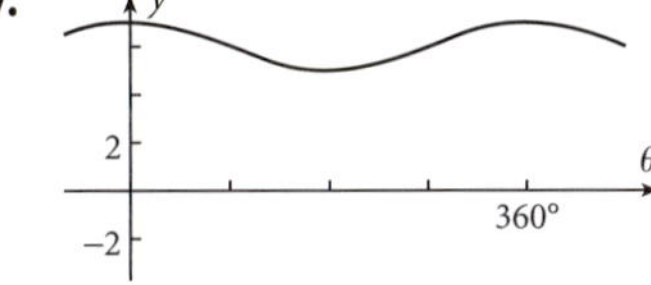

Vertical translation by +6

9. Amplitude = 5; Period = 120°; Phase displacement = 60°; Sinusoidal axis = 6

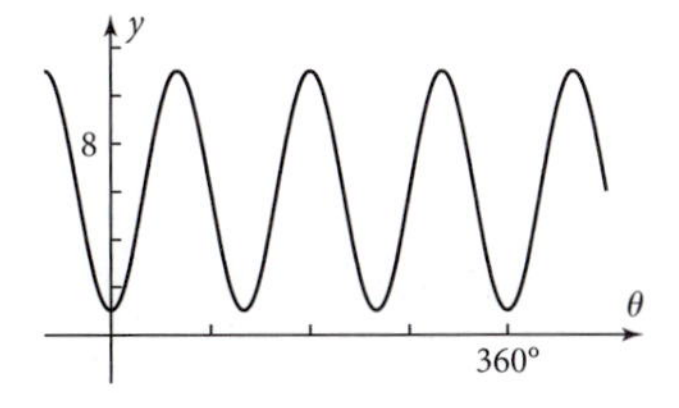

Problem Set 6-2

1. Amplitude = 4; Period = 120°;
Phase displacement = −10°; Sinusoidal axis = 7

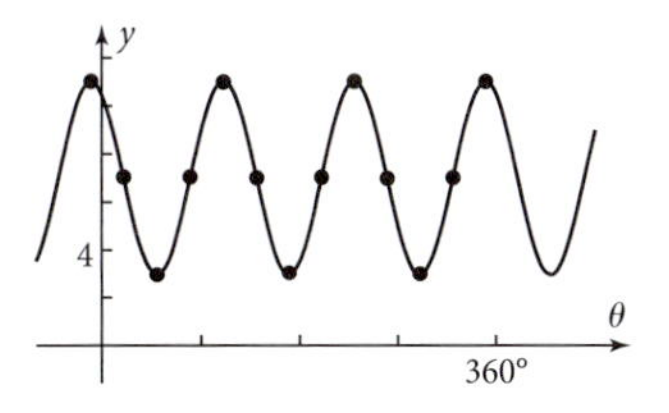

3. Amplitude = 20; Period = 720°; Phase displacement = 120°; Sinusoidal axis = −10

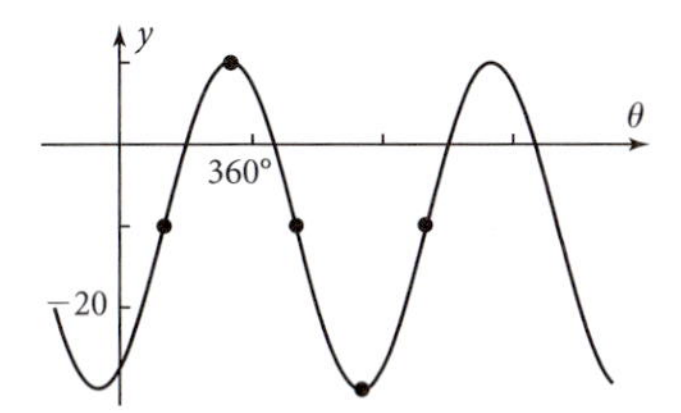

5a. $y = 9 + 6\cos 2(\theta - 20°)$

5b. Amplitude = 6; Period = 180°;
Frequency = $\frac{1}{180}$ cycle/deg; Phase displacement = 20°;
Sinusoidal axis = 9

5c. $y = 10.0418...$ at $\theta = 60°$; $y = 8.7906...$ at $\theta = 1234°$

7a. $y = -3 + 5\cos 3(\theta - 10°)$

7b. Amplitude = 5; Period = 120°;
Frequency = $\frac{1}{120}$ cycle/deg; Phase displacement = 10°;
Sinusoidal axis = −3

7c. $y = -8$ at $\theta = 70°$; $y = 1.9931...$ at $\theta = 491°$

9. $y = 1.45 + 1.11\sin 10(\theta + 16°)$

11. $y = 1.7\cos(\theta - 30°)$

13. $r = 7\cos 3\alpha$

15. $y = 35 + 15\sin 90\theta$

17. $y = 4 - 9\sin\frac{9}{13}(\theta + 60°)$

19. $y = 12.4151...$ at $\theta = 300$; $y = 2.0579...$, $1.9420...$ below the sinusoidal axis at $\theta = 5678°$

21. $y = 4 + 3\cos 5(\theta - 6°)$

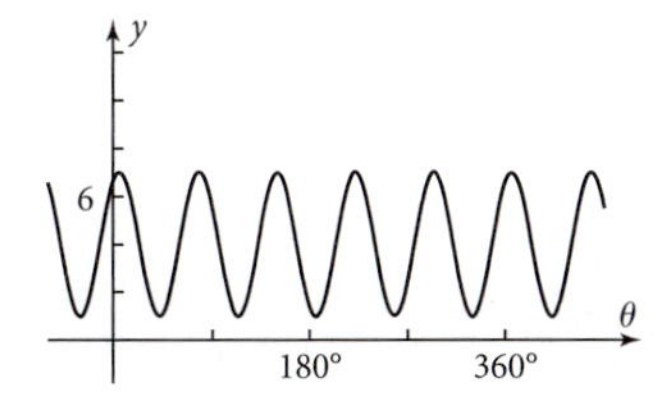

23a. $y = 6 + 4\cos 3(\theta + 10°)$

23b. $y = 6 - 4\cos 3(\theta - 50°)$

23c. $y = 6 + 4\sin 3(\theta + 40°)$

23d. $y = 6 - 4\sin 3(\theta - 20°)$

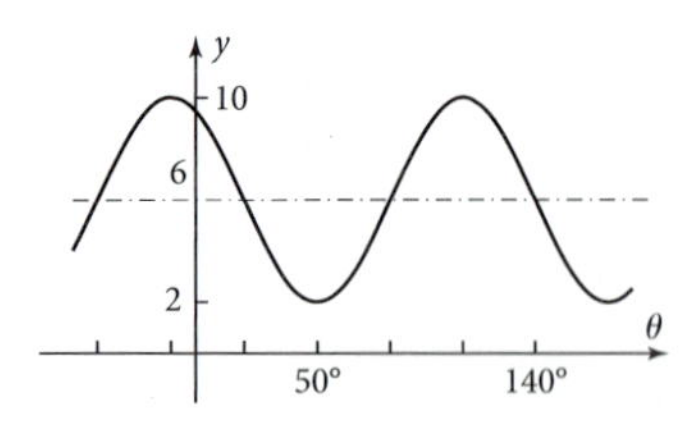

25a. 60 cycles/deg. Thinking in terms of complete cycles (60 of them) gives a clearer mental picture than thinking in terms of fractions.

25b. Period = 1.2°; Frequency = $\frac{5}{6}$ cycle/deg. The frequency is 300 divided by 360°.

27a. Subtract 3 from both sides, then divide both sides by 4. Also, $2(\theta - 5°) = \dfrac{\theta - 5°}{1/2}$

27b. Now the 4 and the $\frac{1}{2}$ are the dilations, and the -3 and the $-5°$ are the opposites of the translations. You could also say that $\frac{1}{4}$ and $\frac{1}{1/2}$ are the reciprocals of the dilations.

27c. It isolates y as a function of θ, making it easier to calculate y given θ.

Problem Set 6-3

1a.

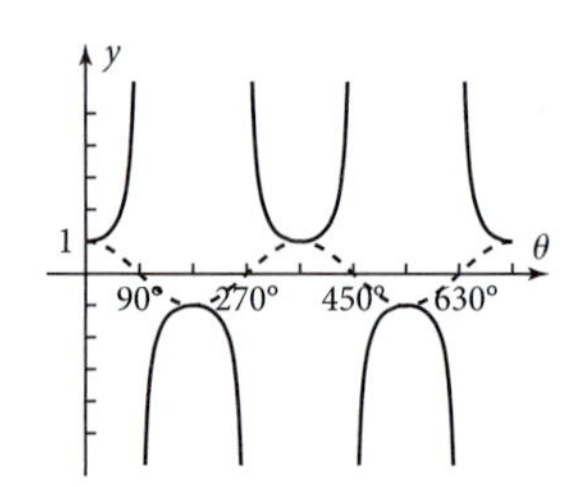

1b. Asymptotes occur where $\theta = 0$ on the cosine graph, at $\theta = 90° + 180°n$

1c. Yes, where $\cos\theta = \pm 1$, at $\theta = +180°n$

1d. No, the concavity changes at the asymptotes, not at points on the graph.

3.

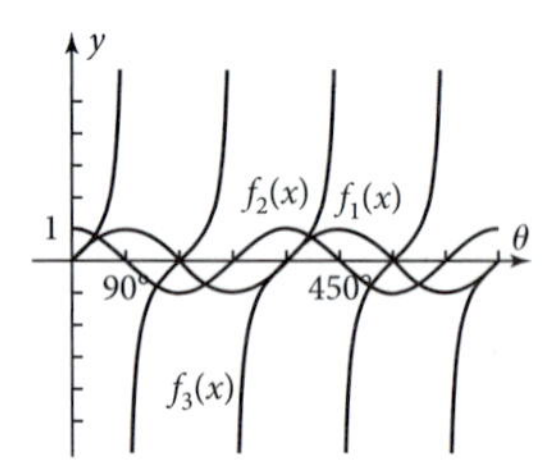

The graph of $f_3(x)$ is the same as the graph of $\tan\theta$.

5. See Figure 6-3a.

7. For all θ, where n is an integer $\sin(\theta + 180°n) = -\sin\theta$ and $\cos(\theta + 180°n) = -\cos\theta$. Therefore, for all θ: $\tan(\theta + 180°n) = \dfrac{\sin(\theta + 180°n)}{\cos(\theta + 180°n)} = \dfrac{-\sin\theta}{-\cos\theta} = \dfrac{\sin\theta}{\cos\theta} = \tan\theta$. So the period of the tangent function is 180°. Furthermore, for all θ: $\cot(\theta + 180°n) = \dfrac{\cos(\theta + 180°n)}{\sin(\theta + 180°n)} = \dfrac{-\cos\theta}{-\sin\theta} = \dfrac{\cos\theta}{\sin\theta} = \cot\theta$.

9. The domain of $\sec\theta$ is where $\frac{1}{\cos\theta}$ is defined, i.e., all $\theta \neq 90° + 180°n$. The range is $|y| \geq 1$, i.e., $(-\infty, -1] \cup [1, \infty)$.

11. θ-translation of $+5°$, θ-dilation of $\frac{1}{3}$, y-dilation of 5, y-translation of $+2$

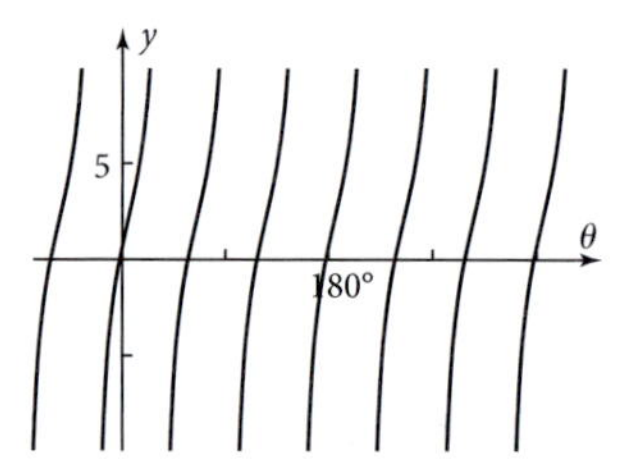

13. θ-translation of $-50°$, θ-dilation of 2, y-dilation of 6, y-translation of $+4$

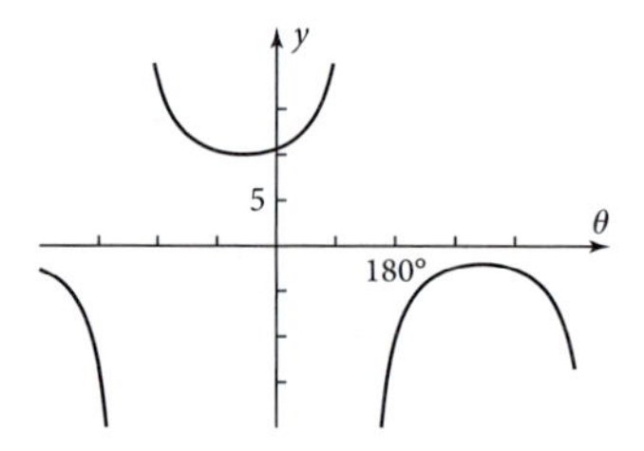

15a.

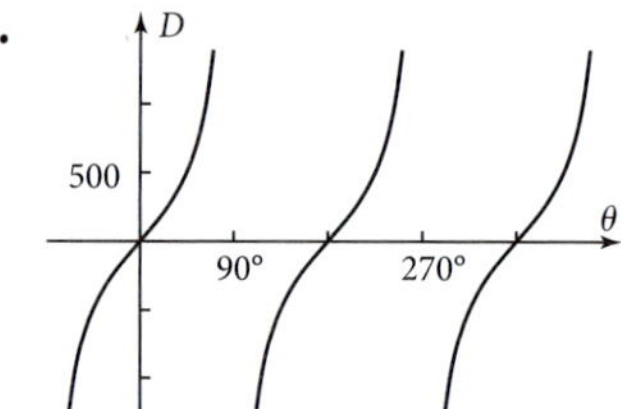

15b. ≈714 m to the right of the lighthouse; ≈28,645 m to the left of the lighthouse

15c. ≈76.0°; ≈116.6°

15d. When $\theta = 90°$, the beam of light is parallel to the shore.

Problem Set 6-4

1a.

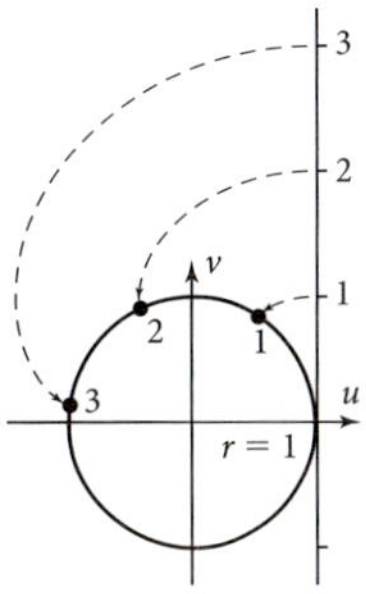

1b.

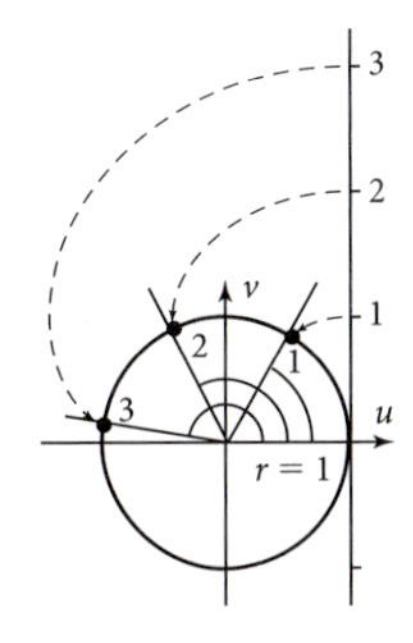

1c. The arc length on the unit circle equals the radian measure.

3. $\dfrac{\pi}{3}$

5. $\dfrac{\pi}{6}$

7. $\dfrac{2}{3}\pi$

9. $-\dfrac{5}{4}\pi$

11. 0.6457...

13. 2.1467...

15. 18°

17. 30°

19. 15°

21. 135°

23. 270°

25. 19.4805...°

27. 72.1926...°

29. 57.2957...°

31. −0.9589...

33. 1.1192...

35. 0.3046...

37. 0.3217...

39. $\dfrac{\sqrt{3}}{2}$

41. $\dfrac{1}{\sqrt{3}}$

43. 1

45. 4

47. 1

49. $y = 5 + 7\cos 30(\theta - 2°)$

50. $y = 5.5 + 0.5\cos \frac{36}{17}(\theta - 15°)$

51. 13.9255... cm

53. 64.6230...°

Problem Set 6-5

1. $\dfrac{\pi}{6}$ units

3. $\dfrac{\pi}{2}$ units

5. 60°

7. 45°

9. $\dfrac{\pi}{2}$ units

11. 2 units

13. 1.5574...

15. −1.0101...

17. 1.2661...

19. 0.2013...

21. $\dfrac{\sqrt{3}}{2}$

23. $\dfrac{1}{\sqrt{3}}$

25. Period = 10; Amplitude = 2;
Phase displacement = +4; Sinusoidal axis = + 3

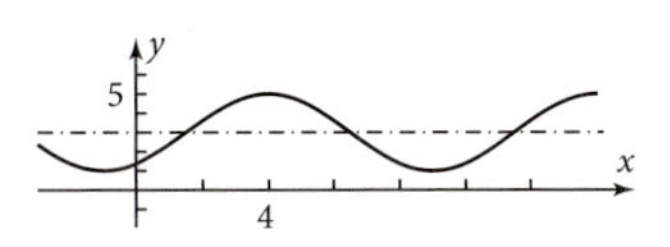

27. Period = 8; Amplitude = 6;
Phase displacement = −1; Sinusoidal axis = +2

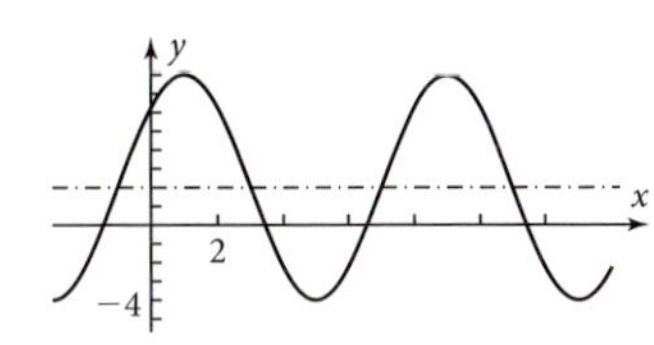

29. Period = 4; Asymptotes at $4n$;
Points of inflection at $2 + 4n$

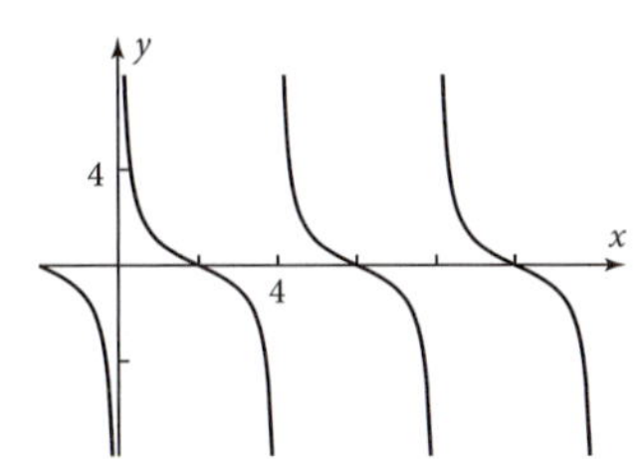

31. Period = 2π; Asymptotes at $\frac{\pi}{2} + n\pi$; Critical points at $n\pi$, specifically $(+2n\pi, 3)$ and $(+(2n + 1)\pi, 1)$

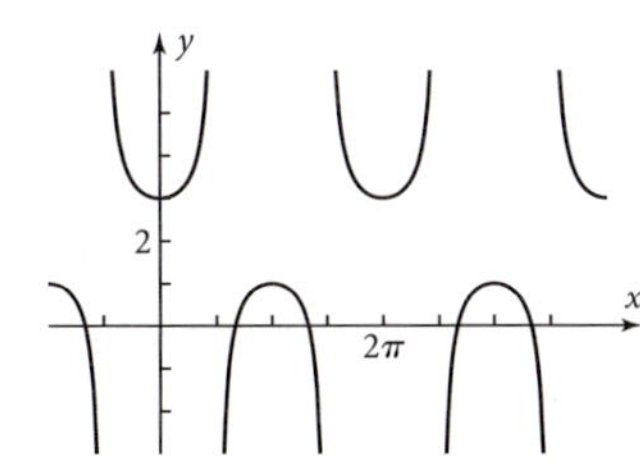

33. $y = 5 + 2\cos\frac{\pi}{3}(x - 1)$

35. $y = -2 + 5\cos\frac{\pi}{15}(x + 5)$

37. $y = \csc\frac{\pi}{6}x$

39. $y = 3\tan x$

41. $z = -8 + 2\sin 5\pi(t - 0.17)$

43. $z(0.4) = -8.9079...; z(50) = -8.9079...;$ $z(50)$ is 0.9079... below the sinusoidal axis.

45a. Horizontal translation of $+\frac{\pi}{2}$; $\sin x = \cos\left(x - \frac{\pi}{2}\right)$

45b. Horizontal translation of $+2\pi$; the graph would coincide with itself and appear unchanged.

45c. $+2\pi$ or -2π, or any multiple of $\pm 2\pi$

45d. A horizontal translation by a multiple of 2π results in a graph that coincides with itself. The period of the sine function is 2π.

45e. Answers will vary. As k increases, the graph moves to the right.

47a. Wrapping the x-axis around the unit circle converts distances along the x-axis to arc lengths, and vice-versa. In particular, it shows that a circular function's independent variable (arc length) is the same as a distance along the x-axis. So for both types of functions, the independent variable is a distance along the x-axis.

47b. A radian measure corresponds to an angle measure, using $m^R(\theta) = m^\circ(\theta) \cdot \frac{\pi}{180^\circ}$, but because a radian measure is a pure number, it can represent something other than an angle in an application problem.

49. Journal entries will vary.

Problem Set 6-6

1. 0.4510..., 5.8321..., 6.7342..., 12.1153..., 13.0173...

2. 1.1592..., 5.1239..., 7.4424..., 11.4070..., 13.7256...

3. 1.7721..., 4.5110..., 8.0553..., 10.7942..., 14.3385...

4. 2.0943..., 4.1887..., 8.3775..., 10.4719..., 14.6607...

5a. $x \approx 1, 5, 21, 25$

5b. $y = 2 + 5\cos\frac{\pi}{10}(x - 3)$

5c. $x \approx 0.9516..., 5.0483..., 20.9516..., 25.0483...$

5d. $x = 0.9516..., 5.0483..., 20.9516..., 25.0483...$

5e. 100.9516...

7a. $x \approx -2.9, -0.5, 1.1, 3.5, 5.1$

7b. $y = -2 + 4\cos\frac{\pi}{2}(x - 0.3)$

7c. $x \approx -2.8608..., -0.5391..., 1.1391..., 3.4608...,$ 5.1391...

7d. $x = -2.8608..., -0.5391..., 1.1391..., 3.4608...,$ 5.1391...

7e. $x = 101.1391...$

9a. $x \approx -10.6, -3.4, 5.4, 12.6, 21.4$

9b. $y = 1 - 3\cos\frac{\pi}{8}(x - 1)$

9c. $x \approx -10.5735..., -3.4264..., 5.4264..., 12.5735...,$ 21.4264...

9d. $x = -10.5735..., -3.4264..., 5.4264..., 12.5735...,$ 21.4264...

9e. $x = 101.4264...$

11a. $\theta \approx 130^\circ, 170^\circ, 310^\circ$

11b. $y = 6 + 4\cos 2(\theta - 60^\circ)$

11c. $\theta \approx 129.2951...^\circ, 170.7048...^\circ, 309.2951...^\circ$

11d. $\theta = 129.2951...^\circ, 170.7048...^\circ, 309.2951...^\circ$

13a. $x = 2.6905..., 3.5926..., 8.9737..., 9.8758...,$ 15.2569..., 16.1589...

13b. $x = 203.7524....$

Problem Set 6-7

1a.

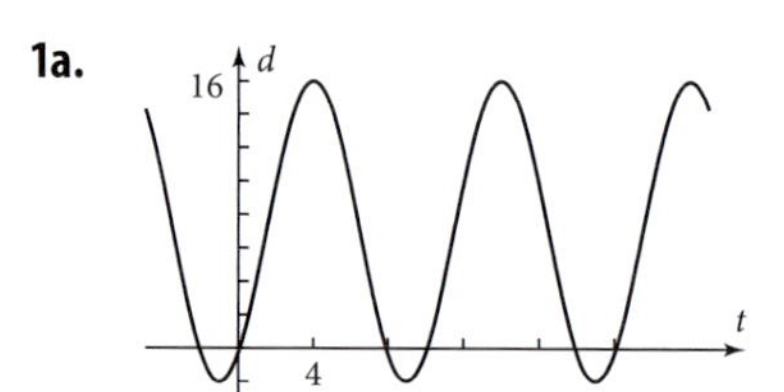

1b. −2 ft; part of the wheel is underwater.

1c. $d = 7 + 9\cos\frac{\pi}{5}(t - 4)$ **1d.** ≈ 4.2 ft

1e. $t = 0.0817...$ s. Because the period is 10 s and the point reaches the top for the first time at $t = 4$ s, the point could not have been at the bottom for any $1 < t \leq 4$. Thus, the wheel must have been coming out of the water at $t = 0.0817...$ s.

1f. "Mark Twain" was riverboat terminology meaning that the water was 2 fathoms deep.

3a.

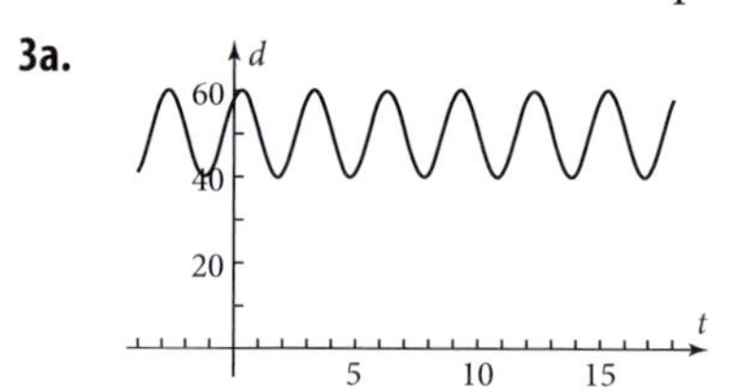

3b. $d = 50 + 10\cos\frac{\pi}{1.5}(t - 0.3)$

3c. $d(17.2) = 43.3086...$ cm

3d. $d(0) = 58.0901...$ cm

3e. $t = 0.0846...$ s

5a. $y = 12 + 15\cos\frac{\pi}{50}x$

5b.

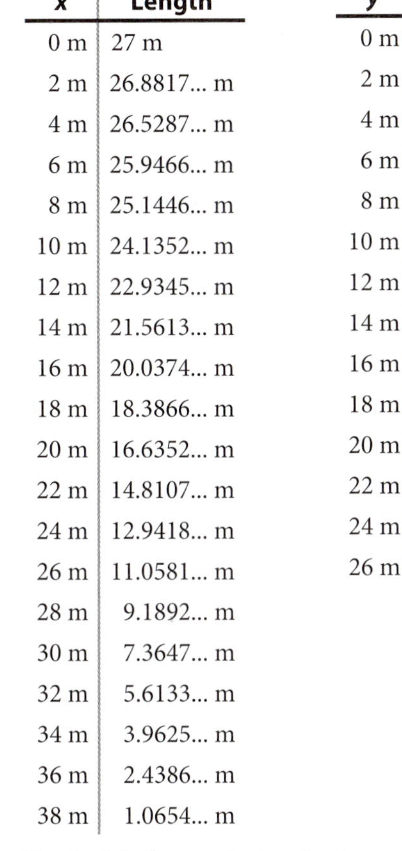

x	Length
0 m	27 m
2 m	26.8817... m
4 m	26.5287... m
6 m	25.9466... m
8 m	25.1446... m
10 m	24.1352... m
12 m	22.9345... m
14 m	21.5613... m
16 m	20.0374... m
18 m	18.3866... m
20 m	16.6352... m
22 m	14.8107... m
24 m	12.9418... m
26 m	11.0581... m
28 m	9.1892... m
30 m	7.3647... m
32 m	5.6133... m
34 m	3.9625... m
36 m	2.4386... m
38 m	1.0654... m

5c.

y	Length
0 m	39.7583... m
2 m	36.6139... m
4 m	33.9530... m
6 m	31.5494... m
8 m	29.2961... m
10 m	27.1284... m
12 m	25 m
14 m	22.8715... m
16 m	20.7038... m
18 m	18.4505... m
20 m	16.0469... m
22 m	13.3860... m
24 m	10.2416... m
26 m	5.8442... m

5d. Vertical timbers ≈ 324 m; horizontal timbers ≈ 331 m

7a. 11 yr

7b. $S = 60 + 50\cos\frac{2\pi}{11}(t - 1948)$

7c. $S(2020) \approx 12$ sunspots

7d. $S(2021) \approx 27$ sunspots; $S(2022) \approx 53$ sunspots; maximum in 2025.

7e. The sunspot cycle resembles a sinusoid slightly but is not one.

9a.

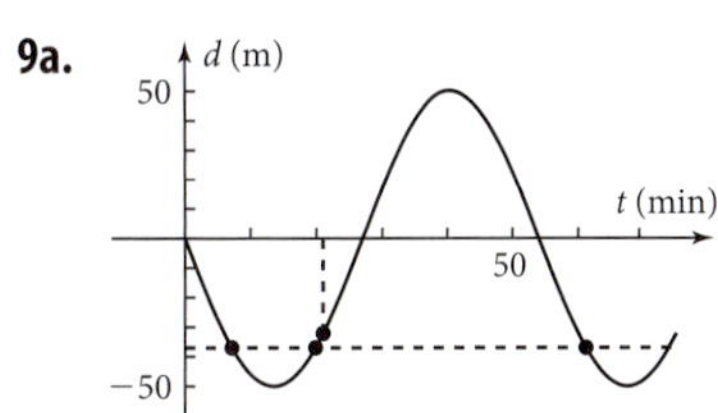

9b. $t = 40.5$ min

9c. $d = 50\cos\frac{\pi}{27}(t - 40.5)$

9d. $d \approx -32.1$ m

9e. $t = 7.1597...$ min, 19.8402... min, 61.1597... min

11a.

11b. $y = 12 - 12\cos\frac{x}{12}$

11c. ≈8.2 in.

11d. $x = 17.8483...$ in. and 57.5498... in.

13a. Frequency = 60 cycles/s; period = $\frac{1}{60}$ s

13b. Wavelength = 220 in.

13c. 34 ft 4.5 in.

15. Answers will vary.

Problem Set 6-8

1a. 15 rad/s

1b. $\frac{450}{\pi}$ rev/min = 143.2394... rev/min

3a. $\omega = 150$ rad/s; $\omega = 1432.3944...$ rev/min

3b. 1425 in./s

3c. If the stone were hurled with the same velocity as the tip of the blade, it would travel at $v = 80.9659...$ mi/h.

5a. $\frac{2\pi}{3}$ radians

5b. 8.3775... cm

5c. $\omega = \frac{4\pi}{3}$ rad/s = 4.1887... rad/s; $v = 16.7551...$ cm/s

7a. $\omega = \frac{2}{3}$ rev/s

7b. $\omega = \frac{4}{3}\pi$ rad/s = 4.1887... rad/s

7c. $v = \frac{280\pi}{3}$ cm/s = 293.2153... cm/s

7d. Points of Ima's body along the axis have $v = 0$ cm/s, but $\omega = \frac{4}{3}\pi$ rad/s = 4.1887... rad/s.

7e. $v = 150$ cm/s; fingertips are moving slower.

9a. $\omega = \frac{10\pi}{3}$ rad/s = 10.4719... rad/s

9b. $v = \frac{50\pi}{3}$ cm/s = 52.3598... cm/s

9c. $v = \frac{50\pi}{3}$ cm/s = 52.3598... cm/s

9d. $v = \frac{50\pi}{3}$ cm/s = 52.3598... cm/s

9e. $\omega = \frac{10\pi}{9}$ rad/s = 3.4906... rad/s

9f. $\omega = \frac{10\pi}{9}$ rad/s = 3.4906... rad/s; $v = 0$ cm/s

11a. $v = \frac{4\pi}{3}$ ft/s = 4.1887... ft/s; $v = \frac{10\pi}{11}$ mi/h = 2.8559... mi/h

11b. $v = \frac{4\pi}{3}$ ft/s $= 4.1887...$ ft/s; linear velocity

11c. $\omega = 44.4444...$ rev/min; angular velocity

13a. $\omega = 10\pi$ rad/s $= 31.4159...$ rad/s

13b. $\omega = 10\pi$ rad/s $= 31.4159...$ rad/s

13c. $v = 20\pi$ in./s $= 62.8318...$ in./s

13d. $v = 20\pi$ in./s $= 62.8318...$ in./s

13e. $\omega = \frac{10\pi}{9}$ rad/s $= 3.4906...$ rad/s

13f. $v = 0$ in./s; $\omega = \frac{10\pi}{9}$ rad/s $= 3.4906...$ rad/s

15a. $\omega = \frac{5 \text{ ft}}{\text{s}} \cdot \frac{1 \text{ rad}}{25 \text{ ft}} = 0.2$ rad/s

15b. $\omega = \frac{6}{\pi}$ rev/min $= 1.9098...$ rev/min

15c. The linear velocities are the same—members all march at the same speed.

15d. $\omega = 0.5$ rad/s

15e. 2.5 times as fast; $2.5 = \frac{\text{diameter of large circle}}{\text{diameter of small circle}}$

17a. $\omega = 90\pi$ rad/s $= 282.7433...$ rad/s

17b. Gear 2: $v = 180\pi$ cm/s $= 565.4866...$ cm/s; $\omega = 12\pi$ rad/s $= 37.6991...$ rad/s

Gear 3: $\omega = 12\pi$ rad/s $= 37.6991...$ rad/s; $v = 36\pi$ cm/s $= 113.0973...$ cm/s

Gear 4: $v = 36\pi$ cm/s $= 113.0973...$ cm/s; $\omega = 2\pi$ rad/s $= 6.2831...$ rad/s

17c. $v = 0$ cm/s; $\omega = 2\pi$ rad/s $= 6.2831...$ rad/s

17d. $\omega = 60$ rev/min

17e. ratio $= 45 = \frac{15 \text{ cm}}{2 \text{ cm}} \cdot \frac{18 \text{ cm}}{3 \text{ cm}}$

Problem Set 6-9

R1a.

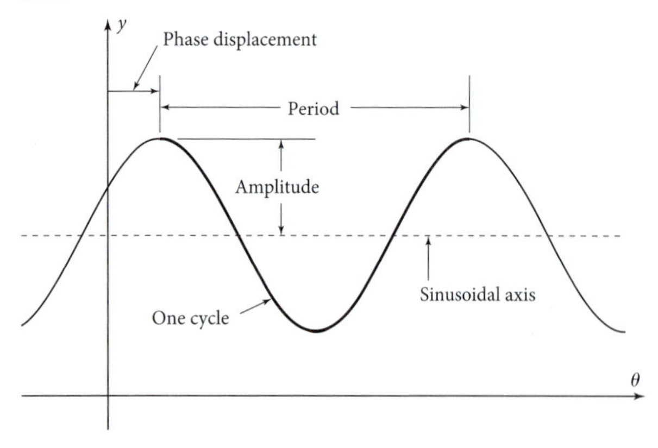

R1b. Argument

R3a.

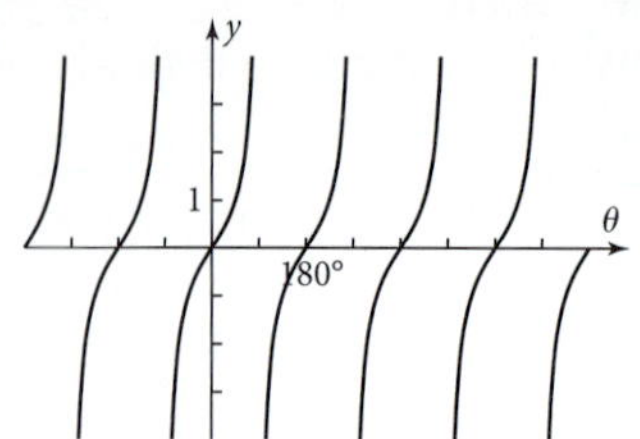

R3b. For all θ, $\sin(\theta + 180°) = -\sin\theta$ and $\cos(\theta + 180°) = -\cos\theta$. Therefore, for all θ: $\tan(\theta + 180°) = \frac{\sin(\theta + 180°)}{\cos(\theta + 180°)} = \frac{-\sin\theta}{-\cos\theta} = \frac{\sin\theta}{\cos\theta}$ $= \tan\theta$. So the period of the tangent function is 180°.

R3c.

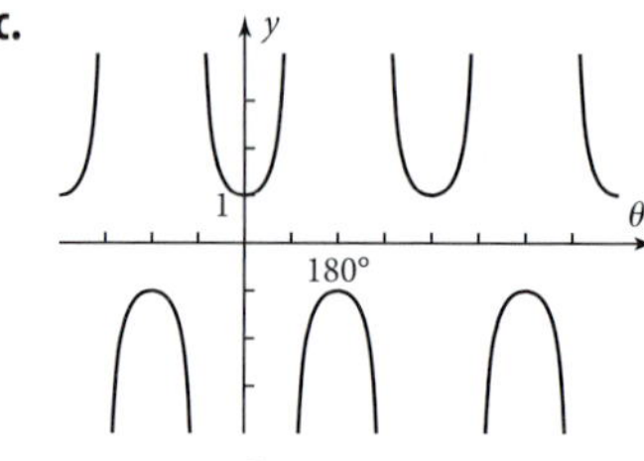

$y = \sec\theta = \frac{1}{\cos\theta}$

R3d. $\sin\theta = 0$ at $\theta = 0°, 180°, 360°$, etc., so $\csc\theta = \frac{1}{\sin\theta}$ goes to infinity at these points.

R3e. The cosecant graph changes concavity only at the asymptotes, not at any points that are actually on the graph, so it has no points of inflection. Within each region of concavity, it has a critical point. The cotangent graph is always decreasing, so it has no critical points. It is concave up to the left of $(90° + 180°n)$ and concave down to the right, so $(90° + 180°n)$ are the points of inflection.

R3f. θ-translation of $+40°$, θ-dilation of 3, y-dilation of 0.4, y-translation of $+2$

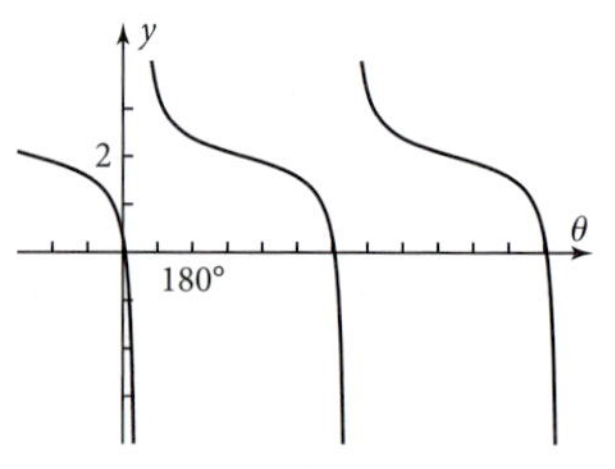

Period $= 540°$; the value of y is unbounded, so the "amplitude" is infinite.

R5a.

R5b. $\frac{\pi}{3}$ units; 2.3 units

R5c. Major radius, $a = 90$ cm; Minor radius, $b = 39.2300...$ cm; $x = 90 \cos t$, $y = 39.2300... \sin t$

The grapher agrees with the figure.

Both angles measure about 30° (precisely 29.92...°). A ball shot straight from one focus to the point where $t = 1$ will bounce off of the tangent line at the same angle at which it approached, thus going straight through the other focus. (This is true no matter which point on the ellipse is aimed for!)

R5d. 53.1301...°

R5e. $\cos \frac{\pi}{6} = \frac{\sqrt{3}}{2}$; $\sec \frac{\pi}{4} = \sqrt{2}$; $\tan \frac{\pi}{2}$ is undefined.

R5f.

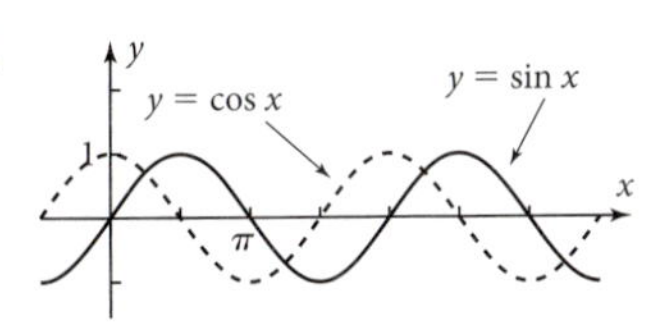

R5g. Period $= 2\pi \div \frac{\pi}{10} = 20$

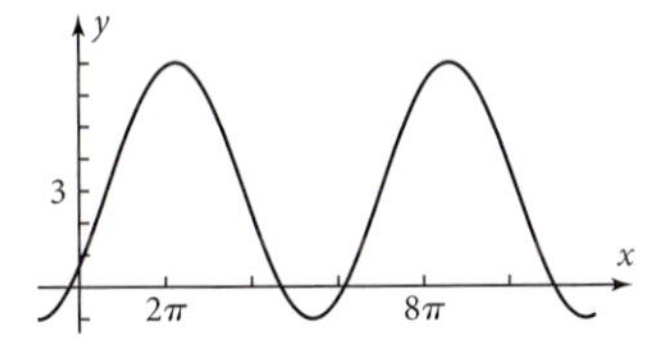

R5h. $y = -10 - 35 \sin \frac{\pi}{20}(x - 13)$

R7a.

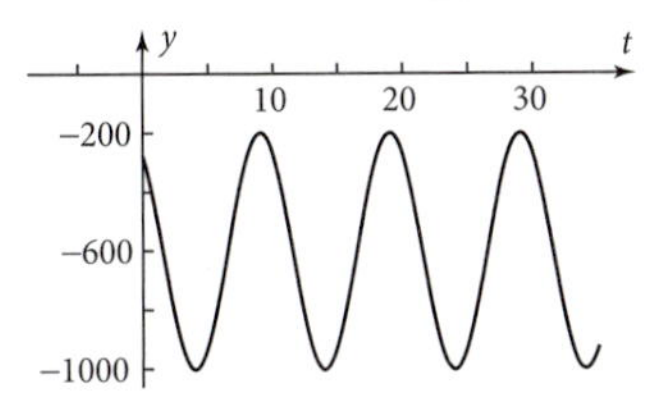

R7b. $y = -600 + 400 \cos \frac{\pi}{5}(t - 9)$

R7c. At $t = 0$, $y = -276.3932...$ m; submarine could communicate.

R7d. Roughly 9 s $< t <$ 7 min 51 s.

C1a. 22.9183...°

C1b. 3.2 ft

C1c. 1.6 ft

C1d.

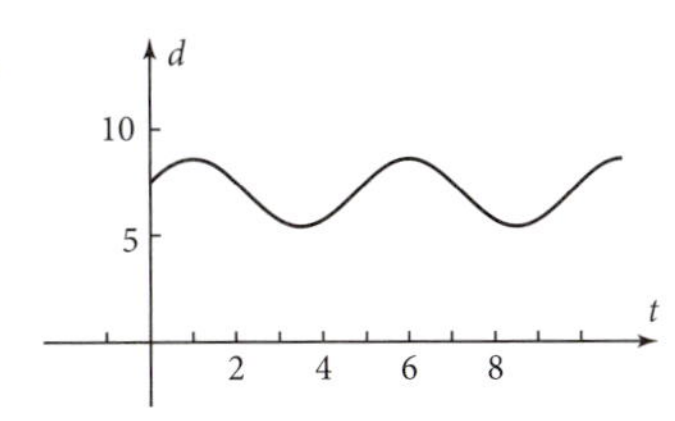

C1e. $d = 7 + 1.6 \cos \frac{2\pi}{5}(t - 1)$

C1f. 5.7055... ft

C1g. 1.9941... s

C1h. False

C3a. $v_{\text{farthest}} = 6.4\pi$ ft/s $= 20.1061...$ ft/s; $v_{\text{closest}} = -1.6\pi$ ft/s $= -5.0265...$ ft/s

C3b. Clockwise with respect to the center of the merry-go-round

C3c. $d(t) = 13 + 6 \cos \frac{2\pi t}{5} + 4 \cos \pi t$, where d is in feet and t is in seconds.

C3d. One cycle is shown here.

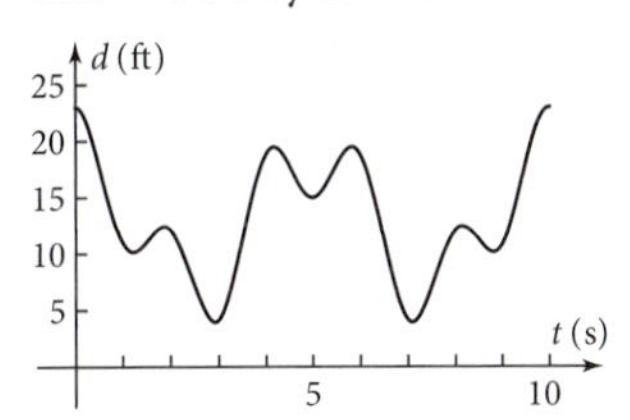

C3e. Answers will vary. There are frequent and rapid shifts of direction, at high velocity.

T1.

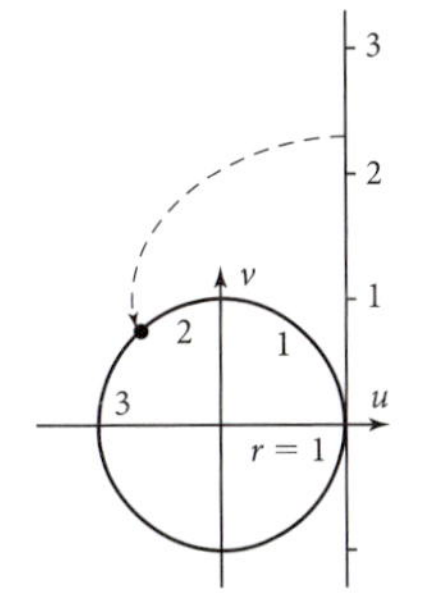

T3. $2.3 \cdot \frac{180°}{\pi} = 131.7802...°$; in Quadrant II.

T5. 36°

T7.

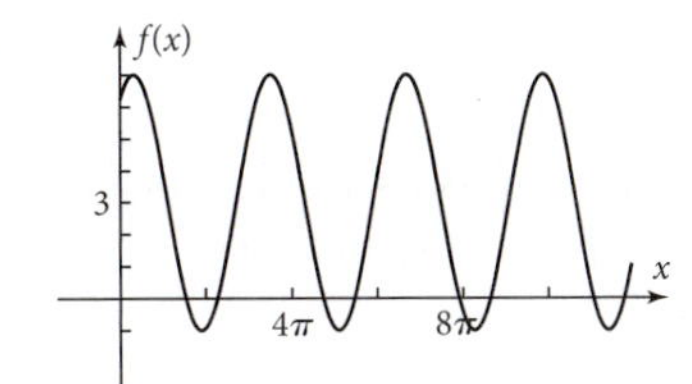

T9. $\omega = 4$ rad/s

T11. $y = -20 + 30 \sin \frac{\pi}{8}(x - 3)$

T13. $d = 1$ ft at $t = 9.6 + 11.2n$ h

T15.

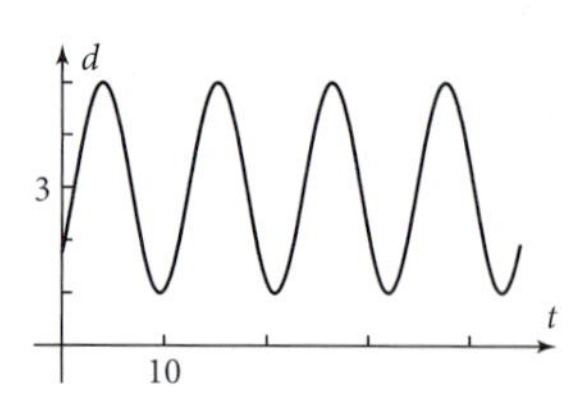

T17. $t \approx 2.71$ h

T19. $\omega = 4\pi$ rad/s $= 12.5663...$ rad/s

T21. $\omega = 10\pi$ rad/s $= 31.4159...$ rad/s

T23. $d(t) = 28 + 24 \cos 4\pi t$

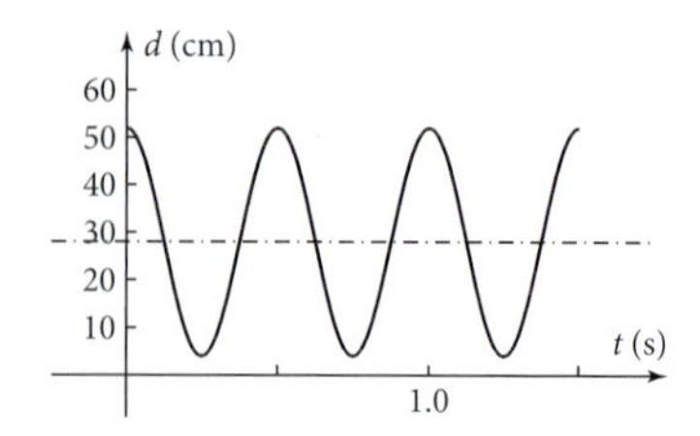

Chapter 7

Problem Set 7-1

1. The sum is 1.

3.

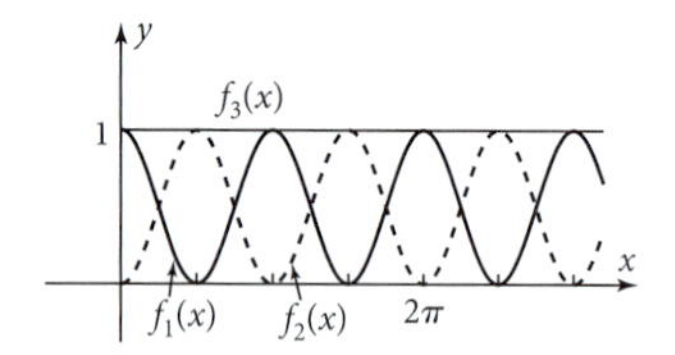

$f_1(x)$ and $f_2(x)$ are symmetrical with respect to each other across the $y = \frac{1}{2}$ line, so the amount, a, that one graph is above $\frac{1}{2}$ is the same as the amount the other graph is below $\frac{1}{2}$. When added: $\left(\frac{1}{2} - a\right) + \left(\frac{1}{2} + a\right) = 1$

5. Because $r = 1$, $\sin 50° = \frac{v}{r} = v$ and $\cos 50° = \frac{u}{r} = u$.

Problem Set 7-2

1. $\sec x = \frac{1}{\cos x}$

3. $\frac{\sin x}{\cos x}$

5. Cosine x and sine x are the lengths of the legs of a right triangle whose hypotenuse is a radius of a unit circle. The Pythagorean theorem then says $(\cos x)^2 + (\sin x)^2 = 1^2$.

7.

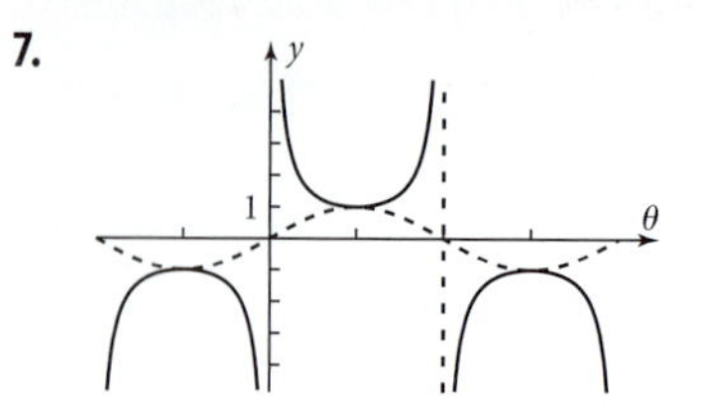

Asymptotes at $\theta = 0°, 180°, 360°, ...$

9. $\cos^2 x + \sin^2 x = 1 \Rightarrow \cos^2 x + \sin^2 x - \cos^2 x$
$= 1 - \cos^2 x \Rightarrow \sin^2 x = 1 - \cos^2 x$

11. $\cos^2 x + \sin^2 x = 1$ $\quad \cos x \cdot \sec x = 1$
$\csc^2 x - \cot^2 x = 1$ $\quad \sin x \cdot \csc x = 1$
$\sec^2 x - \tan^2 x = 1$ $\quad \tan x \cdot \cot x = 1$

13. $\sec x = \frac{1}{\cos x} \Leftrightarrow \csc x = \frac{1}{\sin x}$

$\cot x = \frac{1}{\tan x} \Leftrightarrow \tan x = \frac{1}{\cot x}$

$\tan x = \frac{\sin x}{\cos x} = \frac{\sec x}{\csc x} \Leftrightarrow \cot x = \frac{\cos x}{\sin x} = \frac{\csc x}{\sec x}$

$\cos^2 x + \sin^2 x = 1 \Leftrightarrow \sin^2 x + \cos^2 x = 1$

$1 + \tan^2 x = \sec^2 x \Leftrightarrow 1 + \cot^2 x = \csc^2 x$

15. Answers will vary.

Problem Set 7-3

1. $\cos x \tan x = \cos x \cdot \frac{\sin x}{\cos x} = \left(\cos x \cdot \frac{1}{\cos x}\right) \cdot \sin x$
$= \sin x$

3. $\sec A \cot A \sin A = \cot A \sin A \sec A$
$= \cot A \sin A \frac{1}{\cos A} = \cot A \frac{\sin A}{\cos A} = \cot A \tan A = 1$

5. $\sin^2 \theta \sec \theta \csc \theta = \sin^2 \theta \frac{1}{\cos \theta} \frac{1}{\sin \theta}$
$= \frac{\sin^2 \theta}{\cos \theta \sin \theta} = \frac{\sin \theta}{\cos \theta} = \tan \theta$

7. $\cot R + \tan R = \frac{\cos R}{\sin R} + \frac{\sin R}{\cos R}$

$= \frac{\cos R}{\sin R} \cdot \frac{\cos R}{\cos R} + \frac{\sin R}{\sin R} \cdot \frac{\sin R}{\cos R}$

$= \frac{\cos^2 R}{\sin R \cos R} + \frac{\sin^2 R}{\sin R \cos R} = \frac{\cos^2 R + \sin^2 R}{\sin R \cos R}$

$= \frac{1}{\sin R \cos R} = \frac{1}{\sin R} \cdot \frac{1}{\cos R} = \csc R \cdot \sec R$

9. $\csc x - \sin x = \frac{1}{\sin x} - \frac{\sin x}{\sin x} \cdot \sin x$

$= \frac{1}{\sin x} - \frac{\sin^2 x}{\sin x} = \frac{1 - \sin^2 x}{\sin x} = \frac{\cos^2 x}{\sin x} = \frac{\cos x}{\sin x} \cdot \cos x$

$= \cot x \cos x$

11. $(\tan x)(\cot x \cos x + \sin x)$

$= \left(\frac{\sin x}{\cos x}\right)\left(\frac{\cos x}{\sin x}\cos x + \sin x\right)$

$= \frac{1}{\cos x}\left(\sin x \cdot \frac{\cos x}{\sin x} \cdot \cos x + \sin^2 x\right)$

$= \sec x\,(\cos^2 x + \sin^2 x) = \sec x$

13. $(1 + \sin B)(1 - \sin B) = 1 - \sin^2 B = \cos^2 B$

15. $(\cos\phi - \sin\phi)^2 = \cos^2\phi - 2\cos\phi\sin\phi$
$+ \sin^2\phi = \cos^2\phi + \sin^2\phi - 2\cos\phi\sin\phi$
$= 1 - 2\cos\phi\sin\phi$

17. $(\tan n + \cot n)^2 = \tan^2 n + 2\tan n\cot n + \cot^2 n$

$= \tan^2 n + 2 + \cot^2 n = (\tan^2 n + 1) + (1 + \cot^2 n)$

$= \sec^2 n + \csc^2 n$

19. $\frac{\csc^2 x - 1}{\cos x} = \frac{\cot^2 x}{\cos x} = \cot^2 x\frac{1}{\cos x}$

$= \cot x \cdot \frac{\cos x}{\sin x} \cdot \frac{1}{\cos x} = \cot x \cdot \frac{1}{\sin x} = \cot x \csc x$

21. $\frac{\sec^2\theta - 1}{\sin\theta} = \frac{\tan^2\theta}{\sin\theta} = \tan^2\theta\frac{1}{\sin\theta}$

$= \tan\theta\frac{\sin\theta}{\cos\theta}\frac{1}{\sin\theta} = \tan\theta\frac{1}{\cos\theta} = \tan\theta\sec\theta$

23. $\frac{\sec A}{\sin A} - \frac{\sin A}{\cos A} = \frac{1}{\cos A\sin A} - \frac{\sin A}{\cos A}$

$= \frac{1}{\cos A\sin A} - \frac{\sin^2 A}{\cos A\sin A} = \frac{1 - \sin^2 A}{\cos A\sin A}$

$= \frac{\cos^2 A}{\cos A\sin A} = \frac{\cos A}{\sin A} = \cot A$

25. $\frac{1}{1 - \cos x} + \frac{1}{1 + \cos x}$

$= \frac{1 + \cos x}{(1 + \cos x)(1 - \cos x)} + \frac{1 - \cos x}{(1 - \cos x)(1 + \cos x)}$

$= \frac{1 + \cos x}{1 - \cos^2 x} + \frac{1 - \cos x}{1 - \cos^2 x} = \frac{1 + \cos x + 1 - \cos x}{1 - \cos^2 x}$

$= \frac{2}{\sin^2 x} = 2\csc^2 x$

27. $(\sec x)(\sec x - \cos x) = \sec^2 x - \sec x\cos x$

$= \sec^2 x - 1 = \tan^2 x$

29. $(\sin x)(\csc x - \sin x) = \sin x\csc x - \sin^2 x$

$= 1 - \sin^2 x = \cos^2 x$

31. $\csc^2\theta - \cos^2\theta\csc^2\theta = (\csc^2\theta)(1 - \cos^2\theta)$

$= (\csc^2\theta)(\sin^2\theta) = 1$

33. $(\sec\theta + 1)(\sec\theta - 1) = \sec^2\theta - 1 = \tan^2\theta$

35. $(2\cos x + 3\sin x)^2 + (3\cos x - 2\sin x)^2$
$= 4\cos^2 x + 12\cos x\sin x + 9\sin^2 x$
$+ 9\cos^2 x - 12\cos x\sin x + 4\sin^2 x$
$= 4\cos^2 x + 4\sin^2 x + 9\sin^2 x + 9\cos^2 x = 4 + 9 = 13$

37.

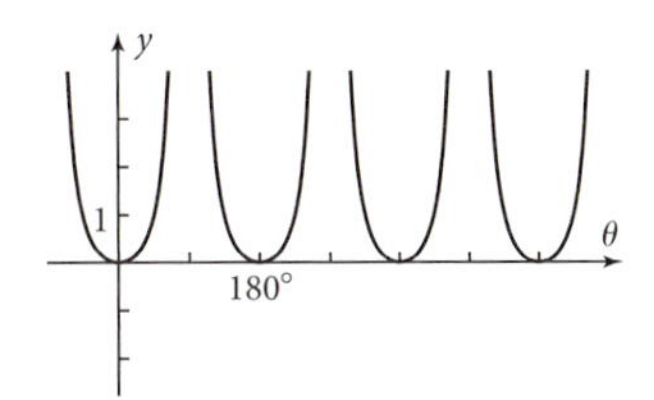

39. $y = (2\cos x + 3\sin x)^2 + (3\cos x - 2\sin x)^2$

x	y
0	13
1	13
2	13
3	13
4	13
5	13

41. For example, $\cos\frac{\pi}{4} = \frac{1}{\sqrt{2}} = 0.7071...,$

but $1 - \sin\frac{\pi}{4} = 1 - \frac{1}{\sqrt{2}} = 0.2928....$

43. $\sec^2 A + \tan^2 A\sec^2 A = (\sec^2 A)(1 + \tan^2 A)$
$= (\sec^2 A)(\sec^2 A) = \sec^4 A$

45. $\frac{1}{\sin x\cos x} - \frac{\cos x}{\sin x}$

$= \frac{1}{\sin x\cos x} - \frac{\cos x}{\sin x}\cdot\frac{\cos x}{\cos x} = \frac{1 - \cos^2 x}{\sin x\cos x} = \frac{\sin^2 x}{\sin x\cos x}$

$= \frac{\sin x}{\cos x} = \tan x$

47. $\frac{1}{1 + \cos p} = \frac{1 - \cos p}{(1 + \cos p)(1 - \cos p)} = \frac{1 - \cos p}{1 - \cos^2 p}$

$= \frac{1 - \cos p}{\sin^2 p} = \frac{1}{\sin^2 p} - \frac{\cos p}{\sin^2 p} = \csc^2 p - \frac{1}{\sin p}\cdot\frac{\cos p}{\sin p}$

$= \csc^2 p - \csc p\cot p$

49. $\frac{1 + \sin x}{1 - \sin x} = \frac{(1 + \sin x)(1 + \sin x)}{(1 - \sin x)(1 + \sin x)}$

$= \frac{1 + 2\sin x + \sin^2 x}{1 - \sin^2 x} = \frac{1 + 2\sin x + \sin^2 x}{\cos^2 x}$

$= \frac{1}{\cos^2 x} + \frac{2\sin x}{\cos^2 x} + \frac{\sin^2 x}{\cos^2 x}$

$= \sec^2 x + 2\cdot\frac{1}{\cos x}\cdot\frac{\sin x}{\cos x} + \tan^2 x$

$= \sec^2 x + 2\sec x\tan x + \tan^2 x$

$= \sec^2 x + 2\sec x\tan x + (\sec^2 x - 1)$

$= 2\sec^2 x + 2\sec x\tan x - 1$

51. $\sec^2\theta + \csc^2\theta = \frac{1}{\cos^2\theta} + \frac{1}{\sin^2\theta}$

$= \frac{\sin^2\theta}{\cos^2\theta\sin^2\theta} + \frac{\cos^2\theta}{\cos^2\theta\sin^2\theta} = \frac{\sin^2\theta + \cos^2\theta}{\cos^2\theta\sin^2\theta}$

$= \frac{1}{\cos^2\theta\sin^2\theta} = \sec^2\theta\csc^2\theta$

53. $\dfrac{1 - 3\cos x - 4\cos^2 x}{\sin^2 x} = \dfrac{(1 - 4\cos x)(1 + \cos x)}{(1 - \cos^2 x)}$

$= \dfrac{(1 - 4\cos x)(1 + \cos x)}{(1 - \cos x)(1 + \cos x)} = \dfrac{1 - 4\cos x}{1 - \cos x}$

55. Journal entries will vary.

Problem Set 7-4

1a. $\theta = 44.4270...^\circ + 360^\circ n$ or $135.5729...^\circ + 360^\circ n$

1b. $\theta = 44.4270...^\circ, 135.5729...^\circ, 404.4270...^\circ, 495.5729...^\circ$

3a. $x = -0.2013... + 2\pi n$ or $3.3429... + 2\pi n$

3b. $x = 3.3429..., 6.0818..., 9.6261..., 12.3650...$

5a. $\theta = -75.9637...^\circ + 180^\circ n$

5b. $\theta = 104.0362...^\circ, 284.0362...^\circ, 464.0362...^\circ, 644.0362...^\circ$

7a. $x = 1.4711... + \pi n$

7b. $x = 1.4711..., 4.6127..., 7.7543..., 10.8959...$

9a. $\theta = \pm 78.4630...^\circ + 360^\circ n$

9b. $\theta = 78.4630...^\circ, 281.5369...^\circ, 438.4630...^\circ, 641.5369...^\circ$

11.

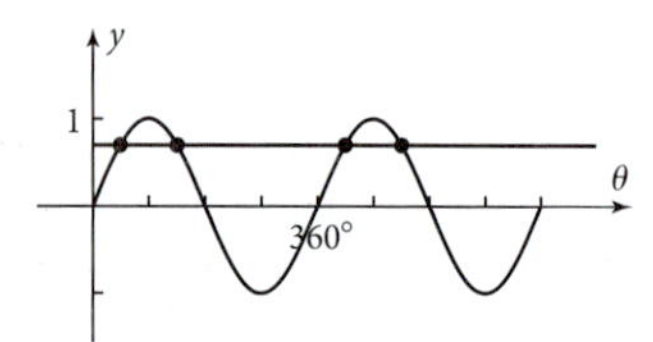

13. No angle has a cosine of 2 ($-1 \le \cos x \le 1$ for all x, because the adjacent side in a right triangle can never be greater in absolute value than the hypotenuse), but there are infinitely many angles whose tangent is 2—the opposite side can be twice the adjacent side, and $\tan(1.1071... + \pi n) = 2$.

15. 217°

17. $\dfrac{\pi}{6}$

19. $2\pi - 2$

21. $\theta = 120^\circ, 300^\circ, 480^\circ, 660^\circ$

23. $\theta = -257^\circ, -17^\circ, 103^\circ, 343^\circ$

25. $x = 0.3918..., 1.6081..., 2.3918..., 3.6081..., 4.3918..., 5.6081...$

27.

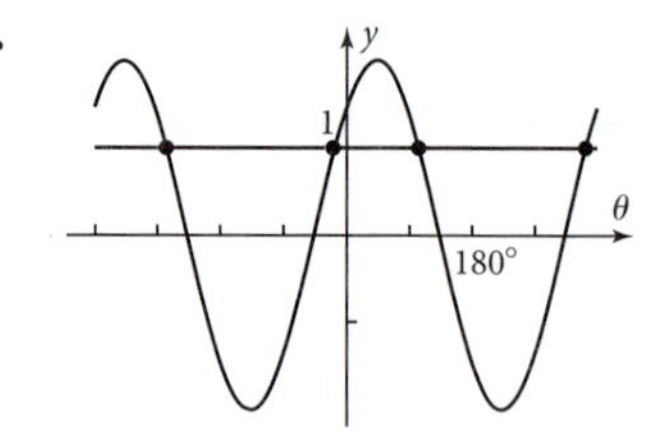

29. $\theta = 0^\circ, 120^\circ, 240^\circ, 360^\circ, 480^\circ, 600^\circ, 720^\circ$

31. $\theta = 210^\circ, 330^\circ, 570^\circ, 690^\circ$

33a.

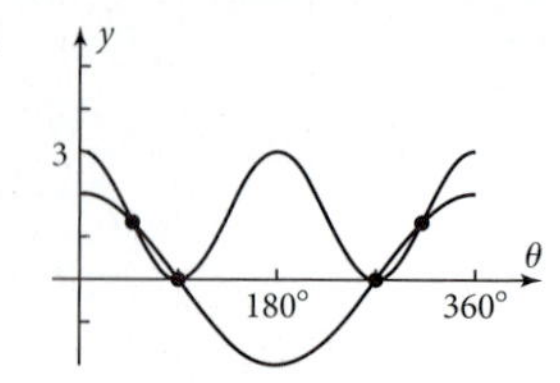

$\theta = 48.1896...^\circ, 90^\circ, 270^\circ, 311.8103...^\circ$

33b. $3\cos^2\theta = 2\cos\theta \Rightarrow (\cos\theta)(3\cos\theta - 2) = 0$
$\Rightarrow \cos\theta = 0$ or $\cos\theta = \dfrac{2}{3}$
$\theta = 48.1896...^\circ, 90^\circ, 270^\circ, 311.8103...^\circ$

35a.

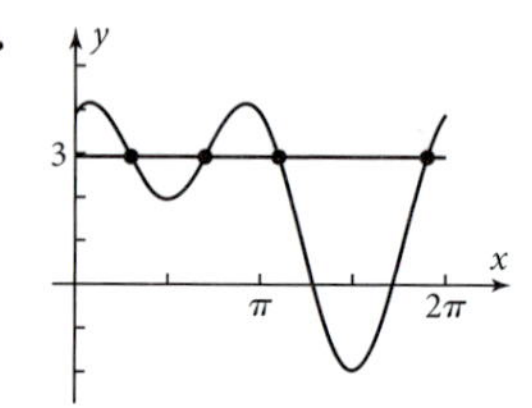

$x = \dfrac{3\pi}{10}, \dfrac{7\pi}{10}, \dfrac{11\pi}{10}, \dfrac{19\pi}{10}$

35b. $4\cos^2 x + 2\sin x = 3 \Rightarrow \sin x = \dfrac{1 \pm \sqrt{5}}{4}$

$\Rightarrow x = \arcsin\dfrac{1 + \sqrt{5}}{4}$ or $\arcsin\dfrac{1 - \sqrt{5}}{4}$

$x = \dfrac{3\pi}{10}, \dfrac{7\pi}{10}, \dfrac{11\pi}{10}, \dfrac{19\pi}{10}$

37a. $y = 500\tan\theta$

37b. $\theta = 5t$; $y = 500\tan 5t$

37c. $t = 10.0388...$ s, $46.0388...$ s, $82.0388...$ s, $118.0388...$ s

39a. $x = 0, x \approx 1.3, 2.4$

39b. $x = 0, 1.2901..., 2.3730...$

39c. x appears both algebraically (as x) and transcendentally (in the argument of tangent).

41a. Graph should match Figure 7-4o.

41b.

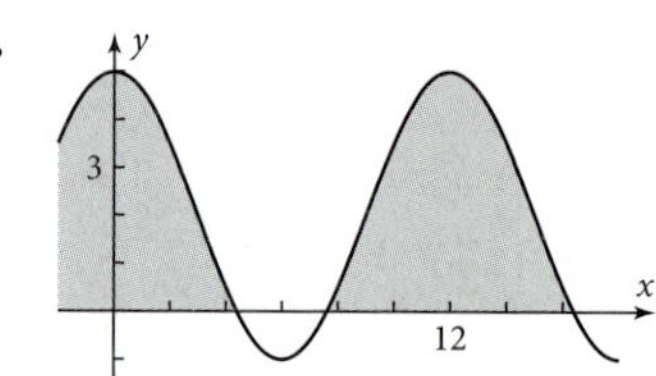

41c. $x = 7.6063..., 16.3936...$

43. The result is $\cos^2 x = \cos^2 x$, which is identically true for all x. The two graphs coincide, and each looks like $y = \cot\left(\frac{1}{2}x + \frac{\pi}{4}\right)$. Evidently, $\frac{1 - \sin x}{\cos x} = \frac{\cos x}{1 + \sin x}$ $= \cot\left(\frac{1}{2}x + \frac{\pi}{4}\right)$ is an identity.

Problem Set 7-5

1a.

t	x	y
−2.0	−5	−5
−1.5	−3.5	−4
−1.0	−2	−3
−0.5	−0.5	−2
0	1	−1
0.5	2.5	0
1.0	4	1
1.5	5.5	2
2.0	7	3

1b.

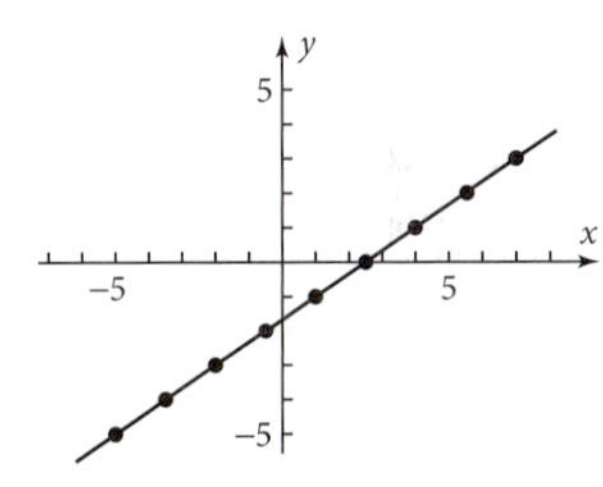

1c. The graphs match.

3a.

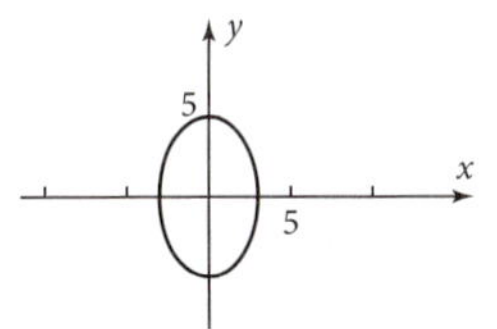

3b. $\left(\frac{x}{3}\right)^2 + \left(\frac{y}{5}\right)^2 = 1$

3c. The x- and y-radii are different, so the graph is an ellipse.

5a.

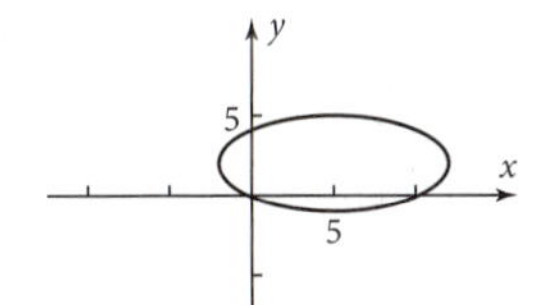

5b. $\left(\frac{x-5}{7}\right)^2 + \left(\frac{y-2}{3}\right)^2 = 1$

5c. The x- and y-radii are different, so the graph is an ellipse.

7a. $x = 6 + 5\cos t,\ y = 9 + \sin t;\ 0° \le t \le 360°$

7b. The graph matches the figure.

9a. $x = 1 + 0.4\cos 0.5t,\ y = 4 + 2\sin 0.5t;$
$x = 1 - 0.4\cos 0.5t,\ y = 4 + 2\sin 0.5t;$
$x = 14 + 0.4\cos t,\ y = 4 + 2\sin t;$
$180° \le t \le 540°$

9b. The graph matches the figure.

11a. $x = 8 + 5\cos 0.5t,\ y = 2 + \sin 0.5t;$
$x = 8 + 5\cos 0.5t,\ y = 2 - \sin 0.5t;$
$x = 8 + 3\cos t,\ y = 9 + 0.6\sin t;$
$360° \le t \le 720°$

11b. The graph matches the figure.

13a. $x = 5 + 0.8\cos(t + 180°);$
$y = 6 + 4\sin(t + 180°);$
$x = 5 + 0.8\cos(t);\ y = 6 + 4\sin(t);$
$x = 5 + 4\cos t,\ y = 6 + 4\sin t;$
$-90° \le t \le 90°$

Note the method to show a different half of each curve.

13b. The graph matches the figure.

15a. (60 m, 75.9 m)

15b. $t = 5$ s; $y(5) = 77.5$ m

15c. $t_1 = 0.8355...$ s, $t_2 = 7.3277...$ s
$x(t_1) = 16.7103...$ m
$x(t_2) = 146.5549...$ m

15d. $x = 160$ m $\Leftrightarrow t = 8$ s $\Rightarrow y = 6.4$ m; the ball will go over the fence.

15e. $y = 2x - \frac{4.9}{400}x^2$

17. $-2\pi \le t \le 2\pi$

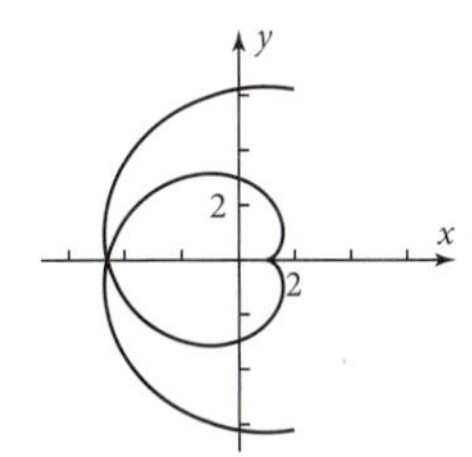

19. $-2\pi \le t \le 2\pi$

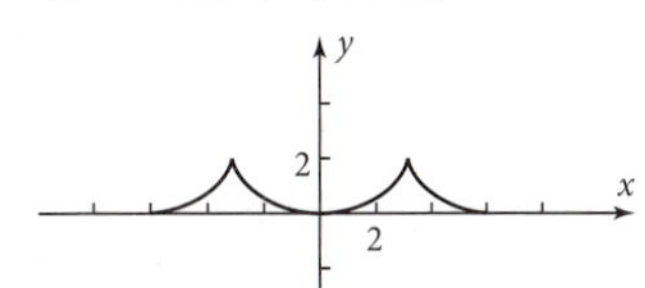

21a. The graph should resemble Figure 7-5f.

21b. Answers will vary. When the graph is plotted slowly, we can see that the x_2-values proceed along the x-axis as the x_1-values proceed along the circle, and that the y_1- and y_2-values are the same.

21c. Change the second graph to
$x_2(t) = \cos t,\ y_2(t) = t.$

Problem Set 7-6

1. Graphs should match the darker portion of the corresponding graphs in Figure 7-6d.

3.

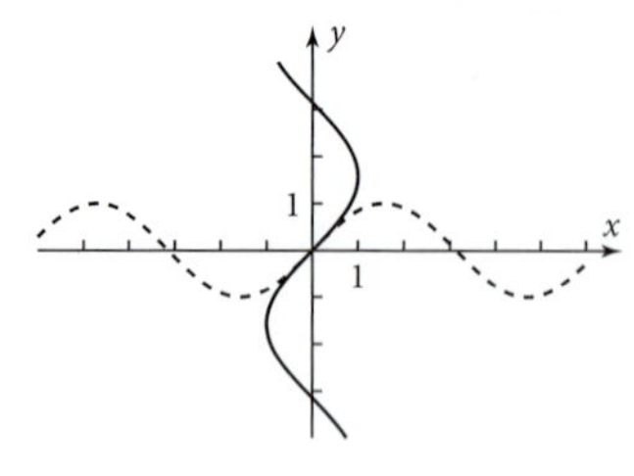

You can make a table of values and show that the (x, y) pairs of one graph are the same as the (y, x) values of the other.

5. $\frac{3}{4}$

7. $\frac{5}{13}$

9. $\frac{15}{17}$

11. $\frac{3}{2}$

13. Undefined

15. 3 is not in the domain of $\cos^{-1}x$.

17. $\cos(\sin^{-1} x) = \sqrt{1 - x^2}$, $-1 \le x \le 1$.
The graphs match.

19. $\sin(\tan^{-1} x) = \dfrac{x}{\sqrt{x^2 + 1}}$, all real x.
The graphs match.

21. $\sin(\sin^{-1} x) = x$, $-1 \le x \le 1$.
The graphs match.

23a. $y = f(x) \Leftrightarrow x = f^{-1}(y)$, so $x = f^{-1}(y) = f^{-1}(f(x))$

23b. $y = f^{-1}(x) \Leftrightarrow x = f(y)$, so $x = f(y) = f(f^{-1}(x))$

25a. $y = 100 + 150 \sin \frac{\pi}{700}(x - 162.5956...)$.

25b. Tunnel: about 1025.2 m; bridge: about 374.8 m

25c. Tunnel: about 950.7 m; bridge: about 449.3 m

Problem Set 7-7

R1a. u and v are the legs of a right triangle with hypotenuse 1.

R1b. $\cos\theta = \dfrac{\text{horizontal coordinate}}{\text{radius}} = \dfrac{u}{1} = u$;
$\sin\theta = \dfrac{\text{vertical coordinate}}{\text{radius}} = \dfrac{v}{1} = v$

R1c. $u^2 + v^2 = 1$ and $u = \cos\theta$ and
$v = \sin\theta \Rightarrow (\cos\theta)^2 + (\sin\theta)^2 = 1$

R1d. For $\theta = 30°$, $\sin\theta = \dfrac{1}{2}$ and
$\cos\theta = \dfrac{\sqrt{3}}{2} \Rightarrow \sin^2\theta + \cos^2\theta = \dfrac{1}{4} + \dfrac{3}{4} = 1$

R1e.

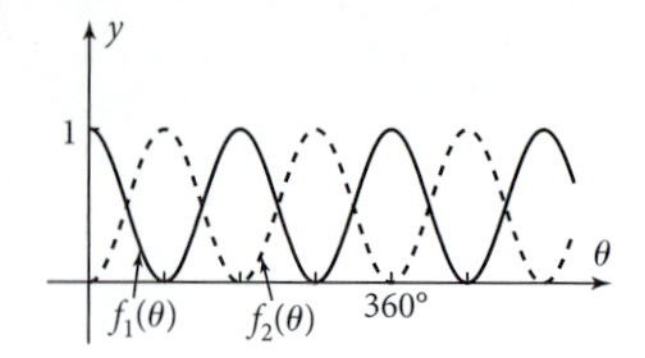

The graphs are symmetric across the line $y = \frac{1}{2}$, where one graph is above the line by the same amount as the other is below it.

R3a. $\tan A \sin A + \cos A = \dfrac{\sin A}{\cos A}\sin A + \dfrac{\cos A}{\cos A}\cos A$
$= \dfrac{\sin^2 A + \cos^2 A}{\cos A} = \dfrac{1}{\cos A} = \sec A$ for $A \ne \dfrac{\pi}{2} + \pi n$

R3b. $(\cos B + \sin B)^2$
$= \cos^2 B + 2\cos B \sin B + \sin^2 B$
$= (\cos^2 B + \sin^2 B) + 2\cos B \sin B$
$= 1 + 2\cos B \sin B$ for all real B

R3c. $\dfrac{1}{1 + \sin C} + \dfrac{1}{1 - \sin C}$
$= \dfrac{1 - \sin C}{(1 + \sin C)(1 - \sin C)} + \dfrac{1 + \sin C}{(1 - \sin C)(1 + \sin C)}$
$= \dfrac{1 - \sin C}{1 - \sin^2 C} + \dfrac{1 + \sin C}{1 - \sin^2 C} = \dfrac{1 - \sin C}{\cos^2 C} + \dfrac{1 + \sin C}{\cos^2 C}$
$= \dfrac{2}{\cos^2 C} = 2\sec^2 C$ for $C \ne \dfrac{n}{2} + \pi n$

R3d. $\csc D(\csc D - \sin D)$
$= \csc^2 D - \csc D \sin D = \csc^2 D - 1$
$= \cot^2 D$ for $D \ne n\pi$

R3e. $(3\cos E + 5\sin E)^2 + (5\cos E - 3\sin E)^2$
$= 9\cos^2 E + 30\cos E \sin E + 25\sin^2 E + 25\cos^2 E$
$- 30\cos E \sin E + 9\sin^2 E = 9\cos^2 E + 9\sin^2 E$
$+ 25\sin^2 E + 25\cos^2 E = 9 + 25 = 34$

R3f.

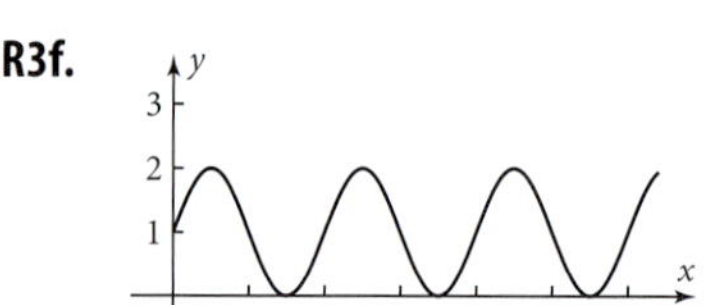

R3g.

E	$(3\cos E + 5\sin E)^2 + (5\cos E - 3\sin E)^2$
0	34
1	34
2	34
3	34
4	34
5	34

R5a.

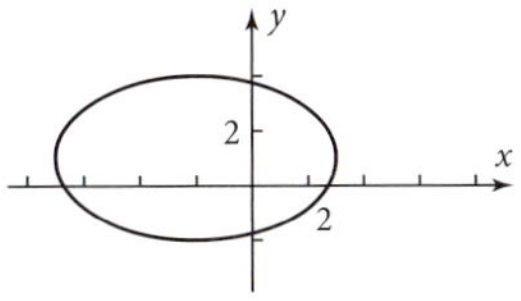

R5b. $\left(\frac{x+2}{5}\right)^2 + \left(\frac{y-1}{3}\right)^2 = 1$

R5c. The equation in part b is the formula for an ellipse centered at $(-2, 1)$ with x-radius 5 and y-radius 3.

R5d. $x_1(t) = 7 + 5\cos t$, $y = 2 + 0.8\sin t$, $180° \le t \le 360°$

Dashed portion:

$x_2(t) = 7 + 5\cos(t - 180°)$,

$y_2(t) = 2 + 0.8\sin(t - 180°)$,

$180° \le t \le 360°$

C1a. $x = 20\cos\frac{2\pi}{3}t$, $y = 20\sin\frac{2\pi}{3}t$

C1b. $x = -10$ cm, $y = -17.3205...$ cm

C1c. $t = 0.25$ s, 1.25 s, 3.25 s; $x = 17.3205...$ cm, $-17.3205...$ cm, $17.3205...$ cm

C1d. Answers will vary.

C3a. Period $= \pi$; Amplitude $= \frac{1}{2}$; Sinusoidal axis: $y = \frac{1}{2}$; Phase displacement $= 0$

C3b. $y = \frac{1}{2} + \frac{1}{2}\cos 2x$

C3c.

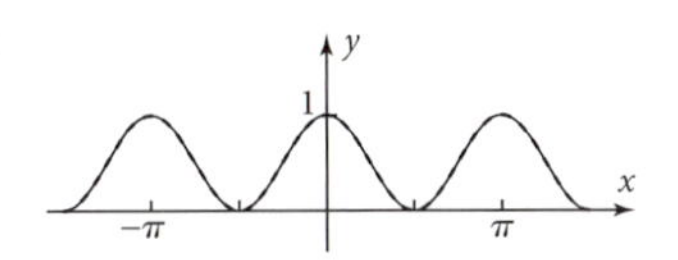

Some numerical confirmation:

x	$\cos^2 x$	$\frac{1}{2} + \frac{1}{2}\cos 2x$
0	$(1)^2 = 1$	$\frac{1}{2} + \frac{1}{2}(1) = 1$
$\frac{\pi}{3}$	$\left(\frac{1}{2}\right)^2 = \frac{1}{4}$	$\frac{1}{2} + \frac{1}{2}\left(-\frac{1}{2}\right) = \frac{1}{4}$
$\frac{\pi}{2}$	$(0)^2 = 0$	$\frac{1}{2} + \frac{1}{2}(-1) = 0$
π	$(-1)^2 = 1$	$\frac{1}{2} + \frac{1}{2}(1) = 1$

T1. $\cos^2 x + \sin^2 x = 1$

T3. $\cot x = \frac{1}{\tan x}$

T5. $\theta = 30° + 360°n$ or $150° + 360°n$

T7. Both $y = \cos^{-1} x$ and $y = \sin^{-1} x$ must be functions, centrally located near the origin, and continuous on their domains.

T9. $(1 + \sin A)(1 - \sin A) = 1 - \sin^2 A = \cos^2 A$ for all real A

T11. $\frac{\sin C}{1 + \cos C} \cdot \frac{1 - \cos C}{1 - \cos C}$

$= \frac{\sin C(1 - \cos C)}{(1 + \cos C)(1 - \cos C)} = \frac{\sin C(1 - \cos C)}{1 - \cos^2 C}$

$= \frac{\sin C(1 - \cos C)}{\sin^2 C} = \frac{1 - \cos C}{\sin C}$

T13. $\theta = 53.1301...°$

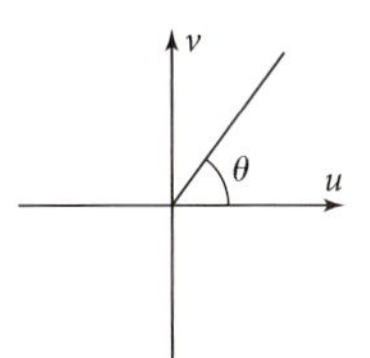

T15. $\theta = \pm 53.1301...° + 360°n$

T17. $\theta = 54.7448...° + 180°n$

T19. $x = \tan t$, $y = t$, $-7 \le t \le 7$ (or whatever are the y-limits of your graph)

Chapter 8

Problem Set 8-1

1. The graphs match.
3. Amplitude $= 5$; Phase displacement $= 53.1301...°$; Answers are reasonably close to estimates.
5. $f_3(\theta) = f_4(\theta)$ for all θ
7. Let $\theta = \pi$ and $D = \frac{\pi}{2}$. Then $\cos(\theta - D) = \cos\frac{\pi}{2} = 0$, but $\cos\theta - \cos D = \cos\pi - \cos\frac{\pi}{2} = -1$.

Problem Set 8-2

1. $y = 13\cos(\theta - 22.6198...°)$
3. $y = 25\cos(\theta - 106.2602...°)$
5. $y = \sqrt{185}\cos(\theta - 233.9726...°)$
7. $y = 6\sqrt{2}\cos(\theta - 315°)$
9. $y = 2\cos(\theta - 30°)$
11. $y = 5\cos(x - 2.2142...)$
13.

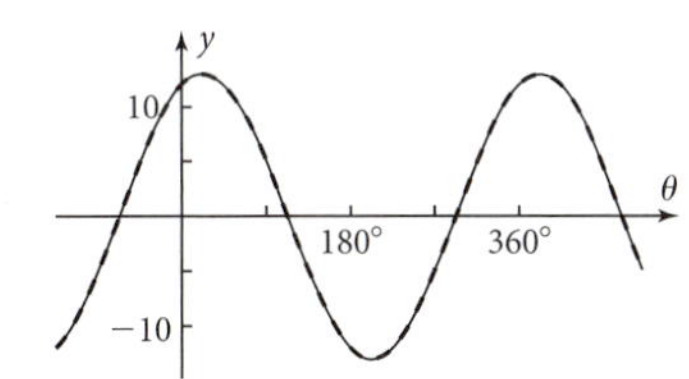

15. $y = \sqrt{2}\cos\left(3x - \frac{\pi}{4}\right)$
The 3 has no effect on the work.

17. Consider $A = \pi$ and $B = \frac{\pi}{2}$. Then $\cos(A - B) = \cos\frac{\pi}{2} = 0$, but $\cos A - \cos B = \cos\pi - \cos\frac{\pi}{2} = -1$.

$y = 5\sqrt{3}\cos\theta + 5\sin\theta$

21. $y = -2.5\sqrt{3}\cos 3\theta + 2.5\sin 3\theta$

23. $\theta = 124.0518...^\circ$ or $344.8727...^\circ$

25. $x = 4.4460...$ or $2.5547...$

27. $\cos 2\theta = \cos(5\theta - 3\theta)$
$= \cos 5\theta \cos 3\theta + \sin 5\theta \sin 3\theta$;
$\theta = 36.2711...^\circ, 143.7288...^\circ, 216.2711...^\circ, 323.7288...^\circ$

29a. $\cos 70^\circ = 0.3420... = \sin 20^\circ$

29b. $\cos(90^\circ - \theta) = \cos 90^\circ \cos\theta + \sin 90^\circ \sin\theta$
$= 0 \cdot \cos\theta + 1 \cdot \sin\theta = \sin\theta$

29c. *Co-* means "complementary."

31. See the derivation on pages 391–392.

33. Journal entries will vary.

Problem Set 8-3

1. Let $A = B = 90^\circ$. Then $\sin(A + B) = \sin 180^\circ$
$= 0 \neq 2 = \sin 90^\circ + \sin 90^\circ = \sin A + \sin B$.

3. $\tan(60^\circ - 30^\circ) = \tan 30^\circ = \dfrac{1}{\sqrt{3}}$

and $\dfrac{\tan 60^\circ - \tan 30^\circ}{1 + \tan 60^\circ \tan 30^\circ} = \dfrac{\sqrt{3} - \frac{1}{\sqrt{3}}}{1 + \sqrt{3} \cdot \frac{1}{\sqrt{3}}} = \dfrac{1}{\sqrt{3}}$

5. Tables will vary but should confirm that $\cos(-x) = \cos x$.

7. The graphs are the same.

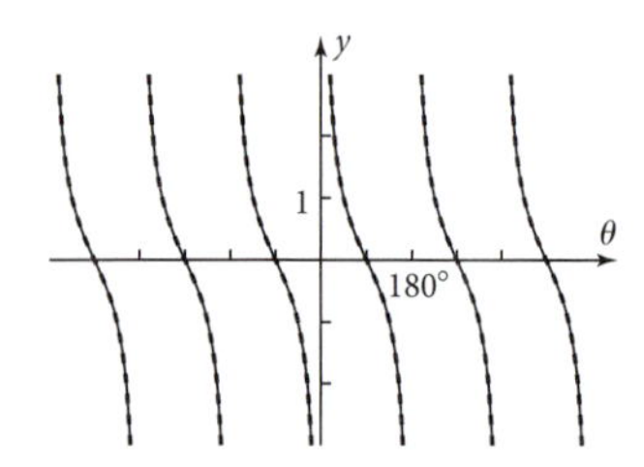

9a. The v-coordinates are opposite for θ and $-\theta$.

9b. The u-coordinates are the same for θ and $-\theta$.

9c. The slopes of the rays for θ and $-\theta$ are opposites.

9d. For any function $f(x)$, if $f(-x) = f(x)$, then $\frac{1}{f(-x)} = \frac{1}{f(x)}$ when defined. If $f(-x) = -f(x)$, then $\frac{1}{f(-x)} = \frac{-1}{f(x)}$ when defined.

11. $\cos(\theta - 90^\circ) = \cos\theta \cos 90^\circ + \sin\theta \sin 90^\circ$
$= \cos\theta \cdot 0 + \sin\theta \cdot 1 = \sin\theta$

13. $\sin\left(x - \frac{\pi}{2}\right) = \sin x \cos\frac{\pi}{2} - \cos x \sin\frac{\pi}{2}$
$= \sin x \cdot 0 - \cos x \cdot 1 = -\cos x$

15. $\sin(\theta + 60^\circ) - \cos(\theta + 30^\circ)$
$= (\sin\theta \cos 60^\circ + \cos\theta \sin 60^\circ)$
$- (\cos\theta \cos 30^\circ - \sin\theta \sin 30^\circ)$
$= \frac{1}{2}\sin\theta + \frac{\sqrt{3}}{2}\cos\theta - \frac{\sqrt{3}}{2}\cos\theta + \frac{1}{2}\sin\theta = \sin\theta$

17. $\sqrt{2}\cos\left(x - \frac{\pi}{4}\right) = \sqrt{2}\cos x \cos\frac{\pi}{4} + \sqrt{2}\sin x \sin\frac{\pi}{4}$
$= \sqrt{2}\cos x \cdot \frac{1}{\sqrt{2}} + \sqrt{2}\sin x \cdot \frac{1}{\sqrt{2}} = \cos x + \sin x$

19. $\sin 3x \cos 4x + \cos 3x \sin 4x = \sin(3x + 4x)$
$= \sin 7x$

21a. $\cos(x + 0.6) = 0.9 \Rightarrow x = 2\pi n \pm 0.4510... - 0.6$

21b. $x = 5.2321...$ or $6.1342...$

23a. $\sin 2\theta = 0.5\sqrt{2} \Rightarrow 2\theta = 45^\circ + 360^\circ n$ or $135^\circ + 360^\circ n$

23b. $\theta = 22.5^\circ, 67.5^\circ, 202.5^\circ$, or 247.5°

25a. $\tan x = \sqrt{3} \Rightarrow x = \frac{\pi}{3} + \pi n$

25b. $x = \frac{\pi}{3}$ or $\frac{4\pi}{3}$

27. $\dfrac{297}{425}$

29. $\dfrac{304}{425}$

31. $\dfrac{304}{297}$

33. $\sin 15^\circ = \sin(45^\circ - 30^\circ) = \sin 45^\circ \cos 30^\circ$
$- \cos 45^\circ \sin 30^\circ = \frac{\sqrt{2}}{2} \cdot \frac{\sqrt{3}}{2} - \frac{\sqrt{2}}{2} \cdot \frac{1}{2}$
$= \dfrac{\sqrt{6} - \sqrt{2}}{4} = 0.2588...$

35. $\sin 75^\circ = \cos(90^\circ - 75^\circ) = \cos 15^\circ = \dfrac{\sqrt{6} + \sqrt{2}}{4}$
$= 0.9659...$

37. $\tan 15^\circ = \dfrac{\sin 15^\circ}{\cos 15^\circ} = \dfrac{\left(\frac{\sqrt{6} - \sqrt{2}}{4}\right)}{\left(\frac{\sqrt{6} + \sqrt{2}}{4}\right)} = \dfrac{\sqrt{6} - \sqrt{2}}{\sqrt{6} + \sqrt{2}}$
$= \dfrac{8 - 4\sqrt{3}}{3} = 2 - \sqrt{3}$

39a. $\theta = 90^\circ - \sin^{-1} x \Rightarrow \sin^{-1} x = 90^\circ - \theta$
$\Rightarrow x = \sin(90^\circ - \theta) = \cos\theta$

39b. $-90^\circ \leq \sin^{-1} x \leq 90^\circ \Rightarrow -90^\circ \leq 90^\circ - \theta \leq 90^\circ$
$\Rightarrow -180^\circ \leq -\theta \leq 0^\circ \Rightarrow 0^\circ \leq \theta \leq 180^\circ$

39c. θ is the unique value of x such that $0 \leq \theta \leq 180^\circ$, which is the definition of $\cos^{-1} x$.

41. $\cos(A + B + C) = \cos A \cos B \cos C$
$- \cos A \sin B \sin C - \sin A \cos B \sin C$
$- \sin A \sin B \cos C$

Problem Set 8-4

1a. $f_1(\theta)$ is the tall single arch and trough, $f_2(\theta)$ the narrow wiggly curve.

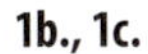

1b., 1c.

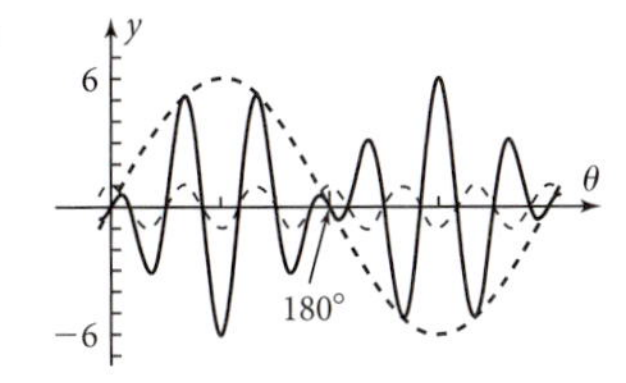

1d. Sinusoid with variable amplitude $6 \sin \theta$

3. $y = 3 \cos \theta + 2 \sin 13\theta$

5. $y = 5 \sin \theta \cos 11\theta$

7. $y = 2 \cos \frac{\pi}{3}x + 4 \sin 4\pi x$

9. $y = 4 \sin \frac{\pi}{6}x \sin 3\pi x$

11. $y = 2 \sin 2x \sin 28x$

13a. $y = 3 \cos 120\pi x + \cos 800\pi x$

13b. $\frac{1}{60}$ s; $\frac{1}{400}$ s

13c. 60 cycles/s = 60 Hz; 400 cycles/s = 400 Hz

13d. Yes

15. Journal entries will vary.

Problem Set 8-5

1. $\sin 65° + \sin 17°$

3. $\cos 102° + \cos 4°$

5. $\sin 7.9 + \sin 0.3$

7. $\cos(3x - 7.2) - \cos(3x + 7.2)$

9. $2 \cos 29° \cos 17°$

11. $2 \sin 4 \cos 2$

13. $2 \sin 3.4 \sin 1$

15. $-2 \cos 5.5x \sin 2.5x$

17. $y = 2 \cos \theta \cos 9\theta = \cos 10\theta + \cos 8\theta$

19. $y = \cos x + \cos 15x = 2 \cos 8x \cos 7x$

21. $2 \cos 2x \sin x = 0 \Rightarrow \sin x = 0$ or $\cos 2x = 0 \Rightarrow x = 0, \pi, 2\pi, \frac{\pi}{4}, \frac{3\pi}{4}, \frac{5\pi}{4}, \frac{7\pi}{4}$

23. $2 \cos 4\theta \cos \theta = 0 \Rightarrow \cos 4\theta = 0$ or $\cos \theta = 0 \Rightarrow \theta = 22.5°, 67.5°, 112.5°, 157.5°, 202.5°, 247.5°, 292.5°, 337.5°, 90°, 270°$

25. $\cos x - \cos 5x = 2 \sin 3x \sin 2x = 2(\sin 3x)(2 \sin x \cos x) = 4 \sin 3x \sin x \cos x$

27. $\cos x + \cos 2x + \cos 3x = \cos 2x + (\cos x + \cos 3x) = \cos 2x + (2 \cos 2x \cos x) = (\cos 2x)(1 + 2 \cos x)$

29. $\cos(x + y) \cos(x - y)$
$= \frac{1}{2} \cos((x + y) + (x - y)) + \frac{1}{2} \cos((x + y) - (x - y))$
$= \frac{1}{2} \cos 2x + \frac{1}{2} \cos 2y = \frac{1}{2}(\cos 2x + 1) + \frac{1}{2}(\cos 2y - 1)$
$= \frac{1}{2}[\cos(x + x) + \cos(x - x)] + \frac{1}{2}[\cos(y + y)$
$- \cos(y - y)] = \frac{1}{2}(2 \cos x \cos x) + \frac{1}{2}(-2 \sin y \sin y)$
$= \cos^2 x - \sin^2 y$

31a. $f(t) = \cos 442\pi t + \cos 438\pi t$

31b. $f(t) = 2 \cos 440\pi t \cos 2\pi t$

31c. The smaller-period sinusoid has frequency 220 cycles/s, and the larger-period sinusoid has frequency 1 cycle/s, so the combined note will sound like A220 getting louder and softer twice per second.

33a. $k = 0.5$: periodic function with a variable amplitude; $k = 5$: periodic function with a variable sinusoidal axis; $k - 6$: sinusoid (with amplitude $\frac{1}{2}$ and period $\frac{\pi}{6}$)

33b. $k = 0.5$: periodic function with a variable sinusoidal axis;
$k = 5$: periodic function with a variable amplitude;
$k = 6$: sinusoid (with amplitude $\sqrt{2}$ and period $\frac{\pi}{3}$)

33c. Answers will vary.

35. $\sin^2 x = \sin x \sin x = -\frac{1}{2}[\cos(x + x) - \cos(x - x)]$
$= -\frac{1}{2}\cos 2x + \frac{1}{2}\cos 0 = -\frac{1}{2}\cos 2x + \frac{1}{2} \cdot 1 = \frac{1 - \cos 2x}{2}$;
$\cos^2 x = \cos x \cos x = \frac{1}{2}[\cos (x + x) + \cos (x - x)]$
$= \frac{1}{2}\cos 2x + \frac{1}{2}\cos 0 = \frac{1}{2}\cos 2x + \frac{1}{2} \cdot 1 = \frac{1 + \cos 2x}{2}$;
the formulas reduce the square of a trig function to the first power of a trig function.

Problem Set 8-6

1. $\cos 2x$ has period π and amplitude 1, but $2 \cos x$ has period 2π and amplitude 2.

3. Tables will vary but should confirm that $\sin 2x = 2 \sin x \cos x$.

5. Tables will vary but should confirm that $\cos 2x = 2 \cos^2 x - 1$.

7. Tables will vary but should confirm that $\sin \frac{1}{2}A = \sqrt{0.5(1 - \cos A)}$.

9. Tables will vary but should confirm that $\cos \frac{1}{2}A = -\sqrt{0.5(1 + \cos A)}$

11a.

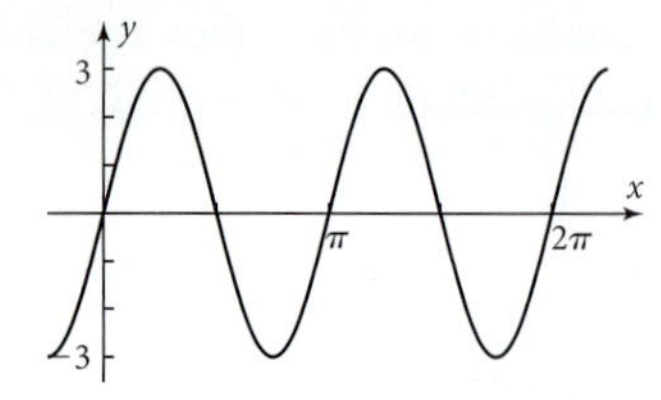

11b. $y = 3 \sin 2x$

11c. By the double argument property, $y = 6 \sin x \cos x = 3 \sin 2x$.

13a.

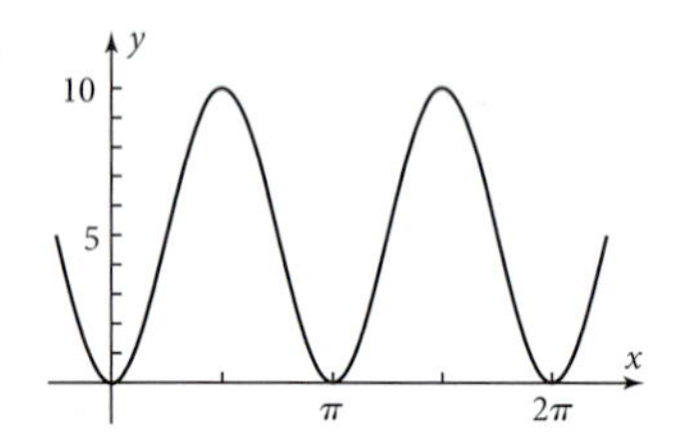

13b. $y = 5 - 5 \cos 2x$

13c. By the double argument property,
$y = 10 \sin^2$
$x = 5 - 5 \cos 2x$.

15a.

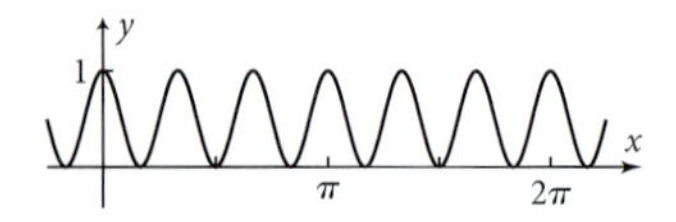

15b. $y = \frac{1}{2} + \frac{1}{2}\cos 6x$

15c. By the double argument property, $y = \cos^2 3x = \frac{1}{2} + \frac{1}{2}\cos 6x$.

17. Not a sinusoid: The graph is not symmetrical across any horizontal central axis.

19a. The dotted curve represents $f_1(\theta)$ and the solid curve represents $f_2(\theta)$.

19b. The + should be used on $[720°n - 180°, 720°n + 180°]$ and the − on $[720°n + 180°, 720°n + 540°]$.

19c. $|f_1(\theta)| = f_2(\theta)$

19d. By definition, $\sqrt{x^2}$ means the *positive* value, so

$$\sqrt{x^2} = \begin{cases} x, & x \geq 0 \\ -x, & x < 0 \end{cases} = |x|.$$

The derivation of the half argument properties takes the square root of the squares of sines and cosines.

21a. $\sin 2A = \frac{24}{25}$; $\cos \frac{1}{2}A = \sqrt{0.8}$

21b. $A = 53.1301...°$; $\sin 2A = \sin 106.2602...° = 0.96 = \frac{24}{25}$; $\cos \frac{1}{2}A = \cos 26.5650...° = 0.8944... = \sqrt{0.8}$

23a. $\sin 2A = \frac{24}{25}$; $\cos \frac{1}{2}A = -\sqrt{0.2}$

23b. $A = 233.1301...°$; $\sin 2A = \sin 466.2602...° = 0.96 = \frac{24}{25}$; $\cos \frac{1}{2}A = \cos 116.5650...° = -0.4472... = -\sqrt{0.2}$

25a. $\sin 2A = -\frac{24}{25}$; $\cos \frac{1}{2}A = \sqrt{0.8}$

25b. $A = 666.8698...°$; $\sin 2A = \sin 1333.7397...° = -0.96 = -\frac{24}{25}$; $\cos \frac{1}{2}A = \cos 333.4349...° = 0.8944... = \sqrt{0.8}$

27. $\sin 2x = \sin(x + x) = \sin x \cos x + \cos x \sin x = 2 \sin x \cos x$

29. $\sin x \cos x = \frac{1}{2} \sin 2x$

31a. Graphs are the same for both expressions.
$A \neq 45° + 90°n, A \neq 90° + 180°n$

31b. $\tan 2A = \tan(A + A) = \dfrac{\tan A + \tan A}{1 - \tan A \tan A} = \dfrac{2 \tan A}{1 - \tan^2 A}$

31c. $\tan 2A = \dfrac{\sin 2A}{\cos 2A} = \dfrac{2 \sin A \cos A}{\cos^2 A - \sin^2 A} \cdot \dfrac{\left(\frac{1}{\cos^2 A}\right)}{\left(\frac{1}{\cos^2 A}\right)} = \dfrac{\left(\frac{2 \sin A}{\cos A}\right)}{1 - \frac{\sin^2 A}{\cos^2 A}} = \dfrac{2 \tan A}{1 - \tan^2 A}$

33. $x = \frac{\pi}{6}, \frac{\pi}{3}, \frac{7\pi}{6}, \frac{4\pi}{3}$

35. $\theta = 45°, 135°, 225°, 315°$

37. $x = \frac{\pi}{3}, \frac{5\pi}{3}, \frac{7\pi}{3}, \frac{11\pi}{3}$

39. $\sin 2x = 2 \sin x \cos x \cdot \dfrac{\sec^2 x}{\sec^2 x} = \dfrac{2 \sin x \sec x}{\sec^2 x} = \dfrac{2\frac{\sin x}{\cos x}}{\sec^2 x} = \dfrac{2 \tan x}{1 + \tan^2 x}$

41. $\sin 2\phi = 2 \sin \phi \cos \phi \cdot \dfrac{\sin \phi}{\sin \phi} = 2 \sin^2 \phi \dfrac{\cos \phi}{\sin \phi} = 2 \cot \phi \sin^2 \phi$

43. $\sin^2 5\theta = \frac{1}{2}[1 - \cos(2 \cdot 5\theta)] = \frac{1}{2}(1 - \cos 10\theta)$

Problem Set 8-7

R1a. Amplitude = 13;
Phase displacement = $\cos^{-1}\frac{5}{13} = 67.3801...°$;
$y = 13 \cos(\theta - 67.3801...°)$

R1b. The graphs match.

R3a. $\sin(-x) = -\sin x$, $\cos(-x) = \cos x$, $\tan(-x) = -\tan x$

R3b. $\sin(x + y) = \sin x \cos y + \cos x \sin y$

R3c. $\cos(x + y) = \cos x \cos y - \sin x \sin y$

R3d. $\tan(x - y) = \dfrac{\tan x - \tan y}{1 + \tan x \tan y}$

R3e. $\cos(90° - \theta) = \sin \theta$

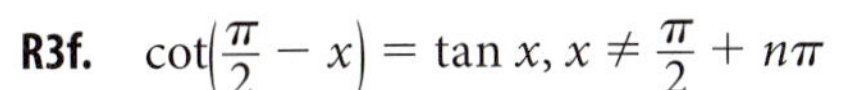

R3f. $\cot\left(\frac{\pi}{2} - x\right) = \tan x, x \neq \frac{\pi}{2} + n\pi$

R3g. $\csc\left(\frac{\pi}{2} - x\right) = \sec x, x \neq \frac{\pi}{2} + n\pi$

R3h.

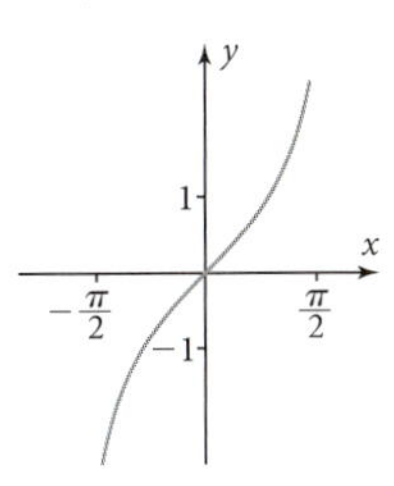

The graph is symmetrical about the origin; $x \neq \frac{\pi}{2} + n\pi$

R3i. $x = 0.5796..., 2.5619..., 3.7212..., 5.7035...$; These are the x-coordinates of the intersection of the two graphs.

R5a. $\frac{1}{2}\cos 41° + \frac{1}{2}\cos 15°$

R5b. $-2\cos\frac{13}{2}\sin\frac{3}{2}$

R5c. $-2\cos 12x + 2\cos 10x$

R5d. $\theta = 0°, 90°, 180°, 270°, 360°$

R5e. $\cos\left(x + \frac{\pi}{3}\right)\cos\left(x - \frac{\pi}{3}\right)$

$= \frac{1}{2}\cos\left(\left(x + \frac{\pi}{3}\right) + \left(x - \frac{\pi}{3}\right)\right) + \frac{1}{2}\cos\left(\left(x + \frac{\pi}{3}\right) - \left(x - \frac{\pi}{3}\right)\right)$

$= \frac{1}{2}\cos 2x + \frac{1}{2}\cos\frac{2\pi}{3}$

$= \frac{1}{2}(2\cos^2 x - 1) + \frac{1}{2}\left(-\frac{1}{2}\right) = \cos^2 x - \frac{3}{4}$

C1a. $\sin 72° = 2\sin 36°\cos 36°$

C1b. $\sin 72° = 4\sin 18°\cos 18°(1 - 2\sin^2 18°)$

C1c. $\cos 18° \Rightarrow 8\sin^3 18° - 4\sin 18° + 1 = 0$

C1d. $\sin 18° = \frac{1}{2}$ or $\frac{-1 + \sqrt{5}}{4}$

C1e. Because $0 < \sin 18° < \sin 30° = 0.5$, $\sin 18° = \frac{-1 + \sqrt{5}}{4}$.

C1f. Answers will vary.

T1. Amplitude $= \sqrt{85}$, period $= 2\pi$

T3. Amplitude $= 1$, period $= 180°$

T5. Harmonic analysis

T7. $\sin 77°$

T9. $\cos x = 1 - 2\sin^2\frac{1}{2}x \Rightarrow \sin^2\frac{1}{2}x = \frac{1}{2}(1 - \cos x)$

$\Rightarrow \sin\frac{1}{2}x = \pm\sqrt{\frac{1}{2}(1 - \cos x)}$

T11. $\theta = 76.7082...°, 209.5519...°, 436.7082...°, 569.5519...°$

T13. $\cos^2\theta = \frac{1}{2} + \frac{1}{2}\cos 2\theta$

T15. $y = 4\cos x\cos 13x$

T17. $\sin(90° - A) = \cos A = \frac{15}{17}$; cofunction property

T19. Answers will vary.

Chapter 9

Problem Set 9-1

1. All measurements seem correct.

3.

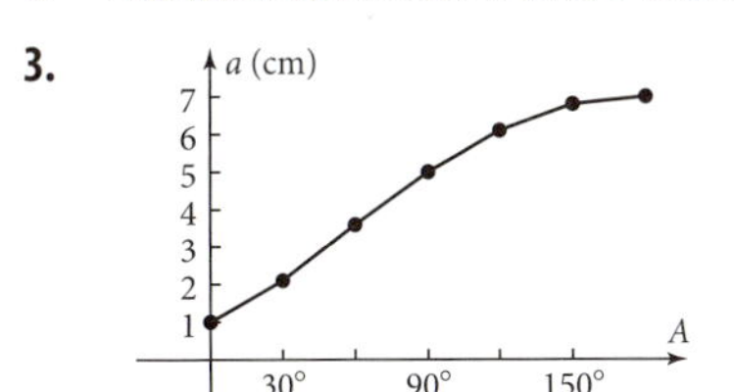

5. The formula is $a^2 = 3^2 + 4^2 - 2 \cdot 3 \cdot 4\cos A$, that is, $a^2 = b^2 + c^2 - 2bc\cos A$.

Problem Set 9-2

1. $r \approx 3.98$ cm

3. $r \approx 4.68$ ft

5. $U \approx 28.96°$

7. $T \approx 134.62°$

9. This is not a possible triangle, because $7 + 5 < 13$.

11. $O = 90°$

13a. $r \approx 4.0$ cm, $p = 4.0$ cm, $m = 5.0$ cm, $R = 51°$

13b. $m = 5.0$ cm, $e = 6.0$ cm, $g = 8.0$ cm, $G \approx 93°$

15. $\cos^{-1}\frac{15^2 + 21^2 - 33^2}{2 \cdot 15 \cdot 21} \approx 132.2°$

17. $X = (4\cos Z, 4\sin Z)$, $Y = (5, 0)$, so z^2

$= (4\cos Z - 5)^2 + (4\sin Z - 0)^2$

$= 4^2\cos^2 Z - 2 \cdot 4 \cdot 5\cos Z + 25 + 4^2\sin^2 Z$

$= 4^2(\sin^2 Z + \cos^2 Z) + 5^2 - 2 \cdot 4 \cdot 5\cos Z$

$= 4^2 + 5^2 - 2 \cdot 4 \cdot 5\cos Z$

Problem Set 9-3

1. ≈ 5.44 ft^2

3. ≈ 6.06 cm^2

5. $s = 26.9814...$ cm^2

7. $s = 4.3906...$ in.^{2}

9a. $5 + 6 < 13$, so the triangle inequality shows that no triangle can have these three sides.

9b. Area $= \sqrt{-504}$

According to Hero's formula, the triangle would have to have an impossible area. So no such triangle exists.

11a. $A = 6\sin\theta$

θ	A
0°	0.0000
15°	1.5529...
30°	3.0000
45°	4.2426...
60°	5.1961...
75°	5.7955...
90°	6.0000
105°	5.7955...
120°	5.1961...
135°	4.2426...
150°	3.0000
165°	1.5529...
180°	0.0000

11c. False. The function increases from 0° to 90°, then decreases from 90° to 180°.

11d. The figure is only a triangle with positive area for $0° < \theta < 180°$, so that is the domain.

13. $A = \frac{1}{2}bc \sin A = \frac{1}{2} \cdot \sqrt{3} \cdot 2 \cdot \frac{1}{2} = \frac{\sqrt{3}}{2}$.

$A = \sqrt{s(s-a)(s-b)(s-c)}$

$= \sqrt{\frac{3+\sqrt{3}}{2}\left(\frac{3+\sqrt{3}}{2} - 1\right)\left(\frac{3+\sqrt{3}}{2} - \sqrt{3}\right)\left(\frac{3+\sqrt{3}}{2} - 2\right)}$

$= \frac{\sqrt{3}}{2}$

Problem Set 9-4

1. $b \approx 5.23$ cm; $c \approx 10.08$ cm

3. $h \approx 249.92$ yd; $s \approx 183.61$ yd

5. $a \approx 9.32$ m; $p \approx 4.91$ m

7. $a \approx 214.74$ ft; $l \approx 215.26$ ft

9a. $x = 679.4530...$ m; $y = 861.1306...$ m

9b. $67,220.70.

9c. $105,421.05 over y and $38,200.35 over x.

11a. $A = 33.1229...°$

11b. $C = 51.3178...°$

11c. $C = 128.6821...°$

11d. This is the complement of 51.3178...° and one of the *general* values of arcsin $\frac{10 \sin A}{7}$.

11e. The principal values of arccos x go from 0° to 180°; a negative argument will give an obtuse angle and a positive argument will give an acute angle, always the actual angle in the triangle. But the principal values of arcsin x go from $-90°$ to 90°; a negative argument will never happen in a triangle problem, but a positive argument will only give an acute angle, whereas the actual angle in the triangle may be the obtuse complement of the acute angle.

13. Answers will vary.

Problem Set 9-5

1. $c \approx 5.32...$ cm or 1.32... cm

3. $c \approx 7.79$ cm

5. No solution.

7. $\approx$5.52 in.

9a. $\approx$38.65 mi.

9b. The other answer is ≈ -12.94 mi. This means 12.94 miles to the west of Ocean City.

9c. 99.29°

11. $S \approx 141.18°$

13. $G \approx 57.75°$

Problem Set 9-6

1. $|\vec{a} + \vec{b}| \approx 14.66$ cm; $\alpha \approx 45.84°$

3. $|\vec{a} + \vec{b}| \approx 11.69$ in.; $\alpha \approx 150.00°$

5a. Lucy's bearing is 150.9453...°

5b. The starting point's bearing from Lucy is 330.9453...°.

5c. 205.9126... m

7a. $|\vec{r}| \approx 329.3$ lb; $\alpha \approx 17.02°$

7b. Abe and Bill neglected the fact that the magnitude of the sum of two vectors does not equal the sum of the magnitudes if the vectors do not point in the same direction.

9. $6.0376...\vec{i} - 5.2484...\vec{j}$

11. $-6.1344...\vec{i} + 14.4519...\vec{j}$

13a. |horizontal component| = 409.5760...; Ground speed is about 410 mi/h.

13b. |vertical component| = 286.7882...; Climb rate $\approx$ 421 ft/s

15a. $-12.8175...\vec{i} + 54.3745...\vec{j}$

15b. $|\vec{r}| = 55.8648...$ units; $\theta = 103.2640...°$

17. $|\vec{r}| = 70$ mi at a bearing of 41.7867...°

19. $|\vec{r}| = 167.8484...$ mi/h at a bearing of 304.1074...°

21. $\vec{r} = 6.8478...\vec{i} + 51.2124...\vec{j}$;

$|\vec{r}| = 51.6682...$ newtons; $\theta = 82.3838...°$

23.

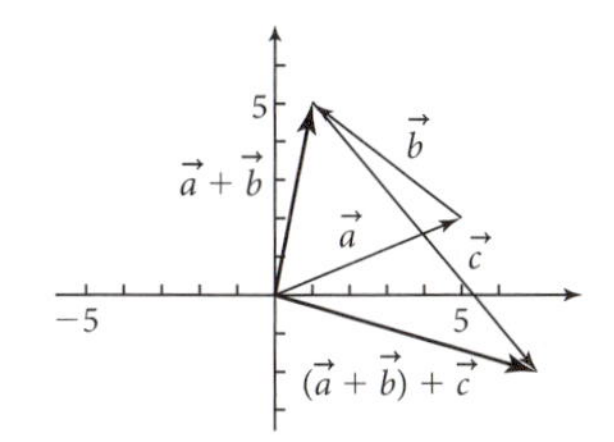

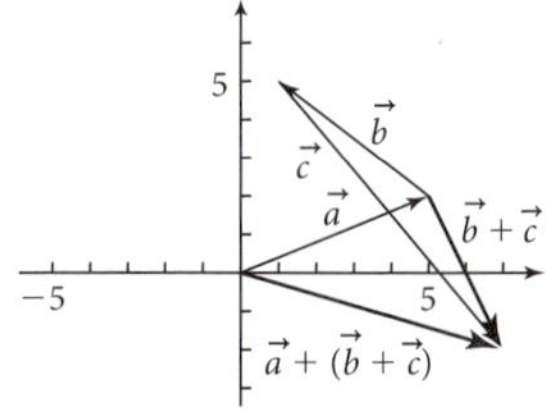

25. If $a\vec{i} + b\vec{j}$ and $c\vec{i} + d\vec{j}$ are any two vectors, then a, b, c, and d are real numbers. So $a + c$ and $b + d$ are also real numbers, because the real numbers are closed under addition. Therefore, the sum $(a + c)\vec{i} + (b + d)\vec{j}$ exists and is a vector, so the set of vectors is closed under addition. The zero vector is necessary so that the sum of any vector $a\vec{i} + b\vec{j}$ and its opposite, $-a\vec{i} + b\vec{j}$, will exist.

27. *Scalar* is from the Latin *scālae,* meaning "ladder."

Problem Set 9-7

1. $CD \approx 445.1$ m

3a. $\approx$8.7 km

3b. $A \approx 607.5$ km^2

5. $\approx$1380.6 yd or 608.4 yd

7a. 0.6326... km

7b. 0.1265... km/s

7c. 53.1210...°

9a. $\approx$463.4 km/h

9b. $\approx$537.0 km/h

11. Let F = the other person's force. $F = 66.0732... \approx 66$ lb. Then the magnitude of the resultant force vector is about 110.8 lb.

13a. |normal component| = 38,974.8025... $\approx$ 39,000 lb, which is not much less than the weight of the truck.

13b. |parallel component| = 8998.0421... $\approx$ 9000 lb, which is surprisingly large!

15a. $x = 5.7735... \approx 5.77$ lb;
|resultant force| = 11.5470... $\approx$ 11.55 lb

15b. $\theta = \tan^{-1}\dfrac{x}{10}$
The graph shows that θ approaches a horizontal asymptote at 90° as x gets larger.

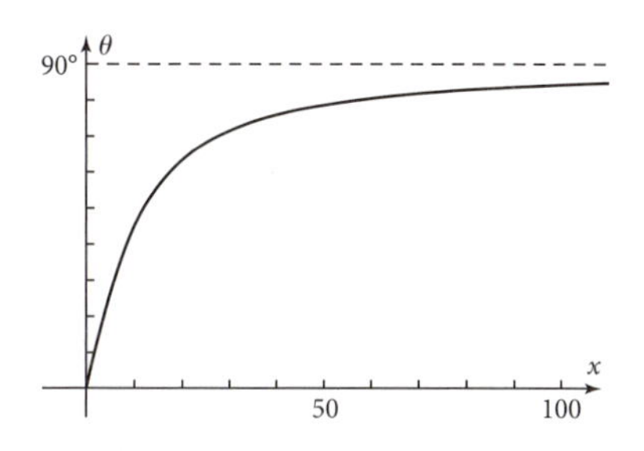

15c. String tension = |resultant force| = $\sqrt{10^2 + x^2}$
The graph shows that the tension approaches x asymptotically as x gets larger.

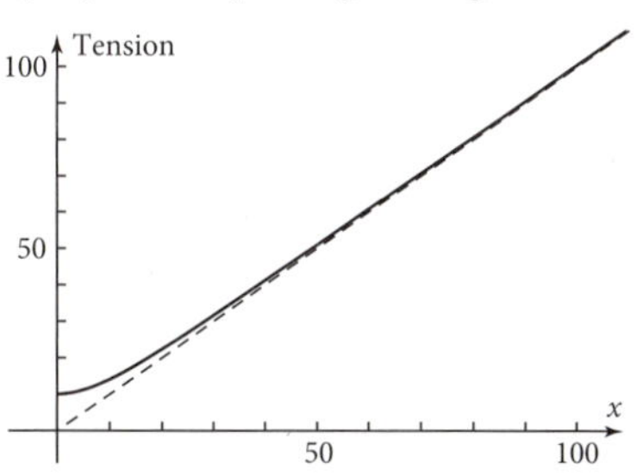

17. $|\vec{t}_1| = 44.2275...$; $|\vec{t}_2| = 54.2531...$,

19. $|\vec{r}| = 290.7331...$ km/h at a bearing of 50.1263...°

21. $\approx$357.5 km

23a. $\approx$133.4°

23b. Area $\approx$ 6838.2 m^2

25a. Answers will vary.

25b. The program should give the expected answer.

25c. Label the 95° angle A, and label the rest of the vertices clockwise as B through F.

$AC = \sqrt{20^2 + 22^2 - 2 \cdot 20 \cdot 22 \cos 114°}$

$= 35.2410...$ m

$\angle ACB = \sin^{-1} \dfrac{20 \sin 114°}{AC} = 31.2287...°$

$\angle ACD = 147° - \angle ACB = 115.7712...°$

$AD = \sqrt{AC^2 + 15^2 - 2 \cdot AC \cdot 15 \cos \angle ACD}$

$= 43.8929...$ m

$\angle ADC = \sin^{-1} \dfrac{AC \sin \angle ACD}{AD} = 46.3050...°$

$\angle ADE = 122° - \angle ADC = 75.6949...°$

$AE = \sqrt{AD^2 + 18^2 - 2 \cdot AD \cdot 18 \cos \angle ADE}$

$= 43.1295...$ m

$\angle AED = \sin^{-1} \dfrac{AD \sin \angle ADE}{AE} = 80.4510...°$

$\angle AEF = 115° - \angle AED = 34.5489...°$

$AF = \sqrt{AE^2 + 17^2 - 2 \cdot AE \cdot 17 \cos \angle AEF}$

$= 30.6817...$ m

25d. For a nonconvex polygon, you might not be able to divide it into triangles that fan out radially from a single vertex.

Problem Set 9-8

R1a. Answers may vary slightly.

θ	Third side (cm)
30°	2.5
60°	4.6
90°	6.4
120°	7.8
150°	8.7

R1b. $5 + 4 = 9$; $5 - 4 = 1$

R1c. $\sqrt{5^2 + 4^2} \approx 6.4$; Yes

R1d.

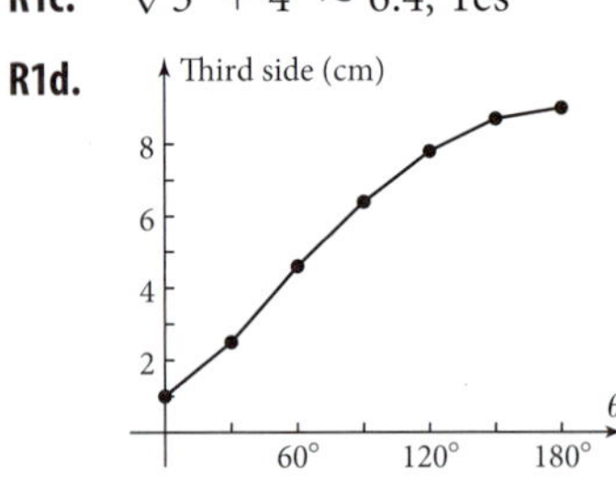

No, the shape is not a sinusoid.

R3a. $A \approx 340.5$ ft²

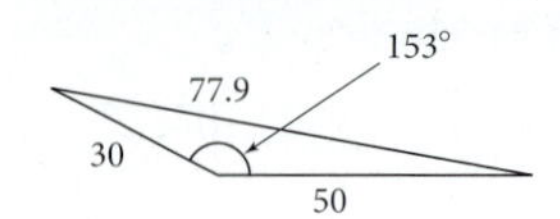

R3b.

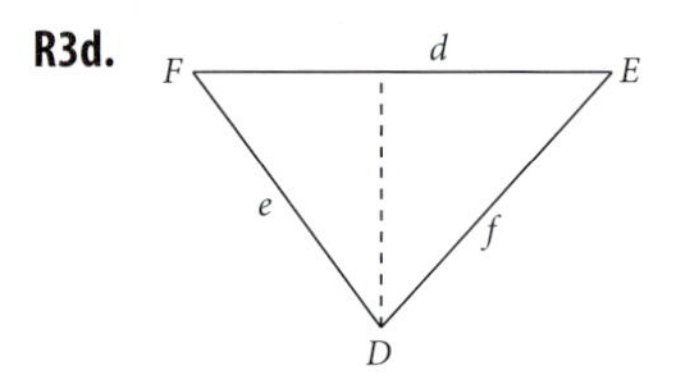

$\theta = 103.1365...°$; $A = 42.8485...$ mi²; $s = 17$;
$A = 42.8485...$ mi²

R3c. $\theta \approx 41.8°$ or $138.2°$

R3d.

altitude $= e \sin F$; $A = \frac{1}{2}bh = \frac{1}{2}de \sin F$

R5a. ≈11.4 cm or 3.4 cm

R5b. $\sin \phi \approx 1.6$, which is not the sine of any angle.

R5c. The 5-cm side must be perpendicular to the third side, making the 8-cm side the hypotenuse of a right triangle. Then $\theta = \sin^{-1}\frac{5}{8} \approx 38.7°$.

R5d. ≈10.5 cm

R7a. ≈137.8 km, so it is out of range.

R7b. 177.1700... km or 325.1113... km

R7c. $(-520 \cos 40°)^2 - 4 \cdot 1 \cdot 57{,}600 = -71{,}722.7663...$, so x is undefined.

R7d. The line from the plane to Tokyo Airport must be perpendicular to the flight path, so $x \approx 22.6°$

R7e. $240^2 = 57{,}600 = (177.1700...)(325.111...)$
The theorem states that if P is a point exterior to circle C, PR cuts C at Q and R, and PS is tangent to C at S, then $PQ \cdot PR = PS^2$.

R7f. Nagoya Airport is closer by about 35.2 km.

R7g. ≈7.6°

R7h. $\sqrt{3000^2 + 400^2} \approx 3026.5$ lb

R7i. The helicopter can tilt so that the thrust vector exactly cancels the wind vector.

C1. Student essay

C3. Student project

C5. $\theta = 37.1847...°$; $\vec{a} \cdot \vec{b} = 29$

The dot product can also be calculated by finding the sum of the products of the $\vec{i}$ coefficients and the $\vec{j}$ coefficients: $\vec{a} \cdot \vec{b} = 3 \cdot 7 + 4 \cdot 2 = 29$. This method is covered in Chapter 12. The dot product is called the scalar product because the answer is a scalar, not a vector.

T1. $d^2 = c^2 + e^2 - 2ce \cos D$

T3. $A = \frac{1}{2}de \sin C$

T5. SAS is shown, but the law of sines works only for ASA, SAA, and SSA.

T7. The range of $\cos^{-1}$ is $0° \le \theta \le 180°$, which includes every possible angle measure for a triangle. But the range of $\sin^{-1}$ is $-90° \le \theta \le 90°$, so the function $\sin^{-1}$ cannot find obtuse angles.

T9.

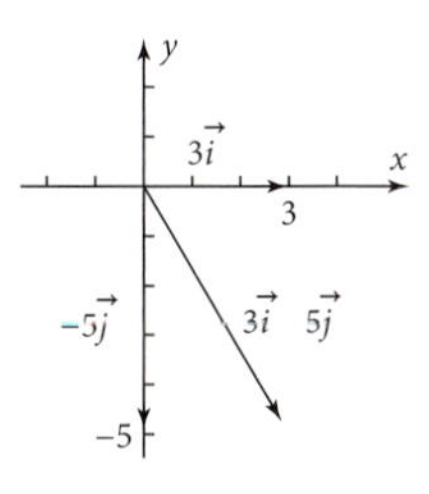

T11. ≈ 3.2 cm

T13. ≈ 30.9 ft

T15. Answers will vary.

T17. 6.5423... cm or 2.4456... cm

T19. $(6.5423...)(2.4456...) = 16 = 4^2$

T21. $-4\vec{i} + 2\vec{j}$

The graph shows that $\vec{d}$ equals the displacement from the head of $\vec{b}$ to the head of $\vec{a}$, analogous to "where you end minus where you began."

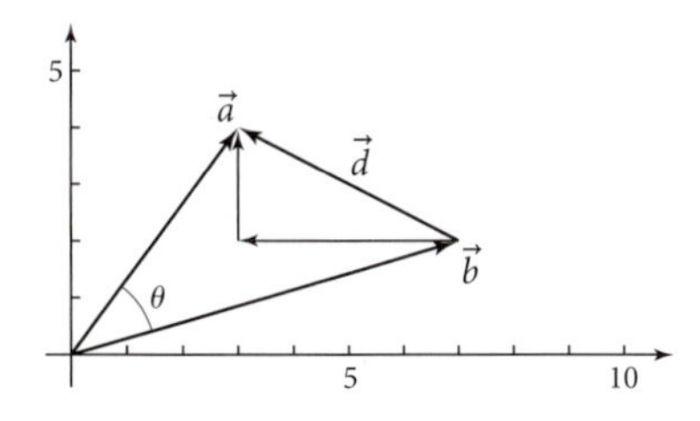

Problem Set 9-9

1.

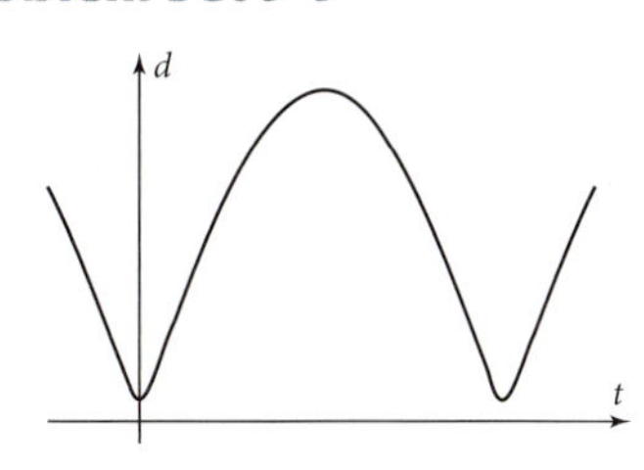

3. $\sin\theta = -\frac{5}{13}$, $\cos\theta = \frac{12}{13}$, $\tan\theta = -\frac{5}{12}$, $\cot\theta = -\frac{12}{5}$, $\sec\theta = \frac{13}{12}$, $\csc\theta = -\frac{13}{5}$

5.

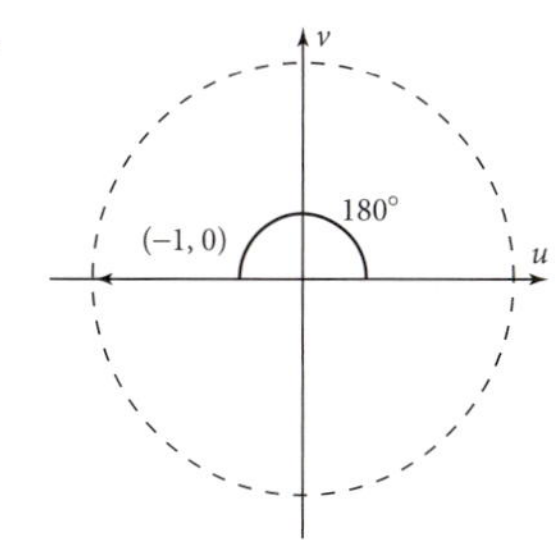

$\cos 180° =$ the u-coordinate on the x-axis $= -1$

7. Sinusoidal

9. 114.5915...°

11.

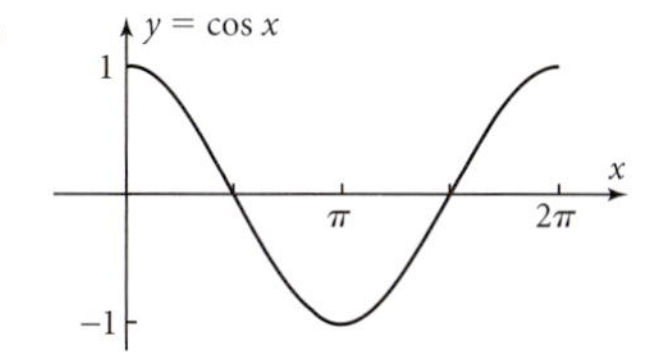

13a. 2π or 360° times horizontal dilation is the period.

13b. Amplitude

13c. Phase displacement or phase shift

13d. Sinusoidal axis

15. $y = 3.4452...$

17.

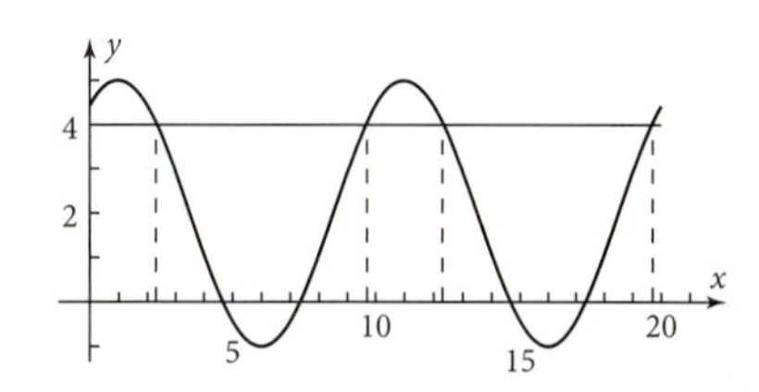

19a. 24 rad/s; 229.1831... rev/min

19b. 120 cm/s

19c. 95.4929... rev/min

21. Reciprocal properties: $\sec\theta = \frac{1}{\cos\theta}$, $\csc\theta = \frac{1}{\sin\theta}$, $\cot\theta = \frac{1}{\tan\theta}$

Quotient properties: $\tan\theta = \frac{\sin\theta}{\cos\theta}$, $\cot\theta = \frac{\cos\theta}{\sin\theta}$

Pythagorean properties: $\sin^2\theta + \cos^2\theta = 1$, $\tan^2\theta + 1 = \sec^2\theta$, $1 + \cot^2\theta = \csc^2\theta$

23. $\cos(x - y) = \cos x \cos y + \sin x \sin y$

Cosine of first, cosine of second, plus sine of first, sine of second

25. $\cos(90° - \theta) = \cos 90° \cos\theta + \sin 90° \sin\theta$
$= 0 \cdot \cos\theta + 1 \cdot \sin\theta = \sin\theta$;
$\cos(34°) = \cos(90° - 56°) = \sin 56°$

27. $6 \sin 2\theta = 6 \sin(\theta + \theta)$
$= 6(\sin\theta \cos\theta + \cos\theta \sin\theta)$
$= 6 \cdot 2 \sin\theta \cos\theta = 12 \sin\theta \cos\theta$

29. Larger sinusoid: $y = 3 \cos 6\theta$; Smaller sinusoid: $y = 2 \sin 30\theta$; Combined: $y = 3 \cos 6\theta + 2 \sin 30\theta$

31. $y = \cos 21\theta + \cos 19\theta$

33. $\theta = 78.6900...°$

35. See Figure 9-9f in the student text.
Possible parametric equations: $x = \cos t$, $y = t$

37. $\theta \approx 63.4°, 243.4°, 423.4°, 603.4°$

39. In $\triangle ABC$, $\dfrac{a}{\sin A} = \dfrac{b}{\sin B} = \dfrac{c}{\sin C}$. The length of one side of a triangle is to the sine of the angle opposite it as the length of any other side is to the sine of the angle opposite that side.

41. $\approx 134.6°$

43a. $2\vec{i} + 16\vec{j}$

43b. $|\vec{r}\,| \approx 16.1$; $\theta \approx 82.9°$

43c.

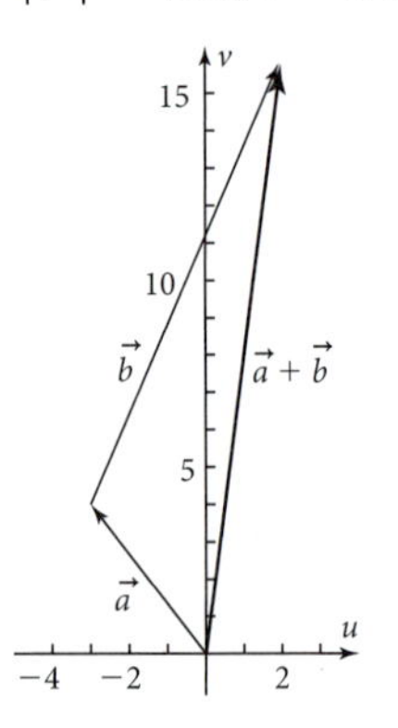

43d. False. This is true only if $\vec{a}$ and $\vec{b}$ are at the same angle.

45a. $\vec{a} = 4\vec{i} + 3\vec{j}$; $\vec{b} = -9\vec{i} + 4\vec{j}$; $\vec{c} = 2\vec{i} - 5\vec{j}$

45b. $\vec{d} = -3\vec{i} + 2\vec{j}$

45c. $|\vec{d}\,| = \sqrt{13} \approx 3.6$ newtons; $\theta \approx 146.3°$

45d. Using the graph, the measured length of $\vec{d}$ is approximately 3.6 units and the measured angle is approximately 146°, which agree with part c.

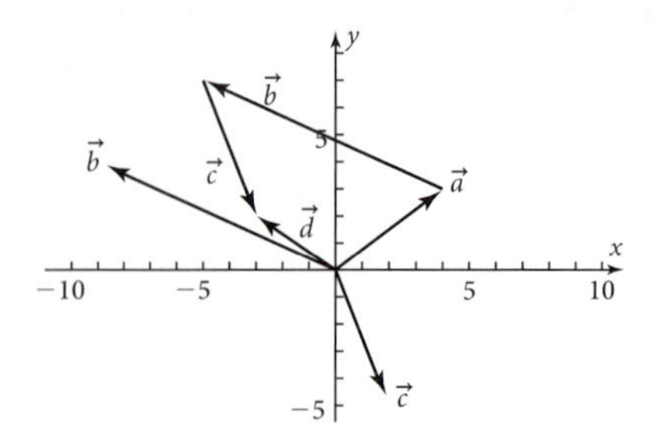

Chapter 10

Problem Set 10-1

1. $y = \pm\sqrt{1 - x^2}$. The graph should look like Figure 10-1a. The graph consists of all points whose distance to the origin is 1: $\sqrt{(x - 0)^2 + (y - 0)^2} = \sqrt{x^2 + y^2} = 1$

3. $\dfrac{4}{36}x^2 + \dfrac{9}{36}y^2 = 1 \Rightarrow \dfrac{x^2}{9} + \dfrac{y^2}{4} = 1 \Rightarrow \left(\dfrac{x}{3}\right)^2 + \left(\dfrac{y}{2}\right)^2 = 1$
The dilations 3 and 2 appear as the denominators of x- and y-terms, respectively.

5.

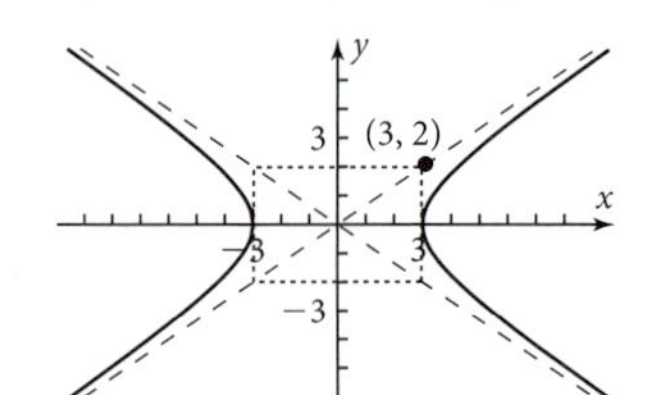

The asymptotes are $y = \dfrac{2x}{3}$ and $y = -\dfrac{2x}{3}$, which have slopes $\pm\dfrac{2}{3}$.

7. $y = \pm\sqrt{1 - x}$

Problem Set 10-2

1. See Figure 10-2c.

3. See Figure 10-2c.

5a. Ellipse

5b.

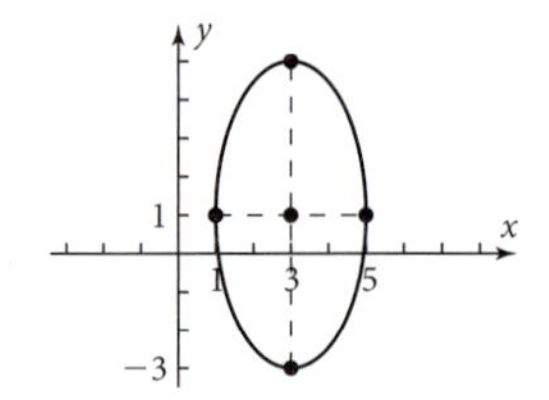

5c. $4x^2 + y^2 - 24x - 2y + 21 = 0$

5d. The graphs match.

7a. Hyperbola

7b.

7c. $-9x^2 + 25y^2 + 36x + 50y - 236 = 0$

7d. The graphs match.

9a. Circle

9b.

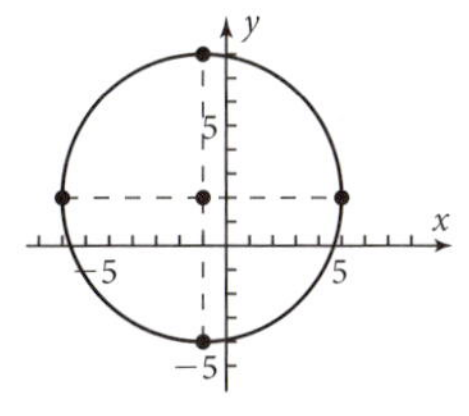

9c. $x^2 + y^2 + 2x - 4y - 31 = 0$

9d. The graphs match.

11a. Parabola

11b.

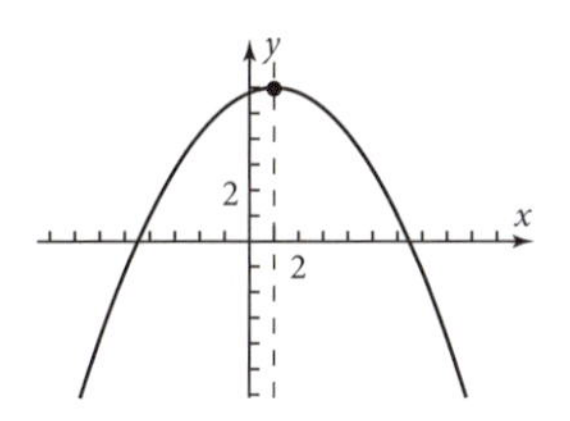

11c. $0.2x^2 - 0.4x + y - 5.8 = 0$

11d. The graphs match.

13. Ellipse; $\left(\dfrac{x+4}{6}\right)^2 + \left(\dfrac{y-3}{4}\right)^2 = 1$

15. Hyperbola; $-\left(\dfrac{x-4}{5}\right)^2 + \left(\dfrac{y+2}{3}\right)^2 = 1$

17. Parabola; $x = 4 - \dfrac{1}{4}(y+2)^2$

19a. Circle

19b. $(x-5)^2 + (y+4)^2 = 36$; Center $(5, -4)$, radius 6

19c. The graph agrees.

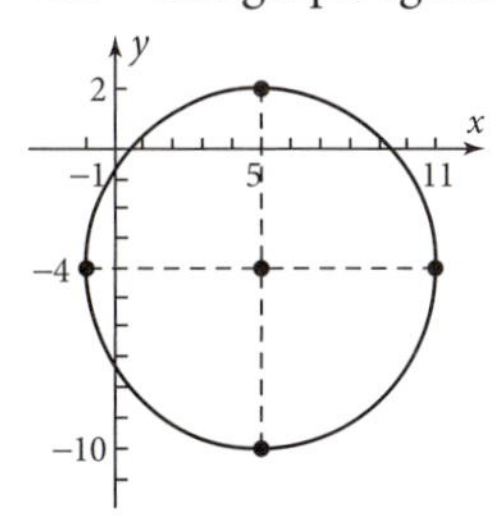

21a. Ellipse

21b. $\left(\dfrac{x-2}{3}\right)^2 + \left(\dfrac{y+5}{2}\right)^2 = 1$; Center $(2, -5)$, x-dilation 3, y-dilation 2

21c. The graph agrees.

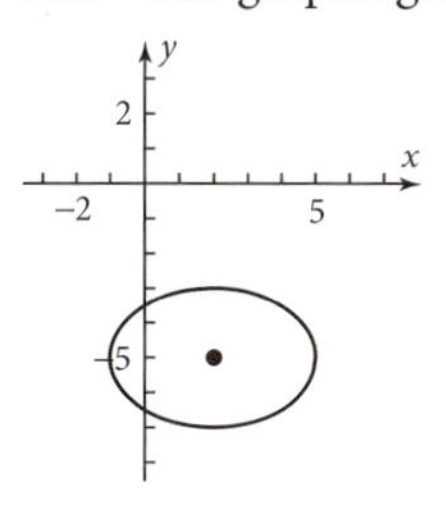

23a. Ellipse

23b. $\left(\dfrac{x+1}{3}\right)^2 + \left(\dfrac{y-2}{5}\right)^2 = 1$; Center $(-1, 2)$, x-dilation 3, y-dilation 5

23c. The graph agrees.

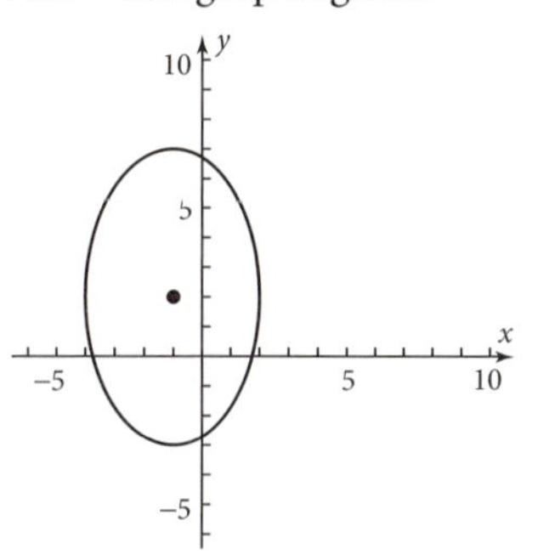

25a. Ellipse

25b. $\left(\dfrac{x}{5}\right)^2 + \left(\dfrac{y-6}{4}\right)^2 = 1$; Center $(0, 6)$, x-dilation 5, y-dilation 4

25c. The graph agrees.

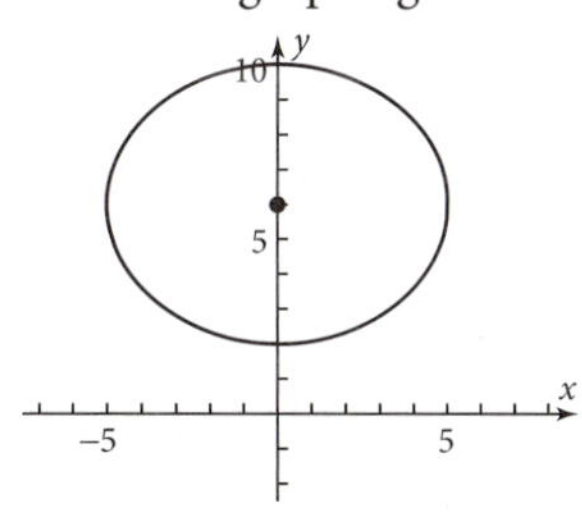

27a. Hyperbola

27b. $\left(\dfrac{x-2}{4}\right)^2 - \left(\dfrac{y+3}{5}\right)^2 = 1$; Center $(2, -3)$, x-dilation 4, y-dilation 5, opening in the x-direction

27c. The graph agrees.

29a. Hyperbola

29b. $-\left(\frac{x+6}{3}\right)^2 + \left(\frac{y+7}{5}\right)^2 = 1$; Center $(-6, -7)$, x-dilation 3, y-dilation 5, opening in the y-direction

29c. The graph agrees.

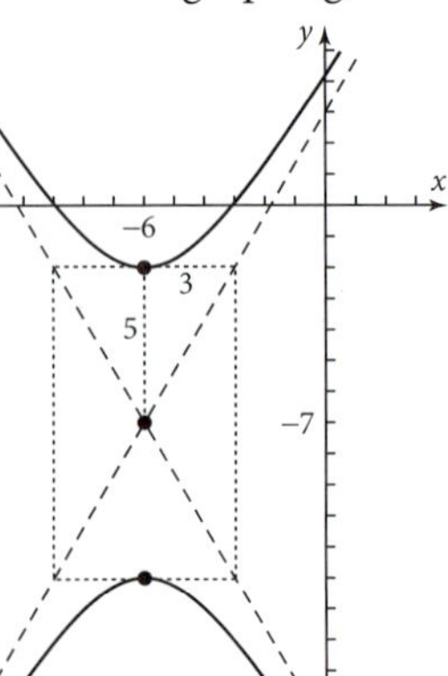

31a. Parabola

31b. $x = -3(y+2)^2 + 7$; Vertex $(7, -2)$, x-intercept -5

31c. The graph agrees.

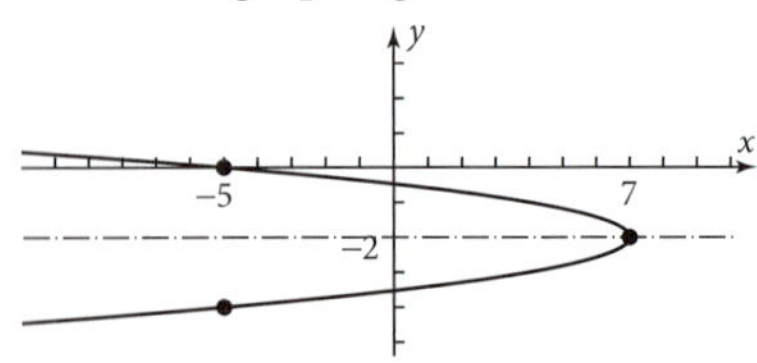

33a. Parabola

33b. $x = 0.5(y+3)^2 - 0.5$; Vertex $(-0.5, -3)$, x-intercept 4

33c. The graph agrees.

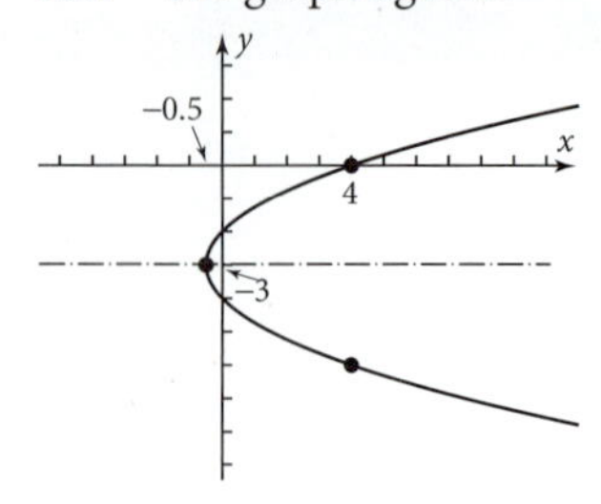

35a.

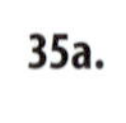

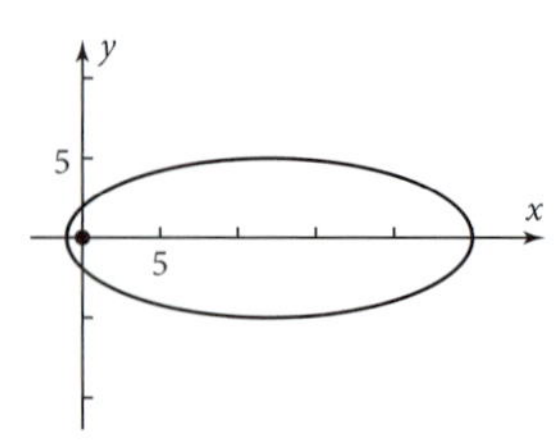

35b. The spaceship is closest to the Sun (at perihelion) at $(-1, 0)$, 1 million mi away. The spaceship is farthest from the Sun (at aphelion) at $(25, 0)$, 25 million mi away.

35c. $y = \pm\frac{5}{13}\sqrt{105} = \pm 3.9411...$ million mi

35d. 20.3846... million mi

37a. See Figure 10-2m.

37b. The type of figure remains the same, but the figure may be rotated, and the shape may be distorted.

Problem Set 10-3

1a. Circle

1b.

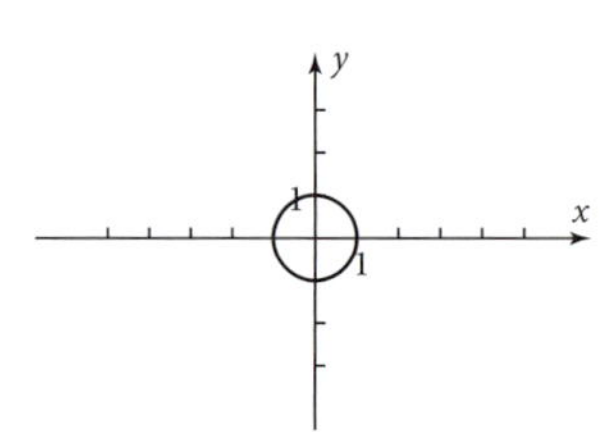

1c. The graphs match.

3a. Hyperbola

3b.

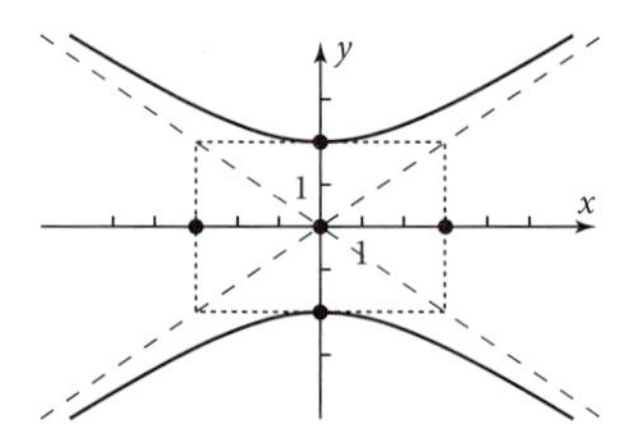

3c. The graphs match.

5a. Hyperbola

5b.

5c. The graphs match.

7a. Circle

7b.

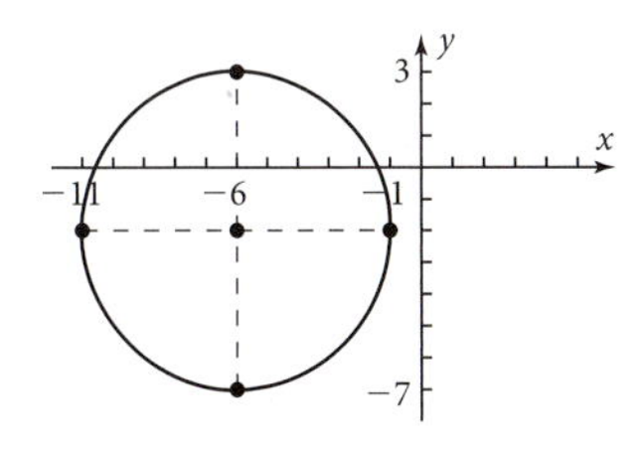

7c. The graphs match.

9. $x^2 + y^2 = 1$.
Agrees with *circle* because x^2 and y^2 have equal coefficients (and no *xy*-term).

11. $\left(\frac{x-4}{5}\right)^2 - \left(\frac{y-3}{2}\right)^2 = 1$.
Agrees with *hyperbola* because x^2 and y^2 have opposite signs (and no *xy*-term).

13. $x = -4 + 6\cos t,\ y = 3 + 4\sin t$

15. $x = 4 + 5\tan t,\ y = -2 + 3\sec t$

17. $x = 4 - \frac{1}{4}(t+2)^2,\ y = t$

19a.

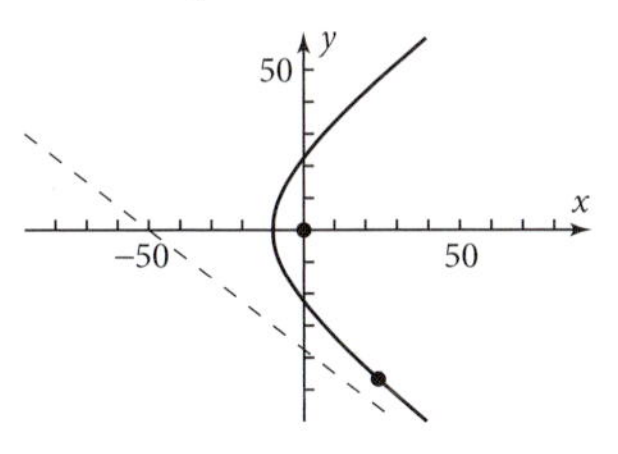

19b. $(x, y) = (24.0326..., -46.7222...)$;
$d = 52.5407...$ thousand mi

19c. The meteor is closest to Earth (at perigee) at $(-10, 0)$ for $t = 0$, which is 10,000 mi from the center of Earth, or 6,040 mi from Earth's surface.

19d. $y = -\frac{3}{4}x - 37.5$

19e. This is the path the meteorite will follow after it escapes Earth's gravity.

Problem Set 10-4

1.

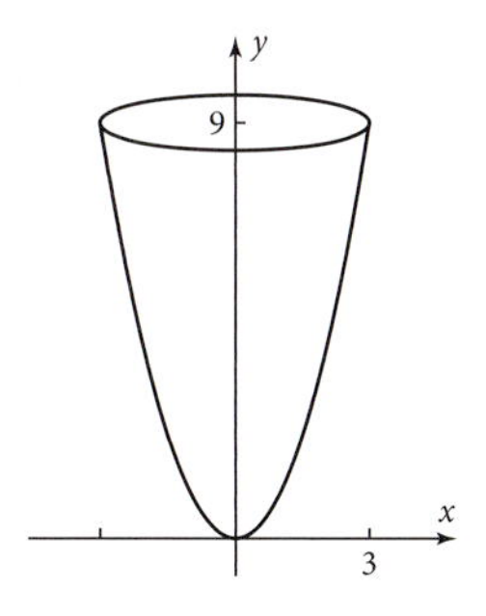

3.

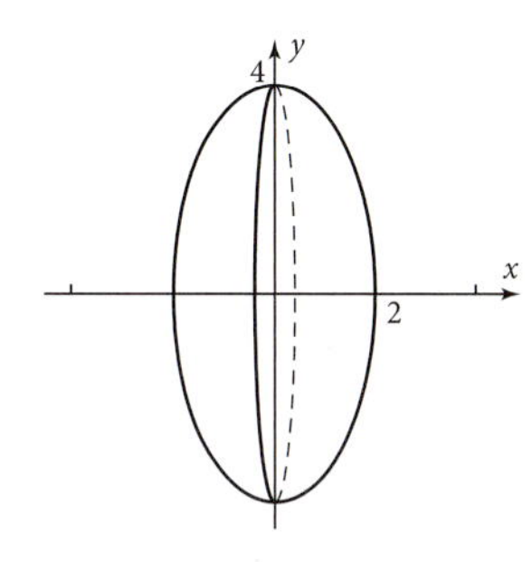

5.

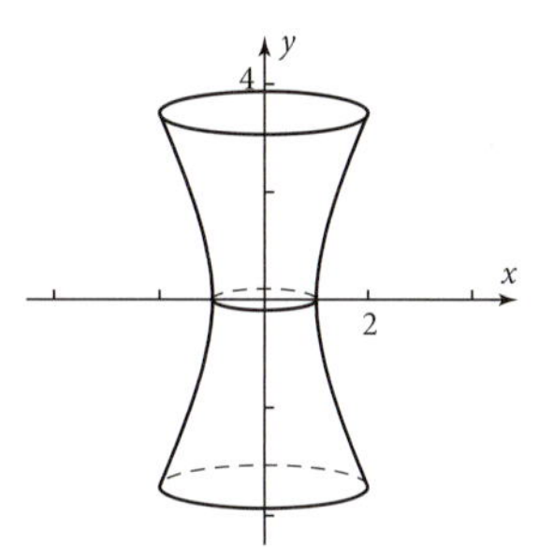

7.

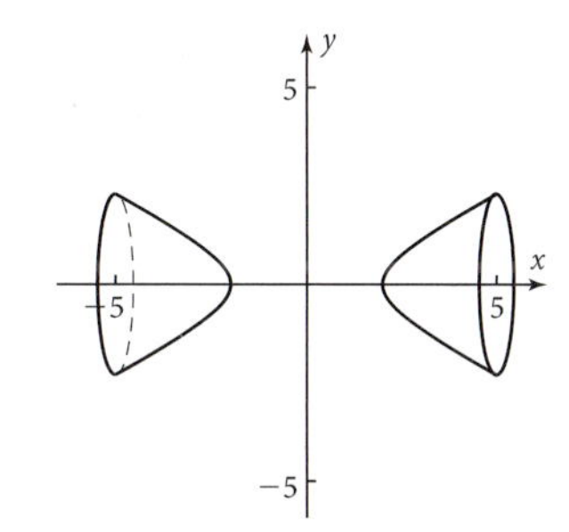

9.

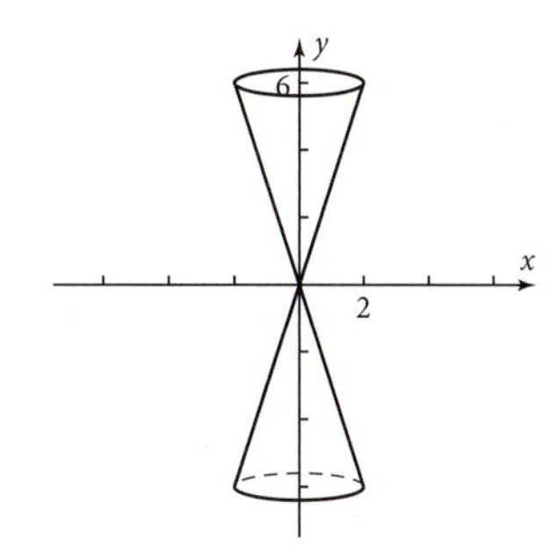

11. Answers will vary.

13. $A = \frac{1}{2} \cdot 2x \cdot y = xy = x(9 - x^2) = 9x - x^3$

Maximum area at $x = 1.7320...$; $A = 10.3923...$

15. $V = 2\pi x^2\sqrt{25 - x^2}$

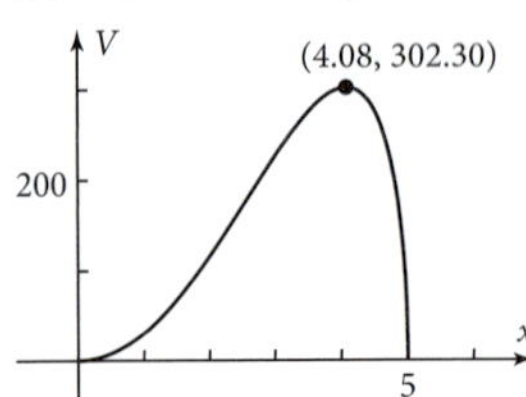

Maximum volume at $x = 4.0824...$; $V = 302.2998...$

17a. $r = 3.5355...$; $h = 7.0710...$

17b. $r = 4.2532...$; $h = 5.2573...$

17c. No

17d. No

19. The cylinder should be constructed with radius 12.2474... ft and altitude 17.3205... ft. 1332.8648... ft^2 of the heavy steel plate will be needed.

Problem Set 10-5

1a. $y = 8.4327...$; This agrees with the graph.

1b. $d_1 = 5.2$; $d_2 = \frac{26}{3}$; $d_3 = \frac{44}{3}$

1c. $d_2 = \frac{26}{3} = \frac{5}{3} \cdot 5.2 = ed_1$

1d. $|d_2 - d_3| = \left|\frac{26}{3} - \frac{44}{3}\right| = \left|\frac{-18}{3}\right| = 6$

1e. x-dilation is $3 = a$, the transverse radius; y-dilation is $4 = b$, the conjugate radius

1f. $c^2 = 5^2 = 3^2 + 4^2 = a^2 + b^2$

1g. $a = 3 = \frac{5}{3} \cdot 1.8 = ed$; $c = 5 = \frac{5}{3} \cdot 3 = ea$

3a. The vertex is equidistant from the focus and the directrix. The eccentricity is the ratio of the distances from a point on the curve to the focus and to the directrix, so $e = 1$.

3b. $y = 4.5$; This agrees with the graph.

3c. $d_1 = 6.5$; $d_2 = 6.5$

3d. $d_1 = 6.5 = d_2$;
The eccentricity is $e = \frac{d_2}{d_1} = \frac{6.5}{6.5} = 1$.

5.

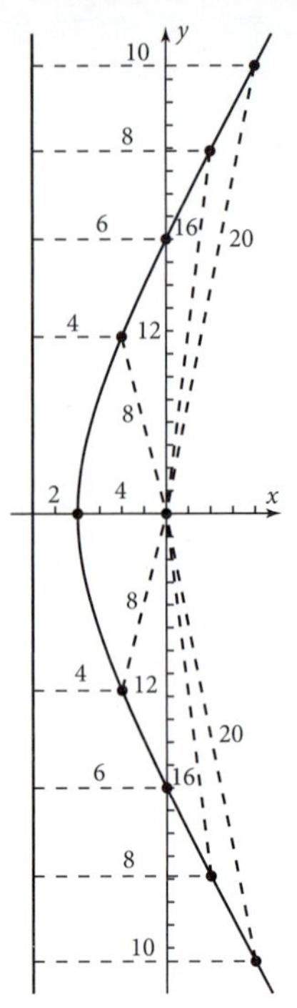

7. The graphs look the same as in Problems 5 and 6.

9a. The major axis is 283 million mi long. The major radius is 141.5 million mi.

9b. $c = 13.5$ million mi; $b = 140.8545...$ million mi

9c. $\frac{x^2}{20{,}022.25} + \frac{y^2}{19{,}840} = 1$

9d. $x = c = 13.5$ million mi; The distance to the Sun is 140.2120... million mi.

9e. $e = 0.0954...$

9f. 1469.6296... million mi.

9g. $x = 141.5 \cos t$; $y = 140.8545... \sin t$

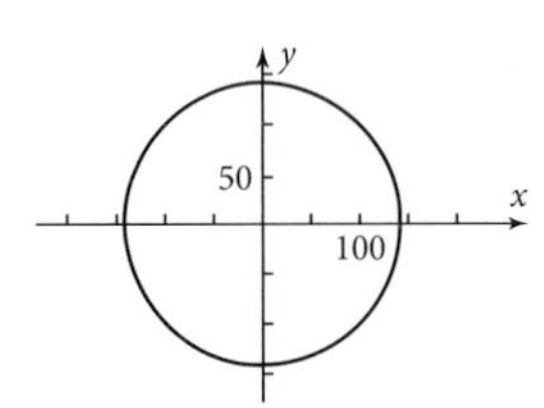

9h. The major and minor radii are nearly equal, and the eccentricity is close to zero.

11a. Ellipse

11b. Major radius, $a = 5$; minor radius, $b = 3$; $c = 4$; $d = 6\frac{1}{4}$; $e = \frac{4}{5}$

11c.

13a. Hyperbola opening vertically

13b. transverse radius, $a = 3$; conjugate radius, $b = 6$; $c = 6.7082...$; $d = \sqrt{1.8} = 1.3416...$; $e = \sqrt{5} = 2.2360...$ slope of asymptotes, $m = \pm\frac{a}{b} = \pm\frac{1}{2}$

13c.

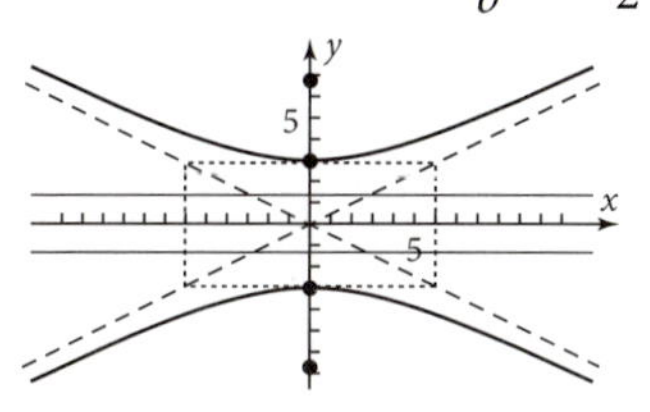

15a. Ellipse

15b. major radius, $a = 4$; minor radius, $b = 3$; $c = \sqrt{7} = 2.6457...$; $d = \dfrac{16}{\sqrt{7}} = 6.0474...$; $e = \dfrac{\sqrt{7}}{4}$ $= 0.6614...$

15c.

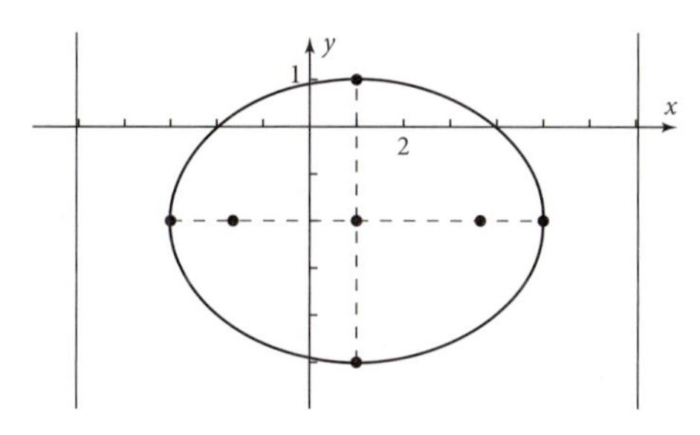

17a. $5x^2 - 3y^2 = -30 \Rightarrow -\left(\dfrac{x}{\sqrt{6}}\right)^2 + \left(\dfrac{y}{\sqrt{10}}\right)^2 = 1$
Hyperbola opening vertically

17b. transverse radius, $a = \sqrt{10} = 3.1622...$; conjugate radius, $b = \sqrt{6} = 2.4494...$; $c = 4$; $d = 2.5$; $e = \frac{2\sqrt{10}}{5} = 1.2649...$

17c.

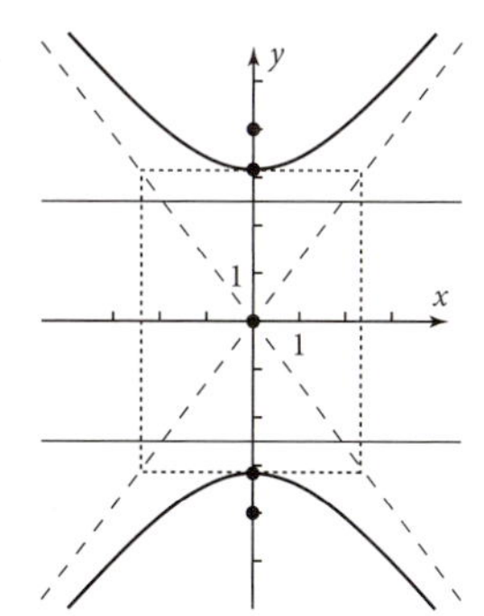

19a. $x = -\frac{1}{4}y^3 + 3 \Rightarrow x - 3 = -\frac{1}{4}y^2$;
Parabola opening left

19b. $p = -1$; $e = 1$

19c.

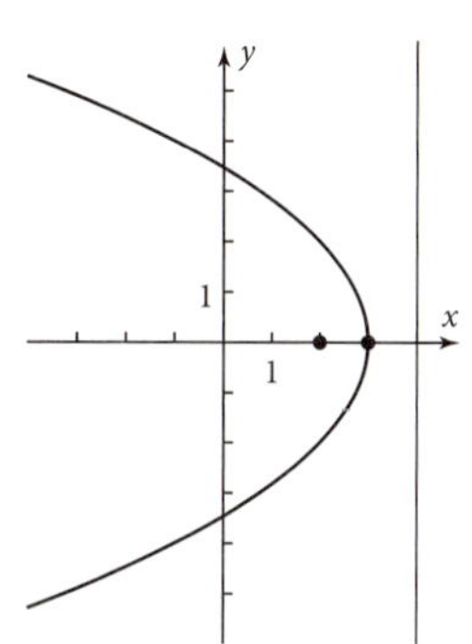

21a.

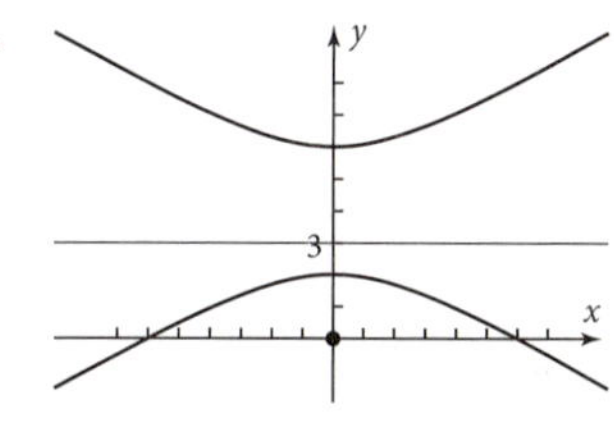

21b. $x^2 - 3y^2 + 24y - 36 = 0$ or
$-\left(\dfrac{x}{2\sqrt{3}}\right)^2 + \left(\dfrac{y-4}{2}\right)^2 = 1$
Hyperbola opening vertically

21c. The graphs match.

23a.

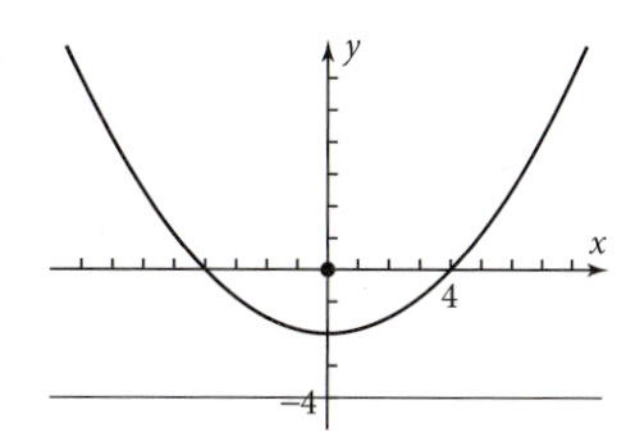

23b. $x^2 - 8y - 16 = 0$ or $y = \frac{1}{8}x^2 - 2$
Parabola opening vertically

23c. The graphs match.

25a.

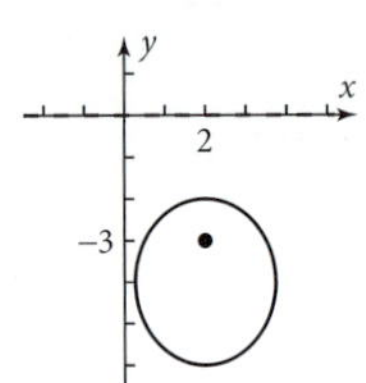

25b. $4x^2 + 3y^2 - 16x + 24y + 52 = 0$ or $\left(\frac{x-2}{\sqrt{3}}\right)^2 + \left(\frac{y+4}{2}\right)^2 = 1$. Ellipse.

25c. The graphs match.

27a.

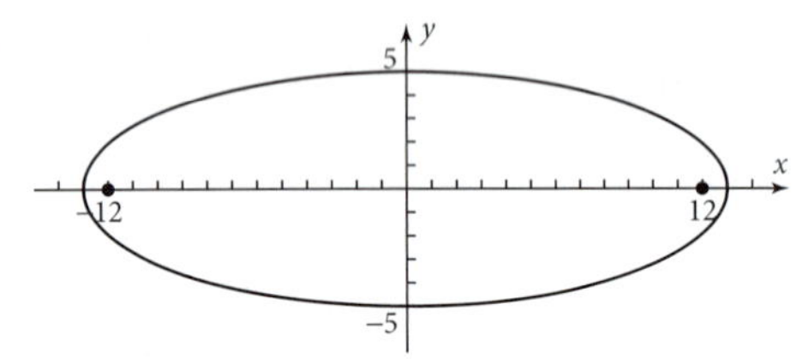

27b. $25x^2 + 169y^2 - 4225 = 0$ or $\left(\frac{x}{13}\right)^2 + \left(\frac{y}{5}\right)^2 = 1$

27c. The graphs match.

29a.

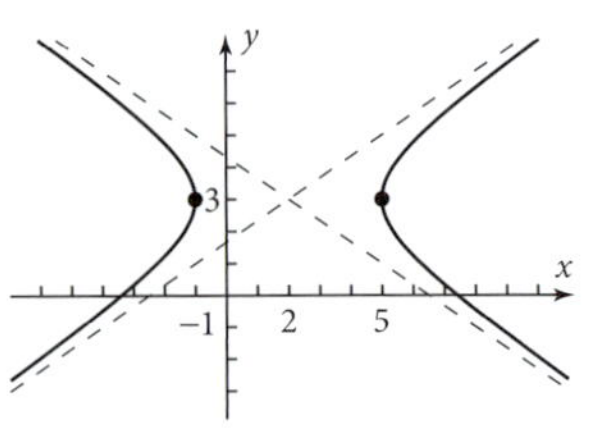

29b. $\left(\frac{x-2}{3}\right)^2 - \left(\frac{y-3}{2}\right)^2 = 1$

29c. The graphs match.

31a.

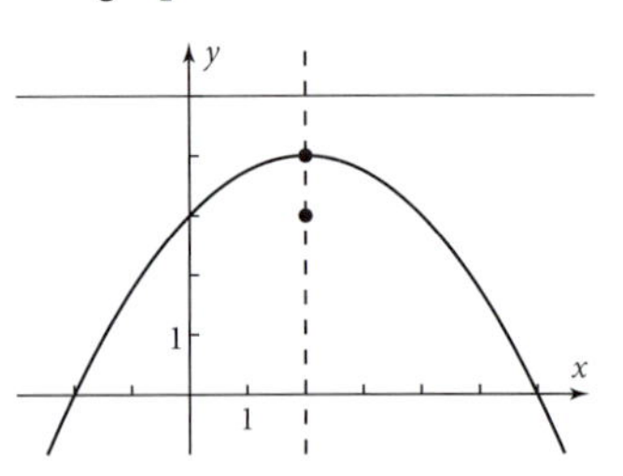

31b. $y - 4 = -\frac{1}{4}(x-2)^2$ or $x^2 - 4x + 4y - 12 = 0$

31c. The graphs match.

33. For Problem 9, $LR = 280.4240...$ million mi. For the ellipse of Problem 11, $LR = \frac{18}{5} = 3.6$. For the hyperbola of Problem 17, $LR = \frac{6\sqrt{10}}{5}$ $= 3.7947....$

Problem Set 10-6

1a. $\begin{bmatrix} \cos 50° & \cos 140° \\ \sin 50° & \sin 140° \end{bmatrix} \cdot \begin{bmatrix} 6 \\ 8 \end{bmatrix} = \begin{bmatrix} -2.2716... \\ 9.7385... \end{bmatrix}$

Image point $(-2.2716..., 9.7385...)$, the rotation angle measures 50°.

1b. By the Pythagorean theorem, $\sqrt{6^2 + 8^2} = \sqrt{100}$ $= 10$, and $\sqrt{(-2.2716...)^2 + 9.7385...^2} = \sqrt{100}$ $= 10$, which is the same.

1c. $\begin{bmatrix} \cos(-90°) & \cos 0° \\ \sin(-90°) & \sin 0° \end{bmatrix} \cdot \begin{bmatrix} 6 \\ 8 \end{bmatrix} = \begin{bmatrix} 0 & 1 \\ -1 & 0 \end{bmatrix} \cdot \begin{bmatrix} 6 \\ 8 \end{bmatrix} = \begin{bmatrix} 8 \\ -6 \end{bmatrix}$

Line segments are perpendicular as shown by slopes $\frac{8}{6}$ and $\frac{-6}{8}$, which are opposite reciprocals.

3. $x = -4 + 3\cos 20° \cos t + 5\cos 110° \sin t$
$y = 2 + 3\sin 20° \cos t + 5\sin 110° \sin t$

5. $x = -2 + 7\cos(-10°) \tan t + 3\cos 80° \sec t$
$y = -1 + 7\sin(-10°) \tan t + 3\sin 80° \sec t$

7. $x = 2 + \frac{1}{32}t^2 \cos 25° + t\cos 115°$
$y = -1 + \frac{1}{32}t^2 \sin 25° + t\sin 115°$

9. $x = 8 + 25\cos(-30°) \cos t + 7\cos 60° \sin t$
$y = 5 + 25\sin(-30°) \cos t + 7\sin 60° \sin t$

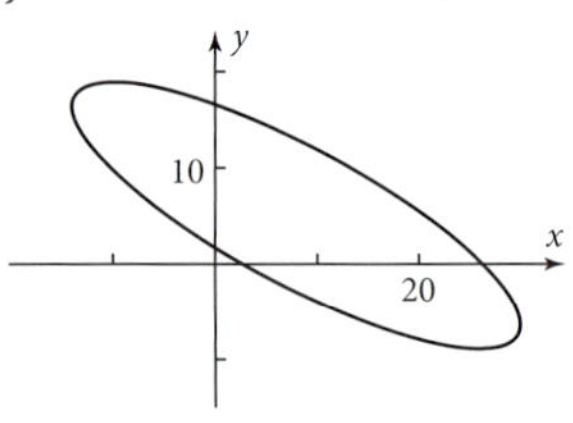

11. $x = 4\cos 15° \sec t + 3\cos 105° \tan t$
$y = 4\sin 15° \sec t + 3\sin 105° \tan t$

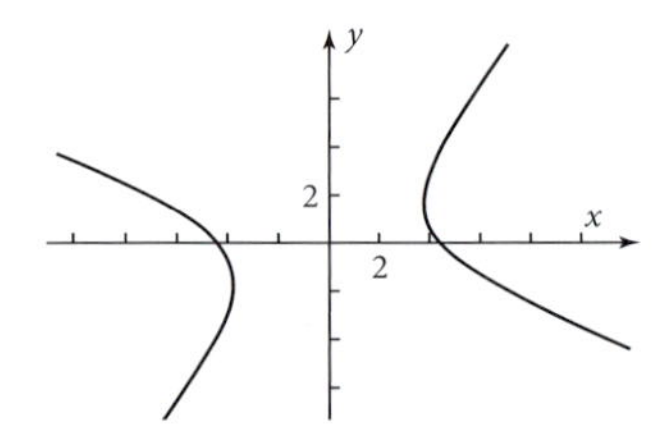

13. $x = -8 + t^2\cos(-30°) + t\cos 60°$
$y = 5 + t^2\sin(-30°) + t\sin 60°$

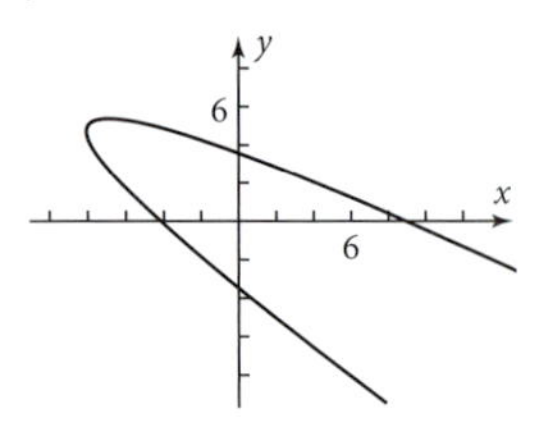

15. -83; Ellipse

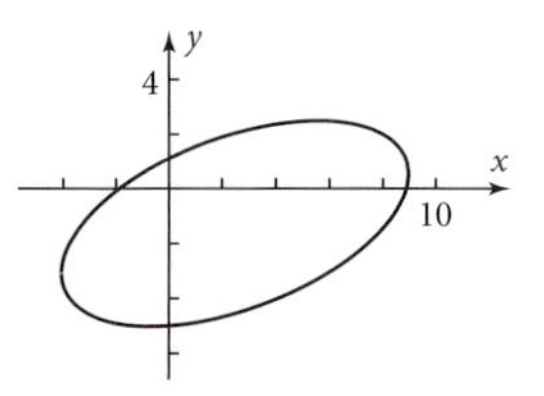

17. 28; Hyperbola

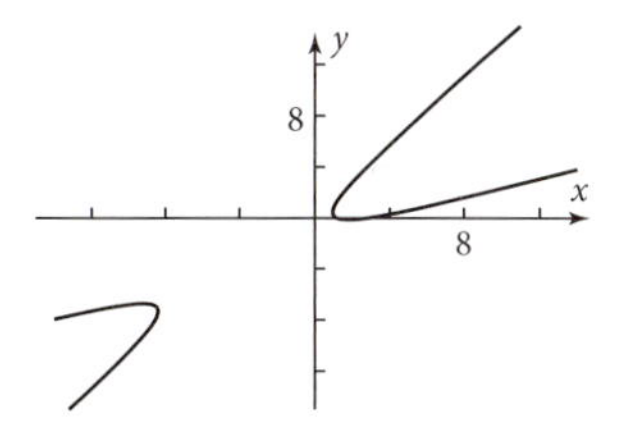

19. 0; Parabola

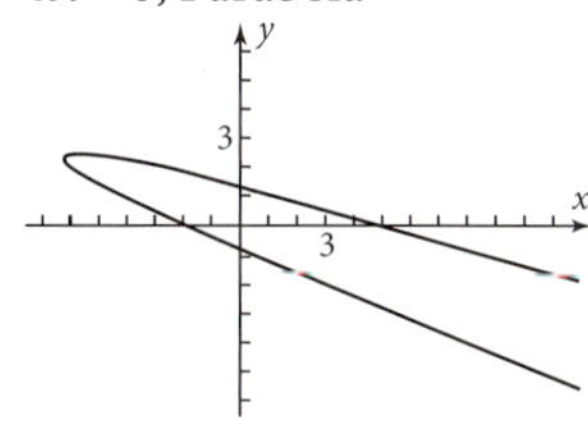

21. If $B = 0$, then $B^2 - 4AC = -4AC$.
A and C have the same sign (ellipse) $\Leftrightarrow -4AC < 0$.
A and C have opposite signs (hyperbola) $\Leftrightarrow -4AC > 0$.
Either $A = 0$ or $C = 0$ but not both (parabola)
$\Rightarrow -4AC = 0$.

23a. $\begin{bmatrix} \cos v & \cos(v + \frac{\pi}{2}) \\ \sin v & \sin(v + \frac{\pi}{2}) \end{bmatrix} \begin{bmatrix} a \cos t \\ b \sin t \end{bmatrix}$

$$= \begin{bmatrix} \cos v \cdot a \cos t + \cos\left(a + \frac{\pi}{2}\right) \cdot b \sin t \\ \sin v \cdot a \cos t + \sin\left(v + \frac{\pi}{2}\right) \cdot b \sin t \end{bmatrix}$$

$$= \begin{bmatrix} (a \cos v) \cos t + \left[b \cos\left(v + \frac{\pi}{2}\right)\right] \sin t \\ (a \sin v) \cos t + \left[b \sin\left(v + \frac{\pi}{2}\right)\right] \sin t \end{bmatrix}$$

23b. $a = \sqrt{34} = 5.8309...$, $b = \dfrac{2\sqrt{34}}{5} = 2.3323...$, $v = 59.0362...°$

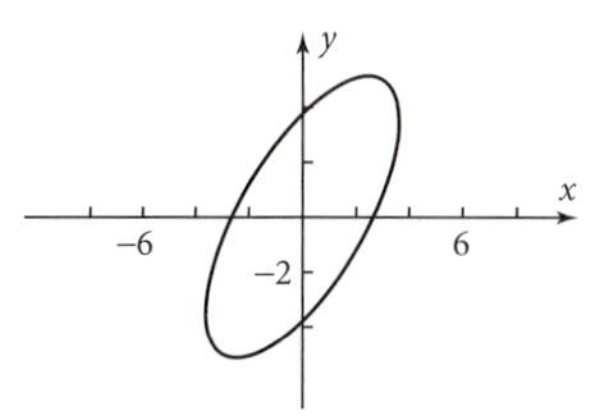

25. Answers will vary.

Problem Set 10-7

1. Assume the major (52-in.) and minor (26-in.) axes are already drawn, perpendicularly bisecting each other. Drive a nail at one end of the minor axis, tie a string to it, and tie a pencil to the other end of the string so that the pencil is 26 (half of 52) in. from the nail. Use this as a compass to draw an arc of a circle with radius 26 in., intersecting the major axis in two points, which will be the foci. Now drive nails at the two foci and tie a 52-in. string between them. Use a pencil to pull the string taut, and slide the pencil back and forth, always keeping the string taut, to draw the ellipse. (Actually, this will draw *half* the ellipse; the string and pencil will have to be flipped to the other side of the nails to draw the other half.)

$e = \dfrac{\sqrt{3}}{2} = 0.8660...$

3a. $y = -\dfrac{1}{1000}x^2$

3b.

x	y
0	20
± 50	22.5
± 100	30
± 150	42.5
± 200	60
± 250	82.5
± 300	110
± 350	142.5
± 400	180
± 450	222.5
± 500	270

3c. 4690 ft

5a. Parabola. The meteorite's path does not intersect Earth's surface.

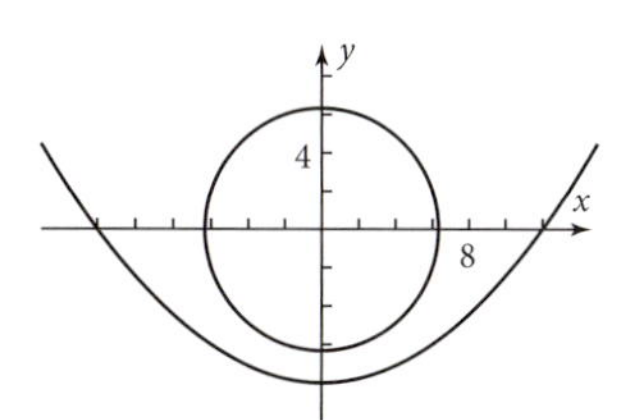

5b. No real solution.

5c. Hyperbola. The branch with the positive square root does not intersect, but the branch with the negative square root does:

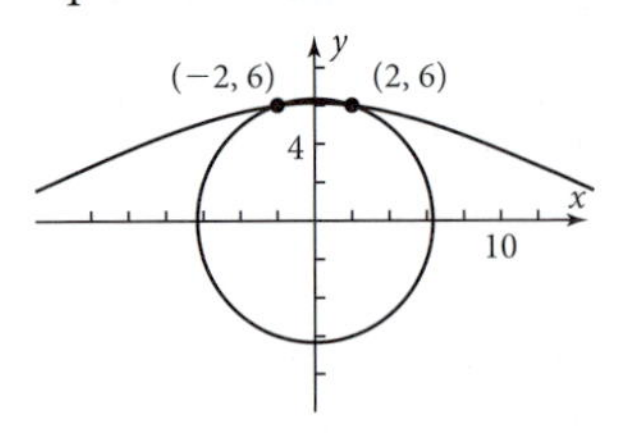

5d. The meteorite strikes at (−2000 km, 6000 km) or (2000 km, 6000 km), depending on which way it is traveling.

7a. $\left(\frac{x - 200}{100}\right)^2 + \left(\frac{y}{28.0056...}\right)^2 = 1$

7b. 28.0056... ft at the center.

7c. $x = 245.0693...$ or $154.9306...$; Truck should stay in the interval 154 ft $\le x \le$ 245 ft.

9a. $(x - 8)^2 + (y - 0)^2 = 4^2$, a circle with center (8, 0) and radius 4

9b.

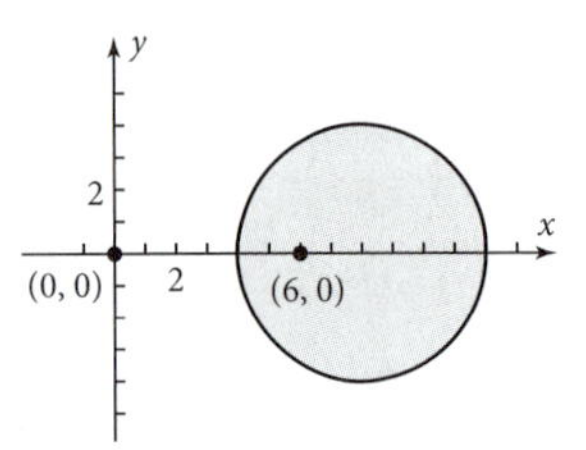

9c. (15, 0) is closer to Supplier 2 but outside the shaded region, so Supplier 1 is less expensive.

11. Hyperbola: $x = 10 + 4 \sec t$, $y = \frac{23}{2} + \frac{26}{3} \tan t$, $-47° \le t \le 37°$.

Top ellipse: $x = 10 + 5 \cos \frac{360}{84} t$, $y = 18 + \sin \frac{360}{84} t$.

Bottom ellipse: $x = 10 + 6 \cos \frac{360}{84} t$, $y = 2 + \frac{6}{5} \sin \frac{360}{84} t$

Problem Set 10-8

R1a. Circle **R1b.** Ellipse

R1c. Hyperbola **R1d.** Hyperbola

R1e. Parabola

R3a. i. $x = \cos t$, $y = \sin t$

R3a. ii. $x = \sec t$, $y = \tan t$

R3a. iii. $x = t$, $y = t^2$

R3b. i. $x = 2 + 7 \cos t$, $y = -3 + 4 \sin t$

R3b. ii. The graph matches the sketch.

R3c. i. $x = -4 + 2 \tan t$, $y = 1 + 3 \sec t$

R3c. ii. The graph matches the figure.

R3d. i. $x = 8 - 3(t + 2)^2$, $y = t$

R3d. ii. The grapher graph matches the given figure.

R3e. $\left(\frac{x}{3}\right)^2 - \left(\frac{y}{5}\right)^2 = 1$

R5a. $d_1 = 7$, $d_2 = 5.6$, $d_3 = 10.4$; $5.6 = 0.8 \cdot 7$; $5.6 + 10.4 = 16$

R5b. $\left(\frac{x}{8}\right)^2 + \left(\frac{y}{4.8}\right)^2 = 1$; $y = \sqrt{19.8}$; $d_2 = 5.6$; $d_3 = 10.4$
The measurements agree.

R5c. Major radius, $a = 90$ cm;
Minor radius, $b = 39.2300...$ cm;
$x = 90 \cos t$, $y = 39.2300... \sin t$
The grapher agrees with the figure.
Both angles measure about 30° (precisely 29.9232...°). A ball shot straight from one focus to the point where $t = 1$ will bounce off of the tangent line at the same angle at which it approached, thus going straight through the other focus. (This is true no matter which point on the ellipse is aimed for!)

R5d. Hyperbola opening horizontally

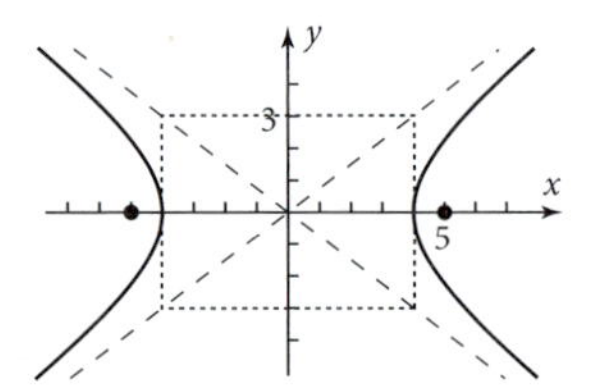

Major radius, $a = 4$; Minor radius, $b = 3$;
Focal radius, $c = \sqrt{a^2 + b^2} = \sqrt{25} = 5$;
Eccentricity, $e = \frac{c}{a} = 1.25$; Directrix radius, $d = \frac{a}{e} = 3.2$

R5e.

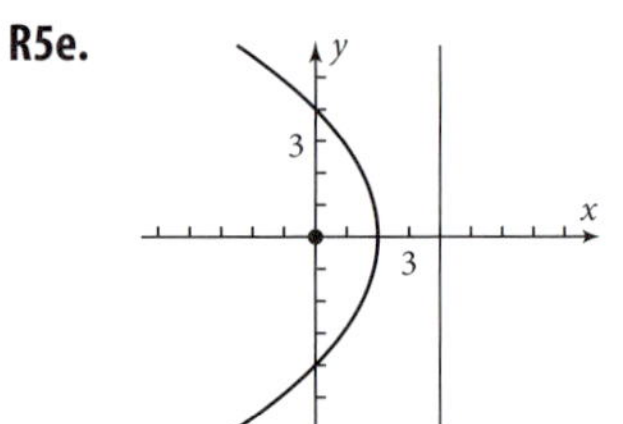

$x = 2 - \frac{1}{8}y^2$

R5f. $x = 2 + 2.4 \tan t$, $y = 4 + 1.8 \sec t$

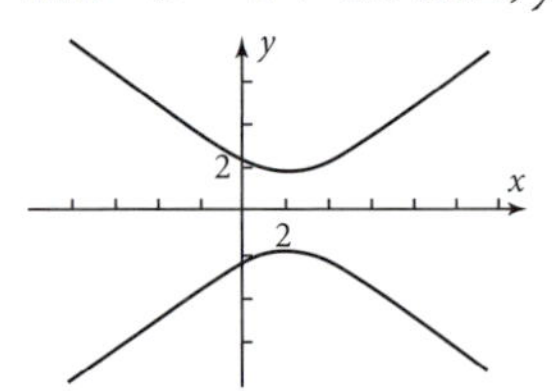

R7a. $x = \frac{1}{100} y^2$; Domain: 0 in. $\le x \le$ 16 in.

Answers to Selected Problems

R7b.

R7c. Yes. As discovered in part a and shown in the graph of part b, the focus is at $x = 25$, while the dish extends only to $x = 16$.

C1a. $a = 0.01 \Rightarrow p = \dfrac{1}{4a} = 25$

C1b. (4, 20)

C1c. $A = 43.6028...°$

C1d. $B = C = 68.1985...°$; The angles in the figure match this.

C1e. $y = \dfrac{5}{2}x + 10$
Successive zooms centered at (4, 20) show that the parabola "looks like" the line more and more as we zoom in, indicating that the line really is tangent.

C1f. Answers will vary.

C1g. Conjecture: An incoming ray that is parallel to the axis of the parabola is always reflected to pass through the focus.

T1. Ellipse

T3. Parabola

T5. Parabola

T7.

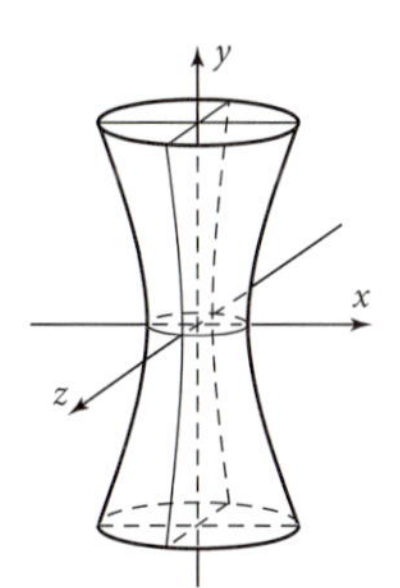

T9. The ratio of the two distances, $\dfrac{d_1}{d_2}$ is the eccentricity.

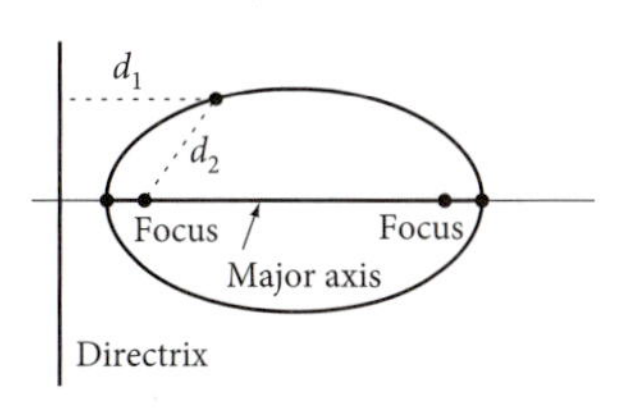

T11.

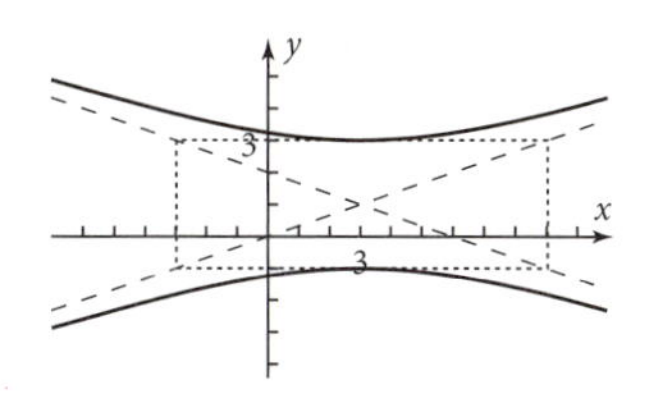

T13. $-x^2 + 9y^2 + 6x - 18y - 36 = 0$

T15. The graph will be a unit circle because the x- and y-dilations both equal 1. Its center is translated to the point $(5, -3)$. $(x - 5)^2 + (y + 3)^2 = 1$

T17a. Center (45, 0); $b = 24$ thousand mi; $e = 0.8823...$

T17b. The cylinder with maximum volume has radius $x = 3.2659...$, height $2y = 2.3094...$, and volume $77.3887...$ The y-intercepts are $(0, \pm 11.2941...)$.

T17c. The satellite comes within 2000 mi of Earth's surface.

T19. $x = 1 + 7\cos(-25°)\cos t + 3\cos 65° \sin t$,
$y = -2 + 7\sin(-25°)\cos t + 3\sin 65° \sin t$,
$0° \le t \le 360°$.

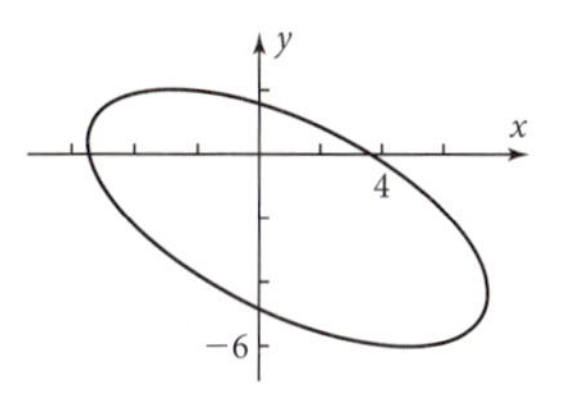

Chapter 11

Problem Set 11-1

1.

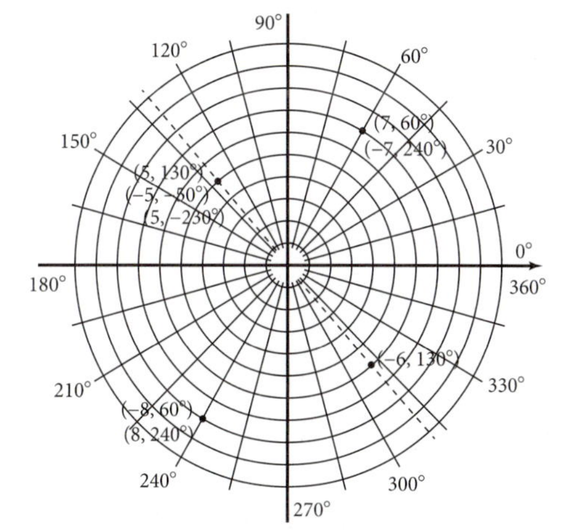

3. $(8, 240°)$ is at the same place as $(-8, 60°)$. $(-7, 240°)$ is at the same place as $(7, 60°)$.

5. $(-5, -50°)$

7. θ is the independent variable. This disagrees with the custom of putting the independent variable first in an ordered pair.

Problem Set 11-2

1.

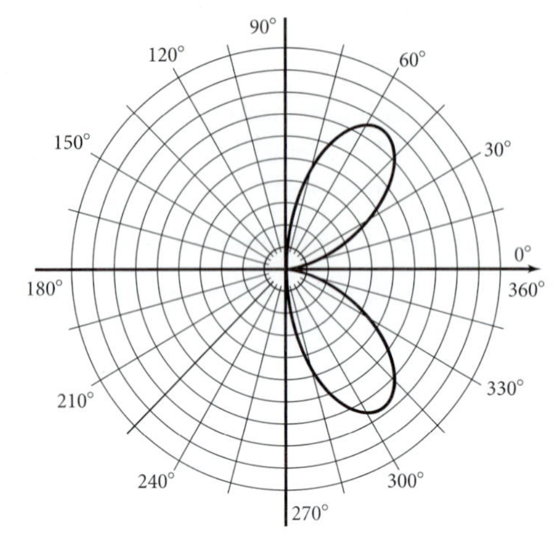

3.

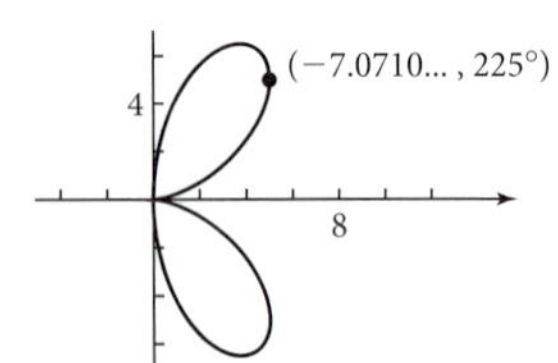

The graph is being retraced between 180° and 360°. The figure has two (*bi-*) "leaves" (*folium*).

5.

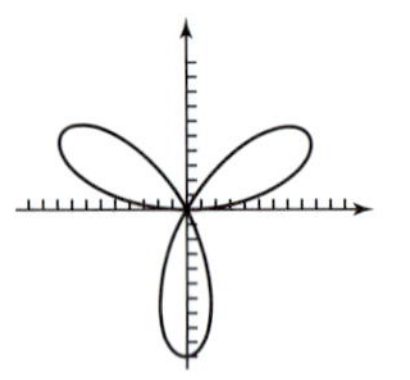

$r < 0$ for $60° < \theta < 120°$

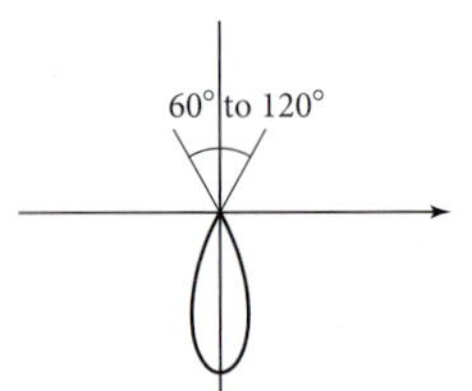

7.

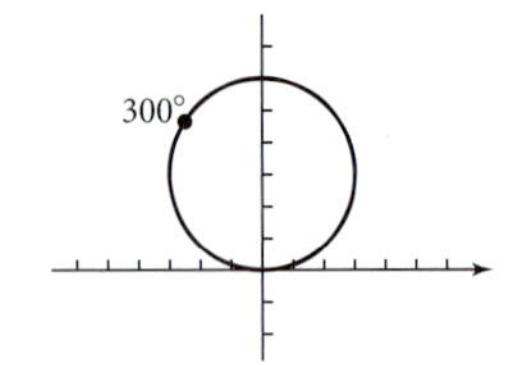

When $\sin\theta$ is negative, that is, $180° < \theta < 360°$, then r is negative, so the points are plotted in the opposite direction.

9. $r = 6\sin\theta$; $r^2 = 6r\sin\theta$; $x^2 + y^2 = 6y$; $y^2 - 6y + 9 + x^2 = 9$; $(y - 3)^2 + x^2 = 3^2$; A circle with center (0, 3) and radius 3

11.

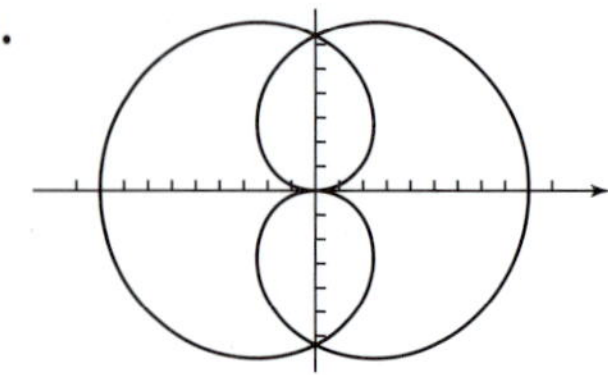

$r < 0$ for $180° < \theta < 540°$

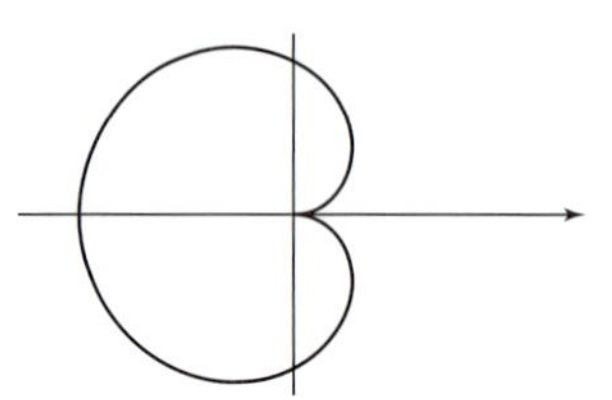

13.

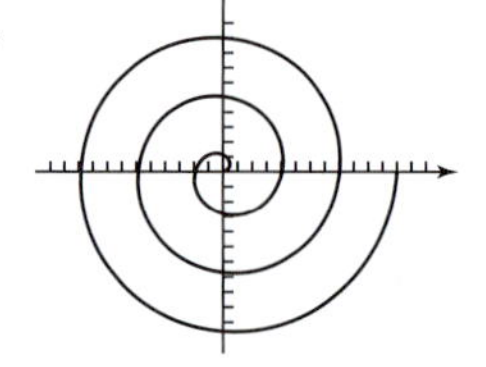

4, 8, 12, . . .

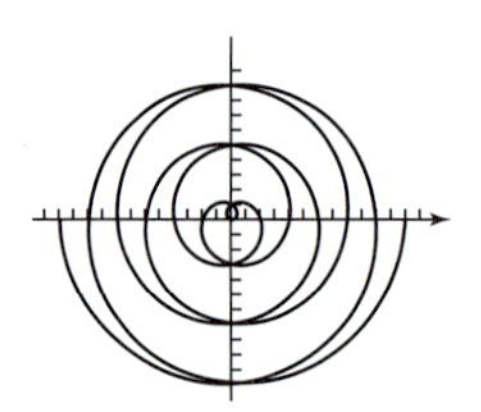

The graph is the original spiral with its mirror image across the y-axis.

15a. $r = \cos\theta$

15b. $r = \sin\theta$

15c. $r = -\cos\theta$

15d. $r = -\sin\theta$

17a.

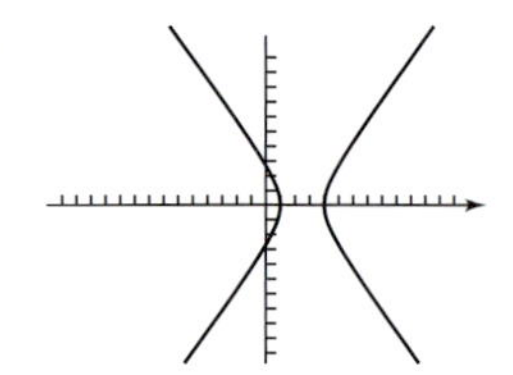

17b. $r = \dfrac{8}{3 + 5\cos\theta}$

$3r = 8 - 5r\cos\theta = 8 - 5x$

$9r^2 = 9x^2 + 9y^2 = 64 - 80x + 25x^2$

$16x^2 - 80x - 9y^2 = -64$

$16\left(x^2 - 5x + \dfrac{25}{4}\right) - 9y^2 = -64 + 16 \cdot \dfrac{25}{4} = 36$

$\dfrac{\left(x - \frac{5}{2}\right)^2}{\frac{9}{4}} - \dfrac{y^2}{4} = \left(\dfrac{x - \frac{5}{2}}{\frac{3}{2}}\right)^2 - \left(\dfrac{y - 0}{2}\right)^2 = 1$

Hyperbola with center $\left(\frac{5}{2}, 0\right)$, horizontal transverse radius $\frac{3}{2}$, and vertical conjugate radius 2

17c. One focus is at the pole. $e = \dfrac{5}{3}$.

19a.

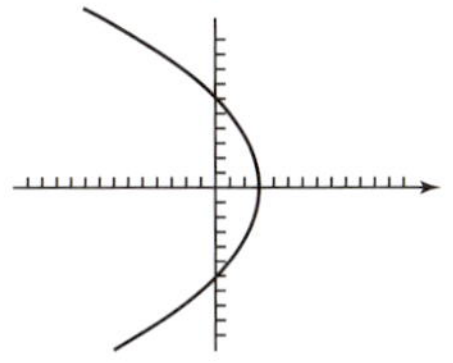

19b. $r = \dfrac{6}{1 + \cos\theta}$, $r = 6 - r\cos\theta = 6 - x$,

$r^2 = x^2 + y^2 = 36 - 12x + x^2$, $12x = 36 - y^2$,

$x = -\dfrac{1}{12}y^2 + 3$

Parabola opening left, with vertex (0, 3)

19c. The focus is at the pole. $e = 1$.

21. Let the polar axis be where the two branches of the loop cross. Then $r = 75 - 135\sin\theta$.

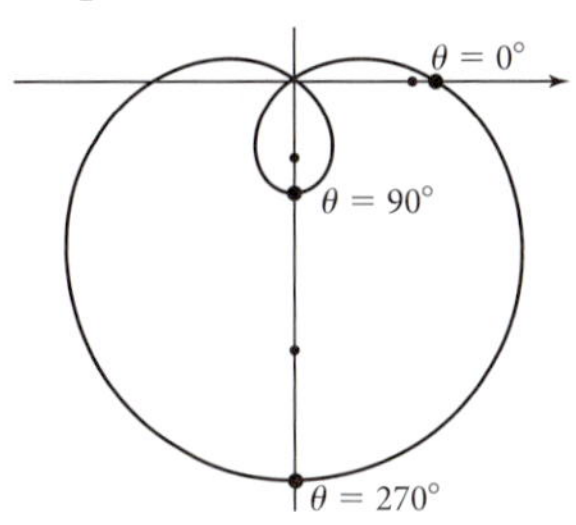

23a. $(x, y) = (0°, 2), (90°, 6), (180°, 2), (270°, -2), (360°, 2)$

23b. Answers may vary. Sample answer: $(r, \theta) = (2, 0°), (6, 90°), (2, 180°), (2, 90°)$

23c. $(x, y) = (0°, 2)$ and $(x, y) = (360°, 2)$. Both correspond to $(r, \theta) = (2, 0°)$;
$(x, y) = (90°, 6)$ corresponds to $(r, \theta) = (6, 90°)$;
$(x, y) = (180°, 2)$ corresponds to $(r, \theta) = (2, 180°)$;
$(x, y) = (270°, -2)$ corresponds to $(r, \theta) = (2, 90°)$.

23d. The inner loop of the limaçon corresponds to the part of the rectangular graph that is below the x-axis.

23e. For any given value of θ, there is only one value of r.

Problem Set 11-3

1. (1.9230..., 78.4630...°); (1.9230..., −78.4630...°)

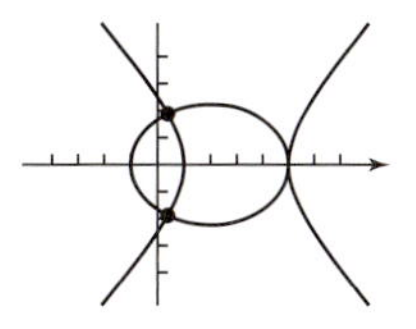

3. No solution

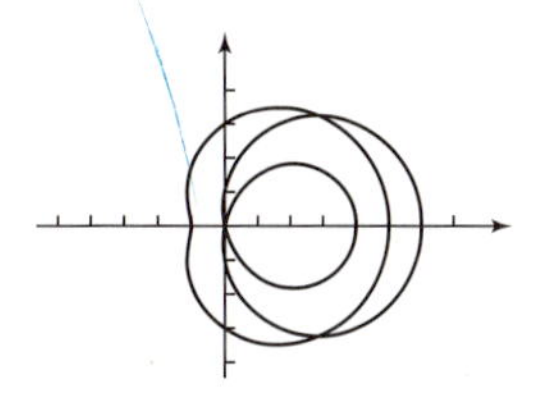

5. (4.1196..., 63.2717...°); (4.1196..., 116.7282...°); (1.2137..., −34.0431...°); (1.2137..., 214.0431...°)

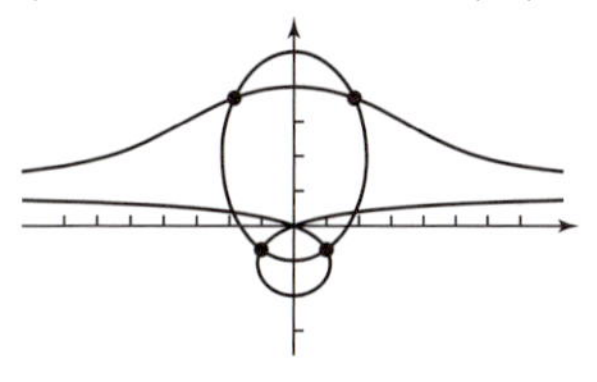

7. (0.3739..., 0.7479...); (1.2407..., 2.4815...); (1.8677..., 3.7355...); (2.9038..., 5.8076...); (3.3506..., 6.7013...); (4.6134..., 9.2268...); (4.7858..., 9.5716...)

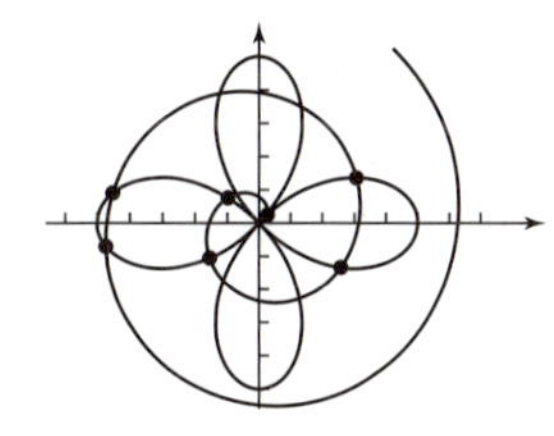

9.

For a given θ, $r_2(\theta)$ is the opposite of the value of $r_1(\theta)$ at $\theta + 180°$.

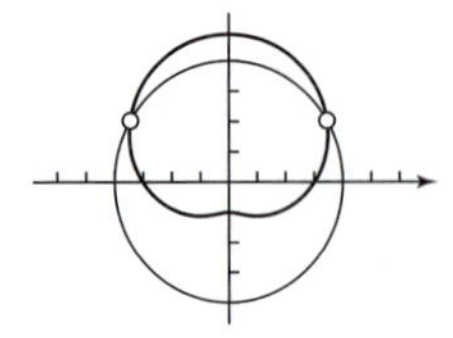

$r_1(\theta) = r_3(\theta) \Rightarrow (r, \theta) = (4, 30°)$ or $(4, 150°)$ while $r_2(\theta) = r_3(\theta) \Rightarrow \sin\theta = \frac{7}{2}$, which is impossible. However, the graphs of $r_1(\theta)$ and $r_2(\theta)$ coincide, so for $r_2(\theta)$ those same two points are intersections with $r_3(\theta)$, but not solutions to the system. To find the intersections of $r_1(\theta) = f(\theta)$ and $r_2(\theta) = g(\theta)$, first set $f(\theta) = g(\theta)$ and solve. To find the additional intersections, set $f(\theta) = -g(180° - \theta)$ and solve.

Problem Set 11-4

1. $\sqrt{2}$ cis 135°

3. 2 cis 330°

5. 5 cis 216.8698...°

7. 8.6023... cis 54.4623...°

9. 1 cis 0°

11. 1 cis 270°

13. 6.6323... + 4.4735... i

15. $-3 + 3i\sqrt{3}$

17. $-1 - i$

19. -5

21. $-3i$

23a. 15 cis 83°

23b. 0.6 cis 11°

23c. 9 cis 94°

23d. 125 cis 108°

25a. 8 cis 289°

25b. 2 cis 187°

25c. 16 cis 116°

25d. 8 cis 153°

27. 3 cis 40°, 3 cis 160°, 3 cis 280°

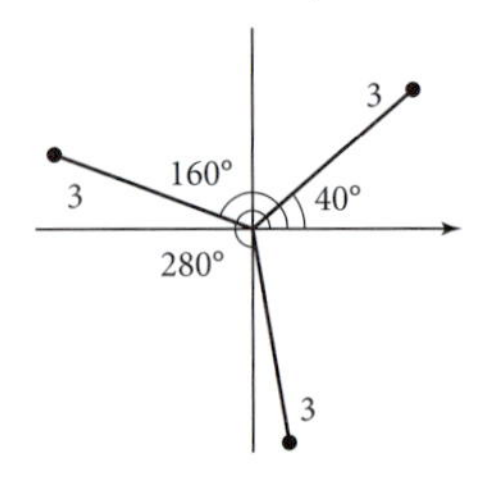

29. 2 cis 20°, 2 cis 110°, 2 cis 200°, 2 cis 290°

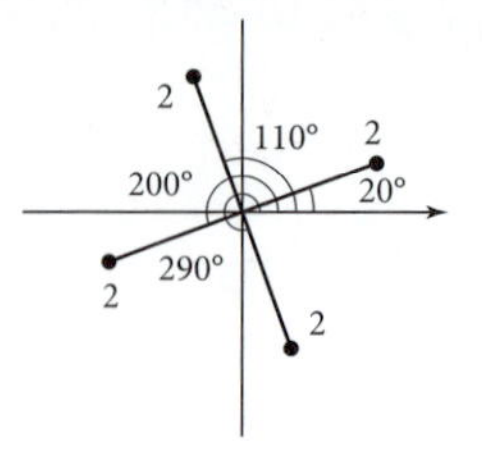

31. cis 45°, cis 225°

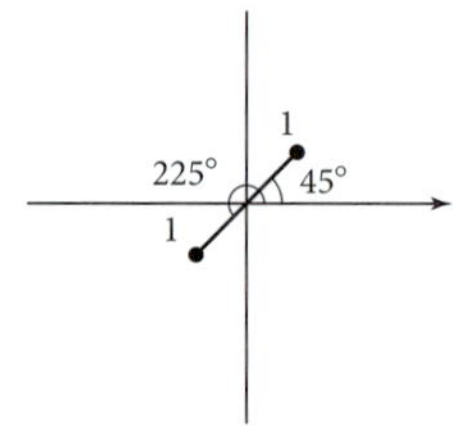

33. 2, 2 cis 120°, 2 cis 240°

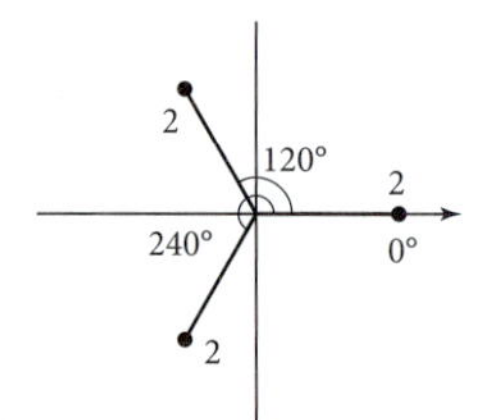

35. cis 30°, i, cis 150°, cis 210°, $-i$, cis 330°

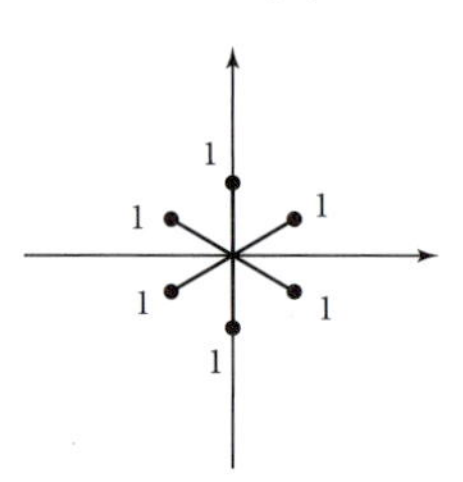

37. $(\cos\theta + i\sin\theta)^3 = \cos^3\theta + 3i\cos^2\theta\sin\theta + 3i^2\cos\theta\sin^2\theta + i^3\sin^3\theta = \cos^3\theta + 3i\cos^2\theta\sin\theta - 3\cos\theta\sin^2\theta - i\sin^3\theta = \cos^3\theta - 3\cos\theta\sin^2\theta + i(3\cos^2\theta\sin\theta - \sin^3\theta)$. But by De Moivre's theorem, $(\text{cis}\,\theta)^3 = \text{cis}\,3\theta = \cos 3\theta + i\sin 3\theta$. Equating real parts gives $\cos 3\theta = \cos^3\theta - 3\cos\theta\sin^2\theta = \cos\theta(\cos^2\theta - 3\sin^2\theta)$.

Equating imaginary parts gives $\sin 3\theta = 3\cos^2\theta\sin\theta - \sin^3\theta = \sin\theta(3\cos^2\theta - \sin^2\theta)$.

39. Journal entries will vary.

Problem Set 11-5

1a. $x = 473 - 300t$; $y = 155 + 100t$

1b. $\dfrac{473}{300}$ h; 312.6666... km

1c. 316.2277... km/h

3a. $x = 187.9385...t$; $y = 10 + 68.4040...t - 16t^2$

3b.

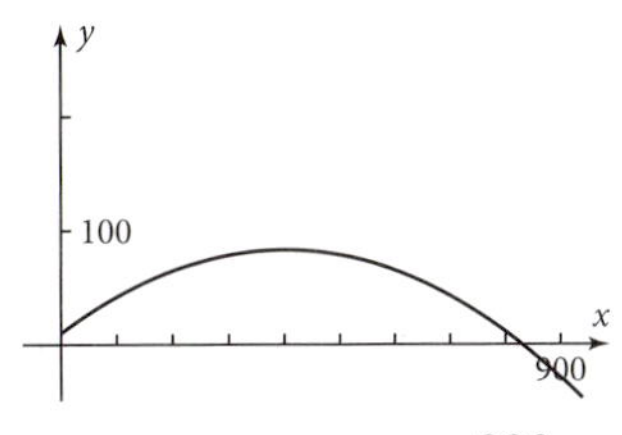

3c. $x = 900 \text{ ft} \Leftrightarrow t = \dfrac{900}{200 \cos 20^\circ} = 4.7887... \text{ s}$
At this time,
$y = 10 + 200\dfrac{900}{200 \cos 20^\circ} \sin 20^\circ - 16\left(\dfrac{900}{200 \cos 20^\circ}\right)^2$
$= -29.3484...$ ft.
The cannonball will fall short.

3d. 67.1111...° or 22.2522...°

3e. Answers may vary.

5a. The graph appears to be an ellipse (see 5c).

5b. The horizontal coordinate of P is the same as that of B, and the vertical coordinate of P is the same as that of A. $x = 5 \cos$, $y = 3 \sin t$. This is the parametric description of an ellipse.

5c.

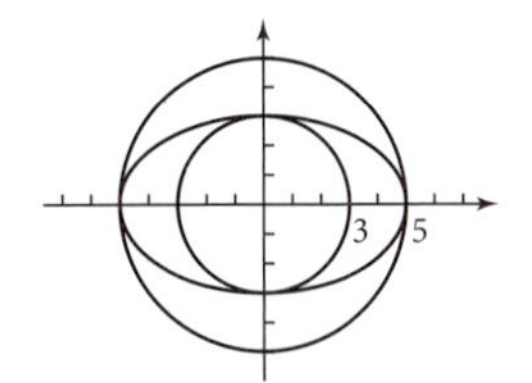

Yes, the circles have the same relationship.

5d. $\left(\dfrac{5 \cos t}{5}\right)^2 + \left(\dfrac{3 \sin t}{3}\right)^2 = 1$

The variables have unequal coefficients, but the same sign.

7a. $\vec{v}_1, \vec{v}_2$, and $\vec{v}_3$ placed head-to-tail connect the origin with $P(x, y)$.

7b. $\vec{v}_1 = (50t)\vec{i} + 0\vec{j}$; $\vec{v}_2 = 0\vec{i} + 50\vec{j}$

7c. $\vec{v}_3 = -(70 \sin t)\vec{i} - (70 \cos t)\vec{j}$;
$\vec{r} = (50t - 70 \sin t)\vec{i} + (50t - 70 \cos t)\vec{j}$

7d. The graph is correct.

7e. $-1.9883...$ cm
Because P is below the track, as the wheel rotates clockwise, P moves backward.

9a. $\vec{r} = (5 \cos t + 5t \sin t) \cdot \vec{i} + (5 \sin t - 5t \cos t) \cdot \vec{j}$

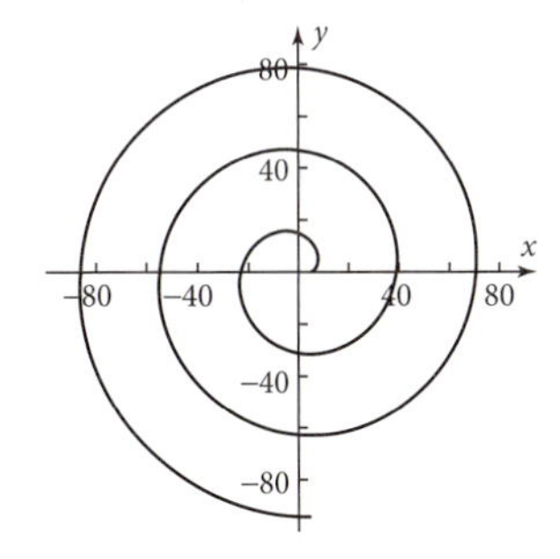

9b. Student project. The drawing should resemble the graph.

11a. The equation is correct.

11b. $x = r \cos t = \dfrac{9 \cos t}{5 - 4 \cos t}$; $y = r \sin t = \dfrac{9 \sin t}{5 - 4 \cos t}$
The equations give the same ellipse.

11c. $x = 4 + 5 \cos t$; $y = 3 \sin t$

11d. The two finished graphs coincide. The t in 11b refers to the angle with the positive x-axis, measured from the origin, but the angle t in 11c is measured from the center (4, 0) of the ellipse.

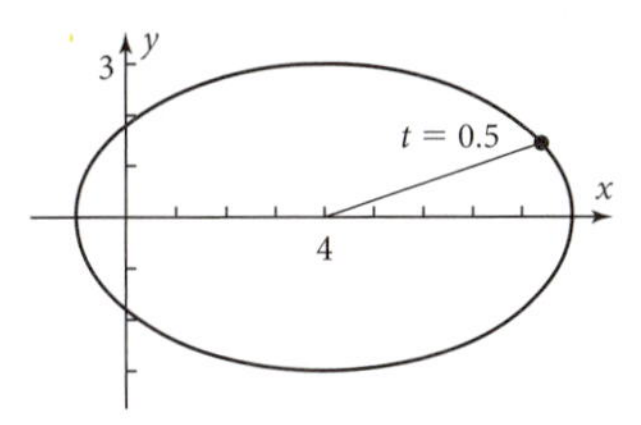

Problem Set 11-6

R1.

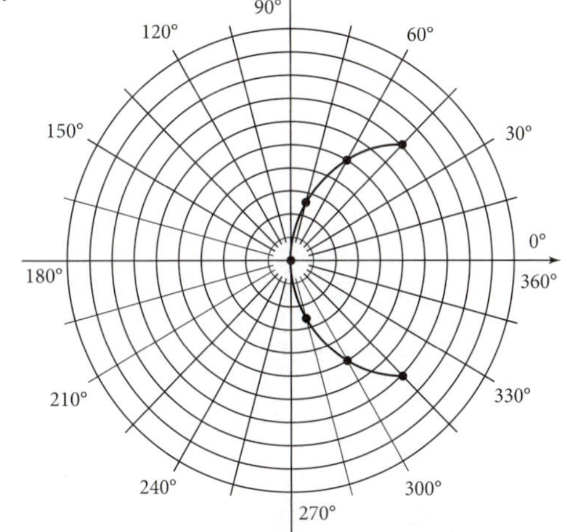

Answers to Selected Problems

R3. (4.6666..., 83.6206...), (4.6666..., −83.6206...°); The graphs also cross at (−2, 180°), (2, 0°).

R5a. $x = 60t \cos 24° = 54.8127...t$
$y = 60t \sin 24° = 24.4041...t$
22.2614... mi

R5b. i. The drawing seems to match the description.

R5b. ii. $\vec{r} = \left(21 \cos t + 9 \cos \frac{7}{3}t\right) \cdot \vec{i} + \left(21 \sin t + 9 \sin \frac{7}{3}t\right) \cdot \vec{j}$

R5b. iii. $0 \le t \le 6\pi$

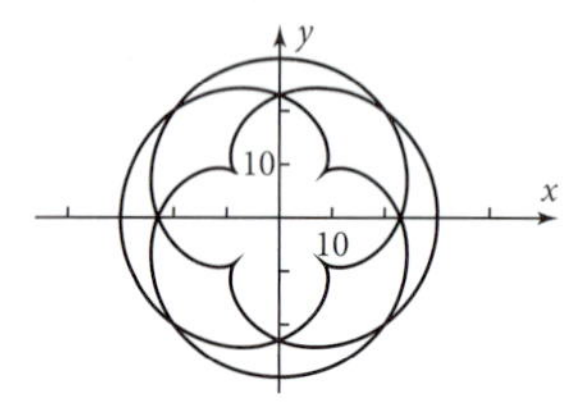

C1a. $\vec{r} = (10 \cos A + 3 \cos B) \cdot \vec{i} + (10 \sin A + 3 \sin B) \cdot \vec{j}$

C1b. $A = 2t$; $B = 12t$;
$\vec{r} = (10 \cos 2t + 3 \cos 12t) \cdot \vec{i} + (10 \sin 2t + 3 \sin 12t) \cdot \vec{j}$

C1c.

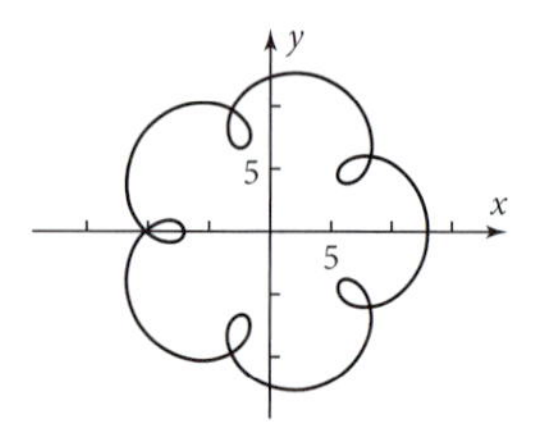

C1d. As the planet circles the center once, the point does go left to right and back six times (as viewed from above), but the point moves toward and away from the center only five times, and this type of motion is what generates the loops.

C1e. i.

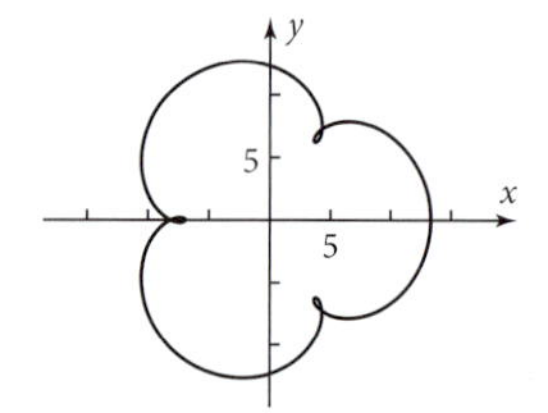

C1e. ii.

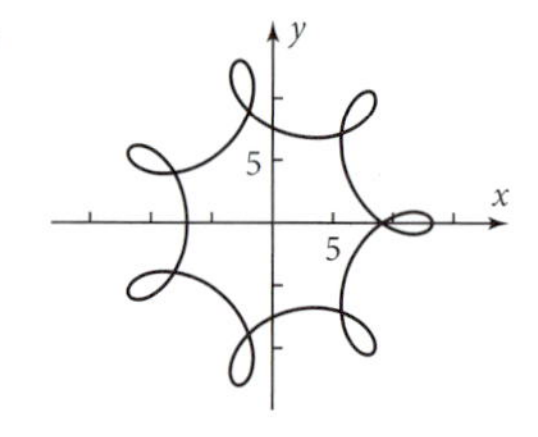

C1e. iii.

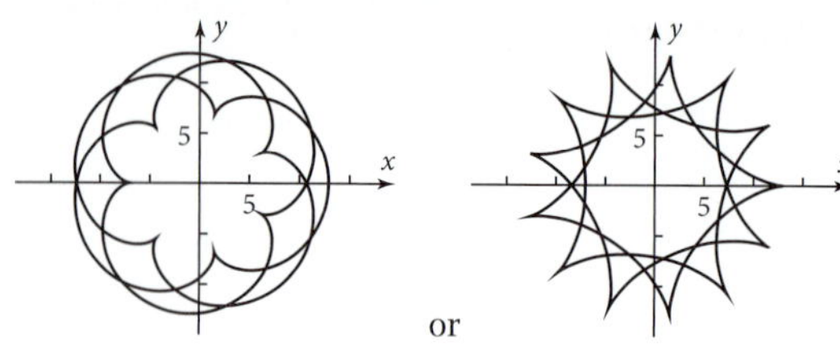

C3a. $a^2 = r^2 + k^2 - 2rk \cos(\theta - \alpha)$
$r = k \cos(\theta - \alpha) \pm \sqrt{k^2 \cos^2(\theta - \alpha) - k^2 + a^2}$

C3b.

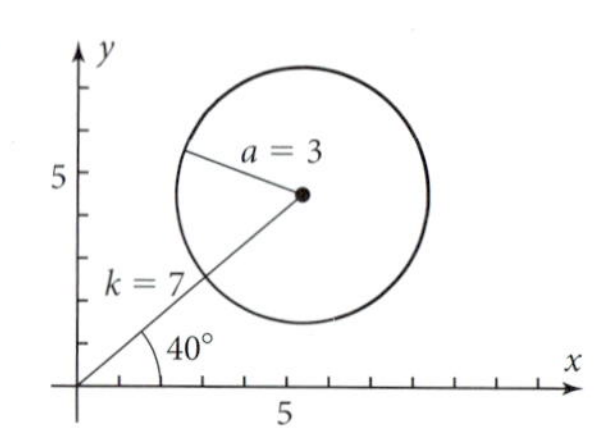

The grapher first plots the arc of the circle farther from the origin. Then the grapher plots the arc of the circle nearer the origin.

C3c. The graph is the same, except that the grapher plots the nearer part first with $r > 0$ and then plots the farther part with $r < 0$.

C3d. $r(50°) = 7 \cos 10° \pm \sqrt{49\cos^2 10° - 40}$
$= 9.6363..., 4.1509...$

C3e. $r(90°) = 7 \cos 50° \pm \sqrt{49 \cos^2 50° - 40}$
$= 4.4995... \pm \sqrt{-19.7543...}$
There are no real solutions.

T1.

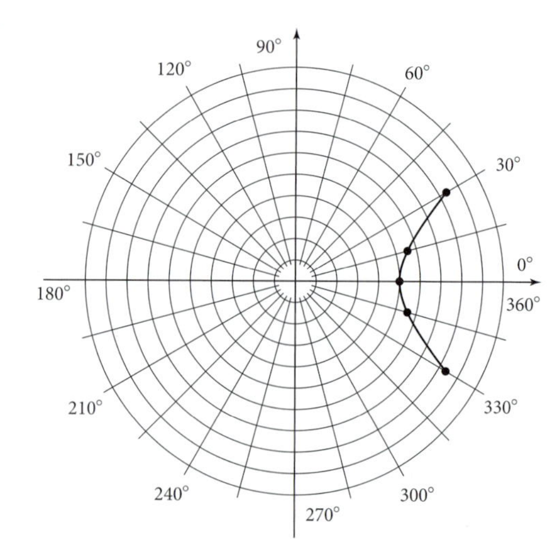

T3. $r = 4 \sec(\theta - 50°)$

T5. 15 cis 91°

T7. 64 cis 150°

T9. (7, 390°), (−7, 210°)

T11. For $r = 3 \sin 4\theta$, the coefficient of θ is even, so there are $4 \cdot 2 = 8$ leaves.

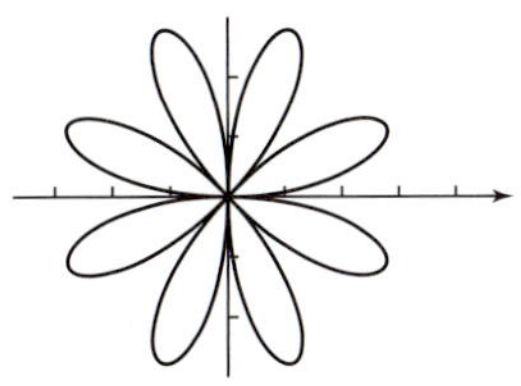

For $r = 3 \sin 5\theta$, the coefficient of θ is odd, so there are 5 leaves.

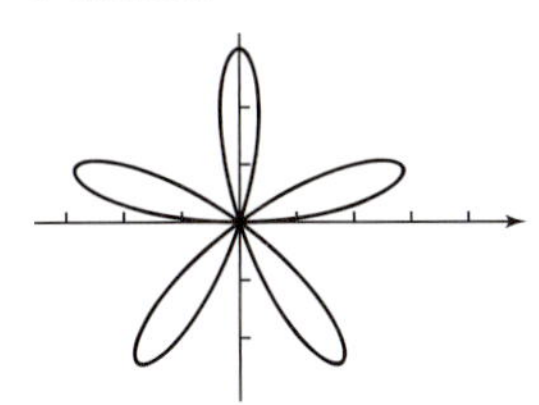

T13. $3 - 5.1961...i$

T15. $1033.7131...\text{ cis } 156.1570...°$

T17. Answers will vary.

Chapter 12

Problem Set 12-1

1.

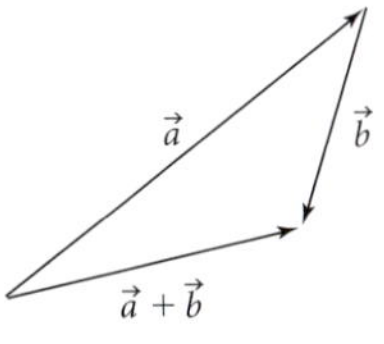
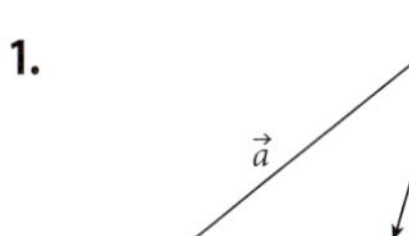

3.

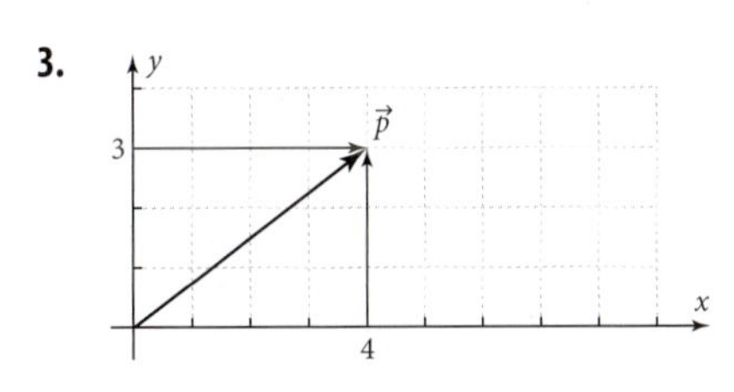

5.

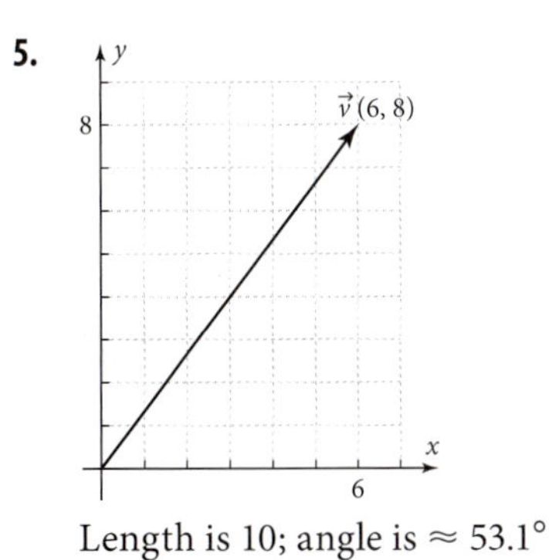

Length is 10; angle is $\approx 53.1°$

7.

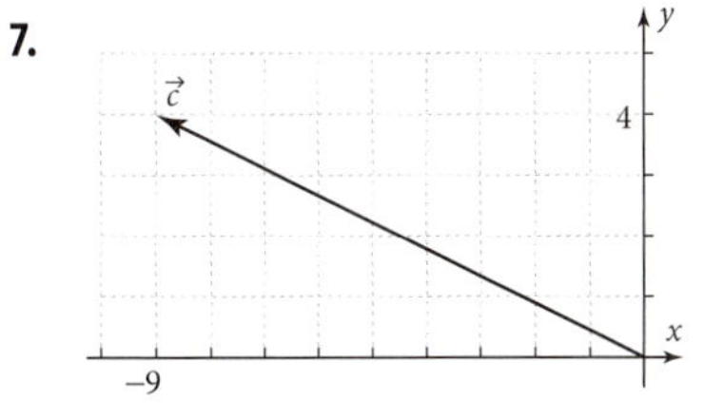

Problem Set 12-2

1a. $\vec{a} = 5\vec{i} + 3\vec{j}$; $\vec{b} = 2\vec{i} + 4\vec{j}$

1b.

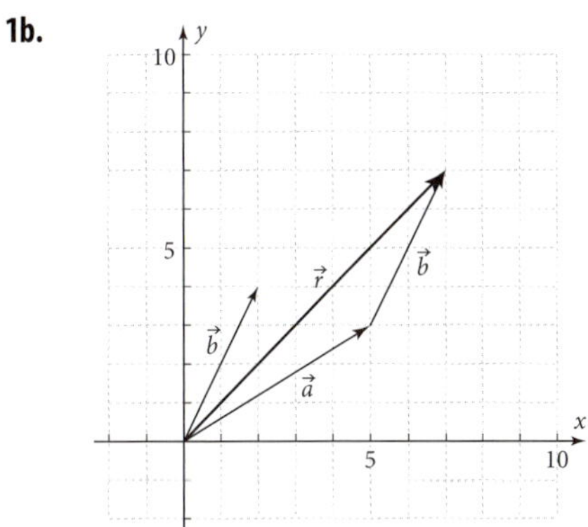

$\vec{r} = \vec{a} + \vec{b} = (5\vec{i} + 3\vec{j}) + (2\vec{i} + 4\vec{j}) = 7\vec{i} + 7\vec{j}$.

1c. Translate $\vec{a}$ so that its tail is at the head of $\vec{b}$, then draw the resultant vector $\vec{r} = \vec{b} + \vec{a}$.
$\vec{b} + \vec{a} = (2\vec{i} + 4\vec{j}) + (5\vec{i} + 3\vec{j}) = (7\vec{i} + 7\vec{j}) = \vec{a} + \vec{b}$.

1d. $|\vec{a}| = 5.8309...$; $|\vec{b}| = \sqrt{2^2 + 4^2} = \sqrt{20}$ $= 4.4721...$; $|\vec{a} + \vec{b}| = 9.8994...$; $|\vec{a}| + |\vec{b}| = 10.3030...$; $|\vec{a} + \vec{b}| < |\vec{a}| + |\vec{b}|$ because the length of any side of a triangle is less than the sum of the lengths of the other two sides.

3a.

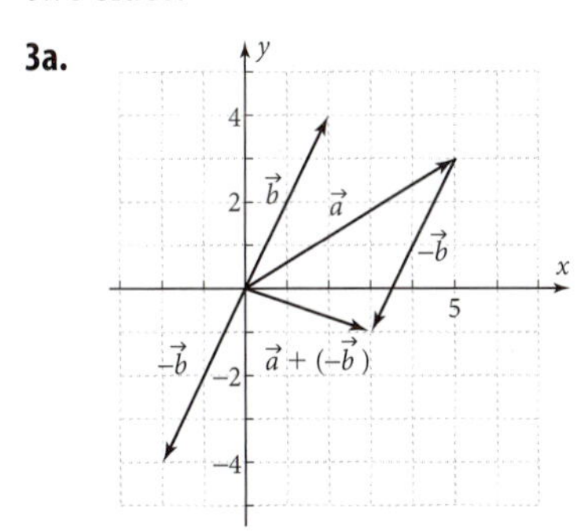

In short, subtracting a vector is the same as adding its opposite because vector arithmetic is defined in terms of components, which are *scalars* (numbers), and subtracting a number is the same as adding its opposite.

3b.

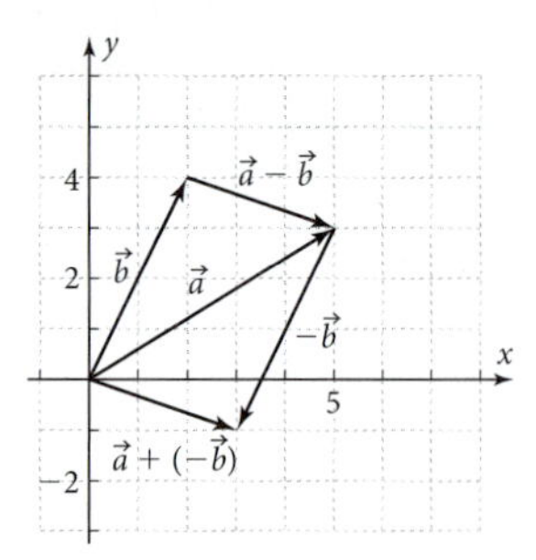

On the graph, the new vector can be seen to have the same magnitude and direction as $\vec{a}-\vec{b}$, found in 3a.

3c. $\vec{a}-\vec{b}=(5\vec{i}+3\vec{j})-(2\vec{i}+4\vec{j})=3\vec{i}-\vec{j}$, which matches the vectors found in 3a and 3b.

5a. $47\vec{i}+45\vec{j}$

5b. $u=0.8574...\vec{i}+0.5144...\vec{j}$; $10\vec{u}=8.5749...\vec{i}+5.1449...\vec{j}$

5c. $243.4349...°$

7. $-\vec{i}+3\vec{j}$

9. $2\vec{i}+4\vec{j}$

11a. $\vec{A}=20\vec{i}+73\vec{j}, \vec{B}=45\vec{i}+10\vec{j}$

11b. $25\vec{i}-63\vec{j}$

11c. $10\vec{i}-25.2\vec{j}$

11d. $30\vec{i}+47.8\vec{j}$

11e. ≈ 56.43 km; $\theta \approx 57.9°$

11f. $|\overrightarrow{AR}| \approx 27.1$ km to Artesia

13. $6\vec{i}+\frac{19}{3}\vec{j}$

15. $\frac{1}{2}\overrightarrow{EF}=2\vec{i}-3\vec{j}$, but this is just $\frac{1}{2}(16\vec{i}-2\vec{j}) = \frac{1}{2}\left[(6\vec{i}+2\vec{j})+(10\vec{i}-4\vec{j})\right]=\frac{1}{2}(\vec{E}+\vec{F})$.
The position vector of the midpoint of two points is the average of their position vectors.

17a.

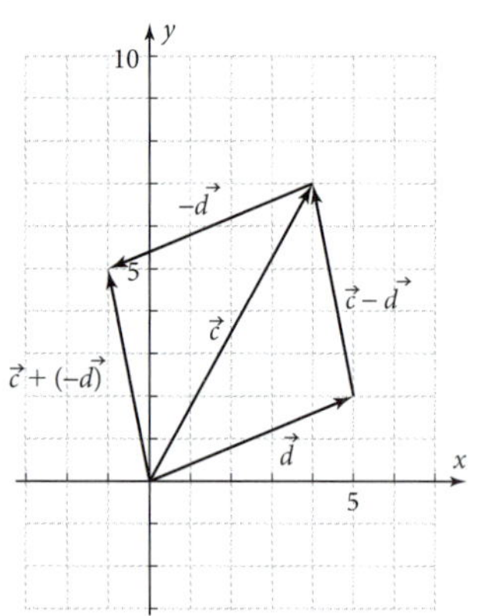

$\vec{a}+\vec{b}=\vec{b}+\vec{a}$

17b. $\vec{v}+(\vec{w}+\vec{x})$
$=(a\vec{i}+b\vec{j})+\left[(c\vec{i}+d\vec{j})+(e\vec{i}+f\vec{j})\right]$
$=(a\vec{i}+b\vec{j})+\left[(c+e)\vec{i}+(d+f)\vec{j}\right]$ by the definition of vector addition, $=[a+(c+e)]\vec{i}+[b+(d+f)]\vec{j}$ by the definition of vector addition, $=[(a+c)+e]\vec{i}+[(b+d)+f]\vec{j}$ by the associativity of addition for scalars, $=\left[(a+c)\vec{i}+(b+d)\vec{j}\right]+(e\vec{i}+f\vec{j})$ by the definition of vector addition, $=\left[(a\vec{i}+b\vec{j})+(c\vec{i}+d\vec{j})\right]+(e\vec{i}+f\vec{j})$ by the definition of vector addition, $=(\vec{v}+\vec{w})+\vec{x}$, so vector addition is associative.

17c.

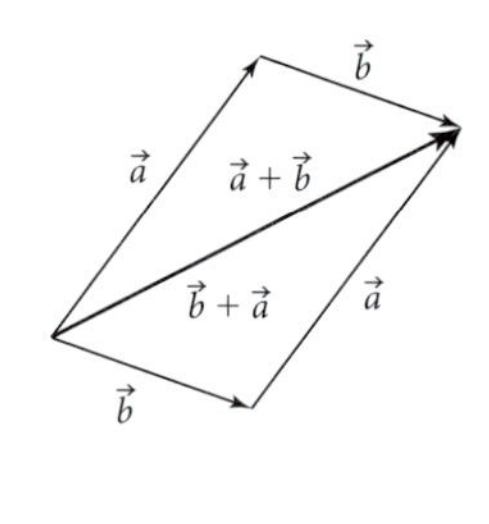

$\vec{a}-\vec{b}\neq\vec{b}-\vec{a}$

17d. $s(\vec{v}+\vec{w})=s\left[(a\vec{i}+b\vec{j})+(c\vec{i}+d\vec{j})\right]$
$=\left[s(a+c)\vec{i}+s(b+d)\vec{j}\right]$ by the definition of vector addition, $=[s(a+c)]\vec{i}+[s(b+d)]\vec{j}$ by the definition of scalar multiplication, $=(sa+sc)\vec{i}+(sb+sd)\vec{j}$ because real-number multiplication distributes over real-number addition, $=s(a\vec{i}+b\vec{j})+s(c\vec{i}+d\vec{j})$ by the definition of scalar multiplication, $=s\vec{v}+s\vec{w}$, so scalar multiplication distributes over vector addition.
There is actually a second distributive property for vectors: $(s+t)\vec{v}=(s+t)(a\vec{i}+b\vec{j})=[(s+t)a]\vec{i}+[(s+t)b]\vec{j}$ by the definition of scalar multiplication, $=(sa+ta)\vec{i}+(sb+tb)\vec{j}$ because real-number multiplication distributes over real-number addition, $=\left[(sa)\vec{i}+(sb)\vec{j}\right]+\left[(ta)\vec{i}+(tb)\vec{j}\right]$ by the definition of vector addition, $=s(a\vec{i}+b\vec{j})+t(a\vec{i}+b\vec{j})$ by the definition of scalar multiplication, $=s\vec{v}+t\vec{v}$, so $(s+t)\vec{v}=s\vec{v}+t\vec{v}$.

17e. You must show that the sum of any two vectors is also a vector. $\vec{v} + \vec{w} = \left(a\vec{i} + b\vec{j}\right) + \left(c\vec{i} + d\vec{j}\right) = (a + c)\vec{i} + (b + d)\vec{j}$ by the definition of vector addition, where a, b, c, and d are real numbers. But the real numbers are closed under addition, so $a + c$ and $b + d$ are real numbers. Therefore, $(a + c)\vec{i} + (b + d)\vec{j}$ or, $\vec{v} + \vec{w}$, is a vector.

If there were no zero-vector, a sum of the form $\left(a\vec{i} + b\vec{j}\right) + \left[(-a)\vec{i} + (-b)\vec{j}\right] = [a + (-a)]\vec{i} + \left[b + (-b)\vec{j}\right] = 0\vec{i} + 0\vec{j}$ would not yield a vector.

Problem Set 12-3

1.

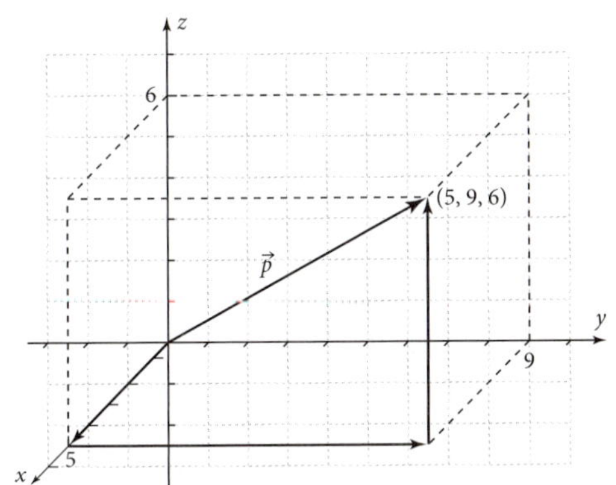

3.

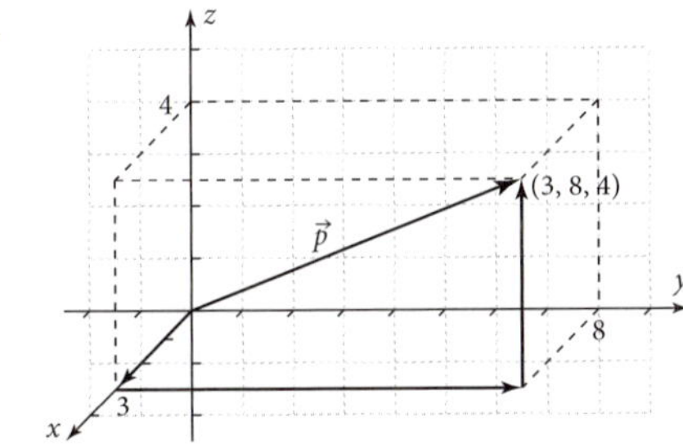

5a. $\vec{a} + \vec{b} = 11\vec{i} - 3\vec{j} - 2\vec{k}$; $\vec{a} - \vec{b} = -3\vec{i} + 7\vec{j} - 4\vec{k}$; $\vec{b} - \vec{a} = 3\vec{i} - 7\vec{j} + 4\vec{k}$

5b. $3\vec{a} = 12\vec{i} + 6\vec{j} - 9\vec{k}$;
$6\vec{a} - 5\vec{b} = -11\vec{i} + 37\vec{j} - 23\vec{k}$

5c. $|\vec{a} + \vec{b}| = 11.5758...$; $|\vec{a}| + |\vec{b}| = 14.0454...$; No

5d. $\vec{u} = (0.8082...)\vec{i} - (0.5773...)\vec{j} + (0.1154...)\vec{k}$;
$20\vec{u} = (16.1658...)\vec{i} - (11.5470...)\vec{j} + (2.3094...)\vec{k}$

7. $\overrightarrow{RS} = 3\vec{i} + 7\vec{j} - 6\vec{k}$; $|\overrightarrow{RS}| = \sqrt{94}$

9. $\overrightarrow{BA} = 6\vec{i} + 7\vec{j} - 6\vec{k}$; $|\overrightarrow{BA}| = \sqrt{121} = 11$

11a.

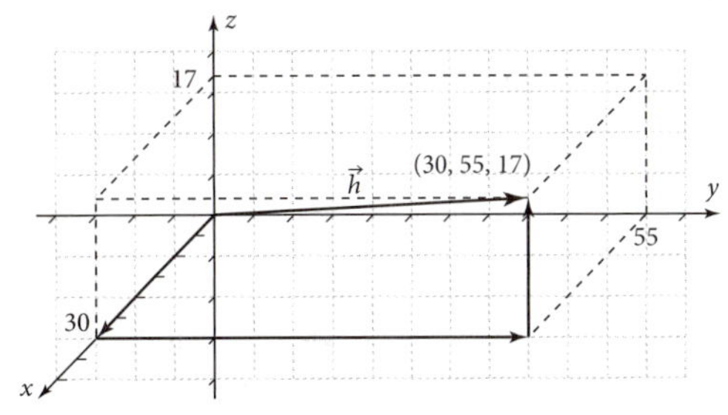

11b. $\vec{h} = 30\vec{i} + 55\vec{j} + 17\vec{k}$; 17 ft above ground; $\approx$64.9 ft from the origin

11c. $-20\vec{i} - 55\vec{j} - 9\vec{k}$

11d. $\approx$59.2 ft

11e. $\approx$17.8 ft

11f. $24\vec{i} + 38.5\vec{j} + 14.3\vec{k}$; 14.3 ft

13. $25\vec{i} + 24\vec{j} + 13\vec{k}$

15. $4.6\vec{i} - 6.6\vec{j} - 0.8\vec{k}$

17.

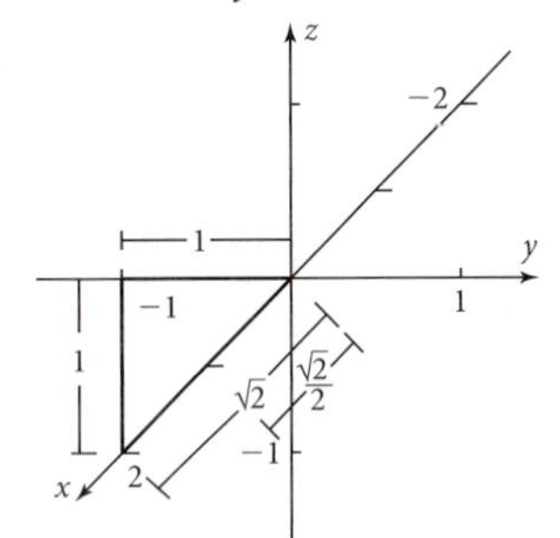

By the Pythagorean theorem, the length of 2 x-units is equal to $\sqrt{2}$ y- or z-units. So each x-unit is equal to $\frac{\sqrt{2}}{2}$, or about 70% of the y- or z-units.

19a. $|\vec{a}| = 9.3273...$; $|\vec{b}| = 14$

19b. $\vec{a} + \vec{b} = (8, 16, 9, 8)$

19c. $\vec{a} - \vec{b} = (-2, -6, -5, 6)$

19d. $\overrightarrow{AB} = (2, 6, 5, -6)$

19e. (3.8, 7.4, 4, 4.6)

Problem Set 12-4

1. 598.9766...

3. −872.6317...

5. 0

7. 80.4095...°

9. 131.1288...°

11. 0°

13. $\vec{a} \cdot \vec{b} = 21$; $\theta = 65.2077...°$

15. $\vec{a} \cdot \vec{b} = -31$; $\theta = 137.8564...°$

17. $\vec{a} \cdot \vec{b} = 0; \theta = 90°$

19a. $\vec{F}_{res} = 45\vec{i} + 120\vec{j} + 15\vec{k}$

19b. 120 lb

19c. 45 lb

19d. 15 lb

19e. $\approx$129.0 lb

19f. $|\vec{F}_1| \approx 72.3$ lb; $|\vec{F}_2| \approx 58.5$ lb; No

19g. $\vec{F}_1 \cdot \vec{F}_2 = 4000; \theta \approx 19.0°$

21. Answers will vary.

23. $-(46.6755...)\vec{i} + (56.0106...)\vec{j} - (37.3404...)\vec{k}$

25a. 2.9199...

25b. $(2.3684...)\vec{i} - (0.9473...)\vec{j} - (1.4210...)\vec{k}$

27a. $-1.6903...$

27b. $-(0.2857...)\vec{i} + (1.4285...)\vec{j} - (0.8571...)\vec{k}$

29. Scalar product, inner product, dot product

31. Journal entries will vary.

Problem Set 12-5

1. $3\vec{i} + 5\vec{j} - 7\vec{k}, -3\vec{i} - 5\vec{j} + 7\vec{k}$

3. $3x - 5y + 4z = 45$

5. $8x - 6y - 8z = 18$

7. $5x - 3y - z = 37$

9. $z_1 = 10;\ z_2 = \frac{21}{5};\ d = 7.914...;\ y = \frac{-54}{7}$

11a. $30x - 17y + 11z = 900$

11b. The x-axis; at $x = 30$ m

11c. $x = -31.8181...$ m, about 31.8 m deep

11d. $\approx 72.3°$

13. $\vec{n}_1 = 2\vec{i} - 5\vec{j} + 3\vec{k}, \vec{n}_2 = 7\vec{i} + 4\vec{j} + 2\vec{k}$
$\Rightarrow \vec{n}_1 \cdot \vec{n}_2 = 2 \cdot 7 - 5 \cdot 4 + 3 \cdot 2 = 0$

15. Let $P_0(x_0, y_0, z_0)$ be a fixed point in the plane, and let $P(x, y, z)$ be an arbitrary different point in the plane, so that the displacement vector from P_0 to P is

$\vec{d} = (x - x_0)\vec{i} + (y - y_0)\vec{j} + (z - z_0)\vec{k}$.

Then $\vec{n} \cdot \vec{d} = 0$

$\Leftrightarrow A(x - x_0) + B(y - y_0) + C(z - z_0) = 0$
$\Leftrightarrow Ax + By + Cz - Ax_0 - By_0 - Cz_0 = 0$
$\Leftrightarrow Ax + By + Cz + D = 0$, where D
$= -Ax_0 - By_0 - Cz_0$.

Problem Set 12-6

1. $-8\vec{i} + 7\vec{j} - 2\vec{k}$

3. $-4\vec{i} - 6\vec{j} + 2\vec{k}$

5. 34

7. Student program. See **www.keymath.com/precalc** for an example program.

9. $19x + 24y - 17z = 41$

11. $12x + 14y - 3z = 63$

13. 62.6498...

15. 33.7268...

17a. $\vec{d}_{z,x} = 10\vec{i} - 5\vec{k};\ \vec{d}_{z,y} = 15\vec{j} - 5\vec{k}$

17b. $75\vec{i} + 50\vec{j} + 150\vec{k}$

17c. 87.5 ft^2

17d. $|\vec{d}_{z,x}| \approx 11.2$ ft; $|\vec{d}_{z,y}| \approx 15.8$ ft; $|\vec{d}_{x,y}| \approx 18.0$ ft;
$\theta_x \approx 60.3°;\ \theta_y \approx 37.9°;\ \theta_z \approx 81.9°$

17e. $3x + 2y + 6z = 78$; the point is above the awning.

19a. $-9\vec{i} + 42\vec{j} + 43\vec{k}$

19c. $\vec{a} \cdot \vec{b} = -12; \theta = 101.1687...°$

19d. $|\vec{a} \times \vec{b}| = \sqrt{9^2 + 42^2 + 43^2} = \sqrt{3694}$
$= 60.7782...;\ |\vec{a}| \cdot |\vec{b}| \cdot \sin\theta$
$= \sqrt{38} \cdot \sqrt{101} \cdot \sin 101.1687...° = 60.7782...$

21a. $\vec{h} = 5\vec{i} - 10\vec{j} + 20\vec{k} = -\frac{5}{3}(-3\vec{i} + 6\vec{j} - 12\vec{k})$
$= -\frac{5}{3}\vec{g}$

21b. $|\vec{g} \times \vec{h}| =$
$|(-3\vec{i} + 6\vec{j} - 12\vec{k}) \times (5\vec{i} - 10\vec{j} + 20\vec{k})|$
$= |0\vec{i} + 0\vec{j} + 0\vec{k}| = 0$

21c. $|\vec{g} \times \vec{h}| = |\vec{g}| \cdot |\vec{h}| \cdot \sin 180° = |\vec{g}| \cdot |\vec{h}| \cdot 0 = 0$

23. $z = 2; x = 5$

Problem Set 12-7

1.

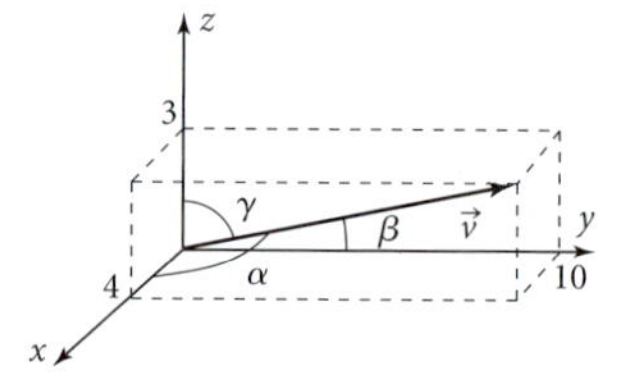

3. $c_1 = \frac{2}{\sqrt{38}};\ \alpha = 71.0681...°;\ c_2 = \frac{-5}{\sqrt{38}};$
$\beta = 144.2042...°;\ c_3 = \frac{3}{\sqrt{38}};\ \gamma = 60.8784...°;$

5. $c_1 = \frac{-4}{21};\ \alpha = 100.9805...°;\ c_2 = \frac{8}{21};\ \beta = 67.6073...°;$
$c_3 = \frac{19}{21};\ \gamma = 25.2087...°$

7. $c_1 = \dfrac{7}{\sqrt{59}}$; $c_2 = \dfrac{1}{\sqrt{59}}$; $c_3 = \dfrac{-3}{\sqrt{59}}$

9. $c_1 = \dfrac{9}{17}$; $c_2 = \dfrac{-8}{17}$; $c_3 = \dfrac{12}{17}$

11. $\vec{u} = \dfrac{8}{3\sqrt{41}}\vec{i} + \dfrac{16}{3\sqrt{41}}\vec{j} - \dfrac{7}{3\sqrt{41}}\vec{k}$

$c_1 = \dfrac{8}{3\sqrt{41}}$; $c_2 = \dfrac{16}{3\sqrt{41}}$; $c_3 = \dfrac{-7}{3\sqrt{41}}$

13. $\left(\frac{7}{9}\right)^2 + \left(\frac{4}{9}\right)^2 + \left(\frac{4}{9}\right)^2 = \dfrac{49}{81} + \dfrac{16}{81} + \dfrac{16}{81} = \dfrac{81}{81} = 1$;

$c_1 = \frac{7}{9}$; $\alpha = 38.9424...°$; $c_2 = \frac{4}{9}$; $\beta = 63.6122...°$;

$c_3 = \frac{-4}{9}$; $\gamma = 116.3877...°$

15. $c_3 = \pm\frac{6}{23}$; $\gamma = 74.8783...°$ or $105.1216...°$

17. $\gamma = 45°$ or $135°$

19. No possible value.

21a. ≈ 14.0 ft/s

21b. $\alpha = 69.0190...°$; $\beta = 38.0264...°$; $\gamma = 59.9152...°$

21c. $30.0847...°$

21d. $5\vec{i} + 11\vec{j}$

21e. $65.5560...°$

21f. The two direction angles would change. Because $\cos^2\alpha + \cos^2\beta + \cos^2\gamma = 1$, the change in γ has to affect at least one of α and β. The azimuth angle would not change.

23. $c_1{}^2 + c_2{}^2 + c_3{}^2 = \left(\dfrac{A}{|\vec{v}|}\right)^2 + \left(\dfrac{B}{|v|}\right)^2 + \left(\dfrac{C}{|v|}\right)^2$

$= \left(\dfrac{A}{\sqrt{A^2 + B^2 + C^2}}\right)^2 + \left(\dfrac{B}{\sqrt{A^2 + B^2 + C^2}}\right)^2$

$+ \left(\dfrac{C}{\sqrt{A^2 + B^2 + C^2}}\right)^2 = \dfrac{A^2 + B^2 + C^2}{A^2 + B^2 + C^2} = 1$

Problem Set 12-8

1a. $(5, -3, 4)$

1b. $\frac{9}{17}\vec{i} + \frac{12}{17}\vec{j} + \frac{8}{17}\vec{k}$

1c. $\left(\frac{9}{17}\right)^2 + \left(\frac{12}{17}\right)^2 + \left(\frac{8}{17}\right)^2$

$= \dfrac{81}{289} + \dfrac{144}{289} + \dfrac{64}{289} = \dfrac{289}{289} = 1$

3. $(23, 21, 20)$

5. $\left(\frac{1}{2}, -9, 0\right)$

7. $\left(\frac{2}{7}\right)^2 + \left(\frac{6}{7}\right)^2 + \left(-\frac{3}{7}\right)^2 = \dfrac{4}{49} + \dfrac{36}{49} + \dfrac{9}{49} = \dfrac{49}{49} = 1$;

$\vec{r} = \left(5 + \frac{2}{7}d\right)\vec{i} + \left(-1 + \frac{6}{7}d\right)\vec{j} + \left(4 - \frac{3}{7}d\right)\vec{k}$

9. $c_1 = \dfrac{2}{\sqrt{29}}$; $c_2 = \dfrac{-3}{\sqrt{29}}$; $c_3 = \dfrac{4}{\sqrt{29}}$;

$\vec{r} = \left(1 + \dfrac{2}{\sqrt{29}}d\right)\vec{i} + \left(-8 - \dfrac{3}{\sqrt{29}}d\right)\vec{j} + \left(-5 + \dfrac{4}{\sqrt{29}}d\right)\vec{k}$

11. $\vec{r} = \left(5 + \frac{9}{25}d\right)\vec{i} + \left(1 + \frac{4}{5}d\right)\vec{j} + \left(-4 + \frac{12}{25}d\right)\vec{k}$

13. $\left(-\frac{79}{3}, -\frac{76}{3}, \frac{53}{3}\right)$

15a. $\vec{u} = -\dfrac{3\sqrt{2}}{10}\vec{i} + \dfrac{2\sqrt{2}}{5}\vec{j} + \dfrac{\sqrt{2}}{2}\vec{k}$

15b. $\left(10 - \frac{3\sqrt{2}}{10}d\right)\vec{i} + \left(14 + \frac{2\sqrt{2}}{5}d\right)\vec{j} + \left(3 + \frac{\sqrt{2}}{2}d\right)\vec{k}$

15c. 8 ft, because you know the floor is at $z = 0$ and you know a point on the ceiling is at $z = 8$.

15d. The roof is perpendicular to the xz-plane, so the normal to the roof is parallel to the xz-plane and therefore has $0\vec{j}$ as its y-component. Because the triangular part of the wall is a 45°-45° right triangle, simple geometry shows that the line through (6, 0, 8) and (0, 0, 14) is a normal. This vector is $(0 - 6)\vec{i} + (0 - 0)\vec{j} + (14 - 8)\vec{k} = -6\vec{i} + 0\vec{j} + 6\vec{k}$.

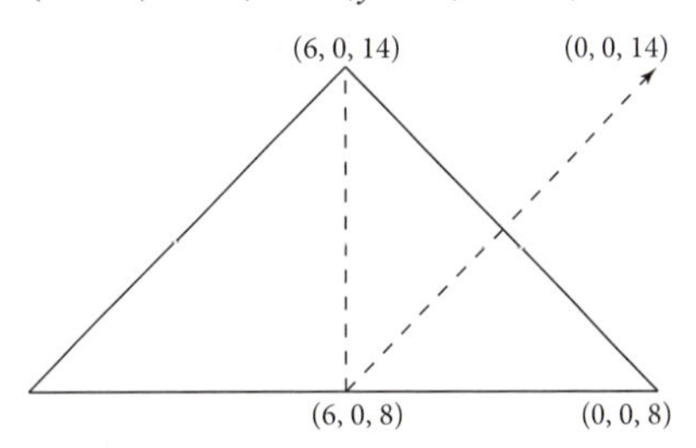

15e. $-6x + 6z = 48$

15f. (4.375 ft, 21.5 ft, 12.375 ft)

15g. *Forensic* means "belonging to, used in, or suitable to public discussion and debate." The evidence about the bullet and its path could be used in a trial.

Problem Set 12-9

R1a. They start at the origin and go to a point.

R1b.

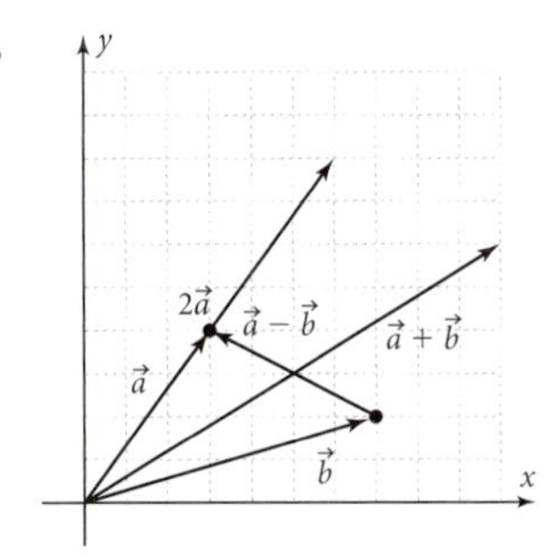

R1c. $-4\vec{i} + 2\vec{j}$

R1d. 11.6619...

R1e. $\theta = 53.1301...°$

R3a.

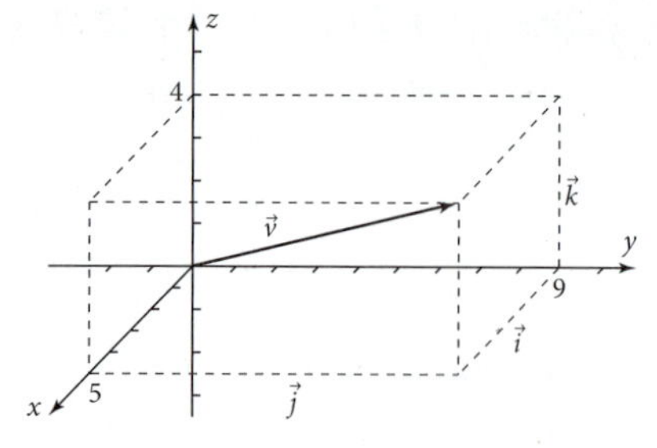

R3b. $12\vec{i} - 23\vec{j} + 20\vec{k}$

R3c. 8.0622...

R3d. $\vec{u} = \frac{6}{\sqrt{65}}\vec{i} - \frac{5}{\sqrt{65}}\vec{j} + \frac{2}{\sqrt{65}}\vec{k}$

R3e. $-3\vec{i} + 9\vec{j} - 9\vec{k}$

R3f. $3.9\vec{i} + 1.3\vec{j} - 4.3\vec{k}$

R5a. Let $P_0 = (x_0, y_0, z_0)$ be a fixed point on the plane, and let $P = (x, y, z)$ be an arbitrary point on the plane. Then $\overrightarrow{P_0P} = (x - x_0)\vec{i} + (y - y_0)\vec{j} + (z - z_0)\vec{k}$ is contained within the plane and so is normal to $\vec{n}$.
Thus: $\overrightarrow{P_0P} \cdot \vec{n} = 0$
$A(x - x_0) + B(y - y_0) + C(z - z_0) = 0$
$Ax + By + Cz = Ax_0 + By_0 + Cz_0$
$Ax + By + Cz = D$, where $D = Ax_0 + By_0 + Cz_0$.

R5b. $\vec{n}_1 = 3\vec{i} - 7\vec{j} + \vec{k}$; $\vec{n}_2 = -3\vec{i} + 7\vec{j} - \vec{k}$

R5c. $2x - 7y - 3z = 1$; $z = \frac{-121}{3}$

R5d. $3x + 6y + 9z = 45$.

R7a.

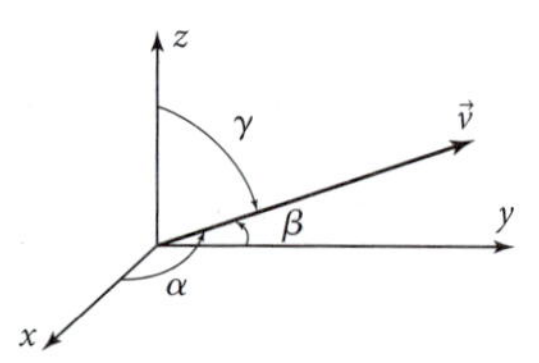

R7b. $c_1 = \frac{6}{5\sqrt{5}} = 0.5366...$; $c_2 = \frac{-8}{5\sqrt{5}} = -0.7155...$;
$c_3 = \frac{1}{\sqrt{5}} = 0.4472...$; $\alpha = \cos^{-1} c_1 = 57.5436...°$;
$\beta = \cos^{-1} c_2 = 135.6876...°$; $\gamma = \cos^{-1} c_3 = 63.4349...°$

R7c. $c_3 = \pm 0.9327...$; $\gamma = 21.1342...°$ or $158.8657...°$

R7d. $c_3 = \pm\sqrt{1 - \cos^2 30° - \cos^2 40°}$
$= \pm\sqrt{-0.3368...}$, which is imaginary. The sum of the angles between any two axes and the vector must be at least 90°, the angle between the two axes.

C1a. 6.7938...

C1b. $\vec{r} = \left(7 - \frac{7}{25}t\right)\vec{i} + 9\vec{j} + \frac{24}{25}t\vec{k}$; $d = 4.3384...$ ft

T1.

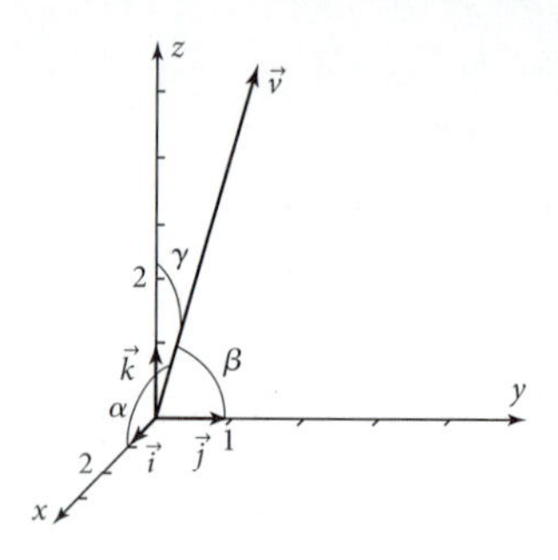

T3. $\vec{a} \cdot \vec{b}$ is a scalar quantity, while $\vec{a} \times \vec{b}$ is a vector.

T5a. $\vec{a} \cdot \vec{b} = 21$

T5b. $\vec{a} \times \vec{b} = -16\vec{i} - 2\vec{j} + 13\vec{k}$

T7. $\vec{v}_1$ and $\vec{v}_2$ are perpendicular if and only if $\vec{v}_1 \cdot \vec{v}_2 = 0$.

T9. The length is one unit.

T11. $2\vec{i} - 6\vec{j} + 5\vec{k}$

T13. $\frac{3}{\sqrt{89}}\vec{i} + \frac{8}{\sqrt{89}}\vec{j} + \frac{4}{\sqrt{89}}\vec{k}$

T15. $-64\vec{i} + 7\vec{j} + 34\vec{k}$

T17. 7.1019...

T19. $\left(\frac{8}{9}\right)^2 + \left(\frac{1}{9}\right)^2 + \left(\frac{4}{9}\right)^2 = \frac{64}{81} + \frac{1}{81} + \frac{16}{81} = \frac{81}{81} = 1$

T21. -18

T23. $7x + 5y + 10z = 140$

T25. 14 ft

T27. Answers will vary.

Chapter 13

Problem Set 13-1

1.

Iteration Number	Side Length	Total Perimeter	Total Area
0	10 cm	40 cm	100 cm^2
1	4 cm	64 cm	64 cm^2
2	1.6 cm	102.4 cm	40.96 cm^2

3. $P(3) = 163.84$ cm; $A(3) = 26.2144$ cm^2;
$P(4) = 262.144$ cm; $A(4) = 16.777216$ cm^2

5. The perimeter approaches infinity, whereas the area approaches zero.

Problem Set 13-2

1. $\begin{bmatrix} -2 & 13 \\ 0 & 10 \\ 0 & 11 \end{bmatrix}$

3. $[-22 \quad 22 \quad -2]$

5. $[14 \quad -42]$

7. Undefined

9. $\begin{bmatrix} 4 & 4 \\ 5 & 3 \end{bmatrix}$

11a. $[4.81 \quad 4.70 \quad 4.96 \quad 1.70]$
The company's total annual income from Texas is \$4.96 million. Of this, \$1.53 million is earned annually from mortgages.

11b. As written, the product cannot be found because the number of columns in the first matrix does not equal the number of rows in the second. However, there is no reason not to have the yield matrix after the investment matrix because analysts can write $[M]$ as a 4×3 matrix and $[Y]$ as a 3×1 matrix.

11c. The number of columns in the first matrix does not equal the number of rows in the second.

13a. $[M]^{-1} = \begin{bmatrix} 0.7 & -2.9 & 2.1 \\ -0.1 & 1.7 & -1.3 \\ -0.3 & 0.1 & 0.1 \end{bmatrix}$

$[M]^{-1}[M] = [M][M]^{-1} = \begin{bmatrix} 1 & 0 & 0 \\ 0 & 1 & 0 \\ 0 & 0 & 1 \end{bmatrix} = [I]$

13b. $\det[M] = 10$;

$\text{adj}[M] = [M]^{-1} \cdot \det[M] = \begin{bmatrix} 7 & -29 & 21 \\ -1 & 17 & -13 \\ 3 & 1 & 1 \end{bmatrix}$

15. Finding the inverse requires dividing by the determinant, but $\det\begin{bmatrix} 6 & 3 \\ 8 & 4 \end{bmatrix} = 0$. The determinant "determines" whether a matrix can be inverted.

17. $x = 1.22, y = 6.9, z = 3.4$

19. $y = 0.5x^2 + 3x - 7$; $y(20) = 253$

21. $\begin{bmatrix} 2 & 3 \\ 4 & 5 \end{bmatrix}\begin{bmatrix} 6 & 7 \\ 8 & 9 \end{bmatrix} = \begin{bmatrix} 36 & 41 \\ 64 & 73 \end{bmatrix}$

$\begin{bmatrix} 6 & 7 \\ 8 & 9 \end{bmatrix}\begin{bmatrix} 2 & 3 \\ 4 & 5 \end{bmatrix} = \begin{bmatrix} 40 & 53 \\ 52 & 69 \end{bmatrix}$

Problem Set 13-3

1. $\begin{bmatrix} 4 & 6 & 2 \\ 2 & 4 & 6 \end{bmatrix}$; Dilation by 2

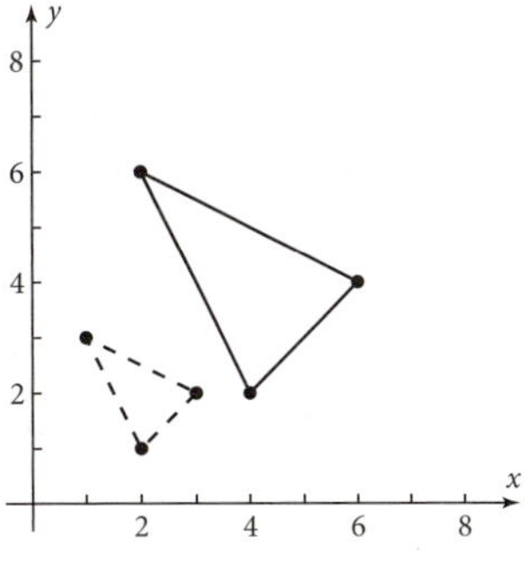

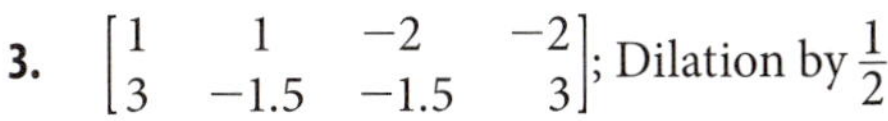

3. $\begin{bmatrix} 1 & 1 & -2 & -2 \\ 3 & -1.5 & -1.5 & 3 \end{bmatrix}$; Dilation by $\frac{1}{2}$

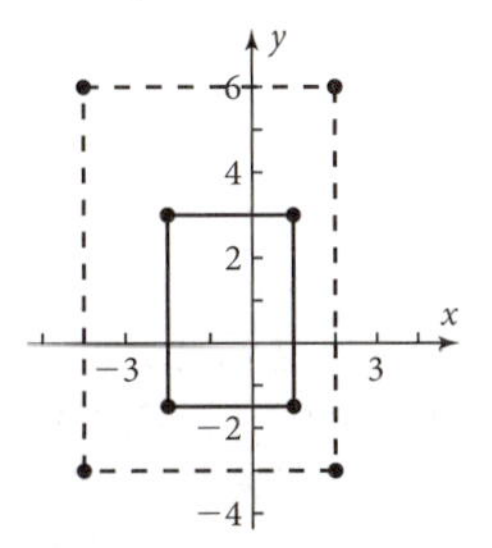

5. $\begin{bmatrix} 1.8 & 1.2 & 3.6 & 4.2 \\ 2.6 & 3.4 & 5.2 & 4.4 \end{bmatrix}$; Rotation by $\cos^{-1} 0.8$ $= 36.8698...°$ counterclockwise

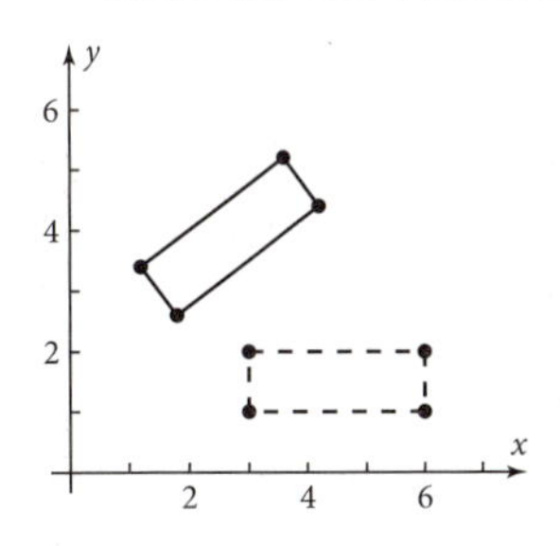

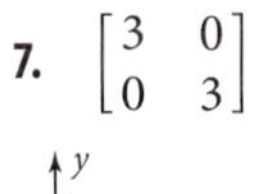

7. $\begin{bmatrix} 3 & 0 \\ 0 & 3 \end{bmatrix}$

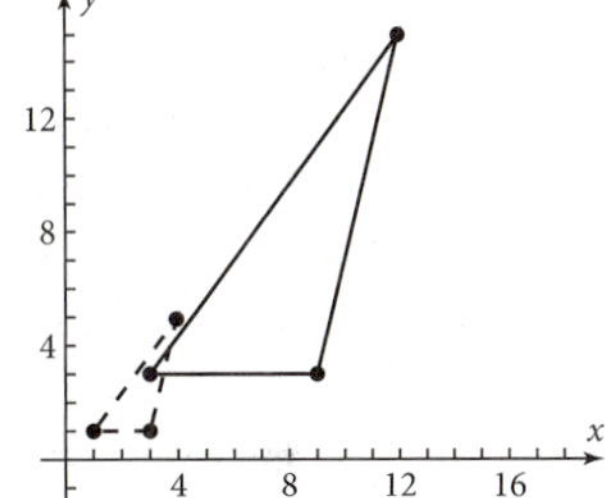

9. $\begin{bmatrix} \cos(-50°) & \cos 40° \\ \sin(-50°) & \sin 40° \end{bmatrix}$

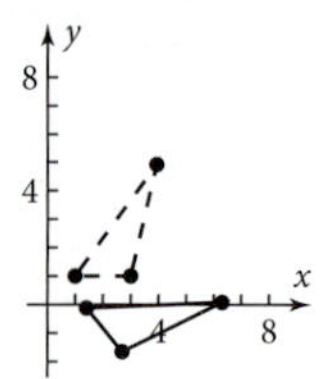

11. $\begin{bmatrix} \cos(-50°) & \cos 40° \\ \sin(-50°) & \sin 40° \end{bmatrix} \begin{bmatrix} 3 & 0 \\ 0 & 3 \end{bmatrix}$

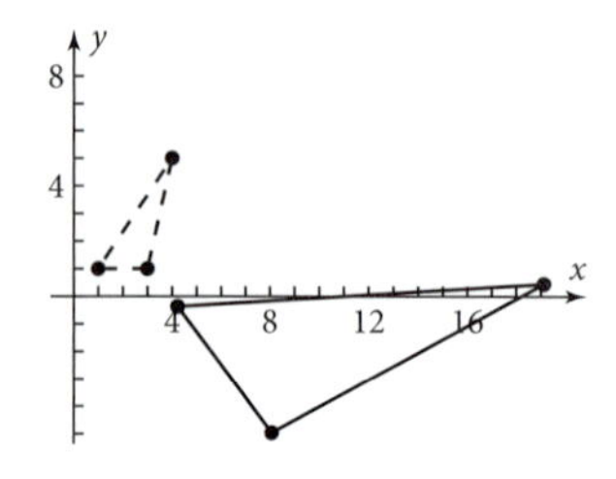

13. $[A]$ will dilate the image by 0.8 and rotate it counterclockwise by 20°. The pre-image matrix is $\begin{bmatrix} 5 & 7 & 9 & 9 \\ 2 & 2 & 4 & 6 \end{bmatrix}$.

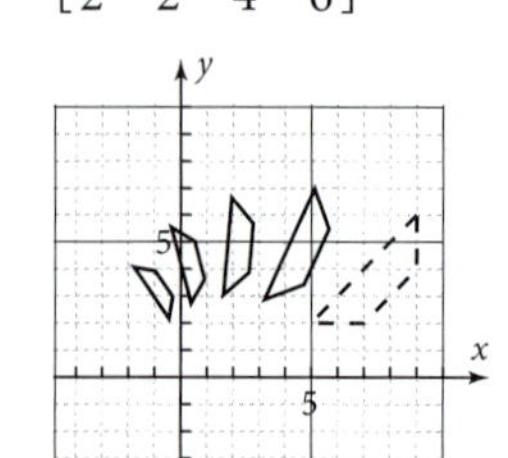

15. $[A] = \begin{bmatrix} \cos 90° & \cos 180° \\ \sin 90° & \sin 180° \end{bmatrix} = \begin{bmatrix} 0 & -1 \\ 1 & 0 \end{bmatrix}$

17. $[A] = \begin{bmatrix} 5 & 0 \\ 0 & 5 \end{bmatrix}$

19. Student program. See **www.keymath.com/precalc** for an example program.

21. Answers will vary.

Problem Set 13-4

1a. $[A] = \begin{bmatrix} 0.9 \cos 20° & 0.9 \cos 110° & 6 \\ 0.9 \sin 20° & 0.9 \sin 110° & -1 \\ 0 & 0 & 1 \end{bmatrix}$;

$[M] = \begin{bmatrix} 3 & 7 & 7 & 3 \\ 2 & 2 & 4 & 4 \\ 1 & 1 & 1 & 1 \end{bmatrix}$

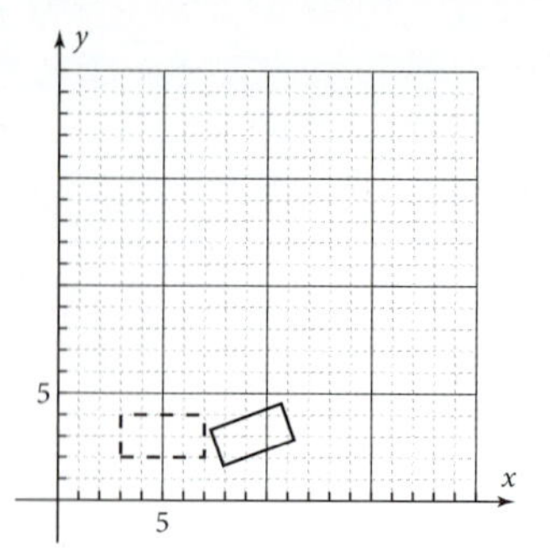

1b.

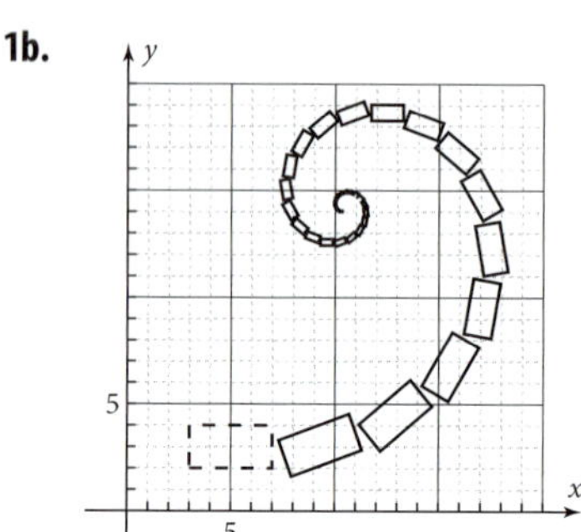

The images seem to be attracted to (10, 14).

1c. (10.4045..., 14.2777...) ≈ (10, 14) from part b.

1d. $\begin{bmatrix} X \\ Y \end{bmatrix} = \begin{bmatrix} 10.4044... \\ 14.2773... \end{bmatrix} \approx \begin{bmatrix} 10.4045... \\ 14.2777... \end{bmatrix}$ from part c.

3a. The figure will be rotated 20° clockwise and dilated by a factor of 0.8, then translated 2 units horizontally and 6 units vertically.

3b. $[M_1] = \begin{bmatrix} 25 & 23 & 23 & 25 \\ 15 & 15 & 12 & 12 \\ 1 & 1 & 1 & 1 \end{bmatrix}$

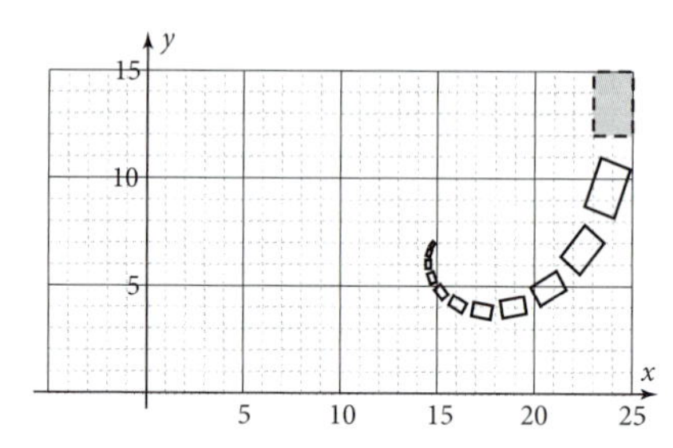

The rectangles converge to approximately (16, 7).

3c.

The images appear to be attracted to the same fixed point.

3d. $[T_2] = \begin{bmatrix} 0.7\cos 35° & 0.7\cos 125° & 7 \\ 0.7\sin 35° & 0.7\sin 125° & -3 \\ 0 & 0 & 1 \end{bmatrix}$

$= \begin{bmatrix} 0.5734... & -0.4015... & 7 \\ 0.4015... & 0.5734... & -3 \\ 0 & 0 & 1 \end{bmatrix}$

Applying $[T_2]$ iteratively to $[M_1]$:

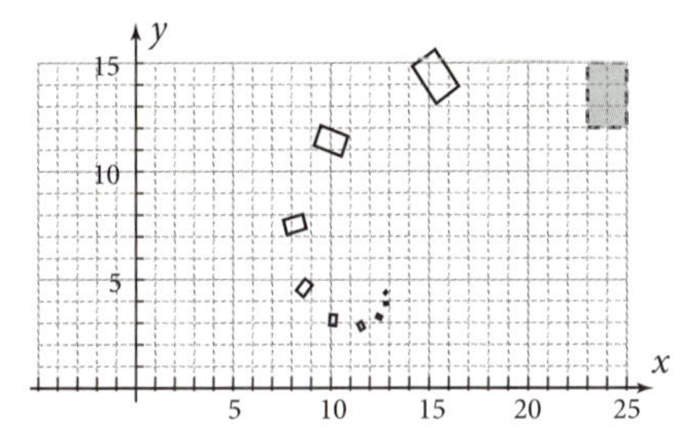

The rectangles converge to approximately (12, 4.5).

3e. Only the transformation matrix determines the fixed point attractor, independent of the starting pre-image matrix.

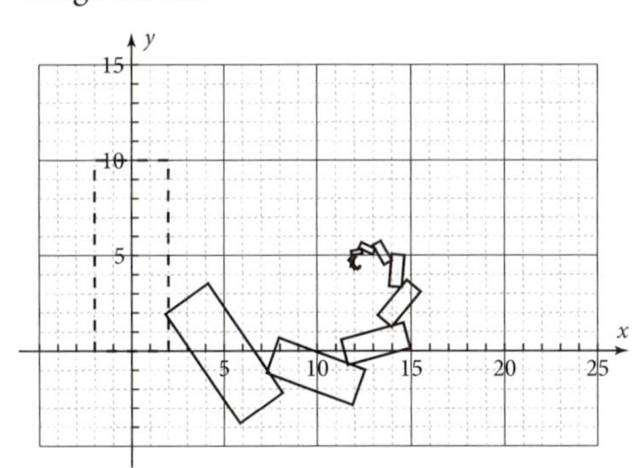

The rectangles still converge to approximately (12, 4.5).

Problem Set 13-5

1a. $[A] = \begin{bmatrix} 0.5 & 0 & 7.5 \\ 0 & 0.5 & -5 \\ 0 & 0 & 1 \end{bmatrix}$;

$[B] = \begin{bmatrix} 0.5 & 0 & -7.5 \\ 0 & 0.5 & -5 \\ 0 & 0 & 1 \end{bmatrix}$;

$[C] = \begin{bmatrix} 0.5 & 0 & 0 \\ 0 & 0.5 & 10 \\ 0 & 0 & 1 \end{bmatrix}$

1b. $[A][A][M] = \begin{bmatrix} 15 & 11.25 & 7.5 \\ -10 & -2.5 & -10 \\ 1 & 1 & 1 \end{bmatrix}$

$[A][B][M] = \begin{bmatrix} 7.5 & 3.75 & 0 \\ -10 & -2.5 & -10 \\ 1 & 1 & 1 \end{bmatrix}$

$[A][C][M] = \begin{bmatrix} 11.25 & 7.5 & 3.75 \\ -2.5 & 5 & -2.5 \\ 1 & 1 & 1 \end{bmatrix}$

$[B][A][M] = \begin{bmatrix} 0 & -3.75 & -7.5 \\ -10 & -2.5 & -10 \\ 1 & 1 & 1 \end{bmatrix}$

$[B][B][M] = \begin{bmatrix} -7.5 & -11.25 & -15 \\ -10 & -2.5 & -10 \\ 1 & 1 & 1 \end{bmatrix}$

$[B][C][M] = \begin{bmatrix} -3.75 & -7.5 & -11.25 \\ -2.5 & 5 & -2.5 \\ 1 & 1 & 1 \end{bmatrix}$

$[C][A][M] = \begin{bmatrix} 7.5 & 3.75 & 0 \\ 5 & 12.5 & 5 \\ 1 & 1 & 1 \end{bmatrix}$

$[C][B][M] = \begin{bmatrix} 0 & -3.75 & -7.5 \\ 5 & 12.5 & 5 \\ 1 & 1 & 1 \end{bmatrix}$

$[C][C][M] = \begin{bmatrix} 3.75 & 0 & -3.75 \\ 12.5 & 20 & 12.5 \\ 1 & 1 & 1 \end{bmatrix}$

1c.

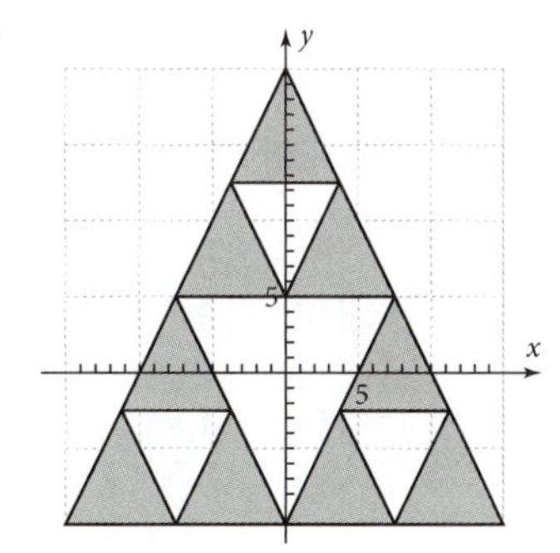

1d. 3rd iteration: 27 images; 20th iteration: 3,486,784,401 images

1e. Pre-image area: 450; 1st iteration area: $450 \cdot \frac{3}{4}$; nth iteration area: $450 \cdot \left(\frac{3}{4}\right)^n$

1f. As $n \to \infty$, the area approaches zero. Sierpiński's triangle has zero area!

3. Student program. See **www.keymath.com/precalc** for an example program.

5. The results should resemble the figure.

7. $[A] = \begin{bmatrix} 0.5 & 0 & 10 \\ 0 & 0.5 & 10 \\ 0 & 0 & 1 \end{bmatrix}$; $[B] = \begin{bmatrix} 0.5 & 0 & 0 \\ 0 & 0.5 & 10 \\ 0 & 0 & 1 \end{bmatrix}$;

$[C] = \begin{bmatrix} 0.5 & 0 & 0 \\ 0 & 0.5 & 0 \\ 0 & 0 & 1 \end{bmatrix}$; $[D] = \begin{bmatrix} 0.5 & 0 & 10 \\ 0 & 0.5 & 0 \\ 0 & 0 & 1 \end{bmatrix}$

The combined image space of all four transformations now covers the entire square.

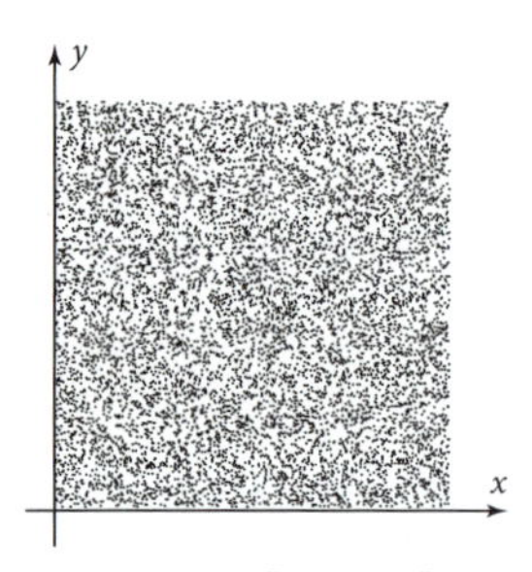

9a. $[D] = \begin{bmatrix} 0 & 0 \\ 0 & 10 \\ 1 & 1 \end{bmatrix}$

9b. $[A] = \begin{bmatrix} 0.6\cos 30° & 0.6\cos 120° & 0 \\ 0.6\sin 30° & 0.6\sin 120° & 5 \\ 0 & 0 & 1 \end{bmatrix}$;

$[B] = \begin{bmatrix} 0.6 & 0 & 0 \\ 0 & 0.6 & 0 \\ 0 & 0 & 1 \end{bmatrix}$;

$[C] = \begin{bmatrix} 0.6\cos(-30°) & 0.6\cos 60° & 0 \\ 0.6\sin(-30°) & 0.6\sin 60° & 4 \\ 0 & 0 & 1 \end{bmatrix}$

9c. $[A][A][D] \approx \begin{bmatrix} -1.5 & -4.6 \\ 7.6 & 9.4 \\ 1 & 1 \end{bmatrix}$; $[A][B][D] \approx \begin{bmatrix} 0 & -1.8 \\ 5 & 8.1 \\ 1 & 1 \end{bmatrix}$;

$[A][C][D] \approx \begin{bmatrix} -1.2 & -1.2 \\ 7.1 & 10.7 \\ 1 & 1 \end{bmatrix}$; $[B][A][D] \approx \begin{bmatrix} 0 & -1.8 \\ 3 & 6.1 \\ 1 & 1 \end{bmatrix}$;

$[B][B][D] = \begin{bmatrix} 0 & 0 \\ 0 & 3.6 \\ 1 & 1 \end{bmatrix}$; $[B][C][D] \approx \begin{bmatrix} 0 & 1.8 \\ 2.4 & 5.5 \\ 1 & 1 \end{bmatrix}$;

$[C][A][D] \approx \begin{bmatrix} 1.5 & 1.5 \\ 6.6 & 10.2 \\ 1 & 1 \end{bmatrix}$; $[C][B][D] \approx \begin{bmatrix} 0 & 1.8 \\ 4 & 7.1 \\ 1 & 1 \end{bmatrix}$;

$[C][C][D] \approx \begin{bmatrix} 1.2 & 4.3 \\ 6.1 & 7.9 \\ 1 & 1 \end{bmatrix}$

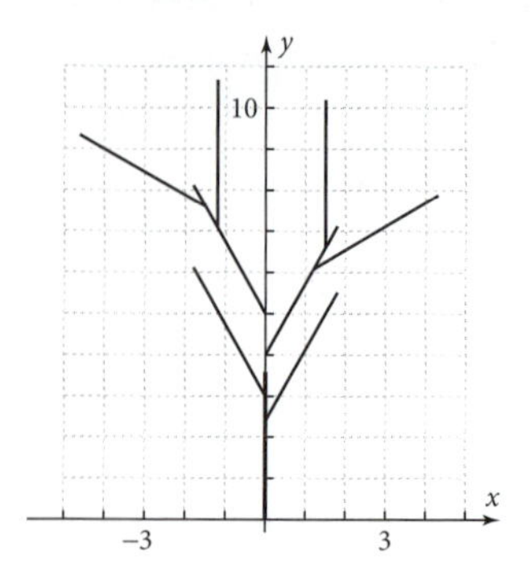

9d. 5000 iterations:

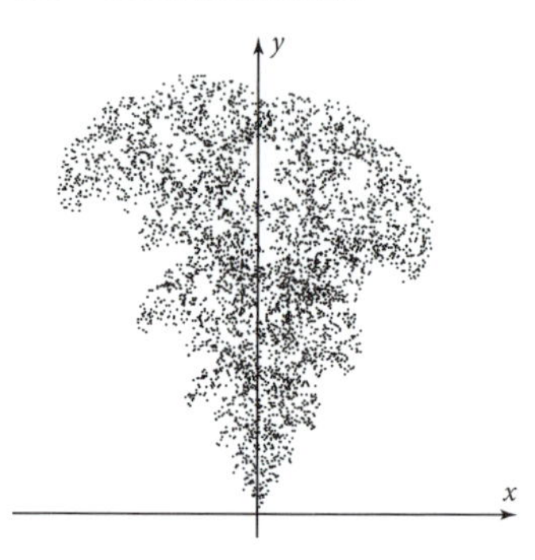

9e. "Strange attractor"

9f. 0th: 10 units; 1st: 18 units; 2nd: 32.4 units; 3rd: 58.32 units; 100th: $3.3670... \times 10^{26}$ units

If the iterations were done forever, the length would become infinite.

9g. $[A] = \begin{bmatrix} -4.6762... \\ 7.4880... \end{bmatrix}$ appears to be in the "foliage" of the left "branch" of the tree.

$[B] = (0, 0)$ is the "root" of the tree, i.e., the base of the "trunk." $[C] = \begin{bmatrix} 3.7410... \\ 5.9904... \end{bmatrix}$ is in the "foliage" of the right "branch" of the tree.

11a. $\begin{bmatrix} -2.9769... \\ 14.2984... \end{bmatrix}$

This is the topmost point of the "fern."

11b. (X_B, Y_B) is the point on the bottom-left leaf that's in the same relation to both that leaf and the entire fern. (X_C, Y_C) is the point on the bottom-right leaf that's in the same relation to both that leaf and the entire fern, and (X_D, Y_D) is the base of the "stem."

11c. $(X_B, Y_B) = (-0.6561..., 2.2628...)$.

$(X_C, Y_C) = (0.9616..., 3.5276...)$.

$(X_D, Y_D) = (0, 0)$. These results confirm the conjecture.

Problem Set 13-6

1a. $r = \frac{1}{5}, N = 25 = 5^2$

$$D = \frac{\log N}{\log \frac{1}{r}} = \frac{\log 25}{\log 5} = \frac{2 \log 5}{\log 5} = 2$$

1b. $r = 0.01 = \frac{1}{100}, N = 10{,}000 = 100^2$

$$D = \frac{\log N}{\log \frac{1}{r}} = \frac{\log 10{,}000}{\log 100} = \frac{2 \log 100}{\log 100} = 2$$

1c. The smaller squares are identical and self-similar, and you can carry out the division process infinitely.

3a. 1.5129...

3b. The area approaches zero. This is consistent with the dimension being less than 2.

3c. The perimeter approaches infinity. This is consistent with the dimension being greater than 1.

3d. $r = 0.5 = \frac{1}{2}, N = 4$

$$D = \frac{\log N}{\log \frac{1}{r}} = \frac{\log 4}{\log 2} = \frac{2 \log 2}{\log 2} = 2$$

The dimension is a whole number, not a fraction. Therefore the image is not a fractal.

3e. $r = 0.6 = \frac{3}{5}, N = 4$

$$D = \frac{\log N}{\log \frac{1}{r}} = \frac{\log 4}{\log \frac{5}{3}} = 2.7138...$$

As n approaches infinity, the sum of the areas of the smaller squares, $A = A_0 \cdot (4 \cdot 0.6^2)^n = A_0 \cdot 1.44^n$, also approaches infinity. This is consistent with the dimension being greater than 2. However, the actual total area of the figure is the same as the area of the original square, because the smaller squares overlap.

5. If the fractal dimension is less than 1, then the length will be zero; if the dimension is exactly 1, then the length will be finite; if the fractal dimension is greater than 1, then the length may be infinite. If the fractal dimension is less than 2, then the area will be zero; if the dimension is exactly 2, then the area will be finite; if the fractal dimension is greater than 2, then the area may be infinite.

Problem Set 13-7

R1. 1024 segments; 0.0173... units; $\left(\frac{2}{3}\right)^n \to 0$ units as $n \to \infty$

R3a. Dilation by 0.6 and rotation about the origin by 30° counterclockwise

R3b.

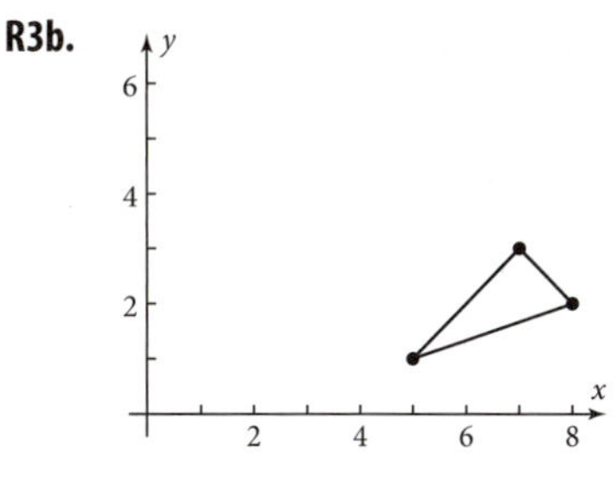

R3c.

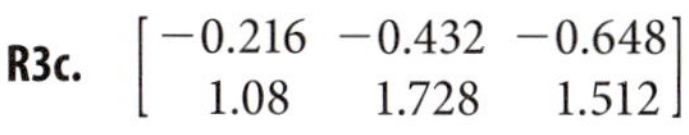

$$\begin{bmatrix} -0.216 & -0.432 & -0.648 \\ 1.08 & 1.728 & 1.512 \end{bmatrix}$$

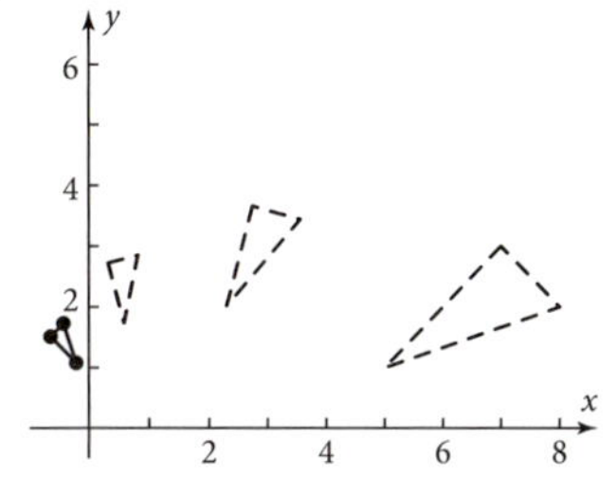

R3d. 90°

R3e. Distance of pre-image vertex: $d = \sqrt{5^2 + 1^2} = \sqrt{26}$; Distance of image vertex: $d = \sqrt{(-0.216)^2 + 1.08^2} = \sqrt{1.213056} = \sqrt{26 \cdot 0.6^6}$ $= 0.6^3 \cdot \sqrt{26}$

R5a. $[B] = \begin{bmatrix} 0.6 \cos 30° & 0.6 \cos 120° & -5 \\ 0.6 \sin 30° & 0.6 \sin 120° & -2 \\ 0 & 0 & 1 \end{bmatrix}$

R5b. $[A][A][M] =$

$$\begin{bmatrix} 6.7698... & 7.8498... & 7.2263... & 6.1463... \\ 3.4239... & 5.2945... & 5.6545... & 3.7839... \\ 1 & 1 & 1 & 1 \end{bmatrix}$$

$[A][B][M] =$

$$\begin{bmatrix} 2.7736... & 3.8536... & 3.2301... & 2.1501... \\ -1.6545... & 0.2160... & 0.5760... & -1.2945... \\ 1 & 1 & 1 & 1 \end{bmatrix}$$

$[B][A][M] =$

$$\begin{bmatrix} -3.2301... & -2.1501... & -2.7736... & -3.8536... \\ -0.5760... & 1.2945... & 1.6545... & -0.2160... \\ 1 & 1 & 1 & 1 \end{bmatrix}$$

$[B][B][M] =$

$$\begin{bmatrix} -7.2263... & -6.1463... & -6.7698... & -7.8498... \\ -5.6545... & -3.7839... & -3.4239... & -5.2945... \\ 1 & 1 & 1 & 1 \end{bmatrix}$$

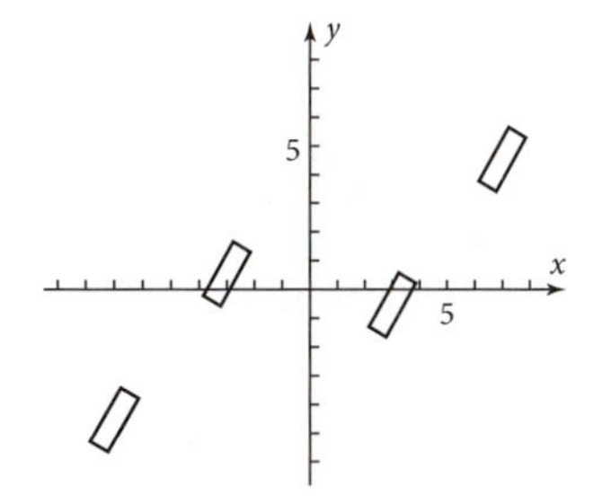

R5c. The attractor is correct. Probabilities are assigned to the transformations (in this case, equal probabilities of 0.5 each). An initial point is chosen. Then a random value between zero and 1 determines which transformation is performed on that point. Then the procedure is repeated on the resulting point. As many iterations as desired are performed.

R5d. Transformation $[A]$, when iterated, attracts each point of the fractal image to the fixed point (5.6175..., 7.6714...) shown on the figure.

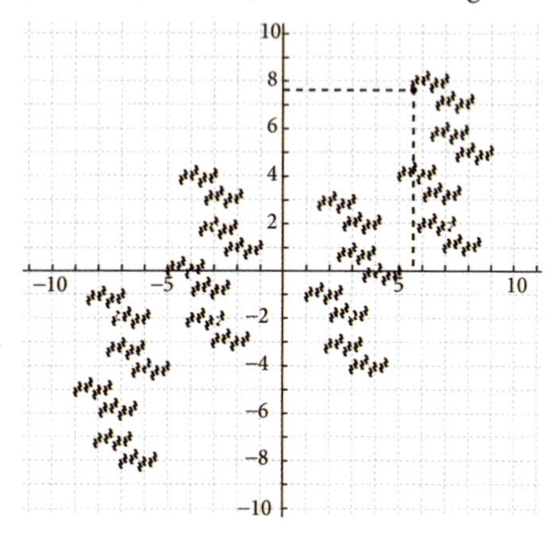

R5e. Let p be the perimeter of one rectangle and P be the total perimeter. $N = 2^n$; $p = 0.6^n \cdot 16$; $P = N_p = 1.2^n \cdot 16$

n	N	p	P
0	1	16	16
1	2	9.6	19.2
2	4	5.76	23.04
3	8	3.456	27.648

$P_{50} = 145{,}607.0104...$

The total perimeter is infinite, because $1.2^n \cdot 16$ approaches infinity as n approaches infinity.

C1a. Sources will vary.

C1b. 20

C1c. 33; If $e_{i,j}$ is the element in row i, column j, then the sign of the cofactor of $e_{i,j}$ is $(-1)^{i+j}$.

C1d. $$\text{adj}[A] = \begin{bmatrix} 20 & 20 & -34 & 13 \\ -6 & -47 & 43 & -8 \\ -44 & 38 & 1 & -4 \\ 33 & -8 & -11 & 3 \end{bmatrix}$$

The 20 and 33 appear at the correct places in the adjugate matrix because the inverse involves taking the *transpose* of the matrix of cofactors. So 33 appears in row 4, column 1 rather than row 1, column 4.

C1e. The number of calculations is thus tedious!

T1. $$\begin{bmatrix} 13 & 63 \\ 23 & 62 \end{bmatrix}$$

T3. Multiplying a given row in the first matrix by a given column in the second matrix is like vector multiplication—the corresponding entries are multiplied and their products are added.

T5. Dilation by 0.9, rotation about the origin by 15° counterclockwise, x-translation by 3, and y-translation by 2

T7. $$[A] = \begin{bmatrix} 0.5\cos 20° & 0.5\cos 110° & 0 \\ 0.5\sin 20° & 0.5\sin 110° & 5 \\ 0 & 0 & 1 \end{bmatrix}$$

$$[B] = \begin{bmatrix} 0.5 & 0 & 0 \\ 0 & 0.5 & 0 \\ 0 & 0 & 1 \end{bmatrix}$$

$$[C] = \begin{bmatrix} 0.5\cos(-30°) & 0.5\cos 60° & 0 \\ 0.5\sin(-30°) & 0.5\sin 60° & 5 \\ 0 & 0 & 1 \end{bmatrix}$$

T9. $$[A][C][D] = \begin{bmatrix} -0.8550... & -0.4209... \\ 7.3492... & 9.8112... \\ 1 & 1 \end{bmatrix}$$

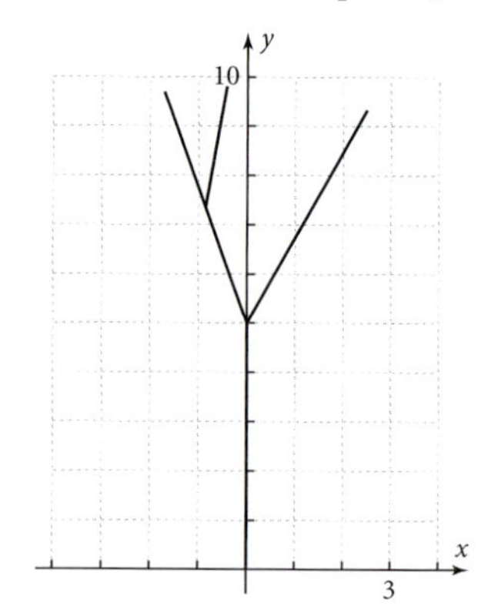

T11. Answers will vary.

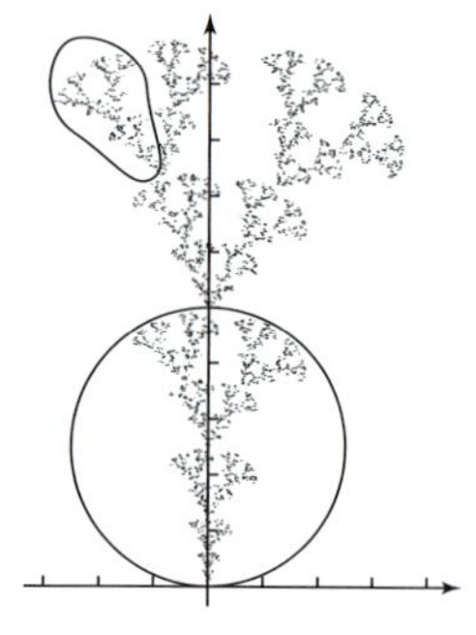

T13. As n approaches infinity, the total length, $L = 1.5^n \cdot 10$, also approaches infinity.

T15. 3

T17. 1.5846...

T19. Answers will vary.

Problem Set 13-8

1. It is a transformation of the unit hyperbola $-x^2 + y^2 = 1$, where x^2 and y^2 have opposite signs and there is no xy-term. Opens in the y-direction because the sign of the y^2-term is positive. Dilation factors of 3 for x and 5 for y, horizontal translation of 4 and vertical translation of 1.

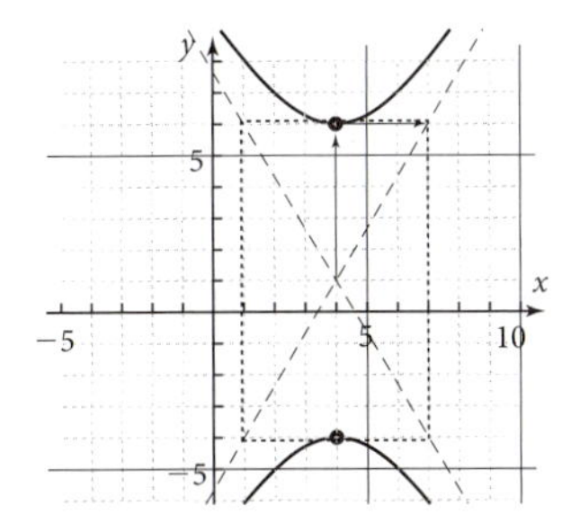

3.

$$\left(\frac{x-4}{3}\right)^2 - \left(\frac{y-1}{5}\right)^2 = 0$$

The right side is zero instead of 1, indicating a degenerate hyperbola.

5.

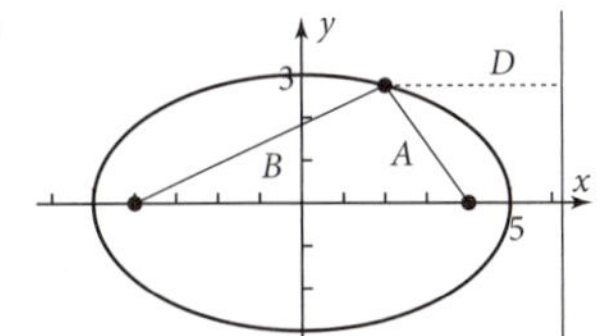

$A + B =$ length of major axis; $D = 6.25A$

7. $\begin{cases} x = 5\cos t \\ y = 3\sin t \end{cases}$; $\begin{cases} x = 5\cos t + 2 \\ y = 3\sin t + 1 \end{cases}$

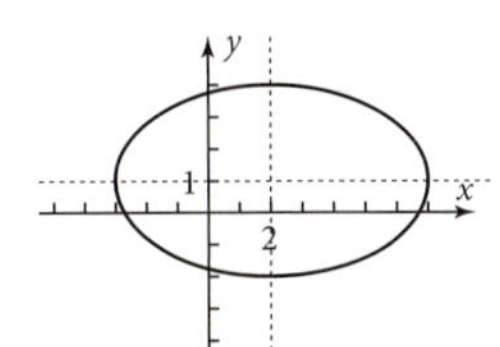

9. $r = -5.0621...$ The point is in QIII because r is negative, and thus the point is on the negative part of the rotating number line.

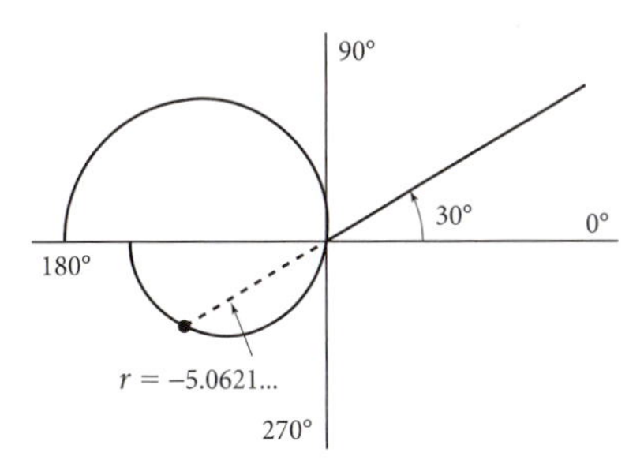

11. The crash occurs at $(r, \theta) = (3960, 325.9227...°)$.

13. $-13.6808... + 37.5877...i$

15. $y = x^{2/3}$. The graph agrees with the figure.

17.

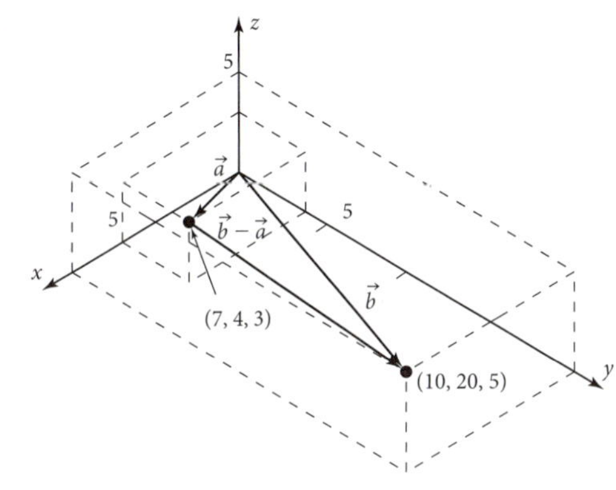

19. 1640.1219... m

21. $\theta = 33.1626...°$

23. $(\vec{a} \times \vec{b}) \cdot \vec{a} = -40 \cdot 7 - 5 \cdot 4 + 100 \cdot 3 = 0$

$$\cos\theta = \frac{(\vec{a} \times \vec{b}) \cdot \vec{a}}{|\vec{a} \times \vec{b}| \cdot |\vec{a}|} = 0 \Rightarrow (\vec{a} \times \vec{b}) \perp \vec{a}$$

25. $(2)(2) + (7)(6) + (4)(1) = 50$

27. $\begin{bmatrix} 70 & 63 \\ 57 & 50 \end{bmatrix}^{-1} = \frac{1}{-91}\begin{bmatrix} 50 & -63 \\ -57 & 70 \end{bmatrix}$

$$= \begin{bmatrix} -0.5494... & 0.6923... \\ 0.6263... & -0.7692... \end{bmatrix}$$

29. The elements in any row of the left matrix in a product and the elements in any column in the right matrix can be considered to be the components of vectors. Multiplying and adding the respective elements in a row and column follows the same pattern as multiplying and adding the coefficients of the components of two vectors.

31. (3.5, 23.2)

33. (3.5130..., 23.1639...)

35. $\approx(6.6386..., 6.2619...)$, at the extreme upper right of the fractal image.

37. $\begin{bmatrix} 0.4 & 0 & 0 \\ 0 & 0.4 & 0 \\ 0 & 0 & 1 \end{bmatrix}$

39.

n	*N*	Perimeter of Each	Total *P*
0	1	36	36
1	4	12	48
2	16	4	64
3	64	$\frac{4}{3}$	$\frac{256}{3}$

41.

n	*A*
0	81
1	36
2	16
3	$\frac{64}{9} = 7.1111...$

43. 0; ∞
This makes sense because an object of dimension less than two has zero area, while an object of dimenstion greater than one has infinite length.

Chapter 14

Problem Set 14-1

1. $\frac{1}{12}$
3. $\frac{5}{6}$
5. $\frac{1}{6}$
7. $\frac{5}{9}$
9. 1
11. $\frac{1}{18}$
13. $\frac{1}{12}$

Problem Set 14-2

1a. A random experiment
1b. 52
1c. 12
1d. $\frac{3}{13}$
1e. $\frac{1}{2}$
1f. $\frac{1}{13}$
1g. $\frac{5}{13}$
1h. $\frac{1}{52}$
1i. 1
1j. 0
3. Student research problem

Problem Set 14-3

1a. 91
1b. 20
3a. 16
3b. 55
3c. 20
5. 693
7a. 9
7b. 18
7c. 504
9. 20
11. 1018
13a. 10
13b. 9
13c. 90
13d. 8
13e. 720
13f. 3,628,800
15a. The second plan gives 10,799,100 more plates.
15b. 24,317,748
15c. No, there would not be enough plates.
17. Journal entries will vary.

Problem Set 14-4

1a. 132
1b. 11,880
1c. 479,001,600
3. 15,600
5. 210
7a. 720
7b. 120
7c. $\frac{1}{6}$
7d. $16\frac{2}{3}\%$
7e. $\frac{1}{720}$
9a. 362,880
9b. 40,320
9c. $\frac{1}{9}$
9d. $11\frac{1}{9}\%$
11a. $\frac{1}{3}$
11b. $\frac{1}{12}$
11c. $\frac{1}{84}$
13a. 40,320
13b. 10,080
13c. $\frac{1}{4}$
15a. 360
15b. 840
15c. 20
15d. 415,800
15e. 5040
15f. 3360
17a. 24
17b. 120
17c. 40,320
19a. 144
19b. $\frac{1}{35}$

Problem Set 14-5

1. 10

3. 2,220,075

5. 1

7. 1

9. 360

11. $7.2710... \times 10^{44}$

13. 792; *Group* tells us the order is not important. We are not asked who sits in which seat.

15. 1,344,904; no

17a. 20 **17b.** 15

17c. 35 **17d.** 1

19a. 2,598,960 **19b.** 635,013,559,600

19c. The order of the cards in a hand is not important.

21a. 45

21b. 252

21c. 45. Choosing eight elements to include is the same as choosing two elements *not* to include.

23a. 120

23b. 2,598,960

23c. 311,875,200

23d. Permutation: parts a and c; combination: part b

25a. $\frac{140}{429} \approx 33\%$ **25b.** $\frac{175}{429} \approx 41\%$

25c. $\frac{105}{143} \approx 73\%$ **25d.** $\frac{5}{13} \approx 38\%$

27a. $\frac{4}{7} \approx 57\%$ **27b.** $\frac{5}{7} \approx 71\%$

27c. $\frac{2}{7} \approx 29\%$

27d. $P(\text{at least one rookie card})$ $= 1 - P(\text{no rookie cards})$, because part b is the opposite (complement) of part c.

29a. 75,287,520

29b. 7,376,656

29c. $\frac{97}{990} \approx 9.8\%$

29d. No, because it is not likely that the defective bulbs will be found. To be at least 50% sure of finding the defective bulbs, you must have a sample of at least size 30.

31. Journal entries will vary.

Problem Set 14-6

1a. $0.56 = 56\%$ **1b.** $0.3 = 30\%$

1c. $0.2 = 20\%$ **1d.** $0.06 = 6\%$

1e. $0.94 = 94\%$

3a. $0.28 = 28\%$ **3b.** $0.18 = 18\%$

3c. $0.12 = 12\%$ **3d.** $0.42 = 42\%$

3e. $0.54 = 54\%$

5. $0.88 = 88\%$

7a. $0.336 = 33.6\%$ **7b.** $0.024 = 2.4\%$

7c. $.976 = 97.6\%$ **7d.** $0.056 = 5.6\%$

9a. $0.3676... \approx 36.77\%$

9b. $0.9998... \approx 99.99\%$

11a. i. 12%

11a. ii. 28%

11a. iii. 48%

11a. iv. 12%

11b. $12\% + 28\% + 48\% + 12\% = 100\%$. These are all the possibilities.

13a. $0.0009 = 0.09\%$

13b. $0.2 = 20\%$

13c. They are not independent. An engine is more likely to fail if the other one has already failed.

Problem Set 14-7

1a. $\frac{3}{4}$

1b. $P(0) = \frac{1}{64} \approx 1.6\%$; $P(1) = \frac{9}{64} \approx 14\%$; $P(2) = \frac{27}{64} \approx 42\%$; $P(3) = \frac{27}{64} \approx 42\%$

1c. $\frac{1}{64} + \frac{9}{64} + \frac{27}{64} + \frac{27}{64} = \frac{64}{64} = 1 = 100\%$. These are all the possibilities.

1d.

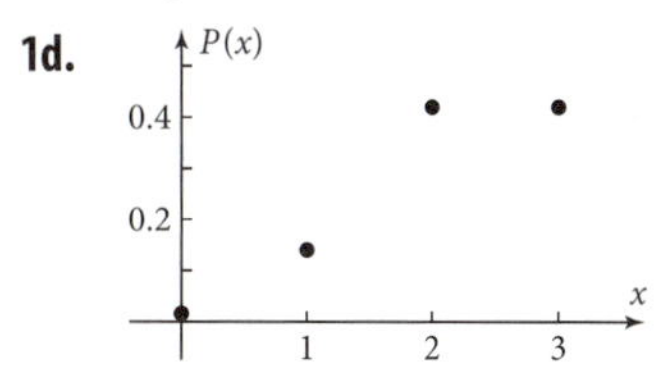

1e. Binomial

3a. $P(3) = {}_{10}C_3(0.3)^7(0.7)^3 = 0.0090...$

3b.

x	$P(x)$
0	0.000005...
1	0.0001...
2	0.0014...
3	0.0090...
4	0.0367...
5	0.1029...
6	0.2001...
7	0.2668...
8	0.2334...
9	0.1210...
10	0.0282...

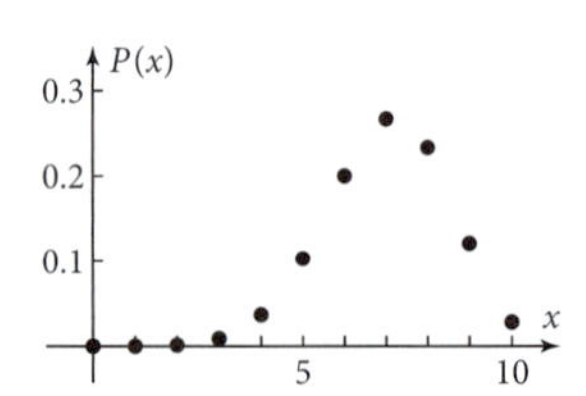

3c. $P(\text{more than } 5) = 0.8497...$; $P(\text{at most } 5) = 0.1502...$; "More than 5" is more likely.

5a. $P(x) = {}_{20}C_x(0.92)^{20-x}(0.08)^x$

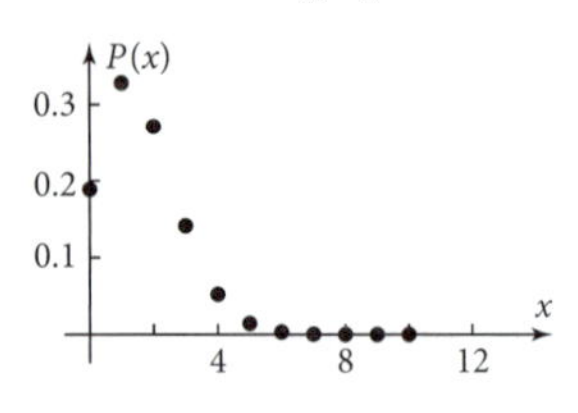

5b.

x	$P(x)$
0	0.1886...
1	0.3281...
2	0.2710...
3	0.1414...

5c. $1 - [P(0) + P(1) + P(2) + P(3)] = 0.0706...$

7a. i. 0.5471... **7a. ii.** 0.3316...

7a. iii. 0.0988... **7a. iv.** 0.1212...

7a. v. 0.0224...

7b. $0.9991... \approx 99.91\%$

9a. 0.1296 **9b.** 0.0256

9c. 0.2073... **9d.** 0.0614...

9e. i. 0.2073...

9e. ii. 0.0921...

9e. iii. 0.1658...

9e. iv. 0.1105...

9f. $\approx 71\%$

9g. A six-game series is most likely, followed by seven, five, and then four games.

11a.

n	PDs of n	x
1	—	0
2	1	1
3	1	1
4	1, 2	2
5	1	1
6	1, 2, 3	3
7	1	1
8	1, 2, 4	3
9	1, 3	2
10	1, 2, 5	3

11b. Let N be the list of numbers that have x proper divisors. Let M be the number of elements in N.

x	N	M
0	1	1
1	2, 3, 5, 7	4
2	4, 9	2
3	6, 8, 10	3

11c.

x	$P(x)$
0	$\frac{1}{10}$
1	$\frac{4}{10}$
2	$\frac{2}{10}$
3	$\frac{3}{10}$

11d.

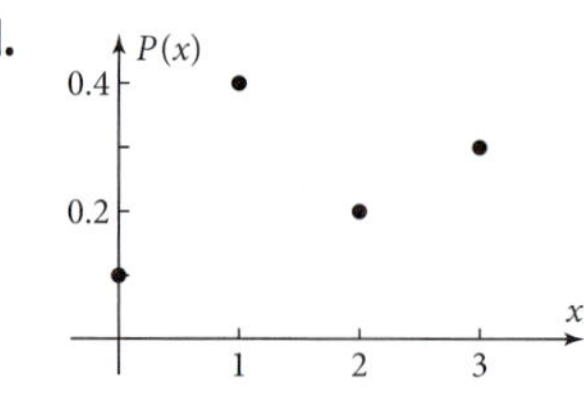

No pattern is evident.

13a. 0.9972... **13b.** 0.9945...

13c. 0.9917... **13d.** 0.8830...

13e. 0.1169...

13f. Here are a few selected values. The formula used is $f(x) = 1 - \frac{_{365}P_x}{365^x}$.

x	P(x)
10	0.1169...
20	0.4114...
30	0.7063...
40	0.8912...
50	0.9703...
60	0.9941...

13g.

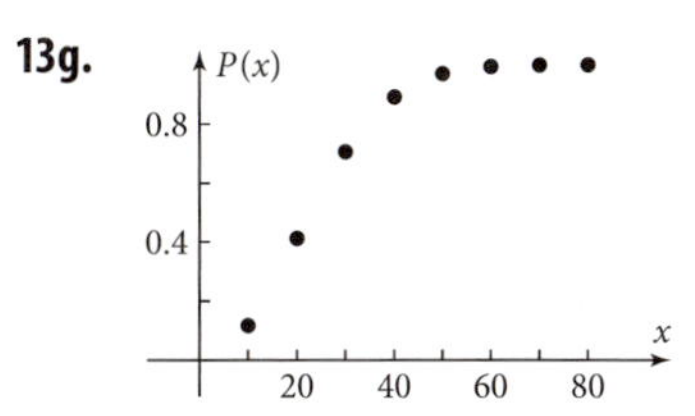

13h. i. $n = 23$ **13h. ii.** $n = 57$

Problem Set 14-8

1a. 2.3 neutrons per fission

1b. Mathematical expectation is a mathematical abstraction, meaning, for example, that you would expect 10 fissions to produce 23 neutrons, 100 to produce 230, and so on.

3a. $510 **3b.** 0.252

3c. $504

3d. He should choose Option A.

5a.

x	P(x)	x · P(x)
0	0.1233...	0
1	0.3205...	0.3205...
2	0.3332...	0.6664...
3	0.1731...	0.5195...
4	0.0450...	0.1800...
5	0.0046...	0.0233...

5b. 1.71

7a. $\frac{1}{5}, \frac{4}{5}$ **7b.** 0

7c. $\frac{1}{16}$ **7d.** $\frac{1}{6}, \frac{3}{8}$

7e. If you can eliminate at least one answer, you have at least a small fractional expected number of points per question.

9a. ≈15 **9b.** 9,985

9c. ≈15

9d.

x	P(x)	A(x)	D(x)
15	0.00146	10000	15
16	0.00154	9985	15
17	0.00162	9970	16
18	0.00169	9954	17
19	0.00174	9937	17
20	0.00179	9920	18

9e.

x	I(x)	O(x)
15	$400,000	$300,000
16	$399,400	$300,000
17	$398,800	$320,000
18	$398,160	$340,000
19	$397,480	$340,000
20	$396,800	$360,000

9f.

x	NI(x) = I(x) − O(x)
15	$100,000
16	$99,400
17	$78,800
18	$58,160
19	$57,480
20	$36,800

The company has .less income each year because (1) there are fewer people still alive to pay and (2) it has a higher payout because there are more people dying.

9g. It would expect to make about $71,773 per year. This is enough to pay a full-time employee.

Problem Set 14-9

R1a. $\frac{3}{8}$ **R1b.** $\frac{3}{8}$

R1c. $\frac{1}{8}$ **R1d.** 0

R1e. $\frac{4}{8} = \frac{1}{2}$ **R1f.** $\frac{7}{8}$

R1g. $\frac{8}{8} = 1$ **R1h.** $\frac{2}{8} = \frac{1}{4}$

R1i. $\frac{6}{8} = \frac{3}{4}$

R3a. i. 220 **R3a. ii.** 31

R3b. i. 15 **R3b. ii.** 20

R5a. 35

R5b. In a permutation, order is important; in a combination, it is not. That is, rearrangements of the same choice of objects are considered different permutations but the same combination.

R5c. i. 3,921,225

R5c. ii. 94,109,400

R5c. iii. 4,178,378,490

R5d. 0.2610... ≈ 26.1%

R7a. $P(4) = {}_6C_4(0.4)^2(0.6)^4 = 0.31104$

R7b. $P(x) = {}_6C_x(0.4)^{6-x}(0.6)^x$

x	P(x)
0	0.004096
1	0.036864
2	0.138240
3	0.276480
4	0.311040
5	0.186624
6	0.046656

R7c.

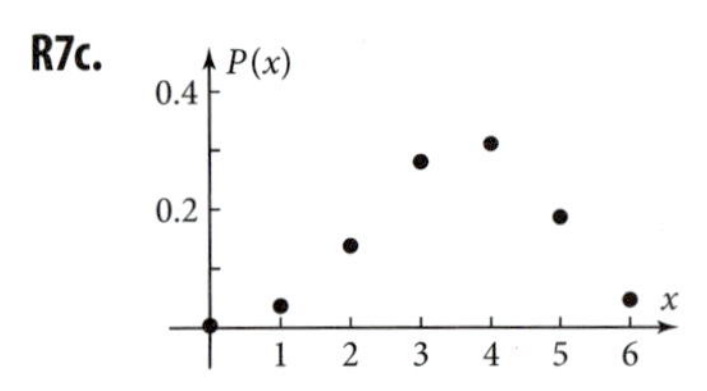

R7d. 0.54432

R7e. The probabilities are the terms in the expansion of the binomial 0.6 + 0.4 raised to a power: $(0.6 + 0.4)^6$

C1a. 230

C1b. two: 529; three: ≈1217; four: ≈2799; geometric

C1c. $\approx 5.34 \times 10^{363}$

C1d. 0.29

C1e. 0.667

C1f. ≈0; no

C1g. Because the expression for the number of active neutrons, $100(2.3)^{1000t}$, if t is in seconds, is an exponential equation

C1h. 0.0552...

T1. ${}_7P_3 = \dfrac{7!}{(7-3)!} = \dfrac{7!}{4!} = \dfrac{7 \cdot 6 \cdot 5 \cdot 4 \cdot 3 \cdot 2 \cdot 1}{4 \cdot 3 \cdot 2 \cdot 1}$
$= 7 \cdot 6 \cdot 5 = 210$

T3. In a permutation, order is important; in a combination, it is not. That is, rearrangements of the same choice of objects are considered different permutations but the same combination.

T5. 0.98

T7. The probabilities are the terms in a binomial series, in this case, the expansion of $(0.8 + 0.2)^3$.

T9. $\dfrac{1}{120} = \dfrac{5}{6}\%$

T11. −\$99; \$1

T13. ≈\$166.67

T15. 0

T17. $\dfrac{1}{9}$ point

T19. 0.3087

T21. $0.16807 + 0.36015 + \cdots + 0.00243 = 1$

T23.

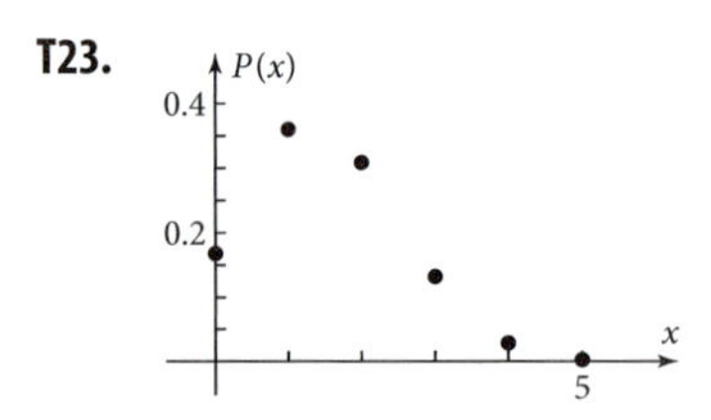

T25. 0.0752

T27. 0.8648

Chapter 15

Problem Set 15-1

1. 23; You would have to add nine 2s; $5 + 9 \cdot 2 = 23$; $5 + 99 \cdot 2 = 203$

3. Linear

5. 14; $14 \cdot 10 = 140$ = the partial sum; $\dfrac{5 + 203}{2} \cdot 100 = 10{,}400$

7. Quadratic regression fits exactly:
$y = x^2 + 4x = 100^2 + 4(100) = 10{,}400$

9. 6138

Problem Set 15-2

1a. Geometric **1b.** 64, $85.\overline{3}$

1b. 64, $85.\overline{3}$

1c. $(6.3139...) \times 10^{13}$

1d. The 37th term

3a. Arithmetic **3b.** 19; 6

3c. −1229 **3d.** The 50th term

5a. Geometric **5b.** 856.25; 2140.625

5c. $1.3640... \times 10^{41}$ **5d.** The 13th term

7a. Geometric **7b.** −36.45; 32.805

7c. −0.0014... **7d.** The 12th term

9a. Neither **9b.** 120; 143

9c. 9999 **9d.** The 57th term

11a. Arithmetic **11b.** $4x - 3a$; $5x - 4a$

11c. $100x - 99a$ **11d.** The 240th term

13. Geometric; the 11th square; $9.2233... \times 10^{18}$ grains. The king was upset because the number of grains of rice got so large so quickly.

15a. $1,267,500, $1,235,000, $1,202,500, $1,170,000, . . .; arithmetic sequence; $325,000; 40 yr; The depreciation function is linear and the scatter plot points lie on a straight line.

15b. $1,170,000, $1,053,000, $947,700, $852,930, . . .; $130,000 the first year, $117,000 the second year, $105,300 the third year; 15 yr.

17. 91%; ≈82.8%; ≈75.4%; geometric; ≈15.2%; 25 washings.

19a. $t_n = t_{n-2} + t_{n-1}$, $t_1 = t_2 = 1$; 89; 144; 6765

19b. 1, 2, 1.5, $1.\overline{6}$, 1.6, 1.625, 1.6153..., 1.6190..., 1.6176..., $1.\overline{618}$

19c. Answers will vary. The spirals in each direction usually are consecutive Fibonacci numbers.

19d. Answers will vary. Leonardo Fibonacci was an Italian mathematician of the late 12th and early 13th centuries. The Fibonacci term t_n is the number of pairs of rabbits there will be in the nth month if you start with one pair and if every pair produces another pair every month but not starting until they are two months old.

21a. To get to step 3, she can take one step from step 2 or two steps from step 1. So, the number of ways to get to step 3 is the number of ways to get to step 1 plus the number of ways to get to step 2. Similarly, the number of ways to get to step 4 is the number of ways to get to step 2 plus the number of ways to get to step 3.

21b. $t_n = t_{n-1} + t_{n-2}$ where $t_2 = 2$ and $t_1 = 1$; 10,946

21c. If you let $t_0 = 1$, then this is the same sequence.

21d. $7.5401... \times 10^{18}$

23. In the 41st month, after 40 payments. $4015.49.

25a. $1275, $1350, $1425; these are called *arithmetic means.*

25b. $4,725,000

27a. ln 48, ln 96, ln 192

27b. Geometric

27c. The logarithms have a common difference of 0.6931..., so they form an arithmetic sequence.

Problem Set 15-3

1a. $3 + 8 + 13 + \cdots + 48$; There is a common difference of 5.

1b. 255; Yes, the answers are the same.

1c. 25,050; Answers will vary.

3a. 13.4217... cm; 446.3129... cm

3b. The answers are getting closer and closer to 500. You can make the answer as close to 500 as you want by taking the sum of enough terms. The maximum depth the pile will attain is 500 cm, regardless of how long it is pounded.

3c. $\lim_{n\to\infty} S_n = 100 \cdot \frac{1}{1 - 0.8} = 500$ cm

5a. The series is geometric because there is a common ratio, 1.01. The amount at the fifth month is the fifth partial sum because it is the sum of the first five terms of the series; ≈$510.10.

5b. ≈$23,003.87; ≈$11,003.87

5c. 241 months

7a.

n	P	I_n	P_n	B_n
4	1,050.00	999.25	50.75	199,798.49

Answers to Selected Problems

7b. $B_1 = B_0(1 + I) - P$
$B_2 = B_1(1 + I) - P$
$= (B_0(1 + I) - P)(1 + I) - P$
$= B_0(1 + I)^2 - P(1 + I) - P$
$B_3 = B_2(1 + I) - P$
$= (B_0(1 + I)^2 - P(1 + I) - P)(1 + I) - P$
$= B_0(1 + I)^3 - P(1 + I)^2 - P(1 + I) - P$
$B_4 = B_3(1 + I) - P$
$= B_0(1 + I)^4 - P(1 + I)^3 - P(1 + I)^2 - P(1 + I) - P$
$B_4 = B_0(1 + I)^4 - P(1 + I)^3 - P(1 + I)^2 - P(1 + I) - P$
$= B_0(1 + I)^4 - P(1 + (1 + I) + (1 + I)^2 + (1 + I)^3)$
$= B_0(1 + I)^4 - P\dfrac{1 - (1 + I)^4}{1 - (1 + I)}$
$= B_0(1 + I)^4 - P\dfrac{1 - (1 + I)^4}{-I}$
$= B_0(1 + I)^4 + \dfrac{P}{I}(1 - (1 + I)^4)$

7c. 611 months

7d.

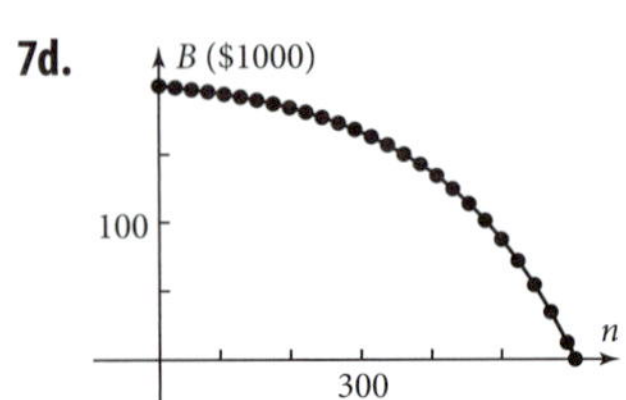

False: Halfway through the mortgage (month 305), the balance is still $164,222.84, about 82% of the original amount.

9a. 0.01024; 0.07680; 0.23040; 0.34560; 0.25920; 0.07776

9b. About 34.6%

9c. Letting X = the number of "point-ups" out of five flips, the first four terms of the series,
$P(X \le 3) = P(X = 0) + P(X = 1) + P(X = 2)$
$+ P(X = 3) = 0.01024 + 0.07680 + 0.23040 + 0.34560$
$= 0.66304$, about 66.3%

9d. About 61.8%, not the same probability

11. $9 + 11 + 13 + 15 + 17 = 65$

13. $3 + 9 + 27 + 81 + 243 + 729 = 1092$

15. 24,414,062

17. 12,171.6

19. −42,120

21. −738,100

23. 26

25. 19

27. 31

29. Converges; $\lim_{n\to\infty} S_n = 1000$

31. Diverges

33. Converges; $\lim_{n\to\infty} S_n = 428\frac{4}{7}$

35. Converges; $\lim_{n\to\infty} S_n = 512\frac{32}{39}$

37. $x^3 - 3x^2y + 3xy^2 - y^3$

39. $32x^5 - 240x^4 + 720x^3 - 1080x^2 + 810x - 243$

41. $x^{12} + 6x^{10}y^3 + 15x^8y^6 + 20x^6y^9 + 15x^4y^{12} + 6x^2y^{15} + y^{18}$

43. $56x^3y^5$

45. $-1365p^4j^{11}$

47. $-1716x^{18}y^{14}$

49. $48{,}384x^3y^5$

51. $-1365r^4q^{11}$

53. Answers will vary.

55. Journal entries will vary.

Problem Set 15-4

R1a. 17; 20

R1b. 75

R1c. $75 = 6 \cdot \dfrac{5 + 20}{2}$

R1d. 80; 160

R1e. 315

R3a. $2 + 5 + 8 + 11 + 14 + 17$

R3b. There is a common difference, 3.

R3c. $6 \cdot \dfrac{2 + 17}{2} = 57$

R3d. The 528th partial sum

R3e. sum(seq(4000·0.95^(N−1),N,1,200,1)) = 79,997.1957....

R3f. $4{,}000 \cdot \dfrac{1 - 0.95^{200}}{1 - 0.95} = 79{,}997.1957...$

R3g. The 76th term

R3h. 80,000

R3i. $1716a^6b^7$

R3j. i. Maya: arithmetic; Vincent: geometric

R3j. ii. Maya: 22 in.; Vincent: ≈ 4.38 in.

R3j. iii. Maya: $S_{21} = 21 \cdot \dfrac{12 + 22}{2} = 357$ in.;
Vincent: $S_{21} = 36 \cdot \dfrac{1 - 0.9^{21}}{1 - 0.9} = 320.6091... \approx 320.6$ in.;
Vincent is behind by 36.3908... ≈ 36.4 in.

R3j. iv. Yes. For Vincent,
$\lim_{x\to\infty} S_n = 36 \cdot \dfrac{1}{1 - 0.9} = 360 > 357.$

R3k. Second dose: 800 mg; third dose: 980 mg; nth dose: $\sum_{x=1}^{n} 500 \cdot \frac{1-0.6^n}{1-0.6}$. The total after n doses is the nth partial sum of a geometric series with first term 500 and common ratio 0.6. Sixteenth dose: $\approx$ 1249.6 mg $\lim_{x\to\infty} S_n = 500 \cdot \frac{1}{1-0.6} = 1250$ mg

C1a. $2, 1, \frac{1}{2}, \frac{1}{4}, \ldots; t_1 = 2, r = \frac{1}{2}$

C1b. $H_2 = 2.7071\ldots$ m; $H_3 = 3.2071\ldots$ m; $H_4 = 3.3838\ldots$ m

C1c. $2 + \frac{\sqrt{2}}{2} + \frac{1}{2} + \frac{1}{4} \cdot \frac{\sqrt{2}}{2} + \frac{1}{8} + \frac{1}{16} \cdot \frac{\sqrt{2}}{2} + \cdots =$ $\left(2 + \frac{1}{2} + \frac{1}{8} + \cdots\right) + \left(\frac{\sqrt{2}}{2} + \frac{1}{4} \cdot \frac{\sqrt{2}}{2} + \frac{1}{16} \cdot \frac{\sqrt{2}}{2} + \cdots\right)$; in both subsequences, $r = \frac{1}{4}$.

C1d. i. 3.6094... m

C1d. ii. 3.2189... m

C1d. iii. 0

C1d. iv. ∞

C1d. v. 2.3047... m above the ground.

C3a. $a^{1.8}$; $1.8a^{0.8}b$; $0.72a^{-0.2}b^2$; $-0.048a^{-1.2}b^3$; $0.0144a^{-2.2}b^4$

C3b.

n	c_n	x	c_{n+1}
1	1	5	5
2	5	4	10
3	10	3	10
4	10	2	5
5	5	1	1
6	1	0	0
7	0	−1	0
8	0	−2	0

From the seventh term on, all coefficients are zero.

T1. 6, 12, 24, 48, 96; $S_5 = 186$

T3. $S_n = \frac{n}{2}(t_1 + t_n)$

T5. $7 \cdot 1 + 7 \cdot 3 + 7 \cdot 9 + 7 \cdot 27 = 280$

T7. $\frac{15!}{6! \cdot 9!} a^6 b^9$

T9. $t_n = 7 \cdot 2^{n-1}$

T11. Geometric; 39.6312... ft

T13. $\lim_{n\to\infty} S_n = \lim_{n\to\infty} 5 + 2\left(4 \cdot \frac{1}{1-0.8}\right) = 45$ ft. As n grows larger, 0.8^{n-1} goes to zero.

T15. $n = 33$

T17. 249.9667... mg

T19. It is approaching 250 mg. $50 \cdot \frac{1}{1-0.8} = 250$ mg.

T21. $\$200(0.9)^n < \$5 \Rightarrow n > \frac{\log \frac{5}{200}}{\log 0.9} = 35.0119\ldots$ or 36 months.

T23. 0.0878...

Chapter 16

Problem Set 16-1

1. Distance = 0.5157... ft; average rate = 0.5157... ft/s

3. $\approx$0.41 ft/s

5. $\approx$22.2 cm

7. Approximate the region with n rectangles, each of width $\frac{7-2}{n}$ and height $g(x)$ evaluated at the midpoint of the subinterval. The exact area is the limit of the sum of the areas of the rectangles as the number of rectangles becomes infinite.

Problem Set 16-2

1. See page 791 for the complete definition.

3. $\lim_{x\to a} f(x) = 3$; $\lim_{x\to b} f(x)$ does not exist because $f(x)$ approaches different numbers from left and from right. $\lim_{x\to c} f(x) = 6$

5. $\lim_{x\to a} f(x) = 7$; $\lim_{x\to b} f(x) = 5$; $\lim_{x\to c} f(x) = 2$

7. $\lim_{x\to a} f(x) = 4$; $\lim_{x\to b} f(x)$ does not exist because $f(x)$ becomes infinite. $\lim_{x\to c} f(x) = 6$

9. $f(x) = -2x + 13$, if $x \neq 4$; $\lim_{x\to a} f(x) = 5$ There is a removable discontinuity at (4, 5).

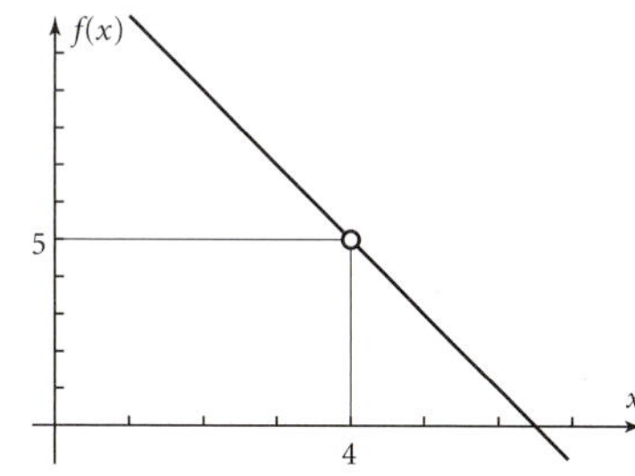

11. $f(x) = -x^2 + 7x - 9$, if $x \neq 4$; $\lim_{x\to4} f(x) = 3$
There is a removable discontinuity at (4, 3).

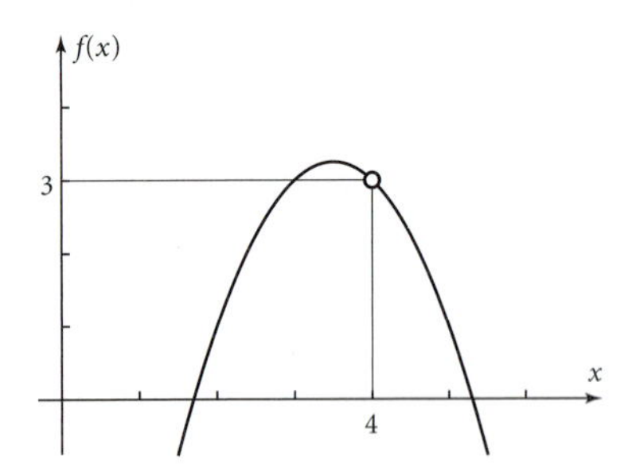

13a. $3.65 < x < 4.35, x \neq 4$

13b. Sample answer: $f(4.1) = -2(4.1) + 13 = 4.8$, which is within 0.7 unit of 5.

13c. The reasoning from parts a and b can be repeated by replacing 0.7 with any arbitrarily small positive number and finding the corresponding interval of x-values, thus showing that 5 really is the limit of $f(x)$ as x approaches 4.

15. $3.7236... < x < 4.1708...$.
The reasoning can be repeated by replacing 0.2 with any arbitrarily small positive number and finding the corresponding interval of x-values, thus showing that 3 really is the limit of $f(x)$ as x approaches 4.

17. $f(x) = -x^2 + 8x - 7 + \dfrac{1}{x-2}$.
The discontinuity is not removed because the remainder does not equal zero. So there is a vertical asymptote at $x = 2$, and thus no limit of $f(x)$ as x approaches 2. The graph of the polynomial part, $y = -x^2 + 8x - 7$, is a curved asymptote for the graph of function f.

Problem Set 16-3

1a. Average rate = 2 ft/min.
The line is a secant line passing through points (4, 31) and (6, 35) with slope 2. See part e for the graph.

1b. At $c = 4$, $r(x) = \dfrac{f(x) - f(4)}{x - 4} = -x^2 + 6x + 2$, if $x \neq 4$. $r(6) = -6^2 + 6(6) + 2 = 2$, which agrees with part a.

1c. $\lim_{x\to4} r(x) = 10$ ft/min;
The line $y = 10x - 9$ is tangent to the graph.

1d. $f'(x) = -3x^2 + 20x - 22$;
$f'(4) = 10$, which agrees with part c.

1e. $f'(1) = -3(1^2) + 20(1) - 22 = -5$;
The line $y = -5x + 15$ is tangent to the graph.

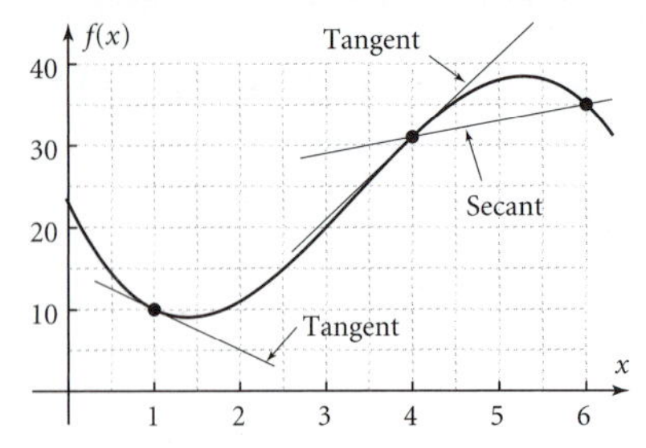

3a.

$$\begin{array}{r|rrrr} 1 & 1 & -17 & 105 & -89 \\ & & 1 & -16 & 89 \\ \hline & 1 & -16 & 89 & 0 \end{array}$$

Thus $x = 1$ is a zero of function f, meaning that a 1-ft board is free! $x = 8 \pm 5i$, which are nonreal complex numbers. Thus there are no other real zeros of function f.

3b. \$32.11; $\approx$14.9 ft.

3c. $r(10.1) = 66.31$; $r(10.01) = 65.1301$; $r(10.001) = 65.013001$; $\lim_{x\to10} r(x)$ appears to equal 65 cents per foot.

3d. $f'(x) = 3x^2 - 34x + 105$; $f'(10) = 65$; $y = 65x - 389$. The graph shows that the line is tangent to the graph of function f at $x = 10$.

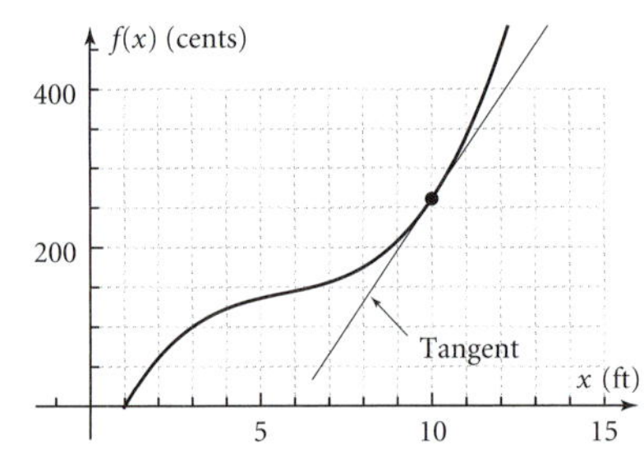

3e. $f'(5) = 10$ ¢/ft, which is lower that the 65 ¢/ft for a 10-ft board. For longer boards, the price per foot increases because taller trees are harder to find.

5. $f'(x) = 7x^6$

7. $f'(x) = 48x^5$

9. $f'(x) = 27x^2 - 10x + 2$

11. $f'(x) = 6x^5$

13. $f'(x) = 3x^2 - 24x + 36$; vertices at $x = 2, 6$

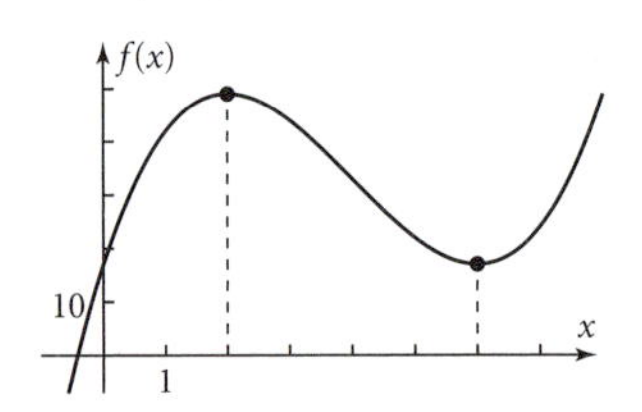

15. $f(x) = (2x - 7)^3 = 8x^3 - 84x^2 + 294x - 343$
$f'(x) = 24x^2 - 168x + 294$
Amos's answer is equivalent to
$3(4x^2 - 28x + 49) = 12x^2 - 84x + 147$, which is exactly half of the correct answer.

17a. Average rate = 0.9999...
Conjecture: $\lim_{x \to 1} \frac{\sin x - \sin 0}{x - 0} = 1$.

17b. Average rate = 0.5402...; cos 1 = 0.5403...

17c. Conjecture: If $f(x) = \sin x$, then $f'(x) = \cos x$. The conjecture is consistent with the graph of $f(x) = \sin x$:
At $x = \frac{\pi}{2}$, the slope appears to be 0, and $\cos \frac{\pi}{2} = 0$.
At $x = \pi$, the slope appears to be -1, and $\cos \pi = -1$.
At $x = 2\pi$, the slope appears to be 1, the same as at $x = 0$, and $\cos 2\pi = 1$.

Problem Set 16-4

1. Integral ≈ 331

3a. (rate)(time) cannot be used because the rate varies.
Distance = $\int_0^{10} 11x \cdot 1.4^{-x}\, dx$

3b.

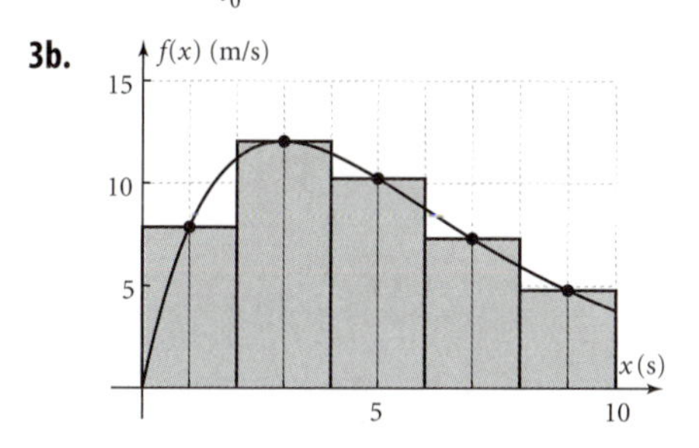

$R_5 = 2 \cdot f(1) + 2 \cdot f(3) + 2 \cdot f(5) + 2 \cdot f(7) + 2 \cdot f(9) = 84.4119...$

3c. $R_5 = 84.4119...$, which agrees with part b.
$R_{10} = 82.9915...$; $R_{20} = 82.6239...$; $R_{100} = 82.5052...$

3d. About 82.5 m

5a.

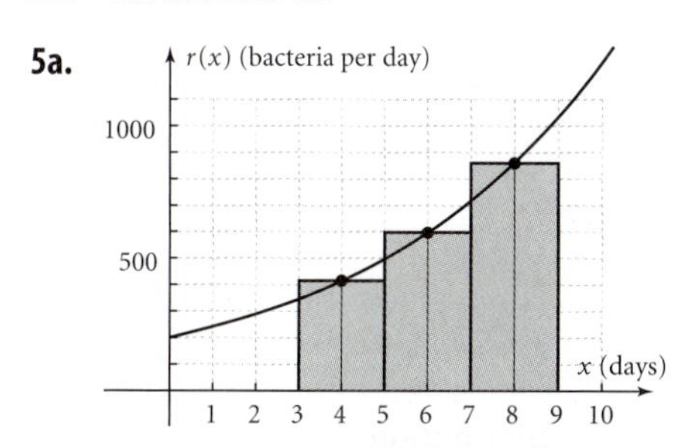

$R_3 \approx 3744$ bacteria; the units are $\frac{\text{bacteria}}{\text{day}} \cdot$ days, which gives bacteria.

5b. By program:
$R_3 = 3743.7603...$, which agrees with part a.
$R_{10} = 3762.6506...$
$R_{20} = 3764.0668...$
$R_{100} = 3764.5172...$
Conjecture: Riemann sums approach about 3765 bacteria, so the number of bacteria grows by about 3765 from day 3 to day 9.

7. Answers will vary.

Problem Set 16-5

R1a. $f(x)$ changes from $\sin(\frac{\pi}{3})$ to $\sin(\frac{\pi}{3} + 0.01)$, or 0.004956.... Rate is $\frac{0.004956...}{0.01} = 0.4956...$, which is close to 0.5. $f(x)$ changes from $\sin(\frac{\pi}{3})$ to $\sin(\frac{\pi}{3} + 0.001)$, or 0.0004995.... Rate is $\frac{0.0004995...}{0.001} = 0.4995...$, which is closer to 0.5.

R1b. The line with slope 0.5 containing the point $\left(\frac{\pi}{3}, \left(\sin \frac{\pi}{3}\right)\right)$ is tangent to the graph of function f.

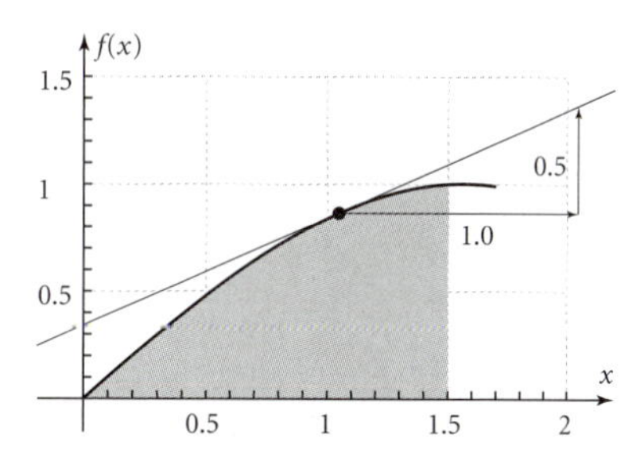

R1c. See graph for part b. There are approximately 3.7 squares, and the area of each square is (0.5)(0.5) = 0.25. Area of region ≈ (3.7)(0.25) = 0.925.
Exact area is (1 − cos 1.5) = 0.9292..., so the approximate value is close.

R3a. $r(2.01) = 115.212$, $r(2.001) = 119.52012$;
Conjecture: Instantaneous velocity at $x = 2$ is 120 ft/min.

R3b. $r(x) = \frac{(120x^3 - 1200x^2 + 3480x - 2400) - 720}{x - 2}$

$r(x) = \frac{120x^3 - 1200x^2 + 3480x - 3120}{x - 2}$

2	120	−1200	3480	−3120
		240	−1920	3120
	120	−960	1560	0

$r(x) = 120x^2 - 960x + 1560, x \neq 2$
$\lim_{x \to 2} r(x) = 120(2^2) - 960(2) + 1560 = 120$ ft/min, confirming the conjecture in part a. The train is going north because $d(x)$ is the number of feet north of the crossing, and $d(x)$ is increasing (positive rate of change) when $x = 2$ min.

R3c. $d'(x) = 360x^2 - 2400x + 3480$
$d'(2) = 120$ ft/min, which confirms the result of part b.

R3d. The vertical coordinate at $x = 2$ is 120, the instantaneous velocity.

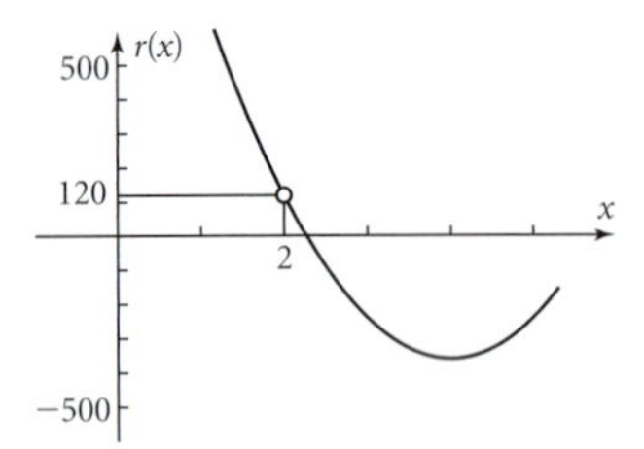

R3e. The line containing the point $(2, d(2))$ with slope 120 is tangent to the graph of function d at that point.

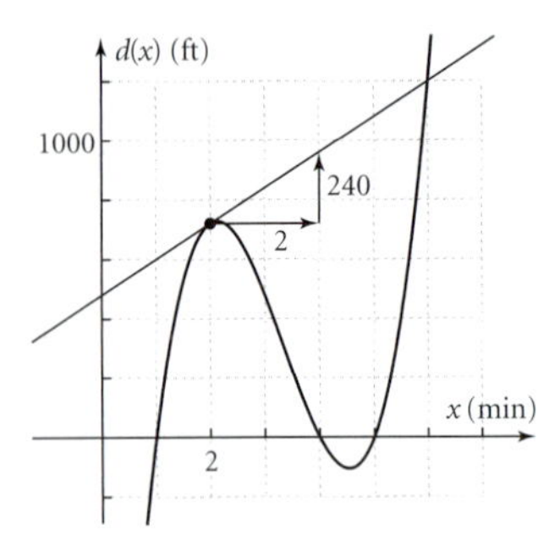

R3f. $360x^2 - 2400x + 3480 = 0 \Rightarrow x = 4.5351...$
Train is stopped at about 4.54 minutes.

C1a. $R_{10} = 55.98$; $R_{100} = 55.9998$
Conjecture: The integer is 56.

C1b. Number of squares ≈ 11.2; Each square represents $(0.5)(10) = 5$ units. Integral $\approx (11.2)(5) = 56.0$, confirming part a.

C1c. $g(4) - g(2) = 4^3 - 2^3 = 56$, the value of the integral from part a. The derivative of function g is $g'(x) = 3x^2$, the value of $f(x)$ in part a.

C1d. 2; $R_{100} = 2.00008224...$, which is close to the exact value, 2.

T1. $L = \lim_{x \to c} f(x)$ if and only if $f(x)$ stays arbitrarily close to L wherever x is kept close enough to c on both sides, but $x \neq c$.

T3. The slope of the solid secant line is the average rate of change, approximately $\frac{3.7 - 2.8}{5 - 2} = 0.3$.

T5. 11.1

T7. There is no value of $f(4)$ because division by zero is undefined.

T9. $3.7236... < x < 4.1708...$, and $x \neq 4$.
The restriction $x \neq 4$ avoids division by zero.

T11. $f(x) = 0.5x + 0.5 - \frac{2}{x - 5}$
The discontinuity at $x = 5$ is not removable because the simplified expression still has a zero denominator if $x = 5$.

T13. The graph of $y = 0.5x + 0.5$ is a slant asymptote to the graph of function f.

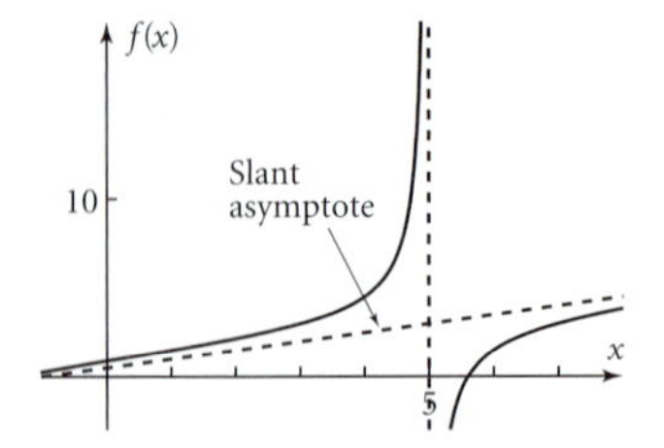

T15. $R_3 = 1 \cdot f(2.5) + 1 \cdot f(3.5) + 1 \cdot f(4.5) = 11.175$

T17. Answers will vary.

Problem Set 16-6

1. 0.72

3. 0.2

5. $\frac{10!}{4!\,6!}$

7. 0.9

9. 1000

11. 270

13. Removable discontinuity

15. 1

17. The instantaneous rate of change of $f(x)$ with respect to x

19. "The definite integral of $f(x)$ with respect to x from 2 to 5"

21. $P(1) = 0.4$; $P(2) = 0.24$; $P(3) = 0.144$; $P(4) = 0.0864$; $P(5) = 0.05184$

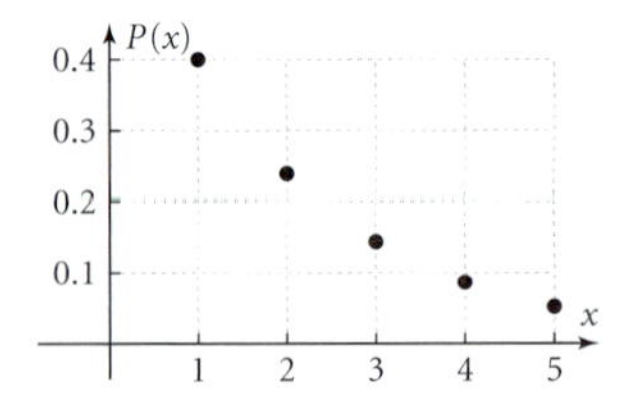

23. 0.9939...

25. 0.31104; binomial distribution

27. 0.56; we must assume the probabilities are independent.

29. 0.24

31. \$5860

33. 320

35. $\approx$\$547.03; a sequence is more reasonable than a continuous function because the value increases by jumps, one jump each compounding period.

37. The sequence is $1 + 1, 4 + 1, 9 + 1, \ldots, 49 + 1$;
$t_n = n^2 + 1$; $t_{100} = 100^2 + 1 = 10{,}001$.

39. 1; 0.4352...; 0.2675...; 0.1894...; 3.6030...

41. 99 \$/ft

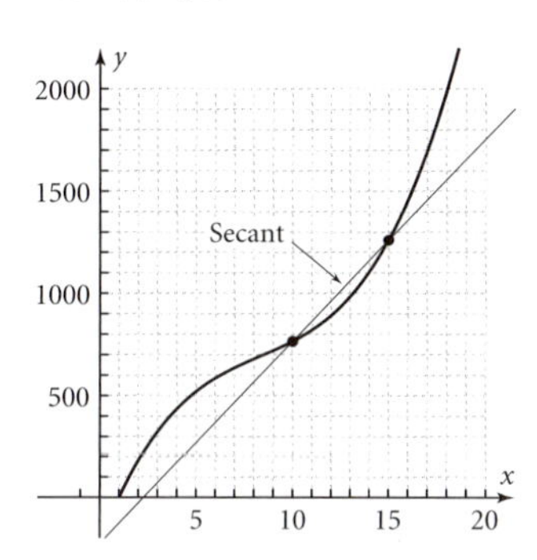

43. $f'(x) = 3x^2 - 50x + 249$;
$f'(10) = 3(10^2) - 50(10) + 249 = 49$ \$/ft, which agrees with Problem 42.
The graph shows that the line containing the point $(10, f(10))$ with slope 49 is tangent to the graph at that point.

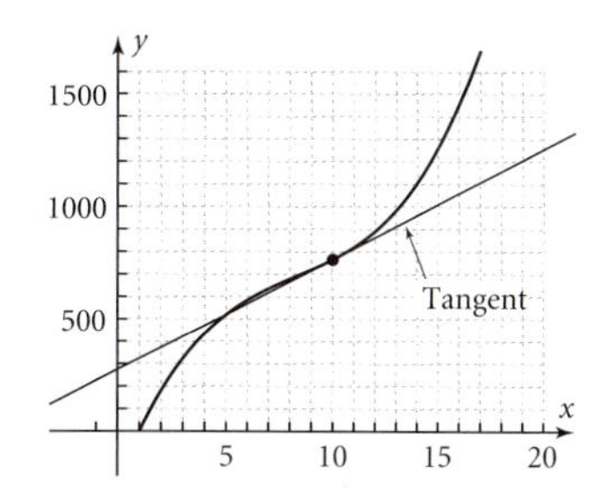

45. Number of grid squares $\approx$ 10.8 (exactly 10.8)
Each square represents $(20)(1) = 20$ liters.
$\int_0^6 r(x)\,dx \approx (10.8)(20) = 216$ liters

47. $R_{10} = 215.82$; $R_{100} = 215.9982$
Conjecture: The integer is 216.

49. Answers will vary.

Glossary

A

Add–add property (of linear functions) (pp. 77, 80): If f is a linear function, adding a constant to x results in adding a constant to the corresponding $f(x)$-value.

Add–multiply property (of exponential functions) (pp. 78, 80): If f is an exponential function, adding a constant to x results in multiplying the corresponding $f(x)$-value by a constant.

Algebraic (p. 4): Involving only the operations of algebra performed on the variable—namely, addition, subtraction, multiplication (including integer powers), division, and roots.

Amplitude (of a sinusoidal graph) (p. 283): The vertical distance from the sinusoidal axis of a graph to its maximum or minimum.

Analytic geometry (p. 522): The study of properties of geometric figures using coordinate systems.

Angular velocity (pp. 328, 329): The number of degrees (or radians or revolutions) per unit of time through which a point on a rotating object turns.

Argument (p. 9): In $f(x)$, the variable x or any expression substituted for x.

Arithmetic mean (p. 765): The term(s) that can be inserted between two given numbers to form an arithmetic sequence.

Arithmetic sequence (pp. 755, 757, 761): A sequence in which each term is formed recursively by adding a constant to the previous term.

Arithmetic series (pp. 755, 775): The sum of the terms of an arithmetic sequence.

Asymptote (p. 4): A line or curve that the graph of a function stays arbitrarily close to as x or y approaches positive or negative infinity.

Auxiliary Cartesian graph (p. 565): A graph of a polar equation made using Cartesian coordinates (θ, y) and a Cartesian coordinate system, rather than a polar graph using (r, θ).

Average rate of change (pp. 244, 801): The average rate of change of a function f over an interval from $x = a$ to $x = b$ is the change in the y-values of the function divided by the corresponding change in the x-values, $\dfrac{f(b) - f(a)}{b - a}$.

Axis of symmetry (p. 501): A line that divides a graph into two symmetric halves.

B

Barnsley's method (pp. 681–682): A means of generating fractal figures that involve iterations of multiple transformations; at each stage, one of a given set of transformations is selected at random and performed on the point that is the image of the preceding transformation.

Bearing (p. 468): An angle measured clockwise from north; used by navigators for a velocity or displacement vector.

Binomial coefficients (p. 774): The coefficients of the terms in a binomial series.

Binomial distribution (p. 734): Probabilities that are associated with r successes in n repeated trials for an event with two outcomes.

Binomial expansion (p. 773): See **Binomial series**.

Binomial experiment (p. 735): A random experiment in which the same action is repeated a number of times and in which only two results of the action are possible.

Binomial formula (p. 774): For any positive integer n, $(a + b)^n = a^n + \binom{1}{n}a^{n-1}b + \binom{2}{n}a^{n-2}b^2 + \cdots + \binom{n}{n-2}a^2b^{n-2} + \binom{n}{n-1}ab^{n-1} + b^n$.
The binomial formula may also be written as $(a + b)^n = \sum_{r=0}^{n}\binom{n}{r}a^{n-r}b^r$.

Binomial series (pp. 734, 767, 773): A series that comes from expanding a positive integer power of a binomial expression.

Boolean variable (pp. 11, 12): A variable that equals 1 if a condition is true and 0 if the condition is false.

C

Cardioid (p. 565): A type of limaçon that has a cusp at its pole. This occurs if, for $r = a + b \cos \theta$ or $r = a + b \sin \theta$, $a \neq 0$, $b \neq 0$, and $|a| = |b|$.

Center-radius form (pp. 495, 499): The form of an equation for a circle, ellipse, or hyperbola that indicates the center and radii.

Circular functions (pp. 307, 308): Periodic functions whose independent variable is an arc length around a unit circle (or angle in radians, a real number without units).

Closed interval (p. 359): An interval of values that includes the endpoints. Such an interval can be written as $a \leq x \leq b$, $[a, b]$, or $x \in [a, b]$.

Coefficient of determination (pp. 132, 133): The fraction of SS_{dev} that has been removed by the linear regression $r^2 = \dfrac{SS_{dev} - SS_{res}}{SS_{dev}}$

Cofunction (pp. 350, 399): Two trigonometric functions that have equal values for complementary angles. For example, $\tan(A) = \cot(90° - A)$, so tangent and cotangent are cofunctions.

Combination (pp. 719, 720): A subset of elements selected from a set without regard to the order in which the selected elements are arranged.

Common difference (pp. 757, 761): The constant added to a term to get the next term in an arithmetic sequence or series.

Common logarithm (p. 96): A logarithm with base 10.

Common ratio (pp. 759, 761): The constant multiplied by a term to get the next term in a geometric sequence or series.

Complementary events (p. 730): For any event A, the event "not A" is the complementary event. The probability that event A will not occur is $P(\text{not } A) = 1 - P(A)$.

Completing the square (pp. 184, 185): Reversing the process of squaring a binomial to go from $y = ax^2 + bx + c$ to the form $y = a(x - h)^2 + k$, which shows translations and dilations.

Complex conjugates (pp. 182, 579): Two complex numbers of the forms $a + bi$ and $a - bi$, where a and b are real numbers and $i = \sqrt{-1}$.

Complex number (pp. 177, 578): The sum of a real number and an imaginary number.

Complex plane (pp. 181, 578): The representation of complex numbers by points in a Cartesian coordinate system.

Components (of a vector) (pp. 466, 467): The horizontal and vertical elements of a vector. Vector $\vec{v} = x\vec{i} + y\vec{j}$ has horizontal component $x\vec{i}$ and vertical component $y\vec{j}$.

Composite argument property (p. 388): A property that reexpresses a trigonometric function of the sum or difference of two angles. For example, $\cos(A - B) = \cos A \cos B + \sin A \sin B$.

Composite function (pp. 23, 30): A function of the form $f(g(x))$, where function g is performed on x and then function f is performed on $g(x)$. The domain of the composite function is those x-values in the domain of g for which $g(x)$ is an element of the domain of f.

Composition of ordinates (p. 405): The method by which sinusoids or other functions are combined by adding or multiplying the corresponding ordinates for each value of x.

Concave up (or **concave down**) (p. 65): The graph of a function (or a portion of the graph between two asymptotes or two points of inflection) is called concave up when its "hollowed out" side faces upward; it is called concave down when its "hollowed out" side faces downward.

Conic section (p. 495): The curve formed by the intersection of a plane with a cone.

Conjugate axis (p. 514): For a hyperbola, the name given to the axis of symmetry perpendicular to the transverse axis.

Conjugate binomials (p. 354): Two binomials that are multiplied to get a difference of two squares. For example, the conjugate of $1 + \cos B$ is $1 - \cos B$.

Conjugate radius (p. 525): The radius from the center of a hyperbola to one endpoint of the conjugate axis.

Constant (p. 9): A number with a fixed value, as opposed to a *variable,* which can take on different values.

Constant function (p. 66): A function of the form $y = b$, for any real constant b.

Continuous (function) (pp. 189, 755, 756): A function for which the limit of $f(x)$ as x approaches c equals the value of $f(c)$ for all $x = c$ in the domain, and whose graph has no gaps or discontinuities.

Converge (p. 772): A series converges when its partial sums approach a finite limit as the number of terms becomes infinite.

Correlation coefficient (pp. 132, 133): The value r, which is the positive or negative square root of the coefficient of determination.

Cosecant (pp. 262, 263): One of the trigonometric functions, equal to the reciprocal of the sine function; abbreviated "csc."

Cosine (pp. 256, 263): One of the trigonometric functions; abbreviated "cos." If θ is an angle of rotation of the positive u-axis terminating in a position containing the point (u, v), then $\cos \theta = \frac{u}{r}$, where r is the distance from the point (u, v) to the origin.

Cotangent (pp. 262, 263): One of the trigonometric functions; equal to the reciprocal of the tangent function; abbreviated "cot."

Coterminal angles (p. 249): Two angles in standard position are coterminal if their degree measures differ by a multiple of 360°.

Critical point (p. 190): A point where the tangent line to the graph of a function is either horizontal or vertical.

Cross product (p. 633): The cross product of two vectors is a vector with the following properties: $\vec{a} \times \vec{b}$ is perpendicular to the plane containing $\vec{a}$ and $\vec{b}$; the magnitude of $\vec{a} \times \vec{b}$ is $|\vec{a} \times \vec{b}| = |\vec{a}| \cdot |\vec{b}| \sin \theta$, where θ is the angle between the two vectors when they are placed tail-to-tail; the direction of $\vec{a} \times \vec{b}$ is determined by the right-hand rule.

Cusp (p. 565): A point at which the graph of a relation changes direction abruptly.

Cycle (p. 254): The part of a graph from any point to the point where the graph first starts repeating itself.

D

De Moivre's theorem (p. 583): The "short way" to raise a complex number to a power: raise the modulus to the power and multiply the argument by the exponent. For example, if $z = r(\cos \theta + i \sin \theta) = r \text{ cis } \theta$, then $z^n = r^n \text{ cis } n\theta$.

Decreasing (function) (p. 38): A function for which $f(x_2) < f(x_1)$ for all $x_2 > x_1$ in the domain.

Definite integral (pp. 809, 812): The area under a curve, found by Riemann sums. Also, the product of two factors, when one factor may vary.

Degree (of a polynomial) (p. 189): For a polynomial expression, the greatest number of variables multiplied together in any one term.

Dependent events (p. 709): Events are called dependent if the outcome of one event is affected by the outcome of the other event.

Dependent variable (p. 4): The output of a function, that is, the $f(x)$-variable.

Derivative (pp. 798, 800, 801): The instantaneous rate of change of a function.

Derivative function (p. 808): The function, written $f'(x)$, that gives the instantaneous rate of change of $f(x)$ at any x-value.

Determinant (pp. 635, 660): A particular square array of numbers, used in inverting a matrix, or in determining the cross product of two vectors.

Deviation (pp. 131, 133): For a data point (x, y), the directed distance of its y-value from $\bar{y}$, $y - \bar{y}$ where $\bar{y}$ is the average of the y-values.

Dilation (p. 18): A stretch or shrink of a function. The function $g(x) = a \cdot f\left(\frac{1}{b}x\right)$ represents a dilation of the function f by a factor of a in the y-direction and b in the x-direction.

Dimension (of a fractal, the Hausdorff dimension) (p. 688): If an object is transformed into N self-similar pieces, the ratio of the length of each piece to the length of the original object is r, and the subdivisions can be carried on infinitely, then the dimension D of the object is $D = \dfrac{\log N}{\log 1/r}$.

Direction angles (of a position vector) (p. 640): α, angle from the x-axis to the vector; β, angle from the y-axis to the vector; and γ, angle from the z-axis to the vector.

Direction cosines (pp. 640, 642): The cosines of the direction angles of a position vector.

Direct variation function (p. 10): A function with the general equation $f(x) = ax$, where a is a constant.

Directrix (pp. 523, 524): A line associated with a conic section.

Directrix radius (pp. 524, 525): The radius from the center of an ellipse or hyperbola to a directrix along the major or transverse axis.

Discrete function (p. 756): A function whose domain is a set of disconnected values.

Discriminant (of a general quadratic relation or conic section) (p. 543): For the general quadratic relation equation $Ax^2 + Bxy + Cy^2 + Dx + Ey + F = 0$, the discriminant is $B^2 - 4AC$. The value of the discriminant determines what type of conic section the equation represents.

Discriminant (of a quadratic equation) (p. 180): For a quadratic equation $y = ax^2 + bx + c$, the discriminant is $b^2 - 4ac$. The discriminant discriminates between quadratic equations that have real-number solutions and those that have nonreal complex-number solutions.

Displacement (pp. 48, 256, 463): The difference between an object's initial and final position.

Displacement vector (pp. 487, 605): The vector that indicates the difference between an object's initial and final positions.

Distance formula (p. 391): A form of the Pythagorean theorem giving the distance D between two points (x_1, y_1) and (x_2, y_2) in a plane in terms of their coordinates:
$D = \sqrt{(x_2 - x_1)^2 + (y_2 - y_1)^2}$.

Diverge (p. 772): A series diverges if the partial sums do not approach a finite limit as the number of terms becomes infinite.

Domain (p. 4): The set of values that the independent variable of a function can have.

Dot product (pp. 485, 619, 620, 621): $|\vec{a}| \cdot |\vec{b}| = |\vec{a}||\vec{b}| \cos \theta$, where θ is the angle between the two vectors when they are translated tail-to-tail. Also called the *scalar product* or *inner product*.

Double argument property (pp. 423, 425): A property that reexpresses a trigonometric function of two times an angle. For example, $\sin 2A = 2 \sin A \cos A$.

E

Eccentricity (pp. 523, 524): Eccentricity, e, is the ratio of the distance between a point on a conic section and the focus, d_2, and the distance between the same point and the directrix, d_1:
$e = \dfrac{d_2}{d_1}$

Elements (p. 658): The numbers in a matrix.

Ellipse (p. 367): The set of all points P in a plane for which the sum of the distances from point P to two fixed points (the foci) is constant.

Ellipsis format (p. 261): A three-dot (...) notation that indicates something has been left out.

Ellipsoid (p. 515): The three-dimensional surface generated by rotating an ellipse about one of its axes.

End behavior (pp. 137, 190, 217): The behavior of a function $f(x)$ as x approaches large numbers in either the positive or negative direction.

Even function (p. 50): A function f is an even function if and only if $f(-x) = f(x)$ for all x in the domain.

Event (p. 706): In probability, a set of outcomes.

Explicit formula (pp. 758, 760): For a sequence, specifies t_n as a function of n.

Exponential function (pp. 10, 296–297): A function in which the independent variable appears as an exponent.

Exponential regression (p. 138): A method of finding the best-fitting exponential function for a set of data.

Extrapolation (p. 6): Using a function to estimate a value *outside* the range of the given data.

Extreme point (p. 190): A point at which a function changes direction.

F

Factorial (p. 715): For any positive integer n, $n! = 1 \cdot 2 \cdot 3 \cdot \cdots \cdot n$ or, equivalently, $n! = n \cdot (n-1) \cdot (n-2) \cdot \cdots \cdot 2 \cdot 1$. 0! is defined to be equal to 1.

Fibonacci sequence (p. 764): The sequence of integers 1, 1, 2, 3, 5, 8, 13, 21, 34, 55, 89, 144, 233, ... , formed according to the rule that each integer is the sum of the preceding two.

Fixed point attractor (or **fixed point**) (p. 674): When a figure is transformed, the point that the transformations approach as the number of iterations approaches infinity.

Focal distance (p. 532): The distance from the center of a conic section to either focus.

Focal radius (pp. 524, 525): The radius from the center of an ellipse or hyperbola to a focus along the major or transverse axis.

Focus (pp. 523, 524): A point associated with a conic section.

Fractal (p. 681): A geometric figure that is composed of self-similar parts.

Frequency (pp. 285, 286): The reciprocal of the period of a sinusoidal function.

Function (p. 8): A relationship between two variable quantities for which there is exactly one value of the dependent variable for each value of the independent variable in the domain.

G

Geometric mean (p. 765): The term(s) that can be inserted between two given numbers to form a geometric sequence.

Geometric sequence (pp. 755, 759, 761): A sequence in which each term is formed recursively by multiplying the previous term by a constant.

Geometric series (p. 755): The sum of the terms of a geometric sequence.

Greatest integer function (p. 53): The greatest integer function $f(x) = \lfloor x \rfloor$ returns the greatest integer less than x.

H

Half argument property (pp. 423, 427): A property that reexpresses a trigonometric function of half an angle. For example, $\sin \frac{1}{2}A = \pm\sqrt{\frac{1}{2}(1 - \cos A)}$.

Harmonic analysis (p. 405): The reverse of the composition of ordinates; a method by which parent sinusoid functions are found from the resultant function.

Harmonic series (p. 777): A series such as $1 + \frac{1}{2} + \frac{1}{3} + \frac{1}{4} + \cdots + \frac{1}{n}$ in which successive terms are the reciprocals of the terms in an arithmetic sequence.

Hausdorff dimension (p. 688): See **Dimension.**

Hero's formula (p. 452): A formula for finding the area of any triangle with sides a, b, and c:
$\text{Area} = \sqrt{s(s-a)(s-b)(s-c)}$
where s is the semiperimeter (half the perimeter).

Hyperbola (pp. 495, 525): The set of all points P in a plane for which the absolute value of the difference of the distances from point P to two fixed points (the foci) is constant.

Hyperboloid (p. 515): The three-dimensional surface generated by rotating a hyperbola about one of its axes.

I

Identity (p. 351): An equation that is true for all values of the variable in the domain.

Identity matrix (p. 660): A square matrix with 1's along the main diagonal (the diagonal from the upper-left corner to the lower-right corner) and 0's everywhere else. Multiplying a square matrix [*A*] by an identity matrix leaves [*A*] unchanged.

Image (pp. 540, 666): The point, figure, or matrix that results when a transformation is applied to an original point, figure, or matrix.

Imaginary number (pp. 178, 578): A number that is the square root of a negative number.

Increasing (function) (p. 38): A function for which $f(x_2) > f(x_1)$ for all $x_2 > x_1$ in the domain.

Independent (events) (p. 709): Events are called independent if the outcome of one event does not affect the outcome of the other event(s).

Independent variable (p. 4): The input of a function, that is, the *x*-variable in $f(x)$.

Indeterminate form (p. 798): An expression that has no direct meaning as a number, for example, $\frac{0}{0}$ or $0 \cdot \infty$.

Inner product (p. 619): See **Dot product.**

Inside transformation (p. 17): A transformation applied to the argument of a function, which affects the graph of the function in the horizontal direction.

Instantaneous rate of change (pp. 244, 798, 801): The instantaneous rate of change of $f(x)$ at $x = c$, called the derivative of function *f* at $x = c$, is the limit of the average rates of change as *x* approaches *c*:

$$f'(c) = \lim_{x \to c} \frac{f(x) - f(c)}{x - c}$$

Interpolation (p. 6): Using a function to estimate a value *within* the range of given data.

Intersection (p. 727): If *A* and *B* are two events, then the intersection of *A* and *B*, $A \cap B$, is the set of all outcomes that are found in both event *A and* event *B*.

Inverse (of a matrix) (p. 660): If the product of two square matrices is the identity matrix, then the two matrices are inverses of each other. The inverse of a matrix [*M*] is denoted $[M]^{-1}$.

Inverse (of a function) (p. 35): The inverse of a function is the relation formed by interchanging the variables of the function. If the inverse relation is a function, then it is called the *inverse function* and is denoted $f^{-1}(x)$.

Inverse variation function (p. 10): A function with the general equation $f(x) = \frac{a}{x}$ or $f(x) = \frac{a}{x^n}$, where *a* is a constant and $x \neq 0$.

Invertible (function) (pp. 38, 42): A function whose inverse is also a function.

Iteration (p. 668): The process of performing an operation over and over again, each time operating on the image from the preceding step.

L

Law of cosines (pp. 444, 446): In triangle *ABC*, with sides *a*, *b*, and *c*, and opposite angles *A*, *B*, and *C*, respectively, $a^2 = b^2 + c^2 - 2bc \cos A$.

Law of sines (p. 455): In triangle *ABC*, with sides *a*, *b*, and *c*, and opposite angles *A*, *B*, and *C*, respectively,

$$\frac{a}{\sin A} = \frac{b}{\sin B} = \frac{c}{\sin C}.$$

Leading coefficient (p. 189): In a polynomial, the coefficient of the highest-degree term.

Limaçon of Pascal (or **limaçon**) (pp. 562, 565): A type of polar graph with the equation $r = a + b\cos\theta$ or $r = a + b\sin\theta$, where $a \neq 0$ and $b \neq 0$. Limaçon is a French word for "snail."

Limit (p. 791): A number that $f(x)$ stays arbitrarily close to if *x* is kept close enough to *c*, but not equal to *c* (or if *x* is kept large enough).

Linear combination (p. 388): Of two expressions *u* and *v*, a sum of the form $au + bv$, where *a* and *b* are constants.

Linear function (p. 10): A function with the general equation $y = ax + b$, where *a* and *b* are constants.

Linear regression (pp. 129, 138): The process of finding the best-fitting linear function for a given set of data.

Linear velocity (pp. 328, 329): The distance per unit of time that a point on a rotating object travels along its circular path.

Linearizing data (p. 144): The process of using logarithms to transform data so that a linear function fits the transformed data.

Logarithm (pp. 88, 90, 321, 322): $y = \log_a x$ is an exponent for which $a^y = x$.

Logarithmic function (pp. 102, 103): A function that is the inverse of an exponential function.

Logarithmic regression (p. 140): A method of finding the best-fitting logarithmic function for a set of data.

Logistic function (pp. 110, 111, 112): A function of the form $f(x) = \dfrac{c}{1 + ae^{-bx}}$ or $f(x) = \dfrac{a}{1 + ab^{-x}}$, where a, b, and c are constants, e is the base of the natural logarithm, and the domain is all real numbers.

Logistic regression (p. 141): A method of finding the best-fitting logistic function for a set of data.

M

Major axis (p. 500): For an ellipse, the longer of its two perpendicular axes; for an ellipsoid, the longest of its three axes.

Major radius (p. 524): The radius from the center of an ellipse or hyperbola to a vertex along the major axis.

Mathematical expectation (pp. 740, 742): The weighted average of the values for a random experiment each time it is run. The "weights" are the probabilities of each outcome.

Mathematical models (pp. 4, 489): Functions that are used to make predictions and interpretations about a phenomenon in the real world.

Matrix (p. 658): A rectangular array of terms called *elements.*

Minor axis (p. 500): For an ellipse, the shorter of its two perpendicular axes; for an ellipsoid, the smallest of its three axes.

Minor radius (p. 524): The radius from the center of an ellipse or hyperbola to one end of the minor axis.

Modulus (p. 579): The magnitude, or absolute value, of a complex number.

Multiply–add property (of logarithmic functions) (pp. 102, 103): If f is a logarithmic function, multiplying x by a constant results in adding a constant to the corresponding $f(x)$-value.

Multiply–multiply property (of power functions) (pp. 79, 80): If f is a power function, multiplying x by a constant results in multiplying the corresponding $f(x)$-value by a constant.

Mutually exclusive (p. 709): When the occurrence of an event excludes the possibility that another event will also occur.

N

Natural exponential function (p. 69): An exponential function with base e, where $e = 2.71828\ldots$.

Natural logarithm (p. 96): A logarithm with base e, where $e = 2.71828\ldots$.

Normal (vector) (p. 627): A vector is normal to a plane or another vector if it is perpendicular to that plane or vector.

Numerical (p. 4): Referring to constants, such as 2 or π, rather than to parameters or variables.

O

Oblique triangle (p. 443): A triangle with no right angles.

Odd function (p. 50): A function f is an odd function if and only if $f(-x) = -f(x)$ for all x in the domain.

One-to-one function (p. 38): A function f is a one-to-one function if there are no y-values that correspond to more than one x-value.

Open interval (p. 359): An interval of values that does not include the endpoints. Such an interval can be written as $a < x < b$, (a, b), or $x \in (a, b)$.

Outcome (p. 706): A result of a random experiment.

Outside transformation (p. 17): A transformation applied to the value of a function, which affects the graph of the function in the vertical direction.

P

Parabola (pp. 67, 532): The set of all points P in a plane for which P's distance to a fixed point (the focus) is equal to its distance to a fixed line (the directrix).

Paraboloid (p. 514): The three-dimensional surface generated by rotating a parabola about its axis of symmetry.

Parameter (pp. 40, 366): A variable used as an arbitrary constant, or an independent variable on which two or more other variables depend.

Parametric equations (pp. 35, 39): A pair of equations that express the coordinates x and y of points on the curve as separate functions of a common variable (the parameter).

Parametric function (p. 366): A function specified by parametric equations.

Partial sum (of a series) (pp. 436, 755): The nth partial sum of a series is the sum of the first n terms of the series.

Period (pp. 59, 254): For a periodic function, the difference between the horizontal coordinates of points at the ends of a single cycle.

Periodic function (pp. 59, 254–255): A function whose values repeat at regular intervals.

Permutation (p. 714): An arrangement in a definite order of some or all of the elements in a set.

Phase displacement (p. 283): The directed horizontal distance from the vertical axis to the point where the argument of a periodic function equals zero. For example, if $y = \cos(x - 2)$, the phase displacement is 2; or if $y = \sin\left(x + \frac{\pi}{3}\right)$, the phase displacement is $-\frac{\pi}{3}$.

Piecewise function (p. 49): A function defined by different rules for different parts of its domain.

Point of inflection (p. 202): The point where a graph switches its direction of concavity.

Point-slope form (p. 66): $y = y_1 + a(x - x_1)$, where (x_1, y_1) is a fixed point on the line.

Polar axis (p. 563): In the polar coordinate system, a fixed ray, usually in a horizontal position.

Polar coordinates (pp. 561, 564): A method of representing points in a plane by ordered pairs of (r, θ). The value r represents the distance of the point from a fixed point (pole), and θ represents the angle of rotation from the polar axis to a position that contains the point.

Polar equation (p. 562): An equation $r = f(\theta)$, where θ is a rotation from the positive axis, and r is the distance from the origin.

Pole (p. 561): In the polar coordinate system, the origin.

Polynomial (p. 189): An expression involving only the operations of addition, subtraction, and multiplication performed on the variable (including nonnegative integer powers).

Polynomial function (p. 9): A function of the form $y = p(x)$, where $p(x)$ is a polynomial expression. The general equation is $p(x) = a_n x^n + a_{n-1} x^{n-1} + \cdots + a_2 x^2 + a_1 x + a_0$.

Polynomial operations (p. 189): The operations of addition, subtraction, and multiplication.

Position vector (pp. 603, 605): For a point (x, y), its position vector starts at the origin and ends at the point.

Power function (p. 10): A function of the form $y = ax^n$, where $a \neq 0$.

Power regression (p. 137): A method of finding the best-fitting power function for a set of data.

Power rule for derivatives (p. 804): A rule for finding the derivative of a power function. If $f(x) = kx^n$, where n is a nonnegative integer and k is a constant, then $f'(x) = k \cdot nx^{n-1}$

Pre-image (p. 666): The point, figure, or matrix to which a transformation is applied.

Principal branch (pp. 269, 374): A defined portion of a trigonometric function's graph that is invertible. Within the principal branch, each output value has only one input value.

Principal value (p. 315): The value along a chosen branch of a multi-valued function, such as an inverse trigonometric function.

Probability (pp. 705, 706): If the outcomes of a random experiment are equally likely, then the probability that a particular event will occur is equal to the number of outcomes in the event divided by the number of outcomes in the sample space.

Probability distribution (p. 734): A probability function that tells how the 100% probability is distributed among the various possible events.

Proper divisor (p. 739): A proper divisor of an integer n is a positive integer less than n that divides n exactly.

Proportionality constant (p. 67): The constant a in the equation for a power function $y = ax^n$. In general, a constant that equals the ratio of two quantities.

Pythagorean property (pp. 345, 526): A property that is the result of the Pythagorean theorem and the definitions of the trigonometric functions. For example, $\cos^2 \theta + \sin^2 \theta = 1$.

Q

Quadratic function (p. 9): A function of the form $y = ax^2 + bx + c$, where $a \neq 0$.

Quadratic regression (p. 142): A method of finding the best-fitting quadratic function for a set of data.

Quadric surface (p. 514): A three-dimensional surface defined by a quadratic equation in three variables. Often they are conic sections which have been rotated about an axis.

R

Radian (p. 300): A central angle of one radian intercepts an arc of the corresponding circle equal in length to the radius of the circle.

Radian measure (pp. 301, 302): The angle measure of a central angle in a circle, defined as the arc length corresponding to the angle divided by the radius of the circle. Radian angle measures are real numbers with no units.

Random experiment (p. 706): In probability, an experiment in which there is no way of telling the outcome beforehand.

Range (p. 4): The set of all values of the dependent variable that correspond to values of the independent variable in the domain.

Rational algebraic function (p. 11): A function that can be expressed as the quotient of two polynomial expressions.

Rational function (p. 177): See **Rational algebraic function.**

Recursion formula (pp. 757, 760): A formula for the terms of a sequence that specifies t_n as a function of the preceding term, t_{n-1}.

Reciprocal function (p. 75): The parent reciprocal function is $f(x) = \frac{1}{x}$.

Reciprocal property (pp. 262, 347): For certain pairs of trigonometric functions, it is the property where the values of one trigonometric function are the reciprocals of the values of the other trigonometric function. For example, $\sec x = \frac{1}{\cos x}$.

Reference angle (pp. 249–250): For an angle in standard position, the positive acute angle (measured counterclockwise) between the horizontal axis and the terminal side.

Reflection (pp. 36, 42, 48): The reflection of a plane figure across a given line in the plane is a transformation for which each point in the image is the same distance from the line on the opposite side as the corresponding point in the original figure.

Regression line (p. 129): The line that makes the sum of the squares of the residuals, SS_{res}, a minimum.

Relation (p. 8): Any set of ordered pairs.

Residual (pp. 129, 133): The residual deviation of a data point from the line $\hat{y} = mx + b$ is $y - \hat{y}$, the vertical directed distance of its y-value from the line.

Residual plot (p. 159): A scatter plot of residuals.

Resultant vector (pp. 465, 605): The sum of two or more vectors.

S

Sample space (p. 706): The set of all outcomes of a random experiment.

Scalar (pp. 463, 603, 605): A quantity, such as time, speed, or volume, that has magnitude but no direction.

Scalar product (pp. 485, 619): See **Dot product.**

Scalar projection (pp. 622, 623): If θ is the angle between $\vec{a}$ and $\vec{b}$ when they are placed tail-to-tail, then the scalar projection of $\vec{a}$ on $\vec{b}$ is $p = |\vec{a}| \cos \theta$.

Scalar quantity (p. 463): A quantity, such as distance, time, or volume, that has magnitude but no direction.

Secant (pp. 262, 263): One of the trigonometric functions, equal to the reciprocal of the cosine function; abbreviated "sec."

Self-similar (p. 681): When an object or figure can be broken into smaller parts that are similar to one another and to the original object or figure.

Sequence (pp. 755, 756, 760): A function whose domain is the set of positive integers. The independent variable is the term number, n, and the dependent variable is the term value, t_n.

Series (p. 755): The indicated sum of the terms of a sequence.

Simple event (p. 706): A single outcome of an experiment.

Sine (pp. 256, 263): One of the trigonometric functions; abbreviated "sin." If θ is an angle of rotation of the positive u-axis terminating in a position containing the point (u, v), then $\sin \theta = \frac{v}{r}$, where r is the distance of the point (u, v) from the origin.

Sinusoid (p. 247): Any translation or dilation of the parent function $y = \sin x$ or $y = \cos x$.

Sinusoidal axis (p. 283): The axis that runs along the middle of the graph of a sinusoid.

Sinusoidal regression (p. 171): A method of finding the best-fitting sinusoidal function for a set of data.

Slope (p. 66): The tilt of a linear function calculated by the change in y divided by the change in x and represented by a in the equation $y = ax + b$.

Slope-intercept form (p. 66): The general equation $y = ax + b$, where a and b are constants and the domain is all real numbers.

Standard position (of an angle) (p. 248): The position of an angle in a coordinate system with the vertex at the origin and the initial side along the positive horizontal axis.

Standard position (of an arc) (p. 308): The position of an arc in a coordinate system with one endpoint at the positive horizontal axis, and extending counterclockwise.

Strange attractors (p. 679): A figure generated by the iterative application of several different transformations.

Sum and product properties (pp. 417, 419): Properties that reexpress the sum or difference of two trigonometric functions as a product of two trigonometric functions, or vice versa. For example, $\sin x + \sin y = 2 \sin \frac{1}{2}(x + y) \cos \frac{1}{2}(x + y)$.

Sum of the squares of the deviations (pp. 131, 133): A measurement of a data set obtained by squaring each point's deviation from the mean, thereby making each value positive or zero, and then summing the results. Abbreviated SS_{dev}.

Sum of the squares of the residuals (pp. 131, 133): A measure of how well a line fits a data set, obtained by squaring each residual, thereby making each value positive or zero, and then summing the results. Abbreviated SS_{res}.

T

Tangent (function) (pp. 262, 263): One of the trigonometric functions; abbreviated "tan." If θ is an angle of rotation of the positive u-axis terminating in a position containing the point (u, v), then $\tan \theta = \frac{v}{u}$.

Tangent line (p. 800): The tangent line of a curve at a point is the line whose slope equals the slope of the curve at that point.

Term index (p. 769): The number of a term in a sequence or series. For example, for $t_2 = 75$, the term index is 2.

Transformation (p. 15): An operation that maps points in the plane or space uniquely. Some examples are dilation, translation, or rotation, which can change the shape or the position of the graph of a relation or figure.

Transformation matrix (pp. 666, 676): A matrix $[T]$ that, when multiplied by a matrix $[M]$, performs a rotation, dilation, and/or translation on matrix $[M]$.

Translation (pp. 18, 672): A shift of a function, graph, or image in the vertical and/or horizontal direction.

Transverse axis (p. 514): For a hyperbola, the name given to the axis of symmetry that runs from vertex to vertex.

Transverse radius (p. 525): The radius from the center of a hyperbola to a vertex along the transverse axis.

Trial (p. 706): One occurrence of a random experiment.

Trigonometric functions (p. 255): The six functions sine, cosine, tangent, cotangent, secant, and cosecant.

U

Union (p. 728): If A and B are two events, then the union of A and B, $A \cup B$, is the set of all outcomes that are in event A *or* event B (or both).

Unit circle (p. 263): A circle of radius 1.

Unit vector (pp. 603, 605, 614): A vector that is one unit long.

V

Variable (p. 9): A symbol used to stand for any one of a set of numbers, points, or other entities; the set is called the *domain* of the variable, and any member of the set is called a *value* of the variable.

Vector (pp. 463, 605): A directed line segment $\vec{v}$.

Vector projection (p. 621): If $\vec{u}$ is a unit vector in the direction of $\vec{b}$, then the vector projection of $\vec{a}$ on $\vec{b}$ is $\vec{p} = |\vec{a}| \cos \theta \, \vec{u}$, where θ is the angle between $\vec{a}$ and $\vec{b}$.

Vector quantity (pp. 463, 605): A quantity, such as force, velocity, or displacement, that has both magnitude (size) and direction.

Vertex (pp. 67, 500): For a polynomial function (including a parabola), a point where the graph changes direction. For an ellipse or hyperbola, an endpoint of the major axis or transverse axis.

Vertex form (pp. 67, 499): The form of the equation for a parabola, ellipse, or hyperbola that indicates the vertex or vertices. For example, for a quadratic function, the form $y = k + a(x - h)^2$, indicates that the vertex of the parabola is (h, k).

X

***x*-intercept** (p. 8): The x-coordinate of a point where a graph meets the x-axis. For this value of x, $y = 0$.

Y

***y*-intercept** (p. 8): The y-coordinate of a point where a graph meets the y-axis. For this value of y, $x = 0$.

Z

Zero (of a function) (pp. 177, 178): A real or complex nonreal value of $x = c$ for which $f(c) = 0$.

Index of Problem Titles

D

E

F

U

V

W

X

Z

General Index

D

E

F

J

K

L

M

S

T

U

V

Photograph Credits

Chapter 1

1: Fotosearch; **4:** © iStockphoto.com/Irina Tischenko; **6:** Library of Congress/Science Photo Library; **7:** Henrik Weis/Photolibrary; **9:** AP Photo/Dale Atkins; **23:** © John Gillmoure/CORBIS; **35:** © Robert W. Ginn/PhotoEdit; **43:** Andre Goncalves/Shutterstock.com; **46:** Jeremy Woodhouse/Getty Images; **48:** © Jacksonville Journal Courier/The Image Works; **56:** Doug Menuez/Photolibrary; **57:** © Laura Dwight/PhotoEdit; **58:** NASA/courtesy of nasaimages.org; **61:** Durden Images/Shutterstock.com

Chapter 2

63: Erwin & Peggy Bauer/Wildstock; **67:** Michael L. Smith/Photographic Reflections; **68:** Radium Institute, courtesy of AIP Emilio Segre Archives; **77:** Ken Karp Photography; **82:** SPL/Photo Researchers, Inc.; **85:** The Granger Collection, NYC—All rights reserved.; **86:** William Milner/Shutterstock.com; **89:** © Michael Newman/PhotoEdit; **90:** Courtesy of Eric Marcotte, Ph.D., www.sliderule.ca; **96:** Kevin Schafer/Getty Images; **98:** NASA/courtesy of nasaimages.org; **107 (left):** Sinclair Stammers/Photo Researchers, Inc.; **107 (right):** © Reuters/CORBIS; **112:** AP Photo/David Zalubowski; **116:** AP Photo/Sayyid Azim; **117:** © Steve & Ann Toon/Robert Harding World Imagery/Corbis; **120:** Ron Dahlquist/SuperStock; **124:** © Braud, Dominique/Animals Animals–Earth Scenes

Chapter 3

127: Kwame Zikomo/Purestock/SuperStock; **132:** Courtesy of the Computer History Museum; **137:** Mike Robinson/SuperStock; **141:** NASA/courtesy of nasaimages.org; **142:** Tom Bean/Getty Images; **149:** © John Cancalosi/Alamy; **159:** Barry Runk/Grant Heilman Photography, Inc.; **161:** © Graham Neden; Ecoscene/CORBIS; **165:** AP Photo/Michael Probst; **170:** Manfred Steinbach/Shutterstock.com; **173:** © Dan Suzio

Chapter 4

175: © Richard Cummins/CORBIS; **201:** Gerald Nowak/Photolibrary; **202:** © Elizabeth Kreutz/NewSport/Corbis; **209:** The Granger Collection, NYC—All rights reserved.; **231:** Image Source/Getty Images; **234:** © Brendan McDermid/Reuters/Corbis; **244:** © iStockphoto.com/Sreedhar Yedlapati

Chapter 5

245: Brian Bahr/Getty Images; **247:** © Ben Wood/CORBIS; **249:** © Jon Hicks/CORBIS; **253:** age fotostock/SuperStock; **255:** Jason Reed/Getty Images; **259:** © Tim De Waele/Corbis; **273 (left):** © Peter TurnleyCORBIS; **273 (right):** © iStockphoto/Holger Mette; **274:** Stocktrek/Jupiter Images; **276:** Paul Foerster; **277:** © Michael Newman/PhotoEdit; **278:** Ian Cartwright/Getty Images

Chapter 6

281: Severin/Getty Images; **283:** Ken Karp Photography; **289:** Ken Karp Photography; **292:** Ken Karp Photography; **298:** Chad Ehlers/Getty Images; **301:** © The Trustees of the British Museum; **307:** thumb/Shutterstock.com; **314:** Dave L. Ryan/Photolibrary; **319:** Image Source/Jupiter Images; **322:** David W. Hamilton/Getty Images; **323:** © Jose Carillo/PhotoEdit; **324 (left):** NASA/courtesy of nasaimages.org; **324 (right):** Ryan McVay/Getty Images; **326:** Geostock/Getty Images

Chapter 7

343: © Inga Spence/Alamy; **352:** Christine Osborne; **354:** Ken Karp Photography; **355:** David Benton/Shutterstock.com; **367:** © CUMMINS/UNEP/Still Pictures/The Image Works; **369:** William S. Helsel/Getty Images; **376:** © eye35.com/Alamy; **380:** J. A. Hampton/Getty Images; **383:** © Bettmann/CORBIS

Chapter 8

385: Fotosearch; **389:** © Kevin Nicholson/Alamy; **394:** M.C. Escher's "Rippled Surface" © 2010 The M.C. Escher Company-Holland. All rights reserved.; **395:** Ken Karp Photography; **409:** © Cranston, Bob/Animals Animals—Earth Scenes. All rights reserved.; **412:** AP Photo; **415:** © Topham/The Image Works; **436:** Courtesy of B&K Precision Corp

Chapter 9

441: © Bill Bachmann/The Image Works; **451:** Wikimedia Commons; **464:** © Carl & Ann Purcell/CORBIS; **465:** © Richard Bickel/CORBIS; **466:** AP Photo/Reed Saxon; **468:** © Philipp Hympendahl/Alamy; **470:** © iStockphoto.com/Allen Krughoff; **473:** © Bettmann/CORBIS; **475 (left):** Erik Simonsen/Getty Images; **475 (right):** NASA/courtesy of nasaimages.org; **477:** Image Source/Photolibrary; **479:** PureStock/Photolibrary; **480:** © iStockphoto.com/Michael Fuller; **484:** Johan Knelsen/Shutterstock.com; **485:** Wikimedia Commons; **488:** Space Frontiers/Getty Images; **491:** NASA/courtesy of nasaimages.org

Chapter 10

493: Yoshikazu Onishi/Sebun Photo/Getty Images; **513:** AP Photo/Keith Bedford; **514:** AP Photo/NASA Jet Propulsion Laboratory; **515:** Ken Karp Photography; **516 (left):** Martin Mette/Shutterstock.com; **516 (right):** Mark Winfrey/Shutterstock.com; **520:** Kzenon/Shutterstock.com; **523:** © Kelly-Mooney Photography/Corbis; **535 (left):** © Mark Antman/The Image Works; **535 (right):** Mark Keller/SuperStock; **537:** AP Photo/Billings Gazette, Larry Mayer; **543:** Courtesy of the Kepler-Gesellschaft Society; **548:** © Morton Beebe/CORBIS; **549:** © iStockphoto.com/Gordon Donovan; **550:** Scott Kardel/Palomar Observatory/Caltech; **556:** © iStockphoto/cbpix

Chapter 11

559: © Royalty-Free/Corbis; **563:** Francisco Amaral Leitao/Shutterstock.com; **570:** © Mike Zens/CORBIS; **572:** © iStockphoto/cjmckendry; **584 (top):** © Tek Image/Science Photo Library/Corbis; **584 (bottom):** The Granger Collection, NYC—All rights reserved.; **585:** © David Young-Wolff/PhotoEdit; **586:** © Tony Freeman/PhotoEdit; **589:** Stock Connection/SuperStock; **591:** Comstock/Jupiter Images; **593:** © Bill Aron/PhotoEdit; **600:** Michael Melford/Getty Images

Chapter 12

601: © Donald Corner & Jenny Young /Artifice Images; **604:** © Library of Congress–digital ve/Science Faction/Corbis; **612:** © James L. Amos/CORBIS; **618 (left):** © CORBIS; **618 (right):** AIP Emilio Segre Visual Archives, Born Collection; **621:** Ken Karp Photography; **624:** © IMAGES, Agence Photographique/eStock Photo—All rights reserved.; **625:** Motacilla/Wikimedia Commons; **639:** Laura Murray; **645:** © Reuters/CORBIS

Chapter 13

655: Kevin Schafer/Stone/Getty Images; **659:** Scott Smith/Photolibrary; **661:** © Bettmann/CORBIS; **669:** Courtesy of National Archives (127-GR-137-57875); **678:** © Oleksiy Maksymenko/Alamy; **681:** (left): © Jack Fields/CORBIS; **681 (right):** © Eric Crichton/CORBIS; **686:** © Jim Zuckerman/CORBIS; **689:** NASA/courtesy of nasaimages.org; **692:** Courtesy of the University of Texas Libraries, The University of Texas at Austin.; **693:** © John Lund/Blend Images/Corbis; **700 (left):** United States Mint image; **700 (right):** © David Wall/Alamy; **702:** Art Resource, NY

Chapter 14

703: Vince Streano/Getty Images; **705:** Laura Murray; **707:** National Weather Service; **710:** © iStockphoto/tomograf; **711:** © iStockphoto/Holger Mette; **712:** © iStockphoto/Ian Hamilton; **713:** Platehut.com; **717 (left):** © David Young-Wolff/PhotoEdit; **717 (right):** Corbis/SuperStock; **722:** Jupiter Images/Getty Images; **726:** Lonny Kalfus/Getty Images; **730:** © Bettmann/CORBIS; **732:** © Richard Gross/Corbis; **733:** David Sanger/Getty Images; **734:** © iStockphoto/Brenda McEwan; **737:** Laura Murray; **738:** Exactostock/SuperStock; **742:** © Image Source/Alamy; **744:** © Bettmann/CORBIS; **747:** © iStockphoto/Gene Chutka; **748:** © Keoki Stender

Chapter 15

753: NASA/courtesy of nasaimages.org; **756:** The Granger Collection, NYC—All rights reserved.; **764:** Amanda Vivan/Getty Images; **765:** Richard A. Cooke III/Getty Images; **767:** S. Lowry/Univ Ulster/Getty Images; **772:** Ken Karp Photography; **781:** © iStockphoto/SweetyMommy; **785:** © James Pickerell/The Image Works

Chapter 16

787: Pernilla Bergdahl/Getty Images; **798:** FloridaStock/Shutterstock.com; **806:** © Kevin Fleming/CORBIS; **814:** The Granger Collection, NYC—All rights reserved.; **816:** © iStockphoto.com/Steve Krull

CONIC SECTION EQUATIONS

Circle

$$\left(\frac{x-h}{r}\right)^2 + \left(\frac{y-k}{r}\right)^2 = 1$$

Center (h, k), radius r

Ellipse

$$\left(\frac{x-h}{d_x}\right)^2 + \left(\frac{y-k}{d_y}\right)^2 = 1$$

Center (h, k), x- and y-dilations d_x and d_y

Hyperbola

Opening in x-direction:

$$\left(\frac{x-h}{d_x}\right)^2 - \left(\frac{y-k}{d_y}\right)^2 = 1$$

Center (h, k), x- and y-dilations d_x and d_y

Asymptote slopes are $\pm\dfrac{y\text{-dilation}}{x\text{-dilation}}$

Opening in y-direction:

$$-\left(\frac{x-h}{d_x}\right)^2 + \left(\frac{y-k}{d_y}\right)^2 = 1$$

Center (h, k), x- and y-dilations d_x and d_y

Asymptote slopes are $\pm\dfrac{y\text{-dilation}}{x\text{-dilation}}$

Parabola

Opening in y-direction:

$$y = k + a(x-h)^2$$

Vertex (h, k), dilation factor a

Opening in x-direction:

$$x = h + a(y-k)^2$$

Vertex (h, k), dilation factor a

LOGARITHMIC FUNCTION PROPERTIES

$$\log_b xy = \log_b x + \log_b y$$

$$\log_b \frac{x}{y} = \log_b x - \log_b y$$

$$\log_b x^y = y \log_b x$$

$$\log_a x = \frac{\log_b x}{\log_b a}$$

MEASURES OF FIT OF FUNCTIONS TO DATA

Sum of the squares of the deviations

$$SS_{\text{dev}} = \Sigma(y - \bar{y})^2$$

Sum of the squares of the residuals

$$SS_{\text{res}} = \Sigma(y - \hat{y})^2$$

Coefficient of determination

$$r^2 = \frac{SS_{\text{dev}} - SS_{\text{res}}}{SS_{\text{dev}}}$$

SEQUENCES AND SERIES

Arithmetic Sequences and Series

For an arithmetic sequence with first term t_1 and common difference d,

$$t_n = t_1 + (n-1)d$$

The nth partial sum of an arithmetic series is

$$S_n = \frac{n}{2}(t_1 + t_n)$$

Geometric Sequences and Series

For a geometric sequence with first term t_1 and common ratio r,

$$t_n = t_1 \cdot r^{n-1}$$

The nth partial sum of a geometric series is

$$S_n = t_1 \cdot \frac{1 - r^n}{1 - r}$$

The sum of an infinite geometric series is

$$\lim_{n\to\infty} S_n = t_1 \cdot \frac{1}{1-r} \qquad (\text{if } |r| < 1)$$

Binomial Formula

The terms of a binomial series come from expanding a binomial $(a + b)^n$:

$$\text{Term with } b^r = \frac{n!}{(n-r)!r!}a^{n-r}b^r = \binom{n}{r}a^{n-r}b^r$$

The sum of the binomial series, or the binomial formula, is

$$(a+b)^n = a^n + \binom{n}{1}a^{n-1}b + \binom{n}{2}a^{n-2}b^2 + \ldots + \binom{n}{n-2}a^2b^{n-2} + \binom{n}{n-1}ab^{n-1} + b^n$$